COMPUTER INTEGRATED MANUFACTURING

International Conference on Computer Integrated Manufacturing (ICCIM)

Proceedings of the ICCIM'91
Ed. B.S. Lim
ISBN 981-02-0684-4

Proceedings of the ICCIM'93
Eds. A. Sen, J. Winsor and R. Gay
ISBN 981-02-1568-1 (set)
ISBN 981-02-1946-6 (vol. 1)
ISBN 981-02-1947-4 (vol. 2)

Proceedings of the ICCIM'95
Eds. J. Winsor, A.I. Sivakumar and R. Gay
ISBN 981-02-2376-5 (set)
ISBN 981-02-2973-9 (vol. 1)
ISBN 981-02-2974-7 (vol. 2)

ICCIM'91 and ICCIM'95 were published by World Scientific Pub. Co. ICCIM'93 was jointly published by World Scientific Pub. Co. and Global Pub.

Proceedings of the 4th International Conference

COMPUTER INTEGRATED MANUFACTURING

21 – 24 October 1997
Singapore

Volume 1

Editors

A. Sen, A.I. Sivakumar and R. Gay
Gintic Institute of Manufacturing Technology
Nanyang Avenue
Singapore

Springer

A. Sen, A.I. Sivakumar and R. Gay
Gintic Institute of Manufacturing Technology
71 Nanyang Drive
Singapore 638075

Computer Integrated Manufacturing

ISBN 981-3083-66-2 (vol. 1)
ISBN 981-3083-67-0 (vol. 2)
ISBN 981-3083-68-9 (set)

Printed in Singapore

Typesetting: Camera-ready by authors
5 4 3 2 1 0

ORGANISING COMMITTEE

Chairman : Dr. Robert Gay, Gintic

Members :
Mr. Goh Kiah Mok, Gintic
Mr. Lew Sin Chye, Nanyang Technological University
Dr. Lim Kah Bin, National University of Singapore
Ms. Lim Lai Lee, Gintic
Ms. Jacqueline Ng, Miller Freeman
Dr. Anirudha Sen, Gintic
Dr. AI Sivakumar, Gintic
Dr. Soh Yeng Chai, William, Nanyang Technological University
Dr. V. Subramaniam, National University of Singapore
Dr. Tay Kiang Meng, Gintic
Mr. Tay Meng Leong, Nanyang Technological University
Mr. Stephen Teng, Singapore Industrial Automation Association

TECHNICAL COMMITTEE

Chairman : Dr. AI Sivakumar, Gintic
Co-Chairman : Dr. Anirudha Sen, Gintic
(Editorial Com)

Members :
Mr. Henry Chang , Productivity and Standards Board
Dr. Opas Chutatape, Nanyang Technological University
Mr. Goh Kiah Mok, Gintic
Mr. Lew Sin Chye, Nanyang Technological University
Dr. Lim Kah Bin, National University of Singapore
Mr. Jasbir Singh, Gintic
Dr. V. Subramaniam, National University of Singapore
Dr. Tay Kiang Meng, Gintic
Mr. Tay Meng Leong, Nanyang Technological University

INTERNATIONAL ADVISORY COMMITTEE

PREFACE

ICCIM '97 is the fourth international conference we have organised in the field of computer integrated manufacturing. It is gratifying to see that since its inception in 1991, ICCIM has established a track record of bringing to the public what latest technology is today and what will be available tomorrow.

This year our theme for ICCIM is "Manufacturing into the next Millennium" - an appropriate title as we approach the end of the twentieth century. In this age of rapid change, it is difficult to see beyond a few years what new markets and technologies will emerge. However, at ICCIM'97 we believe that many of the ideas presented and some of the research work reported will become reality tomorrow. For example, just since the last ICCIM conference in 1995 the growth of Internet-based business has suddenly exploded and made fortunes for many people, very often, unexpectedly. While the scale was beyond expectations, the trend as related to the manufacturing industry was anticipated and discussed in a number of papers from previous ICCIM conferences.

We have divided the proceedings into two volumes. In the first volume, we gathered the papers dealing with systems level management and planning of manufacturing activities, whilst the second volume concentrates more on product related and implementation related issues.

In this volume the following topics are covered :

- Management Aspects
- Manufacturing Planning and Control
- Advanced Manufacturing Processes

On behalf of the Organising and Technical Committees of ICCIM '97, we would like to thank all those who have made this conference what it is today - the authors, International Advisory Committee , reviewers and participants.

A special thanks to Ms. Lim Lai Lee for her contribution to the numerous details in organising this event and for collating the articles for these two volumes.

October 1997

A.Sen
A.I. Sivakumar
R.Gay

CONTENTS

Volume 1

Flexibility and Change Management

Business Process Reengineering

CIM Design and Planning

Integrated Databases

Internet and the WWW

CIM Justification

CIM Implementations

MANUFACTURING PLANNING AND CONTROL

Aggregate Planning

Volume 2

PRODUCT DESIGN AND CONCURRENT ENGINEERING

Geometric Modelling

Product Design

Product Analysis

Process Planning

Inspection and Diagnostic Systems

Robotics and AGV Systems

ADVANCED MANUFACTURING TECHNOLOGY

KEYNOTE SPEECHES

KEYNOTE SPEECHES

House of Business Engineering
- The Universal Language for Process Management -

Prof. Dr. August-Wilhelm Scheer

Institute for Information Systems (IWi), University of the Saarland, D-66123 Saarbrücken

Abstract. In this paper "ARIS - House of Business Engineering", an architecture of business processes is presented. It consists of the four levels 'Process Design', 'Process Management', 'Process Workflow' and 'Process Application', which encompass the whole life-cycle range: from business strategy to the execution of the processes and the implementation of the information technology. Thus, it leads to a completely new process-oriented software concept.
The "ARIS - House of Business Engineering" is a universal language for process management, which means it is applicable for every type of business process: in manufacturing, in the service industry and in the public services. At the same time the architecture bridges the gap between business process modeling and workflow-driven applications, between business process re-engineering and continuous process improvement.

1 Process Design and Control as Key Management Tasks

The term "business process" is defined universally. A business process is described as a procedure relevant for adding value to an organization. It is viewed in its entirety, from beginning to end.
Fig. 1 illustrates the business process of order entry processing. The initial requirements of the customer lead to order acceptance by the manufacturer's sales department. Sales then relays information to purchasing, in order for them to supply bought-in parts. Finally, production plans and executes the work-order. Fig. 1 illustrates this procedure by a series of events triggering functions. The initial event of the process is „demand occured". The final event is the completion of the product in manufacturing. Events do not only trigger functions, they are themselves the results of functions. Processes can be split into sub-processes. Conversely, sub-processes can be joined together. By introducing logical operators, the control structure with its event-driven process chain (EPC) can be expanded to accommodate variously complex procedures, as shown in fig. 1 [5]. Besides describing the procedural structure of events and functions, there must also be a focus on describing the organizational units assigned to the functions. Many re-engineering projects are actually directed at re-allocating functions to organizational units.

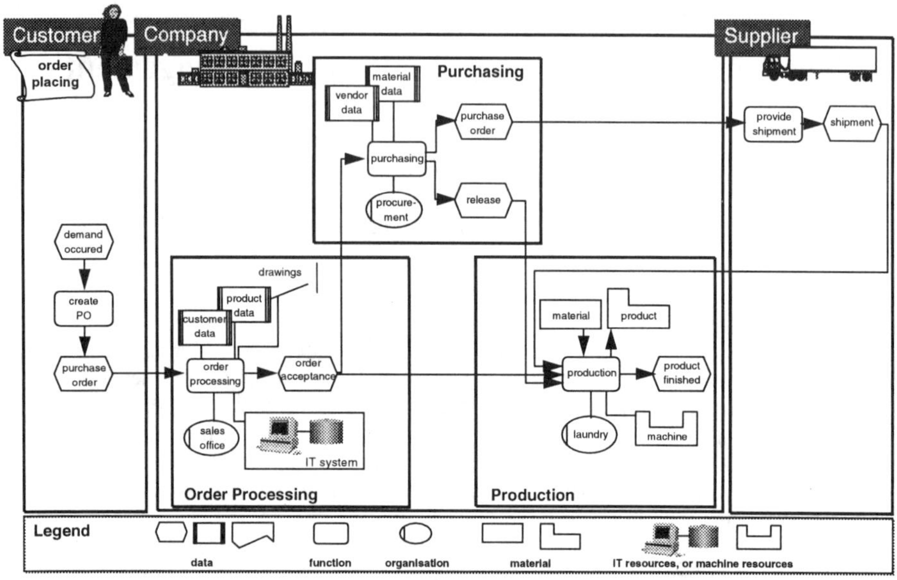

Fig. 1. business process model

A business process consists of two function classes. The first function class describes how processing rules transform input data into output data. Functions of this kind are executed in the "office area". In addition to the transformation of data, a second type of transformation can be carried out in a business process: the transformation from input material to output material. This process is called manufacturing. Material transformation comprises physical change, but can also involve a change in location, that is the function of material handling.

For many years now, the process of material transformation in industrial enterprises has been mastered quite well. This process can be described minutely and is usually controlled precisely regarding its scheduling and costs. On the other hand, management's knowledge regarding administrative processes is usually quite scant. Whereas the process of a production order is described minutely by the routing, descriptions regarding the business process in sales, purchasing or accounting are rare. Finally, to make matters even worse, in many industrial enterprises, scheduling and cost shortcomings are more frequently found in administrative rather than in production departments. Therefore, it seems appropriate to examine whether and how procedures which have proven to be most successful in controlling manufacturing processes might also be applicable in the back office.

2 An Architecture Supporting Business Processes

When characterizing the methods used for process control in manufacturing, we deduce the following four main tasks:

1. Describing and optimizing the process structure, based on routings,
2. Optimization planning of current business processes with regard to capacity, time, and costs (production scheduling control),
3. Controlling the execution of individual processes (material flow control),
4. Supporting the function execution, that is, material or data transformation rules.

On the whole, these tasks can also be translated to processes in the service sector, where data transformation has a high priority. In fig. 2 these tasks are allocated to a Four Level Model, the "ARIS - House of Business Engineering". The figure sums up the individual levels of the architecture and depicts their correlation. In the following paragraphs each level will be described in detail.

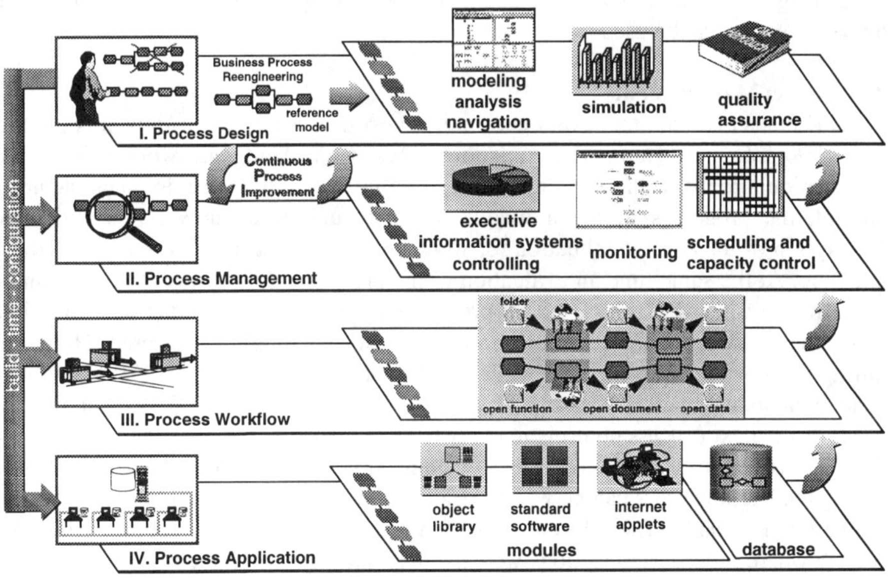

Fig. 2. ARIS - House of Business Engineering

2.1 Process Design

Level I „Process Design" describes business processes according to the routing. Therefore, the ARIS (Architecture of Integrated Information Systems) concept provides a method to cover every aspect of a business process [cf. 4]. ARIS consolidates various views into one business process (e.g. Function View and Data View). The Control View records the relationships between the other views; it also utilizes the event-driven process chain (EPC) -method, as illustrated in fig. 1.

On the one hand the ARIS concept can be employed as a front end for managing manufacturing routings and material flows. This offers more variations for representing alternate procedures. Obviously, presenting the manufacturing methods in a graphical form is much more user-friendly than listing routings in tables. By including the material flow in the ARIS concept, relationships between product and business process models can also be addressed. On the other hand, concepts designed to manage different versions of manufacturing routings (even expert systems) are well suited for describing business processes in the service sector. In addition to modeling methods the ARIS TOOLSET on Level 1 offers tools for analyzing and simulating processes [2].
As an aid to modeling business processes, existing information on useful structures of the business processes can be included in the basic solution. These reference models, empirically derived from best practice cases or theoretical considerations, lead to substantial time savings in designing optimal procedures. Reference models can be described according to the ARIS concept and stored in the ARIS TOOLSET. When structuring processes, every function, e.g. analysis, comparison, model adjustment and model changes, can be used.

2.2 Process Management

Level II plans and monitors every current business process from the "business process owner's" point of view. The functions are assigned to the workstations and workstation groups with respect to time, so that, as is the case in manufacturing control, the processes in the individual capacity units are known. A compressed description of the capacity situation for workstation groups can be derived from these data, too. At the same time an evaluation of the processes with respect to costs can be made if there is a description of the processes by functions and functional duration. Here, the duration of the functions can first of all be roughly estimated. However, during the next steps it becomes clear that by means of access to workflow systems exact data about actual values of processing times can be obtained. These can then be used for the next process controls in order to better estimate the duration of functions. The intensive discussion in business administration about activity-based costing during the last few years can be widely resolved if such a generalized business process deliberation is made. There has always been activity-based costing, but only in those sectors where process descriptions were available, this means in the field of evaluating manufacturing processes. This is the reason why there are terms like concurrent costing, where the actual costs of a manufacturing order and thus of a manufacturing process can be determined in the course of processing.
The information on "to-be" and "as-is" deviations from the schedule and the costs is utilized to continuously improve business processes. This creates a closed loop between Level I (Process Design) and Level II (Process Management), leading to continuous process improvement (CPI). Every method used in describing Level I, such as process analysis, model comparison or simulation, can be employed for continuous process improvement.
BPR and CPI should be regarded in the same vein. When a certain situation arises, causing a company to reflect on its structures, this in turn can lead to a BPR project. However, even after resolving the problem, processes still change. New organizational

concepts can arise. New best practice cases become available as reference models. New technologies are invented. New knowledge is obtained from processes, which have just been implemented, leading to an adjustment of the process. Hence, process design is a continuous process. Frequently, conflicts of interest lead to apparent disparities between BPR and CPI: applications vendors are sometimes blamed for the lengthy procedure occasionally necessary to implement their software. They are concerned that their product could be held responsible for any additional delay if they are connected with a BPR project. Therefore, they oppose BPR strategies and recommend rapid installation of their software and subsequent continuous process improvement. Due to their interest in selling consulting services, consulting companies, on the other hand, recommend the opposite approach: first, develop a new engineering (organizational) concept and then support it with the new software. This prevents unnecessary and awkward procedures from being carried over into the new software concept. The contradictions of these two approaches are resolved in the "ARIS - House of Business Engineering" because BPR and CPI are so closely intertwined.

2.3 Process Workflow

The entire business process procedure (cf. fig. 1) is generally not handled by a single software application system. Moreover, several function-oriented systems are usually used by Sales, Purchasing, Production or Finance. None of these systems is capable of providing information on the entire process, such as the processing state of an order. Therefore, is seems obvious that we should not hand over the responsibility for the whole process control to a single function, but rather to a separate system level. This level is known as workflow.

Workflow systems transport the objects to be processed (documents) from one workplace to the next. Better yet, they send them from the computer system at one workplace to the computer system at the next work step. Therefore, the procedure must be described in detail and must include each individual type of process or business user, respectively [1]. In fig. 2 the document flow is illustrated by a "folder", which is transported from one workplace to the next. This folder contains electronic references regarding the data required for processing and the function elements that need to be called up.

The workflow system is informed of the process status, execution time and the appropriate business user of every business process. Thus, the workflow system is also the foundation for Process Management at Level II. It reports the data for cost and scheduling evaluations and provides process information for process monitoring. An agreement by the Workflow Management Coalition, a group of Workflow vendors, has standardized interfaces [6]. Now, various workflow systems can be linked with one another.

The process representation of workflow systems can also be used to guide business users. This increases their knowledge of the interrelationship of organizational business processes. The specific procedure in fig. 3 (right box) follows from the general business process procedure. You create a specific procedure by giving information on particular business users and by selecting a certain path outlined in the general business process description. Thus, business users can always see how their activity is embedded in the process, who will precede and who will succeed them within the process. For example, they can also see that only the left branch of a business process is relevant for them; the control flow of the right branch might be deleted. Since a particular process has not been created for the business user of the succeeding activity, only the department name, "Warehouse", is listed. Depending on the capacity situation at that time, the business user of the next operation is not determined until the conclusion of the task. During 'Process Workflow' processes with precisely defined procedural structures can be differentiated from processes with only roughly defined procedure steps.

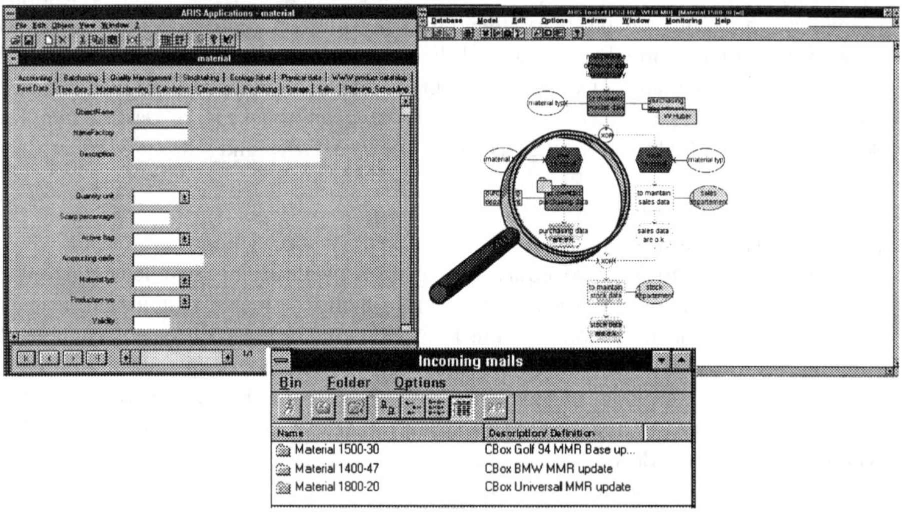

Fig. 3. process-oriented user interface

2.4 Process Application

Finally, at the fourth level of the "ARIS - House of Business Engineering", the functions in the process are executed. This computer-aided task is carried out by function elements. These can either be processing modules of classic application software or processing objects from object libraries. In Internet applications, applets correspond to these encapsulated program modules.

Current vendors of integrated software systems are splitting their systems into smaller modules. Many of them are now just loosely coupled. This makes it possible to release upgrades for each individual module and not across-the-board for the entire system. On the whole, there is a strong tendency today towards splitting application software into individual components (componentware). These modules are re-assembled into complete solutions according to process models. The operational data in these applications are managed by database systems.

In the object-oriented approach, data and functions are encapsulated and communicate via a messaging system, which performs material handling for the workflow system. The objects correspond to the "folder" and provide references to data and functions. It is important to note that Level III is responsible for the entire process of the operation.

It calls up objects to be processed, such as electronic forms for filing insurance claims, loan application forms for loan processing operations or customer orders for customer order processing. It then passes them on to the appropriate processing station and calls up the program modules. This separation of the control flow of programs and function execution is bringing about tremendous changes in the software market. Vendors of conventional application software will have to decide whether they want to be "module brokers" at Level IV and just provide "componentware" with some editing functionality - or if they want to move up to the rapidly growing workflow systems market. Conversely, software manufacturers without much experience in applications are reaching a new point of departure, now that workflow systems are being developed. Particularly in service applications, the processing rules at Level IV can be so simple that they only involve data entry or document editing. Many functions could therefore be executed at this level, such as calling up a spreadsheet or a word processing program. This makes workflow systems which control the coherence of a procedure all the more important.

This means for users that a new architecture for application software is on its way. Service providers, such as banks and insurance companies, do not have a large selection of standard applications at their disposal to support their operational procedures. Now they can document (model) their business procedures at Level I and can control their procedures by implementing a workflow system at Level III. At Level IV, they can still use their existing software to support the processing rules. Nevertheless, today it is necessary to split software at Level IV and make it accessible to workflow control. By separating procedure control from function execution statements, information systems are split into data management, procedure control and function execution.

2.5 Interaction between the Levels

When supporting business processes in their entirety, it is not sufficient to simply split the whole process into the four parts intellectually or as a physical system, as described above. We must also separate their links with one another. We have already noted that the individual business events in the Process Workflow Level are generated by copying the business process design at Level I. The generation of this business design is thus a link between the business process modeling tool and the workflow system. In the Workflow Management Coalition, experts are working on creating accepted standards for this link. The same applies to delivering workflow results to Level II, for example, by delivering details regarding as-is schedules or as-is amounts to Level II for evaluation purposes. These two links make it possible to immediately update a business process procedure, even in execution and evaluation levels. This occurs without having to manipulate any computer programs [cf. 3]. Furthermore linking Level I with Level IV will enable model based customizing: changing the attributes of the data model at Level I alters the data tables in Level IV. Modifying process models, in turn, varies the sequence of function procedures. Changing function models either switches off or activates functions. Finally, employing the organizational model allocates functions to certain organizational units and determines the screen sequence. Thus, organizational Design Level I plays a tremendous role within the whole architecture.

References

1. Galler, J., Scheer, A.-W.: Workflow-Projekte: Vom Geschäftsprozeßmodell zur unternehmensspezifischen Workflow-Anwendung. Information Management 1/95, S. 20-27.
2. IDS Prof. Scheer GmbH (ed.): ARIS-Toolset - Business Reengineering mit dem ARIS-Toolset. Saarbrücken 1994.
3. Loos, P.; Scheer, A.-W.: Vom Informationsmodell zum Anwendungssystem - Nutzenpotentiale für den effizienten Einsatz von Informationssystemen. In: König, W. (Ed.): Wirtschaftsinformatik '95 - Wettbewerbsfähigkeit, Innovation, Wirtschaftlichkeit. Heidelberg 1995, S. 185-201.
4. Scheer, A.-W.: Architecture of Integrated Information Systems - Foundations of Enterprise Modelling. Berlin et. al. 1992.
5. Scheer, A.-W.: Business Process Engineering - Reference Models for Industrial Enterprises. 2nd ed., Berlin et. al. 1994.
6. Hollingsworth, D.: The Workflow Reference Model. In: Workflow Management Coalition (Ed.): Document TC00-1003, Draft 1.1 3-Jan-95.

Semiconductor Virtual Enterprises: Requirements and Solution Proposals

C. Makatsoris and C.B.Besant

Department of Mechanical Engineering,
Imperial College, London SW7 2BX

Abstract. Semiconductor manufacturing enterprises are global businesses covering multiple sites around the world and consist of a number of shop floors, service providers and suppliers of materials and components. Because of the distribution of activities such enterprises are termed Virtual Enterprises. Due to the dynamic nature of this type of industry constrained by short product life cycles and the fast changing pace of technology, such manufacturing organisations are faced with challenges in an effort to satisfy customer needs but also the enterprise's own goals. It is well known that these are two conflicting targets. Currently state-of-the-art systems are not capable of offering adequate solutions to address such challenges. This has lead to the conception and work towards a solution that aims to fill the gaps that arise due to the current state-of-the-art solutions. The X-CITTIC system is a distributed toolset that aims to address those challenges. X-CITTIC will supply solutions to the challenges of the Virtual Enterprise by providing a fast planner for promising reliable due dates to the customers, by optimising the order flow through the Virtual Enterprise with the creation of robust and optimised order flow plans, by 'repairing' those plans in the occurrence of unplanned events and finally, by providing accurate, relevant and timely information for fast decision making. The X-CITTIC system will be presented in this paper.

1 Introduction

The virtual enterprise can be defined as a subset of the units and processes within the demand chain network, consisting of a matrix of largely co-operating manufacturing, stores and transport units of mixed ownership which behave like a single company through strong co-ordination and co-operation towards mutual goals. Semiconductor manufacturing is a global activity covering multiple manufacturing sites consisting of shop floors, service providers and suppliers of material and components. For that reason semiconductor manufacturing enterprises can be termed Virtual Enterprises.

Due to the fast pace of change of technology, such enterprises are faced with challenges when it come to the planning and control activities. Existing state-of-the-art systems are not capable of addressing those challenges, leaving gaps limiting flexibility for proactive decision making, and reaction to unexpected events. This has lead to the conception and undertaking of the development of a distributed toolset, termed X-CITTIC, which aims to will address those challenges. X-CITTIC will provide planning, scheduling and control solutions to improve manufacturing performance in a microelectronics virtual enterprise, by integrating with the heterogeneous legacy manufacturing systems and by bridging the gap between the higher enterprise wide planning levels and the globally distributed lower shop floor manufacturing execution systems. This work is currently being tackled by a balanced consortium comprising

- the users GEC-Plessey-Semiconductors, Telefunken-Microelectronics who will provide the reference cases for pilot systems, and also ALCATEL-Mietec.
- Nimble, an experienced IT supplier and,
- the academic institutions Imperial College, London, INESC, Porto, FhG-IPA Stuttgart.

The aim of this paper is to describe the problems semiconductor Virtual Enterprises are currently faced with. It also aims to address the requirements of Virtual Enterprises by proposing solutions explored under the European Esprit project entitled X-CITTIC. The paper is organised as follows. In section 2 a brief overview of the semiconductor manufacturing process is given. In section 3 the challenges of the Virtual Enterprise are presented in the context of the semiconductor industry. In section 4 an outline of the X-CITTIC system is given where each major component is described in more detail in the subsequent sections.

2 Semiconductor Manufacturing

A rough overview of the basic product flow in a semiconductor production chain [Frederix 96] is shown in figure 1. Chip production starts in a shop floor, termed wafer fab, with a virgin disk of pure silicon, or a wafer. A wafer, in the course of integrated circuit manufacture undergoes a transformation through a series of time consuming and complex processes. Wafers are then tested to determine which die meet the standards required at a *probe test unit*. Good circuit dies are then extracted from a wafer through special processes and are bonded and wired into its product package at the *assembly unit*. Finally tests on packaged integrated circuits take place at the *test unit*. Some of these units, and even the whole chain, may be located within the same factory. However, there are many such units available in a virtual enterprise, with different factory combiniations, product and manufacturing capabilities, as well as transport services and stores and warehousing.

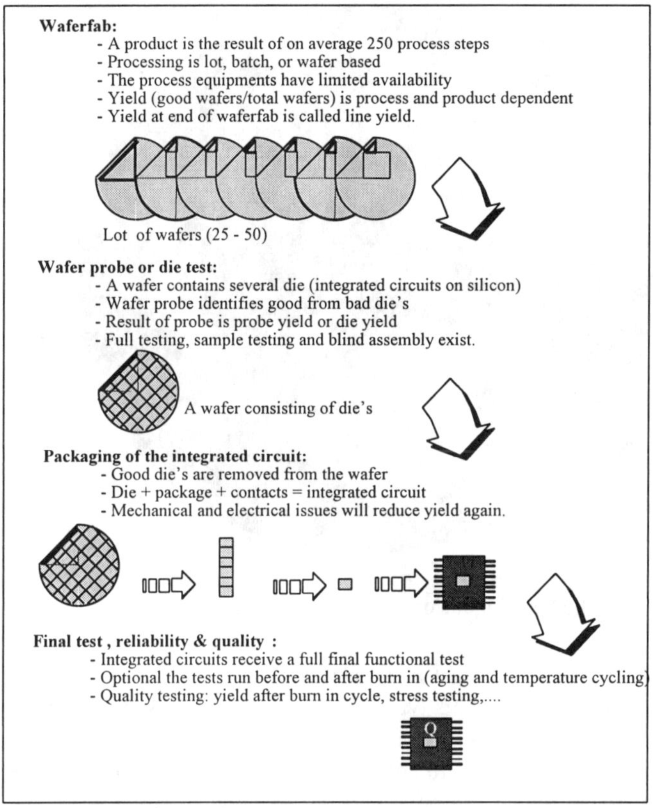

Fig. 1 : Overview of the Semiconductor fabrication process[Frederix, 1996].

A vast number of different technologies are involved in the production of a vast number of products through the number of processes and shop floors involved. For that reason enterprises consist of a number of different facilities for wafer fabrication, assembly and final testing. Such enterprises often cross national boundaries and are closely coupled in partnerships with suppliers, subcontractors and customers (fig. 2). There are many examples of such enterprises which have been formed and grown by acquiring a number of smaller enterprises in the field. Also a semiconductor manufacturing enterprise may be part of a larger corporation. This is certainly the case for all the industrial partners in the X-CITTIC project.

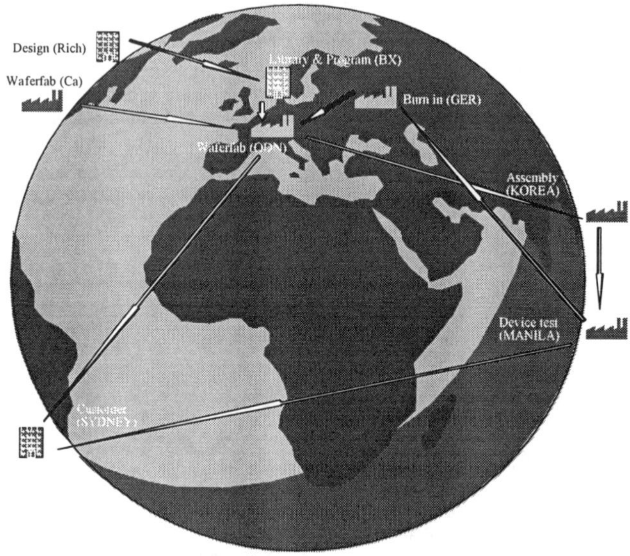

Fig. 2: Example of a Semiconductor Manufacturing Virtual Enterprise [Frederix, 1996].

3 Challenges of the Virtual Enterprise

A distinctive characteristic of this type of industry is the very high initial investment required, which is ever increasing with the introduction of new technologies. For that reason enterprises have resorted to another tactic to minimise the investments in new technologies made fully by the enterprise. This tactic involves shared investments and facilities by two or more competitive enterprises. It is also common to employ outsourcing. Outsourcing may involve any of the stages for integrated circuit manufacturing including design, wafer fabrication, assembly, final test, sales and distribution etc. It can be over 30% of the value added, and it is considered as a cost and not capital.

It is, therefore, obvious that semiconductor manufacturing enterprises comprise of a network crossing national boundaries, and the nodes of that network are the different stages involved in the production. This leads to different routing options within this network depending on technology, product, resource availability and costs. In addition to those 'processing stages', where manufacturing is involved, the other stages in the virtual enterprise network include logistics stages and stocking points. Therefore, there are three main node types which constitute the enterprise network, processing nodes, stocking nodes and logistics nodes and there are different options in how orders can be routed within this network. For example, a customer order may consist of a number of line items to be delivered either in a set, in a so called kit-of-parts, or assembled within the same package, termed multi-chip-module (MCM). The order components may, thus have to be manufactured in different processing stages, depending on the technology involved. In addition, they have to spend some time in a number of stocking points and logistics stages. The customer wants all of the order components to be delivered at the same time. However it is easier to plan each line item separately and consequently order components may be delivered at different times.

This distribution of activities can cause such enterprises to encounter problems in realising global goals [Makatsoris, 1996]. Other problems are caused by the gaps that exist between higher level planning systems and local execution of tasks and the fact that every unit in this network has its own control systems and infrastructures. Furthermore responsive control to cope with the highly dynamic nature of semiconductor manufacturing is virtually non-existent, across the enterprise network. A lack of co-ordination and correct information flowing between and across the levels of the organisation makes global control difficult to realise.

Order entry is one of the processes which is performed in an uncoordinated fashion. Sales departments are trying to satisfy the customers as much as they can and therefore, they prefer to take customer request dates as the date they want the order to be delivered. However, capacities and the current inventory state are not been taken into account at this stage. Thus, these dates are then modified but in most cases they differ from the original customer request date. Hence to meet the request dates of key customers, strategic stocks are held, possibly to the detriment of other customers. On the other hand, inventories have to be kept low to satisfy financial control. Therefore, the planning department is left to satisfy objectives which include satisfying customer needs, business needs and the needs of the enterprise itself. These conflicting objectives lead to compromises which are results of the uncoordinated flow of information within the organisation and thus lead to reduction in customer satisfaction or that of the enterprise.

To improve customer service, enterprises are aiming at reducing their lead times by keeping strategic products in inventory. These products are either determined by forecasts or by business agreements with strategic customers, which take place before the placement of the firm order. Also in the semiconductor industry due to yield variations of final products, a limited amount of excess is produced. This excess can allow shorter lead times. However, such techniques are subject to many inaccuracies, and this leads to increased risks of high inventory costs.

A key challenge for supply chain management is to balance customers needs with the cost of inventory. Inventory must be optimised for both customer service and enterprise economics, not only at the end customer but also at the intermediate ones in the supply chain to attain a fair but balanced distribution of inventory cost between customers and suppliers as well as its overall minimisation. The approach must not be obscured and should ensure that the supply chain is assessed as a whole. Seeing the large picture is of paramount importance. Any system has to be capable of responding to its inherent dynamics and static rules and policies have little validity here. Success may be measured by how well it responds to input in an efficient, co-ordinated and co-operative manner to provide an output which is consistent, coherent and effective in maximising returns to the enterprise.

The shrinking of organisations and pulling back to core competencies inevitably leads to utilising further outsourced services. These are now used for most enterprise functions from design to manufacturing. They are used not without extra cost. Effort is required to monitor quality and supply. Good relationships are required to ensure good co-operation, but at the same time it is realised that the outsourced supplier has other customer enterprises to serve, so that there is a limit to what can be achieved, particularly when something goes wrong in the supply flow, into or out of the supplier. Lowest price alone cannot be the negotiating gambit with the supplier to retain services but a price based on an acceptable level of co-operation to reduce risk in the supply flow. Such co-operative arrangements should have heavy penalties for any defection from an agreed strategy.
Demand swings are another problem, and one which is highly probable in a fluid market place. Enterprises must make investment decisions at the right time to ensure new capacity is available on a demand upswings. Extra outsourced capacity may be difficult to come by when the economic cycle changes and this leads to longer order lead-times. Conversely if extra capacity is secured, through greenfield investments for instance, and demand shows no sign of increasing this extra cost is an economic burden to an enterprise. Enterprise alliances are becoming more common to offset this risk.
Inside the core enterprise management of the supply chain has gained increased importance, elevating the logistics manger to director levels in some organisations. No plan can be accurate as it is about anticipating the future in a highly dynamic environment. But, customer delivery date promises, and logistic supply plans should reflect in a much better way, than most current systems provide, the detail of the information available, and predicted, to provide good control possibilities. Such controls need to have intelligent decision making as the basis and take into account the influence on flow through the supply network. Local autonomies, as manufacturing shop floors and factories, warehouses, and transport providers will have to be constrained to work within specific guidelines to ensure co-ordinated actions throughout the supply network. A decision making process may be automatic in certain instances, in others a rich choice of information and decision possibilities should be presented interactively to the assigned decision maker(s), with an ability to check consequential effects before any real decision is made, and in other instances an early warning may be sent up stream through the supply network to initiate possible control adjustments which may need prior negotiation. The way the decision making processes are formulated and organised for control actions is one of the challenges to improved system design.
The trends are clear. Industrial enterprises have to change in both their internal and external perceptions in order to secure a confident future and competitive edge, but there is no easy solution to supply chain control, trade-offs between conflicting requirements are necessary. The way these trade-offs are quantified will depend upon each enterprise involved and may be characterised by the industry to which it belongs to.

As a result today's customer needs may be summed up as,

- Fast interactive transactions to agree an order delivery date-time

- A short lead time from order placement to delivery to the site specified.
- Delivery on time precision of the full order.
- The product(s) to have the right price and quality
- Flexibility to adopt to specific customer requests.

However, those needs cannot be addressed simultaneously and fully since there are three major requirements that are currently missing from or are inadequate in the state-of-the-art systems:

Planning: Production Plans are not feasible nor optimised with respect to the whole virtual enterprise network due to the lack of adequate *Enterprise Control.*

Enterprise control: Feedback of actual state (e.g. capacity, inventory levels etc.) of the Virtual Enterprise and comparison against requirements is inadequate, if any, and it is based on static calculations. Hence, there is currently a lack of timely and adequate feedback of the relevant information for required for planning.

Plan control: Reactive actions performed on an *ad hoc* basis and without consideration of the effects on the other nodes in the virtual enterprise network. Therefore, there is a lack minimising deviations from the initial plan with the occurrence of unplanned events by suggesting actions that can alleviate problems that occurred by such events with the minimum disruption.

4 X-CITTIC - A Planning and Control System for Virtual Enterprises

X-CITTIC is a distributed system for the operation of Virtual Enterprises geared to the needs of the semiconductor industry. X-CITTIC will address the previously mentioned needs by filling the gaps of existing systems with the provision of a configurable toolset capable of:

- promising reliable due dates to customers while they are being captive when placing the order,
- optimising the Virtual Enterprise network both in terms of customer satisfaction but also in terms of the enterprise's internal objectives,
- reacting to unplanned events by minimising any deviations from the predicted Virtual Enterprise plan,
- providing all the necessary information, presented in the right format for fast decision making.

More specifically, the X-CITTIC capabilities will be offered by the following functionalities:

- Integration of feasibility checked orders in the virtual enterprise in real time.

- Virtual enterprise capability and capacity-checked determination of customer delivery dates.
- Scheduling of an order flow plan for the optimised and selected virtual enterprise routes.
- Supervision of the production progress including monitoring, reporting, alerts and early warnings
- Reactive localised plan repair triggered by out of control signals
- Networking of all virtual enterprise units on an information technology level based on CORBA
- Integration with legacy systems (e.g. there are interfaces to business systems like SAP R/3 and to Manufacturing Execution Systems like WorkStream, Promis etc.)

Therefore, the X-CITTIC toolset will consist of the following components:

- A fast pseudo real-time multiple request handling component for due date fixing, termed *Order Promise (fig. 3)*.
- A fast responsive order request driven rough-cut planner, termed *Rough Planner (fig. 3)* [Dudenhausen, 1997; Leach 1997].
- A Virtual Enterprise order flow planning and scheduling system for optimising the order flow, termed *Fine Planner (fig. 4)*.
- A Distributed virtual enterprise event-driven intelligent *Reactive Controller (fig. 5)*.
- And finally, a comprehensive configurable *Information Manager* [Zhou, 1997].

5 Order Promise and Rough Planning

Customers, through the core enterprise, see the virtual enterprise as a closed unit. In a so-called 'one-face-to-the-customer' relation, a customer can consult an order taker in the core enterprise at any time and global location suitable to them. The *order promise* and *rough planning* components determine whether customer order requests can be met and if not make offers on-line through the order taker for realistic delivery dates for the customer to consider and accept there and then, if not the process will have to be repeated for any subsequent call for the same request.

Order Promise has to be able to deal with customer requests from around the world through the world wide sales network of the core enterprise. However, orders cannot be processed simultaneously as each order will require capacity to be committed. For that reason, as orders are entered, sometimes simultaneously, they are sequenced based on priorities given by marketing. Such priorities are user defined. Once orders have been given priorities, are then processed by the X-CITTIC system.

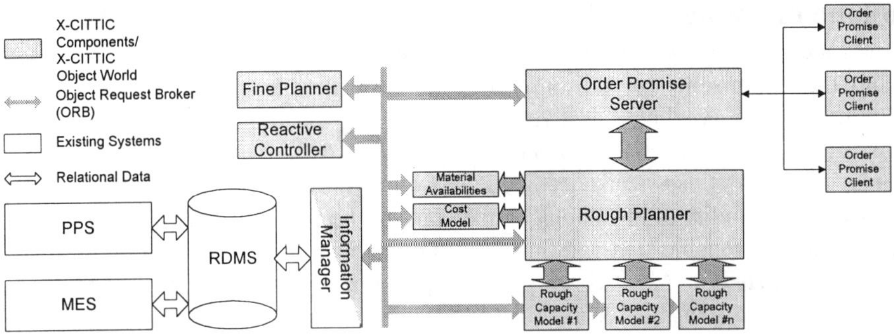

Fig. 3: Order Promise and Rough Planner

For processing the orders entered into the X-CITTIC system, the following two scenarios have to be distinguished. Firstly, order requests, may be wholly or in part satisfied from 'available-to-promise' through the *order promise* component. Secondly, any shortfalls are presented as order line items (products) automatically to the *rough planner* to find an acceptable solution. This is by means of concurrent negotiation with the virtual enterprise units, through their capacity and cost models.

The mechanism the *rough planner* uses is to divide the *order promise* component request into several suborders according to the virtual enterprise network and the capabilities of the units. The suborders offered to the VE unit models, which are set to their current capacity time status, and will be processed by exploring the options. There are a number of possibilities to fulfil an order utilising different intermediate virtual enterprise start and finish times, splitting order quantities over time and/or virtual enterprise unit. With the help of co-ordinated automated negotiations, where the general objectives of the customer as well as the virtual enterprises own interests are taken into account, *rough planning* produces capacity-checked solutions for an order request as well as optimising the virtual enterprise route options. A utility function parameterised for several criteria, is employed by *rough planning* in order to select the best solution.

Rough planning has to take place within a few minutes at the maximum, since customers through the order taker wait for a response while still on-line. At the same time, planning results have to meet the highest standards, so that delivery can be kept and the production cost minimised. Contemporary PPC-systems are not able to solve this dilemma. X-CITTIC *rough planning* is performed on the basis of detailed and capacity-checked fine planning carried out at an earlier time and the order promise component automatically integrates any new accepted customer orders into the existing plan. Thus, *rough planning*, although its main function is to determine reliably the customer order delivery date promise, it is in fact only updating the original plan by subsequent action of the *fine planner*. Now it is possible to promise a dynamic and capacity-checked delivery date within seconds after receiving a customer request.

6 Fine Planning

The *Fine Planner* is a distributed multi-agent based planning application which deals with the optimisation of the order flow through the Virtual Enterprise network. Optimisation is based on objectives aiming to satisfy customer needs. It is also based on objectives which satisfy the enterprise's targets. To optimise the order flow, the *Fine Planner* utilises capacity models which allow accurate determination of capacity availability of the wholly owned units of the core enterprise. This allows the assignment of reliable start and finish times to each wholly owned unit in the Virtual Enterprise network. Also, as a result of capacity analysis, lot release schedules for those units are generated concurrently within the same planning cycle [Teixeira, 1997].

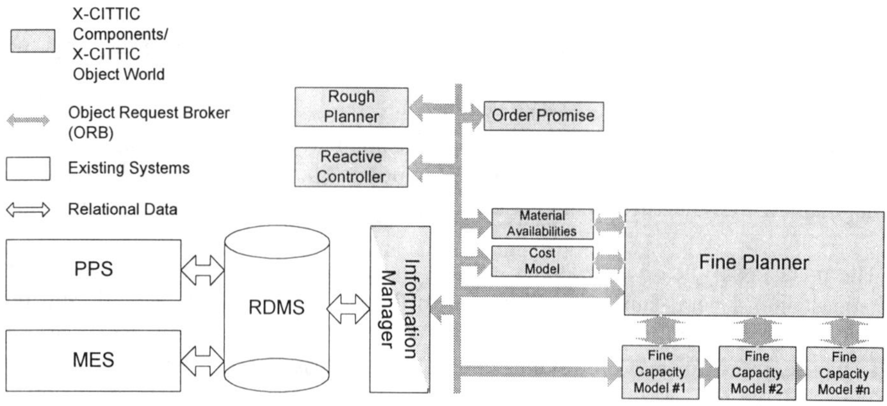

Fig. 4: Fine Planner

It is obvious that a core enterprise cannot impose a detailed fine plan for order flow for units which are service providers to the core enterprise, as these are autonomous units and may belong to other virtual enterprises not involving the core enterprise. However, reliable start and finish times are assigned to such units, based on more aggregate models based on contractual commitments between the core enterprise and the service providers. Such inputs, are binding in the virtual enterprise, and take the form of time-date windows for virtual enterprise unit order move-in and move-out as well as minimum quantity level.

The *fine planner* exercises a user defined planning horizon. The optimisation process aims to maximise revenues and minimise total cost across the network through a number of concurrent calculations with exploration of alternatives. It takes into account all demand classes, including firm and engineering orders, forecasted demand and inventory replenishment orders. Optimisation is subject to capacity and material and tooling availability constraints and includes transportation cost optimisation as well.

Inputs to the *fine planner* come from a comprehensive, Activity Based Costing cost model, for revenue maximisation, the bill-of-materials system, the Material-Requirements-Planning system and also the demand forecasts. The resulting plans are issued to the relevant units in the Virtual Enterprise network and are also used by the *order promise* component, for its 'available-to-promise' calculation.

The plans generated by the *fine planner* need to be robust so that the virtual enterprise can experience little change of input once confirmed. However semiconductor virtual enterprises are very dynamic, and experience regular changes in customer's needs, yield losses, and capacity fluctuations due to unforeseen events. So, re-planning is highly necessary for the control of order flow and must be responsive to out of control events or predictive failure to conform. Such predictions may be flagged by the local Manufacturing Execution System and dealt by the reactive control component.

7 Reactive Control

If order flow is not properly controlled, resulting delays can cause major discrepancies between planned schedules and actual order flow and will be experienced by all virtual enterprise units affected by the delinquent order. As a customer order is very complex in its makeup, comprising line items, virtual enterprise unit orders, and at the lowest level a product lot, the out of control of one lot can prejudice the whole order. It is the task of reactive planning and control to minimise the impact of disruptions on virtual enterprise units not directly affected as well as to bring order flow into control once more. If conflicts arise, fast decisions have to be made, so that the period without a feasible plan to correct the disruption remains short.

It is the task of a monitoring function within the *information manger* to immediately recognise any disruptions down to the lot level, as well as to inform the control component. The controller then reacts by selecting an appropriate strategy. Within X-CITTIC systems, the following four ways of reacting to disruptions are considered:

- **intra-local plan repair**
  Through minor local changes to the virtual enterprise schedule involving one or two virtual enterprise units by,
  a)reprioritising
  b)utilise any available-to-promise local stock
- **negotiation-based re-planning**
  If local changes alone cannot guarantee the meeting of the order in full at the promised delivery dates then by means of negotiation with the units in question the schedule is changed so that maintaining the global virtual enterprise schedule as much as possible by,

a) reallocate orders to other units
b) change schedule date dates but not the order delivery promise date

- **early warnings**
  Any relevant unit will be informed about potential conflicts at the earliest possible moment.
  a) allowing early negotiation to take place for re-planning
- **global re-planning**
  If disruptions cannot be compensated by scheduling changes, a new global *rough planning* that includes the whole VE, becomes necessary and customer immediately informed.

8 Information Management and IT-Architecture

The *information management* component [Zhou, 1997] provides in suitable form the necessary information available to other X-CITTIC components. It also provides the IT infrastructure for the integration of the X-CITTIC components both within the system and also with the outside world, based on state-of-the-art information technology. The information manager offers the following,

- full object-orientation
- distributed application according to the CORBA standard
- migration strategies for existing systems

9 Discussion

In this paper the challenges of the Virtual Enterprise and more specifically the challenges of a semiconductor Virtual Enterprise have been discussed. A system, entitled X-CITTIC, which is currently under development under the European Esprit work programme has also been described.

The X-CITTIC project will generate a versatile, adaptable and flexible toolset which will provide customer focused decision support for the virtual enterprise. This toolset fills the gap between the distributed resource planning systems and local planning and execution systems providing a tactical planning and control tool. Such a toolset will achieve enterprise goals and customer expectations through better capacity models at the enterprise level, and dynamic response to events in real time. The toolset, will also be tuneable to different enterprise policies. Its flexibility and adaptability will allow it to be configured for different virtual enterprises within the discrete manufacturing sector.

Currently, the project is in its prototyping phase but with the majority of the main components and algorithms being benchmarked. However, there is still development work to be carried out for the full on-line prototype. Finally, the important features of the system will be demonstrated in reference cases at GEC Plessey Semiconductors and Telefunken Microelectronics.

10 References and Bibliography

1. Barnett, W., Presley, A., Johnson, M. and Liles, D.H., 1994, *An architecture for the virtual enterprise*, Proceedings of the IEEE International Conference on Systems, Man and Cybernetics, Vol. 1, (1994)
2. Chen, H., Harrison, M., Mandelbaum, A. Van Ackere, A. and Wein, L.M., *Empirical evaluation of a queuing network model for semiconductor wafer fabrication*, Operations Research, Vol.36 No.2, (1988)
3. DuBois, F. L. and Oliff, M.D., *Aggregate Production Planning in practice*, Production and Inventory Management Journal, Third Quarter, (1991)
4. Dudenhausen H-M, Halmosi H., Loffler B., *Real-Time Order Planning in Semiconductor Virtual Enterprises*, submitted to IFAC/IFIP Conference on Management and Control of Production and Logistics, Campinas, Brazil, (1997).
5. Frederix F., *Planning and Scheduling multi-site Semiconductor Production Chains*, Esprit Integration in Manufacturing Conference, Galway, Ireland, 2-4 October 1996.
6. Leach N. P., Makatsoris C., Richards H. D.,*Supply Chain Control: Trade-Offs and System Requirements*, Quality, Logistics and Technology Management, ESPRIT-COPERNICUS Symposium, Budapest, Hungary, (1996).
7. Leach N., Ristic M., *Capacity Models for Scheduling the Virtual Enterprise*, submitted to IFAC/IFIP Conference on Management and Control of Production and Logistics, Campinas, Brazil, (1997).
8. Leachman, R.C., Benson, R.F., Liu, C. and Raar, D.J., *IMPReSS: an automated production planning and delivery quotation system at Harris Corporation-semiconductor sector*, Interfaces, Vol.26, Jan/Feb 1996
9. Makatsoris, H., Leach, N.P., Richards, H.D., Besant, C.B. and Ristic, M., *Addressing the planning and control gaps in semiconductor virtual enterprises*, Esprit Integration in Manufacturing Conference, Galway, Ireland, 2-4 October 1996.
10. Teixeira E. M. A., Makatsoris C., Besant C.B., *Distributed Capacity Analysis for Proactive Planning in Semiconductor Virtual Enterprises*, submitted to IFAC/IFIP Conference on Management and Control of Production and Logistics, Campinas, Brazil, 1997.
11. Wortman, J.C., Euwe, M.J., Taal, M., and Weiss, V.C.S., *A review of capacity planning techniques within standard software packages.*, Production Planning and Control, Vol.7 No.2, (1996).
12. Zhou Q., Besant C. B., *An information Management Architecture for Multi-Site Production Planning and Control*, submitted to IFAC/IFIP Conference on Management and Control of Production and Logistics, Campinas, Brazil, 1997.

The Latest Research Trends in CIM

Grier C I Lin

Centre for Advanced Manufacturing Research
School of Engineering, University of South Australia, The Levels, SA 5095
Australia

Abstract. Manufacturing enterprises play an important role in improving the economic environment of a country. The economic environment of many developed countries is under threat from emerging industrialised countries. Today the capability to produce high quality products with shorter delivery times and the ability to produce according to the diverse customer requirements has become the characteristics to be order-qualifiers for manufacturing industries. Application of intelligent manufacturing systems and CIM is the panacea for overcoming the above issues while retaining the employment level and revenue of a country in today's high competitive global market. With the developments taking place in CIM and its related technologies, the application of CIM in manufacturing enterprises has become a reality from the dream.

1. Introduction

The characteristics of the present world market includes higher competition, shorter product life cycle, and smaller batch sizes to satisfy variety of customer profiles. To fulfil these requirements, manufacturing companies need to be flexible, adaptable, responsive to changes, reactive and able to produce variety of products in a short time at a lower cost. In addition, they should be able to address new environmental requirements and concerns. Due to these needs, manufacturing companies are compelled to seek advanced technologies as a panacea for all of these issues. The most significant outcome of this search resulted in the concept of Computer Integrated Manufacturing (CIM) in the early 70s. The concept of CIM was initially coined by Dr. Joseph Harrington in 1973 in a book published by the name "Computer Integrated Manufacturing" [1]. However, until the early 1980s, CIM didn't become a commonly known acronym as it exists today.

CIM promulgates a fundamental strategy of integrating manufacturing facilities and systems in an enterprise through computers and its peripherals. The definition of CIM has various connotations depending on the application of it. CIM is a process of using computers and communication networks to transform islands of enabling technologies into a highly interconnected manufacturing system. CIM involves integration of advanced technologies in various functional units of an enterprise in an effective manner to achieve the corporate objective of the manufacturing enterprise. Effective integration requires an in-depth understanding of all the technologies and comprehensive knowledge of all activities in all functional units of a company. To be successful in the application of CIM, first the integration of advanced manufacturing technologies should be achieved, the computers act only as subordinates to the technologies. However, without computers the integration cannot be effective and efficient since computers help to organise, retrieve and store the information in an orderly manner with high accuracy and speed.

The proposal of CIM in the early 70s might have appeared to be too futuristic, but in today's technologies it is possible to reach the goal of CIM, without much difficulty. In 1976 Lin [2] has stated that the future of engineering is inextricably bound with the application of computer integrated technology. He has suggested implementing integrated advanced technologies is an effective approach towards solving the problems of decreased productivity relative to labour costs and consequent rise in unit costs which are continually plaguing present day manufacturing managers.

CIM is a management and manufacturing strategy. While the concept of CIM is broadening its application in manufacturing industries, it is interesting to note that some innovative manufacturing and management strategies are also beginning to surface. Concurrent Engineering (CE), Lean Manufacturing (LM), Responsive Manufacturing and Agile Manufacturing are some of the new terms evolving today to reflect the dynamic nature of improvements in manufacturing applications. These concepts are proposed by various academia. The induction of new terms may have resulted due to the result of CIM being an intricate concept to be understandable by manufacturing managers and to be implemented. Nevertheless, the concept of CIM is far broader than these new terms and it is still able to embrace all the new features of similar strategies. The CIM stands as the innovative application for yesterday's proposals and newer applications required today and for the future.

2. Research Trends in CIM related Strategies

In today's competitive market which requires short product delivery times with high quality and mass customisation, research should be further strengthened towards application of CIM in manufacturing industries. The realisation of CIM requires effective integration of a number of available advanced technologies. The technologies could be: office automation, computer aided design (CAD), computer aided manufacturing (CAM), computer aided process planning, computer aided quality control, automated storage and retrieval systems, robotics, group technology, cellular manufacturing, flexible manufacturing systems, computer aided engineering, automated guided vehicle systems (AGVS), automated assembly systems, automated material handling, automatic inspection, statistical process control, supplier qualification standards, automatic order processing, integrated production scheduling, automated shop floor data collection, computer numerically controlled machines etc.

To revitalise the manufacturing industries, many variations in manufacturing methodologies were developed and proposed by researchers. The main focus for revitalising the industry and the paramount importance of reducing the product delivery time can be understood by the Figure 1. It depicts the trend of increasing product development time due to complexity of a product and the shortening of the product life time, due to change in life styles.

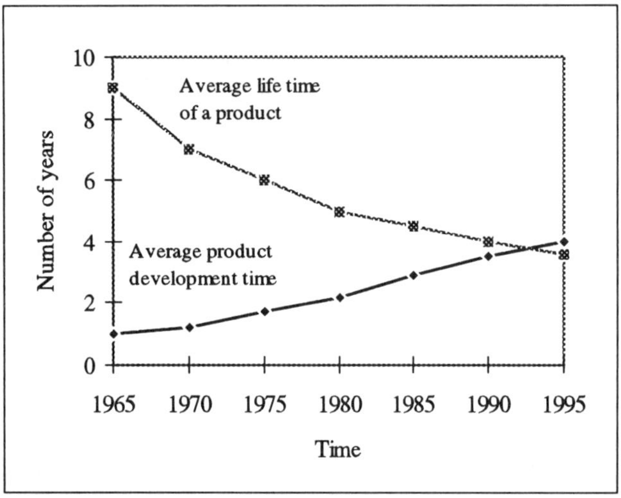

Figure 1. Product development time against life time of product (adopted from Syan [2])

The concept of Concurrent Engineering (CE) has been around the manufacturing circles from early 1960s in various forms requesting the use of multi-disciplinary teams to accelerate product introduction. However, the concept was given the name Concurrent Engineering with an appropriate definition by US Defence Advanced Research Projects Agency in 1987. CE involves, a systematic and simultaneous approach to the integrated design of products and their related processes including marketing, manufacturing, sales and purchasing [3]. Further, it involves formation of multi-disciplinary teams for the rapid product development and introduction of the product into the market. CE could be considered as a management strategy rather than the manufacturing strategy.

In late 1950s, Japanese automobile manufacturers realised that mass production did not fit into the production and management strategy required to provide the product differentiation which was evolving as a fundamental market feature [4]. This realisation resulted in the formation of the concept, Lean Manufacturing (LM). LM involves addressing the product strategy, product development, supply chain, manufacturing and product distributions for the production of diverse products in small batches. It may further consist of out-sourcing and effective integration between the suppliers and subsidiaries. From the early 1980s, the impact of LM on mass producers of North American automobile manufacturers had started to unfold and the American companies realised that this concept of LM can be utilised in their work force as well.

The concept of Agile Manufacturing is similar to the concept of LM by emphasising on small batch sizes. However, agility requires: reduction in product development time; allowance for considerable customisation of product features; and incorporating highly adaptive, flexible and efficient manufacturing practices in the product development and manufacturing cycle [5]. It further involves effective integration of technology, business activities, enterprise and people.

The Business Process Re-engineering (BPR) involves identifying each and every business activity, evaluate the importance and relevance of the activities towards achieving the business goal and redesigning all the activities in an efficient and economical manner to achieve the business objectives. BPR achieves these objectives through business modelling and analysing techniques. The use of BPR eliminates unproductive and unnecessary business activities and operations in an enterprise and actuate process simplification and if necessary out-sourcing.

As we could see from these explanations for the new manufacturing strategies and terms, which are evolving simultaneous to CIM, CIM could still provide all the features of lean, agile and concurrent manufacturing in an integrated approach. The potential for CIM in manufacturing industries is considerably large. However, there is often a mismatch between the hypothesis of CIM and the application in a real world. Today, research in CIM is carried out with increased vigour to meet the requirements of industry with the latest facilities and tools available in the market to shorten the mismatch. Further, the trend in CIM research is moving towards finding new ways and tools to make the futuristic factory into a realistic factory.

3. Latest Research Trends in CIM

Today, we are at the threshold of seeing rapid changes in information and communication processing equipment. While the mass production is still needed for certain sectors of manufacturing, it is no longer a competitive weapon in a rapidly changing global economy. The challenge in reviving the manufacturing industry to face the competitive threat depends on the ability to respond to all the requirements of the global market. However, many of the manufacturing industries in the developed world are still operating islands of automation or age old technologies. This trend of clinging to the past is not going to lead the manufacturing industries to face the growing competition in the global market. We can see today the companies are relocating offshore to reduce labour costs. Nevertheless, this phenomenon can be avoided if manufacturing companies invest in CIM related technologies and become globally competitive. The efficient application of CIM in a fully integrated manner will help the industries to face the global and local competition with high confidence.

With the advent of new network technology and internet, the application of CIM has transcended geographical boundaries and embraced the global application of CIM [6]. To facilitate the investment in CIM and to help manufacturing industries, many researchers are seeking solutions in application of CIM. Some of the research trends in CIM which are evolving today can be categorised as follows:

- Justification of CIM and management strategies for CIM
- Enterprise integration for CIM beyond and within geographical boundaries
- Network communications for the implementation of CIM
- Advanced tools and technologies for the application of CIM
- Manufacturing system modelling
- Application of artificial intelligence (AI) such as fuzzy logic, neural network, genetic algorithms and intelligent agents for fully integrated intelligent manufacturing systems.

Each of this categorisation are diverse in nature and could consist of many sub groupings. In addition, few research directions may overlap in more than one grouping due to the complexity of the research in CIM.

Justification and management strategies for CIM involves issues including: providing managers with principles and guidelines for the application of CIM in their working environment; benchmarking the features of the manufacturing supply chain; overcoming the resistance to change towards integrated manufacturing; manufacturing strategy implementation through real-time systems; organisational and social issues in an enterprise; optimised enterprise integration and decision support systems and expert systems for manufacturing managers. Decision support systems and expert systems may further include expert systems for quality inspection, production planning and selecting manufacturing parameters; decision support systems to evaluate designing decisions, financial decisions; and educational and informative systems.

Enterprise integration for CIM beyond and within geographical boundaries consist of research diversity including: architectures and modelling formalism for enterprise integration; evaluation methodologies for enterprise integration; international CAD/ CAM collaboration for CE and CIM implementation through sub-systems integration. While, network communications for the implementation of CIM consist of: application of wide-area networks and internet for CIM; information enhancement by data integration; issues related to integration of client and server for manufacturing shop-floor automation; application of multimedia and hypermedia in CIM environment; data management in CIM systems etc.

The categorisation of advanced tools and technologies for the application of CIM is one of the thriving areas in CIM related research. It consist of robotics and automation for a CIM framework, vision based manufacturing systems and intelligent AGVS; intelligent and knowledge-based enabling systems for CIM etc.

Manufacturing system modelling consist of research variations including: integration of information models with functional models for CIM; object oriented resource modelling; integrated simulation modelling approaches for CIM and modelling approaches and methodologies for designing CIM systems.

Application of AI for fully integrated intelligent manufacturing systems consist of a research direction that includes the application of neural network on manufacturing automation and technologies, intelligent scheduling systems with genetic algorithms; fuzzy scheduling systems, hybrid systems of CIM components, adaptive models for CIM architectures and embedding AI in CIM software.

The Centre for Advanced Manufacturing Research (CAMR) at University of South Australia is proud to be a part of a international research team, carrying out research on CIM related technologies to help industries particularly the manufacturing industries in Australasia and to cater for the present and future requirements. Some of the leading research undertaken by CAMR are:

- CAD and Sensor Based Intelligent Robot Server
- International CAD/CAM Collaboration
- Intelligent Design for Automatic Assembly
- Intelligent Integrated Reverse Engineering
- Automatic Inspection System
- Intelligent and Flexible Inspection System
- Flexible Fixturing
- Decision Support System for CIM Investment Analysis
- Integration of Reverse Engineering and Two Robots for Rapid Prototype Machining
- Flexible Robot Measuring System

The complete list of research activities carried out by CAMR and the relevant list of publications can be obtained from http://www.camr.unisa.edu.au/.

In CAD and sensor based intelligent robot server research, an intelligent robot server has been developed to provide a solution to overcome the issues such as interruption in the entire workcell when a machine is malfunctioning or when a machine is introduced into the workcell or moved out of it for its installation or when accurate calibration procedures, and programming are required to be carried out due to change in the working environment in relation to product changes and modifications. The change over time, accurate calibration, and re-setup ultimately lead to low machine utilisation. The above research minimises all these avoidable activities and improves the productivity of manufacturing enterprise.

International CAD/ CAM collaboration involves the development of a system for international concurrent product design and manufacturing. The system can be utilised to co-edit CAD/ CAM data by geographically separated CAD/ CAM designers for co-editing tasks through their CAD/ CAM applications. The future directions of this project is to extend the international concurrent product design research to accommodate for faster and more reliable communication mediums for audio, video and three dimensional graphic transmission. The result will be the development of a remote manufacturing cell for fully operational CIM.

Intelligent Design for Automatic Assembly involves the development of a design evaluation system on the basis of design for assembly principle. The system evaluates a design similar to a human evaluation in an intelligent way. The research project consist of four modules. They are modules for design simplification, design for automatic assembly evaluation, redesign approach and assembly planning.

Research in intelligent integrated reverse engineering involves the development of an intelligent approach which reduces the time taken to digitise objects, improves the digitisation accuracy and generates free-form surface models.

Automatic inspection system research project presents an automated inspection system using computer vision technology. The inspection task is mainly concerned with verifying the features of an object and two-dimensional measurement. Emphasis is placed on image segmentation and inspection speed.

Research in intelligent and flexible inspection system involves integration of a coordinate measuring machine with a machine vision system and a computer aided design system, to achieve dimensional inspection of parts that are randomly located on a worktable. The vision system in conjunction with the CAD system will recognise the position and orientation of parts and generate the measuring path automatically.

Research in the flexible fixturing involves the development of an automated flexible fixturing system. The system consists of a single device to hold parts or assemblies of different shapes and sizes while they are being subjected to the wide variety of external force-fields and torques associated with conventional manufacturing operations. By employing such a device lead time, overhead cost associated with storing a multiplicity of fixtures required for manufacturing procedures and effort required to design special-purpose fixtures are reduced.

The research in decision support system for CIM investment analysis, provides a simplified and integrated decision support system which comprehensively analyses most of the relevant issues related to investment evaluation. Further, it provides approaches to quantify intangible benefits, incorporation of synergy, impact evaluation in functional units and multi-criteria optimisation.

The research in integration of reverse engineering and two robots for rapid prototype machining investigates the possibility of integrating reverse engineering technologies with two coordinated robot arms for rapid prototype machining. The research project of flexible robot measuring system involves developing a flexible robot measuring system for large and complicated components that can not be handled by traditional measuring methods or coordinate measuring machines.

Today, many manufacturing industries are under pressure to achieve the goal of the objectives within limited capital available for new investments. Optimisation of resource allocation for strategic investments could provide a mechanism to achieve the goal. As Hess [7] has pointed out, the future of CIM depends on optimised integrated systems. Traditionally, to improve the effectiveness and competitiveness of an organisation, all the efforts are focused on achieving a single goal or achieving the objective within a functional unit. This type of single goal seeking mechanism or optimising the results of an individual unit, do not satisfy the overall objective of a manufacturing company. Therefore multiple goal seeking or multi-criteria optimisation which considers all the relevant factors and overall benefit of a company as a single unit, should be utilised to make CIM as the effective application in the present global economic circumstances.

4. Conclusion

CIM is an innovative and expansive concept to provide the solutions manufacturing industries are seeking to survive in the competitive global market. Today, we are at the threshold of utilising innovative and improved computer related technologies for the betterment of manufacturing industries, compared to the situation prevailed a decade ago. The successful future of manufacturing industry is inextricably involved in the efficient and effective utilisation of CIM and its components. The research in CIM and related technologies and the application of it in manufacturing industries are progressing towards a better future. The research should be further strengthened towards developing optimised CIM systems to control the scare resources we are having today to meet the competitive and agility requirements of today's world.

5. References

1. J. Harrington, Computer integrated manufacturing, New York, Industrial Press, (1973).
2. G. C. I. Lin, Integrated Compute- Aided Design, Manufacture and Management, Proce. Annual Engineering Conf., Townsville, Australia, (1976), pp 68-70.
3. C. Syan, and U. Menon, Concurrent Engineering: concepts, implementation and practice, Chapman & Hall, London, (1994).
4. R. Lamming, Beyond Partnership: Strategies for innovation and Lean supply, Prentice Hall, Uk, (1993).
5. C. Montgomery, and L. O. Levine, The Transition to Agile Manufacturing, ASQC Quality Press, Milwaukee, (1996).
6. G. R. Liang, Advent of network-integrated manufacturing age in Taiwan, Proce. of 4 th Int Conf. on Automation Technology, (1996), pp 29-36
7. J. G. Hess, Beyond CIM - Optimized Enterprise Integration, proceedings of Autofact '92 Conf., USA, September, (1992), pp 2-9 to 2-24.

Mass Customization- Opportunities and Challenges for High Value Added Products and Services

Mitchell M. Tseng

Department of Industrial Engineering & Engineering Management
The Hong Kong University of Science and Technology

Abstract. Mass customization, providing products and services that best serve customer needs with mass production efficiency, has become an important topic for manufacturing industries and service providers. In this paper, the opportunities and challenges of mass customization are discussed. Also discussed are approaches to implementing mass customization based on building block identification, product platform formulation, and product life-cycle integration with product family structures.

1 Introduction

The aim of product development is to create a product that satisfies customer needs, within the constraints of the customer's budgets and schedule as well as within the boundaries of a firm's existing or soon to be acquired capabilities. For years, the process of matching needs to capabilities was carried out via a set of discrete product offerings. One of the direct consequences of this approach was an explosive growth of product variety. For instance, the number of store keeping units (SKU) in the average supermarket stores doubled to more than 30,000 between 1980 and 1940. Meanwhile, because of shortened product lead times, more and more products were being created each year and all these products were fighting for shelf space. However, the hidden cost of complexity, from Sales to Logistics, have been largely ignored and the traditional process of product development remained geared toward the "mass production" style of manufacturing. In essence, there was an underlying assumption that the features and quantity of a product which customers would buy could all be forecasted with an adequate amount of lead-time. Thus, product designers could contrive and package certain features with confidence and the manufacturing department could organize to produce new products. The success or failure of a product often was attributed to the quality of forecast, which often was "educated guessing". In reality, deviations from forecasts resulted in the occurrence of simultaneous inventory excesses and shortages, in addition to wasteful uses of resources and ultimate end-user dissatisfaction

In this paper, a different approach in designing the life-cycle of products through, mass customization, will be discussed. The objective of this approach is to proactively connect customer needs to the capabilities of a company in the design process. Some major technical challenges in realizing this approach are then presented along with a discussion of some of the progress already made in the industry and academia.

Essentially, the attempt is to include customers in the product life-cycle, particularly in the design phase. With the information technology we have today, customers can convey their requirements and make trade-offs among to their needs, constraints, and the available capabilities of the manufacturer or service provider.

2 Mass customization

Mass customization is a new paradigm for industries to provide products and services which best serve the customers' needs while maintaining mass production efficiency. The basis of this new approach is a recognition of each customer as an individual and the subsequent production of products with "tailor-made" features under constraints such as cost and schedule [Toffler, 1971]. Mass customization begins by seeking out the customers' needs and constraints and then produces products that best fit each customer's needs. The inherent repetition in product marketing, design, and manufacturing can be recognized through the establishment of patterns. Once patterns are identified and utilized, scale of economy can then be applied for efficiency. In the meantime, customers are also willing to pay more for products which better satisfy their special needs. Figure 1 illustrates how mass production has an advantage in high volume production where the actual volume can defray the cost of huge investments in equipment, tooling, engineering, and training. However, satisfying each individual customer's needs often can be translated into higher value, and where lower production volume cannot justify the large investment. Because mass customization allows companies to garner scale of economy through repetition, it is capable of reducing costs and lead time. Hence, mass customization achieves a higher margin and is more advantageous. With the increasing flexibility built into modern manufacturing systems and programmability in computing and communication technology, companies with low to middle production volumes can gain an edge over competitors by implementing mass customization [Tseng and Jiao, 1996].

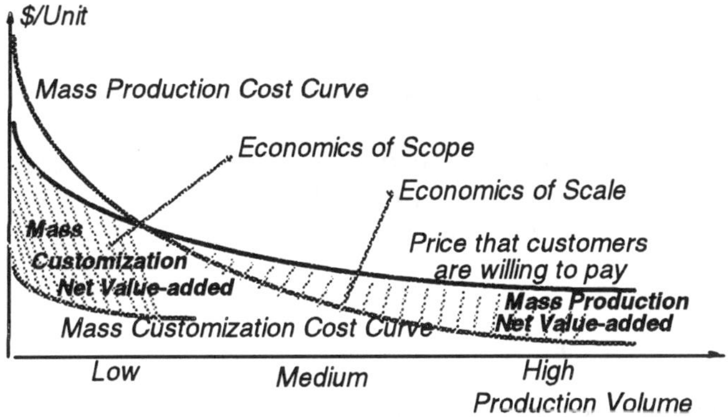

Figure 1: Cost and volume relationships for different styles of production

Besides the tremendous benefits for business, mass customization also raises many issues for theory development. It is also an attempt to summarize many recent trends towards manufacturing flexibility. However, although the notion of pursuing a customization strategy has great appeal, the current literature on state-of-the-art mass customization has only highlighted the benefits while neglecting the means. As a result, mass customization has drawn criticisms that it has the look and feel of a revolutionary paradigm but lacks a coherent framework and thus remains a repackaging of pre-existing ideas with only limited synthesis [Kotha, 1995]. Moreover, competitive strategies in the 90's are many and varied, including manufacturing flexibility, time-based competition, lean production, re-engineering, and continuous improvement. Infrastructure issues are critical to the successful implementation of mass customization so that it can be compatible with other strategies [Lau, 1995]. Considering the aforementioned criticisms and the desire of businesses to become customer driven, different types of mass customization have been identified [Pine, 1997]. In this paper, mass customization is approached from the engineering perspective, namely through form and function.

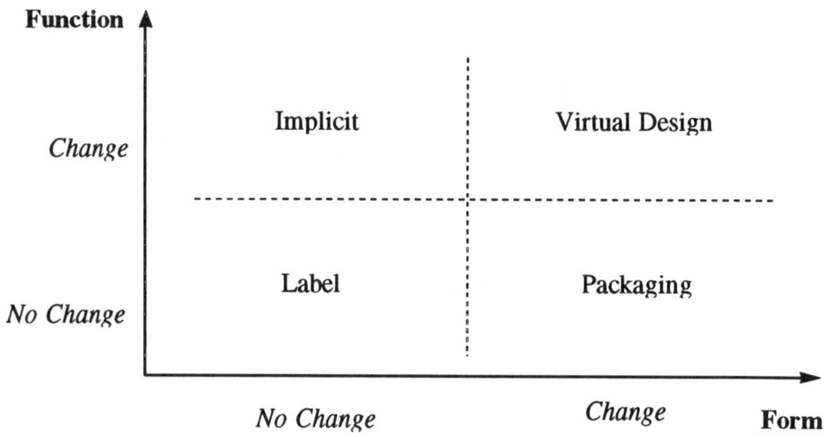

Figure 2: Four Modes of Mass Customization

Based on the effect of individual customers' requirements on form and function, four different modes of customization can be identified, specifically, labeling, implicit, packaging, and virtual design, as shown in Figure 2. The four modes are explained as follows.

2.1 Labeling

Labeling provides superficial changes in outlook that appeal to individual customers. Basically, a standard product is offered. However, depending on the customers, an amount of possible "personal touches", though they are mostly cosmetic, can be added. These additions could be instructions in different languages, form letters to customers, or cosmetic changes in color or decoration. These changes do not change the product form or function, they merely appeal to the customer's liking.

2.2 Implicit

Inherent customization has built-in functionality that can satisfy a wide spectrum of customers without adding to product variety. A common example is the universal power supply which was designed to implicitly satisfy different voltage requirements without the customer having to specify the variety. Implicit customization anticipates the advent of customer needs and attempts to satisfy them with a limited number of flexible products that are adaptive to their need, without changes in the product's outer appearance. Customers are thus satisfied with varying function though they sometimes are not aware of the differences in the product's functionality.

2.3 Packaging Customization

By adjusting the size, volume, or product appearance, the customers' uniqueness can be addressed. OEM manufacturers often cater to their customers' needs and distinguish their products by including different enclosures with their products. Another common practice is to change the box or packaging materials of a product according to the needs of customers. Here, the function remains the same, but the form has been changed. It can be an illusion created so that the customer will mistake a standard product as a custom product.

2.4 Virtual Design

Virtual Design further puts the customer in the center of the design life-cycle. By presenting the capabilities of the firm, particularly different options for product features, customers can then balance their needs with constraints such as schedule, cost, and product features. Thus, the customers are able to express their own implicit or explicit needs in the form of functional requirements or even design parameters [Suh, 1990. The key in this mode of customization is not only well-informed customers but also the ability to present well organized choices so that the customers can made the necessary trade-offs and then specify their needs. The availability of high performance computing and communication technology further enables the customers to articulate their needs by allowing the customers to "immerse" themselves in the environment of a product. Customers will then be in a position to configure the product or even "test" the product to ensure their needs are met. Furthermore, because of communication networks, specific sets of functional requirements can be connected and reconfigured at different geographical locations. Virtual Design remains the most challenging mode of mass customization. Because of its scope of changes in both form and function, it requires a significant amount of energy and effort.

3 Technical Challenges

The essence of mass customization lies in the product and service developers being able to perceive and capture latent market niches and subsequently develop its technical capabilities to meet the diverse needs of their target customers. Successful mass customization from the customers' perspective depends on the balance of three elements: features, cost, and delivery, as shown in Figure 3.

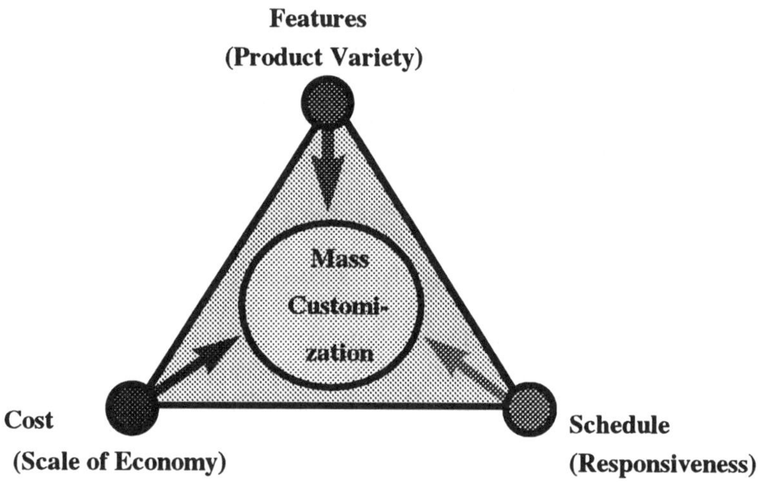

Figure 3: Balancing customers' needs in three different dimensions-
features, cost, and schedule

Thus, the main technical challenge in implementing mass customization is to develop the necessary infrastructure as to facilitate the choice of the best design alternative, i.e. the alternative that simultaneously satisfies these requirements and the customers' constraints [Tseng and Jiao, 1996]. In order to accomplish this balance, three major technical challenges have been identified:

3.1 Maximizing Commonality

Maximal amounts of repetition are essential to achieving the efficiency of mass production, as well as in sales, marketing, and logistics. This can be attained through commonality in design which leads to reusable tools, equipment, and expertise. Intuitively, this principle directly conflicts with the uniqueness of products which promise to satisfy the diverse needs of different customers. The continuous development of customized features results in a continuous accretion of varieties which in turn necessitates constant process changeovers and inventory and logistic complexity. Thus, there is a seeming contradiction towards the main characteristics of mass production, low-cost and high efficiency. Manufacturers are thus faced with the challenge of ensuring a "dynamic stability" [Boynton and Bart, 1991] in which they can serve the widest range of customers and their changing product demands while building on existing process capabilities, experience, and knowledge.

This dynamic stability comes from the proper selection of building blocks of maximal commonality. These building blocks are selected based on (a) current and future customer needs, (b) repeatability in design and fulfillment, (c) ease of configuration, and (d) an appropriate level of aggregation. For instance, if the level of the building blocks is too low, such as at the nuts and bolts level, then there will be too many blocks and configuration of the blocks will become difficult. On the other hand, if the level is too high, such as at the level of complete modules or products, then the blocks will not be reusable or have enough commonality to other products. However, if these blocks are defined and selected appropriately, then due to the similarity over product lines or among a group of customized products, reusability will naturally facilitate increasingly efficient and cost effective product delivery. By optimizing commonality among internal modules, tools, knowledge, processes, and components, low cost efficiency, product portfolio integrity, and continuity of infrastructure can all be achieved. This is particularly true in savings resulting from leveraging downstream investments in the product life-cycles, such as existing design capabilities and manufacturing facilities.

Although commonality and modularity have been important design practices, an emphasis on commonality is usually employed for the purposes of physical design or manufacturing convenience. To achieve mass customization, commonality needs to be approached from the perspective of the customers' needs or functional requirements. By grouping customers' needs by their commonality, a set of designs can be created for the establishment of a series of product families, thus facilitating the mapping between customer needs and the capabilities of the company. Techniques such as inductive learning, clustering, and Group Technology have been applied in the past to maximize commonality [Tseng and Jiao, 1996].

3.2 Product Platform

A product platform provides the logic for connecting modules and contains three major elements:

a) the network structure that describes how different modules are connected
b) the configuration rules that describes the compatibility of modules
c) the inter-relationships between the customers' needs and product families

The underlying architecture of a product platform is in turn called the Product Family Architecture (PFA), from which product variants can be derived to satisfy a spectrum of customer needs related to various market niches. Essentially, the PFA captures and utilizes the commonality of a product family and serves as a repertoire of the knowledge bases of different products. It also prevents variant product proliferation for the same set of customer requirements.

The formulation of product platforms is shown in Figure 4. It integrates building blocks with inputs based on design concepts, process capabilities, skills, technology trends, and competitive direction. It is important that the integrity of a product's topological structure is maintained in order to avoid misinterpretation among customers, designers, and production. For this purpose, tree-structure representation is commonly used for presenting the relationships among building blocks. Other structures, such as bus types and sectional types, are also commonly used in industry.

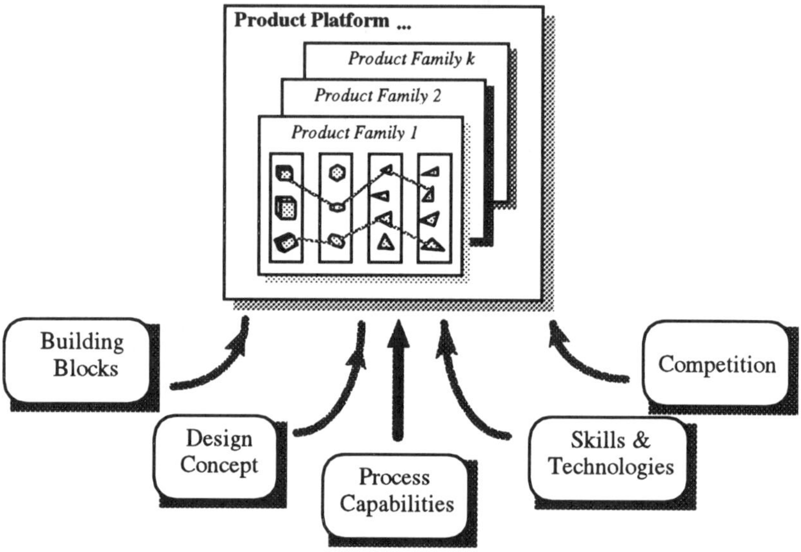

Figure 4: Product Platform Formulation

3.3 Integrated Product Life-cycle

Product development cycles provide the mechanism for bringing the product building blocks and product platform into a coherent process, integrating customers, sales, engineering, manufacturing, and service. To this end, the scope of the design process is extended to include sales and service and is shifted from point product design to product family design, as depicted in Figure 5.

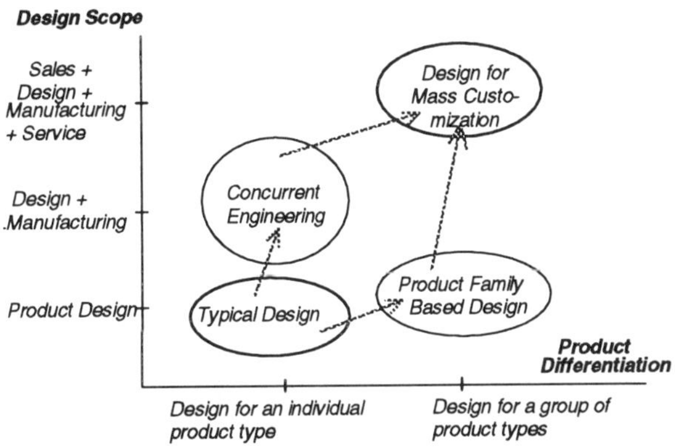

Figure 5: Design Scope Differentiation and Product

Furthermore, this shift and extension has implications for different functions across traditional product design. For example:

- Sales should assist the customers to select the product platform, configure product features, ensure compatibility, and make design trade-offs, in addition to the conventional tasks of acquiring customers and supporting customer relations.
- Marketing should not only focus on the positioning of product families that best represent the strengths of the company but should also proactively consider technology and business trends with competition in mind. In addition, communicating the strengths of the product families and connecting them with customers need to remain top priorities.
- Engineering should focus on product family development, including the structures within families and differentiation among families of products.
- Besides the fundamental operation responsibilities regarding product quality and controlling cost, Manufacturing should gear its operations to take advantage of repetition in order to re-use capabilities and build operational strengths through continuous improvement and learning. Furthermore, flexibility in resources should be emphasized so that responsiveness to customer needs can be enhanced.

4 Conclusion

Faced with growing variety and a demand for rapid response, manufacturing industries and service providers strive to balance customer satisfaction and cost containment. Mass customization recognizes each customer as an individual, meanwhile extracting maximal commonality to achieve scale of economy. Since customers are generally willing to pay more if their needs are better satisfied, the value added in product and service operation can then be enhanced.

However, mass customization must be carefully planned. A framework is developed for identifying different types of customization opportunities. A systematic approach to maximize commonality, develop coherent product family architectures and integrate product life-cycles is proposed. Preliminary results for the implementation of mass customization in local companies have shown some promising benefits.

References

1. Andrew C. Boynton and Victor Bart, Beyond flexibility, building and managing the dynamically stable organization, *California Management Review* (1991).
2. Suresh Kotha, Mass customization: Implementing the emerging paradigm for competitive advantage, *Strategic Management Journal* **16** (1995) 21-42.
3. Ronald S. Lau, Mass customization: the next industrial revolution, *Industrial Management* **37** (1995) 18-19.
4. B. J. Pine II, Mass Customization: *The New Frontier in Business Competition*, Harvard Business School Press (1993).
5. N. P. Suh, *The Principles of Design*, Oxford University Press (1990).
6. Alvin Toffler, *Future Shock*, Bantam Books (1971).
7. Mitchell M. Tseng and Jianxin Jiao, Design for mass customization, *CIRP Annals* **45** (1996) 153-156.
8. Mitchell M. Tseng and Jianxin Jiao, A Variant Approach to Product Definition by Recognizing Functional Requirement Patterns, *Journal of Engineering Design* **8** (1997).

MANAGEMENT ASPECTS

Global Perspectives

Manufacturing in Global Perspective - The Impact of Intranet/Internet Development

[1]Henry Lau, [2]Felix T.S. Chan, [3]Ralph W.L. Ip

[1]Advanced Technology Education Centre, Regency Institute, Australia
[2]Dept. of Ind. & Manuf. Systems Engg., University of Hong Kong
[3]Dept. of Manuf. Engg. & Engg. Mgt., City University of Hong Kong

Abstract. Internet - the information superhighway - which connects more than 50 million individuals and companies worldwide through a hugh, stable and inexpensive network accessed by computers, is continually growing at an exponential rate. However, not all the people are aware that Intranet, which is the use of Internet technologies within a company deployed on an internal network based on open Web technology, is now growing even faster than the external Internet. The growing acceptance of Intranet in enterprises has a significant impact on the future paradigm of manufacturing especially the global perpective of manufacturing. This paper describes the development of such an global network in an manufacturing organsiation with practical example. The impact of such a development on global manufacturing is also covered.

1. Introduction

Internet - the information superhighway - which connects more than 50 million individuals and companies worldwide through a hugh, stable and inexpensive network accessed by computers, is continually growing at an exponential rate [1]. However, not all the people are aware that Intranet, which is the use of Internet technology within a company deployed on an internal network based on open Web technology, is now growing even faster than the external Internet [2-4]. There are three main reasons why Intranet is so well received by the public :

- Intranets are much easier to set up and expand and require minimum training
- Intranets can be implemented quickly, are standards based, have broad vendor support and a range of product offerings
- Intranets integrate electronically to corporate data stored in servers such as product data, cost data, sales and marketing data

The growing acceptance of Intranet in enterprises has a significant impact on the future paradigm of manufacturing especially in terms of global perpectives. Manufacturing is very much related to "information processing" [5] and due to the increasing competitiveness in manufacturing locally and globally, a fundamental shift of paradigm in organisational operations is necessary if competitive edge is to be retained. An "organisational change" is needed and the underlying foundation for this change is efficient information processing. An organisation without information cannot function or adapt. To meet these requirements, an information flow network, which aims at the enhancement of responsiveness and flexibility of manufacturing companies is crucial. Intranet aims at achieving this objective with the scope for future global information interconnection. With the implementation of Intranet, the change of paradigm of manufacturing from traditional type of sequential, function-oriented to simultaneous and integrated approach in dealing operational tasks such as product development is realised.

A manufacturing organisation can be divided into three main groups namely Marketing/Sales, Design and Manufacturing (Fig. 1). Each group may have its own server to store the relevant information. The reality is that the company's corporate data, which includes customers data, sales and product data, is normally stored in the company's main-frame system. However, it is likely that most staff are using Personal Computers (PCs) and/or Macintoshes running DOS/Windows/Apples applications and due to the difference of formats and standards of the involved computer systems, it is difficult, if not impossible, to access data from the enterprise server. It should be noted that the pre-requisite of implementing Intranet is that the corporate data must be transferred to the Web server with standards supported by Internet [3,4]. Once that is done, most of the data from the server can be accessed by clients using Web browsers as the front-end client interface - a client/server system. With this setup, various manufacturing groups can transfer information among themselves via a common interface. In addition, data communication with other groups such as design, marketing can also be achieved as most of the data is now stored in a Web server which can be accessed company-wide. Apart from internal data communication, Intranet can easily combine with an Internet connection so that information can be transferred with local or global customers and suppliers [1-4]. In general, Intranet is "fenced off" from outside Internet by "firewalls" that allows employees to look out but keep security control of the data that is not supposed to be accessed by outsiders.

2. Change of Paradigm of Information Exchange for Manufacturing

The paradigm of information exchange within a manufacturing organisation is facing change due to the impact of Intranet/Internet technology. In the existing practice, the information exchange is carried out in a way as shown in Fig. 1. Various systems may need specific data from other systems e.g. Bill of Material data from a CAD system to a Manufacturing Resouce Planning (MRP-II) system. Problems may arise during the process of data exchange due to various formats and standards adopted by various systems. To overcome this problem, a number of standards such as PDES, STEP and IGES have been established with the aim to achieve "flawless" data transfer among various computer systems [6-9]. In particular, PDES and STEP are quite well received in the manufacturing community for the exchange of product data among different systems such as Computer Aided Design (CAD), Computer Aided Manufacturing (CAM), Computer Aided Process Planning (CAPP) [10-12].

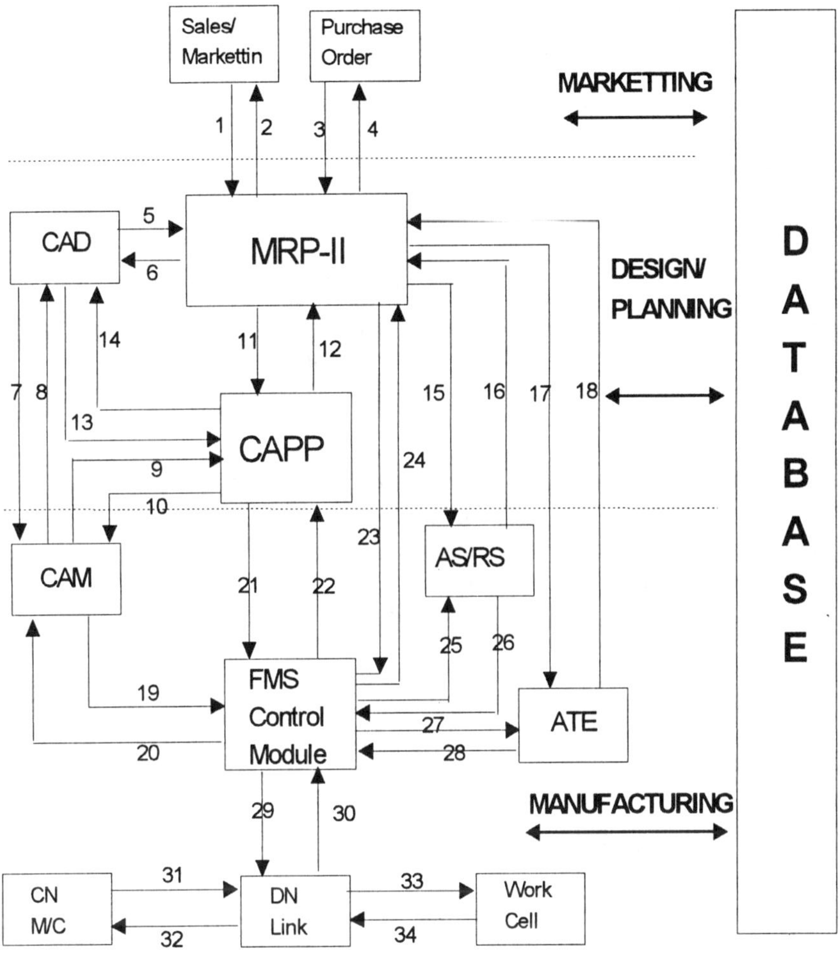

Figure 1 : Information Exchange among Systems

Note : The input and output information for the manufacturing systems is shown in Table 1.

As shown in Table 1, various manufacturing systems require specific data from other systems and basically this data transfer can be accomplished by one of the two ways namely, peer-to-peer communication and client/server communication [13]. In peer-to-peer communication, data is directly sent from a system to another e.g. BOM is directly sent from a CAD system to an MRP-II system through a connection between the two involved systems such as Windows for Workgroup. In this case, there is no separate server to store data whereas in client/server connection, data from various systems is stored in a server which acts as the central database and therefore there is no direct exchange of data between systems. For exmple, the BOM data from CAD system is sent to the server first and the MRP-II system then retrieves data from the server. A server program, such as Netware series and Windows NT, is required to provide a centralised control of data access and data security.

The Intranet approach is essentially a client/server communication with the Web server as the the central database which stores information from various sources such as marketing/sales, design and manufacturing. Web server brings diverse kinds of information to employees via a single front end, the familiar Web browser e.g. Netscape Navigator. Employees can make use of their Web browser to access a series of Web pages with links to design and manufacturing databases. In addition, Intranet also allows group discussion, electronic mails and files transfer within the company. The relevant information from the existing servers is stored in the Web server and in order to be viewed by the Web browsers from client sites, the data must be put in HTML pages. HTML (HyperText Transfer Protocol) is the protocol which negotiates document delivery to a Web browser from a Web server. Relational databases, such as database tables created by Microsoft Access and dBASE series, can be put to Web server and displayed in Web browsers using ODBC facility which is an interface developed by Microsoft to let clients e.g. Web browser sites, access multiple database engines such as Oracle and Ingress databases [14,15]. Some software, such as dbWeb version 1.1, allows users to put databases to Web server using ODBC without the need to write any codes.

3. Benefits and Consideration for Intranet/Internet Approach

As mentioned in the previous context, the traditional paradigm of manufacturing is based on sequential and function-oriented approach. For example, in the product development process, the product design department passes information to the manufacturing department which in turn passes information to the production shops. With the Intranet approach, every department, using the common front-end interface, can access information from the Web server simultaneously. The server program can be configured to determine which department can access which groups of information. With such approach, more efficient and direct communication between departments is achieved and lead time for product development can be shortened significantly [16,17]. Groups discussions, electronic mails and file transfers between departments are also available within the Intranet connection with security control.

Table 1 : Data Exchange for Fig. 1

Line No	Output Information	Source System	Destination System
1	Data for setting master production schedule	Sales/Marketing	MRP-II
2	Actual production data for study and analysis	MRP-II	Sales/ Marketing
3	Place orders	Purchasing	MRP-II
4	Delivery Progress, vendor quality	MRP-II	Purchasing
5	Bill of Material	CAD	MRP-II
6	Inventory records, cost information	MRP-II	CAD
7	Geometric data of design	CAD	CAM
8	Information feedback	CAM	CAD
9	Feedback status and problems found	CAM	CAPP
10	Process data	CAPP	CAM
11	Product information/ cost data	MRP-II	CAPP
12	Process planning summary	CAPP	MRP-II
13	Geometric data	CAD	CAPP
14	Status feedback	CAPP	CAD
15	Component data retrieval, lists of parts to be drawn	MRP-II	AS/RS
16	Inventory situation	AS/RS	MRP-II
17	Retrieval of quality analysis	MRP-II	ATE
18	Quality/measurement results	ATE	MRP-II
19	Status feedback	CAM	FMS C. M.
20	NC programming data	FMS C.M.	CAM
21	Process operations information	CAPP	FMS C.M.
22	Data enquires	FMS C.M.	CAPP
23	Production schedule for FMS	MRP-II	FMS C.M.
24	Orders completed	FMS C.M.	MRP-II
25	Storage and material handling monitoring	FMS C.M.	AS/RS
26	Feedback data as requested	AS/RS	FMS C.M.
27	Quality/dimensional checking	FMS C.M.	ATE
28	Results feedback	ATE	FMS C.M.
29	Monitor machining sequence	FMS C.M.	DNC Link
30	Report on machining status	DNC Link	FMS C.M.
31	Feedback current status	CNC	DNC Link
32	Download programs	DNC Link	CNC
33	Download programs	DNC Link	Work Cell
34	Feedback current status	Work Cell	DNC Link

The other incentive for adopting an Intranet approach is the easy expansion to global manufacture through Internet. As Intranet shares the same software and network equipment as well as speak the same computer language as Internet, it can easily connect to the Internet which is the world wide information superhighway. This can enhance the development of global manufacture for the company. A local car manufacturing plant in Australia, e.g, Mitshubishi, with a number of car component suppliers locally and overseas and the mother company in Japan, may need to send component information to other companies in other States and receive quality and product data from Japan. At the same time, customers can access commercial and sales information from the Internet regarding the new car models, client feedbacks etc. This "information at your finger tips" approach through Intranet/Internet approach can greatly enhance the global competitive edge of the company. Customers can access Web pages specially designed for customers service; employees in other branches can access product, design, sales and other corporate information. Security access is controlled by the Web server programs to set up various access level for various clients, company employees and executives With this set up, Intranet/Internet connection can serve as a highly secure forum from which design and manufacturing engineers from various branches, locally and globally, can share confidential knowledge and ideas with their colleagues.

The other advantages include :

- the ubiquitous Web browser is available for virtually every client Operating System (OS) - unlike proprietary groupware clients e.g Lotus Notes, which may not be available for some platforms
- The market for Web server software is competitive - not a single-vendor solution but the products interoperate well

The essential point for setting up an Intranet in a company is that the full support from the top management to the lower level staff is of utmost importance. As every department is involved in the change process, a clear understanding of the steps required to implement this technology is critical. The other considerations include:

- There are still limited tools exist for linking a Web server to a wide variety of databases or other back-end application though ODBC has been quite well supported in this area.
- Intranet/Internet require TCP/IP (Transmission Control Protocol/Internet Protocol), which is the collection of transport and application protocols used to communicate on the Internet and other network, whereas the traditional Local Area Network (LAN) works with just LAN tansmission protocols and this means that the entire protocol connection for the company need to be changed.
- Although programming standards for the Web, such as Java, Visual Basic Scripts are emerging, HTML is not powerful enough to develop a complex client/server applications. However, this will definitely change in the coming years as more and more Web-related software will be released.

4. Examples of Intranet/Internet Connections for Manufacturing

In a global perspective, a number of Intranet/Internet developments for global manufacture have been implemented. Three examples of these developments are summarised below :

1. In Hong Kong, the Industrial Engineering and Engineering Management Department of Hong Kong University of Science and Technology and the Clothing Industry Training Authority (CITA) has been working together on an industry-project named Hong Kong Accessory Information Network System (HKCAINS)[18]. This project aims to develop a global communication network for Hong Kong's apparel industry in the economic viability of establishing such a network. The main objectives of this initiative are to implement an industry-wide communication infrastructive to support the global business application of Hong Kong's textile and apparel industry; and by preparing and building concensus on an implementation plan, to design, develop and implement a startup site for demonstrating its specialized communication requirements. The major target users are the executive managers, merchandisers, engineers and designers in the textile and apparel industry. They are able to retrieve textile and apparel related information through the communication infrastructure. They can thereby take advantage of current trends in fashion, up-to-date business information, and other worldwide resources [x]. This project is proved to be quite successful and further development is underway to enhance the sophistication of this network.

2. National Semiconductor has developed an internal Web site called Community of Practice which can serve as a highly secure forum from which engineers, designers can share confidential knowledge and ideas with their colleagues who may be located in other regions of the world. With this setup, designers, engineers and workshop personnels can share information which can be restricted to be accessed by certain group of authorised staff. This greatly helps the communication of staff from all over the world and that is important for a multi-national company such as National Semiconductor.

3. 3M, an innovator in manufacturing since 1902, has developed a global network called Global Economic Overview. At present, 3M has 60 subsidiaries in about 53 countries all over the world. So 3M has all the reason for giving the highest priority to timely communications that its executives and engineers can securely access from anywhere in the world. With this communication network, authorised staff can access 3M's Global Economic Overview using Netscope Navigator. Staff can review timely economic data, submit research requests to 3M's Economic Services division, read current economic news, or check a schedule of upcoming videoconferences.

The above there examples demonstrate the impact of the Intranet/Internet development on global manufacture. This development proves to be effective in leveraging the competitive edge of companies in this vibrant global market.

5. Conclusion

The rapid growth of Intranet/Internet technology has a significant impact on the future paradigm of Manufacturing. In order to maintain the competitive edge in the manufacturing market, it is imperative that information can be accessed by employees, customers, suppliers and business partners electronically. This electronic information access lowers the cost of production internally and externally and it enables employees in your company to work more effectively together with customers, suppliers & business partners. However, Intranet is still in its early stage in terms of global manufacturing but it is an appealing and economical solution for companies with the aim to achieve better responsiveness and flexibility.

References

1. The Intranet Explained. (1996) Intranet Australia Pty Ltd.
2. Ayre, Rick. (1996) Setting Up Shop. *PC Magazine*. April, 151-158.
3. Pompili, Tony. (1996) Content and Collaboration. *PC Magazine*. April, 115-149.
4. Derfler, F.J.J. (1996) A universal Client. *PC Magazine*. April, 105-113.
5. Lau, Henry. (1995) Generic Methodology for the Design and Implementation of a Non-standard Specific Information Flow System. PhD Thesis. Adelaide University.
6. Bohn, J.H. (1996) Introduction to Rapid Prototyping (ISO-10303 STEP). Randolph Hall.
7. Mult, H,C. & Hans, I. (1991) Advanced Manufacturing to Achieve World Competitiveness. *Paper for Workshops Manufacturing Australia*, July.
8. Luca, A.D. (1993) A Step by Step Effective Approach to Computer Integrated Enterprise. *Australian Conference on Manufacturing Engineeering*. P109-113.
9. Millett, B.C. (1990) Neutral Manufacturing Data: The Gateway to CIM, *Pacific Conference on Manufacturing Proceeding*, Vol 1, 534-541.
10. EDS Unigraphics. (1995) EDS (http://www.ug.eds.com/ug/translator/step.html).
11. Everitt, P.B. (1988) Product Data Exchange Specification: RAMP's Unique Implementation of the PDES Specification. *SME Technical Paper*, MS88-660.
12. Sargent, P.M. (1990) Materials Data Interchange for Component Manufacture. *Engineering with Computers*, Vol. 6, 237-247.
13. Nance, Barry. (1992) Introduction to Networking. QUE.
14. Microsoft Corporation. (1993) *Professional Features* Book1. Microsoft Visual Basic.
15. Microsoft Corporation. (1994) Windows for Workgroup Manual.
16. Vasilash, G.S. (1987) Simultaneous Engineering, *SME*, P50-55.
17. Zockel, M. (1993) Achieving Simultaneous Engineering. *Australian Conference on Manufacturing Engineering*. Adelaide, 22-24 November.
18. HKCAINS. (1997) http://143.89.48.174/introduction.html

Competences and Capabilities: An Analysis of the Hard Disk Drive Industry

N. Viswanadham*
Yap Chee Meng**
Phun Kim Ming*

*Mechanical & Production Engineering
**Industrial and Systems Engineering
National University of Singapore
10 Kent Ridge Crescent, Singapore 119260

Abstract. In this paper, we use the core competence and core capability framework to analyze the hard disk drive industry and develop various scenarios for this industry. The hard disk drive industry is a highly dynamic industry, manufacturing an intermediate product for big customers who are mostly PC manufacturers. It has a staged manufacturing process and a global supply chain. Price and delivery reliability are the differentiating factors. Based on these characteristics, we identify two operational core business processes: the supply chain process and the new product design and development process; four managerial processes: supplier management, customer engagement, risk evaluation and strategy development, and human resource management and infrastructure development as the two support processes. Using patent analysis on six key players in the disk drive industry, we found the leaders in disk drive component technologies. To create core capabilities, it is important to link these core technological competencies with the two operational processes and create a competitive advantage through low cost and reliable delivery. By developing the supply chain and the new product development processes as core capabilities, a company can certainly keep ahead of the competition. We also identify the possible growth strategies for the disk drive companies.

1 INTRODUCTION

During the 1980's, a promising new approach to strategic management emerged based on the notion of core competence. According to this, there are specific competences that are closely related to the firm's intangible assets, know-how and its accumulated technological resources. The origins of this approach can be traced back to the early work of Penrose [1]. Firms are viewed as bundles of competences rather than bundles of facilities or products or functions. This resource based approach focuses on accumulating capabilities over time and different firms have different competences. The notions were put on a firm foundation by Prahalad and Hamel [2,3]. They defined a core competence as an integration of diverse skills, technologies and know-how. It is the integration of the diverse entities that is distinguishing hallmark rather than a single discrete skill or technology. They are several types of competences : infrastructural, marketing, production and technological. One needs also the organizational competence to provide an environment for people of different skills work together. The characteristic of core competences are that they are not imitatable easily, they contribute to the customer benefit, they do not diminish with time if nurtured continuously and finally they are gateways to new markets.

A firm's core technologies are those underlying, enabling technologies that can lead to multiple businesses and products. In the data storage industry, different products such as hard disk drives, optical drives, tape drives will have their own core technologies but some of these technologies could be common. Disk drives are based on cluster of core and borrowed technologies such as ICs, small motor, digital signal processing, head and media. In fact, small motor technologies, ICs technologies and miniaturization skills are also needed to build other products such as optical drives. From core technologies, a range of key components are made, which are used to develop systems. For a firm to be successful, in addition to the technological competences mentioned above, it needs capabilities to produce generations of products and market them to the customers. That is why new product development and supply chain process are very important (core) business process in the hard disk drive industry.

Prahalad proposed an opportunity gap management framework for resource based strategic management[4]. In this framework, the organizations need to first define their strategic intents, then formulate a strategic architecture, identify the core competences needed to achieve the strategic intents and deploy the competences to exploit new market opportunities. Carl Long and Mary Vickers-Koch [7] stressed that it is important to link core competence to strategic business processes for an organization to create core capabilities.

We present here a framework for developing a capability based organization (See Figure 1). First an organization has to analyze the uncertainties as well as the future challenges in terms of market or technology trend. Understanding the future uncertainties and trends will help an organization to formulate the best strategic intent and develop a strategic architecture to guide the organization to achieve its aspiration. Next, it is necessary to determine the core competences needed and the core business processes, and link them together to create unique core capabilities. With proper resource allocation and development, the core capabilities created can be nurtured and exploited to explore new market opportunities through new product development or new business development.

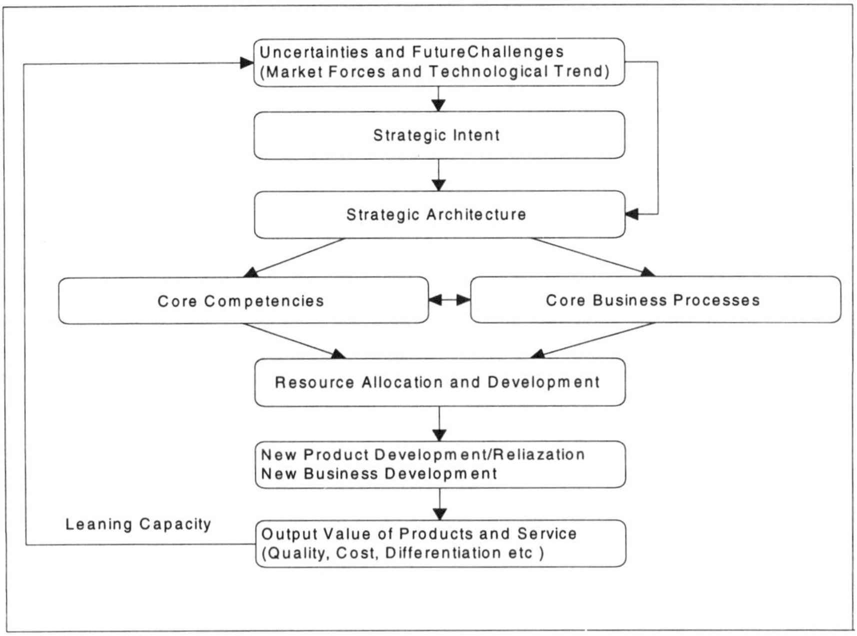

Fig. 1. Framework for developing a capabilities based organizations

In [17], the authors identified that materials core capability are critical to the semiconductor silicon industry and also concluded that firms which build strategy around this core capability are more likely to succeed that firms that do not. Miyazaki [14] has studied the technological competences of many firms in the areas of optoelectronics using bibliometric INSPEC data, US patent data and interviews. In the study, Miyazaki concentrated on the various main factors that affect the rate and direction of competence building and they are previous core business activities and links with customers.

In Section 2, we describe the history of the hard disk drive industry and discuss the generic technologies in the hard disk drive industry. We then identify the five major characteristics of the hard disk drive industry. In Section 3, we identifies the core business processes in the hard disk drive industry based on the five characteristics. In Section 4, we identify the technological competences of six leading companies through patent analysis on them. We analyze the three possible new markets for the hard disk drive industry in the final section of the paper.

2 THE HARD DISK DRIVE INDUSTRY

2.1 Background of The Hard Disk Drive Industry

The history of the hard disk drive industry started in 1956 when IBM introduced its 24", 4.4 Megabyte (Mb) RAMAC 350. In the 60's and 70's, hard disk drives were serving the mainframe and minicomputer market with form factors of 14", 10.8", 9.5" and 8". Then, in the late 70's and early 80's, disk drives of smaller form factors were in demand due to the introduction of personal computer by IBM and Apple. In 1979, Seagate Technology, founded by Al Shugart, shipped the first 5.25" disk drive with a capacity of 5 Mb. Since then, the storage capacity started to grow exponentially, doubling every 18 months. Then, Maxtor Corporation introduced the 5.25" disk drive with 300 Mb in 1983, with 760Mb in 1986 and 1.5Gb in 1989. However, Maxtor was later acquired by Hyundai.

Conner Peripherals, started in 1986, became the main supplier of 3.5" drives for Compaq laptops. In the late 80s and early 90s, there was a shift in the market structure when the 3.5" drives started to replace 5.25" drives in the PC market. Also the 2.5" drives emerged as the main storage medium for the mobile computer and laptop market. All these developments were due to the tremendous technological improvements in the storage capacity of hard disk drives. Also the cost per Mb of storage is reducing continuously at a rate of 50% for every 18 months[13].

Hard disk drives have become mature products (product and process technologies stabilized) after 40 years of history. Its rapid evolution has caused a large number of firms to exit. More than 130 firms entered the hard disk drive industry but less than 30 firms survived [9]. The hard disk drive industry has reached a state of concentration and the top five players (Seagate, Quantum, Western Digital, IBM and Fujitsu) own more than 80% of the market share [11]. Its market is still growing strongly and is expected to rule the market for at least another ten years. "Potential" substitute such as optical disk drives and flash memories could not threaten the hard disk drive market. High storage capacity, high reliability, high accessing rate, small form factor and improved shock resistance are the characteristics not found in the substitutes. Optical disks have very high capacity but have very slow access rates. Flash memories have very low storage capacity and their cost per Gb is very high. The maximum capacity ever achieved is about 80 Mb and costs about $1000.

2.2 A Hard Disk Drive: Core Products and Generic Technologies

A hard disk drive is an electromechanical magnetic recording device whose function is to store data over a long period of time and retrieve it reliably. Disk drive parts and their functions are given below [14]:

- Read Write Heads - These are tiny devices that write and read data from the disk. The head technology has undergone 2 major transitions, that are from ferrite head to thin film inductive head and then to magnetoresistive technology. Since 1991, areal density is growing at a compound growth rate (CGR) of 60%

with the magnetoresistive head technology as compared to only 40% with thin film heads before 1991. Besides MR heads, rapid advances in low-noise thin film technology and PRML sampling channels are also the major reasons for this progress. Beyond year 2000, head technology will evolve into spin valve and giant MR structure.

- Disks/Media - Media is coated with a magnetic layer which allows information to be stored as tiny magnetization patterns. The projected compound growth rate for data rate is about 40% per year.
- Motors - The purpose of spindle motor is to rotate the disk at a constant speed. Stepper motors or voice coil motors are used to position heads in the armature assembly.
- Head Gimbal Assembly - This includes mechanical parts such as the actuator arm, slider and so on. The head is mounted onto the slider and the actuator arm driven by a servo motor is responsible for controlling the position of head across the surface of the media.
- Electronics - The electronics part consists of all the electronic and channel circuitry. The servo circuits control the actuator speed and accuracy in positioning heads.
- Software - The software such as microcodes ensure data are correctly written to or read from the disks and interface with host computer. The code also provides for error detection and correction during data transfers, and error recovery function.
- Other mechanical parts - This includes the bearings, casing, gaskets, air strippers, disk clamps and others.

Evolution of smaller form factor is driven by space constraint of desktop systems such as PCs, workstations, laptops and hand held computers, etc. Advances in heads, disks interface and channel technologies have helped in miniaturization while simultaneously improving the storage capacity and performance. By the year 2000, it is expected that 3.5" drives can provide a maximum storage capacity of 90 Gb, 2.5" drives with a maximum of 10 Gb and 1.8" drives with a maximum of 3.4 Gb. Compared to 1990, reliability of hard disk drives produced today has improved by five times. Improvement in reliability is due to minimization of parts count which includes heads and disks, more integrated electronics, better tolerance control and limiting operation temperature rise. MR head technology would be the critical technology because it requires less spacing between the head and media due to its advantage over thin film technology in resolution, signal amplitude, and write-wide but read-narrow phenomenon. Figure 2 shows the core products and the important generic technologies for a hard disk drive. The core products are the main components of a hard disk drive such as heads, disks, electronics and mechanical parts, and also the integration of these components. For each core product, there are several important generic technologies.

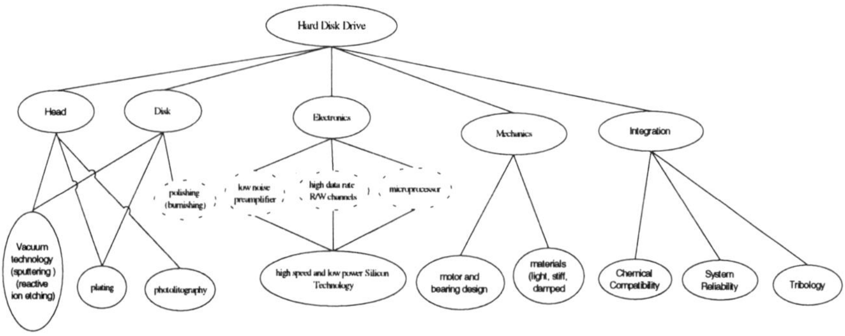

Fig. 2. Core products and generic technologies for a hard disk drive

From Figure 2, it can be seen that heads and disks have 2 common generic technologies: vacuum technology and plating. Electronics can be further divided into 3 sub core products: low noise amplifiers, high data rate read write channel and microprocessors. The important generic technology behind these core products is the high speed and low power silicon technology. For mechanical parts, motor and bearing designs and material science technology are important. For the integration of different components, the important technologies are chemical compatibility, system reliability and tribology. All the generic technologies are the technical core competencies for building a hard disk drive. The core products are the physical embodiments of these competencies.

2.3 Characteristics of the Hard Disk Drive Industry

As we mentioned before, HDD industry is an intermediate product industry with big customers, staged manufacturing and global supply chain. It is very important to realize that these characteristics determine the business processes. The five main characteristics identified for the hard disk drive industry are:

Global supply chain - The hard disk drive industry has a global supply chain. Various components are manufactured in different countries and are shipped to the hard disk drive manufacturers who in turn export the hard disk drives to other countries. All the hard disk drive firms including IBM and Seagate have to obtain their components externally from suppliers. With global sourcing of components, inventory costs and logistics costs are to be kept in control by having an efficient supply chain process (inventory transit as well as safety stocks) with good delivery reliability.

Intermediate product industry - Hard disk drives are not the end products but are critical components in PCs, minicomputers, laptops and mainframe computers. Also, a hard disk drive is made up of several components such the head, media, motors, electronics and PCBs. The hard disk drive companies are intermediate product companies with big customers (OEMs) which acquire components for its products from suppliers.

Highly Dynamic Industry - Rapid Changes in Technology - The hard disk drive industry has reached a state of maturity in terms of its product configuration and use. However, the components of hard disk drives are being continuously improved to provide higher capacity, reliability and access rates. Rapid advances in component technologies put pressure on disk drive manufacturers to change their designs often and introduce new products frequently. Due to these technological advances, and also due to the changes in the PC industry, life cycle of hard disk drives has decreased tremendously, from 2 to 3 years in the early 90s to 3 to 6 months in recent months. This means a new product will probably become obsolete in 3 months. Hence, capabilities in new product design and development are highly desired and are vital for survival.

Staged manufacturing process - The manufacturing of hard disk drives is done in stages and each stage in a different location. The first stage is the manufacture of the components. The second stage is the disk drive assembly. Some companies perform the assembly in two stages; customization is done in the second stage. Logistics plays an important role for shipments between the stages.

Big Customers - Hard disk drives are storage devices in mainframe computers, minicomputers, personal computers, laptop PCs, photocopying machines, fax machines, video editors and so on. Hence, the important customers for the hard disk drive industry are big OEMs such as Acer Computer, Compaq, Hewlett Packard, NEC, Canon, Xerox, and other OEMs or distributors. Besides high bargaining power of the customers, cost for losing a customer is very high. Thus, customer acquisition and retention is a very important business process. Also, price and on time delivery are differentiating factors and thus efficient supply chain and order to delivery processes are essential.

3 CORE BUSINESS PROCESSES IN A HARD DISK DRIVE COMPANY

From the five characteristics of the hard disk drive industry described in Section 2, we can identify eight core business processes for a company manufacturing hard disk drives. Two of them are operational processes, four are managerial processes and two are support processes.

3.1 Operational Processes

The two operational processes are the supply chain and the new product design and development processes.

3.1.1 Supply Chain Process

The supply chain process is a horizontal process that starts with the procurement of the raw materials and ends with the delivery to the OEM. The hard disk drive industry has global operations and this increases the complexity of its supply chain process and also the inventory and logistics costs. Due to the rapid changes in the demand for the PCs, and the technology of the disk drive components, the disk drive companies encounter frequent over- and under-capacity conditions for the components and finished products. Most disk drive companies follow just-in-time manufacturing techniques and carry minimal inventories. Transport of components and disk drives are done by air freight. Some companies use overnight courier services. To maintain a lean supply chain process, one has to manage the interfaces between suppliers and logistics, and disk drive manufacturers and logistics and OEMs. Thus, supplier management and customer engagement processes become critical for effective supply chain management (see Sections 3.2.1 and 3.2.2). The performance measures of the supply chain process include lead time, cost, flexibility and delivery reliability. Best practices include supply of batch of disk drive directly to the factory site of the OEMs, testing thoroughly before shipping on time, everytime in right quantities, alliance with overnight couriers for logistics and customs clearance, etc.

3.1.2 New Product Design and Development

Since the hard disk drive industry is highly dynamic with very short product life cycles and high technological changes, the new product design and development process is critical for survival. Time to market is a critical factor for a disk drive manufacturer to gain competitive advantage. Table 1 shows all the new hard disk drives models introduced from April 1996 till December 1996 by various companies[12]. Within the nine-month interval, 12 models of 2.5" disk drives, 4 models of 3" disk drives, 41 models of 3.5" desktop disk drives, 10 models of 3.5" high capacity disk drives and 1 model of 5.25" high capacity disk drive were introduced and they are mostly from the key players in the industry. The large variety of new products introduced within a short period shows the importance of planning for generations of new products. Besides, most of the new products have improved features and they are integrated with latest technology such as the magnetoresistive (MR) heads and Partial Response Maximum Likelihood (PRML) read write channels. Companies which introduce products incorporating the latest technologies, will have competitive advantage.

Table 1. New products introduced from April 1996 till December 1996 (Source: Computer Data Storage Newsletter, Micro-Journal SARL de presse, France, April 1996 - Jan 1997)

Company	**No. of models**	**Comments**
2.5 " HDDs		
Hitachi	5	4 (PRML + MIG head + glass media) 1 (PRML + MR head + glass media)
Fujitsu	2	MR head + glass media
IBM	2	MR head + PRML
Toshiba	2	MR head + PRML
Seagate	1	MR head + PRML
3" HDDs		
Integral	2	1 thin film head and 1 MR head
JTS	1	-
Western Digital	1	-
3.5 " desktop HDDs		
Samsung	9	7 thin films head and 2 MR heads
Western Digital	8	proximity recording head
Maxtor	7	-
NEC	6	thin film head +PRML
Seagate	5	2 (Proximity recording head), 2 (MR head + PRML) 1(thin film + PRML)
Fujitsu	3	MR head + PRML
IBM	2	MR head
JTS	1	-
3.5 " high capacity HDDs		
Fujitsu	8	MR head + PRML
Hitachi	2	MR head + PRML
5.25 " high capacity		
Cybernectics	1	-

3.2 Managerial Processes

The important core managerial processes are:

3.2.1 Supplier Management

It is essential to build close relationship with suppliers to maintain quality and product delivery times. For the disk drive manufacturers to be flexible, it is essential that suppliers are also flexible. To develop new products as desired by the customers, it is essential that the component manufacturers are innovative. Supplier participation in the design of new products is a prerequisite to short times to market and also to ensure that design of a components for ease in assembly, etc. In all the disk drive companies that were interviewed, supplier management was given priority. Suppliers' engineers are resident in the disk drive plants. Meetings are held every three months to review the relationship.

3.2.2 Customer Engagement

Customer engagement is another core business process. The customers for the disk drive industry are big and powerful companies in both financial and technical terms. Different customers have different specifications and requirements for their computer systems. Besides that, OEMs usually rely on one main supplier for all the disk drives. Hence, it is critical to understand the customer requirements and their products specifications to avoid manufacturability and assembly problems for the customers. Maintenance of close contact with the customers by participating in their design and maintaining their inventories are important for brighter common destiny. All the companies we interviewed have customer engagement as the core process. In some companies, data on MTTF and testing is shared with customers. Also customers' engineers are resident in the factory for consultations.

3.2.3 Risk Evaluation and Management

Bower and Christensen [8, 9, 10]have shown how disruptive technology has changed the market structure of the hard disk drive industry several times and forced a lot of companies out of the market. They trace the reasons for failure of leading companies as being , (i) too closely locked in with the current customers and not being able to forecast correctly the requirements of the future customers, and (ii) not investing in new technologies to manufacture smaller form factors disk drives. If the failed companies had capability in risk evaluation and management, and invested in production in the new form factor disk drives earlier than their competitors , the story of the disk drive industry would have been different.

3.2.4 Strategy Development

Strategy development is critical because the disk drive industry is fast changing and the industry has global operations. Good strategy facilitates effective decision making in organization structure changes, capital investments, strategic alliances or partnerships, acquisitions or joint ventures, diversification into upstream or downstream systems areas, entering new countries with manufacturing faculties along with OEMs and several other long term decisions.

3.3 Support Processes

3.3.1 Human Resource Development

Human resource development is critical to preserve and improve the capability and competencies of the employees so that they can support all the core business processes.

3.3.2 Infrastructural Development

It is essential to use new technologies in IT such as Intranet, EDI, data warehousing, software agents to enhance the capabilities of the company, and also to improve the communication between the stakeholders. This will also improve the quality and timeliness of the data and would reduce the lead times. Reducing the manufacturing lead time using automation, sophisticated testing facilities to increase MTBF are also needed.

4 PATENT ANALYSIS

A patent analysis was carried out to identify technological competencies of six leading companies (A-F) in the hard disk drive industry. A search was done on all the US patents of these companies from year 1991 to 1996. *Intra-firm technology indices* (IFTI) were calculated to measure the R&D strength of a particular firm in a particular technological field. IFTI is defined as the firm's share of patents within a field divided by the firm's share of total patents. The bands for IFTI values are between 0 to infinity. IFTI values of greater than unity indicate areas of relative strength within a field, and values less than one indicate areas of relative weakness. We define $IFTI_i$ as the IFTI value for a particular group of technology if r_i is the firm's share of patents within the group of technology and R_i is the firm's share of total disk drive related technology, then the ratio of r_i over R_i yields $IFTI_i$.

4.1 Intra-firm Comparison

The IFTI values for all the field were calculated for each companies and they are represented by Figure 3a, 3b, 3c, 3d, 3e and 3f. From the IFTI values, we obtained the following observation:

A - IFTI values higher than unity (Figure 3a) for almost all technologies shows that A has technological competence in almost all the main components (the heads, media, motor and control, bearing designs, head gimbal mechanical parts and other mechanical parts).

B - For year 1991-1992, B has patents related to electronics/software, HGA mechanics and other mechanical parts only. Since then, B patents became more diversified with new patents in motor, bearing and head. The high IFTI values in head and bearing for 95-96 shows that head and bearing design has become technological strengths of B (Figure 3b).

C – C's patent is relatively less and it has been decreasing from year 1991 to 1996 This shows that it has become less R&D focused. Head gimbal mechanics, motor design, and electronics are the relative strengths of C as indicated by the IFTI values in Figure 3c.

D - The number of patents owned by D are increasing from 1991 to 1996. This shows that D places more emphasis in R&D now compared to the past. IFTI values in Figure 3d show that bearing design , motor and control technologies, head gimbal mechanical parts and other mechanical parts design are the relative strengths of D.

E - E has the least patents among the six companies. This shows that E does not have strong R&D capability. However, head gimbal mechanical design could be its relative strength because of the high IFTI values (Figure 3e).

F - The IFTI values in Figure 3f shows that F's R&D strength is quite diversified in the components such as motors, bearings, HGA mechanics and other mechanical parts.

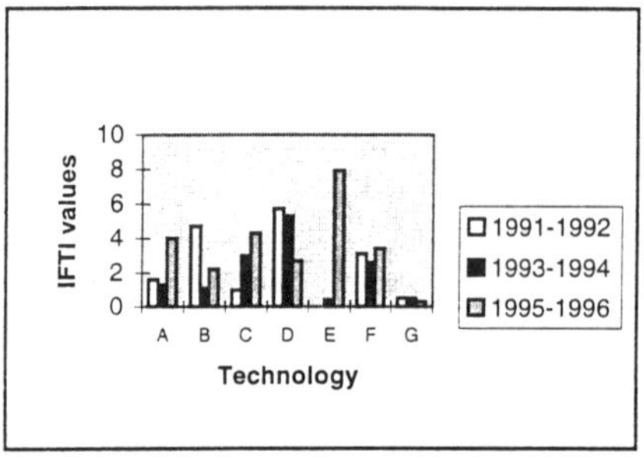

Fig. 3a: IFTI values for A's patents

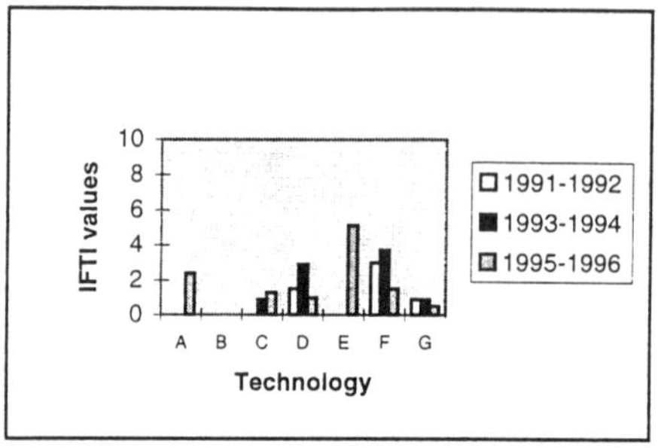

Fig. 3b: IFTI values for B's patents

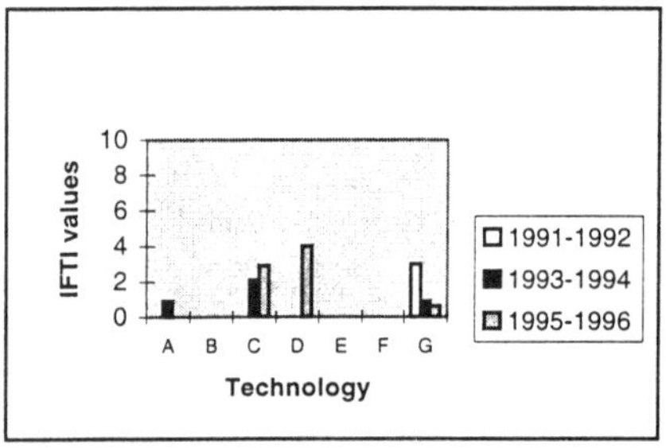

Fig. 3c: IFTI values for C's patent

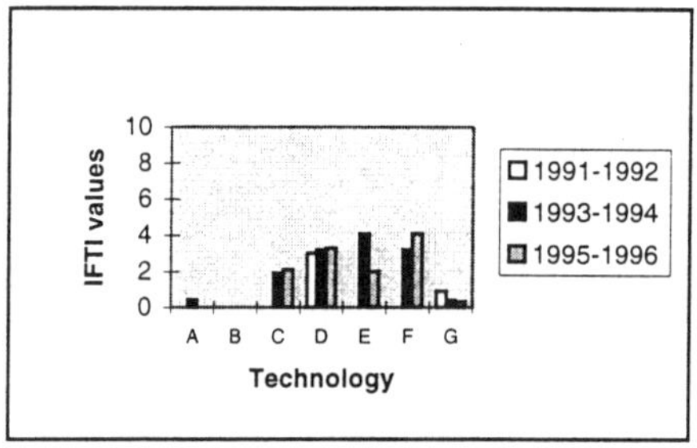

Fig. 3d: IFTI values for D's patents

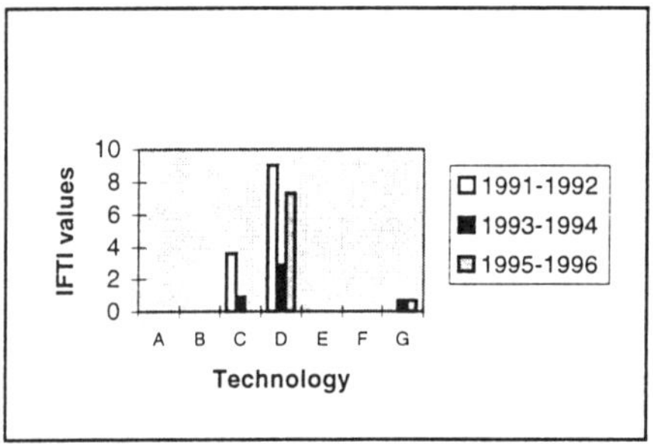

Fig. 3e: IFTI values for E's patents

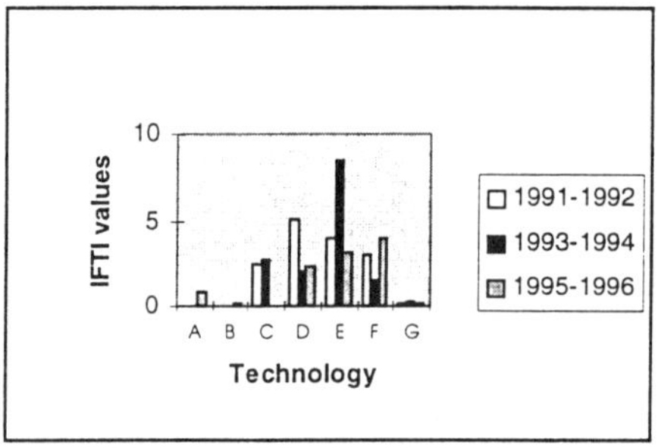

Fig. 3f: IFTI for F's patents

A: Head Related Technology
B: Media Related Technology
C: Motor and Control Technologies
D: Head Gimbal Assembly Related Technologies
E: Bearing Related Technologies
F: Other Mechanical parts
G: Electronic and Software Related Technologies

4.2 Inter-firm Comparison

- A has the highest number of patents for years 1991-1992 and 1993-1994 whereas B has the highest number of patents for year 1995-1996 among the six companies. This indicates that A and B are stronger and more active in R&D compared to the other four players.
- A has the highest percentage of patents in head, media and head gimbal mechanics for the three period. B has the highest percentage in electronics/software technologies for the year 1995-1996. This shows that A and B are the strongest in the respective technologies among the six players.
- From Tables 2a and 2b, only A and B have patents in MR head and PRML technologies among the six players. It is also noted that the number of patents for these two technologies owned by A and B are increasing over the three periods. This fits into the trend in the disk drive technologies whereby the two technologies are the future leading technologies.

Table 2a Number of patents for MR technology

	91-92	93-94	95-96
A	1	7	11
B	0	0	5
C	0	0	0
D	0	0	0
E	0	0	0
F	0	0	0
Others	65	101	173
Total	66	108	189

Table 2b Number of patents for PRML technology

	91-92	93-94	95-96
A	0	0	3
B	0	7	5
C	0	0	0
D	0	0	0
E	0	0	0
F	0	0	0
Others	1	12	6
Total	1	19	14

From the patent analysis, it is clear that A and B emphasize on developing new technologies (MR and PRML). The success of these companies (they are the top two companies in terms of market share from 1991 till 1996) may be partly due to their technological competence.

5 NEW MARKET FOR HARD DISK DRIVES

The growth of the Internet, Intranet and the emergence of the network computers has increased the demand for high capacity, low cost disk drives. Analysts predict that the personal computer sales will outgrow the TV sales by the year 2000. Although network computers are initially planned to be diskless, The computer data storage newsletter, July 96 argues that they must have low cost low capacity drives to satisfy most of the Internet users. Network servers will be managing large databases of text, video and data. Capacities of the order of 1 or 2 Terabytes is considered minimal for video editing and providing movies on demand. The low power consumption, high capacity, low cost per Gb, light weight, small size and shock resistance has made disk drives very popular in consumer electronic applications as well.

With the growth of secure communications via the Internet and Intranet, there is a market for distributed data storage systems where in information in the form of data, video and text is stored and accessed from remote terminals. Applications include E-mail, entertainment networks such as video on demand, education networks where all the students can access lecture notes from remote sites, corporate networks where information about order status, inventory levels, designs is stored on the network and accessed by authorized persons. Such distributed data storage systems will become popular in various contexts. They are basically a cluster of disk drives interconnected by high speed backbone communication networks and accessed from remote sites. Such systems exist even today for specialist banking and other applications. They will become more commonplace and create a market of their own. Capabilities to manufacture data storage devices, assemble them as clusters and interconnect them via secure communications for access by remote users for multimedia applications, and to manage multimedia information and the communication network are required in this context. Disk drive manufacturers can leverage the above capabilities to become data storage systems manufacturers. From the above discussion, it is clear the disk drive manufacturers have 3 options:

To remain as a data storage device manufacturer of various form factors for PCs and workstation and move into rapid optical disk drives if need be. With the advances in components technologies and new emerging markets, this would require core new product development and supply chain capabilities.

(i) To forward integrate into the manufacture of personal computer systems.

(ii) To manufacture data storage systems, which are networks and servers for variety of applications such as video parlors, universities, banks, etc. The three scenarios are outlined in Figure 4.

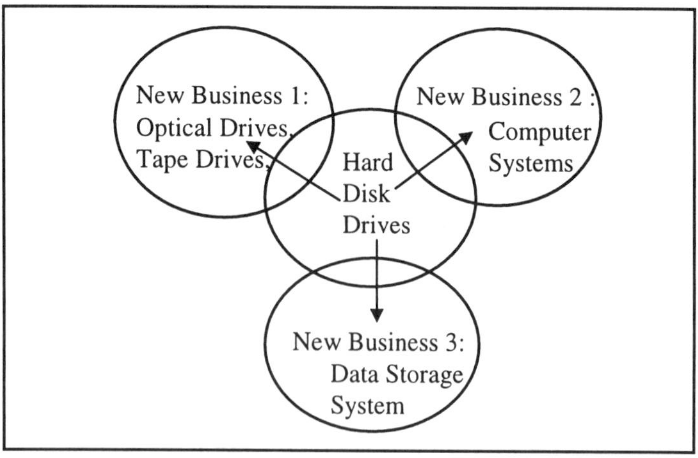

Fig. 4. Possible business opportunities for a hard disk drive company

6 CONCLUSIONS

In this paper, we have taken a different approach towards the analysis of the hard disk drive industry for operational performance management and the opportunity gap management. From the five characteristics of the hard disk drive industry, we have identified the eight important business processes. From the patent analysis and product analysis, we identified the core technologies that provide competitive advantage for a disk drive manufacturers. We have reinforced our ideas and checked our results by collecting information from six leading players in the hard disk drive industry via the Internet as well as through personal interviews with the managers. Core capabilities for most successful hard disk drive manufacturers include supply chain management and the new product design and development. We also argue that a disk drive manufacturer, by acquiring competences in multimedia information management, distributed data storage system architectures, and multimedia information network management, can manufacture data storage systems for various applications such as entertainment, mail service, education and business. We believe that the hard disk drive manufacturers will gain a lot if they think in terms of the processes rather than products. Further work includes performance analysis of each the business processes for their operational efficiency and effectiveness.

References

1. E. Penrose, *The Theory of Growth of the Firm, London*, Basil Blackwell (1959).
2. C.K. Prahalad, and Gary Hamel., Strategy intent, *Harvard Business Review* (May-June 1989) 63-76.
3. C.K. Prahalad, and Gary Hamel., The core competence of the corporation, *Harvard Business Review* (May-June 1990) 79-91.
4. C.K. Prahalad., The role of core competencies in the corporation, *Research Technology Management* (1995) 40-47.
5. Mahen Tampoe, Exploiting the core competencies of your organization, *Long Range Planning* **27** (1994) 66-77.
6. George Stalk, Evans, and Lawrence E. Shulman, Competing on capability: The new rules of corporate strategy, *Harvard Business Review* (March-April 1992) 57-69.
7. Carl Long, and Mary Vickers-Koch, Using core capabilities to create competitive advantage, *Organization Dynamics* (Summer 1995) 7-22.
8. Clayton M. Christensen, and Richard S. Rosenbloom, Explaining the attacker's advantage: Paradigms, organizational dynamics, and the value network, *Research Policy* **24** (1995) 233-257.
9. Joseph L. Bower, and Clayton M. Christensen, Disruptive technologies: catching the wave, *Harvard Business Review* (Jan-Feb 1995) 43-53.
10. Joseph L. Bower, and Clayton M. Christensen, Customer power, strategic investment, and the failure of leading firms, *Strategic Management Journal* **17** (1996) 197-218.
11. Disk Trend Report (1996).
12. Computer Data Storage Newsletter, Micro- Journal SARL de presse, France, Vol. #9-4 to #10-1 (April 1996 - Jan 1997).

13. Edward Grochowsk, and Roger F. Hoyt, Future Trends in the Hard Disk Drives, *IEEE Transactions on Magnetics* **32** (May 1996) 1850-1853

14. Kumiko Miyazaki, *Building Competence in The Firm*, St. Martin's Press Ltd, New York (1995) 105-205.

15. Gary Hamel, and Aime Heene, *Competence-based Competition*, John Wiley and Sons, England (1994).

16. Kanu G.Ashar, *Magnetic Disk Drives Technology: Head, Media, Channels, Interfaces and Integration*, IEEE Press (1997).

17. Steven T. Warsh, Robert L. Boylan, Joseph Morone and Albert Paulson, Core *Capabilities and Strategy: Empirical Evidence for the Silicon Semiconductor Industry*, John Wiley & Sons (1996).

Systems Integration in Manufacturing Firm : Some Empirical Observations in Japan

Deok-Joo Lee*, Hiroshi Katayama* and Hyung-Sik Oh**

*Department of Industrial and Management Systems Engineering, Waseda University, Ohkubo 3-4-1, Shinjuku Ward, Tokyo, Japan.

**Department of Industrial Engineering, Seoul National University, Seoul, 151-742, Korea.

Abstract. The main theme of this study is a comparative study of manufacturing strategy process of firms with respect to the level of systems integration they attained. It was attempted to find the differences in competitive priorities and action programmes between the firms with high and low level integrated systems through statistical analysis of survey data from Japanese manufacturers. It is observed that most of recent manufacturers are much concerned with integrating information system. It is also found that highly integrated firms put stronger emphasis on the manufacturing strategy pertaining to new product development.

1 Introduction

At once driven by volatile market situations and aided by the advancement of information technology, manufacturers have paid attention to the importance and possibility of systems integration within company from the earlier 1980s. Especially the introduction of computers in manufacturing has impelled the technological innovations for the efficient integration of manufacturing systems in the name of CIM. Actually it is observed that the greater part of manufacturing companies already initiated significant efforts to implement CIM in their production spots and attained integration to some extent. And it is expected firmly that, when fully implemented, CIM provides a broad range of flexibility to accommodate various product design, agile system change to meet shifts in demand, as well as greater control over manufacturing processes.[1]

De Meyer[2] showed empirically that companies which intended to implement CIM considered this project not as turnkey-type but as evolutionary one towards a fully integrated manufacturing information system. From this finding, it can be thought that there might exist some levels in the integration of manufacturing system ranging from one in technical sense, computerized links in hardware and software, to one in organizational sense, where information is shared among different functional areas within organizations. In addition, ISO[1] recently developed the CIM reference model in which the 6 levels of integration were defined such that equipment level, station level, cell level, area level, plant level and enterprise level. Along with this rationale on systems integration, we make the level of systems integration in manufacturing firms a subject of this empirical research.

From the Skinner's seminal work[3] which was motivated by the observations that the frustration of manufacturing managers in increasingly competitive environments came from the missing link between manufacturing and corporate strategy, the concept of manufacturing strategy has been paid much attention to and studied by many researchers in both theoretical and empirical ways. As the constitution of modern manufacturing systems have been changed with the advent of advanced technologies such as CAD, CAM, FMS etc., it is called for new perspectives in this line of inquiry. Ferdows & Lindberg[4] examined the link between the efforts of developing FMS and the strategic role of its manufacturing function. Lei & Goldhar[1] discussed the impact of CIM in modern manufacturing firms and concluded that strategy and organization design should be redefined to attain the intrinsic capabilities of CIM. Recently Parthasarthy & Yin[5] studied empirically the conditions under which the higher level of integration could enjoy the higher competitive performance. They showed that integration of parts in one subsystem should have a corresponding level of integration in other subsystem in order that the overall system is to be efficient. However, despite the role of systems integration is becoming important in modern manufacturing systems, the empirical efforts to figure out the strategic role of systems integration in manufacturing firms are hardly found in the present literature.

The purpose of this paper is a comparative study of manufacturing strategy process of firms with respect to the level of systems integration they attained. In more detail, it is attempted to explore the differences in competitive priorities and action programmes with respect to the level of systems integration through statistical analysis using the survey data. Thus the question this paper addresses on is whether there are any differences in manufacturing strategy and action plans between high and low level integrated firms and, if any, what and why.

2 Data and Method

The data used in this study are collected from the 1996 Japan Manufacturing Future Project (MFP) conducted by Waseda University s Manufacturing Roundtable. MFP is a world- wide collaborative research project initiated by Boston Univ. in U.S.A., INSEAD in Europe and Waseda Univ. in Japan since 1983, and now researchers from eleven countries are involved in this project. The MFP is specifically designed to examine the strategic aspects of operations management pertinent to leading manufacturing business units (MBU). The research framework of MFP is based on a behavioural model presented by Fig. 1.

---

[1] ISO / TC–184 / SC–5 / WG–1

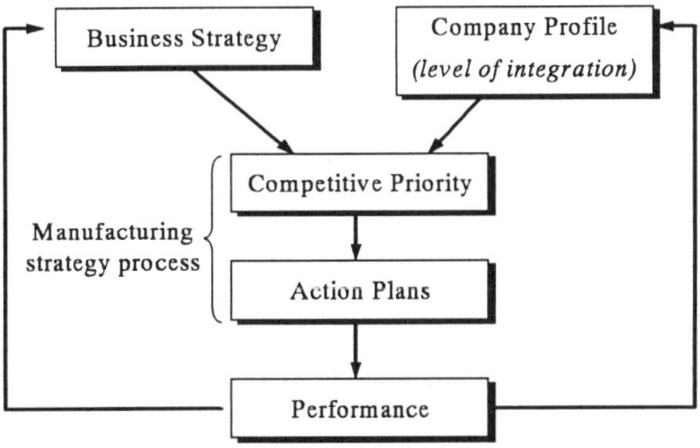

Fig. 1. Behavioural model of MFP

This study views the level of integration as one facet of company profile and discussion is centred around the disparity of manufacturing strategy process in the level of systems integration. Among two constructs representing distinct stages of the process of manufacturing strategy, competitive priorities describe what an MBU should achieve in order to support the business strategy effectively, of which dimensions cover cost, quality, flexibility, delivery and service. Action plans are the manufacturing improvement programmes which the MBU intends to implement in order to achieve its competitive priorities once determined.[6] A list of 16 competitive priorities and of 44 action plans were presented with a 7-point scale answer to each question in the questionnaire.

In Japan, more than a thousand of questionnaires had been mailed to relatively large manufacturing companies or business units and finally 182 companies responded to the survey, which represents a response rate of about 16 %. The sample was divided into two groups depending on the level of systems integration in a manufacturing business unit. A question from the survey used as a basis of sample division is as follows[2] :

'What do you think about the present level of systems integration in your company?'
1. Under contemplation, 2. Under development, 3. Some level, 4. High level

Those indicating an answer 1 or 2 were classified into a *Low-level* group and those indicating 3 or 4 into a *High-level* one. As a result, the *Low-level* group consisted of 123 respondents, the *High-level* group of 34, and there were 25 nil responses. Underlying proposition studied throughout this paper is that there are differences in manufacturing strategy between *High-level* and *Low-level* group. Both groups were then compared along a multitude of dimensions and variables of manufacturing

[2] In questionnaire of Japan, we added a special domestic section including this question to ask about how Japanese firms try to overcome the recent recession after the collapse of bubble economy.

strategy process. Toward this comparative analysis, we performed the statistical t-tests with the null hypotheses that there were no differences of each variable between the two groups[3].

3 Results and Discussion

3.1 Competitive Priorities

The section of questionnaire pertaining to competitive priorities asked companies the degree of importance for each item over the next five years. Table 1. shows the average scores and rankings for the samples of both groups.

As a whole, it is found that Japanese companies tend to place more importance on price and delivery rather than flexibility and service factors. Particularly the competitiveness based on price was revealed as one both groups gave the highest or the second highest priority. This tendency may result from at once the long recession in Japan and more importantly the globalized competition threatened by new competitors such as China and ASEAN countries.[7]

Table 1. Result of competitive priority

(*) : 5% significance, (**) : 1% significance

Competitive priorities	Low		High		t-value
	mean (std. dev.)	rank	mean (std. dev.)	rank	
PRICE					
Ability to profit in price	5.85(1.38)	1	5.97(1.22)	2	−0.46
FLEXIBILITY					
Make rapid design changes(*)	5.12(1.26)	12	5.53(0.93)	6	−2.07
Introduce new products quickly(**)	5.50(1.39)	8	6.21(0.81)	1	−3.77
Make rapid volume changes	4.87(1.24)	15	4.88(1.25)	14	−0.03
Make rapid product mix changes	4.97(1.23)	13	5.03(1.14)	13	−0.27
Offer a broad product line	5.31(1.30)	10	5.38(1.13)	10	−0.29
QUALITY					
Offer consistent quality	5.70(1.20)	4	5.47(1.29)	8	0.98
Provide high−performance product	5.58(1.21)	7	5.50(1.19)	7	0.32
Provide reliable product	5.83(1.10)	2	5.68(1.22)	4	0.73
Provide durable product	4.90(1.20)	14	4.88(1.18)	14	0.07
DELIVERY					
Provide fast deliveries	5.62(1.17)	5	5.74(1.14)	3	−0.51
Provide dependable deliveries	5.77(1.05)	3	5.59(1.21)	5	0.86
SERVICE					
Provide effective after−sales service	5.39(1.19)	9	5.18(1.40)	12	0.88
Provide effective product support	5.18(1.19)	11	5.41(0.89)	9	−1.07
Make a product easily available	4.83(1.46)	16	4.88(1.27)	14	−0.18
Customize products	5.59(1.11)	6	5.35(1.15)	11	1.09

3 In order to avoid methodological bias, a nonparametric statistical test (Mann−Whitney test) was also done and there were no significant differences between the result of both methods.

It is found that there are two variables on which two groups gave significantly different degree of importance in competitive priorities ; *ability to make rapid design change* and *ability to introduce new products quickly*. It can be paraphrased that the *High-level* integration group place a greater emphasis on the manufacturing strategies pertinent to the new product development process. This implies that once firms have achieved the integration of manufacturing systems, they endeavour to improve the capabilities of new product design and development which are considered as one of the most critical competitiveness in the recent capricious market environments.
One more intriguing interpretation can be applied to our result. Miller & Roth[8] suggested a manufacturing taxonomy of which contents are three distinct types of manufacturer such as innovators, marketeers and caretakers and they characterized innovators as those who are differentiated by their ability to make changes in design and to introduce new products. From this context, it can be figured out that the *High-level* group show the characteristics of *innovators* strongly.
It has been extensively stressed that one of the most viable competitive capabilities which can be exploited from the intrinsic technological benefits of the integration of manufacturing systems, in other words CIM, would be flexibility.[5] Seeing that all of flexibility items were given higher scores by the *High-level* group, the strategic perspectives of the firms which already reached systems integration to some degree are focused on attaining the flexibility of manufacturing functions by exerting integrated system to edge their competitiveness. On the other hand, all the items of quality were placed on a greater importance by the *Low-level* group.
From the mid-80s, some researchers proposed a new model of the relationships among competitive capabilities in manufacturing strategy against with the trade-off theory evoked by Skinner[3], called by the cumulative or sand-cone model.[9] In the model, it was asserted that in order to last improvements in manufacturing performances, amongst the factors of competitive capabilities, a company should concentrate, at first, on the quality improvements which would be a cornerstone for pursuing further capabilities such as dependability, speed and cost efficiency.[4] Relying on the cumulative model, it might be explained that the *Low-level* group are more concerned with relatively elementary competitive capability, quality, in comparison with the *High-level* group. Thus the results of the comparative analysis on competitive capabilities imply that the systems integration is one of the critical infrastructure of technology without which modern companies confronted with severe competition cannot enhance their manufacturing capabilities to higher level.

3.2 Action Plans

In another part of the questionnaire, the respondents were asked to indicate the degree of emphasis over the next two years for each action plan items in order to facilitate their goal of manufacturing strategy. In this section we performed the

---

4 There being divergent opinions on the priority order among the factors of competitive capabilities, it is widely presumed that the most basic one is the quality.

comparative analysis on the action plans between two groups in order to understand which programmes are planned to implement by the both groups to achieve their own manufacturing strategies. Table. 2 shows the average scores and rankings of action plans for the both groups.

Except for two items, *Integrating information systems within manufacturing* and *Supplier partnerships*, all of the items were given higher scores by the *High-level* group. That is, the *High- level* group are generally pursuing improvement programmes in manufacturing more aggressively. If we examine at the rankings of action plans, there are more or less differences in the both groups. The *Low-level* group are going to heavily investing in the integration of information technology not only in manufacturing systems but also across functions within business units. In other words, they want to improve systems integration above all. On the other hand, the *High-level* group are concentrating more on the product redesign and new product processes than integrating information systems. As observed earlier, the firms with highly integrated systems seek to improve flexibility, more concretely the ability of design change and new product introduction as a competitive leverage. Accordingly it is observed that they are paying the most attention to the action programmes which have something to do with such competitive priorities. Thus manufacturing strategy process of the *High- level* group show considerable consistency between competitive priority and action plans.

The statistical t-test result reinforces the foregoing contention. As a result of t-test, we obtained ten items which were given significantly higher emphasis score by the *High-level* group. We grouped them into four categories as follows ; 1) programmes for *product design* such as VA/product redesign and design for manufacture, 2) programmes for *process improvement* such as developing new processes for new products, developing new processes for old products, continuous improvement of current processes and complex robotics system, 3) programmes for *quality* such as automatic inspection/testing, SQC/SPC and ISO 9000, 4) programmes for *protecting environments* such as increasing the use of recycling materials. As a conclusion, regardless of systems integration level, most of recent manufacturers are concerned with integrating information system and moreover manufacturers might not engage in new product development, process improvement and quality up unless they achieve integrated manufacturing system to some extent.

In order to get some more compact implication from the analysis of action plans, the factor analysis was conducted to reduce the large set of action plan variables into the smaller number of underlying factors. The procedure of factor analysis adopted in this paper is as follows. Basically factors were extracted by running the least squares extraction method with the varimax rotation. After each run all the items that showed loadings of less than 0.35 on all factors, and items whose loadings were greater than 0.35 on more than two factors were dropped. This procedure was repeated until no such item was found. The result of factor analysis is shown by Table 3[5].

---

[5] As mentioned by Kim & Arnold[6], this procedure increases the parsimony of the subsequent analysis at the expense of the completeness of the study.

Table 2. Result of action plan

(*) : 5% significance, (**) : 1% significance

Action plans	Low		High		t-value
	mean (std. dev.)	rank	mean (std. dev.)	rank	
Empowerment	5.02(1.26)	13	5.09(1.36)	21	−0.25
Developing new performance measures	4.82(1.28)	17	4.88(1.51)	28	−0.22
Reorganizing plant networks	5.33(1.35)	4	5.42(1.14)	7	−0.35
Reconfiguring plant layouts	4.81(1.59)	18	4.88(1.47)	28	−0.24
Worker training	4.78(1.19)	19	5.11(1.12)	20	−1.47
Management training	5.01(1.21)	14	5.38(1.34)	9	−1.51
Supervisor training	4.96(1.27)	16	5.47(1.23)	5	−1.81
CAM	4.27(1.47)	36	4.44(1.52)	40	−0.58
CAD and/or CAE	4.70(1.54)	23	4.91(1.40)	27	−0.71
CIM	4.28(1.61)	35	4.70(1.42)	33	−1.36
Automated inspection / testing(**)	4.58(1.41)	27	5.33(1.29)	12	−2.73
Value analysis/product redesign(**)	5.12(1.23)	8	5.75(1.00)	1	−2.70
Cross-functional teams	5.08(1.10)	9	5.30(1.40)	14	−0.95
Functional teamwork	4.65(1.23)	24	5.03(1.19)	22	−1.56
New processes for new products(**)	5.14(1.35)	7	5.73(0.99)	2	−2.78
New processes for old products(*)	4.77(1.34)	20	5.29(1.03)	15	−2.07
Continuous improvement(*)	5.05(1.04)	11	5.47(1.18)	5	−2.00
Integrating IS within manufacturing	5.77(1.01)	1	5.69(1.16)	3	0.37
Integrating IS across the business unit	5.41(1.28)	2	5.41(1.21)	8	−0.01
Integrating IS in logistics	5.05(1.29)	11	5.36(1.27)	11	−1.23
Reconditioning physical plants	4.55(1.45)	29	4.84(1.29)	30	−1.02
Just-in-time materials management	4.23(1.45)	37	4.69(1.42)	34	−1.62
Simple pick & place robots	3.98(1.61)	41	4.48(1.60)	39	−1.58
Complex robotics systems(*)	3.05(1.71)	44	3.81(1.84)	43	−2.25
FMC and/or FMS	3.70(1.70)	42	4.18(1.67)	42	−1.43
DFM(*)	4.46(1.46)	30	5.03(1.24)	22	−2.25
Concurrent engineering	5.00(1.54)	15	5.32(1.14)	13	−1.13
SQC / SPC(*)	4.03(1.41)	40	4.56(1.04)	37	−1.99
Total Quality Management (TQM)	4.41(1.50)	31	4.76(1.46)	32	−1.18
Total productive maintenance (TPM)	4.41(1.46)	31	4.97(1.51)	25	−1.90
Closing or relocating plants	3.35(2.09)	43	3.57(2.09)	44	−0.55
Production & inventory system	4.14(1.56)	39	4.54(1.52)	38	−1.30
BPR	4.57(1.59)	28	5.00(1.60)	24	−1.37
Outsourcing manufacturing	4.21(1.60)	38	4.39(1.43)	41	−0.59
Supplier partnerships	4.74(1.20)	22	4.66(1.38)	35	0.32
Customer partnerships	5.34(1.38)	3	5.38(1.28)	9	−1.13
Improving the quality of work life	4.62(1.25)	26	4.94(1.09)	26	−1.35
Benchmarking	4.38(1.54)	33	4.61(1.23)	36	−0.77
Certification to quality standard	5.17(1.31)	6	5.29(1.24)	15	−0.47
ISO 9000(*)	5.07(1.48)	10	5.65(1.23)	4	−2.06
Tacking back products from customers	4.38(1.79)	33	4.84(1.60)	30	−1.34
Processes to protect environment	4.77(1.52)	20	5.12(1.41)	19	−1.20
Increasing the recyclable materials(*)	4.65(1.60)	24	5.29(1.50)	15	−2.09
Developing manufacturing strategy	5.24(1.28)	5	5.26(1.28)	18	−0.11

Table 3. Result of factor analysis for action plans

	1	2	3	4	5	6	7
Factor 1 ; *Human-related*							
Supervisor training	.842						
Worker training	.825						
Management training	.756						
Empowerment	.428						
Factor 2 ; *Environments-related*							
Increased use of recyclable materials		.861					
Taking back products from customers		.790					
Processes to protect environments		.756					
Factor 3 ; *Computer application-related*							
CAM			.903				
CAD/CAE			.682				
CIM			.647				
Factor 4 ; *Design/KAIZEN-related*							
Continuous improvement				.639			
Reconditioning physical plants				.578			
Value analysis/product redesign				.519			
Simple robots				.515			
Design for manufacture				.486			
Factor 5 ; *Integration-related*							
Integrating IS across the BU					.814		
Integrating IS within manufacturing					.668		
Integrating IS in logistic system					.666		
Factor 6 ; *Supply chain/restructuring -related*							
Outsourcing manufacturing						.524	
Supplier partnership						.506	
BPR						.468	
Bench-marking						.426	
Factor 7 ; *Quality-related*							
ISO 9000							.526
Certification to quality standard							.506
Processes for old products							.452
Eigenvalue	7.78	2.33	1.82	1.62	1.48	1.21	1.03
% Variance explained	33.1	9.3	7.3	6.5	5.9	4.8	4.1

As illustrated by Table 3, the seven distinct factors which have considerable explanatory power (71%) for the variances were obtained. From the forty-four variables presented originally in the questionnaire, nineteen variables were eliminated by the procedure described earlier, leaving only twenty-five variables in the final seven factors. The title of each factor was determined to represent the included variables as closely as possible. As a result of factor analysis, the seven constructs of action plans, *Human-related* action plan, *Environments-related* action plan, *Computer application-related* action plan, *Design/KAIZEN-related* action plan, *Integration-related* action plan, *Supply chain/restructuring-related* action plan and *Quality-related* action plan, were suggested in this study.

In the sequel, we conducted the comparative analysis on the above 7 constructs of action plans. In the procedure of t–test, the scale value of each factor was determined by a simple average of the included variables. Table 4. shows the results of t–test.

Table 4. Result of action plan factors

(!) : 10% significance, (*) : 5% significance, (**) : 1% significance

Factors of action plans	Low		High		t-value
	mean (std. dev.)	rank	mean (std. dev.)	rank	
Human–related	4.95(1.01)	3	5.25(1.12)	3	–1.47
Environments–related (!)	4.60(1.48)	5	5.11(1.39)	5	–1.78
Computer application–related	4.43(1.33)	7	4.69(1.31)	7	–1.01
Design/KAIZEN–related(**)	4.64(0.94)	4	5.13(0.96)	4	–2.71
Integration–related	5.41(1.02)	1	5.46(1.17)	1	–0.26
Supply chain/restructuring–related	4.49(1.08)	6	4.70(1.05)	6	–0.98
Quality–related(*)	5.02(1.00)	2	5.41(0.82)	2	–2.08

At first it is interesting to note that the rankings were not changed by the level of integration ; both groups were most interested in the integration– related programmes. In another research based on 1996 MFP data[7], it is discussed that Japanese manufacturers are in the midst of redefining their manufacturing strategy, from what is often described as 'Japanese manufacturing methods' such as JIT, SQC, TQM etc. to the investment in managerial and product development related information technology. Along with this structural change, they may well invest heavily in integration of information systems ranging over their entire organization as it is witnessed in this study.

As expected from the result of the item by item analysis presented by Table 3., the comparative analysis on factors also gives a result that the *High–level* group is placing greater emphasis on Design/KAIZEN and Quality– related constructs, without at the expense of putting less emphasis on any of the other 5 constructs. This result ascertains our contention that the companies with high–level integration are working on operationalizing their manufacturing strategy to sharpen the competitive capability of new product development through heavily investing in the action programmes pertinent to design, process and quality.

Moreover it is found that Environments– related factor was also considered as an important action plan by the *High– level* group. Environmental problem is now regarded as an inevitable one and manufacturers also should try to convert their manufacturing systems toward the nature–friendly direction. Considering the Green Round agreement on discussing now, we can conceive that green manufacturing system would be an important competitive weapon which a leading company should be armed with. Thus our result shows that the *High–level* group are being prepared to the circumstances of the coming market in advance and want to be a forerunner by exploiting the advantages of highly integrated manufacturing system.

4 Conclusion

Throughout this paper, we have explored to look for differences of manufacturing strategy process between manufacturers of highly integrated system and those not. The subject of this study came from the conjecture that the integration of manufacturing system would influence the manufacturing function in a company whereby the strategic role of it might be altered.

The differences observed in this study were quite outstanding, as well as intuitive. Our empirical data reveals that the manufacturers with highly integrated system are focusing their competitive capabilities on new product development strategy. They give the competitive priority to the rapid design change and new product introduction, and accordingly they place a greater emphasis on the action programmes related to design, process and quality improvement. This finding is not surprising if we remember it is flexibility that integration provides the firm's manufacturing function with.

There is no manufacturer but has interests in systems integration in these days. The results of this study explain in part for what the companies try to implement the integrated manufacturing system. Highly integrated firms are likely to want to be a leader, not a follower, in market and seem to act as such. They pursue to improve the capabilities of new product design and introduction and to be a preemptor in the new market by exploiting the advantages of integrated manufacturing systems. To think conversely, a manufacturer probably can not develop a new product timely as well as successfully unless its manufacturing system should be integrated to some extent. Hence the systems integration might be deemed as a technological avenue to cope with the challenges of competitive market in manufacturing.

References

1. D. Lei and J. D. Goldhar, Computer-integrated Manufacturing : Redefining the Manufacturing Firm into a Global Service Business, *Int. J. of Operations & Production Management*, **11**(1991) 5–18
2. A. De Meyer, The Integration of Information Systems in Manufacturing, *OMEGA*, **15**(1987) 229–238
3. W. Skinner, *Manufacturing in the Corporate Strategy*, New York: Wiley(1978)
4. K. Ferdows and P. Lindberg, FMS as a Indicator of the Strategic Role of Manufacturing, *Int. J. of Production Research*, **25**(1987) 1563–1571
5. R. Parthasarthy and J. Z. Yin, CIM and Competitive Performance: Moderating Effects of Organization-wide Integration, *J. of Eng. & Tech. Management,* **13**(1996) 83–110
6. J. S. Kim and P. Arnold, Operationalizing Manufacturing Strategy : An exploratory study of constructs and linkage, *Int. J. of Oprns. & Prodn. Mgmt.*, **16**(1996) 45–73
7. A. Meyer, H. Katayama & J. S. Kim, Building Customer Partnerships as a Competitive Weapon?, *Proc. of the Int. Sympo. of Manufacturing Roundtable*, INEAD, France, 1996
8. J. G. Miller and A. V. Roth, A Taxonomy of Manufacturing Strategy, *Management Science*, **40**(1994) 285–304
9. M. A. Noble, Manufacturing Strategy :Testing the Cumulative Model in a Multiple Country Context, *Decision Science*, **26**(1995) 693–721

Flexibility and Change Management

Flexibility in Supply Chain Processes

N. Viswanadham Y.S. Wong C.K. Chan

Department of Mechanical and Production Engineering
National University of Singapore
Singapore

Abstract. In this paper, we view a manufacturing enterprise as a collection of value delivery processes such as new product development process, supply chain process, order to delivery process etc. We are interested in the flexibility issue of such an enterprise and in particular the supply chain process (SCP). We define the delivery time, volume, mix and new product flexibility for a supply chain process and discuss the source factors for designing these flexibility types into SCP. We propose measures for all the above four types of flexibility.

1 Introduction

Traditionally, manufacturing systems are viewed as a sequential arrangement of functions such as design, manufacture, marketing and finance etc. The recent trend is to consider manufacturing enterprises, which consist of suppliers, distributors, retailers and customers. We view such an enterprise as a bundle of value delivering business processes such as new product development process, supply chain process, order to delivery process etc, and study the issue of the performance analysis of the manufacturing enterprises. We focus in particular on the supply chain process, which is the most important mega-process for a manufacturing company.

A supply chain process starts with the procurement of raw materials and ends with the product delivery to the retailer or customer as the case maybe. We define the cycle time, cost, quality and flexibility as the important performance measures. In this paper, we concentrate on developing definitions and measures for flexibility in supply chain processes. Flexibility is fast becoming the next sought after competitive capability. It can help to shorten manufacturing lead times, reduce batch sizes and inventory levels, introduce new products more frequently; it also helps to support market segmentation and product customization.

The existing literature on supply chain management (Bowersox *et al* 1992 and Schary 1995) fails to give a systematic and rigorous treatment of the flexibility issue, especially in terms of empirical examples and flexibility measurement. Lead time and cost are the more commonly used performance measures for a supply chain process as they are easily quantifiable. Hence, there exists great motivation to develop flexibility measures for a supply chain process. In this paper, we attempt to develop measures for *delivery time, volume, new product* and *mix* flexibility.

This paper is organized as follows. In Section 2, we define a manufacturing enterprise as a collection of business processes and consider issues of analysis using this approach. In Section 3, we define the various types of flexibility. We also present the various ways to achieve flexibility throughout the supply chain. In Section 4, we propose flexibility measures.

2. The Manufacturing Enterprise

The manufacturing enterprise consists of several internal and external subsystems. The internal subsystems include factory floor, integrated logistics, organization structure, and product development. The external subsystems include suppliers, distributors, customers, competition, and technology. All these have to act in cohesion so that customer desired products are delivered on time.

Traditionally, manufacturing systems are viewed as a sequential arrangement of functions such as design, manufacture, R & D, marketing, finance and so on. The recent trend is to view a manufacturing system not in terms of functions, divisions, or products but as a collection of value delivering processes. Functional or hierarchical structures typically present responsibilities and reporting relationships whereas process structure is a dynamic view of how the organization delivers value to the customer. We now present a formal definition of a business process.

2.1 The Business Process

A process is a structured, measured set of activities ordered in time and space, designed to produce a customer desired output [Davenport 1993].

A process perspective is a horizontal view of the manufacturing system that cuts across the enterprise with product inputs at the beginning and customers at the end. This approach emphasizes on how work is done rather than focus on specific products. Clearly, structured processes are amenable to measurement in a variety of dimensions: cycle time, defects, variability, flexibility etc. Examples of some typical processes include: supply chain process, order to delivery process, new product development process, customer acquisition process, etc.

In this paper, we are concerned about the flexibility issues in the supply chain process. Before proceeding further, we define a supply chain process (SCP).

2.2 The Supply Chain Process

A supply chain process encompasses the full range of intra- company and inter-company activities beginning with raw material procurement by independent suppliers, through manufacturing and distribution, and concluding with successful delivery of the product to the customer (Cohen and Lee 1986).

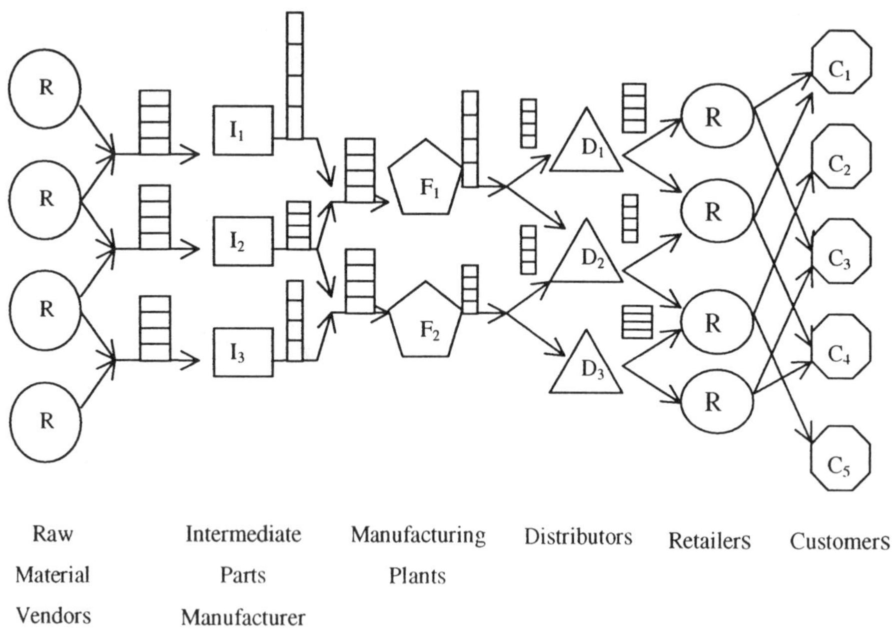

Figure 1 The Supply Chain Network

This is the most important mega process involving the customer for a manufacturing company (See Figure 1). It is triggered by an order from the customer. It starts with procurement and ends with the use of the product by the customer. In the literature, supply chain management (SCM) is the word used for managing the supply chain process.

2.3 Performance Measures for a Supply Chain Process

Performance measures provide the basis for determining the effectiveness and efficiency of the supply chain process as well as for comparing the performance with competing manufacturing enterprises. The critical measures for a supply chain process are: cycle time, quality, process flexibility, cost and process capability. Cost is included as a measure although properly done cycle time reduction will automatically result in cost advantage.

2.3.1 Cycle Time

We define the cycle time of the supply chain process as the length of the longest path from procuring raw materials till delivery to the customer. In order to improve the cycle time of the entire business process, we need to simplify the process by: eliminating non-value adding activities, grouping some of the activities together, using new technologies such as databases, expert systems, electronic data interchange, and reducing the waiting time between organizations and functions by managing the interfaces between them using process owner or case worker kind of coordination. Alliances between organizations along the supply chain also help to cut down the cycle time.

2.3.2 Quality

Quality is the hallmark of a competitive product. It is directly related to customer satisfaction, which involves the delivery of high quality products at the right time, right place and in the right quantities. Thus, quality is also translated to mean *delivery reliability* both in terms of order fill and time of delivery. This requires the defect free working of all subsystems within the supply chain from procurement to delivery.

2.3.3 Process Flexibility

Here we look at flexibility of the entire supply chain process. It is necessary to create flexibility throughout the end-to-end process i.e.: suppliers to retailers to logistics. This is because creating highly flexible subsystems such as the factory floor or automated warehouse with others being rigid may not be beneficial at all. For example, if the supply chain is to be new product flexible, i.e. able to design, prototype and bring to market new products, it is necessary for both the suppliers and manufacturers to be new product flexible, the logistics to transport new products, and the distributors and retailers to sell them to the customers. The four flexibility types that are relevant to the SCP are *delivery time, volume, new product* and *mix*. The flexibility measures are presented in Section 5.

3 Flexibility in the Supply Chain Process

Flexibility is the ability to react effectively to change with little penalty in time, effort, cost or performance. Its effective management involves devising ways and means to cope up with uncertainties, which could be operational (failures, rush orders, volume and variety changes) or tactical (component improvements and new product development in the same range) or strategic (disruptive technology changes) in nature. This section discusses the delivery time, volume, new product and mix flexibility in supply chain processes.

3.1 Delivery Time Flexibility

A supply chain process is ***delivery time flexible*** *if it possesses the capability to reduce or increase the cycle time from procurement of raw materials to delivery of the product to the* *customer.*

The supply chain process is the most important business process involving the customer for any manufacturing company. Today, the customers are demanding shorter and shorter lead times. Some customers demand daily delivery in small lot directly to their factory floor. There may also be situations when customers demand that a delivery be postponed for one reason or another e.g.: due to lack of warehouse space, delays in production schedules, etc.

The delivery time is affected by several programs currently in practice along the supply chain such as supplier management, strategic alliances between manufacturers and logistics operators, formation of business networks based on core competencies, etc. The key issue is time-based competition. All the companies in the supply chain process share a common destiny and the customer sees only the benefit/costs accrued at the end of the chain and not in between. Each one is an internal customer to the other and finally the end user is a customer to all of them. Defect free, on-time delivery to the final customer requires that each partner operates without defects and keeps to the delivery schedules. This is the essence of time based competition (TBC).

Operational uncertainties will lead to variation in delivery time performance. Variation refers to the tendency of a process to produce different results under almost identical conditions. The level of variation will directly affect the level of customer satisfaction. Hence, achieving short mean delivery time alone is insufficient. It is equally important to deliver such results with consistently low variability. The goal should then be to achieve the delivery time target all the time. Hence, there exists ample motivation to understand the possible causes that will lead to a failure in meeting delivery time target. The cause and effect diagram is a good tool to achieve this purpose.

Since the supply chain process is an interconnection of subprocesses, activities and tasks, the variation in the chain is the sum total of the individual variation in the various tasks and interfaces (Montgomery 1985). In order to control the variation, it is essential to control each of the individual tasks and their interfaces. From the control limits of the overall supply chain process (also called the delivery window), we have to determine the allowable limits on processing time for each of the subprocesses, individual activities and tasks. The next step is to devise mechanisms to perform these tasks within those time limits.

3.2 Volume Flexibility

A supply chain process is **volume flexible** *if the volume level and the composition of the product mix can be changed within a short time frame.*

Volume flexibility is an important property of supply chain process in a high technology industry. Swings in final product demands get reflected with a time lag as changes in volumes of intermediate products and components. Good forecasting techniques for the end products and related products along with the product in consideration will help reduce the variability in demand.

It is essential that small batches of products be produced and delivered for a process to be volume flexible. It is thus important to achieve low set-up times at each stage of the supply chain. Human resource management also has a role in volume flexibility. Plants that use temporary workers should have an advantage in adjusting to volume changes. Also, plants that have wage structures linked to plant or division performance would also be better in adjusting to volume fluctuations. The fact that wages are linked to performance automatically reduces the payroll in periods of low sales volume or crisis.

3.3 New Product Flexibility

A supply chain process is **new product flexible** *if generations of new products can be manufactured and marketed rapidly.*

With shrinking product life cycles, it is very important that each generation of products incorporates the latest product technology and reaches the market fast to maximize first mover's advantage. This flexibility will require capabilities like rapid tooling, fast process engineering, fast commissioning of new product lines and highly responsive logistics operations to cope with the changes in volume (Wong 1996). It is also essential that the suppliers and distributors possess capabilities for new product design and manufacture and new product launch respectively.

3.4 Mix Flexibility

A supply chain process is **mix flexible** *if a number of different products can be simultaneously produced and delivered to the customers.*

Mix flexibility is important when a firm has a broad product line and caters to different market segments. Suppliers play a very important role in contributing to mix flexibility. The suppliers must possess the agility to offer a wide array of component parts i.e.: suppliers' flexibility in several dimensions including delivery time, mix, volume and new product. The alternative is to have a great number of suppliers but this will increase the management overhead enormously. In addition, the warehouse and transportation facilities should be able to handle different sizes, shapes and installation procedures.

Distributors form another integral force to achieve mix flexibility. They should possess the ability to sell the varieties onto the customer through effective and efficient marketing and distribution, and by maintaining the appropriate stock level for each product. Also, they can gather first-hand information about the needs, preferences and problems of customers. By maintaining a close relationship with distributors, such vital information is constantly and promptly feedback to the mañufacturer to ensure that the company produces only products that sell rather than trying to sell what are produced.

4 Flexibility Measures

Measures are important because they provide the basis for comparing competing systems. In this section, we propose measures for mix, volume, new product and delivery time flexibility.

4.1 Delivery Time Flexibility

The concept of process capability index can be used to measure the delivery time flexibility. There are two key conditions that must be satisfied before process capability measurements are meaningful:

i. The process must be in a state of statistical control i.e.: all special causes are removed from the process.

ii. The process is normally distributed.

There are two possible scenarios in meeting a customer's delivery time target. The customers can either demand that a delivery be made by a certain date or they can ask that a delivery be postponed to a later date. The first scenario will have an upper delivery time limit to meet and no lower limit (time compression). The second scenario will have a lower delivery time limit to meet and no upper limit (time extension).

The delivery time flexibility measure for each case is as follows:

<u>*Time compression*</u>

$$\textit{Delivery}\ \text{Time Flexibility} = \left(\frac{USL-\mu}{3\sigma}\right) \tag{1}$$

Time extension

$$\textit{Delivery}\,\text{Time Flexibility} = \left(\frac{\mu - LSL}{3\sigma}\right) \quad (2)$$

where, σ is the standard deviation of the delivery process
USL is the upper delivery time limit
LSL is the lower delivery time limit

4.2 Volume Flexibility Measure

Before we propose a measure for volume flexibility, it is important to emphasize the distinction between volume fluctuation and volume flexibility. Only the latter is a desirable attribute of a manufacturing system. The volume fluctuation measure is based on data about the ratio of highest monthly production to lowest monthly production for three years.

The volume fluctuation is given by:

$$\text{Volume Fluctuation} = [((R1)^2 + (R2)^2 + (R3)^2)/3]^{1/2} \quad (3)$$

where R1 is the ratio of the highest monthly production to lowest monthly drive production for year 1. (Note that, a simple variance measure for 36-month production would have been more desirable if the data is available.) (Suraez *et al* 1996)

Volume fluctuation does give an estimate for the volume flexibility level, but it is certainly not the complete picture. There may be situations when volume fluctuation brings with it a cost hike or quality level deprovement. Thus, the flexibility implied by larger volume fluctuation must be weighted against the effects on cost and quality. The reasoning is as follows: a plant that is able to shrink and expand its production volume widely and still keep its cost low and quality level high, is more volume flexible than a plant that presents the same volume fluctuation but has a much higher cost and lower quality. Thus,

$$\text{Volume Flexibility} = \frac{\text{production volume fluctuation}}{\text{(avg cost per unit) (fraction of output with repair)}} \quad (4)$$

Higher scores indicate more volume flexibility. Other things being equal, the measure penalizes plants with higher cost or lower quality levels.

4.3 New Product Flexibility Measure

New product flexibility is proportional to the amount of time that is required to ramp up production of a new product model for high volume production. However, the cost to achieve this ramp-up production must also be taken into consideration. A drive company may be able to achieve very superior ramp-up capability by paying a very high price to the suppliers for their services. In this case, the new product flexibility is achieved at the expense of competitive pricing. Many companies also tend to rush into volume production before the products are fully tested, with significant risks of quality and yield. This is certainly undesirable. Hence, it is very important to take into account the cost and quality issues for the ramp-up. Thus,

$$\text{New Product Flexibility} = \frac{\text{time to set up one new product line}}{\text{(cost of new product line)(fraction of output with repair)}} \quad (5)$$

4.4 Mix Flexibility Measure

The most common measure for mix flexibility is the number of products that a system produces at any point in time. It gives a good indication of the breadth of the product line. However, such a measure is too simplistic. Mix flexibility cannot be simply measured by the simple count of the different products. Rather, one must measure the heterogeneity of the product set (Suarez *et al*

1996). A simple example will illustrate this. Two automobile manufacturing plants, A and B, each producing two types of products. Plant A produces A1 and A2, while plant B produces B1 and B2. Both plants are equally mix flexible if they are assessed according to the number of products they produce. However, A1 may be significantly different from A2 (say A1 is a family sedan and A2 is a jeep that comes with special features like *four –wheel-drive*, 4WD) whereas B1 and B2 are both ordinary family sedans that differ merely in terms of color and engine. Hence, a rigorous analysis is needed before we can decide whether plant A or B is more flexible.

In this paper, we will demonstrate the development of the mix flexibility measure using disk drives as an example. Disk drives differ in terms of form factor, storage capacity, interface type etc. Simply counting the number of different drive models will not give us a good measure for mix flexibility. Thus, we use the following 6 variables to give a more accurate measure. These variables capture physical, technological, quality and market differences: *drive model, form factor, interface type, spindle speed, storage capacity* and *MTBF (Mean Time Between Failure) ratings*.

We will develop an aggregate measure that includes each of these variables. The mix flexibility measure is developed as follow. First, rank each variable based on its importance in differentiating products. A 10 point scale is used, where 1 = not important and 10 = most important. The next step is to score the actual performance of the company for each variable. The company that performs the best for that variable would be assigned a value of 10. The score of the other company's performance for that variable is then compared to the "10" and ranked based on its relative value and impact. For example, company A produces drives with 5 different types of interfaces while company B only uses 4 different interfaces. Hence, the score for company A will be 10 and that for company B will be 8 (4/5 x 10 = 8).

Table 1 Product Mix Summary

Decision Variables	**Importance Level**	*Seagate*	*Western Digital*	*Maxtor*
No. of drive models	10	11	3	4
No. of storage capacity groups	9	7	5	5
No. of spindle speed groups	7	5	3	2
No. of interface types	6	5	2	1
No. of form factor groups	4	4	2	2
No. of MTBF groups	3	4	3	3

The product mix for Seagate, Maxtor and Western Digital is found in Table 1. The mix flexibility measure for each company is computed in Table 2. We conclude that Seagate is the most mix flexible among them, followed by Western Digital and finally Maxtor.

Table 2 Mix Flexibility Computation

Decision Variable	Importance Level	**Seagate**		**Western Digital**		**Maxtor**	
		Performance	*Score*	*Performance*	*Score*	*Performance*	*Score*
Drive Models	10	10	100	2.73	27.3	3.63	36.3
Storage Capacity	9	10	90	7.14	64.3	7.14	64.3
Spindle Speed	7	10	70	6	42	4	28
Interface	6	10	60	4	24	2	12
Form Factor	4	10	40	5	20	5	20
MTBF	3	10	30	7.5	22.5	7.5	22.5
Mix Flexibility		***390***		***200.1***		***183.1***	

5 Conclusion

In this paper, we have provided a complete definition of flexibility in an integrated supply chain process. The point we emphasized here is that flexibility efforts should not be confined to the factory floor only. The entire supply chain must come into play in order to achieve global flexibility. Flexibility management is a competence that involves skillfully managing several resources of the system including suppliers, distributor channels, customers, factory floor control systems, automation hardware, software, people, organization structure and even product design.

Acknowledgements

We would like to acknowledge Mr. Lau Teck Seng (Director DSI), Dr. Chang Kuan Teck and Mr. Soh Swee Hock of Data Storage Institute, for their keen interest in our research, and the time they spent in discussions and in helping us with the visits to various disk drive companies. Finally, we are also very grateful to the managers from Seagate, Maxtor and Western Digital for their participation in our discussions.

References

1. Bowersox, and Gloss, D. J., 1996, *Logistical Management* (McGraw Hill).
2. Cohen, M. A., Lee, H. L., 1986, Strategic Analysis of Integrated Production Distribution Systems: Models and Methods, *Operations Research*, Vol. 36, No. 2, 216-228.
3. Davenport, T. H., 1993, *Process Innovation* (Harvard Business School Press).
4. Maxtor, Homepage, http://www.maxtor.com/products.html dated 22 Jan 1997.
5. Montgomery, D. C., 1985, *Introduction to Statistical Control* (USA: John Wiley and Sons).
6. Phun, K. M., 1997, Competence Based Competition versus Product Centric Based Competition, Technical Report, Department of Mechanical and Production Engineering, National University of Singapore.
7. Schary, P. B., 1995, *Managing The Global Supply Chain* (Copenhagen, Denmark: Handelshojskolem Forlag).
8. Seagate, Homepage, http://www.seagate.com/disc/prodmatrix.shtml dated Mar 1997.
9. Suarez, F. F., Cusumano, M. A., Fine, C. H., 1996, An Empirical Study of Manufacturing Flexibility In Printed Circuit Board Assembly, *Operations Research*, Vol. 44, No. 1, Jan – Feb, 223 – 240.
10. Western Digital, Homepage, http://www.wdc.com/products/drives/general.html dated 1997.
11. Wong, P. K., 1996, Supply Infrastructure for Flexible Production: The Case of Hard Disk Drive Industry In Singapore, Technical Report, Faculty of Business Administration, National University of Singapore.

Flexibility: Its Meaning for Japanese Manufacturers

Hiroshi Katayama, Yoshikazu Azami and Deok-Joo Lee

Department of Industrial and Management Systems Engineering,
School of Science and Engineering, Waseda University, Japan

Abstract. In this paper, in order to clarify the essential factors of manufacturers' understanding about flexibility, which is widely noticed as a key concept by leading firms, a survey was performed and its result was analysed. The data collection was performed by sending a set of questionnaires to senior manufacturing executives. Number of respondent manufacturers of this survey is up to 182 business units of companies and the basic methodology of flexibility structure analysis adopted in this paper is discriminant analysis as well as Hayashi's qualitative analysis model II, which enable to extract some significant factors to differentiate ideas and/or considerations on flexibility issues of two categories of firms, *i.e.* flexibility focused firms and flexibility non-focused firms.

1 Introduction

These years, Japanese manufacturers have been struggling with many difficulties in their business such as shrinking sales revenue, cost inflexibility, which suddenly emerge as a result of explosion of bubble economy in early '91. They have suffered from high Yen exchange rate, low level of domestic demand in industrial as well as consumers' products and in rush of cheaper products from developing countries. To cope with such tough problems, manufacturers try to launch new activities such as offshore operations, outsourcing, reengineering by implementing information technologies. However, these activities cause, on the other hand, other new problems they have to deal with, *e.g.* management technology transfer, global logistics, cavity of industrial structure.

In this situation, one of the fundamental problems manufacturers have to consider is the existence of confusion in the strategy making and implementation of action programmes to accomplish their targets. Actually, competition in the market of shrinking demand and shorter life cycle of products compelled manufacturers to improve their competitive capability on flexibility. Therefore, this paper focuses to discuss actual meaning of flexibility for manufacturers and future directions, where they are pursuing with flexibility enhancement, through identifying the characteristics of flexibility focused firms.

As one of the related studies on manufacturing strategy [3], [4], Ferdows and De Meyer [1] proposed a new model that represents the time-based relationship among the issues of manufacturers strategic directions, called "the sand-cone model". In this model, it was asserted that there is a specific priority among manufacturing strategy such as quality, dependability, speed and cost efficiency, for lasting improvement of manufacturing capabilities. In addition, especially centring on the flexibility, Tombak and De Meyer [5] discussed about how the manufacturers' concerns with this issue, which are planning to implement FMS, differ from those of other manufacturing firms. They showed empirically that managers were implementing FMS in order to achieve the flexibility not of changing the product design, but of dealing with variety in the production systems.

In this paper, in order to clarify the essential factors of manufacturers' understanding about flexibility, which is widely noticed as one of the key concept by leading firms, the result of a survey is analysed and reported. The data collection is performed by sending a set of questionnaires to senior manufacturing executives. The number of responded manufacturers is up to 182 business units of companies and the basic methodology of flexibility structure analysis adopted in this paper is discriminant analysis as well as Hayashi's qualitative analysis model II [2], both of which enable to extract some significant factors to differentiate ideas and/or considerations on flexibility issues of two categories of firms, *i.e.* flexibility focused firms and flexibility non-focused firms.

2 Methodology

2.1 Flow of the procedure

Outline of the procedure applied to perform this study consists of three major steps illustrated in Figure 1, *i.e.* classification of the firms (*Step 1*) based on their responded scores on the strength of flexibility concerned strategy, extraction of sub- consciousness about each group's thought on flexibility (*Step 2*) and extraction of firms' trend or direction of the flexibility issues moving toward (*Step 3*). Details of each step and result are described in the following sections.

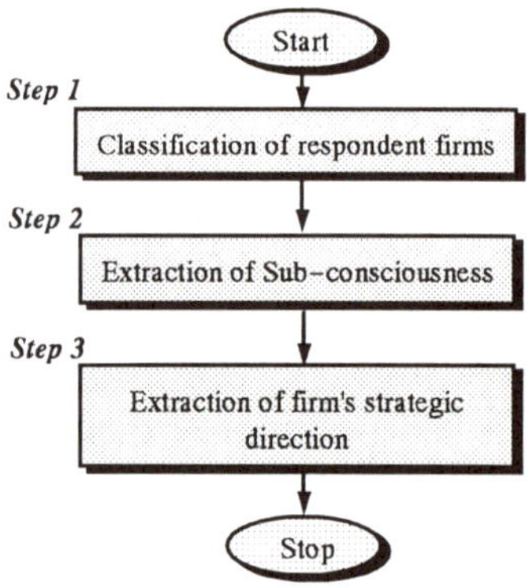

Fig. 1. The proposed procedure

2.2 Data set from a survey

For the purpose to clarify the relation between strategy and action programmes of Japanese manufacturers, a large scale survey project was launched in late 1996. Relatively large manufacturing companies or business units are selected because of their capability to devote in their strategy making, execution of action programmes and their scale of influence in the society.

This survey was performed as Japanese domestic part of international-wide survey called Manufacturing Futures Survey Projects International (MFP- International), which was initiated in 1983 in order to develop a comprehensive data set that can be used for studying the strategic aspects of operations management, and was initially conducted by Boston University in USA., INSEAD in France and Waseda University in Japan. Now its coverage is expanded by involving emerging industrialising countries such as Australia, China, Korea, Mexico, New Zealand, Singapore, Taiwan, Thailand *etc*. The purpose of this survey is to recognise the essential understanding, critical problems as well as strategic direction and action programmes of Japanese domestic manufacturers, and among these problems, this paper especially focuses on flexibility issues, because manufacturers can not avoid to consider the enhancement of such function in order to cope with volatile customer demands.
The behavioural model of manufacturing companies developed as MFP scheme is the set of directions, activities and performance in terms of their competitive priorities based on business strategies and their profile, their action programmes based on their settled competitive priorities and their performance results based on their adopted action programmes. Competitive priorities mean the pattern of priority assignment to each manufacturing strategy, *e.g.* cost/price competitiveness(C) , quality competitiveness(Q), delivery competitiveness(D), flexibility competitiveness(F) and service competitiveness(S). Within these, six issues given in Table 1 are considered as the major possible flexibility concerned competitive priorities. Action programmes mean companies' activities to accomplish the objectives settled by manufacturing strategy, and this category includes forty-four issues in total listed in Table 3. Performances are measured by up to twenty-three indicators listed in Table 7. The data of these issues are analysed in the subsequent sections to extract flexibility-related sub-conscious factors which are hidden in these.

2.3 Classification of respondent companies

For the survey of MFP-Japan in 1996, about 1300 questionnaire booklets had been mailed, and 182 business units of respondent companies as a result, which indicates about 14% response rate. In this survey questionnaires, 6 categories of manufacturing competitive priorities, as shown in Table 1, are settled, which are pertinent to flexibility and asked companies the degree of importance of each ability over the next 5 years on a 7-point scale (1 = least important, 7 = most important). The average value of company' s indication for these six items was calculated and taken as an aggregated flexibility score of each company.
Classification of responded companies was based on this score, *i.e.* the upper one-third and lower one-third percentile firms were classified as flexibility focused and flexibility non-focused firms respectively. Figure 2 shows the distribution of aggregated flexibility scores over the sample. Based on the MFP framework represented by Figure 3, in comparison with flexibility non-focused firms, it is examined what kind of action programmes the flexibility focused firms have invested in the past, how well their performance have been improved by those investments and, henceforth what kind of improvement programmes they are placing an emphasis on.

Table 1. Flexibility concerned competitive priority issues (*Source*: MFP–International)

Items of competitive priority
1. Ability to make rapid design changes
2. Ability to introduce new products quickly
3. Ability to make rapid volume changes
4. Ability to make rapid product mix changes
5. Ability to offer a broad product line
6. Ability to customise products to individual customer's needs

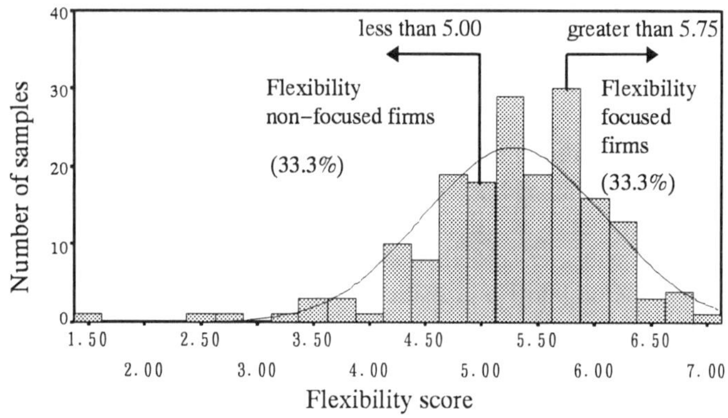

Fig. 2. Distribution of aggregated flexibility score

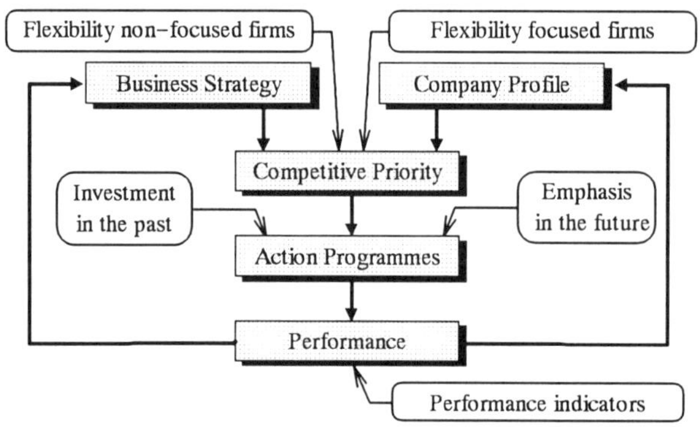

Fig. 3. Behavioural scheme of manufacturing company (*Source*: MFP–International)

3 Results of Analysis

Some important results from the analysis, which try to reveal companies' sub–consciousness on flexibility issues imbedded in their action programmes and manufacturing performance, are described in the following sections.

3.1 Overall tendency about competitive priority on the flexibility issues

Figure 4 represents the recent trend of each macroscopic competitive priority meas–

ure obtained by calculating average of whole scores of concerned competitive priority issues. From this graph it can be noticed that the flexibility score has a precise decreasing trend with, even, the least scores among the other manufacturing strategy factors throughout the last eight years. This indicates that the concept of flexibility is still not concretely understood and popular at all, although it is widely discussed among manufacturers.

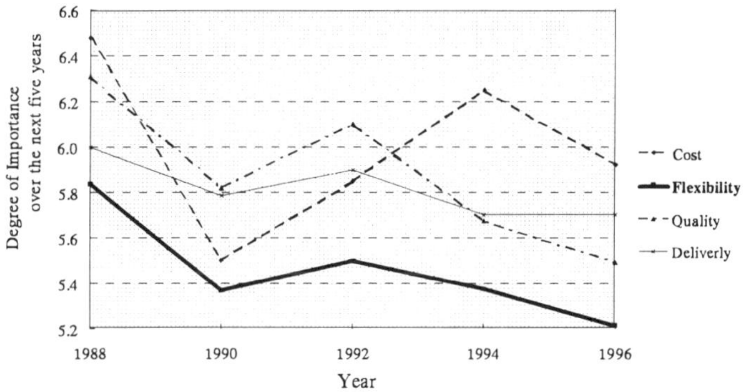

Fig. 4. Time series of competitive priority scores (*Source data*: MFP–Japan)

3.2 Discriminant analysis of action programmes

Figure 5 and 6 illustrate score dispersion of flexibility concerned action programmes together with others on current–future two dimensional plane, of which the former is precisely denoted by their names. In these figures, called gap analysis between current and future scores, the solid line represents a line with the slope of 45° which means the set of the same score of the past and the future emphasis, and the dotted line passes through the average point of all plots with the same slope.

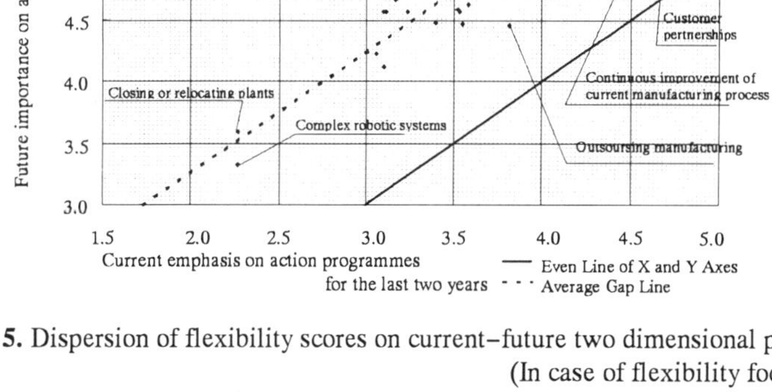

Fig. 5. Dispersion of flexibility scores on current–future two dimensional plane (In case of flexibility focused firms)

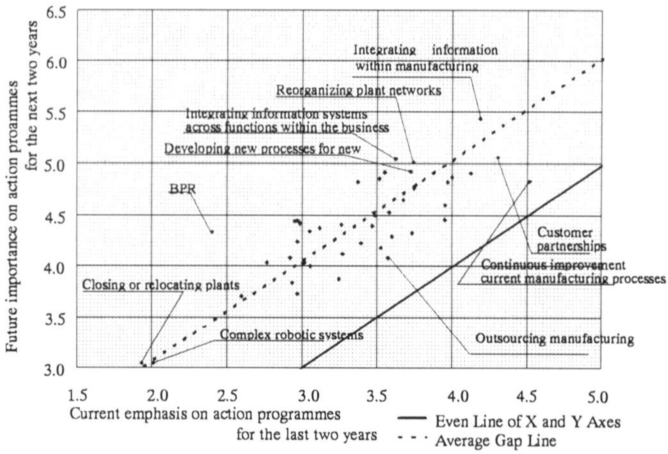

Fig. 6. Dispersion of flexibility scores on current–future two dimensional plane (In case of flexibility non–focused firms)

Table 2 is summary of the result of statistical test for difference of two averages on each action programme, *i.e.* average score of flexibility focused firms and that of flexibility non–focused firms, by *t*–test with significance level of 5 %.

It can be concluded from above gap analysis figures and the result of *t*–test that there are very few significant difference in action programmes between the two categories of firms, thereby it seems that two groups have emphasised similar improvement programmes superficially. In this place, another statistical analysis, the discriminant analysis, is conducted in order to discriminate the characteristics of two groups by paying attention to more sub–conscious thoughts of these firms.

Table 2. Significantly different action programmes between two categories of firms (*t*–test of average scores)

Significantly different action programmes (5%)		Score of F. F.	Score of F. non–F.
Current	• Simple pick & place robots	3.05	3.24
	• Improvement the quality of work life	4.20	3.52
Future	• Continuous improvement of current manufacturing processes	5.51	4.83
	• Integrating information systems with suppliers and distributors	5.58	4.49
	• Simple pick & place robots	4.23	3.87

Where, F.F.: Flexibility focused firms F. non–F.: Flexibility non–focused firms

Table 3 summarises the obtained standardised coefficients of canonical discriminant functions of current and future action programmes, which differentiates both groups with accuracy of obtained discriminant rate. Figures 8 and 9 are their visual representations illustrating how these two groups are being discriminated. The essential action programmes discriminating two groups figured out by this analysis are summarised in Table 4.

It can be noticed, from these facts, that flexibility focused firms consider information integration, CAD/CAE implementation for enhancing design functions, concurrent engineering in design activities and future team working not only restricted within

design function, hardware facilitation such as FMC/FMS and CAM. Therefore, this group tacitly considers information integration through reinforcing critical functions such as design activity and manufacturing facilities to realise their flexibility.
On the other hand, flexibility non-focused firms consider supplier partnership, supervisor training and worker's job enrichment, recycling, quality issues such as ISO 9000, TQM and automated inspection/testing, which seem to be toward relatively diverging directions and, essentially, trying to improve performance of each function by regarding it as island.

Table 3. Standardised coefficients of canonical discriminant functions and discriminant rates (In case of action programmes)

	Action programmes	**Current**	**Future**
1	Giving workers a broad range of tasks	-0.02860	-0.41804
2	Developing new performance measures	0.90361	0.33106
3	Reorganising plant networks	-0.31075	0.24813
4	Reconfiguration plant lay-outs	-0.18503	0.52603
5	Worker training	0.26709	-0.20202
6	Management training	-0.52873	0.52603
7	Supervisor training	-0.41676	-1.25821
8	CAM	0.04098	0.45997
9	CAD / CAE	0.63225	0.81642
10	CIM	-0.55444	-0.29933
11	Automated inspection / testing	0.07475	-0.34544
12	VA / Product redesign	0.16676	0.16541
13	Cross-functional teams	0.22983	-0.05443
14	Functional teamwork	-0.39408	0.69735
15	Developing new processes for new products	-0.57749	-0.14307
16	Developing new processes for old products	0.10345	0.08314
17	Continuous improvement of current manufacturing processes	-0.03910	0.74136
18	Integrating information systems within manufacturing	0.14985	0.65103
19	Integrating information systems across functions within the BU	0.85100	-0.59686
20	Integrating information systems with suppliers and distributors	-0.25070	0.06787
21	Reconditioning physical plants	-0.17800	0.00202
22	JIT	-0.36580	-0.02007
23	Simple pick & place robots	-0.17778	-0.24535
24	Complex robotics systems	0.13766	-0.04828
25	FMC / FMS	1.13354	-0.88427
26	DFM	-1.48008	-0.45572
27	CE	0.77394	0.03468
28	SQC / SPC	-0.54400	0.84733
29	TQM	0.05108	-0.61447
30	TPM	0.17103	-0.14742
31	Closing or relocating plants	-0.06215	0.58399
32	Investing in improved production-inventory control systems	1.17609	0.26186
33	Reengineering business processes	0.10505	-0.45379
34	Outsourcing manufacturing	0.04328	0.38925
35	Supplier partnerships	-1.01562	-0.59392
36	Customer partnerships	0.52218	0.30676
37	Improving the quality of work life	0.63737	0.38155
38	Benchmarking	0.53956	0.37028
39	Certification to customers' quality standard systems	-0.82216	-0.15995
40	ISO 9000	-0.55805	0.18515
41	Taking back products from customers to recycle or destroy	-1.53073	0.36859
42	Improving manufacturing processes to protect the environment	-0.11752	-0.26544
43	Increasing the use of recyclable materials	0.74970	-0.08979
44	Developing manufacturing strategy to support business strategy	0.78430	-0.19983
	Discriminant rate	**83.33%**	**88.51%**

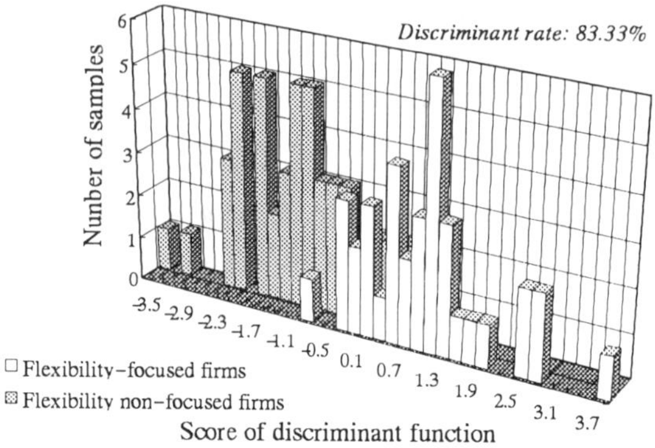

Fig. 8. Score distributions of both categories and discriminant rate (In case of current emphasis on action programmes)

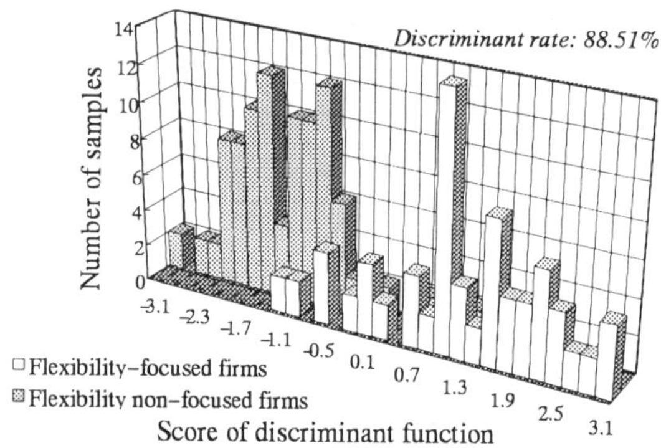

Fig. 9. Score distributions of both categories and discriminant rate (In case of future importance of action programmes)

Table 4. Discriminative action programmes between two categories of firms

Action Programmes	Flexibility focused	Flexibility non-focused
Current	FMC/FMS, CAD/CAE, MRP, Integrating information systems across functions within the business unit, Concurrent engineering	Taking back products from customers to recycle or destroy, Supplier partnership, Certification to customer's quality standard systems, ISO9000, Supervisor training
Future	CAM, CAD/CAE, Reconfiguring plant layouts, Integrating information systems within manufacturing, Functional teamwork	Supervisor training, TQM, Supplier partnership, Giving workers a broad range of task and/or more responsibility, Automated inspection/testing

In order to analyse the foregoing discriminatory factors in more detail, the Hayashi's quantification analysis model II [2], which is a similar statistical model as discriminant analysis except dealing with qualitative data such as ranked scores, was conducted and Table 5 is obtained as a result. Paying attention to rationality of interpretation that positive quantified value contributes to explain flexibility focused firms and negative value for flexibility non-focused firms, it can be noticed that flexibility focused firms relatively move from each hardware and software flexibility such as FMC/FMS, MRP to integration or specific advanced technology such as CIM, DFM.

Table 5. Result of Hayashi's quantification analysis model II

(In case of action programmes)

Action programmes	Current score		Future score		Action programmes	Current score		Future score	
Investing in improved production-inventory control systems (MRP)	0	—	-	—	FMC / FMS	0	-0.641	-	—
	1	—	1	-0.769		1	1.63	1	-0.285
	2	7.75	2	6.99		2	-1.335	2	1.26
	3	8.93	3	6.81		3	0.27	3	0.80
	4	8.42	4	1.29		4	0.55	4	-0.806
	5	7.16	5	-1.152		5	1.80	5	-0.094
	6	9.21	6	-2.491		6	-1.044	6	0.26
	7	-150.696	7	-0.326		7	-7.685	7	0.52
CIM	0	-0.210	-	—	Design for manufacturing (DFM)	0	-3.644	-	—
	1	-2.307	1	7.02		1	-2.847	1	1.77
	2	1.26	2	-314.29		2	-4.290	2	-1.598
	3	1.54	3	6.34		3	-2.898	3	-1.780
	4	-1.259	4	6.96		4	-3.110	4	-0.051
	5	0.76	5	6.78		5	-3.239	5	-0.250
	6	0.33	6	6.85		6	-8.170	6	1.03
	7	-0.289	7	5.87		7	319.84	7	2.29

Discriminant rate of this result is 100%

3.3 Discriminant analysis of the performance

The same analysis as the former section was executed for performance data to clarify their direction more directly and obtained Table 6, 7, 8 as results. Main facts obtained from these analyses are that flexibility focused firms keen on lead time reduction, whereas non-focused firms' directions are, again, widely varied.

Table 6. Significantly different performance indices between two categories of firms

(*t*-test of average scores)

Significantly different performance indices (5%)	Score of F. F.	Score of F. non-F.
· Variety of products producible by manufacturing	136.09	114.21
· Procured materials inventory turnover	110.58	103.90
· Procurement lead time	113.46	105.20
· Average defect rates in customer's hands	109.35	104.12
· Market share	113.91	99.53

Where, F.F.: Flexibility focused firms   F. non-F.: Flexibility non-focused firms

Table 7. Standardised coefficients of canonical discriminant function and discriminant rate

(In case of performance indices)   (*Continued*)

	Performance indicators	coefficient
1	Average defect rates at the end of manufacturing	-0.38312
2	Average unit production costs for typical product	-1.87987
3	Work-in-process inventory turnover	0.22818
4	Equipment changeover time	0.61160
5	Manufacturing cycle time	-0.59782
6	Variety of products producible by manufacturing	-0.33999
7	Productivity of direct production workers	-0.08209
8	Procured materials inventory turnover	-0.04433
9	Procurement lead time	1.36915
10	Average defect rates of the procured materials	1.13082
11	Overall quality as perceived by customers	-0.42096
12	Average defect rates in customer's hands	-0.20482
13	Finished goods inventory turnover	0.44143
14	On-time delivery to customers	-0.07929
15	Delivery lead time	1.12025
16	Speed of new product development	0.73532

Table 7. Standardised coefficients of canonical discriminant function and discriminant rate
(In case of performance indices)

	Performance indicators	Coefficient
17	On-time completion of new product development projects	-0.58605
18	Speed of introducing product design change to manufacturing	-0.75057
19	Extent to which new products meet customer needs	0.83060
20	Extent to which new products meet initial unit cost target	0.32658
21	Number of avoidable engineering change orders for new product	0.11275
22	Market share	0.14202
23	Profitability	0.33401
	Discriminant rate	**72.60%**

Table 8. Discriminative performance measures between two categories of firms

Performance	Flexibility-focused firms	Flexibility non-focused firms
Current performance	Procurement lead time, Delivery lead time, Speed of new product development	Average unit production costs for typical product, On-time completion of new product development projects, Overall quality as perceived by customers

4 Concluding Remarks

Obtained main results from this study are that the remarkable performance difference between both categories of firms is revealed and the factors of these distinctive differences on the direction of action programmes and performance are summarised as follows.

1) Flexibility focused firms seem to be making efforts of lead time reduction or realising quicker response through information system integration with improvement of design and operation functions.
2) Direction of flexibility non-focused firms is diverging.

Acknowledgement

This research is accomplished under the support of Science Research Fund of Japanese Ministry of Education for 1995-1996 and Waseda University Grant of Special Projects 1996-1997. We, the authors, sincerely appreciate their kind supports.

References

1. K. Ferdows, and A. De Meyer, Lasting improvements in manufacturing performance : In search of a new theory, *Journal of Operation Management*, **9**, 2 (1990) 168-184.
2. C. Hayashi, *Information Processing and Mathematical Statistics*, Sangyo Tosho Publishing Co. (1970).
3. J. G. Miller, A. De Meyer and J. Nakane, *Benchmarking Global Manufacturing*, R. D. Irwin (1992).
4. W. Skinner, Manufacturing-missing link in corporate strategy, *Harvard Business Review*, May-June (1969) 136-145.
5. M. Tombak and A. De Meyer, Flexibility and FMS : An Empirical Analysis, *IEEE Transactions on Engineering Management*, **35**, 2 (1988) 101-107.

Structural Fitness to Manage Market's Turbulence

- Create Dynamic Structures to Master the Next Millennium -

Nick Brehmer, Dirk Markfort, Hermann Kühnle

Fraunhofer-Institute
Factory Operation and Automation
Steinfeldstrasse 3
39179 Barleben
Germany
brehmer@iff.fhg.de
markfort@iff.fhg.de
kuehnle@iff.fhg.de
http://wwwue.iff.fhg.de

Abstract. The environment gets more and more turbulent. Enterprises are not able to predict the environment over a longer time and notice the drastic changes on markets. To ensure the long time capability of survival within turbulent markets, enterprises have to generate flexible, adaptable and living organizational structures, best known as dynamic structures. To meet this demand, enterprises need instruments and methods to point out the current organizational structure and to highlight change requirements. These are prerequisites to set goals and to develop strategies to reach them.

The focal point of this paper is the description of the *Structural Behavior* of an organization. The structural behavior describes the ability of an enterprise to react on market changes. Consequently, the structural behavior can be divided in rigid, chaotic and dynamic. This paper describes the characteristics of structural behavior and their degrees. To ensure an holistic view on the entire social-technical system enterprise, we used the Six-Level Model. This model also helps to structure the characteristics of the Structural Behavior and their degrees.

Keywords. turbulent environment, structural behavior, strategic planning, dynamic production and organizational structures, change process, socio-technical system, Six-Level Model

1 Initial conditions and problems

In the past, enterprises operated within a stable and less changing environment, because the vendor markets were predictable. The high efficiency of enterprises based on setted hierarchical structures and the separation of planning and working.

Since the early 80ies, markets getting less predictable and the complexity of enterprises and markets increases. The reaction time to respond to market changes are also increasing. These trends have the following causes:

- *Fast changing markets* - Market's behavior is less predictable and insecure. Long-time planning is not possible and therefore entrepreneurial mistakes are increasing.
- *World wide availability of resources* - Resources are available in shorter times and world wide. The relationships between vendor and buyer are valid over longer times.
- *Rapid development of technology* - Especially the information and communication technology is changing fast and allows world wide contacts in shorter times.
- *Individualization of customer demands* - Customer request individual solutions for individual problems. Enterprises have to adjust to this development.

Traditional grown structures are not able to manage these environmental demands. They are not able to follow market changes in an adequate time.

Therefore, new management concepts were created in the past 10 years. There are concepts like Total Quality Management, Lean Production, Business Reengineering, etc. Lots of enterprises realized these concepts. They minimized costs and waste and manufactured more efficiently. But they changed only ones or twice up to a higher level. These single change processes are not enough to follow the rapid environmental changes. Management plans strategically (5 to 7 years), i.e., they are planning changes within that time frame as well. But the environment will change in the future even more unpredictably. Management concepts with the goal to change enterprise structure once up to date are only partly applicable.

For surviving in a turbulent environment, today's companies need to change to dynamic structures which fulfill the requirements of adaptability, capability of development, management of complexity, robustness, self-similarity and goal-orientation, and optimized usage of resources. Most important is a holistic approach by addressing changes to the company structure.

These dynamic structures are combining elements of mechanical concepts (Lean Production, TQM, etc.) with elements of organic concepts (Fractal Company, Agile Manufacturing, Learning Organization, etc.) in the right way to meet market demands.

2 Structural Behavior

How can a dynamic structure be described? This is a very complex issue that needs to be addressed from several points of view. We are also convinced, that different organizational structures are able to fulfill the mentioned demands.

This aware, the question what dynamic structures should fulfill raises. Answering this question, the scientist has a basis to compare different enterprises and their structures. One does not has to compare the organizational elements and their relationships or the amount of interfaces an order has to pass. The investigation is not only focused on process and the hierarchical structure of an enterprise.

The following abilities of dynamic structures were identified (Warnecke, 1993). Dynamic structures are:

- *Able to Reconfigure* - they are self-organizing, self-optimizing, adaptable and capable to develop
- *Self-Similar* - they are oriented towards overall objectives and similar to each other in goal orientation.
- *Stable* - they are successful and viable in the long run.
- *Capable to Manage Complexity* - they are designed from the holistic view of the company and along the processes. Organizational units are part of these dynamic structures, which are able to manage complexity through breaking down the global tasks in smaller parts that concern the unit.

Considering that, structural behavior shows the ability of the organization to follow market changes. This is independent from the actual organizational structure.

2.1 Six-Level Model

The description of structural behavior bases on the Six-Level Model (Kühnle, 1994). Fig. 1 shows the Model and its six horizontal levels.

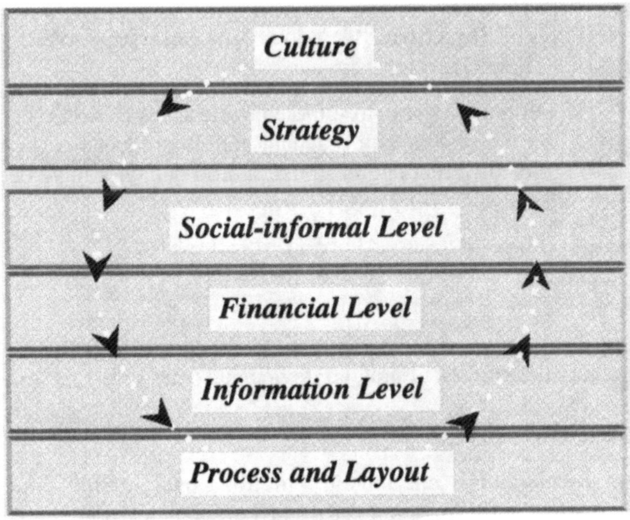

Fig. 1: Six-Level Model

This level model ensures the integral view on the enterprise. It decreases the complexity of the organization, because it subdivides the complex *social-technical system* enterprise into six horizontal levels. The levels culture, strategy, social-informal level, financial and information level as well as process and layout can be searched separately but only the integral view on all levels ensures an integral analysis of the entire organization.

The level culture describes the organization as social system as well as the value and thinking pattern within the organization. Thereby, the organization is seen as culture, as partnership with defined value and orientation patterns.

The level strategy describes the ability of enterprises to serve the market with all available resources. The enterprise has to maximize customer benefits. This level describes the efficiency of organizations.

The focus of the social-informal level are the employees. It describes the relationship of all employees within the organization. The culture of the social-technical system enterprise builds the frame of this level.

The financial level analyses the economic efficiency. Focus of that level are financial processes and their individual adaptations to different organizational forms.

All technical processes and material flows within the value chain are analyzed in the level process and layout. Therefore, it is also called technical level.

To sum it up, one can say that the levels culture and strategy are building the frame for the activities within the other 4 process-levels. The culture defines the organization as social system and with the help of strategies, enterprise is able to define action patterns to ensure the efficacy of resource utilization. Therefore, this model is also called Two+Four-Level Model.

2.2 Characteristics of Structural Behavior of an Enterprise Within Turbulent Markets and Their Degrees

This Six-Level Model helps to structure the characteristics of the structural behavior and their degrees. As the results of a practical and theoretical multi-company study (Hartmann, 1997) three different degrees of structural behavior have been identified:

- rigid,
- chaotic, and
- dynamic.

The following description of these degrees of structural behavior highlights some selected characteristics of the cultural, strategic and social-informal level.

2.2.1 Rigid Structural Behavior

Rigid structural behavior occurs mainly in traditional grown Enterprises. Main characteristics are the over-organized and strange hierarchical organizational structure. Management sets goals and tasks without any employee integration. The way to achieve them is dictated by management. Another characteristic of the cultural level is the degree of informal social systems. There are strongly developed informal systems within organizations with rigid structural behavior. This informal structure is not allowed but plays an important role within all organizational processes.

An important characteristic of rigid structures is the decreasing capability to react and respond to changing environmental constraints. Strategies and goals to achieve are also dictated. Because of the turbulent environment, the capability of management to consider and decide about strategic tasks decreases constantly.

A distinctive sign of rigid structural behavior on the social-informal level is e.g. the separation of competence and responsibility, known as principle of congruence (Warnecke, 1995). To master tasks, an customer order has to overcome hierarchical, horizontal and vertical obstacles and barriers.

One symptom of rigid structures are the high amount of specialists. The causes can be seen in qualifications arranged by management. Qualifications are tasks-oriented, without strategic long-sightedness. There are less or no participation of employees.

2.2.2 Chaotic Structural Behavior

The counterpart of rigid is the chaotic structural behavior. Young enterprises or rigid factories within a fast changing environment are tending to chaos, that means instable structures.

Chaotic structural behavior is characterized through a low organizational degree that causes an uncoordinated performance of management and no or less strategic work. Management is not able to control the increasing environmental changes. Because of a weak leadership, there are no fixed organizational rules or guidelines. Single person starting activities uncoordinated and without a holistic view on the whole enterprise. The mainly small and autonomous units within young organizations are causing lots of frictional losses, redundant work and therefore they waste time.

Management also order adaptations and changes without operational and strategic necessity. They do not have enough information in the right quality to analyze the environment. Because of that, their behavior is reflexive and the enterprise tends to under- resp. overreactions.

The enterprise does not have a fixed vision or goals and their strategies are fuzzy and not clearly defined. Therefore, they are changing faster as the environment requires.

On the social-informal level, enterprise is characterized through a fuzzy assignment of tasks. Management does not dedicate the business to the employees or to a proper department. This destroys systematically the ability of employees to solve problems independently, because nobody feels responsible for the tasks to do. Every employee is waiting for the specification of the tasks or further instructions. Therefore, the fuzzy assignment also causes a decreasing competence and staff motivation.

However, only the comprehensive employee participation guarantees the survival of the enterprise. Management does not have a decisive influence on employee qualifications or the topics of them. The employees decide about and organize their own qualification. Results are past-oriented qualified employees.

2.2.3 Dynamic Structural Behavior

Creating dynamic structures management has to speed up rigid structures or has to slow down chaotic structures (refer fig. 2).

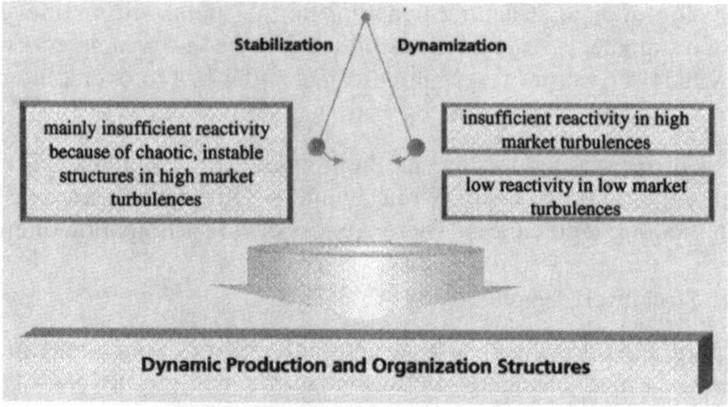

Fig. 2: Creating Dynamic Structures

Main characteristic of dynamic structural behavior on the cultural level is the arrangement of goals and tasks between management and staff. The employees participate in the way how to reach the agreed goals. The team has enough decision competencies to initiate and complete changes. The changes to do are transparent for everyone and the staff is able to influence them. Management and employees are aware of the changing environment and the need to change themselves. They are „living with the change“ actively.

Organizational restrictions and rules are broad, so the employees are having enough space to act independently. This guaranties fast reactions on market demands. The employees do not have to consider high bureaucratically rules and processes that take too much time to overcome. This is one aspect that shows, that dynamic organization are operating close to the market.

The management behavior is characterized by a clear goal orientation. Strategies, goals and the vision are defined under the consideration of internal and external circumstances. The organizational vision, goals, strategies will be permanent adapted to the changing, turbulent environment. The decision making process on all hierarchical levels bases on broadly communicated information within the organization.

The dynamic structural behavior also bases on self-organization and-optimization and self-similarity of organizational units and therefore of the whole enterprise too. These are basic elements of living, flexible and adaptable - best known as re-configurable - structures to satisfy customer demands.

On the social-informal level, dynamic behavior bases on the congruency of task, competence and responsibility within an autonomous, decentralized organizational unit. These units are able to react fast, They are acting close together with the customer, so they also know their demands and requests. This is the basis to manufacture individual solutions for customer in a high quality and shorter times.

Management and employees arrange the qualifications. There are not only task-oriented qualification, but also methodical and social qualifications as well as the training of thinking in contexts and over the entire value chain. Continuos learning is self-evident for all members of the organization.

The above mentioned agreements in qualification, goal setting, etc. show the high level of employee participation. That is also one of the most important instruments to use the existing high potentials of the employees and to develop and extend these potentials.

3 Results

This paper highlights some, selected characteristics of structural behavior and there degrees on the levels culture, strategy and on the social-informal level as well. The characteristics of structural behavior and their degrees are also defined on the other three levels of the Six-Level Model. Therefore, there is an complete instrument to evaluate organizational structures.

Main result is the development of a management instrument, so management is able to analyze the current structural behavior of their own organization. The analysis points not only the structural behavior out, it also highlights the requirements to change. The goal must be the implementation of dynamic structural behavior on all six levels.

Enterprises acting within turbulent markets have to ensure dynamic structural behavior. This guarantees a high level of adaptability and flexibility. Because of these characteristics, enterprises are getting more robust and increase their own competitiveness.

With the implementation of dynamic structural behavior, enterprises are only able to *react* on the market. Being able to *act* on the market, management has to implement dynamic structures. Prerequisite is, beside dynamic structural behavior, the ability to act chance driven. Only the combination of dynamic structural behavior and the ability to act chance driven ensures long-term success.

Like it is mentioned above, the generated management instrument is not only worked out theoretically. This instrument was used in ten several German enterprises. Management was able to identify the current structural behavior of their organizations. The practical usage underlines the fast and simple applicability for management. It was seen that the instrument is easy to handle and good understandable.

Next Steps

For further research activities it is important to work completely out the change driver as the second characteristic to describe organization within a turbulent environment. The combination of these two basic characteristics of enterprises within turbulent markets opens a positioning matrix (refer fig. 3).

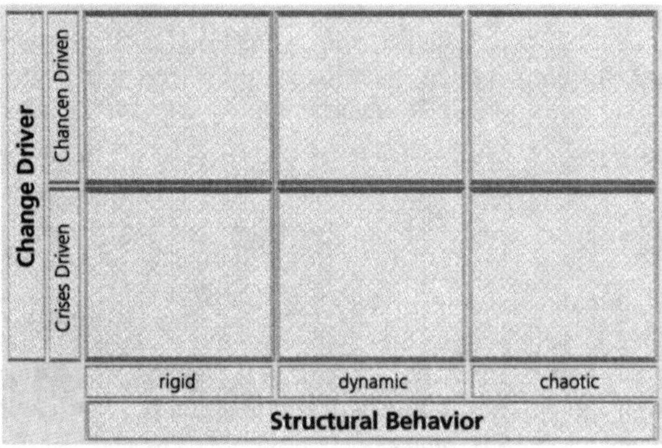

Fig. 3: The Positioning Matrix

Creating this matrix, organizations are able to position themselves within the turbulent environment. Management has an instrument to visualize the current and the goal position. Checking the current individual structural behavior and the change driver of an organization, management is able to derive requirements to change to a dynamic organization.

Considering the possible six fields within that matrix, one can create guidelines to reach the dynamic organization depending on the individual starting position.

References

1. H.-J. Warnecke. (1995). *Aufbruch zum fraktalen Unternehmen*, Springer, Berlin, Heidelberg, New York

2. Kühnle, H. (1994). Die Fraktale Fabrik - Neue Organisationsformen. In: *Menschen, Maschinen, Märkte* (W. Kunerth). Springer, Berlin a.o.

3. Hartmann, M. (1996). *DYNAPRO - Erfolgreich produzieren in turbulenten Märkten. Anforderungen und industrielle Lösungsansätze.* LOGIS, Stuttgart

4. Hartmann, M. (1997). DYNAPRO - *Erfolgreich produzieren in turbulenten Märkten. Methoden für dynamisches Management.* LOGIS, Stuttgart

5. Warnecke, H.J. (1993). *The Fractal Company. A Revolution in Corporate Culture.* Springer-Verlag, Berlin, Heidelberg, New York

6. Förster, T.; Schmelzer, S.F. (1996): *A New Paradigm: The Fractal View of the Company.* In: NGMS International Conference, Irvine, USA, February 7, 1996

The Human Side Of Virtual Enterprise

Jeff Y T Ang
Monash University, Churchill 3842, Australia

Abstract. This paper analyses the factors that caused a transformation in executive attitudes, organisation structures and the human side of an enterprise. It then proposes a framework for emerging Virtual Enterprise (VE).

1. Introduction

We are very much aware that the information age is upon us and that our recent fore-fathers have gone through the Industrial and Agricultural Ages. During the Agricultural Age the human race has mastered the production of food to satisfy the need for food. The Industrial Age has seen the mechanisation of labour and the human race has amplified its physical capabilities, eg going to the moon at will, etc. During this Information Age the human race will endeavour to enhance its intellectual capabilities. To achieve this some important changes are required.

There are few periods in history that deserve the label of "transforming eras" when circumstances sufficiently change to warrant a major shift of assumptions. Major change takes place only occasionally when the working assumptions on which people have depended become so inappropriate that they break down to be replaced by a more appropriate set. This is exemplified by the movement of majority of people from farms to factories to offices.

We are living in one of those rare times in history when two crucial elements for social change are present – new values and economic necessity as detailed by Kanter[6]. These elements are pushing us towards the concept of Virtual Enterprise (VE).

For VE to be successful, change needs to be managed. Let us examine the evolving patterns of organisation during these preceding periods so that new organisation structure and executive concerns can be complemented under VE.

2. The Evolving Patterns Of Organisations

The management style may be broadly classified into rational/scientific and qualitative/humanistic management as depicted in Tables 1 and 2 below.

Table 1: The three eras of rational/scientific management (see Hickman [5])

Era	Executive Concerns	Organisational Response
1910-1935	Well functioning multiunit organisations and management hierarchy	Structure building
1935-1955	Maximisation of production and operations output and efficiency	Productivity enhancement
1955-1970	Increased management control over all business variables	System design

Table 2: The eras of qualitative/humanistic management (adapted from Hickman[5])

Era	Executive Concerns	Organisational Response
1970-1980	Successful competitive positioning in the marketplace	Strategic planning
1980-1985	Superior performance and execution to achieve excellence	Culture shaping
1985-1990	Adaptation to the increasing and accelerating of change	Innovation management
1990-1995	Exploitation of information technology	Strategic alliance
1995 to 2000s	Globalisation	Virtual enterprise

As shown by Kanter[6], the factors bearing on the design of an organisation in the 1890s-1920s forms the traditional industrial corporation as shown in Table 3. This table contains additional factors that future organisation design will be influenced.

Table 3: Organisation design factors (adapted from Kanter[6])

Traditional 1890s to 1920s	Emerging 1960s to 1980s	Virtual Enterprise 1990s to 2000s
Uneducated, unskilled temporary workers	Educated, sophisticated career employees	Well trained masterminds
Simple and physical tasks	Complex and intellectual tasks	Tasks spanning across organisation boundaries
Mechanical technology	Electronic and biological technologies	Information technologies
Mechanistic views, direct cause and effect	Organic view, multiple causes and effects	Organic view, multiple causes and effects
Stable markets and supplies	Fluid market and suppliers	Global market and suppliers
Sharp distinction between workers and managers	Overlap between workers and managers	Self managing teams

Peters[9] pointed out the following factors that forces corporation to re-organise:

1. The shift in strategic resource from an industrial to an information society
2. The coming seller's market and the new competition for the best employees
3. The whittling away of middle management
4. The continuing entrepreneurial revolution
5. The emergence of the new variegated work force
6. The demographic revolution of working women
7. The growing use of intuition and vision
8. The mismatch between our education system and the needs of the new information society
9. The rising importance of corporate health issues
10. The values of the baby boomers, those born between 1946 and 1964, who are now populating the ranks of management

while Naisbitt[7] predicted the following overarching trends that influence our lives:

1. The booming global economy of the 1990s
2. A renaissance in the arts
3. The emergence of free market socialism
4. Global lifestyles and cultural nationalism
5. The privatization of the welfare state
6. The rise of the pacific rim
7. The decade of women in leadership
8. The age of biology
9. The religious revival of the new millennium
10. The triumph of the individual

3. The formulation of Virtual Enterprise

Taking all the above evidence together and a clear picture of the successful firm (Peters[9]) in the future emerges. It will be:

- flatter (have fewer layers of organisational structure)
- populated by more autonomous units (have fewer central staff second-guessers, more local authority to introduce and price products)
- oriented toward differentiation, producing high value added goods and services, creating niche markets
- quality conscious
- service conscious
- more responsive
- much faster at innovation
- a user of highly trained, flexible people as the principal means of adding value

In order to meet the requirements of the successful firm, management will have to be proactive. The essence of proactive management performance are:

1. an obsession with responsiveness to customers
2. constant innovation in all areas of the firm
3. partnership (the wholesale participation of and gain sharing with all people connected with the organisation
4. leadership that loves change (instead of fighting it) and instils and shares an inspiring vision, and
5. control by means of simple support systems aimed at measuring the "right stuff" for today's environment

Furthermore, the nature of day-to-day supervision will need to be changed. Table 4 depicts the new requirements of supervision.

Table 4: Changing nature of supervision

Old	**New**
10 direct reports	50 to 70 direct reports
Scheduler of work	Coach and sounding board for self managing team leaders/coordinators, working on training to emphasize skill development
Rule enforcer	Facilitator, getting experts to help the team as needed
Lots of planning	Lots of wandering
Focused `down' or `up' the structure	Focused `horizontally', working with other functions to speed action-taking
Transmitting middle/top management's needs `down'	Selling teams' ideas/needs `up'
Providing new ideas for workers	Helping workers/teams develop their own ideas, providing ideas for cross-functional systems

As a responsive corporation, its structure must rearrange to meet the faster speed of innovation, that is to turn the organisation structure from mechanistic to organic as explained by Table 5.

Table 5: Mechanistic versus Organic organisational structures

Mechanistic	**Organic**
Channels of communication highly structured, restricted information flow	Channels of communication open with free flow of information throughout the organisation
Operating styles must be uniform and restricted	Operating styles allowed to vary freely
Authority for decisions based on formal line management position	Authority for decisions based on expertise of the individual
Reluctant adaptation with insistence on	Free adaptation by the organisation to

holding fast to tried and true management principles despite changes in business conditions	changing circumstances
Emphasis on formally laid down procedures. Reliance on tried and true management	Emphasis on getting things done unconstrained by laid out procedures
Tight control through sophisticated control systems	Loose, informal control with emphasis on norm of cooperation
Constrained on-job behaviour required to conform to job descriptions	Flexible, on-job behaviour permitted to be shaped by requirements of the situation and personality of the person doing the job
Superiors make decisions with minimum consultation and involvement of subordinates	Participation and group consensus used frequently

4. Implementing the components of Virtual Enterprise

In an information society, human resources are any organisation's competitive edge. Business has given this concept a kind of obligatory lip service in the past. Now it is time to understand the practical side. We will not see profits grow if we do not learn how to grow the people (Naisbitt[8]). This creates a need for human resource management (HRM) to implement the following activities:

1. Planning for human resource
2. Staffing the organisation
3. Appraising & rewarding employees
4. Improving work environments
5. Training employees
6. Establishing & maintaining effective working relationships

In technology, we expect bold experiments that test ideas, obtain new knowledge, and lead to major advances. But in matters of social organisation, we usually propose only timid modifications of conventional practice and balk at daring experimentation and innovation. Why? Forrester[4]. We must be bold to excel the triumph of the individual.

As Toffler[11] pointed out, big organisations, as a rule, only change significantly when certain preconditions are met:

- First, there must be enormous external pressures
- Second, there must be people inside who are strongly dissatisfied with the existing order, and
- Third, there must be a coherent alternative embodied in a plan, a model, or a vision.

Unless there is a vision of a workable alternative to the existing structure, the resistance to change will usually triumph. Even if this plan or vision is ultimately rejected, it can still play a fundamental role in crystallising the issues, mobilising support for change, and accelerating adaptation.

Without some explicit assumptions about the long-term future, and strategic guidelines for dealing with them, without a vision of its own future form, even the largest and seemingly most secure organisation face disaster in a period of revolutionary, technological and economic turbulence.

Bennis[1] painted the following issues that a successful firm needs to address:

"The problems with many organisations ... is that they tend to be overmanaged and underled.

There is a profound difference between management and leadership, and both are important.

`To manage' means `to bring about, to accomplish, to have charge of or responsibility for, to conduct'

`Leading' is `influencing, guiding in direction, course, action and opinion'

Leadership is what gives an organisation its vision and its ability to translate that vision into reality"

5. Management tomorrow – dealing with complexity

The management style of the 2000s will be based on:

- Perspective management through unifying principles
- Power management through freedom to act
- Pivot management through individual fulfilment

The traditional vertical model (Toffler[10]) seeks to solve problems by adding another echelon to the power hierarchy.

The alternative model, congruent with the emerging forms of organisation in the business world and the advanced economies, flattens the hierarchy rather than extending upward. It will be based on networks of alliances, consortia, specialised regulatory agencies, to accomplish ends too large for any single state (please see CALS[1,2]) leading to the birth of Virtual Enterprise.

6. References

1. Bennis, Warren and Nanus, Burt, *Leaders*, 1985
2. CALS Pacific Korea 1996, Conference proceedings
3. CALS-Singapore 1996, Conference proceedings
4. Forrester, Professor Jay, Sloan School of Management – MIT, *A New Corporate Design*, 1965
5. Hickman, Craig & Silva, Michael, *The Future 500*, 1988
6. Kanter, Rosabeth, *The Change Masters*, 1983
7. Naisbitt, John and Aburdene, Patricia, *Megatrends 2000*, pp2-3, 1990
8. Naisbitt, John and Aburdene, Patricia, *Re-inventing the corporation*, 1985
9. Peters, Tom, *Thriving on chaos*, 1987
10. Toffler, Alvin, *Power Shift*, pp 463-464, 1990
11. Toffler, Alvin, *The Adaptive Corporation*, 1985

Business Process Reengineering

BPR and Benchmarking with ARIS-based industry-specific reference models

Dr. Christian Aichele / Sebastian Böhning

IDS Prof. Scheer GmbH, Postfach 10 15 34, 66015 Saarbruecken, Germany

Abstract. Industry-specific reference models are the basis of quality and quantity-related business process reengineering and benchmarking projects. This paper points out the use of such reference models. A case study shows how logistics processes in a company can be optimized through industry-specific reference models.

1. Demands placed on industry-specific reference models and definition of the concept

The increased opening up of markets has, in almost all branches of industry, led to heightened competition. Cost reduction, shorter order processing times and faster product development are, with quality remaining the same, important factors for success in achieving the survival of a company. The industrial nations in the West in particular should take up these challenges to a greater extent than they have so far. The great importance of time (shorter, faster) as a factor of performance has led to a more detailed analysis of the time-consuming cross-departmental business processes in enterprises. The concept of "Business Process Optimization" is regarded as a recipe for success.

The overall objective of industry-specific reference models is to show the optimum situation of a company in a standardized, consistent and, above all, comprehensible way. Only reference models which fulfill these requirements can be applied to the optimization of business processes.

Recent results of scientific research in the field of information modeling and reference models[1] have led many manufacturers and service providers to the realization that the use of industry-specific reference models is a way or a possible method for the optimization of the business processes in a particular company. There is, however, still a discrepancy between manufacturers and service providers asking for industry-specific reference models for the optimization of either their own company as a whole or of particular business processes and the knowledge of how to use industry-specific reference models for optimization. In practice, there are, in this context, even widely divergent ideas about the content of industry-specific reference models and the concept of "reference model". The combination of the words "reference" (in the sense of "recommendation" or of "authority which can be quoted and from which information can be obtained") and "model"[2] (in the sense of "standard" or "pattern" or "abstract representation of a complex matter"[3]) which form the compound "reference model" is the basis for the definition of the concept of "reference model".

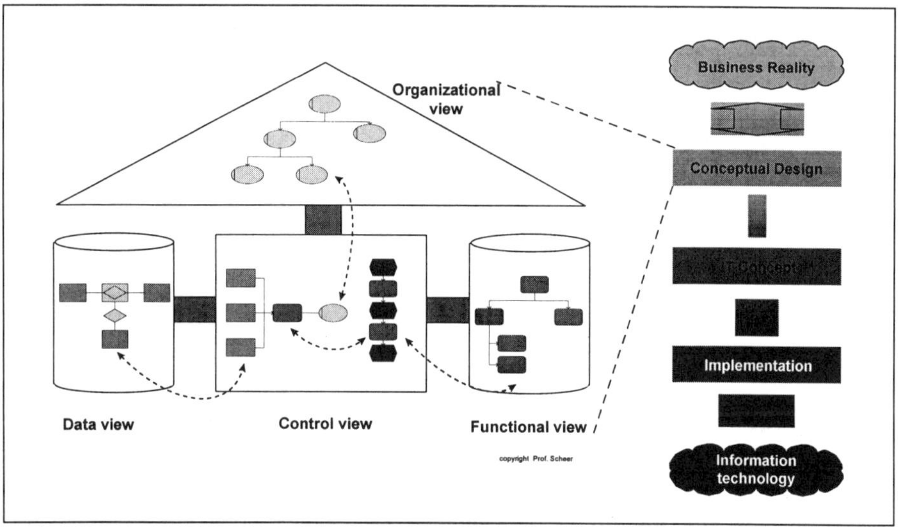

Fig. 1. Structure of industry-specific reference models created according to the ARIS Method

An industry-specific reference model is, on the one hand, the graphic representation of the flows of information and material in a company (including the coordination and control processes of reference companies in the form of process and function models) and, on the other hand, the graphic representation of the information and

---

1 compare Scheer, A.-W.,Business process engineering

2 compare Stachowiak, H., Modelle, page 172

3 compare Eichhorn, W., Modelle

data structures of reference companies in the form of data models[4]. Moreover, industry-specific reference models contain, in the form of representations of organizational structures or organizational charts, the units which make up the company organization structure. The overall objective of industry-specific reference models is the representation of an optimized situation in a company in a standardized, consistent and, above all, comprehensible way. A very important aspect of industry-specific reference models is that besides representing the quality-oriented structure of the industry through graphic models, they contain best practice and/or branch-average indicators[5].

In practice, the following demands are placed on industry-specific reference models:

- Comprehensible representation of optimum business processes for commercial operations (in the fields of general accounting, cost accounting etc.) for logistics processes (as e. g. sales, sales planning, production planning and scheduling, production/operations management, production) and for the technical processes (in the fields of product development, quality planning and quality control, maintenance etc.). These business processes are, on the one hand, supposed to represent the material flow and, on the other hand, show the information flow.
- Comprehensible representation of optimum information and data structures in a company for the processes and functions listed above.
- Comprehensible representation of an optimum company organization structure.
- Instructions for the use of industry-specific reference models and for the optimization of business processes on the basis of industry-specific reference models[6].
- Instructions on the best way of supporting the industry-specific reference models through data processing systems.

2. Structure of industry-specific reference models

Industry-specific reference models contain:

- Function decompositions or function hierarchies which represent all functions that are necessary in certain function areas in their hierarchical assignment;
- Organizational charts which show the organizational units in a company in their hierarchical structure;

---

4 compare ARIS method, ARIS manual

5 compare C. Aichele et al., Optimierung von Logistikprozessen auf Basis von Referenzmodellem

6 compare ARIS procedure models for BPR, benchmarking and process-oriented implementation of ERP

- Data models which position the company information and data in such a way that the interdependent structures in the respective processes become apparent;
- Process models which position functions, data and organizational units in their dynamic sequence;
- and best-practice and/or branch-average indicators which are function and business process-oriented.

The Architecture of Integrated Information Systems (ARIS) developed by Professor A.-W. Scheer can be a methodical basis for the rendering of industry-specific reference models[7]. ARIS can represent all the components of industry-specific reference models which have been listed above. Moreover, the extended Event-Driven Process Chain (eEPC) modeling method contained in ARIS provides the possibility to represent functions, organizational units and data (data clusters, entities, attributes) in a dynamic process in a structured way (see Figure 1).

In the ARIS model, the components which are relevant for the description of the situation in a company and for the transfer of this information into information systems are broken up into the so-called views. In doing so, the individual functions are described through a function view, the individual events, results and environment situations through a data view, and the organizational structure through an organization view. In order to maintain and represent the connection between these static views, a control or process view is introduced, which describes business processes as a dynamic connection of the static views.

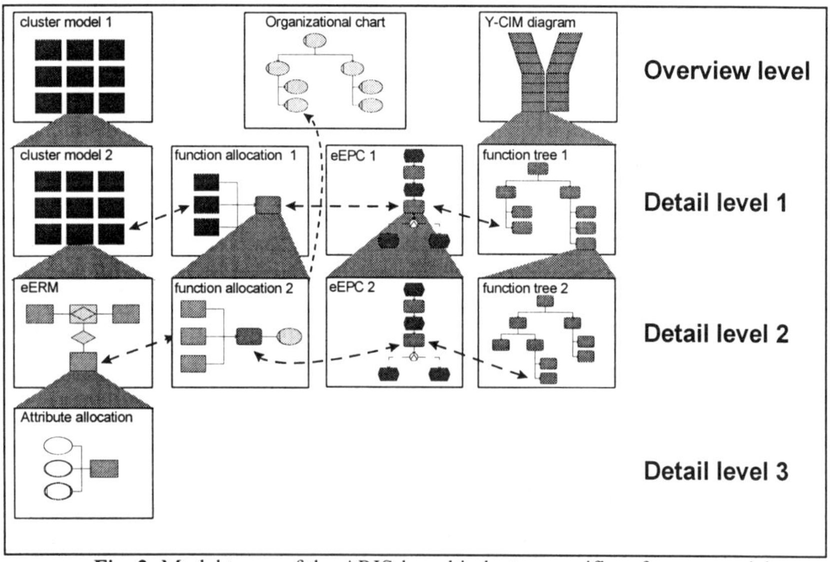

Fig. 2. Model types of the ARIS-based industry-specific reference models

7 see Scheer, A.-W., Architektur integrierter Informationssysteme

Well-established graphic forms of description are used for the representation of the views. Functions are hierarchically grouped in function trees, organization charts are used for the organization view. To describe the data view, the extended Entity Relationship Model (eERM) or the Structured Entity Relationship Model (SERM) is utilized. The ERM is regarded as the description method which is, at present, best suited for describing data structures, as its graphic form of presentation and its clear definitions make it especially user-friendly. The representation of the control and process views is mainly done through the extended Event-Driven Process Chain (eEPC).

It is possible to create industry-specific reference models in all phases of ARIS (Conceptual Design, IT Concept, Implementation). The example of practical application which is given in Chapter 4 refers to the Conceptual Design phase).

3. Reference indicators

3.1 Indicator systems for business process optimization

Information, criteria, indicators and points of orientation will play an important role in economic life in the years to come, i. e.:

- Information about the environment, the market and the competitors;
- Criteria and indicators for a critical assessment of the position of the own company vis-a-vis the competitors;
- Points of orientation for business objectives and strategic plans.

In an economic system based on competition, it is vital to make out the future economic situation as early and as clearly as possible in order to take one's chance as soon as it presents itself. Taking part in business life is a future-oriented activity which requires a great degree of awareness in the planning of each step. It is therefore a matter of great concern for companies to have methods and systems at their disposal with the help of which statements about future combinations of important factors can be made.
Those who are better informed in less time and have the best instruments of management at their disposal, have better chances in business life. If in doubt, they can make analyses on a broader and sounder basis and with greater care, can think in more complex contexts, make more targeted arrangements and be ready for action in less time. Success will be on their side. The question of relevant data or figures which serve as indicators and of an instrument of management or a system of indicators which can be derived from them is of importance in large as well as in medium-sized and small business units.

The management of a company needs indicators

- to be able to distinguish - with the number and complexity of business processes being high in each enterprise - what is important and relevant from what is unimportant and irrelevant, and to be able to obtain aggregated information from a mass of disordered data;
- in order to think in a relational way rather than according to the monetary quantities of profit and costs and to pay more attention to productivity than to production;
- to see causal connections, cause and effect, and the interplay of positive and negative factors;
- to know the company's own position as a competitor on the market as well as the strengths and weaknesses of the company in comparison with the competitors', so that it is in a position to react timely to changes and new developments on the market, and
- to be equipped with an instrument of management for targeted thinking and action.

Indicators are, on the one hand, ratios of business quantities like e. g. output, yield, sales volume, capital, assets and costs compared with each other and with other basic figures and valuation figures within the economic framework. These ratios make it possible for the management of a company and for the staff in charge to find out quickly and without having to go through extensive statistics if their company works in an efficient and cost-effective way (in its own development from one period to the next as well as in comparison with other companies of the same kind). Indicators provide ideas and impulses for improvements and are a valuable, if not indispensable means for successful management.

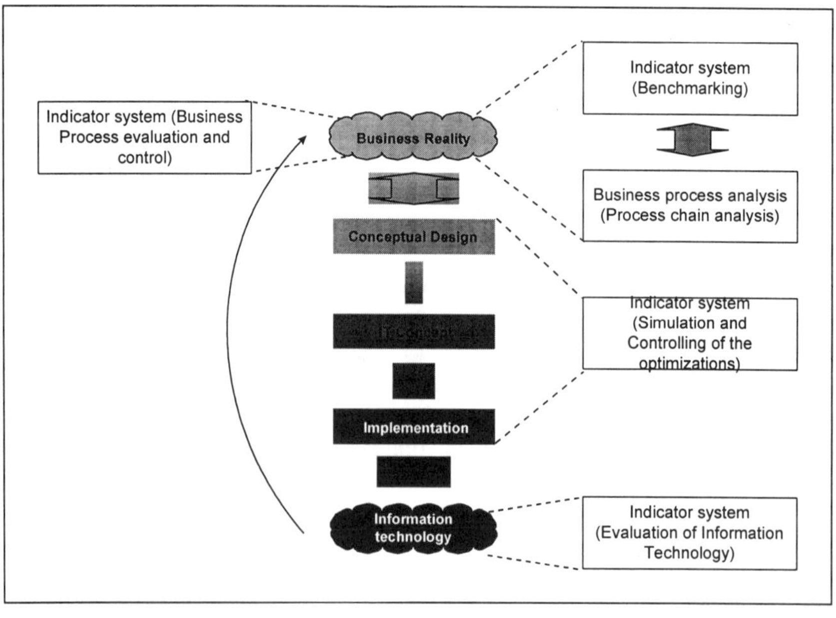

Fig. 3. Indicator systems in the project phase model

The indicator systems which have been developed so far are either incomplete, i. e. quantitatively important task areas of the management are left out of consideration, or their structure is not sufficiently logical. In the past, the management normally used financial quantities like results and costs in order to measure the performance of their company, whereas non-financial quantities like e. g. service level, market shares, order processing times, etc. were hardly considered at all[8].

On the basis of the process-related indicator system developed by C. Aichele, processes can be speedily analyzed and evaluated taking into consideration the company vision, business strategies and corporate objectives as well as the factors of performance connected with them[9]. The correlation of the individual indicator analyses with the business processes makes it possible to proceed to a subsequent targeted process optimization. The relevant indicators of critical areas or processes allow a permanent efficiency control of the individual project phases (Indicator Analysis, Business Process Analysis or Conceptual Design, IT Concept and Implementation). The changes in individual indicators make it also possible to prove the quantitative impact of optimizing the sequence of operations and the company organization structure (see Figure 3).

Through indicator systems which are installed on a permanent basis it is also possible to show, at an early stage, the tendencies of company development on the basis of management decisions, market changes and structural changes, technological innovation and swings in the economic cycle; thus, they timely support reactions by the management (see Figure 3). Timely and effective process analyses have to be an integral part of the management process on all corporate levels.

Indicator analyses establish, in the form of easily comprehensible figures, a meaningful relation between different quantities. On the one hand, they serve as an instrument of allocation and controlling in different functional areas: in the commercial sector (e. g. in personnel management), in the fields of logistics (e. g. in materials management and control, production, distribution) and technology (e. g. in product development and quality assurance); on the other hand, they are an instrument of analysis for the optimization of the sequence of operations and the company organization structure. The integration of the process-related indicator system into the ARIS method developed by Professor Scheer opens up the possibility of increasing the efficiency of the analysis, and, by doing so, of enhancing the efficiency of the subsequent phases (IT Concept and Implementation) as well. Moreover, the quantitative benefit can be ascertained and reviewed in the course of all phases.

---

8 see C. Aichele, J. Kirsch, Geschaeftsprozessanalyse aus Basis von Kennzahlensystemen

9 see C. Aichele, Kennzahlenbasierte Geschaeftsprozessanalyse

3.2 Tasks and benefit of indicator systems

Companies, and above all, their management must be in a position to obtain rapidly, from few, but instructive pieces of information, a precise picture of the situation within and outside the company. Information about cost structures, capacity utilization, production programs, volume of orders, operating position and liquidity is essential for the survival of a company. The strengths and weaknesses of a company in comparison with competitors in the same branch of industry have to be synthesized. Indicators which are well thought-out and carefully evaluated supply the tools for rapid answers to these complex questions. The formation of indicators is the most important way of condensing information.

Indicators or company indicators are reference values for inter-company (company-specific indicators) and intra-company (industry-specific indicators) comparison. Indicators establish, in the form of easily comprehensible figures, a meaningful relation between different quantities.

The amount, preciseness and availability within an adequate period of time of information on alternatives, their consequences and factors which restrict action determine the quality of a decision. It is therefore necessary to provide information systems in a company which supply relevant information for decision-making in a concentrated, aggregated and concrete form. Indicator systems or reference models with indicator systems are such information systems.

4. Case study on the optimization of logistics processes on the basis of industry-specific reference models

The company analyzed in the following case study is a group in the beverage and brewing industry with the following benchmark figures: turnover: DEM 800 million (US$ 500 million); output: 4 million hectoliters/year; employees: 2,500. The project task consisted in analyzing and optimizing the business processes in the operating area of logistics in the main production facility (see Figure 4).

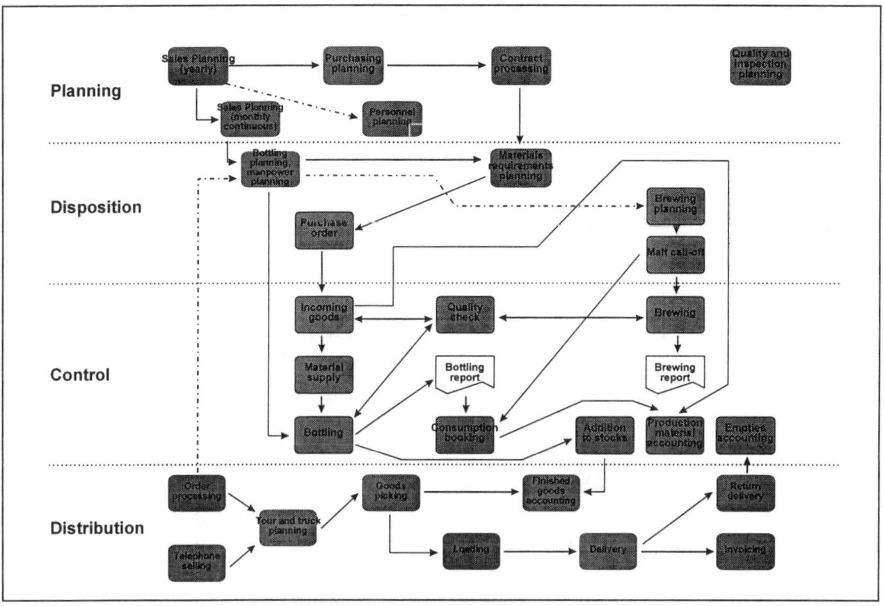

Fig. 4. Processes to be analyzed

In the first phase - Business Process Analysis -, data about the sequence of business operations in the area of logistics were collected in cooperation with the departments in charge and represented in the form of Process Chain Diagrams or Event-Driven Process Chains (see Figure 5).

These processes were subsequently checked for improvement potentials. In the case of the „Inventory Taking“ process (see figure 5), for example, the following weak points or potentials for improvement were discovered:

- inventory data were taken more than once (after bottling and before storage);
- daily physical inventory taken by two employees;
- weekly inventory;
- actual process time was twice as long as reference process time.

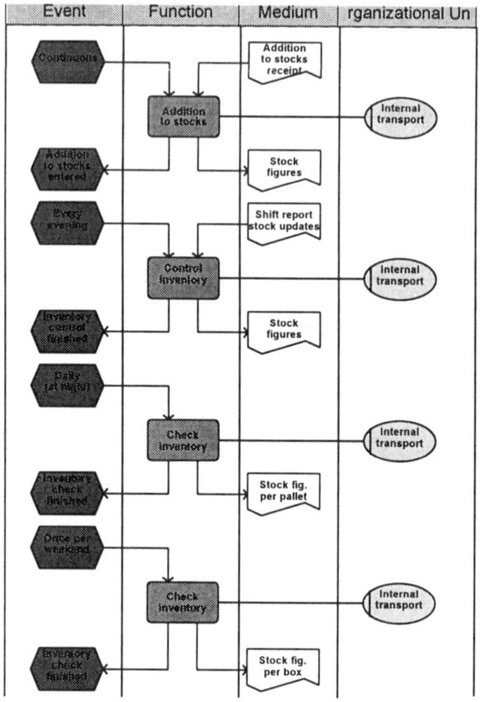

Fig. 5. Process Chain Diagram „Inventory Taking“

The objective of the second project phase of the Business Process Optimization (BPR or Conceptual Design) was to eliminate the weak points detected and to develop improved processes for the future. In cooperation with the departments in charge, the "as-is" processes were compared with processes from the Reference Model for the Beverage Industry and the Reference Model for Breweries developed by IDS Prof. Scheer GmbH. Figure 6 shows the result of this industry-specific comparison.

Very soon, it became apparent that the functions „Sales Planning“ and „Long and Medium-Term Bottling Planning“ were missing in the „As-is“ process whereas there was no optimum dynamic connection between the existing functions „Brewing Planning“, „Bottling Planning“, „Sales“ and „Distribution“ and the process time was far too long compared with the reference process time.

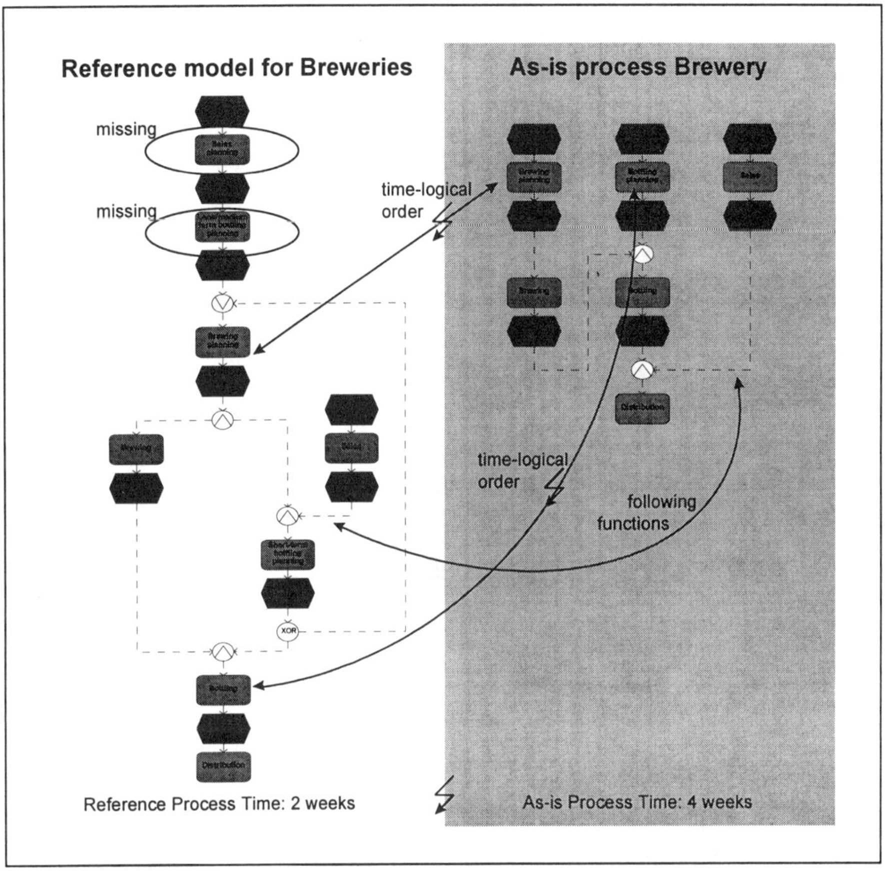

Fig. 6. Comparison of the logistics process

On the basis of these results, the logistics process presented was optimized and a new concept for it was developed (see Figure 7).

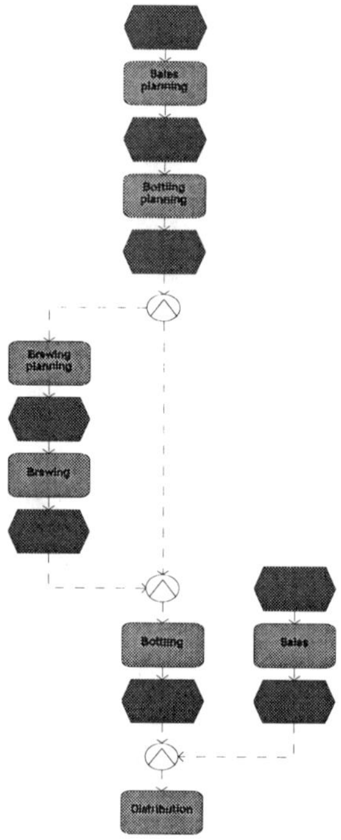

Fig. 7. Optimized logistics process

In the last project phase - the Implementation phase -, the newly developed concepts were introduced in the organization process. In the case presented above, the use of industry-specific reference models for the optimization of logistics processes contained the following advantages:

- clear representation of typical processes and structures in the industry, which provides the project team with a guideline for the "As-is" analysis, for the formulation of industry-specific questions in the "as-is" analysis, for the assessment of weak points of the "As-is" sequences of operations through a comparison with the reference model and for the planning and control of the "As-is" analysis.
- conception of new structures or new processes on the basis of the industry-specific reference models, and, above all,
- shortening of the usual project time and decrease in the usual project cost by approximately 50 - 60%.

From the experience made in the project described above and in numerous other projects with the objective of optimizing business processes in the area of logistics which were carried out on the basis of industry-specific reference models, the following realization can be derived: on the one hand, business process optimization projects can only be carried out in an optimum way if this is done on the basis of consistently used methods and industry-specific reference models. On the other hand, it was shown that a shortening of the project processing time and a decrease in project cost can only be achieved through integrated methods like e. g. ARIS and through the use of industry-specific reference models containing quality-oriented and quantity-oriented information.

References

1. A.-W. Scheer, Business Process Reengineering, Springer Verlag (1996)
2. H. Stachowiak, Modelle - Konstruktion der Wirklichkeit, (1983)
3. Eichhorn, W., Modelle und Theorien der Wirtschaftswissenschaften, in Abel, B. (publisher): Wissenschaftliche Grundfragen der Wirtschaftswissenschaften, Muenchen (19979) 60-140
4. ARIS method, ARIS toolset manual (1997), IDS Prof. Scheer GmbH
5. C. Aichele, T. Elsner, K. Thewes, Otpimierung von Logistikprozessen auf Basis von Referenzmodellen, in Management & Computer, No. 4 (1994)
6. ARIS procedure models, databases for the ARIS toolset, IDS Prof. Scheer GmbH
7. A.-W. Scheer, Architektur integrierter Informationssysteme, Springer (1992)
8. Aichele, Christian, Kirsch, J gen, Geschaeftsprozessanalyse auf Basis von Kennzahlensysteme, in Management & Computer, No. 2 (1995)
9. Christian Aichele, Kennzahlenbasierte Geschaeftsprozessanalyse, Gabler (1997)

A Case Study on BPR to Manufacturing Industry

Felix T.S. Chan[1], Ralph W.L. Ip[2], Henry Lau[3], Sev. V. Nagalingam[4] and Grier Lin[4]

[1]Dept. of Industrial and Manufacturing Systems Engineering, The Uni. of Hong Kong.
[2]Dept. of Manufacturing Engineering and Engineering Management, City Uni. of Hong Kong.
[3]Advanced Technology Education Centre, Regency Institute, Australia.
[4]School of Engineering, Centre for Advanced Manufacturing Research, Uni. of South Australia.

Abstract. An Business Process Reengineering(BPR) effort was initiated at a manufacturing industry which had recently performed poorly. This paper describes the approach of a BPR effort at the company and how the analysis of specific projects related to the proposed changes were considered critical. Also identified are the areas for this particular company that must be improved if the BPR attempt is to be successful. It is believed that by following this approach to BPR, more companies would be successful than the current situation.

1. Introduction

The international organisation studied has service centres distributed throughout Australia, which are responsible for storing, processing, and delivering goods to customers. An idea to have a centralised centre was obtained from a case study of a very successful BPR attempt by BOC Gases Australia[1]. There were many variables that would be effected by such a proposal, and these were listed as projects. On detailed analysis of several of these projects, in particular of the slitter 1 machine, it was found that changes to the model needed to be made and another model was developed. It became apparent that if the BPR model had been implemented before these projects were studied, then wrong decisions could have been made and the BPR effort would have failed. The analysis of the projects also allowed for other areas of concern to be identified which must be improved if the BPR effort is to succeed.

The company selected to apply BPR is an international business which consists of several divisions. The main reason why this company was selected was its poor performance over the last financial year, which had prompted the managing director to call for drastic changes, i.e. the company has signalled major cost cutting in the group's steel division and........that big changes would have to be made.

This commitment from senior management to cost cut and make radical changes made it an ideal company to apply BPR. It was seen that the service centre had a clear vision;

> "We will maintain market leadership and expand our processing and support services for Flat Products to achieve its growth objectives by enabling market offers tailored to customer needs."

This vision emphasises market leadership, growth, and customer satisfaction as its primary goals.

2. Overview Of BPR

BPR is the latest idea found in management literature. It is concerned with making significant, radical changes to a company based on the business processes.

BPR has been defined by Hammer and Champy[2] as...

> " ...the fundamental rethinking and radical redesign of processes to achieve dramatic improvements in critical contemporary measures of performance, such as cost, quality, service and speed."

The need for radical changes has occurred recently as companies were discovering that they could no longer afford to operate as they currently did. The need to reengineering was often realised when the business was in a critical situation and had no other option but to take drastic action. These critical problems arose due to changes in the market, which were highlighted by Katz et al[1];

- Customers expecting more from the product or service, at a lower cost.
- Competition on a global scale, and thus more intense.
- New technology.

It appears that BPR is still evolving as it is not just done when an organisation has reached crisis point. Skinner[3] mentioned two types of reengineering efforts: proactive and reactive. Proactive reengineering is when the company pursues projects for business growth, whereas reactive reengineering is done when the business is experiencing major difficulties.

The main characteristic of BPR is that it relies on radical changes, Hammer and Champy[2] suggest a clean slate approach as the best way to do this. This is because the people involved in the redesigning are not limited to past processes and procedures, which allows for a greater diversity of ideas to be generated.

Radical changes are also achieved as BPR focuses on the entire organisation. This is not only including the company's direct employees and processes, but also the customers and suppliers. Parker[4] referred to three main areas that are affected by BPR: people, processes, and technology.

However, most companies failed in applying BPR, it has been found that only 5% of such attempts succeeded[2]. There were seen to be several problems responsible for a

BPR attempt. One such problem was that processes were often fixed, and not eliminated or changed significantly. Specific reference is made to Hammer[5] who describes many businesses have spent considerable amounts to use technology to mechanise old business methods. In other words, the process was merely made faster, but ignored the possibility to eliminate, combine, and streamline processes. Such mechanisation would require an extremely high expenditure, but gave disappointing results.

Since BPR is concerned with radical changes to the entire organisation, it can be expected that employees will be resistant to change. In fact, Brittain[6] stated that "overcoming the human resistance to change is the biggest single issue in reengineering."

In particular, BPR tends to be used only when the company is in severe financial trouble, and so the radical changes are tried to be made fast. This can lead to factors not being considered or analysed in depth, and as a result, wrong decisions being made.

3. Overview Of The Reengineering Effort

3.1 Understanding The Company

In order to obtain a better understanding of the company, information such as organisational charts, customers, suppliers, and products were studied. It was seen that there were two similar service centres distributed around Australia, which process steel supplied from two different sites, and then distribute the finished products to customers (see Figure 1).

In addition to supplying the local customers, it was seen that the South Australia(SA) and Victoria(VIC) centres supplied NT and Tasmania respectively. Obviously, there existed a degree of centralisation, and it was also interesting to note that the SA and VIC plants were also similar in product volume and mix.

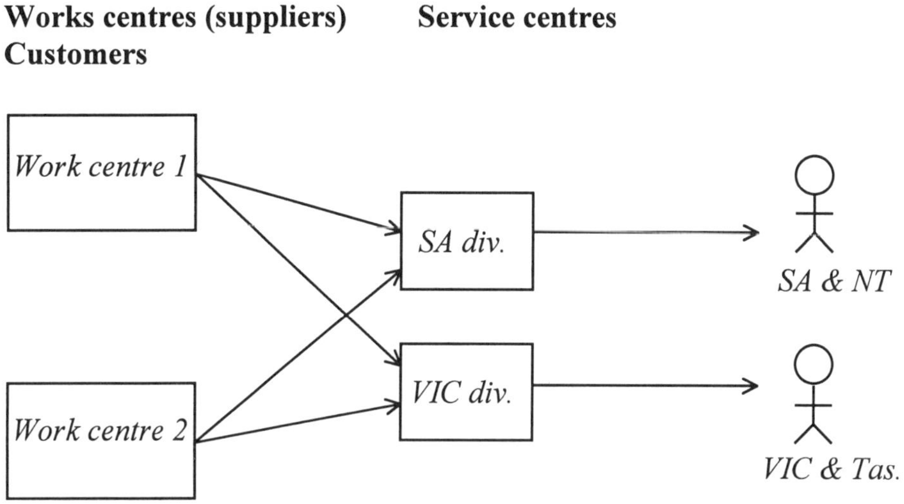

Fig. 1 Model 1: Current situation

3.2 Developing The BPR Model

Mention has been previously made that BPR only succeeds in 5% of applications. One very successful application has been described by Katz et al[1], where BOC gas combined their service centres into one centralised centre.

From this observation and the above model of the service centres, the concept of a large, centralised service centre, obtained by combining the SA and VIC sites, was developed. This allowed a model showing these radical changes to be developed (see Figure 2).

Some of the benefits from a centralised service centre would be increased productivity, reduction in employees, and released assets. The extent of the effects would need to be studied and quantified in order to determine if it is feasible or not. Katz et al[1] discussed how over 100 hundred projects relating to such a radical change were identified. These projects covered all areas and variables that would be changed from the concept of a centralised service centre. These projects were seen as fundamental in determining the affects of the BPR model and thus if the model was feasible or if improvements could be made to the model.

Works centres (suppliers) **Service centre**
Customers

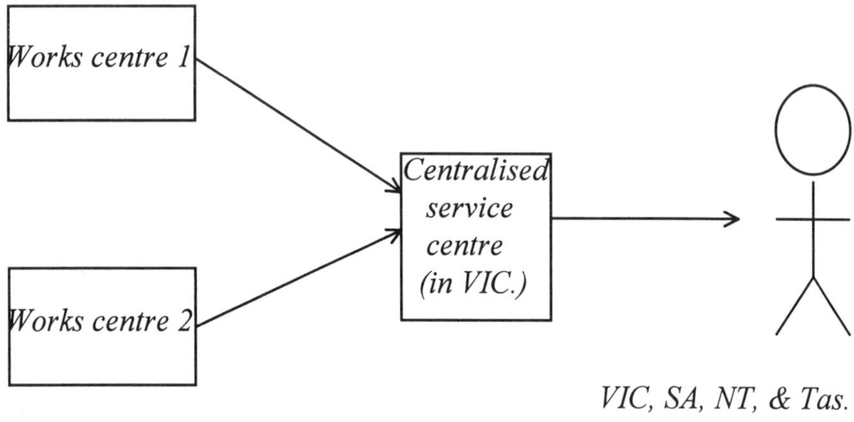

Fig. 2 Model 2: Centralised service centre

A similar approach was followed for the BPR model suggested above, and the specific projects are listed below:

- Duplicated positions
- Customer satisfaction
- Machine relocation
- Cost of relocation
- Communication network
- Scheduling
- Business processes
- Expansion capacity of VIC plant

However, time restrictions prevented a detailed analysis of each of the projects, though several were initiated and the results of others were estimated.

Duplicated position

The analysis into the duplicated positions was commenced by studying the organisational charts for both the SA and VIC sites. From this it was seen that many of the positions were the same, which allowed for the reduction of employees. Thus, on moving the SA centre to VIC, there would be no need for say, two state managers, allowing for one to be eliminated. In all, approximately 20 staff people could be eliminated, and reductions could also be made in the number of operators (approximately 20), since many of the tasks and responsibilities would also overlap.

Customer satisfaction

A project concerned with customer satisfaction was listed. This is a very important factor to consider, and should be conducted by surveys, interviews, and the customer actually being involved in decision making. From this analysis, the organisation can try to obtain high customer satisfaction and avoid any problems that may arise from the radical changes.

To maintain customer satisfaction, it was decided that there would need to be some people to remain in SA as representatives. These representatives would be responsible for meeting clients about any concerns that they might have, and as the customers often enquire into areas which require the help from say the accountant, they would need to obtain additional skills.

It would also be necessary to determine which customers rely on Just-In-Time(JIT) or are considering using this method. This would have a significant effect on the amount of finished stock that is to be held in SA.

Machine relocation and Cost of relocation

The analysis then focused on the necessity of relocating the machines to VIC. Such an exercise would be extremely expensive, especially due to the downtime of machines and dissatisfied customers. However, on relocating the SA machines, it was expected that the productivity would increase since two identical ones would be working together at the same site. This is because similar products are processed and intelligent scheduling would reduce setups.

Based on the low operating time of all machines at the SA site, it was decided to see if radical improvements to machine productivity could be obtained some other way besides relocating. Since the manufacturing aspect is a major component of the business, it is fundamental that it should also be studied. It is expected that if such an analysis is not conducted in manufacturing organisations that wish to introduce BPR, wrong decisions can be made and the attempt will fail or under perform.

Detailed analysis of a particular machine, slitter 1, showed that significant improvements could be achieved. Similar reductions were expected to be able to be made with other machines, not only at the SA site, but also in VIC.

Thus, it was felt that due to such significant improvements, the SA machines may not have to move in order to obtain large improvements. The analysis of slitter 1, therefore, could in fact save the company considerable time and expense and thus prove to be the most important project completed. More importantly, the considerable savings generated were obtained from relatively simple tasks which would be proved easy to implement. The above analysis allowed the concept of relocating just the sales and administration division to arise.

3.3 Refining The Model

The changes resulting from the customer satisfaction analysis, in relation to local customer representatives and inventory, required that the model be changed to incorporate this matter (see Figure 3). One point which is worth to mention here is that, the key operation remaining at the SA site is managing the inventory.

However, the results obtained from the analysis of the projects, in particular, the machine relocation and the cost of relocation, have suggested that further refinements should be made to the model. It was felt that due to such significant improvements, the SA machines may not have to move in order to achieve higher productivity. Thus, the impact of this analysis would have a significant effect on the decision making. The slitter 1 analysis enabled the model to be refined so that another model could be considered (see Figure 4), whereas the previous changes, such as local customer representatives, were seen as essential.

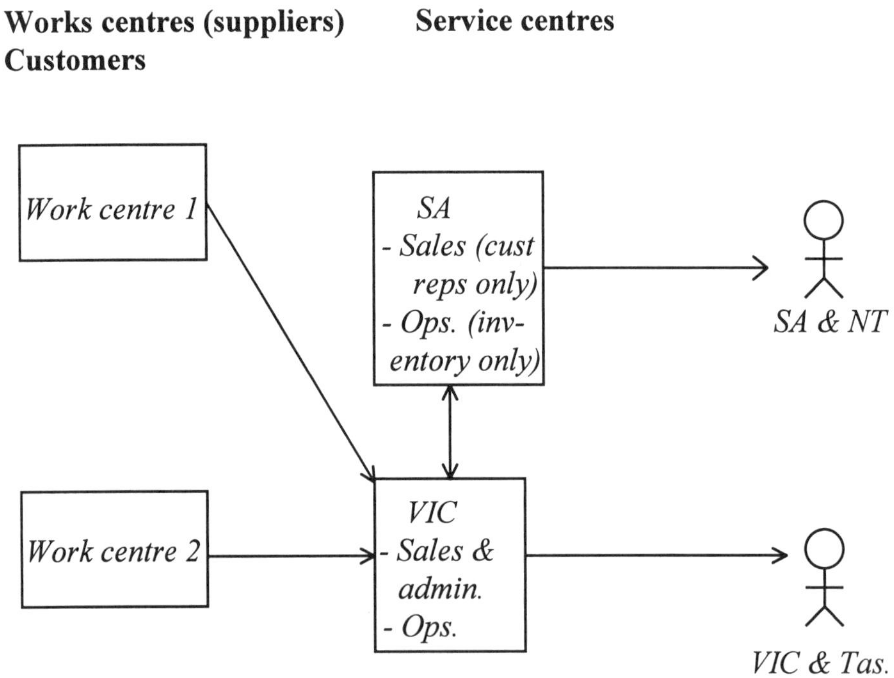

Fig. 3 Model 3: Centralised service centre Mk 1

Works centres (suppliers) **Service centres** **Customers**

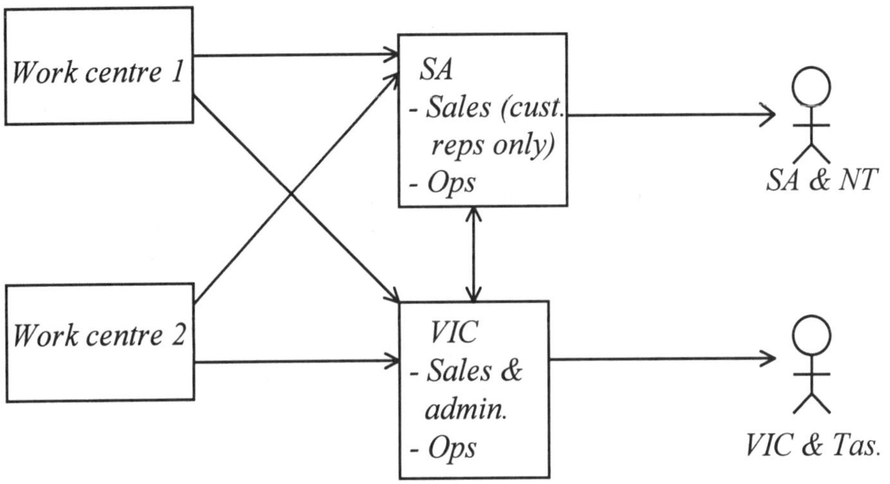

Fig. 4 Model 4: Centralised service station Mk 11

4. Comparing The Models

It was necessary to conduct a financial simulation for both models, which can only be done sufficiently when all of the specific projects have been completed. Due to time restraints, the results of projects have been assumed, but are considered to give a reasonable valuation of the proposed models.

The financial simulation was completed using the Profit and Loss Statement and Balance Sheet for the current model, and then making the changes for the two proposed models, i.e. Model 3 and Model 4. This was done using MS Excel, which was found to be a powerful tool in the simulation since such changes could easily be made using formulae. The results of the financial simulation are seen in Table 1 below and clearly show Model 3 as the best.

Table 1 Financial simulation

Model	Net Assets	Net Profit (Loss)	Shareholders' Equity	Released Assets	Ranking
1	$50,525,189	$(4,107,903)	$46,417,286	0	3
3	$50,525,189	$357,937	$50,883,126	$2,042,451	1
4	$50,525,189	$(443,273)	$50,081,916	$1,000,000	2

There were some considerable changes to the assets as fixed assets, such as buildings, were converted to cash. This is seen in the 'released assets' column and is important to consider since these funds can be invested profitably, such as in shares.

It should be noted that abnormal expenses were ignored for the purpose of the financial feasibility analysis since they are 'once-off' costs. There were two types of these costs, which were the relocation cost, and the remuneration packages paid to employees. These abnormal items have been considered for the payback period in the comparative analysis below.

The financial simulation has limitations to identifying the best model since it completely ignores qualitative factors, which have been considered in the analysis below.

A comparative analysis enabled the qualitative factors to be considered and converted into a quantitative value, ie. the probability of success. The factors considered to reduce the probability of success were;

- High degree of change
- Big effort
- Many people/divisions involved
- Long lead time
- External commitment

The analysis ignored customer satisfaction and employee morale since these were considered as being almost the same for each model.

Weightings were applied to each of the factors(see Table 2), and each model was given a score as to how it performed, which were then totalled. The resulting probability of success factor was found to be extremely low for each model, but this was considered acceptable since the technique used was consistent.

Model 1 is not an option since the organisation would not remain competitive if it continued to operate as it currently does. Calculated risks need to be taken in any business in order to achieve outstanding results.

Although Model 3 performs better than Model 4, it is less likely to succeed and so the expected value is considerably lower. The payback period from implementing Model 3 is also significantly longer than Model 4 since it has a very high relocation cost and remuneration packages to pay. Thus, Model 4 is quite clearly the best option.

Table 2 Comparative analysis

Criteria	Weighting	Model 3		Model 4	
		Score	Total	Score	Total
High degree of change	9	9	81	6	54
Big effort	8	9	72	6	48
Many people/div. involved	7	9	63	5	35
Long lead time	7	8	56	6	42
External commitment	6	6	36	3	18
Total			308		197
Probability of success			0.1676		**0.4676**
Net benefit			$4,465,840		$3,664,630
Expected value			$ 748,500		**$1,713,600**
Payback period			1.69 yrs		**0.26 yrs**

Depending on the number of models analysed, further refinements may be able to be made so that eventually the best model is found and selected. Similarly, if Model 4 was implemented successfully, then the idea of relocating the operations division, as seen in Model 3, would be less risky.

5. Why BPR Could Fail At The Company

There were several major areas of concern specific for the company studied that would prevent BPR to be applied successfully. Several instances observed during this case study showed that there was no management commitment to change. The first of these was evident in a confidential report from the state manager to the executive general manager in March of last year, in which the following was stated;.

> "A comparison with other Service Centres with a similar volume and mix of product has shown that they also produce similar results...if the effects of the Works commission is eliminated."

In fairness to the state manager, he did raise some viable concerns to the performance evaluation of the service centres, but it was his acceptance of the satisfactory performance that caused concern. There was also no commitment to productivity improvement.

The management commitment problem was further displayed on discussion with the operations manager, who had set a goal of increasing the average tonnes produced per shift by 3 tonnes for a particular machine. The manager, however, was unsure how this target would be achieved. Obviously, the manager knows what has to be done, but is unsure of how to get there.

The above findings shows that the concept of change is not present and that the commitment to change expressed by senior management has not filtered through the divisions of the organisation. To reach a goal, it is fundamental to study the variables that need to be improved and the action that needs to be taken. If radical changes are to occur at this company, management must show commitment in such projects or otherwise poor decisions may be made and the BPR effort will fail.

The company displayed little sign of continuous improvement techniques and teams, and this problem is directly related to management commitment discussed above. Only when there is a strong culture of teams dedicated to continuous improvement can such radical changes be made. These teams would be able to carry out the specific projects, but can only be established with a full support from the senior management.

Another area of concern was that of communication. There was seen to be little communication between divisions and different levels of employees. This often led to problems, but more importantly, prevented these problems from being corrected and avoided in the future.

For any change in project management, especially the one such as BPR which involves radical changes, it is critical that there is effective communication throughout the entire organisation. This can eliminate many unnecessary problems and enable the change to be implemented successfully by relieving the human resistance to change.

The management information supplied needs to be improved if the BPR project is to succeed. There were several instances where information supplied to management was misleading and sometimes even inaccurate, and this can lead to wrong decisions being made. For a BPR application, which involves radical changes, it is essential that this information be correct.

The major concern that arose, however, was based upon the specific analysis of the slitter 1 machine, and was involved with how the BPR effort should be handled by any company. An analysis of this machine was conducted for it was found to be operating less than 4 hours per day. The idea of a centralised service centre appeared to solve this productivity problem as the two same machines working together would enable intelligent scheduling to reduce setups. However, on conducting detailed analysis, it was shown that the machine delays could be reduced significantly to achieve a 40% increased in machine utilisation. This significant improvement could be made some other way than relocating the machines. As a result, the analysis provided an option that wasn't even considered before. It is expected that detailed analysis of the other projects would also lead to model refinements.

According to this research findings, it is believed that it is absolutely critical that all variables that are affected by the BPR concept should be analysed in depth. This may take considerable time, but in the long run, will enable the best decisions to be made. It would also be possible, as in the case of the slitter 1 analysis, to apply these improvements immediately since they do not involve radical changes. In this way, more information concerning the project can be obtained, such as the performance improvement of other machines and the capacity of the VIC centre. This may show that only a few machines need to be moved. Whatever the outcome, the study of specific projects in a BPR effort should be done by companies as it prevents radical decisions being made hastily and thus wrong decisions.

6. Conclusion

The project completely highlighted the significant importance of analysing in details of all the variables that are affected by a BPR model. These variables, listed as specific projects, enable the outcomes of the model to be determined as well as any refinements or additions to the model. This may take considerable amount of time and result in many projects completed and models developed, but this thorough investigation will allow the best model to be selected, both in terms of profitability, customer satisfaction, and likelihood to succeed. It is believed that by following this approach to BPR, considerably more companies would be successful than the 5% that are currently exist.

7. References

1. Katz, M., Katz, L., Morien, R. and Mitchinson, S., "Reengineering customer services at BOC Gases Australia", *Business Change & Reengineering, Vol. 2, No. 3*, 1995, pp. 4-17.
2. Hammer, M. and Champy, J., *Reengineering the corporation*, Australia, Harper Business, 1993.
3. Skinner, L., "BPR: Current issues and applications- BPE executive summary", *Georgia, Institute of Industrial Engineers*, 1993.
4. Parker, J., "BPR: Current issues and applications- An ABC guide to BPR", *Georgia, Institute of Industrial Engineers*, 1993.
5. Hammer, M., "BPR: Current issues and applications- Reengineering work: don't automate, obliterate", *Georgia, Institute of Industrial Engineers*, 1993.
6. Brittain, C., "Reengineering complements Bellsouth's major business strategies", *Industrial Engineering*, February, 1994.

Enterprise Modelling with Intensive Knowledge

Zhao ZhenYing and Robert de Souza

Centre for Engineering and Technology Management
School of Mechanical & Production Engineering
Nanyang Technological University, Nanyang Avenue,
Singapore 639798
M96001B61@ntuvax.ntu.ac.sg

Abstract. As manufacturing systems become more sophisticated and complicated, effective manager know how plays a crucial role in managing an enterprise and in dealing with its dynamics and uncertainty. In this paper, a formalism based on the CIM-OSA reference model is presented to specify the business processes and enterprise activities at the knowledge level.

1 Introduction

The design and improvement of modern manufacturing enterprises is an extremely complex process, for it involves a non-trivial combination of technological, human, machine and organizational issues. To cope with such a challenging problem requires the model of the enterprise not only at shop floor level but also at organizational and business level, and manipulating the model (such as simulation) to produce guidelines, decision, suggestion or knowledge to the design and improvement of the enterprise.

There is a vast body of well-developed enterprise models in the context of Computer Integrated Manufacturing (CIM) which are ready to be used either in the design or in the implementation phases of enterprise integration. For instance, the ISO Technical Committee TC 184 on enterprise modelling and integration has proposed a Reference Model for shop floor production standards (ISO 1990). This model provide a conceptual framework for understanding discrete parts manufacturing; and can be used to identify areas of standards necessary to integrate manufacturing systems. The Purdue Consortium has developed an engineering-oriented architecture and associated implementation methodology. The architecture was published as the Purdue Enterprise Reference Architecture (PERA) (Williams 1994). The GRAI/LAP laboratory of the university of Bordeaux has developed a framework and modelling tools to support enterprise integration (Doumeingts 1987). ESPRIT projects have also developed modelling reference architectures and industrial applications. CIMOSA projects are, however, the most significant results. In the CIMOSA scenario, the business process is distinguished from the enterprise activities. An enterprise is viewed as a large collection of concurrent business processes executed by a set of functional entities that contribute to business objectives or goals (Vernadat 1995, Vernadat 1996). The architecture for integrated Information System (ARIS) has been developed at the University of Saarbrüchen in Germany (Scheer 1994). Its overall structure is similar to CIM-OSA, but it deals with more traditional business-oriented issues of enterprises such as order processing, inventory control and so on.

All these aforementioned architectures have produced many fine results, however, most of them are focused on formulating the data and information aspects of an enterprise. The knowledge is mixed with information and treated as part of the information in processing and utility. The modelling language employed by the above schemes is also suitable for representing data and information but not for expressing various knowledge. Meanwhile, these modelling schemes focus on precise and complete knowledge, or complete unknown random factors in the model. Little attention is paid to the imprecise and incomplete knowledge that persists in the enterprise. Actually, the enterprise paradigm has shifted with advances in technology, from the original labour and physical facility intensive mode through a data and information intensive mode, and onto the present knowledge intensive format. This knowledge embodied in an enterprise (common sense, experience, rules etc.) plays a significant role and often becomes the most important resource for the success of the modern enterprise (Paul Gosset, 1994). The functionality and the behavior of the enterprise are increasingly governed by the knowledge of these "intelligent" enterprises. Therefore, in order to formulate dominant and representative features of real-life enterprises, it can be argued that more emphasis ought to be placed on the knowledge instead of on detailed data and information representation during modelling.

2 Knowledge Integration of an Enterprise

Modelling an enterprise requires sufficient understanding to represent its behavior and functionality effectively. If the behaviour and functionality of the enterprise are completely known, namely, the knowledge of the enterprise is precise knowledge, then conventional precise knowledge representational methods such as mathematical equations, rules, facts, frames, logic and the like can be used. However, if the functionality and behaviour of the enterprise are completely unknown, namely, the knowledge of the enterprise belongs to random knowledge, one feasible modelling approach is the random representation on the basis of probability theory or stochastic theory. In reality, the understanding of the enterprise functionality and behavior usually fall between these two polar states. The enterprise behavior and functionality are either partially known or undergo change. In other words, the knowledge of the enterprise is muddy knowledge, i. e. neither precise nor random. Hence, vague, imprecise, ambiguous knowledge representational methods such as fuzzy logic, Bayesian probability, certainty factor model, etc. should be adopted in modelling. Conventional modelling methods in the CIM scenario reveal very little, if anything, about this imprecise, vague ambiguous knowledge.

Furthermore, these knowledge may manifest itself in two broad forms, either as numerical knowledge or symbolic knowledge. Numerical knowledge may appear in mathematical equations, regression models, stochastic models, or as a data set. A trained neural network can also be employed to represent input-output numerical relationships whose transformation process cannot be easily described by obvious mathematical formulae. On the contrary, symbolic knowledge can be presented in linguistic, symbolic form such as facts, frames, semantics, production rules, logic, natural language and so on. The above-mentioned precise, muddy or random knowledge that appear in either symbolic or numerical form can formalize six categories of knowledge: precise numerical knowledge (such as mathematical function), precise symbolic knowledge (such as rules, facts), muddy numerical knowledge (such as Bayesian probability, certain factor theory), muddy symbolic knowledge (such as fuzzy knowledge), random numerical knowledge (such as stochastic and chaotic process) or random symbolic knowledge (such as random symbolic generation).

Normally, the behaviour and functionality of an enterprise contains varying degrees of these six types of knowledge and hence, modelling needs an integration of them. The integration should not be simply the lump sum to make an accumulated whole, but be carried out in a hybridisation and coupling fashion to form a synergistic whole. One promising way to accomplish thus action is to integrate the multiple types of knowledge at the knowledge level. In other words, using knowledge about the domain knowledge to accomplish "deep" level integration. The knowledge about the domain knowledge is meta knowledge, namely, the knowledge concerning the propriety, characteristic and utility of the domain knowledge. The integrated structure is shown in Fig. 1. The function of meta knowledge is to use, control, manage and coordinate the domain knowledge.

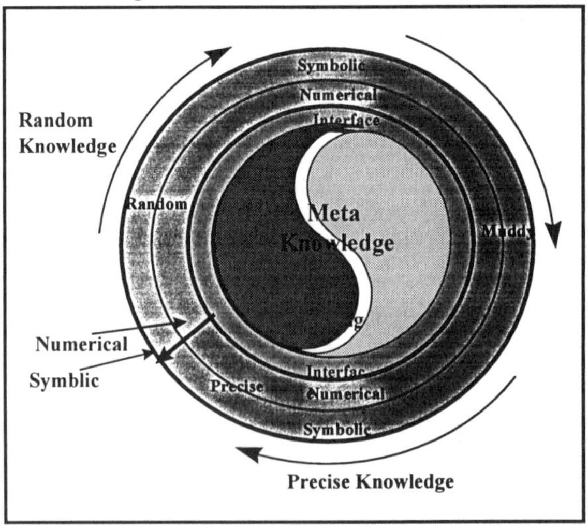

Fig. 1. The integration using meta system as a kernel

3 Activity Formalism

3.1 Definition and Graphic Representation

In essence, an activity is defined as a transformation function which transforms the input state into an output state under some condition. It can be written as equation 1.

$$\text{Output state} = f(\text{Input state}) \text{ subject to guard}(f) = \text{C}. \qquad (1)$$

Several formalisms for graphical representations of an activity have been proposed as shown in Fig. 2. A Generic Activity Model (GAM) has been proposed by ISO TC 184(ISO 1990). It makes use of a box with 10 legs as illustrated by Fig. 2(a). It differentiates between information flow and the material flow and makes for explicit resource input and output. However, the input and the output have no precise semantics, and the model fails, in the authors' opinion, to describe the triggering condition as well as the ending status of the activity. Furthermore, the representation is deemed too complex. Another simpler representation is the IDEF0 ICOM box depicted by Fig. 2(b) based on actigrams of SADT(Vallespir et al. 1991). It is often used in practice to represent either processes or activities. However, it may be considered too simple. Another candidate for activity representation is the CIMOSA enterprise activity box. It has a graphical representation compliant to the ISO GAM representation and has semantically well-defined inputs and outputs (Fig. 2(c)) (Vernadat, 1995). However, it does not clearly represent the triggering condition and the control input, as well as the knowledge characteristics and function in the activity is not explicitly specified, hence, the authors have proposed a revised activity box as shown in Fig. 2(d) which is modified from CIM-OSA representation. The difference between the proposed activity and that of CIM-OSA lies in the fact that the control output is modified to express the triggering condition together with facilitate the recording of the termination status and subsequent post processing. The function input and output also contains knowledge views beyond the physical and information flows.

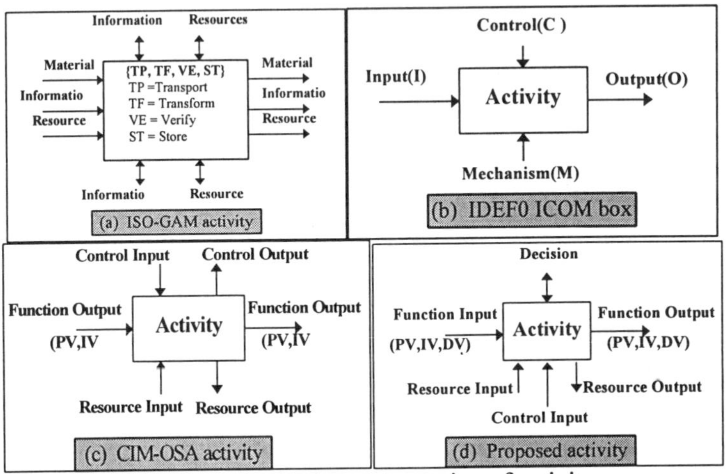

Fig. 2. The graphic representation of activity

In Fig. 2(d), the Function Input (FI) refers to a set of object views to be transformed by the activity; whilst the Function Output (FO) is the one transformed by the activity. Generally, the object views offered by CIM-OSA are referred to as the physical and the information views. The knowledge is mixed with information and data in the information view (Vernadat, 1996). In order to emphasize the function of the knowledge, a knowledge view is added to separate the knowledge from the information in the object view, and to modify the function input and output as the Physical View (PV), Information View (IV) and Knowledge View (KV). The Resource Input (RI) refers to a set of object views used as the resource to support the execution of the activity; it will not affect the transformation of the FI, FO; whilst the Resource Output (RO) is a set of object views used as resources to be transformed after execution of the activity. The resource object view encompasses the physical and information views. Control Input (CI) refers to a set of information views used to support or constrain the execution but is not modified by the activity. The Decision (D) contains the triggering condition for execution of the activity and is post processed after the activity runs. It will not be affected by execution of the activity. It is mainly concerned with conditions, the constraints to perform the activity, and with the termination status of the execution.

3.2 Mathematical Description

Although graphical notations are useful to understand, the mathematical expression can give more precise and deep insight about the nature of activities. Essentially, an activity can be expressed as follows:

$$\boldsymbol{O}_A = \delta(D.\text{trigger\_port})\boldsymbol{f}(\boldsymbol{I}_A, \boldsymbol{fe}, t) \tag{2}$$

where $\boldsymbol{O}_A \in \boldsymbol{R}^{2\times3}$ is the output vector of the activity $\boldsymbol{O}_A=(\boldsymbol{F}_O,\boldsymbol{R}_O)^{\mathrm{T}}$. $\boldsymbol{I}_A \in \boldsymbol{R}^{3\times3}$ is an input vector of activity $\boldsymbol{I}_A=(\boldsymbol{F}_I,\boldsymbol{R}_I,\boldsymbol{C}_I)^{\mathrm{T}}$; $t\in\boldsymbol{R}$ is time; $\boldsymbol{f}\in\boldsymbol{R}^{3\times3}\times\boldsymbol{R}^{n}\times\boldsymbol{R} \rightarrow\boldsymbol{R}^{2\times3}$ is a mapping field of activity; D is decision and

$$\mathrm{D} = \{\text{Trigger\_port; Post\_processor}\} \tag{3}$$

$\boldsymbol{fe}$ is a user-defined feature set. It is intermediate variable $\boldsymbol{fe}\in\boldsymbol{R}^{n} = (feature_1, \ldots, feature_n)$,

$$\boldsymbol{fe} = \boldsymbol{g}(\boldsymbol{I}_{\mathrm{A}}, t) \tag{4}$$

in which $\boldsymbol{g}\in\boldsymbol{R}^3\times\boldsymbol{R} \rightarrow\boldsymbol{R}^{n}$ is a mapping from the FI to the defined feature set.

$$\boldsymbol{F} = (F_{\mathrm{PV}}, F_{\mathrm{IV}}, F_{\mathrm{KV}})^{\mathrm{T}} \tag{5}$$
$$\boldsymbol{R} = (R_{\mathrm{PV}}, R_{\mathrm{IV}}, R_{\mathrm{KV}})^{\mathrm{T}} \tag{6}$$
$$\boldsymbol{C} = (C_{\mathrm{IV}}, C_{\mathrm{KV}})^{\mathrm{T}} \tag{7}$$

where F_{PV}, F_{IV} and F_{KV} denote the functional physical view, information view and the knowledge view, respectively. R_{PV}, R_{IV} and R_{KV} is resource physical, information and knowledge view respectively. The mapping is conducted by the integrated knowledge shown in Fig. 1. $\delta(\cdot)$ is delta function in term of triggering condition in decision,

$$\delta_{\text{D}.Trigger\_port} = \begin{cases} 1 & \text{if D.Trigger\_port is true} \\ 0 & \text{if D.Trigger\_port is false} \end{cases} \quad (9)$$

In the case that mapping field $f(\cdot)$ is an unit mapping field, namely, $f(\cdot) = \mathbf{I}$, the resource will never be changed and the function input equals the function output. This means the activity is just a delay. If $f(\cdot)$ can be well constructed in either symbolic or numerical way, then the activity is called structured, otherwise, the activity is non-constructed.

3.3 Activity Specification

Enterprise activity is used to describe enterprise functionality with the allocation of resource and decision support. In CIM-OSA, the representation employs Pascal-like program template. This representation is complex and incomplete with lack of systematisation (Nager et al., 1995). Hence, the template representation of activities is enhanced. Its specification is characterized by three components as shown in equation 12.

$$\text{Activity} := \{ \text{Header; Decision; Body} \}. \quad (10)$$

The Header comprises the description of the activity including activity name, activity type, design authority and illustration to the activity. The detail representation for the integrated knowledge can refer to De Souza (1997).

4 The Business Process

The business process is used to describe the enterprise behavior. It is aggregated by a series of activities and only triggered by events such as order arrival, machine breakdown etc. In computer theory, it has been shown that activities can be combined into processes by using a parallel operator denoted as "||" to describe any flow of control assuming that this operator is not commutative(Curtis et al., 1992). In CIM-OSA, a process is aggregated by a control structure rule set (procedural rule set) such as sequence, parallelism, rendez-vous, choice or loop to connect activities to a network fashion(Vernadat, 1995). However, many business process in manufacturing enterprise go hand in hand with a vast body of incomplete and vague fashion. Furthermore, some connections among the activities may be based on stochastic relations. In short, it is strongly dependent on multiple types of knowledge. Therefore, its representational formalism should adopt a similar structure as shown in Fig. 1, although the business process and enterprise activity are at different layers within the enterprise. In other words, although the meta system and domain knowledge are different, the integrated structure is same.

In a general mathematical form, the process can be expressed as

$$\text{P} = \delta(\text{D.condition}) f(\boldsymbol{EA}, t) \quad (11)$$

where D denotes the Decision which shown in equation 3. The Triggering condition is event driven, and the End_processor performs the post operations after the business process such as results storing and terminating status recording. $\boldsymbol{EA} \in \boldsymbol{R}^n$ is the enterprise activity vector and $\boldsymbol{EA}=(EA_1, \ldots, EA_n)^T$. EAi represents the *ith* enterprise activity. $f(\bullet) \in \boldsymbol{R}^n \times \boldsymbol{R} \rightarrow \boldsymbol{R}$ is the mapping function which is constructed by an integration of the multiple categories of knowledge as shown in Fig. 1. The $\delta(\cdot)$ is defined by equation 9.

In a more detailed form, a business process also consists of three components as shown in equation 11. The header defines the name, type and illustration of the process. The Decision contains the triggering condition for execution of the business process and post processes sequences. Its structure is the same as a equation 4 except that it is for a process. The body specifies the behavior of the business process and is constructed of an integration of multiple types of knowledge as Fig. 1. The process is at a high level and operates on the activities. Correspondingly, the knowledge is also different. For example, the precise symbolic knowledge usually contains various rules such as production, sequential, conditional or spawning rule, loop, termination rule or constraint.

5 Example

Let us consider an activity performing machine maintenance depicted by Fig. 3.

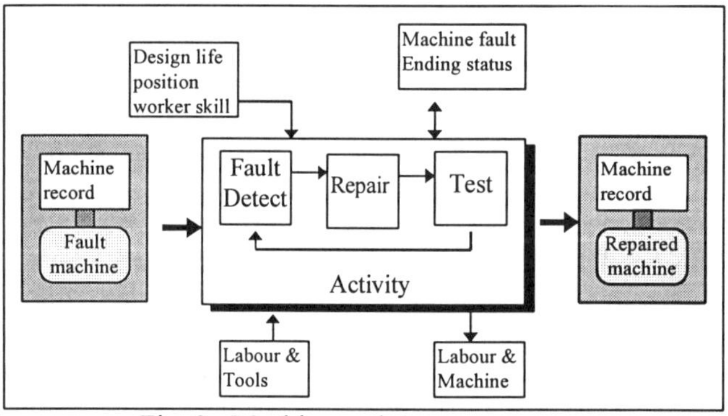

Fig. 3. Machine maintenance process

The machine maintenance activity is performed by a team of workers and consists of three basic steps. The Function Input consists of the faulty machine and the associated information and knowledge about the faulty machine such as the machine fault times, composition of the machine etc.. The Function Output contains the repaired machine and the corresponding knowledge and information about the machine. This activity needs a resource support such as labour, the fault analysis equipment and software, maintenance tools. The Control Input is necessary for activity execution but will not be changed by the activity. For example, the machine designed life, the function of the machine in the production line, machine worked time are Control Inputs. The

Decision includes the triggering rules such as when the machine is at fault and needs to be repaired, the activity start, and the post processor activated for such tasks as recording the terminating status, storing the results and so on. The mapping field consists of fuzzy knowledge such as maintenance time which has a fuzzy relationship with the worker skill, repaired results and testing; the stochastic knowledge includes the processing time that exists randomly to some degree; the deterministic numerical knowledge includes the signal processing, fault tree reasoning; the symbolic knowledge includes fault position reasoning based on the fault tree, and the maintenance results analyzed by reasoning. The following is part description of this maintenance activity at the specific level.

```
ENTERPRISE ACTIVITY i
   /*Header: the name, type, identifier and explanation of this activity */
Name: machine-maintenance
Type: man-machine
Identifier: EAi
Description: This activity is used to represent a maintenance process
Decision: {/* Decision: the activity to decide the execution of this activity */
   IF machine is fault and worth repairing THEN START ACTIVITY;
   ELSE  RETURN Activity cannot be executed;
                        ⋮
WHEN   repaired_results = FINISH  DO {   /* Postprocess */
              Termination_status = repaired_status;
              SAVE(results);  ...;}
              }
      /* Body: inputs and outputs of this activity */
Function Input:    FI.PV, FI.IV, FI.KV
Function Output: FO.PV, FO.IV, FO.KV
Resource Input:   RI.PV, RI.IV
Resource Output: RO.PV, RO.IV
Control Input:     CI.IV
Features:  actual_processing_time, fault_rate;
   /*Body: the integration of knowledge to map the inputs to outputs */
Activity Knowledge:{
   Precise Symbolic Knowledge { /* knowledge precisely defined */
      RO.PV.workers = RI.PV.workers;
                        ⋮
      FO.PV  SEQDO { fault_identify; repair; test };
                                  ⋮
      IF test.status = OK THEN {
              FO.PV.machine = repaired machine;
              repaired_status = OK;}
      ELSE repaired_status = FALSE;
                        ... }
   Precise Numerical Knowledge {
      FO.KV.fault_times = FI.KV.fault_times +1;
      FI.IV.Max_processing_time = 50.0;
      FI.IV.Min_processing_time =  19.0;
```

```
                              :
                               }
Muddy Symbolic Knowledge: {
   FO.KV.Fuzzification = FI.KV.triangular;
   FO.KV.Defuzzification_method = FI.KV.Centroid;
                              :
        }
Random Knowledge {
   normal_distribution = normal(FI.IV.mean, FI.IV.variant);
   FO.IV.pt_random = GENERATE(normal_distribution);
                                        :
   }
Meta Knowledge {  /*The function of each types of knowledge */
   PK : { fault_identify,  maintenance} /* precise knowledge */
   MK : { processing_time, fault_times, ...}
   RK: { pt_random}  /* random knowledge */
   CONSTRAINT: { /* some constraint for operation order */
    Decision.repaired_status NEED {fault_detect & test}
          fault_detect AFTER signal_processing;
          fault_detect USE ( diagnosis_method 1 );
                              :
                     }
   Conflict_Resolver: { diagnosis_results =
      OR(diagnosis_method1, diagnosis_method2};
      WHEN activity = START {
      Fe = MAPPING(RI, FI, CI);
      (FO,RO) = MAPPING(FI, RI, CI, Fe);
                              :
                              }
}
END ACTIVITY
```

6 Conclusion

Business process and enterprise activity serve as a basis for modelling enterprise behavior and functionality in the context of CIM. As knowledge becomes the most important resource of today's enterprises, modelling at the knowledge level can be simplified. The modelling process made systematic and capable enough to capture the dominant features of enterprises in the model. In this paper, a formalism to represent enterprise activities and business processes at the knowledge level has been specified based on the CIM-OSA results. The representation involves the integration of multiple types knowledge. The meta system as a kernel to accomplish the integration is an efficient structure to achieve "deep" level integration.

References

1. F. B. Vernadat, CIM business process and enterprise activity modelling, *Proceedings of the IFIP TC5 working Conference on model and methodologies for enterprise integration*, Queensland, Australia, November 1995.
2. F. B. Vernadat, *Enterprise modelling and integration*, Chapman & Hall, London, UK, 1996.
3. G. Nager, et al., Structuring and configuration of CIM systems for branch-specific medium-sized enterprise, *Journal of Intelligent Manufacturing*, Vol. 6 Number 3 June 1995.
4. ISO, 1990, Reference model for shop floor production standards. Technical Report 10314, Part 1. ISO TC 184/SC5/WG1 N126 and ISO TC 184/SC5/WG1 N160.
5. P. Gosset and P. Massotte, Knowledge Systematisation and sharing - GNOSIS(IMS) impact on industry, *Proceedings of the Tenth CIM-Europe Annual Conference*, 5-7 October, 1994, Copenhagen, Denmark.
6. T. J. Williams et al., Architecture for Integrating Manufacturing Activities and Enterprises, *Computer in Industry*, 24, (2-3), 1994, pp. 111 -140.
7. De Souza, Robert and Zhao, Zhenying, "A Case For Intelligent Representation of Dynamic Resources in Simulation," accepted by *international Journal of Production Operation*, 1997.
8. Curtis, B., Kellner, M. I., Over, J., 1992, Processing modelling, Community of the ACM, 35 (9), pp. 75-90.
9. Vallespir, B., Chen, D., Zanettin, M., Documeingts, G., 1991, Definition of a CIM architecture within the ESPRIT project IMPACS. Computer Applications in Production Engineering:Integration Aspects, Elsevier, Amsterdam. Pp. 731-738.
10. Doumeingts, G., Vallespir, B., Darracar, D. and Roboam, M., Design methodology for advanced manufacturing systems, Computer in industry, 9(4), 271-96, 1987.
11. Sheer, W-A, and Kruse, C., ARIS-Framework and toolset: A comprehensive business process re-engineering methodology. Proceedings of the Third International Conference on Automation, Robotics and Computer Vision (ICARCV'94), Singapore, 8-11 November, pp. 327-31.

CIM Design and Planning

Comparison Between CIMS and CIPS

G. Xiong

Institute of Industrial Process Control, Zhejiang University, HangZhou, 310027,China

X. M. Xu

Department of Automation, shanghai Jiao Tong University,200030,China

Y. X. Sun

Institute of Industrial Process Control, Zhejiang University, HangZhou, 310027 , China

Abstract. In this paper, the authors firstly define different kind of industries . Then, process industry is compared with discrete industry in automation 、 process character and history. Next , computer integrated manufacturing systems and computer integrated process systems are defined and compared . The authors also summarize the theories and technologies included in computer integrated manufacturing systems and computer integrated process systems. At last, conclusions are drawn.

1 Introduction

The authors firstly define different kind of industries. Then, process industry is compared with discrete industry in automation 、 process character and history. Next, computer integrated manufacturing systems and computer integrated process systems are defined and compared. The authors also summarize the theories and technologies included in CIMS (computer integrated manufacturing systems) and CIPS (computer integrated process systems) . The comparison can help researchers and builder of CIPS and CIMS learn from each other, make them research and build CIMS or CIPS more successfully and quickly with lower cost .

2 Type of Manufacturing Enterprise

In discussing manufacturing unit operations, one that has proven useful for discussing process control and communication application is segmenting the major industries into continuous, discontinuous, batch and discrete unit operations.

Definition 1:(Continuous operations) are those where the raw materials, intermediates, and products are normally fluids and are processed in a continuous fashion for long periods of time, often for years, without shutdowns. The process technology is often the key factor for economic success rather than the product technology. These operations are typified by the chemical, petroleum, and power industries. A typical example would be a large distillation operation as used in petroleum and chemical separations.

Definition 2:(Discontinuous operations) when running, are the same as continuous operations with the exception that they frequently change from one product to another. This either requires shutdown and start-up at frequent intervals or moving from one operating condition to another in order to make a similar product. Such changes may occur once a few hours or days. To make such processes economical to operate, additional automation is used to make these changes in operating conditions. Examples of such operations are specialty chemical processes, paper-making machines, and steel rolling mills.

Definition 3:(Batch operations) although run in discontinuous fashion, are somewhat different in that the processing is done in a specified sequence or recipe. The raw materials are mixed together
and then processed over a given trajectory of operating conditions, such as temperature and pressure. The oldest form of unit operations is still used extensively in many industries, such as chemical, food, minerals, pharmaceutical, leather, textile, lumber, electronic and metal part processing. All batch processes include the control requirement of following sequences of operations according to a specific recipe for each desired product.

Definition 4:(Discrete operations) are those that produce one product at a time ,such as automobiles, refrigerators, or light bulbs. Such processes may use an assembly line where the product moves through the various unit operations ,or the product may be relatively stationary. With different processing steps have been done at the same location. Products may be produced in a quantity of one or by the thousands in a mass production line. These include computers, appliances, furniture, pencils and toys. The introduction of robots, vision system ,and other handing devices, as well as computer-controlled machine tools and automated inspection systems ,is providing the means for automating industrial discrete manufacturing operations .Many of these control systems require complex models and control algorithms.

Definition 5:(Process Industry) Generally, petroleum 、 electric power 、 metallurgy 、 chemicals engineering 、 pulp & paper 、 pharmaceutical 、 food 、 rubber & plastic et. al ,belong to process industry whose operations mainly include continuous operation and discontinuous operation and discontinuous operation.

Definition 6: (Discrete Industry) Generally, food 、 minerals 、 pharmaceutical 、 leather、 textile、 lumber、 automobiles、 refrigerator and light bulbs et. al. belong to discrete industry whose operations mainly include batch operation and discrete operation.

3 Comparison of Automation between Process Industry and Discrete Industry

3.1 Level of Automation

The level of process automation in the fluid process industries is higher than in the discrete industries due to the nature of the manufacturing processes and to the development of suitable devices to measure the process state in real time .

In fluid process manufacturing it is possible to reliably measure the basic parameters (pressure, level and temperature are included) of flow, and more elaborate composition measurements have been used in real-time operation for the last four decades .As these industries produced primarily large volumes of commodity products, the economic advantage was gained by those with the more efficient process operations. This incentive led to significant investments in improving process designs and their control systems. Today these industries are highly automated and it would be impossible to operate them manually.

In contrast it has been much more difficult to measure the state of the discrete processes and to automate the handling of discrete parts. Generally, economic advantage is given with the best product design, so most development efforts are in improved products rather than processes.

The discrete industries have led the fluid process industries in the application of business automation. They developed many of the techniques for such things as improved product design, production planning, and order handling ,and computers were increased investment in discrete process and business automation. They are now being coupled together to achieve complete computer-integrated manufacturing[1-2].

3.2 Use of Digital Technology

Digital computers were first applied in fluid process operations to gather real-time data for use in process models to optimize the operating conditions. Such applications have now been developed for many processes ,but with considerably more investment than had been originally estimated.

The first supervisory computers adjusted the set points of the centralized of analog controllers being used for the feedback control of the processes. The next major use of the digital computers was in the development of DDC (Direct Digit Control). Centralized computers replaced the individual analog controllers, and CRTs (Cathode-ray Tubes) replaced analog operator panels. Often two computers were used in parallel to achieve the desired availability. With the rapid infusion of economic and reliable microprocessor technology, modern fluid process control systems now are designed with a distributed architecture incorporating many types

of digital computers.
In the discrete industries, computers in manufacturing were first applied in the business operations and in the control of individual machine tools. Centralized control of production processed was not developed as in the fluid process operations, and when computers were first applied on the factory floor, technology had advanced to the point that small reliable computers were available. These were first used with NC, DNC, robots, and PLCs, but now are widely applied in distributed architectures for all aspects of production control.
In discrete operations, process control has always been distributed in the plants near the operating equipment. In contrast, fluid process control was centralized using analog technology and only now it is being distributed near the plant unit operations with digital technology.
Also, in discrete operations the operator communication devices have always been located on the plant floor at the individual production units. As more digital technology is applied, this is changing with centralized operator and supervisory control rooms being installed. In the fluid process operations, the operator communication devices have been centralized for decades, and this is continuing with the application of digital technology.

3.3 Human Involvement in Production Operations

At the turn of the century, production operations in almost all industries were carried out on the plant floor with manual labor. As the fluid processes applied increasing levels of automation, this situation changed dramatically to where only one or two people are now required to operate multimillion dollar production plants.
In the discrete industries most production operations are still done today with manual labor. However, some major manufacturers have invested heavily in the last decade to automate many of these production operations, and some are now completely automate many of these production operations, and others are now completely automated. These require significantly fewer production workers who are located in centralized operating areas[1-2].

3.4 Convergence of Process Control into CIM

The technical and office operations of both discrete and fluid processes are rapidly becoming more automated through the application of digital technology. As discussed above, the production operations in both industries are going toward distributed process control and centralized operator communications .As these developments continue, and are combined together, CIM will be achieved in both industrial areas.
This is not to imply that CIM will be the same in all industries ,as there will be significant differences between systems and the detailed technology applied. This will be most apparent at the immediate production operations where different types of measurements and actuation are required due to the physical nature of the processes. However, at the level of operator communication, many common technologies are starting to be used for graphic displays ,process diagnostics, and

other similar functions. As higher-level supervisory and business functions are incorporated in CIM, many of these technologies could be used in all types of industrial applications.

4 Comparison of Production Process between Process Industry and Discrete Industry

Object	Process industry	Discrete industry
Material Flow:		
(1) Processed material quantity	Big and centralized	Big but decentralized
(2) Manufacturing process	Continuous	Can be discontinuous
(3) Material transfer	Through sealed chamber and pipeline	Through transport
(4) Character of physical chemistry	Change continuously	Unchanged
(5) Energy change	Consumed and transformed	None
Technology:		
(1) Technological process	Unchanged	Be changed often
(2) On-line technological parameter	Often changed	Unchanged
(3) Relationship between neighboring technologies	Restrained continuously, synchronism	Restrained dependently, loose
(4) Product design	Fixed in the first time	Changed to meet various need
(5) Flexibility	Weak	Strong
Production Control :		
(1) Production mode	Large batch	Small batch or signal piece
(2) Requirement in real-time	Real-time on-line monitor	No need
(3) Measurement of technological parameter	On-line measurement is difficult, low precision	Measurement is easy, high precision
(4) Product quality	Guaranteed by control system	Effected by many factors
(5) Mathematical model	Reaction dynamics; thermodynamics equation; experience equation; regression model et al.	Discrete event dynamic model; maximum and minimum algebra; Petri net; Stochastic service model et al.
(6) Execution tool	DCS, PLC, Heat-exchanger, Reaction tower, valve	Numerical controlled machine tool; FMS
Other:		
(1) Device	Many type ,huge volume, complex structure	Less type, compact volume, simple structure
(2)Optimization object	Manufacture steadly, safe, low cost, high output, high product quality	Shorten product's life period; improve the utilization ratio of device
(3) Information integration	Easy because of less information type	Difficult because of many symbol and graph types
(4) Work environment	High temperature, high pressure, easy ignited, explode easily, corrupted easily	Normal temperature and pressure, relatively safe

Fig. 1. Comparison of production process between process industry and discrete industry

Comparison of production process between process industry and discrete industry is as fig.1.[3-6].

5 The Definitions of CIMS and CIPS

In 1974, the concept of CIM was first mentioned by Dr. Joseph Harrington in his book 《Computer Integrated Manufacturing》[7].In the book, he put forward two basic points of view:
(1). All manufacturing activities ,which includes market analysis, product design, product manufacturing, management, administration and service after sale , are an unseparated whole. So, these activities must be
connected together and be considered as a whole.
(2). In essence ,the whole manufacturing process is a process of collecting data, transmitting data and processing data, the end product can be considered as matter form of the data.

Definition 7: (CIMS) CIMS(Computer Integrated Manufacturing Systems) is defined by the expert group of Chinese 863/CIMS as: CIMS is a system based CIM philosophy. Through computer hardware and software, it comprehensively uses modern management technology, manufacturing technology, information technology, automation technology and system engineering technology. It integrates three factors that is connected with all manufacturing activities, such as human, technology , administration & management with their information flow and matter flow so it becomes an organic large scale system that can run optimistically.

Definition 8: (CIPS) Now ,the term CIMS is widely used in all types of enterprises .We call CIMS used in process industry CIPS(Computer Integrated Process Systems).

6 History of CIMS and CIPS

In 1950's, computer began to be applied in control system of process industry .From then ,the history of process industry automation can be divided into four stages:
(1) 50's-60's: Use DDC(Direct Digit Control);(2) 70's-early of 80's: Use DCS (Distributed Control System);(3)early of 80's - end of 80's :Use DCS + optimization ; (4) 90's: Use CIPS（Computer Integrated Process（or Production）Systems）.From 90's ,production dispatch、plan optimization、administration、management & decision are used in computer control system, this connects market consciousness with optimized control、accelerates the development of computer control system to CIPS, makes enterprise gain more economic profit and better social benefit. Clearly, CIPS is the inevitable direction of process industry automation [8].

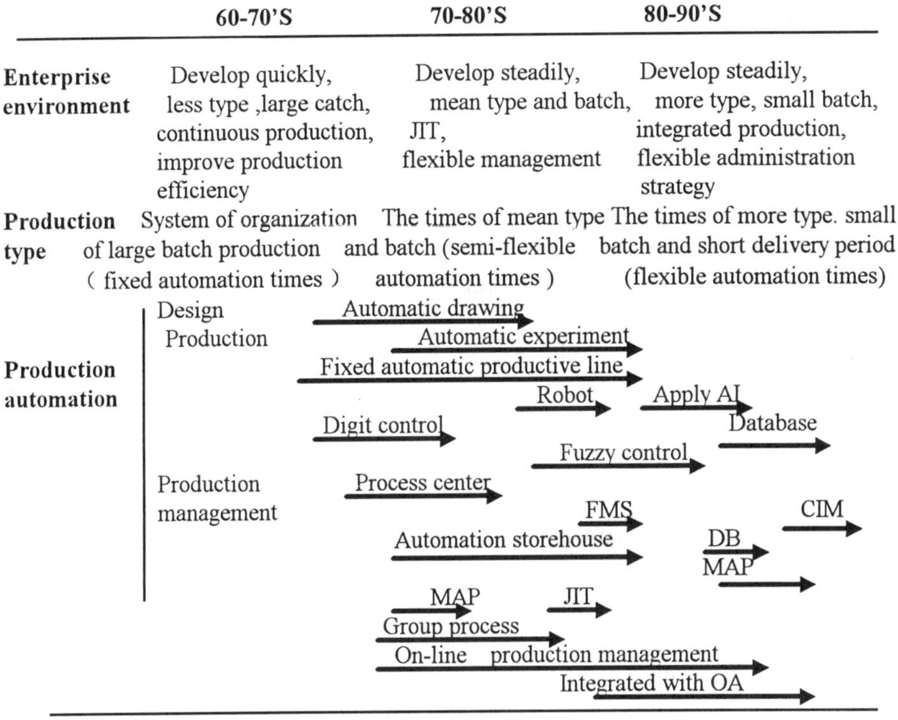

Fig.2. History of discrete industry

History of CIMS can be summarized as fig.2.

7 Comparison Between CIMS and CIPS

(1)CIPS is used in process industry ,CIMS is used in discrete industry.
(2)They have common essence: Through integrating human with technology and management ,an enterprise becomes an organic system that has good internal quality and can react quickly to external change.
(3)They both realize information integration and knowledge integration through computer technology、communication network technology and distributed database technology. This is the inevitable requirement on production enterprise during information times .
(4)They have different starting point. Computer is applied in discrete industry more lately. Discrete industry has lower basic automation and it's automatic hardware is less. Specially in China, to realize " island automation " or " low level automation " is more urgent. At the same time, we should draw a lesson from advanced countries in the development of CIMS . Because we develop CIMS from zero, we should put forward a best overall plan about CIMS when partial automation system is built in

an enterprise .

Computer was applied in process industry more early because they require automation more urgent. Even in China, low level automation is generally realized in process industry .They have better base of realizing CIPS.

(5) They have different means to realize the object. Discrete enterprise can reduce single product cost, improve product quality, increase product's type, meet changing market's requirement and raise production efficiency in different degree through unit automation, enterprise flexibility, building CIMS et al., and it's key is " flexibility". Process enterprise can raise production efficiency, improve product quality, increase product quantity, control environment pollution and save energy through advanced control technology; Through CIPS, process enterprise can improve it's adaptability to market, improve management level and production level, increase utilization ratio of raw material and reduce waste material et al. .It's key is the high quality of control and management system.

8 Theories and Technologies Included in CIMS and CIPS

CIMS and CIPS have common points and different points in theories and technologies because their controlled objects and building means are different and their aims are the same .When we research or build CIMS or CIPS ,we should compare them ,make them complement each other. Thus, CIMS and CIPS can become ripe and perfect more quickly.

In the concrete, main theories and technologies included in CIMS are [9-11]:

(1) The model and reference architecture of CIMS: CIMS-OSA 、 Purdue enterprise reference architecture 、 and GRAI-GIM are famous, but they still need to be developed. (2) Computer networks in CIMS: To develope the computer networks suitable for CIMS environment. (3) CAD/CAPP/CAM: To realize information integration. (4) Computer graphics: To develops IGES and STEP et.al. (5) CIMS/MIS(Management Information System) (6) CIMS/DDBS (Distributed Database System) (7) EDI(Electronic Data Interchange) (8) QFD (Quality Function Deployment).

Main theories and technologies included in CIPS are [4,12]:

(1) The standard model and reference architecture suitable for CIPS. (2) The standard computer networks suitable for CIPS. (3) CIPS/DDBS. (4) The integration of business system, human system and technology system in CIPS. (5) The theory and technology of integrated control and comprehensive optimization of complex process industry. (6) Advanced control and optimization method. (7)CIPS/MIS. (8) Simulation in CIPS. (9) Reliability analysis of CIPS

Conclusions In the paper, CIMS and CIPS are compared in detail. To understand their common and difference can help researchers and builder of CIPS and CIMS learn from each other, mark them research and build CIMS or CIPS more successfully and quickly with lower cost. Related work needs go on ,related researchers need work together, related organization need be found early.

References

1. S. B. Gershwin , R. R. Hildebrant , R. Suri , and S. K. Mitter. A Control Perspective on Recent Trends in Manufacturing Systems. *IEEE Control Systems*, (1986) pp3-15
2. John W. Bernard. CIM in the Process Industry. *Instrument Society of American*. (1989)
3. G. L. Song , X. S. Jiang , and B. Y. Liu. The Overall Architecture of Large Process Industry. *Information and Control*. Vol.23, No.2 (1994)pp65-70
4. B. Y. Liu. CIM and The General Technologies for Industrial Automation -- Goals , Methods and Practices . *Information and Control*. Vol. 22 , No.1 (1993) pp29-35
5. G. L. Song ,X. S. Jiang .CIMS in Continuous Production Process. *Computer Integrated Manufacturing Systems*. Vol 1. No.1 (1995)pp15-20
6. G. Xiong, X. M. Xu , and ZH. J. Zhang. To Accelerate the Theory Research and Practice of CIMS. *Computer Integrated Manufacturing Systems*. Vol 1. No.3 (1995) pp35-38
7. Joseph Harrington . Computer Integrated Manufacturing. *INDUSTRIAL PRESS* . Inc, 200 Madison Avence, New York (1973)
8. G. Xiong , X. M. Xu , and ZH. J. Zhang. et al.. The Theories and Technologies of CIPS . *Information and Control* . Vol.25 , No.5 (1996) pp65-69.
9. T. J. Williams. Architecture for Integrating Manufacturing Activities and Enterprises . *Proceedings of 12th World Congress IFAC International Federation of Automatic Control*, (1993)10:273-283
10. Clydl R. Van Haren . A Reference Model for Computer Integrated Manufacturing from the View Point of Industrial Automation. *Proceeding of CIMCON 90*,(1990) pp42-61
11. B. H. Li .The Provisional Constitution 、Standard、Building Guide of CIMS. *Chinese Weapon Industry Press*. 1994.9
12. G. Xiong, X. M. Xu, and ZH. J. Zhang .The Architecture of CIM/CIP and its Development . *Systems Engineering - Theory 、Methodology 、Applications*. Vol.4, No.4 (1995) pp11-16

Expert System For Pre-Determining CIM Levels
(WITH PARTICULAR REFERENCE TO WAFER FABRICATION INDUSTRY)

Ranjith Nayar

III Sem, MSc. (CIM), Nanyang Technological University, Singapore

"It is possible to get totally consumed by computers"
- Dr Allan Newall, AI pioneer

Abstract It is widely recognised that to remain competitive in today's world, electronics manufacturing industries need a fairly large number of CIM solutions. However, CIM is usually a major investment decision -- and it is difficult to pre-determine its effectiveness and optimum level for a given situation.

Industry, therefore, increasingly feels the need for an Expert System to pre-determine (at least, guide) the extent of CIM, the choice of technologies and to help solve the perennially sticky problem of hardware sizing.

This paper attempts to discuss such an Expert System to help CIM managers' decision making. It further discusses the need to develop criteria to measure the effectiveness of CIM solutions and examines the inputs and questions that will need to be analysed by an appropriate Expert System.

Introduction
Perhaps the greatest problem a modern business enterprise faces in today's world is the sharp increase in competition. The old days were simple -- healthy companies survived, while the ones not up-to-the-mark went down. Today, competition is so intense that there is little, in terms of quality, to distinguish one competitor from the other. This is especially true of electronics manufacturing companies.

Increasingly, what determines the success or failure of a company is the competitive edge it determines from factors like greater product range or variations, faster turn-around time and highly competitive pricing. Electronics manufacturers are aware of this and constantly look for means to increase their competitiveness.

Attempt is being made in this paper to develop on the following issues:

a) decision making and analysis
b) questions that need to be answered for a knowledge base/ expert system
c) systems design issues
d) criteria to measure the effectiveness of (applying) CIM

Multi-layered decision making and analysis is illustrated in a case study on choice of CIM solutions.

Background

Classical Industrial Engineering techniques of Methods Improvements and productivity enhancement have been applied with great success in electronics industries. Techniques like Line Balancing and Work Study have tremendously boosted capacity utilisation and helped reduce overheads component in product costing.

However, it is increasingly recognised that these techniques have already been used to saturation and, while it is important to maintain (and even fine tune) these methods, other means have to be used now to post any new quantum gains. Moreover, with the globalization of the market, competition is no longer with the 'friendly, neighbourhood' competitors, but with the multinational 'big boys'. Increasingly, therefore, manufacturing facilities are turning to CIM technology as the panacea to address the competitiveness issue.

1. The Problem

With impressive strides made in hardware, software and middleware technologies in recent years, greater number of 'solutions' are available in the market for relatively low costs -- Manufacturing Execution Systems (MES) are increasingly common and getting more sophisticated, shop-floor simulation, scheduling and factory automation have become buzzwords and it has become quite tempting to "buy" productivity!

However, as the sophistication -- and complexity -- of the solutions increases, quite often managers feel that the productivity gains thereof are not commensurate with the implementational efforts that need to be put in. Thus, while the tools themselves are useful, even necessary, the complexity of implementation and the breadth and depth of decisions to be made, make it necessary to carefully pick and choose from among the CIM technologies available in the market. Consequently, there is need for a proper evaluation of effectiveness of the different technologies available. Conventional decision analysis techniques become insufficient here, especially in view of the paucity of data usually available, the high risk of wrong decisions and the fact that many of the situations and choices cannot be

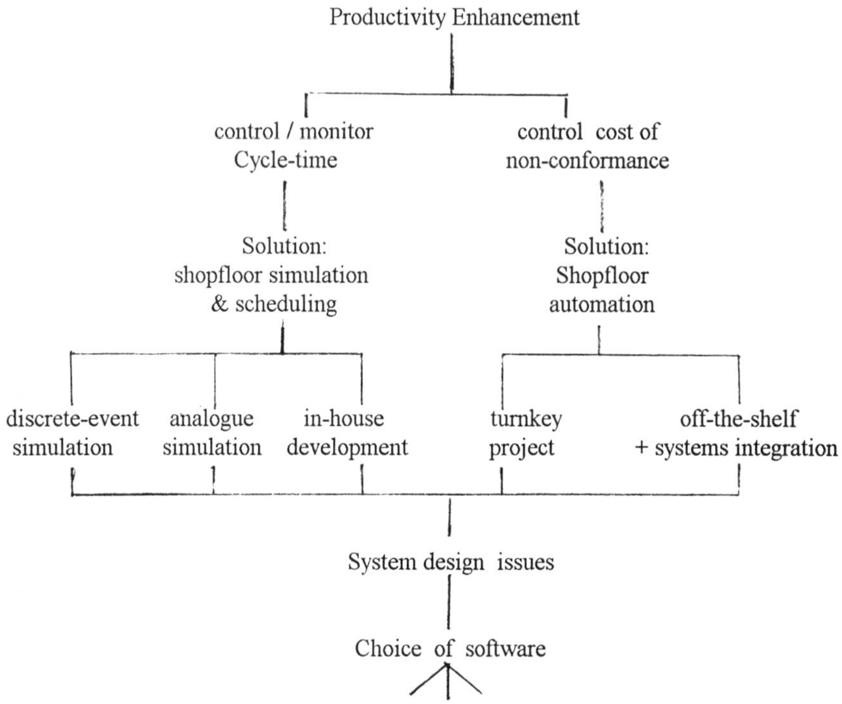

Fig. 1: Multi-layered decision requirement

quantified because, in the real world, they are based on the practitioner's 'gut feeling'.

Moreover, most of these decisions are 'multi-layered', each alternative containing two or more sub-alternatives. A typical productivity enhancement strategy, for instance, can be decomposed as shown in Fig. 1 above.

2. Designing An Appropriate Solution

Often, the knowledge base needed to make these decisions is not available within the organisation. Therefore, a CIM manager needs to have a formal Decision Support System. A typical Management Decision Support System needs to perform three levels of activities (Dyer, 1987) -

i) Information management : storage, organisation and retrieval of data

ii) Information analysis : search, analysis and display of variables, trends, relationships

iii) Decision analysis : prioritization and selection from among alternatives

While there is a fairly large body of work / opinion on the first two (viz. information management and analysis), decision analysis is a more fuzzy area, especially where decisions lie outside the quantifiable domain.

For instance, given that two options can be evaluated by quantifiable parameters, the choice between two equally good options may appear to be either too complex or too discretionary (as, typically, in the choice between discrete event simulation and continuous simulation). The choice may often be felt to lie outside the domain of quantifiable decisions. If so, how can a decision be taken here?

There are two solutions to this problem. The first of these is to make use of the Analytical Hierarchy Process, or AHP. The other is to employ Case Based Reasoning, or CBR.

Analytical Hierarchy Process

AHP has seen wide application and is a proven decision making methodology (Golden LB; Wadil EA & Harker PT; 1989). Its usefulness in the evaluation of advanced manufacturing technologies has been discussed by Lim KH (1994). The process basically involves breaking down the objective into n-level sub-objectives and then applying a qualitative analysis, bottom up, level by level (Fig 2 below). The lines branching out from options represent the contribution of that option on all the sub-objectives of the hierarchy[5]. AHP, however, has a major disadvantage in that it involves tremendous amount of calculations to be carried out by the decision maker.

Case-Based Reasoning

CBR, on the other hand, is a technique of decision analysis that involves reasoning from precedence, adapting previous solutions to suit current problems, highlighting aspects of the present situation and then designing the appropriate solution. (This contrasts with rule based reasoning, which solves problems by chaining together rules of inference.) The procedure is summarised in Fig. 3 below[9].

In building Expert Systems, reasoning ability can be effectively developed by incorporating case-based reasoning. Computers can be programmed to access and utilise stored events relevant to the problem being studied (Kolodner & Reisbeck, 1986; Kolodner, 1988; Reisbeck and Schank 1989), and hence the CBR technique can even be computerised.

The major advantage of CBR is that the analysis is qualitative and, while it may not give the kind of best-fit that an intense analysis like AHP would, it is more easy to understand and implement.

Therefore, among the first steps in building the Knowledge Base for the Expert System is to build a "case bank", based on past cases from within the company (including other divisions and locations) and other cases from industry, obtained through literature survey. These cases then have to be categorised for analysis.

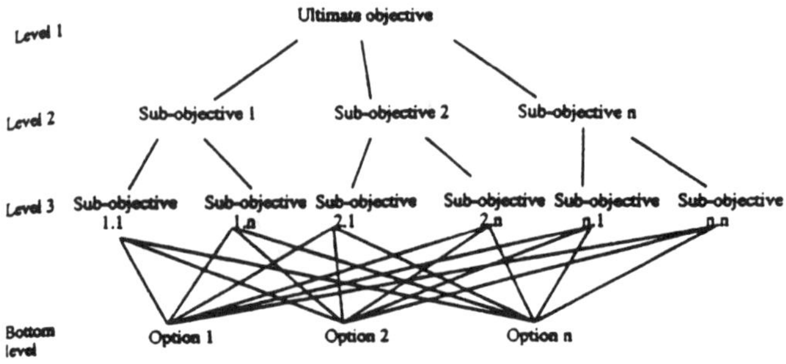

Fig 2 : The analytical hierarchy

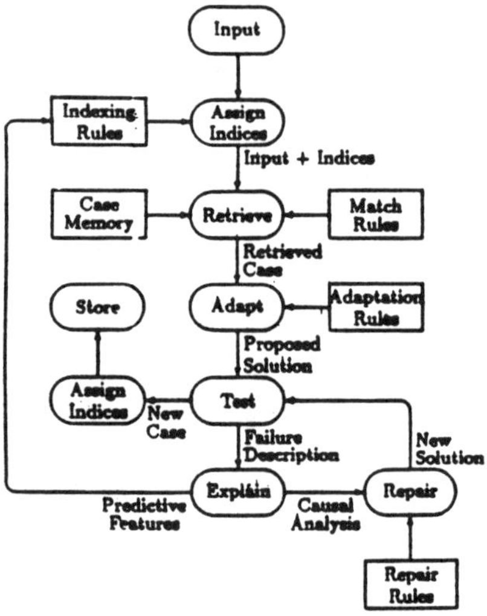

Fig. 3 : Case-Based Reasoning - Flow chart

Building The Knowledge Base

Much of the rules and procedures of manufacturing are informal and hence, can be the basic building blocks for the case bank. These informal rules usually have their basis in past experience and are fairly reliable (Hayeck, 1973). In the field of Expert Systems, rules are formalisms intended to encode knowledge (Chandrashekaran, C. & R. Ramesh; 1987).

The process of CIM-related decision making has to be understood first, before starting on the actual design of the Expert System and its Knowledge Base. We need to frame questions to understand the key issues involved, some of which are :

- defining primary manufacturing / productivity parameter(s) for focus

  This may be lead-time, cycle-time, process turn-around time, new product introduction time or cost parameters including cost of bad quality. The choice of focus parameter will vary from company to company.

- establishing the current value of the selected parameter(s)

  Also establish the target value for these parameters, sought to be achieved through the CIM element under review (along with the time-frame within which this achievement is desired). Quite often the primary parameter the company seeks to improve is a dependent variable and either has an impact on, or is impacted by, other variables. This dependence has to be examined in order to quantify the range of primary parameter's improvement expected. For instance, the reduction of cycle time primary variable may necessitate de-stressing of equipment utilisation.

  The knowledge base should also contain a mathematical model which converts the gain in these manufacturing parameters into financial parameters incorporating also the second dependent variable.

- efficiencies and bottlenecks of the current method of operation

- cost of non-conformance or scrap

  Dollar value, however, is not the only measure. Of equal concern is the productivity improvement that can be accrued / effected, which result directly in faster, more efficient production and better turns ratios. Productivity improvements are usually two-fold, in that operator productivity is enhanced while simultaneously improving machine utilisation through avoidance of stepping back-and-forth between first-round production and rework.

- existing / desired levels of interaction with other CIM components like the MES and CAM software

  This is also a good time to examine the shortfalls or drawbacks thereof.

- other CIM solutions that have been tried for similar problems or areas

  A scale needs to be developed to quantify the similarity of two situations.

- non-CIM alternatives for addressing the problem / production area in focus

  The cost of that effort and the quantifiable and qualifiable results thereof would become the benchmark for measuring the results expected from implementing CIM as the solution.

- the costs of an equivalent CIM solution (equivalent in terms of benefits)

- defining an appropriate financial tool
  Tools like Internal Rate of Return (IRR) or Net Present Value (NPV) test the soundness of the proposal under consideration.

- significant factors that weigh in favour of (and against) the introduction of the CIM technology.

Expert System for System Design
The role of the proposed Expert System also includes sizing of memory and disk space. Sizing is usually an ambiguous and tricky exercise. Case based reasoning is a good technique for use here, being based on reasoning from precedence.

Various issues need to be addressed to evolve a comprehensive necessary-and-sufficient design of hardware to meet current and anticipated requirements, including possibilities for future expansions and upgrading. These include issues like memory and swap area requirement, system response, cross-platform compatibility, cross-version compatibility with other software (existing/ in the pipeline), compatibility with MES and databases in use by different applications, and the number of child processes spawned / run by the software (which may vary non-linearly with the addition of new users / equipment).

Other important issues pertain to scalability and portability of the software, and non-technical issues like costs (up-front, user licences, training, maintenance contracts), availability of local support / help-line, ease of use and training requirements / provided).

3. Measuring Effectiveness of CIM Solutions

While application of CIM technologies (in part or in full) does lead to tremendous gains in productivity (especially in terms of cycle-time reduction, a more efficient management of WIP, etc.), and consequent enhancement of competitiveness, few companies will commit huge financial resources towards it. This is especially so because many of the gains achieved in these manufacturing parameters are difficult to intuitively covert to financial gains. Consequently, there is need to develop appropriate indices to measure and present the gains of applying CIM.

The only logical measure of effectiveness of CIM technologies is to measure the gains accrued, in terms of three parameters of productivity - cost, quality and time - and then add to that an analysis of the intangibles. Taken together, these sum up the quantifiable and non-quantifiable (viz., the qualifiable) parameters that are necessary and sufficient for a proper evaluation. Simulation of the scenario planned helps in predetermining these gains.

- Cost expended and dollar returns. This is, effectively, the approval considerations for budget outlay and ROI. The first step for this is for the company to set its long- and medium-term objectives. Few CIM projects are "get-rich-quick" schemes.

- Quality improvement, for instance quantifiable sub-parameters like yield, % rework, increased output / moves, defect density, cost of bad quality, etc.

- Time parameters, such as reduced information access time (and consequently, less waiting / queuing / down time). This can be quantified by a comparison between previous and expected / simulated (post-CIM) information access time. Information to be accessed on the shop-floor includes specifications, control limits, equipment status, process recipes and on-line assistance. The spin-off consequence of all this is greater productivity that can be expected from the operators and engineers who can now concentrate on getting more production done -- which is their primary task.

Intangibles

Conventional calculations of financial indices like return on investment (ROI) and expected net present value (EPNV) make financial sense, but major decisions on business process re-engineering and process redesign should not be based on these indices alone. Quite often, these indices, in fact do the opposite, by making it appear that it makes more sense to be expanding and improving upon present practices. This is because re-engineering is usually far more expensive and its returns more difficult to forecast. If these indices are relied on totally for decision making, managers would go on expanding on current facility and would soon be carrying massive outdated production technology.

There is a need to move beyond these indices and see re-engineering investment as a long-term strategic choice. Product quality improvement, cycle-time reduction, greater product flexibility and improved competitiveness will have to be considered on par with financial parameters and indices.

It is a truism that companies rarely implement all available CIM technologies together. Rather, CIM application is more evolutionary in nature. A couple of components may be implemented at a time[6]. (One notable exception is the setting up of a new wafer fabrication plant, where upfront application of state-of-the-art CIM technologies is seen as critical). However, investment in advanced manufacturing technologies has to be thought of as a way of keeping the company competitive so as to maintain and improve market share. This has to be a strategic objective for any company. That being the trend, today companies face the threat of getting left behind if they do not move fast in this direction.

Project Gestation Period Issues

With higher employment opportunities available, especially in the East, the turnover rate of managers has been so high that long-term product enhancement and production process changes tend to be avoided. On an average, the gestation (and maturity) period for automation-type projects is three to five years. Thus, decision makers who do not have long-term career plans within the same company, are not motivated to push for such projects.

Selection Criteria

In comparing equivalent CIM technologies from different vendors, the CIM manager needs a selection criterion. This can be a fairly classical rule-based methodology which can easily be handled by the Expert System. Groover and Zimmers[4] discuss such a measure, with a point allocation system. An example of the criteria and the relative weights is given below. Weights have to be assigned based on the relative importance of the factor to the particular user company, and is only suggested here as a guideline for a wafer fab.

A. Systems & General considerations	60
B. Applications - Hardware (Electrical)	25
C. Applications - Hardware (Mechanical)	15
	100

A. Systems & General considerations

1. Cost (hardware / software, added specialities for the project, maintenance, ...)
2. System Management (file management procedures, user diagnostics, ...)
3. Peripherals (source code availability, clear documentation, efficient interfaces,)
4. Service Contract (warranty, discontinuance clauses, on-line & on-site service, ...)
5. Quality (user group existence and support, vendor quality responsiveness and financial stability, number of installations, growth, reliability, simultaneous operation, crash recovery, power loss recovery, environmental sensitivity, human factors considerations, hardware / software response time, output device speed, ...)
6. Delivery and logistics support (staging in-plant benchmarks, installation, pre-delivery inspection, installation guide, supply revisions, billing, proprietary agreements, ...)
7. Miscellaneous considerations (compatibility with existing equipment, graphic terminal control, operation, input methods, prompting, plotting control, ...).

B. Applications - Hardware (Electrical)

1. Electrical design applications : logic diagrams, circuit diagrams, schematics
2. Electrical Interfaces : Circuit analysis logic
3. Documentation of electrical systems

C. Applications - Hardware (Mechanical)

1. Mechanical design, logic diagrams, process diagrams, schematics, architectural drawings, structural design, piping from/to lists and mapping.
2. Dimensional mechanical design and footprint.

4. Case Study : Case-Based Reasoning

The local unit of a wafer fab manufactures high quality, competitively priced ICs for the local market (around 25%) and the US and Japanese markets. In line with its strategic objective of being among the top IC manufacturers of the world, two years ago the company took a strategic decision to ramp-up its local manufacturing base to 200%, with the objective of capitalising on lower production costs and faster process turn-around factors. A technological audit revealed that the CIM technologies it currently had would be the primary bottleneck in the ramp-up effort.

Management first identified the critical manufacturing parameters which it aimed to improve. The list was narrowed down for focus purposes (and also because improvement of these focus parameters would have a cascade effect on other parameters as well). The final shortlist identified cycle-time and cost of misprocessing as the key focus parameters to be improved. An analysis of the technological audit suggested CIM upgrades -- full shop-floor automation, an executive information system and a simulation package for shop-floor scheduling. It was recognised that the gains of this strategy are difficult to calculate in dollar terms and focus should be on competitive advantages derived from achieving its strategic objective.

Two broad strategies were available -- either to develop and implement the required CIM interfaces in-house, or to farm out the job to consultants/ system integrators.

Case Based Reasoning (CBR) was applied here, to choose between the two alternatives. Prior examples were available from the other units of the company and from known cases of competitors' efforts in the same direction. The analysis followed the steps described in Fig.1.

The cases retrieved for study were first clustered into two classes. The first class of cases involved development of home-grown equipment integration packages for various classes of equipment. Such home-grown packages have advantages -- near-perfect fit for the objectives of the exercise and availability of hands-on developmental experience. However, the in-house software development strategy has a major weakness -- over the years, little software enhancement is usually possible, because of constraints of manpower resources and limitation of skills.

The second class of cases retrieved and analysed involved implementation of 'turnkey' solutions by software consultancy houses. Advantages of this approach included hassle-free implementation, tighter deadlines since more resources can be applied for the short time (as the company does not have to carry human resources beyond the project gestation period). Major disadvantages are however present, most notably the limited expertise on the 'nuts and bolts' aspects among the company's internal resources and the higher costs involved.

After analysing in-depth the advantages and disadvantages from case-studies, a new strategy was evolved. This was to adopt the via-media - viz. to buy commercially available software packages, followed by developing systems integration in-house. This strategy combines the advantages of both the classes of cases studied -

- far fewer systems integrators are required to customise state-of-the-art software packages and could double as systems maintenance personnel;
- development time is minimised (being limited to integration and testing time);
- virtually no overheads need be carried; and
- by selecting financially viable vendors, the company is assured of future software enhancement and upgrades.

The decision making process adopted in designing the appropriate solution started with the calculation of the cost of buying off-the-shelf packages and licences. Payback period was computed on the basis of quality and was found to be in the range of a few days, calculating on the basis of cost-saving from the avoidance of misprocessing on batch processing equipment (which could result in scrapping of upto 200 wafers). It was further decided to implement equipment integration in an evolutionary manner, starting with batch processing equipment which would offer the shotest payback period.

Finally, thumb-rule guidelines were worked out for sizing of the memory requirements for the company's standard platform. Sizing is a tricky issue, involving trade-offs between speed and extensibility. It also depended on the nature of the equipment being interfaced. Highly communicative equipment which have large numbers of alarms and parameters which require monitoring would necessitate larger memory resources. System configuration also depends on the platform chosen to run the application, which in this case was a company-wide standard.

For evolving selection criteria to chose from among the alternative 'solutions' (software and hardware), the CIM department conducted a survey among CIM managers across other units of the company and, based on their inputs and experiences, developed a point allocation system. The final selection was then based on this evaluation.

5. Conclusions

It is widely accepted that to remain competitive in today's market a company has to adopt some form of CIM. Particularly so in the case of a wafer fab, where the technology itself is so complex that it is virtually impossible to engage in manufacturing without quite an intense application of CIM technologies.

This gives rise to some major issues. Perhaps the most important of these is the purchasing decision, because, obviously, there is a need to pick and choose the technologies/ components that the CIM manager wants to have. We have examined here the questions of decision making and CIM effectiveness. The knowledge base rcquircd to make such complex decisions is usually not available within the company. Thus, there is a need for a good decision support system to help the CIM manager make the right choices.

We suggest development of an Expert System for this. This paper also examines the inputs expected for developing the Knowledge Base for such an Expert System.

***

REFERENCES

1. Bazerman, Max H., Judgement in Managerial Decision Making, John Wiley, (1986)

2. Bortz, F., Mind Tools - The Science of Artificial Intelligence

3. Golden LB; Wadil EA & Harker PT; "The Analytical Hierarchical Process - Applications and Studies", Springer-Verlag, New York, (1989).

4. Groover, Meikell P. & Zimmers Emory W.; CAD/CAM : Computer-Aided Design and Manufacturing, Prentice-Hall of India, (1992).

5. Lim Kian Hin; "Evaluation and Justification of Advanced Manufacturing Technologies" (dissertation work, 1994).

6. Luong, Lee & Spedding, Trevor: A Methodology For Optimising Technological Applications in Manufacturing Industry.

7. Mitra, Gautam, *Ed.* Computer Assisted Decision Making, North Holland, (1986).

8. Rembold U., Nnaji B.O., Storr A., Computer Integrated Manufacturing & Engineering, Addison-Wesley, (1993).

9. Riesbeck C.K. & Schank R.C.; Inside Case-Based Reasoning, Lawrence Erlbaum Associates, (1989).

10. Tse A., How Do We Evaluate a Knowledge Based System in Marketing? *Expert Systems for Information Management*, 5(2),(1992).

The Architecture of CIPS

G. Xiong

Institute of Industrial Process Control, Zhejiang University,Hangzhou,310027,China

X. M. Xu

Department of Automation ,Shanghai Jiao Tong University, Shanghai ,200030,China

Y. X. Sun
Institute of Industrial Process Control, Zhejiang University,Hangzhou,310027,China

Abstract. In this paper, the authors firstly give definitions of CIMS、CIPS、history of process industry automation. Then the authors analyze the necessity to research the architecture of CIPS, the common principles of building a architecture, the common points of different architectures. Next, the authors summarize the present situation about the architecture of CIPS. At last, the authors introduce a new architecture of CIPS and put forward a new merged architecture .

1 Introduction

1.1 Definitions of CIMS and CIPS

Definition 1 (CIMS): CIMS(Computer Integrated Manufacturing Systems) is a system based on CIM philosophy. Through computer hardware and software, it comprehensively uses modern management technology、manufacturing technology、information technology、automation technology and system engineering technology. It integrates three factors that is connected with all manufacturing activities, such as human、technology 、administration & management with their information flow and matter flow ,so it becomes an organic large scale system that can run optimistically[1].
The definition given by the expert group of Chinese 863/CIMS is mainly based on discrete industry, such as machine industry. In fact ,CIMS is also the direction of process industry, such as petroleum、power、metallurgy、chemical engineering、pulp &paper、pharmaceutical food、rubber & plastic et al..

Definition 2 (CIPS): CIMS used in process industry is called CIPS（Computer Integrated Process（or Production） Systems）[2].

1.2 History of Process Industry Automation

In 1950's,computer began to be applied in control system of process industry. From then, the history of process industry automation can be divided into four stages: (1) 50's-60's:Use DDC(Direct Digit Control).Computer is used to replace normal meter, control algorithms mainly include PID and feed forward. DDC didn't popularize because of hardware's low reliability; (2) 70's-early of 80's:Use DCS (Distributed Control System). Hardware's reliability is enhanced by its decentralized control loops and unified display and supervision . But its control algorithms aren't enhanced ,process operation condition is still given by man ; (3)early of 80's - end of 80's :Use DCS + optimization . Optimal control and advanced process control algorithms , such as multi-variable decoupling, multi-variable bounded control, predictive control , deductive control, ANN and many model-based control, began to be used successfully; (4) 90's: Use CIPS （Computer Integrated Process（or Production） Systems）.From 90's,production dispatch、plan optimization、administration、management & decision are used in computer control system, this connects market consciousness with optimized control、accelerates the development of computer control system to CIPS ,makes enterprise gain more economic profit and better social benefit. Clearly, CIPS is the inevitable direction of process industry automation.

2 The Definition, the Necessity ,the Building Principles and Main Content of CIPS Architecture

2.1 The Definitions of CIPS Architecture

CIPS is complex, it consists of many sub-projects which coordinate to realize its object: integrate plant's all departments and operate plant optimally. So, it is necessary to help CIPS developer understand detailed task, realize system analysis、system design 、 system developing and system running . Recently, some people are developing graphical or narrative tools to describe CIPS's developing and building process , describe the framework of necessary process, detailed content of each step in clear form; The tools also assure all the present manual 、 figure and computer aided tool cam still be used by CIPS's developer. Those tools is called the architecture of CIPS.

2.2 The Necessity of CIPS Architecture

Architecture can broadly research that how an enterprise be changed from present administration mode from system's angle[3-4];Architecture can clearly express the development process of CIM/CIP project or other projects of enterprise from forming initial idea to that actual system can run, and it also explain a series of related problem included in enterprise from forming system idea to system's running. At the same time, architecture provides method to compare different schemes and select the best scheme and technology; It also help and direct enterprise to gain the technology and ability needed in building CIMS/CIPS; The architectures of CIMS and CIPS have common points, so an enterprise can form particular architecture of its own through the common architecture ,they need not do everything from zero or repeatedly, and this can shorten the development period of CIMS/CIPS and save product cost remarkably.

2.3 The Building Principles of CIPS Architecture

(1) Abstraction: overlook some details about system character and system behavior; (2)Modularity: System is divided into single and independent unit, those unit can be freely changed with the same unit included in other part and they don't effect overall function of the system;(3)Openness: In space, every parts included in CIMS/CIPS can be integrated efficiently ,it can add new function, new technology or new part can be integrated with it easy and efficiently. In time, architecture can prolong the life period of CIMS/CIPS greatly;(4)User's requirement and system building are divided :Separate system plan and design from system running ,allow special user interface coordinate with control system, don't consider the concrete method used in control system to meet user's requirement. It is proved that no one CIM/CIP architecture conform to all the principles.

2.4 Main content of CIPS Architecture

(1)Model set: In narrow sense, enterprise reference architecture is the set of describing integrated enterprise from different angles on general level; In a broad sense, enterprise reference architecture describe enterprise at different development stages.

(2)Set of methodology :It is the effective set of guide, technology and process.

(a). reference model: describe engineering structure of building integrated enterprise broadly and generally. (b).Modeling mechanism: help to build, research and judge model . (c) Structural method: Point out overall stages about how to change old system into future system according to the given object and restraint. (d) Function judgment standard: system should be judged from several angles(economy, reliability et al).

(3)Modeling method and tool: used on forming ,developing and analyzing the model. Tool is subordinated to special methodology architecture or reference architecture, or tool is independent.

(4)Basic architecture: a system that support processing and communication functions needed by enterprise information integration.

At present ,no architecture includes all the content .

3 Development of CIPS Architecture

Many kinds of CIM/CIP architecture energetically explain and direct the task included in production activity and enterprise integration, only several kinds of them clearly explain how an enterprise to integrate and what is needed. All the CIM/CIP architectures put forward now can be divided into two types: oriented to the part and oriented to the whole. It is necessary to point out ,CIMS and CIPS have common points in the necessity of researching architecture 、 its forming principles and its main content. Their difference lies in that they have different basic structural on low level ,the difference is caused by the different characters included in the two types of industries. At present ,many architectures are put forward according to the requirement of discrete industry ,to summarize their characters is helpful for researching CIPS architecture.

3.1 CIM/CIP that is Oriented to the Part

From system design to system building ,CIMS/CIPS are always a complex and large scale system. A effective method can divide it into CIMS/CIPS again at system building stage. CIMS/CIPS can be divided vertically and horizontally.
Vertical division: To divide CIMS/CIPS correctly 、 suitably 、 simply and comprehensively, the architecture shouldn't include unnecessary branch system and redundancy. It reflects the CIM/CIP architecture that is oriented to functional architecture.
Horizontal division: It reflects the architecture that is oriented to control structural. The principle of defining suitable levels of hierarchical control system is to reduce the number of hierarchical control levels and to ensure effective control[5].

3.1.1 Control Structure that is Oriented to Functional Structure

Its typical is the wheel structure of SME in American. The wheel structure in 80's mainly show the function of design and manufacturing, and it doesn't emphasize the simplification before building automatic system and the interaction among enterprise 、 consumer and supplier. The wheel structure of CIM/CIP in 90's put consumer in the kernel, this indicates that the main task of building CIMS/CIPS is to make enterprise attract consumer and win competition.

3.1.2 The Architecture that is Oriented to Control Function of System

Its typical example is the hierarchical control model developed by NBS in 1980,the model includes five levels. The model developed by ISO includes six levels. Because of its many advantages ,hierarchical control is applied in most enterprise .But, it has defects and many special control structure are put forward.

3.2 CIM/CIP Architecture that is Oriented to the Whole

At present ,many kinds of CIM/CIP architecture that is oriented to the whole are developed or are being developed ,such as: CIM-OSA (ESPRIT project developed by AMICE research team) 、 GRAI-GIM and Prudue enterprise reference architecture (developed by the CIM committee of Prudue university) et al.. But ,it is proved that no one architecture is perfect in necessary ability.

3.2.1 CIM-OSA

CIM-OSA (Computer Integrated Manufacturing -Open System Architecture) have several characters: (1)It is oriented to overall life period of system;(2)system's plan and design is separated from system's running;(3)Standardization and Openness. The conception of general level is favorable to the standardization of information unit and manufacturing unit. Partial general level is favorable to CIMS/CIPS standardization in a concrete industry; (4)Structure method of planning and designing ;(5)It is executable in computer .It's ultimate aim is to make all the structure 、 model 、 tool and technology related to architecture be executable in computer;(6)It can be used in production monitor and factory management;(7)CIM-OSA is one of the architectures that are researched most fully and broadly. It becomes one of initial standard of ISO[5-6].

Demerits: No effective method system. CIM-OSA is mainly oriented to structure, it doesn't put forward a method about how to build architecture. CIM-OSA stress how to describe and get CIMS, so it is difficult for developers to use CIM-OSA in researching factory integration and building CIMS; It is difficult for the user who aren't expert in computer to master CIM-OSA.

CIM-OSA needs be perfected in such areas: Make it more executable; Provide a method system that is more useful; Eliminate the restriction that CIM-OSA can be used only in discrete industry.

3.2.2 Prudue Reference Architecture

The architecture was developed by the group of American-Prudue university during 1989-1992.Based on the experience of building CIMS/CIPS in discrete and process manufacturing enterprises, Prudue reference architecture clearly stress the import role of human and organization in building CIMS/CIPS, it more completely discuss about the function 、 task 、 design and analysis method in every stage of CIMS/CIPS overall life period.

Its advantages:(1)It is oriented to system's overall life period;(2)Because it is described informally, it can clearly be understood by the users who have no computer education ;(3)It takes the integrated relationship between human and enterprise into account;(4)System's plan and design is separated from system's running; (5) Structure design method;(6)It has strong expandability .It is suitable to many kinds of enterprises that are different .It's method system (expressed in "Prudue CIMS building manual") is more perfect than CIM-OSA and GRAI-GIM.

Demerits: No necessary modeling technology that is used in computer modeling.

Prudue reference architecture needs be perfected in normalization: (1)At present ,main task is the normalization in describing system and developer's organization;(2)Research the necessary modeling technology that is added in computer modeling;(3) The normalization of the tasks related to manufacturing or consumer service is simple and clear in conception but in practice ,it needs many kinds of devices, so the information in the area need more practice.

3.2.3 Architecture of GRAI-GIM

Its advantages:(1)It has a good method system about using architecture ,it also includes the technology and tool used in developing integrated program on illustration and describing stage;(2)Its describing file only discuss about how to develop computer system and related hardware to realize predefined factory integration;(3)It is mainly used in discrete industry ;(4)It has expandability. The method system about overall life period of GRAI-GIM now is limited in detail design stages about concept 、 analysis 、 describing and developing of enterprise integrated system ,it doesn't discuss about related problems in system's structure 、 system's check and deliver 、 system's use stages;(5)It is easily be understood by common people;(6)It has structural plan and design method;(7)It takes economic and technical judgment of future system into account .

Demerits: (1)It doesn't discuss about the relationship among human 、 human organization et al..(2)GRAI-GIM is the same as CIM-OSA, they consider human as resource about required technique or physical ability, so they don't discuss human training plan 、 overall requirement 、 human organization and detail organization et al..

GRAI-CIM need expand its research range, human relationship should be included; Its content need be augmented ,the detail methods of modeling and application about structure 、 check and deliver of enterprise system should be included.

4 The Architecture of CIPS

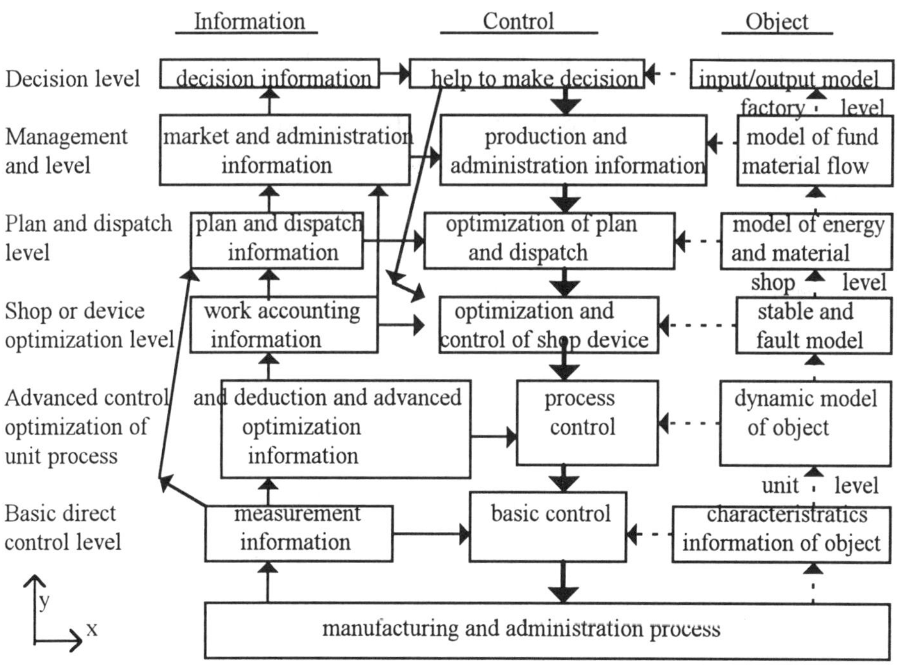

Fig.1. A architecture about overall process of CIPS

4.1 The Overall Process Architecture and the General Technology of CIPS

The key project met by automation technology is how to realize information integration and its application. It provides assisting decision optimization means for every leading 、 management and production department (such as administration decision, optimized dispatch and optimized production) and connects those optimization plans with actual production control system, this is a composite technology about automation that is called "General technologies"(Figure 1)[7-9]. Its main objects are to provide a series of reference models (included modeling method and related specifications) for large scale process industry ,a platform that can be used in different developing stages 、 test and verify the reference model and support platform that are used in demonstrative factories. To describe the relationship among reference models clearly, to verify that reference model is dependent on support platform ,the architecture of automatic system should be built firstly ,because it can make every relationship clear and direct developing demonstrative project from beginning to end. It is clear that the architecture is a very important project included in general technologies ,it directly effects the whole system's success or failure.

From the point of whole life period view, a large scale complex automatic system should undergo several stages, such as feasibility demonstration 、 requirement analysis 、 initial design 、 detail design 、 developing and building 、 running and maintenance et al.. Based on the idea of CIM-OSA that connect every stage in whole

life period together ,they put forward a architecture of a large scale process industry that is oriented to the whole life period together ,they put forward a architecture of a large scale process industry that is oriented to the whole life period of automatic system. Thus, a special task can be finished on one stage, every stage can also be connected. In main stage of system's developing process, there are related hardware and software platform to support developing, this can accelerate developing process and assure project's quality.

According to vertical relationship of enterprise's organization mode, a chemical enterprise is divided into five levels by NBS in American(Figure 2).It is divided into six levels by ISO: enterprise level; factory level; project level; unit level; control station level and device level. Song et al. put forward a three level structure for a chemical enterprise. Horizontally ,every level can be considered as generalized control system because they include three factors of control system: information 、 control and object.

General technologies can help building CIPS, it need a series of reference models that serve as sample plate of new project; use a series of developing tool to develop different module; Those modules can run coordinately under unite integration frame. At present ,many achievements are gained about general technologies and architecture.

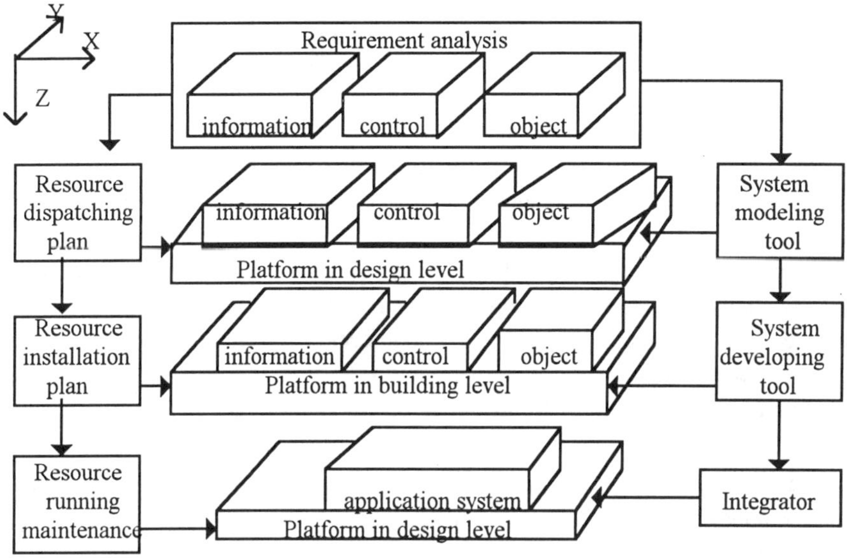

Fig .2. A sectional drawing of architecture model of CIPS

4.2 Discuss about Architecture of CIPS

4.2.1 Natures Included in the Ideal Architecture that is Oriented to the Whole

It is insufficient that only give out the description and design of computer control system, user need the direction about how to design 、 develop 、 build and use enterprise integration, So the architecture that is oriented to the whole is most valuable.

The ideal architecture have several characters:

(1)The architecture not only provides model about information integration needed in decision 、 dispatch and control ,it also provides model about project and plan that is needed in enterprise or factory integration.

(2) The architecture is connected with method system. The method system direct the design and building about information 、 matter flow and product integration et al..

(3) The method system should include computer or other support tool used in describing and developing enterprise integration, this can help user and building team complete predefined integration task.

(4) The architecture and its related model 、 technology 、 grammar and mean of tool's formative definition should be clear , the architecture developed should be suit to the trend of open computer system, the architecture has the form that can be executed by computer.

4.2.2 A Compound Architecture

At present, there is no perfect architecture, to combine CIM-OSA 、 GRAI-GIM and Prudue reference architecture absorb their merits and form a new compound architecture is a suitable way.

The compound architecture have such content:

(1)Use Prudue reference architecture to direct the whole plan and design. Because it can easily be understood and accepted by those designer who have no computer background.

(2)Use the model frame of CIM-OSA to provide analysis method for the data flow 、 matter flow and energy flow included in Prudue reference architecture.

(3)Use the model frame of GRAI-GIM as necessary reference to assure the perfection of requirement.

The existing methods and tools can help each other.

(1)On functional data or description level ,GRAI-GIM 、 GRAI-GRID 、 GRAI-NET are adopted ,formalized system of integrated basic structure of CIM-OSA 、 integrated engineering environment and integrated running environment are used as design model.

(2)The human 、 organization 、 manufacturing and service devices included in Prudue reference architecture are formalized.

(3)Use the integrated basic structure of CIM-OSA or express and operation level of Prudue reference architecture and they are perfected by necessary normalization.

Conclusions . Many countries of the world have understood that research the architecture of CIPS is important .CIPS is becoming useful in practice. Even through there is no perfect architecture ,it development direction is clear ,those existing architectures, such as CIM-OSA 、 GRAI-GIM and Prudue reference architecture ,will become a perfect and useful architecture through complement and expansion; Future perfect and useful architecture through complement and expansion; Future perfect architecture will be based on existing architectures ,expanding their connotation and extension; It is also important to research a standardization of judging if a architecture is useful, this method system can direct an enterprise to select CIPS architecture.

References

1. B. H. Li .The Provisional Constitution 、 Standard、 Building Guide of CIMS. *Chinese Weapon Industry Press.* 1994.9
2. G. Xiong, X. M. Xu , and ZH. J. Zhang. To Accelerate the Theory Research and Practice of CIMS. *Computer Integrated Manufacturing Systems.* Vol 1. No.3 (1995) pp35-38
3. T.J.Williams. Architecture for Integrating Manufacturing Activities and

Enterprises . *Proc. of 12th World Congress IFAC International Federation of Automatic Control*, (1993)10:273-283

4. T.J.Williams. The Purdue Enterprise Reference Architecture-A Technical Guide for CIM Planning and Implementation;1992
5. Michael Kaskowitz . CIM at Apple Computer, A Case - study. Institute of Industrial Engineers, *International Industrial Engineering Conference Proceedings* ,(1991)97-103
6. Matt Johnson. Towards a distributed control architecture for CIM .*Proceedings of CIMCON 90*,(1990)166-176
7 G. L. Song , X. S. Jiang , and B.Y. Liu . The Overall Architecture of Large Process Industry. *Information and Control.* Vol.23, No.2 (1994) pp65-70
8. B.Y. Liu. CIM and The General Technologies for Industrial Automation - Goals, Methods and Practices. *Information and Control.* Vol.22, No.1(1993) pp29-35
9. G. L. Song ,X. S. Jiang .CIMS in Continuous Production Process. *Computer Integrated Manufacturing Systems.* Vol 1. No.1 (1995)pp15-20

The Resource Flowing Model of Green Manufacturing Systems*

Fei Liu Xiaohui Chen

The First Department Of Mechanical Engineering,
Chongqing University, Chongqing, China

and

Hua Zhang

Automobile Department , Wuhan Metallurgical University
Of Science & Technology ,Wuhan, China

Abstract. Green manufacturing is a new concept occurred in recent years. Its goal is to minimize the negative effects on the environment through products life cycle. In this paper, it is pointed out that the root cause of environment pollution of manufacturing systems is the irrational consumption of resources. Therefore, the resources problem and environment problem should be considered together when the green manufacturing is studied. From this view point, the concept and connotation of green manufacturing and green manufacturing system is studied further. And then, the factors influencing resources consumption of manufacturing systems are presented by analyzing the operation processes and the materials flowing. Finally, the resources flowing model of green manufacturing systems is set up.

1 Introduction

Environment, resources and population are the main problems which human beings must be faced with at the present time. In particular, the environment problem is imperiling the existence and development of human society severely. With the global environment deteriorating, people have to pay more attention to the solution of the environment problem and have recognized by long term's researches and practices that environment problem should never be isolated, which has close relations with resources problem and population problem. Especially, the resources problem not only is involved in how to use the limited resources in the world, but also is the main origin of environment problem. Therefore in recent years a new concept has been put

* This project is supported by National Natural Science Foundation of China and National 863/CIMS Program of China.

forward: to use resources effectively at most and discharge waste at least is the basic solution of environment problem.

Manufacturing is the industry which transfers the useful resources (including energy) into industrial products or living products. It is not only the fundament industry which creates wealth for people, but also it is the important origin of environment pollution by discarding wastes. Since this, how to make the environment pollution of manufacturing industry minimize has become more and more important. Therefore the new concept----green manufacturing[1] has been put forward.

2 The concept and connotation of Green Manufacturing Systems

In reference [1], green manufacturing is defined as " a system that integrates product and process design issues with issues of manufacturing production planning and control in such a manner as to identify, quantify, assess, and manage the flow of environmental waste with goal of reducing and ultimately minimizing environmental impact while also trying to maximize resource efficiency " .

In order to well understand the concept and connotation of green manufacturing, We can consider two aspects as following:

(1) The concept of green manufacturing should reflect the view point that the environment problems of manufacturing industry should not be considered isolatedly without considering the resources consumption problems, because there are close relationships between resources consumption and environment pollution of manufacturing industry. In fact, green manufacturing is a system engineering issue considering the resources and environment problem synthetically, and its objective is to minimize negative impacts on environment within the whole life cycle of products.

(2) Any manufacturing process is fulfilled in manufacturing systems, so is the green manufacturing. In another way, the green manufacturing and manufacturing systems should be considered together. Since then we put forward the concept of green manufacturing systems: Green manufacturing system, just as its name implies, is a manufacturing system to carry out green manufacturing. We can explain this further as following:

It can be inferred from the function definition of manufacturing systems[2] that a manufacturing system is an input and output system in which the manufacturing resources (like raw materials, energy) are input, and go through manufacturing processes (Here manufacturing processes are general which include designing, manufacturing, assembling , etc.) ,and the products (including semifinished products) are output; meanwhile the wastes are discharged and give rise to environment pollution. This process is shown as Fig.1.

In the figure, the dash line presents that under some particular conditions the environment pollution is not caused by discarded wastes but directly by the process of manufacturing and use of products. This kind of pollution is much less than the waste pollution.

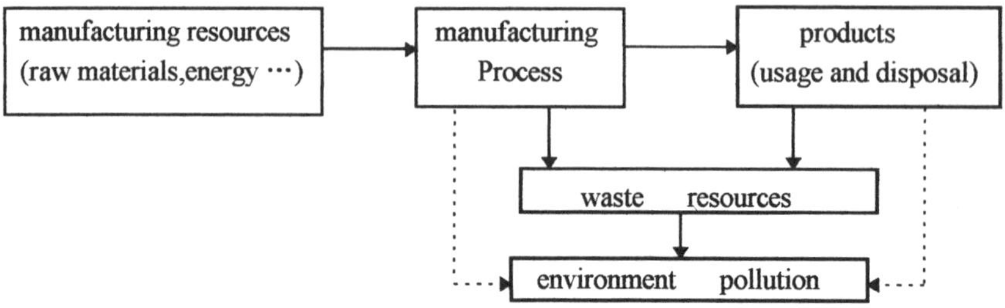

Fig.1. The negative effects of manufacturing system to environment

Compared with traditional manufacturing system, the green manufacturing system must ensure the least environment pollution. This characteristic is indicated in Fig.2.

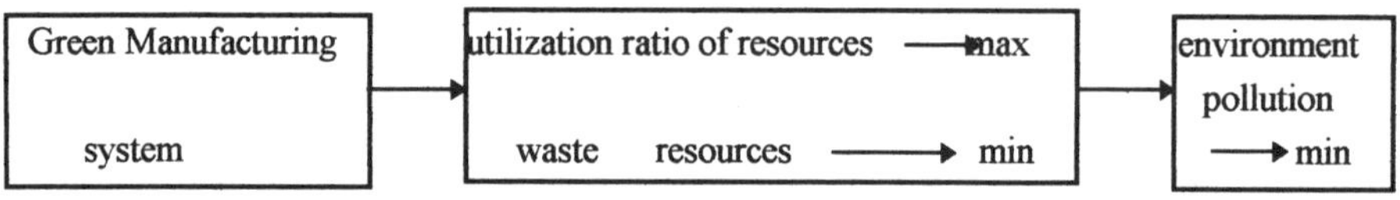

Fig.2. The characteristics of Green Manufacturing Systems

3 The resources flowing model of Green Manufacturing systems

3.1 The resources in a manufacturing systems

The resources in a manufacturing systems are called manufacturing resources. Manufacturing resources are the substance basis on which manufacturing system exits and works. The definition of manufacture resources can be divided into two kinds. One is the narrow sense definition, that is substance resources; the other is wide sense definition which means besides the materials and energy, it also includes fund, technology, information, labor and so on. The constitution of manufacturing resources system is shown as Fig.3.

For the resources which directly influence the green manufacturing systems are the substance resources, mainly include materials and energy resources, so the resources referenced in this paper are the substance resources, especially the materials and energy resources.

3.2 The factors influencing the resources consumption state in manufacturing systems

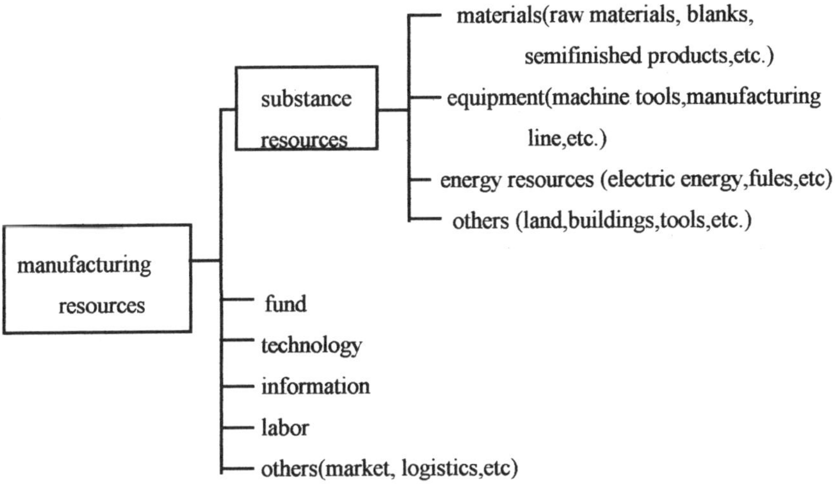

Fig.3. Constitution of manufacture resource systems

(1) Analysis of the factors influencing the resources consumption from the working processes of manufacturing systems

From the process definition of manufacturing systems[2], we know that a manufacturing system may be considered as a process which includes market analyzing, product designing, process planning, manufacturing, assembling, packing, product sale and service after sales, etc. Figure 4 illustrates the main processes and information flowing of a manufacturing system.

The researches and analysis show that all processes of manufacturing systems directly or indirectly affect the resources consumption. For example, market information and market analysis will determine the kind and quantity of products; products design will determine the specific type, shape and characteristic of products, and the products kind, quantity, type, shape and characteristic will affect the kind, quantity and utilization ratio of resources. Different process schemes and different process routes result in different materials and energy consumption while manufacturing the same products. Moreover, manufacturing (including assembling) is the main process in which the resources are transferred and consumed directly. The package pattern, the conditions of transportation, sale and service will directly or indirectly influence the state of resources consumption.

(2) Analysis of factors influencing the resources consumption from materials flowing of manufacturing systems

As above mentioned, a manufacturing system is an input ---- output system which transfers the manufacturing resources into products or semifinished products. In a sense, the manufacturing process which transfers the manufacturing resources into products or parts within manufacturing system is a dynamic process of materials flowing. This dynamic process can be explained roughly by Fig.5.

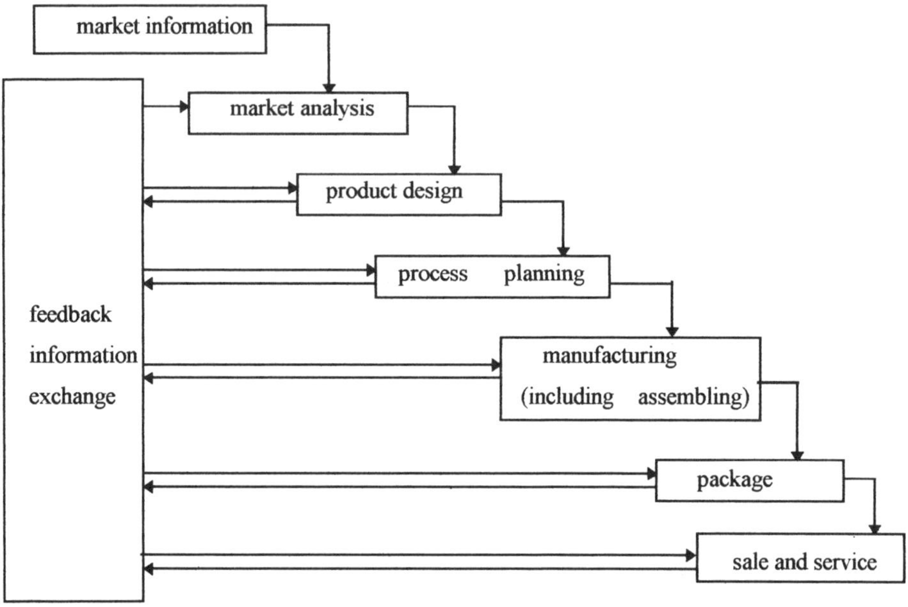

Fig.4. The main processes and information flowing of manufacturing system

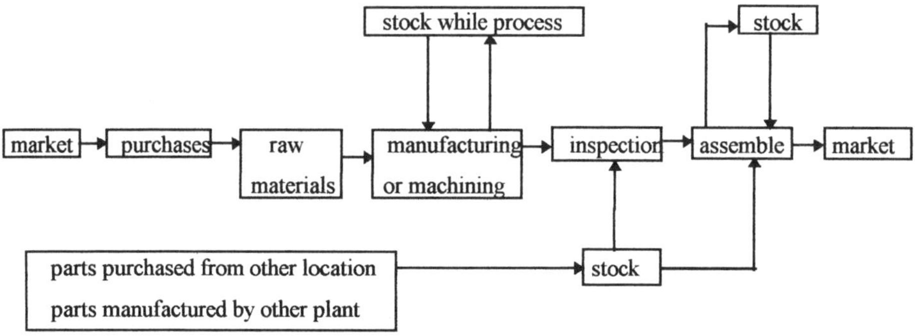

Fig.5. Material flowing of manufacturing system

Because the materials flowing needs energy to drive or transfer, the material flowing process in manufacturing systems is not only the main process of materials transferring and consuming, but also the main process of energy consuming.

The analyses and researches show that all the processes of materials flowing have direct or indirect influence on the resources consumption of manufacturing systems. For example , different specification of raw materials purchased at market will results in different raw materials utilization ratio. Too many spare parts storage may cause overstock and finally have to give up, which means wasting resources.

3.3 The resources flowing model of green manufacturing systems

By synthetically considering the connotation of green manufacturing and the factors influencing the resources consumption within manufacturing systems, a resources flowing model of green manufacturing systems is created as illustrated in figure 6.

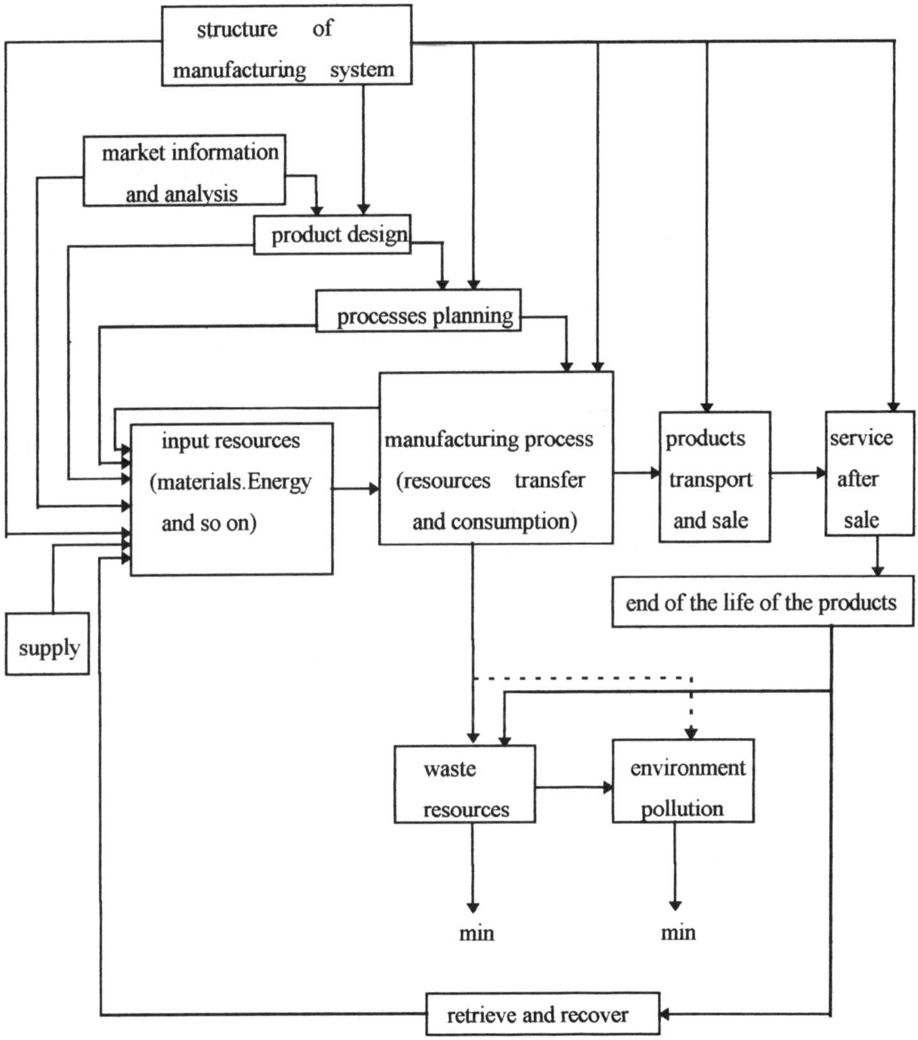

Fig.6. Resources flowing model of Green Manufacturing systems

In the Fig.6, there are some viewpoints which are explained as follows:

(1) The characteristic objectives of green manufacturing systems are to make the minimum wastes and environment pollution. The fundamental factors which decide the

two objectives are the resources flowing process.

(2) The factors influencing the resources flowing (the input quantity of resources, process of resources consumption, quantity of waste resources and so on) of manufacturing systems are various, which include the structure of manufacturing systems (such as components of equipment, workshop layout), product design, process planning, manufacture process, product transportation, disposal after use.

(3) The strategic way to carry out green manufacturing has been point out in the model ,whose objective is to make the resources consumption and wastes least, considering fully processes and optimizing each of the processes which influence the resources consumption and environment pollution in the whole product life cycle, to optimum the process of resources flowing and achieve the goal of green manufacturing.

4 Conclusions

Green manufacturing is a manufacturing system engineering issue about making full use of resources and minimizing environment pollution during the full product life cycle.

The root cause of environment pollution caused by manufacturing systems is the irrational resources consumption. A resources flowing model of green manufacturing systems has been put forward in this paper. The model presents that the characteristic objective of green manufacturing system is to pursue the least wastes and environment pollution, and the determinant factor of these two objectives is the resources flowing. There are many factors influencing the resources flowing of manufacturing system, including manufacturing systems structure, product design scheme, process planning, manufacturing process and the dispose method of product after use, etc. The strategic way of implementing green manufacturing is to optimum these factors and processes with the goal of the least resources consumption and wastes .

References

1. S.A.Melnyk, and R.T.Smith, *Green Manufacturing,* Society of Manufacturing Engineers (1996) pp. 1 - 8.
2. F.Liu , D.Yang , J.Chen, *Manufacturing System Engineering,* China National Defense Publishing House Press (1995) pp. 3 - 9.
3. F.Liu, The significance and main contents of the studies on Manufacturing resources, *Proceedings of development strategic of Advanced Manufacturing Technology, Beijing, China* (1995) pp. 225 -228.
4. H. Zhang, F. Liu, A Study on Architecture of Green Manufacturing and Some Strategic Problem in Its Implement, *Computer Integrated Manufacturing Systems, Vol.3,No.2* (1987) pp. 11 - 14.

Impact of Factory Size on Workflow and Choice of Layouts

Lee Pui Mun
Nanyang Business School
Nanyang Technological University

ABSTRACT

This paper presents a study that compares the effectiveness of a manufacturing system employing different layout concepts (functional versus cellular) in relation to factory size. The purpose of this study is to analyze relevant performance statistics of the two different layouts under the effect of varying factory sizes. Results from such a study will provide practitioners with comparative information to assist in *economic* decision-making related to factory layouts. The findings in this study indicate that at large factory size, the percentage of move time in relation to throughput time in functional layout stayed relatively unchanged while the percentage of wait time in a cellular layout also stayed relatively unchanged. While previous studies had shown that there is an increase in move time for functional layout and an increase in wait time for cellular layout whenever factory size increases, this research has also shown that the increase in percentage move time or wait time tapers off as factory size increases further. Thus, in a large factory, justification for cellular or functional layouts should not focus entirely on productivity as it had been shown that both systems are producing at about the same rate, but the economic implications of large amount of wait time versus move time should also be analyzed to select a cost-effective layout.

1 INTRODUCTION

As global competition becomes more intense, manufacturing companies are forced to improve their operation efficiency. Among the numerous improvement activities that many firms are undertaking, employing alternative layout concept is definitely a popular one. Companies have increasingly considered replacing the traditional functional layout, also known as the job shop layout, by cellular layout in order to improve the efficiency of their manufacturing operations. This is part of the ongoing trend in fulfilling the goal of Just-in-Time (JIT) production and Total Quality Management (TQM).

Cellular layout, a concept first proposed by Ralph Flanders in the 1920s [4], has been successfully implemented in Japanese factories. This layout concept enables a family of products that shared processing similarities to be manufactured on a group of machines placed in a single cell. In contrast, the traditional functional layout requires each batch of job units to travel to different types of machines located at different parts of the factory floor.

As compared to the traditional functional layout, cellular layout is supposed to generate superior production advantages such as throughput time reduction, reduced setup times, and shorter traveling distances [1, 6, 8, 9]. These advantages will enable efficient production control, increase operating cash-flow with less investment in work-in-process inventory, and a superior customer service level [7].

However, the potential increase in production efficiency by using cellular layout has always been a source of contention when compared to the traditional functional layout. Many practitioners who implemented the cellular layout had often made extensive prior modifications to their shop floor operations. These modifications include optimizing machine dedication, improving set-up times, making changes in the areas of production planning, process control, and operations management [2]. It is unclear whether these extensive modifications and improvements could have by its own accord contributed to the net improvement in production efficiency. In every sense, these prior modifications would have made it more difficult to assess the true benefits of implementing cellular layout over a previous functional layout. Besides, every manufacturing facility has its own unique production characteristics. Implementing a cellular layout without conducting a feasibility study is both risky and unsound. A feasibility study should focus on the difference in performance when using cellular layout in replacement of functional layout, without considering any prior modifications to the shop floor operations.

However, because of the complex job shop interactions, neither mathematical modeling nor physical experimentation could be efficiently and effectively used to conduct a comparative study between cellular and functional layouts. Computer simulation provides the only possible alternative for a quick and sufficiently accurate solution to the feasibility study. It can replicate the existing conditions of a functional layout and transplant these conditions to a proposed cellular layout. In this manner, one can easily detect any difference in performance between the two systems. Also, those factors of interest can easily be adjusted in order to gauge the sensitivity of both systems.

This paper illustrates the use of a computer simulation to create models for the comparative study between functional layout and cellular layout. The purpose of this study is to compare the performance of both layout concepts and to evaluate the usefulness of employing computer simulation to create models, perform simulated runs, and obtain relevant data that can facilitate feasibility analyses. The objective of this study is to analyze production efficiency between using functional and cellular layouts in relation to factory size.

2 RESEARCH BACKGROUND

Simulation uses abstract models to represent a real world situation. The main objective of a simulation study is to understand the behavioral dynamics of the real world elements based upon the results generated from the simulated behavior of the abstract model. Computer simulation requires the model to be created and tested on a computer. The experiments using computer simulation can be conducted with the use of normal programming languages, simulation languages, or simulation software packages.

A simulation software package is a user friendly software tool that helps to reduce modeling effort, provides comprehensive statistical analysis, enables sensitivity analysis, and generates visual results via animation. Many firms are increasingly turning to the use of these simulation packages to conduct quick analyses and to gain computer visualization of the interactions within the simulated system.

Several prior simulation studies had been conducted to compare the performances of cellular layout and functional layout. However, the design variables used in those studies were substantially different from this comparative study. In Flynn's [5] study, the main design variables were different types of layout configuration and different distributions of incoming demand for parts within the shop. In Cummings [3] study, the main design variables were a range of workload conditions and the effect of absenteeism on the shop floor. In this study, the design variable is traveling distance per batch of job units.

3 RESEARCH MODEL

3.1 The Work System

The simulation model used in this research represents hypothetical factory layouts and production conditions. The objective of the model is to act as a platform to test impact of factory size on different types of layouts. The model attempts to replicate normal working conditions in a general manufacturing industry. The functional layout and the U-shaped cellular layout are presented in Figures 1 and 2, respectively. Similar machines in the functional layout are grouped into four work centers. Transporters such as forklift trucks are used to deliver and collect batches of products to and from each work center according to the process plans. Movement of batches of products within the work center, however, is handled by the machine operator. For the cellular system, there are three designated manufacturing cell processing a specific product family. Cell 1 serves Products 1, 2 and 3. Cell 2 processes Products 4, 5, 6, and 7. Lastly, Cell 3 handles Products 8, 9, and 10. Again, transporters are used to move batches of products to and from the cell. Both layout systems are equipped with four forklift trucks. The loading/unloading times are assumed to be constant. Transporter capacity is assumed as infinite, which means different job batches waiting for the truck can all be served at once. However, multiple batches will only be moved by the transporters when they are all moving from the same pickup point to the same drop-off point. The transporter moving speed is modeled at 4.5 miles per hour, which is the average speed found in industry practice. The paths of the transporter are also shown in Figures 1 and 2.

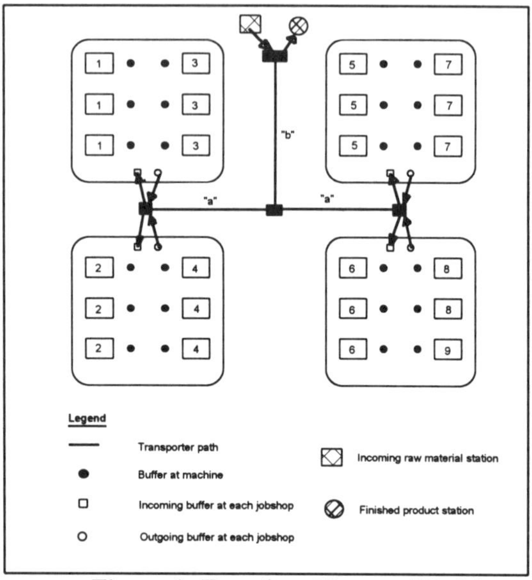

Figure 1: Functional Layout Model

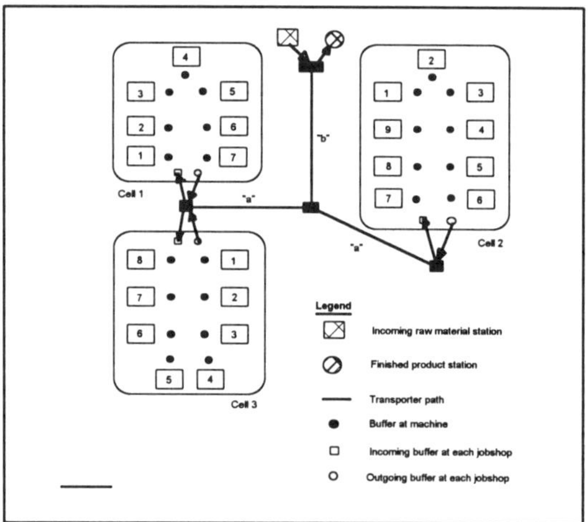

Figure 2: Cellular Layout Model

3.2 The Production Characteristics

There are 10 different types of machines and ten different types of products. Products are processed in batch-mode, i.e. all products in a single batch will move as a unit from one machine to the next irrespective of the layout system employed. Buffers are used at every machine as well as at staging point of each work center and cell. Buffer capacity is assumed as infinite. Job routings are randomly generated with a flow that can best be described as "jumbled". The number of operations per product varies from three to six, and each operation is performed at a different type of machine. No machine can be revisited and no backtracking is allowed. Batches of products are processed at the machine in the order of their arrival at the machine's buffer, i.e. *first come first served* (FCFS). This priority rule is chosen because it is a frequently used sequencing rule in practice. Key machine utilization rates average at about 70% for the functional layout and at about 90% for the cellular layout.

Transfer times between machines within work center or manufacturing cells are assumed as negligible due to their short distances. Setup times for each type of machines are set at a constant of 0.5 time unit, which is about 10% of the average processing time for each machine. The number and type of machines, the mean processing times, and sequencing of jobs are identical for both layout systems. Processing times at each machine are exponentially distributed. Each machine will work on only one batch at a time.

Batches of raw materials are scheduled to arrive at a receiving buffer station with an infinite capacity. The order of pickup at this receiving station by the transporters is again FCFS. The first arrival of each batch is assumed to occur at different but predetermined point in time. However, the time between arrivals for each subsequent batch is assumed to follow an exponential distribution with a pre specified Mean Time Between Arrival (MTBA). Only one batch is scheduled to arrive at any period. Finished batches of products are routed to a shipment station, which will not be modeled and therefore assumed to have zero processing time.

4 DESIGN METHODOLOGY

The main objective of this comparative study is to analyze production efficiency between using functional and cellular layouts in relation to factory size. To achieve this objective, the overall average throughput time of each different batch of products in the system must be measured. Average overall throughput time includes average overall waiting time in buffers, average move time in the system, and average processing time at each machine. The experimental study used traveling distance as the independent factor. Travel distance means the total distance traveled by each batch of products throughout its processing sequence. The treatment factor in this simulation experiment is the length of the transporter's path, which is formed by an inverted **T** shape as shown in Figures 1 and 2. The various configurations of path lengths used in this study are: 100/200, 200/400, 300/600, 400/800 (all distances are measured in feet), with the first number corresponding to "a" on the travel path and the second number corresponding to "b" on the travel path. These distances are used to represent various factory sizes, ranging from small to very large. For example, a travel path of 100/200 represents a manufacturing area of 160,000 sq. feet, while 400/800 will represent an area of 2,560,000 sq. feet. There are altogether (2x4) 8 different simulation runs, four for the functional layout and four for cellular layout. The performance indicator is the average number of batches completed for all product models within the study period at various travel distances.

The simulation cycle (study period) in the research is set at 480 time units (1 time unit = 1 minute) or equivalent to an 8-hour shift working day. In this model, the manufacturing plant is assumed to operate on 1-shift (8 hours) per day only. No break time or delays are considered. Each different type of simulation cycle/run is replicated for five times. The performance results reported represent the overall mean of a set of five runs for each different simulation configurations.

5 DISCUSSION OF SIMULATION RESULTS

In this research, the performance measurement between functional and cellular layouts is the average number of batches made for all product models in the system within the study period. To obtain this measure, the sum of the average number of batches made for each product model (there are 10 product models in this research) is computed. This sum is calculated for each different travel distance configuration. Table 2 shows the simulation results obtained in this research. When factory size is large and the travel distance is long, the superiority in output volume by cellular layout against the functional layout decreases.

Table 2: Total Number of Batches Processed Within Study Period at Various Travel Distances

	Travel Distances			
Layout Configurations	100/200	200/400	300/600	400/800
Function	*347.60*	*172.40*	*114.00*	*86.40*
Cellular	*377.80*	*202.25*	*130.75*	*94.00*
Difference:	***30.20***	***29.85***	***16.75***	***7.6***

Italics values represent the total number of batches made for all ten product models.

To gather more information on the performances of functional and cellular layouts, further analysis was carried out. There are three components that made up overall throughput time. These are wait time, move time, and processing time. To measure the mean wait time per batch, the mean average wait time at each buffer was first computed. This mean wait time per buffer will then be multiplied by the estimated average number of buffers that each type of product would be encountering throughout its entire manufacturing process. Both functional and cellular layouts generated a different average number of buffers. To estimate the mean move time per batch, the sum of the average wait time per batch and the average machine process time per batch is subtracted from the mean throughput time per batch. The percentage breakdown of the three components of mean throughput time for both the functional and cellular layouts are presented in Figures 4 and 5 respectively.

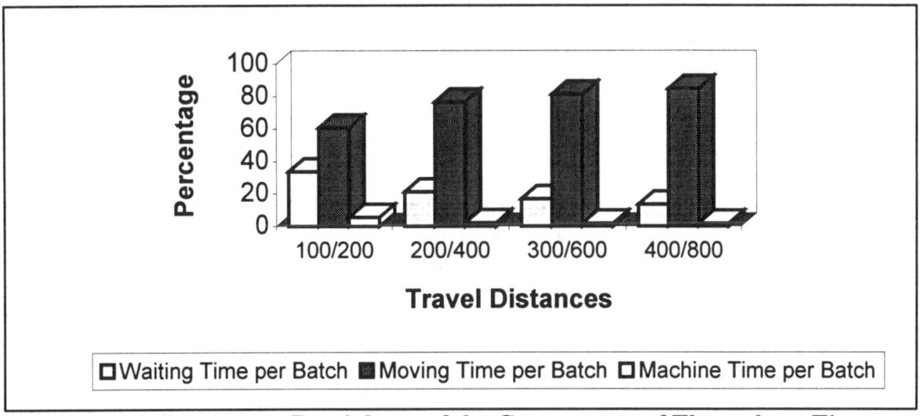

Figure 4: Percentage Breakdown of the Components of Throughout Time [Functional Layout]

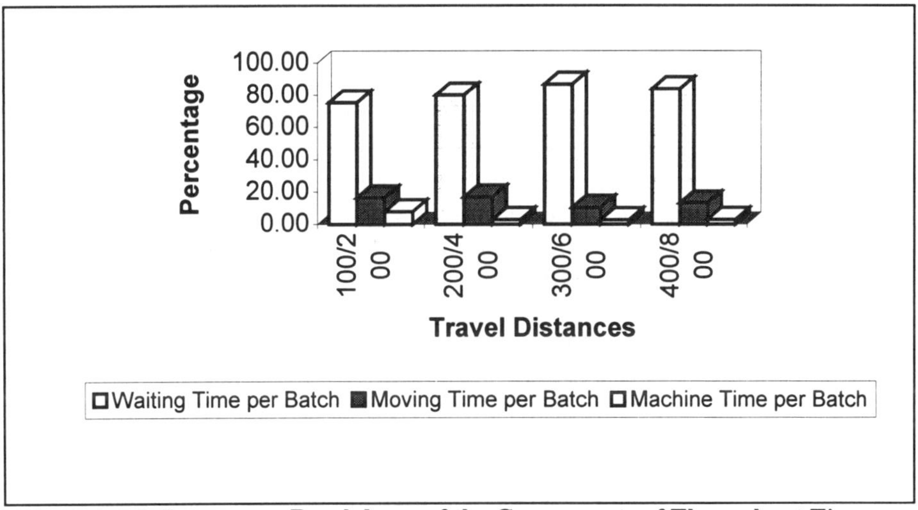

Figure 5: Percentage Breakdown of the Components of Throughout Time [Cellular layout]

Results from this simulation study indicate that using cellular layout produces larger number of batches at all different configurations of travel distance when compared to a system employing functional layout. Further analysis of the simulation data reveals that in functional layout system, the majority part of the overall throughput time is spent in moving the batches (high move time). However, in the cellular layout system, wait time at the buffers constitutes the largest portion of the overall throughput time.

Factory size has a significant impact on the efficiency of cellular layout. When factory size is small, cellular layout has clear advantage over functional layout in terms of output performance. When factory size is large, the superiority in output volume by cellular layout against the functional layout decreases dramatically. This can be attributed to the fact that at longer traveling distances, the rate of gain in shorter move time for cellular layout over functional layout is much less than the rate of increase in wait time for cellular layout. Thus, there is a trade-off between having shorter move times but longer wait times when considering the use of cellular layout in large factories.

The performance of cellular layout is also severely affected by bottlenecks created when a single machine is now used to process the product family. Under the traditional functional layout, the workload can be spread among those similar machines equipped in a process center. Simulation results showed that long queues were consistently generated at the most heavily utilized machine in each of the three cells. The capacity utilization at these bottleneck machines reached as high as 92%. Conversely, the same machines in functional layout had capacity utilization rate of about 70% with almost no queues.

Another observation is that in functional layout, percentage wait time (which is basically idle time) reduces while in cellular layout, percentage wait time increases initially but became relatively constant, as factory size increases. Also, percentage move time in functional layout increases initially but became relatively constant at large factory size while in cellular layout, percentage move time reduces slightly as factory size increases. The implication from this observation is that when factory size is large, the rate of increase in percentage wait time for cellular layout and the rate of increase in percentage move time become minimal. Thus, the decision to use either cellular or functional layout should not be based entirely on output gain. There should be a focus on the economic comparison of cost incurred in moving parts over large distance versus opportunity cost being lost in having excessive wait time. Move time can be associated with non-value added activities while wait time is idle time. Moving parts entail control, coordination and resources (equipment and manpower) as well as unnecessary time spent while wait time (as idle time) only represents foregoing opportunity cost. Taking these considerations into perspective, one can calculate the cost incurred by per unit move time and the cost per unit of idle time, and then using simulation to generate results such as those shown in this paper, and proceed to perform an engineering economy study to estimate the true monetary benefit of using cellular layout over functional layout. This is the type of economic justification that will greatly aid in decision-making.

6 CONCLUSION

This comparative study has found that the difference in output performance between cellular and functional layouts is very much affected by factory size. The study also indicates that at large factory size, the percentage of move time in relation to throughput time in functional layout stayed relatively unchanged while the percentage of wait time in a cellular layout also stayed relatively unchanged. While previous studies had shown that there is an increase in move time for functional layout and an increase in wait time for cellular layout whenever factory size increases, this research has also shown that the increase in percentage move time or wait time tapers off as factory size increases further. Thus, in a large factory, justification for cellular or functional layouts should not focus entirely on output volume as it had been shown that both systems are producing at about the same rate, but emphasis should also be placed on the economic implications of large amount of wait time versus move time. The conclusion drawn from this research is that one must always look into the economic comparison of cost incurred in moving parts over large distance versus opportunity cost incurred in excessive wait time when faced with the choice of either using functional or cellular layout in large factory.

The generalization of the findings obtained in this study is limited by the models used and the assumptions on which the models were based. In this comparative study, all resources were assumed fixed and no other changes or improvements were assumed when functional layout is converted to cellular layout. The design parameters used in this study are chosen to closely represent actual manufacturing environment. Different results may surface if different model configurations, product mix, processing sequence, batch sizes and assumptions were constructed. However, the fact remains that cellular layout may not always be superior to functional layout. The decision to implement either of the two layout concepts should depend on the nature of the production characteristics, the size of the manufacturing plant as well as a comparison in costs incurred in moving versus waiting.

In summary, this study has shown that there exists a need to conduct feasibility study whenever cellular layout is recommended for replacing existing functional layout. Because of the complex interactions in a manufacturing shop floor, computer simulation is an efficient tool to provide a quick and insightful comparative analysis for decision-makers who are wrestling with the problem of considering large amount of investment in a relatively short time period. Besides projecting manufacturing performances, computer simulation also helps to identify potential bottleneck machines and allows decision-makers to identify resource deficiencies.

REFERENCES

[1] Black, J.T., "Cellular Manufacturing Systems Reduce Setup Time, Make Small Lot
Production Economical," ***Industrial Engineering***, Vol.15, No.11, 1983, p.35.

[2] Burbidge, J.L., "Whatever Happened to GT?", ***Management Science***, 1978, p.86.

[3] Cummings, G.F., "Simulation Model to Compare Group Technology and Functional
Layout," ***Proceedings of the Summer Computer Simulation Conference***, 1980, pp.626-630.

[4] Flanders, R. E.," Design Manufacturing and Production Control of a Standard Machine," ***Transactions of ASME,*** Vol. 46, 1925.

[5] Flynn, B.B., and Jacobs, F. Robert, "An Experimental Comparison of Cellular (Group Technology) Layout with Process Layout," ***Decision Sciences***, Vol 18, No.3, 1987, pp.562-581.

[6] Hyer, N.L. and Wemmerlov, U. "MRP/GT: A Framework for Production Planning and Control of Cellular Manufacturing," ***Decision Sciences***, 1982, p681.

[7] Jackson, D., ***Cell System of Production, Business Books Limited, London***, 1978.

[8] Opitz, H. and Wiendahl, H.P., "Group Technology and Manufacturing Systems for Small and Medium Quantity Production," ***International Journal of Production Research***, 1971, p181.

[9] Suresh, N.C., "Optimizing Intermittent Production Systems through Group Technology and an MRP System," ***Production and Inventory Management***, 1979, p.76.

Integrated Databases

A Study of SDAI Implementation on Relational Database

*Yan Kai-tao *Deng Jia-ti **Zeng Qing-hui ***He Yue

* Beijing University of Aeronautics &Astronautics 720 Lab. Beijing 100083
**Linpan Oil Production Plant, Shengli Petroleum Administration, Shandong 251507
*** CIMS Reserch Center, Xi'an JianTong Uninversity, Xi'an 710049

Abstract: SDAI is the standard data access interface of STEP. It defines a functional interface between an engineering application and a data model described by EXPRESS. This paper discusses a method of SDAI implementation on relational database, expounds a conversion mechanism between STEP model and relational model, and develops a SDAI interface function library of data conversion on relational database, thus provides the basis for engineering applications of STEP.

Keywords: STEP EXPRESS SDAI Relational Database

1 INTRODUCTION

Database was initially used to support application programs of large quantities of structured data. The hierarchical database come out in 1970's first brought about the separation of applicatoin program from data, established a record-oriented model. Taking the relational algebra theory as its solid basis, the relational database emerged in the late 1970's set up a relational data model, making further improvement on data independence. Since 1990's, Object-oriented database has been providing the complex data structure and access of new application field with high quality support. Although the Object-oriented theory has not been perfected so far, and the problem that how to integrate perfectly the concept of database with Object-oriented model or Object-oriented concept has not been solved as well, thedatabase shows great vitality with it's advanced mechanism and good adaptability[1].

The task of SDAI(Standard Data Access Interface)is to provide standard service for data sharing, and to establish unified format and access interface, so that various data expressions and organizations have the unified format to exchange. The composition of multi-type and is omerized system has almost been a general phenomena in modern manufacturing, and especially needs data management function able to process complex data and control complex data structure, in which the database technology is one of the major support technologies. Relational database holds a primary ratio and important position in the database composition in the engineering applications proceeded and be in progress in quite time. This paper studies the SDAI implementation on relational database, having great significance.

2 THE OBJECT AND IMPLEMENTATION BACKGROUND OF SDAI

STEP is an international standard of data conversion published by ISO,whose object is to summarize the information needed by computerized products models possessing integrity and perfectibility during the lifetime of the products. It's standard data access interface SDAI is attempt to provide a mechanism for product data exchanging and sharing, and to establish a standard application program interface for STEP model to separate data applicating from data storing technology, thus to give applicators a unanimous data access enviroment, having users felt that, it appears that the core level data description managed by interface be consistant with the data description defined in the application model[2].

The application background of SDAI is the various of applications model established by EXPRESS. The applications requires to exchang the meeded information from database system, with no concern for the data orgnization, storage and management structure. SDAI defines a series of service functions of data access operation of data structure of building model by EXPRESS language to meet those requirements. On one hand it provides the requirements of enviroment and status and service function for implementation of operating EXPERSS model data, on the other hand it sets up a bridge between EXPRESS model data and application program throught the implementation forms of different programming languages(language binding , to have the product information model of STEP realized in application.

Taking SDAI model as a data cell, SDAI makes up SDAI implementation function with language binding of SDAI commands and specific program, and provides the functions of data reading, writing and storing of EXPRESS application model by operathing those functions.

3 FUNCTIONAL REQUIREMENTS OF SDAI IMPLEMENTATION ON RELATIONAL DATABASE

The functional requirements of SDAI include five aspects: data creating, deleting, amending, inquiring and effectiveness checking, not including contents such as range of support activity, data management, mode plan and performing development.

Data creating includes that of mode, entity, entity instance and partial entity instance, correspondingly, the deleting operation of those creatings as well. Amending may be understanded as the integrated operation of deleting and creating.

Data query requires the function of SDAI to support retrieving specific information, including the retrieval of generated data, detailed as follows:

(1) Model structure query: queries the integrated information of relevant entity and it's attribute in the model.

(2)Entity structure query: queries concerned entity structures, such as entity, type, entity attribute, attribute definition, dependency of related entity and attribute structure,etc.

(3)Entity information query: queries the information of entity instance, including the retrieval of specific attribute and the whole attribute, the retrieval of binded item, and entity types etc.

The effective checking function of SDAI is used to ensure the integrity, accuracy and security of data in practical application. It's operation set includes that if the citation of newly created model from the data structre of other model is standard, if the references needed by newly found entity exists, meets inheritance bind, if the newly found super/sub. type entity instance abides by the model definition, has effective physical meaning in data creation; In data amending and deleting, if there is a uniformity principle between the related entities and if the correspond ingly link processing is completed; in query the self-restrained checking of input conditions, the friendly interface indication of retrieval process and results.

4 APPLICATION BASIS OF SDAI IMPLEMENTATION ON REALATIONAL DATABASE

The implementation of SDAI on relational database requires basic support of database management system to accomplish the expression of all the complex objects, relational binding and closing of application engineering on the realtional database.

(1) The store cell of relational database is a two-dimensional table. In brief, a relational basic table can independently be mapped into a entity in the EXPRESS Model with it's complex entity be decomposed into simple entities. The link or bind between complex entities are reflected by the relation between basic tables in the relational database management system. The more complex the types of data and the more the cell classes sets made up by data and the richer the way of binding, the broader the range of application quantity of work needed by the flexibility and maintainability of data conversion interface.

(2) It's realized by keywords for relational database to support the uniqueness function of record recogntion. Every basic table identifies various records by the master keyword, master keyword attribute or the uniqueness of combination of the two. Those unique identifiers are no allowed to be changed during data access operation, and naturally, they correspond with the UNIQUE bind within the entity definition of EXPRESS model.Moreover, the uniqueness bind of basic table name assures the uniqueness recognition of basic table and EXPRESS entity corresponded with the E/R medium. The instanced of basic table and it's attribute and that by indispensable relation and other relative basic table give guarantee for the smooth proceeding of entity instanced in the EXPRESS model, and the data access process completely meets the requirements of UNIQUE RULE BIND OF EXPRESS model.

(3) Relational database management system provides security limits of basic table creating and using and the general operation of data eternal access.

(4) The creation of partial content of record of relational database management system has two cases: the one is that without viloating the unique bind of keywords, the newly corresponding basic table record can be directly produced; the other is that with the unique bind of keywords disobeyed, the correspinding record on the basic table is first required to retrieve and to be deleted according to the record data to be input, then the new data is stored in the new record produced by basic table.

(5) Relational database management system shows the undefined value, such as the void value in SQL, by defining the performance characteristics of basic table in the basic table creation. This function meets the requirements of attribute optional and EXISTS binds of EXPRESS.

(6) With some restrictions, the binding deleting function of general relational database management , can only bindly delete some simple binding relations indicated clearly in application, while the complex ones require programmer delete artificially with programming.

(7) The improved relational database management system supports rule management, eg has strong adaptability to the requirements of data effectiveness function.

(8) SDAI doesn't support dynamic user definition entity. Most retional database system provides dynamic interfaces, performs extending or amending of application

mode by binding with advanced languages. However the suppor to mode dynamic changing is too limited to remedy the function defect of SDAI, and because of the restrictions of the fundemental principle of relational database structure mode, is almost impossible to meet the requirements of SDAI concerning in this aspect.

5 CONVERSION OF EXPRESS MODEL MAPPED TO RELATIONAL MODEL

The key technology of SDAI implementation is to implement the mechanism of interface conversion between the STEP data model and various application enviroment data models, with the basis of SDAI implementation on relational database be the conversion between EXPRESS model and relational model. In order to express data structure clearly, the STEP data management model is first described formatedly, whose WSN expressed as follows:

<model>::={<declaration set><relation set>}

<declaration set>::=[{<type declaration>kentity declaration>
|<mode declaration>|<constant declaration>|
|< algorithm declaration><rule declaration>}]

<type declaration>::={<data type ><declaration bind>}

<relation set>::={<relation identification><relation definition>}

<relation definition>::=[{<relation definition among modes>|<declaration relation definition of mode and the other declaration relation definition within the mode>|<relation definition entities >|<declaration relation definition of entity and the other declaration relation definition>|<relation definitoin of entity and attribute>}]

......

whose abstractive logical expression is:

$$\text{Model}= \{\zeta , \mathfrak{R}\} \qquad (1)$$

$$\zeta \leq R , \tau \qquad (2)$$

where (1) ζ is declaration set, whose element is declaration identification.

(2) $\mathfrak{R}$ is relational set,whose element is relational identification

(3)<= indicating the binding components among the declarations in the model.

(4) τ is type identification set, which assigns each declaration relational identification with type value. $\tau\subseteq\Im$,$\Im$ is acquired in this way: $\zeta \cup \mathfrak{R}\rightarrow\Im$.

Type set $\Im$ contains the types of data, abstractive data ,tuple data and set data. Supposing that U is attribute set. If$\{A_1 ,A_2,...,A_n\}\subset U$, $T_1,T_2, ...,T_n$ are the non-tulpe data types in the $\Im$, and type T_j deosn't contain attribute A_i, then $[A_1: T_1,...,A_n: Tn] \in\Im$.

The range of type is defined by declaration and relation. The content included by data management is declaration, the staus of declaration object and relational set, in which the relational definition provides operation mechanism of algorithm involved by various functions and processes.

The EXPRESS data management usually makes up a logical unit in forms of several mode sets, when it is implemented on the database. So it should first be considered to express the data structure of EXPRESS to express the binding among the mode declaration the other declarations and various relations, especially the binding among various entities declarations.

The data management instance of STEP data management mode may be logically expressed as follows:

$$(\text{decl}, \quad \text{st}, \quad \text{rel}) \tag{3}$$

where (1) decl is a declaration identification;

(2) st is a mapping, which identifies each pair of declarations having connections with each other in the mode, for example, the mapping (S,e), between declarations of mode and entity, in which $S \in \zeta$,e=Schema(S),The specific value in the $\| \tau (S) \|$ is called status in place of e in the S, signed as $st_s(e)$.

(3) rel is a mapping, indicating every relational identification. $R_1 \in \Re$ is a relationship, expressed as: $rel(R_1) \subseteq \| \tau (R_1) \|$.

The data conversion between the STEP model implemented by SDAI and application system is convertible, so is the conversion function of conversion mechanism for interface implementation.Reference [3] expounds the conversion of IDEF model into the EXPRESS model, possessing typical E_R significance, provides a conversion method of relational model into the EXPRESS model.

6 DEVELOPMENT OF SDAI INTERFACE FUNCTION BANK ON THE RELATIONAL DATABASE

The SDAI standard has drafted SDAI operation and operation function bank of the system, requiring that each SDAI language binding implement all the functions given by the standard. The standard provides the standard format of these function implementation of C language binding , but the concrete content of its function is empty.

Most large relational database management systems have the function of language binding,in which the binding with C language makes data processing more simple, convenient and effective.Discussing SDAI interface function bank on the relational database management system,this paper selects ORACLE database and C language binding as the object of the studying.

The unique identification of entity instance is very important and crux when data is stored and mantained. System programmer controls and re solve thos problem on the relational database through nominating the relations among the basic table name, basic table keywords, keywords unique index and the inheritance among basic tables when the database structure designning is established.

Because the UNIQUE limit of EXPRESS model and the basic table keywords and their unique indexes of relational database have the effectiveness of different in approach but equally satisfactory in result, the UNIQUE limit can be conveniently converted into the keywouds unique index of relational database. The complex entity and supertype or subtype inheritance are alse integratedly mapped through relation appointment among basic tables on the relational database.

The core layer support of SDAI operation interface database access is implemented by several functions, whose value is directcdly database-operated by language binding. The corresponding function is called according to the condition information and the operation requirements input by users on the SDAI interface, and completes the processing of database basic table record in the light of entity instance identifier.

The difficult degree of query operation is depended upon the structure complexity of processed object.The query of simple entity attribute definition and attribute value can get results by separatly processing on the single basic table, but the query of complex object can be completed only through entity instance identifier, keywords index, medium model, multi-table bind operation and the corresponding basic information retrieving in the data dictionary of database management system etc.Followong is a illustration of taking the retrieving the implementation of SDAI function of attribute value of a certain attribute in entity as an examlpe.

The SDAI function retrivaling the attribute value of entity attribute is:

void*sdai Get Attr(SdaiInstance instance, Sdai Attr attribute,
Sdai Primitive Type valueType, void* value)

This function is detail analysed as follows:

(1) The format of SDAI operation corresponded to the function is described as follows:

Input: Entity Instance

/*Using specified entity instance ,the designed attribute value will be queried from the entity instance*/

Attribute Definition attribute-name

/*using the attribute name of the attribute whose value was queried */

Output: primitive /*the return value of attribute to query provided by SDAI enviroment.

Possible error status: the possible error status of SDAI error code set when the function is performed is{3.7.11.12.17.24.25.27.99}.The content description of error code confers reference[4].

(2) The entity information intended to be retrieved by the parameter variable table(pvt)of SadiInstance instance can be matched onto the correspoding basic table record of relational database from the Entity Definition item of SDAI dictionary model, thus determines the basic table to be operated. The keywords of Entity Defintion entity are entity-name and parent-schema. This parameter value(EntityInatance)must first be input as one of retrieval terms.

(3) The parameter of SdaiAttr attribute is another retrieval factor in function performing, which indicates the attribute name of the attriute to be retrieved in the entity(AttributeDefintion-attribute-name).

(4) The parameter of Sdai PrimitiveType value Type is one of queried information return values during function performing. It's domain type information of the attribute queried in terms of attribute name described in (3) by the function, through the operation of Attribute Definition of SDAI Dictionary madel.

(5) The void* value parameter is the return information(primitive) in function performing. After successfully performing of the function, the attribute value retrieved will be assigned to the value varible.

(6) void* sdai Get Attr function program includes 3 subprogrames. The subprogram lookcheck retrieves entity to be queried,the existance of attribute, and gets the identifiers of the entity and attribute; The subprogram look for finds out the corresponding basic table name and attribute domain name of the database with madel conversion mechanism; The last subprogram getvalue acquires the attribute domain value from the basic table and assign it to the function varible void* value. The function source program is refered to the appenedix.

(7) Application cases: The function call for the void* sdai Get Attr function application to retrieve material trademark attribute value of material entity in the material quality standard mode whose STEP modelling[5] has been set up is:

Material trademark=Void* Sdai GetAttr (material, material trademark, Sdaistring, void* value).

7 CONCLUSIONS

The information-based and integrated development of modern manufacturing imperatively requires the separation between the managements and usages of data and the database technology. The methodology of STEP points out the implementation way for the separation , in which the SDAI is one of the kernel problems and the key technologies. Studies on the SDAI on Object-Oriented. Database has been published[6], the study of SDAIimplementation on relational database is being made. This paper expounds a principle and method of SDAI implementation on the relational database, establishes a conversion mechanism between the STEP model and the relational model , and Based on that, finishes the development of SDAI interface function bank for data conversion between the two systems of engineering application and data management, thus lays a foundation for the development of application softeware platform of information conversion and sharing application based on STEP.

REFERENCES

(1) David L.Wells et al.Architecture of an Open Object-Oriented Database Management System,Cimputer,Vol.25,No.10,pp.74-81,1992

(2) Frank Leymann,Towards the STEP Neutral Reposioory,Computer Standards& Interfaces, Vol.16,No.4,pp 299-319,1994

(3) Yaikai-tao et al Standard Stratagem of Manfacturing Quality Management Data and Data Conversion Interface,Computer Engineering and Application. Vol.38, No.3, 1997

(4) James E.Fowler, ISO TC 184/SC4/WG T N 262, ISO10303-22. Industrial Automation System-Product Data Access INterface Specitication, 1992

(5) Yan Kai-tao,He Yue,Quality Control Information System Based on STEP, Computer Engineering and Application, Vol.32,No.6,1996

(6) A.Goh et al.Astudy of SDAI Implementation on Object-Oriented Database, Computer standards Interfaces. Vol.16,No.33-43,1994

DATABASES FOR COMPUTER INTEGRATED MANUFACTURING(CIM): A SURVEY

Robiah Ahmad and Abd. Rahman Abdul Rahim

Faculty of Mechanical Engineering, Universiti Teknologi Malaysia
K.B. 791, 80990 Johor Bahru, Johor, Malaysia
Fax: 607-5566159

Abstract. Manufacturing's industry effort to be competitive in a worldwide marketplace are placing a strong emphasis on improving its flexibility and responsiveness to customer demands. This has led to a fundamental reappraisals of manufacturing strategies and the realization that the key to success lies not on improving performance but in improving manufacturing capability. Manufacturing capability deals with production capacity, ability and flexibility. These are the essential elements needed to meet customer needs and remain responsive to change. To remain competitive, firms must operate positive control over the manufacturing process at every stage - from order receipt to customer acceptance and after sales service.

In recent years, many advanced manufacturing companies are moving towards Computer Integrated Manufacturing or CIM. Factors which have influenced this move include the need to increase productivity, reduce cost of material and direct and indirect labor, reduce scrap, rework and inventory, reduce cycle time, increase capacity, improve quality of products and customer service, greater production flexibility and improved market position.

This paper investigates the most widely used database options in manufacturing companies and their strategy in using them in moving towards CIM. In order to carry out the analysis a survey was conducted. The survey was conducted in the United Kingdom. The result of the survey can be used as an indicator of companies' strategy in implementing CIM. Some of the issues that are asked in the questionnaires are the amount spent on Information Technology (IT), database strategy, database model currently used, application of CIM technologies, use of computer systems and data communication architecture and other relevant issues.

Keywords : Computer Integrated Manufacturing (CIM), Information Technology (IT), database

Introduction

The manufacturing process starts with the design of the product. At this stage, component drawings, specifications and bills of materials are documented. The designer must have a clear understanding of the functions and performance expected of that product. The next stage is manufacturing planning. Information processing activities required at this stage include process planning, master production schedule, material requirement planning and capacity planning. Manufacturing control such as quality control or inventory control on the production floor is necessary in order to manage the physical activities to achieve the manufacturing plan. Business functions such as marketing, forecasting, purchase order, invoicing etc. act as an interfacing for communication between customer and manufacturer. CIM therefore, deals with the automation of all information processing activities described above through the use of computer systems.

CIM is a philosophy aiming at achieving a more efficient and effective product design and production processes by integrating the control and distribution of information throughout the entire process. The focus of CIM is on the sharing of information by engineering, production, business and various support groups. CIM automates the flow of information between the operations and activities of the manufacturing firms. It is an approach to manufacturing which uses computer technology through the availability and processing of information from all phases of manufacturing.

Computer Integrated Manufacturing refers to the use of computerized information systems and management philosophies to totally integrate all production related activities from customers' perceived need for a product through product design, process design, production, and after sales support [1].

CIM systems consist of subsystems that are integrated into a whole. These subsystems consist of business planning and support, product design, manufacturing process planning, process control, shopfloor monitoring systems and process automation. The subsystems are designed, developed and applied in such a manner that the output of one subsystems serve as the input of another subsystem. Organizationally, these subsystems are usually divided into business planning and business execution functions, respectively. Planning functions include activities such as forecasting, scheduling, material requirements planning, invoicing and accounting. Execution functions include production and process control, material handling, testing and inspection [2].

One of the important elements to consider as part of CIM implementation is the database strategy. In CIM, databases play an important role throughout the company for information storage and retrieval. Databases for CIM must be extensive and comprehensive since it involves total operation of the organization. This database can provide on-time and accurate information to all parts of the organization. As the number of computer applications increase, it is important to have a database system which is capable of integrating all the information with the rest of the systems throughout the company. Since the emergence of CIM, numerous articles have been written on the issues and success of companies implementing CIM. While many articles and papers focus on individual results, there has been little discussion concerning the overall implementation of CIM. A survey was conducted in the United Kingdom to obtain overall picture of CIM and to investigate their database options and strategies in using them towards CIM.

Developing the Survey

A questionnaire was developed and sent to 250 manufacturing companies to survey their approach in implementing CIM regarding computer application software, database options, information technology expenditure and integration strategy. The results should reflect the trends of those companies in implementing CIM. The questionnaire was divided into four sections. Part A is about background of the company. Part B is about computer application software, Part C deals with the company's database strategy and part D is about integration strategy. Multiple-choice rather than open-ended questions were used to minimize the time required to answer the questionnaire. The survey is a six-page questionnaire that could be completed within 20 to 25 minutes.

Results of the Survey

Of the 250 surveys that were mailed, 68 were completed and returned. The respond rate was 27.2%. The results were compiled from the 68 surveys that were returned. Some surveys were returned incomplete, particularly in Part C (database strategy) so the sample size is not consistent at n=68 for all questions. For mathematical simplicity, responses are converted to percentages. Due to rounding, some results may not total 100%. Following are some of the results of the survey.

1. The respondents were from:

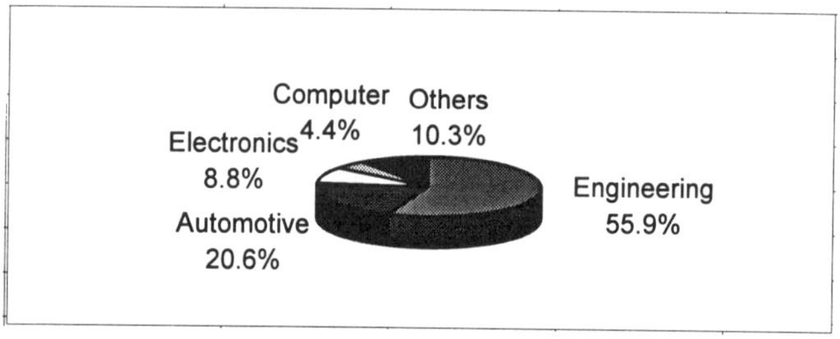

Fig. 1. Respondents of the Survey

2. Number of employees:

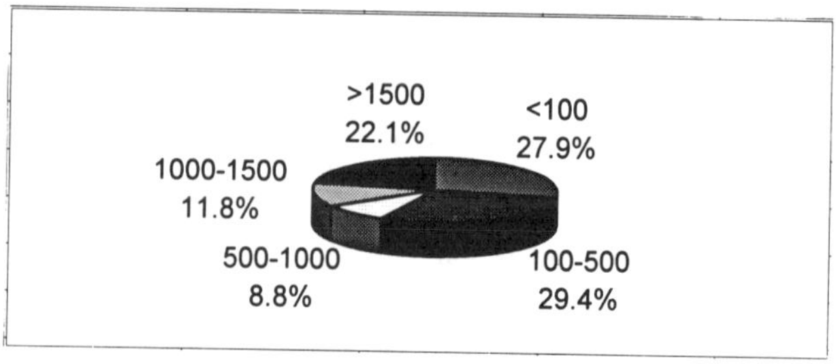

Fig. 2. Number of employees

3. Number of different types of products produced:

- 72% more than 10
- 11.8% 6 to 10
- 10.3% 1 to 5
- 5.9% 1 product only

4. IT expenditure (Table 1):

Table 1: Amount spent on IT (number of companies)

Amount (£)	Last 5 years	Next 5 years
< 1,000	2 (2.9%)	1 (1.5%)
1,000 - 10,000	13 (19.1%)	15 (22.4%)
10,000 -	21 (30.9%)	21 (31.3%)
1,000,000	32 (47.1%)	30 (44.8%)
> 1,000,000	n=68	n=67

Almost half of the companies surveyed claimed that they spent more than £ 1 million on IT expenditure. This shows that companies are willing to make large investment in computer technology. Companies that have short product life cycle have to respond to changing customer demand. Investment in IT is necessary to enable them to produce good quality products at the right time and at the right cost. IT has offered many advantages in the way manufacturing process can be organized.

5. Types of computers used (Table 2):

Table 2: Types of computers used

Types	No. of companies
Supercomputer	6
Mainframe	24
Workstation	44
Minicomputer	50
Microcomputer	60

It is not surprising that all the six companies that used supercomputers spent more than 1,000,000 for the last five years in IT expenditure. Of the companies that used supercomputers, 2 are engineering companies and 3 are automotive companies. 29 of the companies that spent more than £1 million on IT used microcomputer and mini computer, 27 used workstation and 20 used the mainframe. Most companies surveyed give more than one response for this question. Mainframe computers are required to handle massive transactions especially during product change. Large processing capability is needed for data retrieval and information flow throughout the company. On the other hand, microcomputers have emerged as low cost alternatives which provide users the ability to automate an increasing number of applications and functions.

6. Data communication architecture (Table 3):

Table 3: Data communication architecture

Types	No. of companies
Ad-hoc	28
LAN	36
WAN	14
LAN/WAN	22
None	6
Proprietary	1

The most popular data communication architecture is Local Area Network (LAN). Surprisingly, quite a number of companies use ad-hoc communication architecture. Wide Area Network (WAN) is important for companies that are widely distributed. Fast and accurate services for communications is important for these companies so that they can respond quickly to the market demands of their customers and fast access to their suppliers.

7. Computer application software (Fig.3), n=68:

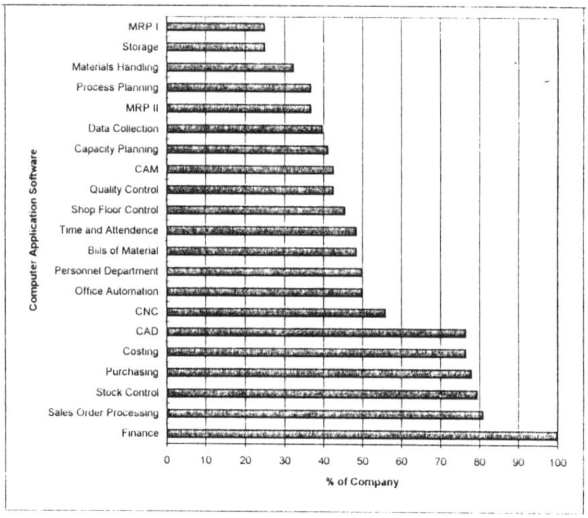

Fig. 3. % of company used computer application software

All the companies surveyed use application software in finance. This is probably due to wide off the shelf availability of this software, the cost is much lower compared to engineering software and easy to use to avoid repetitive simple calculation. Majority of the software used is for business functions. Engineering software which is highly used is CAD. There are a lot of CAD commercial packages available ranging from two dimensional drafting to three dimensional solid modeling with color shading. The slow pace of computer based technologies' entrance in many companies are due to either technical inadequacies or problems in performance. Usually, large companies with strong financial standing and resources are the one who have early applications of these new technologies to improve efficiency and productivity to gain competitive advantage over other competitors.

8. Use of Database Management System (DBMS), n=68:

- 82.4% Yes 17.6% No

9. Database Strategy (Table 4):

Table 4: Database Strategy Applied

Database	No. of Company
Centralized	34
Decentralized	15
Distributed	15
N/A	9

The most popular database strategy is centralized database, where all departments share the same database. Some companies give more than one response to this question. In reality, manufacturing companies use a large number of dissimilar computer systems, data systems and databases. The need for sharing data among these systems results in overlapping databases. A common database is essential so that data sharing throughout the organization can be accomplished. In centralized database, a common database on a central computer system is used with common paths to client systems. With current computer technology, it is possible to create a centralized system with sufficient redundancy to avoid lost of data in the case of failure or breakdown.

10. Database product used in the company, n=59:

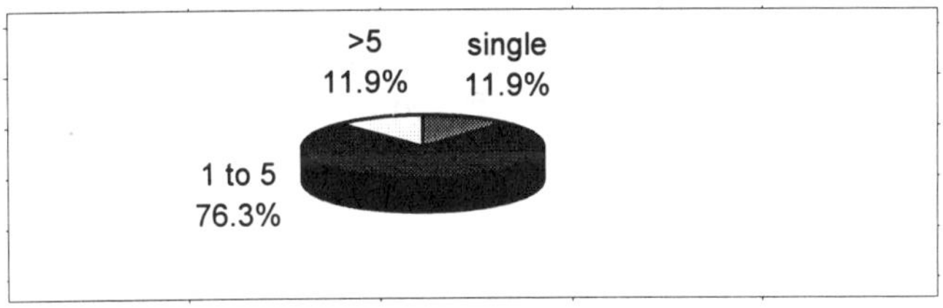

Fig. 4. Database product used in the company

11. Data models applied (Table 5):

Table 5: Data models

Model	No. of Company
Hierarchical	17
Network	9
Relational	43
Object-oriented	3
N/A	18

The most highly used data model is relational database, followed by hierarchical and network. The least popular is object oriented. This is probably due to the complexity of the database and it is still relatively new in the market. Relational database is more flexible, simple to use and maintain compared to earlier type of databases. Structured Query Language (SQL) is used with relational database models to make it more user friendly. Hierarchical model is not widely used because it does not allow direct representation of many-to-many relationship which is an important aspect in manufacturing environment where a lot of information may be interrelated and has many-to-many relationships among each other. The long term solution to manage more complex data with relational data model has problems with could be the object oriented data model which can handle a lot of complex data more efficiently.

12. Master Coordinating Database, n=66:

- 30.3% Yes 69.7% No

13. Implementation of integration strategy, n=66:

- 75.8% Yes 24.2% No

Three quarter of the companies surveyed said that they have strategy for integration of various systems currently used in day-to-day operation.

14. Number of computer suppliers, n=64:

- 21.9% Single 78.1% Multiple

Most companies have multiple computer suppliers. One of the main reasons is company's policy of buying the cheapest computer in the market. Another reason is that not all computer suppliers are good in any one area. One supplier might be reliable in supplying microcomputer but on the other hand he cannot supply a plotter or a system integrator. Companies are forced to find another supplier. However, the practice of using multiple suppliers will create problem in integration at a later stage. Some computer systems are incompatible and cannot 'talk' to each other.

15. Planning for CIM, n=67:

- 41.8% Yes 58.2% No

16. Application of Electronic Data Interchange (EDI), n=68:

- 39.7% Yes 60.3% No

17. Same computer system with the supplier, n=68:

- 85.3% Yes 14.7% No

Most companies have the same computer system as their suppliers. This will result in less problem in manufacturing stage. For instance, engineering drawings from the main company can be easily accessed from the supplier's site if both systems are compatible.

Companies Spent More Than RM 1 million on IT

32 companies surveyed spent more than RM1 million on IT expenditure for the last five years. Further analysis is carried out to determine the trend of these companies in moving towards CIM. The amount of capital invested in computer technology is usually directly related to the move towards CIM.

1. The respondents were from:

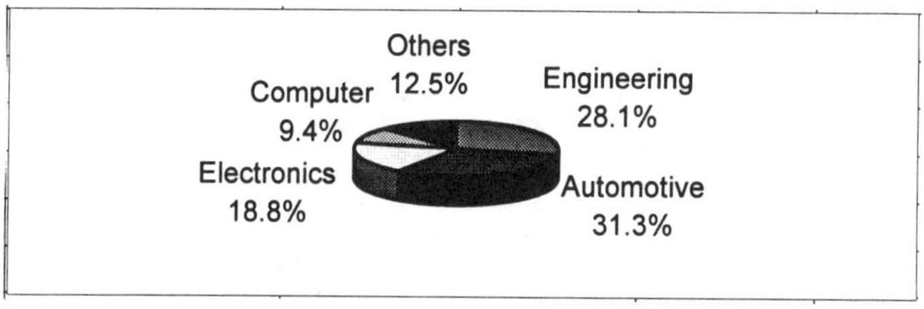

Fig.5. Companies which spent more than RM 1 million on IT

Automotive companies invest heavily on IT followed by engineering and others. One of the main reasons could be due to stiff competition from foreign manufacturers that forced these companies to automate the information flow to reduce manufacturing lead time as well as overall production costs.

2. Types of computers used (Table 6):

Table 6: Types of computers used

Types	No. of companies
Supercomputer	6
Mainframe	20
Workstation	29
Minicomputer	29
Microcomputer	27

It is not surprising that all the six companies that used supercomputers spent more than RM 1,000,000 for the last five years in IT expenditure. Most companies surveyed give more than one response for this question. Three of the companies that used supercomputers are automotive companies.

3. Data communication architecture (Table 7):

Table 7 : Data communication architecture

Types	No. of companies
Ad-hoc	10
LAN	18
WAN	12
LAN/WAN	19

Most companies used LAN/WAN architecture. However, there are some companies that give more than one response to this question because they used a combination of data communication architecture.

4. Computer application software

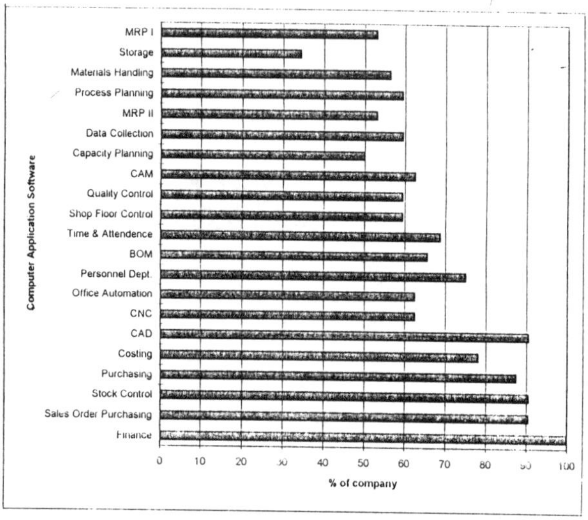

Fig. 6. % of company used computer software

Most companies used computer to handle the business function. CAD is the most widely used engineering software.

5. Database Strategy (Table 8):

Table 8: Database strategy applied

Database	No. of Company
Centralized	18
Decentralized	9
Distributed	12
N/A	1

6. Data models applied (Table 9):

Table 9 : Data models

Model	No. of Company
Hierarchical	13
Network	3
Relational	29
Object-oriented	3
N/A	1

7. Planning for CIM:
- 40.6% Yes
- 46.9% No
- 9.3% Currently implementing CIM
- 3.1% N/A

Three companies claimed that they are currently implementing CIM. Two of these companies are Japanese multinationals and the other one is an American multinational.

Analysis of Engineering Company

Since engineering companies are most likely to implement CIM compared to the other manufacturing companies, further analysis was carried out. The total number of engineering companies responding to the survey is 38 which is about 56% of the response rate.

1. Number of employees:

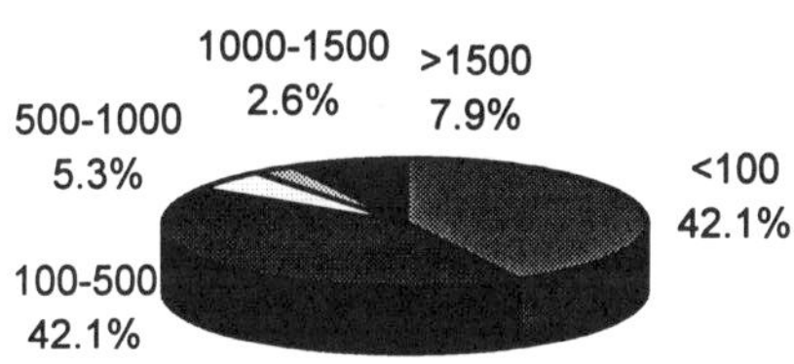

Fig. 7. Number of employees

Majority of the engineering companies have employees fewer than 500. These companies can be classified as either small or medium size enterprise. The level of computer application cannot be judged from the number of employees worked in the company. Fewer employees could mean higher level of automation.

2. Number of different types of products produced, n=37:
- 70.3% more than 10
- 13.5% 6 to 10
- 8.1% 1 to 5
- 8.1% 1 product only

3. IT expenditure (Table 10):

Table 10: Amount spent on IT (number of companies)

Amount	Last 5 years	Next 5 years
< 1,000	1 (2.7%)	0 (0.0%)
1,000 - 10,000	10 (27.0%)	14 (36.8%)
10,000 - 1,000,000	17 (45.9%)	15 (39.5%)
> 1,000,000	9 (24.3%)	9 (23.7%)
	n=37	n=38

4. Types of computers used (Table 11):

Table 11: Types of computers used

Types	No. of companies
Supercomputer	2
Mainframe	9
Workstation	24
Minicomputer	31
Microcomputer	20

Most companies surveyed give more than one response for this question.

5. Data communication architecture (Table 12):

Table 12: Data communication architecture

Types	No. of companies
Ad-hoc	5
LAN	20
WAN	22
LAN/WAN	6
None	3
Proprietary	1

6. Computer application software :

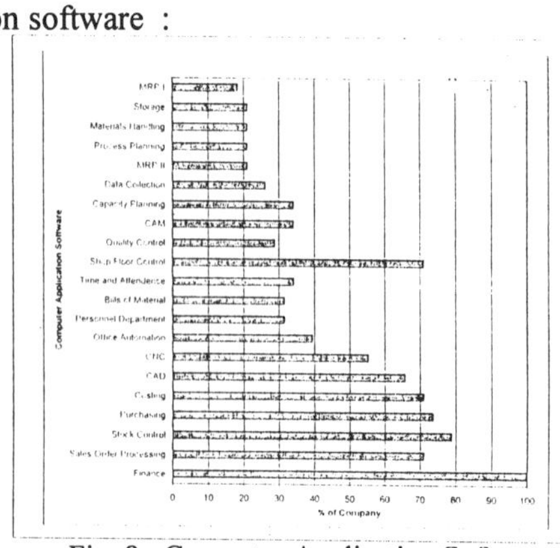

Fig. 8 : Computer Application Software

8. Use of Database Management System (DBMS), n=38:

- 76.3% Yes 23.7% No

9. Database Strategy (Table 13):

Table 13: Database strategy applied

Database	No. of Company
Centralized	19
Decentralized	12
Distributed	8
N/A	5

10. Database product used in the company, n=33:

- 18.2% single database
- 72.7% 1 to 5 database
- 9.1% more than 5

11. Database models applied (Table 14):

Table 14: Database models

Model	No. of Company
Hierarchical	6
Network	4
Relational	16
Object-oriented	1
N/A	15

12. Master Coordinating Database, n=35:

- 25.7% Yes 74.3% No

13. Implementation of integration strategy, n=36:

- 72.2% Yes 27.8% No

14. Number of computer suppliers, n=35:

- 25.7% Single 74.3% Multiple

15. Planning for CIM, n=38:

- 36.8% Yes 63.2% No

16. Application of Electronic Data Interchange (EDI), n=38:

- 28.9% Yes
- 71.1% No

17. Same computer system with the suppliers, n=38:

- 15.8% Yes 84.2% No

Conclusion. Majority of computer applications in place are for simple applications such as finance, purchasing, costing and other business functions. The use of applications in more strategic areas such as process planning, capacity planning, MRPII and other areas are lagging behind. This is probably due to the limited number of applications available in the marketplace and some of the software are prohibitively expensive for smaller companies. In the future, when there are more software available and at a lower cost, it is expected that the trend will reverse where more small and medium size companies can afford to buy these software. The centralized database is the traditional design, the simplest and the most workable. However, it is questionable whether this is a feasible architecture for CIM environment in the long run where factory automation will require high level of data sharing. The heterogeneous, physically distributed model of CIM systems require support for large volumes of data distributed over a wide variety of computing systems with a very high level of accessibility among various subsystems.

From the result of the survey, the relational data model is the most widely used in companies. However, there are some complicated tasks which relational database cannot handle as efficiently as object oriented database. Relational database is better suited for report intensive application. Object oriented database, on the other hand, have more potential for workstation applications such as CAD which requires high level of performance and computation capabilities. Even though many of the companies have CIM in part of their organization and many of them said that they have an integration strategy, there is little evidence of integration between these parts.

References:

1. H. Noori and R. W. Radford, *Production and Operations Management: Total Quality and Responsiveness,* McGraw Hills Inc., (1995).
2. S. Kalpakjian, *Manufacturing Engineering and Technology*, 3rd. Edition, Addison Wesley Publishing Company Inc., (1995).
3. M. P. Groover, *Automation, Production Systems and Computer Integrated Manufacturing*, Prentice Hall International Edition (1987).
4. Anon, Database Management: Gate to CIM, *American Machinist*, October 1987, pp. 81-88.
5. B. Harris, Managing the Database, *Professional Engineering*, February, 1991.
6. K. McIntosh, Shaping Up to the Information Age, *Manufacturing Engineer*, October 1991.
7. K. McIntosh, Data Direkt, *Manufacturing Engineer*, February 1992.
8. C. K. Wassk, Know How to Choose the Right CIM System Integrator, *Industrial Engineering*, July 1990.
9. J. Weston, Bringing CIM to Fabrication, *Professional Engineer*, April 1991.
10. J. Gomsi and M. Desanti, Relational and Object Oriented: Understanding the Database Debate, *International Journal on CIM*, Fall 1990.
11. N. W. Hopwood Jr., Managing Information as a Resource, *CIM Technology*, August 1987.
12. J. E. Cole, Design-to-Manufacture - an Integrated Information System, *Computer Aided Engineering Journal*, October 1989.

Automated Knowledge Acquisition for Multiexpertsystems in Manufacturing

André Schekelmann* and Wilhelm Dangelmaier

Heinz Nixdorf Institute and University of Paderborn
Fürstenallee 11, D-33102 Paderborn, Germany
{asm,whd}@hni.uni-paderborn.de

Abstract. Knowledge acquisition is known as a major "bottleneck" in the development of expert systems. Problems increase, if multiple experts have to be considered, because they usually have different views of the domain, e.g. a manufacturing system, i.e. the usually use different terminologies to describe the domain and do not agree in their decisions. In the proposed approach, terminological problems are overcome by relating the terminologies of the experts to the terminology of a basic model using machine learning techniques. Based on this relationship, a translation among the terminologies of the experts using a two–step procedure becomes possible. To acquire the knowledge of the experts, using machine learning, especially using Inductive Logic Programming is discussed. Finally, approaches for discovering and overcoming of conflicts in the behaviour of the experts are presented.

1 Introduction

In a manufacturing system, the different resources are connected by the flow of material. Controlling this flow has to be considered as an *on–line problem* because neither the amount of jobs nor the amount of resources is fixed for a certain period of time. This is true, even if a complete production schedule for a certain period is known in advance because of special orders of customers or breakdowns of resources, or because of changes in the predetermined sequence of jobs due to quality control and possible reworks.

Therefore, traditional *optimizing techniques* are not suitable to solve such problems, because they always assume fixed amounts of jobs and resources in order to compute an optimal allocation of jobs to resources according to a given target function. *Priority rules*, a class of well–known algorithms for on–line controlling the flow of material in manufacturing, are not suitable, too, because they are often to general and do not take into account special properties of the particular manufacturing system. In addition, priority rules always implement fixed strategies that do not consider the particular state of the manufacturing system.

* This work is supported by the DFG Graduiertenkolleg "Parallele Rechnernetzwerke in der Produktionstechnik", GRK 124/2-96.

A promising approach for the development of problem-specific strategies, able to react fast to unpredictable changes in the manufacturing system, are knowledge–based techniques (see [9] for an overview), especially if knowledge and experiences of several domain experts, e.g. schedulers working in different shifts, are taken into account. Such human schedulers have detailed knowledge about special properties of the manufacturing system. In addition, human beings are usually able to react flexible in a changing world. Integrating the knowledge of experts into on–line strategies for controlling the flow of material should therefore lead to strategies that take into account the special properties of the manufacturing system and its changing states.

Unfortunately, the acquisition of knowledge for such expert systems is known as a bottleneck, because experts often act intuitive and are not able to express all their knowledge in an explicit way, e.g. as rules. To overcome this problem, approaches for machine learning have been applied to problems in manufacturing (see [1] for a survey). Problems in knowledge acquisition increase, if several experts are taken into account: "If knowledge acquisition for an expert system with a single expert can be described as a bottleneck, acquisition from multiple experts [...] has the potential to become a log jam" [8].

This acquisition of knowledge from multiple experts in manufacturing leads to two main problems. Firstly, experts usually do not use the same terminology, because they perceive the domain differently and use therefore different terminologies to describe it. E.g., one expert uses the attribute *priority* to describe the importance of a job, allowing the values *small* and *great*, another expert uses the attribute *importance*, allowing the values *low, medium* and *high*. Secondly, experts usually do not agree in their decisions, i.e. in the same situations, they react differently. E.g., one expert may chose a job with the highest importance, another expert a job with the shortest slack time. These differences have to be analyzed and resolved.

The paper is organized as follows: In the next section, an approach for overcoming terminological problems is proposed. Thereto, the need of a common model as a preposition for relating terminologies is motivated and a relationship between the attributes of the individual experts and the attributes of the model is introduced. Based on this relationship, that can be discovered using machine learning techniques, a translation between the terminologies of the experts becomes possible. In the following section, the acquisition of the knowledge of the experts using machine learning techniques is discussed; the need of approaches for symbolic learning is motivated and the problem of choosing the appropriate attributes for expressing the examples when using a machine learning algorithm for learning from examples is discussed. Afterwards, approaches for analyzing and integrating the knowledge of the experts are given. In the final section, the main results of the paper are summarized. Throughout the paper, the presented approaches are illustrated by an example taken from the domain of controlling the flow of material in a manufacturing system; in the example, jobs described by different attributes have to be scheduled on resources by human experts. Furthermore, the approaches are compared with other approaches in the literature.

2 Overcoming Terminological Problems

As have been stated in the previous section, different experts use different terminologies for describing a manufacturing system. Generally, among the attributes of such terminologies, *consensus, conflict, correspondence* and *contrast* are possible (see [12]). Consensus holds, if the *same* attributes are assigned to the *same* distinction, conflict, if they are assigned to *different* distinctions. Correspondence holds, if *different attributes* are assigned to the *same* distinction, and, finally, contrast holds, if they are assigned to *different* distinctions.

2.1 The Necessity of a Simulation Model

Usually, such terminological problems between humans are overcome by describing a concept using an agreed terminology or by naming the same concrete object in different terminologies. E.g., an Englishman may describe a pencil as a drawing utensil, suitable for writing on paper that can be erased easily and a German may recognize it as a "Bleistift" based on this description. Or the Englishman may name a certain writing utensil that can be seen in reality by both the Englishman and the German a "pencil" whereas the German name it a "Bleistift".

Overcoming terminological problems by describing a concept in different terminologies requires intelligence and intuition, and is therefore difficult to formalize and to implement in a system (see [11] for details). For an automated overcoming of terminological problems as intended here, the overcoming of terminological problems by naming a concrete object in different terminologies seems easier to realize. When applying the approach to manufacturing systems, however, it reveals its difficulties, because it is only applicable to physical existing objects, e.g. machines. It is not directly applicable to abstract concepts as the *urgency* of a job that can not be seen in real life. In the context of material flow systems, instead of a concrete physical object, a *state of the manufacturing system* forms the basis for the experts. Therefore, it must be possible, that this state can be seen and evaluated by both experts. An opportunity to guarantee this might be, that the experts observe the manufacturing system together, i.e. they see the same concrete state and describe it using their special attributes and values. This easy opportunity can not be used, because the experts are usually not available at the same time at the same location.

An opportunity to overcome these difficulties is the use of a common model using a basic terminology that is understandable to all experts. Using the attributes of this model, a concrete state of the manufacturing system can be described, the description can be presented to each expert, each expert can name this state using his terminology and the resulting descriptions can be compared. Thereto, the model uses detailed data from the manufacturing system, e.g. specifications of jobs or machines, or feedback from the manufacturing system. A formal framework for such a model building is given by the information model for manufacturing process monitoring and control (see [5]), details for applying this modeling approach are described in [11].

It might be argued, if a terminology like the one of the model exists that is understandable to each expert, that this terminology could be used as a communication language of the experts, and there is no reason for using different terminologies. This view should be contradicted here, because of the different advantages of the use of individual terminologies. Firstly, the experts usually use a higher level of abstraction which is based on their individual view of the domain. Therefore, their knowledge is best expressed using their individual terminology (section 3). Secondly, individually represented knowledge is more easily understandable to an expert, so the expert can evaluate and eventually alter it. Thirdly, if an expert is able to express parts of his knowledge explicitly at all, e.g. as rules, it will certainly be easier for the expert to use his "own words". Allowing individual terminologies, such rules can be integrated easily.

2.2 Reducibility and FOIL–Reducibility

In the previous section, the necessity of a common model based on a basic terminology has been motivated. The relationships between the terminology of an expert and the terminology of such a model can be formalized as follows:

Definition 1 (Reducibility). Assume an attribute A of an expert and the set of attributes $\mathcal{M}$ of the model. Assume further a set of operators $\mathcal{V}$ on $\mathcal{M}$. A is called *reducible to* $\mathcal{M}$ *using* $\mathcal{V}$, if there are attributes $M_1, \ldots, M_n$ from $\mathcal{M}$, that can be combined using operators from $\mathcal{V}$ to a formula that can be assigned to the same distinction as A. A is called *reducible to* $\mathcal{M}$, if there is a set of operators $\mathcal{V}$, so that A is reducible to $\mathcal{M}$ using $\mathcal{V}$.

E.g., if the set of operators $\mathcal{V}$ includes subtraction, the attribute *slack_time* is reducible to a set $\mathcal{M}$ = {*work_remaining, time_till_due_date, priority, waiting_time*} of attributes of a model using $\mathcal{V}$, because it holds: *slack_time* = *time_until_due_date* − *work_remaining*, i.e. *slack_time* is reducible to $\mathcal{M}$.

The concept of reducibility is an extension of the concepts consensus, conflict, correspondence, and contrast as defined in [12], because it establishes a relationship between a single attribute on the one hand, and an *expression over various attributes* on the other hand, whereas the concepts defined in [12] only establish relationships between *two attributes*. Besides, these concepts can be easily expressed in the presented formalism (see [11]).

For testing reducibility, FOIL, a tool for *Inductive Logic Programming* is used. Thereto, examples are used that describe a situation using the attribute A on the one hand, and the attributes from $\mathcal{M}$ on the other hand. Inductive Logic Programming and FOIL are discussed in section 3 in more detail. Using FOIL leads to the following definition:

Definition 2 (FOIL–Reducibility). An attribute A is called *FOIL–reducible* to a set of attributes $\mathcal{M}$ using a set of operators $\mathcal{V}$, if FOIL is able to induce a theory with high classification accuracy for the target relation A given the attributes from $\mathcal{M}$ and the operators from $\mathcal{V}$ as background knowledge.

It should be mentioned that it is possible that a theory with a high classification accuracy exists but FOIL fails to induce such a theory; testing for reducibility using FOIL leads therefore to a stronger notion of reducibility.

The approach presented here is more powerful than other approaches in the literature: In the approach presented in [12], *knowledge grids* are used to discover consensus, conflict, correspondence, and contrast in the terminological systems of experts. These knowledge grids have to be filled in by the experts and traded between them, e.g. an expert has to fill in values for attributes used by another expert to describe an agreed set of entities. In the approach presented here, reducibility between an attribute and a set of attributes is discovered; this concepts goes beyond the concepts in [12]. In addition, the approach presented here uses machine learning techniques and is therefore quasi–automatic and involves the experts to a degree that is by far lower than in the approach in [12]. The approach in [3] is more similar to our approach, as it uses Inductive Logic Programming for discovering relationships between terminologies, too. In contrast to our approach, it only discovers correspondences between attributes, though.

2.3 Relating Terminologies

Based on the results of the previous section, for the terminology of each expert a set of rules can be assumed that maps attributes of the model to attributes of the respective terminology. Using these rules, a translation between the terminologies of the experts becomes possible. This translation from terminology $\mathcal{S}$ of an expert S into terminology $\mathcal{T}$ of an expert T is done in two steps: In the first step, a description in terminology $\mathcal{S}$ is transformed into an equivalent description in the terminology $\mathcal{M}$ of the model, and in the second step into an equivalent description in terminology $\mathcal{T}$ (see [11] for details).

The translation step from terminology $\mathcal{S}$ to terminology $\mathcal{M}$ is based on the rules $h_1^S :- b_1^S, \ldots h_k^S :- b_k^S$ that establish the connection between the attributes from $\mathcal{M}$ and $\mathcal{S}$. Thereto, these rules that map attributes from $\mathcal{M}$ to $\mathcal{S}$ are used in an *abductive* way to map attributes from $\mathcal{S}$ to $\mathcal{M}$. E.g., based on a rule $urgency(A, high) :- priority(A, 4),\ work\_remaining(A, B), B > 3$, from *urgency is high*, *priority is 4 and work_remaining is greater than 3* is inferred. This inference is logically not justified, because *priority is 4 and work_remaining is greater 3* is only a sufficient condition for *urgency is high*, but not a necessary one.

Here, the use of abduction can be justified, because it can be assumed that all implications between $\mathcal{M}$ and $\mathcal{S}$ are known, i.e. there is no description other than *priority is 4 and work_remaining is greater than 3* and that implies *urgency is high*. In this way, *priority is 4 and work_remaining is greater than 3* can be regarded as sufficient conditions for *urgency is high*, too. Under this assumption, the abductive inference can be justified.

In the second step, the obtained description in terminology $\mathcal{M}$ is further translated into an equivalent description in terminology $\mathcal{T}$ using the rules $h_1^T :- b_1^T, \ldots, h_l^T :- b_l^T$ that map attributes from $\mathcal{M}$ to attributes from $\mathcal{T}$.

3 Acquiring Knowledge using Machine Learning

After relating the terminologies used by different experts, we are now concerned with the problem of acquiring rules underlying the decisions of the experts. To overcome the aforementioned problems of direct or indirect knowledge acquisition, approaches for machine learning, namely *learning from examples* should be used. In these approaches, the expert does not give his knowledge explicitly, but gives examples of his behavior, i.e. the expert describes situations and chosen decisions in this situations. Machine learning algorithm then try to discover regularities underlying these examples. These algorithms can be divided into approaches for *subsymbolic* and *symbolic* learning (see e.g. [14]).

Approaches for *subsymbolic* learning like *Case–Based Reasoning* or *Neural Nets* represent the knowledge in an uninspectable way. Such approaches can therefore be considered as a *black box*, i.e. either they solve a problem or they do not. If they solve a problem, they do not offer any kind of explanation for this solution. If they fail to solve a problem, they do not offer an explanation either, i.e. they do not offer a starting point for changing the algorithm, the representation of the knowledge or the knowledge itself (see [14]).

For controlling the material flow in manufacturing systems, such approaches are inappropriate, because an *inspectable* representation of the knowledge is needed: In practice, experts learn through their own experience, i.e. they learn by *individual* [16] learning. The possibilities for *learning through communication* [16] are limited, because the experts often work in different shifts, so it is difficult for them to share their knowledge with their co–workers. Therefore it is important that the knowledge of the experts is represented in a way *inspectable for humans*: To support individual learning it is useful that an expert can inspect the knowledge that was acquired from him and gain insight in his behaviour; to support learning through communication, it is useful that an expert can inspect the knowledge acquired from other experts and compare it with his own experiences to profit from the experiences of other experts.

Such inspectable knowledge representations, e.g. decision trees or rules, are induced by *symbolic approaches* for machine learning. These representations can be easily understood by the experts, because they are used to them as they occur often in manufacturing as decision tables or priority rules. Besides, when a solution is determined based on such a tree or such rules, the used part of the tree or the used rule(s) can serve as an explanation for the solution.

To express these rules, *propositional* languages are not sufficient, but *first–order* languages are needed, because the decisions of experts are often based on heuristics like "*if a job a has a higher priority than another job b, then prefer job a*" or "*if a job has waited more than 10 minutes, then always prefer this job*", although the experts are usually unaware of this. To express the first heuristics, the first–order rule IF $priority(a) > priority(b)$ THEN $process(a)$ is sufficient. To express this heuristics in propositional logic however, $n(n-1)/2$ rules are necessary, if the priority of the jobs ranges from 1 to n, because an attribute can only be compared with a value, but not with another attribute. To express the second heuristics, the first–order rule IF $waiting\_time(JobX) > 10$

THEN $process(JobX)$ is sufficient, where $JobX$ is a *variable* which can be bound by every concrete job. To express this rule in propositional logic, m rules are necessary, if m jobs are to be considered, because variables can not be used.

Approaches for inducing theories in first–order logic belong to the field of *Inductive Logic Programming* (see e.g. [6]). Such approaches are able to determine regularities underlying a set of examples of a *target relation* on the basis on background knowledge given as *background relations*, and to express this regularities as a theory in first–order logic. In our approach, FOIL (see e.g. [10]), i.e. the same tool that has been used to determine relationships between terminologies (see section 2.2) is used.

When using Inductive Logic Programming, the choice of the attributes for expressing the examples is always a crucial point, because the quality of the induced theory naturally strongly relates to the examples and the attributes that are used to characterize them. If too many attributes are used, the machine learning algorithm may have difficulties to distinguish between attributes that really correlate with the regularities underlying the examples and those that only seem to do so. The significance of this problem usually decreases with the number of presented examples, however. Severe problems arise, if too few attributes are used, or if the attributes are very detailed, and so the information is distributed over many attributes. In the optimal case, only such attributes would be used, that have a strong correlation with the regularities underlying the examples. Unfortunately, these attributes are not known in the beginning.

In the domain regarded here, when acquiring the knowledge of an expert X, there are three possible sets of attributes: The set of attributes $\mathcal{X}$ used individually by the expert X, the set $\mathcal{M}$ of attributes of the model, or both sets together, that means the set $\mathcal{X} \cup \mathcal{M}$. Only using the attributes from $\mathcal{X}$ might not be sufficient, because it can not be guaranteed that the expert really knows all the attributes that correlate with his decisions and is able to give them explicitly. When using the attributes of $\mathcal{M}$, there should normally no information missing; the set can become quite large, however. A more severe disadvantage is, that the attributes of $\mathcal{M}$ are on a low level of abstraction, whereas the expert uses attributes on a higher level of abstraction and his knowledge could therefore be better expressed on this higher level of abstraction, too. Therefore, the set $\mathcal{X} \cup \mathcal{M}$ should be used, because there is no information missing and attributes on a high level of abstraction are provided that may better explain the decisions of the expert than the attributes of $\mathcal{M}$. This choice has been confirmed by experiments using FOIL 6.3 and different sets of attributes.

4 Analyzing and Integrating Knowledge

Experts do not only use different terminologies, they also usually do not agree in their decisions, i.e. in some situations, experts will react different than other experts. In this section, approaches for *discovering* and *overcoming* of such differences are discussed.

Before analyzing and integrating the knowledge of different experts, the rules of the knowledge bases have to be translated into a common terminology, i.e. in the terminology of the common model (see section 3). The analyzing of the knowledge then consists of determining the situations where conflicts arise, i.e. where the experts draw different conclusions. The problem of determining whether such situations exist or not is called *consistency problem.*

In real world applications, the finiteness of the domains of all attributes can be assumed. E.g., as values for the attribute *waiting_time*, not all real numbers are needed but only numbers restricted by a given lower and upper bound; besides, not any precision is needed. E.g., for *waiting_time*, the domain $\{1, \ldots, 20\}$ can be assumed, i.e. the *waiting_time* can always be assumed as a natural number between 1 and 20 (e.g. hours).

Under the assumption of finiteness of the domain of each attribute, the consistency problem can be shown as *decidable*, but *Co–NP–complete.* Nevertheless, this problem has to be tackled. In our approach, the problem is reduced to a number of satisfiability problems of formal logic. In doing so, it is possible to use standard approaches for solving such satisfiability problems and therewith the consistency problem.

The approaches for integrating the knowledge of different experts can be subdivided into *compiling* and *interpretative ones* (see [2]). In interpretative approaches, the rulebases of the experts stay separate. When in a concrete situation a decision is pending, all experts are asked one after another. If they agree in their decisions, this decision will be chosen, if they do not agree, this conflict is handled using a special strategy. In compiling approaches, the rulebases of the different experts are integrated into one rulebase, comprising all the knowledge.

Both approaches have different advantages and disadvantages: Interpretative approaches preserve the individual knowledge and therefore individual differences, whereas compiling approaches replace individual knowledge by one integrated theory, neglecting individual differences (see also [7]). However, the compiling approach offers a compact representation of the knowledge, easily understandable by the experts, whereas the knowledge in an interpretative approach is distributed over several rulebases. If an individual theory changes, this change has an immediate effect in the interpretative approach, however, whereas in the compiling approach the integrated theory has to be updated what can cause complex problems. As can be seen, both approaches have different advantages and disadvantages and the choice of an approach strongly depends on the particular domain and the application. When it is important to have an integrated theory that can be inspected by the experts, e.g. because the acceptance of the system by the experts would be higher, the compiling approach should be used. If such a theory is not necessary, the interpretative approach should be used because of its computational advantages and because it preserves the individual lines of reasoning of each expert.

In our approach, the knowledge is integrated based on confidence factors that quantify the quality of the rulebases. Thereto, the rulebases of the experts are evaluated in a preprocessing phase by using the rules of the experts in a simulation environment for controlling the flow of material and measuring their performance in regard to given evaluation functions, e.g. the *mean utilization of the resources* or the *mean tardiness of the jobs*. To each rulebase, a *confidence factor* can then be assigned, quantifying its performance. These confidence factors are used to resolve conflicts. In an interpretative approach, when a conflict arises, the decisions of the experts are rated with this confidence factor and the decisions with the highest culminated confidence factor is chosen. In a compiling approach, the knowledge bases are merged e.g. by choosing the rule with the highest confidence factor at a time one after each other, until all situations can be handled. By using such confidence factors, "better" experts have a higher influence on the decisions than "worse" experts. They do not dominate totally, though, because if a lot of the "worse" experts agree on a decision, they can overrule a "better" expert. This takes into account, that even the best expert may fail sometimes and should not be assumed as infallible.

The approaches presented here seem to be similar to those presented in the field of multi-agent learning in [15], [4] or [13]. In these approaches, however, the examples are always taken from one source and are distributed to several learning agents, i.e. each set of examples describes the *same concept*. The conflicts in the resulting theories are then handled using an interpretative ([15]) or a compiling approach ([4], [13]) based on the classification accuracy of the theories in regard to the examples. In the approach presented here, each expert forms his own examples by choosing decisions for situations, i.e. the sets of examples describe *different concepts*. The use of the classification accuracy as a criteria for resolving conflicts is therefore not appropriate here, because the sets of examples describe different concepts and it is senseless to apply rules induced from examples of a concept to examples of another concept. The problem of integrating different concepts also considered in [7], in that approach, however, conflicts are resolved by the user. In the approach presented here, conflicts are resolved automatically in regard to the quality of the rules for the domain they should be used in.

5 Conclusion

The presented approach for acquiring knowledge from several experts for controlling the flow of material in manufacturing systems supports learning of the experts and offers an opportunity for the construction of a corporate memory (see [16]). *Individual learning* is supported by the use of a symbolic approach for machine learning, as it allows an expert to inspect the knowledge that was acquired from him. This inspectable knowledge representation and the possibility of translation between the terminologies of the experts also supports *learning through communication*. This kind of learning is further supported by the presented approach for discovering conflicts in the knowledge of different experts, because it allows an expert to compare his behaviour with the behaviour of other experts. Finally, the integration of the knowledge allows the construction of a *corporate memory* that comprises the knowledge of all experts and serves as a storage device for knowledge available in an enterprise.

References

1. Haldun Aytug, Siddhartha Bhattacharyya, Gary J. Koehler, and L. Snowdon. A Review of Machine Learning in Scheduling. *IEEE Transactions on Engineering Management*, 41(2):165–171, 1994.
2. P. Brazdil, M. Gams, S. Sian, L. Torgo, and W. van der Velde. Learning in Distributed Systems and Multi-Agent Environments. In Yves Kodratoff, editor, *EWSL 91. European Working Session on Learning*, Berlin, Heidelberg, 1991. Springer–Verlag.
3. Pavel Brazdil and Stephen Muggleton. Learning to Relate Terms in a Multiple Agent Environment. In Yves Kodratoff, editor, *EWSL 91. European Working Session on Learning*, Berlin, Heidelberg, 1991. Springer-Verlag.
4. Pavel Brazdil and Luís Torgo. Knowledge Integration and Learning. In B. Wielinga et al., editors, *Current Trends in Knowledge Acquisition*. IOS Press, 1990.
5. Wilhelm Dangelmaier and Karin Geck-Mügge. Information Model for Manufacturing Process Monitoring and Control. ISO TC184/SC4/WG8-P3.
6. Nada Lavrač and Sašo Džeroski. *Inductive Logic Programming. Techniques and Applications*. Ellis Horwood, 1994.
7. Steven R. LeClair. Interactive Learning: A Multiexpert Paradigm for Acquiring New Knowledge. *SIGART Newsletter. Knowledge Acquisition Special Issue*, (108):34–44, 1989.
8. Karen L. Mc Graw and Karan Harbison-Briggs. *Knowledge Acquisition: Principles and Guidelines*. Prentice Hall, Englewood Cliffs, New Jersey, 1989.
9. S.J. Noronha and V.V. Sarma. Knowledge-Based Approaches for Scheduling Problems: A Survey. *IEEE Transactions on Knowledge and Data Engineering*, 3(2):160–171, 1991.
10. John Ross Quinlan and Richard Michael Cameron-Jones. Induction of Logic Programs: FOIL and related Systems. *New Generation Computing*, 13:287–312, 1995.
11. André Schekelmann and Wilhelm Dangelmaier. Agents as Experts: Overcoming Terminological Problems (to appear). In *SCAI 97. Proceedings of the Scandinavian Conference on Artificial Intelligence*. IOS Press, 1997.
12. Mildred L.G. Shaw and Brian R. Gaines. Comparing conceptual structures: consensus, conflict, correspondence and contrast. *Knowledge Acquisition*, 1:341–363, 1989.
13. Riyaz Sikora and Michael J. Shaw. The Evolutionary Model of Group Problem–solving: A Computational Study of Distributed Rule Learning. Technical report, Beckman Institute for Advanced Science and Technology, The University of Illinois at Urbana-Champaign, 1993.
14. Edgar Sommer. An Approach to Quantifying the Quality of Induced Theories. In C. Nedellec, editor, *IJCAI 95 Workshop on Machine Learning and Comprehensibility*, 1995.
15. Luís Torgo. Rule Combination in Inductive Learning. In Pavel Brazdil, editor, *ECML 93. Proceedings of the European Conference on Machine Learning*, Berlin, Heidelberg, 1993. Springer–Verlag.
16. Gertjan van Heist, Rob van der Spek, and Eelco Kruizinga. Organizing Corporate Memories. In *Proceedings of the 10th Knowledge Acquisition for Knowledge–Based Systems Workshop*, 1996.

A Hybrid Information Manufacturing System with an Integrated Rule Knowledge Database

[1]Henry Lau, [2]Felix T.S. Chan, [3]Ralph W.L. Ip

[1]Advanced Technology Education Centre, Regency Institute, Australia
[2]Dept. of Ind. & Manuf. Systems Engg., University of Hong Kong
[3]Dept. of Manuf. Engg. & Engg. Mgt., City University of Hong Kong

Abstract. The concept of a Generic Non Standard-specific Information Flow System (GNSIFS) has been discussed in a previous paper [1]. The main feature of this system is that it can perform data exchange among a group of manufacturing systems regardless of the operating systems, standards being used by them. Further research in this area shows that the database system of GNSIFS should consist of a relational data model and a free format data model and that the capability of this information flow system can be much enhanced if a Rule Knowledge DataBase (RKDB), which contains a large collection of rules and facts, is incorporated with the database system of GNSIFS. This paper describes the methodology for the development of this Hybrid Manufacturing Information System (HMIS) and the test of this system in a real company to evaluate its viability in an industrial environment.

1. Introduction

For a manufacturing company to become world competitive, it must, among other things, have an effective information processing capability. In today's environment, where there are a great variety of computer systems which an even greater variety of standards, the linking of computers to enable electronic data interchange is difficult and costly. This problem is addressed in two previous papers [1,2] with the development of a Generic, Non-standard Specific Information Flow System (GNSIFS) which enables computer users to select data to be shared in a company and store it in a common database from which it can be accessed by any other computer systems in the company (See Fig. 1). The information in the common database, e.g. Bill-Of-Material (BOM), tool data, is stored in a neutral format and the translation to and from this neutral format is done in the background by a Generic Format Translator which converts the formats of files to enable efficient data transfer [2]. It is proposed that the information in the database is carefully selected to minimize memory requirements and speed up access time.

Since the detailed description of this proposed information flow system is covered in previous papers, it is not to be discussed here but a short description follows. The two main features of GNSIFS are:

(a) All the data is converted to a neutral format prior to being sent to the common database. The format conversion, which is a built-in feature of this system, is done transparently. In other words, users do not need to take into consideration the formats being used in different computer systems.

(b) This generic information system rationalises the storage system by selectively storing the relevant text data in the common database e.g. BOM, inventory data. This is cost effective as it greatly reduces the size of the storage system and the speed of access.

Further research shows that the critical constituent of GNSIFS is the DataBase Management System (DBMS) which handles the storage and retrieval of data between the various systems and the common database. Such a database system must be flexible in order to cope with data transfer among different combinations of computing systems.

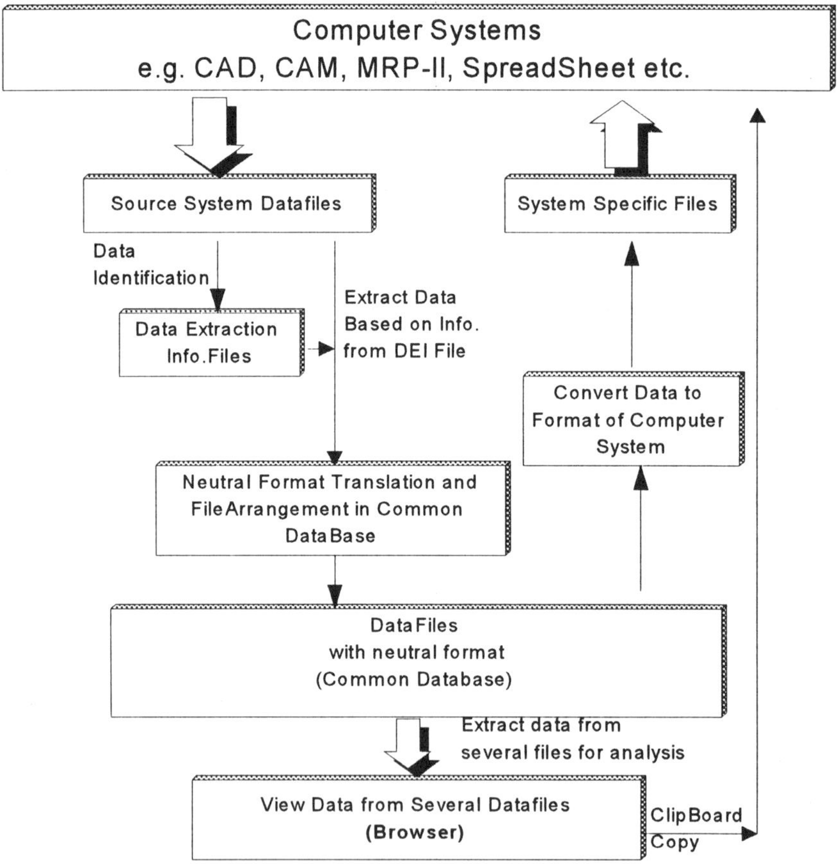

Figure 1 : Schematic Diagram of GNSIFS

Before starting the design of this DBMS, a survey of the existing data models was conducted with the aim to determine the appropriate data models for building the DBMS. The study shows that there are various types of database models including relational, hierarchical, network, free format, multi-user distributed structures and object-oriented data structure [3,4,5,6]. These models are summarised below :

(a) Relational - Relational databases are the most common data models [7,8]. House [9] states the advantages of the relational database as :

  (i) it is well-developed with a large number of supporting tools and publications for those who want to develop a relational data model for special application,
  (ii) it requires only 4 Megabytes RAM memory or more , and
  (iii) it is reliable to handle alphanumeric files.

The relational database files have no related parent files i.e. all files are independent. However, relationships between files can be created, based on common data fields in the files, e.g. Employee Identification Number. For example, there are two files in the database; an Employees File, which records the details of employees such as Name, Address, Position, Employee Identification Number, and a Sales File, which records the details of sales data such as Sale Amount, Sale Description, Employee Identification Number responsible for the sale. When the application program requests the DBMS to look for the sales amount for certain employee with name, e.g. George Smith, the DBMS will get information from the two files i.e. Employees File and Sales File because the Employee Identity Number is in both files.

(b) Hierarchical - A hierarchical database is one where the relationships between the files form a hierarchy where the files are ranked from top to bottom, with higher level files being the parents of lower files. In a hierarchy, a parent file may have several lower-level files under it but the lower-level file may have only one parent file e.g. directories and subdirectories.

(c) Network - A network database is similar to a hierarchical database except that a file can have multiple parents.

(d) Free Format - Free format database is in the form of text files, like the text part of this report. Any character, word, or combination of these can be found changed, deleted, inserted or replaced using the text editor.

(e) Multi-user Distributed Structures - Multi-user, distributed structures databases, as the name implies, can have any of the above structures. The main feature of this model is that they allow the database to be read by many users at the same time. However, only one user is allowed to write into a record at a given time to avoid data inconsistency.

(f) Distributed Data Model - This model is designed to be used in "heterogeneous" environment where data is distributed in different levels such as corporate data in factory level, design data in center level. The data in various levels is separated from the others by a security "wall" but particular data can be interchanged between specified levels provided that there is authorized permission by top management [10]. This model is appropriate for companies where there is a complex data structure and data security at different levels is essential as in a manufacturing company. Although data is stored in various "levels", it is accessible to all and therefore from the user's perspective, this data model is no different from a central database and basically whole files are transferred to the user without any format conversion.

(g) Object-oriented - In an object-oriented database model, an "object" is a representation of a complex data type. Barry [11] points out that DBMS based on object-oriented data model can maintain the integrity of complex entities by providing a way to manage complex data types through "inheritance", "encapsulation", and "polymorphism". However, Khoshafian [12] cautions that object-oriented databases have their own problems. Object-oriented database technologies and products are still in their infancy and lack a consistent, agreed-upon conceptual model. In addition, it requires computers of at least 8 Mbytes RAM to achieve efficient operation [13,7,14,15]

Based on the concept of GNSIFS [1,2], there should be two groups of information in the common database, one group (Group A) dealing with fixed information (fixed data structure) such as costs, overhead rates, process times and machine availability while the other group (Group B) contains information relating to the products. Group B data can be arranged as an alphanumeric component string which contains data such as Bill of Materials, tool data, quality data etc.

Study in this area shows that the capability of this database system can be much enhanced if a Rule Knowledge DataBase (RKDB), which contains rules, facts and the inference engine, is used in conjunction with the traditional DBMS to form a Hybrid Manufacturing Information System (HMIS). This RKDB is of the "dynamic" nature which grows with experience represented by the rule base. As a result, this system allows users to build up knowledge by inputting new data to the knowledge database.

2. Traditional DBMS for GNSIFS

The DBMS is responsible for efficient data exchange of GNSIFS. Such a database system must be flexible in order to cope with data transfer among different combinations of computing systems. To get a better understanding of the development of database management systems and the reasons why a special designed DBMS is necessary for GNSIFS, database systems development was studied in some depth.

The proposed DBMS for GNSIFS is mainly concerned with the efficient transfer of alphanumeric data among computer systems and a relational database model "looks" appropriate for this task (it is shown in the later context that a flexible database system is also required) because :

- alphanumeric data transfer is reliable
- any data fields can be searched for quickly
- relational DBMS development tools are available
- it is well developed data model

Analysis of Database Development Tools on the market indicates that most of the popular development tools are for relational databases. Browning [16] states that the popular relational database development packages on the market include dBase-series, Advanced Revelation, DataEase, KnowledgeMan, Microsoft Access, Microsoft FoxPro,

Paradox, R:Base and SuperBase. Most of these packages have the built-in capabilities to create customised database structures which greatly reduce the time to develop the database management system. The main feature of this package is that the user has to define the lengths of fields as this allows the software to produce records of fixed lengths so that these records can be accessed very quickly with the help of index files. An index file contains the specific fields of the database followed by the record number of the record which contains that specific field. Since the field data of index file is sorted in order, the search for data is fast [17].

As discussed before, there are two groups of information in the common database, one dealing with fixed information (Group A) such as costs, overhead rates, process times and machine availability while the other group (Group B) contains "flexible" information relating to the products, such as Bill of Materials, tool data, quality data etc.

For the fixed information of Group A data, the Relational DataBase Management System (RDBMS) is ideal since the length of fields can be easily assigned, e.g. the scrap rate field should not need more than 2 digits before decimal and 2 digits after decimal. The relational database development tools are used to create the database system for this group. However, the relational database development tools are not appropriate for Group B data for which the length is not easily pre-determined. For example, the tool data for one component can contain 5 to 10 pages of text information while other components may have no tool data. A DBMS for this group must therefore have the flexibility to allow lengths from zero to unlimited characters. Consequently, a Flexible DataBase Management System (FDBMS) with a free format capability is required (Fig. 2). The main limitation of such a free format database is the time to locate a particular field data because the search must be done character by character. However, if the data in Group B is related to components, then, as long as the location of component string is available, the information can still be retrieved relatively quickly.

The features of the FDBMS must include :

- no limit on size of data fields;
- size of data fields to be variable during running time i.e. users can alter the size of data fields when actually running the program;
- number of fields to be variable during running time;
- search of records (component alphanumeric string) is done based on a unique key
  field, normally the component number.

Therefore, it is proposed that the DBMS for GNSIFS should consist of a RDBMS which handles fixed information and a FDBMS which handles a range of components data.

3. Rule Knowledge Database

The GNSIFS will be more complete if it is also incorporated with RKDB which contains mainly rules and facts for expert systems. Expert systems with moderately large number of rules, can be compiled in LISP or Prolog [18] and then held in memory. In addition, the comparatively small number of facts can also be represented as lists (sequences represented in the form of linked cells) in memory. The problem is that these traditional techniques are no longer enough when large collection of data, e.g. a few thousand of rules, are handled. In this respect, an appropriate data model is required to tackle this problem.

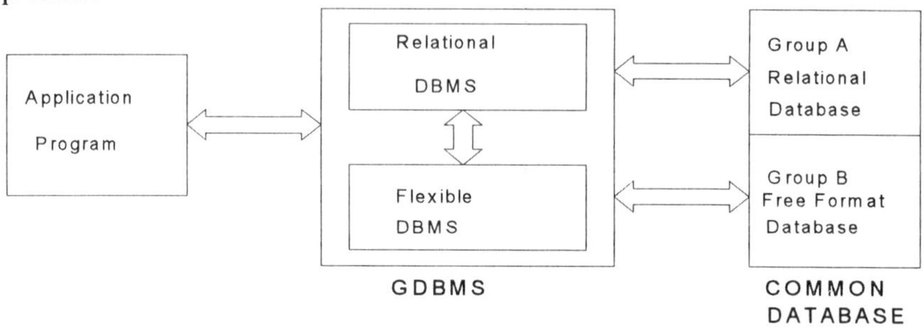

Figure 2 : A Proposed Generic DBMS

A literature survey was conducted to examine the type of data model which can fit in to GNSIFS and is able to manage a large collection of rules and data. It was found that the first attempt was carried out by Jarke and Vassiliou [19] who tried a simple connection to an existing relational database by passing the string representation of an SQL query in one direction and retrieving a set of answer in the other. This technique was basically concerned with the use of relational data model to store the rules and facts. It was simple to implement, but due to the limitation of relational data model, particularly the awkwardness of the standard relational query language Structured Query Language (SQL) and the lack of data-modelling constructs in a relational schema, it was not considered to be a good approach. In the research of new models, Shipman [20] developed the Functional Data Model (FDM), together with the associated Daplex programming language (similar to Access Basic in Microsoft Access). This model is based on the fundamental ideas of modelling entities and relationships developed by Chen [21], which are now adopted in System Analysis. Poulovassilis [22] developed Functional Data Language (FDL), which is based entirely on functional programming and list abstraction but otherwise looks similar to FDM. Another approach was carried out by Tsur and Zaniolo [23] to develop the Logic Data Language {LDL) which handles complex term structures involving function symbols as though they were proper data structure. Another study was carried out by Dittrich[24] and Bancilhon [25] with the aim to explore the features of object-oriented database as the basis for the new data model. Object-Oriented DataBase (OODB) model breaks away from the relational data model and takes seriously the handling of complex data structures and lists, with procedures (can be rules) being stored in the database instead of monolithic user application programs (like the application programs developed by Microsoft Access [26]).

In view of the survey, it is correct to say that a number of data models have been developed or under development for the construction of a RKDB. However, many of these models are still not yet fully developed into a commercially viable product like dBase-series tools, and that means specially-written programs are still required if any of these models, which have been described, is to be used in an industrial or commercial environment. This is not practical particularly for a medium to small size company as the investment outlay for such a task is unpredictable.

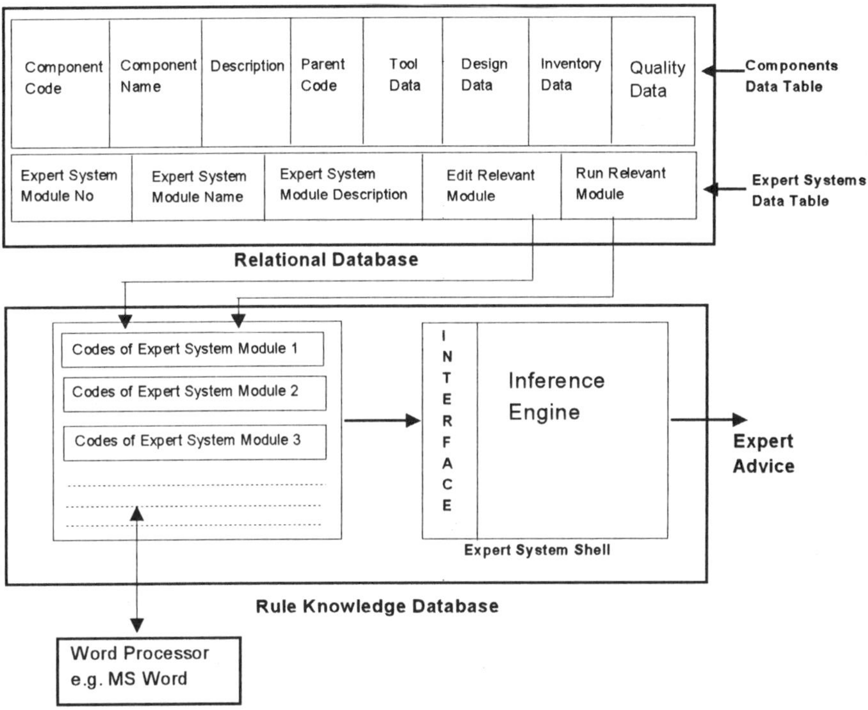

Figure 3 : The Hybrid Manufacturing Information Database System

A Hybrid Manufacturing Information System (HMIS) has been developed as a result of the research in this area, which makes use of the well-developed relational DBMS developing tools in conjunction with a expert system shell, e.g. CLIPS [27,28] or ART-IM [30], to form the framework of this system.

Referring to Fig. 3, the main features of this HMIS are described below :

- The relational database system is used to form a data table with the data fields, e.g. expert system module no, for the expert system modules. As RDBMS is characterised by its efficient searching mechanisms for data stored in data fields of a data table, the data under these data fields can be managed efficiently. With this table, users can locate the required expert system module quickly for further processing. For example, if users want to find the expert system module for the automatic process planning such as in a Computer Aided Process Planning (CAPP) [29], the *search* function can be used to locate the relevant expert system module for this purpose. Relational DBMS tools such as Microsoft Access provides this *search* facility.

- The data field "Edit Relevant Module" contains a macro routine (a sequence of commands of operations) which enables the calling up of the associated word processor (can be configured) and loads the codes (normally in text form) of the relevant expert system module to the word processor program. This enables users to view or edit the source codes of the module. Most commercial RDBMS developing tools have the capability to create this macro function.
- The data field "Run Relevant Module" contains the macro function which enables the loading of the relevant module to the expert system shell. The expert system modules can be associated with the particular expert system shell by specifying the "extension" of the file names. For example, the file "exp_mod1.clp" will associate with the CLIPS expert system shell by identifying the file extension ".clp". Once the module is loaded to the expert system shell, users can run the program to get the required result.

With this technique, the large collection of rules and facts can be stored in separate text files with unique file-names. The relevant file can be loaded to the associated Word Processor or Expert System Shell for further processing. This methodology takes advantage of the well-developed data fields searching engine of RDBMS for handling the large collection of rules which are used in expert system operations. Strictly speaking, this HMIS cannot be considered as a new data model but it can certainly provide a practical and cost-effective solution to the handling of data for expert system routines and because it uses existing development tools and application software to carry out the tasks, no extra programs need to be written. This can be considered as a simple but workable approach although the combination of the relational and the rule knowledge database may be relatively slow to run when the rules to handle are large due to the requirement of steps to call various commands to load files to different software programs.

This HMIS has been tested in a company in Australia to evaluate its viability and results are promising. The following section describes the test of this system.

4. Evaluation of HMIS

In order to prove that the HMIS works, it was tested in a local organisations in Australia. This company operates several advanced machines including :

- a Sintering Laser Station (SLS-2000) rapid prototyping machine, model Sintering Station 2000, which runs on UNIX,
- a laminated object rapid prototyping machine, model LOMM 2030, which runs on Windows NT, and
- a digitbotic, non-contacts, laser scanning, coordinate measuring machine, which also runs on Windows NT.

Rapid prototyping machines can produce a working model directly from a Computer Aided Design (CAD) model in a few hours. The SLS-2000, uses a laser to fuse layers of wax or nylon while in the LOMM layers of self-adhesive paper are laser cut and stuck together to form the object. The Laser Scanning Machine is able to check dimensions accurately and quickly without physical contact with components. Apart from these machines, the company is also running CAD packages and Microsoft Office programs.

In this test, it is required to:

(i) extract "selected" technical data from the three machines electronically,

(ii) transfer the extracted data to Excel for graph plotting and analysis, and

(iii) store the relevant technical data extracted from the three machines together with key dimensions of components, which comes from the CAD system, and data analysis results from Excel, to a common database.

More importantly, this test is to assess the feasibility of the proposed RKDB system which works in conjunction with the relational database. An expert system module has been specifically written for providing advice on actions to be taken when there is some dimensional discrepancy with the components from the SLS-2000 machine. For example, a certain dimension of a component is oversized and expert advice is required regarding the adjustment of process parameters such as set-up time, operation time, which may influence the dimensional outcome of the component. User starts up the relational database table about expert system modules and the correct module is found using the *search* command. When the "Run Relevant Module" macro is activated, the relevant expert system module, which is stored in the RKDB, was retrieved and loaded to the CLIPS expert system shell which runs the program. Through the inference engine, expert advice regarding necessary actions to be taken, e.g. reduce operation pressure, was provided.

The computer systems in the Company run on various operating systems. Table 1 shows the input/output information for these systems.

Table 1 Input/output Information for the Computer Systems

Name of Computer Systems	**Input Information**	**Output Information**	**Operating System**
Sintering Station 2000	Data Analysis Results	Technical Data	UNIX
LOMM 2030	Data Analysis Results	Technical Data	Windows NT
Laser Scanning Machine	Data Analysis Results	Data Analysis Results	Windows NT
ComputerVision Cadds 5	None	Key Dimensions of Components	UNIX
Excel 5.0	Technical Information, Quality Results	Data Analysis Results	Windows 3.x
Dimensional Checking Machine	None	Dimensional Checking Results	DOS

As can be seen from the table the systems have a range of operating systems such as UNIX, DOS/Windows and Windows NT. These are all common operating systems and a commercial network operating system, such as Netware, is able to handle the "networking" of these computer systems. The schematic diagram of the integration of the systems is as shown in Fig. 4.

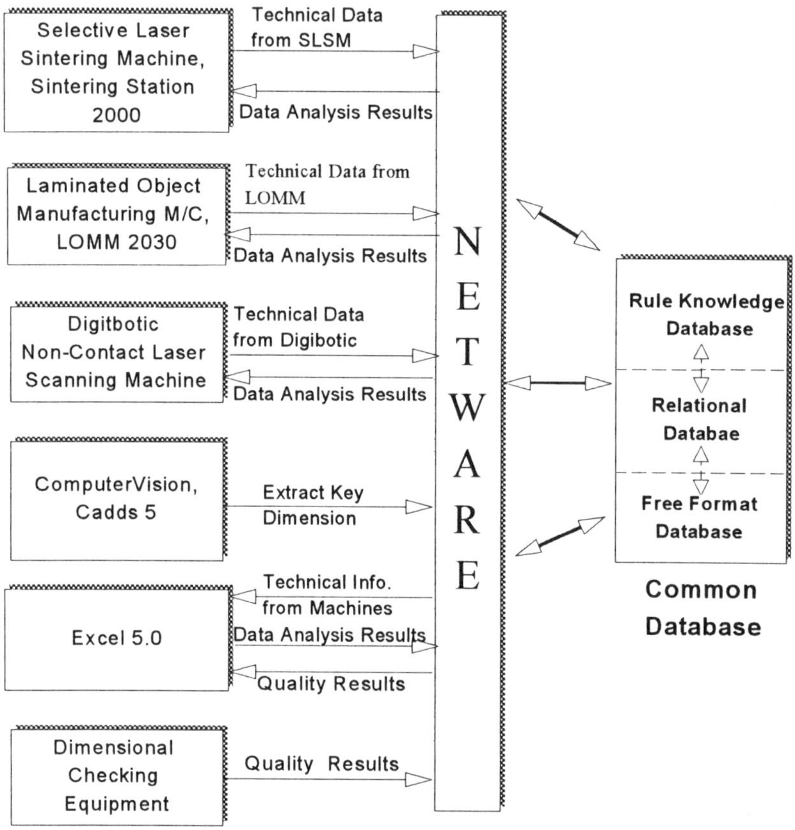

Figure 4 : Integration of Computer Systems in SACFM

5. Results of Evaluation

In general, the test results are quite successfully. Data exchange among various systems was achieved with data from the common database which consists of rule knowledge database, relational database and the free format database. The following is the summary of the test :

- The electronic data transfer between computer systems and common database saves time. As the required data of components is stored in the free format database under different categories, the storage and retrieval of component data can be carried out in a more efficient way.

- Machine utilisation is increased due to the fast data collection and analysis.
- Macro functions from relational database enable the loading of expert system modules, which contain rules and facts, from the RKDB to the associated expert system shell (CLIPS is used in this test) for the inferencing operations. This makes the search of relevant expert system modules much easier.
- Macro functions from relational database also enables the loading of relevant expert system modules to an associated word processor so that editing of codes can be carried out in an easy way under a familiar user-interface environment.

6. Conclusion

The principles of Hybrid Manufacturing Information System (HMIS) has been described in this paper. Instead of developing a new data model for handling the large collection of rules and facts, a methodology using the existing well-developed tools to build the "hybrid" common database which consists of rule knowledge database, relational database and free format database is elaborated. This HMIS can be used in companies without significant change of the existing equipment and software being used in the companies. Popular database developing tools such as Microsoft Access, dBase-series and Oracle can be used to develop HMIS. Familiar Word Processor such as Word, WordPerfect as well as Expert System Shells such as CLIPS, ART-IM can be set up as associated software for the HMIS. Test of this system in a real company indicates that this methodology is workable. Further refinement and review of this system is underway and more trials of this system are carried out to ensure that HMIS is industrially viable and can be used under various manufacturing environments.

References

1. Lau, Henry & Zockel, Fred, A Generic Non Standard-Specific Rationalised Information Flow and Storage System for Computer-based Manufacturing, *Asia Pacific Conference on Industrial Automation*, Singapore, 1994.
2. Lau, Henry, Chan, Felix & Pun , Frankey, Constituents of Generic Information Flow System for Computer-based Manufacturing, *12th International Conference on CAD/CAM, Robotics & Factories of the Future, Technologies* (Proc. PP:426-431), U.K., August, 1996
3. Buddenhagen, David. Designing a Production System Database. *Synergy '86 Conference Proceedings*, California.
4. Coombes, Grahan. 1987. An Integrated Data Management Environment for CIM. *Autofact '87 Conference Proceedings*, Michigan.
5. Dcrus, Kenneth . 1986. Integrated Database: A Necessary Component of CIM. *Synergy '86 Conference Proceedings*, California.
6. Navathe, Shamkant B. 1992. Evolution of data modeling for databases. *Communications of the ACM*, Sept, V35, p.112-124.
7. Glen, Ron. 1992. The great new database debate. *Canadian Datasystems*, April, p.23.

8. Markowitz, V.M. 1994. Safe Referential Integrity and Null Constraint Structures in Relational Databases. *Information systems.*
9. House, Bill. 1991. Comparing the greats (comparing database management systems. *Data Based Advisor*, p.170.
10. Bell, David & Grimson, Jane. 1992. *Distributed Database Systems*. Addison-Wesley Computer Science
11. Barry, John. 1992. The world of object-oriented DBMSs. *DBMS*, June,,v5,p.53.
12. Khoshafian, Setrag. 1991. Modeling with Onject-oriented databases. *AI Expert*, Oct, p.26.
13. Price, Gwyn & Gray, Alec. 1994. *Object Oriented Databases & their Application to Historical Data.* University of Wales College of Cardiff
14. Ahmed, Shamin; Wong, Albert; Sriram, Duvvuru; Logcher, Robert. 1992. Object-oriented database management systems for engineering; a comparison.. *Journal of Object-oriented programming*, June, v5, P.27.
15. Loomis, Mary. 1992. Object versioning (ODBMS). *Journal of Object-oriented programming*, Jan, p.40.
16. Browning, Dave. Relational Databases - New Blood New Power. 1993. *PC Magazine*, May.
17. Simpson, Alan. 1990. *Understanding dBase IV 1.1.* Sybex.
18. Winston, P.H., *Artificial Intelligence*, Addison-Wesley Publishing Company, 1979.
19. Jarke, M nd Vassiliou, Y. 1984, Coupling Expert Systems and Database Management Systems, *Artificial Intelligence Applications for Business*, W. Reitman (ed.), Ablex, pp.65-86.
20. Shipman, D, 1979, The Functional Data Model and the Data Language DAPLEX, Proc. ACM SIGMOD79 Conference (Boston), revised and printed in *ACM Trans. Database Systems* (6), 140-173.
21. Chen P.S., 1976, The Entity-Relational Model - Towards a Unified View of Data, *ACM Trans. on Database Systems* (1), 9-36.
22. Poulovassilis, A, 1988, FDL: An Integration of the Functional Data Model and the Functional Computational Model, Proc. *Sixth British National Conference on Databases* (BNCOD6), W.A. Gray (ed.), Cambridge Univ. Press, pp. 215-236.
23. Tsur. S. and Zaniolo, C., 1986, LDL: A Logic-Based Data Language, Proc. *12th International VLDB Conference* (Tokyo), Morgan Kauffman.
24. Dittrich, K, 1988, *Advances in Object-Oriented Database Systems*, Springer-Verlag.
25. Bancilhon, F, 1988, Object-Oriented Database Systems, Proc. *7th ACM SIGART-SIGMOD-SIGACT Symposium on Principles of Database Systems (Austin),* ACM Publications, pp. 1-11.
26. Microsoft Access, Users Guide, 1996, Microsoft Corporation.
27. CLIPS Reference Manual, Version 6.0. (1993)
28. Giarratano & Riley. (1993). *Expert Systems: Principles and Programming.* International Thompson Publishing.
29. Alting L., Zhang, H., 1989, Computer Aided Process Planning : the State of the Art Survey, *International Journal of Production Research*, Vol. 27, No. 4, pp 553-583.
30. Liao, T., Warren and Chen, L. J., 1993, An Evaluation of ART1 Neural Modes for GT Part Family and Machine Cell Forming, *Journal of Manufacturing Systems*, Vol 12, No 4, pp282-290.

A CORBA Based Information Integration Platform System for CIMS Environments[1]

Ge Yu, Bin Zhang, Guoren Wang, Jin Shi, Xiaochun Yang, Huaiyuan Zheng

Department of Computer Science and Engineering
Northeastern University, Shenyang, 110006 China
{yuge, zhangbin, wanggr, zhenghy}@mail.neu.edu.cn

Abstract This paper presents the design of a new generation information integration platform system for supporting the CIMS integration framework, by exploiting the novel techniques of software bus paradigm. The paper tries to explore deeply the related key techniques and propose new implementation methods, including the database integration mechanism and the integration model based on CORBA specification, the multidatabase query processing, the transaction management and distributed processing on top of object brokers, the object wrapping and adapting techniques, as well as the non-traditional database integration techniques.

1 Introduction

The essence and purpose of a CIM system is the integrated optimization for an enterprise, by integrating the discrete and diverse production elements like people, machines, and resources into an optimizable integrated entity to effectively support whole life cycle of a product development and manufacturing thus to greatly raise the competence capability of the enterprise. The integration in an enterprise includes from information integration, functional integration to application integration, from procedural integration, domain integration to enterprise integration. Generally, the enterprise integration is constructed in terms of an integration framework[3]. From the viewpoint of CIMS, the integration is to integrate all data sources and the applications into a unified framework. The foundation of implementing the enterprise integration and domain integration is the information integration, which resolves the exchange, conversion, and sharing among complex, distributed and heterogeneous data in a CIM system, where some data are conventional (structural) and some data are non-conventional (unstructural), such as multimedia engineering data. An effective approach for in-

[1]This work has been partially supported by the National 863 High-Tech Programs of China with grant 863-511-9611-006.

formation integration is to establish information integration platform, to provide a unified infrastructure for integrating various information sources.

From the end of 1980s to the early of 1990s, many researchers and organizations have been making research on the techniques of information integration framework and integration platform techniques, and many standards and systems have been proposed, for example, DEC Inc.'s FBE (Framework Based Enterprise)[3], IBM Inc.'s CDF (Common Data Facility)[5]. In China, The Expert group of the 863/CIMS Subject also worked out an integration framework architecture and made the goal of information integration.development of an information integration platform relies on multidatabase techniques. A multidatabase integrating system provides users a single data definition and manipulation language to access different independent databases and files, which is realized by integrating related parts of participant databases to provide a global schemas and views. The structure of a multidatabase system consists of three layers: local, communication and global. Local layers transform local data into common formats, global layers provide global user interfaces, data mapping, global execution control and management, and communication layers transfer data/commands between local layers and global layers. Some representative multidatabase systems are Interbase developed at Purdue University[2], Pegasus developed at HP Lab.[1]. We also developed two multidatabase systems – PolyBase[10] and SuperBase[15] to support database integration in CIMS. These systems are able to support integrations of different types of databases like relational, network, hierarchical and object-oriented (OO), and engineering files like CAD. However, with the advance of research on the interoperability in distributed, heterogeneous computing environment and new software techniques like OO analysis methodology, three main weakness emerged in these systems:

- System-oriented, not application- oriented on the implementation of data integration. For example, the data mapping from the local format to the common format is performed by specially-designed translator. Whenever a new database is added into the system, the corresponding translator must be made and added, which would leads to great changes and adjustment in the system. Moreover, global application interface is predefined by the system, and adding a new interface needs an expert to do. These limitations will restrict the flexibility, interconnectivity, and expandability of the system.
- Lack of an uniform specification on integration model and languages. For instance, Pegasus adopts functional object model and functional object language, while PolyBase adopts non-1NF relation model and nested SQL. These would leads to new heterogeneity between different integration systems, and make the integration processing more complex.
- Not support OO paradigms sufficiently. More and more applications tend to adopt OO model and OO design method because OO model has a lot of better features,such as encapsulation and inheritance, than traditional methods. However, previous integration systems can only support traditional database systems like the relational model and operations.

Therefore, it is a very urgent task to develop a new generation information integration platform (system) for CIMS environments. The goal of this paper is try to explore the related critic issues. The rest of the paper is organized as follows. Section 2 introduces the concepts of software bus used for information integration, and the CORBA specifications. Section 3 presents the design ideas of the information integration platform system - SCOPE, and its architecture and construct. Section 4 describes the design of the functional classes with related implementation issues. Finally, Section 5 summarizes the paper.

2 Software Bus and Multidatabase System

In order to support interoperability under heterogeneous, distributed and autonomous computing environments, and to solve information integration problems, the international organization OMG (Object Management Group) made a new specification CORBA (Common Object Request Broker Architecture), and released Version 2.0 in 1994[6]. The main idea of CORBA is the common software bus structure, which has functions similar to a hardware bus in a computer system. By using a software bus, any application module can be plugged into the bus easily like an integrated circuit (we call it software IC) as long as it is complaint with the bus specification and thus different software ICs can be connected and integrated with each other. Although this idea has been proposed for many years, it became available in recent years after advance of the object-oriented techniques and open client/server techniques.

According to the CORBA specification, the core of Software Bus implementation is ORB (Object Request Broker). Not only does ORB support standard OMG object model, but also have functions of distributed process management and communication management. It can replace the corresponding parts (e.g. RPC, sockets) in the traditional client/server architecture and provides good location transparency and distribution transparency. In order to describe plugs and plug-ins on software bus, IDL (Interface Definition Language) is defined in CORBA, which supports encapsulation and isolation among component systems. A component system can be plugged into the ORB software bus as an object by defining and describing interface parameters with IDL, to provide services for other systems and/or send requests to other systems, that is, plug and play style. Many computer makers have provided ORB products, such as DEC's ObjectBroker, HP's ORB Plus, IBM's DSOM, and Iona's Orbix.

In a distributed system, heterogeneity exists at three basic levels. The first is the platform level, where database systems reside on different hardware, use different operating systems and network with different communications protocols. The second is the system level, where data are managed by different database management systems, even file systems, moreover, they might be based on different data models and languages. The third is semantic level. Since different databases or files were designed independently, semantic conflicts among them exist inevitably, which include schema conflicts and data conflicts. CORBA handles the heterogeneity at the platform level and provides an infrastructure for handling heterogeneity in database systems. The semantic interoperability remains to be solved at the application programming level. To do it, a widely accepted standard data model is needed to choose as common data model at first. Fortunately, the international organization ODMG (Object data Manager Group) specified a standard OO model – ODMG-93 for OO databases in 1993[7], which consists of the object model and query language OQL. Now, many famous OO database products, such as O2, ObjectStore, have adopted this standard. Therefore, it is possible to build an object-oriented CIMS information integration platform that supports database interoperability well and conform to the international standard, on the basis of new interoperability and integration mechanism.

To build information integration platform based on CORBA software bus, we regard the following key techniques to be considered:

- Object-oriented integration model and query language implementation technique, including query processing, query optimization, and binding the global execution language to the integrated object methods.
- Object-oriented global transaction management and execution technique. ORB system provides object reference, invoking and binding facility, by which the global transaction management and distributed execution mechanism performs transaction creation, scheduling, execution, committing or abort.
- Object-oriented integration technique, including object wrapping for local data, transformation from local schemas to common model, capturing and refinement of semantics, resolution of semantic conflicts, and management of integration repository .
- Integration architecture and methodology on software bus. Software bus can support distributed object implementation transparency. With the new features, multidatabase integration architecture, functional level dividing, integration schema structure and supporting environment associated with OO development are new research topics.

Recently, people pay more attention on multidatabase systems developing based on CORBA. A successful system is MIND system implemented by METU in Turkey[4], which implements interoperations of relational databases and OO data-bases. However, some critical problems as follows haven't be solved:

- Only supporting interoperations among databases, not providing data semantic integration.
- Not designed for CIMS environments, and no consideration on supporting engineering files even ordinary file systems.
- Not adopting standard OQL language as global interface. There also exist some problems in the implementation of query processing and transaction management.

Therefore, we designed an information integrated platform system which conforms CORBA specification, called SCOPE (System of CIMS Object Platform

Environments). This platform system intends to realizes integration of heterogeneous, distributed and autonomous data source (databases and files), by using the new Software Bus integration paradigm of request/service object mechanism. We have main goals as follows.

- To provide single interface to support various applications on multiple platforms including different hardwares and softwares. In detail, to provide a common API for data sharing and exchanging of relational databases, OO databases and engineering data files in CIMS environments.
- To support a good openness and standardization in order to reduce integration complexity and optimize integration procedure.
- To guarantee reuse of legacy resources (mainly, software and data), with "plug and play" style to decrease costs of system maintenance and evolving.

Above all, in the course of the design and implementation of SCOPE, we will try to resolve key integration techniques based on CORBA Software Bus under complex data resource environments of CIMS.

3 Architecture Design

The role of an information integration platform is to provide data sharing among heterogeneous, distributed, and autonomous component systems for higher-level application integration platform. Thus, we adopt the multi-level federated schema structure[13]. Schemas are classified into local schemas, export schemas, global schemas, and external schemas. A common data model complaint with the ODMG object model is defined to represent export schemas and global schemas. The schemas are generated with 3-layers schema transformation method[11]:

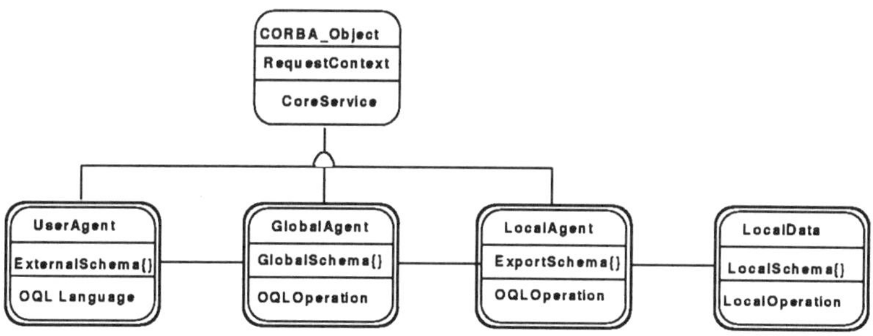

Figure 1: Object Class of SCOPE

- Local schemas (descriptions) are wrapped and transformed to export schemas that use the common data model;
- Related export schemas are integrated to the global schema after the conflicts are resolved;
- External schemas that meet the need of applications are defined on the global schema.

As for the system analysis and design of SCOPE, we adopt OMT (Object Modeling Technique) method, by which, the main system components of SCOPE are abstracted as four object classes: User Agent Object (UAO), Global Agent Object (GAO), Local Agent Object(LAO), and Data Source Object (DSO). The implementation details of each object are encapsulated and only interfaces for services are provided . The interface consists of schemas and operations, defined with the extended IDL language. In OMT diagrams, an object is represented as a box which is divided as three parts. The first part contains the class's name, the second part contains the interface, and the third part contains the attributes. The direct line between objects means communication. Fig.1 shows the structure, interface, and their relationship of the four classes in terms of OMT. Three of them are sub-classes of the CORBA object class. Their interfaces correspond to 3-layers federate schema structure, respectively.

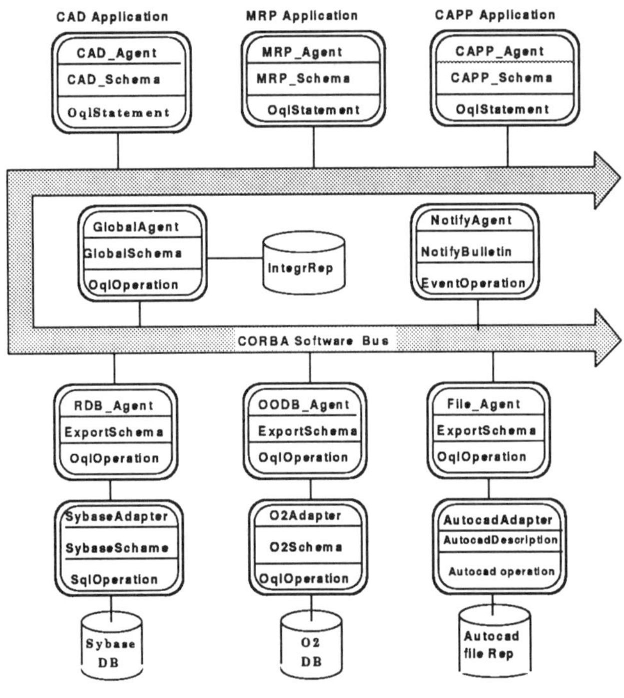

Figure 2: Architecture of SCOPE

Fig.2 shows the architecture of SCOPE, in which three typical data sources (Sybase, O2, and Autocad file) and three typical applications (CAD, MRP and CAPP) are illustrated. In terms of their roles, GAO provides service for UAO, and LAO provides service for GAO. GAO and LAO act as the middleware of the whole system. In addition, we create a Notification Agent Object (NAO) to handle particular events occurring during the system running.

4 System Functional Class

In this section, we describe the design ideas and the implementation issues of major functional object classes in SCOPE. Each class of object is decomposed into several sub-classes, and is designed as a software IC that can be plugged in CORBA software bus directly. These software ICs are used to build the whole platform system with multi-level client/server architecture. Due to the limitation of the space, not all details are given. The interested reader can refer to the related document[8].

4.1 User Agent Object

User Agent Object (UAO) handles API processing, interactive interface processing, and language processing. The definition of the UAO class is shown in Fig.3, where API interface provides useful sub-libraries with some popular language,such as C,C++ to support subroutine calling and binding; Interactive interface (CmdProc) supports user's friendly operations, such as visual interface; and Language Processor (InterfaceProc) performs pre-processing of database's operations to reduce the workload of the global query processing, for example, to do lex, parse, and iteration processing.

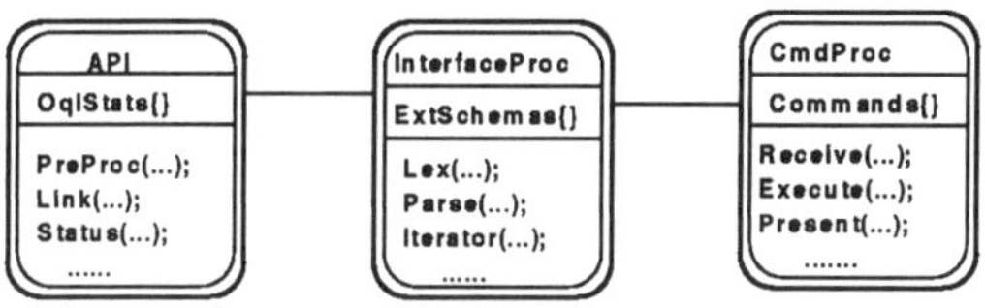

Figure 3: User Agent Object

4.2 Global Agent

Global Agent Object (GAO) runs as a server to perform query processing and optimization, result processing, transaction management, concurrency control, integration processing, and global dictionary management. In particular, the parallel query execution is supported to get high performance. Fig.4 gives the definition of the GAO class.

1. Schema Integrator and Integration Repository

We adopts the ODMG93 OO model as the integration model[7], and define new data integration language – Object Definition Language (ODL) which based on ODMG-93 syntax extended with object integration clauses and integrated object management clauses. A semi-automated integration method is proposed, by using the description assertions of corresponding relationships among export schemas[14]. In detail, schema integration consist of two phases :

- Definition phase : Integrator uses ODL to define all corresponding relationship assertions among the export schemas to be integrated.
- Integration phase: According to the relationship assertions, the integration schemas are generated automatically.

2. Query Processor and Result Processor

We adopt OQL (Object Query Language) as the global language complaint with the ODMG93 specification. When a client sends an OQL query to access integration database, the query is decomposed into several local queries which can be executed in parallel way by using different threads at the same site or different processes at the different sites. By ORB, the sub-queries are delivered to corresponding LAO servers to execute, and local results are returned to GAO to do merging and formatting.

3. Transaction Manager and Lock Manager

Transaction manager (TransactionMGR) is designed to support multidatabase transaction management and distributed execution, by using flexible transaction model and strategy[12], which extends the ordinary 2PC protocol to support flexible atomicity of multidatabase transactions. Lock Manager is designed to support multidatabase concurrency control, by using an extended 2PL protocol, which supports strong database consistency as well as weak database consistency. The Global Log system is maintained to ensure the persistence of the committed global transaction. In addition, the stub-style interface is used to start up local transaction.

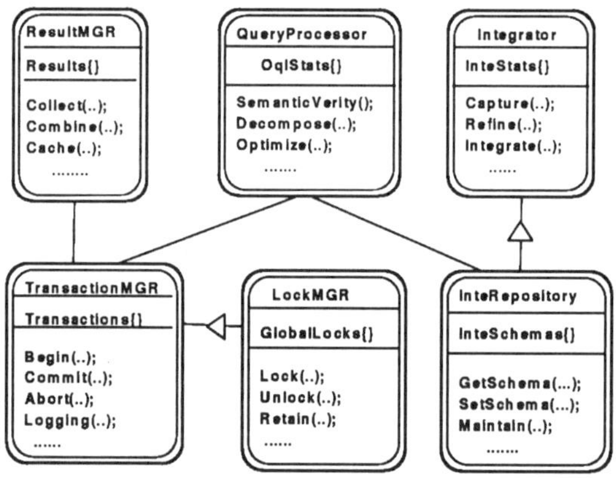

Figure 4: Global Agent Object

4.3 Local Agent Object

Local Agent Object (LAO) runs as a daemon process, which includes QueryExecuter, Object Wrapper, Object Repository and Object Dictionary repository, as shown in Fig.5.

1. Object Wrapper

Object Wrapper performs object wrapping for data from local data sources by the following two steps:

- Export schema generation by transforming local schema (descriptions) into export schemas with the common OO model and save them into Object Repository (ObjRepository). To cope with engineering data files, a special operation mapping technique is exploited[9].

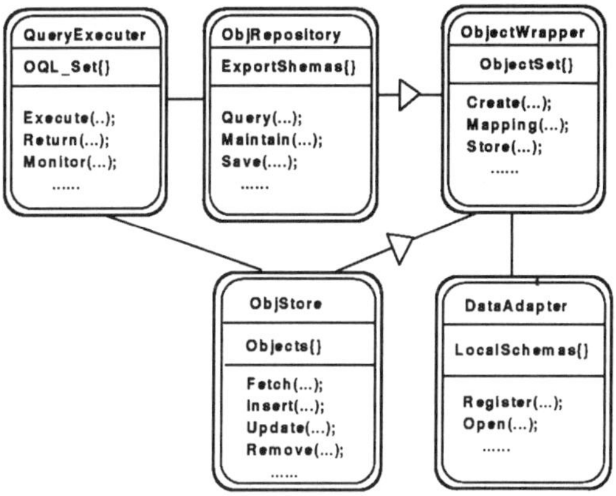

Figure 5: Local Agent Object and Data Adapter

- Object generation by extracting, converting and creating objects for local data and store them into Object Store (ObjStore).

2. Query Executer

Query Executer receives query operations from GAO, performs local query on ObjStore and returns the results to the requesting GAO.

3. DataAdapter

A data adapter is an interface to the local data source that establishes connections with local data source, sends requests to data source and accesses local data. To connect a local data source, it must be registered into the system in advance.

4. Notification Agent Object

To deal with special events occurred in the system such as, schema change, system crash, transaction abort, and alarm, Notification Agent Object is needed to monitor, receive, trigger, and process events, for example, to activate and disactivate an object, detect failures, etc.

5 Conclusions

In this paper, we have presented the design of a CIMS information integration platform system based on CORBA software bus mechanism. It has the main features as follows.

- New multidatabase integration techniques, including the architecture of integration platform, non-traditional database integration, parallel processing and optimization, flexible transaction management and concurrency control.
- Better heterogeneous transparency, distribution transparency, and integration transparency. The implementation environment is based on CIMS factory applications.

- Conform to international standard and trends. In particular, the standardization of the integration model and language is emphasized.

So far, the detailed design have been finished. The prototype system is being developed, and is expected to be finished by the end of 1997. The implementation environment of SCOPE is SUN workstations, Pentium PCs, UNIX, Windows, and Windows/NT OS, DEC's ObjectBroker and CORBA development toolkits, O2 Object-Orient DBMS, Sybase relational DBMS, and Autocad engineering data files. Our future work will include to improve the performance and scalability of the system.

Reference

[1] R.Ahmed,J.Albert, et al., "An overview of Pegasus," *Proc. of RIDE-IMS'93*, pp.273-277.

[2] O.A.Burkres,J.Chen, "InterBase, an execution environment for heterogeneous software system," *IEEE Computer*, 26(8), Aug.1993, pp.57-69.

[3] FBE:Framework based Environment ver.1.0, *DEC Corp.*, Aug.1994.

[4] A.Dogac, "A Multidatabase system implementation on CORBA," *Proc. of RIDE'96*, pp.2-12.

[5] CIM Communication and Data Facility Tech. Manual, *IBM Corp.*, 1992.

[6] T.J.Mowbray, and R.Zahavi, "The essential CORBA: Systems integration using distributed objects," *John Willey,and Sons, Inc.*, 1995.

[7] The Object Database Standard: ODMG-93, edited by R.Cattell, *Morgan Kaufmann Pub.*, 1996.

[8] "Design Document of SCOPE/CIMS," , *Tech. Report, of Database Lab. of NEU*, Shenyang, Jan.1997.

[9] Z.Song, "Studies on unstructural data integration technique," *Matser thesis of NEU*, Jan.1997.

[10] G.Wang, G.Yu, "Query Processing Technique in PolyBase system," *Software Journal*, Vol.7, Beijing, 1996,pp.140-146.

[11] G.Wang, G.Yu, "A data model in CIMS information integrated platform," *Computer Journal* , Vol.1, Mar.1997.

[12] G.Yu, S.Wang, "Flexible transaction management for CIMS multidatabases," *Proc. of the 7th China Advanced Database System Conf.*, Xi'an, Sep.1993, pp.355-362.

[13] G.Yu, H.Zheng, "Research on CIMS information integrated platform multi-level integrated mechanism," *Proc. of The 3rd National CIMS Conf.*, 1994, Wuhan, pp.72-76.

[14] B.Zhang, X.Shi, "Classification and resolving of conflicts in Object-Oriented multidatabase systems," *Computing Researching and Developing*, Vol.2,1997.

[15] B.Zhang, G.Wang, "Design and implementation of a multidatabase integrated system Superbase on CIMS environments," *Journal of CIMS*, Vol.2, Beijing, Feb.1997.

A Specification of CIM Information System*

Luo Junzhou Gu Guanqun Fei Xiang

Department of Computer Science , Southeast University, Nanjing 210096, P. R. China

Abstract. This paper presents a Petri net based formal method for CIM information system modeling and analysis. An extended predicate/transition net is introduced briefly, and the CIM integrated framework, which is composed of a set of high-level net models, is given. Finally, the paper focuses on the process models that are used in the conceptual design and analysis.

1 INTRODUCTION

Computer-integrated manufacturing(CIM) is a new paradigm developed in the 1980s to improve productivity, efficiency and competitiveness of manufacturing systems. To complement the research on manufacturing automation, which has resulted in many so-called 'islands of automation', research now focuses more on tools and methods aimed at tying together locally integrated modules such as CAD, computer-aided process planning(CAPP), FMS, MRP II etc. Current solutions rely on factory analysis methods based on reference models(e.g. GRAI, IDEF0, CIMOSA), structured design methodologies, database design, computer communication support(e.g. MAP and TOP), and data exchange protocols(e.g. PDES, SET, STEP).

The fundamental aims of an information system design methodology is to provide data analysis with a set of representative models, methods, software tools, and validation tools to transform a user-oriented verbal description of a real-world object organization or enterprise into a conceptual model (or theoretical representation), and then into a computerized model, such as the database-oriented description (or computer representation). In CIM environments, it is now crucial that information system design and analysis, as well as database design, are performed in synergy with factory analysis. These design tasks can no longer be performed on the basis of more or less precise user requirements formulation, but must take into account the full reality of the manufacturing enterprise. The purpose is to build fully integrated systems. Therefore, relationships of one part of the organization to other parts must be well understood, and the impact of automating some parts of the company must be evaluated and must remain coherent with the rest of the company.

Nowadays in the research on the architecture of information integration (e.g.

* This work is supported by national natural science foundation of China

CIMOSA), the research models for the design methodology (e.g. IDEF0) and the models and technologies in software engineering, researchers are paying much more attention to the establishment of the functional data models, which is a kind of conceptual modeling, and ignore the importance of behavior models in the information system design. Behavior represents when process elements are performed (e.g. sequencing), as well as aspects of how they are performed through feedback loops, iteration, complex decision-making conditions, entry an exit criteria, and so on.

Gradually, Petri Nets are showing their strong abilities of modeling and analysis in the manufacturing system design because of their special advantages. They are promising tools for describing and studying information processing systems which are characterized as being concurrent, distributed, parallel, asynchronous, non-deterministic and stochastic. Petri nets have several advantages: they capture the precedence relations and structural interactions of unpredictable, concurrent and asynchronous events; deadlocks, conflicts and buffer sizes can be modeled easily and efficiently; Petri nets have a well developed mathematical foundation that the various extensions provide a rich capability of modeling system at different levels of abstraction for system design. Thus, it is the conviction that at the organization and at the conceptual modeling levels Petri nets provide a more suitable framework to analyze dynamic properties of data and information and to implement company policies. Petri Nets can be profitable coupled to existing data modeling approaches to extend their analytical power.

This paper presents an approach to modeling and analyzing CIM information system in terms of Petri nets theory in order to set up a series of high-level integrated models based on Petri nets. In these modes the system integration relationships between logical and physical information accesses can be generated, and the communication and data flow between the manufacturing customers (clients) and the computer resources (servers) can be analyzed. In one word, an integrated framework for CIM information design and analysis will be well set up in the paper.

2 A PETRI NET SYSTEM

A kind of Extended Predicate/Transition Nets, EPr/TN net system, based on standard predicate net, is developed as an adequate tool for the specification and analysis of a distributed system. In this high-level Petri net, the inhibitor arc and the erase arc are added to a common Place/Transition net, and then markings, arc labels and transition inscriptions as well as transition occurrence conditions and transition rules are well defined. EPr/TN net system has great ability to specify the distributed system.

2.1 Net System Definition

A EPr/TN net system, Σ, is a nine-tuple, $\Sigma=(S,T;F,V,A,D,L,I,M_0)$, where

1. S is a place set, and T is a transition set;
2. F is a set of input arcs and output arcs, $F\subseteq S\times T\cup T\times S$; V is a set of inhibitor arcs, $V\subset S\times T$; A is a set of erase arcs, $A\subset S\times T$; $F\cap V=\phi$, $F\cap A=\phi$, $V\cap A=\phi$;

3. $D=\{D_s$ is the domain of a place $\mid s \in S\}$;
4. $L=\{K(f)$ is the label of a arc $f \mid f \in F \cup V \cup A\}$;
5. $I=\{P(t)$ is the predicate in the bound variable of a transition $t \mid t \in T\}$;
6. $M=\{M_s$ is the marking of a place $s \mid s \in S\}$, M_0 is a initial marking.

2.2. Transition Rule

A transition t is enabled if $\Sigma=(S,T;F,V,A,D,L,I,M_0)$ is a EPr/TN net system, and under a marking M and a permissible interpretation ∂t of its bound variables the following holds:

1. $M \geqslant M^{\partial t}$;
2. $\partial t(P(t))=$ true if t contains a transition inscription P(t); and
3. $\forall s \in {}^{V}t \wedge \forall y \in N_s^{\partial t} \Rightarrow M_s(y)=0$, where $N_s^{\partial t}$ is the inhibitor set of a place s.

A transition under a marking M which is enabled by a permissible interpretation ∂t, generates through its occurrence a successor marking M′ given by the following:

1. for $s \in {}^{\bullet}t - t^{\bullet}$, $M'_s = M_s - \partial t(K(s,t))$
2. for $s \in {}^{\bullet}t \cap t^{\bullet}$, $M'_s = M_s - \partial t(K(s,t)) + \partial t(K(t,s))$
3. for $s \in {}^{A}t - t^{\bullet}$, if $y \in D_s - N_s^{\partial t}$ then $M'_s(y) = M_s(y)$ else $M'_s(y)=0$
4. for $s \in {}^{A}t \cap t^{\bullet}$, if $y \in D_s - N_s^{\partial t}$ then $M'_s(y) = M_s(y) + [\partial t(K(t,s))](y)$ else $M'_s(y) = [\partial t(K(t,s))](y)$
5. for $s \in t^{\bullet} - ({}^{\bullet}t \cup {}^{A}t)$, $M'_s = M_s + \partial t(K(t,s))$
6. for $s \in S - ({}^{\bullet}t \cup {}^{A}t \cup t^{\bullet})$, $M'_s = M_s$

3 INFORMATION SYSTEM FRAMEWORKS AND MODELS

The information system of a manufacturing enterprise is made of all pieces of data and information used/stored/processed for the needs of users and applications of the enterprise. The information system is the memory of the manufacturing system, containing its valuable data, know-how and expertise. It is made of data and information as well as constraints and rules. It is usually computerized in the form of one or more databases.

Manufacturing environments are characterized by the fact that all kinds of data and information are present, from static to very dynamic types of information, from simple to highly structured types of information, from very short to long duration types of transaction processing. Major types of information in manufacturing are from product, process, resource, planning, production and administrative. Manufacturing environments are also characterized by typical functional areas which can be the sources of structural, declarative or procedural knowledge for the information system. Some common application systems are from CAD/CAE, Process Planning(CAPP), Production Planning(MRP II), Shop-Floor Control(SFC), and so on.

In order to design and analyze CIM information system, to completely describe database and to support the CIMS implementation, a structured approach, based on the modeling and analysis architecture presented by F. Vernadat, can be used for the establishment of CIM information system by using the Petri net system given in the above section. The methodology is structured into the following three major phases.

3.1 Organization Analysis

This phase is concerned with functional analysis of the organization (or a part of the enterprise) in terms of objects, functions and rules (or policies) of the enterprise, based on the analysis of business objectives and constraints as defined by users. It is usually based on high-level models and always involves a functional decomposition principle. The main objective is to build an organization schema defining the organization functionality, the enterprise entities and the organization behavior, which can be represented by some certain Petri nets.

An organization network can be used in Petri net models as a description of an organization process, which is oriented at specifying the modifications produced in the context in which its component functions are executed, the relationships between its component functions, and communications with other networks. An organization network is then a set of functions and object flows, and precedence relationships between them; the component functions (represented by transitions) can be activated only when their input objects (represented by places) are present and when their occurrences generate the output objects.

Basically, the organization analysis is aimed at describing the enterprise structure according to a top-down approach from three complementary viewpoints: functional modeling (describing the organizational structure), information modeling (describing the information system), and dynamic modeling (describing the organizational behavior).

3.2 Conceptual Design

This phase is concerned with the detailed design of the information model which will have to be implemented to support the activities of the physical system and the decision system of the organization. It is concerned with modeling and analysis of the static and dynamic properties of the information system using semantic or object-oriented data models. This is achieved on the basis of data processing requirements as expressed by users or applications during organization analysis. The objective is to build a conceptual schema of the information system, expressed in an implementation independent, computer processable, language subject to formal verifications and consistency checking.

Dynamic features concern both the operations on the data and the causal relationships existing between the organization operations. They are expressed in the form of process nets, which represent the detailed logic of application programs or transactions implementing enterprise activity execution. Manipulations of data are specified by means of a transition definition language.

The conceptual model is aimed at supporting the conceptual design phase, which can be a bottom-up approach for designing the enterprise database system of the CIM system. This phase involves four main tasks: forms analysis, user requirements analysis, database view design and view integration. Therefore, the conceptual model can be composed of a data description model, a process description and a operation model. It is structurally related to the previous organization model but concentrates on information object modeling and processing at a very detailed level.

3.3 Implementation Design

This phase is concerned with the implementation description of the internal schema of the information system to be automated. The internal schema is made of the following 2 points: (1) an implemented information model, and (2) implemented database transactions performing units of processing on data of the implemented information model. Database transactions can be used by application programs or users of the information system. The implemented information model can be made of one or several physical database schemata (data structure, indexes, integrity constraints and access rights) to be implemented on commercial database management systems(DBMSs) used by the manufacturing system. Thus, one objective of this task is to transform the conceptual schema of the previous phase into one or several implementation data models (using the relational model, the hierarchical model, the network model, object-oriented models or even flat-files).

Three languages of data definition language, data manipulation language and process definition language can be defined in the phase in order to complete implementation design.

4 A TYPICAL EXAMPLE

A Flexible Manufacturing Cell(FMC) is composed of a number of machine tools (including NC machines, robots, measuring machines etc.), some clamping devices and pallets, cutting and measuring tools, and devices. The different kinds of parts can be produced in terms of a given working flow. A process plan indicates the sequence of operations to be performed for a certain part as well as the machines and tools to be used. Suppose that there are some main devices such as machine tools, fixtures and cutting and measuring tools in the FMC, in which by means of fixture a part must be mounted on a pallet to be processed . The sequence of machining operations to be applied to a part is given in the form of a process plan.

The supposed process plan can be specified in the EPr/TN net system, in where places indicate the acted object classes such as parts, machine tools (machines, for short), fixtures, and cutting and measuring tools (tools, for short), and transition indicates the action classes such as process operations. The predicate/transition net system describes the process of the action (function) execution in details. The net also describe the effect on the related objects to be produced at each execution step and the implementation process in each action of the action class. An action is an instance of the action class. Figure 1 gives a process net for the FMC process operation, which is simplified EPr/TN net system, a five-tuple, Σ=(O,T,Tg,IN,OUT).

O (Object) is a set of object classes to be operated in the process. T (Transition) is a set of state transitions such as database transactions and basic mechanical operations, which make system state change. The database manipulation can be expressed by SQL-like language, and the mechanical or manual operations can be expressed by natural language. Tg (Trigger) is a trigger of the state transition with a

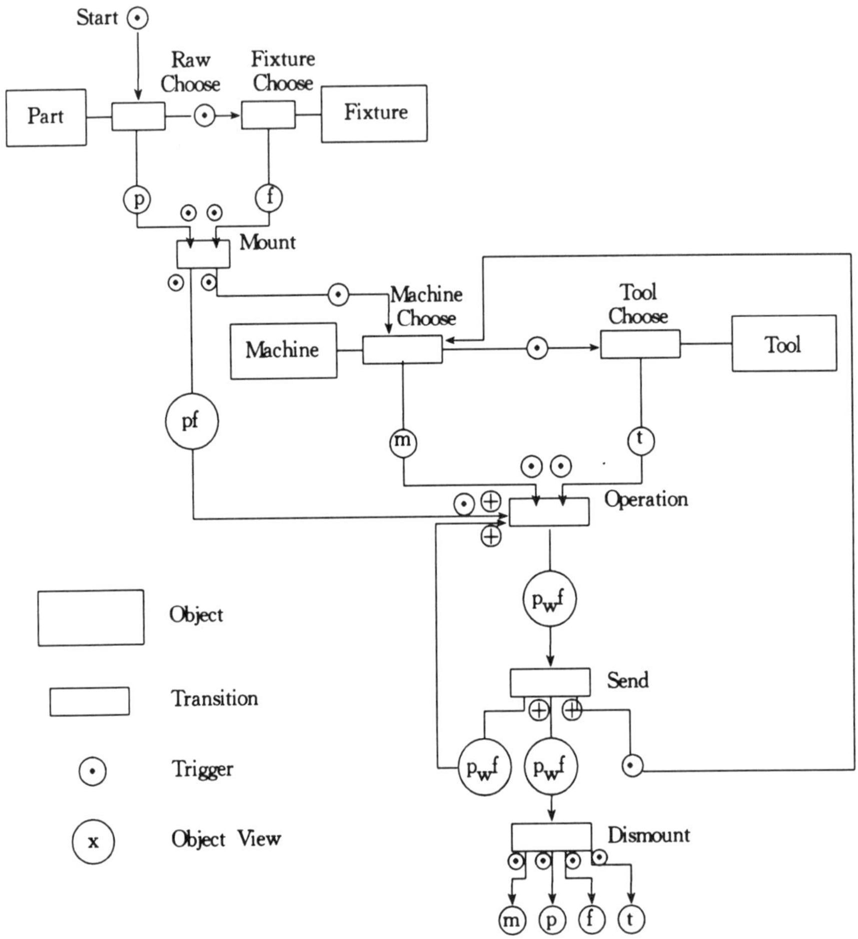

Fig. 1. Process Net Model for FMC

command or request, which controls the transition execution. IN (Input) is the transition input of object instances. (p) $\in$Part, (f) $\in$Fixture, (m) $\in$Machine, (t) $\in$Tool. (pf) indicates the mounted part and fixture, and (p_wf) indicates the worked (processed) part. When a transition has several inputs, (·) on the arc indicates 'AND' of the multi-inputs and enables the transition, and (+) on the arc indicates 'XOR' of the multi-inputs and enables the transition. OUT (OUTput) is the transition output of object instances. When a transition has several outputs, (·) on the arc indicates 'AND' of the multi-objects, and (+) on the arc indicates 'XOR' of the multi-objects.

The basic database manipulations or the basic mechanical operations, which correspond to state transitions in figure 1, can be given as follows.

Raw Choose:

Select raw material from Part. DBF where <predication>;
Update the Part. DBF —— the selected raw material state as 'ready';
Take the selected raw material (p) out from the storage;
Fire the trigger to start Fixture Choose.

Fixture Choose:

Select fixture from Fixture. DBF where <predication>;
Update the Fixture. DBF —— the selected fixture state as 'ready';
Take the selected fixture (f) out from the storage;

Mount:

If (p) $\wedge$ (f), mount (pf);
Update Part. DBF —— the (p) state as 'used';
Update Fixture. DBF —— the (f) state as 'used';
Fire the trigger to start Machine Choose.

Machine Choose and Tool Choose are much similar to Part Choose and Fixture Choose.

Operation:

If (pf) $\wedge$ (m) $\wedge$ (t), machining as the operation step;
Update (p) state as 'worked';

Send:

If there are other operation steps in the process plan, return back to the transition 'Operation', otherwise send (pwf) to the next transition.

Dismount:

Update Part. DBF (because of (p) state as product) to change the quantity of this raw material as (quantity-1), and add one to the quantity of the product;
Update Fixture. DBF, Machine. DBF and Tool. DBF to change the selected fixture, machine tool and tool state as 'free', respectively.

The process net defines a specific process, including process execution, data collection and analysis and process control. Process is a combination of the ordered steps, in which each step is called as process element. The process element is an atomic action and can not be divided into smaller one.

The process model is used to specify the dynamics of the data processing system being analyzed, i.e. to specify how the transactions and the views are related to one another and used in a given application system. In fact the process model is used to specify at the conceptual design level the internal behavior of activities and functions defined at the organization analysis level. The model can also serve as a validation tool of data schemata obtained by simulating the execution of the transactions in their application context.

5 CONCLUSIONS

The purpose of this paper is to illustrate the very rough use of Petri Nets for the design, analysis and formal specification of various aspects of manufacturing information systems. Obviously, Petri nets, and especially high-level Petri nets, are efficient models to represent and analyze company policies and the behavior of information systems in addition to classical data models used to represent the static properties of information systems.

The method presented covers the analysis of an organization, the modeling of its information system, and the design of the database environment which will have to support the operation of the organizational activities. There are three major advantages of the method. The first is that high-level Petri nets are completely used as the formal techniques for establishing models of the CIM information system, which become an integrated model. The second is the layered approach (organization level, conceptual level and implementation level), which provides various categories of users (managers, engineers, data analysts and programmers) with appropriate tools for their modeling and analysis needs. The last is the use of computer-aided tools, by means of which the modeling and implementation can completed rather fast, and the specific models of a certain enterprise can be set up and the simulating behavior can be represented in the computer.

REFERENCES

1. F. DiCesare, et al., *Practice of Petri Nets in Manufacturing*, England: Chapman&Hall, 1993
2. F. Vernadat, A. Di Leva, P. Giolito, Organization and Information System Design of Manufacturing Environments: the New M* Approach, *Computer-Integrated Manufacturing Systems*, Vol. 2, No. 2, May 1989
3. B. Curtis, M. I. Kellner, J. Over, Process Modeling, Communication of the ACM, Vol. 35, No. 9, September 1992
4. Gu Guanqun, Luo Junzhou, EPr/TN Net System and the Formal Description Technique of Network Protocols, *Journal of Computer* (in Chinese), Vol. 17 Supplement, 1994
5. S. Joshi, A. A. Desrochers, Performance Analysis of Network and Database Transactions Flow in a CIM System, *Computer-Integrated Manufacturing Systems*, Vol.5, No.2, May 1992

Internet and the WWW

Internet for Higher Productivity in Manufacturing

Syed M Rahman, Baikunth Nath and Ruhul Sarker

Gippsland School of Computing and Information Technology
Monash University, Churchill, VIC, Australia 3842
{syed.rahman,Baikunth.Nath, Ruhul.Sarker}@fcit.monash.edu.au

Abstract. *Internet is proving itself as a powerful media for business communications. Internet, incorporating computers and multimedia, has provided tremendous potential for remote integration and collaboration in business and manufacturing applications. In this paper the Internet functionality related to applications in manufacturing environment is discussed, critically analysing the limitations and issues of existing Internet technology with proposals to overcome them are presented.*

1 Introduction

The Internet appears as a shimmering ocean of digital information. Professional journals, literary classics, national news services, stock reports, virtual communities, image databases, and digital libraries are just some of the wonders awaiting the uninitiated. Graphical browsers provide intuitive point-and-click access. Powerful search tools combine with intelligent software agents to put a global network of information at fingertips. In the past few years, a diverse group of tools for resource discovery on the Internet has become available. Each tool serves a slightly different purpose. These tools carry the searcher into the brave new information society for tomorrow.

The use of computers in manufacturing organisations has been proved essential for improving productivity and quality. Computers have been employed in almost all aspects of manufacturing organisations starting with marketing to delivery of goods to customers, including after sales services. A number of new technologies and planning processes such as computer-aided design (CAD), computer-aided manufacturing (CAM), computer-aided engineering (CAE), computer-aided process planning (CAPP), computer integrated manufacturing (CIM), flexible manufacturing systems (FMS), manufacturing resource planning (MRP-II) and computer numerical control (CNC) machines, etc. are based on computers. Information systems have an important role in enterprise integration and management strategies for improving productivity and quality. With the increasing role of information technology in manufacturing, the areas suitable for LAN and WAN use needs to be investigated very carefully taking into account the operational characteristics of manufacturing. There is potential for Internet use, combined with multimedia, in manufacturing applications such as CAD, CAM, CAPP, CAE, MRP and training.

Multimedia works well in the computerised manufacturing environment, mainly because it is cost-effective, convenient and the results can be measured (Gunasekaran et al, 1996). There is a need for the enterprise integration due to the fact that modern manufacturing operations are more decentralised in nature. This may be easily achieved using multimedia capable computer network, especially in the process of integrating various functional areas, such as marketing, design and planning, production and distribution within the enterprise.

With the advancement of production automation, the involvement of digital computers in the design and manufacture of products is increasing rapidly. The use of computers in design and manufacturing constitutes the most significant opportunity for substantial productivity gains in industry today. CAD and CAM play an important role in this context. Computer-aided design (CAD) can be described as any design activity that involves the effective use of the computer to create or modify an engineering design. There are two basic reasons for using the computer in the design of a product: (i) to increase the productivity of the designer and (ii) to create a data base for manufacturing. The product designer uses the interactive graphics system to establish the geometry, dimensions, and tolerances for the various parts. These same design data can be displayed for the process planner to use in preparing/generating the NC part program. Some CAD/CAM systems simulate the tool path on the monitor from which necessary changes or modifications can be decided. The NC part programs direct the CNC machine to fabricate a product (Groover, 1987). The establishment of CAD and CAM facilities in one organisation, it involves huge amount of capital investment to purchase CNC machines, computers and relevant software and subsequently high running cost is also required. To operate and maintain these facilities, it requires:

- expert designer, NC part programmer and CNC machine operator.
- technically sound management and qualified maintenance team.

It is quite implied that substantial amount of money in maintenance and operation is also involved. Though there are many advantages of computer integrated manufacturing, but it is evident that the unit cost will be much higher in this system, if the capacity is not properly utilised. To achieve proper utilisation of the facilities, sufficient and continuous demand from the customer is a necessary condition. In such situations the Internet can help both users/retailers and manufacturer, even without any knowledge of the facility types and locations by the users. Based on their long term plan, external demand and cost-benefit analysis the manufacturer may have the option to purchase the facility and establish it in their own site or fabricate from elsewhere using Internet. It's like a make or buy decision. In addition, the users and manufacturing companies wishing to fabricate their products using distant facilities may also search for the fabrication service providers through the Internet according to their requirements. Considering the unit cost and capacity utilisation in computer integrated manufacturing, an attempt has been made to study the uses of Internet in manufacturing and the scope for future research. In this paper, the application of Internet in design and fabrication using CAD/ CAM will be considered. It is expected that the use of Internet in manufacturing will increase the productivity of this sector.

The organisation of the paper is as follows: Section 2 presents the concepts of multimedia. The possible application areas in manufacturing are described in section 3. Section 4 discusses the prior research on Internet in manufacturing. Section 5 provides the possible applications of Internet in manufacturing. The scope of future research is discussed in section 6 and section 7 concludes the paper.

2 Multimedia

Multimedia has different meanings to different people and businesses. The most common definition is the merging of graphics, text, sound, and video in various combinations with computer and with which the user can often interact. The category of graphics includes art, photos, and animation; sound includes music, speech, and sound effects. Burger (1995) made a strong statement about multimedia in his book that "multimedia is not about technology, but about effective communication". This is true for most of the users of multimedia.

At first glance, the primary benefit of multimedia is the combination of several traditional media. Text offers clarity and self-pacing. The spoken word can be even more accessible. Graphics provide visualisation and communicate style. Music imparts mood and character. Video captures the moving events of the world around us. Computers store and access vast amount of information. Most of the additional benefits of multimedia stem from the digital nature of most related media. Digital information suffers no generation loss when replicated or transmitted. It is simple to edit, copy and create. Random access capability of digital systems helps in instantaneous locating of any point or parts and making the systems interactive in nature.

Till today, the application areas of multimedia are business and industry, government, education, medicine and nursing, encyclopedic resources, entertainment and communication. Business applications include sales/ marketing presentations, trade-show productions, employee training, direct marketing, retail vending, and point-of-sale information. Education is an extremely powerful application for multimedia. The value of supplementing dry textbooks with film and video has been recognised for years. Multimedia not only offers the ability to integrate this audio-visual experience directly with text, but adds the power of interactivity. Students can learn at their own pace. Multimedia extends education into the home.

3 Possible Applications in Manufacturing

Some functional areas of a manufacturing organisation which can be considered for Internet uses are shown in table 1.

Table 1. Functional areas in manufacturing organisation for use of Internet

Functional Areas	*Possible Areas for Internet Use*
Marketing	Sales, customer service, advertising.
Engineering Design and Planning	CAD, CAPP, CAE, MRP, PPC, Scheduling
Manufacturing	CAM, CIM, FMS
Inventory Control and Warehousing	ASRS, Stock Control, Purchasing
Maintenance	Scheduling, Spare Provisioning
Personnel	Training, Demonstration

Marketing requires a very general interactive information system that improves the communication between the company and customers. For example, the Internet supported by multimedia can be used to support advertising, sales, customer service and feedback, in a way that also improves the image of the company amongst its customers. In production subsystems, computerised and automated operations require fast retrieval of information from different sources and processing of it on-line for the purpose of making decisions interactively about scheduling, routing, batch sizes, etc.

It is necessary to study the media requirements for various manufacturing activities for making effective use of the Internet. Different operational aspects in manufacturing require different combinations of media (like text, graphics, audio, video, etc.). These requirements are shown in table 2. For example, graphics and text may be used in forecasting. Considering that a production manager would like to decide the schedule for a future period, then the demand pattern, using graphics & text, for that period could be extracted from the database by connecting to marketing that may be situated at a distant place. Similarly, the purchasing section can retrieve the required information (images & text) from marketing so that it can organise its purchasing activities taking into account the lead time involved. This implies that any department can access marketing information in order to have relevant inputs for their operational planning and activities. People in different parts of an organisation that are geographically separated may need to access information from other areas and their use of this information can be enhanced with the help of LAN or Internet. This sort of open communication will not only improve their understanding of the operation of the whole system, but also generates a sense of pride, a greater commitment to team work and encourages co-operation with people in different functional areas. This can lead to increased productivity and quality within the organisation.

Table 2. Possible media requirements for manufacturing operations

Operational Aspects	***Possible Media Requirements***
Sales Forecasting	text, graphics
Product Marketing	text, graphics, video, audio, images
Raw material Purchasing	text, images, audio, video
Production Planning	text, graphics, video, audio, images
Production Control and Scheduling	text, graphics, video, animation, images
Supply Chain Management	text, audio, video, graphics
Stock Control	text, graphics, video
Facility Planning and Design	text, graphics, video, animation, images
Inspection & Quality Assurance	text, audio, video, graphics
Process Control	text, graphics, video
Training and Demonstration	text, graphics, audio, video, animation, images
Maintenance	text, graphics, video, animation

4 Internet in Manufacturing: Prior Research

The researchers are trying to take the advantages of Internet, supported by multimedia, in manufacturing. However, very few attempts have been made in this case. Whang et al (1996) present the concepts of an Internet assisted manufacturing system for agile manufacturing practice. It consists of a CAD/CAPP/CAM/CAA integrated Central Network Server (CNS) which links to local FMS or CNC machines by means of cable connections. After a local user inputs the product information, the CNS can generate complete CAD/CAPP/CAM/CAA files and control the remote FMS or CNC machines to accomplish the whole production process. This system uses

Internet as an interface between a user and the CNS and allows a local user to operate remote machines connected to the Internet.

The ARPA Manufacturing Automation and Design Engineering (MADE) program has been developing Internet-based tools, services, protocols and design methodologies that would allow contractors to compose teams of specialists from different locations and organisations as project needs arise (Cutkosky et al, 1996). As a practical test of what the MADE program has achieved, members of the MADE community undertook an ambitious exercise in geographically distributed design and prototyping called Madefast. The exercise tested current MADE technology, identified future research directions and established a network of tools, services, documentation and contacts for the MADE community. Madefast is an early example of a new and rapidly growing genre of projects that use the World Wide Web (WWW) extensively for collaborating and archiving results. Accordingly, the WWW is the best primary resource for documenting the project.

In today's global marketplace, industries try to compete with cost-effective products engineered using emerging techniques, which are highly flexible and rely heavily on computers and information interchange among them. The active involvement of various automation technologies in manufacturing has had a positive effect, but has also led to the creation of various incompatibilities and given rise to situations commonly known as 'islands of automation'. The implementation of MAP (manufacturing automation protocol) and TOP (technical and office protocol) has started to bridge the communication gap, but there still exists an 'island of automation' between the controllers and host computers. Jagdale and Merchant (1996) presented a direction towards connecting these different islands of automation. The objective is to create a generic Internet-based utility for hooking up host computers, which handle high-level decisions and dedicated process controllers. This gives the controllers various facilities to interact, query, report and co-ordinate activities with computers located anywhere on the Internet.

The World Wide Web (WWW) is increasingly being used in companies. Currently the WWW provides new potential concerning the integration of external partners into the product development. New concepts can be realised by web tools especially regarding the integrated data and process management. Reinhart and Brandner(1996) introduces some possible solutions of such situations.

The US administration envisions the information super highway as a seamless web of communications networks, computers, databases, and consumer electronics that will put vast amount of information at users' fingertips. A solid grasp of the fundamentals would allow US manufacturing companies to fully appreciate and utilise the information super highway to broaden their global competitive advantage. Liu (1995) provides a conceptual framework of data communication and networking for understanding the information super highway.

The concept of computer supported collaborative work (CSCW) is achievable using Internet as a very potential tool. Since the 1980s, the need in developing the CSCW environment has been recognised by various researchers. However, most of them focused on the technical citing, text editing, and electronic mail message transmission, i.e. two dimensional graphics and text. Use of media on demand multimedia electronic mail as a tool of collaboration on the web has been proposed by

Tsoi and Rahman (1996). To enable the CSCW to accommodate 3D CAD/CAM, Lin et al (1995) have explored an international collaborative environment that links geographically distributed locations: Taiwan and Australia. They used two examples to test the proposed environment internationally and domestically. UNIX, LAN and Internet are adopted in this international collaborative integration.

The increasing amounts of information available on the Internet are also relevant to manufacturers. These include abstracts and texts of technical reports from manufacturing research centers; personnel directories of national laboratories, engineering departments, and private companies; government reports, policy papers, statistics, and regulations; commercial sites for parts distribution; and engineering tools such as models and simulations. McGregor and Pacheco (1995) gave an overview of technical information of interest to manufacturers that can be accessed through the Internet, and described the modes of access by which this information is made available.

The use of IT in manufacturing is changing, mainly towards supplying greater customer service and choice. To achieve the necessary flexibility it is essential to have access to all available information. Decision support is becoming a key element of MRP systems (Montgomery, 1996).

Building an information strategy for the supply chain has now become a critical element for success. Increasing customer demand for responsiveness and choice is leading to change, and advances in Efficient Consumer Response (ECR) are bringing about a revolution in the way retailers, suppliers, distributors and manufacturers interact (Hinkins, 1996).

Epstein (1996) indicated that equipment outsourcers must be checked out when looking for the hottest undiscovered area of technology. As competition intensifies for technology in general, more and more equipment makers are turning to contract manufacturers-outsourcers to cut costs and manage inventory so that they can concentrate on their core needs: product design, marketing and sales. The outsourcers are not just being called on to satisfy incremental demand, as they were in the Eighties. Instead companies are looking to kick out their manufacturing operations altogether. They don't want to deal with them any more. It is a headache to them but a gold mine to outsourcers like Solectron and SCI Systems.

5 Internet in Manufacturing: Proposed Applications

Now days, engineering and manufacturing groups mostly depend on CAD and CAM packages for drafting and design, manufacturing and maintaining their products. Many designers are drafting and designing similar parts/ products, developing manufacturing processes and schedules using similar software in different corners of the country or in different countries in the world. That means, people are wasting their valuable time simply repeating the same tasks. The design specifications and technical documentation, for standard parts and products, can be created and accessed on-line through the Internet. The introduction of such a system will reduce the man-hour in manufacturing organisation and hence increases the productivity.

Internet already proved as an unparalleled media for business communications. Internet, incorporating communication, computers and multimedia, has provided tremendous potential for remote integration and collaboration in business and manufacturing applications. The uses of Internet for manufacturing applications is nothing closer compared to its growth and are very few. The reasons contributing to this slow utilisation may be accounted for lack of experience in the degree of integration that the Internet is capable of providing to enable a heterogeneous computerised environment to share data, resources, and programs as unrestrictedly as possible incorporating multimedia in them.

It is possible that a designer working at a CAD station can remotely communicate with a remote CAM facility to obtain a fabricated part with rapid turnaround using the Internet in a cost-effective and efficient manner. An integrated environment enables staff to communicate more effectively and thereby improve the overall productivity of their companies. Internet may be used in CAD, CAE and CAM by providing users with tools to reengineer and integrate product and process information in a distributed environment. Internet enables cost effective use of multimedia. It is expected that the marriage of communication and multimedia over the Internet will accelerate the use of Internet in manufacturing environment. The use of Internet in the design and engineering process is tremendous taking into account the role of text, drawings, full-vector graphics, and full-motion video on one hand and Internet as an interface for collaborative design with access to manufacturing database and research database on the other hand. Realising the importance of the application of Internet and multimedia in manufacturing, especially in the process of integrating different manufacturing facilities, geographically distributed in different locations, the possible options for utilisation of the facilities and the resulting linkages among them have been studied. The major activities involved in the process of fabrication are identified as (i) Drafting and design, (ii) Developing NC codes and (iii) Automated CNC machining.

Depending on the location of these activities (local or remote) the choices are:

1. A service provider manufactures parts/products. The search for the providers, their evaluation including cost benefit analysis and finally supplying them with the technical specifications is achieved through the Internet.
2. The designing of the part/product is done on a CAD system at the local site. The search for an appropriate CAM/CNC service provider to complete fabrication using the design instructions is achieved through the Internet.
3. Local facilities are used to design the part/product on a CAD system, to develop NC codes for CNC machining. The search for an appropriate CNC machining service provider to complete fabrication is done on the Internet. The instructions generated for the fabrication process may be provided off-line or on-line incorporating interactive multimedia.
4. Supports for all the activities in the process of fabrication are taken through the Internet from the service providers each of which are geographically separated.
5. Supports for design and programming are taken from one service provider, and support for fabrication services are taken from another company.
6. Support for the design services are taken from one service provider, while the programming and fabrication services are taken from another company.

In cases of options 1 to 3 the interaction between the customer and the service providers may be classified as 1:1, while in cases of options between 4 to 6 the interaction is 1:many i.e. 1:2 or 1:3. The 1:1 interactions are shown in figure 1. Only one of the Internet connections depending on the option chosen, shown by a double arrow, exists. The 1:many interactions are shown in figure 2. All three Internet connections (i.e. conn. 1, 2 & 3) exist for option 4 representing a 1:3 interactions. For options 5 and 6 the Internet connections (1 & 2) or (2 & 3) are 1:2 interactions. However, some options may exist with all activities will be completed at the customer premises. But that does not include the Internet and no interaction exists. A feasibility study needs to be conducted depending on the short and long term strategic planning for each of the above options. In the process of the study, the following tangible cost may be taken into consideration for each of the above options.

Option-1: Costs for (product + handling and delivery + Internet uses, if any)

Option-2: (unit cost for using CAD system) x (time required to design) + Costs for (CAM/ CNC services + handling and delivery + Internet uses, if any)

Option-3: (unit cost for using CAD system) x (time required to design) + (unit cost for program development) x (time required to develop) + Costs for (CNC machining services + handling and delivery + Internet uses, if any)

Option-4: Costs for (design + coding + fabrication + handling and delivery + Internet uses, if any)

Option-5: Costs for (design and coding + fabrication + handling and delivery + Internet uses, if any)

Option-6: Costs for (design + coding and fabrication + handling and delivery + Internet uses, if any)

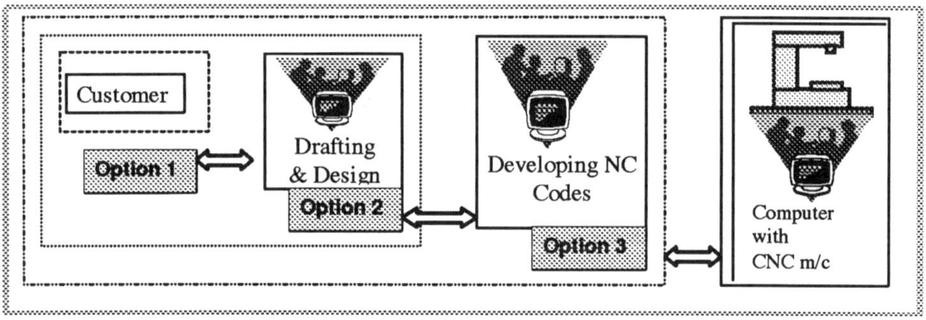

Figure 1: Proposed manufacturing system for option 1-3 with 1:1 interaction

For in-house development of CAD and CAM facilities, it is necessary to conduct a feasibility analysis incorporating the cost estimates involved with each of the options, their expected life and other constrains such as due dates, quality, service level etc. This would enable to reduce investment and product cost and improve productivity in terms of money. On the other hand the dedicated design and manufacturing service providers are able to use their capacity at a higher level. Higher capacity usage reduces the production cost, since the fixed cost is distributed among more products.

Though the design and fabrication can be performed from a distant location with the help of Internet, the delivery of products may take longer than the expected delivery time or the delivery cost could be higher. In that case, the reliability and service level of the service provider needs to be analysed. For a longer term decision, a detailed financial analysis should be performed considering all the cost parameters.

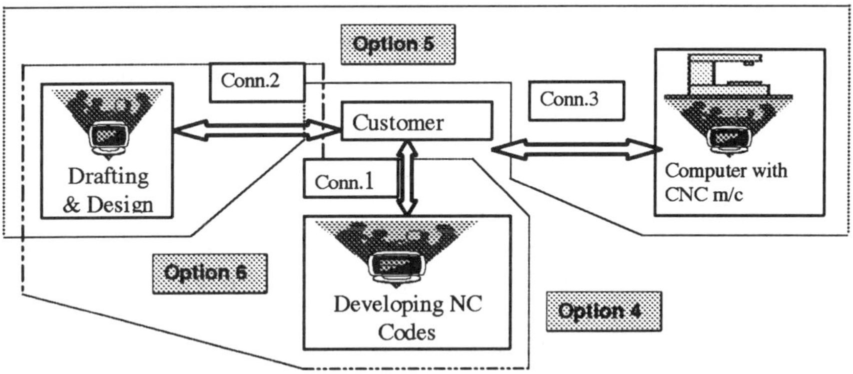

Figure 2: Proposed manufacturing system for option 4-6 with 1:many interaction

Sometimes, it may be feasible to take consultancy services on a short-term contract or job basis than hiring an expert. Internet can help in searching individual expert and consulting service providers in the intended area. It is also possible to get the work completed without going to the production site.

National and multinational companies having offices, design facilities, manufacturing facilities, research facilities, etc. in different geographical locations may use Internet to make their facilities available or to access facilities from others on mutual agreement to advise/ determine/ assign product and activities. Some of these are listed below.

- design project to be carried out in a design facility,
- product to be produced in a manufacturing facility,
- production level in each manufacturing facility,
- distribution plan from each manufacturing facility to each market region,
- raw material purchasing from different locations/ suppliers,
- allocation/ receiving plan of raw material to/from each manufacturing facilities, etc.

Alternatively, these can be dictated by the management/ decision-maker from one central location to all facilities in the company. The decision making process for multiple products involving a number of raw materials, manufacturing facilities, suppliers, market areas and alternative transportation is very complex. The uncertainty in demand and raw material delivery even make the problem more complex in nature. An expert can formulate the decision problem and solve the problem accessing the databases of all remote facilities from a central location. In this case, one team in a central location is enough for the decision making process, which is expected to improve the overall productivity.

6 Issues in Internet Connectivity

New technologies, specially with open architectures, always open new opportunities for litigation and carries inherent security risks. In absence of sufficient security measures, the information transferred over the network is vulnerable to encroachment to one degree or another. Security risk is one of the main issues in using Internet for manufacturing productivity improvement. Much research needs to be done to solve the problem in this field. The other issues include the bandwidth requirements for real time control and synchronisation between the task of the CNC machine. The consideration of network delay and jitter needs to be taken into consideration in the synchronisation process.

7 Conclusions

In this paper the applications of the Internet in manufacturing have been studied. The possible alternatives of using Internet in complete product fabrication with or without modern manufacturing facilities are also discussed. The central theme of this paper is the effective use of Internet in manufacturing, especially in fabrication. The main emphasis for improving productivity is to bring various elements of the systems together. Multimedia should be an excellent tool for this. Multimedia can also improve the effectiveness of CAD and CAM in manufacturing systems. The lead time for transferring information, material and services can be reduced by an interactive multimedia application. Though the design and fabrication can be performed remotely using the information provided by Internet, the delivery of products may take longer than expected delivery time which must be considered at the planning phase.

References

1. Burger, J. *Multimedia for Decision Makers*, Addison-Wesley Publishing Company, 1995
2. Cutkosky, M.R.; Tenenbaum, J.M.and Glucksman, J. Madefast: collaborative engineering over the Internet, *Communications of the ACM*, Vol: 39 Iss: 9, 1996, pp. 78-87.
3. Epstein, J. Non-tech tech [outsourcing], *Financial World*, Vol: 165 Iss: 9, 1996, pp. 47-9
4. Gunasekaran, A.; Bignall, R. J. and Syed M Rahman, Multimedia in manufacturing, *Production Planning and Control*, Vol 7, No. 5, 1996, pp440 - 451.
5. Groover, M. P., *Automation, Production Systems, and Computer Integrated Manufacturing*, Printice-Hall, 1987.
6. Hinkins, T. Chain of command [manufacturer-retailer supply chain management], *Information Age*, Vol: 1 Iss: 10, 1996, pp. 16-17
7. Jagdale, S.S.and Merchant, N. Implementing distributed controls for FMCs using Internet utilities, *Computers & Industrial Engineering*, Vol: 31 Iss: 1-2 , 1996, pp. 87-90
8. Liu, D. Information superhighway: essential fundamentals, *AUTOFACT '95. Conference Proceedings*, vol.2, 1995, pp. 179-92
9. Lin, G.C.I.; Yung-Chou Kao; Hon-Chy Liaw; Lee, R.S.; Chen, L.S.and Dayong Zhang, Application of an environment for international collaborative CAD/CAM, *Proc. of the 3rd International Conference, Computer Integrated Manufacturing*, vol.1, 1995, pp. 323-30
10. McGregor, B. and Pacheco, J., Manufacturing resources on the Internet, *AUTOFACT '95. Conference Proceedings*, vol.1, 1995, pp. 393-401
11. Montgomery, S., Industrial strength [Manufacturing firms IT systems], *Information Age*, Vol: 1 Iss: 10, 1996, pp. 36-8, 40,
12. Reinhart, G.and Brandner, S., Integration of suppliers into the data and process management, *ZWF Zeitschrift fur Wirtschaftlichen Fabrikbetrieb*, 91, 9, 1996, pp. 391-4
13. Tsoi K N and Syed M Rahman: Media-On-Demand Multimedia Electronic Mail: A Tool for Collaboration on the Web. *Proc. of the 5th IEEE Int. Symposium on High Performance Distributed computing HPDC' 96*, August 6-9, New York, USA. 1996. pp. 121-126
14. Wang, Z.Y.; Rajurkar, K.P. and Kapoor, A. Architecture for agile manufacturing and its interface with computer integrated manufacturing, *Journal of Materials Processing Technology*, Vol: 61 Iss: 1-2 1996, pp. 99-103.

Concepts for Improving the Network-ability of Small and Medium Sized Companies

Claus Schmid, Ulf Zillig, Fritz Huber

Institute for Industrial Engineering and Management,
Swiss Federal Institute of Technology ETH Zürich,
Switzerland

Abstract. The following paper initially describes the evolution of the supply structures in Switzerland which are basically consisting of small and medium sized companies. Concluding that manufacturing networks of increasing density and complexity are to be the outcome of this process the problems brought about are derived. Finally a conceptual framework embracing six dimensions is developed to tackle these problems.

1 Evolution of Supply Structures

Improving the competitiveness of companies belonging to the so-called supporting industry was focused so far mainly on streamlining the processes within each single company [1]. Especially trends and efforts such as Lean Production and Total Quality Management where concentrated on the three leading key factors of success: costs, door-to-door time and product quality and their relevant parameters within the production processes [2]. Having optimized processes within each company there is still considerable space for further improvement. Activating this potential it is necessary to leave the internal stage of activities. In modern industrial manufacturing even a company accordingly organized, using adequate technologies will not be able to compete successfully unless the business surroundings are taken into consideration. Figure 1 illustrates the evolution from rather simple arrangements to chain shaped structures to contemporary supply networks. Producing especially investment goods the company at the end of the supply structures offers its products to the market while receiving sub-assemblies, parts and materials from suppliers on various stages. Each company along the vertical paths provides the described elements for the following.

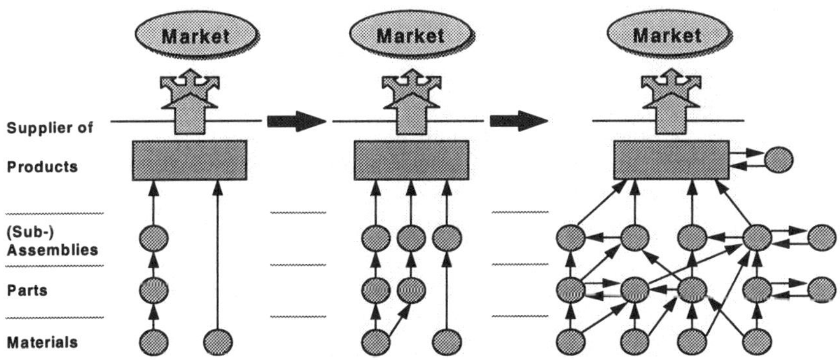

Fig. 1. Phases of supply structure evolution

Initially only a few suppliers provided mainly low value materials and parts. The supply chain resulted from a more assembly oriented organization within the product suppliers. Today lean production supports the concentration on core competences more than ever [3]. The number of processes being outsourced is continuously rising while the assembly remains the main value adding in-house process. As a reaction to the growing demand for cost reduction companies on lower stages of the supply hierarchy are starting outsourcing initiatives themselves while reducing their level of value adding activities [4]. This carries the evolution one step further towards a manufacturing network with increasing density. Along with this evolution, efforts to control and optimize the business surroundings become more and more important. Internal complexity is reduced drastically while in return external coordination and management is getting more sophisticated and will become one of the key factors of success [5]. The future form of manufacturing organization therefore tends to be an external one.

The so-called supply chain approach is reflecting this realization, although the environment is considered to be organized mainly in chain-shaped structures. Although this assumption already is a quite dramatic simplification, most supply chain approaches are intended to provide rules for the management of the direct suppliers only. Nevertheless each company has a highly variable and flexible spectrum of interacting partners which is described as manufacturing network [6]. Manufacturing networks therefore are temporary and non-temporary arrangements of more than two legally independent companies interfacing and exchanging materials, information or services. In a first approach this model seems to be a suitable description for the complex structures in which today's manufacturing companies are acting, but it is neither possible nor realistic to determine an exact and persistent borderline containing one distinctive manufacturing network in a top-down approach [7]. One company can be a part of various manufacturing networks occupying different positions in each structure. Although manufacturing networks are rather fuzzy to identify, there is at least the possibility for a conceptional classification. Starting from the main components of networks, elements and links it is appropriate to distinguish between dominating and non-dominating elements on the one hand and homogeneous and heterogeneous links on the other hand.

A manufacturing network with one dominating element on top of the hierarchy usually tends to be an homogeneous network as well because the dominating element defines suitable standards for the relevant links in terms of material and information flows including the services. Typical examples for such a type of manufacturing network can be found in the car industry. Large companies are in a well defined state of control which is exercised by standards for transport means or transport arrival times (material flow) and standards for data exchange or bar coding (information flow). This type of manufacturing network can be controlled with the help of resources provided by large companies. The structure itself is tailored to the requirements of the dominant element which is also setting the standards. A system of universal and case by case rules is created, installed, updated and overseen by the dominating element.

In contrast to this dominated, hierarchical type a completely different manufacturing network comprises about 80% of the Swiss manufacturing industry [8]. The elements of which are throughout small and medium sized companies without any sort of network-dominating partners. Consequently the links between these elements are not conformed to any kind of general standard. Mainly different means of transport, process descriptions, data formats, material specifications or forms are used in each link of the network. Compatibility was never an argument because the heterogeneous, non-dominant manufacturing networks are self driven and self controlled [9]. The evolution illustrated in Figure 1 set a pace which made it impossible to negotiate and adopt standards. Furthermore small and medium sized companies do not have the resources for a comprehensive manufacturing network management. The solutions to establish links to customers and suppliers are created when necessary. Finally flexibility often is a key criteria for being competitive in a heterogeneous manufacturing network and prevents the definite arrangement of standards.

In conclusion it can be stated that on the one hand the management of the manufacturing network which surrounds the companies becomes more and more important. On the other hand an evolution can be identified concerning especially small and medium sized companies which are parts of increasingly complex heterogeneous structures.

2 Problems of Networking Companies

Starting from the general situation described above, a research project focused upon the specific situation of small and medium sized companies being part of a rather non-dominated, heterogeneous manufacturing network was launched. Based on 8 Swiss companies supplying materials, parts, sub-assemblies, assemblies and complete products the entire spectrum of the supply industry is covered within the project. In order to manifest the described stage of evolution the companies themselves as active elements of the network and their links into the surrounding manufacturing network were under scrutiny. A variety of analysis activities were carried out in order to identify potential for improvement within the manufacturing network. As far as the elements of the network are concerned, the interfacing processes were analysed. Departments involved and duration times were systematically collected and prepared. All types of interfaces with other elements of the manufacturing network were taken into consideration, beginning with material flows and information flows to the organization of services. In consequence the complete infrastructure providing the resources for interfacing processes was the focus of the analysis.
Analyzing the links between the elements of the manufacturing network both customer and supplier orders were collected and analyzed with statistical tools. Number and location of customer or supplier, type of order, order value, number of order positions, type of product, product quantity exchanged, means of transport and date are aggregated for a period of one year. In the following some results are to be discussed in order to work out the potentials which should be activated by new concepts.

The usage of distribution systems by different companies can be summarized as follows. The suppliers in majority deliver their products with the help of their own distribution systems. Mail is also used frequently but not in larger quantities, especially by smaller companies. The other main distribution system used abundantly are external distribution systems, i.e. professional carriers. In conclusion it becomes obvious that it is possible to activate inherent potential within the material flows because the

distribution systems are open for reengineering alternatives. Rather static procedures such as distribution by public mail are clearly the minority.
Another interesting aspect of material flows in manufacturing networks is illustrated in Figure 2. The form of integration between customers and suppliers is structured into 6 stages. Beginning with step 1 which is the most simple form of integration at the material flow interface the possible integration stages climb up to step 6 in which even the procurement itself is taken over by the partner. In-between there are various degrees of integration such as transportation of incoming products right to the inspection department or directly to the production line of the customer.

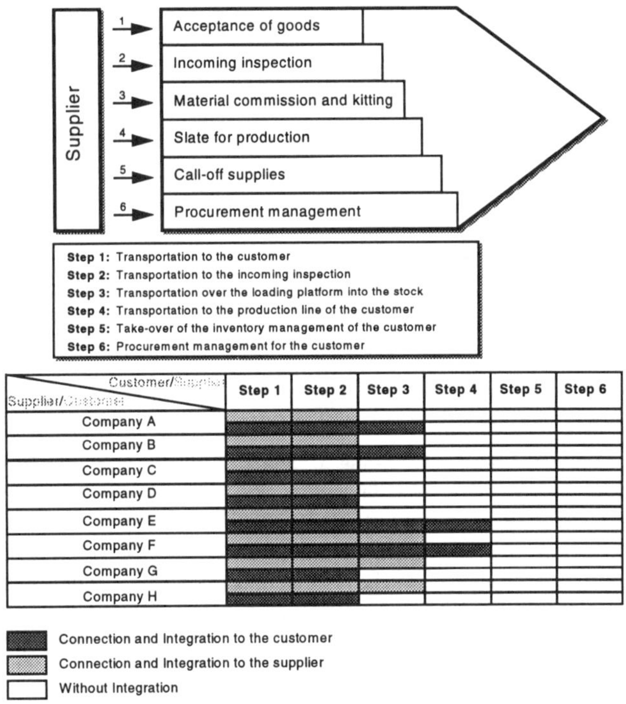

Fig. 2. Integration of material flows

The interface process analysis shows that all participating companies are carrying out an final quality check before the products are packed up and shipped to customers. After the transport the goods-inwards inspection is performed by the customer before the products are repacked into the internal means of transport [10]. It is also visible in Figure 2 that in all companies the customer side of the business has reached a higher stage of integration than the supplier side. Nevertheless there is still a considerable potential for improvement because the higher stages of integration on both sides are not activated yet. Furthermore, there is no set of rules available which stage of integration might be optimal with which partner in a manufacturing network.

Analyzing the network's elements and links in terms of information flow and processing there are also a number of interesting conclusions to be drawn. Figure 3 shows the different media used in the companies for information transfer to customers. The columns represent how often one type of media is used in how many companies within the network. The profile of information transfer to suppliers is almost identical. It is remarkable that despite the fact that all companies do have internal computer networks in operation none of them is using the opportunities of electronic data transfer. The same applies to the use of any kind of new electronic media such as internet-based software or teleconferencing. Information transfer with the help of electronic media is not used at all.

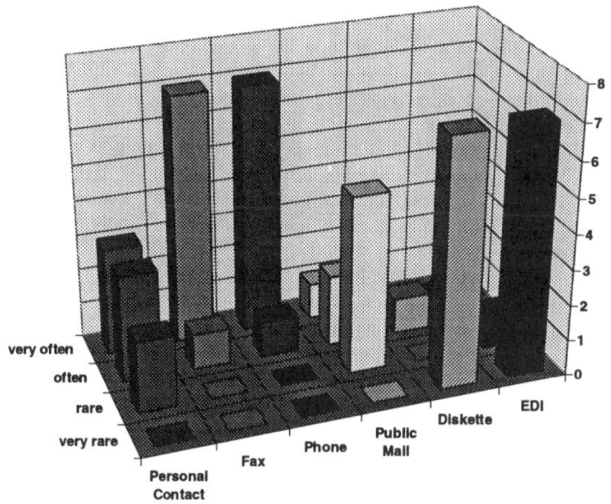

Fig. 3. Usage of media for information transfer

In return the technical preparations in terms of infrastructure are available. All companies are equipped with LANs of various topologies. Most of the companies are already running an internal client server architecture which would be capable of using new media such as intelligent internet applications or EDI. Nevertheless it can be stated that the potential of new information flow technologies [11] is not appreciated. Although the technological prerequisites are available the managerial tools are not at hand to develop the concepts for making use of these potentials in a sophisticated way [12]. In return an analysis of the information transferred in context with the manufacturing processes showed that over 85% of the contents is suitable for formalized coding. Therefore it would be possible to realize electronic exchange procedures.

An interesting aspect is revealed when looking at the values transferred between the companies and their customers. All sorts of exchange processes between partners within the manufacturing network have been structured by classes according to the value of the transfer in Swiss Franks for each participating company. In general it can be said that a large number of transfers do have a value of under 500 Swiss Franks and that more than half of all transfer values are well under 10'000 Swiss Franks. In parallel with these relatively low transfer values goes a significant increase in transfer frequencies. In addition the figures analyzed show that more than 80% of all orders placed by customers only have 1 or 2 positions. Figure 4 gives an overview of the order structure.

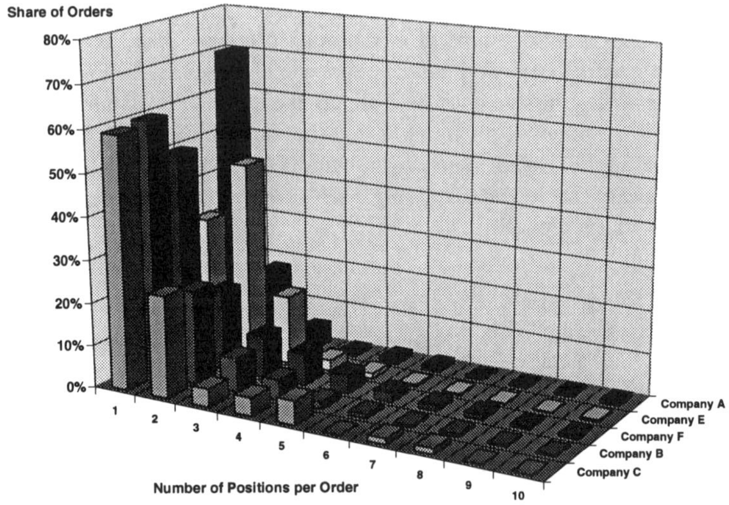

Fig. 4. Structure of Orders placed

These results are to be interpreted as visible signs of successful implementation of various Just-in-Time concepts in small and medium sized companies. Placing a higher amount of orders for small quantities of products which reach the supplier with a relatively high frequency has become practice throughout the manufacturing network as the figures clearly show. Implementing Just-in-Time principles in an isolated way, just focusing on the benefit for each single element of the network only offers limited potential for improvement. Pushing the transaction frequency to constantly increasing numbers with less positions per order and reduced value transferred is economically sensible only to a certain degree.
In terms of information and material flows these results clearly show that the resources needed to process and handle information are increasing. Due to higher transaction frequencies the expenses for handling, processing and storing of information are about to reach a critical limit. In return the value of the transactions carried out are about to decline further and further. Preserving and improving the competitiveness of the entire network therefore demands the ability to cope with these phenomena.

In conclusion there a three main aspects to be taken into consideration. Firstly the manufacturing networks become increasingly complex and difficult to manage. In addition the internal optimization demands an increasing capacity in terms of material and information flows throughout the network. This capacity required and the simultaneously growing complexity of the network demands new solutions. Thirdly there is still potential available which is ready to be activated by restructuring material and information flows through the network. Due to complexity and fuzzy as well as dynamic structures of manufacturing networks all types of algorithm-based superior control functions are bound to fail. Within this tension between growing complexity and the need for control and management, network-ability is becoming the core competence of the future. Network-ability is not a one dimensional term but describes the ability and competence of a company to cooperate with partners on all process levels [13]. The future form of manufacturing organization therefore is mainly an external one. Heterogeneous and highly interlaced manufacturing networks consisting of small and medium sized companies will represent the backbone of the European manufacturing industries [8]. The key question for small and medium sized companies being part of a heterogeneous manufacturing network is how to improve their network-ability while taking all possible solutions into consideration and still paying attention to the questions of a networked cooperation system.

In the following a framework of concepts for improving the network-ability of small and medium sized companies in heterogeneous manufacturing networks is presented.

3 Conceptual Structure for Improving Network-ability

Ameliorating the competitiveness of the manufacturing network means improving the network-ability of its elements. As discussed above manufacturing networks themselves are rather complex structures, fuzzy to differentiate with a high degree of interdependencies. Optimization requirements and potentials for improvement are difficult to locate and variable in their particular abundance. Working on significant improvements in the network-ability of a manufacturing company therefore can only be based on a variety of basic concepts as shown in Figure 5 which have to be chosen and implemented with maximum benefit for the network and the company itself.

First of all the main lever which offers the best expectation has to be chosen. This first choice has to be carried out in accordance with the strategy, provided the fact that small and medium sized companies do have a formulated strategy. If this might not be the case usually there is at least an implicitly formulated strategy available which provides the background for choosing the appropriate levers. A selection of possible levers is presented in Figure 5. In conformity with the current business strategy or manufacturing strategy an expanding cooperation with various network partners might be an interesting option for example. In addition the minimization of non-value-adding processes at the networking interfaces offers a suitable lever to expand the number of links without inflicting high operational costs. After choosing the main levers, the formal principles have to be checked ensuring that the principles of network management are carried forward as a first step for the basic concepts. The main task is to remind and focus on the fact that integrated manufacturing network management is depending on a holistic approach, i.e. aspects of the customer's customer and the supplier of the supplier should be taken into consideration while customer focus and time to market as direct key factors of success are still to be obeyed.

Moreover the flow system as the backbone of the network is another formal principle which has to be taken into consideration. The main objective of this formal principle is to keep material and information in a condition of directed flow. All unnecessary stocks and stops are to be prevented combined with the holistic approach. Finally the applied system theory should be taken into consideration as well before the basic concepts are worked out. Manufacturing networks in reality are as discussed above far too complex and fuzzy to be interpreted in with the help of mathematical formulas. Nevertheless it is vital to keep in mind the general assumptions and basics of systems theory.

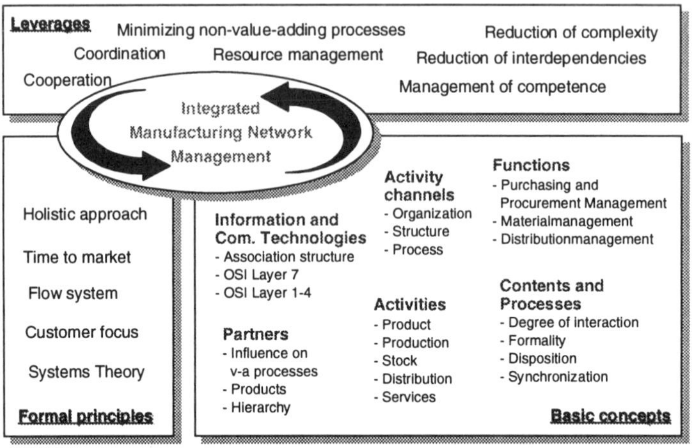

Fig. 5: The Manufacturing Network Management Circle

After identifying appropriate levers and equipping them with the relevant formal principles, the next step is the elaboration of basic concepts. The implementation of these concepts is intended to improve the network-ability of small and medium sized companies directly and to prepare them for future requirements in network-manufacturing. As illustrated in Figure 6 the basic concepts are to be structured in two complementary aspects. The objective is the construction of a tangible framework for developing concepts to improve the network-ability of small and medium sized companies. The concept cube structure as displayed in Figure 6 is intended to be a normative model, composing the dimensions in which optional solutions can be placed. Moreover all 6 dimensions are designed to be completely independent with a distinctive assignment to either material flow or information flow solutions. Both groups of dimensions are shown in Figure 6 as cubes. In-between these two main cubes there are six more cubes representing miscellaneous concepts integrating the aspects of both material and information flow concepts. The upper right concept cube covers the material flow solutions whereas the lower left cube concentrates on information flow solutions. The remaining quadrants between these two are used to develop integrated solutions. While Figure 6 shows the two main aspects in relation to each other including the transition cubes for basic concepts with mixed elements, the structure of each dimension is listed in Figure 5.

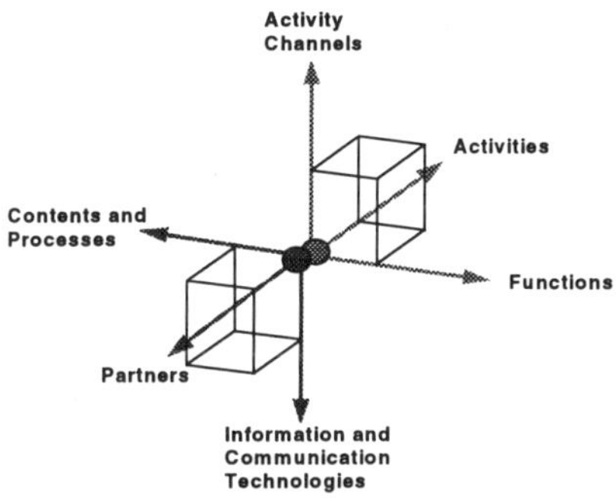

Fig. 6. Network-ability Concept Cubes

In terms of material flow the three main dimensions composing the concept cube are the activity channels, the activities themselves and the functions.

- *Activity channels* describe where the solution is based initially. The dimension contains organizational structural and processual stages. Using the organizational stage as a basis for concepts means that the complete organizational arrangement of the company is inflicted. Structure based concepts are grounding on compositions of processes with one common purpose whereas single processes are the most detailed stage. Concepts with an organizational focus have a higher potential to achieve an direct effect on the network-ability. In contrast process based concepts do have rather limited impact on the network-ability of the company but on the degree of autonomy. Concepts which focus on the reengineering or redesign of single processes are far more suitable to achieve an appropriate adjustment when a detailed match of interfaces is necessary to reach a higher network-ability.
- *Activities* as such are represented by their assignment to the product itself, the production, storage and internal logistics, transport and external logistics and finally service. This dimension of the material flow oriented cube helps to structure basic concepts according to objective of activities. In case the basic concepts are intended to have a direct impact on value-adding activities, they should be planned with a clear focus on product and production while non-value adding processes are rather likely to be influenced when addressing activities in the logistics and service area. The same applies to activity based costing initiatives. The decision on which sort of activities the basic concepts are to be focused upon is vital for a determined implementation of the concept itself.
- *Functions* are the third dimension of the material flow cube containing purchasing and procurement management, material management and distribution management as three possible elements of basic concepts for improving the network-ability. Especially when the concepts are intended to improve material flows, these three functions are to be the integrated. Nevertheless they could be a part of the structure of concepts focusing on information flows as well.

The front cube in Figure 6 represents the main dimensions of concepts with the objective to improve the information flows. In the following the information flow aspect of the concept cube model is to be discussed.

- *Information and communication technologies* are the media for information flows. Especially the potential of the new information and communication technologies is not recognized or used properly in the small and medium sized companies under investigation. The structure of this dimension concerning the information flow is mainly based on the Open Systems Interconnection (OSI) Model. Furthermore the association structure has to be taken into consideration. It is important whether a concept is designed for 1:1, 1:m or m:m associations. When formulating a concept for improving the network-ability by optimizing information flows, the physical network should be determined as well as the applications used and the association structure.
- *Partners* are to be considered in terms of hierarchy, products and influence on the value-adding processes. This dimension therefore represents the external aspects. As far as the hierarchy in the relevant area of a manufacturing network is concerned the main issue is the balance of bargaining power based on information. A possible shift in powers can be initiated by introducing and implementing new information flow concepts. Therefore this strategic aspect is integrated in the structure of this dimension. It has to be clear which partner within the network has to be supplied with which information. The concepts including information flow aspects are supposed to take into consideration what type of products in general are exchanged. Information flow concepts (and material flow concepts as well) depend on the question whether these products are mainly according to standards or whether the products are highly differentiated variants made on customer specification. The concluding element of this dimension is the degree of influence on the value-adding processes that a partner within the network has. Each concept worked out, has to specify this potential given to a partner.
- *Contents and processes* representing the third dimension of the information flow concept cube is structured with the degree of interaction, formality synchronicity and the disposition system. The degree of interacting is intended to define the level of co-operation based on the information flow. It is necessary to specify if information is simply dumped and can be collected by the partner with the help of a mail-box system, if there are search routines included, or if the partner is able to generate own routines. In terms synchronicity as the second element of the structure, the concept is to define whether the cooperation via information flows has to be synchronous, intermittent or asynchronous. The disposition includes the specification which partner in a manufacturing network is responsible for initiating the information flow. Finally the formality has to define the possibility to structure the contents of the information itself.

The cubes with the objective to provide a framework for the conceptual modelling of solutions to optimize the network-ability, with the two main cubes concerning the material flow and the information flow. As shown in Figure 6 the two main cubes are positioned and oriented diametrically. The network-ability of a company can also be improved by concepts integrating both aspects of material and information flows. These concepts are to be located in the remaining six cubes. The model therefore offers a structural framework for all possible combinations of both aspects.

4 Summary

Having not much in common with the rather linear or even punctual structures manufacturing networks with fuzzy structures, and a high degree of complexity are the future forms of manufacturing cooperation. Therefore mainly small and medium sized companies in non-dominated, heterogeneous networks do require concepts to improve their network-ability. Developing concepts which are intended to improve the network-ability does depend on a framework. Otherwise possible concepts might be based on low potential levers or they might be inadequate because some important aspects might be disregarded due to the complexity of the network and the large variety of possible solutions. In order to structure the variety of possible solutions a conceptual framework is presented. Being based on two cubes with three dimensions, each structure for the material flow and the information flow is described. The adjacent cubes are representing integrated concepts comprising material flow elements as well as information flow based solutions.

5 References

1. R. Goffee and G. Jones, What Holds the Modern Company Together?, *Harvard Business Review* **11/12**, (1996) 133-148.
2. L. A. Kappelmann and J. Cappel, Confronting the Year 2000 Issue, *Journal of Systems Management* **4**, (1996) 4-3.
3. R. Gremir, Going Virtual, Prentice Hall (1995) pp. 200-215.
4. H. W. Chesbrough and D. J. Teece, When Is Virtual Virtuous ? - Organizing for Innovation, *Harvard Business Review* **1**, (1996) 65-73.
5. F. Huber, Kooperative Entwicklungs- und Produktionsnetzwerke, *TBI '96 Strategien für innovative kleine und mittlere Unternehmen*, (1996) 53-60.
6. X. De Groote, The Flexibility of Production Processes: A General Framework, *Management Science* **7**, (1994) 933-945.
7. P. Pekar, and R. Allio, Making Alliances Work - Guidelines for Success, *Long Range Planning* **4**, (1994) 54-65.
8. P. H. Dembinski and H. L. Unterlerchner, L'internationalisation des PME suisses à l'horizon de l'an 2000, Georg et Eco'Diagnostic, (1994) pp. 5-8.
9. J. Layden, Supply Chain Management Creates New Roles for IEs, *IIE-Solutions* **7**, (1996) 35-39.
10. Th. H. Davenport, S. L. Jarvenpaa and M. C. Beers, Improving Knowledge Work Processes, *Sloan Management Review* **8**, (1996) 53-65.
11. R. Benjamin and R. Wigand, Electronic Markets and Virtual Value Chains on the Information Superhighway, *Sloan Management Review*, (1995) 63-72.
12. L. Thach and R. Woodman, Organizational Change and Technology: Managing on the Edge of Cyberspace, *Organizational Dynamics* **23**, (1994) 30-46.
13. K. Fischer, C. Schmid and U. Zillig, Sind Schweizer KMU netzwerkfähig?, *Technische Rundschau* **46**, (1996) 96-100.

Network Middleware Illustrated for Enterprise Enhanced Operation

W. Loeve, M.E.S. Vogels, H. van der Ven, E.H. Baalbergen

National Aerospace Laboratory (NLR), P.O. Box 90502,
1006 BM Amsterdam, The Netherlands

Abstract. Enterprises are confronted with the challenge to lower cost of product engineering. Engineers enter and leave enterprises. Much engineering know-how can be conserved for use by successive generations of engineers in documented engineering procedures, in software for simulation and in the supporting documents and simulation related data. In the paper the considerations and the technical principles are illustrated by examples from two applications: a national virtual centre for Computational Fluid Dynamics for engineering and an IT framework for distributed multi-discipline design, analysis and optimization of industry products.

Keywords. Network middleware, enterprise automation, engineering.

1 Introduction

Enterprises are confronted with the challenge to lower cost of product engineering. In this context engineering staff is minimized. Increasing the efficiency of enterprises by minimizing staff requires increased attention for management and control of engineering know-how. In view of this engineering information has to be conserved in enterprises in the form of software such as for simulation, electronic documents (engineering procedures and manuals for simulation software) and data on electronic carriers. For engineering know-how in these forms it is possible to make use of Information Technology (IT) for management, development control, access and application.

Enhancing enterprise operations in engineering is possible by enhancing accessibility and applicability of enterprise know-how. Minimizing staff to lower enterprise cost nowadays very often means that engineers are only hired to perform specific jobs. This means that it is required to enhance accessibility and applicability of enterprise know-how for engineering with partly temporary staff members. IT makes it possible to transfer more efficiently know-how between (generations of) engineers at various geographical locations.

It is a task of the enterprise itself to control the use of IT for enhancing operations. This makes that IT is at least a required core competence of the enterprise. In chapter 2 a UNIX network and network middleware is described by which enterprise enhanced operation is possible. In chapter 3 and chapter 4 the application of the network and the middleware is illustrated for application of computational fluid mechanics and for application of a combination of computational fluid mechanics and computational solid mechanics respectively.

2 A UNIX Network for Optimal Engineering Support

2.1 Computer Network Required for Engineering

The most advanced industrially applicable computational fluid dynamics (CFD) software for aerodynamic analysis in aircraft engineering at the moment is based on the Reynolds-Averaged Navier-Stokes equations (Rubbert, 1994). One of the interesting applications of CFD for the near future is support of aerodynamic optimization (Borland, 1994). Optimization requires more computer performance than analysis, because for optimization for instance about 40 times solution of the flow equations in combination with an optimization algorithm is required. For NLR aerodynamic optimization is the most demanding engineering application for which it was decided to procure a NEC SX-4/16. This computer is meant to be applied in support of engineering and scientific simulation in The Netherlands (Loeve 1996). With a NEC SX-4/16 (a vector computer with 16 processors and shared memory) it is easy to realize for CFD in engineering a computational performance on 14 processors of 8,500MFLOP/s (Ven van der, 1997). This is realized with relatively little effort (only DO-loop parallelization). One aerodynamic optimization of 40 analysis runs then takes about 10 hours.

The process of engineering an aircraft is characterised by a large number of activities executed by engineers from various disciplines mostly located at different geographical locations. The engineers use a variety of simulation software tools. Engineers need workstations for graphical interaction with their tools. Sharing of know-how between specialists has to be safeguarded which means that information servers and intranet servers are required in addition to compute servers and workstations. The network has to give access, for all involved persons and groups, to the computers as well as to the software, data and electronic documents that are stored on the computers. Computers, workstations and information servers such as data management servers with tape robots, are commonly equipped with UNIX operating systems. Complexity of the computer networks and of the collections of software systems used in engineering processes both increase continuously in industry.

The computer network that serves the two establishments of NLR is presented in figure 1. The external connections indicated in the figure concern primarily the use of the SX-4. This computer is by far the most powerful computer in the Netherlands as can be seen in (Dongarra 1996).

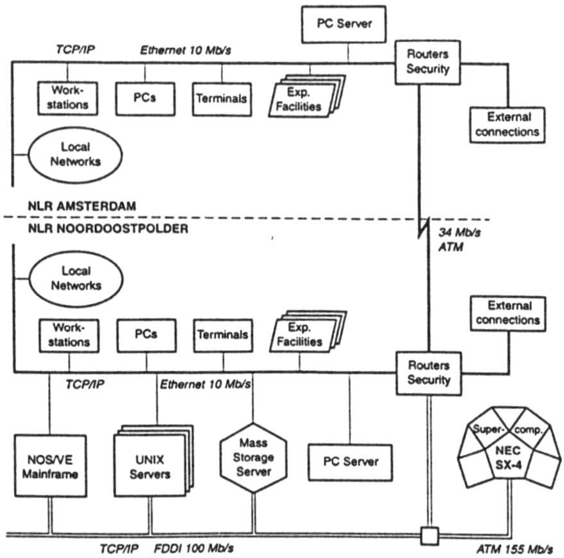

Fig.1 The NLR high performance computer network

2.2 The need to Present the Network to Engineers as a Unity

Many problems occur in practice when engineers try to use a network:

- Remote access to computers is difficult to organize.
- Programs lack uniformity in user interface and data interface. Incidental use is practically impossible because of time required to familiarize.
- Problems with programs on different computers can be:
  - finding out on which computer a specific program may run,
  - exchange of files between computers,
  - incompatible data formats due to different byte ordering and representation of real numbers in different computers.
- Users have to organize, manage, and find a way through more on-line information than that is accessible on a single computer.

Removal of the technical and organizational shortcomings mentioned above will enhance the enterprise engineering. Information about know-how (data, documents and software) shall be made available on-line. Information shall be managed and controlled in such a way that reconstruction of old information is possible and information can be exchanged between engineers in the same way as if they all make use of one computer. Engineers have to be assisted in organizing, re-using and integrating a continuously growing amount of software, data and supporting help information in networks.

Only by making high performance computing a natural part of desktop workstation applications, the use of high performance computing in engineering will spread according to the need for competitive enterprise operation.

2.3 Network Middleware for Enterprise Enhanced Operation

Since computers were introduced at NLR in the late fifties, this institute has applied a centralized IT approach. The main reasons were economy of scale and the need to be able to procure the computing power that is required for the most demanding applications (Loeve 1976). In addition productizing of software developed by NLR for CFD in industry is performed by IT professionals. In this development much attention was always paid to quality management. This resulted in an ISO 9001 and AQAP-110 certificate for development of simulation software. It also resulted in development of architecture and construction principles for development of CFD software according to industry needs (Vogels 1997).

Finally, quality management resulted in the development of a software facility called SPINE that supports the construction of general as well as application-oriented working environments (Loeve 1997). A SPINE- based working environment gives access to local and remote computing resources and information in a network. The network presents itself with SPINE as one single computer (a "metacomputer") to its users. In addition to hiding the networking details, SPINE provides mechanisms for easy customization of the metacomputer (to the needs of a particular user, group of users, or application area), for easy operation of the metacomputer by the user, and for management and control of information. To support the operational use of a working environment, SPINE comprises technical solutions for the key elements of a metacomputer: an easy-to-use desk top, facilities for information management and middleware.

The desk top element of SPINE provides the user of a working environment with a graphical integrated user shell. The shell provides a uniform and consistent view with respect to presentation and manipulation of data files, documents, and source code, and with respect to execution of tools. The data management and processing operations are available in terms of point-and-click and drag-and-drop operations on windows and icons. The shell provides control over multiple windows in combination with multi-tasking. This makes it possible to have several tools running simultaneously. The shell also provides the user with contextual on-line help, both on the operational aspects of the working environment as well as on the applications.

In addition to the Windows like look-and-feel of SPINE, the shell supports data-oriented, tool-oriented, as well as work-flow driven modes of working. The work-flow capability of the shell supports work-flow management, in which data and tools can be configured in a network that supports, or even may enforce, certain scenarios or activities. Work flows are considered to resemble an engineer's way of working, and to provide an intuitive means of specifying jobs and controlling the execution of the jobs. Work-flow management also enables transparent and automated information management, e.g., to enforce synchronization and reconstruction of activities and results. SPINE supports the Computer Supported Cooperative Working aspects of information management by providing tools for managing storage, access to, and modifications of data, documents, software and any information related to this, and for the exchange of information among users.

The task of SPINE's middleware is to manage the available resources in order to accomplish the single computer look-and-feel, and to exploit the potentials of the underlying computer network. The middleware also takes care of the application of data conversions required to deal with hardware and data-format heterogeneity.

SPINE, by being based on standard UNIX and TCP/IP software, has proved to be applicable for realization of environments that span both local-area and wide-area networks. For example, the NICE environment (see chapter 3) spans the local area networks of the NICE partners, but also gives access from remote workstations to NLR's SX-4 vector computer and CFD software as well as the compute power and software in other partners in the NICE project, over the Internet. In this way, a Netherlands-wide CFD working environment has been realized.

SPINE is being further productized jointly by NEC and NLR. The working environments constructed using SPINE are highly portable and easy to customize. Moreover, the working environments are open and extendible, in that any UNIX and TCP/IP based computing system can easily be integrated, and that existing software and other forms of electronically available information can easily be integrated. At the moment SPINE is implemented in about 20 organisations for integration of local workstations in the organisations with the SX-4 at NLR. In some cases SPINE is just as in NLR used by other organisations for integration of not only user-workstations with the supercomputer but also with other servers.

3 The use of the UNIX Network and Middleware for Computational Fluid Mechanics

3.1 CFD work Environment in one Organisation

Since the earliest developments of SPINE it has been applied to ISNaS, a working environment for flow simulations (Vogels 1989). At first the ISNaS CFD working environment was aimed to support the use of flow simulation packages across the NLR network. The network consists of workstations and terminals that give access to mainframes acting as compute-, intranet-, internet- and data management servers. A supercomputer is integrated in the network since 1987. Because of its success, more functionality was and is added continuously to the network and to the network middleware of SPINE.

ISNaS now supports both use and development of simulation software. The use of the simulation software is supported by integrating the entire pipeline of simulation analysis: geometry modellers, grid generators, flow solvers, and postprocessors. File transfer and remote logins are hidden from the user. Exchange of different file formats between the various tools is made transparent to the user using the SPINE facility for implicit file conversion. Exchange of documents and data is made possible through easy-interface databases. Feedback to the developers of both the working environment and the simulation software is made easy by the use of electronic forms.

The development of simulation software is supported by providing via SPINE a software version management tool, and a toolpack sx4dev which integrates several development tools on the NEC SX-4 (more about the toolpack in Section 3.2). A framework for the creation of a regression test suite is available within the working environments. For support of development control, electronic forms are used to manage error reports.

For both users and developers project management systems can be set up. Such a project management system consists of an electronic project archive, a search engine on this archive, a drag-and-drop mechanism to submit documents to the archive, and a database for electronic forms. We will see an example of such a project management system in Section 3.2.

The building blocks of the working environment supporting both use and development of CFD software are shown schematically in Figure 2. As is clear from the figure development and use interact and overlap.

The development of a working environment is dominated by two principles:

- an integrated tool has added value to the user,
- the working environment shall adapt to the user and not vice versa.

This also implies that integrated tools do not stand alone in the working environment: they are surrounded by tools that either supply input or use the output of the specific tool. This integration of tools greatly facilitates their use.

The success of the present working environment is probably best exemplified by the following. A CFD trainee was asked to perform an analysis of air heater flow. The trainee had a thorough knowledge of flow physics, and numerical mathematics. But he was less familiar with supercomputers, networks, UNIX and postprocessing. Using the working environment he was able to perform and analyze a specific flow configuration within one week. The analysis consisted of the entire pipeline from preprocessing up to visualization.

3.2 National CFD Working Environment

To enhance the use of HPCN techniques the Dutch government partly finances the project NICE, Netherlands Initiative for CFD in Engineering. NICE aims to promote the use of CFD methods, developed at universities and large technological institutes, to industry and small and medium enterprises. On the one hand, new (parallel) algorithms are developed motivated by the needs and wishes from the industry, on the other hand, these algorithms are used to analyze flow problems from the same industry. Development and use are combined in a single initiative.

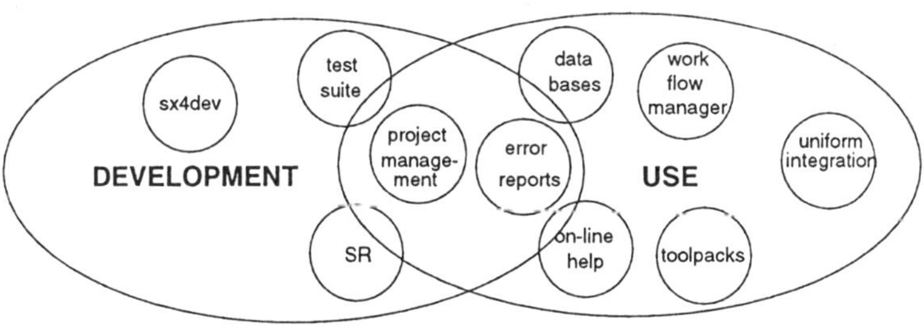

Fig. 2. Building blocks for both use and development of engineering software

In order to facilitate the use of remote resources NICE develops a nation- wide simulation centre HFS based on SPINE. HFS consists of two parts: a common part to all NICE partners, and a part specifically tuned to the needs and wishes of a NICE partner. The network on which HFS operates consists of the local network of the partner and the NEC SX-4 at NLR. HFS is a specialized 'ISNaS' (see Section 3.1) at each partner, extended with a common set of tools for both project management and development on and use of the NEC SX-4.

Management of the project NICE is facilitated through HFS. A project management system is part of the common tools in HFS. The system consists of an electronic project archive, electronic error reports, and search engines to search both the project archive and the error report database. The project archive has a technical report repository. Development on and use of the NEC SX-4 is supported by the toolpack sx4dev that integrates tools for compiling, executing, analysing and debugging source. Automatic makefile generation allows for manipulation of the source code. Options of the tools are based on the experience of experts in supercomputing. Using this tool the computational kernel of a flow solver of TNO-TPD was compiled, executed and analyzed within a quarter of an hour. Before that, TNO-TPD had never worked on an NEC supercomputer.

By supporting both development and use of CFD software HFS is suited for a wide range of customers. From novice users inexperienced with CFD, through experienced users with a need of more powerful resources, up to developers of new algorithms and software.

An example of a customer needing more powerful resources is ESTEC. For a time-accurate simulation using a multi grid, multi block structured solver they required more memory than their IBM SP2 could supply. Moreover, throughput times became prohibitive. Even though the code was originally written for scalar machines, a porting effort of roughly three manweeks increased the megaflop rate from 80 MFLOP/s on the IBM SP2 to 500 MFLOP/s on one single processor of the NEC SX-4.

4 The use of the UNIX Network and Middleware for Multi-Discipline Analysis and Optimization

4.1 Problem Analysis

To effectively resolve cross-discipline trade-offs both to improve aircraft performance and reduce development timescales and costs, concurrent engineering principles are under investigation for the preliminary design stage.

In Europe, the MDO Consortium [1] is addressing integration of design and analy-

[1]The MDO project (Multi-Disciplinary Design, Analysis and Optimization of Aerospace Vehicles) is a collaboration between British Aerospace, Aerospatiale, DASA, Dassault, SAAB, CASA, Alenia, Aermacchi, HAI, NLR, DERA, ONERA, and the Universities of Delft and Cranfield. The project is managed by the British Aerospace and is funded by the CEC under the BRITE-EURAM initiative (Project Ref: BE952056).

sis tools creating a Multi-Disciplinary Optimization (MDO) capability (MDO Consortium, 1996). The purpose is to develop and demonstrate the viability of MDO and validate this for simplified but realistic aircraft preliminary design tasks concerning the design of a large civil aircraft where a non-trivial interaction is expected between the mono-disciplines of aerodynamics and structures. NLR, one of the Consortium members, carries out the work in a multi-disciplinary team with specialists from aerodynamics, flight mechanics, structures, and information technology.

In the MDO problem, different aircraft designs are compared by comparing the Direct Operating Cost (DOC) for the aircraft design. For the problem at hand, the DOC is simplified in a model containing drag and total weight. The problem of DOC evaluation for an aircraft design can be solved in more than one process, each leading to a different approximation of the DOC. Therefore a DOC value for an aircraft design can only be interpreted in the context of the process in which it was achieved.

Process model

The process is modelled in so-called N^2 diagrams where the *(N)* contributing disciplines form the diagonal of an $N*N$ matrix, the off-diagonal entries show the dependencies between the disciplines. An above-diagonal entry denotes a feedforward coupling, a below-diagonal entry denotes a feed backward coupling.

The top-level N^2 diagram for the NLR process for the early problem is given in figure 3. The figure shows the involvement of the disciplines.

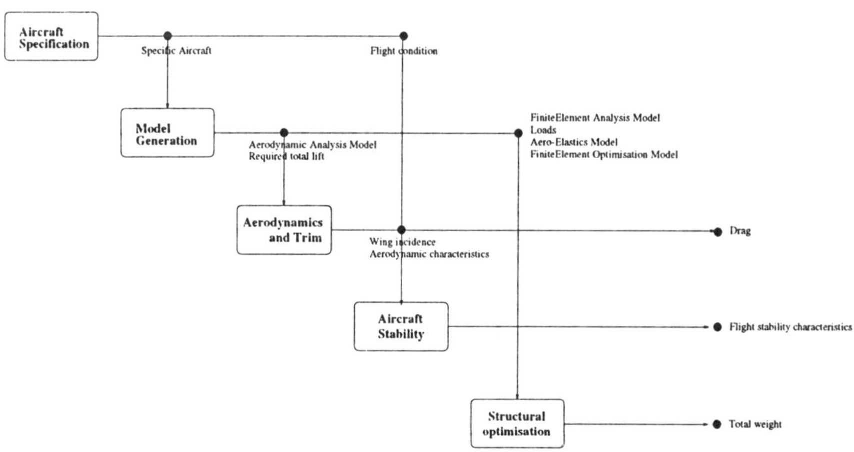

Fig. 3. Multi-discipline analysis process

The diagram is essential to avoid ambiguity in the analysis process. The diagram shows for example that the loads for the structural optimization in the present analysis process are the result of "model generation" and not of "aerodynamics and trim". Such choices for the process should be explicit when interpreting the DOC value for an aircraft design. On lower levels, similar N^2 diagrams show the detailed processes. The level of detail of the N^2 diagrams is dictated by its usefulness for the MDO-process and usually stops at the level of a mono-discipline.

Changes in the mono-discipline software for the multidiscipline context
On the MDO level, it is assumed that all disciplines have their own (mono-disciplinary) application software. The application software runs on the NLR heterogeneous UNIX network, involving e.g. SGI work stations, CD4810 scalar compute server, and the NEC SX-4 parallel vector machine. Data, documents and software are stored via a data management server on a tape robot. Most software is home--brew, implemented in Fortran and C. The software frequently changes to satisfy additional requirements.

Three types of changes are required for the mono-discipline software to function in the multi-discipline context. First, the mono-discipline software has to accommodate for the communication with the other disciplines, on the level of detail and in the terminology specified in the N^2 diagrams. Whenever possible, this accommodation consists of interface programs which translate between the MDO-parameters/results and the mono-discipline's parameters/results. In this way, the mono--discipline application software can remain unchanged, and can be used for solving mono- and multi-discipline problems.

Second, whereas in mono-disciplinary work it may be acceptable that the work halts when a specialist is unavailable, in multi-disciplinary work carried out in a team, it is not acceptable. The work may be taken over by a colleague less familiar with the MDO problem, or by a multi-discipline team member of a different discipline. A prerequisite for enabling someone else to perform the absent specialist's work, is accessibility to the mono-discipline software. This is organised in the MDO environment "ismo" (Section 4.2).

And, finally, again because of the team character of the multi-disciplinary work, it is required that the application software, used for the MDO work, is identified and reproducible.

4.2 Set-up of MDO Environment

One of the purposes of the MDO environment "ismo" is, as explained above to provide access to all mono- and multi-discipline application software of the MDO team. This includes the shielding of the computer network, and providing the virtual computer feel. For this purpose, the mono- and multi-discipline application software is integrated into an MDO environment based on NEC-SPINE.

Application software which is unchanged by the multi-discipline context, and which is already under version management, is considered to be identified and reproducible. Newly developed application software (such as the interface programs with the MDO-level), and changes in existing application software are version managed in the MDO environment.

The challenge for information technology is not only to capture the MDO process model as in N^2 diagrams, but also to supply an MDO environment which supports the work according to the MDO process view of the disciplines involved. This is the leading principle for the information engineering in the MDO environment. The application software in the MDO environment is structured (in levels and groups) according to the N^2 diagrams. In this way, the MDO environment reflects the MDO team's view on the MDO process. In Figure 4, three levels of the MDO environment are given.

From top to bottom we see in figure 4 first the highest level window which gives access to the environment including the tools. In the second window in figure 4 tool access is presented for version management as well as access to the specific MDO programs. The latter programs are shown in the bottom window with the elements (in alphabetical order): 1 ac-performance; 2 ac-specification; 3 ac-stability; 4 aerodynamics-and-trim; 5 model-generation; 6 multi-inspection; 7 structural-optimization.

The diagonal elements in the MDO process (Figure 3) correspond to the "ismo" elements 2, 5, 4, 3, 7 respectively. The "ismo" elements 1 and 6 are additional

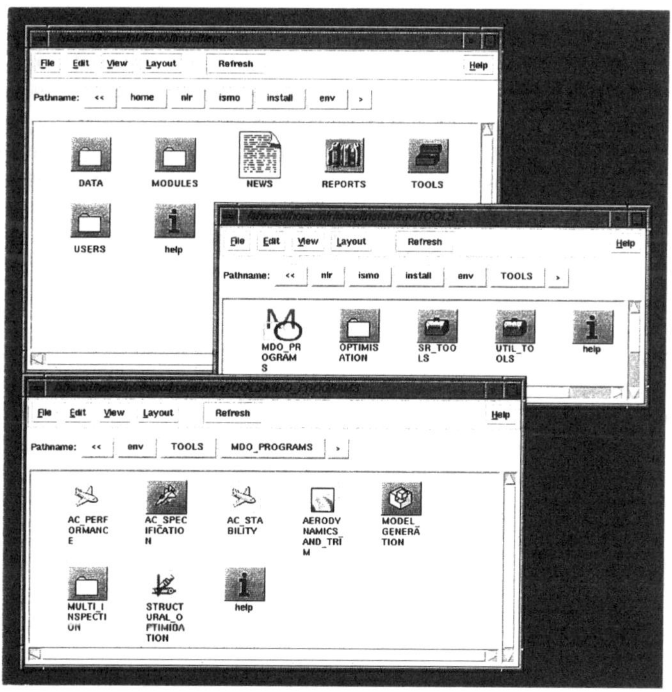

Fig. 4. Screendump of three upperlevels of MDO environment

compared to the diagonal elements in the MDO process; they are included in the environment "ismo" because:

- "ac-performance" has been identified as a contributing discipline, but has been left empty for the problem at hand,
- "multi-inspection" is a passive process, not producing any information to the DOC formula.

At each level engineers are supported by help information that is relevant for the concerning level. Manuals for instance can be accessed by dropping the icon representing a program on the help icon.

5 Concluding Remarks

Hiring engineering staff when needed requires that enterprise know-how is conserved in software for simulation, electronic documents and data.

For engineering in industry it is essential that cooperating engineers can use the computer network of the enterprise as well as remote high performance computer as one single virtual computer for management, control, exchanging, accessing and using enterprise know-how.

6 References

1. C.J. Borland, J.R. Benton, P.D. Frank, T.J. Kao, R.A. Mastro and J.F.M. Barthelemy, (1994) Multidisciplinary design optimization of a Commercial Aircraft Wing - An Exploratory Study. 5th Symposium on Multidisciplinary Analysis and Optimization, Panama City Beach, Sept. 7-9, 1994. AIAA-94-4305-CP, pp 505-519.
2. J.J. Dongarra, H.W. Meuer and E. Strohmaier, E (1996) TOP500 Supercomputer Sites November 1996; retrievable from internet by anonymous ftp to www.netlib.org/-benchmark/top500.ps.
3. W. Loeve, (1976) A hierarchical network linking two research laboratories; Computer Networks 1 North-Holland Publishing Company.
4. W. Loeve, (1996) From parallel CFD to a tool for computer aided engineering; in *Parallel Computational Fluid Dynamics* Elsevier.
5. W. Loeve, E.H. Baalbergen, M.E.S Vogels and H. van der Ven, (1997) High performance computing in simulation that fulfils user-driven quality criteria; in *DASIA '97 Conference* ESA Publications Division.
6. MDO Consortium (1996), http://www.telecall.co.uk/~srcbae/mdo.
7. P.E Rubbert, (1994) CFD and the Changing World of Airplane Design; AIAA Wright Brothers Lecture, Anaheim California; September 18-23.
8. H. van der Ven and J.J.W. van der Vegt, (1997) Experience with advanced CFD algorithms on NEC SX-4; to be published in *VECPAR '96 selected papers from the 2nd International Meeting on Vector and Parallel Processing (Systems and Applications)* (ed Laginha M. Palma et al), Springer.
9. M.E.S. Vogels and W. Loeve, (1989) Development of ISNaS: An Information System for Flow Simulation in Design, in *Computer Applications in Production and Engineering* (ed F. Kimural) North-Holland ISBN 0 444 88089 5.
10. M.E.S. Vogels, (1997) Architecture and construction principles for computational fluid dynamics software for engineering in industry; Thesis, ISBN 90-9010295-7; M.E.S Vogels, National Aerospace Laboratory Amsterdam.

Impacts of Network-Based Commerce on Asian Industry

A Lee Gilbert, Director

MBA in International Business Programme, Nanyang Business School, Singapore

Abstract

As an information technology-enabled innovation, the consequences of the rapid diffusion of electronic commerce to Asia are far from clear. Following a brief map of the evolutionary path of money, the paper views the strategic interests of the actors in electronic commerce, applies multiple scenario technique to reveal the likely impacts, and closes by applying lessons from the financial services industry to manufacturing firms and their stakeholders.

1 Introduction

Technology has a powerful impact on industry structure. Smaller enterprises use the Internet as a channel for one-time EDI transactions. Suppliers now search the Web for new customers. In New York City, Citibank and Chase Manhattan recently announced re-loadable, stored value cards to be used for retail transactions [1]. This trend is diffusing rapidly to Asia, where HongKong banks have more than 200,000 smart card-based electronic purses in circulation; and the Singapore National Computer Board is carefully field testing a new national infrastructure designed to support global electronic commerce. In North America, one Kentucky-based bank already offers virtually every retail service on the Internet [2]. The contrast raises strategic issues for Asian producers: how can they access new opportunities created by electronic commerce, and what decisions and actions will lead to sustainable competitive advantage in this arena?

The trend toward conducting business through electronic media integrates mature technologies such as EDI, emerging technologies such as smart cards, and the growing commercial use of the Internet. The popularity of this public network ballooned only after colorful graphics, audio/video capabilities, and "point and click" navigation replaced UNIX commands. With the arrival of adequate transaction security software, the Internet became a global market, with a wide variety of goods and services now only a mouse click from its many users. Asian firms are exploring technology-based global services and alliances to exploit these new opportunities. Because these alter the nature of money, they may lead to new technical and legal standards, awkward governance issues, and strategic shifts in the structure of the value systems linked to public networks.

---

1 Business Times, Singapore, May, 1996

2 "Security Issues: Security First Network Bank," http://www.sfnb.com/security.html

2 Electronic Commerce and Industrial Structure

Electronic Data Interchange (EDI) is the inter-process (computer application to application) communication of business information in a standardized electronic form. This includes EDI, content and process related to production and distribution, the transfer of moneys, and sharing common data bases to aid in the efficient conduct of business. Electronic Funds Transfer (EFT) describes direct payment transfers, for example to a creditor's bank account. First Virtual applies EFT to make payouts to merchants who make sales through the Internet. They collect numbers from the bottom of the check merchants use to pay for application to the service, then store these offline. Payments apply a "Virtual PIN" process to conceal sensitive information, and all transactions are positively acknowleged via return email. This process is called electronic reconciliation (ERS), and requires transmission of statements by Internet. These processes represent the inter-organisational flows of information required to coordinate (and often to replace) the physical activity which creates product value [3].

Such flows are not static. The structure of a value chain and its external value system must adapt to changes in three contextual factors: physical links among organisations, the information flows among them, and the drivers for production, distribution, and transactions costs which determine economic performance [4].

The traditional value chain for women's jeans is to produce, distribute, then sell. In a reconfigured system, a customer is carefully measured after selecting a style. The retailer transmits this style and measurement information to a fabric cutter in another firm. They, in turn, forward laser-cut fabric pieces to stitchers at a third company. The sewn jeans are washed by a laundry at a fourth site, which ships the finished pair to the retailer for delivery to the customer. This buying-manufacturing-delivery cycle alters the sequence of activity in the value chain to maximize product differentiation, and is one model for a new breed of emerging global, virtual companies [5].

But electronic commerce is also relevant to cost leadership. To be the lowest cost producer demands the best product development, design, manufacturing, and distribution links over the life-cycle of a product or product family. This requires digital connections to markets and suppliers. For example, to sell Proton automobiles in the Philippines, Malaysia must compete with Korean manufacturers not only in labor costs, but in implementing computer assisted design and manufacturing software and in managing its international network [6].

---

[3] Porter and Millar 1985.
[4] Arrow 1974.
[5] Interactive Custom Jeans, at http://206.65.84.96/
[6] Darby 1995

2.1 The Actors and their Interests

To fully understand long-term impacts, it is first necessary to examine the interests of the main parties in the electronic commerce trend:

Parties to transactions are the driving force for the diffusion of electronic commerce. Their needs vary by transaction type: obviously, security and authentication are critical for high-value items, while small convenience exchanges (e.g., on-line news) demand very low costs and easy execution, and other transactions may require privacy. Manufacturers may be more sensitive to speed, security, and privacy, while suppliers will probably be concerned with added fixed costs, difficulties in managing their interfaces with multiple payment channels, and liability issues associated with performance failures.

Manufacturers design, produce, and distribute their goods through channels, thus creating a series of transactions along the value chain. In many segments, the basis of competition is shifting toward shorter product cycle times and more cooperative links with customers, which demands global telecommunications network access.

Industrial park operators provide the infrastructure necessary for production and distribution. In a high-velocity environment, their tenants need access to information-related functionality such as logistics networks, international standards, and telecommunications.

Fiduciaries are entrusted with the transfer of payment among the parties to the transactions. They include banks, various types of card operators, government agencies, and POS network operators.

Governments have legitimate interests in regulating money flows within a national economy, and may intervene by limiting consumer credit or regulating transborder currency flows. Central banks worry (perhaps uneccessarily) about loss of revenue from the float associated with seignorage. Open policy issues include controls over the right to issue electronic money, reserve requirements for issuers, the definition of money from an economic policy perspective, and interest and tax payments on value held and remitted on the network [7].

Suppliers of information technology see the rapid evolution of electronic commerce as a market opportunity. Transaction security is a critical success factor: contneders include the Visa- and Microsoft-backed Secure Transactions Technology (STT), or the Secure Electronic Payment Protocol (SEPP) standard proposed by MasterCard, GTE, IBM, Netscape, and CyberCash. Both standards were viable from a technical perspective, but suppliers and fiduciaries soon realized only one was likely to attract the necessary critical mass of merchants, so they joined the Secure Electronic Transaction (SET) standard, which now looks sure to dominate over the near term.

---

[7] Laster, D and J Wenninger, "Policy Issues Raised by Electronic Money," Federal Reserve Bank, USA, http://startide.ctr.columbia.edu/citi/history/emon.html

2.2 The Evolution of Money

Barter was an effective institutional solution to the problem of exchanging an excess of one commodity (e.g., your labour) to fill a shortage of another (rice or fish), and improved local trading efficiency. However, as barter markets are inherently inefficient, especially for goods that cannot be stored over time, cash gradually came into use.

Cash solves the problem of storing value over time, and by enabling price comparisons, reduces costs for negotiating transactions. Portable and largely anonymous, cash is convenient, yet its owners are vulnerable to loss, theft, and damage. Cash also has drawbacks: it costs billions to manually count, store, move, and protect printed cash, while the war against counterfeiting, necessary to stabilize its value, escalates with each advance in reprographics technology.

Checks are essentially a personal promise of cash, while *Commercial instruments*, with terms and conditions of payment tailored to fit specific needs, represent the institutional evolution of the check. A letter of credit, for example, authorizes a bank to pay a previously negotiated amount to a seller on behalf of a buyer upon the satisfaction of specific terms and conditions, normally associated with delivery [8]. The intended effect of most commercial instruments is to mitigate risk, and hence reduce the transaction costs, for all parties.

Credit card transactions require the card service operator (e.g., Visa) to promise to pay the merchant, and obligate the card owner to repay the operator. As a substitute for cash, cards reduce the owner's risk of loss, and as a substitute for checks, they reduce a merchant's cost for collections. Credit card operators help merchants verify cards and assume some responsibility for monitoring fraud, thus reducing liability for both the card holder and merchant [9]. According to Scott Cook of Intuit, with few changes in the system, the standard credit card is a secure and reliable medium for payment over a network [10].

Stored-value cards, which also contain a small magnetic memory, have been in widespread use for more than a decade as a convenient exchange medium for fairly small transactions such as telephone tolls and public transport fees. Stored-value card service operators are paid in advance for services which will be delivered only in the future, and earn significant revenues by investing the resulting "float" in the money markets. Dimensionally similar to credit cards, a proprietary stored value card is less bulky than coins, eliminates the need to make change, and generates no records to link card owners with their transactions. However, its widespread use necessitates a reader at each point of sale, and outlets where cards can be obtained. Also, despite initiatives to increase security (such as the Oscard), the

---

8 Bannock and Manser, 1995.

9 Richards, S; "Electronic Money/Internet Payment Systems," Ohio State University, 1996, http://www2.cob.ohio-state.edu/~richards/bankpay.htm

10 Flohr 1996, p. 74-84.

underlying technology is highly vulnerable to counterfeiting and duplication, which inhibits applications for larger transactions [11].

Smart cards contain a single-chip microprocessor with onboard memory which can be access-protected and are more expensive than earlier card technologies. Downward-compatible with credit cards, and providing a secure interface to ATM cash dispensers, they will rapidly displace credit and stored-value cards. Smart card processors provide sufficient security for use in high-value transactions, although users must now enter a PIN during payment. Widely used in Europe for small transactions, smart cards linked to DTMF transceivers can access ATM and telephone networks, while the CAFE field trials take the infrared path to wireless secure transactions [12].

Electronic Payment requires "using networks to conduct business transactions with one another electronically" [13]. A payer moves funds to the recipient's account, and both receive verification of the transfer. Large institutions have used electronic funds transfers, for a variety of applications ranging from payroll to inter bank transfers, for decades. With the diffusion of the World Wide Web and new secure protocols for use on public networks (e.g., STT, SET, and SEPP), smaller producers can safely movefunds to suppliers, mortgage company, or data services, who can immediately use inflows of electronic money to cover their business expenses, or convert it to other forms of cash. These new systems include electronic checking services and digital cash.

Digital Cash now appears in a variety of competing forms. The key differences lie in the capacity to conduct *off-line* transaction (without the presence of the buyer or on-line links to a bank), and whether or not the e-money leaves a trail of information to *identify* the original purchaser with its use as it flows through the economy [14].

Analysis: The shift to e-money is strategic because it solves time and space problems, while its digital cash variant is significant chiefly because it also enables protecting the anonymity of participants to transactions. Today, it is quite feasible to design a secure payment system to prevent duplicating large amounts stored in a computer or smart card and ensure payment to sellers, while masking the identity of buyers. Network-based e-money thus appears ideal for global commerce [15]. It complements EDI and other network-based forms of commercial cooperation, and is especially relevant to small suppliers.

---

11 Jack 1996, p. 82.
12 Flohr 1990.
13 Network World, Feb. 1994
14 Miller, J; "E-money mini-FAQ (release 2.0)," 5 2 96. http://www.ex.ac.uk/~RDavies/arian/emoneyfaq.html
15 Economist, Vol. 333 n 7891, 26 Nov. 1994, p. 21.

2.3 Emerging Trends in Technology

Many technologies could play an important role. The National Industrial Information Infrastructure Protocol (NIIIP) Consortium is a group of leading U.S. information technology suppliers and users with a common interest in deploying computing solutions to make manufacturing industry more efficient and globally competitive through enabling the "virtual enterprise". The project uses intelligent agents to build virtual "office buildings" which house agencies and agents that can pursue a variety of goals, such as information search and retrieval, change notification, and product pricing [16].

Digital Equipment Corporation was a pioneer in communicating information over Internet newsgroups and bulletin boards, because "This is a technology that makes us look attractive by allowing us to project ourselves all over the world -directly to the screens of our customers." Customers around the world had access to over 3,000 documents, including technical manuals and research reports. Digital was also the first computer industry vendor to offer online ordering of its full product line over the Internet. Customers could obtain descriptions of hardware, software, and services, read Digital's announcements, configure systems, generate price quotations, and place orders. Potential customers could also test run their applications in the new Alpha AXP environment - a computer that was sold as a server for use with the Web and other internet applications. By early April 1994, more than 2,500 potential customers from 27 countries had logged onto these systems for a test drive [17].

Of course, innovative manufacturers will soon discover new applications of E-commerce technologies, as outlined in Figure 1:

Figure 1: Typical Manufacturing Applications of E-Commerce Technologies

TECHNOLOGY/function	Design	Production	End Use
1. **Web Publishing Technology (HTML)**	Global Intranet to support CAD	Standards Manuals Intranet for CAM	Customer Support Marketing & Sales
2. **Global Positioning Systems (GPS)**	-	Tracking Inbound Shipment	Tracking Outbound Shipment
3. **Secure Transactions**	Confidential design data	-	Confidential customer data
4. **Payment Systems**	Worldwide design teams	Virtual value chain	More responsive customer support.
5. **Wireless PDA**	Anywhere, anytime designs	MRP tracking and expediting	Customer Support

The following section presents the technique used to study the potential impacts of these applications.

---

[16] Gilbert 1996

[17] ISworld, http://www.cox.smu.edu/mis/cases/dec/internet.html#interprod

3 The Future of Network Commerce

Forecasting is awkward when the future is likely to be determined by trends that are not extrapolations of the present, or events for which there is no precedent [18]. A scenario is "a hypothetical sequence of events constructed for the purpose of focusing attention on causal processes and decision points [19]" and prsents logically coherent descriptions of activity within fundamentally different futures. The technique identifies major opportunities, threats, and uncertainties, and facilitates strategic learning at low real and opportunity costs.

Building Multiple Scenarios requires special knowledge and skills, and cooperation between functional specialists and external experts. Useful scenarios represent the few dimensions of an alternative future relevant to decisions actually at hand. The first task is to identify the critical decisions which are most sensitive to environmental factors.

Selecting scenario logics (two to four are enough) is next. For example, a manufacturing firm evaluating a major new EDI system would first identify key variables in its task environment. These might include national standards, internal procedures governing transactions, and vendor relationships. Factors in its remote environment include economic growth, local and international policy governing currency flows, user perceptions of service value, plus performance, cost, and access to national and regional telecommunications infrastructure. Three scenario dimensions, each with only two states, may economically represent the key dimensions:

Figure 2: Three E-Commerce Scenarios

SCENARIO/dimension	COMPUTE	CONNECT	CONTENT
6. **Baseline**	Commodity	Bandwidth limited	Local asset
7. **MNC Domination**	Scale economies	Private networks	Data mining
8. **Spherical Hub**	Declining asset	Global commodity	Sustainable asset

Such draft scenarios are merely the basis for more elaborate treatment and analysis. The goal is to find the critical issues, threats, and opportunities emerging from each scenario, and to identify those policies which will be the most robust in the more likely outcomes. Once scenarios are complete, the focus shifts to task-level decision factors that may affect the outcome of key decisions. At each step, multiple scenarios highlight specific needs for domain knowledge, define boundaries for analysis, and differentiate critical variables.

[18] Kahn 1974.

[19] Kahn and Weiner 1967, p. 6.

4 Analysis: *lessons for manufacturers from banking*

The highly IT-dependent banking industry commands a central role in the commercial world. Banks see a need for secure and convenient systems for use on public networks [20]. One leading survey predicts 16 percent annual growth for electronic commerce and payments, with projected global flows of US $800 billion by the turn of the century [21]. However, this will demand links between fiduciaries and merchants, which will happen only if they can agree on technical and business standards. With these, any networked computer could be an automatic teller machine, from which consumers can download the electronic equivalent of cash in the privacy of their homes, hotel rooms, or (with wireless links) even rental cars. Customers can perform routine personal banking -opening accounts, making deposits and securing overdrafts, loans or guarantees- on-line.

While many local banks have an Internet presence (in Singapore, DBS, OCBC, OUB, and UOB have sites providing information on their banks, products and services, and smaller banks are set to follow suit), US banks lead in the actual delivery of interactive services to customers. Tiny Kentucky-based Security First Network Bank (SFNB), a pioneer in the Internet banking arena, typifies the dramatic impact of public network access. The world's first insured bank to operate on the Internet [22], SFNB provides the same services as any "brick and mortar" bank, plus added convenience to customers on the Internet.

SFNB fees compare favorably with traditional services, because its virtual banking business model is more efficient. SNFB portrays its customer coverage as a giant "footprint" spanning the nation through the Internet. However, its only physical infrastructure is the "control hub" for Internet operations, located in and controlled from Pineville, Kentucky. SFNB not only offers its customers a vast array of on-line services, but actually cut costs, mainly by reducing operating expenses and leveraging its low-cost physical infrastructure. [23].

Network operators and banks address security at several levels and points of interaction among the Bank and its customers over the Internet. Critical issues are: (1) end-to-end transaction integrity across the Internet, and (2) protecting business processes against intrusion. As a precaution, SFNB regulates money outflow (as for withdrawal limits for ATM transactions), with limits tailored to customer needs.

Although network-based competition may erode the value of branches, established firms may be able to retain the residual value of their knowledge and expertise, and use these assets to retain their customer base. If so, they may continue to dominate specific segments

---

[20] Bogolin, 1995)
[21] Business Times, May, 1996
[22] SFNB ; http://www.sfnb.com.
[23] ibid.

of markets, even if Internet banking becomes the dominant model in the future. However, it is equally likely that the structure of the banking industry, the firm's pattern of interaction with its customers, and the critical measures for success will change dramatically [24].

As the world's first Internet-based bank, Security First Network Bank (SFNB) provides a window to the future. Its efficient online business model enabled the bank to cut operating costs while delivering superior customer service. However, we have only a hazy view of its long term viability. In response, global banking firms and specialized software companies are likely to forge strategic alliances to seek the potential synergy between the two business types. The three-way joint venture between Windows Corporation, the personal finance software firm Intuit, and Visa International is an early example.

Scenario analysis shows that service firms are likely to face intense competition from non-traditional competitors. Thus, strategic alliances among software development firms, Internet service providers and local banks may be required to defend current customer relationships.

On the technology front, we see that firms with software development strengths have a potential source of competitive advantage, because software lies at the heart of both banking and networks, and the future of electronic commerce seems likely to require integrating a wide range of new hardware devices into network applications. For example, wireless terminals would provide both mobility and instant access to financial information and resources, while smart card readers would become standard on every PC and public telephone.

Another key competence will be leveraging content to enhance customer service [25]. By extension, the banking industry suggests that it will be critical for many other firms to maximize the value of customer assets and improve product distribution efficiency [26]: Here banks lead the way for the manufacturing industry.

1. Maximizing the value of the customer asset means each firm must understand its customer, learn more about market segmentation and client interests, then focus all their energy on exploiting the full value of their relationships with customers. The scenario analysis suggests that networks are ideal tools to support such objectives.

2. Cost effective design, production, and distribution of online customer support services requires substantial organizational skills, plus competence in software development and I.T. operations, and analysis suggests that firms without access to these assets will be at an increasingly severe disadvantage under all three scenarios.

---

[24] Prince, 1996
[25] Post et al., 1995, Chia, 1995
[26] Sraeel, 1995

5 Conclusion

In summary, while many important factors remain unclear, network-based electronic commerce appears poised to dynamically alter global relationships between major actors in manufacturing and its support industries. As the e-commerce trend matures, the resulting transformation will create new opportunities for nimble manufacturing firms with well-developed network access skills. The business risk for Asian producers, whose current investments in network technology lag behind their global competitors, is likely to be dramatic. Thus, Asian industrial park operators have an opportunity to invest in network capacity and manufacturing-related functionality, and use this as the basis to form alliances with producers on favorable terms.

Selected References

Arrow, K; The Limits of Organization, Norton, NY, 1974.

Bannock, G and W Manser, International Dictionary of Finance, Economist Books, London, 1995.

Benjamin, R and R Wigand; "Electronic Markets and Virtual Value Chains on the Information Superhighway," Sloan Management Review, Winter 1995, p. 62.

Bogolin, K; Homing in on Home Banking, Consumers are Ready", Bank Management, January/February 1995, p. 60-62.

Chia, P "Citibanking the World", Bank Management, July/August 1995, pp. 30-37.

Darby G, "The Value Added Teleport," Proceedings of PTC 1995, Honolulu., 1995.

Flohr, U; "Electric Money," Byte, June 1996, p. 74-84.

Gilbert, D; "The Role of Intelligent Agents in the Information Infrastructure," Proceedings of PTC 96, Honolulu, 1996.

Hannan, T.H., and J.M McDowell; "The Impact of technology adoption on market structure", Review of Economics and Statistics, February, 1990, (72) 1, pp. 164-168.

Jack, A; "Oscard Makes Phones Pay, " Byte, June 1996, p. 82.

Kahn, H; The Future of the Corporation, Mason and Libscomb, New York NY, 1974.

Kahn, H, and Weiner; The Year 2000, Macmillan, London, 1967, p. 6.

Kobielus, J; "Look before leaping into Internet electronic commerce", Network World, February 28, 1994, pg. 32.

Porter, M and V Millar; "How Information Gives You Competitive Advantage," Harvard Business Review, (63) 4, 1985.

Post, G, V., A Kagan, and KN Lau; "A Modeling Approach to Evaluating Strategic Uses of Information Technology", Journal of Management Information Systems, Fall 1995, (12) 2, p. 161-187.

Prince, C; "Winning the Home Banking War," Bank Systems & Technology, April 1996, pp. 46-9.

Richards, S; "Electronic Money/Internet Payment Systems," OSU, 1996, http://www2.cob.ohio-state.edu/~richards/bankpay.htm

Slater, A;"A Primer on Internet Security: Citibank," Class Presentation, Global Information Technology, October 1996.

Sraeel, H; "Five major Trends in Banking," Bank Systems & Technology, Jan 1996, pg. 38-46.

Tranctincsky, N, and S Jravenpaa; "Information Systems Design Decisions in a Global Versus Domestic Context", MIS Quarterly, December, 1995, pp. 507-528.

A CALS Framework for Electronic Commerce in Virtual Manufacturing Enterprises

Dennis C. H. Sng
(dennis@gintic.gov.sg)

Systems Technology Division, Gintic Institute of Manufacturing Technology, Singapore

Abstract. Today, the increasing popularity of the Internet and intranets has many enterprises embarking on Electronic Commerce (EC) initiatives. The convergence of computer networking and telecommunications technologies has made it possible for virtual enterprises to be formed from previously separate companies. The growing CALS movement is attracting a lot of interest, especially in Asia, with its promise of a strategy for EC. This paper presents a framework for the implementation of CALS to support EC in virtual manufacturing enterprises.

1. Introduction

Global competition has raised customer expectations of product performance, quality, and cost. Enterprises are outsourcing processes in a product value-chain in an effort to meet these challenges, thus forming consortiums of client, supplier, and sub-contractor companies. In such a consortium of enterprises, the challenge is to create a uniform understanding of the product design, manufacturing process, and logistics support required to take a product from concept through production to delivery. Lack of, or outdated information, often result in unnecessary engineering changes, which add to the cycle-time and cost. A 1993 National Productivity Board survey found that Singaporean companies spent S$10 billion a year to produce products, of which S$7 billion went into rectification of avoidable problems [1]. Enterprises are embracing Concurrent Engineering (CE) to address these needs.

For effective CE, the product developers must be able to work from a common product database, in real time, on the design, development, manufacturing, distribution, and servicing of products. This product information must be based on a model of the Product Development Process which spans across the product value-chain organisations. This will facilitate Electronic Commerce (EC) across organisations in the value-chain, resulting in a reduction in overall product time-to-market and life-cycle costs, as well as an improvement in product quality.

For this strategy to succeed, a framework unifying information technology (IT), information interchange standards, and information infrastructure with key business processes, is required. Such a framework will integrate a manufacturing enterprise with its clients and suppliers (which may be distributed globally). They will be able to share and exchange information seamlessly and in real-time. This convergence of computer networking and telecommunications technologies is making it possible for groups of companies to coordinate geographically and institutionally distributed capabilities into a single "virtual enterprise" or VE [2]. The shared information must include both business information (eg. purchase orders, invoices), and technical information (eg. CAD data, technical documents). With such a framework, it would be possible to implement an EC strategy to give companies a competitive edge in the global market place, and turn them into world-class manufacturing enterprises.

Ideally, this framework should be generic enough to be applicable in any industry and over all phases of a product's life-cycle. However, in this paper, the scope will be limited to the Die & Mould sector, and how it supports the Electronics and Precision Engineering clusters. Also, rather than studying the entire product life-cycle, the focus shall be on the development & engineering phases of a product (ie. the phases between industrial design and manufacturing). We will first look at how CALS and EC have evolved. Next, we look at the infrastructure being developed in Singapore to support EC. Finally, we will propose our CALS Framework for Electronic Commerce in Virtual Manufacturing Enterprises.

2. CALS Overview

CALS (Continuous Acquisition & Life-Cycle Support) began in 1985 as a productivity and quality initiative of the US Department of Defense. Based on standards for communicating information, its goal was to integrate the product-oriented systems of various organisations. Originally applied to CAD data and technical manuals, CALS now applies to all data being communicated about the design, development, manufacture, distribution, installation, and servicing of products, ie. the entire product life-cycle. CALS is a strategy designed to move from the current paper-intensive environment to an automated and integrated mode of operation with "minimal" paper requirements. Its goal is *interoperability* [3]. CALS supports CE through the sharing of product data among organisations. The success of CALS relies on two key elements: standards and infrastructure:

- *Standards* are the foundation of CALS as they enable the sharing of product data. The CALS Industry Steering Group supports a set of baseline standards which cover electronic data interchange (EDI), image & raster formats, technical documents, product data modelling & exchange standards, interactive electronic technical manuals, and many more. Standards are also emerging in other areas, such as workflow management and product data management.
- An enterprise or consortium-wide information *infrastructure* should provide an open environment based on common standards for user interaction and information exchange. It provides the basis for information sharing, virtual teaming, and enterprise integration, allowing any member of the consortium to access information anywhere in the world, from anywhere in the world. The best candidate for a world-wide information infrastructure is the *Internet*.

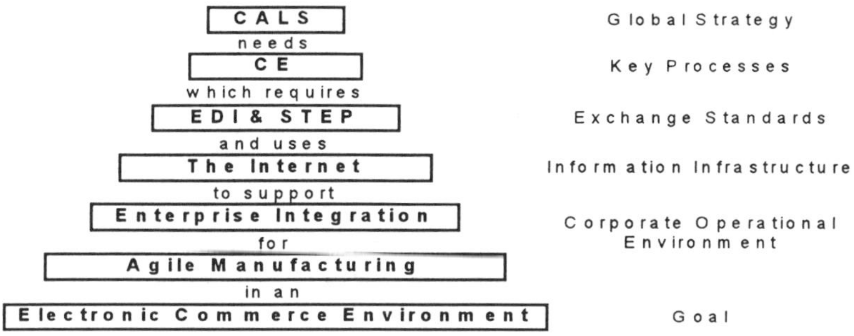

Figure 1: Relationship of CALS to other initiatives

CALS is thus a strategy that can be applied to any industry with a significant volume of business & product information that must be shared within the enterprise and among its suppliers. In the USA, it has been successfully applied in the defence, aerospace, shipbuilding, and automotive industries. In Japan & Korea, leading companies in the automotive and electronics industries have CALS programs. In Singapore, many companies are using EDI for information exchange with their suppliers and are now planning to go beyond EDI to EC.

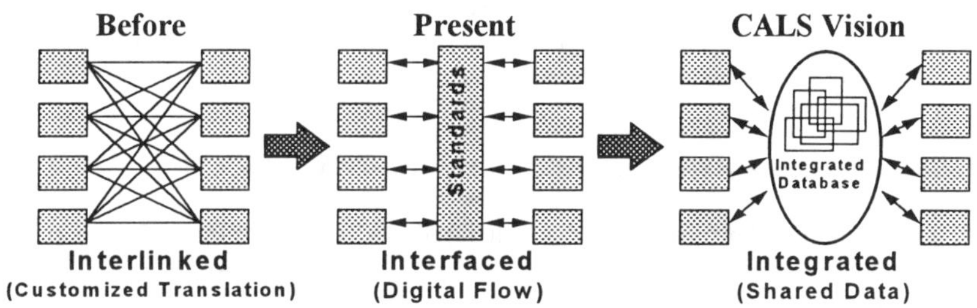

Figure 2: CALS - From Information Exchange to Information Sharing

3. Evolution of Electronic Commerce & Virtual Enterprises

The Gartner Group defines EC as "an evolving set of information technology tools and implementation techniques, as well as the business strategies and practices necessary to 'do business electronically'". Generally, EC covers all forms of business transactions that are conducted electronically using telecommunications networks [4]. In this paper, we shall focus on EC between business partners, rather than business to consumer transactions. Business to business transactions can occur between suppliers & customers, between the private sector & public sector, and between trading partners. It is the opinion of the author that the business-to-business component of EC will represent the more significant portion of transactions, both in terms of value and volume, in the short to medium term.

EC includes the electronic trading of goods & services, and the distribution of electronic material (eg. software, video, images, etc.). Upstream and downstream transactions include the facilitation of contracts between business partners, pre &

post sales support, electronic procurement, and support for shared business processes. These EC supported business processes can span corporate boundaries, with each enterprise acting out its own portion of the shared processes. It thus gives rise to the concept of the VE, where each participating organisation contributes its own core competencies in a collaborative consortium of enterprises.

EC is not only a network, a standard, application software, electronic mail, electronic funds transfer, transaction processing, or the Internet. EC refers most effectively to processes rather than any technology or event. Business processes take place over periods of time, and normally possess a repetitive quality (possibly with variations). Events occur instantaneously and generally reflect isolated behaviour, as is more evident in retail purchasing. It is the emphasis on (repeatable) business processes that will derive benefits from the EC concept [4]. The processes associated with the manufacturing, distribution, and procurement of products and services are the ones most influenced by EC.

Our concept of VEs differs from contemporary definitions like those by Goldman et al. [2]. Those have very dynamic and opportunistic characteristics. Our concept is more strongly influenced by Asian culture where the member organisations maintain long term business partnerships, but yet have a high degree of agility because of the use of IT and information infrastructures. The key feature of our vision is *common access to shared product and business information via IT and information infrastructure*.

4. Information Infrastructure - Singapore ONE

The National Computer Board's (NCB) IT2000 masterplan aims to transform Singapore into an "Intelligent Island" with a broadband information infrastructure playing a key role in delivering applications and services to every sector of society [5]. The means to achieve this is the National Information Infrastructure (NII), "*a well integrated and extensive nation-wide information infrastructure that facilitates the unobstructed flow of information across and within all sectors of Singapore*" [6].

In the near future, the NII will take the form of Singapore ONE (One Network for Everyone) a national broadband network that will deliver a range of multimedia services to the workplace, the home and the school. The key players in this initiative are: the Telecommunications Authority of Singapore (TAS) who are responsible for infrastructure development, the National Science & Technology Board (NSTB) for broadband technology development, and the NCB for application development.

Singapore ONE comprises two levels - an infrastructure level of networks and switches, and a level of applications and multimedia services [7]. The *infrastructure* level consists of a core broadband network connecting several local access networks. The backbone is built, owned, and operated by an industry consortium. Currently, both Singapore Telecom's (SingTel) and Singapore Cable Vision's (SCV) local access networks can be used for the delivery of multimedia services. In the future, technologies such as fibre-to-the-home or broadband wireless may become viable as access networks. The key components of the network are shown in Figure 3:

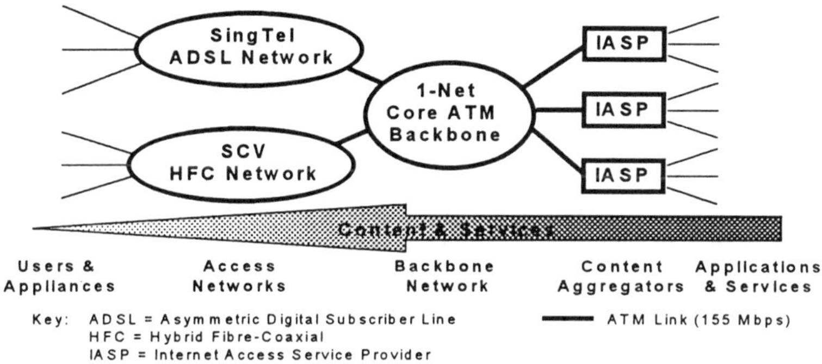

Figure 3: Singapore ONE Network

- The 1-*Net Backbone Network* is a high speed ATM network running at 155-622 Mbps. This network, which has been under construction since 1996, will also link Singapore to the rest of the world via high-speed digital undersea cables.
- *SingTel's* existing telephone network with fibre-to-the-curb and copper twisted-pair cable for the local loops. Asymmetric Digital Subscriber Line (ADSL) technology allows high bandwidth data transmission over the downlink (ie. from access network to the user) over the existing copper twister-pair cable in the local loop. With ADSL, transmission speeds of between 1.5-6.1 Mbps downlink and 64-640 Kbps for the uplink are realisable [8].
- *SCV's* Hybrid Fibre-Coax (HFC) network will provide services via cable modems. The backbone infrastructure is again optical fibre, with copper coaxial cable running up to the door. Cable modems typically allow up to 30 Mbps service for the downlink and up to a few Mbps service for the uplink [9]. However, the user's computer is connected to a cable modem via a shared 10 Mbps LAN which reduces the available bandwidth.

Applications and services are essential to the success of Singapore ONE. Application development is encouraged in four areas - government, home, education, and business. Singapore ONE will make it more convenient for the public to carry out government transactions, through services like multi-function kiosks and one-stop government centres with video conferencing facilities. High speed communications will open up a whole new channel home applications, including entertainment on demand, electronic shopping, home banking, and electronic information services. Singapore ONE will enable easy access to high quality multimedia teaching and learning materials, as well as support distance learning and multi-party collaboration by students at different locations. For businesses, a high speed multimedia network will mean better communication, coordination, and business transactions. Potential business applications include video conferencing, multimedia information services for corporations, tele-commuting, and EC.

Singapore ONE will be an operational pilot network by 1998. In the initial years, the government will be the anchor tenant. But in time, Singapore ONE will take on a commercial flavour as more applications and services ride on it, forming the critical mass which will lead to an explosive growth of activity on the network. Singapore ONE will be implemented in two overlapping phases.

- *Phase 1 (1996 - 2001)* will start with the selective deployment of a pilot core broadband network, comprising several broadband ATM switches linked to offices, public places and homes via existing local access networks. By 1997, pilot services will include access to schools' curriculum materials, the digital library, and commercial services for homes, as well as virtual government services in public sites. Up to 5,000 homes will pilot the various services on Singapore ONE by end 1997.
- *Phase 2 (1999 - 2004)* will see the network grow in capacity as more switches and cables are added, and more applications will come on-stream. Over the years, as cost declines, home connections may be upgraded to allow home users to take full advantage of the high bandwidth of Singapore ONE. Private sector activities will gradually become the main driving force for the network.

5. CALS Framework for Electronic Commerce

5.1 Generic Framework

In this section, we propose a generic CALS framework for EC. This framework consists of sets of applications which are generic to the manufacturing industry. It can then be customised to specific industry clusters and sectors. In this way, the core technologies can be re-used in different application scenarios and domains.

5.1.1 Architecture

The proposed architecture of the generic CALS framework is shown in Figure 4:

- *CALS Systems Architecture*: This provides the specification for the integration of CALS infrastructure with manufacturing application systems with the help of reference models. Reference models are represented by functional, information, and organisational models. The CALS architecture is therefore the foundation on which fully integrated systems can be developed and implemented.
- *Common User Interface*: A common user interface to the various applications in the framework improves usability and portability. This usually takes the form of a Web browser as the primary user interface. A browser's universality and portability makes it ideally suited for our framework of applications.
- *Applications*: The applications for EC are wide ranging and include those shown in Figure 4. The various application sets in the framework are discussed in greater detail below.

- *Leveraging Existing Solutions*: The ability to leverage the existing range of manufacturing solutions is critical to the success of this framework. By being able to interface to or even integrate with the existing base of solutions, the acceptance of the framework by the industry would be improved.
- *Infrastructure*: The CALS infrastructure consist of standards, communication protocols, networks, and interfaces. It should support common protocols for database access, data compression utilities, and security mechanisms.

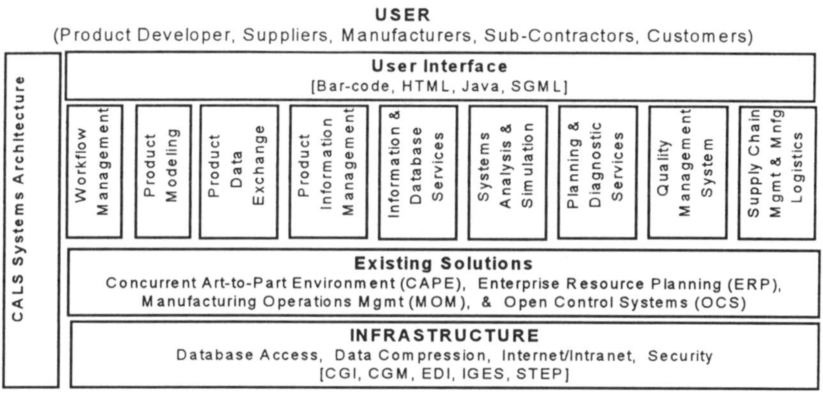

Figure 4: Generic CALS Framework

5.1.2 Applications

Workflow Management: A detailed understanding of workflow is an essential requirement in developing the full CALS framework. Workflow can be highly complex in manufacturing environments and would require effective representation for thorough understanding. The CALS framework will provide the standard format by which the workflow could be represented, analysed, communicated and managed at any required level of detail.

Product Modelling & Data Exchange: Product development teams create data of various types, eg. CAD/CAM models; NC programs; bills of material; schedules; and technical documents. For an organisation to integrate its business processes, it must first determine how product information is represented and then determine how it is communicated to others. The key requirement of complex product manufacturers is robust, stable standards for information. While some companies can tolerate periodic translations between dissimilar CAD systems using ISO STEP, IGES or direct translators, the survival of complex product manufacturers is based on the massive amounts of information moved within and between their internal organisations, and to and from suppliers and key customers.

Product Information Management: What makes CALS different from other EC strategies is the concept of a shared data environment that goes beyond simple digitisation and data exchange. CALS allows companies to extend their enterprise to encompass subcontractors working in full collaboration (ie. CE). *Shared information is the key.* But often CE implementations attempt to interface a number of discrete, specialised applications (CAD, MRP systems, paper-based systems, and so on). Thus, it is often difficult to access or share information.

The solution is a common product database storing all the information needed by all involved, without limiting access based on life-cycle stage. Such an *integrated information environment* supports the execution of streamlined business processes across enterprise boundaries by teams with physically distributed members, enabled by the controlled access to data, effective sharing of data, and the collaborative preparation of project deliverables. What is needed are consistent methodologies for implementing Product Information Management Systems in such a VE.

Information & Database Services: Our vision involves the integration of data from multiple information sources via various information and database services. These services will support data in many diverse formats. From these information sources, information like technical specifications & CAD models, part cost, delivery times, inventory, and other supplier information can be obtained instantaneously. Ultimately, customers will be able to place orders through such a network after the required information is retrieved, and be able to track order status.

Systems Analysis & Simulation: Efficient utilisation of a manufacturing system requires a thorough understanding of its behaviour. This can be achieved through systems analysis and simulation, which are widely used to support decision making during system design and operation. There is a critical need for the enterprise data management system to include detailed information on manufacturing systems to support analysis and simulation tools. The CALS framework will include manufacturing systems information in standardised format and with standard interfaces for systems analysis and simulation tools, resulting in substantial reduction in analysis and simulation times leading to efficient use of resources.

Planning & Diagnostic Services: As more automated control systems are introduced into distributed manufacturing facilities, it will become more difficult to maintain today's centralised process planning systems. A new approach is required to permit distributed manufacturing operations planning via a network of cooperating agents on the Internet. Planning tasks can be distributed across diverse locations, with the information consolidated to form a complete manufacturing plan.

Currently, diagnostic tasks are mainly dependent on the knowledge of repair personnel, or the information in trouble-shooting manuals. By exploiting the Internet, the diagnostic knowledge and information of specific products, processes or equipment can be captured and accessed by users all over the world. Cases of known problems and corrective actions can be stored on the net to ensure the timeliness and accuracy of information required for future diagnosis of similar problems.

Quality Management System: The World-Wide Web can provide a mechanism for authorised users to retrieve relevant quality documents from anywhere in the world which provides Internet facilities. It will also allow regional review and approval of corporate quality documents which may need a wider scope of opinions.

Different world-wide branches of a multi-national company can also have ready and on-line access to corporate policies and documents. Other potential applications include a corporate intranet-based *Qualified Supplier Management System,* which provides a world-wide database of qualified suppliers/vendors. Vendor history can be updated by various world-wide branches of a company which can be used for variance analysis and quality improvements.

Supply Chain Management (SCM) & Manufacturing Logistics: The SCM business model organises the company around the customer and suppliers in a completely integrated manner. Logistics begins where purchasing ends and ends at customer shipments. In between, activities involved include receiving, storage, production, inventory control, distribution, etc. SCM however, encompasses the entire product life-cycle from raw materials and manufacture of parts, to the assembly of the final product and its distribution to the customer, to its product maintenance and disposal or re-cycling. The ability to share business and product information is critical to the SCM concept.

5.2 Framework for Die & Mould Sector

In a typical supply chain scenario, a client will place orders with a manufacturer, say of consumer electronics products. This manufacturer in turn will order tools (die sets and moulds) from tool (die & mould) makers. These tool makers will order standard parts for their die sets and mould base from component suppliers. These tool makers will then design the tools, fabricate the parts, before assembling them into the final tool. These tools can then be used to produce plastic and sheet metal parts for the manufacturer.

The CALS Framework for Die & Mould sector [Figure 5] has been customised from the Generic Framework to support Electronic Commerce in such a supply chain scenario. The applications under this framework have been divided into two categories: Product Development Support and Manufacturing Engineering Support. Projects under the category of Product Development Support are:

- *QuickMold*: This project aims to develop tools to shorten the design life-cycle for the die & mould sector. It captures the knowledge of skilled mould designers in a 3D solid modelling tool.
- *Product Data Exchange Service (PDEX):* This project aims to provide effective CAD/CAM data exchange service between product developers and their suppliers. It also tries to help the companies to upgrade from traditional 2D draughting to 3D solid modelling for their design work.
- *Standard Parts Database Server:* This project allows engineers and purchasers to search for mould base and die set parts, and vendors for these parts, over the Internet via the World-Wide Web. Part specifications & drawings, and vendor information can be obtained conveniently via a Web browser.

The scenario here begins with the product developer at a consumer electronics manufacturer who sends the product model to the mould designer via the PDEX service (assuming that the manufacturer and mould maker are using different CAD systems). The mould designer can then begin designing the mould using QuickMould. In the process of mould design, the mould designer will make use of the Standard Parts Database Server to identify suitable mould base parts, and suppliers for those parts. The part parameters can be downloaded into QuickMould and the mould designer can proceed with the design without having to create models of the parts if they do not already exist in his database.

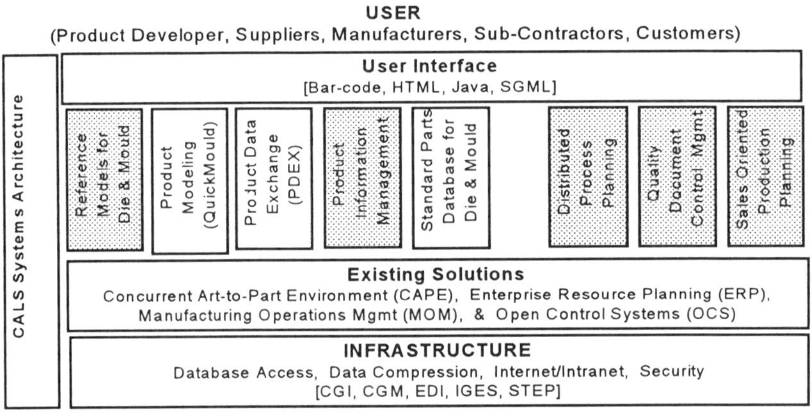

Figure 5: CALS Framework for Die & Mould Sector

The Manufacturing Engineering Support applications will facilitate the integration of the manufacturer with his sub-contractors and suppliers, as well as his clients. Projects under this category are:

- *Methodologies for Product Information Management Implementation* for implementing distributed product development environments which span the VE
- *Development of Reference Models* of best practices in supply chain management in electronics industries
- *Distributed Process Planning Services* for distributed manufacturing operations planning via a network of cooperating agents on the Internet
- *Quality Document Control Management System* using the Internet for better material quality control
- *A Sales-oriented Production Planning System* for on-line inter-organisational sales transaction and suppliers' production capacity booking
- *Third-party Logistics Management* for monitoring and managing suppliers' delivery lead-time via the Internet or intranets

5.3 Deployment

The National Computer Board has adopted the proposed CALS Framework as the strategy for implementing the IT2000 initiative in Singapore's Manufacturing &

Distribution cluster. The CALS Framework will be pilot tested on the Die & Mould sector, a key supporting industry in Singapore. Deployment of the CALS Framework for this sector will be in two phases: Phase 1 will include the Product Development Support applications, while Phase 2 will cover the Manufacturing Engineering Support applications. With the experience of deployment in the Die & Mould sector, frameworks for other industries can be derived.

6. Conclusion

In this paper, we have shown how Electronic Commerce (EC) supported the evolution of virtual enterprises (VE), and why a framework for EC can facilitate the integration of the organisations in the VE. We have proposed a Generic Framework for EC, which can be customised for specific industries. The applications deployed in each framework will facilitate intra- and inter-enterprise business processes over the entire product life-cycles. The Singapore ONE network will be the high-speed infrastructure to realise the transactions within and between VEs.

Acknowledgements

The author would like to acknowledge the contributions of all staff at the Systems Technology Division of Gintic Institute of Manufacturing Technology and the Manufacturing & Distribution Cluster at the National Computer Board who have contributed to formulating the CALS Framework vision.

References

[1] "Singapore firms spend billions to correct avoidable problems", *Straits Times*, Singapore, 2 April 1993.

[2] Goldman, Steven L., Nagel, Roger N., and Preiss, Kenneth, *Agile Competitors and Virtual Organizations: Strategies for Enriching the Customer*, Van Nostrand Reinhold, New York, USA, 1995.

[3] Payne, Robert A., "CALS", *Datapro*, Datapro Information Service Group, New Jersey, USA, June 1994.

[4] Wallace, Stephen, and O'Dowd, Helen, *Electronic Commerce*, Project Report, Information Assets Pty Ltd, Brisbane, Australia, 1 November 1996.

[5] National Computer Board, *The IT2000 Report: A Vision of an Intelligent Island*, SNP Publishers Pte Ltd, Singapore, March 1992.

[6] Yap, Michael K. S., "Singapore NII: Beyond the Information Highway", *Information Technology - Journal of the Singapore Computer Systems*, Vol 6 No 1, Singapore Computer Society, Singapore, September 1994.

[7] NCB, *Singapore ONE,* National Computer Board, Singapore. (Available: http://www.s1.gov.sg/general_ info.html)

[8] Juliussen, Egil., "Technology 1997 Analysis & Forecast - Computers", *IEEE Spectrum,* Institute of Electrical and Electronics Engineers, Inc., New York, USA, January 1997.

[9] Bell, Trudy E., and Riezenman, Michael J., "Technology 1997 Analysis & Forecast - Communications", *IEEE Spectrum,* Institute of Electrical and Electronics Engineers, Inc., New York, USA, January 1997.

CIM Justification

ADVANCED MANUFACTURING TECHNOLOGY: STRATEGIC AND OPERATIONAL CHANGE FOR SMALL-AND-MEDIUM-SIZED MANUFACTURING FIRMS (SMEs)

Kam B. Chung and Yeong W. Kenneth*

School of Mechanical and Production Engineering
Nanyang Technological University

ABSTRACT: Local companies are encouraged and exhorted to utilise advanced manufacturing technology (AMT) with the ultimate objective of achieving complete integration in manufacturing. In the wake of this development, this paper discusses a study of the factors that influence the acquisition of AMT by some local SMEs as a change process for enhancing strategic and operating performance.

INTRODUCTION: Rationale and Background

The management of technology drives marketing, design, manufacturing, and ultimately the overall performance of an enterprise or firm. As this is critical for a small firm, it should be properly integrated with the business strategy to provide effective responses to the challenges and opportunities emerging in industries today. Fred Bucy, former chairman of Texas Instruments, expressed, "Technology is the design and manufacturing know-how to produce goods.....Products.... are the results of technology but are not themselves technology." Gerard H. Gaynor defines technology as a way to do something, a way to perform a value-chain activity. When a product is used to do something, it is that use which defines it as technology for the user. These are the operational issues that condition the strategic posture of a firm.

The acquisition of technology can be regarded as an initiation of technological change experienced by an enterprise that will result in a change in its manufacturing activity. Enterprises use technology to improve performance and to enhance competitive advantage (Maidique & Hayes). Thus, the costs and benefits of technology will have to be studied and examined constantly. In the main, capital investment plays a key role in acquiring technology. However, it is the management of the technology and not the technology that is the key to competing successfully. Essentially, the management of any new technology is actually the practice of integrating technology strategy with business strategy in the company (Steele 1988). Figure (1) which illustrates the role of technology in the total enterprise process, rationalises the development of the study.

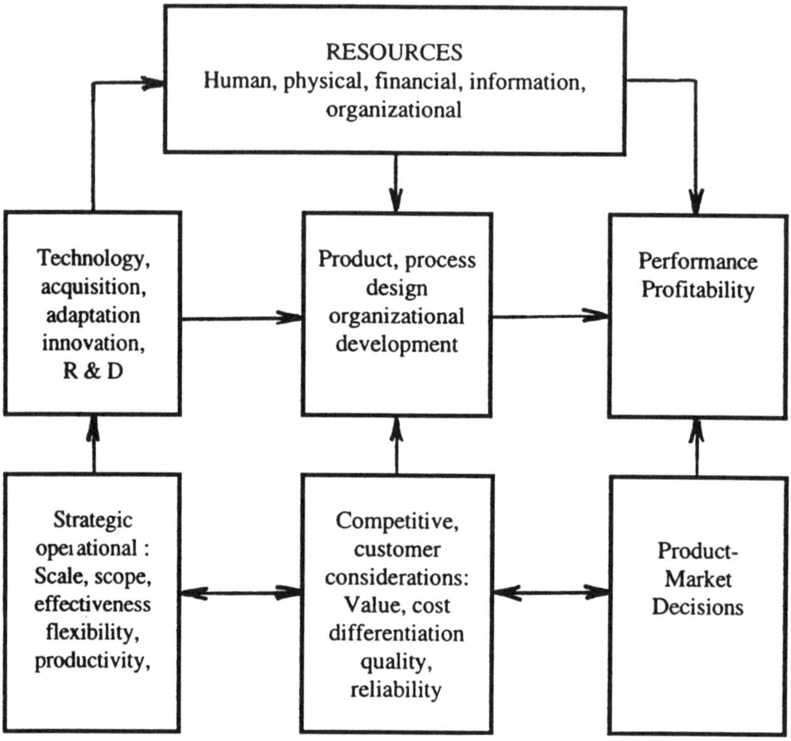

FIG.1 TECHNOLOGY AS A DRIVER OF CHANGE

Advanced Manufacturing Technologies (AMT): The classes of advanced manufacturing technologies (AMT) used in this study is according to Noori (1990).

Definitions of SMEs: Singapore distinguishes between local and overseas SMEs. For this study, the local definition of an SME was used. As such, companies with fixed assets of under $12million in the manufacturing sector and under 50 workers in the services sector are classified under local SMEs. Local companies with fixed assets over $12 million in the manufacturing sector and more than 50 workers in the services sector are considered large. With this definition it is important to note that a local manufacturing company is considered a SME even if it has a workforce of 100 - 200 employees, so long as it meets the above criteria of less than $12 million in fixed assets.

The Manufacturing Sector: SMEs in the manufacturing sector has seen tremendous growth in the last 10 years. In 1987, 2,700 establishments in the manufacturing sector employing more than 10 workers were SMEs. in 1993, this figure grew to 87,700, employing almost 500,000 employees. Their value added grew from $2.6 billion in 1987 to $20.9 billion in 1994 and productivity, as measured by the nominal census value added per worker, grew from $25,700 in 1987to $44,7000 in 1994.

SMEs and Technology: In an APEC survey on small and medium enterprises conducted in 1994, one of the difficulties encountered by SMEs in Singapore is to make significant productive improvements and produce higher value-added products due to global competition and pressure from local MNCs.

Conventional wisdom points out that only large firms have the facilities and expertise to benefit from advanced manufacturing systems. However studies have shown that advanced manufacturing technologies have as much, if not more, to offer to the small firm (Meredith 1987).

THE OBJECTIVE OF THE STUDY was to determine the factors that influence the acquiring of AMT and to determine the technology choice of SMEs. The study on managing advanced technologies in SMEs can include a whole range of topics such as project management, risk management, financial management etc, in the different phases of assimilating these technologies However, this project focused on the acquisition phase of adopting AMT in SMEs.. The study targets only the manufacturing sector of SMEs mainly in the electronics, metal fabrication, food and beverages and plastics industries.

Background of the study: The study was undertaken by a questionnaire survey and case studies of companies. Literature research in this subject was done prior to the designing of the questionnaire. The questionnaire was structured into 5 modules as follows: (A) **General Information and Business Profile** seeks basic information on the main line of business, the ownership pattern, volume of sales, number of employees and their labour structure. (B) **Manufacturing Strategy** seeks to determine the manufacturing profile of the firm and the key factors of the manufacturing function that contributes to the firms strategic planning. (C) **Productivity and Technology Management** seek to determine the factors and considerations of the firm when adopting AMT and the degree of impact it has on the manufacturing function. (D) **Training, Research and Development** seeks to understand the degree of emphasis on human resource training and R&D in line with the acquisition of AMT. (E) **Marketing and Strategic Direction** seeks to understand the effects on marketing due to AMT acquisition and the firm's strategic direction

RESULTS AND ANALYSIS

The questionnaires were forwarded to the managing directors or general managers of small firms. Of the 300 small and medium enterprises in the manufacturing sector that were sent the questionnaire, only 20 responded. This low response of 6.7 % will only yield exploratory results. Not all the issues raised in the questions will be discussed. Many of the questions required a 5 point-scaled response which ranged from (5), very important; to (1), unimportant. The assistance of the Association of Small and Medium Enterprises is acknowledged.

GENERAL INFORMATION AND BUSINESS PROFILE

Business Profile

A good 72% of the firms who responded are from the metals and electronics industries. The 3 firms that did not categorize themselves in the specified industries were involved in subcontracting activities of refrigeration, electroplating and precision valve industries. Majority are locally owned and only 5 had any foreign participation or are foreign affiliate. Half of them are supporting larger firms or MNCs in their respective industries and about 22% are purely end product manufacturers. Another 22% indicated a combination of activities namely as end-product manufacturers and subcontracting activities. Although 2 firms were not willing to specify their sales revenue for the last fiscal years, 8 responded that they were exceeding \$10m in sales while another 5 were in the \$5m -- \$10m category and only 3 were earning less than \$5m in sales. Those with sale exceeding \$10m were mainly in the metals and electronics industries. Results obtained regarding the number of employees were typical of a small and medium manufacturing firm, falling within the range of 51--100 employees, with a significant number of firms exceeding 100 employees.

It is interesting to note that 7 firms indicated that direct labour contributes to 70% or more to the overall number of employees and another 7 firms state that the labour cost was between 30% to 60% of total manufacturing cost. These significantly high figures in labour cost may be due to the labour intensive nature of the industry or the activities they are involved in. Whatever the case, it serves as a motivation to replace labour with technological automation. Of course there are other factors to consider such as technological availability in the industry.

Average Age of Equipment

Seventy percent of the firms that responded has been fully operational for more than 10 years. Only 2 firms indicated that they were in operation for less than 5 years. Of these 2 firms, 1 was a joint venture with 52% foreign participation and the other was a foreign affiliate with 51 to 100 employees. The average age of manufacturing equipment/machinery in use were given as: less than 1 year, 1 firm; 1 to 3 years, 2 firms; 3 to 5 years, 8 firms; 5 to 10 years, 5 firms; more than 10 years, 2 firms. Upgrading of old machinery with new machinery through technological advancements may be crucial to stay competitive. However, the rate of technological advancement in each industry is different. Furthermore firms may remain complacent with existing technology due to lack of pressure from competitors to upgrade. Also a purely sub-contracting firm may find it difficult to justify any investments on machinery when the basic existing machinery can do the required subcontracting job. Thus the activities and the strategic mindset of the firms may cause it to overlook any old and non-competitive machinery

MANUFACTURING STRATEGY

Level of Managerial Involvement

The classification of management levels in a SME is usually: top management: eg. managing director and department managers; middle management: eg. engineers and sales executives; lower management: eg. production supervisors and lead personnel. It would be normal to expect that the larger the company, the lesser the lower management will be involved in the setting of goals and business plans. Lower management like a supervisor will only be required to attend to day to day operations without much understanding of the company's operational strategy. Of the 18 firms that responded, 15 firms with 51 to 100 employees or more indicated that only middle management and above understands the goals and strategies of the company. Only 1 of these firms said that even the supervisors and workers understood its goals and plans. 3 firms with less than 50 employees indicated that every manager, supervisor and worker understands the goals and strategies of the firm. This is a likely indication of the closely-nit working environment typical for a small firm.

Current State of Investment in Capital Equipment Capital intensiveness of the industry plays an important role in the adoption of AMT. The likelihood of a new piece of machinery or technology emerging in a capital intensive industry is greater than that of a more labour intensive industry. The greater the need for machinery in the type of manufacturing activities, the greater the awareness of any new emerging technology. Eight firms are actively investing to upgrade or advance more than 70% of their capital equipment, while three firms indicated about 30-50%, and four firms less than 30% on same.

It is a good sign that the majority of the firms are actively investing or has been investing in machinery and equipment. However, the 8 firms that has 70% or more in capital investments were only investing about 10% on technical expertise. This may be an indication that the capital equipment invested in are of similar or familiar technology and are able to operate efficiently with the current state of expertise. It can be highlighted that the more popular investments in AMT were Computer Aided Design, Computer Aided Manufacturing, Computer Numerical Control and Computer Process and Planning Control.

Technological Advancement is important to the SME, with firms citing the need to keep track of technological trends and developments. It is also important for firms to make technology a priority of business operations, possibly a component of business strategy. Moreover, it is very important for firms to know and understand new technologies adopted by competing firms.

On factors of adoption, 10 were specified in the survey that may affect the firms decision on acquiring advanced technology. They are: reduction of labour cost, increase speed of production, increase quality of products, cash-flow factors, culture of the firm, return on investment, managerial and technical expertise, achieving competitive advantage, market considerations and replacement of old equipment and machinery. The returns of the survey indicated 3 out of these 10 factors are significantly more important. They are; reduction of labour cost (mean 4.2), increasing speed of production (mean 4.1), increasing quality of products (mean 4.7). Interestingly, the factors that were listed as less important with respect to their low mean values were culture, managerial and technical preparedness, and the need to replace old capital equipment.

PRODUCTIVITY AND TECHNOLOGY MANAGEMENT

Choice of Technology This section seeks to determine the type of competitive activities of a firm that eventually lead to the choice of technology that the firm is willing to invest in. Ability to achieve competitive pricing, high flexibility, high quality, quick delivery and good service and support are the five factors listed. The results observed were: competitive pricing (mean 4.4), high flexibility (mean 3.7), high quality (mean 4.3), quick delivery of products (mean 4.6), good service and support (mean, 4.2).

Thus, most of the firms share the common view that the ability to make products fast and offer dependable delivery is a critical issue. Comparatively, the need for manufacturing flexibility was viewed as not so important a factor (mean 3.7). However, high flexibility in manufacturing is significant among 7 firms. The preference remains for AMT that gives high quality and increases speed of production. The type of AMT a SME would invest in are: 1) Precision 2) High batch-lot-size 3) High quality 4) Flexibility in manufacturing 5) Speed of manufacturing 6)

Method of Implementing AMT This relates to the group of people involved in the decision making process of acquiring AMT and the degree of external assistance required during the implementation phase The results indicated that 13 of these firms involves middle management in their decision making process when aquiring any new technology while only 5 firms relies solely on upper management. Also, 13 firms support the idea of seeking external support and services during implementation of any new technology. Of these 13 firms, 10 relied on the supplier of that new technology or machinery, while 3 firms indicated that they tend to seek external expertise and consultants instead of the suppliers themselves. Thus there exists the possibility that there may be a lack of in-house technology expertise or skills or that the situation may involve the of acquiring unfamiliar technology.

Degree of Positive Influence on Manufacturing Factors Based on current acquired AMT in the firms, reducing process manufacturing time, increasing quality level and reducing delivery time were cited as having the most positive influence due to the adoption of AMT. The results on this issue clearly reinforces the technology choice of the firms. The acquired AMT were cited to have the least impact on reducing production floor space (mean 3.4), and reducing inspection time (mean 3.5).

TRAINING, RESEARCH AND DEVELOPMENT

Training In the area of training and upgrading of technical skills, it is not surprising that operator and supervisor training are both top on the list with means of 4.2 and 4.3 respectively. Whenever a new item of machinery is purchased, both the operator and supervisor will be given the required training to operate the machine efficiently. In fact, 12 firms cited that it is important (score 4) for top and middle management to have some technical understanding of the new machinery. This may be a typical characteristic of a small manufacturing firm where an engineering manager or a production manager has lots of 'hands on ' experience on the machinery and are greatly involved in the day to day technical operations and problems on the production floor.

Research & Development: The general result for this function is low as R and D can be interpreted in many ways by different firms.

MARKETING AND STRATEGIC DIRECTION

On the subject of marketing, the firms cited 3 factors as having the most positive impact as a result of acquiring AMT. They are; factor 1) reducing pricing structure of products, 2) strengthening of market position and 3) targeting of market buyers. For each of these 3 factors, 7 firms indicated a score of 5 (very important) and no less than 5 firms indicate a score of 4 (important). The return on investment as a result of AMT adoption leads to a lowering of cost, which allow a more competitive pricing structure. Thus acquiring AMT puts the firms in a better position to compete for a share of the market. Finally, 13 of these firms indicated that their strategic direction will be to build on existing market share in the next 2 years while 5 firms cited that creating new markets for new products was their main objective in the near future. The growth of SMEs is highly dependent on the kind of industry they are in. This is because many supporting SMEs with sub-contracting activities rely heavily on the orders from larger firms in the industry. For example, a SME supporting the electronics industry tend to follow the growth trend of larger companies and MNCs.

CASE STUDIES: In addition to the survey, three case studies were done to reinforce the findings gathered from the 18 companies that responded to the questionnaire. The objective of the case studies is to; 1) reinforce the findings from the survey conducted with respect to factors that influence the acquiring of AMT ,2) identify additional factors that influence the adoption of AMT, 3) gain a better understanding of SME's management of new technology by looking at individual company. Two general criteria were used in the selection process; 1) The candidate organisation must be a manufacturing enterprise that fit the definition of an SME in Singapore, 2) The candidate organisation must have already acquired some form of high technology machinery for their manufacturing function or has the intention to do so. The three companies are designated CL, CM, CN. An informal interview guide was prepared for the interview sessions, consisting of questions about the characteristics of the organisation, the factors that influence their decision to acquire new technology and the type of technology they are willing to invest in. The interviews were conducted with senior corporate staff that are involved with the manufacturing function of the organisation. These interview sessions were conducted at the site of the organisation and include the answering of the questionnaire at the end of the session. When possible, other individuals of the organisation were also interviewed to obtain additional perspectives on the process of acquiring new technology.

DISCUSSION AND CONCLUSION

Core Factors that Influence the Acquiring AMT in SMEs

The factors that influence the adoption of new or advanced manufacturing technology can be quite unique and varies with different SMEs in different industries. The size of the operations of SMEs can also lead to different considerations in the adoption strategy. For example in the process of investing in AMT, CM may placed greater considerations on client's manufacturing direction and less emphasis on flexibility of manufacturing than CN which has a larger operation and more diversified market. Similarly, a SME like CL in the wire harness industry may view achieving high quality and precision as its priority due to the processing of very fine wires. However, this does not mean that SMEs in different industries does not share the same investment concerns. This section highlights several core factors that were deduced form the survey and case studies done. The core factors discussed are by no means the only concerns of SMEs in acquiring AMT. Other factors such as culture of the firm, degree of adoption by competitors, business strategy etc. can influence the decision to acquire AMT. However it is hopeful that the study done will be able to highlight certain factors that may be general to SMEs in Singapore. It is also important to note that most of the firms being studied are not end-product manufacturers. Their manufacturing operations serves as a supporting function to larger firms in the industries.

The factors of influence will be divided into internal factors and external factors. Internal factors concerns with the firms current manufacturing and strategic position in adopting AMT. For example, the financial position of the firm is an internal factor of consideration in acquiring new technology. External factors on the other hand deals with customer and supplier issues for which the firm may have or may not have control over. The quantity of orders set by the customer is an external factor in which the firm has no control over.

Financial Considerations; All three firms cited financial funding as a major consideration in acquiring a new technology or machinery. More accurately, the lack of proper financial support and funding causes SMEs to be conscious in their effort to justify the investment. The financing alternatives are fundamentally different for large versus small firms. Even if they did manage to raise the funds with the help of government assistance and financial schemes, they may find that it has a dramatic effect on their balance sheet. Financing the new technology after the acquisition such as training and maintenance cost and the risk on the return on investment may prove too much for a small firm. On the other hand, larger firms have a greater pool of resources to cushion them against any unsuccessful new technology venture. In any case, if the business forces them to invest, then the investment must be justified strongly. In the case of CM and CL, the decision to acquire new technology must have the support of the customer and the security that the size of orders will be worth the investment.

For SMEs, the method of acquisition plays an important role in the decision making process. Purchasing a new technology or machinery with a given trial period can minimize the financial risks involved. The trial period allows the firms to familiarize themselves to the new technology and make better feasibility studies before acquiring it. Of course, this highly depends on the conditions laid down by the supplier of the technology.

Quality: One way for small firms to remain competitive with larger firms and having a competitive advantage over other smaller firms is to focus on quality. Having higher quality can convey two different meanings:1) The reliability of the product or service is higher -a production-process-related phenomenon. 2) The performance of the product or service is higher -a design-related phenomenon.

In the study, we are more concerned with the first meaning of quality related to the process technology since most of the firms studied are involved with supporting larger firms in the manufacturing of end-products. More accurately the manufacturing-base definition of Garvin (1984): quality means conformance to the required specifications. Of course, it is up to the firm to determine what the customers truly want (value-based and user-base quality considerations). In the case of CM building reliability into service by having fast and dependable delivery is a way of improving quality.

The results of Module B of the survey justify the importance of producing high quality products in SMEs Also the results of Module C indicate that achieving high quality and, achieving quick and dependable delivery of products were core competitive activities that lead to the technology choice of the firm. Despite this believe, many small firms seem reluctant to replace old production equipment with newer technology until the old equipment becomes physically obsolete. Not only does the older equipment mean higher cost and lower product quality, it will prove totally unfit for production. Improving quality through new technology can seriously minimize future costs and quality problems. For example, costs incurred through scrap, rework, repair and retest can be greatly reduced or prevented completely. Interestingly, one of the barriers to technology adoption in SMEs is mindset of managers of ignoring the dynamic nature of new operations technology and its enormous potential for quality improvements.

Replacement of labour: For the survey conducted and the interviews with SMEs, there are two motives for acquiring new technology to replace labour. They are; 1) to reduce labour cost and 2) to improve quality through automation. Due to the rising labour costs in Singapore, it is inevitable that the consideration of replacing labour with machinery as far as possible becomes inevitable. Many SMEs started out having a majority of labour operations because of the lack of funds to purchase machines to automate the processes. However as they progress, funds are more readily available and the effort to reduce the rising labour cost is becomes more conscious. The result of Module B question 7a shows the common agreement among the firms surveyed that reducing labour cost is high on the agenda in their decision to acquire new technology. The case study on MEC International highlights the fact that the labour intensive nature of an industry can motivate as much automation as possible by acquiring new technology. For SMEs achieving the high speed of production and high level of quality are vital in order to stay competitive. Replacing labour operations with technology provides tremendous advantages in productivity and quality. Automating through technology not only increases the speed of production, but reduces the error caused by manual work, and hence reduces inspection rework times.

Replacing labour with technology can lead to resistance to change. It is likely that firms have to employ job redesign principles and maybe even as far as compensation policies. It is likely that the demand for technically skilled individuals in machine operation will increase. Correspondingly, the need for unskilled general labour, particularly in the manufacturing industries, will decreases.

Availability of Technical Expertise: There is growing evidence that technology adoption may be subject to the firm's prior experience with technology and the organisational knowledge and skills major determinants of the adoption rationale This is reinforce by the case study on CN where the availability of in-house expertise is a major concern. It has been argued that technology deskills work and destroys jobs. On the other hand, as technology replaces the repetitive operations of manual work, highly skilled workers are required on the shop floor to develop and optimize processes. A previous study on the adoption and implementation of AMT in Singapore cited the need for specialized technical training as one of the critical success factors. The fact that AMT requires highly skilled direct labour, there may be additional costs associated with employee training and higher wages. SMEs in general do not have the formal selection and training practices found in larger firms. Therefore they may have difficulties in upgrading their employees' technical skills which can be achieved through external hiring or through in-house training programs.

Customer-Motivated Factors: There are three points related to the customer that most SMEs are concern with in the acquiring of AMT. They are; 1) The life span and size of orders, 2) The customer base of the AMT acquired, and 3) The manufacturing direction of the customer. Larger firms are committed to stating the size of the orders when contracting manufacturing functions to smaller firms. It is up to the smaller firms to gauge the life of the orders. In any case, the size and the life span of the sub-contracted work must justify the investment of an AMT. However, it is not good enough for the revenue to just cover the cost of investment. SMEs must have the security that future sub-contracting work needs the application of the acquired AMT. In the case of CL, a large base of customer that have the potential to make use of the AMT must be ensured before any acquisition. It is of course necessary that the AMT be flexible enough to adapt to the different requirements of each customer. SMEs also tend to rely on the manufacturing direction of their customers who are usually MNCs or larger manufacturing firms. The strategic direction of larger firms gives them the assurance that the AMT invested will not be left idle. In general, SMEs tend to invest in AMT that are mutually beneficial.

Supplier Motivated Factors: In the same case as customer-motivated factors, SMEs must ensure that the supplier is capable of supplying the required material to sustain the usage of the AMT. It would be pointless for CM to invest in AMT if the supplier has no confidence in supplying the required plastic material for the operation of the AMT. Unlike customer-motivated factors, firms do have some degree of control over these factors. For example, they may resort to keeping stock if a slow down in the following months' supply is anticipated.

Technology Choice of SMEs: The technology choice of SMEs varies for different industries. The choice of which way to perform a value-chain activity or which technology to use should be govern by the competitive advantage that the firm is pursuing in implementing its competitive strategy. In other words, if low-cost -low price is the strategy, then low-cost technologies should be used, consistent with maintaining acceptable levels of quality, attractiveness and so on. Firstly, the decision on the form of technology to focus on should be guided by the manufacturing considerations. Secondly, the choice of technology will also depend mainly on the characteristics of the product: a) Complexity of the product --the greater the complexity, in general, the greater the degree of automation required; b) Precision with which the product must be made; c) Batch or lot-size --the larger the lot size, the less degree of flexibility required; d) Diversity or the number of models in the family --the greater the diversity, the more flexibility is needed. From the study, the form of AMT adopted by SMEs should have the general characteristics of seeking to improve quality, allowing high flexibility and increase speed of production.

Improving Quality. The importance of quality as one of the core factors to SMEs adoption of AMT is accepted. It is not surprising that SMEs main concern with a new technology is its ability to produce the consistently high level of quality needed for them to stay competitive. A number of technology can be considered to improve quality and prevent future quality problems. These include investing in: 1. Computer-aided manufacturing technologies; 2. Process monitoring and control technologies, 3. Automated inspection system; 4. Quality information systems; 5. Improved process and process design techniques.

In CL, the integration of an automated inspection system into one of their crimping machines helped improved the quality control of the product. The automated inspection system was able to detect the poorly crimped wires by analyzing the change in pressure of the crimping operation. The system terminates the operation of the machine which would otherwise allow the faulty operation to continue. Many SMEs in Singapore are often preoccupied with activities such as rework scrap repair, retest operations due to substandard quality products that are produced and discovered before or after shipping to the customer. It is these cost as a result of poor quality that are financially difficult to justify. Reduction of rework operations by through better quality allow for faster and more reliable delivery promises

High Flexibility. SMEs in their respective supporting industries supply manufacturing parts needed by larger firms to produce the end-product. According to Meredith, small firms can benefit through manufacturing flexibility by reacting faster to customer's demand and providing customised products. In the study, 7 firms chose manufacturing flexibility as their priority when investing in AMT. From the interviews conducted, the two kinds of flexibility that SMEs in Singapore are generally interested in is modification flexibility and machine flexibility. Modification flexibility is the ability of a process to implement design changes on a given product. It enables standard products to be modified to meet various customer specifications.. For example, CM produce different plastic components needed by MNCs such as Aiwa and Panasonic for the manufacturing of electrical goods. These plastic components include hi-fi casings, calculator casings, etc. The flexibility in manufacturing needed by CM to produce a whole range of plastic components is absolutely vital. It would be necessary for the in-house technology to be able to welcome the most stringent design requirements laid down by the customer. Failure to do so will result in a loss of confidence by the larger firms in its manufacturing capability. Machine flexibility provide a firm with the capability to switch over its production process very quickly to produce a different set of parts. This is essential because of the different parts required by each customer. Furthermore, with machine flexibility, SMEs are capable of producing the replacement parts required on demand due to diffeerent quality of a set of orders.

Speed of Manufacturing. SMEs usually are in the situation where they overload their production capacity due to the large size of orders from different MNCs. Whenever possible, manufacturing of certain parts may have to be subcontract to vendors. Furthermore, the need to meet delivery datelines add to their woes.. Thus, 11 firms indicated that speed of manufacturing products is an essential characteristics in selecting AMT. High speed of manufacturing can include efficient production and management systems. AMT such as group technology and computer integrated manufacturing can drastically increase the rate of production. On the other hand, a new machine in the market is able to generate several times more units than the older machine.

CONCLUSION: There are two points to note when selecting the type of technology. Firstly, the characteristics of the new technology acquired has to be assessed for tradeoffs. For example, a high degree of flexibility of a new technology may result in a compromise in the quality of the product manufactured. Similarly, high speed of manufacturing can a result in a drop in quality. Secondly, AMT will not live up to their potential without the full management commitment and employee participation. Any AMT geared towards improving quality will not work without the active support of management and positive employee participation.

SMEs realize significant advantages through advanced manufacturing technologies, but that the time and resources required for a successful adoption are often inhibitive. In this study, these constraints are divided into internal and external factors that affects the decision to acquire AMT. Although internal factors vary with each SME depending on its business strategy, factors such as financial considerations, quality, replacement of labour and availability of technical expertise are found to be common among local SMEs. External factors on the other hand tend to be customer-motivated and supplier motivated. The technology choice of a SME directly influence the form of potential AMT that the firm is willing to invest in. The technology choice is in turn influenced by its manufacturing strategy and product characteristics. The ability for the AMT to improve quality, achieve high machine and modification flexibility, and increase speed of production were found to be the essential characteristics of AMT that interest local SMEs.

AMT or any new technology cannot replace good planning and strategy, sound management, and a capable dedicated workforce. There are many firms which do not have any of the new technologies in operation, yet who consistently outperform some of their more rapidly adopting competitors. For those well-managed SMEs that do match their technology applications to the needs of their process and their customers through proper acquisition management, AMT may hold the key to the future success.

References:

Gerard H. Gaynor, Handbook of Technology Management, McGraw-Hill, 1996.

Garvin Managing Quality: the Strategic and Competitive Edge. Free Press Collier Macmillan 1984

D. L. Goetsch, Advanced Manufacturing Technology, Albany, N.Y. Delmar Publishers, 1990.

Lowe, The Management of Technology, Chapman & Hall, 1995.

Johnson A. Edosomwan, Integrating Innovation and Techonology Management, Wiley-Interscience Publication, 1989.

Hamid Noori, Managing the Dynamics of New Technology : Issues in Manufacturing Management, N.J.: Prentice Hall, 1990.

Louis A. Lefebvre, E Lefebvre, and J Harvey, Intangible assets as determinants of advanced manufacturing technology adoption in SMEs, IEEE Engineering Management Journal Vol. 43, No3, 1996

Madique and Hayes. The Art of High Technology Management. Prentice Hall N.J. 1987.

Meredith, J., The Strategic Advantages of New Manufacturing Technologies for Small Firms. Strategic Management Journal Vol. 8, pp249-258. 1987.

From CSF Analysis to Balanced Scorecard Design in Strategic IT Planning

K T Yeo

Division of Systems and Engineering Management
School of Mechanical and Production Engineering
Nanyang Technological University
SINGAPORE

Abstract

The Balanced Scorecard(BSC) design aims to mobilise people to fulfill the company's mission and to translate strategy into action. The BSC is to provide the 'big picture', and IT or any other change programs are the integral and actionable parts. IT strategy at functional level must contribute to the broader corporate business strategy. This paper adapts the BSC design to incorporate the well known Critical Success Factor (CSF) analysis to create an enhanced framework for IT justification, planning and management. The emphases are on using IT to improve company's internal business processes and to stimulate strategic organisational learning and growth.

1 Introduction

"Information technology has not done much to improve companies' productivity but instead has been responsible for increasing costs every year", reported as the concern expressed by a Singapore corporate CEO(1). Many CEOs were similarly wary of "a constant round of replacements and upgrades". The chairman of General Electric, Jack Welch was quoted in his speech at the World Economic Forum, in early 1997, that all the promises made by IT "had come to nought". The same Singapore based CEO was quoted to say that "there is very seldom a proper evaluation of the appropriate technology for the job at hand. Each time, Microsoft comes up with something new, or Intel comes up with a new processor, most IT departments everywhere are busy making a justification for change."(1)

Preoccupation with technology for its own sake will not bring results. Justification for technology without considering corporate mission and strategy is a fallacy. Similarly,

IT justification based on short term financial measures is also inadequate. This is one of the basic ideas of the balanced scorecard(BSC) developed by Kaplan and Norton (2)(3), to go beyond financial measures in performance measurement.

Financial measures are historical in nature and tend to track end results in monetary terms. Justification of technology acquisition initiatives purely in financial terms is not always easy and realistic. Non-financial measures giving emphasis to customer satisfaction, significantly enhancing the internal business process, and promoting strategic learning and innovation, can be more critical for a company's long-term survival and growth.

The aim of this paper is to propose and develop an enhanced framework based on the combination of available strategic planning tools, for better justification, planning and management of technology acquisition. There are two key issues to be addressed in this paper which are concerned with: firstly, how to ensure IT be part of a company's business strategy; secondly, how to go beyond short term financial justification to include measures to improve internal business processes and to stimulate strategic learning, innovations and growth.

The proposed framework for strategic planning of technology links the well practised Critical Success Factor (CSF) analysis with the balanced scorecard (BSC) strategic management system design. The CSF analysis is further enhanced by incorporating the analysis of contradictions. A contradiction surfaces wherever a proposed idea significantly deviating from the realities which are characterised by obstacles or barriers. The proposed critical balanced scorecard system aims at bridging the gap between strategy and implementation with critical analysis. The new design will help a company translates its strategic vision into reality. This can be achieved by building up its core capabilities, especially through appropriate technology applications.

The critical balanced scorecard approach is to ensure that the application of information technology forms an integral part of a company's business strategy. The appropriate technology strategy should go beyond a narrow functional mindset, and contribute directly to the business needs and jobs at hand, stimulate purposeful organisational learning to strengthen internal capabilities and processes; and to satisfy customers' required and latent needs.

2. CSF-Contradiction Analysis

2.1 The CSF Process

The proposed critical balanced scorecard is built on a framework of systemic CSF analysis for strategic IT planning by Yeo(3). The systemic approach gives emphasis to concentrated rethinking of corporate mission, goals, strategies, organisational tasks at hand and the strategic information requirements in IT planning. The aim of CSF analysis is to elicit involvement and views from different levels of personnel to establish those few areas of major concern that are critical to the success of their company.

Yeo's framework aims at separating IT-related CSFs and concerns from those non-related ones. Non-IT related issues may have important impact on the IT successes. Examples of non-IT related issues are leadership, corporate culture and value, resistance to change, senior management support and commitment, human resource management, existing internal capabilities and so on. The strategy in using CSF analysis for IT planning is not to be unduly concerned with the problems of computer technology and systems in the initial stage.

The process of CSF analysis is schematically illustrated in Figure 1. The process works both at the strategic business level and operational solution levels. The CSF framework encourages continual and focused rethinking, almost on a near-real time basis, of one's business and its future direction in relation to the industry's structure and its competitive environment. An indepth analysis looks at the various competitive forces within the industry such as rivalries among the existing competitors, threats of new entries, relative bargaining power of suppliers and buyers and the threats of substitutes of goods or services due to new technological development and innovations(5). The analysis may lead to the formulation of revised or refocused long term objectives and strategies.

Figure 1. The CSF Analysis Process

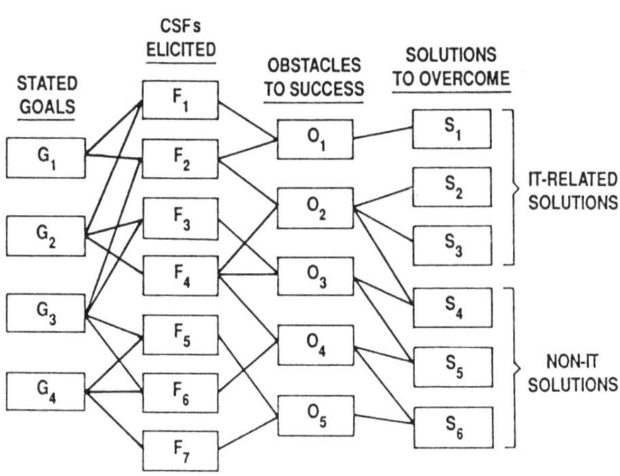

2.2 The Seven-S Framework

A list of CSFs can be elicited from different levels of management and employees, through a series of interviews. The compiled CSF list is a set of practical as well as idealised concepts, ideas and suggestions on how the formulated goals can best be fulfilled. The CSFs can be sorted into a number of categories. A useful reference for grouping CSFs is McKinsey's 7-S framework for achieving corporate excellence. Among these CSFs, some can be classified as hard S's and the others as 'soft'. IT or systems related CSFs and non-related CSFs can also be identified and separated. A seven-S framework is illustrated in Figure 2 (4).

Figure 2. A Seven-S Framework

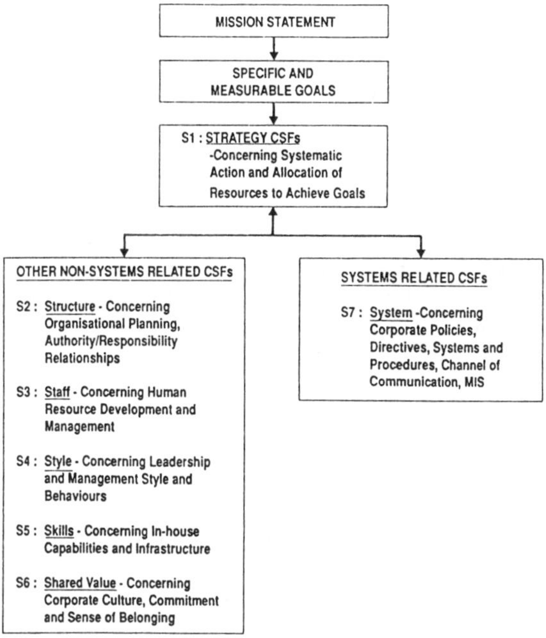

The same 7-S framework can also be used to sort out the multitudes of concerns, perceived barriers and obstacles expressed or elicited in the CSF interviews. The framework is also useful to appreciate the nature of conflicts and contradictions arise from the differences between strategies and action, proposed solutions and initiatives versus the realities of the company and its constraints.

2.3 Analysis of Contradictions

Based on the reality of the industry and internal environment of the company, both senior and functional management are in the position to identify obstacles facing the company, in relation to the desired goals and CSFs. Among the CSFs, some will be idealistic in nature, which are conceptualised as idealised 'models', which may be in conflict with the real-world situations. The gaps or differences between a model and reality are termed as 'contradictions'. These contradictions need resolution in order to achieve progress.

Gaps
as Contradictions

Realities	I	***CSFs***
as Obstacles	<------------------------>	*as Conceptualised Models(tend to be idealised)*

There will be as many 'contradictions' as the number of mismatches between the 'realities' and the 'idealised' CSFs. These contradictions are then analysed and debated, as a process of dialectic. During these debates and discussions, strategic and purposeful organisational learning will take place, and ultimately solutions found.

In proper handling of different types of contradictions, it is useful to separate them into critical contradiction(s) from the subcritical and non-critical ones. By management of exception, greater efforts should be given to the critical or principal contradictions. But one must also be careful about the subcritical ones which may turn into critical contradictions when circumstances changed.

All Contradictions
can be categorised into
I
I
Critical, Sub-critical, Vs Non-critical

&

Each Contradiction
consists of
I
I
Primary Aspects Vs Secondary or Supporting Aspects.

There are various aspects of each contradiction which must be identified and studied. The dominant aspects of a contradiction are the primary influencing factors which are to be identified and separated from the secondary or supporting aspects. These aspects of contradiction can be technology or human factors based which can be seen as 'hard' or 'soft'. The soft aspects are generally referred to as human related problems.

The proper handling and resolution of these contradictions are at the centre of the change initiatives. The implementation of say, a CIM and any other major IT initiatives, can be firmly justified when they represent the principal aspect of a critical contradiction. The planning for and management of this primary aspect, must contribute to the resolution of a critical contradiction in question.

The entire CSF process begins with the formulation of a vision and business strategies. This helps the planner or decision maker to see the 'big picture' from the outset and to prevent a pre-occupation of technology for its own sake. The IT cost has exploded in recent years. IT investment must be in line with corporate business strategy. The CSF approach with the dialectical analysis of contradictions, helps to evaluate the true need for appropriate technological solutions, and in the selection of innovative and cost-effective initiatives. Technology may not be the primary aspect of a critical contradiction. Any IT solution must be business driven, and its performance needs to be measured.

3. The CSF-BSC System

The balanced scorecard(BSC) performance measurement system developed by Kaplan and Norton (2,3) can be incorporated in the CSF analysis for strategic IT planning and justification. The emphasis is to include non-financial perspectives, goals and measures which are related to learning, innovation and growth potentials; higher level of customer knowledge and satisfaction; and improvement in internal business processes. Figure 3 shows Kaplan and Norton's BSC framework to translate a strategy into operational terms.

Figure 3. The Balanced Scorecard Framework (3)

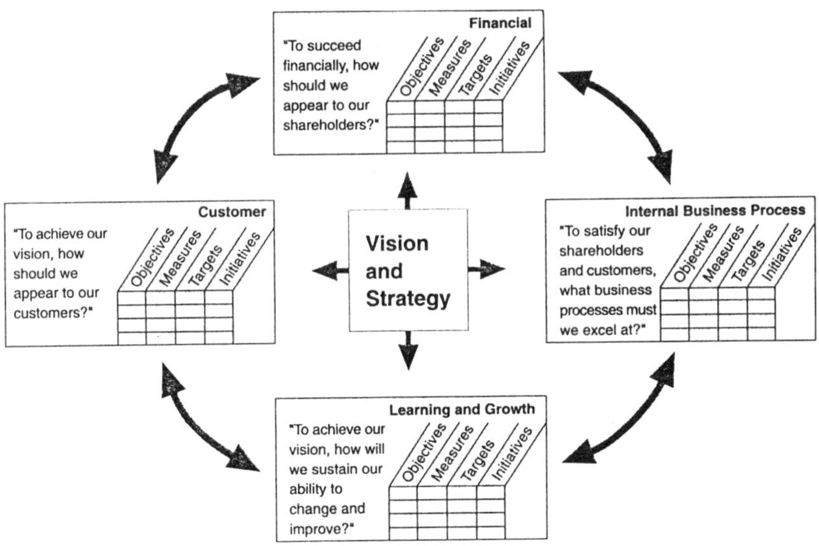

The balanced scorecard design is to give focus on future potential in relation to corporate long-term strategic goals. Such potential is growth oriented; and it is not wholly exogenously given by opportunities in market alone, but more importantly, it is dependent on the company's continual development of internal core capabilities.

This paper looks critically on how such internal capabilities and future performance potentials can be enabled by the introduction of appropriate technologies which of course include IT and computer-aided systems.

The BSC in combination with CSF analysis can be used to improve the justification and evaluation of IT implementation and performance measurement; and ultimately to translate corporate strategy to actions and initiatives. The role of IT in today's business competition can not be under-estimated. Competitive advantages are increasingly derived from information and intelligence, and ability to exploit intangible assets for future innovation and growth.

The substance and process of implementing a critical BSC system design is illustrated in Table 1, which takes into consideration of the CSF process and BSC's four processes of: (i) translating the vision, (ii) communicating and linking, (iii) business planning, and (iv) feedback and learning. The combined, adapted and enhanced processes are described in this section.

Table 1. The Critical Balanced Scorecard Process

			Analysis of contradictions <<---------------------->>		
	(Directional) Perspectives Goals/ Objectives	Perform -ance Measures & Targets	Critical Success Factors (CSFs)	Obstacles or Barriers to Succeed	Solution/ Initiatives
Vision and Business Strategy (Mission and Strategy Statements)	**0. Financial** (Looking backward and focus on shareholders) **1. Customer** (Looking outward to the markets) **2. Internal Business Process** (Looking inward to excel) **3. Learning and growth** (Looking forward to the future with change and improvements).	(Linking measures & targets to strategic goals)	(The few things company must well to meet objectives and performance targets)	(Review internal capabilities and weaknesses)	(Differentiate Hard from soft initiatives Separating IT and non-IT related solutions Can be incremental and radical.)

Process 1: Capturing the vision and stating the mission

Vision is what one wants to become. Mission is what one to do to fulfill that vision. The vision is therefore an image of the corporate future, say, to attain world-class standard, or to be a world leader, or the best in Asia in a particular field of business. There are of course problems in interpreting the vision and strategy. For instance, how one defines 'world-class' or 'the best in Asia'. This process is to help clarifying and building a consensus around a company's vision.

A broad mission statement is what the company aims to accomplish, based on the strategic vision. Corporate strategy is the long-term plan of how to carry out the mission. This could be a difficult concept to practice since, very often, middle managers and other staff do not know how to translate these mission statement into appropriate actions.

There is obviously a need to bridge the gap between 'idealised concepts or models' with reality. The building of 'internal capability' can help to bridge such gap. Strategic and purposeful learning can, in turn, help to build up such capability which goes beyond the narrow knowledge and skills of day-to-day operations. The main focus of the BSC is to give emphasis on strategic learning, innovations and growth, and develop the intangible assets in human capabilities.

Process 2: Defining and setting the goals

This process focuses on developing operational objectives under the four goal perspectives for the BSC namely: financial, customer, international business process and learning and growth (Table 1). The process forces the company's senior executives to clarify the meaning of the mission and strategy statements and their translation into specific objectives. The objectives statement must be tangible and concrete, which can, in turn, be converted into specific measures and targets. This is an important part of a strategic planning process which involves the setting of goals, selection of measures and targets.

Process 3: Communicating and linking

Strategic implementation are the actions that take place at the operational or functional levels. All short and mid-term objectives at implementation levels are to be linked to the corporate strategy and vision. It is important to be aware that there will always be gaps between strategies and implementation. Lofty vision and strategy statements do not translate easily into action at the local or implementation level. This process is to communicate the corporate long-term goals and strategies to all levels, and to link and align objectives from operational to strategic levels.

Kaplan and Norton(2,3) suggest that the corporate senior management may begin the BSC design by formulating the financial and customer objectives. It then mobilises the talent and information in the next levels of managers by getting them to formulate the 'internal-business-processes' and 'learning-and-growth' objectives that would drive the achievement of the financial and customer goals.

The scorecard design serves to inform company employees what the organisation is trying to achieve for shareholders and customers alike, and what key activities are needed to align individual performance with overall strategy. The balanced scorecard is both motivating and obligating.

Process 4: Planning, learning and analysis

Business planning and analysis which go beyond short-term financial measures, help to link change programs and resource allocation, say a proposed major investment in CIM technology and systems, to long-term business strategic priorities. Non-financial goals for improving internal business processes and for strategic learning and growth can be such priorities.

The processes of building a balanced scorecard help to clarify the strategic objectives and identify the few critical drivers or success factors for the company. The BSC system creates a framework for managing an organization's various change programs. The change initiatives may range from reengineering with IT applications, investment in advanced manufacturing technology to the softer aspects of customer-orientation, employee empowerment, teamwork, time-based management, partnering, and continuous improvements under the TQM philosophy. This particular process may include the CSF and contradiction analyses suggested earlier.

CSF and contradiction analysis:

As discussed in the earlier section, CSF analysis is a fact finding and justification step of a proposed change program which may include an IT investment program. The enhanced CSF analysis involves the surfacing of internal contradictions that may stifle growth and innovations and the fulfillment of desired objectives, if they remains unresolved.

Contradictions must be surfaced and creative debates stimulated when the elicited CSFs and perceived realities are in conflict, especially when concerns are expressed and obstacles identified. If purposeful debates are properly conducted, there will be tremendous opportunities for feedback and learning. The learning is likely to be focused and purposeful.

The surfacing and analysis of contradictions can cause the so-called double-loop learning - learning that produces a change in people's assumptions and about true cause-and-effect relationships. The result is to avoid devoting out-of-proportion amount of efforts and investment on resolving secondary contradictions. When dealing with a principal contradiction, the primary or dominant aspects of the contradiction must be identified and separated from the secondary ones; and to be handled and managed accordingly.

Conclusions

The proposed critical balanced scorecard combines the best of CSF analysis and BSC design. The enhanced system promotes strategic learning and innovation for future growth and internal process improvements, in addition to the financial and customer-oriented objectives. The strategic planning processes consist of articulating company's shared vision, mission and strategy, defining goals and objectives, formulating performance measures and expected results; surfacing assumptions and potential internal contradictions, stimulating debates and multi-learning, gathering feedback, testing the hypotheses on which strategy was based, and making the necessary adjustments to proposal of any change program which includes IT implementation.

References

1. Straits Times, "IT does little for productivity: Wuthelam chief", 2 August 1997.
2. Kaplan, R S and Norton S P. Using the balanced scorecard as a strategic management system, *Harvard Business Review*, Jan-Feb(1996) 75-85.
3. Kaplan R S and Norton D P. *The Balanced Scorecard,* Harvard Business School Press,(1996).
4. Yeo, K.T. Systemic CSF Analysis for Strategic IT Planning, *International Journal of Project Management,* 8(4), (1990) 205-313.
5. Porter M E. *Competitive Advantage,* Free Press, (1985).

Quantifying Intangibles in AMT Justification by using the Cost-Value Technique

Sev V Nagalingam, Grier C I Lin

Centre for Advanced Manufacturing Research, School of Engineering
University of South Australia, The Levels, SA 5095, Australia

Felix T S Chan

Department of Industrial and Manufacturing Systems Engineering
University of Hong Kong, Haking Wong Building, Pokfulam Road, Hong Kong

Abstract. Advanced Manufacturing Technologies (AMTs) can offer many tangible and intangible benefits for a company. These benefits are essential for continuous growth of a company and for the company to increase its competitiveness. The intangibles are difficult to quantify and cannot be accounted for by current financial methodologies. These intangibles pose a challenge for a manufacturing manager. This paper presents an effective and simplified approach, which is being applied in a Decision Support System (DSS), to solve the problem created by intangibles. This approach incorporates a Cost-Value technique with current financial methodologies, for the evaluation of intangibles for the justification of Computer Integrated Manufacturing and / or its components.

The Cost-Value technique has been utilised in the past for the selection of a suitable computer hardware and software among the various vendors, who were competing for the tenders. The same methodology can be used to quantify the intangibles by assessing the value of each intangible for the company concerned. When the value of the intangible is established for an AMT, it will overcome the barrier in the justification process. It is expected, having a clear concept in justifying the intangibles related to AMTs and CIM investment will propel the companies towards adopting advanced technologies without any hesitation.

1 Introduction

Justification of Computer Integrated Manufacturing (CIM) is a major challenge to manufacturing managers and decision makers of CIM investment analysis. The benefits and costs associated with CIM are very difficult to quantify due to many tangible and intangible attributes, as well as, considerations of integration the technology brings along with it. CIM is achieved by integrating effectively various Advanced Manufacturing Technologies (AMTs) of an enterprise. AMTs are generally defined as systems providing flexibility as well as data driven computer integration for the manufacturing organisation [1]. The current developments in AMTs are such that the traditional financial methodologies cannot be effectively used in justification of CIM. This arises due to the inability of the traditional methodologies and the immensity of the AMT investment.

Having studied the productivity problems of the early 1970s, Lin [2] found that the demand for products with enhanced quality was increasing while the supply of skilled labour and working hours were decreasing. He suggested implementing integrated advanced technologies such as CIM, as an effective answer to solve the problems of decreased productivity relative to labour costs and the consequent rise in unit costs which are continually plaguing present day manufacturing managers. Recent world competition has been such that performance characteristics such as low prices, high quality levels and short lead times have become requirements for manufacturing viability rather than a feature providing competitive advantage [3]. However, the question arises how to incorporate these measures into the justification process and to evaluate the intangibles which are escaping this process.

To overcome the difficulties associated with the investment in AMT / CIM, a Decision Support System (DSS) for CIM investment is being developed at the Centre for Advanced Manufacturing Research of University of South Australia. This DSS is named the DSSCIM.

A company which invests in AMTs will have a good opportunity to be placed in a more competitive position. In order to invest in AMT, an effective justification procedure is required to justify intangibles. The Cost-Value technique is utilised with necessary modification in DSSCIM to effectively evaluate the intangible attributes associated with investment in AMT.

2 Cost-Value Method

All the qualitative attributes related to an AMT improve the value of that technology. Generally, anything capable of being appreciated could be considered to have a value[4]. Values are naturally intangibles. Many value measurement techniques are used to measure the impact of the values in a decision making process. These techniques includes simple ranking, weighted ranking, pair-comparisons such as regular-pair and scaled-pair comparison, pair-cost-comparison, direct magnitude estimation; criteria analysis and scoring models [5].

None of the above methods assign any monetary value for an attribute except the pair-cost-comparison method. This method utilises monetary units to express the difference between two items in a pair of items being compared. However the Cost-Value method proposed by Joslin [6] provides a framework to assign monetary values to intangibles without resorting to pair comparison among intangibles. When an AMT is considered for investment, comparing the intangible and tangible attributes in a pair is an irrelevant exercise.

The Cost-Value technique was originally proposed to evaluate the selection of computers among vendor proposals. The technique requires the required features of the system considered, to be categorised as mandatory and desirable features. Then the technique enables the decision maker to evaluate only the desirable features of a given system and the cost of it. The mandatory features are not evaluated and these features should be provided in the system for the initial consideration. Later Bhat [7] explained that this cost-value technique can be utilised to evaluate the acquisition of AMTs from vendor proposals. The application of the cost-value technique enables the decision maker to understand the importance of the attributes in the evaluation process and helps to make informed decisions.

The application of cost value technique is effectively utilised in the DSSCIM for the justification of AMT or CIM [8,9]. Conducting an evaluation with the cost-value technique requires adopting the following two steps:

2.1 Identifying the technology considered for the investment.
In any justification issue it is mandatory to identify the technologies which are suitable for consideration based on strategic and tactical forecast. In the DSSCIM, a facility is provided to select the required technology based on the strategic requirements of the company. The required technologies are selected by the DSSCIM by utilising a rule based algorithm and by comparing the expected performance measures of the strategic attributes, which were initially provided by the decision maker [10]. For the selected technologies the DSSCIM provides intangible and tangible attributes associated with it. Providing all the attributes related to an AMT is a significant deviation from the cost-value technique proposed by Joslin in which the attributes of an AMT have to be categorised into desirables and mandatory features. This deviation results in simplifying the technique into an effective tool to evaluate all the values of the attributes related to that technology disregarding the difference of an attribute being desirable or mandatory. To guide the decision maker the attributes associated with every selected AMT for evaluation are automatically provided by the DSSCIM. This provision enables the decision maker to be aware of the tangible and intangible attributes associated with an investment in AMT.

2.2 Identifying the value of an attribute
This is performed by asking the question: What is the value of an attribute of the technology within a functional unit, where the technology is going to be implemented?.

The use of cost-value technique requires identification of the perceived value of each intangible attribute associated with the selected technology. Generally the value of an item is considered less than the cost of the item, when that item is competitively available [6]. This criteria is applicable for attributes sought by a company to achieve its strategic decision. However, there are some exceptions for this criteria where some of the general items will have more sentimental or utility value than its cost. Consider a technology under evaluation has an additional facility to count the number of items processed through it at the additional cost of $5,000. However, if there is no requirement to count the number of items passing through that machine in the manufacturing procedure, the value of having that additional feature is nil for the company for that machine. Nevertheless, if that facility can be used in future operations of the company, a moderate value such as $ 800 can be assigned. As Primrose and Leonard [11] has stated many of the intangible attributes could be broken into tangible attributes. It has been shown by them that the improved product quality could be considered as representing reduced scrap, reduced rework, reduced disruption of production, reduced warrantee payments and increased sales. These representations could be quantified within a reasonable accuracy with little difficulty. On the other hand, if an intangible attribute cannot be divided into number of tangible attributes, the usage of cost-value method will be the only available option to quantify that intangible attribute.

2.3 Application of Cost-Value Method in DSSCIM

Literature in the area of justification methodologies highlights the inability of the financial techniques to evaluate intangibles, which are inherent in CIM systems. Existing cost accounting systems and cost management practices do not adequately support the objectives of automated manufacturing. Due to this impasse, manufacturing managers and decision makers do not have a clear concept in evaluating the potential intangible and indirect costs / benefits.

DSSCIM provides two options to analyse the financial implication of an investment by AMT [10]. One option offers a facility to incorporate cost-value technique to evaluate the intangibles. The other is a standard approach preferred by managers, whereby the intangibles are not considered in the evaluation. The effective use of the cost-value technique requires assigning a monetary value, for each important attribute, depending on a company's requirement. During the financial analysis stage, the DSSCIM guides the decision maker to select appropriate financial-analysis-sheets for a sub-system, which is initially suggested and selected for the analysis. The cost-value option offers guidelines to quantify attributes directly related to the sub-system. At this option, the DSSCIM guides the decision maker to estimate both the benefit from an attribute and the value of an attribute for the company. In the financial analysis module, the benefit from an attribute is incorporated with other benefits. The value of an attribute related to the sub-system, where the AMT has been installed, is deducted from the capital cost according to the cost-value method for comparison against investing in other alternative technologies. A caution should be exercised to avoid counting the benefits twice as a value and an improved savings due the benefit of an attribute. Nevertheless, the value and benefits are considered separately, since both are different entities in an evaluation procedure.

For instance, in the case of the cost-value option, if the organisation is willing to spend about $7,000 for improving quality of the products, or it perceives the value of improved quality in the organisation as worth $7,000, then the decision maker can assign that amount of money to that particular attribute. In other words this technique is estimating a value for an attribute and this analysis will force the decision maker to establish a reliable monetary value for an attribute. Assigning a value is a complex and abstract method. The subjective measurement of value of an attribute may change or vary over time depending on the importance of that attribute to the company. By deducting the value of the desirable attributes from the total capital cost of the investment, the company will be in a better position to justify an investment in AMT compared to investing in alternative technologies using existing financial techniques. Manufacturing companies invest money to make a return on that investment. Hence all the value assignments should be further analysed to verify whether the perceived value on the strategic attributes directly or indirectly improves the financial position of the company within a reasonable time frame. For example, when the product quality is improved, it may increase the sales of the items and lead to capturing more market share compared to competitors or the improved product quality can be offered to the customers on the higher price, whereby increasing the profit margin.

Manufacturing managers determine after-tax-cash-flow analysis using financial techniques such as net present value, internal rate of return and pay back period of the investment. These preferred financial results are generated by DSSCIM, for the benefit of the decision maker. In addition, the DSSCIM offers depreciation methods such as straight line depreciation, sum of years depreciation and fixed declining balance method to calculate cash-flow. These methods can be further enhanced if a requirement arises for alternative methods. Information related to company-tax is captured by providing a dialog box and requesting the managers to enter the appropriate tax rate for the amount calculated by the before-tax-cash-flow analysis. In a comprehensive DSS this information should be automatically calculated by a DSS. Since tax requirements vary in different countries, the automatic tax rate calculation has not been incorporated in the present model. By adding more programming, this feature can be incorporated when the necessity arises.

A typical template utilised in the DSSCIM for the financial analysis is shown in Table1. Using the template for the financial analysis, the following calculations are computed by DSSCIM for each technology selected by the decision maker.

Effective capital investment =
Total capital investment - Investment tax credit -Values of attributes (1)

Total operating savings =
Operating savings (costs) by tangible attributes + Savings by intangible attributes - Total expended investment (2)

Net before tax savings = Total operating savings - Depreciation (3)

Net after-tax savings = Net before tax savings *(1- tax rate) (4)

Net after-tax cash savings = Net after tax savings + Depreciation (5)

Net after-tax cash flow =
Net after tax cash savings - Effective capital investment + investment tax credit (6)

When the cost-value option is not used, the effective capital investment equals the total capital investment, since attributes are not assigned any monetary values. At this option, the template follows a similar template proposed by Canada [12], where the savings by intangible attributes and the value of the attributes in the strategic decision is not considered.

2.4 Illustrative Example for Value Assignment

Let Robotics be considered as one of the technologies for investment. Robots are used in a wide variety of manufacturing applications. Robots are generally regarded as machines that replaces people, and it was justified on that basis in the past. Considering the robot as a replacement of the human labour, limits many opportunities of that technology while creating many labour problems. Whereas applications of the robot can give the establishment a leading edge in manufacturing operations. When the robots are sought for applications where the complexity of the job and very high accuracy is a priority or when the application is in hazardous circumstances, the evaluation of the robot on the financial criteria alone makes justification impossible. Investment in Robotics can generate the following benefits to a company.

The Robot:

1. Provides direct labour cost savings
2. Reduces floor space requirement
3. Increases productivity
4. Improves machine utilization
5. Reduces scrap
6. Reduces reworking costs
7. Improves working condition by eliminating any unpleasant and hazardous tasks
8. Improves moral of the employees
9. Improves and provides more consistent product quality, and
10. Reduces cost of quality control

Attribute 1: Provides direct labour cost savings. The savings of direct labour can be estimated with a very high accuracy. Let this equal to $ 70,000/ annum

Attribute 2: Reduces the requirement of floor space. The savings due the requirement of floor space can also be estimated within a reasonable accuracy. Let this equal to $10,000 /annum

Attribute 3: Increases productivity. The capability to increase the productivity will definitely place the company in a better competitive position compared to its competitors. Hence the value of the productivity improvement can be estimated as $ 15,000

Attribute 4: Improves machine utilization. The capacity to improve the machine utilisation will be beneficial for the company in a near future. Hence a value of $20,000 is assigned to that attribute.

Attribute 5: Reduces scrap. The reduction in scrap due the investment in robotics can be estimated with a fair degree of accuracy. Let this is equal to $5,000 / annum.

Attribute 6: Reduces reworking costs. This attribute could also be objectively estimated within a fair degree of accuracy. Let this is equal to $10,000/ annum.

Attribute 7: Improves the working condition by eliminating any unpleasant and hazardous tasks. Hazardous tasks increase the opportunities of more accidents in the working environment and illness for the employees. Hence an improvement in the working condition will result in more productivity and less interruption to work. This feature can be added value for the manufacturing operations. Let the value of the feature is $ 10,000. This value could be recovered by reduced labour compensations.

Attribute 8: Improves the moral of the employees. This feature will facilitate the retention of the workers, thus avoiding higher worker turnover and retraining costs. Furthermore, it will help to attract more qualified employees for that company. Therefore let the value of the feature is approximately equal to $20,000.

Attribute 9: Improves and provides more consistent product quality. Improvement in the product quality and having consistent quality through out the range of products will in help the company to compete with its competitors. Therefore Let the value of this feature is $20,000

Attribute 10: Reduces cost of quality control. Reduction in the cost of quality control can be roughly estimated to be equal to $5,000 / annum

From these estimations:
The total benefits which can be directly or indirectly measured = $100,000 / annum
The total value of the attributes which are difficult to quantify = $85,000
Hence for a typical example shown in table 1,
The Effective capital investment = $ 293,000-29,300-85,000 = $ 178,700

However, while utilising the Effective capital investment is an index to compare the capital costs related to investing in conventional technologies and to calculate Net Present Value (NPV), Internal Rate of Return (IRR) and Pay Back (PB) Period, the Total capital investment is utilised to perform depreciation related calculations. Since the cost of acquiring asset and the depreciation will not be changed by the value of attributes the new technology offers to the company.

Table 1. Template for financial analysis

Technology	*Robotics*					
Prime Functional Unit	*Assembly*					
Year	0	1	2	3	4	5
Cost- Value analysis incorporated						
	(in1000s)					
Investment, Capitalised						
Equipment Cost	280					
Accessories & Tooling Cost	11					
other---	2					
Total Capital Investment	293					
Tax Credit for the Investment	10.00%					
Investment Tax Credit	29.3					
Effective Capital Investment	178.7					
Investment, Expended						
Engineering	5					
Installation	5					
Startup	2					
Other	0.5					
Total Expensed Inv.	12.5					
Operating Savings (Costs)						
From Attributes	0	100	100	120	130	130
other Savings	0	15	15	17	15	15
Other Costs	2	3	3	2.5	2.5	2.5
Total Operating Savings	-14.5	112	112	134.5	142.5	142.5

By utilising this template, the DSSCIM generates the result for investment in Robotics as shown in Table 2.

Table 2. Display of Results

Sub System :	Robotics		
Department :	Assembly		
Ratios			
NPV : 318.18		Chosen	
IRR : 65.4 %		Discount Rate : 20	%
PayBack Period 2.17	Y		43
(within 10 years)			
Annual Aftrer -Tax NetCash Flows			
	Year 0 :	-163.9	
Year 1	99.45	Year 6 :	125.37
Year 2 :	99.45	Year 7 :	125.37
Year 3 :	118.57	Year 8 :	125.37
Year 4 :	125.37	Year 9 :	125.37
Year 5 :	125.37	Year 10 :	125.37

The parameters for the calculations are considered as follows :

Straight line depreciation method with salvage value of $10,000
Life of the asset is 10 years
Tax rate for the taxable income through out the study period is 15 %
Minimum attractive rate of return (MARR) for the computation of NPV is 20 %

3 Conclusion

Advanced manufacturing technologies have an inherent difficulty in justification due to the many intangible attributes associated with them. Due to this difficulty managers tend to avoid the intangible attributes within the justification process [13]. This avoidance imposes a severe limitation on the justification procedure. Any justification should be made on complete information without ignoring any features which are difficult to incorporate into this procedure. The Cost-Value technique has a disadvantage of being a subjective estimation procedure. However, the consideration of intangible attributes within the justification framework enables a decision maker to derive decisions based on complete information. In addition, incorporating the values of qualitative attributes into the justification procedure will definitely justify the costs involved in investing in AMTs. The results shows by incorporating the benefits due to the intangibles and by value estimation the investment produces favourable financial indices. To reduce the subjectiveness of the cost value technique, further research should be carried our in redefining the intangibles associated with an investment into tangible attributes. Few companies have become leaders in the field of CIM. Learning from them would be a valuable experience for a manufacturing managers who are interested in implementing CIM for their companies. Hence establishing a comprehensive database, which provides benefits accrued for each AMT at the leading companies will effectively help to reduce the subjectiveness associated with the estimations.

4 Acknowledgments

The authors of this paper would like to thank Ms Bonnie Hart, study adviser at University of South Australia for her helpful comments towards editing this paper.

5 References

1. D. V. Hunt, *Dictionary of Advanced Manufacturing Technology* , Elsevier Science, New York, (1987).
2. G. C. I. Lin, *Integrated Computer Aided Design, Manufacture and Management*, Annual engineering conference 1976, Australia, May, (1976), pp 68-70.
3. P. Gilmour, R. U. Hunt, and N. J. Harrison, Implementation of CIM in Australian manufacturing, *Int. J. Computer Integrated Manufacturing Systems*, Vol 4 No 2, (1991) pp 26-30.
4. N. Rescher, *Introduction to Value Theory*, University Press of America, USA, (1982).
5. M. L. Shillito, and D. J. De Marle, *Value, Its Measurement Design and Management*, John Wiley and sons Inc, New York, (1992).
6. E. O. Joslin, *Computer Selection*, Addison-Wesley Publishing, USA , (1968).
7. V. N. Bhat, *Applications of the cost value technique for the AMT acquisition*, Int. Engineering management Conference; Managing in a Global Environment, USA, October (1992), pp 179-83.
8. G. C. I. Lin, and S. V. Nagalingam, *Justification of CIM : Overcoming implementation* barriers, 2nd Int. Conference on Managing Integrated Manufacturing, Leicester, U.K, June (1996), pp 13-18.
9. G. C. I. Lin , and S. V. Nagalingam, *Justification of CIM: A Guided Approach, 4 th Int. Conference on Automation Technology*, Taiwan, July, (1996), pp 3-10.
10. S. V. Nagalingam,. and G. C. I. Lin, A Unified Approach towards CIM Justification, *The Int Journal of Computer Integrated Manufacturing Systems*, in press.
11. P. L. Primrose, and R. Leonard, The financial evaluation and economic application of advanced manufacturing technology, *Planning for automated manufacture*, IMechE, September, (1986),pp 49-54.
12. J. R. Canada, Non - Traditional Method for Evaluating CIM Opportunities Assign Weights to Intangibles, *J. Industrial Engineering*, Vol 18 No 3, (1986), pp 66-71.
13. F. Lefley, Capital investment appraisal of advanced manufacturing technology, *Int. J. Production Research*, Vol 32 No 12, (1994), pp 2751-2776.

CIM Implementations

A new Modelling Approach to Support Identification and Development of Secondary Processes in an Industrial AMT Implementation

Torbjörn Ilar & Prof. Anders Kinnander

Division of Manufacturing Engineering,
Luleå University of Technology, Sweden.

Abstract Implementation of semi-automated manufacturing cells always requires development of both the technical system and also the organisation required to manage and support this system. Overall system performance is highly dependant on the interaction between the technical and organisational systems and it is thus of great importance to consider both of these during the different phase of the design and implementation of such a system.

This paper is a result of a co-operative project between Luleå University of Technical and SCANIA Truck Components where different types of production modelling and simulation tool have been applied to support development and implementation of an advanced press line consisting of three 1200 tons presses and five industrial robots. The project has used two different production modelling tools to analyse different development strategies. An extended process mapping and production simulation approach has been used where both the main and supporting processes are modelled and where the use of the simulation model is focused more towards communication of information about the system than towards generating performance measures for the systems. This model approach is compared with the more common industrial application of simulation as purely a decision support tool; an approach applied during the earlier phases of this AMT project.

Key words
AMT, system implementation, process mapping, $IDEF_0$, discrete event simulation.

1 Introduction

1.1 Problem definition

Industrial implementation of advanced manufacturing technology (AMT) almost invariably involves a high degree of technical as well as organisational complexity. Traditional project management is very much focused on the technical aspects of an AMT implementation and often neglects issues concerning man-machine interaction. Nevertheless, the degree of success of this type of project is often dependant upon how well this interaction is manage which implies that developing understanding and acceptance among personal is important for overall project success.

The use of production simulation for decision support in this kind of project is now quite commonplace. However, its use is usually limited to evaluating alternative technical configuration and seldom takes supporting processes into account. In many cases, even the gains from the technical simulations are of limited benefit to the project as a whole due to poor correlation between the model and the real system where both system complexity as well as limited modelling time are significant constraining factors.

Ongoing research (Gaines & Norrie et al, 1995) into the next generation of manufacturing system stresses the need for semi-automated systems where an organisation's ability to handle rapid changes caused by short product life cycles and customer driven production will become the most important productivity factors. The development of this kind of AMT system must therefore be aimed at addressing continuous change rather than a steady state situation which never will occur.

1.2 Scope of the Paper

This paper investigates the importance of modelling both the main as well as supporting process in order to obtain a realistic model of an AMT system. The work is based on an investigation carried out in co-operation with Scania AB in Luleå to evaluate an extended production simulation and process mapping approach where both the main and supporting processes are modelled and where the used of simulation models is for communication rather then evaluation.

This approach is compared with the more common industrial application of simulation as purely as decision support tool and which was used in an earlier phase of the AMT implementation studied. One important feature of this study was the implementation of an individual learning model to capture the knowledge build up process.

1.3 Learning Curve

One approach to modelling and predicting competence development within an industrial organisation is the use of a theoretical learning curve. Learning curve theory has been applied to a wide range of industrial applications (Naime & Towill et al, 1994) that include human operators in complex tasks such as assembly.

A number of different mathematical equations has been proposed for modelling learning. However, the most widely utilised is probably still the time constant model.

$$Y_M(t) = \left[Y_c + Y_f(1 - e^{t/\tau})\right]$$

where: $Y_M(t)$ is the model output at time t,

Y_c is the model output at time t=0,

$(Y_c + Y_f)$ is the model output at time $t=\infty$,

τ is the model time constant

It is this model that is used in this study.

2 Project History

At the end of 1994 Scania decided to invest in a new production line for the manufacture of sheet metal components from cut blanks to finished pressed parts. This was to be a turn-key installation and the press line consist of three 1250 tons presses and five ABB 6400 robots for material handling between the presses (see figure 1). One of the objectives of the project was to introduce a flow oriented layout with fully automated material handling with a group oriented organisational structure.

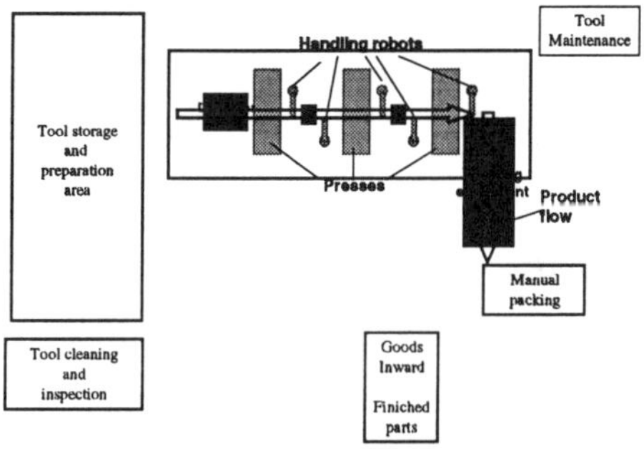

Fig. 1. General System Lay-Out

From a technical point of view, this AMT implementation can be consider as having a low technical risk as it used existing technology bought as a turn key system from a known system supplier.

Scania also has a long experience of sheet metal forming processes as well as material handling and there are a large number of similar references installations. The main technical concern was the ability of the press tools to handle the high press stroke rate required and also the system's ability to handle the scrap material resulting from the clipping of blanks.

Whilst the technical risk was fairly low, the uncertainties and risk associated with the organisation and supporting process was much higher. The new group oriented organisational approach represents a significant change from current management style and placed new demands on the workforce ability to handle support processes such as planning, press tool handling and preparation, quality control and system maintenance. It was these supporting processes that were initially expected to be the limiting factor on system performance.

2.1 Decision support

Scania was very much reliant upon the systems supplier's knowledge of the overall system configuration. However, production simulation was used to support the Scania project management group (Bengtsson et al, 1995). The main objectives of this simulation was to provide the project management team with information and guidance concerning maximum capacity, appropriate batch size, bottle-necks, material flow; all technical aspects of the system (and thus part of the main process).

The only support processes modelled in this simulation study was material handling and, to a limited extent, direct tool set-up in the press line. A retrospective analysis showed that the omission of supporting process in the simulation model used by the management group resulted in poor correlation between the simulation results and the output from the real system was achieved. In fact, the results from this study were of very little use in reducing the uncertainties involved in this project.

3 Supporting Process Modelling

The main objectives with the present study was to identify and model supporting process using $IDEF_0$ schema for process mapping. This was followed by an investigation where production simulation was used to evaluate the robustness of these process and to communicate the results to management and shop floor personnel. This communication was seen as being very important for providing a common understanding of the supporting processes. A more general objective was to develop a modelling approach which supported analyses of both main and supporting process.

An important factor which dictates the usefulness of any methodology is the acceptance of the model by the staff involved in the project. There are many different views as to what is required to achieve such user acceptance. One option is to use "high end" simulation tools which offer a high level of visual fidelity. A second option is to reduce the perceived system complexity by using a hierarchical modelling approach to model processes, sub-processes and activities. The former technique is dependent upon access to technical documentation about the system and its component parts. The latter approach requires information which is not always clearly documented and thus requires frequent meetings between the model builder and the project group (including shop floor personnel) to allow incremental development of a model.

A hierarchical modelling approach using two different modelling tools was used in this investigation. $IDEF_0$ was used for process mapping and discrete event simulation was used for communication of the dynamic aspects of the system.

3.1 Process Mapping

The process mapping investigation consisted of the following steps:

1. Identification of supporting processes,
2. Dividing the support process into sub processes,
3. Identification of activities and activity chains,
4. Identification of inputs, outputs, constrains and resources for each activity.

The processes identified in step 1 are presented in figure 2. These steps were closely followed in sequence; although there was some overlap between steps 3 and 4 due to an increase in common understanding within the project group. An initial model created by the research group was presented to the project group whose reactions formed the input to the next phase of development of the model. The $IDEF_0$ modelling software used allowed export of models in an 'open' format for transfer to other software; in this case a discrete event simulation system.

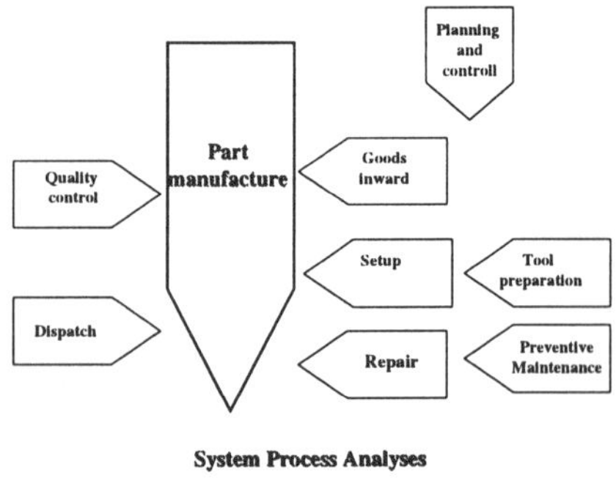

Fig. 2. Identified main and supporting processes

The two level IDEF$_0$ schema for the tool preparation process is presented in Figure 3. One unexpected result from this highly abstract model was gaining an insight into the importance of the conveyor, which is a resource in almost every activity in the tool preparation process and which potentially acts as a bottleneck. Should this information have been available earlier in the project, it is likely that the conveyor system would have been re-designed.

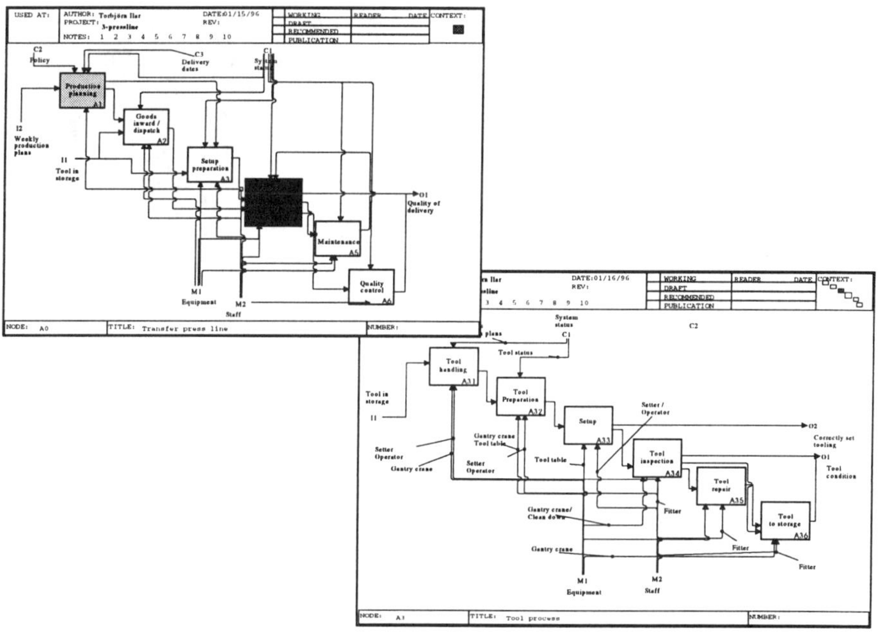

__Fig. 3. Two levels of the IDEF schema for Tool Preparation.__

4 The Simulation Model

As mentioned earlier in this paper, the objectives of the simulation model was not to obtain data which could act directly as a decision support for the project leader but rather to communicate the behaviour and complexity of the system and thereby increase the insight and awareness of the production group manager and operators.

The next step was to identify appropriate constraints for the simulation model i.e. processes to be included. The decision was made to include the main process (sheet metal forming), tool preparation, set-up and maintenance processes. These processes were considered to be the ones with the highest impact on the overall system performance.

The simulation model was built in the commercial production simulation software *Taylor II*. A high level of correlation between the IDEF$_0$ structure and the simulation software was aimed for and the same activities and activity chains were used.

This model is presented in Figure 4. Unfortunately, this figure cannot convey the logical behaviour of the model since this only becomes apparent when the model is run. This is a common drawback discrete event software, and is an important constraint when communicating using simulation models.

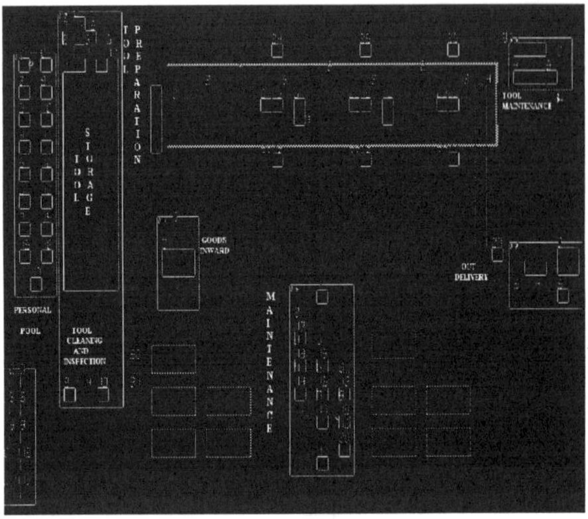

Fig. 4. Discrete event simulation model

4.1 Modelling of competence profile

To be able to model 'learning' associated with the support processes, a competence profile and learning curve were included in the model for two tasks; the tool related process and the maintenance related process. Two values (which change as the model is run) describes a unique competence profile for each operator. These were initially set by the group manager.

The competence index for each operator is updated when s/he performs an activity following the time constant learning model described earlier. For the tooling processes this is:

$$x = x + 1$$

$$tu_{ij} = tu_{ijopt} + \Delta tu_{it} \times e^{-\frac{x}{C_t}}$$

where: x is learning curve index,

tu_{ij} is the operation time in the updating matrix for operation i and operator j,

tu_{ijopt} is the optimum operation time,

Δtu_{ij} is the initial activity time and

c_t is a learning curve factor.

A similar algorithm is used in the maintenance process.

5 Designing and analysing the simulation experiments

Factorial design was used when planning the simulation runs. Four factors were of interest which gives a total of 16 combinations to achieve a four factor, two level, full factorial experiment (Law & Kelton, 1991). The following factors were included in the simulation experiment:

1. *The constant in the learning curve equation.* This is mainly for sensitivity analysis since an exact value for this constant is very difficult to find. Level 0 equates to 5 and level 1 to 15.
2. *The level of operator availability.* Level 0 represents 90% and level 1 70%.
3. *Batch size.* Level 1 equals 200 and level 0 equals 50.

4. *Two different human management philosophies.* Level 0 represents an expert concentration philosophy where the operator who will complete the activity in the shortest time is given the highest priority. Level 1 is for a competence spread philosophy where the operator with the longest activity time has the highest priority.

Factor 3 and 4 of this study can be controlled, so that data may be obtained about how these factors influence another variables referred to as the response variable or performance measure.

The experimental study was conducted in two steps. The first step focused on knowledge build up including factors 1 and 4. In the second step, the best knowledge build up result for each human management philosophy was used as an input and factors 2 and 3 were introduced. The 4 factor composition for the first step and 8 factor compositions for the second step were simulated 5 times to achieve four replicates. This is known as a full experiment run. The mean and the variations could then be calculated for each response. Statistical analysis also generated confidence intervals for each response.

5.1 Model responses

The model responses used were selected with the aim of supporting communication of the behaviour of the system under these different scenarios rather then for providing direct measures of the performance of the real system. The following responses were selected:

- *Relative production rate,* which was defined as the actual production rate verses the production rate from the previous decision support simulation run.
- *Performances of supporting processes,* which was the mean time and variations associated with carrying out tool preparation and maintenance processes.
- *Group competence build up,* which including the current competence index divided by the optimum competence (SDO), variation in competence (VIC) and the competence build up time. The latter response was defined as when operator group reached a competence index of 50% of optimum and the first two responses were calculated as:

$$SDO = \frac{\sum_{1}^{i}\sum_{1}^{j} t_{uij}}{\sum_{1}^{i}\sum_{1}^{j} t_{optij}} \quad \text{and}$$

$$VIC = \sum_{1}^{i}\sum_{1}^{j}(t_{uij} - \bar{t})^2 \qquad \text{(Box \& Hunter et al)}$$

Where: tu_{ij} is the operation time in the updating matrix for operation i and operator j, tu_{ijopt} is the optimum operation time.
The responses were "measured" after a completed simulation run (50 days).

5.2 Results

The most significant results from the simulation study were that;

- The highest initial relative production rate will be achieved with the expert personal management approach and high operator availability, independent of all other factors. It is important to note that this production rate is still less then 75% of the prediction from the earlier simulation study. This shows the importance of expanding production simulation to include both main and supporting process.
- The shortest tool preparation and maintenance process times will be achieved with the expert management approach. However, these times are subject to very high variations due to the inability to handle personnel disturbances.
- The learning curve factor, c_{l}, has no significant effect on the 'best' personal management approach. The lower factor value will, of course, lead to a slower group competence build up, but still offer better performance than the competence spread approach.
- A low operator availability combined with small batch sizes is preferable with the management approach of spreading competence. This indicates the group's ability to handle rapid change, which is expected to be of great importance in the next generation of manufacturing systems.

These results where validated through a presentation and discussions with the project group, rather then a direct compare with the actual system performances.

6 Discussion

Concurrent engineering has been used in industry for many years and also investigated in the academic world. This research has mainly focused on traditional product development (Jeppsson, 1997) neglecting manufacturing, or focuses on an expert oriented project group (Nelsson, 1996). It is often suggested that concurrent engineering requires all functions within the company to share and communicate information (Jeppsson, 1997 and Nelson, 1996) and that the use of different computer based support tools is of great importance.

Development of advanced manufacturing systems has very much in common with traditional product development with similar opportunities to avoid costly mistakes by putting in more resources and utilising computer based modelling tools during the early development phases.

Organisational issues in manufacturing have been described generally in numerous papers (Lei & Goldhar, 1993; Tie & Lin, 1992) as well as specifically for AMT implementations (Towill & Cherington, 1994). All these papers indicate that organisational learning, and thus the development of the supporting process, can only start after the installation and commissioning of the system is complete. The main drawback with developing supporting process after installation is complete is the high negative effect on systems productivity during this development period.

The implementation approach outlined in this paper addresses the importance of the supporting process with the aim being able to carry out parallel development of both main and supporting processes. Two important factors are the close interaction between the end user and the system modeller and the use of models with different levels of abstraction. This approach also requires utilisation of high end visualisation and interactive modelling tools in the early phases of the project. An example of such a model is presented in figure 5. This model was built in the commercial VR software *Superscape*.

This methodology is based upon the concept of *Rapid Factory Modelling* (Legge & Ilar, 1997) which addresses the need to quickly develop a digital factory model based upon the information available during pre-operative stages of the development of the manufacturing system. Current methods of model building are so time consuming that it is often only realistic to build a model once the system design is stable.

The ability to quickly develop a complete model of a proposed manufacturing system is considered important for both testing and user involvement.

Fig. 5. Virtual reality model of the three press line

At the early stages of a project, the level of detail available will be low. However, it is important that a detailed 'generic' (i.e. non specific) model can be created quickly to allow early user involvement and feedback.

This approach must also contain support for the extraction of important variables (responses), which has been addressed by other researchers utilising discrete-event simulation in combination with PLS-modelling (Eriksson & Kinnander, 1997).

7 Conclusions

This paper discusses the importance of modelling the supporting processes in order to obtain a realistic evaluation of advanced semi-automated manufacturing systems and introduce a two level modelling techniques to support this evaluation. The paper also outlines a general approach for implementation of advanced manufacturing systems using computer modelling. Some of the key features of this approach have been investigated in an industrial case study carried out in co-operation with Scania.

The main conclusion from this work are:

- The two level modelling method applied in this investigation offers excellent opportunities to support deeper understanding of the interaction between the main and supporting processes in an AMT system. These interactions are often neglected but have a great impact on the total system efficiency. Using process mapping tools together with production simulation offers the opportunity to identify the most important process and activities, and to use dynamic analyses to communicate the important system characteristics to a broader audience within a company.
- The best source of supporting processes knowledge is from the work force and it is therefore of great importance to obtain models which can be easily communicated to this group. This does not necessarily require the use of high end 3D visualisation tools but rather demands a close working relationship between the "modeller" and the project group.

- Investigation and active development of supporting processes can drastically improved the quality and scope of the actual system behaviour. Modelling this behaviour allows an increase in the understanding by the staff and operators.

The next step in this industrial investigation is to apply high end visualisation and interactive modelling tools to support a continues process and competence development within the system operators.

8 Acknowledgements

The work presented here is part of the research project PIDUS, which addresses the design, implementation, commissioning, management and maintenance of production system. The authors would like to thank: David Legge and other co-workers in the Department of Manufacturing Engineering; Jonas Rindeskog and his staff at Scania and especially the Swedish National Board for Industrial and Technical Development (NUTEK) for funding this work.

9 References

1. Gaines B.R., Norrie D.H. (1995). *Knowledge Systemanization in the International IMS Research Program.* http://ksi.cpsc.ucalgary.ca/KSI/KSI.html
2. Naim M.M., Towill D.R.. *Modelling and Forecasting Industrial Innovations via the Transfer Function S-Shaped Learning Curve.* International Journal of Advanced Manufacturing Technology (1993) no 8 pp329-343
3. Bengtsson J, Hvedng B, Matiasson L (1995). *Production Simulation of the Three Press Lina at Scania.* Luleå University of Technology. *Technical Report*
4. Law, A. M. & Kelton, W. D. (1991). *SIMULATION MODELLING & ANALYSIS.* New York: McGraw-Hill
5. Box, Hunter & Hunter (1978). *STATISTICS FOR EXPERIMENTS.* New York:John Wiley&Sons
6. Jeppsson P. (1997). (In Swedish) *Computer integrated Design Systems in Concurrent Engineering.* Luleå University of Technology. PhD. Dissertation No. 1997:08
7. Nelson J (1996). *On Co-operation in Manufacturing Systems Engineering.* Chalmers University of Technology. Licentiate Thesis PTA 96:06
8. Lei D., Golghar J.D. (1993). *Implementation of CIM Technology: The Key Role of Organizational Learning.* International Journal of Human Factors in Manufacturing. vol 3, pp217-230
9. Tie H. Y., Lin G. C. I. (1992). *Incorporating CIM optimisation in manufacturing management strategy.* Computer Integrated Manufacturing Systems. vol 5 no 4 pp311-325
10. Towill D.R., Cherrington J.E. (1994). *Learning Curve Models for Prediction the Performance of AMT.* International Journal of Advanced Manufacturing Technology (1994) no 9 pp195-203
11. Legge D. & Ilar T. (1997). *The Virtual Factory - Manufacturing System Development and Use - A General Framework.* Working document presented at the GNOSIS IRM 1997
12. Eriksson U & Kinnander A (1997). *Challenges and Opportunities when using Simulation Models of Manufacturing Systems.* Proceeding at the 32nd Int. MATADOR Conference.

A Prototype System for CIMS Information Integration Platform[1]

Shi Jing, Zhang Bin, Yu Ge, Wang Guoren, Shong Baoyan, Jin Lili, Zheng Huaiyuan

Department of Computer Science and Engineering
Northeastern University, Shenyang, 110006 China

Abstract The goals of CIMS information integration platform are to support users efficiently to access the integrated information from multiple, heterogeneous, autonomous, distributed information sources. In this paper, we describe the schema integration mechanism, and present the architecture of CIM-IIPS prototype system and its flexible integration strategies for CIMS complex environments. We also introduce an object-oriented integration data model.In particular,the non-traditional data source integration technique is discussed,and an object-oriented integrated interface to the file system integration is proposed.

1 Introduction

The integration of different application environments is one of the major challenges in the development of integrated information systems. Especially in the Computer Integrated Manufacturing System CIMS , there exist large numbers of the heterogeneous, autonomous and possibly distributed data sources, they lack of efficiently information sharing and data exchanging, and the information inconsistency between data sources are the serious drawbacks of the current CIMS. One of the solution for that *information isolated islands* is to create the integrated data management system to manage the heterogeneous data[1], and support the dynamic data access among the data sources, on the prerequisite

[1]This work has been partially supported by the National 863 High-Tech Programs of China with grant 863-09-2141

of maintaining the existing database systems and applications. In the CIMS information integration field, many research works are made and there appear lots of prototype systems[2-6]. However, those prototype systems and relevant researches have only resolved the traditional database integration, but not meet modern CIMS applications, therefore, it is very urgent to study the techniques about the object-oriented CIMS information integration platform.

Most current commercial information integration platform systems, such as [6], mainly focus on some special hardware environments, using non-object-oriented methods, providing the information exchange but lacking of the information integration. Thus, we design and develop an object-oriented information integration platform system for CIMS application, called CIM-IIPS, which chiefly aims at integrating the information distributed in the existing autonomous and heterogeneous multiple data sources in CIMS environment, and supporting information sharing among systems. We designed the CIM-IIPS architecture to fulfill several important goals, all interrelated, as follows:

- *Flexible Integration Strategies.* For CIMS complex integration requirements, we provide two kinds of integration strategies:*the logical integration approach* and *the physical integration approach*, the former means that the local databases are logically integrated into the integration database , and the latter means that the local databases are physically integrated into the integration database. These two kinds of approaches are suitable for different situations. For example, the physical integration approach is suitable for the top administration level, because the administration level need the comprehensive management information, which are less and more stable, and on the other hand, the logical integration approach is more suitable for the factory level, because the factory level need handle the quantity of dynamic data , and the local data may be frequently changed. So the user may choose the appropriate strategy for different integration requirement.

- *Support Integration for Different Data Source Types.* In CIMS environment, the data sources to be integrated include not only the traditional database systems (e.g., relational database, network database, hierarchical database), but also the non-traditional database systems, such as the object-oriented database and file system in which the schema does not exist. Note that the non-traditional data source integration is increasingly urgent, specifically for CIMS,so we deeply study the file system integration techniques on the basis of the traditional database integration techniques.

- *Data Consistency.* When data is integrated from local autonomous sources, the integrated views data should be consistent with the corresponding local data. In particular to the physical integration strategy, the integrated views are materialized as the integrated database, the integrated system should support the data consistency maintenance for the different integration strategies.

In CIM-IIPS, the integration data model is based on the object-oriented data model, it can integrated the traditional database (e.g. ,ORACLE ,SYBASE), the object-oriented database (e.g., O2)and file system (e.g., DXF file in AUTOCAD). Furthermore, there are two kinds of integrators, *the domain integrator* mainly process the information integration inside each application field, and *the main integrator* resolve the information integration between the application fields.

The remainder of paper is organized as follows. In Section 2, we describe the schema integration mechanism. In Section 3, we first overview the CIM-IIPS system architecture and integration strategy, and present the design and implementation details of Main integrated System, furthermore discuss the data duplicate copy between the global database and the local database and corresponding solution—Bulletin system. In Section 4, we give the integration data model and integration data language. Finally, we discuss about file integration and relevant resolution approach in Section 5 and concluded in Section 6.

2 Schema Integration Mechanism

In CIMS environment, the information to be integrated are generally heterogeneous, so, we adopt the*schema integration mechanism* in CIM-IIPS to support information integration, namely the integrated process is the schema-driven style. We describe the integration as the*four-level schema and three-level mapping* integration mechanism, the four-level schema include the local schema, the mapping schema, the global schema and the global subschema,and while the three-level mapping means Schema Mapping ,Schema Merging and restructuring,and Global Schema Definition, just as shown in Figure 1. The local schema, represented by the local data model, must cover all of the local schemes that are willing to be integrated in all component databases in order to support the "Completeness and Correctness Principle" of the global schema. All local schemes that are willing to be integrated will be mapped to the corresponding mapping schema, which is represented by the single common data model(CDM). Creating the mapping schema is mainly for removing the semantic heterogeneity. Furthermore, merging and reconstructing the mapping schema will produce the global schema, which is also represented by CDM. The global schema will remove all semantic conflicts including all that were not resolved yet during the schema mapping phase. In detail, the schema merging eliminate the redundant information existing in all mapping schemes so that support the "Minimality Principle", and the schema reconstruction mainly aims at removing the semantic conflicts of the schema structure level to ensure the "Semantic Consistency Principle" of the global schema, and meanwhile building the global schema that is easily understood for the users is to meet the "Understansability Principle" of the global schema. Based on the global schema, the global database manager can define the global subschema required for the global user by the global subschema definition language.

3 System Architecture

In CIMS, the information transformation and information share are very frequent, so it is the important task to decide the system architecture for information integration platform. Our CIM-IIPS comprise three levels of system

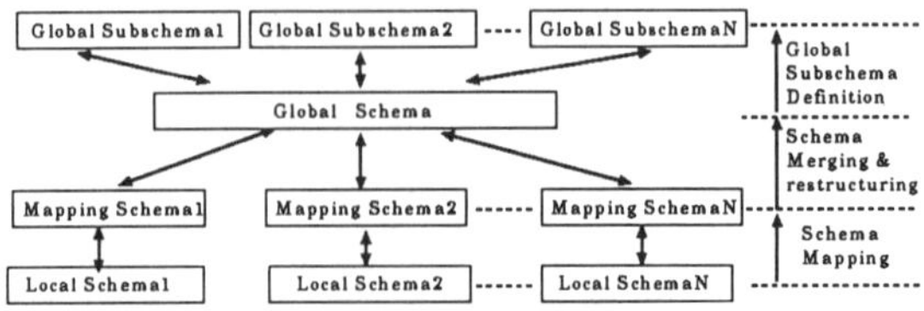

Figure 1: Schema Integration Mechanism

architecture, namely main integration system level,domain integration system level and subsystem integration level. In it, main integration system level specially support the global information exchanging and sharing among each application fields in CIMS. Domain integration system , on the other hand, support the information integration inside each application field(e.g., MRPII, mechanical CAD). The main function of the subsystem is to resolve the heterogeneity of the data models and support the autonomy of the local systems. The subsystem may directly connect to the domain system as well as the main integration system .

For above system architecture, we propose the different integration strategies in our platform system. On the top global level, the main integration system use the physical integration approach in order to improve the system performance of the information transformation and reduce system costs, in contrast with it, the domain integration level adopts the logical integration approach to support frequent data transformation and highly data consistency requirement. In CIM-IIPS, the main integration system is the kernel part, we will given the more details about it in following aspect.

3.1 Main Integration System

Figure 2 shows the architecture of main integration system, which is based on open client/server architecture, and the whole system is divided into two parts:client and server, as suggested. Now, we will describe the modules of the main integration system in more detail in turn as follows.

- Client. In CIMS information integration platform, the client is over-platform, the global users may access the sharing information on any devices, even microcomputer. The function modules of client include:

  (1)Application Interface, which contains Interactive Application Interface, Batch Process Interface and Application Programming Interface(API).

  (2)Middleware Processor, including EXPRESS converting interface and Graphic Programming Interface, can meet various application requests more conveniently and perfectly than main system application interface.

  (3)Language Processor, which is responsible for syntax analysis, semantic check and operation translation to operation request for the global user and then generates Common Operation Quest(COQ) based on middle language code.

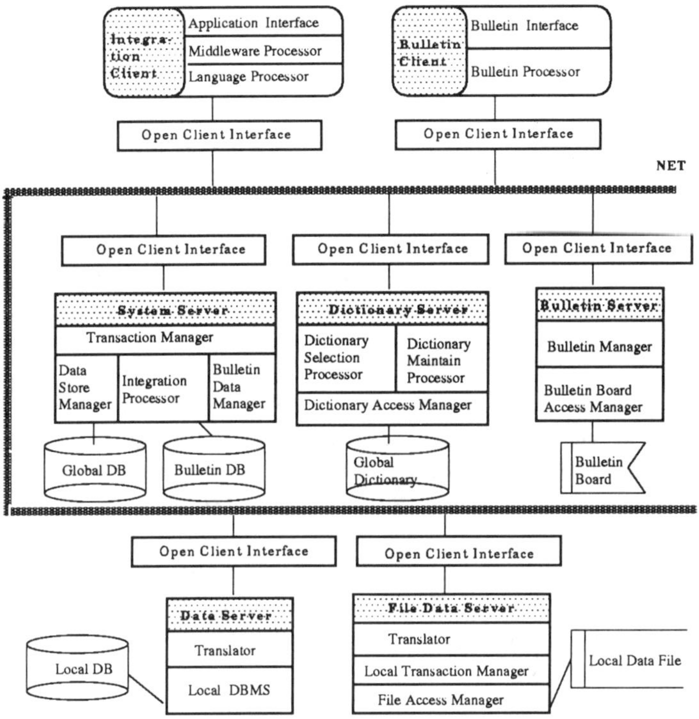

Figure 2: The architecture of Main Integration System

(4)Open Client Interface, which makes use of the open computing environment RDA/DBRPC and heterogeneous network integration functions to correctly send the Common Operation Quest to corresponding servers.

- System Server. The related COQs are sent to system server through Open Server Interface. The global database stores the global sharing information, which are managed by Data Store Manager. The Data Store Manager is chiefly responsible for accessing the global database, translating from COQ to access request of global database, and extracting global data. The bulletin database stores the temporary data announced by each local application, which is supported by Bulletin Data Manager. The Bulletin Data Manager deals with access to bulletin database, extracts the global sharing information from the local data source to bulletin database, and then sends them from bulletin database to Data Store Manager. The major task of Integration Processor is to manage the whole schema integration process, while System Transaction Manager is responsible for the transaction management of main integration system in order to ensure the data consistency in main integration system. In our platform system, both the global database and bulletin database are based on the object-oriented database system O2 .

- Global Dictionary Server. It mainly supports the operation requirement for global dictionary, these requirements are sent to the dictionary selection module or dictionary maintaining module, which is responsible for translating the operation request of global dictionary into access operation for global dictionary, and completing the real access for global dictionary by Dictionary access manager.
- Data Server. It serves for both the global database and bulletin database in System Server, namely, each data server first receives the operation request based on middle operation language from Main System Server through the open server interface, and then the corresponding translator will translate the common operation request into local operation request supported the local database system and proceeded to commit them to the local data source to execute, finally the results will be returned. As shown in Figure 2, the data server comprise two kinds: traditional database system (e.g., ORACLE, SYBASE) and non-traditional data source system (e.g., FILE system), the implementation of corresponding translator are very different from each other.
- Bulletin Server. It is the important part in bulletin system, it chiefly supports the physical integration strategy in main integration level through maintaining the consistency of data duplicate copies between the global database and the local databases. More details will be given in Section 3.2.

3.2 Bulletin System

As above discussion, the physical integration strategy adopted in the main integration level will produce the duplicate copies of the share information between the global integration database and the local databases, so we put forward the method called *"main copy asynchronous update"* to maintain the data consistence between data duplicate copies. According to this method, the local information that are provided for the global sharing are viewed as Main duplication, while the corresponding copy in the global database is called Vice-copy or copy of Main duplicate, when the local data are updated, only the Main duplication allow to be updated but not directly change the Vice-copy. So-called asynchronous update means that when Main duplicate is updated, the vice-copy in the global database may be modified during the range of the allowing time and don't need to update in the mean time.

In CIM-IIPS, we pose the*bulletin facility* for implementing the asynchronous update. After schema merging and reconstructing, the integration processor will automatically notify bulletin server to create the bulletin board file, which is the bright associating the global database with local databases. On the other hand, if local applications have updated the Main duplication, it should announce the related updated content to the Bulletin server using Bulletin client at once. Finally, system server will perform the data consistence maintenance between the integration database and the local databases, according to the messages updating bulletin which came from the bulletin server .

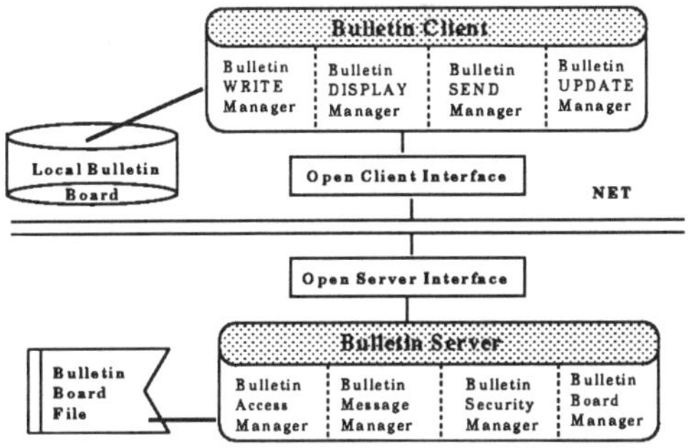

Figure 3: The Architecture of Bulletin System

Fig. 3 shows the bulletin system architecture, which is based on client/server architecture, it consists of several clients and one bulletin server.

- Bulletin client. It includes several modules such as bulletin WRITE manager, bulletin DISPLAY manager, bulletin SEND manager and bulletin UPDATE manager. In some local database, when the sharing information of the global system are ready, or , are changed and hope to update the corresponding global information, the local user will notify the bulletin server about the prepared information or changed data using bulletin client, and then, the bulletin server performs the asynchronous updating operation to global database.
- Bulletin server. It includes four parts: Bulletin Access Manager, Bulletin Message Manger, Bulletin Security Manager and Bulletin Board Manager. Once the global schema is available, the system server will notify the bulletin server to create the corresponding bulletin broads for that schema, so that the local application may announce all of the related local information on them. When the Bulletin Message Manager receives the bulletin information from the bulletin client, it will send those information to the Bulletin Security Manager to process, and furthermore, the legal bulletin information will be sent to the Bulletin Board Manager, which is responsible for the maintaining of the bulletin board and access control. Above functions are all based on the Bulletin Access Manager.

4 Integration Data Model(IDM)

In CIMS environment, the data exchanging between the application fields are very complex, the integration involves in quantities of commonly structural data as well as the non-structural engineering data, so CIMS information integration platform must choose the object-oriented data model so as to easily integrate and transform the engineering data. However, only object-oriented data model can't meet the integration request, so we expend Source Tag Facility[7] and Semantic Context Mechanism[8] into object-oriented data model, summarily, our integration data model(IDM) may be described as follows:

IDM = object-oriented data model + source tag + semantic context

In CIM-IIPS, the notion of object orientation is based on the object-oriented data model O2 , because we use the O2 as the common data model(CDM) to support mapping schema and global integration schema. And the Source Tag Facility means that the schema mapping process not only should describe the mapping relationship between source schema and target schema, but also record where the data in source schema or target schema came from, and a source tag corresponds to a source schema or a target schema . This facility is very important for operating the global integration schema. The Semantic Context Facility is posed for resolving semantic heterogeneity, its main idea is to extract the semantic from data in each integration database and described as a part of schema, namely the attributes in schema may possess own description attribute called context, for example, the attribute SALARY in schema PERSON may have the two semantic attributes CURRENCY and SCALEFACTOR. The context-added data is called semantic value , which are comparable, thus ensure the semantic interoperatability. In our system, the semantic value is the basic unit of information exchanging among heterogeneous information systems.

According to above integration data model, we design an integration data language , it is used to map and reconstruct the integration data schema, perform the physical extraction and reorganization of the global data, and supports the retrieval involving semantic context to global data schema. It is made up of integration data definition language and integration data operation language. Integration data definition language is responsible for defining integration data schema, resolving semantic conflicts between integration schemes, finishing schema merging and reconstructing, and managing integrated data schema. Integration data operation language is mainly to support the retrieval for global schema, particularly it can directly describe the semantic context of retrieval results in SELECT sentence. Note that the main integration system don't allow global user to update the global database, if the local data are updated, corresponding global data modification will be performed by bulletin system , just as discussed in Section 3.2.

5 File Integration

File system integration is a challenge for the traditional integration approach.The file system doesn't support the notion of the*schema*, nor be managed by traditional DBMS, the semantics of data if open deeply dependent on the way in which programs manipulate the data, so the file system integration obviously differ from the traditional database system integration. Traditional approaches to the integration of heterogeneous databases, called Structural Mapping[9], are not able to deal with file system integration, because they are based on the definition of correspondence between data elements of the schemes such as CDM and local data schema , just as described above. In CIM-IIPS, we propose to use the Operational Mapping approach to support the file system integration. The Operational Mapping approach, which overcomes the limitations of Structural Mapping , consists of defining the correspondence between operations at different schemes instead of defining correspondences between data elements. The

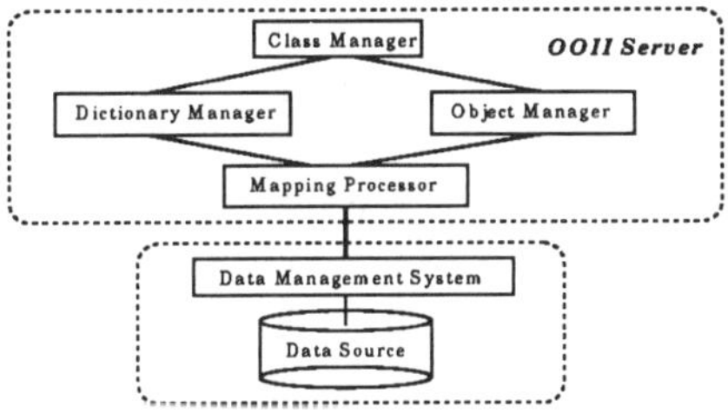

Figure 4: OOII'architecture

main ideas are as follows: some *primitive operations* will first be extracted from the local system, which are viewed as the local schema, and then the mapping schema is defined as a set of *basic operations* on a set of abstract objects, finally, the Operational Mapping is defined as the *implementation* of these basic operations in terms of the primitive operations of the local system, furthermore, more powerful *abstract operations* can be implemented automatically by integration system. At run-time the abstract operations issued by integrated application are translated using the basic operations, and the implementation of the basic operations is invoked.

Obviously, the idea of Operation Mapping need be supported with the use of the object-oriented(OO) technique. In order to define the mapping schema as a set of operations it is convenient to use the object-oriented Data Model, since one of the main characteristics of object orientation is to view data through a set of operations, thus encapsulating the internal representation. The combination of OO techniques and the Operation Mapping approach can perfectly deal with the file system integration problem. In CIM-IIPS, to implement the Operation Mapping approach, we define two kinds of object classes:ObjectCLASS and mappingCLASS. The objectCLASS describes the structural and behavioral characteristic of the object set, while the mappingCLASS supplies the implementation of those behavioral characteristic, that means, the objectCLASS is the abstract of the local systems on the integration level, and the mappingCLASS defines both the implementing code of the basic operations on the corresponding objectCLASS and necessary data structure for supporting the implementation of the basic operations.

To support the file integration, we design and implement an Object-oriented Integrated Interface OOII(which is called Translator in Main Integration System in Fig.2) for local file system in File Data Server , the architecture of OOII is shown in Fig.4. OOII is made up of four parts:Class Manager CM, Mapping Processor MP, Object Manager OM and Dictionary Manager DM, in which CM is the schema manager of the objectCLASS, OM manages the volatile integrated objects, and MP includes the implementation of all basic operations of the corresponding mappingCLASS, it is invoked to create the volatile object by OM, and meanwhile it can call the local systems for accessing the local data. The OOII's design need take several key techniques into consideration, such as the problem about Object identifier(Oid) in MAP, the class and object management, the design of the mapping dictionary and so forth.

6 Conclusions

In this paper, we deeply studied the information integration technology in CIMS environment, described the design and implementation of the prototype system of CIMS information integration . In detail, first, we presented a schema integration mechanism with the four-level schema and three-level mapping, Secondly, we described a three-level system architecture and the design details of main integrator, and we proposed the asynchrony data update mechanism based on the bulletin facility for supporting data consistence maintenance in physical integration. Thirdly, we described the object-oriented integration data model with the notions of Source Tag and Semantic Context. Finally, we introduced the Operational Mapping approach to integrate various of the data sources, and particularly gave the solution for file system integration. Our future work will focus on studying the multiple data source integration technology based on the Common Object Request Broker Architecture CORBA, We believed that this new mechanism will make the integration system more flexible and more scalable.

Reference

[1] Rattner, Laurie, "Information Requirements for Integrated Manufacturing Planning and Control: A Theoretical model",*Ph.D. Thesis, Decision Sciences and Engineering System*, Rensselaer Polytechnic Institute, December 1990.

[2] Krishnamurthy, V. S. Su, H. Lam, and M. Mitchell, IMDAS," An Integrated Manufacturing Data Administration System",*Journal of Data and Knowledge Engineering*, Vol.3,No.2,1988.

[3] Hsu, C. and C. Skevington," Integration of Data ang Knowledge in Manufacturing Enterprises: A Conceptual Framework",*Manufacturing Systems*, Vol. 6, No. 4, 1987.

[4] Smith,J.M.,et al., "Multibase-Integrating heterogeneous distributed database system", *in Proc. AFIPS*, 1981, Vol. 50.

[5] Wong,K.K. and Bazex,P.,MRDSM: "A relational multidatabase management system",*in Proc. Third Int'l Seminar Distributed Data Sharing Syst.*, Mar. 1984.

[6] "System Enabler CDF",*IBM Corp.*.

[7] Wang,Y.R.,Madnick,S.E.,"A Polygen Model for Heterogeneous Database System:The Source Tagging Perspective".*In Proceedings of the 16th Internation Conference on VLDB*,Brisbane, Austrilia, 1990.

[8] Wang Guoren,Yu Ge,Shan Jidi,and Zheng Huaiyuan,"The Data Model of CIMS Information Integration Platform",*Computer Journal*,China,1995.

[9] E.Bertino,M.Negri,G.Pelagatti,and L.Sbattella," Integration of Heterogeneous Databas Applications Through An Object-Oriented Interface",*Information Systems* ,Vol,14,No 5,pp.407-420,1989.

Methods for Design and Implementation of CIM Systems

M.H. Xiong   Z.Y. Han   Y.F. Zhong   J. Zhou

CAD Center, Huazhong University of Science and Technology, P.R.China

Abstract The design and implementation of CIM systems is a complex process involving the integration of multiple systems as well as people, tools and methods. In this paper, in order to help people involved in CIM system projects to design and implement CIM systems, the process of CIM system design and implementation is analyzed thoroughly, and new prototype process, which is practical and flexible, is suggested. Besides, to meet the needs and special requirements of the design process for CIM systems, the CIM system Integrated Design Environment is proposed and its characteristics, structure and components are also particularly defined in this paper.

Keywords Computer-integrated Manufacturing, Development and Implementation, Reference Model, Architecture, Design Environment

1 Introduction

Over a decade or so, computer integrated manufacturing (CIM) has been widely accepted by both industry and academia as the new generation manufacturing strategy. CIM system (CIMS) is considered as having the ability to improve product quality and enhance competitive power in the globe economic market. So, it is recognized as a 'Factory of the future' model or a manufacturing enterprise model in the next century. The CIM and CIMS are now the popular theme for most of the manufacturing enterprises in P.R. China. Some enterprises have been developing this new technology to meet their increasing demands for the best benefits and their production flexibility.

However, because CIMS is a complex organization of people, machines, materials, procedures and information, the enterprises faced with design and implementation of their CIMS usually find themselves thrown into an environment in which there are little appropriate specifications to guide them to develop this project. So, while some enterprises have won honor for their CIMS implementation, others have failed to develop their CIMS. In west countries an estimated more than 50 ~ 75% of CIMS projects fail to meet their expectations[1], and in P.R. China, some enterprises have got into various troubles in the process of CIMS development.

In author's opinion, one of the means to solve this problems mentioned above is to study and put forward some efficient, practical guidelines, tools and methods which are suitable for an enterprise to practice. In this paper, to meet this requirements, methods for design and implementation of CIMS are studied thoroughly, and a new prototype process of CIMS development and a new framework of CIMS integrated design environment are proposed in terms of several CIMS development projects in which the authors have been involved.

2 The Process of CIMS Development

Recently, much effort has been devoted to guide the enterprises how to work when they decide to begin designing and implementing their CIMS. The aim of such effort is to provide a serious of methods, or a framework for these enterprises. The so-called CIM Methodology contains the achievements of this research works. Obviously, it must be paid attention that the invention of every methods of CIMS development is not to impose a strict discipline on the CIMS design and implementation. Every enterprises should use different methods which are suitable for them to develop their CIMS in terms of their own perspectives, conditions and their goals to develop this systems.

The process of CIMS design and implementation (CIMS-DI) includes several phases, and it can be generally divided into such phases as feasibility study, preliminary design, detailed design, system implementation, and system operation and maintenance. This kind of division has a strict structure but little flexibility because it's essentially a structured life-cycle development approach[2]. On the basis of analyzing several CIMS-DI projects in P.R.China which the authors have been involved in, it has been discovered that the usual structured development methods usually can not satisfy all the requirements of CIMS-DI projects, and hence need to be improved to suit CIMS-DI.

2.1 Proposed Process of CIMS-DI

Because it is quit difficult for an enterprise to determine completely and correctly the enterprise demands and the system functions in the phase of feasibility study or system design, the modifying and repeating works are usually unavoidable, especially as the enterprise business objectives and environment have been changed. Practically, this modifying and repeating works are helpful for enterprises to design and implement CIMS of good quality and achieve the promising best benefit.

Therefore, a new prototype process of CIMS-DI is proposed in the paper, which is mainly divided into several phases such as CIMS master planning, implementation sequence determining, detailed design and system implementing. The relationship between these main phases are illustrated in figure 1, and the conditions of input , output and environment (control and mechanism) of each phases are showed in figure 2 by means of the structured design tool IDEFo.

The system detailed design, implementation and integration are completed in accordance with determined implementation sequence, and so, having been meeting the demands of follow-up processes, the amounts of design work are decreased and some repeated useless works are sometimes avoided. Therefore, this process of design and implementation is a 'process for reality' or a 'process for application'.

Compared with above proposed process with the usual development process, except the proposed process breaks the traditional strict linear development mode, the main difference between them is that in the proposed process there are the presentation of CIMS master planning and the feedback coming from the follow-up phases of CIMS master planing.

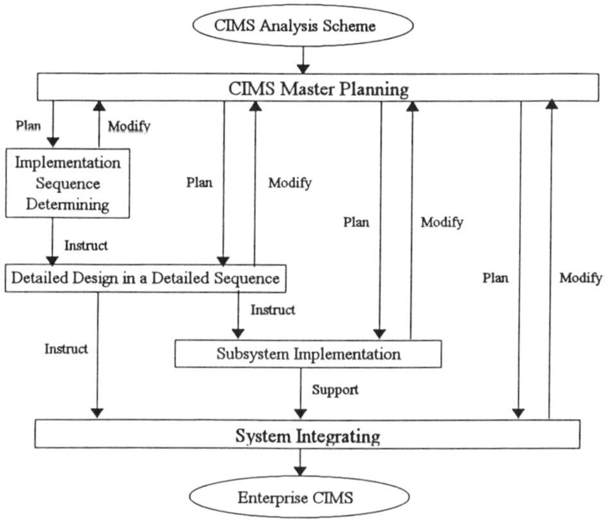

Fig. 1. Relationships between the main phases of CIMS-DI

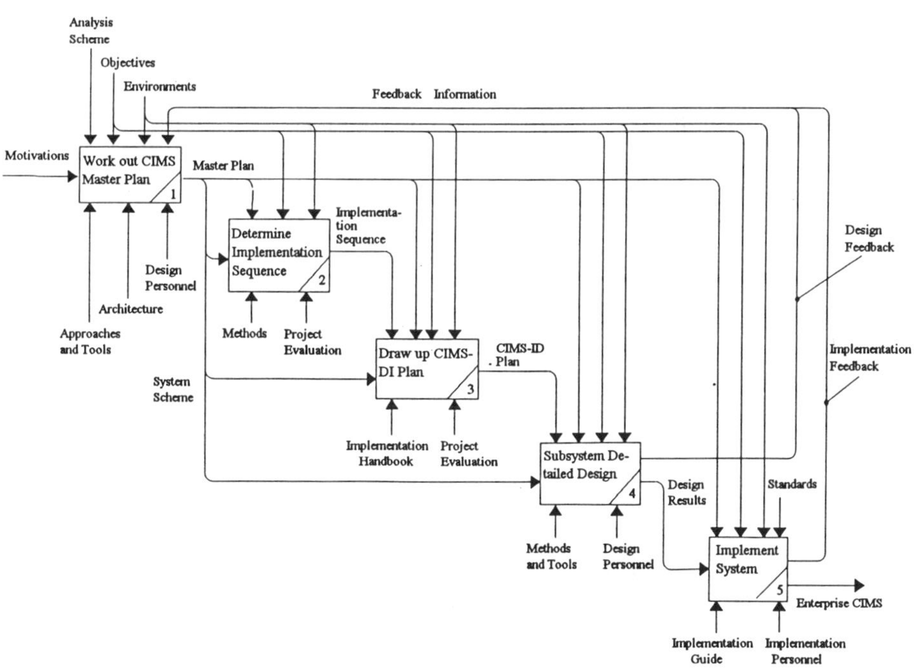

Fig. 2. A new prototype process of CIMS-DI

2.2 CIMS Master Planning

The CIMS master planning (CIMS-MP) is the major phase of this new proposed CIMS-DI process because the CIMS-MP establishes the overall framework of a CIMS and its CIMS-DI. Although the preliminary design in the traditional process also plans the CIMS projects, the CIMS-MP is more important than the preliminary design in the respects of planning a CIMS and guiding an enterprise to develop its CIMS. In other words, the CIMS-MP is an extended function about the CIMS preliminary design.

The contents of CIMS-MP mainly includes the overall framework of CIMS (function model, information model, etc.) as well as some important relationships between some important subsystems. The people who plan CIMS-MP must be organized by system analyzers, system designers and the people implementing CIMS. The CIMS-MP has been corrected and improved along with the process of CIMS-DI, and so the validity and practicality of CIMS-MP are guaranteed.

2.3 The Feedback

The feedback discussed here implies the information coming from the follow-up phases of the CIMS master planing, and according to this feedback the CIMS master plan is renewed. The information included in the feedback can be roughly divided into two parts of the design feedback and the implementation feedback. The design feedback represents the modifying requirements about CIMS-MP from the CIMS designers, and the implementation feedback represents the modifying requirements about CIMS-MP from the people implementing CIMS. The both can decrease the contradictory elements between the CIMS-MP and the plan of detailed design and between the CIMS-MP and the people implementing CIMS.

Actually, with the feedback from the follow-up phases the proposed CIMS-DI process has an ability of 'self-modification' or 'self-adjustment'. That is to say, this proposed process could be more reliable and practicable to guide enterprises to design and implement their CIMS.

3 The Design of CIMS

Of the process of CIMS-DI, the CIMS design is a process involving the integration of a lot of system designers who may use different tools and methods, and be in different design places and environment. This may lead to long development times with requirements that are not well understood and often changed during the design process. In view of these condition, a new method based on the reference models is proposed and the architecture of CIMS integrated design environment is put forward.

3.1 CIMS Design Method Based on Reference Models

Because of the complexity in modelling a CIMS, several reference models are used to view and understand the CIMS from different perspectives. A reference model (RM) presents the information and structure that are previously agree upon and can be treated as a standard or guideline[3]. Each RM corresponds to the conceptual representation of a system from a certain or some perspective(s). Examples of RMs include the function model, information model, hierarchical control structure, operation model and network communication protocols.

Therefore, in order to standardize the design of CIMS and simplify the design work, CIMS design could be transformed to another way that the design process is to rebuild from the components derived from CIMS RMs. This design method is hence called the CIMS design method based on RMs.

The process that the structure components of CIMS are derived and rebuilt from the RMs is very complicated. The effective design of CIMS depends on the reliability and the amounts of the established RMs, as well as on the skills and experiences of designers. For every designers, the RMs actually play a much important part in the process of CIMS design. Also, it should be in many ways that the RMs support the design of CIMS, and this support determines the quality and efficiency of CIMS design to a great extent.

3.2 Integrated Environment for Design of CIMS

In order to meet the needs and special requirements placed upon the design of CIMS discussed above, an architecture, CIMS Integrated Design Environment (CIMS-IDE), is presented in this paper, which enables the systematic integration of design process, design tools and methods and makes CIMS design effective. The main characteristics of the CIMS-IDE are as follows:

(1) The CIMS-IDE integrates every stages of CIMS design into an unifying environment that allows designers to operate with familiar tools and methods, such as IDEF[4], OOD (object orient design), GRAI[5], etc.

(2) The CIMS is designed within the CIMS-IDE in an automatic and mutual way on the basis of several consistent databases.

(3) The CIMS-IDE architecture allows CIMS RMs, regardless of drawing or word, to be stored in a database in certain format/structure, and enables use and maintenance of this database.

(4) A full set of design documents can be delivered in this unifying environment through the design process, which conforms the standards of CIMS and satisfies the requirements of CIMS design.

(5) A standardized and unified platform of the application modules makes it possible and convenient to add to and remove from the CIMS-IDE.

The CIMS-IDE comprises the following three parts: application program (AP), design process management (DPM) and database (DB). The AP can generally include, not constantly, the application modules of requirement analysis, modelling, evaluation, decision support, and file/document output. It serves the application function of the CIMS-IDE. The DPM supplies the functions of basic data processing and design process managing, and it provides the interfaces with the AP and the platform to access DB. The DB, the core of the CIMS-IDE, comprises logically a basic database (BDB) and a working database (WDB). The BDB stores the data and structure of design tools (DT), design methods (DM), RM and integrated infrastructures (II), and the WDB stores the executing results of the AP. The structure of the CIMS-IDE is illustrated in figure 3.

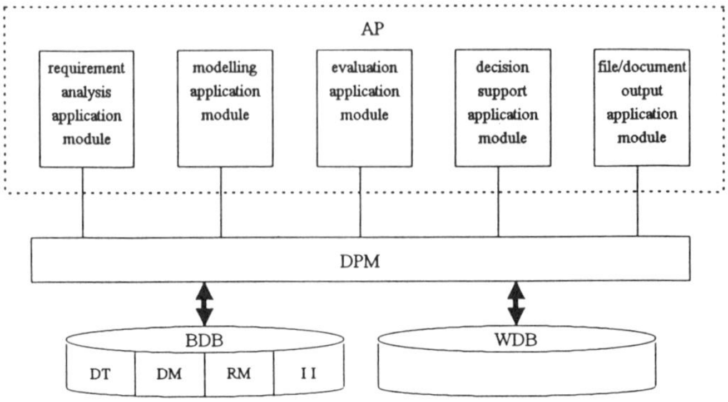

Fig. 3. The structure of CIMS-IDE

The CIMS-IDE is a complex design environment, and its two important parts are the environment, which integrated design tools and methods of CIMS used recently, and the mechanism of CIMS design process management. Actually, the layout plan of CIMS-IDE has been accomplished by a team of researchers organized in CAD Center, Huazhong University of Science and Technology (HUST), P.R.China, and the CIMS-IDE will be implemented in an environment of microcomputer (486/586) and Client/Server architecture. The software and tools used in this development include PowerBuilder, Word for Windows, MS SQL Server, Visual C^{++}, etc.

4 Conclusions

The design and implementation of CIMS is a difficult task in which the people of design and implementation of CIMS have little design experiences and are under tremendous pressure to do well right first time. So, it is important for those people to have the correct and practical 'roadmap' or guidelines as well as the design tools or environment.

In this paper, the issues of CIMS development have been firstly discussed, and then a practical prototype process of CIMS design and implementation is suggested. This proposed process breaks the traditional strict linear development mode, and allows the information feedback from the follow-up phases. Besides, the presentation of CIMS master plan enhances the abilities to plan the CIMS and improves the adaptability of CIMS design. Thus, this new process of CIMS development is more suitable for designing and implementing CIMS.

In order to simplify the design of CIMS, the architecture of CIMS integrated design environment is proposed and defined. Such an environment addresses the needs and special requirements of the design process for CIMS. It integrates every design tools and methods to an unified platform, and the three important parts of this environment are the CIMS design process management supplying an application platform, the application program which can be conveniently added to and removed from the environment, and the database which stores the data and structures of design references as well as the executing results of application program. Implementing the

CIMS-IDE will require a considerable research effort focused primarily on developing a more fundamental understanding of the design process of CIMS. However, this effort should advance the research work for CIMS, and will eventually supply the designers the faster, more effective tools and methods for CIMS design.

5 References

1. D. O'Sullivan, Development of integrated manufacturing systems, Computer-Integrated Manufacturing Systems, 1 (1992) 39-53.
2. Xiong Mohua, Research on CIMS Methodology based on Optimization and Step Ideology, Doctor dissertation of HUST, 1996.
3. Dah-Chun Gong and Leon F. McGinnis, Towards a manufacturing metamodel, Int. J. Computer Integrated Manufacturing, 1 (1996) 32-47.
4. Mayer, R. J. and Painter, M. K., Roadmap for enterprise integration, Proceedings of the AutoFact91 conference (SME, Chicago), 1991.
5. Maloubier, H., Brueil, D., Doumeingts, G., And Gavard, J., Use of GARI method to analyze and design production management systems, Proceedings of the Conference on Advanced Production, Management Systems (Elsevier, Amsterdam), (1984) 127-140.
6. Ebers, M and Lieb, M, Computer integrated manufacturing as a two-edged sword, Int. J. Operations and Production Manage, 2 (1989) 69-92.
7. R. J. Borden, Organizing for CIM: project management, technology and integration, Computer-Integrated Manufacturing System, 2 (1991) 60-69.
8. David A. Koonce, Robert P. Judd and Charles M. Parks, Manufacturing Systems engineering and design: an intelligent multi-model, integration architecture, Int. J. Computer Integrated Manufacturing, 6 (1996) 443-453.
9. J. G. Demel, R. G. Askin, A Multiple-Objective Decision Model for the Evaluation of Advanced Manufacturing System Technologies, Journal of Manufacturing Systems, 3 (1992) 179-194.
10. L. M. Camarinha-Mates, H. Pinheiro-Pita, R. Rabelo and J. Barata, Towards a taxonomy of CIM activities, Int. J. Computer Integrated Manufacturing, 3 (1995) 160-176.
11. C.A. Snyder, J. F. Cox, Development computer integrated manufacturing: major issues and problem areas, Costs and Economics, 17 (1989) 197-204.
12. C. A. Voss, The managerial challenges of integrated manufacturing, Int. J. Operations and Production Manage, 5 (1989) 33-38.
13. Zehenovic, M. Dragutin, Methodology for design of effective computer integrated manufacturing systems, Robotics and Computer-integrated Manufacturing, 3/4 (1990) 279-290.

Computer Integrated Manufacturing System for training in Ngee Ann Polytechnic, Singapore

Wong Chun Chong: wcc@np.ac.sg/tel:4606449/fax:4677384
Lek Pheow Hwa: lph1@np.ac.sg/tel:4606457/fax:4677384

Ngee Ann Polytechnic
Mechanical Engineering Department
535, Clementi Road, Singapore 599489

Abstract. This paper presents the design configuration and implementation of a CIM system completed towards the end of 1995 in the Mechanical Engineering Department in Ngee Ann Polytechnic. The system which links the CAD/CAM facilities with the advanced manufacturing workshop is designed to provide practical training for the students and demonstrate the potential of CIM technology to the manufacturing industries in Singapore. One of the main focus area of the CIM implementation is the issue of information management in a highly automated manufacturing enterprise. Brief introduction to the CIM curriculum in Ngee Ann Polytechnic and the use of the CIM system for training will be presented.

List of acronyms:
AGV-Automatic Guided Vehicle; AI-Artificial Intelligence; ASRS-Automated Storage & Retrieval System; BOM-Bills of material; CAD -Computer Aided Design and Drafting; CAM-Computer Aided Manufacturing; CAPP-Computer Aided Process Planning; CIM-Computer Integrated Manufacturing; CMM-Coordinate Measuring Machine; CNC-Computer Numerical Control; CV-Computervision; CVMac-CV Macro; CVNC-CV Numerical Control; DNC-Direct Numerical Control; EDM-Enterprise Data Manager; EPD-Electronic Product Definition ; FAL-Flexible Assembly Line; FMS-Flexible Manufacturing System ; HMC - Horizontal Machining Center ; JIT-Just In Time; JCF-Job Control Files; KBS-Knowledge Based System; LAN-Local Area Network; MPC-Manufacturing Planning & Control; MRP - Material Requirement Planning ; MSS - Manual Setup Station; NC - Numerial Control; NPCIM - Ngee Ann CIM Kongsi; PC - Personnel Computer; QC - Quality Control; PLC - Programmable Logic Controller; RDBMS - Relational Database Management System; SFC - Shop Floor Control

1. Introduction

Over the last 10 years, Singapore manufacturing industries has shifted towards producing high quality, high value-added products for export to the world market. In response to the emphasis placed on new technology and the changing human resource requirements, local institutions of higher learning embarked upon restructuring their curriculum and acquired new facilities to train the manpower needed to operate modern manufacturing systems. Ngee Ann Polytechnic's efforts in incorporating the latest technology to the curriculum follows closely with the phases of technological changes in Singapore. Between 1980-1990, industrial automation, CAD/CAM, CNC, Robotics, FMS technology were introduced to the curriculum. In the mid 90's, the curriculum is further integrated towards Computer Integrated Manufacturing. During this period, existing and new facilities are integrated to better support the new curriculum. The CIM system was designed and implemented between 1993-1995.

The main objectives of the project are:

1) To set up an integrated CAD/CAM-CIM environment to provide training on the latest techniques in automated manufacture.
2) To promote and support continuing development for areas related to CIM.
3) To demonstrate and promote the potential of CIM and related technologies to the local industries.

2. CIM Architecture and System Overview

The CIM system is modeled on a fictitious company known as the Ngee Ann CIM Kongsi (NPCIM) [1] which produces a variety of discrete engineering components and products. The system architecture of the NPCIM is shown in figure 1. It has two main levels, separately identified as the enterprise level and the shop floor level. A local area network (LAN) operating on the Ethernet TCP/IP protocol, provides the communication backbone for the system.

At the enterprise level, the functional departments are represented by six SUN SPARC-10 Unix workstations loaded with application software for performing the various manufacturing activities. Major commercially available application software used in the project are: Computer Aided Design & Manufacturing (CAD/CAM- CADDS5 & CVNC) , Computer Aided Process Planning (CAPP-METCAPP), Robot/Inspection Cell Design and Simulation (SILMA-ROBOTICS/INSPECTION), Engineering Analysis (ANSYS), and Manufacturing Planning & Control (MPC-EZMRP). All the product design, engineering, manufacturing, and other data are stored in the Enterprise Data Manager (EDM) Vault.

The shop floor level of the CIM is configured in a hierarchical and distributed manner. The main shop floor controller (SFC) takes the role of an area manager that coordinates the activities of the shop floor equipment through the various cell controllers (PC QNX, Macintosh PCs) connected together by a local area network (LAN) running on Ethernet protocol. The cell controllers supervise and control the cell level equipment. The SFC communicates with a sub-network consisting of a Macintosh server and three other clients. The Macintosh clients act separately as ASRS controller and AGV controller. Interaction between the SFC and the ASRS/AGV subsystem is through a Oracle relational database. A level below the cell controllers are the real-time device controllers such as the PLC for conveyors/docking station operations, and other controllers for the CNC machine, the inspection machine, the robots , the AGVs and the ASRS crane.

Figure 2 shows a schematic layout diagram of the entire CIM Shopfloor. Essentially, the facility consists of an automated storage/retrieval system (ASRS), a transportation system using two free ranging Automatic Guided Vehicles (AGVs), a Manual Setup Station (MSS) for material /fixture setup/removal, a CNC Horizontal Machining Center (HMC) for the metal cutting operation, a Coordinate Measuring Machine(CMM) for QC inspection , a Flexible Assembly Line (FAL) for automated assembly using two SCARA type robots , shuttle conveyors, docking stations, feeding/orientation devices and the part/pallet identifications system.

Figure 1: Ngee Ann CIM System Architecture

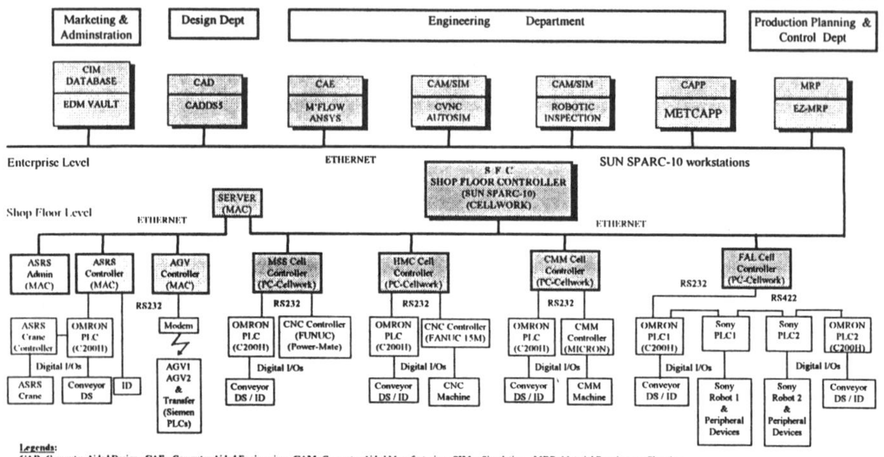

Legends:
CAD=Computer Aided Design CAE=Computer Aided Engineering CAM=Computer Aided Manufacturing SIM = Simulation MRP=Material Requirement Planning
EDM=Enterprise Data Manager MSS=Manual Setup Station HMC=Horizontal Machining Centre CMM=Coordinate Measuring Machine FAL=Flexible Assembly Line
ASRS=Automated Storage and Retrieval System AGV=Automated Guided Vehicle PC=Personal computer PLC=Programmable Logic Controller CNC=Computer Numerical Control
DS=Docking Station ID=Magnetic Read/Write Head I/Os=Inputs/Output

Figure 2 : Physical Layout of the CIM Shop Floor Facility

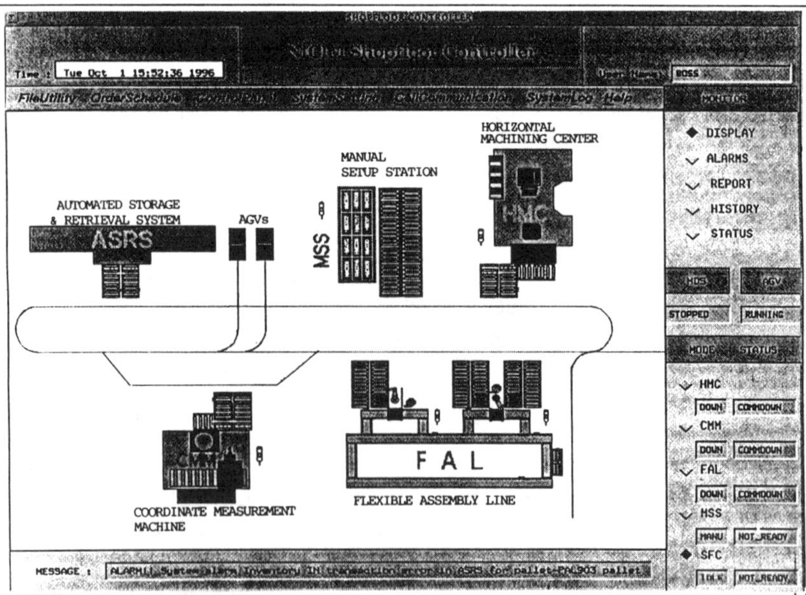

3. Manufacturing Activities and the Information Processing Cycles

The NPCIM was developed to include the following major manufacturing activity cycles as shown in figure 3.:

1) Market Input Cycle
2) Product Design and Development Cycle (PDDC)
3) Detail Design Cycle (DDC)
4) Production Engineering Cycle (PEC)
5) Production Processing Cycle (PPC)
   5.1) Production Planning Cycle
   5.2) Production Execution Cycle

3.1) Market Input Cycle
The market input cycle involves the gathering of information on products and customer needs. In addition, the corporate management may issue directives for new products to be developed and the marketing personnel may also prepare forecasts of sales of existing products. The departmental managers review all the relevant information before releasing them for further activities.

3.2) Product Design and Development Cycle (PDDC)
The cycles involve the design of new products starting from the concept to the preliminary design stage. The resulting preliminary design may be subject to several iterative rounds of analysis, modifications before it is accepted. During this stage, key personnel from the various departments are required to consider various factors related tot he product, such as manufacturability, compliance to the customer requirements,etc. Prototypes may be built to further evaluate and test the products. In the prototyping process, programs for the selected prototyping process (CNC machining, Stereolithography, manual crafting) are generated to produce the prototype for testing, review and final approval by the concerned parties including possibly the customer.

3.3) Detail Design Cycle (DDC)
In this cycle, all the details necessary for the final product are incorporated, documented and released. If the design needs to be modified after this stage , a formal engineering change notification (ECN) procedure may be activated. Once the proposed changes are approved, this cycle is repeated and the changes are incorporated as a new revision.

3.4) Production Engineering Cycle (PEC)
This cycle takes the completed product design and determines the best way to manufacture it, either as designed or with changes to improve its manufacturability. This cycle, therefore, determines the manufacturing processes, the desired sequence of operations, the specific machine tools, fixtures and materials to be used. Manufacturing data such as the CNC/Inspection/Robot programs, routing plans, instructions for setting up machine tools , fixtures and other peripheral are generated.

Figure 3 : Main CIM Activity Cycles

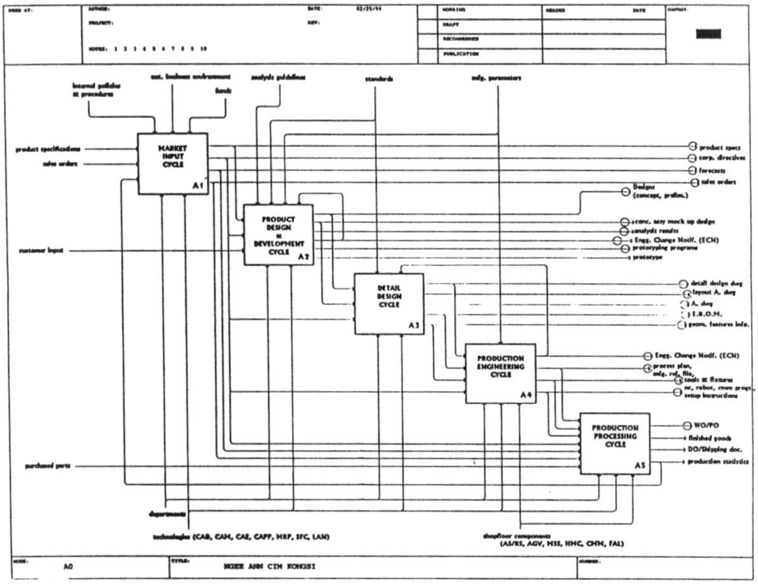

Figure 4 : CIM Information Flow

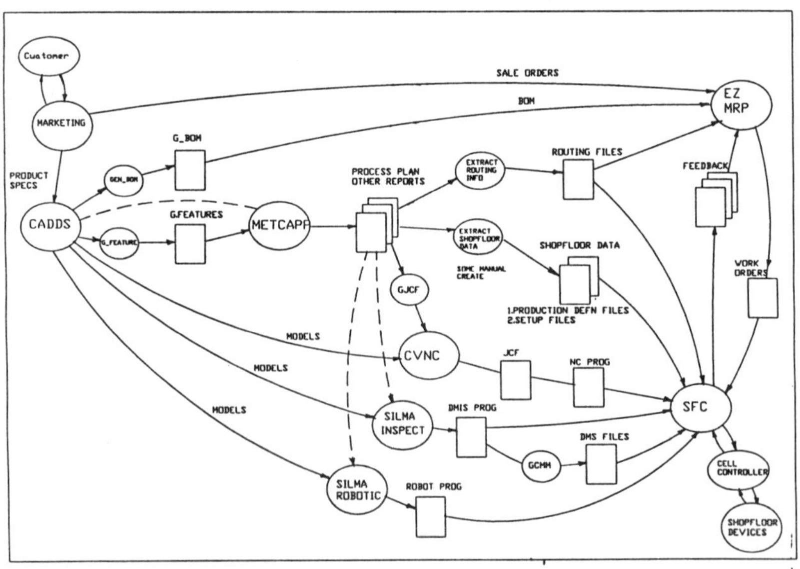

3.5) Production Processing Cycle (PPC)

The production processing cycle consists of two main activities, namely the production planning and the production execution. Production planning involves the creation of the master production schedule and the generation of the necessary work orders and purchase orders. Production execution is the actual approval and issue of work orders for all the make parts to the shop floor. Production starts according to the schedule worked out based on due dates and /or other criteria. The production execution activities are highly automated. The shop floor controller controls and coordinates the shopfloor automated processing stations and material transportation system. It keeps track of all the production status and feeds back to the enterprise level database periodically until all the orders are completed.

4. Information flow

Each of the activity cycles in the NPCIM receives inputs in the form of electronic data. Additional work will be performed at each of the activity and the result is output as electronic data for the subsequent activities. Figure 4 depicts the main information flow across the entire CIM system. The list below summarises the information creation, transaction, conversion and the tools or customised programs developed.

1) Product Design/Development Activities
    Part design/Assembly Structure
    CADDS5/CAMU + Gen BOM
    -CVMac program to extract BOM information & generate BOM file
    CADDS5 + Gen Feature
    -CVMac program to extract features information for processing planning.
2) Design/Process Planning Activities
    Semi-auto Feature extraction for process planning
    Manual process planning
    METCAPP + Reports + Gen Route
    -CVMac programs to generate Routing files
    -CVMac programs to generate CVNC /JCF(NC-Job Control File) files
    Generation of Production Data:
    -CVMac programs to generate Production data files
    Setup instruction files for preproduction operation and MSS tasks
    -Oracle application programs for generation of setup instructions files
3) Design/Engineering Activities
    Generation of Machining/Inspection/Robot programs:
    -CVNC/JCF + Post Process to NC Machining programs
    -SILMA Inspection + DMIS-CMM Inspection programs
    -SILMA Robotics + SONY Robot programs
4) Production Planning/Processing Activities (MRP Processing)
    Load BOM(Bills of Material) + Maintain other Product Information
    Load Routing + Maintain other routing/workstation information
    Maintain inventory information
    Maintain customer/sale/vendor information
    Generate MPS (Master Production Schedule)

Generate production plans and commit plans
Generate work orders/purchase orders
Release Purchase Orders and Work orders:
Production manager initiate downloading of Work orders to SFC
and triggers automatic production process
SFC send necessary production files to cells, check availability of resource
before starting auto-mode operation
Feedback from SFC: Production reports; Production time; Purchasing receipts

5. Concurrent Engineering and Product Data Management

In order to reduce the lead time from concept design to finished product, the concept of concurrent engineering is being increasingly practiced by manufacturing companies around the world since the 80's. Concurrent engineering, by definition, implies the need to encompass all aspects of design, development, manufacturing and marketing. To practice concurrent engineering effectively, it is highly critical that secure concurrent access to product and process information as well as other enterprise data is made possible.

The concept of Electronic Product Definition (EPD) [2] involves defining a new product digitally, via a common electronic model of the entire product assembly. With EPD, product data and all the associated attributes are generated only once and they are stored electronically in a common format accessible by all participants in the product development process throughout the life cycle of the product. The electronic system allows product data to be stored, retrieved and transmitted to any part of the organization as required. This new electronic data management approach reduces significantly the amount of paper documents generated, transferred and maintained between the different stages of product development. Authorized users involved in the product development team have easy access to the up-to-date version of the master documents stored in the electronic vault. The 'virtual' product provides full visibility to the process, resources and changes that occur during the product life cycle. As a result, team members have a better understanding of the whole product and the status of its development. This should reduce non-value-added activities, costly mistake and engineering changes. Over the last few years, better and more user friendly product data management tools are developed to help enterprises manage their product and corporate data. These product data management packages are set to give manufacturers the ability to reduce the time-to-market by providing information and tools to help engineers concurrently develop the products and then share that product information with the rest of the organization in a timely manner.

6. Role of Enterprise Data Manager (EDM) in Information Management

In a CIM enterprise, pervasive use of computers and the related software tools generates large amount of product and processing data. The management, control and maintenance of such data is critical to the successful operation of a CIM system. In the case of NPCIM, a software known as the Enterprise Data Manager (EDM) is used for these purpose[3]. Figure 5 shows the role of EDM in configuring the environment for effective communication, control and management of the data transfer between and within the various work groups in the CIM organization through a client/server network architecture. The clients are made up of members from various functional departments.

The core modules of the EDM software consists of the **EDMVault** host/server software, the **EDMClient** software and the **EDMProjects** software. EDMVault provides distributed data storage areas for product and associated manufacturing processing information. The Members of the individual departments perform various information processing activities and stores the information into the corporate vault. This information may be accessed by authorised members within the department or other departments. EDMClient allows a user to request EDM services, such as data access, data sharing, check-in/check-out, electronic notification via on-line mail facility or via automatic notification schemes. [The EDM server first checks control metadata in the form of RDBMS tables to confirm that the request can be honoured, and then processes the request. If the user has the proper authority, the EDMVault server delivers a copy of the requested EDM data object and locks it from modification by other users. Other users can obtain a copy of the object and read it, but no one else can modify the object until it is checked back into the vault.]The EDMProjects software module provides the system designer with a tool for organising people and their work into projects for on-line tracking and control. With EDMProject, the system designer may define the project tasks, milestone specifications and the appropriate codes for security access for project team members and finally control the user's data flow.

7. Configuration of the Shop Floor Control Functions

The control of the CIM shop floor activities is configured with a cell control application enabler known as **Cellworks**. The core modules of the enabler consists of the **CELLman**, the **CELLguide**, the **MBX Message Bus** and a number of **Servers**. Cellman is used to develop the application logic. CELLguide provides the integrated environment for the configuration of the applications. In a typical application, the CELLman logic will receive messages from device server, perform the logical operations and send messages to the servers. The MBX Message Bus forms the backbone for providing object oriented transparent inter-process communication for information exchange between CELLman applications and servers across a local area network (Ethernet).

Figure 6 shows the configuration and the main functions of the SFC modules which control and monitor the whole shop floor operations in NPCIM. The SFC communicates with the cell level controllers through the TCP/IP server across an Ethernet LAN. It also exchanges information with the enterprise level computers through the EDM vault and MRP. It receives manufacturing data from the EDM vault and report production results to the MRP system

Figure 5 : Detail Design of EDM Environment

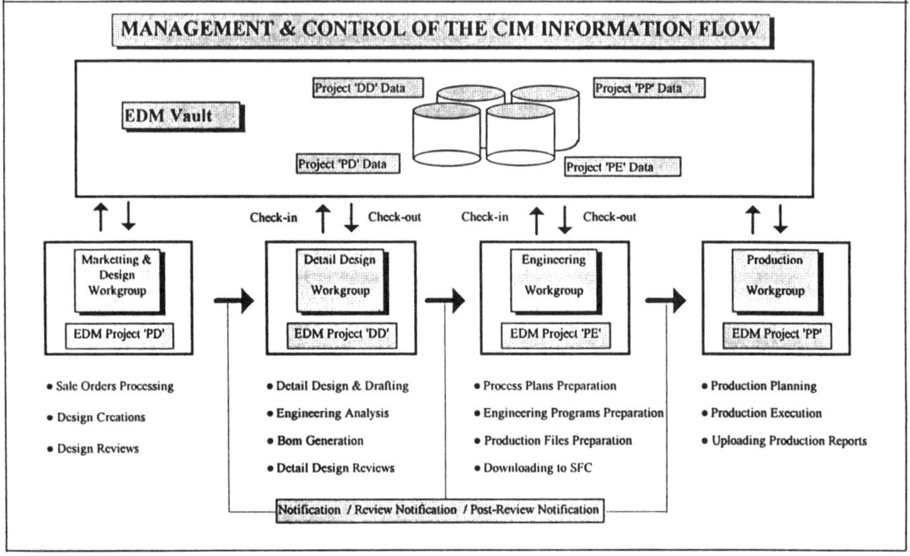

Figure: 6 Configuration of the Shop Floor Controller

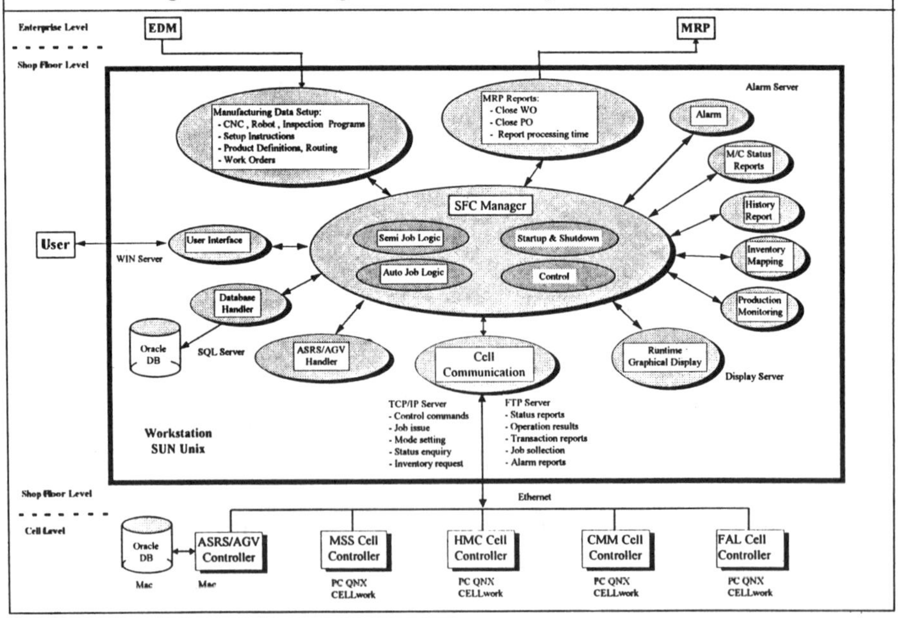

8. Operations and Material Flow in the CIM Shop Floor

Upon receiving the work orders and all the associated manufacturing data from the enterprise level activities, the SFC will initiate the production operation after checking that the material and processing stations are setup and ready for operation. Raw material are retrieved from the ASRS and send to the MSS for setup operation. The material and the fixture are then sent to the HMC cell for machining operations. When all the machining operations are completed, the machined parts are cleaned and sent for inspection at the CMM cell. After inspection, good parts are removed from the fixture and returned to the ASRS for storage and subsequent assembly operation. Defective parts are rejected at this point. When all the necessary component parts of a product are available, the SFC will initiate dispatching of components for the the assembly operation at the FAL. Assembled products are returned to the ASRS. The last process of the shop floor operations is the packaging. All packaged products are stored at the ASRS for shipment.

Each of the cells at the shop floor may be operated in a stand-alone manual mode. This mode of operation is essential for our training environment where students are divided into several small groups so that they can study the operation of the cells in greater detail prior to the integrated operations.

9. Current Status and the CIM Curriculum

The CIM system is being used as part of laboratory practice and case study for the Mechanical Engineering and Mechatronic Diploma course students in the department. Students gain an overview of the key components, configuration and characteristics of the whole CIM system. They also work on assigment topics related to CIM. The Factory Automation System (FAS) module is experimenting with a role play approach in conducting the CIM laboratory practice. Students will role-play as designers and design manager (or other roles) and work through the design activity cycles to understand the process of information creation, electronic review, approval for release and engineering change as well as concurrent engineering in a CIM environment.

Over the past decades, course curriculum in the Mechanical Engineering has undergone several major review and changes so as to keep up with the industrial requirements for trained man-power in manufacturing automation. A CIM option has been introduced and will be offered to the final year students in 1998. Students will gain deeper understanding of the building blocks of CIM including technology such as CAD/CAM, Manufacturing Planning and Control, Robotics, Factory Automation and Integrated Manufacturing System.

10. Future Development

The main focus of the CIM project was to put together an integrated CAD/CAM-CIM environment which can serve the current training requirements in manufacturing automation as well as to create a platform for further development. Plans for further development and enhancement lie in the area of total electronic product definition involving the creation and integration of more comprehensive product and manufacturing process databases. Currently, a software known as Navigator has been installed and configured to allow the CIM project team members and users to 'navigate' across the product structure and all the related attribute information through user friendly graphical user interface. The project manager can track the status of the whole product easily. In this way bottle-necks can be identified early and necessary timely actions taken to reduce the lead time. Another interesting development area which has potential application for our training environment is in the use of Internet for 'network integrated manufacturing' [4]. Users may share design information and conduct design review remotely via the internet. When equipment malfunctions or breaks down, specialists from equipment suppliers may zoom in on the problem areas via video images from on-site video cameras connected to Internet. The specialist can perform real-time trouble shooting and even problem rectification through the Internet access.

11. Conclusion

Despite the demanding specifications of the CIM system, the project has been successfully completed and tested vigorously after nearly two years of hard work by a team of lecturers from the Mechanical Engineering Department and their industrial partners. The main objective of building an integrated manufacturing system which makes use of computer and information technology to bind together the business, design, engineering, planning and production activities in a manufacturing enterprise has been achieved. The project provided a valuable learning opportunity for the lecturers in designing and building our own CIM system. With this facility, the students are able to gain hands-on experience on many aspects of automated manufacturing technology within one site. Local industries can witness the operation of a CIM system and assess the potential benefits they can harness through automation and computerization of their manufacturing activities.

References

1. CIM Task Force Team & Computervision, "CIM Enterprise Level and Shop Floor Level Acceptance Test Plan", Vol 1 and Vol 2 1995.
2. Trevor Broughton, Dr.Stan Manton, John Puttick, "The Time-To-Market Machine: How EPD is Changing the Face of 1995", Computervision Publication.
3. Lek PH, Wong CC, "Developing an Integrated Work Group Environment for a CIM Enterprise" Proceedings of the International Conference on CIM, ICCIM 1995.
4. Gau Rong Liang, "Advent of Network-Integrated Manufacturing Age in Taiwan" Institute of Industrial Engineering, National Chiao Tung University, Taiwan.

Acknowledgments

This paper is a culmination of the work contributed by a team of staff members from the Mechanical Engineering Department. The authors wish to record his appreciation to the contribution and support from the members of the CIM Task force, the workshop staff in Advanced Manufacturing Workshop and the staff members from CAD-CAM Center II .

MANUFACTURING PLANNING AND CONTROL

Aggregate Planning

A MODIFIED p-MEDIAN FORMULATION FOR GENERALIZED GROUPING

Kripa Shanker* and Murari P Barnwal

Industrial and Management Engineering Department
Indian Institute of Technology Kanpur, 208016, INDIA
(*corresponding author, email : ks@iitk.ernet.in)

Abstract. This paper presents a modified p-median formulation for generalized grouping problems where parts can have multiple process plans. The formulation accommodates a specified number of families to be formed (imposed grouping) as well as allows any number of families to be formed (natural grouping) while maximizing similarity amongst the process plans. A heuristic procedure is proposed to form the process families after the solutions of the resulting integer programming formulations have been obtained. The proposed formulation is compared with other p-median formulations and is shown to possess better computational characteristics.

1 INTRODUCTION

We consider the problem of part-machine grouping in a flexible manufacturing environment where an operation can be performed on more than one machine types or where a multi-functional machine can perform more than one operation types. The resulting grouping problem where a part may have multiple process plans (or process routes) is popularly known as *generalized grouping* to differentiate it from the traditional *simple grouping* problem where a part has only one process plan. The binary (0-1) machine-part matrix used as the input in the simple grouping is replaced by binary machine-process plan matrix in generalized grouping. The objective is to form groups of the process plans by selecting only one process plan from the multiple process plans of each part such that some measure of similarity is maximized.

For generalized grouping problem, Kusiak (1987) introduced a p-median model (PMM) in which the desired number of part families to be formed (p) is specified in advance. The present formulation is a modification over the PMM. The PMM is briefly presented below. Following notation and definitions are used.

Notation and Definition :

Indices	i, j :	process plans	k :	part
	m :	machine		

Parameters N : total number of parts

F_k : set of process plans for part k, k=1,...,N

P : total number of process plans over all parts (=$\Sigma | F_k |$)

M : total number of machines

p : the maximum number of process families required

Input Matrix $[a_{i,m}]_{PXM}$: matrix of requirement for machines by process plans

$$\text{where } a_{i,m} = \begin{cases} 1, \text{ if process plan } i \text{ requires machine } m; \\ 0, \text{ otherwise.} \end{cases}$$

Similarity Coefficient Matrix $[s_{ij}]_{PXP}$: The similarity coefficient s_{ij} between process plans i and j can be computed as follows.

(a) similarity between two process plans of the same part, $s_{ij} = -\infty$, $i,j \in F_k$, k=1,...,N.

(b) similarity of a process plan with itself, $s_{ii} = 0$, i=1,...,P.

(c) similarity between two process plans of different parts,

$$s_{ij} = \sum_{i=1}^{P}\sum_{m=1}^{M} \delta\left(a_{im}, a_{jm}\right), \ \forall\ i \in F_k;\ j \notin F_k \text{ and } k=1,\ldots,N.$$

$$\text{where } \delta(a_{im}, a_{jm}) = \begin{cases} 1, \text{ if } a_{im} = a_{jm}; \text{ and} \\ 0, \text{ otherwise.} \end{cases}$$

Decision Variables : All the P process plans are indexed serially from 1 to P. A process family is defined by the index of one of its member plans. The p required families, therefore, can assume their indices from the set {1,..,P}. For plans i and j,

$$x_{ij} = \begin{cases} 1, \text{ if process plan } i \text{ belongs to process family } j; \\ 0, \text{ otherwise.} \end{cases}$$

For example, if a process family comprises of plan 1 (of part 1), plan 4 (of part 2) and plan 9 (of part 4), then the plan family can have the index 1, 4 or 13. Say it is indexed as 4, Then x_{14}=1, x_{44}=1, and x_{94}=1 all other x_{ij}. = 0.

Objective Function : The objective of the generalized p-median model is to form the process plan families by maximizing the value of similarity measure among them and is given as

$$\text{maximize} \sum_{i=1}^{P}\sum_{j=1}^{P} s_{ij} x_{ij} \tag{1}$$

subject to the following constraints:

(a) only one process plan of a part is selected

$$\sum_{i \in F_k}\sum_{j=1}^{P} x_{ij} = 1 \qquad k=1,\ldots,N \tag{2}$$

(b) at most p process plan families are formed

$$\sum_{j=1}^{P} x_{jj} \leq p \qquad (3)$$

(c) process plans are assigned to a family

$$x_{ij} \leq x_{jj} \qquad i,j = 1,...,P \qquad (4)$$

(d) binary decision variables,

$$x_{ij} = 0 \text{ or } 1 \qquad (5)$$

Several attempts have been made over the last decade to generalize the problem environment and also to propose solution methodologies. Shtub [1989] formulated the generalized grouping problem as a generalized assignment problem.

Agrawal [1990] modified Kusiak's PMM with respect to the constraint (b) in eqn. (3). He observed that in view of this constraint, the number of assignments along the diagonal of the dissimilarity coefficient matrix $[s_{ij}]$ is at the most equal to p, and thus the remaining $(N\text{-}p)$ assignments have to be made corresponding to some non-diagonal element of the matrix in a manner such that these $(N\text{-}p)$ assignments appear in only those p columns or p rows that represent the process families and have assignments at their cells corresponding to the diagonal matrix. He modified the constraint (b) in eqn. (3) as

$$\sum_{i=1}^{P} x_{ij} \leq (N - p + 1) x_{jj} \qquad j = 1,...,P$$

(6)

The idea behind this constraint is that out of assignments of P plans, $(N\text{-}p)$ assignments have to be made from the non-diagonal cells. Further, if only one plan (i.e. the index plan) is assigned to each of some $(p\text{-}1)$ families, then at the most $(N\text{-}p+1)$ plans can be assigned to the pth family.

Kusiak and Cho [1992] proposed a maximum clique algorithm where the similarity matrix is represented by a transition graph. The maximum clique of the resulting graph represents a family. The part families are generated by branch and bound procedure.

2 PROPOSED p-MEDIAN FORMULATION

In the Kusiak's PMM, the constraint (c) in eqn. (4) postulates that *at least one* plan should be selected for the process family formation. This implies that even one process plan can make a process family. From the point of view of grouping, this is not a very practical assumption. In fact, such a process plan which is disjoint of all other plans should not even be considered since there is no similarity of the manufacturing attributes of this plan with any other plans. Hence, it is argued that to accord a purposeful meaning to the grouping, *at least two* process plans of two different parts should be selected for process family formation. With this in view the p-median formulation is modified as follows (Barnwal, 1996).

The constraint (c) is modified as:

$$\sum_{i=1}^{P} x_{ij} = 0 \ \ if x_{jj} = 0$$
$$> 1 \ \ if x_{jj} = 1.$$

The above constraint in terms of inequalities is

$$\sum_{i=1}^{P} x_{ij} \leq Bx_{jj}, \text{ and } \sum_{i=1}^{P} x_{ij} \geq 2x_{jj} \ for\ all\ i, j. \qquad (7)$$

where B is limit on the cell size, i.e., the maximum number of plans (or equivalently, the number of parts) desired to be kept in a family.

The advantages of the constraint (c) given by eqn. (7) in place of eqn. (4) are three fold:

- while maintaining all the features of grouping, this modification does not require a prespecified number of families to be formed, and thus would result into *natural* grouping,
- if for some reason, it is required to prespecify the number of families to be formed, this can also be included - thus it provides *flexibility* .
- the number of constraints is less than that in PMM.

Based on the above observations and discussions, the following two models are proposed.

Imposed Grouping Model (IMPGRP)
This model is similar to the PMM. The number of parts to be included in a cell is at least two and upto a maximum of the cell size limit (B). The number of groups to be formed (p), as in case of PMM, is prespecified.

Natural Grouping Model (NATGRP)
This model possesses more flexibility than IMPGRP. There is no prespecified number of groups to be formed. However, the number of parts to be included in a cell is at least two and upto a maximum of the cell size limit (B).

The two proposed models are summarized in table 1.

3 SOLUTION METHODOLOGY

The modified p-median models are integer programming formulation and can be solved by any 0-1 linear integer programming solving package e.g. LINDO or CPLEX. The NATGRP model may result into a large number of process families for larger problems. So a heuristic procedure is applied for merging some of the families

Table 1. Summary of proposed IMPGRP and NATGRP models

<table>
<tr><td>Imposed Grouping Model (IMPGRP)
The number of groups to be formed (p), as in case of PMM, is prespecified. The number of parts to be included in a cell is at least two and upto a maximum of the cell size limit (B).</td><td>Natural Grouping Model (NATGRP)
There is no prespecified value for the number of groups to be formed. The number of parts to be included in a cell is at least two and upto a maximum of the cell size limit (B).</td></tr>
<tr><td colspan="2">maximize $\sum_{i=1}^{P}\sum_{j=1}^{P} s_{ij}x_{ij}$ (1)
subject to
(a) only one process plan of a part is selected : $\sum_{i \in F_k}\sum_{j=1}^{P} x_{ij} = 1 \quad k=1,\ldots N$ (2)
(d) binary decision variables : $x_{ij} = 0$ or 1 (5)</td></tr>
<tr><td>(b) at most p process plan families are formed: $\sum_{j=1}^{P} x_{jj} \le p$ (3)
(c) assignment of plans to families:
$\sum_{i=1}^{P} x_{ij} \le Bx_{jj}$, and $\sum_{i=1}^{P} x_{ij} \ge 2x_{jj}$
for all i, j.
(7)</td><td>-
(c) assignment of plans to families:
$\sum_{i=1}^{P} x_{ij} \le Bx_{jj}$, and $\sum_{i=1}^{P} x_{ij} \ge 2x_{jj}$
for all i, j.
(7)</td></tr>
</table>

which use the same machines or the difference is only due to alternative machines. Subsequently, the assignment of machines to the families is done while minimizing the intercell movements. A refinement step is applied which reassigns bottleneck parts to further reduce the intercell movements. With these in view, the following heuristic is proposed.

Step 1 : *Identification of sets of alternative machines*
Make the list of alternative machines for each type of operation (irrespective of parts).

Step 2 : *Merging the part families*
If the machines used by a process family are a subset or superset of machines used by any other process family, merge the two families. If the subset or superset criteria fails between the two families only due to the presence of machines of the same set of alternative machines, then merge them by assigning either of the two machines. Restructure the process family accordingly. Repeat for every pair of families.

Step 3 : *Assigning the machines to the process families.*
(i) Compute for each machine the usage factor for each process family.

(ii) Assign the machine to the family where the usage factor is maximum, ties can be broken arbitrarily.

(iii) If the sum of usage factors for an alternative machine is zero, i.e. no operation has been assigned to it, then let it replace the machine of the same set which has a positive usage factor for more than one family. It may be noted that such a situation arises when number of alternative machines for a type of operation is relatively large.

Step 4 :*Merging the families and machine cells*
Merge two families and their corresponding machine cells if it eliminates intercell movement and does not violate the cell size limit imposed. Stop when no merging is possible or the groups formed are disjoint.

Step 5: *Refinement*
Go to step 3. Assign the bottleneck parts to that cell which has maximum number of its operations, ties broken arbitrarily.

4 COMPARISON OF MODELS

A comparison of Kusiak's PMM, Agrawal's modified model and the proposed IMPGRP and NATGRP models can be made on the basis of the number of constraints, number of variables and order of complexity. The order of complexity is calculated on the basis of all possible combinations which are likely to be explored while solving the integer programming model. Table 2 gives the comparison of various models.

The three models were implemented on HP 9000 machines in C language. The integer programming models were solved using CPLEX. The proposed heuristic is then applied to merge the process plans and to form part families and machine groups. Several numerical examples were experimented upon (Barnwal, 1996) using the PMM, IMPGRP and NATGRP models. It was observed that

- with NATGRP model, the CPU time varies quadratically with the total number of plans (*P*) while that with Kusiak's model varies exponentially.
- NATGRP takes less time than IMPGRP. In fact, IMPGRP is not able to solve the problem with larger number of plans within a reasonable time (50 secs).
- the number *B*, the maximum parts to be included in a family, has a major effect on IMPGRP model for the solution to result in the lower computational time, but NATGRP is almost insensitive to *B*.

Figure 1 shows the variation in CPU time for CPLEX with the number of parts for representative examples.

Table 2. Comparison of Kusiak's PMM, Agrawal's modified and proposed IMPGROUP and NATGROUP models

Criteria	Kusiak's PMM	Agrawal' modification	IMPGRP model	NATGRP model
GT concept	violates	violates	agrees	agrees
Type of grouping	imposed	imposed	imposed	natural
Number of variables	P^2	P^2	P^2	P^2
Number of constraints	P^2+N+1	$P+N+1$	$2P+N+1$	$2P+N$
Heuristic steps	steps 1, 3 &5	steps 1, 3 &5	steps 1, 3 &5	all steps
Total number of enumeration	$\prod_{k=1}^{N} \lvert F_k \rvert . S(N,p)$	$\prod_{k=1}^{N} \lvert F_k \rvert . {}^{N}C_p . S(N-p,p)$	$\prod_{k=1}^{N} \lvert F_k \rvert . {}^{N}C_p . S(N-p,p)$	$\prod_{k=1}^{N} \lvert F_k \rvert . N^2$

Note : The number of ways in which n different objects can be partitioned into p non-empty subsets is given by Stirling number $S(n,p) \cong p^n/p!$.

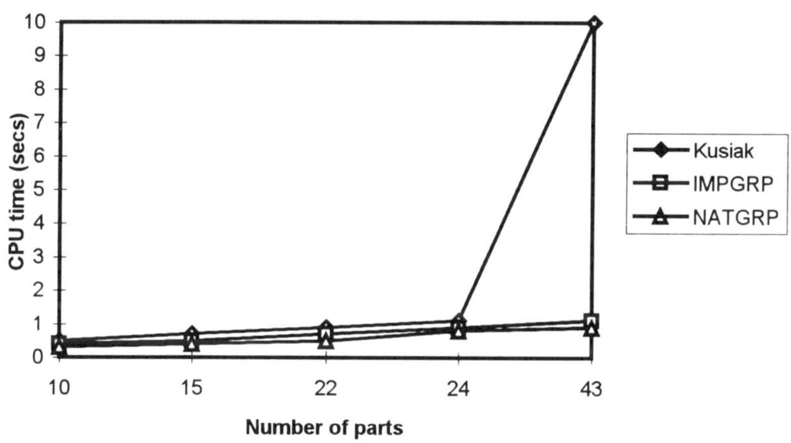

Fig. 1 : Variation of CPU time with number of parts

5 CONCLUSIONS

A modified p-median formulation with two variations - imposed grouping and natural grouping - has been proposed for the generalized grouping problem ensuring that a family contains at least two parts. However, a number B, the maximum parts to be included in a family, is to be specified, instead. The integer programming solution obtained is further refined through a heuristic to yield a more purposeful final solution. The computation efficiency is found to be better than the earlier formulation - PMM.

REFERENCES

1. Agrawal, A. K., Generalized grouping and loading problems in FMS : models and methodologies, Ph. D. dissertation (1990, unpublished), Industrial and Management Engineering Department, Indian Institute of Technology, Kanpur, India.
2. Barnwal, M. P., Models and methodologies for generalized grouping problem, Master's thesis (1996, unpublished), Industrial and Management Engineering Department, Indian Institute of Technology, Kanpur, India.
3. Kusiak, A., The generalized group technology concept, *Int. J. Prod. Res.*, **25** (1987) 561-569.
4. Kusiak, A. and Cho, M., Similarity coefficient algorithms for solving the group technology problem, *Int. J. Prod. Res.*, **30** (1992) 2633-2644.
5. Shtub, A., Modelling group technology cell formation as a generalized assignment problem, *Int. J. Prod. Res.*, **27** (1989) 775-782.

Modeling of Production Planning and Control for the Integration of Information Systems

W. Dangelmaier, R. Holtkamp, T. Langemann

Heinz Nixdorf Institute
University of Paderborn, Germany

H.-J. Sobiech, F. Ziersch

Volkswagen AG, Germany

Abstract. The grown complexity of legacy systems in manufacturing companies and the need to improve the performance of the manufacturing process lead to the necessity to gain detailed insight in production processes and the supporting information systems. In this paper a four step modeling and analyzing approach for production planning and control systems is presented that allows to model structures and dynamics of these systems. With this approach existing production planning and control systems can be analyzed and improved and new ones can be build and integrated. The approach is based on a sound modeling method (MFERT) and supported by an appropriate software tool (OOPUS). The paper presents the approach with a running example from the automotive industry.

1 Introduction

The growing dynamics in logistics and production planning and control (PPC) in industry call for a good understanding of structures and processes. This understanding can be reached by modeling the production planning and control processes and a following analysis based on the model. Especially the interconnection of information systems with the production process and with each other must be well understood to get the best results from these systems. With our method MFERT and the supporting tool OOPUS we have developed a basis for the model based building of complex production planning and control systems.

In a common project with the Volkswagen AG in Braunschweig, Germany, the rear axle production process was modeled. During this phase deficiencies in the integration of several PPC systems were recognized and possibilities to improve existing planning processes where detected. System modeling and analysis lead to a concept for the integration of several legacy systems. In a next step the model based system will be connected to the legacy systems and executed for PPC purposes.

2 Approaches for the Construction of PPC-Systems

In companies of the automotive industry planning processes (e.g. capacity or material planning) were early supported by information systems. As a result of this evolutionary process today's information systems have a low degree of integration. IS[1] are often a solution for only one specific domain without handling the complexity at the interfaces between the systems [4].

Modern IT concepts, like model-based systems for example, are rarely found in automotive industry yet. Theses concepts enable the logical integration of existing applications and data [3]. Therefore redundancy can be avoided and the inherent complexity of this domain can be handled better than in systems which are not model-based. Hence integrated IS in the production planning and control field can be provided to the end user.

To integrate the relevant aspects of the IT environment as well as the relevant planning and control processes in an IT solution a transparent and well suited way of describing such aspects is necessary. New approaches are using methods for *modeling the planning processes and the existing information systems* for the purpose of increasing transparency and building such systems. The models created in such a way can be used as a basis for the simulation and evaluation of the focused processes.

To reduce complexity in a first step often only partial models of the focused system are created (for instance only the manufacturing process of a sub-module). To build a model of the whole system in only one step would be much too complex. In a second step partial models can be integrated to a global one. This paper uses the computer supported planning and control of the rear axle manufacturing process at the Volkswagen AG at the plant in Braunschweig by a model-based approach as a running example that stems from a practical application of this approach.

As production planning and control is to remain a labor-sharing process even as tasks become increasingly concentrated, the primary requirement for the interaction of tools for developing PPC-systems is a common language, a basis for communication which can be used to describe whole production systems and their procedures. To represent this common language, a *dynamic model of the production process* can be used as a basis for all production planning and control processes and for defining the required interfaces. In an object-oriented approach, a set of methods is then assigned to each modeling component. This set of methods

- is provided with data from the model of the production process and plays back its results in the model itself or in reality,
- derives, processes and stores knowledge in a decentralized way and incorporates it as a totality in production planning and control to make it generally available,
- carries out a step-by-step completion and constant adaptation of the model to environmental conditions, thereby making stipulations for optimizing the operation of a manufacturing process according to the current situation.

Using this concept each company should be able to model its individual manufacturing process and derive functions and data for simulation, planning, scheduling and monitoring with a maximum of flexibility relative to changing situations and the possibility

[1] In this paper the term *information systems* (IS) is used as a synonym for IS in the PPC environment.

of adapting to changing organizational circumstances. Because of the complexity of the resulting models and because the running production planning and control system shall be automatically derived from the developed model, the modeling process must be supported by appropriate software tools.

3 Modeling of PPC-Processes and -Systems

3.1 An Approach for Modeling PPC-Systems

This chapter briefly describes the modeling method MFERT[2]. It covers only the main elements of the method which are necessary to understand the described modeling process. For more detailed information see [1].

The main elements of the method are:

- *element_in_state*: An element_in_state is an element in a certain state. An element is a static representation of a part of the universe of discourse (i.e. material, resource or information). It is characterized by its attributes.
- *operation*: An operation is a transformation of elements from one state into another. It is characterized by its input and output elements and by the relative time of the inputs and the outputs to a reference element.

Both model elements can be grouped together into categories which contain element_in_states or operations of the same class. These categories can be hierarchically organized and are called *element_in_state_category* and *operation_category* respectively. This is based on the object oriented paradigm for data modeling.

These model elements cover all informations to describe the production process except its structure. To describe this structure a graph based model is used. In this graph there are two kinds of nodes corresponding to the two kinds of data:

- *element_in_state_category_node*: A node of this kind represents elements in state of a certain category and its subcategories. Graphically it is represented by a triangle.
- *operation_category_node*: A node of this kind represents operations of a certain category and its subcategories. Graphically it is represented by a rectangle.

These two types of nodes can be linked together by *directed edges* representing the flow of material and information in the production process. Only nodes of different kind may be connected by an edge. The result is a *bipartite, directed graph* which represents a model of the production process or at least a part of it. The *functionality* of the system is realized by allocating functionality to each node of the model.

A model of the enterprise can consist of several partial models of different levels of abstraction. They can be linked together by interfaces representing the flow of data between these levels. The interfaces of the lowest level can be linked to the reality[3] or to simulation modules representing the reality. In figure 1 a reference model for a MRP II based PPC-System is shown as an example. For each level the user can select appropriate reference models from a set of predefined building blocks. If there are no suit-

[2] MFERT is a german acronym for „modeling of manufacturing systems"

[3] I.e. directly to machines or to terminals for human interaction with the system.

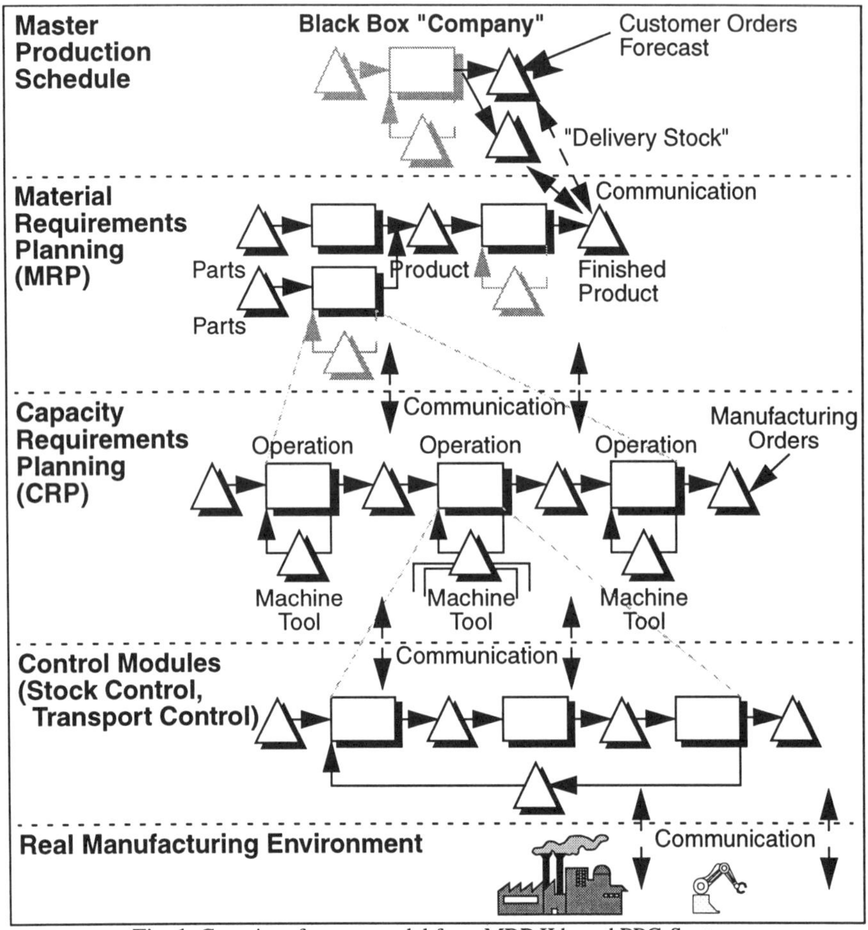

Fig. 1. Generic reference model for a MRP II based PPC-System.

able models available, new models have to be created by deducing them from existing ones.

3.2 A Modeling Framework for Creating PPC-Systems

The OOPUS[4] system serves as an *integrated framework* for the purposeful *modeling, simulation* and *generation* of PPC-systems (see figure 2) based on the concepts of MFERT [1,2]. The functionality is defined by managers for each node of the process graph. This allows the model to become executable. OOPUS offers several graphical editors to support *creation* of PPC-models and a run-time environment for the *execution* of the described models.

In OOPUS the concept of software reuse is strongly emphasized. This is done by the use of building blocks, which are arranged in so called structural prototypes (see

[4] OOPUS is a german acronym for „Object oriented framework for the generation and integration of individual production planning and control systems"

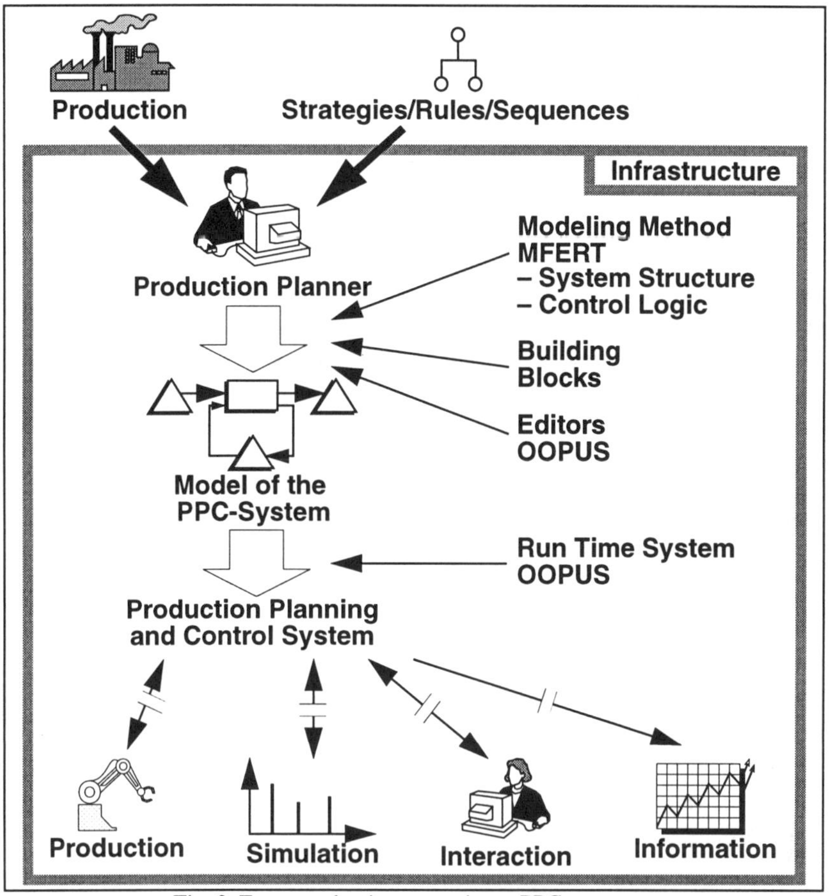

Fig. 2. From production strategies to PPC systems.

figure 4). A system generated by OOPUS can be applied to plan, to document, to control, to monitor and to simulate a production process. Therefore many useful computer supported applications in the manufacturing environment can be integrated by OOPUS.

4 Model Based Analysis and Design of PPC- Systems

4.1 Overview

Within the context of optimizing the in-plant logistics of the plant in Braunschweig an integration of the IS in the rear axle manufacturing process was planned. In particular it was discovered that the logistical interdependencies between several cost centers in the rear axle subdivision are not captured completely [6]. At present they are handled by experienced employees. The information flow between these production planners is mostly verbal and it depends on experience of the single employee to consider all affecting factors to make a proper decision. Due to the necessity to noticeable increase the transparency of the complex structures in the focused area the use of a model-based system was determined [5].

4.1.1 The Aims of the Project

The following potentials for optimization in the use of computer supported PPC systems had been identified in the rear axle production division:

- Providing a uniform and integrated access to all relevant information of existing information systems for the production planners,
- improving transparency of decentralized stock amounts,
- improving transparency of the impact of local decisions to the whole production process,
- IS support for short time cycle control (e.g. shortly changed material demands),
- simulation capabilities and the calculation of alternative production strategies.

The project aims at providing a powerful and model-based PPC-system for the rear axle production division. Therefore the progress figures concept - as the current PPC strategy - has to be adapted as well.

4.1.2 The Steps of the Project

The analysis of the PPC problems and the design of a new system is done in four steps:

1. Building a structural model[5] of the manufacturing process which includes all necessary information about interfaces to existing IS (chapter 4.2).
2. Analysis of the constructed model and comparison with the formulated goals of the project. Identification of weak points and potentials of optimization in the current PPC-system (chapter 4.3).
3. Enhancing the model with control logic for the purpose of execution and simulation (chapter 4.4).
4. Analysis of the dynamic behavior by simulation. Identification of weak points in the dynamic model and iterative improvement (chapter 4.5).

4.2 Modeling the Structure of the PPC-System

The process model of the rear axle production is configured such that production control can be done in consideration of the progress figures concept. In the sense of process control the different nodes of the model can be accounted as single control blocks. Following the individual model elements will be explained and some particularities of them will be exposed. However our model is much more detailed than explained in this text.

4.2.1 Elements in State

Two different kinds of elements in state have to be distinguished: The *applicable material* that is set in the material flow and the *resources* that are assigned to a specific operation and so are not set in the material flow. In the model elements in state are grouped together with regard to their properties. They differ from each other in their information relevant to production control in terms of material transport requirements

[5] In this context *structural* means that there is no functionality included in the model in this step.

in the hall. Different information relevant for manufacturing control are the specific transportation routes of the elements and the adjunctive different transportation times for supply at the respective place of demand.

4.2.2 Operations

In chain production products are handled sequentially by each machine. Products pass through the processing equipment and are assigned to shop orders. A shop order is distinguished by a defined lot size and a non ambiguous product specification. As a consequence in the considered operations that are tied to the manufacturing process (welding, lacquering, mechanical processing and assembling) all process oriented shop orders that require the same operation at the same time compete for that operation.

The operations for transport and shipping can work on several shop orders in parallel, because these operations are not subject to the characteristics of chain production. Consequently a parallel processing is possible in these cases.

4.2.3 Model of Information Systems

The described data relevant for the production of rear axles are merged to a complete model of the rear axle production process. In this model the links to existing IS from which this information can be taken have been integrated. Parts of the resulting model are depicted in figure 3. Regarding data protection and the limited amount of space in this article the complete, much more detailed model is not included.

4.3 Analyzing the model of the PPC-System

The build model can be further enhanced with the demands for information at the particular nodes (figure 3). Using the structural connections of the model the demanded information can be grouped in three categories:

- Information that is available from existing IS without any modifications,
- information that can be made available by simple modifications or additional interconnections of the existing IS and
- information that cannot be supplied by the existing IS.

The second category can be used to improve the existing information systems. All three and especially the third category can guide the building of new information systems. This analysis was done for the rear axle production and several proposals for improvement could be made on this basis.

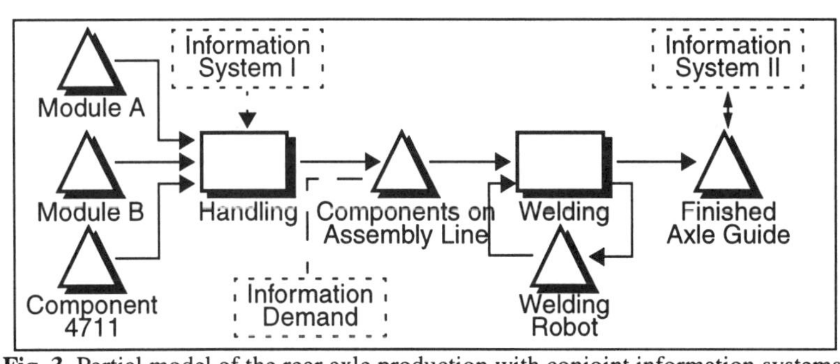

Fig. 3. Partial model of the rear axle production with conjoint information systems.

4.4 Enhancing the Model with Planning and Control Logic

The Planning and Control Logics are analyzed and translated to an algorithm that is described by the functionality of each single node in a so called handler.

4.4.1 Planning Logic

Based on the orders of the car building plants a daily calculation of material requirements is done in the model by means of a bill explosion [7]. By a retrograde calculation the requirements of elements in state at the operation preceding an element in state node is calculated. These requirements are terminated by the given due times. The termination needs a determination of throughput times and set-up times in the model.

Because stock maintenance at each node is done cumulatively with progress figures, the following information for operation planning can be accessed at each node:

- Decentral priority of the shop orders,
- availability at the control block,
- orders still to be planned at the control block,
- increased requirements.

4.4.2 Control Logic

The decentral prioritization of shop orders follows the aim to hold the completion time of the production orders. The retrograde calculation assures all interdependencies related to the sequence of operations to be considered. The dispatching of orders is done by a document for each cost centre which contains:

- the order of production for each production line,
- fixing the type of manufacture and finish date,
- calculated stock.

4.4.3 Functionality at operation nodes

At all operations of the model identical attributes relevant for production control can be found. With regard to production control predicates about

- the availability of the operation,
- the total occupancy of the operation,
- the calculated priority of orders to be done,
- the duration of each order at the machine

must be made.

4.4.4 Functionality at element in state nodes

The model of the rear axle production must enable a stock maintenance. The fundamental element of stock maintenance are time axes at each element in state node.

4.4.5 Algorithm

Production planning and control is done in two steps[7]. In the first step a material requirements planning over the considered time horizon without termination is done. In the second step the demands are terminated with regard to the capacities of the resources. In figure 5 the schematic instruction flow of the algorithm of the second step is shown. This algorithm is related to the structural prototype shown in figure 4. Each node of the prototype represents all nodes of the same kind in the model. In this case

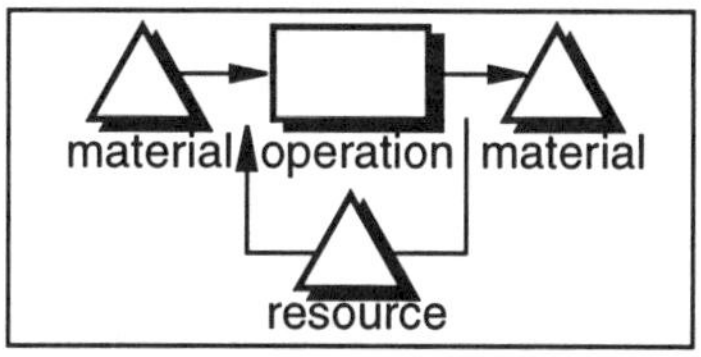

Fig. 4. Structural prototype for the control algorithm.

there are three kinds of nodes: material, resource and operation. The representation of the algorithm in figure 5 is a kind of interaction diagram, where the interactions between the functionality of the nodes of the prototype are shown. At the same time the instruction flow of the algorithm at each node is shown. Because all nodes of the model can be mapped to one of the nodes of the structural prototype from figure 4 no other functionality than that of figure 5 is needed for the model. So a high degree of reuse is reached. Only some parameters of the handlers must be adapted for each of the more than 300 nodes of the model.

4.5 Execution of the PPC-System

The constructed model can be carried out with the OOPUS system. In combination with the run time environment of OOPUS it can be regarded as a PPC system. By carrying out the system with simulated data the performance of the production control algorithm shall be evaluated for the present production. The algorithm can be adapted

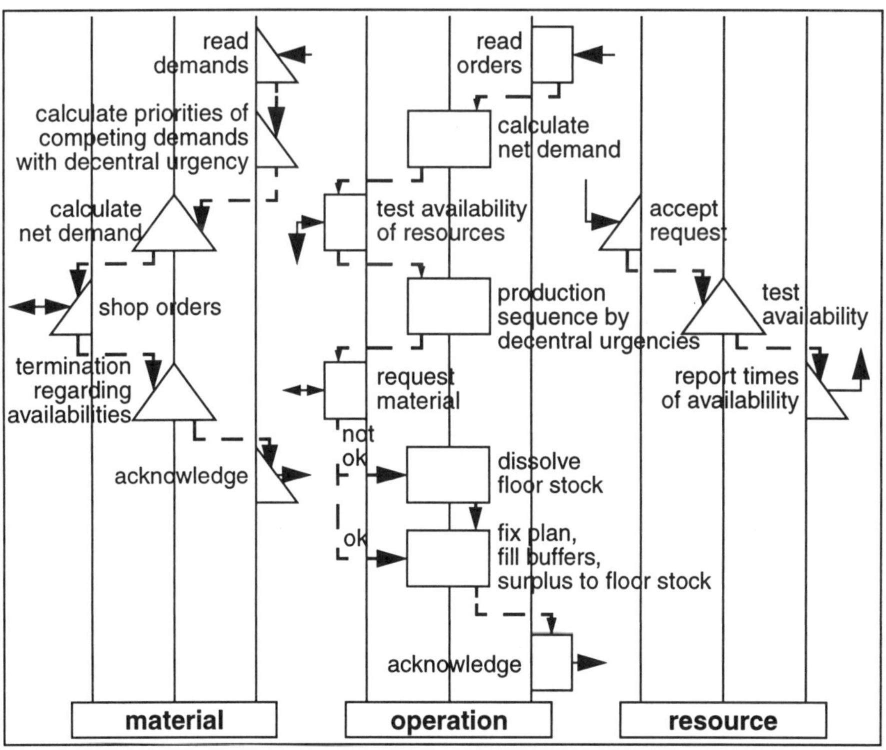

Fig. 5. Coarse flow of the control algorithm (simplified).

until satisfactory performance is reached. Then the same system can work with real data that stems from the production process without further needs for adaptation.

5 Conclusions

In this paper a four stage modeling approach for production planning and control systems was presented. In the first two steps only the structural aspects of the system are modeled and analyzed. The third step enhances the model with dynamic control functionality. The resulting dynamic model is then analyzed by simulation in the fourth step and can afterwards be converted to a running production planning and control system. The modeling is based on a sound modeling method (MFERT) and the whole process is supported by an appropriate tool (OOPUS).

However, even by focussing on the structural aspects of the model in the first two steps many potentials of optimization in the plant of the Volkswagen AG in Braunschweig were detected. Now the following two steps are taken and a prototype of the resulting production planning and control system is installed at the Heinz Nixdorf Institute in Paderborn in order to gain more insight in the relevant information and material flows of the production process by simulation.

6 References

1. W. Dangelmaier: *Distributed production planning and control in adaptive manufacturing processes - THE MANDATE APPROACH.* In: ESPRIT Working Group 9245 (Hrsg.): Improving Manufacturing Performance in a Distributed Enterprise: S. 63-70. Edinburgh (U.K.), 1995.
2. W. Dangelmaier, W. Felser, S. Henkel and R. Holtkamp: *OOPUS - a distributed system for modelling, simulation and production planning and control.* In: ESPRIT Working Group 9245 (Hrsg.): Improving Manufacturing Performance in a Distributed Enterprise: Advanced Systems and tools, S. 79-88. Edinburgh (U.K.), 1995.
3. W. Dangelmaier, T. Langemann and R. Wenski: *Model-oriented Construction of Manufacturing Management Information Systems,* In: CESA'96: Symposium on Discrete Events and Manufacturing Systems, Lille - France, July 9-12, S. 346-351, IEEE, 1996
4. N. Gaither: Production and Operations Management, Fifth Edition, The Dryden Press, Fort Worth et al., 1992
5. K. Müer.: *Optimierung der Ablaufplanung und Ablaufsteuerung der Hinterachsfertigung bei der VW AG, Werk Braunschweig - Sollkonzept eines Benutzerkonzeptes für die Fertigungsplanung der Hinterachsfertigung,* Heinz Nixdorf Institut Paderborn, 1996
6. R. Truderung: *Fertigungssteuerung mit Fortschrittzahlen – Analyse und Möglichkeiten,* Heinz Nixdorf Institut Paderborn, 1994.
7. R. Wallbruch.: *Optimierung der Ablaufplanung und Ablaufsteuerung der Hinterachsfertigung bei der VW AG, Werk Braunschweig – Modellierung und Erstellung eines Fertigungssteuerungsmodells für die Hinterachsfertigung,* Heinz Nixdorf Institut Paderborn 1996

Design of Agile Multi-Line Production System for Volatile Market Environment

Hiroshi Katayama[(1)] **and Masayuki Tanaka**[(2)]

[(1)]Department of Industrial and Management Systems Engineering,
School of Science and Engineering, Waseda University, Japan

[(2)]Sankyo Company Ltd., Japan

Abstract. This paper proposes an agile line balancing procedure of multi-line multi-item lot-wise production system in a volatile market environment. The procedure is composed of three steps, *i.e.* database generation of production formation for each product (Step 1), design of balanced multi-line multi-item lot-wise production formation (Step 2) and product item sequencing for each line (Step 3). Performance of the proposed procedure is evaluated through simulation experiments based on the actual data from a case factory, and the results indicate that stable as well as robust work formation enabling to cope with uncertain market scene is quickly obtained.

1 Introduction

As a recent indication, manufacturing enterprises all over the world are confronted with dramatic changes on business environment such as globalisation, market volatility, information intensive technology development *etc.* and many leading manufacturers are trying to reform their production systems with new production and management technologies. Agile manufacturing is one of the classes of such technological issues and researches concerning to this topic are becoming vital further than its initial stage [2]. Looking at the field of line production system management, efforts for capability enhancement on product variety have been paid by increasing product mixture rate, set up time reduction *etc.* through implementing computer supported lean production system such as electronic KANBAN system, mixed-item line production system refinement by KAIZEN, job sequencing by Target Chasing Method (TCM) for smoothing parts consumption rates, which is called Goal Chasing Method (GCM) in Monden [4]. However, performance of such courageous efforts is still inadequate due to further ambiguity increase in the recent market.
Understanding these background, this paper discusses about an agile design procedure of multi-item lot-wise production system for multi-line facility to offer a relevant weapon to cope with volatile market environment. The proposed procedure is

composed of database generation of line production work formation for each product (Step 1), design of balanced production formation of multi–item lot–wise production formation in multi–line facility configuration (Step 2) and product item sequencing method for each line (Step 3).

The kernel logic of Step 1 is conventional single–item line balancing algorithm [5], and in this step, various balanced work formations of each item are generated for various cycle time values and stacked as line balancing database. The logic of Step 2 is given in terms of quadratic programming model, which aims to obtain stable work formation in a static sense. That is, considered criterion is the minimum variance formation of necessary number of stations or workers for a given product–mix under the constraints of satisfactory resource utilisation. The logic of Step 3 is given as item sequencing problem solver, which tries to seek stable work formation in a dynamic sense. That is, to realise even pace resource utilisation in the actual time-based production operation, and considered criterion is the minimum time–based fluctuation of necessary number of stations or workers of the considered whole production lines under the restriction of resultant solution of Step 2. This module is developed in terms of heuristic approach.

2 Proposed Procedure

2.1 General flow of the procedure

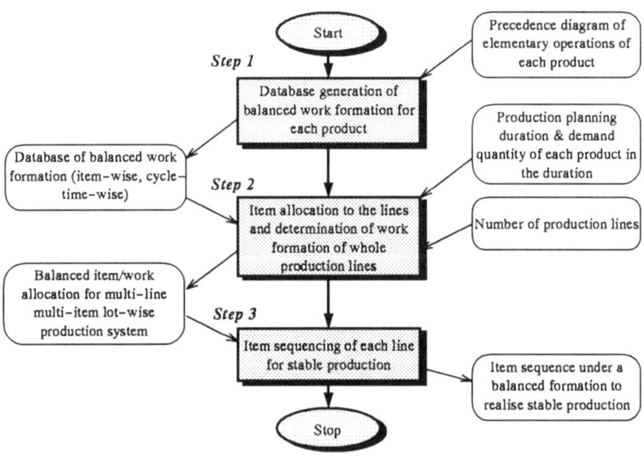

Fig. 1. General flow of the proposed procedure

The proposed prototype procedure was constructed through investigation of actual modification procedures of line production system managed in several production management division of factories, and is composed of three principal steps, *i.e.* generation of database of balanced work formation for each product (Step 1), item allocation to the considered lines together with determination of work formation of whole production lines (Step 2) and item sequencing of each line for stable production (Step 3). Main structure and its components, their input and output information are illustrated in Figure 1. Its final output image, *i.e.* item sequences under a balanced formation realising stable production are also illustrated in Figure 2.

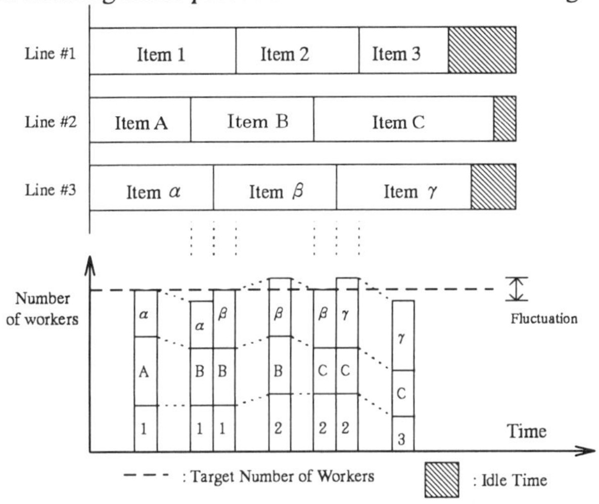

Fig. 2. Output image of proposed multi–line multi–item production formation and item sequencing system

2.2 Generation of work formation database (Step 1)

This step is the preparatory phase of the entire procedure facilitated by the conventional single–item line balancing algorithm, and this kernel technology is to generate and maintain various balanced production formations of each item for various cycle time values. This database is referred by Step 2 as technological asset of management. The input/output information and outline of this step are summarised as follows.

1) Input information : precedence diagram of each item's elementary operations
2) Output information : database of balanced production formation of single– item line production system (item–wise, cycle–time–wise)
3) Procedure of the step : The conventional algorithm used in the single– item line production system design utilised here is formulated as follows [3], [5].

< Mathematical formulation >

Objective function : Max. ϕ_i

$$\phi_i = \frac{T_i}{J_i \cdot C_i} \tag{1}$$

Where,

ϕ_i : balancing efficiency of considered product item i

J_i : number of stations for considered product item i

C_i : cycle time for considered product item i $\quad C_i = \frac{P_i}{Q_i}$

T_i : total operation time of considered product item i defined by the sum of Ot_{ij}

$$T_i = \sum_{j=1}^{J_i} OT_{ij}$$

OT_{ij} : operation time of station j for considered product item i

Q_i : demand quantity of considered product item i in the considered planning duration

P_i : considered length of production period for product item i

Constraints :

① All the precedence relations among elementary operations should be satisfied.

② Each OT_{ij} must not exceed the cycle time of regarding product item i.

$$OT_{ij} \le C_i \quad j = 1,2,\cdots,J \tag{2}$$

This problem formulated above can be solved by branch and bound method, in which lower bound function is defined as Eq.(3).

$$LB_i = r_i \cdot C_i + q_i \cdot C_i \tag{3}$$

$$q_i = \left\lceil \left\{ T_i - \sum_{j=1}^{r} OT_{ij} \right\} \Big/ C_i \right\rceil \tag{4}$$

So, optimistic estimation of balancing efficiency of item i, $\hat{\phi}_i$ at each branching node is calculated by Eq.(5).

$$\hat{\phi}_i = T_i \Big/ LB_i \tag{5}$$

Where,

LB_i : the function expressing minimum possible number of stations which is necessary to maintain for the production duration of product item i

r_i : number of stations that elementary operations of product item i are already assigned to some station

q_i : minimum necessary number of additional stations for product item i at the current branching node

$\lceil x \rceil$: minimum integer greater than or equal to x

$\hat{\phi}_i$: estimated maximum balancing efficiency of product item i

2.3 Item allocation and determination of work formation for each line (Step 2)

This step is the major part of the entire procedure for desirable multi–line multi–item switching production formation represented by quadratic programming model, which actualises balanced allocation of proper product items to considered lines together with stable and higher average resource utilisation under given demand quantities in the specified production duration. The input/output of this step and the mathematical representation of the contents are summarised as follows.

1) Input information

① the length of production duration and demand quantities of each product item

② database of balanced production formation (item–wise, cycle–time–wise)

2) Output information

multi–line multi–item lot–wise production formation actualising stable operation, reasonable resource (number of workers) allocation and higher average resource utilisation during the production duration

3) Procedure of the step

As a design procedure of considered multi–line multi–item production formation, the following five step procedure is developed, which includes quadratic programming sub–procedure illustrated in Figure 3 and is for stable and higher operation resource (number of workers) utilisation between periods of specified item production.

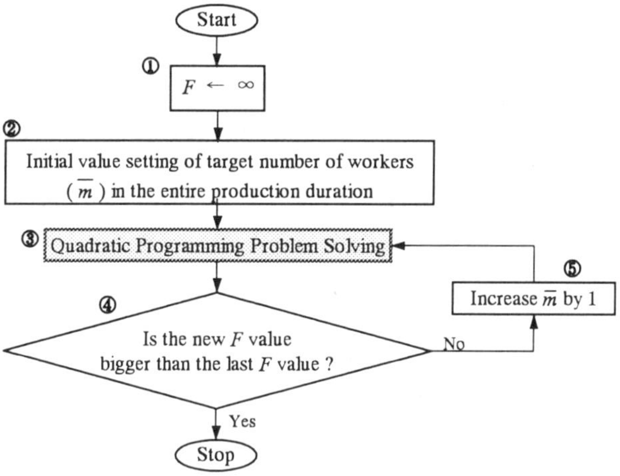

Fig. 3. Flow chart of Step 2

① initial value setting of the objective function of quadratic programming problem, *i.e.* F, in terms of sufficient large value substitution to obtain the optimal solution through the iterative procedure.

② initial value setting of target number of workers (or stations $\bar{m}$) in the entire production duration in terms of proper small number to obtain the optimal solution by increasing its value

③ quadratic programming problem solving

Following mathematical model can be solved, *e.g.* by "Industrial LINGO" [1], a generalised non–linear mathematical programming package for problems of actual business scale.

< Mathematical formulation >

Objective function :

$$Min. \quad F = \sum_{k}^{K} f_k \tag{6a}$$

$$f_k = \sum_{i \in I} \sum_{j \in J_i} \left(m_{ijk} x_{ijk} - \overline{m} \right)^2 \tag{6b}$$

Constraints :

$$\underline{\eta} \le \sum_{j \in J_i} \eta_{ijk} x_{ijk} \le \overline{\eta}, \quad i \in I,\ k \in K \tag{7}$$

$$x_{ijk} = 0 \ or \ 1 \tag{8}$$

$$\sum_{j \in J_i} \sum_{k} x_{ijk} = 1, \quad i \in I \tag{9}$$

$$\underline{w} H_k \le \sum_{i \in I} \sum_{j \in J_i} c_{ijk} Q_i x_{ijk} \le \overline{w} H_k \quad k \in K \tag{10}$$

Where,

I :set of product items

K :number of lines

J_i :set of alternatives of balanced production formation for product item i

$\overline{m}$:target number of workers to be assigned in the considered production duration

m_{ijk} :necessary number of workers for the j-th alternative formation of product item i at line k

η_{ijk} :balancing efficiency of the j-th alternative formation of product item i at line k

$\underline{\eta}$:admissible minimum balancing efficiency

$\overline{\eta}$:admissible maximum balancing efficiency

c_{ijk} :cycle time of the j-th alternative formation of product item i at line k

Q_i :demand quantity of product item i during the production duration

H_k :available production time during the considered production duration at line k

$\underline{w}$:admissible minimum utilisation rate of the production system

$\overline{w}$:admissible maximum utilisation rate of the production system

x_{ijk} :0–1 decision variables indicating adoption of the j-th alternative formation of product item i at line k

If the alternative is adopted $x_{ijk} = 1$

If the alternative is not adopted $x_{ijk} = 0$

④ Is the new F value bigger than the last F value ?

If true Terminate the procedure, as optimal solution is obtained.

If false Go to Step ⑤, as result is not attain the optimal solution yet.

⑤ Increase $\overline{m}$ by 1 and return to ③ for further iteration.

4) Criteria

Following criteria functions are adopted to evaluate the result.

① F :[See Eq.(6a)]

② w_k :

$$w_k = \frac{\sum_i c^{*}_{i \cdot k} Q_i}{H_k} \times 100 \qquad k \in K \tag{11}$$

③ g_k :

$$g_k = \frac{\sum_i m^{\bullet}_{i\bullet k} Q_i c^{\bullet}_{i\bullet k} \eta^{\bullet}_{i\bullet k}}{\sum_i m^{\bullet}_{i\bullet k} Q_i c^{\bullet}_{i\bullet k}} \times 100 \qquad k \in K \qquad (12)$$

Where,

$m^{\bullet}_{i\bullet k}$: resultant necessary number of workers of product item i at line k

Q_i : resultant demand quantity of product item i during the overall production duration

$c^{\bullet}_{i\bullet k}$: resultant cycle time of product item i at line k

$\eta^{\bullet}_{i\bullet k}$: resultant balancing efficiency of product item i at line k

F, the first issue, is the resultant value of objective function of proposed mathematical procedure, that is, attained minimum level of fluctuation of number of workers. w_k, the second issue, is reserved working time ratio at line k and g_k, the third issue, is net utilisation rate of whole production lines.

2.4 Item sequencing at each line for stable production (Step 3)

This step consists of five sub-procedures enabling to attain minimum sum of absolute difference between target and assigned work man-hours to the stations of entire production lines, which could be realised through a heuristic approach, *i.e.* pair-wise item sequencing of two critical production lines as described in Figure 4.

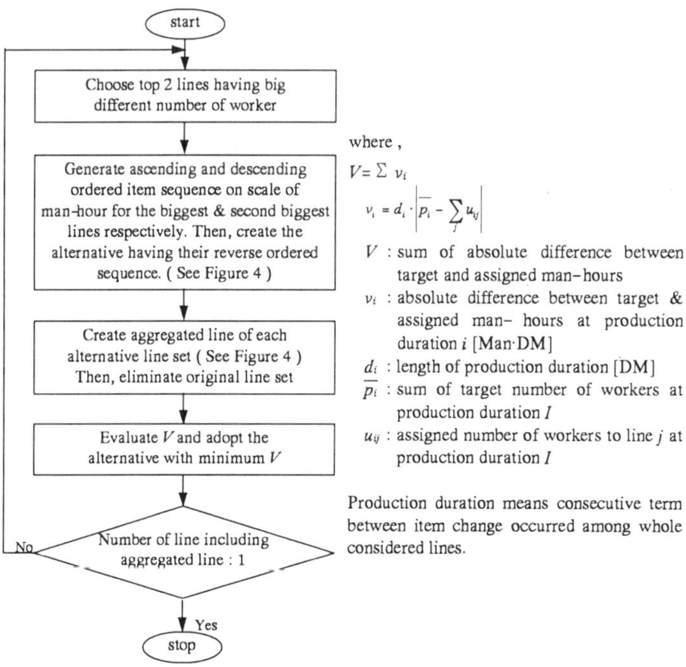

Fig. 4. Heuristic rule for item sequencing

3 Case Study

3.1 Assumptions/Conditions

Simulation experiments are executed under the following assumptions/conditions to validate the feasibility and effectiveness of the proposed procedure.

1) Precondition

Number of considered production lines : 2

Number of product items : 6

Cycle time : several possible values for each product item settled by historical data

Demand quantities : several possible values for each item settled by historical data

2) Feature of each product

Item 1 : product item that requires typical moderate operation–hours

Item 2 : product item that requires the shortest operation–hours (the easiest product)

Item 3 : product item whose process includes operations of basic item 1 and a few additional operations

Item 4 : product item that requires the longest operation–hours (the most difficult item to produce)

Item 5 : family product of item 4 (the same number of elementary operations with different work man·DM)

Item 6 : family product of item 4 (the same number of elementary operations with different work man·DM)

3.2 Precedence diagram of each product's elementary operations

An example of precedence diagram of elementary operations necessary to produce specific product item, which is collected and summarised through factory observation, is illustrated in Figure 5.

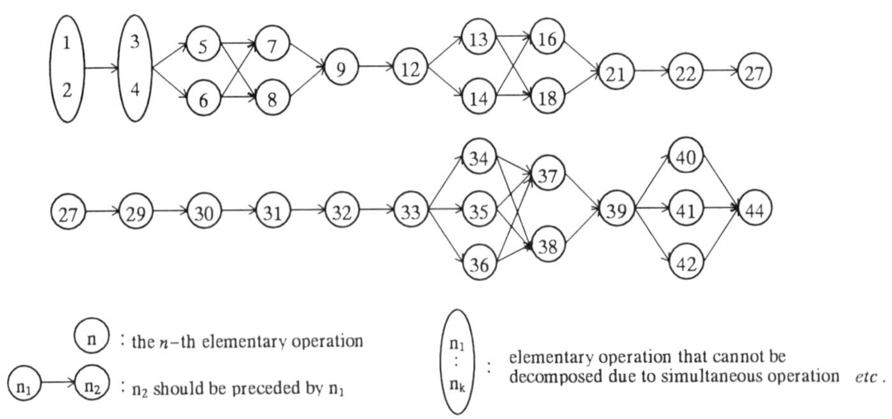

Fig. 5. Example of precedence diagram (Item 2)

3.3 Computational results

Variables and criteria values obtained from Step 2 (item allocation) are summarised in Table 1. Figure 6 demonstrates the result of Step 3 (item sequencing) for both lines. Other characteristics of considered variables and criteria are also summarised in Table 2. These tables indicate that proposed procedure is not only feasible but also able to be qualified, as F and g values are sufficiently small and over 80% respectively.

Table 1. Example result of Step 2

Issues \ Line No.	1	2
Target number of workers ($\overline{m}$)	15	
Objective function (F)	3	
(f_k)	1	2
Line utilisation (w_k)[%]	96.36	93.58
Net line utilisation (g_k)[%]	86.15	85.33
Production item number	1(15) 3(14)	2(14) 4(15) 5(15) 6(16)

*Figures in the parentheses are the number of workers for each item.

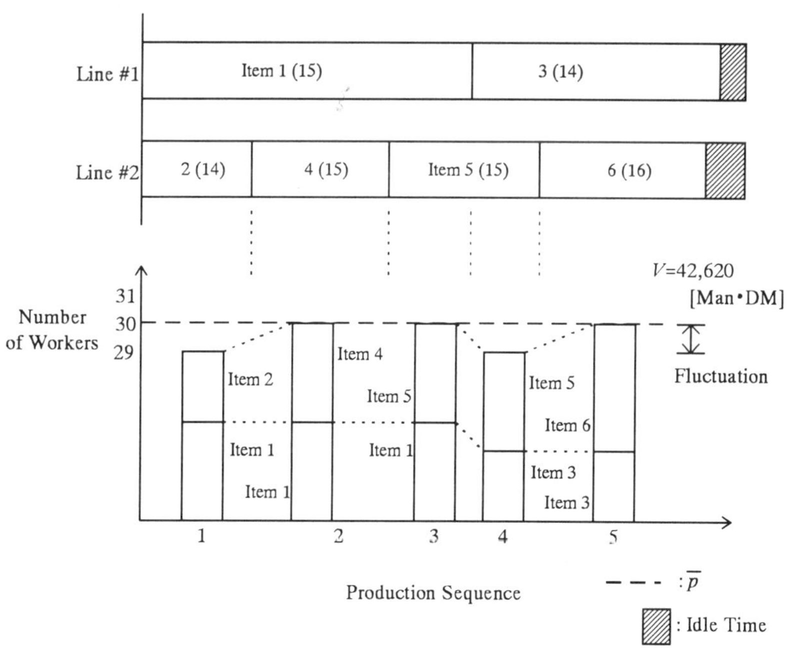

Fig. 6. Production Sequence Obtained by Heuristic Rule

Table 2. Variables and criteria of examined formation

Criteria & Variables \ Item		Item 1	Item 2	Item 3	Item 4	Item 5	Item 6
Two Lines	Number of workers (m)	15	14	14	15	15	16
	Cycle time (C)	155.2	108.6	178.5	189.1	189.1	169.2
	Balancing effi-ciency (η)	87.43	84.13	84.55	86.82	86.82	83.65

4 Concluding Remarks

In this paper, an agile balancing procedure of multi-line multi-item lot-wise line production system is proposed and its performance is evaluated through simulation experiments based on the actual data from a case factory. Noteworthy results obtained is summarised as follows.

1) A relevant agile procedure of production formation generation for multi-line multi-item line production system was developed, which consists of three principal steps, generation of balanced production formation database, multi-item production formation design and product item sequencing.
2) It could be recognised from performance analysis that proposed procedure can offer satisfactory results in actual business scene, *i.e.* allowable fluctuation of necessary work force as well as higher rate of production system utilisation.

It is also recognised that relevancy of the formation database is considered to be significant, and further improvement of the criteria values could be realised through enrichment of this database.

Acknowledgement

This research is accomplished under the support of Science Research Fund of Japanese Ministry of Education for 1995-1996 and Waseda University Grant of Special Projects 1996-1997. We, the authors, sincerely appreciate the kind support by the funds to prepare this paper.

References

1. Cunningham, K. and L. Schrage, *LINGO User's Manual* (LINDO SYSTEM INC., 1988).
2. Gerhardt, L., *Adaptive / Agile Integrated Manufacturing Enterprises (AIME) -An Evolving CIM Strategy for the 21 Century* (Key Note Lecture in the 2nd ICCIM, 1993).
3. Katayama, H., "An agile design procedure for a line production system in a versatile market environment", *International Journal of Computer Integrated Manufacturing*, Vol. 9, No. 4, pp. 299–305, 1996.
4. Monden, Y., *Toyota Production System*, Institute of Industrial Engineers, 1993.
5. Muramatsu, R., *Basic Production Management* (Kunimoto Publishing Co. Ltd., Tokyo, Japan, 1979), p. 169–219.

A Knowledge-Based Selection System for the Multi-Product Assembly Line

A Khan and M K Khan

Department of Mechanical and Manufacturing Engineering,
University of Bradford, Bradford, England (UK)
Tel: 0044-1274-384523, Fax: 0044-384525
Email: M.K.Khan@Bradford.ac.uk

Abstract
In a multi-product manufacturing environment, products are grouped together into families. Production demand for a group of products must be sufficiently large for assembly line justification. In addition, products in the same family must have justified operational similarity. This paper presents Knowledge-Based production selection criteria for the assembly line justification, on the basis of demand volume, operational similarity and operational divisibility. The knowledge-based system is developed for the analysis of these parameters through the use of an Expert System called AM: a computing software. The developed Knowledge-based System for assembly line design and selection also uses analytical techniques for the final results.

Introduction

The objective of this paper is to develop preliminary design principles for selecting an assembly system. Production system are mainly divided into the following three categories based on the rate of demand[1,2] . For very low demand jobbing production is recommended. Batch-type production is used for medium demand volume while for high demand volume mass production is recommended. This classification is limited in that demand volume is not the only criteria for selecting an assembly system. Also demand volume may be very high but due to small work content of production, the assembly system may still not be economical.

Assembly system may also be recommended on the basis of material flow in the system[3] . In multi-product environment, material flow also play a vital role in the system selection. Four type of material flow are generally observed on shopfloor, they are repeated operation, in-sequence, by pass movement and back tracking[4] . In-sequence flow is considered as the most desirable flow for an assembly system due to its unidirection in the system. Products having more similarity in their operations will automatically increase the over-all percentage of such operations on the shopfloor.

In an assembly system, work load on shopfloor are performed simultaneously at all stations. For a perfect balance, the work load must be distributed as evenly as possible among the stations. This will only be possible if the task times of each individual product is divisible into tiny task times[5,6]. Increasing the difference of task times with the average task time will increase the possibility of a parallel station or over-all line inefficiencies. This statement gives an idea for experts in this field to select a product or a group of products for an assembly system.

A Knowledge-Based Assembly System Selection (KBASS) is designed which selects a product or a group of products suitable for an appropriate assembly system. The KBASS gives a clear cut advice on low, medium or high demand volume in relation with total work content, product similarity for increasing the possibility of in-sequence operations on shopfloor and the allowable task time divisibility for even distribution of tasks among stations in an assembly system. The KBASS can be operated even by a computer novice. The user must sit by the computer terminal and type in the description of the problem. The KBASS asks questions, the user answer them, more questions may be asked and answered and finally, a conclusion is drawn from these information.

Literature Review

In literature review, the classification of such systems is mainly based on the rate of demand[1,2]. One of the methodologies available in the literature for the selection of assembly system provides quantitative as well as qualitative factors[6]. Wild's methodology was a preliminary study and much further work is needed for sharpening the selection process. An other classification of assembly systems on the basis of cost effectiveness include manual, special purpose automatic and programmable assembly systems[7]. Assembly system are compared by analysing the effect of parts quality, number of parts in the assembly system, annual production volume per shift, production style variations, design changes, number of products to be assembled and economic climate. The candidate systems are also analysed on the basis of their cost and benefits[8]. In the first step, their cost and benefits are estimated and in the second step, their life cycle cost analysis is performed based on cost breakdown structures. The combined score in both steps, gives a cost / effectiveness measure for selection.

Since, then, a lot of research is conducted on the design and performance of assembly systems. A comprehensive literature review was carried out on the design and analysis of these systems. Quantitative development and qualitative issues are addressed at both the strategic and tactical levels[9]. Assembly systems are divided into two groups i.e., serial configuration and parallel configuration[10]. An analytical model is developed to identify when each configuration is cost effective.

In majority of cases, the researchers have addressed the design, balancing, performance and economy of the assembly systems. The comparison is conducted only within various assembly systems. The selection of an appropriate system which may possibly be an assembly system or non-assembly system have got very less attention from the researchers while a systematic quantitative Knowledge-Based selection for an assembly system is absolutely out of seen.

Selection Structure
A Knowledge-Based Assembly System Selection (KBASS) is designed for a single or multi- product environment which systematically selects assembly system for a product or a group of products under consideration.

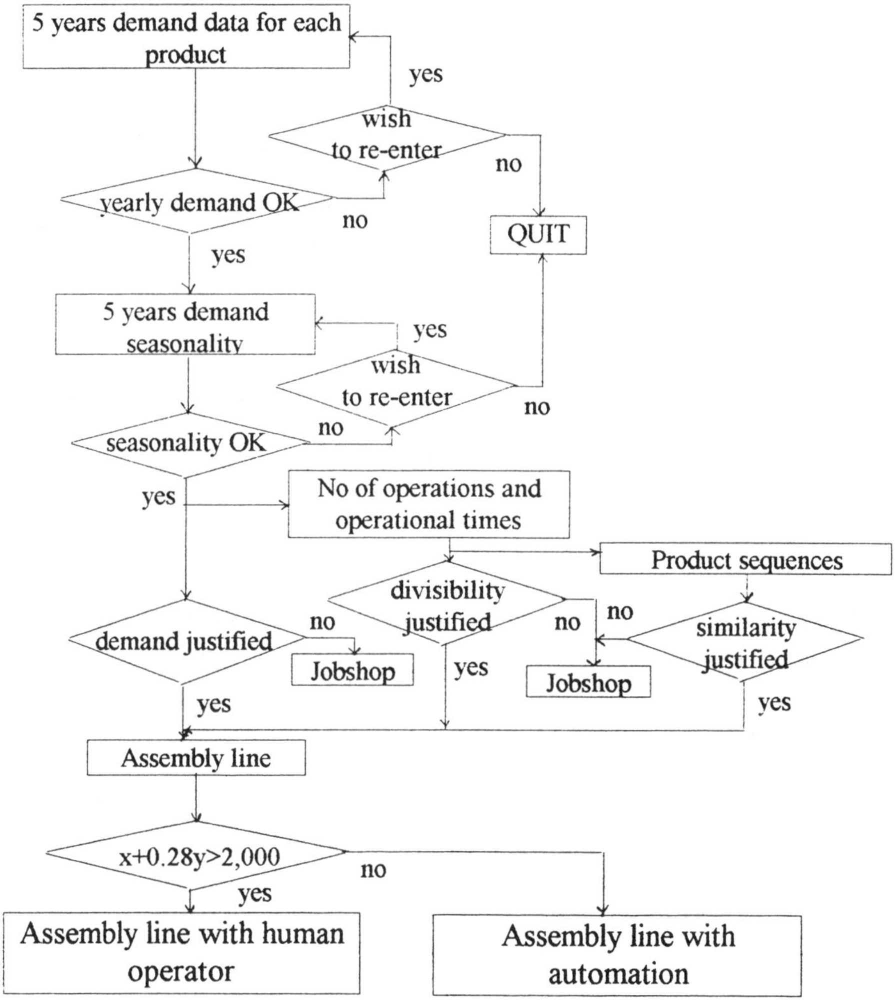

figure 1

The basic flow-chart for the KBASS is shown in the figure 1 which is divided into the following three sections. They are:

1. demand justification in relation with total work content for an assembly system
2. operations similarity in products

3. task times divisibility in products

These selection parameters are extracted from the principles of assembly systems[1,2,8]. The KBASS is developed for the analysis on these parameters through the use of Expert System, computer soft ware, called AM. It is a ruled Based development tool. A set of Knowledge-Based questions are prepared for each section. The user is asked to answer each question which are then compared by IF-THEN production rules with in the KBASS and the program attempts to derive an expert conclusion

Selection Procedure

To select a product or a group of products for assembly systems, the Knowledge-Based system verify demand volume in relation with total work content, operational similarity of all products with the design product and operational divisibility for the mass production on the assembly system. For the analysis on these parameters, the system will require details of the demand volume and operational involvement for each individual product.

Demand volume

The data collection is started with the number of products involved in whole production and the expected demand volume of first year for each product. To make sure that the production will have a minimum life cycle of five years, the KBASS will require percentage increase or decrease for each product in the next five years in the same display.

The KBASS can accept percentage decrease in the demand of any number of products, but, over-all expected demand for the following years must be greater than the demand volume of preceding year. The system will check all these entries and will go ahead for analysis only if all these entries are valid for the system. In case of invalid entries, the system will give corresponding error massage and will bring the same display again.

Demand seasonality

After entering justified yearly demand data for the demand display, the system will need to justify monthly demand for each and every product. Monthly demand share for each product should lie between the range from 5 to 20 percent of the total demand volume for that product. Any entry beyond this range will not be accepted by the system as a valid entry. The reason for this is that each and every month must have a reasonable share of the total demand volume so as to justify a continuous production on assembly line.

To check any growth or replenishment in the product life cycle, the system will provide option for the user to enter monthly demand volume for year after year. In case if demand variation remains the same, or similar to demand variation of any preceding year, then the system will assign monthly demand variation either on the basis of year one if data for demand variation is entered only for year one or on the basis of user choice if data for demand variation is entered for more than one year.

As soon as, the user enters evenly distributed monthly demand data, the system displays total demand sheet which contains all necessary information about demand in the next five years, i.e., individual product monthly and yearly demands, over-all production monthly and yearly demands and their grand total for five years.

Operational involvement

Now the system will need operational involvement in each product so as to find out work content for each product and total work content for whole production. It is considered that the number assigned to any operation is a unique number with its own unique operational time. Each operation will maintain their own number and operational time in any other product if needed. It is therefore necessary that the Knowledge-Based system wants to know about the number of different operations which are involved in whole production. This will sets maximum range for the number of different operations required. Time needed for the assembly of each operation is asked in the next display which is followed by operational involvement for the assembly of each product. From this data, the system will find out individual time for each product and total time required for whole production.

System analysis

Products under consideration are checked for assembly line installation by verifying the following three criterion. They are product similarity, product divisibility and justified expected demand volume. The KBASS verifies each criteria for assembly line installation and does not stop its analysis in the middle even if any criteria is unsatisfactory. The KBASS selects maximum demand product as a design product for comparing with remaining products. Operations involved in the assembly of each individual product are compared with the operations of the design product and their percentage operational similarities with the design product are recorded. The KBASS will verify that each and every product must have at least 50 % similarity with the design product. This statement is based on the practical examples available in the literature. The KBASS displays comparison of these products with the design product and give for each individual product the number of operations which are similar, the number of operations which are dissimilar and their percentage similarity with the design product.

The KBASS also verify operational divisibility for tiny task times so as to justify assembly line installation for production. From literature, 23 examples of assembly lines are selected and are analysed for product divisibility. Number of tasks are plotted against a factor which is equal to maximum task time divided by average task time. A line is drawn between two highest points which covers almost all remaining points. It is therefore derived that the product task times considered for assembly lines must satisfy the following relation:

$$b \leq a / 1.8 + 2.1 \quad (1)$$

where

a = number of task considered
b = maximum task time / average task time

It is also observed that the minimum number of operations and operational times for the assembly line justification is 6 and 12 seconds respectively.

Demand volume in relation with total work content is considered as the third assembly line selection criteria for products under consideration. Generally jobshops are suitable for low demand volume. Assembly line with human operator are suitable for medium demand volume where cycle time is always set for more than 30 seconds so as to maximise line efficiency. When demand volume is very high, then short cycle times are required for the stations on the line. This can only be achieved by the assembly line with automation. The minimum cycle time which can be practically achieved with automated line is 2 seconds . Total yearly demand volume in relation with total work content is justified for the assembly line by the following relation[10] ,

If $x + 0.28y > 2000$, then {Assembly Line with Automation} (2)
or $1000 \leq x + 0.28y \leq 2000$, then {Manual Assembly Line} (3)
or $x < (14{,}400 / y)$, then {Jobshop} (4)

where

x = Demand volume / 1000
&
y = Total work content (in hours)

The KBASS needs justification for all three criterion of assembly line installation. If any condition is failed, the solution is then jobshop with human operator. The system will stop its further analysis and will inform the user that the system has no facility for further analysis in jobshop production.

After successfully checking the three criterion for the assembly line installation, the system give the suggested line with either human operators or automation and asks for the full title of the line

KBASS Description

The KBASS rules are developed in AM: a computer software. The KBASS rules are expressed in the form of 'IF-THEN-OR' and consists of:

(a) the name of the rule which instructs development engine to insert a new rule in the KBASS.
(b) the premise 'IF' of the rule
(c) the action 'THEN' of the rule in which the KBASS reaches the appropriate decision or direction

If all the premise are true, the rule is considered true and action takes place in the desired direction. If any of the premise is false, the rule is considered false, and any statement in the 'OR' part is then taken to be true.

Knowledge-Based

A number of questions are asked from the user in various stages of KBASS which must be answered by the user. The KBASS activate production rules on each answer of the user and derive a conclusion or decision for further processing in the system. The main sets of KBASS production rules are applied in the following manner.

Set 1

Rules (1-5) check demand justification for assembly system

Set 2

Rule (6) checks operational similarity justification for assembly system

Set 3

Rule (7) checks operational divisibility for assembly system

Set 4

Rules (8-9) check type of manufacturing system for the user

Data required

Rules of KBASS are capable of selecting appropriate assembly system for the user. The selection process must, first, obtain enough information from the user to allow the application of sufficient number of rules to develop a specific recommendation. The questioning sequence include inquires about product description and related information. The KBASS asks the user questions about the problem to be solved. The user responds by entering the exact format in which the data are required. The following questions are required for the processing of above set of rules:

Questions required for set 1 rules

number of product required
demand volume for each product in year 1
percentage increase or decrease in demand volume for year 2 to year 5
monthly seasonality for each product in year 1
seasonality for each product in remaining years if different from year 1

Questions required for set 2 and set 3 rules

total number of different operations involved in whole production
operational times for each operation
list of operations involved in each product

Creating KBASS rules

Each set of KBASS rule reaches to the roots of each section and starts justification from the lower level upto the end rule. The main rules of each set are briefly explained as follows:

Set 1 Rules (Demand data justification)

Rule 1:

IF	yearly demand data entered correctly for each product
AND	monthly demand data entered correctly for each product
THEN	check demand data justification
OR	QUIT

Rule 2:

IF	year 1 demand entered correctly for each product (i.e.>0)
AND	year 2-5 demand %age entered for each product
AND	year 1-5 total annual demand > preceding year demand
THEN	yearly demand data entered correctly for each product
OR	Please review your yearly demand data

Rule 3:

IF	monthly demand occurs between (5-20)% of total demand
AND	total entry for 12 months = 100
AND	monthly demand entered for each product in year 1
THEN	check for monthly demand data for remaining years
OR	Please review your monthly demand data

Rule 4:

IF	monthly demand data for remaining years not entered
AND	monthly demand data entered for more than one year
THEN	monthly demand data for remaining years will be based on user choice
OR	monthly demand for remaining years will be based on year one monthly demand data

Rule 5:

IF	demand volume /1000 > 14,400 / total work content (in hrs) by (4)
THEN	conclusion: demand justified for assembly system
OR	jobshop selected

Rule 6:

IF	number of different operations > 5
AND	total operational time > 12 seconds
AND	b < a / 1.8 + 2.1 by (1)
THEN	conclusion: divisibility justified for assembly system

OR jobshop selected

Rule 7:

IF number of products > 1
AND design product selected on basis of higher demand
AND operational similarity with design product > 50% for each product
THEN conclusion: similarity justified for assembly system

OR jobshop selected

Rule 8:

IF demand justified for assembly system
AND divisibility justified for assembly system
AND similarity justified for assembly system
THEN check assembly system selection

OR jobshop selected

Rule 9:

IF demand volume /1000 + 0.28*WC(in hrs) >2,000 by (2)
THEN selection: Assembly Line with Automation

OR selection: Assembly Line with Human Operator

Conclusion

Generally, the selection process defines a system with in various type of assembly systems. Comparison of assembly systems with other manufacturing systems is available in a very crude form. The selection process is purified with the help of design principles for assembly systems. Assembly system installation is justified by checking their high demand in relation with total work content, divisibility of task times for even distribution of work load among stations on line and required similarity in operations of various products. These design parameters are chosen for maximising line utilisation with justified demand volume and minimising line losses which may created due to imperfect load distribution among stations and due to huge set-up and change-over times. The KBASS further helped by selecting appropriate assembly system in jobshop, assembly line with human operator, or assembly line with automation, in a systematic and step by step approach. The data collected will be stored in KBASS and can be retrieved in the planning and design stages for assembly lines.

References

1. Wild, Mass-production Management, New York: Wiley & Sons (1972).
2. Johnson, et. al., Operation Research in Production Planning, Scheduling and Inventory Control New York: Wiley (1974).

3. Ying, et. al., Two sequence-pattern, matching-based, flow analysis methods for multi-flowlines layout design. Int. J. Prod. Res., Vol. 31, No. 7, (1993) 1557-1578.
4. G. Aneke, et. al., A design technique for the layout of multi-product flowlines. Int. J. Prod. Res., Vol. 24, No. 3, (1986) 471-481
5. F. Sawyer, Line Balancing. The machinery publishing Co. Ltd (1970).
6. Wild, On the selection of mass production systems. International Journal of Production Research 13, 443 (1975).
7. Boothroyd, Economics of Assembly Systems. Journal of Manufacturing Systems, Vol. 1, No. 1, (1982) 111-127.
8. Shtub, et. al., A methodology for the selection of assembly systems. Int. J. Prod. Res., Vol. 27, No. 1, (1989) 175-186.
9. Ghosh, et. al., A comprehensive literature review and analysis of the design, balancing and scheduling of assembly systems. Int. J. Prod. Res., Vol. 27, No. 4, (1989) 637-670.
10. Young, Automatic sequence generation and selection for mechanical assembly and part replacement, PhD Thesis, Pennsylvania State University, USA (1993).
11. Daganzo, et. al., Assembly system design principles and trade-offs. Int. J. Prod. Res., Vol. 32, No. 3, (1994) 669-681.
12. Milas, Assembly Line Balancing... Let's Remove The Mystery. Production & Inventory Control, may,IE, (1990) 31-36.
13. Abdou, An integrated approach to facilitates layout using expert systems. Int. J. Prod. Res., Vol. 28, No. 4, (1990) 685-708.
14. Lotter, Manufacturing assembly Hand Book, Butterworths (1989).

Knowledge Based Methodology For Selection Planning And Design Of Flexible Manufacturing System

I. Hussain and M. K. Khan

Department of Mechanical and Manufacturing Engineering,
University of Bradford, Bradford, England (UK)
Tel: 0044-1274-384523, Fax: 0044-1274-384525
Email: M.K.Khan@Bradford.ac.uk

Abstract

Computer Integrated Manufacturing plays a vital role in today's industries. Due to increasing demand in products variety, quality, reliability and comparatively lower production, the manufacturing costs have made the conventional manufacturing enterprises to move towards automation, especially Flexible Manufacturing System (FMS). FMSs are quite expensive and complicated and for this reason are attracting a large number of researchers. This paper presents a hybrid FMS system incorporating a knowledge base (KB) and an analytical technique for the selection, planning and design of FMS. The methodology uses demand forecast, related information about part types, machines calculation and material handling system. The selection of machines and material handling system is based on flexibility and cost optimisation.. The methodology is developed in ***AM*** : an expert system shell. The application of the system is demonstrated with the help of an example.

Introduction

An FMS is a production facility consisting of flexible machining centres and automated Material Handling System (MHS) which are logically controlled by computers. Many supporting workstations such as loading/unload station, deburring station, washing station, inspection station, storage, tool and fixture stations can also be added to FMSs[1]. FMSs offer several benefits over conventional manufacturing, namely, increased productivity quality, uniformity, shorter manufacturing cycle times, reduced labour requirement, and better machine utilisation[2]. FMS can be made a part of computer integrated manufacturing (CIM). CIM as a whole covers computer -aided integration of all functional areas from marketing to manufacturing, while FMS focuses on shop floor organisation including functions like, manufacturing, material handling and tooling[3]. By integrating a FMS into CIM one can achieve tremendous manufacturing advantages[4].

FMSs are quite complex systems and are made up of a number of components which interact with one another[5]. A careful analysis is required before investing a huge amount of capital on FMS installation. Therefore, efficient planning methodologies and control policies must be chosen for the successful implementation of an FMS[1]. There are various modelling techniques which are used to plan, design and evaluate the performance of FMSs, such as: mathematical models, simulation models and (KB) models. Mathematical and simulation models are widely used to investigate the performance of FMSs. Mathematical models are formulated by using integer or mixed integer programming and are solved by using algorithms, heuristics, and queuing networks. Such models are good for small problems but unfortunately for large scale problems, these models are computationally not feasible. Simulation is considered as one of the best tool for analysing the performance of complex manufacturing systems. Simulation is essentially a trial and error process; even an experienced designer would use various guidelines and rules of thumb in the analysis[6]. On the other hand KB systems are found to be quick, realistic and more user friendly for design problems[2,6,7,8].

This paper presents a hybrid (KB/analytical) methodology for selection, planning, and design of an FMS. The paper briefly covers the demand information, part types information, material handling information, machines calculation and MHS selection. The KB helps the user in selecting a FMS and suitable MHS for the selected system. The analytical approach helps in calculating quantity of part types to be produced, required number of machining centres, and finding design parameters of a selected MHS. Two kinds of part types are considered, i.e. cylindrical parts and prismatic parts. Two types of machining centres, i.e. milling and turning machining centres are considered. Both types of machining centres are assumed to be highly automated capable of accommodating maximum number of processes and operations involved on part types. However, the milling and shaping processes on a cylindrical part can be performed by a milling machining centre.

Notations

D_m	= monthly demand
N	= number of operations per part type
T	= total operating time per part type (min)
N_s	= number of simultaneous operations per part type
P_t	= total operations per part type
P_{all}	= personal allowance (%age)
S_{all}	= shrinkage allowance (%age)
T_{all}	= time allowed per part type (hrs)
T_{oc}	= total operation time per month for cylindrical parts (hrs)
T_{op}	= total operation time per month for prismatic parts (hrs)
M_t	= monthly allowable time per part type (hrs)
S_t	= setup time (min)
S_{av}	= system availability (hrs)
T_m	= turning machining centres
M	= total number of machining centres
M_m	= milling machining centres
TTW_d	= total transport work for delivery moves

D = distance matrix
d = delivery matrix
D_N = number of deliveries per hour per vehicle
A_d = average distance per delivery (m)
N_{agv} = number of AGV(s)
C_{agv1} = cost of an AGV (£s)
C_{agv} = total cost of AGV(s) (£s)
T_f = traffic congestion factor
E_{mhs} = efficiency of MHS

Knowledge Based System

The reason for incorporating a KB approach in the hybrid methodology is that the KB solves problems in a given area much more efficiently and effectively than other conventional decision making approaches[8].

The KB is developed and implemented on a Pentium P100 PC using *AM*, an expert system shell. AM uses highly visual interactive interface which helps users to produce applications quickly and easily.

Structure of KBFMS

The structure of the KBFMS is shown in Figure 1.

Information Base

The KBFMS requests information from the user. The information consists of facts about FMS such as number of part types (each part type is considered different from the others), quantity of each part type, part types information regarding number of processes and number of operations, machining centres information, machining centres layouts on shop floor (single row, double row, loop layout), floor space dimensions, MHS information, cost of machining centres and MHS. The KBFMS can consider both cylindrical and prismatic kind of part types. An incorrect piece of information is notified to the user and correct information entry is requested (man-machine interface).

Optimisation

The analytical part of the hybrid methodology calculates the required number of machining centres for the number of units to produce per month. The floor space dimensions are calculated for each layout by considering two assumptions. The design factors for the MHS are calculated for each layout and assumption. The output of the analytical approach forms a base for selecting the optimal number of turning and milling machining centres, the suitable layout and the MHS in the KB.

Knowledge Base

The KB part of the hybrid methodology consists of production rules related to manufacturing systems, machining centres and MHSs. Based on the output from the analytical approach, the KB helps in selecting the FMS, optimal number of machining centres for the FMS, suitable layout and MHS.

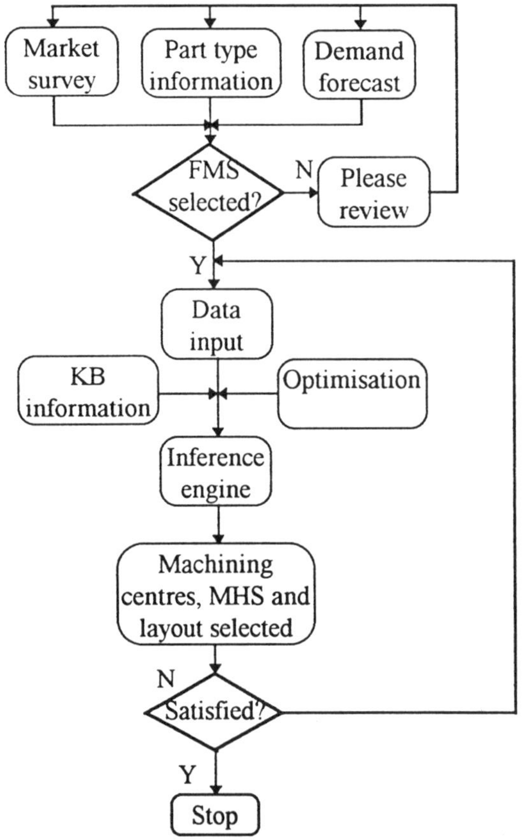

Figure 1

Methodology

The KBFMS makes use of a hybrid technique of KB and optimisation to help user select an FMS, required machining centres, suitable layout and appropriate MHS. The KB part of the methodology not only includes facts and data but also production rules which help management select a suitable manufacturing system, layout and MHS, while the optimisation techniques help in calculating number of machining centres, floor space dimensions and design parameters of the MHS.

FMS Selection, Planning And Design Approach

Based on facts and data the KBFMS checks whether an FMS is required or not before going in to the details of planning and designing. An example is given below which shows how the KBFMS selects an FMS.

If	variation in the part types is more than 7
and	demand for the part types falls in the range of mid volume
and	system is able to react to demand changes
and	system is able to react to design changes
Then	an FMS is required

After selecting a FMS, the KBFMS conducts a detailed planning for the system. In the planning stage the number of processes and number of operations in each process are defined. The maximum number of processes considered on cylindrical and prismatic part types are five and four respectively. Each process can have maximum number of nine operations. The setup time for each batch of part types is found. The allowances such as shrinkage and personal[7] are included to find the exact quantity of parts and the available time in which those parts are to be produced. The cost related to part types such as labour cost per hour and material cost per kg is also defined. The capability of the machining centres and their estimated cost is taken into consideration. Local buffers are considered with each machining centre. The two major MHSs i.e. AGV-robot based and Conveyor-robot based are included in the planning stage. Selection of the suitable MHS is based on the KB. The design parameters are calculated in the design stage. In this paper for the sake of brevity, it is assumed that an AGV with robots is selected as a MHS. As mentioned earlier, a hybrid methodology is developed which makes use of declarative data, facts and rules. An example shows the rules used in the planning stage.

◊ how many processes are considered on a part type?
◊ how many operations are considered on a part type?
◊ can simultaneous operations be performed?
◊ will machining centres be capable to handle simultaneous operations?
◊ will the part travel path be fixed or variable?
◊ can new routes be added easily, in case of machine breakdown?
◊ how many shifts to operate the system per day?
◊ how many days to operate the system per month?

Rules similar to the above help in identifying the system and provide a base for the design stage.

In the design stage, different designing factors which result in the calculation of machining centres are calculated. This includes monthly demand, total operations per part type, time allowed per part type, monthly allowable time per part type, total operation time per month for all part types and availability of the system per month. The calculation of machining centres is based on monthly demand. As mentioned previously two kinds of part types are considered which cover a wide range of rotational and prismatic parts, therefore two types of machining centres are recommended for the FMS, i.e. turning machining centres and milling machining centres.

In the planning stage a suitable MHS is selected, which for this paper is assumed to be AGV-robot based MHS. In the design stage a careful analysis of the MHS is performed by considering three types of layouts, i.e. single row, double row and loop layout. The best selection of the layout for machining centres on shop floor greatly optimise the material flow between different stations. Three design factors are considered in this paper, however, the KB can consider upto twenty factors for a real situation. The factors are:

◊ number of AGVs required
◊ cost of AGVs
◊ efficiency of MHS

MHS Description

Two cases are considered to find the delivery and empty moves of an AGV. In the first case the AGV picks up a part from the loading station and delivers it at the machining centre. From the machining centre an already finished part is picked up and delivered to the unloading station. The AGV moves empty from the unloading station to the loading station. So, in this case for each two deliveries there is one empty move. In the second case, the AGV picks up a part from the loading station and delivers it at the unloading station. From the loading station it moves empty back to the loading station as the machining centre remains busy during that period. When the part is finished at the machining centre, the AGV then moves empty to the machining centre to pick up the part. The AGV picks up the part and delivers it to the unloading station. From the unloading station it moves empty to the loading station. In this case for each two deliveries there are three empty moves.

Two assumptions are considered to express robot capability. In the first assumption a robot is considered to serve two machining centres while in the second assumption one robot can handle three machining centres. If a machining centre is left unassigned in both the assumptions, then a robot serving one machining centre will be selected.

Based on both the cases, the number of moves per hour to milling and turning machining centres are calculated. Distance matrices are also constructed for the two cases and assumptions. The floor space dimensions are calculated under each assumption for each type of layout. The shop floor area is assumed to be rectangular in shape. The floor space constraints are checked for each assumption and layout. If the floor space constraints violate the available floor space dimensions then the KB notifies the violation to the user. The design factors are calculated for each layout under each assumption. As each layout for each assumption gives different results for the design factors, weights are therefore assigned to individual design factors under each assumption in each layout. The selection of the best layout is based on the total weight. The layout with the highest weight is given priority. For example,

If	weight of single layout < weight of loop layout
and	weight of double layout < weight of loop layout
Then	select loop layout

As there are two assumptions which are considered in this paper, while one of the two assumptions is required to be selected for the manufacturing scenario considered. Therefore the selection of an assumption is based on the optimal cost associated with each assumption. The cost associated with the assumptions include operating costs and cost of robots (not discussed in this paper). The robots with optimal cost get selected. For example,

If	total cost associated with 2nd assumption < total cost associated with 1st assumption
The	one robot to serve three machining centres

The KBFMS is illustrated by using a numerical example. The example considers 22 different part types. Out of 22 part types, 12 are cylindrical and 10 are prismatic part types. The demand per year for cylindrical and prismatic part types is 10735 and 10135 respectively. The shrinkage and personal allowance are considered same for all part types as 1% and 5% respectively. Simultaneous operations are considered on the part types. The system operates 2 shifts per day and 24 days per month. The efficiency of the system is considered as 60%. The cost of an AGV is assumed as £10000. The velocity of an AGV and traffic congestion factor are considered as 1 m/sec and 0.95 respectively. The loading and unloading of parts are assumed as same. The robots are assumed to handle maximum weights of part types. The data input values are made as realistic as possible.

Discussion

The number of machining centres and design factors of MHS are calculated using Equations (1-12) in Appendix A. The average operation times per month of cylindrical and prismatic part types are calculated by considering the operation times of individual part types and the simultaneous operations performed on individual part types. Both the individual part type operation times and number of simultaneous operations on an individual part type are not shown in the problem statement. Advancement in technology has made the machine tools capable to operate simultaneously on a part type due to multi spindle movements and availability of tool magazines. Therefore a maximum of three simultaneous operations are considered on a part type. Results in Appendix B show that the average operation times per month of cylindrical and prismatic part types are 657 hours and 773 hours respectively and availability of the system per month is 230 hours. Using Equations (5) and (6) the number of turning and milling machining centres are calculated as 3 and 4. The machining centres are considered capable of accommodating all processes but those

cylindrical parts which have shaping and milling processes will visit the milling machining centres as well.
Table 1 shows the results of the two design factors for the two cases (one robot to serve two machining centres, and one robot to serve three machining centres) under first assumption. The loop layout offers better results for the number of AGVs, cost of AGVs and efficiency of the MHS as compared to other layouts. Table 2 shows the results of the same design factors for the same cases but under the second assumption. In Table 2 the loop layout performs even much better than the first assumption in Table 1 giving optimal number of AGVs, optimal cost and high MHS efficiency.

The two cases (in first case for each two deliveries there is one empty move, while in the second case for each two deliveries there are three empty moves) refer to two extreme situations, while in the practice the problem lies between these two extremes. Table 3 shows the average result of the extreme cases. A comparison of both assumptions is given in this table. The second assumption produces better results in loop type of layout for the average results of both cases.

Conclusion

The KBFMS is hybrid tool of KB and analytical approach, which helps in identifying the system and calculating the designing parameters of the system. After identifying the system the KBFMS requests the user to provide declarative data to conduct detailed planning and design of the machining centres, layouts and MHSs. Three known types of layouts and two MHSs are considered. In this paper an AGV-robot based MHS is considered. A maximum twenty design factors can be considered for a MHS, but only three are discussed in this paper. Even with a limited case example, the KBFMS has shown that it is a valid method for the planning and designing of FMS.

Appendix A

$$D_m = D_m + D_m*(S_{all}/100) \tag{1}$$
$$P_t = N/N_s \tag{2}$$
$$T_{all} = T*(1+P_{all}/100) \tag{3}$$
$$M_t = [(T_{all}*D_m)/60]+[\{S_t/(D_m*60)\}] \tag{4}$$
$$T_m = T_{oc}/S_{av} \tag{5}$$
$$M_m = T_{op}/S_{av} \tag{6}$$
$$M = T_m+M_m \tag{7}$$
$$A_d = TTW/d \tag{8}$$
$$TTW = \sum D*d \tag{9}$$
$$N_{agv} = Round(d/D_N+0.49) \tag{10}$$
$$C_{agv} = N_{agv}*C_{agvl} \tag{11}$$
$$E_{mhs} =[\{(A_d/V_{agv})*T_f/(A_d/V_{agv}+A_e/V_{agv})\}]*100 \tag{12}$$

Appendix B

Results

Average operation time on cylindrical parts = 657 hours
Average operation hours on prismatic parts = 773 hours
] Simultaneous operations are considered
Required number of turning machining centres = 3
Required number of milling machining centres = 4

Table 1: Comparison first assumption

Design Factors	Single row layout		Double row layout		Loop type layout	
	Case 1	Case 2	Case 1	Case 2	Case 1	Case 2
1	10	12	7	9	4	6
2	100000	120000	70000	90000	40000	60000
3	59.16%	48.94%	56.90%	41.35%	95.00%	56.44%

Table 2: Comparison second assumption

Design Factors	Single row layout		Double row layout		Loop type layout	
	Case 1	Case 2	Case 1	Case 2	Case 1	Case 2
1	6	7	4	6	3	4
2	60000	70000	40000	60000	30000	40000
3	56.90%	47.39%	56.90%	41.23%	95.00%	56.52%

Table 3: Comparison (average result of both cases)

Design Factors	Single row layout		Double row layout		Loop type layout	
	1st assumption	2nd assumption	1st assumption	2nd assumption	1st assumption	2nd assumption
1	11	7	8	5	5	4
2	60000	70000	40000	60000	30000	40000
3	54.05%	52.14%	49.12%	49.06%	75.72%	75.76%

References

1. A. A. Aly, and M. Subramaniam, Design of an FMS decision support system, *Int. J. Prod. Res.*, Vol. 31, No. 10, (1993), 2257-2273.
2. J. M. Millichamp, and A. F. A. Wahab, An expert system for FMS design, *Simulation* 48:5, (1987), 201-208.
3. M. P. Groover, Automation, production systems, and computer integrated manufacturing, Prentice-Hall International Editions, (1987).
4. P. G. Ranky, Computer integrated manufacturing, Prentice-Hall International, (1986).

5. H. A. El Maraghy, and T. Ravi, Modern tools for the design modelling and evaluation of flexible manufacturing systems, *Robotics & Computer-Integrated Manufacturing*, Vol. 9, No. 4/5, (1992), 335-340.
6. J. M. Millichamp, Kwon and A. F. A. Wahab, FMS designer: An expert system for flexible manufacturing system design, *Int. J. Prod. Res.*, Vol. 28, No. 11, (1990), 2013-2024.
7. D. R. Sule, Manufacturing Facilities, Location, Planning, and Design, PWS-KENT Publishing Company, (1988).
8. M. Momin, M. K. Khan, and A. K. Kochhar, A knowledge-based approach to reactive scheduling in cellular manufacturing, *Advanced Manufacturing Processes, Systems, and technologies (AMPST'96)*, Mechanical Engineering Publications Limited, (1996), 201-210.

Production Scheduling Systems

Minimizing Mean-Lateness and Numbers of Tardy Jobs in a Job-shop

Sumiharni Batubara, Docki Saraswati, Ihwan Karnoto

Industrial Engineering Department,

Trisakti University, Jakarta, Indonesia

Abstract. Demand's variety on Printed Continuous Form (PCF), such as size, color and number of plies have influence the completion time for producing Printed Continuous Form. First, we calculate the set-up time and test-run time. Next, we determine the time distribution for processing time of PCF order and failure time of unit folder by using SIMAN. Analysis of the results gives that EDD is the best priority rule with mean-lateness 18.2 hours and 4 numbers tardy jobs.

Keywords: scheduling, priority rules, distribution, simulation.

1. Introduction

Scheduling theory has been widely used to model nearly every aspect of plant activities to flexible manufacturing systems. Indeed, not only does one find analytic results, but also one is impressed with a large number of computer simulation packages addressing scheduling problems. Throughout most of the history of scheduling, various independence conditions have been made.

A selected manufacturing plant is designed with five machine lines in Printing Department for producing Printed Continuous Forms (PCF). Each line consist of unwind unit, offset printing unit, processing unit, folding unit and finally, finished goods, which is PCF. The various demand for PCF, characterized by their size, color and number of plies. The purpose of this paper is to demonstrate, how statistical methodology embedded in scheduling may be applied conjunction with SIMAN to improve customer satisfaction by delivering on time.

The completion time for producing PCF, not only depends on set-up time and test-run time, but also depends on process time and failure time, which are uncertain. There are basically two approaches in order to obtain the schedule for producing PCF in Printing Department. Firstly, calculate the standard set-up time and standard run-time, using statistical basis methodology. Secondly, estimate the process-time following with failure time in folding unit, using SIMAN. Finally, model is constructed using SIMAN simulation language to estimate the total completion time, regardless the set-up time. This simulation experiments also examine the effects associated with the order characteristics. The purpose schedule is obtained through the comparison between SPT, EDD and Slack.

This paper is organized in five sections. Literature reviews for approaching the study are shown in section 2. Data collection and data processing is discussed in section 3. Analysis of the results are provided in section 4. Concluding remarks are presented in section 5.

2. Literature Reviews

Productions management decisions are made in the different levels, scheduling problems belong to operational level. The purpose of scheduling is to optimize the use of resources so that the overall production objectives are met. In general scheduling involves the assignments of dates to specific jobs or operation steps through several processors. The problems arise when several jobs can be performed in the same time while the numbers of machines are limited.

Basically, there are two stages in scheduling[7] , first is called loading and the second stage is called dispatching. Loading is assigning a particular job to a particular work center during a period, while dispatching is sequencing the jobs to be worked on at a work center and releasing them after a particular time. Scheduling is prepared for a short period of time, which is designed to implement the Master Production Schedule.

Many scheduling techniques can be employed to schedule a job shop. The type of technique depends on the volume of orders[10] . A job shop is contracted to make to order custom product according to customer's design.

Scheduling techniques can generally be categorized as Forward Scheduling and Backward Scheduling. Forward Scheduling[9] , where jobs are manufactured to customer orders and delivery is requested as soon as possible. In Backward Scheduling, the last operation on the routing is scheduled first. The purpose of Backward Scheduling is to minimize the amount of WIP.

2.1 Notation and Terminology

Tij = is the processing time, time requires to complete a task i from job j

Fij = is the flow time, this is defined to be time that task i spends in the workshop.

Cj = completion time, the time at which processing of job j finished.

$$Cj = Rj + Fj$$

Where, Rj is the Ready time of job j
Fj is the Flow time of job j

---

[7] Krajewski, Lee and Larry P. Ritzman, (1995) Operations Management Strategy and Analysis, Addison Wesley Pub. Co. p. 726.

[10] Stevenson, William J. (1994) Production and Operations Management, Irwin. p. 734.

[9] Riggs, James L. Production System : Planning, Analysis and Control, 4th Ed., John Willey & Sons. p. 544.

dj = due date, the promised delivery date of job j

Lj = lateness, this is to simplify the difference between its completion time and its due date.

$Lj = Cj - dj < 0$, when it completes before its due date.

$Lj = Cj - dj > 0$, when it completes after its due date.

Tj = tardiness, a job is tardy when lateness positive

Sli = slack, the difference between due date of task i and processing time.

$Sli = di - ti$

2.2 N jobs, M machines Scheduling

There are 2 categories in Scheduling with m machines.

1. N jobs, M machines in parallel.
2. N jobs, M machines in series.

Priority decision rules that are used for preparing dispatch list of jobs in this study are: SPT,EDD and Slack.

SPT = shortest processing time, the rule selects the first job with the shortest processing time in the machine, in order to minimize the Mean Flow Time.

EDD = earliest due date, this rule sequences the jobs waiting at the facility based on their due date, in order to reduce the maximum lateness[1].

Slack = this rule selects first jobs with the smallest slack, in order to reduce number of tardy.

2.3 ABC Classifications

A typical manufacturer ranks every product or item, according to their importance or inventory-related cost, is to classify the products or items in three Classes. Class A items, at least 10 to 20 percent of the items fall in this class and about 60 to 80 percent of the total value of annual expenditure. Class B items, is composed of 20 to 30 percent items, that account for about 30 to 40 percent of the total expenditure. The rest is belongs to Class C.

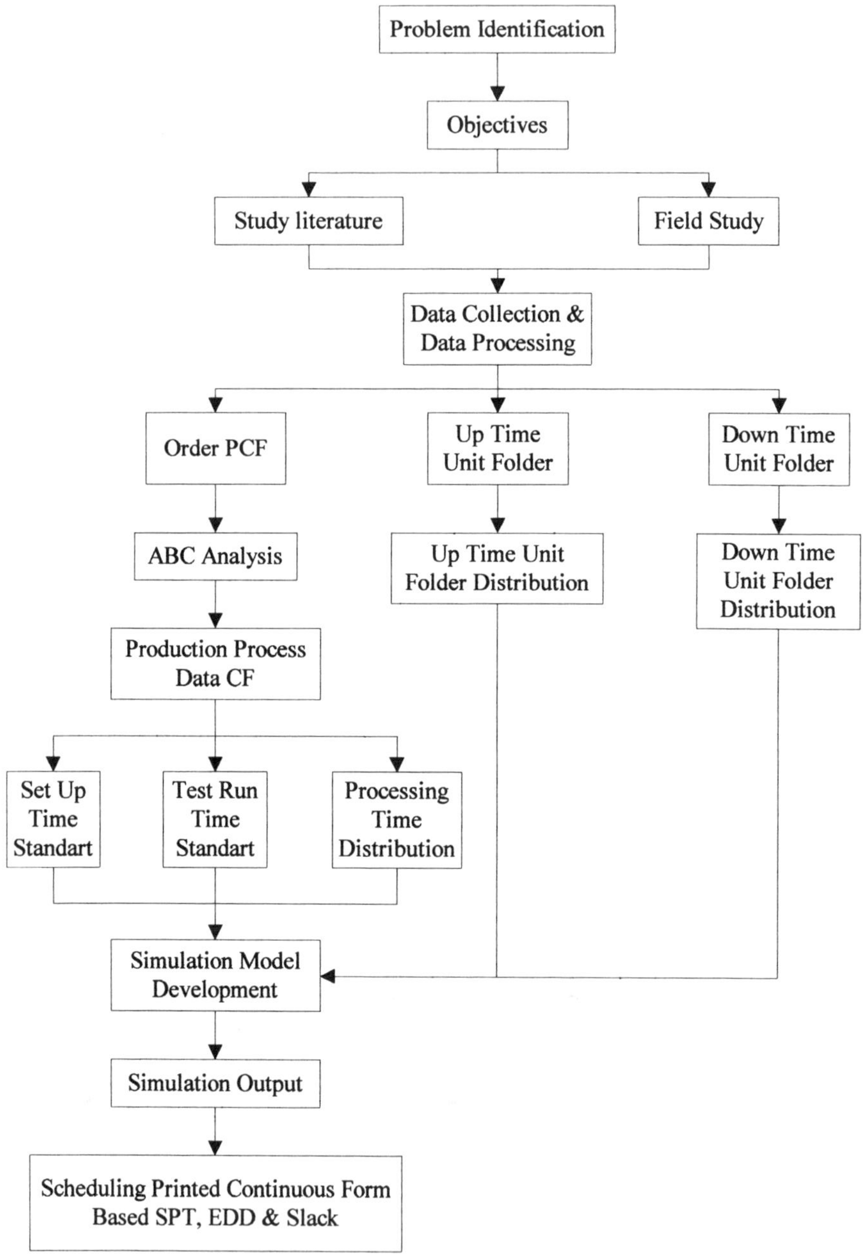

Figure 1. Methodology Research Diagram

2.4 Predetermined Time Standard

In order to obtain a good result, there are several factors should be considered, such as working conditions, operator skill level and method of measurements. Measurement is performed by observing the worker, who will be timed for several times for each element of a set of tasks under observations. In determining a work cycle time, members of data should be collected as a subset of a larger population. The purpose of sampling is to enable to estimate some unknown characteristic of population. Therefore, it is important to test the data collections with test of significance, such as, hypothesis testing of normal distribution, Goodness of Fit test. Finally, the standard time is calculated for each work element and for the complete set of tasks. This standard time will be used as the processing time in scheduling.

3. Research Methodology

Based on ABC analysis, PCF with the paper length of 11" and 12" belongs to Class A. The percentage of these PCF is 79.76% from the total order between January – June 1966. Furthermore, this study will focus on PCF with length 11' and 12', which are to be scheduled in Printing Department.

3.1 The sample and the data collected.

Data used to determine the standard set-up time and test-run time are obtained from observing the process for making PCF with 1, 2, 3 and 4 colors. The stages that had been done to determine the set-up and test-run time will be described as follows:

a. Normalize data test. First, we need to find out whether these data were approximated to normal. Therefore, we have to go through the process of hypothesis testing procedure.
b. Estimate the sample size. We wish to estimate with 95% of confidence level.
c. Determine the standard time to perform a task that the operators were doing under usual working condition and working at a normal pace. Westinghouse is an approach that is used to calculate the standard time after considering the factors for adjustment.

3.2 The result standard times.

In accordance to the above observation, we obtain the results, as follow:

- The set-up standard time are: 40.22 minutes (1 color), 65.17 minutes (2 colors), 90.10 minutes (3 colors) and 110.57 minutes (4 colors).
- The test-run time for every 2000 exemplar are: 28.44 minutes (1color,11"), 34.06 minutes (1 color, 12"), 43.78 minutes (2 colors 11"), 49.37 minutes (2 colors, 12"), 65.59 minutes (3 colors, 11"), 74.80 minutes (3 colors,12"), 96.80 minutes (4 colors,11") and 100.20 minutes (4 colors,12")

3.3 Data collected for PCF 1, 2, 3 and 4 colors with length 11" and 12".

In order to test the data collected, we examine with package programs *Statgraphics* ver.5 with Chi-square Goodness of Fit Test $\alpha = 0.05$. See table 1.

Table 1. Distribution Test Results

Number of colors	Length	N'	Distribution	Least Chi-Square	Mean/ Min.	Deviation/ Max.
1	11"	4.77	Normal	0.2665980	4.57033	0.259036
2	11"	4.17	Uniform	0.6574070	5.12	6.17
3	11"	5.13	Uniform	0.8279620	6.36	7.65
4	11"	2.25	Normal	0.1578710	8.27033	0.327124
1	12"	7.73	Uniform	0.3740360	4.55	5.83
2	12"	9.04	Uniform	0.2533480	6.15	7.74
3	12"	2.44	Normal	0.1755110	7.92033	0.320963
4	12"	2.07	Normal	0.0766727	8.59133	0.320622

3.4 Data collected on failure of Folding Unit .

Up-time and Down-time data was taken from the Maintenance machines. Up-time Folding unit is time interval time between two down-time in folding unit. From the up-time distribution in Folding unit , it is found that N' = 28.35 (enough data) and resulted Lognornal Distribution which has least Chi-square of 0.179279, and has d.o.f. = 2, with mean 9.3610611 and Standard Deviation = 0.112163496.

From the test of Down-time distribution in Folding Unit, it was found that N' = 19.75 (enough data) and resulted Lognormal Distribution with least Chi-square 0.3348750 , it has d.o.f. = 2 with mean = 4.3207955 and standard deviation = 0.115486969.

3.5 Simulation Model.

From the manufacturing and Printing of PCF, it was found that the parameters of processing and printing PCF on different length, number of colors, up-time and down-time in folding unit and variables were influence the printing process. Therefore, the simulation model can be developed. The simulation model has to describe the real system in detail, but simple. In this study, we use SIMAN as the simulation language.

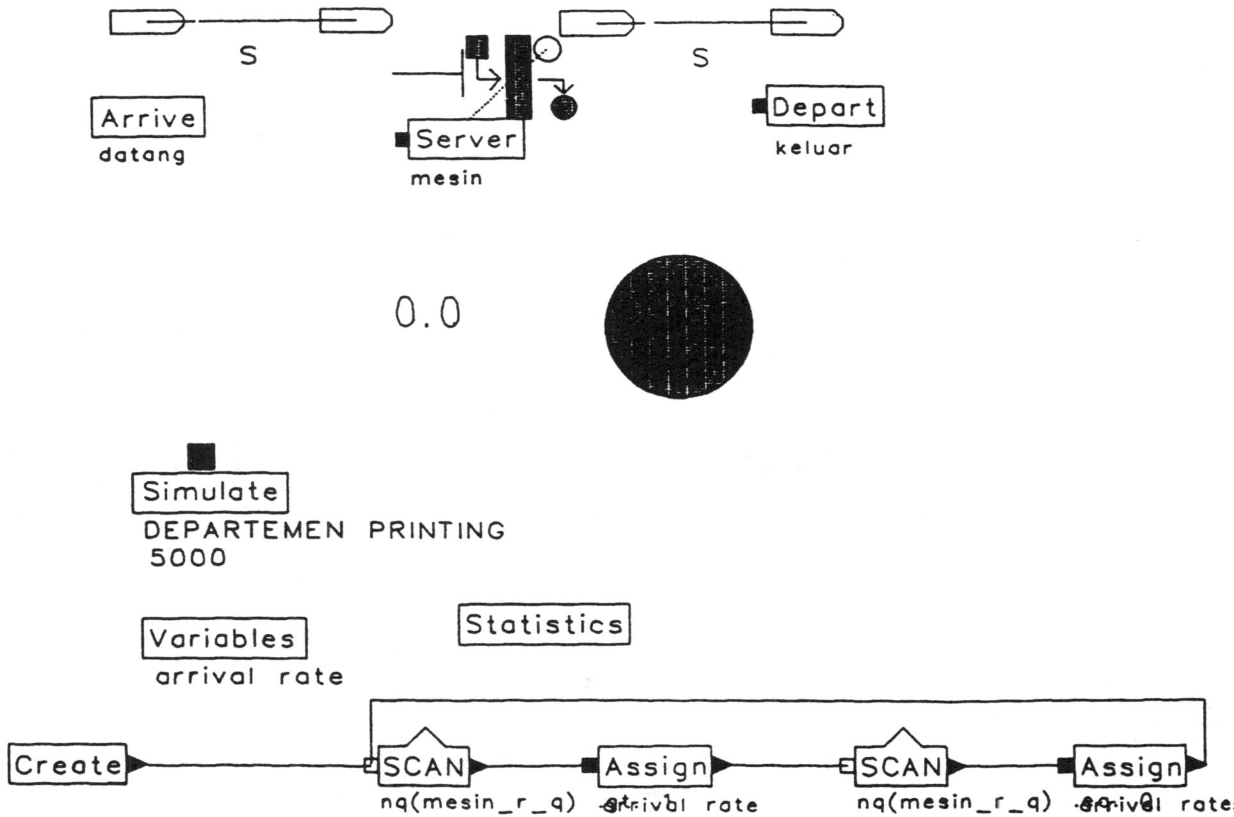

Figure 2. Simulation Diagram Model

3.6 Simulation Results.

Simulation program was replicated for 10 times to obtain the average values for each item. See Table 2 for replications.

Table 2. Replication Results by SIMAN

Length	Number of colors	Average value (minutes)
11"	1	5.16
11"	2	6.84
11"	3	7.25
11"	4	9.12
12"	1	6.75
12"	2	7.15
12"	3	8.97
12"	4	9.13

3.7 Scheduling

Based on the above process time for PCF 1,2,3 and 4 colors with length 11" and 12" plus standard set-up time and test-run time for each item, we schedule these item into line 1, 2, 3, 4 and 5. The propose schedules were based on the three priority rules: SPT, EDD and SLACK. The duration time for the schedules will be taken on 1 July to 6 July 1996.

4. Scheduling Analysis

The criteria that we implement to solve the above problems are based on Mean-Lateness, Mean-Tardiness, Maximize Tardy and Number of Tardy. See table 3.

Table 3. Scheduling Criteria Results

Rule	Mean-Lateness (hours)	Mean-Tardiness (hours)	Maximize Tardy (hours)	Number of Tardy
SPT	-17.86	4.09	39.76	14
EDD	-22.28	0.46	18.2	4
SLACK	-3.19	2.09	14.22	16

According to the objective, this study is to minimize lateness. Therefore, from the above comparison between three priority rules we analyst that EDD and SLACK will give maximize tardy relatively low compare with SPT.

In comparing EDD and SLACK, we obtain that SLACK has the least maximize tardy with 14.22 hours, but if we observe more detail, SLACK will have the most number of tardy, which is 16 orders, as well as with the value of Mean-Lateness and Mean Tardiness SLACK has higher values.

At the end of the scheduling analysis, we can select that EDD is the appropriate rule for this model of scheduling problem. The reasons are:

- Having lower mean-lateness, mean-tardiness and number of tardy compared with SLACK
- Having relatively higher maximize tardy compared with SLACK, but it still can be tolerate.

5. Concluding Remarks

In accordance to the observation during the study, we attempt to make some conclusions:

1) Simulation model is being used to simulate the up-time and down-time in folding unit, to determine its average values with 10 replications.
2) From scheduling analysis we know that the capacity in Printing Department in not enough to fulfill the customer's orders. Therefore, the company should anticipate this situation, at least with the information about the exact time their order will be completed.
3) Scheduling with SPT will give Mean-Lateness 39.76 hour, it happens since SPT ignores the due-date time.
4) Scheduling with SLACK will have the least Mean-Lateness, which is 14.22 hours, on the contrary this rule will give the number of tardy with 16 orders.
5) Finally, we select EDD as the appropriate rule to schedule PCF with Mean-Lateness 18.2 hours and the number of tardy is 4 orders.

References

1. Bedworth, David D. and James E. Baley, (1987) Integrated Production Control System: Management, Analysis, Design. 2 ed., John Wiley & Sons Inc.
2. Dilworth, James B., (1993) Production and Operations Management , Manufacturing and Services. 5th ed. MGH .
3. Enns S.T., An Integrated system for controlling shop loading and work flow. *Int.J. Prod. Res* 33 (1995) 2801-2820
4. Handfield, R.B., R.T. Pannesi, Antecedents of lead-time competitiveness in make-to-order manufacturing firm. *Int. J. Prod. Res.* 33 (1995) 511-537.
5. Hall, Nicholas J., C. Iskandarajah, A survey of Machine Scheduling Problems with Blocking and No-wait in Process. *Operations Research* 44 (1996) 510-525.
6. Keller G., B.Warrack, H.Bartel, (1994) Statistics for Management and Economics,3rd ed., Duxbury Press.
7. Krajewski, Lee and Larry P. Ritzman, (1995) Operations Management Strategy and Analysis, Addison Wesley Pub. Co.
8. Pegden, C. Dennis et.al.(1995) Introduction to Simulation Using SIMAN. 2nd Ed., MGH
9. Riggs, James L. Production System : Planning, Analysis and Control, 4th Ed., John Willey & Sons.
10. Stevenson, William J. (1994) Production and Operations Management, Irwin .
11. Sutalaksana, Iftikar et.al., (1989) Teknik Tata Cara Kerja, Departemen Teknik Industri-ITB

A Hybrid GA-SA Algorithm for Flowshop Scheduling Problems

Baikunth Nath

Gippsland School of Computing and Information Technology
Monash University, Churchill, Australia 3842
B.Nath@fcit.monash.edu.au

Abstract. *Flowshop scheduling for minimising makespan is one of the most well-known problems in the area of scheduling. In this paper, we examine various genetic operators such as crossover and mutation to design a genetic algorithm (GA) for the flowshop scheduling problem with the objective of minimising the makespan. By simulation, it is shown that the two-point crossover and the shift change mutation (two positions are randomly selected, a job at one position is removed and put at another position) are effective for this type of problem. The performance of GA is compared between various specifications, and with the search technique simulated annealing (SA). Simulation results show that the genetic algorithm did not perform as well as simulated annealing. In order to improve the performance of the genetic algorithm, we propose and examine a hybridisation of the genetic algorithm with simulated annealing. The improved performance of GA-SA hybridisation is demonstrated via simulation.*

1. Introduction

Flowshop scheduling for minimising makespan is one of the most well-known problems in the area of scheduling. Manual approaches to scheduling have evolved over many centuries, and they can be deep, subtle and highly varied across industries. In the 1950s and 1960s the rise of computers made it possible to represent the structure of the shop, activities, jobs and other constraints in some detail so that, given appropriate input data and simple despatch rules at decision points, the computer could extrapolate a given schedule into the future at relatively low cost. Various approaches to this problem have been proposed since Johnson's pioneer work [6]. While very small problems can be solved exactly using mathematical tools such as integer programming and dynamic programming, it is difficult to find the optimal solution of a flowshop scheduling problem involving many jobs and machines. Various heuristic methods began to be developed for approximately solving these large problems. Recently, several heuristic approaches based on iterative improvement procedures were applied to the flowshop scheduling because of the improved computational power of computers.

Scheduling heuristics employing simulated annealing are described in Kirkpatrick et al. [4], Osman and Potts [10] and Ishibuchi et al. [4]. Recently many authors applied genetic algorithm to combinatorial optimization problems [5] [9] and scheduling problems [10] [11] [12]. Some empirical studies [8] [9] showed that the ability of genetic algorithms to find near optimal solutions was a little inferior to other search algorithms.

In this paper, we apply genetic algorithm to flowshop scheduling problems and examine a hybridisations of genetic algorithm with simulated annealing. First we examine various genetic operators to design the genetic algorithm for minimising the makespan. By simulation, we show that the two-point crossover and the shift change mutation are effective for this type of problems. Next we compare the genetic algorithm with simulated annealing by applying to randomly generated test problems. Finally, we examine a hybridisation of genetic algorithms with simulated annealing to improve the performance of the genetic algorithm. Computer simulations show the good performance of our modified hybrid GA-SA algorithm.

2. Genetic Operators

In this section we examine various genetic operators such as crossover and mutation for the flowshop scheduling problem before comparing the performance of the genetic algorithms.

General assumptions of the flowshop scheduling problem can be described as follows. Jobs are processed on multiple stages sequentially. At each stage there is one machine available continuously. A job is processed on one machine at a time without preemption, and a machine processes only one job at a time. We assume that *n* jobs are processed in the same order on *m* machines. We use makespan (completion time of the last job) as a criteria for the flowshop problem.

2.1 Coding

In genetic algorithms, each solution of an optimisation problem is usually encoded as a bit string. That is, a binary representation is used for the coding of each solution. In this paper, however, a sequence of jobs is represented as a string. For example, the string "ABCD" represents the job sequence where "Job A" is processed first, then "Job B" is processed, and so on. If the string "ACAD" is generated by genetic operators such as crossover and/or mutation, this string is not a feasible solution to the flowshop scheduling problem because the allele "A" appears twice in the string and the allele "B" does not appear. Therefore the string in the flowshop scheduling problem should be a permutation of given jobs. We denote the sequence of *n* jobs by an n-dimensional vector $x = (x_1, x_2, \quad, x_n)$ where x_j represents the jth processing job.

2.2 Selection

Selection is a process to select two parents for generating a new string, called child. Let N be the number of solutions in each population in genetic algorithm, i.e., N is the population size. Denote N population solutions in the tth generation by $\Phi_t = \{x_t^1, x_t^2, \ldots, x_t^N\}$. Each solution x_t^i is selected as a parent string according to some selection probability $P_s(x_t^i)$. We used the following selection probability in our simulation:

$$P_s(x_t^i) = [f_W(\Phi_t) - f(x_t^i)]^2 \,/\, \sum_{x_t^i \in \Phi_t} [f_W(\Phi_t) - f(x_t^i)]^2 \qquad (1)$$

where $f(x_t^i)$ is the objective function to be minimised in the flowshop scheduling problem (i.e., makespan) and $f_W(\Phi_t)$ is the worst value of $f(\cdot)$ in the tth generation:

$$f_W(\Phi_t) = \max\{f(x_t^i) \mid x_t^i \in \Phi_t\}.$$

The selection probability in (1) represents the selection pressure.

2.3 Crossover

Crossover is an operation to generate a new string (i.e., child) from two parent strings. Since our flowshop scheduling problem is a sequencing problem of n jobs, various crossover operators proposed in the literature [1, 2] are applicable. In this paper we examine 6 crossover operators. The following three crossover operators shown in Figure 1 turned out to be effective, in our simulation, for the flowshop scheduling problem.

1. *One-point crossover*: This crossover is illustrated in Fig. 1(a). A point is randomly selected for dividing one parent. The set of jobs is inherited from one parent to the child, and the other jobs are placed in the order of their appearance in the other parent.

2. *Two-point crossover (i)*: This crossover is illustrated in Fig. 1(b). Two points are randomly selected for dividing one parent. The jobs outside the selected points are always inherited from one parent to the child, and the other jobs are placed in the same manner, as in one-point crossover, in order of their appearance in the other parent.

3. *Two-point crossover (ii)*: Two points are randomly selected for dividing one parent. The jobs between the two selected points are always inherited from one parent to the child, and the other jobs are placed in the same manner in order of their appearance in the other parent.

4. *Two-point crossover (iii)*: This crossover is a mixture of the above two versions of the two points crossover. Each version is applied to each pair of selected parents with equal probability.

5. *Position based crossover (i)*: This crossover is illustrated in Fig. 1 (c). First a random number of positions are determined, then the required number of positions are randomly selected and marked by ▼. The jobs at randomly selected positions are inherited from one parent to the child, and the other jobs are placed in the order of their appearance in the other parent.

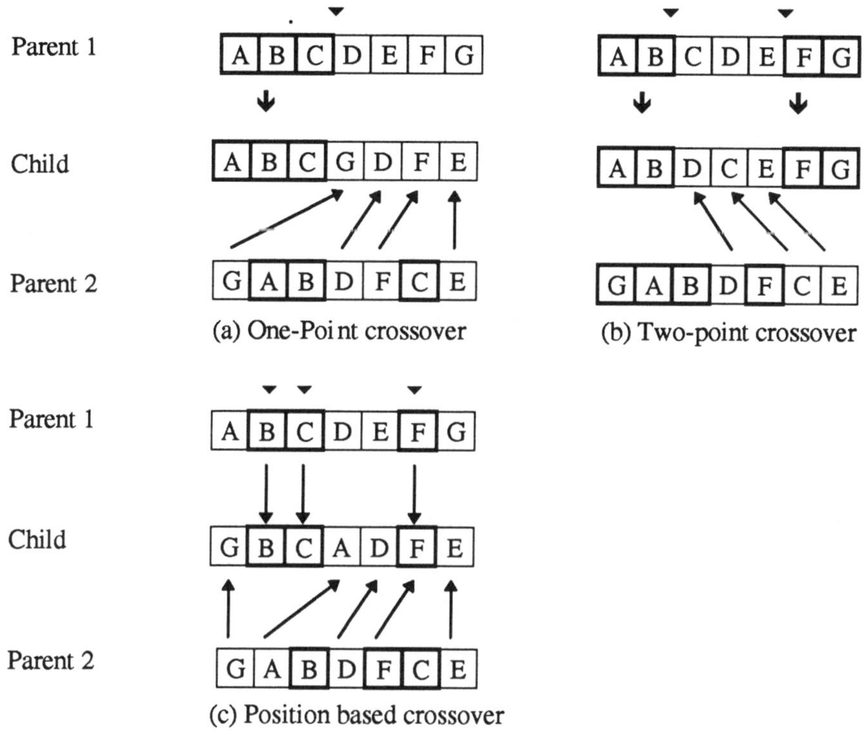

Fig. 1. Crossover Operators

6. *Position based crossover (ii)*: This crossover is almost the same as the position based crossover (i) except for the choice of positions. In this case, each position is independently marked by ▾ with the probability of 0.5. The jobs at selected positions are inherited from one parent to the child, and the other jobs are placed in the order of their appearance in the other parent.

2.4 Mutation

Mutation is an operation to change the order of jobs in each string generated by a crossover operator. Mutation operation can be viewed as a transition from a current solution to its neighbourhood solution in local search algorithm. We consider the following three mutation operators for the flowshop scheduling problem.

1. *Adjacent two-job change*: Adjacent two jobs are randomly selected and interchanged as shown in Fig. 2(a).

2. *Random two-job change*: Two jobs are randomly selected and interchanged as shown in Fig. 2(b). This mutation includes the adjacent two-job change as a special case.

3. *Shift change*: Two positions are randomly selected. A job at one position is removed and inserted at the other position as shown in Fig. 2(c). There are a few

variations to this situation, for example, whether to remove the first job or the second job, etc. This mutation includes the adjacent two-job change as a special case.

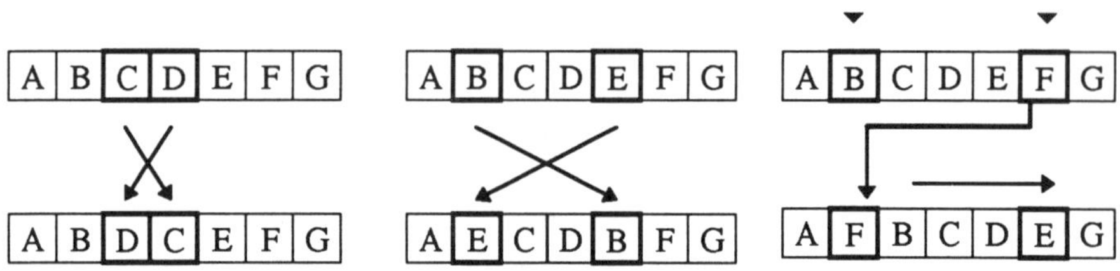

(a) Adjacent two job change (b) Random two job change (c) Shift change

Fig. 2. Mutation Operators

2.5 Genetic Algorithms

By combining the above genetic operators, we can design various genetic algorithms. The outline of the genetic algorithm, for completeness, can be written as follows:

Step1. *Initialization:* Randomly generate an initial population Φ_1 of N strings.

Step2. *Selection:* Select N pairs of strings from a current population according to the selection probability.

Step3. *Crossover:* Apply one of the crossover operators to each of the selected pairs to generate N solutions with the crossover probability P_c . If the crossover operator is not applied to one of the pairs, the selected solution remains as a new string.

Step4. *Mutation:* Apply one of the above mutation operators to each of the generated N strings with mutation probability P_m .

Step5. *Best strategy:* Randomly remove one string from the current population and add the best string in previous population to the current population .

Step6. *Termination:* If a specified stopping rule is satisfied, stop this algorithm. Otherwise return to step 2.

2.6 Comparison between Various Specifications of Genetic Algorithms

To compare various specifications, we randomly generated 50 flowshop scheduling problems with 20 jobs and 10 machines. The processing time of each job at each machine was randomly specified as an integer in the interval [1, 99]. For the flowshop scheduling problem with an objective of minimising the makespan, appropriate parameter values were employed from the above six crossover operators and three mutation operators with the following values of population size N, the crossover probability P_c and the mutation probability P_m :

$N = 5, 10, 20, 30, 50$

$P_c = 0.5, 0.6, 0.7, 0.8, 0.9, 1.0$

$P_m = 0.5, 0.6, 0.7, 0.8, 0.9, 1.0$

By simulation, using genetic algorithms with various parameter specifications, we observed that the following specifications worked best.

Population sizc, $N = 10$

Crossover probability, $P_c = 1.0$

Two-point crossover (i)

Mutation probability, $P_m = 1.0$

Shift change mutation.

As a stopping rule in simulation, we used the total number of evaluations of the objective function in order to compare various parameter specifications under the same computation load. The total number of evaluations of the objective function was specified as 10,000 or 40,000. The total number of generations was specified as to be inversely proportional to the population size N. That is, for the total number of evaluations of 10,000, the total number of generations was specified as $10{,}000 / N$.

Simulation results of various crossover operators are shown in Table 1. Table 1 is the result of a single simulation for each of 50 problems by each crossover operator. Average makespans shown in Table 1 were normalised using the results by the genetic algorithm with the above parameter specifications and the stopping rule of 40,000 evaluations. Further, we iterated the simulation 10 times on the 50 problems for each crossover operator. The average performance of three representative crossover operators, one-point crossover, two-point crossover (i) and position based crossover (i), from the 10 trials and their standard deviation are shown in Table 2. Each value in Table 2 was normalised in the same manner as in Table 1. From Table 2, we observe the high performance of the two-point crossover (i).

Table 1. Comparison of various crossover operators

Crossover operator	Number of evaluations	
	10,000	40,000
One-point crossover	100.82	100.21
Two-point crossover (i)	100.44[+]	100.06[+]
Two-point crossover (ii)	100.86	100.24
Two-point crossover (iii)	100.67	100.13
Position based crossover (i)	101.11	100.41
Position base crossover (ii)	100.80	100.32

[+] the best result

Table 2. Comparison of three crossover operators

Crossover operator	40,000 evaluations Average	Std. deviation
One-point crossover	100.09	0.6738
Two-point crossover (i)	100.01	0.7043
Position based crossover (i)	100.21	0.7218

Tables 3 and 4 show simulation results with various mutation operators. From these results, we observe the high performance of the shift change mutation.

Table 3. Comparison of various mutation operators

Crossover operator	Number of evaluations 10,000	40,000
Adjacent two-job change	106.31	105.27
Random two-job change	101.55	102.52
Shift change	100.45*	100.04*

* the best result

Table 4. Comparison of three mutation operators

Crossover operator	40,000 evaluations Average	Std. deviation
Adjacent two-job change	105.39	2.2734
Random two-job change	102.34	0.7643
Shift change	100.02	0.7052

3. Simulated Annealing Algorithm

Simulated annealing is a heuristic which attempts to overcome the disadvantage inherent in descent methods that no further search for a global minimum is performed after a local minimum is detected (unless the method is repeated using a different starting sequence). A descent method repeatedly attempts to construct an improved sequence (a set of neighbourhoods) of jobs from a current sequence. We used the algorithm proposed by Osman and Potts [10]. First an initial sequence **x** is randomly generated. The transition from a current sequence **x** to a neighbourhood sequence **y** is accepted by the probability

$$P(x \to y) = \min\left\{1, \exp(\frac{f(y) - f(x)}{T_k})\right\}, \quad k = 1, 2, \ldots, K \qquad (2)$$

where T_k is a control parameter called temperature and K is the total number of iterations. In performing simulation, the value of the control parameter is gradually decreased from a large initial value to a small value. In our simulation, we specified the pattern proposed in [8].

$$T_{k+1} = T_k \,/\, (1 + \beta T_k)$$

where β is positive constant.

We determined the positive constant β and the initial values of T_k according [8]. The current sequence **x** is replaced with the candidate sequence **y** with the acceptance probability in (2). This search procedure is iterated until a specified stopping rule (the number of iterations $K = 40{,}000$) is satisfied.

3.1 Comparison between Simulated Annealing and Genetic Algorithms

In this section, we compare the genetic algorithm with simulated annealing. In simulations, the neighbourhood structure based on sift change mutation, and the same stopping rule as in genetic algorithm was used.

We applied the algorithm to randomly generated 50 test problems with 20 jobs and 10 machines. Simulation results are shown in Tables 5 and 6 for 20-job problems and 50-job problems, respectively. In these, tables, average makespan results obtained by the two algorithms are normalised using the results by the simulated annealing with 40,000evaluations. From these tables, we observe that the genetic algorithm is a bit inferior to simulated annealing algorithm.

Table 5. Comparison of algorithms for 20-job problems

Evaluations	10,000	40,000
GA	101.7	101.2
SA	100.9*	100.2*

* the best result

Table 6. Comparison of algorithms for 50-job problems

Evaluations	10,000	40,000
GA	102.6	101.7
SA	101.3*	100.5*

* the best result

4. A Hybrid GA-SA Algorithm

In this section, we propose and examine a hybrid GA-SA, genetic simulated annealing, algorithm to improve the performance of the genetic algorithm. In simulation, we used the following genetic simulated annealing algorithm.

Step 1 *Initialization*

Step 2 *Simulated annealing and termination test:* Apply the simulated annealing algorithm with the constant temperature to each of the N solutions in the current population. If a prescribed stopping rule is satisfied during the search, stop the algorithm. If the simulated annealing search is completed for all the N solutions, let the set of the obtained N solutions be the current population, go to Step 3.

Step 3 *Selection*

Step 4 *Crossover*

Step 5 *Mutation*

Step 6 *Best strategy*

Step 7 *Termination:* If a specified stopping rule is satisfied, stop this algorithm. Otherwise return to Step 2.

In Step 2, the constant temperature is used to avoid extreme deterioration of the current solution during the initial state of annealing with high temperatures. The simulated annealing algorithm was iterated 500 times for each solution of the current population. In order to improve performance of the GA-SA algorithm, we modify the simulated annealing algorithm by randomly selecting s neighbourhood solutions of the current solution and letting the best solution be the candidate solution for the next transition in simulated annealing. Such a modification of simulated annealing is suggested in Ishibuchi [3].

4.1 Simulation results

In a similar manner to the last sections, we applied the genetic simulation annealing to 50 flow shop scheduling test problems with 20 jobs and 10 machines. The simulation results are shown in Table 7. The constant temperature in the GA-SA was specified as $T = 2$. In the genetic simulated annealing search, s (s = 1, 2, 4, 8, 10, 20) candidate solutions were selected in the neighbourhood of the current solution, and the transition to the best solution was examined with the acceptance probability in (2). When s = 1, the simulated annealing in the genetic simulated annealing is the same as the standard algorithm.

Table 7. Performance of the genetic simulated annealing algorithm for 20-job problem

Search algorithms	Number of evaluations 10,000	40,000
GA	101.63	101.21
GA-SA (s=1)	101.33	100.32
GA-SA (s=2)	101.23*	100.24
GA-SA (s=4)	101.27	100.27
GA-SA (s=8)	101.26	100.14*
GA-SA (s=10)	101.28	100.16
GA-SA (s=20)	101.54	100.21

* the best result

Table 7 shows that the introduction of s into the simulated annealing improved the performance of the genetic simulated annealing algorithm.

5. Conclusion

In this paper, we examined the performance of genetic algorithms to identify specific operators and parameters for the flow shop scheduling problem. By simulation, we showed that the genetic algorithm is a bit inferior to simulated annealing. In order to improve the performance of genetic algorithms, we examined a hybrid algorithm: genetic simulated annealing. Again by simulation, we showed that the hybrid GA-SA algorithm produces better results.

References

1. L. Davis, *Handbook of Genetic Algorithms*. Van Nostrand Reinhold, New York, 1991.

2. D. E. Goldberg, *Genetic Algorithms in search, optimisation, and machine learning*. Addition-Wesley, Reading, MA, 1989.

3. H. Ishibuchi, S. Misaki and H. Tanaka, Modified simulated annealing algorithms for flow shop sequencing problem, *EJOR*, **81** (1995) 388-398.

4. H. Ishibuchi, R. Tamura and H. Tanaka, Flow shop scheduling by simulated annealing, *Transactions of the Institute of Systems, Control and Information Engineers*, **4** (1991) 111-117.

5. H. Ishibuchi, N. Yamamoto, T. Murata and H. Tanaka, Genetic Algorithms and neighbourhood search algorithms for fuzzy flowshop scheduling problems. *Fuzzy Sets and Systems*, **67** (1994) 81-100.

6. S. M. Johnson, Optimal two- and three-stage production schedules with setup times included, *Naval Res Logistics Q*. **1** (1954) 61-68.

7. S. Kirkpatrick, C. D. Gelatt Jr. and M. P. Vecchi, Optimisation by simulated annealing, *Science* **220** (1983) 671-680.

8. M. Lundy and A. Mees, *Convergence of an annealing algorithm*, Math. Prog. **34** (1986) 111-124.

9. B. Manderick and P. Spiessens, How to select genetic operators for combinatorial optimisation problems by analysing their fitness landscape, *Computational Intelligence Imitating Life*, IEEE Press, New York (1994) 170-181.

10. I. H. Osman and C. N. Potts, Simulated Annealing for permutation flow shop scheduling, *OMEGA*, **17** (1989) 551-557.

11. T. Starkweather, S. McDaniel, K. Mathias, D. Whitley and C. Whitley, A comparison of genetic sequencing operators, *Proc of the Fourth ICGA*, (1991) 69-76.

12. G. Syswerda, Scheduling optimisation using genetic algorithms, In *Handbook of Genetic Algorithms* (Edited by L Davis), Van Nostrand Reinhold, New York, 1991.

SOLVING FMS LOADING PROBLEM WITH GENETIC ALGORITHM : A STRATEGY FOR SOLUTION GENERATION AND RECOMBINATION

Kripa Shanker* (ks@iitk.ernet.in) and **Neeraj Kumar** (neerk@iitk.ernet.in)

Department of Industrial and Management Engineering
Indian Institute of Technology, Kanpur - 208 016 (U.P.), INDIA.
(*corresponding author)

Abstract. This paper attempts to develop a strategy for using genetic algorithm to solve the part selection and machine loading problem in a flexible manufacturing system (FMS). The strategy is guided by integer programming (IP) formulation of the problem. The solution chromosome consists of part type selection and operation assignment genes. For generation of solutions, constraints are segregated into direct and indirect ones. Direct constraints (large in number) are satisfied by assigning context-dependent values to decision variables. Indirect constraints require the penalty functions. A two-step crossover (part swap followed by route swap) is suggested which retains the solution feasibility across generations. The mutation is allowed only for operation assignment genes of selected part types. Implementation of methodology for small FMSs indicates that with proper penalty scheme and GA parameters, the proposed problem-specific GA can provide optimum solutions.

1 INTRODUCTION

The short to medium term production planning in flexible manufacturing systems (FMSs) is better known as *FMS part selection and loading problem* which involves chiefly the selection of part types from among the total demand, and the assignment of tools and operations (of selected part types) to machines.

The selection and assignment is made so as to optimize one or more of the following objectives – throughput, workload balancing, machining costs, material handling etc. The major constraints that guide the optimization are – assignment of operations incumbent upon part type selection and appropriate tool assignment, tool magazine capacities on machines, machining time availability, tool availability, and tool lives.

Mostly, this problem has been formulated as an integer programming (IP) or mixed integer programming (MIP) problem and solved by MIP software or heuristics[1,2]. While the IP/MIP formulation is capable of incorporating all aspects of the problem, it becomes increasingly difficult and time-consuming to solve as the problem is scaled up. Hence, for efficiently solving this problem, a methodology is required which utilizes the IP formulation and is able to solve large size problems within reasonable time and effort.

The genetic algorithm (GA), an evolution-based non-conventional search method, holds much hope for the FMS loading problem. Gaining popularity as a robust search method, the genetic algorithms have witnessed a rapid growth over the last decade[3,4]. Besides extensive application in unconstrained function optimization and engineering design optimization, they have also been used to solve many intractable operations research and production management problems like the set covering problem, set partitioning problem, quadratic assignment problem, transportation problem, travelling salesperson problem, flow shop scheduling, job shop scheduling, and manufacturing cell design.

The aim of this work is to develop a methodology for solving the FMS part selection and loading problem with the help of genetic algorithm.

The next section describes a typical IP model of the FMS part selection and loading problem. The methodology to use GA for this problem is developed in section 3, elaborating upon the adaptation of GA mechanisms. Section 4 summarizes the current status of methodology implementation and section 5 concludes the paper.

2 FMS PART SELECTION And LOADING PROBLEM

While making decisions regarding the selection of part types and loading of machines, many factors have to be kept in mind:

Production Demand (part type spectrum, operations required, production volumes, due dates, part priorities)

Production Resources (number of machines and tool types, tool-machine compatibilities, their capabilities and flexibilities vis-à-vis various operations)

Production Process (alternative process plans i.e. routes for operations, batching of parts, grouping of machines)

Auxiliary Resources/Activities (tool magazines on machines, local/central buffers, material handling devices such as AGVs and robots for transportation, loading/unloading, machine setup etc., pallets and fixtures)

As is evident from the plenty of research on FMS loading problem, most of these aspects can be quantified and then explicitly considered during the modeling. However, it seems quite difficult, even unnecessary to some extent, to include all aspects in a single model.

The four entities incorporated in the formulation herein are described below. The values in the last column indicate the ranges which will be used in later sections to generate the experimental FMS configurations.

Part type	:	p	=	1, ..., P	(P = number of part types)	6 - 12
Operation	:	o	=	1, ..., O_p	(O_p = number of operations in part type p)	4 - 8
Machine	:	m	=	1, ..., M	(M = number of machines)	3 - 7
Tool type	:	t	=	1, ..., T	(T = number of tool types)	10 - 20

The related problem parameters are listed below along with the ranges used later for problem generation:

BS_p	=	Batch size for part type p	1 - 2
MG_m	=	Tool magazine capacity on machine m	6 - 20
CP_t	=	Number of copies available of tool type t	1 - 3
SL_t	=	Slots required by tool type t	1 - 5
LF_t	=	Life of tool type t	-
$MC(t)$	=	Set of machines compatible for tool type t	-
$TT(m)$	=	Set of tool types compatible for machine m	-
$AR(p, o)$	=	Set of alternative routes (tool-machine combinations) for operation (p, o)	-
	=	$\{a \mid a \equiv (m, t); (m, t)$ is an alternative route$\}$	-
PT_{poa}	=	Processing time for operation o of part type p on alternative route a	5 - 30 min.
A_{po}	=	Total number of alternative routes for operation (p, o) = $\lvert AR(p, o)\rvert$	1 - 3
$OP(m, t)$	=	Set of operations which can be performed by tool-machine combination (m, t)	-
H	=	Scheduling horizon	480 min.

The decision making is reflected in the following binary variables:

$$x_p = \begin{cases} 1 & \text{if part type } p \text{ is selected} \\ 0 & \text{otherwise} \end{cases}$$

$$x_{po} = \begin{cases} a & (= 1, \ldots, \mathrm{A}_{po}) \text{ if operation } (p, o) \text{ is assigned to route } a \\ 0 & \text{otherwise} \end{cases}$$

$$y_{mt} = \begin{cases} 1 & \text{if tool type t is allocated to machine m} \\ 0 & \text{otherwise} \end{cases}$$

The values chosen for the above variables have to satisfy the following constraints:

(a) Uniqueness of an operation assignment and its dependence on selection of corresponding part type (a1) $x_{po} - \text{BIG}.x_p \le 0$ $p = 1,\dots,P$; $o = 1,\dots,O_p$; BIG = a big positive integer.

(a2) $x_{po} - x_p \ge 0$ $p = 1,\dots,P$; $o = 1,\dots,O_p$;

When $x_p = 0$, constraint (*a1*) ensures that $x_{po} \le 0$. Otherwise, when $x_p = 1$, then constraint (*a2*) forces $x_{po} \ge 1$.

(b) Dependence of an operation assignment on appropriate tool-machine assignment

$x_{po} - y_{mt} \le 0$ $p = 1,\dots,P$; $o = 1,\dots,O_p$; (*m*, *t*) corresponds to the alternative route specified by value of x_{po}

(c) Tool magazine capacities on machines

$$\sum_{t \in TT(m)} SL_t . y_{mt} \le MG_m \qquad m = 1,\dots,M$$

(d) Machine capacities

$$\sum_{t \in TT(m)} \sum_{(p,o) \in OP(m,t)} BS_p . x_{po} . PT_{poa} \le H \qquad m = 1,\dots,M \ ; \ a \equiv (m,t)$$

(e) Tool type availability

$$\sum_{m \in MC(t)} y_{mt} \le CP_t \quad t = 1,\dots,T$$

(f) Tool lives

$$\sum_{(p,o) \in OP(m,t)} BS_p . x_{po} . PT_{poa} \le LF_t \qquad m = 1,\dots,M \ ; \ t = 1,\dots,T \ ; \ a \equiv (m,t)$$

(g) Binary nature of all decision variables

$x_{po} = 0$ or 1 $p = 1,\dots,P$; $o = 1,\dots,O_p$; $(m,t) \in AR(p,o)$

$y_{mt} = 0$ or 1 $m = 1,\dots,M$; $t \in TT(m)$

According to the objective being considered, some more variables and constraints may have to be added. For example, if the objective is a function of machine workloads, then the workloads are appended to the list of decision variables and the workload definitions become the additional constraints.

3 METHODOLOGY for APPLYING GA

This section describes how GA can be used to solve the FMS part selection and loading problem, particularly the IP model given in the previous section. Specifically, the solution coding, solution construction, and recombination (crossover/mutation) issues are discussed in detail.

3.1 REPRESENTATION (CODING) of PROBLEM SOLUTION

In order to represent the problem solution in a chromosome, the following genes are required:

- one gene each for x_p variables; number of genes, $G_p = P$.

- one gene each for x_{po} variables; number of genes, $G_{po} = \sum_{p=1}^{P} O_p \approx P.\overline{O}$, where $\overline{O}$ is average number of operations in a part type.

For a problem (*P1*) with 2 part types, 2 operations in each part type, 2 alternative routes for each operation, 2 machines and 2 tool types (both being compatible for both machines), a typical chromosome may look like:

$$CH = \overbrace{10}^{x_p}\ \overbrace{0010}^{x_{po}}$$

The total number of genes required, in other words, the *chromosome length* is:

$$CL \approx P + P.\overline{O}$$

For a 20 part - 10 machine - 20 tool problem (*P2*) with, on an average, 5 operations per part, 2 routes per operation and 5 tools per machine, $CL \approx 120$.

It has been proved that implicit parallelism i.e. the number of schema represented in a chromosome is maximized by the binary alphabet[3]. However, Antonisse[5], based on his new interpretation of schema notation, argues that an *immediately expressive representation* of the problem solution may be better than the forced binary coding. Therefore, even with non-binary alphabet (e.g. x_{po} genes), one may expect a GA performance similar to or better than that with binary alphabet.

3.2 SOLUTION GENERATION by CONSTRAINT SEGREGATION

Even for a moderate-sized FMS loading problem, the variables number around a hundred and the constraints in many hundreds. Due to the large number and interactive nature of variables and constraints, if a solution is generated randomly, the chances are that it would be infeasible. Generating a fully feasible non-trivial solution with a view to optimize the objective function for the FMS loading problem requires extensive heuristic methodology and is a subject in its own right. In short, generating a feasible solution would be too time-consuming and generating a population even more so. And this feasibility will not be guaranteed after the recombination. Moreover, by virtue of containing valuable building blocks, infeasible solutions may actually be helpful for GA. So, despite a strong instinct to use only feasible solutions, it is not appropriate for solving the problem by GA.

On the other hand, if a solution is generated randomly, the number of constraints violated would be very high. Applying penalty functions for all such constraints would make the fitness evaluation a bottleneck.

In order to avoid both the above extremes (consideration of only feasible solutions or heavy use of penalty functions), a technique is required for solution generation which does not insist on fully feasible solutions, yet reduces the level of infeasibility (the number of constraints violated). With this spirit, a quick method is proposed for generation of tolerably good solutions.

First, the constraints are segregated into two categories:

Direct Constraints : A constraint involves only two variables so that one variable can be treated as independent and the other as dependent variable. In other words, one variable is assigned a value randomly and the other one is assigned a value based on the value of the first one.

Indirect Constraints : A constraint involves many variables and is not satisfiable in a straightforward manner.

It is easily seen that the 'operation assignment dependent upon part type selection' (*a1*, *a2*) and 'operation assignment dependent upon appropriate tool assignment' (*b*) constraints belong to the first category (direct constraints). Rest of the constraints (*c* : tool magazine capacities, *d* : machine capacities, *e* : tool type availability, and *f* : tool lives) are indirect constraints.

Notably, the direct constraints form the major chunk of all constraints (nearly 77% for problem *P2*). If these are satisfied while generating the solution, then relatively fewer constraints will remain to be dealt with by penalty functions. Based on this segregation, the solution generation procedure is detailed below:

Step 1. Repeat step 2 for all part types $p = 1, \ldots, P$.
Step 2. Create a gene for variable x_p.
Assign to it value 1 or 0 randomly.
For all operations ($o = 1, \ldots, O_p$) of part type p, repeat step 3.
Step 3. Create a gene for operation (p,o).
If $x_p = 0$ (part p not selected), then go to step 4.
Else (i.e. if $x_p = 1$, part p selected), go to step 5.
Step 4. $x_{po} = 0$ (operation (p,o) not assigned).
Step 5. $x_{po} = a$, random integer a varying from 1 to A_{po}.
From set $AR(p, o)$, assign route a – tool type t to route a – machine m, in other words, $y_{mt} = 1$.

The x_{po} genes are assigned zero or non-zero values based on 0-1 value of corresponding x_p genes, so as to satisfy the (*a1*, *a2*) constraints. Values of y_{mt} variables are completely determined by x_{po} values, so, constraints (*b*) are also satisfied. Also, for the same reason, the y_{mt} variables need not be included in the chromosome.

Relative positions of genes within the chromosome are indicated by the order of steps and are shown below:

$$CH = \mathbf{x_1}\ x_{11}\ x_{12} \ldots x_{1O_1}\ \mathbf{x_2}\ x_{21}\ x_{22} \ldots x_{2O_2} \ldots \mathbf{x_P}\ x_{P1}\ x_{P2} \ldots x_{PO_P}$$

where $O_1, O_2, \ldots, O_P$ are the numbers of operations in part types 1, 2, ..., P respectively. For clarity, x_p genes are shown in bold.

Under this scheme, a solution for a 3-part problem (with 3, 2 and 5 operations respectively) may look like the chromosome given below (part selection i.e. x_p genes are shown in bold):

$CH =$ **1** 2 1 3 **0** 0 0 **1** 3 1 2 2 1

Once the x_p, x_{po} and, consequently, y_{mt} values are assigned, one can check for violation of constraints and, using penalty functions, can evaluate the objective function. The penalty function for a constraint (i) may be simple or complex, linear or non-linear, and may be defined as:

$$\phi(i) = \begin{cases} F(i) & \text{if constraint } i \text{ is violated} \\ 0 & \text{otherwise} \end{cases}$$

If the objective function is to maximize some function Z, then the modified objective function, incorporating the penalty functions, will be:

$$\max \quad Z' = Z - \sum_{i=1}^{CN} \phi(i), \text{ where } CN = \text{total number of constraints.}$$

3.3 RECOMBINATION to MAINTAIN FEASIBILITY LEVEL

3.3.1 TWO-STEP CROSSOVER

In the traditional crossover, crossover points are selected randomly. But in this case, the gene values are context-dependent and if the crossover sites are chosen randomly, then the compatibility between an x_p gene and its corresponding x_{po} genes may vanish and the related constraints will not remain satisfied in the resulting chromosomes. Therefore, specific crossover operator/s have to be designed so that the level of feasibility present in the original chromosomes (solutions) is not lost during crossover.

The whole chromosome can be imagined as being made up of sub-strings, each related to one part type. If the information exchange is made in terms of whole part sub-strings, then the compatibilities within a part sub-string will not be destroyed and feasibility will be intact. In practice, this means that the crossover point/s, though chosen randomly, must always be located only on the x_p genes.

Further, each part type may belong to one of the following three cases:

I. Part is unselected ($x_p = 0$) in both parent chromosomes.
II. Part is selected ($x_p = 1$) in one and unselected ($x_p = 0$) in the other parent chromosome.
III. Part is selected ($x_p = 1$) in both parent chromosomes.

When the part type is unselected in both parent chromosomes (case I), information exchange has no meaning. When the part type is selected in one and unselected in the other parent chromosome, exchange is effective but once the whole part sub-strings have been exchanged, no more information exchange is possible. However, if part type is selected in both parent chromosomes (case III), a meaningful exchange of information can be made (in terms of assignments of operations to different routes) between the part sub-strings pertaining to the same part type in the parent chromosomes. Thus, our crossover operation has two distinct stages:

1. **Part Swap** : Exchange of part type selection information, and
2. **Route Swap** : Exchange of operation assignment information for part type/s which are selected in both the parent chromosomes.

As an example, consider the following parent chromosomes (4 part types with 4, 2, 1 and 3 operations respectively) and *one-point crossover* (crossover point occurring at x_2 i.e. location 6, shown in underline):

$$CH_1 = \mathbf{1}\ 2\ 1\ 3\ 3\ \underline{\mathbf{0}}\ 0\ \mathbf{0}\ 0\ \mathbf{1}\ 3\ 2\ 1$$
$$CH_2 = \mathbf{0}\ 0\ 0\ 0\ 0\ \underline{\mathbf{1}}\ 3\ 2\ \mathbf{0}\ 0\ \mathbf{1}\ 2\ 1\ 3$$

Step 1 of crossover (*part swap*) will exchange the information for part types *2*, *3* and *4* between CH_1 and CH_2. The resultant chromosomes are CH_1' and CH_2':

$$CH_1' = \mathbf{1}\ 2\ 1\ 3\ 3\ \mathbf{0}\ 3\ 2\ \mathbf{0}\ 0\ \mathbf{1}\ \underline{\mathit{2}}\ 1\ 3$$
$$CH_2' = \mathbf{0}\ 0\ 0\ 0\ 0\ \mathbf{1}\ 3\ 2\ \mathbf{0}\ 0\ \mathbf{1}\ \underline{\mathit{3}}\ 2\ 1$$

Since part type *4* is selected ($x_4 = 1$) in both parent chromosomes, step 2 of crossover (*route swap*) would try to exchange information between part *4* sub-strings (locations 12 to 14). Suppose this crossover point occurs at location 13 (shown in italic underline). Then, route information for operations (*4*, *2*) and (*4*, *3*) is exchanged between CH_1' and CH_2'. The resultant chromosomes are:

$$CH_1'' = \mathbf{1}\ 2\ 1\ 3\ 3\ \mathbf{1}\ 3\ 2\ \mathbf{0}\ 0\ \mathbf{1}\ 2\ 2\ 1$$
$$CH_2'' = \mathbf{0}\ 0\ 0\ 0\ 0\ \mathbf{0}\ 0\ 0\ \mathbf{0}\ 0\ \mathbf{1}\ 3\ 1\ 3$$

For ease of visualisation, the part selection genes have been placed along with their operation genes. However, for better understanding of part swap step, one could imagine as if operation genes did not exist and all the part genes stood together. And after this step, all operation genes move with the corresponding part genes.

For both steps of crossover, any type of crossover (*one-point*, *two-point*, *uniform* etc.) can be employed.

3.3.2 MUTATION

Just like crossover, the traditional mutation also can not be applied blindly in our problem. For example, if mutation operates on a part selection gene and if in this process part becomes selected from unselected (x_p goes from 0 to 1) or vice verse (x_p becomes 0 from 1), then the corresponding operation assignment genes will become meaningless i.e. *zero* values for a selected part or *non-zero* values for an unselected part. Similarly, the mutation can not operate on the operation genes of an unselected part because such genes have to retain the value zero. So, the only genes left for mutation to operate upon are the operation genes of selected parts. For each such gene, the mutation can allocate a value from *1* to A_{po}, the number of alternative routes for the corresponding operation (*p*, *o*). As an example, consider the chromosome:

$CH_2'' = \mathbf{0}\ 0\ 0\ 0\ 0\ \mathbf{0}\ 0\ 0\ \mathbf{0}\ 0\ \mathbf{1}\ \underline{3}\ 1\ 3$

Here, mutation can operate only on operation genes of part type *4*. Suppose that gene for operation (*4*, *1*), which has 3 alternative routes, is to be mutated (shown in underline). So, this gene (location 12) is randomly assigned a value from *1* to *3*. If the value assigned is *1*, then after mutation, CH_2'' will become:

$CH_2'' = \mathbf{0}\ 0\ 0\ 0\ 0\ \mathbf{0}\ 0\ 0\ \mathbf{0}\ 0\ \mathbf{1}\ 1\ 1\ 3$

In mutation also, for determining the new value, various mutation operators can be used such as *uniform mutation*, *boundary mutation*, etc.

4 IMPLEMENTATION

In preliminary studies, the methodology outlined is implemented for some small FMSs (generated using the parameter ranges given earlier), with the simplest possible objective functions, and without any parametric optimization. The standard code widely available for simple GA is not useful here because of context-dependent gene-values. The problem specific GA code is written in C++. The GA is implemented with the following parameters:

Population size : 20
Crossover : One - Point Crossover
Crossover Probability : 0.9
Mutation probability : 0.1
Elite fraction : 0.1

Some results are reported below:

4.1. MAXIMIZE No. of PART TYPES SELECTED (max $Z = \sum_{p=1}^{P} x_p$)

The penalty scheme is to simply reduce the objective function value by one for each constraint violated i.e.

$$\phi(i) = \begin{cases} 1 & \text{if constraint } i \text{ is violated} \\ 0 & \text{otherwise} \end{cases}$$

Consequently, the modified objective function will be:

$$\max \quad Z' = \sum_{p=1}^{P} x_p - \sum_{i=1}^{CN} \phi(i) = \sum_{p=1}^{P} x_p - \text{no. of constraints violated}$$

Table 1 compares the GA results with the optimum results given by CPLEX (software for MIP problems).

It is observed that the best GA solution is achieved quite early, within 20 generations. But, out of five problems, GA gives optimal solution only in one (problem 1). However, a slight modification in penalty scheme improves the GA performance considerably. The modified penalty scheme is:

$$\phi(i) = \begin{cases} 2 & \text{if constraint } i \text{ is violated} \\ 0 & \text{otherwise} \end{cases}$$

So that the objective function value becomes:

$$Z' = \sum_{p=1}^{P} x_p - \sum_{i=1}^{CN} \phi(i) = \sum_{p=1}^{P} x_p - 2 \text{ x (no. of constraints violated).}$$

The results obtained with this scheme are shown in table 2 (optimum solutions from GA are shown in bold).

Thus, with the slightest modification in implementation strategy, the GA is able to reach an optimal solution in 60% (6 out of 10) cases. In addition, while the CPLEX gives a unique solution, the GA may turn up the multiple optimal solutions possible (e.g. problem 2).

In another modification to penalty scheme, when the weightage for penalty is reduced to 1/2, the GA ends up with no feasible solutions at all.

4.2 MAXIMIZE TOTAL No. of PARTS SELECTED (max $Z = \sum_{p=1}^{P} BS_p x_p$)

Analogous to the above, the penalty scheme is:

$$\phi(i) = \begin{cases} 2 & \text{if constraint } i \text{ is violated} \\ 0 & \text{otherwise} \end{cases}$$

However, to keep the number of parts selected and the total penalty value of approximately the same order, the total penalty is multiplied by the mean batch size:

$$\max \quad Z' = \sum_{p=1}^{P} x_p - \overline{BS} \sum_{i=1}^{CN} \phi(i) = \sum_{p=1}^{P} x_p - 2 \text{ x } \overline{BS} \text{ x (no. of constraints violated),}$$

where $\overline{BS}$ is the mean batch size.

On observing that GA is not able to arrive at the optimal solution, the population size (PS) is increased from 20 to 40. Results for both levels of population size are shown in table 3 (GA optima indicated in bold):

With population size 20, the GA achieves the optimum solution in only one out of ten problems. But with population size doubled, the optimum is achieved in three problems. This indicates that a bigger population is more effective, which, however, would certainly increase the solution time. An interesting, and counter-intuitive, observation is made in problem 6, namely, that higher population size deteriorates the performance.

Table 1. Comparison of CPLEX and GA results for maximizing the number of selected part types

Problem	No. of Part types (P)	No. of Part types selected (Optimal from CPLEX)	No. of Part types selected (From GA)
1.	10	6 (1, 2, 3, 5, 8, 9)	**6** (1, 2, 3, 5, 8, 10)
2.	9	4 (6, 7, 8, 9)	3 (5, 6, 9)
3.	11	7 (1, 2, 3, 7, 9, 10, 11)	5 (1, 2, 5, 7, 10)
4.	12	8 (1, 3, 4, 5, 6, 7, 9, 11)	6 (1, 5, 6, 9, 11, 12)
5.	9	4 (1, 4, 5, 6)	3 (1, 4, 5)

Table 2. Comparison of CPLEX and GA results for maximizing the number of selected part types (with a modified penalty scheme)

Problem	No. of Part types (P)	No. of Part types selected (Optimal from CPLEX)	No. of Part types selected (From GA)	No. of generations for the best GA solution
1.	11	4 (2, 7, 8, 11)	3 (8, 9, 11)	12
2.	6	3 (1, 2, 4)	**3** (1, 4, 6) **3** (1, 2, 4)	7 24
3.	7	2 (1, 2)	**2** (2, 5)	5
4.	9	3 (3, 6, 8)	**3** (3, 6, 8)	32
5.	9	4 (1, 4, 5, 6)	**4** (1, 4, 5, 6)	44
6.	10	5 (1, 5, 6, 8, 10)	3 (5, 6, 10)	6
7.	11	4 (1, 4, 8, 11)	**4** (1, 8, 9, 11)	17
8.	7	4 (1, 3, 4, 6)	3 (3, 4, 5)	11
9.	6	2 (3, 4)	**2** (1, 4)	4
10.	11	6 (3, 6, 7, 8, 9, 10)	5 (4, 5, 8, 10, 11)	36

Table 3. Comparison of CPLEX and GA results for maximizing the number of selected parts

Prob.	No. of Part types	Mean Batch Size	CPLEX	GA with PS = 20		GA with PS = 40	
	(P)	$\overline{BS}$	No. of Parts selected	No. of Parts selected	No. of generations	No. of Parts selected	No. of generations
1.	11	8	42	39	27	31	8
2.	6	15	28	18	4	**28**	5
3.	8	8	28	16	30	16	6
4.	12	10	38	28	11	**38**	36
5.	11	11	42	37	7	37	11
6.	7	11	32	**32**	12	24	1
7.	11	8	30	18	11	23	30
8.	11	11	34	18	8	29	39
9.	7	11	22	10	15	**22**	32
10.	11	12	46	25	14	38	17

It must be noted that the GA used here is quite primitive and the penalty schemes simplistic, so one can hardly expect desirable results. Nevertheless, the improvement in performance due to modifications in implementation strategy does indicate that the methodology developed is capable of giving good results when applied wisely. It also means that a parametric optimization, assisted by a properly planned design of experiments, *is* mandatory before GA can be forwarded as a valuable tool for the FMS part selection and loading problem.

5 SUMMARY and CONCLUSIONS

In order to develop a GA methodology for FMS part selection and machine loading problem, the large number of constraints involved are segregated into direct and indirect constraints. The direct constraints, which involve only two variables each and form the major portion, are satisfied by generating context-dependent values for those variables. The remaining few indirect constraints are accounted for through the penalty functions. For preserving the feasibility of solutions, a two-stage crossover is outlined – part swap crossover followed by route swap crossover. In mutation, only the operation assignment genes of selected part types are allowed to mutate so that the compatibility between part type selection and operation assignments is not destroyed. Preliminary studies with small FMSs, simplistic penalty schemes, and non-parameterized GA show that, with slight modifications in implementation strategy, the methodology is capable of achieving good results.

Further work is in progress to apply this methodology for medium and large-sized FMS loading problems. The efforts are also being made to optimize the GA parameters involved.

REFERENCES

1. Stecke, K E, Formulation and solution of nonlinear production planning problems for flexible manufacturing systems, *Management Science*, **29** (1983) 273-288.
2. Shanker, K and Tzen, Y J, A loading and dispatching problem in a random flexible manufacturing system, *Int. J. Prod. Res.*, **23** (1985) 579-595.
3. Goldberg, D E, *Genetic Algorithms in Search, Optimization and Machine Learning*, Addison - Wesley (1989).
4. Davis, L, ed., *Handbook of Genetic Algorithms*, Van Nostrand Reinhold (1991).
5. Antonisse, J, A new interpretation of schema notation that overturns the binary coding constraint, *Proc. Third Int. Conf. on Genetic Algorithms*, George Mason University, USA (4-7 July 1989), 86-91.

SCHEDULING HEURISTICS FOR THE m-MACHINE NO-WAIT LOT STREAMING FLOWSHOP

Subodha Kumar
Tapan P. Bagchi
bagchi@iitk.ernet.in

Department of Industrial and Management Engineering
Indian Institute of Technology Kanpur, India 208016

Chelliah Sriskandarajah
chellia@ie.utoronto.ca

Department of Mechanical and Industrial Engineering
University of Toronto, Toronto, Canada M5S 1A4

Abstract. This paper aims at minimizing makespan in m-machine no-wait flowshops with multiple products requiring lot streaming. A "product" here implies many identical items. For the single product case with fixed number of sublots optimal continuous-sized sublots are obtained and then a heuristic is used to find integer-sized sublots. For the multi-product case optimal sequencing of products is attained by solving a traveling salesman problem. Another heuristic yields integer-sized sublots. Finally we evaluate the use of genetic algorithmic meta-heuristics for the interacting decision phases in simultaneous lot streaming and sequencing. We conclude that while GA may deliver makespans similar to those given by heuristic methods that cleverly exploit problem features particular to lot streaming, GA loses out in computational efficiency. On the other hand, GA can find near-optimal number of sublots for each product—a task for which neither an analytical nor a heuristic method presently exists.

Key words and phrases: No-wait flowshop, lot streaming, scheduling, sublots, heuristic algorithms, genetic algorithms.

1 Introduction

This paper addresses the task of lot streaming and scheduling of N products through m-machine no-wait flowshops in order to minimize the total finish time or what is also known as the makespan of the schedule. A "product" or lot consists of many identical items. Lot streaming (lot sizing) is the process of creating sublots (or transfer batches) to move the completed portion of a production sublot to downstream machines so that operations can be overlapped. In flowshops the creation of sublots permits the overlapping of different operations on the same product and may therefore reduce the makespan. However, the planning decisions become more complex when lot streaming is allowed. Then a solution procedure requires the combined optimal creation of sublots as well as the scheduling of these sublots. Baker (1990, 1995) provides overviews of lot streaming in flowshops.

Szendrovits (1975) modeled a single product consisting of equal-sized sublots processed continuously in a multi-stage production system with no idle time between sublots. In general, however, there may be idle times between the production runs of different products. Goyal (1976) found optimal sublot sizes for the m-machine model. Equal-sized sublots were studied also by Baker and Jia (1993).

Presently, processing of sublots is *no-wait*, started as soon as the operation on the preceding machine is finished. This may idle the machines between two sublots. No-wait conditions arise in chemicals and petro-chemical processing and in hot metal rolling where metals must be processed continuously at high temperature. For a survey of no-wait scheduling, see Hall and Sriskandarajah (1996).

The present paper deals with multi-machine lot streaming but with *no-wait* constraint. We study continuous-sized sublots as well as integer-sized sublots. The continuous-sized model is used when the number of items in a product lot is large, so that the lot may be treated as infinitely divisible. This model provides useful insights into the behavior of the no-wait model when sublots are integer-sized.

The next two subsections provide the notations used and the assumptions made in setting up the different models in this paper.

1.1 Notations

m : the number of machines in the flowshop.
$M_1, M_2, ..., M_m$: the m machines in the flowshop.
N : the number of different types of products to be produced.
P_j, $j=1,.....,N$: the different product-types to be produced.
p_{ij} : the processing times of unit product j on machines M_i.
X_{kj} : the total number of items (or units) of product j in the sublot k, where X_{kj} is a positive real number to allow continuous sized sublots.
Y_{kj} : the total number of items of product j in the sublot k, where Y_{kj} is a positive integer number. We use this notation when sublot sizes must be integer-. valued.
p_{ij}^k : the processing times of sublot k of product j on machines M_i.
n_j : the number of sublots for product j.
$\sum_{k=1}^{n_j} X_{kj} = W_j$: the total number of units in product P_j.
s_{ij} : setup time of product P_j on M_i

1.2 Assumptions

1. All products are available at time zero.
2. Each product has a specified number of sublots.
3. The processing of sublots is *no-wait.*
4. *Consistent* sublots are used due to the *no-wait* processing of sublots. A sublot is *consistent* if its size remains the same on all machines.
5. The products are not split, i.e., no *pre-emption* is allowed. This means that all sublots of a product are produced consecutively. Note that it may be optimal to pre-empt a product but we limit our study to the no pre-emption case.
6. The processing time of a sublot is proportional to its size, $p_{ij}^{k} = p_{ij} X_{kj}$
7. *Anticipatory setup times*: Switching from one product to another requires a setup, independent of the processing sequence. Anticipatory setup can be done as soon as a machine becomes free even without the actual arrival of the next sublot.

2. Lot Streaming Single product and Continuous-Sized Sublots

Since we deal here with a single product, the index j denoting product type is dropped. The object is to determine the optimal values of X_i (i = 1, 2,..., n), the *sizes* of the n sublots, that minimize the total makespan. We assume that the number of units in a product lot is large so that the lot can be treated as infinitely divisible. S is defined as the order of the sublot indices (i = 1, 2, ..., n). There is no *loss of generality* in assuming the order of these indices to be the ordered set S = {1, 2, ..., n}. The schedule found from S is denoted as $\widehat{S}$. We use m = 3 to illustrate lot streaming (see Fig. 1).

Let Δ_i be the idle time on the *last* machine M_3 before the processing of sublot i begins and let Π_i be the idle time on the *first* machine M_1 after the processing of sublot i is complete. Makespan C(X) may then be written as follows :

$$C(X) = s_3 + p_3 W + \sum_{i=1}^{n} \Delta_i \, . \qquad (1)$$

Alternatively,

$$C(X) = s_1 + p_1 W + \sum_{i=0}^{n} \Pi_i \, . \qquad (2)$$

where $\Pi_0 = \max\{0, s_2 - s_1 - X_1p_1, s_3 - s_1 - X_1p_1 - X_1p_2\}$, $\Pi_i = \max\{0, X_i p_2 - X_{i+1}p_1, X_i p_2 + X_i p_3 - X_{i+1} p_1 - X_{i+1} p_2\}$, $i = 2,..., n\text{-}1$ and $\Pi_n = X_n (p_2 + p_3)$. Note that in (1), the makespan is given as a function of the idle times $\{\Delta_i\}$ of the last machine M_3, while it is given as a function of the idle times $\{\Pi_i\}$ of the first machine M_1 in (2).

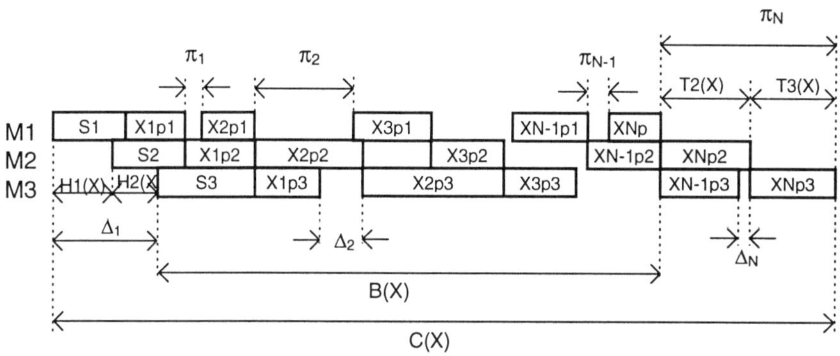

Fig. 1. An Example Showing Sublots of a Single Product Being Processed Through Three Machines

We will first solve the minimum makespan problem with lot size being rational numbers. We will then find the optimal solution in **N** (the set of natural numbers).

Let $Z = \sum_{i=1}^{n} \Delta_i$, where $\Delta_1 = \max\{0,\ s_2 + X_1 p_2 - s_3,\ X_1 p_1 + X_1 p_2 + s_1 - s_3\}$, and $\Delta_i = \max\{0,\ X_i p_2 - X_{i-1} p_3,\ X_i p_1 + X_i p_2 - X_{i-1} p_2 - X_{i-1} p_3\}$, $i = 2, ..., n$. The minimum makespan problem (1) may be then solved by the following LP:

minimize Z, (3)

subject to :

$$\sum_{i=1}^{n} X_i = W. \tag{4}$$

$$\Delta_1 \geq s_2 + X_1 p_2 - s_3.$$

$$\Delta_1 \geq s_1 + X_1 p_1 + X_1 p_2 - s_3.$$

$$\Delta_i \geq X_i p_2 - X_{i-1} p_3, \quad i = 2,..., n.$$

$$\Delta_i \geq X_i p_1 + X_i p_2 - X_{i-1} p_2 - X_{i-1} p_3,\ i = 2, ..., n; \quad X_i \geq 0, \quad i = 1, ..., n. \tag{5}$$

The above formulation may be generalized to any value of m.

3 Lot Streaming of a Single Product: Integer-sized Sublots

The problem of finding optimal integer-sized sublots ($Y = (Y_1, Y_2, ..., Y_n)$) for a single product is NP-hard. Exact exhaustive search type methods here would become prohibitive for large problems. Therefore, we use a heuristic procedure provided by Sriskandarajah and Wagneur (1996) to solve it. This heuristic (called hereafter Heuristic S & W and described below) is used again when we address simultaneous (integer-sized) lot streaming and scheduling of multiple products.

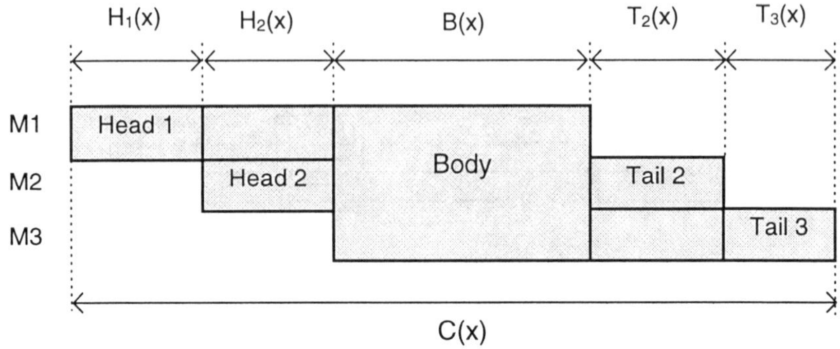

Fig. 2. A Schedule Showing All Five Sections

Heuristic S & W :

$W_0 = W$; begin
for i=1 to n do
begin ; $Y_i = 1$; $W_0 = W_0 - 1$; end
while $W_0 > 0$ do
begin
find $d_i = X_i - Y_i$, $i = 1, ..., n$
find r such that $d_r = \max_{1 \le i \le n} \{d_i\}$
$Y_r = Y_r + 1$; $W_0 = W_0 - 1$
end
end

Heuristic S & W finds integer-sized sublots such that the *maximum absolute deviation* of lot sizes from the optimal continuous solution (obtained by LP in Section 2 above) is minimized. Since each lot must have at least one item in it, the heuristic starts by assigning one item to each lot. The remaining items are assigned to the different sublots one by one. Each such assignment is made to narrow the gap of the maximum deviation of the integer solution from the optimal continuous solution.

Note that the LP illustrated by (3), (4) and (5) for $m = 3$ yields sublot sizes $\{X_i\}$ which may be real numbers. Subsequently Heuristic S & W produces integer-sized sublots. Recall also that W is the total number of units of the single product to be produced and $\{Y_i\}$ are the n integer-sized sublots into which W is divided for lot streaming.

4 The Multiple Product, Continuous-Sized Sublots Case

We describe the procedure using again $m = 3$ as an illustration. The schedule $\hat{S}$ (see Figs. 1 and 2) for S may be divided into five distinct sections or time segments, namely H_1, H_2, B, T_2 and T_3, as indicated below.

Section H_1. It is the beginning portion of $\widehat{S}$, in which machine M_2 is idle throughout. The length of Section H_1 is measured by $H_1(X)= \max\{0, s_1 + X_1p_1 - s_2\}$. We call this section the "Head1" of schedule $\widehat{S}$.

Section H_2. This section follows immediately after section H_1 and occurs in the beginning portion of $\widehat{S}$ during which the machine M_3 is idle. The length of H_2 is measured by $H_2(X) = \max\{0, s_2 + X_1p_2 - s_3\}$. We call this section "Head2" of schedule $\widehat{S}$.

Section T_3. T_3 is the last portion of $\widehat{S}$ in which machine M_2 is idle throughout. The length of Section T_3 is measured by $T_3(X)= X_n\,p_3$. We call this section the "Tail3" of schedule $\widehat{S}$.

Section T_2. It is the portion immediately before the last portion T_3 of $\widehat{S}$, where the machine in M_1 is idle throughout. The length of the Section T_2 is measured by $T_2(X) = X_n\,p_2$. We call this section "Tail2" of schedule $\widehat{S}$.

Section B. B is the middle portion of $\widehat{S}$. The length of section B is measured by $B(X) = C(X) - H_1(X) - H_2(X)- T_2(X) - T_3(X)$. It is the time span between the finish of the last sublot on M_1 and the start of *setup* on M_3. We call this section the "Body" of schedule $\widehat{S}$.

A schedule $\widehat{S}$ containing all five sections—H_1, H_2, B, T_2 and T_3— is shown in Fig. 2. Similarly, one may construct a schedule $\widehat{S}$ for any value of m. The schedule for $m = 4$, for example, will have three heads, one body and three tails.

We define σ as the sequence of product indices, i.e., the N products are scheduled in the flowshop in the order $P_{\sigma(1)}$, $P_{\sigma(2)}$, ..., $P_{\sigma(N)}$. Thus, product $P_{\sigma(i)}$ is scheduled in the ith position in σ. Note that the products are not split, i.e., no pre-emption is allowed once processing begins. This means that *all* the sublots of a given product are produced consecutively. We now state a fundamental result as follows:

Theorem 1 *Suppose that sublot size vector* X(j), $j = 1, 2, ..., N$ *for each product is given, then finding an optimal* σ *that minimizes the makespan of* N *products can be formulated as a traveling salesman problem.*

Proof: First we assume $m = 3$. Since the sublot sizes $X(j)$, $j = 1, 2, ..., N$ for each product are given, the lengths of H_{1j}, H_{2j}, B_j, T_{2j}, T_{3j} are defined precisely. An illustration involving N products is shown in Fig. 3. Makespan F_σ may be now written as follows :

$$F_\sigma = \sum_{i=1}^{N+1} \max\{0, T_{2\sigma(i-1)} - H_{1\sigma(i)}, T_{2\sigma(i-1)} + T_{3\sigma(i-1)} - H_{1\sigma(i)} - H_{2\sigma(i)}\}$$

$$+ \sum_{i=1}^{N+1} (H_{1\sigma(i)} + H_{2\sigma(i)} + B_{\sigma(i)}).$$

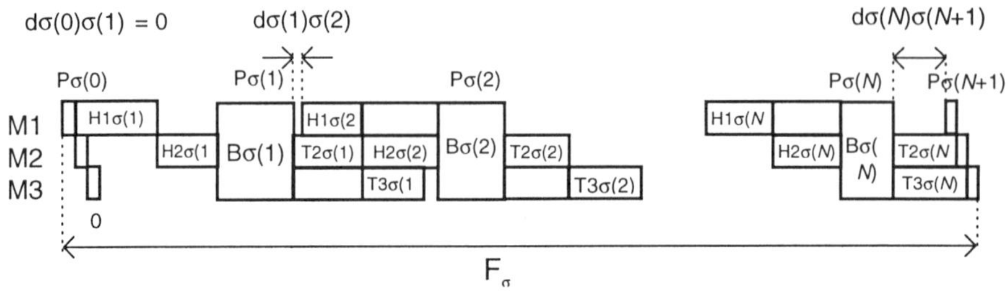

Fig. 3. A Schedule of N Products in the Sequence σ with $m = 3$

Without loss of generality, the schedule starts and ends with dummy products 0, N+1 with $H_{1\sigma(i)} = H_{2\sigma(i)} = B_{\sigma(i)} = T_{2\sigma(i)} = T_{3\sigma(i)} = 0$, where $i = 0$, $i = N+1$ (see Figure 3). Since $\sum_{i=1}^{N+1} (H_{1\sigma(i)} + H_{2\sigma(i)} + B_{\sigma(i)})$ is a constant for given sublot sizes, it is clear that minimizing F_σ is equivalent to minimizing the sum of the terms given by

$$F'_\sigma = \sum_{i=1}^{N+1} \max\{0,\ T_{2\sigma(i-1)} - H_{1\sigma(i)},\ T_{2\sigma(i-1)} + T_{3\sigma(i-1)} - H_{1\sigma(i)} - H_{2\sigma(i)}\}$$

The above expression may be written as

$$F'_\sigma = \sum_{i=1}^{N+1} d_{\sigma(i-1)\sigma(i)},$$

where $d_{\sigma(i-1)\sigma(i)} = \max\{0,\ T_{2\sigma(i-1)} - H_{1\sigma(i)},\ T_{2\sigma(i-1)} + T_{3\sigma(i-1)} - H_{1\sigma(i)} - H_{2\sigma(i)}\}$ may be interpreted as the distance between city i-1 to city i in a traveling salesman problem (TSP) formulation. The above result may be generalized for any $m > 3$. This completes the proof.

The optimization problem formulated as a TSP as above provides significant computational advantages as the result of the numerous optimal and heuristic approaches now available to solve the TSP problem. The size of m does not affect the TSP solution time because of the no-wait constraint. But we do need to calculate the distance matrix given a flow shop problem. The calculation time of the distance matrix depends on both n and m, where n is the number of products. However, the time to calculate the distance matrix is very small even if n and m are large. So this time is negligible.

We next address the multi-product problem employing *integer-sized* sublots.

5 The Multiple Product, Integer-Sized Sublots Case

In Section 1.4 we obtained only a heuristic integer solution for the single product problem. However, even if one obtains an integer solution for the problem, it does not necessarily provide optimal sublot sizes for the problem of simultaneously lot streaming (finding optimal integer-sized sublots) and scheduling of *multiple* products. An example with $m = 2$ is given by Sriskandarajah and Wagneur (1996) to illustrate this fact. To solve the multiple product problem, therefore, other methods must be developed. To this end we propose a different heuristic as follows.

Heuristic Algorithm MHEU

Step 1. Find real valued lot sizes by LP and then round these off byHeuristic S & W.

Step 2. Given these integer solutions for lot sizing, obtain the "optimal" product sequence using TSP, on the strength of Theorem 1.

This procedure generalizes the heuristic method for $m = 2$ provided by Sriskandarajah and Wagneur (1996).

6 GENETIC ALGORITHMS

Simultaneous lot streaming and scheduling involves three interdependent decision stages, the first being the number of sublots n_j in which the total demand W_j for each product must be divided. The second stage optimally sizes these sublots, i.e., it distributes the total W_j into the different sublots to be produced. The third decision optimally sequences the the N products. Genetic Algorithms (GAs) may be used in each of these stages. The information to coded in the solutions has three elements: (1) The sequence in which the N products will be processed. (2) Product j is processed through the machines in n_j sublots. And lastly (3), each of the n_j sublots has a specified (integer) size, Y_{kj} units of product j. For each sublot of product j a *chromosome* encoding this information may be constructed as follows:

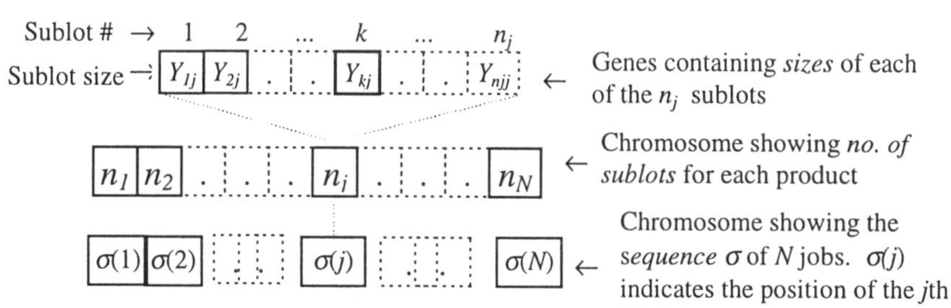

The GA starts with a population of initial solutions. It is conventional to randomize this initial condition. For lot sizing, the n_j sublots for the jth product are randomized as follows. The total demand W_j for product j must be distributed into n_j sublots. Thus, we must have

$$Wj = \sum_{i=1}^{nj} Yij$$

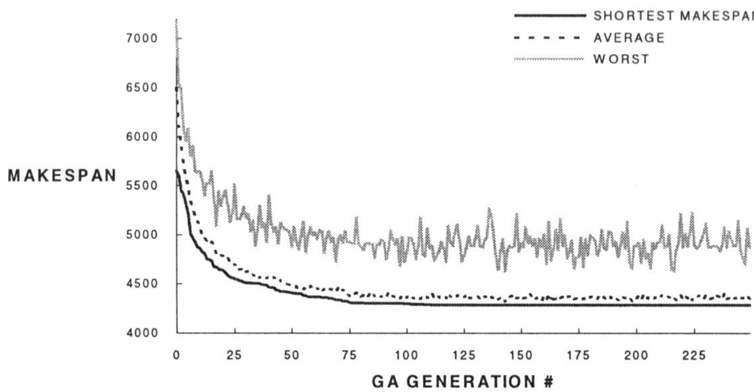

Fig. 4. Convergence of the GA as it searches for the Shortest Makespan Lot Streaming Solution

Therefore, initially W_j is distributed among the n_j sublots, randomly, ensuring that each sublot receives at least *one* unit. Similar constraints are used in crossover and mutation operations during the iteration of the GA.

Fig. 4 displays the typical convergence profile of the GA as it seeks the minimum makespan, using parameters $PS = 50$, $Pc = 0.9$, $Pm = 0.05$ and elite frection $\varepsilon = 0.35$. The GA was parameterized using a statistical design-of-experiments framework.

7 Computational Studies and Conclusions

The methods described in Sections 2, 3, 4 and 5 do not optimize the *number* of sublots. Consequently, in the different heuristics-to-GA comparisons done, the number of sublots for each product was "fixed." Later, we separately addressed the *optimization* of the number of sublots to minimize makespan. Typical results are shown in Table 7.1, the average percent reduction in makespan being calculated using the best makespan found by GA. "Sublot size by LP, Sequence by GA" appears to consistently outperform MHEU with makespans becoming shorter by 15-20% for problems with 10 or more machines and 30 or more jobs. But, CPU time for GA grows rapidly. When not assisted by LP, the GA has to work much harder, highlighting the value of "hybridizing" the GA (Goldberg, 1989). Sizing sublots by GA (while keeping the number of sublots and the job sequence fixed) appears to be comparable in performance to LP + SW heuristic, though the CPU time required is much higher.

The GA re-designed to seek an *optimal number of sublots* produced results that were most insightful (Table 7.2). The GA produced *shorter* makespans than all arbitrarily set "fixed" number of sublots. This reveals the resident opportunity for management to reduce makespan (thus improve shop productivity) if the *number of sublots* is optimized by GA. Recall that no method, heuristic or analytical, presently optimizes the number of sublots in a lot streaming no-wait flowshop.

We conclude that meta-heuristic methods such as GA can easily outperform purely random search in simultaneous lot streaming and sequencing. But GA does not appear to be computationally competitive with heuristics such as MHEU and S-W, which cleverly exploit the structure particular to such scheduling problems.

Table 7.1 MHEU and the "Sublot Size by LP, Sequence by GA" Algorithm

No. of Sublots	m/c	Prods.	Prob. #	Makespan by MHEU Heuristic CPU 88 Sec.	"Size by LP, Sequence by GA" CPU 817 Sec.	% Improvement in Makespan
Fixed	10	30	1	12589	10080	19.93
Fixed	10	30	2	11329	9503	16.12
Fixed	10	30	3	11869	9577	19.31
Fixed	10	30	4	12038	10201	15.26
Fixed	10	30	5	11668	9428	19.20

Table 7.2 "No, of Sublot by GA, Size by LP, Sequence by GA" Algorithm vs. "Fixed No. Of Sublot, Size by LP, Sequence by GA" Algorithm

				FIXED + LP + GA		GA + LP + GA	
Problem #	Setup Instance	m/ c	Prods.	MAKESPAN found	CPU (Sec) on HP 9000/850	MAKESPAN found	CPU (Sec) on HP 9000/850
1	I_1	3	5	1170	129.3	1144	277.0
2	I_1	5	5	1579	835.7	1540	1288.3

References

1. Baker K.R. (1990). "Solution Procedures for the Lot-streaming Problem", *Decision Sciences*, 21, 475-491.
2. Baker, K.R. and Jia D. (1993). "A Comparative Study of Lot Streaming Procedure", *OMEGA*, 21, 561-566.
3. Baker K.R. (1995). *Elements of Sequencing and Scheduling*, book published by The Amos Tuck School, Dartmouth College, Hanover, NH.
4. Goldberg D.E. (1989). *Genetic Algorithm in Search, Optimization and Machine Learning*, Addison-Wesley, Reading, Mass.
5. Goyal S.K. (1976). "Note on Manufacturing Cycle Time Determination for a Multi-Stage Economic Production Quantity Model", *Management Science*, 23, 332-333.
6. Hall, N.G. and Sriskandarajah, C. (1996). A Survey of Machine Scheduling Problems with Blocking and No-wait in Process, *Operations Research*, 44, 510-525.
7. Sriskandarajah C. and Wagneur E. (1996). "Lot streaming and streaming and scheduling multiple products in a two-machine no-wait flowshops", *Working Paper*, Department of Mechanical and Industrial Engineering, University of Toronto.
8. Szendrovits, A.Z. (1975). "Manufacturing Cycle Time Determination for a Multi-Stage Economic Production Quantity Model, *Management Science*, 22, 298-308.

Performance Evaluation and Simulation

Simulation of Complex Production Systems with Different Levels of Detail

Prof. Dr.-Ing. habil. Wilhelm Dangelmaier,
Norbert Holthöfer, Stephan Kress, Ulrich Schäfermeier

Heinz Nixdorf Institute, University of Paderborn, Germany

Abstract. The simulation and on-line animation of complex production systems is still a challenge despite the supercomputers which are available today. The main reason is the exponential growth of the computational effort with the level of detail the simulation and animation are carried out. Therefore we have developed a new hierarchical modelling concept which transfers the level of detail technique of Virtual Reality animations to discrete and continuous simulations. The convergence tendencies of discrete simulation and VR animation in production technology applications encourage to this approach. Based on the hierarchical model the concept of an integration tool for simulation and animation in a virtual environment is developed which re-uses existing simulation and VR system components as far as possible.

1 Introduction

Discrete simulators are used to plan all flows of material in a production system in advance to achieve effective and efficient production and production arrangements [5], [13], [22]. But as the simulation is based on an abstract model of the plant which is usually developed once and before the simulation only certain questions from a previously determined point of view can be examined. On the other hand there are often no alternatives to simulators to avoid time-consuming and expensive running-ins for new products.

Virtual Reality (VR) is a new technique to model plants which gives a vivid and realistic picture of the processes in a plant. The 3D representation even allows the inexperienced user to retrace and understand complex production sequences. To achieve fine granular simulations and real-time animations of sophisticated production systems in Virtual Reality present and foreseeable computer technology is not efficient enough however due to the enormous increase in the necessary computing effort.

Based on the present convergence tendencies of discrete simulation and VR animation, the concept of a situation dependent detailing of simulation and an appropriate integration tool is developed in this paper. The tool allows the user to examine complex production plants under different aspects in real-time. The main idea of this concept is the transfer of the level of detail technique[1] used for fast VR animation to simulators which are coupled on-line to a virtual environment.

2 Convergence Tendencies of Discrete Simulation and VR Animation in Production Technology Applications

Simulation technology is an approved decision support instrument for planning and optimizing production systems, especially in the context of discontinuous piece good production [16]. The first step of a simulation study is the creation of a model. The model focuses on the important aspects (according to the aim of the study) of the real system and is as an abstract representation prerequisite both to grasp and solve the planning task. Three major disadvantages of this traditional method of executing simulation studies can be formulated as:

1. The models are abstract and do not give a vivid picture to the mind.
2. Each question needs an own simulation model (one-way model). New questions require new simulation studies.
3. The questions concerning the production system are to be formulated before and not during the simulation. As a result a solution can only be found within the frame and abstraction the simulation specialist (and not the user!) has chosen. Therefore, new questions cannot be found and formulated.

The lack of vividness and applicability of simulation led to the effect that just 10 % of German manufactures use simulations (for comparison: CAD 60 %, PPC 70 %) despite yearly savings of about 1.000 million DM [16]. To overcome these disadvantages, i. e. to improve the vividness and to make simulation usable for more deliberations and user groups, animation functions are added to simulation tools. Animations which are based on the abstract simulation model show the processes in an iconized way and thus serve for debugging, analysis and presentation of the model. [13], [11]

2.1 Animation - State of the Art

Animations are used to visualize simulations. Today, nearly all simulation systems offer animation components which are permanently improved. Basically, there are three different procedures for generating and executing animations:

1. The graphical animation elements are part of the building blocks of the simulator. Creation of the model and preparation of the animation are done in one step (e.g. SIMPLE++, Simfactory II.5).

---

[1] The idea of this technique is to represent each visible object through a series of 3D models with decreasing level of detail. The model which is actually taken when generating a picture depends on the size of the object on the screen.

2. Model creation and animation generation are done in two steps. Layout and animation elements are designed with a graphical editor and then imported and executed in the simulation system.
3. Animation generation as in 2., but the execution takes place in an external animation system. [13]

The demand to present complex processes more realistically leads to a transition from 2D to 3D representation today. The use of CAD drawings according to procedure 2 is an important intermediate step to achieve this more realistic representation. But solutions following 2 are often limited to one-way data exchange: The CAD drawings are imported by the simulation system and then used for volume models of the simulation elements and for background layout. Examples of this class are Arena and Siman V / CINEMA V (2D) and Automod II (3D) [18]. Further considerations for the coupling of 3D animation with event simulations favour a more modular structure as proposed in 3.

But present 3D animation concepts as in 3 still use a trace file[2] as the central *and* unidirectional interface between simulator and animation components. Such a concept prevents both user interaction in a 3D environment, which is offered already by VR systems, and a feedback of these external events to the simulator. Therefore, the possibility of interaction demands a bidirectional coupling of animation and simulation [6].

2.2 Use of Virtual Reality in Production Technology Applications

The purpose of Virtual Reality is a closer dovetailing of man and computer. In contrast to classical animation, VR enables the user to interact in and manipulate the computer generated 3D models he has conceived [3]. Research efforts in the field of VR are presently characterized by two main tendencies:

- Improvement of immersion and interaction in VR systems ([7], [10]) and
- Revealing meaningful applications based on this technology.

The achieved state of the art and the permanent research efforts to better the man-machine-interface already allow meaningful applications. Present application fields in production technology comprise:

- Off-line programming / teaching of robots in industry [10]
- Simulation of a production cell with several robots [8]
- Development of CAD to a complete virtual prototyping environment (CyberCAD) [3], [4] for all life phases of a product: conception, design, and presentation of ideas [9], design for manufacturability/assembly [21], and validation of functionality and ergonomics [19]

[2] This file includes a chronological list of all events that took place during the simulation with all relevant features for the dynamic animation such as time stamp, object type, and event type.

- Extension of event simulation by VR animation for piece good production systems

The multitude of implemented VR applications both in research institutes[3] and in international industries[4] proves the benefits of VR and refutes the often repeated reproach that VR is just an expensive computer game [1], [10], [14].

2.3 Development Tendencies

The change from 2D to 3D representation shows the increasing importance of animation for simulations. The use of realistic CAD models instead of icons helps to overcome one main disadvantage of simulators, the lacking or just symbolic and therefore unrealistic presentation for the user.

The next stage of development is the combination of VR technology and event simulation because VR allows the user to become an active part of the simulated processes (no bird's-eye view). The created model of the production system is no longer a one-way model with the details for only one question but can be seen as an open model able to accumulate as much information as necessary or possible. Furthermore, the formulation of questions takes place not only before but during the simulation and by the user himself dependent on the dynamic linkage of single objects at run-time.

Presently, the simulation functionality of VR applications is limited to motion simulation or the control of single or isolated groups of robots for example. A generic planning and control functionality to simulate a complete plant in VR has not been realized yet. VR applications for larger environments do not offer simulations which depend on different inputs and events but reproduce „hard coded" scenes which have been generated earlier. Yet, a true VR factory simulation demands modelling and event components as in traditional simulators.

Hence, convergence tendencies of both technologies are necessary but also apparent. A single and simple combination of existing tools however is not sufficient [5]. For a satisfactory solution, a framework is needed which enables a consistent join of both simulation models. Therefore, a hierarchical modelling method is proposed which allows simulation and animation of complex production plants in a virtual environment in real-time.

3 Concept of a Factory Simulator with Situation Dependent Detailing in a Virtual Environment

For complex simulations computer hardware still represents a bottleneck. To avoid this disadvantage a tool is needed which details the simulation due to the user's

---

[3] e.g. several Fraunhofer Institutes (IPA, IAO, IGB, IBP) and the HNI (Virtual Enterprise, VR laboratory)

[4] e.g. Caterpillar, VW, BMW, Daimler Benz, GM, Matsuhita, Boeing, ...

position and angle of vision in the virtual environment. For example, if the user moves to a lathe he will see the processing which takes place at that moment in the minutest details. In the other case (the machine is not visible) the process is described by its duration only (cf. fig. 1).

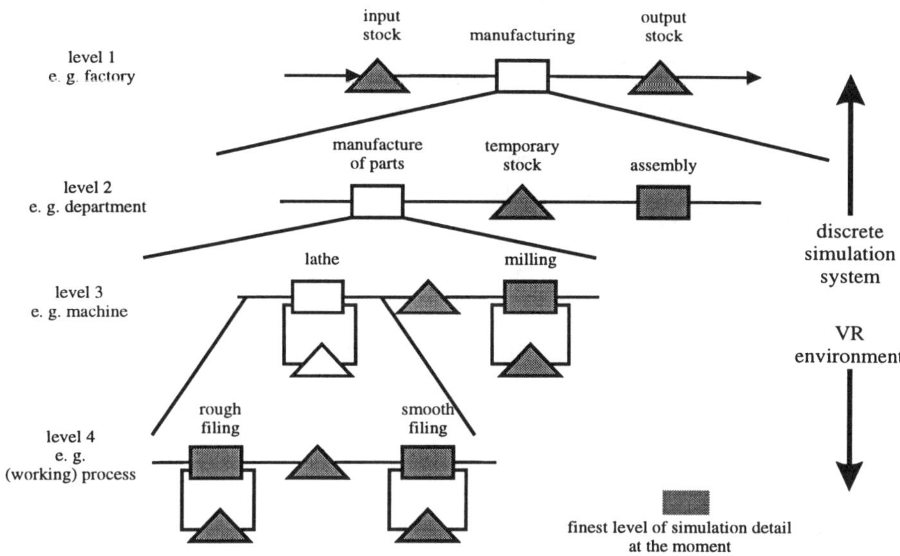

Fig. 1. Hierarchical simulation of manufacturing

The following example will show the achievable reduction: Under the assumption that a factory is represented by a model with n hierarchy levels and the number of objects rises by factor 10 on each level, a complete simulation of the factory necessitates the simulation of about 10^n objects on the deepest level. On the other hand, if just one object is detailed on each level just 9n objects will have to be considered (cf. fig. 2). Whilst complete simulation means an exponential rise of computing effort with each level, a situation dependent simulation causes just a linear rise; therefore it is possible to simulate much more complex plants than before.

As neither (discrete) simulation systems provide this possibility of detailing nor VR environments offer a true simulation of production systems, a tool is necessary that integrates and completes both systems in a framework. Additionally, a (modelling) method[5] is required which allows both a consistent distribution of the simulation on different levels and transitions between these levels.

[5] A modelling method comprises (formal) language constructs and (less formal) proceedings how to use these constructs efficiently to build the model [17].

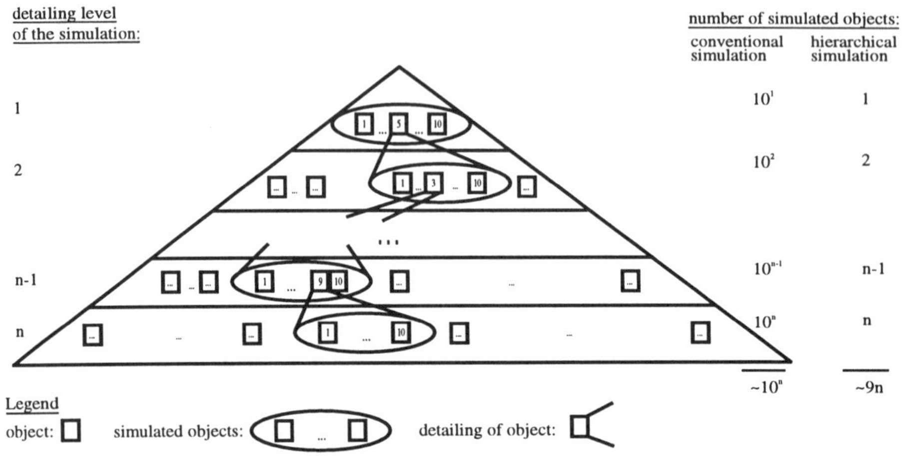

Fig. 2. Reduction of computing effort by hierarchical simulation

3.1 Necessity of Transition Methods for Object-Related Detailing of Simulation

The degree (and level) of simulation has to be determined by the user‘s angle of vision. Therefore, the event combination „start of work“ - „end of work after 3 minutes“ on machine M must be replaceable by a detailed motion simulation when the user approaches M: Whilst an event simulation just considers the average duration (with a certain deviation and error probability), a motion simulation gives a quite more realistic mapping of processes. Consequently, for the transition between different levels of detail methods are needed to prevent the inherent disadvantage of present VR systems (hard coded sequences). If the user‘s interaction (new angle of vision) demands a deeper level of detail additional attributes and finer methods will be evaluated. Thereby, the more detailed simulation is triggered (situation) dependent on the date and the attributes‘ values in the upper level(s). If necessary, parts of the more detailed simulation are re-simulated to assure a consistent simulation in and between the levels (cf. fig. 3). Of course, a complete consistency cannot be achieved by a simulation on different levels because accuracy is lost on higher levels. But usually aggregate or less detailed data of one area of manufacturing is sufficient for other areas to execute a meaningful simulation.

Due to the diversity of detailing possibilities, transition methods must be inheritable between classes as well as definable for individual objects. Especially, the transition from discrete to dynamic simulation demands suitable methods which allow a dynamic immersion in the motion processes dependent on the assigned attributes of the discrete simulation (e.g., a part has been assembled halfway when the user approaches).

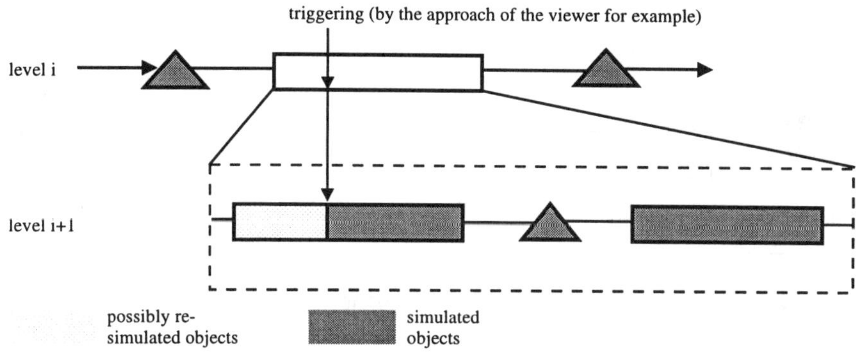

Fig. 3. Detailing of the simulation triggered by the viewer

3.2 Structure of the Concept for Coupling Discrete and Dynamic Motion Simulation

The integrated system must comprise a discrete (material flow) simulation system for event control and management as well as a VR component for visualisation and simulation of the kinematic properties of the manufacturing devices and their environment. Therefore, the simulation is distributed. The viewer influences the simulation by events which he triggers, a bidirectional communication between the simulation systems is necessary. For example, if the user stops the claw arm of a robot this event means a malfunction and non-execution of the accordant transformation for the discrete system as well.

Therefore an interface has to be defined which allows the communication between the two simulators. This interface has to obey the following requirements:

- Bidirectional, synchronous and asynchronous communication
- Enabling of parallel processes
- Control of different consistency conditions

Communication between both models of discrete and dynamic simulation requires further functionality of the interface. A correlation of time models is necessary. The discrete simulator maps the continuous system time into non equidistant events. Presentation on the other hand demands timely equidistant pictures (no jerks). Therefore, a translator is needed which translates the simulator trace into an animation script (cf. fig. 4, direction 1). Vice versa, the translator has to register user dependent changes of the animation script (external events) and write them into the trace with a (rounded) time stamp (cf. fig. 4, direction 2).

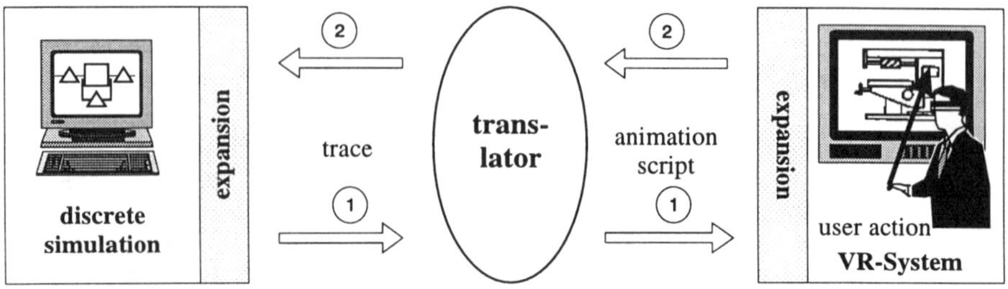

Fig. 4. Interface between discrete simulator and dynamic VR component

3.3 Concept of the Integrated Tool

A concept has to define the functionality of each building block and their integration into one comprehensive tool. Fig. 5 shows an exemplary architecture. The essential building blocks are the discrete simulation system, the VR system, and an integrating framework for both systems. Existing tools for discrete simulation and VR simulation are provided with interfaces (IF) which result from the coupling concept.

The tasks of the integrating framework are:

- Control and translation of the standardized communication (translator).
- Coupling of both submodels in the modelling phase (model generator). Here, checks on semantic consistency base on the hierarchical modelling method. Objects in both models must correspond.
- Debugging in the test phase (debugger).
- Management of references between the distributed objects in the simulation systems and situation dependent control of the objects in their detailing levels at run-time (object manager).
- Synchronization of the system by a common clock pulse.

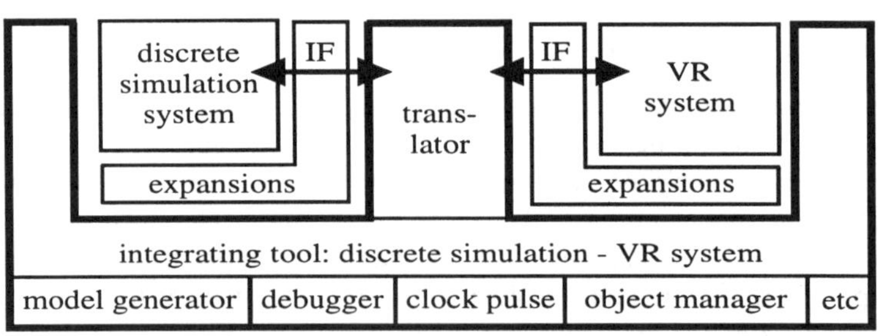

Fig. 5. Exemplary architecture of the integrating framework

The proposed concept of the framework determines the software modules and their requirements. Visualization system and simulation system have to comply the following properties:

- Parallel processes to realize continuous motion and simulation of different objects concurrently.
- Editors for models in both systems with interfaces for queries and manipulation.
- Expandability of events and states.
- Interfaces for coupling to the simulation system / translator.
- 3D graphics and interactive environment in the animation system.

Suitable tools for the integrating framework have to be chosen or developed according to the requirements above. Presumably, there will still be cutbacks on side of the VR systems [15].

4 Conclusions

To achieve fine granular simulations and real-time animations of sophisticated production systems in Virtual Reality with computer hardware which is available today and in foreseeable future a change of modeling concepts is necessary. On the one hand, traditional simulation systems, which are already provided with improved 3D animation using drawings from a CAD system for example, are an approved decision support instrument for planning and optimizing production systems but still lack of applicability, especially interaction, for reformulated deliberations and new, inexperienced user groups. On the other hand, Virtual Reality is a far better technology which allows the user to immerse and interact in a virtual environment but due to the large computing effort of fine granular simulation and animation present VR applications are limited to motion simulation and control of single or isolated groups of machines. The new concept of a hierarchical and situation dependent detailing of simulation reduces the necessary computational expense to a great extent and integrates traditional simulation and VR. The proposed integration tool which re-uses existing simulation and VR components as far as possible enables the user to examine plants to any complexity under different aspects in real-time. The next steps are to elaborate concrete transition methods for the situation dependent detailing and to program the integrating framework.

5 References

1. Ch. Bauer: *Nutzenorientierter Einsatz von Virtual Reality im Unternehmen: Anwendungen, Wirtschaftlichkeit, Anbieter*. München: Computerwoche-Verlag, 1996.
2. K. Büttner et al.: *Effizienzsteigerung bei der rechnerunterstützten Konstruktionstätigkeit durch die Synthese aus CAD und Virtueller Realität.* In: H.-J. Warnecke et al.: (Hrsg.) IPA-/IAO-Forum Virtual Reality '94. Anwendungen und Trends. Stuttgart, Feb. 1994. Berlin etc.: Springer, 1994.
3. K. Carr et al.: *Simulated and Virtual Realities. Elements of Perception*. London: Taylor & Francis, 1995.
4. D. N. Chorafas; H. Steinmann*: Virtual Reality: Practical Applications in Business and Industry*. Englewood Cliffs, New Jersey: Prentice Hall, 1995.

5. W. Dangelmaier; N. Holthöfer; S. Kress; U. Schäfermeier: *Konvergenztendenzen zwischen diskreter Simulation und VR-Animation im Bereich produktionstechnischer Anwendungen.* In: ZWF 91 (1996) 10, S. 486-489.
6. R. A. Earnshaw; Gigante, M. A.; Jones, H.: *Virtual Reality Systems.* London etc.: Academic Press, 1993.
7. Th. Flaig: Virtuelle Welt. *Simulation einer Roboterlinie in Echtzeit mit Hilfe von Virtual Reality.* In: Maschinenmarkt, Heft 25, 1994, S. 62 - 67.
8. Th. Flaig; J.-G. Neugebauer; R. Dänghaus: *Virtual Reality für die Robotersimulation und Off-line Programmierung.* In: CIM-Management Heft 10, 1994, S. 11 - 16.
9. Th. Flaig; Th. Thrainsson: *Virtual Prototyping - New Information Techniques in Product Design. In: Flexible Automation and Intelligent Manufacturing*, Fifth International FAIM Conference, Stuttgart, June 1995.
10. D. Kahaner: *Virtual Reality in Japan.* IEEE Micro, 1993, S. 66-73.
11. H. Kirchner: *Ein Animationswerkzeug für die Simulation.* In: A. Sydow: Fortschritte in der Simulationstechnik, Band 6: Simulationstechnik, 8. Symposium in Berlin, Sep. 1993. Wiesbaden: Vieweg Verlag, 1993.
12. J. Kosturiak; M. Gregor: *Simulation von Produktionssystemen, Teil 2: Marktgängige Simulationssysteme.* In: ZWF 90 (1995) 3, S. 120-124.
13. J. Kosturiak; M. Gregor: *Simulation von Produktionssystemen.* Wien et al.: Springer, 1995.
14. H. Kress: *Integration Aspects within a Virtual Prototyping Environment.* In: Flexible Automation and Intelligent Manufacturing, Fifth International FAIM Conference, Stuttgart, June 1995.
15. J. Lietzau: *RoSi - Ein Modul zur Kopplung des Materialflußsimulationssystem Simple mit dem Bewegungssimulationssystem ROBCAD.* Diplomarbeit, HNI, Paderborn 1995.
16. K.-U. Lindner: *Datengetriebene Simulation in der Fertigung.* Frankfurt am Main; u. a.: Peter Lang GmbH, Europäischer Verlag der Wissenschaften 1994.
17. K. Mertins et al.: *Modellierungsmethoden für rechnerintegrierte Produktionsprozesse: Unternehmensmodellierung, Softwareentwurf, Schnittstellendefinition, Simulation.* München; Wien: Hanser, 1994.
18. R. Meyer: *Kopplung von Simulation mit CAD.* ASIM-Simulationstechnik, 8. Symposium Berlin, Sep. 1993.
19. N. N.: *Jack VR. Analysing the man-machine interface.* Goodwin Marcus Systems, Middlewich, England, 1996.
20. A. Reinhardt et al.: *Simulationsinstrumente - Modellierung und Implementierung.* In: A. Kuhn et al.: (Hrsg.) Fortschritte in der Simulationstechnik, Band 7: Handbuch Simulationsanwendungen in Produktion und Logistik. Braunschweig, Wiesbaden: Verlag Vieweg, 1993.
21. J. Turino: *Managing concurrent engineering: buying time to market.* New York: Van Nostrand Reinhold, 1992.

A Discrete Event Simulation Environment for Statistical Process Control

T.A. Spedding

School of Mechanical and Production Engineering, Nanyang Technological University, Singapore 639798

M. Chandrashekar

Systems Design Engineering, University of Waterloo, Waterloo, Ontario, Canada.

Abstract. This paper describes a discrete event simulation environment for the analysis of Statistical Process Control Systems. The platform for the development is a simulation tool called PROSIM, which permits a user to define components, subsystems, and their interconnections and analyse the resulting system. The components and systems are defined and analysed interactively. A library of standard SPC objects containing models for SPC tools such as charting, capability and costing has been created which help to define the control application. Such a PC based tool is useful at the design as well as trouble shooting stages of a manufacturing system. It is also an effective teaching and research tool.

1. Introduction

This paper explores the potential of using discrete event simulation for analysing and investigating the optimal design of a Statistical Process Control (SPC) system. PROSIM, an object oriented continuous time simulation system based on the Graph Theoretic Method (GTM) is utilised as the basis of the simulation modelling technique. This approach has the advantage that each component or subsystem can be separately programmed, turned into an object, handled independently in time.

The motivation for this work is to develop a *decision support* tool so that the analyst can determine which SPC techniques are the most appropriate under a set of given conditions. Actual conditions may include a complex environment where quality characteristics are subjected to several types of assignable causes. The simulation environment allows the analysis of quality control as a system rather than as specific individual processes so that the performance of the system and its inter and intra relationships can be examined.

After this introduction a brief description of Prosim is presented followed by a discussion of how Prosim can be used to model SPC. The penultimate section presents two examples: a process capability study and an investigation of quality costing, to illustrate the potential of implementing a simulation approach to assist the quality engineering mechanism for a particular system. The findings of the paper are drawn together in the concluding section.

2. The Prosim Simulation Environment

Computer simulation is often an important part of the design and evaluation of systems [1]. It can be used as a prediction tool, to assess the viability of a project; as a diagnostic tool, to study the interactions between various components of a system or reveal inadequate performance of some of them.

In engineering systems analysis, a component forms the basic building block for the development of a system model. Components are interconnected to form subsystems, and subsystems together with components are interconnected to form larger entities called systems. Systems analysis consists of an *equation formulation* phase usually followed by an *equation solution* phase. A typical simulation of a system over time consists of an application of the formulation step followed by repeated application of the solution step over time. A typical parameter sensitivity study involves repeated formulation and simulation steps.

Prosim provides such a flexible environment where a user can define components and subsystems, and interconnect them to create systems, libraries of components and systems for future use, and carry out simulations. Prosim has been developed with the following aims:

1. to provide a single platform which could be used to model phenomena occurring in various engineering fields;
2. to provide a graphical user interface (GUI) through which the system could be defined, assembled, modified, and simulated;
3. to allow the user to model fairly complicated, non-linear systems while at the same time provide automated ways to solve the equations.

The conceptual framework guiding the development of Prosim is the Graph-Theoretic Modelling (GTM) [2] approach. GTM provides a unified methodology for mathematical modelling and analysis of inter-disciplinary systems. The main advantages of the GTM approach are:

a) the capability to model components or subsystems independently, and then interconnect them in any desirable configuration and
b) a matrix based approach which takes advantage of efficient numerical solution techniques.

The GTM approach has been widely used in physical systems modelling, analysis, and design, including electrical and electronic systems, 2D and 3D mechanical systems, hydraulic, and thermal systems. The approach has also been used for probabilistic systems and reliability analysis and discrete event systems applications such as manufacturing systems [3]. The scope of the GTM approach is summarised in Chandrashekar et. al. [2] and presented in Figure 1.

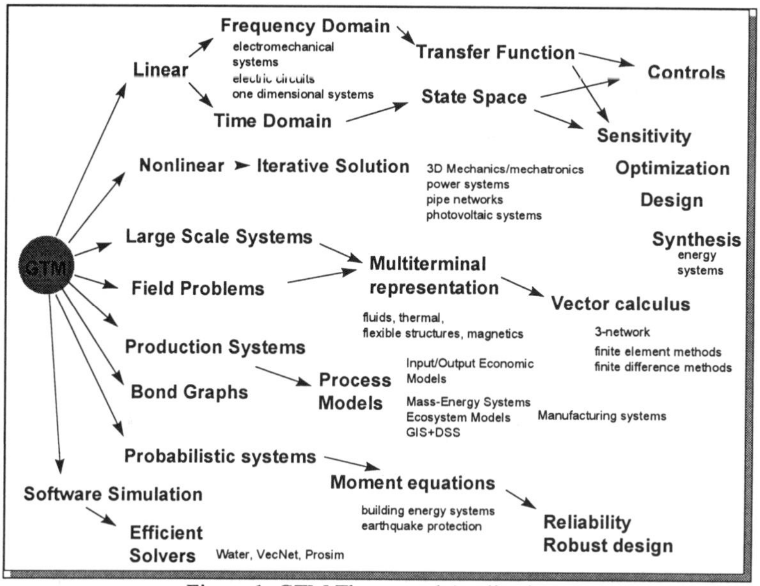

Figure 1. GTM Theory and Applications

Prosim brings together the basic theoretical framework of GTM, symbolic processing and advanced numerical methods, and GUI and object oriented programming methods to create a unique systems modelling and simulation environment.

A key concept in Prosim is that all parts of the system should be modelled independently of each other, then connected together to create the whole system. This creates a *hierarchy* of objects such as the one represented in Figure 2. The elementary building block of the simulation is the *component*, which typically corresponds to a set of parameters and variables related to each other by differential-algebraic equations. Components are then assembled into *systems*, which in turn can be assembled into larger subsystems. The division of a system into subsystems and components is somewhat arbitrary, although it is usually guided by physical considerations.

Once defined, an object (component or system) in Prosim is treated as a black box. The object can be handled by the user as a whole, without any knowledge of encapsulated variables or equations.

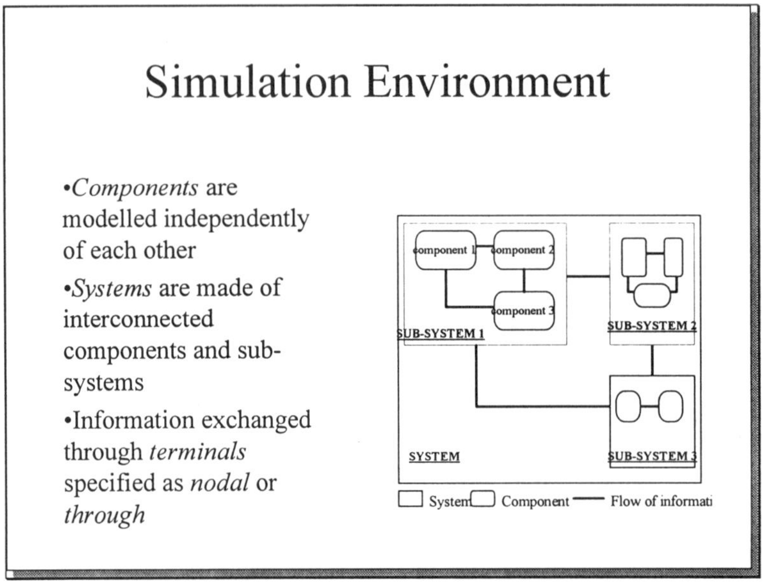

Figure 2 - Hierarchy of Components and Systems.

3 Statistical Process Control Using Prosim

The underlying advantage of simulation techniques or tools is their inherent capability to model variation. Statistical Process Control seeks to control and minimise variation through feedback onto the process thus ensuring a continuous improvement system; based on prevention rather than detection.

Prosim is a time driven system so that events can be naturally programmed in time using a modular programming language. Components, handled as objects, can then be connected together to provide a complete system. Thus components such as random number generation with assignable causes, control charts and feedback operations can be coupled together to provide a complete system of quality operations (Figure 3).

Figure 4, for example shows the coding for an Xbar chart. Once the code for an X bar chart has been developed it can be represented as a single component which can be connected with other components to form a system (also shown in Figure 4).

Once developed, individual components can be stored in a library to form a toolbox of quality components (Figure 5) which, using the copy and paste facility, can be linked together in different system configurations. This approach to simulation offers the quality practitioner several important advantages:

- Simulation models can be used to explore system behaviour
- Modularity of simulation approach provides flexibility
- A Simulation approach is useful in system design

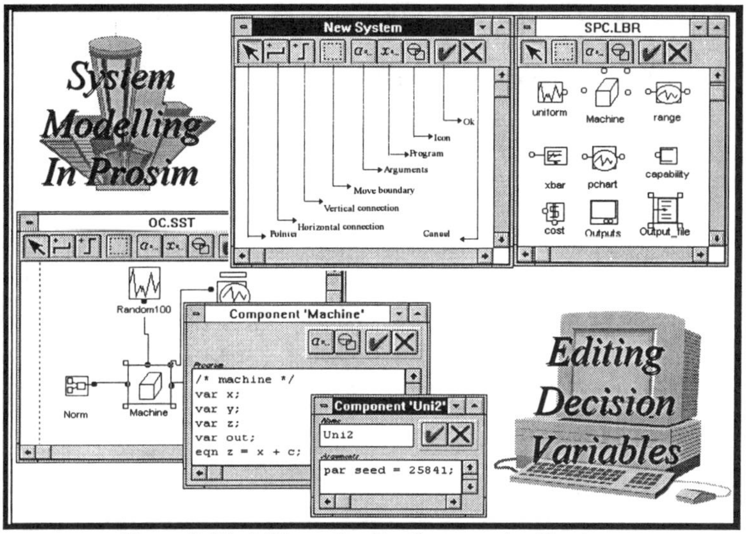

Figure 3 Modelling a Quality System using Prosim

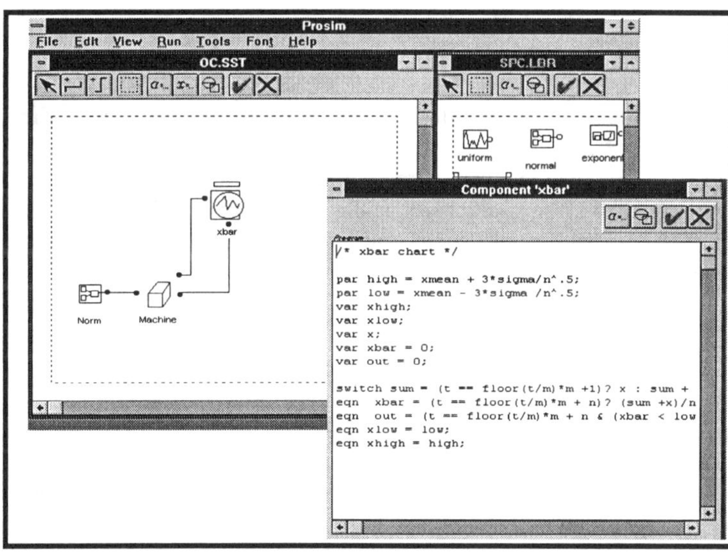

Figure 4 X bar Chart

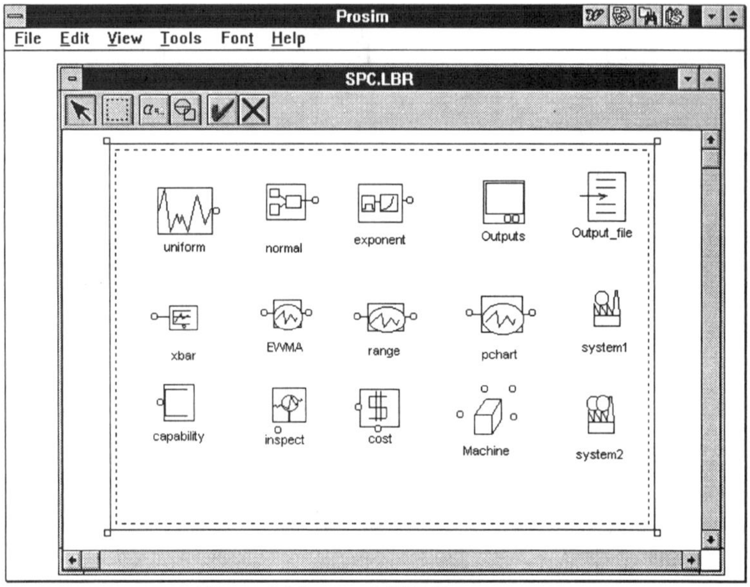

Figure 5 Library of SPC Components

4. Examples and Case Studies

This section illustrates the power and flexibility of using Prosim as a simulation environment to model a quality control system by presenting two case studies. Further examples can be found in Spedding and Chandrashekar [4]. This paper shows illustrations of the power of Prosim to reproduce the operating characteristics and average run lengths of several charts under simple (for validation) and complex (representing a realistic environment) out of control conditions. The paper also presents industrial case studies.

4.1 Process Capability Studies

A study was performed using quality data from a local MNC who produce stamped aluminium and brass components. Here a system was built using Prosim to model the inherent variation of the particular quality characteristic (Figure 6). The out of control characteristics (special causes of variation) were determined by examining control charts from a large amount of data. The out of control conditions were simulated using Prosim by integrating the common (intrinsic variation) and special causes of variation into a single model. Hence the ongoing capability of the system was assessed.

The power of this approach is that the sensitivity of the system's capability can be determined by investigating future situations when the system is not in control. Actions such as process centring can also be assessed or whether management action is required to reduce the variation of the actual process. Such simulation studies can easily be extended to investigate the effects of non-normality and the associated use of either standard Process Capability techniques or more sophisticated analysis (see for example Kotz and Johnson [5]).

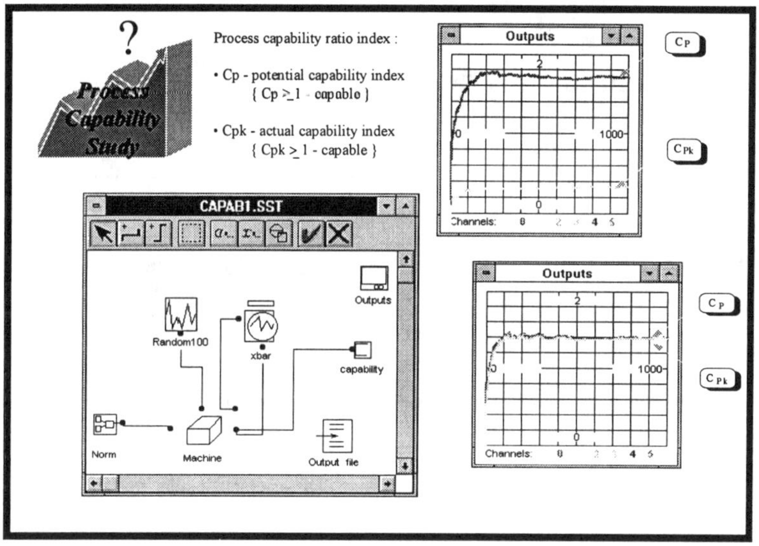

Figure 6 Process Capability Study using Prosim

4.2 Quality Costing

A second example illustrates the potential of using Prosim for quality costing. Here use is made of a component which tracks the out of control situations and the power of a particular charting technique to detect these. Thus the cost of investigating an actual out of control signal and a false alarm can be determined. Also fixed and variable costs for sampling can be evaluated. Results showing the various quality costs can be determined for processes subjected to multiple out of control conditions (Figure 7). The investigation can easily be extended to determine the optimum economic design of the control chart by varying parameters such as the control limits, sample size and sampling frequency.

Theoretical studies have been presented by authors such as Duncan [6] and Montgomery [7]. The advantage of a simulation approach is that it is not constrained by the complexity of the out of control situations (special causes of variation). However use must be made of an optimisation techniques which can be linked to the simulation model results such as Design of Experiments and Response Surface Methodology [8] to determine the optimum configuration for economic control charting.

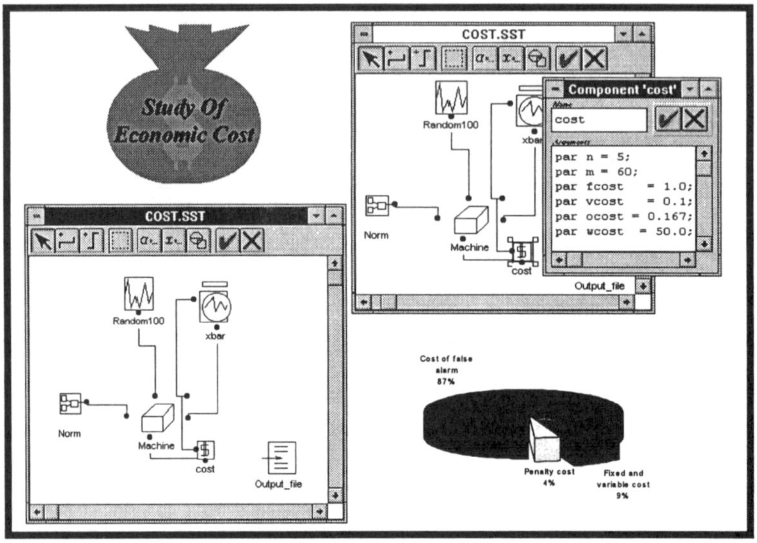

Figure 7 Quality Costing using Prosim

5. Conclusion

This paper has demonstrated how discrete event simulation can be employed for the design and implementation of statistical process control systems. Prosim provides a powerful platform on which to base the simulation. The component based structure of Prosim provides the facility to build rigorous system models with the minimum of effort and the flexibility of the time based programming language allows complex structures to be developed. The tool is based on the concepts of graph theoretic systems modelling approach which has wide acceptance in dynamically systems modelling including manufacturing systems.

Prosim provides a flexible and user friendly environment for creating, testing, and planning different quality control strategies in an interactive manner. The simulation approach presented permits a systematic method for inquiring into different configurations and SPC strategies that are beyond theoretical treatment The engineer is provided with a toolbox of statistical and quality techniques with which to design a simulation model of a system to investigate the optimum configuration.

6. Acknowledgements

The authors would like to thank Seah Yock Peng for help with executing the simulation models and developing the graphical presentation.

7. References

1. F.E. Cellier, *Continuous System Modelling,* Springer Verlag, New York (1991).
2. M. Chandrashekar, P.H. Roe, and G.J. Savage, Graph-Theoretic Models: A Unifying Modelling Approach, *Proc. 23d Annual Pittsburgh Conference on Modelling and Simulation*, Pittsburgh, PA, Part 4 (April 30-May 1, 1992) 213-221.
3. D. Thevenard and M. Chandrashekar Development of a Simulation Platform for Continuous Time Systems Modelling, *IEEE Systems Man Cybernetics Conference*, Beijing, October 14-19, (1996) 2364-2369.
4. T.A. Spedding and M. Chandreshekar, A Component Based Simulation Environment for Statistical Process Control *(In Press).*
5. S. Kotz and N.L. Johnson, *Process Capability Indices*, Chapman & Hall (1993).
6. A.J. Duncan, The Economic Design of X bar Charts when there is a Multiplicity of Assignable Causes, *Journal of American Statistical Association,* 66, (1971) 107-121.
7. D.C Montgomery, The Economic Design of Control Charts: A Renewal and Literature Survey *Journal of Quality Technology*, 12, (1980) 75-87.
8. G.E.P. Box, W.G. Hunter and J.S. Hunter, *Statistics for Experimenters: An Introduction to Design, Data Analysis and Model Building*, John Wiley and Sons, New York (1978)

An Intelligent, Holistic Simulator For A Computer Keycap Assembly Cell

S S G Lee, R B de Souza & T A Spedding

School of Mechanical & Production Engineering
Nanyang Technological University
Nanyang Avenue
SINGAPORE 639798

Abstract. A total simulation environment was developed to model a two-tier, flexible computer keycap assembly cell with multiple servers and a material handling system. It consists of a Windows NT-based graphical user interface created in Microsoft Visual Basic™ ver 3.0, by which the parameters of the cell and other data are input into a simulation engine written in ANSI *C*, and a statistical input/output module, written also in Microsoft Visual Basic™ ver 3.0, for analysing the results of simulation. The user first creates the simulation model by defining six basic simulation blocks: **CREATE**, **SERVE**, **AS/RS**, **PICK-N-PLACE**, **INSPECT**, **DISPOSE**. A *C* routine verifies the simulation model, then passes the model's data to the simulation engine. The simulation engine executes simulation replications after which the statistical module analyses, among other things, the mean time-in-system of the entities and throughput of the servers and displays the mean, standard deviation, skewness and kurtosis on-screen. It can also conduct statistical analyses such as the t, Z, or F-tests to compare the means and variances of two sample data sets. A plot of the moving averages is displayed to assist the user determine the onset of the steady state. An expert system shell written in M.4 can help the novice simulationist analyse the results of simulation and improve the productivity of the cell.

Keywords: System simulation, statistical analyses, expert systems

1. Introduction

In computer simulation, a model of the real world is analysed to give some insight into its dynamic behavior under varying circumstances[1]. Complex real world systems, not easily described by analytical or mathematical models, are ideal candidates for computer simulation. The operations of a real world system and the logical or mathematical relationships among its constituents are formulated in a model based on an underlying set of assumptions[2]. The simulation model is typically replicated over a period of time, and observations recorded and statistically interpreted to estimate the characteristics of the real world system. Simulation is increasingly viewed as an effective, indispensable methodology for analysing and controlling increasingly complex and technical systems[3,4]. Much research has been undertaken to integrate artificial intelligence with simulation [5,6], in order to enhance the advantages of simulation. However, a comprehensive simulation environment with realistic animation and statistical output analysis capabilities is still an important starting point.

This paper reports on the creation of a simulator of a computer keycap assembly cell using Microsoft Visual Basic™[7]. A simulation model of a specific computer keycap assembly cell can be built from icons of pre-defined templates in a Windows®-based graphical user interface (GUI). The simulation engine and the statistical module are then invoked to execute the simulation replications and analyse the output results.

2. The Computer Keycap Insertion Cell

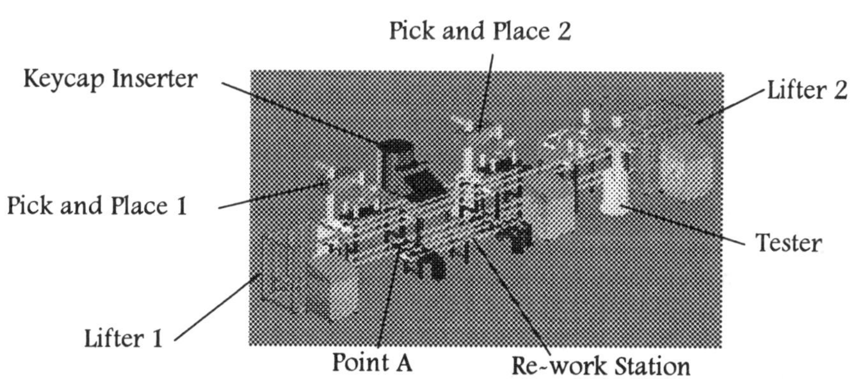

Figure 1. A schematic diagram of the Computer Keycap Assembly Cell

The Computer Keycap Assembly Cell, shown schematically in figure 1, is an experimental assembly cell installed in the School of Mechanical & Production Engineering.

Keyboards are transported on pallets, which are transported from the lower level conveyor to the upper level by Lifter 1. A fresh keyboard, *sans* keycaps, is then manually loaded onto the pallet and conveyed to the "Pick and Place 1" robotic arm which detaches the keyboard from the pallet. The keyboard is conveyed to the keycap inserter station where the keycaps are mechanically-inserted. The finished keyboard is conveyed towards the "Pick and Place 2" robotic arm where it is re-attached to a pallet. However, if a queue forms in front of "Pick and Place 2", the keyboards are temporarily shelved into an automatic storage and retrieval system (AS/RS) located in-between the keycap inserter and the "Pick and Place 2" robotic arm. As soon as the queue clears, the AS/RS retrieves the stored keyboard and places it back onto the conveyor.

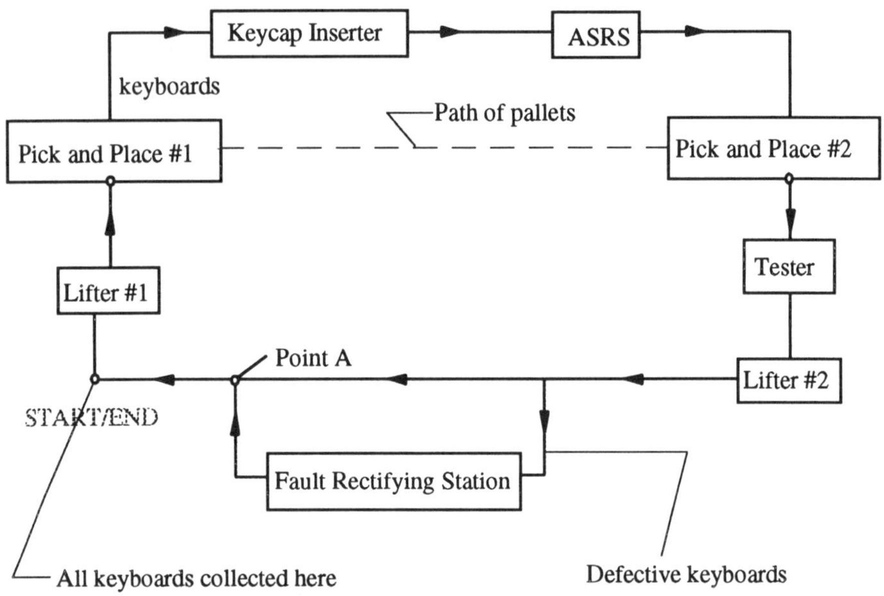

Figure 2. Schematic representation of the Computer Keycap Insertion Cell

The keyboards are inspected at the keyboard tester and the inspection result encoded onto the pallets. Lifter 2 lowers the keyboards, still mounted on their pallets, onto the lower level conveyor. Defective keyboards are re-worked, and the rectified keyboards join the "good" keyboards at point *A* in figure 2 and are conveyed towards Lifter 1, where all keyboards are collected by hand. The free pallets are stacked up in front of Lifter 1, awaiting the arrival of a fresh keyboard. Figure 2 shows a schematic representation of the Cell.

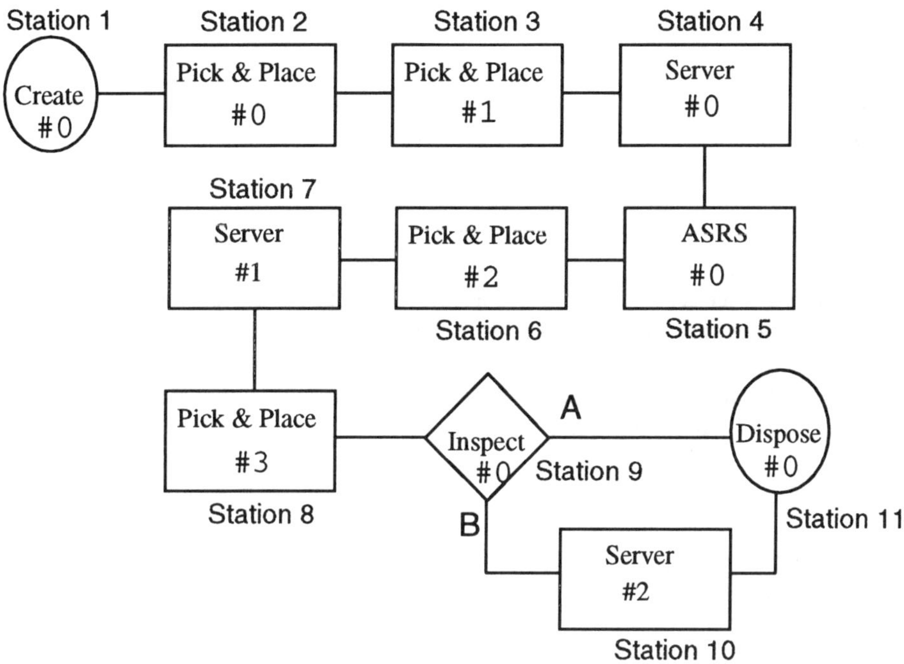

Figure 3. Computer model of the Computer Keycap Insertion Cell.

The computer model of the computer keycap assembly cell is shown in figure 3. The *Create* #0 block is where the keyboards are created or introduced into the assembly cell and wait for an empty pallet to be transported from the lower level conveyor to the upper level. Although the lifter is physically different from the "Pick and Place 1" robotic arm, their operations are similar, so Lifter 1 can be represented by the "Pick & Place 1" block. "Pick and Place #1 and #2 and Lifter 2 are represented by "Pick & Place #1", "Pick & Place #2" and "Pick & Place #3" respectively.

Parts are processed at servers, which can be machines or operators, for a specific duration. *Server #0* corresponds to the keycap insertion station while *Server #1* is the tester and *Server #2* is the re-work station. The *Inspect* block checks if the keycaps are correctly inserted. The good keyboards are conveyed directly to the *Dispose* Block while the defective keyboards are directed to the re-work station, after which, they re-merge with the mainstream to arrive at the *Dispose* block. The *Dispose* block represents the location where the keyboards are removed manually from the assembly cell.

3. Modeling Tools

Microsoft® Visual Basic™ version 3 was adopted as the programming language as it is the quickest and easiest way to create applications for the Microsoft® Windows operating system. It allows the programmer to create attractive, powerful and full-featured applications that exploit the key features of Microsoft® Windows, including object linking and embedding (OLE). Apart from Microsoft Visual Basic™, an off-the-shelf user interface software, *Drag-it™*[8], was used to create natural user interfaces. This software enables the modeler to focus on the design of the graphical user interface without the need to write a complete graphics package. The Watkins *C* simulation library[9] was used in conjunction with ANSI *C* to create the discrete event simulation engine. It can also integrate various simulation libraries into the main program.

4. The Simulation Blocks

Six basic simulation building blocks were created from which the complete simulator may be built. They are *CREATE, PICK-AND-PLACE, SERVER, AS/RS, INSPECT, DISPOSE.* The *CREATE* block creates fresh keyboards or entities and introduces them into the Cell. The *PICK-AND-PLACE* block moves keyboards or entities from one location to another. The time taken to effect the transfer, as well as the time for the robotic arm to return to its initial position, characterise the block. A *SERVER* denotes either a machine or a human operator that works on the keyboard or entity for a specific time duration. The *AS/RS* (Automatic Storage and Retrieval System) block stores keyboards or entities into a storage rack and retrieves them at the appointed time. The *INSPECT* block identifies the defective keyboards and redirects them to the re-work station. The percentage of defective keyboards is pre-determined by the modeler. A random seed is used to generate the desired rejection rate. The *DISPOSE* block figuratively removes the finished keyboards or entities from the simulation model, in order that limited memory space is freed for other simulation tasks.

5. The Simulation Engine

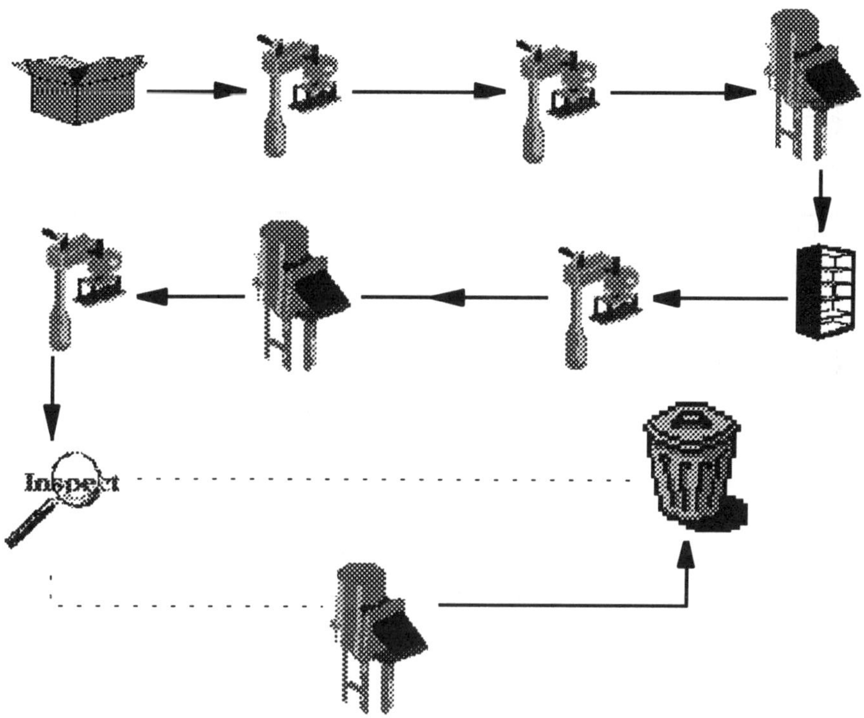

Figure 4. A Schematic of the Simulation Model of the Computer Keycap Insertion Cell

The Computer Keycap Insertion Cell Simulator, a schematic diagram of which is shown in figure 4, consists of 1 *Create* block, 1 *Dispose* block, 3 *Server* blocks, 4 *Pick and Place* blocks, 1 *AS/RS* block and 1 *Inspect* block, making a total of 11 simulation blocks. The model is verified by a "Check" program written in Visual BASIC™.

First, the program tallies the number of *Create, Server, Dispose, Pick and Place, AS/RS* and *Inspect* blocks in the model. It then checks the destination of each block. If no error is detected thus far, the parameters of each block are checked. Whenever an error is detected, further checking is aborted and the errors are displayed in the error dialog box. After checking each and every block in the model, the general conditions for replications are verified.

Only when the model is completely verified will the model parameters, shown in table 1, be defined for the simulation engine.

Table 1. The input parameters for the simulation model.

Code	Comments
3	*Number of replications (integer)*
10000.0	*Simulation time*
123	*Random seed (integer)*
15	*Number of pallets (integer)*
0.1485	*Conveyor speed*
0.8	*Percentage of good keyboards*
2 5.0 0.0 0.0	*Keyboard inter-arrival time*
3 5.0 0.15 0.0	*Server 0 mean processing time*
3 5.0 0.15 0.0	*Server 1mean processing time*
3 5.0 0.15 0.0	*Server 2 mean processing time*
0 2.5 0.0 0.0	*Lifter 0 lifting time*
0 2.5 0.0 0.0	*Lifter 0 lowering time*
0 2.5 0.0 0.0	*Lifter 1 lifting time*
0 2.5 0.0 0.0	*Lifter 1 lowering time*
0 2.5 0.0 0.0	*Pick and Place 0 reach time*
0 2.5 0.0 0.0	*Pick and Place 0 return time*
0 2.5 0.0 0.0	*Pick and Place 1 reach time*
0 2.5 0.0 0.0	*Pick and Place 1 return time*
1 4.0 5.0 0.0	*AS/RS Storage Time*
1 4.0 5.0 0.0	*AS/RS Retrieval Time*
5	*AS/RS Capacity (integer)*

The mean processing times of servers, lifters, pick-and-place manipulators and AS/RS are represented by up to four digits, depending on the probability distribution function used (see Table 1).

Table 2. Coding of the mean processing times of stations

Digit	*1*	*2*	*3*	*4*
Probability Distribution				
Deterministic	0	Single		
Uniform	1	Min	Max	
Exponential	2	Mean		
Normal	3	Mean	Std Dev	
Triangular	4	Min	Mode	Max

The first digit indicates the type of probability distribution used to model the processing time. The remaining three digits are the parameters of the distribution. Thus, for a triangular distribution, the second digit is the minimum and the third digit the modal value, while the fourth digit is the maximum.

The simulation engine is only activated when the input data is verified to be correct. Routines were written for keyboard arrivals, servers, conveyors, lifters, pick-and-place manipulators, and the AS/RS. The simulation engine operates on input parameters such as the mean process times of the servers, the seed value of the random number, and the duration of simulation (see table 1). The program initialises each simulation run, and generates the desired random number. The results of each replication, *viz* the average queue lengths and total processing times of the resources, total production rates and the average time spent by the keyboard in the Cell, are written into ASCII text files and displayed on screen. Also shown are the results of each simulation replication as well as the averages over all the simulation runs and whether the simulation was terminating or non-terminating. With each replication, the display screen is updated. Table 3 shows a typical screen display.

Table 3. The output report produced by the simulation program

Simulation of a Computer KeyCaps Insertion Cell Nanyang Technological University School of Mechanical & Production Engineering					
Simulation Replication : 2			Simulation clock : 16000.00		
Average result across replications			Run : 2		
No. of keyboard processed	:	295.50	No. of keyboard processed	:	292
No. of good keyboards	:	237.00	No. of good keyboards	:	235
No. of bad keyboards	:	58.50	No. of bad keyboards	:	57
			Queue size of Arrival block	:	332
Ave Q size of arrival block	:	161.05	Ave Q size of arrival block	:	164.19
Average time in system	:	4419.71	Average time in system	:	4504.69
Production rate (units/hr)	:	66.69	Production rate (units/hr)	:	66.1
			Simulation started at	:	0.00
			Simulation stopped –>		
			End time reached at	:	16000.00
			Actual time taken	:	17.00 secs
Message : **Simulation completed !** **<Press any key to continue>**					

6. Output Analysis Of Simulation Results

Visual BASIC™ routines were written for the statistical analysis of simulation output for both terminating and non-terminating systems, in particular, the system parameters *time-in-system, time-in-server* (the sum of queue and processing times) and throughput of the servers. Typical statistics for simulation output include; overall mean, overall standard deviation, overall kurtosis, overall skewness, mean of replications, standard deviation of the mean of replications, total number of observations, number of replications, the minimum and maximum number of replications. Statistical tests, such as the t and Z tests for the comparison of two sample means, and the F-test for variance are also provided. To assist the modeler ascertain the on-set of steady state, the moving averages can be plotted. The user selects from a pull-down menu the desired type of statistical test, and the computed statistics are displayed on-screen. The confidence intervals for the tests of significance can also be displayed.

7. Intelligent Analysis

To help the simulationist analyse the performance of the simulated keyboard assembly cell, a rudimentary expert system, incorporating about 120 rules, was built from the M4 expert system shell[10] which runs in the Windows® environment. The simulationist enters the queue lengths, maximum processing times and utilisations of workstations. Since the keycap assembly cell is a serial assembly line, the utilisation of the workstations is a main determinant of throughput. From the experience of operating the Cell, the utilisation rates were categorised as follows:

Table 4. Queue sizes and utilisation rates

Utilisation rate (%)	Workstation is	Avg Q-size/Max Q-size (%)	Workflow is
< 30	Very under-utilised	< 30	Ideal
30 - 50	Under-utilised	30 - 60	Normal
50 - 75	Adequately utilised	60 - 85	Unusual
75 - 90	Fully utilised	>85	Congested
> 90	Over-utilised		

Apart from the queue lengths and utilisations of workstations, the expert system can advise if the average conveyor speed is synchronised with the average processing time of the workstations, or whether the inter-arrival times of keyboards and the number of pallets are sufficient for the keyboards in circulation.

8. Conclusions

A total simulation environment was developed to model a two-tier, flexible assembly cell with multiple servers and a material handling system. It consists of a graphical user interface created in Microsoft Visual Basic™ ver 3.0, by which the parameters of the line and other data are input into a simulation engine written in ANSI *C*, and a statistical input/output module, written also in Microsoft Visual Basic™ ver 3.0, for analysing the results of simulation. The simulation model is created from six basic simulation blocks: CREATE, SERVE, AS/RS, PICK-N-PLACE,

INSPECT, DISPOSE. The simulation engine executes simulation replications for a pre-determined duration of time, after which the statistical module analyses the major performance indices and conducts statistical analysis such as the *t*, *z* or *F*-tests to compare the means and variances of two sample data sets from two different simulation runs.. A plot of the moving averages helps the simulationist determine the onset of the steady state. An expert shell written in M.4 analyses the results of simulation for opportunities for improvement. The authors plan to 0building and simulation, and a facility for virtual reality modeling. The simulation environment can be an invaluable aid for teaching simulation modeling and analysis.

9. Acknowledgements

The authors acknowledge with thanks the support of the School of Mechanical and Production Engineering, NTU and the technical assistance of E K Ong, L W Kee and W K Teo.

10. References

1. Mchaney R, 1994, *Computer Simulation : A Practical Perspective*, Academic Press.
2. Rogers, P.,, & Gordon, R.J., 1993. "Simulation for Real-time Decision Making in Manufacturing Systems", *Proceedings of the 1993 SCS Winter Simulation Conference*, pp 866 - 874.
3. Manivannan S., & Banks, J., 1992. " Design of a Knowledge Based On-line Simulation System to Control a Manufacturing Shop Floor", *IIE Transactions*, **24** : 72 - 83.
4. Pritsker A.B., Grant F.H., & Duket S.D., 1986. "Simulation and Real Time Factory Control", *SME Paper # MS86-212, Proceedings, Real Time Factory Control Conference*, May 13-14, 1986, Michigan, Dearborn.
5. Lee W.L., Spedding T.A., de Souza R.B., & Lee S.S.G., 1995. " Neural Simulation of a Prototype Assembly Line with Adaptive Control", *Proceedings, 3rd International Conference on Computer Integrated Manufacturing*, July 1995, Singapore, vol 2 : 1494- 1498.
6. Lee, W.L., Spedding, T.A., de Souza, R., and Lee, S.S.G. 1996. "Adaptive Simulation of a Keyboard Assembly Cell", *Journal of Computer Integrated Manufacturing*.
7. Microsoft Corporation, 1993, *Microsoft Visual Basic Version 3.0 Programmer's Guide*.
8. Drag-It® / VBX Users' *Manual version 2*, 1994-5, Kelro Software, Calif.

9. Watkins, K. 1993. *Discrete Event Simulation in C.* McGraw Hill Book Company.
10. Cimflex Teknowledge Corporation, 1991. *M.4: A Guide to Developing and Delivering a Knowledge System.*

Collaborative Simulation
Towards a new paradigm

Kees AB & Dr. R. de Souza

CETM, School of Mechanical and Production Engineering
Nanyang Technological Iniversity, Nanyang Avenue, Singapore 639798
Email: mabkees@ntu.edu.sg

Abstract. Modelling of manufacturing and control systems using software tools has traditionally been a serial lifecycle process. Through the different steps of the modelling lifecycle, the user has to gather data, model and test it whereafter it may have to be correct and tested again. This lifecycle or leadtime is long and tedious because of human communication between the parties involved. However, with the explosive growth in the Internet, Email and FTP in specific, parallel and multi communication between people has dramatically changed the face of simulation projects and impacted the simulation lifecycle in particular. New collaborative tools provide the means to substantially reduce lead time by concurrent user modelling using the Internet as a vehicle to communicate with video, audio, chat and whiteboard options.

1. Introduction

Nowadays the use of Internet can speed up the modelling lifecycle tremendously using collaborative windows environments to discuss system models (Grossler,1996) (Fig 1). This means that now we don't have to wait for the system expert to validate the model, all we need is to communicate through the Internet and build and discuss the model concurrently online. Several tools such as Video, Audio, Chat and Whiteboard, which are discussed later, can be used together with the shared modelling application to come to a satisfactory system model. Video Conferencing tools all have the communication features to support a collaborative environment but they miss one feature which is sharing of applications. This is the feature which gives the conferencing tools discussed in this paper the extra functionality useful for collaborative environments.

In simulation studies, the aim is to model and simulate as accurately as possible the real life scenario. This is of utmost importance to provide a acceptable solution to the given problem. The life cycle of a simulation study, as discussed by Osman Balci (1989), provides a guideline for the simulationist in conducting 10 Phases, 10 Processes and 13 Credibility Assessment Stages (CAS). This guideline provided a sequence of processing steps, with a credibility assessment stage after each phase.

Communication and collaboration between parties involved in the simulation study is especially in the CAS a time consuming process

This paper will discuss the impact of collaborative environment tool on the life cycle of a simulation study. In Section 2 the collaborative environment will be discussed with all its features. Section 3 will touch upon Balci's life cycle where after the impact of collaborative environments on simulation studies will be discussed in Section 4. Concluding remarks are given in section 5.

2. Collaborative Environment

Collaboration between people has been common practice for a long time. The ability to collaborate even if people are not at the same place geographically is a new option with emerging collaborative environment tools like Netmeeting. Many software tools, used daily, give the option to start a collaborative environment if discussions with other people is required.

As soon as two or more persons are connected to each other, which can be done either by calling him direct or picking his address from a general directory, one of them has to run the applications which they want to share. In this case it is the modelling tool Arena for system simulation. To share the application the owner (person who runs the program) has to make it shared (Fig. 2) so that the others can see the application and set it to collaborate to allow the others to work online on the application. From this moment onwards anyone in the conference, but one at a time, can make alterations to the model. It is currently impossible for two or more person to make adjustments at the same time because the pointer of the application can be controlled by only one person.

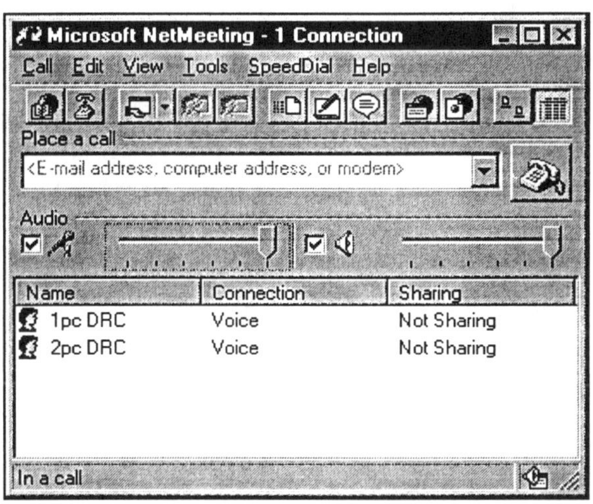

Fig. 1. Two users log-on to Netmeeting and decide on media preferences.

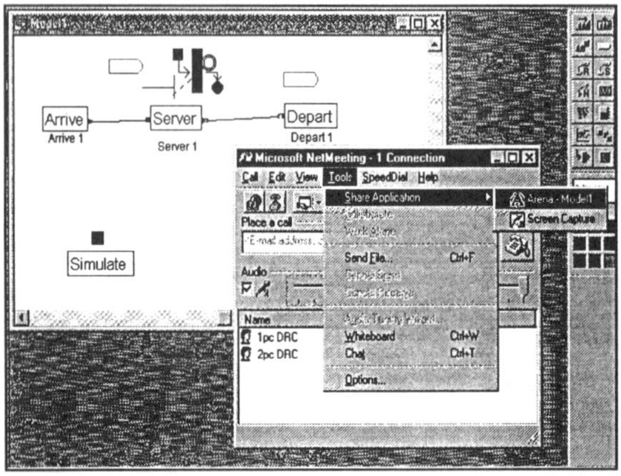

Fig. 2. The simulation software designated as shared and the model built concurrently

Four means to help with communication are Video, Audio, Chat and Whiteboard which will be discussed in the next section.

Video (Fig. 3) gives the conference a more friendly feature because now anyone is able to see the other party on the other end of the line. Just like Audio, Video is only feasible between two parties in the conference, though in ISDN or broadband commincation this may be possible. This is to minimise the amount of data sent through the network.

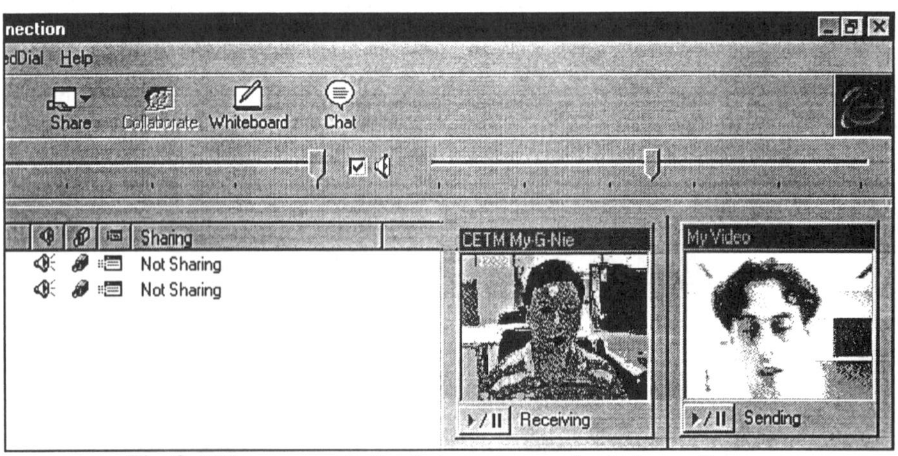

Fig. 3. Video communication between two parties in the conference.

Audio is the most convenient of the two for online discussion but this is only feasible between two persons due to technical half or full duplex constraints of the system. The Chat option (Fig. 4) however allows more than two persons to discuss and manage the collaborative modelling. Chat allows everyone to send types messages to a common window visible to everyone. One particular problem while modelling a system with a group (Who's turn is it?) can easily be solved in this way. Chat can also been used to give online help to the person who is working on the model.

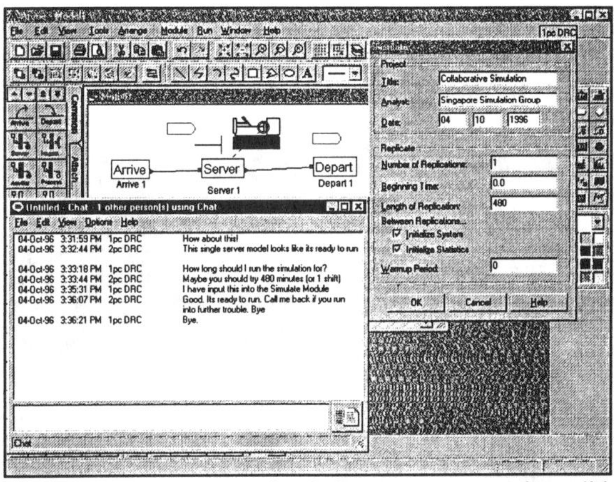

Fig. 4. Chat allows conferencing on simulation parameters and possibly data input.

In case the chat or audio option is not sufficient for the explanation of some modelling problems there is a third way to make things clear, the whiteboard (Fig. 5). The whiteboard gives a range of possibilities for the user to make his point clear. First, he can use text to write a suggestion or explanation. Second, he can cut and paste pictures from any windows application running. Third, he can use drawing tools to indicate important parts. Finally, he can use an indicator hand to stress points in the whiteboard. An important feature is the possibility to save your communication after the collaborative modelling session to verify agreements made during the collaborative simulation modelling.

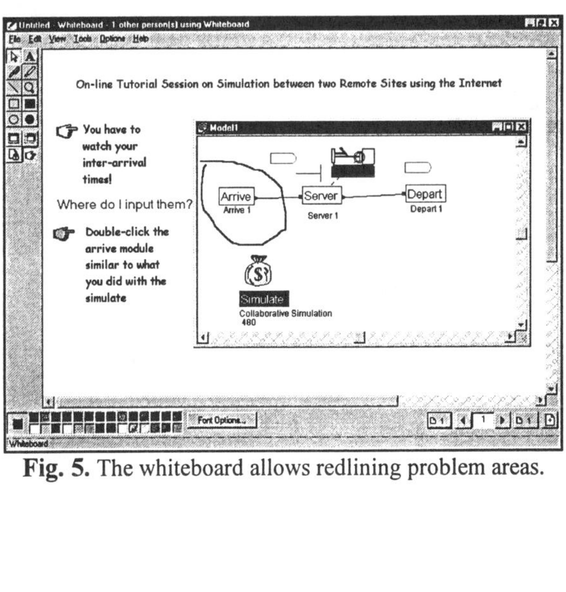

Fig. 5. The whiteboard allows redlining problem areas.

After the collaborative modelling discussion the parties involved can receive a copy of the newly developed model. This can easily be done by selecting the send file option (Fig. 6) whereafter you will be asked to select a file to be sent and also to whom you want to send it. The receiving parties will receive a notice (Fig. 7) that information is sent to them and whether or not they want to accept it. This also allows near real time updating and transfer of simulation data

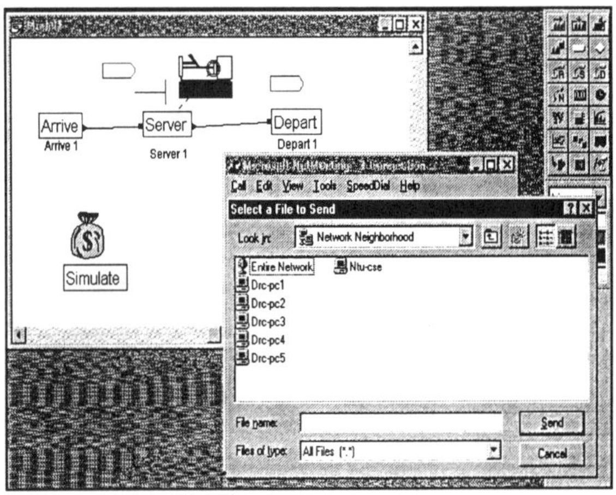

Fig. 6. Select a file to send

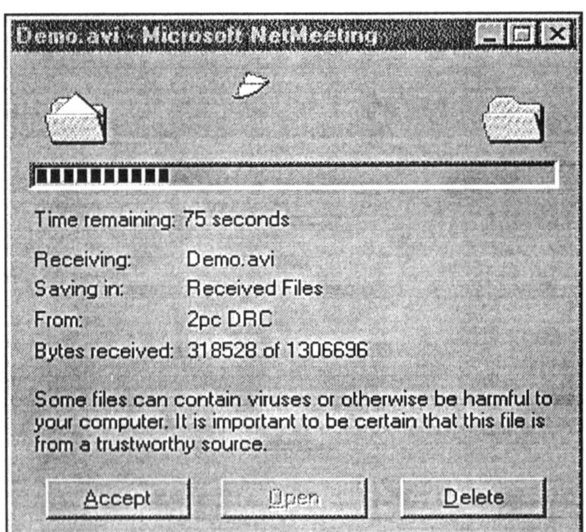

Fig. 7. Receiving a file from a participant

3. The Life Cycle Of A Simulation Study

The Life Cycle of a simulation study (Balci, 1989) consists of ten phases, shown by the boxes, 10 processes, shown by the dashed arrows, and 13 credibility assessment stages which are shown by the solid arrows (Fig. 8). The 10 phases are the corner stones in the life cycle from which one go forward (Processes) or look backwards (Credibility Assessment Stages). In this paper we will not go into detail on describing the phases, processes and credibility assessment stages. This may be found in the Balci paper.

3.1 Phases

The authors believe that the 10 Phases can be divided into 3 parts according to the parties involvement with the phases and their knowledge about the phases. Part 1 will consist of the following phases; Communicated Problem, Formulated Problem and Proposed Solution Techniques.

These 3 phases most likely occur at the location where the problem occurs, i.e. the manufacturing plant, for the knowledge to go through these phases is about the manufacturing process itself. After a Solution Technique is proposed, in this case Simulation, the Life Cycle will go into part 2. This part consist of the following phases; System and Objectives Definition, Conceptual Model, Communicative Model, Programmed Model and Experimental Model. These phases are, in a lot of cases, done by a consultant or research group with expert knowledge of modelling and simulation.

Part 3 consists of the last two phases; Simulation Results and Integrated Decision Support, and will bring the Life Cycle back to the manufacturing site where part 1 started.

3.2 Processes

The ten processes which connect the phases discussed earlier can be, according to the authors, divided into two groups. The first group consists of the processes which link phases within part 1, 2 and 3 and the second group consists of processes which link phases between part 1, 2 and 3. These are the different group divisions:

- Group 1 Processes within part 1 are Problem Formulation and Investigation of Solution Techniques.
- Group 1 Processes within part 2 are Model Formulation, Model Representation, Programming and Design of Experiments.
- Group 1 Process within part 3 Presentation of Simulation results.
- Group 2 Process between part 1 and 2 is System Investigation.
- Group 2 Process between part 2 and 3 is Experimentation.
- Group 2 Process between part 3 and 2 is Redefinition.

The reason for this division is that we think that the communication in Group 1 Processes is of a different level than the communication of Group 2 Processes. Section 3.3 will discussed the difference in the communication levels and its impact on the collaborative environment.

3.3 Credibility Assessment Stages

In the Credibility Assessment Stages (CAS) every phase during the life cycle is verified and validated in relation to its previous phase. According to Bacli (1989), the assessment of the acceptability of the simulation results should be evaluated by a panel of expert peers. This panel should be comprised of people who have expert knowledge of the system under study, expert modellers, expert simulation analysts and people with extensive experience with simulation projects.

This panel examines the overall study based on the project team's presentation at the end of the life cycle. During the life cycle the project team will assess the phases with one or more parties from the panel, what ever knowledge or information is needed. Just like with the processes, we believe that the CAS can be divided into the same two groups. Group 1 with Formulated Problem Verification, Feasibility Assessment of Simulation, Model Qualification, Communicative Model Verification and Validation, Programmed model Verification and Validation, Experimental Design Verification, Presentation Verification and Acceptability of Simulation Results for validation and verification inside each part and Group 2 with System and Objectives Definition Verification for validation and verification between parts 1 and 2.

To improve the life cycle of the simulation study communication between parties has to be improved. It is therefore necessary to identified which processes or CAS take the biggest portion of the time. This changes for every project but in general we can see that communication in Group 2 Processes and CAS, which happen between different parties, can be identified as time consuming. This is where collaborative environments come in useful. Another area where collaborative environments come in useful is in the Processes and CAS in part 2. Using collaborative environments during the 5 phases in part 2 creates a concurrent and iterative process instead of a sequential activity.

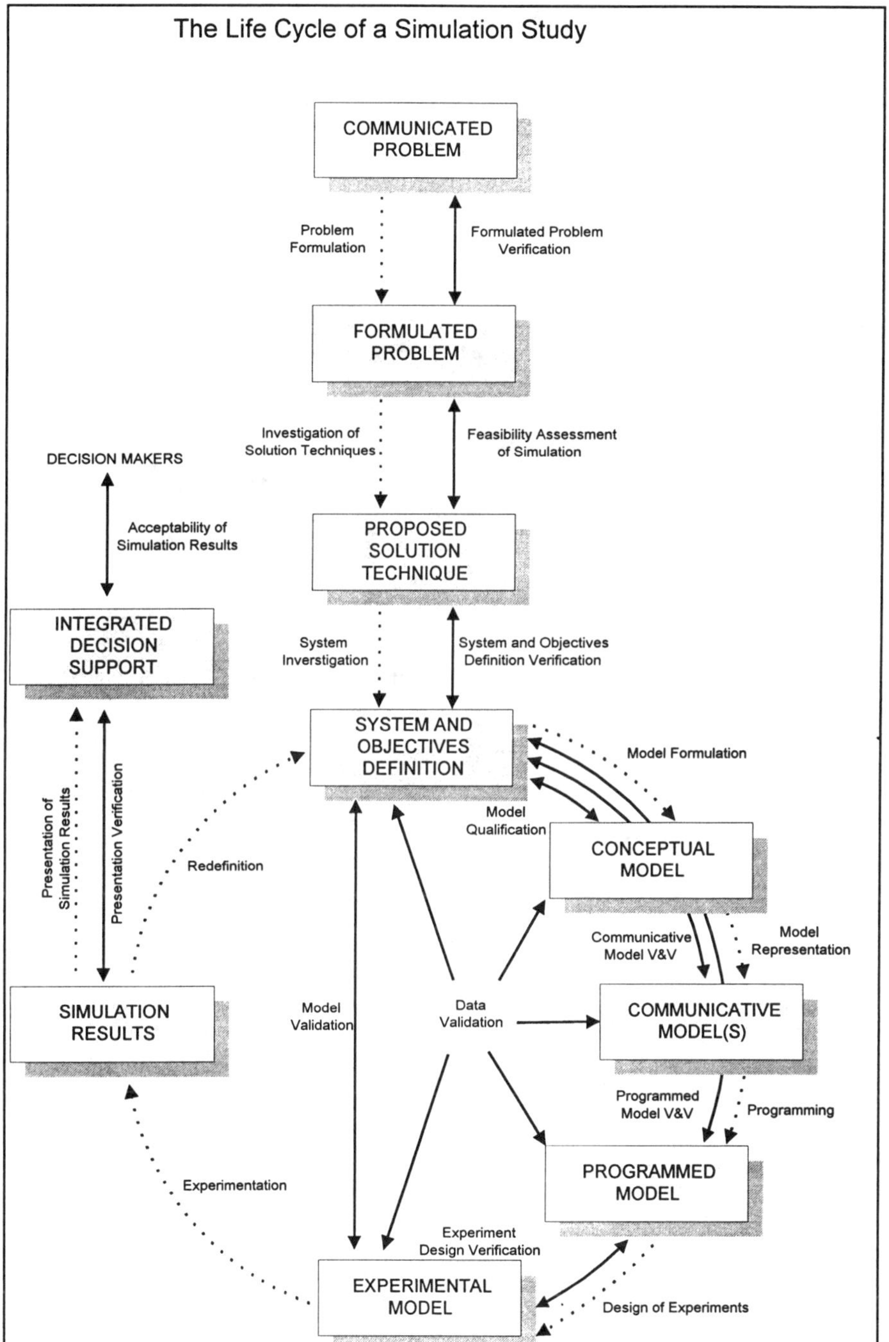

Fig. 8. The Life Cycle of a Simulation Study (Balci, 1989).

4. Improvement In The Life Cycle

As discussed in Section 3, the life cycle can be split up in 3 parts. Part 1 consists of phases inside the company where the problem occurs. Part 2 (Fig. 9) consists of the phases done by the consultant or research group and part 3 returns to the phases in the company to see whether the simulation results are acceptable.

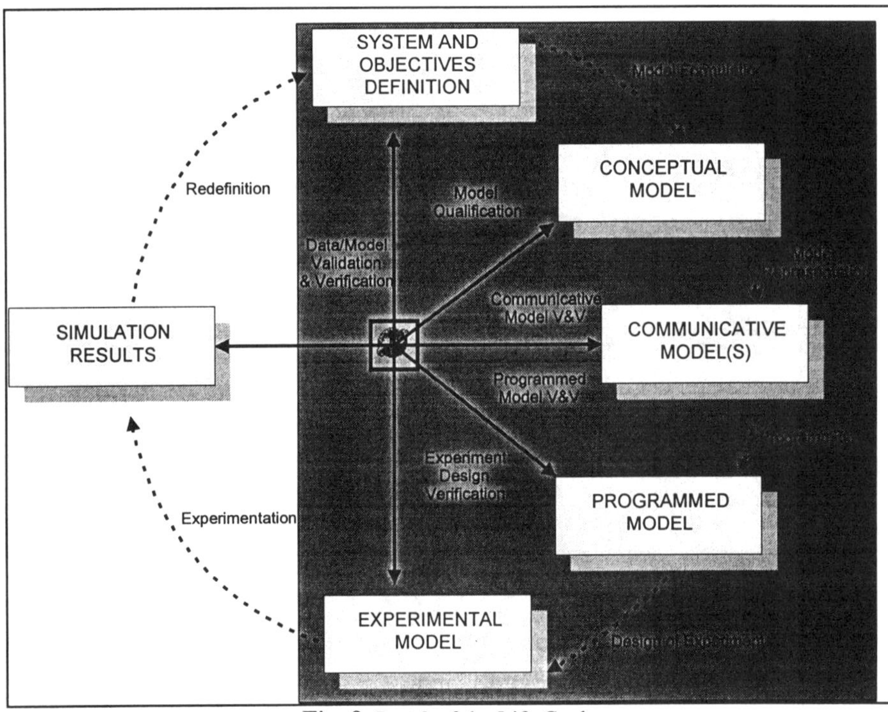

Fig. 9. Part 2 of the Life Cycle.

For part 2, the improved communication with the collaborative tool could result as follows. After the project team calls the system engineers (Fig. 1), The System and Objectives Definition will be achieved via the Collaborative Environment by building a model of the system concurrently (Fig. 2). This improves the speed of work dramatically compared with the old situation when the two parties had to meet somewhere or sent the relevant information to each other. After the System and Objectives Definition is finalised the project team starts formulating the model. During the Model Formulation the team members might need to discuss about the simulation parameters (Fig. 4) to come to a Conceptual Model. Later in the project, after the model has been programmed the Design of Experiment process has to determine the right input variables. The Whiteboard can indicate the location of the input variables, inter-arrival times in this case (Fig. 5), with help of cut and paste pictures and added text explanation.

To create relevant simulation results in the experimentation process real data has to be used. The data has to be sent to the project team. In the collaborative environment this is as easy as copying a file inside your own PC. The only thing which has to be done is to select the file to be send (Fig. 6). The receiving party will be notified that a file is sent and has the change to accept or delete (Fig. 7). As described, only one or two communication features were used during each life cycle step. In real life, more features will be used during each step. As we can see, the process has become concurrent because of the ease of communication and collaboration.

5. Conclusion

As we can see system modelling has grown from a serial user task on a stand alone or local area network (LAN) to a group of concurrent users on the Intranet and nowadays, using collaborative tools, to a global conference on the Internet. Research is ongoing into interpreting how the Balci Life Cycle of a Simulation Study is impacted (Balci, 1989).

References

1. Grossler, A.(1996). Providing simulation models on the Internet. *Industrial Seminar of the Mannheim University, Germany.*
2. Balci, O. (1989). How to assess the acceptability and credibility of simulation results. In *Proc. of the 1989 Winter Conference*. p 62-71.
3. Netmeeting user manual. Internet homepage from Microsoft at http://www.microsoft.com/netmeeting/

Reduction in Product Cost via Manufacturing Simulation

Felix T.S. Chan[1], Ralph W.L. Ip[2], Henry Lau[3], Sev. V. Nagalingam[4] and Grier Lin[4]

[1]Dept. of Industrial and Manufacturing Systems Engineering, The Uni. of Hong Kong.
[2]Dept. of Manufacturing Engineering and Engineering Management, City Uni. of Hong Kong.
[3]Advanced Technology Education Centre, Regency Institute, Australia.
[4]School of Engineering, Centre for Advanced Manufacturing Research, Uni. of South Australia.

Abstract. This paper deals with the simulation study of layout optimisation of an automotive production plant for cost reduction at all levels of manufacturing using Simfactory II.5 simulation package. The case study is an automotive tractor manufacturing company in India. The simulation study conducted shows that the cellular manufacturing system is much better than traditional manufacturing system. This can be seen from the fact that the throughput rate in the cellular system is higher and hence the storing of materials and work-in-process are considerably reduced. Though the operation cost is slightly higher than the traditional layout, we find that other costs have been reduced, such as the resource cost, the product cost and the total layout cost. The changeover from forklifts to conveyors for handling the transportation has also contributed to the above cost reductions. The total quantity of tractors produced at the end of the simulation period showed that the cellular manufacturing system is better than the traditional manufacturing system.

1. Introduction

A manufacturing system is a complex organisation of people, machines, materials, procedures and information. It comprises of a large number of functions, from business planning through materials planning to production. The material handling systems commonly used are fork lifts, conveyors or overhead cranes. Traditional manufacturing systems are characterised by rigid scheduling procedures, heavy machine downtime and non-operative times due to components not available for processing. Cellular manufacturing is used to describe the operations of group technology machining cells. The machining cells may be single machine cell, group machine cell with manual handling or semi-integrated handling systems and flexible manufacturing system (FMS). Cellular manufacturing system (CMS) has been viewed as a link from conventional manufacturing to computer integrated manufacturing (CIM) and the factory of the future. CMS offers the potential to move from inflexible, repetitive batch, mass production to more flexible small-lot production at reasonable costs. Conversion of the conventional production facility into a cellular layout has been made through both the adoption of just-in-time (JIT) production principles and through the advanced manufacturing technology of FMS.

The traditional assembly line manufacturing system for automobiles in the past suffered from throughput and quality problems. Auto manufacturers had to increase the scope of the options being offered, which lead to the further compounding of problems of traditional assembly lines to handle product variation. Efforts to improve the quality through longer cycle times and inspection adversely effected the throughput while lowering cycle times had a negative effect on quality. This lead to the introduction of automated systems like cellular manufacturing and flexible manufacturing systems. By further applying the latest philosophies like JIT it has lead to a considerable reduction in raw material and work-in-process (WIP) inventories. Faster setups and less WIP has shortened the duration of the manufacturing cycle time and also the product lead times.

A major cause for delay in the manufacturing environment is moving materials from one location to another. It is estimated that about 85% of the time in the production system is spent in waiting for transportation, and only about 10% of the time is actually spent in processing[1]. This causes the WIP to swell up and ties up capital that could be used for other purposes within the manufacturing facility. The material handling device could be a forklift, a cart, a worker, or any other form of mover that takes one job from one location to another. The network representation of the model is thus represented as a resource and the material to be moved will have to wait for this resource. Srinivasan[2] is of the opinion that in a trip-based material handling system, the time taken for empty travel and loaded travel must be taken into account for any type of simulation study. Srinivasan[2] considered only specific types of material handling equipment like automated storage/retrieval (AS/AR) systems and automated guided vehicle (AGV) systems when the above conclusion was drawn.

A trend in manufacturing industry today is to focus on lead time reduction as a competitive advantage. Short lead time can be seen as a means to gain a high market flexibility. In the simulation study of a car-body shop Lindau et al.[3] showed that by reducing the time in all aspects of business, it is possible to decrease costs, improve quality and the customer service level. It has been well accepted that simulation can be used to improve production planning and control which is the most complex activity, and any improvements in the efficiency can help in substantial savings. A recent survey on the use of computers and analytical techniques for production planning and control, in the British industry, showed that 17% of the 180 companies which participated in the survey had facilities for testing the implications of the alternative strategies. The aim of this work is to apply simulation technique to reduce the product cost of an automotive tractor manufacturing plant.

2. Overview Of Manufacturing Simulation

Computer simulation is the discipline of designing a model of an actual or theoretical physical system, executing the model on a computer, and analysing the execution output. It is a form of experimentation that involves asking decisions in a simulated environment replicating real world conditions and the main purpose of conducting such experiments is to understand the behaviour of the system and also for evaluating various strategies for the operation of the system. It helps to identify the bottle necks in a system, resource requirements, resource utilisation, operator workload, system throughput, process time and accuracy, cost benefits and training requirements. By using simulation with the help of the best combination of machines, stations, operators and procedures, an increased productivity can be achieved at low cost. By effecting potential changes to an activity or process in the simulation without actually building the prototype, a lot of money can be saved. It can also help in identifying potential safety hazards to humans and thereby increase safety standards. It is useful to build a detailed computer model of a system in order to fully understand its variability and sensitivity to changing conditions. Simulation analysis is a process that requires a considerable amount of user involvement. The complete simulation cycle is given in Figure1.

Three key areas where simulation can be used in manufacturing:

- Testing the system of a new design or part of a system.
- Identifying critical design elements in a system.
- Detecting possible collisions between equipment[4].

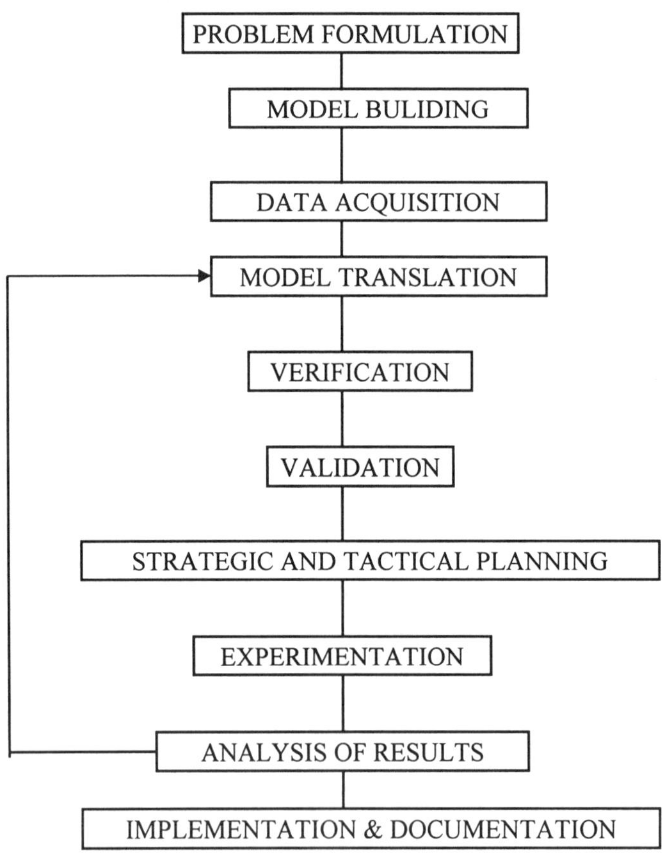

Fig. 1. Simulation Cycle

Just-in-time (JIT) is currently accepted as the philosophy to be followed to match the highly competitive and more open world market. Its idea is the elimination of waste and delay at every stage of manufacturing. The pull production system is followed as a means of production control. The traditional manufacturing system has to be converted to a cellular manufacturing system and the shop floor is arranged as a set of manufacturing cells. To do this, simulation can be used to obtain a design which can take care of the material handling loads, estimate the operator utilisation, determine the reorder levels and also save money by identifying bottlenecks and problem areas and by not investing in the factory directly. Welgama and Mills[5] using SIMAN simulation language in their simulation study of a chemical company in Australia where a change was made from traditional manufacturing system to JIT. The study aimed to show how the material handling load, queue level of orders and lead time varied under normal demand, lesser demand and higher demand. Strategic decision factors and planning decision factors can be easily implemented in a job shop, whereas it is not the same with the shop-level control decision factors. To integrate material requirement planning (MRP) with JIT is not that easy. The most obvious difference between the two is the management philosophy. Ziaul Haq and Faizul Haq[6] developed a simulation model for a job shop, using SIMSCRIPT II.5 simulation language and used the one-card Kanban system. Thus the use of simulation can be justified in the current emerging manufacturing scenario. It has become easier to use, more powerful and more appropriate for regular use by manufacturing mangers. It helps to understand the operations, create accurate models that represent the real life situation, allows it to be tested and evaluated, and all this without making any costly investments and unnecessary disruptions to a facility.

3. Case Study

The case study undertaken for this simulation work is an automotive tractor manufacturing company. It is one of the leading tractor manufacturers in India with its market share alone being about 62%. They manufacture three different types of tractor models - 1035DI, 245DI and 444. They also manufacture tilling and farm equipment. The factory follows the traditional mass production system with different departments processing various components, and they will then be assembled in respective assembly areas, and finally the tractor is assembled in the main assembly section. Figure 2 shows the brief schematic plant layout.

3.1 Traditional Manufacturing Layout

The simulation study for the traditional layout was carried out in two different models. In the first model the layout TRACTOR-1 is as it is seen in the factory today - a functional line layout. The machines are so arranged in the gear shop that they follow the sequence of operation. Buffers are placed between machines to store the items that have been processed by the preceding machine and also for further operations down the line in the succeeding machines. Materials are transported between machines by forklifts. In the second model the layout TRACTOR-2 continues as a functional line layout, but the only difference is that similar machines of the gear shop have been

3.2 Cellular Manufacturing Layout

In the third model, the layout TRACTOR-3 consists of machines arranged into two cells: cell-1 and cell-2. The machines are arranged in such a way that those requiring inter-cell transfer have been placed closer to one another, and transportation of materials in such cases are done by forklifts. Input and Output buffers are used for each cell to store the materials temporary before sending them for their next operations. The drive shaft and the spline shaft are produced by cell-1, and the gear and rear axle are produced by cell-2. In the fourth model the layout TRACTOR-4 is the same as the one discussed in TRACTOR-3 above. The only difference is that no input or output buffers are used to store the materials, and also the inspection department is connected to the two cells and the heat-treatment by means of a conveyor.

3.3 Human Resource Allocation

The allocation of human resources was the responsibility of the shift-in-charge. For layout Tractor-1 the allocation was made on the basis of one-man-one-machine principle. There was one person who has given the responsibility of doing the setting in each machine. They were required to only dismantle the setting done for an earlier operation and to reset the machine for the next one. Overall there was no methodology involved in the allocation of operators. For the allocation of human resources in the second layout Tractor-2, an assumption was made that the allocation could be modified to the extent that one man could operate on two machines simultaneously. This showed that lesser turners, millers, heat-treatment personnel and grinders were required when compared to TRACTOR-1. For the layouts Tractor-3 and Tractor-4, as the machines have been grouped into cells, it was found that the number of operators required was less than the earlier models as one operator could run two machines simultaneously (Table 1).

4. Objectives Of The Simulation Analysis

The simulation for the above layouts were done in order to study the following :

- Layout cost
- Resource cost
- Operation cost
- Raw material consumed cost
- Product cost
- Processing cost
- Current inventory cost
- Number of final products
- Machine utilisation

The models have been constructed and all the assumptions have been validated. Experiments are then executed for all the following factors:

grouped into cells conforming to the operations involved. The lathes have been grouped to produce the blanks, but the milling, gear shaping, gear shaving, tooth rounding and other special spiral tooth generating machines have been grouped to produce the soft gear before heat-treatment. The heat -treatment has been considered as an exclusive cell and the grinding, bench working and inspection have been considered as another machine cell. The assembly area along with the warehouse remains the same as TRACTOR-1. Buffers have been placed only at the exit point of each machine cell.

The other areas of manufacturing are integrated along with the gear shop and the simulation is carried out for the entire manufacturing plant. Conveyors are the mode of material handling system in each of the above machine cells and forklifts have been used sparingly to transport materials between workstations that have not been connected by the conveyors directly.

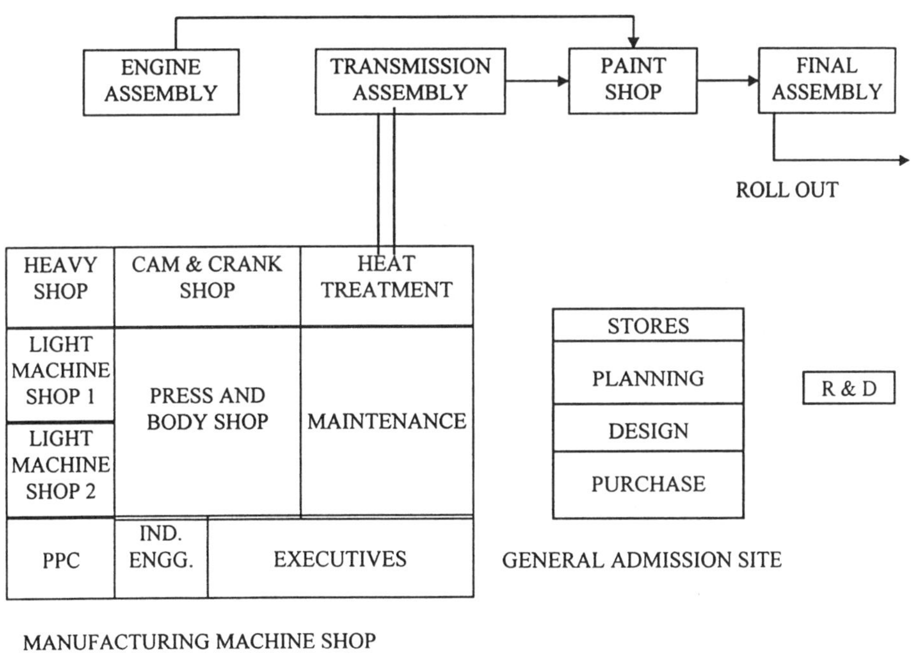

Fig. 2. Schematic plant layout

- length of the simulation run
- number of independent simulation runs
- initial conditions for each simulation run
- length of the warm-up period

Graphs, charts, histograms etc., are used to analyse the simulation results thus obtained.

Table 1. Comparison details of the human resource allocation made in all the four layouts

OPERATOR	DEPARTMENT	TRACTOR 1	TRACTOR 2	TRACTOR 3	TRACTOR 4
TURNERS	GEAR SHOP	14	10	6	6
MILLERS	GEAR SHOP	16	10	6	6
FITTERS	GEAR SHOP	16	10	5	5
HEAT TREATMENT PERSONNEL	HEAT TREATMENT	30	20	10	10
PAINT OPERATORS	PAINT SHOP	20	20	20	20
SHEET METAL WORKERS	PRESS & BODY SHOP	25	25	25	25
ENGINE ASSEMBLERS	ENGINE ASSEMBLY	30	30	30	30
GRINDERS	GEAR SHOP	10	6	3	3
TESTERS	WAREHOUSE	10	10	10	10
FORKLIFT OPERATORS	MACHINE SHOP	30	10	8	7
MACHINISTS	OEC SHOP	25	25	25	25
CLERKS	WAREHOUSE	10	10	10	10
TRANSMISSION ASSEMBLERS	TRANSMISSION ASSEMBLY	30	25	25	25
FINAL ASSEMBLERS	FINAL ASSEMBLY	40	40	40	40
OEC INSPECTORS	OEC SHOP	16	16	16	16
STORES KEEPERS	STORES	12	12	12	12
INSPECTORS	INSPECTION	30	20	10	10
TOTAL		364	299	261	260

5. Results And Discussions

5.1 Number of Parts Produced

The simulation results showed that the four critical components drive shaft, gear, rear axle and spline shaft increased steadily in every layout. Components from the other manufacturing areas also showed a steady increase in the number being produced in each layout. Table 2 given below shows the number of components produced over a given period of time. A comparison for the number of parts produced between traditional and cellular layout using TRACTOR-1 and TRACTOR-3 or TRACTOR-4 values, showed that for all the three types of tractor models the values varied largely. The above results showed that the throughput rate was higher in the cellular layout than in the traditional one and hence more components passing through the cellular system. It can be observed that the quantity of tractors produced increases from 210 in traditional layout to 455 in cellular layout. All these substantial increase is due to the cellular manufacturing and also jointly contributed by the use of conveyors as well as sparse use of forklifts in the latter.

Table 2. Number of Parts produced in all the four layouts

LAYOUT	1035DI MODEL	245DI MODEL	444 MODEL	Total Number
TRACTOR 1	70	70	70	210
TRACTOR 2	70	105	105	280
TRACTOR 3	140	105	140	385
TRACTOR 4	140	175	140	455

5.2 Cost Analysis

Table 3 given below shows the cost details obtained from the final summary cost analysis after the simulation has been completed, these include layout cost, resource cost, operation cost, raw material consumed cost, product cost , the WIP cost and the current inventory cost. A comparison study can be made between traditional and cellular layout using the TRACTOR-1 and TRACTOR-3 OR TRACTOR-4. A graphical representation of product cost for comparison is shown in Figure 3.

Table 3. Comparison of summary cost details for all the four layouts

COST DETAILS	TRACTOR 1	TRACTOR 2	TRACTOR 3	TRACTOR 4
LAYOUT COST	28996604	20896718	21080938	22276024
RESOURCE COST	3.98E+008	4.11E+008	3.99E+008	3.80E+008
OPERATION COST	2.06E+008	2.13E+008	2.18E+008	2.09E+008
RAW MATERIAL COST				
(QTY)	97500	97500	89888	97690
(COST)	18663800	18676720	17319670	18764690
PRODUCT COST				
PRODUCT 1035DI	455324	205329	358851	312947
PRODUCT 245DI	476442	425596	345653	310653
PRODUCT 444	447304	418732	347145	310345
RETESTING	459690	418793	348014	309378
SCRAP	771	921	850	819
CURRENT INVENTORY COST				
(QTY)	244308	244856	243416	241188
(COST)	5.72E+008	5.76E+008	5.15E+008	4.71E+008

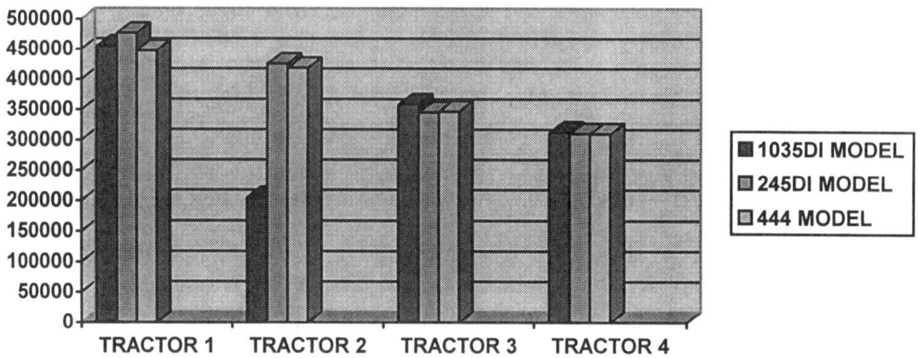

Fig. 3. Graphical representation of product cost for comparison

6. Conclusion

The simulation study conducted shows that the cellular manufacturing system is much better than traditional manufacturing system. This can be seen from the fact that the throughput rate in the cellular system is higher and hence the storing of materials is considerably reduced. The WIP has been reduced in the cellular layout than in the traditional layout. Consequently, the WIP cost and product cost are considerably reduced. Though the operation cost is slightly higher than the traditional one, it has been recovered from other cost saving. The changeover from forklifts to conveyors for transportation has partly contributed to the above reduction. The total quantity of tractors produced at the end of the simulation period showed that the cellular manufacturing system is better than traditional manufacturing system.

7. References

1. Banks, J., " The simulation of material handling system", *Simulation, November*, 1990, pp. 261-269.
2. Srinivasan, M., "Trip-based material handling systems: throughput capacity analysis", *IIE Transactions, Volume 26, Number 1*, 1994, pp. 70-89.
3. Lindau, R. A., Kanflo, T. and Lumsden, K. R., " Impact of real-time information for scheduling a car-body shop - a simulation study", *International Journal of operations and Production Management, Volume 14, Number 3*, 1994, pp. 114-125.
4. Chan, Felix T.S., "Using simulation to predict system performance : a case study of an electro-phoretic deposition plant", *International Journal of Integrated Manufacturing Systems, Volume 6, Number 5*, 1995, pp. 27-37.
5. Welgama, P.S. and Mills, R.J., "Use of simulation in the design of a JIT system", *International Journal of Operations and Production Mangement, Volume 1, Number 9*, 1995, pp. 245-260.
6. Ziaul Haq and Faizul Haq, "Embedding JIT in MRP : The case of job shops", *Journal of Manufacturing Systems, Volume 13, Number3*, 1995, pp. 153-164.

Advanced Methods for Simulation Output Analysis in Manufacturing Systems
Analysis of Variance vs. Artificial Neural Networks

Ulf Eriksson

Department of Manufacturing Engineering, Luleå University of Technology

Abstract. Typical real-world environment contains non-linear, collinear and noisy variables, a properly designed simulation model will reflect that in its operating conditions. Understanding and implementing the sometimes numerous of output data from the model could be a huge challenge in a simulation project, the experiments have to be performed rigorously. This paper presents an exploratory comparison of two methods to use in practical applications for simulation output analysis: the first one is ANOVA commonly used by experimenters in any field, the other one is Artificial Neural Networks a method which gain ground in many statistical applications.

The empirical data was generated by a classical model called "Small Factory Model" controlled by a two-level full factorial design. The result shows that both methods works in combination with factorial design though in this case there was an advantage regarding to artificial neural networks due to it is opportunities to model more than one response at the time.

Keywords. Discrete-event simulation, Factorial Design, ANOVA, Artificial Neural Networks, Decision Support Systems

1 Introduction

Simulation is a technique for developing an understanding of the interaction between the parts of a system, and the system as a whole without having to work with the real system. The level of understanding which can be developed via simulation is difficult to achieve any other way.

A *system* is an entity which maintains its existence through the mutual interaction of its parts. A system exists and operates in time and space [3].

A *model* is a simplified representation of a system, usually over some time period or within spatial limits and is intended to promote understanding of the real system [8].

Simulation is the process of using a model in such a way as to enabling one to perceive interactions that would otherwise not be apparent because of their separation in time or space. This also provides a perspective on what happens within the system, which, because of the complexity of the system, would probably otherwise not be evident [ibid.].

Developing understanding of a system does not mean just building and running a model. To get the model to represent the full range of operating conditions possible in the system, careful experimentation is needed. The more complex the problem the greater the need for good simulation methodology.

A panel discussion at WSC-94, raised the following questions [5]:,

- Are statistical methods used for analysis ?
- How do practitioners view published research on output analysis methods ?
- Is there a need for continued research into the development and investigation of output analysis methods ?

The conclusion was that statistical methods are not often used but that research into improving methods of output analysis should be continued. However, research should be re-focused towards methods that are accessible and understandable by simulation practitioners. A very important concern was that the methods used must be robust and reliable, applicable to a wide range of models/problems and not require specialised knowledge [11].

In the current work, factorial design, Analysis of Variance (ANOVA) and, a method *new* to discrete event simulation, an Artificial Neural Network, were used to control and analyse a discrete event simulation model of a "small factory model" [8]. The ANOVA analysis was performed in Statgraphics, Manugistics Inc. USA, and the Artificial Neural Network analyse performed using a shareware called NNMODEL, Neural Fusion, USA.

1.1 Experimental Method

There are broadly two types of experimental approach used in simulation; interactive and batch. Interactive experimentation allows control of a simulation model while running e.g. by switches and slides. Problems spotted in the model are corrected either in real time, or by changes to the simulation or underlying model [8]. This is sometimes referred to as a "management flight simulator", "learning laboratory" or "micro-world" [10]. It is difficult to examine complex models (systems) with interactive simulation, since the model is only viewed for a relatively short period, however it can give a good explanation of some cause - effect relations in the model.

The batch approach is usually used when comparing alternatives or searching for a target/optimum. Such an approach might involve configuration of two or more parameters which will increase problems associated with planning the experiments. Since experimental factors can be set at many levels there is potentially for a large number of runs to be needed. In general, the use of factorial design is most efficient for this type of experiment. A forum for exchanging experiences and methods for output analysis is the yearly Winter Simulation Conference [11].

1.2 Factorial Design and Analysis

Factorial designs are widely used in experiments involving several factors where it is necessary to study the combined effect of factors on a response.

The most common factorial design is that with k factors, each at one of two levels. The levels could be quantitative, such as batch size, or qualitative, e.g. two different types of machine. A two level "full factorial" replicate requires (see equation 1)

$$2 \times 2 \times \ldots \times 2 = 2^k \tag{1}$$

observations and is known as a 2^k factorial design [9]. A full factorial study is particularly useful in the early stages of experimental work as it requires relatively few runs per factor and can indicate major trends for further experimentation [4].

If the number of factors (k) is high the number of runs in the study increases exponentially. If higher order interactions can be neglected the desired information about the model can be obtained with a fraction of the full factorial design [4]. Also Taguchi design of experiments can be used to reduce the number of experimental runs [6]. However, in the present study, which was performed exactly as the original experiment in [8], two-level full factorial design was used.

1.3 Analysis of Variance (ANOVA)

ANOVA is useful in a range of disciplines. In a One-Way ANOVA analysis each observation is classified or grouped according to one factor. For example, to examine how a particular fertilizer affects the growing rate of corn. If the analyse contains the effects of two or more variables, as is the case in the present work, the technique is called multi-factor ANOVA analysis. The multi-factor ANOVA examines the effects of two or more factors on one variable; typically data collected from a designed experiment.

In ANOVA a test for the significance of each individual factor as well as interaction effects caused by one or more factors acting jointly can be made. The ANOVA table partitions the variability for each chosen response into separate pieces for each of the effects. It then tests the statistical significance of each effect by comparing the mean square against an estimate of the experimental error. A more complete picture of ANOVA is found in [1] and, related to simulation, in [2] and [8]

1.4 Artificial Neural Networks

A neural network is a massively parallel distributed processor that has a natural propensity for storing experiential knowledge and making it available for use. It resembles the brain in two respects [1]:

1. Knowledge is acquired by the network through a learning process.

2. Inter-neuron connection strengths, known as synaptic weights, are used to store the knowledge.

Artificial neural networks are inspired by the human brain but are somewhat primitive compared to the human brain [7].

In the present study a Back Error Propagation (BEP) neural network was used. This type of neural network is constructed from many inner-connected processing elements (neurons) each executing simple non-linear equations. The neurons are partitioned into layers, with each neuron in a layer fully connected to all neurons in the next higher layer, Figure 1. Each connection has a strength or weight associated with it, and a network learns or is trained by modifying these weights. This results in the mapping of inputs to outputs via an abstract hidden layer.

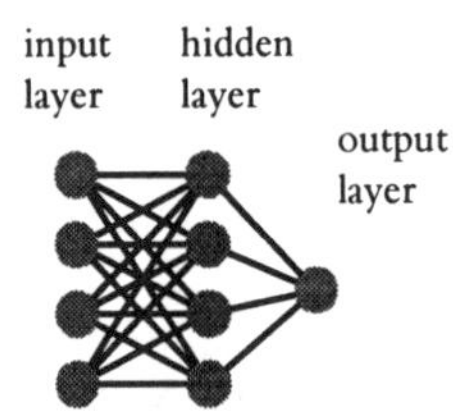

Figure 1. A neural network with four input layers, four hidden layers and one output layer.

After training e.g. with data from a statistically designed experiment, a network can be used to predict the behaviour of multivariate experimental data. Inputs or Independent Variables (X) from the experimental database enter the network at the input layer, and outputs or Dependent Variables (Y) exit at the output layer.

In practice, neural networks are especially useful for classification and function approximation/mapping problems which are tolerant of some imprecision and have lots of training data available. Almost any mapping between vector spaces can be approximated to arbitrary precision [7].

A more comprehensive explanation of Neural Networks can be found in [7].

2 The Simulation Model

The simulation model used in this study originates in Law & Kelton's book Simulation Modeling and Analysis [8] as a tutorial in factorial design ("Small Factory Model"). It is a typical rework process commonly used in both industrial and in informational environment and contains two operations, a machining centre and an inspection station in series, Figure 2.

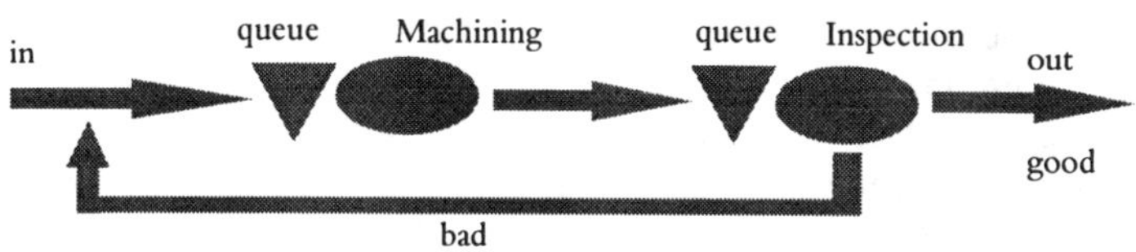

Figure 2. Schematic overview, "Small Factory Model" [8].

Unfinished items arrive, are processed in the machining centre and then checked at the inspection station. Items failing the inspection are returned to the machining station to be re-worked. Since there is only room for one item in each operation any incoming items will have to wait in a queue before the operation if it is occupied.

Parts arrive according to a Poisson process at the rate of one per minute. The performance measurers of interest were the average time that an item spends in the system (makespan) and the average length of the queue for the inspection station (queue length).

Controllable experimental factors, six in total, are:

- machining times,
- inspection times,
- machine uptimes (MTBF),
- machine repair time (MTTR),
- probability of a bad part and
- queue discipline (FIFO and shortest job first).

The purpose of the simulation was to determine which of the production parameters has the greatest effect on the performance measures used. The goal was to reduce both makespan and the average length of the queue for the inspection station. The EXTEND simulator, developed by Imagine That Inc., USA, was used.

2.1 Designing and executing the simulation

The model is assumed to be verified and validated. [8] The experimentation was performed exactly as in the original Law & Kelton [8] tutorial except for the significance level which in this case was increased by 5% to 95%. A two-level full factorial design with 6 factors (2^6) and four replicates (n=5) was used. The complete design matrix is summarised in table 1.

Table 1. Design matrix with "current" system as the lower level (-) and "improved" as the higher level (+).

Factors	Description	current (-)	Improved (+)
1	Machining times (mt)	U(0.65,0.70)	U(0.585,0.630)
2	Inspection times (it)	U(0.75,0.80)	U(0.675,0.720)
3	Machine uptimes (mu)	expo(360)	expo(396)
4	Machine repair times (mrt)	U(7.2,10.8)	U(8,12)
5	Probability of bad part (pbp)	0.10	0.09
6	Queue disciplines (qd)	FIFO	Shortest job first

A warm-up period of 40 hours was used to allow the model to reach a steady state. After the warm up, the model ran for a simulated period of 160 hours. A 2^6 factorial design with four replicates results in 320 runs in total. The result were held in a matrix with 8 columns, six controlled factors and two responses, and 320 rows. As a comparison, the simulation runs in this case took about three hours compared to Law & Keltons (ibid.) 12 hours (Pentium Pro vs. IBM PS/2).

3 Simulation Ouput Analysis

3.1 ANOVA

The results from the ANOVA table are shown in table 2. The value of R^2 (R Square in the table) indicates that the model explains 84% of the variability in makespan and 93% of the variability in queue length. A residual plot which showed no systematic variance indicated that the model was good.

Table 2. ANOVA measures.

Response	Mean	Std Dev
Make span		
Residual	(abs) 0.134298	0.1821
R Square	0.8432	
Queue length		
Residual	(abs) 0.067813	0.1021
R Square	0.9346	

Figure 2 shows standardised effects in a Pareto Chart for the responses makespan and queue length.

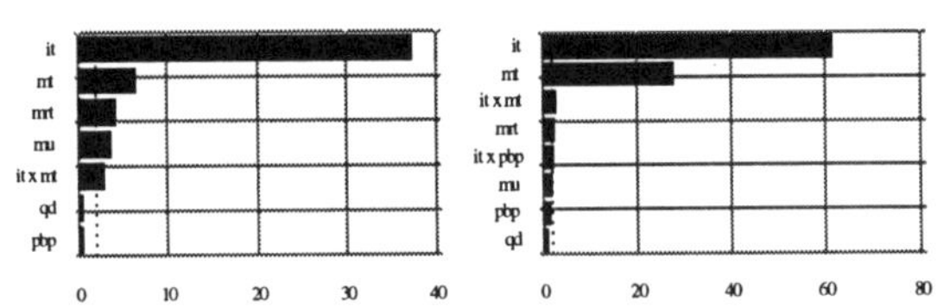

Figure 2. Makespan (left) and queue length effect of significant factors. The dotted line shows the 95% - level.

The most important factors affecting makespan and queue length were found to be the factor it, Inspection time, and the factor mt, machining time. Factors qd and pbp, queue discipline and probability of bad part, had no significant effect on either makespan or queue length and could therefor be regarded as noise. Figure 2. shows that there is a significant interaction between inspection time and machining time (it(mt) but that this is negligible in comparison with the main effects.

A 3d surface plot in figure 3 shows the relative direction of the effects for each response.

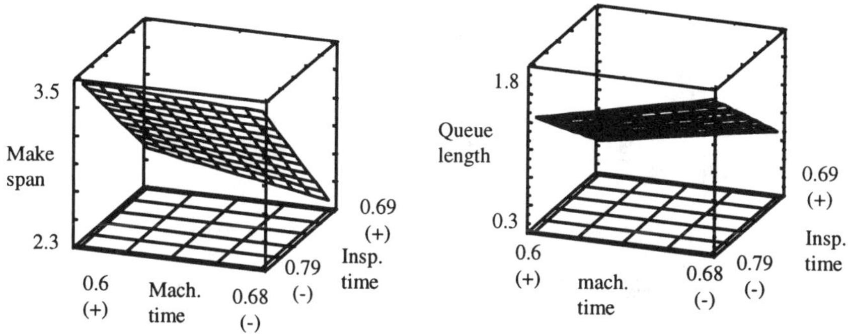

Figure 3. 3d surface plot. The effect of significant factors on the responses makespan (left) and queue length. Note the difference between how much the inspection time affect the responses compared to the effect resulting from machine time.

3.2 Neural Networks

The neural model used was a BEP (Back Error Propagation) with six inputs and two outputs, with four hidden layers, figure 4. The input layer responds to the six input factors and the two output layers responds to the desired performance measurers (makespan and queue length).

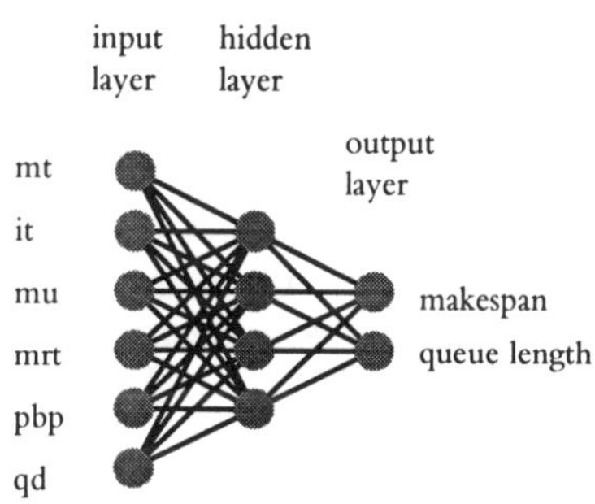

Figure 4. The neural network with six inputs, four hidden layers and two outputs. A diagnostic check of the neural model showed that it gave good predictions for both responses.

Table 3 is a summary of basic performance statistics. A value of R^2 near 1.0 and a rather good residual standard deviation show the model to be reliable. A residual plot was also made to identify any systematic variance.

Table 3. Basic performance statistics of the artificial neural network model, residuals and R^2.

Response	Mean	Std Dev	Min.	Max.	Sum Sq
Make span					
Measured	3.0013	0.4292	2.2635	3.9615	58.7621
Predicted	3.0024	0.3937	2.4709	3.5784	49.4508
Residual	-0.0012	0.1761	-0.4179	0.4401	9.8927
R Square	0.8316				
Queue length					
Measured	1.0412	0.3968	0.5186	2.1402	50.2312
Predicted	1.0406	0.3847	0.5233	1.6595	47.1989
Residual	0.0006	0.0990	-0.3850	0.4807	3.1267
R Square	0.9378				

The taught neural network was felt to be sufficiently good to be used for analysing causality in the model. A sensitivity analysis, table 4, showed that inspection time have the greatest effect on the responses.

Table 4. Sensitivity analysis, shows how the responses are effected by the factors. Minus sign means that the factor correlates in the opposite direction regarding to a response.

Make span		*Queue length*	
Factor	Ave. Sens.	Factor	Ave. Sens.
it	-0.43577	it	-0.44703
mrt	-0.1596	mt	0.32291
mu	-0.12952	mrt	-0.05709
mt	-0.07272	mu	-0.03074
qd	-0.02429	pbp	-0.03824
pbp	-0.04961	qd	-0.01153

Sensitivity is calculated by summing the changes in the output variables caused by moving the input variables by a small amount over the entire training set. The Average Sensitivity variable (Ave. Sens, Table 4) is the average of the values of the change in the output. This value is then divided by the total amount of change for all input variables to normalise the values.

The 3d surface plot in figure 5 gives an overview of how machining time and inspection time are related to the system responses.

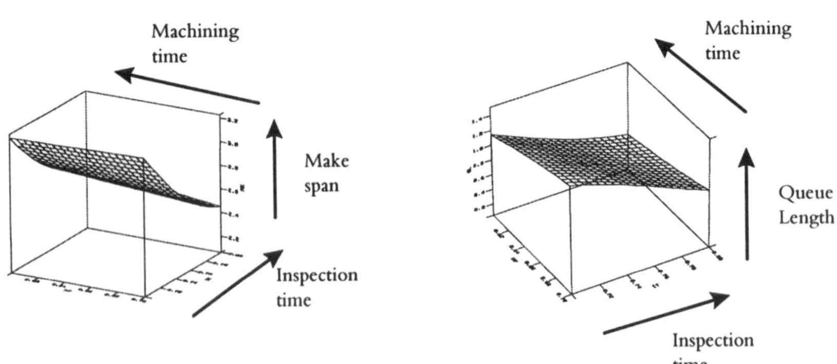

Figure 5. The effect of significant factors on the response makespan and queue length, the arrowheads for machining and inspection points in the "improved" direction.

4 ANOVA vs. Neural Networks

So, which one was the best? As far as the performance measurers of the both methods are concerned, there was no great difference between the methods. The R^2 measure difference was less then 1% (0.097, makespan) and the surface plots are almost identical.

Statisticians fear that neural network "black box" results are impossible to repeat and interpret. While neural networkers ignore distributional assumptions statisticians study the consequences and importance of these assumptions. A further problem is that the jargon associated with statistics and neural networks are different which complicates co-operation between researchers in these two fields [11]. From a practical view point, neural networks are easier to work with because of the possibility to include more than one response in the same time. Another advantage with Artificial Neural Networks is that the underlying design not have to be orthogonal [4].
As far as reliability and variability go, the former is satisfied since repeating this study should give the same results. However validity is impossible to interpret since a generalisation with only this case is doomed to failure. This work indicates that neural networks offer scope for application in simulation, however, more work is required to establish a better theoretical basis for their use in supporting analysis of simulation results.

5 Conclusions

The aim of this exploratory investigation was to compare two alternative methods for analysing output data from a simulation model. The simulation model used to generated the empirical data has its origins in Law & Kelton´s, "Simulation Modelling & Analysis" (1991).[8] The two methods compared were ANOVA and Artificial Neural Networks. ANOVA has a long history in the field of statistical experimentation. Neural networks, on the other hand, have just entered the field but are gaining market share from statistical applications.

The main conclusion of this work is that there is a promising advantage in using neural networks due to their ability to involve more than one response at the same time in the analysis. It was also concluded that the method works well in combination with discrete-event simulation and factorial design. Furthermore access to the world wide web and the availability of easily downloadable shareware applications offers great opportunities for developing new methodologies for performing output data analysis.

Further planned research will explore the use of artificial neural networks in more complex models and other techniques such as genetic algorithms for the analysis and optimisation of manufacturing system models. This will hopefully give simulation practitioners an easier way to present results in a decision making process.

Acknowledgements
This article is a part of research in the PIDUS project (Design, Management and Maintenance of Production system) which is supported by The CIM Institute at Luleå University of Technology and the Swedish National Board for Industrial and Technical Development (NUTEK). The author would like to thank: Professor Anders Kinnander, David Legge and other co-workers in the Department of Manufacturing Engineering; Bertil Pålsson in the Department of Mineral Processing; Carl Bomgardner, Neural Fusion, USA.

References

1. I. Aleksander, and H. Morton, An Introduction to Neural Computing London: Chapman & Hall (1990).
2. J. Banks, and J. S. Carson, II Discrete -event system simulation. Englewood Cliffs, New Jersey: Prentice Hall (1984).
3. L. Bertalanffy, General System Theory: Foundations, Development, Applications, New York: George Braziller (1968).
4. G. E. P. Box, W. G. Hunter and J. S. Hunter. Statistics for experimenters, New York: John Wiley and Sons (1978).
5. J. M. Charnes, and J. S. Carson, et. al. Output Analysis Research: Why bother? A Panel Discussion, Proceedings of the 1994 Winter Simulation Conference, ed. J.D. Tew, et. al.
6. W. Y. Fowlkes and C. M. Crevelling, Engineering Methods for Robust Product Design, Addison-Wesley Publishing Company (1995).
7. S. Haykin, Neural Networks: A Comprehensive Foundation, NY: Macmillan (1994).
8. A. M. Law, and W. D. KELTON, Simulation Modelling & Analysis, New York: McGraw-Hill (1991).
9. D. C. Montgomery, Design and analysis of experiments, New York: John Wiley & Sons (1991).
10. P. M. Senge, The fifth discipline: the art and practice of the learning organization, New ed. - London : Century business (1993).
11. Winter Simulation Conference. http://www.wintersim.org/ (1994, 1995, 1996)

Petri Net Approaches/Soft Modelling Approaches

A Colored Timed Object-Oriented Petri Nets Approach to Modeling Automated Manufacturing Systems

Li-Chih Wang[1] and Shao-Ying Wu[2]

[1] Department of the Industrial Engineering, Tunghai University, Taiwan, R.O.C.
[2] Mechanical Industry Research Laboratories, Industrial Technology Research Institute, Taiwan, R.O.C.

Abstract. This paper presents a new paradigm, called CTOPN (colored timed object-oriented Petri net), to model an automated manufacturing system (AMS). A CTOPN model consists of four major characteristics: (1) the dynamic behavior of an AMS is represented from an object-oriented perspective, (2) the characteristics of material flow (e.g., part routing) and time (e.g., manufacturing time, machine setup time) in an AMS can be fully represented, (3) the control logic/dynamic behavior of an AMS may be explicitly analyzed, (4) an integrated and consistent environment is provided to simultaneously model, schedule and control an AMS. In this paper, we will introduce the definition of a CTOPN and the procedure of developing a CTOPN model.

1 Introduction

An automated manufacturing system (AMS) may be described as consisting of a group of numerically controlled machines and buffers interconnected by the material handling equipments (e.g., robots). Operations of each facility and part/material flows in an AMS are usually controlled by a host computer through a programmable logic controller (PLC). In order to maximize the performance of an AMS, it is necessary to develop an effective and timely mechanism which can not only generate an efficient operations schedule, but also guarantee the validation of the dynamic behavior of that AMS (e.g., deadlock-free, conflict-free).

Due to the characteristics of the graphical representation and mathematical analysis of the control logic of a manufacturing system, Petri nets and their variations, such as colored Petri nets (CPNs) and timed Petri nets (TPNs), have been widely applied to model, control, and schedule a system, such as an AMS, whose activities are concurrent and asynchronous in nature [1,2,3,7]. However, a Petri net-based model is highly system dependent and lacks properties, such as modularity, reusability and a high degree of maintenanability, that are commonly required in modern manufacturing control systems. Although a number of researchers have focused on developing object-oriented manufacturing control softwares[8,9], however, they lack the capability of mathematically analyzing the control logic of an AMS that are owned by Petri net models.

Recently, the development of object-oriented Petri net approaches for manufacturing systems have been studied. Lee and Park [5] develop an object-oriented high-level Petri nets (OPNets) model to increase the maintainability and reusability of objects in Petri net modeling for real-time manufacturing systems. Wang [10,11,12] developed an object-oriented Petri net (OPN) paradigm which not only owns the characteristics of Petri net and OOP, but also incorporates the control/decision knowledge (e.g., scheduling/dispatching knowledge) in the control logic. However, time characteristics of operations/activities in an AMS are not considered in the OPN model, therefore, operations scheduling function cannot be effectively performed by the OPN approach.

In this paper, a new paradigm, called CTOPN (colored timed object-oriented Petri net), which owns the characteristics of colored timed Petri nets and object-oriented programming is introduced to model the dynamic behavior of an AMS. Both the dynamic behaviors inside each facility and among the related facilities are modeled using the OPN approach described in Wang [10,11], then colored tokens, the transform functions and time-delay attributes are added to construct a CTOPN model according to the characteristics of material flow (e.g., part routing) and time constraints (e.g., processing times) in an AMS. After a CTOPN model is analyzed (e.g., deadlock-free, boundedness), it can be used to generate an effective operations schedule for an AMS by modifying some scheduling algorithms, such as the L1-search algorithm described in Lee and DiCesare [6], then the control system may be derived according to the specification (i.e., control logic) of the OPN control model and the near-optimal operations schedule generated by the CTOPN model.

The remainder of this paper is organized as follows. Section 2 describes the definition of a CTOPN model. The procedure of constructing a CTOPN model is presented in section 3. Then, the application of a CTOPN approach in deadlock analyzing 、 scheduling and controlling an AMS is briefly described in section 4. Finally, conclusions are given.

2 The Definition of A CTOPN Model

From an object-oriented perspective, a manufacturing system is composed of a number of related physical objects (i.e., manufacturing facilities); each physical object has its own behavior characterized by the methods and the attributes or status. Mathematically, a system, $\boldsymbol{S}$, may be defined as (see Equation 1):

$$\boldsymbol{S} = (\,\boldsymbol{OPS}, \boldsymbol{R}, \boldsymbol{C}, \boldsymbol{D}, \boldsymbol{I}, \boldsymbol{Inh}, \boldsymbol{O}, \boldsymbol{M_0}\,) \tag{1}$$

in which

i = index of object place (m: total number of object places)

j = index of activity transition (n: total number of activity transitions)

$\boldsymbol{OP_i}$ = object place i which corresponds to physical object i in the system (i = 1, 2, ..., m)

$\boldsymbol{G_j}$ = an activity transition j (j = 1, 2, ..., n)

$\boldsymbol{OPS}$ = a set of object places in the system $=\{\boldsymbol{OP_1},\boldsymbol{OP_2},\ldots,\boldsymbol{OP_m}\}$

$\boldsymbol{R}$ = a set of activity transitions = $\{\boldsymbol{G_1},\boldsymbol{G_2},\ldots,\boldsymbol{G_n}\}$; $\boldsymbol{OPS} \cup \boldsymbol{R} \neq \varphi$, $\boldsymbol{OPS} \cap \boldsymbol{R} = \varphi$, $\boldsymbol{R} = \boldsymbol{R_1} \cup \boldsymbol{R_2}$ in which $\boldsymbol{R_1}$ is a set of *ordinary* activity transitions and $\boldsymbol{R_2}$ is a set of *hierarchical* activity transitions, $\boldsymbol{R_1} \cap \boldsymbol{R_2} = \varphi$.

$\boldsymbol{C(OP_i)}$ = a set of colored tokens associated with object place $i=\{cp_{i1}, cp_{i2},..., cp_{iu_i}\}$ in which u_i represents the types of colored tokens that $\boldsymbol{OP_i}$ may have.

$\boldsymbol{C(G_j)}$ = a set of colored tokens associated with activity transition $j=\{cg_{j1}, cg_{j2},..., cg_{jv_j}\}$ in which v_j represents the types of colored tokens that $\boldsymbol{G_j}$ may have.

$\boldsymbol{D}$: $\boldsymbol{C(OP)}\rightarrow R^+$ is a time-delay attribute transform function for every colored token in each object place in the CTOPN model and R^+ denotes the positive real number.

$\boldsymbol{I(OP, G/cg)}$: $\boldsymbol{C(OP)}\times \boldsymbol{C(G)}\rightarrow \boldsymbol{C(OP)}\cup \varphi$ is a transform function for an arc which connects from object place $\boldsymbol{OP}$ to activity transition $\boldsymbol{G}$ with firing colored token cg ($\boldsymbol{OP} \in \boldsymbol{OPS}$, $\boldsymbol{G} \in \boldsymbol{R}$, $cg \in \{\boldsymbol{C(G_j)}|j=1,2,...,n\}$).

$\boldsymbol{Inh(OP, G)}$: $\boldsymbol{OPS}\times \boldsymbol{R}\rightarrow\{0,1\}$ is the weight function for an *inhibitor* arc which connects from object place $\boldsymbol{OP}$ to activity transition $\boldsymbol{G}$.

$\boldsymbol{O(OP, G/cg)}$: $\boldsymbol{C(OP)}\times \boldsymbol{C(G)}\rightarrow \boldsymbol{C(OP)}\cup \varphi$ is a transform function for an arc which connects from activity transition $\boldsymbol{G}$ with firing colored token cg to object place $\boldsymbol{OP}$.

$\boldsymbol{M_0}$ = the initial marking of the CTOPN model; and $\boldsymbol{M_0}=[\boldsymbol{M(OP_i)}]_{m*1}$ (i=1,2,...,m) in which $\boldsymbol{M(OP)}$: $\boldsymbol{OPS}\rightarrow \boldsymbol{C(OP)}$.

Each object place, $\boldsymbol{OP_i}$, in a CTOPN model corresponds to a physical object (e.g., manufacturing facility), i, in an AMS; each activity transition, $\boldsymbol{G_j}$, and its associated input transform function, $\boldsymbol{I(OP_{i'},G_j/cg_j)}$, colored tokens set, $\boldsymbol{C(G_j)}$, and output transform function, $\boldsymbol{O(OP_{i''},G_j/cg_j)}$, are constructed to represent the message passing relationships between the related object places $\boldsymbol{OP_{i'}}$ and $\boldsymbol{OP_{i''}}$ ($i' \neq i''$; i', i'' = 1, 2, ..., m) according to the part flow characteristics. Furthermore, the colored tokens, $\boldsymbol{C(OP_i)}$, and the associated time-delay attributes, $\boldsymbol{D}(cp_i)$ $cp_i\in \boldsymbol{C(OP_i)}$, for each colored token in object place, $\boldsymbol{OP_i}$, are used to represent the state of that physical object and the duration that state resides, respectively. In a CTOPN model, the internal behavior of a physical object is encapsulated, we may mathematically represent the internal behavior of object place i, $\boldsymbol{OP_i}$, as(see Equation 2):

$$\boldsymbol{OP_i} = (\boldsymbol{SP_i}, \boldsymbol{IM_i}, \boldsymbol{OM_i}, \boldsymbol{AT_i}, \boldsymbol{Pre_i}, \boldsymbol{Post_i}) \quad (2)$$

in which

$\boldsymbol{SP_i}$ = a set of state places in object place i = $\{sp_{i1}, sp_{i2},...,sp_{ia_i}\}$ in which a_i is the total number of state places in $\boldsymbol{OP_i}$

$\boldsymbol{IM_i}$ = a set of input message places for object place i = $\{im_{i1}, im_{i2},...,im_{ib_i}\}$ in which b_i is the total number of input message places of $\boldsymbol{OP_i}$; $\boldsymbol{SP_i}\cap \boldsymbol{IM_i}= \varphi$

$\boldsymbol{OM_i}$ = a set of output message places for object place i = $\{om_{i1}, om_{i2},...,om_{ic_i}\}$ in which c_i is the total number of output message places of $\boldsymbol{OP_i}$; $\boldsymbol{IM_i}\cap \boldsymbol{OM_i}= \varphi$, $\boldsymbol{SP_i}\cap \boldsymbol{OM_i}= \varphi$, $\boldsymbol{SP_i}\cap \boldsymbol{IM_i}\cap \boldsymbol{OM_i}= \varphi$

$\boldsymbol{AT_i}$ = a set of activity transitions in object place i = $\{at_{i1}, at_{i2},...,at_{id_i}\}$ in which d_i is the total number of activity transitions in $\boldsymbol{OP_i}$

$\boldsymbol{Pre_i}$ ($\boldsymbol{Post_i}$): $\boldsymbol{SP_i}\times \boldsymbol{AT_i} \rightarrow\{0,1\}$and $\boldsymbol{IM_i}\times \boldsymbol{AT_i}$ ($\boldsymbol{OM_i}\times \boldsymbol{AT_i}$)$\rightarrow\{0,1\}$ is an input (output) function for each activity transition in object place i

The dynamic behavior inside a physical object i is characterized by the state places $\boldsymbol{SP_i}$ and activity transitions $\boldsymbol{At_i}$, the timing at which an internal activity/method at_{it} (t = 1, 2, ..., d_i) can be performed only when its connected *input* state places (SP_i) and/or message places (IM_i) have the required tokens as specified in its input function Pre_i. Once an activity at_{it} is fired, the suitable colored tokens will be put in its connected *output* state places (SP_i) and/or message places (OM_i) as specified in its output function $Post_i$.

In order to describe the detailed communication (i.e., message passing) process among the distinct physical objects in an AMS, activity transition G_j (j=1,2,...,n) may be further defined as(see Equation 3):

$$G_j = (GOM_j, GIM_j, GH_j) \quad (3)$$

in which

GOM_j = a set of output message places at different object places which are the inputs of G_j = $\{gom_{j1}, gom_{j2},..., gom_{je_j}\}$ in which e_j is the total number of connected output message places; and $gom_{jl} \in \{OM_i | i=1,2,...,m\}$, $l=1, 2,..., e_j$

GIM_j = a set of input message places at different object places which are the outputs of G_j = $\{gim_{j1}, gim_{j2},..., gim_{jf_j}\}$ in which f_j is the total number of connected input message places; and $gim_{jl} \in \{IM_i | i=1,2,...,m\}$, $l=1, 2,..., f_j$

$GH_j = (GP_j, GT_j, GI_j, GO_j)$ is a hierarchical description of activity transition G_j. In other words, if $G_j \in R_2$, then:

GP_j = a set of state places which are comprised in G_j = $\{gp_{j1}, gp_{j2}, ..., gp_{jg_j}\}$ in which g_j is the total number of state places in G_j

GT_j = a set of activity transitions which are comprised in G_j = $\{gt_{j1}, gt_{j2}, ..., gt_{jh_j}\}$ in which h_j is the total number of activity transitions($\in R_1$) in G_j

GI_j (GO_j): $GP_j \times GT_j \rightarrow \{0,1\}$ is an input (output) function for each activity transition comprised in G_j

3 The Development of A CTOPN Model

The development of a CTOPN model for an AMS consists of four phases: (1) requirement analysis, (2) construction of the OPN for each individual physical object in the AMS, (3) construction of the system OPN model, and (4) construction of the CTOPN model. The detailed description of the first three phases are described in Wang [10,11], readers who are interested in the application of these approaches may refer to those papers. In this paper, we will focus on the description of the fourth phase.

Let us first consider an AMS consisting of one lathe, one milling machine, one I/O buffer, one WIP buffer linked by a robot; two different part types, A and B, are processed in the AMS. The routings of type A and type B are I/Obuffer-Lathe-WIPbuffer-Milling_machine-I/Obuffer and I/Obuffer-Milling_machine-WIPbuffer-Lathe-I/Obuffer, respectively. For illustration, the OPN of object 'robot' and the system OPN model are depicted in Figures 1 and 2, respectively [11]. Figure 1 shows that the robot OPN has four activity transitions (at_{31}-start loading, at_{32}-complete loading, at_{33}-start unloading, at_{34}-complete unloading) and three state places (sp_{31}-loading, sp_{32}-idle, sp_{33}-unloading). Currently, the robot is idle and available (i.e., sp_{32} has a token), the robot will start loading a part (at_{31} is fired) if 'a machine request for loading' message sent from the related OPN (i.e., Lathe or

Milling_machine) is received at the input message place im_{31}. Then the robot will be in the state of 'loading a part' (sp_{31}). Once the robot completes loading (at_{32} is fired), the robot will be idle/available (sp_{32}) again and the 'complete loading' message at the output message place om_{31} will be sent to either input place im_{43} of lathe or input place im_{53} of milling machine depending on the firing decision made by gate G_3 (see Figure 2).

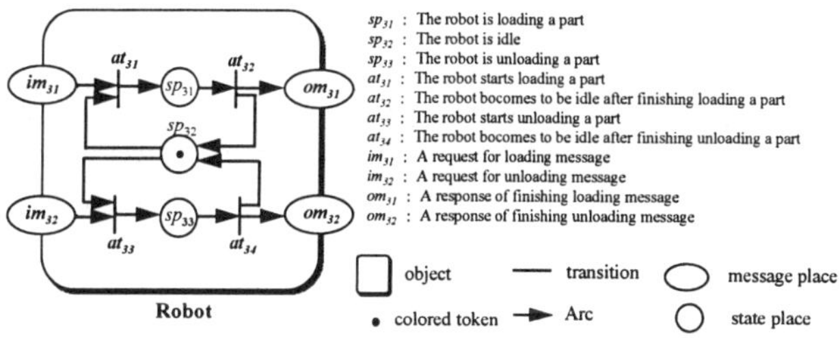

Fig. 1. The individual OPN for object 'robot'.

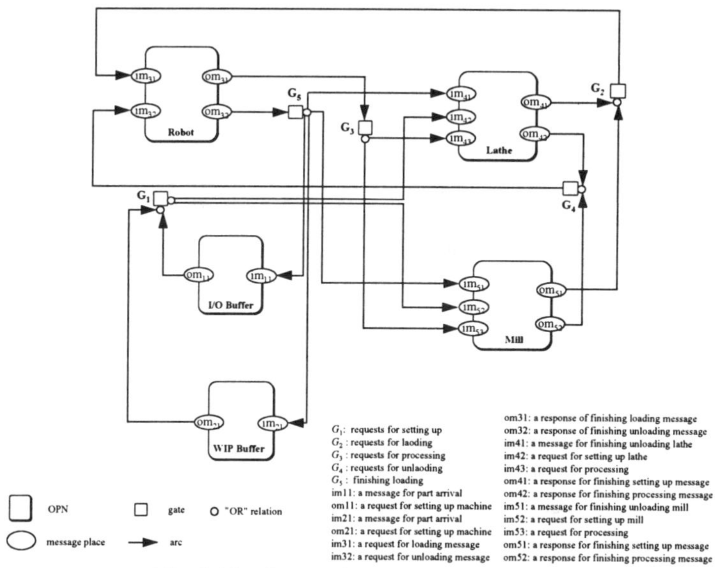

Fig. 2. The System OPN model for the AMS

Construction of a CTOPN model is derived from the system OPN model by first representing each encapsulated OPN as an *object place* and the message passing gate as an *activity transition*, then the internal behavior (i.e., states of the OPN) of each physical object and the routing of each part type may be represented by the associated colored tokens. The procedure of constructing a CTOPN model may be summarized as follows:

Step 1. Represent each encapsulated OPN in the system OPN model as an object place $\boldsymbol{OP_i}$ (i=1,2,...,m).

Step 2. Represent each gate involved in the system OPN model as an activity transition G_j (j=1,2,...,n).

2.1 For any gate whose input-output connection is either a AND-OR or a AND-AND relation, represent it as an *ordinary* activity transition $\in R_1$.

2.2 For any gate whose input-output connection is either an OR-OR or an OR-AND relation, represent it as a *hierarchical* activity transition ($\in R_2$) that may be further decomposed into one state place connected with one output and several input ordinary activity transitions ($\in R_1$) to clarify the firing condition for a gate with "OR" input connections.

Step 3. Represent the [OPN-gate-OPN] message passing relations specified in the system OPN model as [object place-activity transition-object place] relations.

Step 4. For each part type p, assign a colored token ct_{pr} to represent its sub-process r according to the process flow (p = 1, 2,..., P; r = 1, 2,..., w_p).

Step 5. Based on the colored tokens and the allocated resources identified in Setp 4, define the set of colored tokens, $C(OP_i) = \{cp_{i1}, cp_{i2},..., cp_{iu_i}\}$, associated with each object place OP_i, and the time-delay attribute of colored token cp, $D(cp)$. (cp $\in C(OP_i)$; i=1,2,...,m)

Step 6. Define the set of colored tokens, $C(G_j)=\{cg_{j1}, cg_{j2},...,cg_{jv_j}\}$, the input, $I(OP_{i'}, G_j/cg)$, and output, $O(OP_{i''}, G_j/cg)$, transform functions for each activity transition G_j (j=1,2,...,n; $cg \in C(G_j)$):

6.0 Set part type index p =1.

6.1 Set sub-process index r =1.

6.2 Sequentially select each pair of colored tokens, $(ct_{pr}, ct_{p,r+1})$, involved in the process flow of part type p, and identify their allocated object places ($OP_{i'}$, $OP_{i''}$).

6.3 Determine the activity transition, G_j, connecting the two object places ($OP_{i'}$,$OP_{i''}$) according to the [$OP_{i'}/ct_{pr}$ - G_j - $OP_{i''}/ct_{p,r+1}$] message passing relations specified in the partial CTOPN model obtained in Step 3.

6.4 Define the set of colored tokens, $C(G_j)=\{ct_{pr}\}$, the input, $I(OP_{i'}, G_j/ct_{pr}) = ct_{pr}$, and the output, $O(OP_{i''}, G_j/ct_{pr}) = ct_{p,r+1}$, transform functions for activity transition G_j.

6.5 Let $r = r + 1$; If $r < w_p$, go to Step 6.2; Else, go to Step 6.6 (i.e., the last pair of sub-processes has been considered).

6.6 If $p=P$ (i.e., all part types are considered), delete those arcs identified in Step 3 without any transform function associated with and go to Step 7; Else, set $p=p+1$, go to Step 6.0 (i.e., consider for the next part type).

Step 7. Analyze the *bounded* characteristics for each object place, and employ inhibitor arcs, $Inh(OP, G)$, to connect the object place OP ($\in \{OP_i$; i=1,2,...,$m\}$) which must satisfy *1-bounded* constraint and all its input activity transitions G ($\in {}^{\cdot}OP$), so that the safeness of that object place may be maintained.

To illustrate the procedure of constructing a CTOPN model, let us consider the aforementioned AMS example; each encapsulated OPN and gate in the system OPN model shown in Figure 2 corresponds to an object place (OP_1 - I/O Buffer, OP_2 - WIP Buffer, OP_3 - Robot, OP_4 - Lathe, OP_5 - Milling Machine,) and activity transition G_j (j=1,2,...,5) in the CTOPN model (Steps 1 and 2). Since the "OR" relation exists among the inputs to gates G_1, G_2 and G_4, they should be represented as the hierarchical activity transitions whose decompositions are illustrated at the right portion of Figure 3 (Steps 2.1 and 2.2). After the message passing relations are considered (Step 3), the partial CTOPN may be depicted in Figure 3. We can divide the whole process into a sequence of manufacturing and non-manufacturing sub-processes depending on the sequence that the facilities are involved. Then, each sub-process is represented by a colored token (i.e., part type A : $\{ct_{1,1}, ct_{1,2}, ..., ct_{1,11}\}$={a.1, a.2, ..., a.11}; part type B : $\{ct_{2,1}, ct_{2,2}, ..., ct_{2,11}\}$={b.1, b.2, ..., b.11}) (Step 4), whose time-delay attribute is defined as the processing time (see Table 1). Consequently, the colored tokens associated with each object place, $C(OP_i)$ (i=1,2,...,5), may be determined by grouping all the sub-processes performed by that object place's corresponding facility/resource (Step 5; see Table 2).

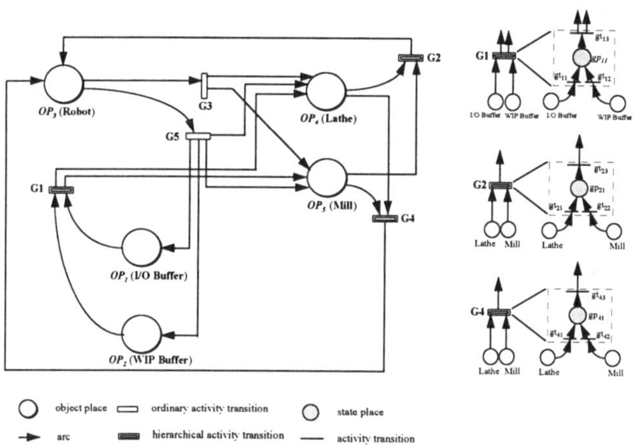

Fig. 3. The intermediate result of constructing a CTOPN model

Table 1. Process flow descriptions for each part type in the AMS

Part A			
sub-process	processing time	colored token	resource allocated
Waiting in I/O Buffer	0	a.1	I/O Buffer
Lathe setup	1.83	a.2	Lathe
Robot loading Lathe	3.28	a.3	Robot
Lathe processing	18.25	a.4	Lathe
Robot unloading Lathe	2.28	a.5	Robot
Waiting in WIP Buffer	0	a.6	WIP Buffer
Mill setup	1.10	a.7	Mill
Robot loading Mill	2.36	a.8	Robot
Mill processing	10.85	a.9	Mill
Robot unloading Mill	1.36	a.10	Robot
Arrived in I/O Buffer	0	a.11	I/O Buffer

Part B			
sub-process	processing time	colored token	resource allocated
Waiting in I/O Buffer	0	b.1	I/O Buffer
Mill setup	1.25	b.2	Mill
Robot loading Mill	2.56	b.3	Robot
Mill processing	12.5	b.4	Mill
Robot unloading Mill	1.56	b.5	Robot
Waiting in WIP Buffer	0	b.6	WIP Buffer
Lathe setup	2.58	b.7	Lathe
Robot loading Lathe	4.22	b.8	Robot
Lathe processing	25.75	b.9	Lathe
Robot unloading Lathe	3.22	b.10	Robot
Arrived in I/O Buffer	0	b.11	I/O Buffer

Now, let us consider the first pair of colored tokens (i.e., a.1 and a.2) in the process flow of part-A (p=1, r=1)(see Table 1), their allocated object places are OP_1 of I/O buffer and OP_4 of lathe, respectively (Step 6.2). We may conclude that activity transition G_1 is employed to connect object place OP_1 with colored token a.1 and object place OP_4 with colored token a.2 from the relation OP_1/a.1-G_1-OP_4/a.2 specified in Figure 3 (Step 6.3). It is clear that activity transition G_1 may be enabled with respect to a.1 if there is a token a.1 in object place OP_1 of I/O buffer (i.e., $M(OP_1) \geqq$ a.1; set $C(G_1)$={a.1}and $I(OP_1,G_1/$a.1)=a.1) and the firing of G_1

will remove a token a.1 from OP_1 and add a token a.2 to object place OP_4 of lathe (i.e., set $O(OP_4,G_1/\text{a.1})$=a.2) since object 'lathe' will start setting up once the request for setting up message sent from object 'I/O buffer' is received as indicated in the sub-processes with colored tokens a.1 and a.2 of Table 1 (Step 6.4). Since there is only one robot, one lathe and one milling machine in the AMS, the inhibitor arcs must be applied to object places OP_3, OP_4 and OP_5 (i.e., $Inh(OP_3,G_2)= Inh(OP_3,G_4)=1$, $Inh(OP_4,G_1)= Inh(OP_4,G_3)=1$, and $Inh(OP_5,G_1)= Inh(OP_5,G_3)=1$), so that the 1-bounded property may be maintained (Step 7). The complete CTOPN model for the AMS is depicted in Figure 4.

Table 2. The colored tokens and the associated time delay assigned for each object place

Object place	*C*(object place) and *D*(colored token) functions
OP_1 (I/O Buffer)	$C(OP_1)$={ a.1, b.1, a.11, b.11 } *D*(a.1)=0; *D*(b.1)=0; *D*(a.11)=0; *D*(b.11)=0
OP_2 (WIP Buffer)	$C(OP_2)$={ a.6, b.6 } *D*(a.6)=0; *D*(b.6)=0
OP_3 (Robot)	$C(OP_3)$={ a.3, a.5, a.8, a.10, b.3, b.5, b.8, b.10 } *D*(a.3)=3.28; *D*(a.5)=2.28; *D*(a.8)=2.36; *D*(a.10)=1.36 *D*(b.3)=2.56; *D*(b.5)=1.56; *D*(b.8)=4.22; *D*(b.10)=3.22
OP_4 (Lathe)	$C(OP_4)$={ a.2, a.4, b.7, b.9 } *D*(a.2)=1.83; *D*(a.4)=18.25; *D*(b.7)=2.58; *D*(b.9)=25.75
OP_5 (Mill)	$C(OP_5)$={ a.7, a.9, b.2, b.4 } *D*(a.7)=1.10; *D*(a.9)=10.85; *D*(b.2)=1.25; *D*(b.4)=12.5

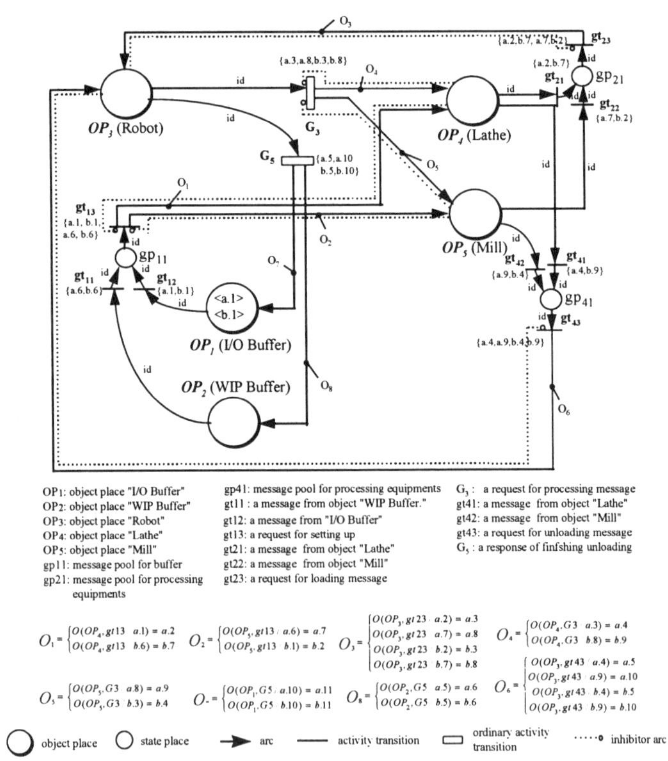

Fig. 4. The CTOPN model for the AMS

4 Applications of The CTOPN Approach

Because of the control logic and dynamic behavior of the AMS can be represented by the CTOPN model. So, the Deadlock Detection Algorithm(DDA) [14] for a CTOPN model can be used to analyze the deadlock property of the AMS.

After the control logic of an AMS represented by the CTOPN model is analyzed and validated, a search algorithm, derived from Lee and DiCesare [6], may be employed to determine the near optimal firing sequence for all gates with specific colored tokens involved in the CTOPN model in terms of minimum accumulated processing time(please refer to [13] for the detailed description of the scheduling search algorithm). This firing sequence implies the operations schedule for each facility in the AMS with nearly minimum makespan. For the CTOPN depicted in Figure 4, the time and sequence of firing each gate and its associated colored tokens may be calculated by CTOPN scheduling search algorithm [13] when two parts (i.e., one A and one B) are processed by the AMS, and we may then construct the operations schedule Gantt chart for each physical object, shown in Figure 5.

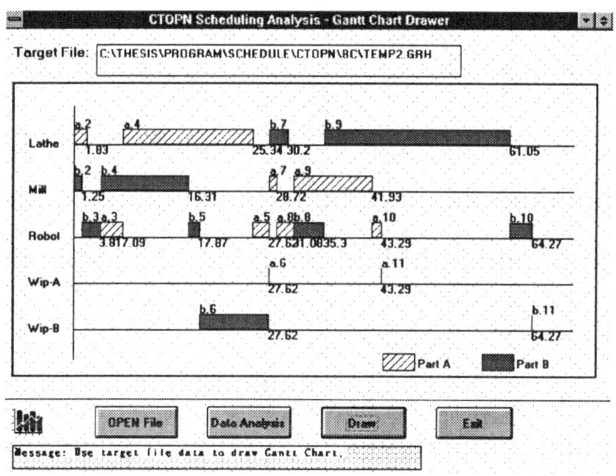

Fig. 5. Gantt chart of operations schedule for the AMS example

Based on the validated CTOPN model and a near optimal operations schedule generated by the scheduling algorithm, We can develop a control model for the AMS. Finally, the CTOPN control model may be directly transformed into the code of a rule-based control software(e.g., *FASTech CELLworks*) by following a number of transformation rules.

5 Conclusion

This paper presents a colored timed object-oriented Petri net (CTOPN) approach, which owns all the characteristics of Petri nets and object-oriented approach, to model and analyze the dynamic behavior of an automated manufacturing system (AMS). Construction of a CTOPN model is based on the individual and system object-oriented Petri nets (OPNs) stored in the OPN class library and including colored tokens (associated with time-delay attributes) and input/output transform functions to represent the characteristics of material flow and time constraints in an AMS, so that the dynamic behavior intra and inter each manufacturing equipment and auxiliary facility in an AMS may be described detailedly. Furthermore, the control logic of an AMS, represented by the CTOPN model, may be analyzed using a deadlock detection algorithm, so that all the possible deadlock situations may be detected and avoided before a control system for that AMS is implemented. Then, an effective near optimal operations schedule and the code of a rule-based control software for an AMS may be generated by employing a search algorithm and a set of transformation rules, respectively. Therefore, the CTOPN approach provides an integrated and consistent environment for modeling, analyzing, scheduling and controlling an AMS simultaneously.

References

1. T. O. Boucher, M. A. Jafrai, and G. A. Meredith, Petri Net Control of An Automated Manufacturing Cell, *Computers & Industrial Engineering*, **17** (1989) pp. 459-463.
2. D. Y. Chao, and D. T. Wang, XPN-FMS: A CAD Tool for FMS Modeling, Analysis, Animation, and Simulation Using Petri Nets and X Window, *The International Journal of Flexible Manufacturing Systems*, **7** (1995) pp. 339-360.
3. K. A. D'Souza, and S. K. Khator, A survey of Petri net applications in modeling controls for automated manufacturing systems, *Computers in Industry*, **24** (1994) pp. 5-16.
4. A. Kusiak, *Intelligent manufacturing systems*, Prentice-Hall (1990) pp. 382-385.
5. Y. K. Lee, and S. J. Park, OPNets : an object-oriented high-level Petri net model for real-time system modelling, *J. Systems Software*, **20** (1993) pp. 69-86.
6. D. Y. Lee, and F. DiCesare, Scheduling flexible manufacturing systems using Petri nets and heuristic search, *IEEE Transactions on Robotics and Automation*, **10** (1994) pp. 123-132.
7 C. M. Liu, and F. C. Wu, Using Petri Nets to Solve FMS Problems, *Int. J. Comput. Integr. Manuf*, **6** (1993) pp. 175-185.
8 P. O'Grady, and R. Seshadri, X-Cell - intelligent cell control using object-oriented programming, *Computer Integrated Manufacturing Systems*, **4** (1991) pp. 157-163.
9 J. S. Smith, and S. B. Joshi, Reusable software concepts applied to the development of FMS control software, *Int. J. Comput. Integr. Manuf*, **5** (1992) pp. 182-196.
10 L. Wang, The development of an object-oriented Petri net cell control model, *Int. J. Adv. Manuf. Technol*, **11** (1996) pp. 59-69.
11 L. Wang, Object-oriented Petri nets for modelling and analysis of automated manufacturing systems, *Computer Integrated Manufacturing Systems*, **26** (1996) pp. 111-125.
12 L. Wang, An integrated object-oriented Petri net paradigm for manufacturing control systems, *Int. J. Comput. Integr. Manuf*, 1996 **(9)** pp. 73-87.
13 L. Wang, and S. Y. Wu, Scheduling automated manufacturing systems using colored timed object-oriented Petri nets, *Journal of the Chinese Institute of Industrial Engineers*, (to appear in 1997).
14 L. Wang, and S. Y. Wu, Colored timed object-oriented Petri nets for modeling and analysis of automated manufacturing systems, *Proceeding of the Fourth International Conference on Automation Technology, July, Taiwan* (1996) pp.431-437.

Coloured-Time, Modular Petri net Approach for the Modelling Zone-Controlled Manufacturing Machines
by
S. Ariffin and R. H. Weston

A coloured-time, Modular Petri net modelling approach to the design and control of manufacturing machines is presented in this paper. Distinct types of zone are defined based on the physical layout of machines systems. The concept of zone control used in conjunction with the modular coloured-time Petri net approach can readily support performance analysis and the selection of candidate machine systems. The method provides a direct mapping between the modelled system and the physical machine which readily support changes to machine task modules. The resultant zone-controlled machine system has well-defined logical properties facilitates the life-cycle engineering of machines, even under condition where high levels of systems complexity are involved.

1. Introduction

Machines in general comprise mechanical mechanisms, and control system elements. Normally a task performed by a machine system is the result of planned actions and the input of control data (Gershwin et al. 1986). When more than one machine is required to co-operate within a manufacturing cell, significant complexities arise with resfect to the design, implementation, operation, maintenance and modification of control strategies (Shin and Epstein 1987, Ben-Arieh 1991). Moreover, in a manner determined by characteristic requirements of the application concerned, the task performanced by an individual machine will be dependent upon those performed by other machines (Pelagagge et al. 1995). Such a multi-machine system can be compared with the mechanical analogue of a distributed mechanical mechanism, requiring synchronism between concurrently operating mechanical mechanisms. Whereas for multi-machine systems computational mechanisms are required to synchronize and achieve overall control of the distributed and concurrent activities involved.

In order to remain competitive many 'world class' industrial companies will need to continue to reduce their dependencies on specialised machines, which have relatively limited capabilities which ultimately will constrain the realisation of business processes. Progressively greater reliance will be placed on ***general-purpose machines*** which can be reprogrammed in an effective way, either centrally or locally as needs dictate (Sweeney 1989). Thus next generation manufacturing machines must allow industry to respond to the demands of variety, delivery and cost. Such machines will need to be readily reconfigured to suit different products (Harrison 1991). This would reduce the lead-times involved in their installation and set-up (following product change) and thereby reduce systems engineering and manufacturing costs (Anon 1992). This implies the need to engineer a new generation of ***computer-controlled machine*** (CCM) systems which more closely meet the needs of their host manufacturing environment, whilst supporting their implmentation within sufficiently short lead-times, commensurate with the falling life-time of the products they are required to produce. Also such machine systems should be easier to modify and maintain within their target manufacturing environment, thereby supporting changing needs and lengthening their useful life-time.

In practice existing industrial approaches to the distribution of machine control functionality leave remaining problems in respect of realising: **modularity** in order to overcome the complexity of machine systems, with associated solution formalisation, including the formal description of communication interfaces and the functional capabilities of modules; **reconfigurability** in terms of enabling the rapid and effective configuration and reconfiguration to handle multi-part production in a multi-machine environment and the rapid reconfiguration of the mix of parts; **concurrency** capability which enables synchronism between distributed and concurrently operating machine systems resulting in high machine utilization and shorter lead-times in meeting customer orders; **software reusability** in terms of enabling future machine control modification and expansion and greater re-use of software modules in different application domains; and **improved interoperability** in terms of more effectively handling multi-machine and multipart schduling problems which can more rapidly fulfil production orders and optimize production orders and optimize production performance. The central question here is how to organise and integrate control systems for individual and groups of machines to yield a new generation of computer-controlled manufacturing machines and systems which are flexible, extendible, and reconfigurable.

In attempting to deal with the above problems, Petri net (PN) modelling methods have gained recognition in academic and industry as a practical tool (Dotan and Ben-Arieh 1991). They have been successfully applied in modelling multi-event synchronization, in respect of different resources in manufacturing systems (Garnousset and Fariness 1989, Cecil et al. 1992, Liu and Wu 1993, Choi et al. 1994 and Wang 1995). PN models can have time parameters attached to them to indicate temporal capabilities, in a manner appropriate to activity modelling in respect of manufacturing machines (Knapp and Wang 1992). Furthermore, the mathematical definitions from PN schema (Ben-Arieh 1991, Knapp and Wang 1992) enables computer programs to be produced for manufacturing control software applications. However, the effort involved in the design and construction of Petri net based models grows rapidly with increase in system complex multi-part multi-machine situations due to the fact that PN modelling is system dependent, and lacks certain inherrant properties (including modular programming, reusability and maintainability) that are commonly required in respect of designing and constructing manufacturing control systems. Thus this paper address the issue of managing complexity as the size of a modelled machine sysem grows.

2. Modular Petri net (MPN) Approach to Modelling Distributed Machines

Thus an approach to creating and using MPN modelling was proposed and anlaysed by the authors (Ariffin *et al.* 1995). The primary design criterion of the MPN approach was to provide means of decomposing conventional PN models into largely self contained modules, i.e. modules which essentially can be decoupled from the remainder of a system in a way which facilitate analysis and promote the reuse of components. The MPN approach conceived facilitates the definition of MPN modules as virtual analogues of the elements of physical systems. Thus enable the reflacement or additional of physical components within complex machine systems. Also task modules can be established based on the physical layout and operating sequences of machines. Thus modules can be added, deleted or modified readily to produce and support the analysis of and selection between alternative candidate machine systems. Also its modular properties facilitate the analysis and change of alterantive operations sequences and the anaysis of breakdown conditions. Thus the study of machine modularity and how it can help handle behavioural complexity in machines (especially as the number of machines grows) can be supported via the MPN approach and supporting simulation tools.

A single piece of equipment or a group of equipment items can be represented by a single task or MPN module. A Petri net graph (which is contained within a task) is represented as a network of communicating processes. This fact can be explained with reference to the simplified example machine system layout shown in **Figure 1**. In this example system a machine cell consist of a robot, an input conveyor and a processing machine. The robot unloads parts from the input conveyor, and loads them onto the machine. Initial modelling of the control function of the machine is broken down (decomposed) with reference to the physical machine layout and operating sequence. In this case the model of the input conveyor corresponds to a model of the operating sequence to be performed (i.e. it takes the form of a control net) whereas the model for the robot relates to the authorized state of the process (i.e. it takes the form of a process net). In the process net, a token is used to represent the availability of a resource (a robot) whilst a taken in the control net models the part to be processed. **Figure 1** shows how the operation of each machine can be represented by a separate MPN module. These modules communicate by a so-called shared-transitioon (ST), which provide a link between the modules. In a real system, a shared-transition will take the form of communication paths (or connecting links) between physical resources or hardware components. In other words, the occurance of a synchronisation event which connects two conditions from different task modules (or machine modules) and can be realized through two PN "places" linking a common transition, is called ST. All *single-transitions* (T) atributed to a ST must be enable for firing before the ST can be enabled. A ST will represent some dynamic property of the system and hence will encode associated state changes between modules.

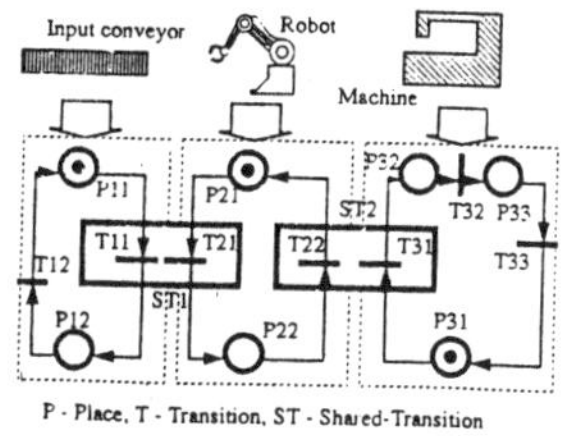

Figure 1: **A Shared-Transition and Modular Petri net Model**

Thus a *shared-transition* (ST) can be used to represent a communication path between at least two hardware components and encodes state changes associated with such a link; e.g. when a job or part is transferred from one machine to another machine. For example, as illustrated by **Figure 1**, ST1 represents the occurance of two conditions (i.e. the arrival of a raw part and the availability of a robot) from different task modules which lead to an occurrance of an event, eg. *start_picking_up_part*. In this case, the part is picked up by the robot from the input conveyor. Another *shared-transition* ST2 represents the *part_being_placed_down* into the machine. In order for a *single-transition* (T) to fire, each of the incoming places must be marked by tokens: a necessary condition for a transition to be enabled. Flow of the tokens (represented by dots) shows the movement of parts and the status of machines in a machine system. After firing, the tokens are transferred to their respective output places.

2.1 Basic Modular Petri nets (MPN) Definitions and Graph

A basic marked Petri net is a five-tuple (Knapp 1992), namely:

PN = (P, T, I, O, m)

for which,

P	=	{Pij} is a set of places j in each task i where i = 1, 2,...n (task number) and j = 1, 2,..n (transition number)
T	=	{Tij} is a set of transition j in each task i, where i is the task number and j is the transition number
I	=	{Pij Tij} is a set of place inputs into a transition (i.e. input function)
O	=	{Tij Pij} is a set of place output from a transition (i.e. output function);
m	=	(mi1, mi2, min} is a set of marked (m) places in each task "i" and token number "n", defining the state of the system and tokens at particular time stamps.

A Petri net graph is a pictorial representation of a Petri net which consists of circle nodes, bar nodes that are linked by directed arcs. The circle nodes and bar nodes represent places and transitions respectively. A marking "min" is the number of tokens in a place Pij. The authors conceived the notion that *shared-transitions* (STs) should be included within distinct modules in conformance with the so-called modular Petri net (MPN) modelling approach. Mathematically, a marked MPN is represented in the following form:

MPN = (P, T, I, O, ST, m)

for which,

ST = {STx} is set of two or more single-transitions (T), where x is the ST number (i.e STx = {Tij} and x = 1, 2,....n))

A *shared-transition* STx will be comprised of at least two *single-transitions* when coordination between events occurring at two or more real machines (which individually are represented by a modular net) the inter operation of which needs to be modelled and controlled.

For the graph corresponding to the example shown in **Figure 1**, the mathematical representation is:

P	=	{(P11, P12), (P21, P22), (P31, P32)}
T	=	{(T11, T12), (T21, T22), (T31, T32)}
I	=	{(P11 T11, P12 T12), (P21 T21, P22 T22), (P31 T31, P32 T32, P33 T33)}
O	=	{(T11 P12, T12 P11), (T21 P22, T22 P21), (T31 P32, T32, P33, T33 P31)}
ST	=	{ST1, ST2} where ST1 = {T11, T21} and ST2 = {T22, T31}
m	=	{m11, m21, m31} = {1, 1, 1}

In **Figure 1(c),** there are three tokens; at the start these will be one each at P11, P21 and P31. The presence of the token at P11 means that the *part_is_ready* and the token at P21 means that the *robot_is_available*, and the token at P31 means that the *output_conveyor_is_ready* to transfer the incoming part. The system state changes by firing transitions. The condition under which *shared-transition* (ST1) will be fired is satisfied when the *single-transitions* T11 and T21 are fired in the first place. Firing a transition means that a specified number of tokens have been removed from each input place of the transition and that a specified number of tokens have been added to each output place of the transition (Mecovsky *et al.* 1990). The physical meaning of this firing is that the robot *picks_up_the_part* and then carries it to the machine for processing. After firing ST1, both P12 and P22 will be marked, which means that the robot is servicing a part and is temporily unavailable (zero token at P11 and P21). The token at P21 will not appear until the *shared-transition* ST2 is fired. When ST2 is fired, the token will transfer to P21 and P32, which indicates that the *robot_is_ready* for its next cycle, and the *part_is_loaded* for machining.

2.2 Coloured-Time, Modular Petri nets

For a system such as a robotic cell, its state varies from time to time (Knapp and Wang 1992). Thus, the MPN structure should be equipped with temporal capabilities for modelling the behaviour of manufacturing machines. A time parameter is necessary to express the temporal behaviour restrictions, associated with each transition (Ben-Arieh 1991). An enabled transition can only fire after the delay time has elapsed. The time could be deterministic and be assigned to each event within a task.

Also, the state of a system modelled with a Petri net is characterized by the markings i.e. by the presence of tokens at places. Tokens are normally divided into two categories: 'availability of resources' and 'actions' where typically different colours are required to distinguish different tokens. Different coloured tokens can be used to indicate different parts which normally compete for the same event (transition). A common question which arises is: which part should be processed first?. This is where a so-called ***colour conflict*** could easily arise. Since a single resource cannot process the multiple parts simultaneously, then a set of control rules must be established to define the order in which operations should occur.

A Coloured-Time MPN (CTMPN) has eight-tuples, namely:

CTMPN = (P, T, I, O, ST, K, π, δ)

for which (i.e. in addition from the above MPN model),

K	=	{(kij)} is a set of coloured token (i is the task number and j is the token number)
I	=	{(Pij, Tij, Nij)}
O	=	{Tij, Pij, Uij)}

π(Tij) is the time delay (tij) for each task I and transition number j

δ(Pij) is a marked function to the power of K in a place for task i and place number j

Nij, Uij = {1, 2,....q} where q is the number of tokens in the net of a task

Nij and Uij record the paths that tokens take to move from place to transition (input) and from transition to place (output), respectively.

Figure 2 shows a pictorial representation of an example modular Petri net (MPN) with coloured tokens defined in the following way:

P	=	{(P11), (P21, P22), (P31, P32)}
T	=	{(T11), (T21, T22), (T31, T32)}
I	=	{(P11, T11, N11), (P21, T21, N21; P22, T22, N22), (P31, T31, N31; P32, T32, N32)}
O	=	{(T21, P22, U21; T22, P21, U22), (T31, P32, U31; T32, P31, U32)}
ST	=	{ST1, ST2} where ST1 = (T11, T21) and ST2 = (T22, T31)
K	=	{(k11), (k21), (k31)}

τ(T11, T12) = (t11, t12)
τ(T21, T22) = (t21, t22)
τ(T31, T32) = (t31, t32)
δ(P11) = {k11}
δ(P21) = {k21}
δ(P31) = {k31}

N11 = {1}			
N21 = {2}	N22 = {1, 2}		
N31 = {3}	N32 = {3}		
U21 = {1, 2}	U22 = {2}	U31 = {1, 3}	U32 = {3}

The token k11 at P11 means a part is ready to be picked up and token k21 at P21 means that the robot is available. When ST1 is fired (as illustrated in **Figure 2**) the tokens k11 and k12 will be at place P22 at a certain time stamp. The time delay for ST1 is taken from the maximum time delay among the enclosed *single-transitions* (in this example ST1 is a combined transition from both T11 and T21). The firing of ST2 will allow the robot to complete its pick and place movements whereby the token k11 and k31 are now placed at P32 while token k21 is returned back to its original place (P21). This means that the part has complete its journey from the input buffer to the output conveyor, while the robot has returned back to its original position. The robot is available to other parts during the next cycle.

Tokens that have passed through a fired *shared-transition* can be divided into two types, so-called 'public' and 'private' types. Token k11 in module M1 is considered public because it needs to be transferred to module M2. Thus represents the movement of a part which is required to be processed along with the processing from start to finish. The token k21 in module M2 and token k31 in module M3 are considered private since each of these tokens is only required to move within its own module. In this case the machines modelled are shared resources and the different types of (coloured) tokens should be initially identified for the whole processing cycle.

The example coloured Petri net of **Figure 2** shows the action of *shared-transitions* ST1 and ST2, each of which have two input places. The activities involved are as follows:

After firing ST1:
If P11 has a public token and P21 has a private token then P22 is active;
If P21 has a private token then P22 is active.

After firing ST2:
If P22 has a public token and P31 has a private token then P32 is active
If P22 has a private token then P21 is active

This means that after the firing of ST1, the token at P11 (k11) and the token at P21 (k21) need to be transferred to P22 in the form of public and private tokens respectively (see **Figure 2 (b)**). In other words, P22 should be composed of both token k11 and k21. On the other hand, after the firing of ST2 a public token k11 has to move across the task module boundary, while a private token (k21 or k31) has to move within the same task module (see **Figure 2(c)**). Yim (1994) explained this phenomena as a colour conflict which can occur when there are several tokens at an input place of a transition. By firing ST2, the token k11 must be selected among the set of tokens of the place P22 and move into module M3. This physically signifies that a part has been placed down on the output conveyor and also indicates that the first cycle is complete.

In the above case, the rules that capture the behaviour of manufacturing operations can be represented as follows:

public token k11: P11 to P22 then to P32 (movement of part)
private token k21: P21 to P22, then back to P21 (activity of robot)
private token k31: P31 to P32, then back to P31 (activity of output conveyor)

The rules reside at the output places which need to receive the tokens. When tokens arrive at a particular place, the *shared-transition* scans all its incoming places and fires the rules that are feasible at that time, provided all the incoming places are occupied by tokens. Each rule, with a designated token type (private or public) consists of a list of tokens which need to be fired. The tokens are passed according to a defined sequence which is specified by the modeller. The time unit delay tij for each *single-transition* (Tij) should be representative of real manufacturing operations and should be incorporated to facilitate an analysis of machining time. The transition always updates its time from the latest time stamp attached to tokens (Ben-Arieh 1991a).

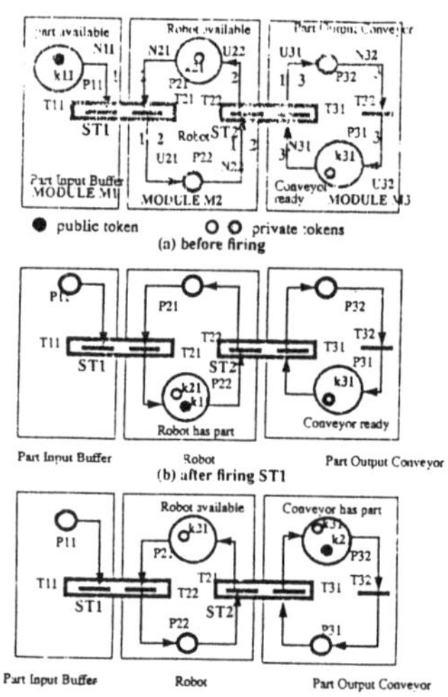

Figure 2: A Simple Coloured Modular Petri net Graph

3. Improving the "Generality of Solutions" by using Zone-Controlled Machine Systems

As described above, modular Petri net (MPN) modules are represented in the form of building blocks. As such they can facilitate modelling and analysis of the operational performance of candidate machine systems. These MPN modules (i.e *machine modules*) can represent equipment functionality (such as conveying, machining, material handling). However, as a modelled machine system increases in complexity (i.e the number and types of machines and parts increase) there is an increased tendency to reuse similar modules. This can create a complex situation at the early stages of design and construction. Thus, the authors concluded that a more generalised method needs to be devised to handle high levels of complexity.

It is realised that MPN modelling does not inherently provide a well-defined predetermined path for material flow, especially when multiple part types need to be processed in the same manufacturing operations, means that it is essential to avoid deadlock and conflict in the system, and care must be taken to detect their possible occurrence at the early stage of any design. Here the purpose is to carry out a logical analysis of functional properties in terms of their liveness, reachability and boundedness of candidate machines systems, in a simple, formal and generalised way.

Moreover, it is not a simple matter to analyse the processing of a given set of parts using different machine types. It is realised that there is a need to handle flexibility in term of producing a part by providing alternate routes through the machine system. In other words "routing flexibility" (i.e. *ability to process a given set of parts on alternative machines* (Raouf 1995) or "path flexibility" (i.e. *possibility of having more than one path from the origin to the destination*) needs to be established in order to process different part types on alternative or different machines.

4. The Design of Zone-Controlled Machine System using the MPN Modelling Approach

The concept of zone-control was introduced by Lin and Lee (Lin and Lee 1992) in which zones were defined with reference to a material flow graphs of a conveyor system. Lin's approach was based on the use of timed PN. The authors have modified and used this concept as a means of improving the way in which MPN models are created to represent the activities of manufacturing machine systems, including: loading, unloading and the conveying of parts. The purpose is to develop and establish use of a formalism for MPN modules in terms of time considerations so that they can be used as building blocks for logical analysis and performance evaluation in resultant machine systems. Logical analysis is necessary to determine certain functional properties of machine systems (i.e. *their liveness, reversibility, reconfigurability, extensibility and so on*).

In order to simplify and enable the design and construction of machines, the authors unified use of the MPN modelling approach and the concept of machine zones. Zones were characterised and categorized into five distinct types (see **Figures 3(a) to (e)**):

(a) **SISO (single-in-single-out) zone** which represents a simple class of activity such as the loading, unloading, simple conveying o a part, a buffer queue or machining of a part.

(b) **SITO (single-in-two-out) zone** which represents a forked-type of activity, such as the conveying of a part which splits into two parts.

(c) **TISO (two-in-single-out) zone** which represent a merge-type of activity such as that realise in parts assembly and the unloading of a number of parts at a same point.

(d) **TITO (two-in-two-out) zone** which represents a mixed-type of activity between SISO-zone type and SITO-zone type. A three-part routing can be established using this class of zone.

(e) **MIMO (multiple-in-multiple-out) zone** with represents a shared-resource activity with multiple input points and multiple output points, but without a mixed type of activity. This can be used to model a robot, AGV, AGV-line conveyor or a machining process.

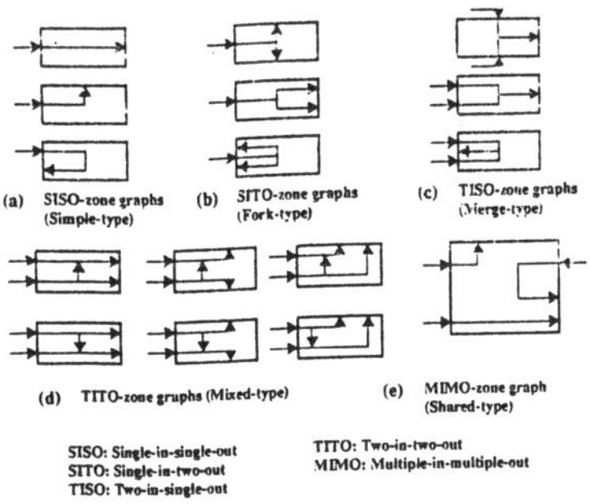

Figure 3: Five Types of Modular Zone Graph

These five types of zone and the equivalent-zone graphs provide basic building blocks for representing a zone-controlled machine system. Using these zone graphs, MPN models and their associated time delays can be constructed to describe the behaviour of resources within machine systems. **Figures 4(a)** to **(e)** present the so-called TMPN (timed modular Petri net) module of the zone graphs shown in **Figures 3(a)** to **(e)** respectively. A *timed-transition* (TT) (represented by a thick bar) in each TMPN module net is used to represent the duration (time delay) of activities associated with each resource. For each TMPN module, a *timed-transition* is used to indicate the time delay associated with given resources. **Figure 4(b)** to **(e)** show the shared resource type, requiring the existence of a shared-resource place in each module. The transitions drawn across each boundary of the module indicate the presence of a shared-transition ST (represented by thin bars). Here *shared-transitions* are used to check input and output conditions to establish communication between modules. In the SITO net (i.e forked-type of net), two *timed-transitions* exist. Conf licts may occur between both *timed-transitions*. In the TISO net (i.e *merge-type of net*), two *timed-transitions* exist and conflicts may occur between both incoming shared-transitions of the net. These conflicts can be resolved by assigning predetermined operation sequences to the input of corresponding *shared-resource transitions*. The MPN simulation approach conceived by the first author has the capability to address this issue.

Consider the example shown in **Figure 6**. Here we assume that the machine system comprises a robot, two identical machines, a loading conveyor and an unloading conveyor. The physical layout is extracted from **Figure 5**. Use of a group of different types of zone graph enables direct mapping between the modelled system and the physical layout of machine elements. The zone-controlled scheme can be used for processing and moving parts along predetermined paths. Obviously the type of zone selected will depend on the types of activity carried out by resources used in the system. Any addition, deletion or modification of real machine types and part routings can be done with reference to a modular layout of the machine system.

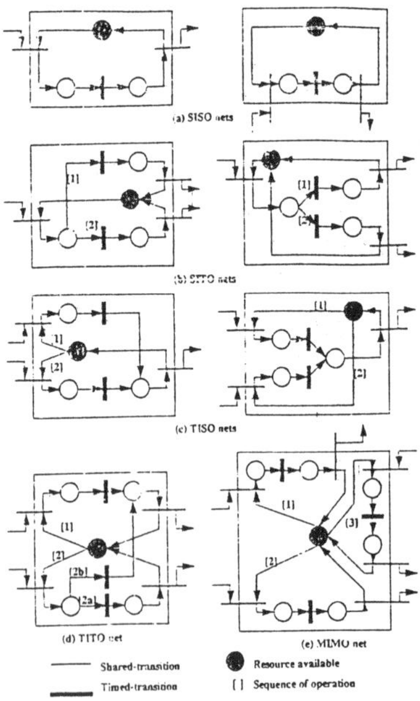

Figure 4: **Timed Modular Petri nets (TMPN) Modules**

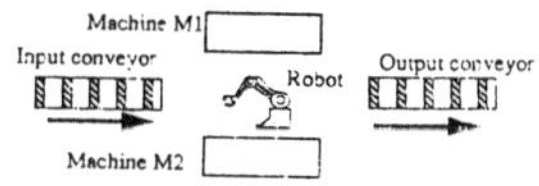

Figure 5: **A Shared Resource Machine Cell Layout**

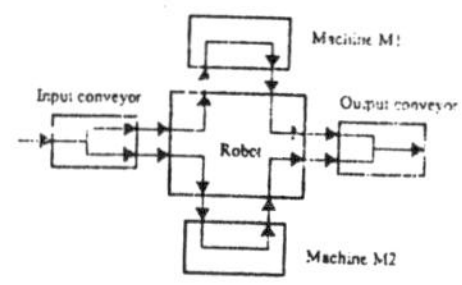

Figure 6: **Modular Layout Graphs for the Machine Cell Figure 5**

TMPN modelling of machine systems can be structured in a more effective way by adopting zone-control concepts is illustrated by **Figure 7**. In this case the TMPN modules organised into zone-controlled units include: two SISO nets representing machining, a SITO net representing the spliting of input parts (which are normally placed in a pallet) that will be "picked-up" by the robot and processed by each machines: a MIMO net representing the resource-sharing activity of the robot; and finally a TISO net representing a collection point for finished parts at the unloading point. This TMPN layout not only indicates possible material flows (parts routing), but also enables status information on the machines to be indicated along with a deterministic sequence of operations for shared resources in the system, through the movement of tokens in the modelled system.

The start of the simulation output derived for this example machine system was as follows:
*** SIMULATION OF TIMED PETRI NET ***
*** Disply_net2(BEFORE FIRED) ***
>>No shared-transition(s) enable!!
>>Single-transition(s) enabled, the process may continue....
FIRING SEQUENCE NO:1
>>Fired Single-Transition: T11 for 1s

TASK	M/C TIME	% UTILIZE
1	0	0.00
2	0	0.00
3	0	0.00
4	0	0.00
5	0	0.00

Parts output place: P54
Current parts completed = 0
*** Display_net3(AFTER FIRED) ***
>>The process may continue....
Then the firing continues until the firing sequence no. 15 where the processing of a part is complete.
This is executed as follows:
*** Display_net2(BEFORE FIRED) ***
>>No shared-transition(s) enable!!
>>Single-transition(s) enable, the process may continue....
FIRING SEQUENCE NO:15
<<Fired Single-Transition: T55 for 5s

TASK	M/C TIME	% UTILIZE
1	5	5.56
2	26	28.89

3	2	2.22
4	2	2.22
5	9	10.00

Parts output place: P54
Total parts complete =2

Thus, the above zone-controlled TMPN modelling approach offers a number of benefits: provides mappings between modelled physical machines; offers a well-structured, distributed and modular model for the control of machine systems where modelled elements correspond directors to physical machine elements. This enables reconfigurability and extensibility; readily supports the additon, delection and modification of modular elements of zones; provides means of planning an appropriate physical machine layout; provides zone-controlled material parts flow and information flow to support the complete machine systems; provides in indication of part routings and also operation sequences for shared resources; and offer means of coping with a growing "size" of machine systems which results in scalable solution.

5. Conclusion

Thus paper reports on the combined use of modular zone-control graphs and modular Petri net (TMPN) to support the design and contruction of machine systems. The approach offers a structured way of simplifying and formalising the representation of various activities and physical objects involved in the operation of machines. This provides a means of defines modular balancing blocks which can be used in different ways and lead to different performance characteristics. Each TMPN module uses two important types of transition, namely: time-transitions, which represent time delays associated with each type of resource; and shared-transitions which check input/output conditions and establish links among modules. The approach has the capacity to cope with routing flexibility at the early stage of design, which is related to the dynamic assignment of machines. This can lead to better designed machine systems, even where high levels of complexity are involved.

6. References

ARIFFIN, S., WESTON, R. H., HARRISON, R. 1995, A modular Petri-nets Approach to the design of distributed machine control systems, *Procs of the 31st Int. MATADOR Conf.*, UMIST, U.K., April, 621-626.

BEN-ARIEH, D. 1991, Concurrent modelling and simulation of reactive manufacturing systems using Petri nets, *Computers Industrial Engineering*, Vol. 20, No. 1, 45-58.

ANON 1992, Integrated machine design and control, *ACME/SERC Application: Case for Support*. Loughborough University of Technology, U.K., December.

CECIL, J.A., SRIHARI, K., EMERSON, C. R. 1992, A review of Petri nets application in manufacturing, *Int. J. Adv. Manuf. Technol*, 7, 168-177.

CHOI, B. W., ICUO, W., JACKMAN, J.K. 1994, Petri net extensions for modelling and validating manufacturing systems, *In. J. Prod. Res.*, Vol. 32, No. 8, 1819-1835.

DOTAN, Y., BEN-ARIEH, D. 1991, Modelling flexible manufacturing systems: the concurrent logic programming approach, *IEEE Trans on Robotics and Automation*, February, Vol. 7, No. 1, 135-149.

HARRISON, R. 1991, A generalised approach to machine control, *Phd Thesis*, Loughborough University of Technology.

GERSHWIN, S. B., HILDERBRANT, R. R., SURI, R., MITTER, S. K. 1986, A control perspective on recent trends in manufacturing systems, *IEE Control System Magazine*, April, 3-18.

KNAPP, G. M., WANG, H. P. 1992, Modelling of automated storage/retrieval systems using Petri nets, *Journal of Manufacturing Systems*, Vol. 11, No. 1, 20-29.

GARNOUSSET, H. E., FARINESS, J. M. 1989, Efficient tools for analysis and implementation of manufacturing system modelled by Petri nets with objects: a production rules compiltion-based approach, *IECON Proceedings IEEE*, Vol. 3, 543-549.

MICOVSKY, A., SESERA, L., VEISHAB, M. 1990. TOR: A Petri net based tool for rapid prototyping of FMS control system and its application to assembly, *Computers in Industry*, 279-292.

LIN, J. T., LEE, C. C. 1992, A modular approach for the modelling of a class of zone-control conveyor system using time Petri net, *Int. J. Computer Integrated Manufacturing*, Vol. 5, No. 4/5, 277-289.

LIU, C. M., WU, F. C. 1993, Using Petri nets to solve FMS problems, *Int. J. Computer Integrated Manufacturing*, Vol. 6, No. 3, 175-185.

PELAGAGGE, P. M., CORDORELLI, G, PALUMBO, M. 1995, Design criteria for cooperating robot assembly cells, *Journal of Manufacturing Systems*, Vol. 14, No. 4, 219-229.

REOUF, A., ANJUM, M. F. 1995, Manufacturing system: Flexibility assessment, *Flexible Manufacturing: recent development*, 69-83.

SHIN, K. G., EPSTEIN, M. E. 1987, Intertask communication in an integrated multirobot system, *IEEE Journal of Robotics and Automation*, April, Vol. RA-3, No. 2, 90-99.

WANG, H. M., WANG, H. B. 1995, Generation of optimal operating strategies for robotics cells: a Petri nets approach, *Int. J. Integrated manufacturing*, June, Vol. 8, No. 1, 32-42.

On Selective Multiple Inheritance in Fuzzy Object Modelling

Zhengkang Lu , K.B. Chuah and Yuejin Zhou

Department of Manufacturing Engineering & Engineering Management
City University of Hong Kong, Kowloon, Hong Kong
(e-mail: 95412030@plink.cityu.edu.hk)

Abstract. Objects as defined in conventional object-oriented modelling require a high degree of crispness. Specially, the values and range of the objects' attributes are stated with a high degree of certainty. This makes it difficult to model realistically problems characterized by imprecision and vagueness. Fuzzy objects provide a solution to the modelling of systems with vaguely defined attribute values. In fuzzy object modelling, we construct the fuzzy objects through their respective inheritance from the superclasses which define their fuzzy attributes and fuzzy reasoning methods. This paper gives the definition and representation of selective inheritance procedure at the conceptual level of the object-oriented methodology and the application in the fuzzy object modelling. It also discusses the two different selective inheritance concepts and the relationship with the name conflict problem of the multiple inheritance in the object-oriented methodology.

1. Introduction

It is generally accepted that objects as defined in conventional object-oriented modelling require a high degree of crispness. The values and range of the objects' attributes need to be stated with a high degree of certainty. This then, makes it difficult to model realistically problems characterized or affected by imprecision and vagueness in information content. Fuzzy objects provide a solution to the modelling of systems with vaguely described attribute values.[1]

In fuzzy object modelling, we construct fuzzy objects through their respective inheritance from the superclasses which define their fuzzy attributes and fuzzy reasoning methods[2]. Owing to the different membership functions and different fuzzy reasoning methods, we must select some suitable fuzzy attribute properties and some reasoning or other methods to construct the fuzzy object class. A class is an abstracted (or superior) object of a number of similar objects. Therefore, there is then a combination problem in the selection of the fuzzy classes as the superclasses.

Some authors have suggested the use of selective inheritance concept – i.e. inheritance of just a subset of attributes and methods from an ancestor class. Without the ability to select subsets, you can rapidly create a combinatorial explosion. It has been shown that having a class inheriting features from all its ancestors is an exponential problem[5]. Dori and Tatcher have also proved theorems about the upper bound on the number of classes one can derive using multiple inheritance[3]. Multiple inheritance may cause a name conflict of a property. Selective inheritance is applicable to resolving such a problem.[5]

This paper gives the definition and representation of the selective inheritance in the conceptual level of the object-oriented methodology and the application in the fuzzy object modelling. It also discusses the two different selective inheritance concepts and the relationship with the name conflict of the multiple inheritance in the object-oriented methodology.

2. The Structure of Fuzzy Object Model

An object-oriented environment is an excellent platform for handling both certain and uncertain information in a uniform way. For every fuzzy class which represents certain fuzzy sets of objects, the subclasses and instances from this class will inherit all of its attributes and methods from this class. A structure will then be constructed and added to the fuzzy sets of objects to represent the actual value of the fuzzy class. The fuzzy objects give us a way of representing knowledge about data and their key abstractions or concepts in the problem domain. A class hierarchy defines the relationship among the different objects in object-oriented modelling. The class hierarchy is linked by the inheritance properties, i.e. we can construct the common properties of an object in its super-class so as to enhance the ability of representing the object in that super class. It also leads to a better organized object-oriented information model.

A typical class hierarchy of the fuzzy objects includes Abstract fuzzy object, Fuzzy class, and Fuzzy object instance[2]. The Abstract fuzzy object defines the fuzzy class creating methods, fuzzy class erasing methods, and the reasoning methods. The Fuzzy class defines the attribute set, rule set, and distribution function set to calculate the membership function. The rule set is a complex attribute set in the fuzzy class. The fuzzy classes permit data to be viewed at different levels of abstraction based on the semantics of the data and their interrelationships. The Fuzzy object instance is an actual instance which inherits the properties of fuzzy class which includes attributes (which may have fuzzy or non-fuzzy domain), rules, and reasoning methods (which are inherited from abstract fuzzy object). A fuzzy object may collect some useful values from different original objects to realize the complex reasoning procedure.

We can create different categories of the fuzzy object classes with respect to the different categories of the fuzzy sets. Different categories of the fuzzy object classes have different properties and membership functions. Some fuzzy object classes may have properties of some special methods such as defuzziness methods. The reasoning methods are widely used concepts in fuzzy objects. They are commonly found in abstract fuzzy objects[2]. The reasoning methods may be related to the **Fuzzy-Set** class or the category of fuzzy objects. The properties of these fuzzy objects are used in the reasoning methods. Another class of the reasoning methods are public reasoning methods which are independent of any fuzzy object class. But they can be selected by several fuzzy objects or fuzzy object classes, and are not used as the common methods which are defined in the **Fuzzy-Set** of all fuzzy objects.

3. The Selective Multiple Inheritance

Instances (usually) inherit only the features of the classes they belong to, but it is also possible in an object-oriented system to allow classes to inherit features from more general superclasses. In this case inherited features can be overridden and extra features added to deal with exceptions. Inheritance is only one of the conceptual structures with which we organize the world. It is a particularly important one and its importance corresponds to the importance of the verb "to be" in a language.

Inheritance is the property that allows an object to receive some characteristics from another superior object (called a class). For example, a lathe and a drill in a production cell have in common many attributes including power, location, and overhead cost in addition to other different attributes. Also, they have some common operations (methods) such as drilling and tapping. A superior object (or class) can be created and called 'machine' with the common attributes and methods. By specifying two objects, 'lathe' and 'drill', as subclasses of 'machine', they will inherit all attributes of 'machine' along with any methods defined for 'machine'. Also, the attributes and methods of a class can be overridden in any of its subclasses. Depending on the implementation language, multiple inheritance could be allowed. In multiple inheritance, a class can inherit attributes and methods from many superior classes.

In the hierarchy of fuzzy objects, a fuzzy object inherits the properties of the super-classes which may be the fuzzy object class or the methods such as reasoning methods[2], see Figure 1. In the creation of the fuzzy object class, we must select one or more other fuzzy object classes as the super-class(es) that form the network model. Sometimes, we can also select one or more methods which are used by the fuzzy object class. So, the methods in the fuzzy object class can be inherited from the super-classes, and can also be inherited from the independent methods especially reasoning methods. In some fuzzy object class, we must define some special methods to manipulate the fuzzy objects. An example of such method is the defuzzification function.

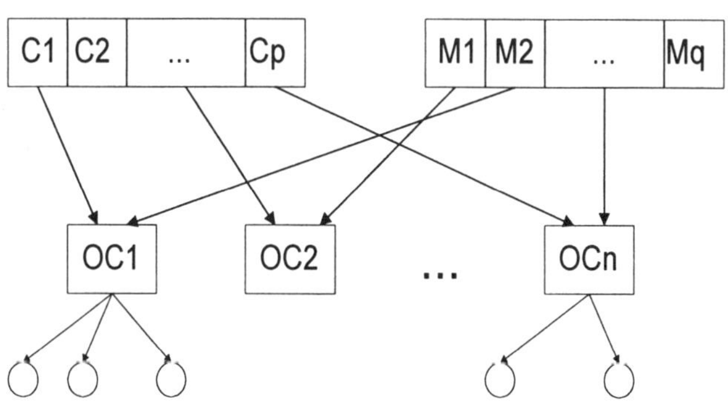

Figure 1 The selective inheritance in fuzzy object modelling

The fuzzy objects (instances) are inherited from their super-class. Let C be a set of classes. If a class A inherits the properties from some C_i and $C_i \in C$ (i = 1, 2, ...), that is, the class A selectively inherit from a set C of classes. We call this inheritance the selective inheritance. The selective inheritance represents the relationship among the class sets or classes which will be used to design the information conceptual model.

There are a couple of disadvantages with this method. The number of classes possible if one explicitly enumerates all the possible combinations using conventional multiple inheritance grows exponentially with the number of ancestors, leading to a combinatorial explosion.[4] Another disadvantage of explicit class enumeration is that it is counter-intuitive to human perception of the real world, and can result in a loss of coherence.[5]

4. The Representation of the Selective Inheritance in Conceptual Level

In conventional object-oriented modelling, the selective inheritance must convert the full inheritance in information modelling by using the additional class. Therefore, the representation of the information model can precisely reflect the real world situation. The conversion procedure itself also results in the combinatorial explosion in the multiple inheritance situation. We therefore introduce the selective inheritance at the conceptual level of the fuzzy object modelling to solve this problem. It will allow us to describe the relationship among class sets more naturally. It also resolves the combinatorial explosion in the construction of the information model.

According to the definitions given by Dori and Tatcher[3] and Suzuki and Ito [4], selective inheritance is inheritance of just a subset of the attributes and methods from an ancestor class. We select some properties as inherited properties and we select one (or more) class(es) in a set of classes as superclass of an object in our definition. In the first situation, the information model can be abstracted further to construct the classes as shown in Figure 2.

Let P be the set of properties in Class A. For Class B we select some properties $P' \in P$ as the inherited properties from Class A. Then we can divide Class A into two parts: Class A1 and Class A2. Class A1 has the properties P'. Class A2 has the properties P-P', and fully inherits the properties from Class A1. Therefore, Class B fully inherits the properties from Class A1. More generally, we can reconstruct Class A from which the other Class B1, B2 ....., Bn selectively inherit the properties, as shown in Figure 2 (c).

With selective multiple inheritance, a class may inherit certain features from any number, m of a given set of, n ancestor classes according to our definition. In the conceptual level of fuzzy object modelling, the selective inheritance is a special relationship among the classes. We need a special symbol to denote this relationship among the object class sets (see Figure 3). In Figure 3, the ellipses represent the sets of classes, and the rectangles represent the class. As shown, Class A selectively inherits the Class Set S which means that Class A inherits the properties of one or more class(es) in the Class Set S.

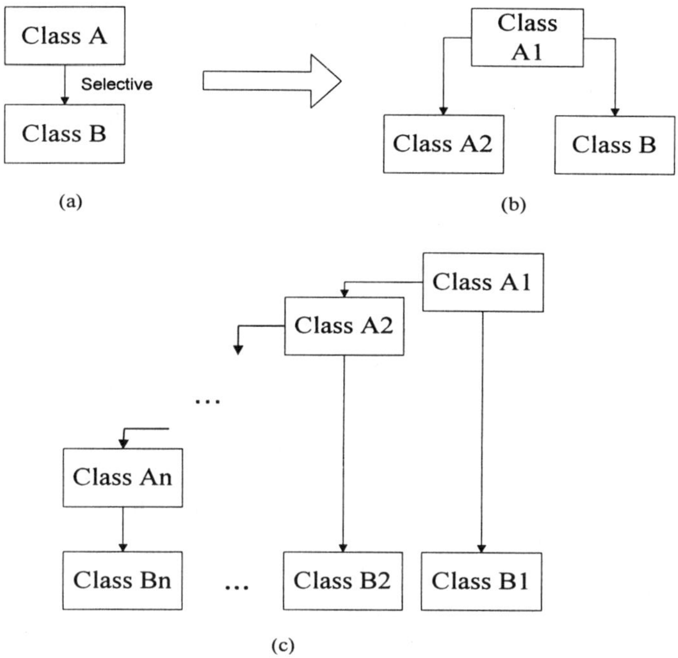

Figure 2 The further abstraction of selective inheritance

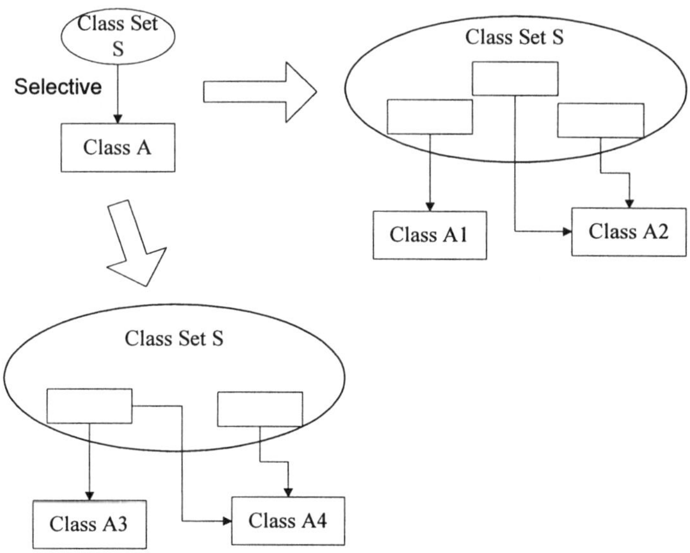

Figure 3 The types of selective inheritance

In selective inheritance, a class can select one or many classes in a set of classes as its superclass. If a class A can only select a class Ci in the set of classes C as the superclass of a class A, we call this selective inheritance, the single selective inheritance. In single selective inheritance, if a class Ci in the set of classes C is selected by a class A, then the class Ci cannot be selected by other classes in this inheritance relation. We describe this inheritance as the exclusively single selective inheritance and call it Class A1. Otherwise, we call the non-exclusive single selective inheritance Class A3. If a class A can select more than a class Ci in the set of classes C as the superclasses of a class A, we call this selective inheritance, the multi-selective inheritance.

Similarly, if some classes {Ci} in the set of classes C are selected by a class A, then the classes {Ci} cannot be selected by other classes in this multi-selective inheritance. We call this exclusively multi-selective inheritance Class A2. Otherwise, we call the non-exclusively multi-selective inheritance Class A4. The exclusively single selective inheritance can be represented by the 1:1 relationship in E-R model, so it is denoted as $S_{1:1}$. The non-exclusively single selective inheritance can be represented by the n:1 relationship in E-R model, so it is denoted as $S_{n:1}$. Similarly, the exclusively multi-selective inheritance can be represented by the 1:n relationship in E-R model, so it is denoted as $S_{1:m}$. The non-exclusively single selective inheritance can be represented by the n:m relationship in E-R model, so it is denoted as $S_{n:m}$.

The fuzzy object class selectively inherits the properties from the Fuzzy Object Category Set and Abstract Methods Set as shown in Figure 4. The selective inheritance from the Fuzzy Object Category Set is the non-exclusively single selective inheritance $S_{n:1}$. That is to say, a fuzzy object class can inherit the properties of one class in the Fuzzy Object Category Set which define the basic properties of the defined fuzzy object. The selective inheritance from the Abstract Methods Set is the non-exclusively multi-selective inheritance $S_{n:m}$, that is, a fuzzy object class can inherit the properties of some classes in the Abstract Methods Set which define the operation methods of the defined fuzzy object such as reasoning methods and other methods.

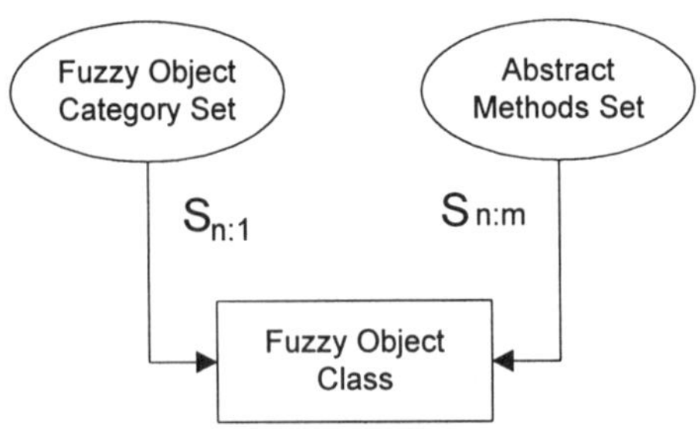

Figure 4 The model of fuzzy objects using selective inheritance

Let {Fi, 1≤i≤m} be the Fuzzy Object Category Set, and {Mj, 1≤j≤n} be the Abstract Methods Set, then a Fuzzy Object Class C can have the set of superclass Ft ∈ {Fi,

$1 \leq i \leq m\}$ and a subset Mc of the $\{Mj, 1 \leq j \leq n\}$. Therefore, the combination of Ft and Mc defines of the fuzzy object class in the fuzzy object modelling. This method provides a the powerful tool for information modelling of the real systems.

5. The Selective Inheritance and the Multiple Inheritance

Multiple inheritance implies the inheritance of features from many ancestors and is often justified as a way to provide reusable descriptions[3]. The aim of the multiple inheritance is to simplify the production and adaptation of code, to improve code sharing among classes and to localize code in the class that might need modification. Multiple inheritance may cause a name conflict of a property, including attributes and methods in the class.

Suzuki and Ito [4] suggest that selective inheritance is applicable to resolving the name conflict in the multiple inheritance. For example, assume that two classes C_1 and C_2 have the same property P and that a class C should inherit P from C_1 but not from C_2. For this, we introduce a "dummy" subclass C' of C_2 from which P is dropped by selective inheritance. To define the class C, it suffices to make C a subclass of C_1 and C'. However, this method will cause another problem.

In the object-oriented methodology, the designing of a class is independent from the other classes except the superclass with which the defined class will share the properties. But, the designing of C' in the above example will depend not only on the class C_2 which is the superclass of the C', but also on the classes C and C_1 to determine the properties P. This will result in creating difficulty in the implementation and maintenance of the class in the information model.

On the other hand, the name conflict problem can be simply solved by combination of defining the inheritance order and redefining properties. In the definition of a class, we can define an order in which the class inherits the properties of the superclasses. This order called inheritance order is only one in a class definition. If the properties which the class inherits are not in the inheritance order, we can redefine the properties in the definition of the class to indicate their superclass. For example, we can define the inheritance order $\{C_1, C_2\}$ in the class C. The class C inherit properties P from C_1, not from C_2, although there are also properties P in the class C_2.

6. Conclusion

In the fuzzy object modelling, there is the combination problem in selecting the fuzzy classes as the superclass. Therefore, we introduce the selective inheritance concept into the fuzzy object modelling in the conceptual level. Then, we distinguish the different definitions of the selective inheritance, and also give the representation of the selective inheritance in conceptual level, and the model of the fuzzy object using the selective inheritance. Finally, we discuss the relationship between the selective inheritance and the multiple inheritance. The selective inheritance plays an important role in the information modelling of the real systems.

References:

1. T.D. Ndousse, Intelligent Systems Modelling with Reusable Fuzzy Objects, *International Journal of Intelligent Systems*, Vol. 12, pp.137-152,1997
2. Z.K. Lu, K.B. Chuah and Y.J. Zhou, The Construction of the Fuzzy Object, *Proc. CIRP International Symposium – Advance Design and Manufacture in the Global Manufacturing Era, Hong Kong*, August 1997
3. E. Dori and E. Tatcher, Selective Multiple Inheritance, *IEEE Software*, Vol. 11, No. 3, pp.77-85, May 1994
4. N. Suzuki and M. Ito, Specialization Constraints for a Complex Object Model Supporting Selective Inheritance, *IEICE Trans. Inf. & Syst.* Vol. E78-D, No 11 pp.1458-1467, November 1995
5. F. Hayes and D. Colman, Coherent Models for Object-Oriented Analysis, *Proc. OOPSLA*, ACM Press, New York, 1991, pp.171-183.

Logistics Management

A Centralised Routing Algorithm for AGVs in Container Ports

Xinguo Yu, Shell Ying Huang

Division of Software Systems
School of Applied Science
Nanyang Technological University
Singapore 639798
email: assyhuang@ntu.edu.sg

Abstract

The work reported here uses a network of time-dependent interedge transit times to model the container port and the traffic conditions. This network is able to represent a container port with complicated track configuration which cannot be represented by a graph which consists of a set of nodes and a set of edges. It is also able to reflect the traffic conditions in the port which changes with time and therefore cannot be modelled by a constant network. A routing algorithm based on Dijkstra's shortest path algorithm is proposed for this time-dependent network with interedge transit times to find the fastest route for AGVs to accomplish their delivery jobs. This centralised routing method is evaluated against another centralised routing method, S. C Daniels' PhD work on real time conflict-free routing algorithm for AGVs. Simulation results showed that the method proposed here outperformed his method.

1. Introduction

Automated Guided Vehicles (AGVs) are driverless vehicles. They are widely used in the manufacturing and shipping industries which incorporate a high degree of automation into their operations, thus reducing their reliance on manual labour while increasing productivity. In the shipping industry the AGVs are employed to handle the containers within a container port. They move along uni-directional or bi-directional tracks within the container storage area and between the container storage area and the loading/unloading bays for ships and lorries.

In this work, the routing of AGVs in container ports is studied. Routing can be centralised or decentralised, . Centralised control employs one computer to perform all functions related to routing and routes planned for AGVs can be optimized in terms of travel time. In a decentralised control system each AGV has the intelligence to work out its own route but there is no guarantee that the route is an optimised one. Needless to say, decentralised systems are more fault tolerant than centralised systems.

Existing methods of routing AGVs adopting either centralised or decentralised approaches use a graph model H = (V, E) to represent the configuration of the AGV tracks. This graph model cannot represent the AGV systems in which it takes different amount of time to pass a junction or a section of tracks due to the layout of the tracks or the different degree of traffic congestion. This work proposed to use a graph with time-dependent interedge transit times to model the AGV track configuration and the traffic conditions in a container port. Centralised routing algorithms based on Dijkstra's shortest path algorithm were proposed to find the fastest routes for AGVs through this network with time-dependent interedge transit times. Extensive simulations were done to evaluate this method of routing AGVs using the proposed model for the traffic system in container ports. Comparison was done with Daniel's Ph.D work, which was another centralised conflict-free routing algorithm for AGVs. Results showed that the proposed method outperformed Daniel's work.

The rest of the paper is organized as follows. Section 2 reviews existing research results on AGV routing. Section 3 presents the approach of modelling of the AGV track configuration and the traffic conditions by a network with time-dependent interedge transit times. After this, the routing algorithms are described in Section 4 and the simulation experiments are reported in Section 5. Section 6 presents the performance results and the comparison with Daniels method. Lastly, Section 7 contains the conclusions.

2. Existing Routing Algorithms for AGVs

In most typical container ports, the traffic lanes for AGVs form a mesh-like topology. The junctions of the lanes are the vertices of the mesh where an edge of the mesh may be (1) a single uni-directional track; or (2) a single bi-directional track; or (3) two uni-directional tracks allowing traffic flow in opposite directions; or (4) any combination of the above. The complexity of the junctions increases when more tracks meet at the junctions.

The problem of routing AGVs is very much simpler if only uni-directional tracks are considered. Routing AGVs in a network which is a simple loop or consists of several loops in the decentralised approach were reported in [1] and [2]. AGVs simply followed some fixed uni-directional loops. The issue of conflicts among AGVs did not exist because tracks were uni-directional.

Directing multiple AGVs over a network with bi-directional pathways was considered "unmanageable" by Young[4]. It is certainly true that to guarantee collision-free, deadlock-free and livelock-free systems is very difficult. However, the benefits of a network with bi-directional links are shown in [3]. Two decentralised routing algorithms, namely the row-first-then-column algorithm and the semi-randomized algorithm, were studied in [8]. In the row-first-then-column algorithm the an AGV always travels along a direction until it reaches the column of the destination in the mesh of AGV lanes. Then it will travel to the row of the destination. In the semi-randomized algorithm, an AGV will make a random choice among the directions it may go under the constraint of monotonic rooting. In both algorithms if a certain move will generate a conflict with other AGVs, then a detour routine will resolve the conflict.

One S. C. Daniels embedded a shortest path algorithm within a branch-and-bound procedure[5] to generate conflict-free AGV routes through bi-directional networks. This is the first centralised conflict-free routing algorithm for bi-directional network. The algorithm is described here:

Step 1: Set Z_u to be infinity. Z_u is the travel time of the route found by this algorithm. Find a shortest path from the origin to destination without regard to any existing traffic. Put the solution into the active list.

Step 2:Remove a solution from the active list and check it for feasibility. If the active list is empty, go to step 3.

a. If no conflict exists, a feasible solution has been found. If the travel time of this solution is less than Z_u, update Z_u and repeat step 2.

b. If a conflict is found, generate two alternative solutions:

b1. Find the shortest path from the origin to the destination over a modified network with the conflicting edge removed.

b2. Determine the minimum delay time necessary to resolve the conflict.

Place the solutions found in the active list and repeat step 2.

Step 3:If any feasible solution was found, Z_u represents the optimal travel time between origin and destination without causing a conflict to occur.

In a network with busy traffic, this system obviously needs numerous re-computation of the shortest paths. This is because the shortest paths found may often be infeasible as the computation for the shortest path does not take the existing traffic into consideration. If the network model of the container port includes traffic conditions of the AGV tracks, only one computation of the shortest path will be required to generate the solution.

3. Modelling of the Tracks and the Traffic Conditions

Because a constant network cannot incorporate the changes in traffic conditions with time, a time dependent network is proposed. The following subsections describe the model and its implementation respectively.

3.1 The Model

Definition 1: Let G = (V, E, F) be a digraph with vertex set V and edge set E. F is a set of functions of time, $d_{ijk}(t)$, which gives the time to travel at time t from the end of edge (i,j) to the end of the edge (j,k). G is called a graph or a network with time-dependent interedge transit times.

A graph with time-dependent interedge transit times may be used to represent AGV traffic systems which cannot be represented by a graph H = (V, E). It is also more general than a network with time-dependent internodal transit times[6]. It is able to represent a network with turn penalties[7]. This means that unequal travelling times across junctions and sections of tracks depending on the current location of a vehicle is reflected by having different values for $d_{ijk}(t)$ and $d_{hjk}(t)$. This is useful, for example, in Figure 1, the horizontal track is a very busy section of the port and the vertical track is less busy. $d_{ijk}(t)$ can be set to the minimum travel time from j to k and $d_{hjk}(t)$ can include the waiting time at j to allow the traffic in the horizontal direction to pass first. Another situation is that there is a side track at a junction so that any AGV trying to turn left can do so in minimum amount of time unless it is blocked by an AGV in front of it, but AGVs trying to turn right or go straight crossing the junction will take a little longer time.

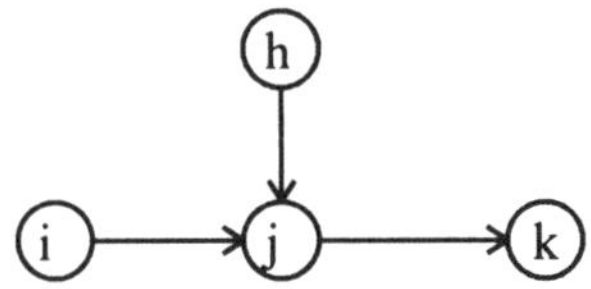

Fig 1. Turn penalty.

This graph with time-dependent interedge transit times is also able to represent various track configurations. If it is a uni-directional track from junction j to k, then along the direction in which the traffic is not allowed, all the $d_{hjk}(t)$ (for any h) will be infinity. If it is a bi-directional track, then all the $d_{hjk}(t)$ (for any h) will be the travelling time for $t = t_1$ when the traffic flow is from j to k at time t_1 and all the $d_{hjk}(t)$ (for any h) will be infinity for $t = t_2$ when the traffic flow is from k to j at time t_2. If there are multiple tracks between j and k, for example, two tracks from j to k in the same direction, then all the $d_{hjk}(t)$ (for any h) will then represent travelling time from end of (h,j) to end of (j,k) using any track allowed by the system at time t.

If a certain link in the AGV tracks is temporarily broken or an AGV breaks down blocking a link from j to k, all the $d_{hjk}(t)$ (for any h) will be set to infinity to prevent attempts to use that link for as long as necessary. All the $d_{hjk}(t)$ (for any h) will be reset to their normal travel times once the blockage is removed.

To summarize, this network with time-dependent interedge transit times is able to represent the changing traffic conditions, the turn penalties and the sudden breakdown of tracks or blockage caused by other reasons.

3.2 Computation of the Travelling Times

If a certain function $d_{ijk}(t)$ in F is not infinity because travelling in the relevant direction is legal in the track topology and the track is operational, then $d_{ijk}(t)$ consists of two parts, $f^1_{ijk}(t)$ and $f^2_{ijk}(t)$ where $f^1_{ijk}(t)$ is the waiting time before an AGV can travel through the junction at j and the edge (j,k) without collision with other AGVs. $f^2_{ijk}(t)$ is the travel time from the end of (i,j) to the end of (j,k) when no other AGV is interfering with the AGV being considered. $f^2_{ijk}(t)$ is a constant for any (i,j) and (j,k). So,

$d_{ijk}(t) = f^1_{ijk}(t) + f^2_{ijk}(t)$.

To find $f^1_{ijk}(t)$, the time for which an AGV will be blocked when it arrives at the end of (i,j) needs to be calculated. Basically, the calculation involves finding out the earliest time an AGV can start at the end of (i,j) without causing collision. This is based on the set of non-departure times as defined below.

Definition 2: Consider a track u which goes from the end of edge (i,j) to the end of edge (j,k). Time t is called a non-departure time through track u if a collision occurs when an AGV starts the travelling along track u at time t.

Let NT_{ijk} be the set of non-departure times from the end of (i,j) to the end of (j,k). When no AGV is travelling in the graph G = (V, E, F), an AGV may travel anywhere in the graph without collision. Therefore NT_{ijk} is empty for all vertices i, j, k in G. After an AGV's route is determined, non-departure times will be inserted into the sets of non-departure times for all the vertices i,j, k which the AGV passes. To illustrate how NT_{ijk} is computed, consider the following example.

Assume that R_1, R_2, ... R_x are the routes scheduled for x AGVs. Route R_{x+1} is the route just computed for AGV x+1. Assume also that there is one single bi-directional track u from (i,j) to (j,k), that is, from end of (i,j) to the end of (j,k).

Case 1: R_{x+1} crosses the junction at j at time t_0 as indicated by the arrows in Figure 2. Assume that R_{x+1} takes one unit of time to cross the junction at j at time t_0. Then t_0 is inserted into NT_{ijk} as another non-departure time.

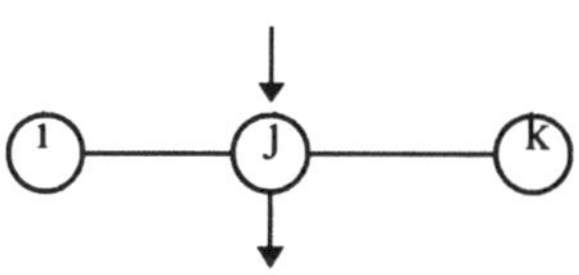

Fig. 2. R_{x+1} crosses junction j.

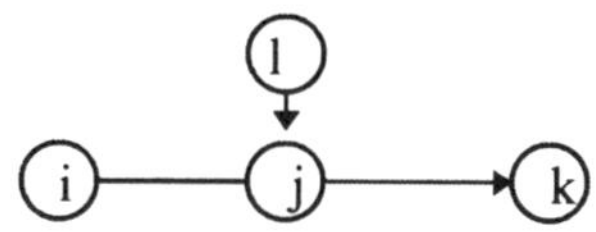

Fig. 3. R_{x+1} turns at junction j.

Case 2: R_{x+1} turns from (l,j) to (j,k) at time t_0 along track u as indicated by the arrows in Figure 3. Assume that an AGV needs one unit of time to go from end j of (l,j) to end k of (j,k). Then t_0 is inserted into NT_{ijk} as another non-departure time.

Case 3: R_{x+1} turns from (k,j) to (j,l) at time t_0 through track u as indicated by the arrows in Figure 4. Assume that an AGV needs one unit of time to go from end j of (k,j) to end l of (j,l). Then t_0 is inserted into NT_{ijk} as another non-departure time.

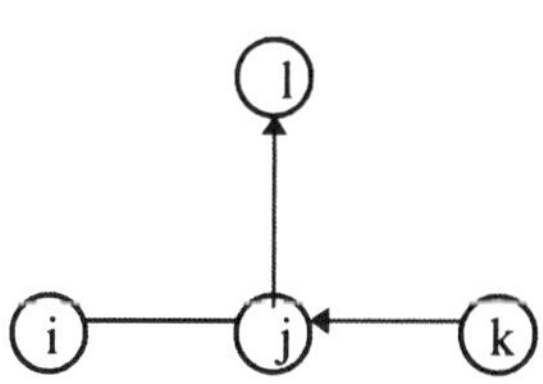

Fig 4. R_{x+1} turns at junction j.

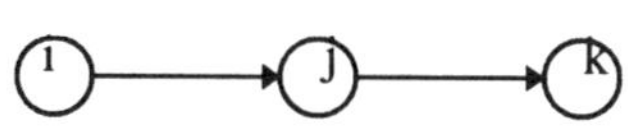

Fig 5. R_{x+1} goes from (i,j) to (j,k).

Fig 6. R_{x+1} goes from(l,k) to (k,j).

Case 4: R_{x+1} goes from (i,j) to (j,k) at time t_0 along track u as indicated by the arrows in Figure 5. Then t_0 is inserted into NT_{ijk} as another non-departure time.

Case 5: R_{x+1} goes from (l,k) to (k,j) at time t_0 along track u as indicated by the arrows in Figure 6. Assume that it takes five units of time to go from (l,k) to (k,j). Then t_0-6, t_0-5, t_0-4, t_0-3, t_0-2, t_0-1, t_0, t_0+1, t_0+2, t_0+3, t_0+4, t_0+5 all become non-departure times in NT_{ijk}.

In the configurations where there are multiple tracks between j and k, NT_{ijk} may or may not be affected by route R_{x+1}. For example, if there are two bi-directional tracks u and v between j and k, then NT_{ijk} is the intersection of NT^u_{ijk} and NT^v_{ijk}, where NT^u_{ijk} is the set of non-departure times with respect to track u and NT^v_{ijk} is the set of non-departure times with respect to track v. So if t_0 is inserted in NT^u_{ijk} but it is not in NT^v_{ijk} then t_0 will not be in NT_{ijk}. On the other hand if t_0 is inserted in NT^u_{ijk} and it is already in NT^v_{ijk} then t_0 will be in NT_{ijk}.

With the values of NT_{ijk}, $f^1_{ijk}(t)$ can be calculated by

$$f^1_{ijk}(t) = \min\{y: y >= t \text{ and } y \text{ is not in } NT_{ijk}\}$$

From this the value of $d_{ijk}(t)$ can be obtained.

4. Routing the AGVs

Routing in the network of time-dependent interedge transit times is a problem of finding the fastest route for AGVs. It is based on Dijkstra's shortest path algorithm. The algorithm is slightly modified as described below:

Introduce a virtual vertex 0 and a virtual edge (0,1) into the network. V is the set of junctions in the port; E is the set of edges linking the junctions. u_{ij} is the shortest time known during the routing process to reach end j of (i,j) from end 1 of (0,1); pred(j) stores the edge in the earliest arrival path to reach j; pred[(i,j)] stores the predecessor edge of (i,j); u_j is the earliest arrival time at junction j from 1; PE is the set of edges with permanent labels and TE is the set of edges with temporary labels.

Step 0: Set $u_{01} = 0$, u_{ij} = infinity for (i,j) in E;
Set $u_j = -1$ for all j in V;
Set pred[(i,j)] = (-1,0) for all (i,j) in E;
PE = {}, TE = E + (0,1);

Step 1: Find (i,j) in E where u_{ij} is the minimum among all (i,j) in TE;
Remove (i,j) from TE and add it to PE;
If $u_j = -1$ then $u_j = u_{ij}$ and pred(j) = (i,j);
If TE is empty, then exit;

Step 2: For all (j,k) in TE:
if $u_{ij} + d_{ijk}(u_{ij}) < u_{jk}$, then $u_{jk} = u_{ij} + d_{ijk}(u_{ij})$ and pred[(j,k)] = (i,j);
Go to step 1;

The example shown in Figure 7 is used to demonstrate the routing process using this algorithm.

Start: $u_{01} = 0$, other u_{ij} set to infinity, all u_i set to -1, PE set to empty, TE = E

Iteration 1: u_{01} is minimal, $u_1 = 0$; pred(1) = (0,1); PE = {(0,1)}; TE = {(1,2), (1,3), (2,4), (2,5), (3,5), (3,6), (4,7), (5,7), (5,8), (6,8), (7,9), (8,9)}; $u_{12} = 4$; pred[(1,2)] = (0,1); $u_{13} = 6$; pred[(1,3)] = (0,1);

Iteration 2: u_{12} is minimal, $u_2 = 4$; pred(2) = (1,2); PE = {(0,1), (1,2)}; TE = {(1,3), (2,4), (2,5), (3,5), (3,6), (4,7), (5,7), (5,8), (6,8), (7,9), (8,9)}; $u_{24} = 24$; pred[(2,4)] = (1,2); $u_{25} = 10$; pred[(2,5)] = (1,2);

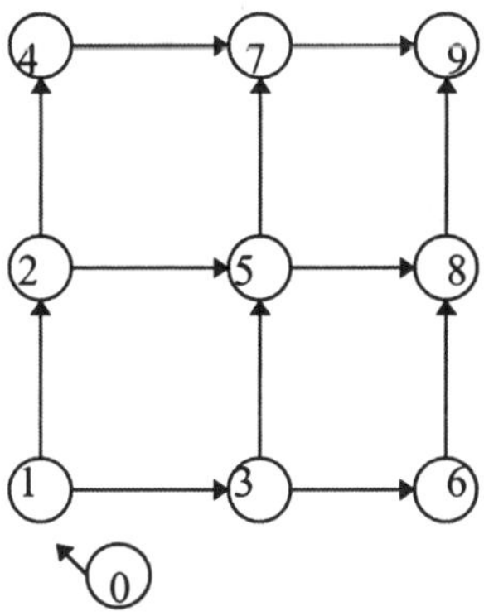

$d_{012}(t) = 4$
$d_{013}(t) = 6$
$d_{124}(t) = 20$
$d_{125}(t) = 6$
$d_{135}(t) = 16$
$d_{136}(t) = 4$
$d_{247}(t) = 20$
$d_{589}(t) = 4 + 2t$
$d_{689}(t) = 20 + t$
$d_{257}(t) = 20$
$d_{258}(t) = 6$
$d_{357}(t) = 4$
$d_{358}(t) = 6$
$d_{368}(t) = 4$
$d_{479}(t) = 12$
$d_{579}(t) = 20$

Fig 7. A network with time-dependent interedge transit times

Iteration 3: u_{13} is minimal, $u_3 = 6$; pred(3) = (1,3); PE = {(0,1), (1,2), (1,3)}; TE = {(2,4), (2,5), (3,5), (3,6), (4,7), (5,7), (5,8), (6,8), (7,9), (8,9)}; $u_{35} = 22$; pred[(3,5)] = (1,3); $u_{36} = 10$; pred[(3,6)] = (1,3);

Iteration 4: u_{25} is minimal, $u_5 = 10$; pred(5) = (2,5); PE = {(0,1), (1,2), (1,3), (2,5)}; TE = {(2,4), (3,5), (3,6), (4,7), (5,7), (5,8), (6,8), (7,9), (8,9)}; $u_{57} = 30$; pred[(5,7)] = (2,5); $u_{58} = 16$; pred[(5,8)] = (2,5);

Iteration 5: u_{36} is minimal, $u_6 = 10$; pred(6) = (3,6); PE = {(0,1), (1,2), (1,3), (2,5), (3,6)}; TE = {(2,4), (3,5), (4,7), (5,7), (5,8), (6,8), (7,9), (8,9)}; $u_{68} = 14$; pred[(6,8)] = (3,6);

Iteration 6: u_{68} is minimal, $u_8 = 14$; pred(8) = (6,8); PE = {(0,1), (1,2), (1,3), (2,5), (3,6), (6,8)}; TE = {(2,4), (3,5), (4,7), (5,7), (5,8), (7,9), (8,9)}; $u_{89} = 34$; pred[(8,9)] = (6,8);

Iteration 7: u_{58} is minimal; PE = {(0,1), (1,2), (1,3), (2,5), (3,6), (5,8), (6,8)}; TE = {(2,4), (3,5), (4,7), (5,7), (7,9), (8,9)};

Iteration 8: u_{35} is minimal; PE = {(0,1), (1,2), (1,3), (2,5), (3,5), (3,6), (5,8), (6,8)}; TE = {(2,4), (4,7), (5,7), (7,9), (8,9)}; $u_{57} = 26$; pred[(5,7)] = (3,5);

Iteration 9: u_{24} is minimal; $u_4 = 24$; pred(4) = (2,4); PE = {(0,1), (1,2), (1,3), (2,4), (2,5), (3,5), (3,6), (5,8), (6,8)}; TE = {(4,7), (5,7), (7,9), (8,9)}; $u_{47} = 44$; pred[(4,7)] = (2,4);

Iteration 10: u_{57} is minimal; $u_7 = 26$; pred(7) = (5,7); PE = {(0,1), (1,2), (1,3), (2,4), (2,5), (3,5), (3,6), (5,8), (5,7), (6,8)}; TE = {(4,7), (7,9), (8,9)}; $u_{79} = 46$; pred[(7,9)] = (5,7);

Iteration 11: u_{89} is minimal; $u_9 = 34$; pred(9) = (8,9); PE = {(0,1), (1,2), (1,3), (2,4), (2,5), (3,5), (3,6), (5,8), (5,7), (6,8), (8,9)}; TE = {(4,7), (7,9)};

Iteration 12: u_{47} is minimal; PE = {(0,1), (1,2), (1,3), (2,4), (2,5), (3,5), (3,6), (5,8), (4,7), (5,7), (6,8), (8,9)}; TE = { (7,9)};

Iteration 13: u_{79} is minimal; PE = {(0,1), (1,2), (1,3), (2,4), (2,5), (3,5), (3,6), (5,8), (4,7), (5,7), (6,8), (7,9), (8,9)}; TE = {};

From the computation above, the fastest route from 1 to 9 is through the edges (1,3), (3,6), (6,8), (8,9). The travel time is 34. It should be pointed out that in iteration 4 it is found that the earliest arrival path to 7 is through junction 5 after coming through the edge (2,5). Then in iteration 8, a faster way to reach junction 7 is known which is through junction 5 after coming through the edge (3,5). This information is indicated by pred(5,7) which is changed to (3,5) from (2,5).

It should be mentioned here that every time after a route is obtained by the algorithm above, the time-dependent network will be updated to reflect the current traffic situation.

5. The Simulation Experiments

Simulation experiments are conducted to evaluate the performance of the routing algorithm using the network with time-dependent interedge transit times as the model for a container port. The layout of the port used in the experiments is shown in Figure 8.

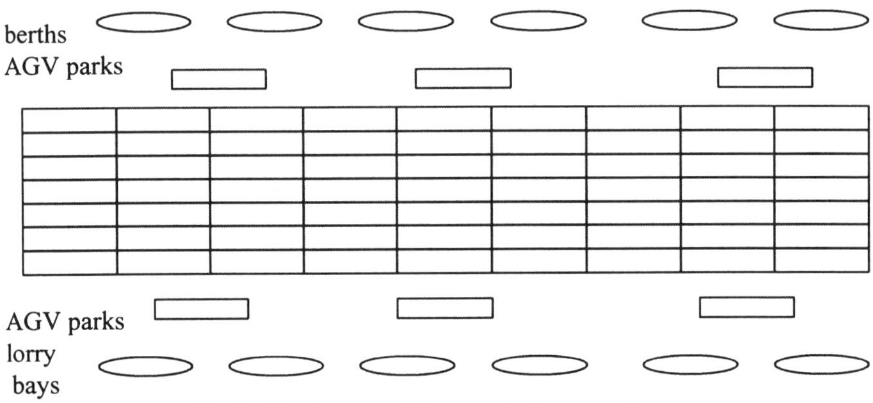

Fig 8. The layout of the port in the experiments.

The port used in the simulation experiments has six ship berths. Next to the ship berths are the AGV parks. Two berths will share one AGV park. The storage area for the containers has a 7 by 9 mesh topology. At the bottom of the layout are the AGV parks and the lorry bays. Each AGV will be allocated to an AGV park. After finishing a delivery job, an AGV returns to its park and waits for the next job assignment. The AGVs in the port are evenly distributed into all the parks.

The simulation assumptions are:

Tracks: When two tracks lie in parallel to each other, assume that AGVs cannot cross from one track to another. An AGV can only surpass the others at junctions with buffers. An AGV is not allowed to do U-turn involving two parallel tracks.
AGV speed: Assume that all road sections in the port have the same quality and that all AGVs have the same speed. Therefore all AGVs travel at the same speed anywhere in the port.
Containers on ships: Assume that the number of containers to be loaded/unloaded to/from ships is randomly distributed.
Ship arrivals: There are always ships waiting to be serviced in the harbour. For each berth one ship will arrive 20 units of time after the previous ship leaves.
Lorry arrivals: For each lorry bay, there is always a lorry to be loaded/unloaded.
Loading/unloading time: Assume that the loading/unloading of a container unto/from a ship or a lorry and the loading/unloading of a container into/from a storage place will take three units of time.
Storage locations for containers: Assume the random policy for choosing locations for storing containers.
Scheduling of AGVs: Each delivery job is given to an AGV which arrives at the nearest AGV park earlier than all other AGVs in the park.

Experiments are conducted for four different track configurations:

(1) Single track without buffer: in the mesh, each edge is a single bi-directional guided track. This is referred to as configuration 1.
(2) Single track with buffer: in the mesh, each edge is a single bi-directional guided track. At both ends of each edge there is one buffer area which can accept one AGV. This area is used to give way to another AGV. This configuration is referred to as configuration 2.
(3) Mixed configuration: Each vertical edge in the mesh has two uni-directional guided tracks with opposite orientation. Each horizontal edge has one bi-directional guided track. This is referred to as configuration 3.
(4) Double track configuration: Each edge in the mesh has two uni-directional guided tracks with opposite orientation. This is configuration 4.

6. Performance Evaluation

Three types of performance figures are collected. (1) The throughput, that is, the number of containers delivered in a fixed amount of time. (2) Mean delay time. This is the average time a delivery request has to wait until an AGV arrives to pick up the container. (3) Mean computation time. This is the average CPU time that is used to compute the route for one delivery job. The results are shown in Figure 9, 10 and 11.

Figure 9 shows that under all the track configurations experimented, the throughput by using the model of the port and the routing algorithm proposed in this work is higher than using a constant network as the model and Daniels' routing algorithm. The improvement in throughput increases as the number of AGVs increases.

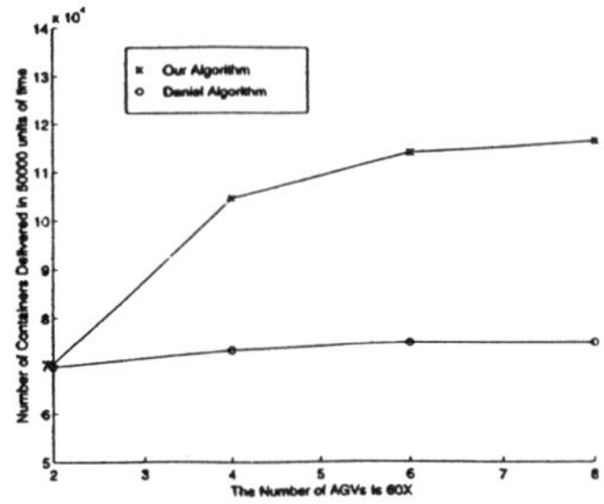

(a) Configuration 1

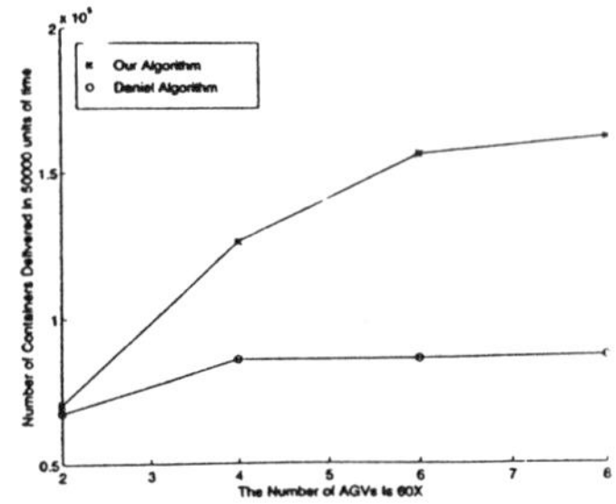

(b) Configuration 2

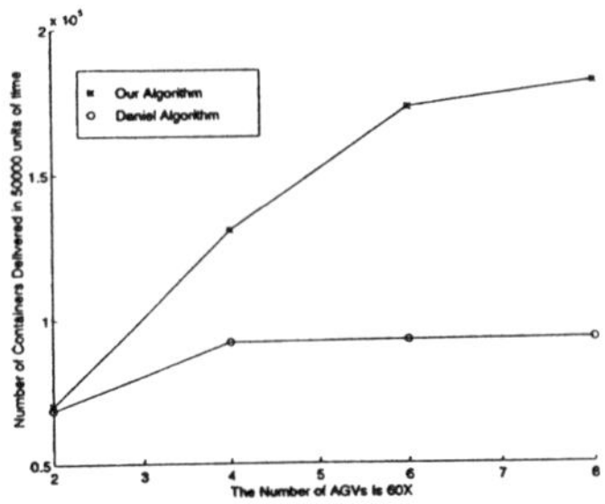

(c) Configuration 3.

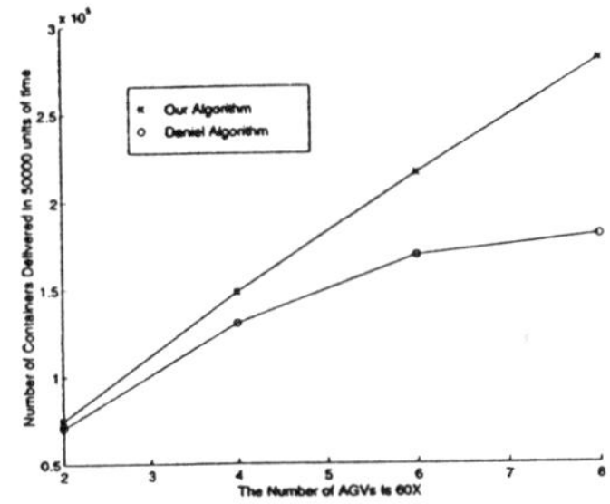

(d) Configuration 4

Fig 9. The throughput.

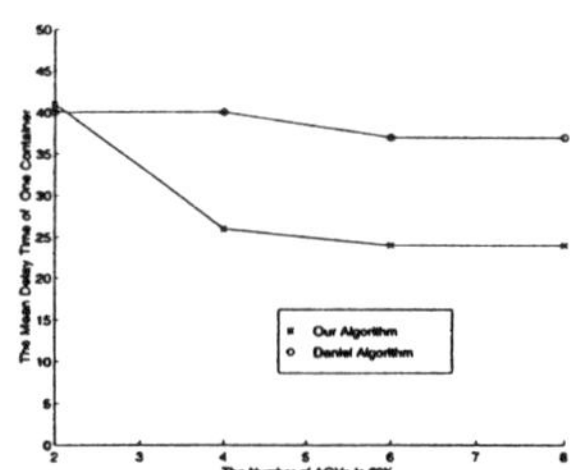

(a) Configuration 1.

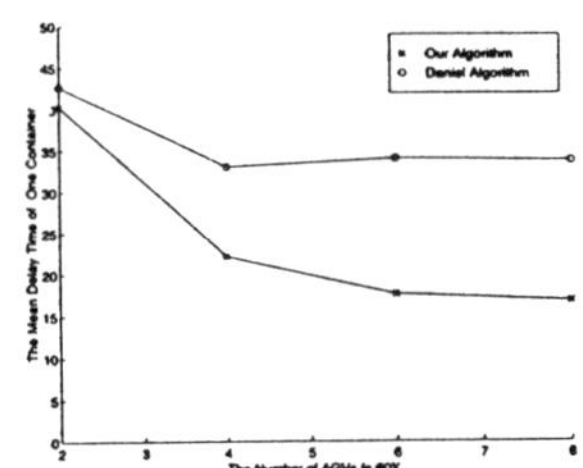

(b) Configuration 2.

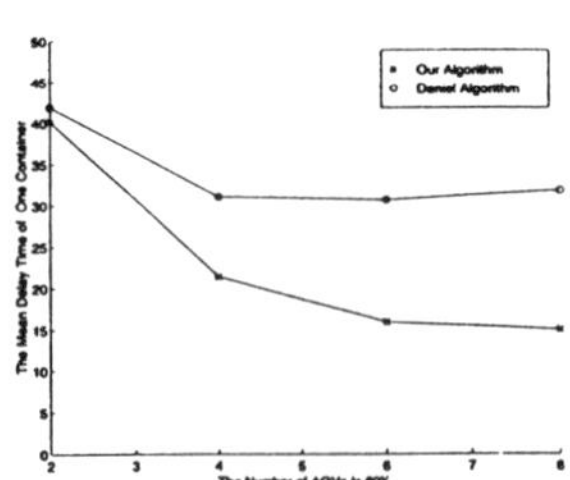

(c) Configuration 3.

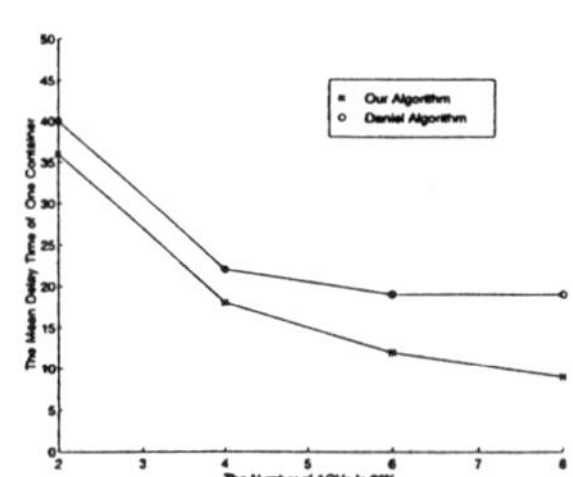

(d) Configuration 4.

Fig 10. Mean delay time for delivery requests.

As shown in Figure 10, there is one case where the mean delay time in our system is longer than Daniels'. This is when each AGV park has 20 AGVs in the single bi-directional track without buffer configuration. In all other cases the delay in our system is at least 8% shorter than Daniels' system.

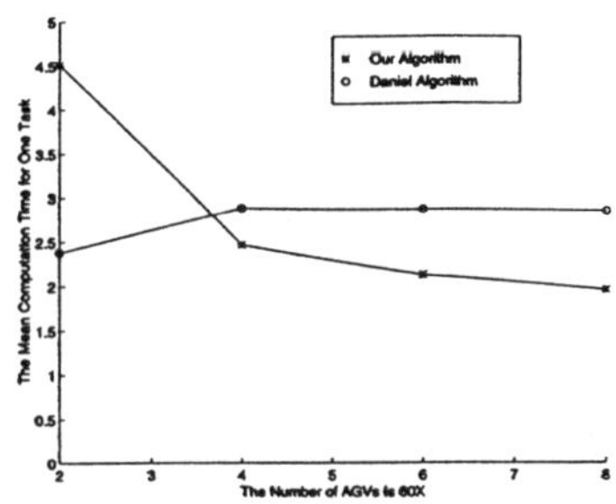

(a) Configuration 1.

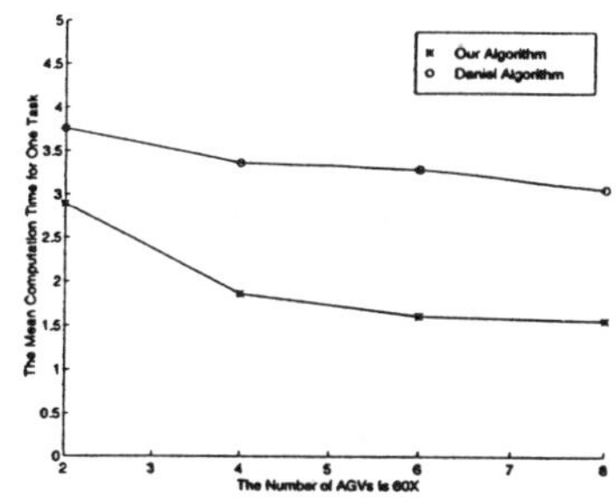

(b) Configuration 2.

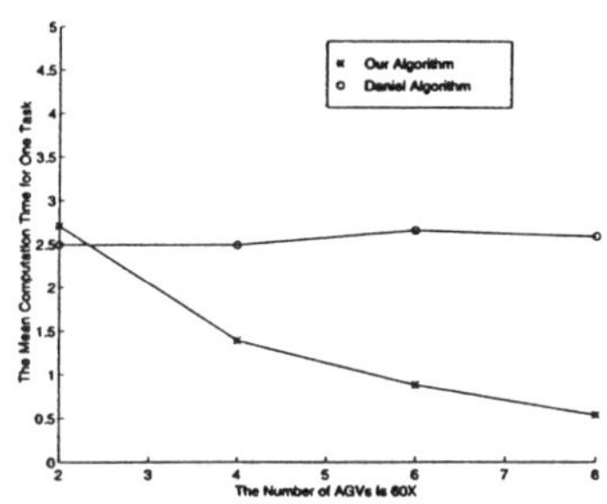

(c) Configuration 3.

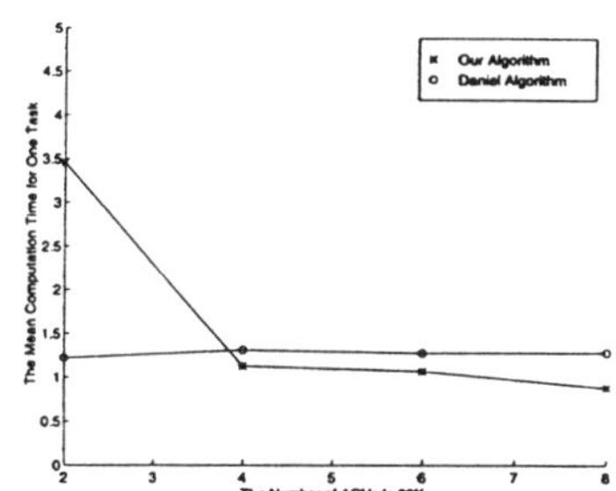

(d) Configuration 4.

Fig 11. Mean computation time to find the route for a delivery.

The following conclusions can be drawn from Figure 11:

(1) The computation time for Daniels' method does not change significantly with the number of AGVs in port.

(2) The computation time of the method proposed here decreases with the number of AGVs in port. This is because when fewer AGVs are in port, an AGV may not be available after a route is found so the computation for a route may be done several times for a delivery job before an AGV is assigned the job. When the number of AGVs in port increases, the chances of recomputation of a route for a delivery is lower and the computation time therefore decreases.

(3) Generally speaking, when the number of AGVs in use is small, the computation time for our method is longer than Daniels' method but when the number of AGVs is high, our method takes shorter time than Daniels'.

The overall conclusion is that our method using a network with time-dependent interedge transit times and the routing algorithm proposed make visible improvement over Daniels's method on the efficiency of moving containers in the port.

7. Conclusion

The work reported here shows that it is beneficial to represent the track configuration and the traffic conditions in container ports by using a network of time-dependent interedge transit times. The routing algorithm modified from Dijkstra's shortest path algorithm works well using this model of the container port. One factor which is not considered in these simulations is the communication time between the central control computer and the individual AGVs. However this factor should have the same effect on our method as on Daniels'. So the comparison between our method and Daniels' done here is fair. Future work may be carried out in the area of evaluation and comparison between a centralised control system and a decentralised control system.

References

[1] Bartholdi, J. J. III and Platzman, L. K. Decentralized Control of Automated Guided Vehicles on a Simple Loop. American Institute of Industrial Engineers Transactions 21(1):76-81, 1989.

[2] Bozer, Y. A. and Srinivasan, M. M. Tandem Configurations for Automated Guided Vehicle Systems and the Analysis of Single Vehicle Loop. American Institute of Industrial Engineers Transactions 23(1):72-82, 1991.

[3] Egbelu, P. J. and Tanchoco, J. M. A. Potentials for Bi-directional Guided Vehicle based Systems. International Journal of Production Research. 24:1075-1079. 1986.

[4] Young, E. L. Private communications, November 1986. Litton Industrial Automation Systems Inc., Automated Vehicles Systems, Zeeland, Michigan.

[5] Daniels, S. C. Real Time Conflict Resolution in Automated Guided Vehicle Scheduling. PhD Thesis. The Pennsylvania State University, Pennsylvania. 1988.

[6] Cooke, K. L. and Halsey, E. The Shortest Route through a Network with Internodal Transit Times. Journal of Mathematical Analysis and Applications. 14:493-498, 1966.

[7] The Minimum Route Problem for Networks with Turn Penalties and Prohibitions. Transportation Research. 3:397-408. 1969.

[8] Jeffery T.L. Soh, Wen Jing Hsu, Shell Ying Huang and Adrian C.Y. Ong. Decentralised Routing Algorithms for Automated Guided Vehicles. Proceedings of 1996 ACM Symposium of Applied Computing. Pp473-479.

Analysis of the Production and Transportation Models for Three Production Bases in ICGCPS

Yuanming Su and Shusaku Hiraki

Faculty of Economics, Hiroshima University,
1-2-1, Kagamiyama, Higashi-hiroshima, 739 Japan

Abstract. In this paper, we analyze the production and transportation models of three production bases in ICGCPS. The types of the transportation are classified and the characteristics of them are clarified. A basic production and transportation ordering model is presented.

1. Introduction

In order to acquire new markets and reduce production costs, a production and transportation system called 'International Co-operative Global Complementary Production System (ICGCPS)' is attracting attention in designing of the production and transportation system.

The ICGCPS is a global production system with several production bases located in several countries. A machining process of parts or components and a final assembly process exist in each production base. In each production base, only specialized kinds of parts or components are produced. Besides been used for producing final products in the production base, the produced parts or components are supplied to other production bases for producing final products. This system is constructed among several countries, especially in developing countries.

In the ICGCPS, because parts or components are supplied among production bases to produce final products, transportation plays a significant role. Consequently, designing a more efficient and rational production and transportation system becomes an very important subject.

For the ICGCPS, some researches have been reported ([1]~[5]). However, the transportations analyzed in them are limited in analyzing the transportations between two production bases. For the transportation of three or more production bases in the ICGCPS under considering transshipment and mixed-loading research is rarely seen until now. From this view point, we examine the transshipment and mixed-loading of an ICGCPS which has three production bases. In this case, the transportations include the transshipment being carried out or not and having mixed-loading or not. In this paper, we investigate the following subjects: (1) Classify the transportations of three production bases in the ICGCPS and clarify the characteristics of them. (2) Propose a basic production and transportation model based on the concept of pull-type ordering system under considering transshipment and mixed-loading. (3) Analyze the variations of ordering, withdrawal and inventory quantities of the model in order to clarify the properties of the model proposed. (4) Compare the model with a case in which components produced by a production base are transported respectively between every two production bases in order to make the characteristics of the model clear.

2. Production and transportation models of the ICGCPS

2.1 Classification of transportations

In this paper, we consider an ICGCPS which has three production bases, production bases 1, 2 and 3. Each production base has a machining process of parts or components (in this paper, we consider as components) and a final assembly process of final products. Components 'a', 'b' and 'c' are produced in the machining process of the production bases 1, 2 and 3, respectively. To produce final products, in each of the final assembly processes, all the components 'a', 'b' and 'c' are necessary. A part of components produced by a machining process is sent to the final assembly process in the same production base to produce final products while the others are transported to the final assembly processes of other production bases to produce final products. Therefore, the components 'a', 'b' and 'c' are supplied mutually among the three production bases and the material flows among the production bases are bi-directions. In order to get higher loading rate, we consider that the transshipment is carried out at production bases. The transshipment means that a part of components needed by stopover production base is unloaded, instead of this, a part of components produced by the stopover production base is loaded into the transportation vessel for the demand of next production base. Based on the transshipment being carried out or not and having mixed-loading or not, we classify the transportations of the ICGCPS in **Table 1**, and describe their main characteristics in **Table 2**.

To describe the transportations, we assume the following conditions:
(1) Make the numbers of transportation vessels as small as possible. (2) No empty transportation vessel occurs in transportation processes. (3) Transportation vessel must go back to the same production base where it starts from before.

Let us define notations as follows:

k: production base (k=1, 2, 3).

t: time period (t=1, 2,⋯).

L_{hk}: the transportation lead time from production base h to k (h=1, 2, 3; k=1, 2, 3; h $\neq$k).

W^k: the capacity of the transportation vessel which starts from the production base k.

Examples of transportation routes, lead times and material flows of three production bases without transhipment in the ICGCPS are shown in **Table 3-1** and **Table 3-2** and with transhipment are shown in **Table 4**.

2.2 Basic production and transportation system

Based on the classification of transportations, let us define a basic production and transportation system with the transshipment mentioned above. **Fig.1.** shows the schematic diagram of material flows and information flows of the system. In this system, components 'a' are produced by production base 1. When the component 'a' is transported to production bases 2 and 3 by a transportation vessel, the vessel starts from the production base 1, makes a brief stop at the production base 2 before arriving at the production base 3. At the stopover production base 2, a part of the component 'a' is unloaded from the vessel, and component 'b' is loaded into the vessel and transported to the production base 3 together with the remainder of the component 'a'. In other word, the transhipment is carried out at the production base 2. Let us define additional notations as follows:

D^k_t : the demand quantity for the final products at production base k in period t.

$F^k_{t:t+j}$: the forecasted demand quantity for the final products in the period (t+j) at production base k, forecasted at the end of period t.

L_k: the production lead time of the components at the machining process at production base k.

Table 1. Classification of the transportations of the ICGCPS

Transportations	Without Transshipment	Respective Piston Transportation		Without Mixed-Loading
		Respective Circuit Transportation		
		Round-Tour's Piston Transportation		
		Round-Tour's Circuit Transportation	Take and Bring Method	
			Take and Collect Method	With Mixed-Loading
	With Transshipment	Round-Tour's Piston Transportation		
		Round-Tour's Circuit and Piston Transportation		

Table 2. Characteristics of the transportations of the ICGCPS

Transportation		Characteristic
Without Transshipment	Respective Piston Transportation	(1) The components produced by a certain production base are transported respectively to other production bases. (2) Transportation vessel brings the components produced by other production bases back to the production base where the vessel started from by a route which is the same as outward route.
	Respective Circuit Transportation	(1) The same as the immediately above (1). (2) But transportation vessel brings the components produced by other production bases back to the production base where the vessel started from by a route which is different from outward route.
	Round-Tour's Piston Transportation	(1) There are Round-Tour in transportation processes when the components produced by a certain production base are transported to other production bases. (2) Transportation vessel brings the components produced by other production bases back to the production base where the vessel started from by a route which is the same as outward route.
	Round-Tour's Circuit Transportation	(1) The same as the immediately above (1). (2) But transportation vessel brings the components produced by other production bases back to the production base where the vessel started from by a route which is different from outward route.
With Transshipment	Round-Tour's Piston Transportation	(1) When the components produced by a certain production base are transported to other production bases, at a stopover production base, a part of the components needed by the stopover production base is unloaded, instead of this, components produced by the stopover production base are loaded into the transportation vessel for demand of the next production base. (2) Transportation vessel brings the components produced by other production bases back to the production base where the vessel started from by a route which is the same as outward route. At the same stopover production base, the laded components are transshipped as well with the components needed by the next production base.
	Round-Tour's Circuit and Piston Transportation	(1) The same as the immediately above (1). (2) But transportation vessel brings the components produced by other production bases back to the production base where the vessel started from by a route which is different from outward route and from the production base, the vessel must go one piston transportation as well to transport the components to three production bases. The transshipment will be carried out at one stopover production base.

Table 3-1. Examples of transportation routes, lead times and material flows of three production bases without transshipment in the ICGCPS

Transportation	Example of transportation routes, lead times and material flows
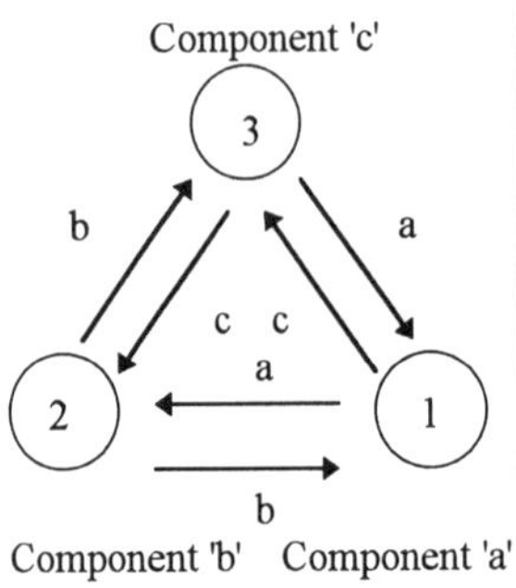 Respective Transportation (→: Material flows)	(1) Respective Piston Transportation between two production bases An example of routes is: a b b c c a W^1: ①⇒②⇒① , W^2:②⇒③⇒②, W^3:③⇒①⇒③ L_{12} L_{21} L_{23} L_{32} L_{31} L_{13} In this case, the characteristics are as follows: a) The components transported in all transportation processes are from one of production bases. b) Without transshipment. c) Transportation lead times are short. d) Three transportation vessels are necessary.
	(2) Respective Piston Transportation among three production bases (I) An example of routes is: a b c b c a W^1:①⇒②⇒③⇒②⇒①, W^3:③⇒①⇒③ L_{12} L_{23} L_{32} L_{21} L_{31} L_{13} In this case, the characteristics are as follows: a) The components transported in all transportation processes are from one of production bases. b) Without transshipment. c) Transportation lead times are longer than case (1). d) Two transportation vessels are necessary.
	(3) Respective Piston Transportation among three production bases (II) An example of routes is: a b c a c b W^1: ①⇒②⇒③⇒①⇒③⇒②⇒① L_{12} L_{23} L_{31} L_{13} L_{32} L_{21} In this case, the characteristics are as follows: a) The components transported in all transportation processes are from one of production bases. b) Without transshipment. c) Transportation lead times are longer than case (1) and (2). d) One transportation vessel is necessary.
	(4) Respective Circuit Transportation among three production bases An example of routes is: a b c a c b W^1:①⇒②⇒③⇒①, W^1:①⇒③⇒②⇒① L_{12} L_{23} L_{31} L_{13} L_{32} L_{21} In this case, the characteristics are as follows: a) The components transported in all transportation processes are from one of production bases. b) Without transshipment. c) Transportation lead times are longer than case (1) and (2), but shorter than case (3). d) Two transportation vessels are necessary.

Table 3-2. Examples of transportation routes, lead times and material flows of three production bases without transshipment in the ICGCPS

Transportation	Example of transportation routes, lead times and material flows
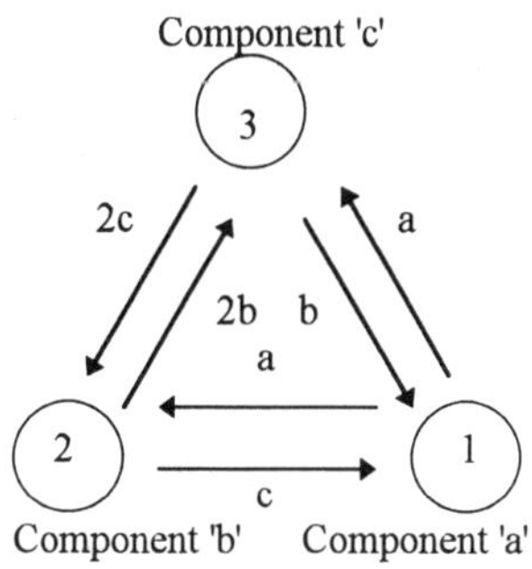 Example of Round-Tour's Transportation	(5) Round-Tour's Piston Transportation a b c a c b An example of routes is: W^1: ①⇒②⇒③⇒①⇒③⇒②⇒① L_{12} L_{23} L_{31} L_{13} L_{32} L_{21} In this case, the characteristics are as follows: a) The components transported in all transportation processes are from one production bases, but there are Round-Tour in transportation processes. b) Without transshipment. c) transportation lead times are longer. d) One transportation vessel is necessary.
	(6) Round-Tour's Circuit transportation (Take and Bring method) An example of routes is: a 2b b b 2c c W^1:①⇒②⇒③⇒①, W^2:②⇒③⇒①⇒② L_{12} L_{23} L_{31} L_{23} L_{31} L_{12} In this case, the characteristics are as follows: a) The components transported in all transportation processes are from one of production bases, but there are Round-Tour in transportation processes. b) Without transshipment. c) Transportation lead times are shorter than case(5). d) Two transportation vessels are necessary.
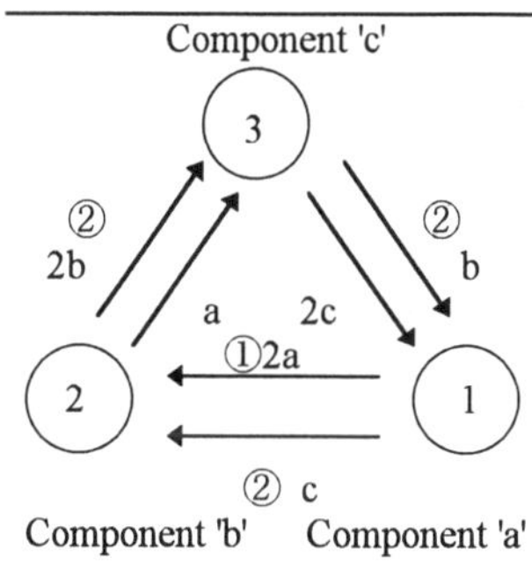 Example of Round-Tour's Circuit Transportation	(7) Round-Tour's Circuit transportation (I) (Take and Bring method) 2a a 2c c 2b b An example of routes is: W^1:①⇒②⇒③⇒①⇒②⇒③⇒① L_{12} L_{23} L_{31} L_{12} L_{23} L_{31} In this case, the characteristics are as follows: a) The components transported in all transportation processes are from one of production bases, but there are Round-Tour in transportation processes. b) Without transshipment. c) Transportation vessel must go two rounds to transport all the components to three production bases, so transportation lead times are long. d) One transportation vessel is necessary.
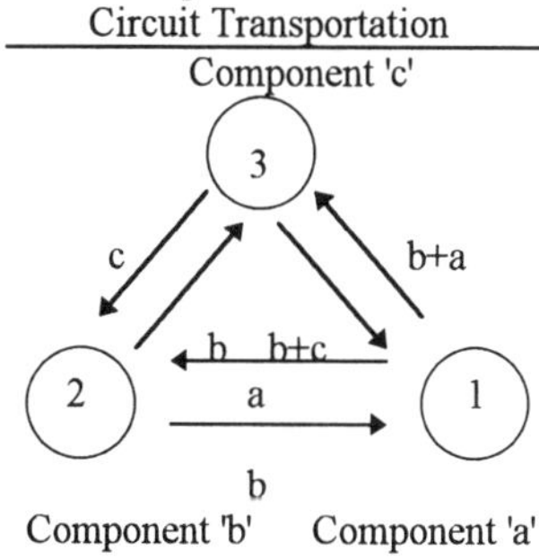 Example of Round-Tour's Circuit Transportation (II)	(8) Round-Tour's Circuit Transportation (II) (Take and Collect method) An example of routes is: a b b+c c b b+a W^1:①⇒②⇒③⇒①, W^3:③⇒②⇒①⇒③ L_{12} L_{23} L_{31} L_{32} L_{21} L_{13} In this case, the characteristics are as follows: a) The components transported in transportation processes are from one of production bases, or from two different production bases. b) With mixed-loading in transportation processes. c) Without transshipment. d) Transportation lead times are shorter than case(7). e) Two transportation vessels are necessary.

Table 4. Examples of transportation routes, lead times and material flows of three production bases with transshipment in the ICGCPS

Transportation	Examples of transportation routes, lead times and material flows
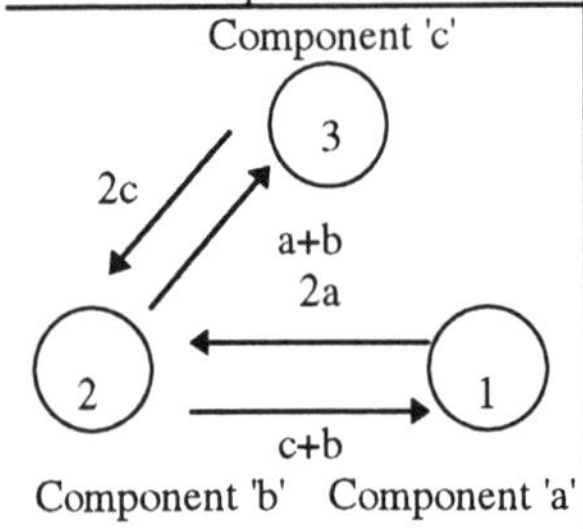 Example of Round-Tour' Piston Transportation	(1) Round-Tour's Piston Transportation 2a a+b 2c c+b An example of routes is: W^1:①⇒②⇒③⇒②⇒① L_{12} L_{23} L_{32} L_{21} In this case, the characteristics are as follows: a) The components transported in transportation processes are from one of production bases, or from two different production bases. b) With mixed-loading in transportation processes. c) Transshipment is carried out at a stopover production base. d) One transportation vessel is necessary.
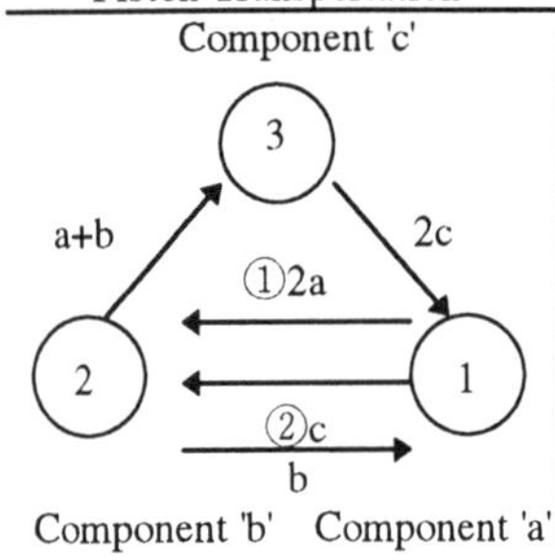 Example of Round-Tour' Circuit and Piston Transportation	(2) Round-Tour's Circuit Transportation 2a a+b 2c c b An example of routes is: W^1:①⇒②⇒③⇒①⇒②⇒① L_{12} L_{23} L_{31} L_{12} L_{21} In this case, the characteristics are as follows: a) The components transported in transportation processes are from one of production bases, or from two different production bases. b) With mixed-loading in transportation processes. c) Transshipment is carried out at one production base. d) Transportation vessel must go one round and one piston transportation to transport all the components to three production bases, so the transportation time are longer than case(1). e) One transportation vessel is necessary

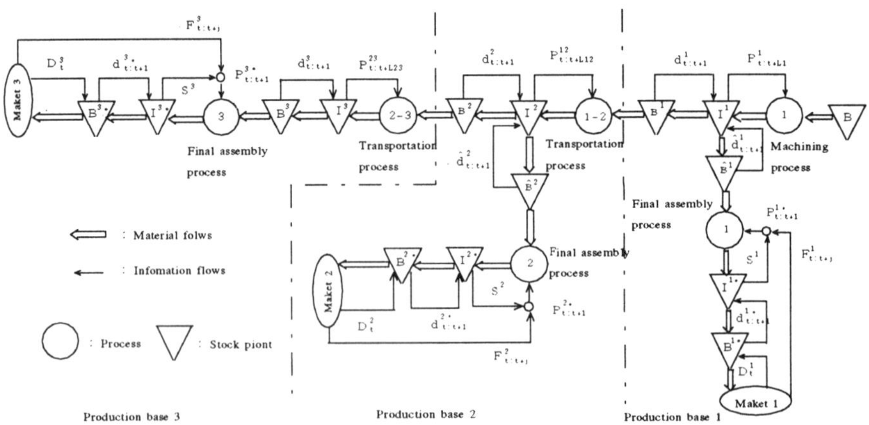

Fig. 1. Schematic diagram of the basic production and transportation system

B^k_t : the on-hand inventory quantity of the components for the transportation process (or final assembly process) at the production base k at the end of period t

B^{k*}_t : the on-hand inventory quantity of the final products for the market at production base k at the end of period t.

$\hat{B}^k_t$: the on-hand inventory quantity of the components for the final assembly process at production base k at the end of period t.

I^k_t: the inventory quantity of the components fabricated by machining process or transported by transportation process at production base k at the end of period t.

I^{k*}_t : the inventory quantity of the products fabricated by final assembly process at production base k at the end of period t.

S^k: the safety stock of the final products at the stock point I^{k*}.

$d^k_{t:t+1}$: the withdrawal ordering quantity of the components from the stock point B^k to I^k, which is required at the end of period t and delivered into the stock point B^k at the beginning of period (t+1).

$\hat{d}^k_{t:t+1}$: the withdrawal ordering quantity of the components from the stock point $\hat{B}^k$ to I^k. which is required at the end of period t and delivered into the stock point $\hat{B}^k$ at the beginning of period (t+1).

$d^{k*}_{t:t+1}$: the withdrawal ordering quantity of the products from the stock point B^{k*} to I^{k*}, which is required at the end of period t and delivered into the stock point B^{k*} at the beginning of period (t+1).

$P^{hk}_{t:t+L_{hk}}$: the transportation ordering quantity of the components from production base h to k, which is calculated at the end of period t, placed into transportation during the period (t+1) and completed at the end of
period ($t+L_{hk}$).

$P^k_{t:t+L_k}$: the production ordering quantity of the components for the machining process at production base k, which is calculated at the end of period t, placed into production during the period (t+1) and completed at the end of period ($t+L_k$).

$P^{k*}_{t:t+1}$: the production ordering quantity of the products for the final assembly process at production base k, which is calculated at the end of period t, placed into production during period (t+1) and completed at the end of period (t+1).

2.3 Formulation

Using the notations defined above, we formulate the basic production and transportation ordering model based on the concept of pull-type ordering system for each production base as follows.

2.3.1 Production base 1

$$I^1_t=I^1_{t-1}-(d^1_{t-1:t}+\hat{d}^1_{t-1:t})+P^1_{t-L_1:t} \quad (1)$$

$$B^1_t=B^1_{t-1}+d^1_{t-1:t}-P^{12}_{t-1:t+L_{12}-1} \quad (2)$$

$$\hat{B}^1_t=\hat{B}^1_{t-1}+\hat{d}^1_{t-1:t}-P^{1*}_{t-1:t} \quad (3)$$

$$P^1_{t:t+L_1}=d^1_{t:t+1}+\hat{d}^1_{t:t+1} \quad (4)$$

$$d^1_{t:t+1}=P^{12}_{t-1:t+L_{12}-1} \quad (5)$$

$$\hat{d}^1_{t:t+1}=P^{1*}_{t-1:t} \quad (6)$$

$$I^{1*}_t=I^{1*}_{t-1}-d^{1*}_{t-1:t}+P^{1*}_{t-2:t-1} \quad (7)$$

$$B^{1*}_t=B^{1*}_{t-1}+d^{1*}_{t-1:t}-D^1_t \quad (8)$$

$$P^{1*}_{t:t+1}=F^1_{t:t+2}+(F^1_{t:t+1}-P^{1*}_{t-1:t})-I^{1*}_t+S^1 \quad (9)$$

$$d^{1*}_{t:t+1}=D^1_t \quad (10)$$

2.3.2 Production base 2

$$I^2_t=I^2_{t-1}-(d^2_{t-1:t}+\hat{d}^2_{t-1:t})+P^{12}_{t-L12:t} \tag{11}$$

$$B^2_t=B^2_{t-1}+d^2_{t-1:t}-P^{23}_{t-1:t+L23-1} \tag{12}$$

$$\hat{B}^2_t=\hat{B}^2_{t-1}+\hat{d}^2_{t-1:t}-P^{2*}_{t-1:t} \tag{13}$$

$$P^{12}_{t:t+L12}=d^2_{t:t+1}+\hat{d}^2_{t:t+1} \tag{14}$$

$$d^2_{t:t+1}=P^{23}_{t-1:t+L23-1} \tag{15}$$

$$\hat{d}^2_{t:t+1}=P^{2*}_{t-1:t} \tag{16}$$

$$I^{2*}_t=I^{2*}_{t-1}-d^{2*}_{t-1:t}+P^{2*}_{t-2:t-1} \tag{17}$$

$$B^{2*}_t=B^{2*}_{t-1}+d^{2*}_{t-1:t}-D^2_t \tag{18}$$

$$P^{2*}_{t:t+1}=F^2_{t:t+2}+(F^2_{t:t+1}-P^{2*}_{t-1:t})-I^{2*}_t+S^2 \tag{19}$$

$$d^{2*}_{t:t+1}=D^2_t \tag{20}$$

2.3.3 Production base 3

$$I^3_t=I^3_{t-1}-d^3_{t-1:t}+P^{23}_{t-L23:t} \tag{21}$$

$$B^3_t=B^3_{t-1}+d^3_{t-1:t}-P^{3*}_{t-1:t} \tag{22}$$

$$P^{23}_{t:t+L23}=d^3_{t:t+1} \tag{23}$$

$$d^3_{t:t+1}=P^{3*}_{t-1:t} \tag{24}$$

$$I^{3*}_t=I^{3*}_{t-1}-d^{3*}_{t-1:t}+P^{3*}_{t-2:t-1} \tag{25}$$

$$B^{3*}_t=B^{3*}_{t-1}+d^{3*}_{t-1:t}-D^3_t \tag{26}$$

$$P^{3*}_{t:t+1}=F^{3*}_{t:t+2}+(F^{3*}_{t:t+1}-P^{3*}_{t-1:t})-I^{3*}_t+S^3 \tag{27}$$

$$d^{3*}_{t:t+1}=D^3_t \tag{28}$$

3. Analysis of the results

In order to clarify the variations of ordering, withdrawal and inventory quantities, we assume the following conditions ([6]), analyze the model which has been formulated in section 2.

(1) The demand quantity for the final products at each production base is a stationary time series with exponential autocorrelation, the time series' mean is μ_{Dk}, variance of demand is σ_{Dk}^2, demands D^h and D^k $(h \neq k)$ are mutually independent.

(2) The forecasted demand for the final products at production base k is calculated by the autocorrelation method:

$$F^k_{t:t+j}=\lambda_k \cdot D^k_t+(1-\lambda_k^{\,j})\mu_{Dk}$$

where, $F^k_{t:t+j}$ is the forecasted demand required in the period (t+j), calculated at the end of period t, λ_k is the autocorrelation coefficient.

The results of the variations of ordering, withdrawal and inventory quantities are summarized in **Table 5**. From this Table, we can see that:

(1) **Fig.2.** shows the relationships between the variance of the stock point I^2 and the transportation lead times L_{12}. The variance of the stock point I^2 is affected by not only the time series of demands of the production bases 2, 3, but also the transportation lead time L_{12}. Therefore, the variance will increase when the transportation lead time L_{12} becomes longer.

(2) **Fig.3.** shows the relationships between the variance of the stock point I^3 and the transportation lead time L_{23}. The variance of the stock point I^3 is affected by the time series of the demand of production base 3 and the transportation lead time L_{23}, but not affected by the transportation lead time L_{12}.

(3) In order to clarify the characteristics, the proposed model (it is called model 2) is compared with a case in which components 'a' are transported from the production base 1 to the production bases 2 and 3 (notations of inventory quantity are $I^{2'}$, $I^{3'}$)

Table 5. Variations of ordering, withdrawal and inventory quantities

Production base (k)		1	2	3
B^k_t	Mean	B^1_0	B^2_0	B^3_0
	Variance	$b_2\sigma_{D2}^2+b_3\sigma_{D3}^2$	$b_3\sigma_{D3}^2$	$b_3\sigma_{D3}^2$
$\hat{B}^k_t$	Mean	$\hat{B}^1_0$	$\hat{B}^2_0$	
B^{k*}_t	Mean	B^{1*}_0	B^{2*}_0	B^{3*}_0
	Variance	σ_{D1}^2	σ_{D2}^2	σ_{D3}^2
I^k_t	Mean	I^1_0	I^2_0	I^3_0
	Variance	0	$c_2\sigma_{D2}^2+c_1\sigma_{D3}^2$	$c_3\sigma_{D3}^2$
I^{k*}_t	Mean	I^{1*}_0	I^{2*}_0	I^{3*}_0
	Variance	$a_1\sigma_{D1}^2$	$a_2\sigma_{D2}^2$	$a_3\sigma_{D3}^2$
$P^{hk}_{t:t+L_{hk}}$	Mean		$\mu_{D2}+\mu_{D3}$	μ_{D3}
	Variance		$b_2\sigma_{D2}^2+b_3\sigma_{D3}^2$	$b_3\sigma_{D3}^2$
$P^k_{t:t+L_k}$	Mean	$\mu_{D1}+\mu_{D2}+\mu_{D3}$		
	Variance	$b_1\sigma_{D1}^2+b_2\sigma_{D2}^2+b_3\sigma_{D3}^2$		
$P^{k*}_{t:t+1}$	Mean	μ_{D1}	μ_{D2}	μ_{D3}
	Variance	$b_1\sigma_{D1}^2$	$b_2\sigma_{D2}^2$	$b_3\sigma_{D3}^2$
$d^k_{t:t+1}$	Mean	$\mu_{D2}+\mu_{D3}$	μ_{D3}	μ_{D3}
	Variance	$b_2\sigma_{D2}^2+b_3\sigma_{D3}^2$	$b_3\sigma_{D3}^2$	$b_3\sigma_{D3}^2$
$\hat{d}^k_{t:t+1}$	Mean	μ_{D1}	μ_{D1}	
	Variance	$b_1\sigma_{D1}^2$	$b_2\sigma_{D2}^2$	
$d^{k*}_{t:t+1}$	Mean	μ_{D1}	μ_{D2}	μ_{D3}
	Variance	σ_{D1}^2	σ_{D2}^2	σ_{D3}^2

$$a_k=(2-3\lambda_k^2+\lambda_k^4),\quad b_k=(1-2\lambda_k+2\lambda_k^2+4\lambda_k^3-2\lambda_k^4-2\lambda_k^5)\qquad (k=1,2,3)$$

$$c_1=[(L_{12}-1)-2\lambda_2+4\lambda_2^3+2\lambda_2^4+2\lambda_2^{L_{12}}-4\lambda_2^{L_{12}+2}-2\lambda_2^{L_{12}+3}+2\sum_{h=2}^{L_{12}-1}(L_{12}-h)\lambda_2^{h-1}]$$

$$c_2=[(L_{12}-1)-2\lambda_1+4\lambda_1^3+2\lambda_1^4+2\lambda_1^{L_{12}}-4\lambda_1^{L_{12}+2}-2\lambda_1^{L_{12}+3}+2\sum_{h=2}^{L_{12}-1}(L_{12}-h)\lambda_1^{h-1}]$$

$$c_3=[(L_{23}-1)-2\lambda_2+4\lambda_2^3+2\lambda_2^4+2\lambda_2^{L_{23}}-4\lambda_2^{L_{23}+2}-2\lambda_2^{L_{23}+3}+2\sum_{h=2}^{L_{23}-1}(L_{23}-h)\lambda_2^{h-1}]$$

base 1 to the production bases 2 and 3 (notations of inventory quantity are I^2, $I^{3'}$) respectively (it is called model 1). **Fig.4.** shows the relationships between the model 1 and the model 2. From the figure, we know that the variance of stock point $I^{2'}$ (model 1) is smaller than the variance of the stock point I^2 (model 2).

The variance of stock point I^3 (model 2) is the same that of as $I^{3'}$ (model 1) if the transportation lead time L_{23} is equal to the transportation lead time L_{13}. This means that when the components 'a' are transported to the production bases 2 and 3 (model 2), if the stopover production base is the production base 2, the variance of the components 'a' at the stock point I^2 will increase. But the variance of the stock point I^3 will decreased because the transportation lead time is shortened. On the other hand, when transshipment is carried out at the stopover production base 2, because the variance of the components 'b' at the stock point I^3 is only affected by transportation lead time L_{23}, so it is the same in the two models mentioned above.

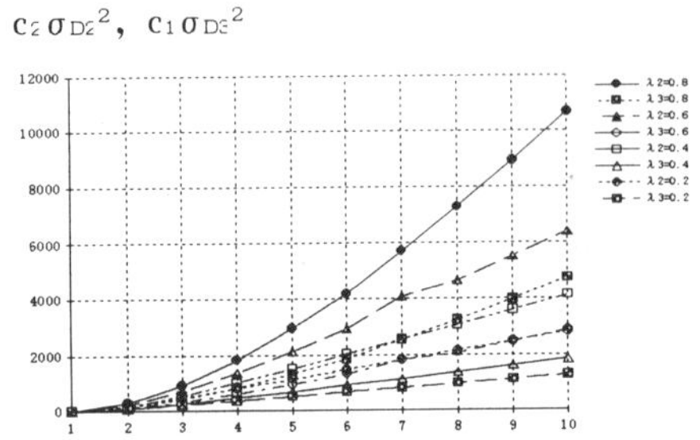

Transportation lead time L_{12}

Fig.2. Relationships between the lead time L_{12} and the variance of the stock point I^2 at the end end period t (σ_{D2}=15, σ_{D3}=10)

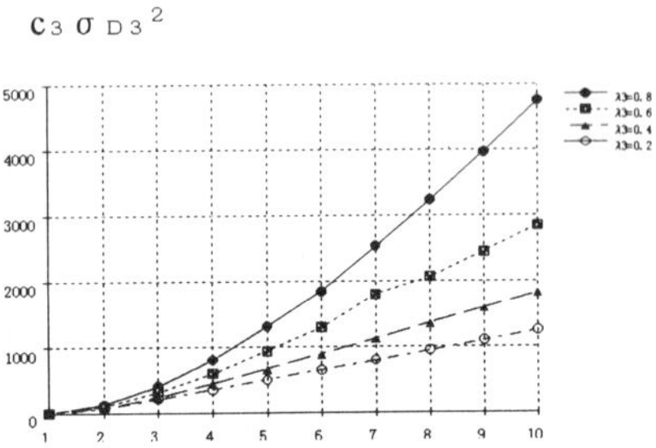

Transportation lead times L_{23}

Fig.3. Relationships of beween the lead time lead times L_{23} and the variance of the stock point I^3 at the end of period t (σ_{D3}=10)

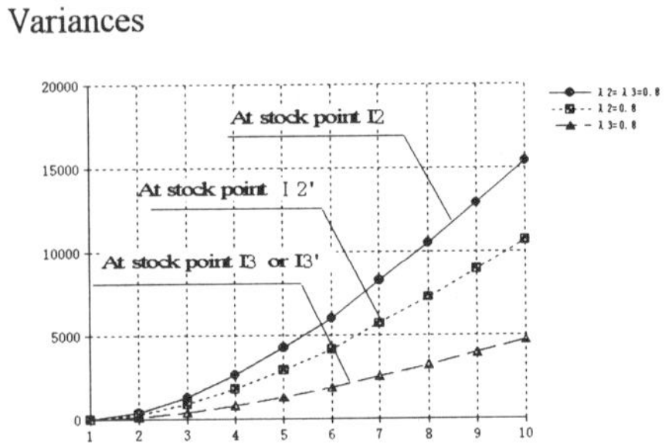

Transportation lead times L_{12}, L_{23} and L_{13}

Fig.4. Relationships of the transportation lead times and the inventory quantities at the stock points I^2, $I^{2'}$ and or I^3 or $I^{3'}$ (σ_{D2}=15, σ_{D3}=10)

4. Conclusions

At last, we can summarize the paper as follows:

(1) Based on the transshipment being carried out or not and having mixed-loading or not, the transportations of three production bases in ICGCPS are classified, and the characteristics of them are described.

(2) A new basic production and transportation model is proposed based on the concept of pull-type ordering system and the variations of ordering, withdrawal and inventory quantities are analyzed to illustrate the properties of the model proposed. (3) The characteristics of the model are clarify by compare it with a case in which components produced by a production base are transported respectively between two production bases.

References

1. H. Katayama, and S. Hiraki, On Design of International Parts Complementation System of Car Industries, *Proc. of International Symposium on Logistics,* (1995) 377-384.
2. S. Hiraki, and H. Katayama, Designing of an International Co-operative Global Complementary Production System, *Proc. of Simulation for Manufacturing and Communications,* (1994) 519-525.
3. S. Hiraki, K. Ishii, and H. Katayama, The Effects of Transportation Interval in Global Complementary Production Systems, *Computer Integrated Manufacturing,* edited by J. Winsor, AI. Sivakumar, and R. Gay, Vol.1, World Science (1995) pp.201-208.
4. S. Hiraki, Comparative Analysis of Ordering Model for an International Co-operative Global Complementary Producduction System, *International Journal of Production Economics,* Vol.44, (1996) 105-117.
5. S. Hiraki, K. Ishii, and H. Katayama, Mathematical Programming Approach to the International Co-operative Global Complementary Production Systems, *Proceeding of the Pacific Conference on Manufacturing,* Vol..II, (1996) 417-422.
6. A. C. Harvey, Time Series Models, *Philip Allan Publishers* (1981).

Supply Chain Management; how it works with JIT, MRPII and SBM

Assoc Prof J R Barker

Associate Dean, School of Information Technology, Bond University, Queensland, Australia.

Abstract. This paper investigates the use of supply chain management (SCM) with the three main manufacturing philosophies, viz JIT, MRPII and SBM.

Introduction.

One of the most crucial components in the manufacturing planning process is to plan for the correct amount of materials to arrive on the shop floor when required by the appropriate resources If this is not done properly, then either excess material will build up or resources will be idle due to non arrival of required materials. Quite often, it may be a mixture of both; too many materials that are not required and not enough materials that are required.

To help companies try to solve this problem, the main manufacturing philosophies that have been used to date are MRPII (Manufacturing Resource Planning), JIT (Just In Time) and OPT (Optimised Production Technology; a philosophy and an associated computer application package).

Companies have implemented these philosophies with varying degrees of success. Authors, such as Aggarwal [1985], Al-Hakim et al [1992], Berger [1987], Browne et al [1988], Fox [1983], Hastings et al [1982], [1985], [1990], Plenert et al [1986] and others, discuss which particular philosophy is relevant and most appropriate in various manufacturing environments.

Supply Chain Management (SCM) extrapolates the materials supply problem to the entire manufacturing (supply) chain. It is concerned with material supply between companies in the chain rather directly addressing the problems *within* an individual company.

Forecasting.

Let :

l be the manufacturing lead time of a product

d be the acceptable delivery time of the same product

The lead time l is the manufacturing lead time within the factory only. It does *not* include the suppliers' lead times (for any necessary raw materials). The following usually determine the necessity or otherwise of forecasting:

$l <= d$, forecasting is not required (1)

$l > d$, forecasting is required (2)

As an example; if it takes only 1 week to make the product (l=1) and the delivery time is 2 weeks (d=2), then no forecasting is required. If, however, it takes 6 weeks to make a product (ie l=6) but the market will only accept a delivery time of 3 weeks (d=3), then forecasting is required.

The lead time l can be made smaller by cutting down set-up times, move times, run times and/or clean-down times. This can be done by buying extra machines, improving manufacturing efficiency, buying newer machines, simplifying product design, using different materials, etc.

JIT and Supply Chain Management.

If $l <= d$, then JIT can be used instead of MRPII, if desired, in a single factory.

The basic JIT philosophy is that nothing is produced until it is needed. In the case of a finished item, the need arises through a customer order. This in turn, creates the need for the components required to make the finished item which creates the need for certain sub-components, etc, all the way down to purchased items (raw material). All of these material 'needs' actually constitute material planning ensuring that raw material items are only purchased when they are required and components (work in process) are only in the factory when they are being processed.

When the purchased items arrive, production commences on the sub-components, then the components etc all the way up to the finished item. The overall affect of this is that, theoretically, inventories (raw material, WIP - work in process, and finished goods) are minimised.

Now let us turn our attention to the complete manufacturing chain instead of just one factory (see figure 1). According to 'pure' JIT, suppliers of a factory should also use JIT. Consequently, if JIT is implemented as per the theory, all manufacturers in the manufacturing chain should be using JIT.

Note that the manufacturing chain is also the supply chain and hence JIT, in a sense, incorporates supply chain management.

For the purposes of the following analysis, it is assumed that the manufacturing (supply) chain is linear, ie each company in the chain has one supplier and one customer, except of course, for the company at the tail of the chain and the company at the head of the chain.

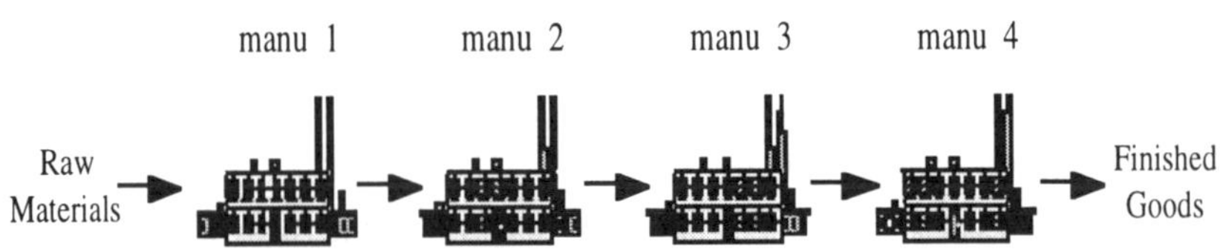

Figure 1: Manufacturing chain.

Assume:

n companies in the chain, where $n >= 2$ (n is 4 in figure 1),

the 'head' company is number 1,

the 'tail' company is number n,

the companies in between (if any) are numbered sequentially from 2 to n-1, for n>2.

Then the lead time and delivery time of company i is $l(i)$ and $d(i)$ respectively. Therefore, the *total* lead time for the whole manufacturing chain is:

$$L = l(1) + l(2) + \ldots + l(n) \quad (3)$$

and the *total* delivery time is:

$$D = d(1) + d(2) + \ldots + l(n) \quad (4)$$

Therefore, the lead time to delivery time comparison (l:d above) becomes, for the manufacturing chain:

$$L:D \quad (5)$$

However, the final customer is only prepared to wait $d(1)$ time units. Hence, the comparison L:D is replaced with:

$$L:d(1) \quad (6)$$

which is a somewhat different to equation (5). L is, of course, the total lead time of company 4 which is its own manufacturing lead time, $l(4)$, plus the suppliers lead times, $l(3) + l(2) + l(1)$.

Equation (6) shows that *unless* the *total* lead time of the whole manufacturing chain is less than or equal to the end customer's delivery time, then every company in the chain cannot use JIT.

For some products, it may be possible for the *total* lead time of the whole manufacturing chain to be less than or equal to the end customer's delivery time, but, for the majority, it is not possible. The lead time of a motor car, for example, should include the time taken to dig the required iron ore out of the ground through to the finished product with its glass, plastic, metal, rubber, electrics, electronics, etc.

Thus, the JIT goal that every JIT company has JIT suppliers, is rarely achievable in practice. However, JIT theory is compatible, by definition, with supply chain management. In practice, the non-JIT companies in the chain usually ensure just-in-time supplies to JIT companies which maintains adequate supply chain management.

As an aside, it is interesting to note that a company can move in & out of JIT. Over a period of time, demand for the company's products may grow to the extent that the lead time for orders (products) becomes greater than the delivery time, ie $l > d$. If this occurs, the company will need to forecast future material requirements and use SBM (or MRPII). At some time in the future, if demand decreases, resources are increased (resulting in shorter lead times) or a mixture of both, the company may again utilise JIT as $l < d$.

To date, I am not aware of this situation being discussed in the literature. I have experienced such a case with at least one company and I intend, in the not too distant future, to write a short paper on the matter.

MRPII and Supply Chain Management.

A simplified schematic of MRPII is given in figure 2. As has been stated previously in the literature, MRPII is, at best, an approximating manufacturing philosophy.

The references given above have documented many problems with MRPII and its implementation. The reasons for these problems are many and include:

- resources, other than machines and labour, were usually not considered,
- constant, rather than variable, lot sizes are used,
- lead times are static in MRPII but not in reality,
- forward scheduling vs backward scheduling,
- unrealistic MPS' do not consider actual job sequencing,
- lack of accurate data,
- sequence dependencies are not considered,
- shop floor loadings are not considered in MRP,
- difficulty with multiple machines per work centre:
  different capacities, speeds and/or capabilities,
- dynamic relationships between resources and materials are not considered (horizontal separation),
- interjob dependencies are not considered ,
- MPS and production schedule are considered independently (vertical separation).

It is beyond the scope of this paper to discuss all the above reasons for the problems with MRPII in detail; that has been done elsewhere in the literature.

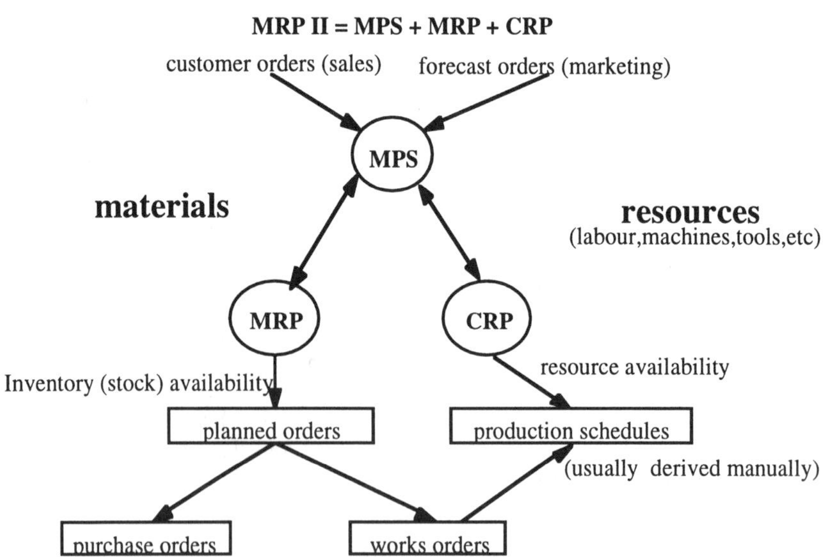

Figure 2: A simple overview of MRPII

Only some of the above reasons will be considered in this discussion. It should be noted, however, that other reasons given above are still relevant and give further insight into why it is difficult to use MRPII with supply chain management.

In MRPII, materials requirements planning (MRP) and capacity resource planning (CRP) are done after a master production schedule (MPS) has been created, including rough cut capacity planning (see figures 2 and 3).

MRP uses a bill of material (BOM) to determine what components and raw materials are required to make the finished goods in the MPS. MRP then uses *backward scheduling* from the due dates in the MPS, to determine what materials are required and when they are required. The static lead times used in the calculations are almost always wrong and hence, the output of the MRP is almost always incorrect.

CRP, on the other hand, typically uses *forward scheduling* which will give different answers to backward scheduling. (Note that the usually incorrect static lead times are again used). Therefore, resources will be scheduled at different times to materials.

As can be seen from figures 2, 3 and 4, the planning functions of materials (MRP) and resources (CRP) are done *separately*. They are treated as if they are *independent* of one another. This is what I have termed *horizontal separation* (see figure 4). They should be planned together, not separately.

From a material supply viewpoint, the resulting discrepancies appear as inaccuracies in quantities and/or due dates. These inaccuracies will be propagated through the supply chain and it is not unusual for them to be magnified as they pass down the supply chain. Constant intervention is therefore required to correct the discrepancies throughout the supply chain.

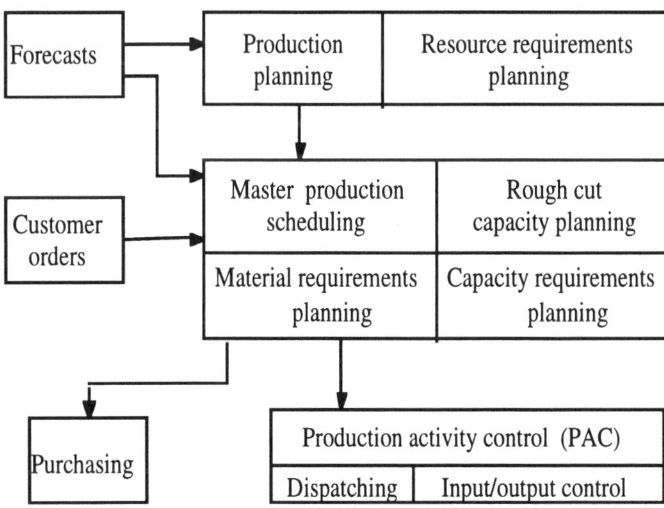

Figure 3: Typical steps involved in MRPII.

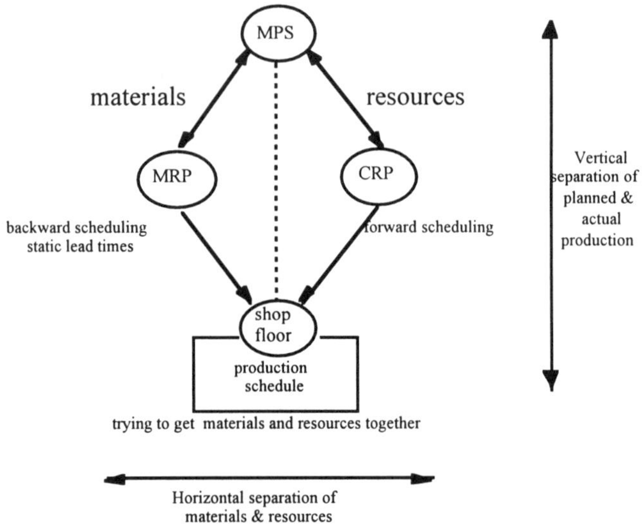

Figure 4: A simple overview of MRPII

Further, the response rate in MRPII is too slow for SCM which expects timely and accurate data. Consider the case where a machine or work centre fails. The processes required, with manual intervention, to adjust inventory requirements in MRPII are so time consuming that they are often not done. Similarly, it is time consuming to alter resource requirements due to unplanned changes in material supply.

Both of the situations in the previous paragraph occur because there is no direct relationship between materials and resources in MRPII. Such a dynamic relationship would allow automatic adjustment of inventory due to unplanned resource changes and/or automatic adjustment of resource allocation due to unplanned material changes.

SBM and Supply Chain Management.

SBM was designed to overcome most, if not all, of the inherent problems of MRPII.

There is no horizontal or vertical separation in SBM. MRP and CRP are joined and, consequently, are done together. Further, the MPS and the actual production schedule are also joined; they are not separate entities as they are in MRPII. Therefore, any changes in what is happening on the shop floor automatically affects the MPS. In MRPII, any such changes must be fed back into MPS; hence the term 'closed loop' MRPII. The MRPII 'diamond' as shown in figures 2 and 4, collapses into a single entity in SBM, enabling the processes required in figure 3 to be replaced by the simple central procedure shown in figure 5.

The material requirements and their associated due dates in SBM are more accurate than those in MRPII because they are derived from actual schedules which incorporate both material and resources. They are also calculated much faster and use more accurate data.

SBM incorporates a global view of manufacturing based on the assumption that a manufacturer is a part of a manufacturing chain and not a single stand-alone entity. SCM is based on the same assumption whereas MRPII is based on the single entity approach.

To clarify this, refer to the manufacturing chain as depicted in figure 1 and consider the following three situations:

- the raw materials to manufacturer 1 are late,
- manufacturer 3 wants material from manufacturer 2 to be delivered later (or earlier),
- manufacturer 4 cancels an order to manufacturer 3 (or changes it's due date).

1) The raw materials to manufacturer 1 are late. Assume that the raw materials to manufacturer 1 are 2 days late. Then it can be shown that, unless extra resources are added somewhere in the manufacturing chain, the delay to a customer at the end of the chain will be at least 2 days. In most cases, it will be greater than 2 days.

All the manufacturers in the chain would have to change their *expected receipt dates* of all affected material.

2) Manufacturer 3 wants material from manufacturer 2 to be delayed. In this situation, the *due dates* of all affected materials on the supply side of manufacturer 3 will need to be altered and the *expected receipt dates* of all affected materials on the customer side of manufacturer 3 will also need to be changed.

3) Manufacturer 4 cancels an order to manufacturer 3. In this situation, all the manufacturers would have to change the *due dates* of all affected materials.

Handling the above situations in MRPII is very difficult because, as has been shown in the previous section, the values used are usually incorrect, the time taken to make the changes is too long and a lot of manual intervention is required.

Recall that the manufacturing chain given in figure 1 is a simple linear chain. In practice, the chain is not nearly so simple. Each manufacturer in the chain usually has multiple suppliers and multiple customers. Consequently, the task is much more complicated than that presented in this simple example.

The assumption has also be made that all the manufacturers in the chain are make-to-order. If one or more manufacturers are make-to-stock, then the above example may also involve increased stock holdings in one or more make-to-stock manufacturers in the chain.

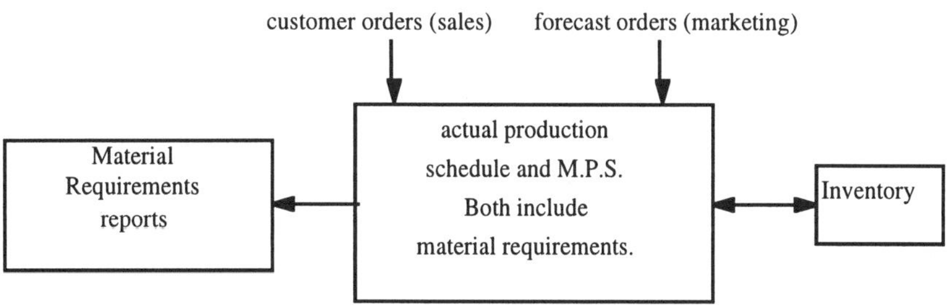

Figure 5: Production Planning in SBM and resulting Material Requirements.

One design criterion of SBM was that its schedules and resulting material requirements could be visible to both a company's suppliers and customers. The increased visibility throughout the supply chain allows all companies in the chain to *synchronise* their material requirements. Any changes in material requirements or supply, by choice or otherwise, of any company in the chain is transmitted quickly to other companies in the chain. Therefore, suppliers are kept up to date with any changes in material requirements from anywhere within the chain and customers are kept up to date with any changes in forthcoming deliveries. This is the very essence of SCM especially with the fast electronic communication that is now available.

Conclusion.

Supply chain management can be used with any of the manufacturing philosophies.

It is inherent in the use of JIT although forward visibility, ie looking at material requirements in the near to distant future, is rather blurred.

MRPII was not designed to be used with supply chain management. The material requirements, as generated by MRP, are usually not accurate enough nor responsive enough to be propagated through the supply chain. Inaccuracies in material requirements and/or due dates may be magnified if introduced directly into the supply chain. Constant corrections are usually required to make MRPII work with SCM.

SBM was designed to use the available power of modern computers to solve complex problems beyond the scope of MRPII. It is the only manufacturing philosophy designed after the introduction of SCM hence its ability to facilitate and encourage the use of SCM.

References.

Aggarwal, S. C., 1985. MRP, JIT, OPT, FMS? Making sense of production operations systems, *Harvard Business Review*, V63 (5), pp 8-16.

Al-Hakim, L.A., Okyar H. and Sohal, A.S., 1992. A Comparative Study of MRP and JIT Production Management Systems (An Australian Case Study), *Unpublished Report*, Monash University, Melbourne.

Barker, J.R., 1993. Schedule Based Manufacturing Philosophy, School of Information Technology, Bond University, *Working Paper 1993-3-088B*.

Barker, J.R., 1993. SBM - Schedule Based Manufacturing, *1993 Australian Conference on Manufacturing Engineering*, Adelaide, November.

Berger, G., 1987. Ten Ways MRP can Defeat You, *APICS Conference Proceedings* , American Production and Inventory Control Society.

Black, W. and Barker, J.R., 1993, SBM - a Worked Example; a Comparison of SBM and MRPII, School of Information Technology, Bond University, *Working Paper 1993-3-106B*.

Browne, J., Harhen, J. and Shivnan, J., 1988. *Production Management Systems: A CIM Perspective*, Addison-Wesley.

Donovan, R.M., 1992. Enhanced Production Scheduling and Capacity Management: Is It Worth It?, *APICS-The Performance Advantage*, pp 42-46, June.

Fox, R. E., 1983. MRP, Kanban, or OPT: What's Best?, *Production & Inventory Management, V24(3).*

Fox, R. E., 1983. OPT vs MRP: Thoughtware vs. Software, *Production & Inventory Management, V24(4).*

Hastings, N.A.J., Marshall, P.H. and Willis, R.J., 1982. Schedule Based MRP: An Integrated Approach to Production Scheduling and Material Requirements Planning, *Journal of the Operational Research Society*, 33, no 11, pp 1021-1029.

Hastings, N.A.J. and Packham, B.J., 1985. Developments in Schedule Based Manufacture Resource Planning, *7th National Conference of the Australian Society for Operations Research*, Adelaide, August.

Hastings, N.A.J. and Yeh, C.-H., 1990. Job oriented production scheduling, *European Journal of Operational Research*, 47, pp 35-48.

Hastings, N.A.J. and Yeh, C.-H., 1992. Bill of Manufacture, *Production and Inventory Management Journal - Fourth Quarter*, pp 27-31.

LaForge, R. and Sturr, V., 1986. MRP practices in a random sample of manufacturing firms, *Production and Inventory Management, V28* (3), pp 129-137.

Plenert, G. and Best, T. D., 1986. MRP, JIT and OPT: What's Best?, *Production and Inventory Management, V27 (2)*, pp 22-29.

Schroder, R., Anderson, J., Tupy, S. and White, E., 1981. A study of MRP benefits and costs, *Journal of Operations Management, V2 (1)*, pp 1-9.

Author's Contact information.

Assoc Prof J R Barker,
Associate Dean
School of Information Technology, Bond University, QLD, 4229, AUSTRALIA.
Tel: 61 7 5595-3344, fax: 61 7 5595-3320, e-mail: jbarker@bond.edu.au.

Quality Management

A FUZZY MULTI-CRITERIA DECISION FRAMEWORK FOR QFD PLANNING

Juite (Ray) Wang

Department of Industrial Engineering, Feng Chia University, Taiwan

Abstract. The objective of Quality Function Deployment is to better understanding of customer needs and appropriately deploy them to the design requirements to maximize the customer satisfaction. However, it is not possible to take all design features into consideration for product improvement, due to the limited product development time and cost. This paper considers the QFD planning as a multi-criteria decision problem and proposes a fuzzy outranking approach to prioritize design requirements recognized in QFD. The fuzzy set is used to represent the imprecise and uncertain product information collected from design teams. The fuzzy outranking relation is used to model the imprecise preference relation between design requirements. A design requirement outranks other requirements if and only if there is sufficient evidence to support that the requirement is superior or at least equal to the others. As a result, the dominant set of design requirements is identified for the focus of design in order to achieve not only the customer satisfaction but also the balanced design of a product. The proposed decision model is more suitable to be used in the imprecise and uncertain early design environment.

1 Introduction

In today's competitive marketplace, companies have to develop products that satisfy customer needs in order to become or remain successful. Poor product definition commonly leads to either failure of that product in the marketplace or extended product development time. Understand of customers' needs, done well and early, leads to successful products and shorter development time. The Quality Function Deployment (QFD) is a tool to help the product development team systematically relate the customer attributes that represent the overall customer concerns to design requirements that represent technical performance specifications of a developing product.[1-2] It has been used successfully by industries in both Japan and U.S.

The QFD planning process is summarized into the following seven steps:

1. Obtaining the customer attributes and their relative importance.
2. Developing design requirements responsive to the customer attributes.
3. Relating design requirements to the customer attributes.
4. Completing the customer competitive survey.
5. Performing the competitive technical benchmarking.
6. Determining the relationships among design requirements.

7. Calculating the design requirements' importance ratings and evaluating their technical difficulties and estimated cost.

Designers need to create or improve a product based on the design requirements recognized in QFD planning process. However, it is not possible to take all design requirements into consideration for product development, because a perfect product may require longer time and higher cost to develop. Product designers need to know how to make tradeoffs in the selection of design requirements that results in higher level of customer satisfaction.

Due to the imprecise and incomplete design information available at the early design stage, it is difficult to assess relative importance of customer attributes, relationships between customer attributes and design requirements, estimated cost, and so on. Most approaches to rank design requirements are to use the numerical value to represent the relative importance of customer attributes and relationship between customer attributes and design requirements.[2-3] For example, a 1-5-9 numerical scale is used to denote weak, medium, and strong relationships between customer attribute and design requirement pairs. Then the technical importance is computed for each design requirement as the weighted sum of the relationships with customer attributes (the relative importance for each customer attribute is used as the weighting factor). As a result, a complete order of all design requirements is built. The design requirements with higher value of technical importance are selected for further development, because they reflect higher degree of customer satisfaction.

The cardinality ranking that provides a complete order of all design requirements does not represent the actual preference structure among them. At the early design stage, design information collected may be too subjective or too incomplete to make a judgment of preferences between two design requirements. Or the designer may be not able to distinguish two requirements and he wishes to make a judgment later until the further information is collected. The incomparability, which exists in practice, between alternatives is completely ignored. Risk must be avoided.

This paper considers the QFD planning as a multi-criteria decision problem[4] and proposes a fuzzy outranking approach to prioritize design requirements recognized in QFD. The fuzzy set[5] is used to represent the imprecise and uncertain product information collected from design teams. The fuzzy outranking relation based on the possibility theory[6] is used to model the imprecise preference relation between design requirements. A design requirement outranks other requirements if and only if there is sufficient evidence to support that the requirement is superior or at least equal to the others. As a result, the dominant set of design requirements that is superior than others are identified for the focus of design in order to achieve not only the customer satisfaction but also the balanced design of a product.

This paper is organized as follows. Section 2 views the QFD planning problem as a multi-criteria decision making problems and develop a fuzzy outranking approach for ranking design requirements. An example to demonstrated the concept developed is presented in Section 3. Section 4 concludes the paper.

2 A Fuzzy Outranking Approach For Ranking Design Requirements

2.1 Preference Modeling For QFD Planning

This paper models the problem of prioritizing design requirements as a multi-criteria decision problem that considers not only their technical importance related to customer needs, but also their estimated cost and technical difficulty. It is expected that the design requirements selected for product development will create a balanced design.[7] Moreover, other life-cycle concerns, such as reliability, maintainability, manufacturability, and so on may be considered.

Due to lack of design information or the available information is too subjective, the outranking relation[8] is used to model the imprecise preference relations existed between design requirements. A requirement r_i outranks r_j (r_i S r_j) if and only if there is a sufficient evidence to believe that the requirement r_i is better than r_j or at least r_i is as good as r_j. The problem of prioritizing design requirements is defined as follows.

Definition

Let A be a finite set of design requirements evaluated in QFD according to a set of criteria C. The result of evaluation for a design requirements is a vector $g(r)=[g_1(r), g_2(r), ..., g_n(r)]$, where the function $g_k(r)$ represents the evaluation result of requirement $r \in A$ for criterion $c_k \in C$. The objective is to discriminate the set A into two subsets: 1) dominant set, S_D, and 2) dominated set, S_E, such that:

$$\forall\, r_j \in S_E, \exists\, r_i \in S_D \Rightarrow r_i \text{ S } r_j \tag{1}$$

The set of criteria includes technical importance, estimated cost, and technical difficulty, and so on. The evaluation result of requirement r with respect to criterion c_k, $g_k(r)$, may be a linguistic term or a fuzzy number.

A fuzzy outranking relation associated with requirements r_i and r_j, denoted by $S(r_i, r_j)$, can be characterized by a degree of outranking in the closed interval [0, 1]. $S(r_i, r_j)$ = 1 implies that r_i outranks r_j with certainty. On the contrary, $S(r_i, r_j)$ = 0 implies that there is no evidence that r_j is outranked by r_i. $S(r_i, r_j) \in (0, 1)$ indicates a credibility of an existing preference of r_i over r_j. Therefore, the imprecise preference structure of design requirements is represented with the fuzzy outranking relations among design requirements.

2.2 Overview of the Possibility Theory

This section provides an overview of the possibility theory[6] used for ranking design requirements. Given two fuzzy numbers M and N. The possibility measure that "M is greater than or equal to N" can be defined as:

$$POSS(M \geq N) = \sup_{u;u \geq v} \min(\mu_M(u), \mu_N(v)) \tag{2}$$

Equation (2) implies that the grade of possibility of the proposition "M is greater than or equal to N." It estimates the maximum chance that an event " $M \geq N$ " will occur.

Similarly, the necessity measure that "M is greater than or equal to N" is defined as:

$$NESS(M \geq N) = \inf_{u} \sup_{v;v \leq u} \max(1 - \mu_M(u), \mu_N(v)) \tag{3}$$

Equation (3) implies that the grade of necessity of the proposition "M is greater than or equal to N." Necessity measure provides an index that estimates the minimum chance that an event " $M \geq N$ " will occur.

2.3 Construction of Fuzzy Outranking Relations

In the QFD planning stage, design information collected may be vague and incomplete. The proposition "requirement r_i is better than r_j " is uncertain. There are some degrees of agreement and disagreement about this proposition. If the degree of agreement is much larger than the degree of disagreement, then r_i may outrank r_j. The idea to construct the fuzzy outranking relation is based on the above concept.

The degree of fuzzy outranking relation is obtained from two types of relation: concordance relation and discordance relation, namely degrees of agreement and disagreement. For two design requirements r_i and r_j, the concordance relation expresses the credibility of the hypothesis that r_i is at least as good as r_j with respect to a certain criterion. The discordance relation is used to express some doubt towards the hypothesis that "r_i is not at least as good as r_j" with respect to some criterion.

Definition: Concordance relation

Concordance relation that r_i is at least as good as r_j with respect to criterion c_k is defined as follows:

$$CI_k(r_i, r_j) = \theta \; POSS_k(r_i \geq r_j) + (1 - \theta) \; NESS_k(r_i \geq r_j) \tag{4}$$

, where θ is the preference ratio, $0 \leq \theta \leq 1$

Both possibility and necessity measures are used to define the concordance relation. The necessity measure $NESS_k(r_i \geq r_j)$ calculates the least chance that r_i is at least as good as r_j with respect to criterion c_k from the conservative viewpoint. On the contrary, the possibility measure of $POSS_k(r_i \geq r_j)$ computes the best opportunity from the aggressive viewpoint. Therefore the concordance index that r_i is at least as good as r_j with respect to criterion c_k should locate between the value of possibility measure $NESS_k(r_i \geq r_j)$ and necessity measure $POSS_k(r_i \geq r_j)$.

A preference ratio θ is defined to incorporate the attitude of a designer into this decision model. If the attitude of a designer is toward to the optimistic, then the value of θ should be greater than 0.5. On the other hand, the value of θ should be less than 0.5, if the attitude is toward to pessimistic.

The global concordance relation, $GCI(r_i, r_j)$, that r_i is greater than or equal to r_j is obtained by aggregating all the single-criterion concordance relation:

$$GCI(r_i, r_j) = \sum_k w_k \bullet CI_k(r_i, r_j), \tag{5}$$

where w_k is the weight used to express the relative importance of criterion c_k.

Definition: Discordance relation

Discordance relation that r_i is at least as good as r_j with respect to criterion c_k is defined as follows:

$$DI_k(r_i, r_j) = NESS_k(r_j \geq r_i) \tag{6}$$

The least chance that requirement r_j is better than or equal to r_i is used to define the degree of doubt that r_i is better than or equal to r_j with respect to criterion c_k.

Next, the concordance relations and discordance relations between r_i and r_j should be aggregated to establish the outranking relation. The aggregation function developed by Sisko et al.[9] is used to obtain degree of the fuzzy outranking relation:

$$S(r_i, r_j) = \begin{cases} GCI(r_i, r_j), & \text{if } GCI(r_i, r_j) \geq DI_k(r_i, r_j), \forall\, c_k \in C \\ \prod_{k^*} \left[1 - DI_{k^*}(r_i, r_j)\right] \dfrac{GCI(r_i, r_j)}{1 - GCI(r_i, r_j)}, & \text{for } \left\{ k \,\middle|\, GCI(r_i, r_j) < DI_{k^*}(r_i, r_j) \right\} \end{cases} \tag{7}$$

If the partial discordance relations for all criteria are not greater than the global concordance relation, then the degree of fuzzy outranking relation between requirements r_i and r_j is equal to the global concordance relation between them. Otherwise, the partial discordance relations that are greater than the global concordance relation for certain criteria will take effect to weaken the belief that r_i outranks r_j. If there is a partial discordance relation $D_{k^*}(r_i, r_j) = 1$ with respect to criterion c_{k^*}, then $S(r_i, r_j) = 0$, i.e. r_j is not outranked by r_i absolutely.

To prioritize a set of design requirements, the outranking relation used to represent the preference structure among design requirements is defined as follows:

(Outranking) $\quad \forall\, r_i, r_j \in A: \; r_i \,\mathrm{S}\, r_j$, if $S(r_i, r_j) \geq \delta \qquad (8)$

where δ is the outranking ratio, $0 \leq \delta < 1$

The outranking threshold δ is used to determine the existence of outranking relation between r_i and r_j. If the degree of outranking relation between r_i and r_j does not exceed outranking threshold, then it is not considered significant and r_j is not outranked by r_i. The value of outranking threshold may range from the smallest value that does not distinguish between two alternatives to the largest value that allows to distinguish between them.

2.4 Constructing the Membership Function of Fuzzy sets

In this paper, the membership function of a fuzzy input data is also evaluated and collected from the interfunctional team. Assume that the evaluation results are considered and represented with a linguistic scale Ω consisting of seven levels. For example, the strongest (weakest) strength of relationship between a customer attribute and design requirement is given the highest (lowest) value in the scale. Otherwise the intermediate value of the scale is given (see Table 1). Based on the linguistic scale, each team member evaluates every criterion and provides the results of his/her evaluation.

Table 1. Linguistic and quantitative scales for each criterion

Quantitative scale	Linguistic scale for relative importance of CAs	Linguistic scale for relationships between CAs and ERs
7	Very important	Strongest
6	important	strong
5	Fairly important	Fairly strong
4	Medium	Medium
3	Fairly unimportant	Fairly weak
2	Unimportant	Weak
1	Very unimportant	Weakest

The obtained linguistic value of an evaluation result is converted into the corresponding quantitative value according to Table 1. Suppose that the evaluation result obtained from the t-th team member by $h(t)$, where $h(t) \in \Omega$ and $t = 1, ..., n$. Then the triangular fuzzy number (l, m, r) is given as follows:

$$m = \frac{1}{n} \sum_{t=1}^{n} h(t) \tag{9}$$

$$l = m - \min_{t} h(t) \tag{10}$$

$$r = \max_{t} h(t) - m \tag{11}$$

3 Illustrated Example

An example of a car design modified from Hauser and Clausing[1] is used to illustrate the concept developed. The set of customer attributes includes "easy to close from outside", "stay open on a hill", "easy to open from the outside", "don't kick back", "doesn't leak in rain", and "no road noise". The corresponding design requirements consist of "energy to close door (r_1)", "check force on level ground (r_2)", "check force on 10° slope (r_3)", "energy to open door (r_4)", "peak closing force (r_5)", "door seal resistance (r_6)", "acoustic transmission window (r_7)", "road noise reduction (r_8)", and "water resistance (r_9)". The criteria involved in the prioritizing process include technical importance (c_1), estimated cost (c_2), and technical difficulty (c_3).

Assume that there are six engineers involved in the QFD planning process. The first step is to determine the relative importance of customer attributes. Based on the linguistic scale defined in Table 1, each engineer assigns the linguistic value to each customer attribute. For example, the linguistic scales of the customer attribute "easy to close from outside" assigned by engineers are {"fairly importance (5)", "fairly importance (5)", "important (6), "very important (7)", "very important (7)", "important (6)"}. The relative importance for the customer attribute "easy to close from outside" is (5, 6, 7) according to the equations (9) - (11). In the similar way, the other input data required for QFD is collected and listed in Appendix.

The following equation is used to calculate the technical importance of design requirement r_j :

$$\hat{t}_j = \left(\hat{y}_1 \otimes \hat{d}_{1j}\right) \oplus \left(\hat{y}_2 \otimes \hat{d}_{1j}\right) \oplus \cdots \oplus \left(\hat{y}_m \otimes \hat{d}_{mj}\right) \tag{12}$$

where:

$\hat{y}_i$: degree of relative importance of customer attribute i, $i = 1, ..., m$

$\hat{d}_{ij}$: degree of relationship between customer attribute i and design requirement r_j ; $i = 1, ..., m, j = 1, ..., n$

$\hat{t}_j$: degree of technical importance of design requirement r_j , $j = 1, ..., n$

Here variables $\hat{t}_j$, $\hat{y}_i$ and $\hat{d}_{ij}$, for all i, j, are fuzzy numbers.

After the required collected input data is converted into fuzzy sets, the possibility measures and necessity measures between design requirements for each criterion are computed based on (2) and (3). Assume that the preference ratio (θ) is equal to 0.5, the individual concordance matrices for criteria of technical importance of design requirement ([cr_{ij}^1]), cost ([cr_{ij}^2]), and technical difficulty ([cr_{ij}^3]) are computed (see Fig. 1(a)). The global concordance matrix ([gcr_{ij}]) that aggregates three individual concordance matrices are shown in Fig. 2(a). The individual discordance matrices for criteria of technical importance of design requirement ([dr_{ij}^1]), cost ([dr_{ij}^2]), and technical difficulty ([dr_{ij}^3]) are calculated (see Fig. 1(b)). Based on (7), the outranking relations among design requirements are constructed (see Fig. 2(b)). The dominant set and dominated set identified are:

$$S_D = \{ r_3, r_6, r_9 \} \text{ and } S_E = \{\{ r_1, r_2, r_4, r_5 \}, \{ r_7, r_8 \}\}$$

It suggests that the design requirements r_3, r_6, and r_9 in the dominant set can be chosen for product development.

It is observed that the value of preference ratio θ influences the degree of outranking among design requirements. The preference ratio reflects the designer's attitude. The larger the preference ratio, the more optimistic a designer is. From the sensitivity analysis shown in Fig. 3, the degree of outranking relation among design requirements increases in most cases, as the value of preference ratio increases. Similarly, the preference structure among design requirements is also influenced by the preference ratio θ and outranking threshold δ. The sensitivity analysis has been performed and it is observed that the larger the value of preference ratio θ, the more the discrimination power is.

(a) Technical importance ($[cr_{ij}^1]$)

(i, j)	1	2	3	4	5	6	7	8	9
1	-	0.57	0.57	0.91	0.87	0.17	1	1	1
2	0.83	-	0.75	0.96	0.93	0.37	1	1	1
3	0.84	0.74	-	0.98	0.95	0.33	1	1	1
4	0.48	0.36	0.34	-	0.67	0.13	1	1	1
5	0.54	0.41	0.39	0.81	-	0.14	1	1	1
6	0.99	0.93	0.94	1	1	-	1	1	1
7	0	0	0	0	0	0	-	0.45	0
8	0	0	0	0	0	0	0.9	-	0.11
9	0	0	0	0	0	0	1	1	-

(b) Technical importance ($[dr_{ij}^1]$)

(i, j)	1	2	3	4	5	6	7	8	9
1	-	0.65	0.68	0.18	0.26	0.97	0	0	0
2	0.35	-	0.51	0.08	0.14	0.86	0	0	0
3	0.32	0.49	-	0.04	0.1	0.87	0	0	0
4	0.82	0.92	0.96	-	0.61	1	0	0	0
5	0.74	0.86	0.9	0.39	-	1	0	0	0
6	0.03	0.14	0.13	0	0	-	0	0	0
7	1	1	1	1	1	1	-	0.8	1
8	1	1	1	1	1	1	0.2	-	1
9	1	1	1	1	1	1	0	0	-

Fig. 1. The concordance matrix and discordance matrix for technical importance

(a) Global concordance relation ($[gcr_{ij}]$)

(i, j)	1	2	3	4	5	6	7	8	9
1	-	0.54	0.29	0.71	0.66	0.32	0.98	0.96	0.5
2	0.66	-	0.56	0.73	0.71	0.43	0.82	0.75	0.58
3	0.92	0.81	-	0.83	0.97	0.52	1	1	0.82
4	0.49	0.43	0.41	-	0.59	0.31	0.82	0.96	0.75
5	0.72	0.45	0.2	0.65	-	0.32	0.98	0.97	0.56
6	0.86	0.71	0.71	0.91	0.81	-	0.97	1	0.75
7	0.17	0.24	0	0.25	0.09	0.12	-	0.63	0
8	0.14	0.25	0	0.12	0.11	0	0.68	-	0.05
9	0.5	0.5	0.32	0.25	0.5	0.25	1	1	-

(b) Outranking relation ($[sr_{ij}]$)

(i, j)	1	2	3	4	5	6	7	8	9
1	-	0	0	0	0	0	0.98	0.96	0
2	0	-	0	0	0	0	0.51	0	0
3	0.92	0.81	-	0.46	0.97	0.0141		1	0
4	0	0	0	-	0	0	0	0.96	0
5	0.68	0	0	0	-	0	0.98	0.97	0
6	0.86	0	0	0.91	0	-	0.97	1	0
7	0	0	0	0	0	0	-	0.07	0
8	0	0	0	0	0	0	0	-	0
9	0	0	0	0	0	0	1	1	-

Fig. 2. The global concordance matrix and the outranking matrix

The value of outranking threshold is dependent on the confidence of designer to the design information collected. If designer believes that the current information is vague and incomplete, s/he can lower the values of thresholds to decrease the discrimination power such that only the dominant requirements are selected for further development. On the contrary, if the information collected is nearly complete, then s/he may want to increase the discrimination power for finding the "best" choice from a set of alternatives.

4 Conclusion

The objective of QFD is to better understanding of customer needs and appropriately deploy them to the design requirements to maximize the customer satisfaction. However, it is not possible to take all design requirements into consideration for product improvement, due to the limited product development time and cost. This paper considers the QFD planning as a multi-criteria decision problem and proposes a fuzzy outranking approach to prioritize design requirements recognized in QFD. The fuzzy set is used to represent the imprecise and uncertain product information collected from design teams. The fuzzy outranking relation is used to model the imprecise preference relation between design requirements. A design requirement outranks other requirements if and only if there is sufficient evidence to support that the requirement is superior or at least equal to the others. As a result, the dominant set of design requirements is identified for the focus of design in order to achieve not only the customer satisfaction but also the balanced design of a product. The proposed decision model is more suitable to be used in the imprecise and uncertain early design environment

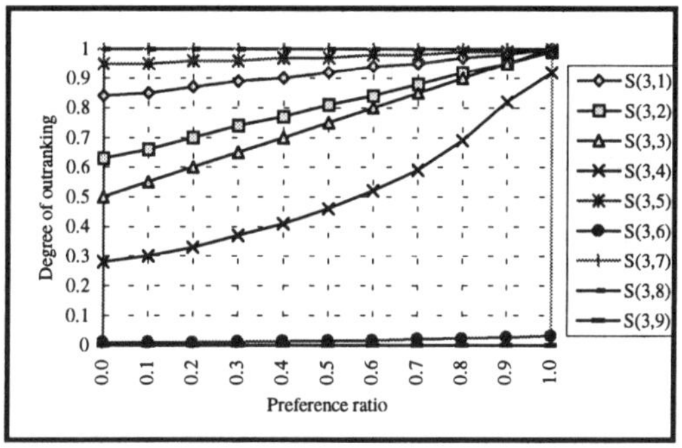

Fig. 3. Relationship between outranking relation ($S(3, j)$, for $j = 1$ to 9) and preference ratio (θ)

Appendix: Sample Data of the Car Design Example

The following shows the input data required for the example of car design. Assume that there are six engineers involved during the QFD planning process. Each engineer assigns linguistic values and then converts them into the corresponding quantitative values with respect to each criterion according to Table 1.

1. Relative importance of customer attributes b_i, $i = 1$ to 6.

i	Customer attribute (b_i)	Relative importance
1	"easy to close from outside"	{5, 5, 6, 7, 7, 6}
2	"stay open on a hill"	{4, 4, 5, 6, 5, 4}
3	"easy to open from the outside"	{3, 4, 2, 3, 4, 3}
4	"don't kick back"	{4, 3, 4, 5, 3, 3}
5	"doesn't leak in rain"	{2, 3, 3, 2, 3, 3}
6	"no road noise"	{1, 2, 2, 1, 1, 2}

2. Relationships between customer attributes c_i and design requirements r_j.

(i, j)	Strength	(i, j)	Strength
(1, 1)	{6, 6, 7, 7, 7, 6}	(1, 5)	{5, 6, 6, 5, 7, 6}

(1, 6)	{3, 2, 3, 2, 4, 3}	(2, 2)	{6, 6, 6, 5, 7, 6}
(2, 3)	{6, 6, 6, 7, 7, 7}	(3, 4)	{6, 7, 6, 7, 7, 7}
(3, 6)	{4, 4, 5, 5, 4, 5}	(4, 2)	{4, 4, 4, 5, 4, 4}
(4, 3)	{3, 4, 4, 3, 3, 3}	(4, 4)	{3, 4, 4, 4, 3, 3}
(4, 6)	{2, 3, 2, 4, 2, 3}	(5, 6)	{6, 5, 6, 5, 7, 6}
(5, 9)	{6, 6, 7, 7, 6, 7}	(6, 6)	{5, 4, 5, 6, 4, 4}
(6, 7)	{5, 6, 5, 5, 4, 5}	(6, 8)	{6, 6, 7, 7, 7, 7}

3. Estimated cost and technical difficulty with respect to each design requirement r_j, $j = 1$ to 9.

j	Design requirement (r_j)	Cost	Technical difficulty
1	"energy to close door"	{3, 4, 4, 5, 3, 4}	{4, 4, 3, 3, 4, 4}
2	"check force on level ground"	{6, 6, 6, 5, 7, 6}	{2, 3, 2, 2, 2, 2}
3	"check force on 10° slope"	{6, 6, 6, 5, 7, 6}	{6, 6, 6, 5, 7, 7}
4	"energy to open door"	{2, 2, 2, 1, 3, 1}	{6, 6, 7, 7, 7, 7}
5	"peak closing force"	{5, 4, 5, 4, 4, 4}	{4, 5, 4, 5, 3, 3}
6	"door seal resistance"	{3, 2, 3, 4, 5, 3}	{7, 7, 7, 6, 6, 6}
7	"acoustic transmission window"	{2, 3, 4, 2, 2, 2}	{3, 4, 3, 2, 4, 4}
8	"road noise reduction"	{1, 1, 1, 2, 1, 2}	{3, 3, 4, 4, 3, 3}
9	"water resistance"	{7, 7, 6, 6, 7, 7}	{6, 5, 5, 6, 4, 5}

4. Weights for technical importance, estimated cost, and technical difficulty.

Technical importance	Cost	Technical difficulty
0.5	0.25	0.25

References

1. J. R. Hauser and D. Clausing, The house of quality, *Harvard Business Review*, 66(3) (1988) 63-73.
2. Y. Akao, *Quality Function Deployment: Integrating Customer Requirements into Product Design*, Productivity Press, Cambridge, Massachusetts (1990).
3. G. S. Wasserman, On how to prioritize design requirements during the QFD planning process, *IIE Transactions*, 25(3) (1993) 59-65.
4. R. L. Keeney and H. Raiffa, *Decisions with Multiple Objectives: Preferences and Value Tradeoffs*, John Wiley, New York (1976).
5. L. A. Zadeh, Fuzzy sets as a basis for a theory of possibility, *Fuzzy Sets and Systems*, 1(1) (1978) 3-28.
6. D. Dubois and H. Prade, Ranking fuzzy numbers in the setting of possibility theory, *Information Sciences*, 30(3) (1983) 183-224.
7. S. Pugh, *Total Design: Integrated Methods for Successful Product Engineering*, Addison-Wesley, Wokingham, UK (1991).
8. B. Roy and PH. Vincke, Relational systems of preference with one or more pseudo-criteria: some new concepts and results, *Management Science*, 30(11) (1984) 1323-1335.
9. J. L.Siskos, J. Lochard and J. Lombard, A multicriteria decision making methodology under fuzziness: application to the evaluation of radiological protection in nuclear power plants, in *TIMS/Studies in the Management Sciences*, edited by H.J. Zimmermann, North-Holland, Amsterdam (1984) pp. 261-283.

Planning-Tools for Quality Management

Dieter H. Müller, Thorsten Tietjen, Thomas Böttcher

Department of Mechanical Engineering, Bremen University
Badgasteiner Straße 1, 28359 Bremen, Germany
bik@bik.uni-bremen.de

Abstract. This paper describes a databased software tool for Quality-Planning. This tool enables companies for a life-cycle orientated use of quality assurance methods during the process of product development. For an adequate use of these activities affecting quality a method for valuation was developed, so-called "Return on Quality".

1 Introduction

The capability to satisfy the customer's needs and requirements concerning the complete life-cycle of industrial products comes more and more to the fore as a factor of competition. Because of the increasing national and international competition in combination with the reduce of the period of development and the heightened requirements on quality and flexibility, preventive actions become more and more decisive importance for products that go off well.
Quality as company's objective can only be achieved on a profitable basis, if the "manufacturing" of quality begins in the phases of specification engineering and product development. For a marketable product have to be considered both the customer's needs and product requirements bearing in mind the technical and the organizatial conditions. Not to pay attention on this requirements can be classified as nonconformity, in the meaning of constant improvement of quality.
As is known a failure that is detected during the phase of product development does carry only small costs. Whereas a defect detected at the customer's causes high costs for repairing or reworking. But the much higher price the company has to pay is the damage of it's image. In consideration of this context there are great expectations of an increasing improvement of the marketability by using a quality-planning that is in a tune with the product. This foresighted planned garantee of the necessary quality features in product development, manufacturing and use is to balance with the resources of the company and the budget costs of the product. A sufficient flexibility of this quality-planning is given by the use of a suitable documentation and software respectively, e.g. a software-tool for the "critical path method".

2 State of the Art

In machine building and other industries it's more and more usual practice that the purchaser gives requirements not only relating to the product itself but also to the process of development and control. Nowadays many companies have established a quality system that is documented in the quality manual. The purpose of this manual is to provide an adequate description of the quality management system. The development and manufacturing processes of the supplier have to comply with these structures, procedures and processes for each manufactured product.

Additional to the check of the quality system, especially in companies out of the field of space technology and aircraft manufacturing, the accordance of the product with the required features and the supplier's ability to meet these requirements is controlled until an order is given. This required disclosure of the company's know-how and the available resources is documented in a textual so-called "Product Assurance Plan" (PA-Plan). This plan fixes early but nearly unalterable all activities for quality assurance. It consists of an enormous number of documents like the management-plan, projekt-plan, test-plan, schedule, and so on.
If it's necessary to change one of these documents, the procedures of the PA-Plan are affected too. This means both the quality activities or methods and the order of events. Inevitable corrections couldn't be included so far, because of he missing transparency of planned quality activities and the life-cycle of products within this PA-Plan. The balance between quality activities and produkt activities happens more from personal experience than in the meaning of a preventive planning. A textual PA-Plan downgrades itself more to a declaration of interest than to a flexible planning-tool. A well suited planning-tool has to react to the existing conditions and changes in the context of development and production.

2.1 Quality Plan

Quality systems are being planned and implemented corresponding to the Industrial Standard EN ISO 9004-1 and in compliance with the Industrial Standard EN ISO 8402. The companies then decide independently which type they are going to use (EN ISO 9001, EN ISO 9002 or EN ISO 9003) as a basis for their quality system. EN ISO 9001 states that there are procedure instructions to be made up for test activities in order to be able to verify that fixed quality demands of a product are also fulfilled. The required examinations and the sketches to be drawn up must be ascertained individually in the quality management plan and a short quality plan as well as in the procedure instructions [19].
In the Industrial Standard EN ISO 9001 however there is no demand made that a company stipulates the drawing up of quality plans. Only if the quality system or the client (customer) demands for it explicitly, is there a necessity for everyone to explain in a well ordered and clear quality plan, the project or product relevant quality document. So the quality plan is a clear summary of the quality system specific to the company and construction then becomes necessary if for example a new product or new process is to be carried out, or significant changes are to be made to existing product or processes. Included within it is the entire seeing through of the development projects by a manufacturer for example, or also the switch of production to a sub contractor. quality systems and also quality plans are based on the maximum 20 elements of the EN ISO 9000ff where every company only has to cover the quality elements that are relevant to them, because not all of the recommended components in the EN ISO 9004 have to be relevant for everyone.
Quality plans change with the quality system or rather its documentation, the quality manual. The individual parts of the quality management system are revised accordingly (e.g. quality elements, procedure or works instructions) as these documents represent the basis of the quality plan on which everything else is built. A contractor will only then draw up a plan if he sees it necessary for purposes of clarity and to ensure the success of the project.

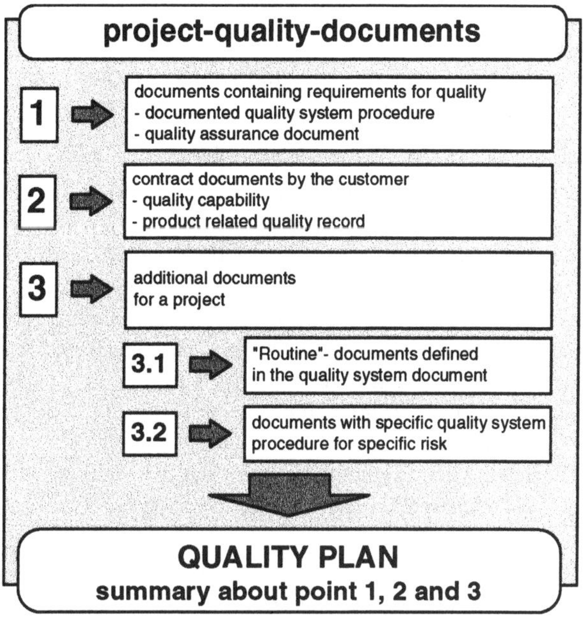

Fig. 1. Quality plan, a part of the documentation of a quality system

A structure plan in contrast to the quality plan is a planning instrument which comes from the project management area. According to DIN 69901 "Project management is the sum of executive duty, organisational hierarchy, organisational techniques and leadership methods needed for the running of a project." A project structure plan is all about the explanation of a project structure. The explanation can follow after the construction, after the start of the project, after basic conditions or from a specific point of view. In the project structure plan the project is hierarchically structured into partial tasks and job lots whereby the partial tasks in the project can further be split up. This then creates project structure levels akin to the structure of a tree. For one thing these specific job lots can explain activities, for the other it can be the sub assembly group. The sum of all the job lots produce a complete picture of all the activities which are necessary for achieving the project goal.

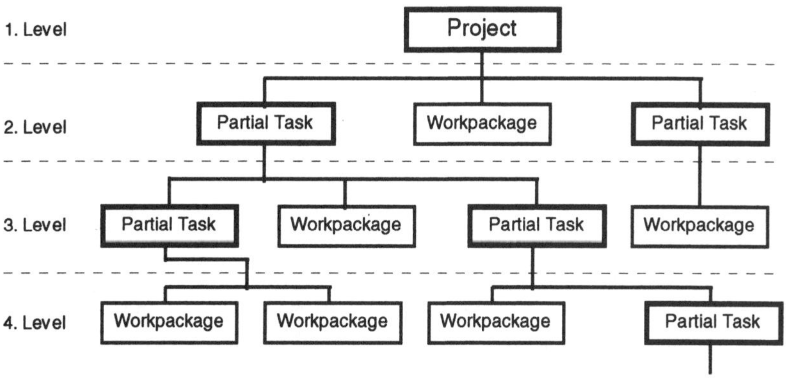

Fig. 2. Work Breakdown Structure [17]

In comparison to the quality management plan which is based on the QM elements, it is possible with the help from a structure plan, to show the whole process of product formation whereby here, the individual partial tasks and processes corresponding to the quality circle can explicitly be examined.

2.2 Implementation of Quality Assurance Methods during Product Formation

Through the description of the concept of quality in the form of the quality circle it was made clear that measures must be grasped in every phase of the life cycle of the product. Fig. 3 contains numerous quality methods which find their use in the individual life phases of the product. These methods have mainly been developed in the last few years and should contribute to the quality assurance of the product. These quality methods are also labelled as tools which lend themselves to fulfil the task of quality assurance. These tasks consist, in the main part, of the basic functions of quality planning, guidance, testing and research and must be fulfilled at every section in the company [1].

Selection, classification and evaluation of the quality methods are understood by quality planning. With this quality methods such as QFD, FMEA or the statistical test plan (DoE) can be used.

During quality guidance the realisation of the product is monitored and corrected and here also methods such as FMEA or QFD are implemented [7].

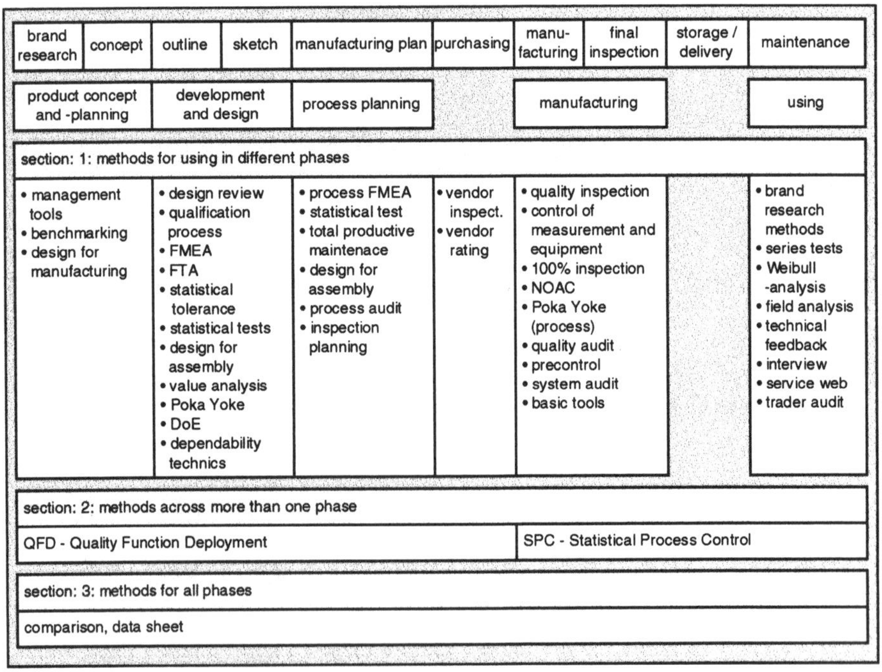

Fig. 3. Quality methods in the parts processes of product formation

During quality testing it is ascertained as to how well the demands on quality have been met [1]. They are sub divided into areas of emphasis: test planning, carrying out of the test and test data registration. With that the test planning is set, what,

when, how and what will be used to carry out the test [7]. This test is then carried out and at the ensuing test date, evaluated. The classical test methods above all find their use here.
Quality demands has as its goal the effect of motivating the workers in a firm thinking and behaving in a quality motivated manner. An example of which is the use of ways of proposing a specific plan of action within a firm or work in quality circles [1].

3 Target Objective of the Planning Instrument

In order to be able to take into consideration all essential aspects of a project/product and to make sense of it, a planning instrument is required especially for quality assurance measures. Preventitive planning for quality assurance measures should be in a position to show, as soon as possible, the processes of product development whilst taking into consideration company resources and also be able to initiate measures so that the previously mentioned quality goals are adhered to.
In order to make the quality assurance methods available for workers, support from a knowledge based system is conceivable. Individual workers can then retrieve information concerning technical topics or also procedures for the use of quality assurance methods.
Quality planning reflects the company's know-how to a certain degree with respect to a product or project. With the drawing out of the quality structure plan all possible requirements should be considered which are in the regulation book and copied into the requirement list. The task of quality assurance involves the planning, carrying out and the monitoring of organisational and technical measures which ensure the demanded quality.
The safeguarding of quality doesn't cover isolated activities, but is an integral part of project activity. The first step for developing a plan lies accordingly in the breakdown of the phases/processes included in the project's course of operation.
Every product phase covers many parallel and sequential processes or activities, which are then best recorded in the form of an ideal plan of action in their chronological order of occurrence.

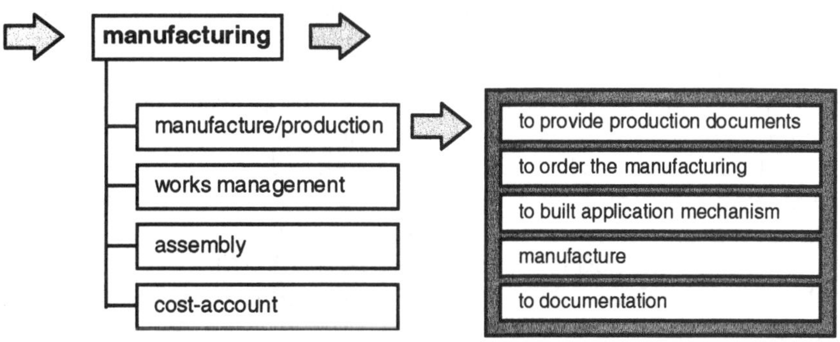

Fig. 4. Degree of sub-divisions; e.g."manufacturing" phase of product formation

The development process of a project's different phases spans the whole project-, beginning with the establishment of market requirements and ending with the standard selling of the product. For every undertaking, the whole process can be split up into many different process parts and each one of these parts consists of a larger number of individual activities or necessary work steps. In support of the quality circle from DIN 55350, the possible internal phases of a job's duration were depicted in order to be able to assign the information requirements that arise to the corresponding structure. The planning of quality assurance measures should be an early task of PA activities, which should be possible to be carried out in the framework of product realisation. From the outset the plan from the organisation's structure must offer the possibility of defining work packages (elements) as individual processes of product creation. Here it must also assign these activities/work and lastly name *PA activities* and appointed *PA methods*.

This model is set out completely by three hierarchies:

- Phase/process parts (The process parts of the product life cycle form the parts phase of the quality circle given in DIN 55350, from marketing to waste management.)
- Elements (The process parts of the product life cycle contain individual work steps e.g. job lots, which are to be carried out in the individual phases.)
- Activities (The process parts are often set together from many individual processes, which can then be described as activities.)

These activities or process parts supply intermediate results which can be checked individually and moreover supply clear instructions about how further operations follow on. Also quality assurance measures can finally be named (e.g. completeness control) and corresponding PA methods (e.g. visual checks, check lists and spot check inspections) can be assigned relatively easily. This finally allows an assignment of quality assurance activities and corresponding PA methods.

3.1 Structuring of the QA Methods

In order to avoid a large reaction of the quality guarantee, it is worthwhile to plan and lay out right at the start of a project's development all necessary quality assurance measures and their accompanying phase in the product life cycle. It is at this point important to ensure a corresponding acceptance of the QA tasks in the individual works areas along with the technical and organisational capabilities. In most firms a number of operational methods already exist which are described as Methods of quality engineering or also Quality methods. Some of the most well known are for example the Failure Mode and Effect Analysis (FMEA), test planning and the Statistical Process Control (SPC). But not only are these economic methods in use, but also some 'simple' quality ensuring measures such as working with checklists or checking by hand of a Should be-Is comparison for example.

The multitude of quality methods is separable into three groups (see also Section 2)

- Methods whose use is only applicable in certain product life phases (Benchmarking, Construction-FMEA, Process-FMEA etc.).
- Methods which influence product quality over more product life phases (e.g. QFD, SPC).
- Methods which can be utilised independent of the product life phase (Checklists, Should be-Is comparisons)

For rough classification of the PA methods to be done, orientation can be achieved around the processes or rather phases of the product's creation corresponding to the organisation according to [7]. This is also dependant upon the tasks within the product life cycle. It is clear with this classification that it is also highly influenced by TQM thinking. Through this, there is then no easy differentiation between project activities and PA activities. Furthermore, a carefully considered task brief of the product life cycle contains many work steps which integrate quality assurance of the product.

3.2 Valuation Method

The suitability of a quality method for a particular product is dependent on the customergiven requirements, the resources and the strategic quality objectives of the respective company.
The first criterion for suitability is that the quality method has to meet the requirements that were given for the method itself. This can be executed with a binary method of excluding. This means that each method is excluded which can't meet a particular requirement. Afterwards all remaining methods are examined more detailed concerning the criteria "time", "costs" and "quality of result".
For this purpose a valuation method was established. This method was recasted from the method "Return on Investment" that is well-known in economics.

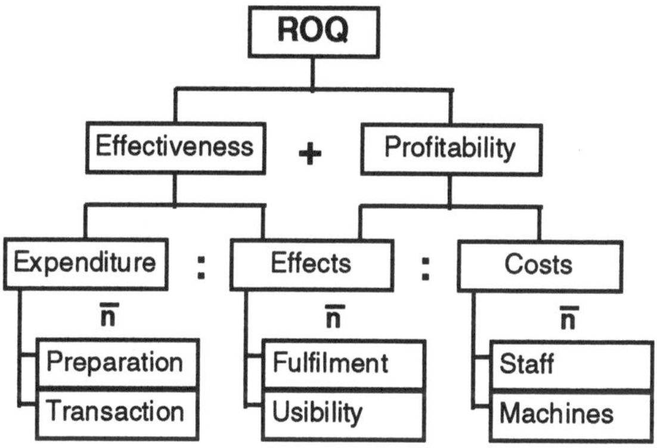

Valuation:

Value	1 - 3	4 - 7	8 - 10
EX / EF / CO	low	middle	high

Weight:

Value	1	2	3
EX / EF / CO	low	middle	high

Fig. 5. ID Value System for Quality Methods

This valuation system we called "Return on Quality" (ROQ). It's a system of characteristic numbers that uses proportional numbers for comparison. The basic numbers of this method are the basic factors "time", "costs" and "quality of result" (effects). This basic factors consist of several components, e.g. the factor "costs" is composed of the components cost of materials, staff expenses, costs for equipment and so on. To valuate these components no absolute values are used, but characteristic numbers in the range from 1 to 10 as known from FMEA. To get a value for the particular factor independent of the number of components, the values of these components are averaged.
The division of the factors "quality of result" and "time" results in the characteristic number "effectiveness", the division of "quality of result" and "costs" results in the characteristic number "profitability". The sum of "effectiveness" and "profitability" is the number "ROQ". For exacter valuation it's possible to weight the included basic factors, e.g. to increase the effect of "costs" or "quality of result".

Using this valuation system it's possible to compare different quality methods to select the one which is most suitable for a particular product, bearing in mind the customer's needs, the product requirements and the technical and the organizatial conditions of the company.

A discussion of the results with reference to the setting up instructions for the ID value method revealed that the order of rank of the examined method's data appears independent of estimate process. The system also responds to changed base criteria, such as other presented goals, where limitations though must be set as the calculated ID values lie very close together. The order of rank of the method is not however influenced.

The method of ID values doesn't give an overall mark for a system, but exclusively allows the comparison with other methods. However a consideration of the table values also allows an estimation of quality methods against one another. Thereby effectiveness and profitability can be obtained through statements about the properties of the examined methods.

3.3 Concept of a Computer Aided Quality Structure Plan

A specific plan of quality assurance measures should be graphically prepared primarily for the different industrial partners (individual/mixed/series) in order to present the PA tasks and their connection to the individual development and manufacturing runs. The quality assurance methods/tasks understood from the analysis phase questionnaire form the basis for the conception of this plan for the partner firms in the project group. However they have the following problem points:

- Graphical presentation of a total project course with a fine subdivision of individual project phases and the assignment of quality assurance measures.
- Identification and presentation of quality assurance measures which will be of standard usage in the partner firms, or rather those which can come into use. Here the separate quality management elements and applied methods are described with respect to the technical usage, the PA organisational expenditure and references to existing operational instructions in firm specific handbooks.
- Investigation and presentation of the effects on the respective product quality and effectiveness of the product development as well as product construction costs by the potential PA method.
- Determination of ID value for the individual PA methods (Time required, consequences, connection of effectiveness/time of usage/performance, estimate of the average cost of operation).
- It is a conceptual yet realisable planning tool which enables an graphical quality assurance plan to be drawn up. This plan is structured by the main elements of the product life cycle and is firm and product specific

The quality structure planning has the goal of setting up a globally valid starting point which allows a full sequence of PA activities right from the start of a project. As such the running phases of the product life cycle should be taken into account in their respective firms. The timing of the implementation of the PA activities should be set out as well as the responsibility for administering the checks and documentation. The basis of the planning should be the phases of the product life cycle and the accompanying activities/work. These phases or rather the work/PA

activities are assigned for quality assurance from several PA methods often at one's disposal. These frequently show serious differences with respect to personnel requirements (in view of qualifications as well as quantity), cost, duration, apparatus and prospects of success. The timing of implementation is also highly variable.
Decisive factors in the decision to use of a certain PA method are frequently criteria such as:

- strategic decision
- PA policy of the firm as a whole
- specified instructions from the client
- the technical ability of the firm
- the financial limitations in the framework of a project/contract
- the efficiency of the method

With this specification, an application data bank is created with which the internal company development can be developed corresponding to the portrayed system. The application is conceived such that the one universal level "Administration", the existing elements and activities in a project can be drawn up. Adaptations to company specific nomenclature, such as the changing of a transitory sequence, are possible without further work. The other level concerns the project or contracts to be carried out. Planning processes for quality assurance measures can then be carried out with the saved information for current projects or also for pre-planned criteria.
Information already respectively exists about the application of phases, elements and activities, such as relevant personnel/departments for example or also globally valid advice on usage. Above all a further table exists which can be described as a method data bank. Therein different quality assurance methods with further information and advice are set out, which can combined with respective individual project activities.
For easy drafting of project activities a pre-structured project scope outline is created from defined criteria, which firstly links the individual project activities to the administration level. This reduces the manual choice of individual plans to the activities and PA methods.

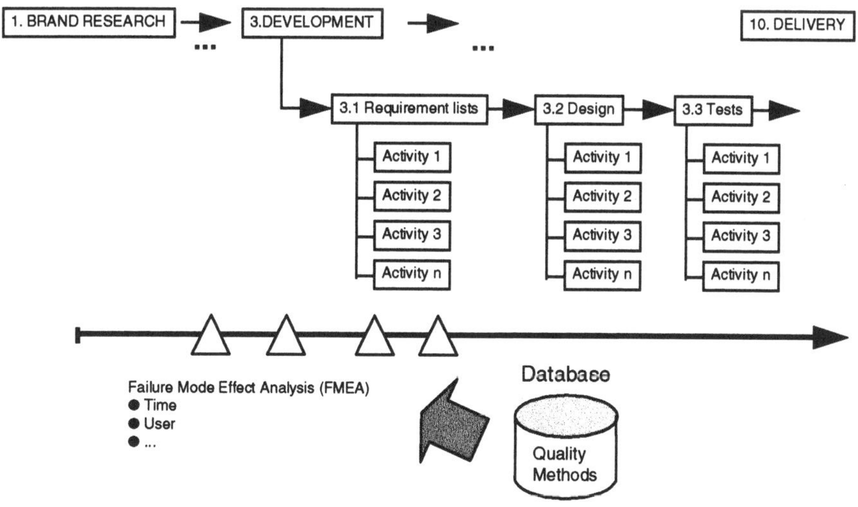

Fig. 6. Assignment of PA activities during the life cycle of a product

The planning level for the project initialisation begins with the laying down of start dates. Additionally, they can be used for the project's pre-structured scope in the application data bank through the linkage with different project criteria. The possibility exists though to manually carry out the choice of project activities. The same goes for the method assignment. Many methods are applicable for different activities in that individual methods offer alternative choices for an activity. Here the integrated ID value system can be fallen back upon, with which a comparison of the possible methods can be carried out.

4 References

1. E. Hering, J. Triemel and H.-P. Blank, Qualitätssicherung für Ingenieure, VDI Verlag, Düsseldorf 1993
2. N.N., DIN EN ISO 8402, Qualitätsmanagement und Qualitätssicherung, Begriffe, Beuth Verlag, Berlin, *Entwurf*,März 1992
3. W. Timischl, Qualitätssicherung, Statistische Methoden, Hanser Verlag, München, 1995
4. H.-P. Wiendahl, Betriebsorganisation für Ingenieure, Hanser Verlag, München, 1989
5. N.N., DIN EN ISO 9004-1, Qualitätsmanagement und Elemente eines Qualitätssicherungssystems - Teil 1: Leitfaden, Beuth Verlag, Berlin, 1990
6. R. Leist and A. Scharnagel, Qualitätsmanagement, Methoden und Werkzeuge zur Planung und Sicherung der Qualität, WEKA-Verlag, Augsburg, 1993
7. T. Pfeifer, Qualitätsmanagement, Carl Hanser Verlag, München 1993
8. N.N., VDI 2221, Methodik zum Entwickeln und Konstruieren technischer Systeme und Produkte, Beuth Verlag, Berlin 1987
9. G. F Kamiske and P. Theden, Qualitätstechniken steigern die Rentabilität, *Qualität und Zuverlässigkeit*, QZ 40 (1995) 11, Carl Hanser Verlag, München, S. 1263-1267
10. A. Bürgi, Führen mit Kennzahlen, Cosmos Verlag, Bern 1978
11. K.-H. Groll, Erfolgssicherung durch Kennzahlensysteme, Rudolf Haufe Verlag, Freiburg i. Br. 1988
12. G. Schott, Kennzahlen, Instrument der Unternehmensführung, Forkel-Verlag, Stuttgart, 1981
13. P. Rinza and H. Schmitz, Nutzwert-Kosten-Analyse, VDI Verlag, Düsseldorf, 1977
14. G. F. Kamiske, Return on Quality RoQ, *Qualität und Zuverlässigkeit*, Carl Hanser Verlag, München, QZ 40 (1995) 10, S. 1122 f.
15. J. Platz and H.-J. Schmelzer, Projektmanagement in der industriellen Forschung und Entwicklung, Springer Verlag, Berlin, 1986
16. RWK Rationalisierungskuratorium der Deutschen Wirtschaft e.V. (editor), Projektmanagement-Fachmann, Volume 1 and 2, RKW e.V., Eschborn, 1996
17. J. M. Juran, Der neue Juran: Qualität von Anfang an, Landsberg/Lech, Moderne Industrie, 1993
18. K. B. Clark and T. Fujimoto, Automobilentwicklung mit System: Strategie, Organisation und Management in Europa, Japan und USA, Campus, Frankfurt, 1992
19. N.N., DIN Taschenbuch 226, Qualitätsmanagement und Statistik, Verfahren, 3: Qualitätsmanagementsysteme, Beuth Verlag, Berlin, 1995

Distributed Quality Assurance System Framework for Agile Manufacturing

Yang Hongpeng, Zeng Xiaoliu, Min xiaoyong, Lin Zhihang

CIMS Research Center, Xi'an Jiaotong University, China

Abstract. In this paper, the quality management requirements for agile manufacturing are firstly analyzed. According to the requirements, a generic quality model and several distributed quality assurance system architectures are proposed which focus on the development of a multi-site approach to quality management. Some reference frameworks such as functional model, elements and modules of distributed quality assurance system(DQAS) are presented. Finally, the implementation issues of a DQAS are also discussed.

Introduction

Agile manufacturing is a new manufacturing philosophy which proposed by the Iacocca Institute at Lehigh University in 1991[1]. The basic idea of agile manufacturing is to rapidly respond the requirement of competitive market and develop new product with high quality, low cost and short lead-time. One of the most important feature of agile manufacturing is multi-site virtual company. How to establish an efficient quality assurance system to control the product quality in multi-site agile manufacturing environment is an important problem need to solve.

The efficient and effective management of quality is acknowledged to be a critical factor for the manufacturing industries in the ongoing battle for competitive advantage within the world market. Without increasing resources, quality management must stay in line with ever increasing market-driven product, process and service demands. The complications of the situation are further multiplied for agile manufacturing enterprise who have more than one industrial site and therefore quality issues arising at both local and global level. Without the integration of quality information, expertise and knowledge existing within each single site, these multi-site enterprises can often duplicate quality efforts without gaining any global benefits from individual improvement.

The reasons for need of distributed quality assurance system for agile manufacturing are found in three main areas: first, the need to accommodate new pattern of manufacturing activity, agile manufacturing, which has meant that accurate information must be immediately available to allow the identification of quality problems; second, the great increase in the volume of quality problems produced in the multi-site agile manufacturing environment demand an efficient distributed quality system to support the quality management in the whole product lifecycle from global level to local level ; third, the need for the evolution of the quality system in time, whose target is here to improve the flexibility required to respond the faster and faster change of the market.

This paper introduces an on-going project "Distributed Computer Quality Assurance System for Agile Manufacturing" which supported by China High-Tech Plan. The objective of the project is to study and develop a generic quality model and several distributed quality system architectures for agile manufacturing, to propose a general quality system reference framework for agile manufacturing which covers the main activities of quality management, and to discuss some application issues of distributed quality assurance system in agile manufacturing environment.

The Generic Quality Model for Agile Manufacturing

The Requirements For Generic Quality Model

According to the requirements of agile manufacturing and the basic quality management principles of TQM, quality management should concern two kinds of quality management activities. From an operational point of view, quality management is control process. Quality system can be looked as a decision and control center which rapidly measuring quality performance, comparing quality performance with the quality objective, providing procedures supporting decision making, and feedbacking the decision to manufacturing system. This aspect of quality management can be shown in Fig. 1.

From a evolution point of view, quality management is a never ending improvement process. Quality system can be looked as a quality evolution center which rapidly measuring the current status of the quality system, comparing performance with the

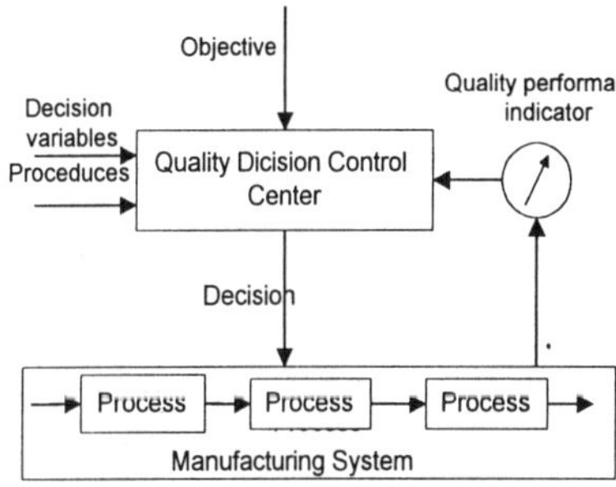

Fig.1. Quality Control Principle

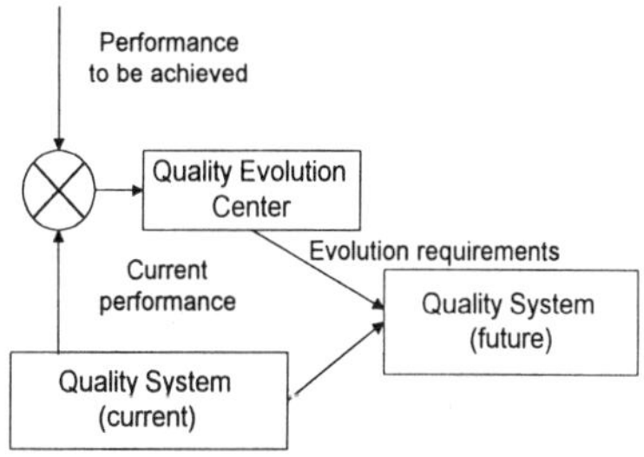

Fig. 2. Quality Evolution Principle

new performance to achieve, and leading to an output which is globally an ask for evolution. This aspect of quality management can be shown in Fig. 2.

As shown above, the generic quality model should have the capability to show the management characteristics of both these activities alone with a strong link with manufacturing system.

The Generic Quality Model

The generic quality model for agile manufacturing should include a quality control subsystem and a quality evolution subsystem linking to the manufacturing system. Fig. 3 summarizes above two aspects within a global model. Quality control subsystem is relevant to the operational point of view while quality improvement subsystem is relevant to the quality evolution point of view.

It must keep in mind that this generic quality model is just conceptual, it is not related to the organization of the enterprise.

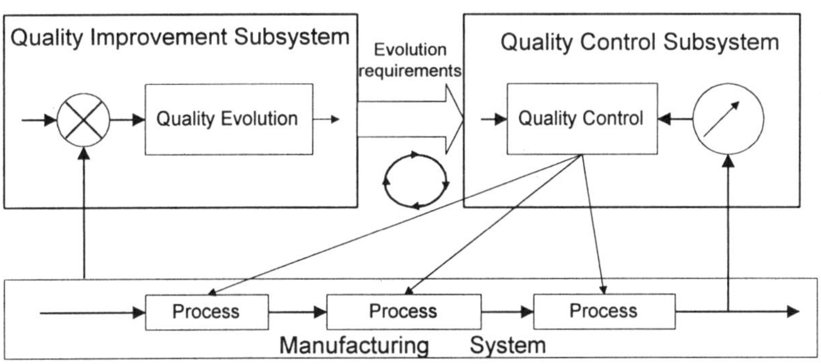

Fig. 3. Generic Quality System model

The Distributed Architectures of DQAS

An architecture is used to develop a system from a set of components. Architectural approaches to quality system development focus on viewing a system from various perspectives and showing their integrative relationships. Having an available set of DQAS architectures will help in the effective and efficient implementation of the DQAS for agile manufacturing and TQM. The DQAS architecture will be identified from physical, logical, functional, and network characteristics and views. These views serve the purpose of aiding both designer and the user of the DQAS.

Physical View of The DQAS Distributed Architecture

The physical attributes of the DQAS architecture are all highlighted in Fig. 4. The architecture illustrated in Fig.4 represents a position distribution structure and information flow structure within a agile manufacturing enterprises. This model provide a road map for quality information flows within the architecture.

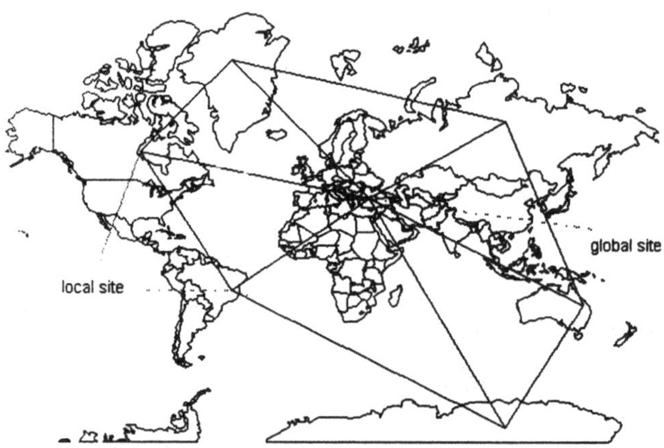

Fig. 4. The Multi-site Distributed Architecture of DQAS

Logical View of the DQAS Distributed Architecture

The logical architecture of DQAS is shown in Fig. 5 which describes how the quality issues will be processed at a multi-site agile manufacturing environment. Logically, this architecture is a top-bottom structure. The top is the global level which provides the instructions and plans to the local level where execution of these plan occur. The bottom is the distributed local level which provide the local level quality information feedback to the global level.

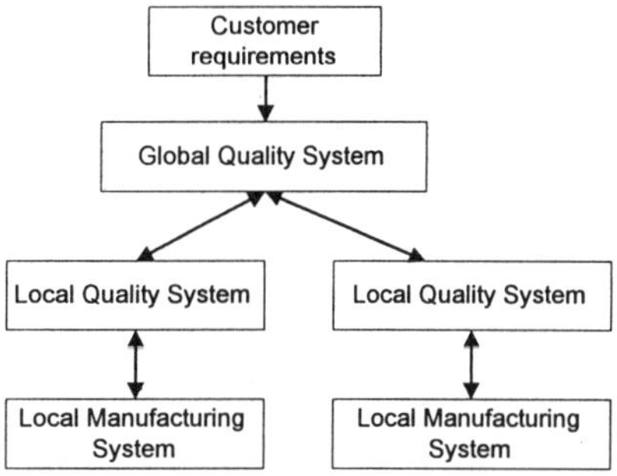

Fig. 5 The Logical Distributed Architecture of DQAS

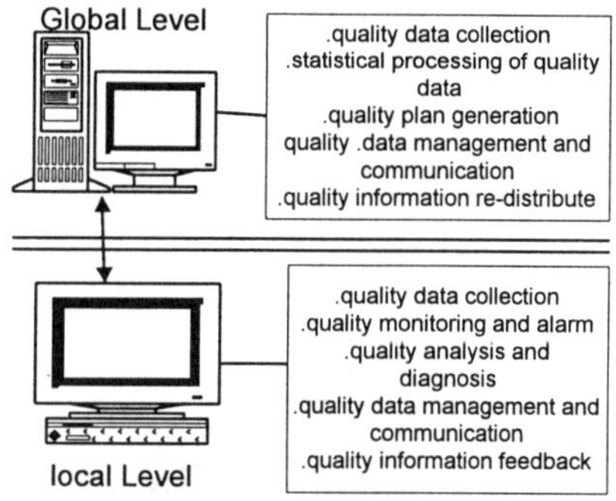

Fig. 6 The Functional Distributed Architecture of DQAS

Functional View of the DQAS Distributed Architecture

According to the logical architecture of DQAS, the functional distributed architecture is illustrated in Fig. 6 which different quality management functions are distributed in the global level and local level. At the global level, quality management rules are planed and guidelines for their application are provided. The main function consists of quality data collection from both customers and the local level, statistical processing of quality data, quality plan generation, quality data management and communication, and quality information re-distribute. The output of the global level consists of the reports on input information after processing and the decision will be re-distributed to the local levels. At the local level, quality management rules are executed and the results are feedback. The main function consists of quality data collection from shop-floor, quality monitoring and alarms, quality analysis and diagnosis, quality data management and communication, and quality information feedback. The output of the local level consists of the quality diagnostic reports and the local quality information will be feedback to the global level.

Network View of the DQAS Distributed Architecture

The network infrastructure supporting the DQAS is shown in Fig.7. It is a multi-site Client/Server interconnect distributed architecture. In each site, there exists a local Client/Server system to support the local quality management. The global network was constructed by all these local Client/Server systems linking to the INTERNET to support the global distributed quality management in a agile manufacturing environment.

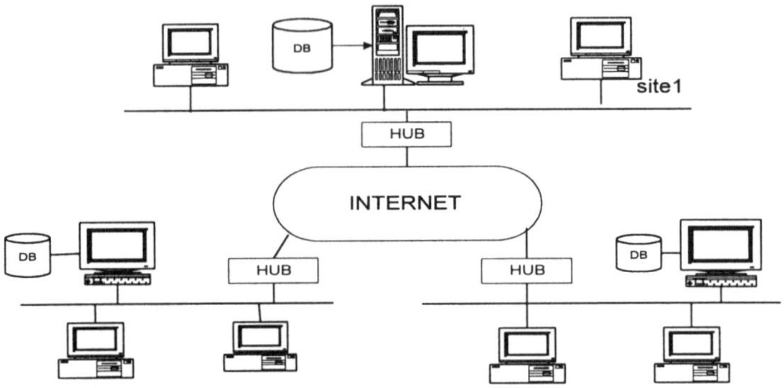

Fig. 7. The Network Distributed Architecture of DQAS

The Reference Framework of DQAS

So far, we have presented various architecture views of the DQAS, these views serve as the skeletal framework for the design of a DQAS. To describe the DQAS more fully, we present various reference models that would function within these architectures. These reference models are the foundation for designing a practical DQAS.

Functional Model of DQAS

The functional model of DQAS is represented by the IDEF0 which is one of the ICAM definition family of enterprise modeling tools. Graphical descriptions of the IDEF0 elements, and the decomposition relationship of activities which is an approach to help in the linkage of the functions at various levels of abstraction are presented in Fig.8. A complete functional model of DQAS would be very complex. Fig.8 is just a high-level example of more than twenty such functional models.

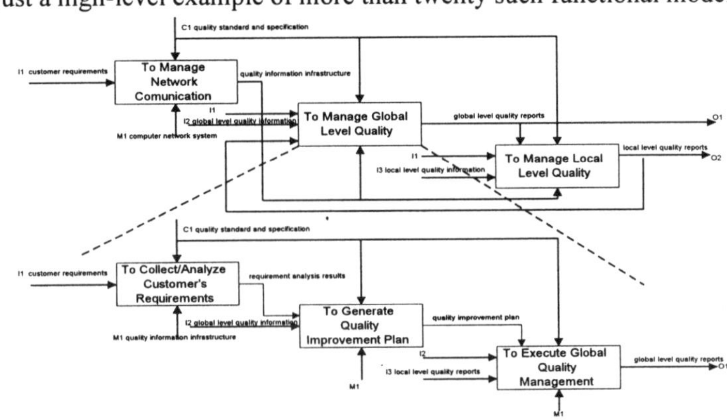

Fig.8. The High-level Functional Model of DQAS

As shown in Fig.8, the DQAS consists of system network&communication management, global level quality management and local level quality management. The global level quality management includes acquiring and understanding customer's requirements, generating quality evolution strategy, executing global quality management.

Elements of DQAS

Functional model represent the function decomposition hierarchical structure of DQAS. To describe the DQAS more fully, we present various elements and modules that would function within these functional models. These elements and modules of DQAS are grouped in table 1, including quality planning module, quality control module, quality evolution module, quality improvement module, quality report generation module, quality documentation management module, quality database management module, system communication module.

Table 1 Elements and Modules of· DQAS

Quality Planning Module	**Quality Control Module**
. Market quality plan	. Statistical process Control
. Procedure quality plan	. Design quality control
. Product assurance plan	.After-sale quality control
. Customer support plan	. Incoming quality control
. Data collection and feedback plan	. Key parts quality trace
. Quality cost plan	
Quality Evolution Module	**Quality Improvement Module**
. Evolution requirement analysis	. Computer quality simulation
. Quality system audit	. System improvement
. Competitive benchmarking	. Corrective action generation
. Quality evolution decision	. Human education and training
Quality Report Generation Module	**Quality Documentation Management**
. Customer requirement reports	. Quality standard documentation
. Vendor performance reports	. Quality system certification documentation
. Quality audit reports	. Material documentation
. Quality index reports	. Process documentation
. Quality cost reports	. Product documentation

Quality Database Management Module

- . Incoming material data
- . In process quality data
- . Key part trace data
- . Finished product data
- . Customer feedback data
- . Quality cost data
- . Competitive data

System Communication Module

- . Network communication protocol
- . Quality data exchange standard
- . Remote database query
- . Remote server register
- . System monitoring
- .Data communication tools

The Implementation Issues of DQAS

The DQAS is a very complex system. This complexity usually causes many quality system implementation to fail. Therefore, to order to develop and implement a practical DQAS in a enterprise, it is very important to take more attention to these issues.

Enterprise Quality Culture

Enterprise quality culture is the soul of enterprise quality management. It consists of quality sense of all staff, the attitude to quality, the attitude to customer requirements, quality objective etc. Good enterprise quality culture is the basis for successfully implementation of DQAS.

Enterprise Quality Standard

Quality standard is the basic criteria of enterprise quality management. Standard requirement should be defined early in the system development phases. This will also allow for a more stable implementation environment. It also be important from the perspective of being able to phase in the system that the standards be determined early on so that future modular subsystem may be implemented with little interface difficulties to existing subsystem.

Enterprise Information Infrastructure

Enterprise information infrastructure is the foundation of a effective DQAS. Without this infrastructure, the DQAS can not be efficient implemented in a enterprise. The main elements of enterprise information infrastructure consists of an interconnect enterprises intranet network within the agile manufacturing enterprises, a distributed network database management system, and the quality data exchange standard.

Human Factor

The people issue is usually the major reason that many implementations do not reach fruition. so, the DQAS development team should include users from all impacted areas of enterprises, from top manager to the shop-floor employees. Education and training is the critical means to improve the human factor.

Conclusion

Quality Assurance system for agile manufacturing is a new research topic, there are a few literature related to this issue. This paper is our initial viewpoint about DQAS for agile manufacturing. A generic quality model and several distributed architectures of DQAS have been proposed. A reference framework which including functional model and elements for designing a practical DQAS in a agile manufacturing enterprises has been described. In the end, the implementation issues of a DQAS are also discussed.

References

1. Roger Nager, Rick dove, 21st Century Manufacturing Enterprises Strategy. Iacocca Institute, Lehigh University, 1991.

2. G. Hohner. Quality Information in Manufacturing: Distributed Processing and Manufacturing Cells. CIM review, Summer,(1990).

3. Giorgio Basaglia et al, Quality Management in Distributed Multi-site Manufacturing Environment. Sharing CIM Solution, IOS Press,(1994)378-388.

4. John KATAKIS et al, A CNMA Pilot Implementation Supporting a Distributed Quality Control System at INTRACOM. Sharing CIM Solution, IOS Press(1994)189-197.

5. Joseph Sarkis et al, Quality Information System in Advancing Manufacturing Environment. Quality Engineering 8(3), (1996)419-431.

Computer Integrated Quality Information System Based on Client/Server Under CIMS Environment

Yang Hongpeng, Lin Zhihang

CIMS Research Center, Xi'an Jiaotong University, China

Abstract. In this paper, a computer integrated quality information system(CIQIS) with Client/Server architecture, based on ORACLE DBMS is described. The developed system consists of the functional subsystems and integration platform. The functional subsystem includes quality inspection planning generation, incoming quality management, statistical process control, quality trace of key parts, after-sale quality management, quality index statistic and analysis, quality system documentation management, and quality information reporting. The integration platform consists of system service layer and application service layer, which including the system administration, distributed quality information management, quality data communication etc. Based on the developed CIQIS prototype system, two practical computer aided quality information systems have been implemented in Shanghai Second Textile Machinery LTD and Chengdu Aircraft Manufacturing Company.

Introduction

Facing the increasing competition from the global market, product quality is recognized as one of the most critical competitive weapon by most enterprises. Total Quality Management (TQM) has been widely accepted as one effective approach to improve product quality. But effective TQM depends on the timely provision of adequate and correct quality information to support a process of never ending improvement. Computer integrated quality information system(CIQIS) is needed to create an environment where the correct quality information is made available to the appropriate person at the right time, and in which automatic feedback of control information to the process may be used to improve the quality.

The reasons for need of computer integrated quality system are found in three main areas: first, the need to accommodate new patterns of manufacturing activity, CIM and JIT, which have meant that accurate information must be immediately available to allow the identification of quality problems; second, the great increase in the volume of quality data provided by the new technologies which demands an effective quality data management strategy; third, the need for timely and accurate information to aid the never ending quality improvement process.

A definition of the Integrated Quality System(IQS) and a method of systematic quality system design have been proposed by Dessouky[1]. A broad concept of CAQ is described by Tuttle[2], who examines requirements for an integrated system incorporating a comprehensive set of quality functions. Another concept of Automating Quality System has been proposed by J.D.Tannock[3], who presents a strategy for the automation and integration of quality system for manufacture, especially emphasis on the systematic structured quality system design and improvement process together with a comprehensive approach to automation in data collection and quality data management. More detail research about Quality Information System in CIM environment has been done by Chia-hao Chang[4], who describes the role and the basic structure of Quality Information System in a CIM system together with the interfaces with other subsystem in a CIM system. Recently, Basaglia[5] presents a generic integrated quality system in distributed multi-site manufacturing environment.

Above all, two important issues about computer integrated quality information system in a CIM environment have been proposed. One is to develop a efficient computer integrated quality information system for collecting, storing, analyzing and providing pertinent quality information to all the activities of quality management in a CIM system. Another is to provide a systematic approach and an integration platform and tools for the evolution of the quality system for never ending quality improvement process.

This paper is the research result of the project "computer integrated quality information system under CIMS environment" which supported by China High-Tech Plan. The objective of the project is to study and develop the general computer integrated quality information system for mechanical manufacturing enterprise, which covers the main activities of quality management and provides the integration platform and tools for the quality information distributed in the different levels of the enterprises.

The System Architecture of CIQIS

According the objective of the project, a computer integrated quality information system with Client/Server architecture which consists of functional subsystems and integration platform is described. The architecture of CIQIS is shown in Fig.1.

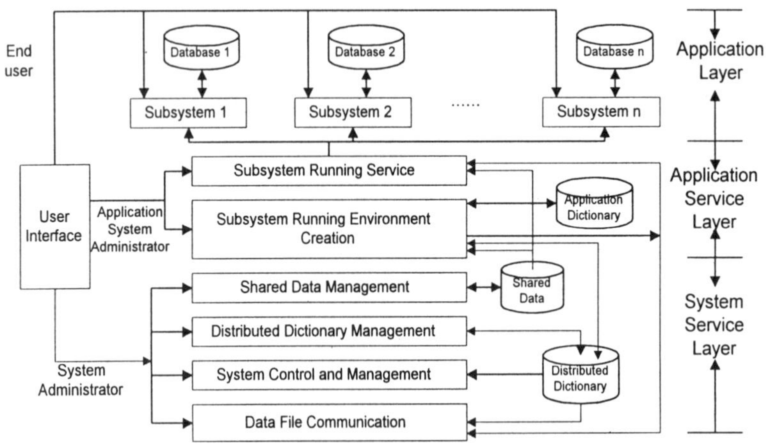

Fig.1. The System Architecture of CIQIS

The developed CIQIS system consists of three layers, that is functional application subsystem layer, application service layer and system service layer. Functional application subsystem layer composes of individual application subsystems which can perform certain quality management function, those are quality inspection planning generation subsystem, incoming quality management subsystem, statistical process control subsystem, quality trace of key parts subsystem, after-sale quality management subsystem, quality index statistic and analysis subsystem, quality system documentation management subsystem, and quality information reporting subsystem. Application service layer is used to integrate above individual subsystems into a global quality information system which including application subsystem running environment creation and application subsystem running service.

System service layer is used to provide common system service for global quality information system which including shared data management, distributed dictionary administration, system control and management, and the communication tools for the quality data.

The CIQIS is used by three kinds of users, that is end user, application system administrator, and system administrator.

The Functional Subsystems of CIQIS

Quality Inspection Planning Generation

Inspection planning generation subsystem consists of inspection instruction generation, inspection diagram drawing, and inspection method library and inspection tool library management. It provides a integrated environment to facilitate the development of inspection instruction and diagram of mechanical parts based on the geometric information and quality information of parts. Users, inspection planner can design the inspection process, including the inspection items, methods, devices, inspection tools and sequence using the inspection method library and inspection tool library of the subsystem.

Incoming Quality Management

Incoming quality management subsystem consists of incoming quality inspection planning generation, incoming quality information management, and supplier quality evaluation. The main function of this subsystem is to control the incoming quality of external products and materials.

Statistical Process Control

Statistical process control subsystem is a general process quality control module which can be used in the area of process quality control. It provides the tools for shop-floor quality control which includes quality data collection, transmit, analysis, generation and selection of quality control charts, on-line monitoring and warning of manufacturing process, analysis of stability of manufacturing process, and quality data management and communication.

Quality Trace of Key Parts

Quality trace of key parts subsystem consists of key parts quality data collection and key parts quality control. The main function of this subsystem is to collect and store the quality data of key parts step by step, and to trace and control quality problems while occurring within the lifecycle of the key parts.

After-sale Quality Management

After-sale quality management subsystem consists of customer information collection, customer information analysis, and customer information feedback. The main function is to collect and process the quality information during the product usage.

Quality Index Statistic and Analysis

Quality index is the integrated reflection of enterprise quality status, and the foundation of quality decision. The function of quality index statistic and analysis subsystem is to realize the comprehensive statistic and analysis of quality index in different levels of enterprise.

Quality System Documentation Management

Quality system documentation management subsystem offers an integrated computerized environment to facilitate the development and implementation of quality system documentation for enterprises working towards ISO9000 certification which providing a easier way for user with generic set of quality manual and procedure templates to develop their quality document of the selected contractual model of ISO9000 and an on-line self-help training module.

Quality Information Reporting

Quality information reporting subsystem offers an rapid quality information feedback system to meet the reporting and feedback requirements of the enterprises and it's customers. It provides a rapid way for leader and user to understand the faults, problems, cost of quality and actions from design, production and service.

The Integrated Platform of CIQIS

The Integrated Environment of CIQIS

System hardware environment is a Client/Server System as shown in Fig. 2. SUN4/75 workstation works as the system server, personnel computer and terminals works as the client, and the TCP/IP local network is used to link the client and server.

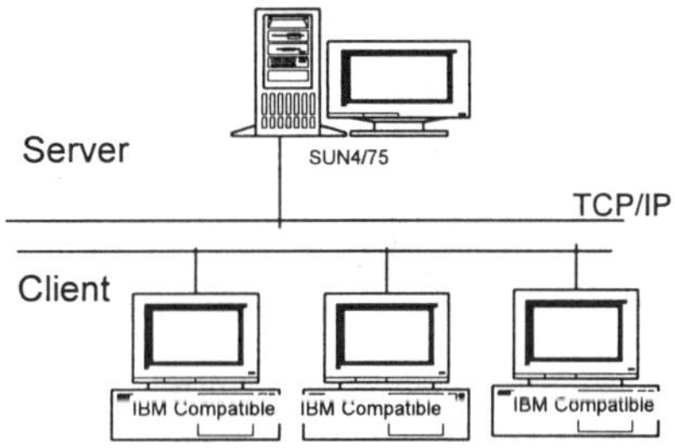

Fig.2. Hardware Enivironment of CIQIS

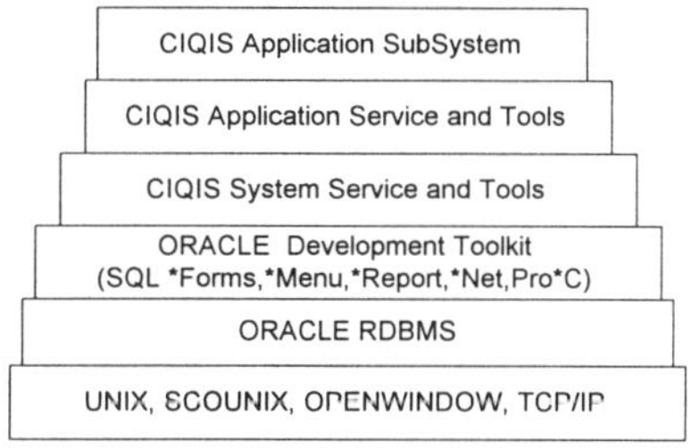

Fig.3. Software Enivironment of CIQIS

System software environment includes UNIX and SCOUNIX operation systems, ORACLE database management system and C programming language. System software structure is shown in Fig.3.

The Function Distribution of CIQIS in a Client/Server System

Client/Server System represent a new computer system architecture. It can make it easier for user to access information, to share data and processing power, to use and develop applications, and to manage a distributed computers environment.

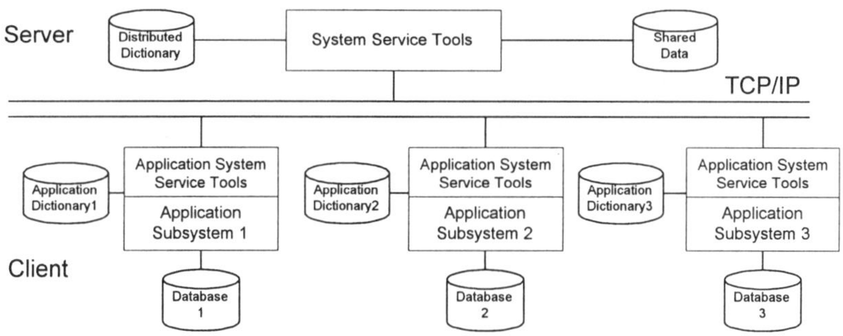

Fig. 4. The Function Distribution of CIQIS in A Client/Server System

A successful Client/Server application system have to distribute the system function in the suitable node in the Client/Server architecture. The function distribution of CIQIS is shown in Fig. 4. The system service tools, together with the shared database and the distributed dictionary are distributed in the server. The application subsystem service tools and application subsystem together with individual subsystem database and application dictionary are distributed in the clients. The application subsystem in one node can use the database query and communication tools to search the shared data in the server and local data in other node. Function distribution of CIQIS in a Client/Server System has improved the system processing efficiency.

Database Remote Query and Communication

One of the important function of integration platform is to provide the database query and communication which enable any node can search data stored in the other nodes in a Client/Server system. Fig. 5 has shown the database chain creation and remote user service process. Based on the subsystem data query requirement and authorization described in the distribution dictionary and application dictionary, database query and communication tools can create the database query chain and remote user register automatically.

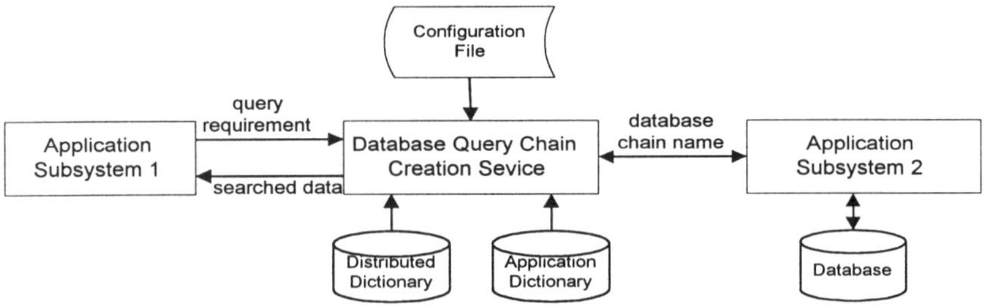

Fig.5. The Remote Database Query Process

The Generation Principle of Application System

The generation principle of application system from subsystem is shown in Fig. 6. The individual subsystems developed by subsystem designer are described in the softcomponent dictionary; application system administrator use application system construction facility to create the application system running service environment

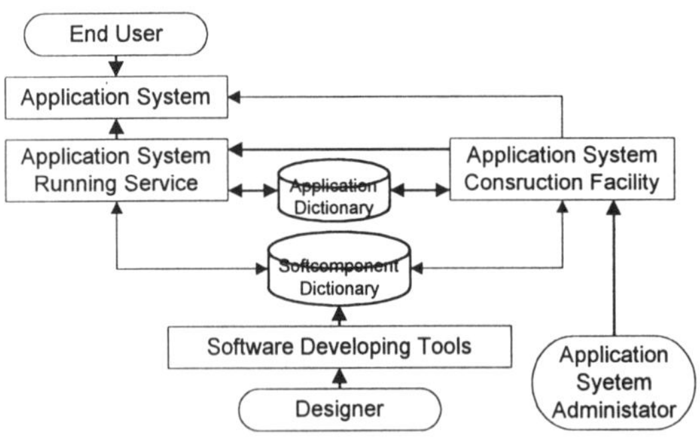

Fig. 6. The Application System Generation Principle

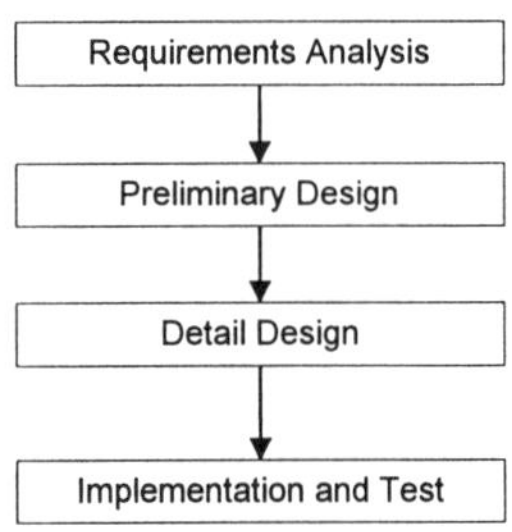

Fig. 7. The Implementation Process of A Practical CIQIS

and application system by using application dictionary and softcomponent dictionary; the end user operates the application system.

The Implementation of a Practical CIQIS

Development Process of a Practical CIQIS

Implementation of a practical CIQIS in Enterprises is a complex system engineering. It is necessary to use structure approach to develop the system step by step. We recommend the developing process for implementing of a practical CIQIS as shown in Fig. 7.

Requirements analysis is to understand the practical quality management requirements of a enterprise. Preliminary design is to represent the CIQIS functional models and information models by using structure system analysis method such as SD, IDEF0, IDEF1x etc. Detail design is to design the functional modules and database structure and physical programming. Implementation and test is to apply the CIQIS in practical field.

Application Examples

Based on the developed CIQIS prototype system, two practical CIQIS have been developed for Shanghai Second Textile Machinery LTD. and Chengdu Aircraft Company which both are CIMS Application Factory supported by China 863/CIMS Programme. Both systems have provided efficient support to the implementation of TQM and CIM, and made good economical benefits for the enterprises.

Conclusion

The role of computer integrated quality information system under CIMS environment has received little emphasis form the practitioner and research literature. In this paper, we have provided an overview that has identified the issues facing CIQIS in a CIM system. This overview includes the requirements, architecture, major subsystems, integration platform, and implementation of a CIQIS.

References

1. Dessouky et al., A methodology for integrated quality system. Trains of ASME. J. of Engng for Ind. 109(1987).

2.Tuttle J, Computer aided quality, ASQC, Quality Congress Transactions, (1987)293-299.

3. J D Tanock et al. Automating quality system. J. of Engng Manufacture, Vol.204, (1990) 231-236.

4. Chia-hao Chang. The structure of quality information system in a computer intergrated manufacturing environment. Computer Ind. Engng, Vol.15,No.1-4,(1988) 338-343.

5. Giorgio Basaglia et al., Quality management in distributed multi-site manufacturing environments. Sharing CIM Solutions, IOS Press, (1994)378-388.

6. Lin Zhihang et al., The research report of computer integrated quality information system under CIM environment. Xi'an Jiaotong University, (1995).

Multivariate Robust Design Optimization

Kwok-Leung Tsui
Department of Industrial Engineering & Engineering Management
Hong Kong University of Science & Technology
Clear Water Bay, Hong Kong

ABSTRACT. Robust Design is an important method for improving product or manufacturing process design by making the output response insensitive (robust) to difficult-to-control variations (noise). Most research in robust design has been focused on problems with a single characteristic or response. This paper concerns the application of robust design method to problems with multiple characteristics. We extend the multivariate loss function considered in Pignatiello (1993) to include the smaller- and larger-the-better type characteristics. We develop appropriate two-step procedures that minimize the average multivariate loss. Similar to the advantages in the single characteristic problem, the proposed two-step procedures substantially reduce the dimension of the design optimization problem and allow for future changes of design requirements without re-optimization. We illustrate the proposed procedures with a polysilicon deposition example.

1 Introduction

Continuous quality improvement has become widely recognized by every industry as critical to maintaining a competitive advantage in the international marketplace. It is also recognized that quality improvement activities are most efficient and cost effective when implemented as part of the design and development stage, through such techniques as statistical design of experiments. In the early 1980's, Genichi Taguchi, a Japanese quality consultant, introduced to several major American industries a systematic method of applying experimental design which has become known as robust design. The widespread acceptance and application of this methodology in America as well as in the world have resulted in significant improvements in product quality, manufacturability and reliability at low cost.

The objective of robust design is to reduce performance variation in products and processes by selecting the settings of easy-to-control factors which provide the best performance and the least sensitivity to noise. Most research in robust design has been focused on problems with a single characteristic or response (see Nair, 1992 for a panel discussion on robust design). Fewer works have been published concerning the application of robust design to problems with multiple characteristics (see Logothetis and Haigh (1988), Pignatiello (1993), and Elsayed and Chen (1993)).

For products with a single characteristic, Taguchi (1986) proposes a quadratic loss function to measure the product quality. He also proposes a two-step procedure to minimize the average loss: first minimize a loss measure, then shift the mean to target. As explained in Leon, Shoemaker, and Kackar (1987) for the single characteristic problem, there are two major advantages of the two-step procedure: (i) it reduces the dimension of the original optimization problem; (ii) it does not require re-optimization for future changes of the target value.

For products with multiple characteristics, the overall product quality depends on how the values of the responses deviate from their ideal values. In general, the quality loss of the product is not always an additive function of the quality losses of individual responses as the responses may interact with each other. For example, Byrne and Taguchi (1987) describe an example with a connector and a tube. The two responses of interest are the force required to insert the tube into the connector and the pull-off force. The two design objectives are: (i) to have the insertion force to be as low as possible for ease of manufacturing; (ii) to have the pull-off force to be as high as possible so that the tube would not fall off from the connector when being used. Obviously, the two responses interact positively (one increases as the other increases or vice versa) and the two objectives conflict with each other. Thus it is not appropriate to consider the individual quality losses of the two responses separately. Pignatiello (1993) considers a multivariate quality loss function to measure the product quality with consideration of the interactions among characteristics.

In this paper we extend the multivariate loss in Pignatiello (1993) to include the smaller- and larger-the-better characteristics. We show that there also exist two-step procedures for the multiple characteristic problem. We derive the appropriate procedures under various models and show that the procedures have the same advantages as those for the single characteristic problem. In Section 2 we discuss the multivariate quality loss for products with multiple characteristics. In Section 3 we derive the two-step procedures under various models. Section 4 illustrates the procedures with an example. Section 5 concludes the paper with a discussion.

2 Multivariate Quality Loss

The overall quality loss of a product depends on how the values of the multiple characteristics deviate from their ideal values. In general, the overall quality loss is not always an additive function of the quality losses of individual responses as the responses may interact with each other. It is assumed that the responses $(Y_1, ..., Y_K)$ are affected by the control and noise factors as follows:

$$Y_i = f_i(\mathbf{C}, \mathbf{N}) + \epsilon_i, \ i = 1, ..., K,$$

where $\mathbf{C} = (C_1, ..., C_p)^T$ is the vector of control factors, $\mathbf{N} = (N_1, ..., N_q)^T$ is the vector of noise factors, and f_i is the transfer function for Y_i. The control factors are assumed to be fixed and the noise factors $\mathbf{N}$ are assumed to be random. The pure error ϵ_i represents the remaining variability of the manufacturing process that is not captured by the noise factors. A loss occurs if any of the responses (Y_i) deviates from its target t_i.

Similar to the single response problem, the robust design objective is to choose the values of the control factors to minimize the average loss over the random noises $\mathbf{N}$ and ϵ_i. The common loss function for a single response is the quadratic loss function. A natural extension of this loss function to multiple correlated responses is the multivariate quadratic function of the deviation vector $(\mathbf{Y}-\tau)$ where $\mathbf{Y}=(Y_1,...,Y_K)^T$ and $\tau=(t_1,...,t_K)^T$, i.e.,

$$ML(\mathbf{Y},\tau) \quad = \quad (\mathbf{Y}-\tau)^T Q(\mathbf{Y}-\tau), \tag{2.1}$$

where Q is a positive definite constant matrix. The values of the constants in Q are related to the repair and scrap cost of the product and can be determined based on the product functional requirements. Pignatiello (1993) considers the loss function in (2.1) to measure the quality of products involving only nominal-the-best (NTB) characteristics. He illustrates how the constants in Q can be determined for problems with two characteristics.

The same loss function can be extended to measure smaller-the-better (STB) and larger-the-better (LTB) characteristics. For a STB characteristic, we only need to replace the fixed target t_i by zero. For a LTB characteristic, we only need to consider the reciprocal of the response, $1/Y_i$, and treat it as a STB characteristic. Ng and Tsui (1992) illustrated this extension for the single characteristic situation. Although it is easy to extend the loss function to cover the STB and LTB characteristics, it is very hard to derive the two-step procedure to minimize the average loss as it is practically impossible to adjust the mean of the STB or LTB characteristic to its target. Section 3 will address this in detail.

It can be shown that, if $\mathbf{Y}$ follows a multivariate normal distribution with mean vector $\mu(\mathbf{C})$ and covariance matrix $\Sigma(\mathbf{C})$, the average (expected) loss can be written as:

$$\begin{aligned} R(\mathbf{C}) &= E_{N,\epsilon}(\mathbf{Y}-\tau)^T Q(\mathbf{Y}-\tau) \\ &= trace[Q\Sigma(\mathbf{C})] + [\mu(\mathbf{C})-\tau]^T Q[\mu(\mathbf{C})-\tau]. \end{aligned} \tag{2.2}$$

The simplest approach to solve the robust design problem is to apply algorithms to directly minimize the average loss function in (2.2). In practice, as the mean vector and covariance matrix are unknown, they can be estimated by the sample mean vector and sample covariance matrix or a fitted model based on a sample of observations of the multivariate responses.

Tsui and Liu (1990) illustrate this approach in the optimization of an IC process with fifteen responses. They first fitted a regression model to each of the responses, then approximated the multivariate quality loss in (2.2) using the coefficients of the fitted model. They applied the random search algorithm to directly minimize the approximated loss. The application resulted in a 30% reduction of variation and 10% increase of yield. Pignatiello (1993) discusses other approaches for direct minimization of the average loss.

The major disadvantage of the direct minimization approach is that the dimension of the optimization problem can be quite large and it may take a long time to locate the optimum. This may cause a long development time during the product realization process. In addition, if there are changes on the targets for some characteristics, re-optimization will be required and thus will further increase the development time. As explained in Leon et al. (1987), a two-step procedure originally proposed by Taguchi (1986) will rectify these problems.

3 Two-Step Optimization

For problems with a single characteristic, Taguchi (1986) proposes to minimize the average loss by two steps: first minimize an appropriate variability measure then move the mean response to target. Box (1988), Nair and Pregibon (1987), and Tsui (1988) discuss the appropriate two-step procedures under additive and multiplicative models and propose data analytic methods for identifying the corresponding variability measure for these problems. For problems with multiple characteristics, it is also possible to minimize the average loss with a two-step procedure. Below we will derive the appropriate procedures under additive and non-additive models.

3.1 Additive Models

Pignatiello (1993) proposes a two-step procedure to minimize the average loss in (2.2) as follows. Suppose the control factor vector can be decomposed into two sub-vectors, $\mathbf{C}^T = (\mathbf{C_1}^T, \mathbf{C_2}^T)$, such that the covariance matrix of $\mathbf{Y}$ is a function of only $\mathbf{C_2}$, i.e., $\Sigma = \Sigma(\mathbf{C_2})$. Then the average loss can be minimized by the following two-step procedure: (i) find values of $\mathbf{C_2}$ that minimize $trace[Q\Sigma(\mathbf{C_2})]$, say $\mathbf{C}_2^*$; (ii) at the values of $\mathbf{C}_2^*$, find values of $\mathbf{C_1}$ that minimize the off-target component in (2.2), $[\mu(\mathbf{C}) - \tau]^T Q[\mu(\mathbf{C}) - \tau]$.

According to Leon and Wu (1992) and Tsui (1996a), a necessary condition for the two-step procedure to minimize the average loss is that the objective function of the second step needs to be minimized to its absolute minimum within the feasible region. Therefore, the procedure proposed in Pignatiello (1993) does not necessary lead to the true optimum unless the off-target component in (2.2) can be reduced to its absolute minimum.

To satisfy the necessary condition described above, a natural scenario is that the off-target component can be reduced to zero (absolute minimum) by bringing all response means to targets. Suppose there exist individual adjustment factors that can be used to shift all means to targets. Then the two-step procedure described above will lead to the true optimum.

In practice, it is not always possible to shift the adjustment factors to bring all response means to targets. A common situation happens when we can only shift the means of some responses, say $\mathbf{Y_1} = (Y_1, ..., Y_{K_1})^T$, to targets while the means of other responses, say $\mathbf{Y_2} = (Y_{K_1+1}, ..., Y_K)^T$, cannot be shifted to targets. This often happens when the response is the STB or LTB characteristic, where the target is zero or infinity. As discussed in Tsui and Li (1994), although there exist adjustment factors that are significant for mean but insignificant for variance, it is not possible to shift the mean to target within the feasible range for these characteristics. Below we derive a two-step procedure that minimizes the average multivariate loss involving the NTB, STB, and LTB characteristics. Note that the situation of having only the NTB characteristics is a special case of our problem.

For $\mathbf{Y}^T = (\mathbf{Y_1}^T, \mathbf{Y_2}^T)$, we denote the corresponding components of the mean vector and target vector as $\mu^T = (\mu_1^T, \mu_2^T)$ and $\tau^T = (\tau_1^T, \tau_2^T)$; and the corresponding components of the Q matrix in (2.2) as Q_{11}, Q_{12}, Q_{21}, and Q_{22}. Then the average loss can be written as:

$$\begin{aligned} R(\mathbf{C}) &= trace[Q\Sigma(\mathbf{C})] + [\mu_1(\mathbf{C}) - \tau_1]^T Q_{11}[\mu_1(\mathbf{C}) - \tau_1] + 2[\mu_1(\mathbf{C}) - \tau_1]^T \\ &\quad Q_{12}[\mu_2(\mathbf{C}) - \tau_2] + [\mu_2(\mathbf{C}) - \tau_2]^T Q_{22}[\mu_2(\mathbf{C}) - \tau_2] \\ &= trace[Q\Sigma(\mathbf{C})] + OT_1(\mathbf{C}) + OT_{12}(\mathbf{C}) + OT_2(\mathbf{C}). \end{aligned} \tag{3.1}$$

Suppose that the covariance matrix of $\mathbf{Y}$ and the third off-target component are functions of the non-adjustment factors $\mathbf{C_2}$ only, i.e., $\Sigma = \Sigma(\mathbf{C_2})$ and $OT_2 = OT_2(\mathbf{C_2})$, and that the adjustment factors $\mathbf{C_1}$ can be used to shift the mean vector μ_1 to its target τ_1. As both OT_1 and OT_{12} drop to zero when $\mu_1 = \tau_1$, it can be easily shown that the average loss in (3.1) can be minimized by the following two-step procedure: (i) find values of $\mathbf{C_2}$ that minimize $trace[Q\Sigma(\mathbf{C_2})] + OT_2(\mathbf{C_2})$, say $\mathbf{C_2^*}$; (ii) at the values of $\mathbf{C_2^*}$, find values of $\mathbf{C_1}$ that shift the mean vector μ_1 to its target τ_1.

Note that the assumptions above are parallel to the assumptions of the single characteristic problem under an additive model. The two-step procedure above is only appropriate when the responses follow an additive model. As explained in Leon et al. (1987) for the single characteristic problem, the two step procedure under a multiplicative model will be different. Below we derive the two-step procedures for the multiple characteristic problem under a general class of models including the multiplicative model.

3.2 Constrained Minimization with Non-additive Models

Since there exist adjustment factors that can be used to shift the first K_1 means without affecting the variance, there are two different robust design objectives that are commonly assumed in the literature: constrained and unconstrained minimization of the average loss. As explained in Leon and Wu (1992) and Tsui (1988), the solutions of these two optimization problems are the same under an additive model but different under a non-additive model. Below we first derive the two-step procedure for the constrained minimization problem under a non-additive model. Our objective is to minimize the average loss under the constraint of bringing the first K_1 means to targets.

Suppose further that for each of the first K_1 responses (whose means can be adjusted to targets), there exists an independent adjustment factor a_i that can be used to shift the ith mean m_i to its target t_i without affecting all other means, i.e., $m_i = m_i(a_i, \mathbf{C_2})$, for $i = 1, ..., K_1$, and $m_i = m_i(\mathbf{C_2})$, for $i = K_1 + 1, ..., K$, and $\mathbf{C}^T = (a_1, ..., a_{K_1}, \mathbf{C_2}^T)$. For simplification purpose, we assume that the K responses are uncorrelated with each other, i.e., Σ is a diagonal matrix with elements σ_i^2, $i = 1, ..., K$. Instead of assuming that the variance is a function of $\mathbf{C_2}$ only as in the additive model case, we follow the more general assumptions for non-additive models described in Box (1988), Nair and Pregibon (1987), and Tsui (1988). We assume that the variance is the product of a power function of mean and a function of $\mathbf{C_2}$ only, i.e.,

$$\sigma_i^2(\mathbf{C}) = \mu_i^{\lambda_i}(\mathbf{C}) P_i(\mathbf{C_2}), \ \lambda_i > 0, \ and \ i = 1, ..., K, \tag{3.2}$$

where P_i is called the performance measure independent of adjustment (PERMIA) in Leon et al. (1987).

Let q_{ij} denotes the $(ij)th$ element of the Q matrix. It follows that the average loss can be expressed as:

$$\begin{aligned} R(\mathbf{C}) &= \sum_{i=1}^{K} q_{ii} m_i^{\lambda_i}(\mathbf{C}) P_i(\mathbf{C_2}) + \sum_{i=1}^{K_1}\sum_{j=1}^{K_1} q_{ij}[m_i(a_i, \mathbf{C_2}) - t_i][m_j(a_j, \mathbf{C_2}) - t_j] \\ &\quad + \sum_{i=1}^{K_1} \sum_{j=K_1+1}^{K} q_{ij}[m_i(a_i, \mathbf{C_2}) - t_i][m_j(\mathbf{C_2}) - t_j] \\ &\quad + \sum_{i=K_1+1}^{K} \sum_{j=K_1+1}^{K} q_{ij}[m_i(\mathbf{C_2}) - t_i][m_j(\mathbf{C_2}) - t_j]. \end{aligned} \tag{3.3}$$

Since there exist adjustment factors for the first K_1 means, under the constraint of shifting the first K_1 means to targets, the second and third terms of the average loss in (3.3) will drop to zero after the adjustment. It follows that the average loss after adjustment (R_d) can be expressed as a function of $\mathbf{C_2}$ only:

$$\begin{aligned} R_d(\mathbf{C_2}) &= \sum_{i=1}^{K_1} q_{ii} t_i^{\lambda_i} P_i(\mathbf{C_2}) + \sum_{i=K_1+1}^{K} q_{ii} m_i^{\lambda_i}(\mathbf{C_2}) P_i(\mathbf{C_2}) \\ &\quad + \sum_{i=K_1+1}^{K} \sum_{j=K_1+1}^{K} q_{ij}[m_i(\mathbf{C_2}) - t_i][m_j(\mathbf{C_2}) - t_j]. \end{aligned} \tag{3.4}$$

Therefore, the minimization of the average loss in (3.3) under the constraint of shifting the first K_1 means to targets can be solved by the following two-step procedure: (i) find values of $\mathbf{C_2}$ that minimize the average loss after adjustment $R_d(\mathbf{C_2})$ in (3.4), say $\mathbf{C_2^*}$; (ii) at the values of $\mathbf{C_2^*}$, for each of the first K_1 means, find value of a_i that shifts the mean m_i to its target t_i.

When there doesn't exist independent adjustment factor, it may not be possible to shift the first K_1 means to target. This happens when the same adjustment factor affects more than one response means. Under this situation, tradeoff among different responses needs to be made. Elsayed and Chen (1993) propose a strategy to solve this problem under the assumption that the possible choices of non-adjustment factors are restricted into the discrete testing values in the experiment.

4 A Polysilicon Deposition Example

Deposition of polysilicon on top of wafers is one of the middle steps in integrated circuit fabrication. The polysilicon layer is very important for defining the gate electrodes for the transistors. The process we consider is from a class project conducted by Peter Hey (an engineer) and Paul Sherry (a facilitator) during a Robust Design workshop offered by the Quality Assurance Center of AT&T Bell Laboratories (1984). The description of the process, planning of the experiment, and analysis of the data were documented as a case study and included as teaching material in the workshop.

The process is briefly described as follows. A reduced-pressure reactor is used to deposit polysilicon on a wafer. Silane and nitrogen gas are pumped in to the reactor and the temperature and pressure are controlled. Fifty wafers mounted on two quartz carriers are placed inside the reactor and a polysilicon layer is deposited on top of the wafer. Two major problems existed at the time of the study and they were non-uniform thickness of the polysilicon layer and presence of surface defects (called pepper). The objective of the experiment is to minimize the thickness variation while maintaining the thickness target at 3600 angstroms and to reduce the pepper count.

There are six control factors of interest in the experiment: deposition temperature (A), deposition pressure (B), nitrogen flow (C), silane flow (C), settling time (E), cleaning method (F). These control factors are studied by an eighteen run orthogonal array. At each of the control factor combination, a nine run noise plan is used to study the variation effect. Thickness is a nominal-the-best characteristic and thus it is possible to identify an adjustment factor to shift the mean to target. Pepper count is a smaller-the-better characteristic and thus it is not possible to shift the mean to zero target even if there exists an adjustment factor. Since it is believed that the variances of both thickness and pepper count increase as their means increase, the non-additive model in (3.2) with $\lambda_i = 2$ is used for both two responses. As pointed out in Section 3, it is possible to have multiplicative model for smaller-the-better responses although the mean cannot be shifted to zero target. Below we will apply the two-step procedure developed in Section 3 to minimize the average loss of the two responses under the constraint of shifting the thickness mean to target.

For simplification, we assume that thickness and pepper count are uncorrelated and carry the equal weight, i.e., the average loss function is given in (3.3) with $K_1 = 1$, $K = 2$, $q_{11} = q_{22}$, $q_{12} \neq 0$, and $\lambda_i = 2$ for $i = 1, 2$. Following the two-step procedure described in Section 3.2, we will first minimize the average loss after adjustment in (3.4), which reduces to:

$$R_d(\mathbf{C_2}) = q_{11}\{t_1^2 P_1(\mathbf{C_2}) + m_2^2(\mathbf{C_2})[P_2(\mathbf{C_2}) + 1]\}, \tag{4.1}$$

then shift the thickness mean to target ($t_1 = 3600$). As $\lambda_i = 2$ for $i = 1, 2$, the mean and the PERMIA functions $P_i = \sigma_i^2/\mu_i^2$ of thickness and pepper count were computed for each control factor combination. As it is believed that there is no significant interaction between control factors, main effect analysis is applied to each of the following measures: thickness mean, log thickness PERMIA, log pepper mean, and log pepper PERMIA. (The logarithm is used as the PERMIA and pepper mean are non-negative and close to zero.) Analysis of variance method is applied to identify the significant main effects for each of the measures and the fitted models are given below. For illustration purpose, for each of the significant main effects, it is assumed that the relationship is a piecewise linear for quantitative factors ($A, B, ..., E$) and discrete for categorical factor (F).

Based on the fitted models, the average loss can be minimized by following the two-step procedure described in Section 3.2. As thinkness is the nominal-the-best response, we need to identify the adjustment factor for thickness and use it to shift the thickness mean to target (3600) at the second step. As factor D is significant for thickness mean but insignificant for thickness PERMIA, pepper mean, and pepper PERMIA, it is identified as the mean adjustment factor for thickness. Therefore, following the first step of the two-step procedure, we first minimize the average loss after adjustment R_d in (4.1) with respect to factors A, B, C, E, and F. According to the fitted models, the optimal values within the ranges are found to be: $x_A = -1$, $x_B = -1$, $x_C = -1$, $x_E = 0$, and $x_F = 1$. At the second step, at the optimal values of these five factors, it is found that $x_D = 2.62$ would shift the thickness mean to target. As this value is out of the range of x_D, an alternative is to change x_C from -1 to 1. As shown in analysis of variance results, an increase of x_C does not significantly increase the pepper PERMIA measure but significantly increases the thickness mean. By changing x_C from -1 to 1, it is found that $x_D = -0.45$ would shift the thickness mean to target and the value is within the range. It follows that the average loss after adjustment in (4.1) for these two solutions are:

$$\begin{aligned} R_d &= q_{11}\{3600^2 * 0.00019 + 0.72^2 * (0.933 + 1) = 2463.4q_{11}, \; \textit{and} \\ R_d' &= q_{11}\{3600^2 * 0.00019 + 0.65^2 * (0.933 + 1) = 2463.2q_{11}, \end{aligned}$$

which are practically the same. Therefore, the optimal solution is chosen to be $x_A = -1$, $x_B = -1$, $x_C = 1$, $x_D = -0.45$, $x_E = 0$, and $x_F = 1$.

5 Discussion

In this paper we extended the multivariate loss considered in Pignatiello (1993) to include the smaller- and larger-the-better characteristics. We developed appropriate two-step procedures that minimize the average multivariate loss. Similar to the advantages in single characteristic problems, the proposed two-step procedures substantially reduce the dimension of the design optimization problem and allow for future changes of design requirements without re-optimization.

In terms of statistical modeling, the proposed two-step procedures are special cases of the loss model approach: first compute estimates of loss measures, then determine the "optimal" factor settings by fitting a model to these loss estimates. An alternative modeling approach is to first model the observed response, and then determine the "optimal" factor settings from the fitted response model. This approach is called the response model approach and was first proposed in Welch et al. (1990). As illustrated in Tsui (1996b), the loss model approach may result in efficiency loss, information loss, and unnecessarily expensive experiments. The response model approach does not suffer these problems but generally requires more work and depends more critically on the adequacy of the fitted model. Recognizing the advantages of the response model approach, Shoemaker, Tsui, and Wu (1991) develop analysis techniques for examining individual control-by-noise interaction plots to reveal control factor levels that dampen the effects of individual noise factors.

Acknowledgements
This research was supported by the National Science Foundation under Grants DDM-9257918 and DDM-9312698.

REFERENCES

1. AT&T, Robust Design Through Experimentation Workshop Notes, Quality Assurance Center, AT&T Bell Laboratories (1984).

2. Box, G.E.P., "Signal to Noise Ratios, Performance Criteria and Transformation", *Technometrics* (1988) Vol. 30, 1-31.

3. Byrne, D.M. and Taguchi, S., "The Taguchi Approach to Parameter Design", *Quality Progress* (1987) 20, December, 19-26.

4. Elsayed, E.A. and Chen, A., "Optimal Levels of Process Parameters for Products with Multiple Characteristics", *International Journal of Production Research* (1993) Vol. 31, No.5, 1117-1132.

5. Leon, R.V., Shoemaker, A.C., and Kackar, R.N., "Performance Measure Independent of Adjustment: An Explanation and Extension of Taguchi's Signal to Noise Ratio", *Technometrics* (1987) Vol. 29, No. 3, 253-285.

6. Leon R.V. and Wu, C.F.J., "A Theory of Performance Measures in Parameter Design", *Statistics Sinica* (1992) 2, 335-358.

7. Logothetis, N. and Haigh, A., "Characterizing and Optimizing Multiresponse Processes by the Taguchi Method", *Quality and Reliability Engineering International* (1988) 4, 159-169.

8. Nair, V. N., "Taguchi's Parameter Design: A Panel Discussion", *Technometrics* (1992) vol. 34, 127-161.

9. Nair, V.N. and Pregibon, D., "A Data Analysis Strategy for Quality Engineering Experiments", *AT&T Technical Journal* (1987) 73-84.

10. Ng, K.K. and Tsui K.-L., "Expressing Variability and Yield with a Focus on the Customer", *Quality Engineering* (1992) Vol.5, No.2, 255-268.

11. Pignatiello, J.J., "Strategies for Robust Multiresponse Quality Engineering", *IIE Transactions* (1993) Vol. 25, No. 3, 5-25.

12. Shoemaker, A.C., Tsui, K.-L., and Wu, C.F.J., "Economical Experimentation Methods for Robust Parameter Design", *Technometrics* (1991) Vol. 33, 415-427.

13. Taguchi, G., *Introduction to Quality Engineering: Designing Quality into Products and Processes*, Asian Productivity Organization, Tokyo, Japan (1986).

14. Tsui, K.-L., "An Empirical Two-Step Procedure for Robust Design", unpublished manuscript (1988).

15. Tsui, K.-L., "A Multi-Step Analysis Procedure for Robust Design", *Statistica Sinica* (1996a) Vol. 6, No.3, 631-648.

16. Tsui, K.-L., "A Critical Look at Taguchi's Modeling Approach for Robust Design", *Journal of Applied Statistics* (1996b) 81-95.

17. Tsui, K.-L. and Liu, L., "Robust Design of a High Dimensional IC Process Through Simulation Experiments", *Proceedings of the SEMICON/KOREA 90 Technical Symposium* (1990).

18. Tsui, K.-L. and Li, A., "Analysis of Smaller- and Larger-the-Better Robust Design Experiments", *International Journal of Industrial Engineering* (1994) 193-202.

19. Welch, W.J., Yu, T.K., Kang, S.M. and Sacks, J., "Computer Experiments for Quality Control by Parameter Design", the *Journal of Quality Technology* (1990) Vol. 22, 15-22.

Computer Aided Design of np Quality Control Chart

Zhang Wu and Xiaolan Zhang

Nanyang Technological University, Singapore 639798

Abstract. This paper presents an optimization algorithm for the design of the np control chart. The objective is to maximize the detecting effectiveness of the np chart, on the condition that the false alarm rate is held at the specified level. The detecting effectiveness is measured by the out-of-control *ATS* (Average Time to Signal). A summation formula for calculating *ATS* has been derived. This formula is more accurate than that used by the economic control chart models. In addition, it allows the fraction nonconforming shift to occur during the taking of a sample. A fractional control limit scheme is also proposed to facilitate the comparative studies of the control charts. The results of a case study have shown that this optimization algorithm does have the potential of improving the detecting effectiveness of the np chart.

Introduction

np control chart has been widely used in the industries to monitor the number d of nonconforming units per sample in a process. A nonconforming unit is a product which fails to meet at least one specified requirement.

In the design of a np chart, four parameters have to be determined, namely, the sample size n, the sampling interval h, the lower control limit LCL and the upper control limit UCL. During the process control, for every h output units, n of them are taken as a sample and examined. If the number d of the nonconforming units found in a sample falls inside the range bracketed by LCL and UCL, the process is thought under control, otherwise out-of-control is signalled. It is common to base a np chart on 100% inspection of all process output. In this case, the sampling interval h is equal to the sample size n. Furthermore, given that the type I error α (or the false alarm rate) has been specified, the control limits are also dependent on the sample size n. Thus, under the 100% inspection, the only independent parameter for the np chart is the sample size n. Our study will focus on this situation.

The detecting effectiveness of a control chart is usually measured by the Average Time to Signal (*ATS*). For the np chart, *ATS* is defined as the average number of inspected units required to signal a fallout (out-of-control) after it occurred. The smaller the *ATS*, the earlier the fallout is signalled, and the more effective the np chart is. The *ATS* of a np chart can be calculated as follows:

$$ATS = h \cdot ARL \qquad (1)$$

Where, *ARL* is the Average Run Length, i.e., the average number of samples required to signal a fallout. For a np chart, if a small sample size n is used, h is also small (h is equal to n in 100% inspection), but *ARL* is large. Conversely, if a large n is used, h is also large, but *ARL* is small. So, the change of n in either direction will cause h and

ARL to compete with each other, because each of them is a factor in equation (1). Logically, there must be an optimum value of n, which will make *ATS* minimum.

Traditionally, the sample size n is rarely determined analytically. Usually, the QA (Quality Assurance) engineer selects a n value based on the rational subgrouping, or the production shift. That is, the sample size is set as the number of units produced in a production shift. However, a sample size determined by this way may be too small or too large, and is unlikely to result in an *ATS* close to the minimum.

Researchers have proposed several alternative approaches for determining the sample size. Three of them are briefly introduced as below:

Least-One-Nonconforming Method [1]

The sample size is chosen so that the probability of finding at least one nonconforming unit, when the process is under control, is very high (say, ≥ 95%).

50%-Power Method [2]

The sample size is set large enough so that the chance of detecting a fraction nonconforming shift in a single sample is approximately equal to 50%.

Economic Design Method

This method is to optimize the sample size, as well as other parameters, so that minimum average hourly cost can be achieved. The economic design is first applied to the $\overline{X}$ chart [3], and later introduced to the np chart [4][5]. However, as attractive as the economic design might be, it has a few weak points:

(1) To minimize the average hourly cost in the economic design, the QA engineer encounters the difficulty of specifying many input costs which are prone to change and sometimes unavailable.

(2) The economic design uses following formula to calculate the average cost per unit time:

$$E(A) = \frac{E(C)}{E(T)} \tag{2}$$

Where, E(.) is the expect operator, and

A	is the cost per unit time.
C	is the total cost incurred during a cycle.
T	is the length of the cycle.

It is well known that the expected value of a ratio is not equal to the ratio of the expected values (even for independent random variables). So, inherent inaccuracy has been introduced by using equation (2).

(3) The economic np chart cannot handle the fraction nonconforming p shift happening during a sample taking. It is assumed that the p shift always takes place immediately after a sample taking. This assumption may incur a serious error when *ARL* is short (or p shift is large).

Recently, adaptive control charts have been explored intensively by many researchers [6][7][8]. When running an adaptive chart, the parameters of the chart are updated on-line. The objective is to minimize *ATS* in bid to achieve the greatest detecting effectiveness.

In this paper, an optimization model for np chart is proposed. As in the adaptive control charts, the objective function to be minimized is the out-of-control *ATS*. This objective is quite tangible. It removes the burden of specifying the input costs as required in the economic models. The minimization of *ATS* immediately leads to the reduction of the nonconforming units, and therefore, directly benefits the industries.

The last two weak points of the economic models have also been overcome by the optimization model. It will be detailed shortly.

For simplicity, the current optimization model only handles the positive p shift (increase of fraction nonconforming), and therefore, only the upper control limit *UCL* is active.

Optimization Formulation

To design a np control chart, the designer has to provide following specifications:

p_0	the value of the fraction nonconforming when the process is in control.
p_d	the value of the fraction nonconforming when the process is considered out of control.
ATS_0	the average time to signal when the process is in control.
λ	the intensity of the fraction nonconforming shift.

p_0 is usually estimated from the data observed in the pilot run. p_d is the barely tolerated fraction nonconforming that degrades the product quality level badly. The value of p_d is determined by the QA engineer, based on the consideration of many factors, such as the quality, manufacturing, cost and the customers' requirements in acceptance sampling.

ATS_0 is also determined by the QA engineer with regard to the false alarm rate. If the cost of handling the false alarms is high, a larger ATS_0 should be used in order to reduce the false alarm rate. But it may affect the effectiveness of the control chart in the meantime. When the effectiveness of different control schemes (e.g., np charts with different sample sizes) is to be compared, the ATS_0 values of the control schemes must be made identical, otherwise the comparison is unfair and the results are meaningless.

The reciprocal of λ is the mean time interval that the process remains in control.

The optimization model is a single variable optimization without explicit constraint function.

$$\text{Objective:} \qquad ATS = \text{minimum}$$
$$\text{Design variable:} \qquad n \qquad (3)$$

Only the sample size n is treated as the independent design variable. Other control chart parameters (i.e., sampling interval and upper control limit) are dependent on n. The optimization formulation is to express ATS in terms of n.

From the specified in-control ATS_0, we can find the in-control average run length ARL_0 which is inversely proportional to the sample size n (note, the sampling interval h is equal to n in 100% inspection).

$$ARL_0 = ATS_0 / h = ATS_0 / n \tag{4}$$

The type I error α is the reciprocal of ARL_0.

$$\alpha = 1 / ARL_0 = n / ATS_0 \tag{5}$$

Suppose the random number d of the nonconforming units found in a sample follows a binomial distribution. The upper control limit UCL is obtained by solving following equation:

$$1-\alpha = \sum_{i=0}^{UCL} C_i^n p_0^i (1-p_0)^{n-i} \tag{6}$$

Where, the in-control fraction nonconforming p_0 is specified by the user.

Then, from UCL, the type II error β under p_d and the out-of-control average run length ARL can be calculated.

$$\beta = \sum_{i=0}^{UCL} C_i^n p_d^i (1-p_d)^{n-i}$$
$$ARL = 1/(1-\beta) \tag{7}$$

Next, let $\Pr(k)$ be the probability that the fraction nonconforming p shift occurs in the time interval between the kth and the $(k+1)$ units inside a sample ($0 \le k \le n-1$), and $ats(k)$ be the average time to signal if the p shift does occur at this interval. The objective function ATS is, therefore, defined as the mean value of $ats(k)$.

$$ATS = \sum_{k=0}^{n-1} ats(k)\Pr(k) \tag{8}$$

As in the economic control chart designs, the p shift is assumed to occur according to a Poisson process with intensity λ. It leads to [3],

$$\Pr(k) = \frac{\lambda e^{-\lambda k}}{1-e^{-\lambda n}} \tag{9}$$

Suppose $P_1(k)$ is the probability of the event E that the p shift occurs in the time interval between the kth and the $(k+1)$ units and it has been signalled at the first sampling after it occurs. Then, $(1 - P_1(k))$ is the complementary probability that the p shift is detected in a following sample, rather than the first one. If the event E is true,

$$ats(k) = h - k = n - k \tag{10}$$

Otherwise,

$$ats(k) = ARL \cdot h + (h-k) = ARL \cdot n + (n-k) \tag{11}$$

So,

$$\begin{aligned} ats(k) &= P_1(k)(n-k) + [1-P_1(k)][ARL \cdot n + (n-k)] \\ &= (n-k) + [1-P_1(k)] \cdot ARL \cdot n \end{aligned} \tag{12}$$

Event E is true when the number d of the nonconforming units found in the first sample is greater than UCL. It will arise under either of the following two circumstances:

(1) i nonconforming units have been found in the first k units and more than (UCL - i) nonconforming units have been found in the remaining (n - k) units. So, the total number of the nonconforming units found in the whole sample is greater than UCL.

(2) More than UCL nonconforming units have been found in the first k units (when $k > UCL$). In this case, the nonconforming units found in the remaining portion of the sample, no matter how many, will make no difference.

Above discussion leads to

$$P_1(k) = \begin{cases} \sum_{i=0}^{k} \Pr(x_1 = i)\Pr(x_2 > UCL - i) & if \quad k \leq UCL \\ \sum_{i=0}^{UCL} \Pr(x_1 = i)\Pr(x_2 > UCL - i) + \sum_{i=UCL+1}^{k} \Pr(x_1 = i) & if \quad k > UCL \end{cases} \tag{13}$$

$$\Pr(x_1 = i) = C_i^k p_0^i (1 - p_0)^{k-i}$$

$$\Pr(x_2 > UCL - i) = 1 - \Pr(x_2 \leq UCL - i) = 1 - \sum_{j=0}^{UCL-i} C_j^{n-k} p_d^j (1 - p_d)^{(n-k)-j} \tag{14}$$

Where, $\Pr(x_1 = i)$ is the probability that the number x_1 of the nonconforming units found in the first k units of the sample (in which the p shift happens) is equal to i, and $\Pr(x_2 > UCL-i)$ is the probability that the number x_2 of the nonconforming units found in the remaining (n - k) units of the same sample is larger than (UCL-i). It is noteworthy that, $\Pr(x_1 = i)$ is obtained under $p = p_0$ (before the p shift), and $\Pr(x_2 > UCL-i)$ is calculated under $p = p_d$ (after the p shift).

To this point, we have completed the derivation of the objective function ATS from the design variable n. To sum up, for a given value of n, the objective function ATS is computed by following steps:

(1) Compute the type I error α by equation (5).
(2) Obtain the upper control limit UCL by solving equation (6).
(3) Calculate the out-of-control average run length ARL by equation (7).
(4) For each k value ($0 \leq k \leq n$-1), calculate $\Pr(k)$ by equation (9) and $ats(k)$ by equations (12) to (14).
(5) Finally, compute the objective function ATS by equation (8).

In the next section, we will discuss a fractional control limit scheme, by which, ATS is generally an unimodal function (function with an unique minimum) of the sample size n (see the solid curve in Figure 1). For this unimodal, single variable and nonconstrained objective function (equation (3)), the search for the optimum n is straightforward. In this study, we use the golden search method [9] to perform the optimization search.

It is noted that equation (8) evaluates *ATS* in accordance with the very definition of mean, and therefore removes the inherent inaccuracy involved in equation (2). Furthermore, equation (8) allows the p shift to take place at any moment during the taking of a sample.

Fractional Control Limit

For the attribute control chart, the exact upper control limit obtained by solving the type I error equation (6) is a fractional one which can be represented as

$$UCL = I + f \qquad 0 \le I < n, \quad 0 \le f < 1 \tag{15}$$

Where, I and f are called the integral and fractional parts of *UCL*, respectively. The fractional *UCL* ensures the in-control ATS_0 (or the false alarm rate) equal to the specified value.

However, in process control practice, rounded *UCL* is customarily used. If I or less units are found nonconforming in a sample, the process is thought in control. On the other hand, if $(I+1)$ or more nonconforming units are found, out-of-control is signalled. Since this round-off procedure is equivalent to rounding *UCL* down to the smaller adjacent integer, it tightens the upper control limit.

In a comparative study, if the rounded control limits are used, they will mask the relative goodness in performance of the different control charts. Since the rounded *UCL* has been tightened, the resultant *ATS* values of all the charts will be reduced. But, different charts experience different degrees of reduction in *ATS*. The in-control ATS_0 values also become smaller, more or less, than the specified value. Suppose the fractional upper control limits of two np charts are 4.01 and 4.99 respectively. While the round-off procedure may have little impact on the 4.01 control limit of the first chart, the round-off of 4.99 to 4 will exaggerate the detecting power of the second chart irrationally, and also increase the false alarm rate to an intolerable degree. This problem is particularly serious when the upper control limit *UCL* is relatively small.

As mentioned earlier, it is vital to make all the control charts involved in a comparative study generate same ATS_0, which should be equal to the specification. Since the rounded control limit cannot make it, the fractional *UCL* scheme is used in this study. The detailed procedure for designing the fractional control limit can be found in a technical report [10].

Figure 1 displays two *ATS-n* curves resulting from the case study to be discussed in the next section. The solid curve is obtained by using the fractional *UCL* and the dashed one by using the rounded *UCL*. A general impression is that the characteristic of the solid curve is much more desirable than that of the dashed curve. While the solid curve is unimodal and smooth, the dashed curve is seriously distorted by the round-off procedure and exhibits a saw-tooth shape. Since the round-off procedure has tightened the *UCL*, the dashed curve is always below the solid one. That is, the rounded *UCL* generates a smaller out-of-control *ATS*. However, it is obtained at the expense of smaller in-control ATS_0 (greater false alarm rate).

The fractional control limits can not only be used in the comparative studies, but can also be applied to the real process control with the help of a Random Number Table (uniform distribution between 0 and 1), where the specified ATS_0 has to be maintained carefully.

Case Study

A np chart is being designed to monitor the number of nonconforming units in an electronic component production line. Following parameters have been specified by the QA engineer:

p_0	= 0.01,	the fraction nonconforming when the process is in control.
p_d	= 0.05,	the fraction nonconforming when the process is considered out of control.
ATS_0	= 50000,	the in-control average time to signal.

Since the intensity λ of the fraction nonconforming shift has not been specified, the p shift is assumed to occur randomly, following an uniform distribution.

The performance of seven different np chart designs is compared, and the results are enumerated at Table 1 (The results have been verified by Monte Carlo simulation).

The first design (exhaustive optimization) finds the optimum sample size by evaluating and comparing the *ATS* values corresponding to n from 2 to 300 (The results are illustrated by the solid curve in Figure 1). The second design (golden optimization) searches the optimum n using the golden search method. These two optimization designs result in exactly the same optimum sample size, control limit, ATS_0 and *ATS*. However, since the golden search method is more ingenious, it obtains the optimum solution much faster, just in a second of CPU time.

The last three designs in Table 1 come in the light of the recommendation from Duncan [2], who concluded that: "for a chart designed to pick up a shift in the process average from 0.01 to 0.06 sample size of about 30 to 70 were optimum". So, the lower bound, upper bound and central point of his range are used.

From Table 1, it can be seen that all of the seven designs result in identical ATS_0, which is equal to the user's specification. It means that the requirement on the false alarm rate is met by all. Consequently, the comparative studies on the detecting effectiveness between the different designs are fair and meaningful.

When looking into the out-of-control *ATS*, it is found that, the optimization design achieves the minimum *ATS* value of 214. It means that the optimum np chart can detect the p shift faster than any other np charts. The *ATS* produced by the least-one-nonconforming design is as large as 136% of the minimum *ATS*. The *ATS* values produced by the recommended sample sizes are also quite large. Especially, when $n = 30$ is used, the *ATS* value is nearly doubled compared to the optimization design. Last, the *ATS* value obtained by the 50%-power design looks quite good in this particular case, even though, it is still larger than that of the optimization design.

Table 1 Designs of the np Charts

No	design method	sample size n	upper control limit CUL $(I+f)$	in-control ATS_0	out-of-control ATS
1	exhaustive optimization	137	4 + 0.996	50000	214
2	golden optimization	137	4 + 0.996	50000	214
3	least-one-nonconforming	299	7 + 0.683	50000	294
4	50%-power	114	4 + 0.752	50000	222
5	recommended sample size I	30	2 + 0.878	50000	399
6	recommended sample size II	50	3 + 0.411	50000	281
7	recommended sample size III	70	3 + 0.853	50000	244

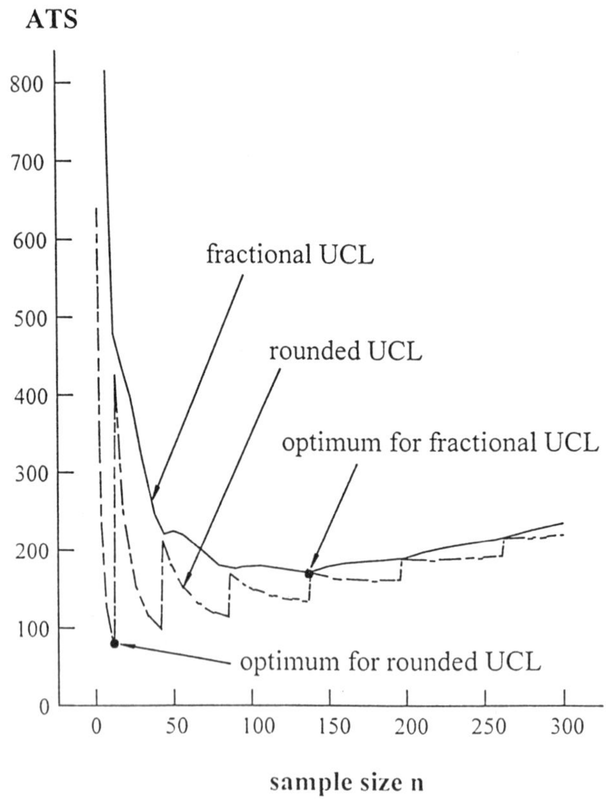

Fig. 1. Out-of-control ATS versus sample size n

Conclusions and Discussions

The aim of this optimization approach is to maximize the detecting effectiveness of the np control charts. It is realized by minimizing the out-of-control *ATS*, using the in-control ATS_0 as an implicit constraint. The sole independent design variable is the sample size *n*. The very immediate effect of this optimization is the reduction of the nonconforming units. This characteristic is definitely useful to the industries. The whole algorithm can be easily implemented in a computer program, and then used in routine designs of the np charts.

The resultant *ATS* of the optimization design is always smaller than that generated by any other designs. It can be seen from Table 1 that fairly significant improvement in detecting effectiveness can be achieved in common situations.

The optimization design is carried out once, and the resultant np chart can be used continuously until the process conditions have changed. No additional inspection instruments and manpower are required. Only the grouping of the output units has to be adjusted according to the optimum sample size.

The optimization model has overcome the three shortcomings found in the economic model of np chart. Particularly, the formulae introduced in equations (8) to (14) remove the inaccuracy inherent in the economic model, and allow the *p* shift to take place at any moment during the sample taking.

A fractional control limit scheme is also introduced. It is not only very useful to the comparative studies of the control charts for attributes, but also can be adopted in practice.

With minor modification, the optimization design can be applied to the non-100% inspection cases.

References

1. W. B. Rice, *Control Charts in Factory Management*, John Wiley & Sons, New York (1947).
2. A. J. Duncan, *Quality Control and Industrial Statistics*, Richard, D. Irwin, Illinois (1986).
3. A. J. Duncan, The Economic Design of X-bar Chart to Maintain Current Control of a Process, *Journal of the American Statistical Association*, **51** (1956) 228-241.
4. W. K. Chiu, Economic Design of Attribute Control Charts, *Technometrics*, **17**, 1 (1975) 81-87.
5. A. J. Duncan, The Economic Design of p-Charts to Maintain Current Control of a Process: Some Numerical Results, *Technometrics*, **20**, 3 (1978) 235-243.
6. M. R. Reynolds, R. W. Amin, J. C. Arnold and J. A. Nachlas, $\overline{X}$ Charts with Variable Sampling Intervals, *Technometrics*, **30** (1988) 181-192.
7. J. J. Flaig, Adaptive Control Charts, *Statistical Process Control in Manufacturing* edited by J. B. Keats and D. C. Montgomery. Marcel Dekker, New York (1991).
8. S. S. Prabhu, D. C. Montgomery and G. C. Runger, A Combined Adaptive Sample Size and Sampling Interval $\overline{X}$ Control Scheme, *Journal of Quality Technology*, **26** (1994) 164-176.
9. J. N. Siddall, *Optimal Engineering Design: Principles and Applications*, M. Dekker, New York (1982).
10. Z. Wu and X. L. Zhang, Fractional Control Limits for np Control Chart, Technical report, Nanyang Technological University, Singapore (1996).

Control Issues

On the Inclusion of Maintenance in Manufacturing Planning and Control Systems

David Sherwin[1], Patrik Jonsson[2] & Magnus Lesshammar[3]

[1]Department of Industrial Engineering, Lund University , Sweden.
[2]Department of Management Economics, Växjö University, Sweden.
[3]Department of Mathematics Statistics & Computer Science, Växjö University, Sweden.

Abstract. Few systems such as MRP(II), JIT, and CIM have integrated software for the maintenance of the machinery. This becomes important with efforts to eliminate redundancy and buffer stores. Unless production and maintenance planning are co-ordinated by the same computer system, incorporating a loss function for omission/delay to PM, managerial control of quality may be lost as demand increases. Only 24 of 284 Swedish firms had integrated maintenance through IT.

Introduction

Both maintenance and production planning systems now use computers to store data and to plan and control the work. The computer has made it possible to automate the planning and recording of maintenance work for later review and analysis of data. In production it is an essential adjunct to planning Just-in-time (JIT) manufacture as well as in the direct control of machinery. However, the battles between Maintenance and Production continue under the extra pressures of more complex machinery and the higher productivity required to make it economical. Complex machinery probably needs more maintenance but seldom gets it because it is being intensively used to drive the JIT-oriented Computer Integrated Manufacturing (CIM) system. CIM now encompasses three-quarters of the whole product cycle from design (CAD) through resource planning (MRP) to production activity control (PAC). By means of systems, such as Enterprise Resource Planning (ERP), CIM also integrates the other organisations in the supply chain.

However, manufacturing planning and control systems (MPCS), do not specifically allow for maintenance, and existing maintenance software does not allow for integration of maintenance work with production plans. MRP II systems are getting more and more integrated with Purchasing, Sales and Accounts, however, this presents the Maintenance manager with the problem of upsetting all those departments as well when the un-integrated preventive maintenance (PM) schedule says that machinery must be maintained, disrupting the production schedule. Clearly integration is desirable. Obvious alternatives are that only breakdown maintenance and/or weekend block preventive maintenance is done between the overhaul periods during which production workers take their annual holidays. There are, however, some less obvious possibilities which use the ability of advanced Maintenance Management and Information Systems, (MMIS), and MRP II systems to plan work very quickly for minimum time or cost. This enables them to take full advantage of maintenance opportunities arising at failures.

The aim of the paper is to examine some of these possibilities in relation to the capability of commercial software for Maintenance and MPCS and their attendant philosophies. It also analyses the extent of use of integrated MMIS approaches in the Swedish industry in general, and in two case studies. The possibilities for integrating PM in MPCS and CIM are examined through a conceptual study of previous research, industry experiences, and logical discussions. In a mail survey of maintenance managers in Swedish manufacturing firms, 284 relevant responses were received, (response rate 48%). This study records the use of integrated MMIS in various manufacturing companies. In two case studies, the objective is to analyse the use of MMIS and its links to manufacturing and production..

Maintenance Planning and Control

Four aspects of planning of preventive maintenance (PM) are emphasised.

1. Characteristics of maintenance planning systems.
2. Integration of maintenance into MPCS.
3. The role of PM in modern maintenance concepts.
4. How modelling of losses due to delay of PM can be conducted.

1. Characteristics of Maintenance Planning Systems

A large variety of maintenance management information system (MMIS) software is available on the market, and they cover a large part of the maintenance management area. Raouf et al [1] identified the following eight basic functions of modern computerised MMIS; namely maintenance reports, inventory management, preventive maintenance management, work order management, craftsmen management, quality control management, vendor evaluation, and other more general considerations. All these functions are separate and are not directly linked to manufacturing or the management information system (MIS). Most extant MMIS are software equivalents of the paper systems of the previous generation and have the following essential features:

1. *Asset Register* - Ideally to serve the needs of both engineers and accountants
2. *Maintenance Schedule* - Details work to be done, resources and periodicity.
3. *Maintenance Planning Facility* - Brings up and records PM and repairs.
4. *Asset History File* - records overhauls and failure repairs as well as PM.
5. *Mechanism for Change* - Alters schedules and reports deficiencies in design.

Some primitive systems omit items 4 and/or 5 and fail to distinguish between schedules and plans. PM achievement may be recorded under 3 or 4 above. Most current software has no modules for using the data to improve schedules. This will almost certainly be a feature of the next generation. As PM Schedules, and therefore plans, change to implement successive optimisations [2], the case for integration with production becomes stronger.

2. Co-Planning of Production & Maintenance - MMIS & MRPII

Before discussing how integration of MMIS with MPCS might be achieved first consider why it would be desirable. It would be more to the point to ask the progenitors of various MPCS schemes why the MMIS was omitted. JIT advocates use of Total Productive Maintenance (TPM), but ignores the need for assured periods of downtime to carry out PM. MRP II has the same fault. Neither contain any way of costing omission/delay to PM. MRP II and JIT schemes seemingly assume PM occurs at weekends and annual shut-downs. It is cheaper to plan PM for free time.

2.1 Managerial Considerations

There are similarities between standard MMIS planning and the business plan, production plan, master production schedule and material requirements plan of MRP II. It is possible to use an MRP II system to plan PM. Most MRP II systems are adjustable to include maintenance planning. The

MMIS planning module is usually easier to use because it was designed especially for maintenance. Given an installation with both systems available, a reasonable compromise might be to plan the jobs on the MMIS to establish time and resource requirements and then include them in the MRP II plan as simple delays. Otherwise all the resources available to the Maintenance department must be entered on the MRP II system, which may overcrowd it.

2.2 Literature

A small number of papers that addressed the idea of using MRP for maintenance planning were presented in the 1980's [e.g. 3-8]. They explained how preventive maintenance plans can work independently of MRP, act as a catalytic base of MRP, or as adjunct to a MRP system. They also emphasised needs to restructure the traditional MRP, since maintenance and manufacturing planning are conducted in somewhat different ways. Newman [3] emphasised that there are two types of inventory in maintenance planning ; the traditional parts inventories and the "human skills inventory". This will increase the complexity in the MPCS. Bruggeman and van Dierdonck [4] identified another shortcoming of the present MRP systems. Total gross spare parts requirements are determined by a planned master maintenance schedule that has a binomial distribution, breakdown maintenance with Poisson distribution and deterministic quantities of spare parts for new products. However, MRP software does not contain probabilistic Bills of Materials (BOM) and does not provide statistical routines for automatic safety stock determination. For proper use of MRP II software for maintenance planning, it would be necessary to develop the BOM file. Other problems with integration of maintenance into MRP are e.g. the difficulties of defining and timing each specific inspection and maintenance sequence; constructing a machine failure probability index; and determining that identical parts are used in different machines produced by different manufacturers. If not using correct data, integration of the two systems might only lead to increased problems of complexity. All these papers, [3-8], advocate doing the maintenance resource planning as well as the time planning on the MRP II system.

2.3 Some Ideas.

After entering maintenance data into the BOM, the PM plans can be generated by a maintenance master schedule or by treating the events as dependent demands. Production is strict to due dates, linked to customer orders. PM has more freedom; model optima are rather flat, [14]. PM can and should be treated like other jobs except where the MRP II is operated to maximise profit in a jobbing shop with too many offers of work. Then PM will be seen as work done at a loss and so delayed repeatedly. This can be overcome by assuming that a failure will occur if the PM is omitted with a probability-cost which increases with the time since the due-date, see below. The probability-cost of the failure will increase with time until the PM is eventually worked into the Production Plan as a job worth doing. This way of thinking about PM is useful when contesting the need to stop a machine with a Production Manager holding a full order book. It puts a price on the delay to or omission of PM and rationalises slack capacity requirements and spares and buffer inventories. The MRP II system could also possibly be used for planning overhauls and annual shutdown periods, however, project planning software is probably better. The ordering of materials and spare parts, and the analysis of job durations by breaking them down into elemental operations could be handled by MRP II.

3. CIM and MMIS

MRP II is usually an integral part of CIM systems so all the same remarks and ideas would apply to CIM. In a true CIM, the computer is central and passes all instructions. Designs are made by CAD and converted by the Computer Aided Process Planning (CAPP) system and the Computer-Aided

Manufacture (CAM) modules into instructions for ordering materials, choosing machines and making components. Robotic devices move materials from station to station under central computer control. Even the sales and purchasing systems are integrated, but significantly not maintenance. CIM generally also includes the design or specification of manufacturing machinery required to make a new product. Integration of the MMIS at this point in the product cycle would be useful if it led to earlier consideration of the maintenance requirements of such new machinery or alterations to those of existing plant to be used in new ways. Both stress levels and intensity of use affect maintenance needs, but this is seldom acknowledged in schedules and cannot be quantified without data collection and analysis. Terotechnological considerations, including costs of ownership and expected life of plant, could usefully be considered as part of the development of products as well as of the machinery to make them. Early Flexible Manufacturing Systems (FMS), for example, suffered from neglect of reliability and maintenance requirements at the planning stage.

In a good CIM system, allowing full simulation of production, the reliability, maintainability and PM of the machines can also be simulated, but this will not be helpful unless the reduction in failures due to PM is properly modelled. In mass production and process plants, this simulation would lead to a cost-based assessment of the need for buffer stores between stages of production, costed as alternatives to redundant machines. Thus we can assess the capacity of existing plant to accommodate different products or increased production rates, or how best to spend a limited sum on improvements. Simulations can also estimate the effects of changes to the workforce, particularly the numbers of skilled maintenance fitters.

Modern Maintenance Concepts (Fads?),CM, TPM, RCM

1. Introduction

Maintenance was almost forgotten in the current generation of production software, but does this matter if it gets done somehow? Most managers still regard maintenance as a fixed overhead, but it is really a profit centre, a major factor in Life-Cycle Profit (LCP) calculations. Advocates of Total Productive Maintenance (TPM) and Reliability-Centred Maintenance (RCM) claim to reduce maintenance costs. In fact, both are incomplete schemes, but beguiling because they purport to offer continuing savings for relatively small one-off expenditures. In the present context, they promise improvement without integrating maintenance into the company's IT system, and so must be examined carefully. They were preceded as the current fashion by Condition Monitoring (CM). Many expensive monitors were installed, sometimes where improved operator training would have saved more money for less investment. All three have good points, though, and can contribute to optimising maintenance in conjunction with production, to achieve the maximum Life-Cycle Profit (LCP).

2. Condition Monitoring (CM)

In a previous paper [13], models for continuous monitoring and inspection at random intervals were presented with the suggestion that care was needed to discover which policy was cheaper and indeed whether CM was the best type of policy. Human monitoring, by operators reporting impending failures or poor performance is a grossly undervalued feature of TPM. Monitors are expensive and subject to failure to warn and to false alarms. All this can be, but seldom is, modelled to explore for the cheapest feasible policy. More usually, CM is seen as a panacea, and exaggerated claims are made regarding the savings. It can be the right policy for items which give warning of costly failures, but it is by no means always the winner over Age or Block renewal policies, [14]. If Age Renewals can be done in system free time, the savings are greater, and may well compete with CM. Continuous CM is not amenable to optimisation, but whether or not interval inspection can offer worthwhile savings depends on how many inspections at cost C_I can be bought for the saving between preventive (C_M) and corrective (C_F) maintenance, relative to the ratio of *MTBF* to mean warning time (*MWT*).

$$(C_F - C_M) / C_I \geq MTBF / MWT \qquad (1)$$

If the inequality is false, inspection is unlikely to be the best policy. This is a simple, rough rule of thumb but it still requires cost and reliability data. If the inequality is not well met then Age and Block policies may be considered, requiring individual failure times to be analysed for distribution.

3. Total Productive Maintenance (TPM)

TPM, [10], was invented to cover the major problem posed by JIT which is getting PM done when the next KANBAN could arrive at any moment. It relies upon the fact that many failures to machines are caused by maloperation, or lack of primary care, (greasing, spannering and cleaning), which can be corrected by training the operator. The problems come later, when failures occur which can be suppressed by skilled PM but not by the operator. TPM does not exclude skilled PM, in fact it is advocated in TPM literature [10], but with no specific strategy to allow it to be planned, it is bound to be neglected if the factory gets too busy. The reports of dramatic falls in failure rates come from increased operator care, especially the operator acting as condition monitor. A smaller return occurs in plants which already have optimised PM and CM.

4. Reliability-Centred Maintenance (RCM)

RCM, [11], purports to be a system for finding the plant's "correct" schedule of maintenance. Its best feature is the way it involves the people closest to the plant in the re-assessment of maintenance needs. Its most dramatic successes have been obtained where the plant was previously grossly over-maintained, (e.g. civil aircraft) or over-maintained in some areas and under-maintained in others, (e.g. oil production platforms). It is also quite effective at highlighting plant which is failing frequently because it is over-stressed or abused. However, its most serious flaw, both in theory and in practice, is to suppose that there is a maximum possible inherent reliability for a machine which can be realised by just the right schedule. Any capable machine can be made more reliable by increased PM intensity, but at increased cost. RCM does not recognise that maintenance is an economic problem. RCM uses many well-established techniques such as FMECA/Design Audit, Brain-storming, and small group dynamics. Its decision trees find feasible rather than best solutions and eschew re-design except as a last resort. It under-emphasises the value of data collection and feedback to designers. By placing the responsibility for judging the seriousness and relative frequency of failures upon the operators and maintainers, it neatly avoids the requirement of maintenance truly based on reliability for accurate and complete data. This makes it attractive to managers who do not wish to invest in better MMIS's, but these subjective judgements can be wrong, and they go unchecked.

Modelling of Losses Due to Delay in Executing PM

1. Effect of Chronic under-maintenance

Chronic under-maintenance results in poor quality products and shortens the life of plant. Ironically, well-maintained plant can be replaced with more modern machinery earlier than ill-maintained, although the latter may be in greater need of renewal earlier than the former. This is because good profits from higher market share due to better quality achieved by superior maintenance allow earlier renewal to sustain and extend the commercial advantage. This would be difficult to model, and subject to dispute about methods, but is a reasonable explanation of Japanese industrial success. It is also an argument against subsidies for plant renewal.

2. Risk Arising from Omission of a Single PM Routine

It is easier to model the effects of a single delay or omission, provided that schedules are initially based upon data analysis and modelling. Typically, there will be an argument between Production and Maintenance as to whether a machine should be shut down for PM or used for an order, delay to which would result in a loss. The problem is to find out which probable loss will be greater. In a simulation of the whole system over the life of the plant, the results would be discounted to net present value (NPV), to evaluate the project, but in real time decisions one would use current costs. Two simplified examples, derived from [12], are given. They cover Block Renewal and CM in the form of delayed action following warning by a continuous monitor, respectively.

2.1 Block Renewal

In Block Renewal, the wearing component would fail, in the absence of PM, according to a time-from-renewal distribution with rising hazard rate average at a cost $C_F > C_M$, the PM cost. PM occurs at fixed intervals and failures are repaired by renewal without disrupting the PM schedule. The expected number of failures between scheduled renewals is then $N(t)$, Cox´s Renewal Function for intervals t. Following Jardine [14 ,p69], the optimal renewal interval, assuming a continuing requirement, is

$$t^* = Min_t\left[C(t) = \{C_M + N(t).C_F\} / (1 - e^{-it})\right] \qquad (2)$$

for interest rate i, where $C(t)$ represents the total cost to infinity, discounted to $x=0$. If the operation is done late, at instant t after $x=nt^*$ having last been done at nt^* then if $0<t<2t^*$, there is a loss in two successive cycles. The total loss, putting $r=1/(i+1)$, is

$$L(nt^*+t) = C(t).r^{nt^*+t} + C(2t^*-t).r^{(n+2)t^*} - C(t^*)\left(r^{(n+1)t^*} + r^{(n+2)t^*}\right) \qquad (3)$$

In a real time problem, the terms in r would be set to unity. In theory, if two or more successive timings are in error, the loss functions can be added, but in practice this would start to become chronic neglect with long-term effects acting to increase the eventual loss. Equation *(3)* breaks down if the total delay exceeds $2t^*$, but disregarding the discounting factors, the total loss for a single delay of any length t is

$$L(t) = C(t) + C(yt^*-t) - yC(t^*), \quad y = int(t/t^*) \qquad (4)$$

This can be discounted as if all occurring at $x+t$ with little error when $x>>t^*$. If the distribution of failure times is very peaky and/or the failure is much more expensive than PM, this loss function can become sufficiently large to compete with the cost of lost or delayed production. It is correct, in this case, to include the cost of lost production in the failure cost and , if appropriate, the PM cost in the Block model. The loss function would then give the "normal" or "ordinarily expected" loss for PM interval variation, which would be offset against the "special" loss, for example the cost of lateness or loss of order or customer. The aegis for the model to be applicable is therefore one where delay to PM is not usual, but occurs only when plans must be changed due to unforeseen circumstances.

2.2 Continuous Monitoring

In a previous paper, [13], a model for continuous monitoring against Poisson failures which give warning signals shortly before failing was discussed. No failures will occur unless action is delayed when the alarm sounds, perhaps until the end of shift or the next weekend. The model assumed constant conditional transition rates between four states as follows

S1	All is well, no warning.	to S2 at rate f	No cost
S2	Warning showing	to S3 at rate w	No cost
		to S4 at rate g	
S3	Under PM	to S1 at rate m	C_M
S4	Failurc Rcpair	to S1 at rate r	C_F

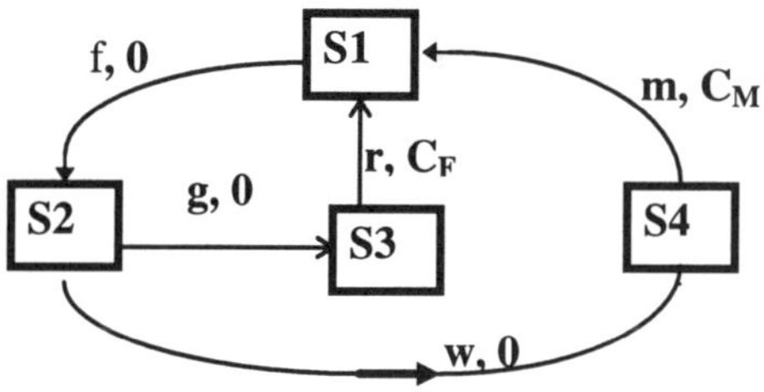

Fig. 1. Markov Model for Continuous Monitoring with Delay, (rate, cost)

impending failure showing. The variable in this case is w , the new value being *1/d* where d is the proposed expected delay in starting PM counting from the time the warning appeared. This is not subject to optimisation unless C_M is a function of d; for constant costs, the loss is minimised by reduction of d. Practically, the costs are often defined only for a few fixed values of d, so that one would calculate to see whether to leave it until the next break. The operating cost rate is given by

$$c(w) = f(gC_F + wC_M)/(f + w + g) + c_{mon} \qquad (5)$$

where the last term is the cost per unit time of owning and operating the monitor.

Suppose that the potential gain from finishing an order on time is G, and the loss expected from making and disposing of the order if it is late due to failure of the equipment after the warning has sounded is S, then we would proceed if

$$G.e^{-gd} \geq c(w|w = 1/d).T(w|w = 1/d) - c(w).T(w) + S\left(1 - e^{-gd}\right) \qquad (6)$$

where
$$T(w) = 1/f + 1/\left(g + w\right) \qquad (7)$$

is the expected up time in a cycle, and *1/w* is the normal delay.

Having seen how integration of the MMIS would help, we examine what is being done in Sweden.

Integration of MMIS - A Survey

A survey was conducted to find how much Swedish manufacturing firms integrated maintenance information systems with manufacturing. There were four characteristic types of MMIS.

1. Manual system, operated separate from other functions and systems,
2. Customer built software, operated from separate computer network,
3. Standard software, operated from separate computer network,
4. Maintenance integrated into the overall MIS of the organisation through a common network.

Table 1 shows the results of the survey. Firms with 50 to 100 employees were considered small, firms with 101 to 500 employees were medium sized and firms with more than 500 employees were considered large. Less than 10 percent of the firms reported that they had any kind of integrated MMIS, and about two thirds used a manual MMIS. The firms with integrated MMIS were mostly large and medium sized paper mills and mechanical engineering companies, with continuous flow or sequential operation + assembly lines. More than half of the medium and about a third of the large companies rely on a separate, manual system. We doubt whether this is the optimum solution for these firms. Most firms should be able to gain advantages through integration of maintenance into MPCS and CIM.

MMIS	Manual	Special	Standard	Integrated	Total
Small	105	7	8	7	127
Medium	72	25	16	11	124
Large	10	11	6	6	33
Job shop	39	5	5	2	51
Batch	93	17	14	8	132
Assembly line	12	7	2	2	23
Continuous flow	43	14	9	12	78

Table 1. MMIS Characteristics from Survey of Swedish Companies

Links between MMIS & Manufacturing - Case Studies

Two different firms were studied. The objective was to see at what level the studied firms used MMIS and how the interaction between Maintenance and Production worked out. The first firm was customer order oriented and produced articulated haulage vehicles. The other was a process firm producing aluminium profiles. Both firms had complex machinery.

1. Case Study No.1 (Articulated Vehicles)

The Manufacturing process has a functional layout. Maintenance is under the Production Department. Unlike Case Study No.2, the Maintenance Manager is also responsible for Quality.

A PC-based system for the planning of Maintenance was introduced recently. The system contains two modules, one for time schedules and another for instructions. The module for time schedules is separated into daily/weekly-based maintenance and Periodic Maintenance of longer periodicity. Daily maintenance is carried out each morning by the operators before starting the equipment. Weekly maintenance is conducted every Friday, if possible, and focuses on direct consequences for operation. Periodic Maintenance is mostly for safety and includes checks for hidden faults. Only Periodic Maintenance is signed off by the fitters.

There is almost no interaction between Maintenance and Production planning, despite the common head. First the production is planned and after that maintenance. Only the bigger items of maintenance work are considered during production planning. According to the Maintenance Policy Statement the objectives are to reduce or eliminate the appearance of faults and failures, reduce or eliminate the need for maintenance and prevent wear, destruction and breakdowns.

2. Case Study No.2 (Aluminium Profiles)

The manufacturing is managed as a demand-led pull-system. Maintenance is organised as a separate function, within the Production department. The area of responsibility for the Maintenance manager is maintenance of the facilities, both machinery and buildings, and Production Technology.

The MMIS is a standard commercial system containing modules for Facilities, Preventive Maintenance, Work Orders, Inventory and Purchasing. Six UNIX-based terminals are located in the manufacturing facility to be used for reporting maintenance. The MMIS is isolated from the other IT systems. There should be a strong interaction between maintenance and production in this plant. It is possible to collect a lot of the needed data through the Programmable Logic Controller (PLC)-system, that controls the presses. So unlike Case Study No.1, most of the necessary data for increased interaction between MMIS and MPC already exist in the IT system, but they do not do it.

In the future, the MMIS will consist of a Client-Server system in Windows NT. There will be 11 clients reporting needs for maintenance. Maintenance conducted by operators will increase and only more advanced maintenance will be done by maintenance staff. The Overall Equipment Effectiveness (OEE) index of Nakajima, [10] will be used to measure "maintenance efficiency".

Conclusions

1. The Case Studies show that even when it comes time to update systems, the opportunity is not being taken to include maintenance in an integrated system.
2. The Survey shows that many companies are still using manual systems.
3. Maintenance planning seems to have been neglected by those who devised the current generation of MRP II, JIT, and CIM systems.
4. The JIT philosophy advocates and requires PM but does not go beyond the somewhat vague concepts of TPM in explaining how it can be planned. It is difficult to see how JIT/KANBAN could be modified to include the planning of maintenance.
5. In contrast, MRP II is eminently suitable for such integration and would probably lead to great benefits by allowing some work now done at premium rates in unsocial hours to be performed in normal working time.
6. MRP II is not entirely suitable for planning annual shut down periods. Critical path analysis software is better for this purpose but MRP II can be useful for detailed planning.
7. In full CIM there is really no excuse for not incorporating the MMIS and other terotechnological functions into the system.
8. Production system simulations should include the reliability, maintainability and PM of machinery and optimisation of redundancy and buffer store capacity.
9. The arguments between Maintenance and Production regarding well-used machinery can be resolved by calculating loss functions, provided that they do not occur so often as to present the danger of chronic under-maintenance. This needs an integrated data system.
10. It is vital that integration of maintenance into the CIM/IT systems includes upgrading of the level of detail in the recorded data, so that full use of existing and new models for optimisation can be made. The words are even more important than the figures in these data, because maintenance is at the component level.
11. The extra capital costs of integrating maintenance into the next generation of CIM software need not be great, because there will be savings for not recording data more than once. The extra capital expenditure is likely to be quickly recouped and to prove a wise investment.

References

1. Raouf, A, Ali, Z, & Duffuaa, S.O. (1993). Evaluating a Computerized Maintenance Management System, *International Journal of Operations and Production Management,* Vol. 13, No. 3, pp. 38-48.
2. Sherwin, D.J. (1984). Improved Schedules by Using Data Collected under Preventive Maintenance, *IEEE Transactions on Reliability,* **R-33**, 4, pp. 315-320.
3. Newman, R.G. (1985). MRP Where M = Preventive Maintenance, Production and Inventory Management Journal, **26**, 2, pp. 21-27.
4. Bruggeman, W, van Dierdonck, R. (1985). Maintenance Resource Planning: An Integrative Approach, *Engineering Costs and Production Economics*, **9**, pp. 147-154.
5. Ettkin, L.P, Jahnig, D.G. (1986). Adapting MRP II For Maintenance Resource Management can Provide a Strategic Advantage, *Industrial Engineering*, **18**, 8, pp. 50-59.
6. Bojanowski, R.S. (1984). Improving Factory Performance with Service Requirements Planning (SRP), *Production and Inventory Management Journal*, **25**, 4, pp. 31-44.
7. Raeker, R.H. (1986). Preventive Maintenance Planning with an MRP II System, *APICS 29th Annual International Conference Proceedings, Falls Church, Va*, pp. 1-3.
8. Russel, S.P. (1985). Capacity Planning for Preventive Maintenance: Pay Me Now or Pay Me Later, *APICS 28th Ann. Int'l Conference Proceedings, Falls Church, Va.*, pp. 101-104.
9. Sherwin, D.J. (1990). Design for a Comprehensive Maintenance Planning Recording and Optimisation System, *Engineering Costs and Production Economics*, **20**, pp. 37-42.
10. Nakajima, S. (1988). *Introduction to TPM*, Productivity Press, Cambridge, USA.
11. Moubray, J. (1991). *Reliability-centred Maintenance*, Butterworth-Heinemann, Oxford.
12. Sherwin, D.J. (1994). Maintenance - A Forgotten Factor in Production Planning Systems, *Proceedings of the O.R. Society Conference, Glasgow, U.K.*
13. Sherwin, D.J. (1990) Inspect or Monitor? *Engineering Costs & Production Economics*, **18**, pp 223-231.
14. Jardine, A.K.S. (1973) *Maintenance Replacement & Reliability,* Pitman, London.

Optimal Periodic Replacement Strategy for Non Self-Announcing Failure Equipment

Anis Chelbi *, Daoud Ait-Kadi *, Moez Limayem **

* Mechanical Engineering Department, Laval University, Quebec, PQ, G1K 7P4, Canada
**Information Systems Department, Laval University, Quebec, PQ, G1K 7P4, Canada

Abstract. This paper addresses the problem of the maintenance of randomly deteriorating equipment whose failure is not self-announcing. In principle, such equipment have to be inspected in order to verify whether or not they are operational. However, there are many situations in which there is neither technological nor human means to inspect the state of these equipment. In those cases, a systematic replacement strategy can be adopted. This strategy consists of replacing the equipment at pre-determined instants T, 2T, 3T, ... regardless of its state. The mathematical model corresponding to this policy is presented and the conditions for the existence and uniqueness of an optimal replacement period are established. This model takes into account the replacements cost, as well as the inactivity cost associated with the period of time between failure and replacement. Due to the complexity of the model, a procedure allowing the computation of the optimal replacement period is proposed. Finally, some examples are treated for given costs and lifetime probability distributions.

Introduction

The failure of an equipment is an undesirable event which usually has heavy consequences at both human and economic levels. Therefore, it is desired to develop maintenance strategies which allow to reduce the frequency and the duration of random interruptions of service, while respecting constraints associated with budget, productivity, space etc. A maintenance strategy is defined as a set of actions that allow to maintain or restore a system in a specified condition or in a position to provide a given service.

In this paper, we focus on the maintenance of equipment whose failure is not self-announcing. As a rule, such equipment are to be inspected according to a specific sequence in order to verify if they are operational or in a failed state. Each inspection consists of taking one or more measurements of operating characteristics which allow to be informed about the deterioration level of the equipment. In situations where that inspection is possible, lot of scholars (Barlow et al. [1], Chelbi et al. [3], Christer et al. [4] ...) have succeeded to establish the instants when the inspection must be performed in such a way to minimize the total average cost, which includes inspections cost and the cost associated with downtime between failure occurrence and its detection.

There exits, however, lot of situations in which, for various reasons, there is not any technological or human mean to inspect those equipment. In such situations, the maintenance managers are usually compelled to replace the equipment periodically no matter its state. Such periodic replacement strategies have largely been treated in the literature. An excellent survey of those strategies is supplied in [8]. The models corresponding to these different strategies have a common assumption which states that the condition of the equipment can be known at all time.

We are unaware of any publications dealing with periodic replacement strategies for systems whose failure cannot be detected as it is the case of a large number of equipment or assemblies which are very difficult to access.

In this paper, we propose a systematic replacement strategy for such equipment. This strategy consists of replacing the equipment at pre-determined instants T,2T, 3T, ...regardless of its state.

In section 1, the definition of the strategy, the working assumptions and the used notations are reminded. Section 2 will be devoted to the elaboration of the mathematical model and the establishment of the conditions of existence and uniqueness of an optimal replacement period T*. These conditions will be made explicit in the case of three lifetime probability distributions commonly used in reliability, namely the Gamma law, the Weibull law, and the truncated Normal law. A numerical algorithm has been developed to generate the optimal replacement period, this algorithm will be validated using numerical examples in section 3. The numerical results will finally be discussed in section 4.

1 Definition of the strategy and working assumptions

Consider an equipment failing in a random manner and whose failure is not self-announcing and cannot be detected. The proposed strategy consists of replacing that equipment by another new identical one at pre-determined instants T, 2T, 3T, ... no matter its state. If a failure happens between consecutive instants kT and (k+1)T, there occurs a penalty due to the equipment downtime between the instant of failure and the replacement time (k+1)T. On the other hand, if the equipment survives throughout the interval [kT, (k+1)T] then preventive replacement is performed at instant (k + 1)T (see figure 1).
The costs associated with this strategy are the cost of replacement of the equipment found out of order, the cost of preventive replacement and the cost associated to the inactivity period.

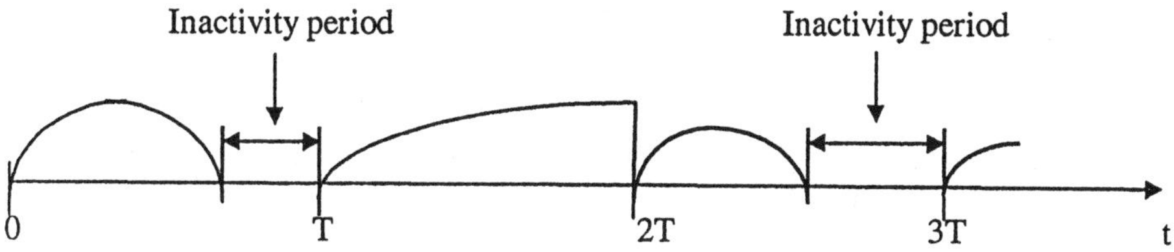

Figure 1. Definition of the strategy.

2 The mathematical model

Suppose the lifetime of the equipment is a random variable, having a density function f(.) and distribution function F(.) of mean μ.
Let's define
C_1 : the replacement cost of the equipment found out of order.
C_2 : the cost of preventive replacement ($C_2 < C_1$).
C_d : the cost of one time unit of inactivity due to failure.
The replacements are performed only at instants kT, k = 1, 2, 3, ...
Let $B_i(T)$ be the average total cost per time unit over an infinite time span. This cost is expressed as follows:

$$B_i(T) = \lim_{t \to \infty} \frac{E[C(t)]}{t} \tag{1}$$

where E[C(t)] represents the average total cost in time interval [0,t]. This cost includes the mean cost of inactivity as well as the average cost of replacement at the instant t. It can be shown, using the renewal theory [5], that the cost $B_i(T)$ can be expressed over a replacement cycle of length T, in the following way:

$$B_i(T) = \frac{1}{T}\left[C_d \int_0^T (T-x)f(x)dx + C_1 F(T) + C_2 R(T)\right] \tag{2}$$

where R(T) represents the equipment reliability function evaluated at the instant T. Replacing R(T) by 1-F(T) in (2), we obtain:

$$B_i(T) = \frac{1}{T}\left[C_d \int_0^T (T-x)f(x)dx + (C_1 - C_2)F(T) + C_2\right] \tag{3}$$

2.1 Graphical representation of the average total cost per time unit

It is possible to have an idea about the behavior of the curve of $B_i(T)$ from its expression (equation (3)). Since the functions F(.) and f(.) are continuous functions, $B_i(T)$ is continuous as well.
On the other hand :

$$B_i(0) = \infty \quad \text{et} \quad \lim_{T \to \infty} B_i(T) = C_d \tag{4}$$

(see reference [2] for the details about the proof of these results).

Thus, the curve for the average total cost per time unit will have one of the behaviors below:

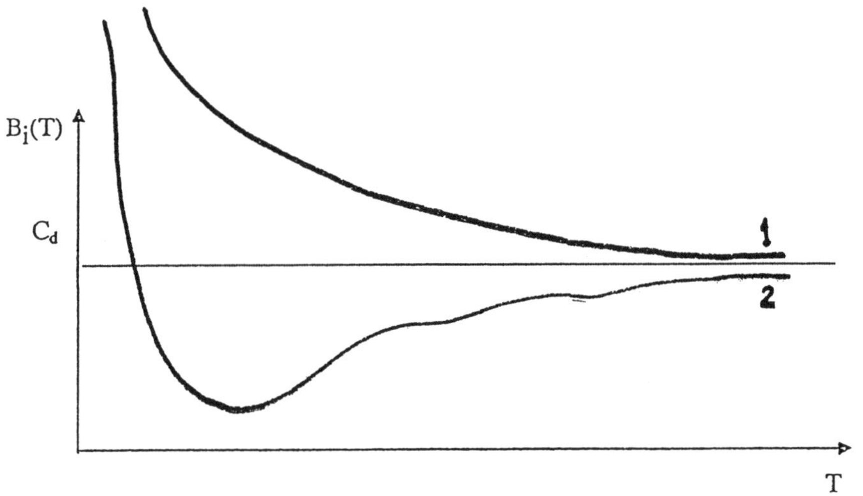

Figure 2. Possible behavior of the cost curve $B_i(T)$.

It is evident that the case of curve 1, for which the minimal cost corresponds to an infinite replacement period, is not to be considered in practice. In this case, the equipment would remain infinitely out of order! However, the behavior of curve 2 indicates that there exists at least one value T* which minimizes the average total cost per time unit.
Obviously, the maintenance manager will only be concerned with the case of the existence of a finite strategy T*, which is the one that corresponds to the minimum of all the local minima of $B_i(T)$.

2.2 The existence and uniqueness of a finite optimal strategy

Theorem

There exists a finite and unique optimal strategy T* if :

$$\frac{C_d}{C_1 - C_2} f(t) + \frac{df(t)}{dt} \rangle 0 \qquad \forall t \rangle 0 \tag{5}$$

and

$$\mu \geq \frac{C_1}{C_d} \tag{6}$$

The complete proof of this theorem is supplied in [2].

Let's denote the left side of equation (5) by: $H(T) = \frac{C_d}{C_1 - C_2} f(T) + \frac{df(T)}{dT}$

Generally speaking, such a combination H(T) of a function f(T) with its derivative does not keep the same sign for all the values of T. The sign of H(T) can change from a certain value of T. We therefore cannot come out with a general condition on the existence and uniqueness of a finite optimal solution. However, it is interesting to express the conditions of existence and uniqueness (equations (5) and (6)) for commonly used lifetime probability distributions. Hence, we could maybe derive one or more conditions on the parameters of these probability distributions so that an optimal strategy might exit, and be unique. We choose the following three distributions: the Gamma law of order α and parameter λ, the truncated normal law of mean μ and standard deviation σ, and the Weibull law of shape parameter α and scale parameter λ.

* *Case 1: Gamma law (α, λ)*

$$f(T) = \frac{\lambda^\alpha}{\Gamma(\alpha)} T^{\alpha-1} \exp(-\lambda T) \quad \text{with } \alpha \geq 0 \text{ and } \lambda \geq 0 \tag{7}$$

where $\Gamma(.)$ represents the gamma function.

Let:

$$\frac{df(T)}{dT} = \frac{\lambda^\alpha}{\Gamma(\alpha)} \left[(\alpha-1)T^{\alpha-2} \exp(-\lambda T) + T^{\alpha-1}(-\lambda \exp(-\lambda T))\right] \tag{8}$$

$$\frac{df(T)}{dT} = \frac{\lambda^\alpha}{\Gamma(\alpha)} \exp(-\lambda T) T^{\alpha-1} \left[\frac{(\alpha-1)}{T} - \lambda\right] \tag{9}$$

$$\frac{df(T)}{dT} = f(T)\left[\frac{(\alpha - 1)}{T} - \lambda\right] \tag{10}$$

The first existence and uniqueness condition (equation (5)) is then written as follows:

$$\frac{C_d}{C_1 - C_2} f(T) + f(T)\left[\frac{(\alpha - 1)}{T} - \lambda\right] \rangle\ 0 \qquad \forall\, T \rangle\ 0 \tag{11}$$

f(T) being a probability density function, $f(T) > 0 \ \ \forall\ T > 0$, we therefore have the following condition:

$$\frac{C_d}{C_1 - C_2} + \frac{(\alpha - 1)}{T} - \lambda \rangle\ 0 \qquad \forall\, T \rangle\ 0 \tag{12}$$

This condition is verified for :

$$\begin{cases} \alpha = 1 \\ \text{and} \\ \lambda \ \langle\ \dfrac{C_d}{C_1 - C_2} \end{cases} \tag{13}$$

Note that $\alpha = 1$ corresponds to an exponential distribution. Its mean μ is then equal to $1/\lambda$.
To satisfy the second condition (equation (6)), it is then required that :

$$\lambda \le \frac{C_d}{C_1} \tag{14}$$

The combination of all these conditions leads to an exponential law whose parameter λ must be less or equal to the ratio $\frac{C_d}{C_1}$. It is therefore certain that the cost $B_i(T)$ of this strategy admits only one minimum $B_i(T^*)$ for any equipment whose lifetime distribution is exponential (constant failure rate) and whose parameter λ is less or equal to $\frac{C_d}{C_1}$.

It is interesting to notice that the case of an exponential distribution of parameter $\lambda > \frac{C_d}{C_1}$ corresponds to an infinite solution. This situation is not conceivable in practice. In this particular case, it is imperative to find the means to inspect the equipment periodically. Indeed, periodic inspection strategies are recommended for equipment with constant failure rate [1]. The optimal inspection period T_i should verify the following relationship (see Rodrigues [7])

$$\exp(\lambda T_i) - \left(1 + \lambda \frac{C_i}{C_d} + \lambda T_i\right) = 0 \tag{15}$$

where C_i represents the cost of one inspection.

* *Case2: Weibull law* (α, λ)

$$f(T) = \alpha\lambda^{\alpha} T^{\alpha-1} \exp\left[-(\lambda T)^{\alpha}\right] \quad \text{with } \alpha > 0 \text{ and } \lambda > 0 \tag{16}$$

$$\frac{df(T)}{dT} = \alpha\lambda^{\alpha}\left[(\alpha-1)T^{\alpha-2}\exp\left[(-\lambda T)^{\alpha}\right] + T^{\alpha-1}\left(-\lambda^{\alpha}\alpha T^{\alpha-1}\right)\exp\left[(-\lambda T)^{\alpha}\right]\right] \tag{17}$$

$$\frac{df(T)}{dT} = \alpha\lambda^{\alpha}T^{\alpha-1}\exp\left[-(\lambda T)^{\alpha}\right]\left[\frac{(\alpha-1)}{T} - \alpha\lambda^{\alpha}T^{\alpha-1}\right] \tag{18}$$

$$\frac{df(T)}{dT} = f(T)\left[\frac{(\alpha-1)}{T} - \alpha\lambda^{\alpha}T^{\alpha-1}\right] \tag{19}$$

The first condition of existence and uniqueness (equation (5)) is then written as follows :

$$\frac{C_d}{C_1 - C_2} f(T) + f(T)\left[\frac{(\alpha-1)}{T} - \alpha\lambda^{\alpha}T^{\alpha-1}\right] > 0 \qquad \forall T > 0 \tag{20}$$

Knowing that f(T) > 0, we therefore have :

$$\alpha\lambda^{\alpha}T^{\alpha-1} + \frac{(1-\alpha)}{T} < \frac{C_d}{C_1 - C_2} \qquad \forall T > 0 \tag{21}$$

For this condition to be verified, we need to have :

$$\begin{cases} \alpha = 1 \\ \text{and} \\ \lambda < \dfrac{C_d}{C_1 - C_2} \end{cases} \tag{22}$$

Considering also the second condition $\mu \geq \dfrac{C_1}{C_d}$ and combining it with the first one, we once again have an exponential law of parameter $\lambda \leq \dfrac{C_d}{C_1}$ and the same conclusion as for case 1 with the Gamma law is applied.

* *Case 3: Truncated Normal law (μ, σ)*

$$f(T) = \frac{1}{\sqrt{2\Pi}\sigma} \exp\left[-\frac{1}{2}\left(\frac{T-\mu}{\sigma}\right)^2\right] \tag{23}$$

$$\frac{df(T)}{dT} = \frac{1}{\sqrt{2\Pi}\sigma}\left[-\frac{(T-\mu)}{\sigma^2}\exp\left[-\frac{1}{2}\left(\frac{T-\mu}{\sigma}\right)^2\right]\right] \tag{24}$$

$$\frac{df(T)}{dT} = \frac{(\mu - T)}{\sigma^2} f(T) \tag{25}$$

The first existence and uniqueness condition can therefore be re-written :

$$\frac{C_d}{C_1 - C_2} f(T) + \left[\frac{(\mu - T)}{\sigma^2}\right] f(T) \ > 0 \qquad \forall\, T > 0 \tag{26}$$

f(T) being positive, this condition to be satisfied becomes :

$$\frac{T-\mu}{\sigma^2} < \frac{C_d}{C_1 - C_2} \qquad \forall\, T > 0 \tag{27}$$

The second condition (equation (6)) is simply written as: $\mu \geq \dfrac{C_1}{C_d}$

In this case, it is impossible to find a condition on the parameters μ and σ which would allow us to assert that the cost $B_i(T)$ allows only one minimum. However, it is certain that this cost allows at least one local minimum. In fact, to affirm that there wouldn't be any finite solution is equivalent to state that the relationship below should be satisfied :

$$\frac{T-\mu}{\sigma^2} \geq \frac{C_d}{C_1 - C_2} \qquad \forall T \rangle 0 \tag{28}$$

whereas there certainly exist at least one value T which does not satisfy this inequality for given μ, σ, C_1, C_2 and C_d.

A computation procedure has been developed to find, when it does exist, the optimal strategy T* which minimizes the average total cost per time unit, $B_i(T)$. The procedure is based on a non-linear equations solving routine (see [6]).
The user has simply to enter the lifetime distribution function of the equipment as well as the costs C_d, C_l and C_2. The program generates then the optimal replacement period T* and the corresponding average total cost per time unit.

3 Numerical results

All the numerical procedures were coded in FORTRAN 77. The computations were performed on a HP - 700 workstation. Depending on the application, the user can utilize the distribution f(.) which best adjusts the lifetime data at his disposal.
Many tests were performed , and four of them are presented below. For these four tests, we used Exponential and Gamma distributions, with different combinations of costs C_d, C_l, C_2 (see table1). It should be noted that the parameters of the lifetime distributions and the costs were arbitrarily chosen.

Table 1. Input data and results.

Test ≠	Probability distribution	C_d (\$/tu)	C_l (\$)	C_2 (\$)	n	m	T* (tu)	$B_i(T^*)$ (\$/tu)
1	Exp. ($\lambda = 0.1$)	60	450	150	1	1	17	54.4
2	Gamma (α=10, λ= 1)	50	400	180	1	1	7.1	32.5
3	Gamma (α=10, λ= 1)	50	600	180	2	1	6.2	36.5
4	Gamma (α=10, λ= 1)	20	400	180	2	0	∞	-

4 Discussion

In the test ≠1, we considered an equipment whose random lifetime follows an exponential distribution of parameter $\lambda = 0.1$ failures per time unit $\left(\mu = \frac{1}{\lambda} = 10\right)$, with a combination of costs as supplied in table 1. According to those values, it is noticed that:

$\frac{C_d}{C_1} = 0.13 \rangle \lambda$, which is a situation where a finite and unique solution T* exists (see equation (14)). The algorithm described in section 3, gives a finite and unique optimal period (T* = 17 units of time).

In test ≠2, we considered another equipment whose lifetime probability density function is a Gamma law of parameters $\alpha = 10$ and $\lambda = 1$, with another combination of costs. The corresponding results are given in table 1.

In the scope of the test ≠3, we supposed that for logistic reasons related, for example, to labor or other maintenance resources, the cost of repair at failure, C_1, is assigned a value of $C_1' = \$600$ instead of \$400. In this case, a decrease in the replacement period is expected. In fact, the results presented in table 1 clearly demonstrate this effect (the value changes from T* = 7.1 to T* = 6.2 units of time).

Finally, in test ≠ 4, it was assumed that with respect to the initial condition with (C_1 = 400 \$), the cost of one time unit of inactivity changes from \$50 to \$20/ time unit. This could be due, for instance, to the existence of stocks of goods manufactured by the equipment in question. This decrease in inactivity cost generated an infinite replacement period. This situation was foreseeable given the fact that the lower the inactivity cost becomes, the more there would be a tendency not to intervene on the equipment . However, it is obvious that the equipment would not be left out of order indefinitely. That is why, in this case, it is imperative to think about and find the necessary resources to be able to inspect the equipment.

On the other hand, if the inactivity cost is very high due, for example, to great production losses following the equipment failure, the decision maker could think of investing in the acquisition and the exploitation of a certain technology such as sensors and other control devices, which would allow him to detect the failure instantaneously and , thus be able to intervene promptly. A Block replacement strategy could then be considered under specific economic conditions (see [2]).

5 Conclusion

A periodic replacement strategy for non self-announcing failure equipment was developed. The performance of this strategy was evaluated in terms of the total

average cost per time unit over an infinite time span. The mathematical model which governs this replacement strategy involves the equipment lifetime probability distribution as well as the costs incurred to perform the replacements, and the cost associated with the inactivity period between failure and replacement. The conditions of existence and uniqueness of an optimal replacement period were established in the general case and were made explicit in the case of frequently used life-time laws such as the Gamma , Weibull, and truncated Normal law.
Considering the complexity of the model, a computation procedure was proposed to generate the optimal replacement period. The elaborated computation modules do not demand particular restrictions on the choice of the parameters and distributions intervening in the model.
The mathematical model corresponding to this strategy can be the object of further possible extensions, namely taking into account the availability of spare parts. This strategy could also be analyzed considering a larger set of possible actions including replacement and minimal repair which restores the failure rate of the equipment to its value just prior to failure.

6 References

1. Barlow, R.E.et Proschan, F., 1965, *Mathematical Theory of Reliability* (New York: John Withey & Sons).
2. Chelbi A., 1996, *Strategies optimales d'inspection et de replacements d'equipments sujets à des défaillances aléatoires,* Ph.D thesis, Department of Mechanical engineering Laval University, Quebec, Canada.
3. Chelbi A., and Ait-Kadi D., 1997, *Inspection and Predictive Maintenance Strategies,* to appear in the International Journal of Computer Integrated Manufacturing, Vol. 11, No 2.
4. Christer, A.H., and Wang, W., 1995, *A Simple Condition Monitoring Model for a Direct monitoring process,* European Journal of Operational Research, Vol 82, pp 258-269.
5. Çinlar E., 1975, *Introduction to Stochastic Processes.*
6. Press. W., Vetterling. W., Teukolsky. S., Flannery. B., 1992, *Numerical Recipes in FORTRAN, The Art of Scientific Computing.* Second Edition. Cambridge University Press.
7. Rodrigues Dias, J., 1983, *Influence de la période d'inspection sur les coûs dans l'inspection périodique des systèmes,* Revue de Statistique Appliquée Vol 31/4, pp 5-15.
8. Valdes-Flores, C., Feldman, R.M., 1989, *A Survey of Preventive Maintenance Models for Stochastically Deteriorating Single-Unit Systems,* Naval Research Logistics, Vol. 36, pp 419-446.

A strategy for real time flow control and breakage compensation of cutting tools in flexible manufacturing systems: Total Tool Management System (TTMS)

Ralph W.L. Ip, Henry Lau, Felix T.S. Chan

Department of Manufacturing Engineering and Engineering Management, City University of Hong Kong, 83 Tat Chee Avenue, Kowloon, Hong Kong

Abstract. Cutting tools are essential and costly components in a flexible manufacturing system (FMS). Tool location scheduling and tool switching control are vital managerial decision functions for tool installation on CNC machines. A well-designed tool management system should include the features of scheduling, transportation and tool information monitoring. The Total Tool Management System (TTMS) as presented in this paper is a computer based tool monitoring system; it consists of off-line tool sequence planning, real time tool flow control and tool break compensation features. The part geometry, machining features and tool availability are the major criteria at the initial tool scheduling process. Fuzzy type part ordering approaches are employed as the central logical algorithm in the tool scheduling process. The results from a batch of part-mix jobs showed that the average machine utilisation rate and the lateness were 0.78 and 13 minutes respectively for a three machines system. These performance are very close to the best results from the LPT and SST dispatching rules.

1 Introduction

Minimum job delay, high machine utilisation, high manufacturing flexibility and reliable adaptive control are key successful factors in an FMS. Some research studies and industrial cases [6, 17] have demonstrated that using job scheduling and tool management concepts could maximise machine utilisation, shorten make-span, minimise tool inventory and save on tooling costs. However, no simple rules can be directly used in real-time cases of an FMS. In fact, a practical tool management system should possess self-adapting decision making abilities, which can respond to different real-time problems arising from different machining operations, e.g. machine breakdown. Although some tool scheduling philosophies and management models for FMSs [11, 12] have been suggested, little research on real-time basis tool scheduling has been done.

The usual practice in tool selection is to match the geometry of the parts with the tool cutting edges. Some studies have shown the methods of choosing appropriate cutting tools for different machining operations. Smith and Stephenson [14] have proposed a mathematical model to select optimum tool sizes for curved surface machining. Although some sophisticated machining strategies [7] have been developed to widen the application of conventional cutting tools in the machining process, in practice, part geometry continues to be the major criterion in tool selection. The approaches in feature recognition are common investigation topics on CAD/CAM technologies [9]. Part feature information, including geometry, orientation and assembly sequence, can be extracted from CAD data files to assist manufacturing process planning. Many feature recognition techniques have been proposed by researchers, such as graph-based approach, convex decomposition, syntactic pattern and Constructive Solid Geometry (CSG) model. Some researchers stated that better machining outputs can be achieved from a well organised tool plan. They also found that machining results are highly dependent upon the chosen machining parameters, e.g. tool diameter, cutting edge geometry and cutter orientation. Some sophisticated machining strategies [7] can broaden the usage of the conventional machining process. However, part geometry is still the major criterion in the tool selection process.

Some tool scheduling theories and models have been proposed, such as rule based tool scheduling [5] and minimising the number of tool switch approaches [4] but the real-time scheduling techniques have not been well established. Using computer simulation would be an effective strategy in studying a real time scheduling problem [1]. Chan and Smith [2, 3] have conducted a series of simulation studies for flexible manufacturing. In the studies, special tasks in flexible manufacturing systems were evaluated by their simulation models. The tasks include workstation utilisation, production rate, lead time control and material handling system management. Chan and Smith [3] also stressed that using simulation techniques would be an economic strategy in process scheduling and planning.

The aim of this paper is to report the principle of a recently developed tool management system, namely Total Tool Management System (TTMS). The system consists of an off-line and real-time tool assignment scheduler, a new tool list will be recommended in the case of tool break and machine breakdown. The tool exchange, transportation and tool data communication are all controlled by the system in real-time fashion through an electronic network.

2 Design concept of the TTMS

The TTMS is a piece of large scale tool scheduling and monitoring software. It is mainly designed to handle part-mix manufacturing problems among a group of identical machines. The system provides off-line tool scheduling and real-time tool monitoring functions. In the off-line tool scheduling mode, part geometrical information, e.g. size and shape, as initially designed by CAD models are extracted by a CAM based feature recognition software tool. The recognised geometrical features are then used to design a hierarchical machining operation sequence. Cutting tools are assigned to each operation based upon the criteria of actual machining process and their availability. Part programmes for each operation are generated from NC code post-processors. The machining time for each operation is also determined. Figure 1 illustrates the overall scheme of off-line tool scheduling strategy in the TTMS.

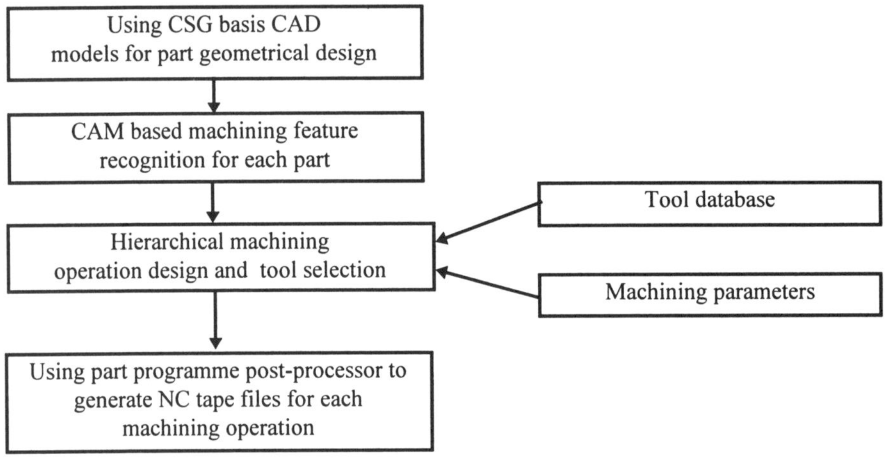

Fig. 1. Tool assignment in off-line mode of the TTMS.

The real-time monitoring of cutting tools in an FMS is a far more complicated task. The TTMS also supports a real-time tool monitoring function to resolve the general operation problems, such as the change of tool / material conditions and machine breakdown. Tool data for each cutter are kept individually in a tool data file and stored in the common tool database of the TTMS. Up-dated tool data are always carried by each cutter using a magnetic type micro-chip system. At the beginning of each batch of job, the scheduled cutters will be loaded into the magazine of the machines. Very often, the capacity of magazines is smaller than the number of the tools required for the entire machining sequence, thus tool sharing strategy is employed in the TTMS. During the machining process, information of tool position and cutter conditions, are continuously checked by the TTMS. When tool break or machine breakdown occur, the pre-programmed tool schedules have to be amended. Information of the broken cutters or halted machines are automatically sent back to the TTMS. In the case of tool break, a new cutter with the same or similar geometry will be selected from the common tool database and used to exchange with the broken one. Due to the reason of some tool geometry change, such as tool length, the original generated part-programmes should be correspondingly re-created. Machine halting may come from many reasons, such as the breakdown of control system or power supply. Thus, the assigned machining operations cannot be finished. When these problems occur, the remaining planned tool schedules have to be modified. All the current information including the unfinished machining operations, machine conditions, tool position on tool magazines and tool data are sent back to the TTMS. A new tool assignment schedule is re-designed in the off-line mode, and a new start of the remaining machining operations will be issued.

3 Principle of part ordering and tool assignment

In Ip et al. [8] work, geometry of a part is designed by a CAD system using the CSG model, and its machining operations are recognised by examining the CSG tree via a CAM model. When the machining features of each part are recognised, the sequence of the operations can be listed in the form of a hierarchical structure as shown in Figure 2.

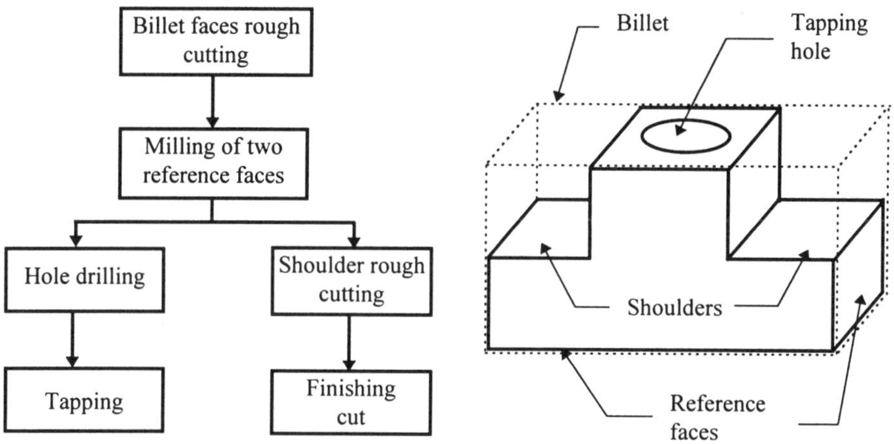

Fig. 2. The recognised machining operations and the machining hierarchy of a tee-bolt.

In the off-line tool scheduling mode for production process planning, the orders of the machining operations in the hierarchy are designed based upon the logical Boolean operation flow in a CSG tree. For example, the two milling operations for producing the shoulders of a tee-bolt are recognised as the two successive subtraction of a cube shape primitive from a rectangular prism. The threaded hole is a combination of drilling and tapping. They are recognised as two subtractions of a cylinder and a helical spring from the rectangular prism.

Cutting tools are chosen from the common tool database in the TTMS using the generated NC machining codes in the hierarchy as shown in Figure 1. In fact, some other factors should also be considered in the tool selection process, such as work-piece material properties, the availability of cutting tools at the current status and machining parameters. The information of the selected tools, including tool type, tool diameter and tool length, for each machining operation will be written into a tool list file.

Parts in a batch of incoming job are needed to be ordered by sorting through their priority weight. Fuzzy approaches [18] are adopted as a tool to plan the part order. The selected cutting tool information in the tool list file and total machining time for each part are the determined results from the off-line tool scheduling mode and the

NC post-processor respectively. These information will be considered together with due date, starting time, quantity of each part in a batch of job in the part selection process. A priority weight which is calculated from fuzzy rules, will be assigned to each part. Using fuzzy logic approach is considered better than other methods, e.g. linear programming [13], queue theory [10] and network models [15, 16], because it can reduce the total computational time and satisfy an aggregated criteria. The schematic diagram of part ordering in the TTMS is given in Figure 3.

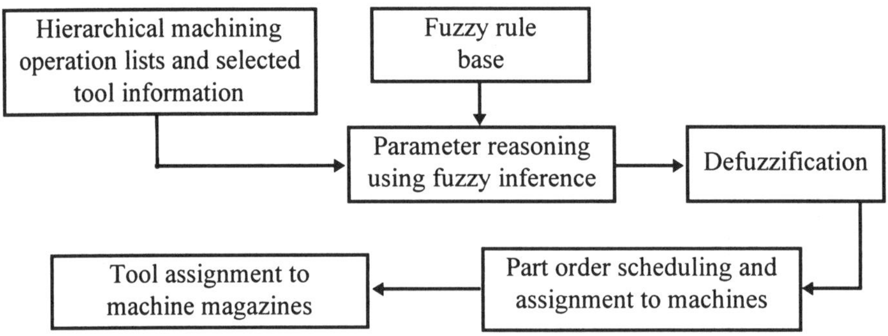

Fig. 3. Fuzzy type part ordering and tool assignment model.

In the part ordering process, the unallocated tools T_a, the number of sharing tools T_s and the slack ratio SR are treated as the input variables to the process. These data can be obtained from the hierarchical machining operation and the tool list file.

T_a : The tools have been scheduled for the specific operations and are waiting to be loaded on machines tool magazine.

T_s : The tools are used for more than one operation, and will be shared between machines.

$$SR = \frac{t_d - t_s - Q\ t_m}{Q\ t_m} \tag{1}$$

where t_d is the part due date, t_s is the starting time of a part, t_m is the total machining time of the part, and Q is the total number of parts. It needs to stress that some other production planning factors will affect the priority in part ordering, however, equation (1) considers the four basic factors only.

The characteristics of the three input variables can be fuzzified, that is to convert them into linguistic variables. The linguistic values for T_a and T_s are labelled as {L, RL, M, RH, H}, and SR is also labelled as {VS, S, M, B, VB}. The abbreviations for the linguistic labels are: 'L' for 'Low', 'H' for 'High', 'R' for 'Rather', 'M' for 'Medium', 'V' for 'Very', 'S' for 'Small', and 'B' for 'Big'. Figure 4 shows the characteristic function of T_a, T_s and SR with fuzzy labels.

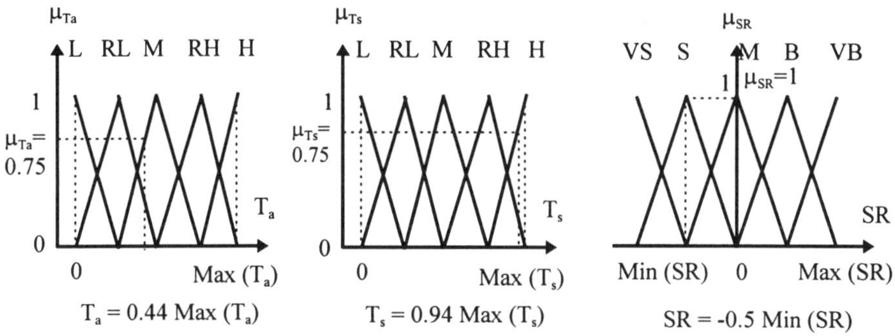

Fig. 4. Characteristic function of T_a, T_S and SR.

Out of all the possible combinations of the linguistic labels from a total of 125 fuzzy rules, i.e. 5 x 5 x 5. The output part variable, namely part priority weight w, can be grouped into 45 different grades (0 to 44), and the characteristic functions of the part priority weight is shown in Figure 5.

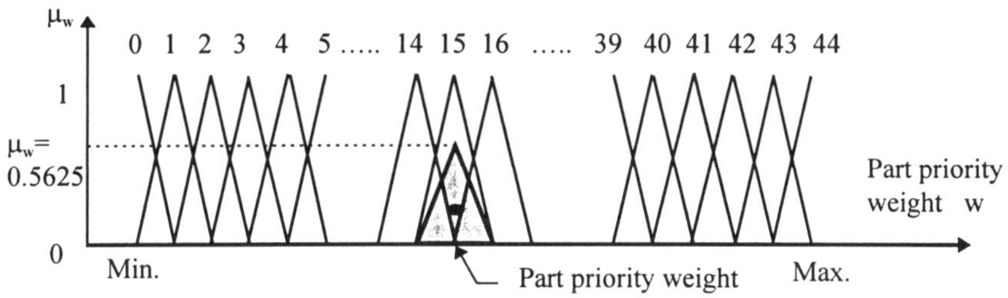

Fig. 5. Characteristic functions of the part priority weight.

Inference engine, by using product-operation method, is used to calculate the grades in the fuzzy characteristic functions of the part priority weight. For example, if T_a = 0.44 Max (T_a), T_s = 0.94 Max (T_s) and SR = -0.5 Min (SR), the corresponding values of μ_{Ta} = 0.75 on characteristic function M, μ_{Ts} = 0.75 on characteristic function H, μ_{SR} = 1 on characteristic function S of the three input variables. For the results of operating the fuzzy rules, the grade of the part priority weight is 15 and the value of μ_W = 0.5625 (0.75 x 0.75 x 1). In the defuzzification process, Centre of Area (COA) method is used to calculate the part priority weight from the corresponding grade. The parts are scheduled in according to their priority order. In fact, high priority weight parts are always scheduled at the top of the operation list.

4 Part-mix scheduling case study

In a batch of jobs, five different parts were designed by a CSG based CAD software. The recognised machining operation hierarchy, selected cutting tools and the machining time for each operation are illustrated in Figure 6.

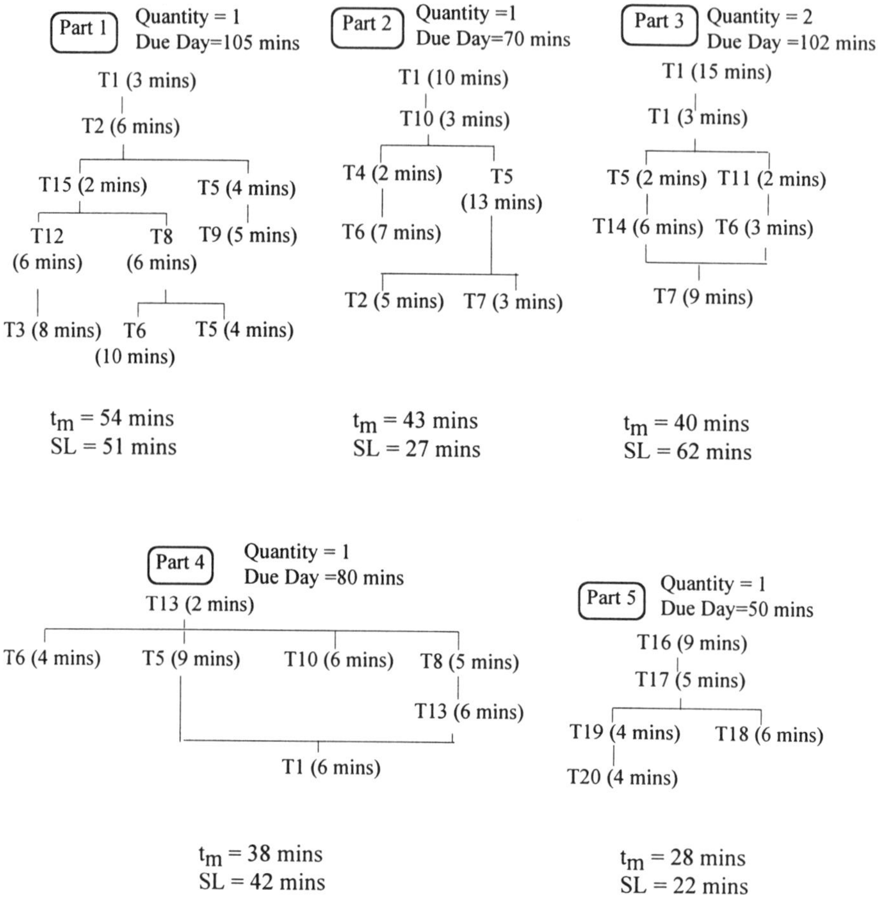

Fig. 6. The hierarchical machining operations and tool information of the five different parts.

Three identical machining centres numbered as 1, 2 and 3 were used and twenty different cutters numbered from T1 to T20 were selected. A tool transportation robot served the machines for the purpose of tool sharing. The tool exchange time between two adjacent machines was 2 minutes, and 3 minutes was required to bypass a machine. The starting time t_s of the jobs was at the 0 minute.

Using the fuzzy basis part ordering strategy, the part 3, 5 and 2 were firstly assigned to machine 1, 2 and 3 respectively. The remaining parts 4 and 1 were scheduled into machines 1 and 2 after the completion of the parts 3 and 5 on the machines. The second part 3 was assigned to machine 3 at last, see Figure 7. The gaps between operations in the parts show that the operation cannot start due to the reason of tool waiting.

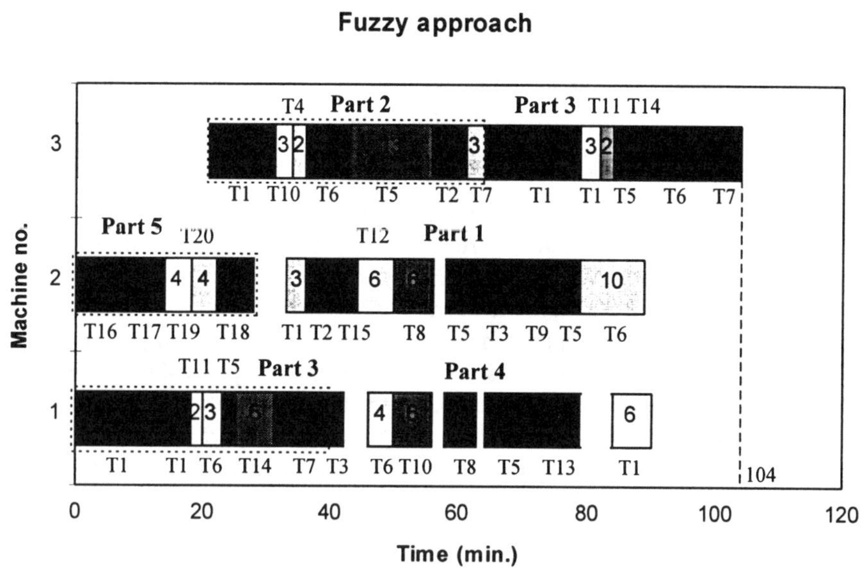

Fig. 7. The scheduled parts and corresponding cutting tool assignment list.

The same batch of job as shown in Figure 6 was assigned to the machines again using different dispatching rules, such as Short Processing Time (SPT), Longest Processing Time (LPT), Earliest Due Date (EDD) and Shortest Slack Time (SST). Table 1 displays a summarised result of the schedules under the rules. In fact, the final completion time as shown in Table 1 means the end time of the batch. Three part ordering performance measuring quantities: machine utilisation, number of tardy jobs and lateness are also listed in Table 1. Machine utilisation is defined as the ratio of the total machining time on the machines to the final completion time, the amount of time delay after the target due date is the lateness and the number of parts which cannot meet the due date is known as the number of tardy jobs.

Table 1. Results for different scheduling strategies.

Scheduling methods	Final completion time (min.)	Machine utilisation 1	2	3	Mean	Tardy job (Part)	Lateness (min.)
SPT	128	0.53	0.63	0.73	0.63	1 & 2	23, 33
LPT	101	0.93	0.80	0.67	0.80	5 & 4	51, 21
EDD	134	0.51	0.62	0.69	0.61	1	29
SST	110	0.75	0.75	0.71	0.74	3	8
Fuzzy	104	0.75	0.79	0.80	0.78	3 & 4	2, 11

5 Discussion

The results as shown in Table 1 are mixed. None of the scheduling methods can offer the best solution for the part ordering. The results indicate that the LPT rule can give the highest mean machine utilisation, i.e. 80%; and only one tardy job is found in the batch when the EDD and SST rules are employed. However, the minimum lateness, i.e. 8 minutes occurs when the parts are scheduled by the SST rule. Although the proposed fuzzy basis part ordering strategy cannot provide the best solution in the three output performance measurement, the results are very close to the best one for all measurements. For example, the average machine utilisation is 2% less than the top one, and the lateness is the second least. It is easily to explain the results that the fuzzy approach considers a group of scheduling criteria instead of just a single constraint, thus, the solution is given by compromising the requirements from each individual rule. This can explain the reason why the minimum lateness result is given by the SST rule, and the highest machine utilisation rate is given by the LPT rule. In fact, the average machine utilisation given from the SST, i.e. 0.74 and the lateness given by the LPT, i.e. 51 and 21 minutes, are not as good as the results given by the fuzzy scheduling method.

6 Conclusion

The investigated TTMS is a tool scheduling and monitoring strategy used in an FMS. Part geometry designed by CAD models are initially extracted from the system database and the machining features are recognised by feature based CAM models. Cutting tools are assigned to each recognised machining operation to form a hierarchy. The parts are scheduled to machines using fuzzy approaches in the off-line mode. Tools can be shared between machines and a closed loop tool monitoring function is designed for real-time control. The scheduled part orders and tool sequence of the remained jobs will be re-ordered when tool break or machine breakdown occur. The part scheduling case study indicated that the proposed fuzzy approach can offer a compromised result for a part-mix schedule. Although the concept of the proposed tool management system was developed, some trials should be carried out to validate its performance and reliability. A real-time test of the re-scheduling ability of the system will be intensively studied.

References

1. Carrie, A.S., *Simulation of manufacturing systems*, John Wiley & Sons (1988).
2. Chan. T.S. and Smith, A.M., Simulation aids JIT assembly line manufacture: A case study, *International Journal of Operations and Production Management*, (1993) Vol. 13, No. 4, 50 - 74.
3. Chan. T.S. and Smith, A.M., Simulation approach to assembly line modification: A case study, *International Journal of Manufacturing Systems*, (1993) Vol. 12, No. 3, 239 - 245.
4. Crama, Y., Kolen, A.W.J., Oerlemans, A.G. and Spieksma, F.C.R., Minimising the number of tool switches on a flexible machine, *International Journal of Flexible Manufacturing Systems*, (1994) Vol. 6, 33 - 54.
5. Grabot, B. and Geneste, L., Dispatching rules in scheduling: a fuzzy approach, *International Journal of Production Research*, (1994) Vol. 32, No. 4, 903 - 915.

6. Han, M.H., Na, Y.K. and Hogg, G.L., Real-time tool control and job dispatching in flexible manufacturing systems, *International Journal of Production Research*, (1989) Vol. 27, No. 8, 1257 - 1267.
7. Ip, W.L.R., The application of an inclined end mill machining strategy on 3-axis machining centre, *International Journal of Machine Tools and Manufacture*, (1993) Vol. 33, No. 2, 115 - 133.
8. Ip. W.L.R., Cheng, K.K., Hui, I.K. and Chan, T.S.F., Dynamic tool management in FMS- A practical approach, *IEEE International Conference on Systems, Man and Cybernetics*, (1995), Vancouver, Canada, 120 - 125
9. Joshi, S. and Chang, T.C., Graph-based heuristics for recognition of machined features from a 3D solid model, *Computer Aided Design*, (1988) Vol. 20, No. 2, 58 - 66.
10. Khator, K. and Leung, L.C., Intermediate Tool Requirement Planning for FMS, *Journal of Manufacturing Systems*, (1994) Vol. 13, No. 1, 9-19.
11. Kokko, T., Total tooling management system (TMS), *Proceedings of 7 th International Conference on Flexible Manufacturing Systems*, (1988) Stuttgart, Germany, 69 - 77.
12. Little, D. Kehoe, D.F. and Al-Maliki, I., Tool management: the key issues for integrated manufacture, *Proceedings of 7 th International Conference on Flexible Manufacturing Systems*, (1988), Stuttgart, Germany, 53 - 68.
13. Sarin, S.C. and Chen, C.S., The machine loading and tool allocation problem in a flexible manufacturing system , *International Journal of Production Research*, (1987) Vol. 25, No. 7, 1081-1094.
14. Smith, G. and Stephenson, H., The selection of optimum tool sizes for three-axis and multi-axis machining of highly curved features, *Proceedings of the 7 th National Conference on Production Research*, (1991) Hatfield, 316 - 320.
15. Sodhi, M.S., Askin, R.G. and Sen, S., Multi-period tool and production assignment in flexible manufacturing systems, *International Journal of Production Research*, Vol. 32, No.6, 1281-1294.
16. Sodhi, M.S., Agnetis, A. and Askin, R.G., Tool Addition Strategies for flexible manufacturing systems, *The international journal of flexible manufacturing systems*, (1994) Vol.6, 287-310.
17. Sodhi, M.S. and Lamond, B.F., Tool life models for flexible manufacturing system, *Proceedings of the 4th International Conference on Data and Knowledge Systems for Manufacturing and Engineering*, (1994) Hong Kong, 399-404.
18. Yan, J., Ruan, M. and Power, J., *Using Fuzzy logic*, Prentice Hall (1994).

ADVANCED MANUFACTURING PROCESSES

Computer-controlled Liquid Bulge Forming of Tubes

S.Thiruvarudchelvan, G.L.Seet and H.Y. Wong,

School of Mechanical and Production Engineering,
Nanyang Technological University,
Nanyang Avenue, Singapore 639798.

Abstract. A computer controlled liquid bulge forming machine has been designed, fabricated and commissioned. It enables internal pressure, axial contraction and axial force control to be automatically monitored and controlled. The system incorporates pressure and displacement sensors to allow the computer to monitor and display the process parameters in real-time processes. The bulging pressure can also be adjusted automatically and the process rates varied accordingly to optimize the overall bulging process.

This machine consists of a horizontal press frame with two opposed hydraulic units at the ends and two sliding units carrying the die halves. Two sets of three hydraulic cylinders open and close the die halves containing the tube to be bulged. Once the die halves are closed two differential pistons, operating in two pressure vessels, attached to the end plates, move in and seal the tube. The hydraulic pressure is then applied to the tube through the central bore of one of the differential pistons. Pistons, actuating under another hydraulic circuit, then apply an axial compressive force on the tube. When the bulging is complete, the tube is de-pressurized, pistons withdrawn, die halves opened and the bulged tube removed.

During the bulging process the hydraulic pressure inside the tube, the axial force on the tube and the axial contraction of the tube are varied independently to produce a bulged tube. The profiles of the pressure and the axial force with the axial contraction are monitored in real-time and displayed on the computer display. Several trials were conducted to determine the best process paths. Once this is determined the computer is programmed to follow these profiles for repeated bulges. Results from the bulging experiments conducted with this machine are reported in this paper.

1 Introduction

Mechanical bulging, bulging with the aid of flexible tools such as urethane, hydraulic bulging [1-8] and electromagnetic bulging [9] are methods used to produce useful engineering components from tubular metal blanks. These methods have their advantages and disadvantages. However, the hydraulic bulging process has the following advantages:

ability to produce larger circumferential expansions of the bulged tube.

flexibility of being able to monitor and control the internal hydraulic pressure, the axial force and the axial compression to achieve the desired circumferential expansion of the tube.

In spite of the advantages and the variety of products that can be produced by the liquid bulge forming process, published information on the process is limited. Large industrial-type hydraulic bulging machines are used in Japan for the manufacture of tubular components. These are elaborate and expensive to be used for research purposes. A small scale experimental machine was, therefore, developed and used for these series of tests.

Some investigations of the hydraulic bulging process have been reported [1-8]. These research findings are of direct relevance to the present work on hydraulic bulging. There are two modes of failure in the hydraulic bulging process:

excessive axial compressive force on the tube causing buckling or collapsing of the tube.

excessive internal pressure causing plastic instability in tension leading to excessive wall thinning and eventual bursting of the tube.

The bulging process has to be controlled so that the pressure and the axial stress on the tube remain within the limits for satisfactory forming. This can also be expressed in terms of the ratio of the meridianal stress, σ_α to the hoop stress, σ_θ in the bulged region. It was found that for satisfactory bulging [7], $0 \geq \beta \geq -1$, where $\beta = \frac{\sigma_\alpha}{\sigma_\theta}$.

Sauer et al [3] have carried out theoretical and experimental studies of the free bulging of thin tubes under internal pressure and axial compressive stress. They obtained a critical stress for buckling during bulging. These authors have also developed a necking criterion to predict the onset of instability in tension under internal pressure and axial compressive stress. They showed that greater expansion and lesser thinning are possible with increased axial compressive stress.

A device for applying an axial compressive force proportional to the hydraulic pressure was developed and bulging tests conducted [4]. The device eliminated the need for an elaborate control system to coordinate the hydraulic pressure and the axial compression of the tube. It is easy to operate, but the experimental results indicated that the circumferential expansion of the tube bulged with the device is limited. Hence, it was decided to design and develop an experimental tube bulging machine where the pressure and the axial compression can be independently controlled . The results of investigations using this machine are reported in this paper.

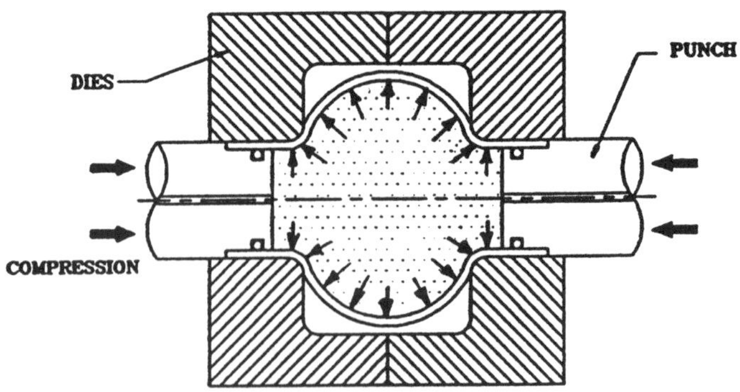

Fig. 1. A schematic diagram of the basic tooling for hydraulic bulging of tubes

2 Operation of the liquid bulge forming machine

The basic tooling needed for hydraulic bulging of tubes is illustrated in Figure 1. The tube to be bulged is placed inside the die halves and the die halves clamped as shown. Thereafter, two punches advance and contact the ends of the tube to seal the space inside. Through the central axial bore of one of the punches hydraulic oil is filled inside the tube. The pressure inside the tube is maintained by a hydraulic pump connected to the end of the punch, that is farther away from the tube. The punches are then moved towards each other at the same speed, and an axial compressive force applied on the tube helps feed it as bulging progresses. The axial movements of the punches, the axial force on the tube and the hydraulic pressure must be coordinated and controlled to stay within a permissible region for successful bulging. When the tube is bulged to the desired circumferential expansion, the axial force is released, the fluid de-pressurized and the punches withdrawn. At this stage the die halves can be opened and the product removed, thus completing the forming cycle.

The experimental tube bulging machine developed is shown in the sectional elevation in Figure 2. Components of the machine are mounted on a horizontal press frame consisting of three round tie bars and two rectangular end plates. The press frame and the tie bars are designed to withstand a central axial force of 1 MN. The two die halves are mounted on two slider plates, which slide freely on the tie bars. Each end plate carries three double acting hydraulic cylinders (each of capacity 40 kN), the plungers of which are attached to the corresponding slider plate. By actuating these hydraulic cylinders the die halves can be opened, closed, or held clamped together. These cylinders are connected to a hydraulic pump through a flow divider, so that the die halves can open or close in synchronism, and close up at the mid-point of the press frame.

The pressurized fluid from a pump enters the punch through a flexible hose connected to the free end of one of the punches, and flows into the tube through the central bore of this punch. The other end of the punch that applies the axial load on the tube enters the die half. The free ends of the punches extending out of the compression cylinders are used to measure the end displacements of the tube being bulged.

By supplying an independent hydraulic pressure to the annular space at the larger diameter of the punch, it is pushed against the tube to generate the axial force needed for bulging. A transducer is used on this line to monitor the hydraulic pressure and thereby control the axial force on the tube. A similar transducer is used for monitoring the internal pressure. The forward displacement of the punch is thus determined by this pressure supplied through a lateral port on the axial compression cylinder. While the punch moves forward, the hydraulic fluid in the annular space between the compression cylinder and the end plate moves out through another port (inclined) on the compression cylinder. To retract the punch, pressurized fluid is admitted through the inclined port into the annular space and the fluid in the opposite annular space at the middle of the punch moves out through the lateral port on the compression cylinder. The axial force on the tube, is thus, generated by a pressure independent of the internal pressure in the tube.

When the liquid bulge forming operation is complete, the axial force and the fluid pressure are reduced to zero. Then the punches are withdrawn from the tube to their initial positions. The clamping cylinders are retracted to open the die halves and to allow the bulged tube to be removed.

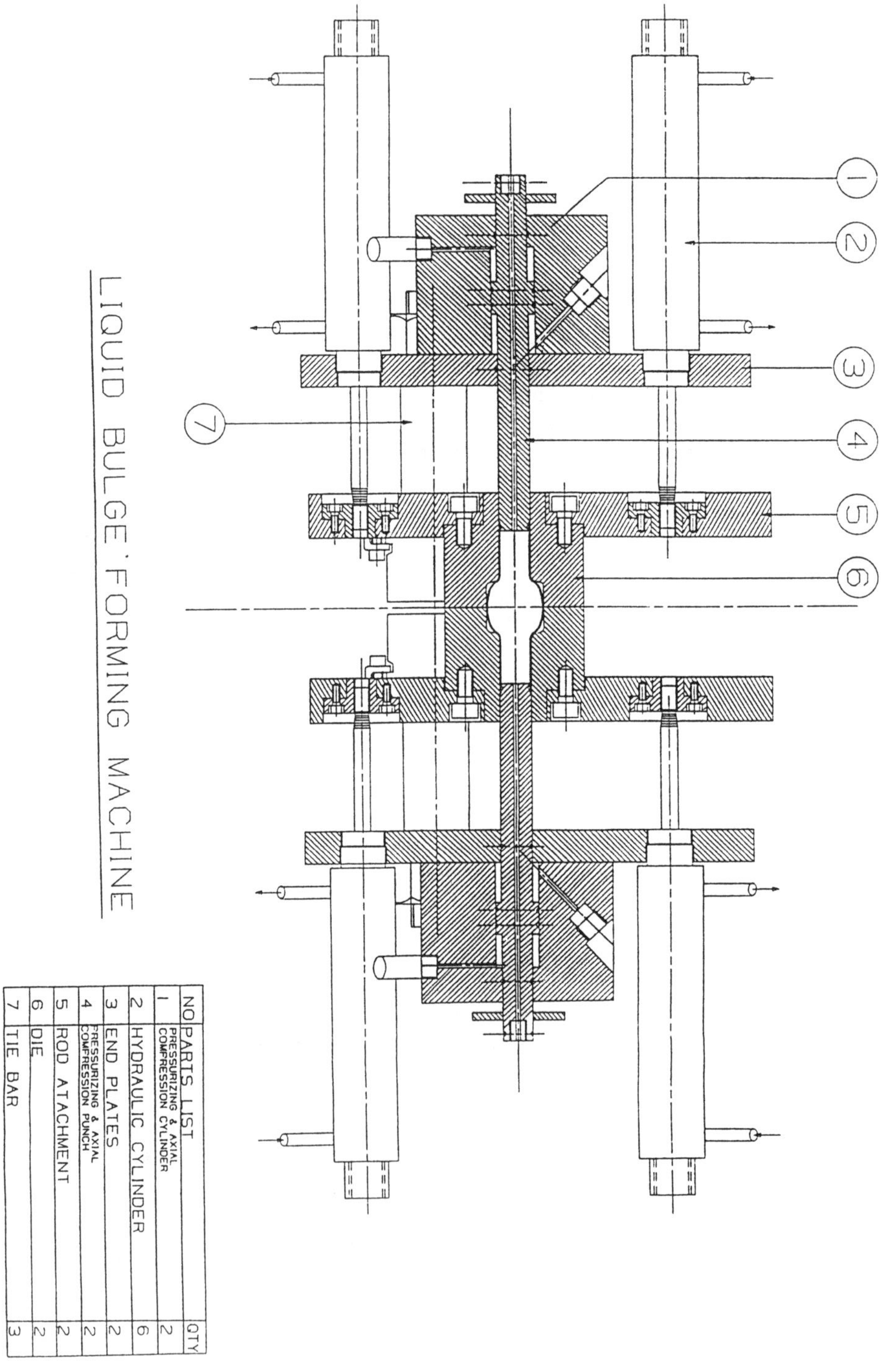

Fig. 2. A sectional view of the tube bulging machine

3 Computer Monitoring and Controlling of Process Parameters

A personal computer was introduced into the system to perform monitoring and controlling of the process. The three important parameters to be monitored and controlled are:

the axial compression of the tubular blank.

the internal fluid pressure.

thc prcssurc that generates the axial force on the tube.

This implies that two types of transducers would be needed; two linear differential voltage transducers (LDVT) and two pressure transducers.

In order to accurately track the total displacement of the blank that is in contact with the two punches, two spring-loaded LDVTs have been installed on the machine; one for each punch at the ends of the machine using the stopper discs as flags. One pressure transducer was connected between the end of one of the punches and the flexible hose connecting the punch to the hydraulic pump. The other pressure transducer was installed on the hydraulic line that supplies the pressure to exert the axial force.

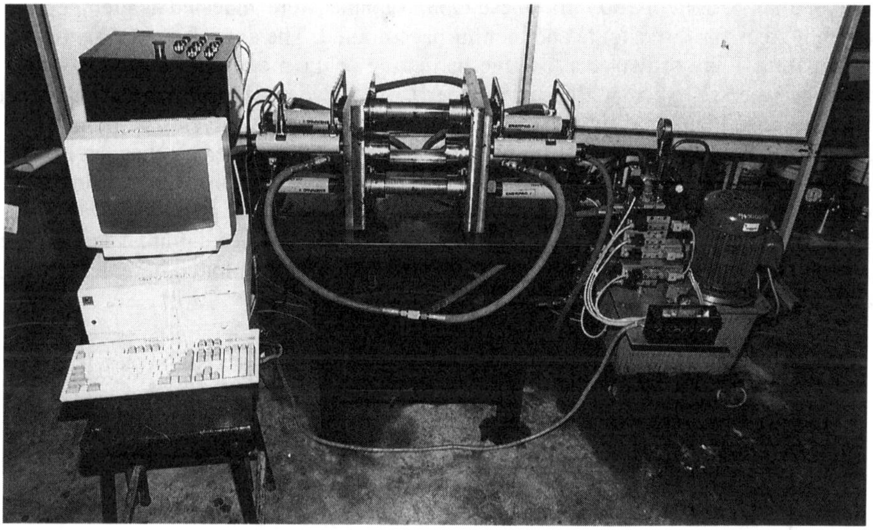

Fig. 3. Photograph of the machine

These transducers are connected to a data acquisition card installed in the computer. It receives the analog signal (displacement or pressure) from the respective transducers and performs an ADC (analog to digital conversion). The raw data, from the two LDVT readings are summed to give the total compression of the tube and the data are subsequently scaled to convert the ADC readings to engineering units. The readings of the transducers are stored and displayed in a graphical form showing axial compression, internal pressure and the pressure that generates the axial force during a bulging operation. The software for the above processes were written in C language.

The closed loop concept is implemented on the present system. A data acquisition card (DAC) and a proportional pressure reducing valve are used. An electrically

actuated pressure reducing valve is used to actuate and control using electrical signal sent out from the computer. The closed feed-back loop, which allows the operator not only to capture the process parameters, but also sends a feed-back rectifying signal to automatically adjust these parameters for better results. In this feed-back system, when the computer receives the transducer input signal from the data logger, it would be programmed to process these data and perform some computation and finally send out a proportional voltage signal to the valve and adjust the hydraulic pressure accordingly. Several different internal pressure-axial compression characteristics that can be preset and applied to the bulging process. The optimum pressure-displacement characteristic can then be selected from the best bulged tube. This characteristic can then be stored in the computer and used again and again to control the process.
The rate of increase of the internal pressure in the tube was controlled by a software written for this purpose to regulate the control valves simultaneously throughout the bulging operation. Controllable pressure build-up rate was achieved by continuous adjustment of the control valves. A photograph of the machine is shown in Figure 3.

4 Experimental and Results on Liquid Bulge Forming

Aluminium (6063T5) tubes of outer diameter 25.4 mm and thickness 1 mm were machined to a length of 150 mm. These tubular blanks were annealed at a temperature of 415 degrees for three hours and then furnace-cooled. The annealed specimens were tested within a day or two to avoid age-hardening of the metal. The outer surfaces of the tubes were polished with emery paper to remove machining marks and other scratches. The blank is then placed in position within the die halves. The punches are advanced to hold the blank firmly in position. The control system is then set up, and the two LDVT displacement transducers are moved into positions for measuring the axial compression of the tube. The spring-loaded plunger of the LDVT is pressed against the punch-end stopper disc that also acts as the reference datum. The software is used to capture and process the data generated, and a graph plotted in real time. The system is now ready to control the process parameters.

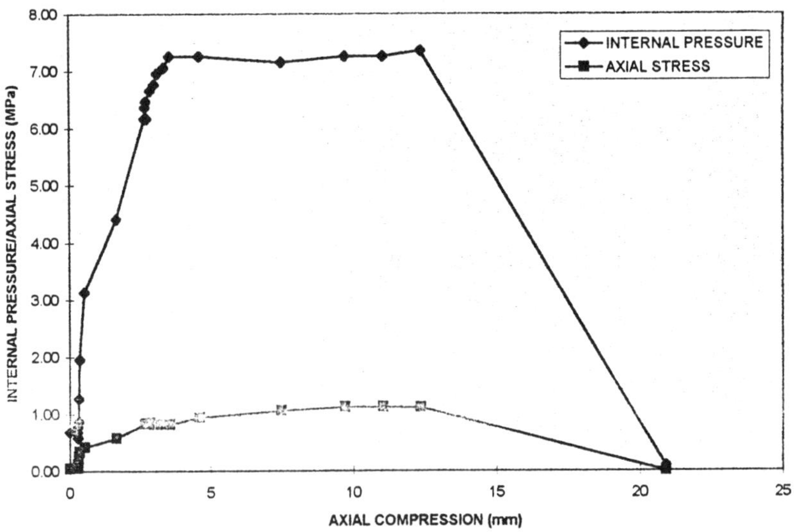

Fig. 4. Pressure versus axial compression for the buckled tube

Several experiments were conducted with different values for the internal pressure, the axial force and the rates of increase of these quantities. Figure 4 shows the internal pressure and the pressure to generate the axial force versus axial compression for a test which resulted in a buckled tube. The internal pressure increased gradually to a value of 75 kg/sq. cm and the pressure generating the axial force reached a steady 191 kg/sq. cm. When the axial compression of the tube reached value of about 12.36 mm the tube buckled quickly and then ruptured.
In another test the final value of the internal pressure was 80 kg/sq. cm and the final value of the axial force generating pressure was 151 kg/sq. cm. when the tube burst. The two pressures measured versus the axial compression for this test is shown in Figure 5. When the total compression of the tube was about 4 mm the tube burst and the pressure dropped rapidly to zero as seen in Figure 5. The failure in this case is attributed to excessive hoop stress and thinning caused by the internal pressure. The pressure-build-up rate was also too high.

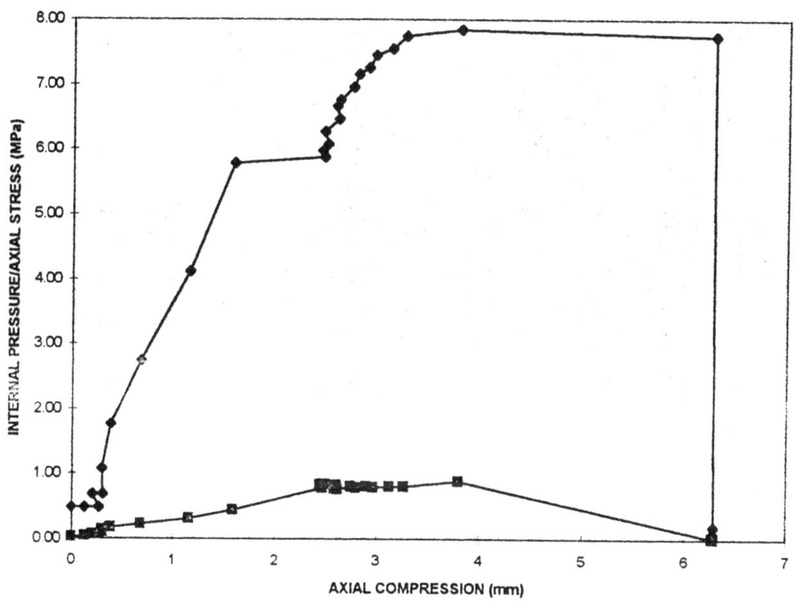

Fig. 5. Pressure versus axial compression of the burst tube

For the successfully bulged tubes, the internal pressure was increased gradually (faster in the initial stages and more gently in the final stages) to a maximum of 76 kg/sq. cm, while the pressure generating the nominal axial stress on the tube was increased to 188 kg/sq. cm. Figure 6 shows a photograph of the bulged tube, while Figure 7 shows the pressures with axial compression for this test. Total axial compression of the tube was 11.15 mm and the bulged diameter was 33.75 mm approximately. During the test the ratio of the internal pressure to the pressure that generated the axial compression on the tube was more or less constant until a compression of about

5 mm. Thereafter the tube bulged steadily while the pressures increases gently to a maximum values of 76 and 188 kg/sq. cm at the final axial compression.

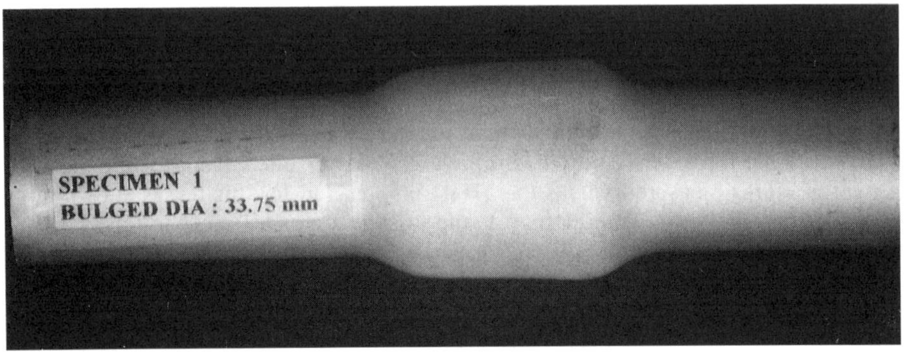

Fig. 6. Photograph of a satisfactorily bulged tube

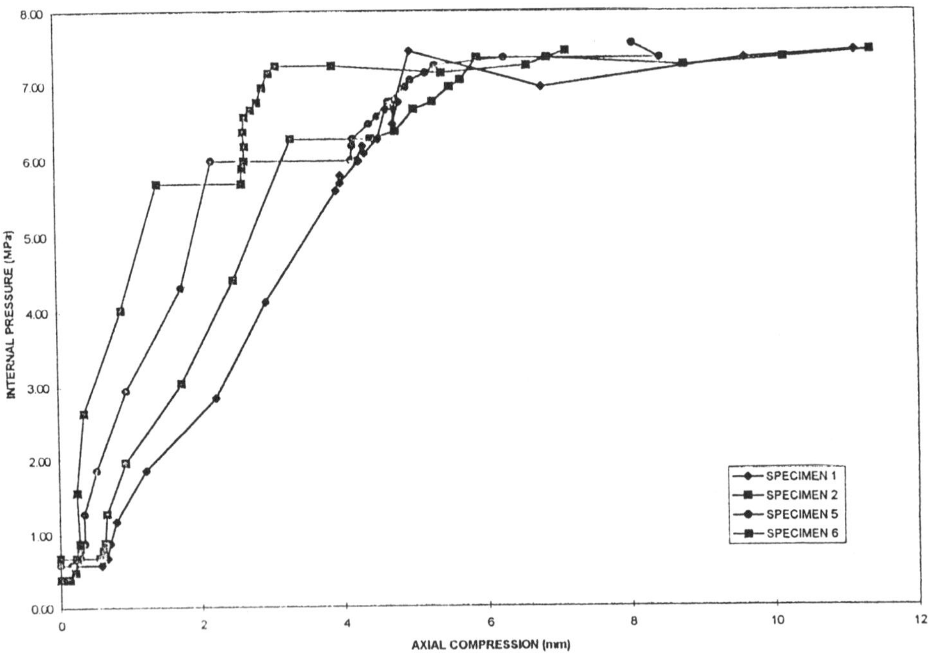

Fig. 7. Pressure versus axial compression of a satisfactorily bulged tubes

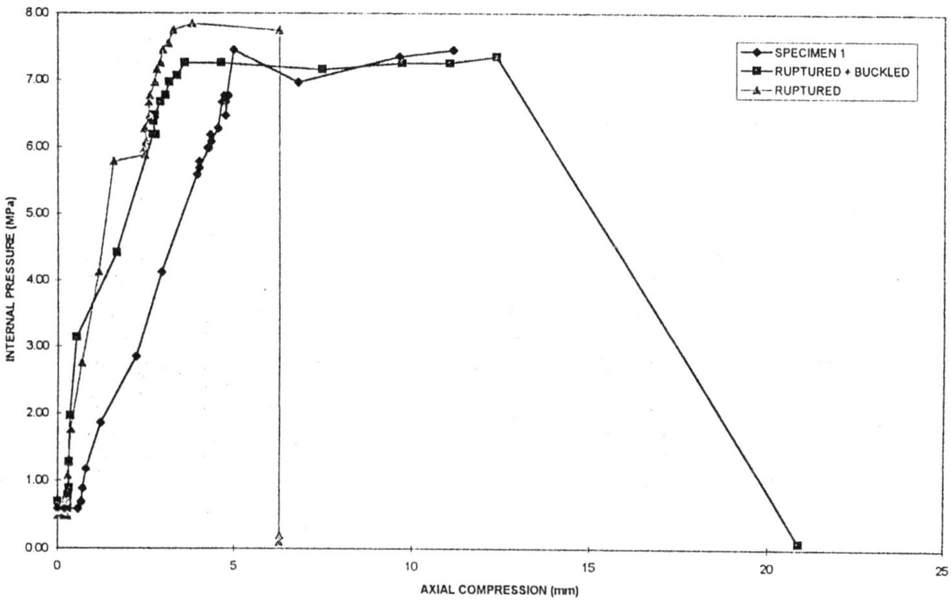

Fig. 8. Comparison of the internal pressure versus axial compression of buckled, burst and satisfactorily bulged tube

Strains on the bulged tubes were calculated from the measured distortions of circles of diameter 1.89 mm imprinted on the tube by photochemical plating before the bulging process. The thickness strains were calculated from measurements of the thickness of the bulged tube cut open longitudinally into two halves. Strains on the bulged diameter of the tube at five positions are tabulated in Table 1.

Table 1. Strains on the bulged tube

Measured Point	1	2	3	4	5
Circumferential strain, ε_θ	0.236	0.227	0.237	0.252	0.235
Meridianal Strain, ε_α	-0.203	-0.174	-0.203	-0.181	-0.188
Thickness Strain, ε_τ	-0.089	-0.094	-0.101	-0.088	-0.078
Strain Ratio, $\rho = \varepsilon_\alpha/\varepsilon_\theta$	-0.857	-0.767	-0.86	-0.72	-0.800
Stress Ratio, $\beta = (2\rho+1)/(2+\rho)$	-0.625	-0.433	-0.622	-0.344	-0.5

5 Discussion

Figure 8 shows the comparison of the internal pressure versus axial compression of the tube in the tree cases of buckled, burst and successfully formed tubes. It can be seen that the satisfactorily bulged tube had the most gentle slope amongst the three, followed by the buckled and the burst specimens, respectively. For the burst specimens the internal pressure build-up was too high for such a low value of axial

compression that both the circumferential and meridianal strains were increasing positively and their sum would reach the value of n in the stress-strain curve $\sigma = a\varepsilon^n$, and necking would occur.
In Table 1 all the successfully bulged tubes have the strain ratio, ρ fall in the region, $-1 < \rho < -1/2$, for the stress ratio as shown in reference [7]. The thickness strains were rather small ($< |0.094|$) in the successfully bulged tubes.

6 Conclusions

Controlling the internal pressure, the axial-force generating pressure and the axial compression of the tube during bulging is found effective for bulging of tubes with substantial circumferential expansions. The maximum circumferential expansion obtained with the present set up is about 32.9 percent. By displaying the process parameters in a graphical form during the bulging process, better characteristics can be chosen for effective bulging of tubes. Feed-back control of the process using a computer ensures the optimum path for the process parameters and bulging can be conducted with preset pressure-displacement characteristics, thereby obtaining larger circumferential expansions of the tube.

References

1. T.R.Barlow, R. Crampton, and M.S.J.Hashmi, The design of an experimental hydraulic bulge forming machine, Proc. 21st Int. MTDR Conf., Manchester, Macmillan, London, (1980) 475-481
2. M.S.J.Hashmi, Radial thickness distribution around a hydraulically bulge formed annealed copper T-joint: Experiments and theoretical predictions, Proc. 22nd Int. MTDR Conf., Birmingham, Macmillan, London, (1981) 507-516.
3. W.J.Sauer, A.Gotera, F. Roff and P.Huang, Free forming of tubes under internal pressure and axial compression, Trans. NAMRC-VI, SME, Michigan, (1978) 228-235
4. S.Thiruvarudchelvan and A.C.Lua, Bulge forming of tubes with axial compressive force proportional to the hydraulic pressure, *J. Mater. Shaping Technol.*, **9** (1991) 133-142
5. S.Thiruvarudchelvan, Ang Hock Eng and Seet Gim Lee, Design and development of a hydraulic tube bulging machine, MANUTECH'92, Manufacturing Symposium, Singapore, (1992) D3/1-D3/9
6. Fritz Dohmann and Christoph Hartl, Liquid-bulge-forming as a flexible production method, *J. Mater. Process Technol.*, **45** (1994) 377-382
7. J Tirosh, A. Neuberger and A. Shirizly, On tube expansion by internal fluid pressure with additional compressive stress, *Int. J. Mech. Sci.*, **38** (1996) 839-851.
8. S.Thiruvarudchelvan, G.L. Seet and H. E. Ang, Computer monitored hydraulic bulging of tubes, *J. Mater. Process Technol.*, **57** (1996) 182-188.
9. H.Zhang, M.Murata and H.Suzuki, Effects of various conditions on tube bulging by electromagnetic forming, *J. Mater. Process Technol.*, **45** (1994) 113-121.

Automatic Mathematical Optimization of Sheet Metal Forming Processes

L.T. Kisielewicz[1], P. Guyon[2], E. Haug[2] K. Ando[3]

[1]ESI Asia Pacific, Japan [2]PSI, France [3]Nihon ESI, Japan

Abstract. The paper describes the first application of mathematical optimization applied to the numerical tryout of dies as applied with sheet metal forming simulation products. This new technique is settled in the industrial environment and needs are described both from a theoretical and a practical standpoint with two application examples.

1.0 Introduction

With the common trend among major automakers to reduce their overall design cycle from the 60 months that use to be the norm a few years ago to the US automakers' target of 30 months by the year 2000, not to mention the 24 months that Honda of Japan is aiming at, the practical solution is to shift the organization from a sequential engineering method, in which each team is working after the upstream team completed its tasks, to a simultaneous engineering organization in which all teams share and exchange data and enable the product engineers and process engineers to work jointly at a very early stage of the design project. In order to make that new process possible, the only practical solution is to rely on Virtual Prototype Testing, i.e. on numerical simulation, provided of course that such simulations can be reliable and predictive. The use of numerical simulation of metal forming at very early project stages make it possible to improve the quality of the products and to reduce both the costs and time schedules. As the movement towards extensive usage of numerical simulation is nowadays well underway, the following target set by the industry is to obtain optimum solution with respect to functional requirements, cost and duration of manufacturing and reliability and maintainability of their products.

Conceptual design is characterized by ingenuity, creativity and engineering judgment of the designer. It is a critical stage of the design process, dealing with the overall planning of a system to serve its functional purposes. At this stage, the designer experiences the greatest challenges as well as chances of success or failure. Selection of the overall technology, type of structure and materials are some of the technical decisions made by the designer at the conceptual design stage. In general, this part of the design cannot be performed by a computer. The designer may, however, use the computer to make systematic investigations into the relative merit of several competing concepts in search of the best one.

Optimization within a selected concept is the stage when the designer applies a trial-and-error procedure to select the best alternate among several possible designs satisfying the functional requirements. The computer is most suitable to carry out that part of the design using methods of automated search for the optimal solutions. Optimization is an automated design procedure giving the optimal values of certain design quantities considering prescribed criteria and constraints.

Numerical optimization methods are usually based on a branch of numerical mathematics called programming methods. The recent developments in this branch are closely related to the rapid growth in computing capacities affected by the development of computers. In the numerical methods, a near optimal solution is automatically generated in an iterative manner. An initial guess is used as starting point for a systematic search for better solutions. The search is terminated when the optimization criteria are satisfied.

This paper present a method for automatic optimization of sheet metal forming processes. It recalls briefly the situation in the industry including a brief presentation of numerical methods applied to the simulation of sheet forming processes. A numerical method for non-linear optimization is then described followed by two practical examples of application including one for an industrial case.

2.0 Numerical Die Tryouts

2.1 Basic Industrial Needs

In order for any numerical simulation based solution to be effective for industrial application of sheet metal forming, it must be capable of predicting all physical phenomena occurring [Fig.1]. More important than this basic criterion is the capability of predicting all possible troubles that can develop during the process (wrinkling, bulging, tearing, etc.) with sufficient accuracy and reliability to support the decisions of the engineers for modifying the geometry of the tools or a process parameter. This prerequisite set of conditions calls for sound physical modeling of the non linear structural dynamics underlying the sheet metal forming apparent phenomena.

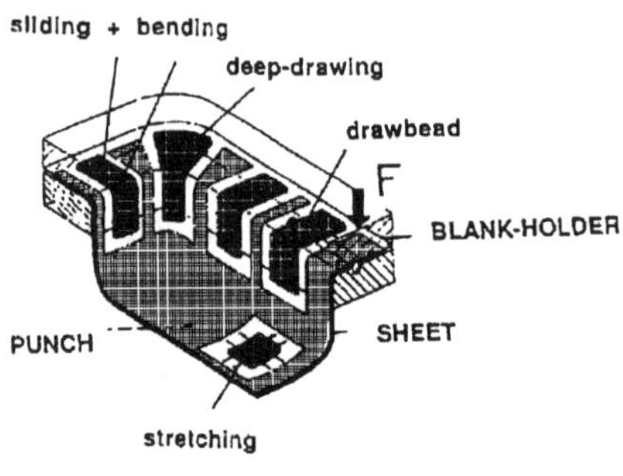

Fig. 1 - Phenomena in Sheet Metal Forming

If proper physical modeling is necessary, it appears not to be sufficient by itself. A number of academic developments have for some time addressed this issue and provide acceptable solutions. However this did not mark the emergence in the industry of numerical simulation for sheet metal forming application. A second set of important criteria relates to the ease of use and acceptance by the process engineers to use the new technology. This requires to insert between the sound modeling of physics and the process engineer an interface that hides all numerical features behind the corresponding real world concepts. Numerical simulation software products need to learn the language of the users for a successful industrial implementation and not the reverse of expecting from the process engineers to learn the language of the computer based solutions.

2.2 Numerical Simulation

PAM_STAMP™ is a finite element based software product specially developed and validated for the simulation of sheet press forming processes [5]. It takes into account large elastic-plastic phenomena with isotropic strain hardening, including considerations for the anisotropy of the material resulting from the rolling of steel, for the strain rate effects and for the friction at the contact surfaces between the blank and the tools.

The software product can handle all phases of the forming process including the deformation under gravity loading when the blank is positioned on the tools, the deformation during blankholding, multiple stroke deformations under the punch, springback after withdrawal of the tools and after material trimming.

A number of advanced numerical features have been implemented in PAM_STAMP™ to enhance its computational performances such as efficient contact simulation and adaptive meshing capabilities. The improved contact simulation alights the need for a well structured mesh of the tools. Automatically generated meshes can be used without modification, thus reducing the time needed to generate the description of the tools' geometry. The adaptive meshing capability which remeshes the blank during the simulation helps the engineer in obtaining accurate results with an originally coarse mesh.

The reliability of the numerical simulation is an essential part of the procedure. No industrial application can be contemplated until the software product and its associated application methodology has been thoroughfully validated against experimental results. In the first years following the first release of PAM_STAMP™ in 1990, several major projects carried out with major automakers in Europe and Japan have targeted such industrial validation. Results of these projects have been extensively reported in the past [1][2][3][6][8][11]. One typical result of these validation projects is depicted on figure 2.

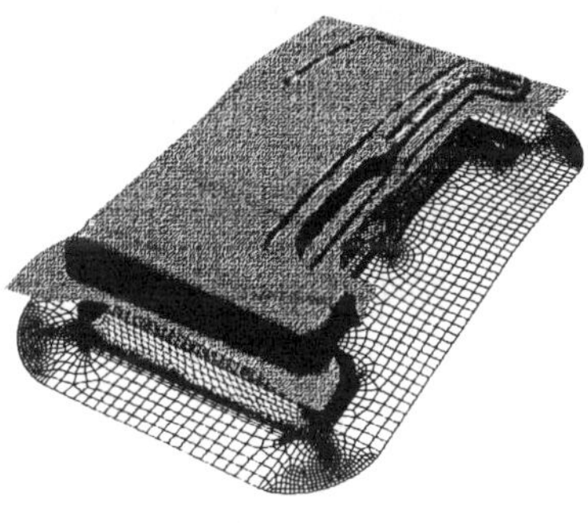

Fig. 2 - Industrial Example

3.0 Mathematical Optimization

3.1 Principles

Any optimization problem can be expressed mathe- matically as follows,

Minimization of objective function	$f(x)$
Design parameters constraints	$g_i(x) \le c_i$
Design parameters boundaries	$a_j \le x_j \le b_j$

The solution consists in finding the value x_k of the design parameters which minimize the objective function f(x) and satisfy the constraint functions $g_i(x)$ within their bounded range of validity [aj,bj].

A possible method to solve such mathematical optimization problem is using polynomial approximations. This method uses approximations of functions to proceed towards the solution by an iterative approach. First some linear or higher order approximations of the objective and constraint functions from a given starting point of the optimization path is calculated. The method uses these approximations of the objective and constraint functions to calculate a first approximate optimum.This first solution is used a a new starting point for the next iteration.

The default approximation of the objective and constraint functions is the linear approximation. However, amplitudes of the parameter variations must be limited in order to warrant the validity of approximated functions. The method using polynomial approximations is the most powerful method to find the solution with the minimum amount of computation.

While the method uses approximated functions to converge towards the optimum solution, it can of course also use the exact values of the objective and/or constraint functions.

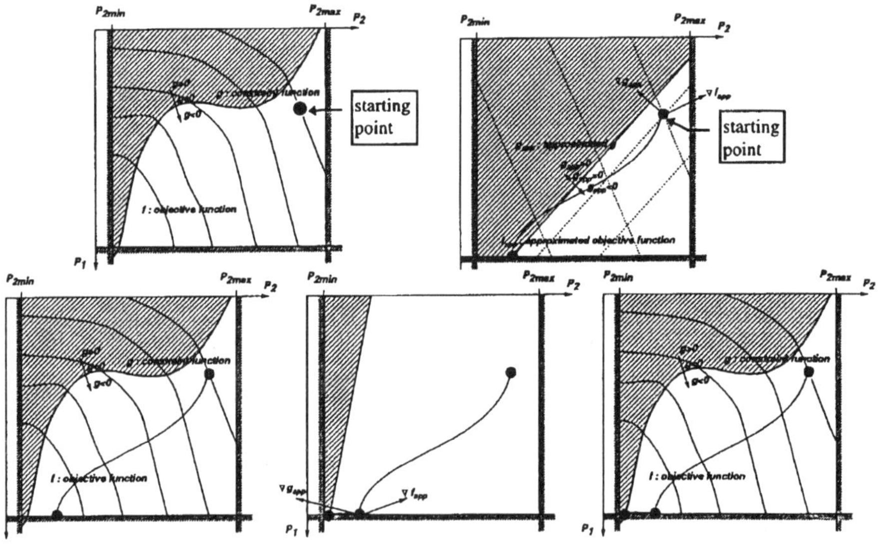

Fig. 3 - Principle of Polynomial Approximation Method

Figure 3 illustrates the way in which the polynomial approximation method works. The example on this figures uses linear approximations for a case with two design parameters and one constraint function with an admissible upper bound equal to zero.

The first graph shows the iso-values of the objective and constraint functions. The optimum solution lies on the left hand lower corner of the diagram.

The second graph shows the first linear approximation of the functions. The gradient and function values are used to calculate the approximation of the functions. The gradient and function values are calculated from the starting point of the optimization path. The basic method finds the solution of the approximated problem.

The third graph shows the approximate solution in the context of the real situation. This approximate solution is not the solution of the real problem, but the method has provided a fair approximation of the optimum solution with the first iteration.

The fourth graph shows the second linear approximation of the functions. The gradient values and function values are calculated on the previous solution point in order to calculate a new approximation of the functions.

The fifth picture shows the new approximate solution in the context of a real problem This solution is identical to the solution of the real problem, i.e., the method has converged.

3.2 Software Implementation

The optimization algorithm must be robust, even in the context of locally non-linear or fast varying objective and constraint functions presenting numerous local minima [Fig. 4]. It must also be computationally economical by reducing as much as possible the number of calls to non-linear solvers like PAM-STAMP™.

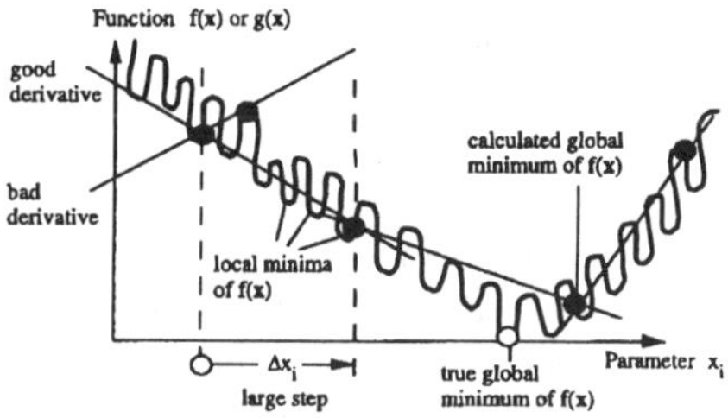

Fig. 4 - Example of Wavy Objective Function

The mathematical optimization software product, PAM-OPT™, satisfies these requirements. The objective function f(x) and the constraint functions gi(x) are replaced by local polynomial approximations. The objective and constraint functions are calculated for some values of the design parameters. The functions are thus known only point wise for specific values of the design parameters which are used for fitting polynomials. The software product loops on approximated functions, and finds approximated optimum solutions. However, if too large increments of the design parameters are used, the polynomial approximations could stray substantially from the real objective constraint functions. In order to avoid numerical oscillations due to such large deviations, the magnitudes of variations must be restricted.

Because of the wide variety of optimization situations to be handled by PAM_OPT™ in conjunction with PAM_STAMP™, a specific Optimization Solver Interface (OSI) matching the specifics of each problem must be prepared [Fig. 8]. This OSI has to read the output of PAM_OPT™ and prepare the input for the solver according to the requests of PAM_OPT™. After the solution is obtained, the OSI needs to provide to PAM_OPT™ the data needed to develop the next step of the optimization strategy.

PAM_OPT™ can be used in conjunction with PAM_STAMP™ as depicted on the flow chart, but also with any other PAM solver. It can also, with proper OSI, call several solvers for treating complex problems and call identical processes in parallel for CPU efficiency.

4.0 Application Examples

4.1 2D Case

A simple 2D case is used to demonstrate the capabilities of the coupled PAM_OPT™ / PAM_STAMP™ solution for optimizing press forming processes. The selected optimization problem is to minimize the mass of the blank under geometrical (minimum thickness and shape tolerance) and mechanical (no rupture) constraints. The design parameters selected are the blank thickness and length (defining the mass of the blank) and the blankholding pressure. The optimization problem is expressed as,

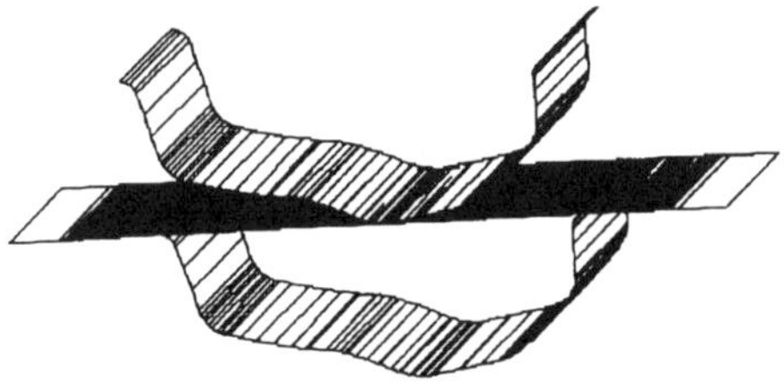

Fig. 5 - 2D Press Forming Case

Minimize (thickness*length)

The 2D blank is modeled with 83 shell elements [Fig. 5] whose final thicknesses and strains are used in the mathematical optimization procedure.

The mechanical constraint, i.e. no rupture, is formulated using the Forming Limit Diagram (FLD) formulation. The ideal FLD used in this exercise is depicted on figure 6. In order to fulfill the mechanical constraints, all points of the blank represented in the major / minor strain plane must lie below the rupture limit. This can be checked by prescribing a negative value for all distances between the points representing the blank and the rupture limit.

For each point representing the blank the distance to the limit of rupture is calculated as follows,

$$d_i = \min\,[d_{i1}, d_{i2}]$$

where d_{i1} is the distance to the left branch of the FLD and d_{i2} the distance to the right branch of the FLD. When d_i is greater than zero, a rupture occurs at that point.

The mechanical constraint is then expressed as follows,

$$\text{if } \max\{d_i\} > 0$$
$$\text{then } Km = \sqrt{[\Sigma_i(\max\{d_i,0\})^2]}$$
$$\text{if } \max\{d_i\} \leq 0$$
$$\text{then } Km = \max\{d_i\}$$

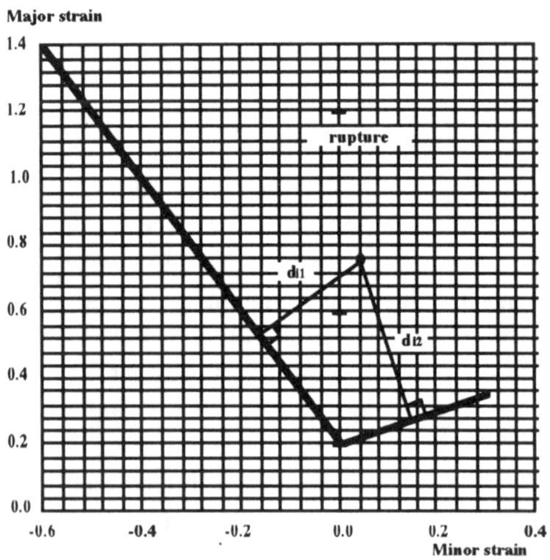

Fig. 6 - FLD used for rupture criterion

and the constraining condition for the optimization problem is written as,

$$Km \leq 0$$

The first geometrical constraint is related to the minimum thickness authorized for the blank, set at 65% of the initial blank thickness. The second geometrical constraint correspond to the shape tolerance of the final part.

In addition the design parameters (blank thickness, blank length, blankholding pressure) are bound to the following normalized values,

$1.6 \leq$ initial blank thickness ≤ 2.8
$0.5 \leq$ initial blank length ≤ 1.2
$3.8 \leq$ blankholding pressure ≤ 96.0

Using as starting point of the optimization procedure a 1.0 long blank, a 2.0 thick blank representing a normalized blank mass of 2.0 and a blankholding pressure of 19.231, the optimum solution found by PAM_OPT™ was a 0.9412 long blank with 1.6 thickness and a blank mass of 1.5, with a blankholding pressure of 9.578. All mechanical and geometrical constraints are satisfied with that optimum solution.

Using a second starting point (corresponding to the lower bounds of the design parameters) the same optimum blank is obtained but with a different blankholding pressure (which indicates that below a threshold value this design parameter has no influence on the optimization procedure).

The automatic optimization required 12 iterations with 64 PAM_STAMP™ simulations requiring a total time of 192 minutes. Using the parallel version of PAM_OPT™, this optimization procedure requires 27 minutes clock time.

4.2 3D Case

The 3D industrial case used is the dashpot suspension cup of the Renault Twingo passenger car. It general shape is relatively simple [Fig. 7] but rather complicated to deep draw. The details of the initial process adjustment for this part has been reported elsewhere [9].

The optimization case applied to the dashpot suspension cup is to minimize the initial thickness of the blank and the blankholding pressure under a number of mechanical constraints.

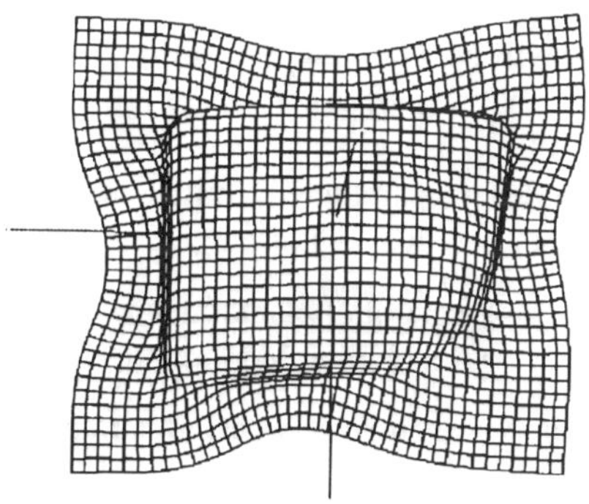

Fig. 7 - Dashpot Cup Geometry

Minimize (thickness*holding pressure)

Several optimization cases are studied. The first one with two constraints (metal rupture and blank thickness) and tight boundaries on the design parameters. The second is similar except that the boundaries on the design parameters are more widely defined. The final case adds an additional constraints on the formation of wrinkles under the blankholder.

The first constraint is to avoid metal rupture which is accounted for using the FLD formalism. The same FLD as for the 2D case is used for the industrial part with a similar expression of the constraints (see section 4.1 above). This constraint is applied both for upper surface strains and for lower surface strains, i.e. for each face of the blank. This double application is somewhat redundant. It is applied to improve the convergence rate, although it can be seen from the detailed results of the optimization that such redundancy is not necessary as the convergence is rather fast.

The second constraint is related to the minimum thickness to be obtained at the end of the process. As for the 2D case, this thickness cannot be less than 65% of the initial blank thickness.

$$t_{fin} \geq 0.65\ t_{ini}$$

The additional constraint applied for the last optimization case is related to the formation of wrinkles underneath the blankholder (to be avoided) [Fig. 8]. This constraint was expressed in terms of displacement of the blankholder during the process, thickening of the blank underneath the blankholder and 'vibrations' of the system. The following constraint is used,

(blankholder displacement) ≤ (thickening of the blank) + (amplitude of vibrations)

This representation of wrinkles formation is rather rudimentary and applicable only to wrinkles forming under the blankholder. A more accurate representation of wrinkle formations valid throughout the blank, i.e. also under the punch(es), is being developed and will be available in a short time.

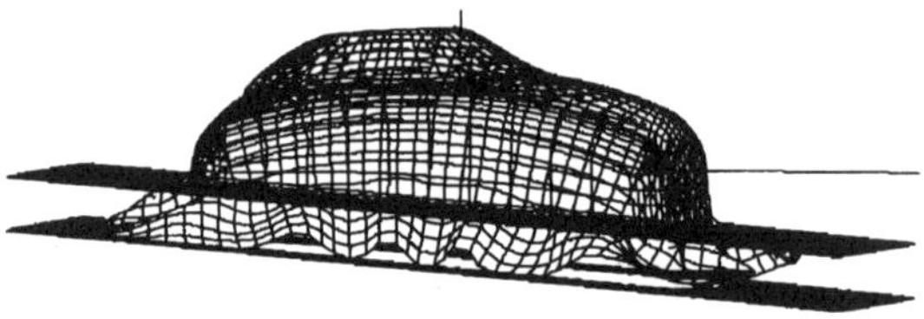

Fig. 8 - Typical Wrinkles Under Blankholder

The design parameters, initial thickness of the blank and blankholder pressure, are bounded to two sets of ranges. In the tight range the following boundaries are used,

1.0 ≤ initial blank thickness ≤ 3.0
0.5 ≤ blankholder pressure ≤ 14.0

In the loose range, the design parameters are bounded to the following values,

0.1 ≤ initial blank thickness ≤ 3.0
0.1 ≤ blankholder pressure ≤ 14.0

Several starting points are considered in the three optimization cases to assess the validity of the final optimum solution.

The optimization with the tight ranges for design parameters converges to the same solution whichever the starting point is. This seems to indicate that the optimization problem expressed in that way has no local solution and that the optimum solution obtained is very likely the absolute minimum. Figure 9 shows the level of improvement that is obtained automatically. For this optimization case, one convergence is studied by considering only the rupture criterion on the upper surface to assess the validity of the convergence. The result is the same as for the cases considering the FLD rupture criterion on both the upper and lower surfaces of the blank, reinforcing the conclusion that solution found is the absolute minimum.

With the loose ranges for design parameters, the optimization paths converge to different solutions. In addition, with the lower value of the blankholding pressure, large scale wrinkles are forming underneath the blankholder as depicted on figure 8.

By introducing the additional constraint on the formation of wrinkles, as explained here above, PAM_OPT™ provides an optimum solution different from the solution obtained with the first case. The final solution however matches all three sets of constraints and provides nevertheless significant improvements with respect to the starting point.

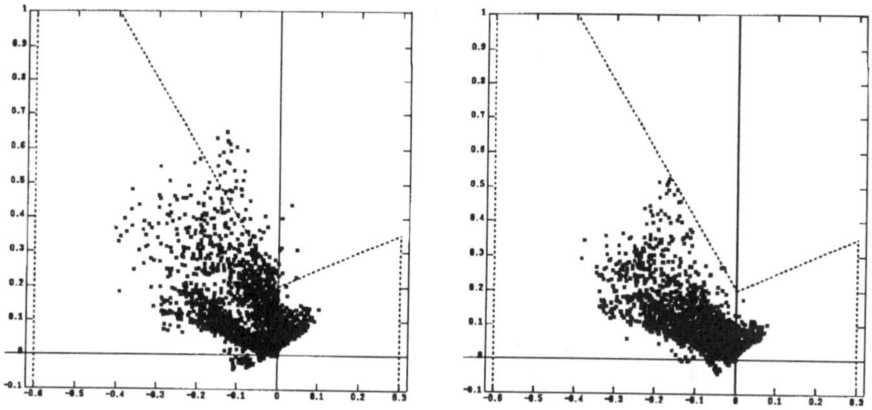

Fig. 9 - Strain Distributions on FLD Diagram

Table 1 hereunder provides the computational costs for each optimization path. The CPU times correspond to a DEC Alpha engineering work station.

Table 1 - Computational costs

Path	Number of iterations	PAM_STAMP™ runs	CPU time
Tight ranges			
1	3	10	241 min
2	4	13	298 min
3	5	16	379 min
4	4	13	316 min
Loose ranges			
1	3	10	171 min
2	3	10	373 min

5.0 Conclusions

The potentials of numerical simulation which is now being recognized by process engineers for steel sheet forming process sees new potential for development by its coupling to mathematical optimization. The paper has recalled briefly the background of that field and has introduced such coupling summarizing the techniques underlying the optimization process and illustrating it with two examples including one with an industrial part.

6.0 References

1 S. Aïta, et al., Numerical Simulation of Stretch Drawn Autobody, *SAE Int. Congr. and Exh.* (1992)

2 S. Aïta, et al., Industrial Sheet Metal Forming Simulations Issues in Code Development, Material Modeling, Validation and Design Methods, *Int. Conf. Comp. Plast.* (1992)

3 F. El Khaldy, et al., Industrial Validation of CAE Finite Element Simulation of Stretch-drawn Autobody Part (Front Fender Case), *17th Int. Deep Drawing Res. Group* (1992)

4 M. El Mouatassim, et al., The Simulation of Multi Operation Deep Drawing Process at Renault with PAM_STAMP™, *J. Mat. Proc. Tech. 45* (1994) 317-322

5 E. Haug, et al., Industrial Sheet Metal Forming Simulation using Explicit Finite Elements, *Int. VDI Conf.* (1991)

6 E. Haug, et al., Investigation for an Innovative Methodology for Sheet Metal Stamping, *Unformtech. Koll. Darmstadt* (1994)

7 E. Haug, et al., Numerical Simulation of Industrial Sheet Forming Processes with PAM_STAMP™, *4th Eur. Cars / Trucks Symp.* (1995)

8 W.C. Hsieh, F.K. Chen, An Optimal Die Design for Stamping a Rear Floor Panel of a Passenger Car, *PAM User's Conf. Asia* (1994)

9 L.T. Kisielewicz, F. Mehrez, Benefits from Computer Aided Tryout, *Int. Sem. Prec. Met. Form.* (1996)

10 T. Mahé, La simulation fait ses preuves en emboutissage, *Industries et Techniques 757* (1995)

11 A. Messina, The Role of Numerical Simulation in Design and Optimization of Car Components Deep Drawing Process, *2nd Eur. Works. Adv. Num. Sim.* (1991)

12 A.K. Pickett, et al., Industrial Press Forming of Continuous Fiber Reinforced Thermoplastic Sheets and the Development of Numerical Simulation Tools, *3rd Int. Conf. Flow Proc. Compos. Mat.* (1994)

Computer - Aided Analysis of Roll Bending in Strip Profile Control on Four High Cold Rolling Mills

K. Srinivasa Murthy, A. S. Deshpande

Department of Mechanical Engineering, Indian Institute of Technology, Powai, Mumbai - 400 076 [INDIA]

Abstract. In the present analysis, a mathematical model is developed on windows environment using C++ to predict the final outgoing strip profile with or without the application of inner, outer or both inner and outer chock bending forces to minimize the variations of the strip thickness distribution in the transverse direction, which would help in achieving a better strip flatness under the given set of rolling conditions. The rolling equations are solved analytically along the length of arc of contact. The pressure distribution at work roll - back-up roll interface and at work roll - strip interface are calculated from beam on elastic foundation theory. The strip profile is calculated by taking into account back-up roll deflection, work roll deflection, work roll flattening and work roll thermal expansion across the width of the roll barrel. Computations are performed for various realistic ranges of rolling conditions.

Nomenclature

2b	Width of the Strip width, mm
2l	Back-up roll barrel length, mm
2L	Bearing center distance of back-up roll, mm
a	Thermal Diffusivity, m^2/sec
B	Width of contact between work roll and back-up roll, mm
Bi	Biot number, dimensionless
E_{BR}	Young's modulus of the back-up roll, Kg/mm^2
$(F_o)_{WB}$	Modulus of the foundation at work roll and back-up roll interface, Kg/mm^3
$(F_o)_{WS}$	Modulus of foundation at work roll -strip interface, Kg/mm^3
h	Cooling Heat Transfer Coefficient, KW/m^2 K
M	Number of parts to which 2l is divided
N	Number of parts to which 2L is divided
k	Thermal Conductivity, W/m K
l_A	Length of the arm at which separator or counter force is applied, mm
p	Average load per unit width at work roll and strip interface, mm
p_J	Load/unit width at work roll and strip interface at J (J=0 to M), Kg / mm
P_S	Roll separator or counter force, Kg / mm^2
q	Average load per unit width at work roll and back-up roll interface, mm
Q	Contact Heat Flux, MW/m^2

q_I Load/width at work roll and back-up interface at I (I = 0 to N), Kg/mm
Pe Peclet number, dimensionless
T Temperature of the roll surface, °C
T_o Coolant temperature, °C
$(Y_{A-C})_I$ Deflection at a point I in the region A–C, mm
$(Y_{C-D})_I$ Deflection at a point I in the region C–D, mm
α Thermal Expansion Coefficient, K^{-1}
ϕ_o Angle subtended at the entry point, radians
ζ Constant used in beams on elastic foundation, dimensionless

1 Introduction

Production of flat rolled strip is one of the main challenges of the present cold rolling industry. Flatness is defined as the ability of the strip to lie on an horizontal surface. Flatness is directly related to the strip crown. Strip crown is defined as the difference between the thickness at the center and at the edges of the strip. For good dimensional tolerances and mill yield, the crown must be maintained at the minimum possible limit. Flatness defect occurs because of longitudinal stresses induced by a mismatch between roll gap profile and the input strip profile. This mismatch leads to non-uniform elongation across the strip width. Hence, if the cross-sectional profile of the strip entering and leaving the roll gap are geometrically dissimilar, stresses are induced. If these stresses exceed the buckling limit of the strip, then bad strip shape occurs. Good quality strip shape can be maintained by ensuring that the incoming and outgoing profiles are geometrically similar i.e., the crowns in each case should be proportional to the mean thicknesses.

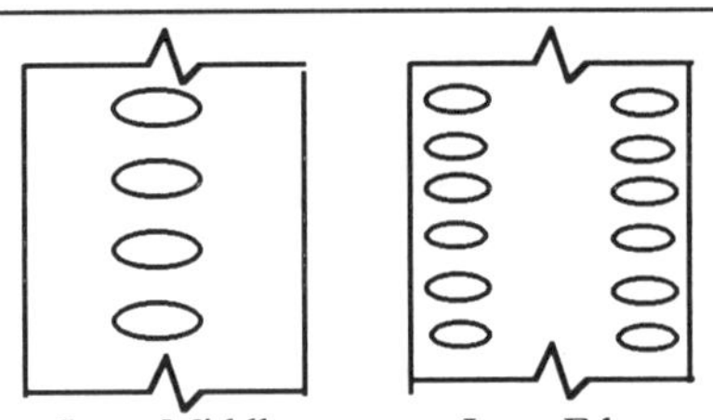
Long Middle (Center Buckles) Long Edge (Side Buckles)
Fig. 1

Most commonly occurring cold rolling defects are long middle and long edge (Fig.1) Long middle shape occurs when the middle portion of the strip is elongated more than the edge, resulting in compressive stresses over this portion. Long edge shape occurs when the edge portion of the strip is elongated more than the middle, resulting in compressive stresses over this portion. These defects can be minimized by controlling the strip crown with the help of roll bending systems. This system alters the crown by regulating the degree of roll flexture. Two types of roll bending systems are possible. If the work roll bending cylinders are in between the work roll chocks, it is called positive roll bending (Roll Bending (Fig.2)). It tries to minimize the roll gap at the center of the work roll barrels. If the work roll bending cylinders are in between work roll and back-up roll chocks, it is called negative bending (Counter Roll Bending). It tries to widen the roll gap at the center of the work roll barrels. To correct the long edge problem, roll bending force is applied. To correct the long middle problem, roll counter bending is applied. Based on the incoming strip crown, to maintain the flatness, roll bending or roll counter bending cylinders are activated.

Stone et. al. [1] by assuming the work roll as a beam resting on an elastic foundation, calculated the roll bending effect. Shohet et. al. [2] describe quantitatively the effect of roll bending on strip profile by compatibility conditions. Yuen et. al. [3] develop a predictive model by considering rolls as cantilevers. The common assumption in the above roll bending methods is that the work and back-up roll barrels are in full contact across the width in loaded condition, which is not true in a cold rolling mill.

In the present work, an attempt is made to develop an off line predictive model to suit rolling of various materials, widths, thicknesses with and without different magnitudes of roll bending forces on inner, on outer or on both inner and outer chocks. Variable contact length between work roll and back-up roll is taken into account.

In addition, a detailed model is developed to incorporate the additional crown generated on the work rolls due to thermal effects based on the prediction of roll temperatures. From the literature, it is found that many investigations are carried out [4] [5] [6] [7] in estimating the work roll crown. In the present analysis, an analytical solution developed by Tseng et. al [7] is used.

2 Analysis

2.1 Rolling Load

A detailed formulation is given in one of our previous papers [8].

2.2 Strip Profile

The calculation of the strip profile by taking into consideration work roll deflection, back-up roll deflection and work roll flattening is explained in one of our papers [8]. An improved method to incorporate roll bending on strip profile, the pressure distribution calculation at both work roll and back-up roll interface and work roll and strip interface is explained in sections 2.3 and 2.4.

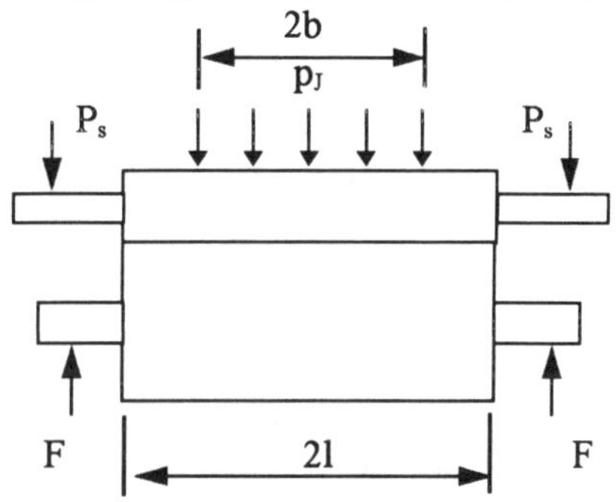

FIG. 2 A Freebody Diagram of the Bottom Work Roll and Backup Roll

2.3 Pressure Acting Between Work Roll and Backup Roll [9]

The free body diagram of bottom work roll and backup roll can be represented as shown in Fig. 2. For the present analysis, the work roll is assumed as the finite beam resting on an elastic foundation. The deflection of work roll from beams on elastic foundation theory can be calculated as the summation of deflections due to

1. Uniformly distributed load at work roll and strip interface (Fig. 3)
2. Bending momentum acting on the work roll due to roll bending forces (Fig. 3)

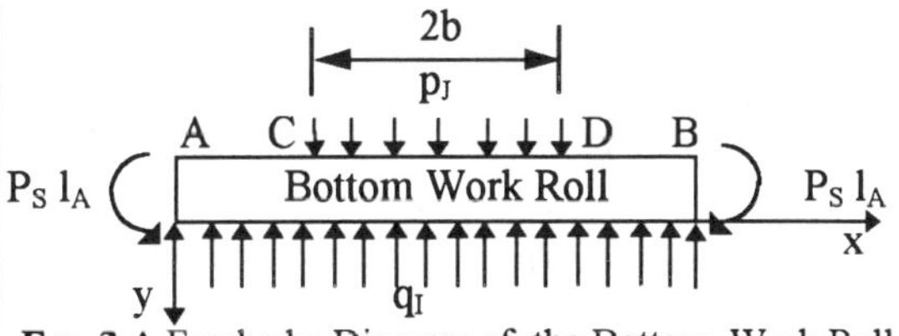

FIG. 3 A Freebody Diagram of the Bottom Work Roll

For case 1, the deflection is calculated as follows

$$(y_{A-C})_I = (p/F_{WB})\,(1/(\mathrm{Sinh}\zeta 2l+\mathrm{Sin}\zeta 2l))\,\{(\mathrm{Cosh}\zeta x\,\mathrm{Cos}\zeta x)[\mathrm{Cosh}\zeta(l-b)\mathrm{Sin}\zeta(l+b)- \mathrm{Sinh}\zeta(l-b)\mathrm{Cos}\zeta(l+b)+\mathrm{Cos}\zeta(l-b)\mathrm{Sinh}\zeta(l+b)-\mathrm{Sin}\zeta(l-b)\,\mathrm{Cosh}\zeta\,(l+b)] + (\mathrm{Cosh}\zeta x\,\mathrm{Sin}\zeta x+\mathrm{Sinh}\zeta x\,\mathrm{Cos}\zeta x)\,[\mathrm{Sin}\zeta\,(l-b)\,\mathrm{Sinh}\zeta\,(l+b)-\mathrm{Sinh}\zeta\,(l-b)\,\mathrm{Sin}\zeta\,(l+b)]\}$$

when x lies A–C (1)

$$(y_{C-D})_I = (y_{A-C})_I + p\,/\,F_{WB}\,[1-\mathrm{Cosh}\zeta\,(x-l+b)\mathrm{Cos}\zeta\,(x-l+b)]$$

when x lies C–D (2)

For case 2, the deflection is calculated as follows

$$(y_2)_I = 2P_s\,l_A\,\zeta^2\,/\,F_{WB}\,(1/(\mathrm{Sinh}\zeta 2l+\mathrm{Sin}\zeta 2l))\{\mathrm{Sinh}\zeta x\,\mathrm{Cos}\zeta(2l-x)-\mathrm{Cosh}\zeta(x)\,\mathrm{Sin}\zeta(2l-x)+\mathrm{Sinh}\zeta(2l-x)\,\mathrm{Cos}\zeta(x)-\mathrm{Cosh}\zeta\,(2l-x)\mathrm{Sin}\zeta x\} \quad (3)$$

where $\zeta = (F_{WB}/E_{WR}I_{WR})^{0.25}$ and $(F)_{WB} = B(F_o)_{WB}$

$B = 2.15\,((q\,D_{WR}\,D_{BR})/E_{WR}\,(D_{WR}+D_{BR}))^{0.5}$

$(F)_{WB} = 0.65\,(E_{BR}\,B^4/(E_{WR}I_{WR}))^{1/12}\,E_{BR}/(1-\nu_{WR})^2$ (4)

The final deflection of work roll can be expressed as

$(y_{WB})_I = (y_{A-C})_I + (y_2)_I$ when x lies A–C

$(y_{WB})_I = (y_{C-D})_I + (y_2)_I$ when x lies C–D (5)

The pressure acting between the work roll and backup roll (pressure in the elastic foundation) can be represented as

$q_I = (F_o)_{WB}\,(y_{WB})_I$ (6)

2.4 Pressure Acting at the Work Roll Strip Interface

The pressure at the work roll and strip interface can be calculated by beam on elastic foundation theory. The work roll is assumed to have a length equal to that of the strip as shown in Fig. 4. The deflection can be derived as

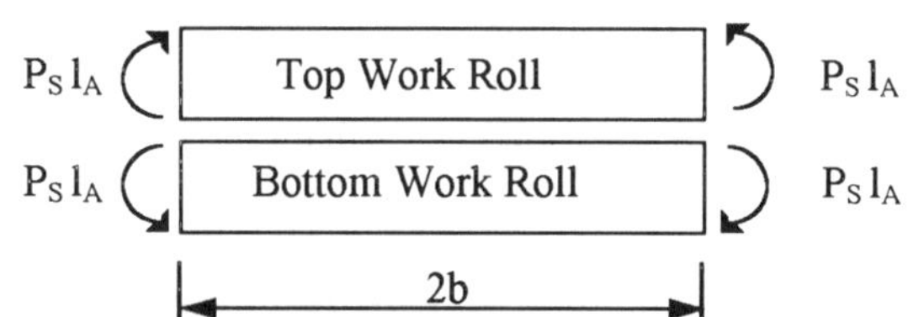

Fig. 4 A schematic diagram of the work rolls and strip

$$(y_{WS})_J = 2P_s\,l_A\,\zeta^2\,/\,F_{WS}\,(1/(\mathrm{Sinh}\zeta 2b+\mathrm{Sin}\zeta 2b))\{\mathrm{Sinh}\zeta x\,\mathrm{Cos}\zeta(2b-x)-\mathrm{Cosh}\zeta(x)\,\mathrm{Sin}\zeta(2b-x)+\mathrm{Sinh}\zeta(2b-x)\,\mathrm{Cos}\zeta(x)-\mathrm{Cosh}\zeta(2b-x)\mathrm{Sin}\zeta x\} \quad (7)$$

where $\zeta = (F_{WS}/E_{WR}I_{WR})^{0.25}$ and $(F)_{WS} = B(F_o)_{WS}$

$B = 2.15\,((p\,D_{WR})/2E_{WR})^{0.5}$

$(F)_{WS} = 0.65\,(B^4/I_{WR})^{1/12}\,E_{WR}/(1-\nu_{WR})^2$

The pressure acting between the work roll and backup roll (pressure in the elastic foundation) can be represented as

$p_J = (F_o)_{WS}\,(y_{WS})_J$ (8)

2.5 Thermal Gradients In The Roll Body

The solution is developed by assuming heat flux over one portion and conductive cooling over the remaining portion of the roll periphery as shown in Fig. 5. The temperature variation are assumed to be cyclic steady state. It is assumed that only a small center portion of the work roll is considered in which there is no temperature variation along the length of the roll body [7].

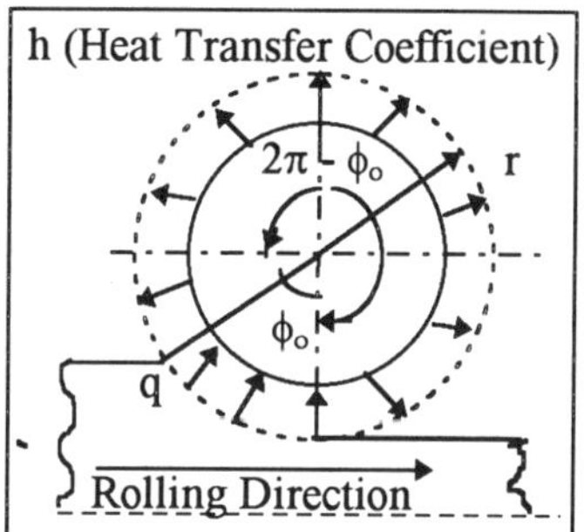

Fig. 5 Roll geometry with conductive and convective boundaries

2.9 Thermal Expansion [7]

The thermal expansion at the center of the work roll can be estimated by

$$\Delta_0 = (1 - \nu_{WR}) / (1 + \nu_{WR}) \alpha D_{WR} (q / 2h)(b / L) U_{max} \tag{9}$$

2.10 Thermal Crown [7]

It is observed in the literature that the axial variation of the thermal expansion or camber is can be approximated to a parabola or a cosine curve. In the present solution, the variation of crown is assumed as a cosine curve with a peak at the center of the strip and zero at the edge of the strip.

$$\Delta(z) = C_T \cos(\Pi z / W) \Delta_0 \tag{10}$$

The thermal crown so determined, is then superimposed on other deflections, to estimate the final strip profile.

3 Method of Execution of the Program

The present software calculates three important sets of results viz.

- Rolling load and pressure distribution
- Elastic deformation of rolls and work roll flattening after taking the roll bending effect into consideration.
- Change in work roll radius due to thermal gradients along the width direction

The software is in two modules namely load module and profile module [8]. The programs are coded in C++ version 4.5 (Borland), which can be executed in IBM PC or its compatibles. The sub tasks are split into meaningful classes (subroutines) in order to avoid memory limitations. The results are stored in data files and they can be shown on the screen as a graphical display. The program is approximately 4500 lines in length and takes 30 seconds for execution of both modules on Pentium. The methodology is explained in the earlier paper [8].

3.1 Solution For Roll Bending

The pressure acting at work roll and backup roll interface and at work roll and strip interface are calculated by equations (6) and (8) respectively. These calculations are repeated, until there exists only positive pressures at work roll and back-up roll interface with a variable length of the roll barrel.

4 Results And Discussion

It is assumed that the roll bending force can be applied as a free choice either at the inner chock, at the outer chock or at both inner and outer chocks simultaneously. It is also assumed that the maximum value of the bending force available in each case is 25000 kg. The independent effect of inner chock bending (0 - 25000 kg) and outer chock bending (0 - 25000 kg) and the combined effect of inner and outer chock bending (0 - 25000 kg) on the strip profile are analyzed. The geometrical parameters and the material parameters of the rolls and the strip used, need to be supplied by the user. However, one set of normal values (default data) used in the present analysis is shown in Table 1. The effect of following principal parameters on the strip profile is studied.

- work roll diameter (300 - 450 mm),
- back-up roll diameter (800 - 1200 mm),
- strip width (800 - 1200 mm) and
- type material to be rolled (mild steel, 0.51 % C steel and Stainless Steel 18 : 8)

The effectiveness of the roll bending for controlling the strip profile will be better with the increased arm length or with the increased magnitude of force on the arm. As the outer chock has more arm length compared to inner chock, the strip profile can be corrected to a greater extent by outer chock bending. The improvement in the strip profile achievable can be further increased by simultaneous utilization of force on both inner and outer chocks. In the present analysis, the outgoing thickness is assumed as 1.6 mm. Because of similarity, only top half of the strip is shown in all Figures. The strain hardening coefficients of different materials is given in Table 2.

Table 1 Default Data (For Fig. 5 to Fig. 34)

Strain Hardening Law, Kg /mm^2 (Mild Steel)	$23.0+3.46\,\varepsilon^{0.2}$
Modulus of elasticity of strip material, Kg /mm^2	21000
Poisson's ratio of strip material	0.28
Work Roll Radius, mm	250
Back-up Roll Radius, mm	540
Coefficient of Friction	0.05
Modulus of Elasticity of Work Roll and Backup Roll Material, Kg/ mm^2	22000
Poisson's ratio of Work Roll and Back-up Roll Material, Kg /mm^2	0.30
Bearing Centre Distance, mm	2700
Roll Barrel Length, mm	1600
Thermal Diffusivity, m^2/sec	$1.068\text{X}10^{-5}$
Thermal Conductivity, W/m K	45.2
Thermal Expansion Coefficient, K^{-1}	$1.3\text{X}10^{-5}$
Cooling Heat Transfer Coefficient, Kw/m^2 K	20.0
Contact Heat Flux, MW/m^2	20.0

Table 2 Material Data

Material	Strain Hardening Law
Mild Steel	$20\,(1+135\,\varepsilon)^{0.295}$
0.51 % C Steel	$34.64\,(1+92\varepsilon)^{0.25}$
Stainless Steel (18:8)	$33\,(1+135\,\varepsilon)^{0.36}$

5.1 Effect of Roll Bending with Default Data

The data utilized in generating Fig. 5, 6 and 7 is shown in Table 1. Figures 5 and 6 show the independent effect of inner and outer chock bending on the strip profile. Fig. 7 shows the combined effect of inner and outer chocks on the strip profile. Without any roll bending force, the crown is of the order of 150 microns. With inner chock bending alone, the minimum crown is 90 microns. With outer chock bending alone, the crown can be reduced to about 60 microns, but with both inner and outer chock bending forces together, it can even go to a negative value of 4 microns. It can be seen that inner chock bending alone can not suffice in achieving a good enough strip profile. It is observed that a force of 25000 kg on outer chock and only a force of 18000 kg on both inner and outer chocks are sufficient for better strip profile.

5.2 Effect of Roll Bending with a Change in Work Roll Diameter

A work roll diameter of 300 mm and all other data as shown in Table 1 is utilized in generating Fig. 8, 9 and 10. It is observed that the application of 25000 kg on inner chock, 20000 kg on outer chock or 10000 kg force on both inner and outer chocks can lead to a better strip profile.

A work roll diameter of 350 mm and all other data as shown in Table 1 is utilized in generating Fig. 11, 12 and 13. Fig. 11 shows that the only inner chock bending is not sufficient in achieving a better strip profile. Fig. 12 and 13 show that the application of 25000 kg on outer chock and 10000 - 20000 kg force on both inner and outer chocks can lead to a better strip profile.

Work roll diameter of 450 mm and all other data as shown in Table 1 is utilized in generating Fig. 14, 15 and 16. Fig. 14 shows that the only inner chock bending is not sufficient in achieving a better strip profile. Fig. 15 and 16 show that the application of 25000 kg on outer chock and 20000 kg force on both inner and outer chocks can lead to a better strip profile.

Hence, it can be seen from the study of the effect of work roll diameter on the strip profile variation, the strip profile can be improved with an increase in the work roll diameter to a certain extent, even without the application of any roll bending forces. This is because of the increase in work roll diameter results in less work roll deflections. It is observed that with an increase in work roll diameter from 300 mm to 400 mm resulted in a decrease of crown from 178 microns to 150 microns without applying any roll bending force. As the work roll diameter increases, it becomes difficult for the chock bending forces to bend the roll, which ultimately cannot help in achieving the better strip profile. However, with the increased work roll diameter, the chock forces required to achieve a particular crown variation are high. For a 300mm work roll diameter, a chock force of 10000 kg on both inner and outer chocks is sufficient to achieve a crown of 60 microns. On the other hand, for a 400mm work diameter, a chock force of 18000 kg on both inner and outer chocks is needed to attain the same crown of 60 microns. Hence, work roll diameter has to be a minimum practical value so that the required crown control is possible with the minimum values of chock forces.

5.3 Effect 0f Roll Bending with a Change in Back-Up Roll Diameter

Back-up roll diameter of 800 mm and all other data as shown in Table 1 is utilized in generating Fig. 17, 18 and 19. The strip profile is good enough without the application of any roll bending forces. It is observed that any magnitude of force on inner or / and outer chock tries to worsen the strip profile. To improve the strip profile in this case, the roll bending force should be applied between work roll and back-up roll (which is called roll counter bending) instead of bending force between work rolls.

Back-up roll diameter of 1200 mm and all other data as shown in Table 1 is utilized in generating Fig. 20, 21 and 22. Fig. 20 and 21 show that the independent application of inner chock or outer chock bending are not sufficient in achieving a better strip profile. Fig. 22 shows that the application of 25000 kg force on both inner and outer chocks can lead to a better strip profile.

Without the application of any roll bending forces, the strip profile worsens with the increased back-up roll diameter. This is because the work roll deflection increases with increased back-up roll diameter at a greater rate than the decrease of back-up roll deflection. The increase in back-up roll diameter is having less effect in altering the range of profile control available with the application of chock bending forces.

5.4 Effect of Roll Bending with a Change of Strip Width

Strip width of 800 mm and all other data as shown in Table 1 is utilized in generating Fig. 23, 24 and 25. It is observed that the application of roll bending force on inner chock, outer chock or on both inner and outer chock are not sufficient in achieving a better strip profile.

Strip width of 1200 mm and all other data as shown in Table 1 is utilized in generating Fig. 26, 27 and 28. Fig. 26, 27 and 28 show that the application of 10000 to 20000 kg on inner chock, 10000 kg on outer chock and below 10000 kg force on both inner and outer chocks respectively, can lead to a better strip profile.

The increase in strip width improves the strip profile. This is because the decrease in work roll deflection rate is more than the increase in back-up roll deflection and increase in work roll flattening. The increase in strip width needs smaller values of roll bending forces in correcting the strip profile.

5.5 Effect of Roll Bending with a Change in Rolling Material

The material to be rolled is taken as 0.51 %C steel and all other data as shown in Table 1 is utilized in generating Fig. 29, 30 and 31. It is observed that the independent application of inner chock and outer chock bending are not adequate to correct the strip profile. The application of 20000 kg force on both inner and outer chocks is sufficient in achieving the better strip profile.

The material to be rolled is taken as Stainless Steel (18: 8) and all other data as shown in Table 1 is utilized in generating Fig. 32, 33 and 34. It can be noted that the 25000 kg of roll bending force on inner or / and outer chock is not sufficient even to improve the strip profile marginally.

The magnitude of the bending forces needed to control the strip profiles seems to be highly dependent on the type of material to be rolled.

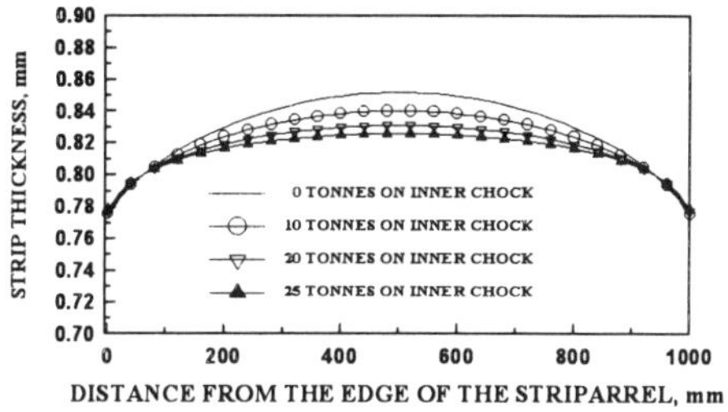

Fig. 5 Strip profile with default data

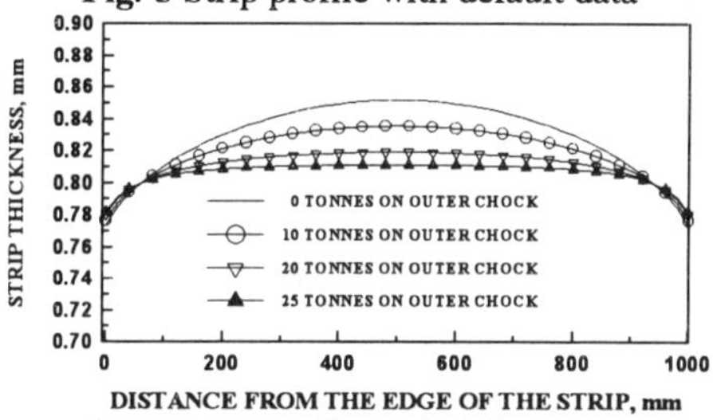

fig. 6 Strip profile with default data

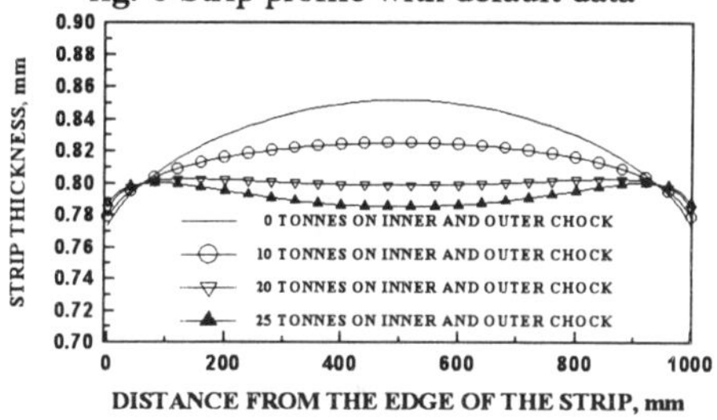

Fig. 7 Strip profile with default data

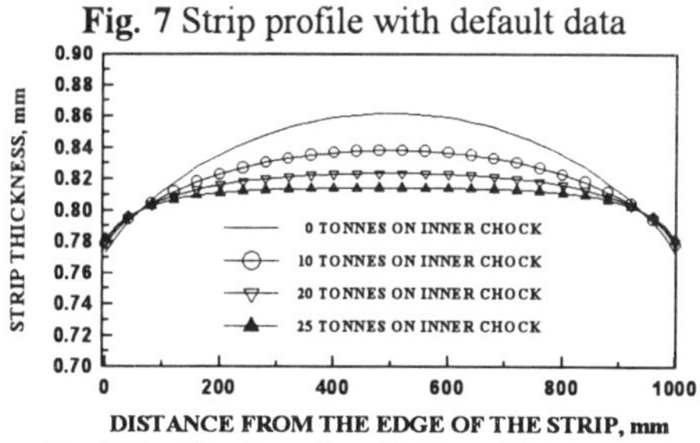

fig. 8 Profile (work roll dia. of 300 mm)

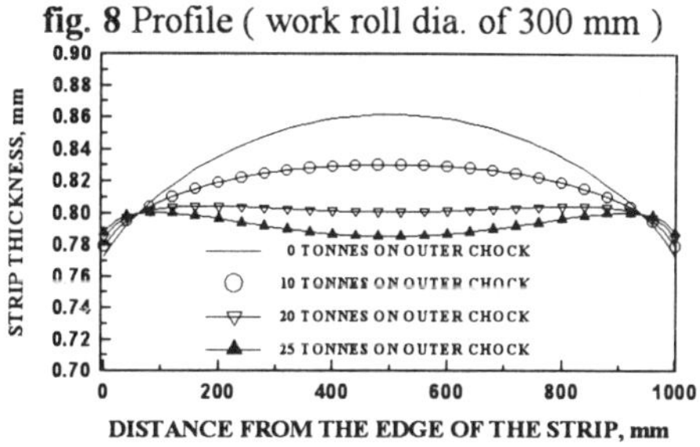

Fig. 9 Profile (work roll diameter = 300 mm)

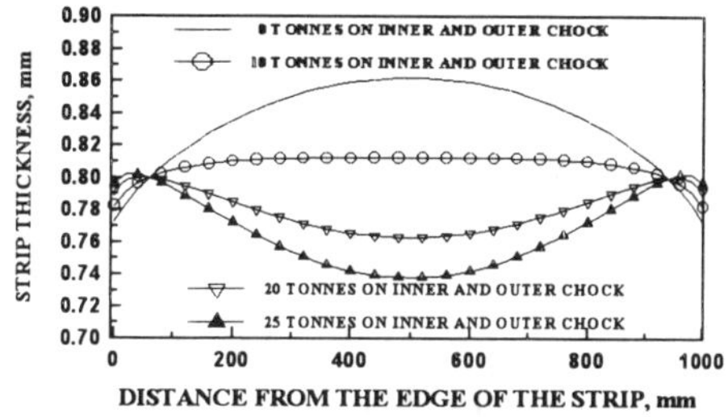

Fig. 10 Profile (work roll diameter=300 mm)

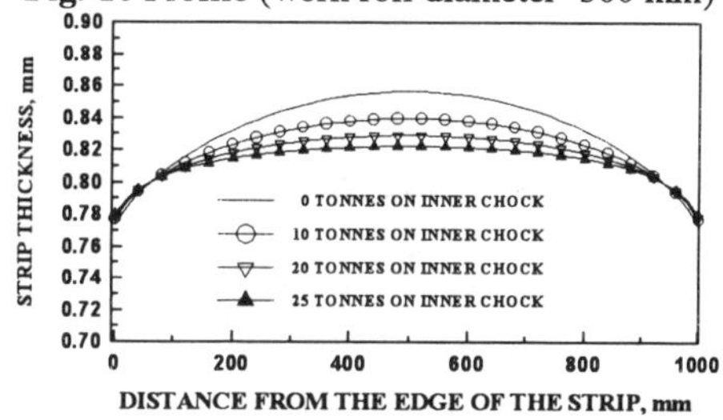

Fig. 11 Profile (work roll diameter=350 mm)

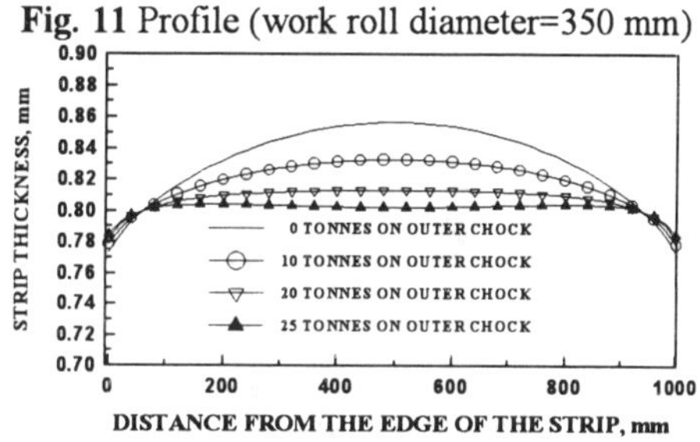

Fig. 12 Profile (work roll diameter=350 mm)

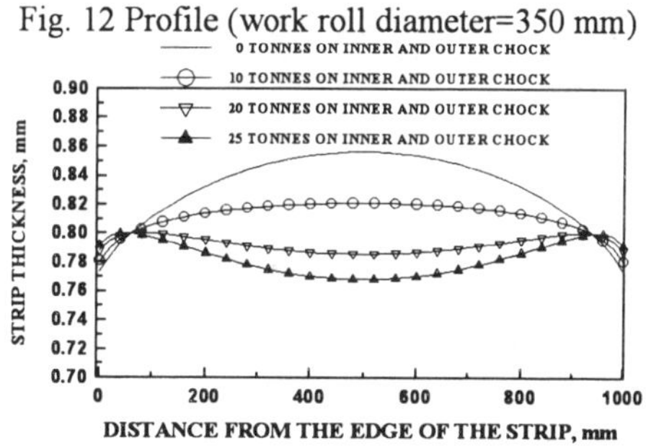

Fig. 13 Profile (work roll diameter=350 mm)

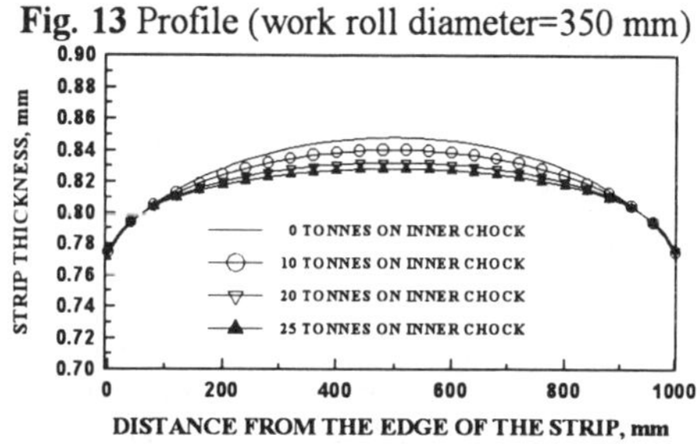

Fig. 14 Profile (work roll diameter =450 mm)

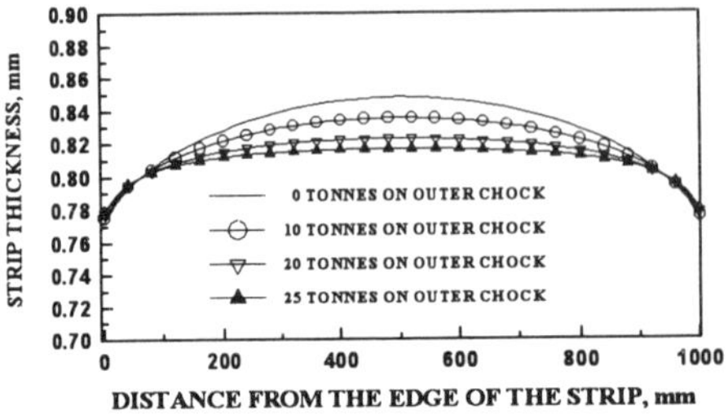

Fig. 15 Profile (work roll diameter=450 mm)

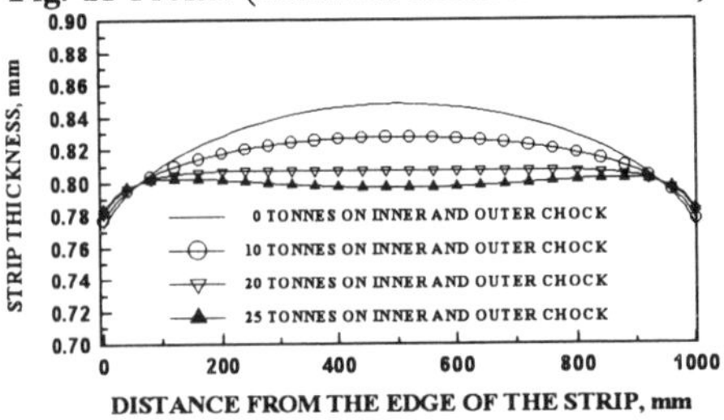

Fig. 16 Profile (work roll diameter=450 mm)

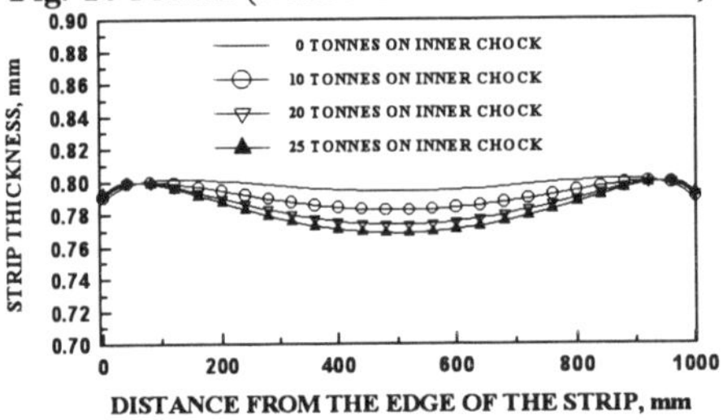

Fig. 17 Profile (backup roll dia. =800 mm)

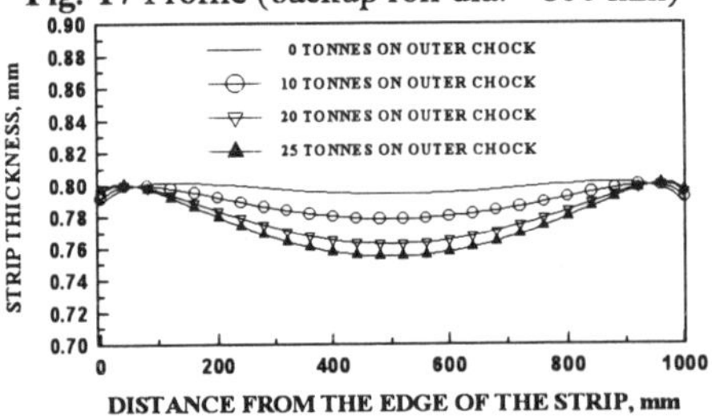

Fig. 18 Profile (back-up roll dia.= 800 mm)

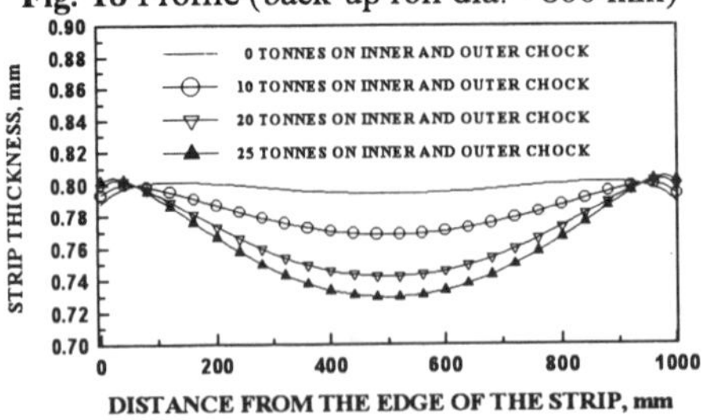

Fig. 19 Profile (back-up roll dia.= 800 mm)

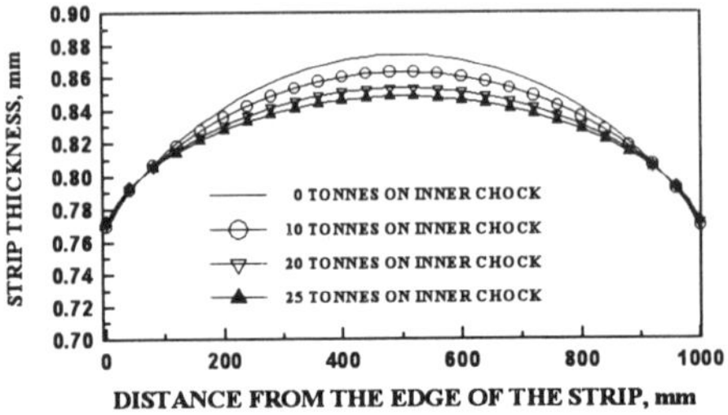

Fig. 20 Profile (back-up roll dia. = 1200 mm)

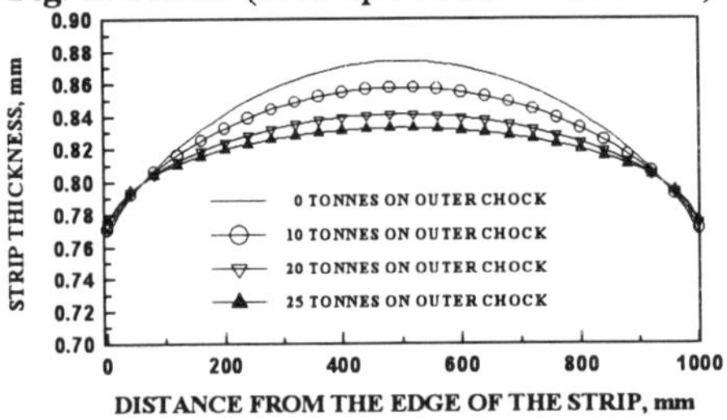

Fig. 21 Profile (back-up roll dia. =1200 mm)

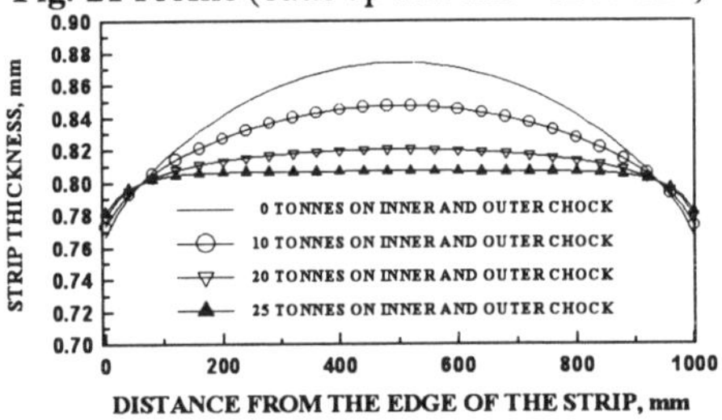

Fig. 22 Profile (back-up roll dia.=1200 mm)

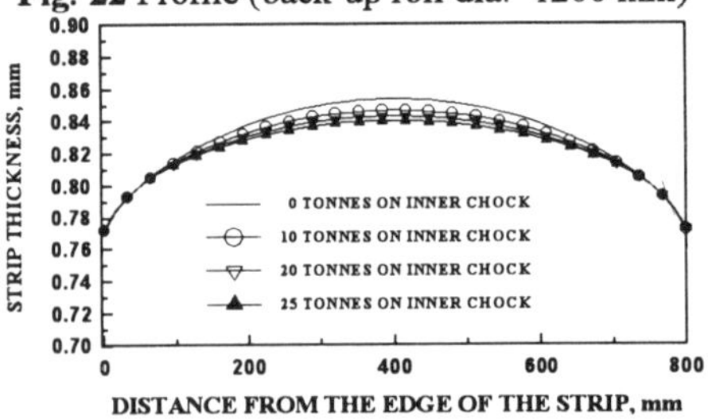

Fig. 23 Profile (initial width = 800 mm)

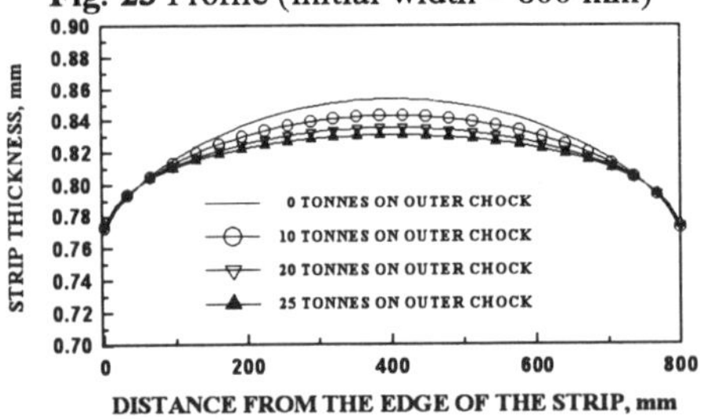

Fig. 24 Profile (initial width=800 mm)

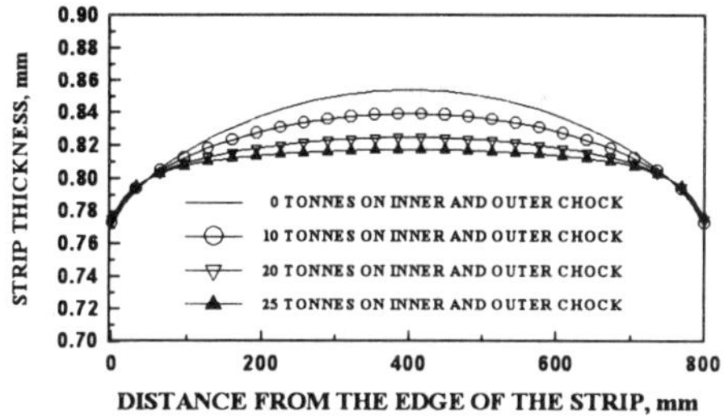

Fig. 25 Profile (initial width=800 mm)

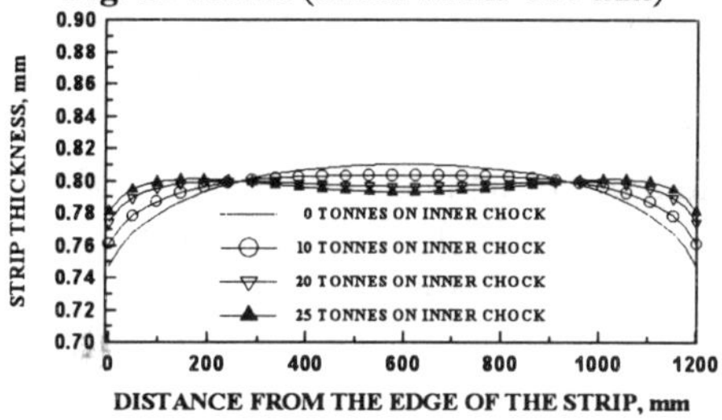

Fig. 26 Profile (initial width = 1200 mm)

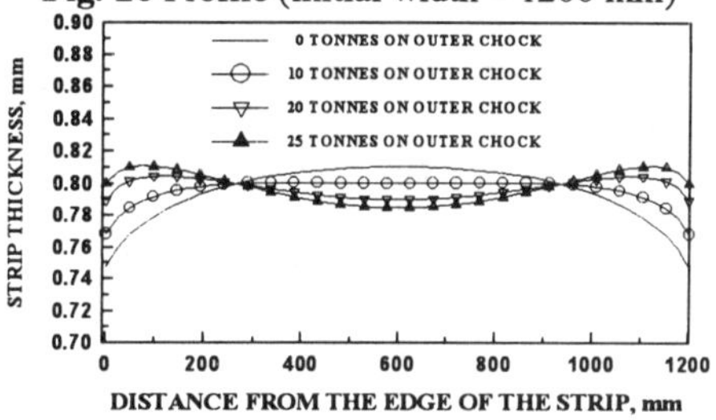

Fig. 27 Profile (initial width = 1200 mm)

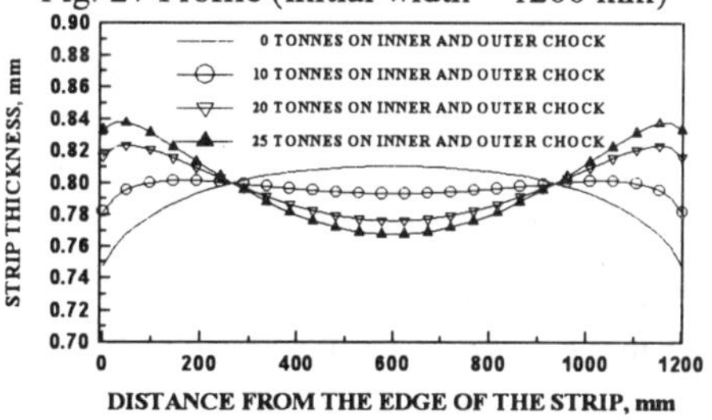

Fig. 28 Profile (initial width = 1200 mm)

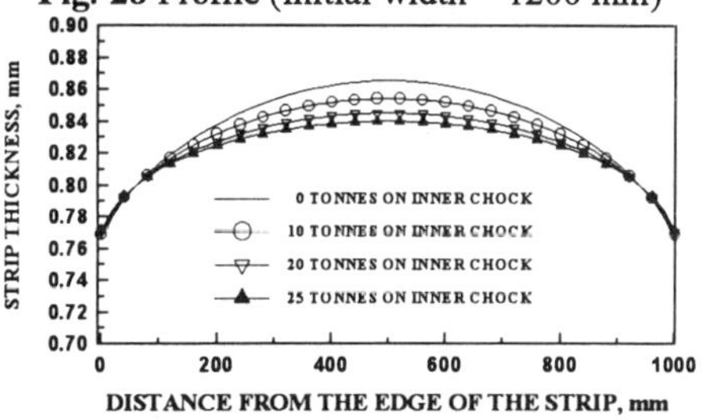

Fig. 29 Profile (0.51% c steel)

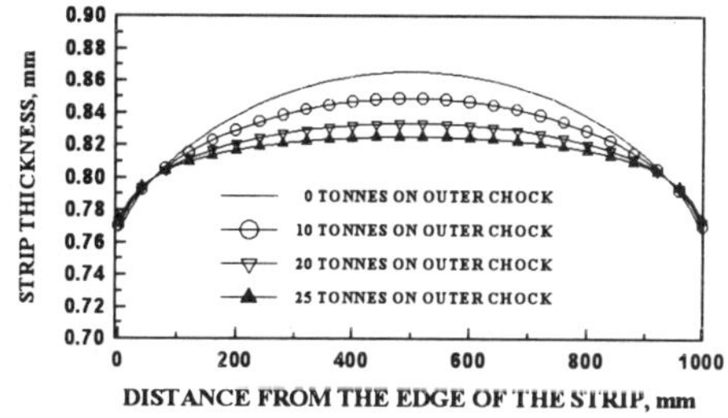

Fig. 30 Profile (0.51% c steel)

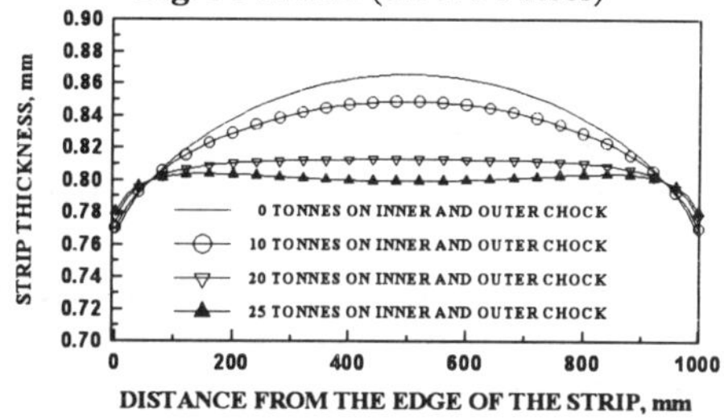

Fig. 31 Profile (0.51% c steel)

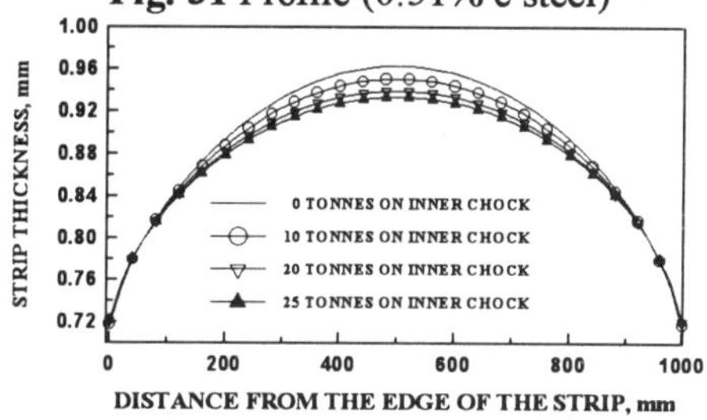

Fig. 32 Profile (Stainless steel (18:8))

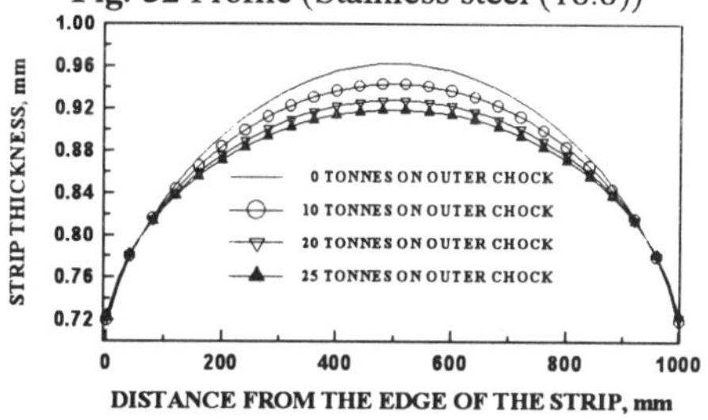

Fig. 33 Profile (Stainless steel (18:8))

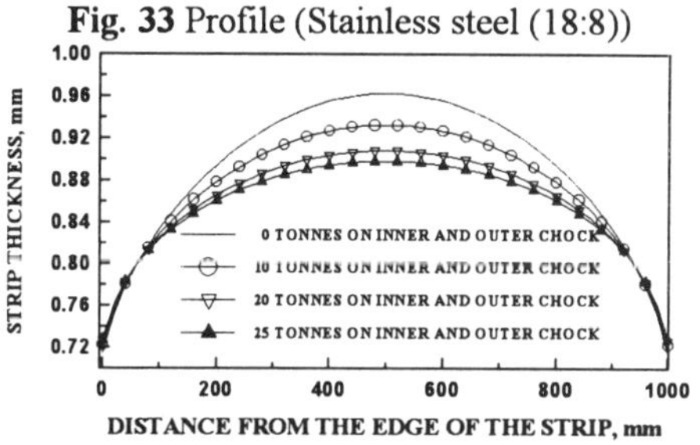

Fig. 34 Profile (Stainless steel (18:8))

6 Conclusions

1. The program generated is capable of suggesting an optimum roll bending force for a variety of combinations of realistic rolling parameters for a 4 high cold rolling mill having facilities for roll bending with inner and outer chocks.
2. For mild steel as well as medium carbon steels using work roll diameter in the range of 300 - 450 mm , back-up roll diameter of 800 - 1200 mm used for strip widths of 800 -1200 mm and a roll bending force within 25000 kg can result in almost zero crown on the outcoming strip without taking the edge drop into consideration.
3. For harder materials, it has been noted that the roll bending alone, within the 25000 kg limit is not sufficient to correct the strip profile fully. In that case, roll cambers have to be resorted to.
4. This program package can be useful to mill designers to decide what type of roll bending facilities should be provided on a practical 4 high mill, if the range of materials to be rolled and other mill parameters are known.

7 References

1. Stone, M.D. and Gray, R., 'Theory and practical aspects in crown control', *Iron and Steel Engineer*, Aug. 1965.
2. Shohet, K.N. and Townsend, N.A., 'Roll bending methods of crown control in four high plate mills', *J. of the Iron and Steel Institute*, Nov. 1968.
3. Yuen, W.Y.D., Wechner, B., 'A physical model for on-line roll bending set up in cold rolling', *The 5 th international rolling conference*, London, 1990, p. 234.
4. Patula, E.J., 'Steady-state temperature distribution in a rotating roll subject to heat fluxes and convective cooling', *ASME J. of heat transfer*, vol. 103, Feb. 1981, p. 36.
5. Bennon, W.D., 'Evaluation of selective coolant application for the control of work roll thermal expansion', *ASME J. of Engg. Industry*, vol. 107, may 1985, p. 146.
6. Sauer, W., 'Thermal camber model for hot and cold rolling', *Steel Research*, vol. 67, no. 1, 1996, p. 18.
7. Tseng, A.A., Tong, S.X., Raudensky, M. and Chen, T.C., 'Thermal expansion and crown evaluations of rolls in rolling processes', *Steel Research*, vol. 67, no. 5, 1996, p. 189.
8. Deshpande, A. S., Srininivasa Murthy, K., 'Computer analysis for the prediction of a strip profile in cold rolling', *J. of Materials Processing Technology*, 63, 1997, p. 712.
9. Hetenyi, M. 'Beam on elastic foundation : *Theory with applications in the field of civil and mechanical engineering*', Univ. of Michigan press, Ann Arbor 1961.
10. Abramowitz, M., Stegun, I.A., 'Handbook of mathematical functions with formulas, graphs and mathematical tables', Dovers publications Inc, New York, 1965.
11. Timeshenko, S., and Goodier, J. N., '*Theory of elasticity*', 3 rd edition, McGraw hill, New York, 1970.

An Application of Repetitive Control to Disturbance Rejection in Servo Systems

ZHAOWEI ZHONG

Gintic Institute of Manufacturing Technology, Singapore 638075

Abstract. This paper reports an application of repetitive control to disturbance rejection in servo systems. At certain frequencies, the characteristic of disturbance rejection becomes worse due to the repetitive control. A new iterative learning compensator is presented to solve the problem. A numerical example is used to confirm the effectiveness of the proposed method. The use of Butterworth band-pass filters, infinite impulse response digital filters, or the fast Fourier transform technique is investigated to realise the proposed compensator. Also, a neural network is applied to the repetitive control system. The results demonstrate that a repetitive control without deteriorating the controlled variable at any frequency can be realised.

1 Introduction

In practice, disturbances on servo systems are often periodic. For example, the torque fluctuation of load causing rotational vibration of the worktable in a CNC gear cutting machine mainly synchronises with both the rotation and cutting edges of the cutting tool. The disturbance causing rotational speed fluctuation in a motor usually synchronises with the rotation of the armature. The frequencies of these disturbances are the same or integral times higher than ω_0, where $\omega_0=2\pi/L$, L is the basic period of the disturbances. To reject such periodic disturbances, repetitive control is useful.

Repetitive control was originally proposed by Inoue et al. [1] to develop a servo system, whose reference commands to be tracked are periodic signals of a fixed period L, using the tracking errors of the preceding periods. There is a similar control method called iterative learning control. It was first developed by Uchiyama [2] to generate a reference input by repeatedly using the error information of the foregoing trials. After that, many studies on these control methods were carried out [3-6]. According to References 5 and 6, repetitive control can be regarded as a simple learning control.

These methods are also useful in rejecting disturbances whose frequencies are the same or integral times higher than ω_0, the basic frequency of the iterative learning compensator. However, it is known that the rejecting effect is small for disturbances whose frequencies are not the same or integral times higher than ω_0. Moreover, at certain frequencies, the influence of disturbances on a repetitive control system which has a basic servo system and the typical iterative learning compensator, on the contrary, is larger than that on the same basic servo system without the iterative learning compensator [7].

In this paper, this problem is solved by means of a new iterative learning compensator. To realise this proposed compensator, the use of Butterworth band-pass filters, infinite impulse response digital filters, or the fast Fourier transform technique is investigated. Also, a neural network is applied to the repetitive control system. A numerical example is used to confirm the effectiveness of the proposed method.

2 Repetitive Control System

The repetitive control is discussed based on a basic SISO servo system shown in Fig. 1 excluding the broken lines and $H_r(s)$. It is assumed that in the basic servo system, the following two basic control purposes have been achieved.

(1) The system is asymptotically stable.
(2) The output y(t) tracks the reference input r(t).

Also, it is assumed that if the disturbance T_L is bounded, the output y is also bounded. In Fig. 1, the symbols represent the following:

$G_p(s)$ = Plant (process, object) to be controlled
Y(s) = Output of the system — R(s) = Reference input
U(s) = Repetitive control — E(s) = Error
$T_L(s)$ = Disturbance — $H_r(s)$ = Iterative learning compensator
$G_1(s)$ = Dynamic compensator 1 — $G_2(s)$ = Dynamic compensator 2.

From Fig. 1, Y(s) is expressed as a linear combination as follows:

$$Y(s)=G_r(s)R(s)+G_L(s)T_L(s), \tag{1}$$

where

$$G_r(s)= G_1(s)G_L(s) \tag{2}$$
$$G_L(s)= G_p(s)/\Delta \tag{3}$$
$$\Delta = 1+ G_p(s)[G_1(s)+ G_2(s)]. \tag{4}$$

For case I of the repetitive control, let

$$U(s)=H_r(s)E(s), \tag{5}$$

and for case II, let

$$U(s)=H_r(s)\, G_1(s)E(s), \tag{6}$$

where $H_r(s)$ is an iterative learning compensator given by

$$H_r(s)=H_u(s)H_0(s)/[1-H_0(s)]. \tag{7}$$

$H_u(s)$ is for compensating the gain and phase of the iterative learning compensator to guarantee the stability of the repetitive control system and convergence of the error. H_0 is the central component of the iterative learning compensator, which satisfies the following condition:

$$\|H_0(s)\|_\infty < \infty. \tag{8}$$

In the closed loop of the repetitive control system, the following equations are obtained:

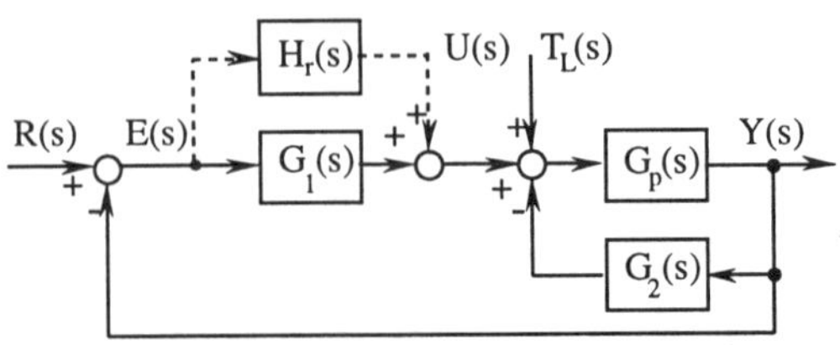

Fig. 1. Scheme of the servo system.

$$E(s) = \frac{[1-G_r(s)]R(s)-G_L(s)T_L(s)}{1+H_0(s)[G_L(s)H_u(s)-1]}[1-H_0(s)] \qquad \text{(for case I)} \qquad (9)$$

$$E(s) = \frac{[\Delta^{-1}+G_2(s)G_L(s)]R(s)-G_L(s)T_L(s)}{1+H_0(s)[G_r(s)H_u(s)-1]}[1-H_0(s)]. \quad \text{(for case II)} \qquad (10)$$

From Eqs. (9) and (10), it is obvious that under the assumptions mentioned above, if $H_0(s)$ satisfies Eq. (11), the influences, on the error e(t), of r(t) and $T_L(t)$ whose frequencies are the same or integral times higher than ω_0, become 0 in the steady state:

$$|1-H_0(j\omega_n)|=0, \qquad (11)$$

where $\omega_n = n\omega_0$, n = 1, 2, 3,... . Based on the assumptions, a sufficient stable condition of this repetitive control system is

$$\|H_0(s)[1-G_m(s)H_u(s)]\|_\infty<1, \qquad (12)$$

where

$$G_m(s) = \begin{cases} G_L(s) & \text{(In case I)} \\ G_r(s) & \text{(In case II).} \end{cases} \qquad (13)$$

Here we have two choices, $G_L(s)$ and $G_r(s)$, to satisfy the sufficient stable condition and to obtain a better characteristic of disturbance rejection.

The typical $H_0(s)$ is

$$H_0(s)=\exp(-Ls) \qquad (14)$$

$$L=2\pi / \omega_0. \qquad (15)$$

It satisfies Eqs.(8) and (11). For the frequencies $\omega=\omega_n$, from Eqs. (9) or (10) and (11), the error e(t) becomes 0 in the steady state. However, for the other frequencies,

$$|1-H_0(j\omega)|=|1-\exp(-j\omega_0 L)|\le 2. \qquad (16)$$

Particularly, when $\omega=\omega_0(2n-1)/2$, n=1, 2, 3,...,

$$1-H_0(j\omega)=2 . \qquad (17)$$

As shown by Eqs. (9) and (10), since the error E(s) is in proportion to $[1-H_0(s)]$, $|1-H_0(j\omega)|$ can be used to approximately estimate the characteristic of disturbance rejection. Equations (16) and (17) mean, for disturbances $\omega\neq\omega_n$, the effect of disturbance rejection is small. Moreover, at certain frequencies, the influence of disturbances on the repetitive control system which has a basic servo system and the typical iterative learning compensator, on the contrary, is larger than that on the same basic servo system without the learning compensator. This is a big problem [7].

3 Iterative Learning Compensator

Instead of Eq. (14), let $H_0(s)$ be

$$H_0(s) = a\,F\exp(-Ls) + b \qquad (18)$$

and

$$F = \begin{cases} 1 & (\omega=\omega_n) \\ 0 & (\omega\neq\omega_n) \end{cases} \qquad (19)$$

$$a+b=1. \qquad (20)$$

The following equation can be obtained:

$$|1-H_0(j\omega)| = \begin{cases} 0 & (\omega=\omega_n) \\ |a| & (\omega\neq\omega_n). \end{cases} \qquad (21)$$

Equation (11) is satisfied and the steady state error becomes 0. Choose a as

$$0<|a|\leq 1. \tag{22}$$

Then for any frequency,

$$|1-H_0(j\omega)| \leq |a| \leq 1. \tag{23}$$

Thus, a repetitive control without deteriorating the controlled variable at any frequency can be realised.

However, the filter given by Eq. (19) may be difficult to realise for some applications. Therefore, instead of Eq.(19), let F be

$$\left.\begin{array}{ll} F(j\omega)=1 & (\omega=\omega_n) \\ |F(j\omega)|\leq 1 & (\omega\neq\omega_n). \end{array}\right\} \tag{24}$$

Equation (11) is also satisfied so that the steady-state error becomes 0. Since

$$|1-F(j\omega)\exp(-jL\omega)|\leq 2 \tag{25}$$

$$|1-H_0(j\omega)|=|a||1-F(j\omega)\exp(-jL\omega)|, \tag{26}$$

choose a as

$$0<|a|\leq 0.5\ ; \tag{27}$$

then for any frequency,

$$|1-H_0(j\omega)|\leq 1. \tag{28}$$

4 Numerical Example

The basic system represented by Eqs. (29), (30) and (31) is used as a numerical example for confirming the effectiveness of the proposed method.

$$G_p(s) = \frac{b_0}{(s + a_0)} \tag{29}$$

$$G_1(s) = \frac{c\,(s + a_0)}{b_0 s} \tag{30}$$

$$G_2(s)=0. \tag{31}$$

Let K_d, K_u be constants and

$$H_u(s)=K_d s+K_u. \tag{32}$$

To realise the filter shown by Eq.(24), Butterworth band-pass filters given as follows are used:

$$\left.\begin{array}{l} F(s)=F_1(s)+F_2(s)+\ldots+F_n(s) \\ F_n(s)=1/(\lambda^3+2\lambda^2+2\lambda+1) \\ \lambda=(s^2+\Omega_{n1}\Omega_{n2})/[s(\Omega_{n2}-\Omega_{n1})] \end{array}\right\}, \tag{33}$$

where Ω_{n1} and Ω_{n2} are, respectively, the lower frequency and upper frequency that define the band-pass width. The central frequencies of the band-pass filters are ω_1, $\omega_2,\ldots,\omega_n$.

In real systems, we often encounter the case in which it is sufficient for practical purposes to reject the influences of some special frequencies of T_L. For example, we can choose n=1, 2, 8 according to practical needs. Therefore, from now on, the computation is carried out assuming n=1, 2 for simplification.

Let the transfer function from the disturbance T_L to the output y be $G_c(s)$. $G_c(s)$ is given by

$$G_c(s) = \frac{Y(s)}{T_L(s)} = \frac{G_L(s)}{1+H_0(s)[G_m(s)H_u(s)-1]}[1-H_0(s)] \,. \tag{34}$$

If the repetitive control is not used, $G_c(s)=G_L(s)$.

In chapters 2 and 3, $|1-H_0(j\omega)|$ is used to approximately estimate the characteristic of disturbance rejection. It is necessary for practical applications to compute $|G_c(j\omega)|$, to exactly evaluate the characteristic of disturbance rejection.

Figures 2, 3 and 4 show examples of the computation of $|G_c(j\omega)|$, where Eq. (5) is used to calculate the repetitive control u, K_d=0,005, K_u=1, L=0.1, a_0=22.1, b_0=68.4 and c=100. When the typical $H_0(s)$ = exp(−Ls) is used for repetitive control, $|G_c(j\omega)|=0$ at $\omega=\omega_n$; however, there are some frequencies $\omega\neq\omega_n$ where $|G_c(j\omega)|>|G_L(j\omega)|$. That is, at these frequencies, the characteristic of disturbance rejection becomes worse due to the repetitive control. When the filter of Eq. (33) or Eq. (19) is used for repetitive control, as shown in Figs. 3 and 4, $|G_c(j\omega)|=0$ at $\omega=\omega_n$; moreover, $|G_c(j\omega)|\leq|G_L(j\omega)|$ at any frequency over the whole frequency range. Thus, a repetitive control without deteriorating the controlled variable at any frequency can be realised.

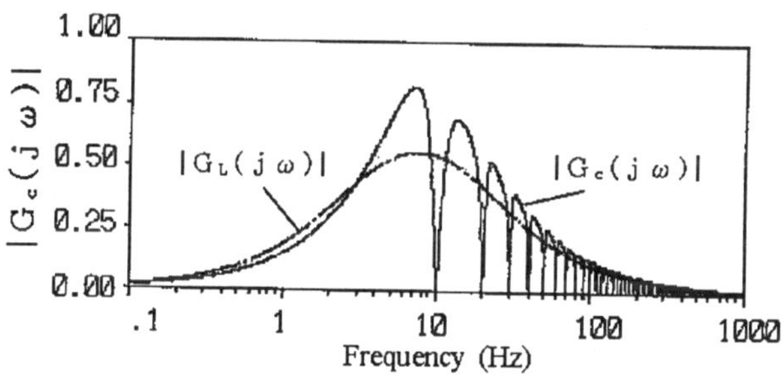

Fig. 2. Computation of $|G_c(j\omega)|$ with the typical compensator (using Eq. (5)).

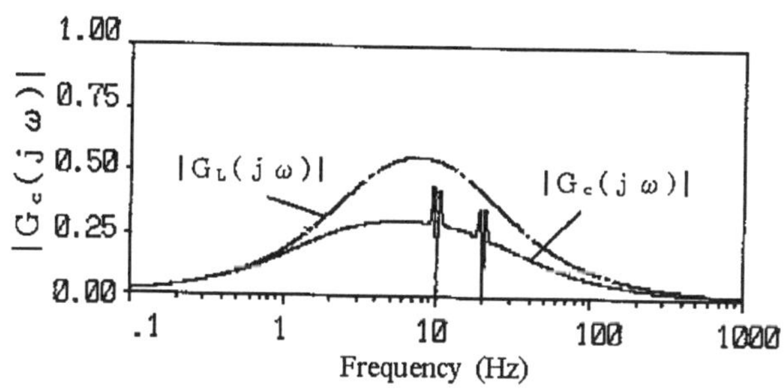

Fig. 3. Computation of $|G_c(j\omega)|$ with the proposed compensator (a=0.4, using Eqs. (5) and (33)).

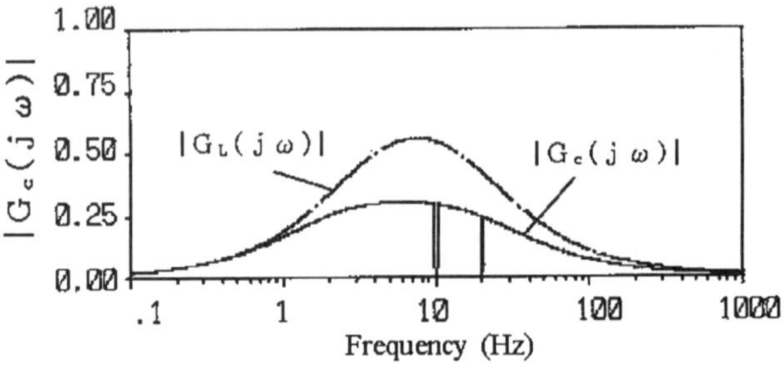

Fig. 4. Computation of $|G_c(j\omega)|$ with the proposed compensator (a=0.4, using Eqs. (5) and (19)).

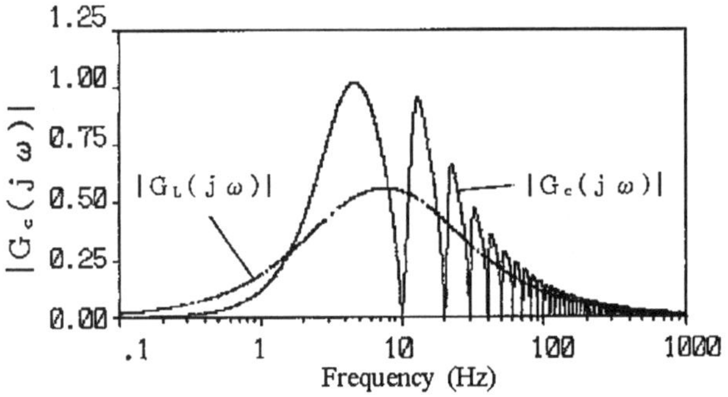

Fig. 5. Computation of $|G_c(j\omega)|$ with the typical compensator (using Eq. (6)).

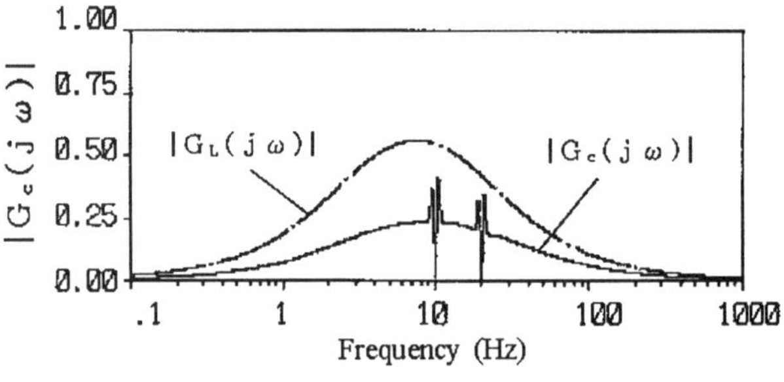

Fig. 6. Computation of $|G_c(j\omega)|$ with the proposed compensator (a=0.4, using Eqs. (6) and (33)).

Figures 5, 6 and 7 show other examples of the computation of $|G_c(j\omega)|$, where also K_d=0,005, K_u=1, L=0.1, a_0=22.1, b_0=68.4 and c=100 but Eq. (6) is used to calculate the repetitive control u. Comparison with Figs. 2, 3 and 4 shows that the statement mentioned above for Figs. 2, 3 and 4 is also true for Figs. 5, 6 and 7. However, the characteristics of disturbance rejection are different. This means that we can choose Eq. (5) or (6) to calculate the repetitive control u to obtain a better characteristic of disturbance rejection. In this example, using Eq. (6) is better than using Eq. (5).

Simulation experiments were carried out. An example of the simulations is shown in Fig. 8, where the same periodic disturbance as in Fig. 8(a) is rejected by using the repetitive control as shown in Fig. 8(b) & (c). In Fig. 8(b), the filter of Eq. (19) was realised by using the fast Fourier transform (FFT) technique. It usually takes time to complete the computation of FFT. However, it is possible for practical applications to realise this filter using FFT by taking advantage of the characteristic of the repetitive control in which the control u is computed using the error information of the preceding periods, or by using hardware and so on. Also, in Fig. 8(c), infinite impulse response digital filters were constructed by transforming the filters of Eq. (33) to the z domain. Since the order of the digital filters is six, the computing time is very short and it is sufficient for practical applications. The results of the simulations also confirmed the effectiveness of the rejection method proposed in this paper.

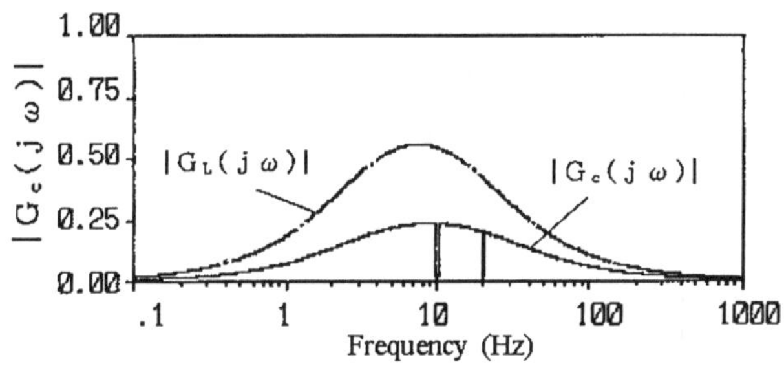

Fig. 7. Computation of $|G_c(j\omega)|$ with the proposed compensator (a=0.4, using Eqs. (6) and (19)).

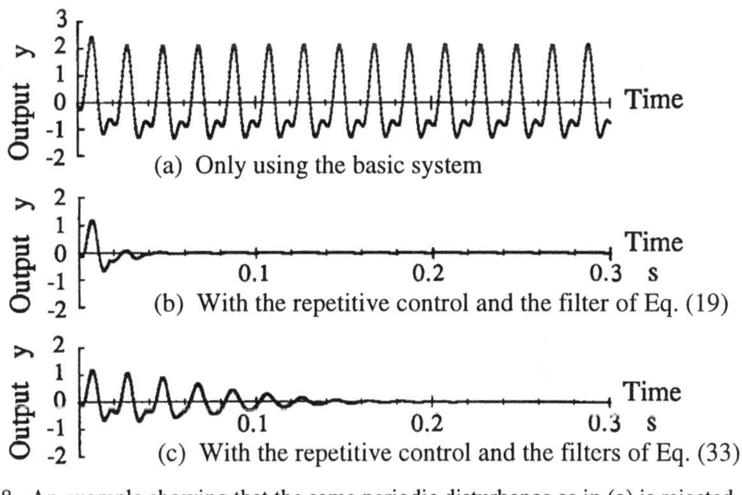

Fig. 8. An example showing that the same periodic disturbance as in (a) is rejected by using the repetitive control (b) & (c).

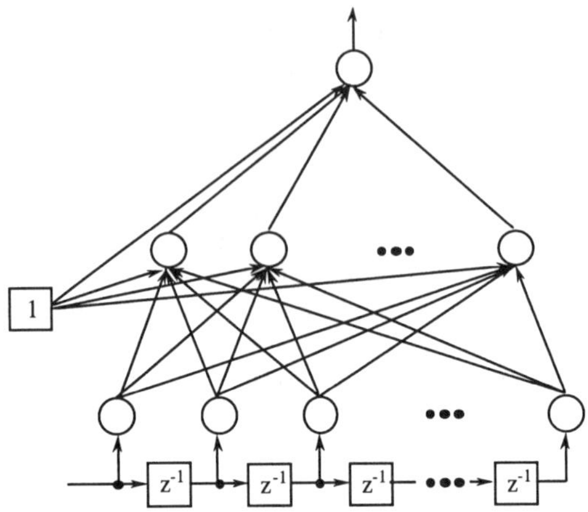

Fig. 9. Neural network.

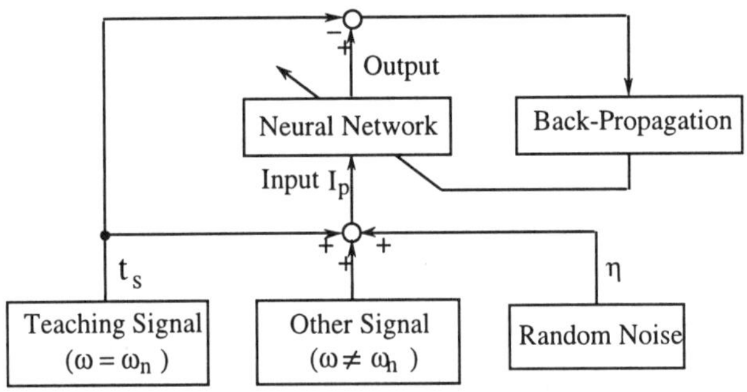

Fig. 10. Training method of the neural network.

Also, it was tried to use a neural network shown by Fig. 9 to realise the filter F. To train the network, the method shown in Fig. 10 was used, where,

$$t_s = \Sigma \sin(\omega_n t + \phi_n) \tag{35}$$

$$I_p = t_s + \Sigma \sin(\omega_k t + \phi_k) + \eta \qquad (k \neq n). \tag{36}$$

Figure 11 shows an example of the training process, where, n=1, 2, L=20ms. After training, the neural network was applied to the repetitive control to reject periodic disturbances.

Figure 12 and 13 show two examples of the simulations, where a_0=8.1, b_0=25.0, c=100, Eq. (6) was used to calculate the repetitive control u. As shown by Fig. 12(b) and (c), the same periodic disturbance as in Fig. 12(a) is rejected by using the typical repetitive control shown by Eq. (14) and the method of Chapter 3.

When the disturbance includes some random noise, the results of using only the basic system, the typical repetitive control, and the method of Chapter 3 are shown by Fig. 13 (a), (b) and (c) respectively.

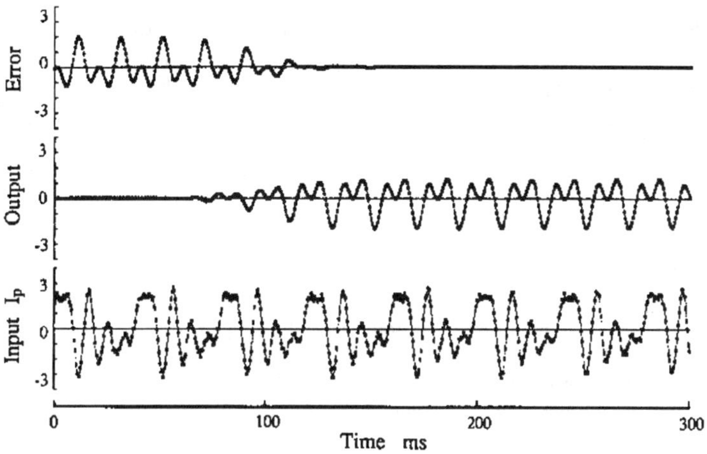

Fig. 11. An example of training process.

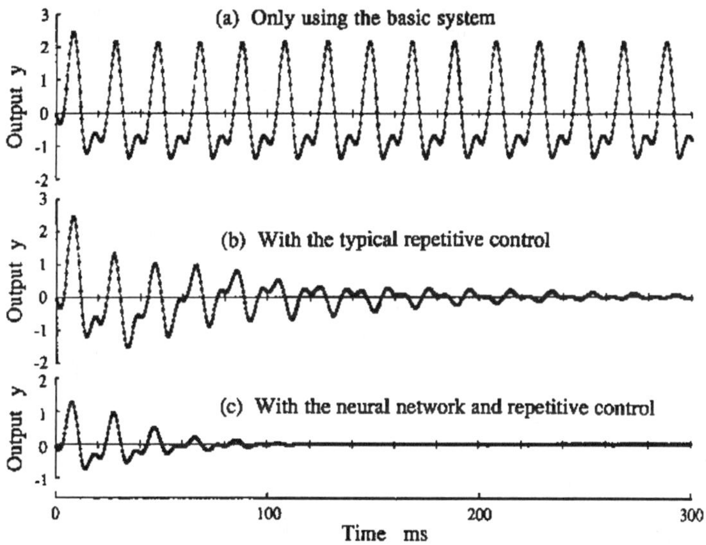

Fig. 12. An example of simulations with a periodic disturbance.

Other results of the simulations also confirmed the effectiveness of the rejection method proposed in Chapter 3. A repetitive control without deteriorating the controlled variable at any frequency can be realised. Moreover, the repetitive control system can be stable even with a larger value K_d when the neural network or the digital filters are used, so that the convergence of the error is faster.

5 Conclusions

Repetitive control is useful to reject periodic disturbances. However, at certain frequencies, the influence of disturbances on a repetitive control system having a basic servo system and the typical iterative learning compensator is larger than that on the same basic servo system without the iterative learning compensator. A new iterative learning compensator has been proposed to solve this problem. A numerical example was used, and numerical simulations were carried out to confirm the effectiveness of the proposed method. The results demonstrated that the repetitive control can reject periodic disturbances without deteriorating the controlled output of the servo system at any frequency.

6 References

1. T. Inoue, M. Nakano, and T. Kubo, High accuracy control for magnet power supply of proton synchrotron in recurrent operation, Trans. IEE Japan, **100**, 7 (1980) pp. 16-22.
2. M. Uchiyama, Formation of high-speed motion pattern of a mechanical arm by trial, Trans. SICE, **14**, 6 (1978) pp. 706-712.
3. S. Kawamura, System theoretic study on learning control method, Trans. SICE,.**21**, 5 (1985) pp. 445-450.
4. T. Mita, Iterative control and its applications to the trajectory control of robot manipulators, Trans. SICE, **21**, 11 (1985) pp. 1223-1230.
5. S. Hara, Repetitive control, Jour. of SICE, **25**, 12 (1986) pp. 1111-1119.
6. S. Hara, Y. Yamamoto, and T. Omata, Repetitive control system: a new type servo system for periodic exogenous signals, IEEE Trans. Automat. Contr., **AC-26** (1988) pp. 659-667.
7. M. Gotou, S. Matsubayashi, and F. Miyazaki, A robust servo system with an iterative learning compensator and a proposal of multi-period learning compensator, Systems and Control, **31**, 5 (1987) pp. 367-374.

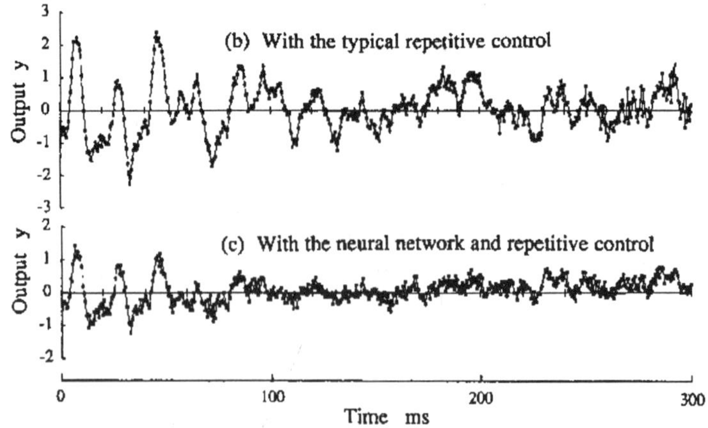

Fig. 13. An example of simulations with a periodic disturbance and random noise.

AUTHOR INDEX

COMPUTER INTEGRATED MANUFACTURING

INTERNATIONAL CONFERENCE ON COMPUTER INTEGRATED MANUFACTURING (ICCIM)

Proceedings of the ICCIM'91
Ed. B.S. Lim
ISBN 981-02-0684-4

Proceedings of the ICCIM'93
Eds. A. Sen, J. Winsor and R. Gay
ISBN 981-02-1568-1 (set)
ISBN 981-02-1946-6 (vol. 1)
ISBN 981-02-1947-4 (vol. 2)

Proceedings of the ICCIM'95
Eds. J. Winsor, A.I. Sivakumar and R. Gay
ISBN 981-02-2376-5 (set)
ISBN 981-02-2973-9 (vol. 1)
ISBN 981-02-2974-7 (vol. 2)

ICCIM'91 and ICCIM'95 were published by World Scientific Pub. Co. ICCIM'93 was jointly published by World Scientific Pub. Co. and Global Pub.

Proceedings of the 4th International Conference

COMPUTER INTEGRATED MANUFACTURING

21 – 24 October 1997
Singapore

Volume 2

Editors

A. Sen, A.I. Sivakumar and R. Gay

Gintic Institute of Manufacturing Technology
Nanyang Avenue
Singapore

Springer

A. Sen, A.I. Sivakumar and R. Gay
Gintic Institute of Manufacturing Technology
71 Nanyang Drive
Singapore 638075

Computer Integrated Manufacturing

ISBN 981-3083-66-2 (vol. 1)
ISBN 981-3083-67-0 (vol. 2)
ISBN 981-3083-68-9 (set)

Printed in Singapore

Typesetting: Camera-ready by authors
5 4 3 2 1 0

ORGANISING COMMITTEE

Chairman : Dr. Robert Gay, Gintic

Members :
Mr. Goh Kiah Mok, Gintic
Mr. Lew Sin Chye, Nanyang Technological University
Dr. Lim Kah Bin, National University of Singapore
Ms. Lim Lai Lee, Gintic
Ms. Jacqueline Ng, Miller Freeman
Dr. Anirudha Sen, Gintic
Dr. AI Sivakumar, Gintic
Dr. Soh Yeng Chai, William, Nanyang Technological University
Dr. V. Subramaniam, National University of Singapore
Dr. Tay Kiang Meng, Gintic
Mr. Tay Meng Leong, Nanyang Technological University
Mr. Stephen Teng, Singapore Industrial Automation Association

TECHNICAL COMMITTEE

Chairman : Dr. AI Sivakumar, Gintic
Co-Chairman : Dr. Anirudha Sen, Gintic
(Editorial Com)

Members :
Mr. Henry Chang , Productivity and Standards Board
Dr. Opas Chutatape, Nanyang Technological University
Mr. Goh Kiah Mok, Gintic
Mr. Lew Sin Chye, Nanyang Technological University
Dr. Lim Kah Bin, National University of Singapore
Mr. Jasbir Singh, Gintic
Dr. V. Subramaniam, National University of Singapore
Dr. Tay Kiang Meng, Gintic
Mr. Tay Meng Leong, Nanyang Technological University

INTERNATIONAL ADVISORY COMMITTEE

PREFACE

ICCIM '97 is the fourth international conference we have organised in the field of computer integrated manufacturing. It is gratifying to see that since its inception in 1991, ICCIM has established a track record of bringing to the public what latest technology is today and what will be available tomorrow.

This year our theme for ICCIM is "Manufacturing into the next Millennium" - an appropriate title as we approach the end of the twentieth century. In this age of rapid change, it is difficult to see beyond a few years what new markets and technologies will emerge. However, at ICCIM'97 we believe that many of the ideas presented and some of the research work reported will become reality tomorrow. For example, just since the last ICCIM conference in 1995 the growth of Internet-based business has suddenly exploded and made fortunes for many people, very often, unexpectedly. While the scale was beyond expectations, the trend as related to the manufacturing industry was anticipated and discussed in a number of papers from previous ICCIM conferences.

We have divided the proceedings into two volumes. In the first volume, we gathered the papers dealing with systems level management and planning of manufacturing activities, whilst the second volume concentrates more on product related and implementation related issues.

In this volume the following topics are covered :

- Product Design and Concurrent Engineering
- Factory Automation

On behalf of the Organising and Technical Committees of ICCIM '97, we would like to thank all those who have made this conference what it is today - the authors, International Advisory Committee , reviewers and participants.

A special thanks to Ms. Lim Lai Lee for her contribution to the numerous details in organising this event and for collating the articles for these two volumes.

October 1997

A.Sen
A.I. Sivakumar
R.Gay

CONTENTS

Volume 1

Flexibility and Change Management

Business Process Reengineering

CIM Design and Planning

Integrated Databases

Internet and the WWW

CIM Justification

CIM Implementations

MANUFACTURING PLANNING AND CONTROL

Aggregate Planning

Production Scheduling Systems

Performance Evaluation and Simulation

Volume 2

PRODUCT DESIGN AND CONCURRENT ENGINEERING

Geometric Modelling

Product Design

Product Analysis

Process Planning

Rapid Prototyping

CE Environments

FACTORY AUTOMATION

Shopfloor Integration

Assembly Automation

Inspection and Diagnostic Systems

Robotics and AGV Systems

ADVANCED MANUFACTURING TECHNOLOGY

PRODUCT DESIGN AND CONCURRENT ENGINEERING

Geometric Modelling

Using Soft Objects to Represent Medical Structures

H.L. Lim, Y.P.Wang, C.K. Chui, Y. Y. Cai

Institute of Systems Science,
National University of Singapore,
Kent Ridge, Singapore 119597
Email: {hllim|ypwang|cheekong|yyycai}@iss.nus.sg

Abstract

Soft object is one of the most effective object representations in 3D computer graphics. Using it, the surface of an object is represented by the summation of the field functions of a set of key points. Soft object has the advantage of being able to represent blobby or tubular objects and their deformations with relatively few primitives and operations. In this paper, we developed an improved rendering method to visualize the surfaces. We compute all pairs of keypoints whose field function interact. For each pair of the points, we compute axial and radial rays from the axis joining them. Newton's method and regula falsi are then used to determine the intersections between the rays and the iso-surface. Based on the intersections, we can subdivide the iso-surface of the soft-object into patches. After computing the normals at the vertices of the patches, the soft objects can be rendered. We explore using soft objects to model blood vessels and catheters for our project for the simulation, visualization and design of catheters.

1. Introduction.

3D graphics has traditionally employed polygonal patches as the basic elements for 3D models. Although the use of polygons has advantages such as speed of rendering and ease of representation, it does have limitations. One of its major drawbacks is that polygons are rigid structures. When they are used to represent organic, soft, or fluid objects, their rigidity makes them difficult to represent the deformation of these objects when they move or deform. An alternative and increasingly popular method to represent these objects is to apply a class of object model called soft objects. Using this representation, each object or its primitives are represented by a set of field functions. Since the field functions within an object and between objects can interactive freely, the surfaces of these objects can be smooth connected. Their deformation can also be realistically displayed.

Because of the advantages of soft objects, we have applied them for representing anatomical models in our project, *daVinci* (Visual Navigation of Catheter

Insertion). The objective of the project is to achieve realistic simulation and visualization of the process of catheter navigation through a network of blood vessels. Through the use of soft objects, we hope to achieve realistic representation of organic tissues in both their original and deformed shapes.

2. The Technology of Soft Objects.

2.1 Representing Surfaces by Field Functions.

Central to the technology of soft objects is its use of field functions to represent iso-surfaces. The idea of using field functions was first proposed by Blinn [1]. He described a technique that approximates atoms by Gaussian distribution functions. Each atom is assigned a Gaussian distribution function:

$$f(r) = ae^{br^2}$$

where the van der Waals potential of the atom is defined as the function of its radius. The potential value at any point is then

$$F = \sum_i f_i(r) = \sum_i a_i e^{b_i r_i^2}$$

At any point where the potential value is equal to or greater than a threshold value, the surface will be visible. If two or more atoms share a common space, their potentials at that point add up, and then this value is compared to the threshold value. Atoms also may have negative potential, creating the effect of repulsion between them.

2.2 Wywill's Representation of Soft Objects.

Wywill et at have improved upon the basic technology of soft objects [2] [3] [4]. They are also the first to use the term *soft objects* for the class of soft, deformable objects. Using their representation, an object contains one or more keypoints. The field function of a key point is of the form

$$f(r) = a\frac{r^2}{R^2} + b\frac{r^4}{R^4} + c\frac{r^6}{R^6} + 1$$

where R is the value at which the field decays to zero. Thus, the isosurface of the field function at that value represents the surface of the object.

In the above equation, a, b and c are constants such that $C(0.0) = 1$, $C'(0.0) = 0$, $C'(R) = 0$, $C'(R) = 0$, and $C(R/2) = 0.5$. Thus, the field value at a point decays from 1 to 0.5 and to 0 when it moves from positions where $R = 0$, $R = 0.5$ and $R =$

1. Compared with Blinn's formulation, Wywill's definition of soft objects can be more quickly computed as there are no exponential functions to be evaluated.

3. The Visualization of Soft Objects.

3.1 Existing Approaches in Rendering Soft Objects.

Blinn displays the soft objects by numerically evaluating the closest surface at each pixel [1]. Each atom in his model is transformed into a standard viewing space, where the axes are parallel or orthogonal to the viewing direction. After the centres and their radii have been transformed to the new coordinate system, the visible field surface of the model at each pixel is computed by tracing a ray from the viewpoint origin through the pixel. If only an atom is present, the ray intersection can be found analytically. However, if more than one atom is present, numerical methods would need to be used. Blinn examined both Newton's method and regula falsi. The former has fast convergence but may not give an answer if the initial guess is not close to the actual solution. The latter is slower but guaranteed to converge. Blinn therefore proposed using a hybrid of Newton's method and regula falsi. Newton's method is first used. If the intermediate result is outside the bracket of solution, regula falsi is then used.

Wyvill instead overlays the space with a 3D grid. All the cubes formed by the grid that are intersected by the surfaces of the soft objects are found. The surfaces within each cube are approximated by polygons. These polygons are then rendered.

To achieve this efficiently, Wywill scans the cubes along one axis to determine some cubes that intersect the surfaces of the soft objects. These cubes are used as seeds, whereby other connected cubes are successively checked to determine whether they intersect the surfaces.

For each intersecting cube, the field values at its eight vertices are found. From these values, the polygons within the cube and part of a surface can be constructed.

Wyvill also describes a method for ray tracing soft objects [6]. He first finds the intersections of each ray with all the bounding spheres of the soft objects. The intersections are ordered in the sequence along the ray. Using these intersections as starting values, the actual intersections between the ray and the surfaces of the soft objects are found numerically.

3.2 A New Method in Rendering Soft Objects.

Amongst the various methods for rendering soft objects, the ray tracing approaches by Blinn and by Wywill can create high quality images but are costly as the surfaces at every pixel need to be found. When the model moves or rotates, the same ray tracing operations need to be carried out from scratch. Wyvill's method of using 3D grid to divide the soft objects into polygons is more general and efficient. However, the 3D grid divides the surfaces into irregular patches. The searching of cubes that intersect the surfaces would also require considerable computations. We therefore propose a new method for computing and rendering the surfaces of soft objects.

3.2.1 Determination of local axes and rays to be traced.

This method takes advantage of the fact that most of the isosurface would usually surround a keypoint if there are no other influencing nearby keypoints. First, we detect all pairs of keypoints whose field functions are likely to interact. For each pair of the points, the axes joining them are first found. This is defined as the local X axis. We also determine another two axes that are non-parallel to each other and the local X axis. They are referred to as the local Y and local Z axes (we will discuss more about the determination of the local Y and Z axes later). The portion of the local X axis within the endpoints is divided into a predefined number of intervals. With each endpoint of the interval as origin, we trace a predefined number of evenly spread-out, axial rays that are parallel to the plane containing the local Y and Z axes. These rays are referred to as *axial rays*. If any of the endpoints is a terminating keypoint (a keypoint that can only be influence by the other keypoint pairs with it), a set of rays would also have to be traced from it and cover the directions that are outside the axial rays. We achieve this by subdividing the hemisphere containing the solid angle not covered by the axial rays and tracing rays in the directions defined by the subdivided angles. These rays are referred to as the *radial rays*. The rays to be traced for a set of terminating keypoints are shown in figure 1.

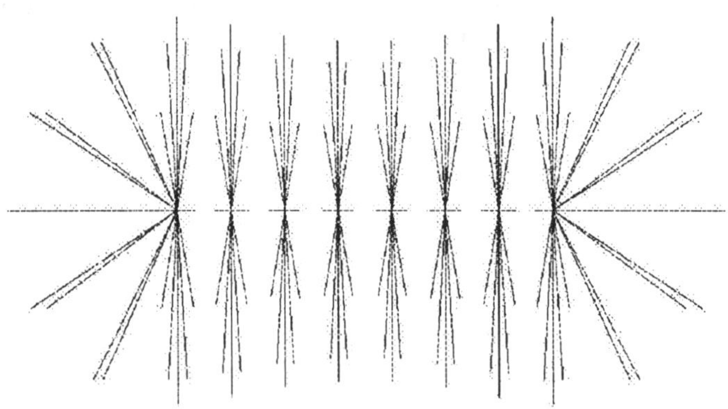

Figure 1: Tracing of rays from a pair of keypoints.

Note that if the field function of a keypoint interacts with the field functions of more than one keypoint, it would belong to more than one keypoint pair. The local-Y and local-Z axes for each pair would have to be selected such that they remain non-parallel to the local-X axis while providing continuity across the pairs.

The ideal situation would be for the same axes to be used for the adjacent pairs. Failing which, we determine the axes that are closest to their counterparts. We then find out if there are solid angles that are not covered by the axial rays of the pairs. Should this be the case, some radial rays would have to be traced to cover these angles.

3.2.2 Computing ray intersections.

After the rays have been defined, each of them is traced. The intersection between each ray and the iso-surface is computed. Like Blinn's method, we first apply Newton's method and then regula falsi to obtain the solution. However, in most situations Newton's iterations would suffice. There may be situations where numerical iterations fail to find a solution. This indicates that there are other nearby keypoints, such that the line is always within the isosurface. In this situation, the point on the line that has the lowest field value is determined and used as its approximation.

The formula of Newton's method is

$$p_{new} = p_{old} - \frac{F(p_{old}) - T}{F'(p_{old})}$$

Where p_{old} is the old guess and p_{new} is the new guess of the iteration, T is the threshold value of the isosurface, $F(p)$ and $F'(p)$ are respectively the field function and the derivative of the field function.

Since we are attempting to find the intersection between a ray and the isosurface, the equation of the ray with direction (ax, ay, az) and origins (x_0, y_0, z_0) can be written as the following parametric equations:

$$x = a_x p + x_0$$

$$y = a_y p + y_0$$

$$z = a_z p + z_0$$

The function of the isosurface can be expressed as a function of r^2:

$$F = \sum_i f_i(r_i^2)$$

As $r_i^2 = (x - x_i)^2 + (y-y_i)^2 + (z - z_i)^2$, where (x_i, y_i, z_i) is the coordinates of the keypoint. By substituting the parametric equation of the ray, r^2 can be expressed as a quadratic equation of p.

Further substituting this quadratic equation to F, we obtain an implicit equation of p. Likewise, the derivative of the iso-surface can also be expressed as a function of r^2, which can be converted to an equation of p:

$$F'(f) = \sum_i f_i'(r_i^2) = \sum_i \frac{df_i(r_i^2)}{d(r_i^2)} \frac{d(r_i^2)}{dp}$$

Using the above substitution, the equation of Newton's method can be expressed totally by p, which can then be evaluated.

3.3.3 Creation and rendering of patches.

After the tracing of rays, patches are defined from vertices that are the surface intersections of adjacent rays. The normals at these vertices are computed. This can be derived from the partial differentiation of the field functions describing the soft objects using the following equations:

$$N = \frac{\partial F}{\partial x} i + \frac{\partial F}{\partial y} j + \frac{\partial F}{\partial z} k$$

Having obtained both the coordinates and normals of the patches, the soft objects can be rendered.

The images of the mesh and shaded surfaces of soft objects generated by the current technique are shown in figures 2 to 9.

3.3.4 Advantages and disadvantages of the current approach.

Compared with Blinn and Wywill's methods, the current approach has the advantage that the computed patches are substantially fewer. It therefore compares favorably with them in terms of speed. Because the patches are directly subdivided from the iso-surface, they tend to have a better fit to the surface topology and their sizes are more uniform. This can improve the quality of image. The disadvantage of the current approach is that an axis needs to be defined for every pair of interacting keypoints. Therefore, the current approach is efficient in objects that are tubular in shape (such as the vascular anatomy), but not in models where the field functions of many keypoints would interact with each other,

4. Using Soft Objects to represent Vascular Anatomy.

4.1 Visualization of Catheter Navigation.

Guidewire and catheter are two major instruments in angiographic vascular catheterization [8][9]. Each of them has multiple variations designed to facilitate a

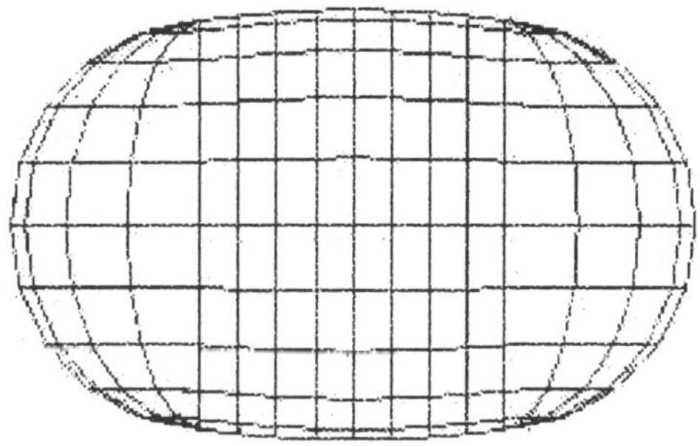

Fig 2: Mesh surface of first sample soft object.

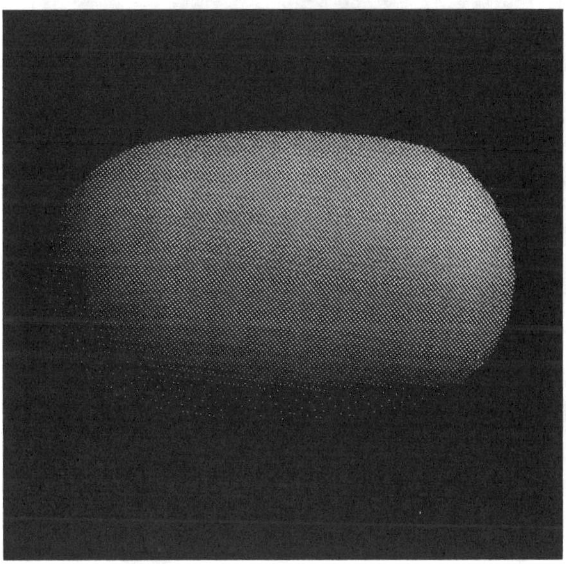

Fig 3: Shaded Image of the first sample soft object.

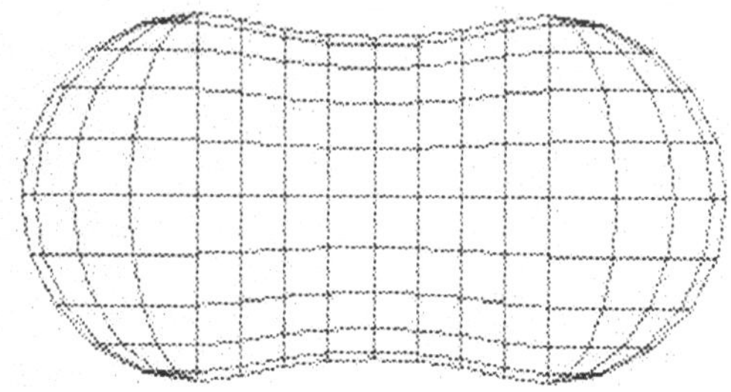

Fig 4 Mesh surface of the second sample soft object.

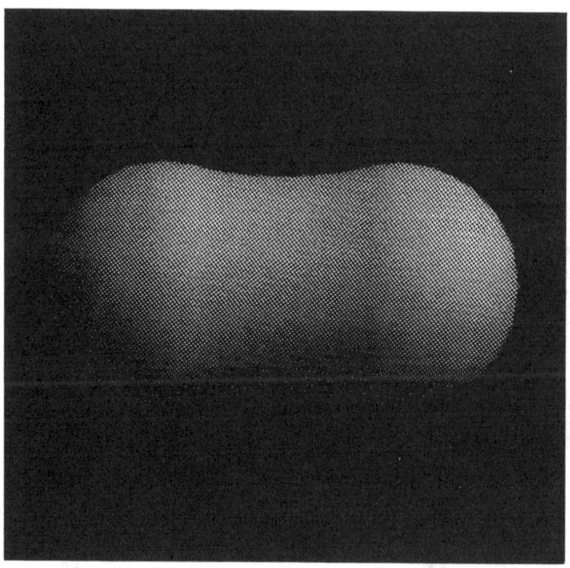

Fig 5 Shaded image of the second sample soft object.

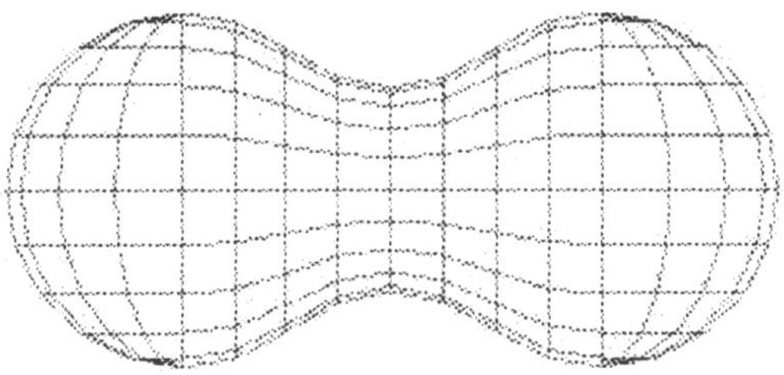

Fig 6: Mesh surface of the third sample soft object.

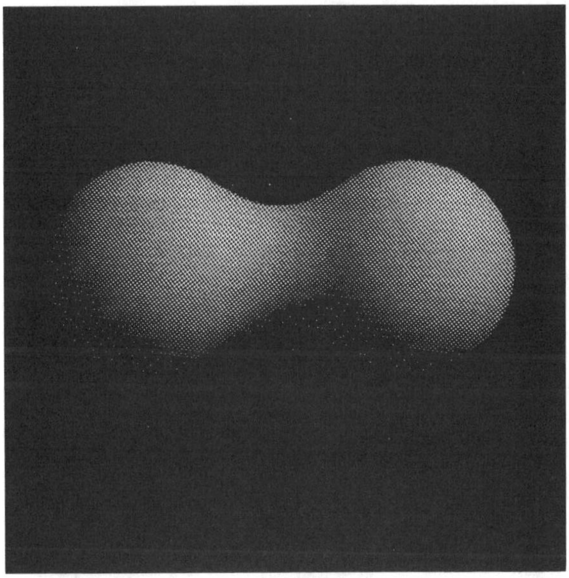

Fig 7: Shaded Image of the third sample soft object.

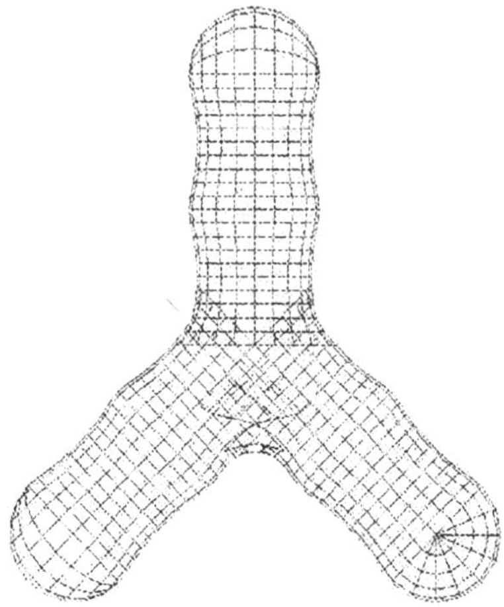

Fig 8: Mesh surface of the fourth sample soft object.

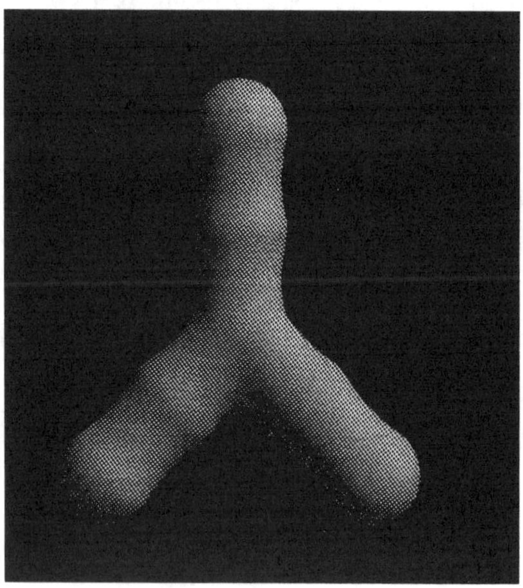

Fig 9: Shaded Image of the fourth sample soft object.

particular function. Catheters are categorized according to their material, shape, length, etc. Some serve multiple purposes in addition to their original intentions. They possess the properties of pliability, torquability, and shape memory. The function of the guidewire is to provide support and guidance for passage of an overlying catheter through the artery and into the lumen of the selected artery.

In the catheterization process, the tip part of pre-shaped catheters usually undergoes various changes. For instance, the insertion of a guidewire into the catheter tip may straighten up the tip shape while the withdrawal of the guidewire would reshape the catheter back to its original configuration. The simulation of the interaction between catheter and guidewire is therefore an important task for the analysis of the deformation in catheterization.

4.2 Basic Information of Vascular Anatomy.

The objective of our system called daVinci (Visual Navigation of Catheter Insertion) [10][11][12][13][14][15] is to achieve realistic simulation and visualization of the process of catheter navigation through a network of blood vessels. In daVinci, the vascular anatomy is represented by the central lines and their corresponding radii of the vessels. Using the marching cube technique, the image data are segmented to obtained coordinates of the central lines and the radii.

This anatomical structure is stored in a file. In the file are blocks which represent the segments of the central line. Each block consists of an index number, a description, a radius and vertices of the segment nodes.

4.3 Representation of Vascular Anatomy Using Soft Objects.

To determine the soft object representation of the vascular anatomy, the anatomical structure file is analysed. Each block of the segment is treated as a soft object, while the vertices of the segment are used as keypoints. Figure 9 and 10 show a model of a bifurcation joint represented by a soft object using different number of keypoints.

5. Conclusion.

In this paper, we have described the use of soft objects to represent medical models such as blood vessels. We have also implemented a new method for displaying these objects. With further development, we hope to improve the efficiency and explore further use of the soft object technology in medical applications.

Acknowledgement:

We would like to thank National Science and Technology Board of Singapore for funding our research.

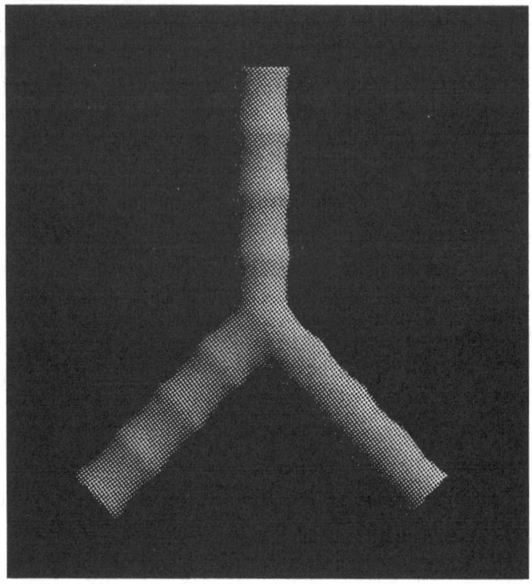

Fig 10: Shaded image of a blood vessel simulated by a soft object with sparse keypoints.

Fig 11: Shaded image of a blood vessel simulated by a soft object with dense keypoints.

References:

[1] Blinn, J. "A generalization of algebraic surface drawing", ACM Transaction of Graphics, 1982, Vol. 1, No. 1.

[2] Wyvill G., McPheeters C., Wyvill B.", "Soft Objects", Advanced Computer Graphics (Proceedings of Computer Graphics Tokyo '86), Tsiyasu L. Kunii (Editor).

[3] Wyvill B., McPheeters C., Wyvill G., "Animating soft objects", The Visual Computer, Vol. 2, No. 4, Aug, 1986.

[4] Wyvill G., McPheeters C., Wyvill B., "Data structure for soft objects", The Visual Computer, Vol. 2, No. 4, Aug, 1986.

[5] Wyvill B., Wyvill G., "Field functions for implicit surfaces", The Visual Computer, Vol. 5, No. 1/2, Mar 1989.

[6] Wyvill G., Trotman A., "Ray-Tracing Soft Objects", New Trends in Computer Graphics (Proceedings of CG International '90), 1990.

[7] Wyvill G., Wyvill B., McPheeters C., "Solid Texturing of Soft Objects", IEEE Computer Graphics and Applications, Vol. 7, No. 12, 1987.

[8] D. Kim and D. E. Orron, "Peripheral Vascular Imaging and Intervention", Mosby-Year Book, 1992, Inc. Missouri, USA.

[9] CORDIS Corporation Diagnostic and Interventional Radiology, Technical Report, Miami, USA 1992.

[10]C. K. Chui, H. T. Nguyen, Y. P. Wang, R. Mullick, R. Raghavan, James H. Anderson, "Potential Field of Vascular Anatomy for Realtime Computation of Catheter Navigation", First Visible Human Conference, Bethesda, MD, USA, October 1996.

[11]Y. P. Wang, C. Chui, Y. Y. Cai, R. Viswanathan and R. Raghavan, "Potential Field Supported Method for Contact Calculation in FEM Analysis of Catheter Navigation", XIXth International Congress of Theoretical and Applied Mechanics, Kyoto, Japan, August 1996.

[12]Y. Y. Cai, R. Viswanathan, Y. P. Wang, C. K. Chui, R. Raghavan, "Simulation of Catheter-guidewire Interaction for Catheterization by Arc Parameterization", Fourth International Conference on Control, Automation, Robotics and Vision, Singapore, December 1996.

[13]J. Anderson, W. Broky, C. Kris, , R. Viswanathan, and R. Raghavan, daVinci: "A vascular Catheterization and Interventional Radiology-based Training and Patient Pretreatment Planning Simulator", Society of Cardiovascular and Interventional Radiology 21st Annual Meeting, March Seattle, USA, 1996.

[14]Y. P. Wang, C. K. Chui, Y. Y. Cai , H. L. Lim "CathViewer: An Interactive System for Catheter Design and Testing", Fourth International Conference on Computer Integrated Manufacturing, Singapore, Oct 1997.

[15]Y. Y. Cai, Y. P. Wang, C. K. Chui, H. L. Lim, "Integration of Geometric Modeling and Physical Modeling for Catheter Navigation", Fourth International Conference on Computer Integrated Manufacturing, Singapore Oct 1997.

Integration of Geometrical Modeling and Physical Modeling For Catheter Navigation

YY Cai, YP Wang, CK Chui, and HL Lim

Institute of Systems Science
National University of Singapore
Heng Mui Keng Terrace, Kent Ridge 119597
Republic of Singapore
Tel: (65)7726724, Fax: (65)7744990
Email: {yycai | ypwang | cheekong | hllim} @iss.nus.sg

Abstract. Catheter manufacturing is a very lucrative business emerging from high-technology development in the minimal invasive surgery. We reports an integration of physical modeling and geometrical modeling for catheterization application. A potential space model of multiple-dimensional information is developed to describe both physical and geometrical properties of the vascular structure. The physical behaviors of object interactions between vascular walls and catheters can be obtained by applying a Finite Element Method (FEM) for the analysis of force, displacement, deformation, etc. Different from the conventional boundary representation that is usually hollow inside a spatial object, the potential space model is defined in any space points not limited to object boundary. This model can be utilized to extract smooth multiple-branch boundary geometry of the complex vascular network by iso-surfacing technique. The interventional device models are designed as soft objects with various geometrical features and physical properties to provide flexibility of object deformation during the catheter navigation. The approach developed is applied in the design of interventional device, the catheter navigation in the simulation of the catheterization, and visualization of the angiographic vascular network.

1 Introduction

Design and manufacturing automation has been a driving force for the development of CAD systems. Geometric modeling represents one of the central components of CAD systems. Many representation schemes have emerged since mid 1970s. A Constructive Solid Geometry (CSG) object is built from the standard primitives, using regularized Boolean operations and rigid motion. The modeling system PADL by Voelcker and Requicha in the University of Rochester [1] represents the earliest CSG work. Figure 1 (a) shows a typical example of a CSG description of an object: the solid C is made from blocks A and B using a union operation; solid E is then made from the solid C and cylinder D using a difference operation. An object can be unambiguously described by its topological and geometrical descriptions. Briefly, the topological information specifies abstractly vertices, edges, and faces, and indicates their incidence and connectivity, while the geometrical representation specifies the equations of the surfaces and orientations. The oldest formalized schema for representing the boundary of a polyhedron and its topology appears to be the winged-edge representation first proposed by Baumgart [2]. BUILD is the first B-Reps (Boundary Representation Scheme) system developed by the CAD Group at the University of Cambridge [3]. As a typical representation scheme, B-reps, has become a de-facto industrial standard for core representation in a geometric modeler nowadays. Figure 1(b) shows the boundary representation of the same solid mentioned above. Slightly different from the Westerners, the Japanese started their modeling research from the half-space concept. Okino *et al* at the Hokkaido University developed the TIPS (Technical Information Processing System) in 1973 [4]. The half-space scheme is sometimes classified as a case of CSG, however, the geometry meaning and the developed motivation of this scheme are distinct from others.

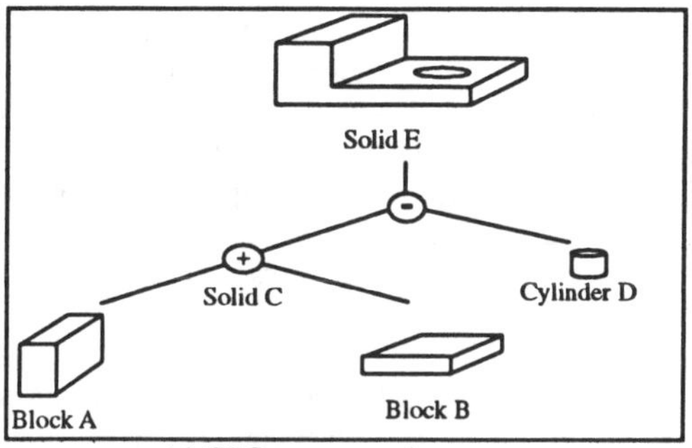

(a) CSG representation

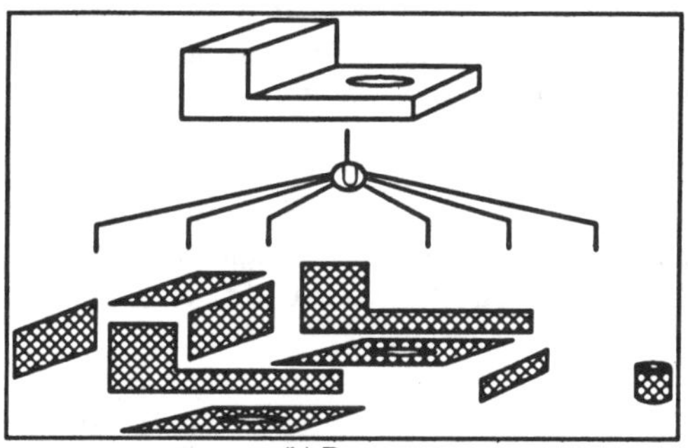

(b) B-reps

Fig. 1. Modeling representations in CSG and B-reps

CSG and B-Reps are two main streams of object modeling. Apart from these, there exist a number of other solid representation schemata which play a peripheral role in solid modeling due to certain limitations. Arbitrary 3D objects can be subdivided into any specified resolution in a hierarchical octal tree structure. Octree was originally proposed by Hunter [5]. Unlike the CSG or B-Reps, the Boolean operation (union, intersection, and difference) using the Octree scheme appears dramatically simple with linear processing time [6]. However, space decomposition is dependent of the coordinate system and this makes it unattractive to rotate objects so represented and also causes inconsistency in congruent objects lying at different orientation. Besides, feature detection and extraction are extremely difficult. Conceptually, Euler operators can be thought of as creating and modeling consistently the topology of manifold object surfaces. By locally deleting or adding vertices, edges, and faces, they can be used to create closed surfaces and to modify these surfaces. As an intermediate language, Euler operators are used in the Geometric WorkBench (GWB), a modeling system developed at the Helsinki University of Technology, incorporating B-Reps [7]. Euler operators can insulate the operations implemented on top of them from details of the data structures used to represent the surface topology. Another advantage is that they ensure topological consistency throughout the modeling process. For quite sometimes, the defined structure of geometry and topology information relating to a solid object, although computationally efficient, was often in terms of low-level details, and was therefore an incomplete definition for product modeling. Ansaldi, Floriani and Falicidieno developed a different graph-based boundary representation called the Face Adjacency Graph (FAG) or Face Adjacency Hypergraph (FAH) in the early 80s [8]. The faces are described as the nodes, whereas the arcs and the hyperarcs represent the relationships among the faces induced by the sets of edges and vertices, respectively. In a feature-based design scheme, the designer can create the part using features directly instead of using lower level geometric entities thus incorporating higher level information into the design. This feature information can then be used for manufacturing planning. The concept of AAG was introduced by Joshi for recognizing features such as slots, pockets, steps for the purpose of automatic process planning [9]. The arc of the graph represents the connectivity between two faces and the attributes are assigned to the arcs according to the geometry of the adjacency relationships.

Romulus (Evans & Sutherland) was a typical example of the early commercialized CAD systems. Modern CAD modelers are becoming more and more capable and versatile. Parametric Technology Corp. (ProEngineer), SDRC (I-DEAS), AutoDesk (AutoCAD), IBM/Dassault (CATIA), CV (CADDS), EDS/UniGraphics (UG), HP(CoCreate) are major players in the current market [17]. Lots of modeling kernels have been developed during the past three decades. Parasolid (EDS), Acis (Spatial) and DesignBase (Ricoh), however, are widely considered as three major geometric modeling kernels nowadays. Traditionally, CAD systems were developed on the Unix platforms. However, with the rapid growing of the Pentium processor, Window platform based CAD packages such as SolidWorks, SolidEdge (Intergraph), PT/Modeler (ProEngineer), Helix (MicroCADAM) etc., are aggressively entering the market.

Geometrical and physical properties are two distinct aspects of real-world objects. They are, however, commonly treated as separate and independent parts in most of the engineering modeling domains. For instance, CAD modeling usually chooses boundary geometry to describe the object shape with little concerns or even complete ignore of the physical behavior of objects. The boundary abstraction of a 3D object is indeed based on a uniformity of the interior of an object. It has been acknowledged that the B-reps is rather suitable for the conventional tasks of design and manufacturing. The hollow assumption inside the object body, however, may be inappropriate for tasks in advanced design, simulation, and concurrent engineering [10]. On the other hand, physically-based modeling has recently become a subject of widespread interest in the graphics and visualization circle due to the advances of medical modeling, virtual reality, etc. Some researchers work on various new methods for representing deformable models while others use physically-based model as a tool to explore the phenomenon of natural motion [11-12]. Computational mechanics techniques such as the finite element method are increasingly used in physically-based modeling [13]. The idea of the FEM is that a continuum can be approximated by dividing it into discrete elements. They make it possible not only to calculate the value of important physical quantities inside a three-dimensional body, but also to analyze the interaction between various objects.

For highly realistic medical/biomedical modeling, simulation, as well as interventional device designing, geometrical and physical properties are equally important. This work reports the integration of physical modeling and geometrical modeling for catheterization application. The multiple-dimensional potential space model and interventional device model are developed. Physical behaviors of object interactions between vascular walls and catheters can be derived based on the potential space model and catheter device model using FEM. Different from the conventional boundary representation that is usually hollow inside an object, the potential space model is defined in any space points not limited to object boundary. The boundary geometry of the vascular object extracted from the potential space model can be very complex with smoothed multiple-branch structure. Besides, physical modeling based on the explicit multiple-dimensional potential space can significantly speed up the analysis of force, displacement, deformation, etc. The interventional devices are designed as soft objects with various geometric features and physical properties. The integration of geometry and material properties within the catheter device provides flexibility of object deformation during the catheter navigation. The approach developed is applied in the design of interventional devices, the catheter insertion in the simulation of the catheterization, and visualization of the angiographic vascular network.

2 Multiple-dimensional Potential Space Model

A central line hierarchical structure is created on the basis of the vascular anatomy. From image data, the cross-sectional vessels segmented are used to obtain the coordinates of the central line assuming that all vessels in 3D space are in circular cylindrical structures with varied radii. With same assumption, it is also possible to calculate the radii of the vessels from the cross-sectional areas of the segmented vessels.

2.1 Vascular Potential Field

The potential space is a multiple-dimensional field where each point defines a scalar quantity and a 3D vector. The scalar factor used is the shortest distance from the space location to the central lines. The 3D vector, however, is the direction of the shortest path from the space location to the central lines (Figure 2). To avoid inconsistency, the potential distances are normalized with respective radii of the vessels. So it is clear that the potential value at a space point tells whether the grid is inside the vessel or on the surface. This is significant to conduct inside/outside testing and boundary testing as well. Similarly, vector unifying is performed.

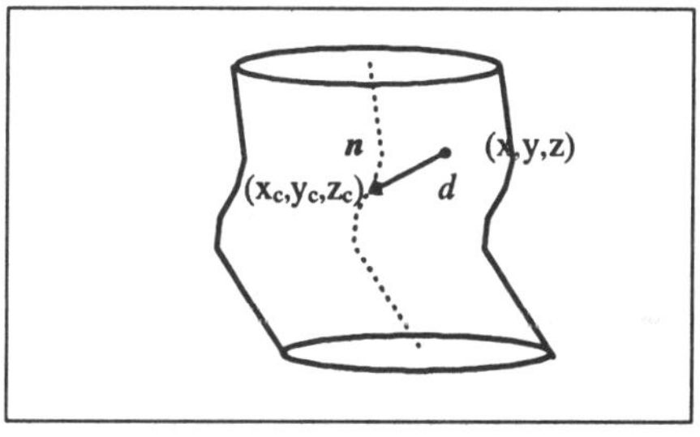

Fig. 2. The potential field consists of a shortest distance (*d*) and its associated direction (***n***) from each point (x,y,z) to the central lines (*dash line*).

Computing the shortest distances and directions sometimes can be time-consuming. A discrete model with pre-computed distances and directions can significantly improve the efficiency of the real-time simulation. For vascular network, a sparse grid space with a given thickness embracing the vascular vessels is used to specify the region of the discrete model. A triple-linear interpolation of the distances and vectors at the adjacent grid points is used to compute the normalized distance and unit vector for an arbitrary point in the region. Beside, a volumetric structure derived from the discrete model can be used for visualizing the object geometry.

2.2 Extraction of Vascular Boundary Geometry by Iso-surfacing

The complexity of vascular geometry is one of the reasons for the hesitation to apply the conventional surface/solid modeling techniques for previous researchers. In this work, a "Marching Cubes" algorithm is applied to modeling the vascular geometry [14]. To achieve this, a discrete volume model is created based on the potential space model. That is first to separate the distance information from the potential field and then convert the value into a characteristic value by a 8-connectivity checking. The grid size is equivalent to the voxel size. As a result, a grid near the vessel boundary usually has a value less than eight while a inner grid has a value of eight. This conversion is able to smooth the boundary surface. With the potential volume, the boundary geometry is subsequently extracted using a lookup table based on binary (above/below threshold) values at the corners of a volume voxel to produce high-speed polygon generation. The local gradients are also calculated to produce visually smooth iso-surfaces across neighboring voxels.

Visualization of the three-dimensional vascular networks is one of the necessities for our catheterization simulation. The geometric model constructed from the central line models (with variable radii) together with the boundary geometry extracted from the potential space model provide a viable way for geometric representation. Figure 3 shows the extracted boundary surface of the vascular structure of the Visible Human.

3 Interventional Device Model

Catheter and guidewire are major interventional devices in angiographic vascular catheterization. They have multiple variations designed to facilitate a particular function. Catheters are categorized according to their material, shape, length, etc. Some serve multiple purposes in addition to their original intentions. They possess the properties of pliability, torquability, and shape memory. Guidewire is another type of interventional device with few variation. The function of the guidewire is to provide support and guidance for passage of an overlying catheter through the arterotomy and into lumen of the selected artery.

Catheter manufacturing is a very lucrative business emerging from high-technology development in the medicine field. However, surgeons who practice the interventional procedures often use personnel preference and prior experience in selecting the devices, for example, the angioplasty balloons and stents for cardiovascular procedures, since the devices are not designed as matching systems. This practice can potentially develop inconsistencies in the performing characteristics of those devices. Such inconsistence can produce damage to the vascular wall tissue.

3.1 Deformable Interventional Device

Traditionally, CAD tools are developed for designing and manufacturing of the hard (or rigid) engineering objects. The interventional medical devices, however, often compose of highly soft material which allows the devices to be deformed during the process of operations due to the interaction of the devices and the human angiographic structures. Furthermore, they can very unlikely have an opportunity to validate the design concept themselves at a realistic checking environment. And the validation process is considered as the utmost crucial step in any design procedure. Therefore, conventional CAD methodology and design tools are not appropriate for the tasks of interventional device designing and evaluating: (1) new approaches of soft object design need to be developed and applied to support high-realistic simulation of catheterization; (2) testing and evaluating of catheter, especially, the new designs, require a very friendly integration of the design and evaluation phases.

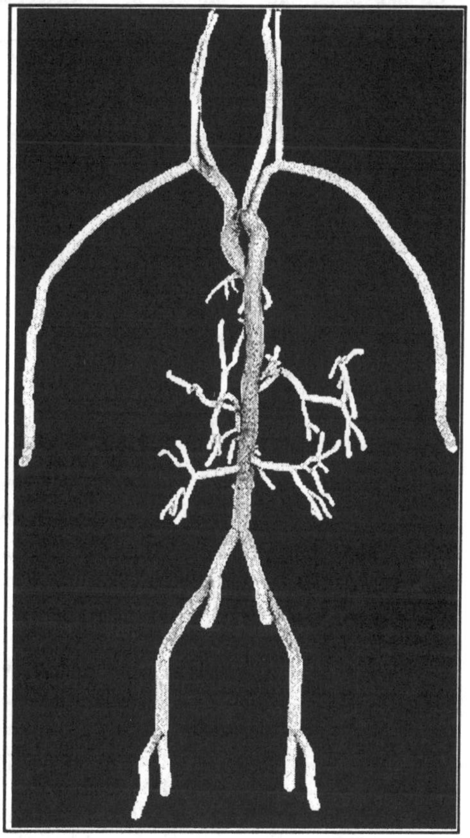

Fig. 3. The boundary surface of the Visible Human vascular structure extracted from the multiple-dimensional potential space model using iso-surfacing technique

3.2 Feature-based Modeling of Interventional Device

A feature-based design approach is applied in this work to model the interventional devices. Typical features of catheter and guidewire include generalized sweeping, tapering, lofting, holing, shelling, threading, filleting, etc. The parametric and variational methods are particularly important to the designing tasks of interventional devices such as catheter and guidewire. We have created varational and parametric based models for catheter, guidewire, balloon and stent, etc. (Fig. 4.). We are corrently developing software packages for presenting, designing and validating of interventional devices on the Intel PC-based Window NT platform using the object-oriented paradigm and OpenGL toolkit. It aims to provide high value-added sales tool, customized design tool and realistic validation tool for interventional products. The work features (1) integration of interactive 3D design and real-time navigation based validation; (2) customized design with the realistic vasculatures models; (3) varational and parametric based product designing.

Many interventional devices are soft objects by nature. In the validation procedure, interventional device may have various deformations due to interaction between vasculature and device. We model the device at various layers of backbone, geometry, material and graphics. The FEM is embedded in the design models to support the deformation analysis. More details will be reported separately.

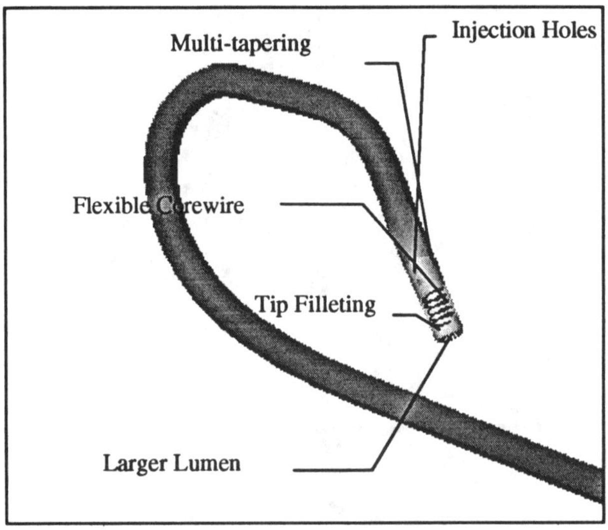

Fig. 4. A catheter designed using the feature-based modeling approach

4 Integration of Geometrical Modeling and Physical Modeling for Catheterization Application

In our integrared model of the vascular potential space and devices, we have a dynamic FEM engine, together with all the material properties of the catheter/guidewire and the vessels. The central line network of the vasculature preserves the anatomical relationship of the vessels. The normalized distance and the unit vector respectively at a grid point in the physical model are used as references to determine the magnitude and direction of the contact force for FEM calculations [15]. The contact force between the vessel wall and catheter/guidewire at a FEM node of the catheter/guidewire is formed when the catheter/guidewire is moved at the grid point outside the vessel. The distance values of the potential space provide explicit information for the inside/outside testing of the catheter/guidewire nodes against the vessels. The central lines play a role similar to gravity centers. This is because when the catheter/guidewire nodes are at the grids outside the original shape of the vessels, these nodes are usually "pulled back" into the vessels by the interaction forces (Figure 5). These forces are pointed to the central lines of vessels model. The FEM engine is applied to solve the deformation of the catheter/guidewire when interaction with the vessels. The model developed provides possible solution for high realistic and real-time catheterization simulation.

An adaptive iterative procedure for the catheter/guidewire FEM analysis is used in this work: (1) inside/outside checking for the catheter/guidewire nodes; (2) generation of the contact forces for the outside nodes; (3) construction of the stiffness matrix; (4) solving the displacement from the stiffness-displace equation; (5) checking the nodes and performing iteration for the optimal displacement update.

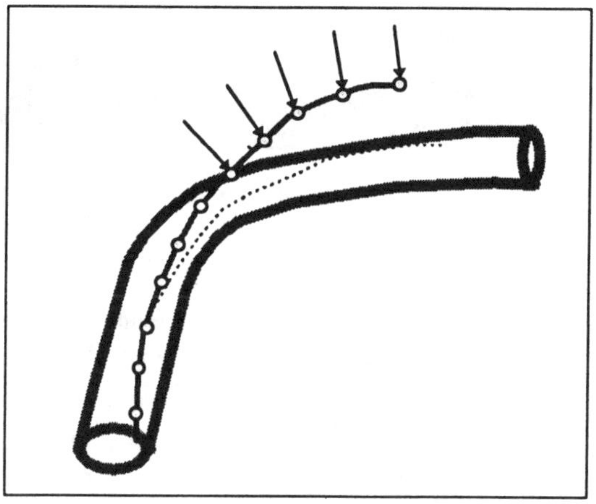

Fig. 5. The contact force pushed the catheter nodes (outside the vessels) back insides the vessels

In order to simulate the vascular catheterization procedures, the research team at ISS has developed a high realistic system for visual catheter navigation [16]. The system consists of hardware device and software system. A mechanical device is designed to manipulate the catheter/guidewire for inserting and moving them inside the vascular tree. The pulse streams generated by the incremental optical encoders are measured and fed to the interface software system.

The navigation is simulated with the visual aides of fluoroscopic and 3D views. The geometry of vascular structure is displayed based on the geometric model. Figure 6 illustrates a three-dimensional view inside the vessel. This internal view displays the geometric structure of the Visible Human simulating a viewpoint at the tip of the catheter during the navigation. The navigation is fully engined by the real time FEM calculation. The integrated model provides both geometrical and physical information for the catheter navigation application (Figure 7).

Conclusions. This work presents an integrated approach of geometric and physical modeling for catheterization application. The geometric and physical properties of the complicated vascular structure is modeled based on the multiple-dimensional potential space. The boundary information extracted is utilized to render the three dimensional structure during the navigation procedure. The moving of the catheter/guidewire inside the 3D vascular networks is determined within the physical model. The integration of the geometrical modeling and physical modeling provides an efficient solution for the high realistic medical simulation.

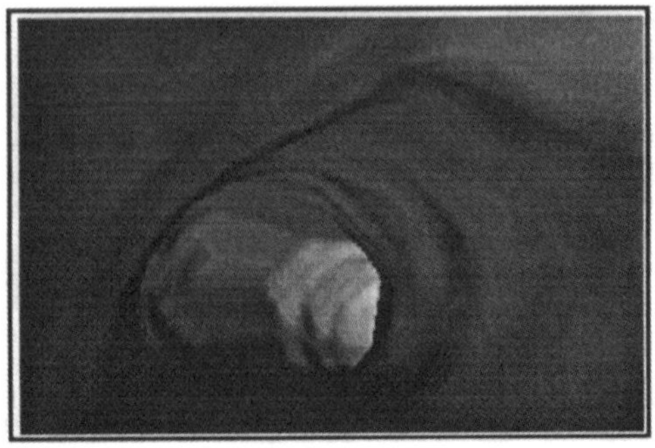

Fig. 6. A fly-through navigation view inside the vascular network

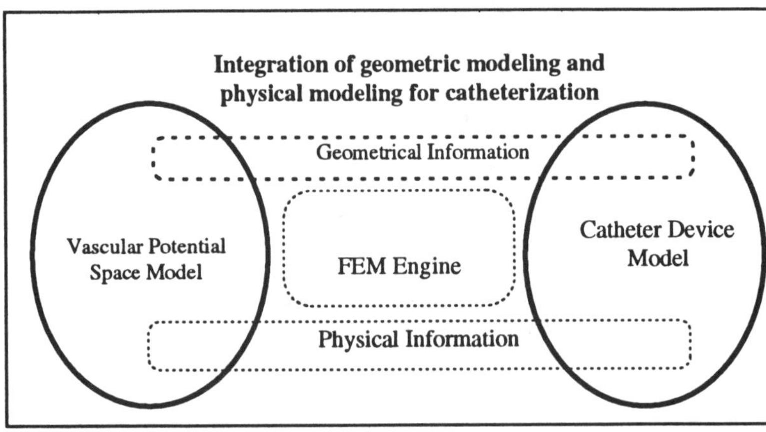

Fig. 7. The integration of geometrical modeling and physical modeling for catheterization.

Acknowledgment. We would like to thank National Science and Technology Board of Singapore for funding this research.

References

1. H. B. Voelcker, An Introduction to PALD: Characteristic, Status, and Rationale, *Technical Memo No. 22,* Production Automation Project, University of Rochester, 1974.
2. B. Baumgart, A Polyhedron Representation for Computer Vision, *Proceedings of the National Computer Conference, AFIPS,* **44** (1975), pp. 589-596.
3. I. C. Braid, *Design with Volumes*, Ph D Thesis, University of Cambridge, 1973.
4. N. Okino, Y. Kakazu, and H. Kubo, TIPS-1, Technical Information Processing System for Computer Aided Design and Manufacturing, in *Computer Languages for Numerical Control* , edited by J. Hatvany, North Holland, Amsterdam, 1973, pp. 141-150.
5. G. M. Hunter, *Efficient Computation and Data Structures for Graphics,* Ph D Dissertation, Princeton University, Princeton, NJ, 1978.
6. D. Meagher, Geometric Modeling Using Octree Encoding, *Computer Vision, Graphics and Image Processing*, **19** (1982), pp. 129-147.
7. M. Mantyla and R. Sulonen, GWB: A Solid Modeler with Euler Operators, *IEEE Computer Graphics Applications,* **2** (1982), pp. 17-31.
8. S. Ansaldi, L. De Floriani, and B. Falicidieno, Geometric Modeling of Solid Objects Using A Face Adjacency Graph Representation, *Computer Graphics,* **19(3)** (1985), pp. 131-139.
9. S. Joshi, *CAD Interface for Automated Process Planning,* Ph D Thesis, Purdue University, 1987.

10. S. C. Lu, A. B. Rebello, D. H. Cui, R. Yagel, R. A. Miller and G. L. Kinzel, A Visualization Tool for Mechanical Design, *Proceedings of Visualization'96*, San Francisco, CA, pp. 401-403.
11. J. Platt, D. Terzopoulos, K. Fleischer and A. Barr, Elastically Deformable Models, *Proceedings of ACM SIGGRAPH'1987*, pp. 205-214.
12. J. Weil, The Synthesis of Cloth Objects, *Proceedings of ACM SIGGRAPH'1986*, pp. 225-232.
13. Y. C. Fung, *Bio-mechanics: Mechanical Properties of Living Tissues*, Second Edition, Springer-Verlag, New York, 1993.
14. W. E. Lorensen and H. E. Cline, Marching Cubes: A High Resolution 3-D Surface Construction Algorithm, *Computer Graphics Vol. 21, No. 4, (Proceedings of SIGGRAPH'87)*, July 1987, pp. 163-169.
15. Y. P. Wang, C. K. Chui, Y. Y. Cai, R. Viswanathan and R. Raghavan, Potential Field Supported Method for Contact Calculation in FEM Analysis of Catheter Navigation, in *Proceedings 19th International Congress of Theoretical and Applied Mechanics,* August 1996, Kyoto, Japan.
16. J. Anderson, W. Brody, C. Kriz, Y. P. Wang, C. K. Chui, Y. Y. Cai, R. Viswanathan and R. Raghavan, daVinci: A Vascular Catheterization and Interventional Radiology-based Training and Patient Pretreatment Planning Simulator, *Society of Cardiovascular and Interventional Radiology 21st Annual Meeting*, March 1996, Seattle, USA.
17. P Marks and K Riley, Aligning Technology For Best Bussiness Results, Design Insight, 1995, USA.

A Mechanism for Coding the Form Feature

Junjun Wu, Tongyang Wang, Minghua Wu and Ji Zhou

CAD Center, HuaZhong University of Science & Technology, P.R.China

Abstract Feature-based modelling is concerned with unambiguous computer representation of the construction and manipulation to the feature-based solid model. This representation should permit us to distinguish different features when manipulating the model interactively and to regenerate the proper model according to the recorded constructive history conveniently. The mechanism of coding the form feature plays an important role in the representation. This paper describes a computational method for fully automating the generation of the feature. In this paper, we propose a coding approach by the nature of the form feature to maintain the design intent while changing the topology of the part, thus to automate feature generation while greatly increasing the flexibility.

1 Introduction

In feature-base modelling, it is greatly required that more flexibilities and convenience should be provided. One of the main embodiments is that the design intents should be maintained while the topology of the part is changed. This is determined by the geometric representation and the maintenance of the feature constructive history. It involves how to record the constructive history and leads to the main purpose of this paper. The constructive history is implemented by referenced relationships among the faces, edges and vertices of the designed object's Breps, that is, the relationships between features.

In reference[1] the author pointed out that the most useful schemes for representing solids have been, and still are, Constructive Solid Geometry(CSG) and Boundary Representations(Breps). Both representations are useful complementary, we believe. Because CSG well supports the parameterized part families and Breps is excellent for graphic user interaction, many feature-based practitioners adopt both CSG and Breps in hybrid form and this trend is likely to continue indefinitely. But the development of a technique for feature generation automatically has turned out to be very difficult when attacked only by pure geometry, especially for arbitrary 3D shapes. The automated feature generation must be implemented at a higher level through attaching some information to the geometric entities.

As we have known, the relationships between features are very crucial in feature-base design. These relations can be expressed explicitly as constraints from which many of the design variants can be derived. It is safe to say that only the relationships between features can be recorded and used to implement the constructive history of the design

object. But the relationships are recorded using the codes of the feature. Now the problem is how to code the feature by greatly decreasing the ambiguities simultaneously.
The remainder of this paper is organized as follows. The related works are discussed, then the role of coding feature face, then representations for coding mechanism for the form feature, at last drawing the conclusion.

2 Related Works

In a commercial system named SolidWorks 96, a problem exists. Consider the example shown in Figure 1. It illustrates a design object with SolidWorks 96. It has been designed in the following three steps:
1. create a block by drawing a rectangular profile on the sketch and extrude it by a specified depth.
2. add a torus on the top of the box by revolving a circular profile around a axis by 360 degree;
3. chamfer the circular edge on the right by specifying a constant radius.
Change the definition of the torus after defining the design object on the left. After regeneration, the result is shown on the right. We can see that the chamfered edge is the one on the left.

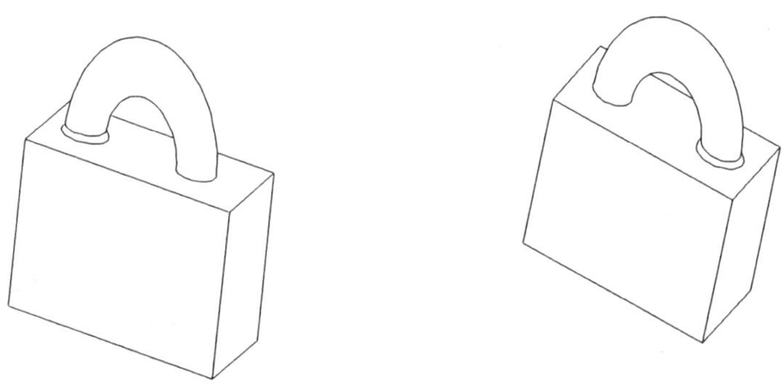

Fig. 1. An example created by SolidWorks'96

Why does it happen? It is because SolidWorks 96 cannot distinguish the 2 edges. The common character of these two edges is that they all are the intersections between the top surface of the block and the torus' surface. This example created by SolidWorks 96 demonstrates that the coding mechanism used by SolidWorks 96 has a limitation in some cases.
In the prototype of feature-based design system called GHCAD, developed by Huzhong Univ. of Sci & Tech and Tsinghua Univ., all the faces are coded according to the sequence of searching the topology of the design object's Breps. The code which includes three identifications is taken as an attribute attached to the face. The three identifications are the feature identification, pattern identification and face identification decided by the searching sequence. Although this is a valid method, there exist some drawbacks. One of the main drawbacks is the existence of ambiguity when executing the same steps as what have been done to SolidWorks 96 above.

Another is that it is assumed the topology of the design object can not be variable, so it cannot derive many design variants, thus, its flexibility is greatly decreased.
To overcome the drawbacks discussed above, it is necessary to find a new method to code the feature. Anyway, the new method for coding feature should enhance the capability of decreasing ambiguity while increasing the flexibility.
In reference[3], Vasilis Capoyleas tried to solve this problem by assigned a name to all the entities and identify it from its representation. Although his naming schema worked well to some.extent, there existed a main drawback. To name an element he had to name many relative elements and he had to handle a graph structure to distinguish the element needed from other elements.

3 The Role Of Coding Feature Face

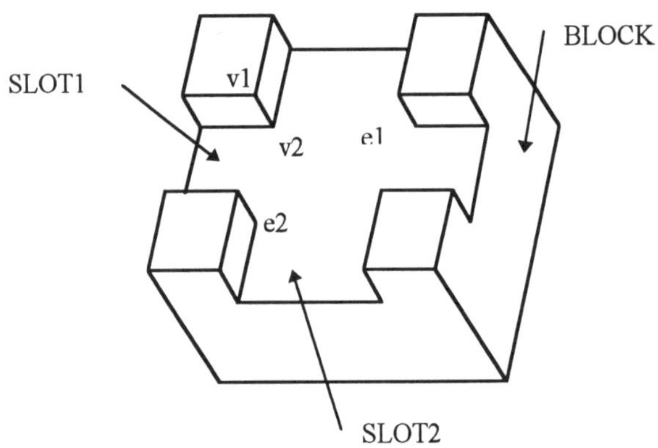

Fig. 2. A part with two slots

In a feature-based model, the basic topological entities are face, edge and vertices. It is obvious that new faces will not be created after boolean operation. As to edges and vertices it is not the same. See the part depicted in figure 2. This part includes 3 features, that is, one block and two intersecting slots. The constructing sequence is from block to slot1, then to slot2. After boolean operations some new edges(eg. e1,e2) and new vertices(eg. v1,v2) are generated, but no new faces are created.
From the above example, it can be concluded that it is the feature face, and only the feature face that can be acted as the role to record the relationships between features. That is to say the parent_son relationships are maintained by the information on feature face. It is difficult to treat the edge and vertex as the fundamental entities to record the relationships between features, for they are dynamically changed. Although edge and vertex are not taken as the fundamental entities, they themselves can be recorded by the feature face. One edge can be recorded by two intersecting feature faces and one vertex by three intersecting feature faces. It is one of the key points in feature-based modelling. So it is pertinent to say that only the feature faces can be taken as the basis of coding.
When designing a part using the feature-based modelling system, usually begin with a base feature. Then add the other features to this base feature. The parent-son

relation is created during the constructing process. The parent-son relations are recorded by recording the feature entities such as faces, edges and vertices. All these feature entities can be recognized by decoding the code attached to the feature faces in advance and the relationships between features can be acquired through the recognized entities. Hence the constructing history of the part can be replayed.

It should be made certain that all the features be coded and the referenced faces, edges and vertices be recorded during the constructing process of the part in a sound feature-based, constraint-based design system.

4 Coding Mechanism For Form Feature

4.1 Feature Classification

Not all the features' forms are generated with the same method. It is useful to classify various features by their natures because every type of features has its special character. In reference[2], C. M. Hoffmann classified feaures into three types. The first type is generated features which are formed by extruding, revolving or sweeping a profile along a trajectory. The second type is datum feature which includes datum point, datum axis and datum plane and the third is modifying feature, that is, fillet, chamfer, shell and so on. This method of classifying form features is quite useful for coding feature.

It should be noted that most of the features belong to the first type and the first feature in a design body is sure to fall into this type. The feature volume of this type can be gotten easily. Datum feature is used to construct the other features easily during the design process and this kind of feature has no feature volume. Modifying feature operates directly on the Breps of the design object. The character of Modifying feature is that the feature volume cannot be gotten easily. We do not consider datum feature and modifying feature here only to limit the scope of this paper.

4.2 Coding Principle

Before presenting the method of coding feature, the coding principle must be provided.

As discussed above, the feature face is the basic element in feature-based design. So we code the feature face to identify it uniquely. We can identify edge with the codes of the faces between which the intersection is the edge and some extra information. The extra information is added here to decrease the ambiguity. As to vertex, it can be identified by 3 faces whose intersection is the vertex itself. The extra information is added to decrease the ambiguity.

4.3 Coding Method For 2d Sketch

Because all the generated features are formed by sweeping a profile along a trajectory, it is natural to consider that the profile is the fundament of the generated features. Every face on the generated feature has a parallel on the profile. We can directly perceive through the sense that the edge on the profile is the parallel. So it is necessary to establish a relation between the elements of the profile and the generated feature's faces.

Let's first see the sketch entities in the 2d sketch. The sketch entities include straight-line, arc, circle and curve. Except for the circle marked by only one point, all the sketch entities can be marked by two points. What we concern about is the identification of the entity, not the location and size. For instance, we can mark a spline curve by its start point and end point. The shape is not the thing that we should concentrate on. So point is the basic entity in 2d sketch. And The other entities such as straight-line, arc and so on can be recorded by points that related to them.

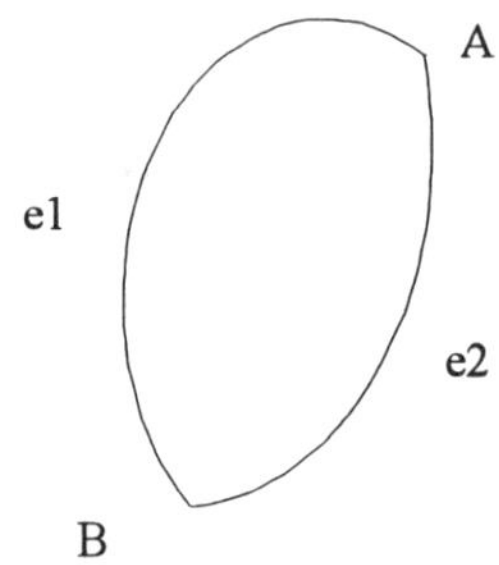

Fig. 3. A 2D profile consisted of 2 arcs

We assign an identification which can be defined by a constant to mark the point. It must be satisfied that the constant is greater than zero or equal to zero. So straight-line, arc, circle and curve can be encoded by the following triples

E = [c1, c2, rn]

In the above triples, E means entity in 2d sketch. c1 and c2 stand for the codes assigned to the points related to E. When E is a circle, the value of c2 is zero. Now pay attention to rn. It means referenced number. The reason it is introduced into the above description is that we should consider the case depicted in figure 3.

In most of the cases, the number of the entities that references the same two points is one. But in figure 3, the profile is closed and the 2 entities both references point A and point B. In order to identify the 2 entities, rn is included in the triples. It records the generating sequence of the entity. We assume that the value of c1 is 1 and c2 is 2. If e1 is sketched first, the representation of e1 and e2 are as the following descriptions

e1 = [1, 2, 1]

e2 = [1, 2, 2]

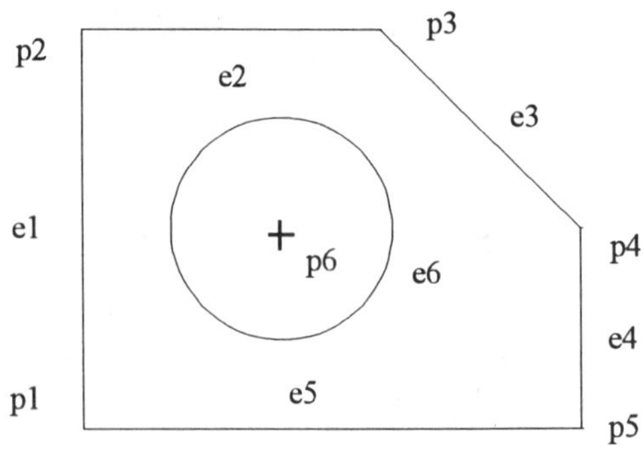

Fig. 4. A 2D profile

A constant is attached to the point while sketching it. The value of the constant is increased by 1 according to the sketching sequence and the value can be reused only if the sketch point to which it attaches does not exist because of deleting it from the sketch. With this mechanism the topology of the profile can have great flexibility. Consider the example depicted in figure 4. All the sketch points are assigned the value varied from 1 to 6. The straight line e1 is recorded by points p1 and p2 and straight line e2 is recorded by the point p2 and point p3, and so on. The circle is recorded by point p6. Every straight line is parallel to a new plane on the feature F1 formed by extruding this profile. As to the circle, a conical face. If another feature F2 referenced the face which is parallel to the entity e1. It can be assured that the profile's shape can vary arbitrarily if the two points related e1 are not deleted, while satisfying the relation between F1 and F2 at the same time.

4.4 Coding Method For Form Feature

The coding method in 2d is the basis of coding form feature. The identification of face has the following description

ffi = [fi, Esp, Est, pi, op]

In the above description, ffi means feature face identification. The following section will discuss every domain in it.

fi stands for feature identification. In order to maintain the parent_son relation and to manipulate the model interactively, it is useful to introduce the feature identification into the structure.

Esp means entity in the sketch in which the profile is sketched and Est stands for entity in the sketch in which the trajectory is sketched. In feature-based design, most of the features are generated features which are formed by extruding, revolve and sweeping a profile along a path. Every side face is parallel to one of the entity in the sketch. This is the reason why the above description includes Esp. It is the same to Est. As to the front face, the value of Esp is -2 and to the back the value of Esp is -1. In the case of revolusion with 360 degree rotating, there are no front and back face and it is unnecessary to code them.

pi means part identification. In assembly environment, more than one part will be inserted into the assembly document. To every feature in a part, the feature identification is unique, but with more than one feature, it does exist that more than one features have the same feature identification. In order to handle this case, add the part identification into the face identification. But it also should be noted that in the process of designing a part, it is unmeaningful to assign a part identification to the part. The part identification is generated when it is inserted into the assembly document. So the part identification is dynamically changed and it is determined by the sequence inserting the assembly document.

Even with the above domains, there exist some special cases that cannot be handled. See the example depicted in figure 5. The Brep is formed by extruding a profile and a draft angle. There exist two methods to handle the corners if the profile is expanded. If the EXTEND method is adopted, the edges in the expanded profile should be extended to their intersection point. The extensions are then swept back along the path and intersected to produce two new faces and a new edge. Or the ROUND method will be adopted. For the edge extensions do not intersect, an arc tangent to the two edges is created and swept to create a closed face in this case. The example belongs to the later. Considering the four cone surfaces in the four corners, they differ from the other faces on the Brep. So the coding mechanism is invalid to this case. This case can be solved by given an additional domain named op in the description. And for the necessity to extend the system, it is useful to append an optional domain to the feature face identification.

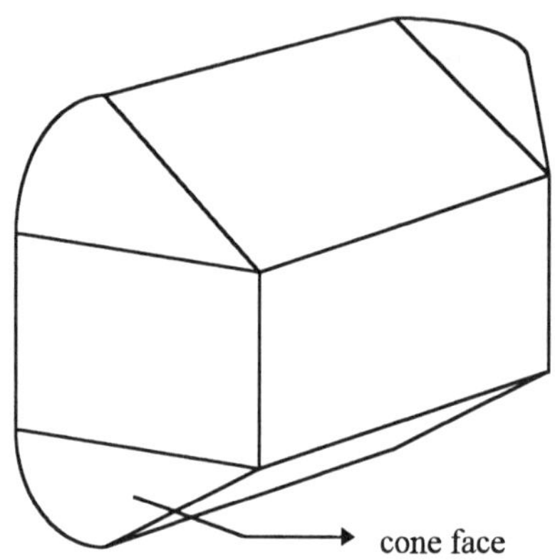

Fig. 5. Extrude with round draft

4.5 Record Edge

Edge can be recorded by the following triples

Ed = [ffi1, ffi2, op]

In it, ffi1 and ffi2 are the two faces with which the edge is formed. And op expressed by duple [u, v] which is used to record the value of the face's parameter space.
Consider the example depicted in Figure 1. Both of the 2 circular edges are formed by intersecting face f1 and f2. We can select a point on the edge respectively, It can be made certain that the duples of the 2 points are different from each other. So with the triples we can distinguish the 2 circular edges easily.

4.6 Record Vertex

Vertex can be recorded by the following description

V = [Ed1, Ed2, Ed3]

In the above triple, Ed1, Ed2 and Ed3 are the codes of the 3 intersected edges. With it, every vertex has a unique expression.

5 Conclusion

With the above coding mechanism, every feature element such as vertex, edge and face can be identified uniquely. It is helpful to decrease ambiguities greatly and automate feature's generation properly. Also, the flexibility of the feature-based model is enhanced.

References

1. Wesley M A, Lozano-Perez T, Lieberman L I, Lavin M A and Grossman D D, A geometric modelling system for automated mechnical assembly, IBM J. Res. DEV., vol.24, No.1(1980), pp64-74
2. C. M. Hoffmann, and R. Juan, Erep, an editable, high level representation for geometric design and analysis, in P. Wilson, M. Wozny, and M Pratt, (Eds.) Geometric and Product Modelling, North Holland(1993), pp129-164.
3. Vasilis CapoYleas, Xiangping Chen and Christoph M Hoffmann, Generic naming in generative constraint-based design, CAD, Vol.28, No.1(1996). pp17-26.

Computation of the Skeleton of 3D Polyhedral Solid

Yujin Hu, M.M.F. Yuen* and Ji Zhou

CAD Center
Huazhong University of Science and Technology
Wuhan, Hubei, P. R. China
*Hong Kong University of Science and Technology
Clear Water Bay, Kowloon, Hong Kong

Abstract. The skeleton of an object, also known as the medial axes transform, is an alternative geometric representation that has great potential in engineering. In this paper, a new method constructing the skeleton of 3D polyhedral solid is present. The basic principle of the method is the abstract Delaunay triangulation of a set of active topological entities(ATE), which are a new classification of the object boundary. In order to conveniently construct Delaunary triangle, the conception of the adjacent connecting graph(ACG) of the ATE is introduced. The proposition and algorithm for determining the topology and geometry of the object skeleton are given. Finally, some examples illustrate the computation properties of the method.

1. Introduction

The skeleton of an object, also known as the medial axes transform, is an alternative geometric representation that has great potential value in engineering. It was first proposed by Blum[1] as a technique for biological shape description. It is defined as the locus of the centers of all inscribed spheres of an object, where an inscribed sphere is a sphere that is not contained by any other interior sphere.

The skeleton is closely related to the Voronoi diagram of an object. Lee[2] says that the skeleton is a subset of the Voronoi diagram. Lavender[3] also says that if the object boundary is suitably defined the object skeleton may be recovered from the Voronoi diagram. So the key constructing the object skeleton is to construct its Voronoi diagram.

Methods generating skeletons can be classified into two main types, namely discrete boundary method and exact tracing method. The discrete boundary method[4][5][6][8] is based on the properties of the Delaunay triangulation of point sets. The object boundary is firstly discretized into a point set, then Voronoi diagram of a point set is computed via its dual Delaunay triangulation. Due to the discrete point sets can't completely represent the continuous object boundary, so this method only gets the approximate skeleton of an object except the boundary of an object is infinitely subdivided. The exact tracing method[7][9][10][11] doesn't need to discritize the object boundary, and it directly solve a system of non-linear equations of the object skeleton. It begins from the intersection of the object boundary, then traces the equidistant between the object boundaries so that all equidistant of the object boundaries are found. However, because it is difficult to recover non-linear equations of the object skeleton into explicit equations, much computation is required to solve the system of non-linear equations.

In this paper, a new method constructing the object skeleton is proposed. The object boundary is represented into a new set of topological entities. At the same time, by means of the relationship of the skeleton and Voronoi diagram, the definition of the active topological entity sets (ATES) of the skeleton, which is a set of the topological entities making a contribution to the skeleton, is introduced. Thus the computation of the skeleton is transferred into the computation of Voronoi diagram of ATES. Because the computation of Voronoi diagram of ATES is very complex, we only compute its dual --- the abstract Delaunay triangle/tetrahedron. Then based on the abstract Delaunay trianglation and some skeleton properties, the topology and geometry of the skeleton entities are separately constructed. In order to compute conveniently abstract Delaunay triangle/tetrahedron a conception of adjacent connecting graph(ACG) is introduced.

2. Basic Definitions

2.1. Voronoi Diagram and Skeleton Entity

Definition 1 Let $P = \{p_1, p_2, \ldots\ldots p_n\} \subset R^m$ be a set, where $2 \le n \le \infty$ and $p_i \ne p_j$ for $i \ne j$, $i, j \in I_n = \{1,2,\ldots\ldots n\}$. We call the region given by

$$V(p_i) = \bigcap_{j \in I_n} H(p_i, p_j) \quad i \ne j$$

the Voronoi domain associated with entity p_i, and the set given by

$$Vd(P) = \{V(p_i), \ldots, V(p_n)\}$$

the Voronoi diagram of P. Where

$$H(p_i, p_j) = \{p \in R^m | d(p_i, p) = d(p_j, p)\}$$

$d(p_i, p)$ is the minimum Euclidean distance between entity p and entity p_i. $H(p_i, p_j)$ is equidistant from p_i and p_j. If we connect all the entities adjacent Voronoi domains, Delaunay triangle sets $DT(P)$, a dual of Voronoi diagram $Vd(P)$, would be formed. Fig. 1a shows $Vd(P)$ and $DT(P)$ of a point set on a plane. If P not only includes points but also other entities such as 3D line, 3D plane, then $Vd(P)$ is Voronoi diagram of mixed entity sets, $DT(P)$ is the abstract Delaunay tetrahedron sets of mixed entity sets. Fig. 1b shows Voronoi diagram of a block. It can be classified into two parts, interior Voronoi diagram(fine lines) and exterior Voronoi diagram(dot lines). Since points on the Voronoi diagram are equidistant from two or more boundary entities of the block, it is easy to see that the interior Voronoi diagram of the block is actually the skeleton of the block. The shade part in Fig. 1b shows the equidistant between top face and bottom face of the block, and the complete skeleton of the block is shown as Fig. 1c.

The skeleton of an object is still represented the structure of vertex, edge and face. We generally call them skeleton vertex, skeleton edge and skeleton face.

Definition 2 Skeleton face is a bounded portion of the equidistant surface of two boundary entities. It is bounded by skeleton edges and skeleton vertices. It is also defined the locus of the center of the sphere tangent to two boundary entities. As shown in Fig. 1c.

Definition 3 Skeleton edge is a bounded portion of the equidistant curve of three or more boundary entities. It is bounded by two skeleton vertices. It is also defined the locus of the center of the sphere tangent to three or more boundary entities. It connects two or more adjoining skeleton faces. See Fig. 1c.

Definition 4 Skeleton vertex is an equidistant point of four or more boundary entities. It is also defined the center of the sphere tangent to four or more boundary entitics. See Fig. 1c.

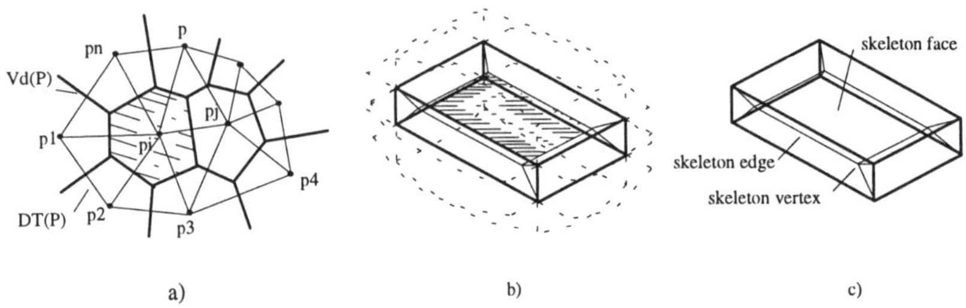

Fig. 1 Voronoi diagram and the skeleton of a block

2.2 Active Topological Entity

Due to the object skeleton is contained in Voronoi diagram of the object boundary topological entities, it motivates us to think if the topological entity creating the object skeleton can directly be separated from the boundary topological entities of the object. Thus we introduce definitions as follow.

Definition 5 Let p_i , p_j be two topological entities of an object O. If it satisfies the following condition

$$\bigcap_{j \in I_n} H(p_i, p_j) \bigcap O \neq \phi \qquad i \neq j$$

then p_i is called active topological entity(ATE), where I_n represent all boundary topological entities.

Definition 6 Let $P = (p_1, p_2, \ldots\ldots p_n)$ be a set of boundary topological entities of an object, then all ATEs in P, written as $\tilde{P} = (\tilde{p}_1, \tilde{p}_2, \ldots\ldots \tilde{p}_m)$, $m \leq n$, is called as active topological entity sets(ATES).

3. Construction of ATES and its Adjacent Connecting Graph

3.1 Classification of the topological entity

In order to construct conveniently the active topological entity and the skeleton of an object, we will further classify the object topological entities, vertices, edges and faces, into new topological entities.

Definition 7 If edge angle $\alpha_e(t)$ between two adjoining faces of an edge E is all greater than 180 degree, then the edge E is called concave edge E_a. Similarly, if edge angle $\alpha_e(t)$ is all less than or equal 180 degree, then the edge E is called convex edge E_v.

Definition 8 If vertex is a intersection of concave edges, then the vertex is called concave vertex V_a. Similarly, if vertex is a intersection of convex edges, then the vertex is called convex vertex V_v. If vertex is a intersection of only one concave edge and two or more than two convex edges, then the vertex is called mixed_I vertex MV_1.

If vertex is a intersection of only one convex edge and two or more than two concave edges, then the vertex is called mix_II vertex MV_2.

3.2 Construction of Active Topological Entity

Proposition 1 Let $P = (p_1, p_2, \ldots\ldots p_n)$ be a set of boundary topological entities of an object, then all faces, concave edges, concave vertex as well as mix_II vertex are ATEs, and others are not ATEs.

This property is not difficult to proof. Only need to proof the intersection of Voronoi region of no ATE p and object O is empty, i.e. $\bigcap_{j \in I_n} H(p, p_j) \bigcap O = \phi$.

3.3 Construction of Adjacent Connecting Graph

Because ATE is a alternative represention of the boundaries topological entity of an object, so among ATEs they still preserve the relationships similar to the topological connections of object boundaries. We call a graph consisting of these connecting relationships as adjacent connecting graph(ACG).

ACG is mainly used to implement the abstract Delaunay triangulation of ATES. Fig. 2a shows a simple 2D L-shaped polygon. ATES consists of one concave vertex and six boundary edges. The adjoining relationship between two topological entities can be represented as a graph with two nodes, one for each ATE, and one edge connecting the nodes. Fig. 2b shows the complete ACG of ATES for the L-shaped polygon. Fig. 2c and Fig. 2d show the example in 3D.

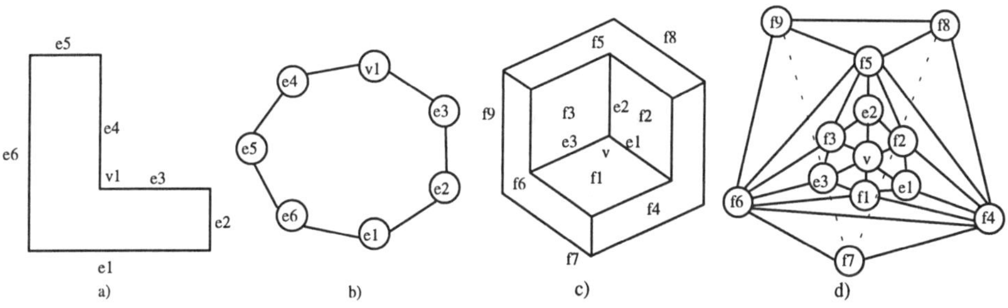

Fig. 2 Adjacent connecting graph of objects

From above construction of the ACG, we discover that the ACG of ATES exists the features as follow:

1). ACG is closed.

2). Each node of ACG connects at least with two nodes (2D) or three nodes (3D).

3). If the ACG is conceived into a domain (2D) or solid (3D), then its boundary generally is segment/triangle. So we often call the boundary face of ACG as boundary triangle.

4. Abstract Delaunay Triangulation

Although the definition of ATES reduces the searching space computing Voronoi diagram, constructing exact Voronoi diagram of ATES is still complex because computing Voronoi diagram of ATES has to involve the computation of surface intersection. Consequently, we don't directly compute the Voronoi diagram of ATES, but we compute its dual abstract Delaunay triangle or tetrahedron by mean of ACG. Once abstract Delaunay triangle or tetrahedron is exactly determined, the topology of the skeleton and the geometry of the skeleton vertex is actually determined.

4.1 The Criterion of Abstract Delaunay Triangulation

Proposition 2 If the radius of inscribed sphere tangent to any four ATEs in ATES is minimum, then the tetrahedron constructed by the four ATEs is an abstract Delaunay tetrahedron.

4.2 The Unit Normal Vector of Abstract Delaunay Triangle

The ACG is an abstract topological graph. It reveals the topology information between the ATEs, but it does not inform any geometry information between the ATEs. Actually, the inscribed sphere corresponding to abstract Delaunay tetrahedron in the ACG is closely related to the geometry of ATEs, thus we define a unit normal vector for each abstract Delaunay triangle face.

Definition 9 Let $B(c,r)$ be the inscribed sphere corresponding the abstract Delaunay tetrahedron with ATEs p_i, p_j, p_k, p_l, where c, r are the center and radius of inscribed sphere respectively. Let q_i, q_j, q_k, q_l be the corresponding four tangent points, $\hat{N}(q_i, q_j, q_k)$ be a unit normal vector of the plane patch constructed by three tangent points q_i, q_j, q_k, which directs the center of $B(c,r)$, then $\hat{N}(q_i, q_j, q_k)$ is called the unit normal vector corresponding to abstract Delaunay triangle $\Delta_{p_i p_j p_k}$, it denotes $\hat{n}(p_i, p_j, p_k)$.

The unit normal vector of abstract Delaunay triangle has two functions. One is to provide the searching direction for solving non-linear system equations. Other is to provide the tangent vector for computing the geometry of skeleton edge.

4.3 Algorithm

Based on above criterion and definition of ATES, we give the algorithm of abstract Delaunay triangulation as follow.

Input: ATES $\tilde{P} = (\tilde{p}_1, \tilde{p}_2, \ldots\ldots \tilde{p}_m)$.

Output: tetrahedron sets $Tetr = (t_1, t_2, \ldots\ldots t_s)$.

Step 1 Initializing tetrahedron list $Tetr$.

Step 2 Construct cyclical list CG to store the ACG of $\tilde{P}$. Each node in CG records the complete information of each boundary segment/triangle in the ACG, which contains the unit normal vector and neighborhoods of each boundary segment/triangle.

Step 3 If CG is not empty, then select a node $T(p_i, p_j, p_k, pt_i, pt_j, pt_k, \hat{n})$ from cyclical list CG. Where p_i, p_j, p_k is three ATEs constructing boundary triangle $\Delta_{p_i p_j p_k}$, pt_i, pt_j, pt_k are three pointers pointing to adjacent boundary triangle, $\hat{n}$ is the unit normal vector of $\Delta_{p_i p_j p_k}$. If CG is empty, then output list $Tetr$.

Step 4 Search fourth ATE p_l from $\tilde{P}$ so that the inscribed sphere of four ATEs corresponding to abstract Delaunay tetrahedron $t(p_i, p_j, p_k, p_l)$ is minimum.

Step 5 Compute exactly the unit normal vector of each triangle of abstract Delaunay tetrahedron $t(p_i, p_j, p_k, p_l)$.

Step 6 Record abstract Delaunay tetrahedron $t(p_i, p_j, p_k, p_l)$ and correspond-ing inscribed sphere $B(c, r)$.

Step 7 Delete the node $T(p_i, p_j, p_k, pt_i, pt_j, pt_k, \hat{n})$ from CG. Then check if abstract Delaunay triangles $\Delta_{p_i p_j p_l}$, $\Delta_{p_j p_k p_l}$, $\Delta_{p_k p_i p_l}$ are nodes of CG respectively. If some of them are nodes in CG, then delete them from CG. If some of them are not nodes of CG, fill them into CG, goto Step 3.

Finding fourth entity above algorithm is very key and important. It requires much more times to solve non-linear system equations which govern the center and radius of the inscribed sphere of any four entities. In general, the construction of the non-linear system equations is not difficult and there is a general method to formulate them[3][11][6]. However, to solve non-linear equations has to consume more times. Newton-Raphson method is adopted to solve non-linear equations in this paper. Initial value $\vec{X}$ of the solution is determined with $\vec{X} = k\hat{n}$, where k is an experienced coefficient. Moreover, because the solution of non-linear equations is not unique, i.e. the inscribed sphere of any four entities may be not unique, so the solved solution has further to be checked. The algorithm computing minimum inscribed sphere describes in detail as follow.

Input: Node $T(p_i, p_j, p_k, pt_i, pt_j, pt_k, \hat{n})$ and ATES $\tilde{P} = (\tilde{p}_1, \tilde{p}_2, \ldots\ldots \tilde{p}_m)$.

Output: Inscribed sphere $B(c, r)$ or null.

Step 1 Let three entities corresponding to triangle nodes be p_i, p_j, p_k, initial minimum radius $\min\_r$ be 1.0e+10. Moreover, initializing $B(c, r)$ is empty.

Step 2 For all ATEs $l = 1, m$.

2.1 Select an entity p_l from P.

2.2 Determine initial value of the solution and solve non-linear system equations to obtain the inscribed sphere $B(c, r)$ corresponding to p_i, p_j, p_k, p_l.

2.3 If $B(c, r)$ is empty, goto Step 2.1

2.4 Check if $B(c, r)$ is inside object, if false, goto Step 2.1

2.5 Check if the projects of the center of $B(c, r)$ on the entities p_i, p_j, p_k, p_l are all inside corresponding entity respectively, if false, goto Step 2.1.

2.6 Compare the radius of $B(c, r)$ with $\min\_r$, and record corresponding tangent points(i.e., the coordinates of the projects in Step 2.5) as well as inscribed sphere $B(c, r)$.

Step 3 output $B(c,r)$.

5. Construction of Topology for Object Skeleton

The constructions of topology for the object skeleton mainly determine the connection among skeleton vertices or skeleton edges. In order to define the topology of object skeleton, two definitions related to the object skeleton are introduced as below.

Definition 10 The trunk of a skeleton is a portion of the skeleton which is not incidence with the object boundary.

Definition 11 The branch of a skeleton is a portion of the skeleton which is incidence with the object boundary.

Abstract Delaunay triangulation exactly determines the geometry of the skeleton vertex. Moreover, it also determines the topology between skeleton vertices or skeleton edges exactly because abstract Delaunay tetrahedron set is dual to Voronoi diagram.

Proposition 3 In abstract Delaunay tetrahedron set $Tetr = (t_1, t_2, \ldots\ldots t_s)$, if any two tetrahedrons share same Delaunay triangle Δ_{ijk}, then the inscribed sphere centers of the two tetrahedrons define a skeleton edge of the trunk of object skeleton.

Proposition 4 In abstract Delaunay tetrahedron set $Tetr = (t_1, t_2, \ldots\ldots t_s)$, if any several tetrahedrons share same Delaunay edge l_{kl}, then the inscribed sphere centers of the several tetrahedrons define a skeleton face of the trunk of object skeleton.

Proposition 5 In abstract Delaunay tetrahedron set $Tetr = (t_1, t_2, \ldots\ldots t_s)$, if the Delaunay triangle of a tetrahedron is boundary triangle Δ_{ijk}, then the inscribed sphere center of the tetrahedron and the convex vertex of the object beside Δ_{ijk} define a skeleton edge of the branch of object skeleton.

Proposition 6 In abstract Delaunay tetrahedron set $Tetr = (t_1, t_2, \ldots\ldots t_s)$, if one boundary Delaunay edge l_{kl} is shared by several tetrahedron, then the inscribed sphere centers of the several tetrahedrons and the convex edges of the object beside l_{kl} define a skeleton face of the branch of object skeleton.

For above properties, we explained as below with the doubly tapered block shown as in Fig. 3a. Fig. 3b shows its ACG with six ATEs, which are described as nodes a, b, c, d, e, f respectively. Fig. 3c shows five abstract Delaunay tetrehadrals. Their inscribed sphere centers are exactly corresponding the skeleton vertices v_1, v_2, v_3, v_4, v_5 of the object skeleton respectively. Due to tetrahedron *abde*, *abfd*, *dbef* share *db* edge, tetrahedron *abde* and *abdf* share triangle Δabd face, tetrahedron *abde*, *dbef* share triangle face Δdbe, tetrahedron *abdf*, *dbef* share Δdbf, so the inscribed sphere centers of three tetrahedrons *abde*, *abfd*, *dbef* construct three skeleton edges v_1v_2, v_1v_3, v_2v_3 and one skeleton face $v_1v_2v_3$. Moreover, due to triangle Δabe is a boundary triangle, vertex B is the convex vertex beside triangle Δabe, so v_2 and B define a branch edge v_2B; due to tetrehadrals *abde* and *abfd* share the boundary edge ab, so v_1, v_2 and the convex edge BC beside ab define a branch face $v_1 v_2 BC$ of the skeleton. Similarly, another three tetrahedrons will construct the skeleton edges v_3v_4, v_3v_5, v_4v_5 skeleton face $v_3v_4v_5$ etc.

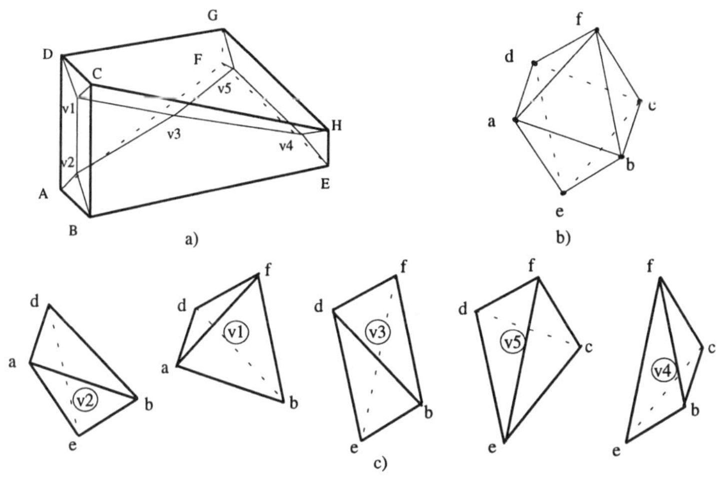

Fig. 3 Topology construction of skeleton for doubly tapered block

6. Computation of Geometry for Object Skeleton

According to the definition of the skeleton edge and skeleton face, we know the geometry of skeleton entities is closely dependent on the geometry of an object boundary. In general, the geometry of skeleton entities of polyhedral solids can describe with simple algebraic equation. Because the object boundaries consist of lines/planes, then the skeleton curve/surface corresponding to skeleton edge/face is a portion of quadrics of the following types: a) point/point (plane), b) plane/plane (plane), c) line/line (hyperparabolic), d) point/line (cylinder parabolic), e) point/plane (circle parabolic), f) line/plane (parabolic cylinder).

Based on the topological construction and geometrical approximate mentioned above , the algorithm generating the trunk and branch of the object skeleton is given as follow.

Input tetrahedron sets $Tetr = (t_1, t_2, \ldots\ldots t_s)$

Output the trunk and branch of the object skeleton

Step 1 For all tetrahedrons do

For all tetrahedrons edge which is interior CG do

1.1.1 searching all tetrahedrons sharing same common edge.

1.1.2 if common edge is interior, then compute exact skeleton trunk.

1.1.3 else (i.e., common edge is boundary edge), then compute exact skeleton branch.

Step 2 Display the skeleton of an object

7. Example and Conclusion

7.1 Example

In this section, we show a few examples of our method. Fig 4a shows a 2D example with curve boundaries. It consists of 29 ATEs, i.e., 21 edges and 8 concave vertices. The trunk of skeleton has 27 skeleton vertices (including 7 pairs degeneracy skeleton vertices) and 26 skeleton edges (including degeneracy skeleton edges). The branch of skeleton has 13 skeleton edges. All the skeleton edges are described as Bezier curves and line. Fig. 4b shows another 2D example with multiple-connection polygon.

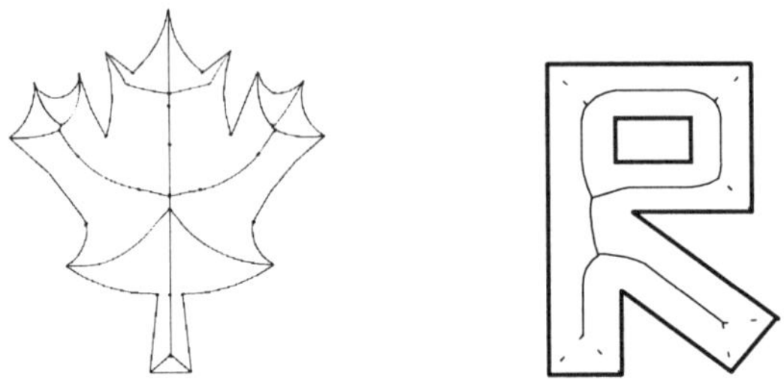

a) b)

Fig. 4 Skeletons of two 2D domain

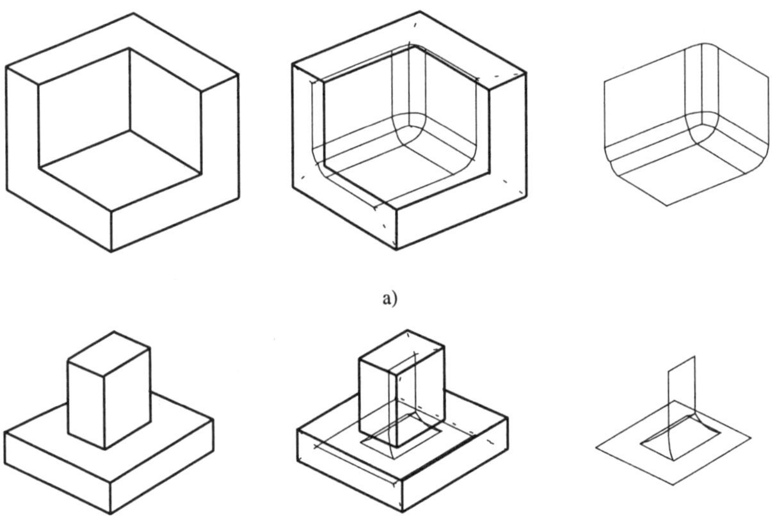

a)

b)

Fig. 5 Two polyhedrals and its skeletons

Fig. 5 shows two 3D polyhedral examples. The total ATEs shown in Fig. 5a are 9 faces, 3 concave edges and one concave vertex. The skeleton in this case consists of 12 trunk faces and 15 branch faces, and on the skeleton trunk there are 19 skeleton vertices and 32 edges. In this example there is no degeneracy skeleton vertex. Fig 5b shows other examples with 19 ATEs, and its skeleton consists of 6 trunk faces and 14 branch faces. In this example there are more degeneracy cases related to concave edge and mixed vertex. In this paper, the tolerance we check coincided point (or degeneracy) is same level tolerance as to solve non-linear system equations.

7.2 Conclusion

In this paper we have presented a method for skeleton construction of an object in 3D. Because the method is based on Voronoi diagram, the topology correctness of the skeleton can be ensured. In addition, we have not to divide the object boundary infinitely and not to trace the object skeleton, so there are less computation complex than discrete boundary method and exact tracing method. Although this method can not directly construct exact skeleton of the object with curve shaped boundary, we believe the same conception and classification introduced in this paper can be lead to an algorithm for skeleton construction of the object with non-intersection curve shaped boundary. Actually we have done for 2D domain with curve shaped boundary as shown in Fig. 4a, but the calculation of non-linear equations for curve shaped boundary is more complicated.

References

1. H. A. Blum Transformation for extracting new descriptors of shape. In W. Wathen-Dunn (Ed.) Models for Perception of Speech and Visual Form MIT Press, USA(1967)pp.362-380
2. Lee, DT. and Drysdale,RL, Generalization of Voronoi Diagrams in the Plane, SIAM J. Comput. 10(1981)pp.73-87
3. David Lavender and Adrian Bowyer et al. Voronoi Diagrams of Set-Theoretic Solid Models, IEEE Computer Graphics and Application 16(1992) pp.69-77
4. T. K. H. Tam and C. G. Armstrong, 2D Finite Element Mesh Generation by Medial Axis Subdivision, AES, 13(1991)pp.313-324
5. J. A. Goldak X. Yu, Constructing Discrete Medial Axis of 3D Objects, International Journal of Computational Geometry & Applications, 1(1991) pp.327- 339.
6. J. M. Reddy Computation of 3D Skeletons Using a Generalized Delaunay Triangulation Technique, Computer Aided Design, 27(1995)pp.677-694
7. E. Sherbrooke and N. Patrikalakis Compution of the medial axis transform of 3D polyhedra, Proc ACM/IEEE Symposium on Solid Modeling and Application, Salt Lake City,Uath, (1995) pp.187-200
8. DJ Sheey,CG Armstrong and DJ Robinson, Computing the Medial Surface of a Solid from a Domain Delaunay Triangulation, Proc ACM/IEEE Symposium on Solid Modeling and Application, Salt Lake City,Uath, (1995) pp.201-212.
9. Niranjan Mayya and V. T. Rajan, An efficient Shape representation Scheme Using Voronoi Skeletons, Pattern Recognition Letters 16(1995) pp.147-160.
10. D. Dutta and C. M. Hoffmann, On the Skeleton of Simple CSG Objects, Journal of Mechanical Design, 115(1993) pp.87-94
11. C. M. Hoffmann, How to Construct the Skeleton of CSG Objects, Technical Report CSD-TR-1014, Computer Sciences Department, Purdue University, Sept. 1990

Creating Blending Surface and N-sided Region Surface with Rational Bezier Patches

Luo Hongzhi, Yi Wensen, Li Shiqi, Zhou Ji

CAD Center, Mechanical Institute at Huazhong University of Science and Technology, Wuhan, P. R. China

Abstract. This paper presents mathematical techniques for constructing G^1 continuous blending surfaces and fill arbitrary N sided regions using rectangular rational Bezier patches. In this paper the derivation of the filling of an N-sided region with the rational Bezier surface is described. An algorithm for generating an N-sided region surface of G^1 continuity with the rational Bezier surface and the test results of the algorithm are presented. This algorithm can be used for any order of rational Bezier patches and its shape can be interactively adjusted by changing the fullness of the surface. Rectangular rational Bezier patches are preferred for the surface because nearly all CAD/CAM system can handle this type of patch quite effectively. In addition, they are supported by STEP.

1 Introduction

The generating Blending and N-sided surface is important in product geometric modeling and automatic manufacturing systems. Many papers published have discussed using non-standard surfaces to building G^1 or nearly G^1 blending surface or N-sided surface[1-3]. It is harmful for application and data exchange to bring non-standard surfaces which do not supported by STEP into CAD/CAM system. They are only applicable in an island of system or in particular circumstance. For the unified treatment of product geometric model, system integration and data exchange of product, we are hoping to build blending surface and N-sided surface with standard surface which is supported by STEP, and can be treaded for the life cycle of the product geometric model in a unified way. There are a few algorithms published to fill a N-sided hole with standard surface, such as bezier patches, but N is limited to three, four and five[4].

This paper puts forward a new method to create blending surface and N-sided region surface with rectangular rational Bezier patches. In this paper, the derivation of the filling of an arbitrary N-sided region with the rational Bezier surface is described. An algorithm for generating an N-sided region surface which has G^1 continuity with the adjacent rational Bezier surface and the test results of the algorithm are presented. The algorithm can be used for any order of rational Bezier patches and its shape can be interactively adjusted by changing the fullness of the surface. The form of the N-sided region surface is in conformity with STEP.

2 Basic description of N-sided region

In an affine coordinate system of space E^3, a rectangular rational Bezier patch $R(u,v)$ has the representation

$$R(u,v)=\frac{S(u,v)}{W(u,v)},\qquad 0\le u,v\le 1, \tag{1}$$

where

$$S(u,v)=\sum_{i=0}^{n}\sum_{j=0}^{m} r_{ij}w_{ij}B_i^n(u)B_j^m(v); \tag{2}$$

$$W(u,v)=\sum_{i=0}^{n}\sum_{j=0}^{m} w_{ij}B_i^n(u)B_j^m(v); \tag{3}$$

in which $\{\vec{r}_{ij}\}$ is net vertices of patch $R(u,v)$, and w_{ij} is corresponding weights of vertex $\vec{r}_{ij}$. $B_i^n(u)=c_k^l t^l(1-t)^{k-1}$ is basic function of Bernstein.

Theorem 1. *Let rectangular rational Bezier patches $R_1(u,v)$ and $R_2(u,v)$ be C^0 continuous along a common boundary curve u=0. Then $R_1(u,v)$ and $R_2(u,v)$ are G^1 continuous if the following conditions hold*

$$R_{2u}(0,t)=\lambda(t)R_{1u}(0,t)+\mu(t)R_{1v}(0,t)\qquad 0\le t\le 1 \tag{4}$$

where $\lambda(t)<0$ for $0\le t\le 1$.

2.1 Boundary condition for N-sided region

Let $\{P_j(u,v)\}, j=1,2,\cdots,n$ are rational patches surrounding an isolate N-sided region D (see Figure 1). Assume that there are four boundary Bezier curves connect to corner vertex of the region, and the tangent vectors at the corner vertices are co-linear. We request that to construct a N-sided rational Bezier surface S in the region, S is G^1 continuous with $\{P_j(u,v)\}, j=1,2,\cdots,n$. and the patches of $S:\{\vec{S}_j(u,v)\}, j=1,2,\cdots,n$ are G^1 continuous along their common boundary curves.

2.2 Consistency condition

When a patch must be smoothly joined with two patches across two adjacent edges at a corner vertex, the inconsistent problem might be encountered.

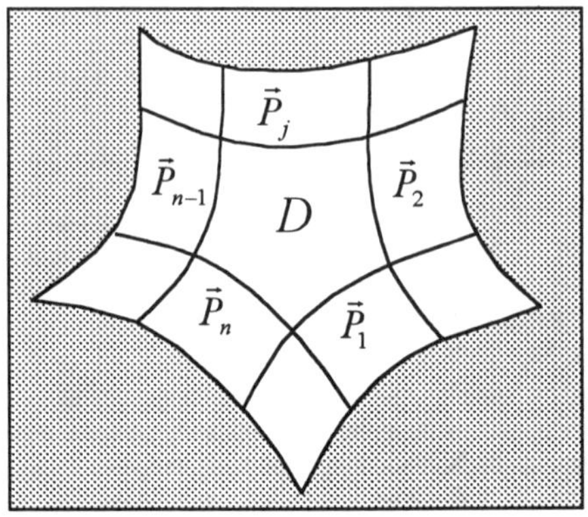

Fig. 1. Boundary condition at a N-sided region

Let $\vec{S}_j(u,v), j=1,2,\cdots,n$ be blending patches, then following condition must be hold at four corner vertices of $\vec{S}_j(u,v)$

$$\frac{\partial^2\vec{S}_j(0,0)}{\partial u\partial v}=\frac{\partial^2\vec{S}_j(0,0)}{\partial v\partial u},\quad \frac{\partial^2\vec{S}_j(1,0)}{\partial u\partial v}=\frac{\partial^2\vec{S}_j(1,0)}{\partial v\partial u}$$

$$\frac{\partial^2\vec{S}_j(0,1)}{\partial u\partial v}=\frac{\partial^2\vec{S}_j(0,1)}{\partial v\partial u},\quad \frac{\partial^2\vec{S}_j(1,1)}{\partial u\partial v}=\frac{\partial^2\vec{S}_j(1,1)}{\partial v\partial u} \tag{5}$$

3. G^1 Blending at the corner vertex of N-sided region

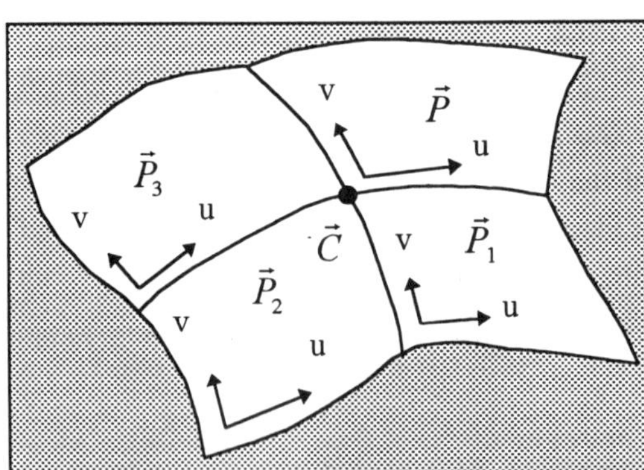

Fig. 2. Blending at corner vertex $\vec{C}$ of N-sided region

As shown in Figure 2, assume that there exist three basic patches $\vec{P}_1(u,v), \vec{P}_2(u,v)$

and $\vec{P}_3(u,v)$ at common vertex $\vec{C}$, and they are satisfied with boundary condition of N-sided region. Now we want to insert a patch $\vec{P}(u,v)$ which has G^1 continuity with $\vec{P}_1(u,v)$ and $\vec{P}_2(u,v)$ simultaneity. Without losing generality, we can assume that at the corner vertex $\vec{C}$, the across boundary tangent vectors of $\vec{P}_1(u,v), \vec{P}_2(u,v)$ and $\vec{P}_3(u,v)$ are satisfied with following conditions:

$$\vec{P}_{1u}(0,1) = \frac{1}{k_1}\vec{P}_{2u}(1,1), \quad \vec{P}_{3v}(1,0) = \frac{1}{k_2}\vec{P}_{2v}(1,1) \tag{6}$$

Due to that there are at least G^1 continuity between $\vec{P}_1(u,v)$, $\vec{P}_2(u,v)$ and $\vec{P}_3(u,v)$, and according to theorem 1, following condition must be hold

$$\begin{aligned}\vec{P}_{1u}(0,1) &= \lambda_{12}(v)\vec{P}_{2u}(1,v) + \mu_{12}(v)\vec{P}_{2v}(1,v)\\ \vec{P}_{3v}(u,0) &= \lambda_{32}(u)\vec{P}_{2v}(u,1) + \mu_{32}(u)\vec{P}_{2u}(u,1)\end{aligned} \tag{7}$$

where, $\lambda_{12}(v), \mu_{12}(v)$ are the blending function between $\vec{P}_1(u,v)$ and $\vec{P}_2(u,v)$, and $\lambda_{32}(v), \mu_{32}(v)$ are the one between $\vec{P}_2(u,v)$ and $\vec{P}_2(u,v)$. Further, we can force that the endpoint condition of them are as follows

$$\begin{aligned}\lambda_{12}(1) &= \frac{1}{k_1}, \lambda'_{12}(1) = 0, \mu_{12}(1) = 0, \mu'_{12}(1) = 0,\\ \lambda_{32}(1) &= \frac{1}{k_2}, \lambda'_{32}(1) = 0, \mu_{32}(1) = 0, \mu'_{32}(1) = 0,\end{aligned} \tag{8}$$

from (6),(7) and (8), we have

$$\vec{P}_{1uv}(0,1) = \vec{P}_{2uv}(1,1) / k_1, \quad \vec{P}_{3vu}(1,0) = \vec{P}_{2vu}(1,1) / k_2 \tag{9}$$

because the patches $\vec{P}_1(u,v)$, $\vec{P}_2(u,v)$ and $\vec{P}_3(u,v)$ exist, there must be

$$\begin{aligned}\vec{P}_{1uv}(u,v) &= \vec{P}_{1vu}(u,v)\\ \vec{P}_{2uv}(u,v) &= \vec{P}_{2vu}(u,v)\\ \vec{P}_{3uv}(u,v) &= \vec{P}_{3vu}(u,v)\end{aligned} \tag{10}$$

from (9) and (10), we can derive

$$k_1\vec{P}_{1uv}(0,1) = k_2\vec{P}_{3vu}(1,0) \tag{11}$$

let $\vec{P}(u,v)$ be G^1 continuity with $\vec{P}_1(u,v)$ and $\vec{P}_3(u,v)$ along their common boundary curve, and the blending function are $\lambda_1(u),\mu_1(u)$ and $\lambda_3(u),\mu_3(u)$ respectively, from theorem 1, we have

$$\vec{P}_v(u,0) = \lambda_1(u)\vec{P}_{1v}(u,1) + \mu_1(u)\vec{P}_{1u}(u,1),$$
$$\vec{P}_u(0,v) = \lambda_3(v)\vec{P}_{3u}(1,v) + \mu_3(v)\vec{P}_{3v}(1,v) \tag{12}$$

then, the twist vector of the patch $\vec{P}(u,v)$ at $\vec{P}(0,0)$ is

$$\vec{P}_{vu}(0,0) = \lambda_1'(0)\vec{P}_{1v}(0,1) + \lambda_1(0)\vec{P}_{1vu}(0,1) + \mu_1'(0)\vec{P}_{1u}(0,1) + \mu_1(0)\vec{P}_{1uu}(0,1)$$
$$\vec{P}_{uv}(0,0) = \lambda_3'(0)\vec{P}_{3u}(1,0) + \lambda_3(0)\vec{P}_{3uv}(1,0) + \mu_3'(0)\vec{P}_{3v}(2,0) + \mu_3(0)\vec{P}_{3vv}(1,0) \tag{13}$$

let

$$\lambda_1(0) = k_1;\ \lambda_1''(0) = 0;\ \mu_1(0) = 0;\ \mu_1'(0) = 0$$
$$\lambda_3(0) = k_2;\ \lambda_3'(0) = 0;\ \mu_3(0) = 0;\ \mu_3'(0) = 0 \tag{14}$$

from (11)-(14), we can easily derive that

$$\vec{P}_{uv}(u,v) = \vec{P}_{vu}(u,v)$$

this means that at the vertex $\vec{C}$, the inserted patch $\vec{P}(u,v)$ is satisfied with consistency condition. So patch $\vec{P}(u,v)$ does exist. Now we have proved following theorem.

Theorem 2 *Suppose $\vec{C}$ is common corner vertex of four rectangular rational Bezier patches $\vec{P}_1,\vec{P}_2,\vec{P}_3$ and $\vec{P}$, at vertex $\vec{C}$ their across boundary tangent vectors are co-linear. If $\vec{P}_1,\vec{P}_2,\vec{P}_3$ does exist, and they are C^0 and G^1 continuity along their common boundary curves, then $\vec{P}$ is satisfied consistency condition if end-point condition of blending function between $\vec{P}$ and its adjacent patches is set as (14).*

4 G^1 Blending at the center vertex of N-sided region

As in Figure 3, $\vec{O}$ is common vertex of N patches in the N-sided region. Given vertex $\vec{O}$ and rational Bezier curves $\{\vec{C}_j(t)\}, j = 1,2,\cdots,n$ which run between center points of each boundary curves and *vertex* $\vec{O}$. We can construct following G^1 blending equations along each $\vec{C}_j(t)$ for correspondent patches according to theorem 1:

$$\begin{cases}\vec{S}_{2u}(0,t)=\lambda_1(t)\vec{S}_{1v}(t,0)+\mu_1(t)\vec{S}_{1u}(t,0)\\ \vec{S}_{3u}(0,t)=\lambda_2(t)\vec{S}_{2v}(t,0)+\mu_2(t)\vec{S}_{2u}(t,0)\\ \cdots \\ \vec{S}_{1u}(0,t)=\lambda_n(t)\vec{S}_{nv}(t,0)+\mu_n(t)\vec{S}_{nu}(t,0)\end{cases} \tag{15}$$

The status of two patches blended along their common boundary curve $\vec{C}_j(t)$ is: the first point of $\vec{C}_j(t)$ is $\vec{O}$, at which the across tangent vector is not co-linear in general; while the end point is the common vertex of four patches, the across tangent vector is co-linear. It is easy to know from the theorem 2 and the relationship of tangent vectors at the first point of $\vec{C}_j(t)$ that the consistence condition will be satisfied at the first and end point of $\vec{C}_j(t)$, if end condition of blending function be set as follows:

$$\begin{cases}\lambda_j(0)=\alpha_j;\lambda'_j(0)=0;\mu_j(0)=\beta_j;\mu'_j(0)=0;\\ \lambda_j(1)=\gamma_j;\lambda'_j(1)=0;\mu_j(1)=0;\mu'_j(1)=0;\end{cases} \tag{16}$$

when t=0, from (15) and (16), we can obtain the equations of tangent vectors

$$\begin{cases}\vec{S}_{2u}(0,0)=\alpha_1\vec{S}_{1v}(0,0)+\beta_1\vec{S}_{1u}(0,0)\\ \vec{S}_{3u}(0,0)=\alpha_2\vec{S}_{2v}(0,0)+\beta_2\vec{S}_{2u}(0,0)\\ \cdots \\ \vec{S}_{1u}(0,0)=\alpha_n\vec{S}_{nv}(0,0)+\beta_n\vec{S}_{nu}(0,0)\end{cases} \tag{17}$$

and the equations of twist vectors

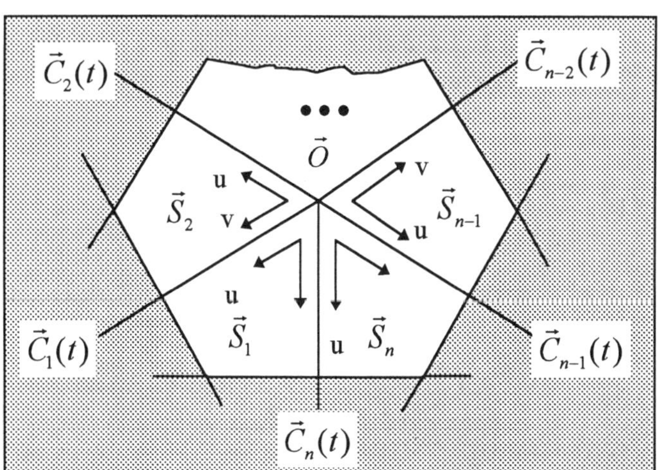

Fig. 3. Blending at center vertex $\vec{O}$ of N-sided region

$$\begin{cases}\vec{S}_{2uv}(0,0)=\alpha_1\vec{S}_{1vu}(0,0)+\beta_1\vec{S}_{1uu}(0,0)\\ \vec{S}_{3uv}(0,0)=\alpha_2\vec{S}_{2vu}(0,0)+\beta_2\vec{S}_{2uu}(0,0)\\ \cdots\\ \vec{S}_{1uv}(0,0)=\alpha_n\vec{S}_{nvu}(0,0)+\beta_n\vec{S}_{nuu}(0,0)\end{cases} \tag{18}$$

by solving equation (17), all $\{\alpha_j,\beta_j\}, j=1,2,\cdots,n$ could be obtained, let

$$\begin{cases}\vec{X}_j=\vec{S}_{juv}(0,0)=\vec{S}_{jvu}(0,0)\\ \vec{b}_j=\beta_j\vec{S}_{juv}(0,0)\end{cases} \tag{19}$$

in (18), then (18) can be rewrite as

$$AX=B \tag{20}$$

by solving equation (12), we can get twist vectors $X=\left[\vec{x}_1,\vec{x}_2,\cdots,\vec{x}_n\right]^T$. And using of equation (21)

$$\vec{x}_j=m_jn_j\left[\frac{w_{j11}}{w_{j00}}(\vec{r}_{j11}-\vec{r}_{j00})+\frac{w_{j01}w_{j10}}{w_{j00}^2}(2\vec{r}_{j00}-\vec{r}_{j01}-\vec{r}_{j10})\right]\quad (j=1,\cdots,n) \tag{21}$$

we can obtain net vertex $\vec{r}_{j11}$ corresponding to $\vec{x}_j$.

5 Solution of the twist vectors equations

In equation (20), $|A|=\alpha_1\alpha_2\cdots\alpha_n-1$. We know that $\alpha_j<0, j=1,2,\cdots,n.$ from theorem 1, so when number n is odd, $|A|\neq 0$, there is only one solution for (20); while as n is even, A may be singularity, due to $\alpha_1\alpha_2\cdots\alpha_{n-1}\neq 0$, the order of A is at least $n-1$. Then we can rewrite (20) as follows

$$\begin{cases}\vec{x}_2=\alpha_1\vec{x}_1-b_1\\ \vec{x}_3=\alpha_1\alpha_2\vec{x}_1-\alpha_2b_1-b_2\\ \cdots\\ \vec{x}_n=\alpha_1\alpha_2\cdots\alpha_{n-1}\vec{x}_1-\alpha_2\alpha_3\cdots\alpha_{n-1}b_1-\cdots-\alpha_{n-1}b_{n-2}-b_{n-1}\end{cases} \tag{22}$$

where

$$\vec{x}_1=\frac{\alpha_2\alpha_3\cdots\alpha_n\vec{b}_1+\alpha_3\alpha_4\cdots\alpha_n\vec{b}_2+\cdots+\alpha_n\vec{b}_{n-1}+\vec{b}_n}{\alpha_1+\alpha_2+\cdots+\alpha_n} \tag{23}$$

There are three cases to solve equation (20) according to the value of the numerator and denominator of $\vec{x}_1$:

a. the denominator of $\vec{x}_1$ is not equal to zero, then $\vec{x}_1$ exist, equation (20) has sole solution.

b. the numerator and denominator of $\vec{x}_1$ are all equal to zero, then $\vec{x}_1$ is not sole, equation (20) has infinite solution. In this case, we can find a suitable $\vec{x}_1$, and get a suitable solution for equation (20).

c. the denominator of $\vec{x}_1$ is equal to zero, but the numerator of $\vec{x}_1$ is not, then $\vec{x}_1$ does not exist, equation (20) has no solution. In this case, we should adjust some value of items in equation (20), and translate case c to case b.

Following is one of the methods to solve the equation (20) in case c: Let the numerator of $\vec{x}_1$ is equal to zero, e.g.

$$\alpha_2\alpha_3\cdots\alpha_n\vec{b}_1+\alpha_3\alpha_4\cdots\alpha_n\vec{b}_2+\cdots+\alpha_n\vec{b}_{n-1}+\vec{b}_n=0 \tag{24}$$

$$\begin{aligned}\vec{b}_j=-\beta_j\Big\{&\left[2m_j w_{j1}/w_{j0}+m_j(m_j-1)w_{j2}/w_{j0}-2m_j^2(w_{j2}/w_{j0})^2\right]\\&(\vec{r}_{j1}-\vec{r}_{j0})+m_j(m_j-1)(\vec{r}_{j2}-\vec{r}_{j1})w_{j2}/w_{j0}\Big\}\\&j=1,2,\cdots,n\end{aligned} \tag{25}$$

where m_j is the order of curve $\vec{C}_j(t)$, $\vec{r}_{jk}$ is control vertex of curve $\vec{C}_j(t)$, w_{jk} is weight of $\vec{r}_{jk}$

Rising the order of curve $\vec{C}_j(t)$ by one, we can get an adjustable vector $\vec{r}_{j2}$ on $\vec{C}_j(t)$. Changing $\vec{r}_{j2}$ will change the value of $\vec{b}_j$ while not affect other value, the result is very important. Solving equation (25) for new $\vec{b}_j', j=1,2,\cdots,n$, we can get $\vec{r}_{j2}', j=1,2,\cdots,n$, then find $\min\left\{\left|\vec{r}_{j2}-\vec{r}_{j2}'\right|, j=1,2,\cdots,n\right\}$ to make change. Now case c has been changed to case b.

By applying all the results mentioned above, we can easily generate a rational Bezier blending surface $\vec{S}:\vec{S}_j(u,v), j=1,2,\cdots,n$ on a N-sided region which is of G^1 continuity with all adjacent patches, also the consistency condition is easy to be satisfied and the inconsistency is easy to be corrected. The algorithm is very powerful and flexible.

6. Examples

Several examples are shown as follow.

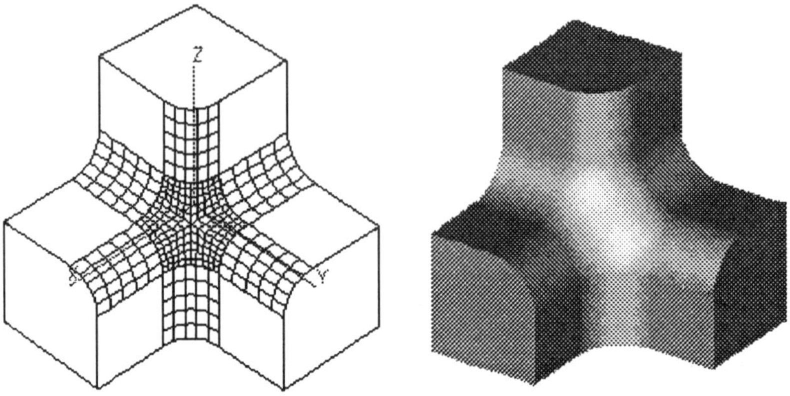

Fig. 4. Wireframe and shaded display of Blending surfaces and a six-sided region surface

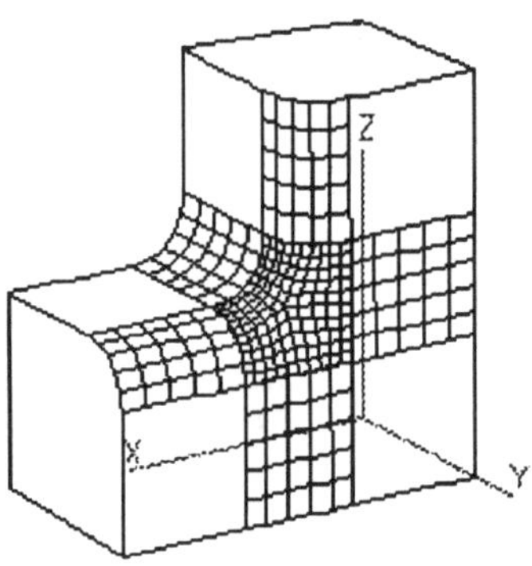

Fig. 5. Blending surfaces and a five-sided region surface

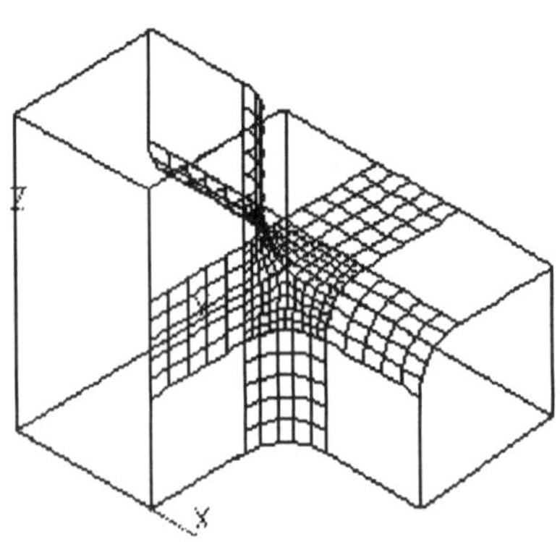

Fig. 6. Blending surfaces and a six-sided region surface

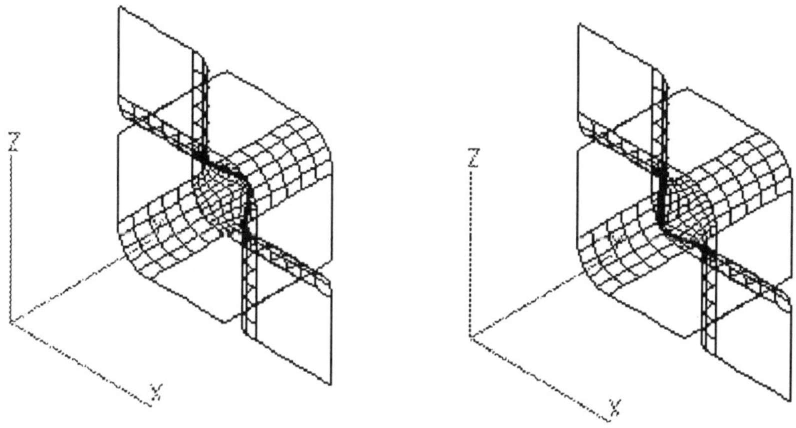

Fig. 7. Adjust fullness of a N-sided surface

7. Conclusion

The preceding sections have shown in details the key techniques that use rectangular rational Bezier patches to construct an arbitrary N-sided region surface to fill a N-sided hole and guarantee G^1 continuity along their common boundary edges. The main contribution of the new method is to build the blending surface and N-sided surface with standard surface form which is widely used and is supported by STEP. So the method will play an important role in geometric modeling and manufacturing field. Due to the length limit, we could not describe how to create center vertex and how to construct $\vec{C}_j(t)$ in N-sided region in detail.

References

1. Calmull E E, Clark J. Recursively Generated B-Spline Surfaces on Arbitrary Topological Meshes, Computer Aided Design, 10(1978) 350-355
2. Chiyokura H, Kimura F. Design of solid with free-form surfaces, in: ACM Computer Graphics, 17(1983) 289-298
3. Wang Lazhu, Zhu Xinxiong. Constructing N-sided patches with NURBS, CADDM, 5(1995) 26-32.
4. Sarrage R F. G^1 interpolation of generally unrestricted cubic bezier curves, Computer Aided Geometric Design., 4(1987) 23-39
5. Luo Hongzhi. Study and Practice for Integral Geometric Modeling System, Doctoral dissertation. Huazhong University of Science and Technology, 1(1992), pp. 78-88.

Optimal Expected-Time Algorithm For Finding Largest Empty Circle And Its Application On PCB Stamping Mould Design

Wang Jia ye, Boey Seng Heng, Lee Eng Wah

Gintic Institute of Manufacturing Technology
71 Nanyang Drive
Singapore 638075

Abstract. The design of PCB stamping mould presents us with a problem of finding the largest empty circle (LEC) between polygons. This paper presents a new algorithm for finding the LEC with the expected time to find a local LEC of ***O(1)*** and preparation time of ***O(N)***, where ***N*** is the total number of holes and sides of profiles.

1 Introduction

A major step in PCB manufacturing process is to cut the profiles and small holes after etching. Stamping is an efficient approach to complete the step for single and double layers PCBs. However, an adequate amount of ejectors pins must be provided on a stamping mould (Fig. 1) to push the PCBs out of the punch in order to prevent the PCBs from breaking. A difficult task in the 2D design of stamping mould is to determine the locations and sizes of the ejectors. The sizes of the ejectors are specified by their diameters, with their values chosen as large as possible. As shown in Fig. 2, the empty space between the holes and profiles must be wide enough to place an ejector. This leaves us with the problem of finding the ***largest empty circle*** (**LEC**) in the space between the holes and profile for placing an ejector. This problem could be solved by creating Voronoi diagram for holes and profiles.

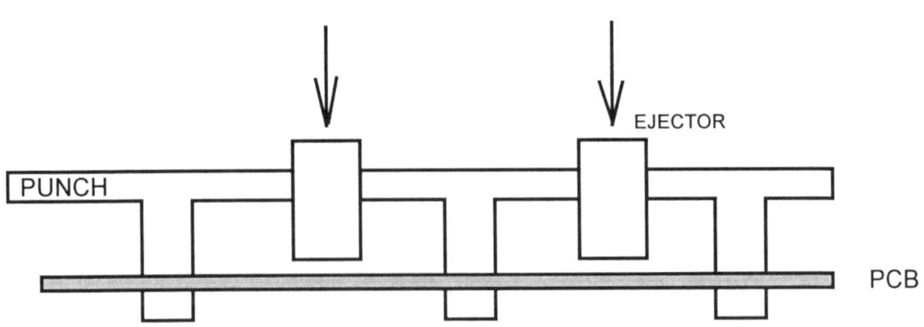

Fig. 1 The PCB Stamping Mould.

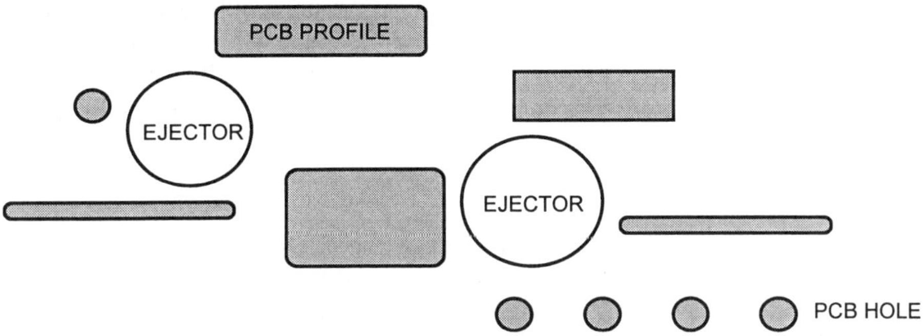

Fig. 2 Design of PCB Stamping Mould.

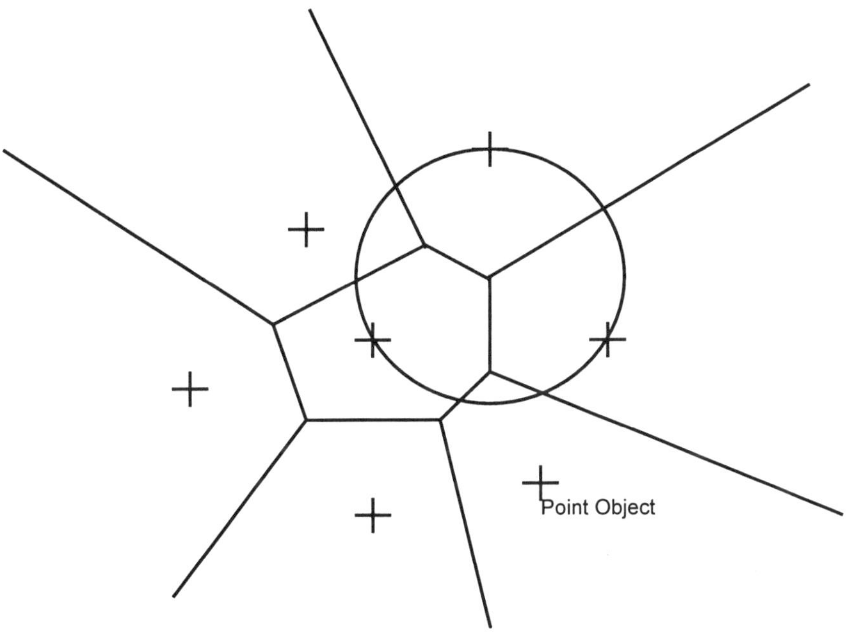

Fig. 3 Voronoi Diagram for a Group of Point Objects.

The Voronoi diagram for a group of point objects is shown in Fig. 3. It is noteworthy that the center of the LEC for a group of points is located at one of the vertices of the Voronoi diagram. It is shown in [1] that these tasks of creating Voronoi diagram and finding the center of LEC can be completed in a total time of *O(NlogN)*, where *N* is the number of the points. It is further proven in [2] that the center of the LEC for a group of polygons is located at one of the vertices of their Voronoi diagram. The time complexity of creating the Voronoi diagram of polygons is also *O(NlogN)*. In this paper, a new algorithm for finding the LEC of a group of holes and profiles for the design of ejectors for PCB stamping mould is presented. Using this new algorithm, the expected time to find a local LEC is *O(1)* with preparation time of *O(N)*, where *N* is the total number of holes and sides of profiles. This new algorithm is proven to be efficient and robust, and it can be extended to create Voronoi diagram of disjoint polygonal object as in [3].

2 Algorithm

The profiles on a PCB contain closed polylines, which often consist of line segments and arcs. The circles, line segments and arcs are called the ***sides*** of closed polylines. These circles and closed polylines, in 2D, are the objects of interest for the new algorithm. Two definitions are needed for the new algorithm:

> **Definition 1:** A ***Voronoi polygon*** **(VP)** associated with a side is a polygon that any point within that polygon is nearer to that side than any other sides and vertices, and any point with the property must be in the polygon.

> **Definition 2:** A ***Voronoi polygon*** **(VP)** associated with a vertex is a polygon that any point within that polygon is nearer to the vertex than any other sides or vertices, and any point with the property must be in the polygon.

The new algorithm requires the 2D plane to be partitioned by VPs to form a net called a ***Voronoi diagram*** (Fig. 4). The segments forming a VP are called the **sides** of the Voronoi diagram. Every side of Voronoi diagram, except those on the objects, is common to two VPs. The distances from any point on the common side of two VPs to the vertices or sides on the objects are same. Voronoi sides can be generated in six cases.

(1) A bisector of two sides (e.g., in Fig. 4, ***P4P5*** is the bisector of sides ***GH*** and ***EF***).

(2) The bisector of two vertices (e.g., in Fig. 4, ***P1P2*** is the bisector of vertices ***A*** and ***G***).

(3) The trace of points with same distances to a vertex and a side (e.g., in Fig. 4, the distance of any point on ***P2P3*** to vertex ***G*** is the same as the perpendicular distance of side ***AB*** to that point).

(4) The trace of points with same distances to an arc and a line segment.

(5) The trace of points with same distances to an arc and a vertex.

(6) The trace of points with same distances to two arcs.

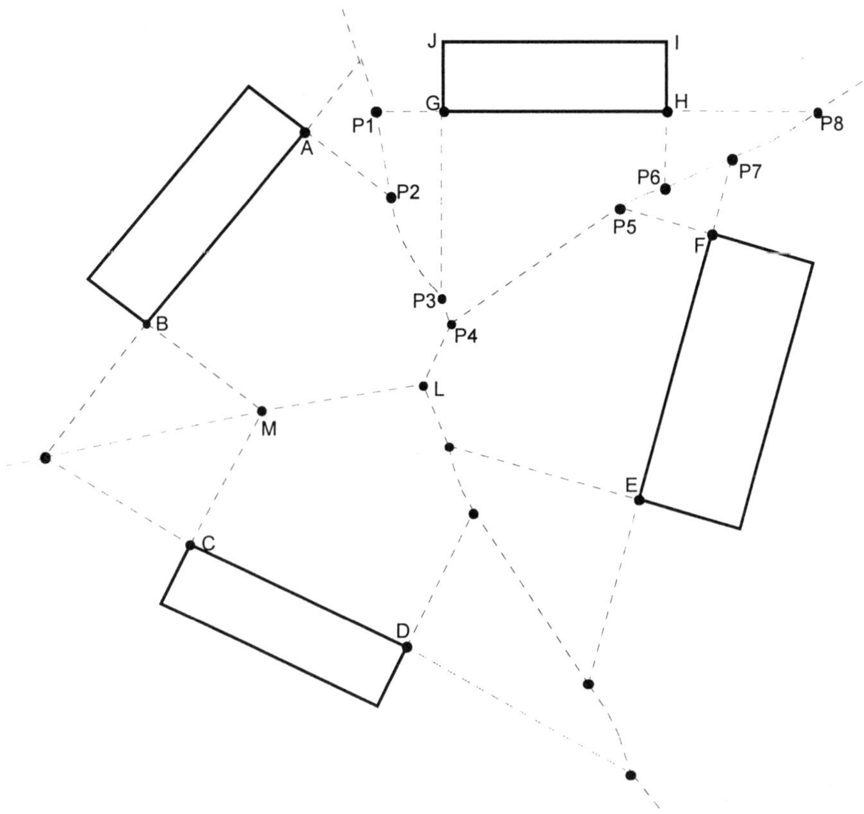

Fig. 4 A Voronoi Diagram.

The traces in cases (3) and (4) are parabolas. The traces in cases (5) and (6) can be parabolas or arcs. A VP associated with a side is bounded by two rays perpendicular to the side from its two end points (e.g., ***GP3*** and ***HP6*** for VP associated with side GH in Fig. 4). A VP associated with a vertex is bounded by two rays perpendicular to the neighbouring sides of that vertex (***GP1*** and ***GP3*** for VP associated with vertex G in Fig. 4). The region bounded by the two rays is called the ***bounded region***.

Our new algorithm is based on an incremental process. For example, when a new object, a circle or a polyline (***GHIJ*** in Fig. 6), is added to an existing Voronoi diagram as shown in Fig. 5. The existing Voronoi diagram is amended to create a new Voronoi diagram (Fig. 4) starting from any vertex of the new object, say ***G*** in Fig. 6, as illustrated below:

- Since ***G*** is in the VP associated with vertex ***A*** and ***A*** is in the bounded region of ***G***, the Voronoi side of the two vertices ***A*** and ***G*** is generated by intersecting their bisector with their bounded region.

- The bisector ***P1P2*** of the two vertices ***A*** and ***G*** intersects the bounded region of ***G*** at ***P1***; ***P1P2*** intersects with ***AK*** at ***P2***. ***AK*** is in the common side of VP associated vertex ***A*** and VP associated with side ***AB***; and ***P2*** is in the bounded region of ***G***.
- The trace of points with same distances to a vertex ***G*** and side ***AB*** is generated as the parabola ***P2P3***, where ***P3*** is the intersection point of ***P2P3*** and the boundary of the bounded region of ***GH***.
- After ***P3***, the trace of points with same distances to line segment ***GH*** and ***AB*** is generated as the side ***P3P4***, which is the bisector of the line segments ***GH*** and ***AB***. This side is in the bounded region of ***GH*** and in the VP associated with ***AB***.
- This tracing process generated polygon ***GP1P2P3*** as the VP associated with ***G*** and polygon ***ABMLP4P3P2A*** as the VP associated with ***AB***.

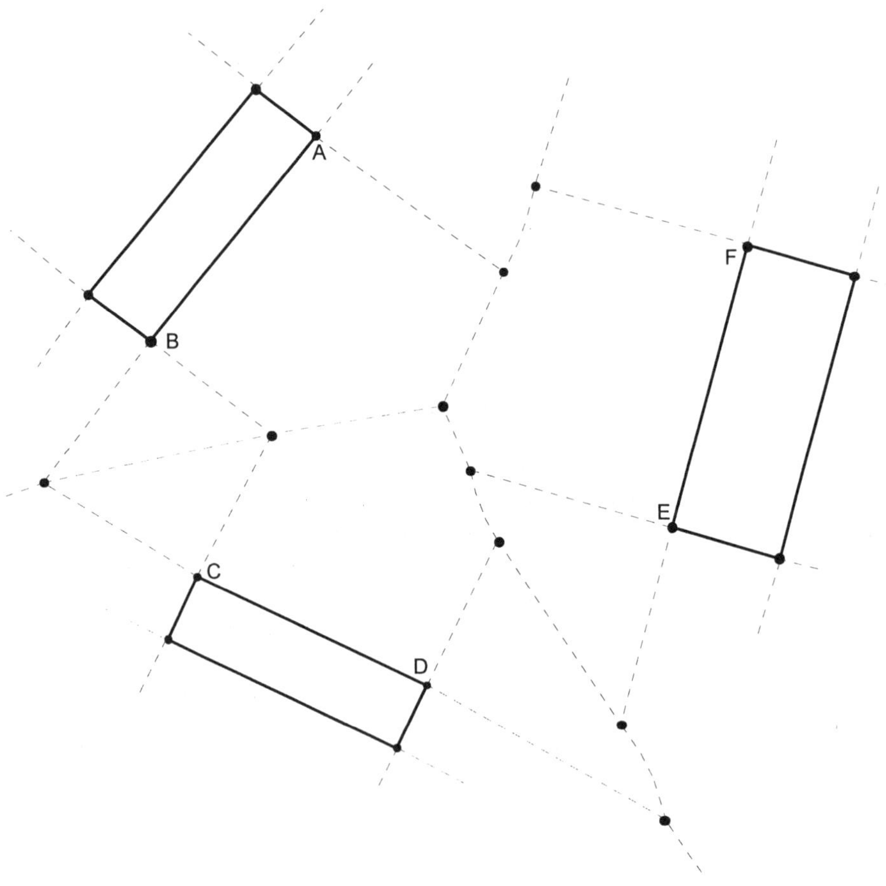

Fig. 5 An Existing Voronoi diagram.

The similar approach can be applied to generate or amend the VPs associated with ***GH***, ***EF***, ***H*** and ***F***. After deleting all the unwanted line segments created during the process of generating new VPs, the new Voronoi diagram as shown in Fig. 4 is formed.

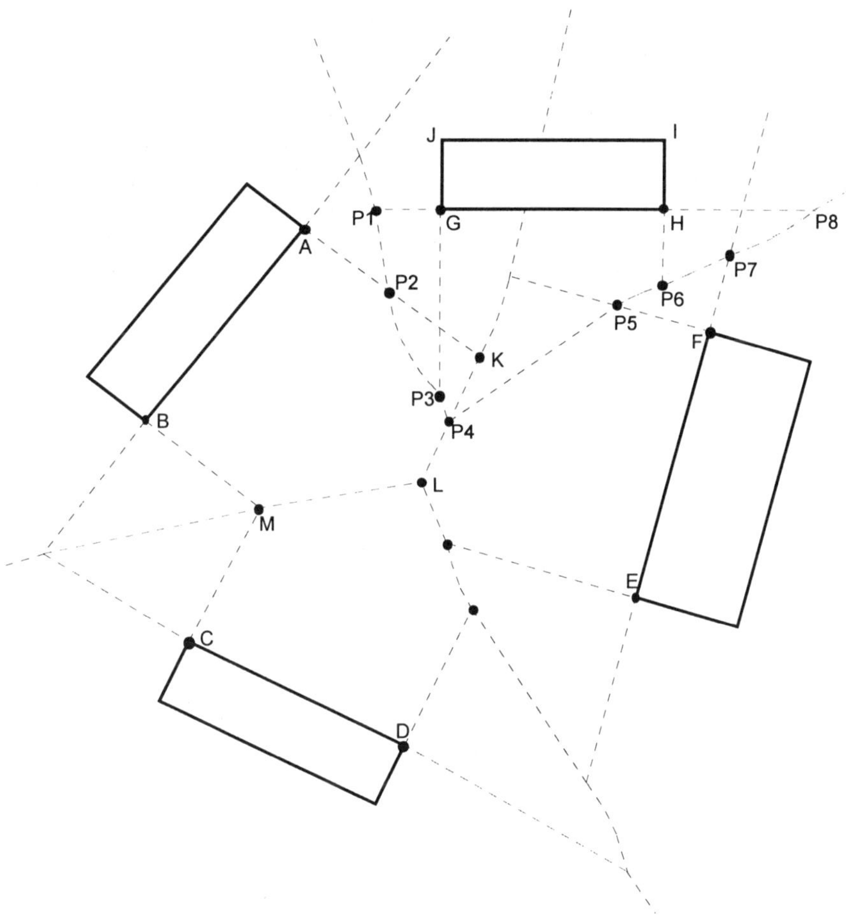

Fig. 6 Creation of a New Voronoi diagram.

From the algorithm described above, it is obvious that a VP is determined by its neighbouring objects.

In order to apply the new algorithm efficiently for ejector design, an estimated **starting point (SP)** must be specified by the designer. This SP is the estimated location of the ejector and it is in the area where the LEC could be found. The incremental process can be carried out by finding objects surrounding the SP using a spiral search method as shown in Fig. 7. This spiral search method partitions the 2D plane into small squares (or cells) with the search sequence as indicated by the numbers in the squares of Fig. 7.

In order to speed up the search for new objects, a ***tolerant circle (TC)*** centered at SP is created in which the LEC is expected. Since the center of the ejector is expected to be within the tolerant circle, the problem of designing an ejector becames the process of locating the center of the largest empty circle within the tolerant circle with the same radius as the LEC. This largest empty circle is local with respect to the tolerant circle and it is called a ***Local LEC***. It is noteworthy that the SP is situated at one of the several VPs created near the SP; and the center of the local LEC is at one of the vertices of the Voronoi diagram within the TC, or at the intersection point of the sides of the Voronoi diagram and the TC. Therefore, a new object can be ignored if its minimum distance to the center of the TC is greater than the maximum distance between the TC and the existing objects. This is because the new object has no effects on the Voronoi sides within the TC under such circumstance. The search will end if the minimum distance from the center of the TC to the square visited is greater than the maximum distance between the TC and the existing objects. That is, when the objects within that square are far enough to have no effects on the Voronoi sides.

		17	16	15	14	13		
		18	5	4	3	12		
		19	6	1	2	11		
		20	7	8	9	10		
		21	22	23	24	25		

Fig. 7 The Spiral Search Method.

3 Conclusion

Since the local LEC is determined by the local objects (sides and vertices) surrounding the SP and the incremental process only searches for objects surrounding the SP as provided by the designer, there is no need to create the Voronoi diagram for the whole PCB. This makes the new algorithm more efficient for locating the Local LECs for ejector design. Although the time complexity of the new algorithm may be ***O(N*N)*** for the worst case, the total optimum expected time to find a local LEC will be ***O(N)*** due to the hugh number of objects with uniform distribution in a single PCB layout. It is therefore efficient to apply the algorithm for the design of ejectors in a PCB stamping mould.

Reference

1. M.I.Shamos and D.Hoey. Closest-Point Problems. In Proc. 16th Annual IEEE Symposium on Foundations of Computer Science, October 1975.
2. Martin Held. "*On the Computational Geometry of Pocket Machining*", Springer-Verlag, 1991.
3. Wang Jia Ye, Wang Wenping, Liu Ding Yuin. Optimal Expected-Time Algorithm for Voronoi Diagram of Disjoint Polygonal Object. Proceedings of ICMA'97,1997.

Product Design

The Research On Design Principles To The CAD/CAM Geometric Platform Based On Shared Standard Semantic Models Of Step/Ap203

Ren, Aihua and Wei, juxia

Dept. of Computer Science and Engineering, Beijing University of Aeronautics and Astronautics, Beijing, China 100083
email: renah@buaa.edu.cn

Abstract. This paper expounds the approaches and the design principles to the CAD/CAM geometric platform based on shared standard semantic models of STEP/AP203[3] in the CIMS environment. general purpose geometric platform frame which is independent of specific graphic systems is given in this paper. The frame is fit for different kinds of specific graphic systems. The advantages for the approaches show that the software design of CAD/CAM is independent from specific graphics systems. Within the scope of some powerful graphics system, the CAD software can be run on these graphic systems by plugging the systems' APIs in the correspondent API(Application Program Interface) sockets.

Key words STEP tools Object-oriented design Geometric modular

1 Introduction

Today, the focus on CAD/CAM developing is the integration of CAD/CAM with many subsystems which have different constructions and various technical functions, based on "Data Base of Product Definitions".

Among the majority of present commercial CAD/CAM systems, the graphic information is transmitted mainly by IGES(Initial Graphic Exchange Specification) or by neutral file between CAPP and CAD or different CAD systems. In the CAD/CAM system based on STEP, product data are exchanged mainly by sharing the STEP product model data base or by the STEP neutral file (ISO10303-21). In STEP based system, not only the graphic information can be transmitted but also the whole product message can be transformed. STEP is more satisfied for information integration of modern manufacturing systems than IGES. But for the developers and programmers of STEP based CAD/CAM， they have to understand both data model definitions of STEP and data model definitions of the specific commercial graphic system. In order to make STEP based CAD programmers can concentrate their jobs on STEP only and make the developed CAD/CAM system be independent from the specific commercial graphic system, it is necessary to develop a generic STEP based object-oriented geometric platform. It makes the CAD/CAM system can use various graphic system as the base support. This platform is more opening. Therefore the flexibility is very important. Based on this framework structure, STEP based data model can be automatically transformed into any data form of the specific commercial graphic system. And the framework can supply many operations to STEP/AP203 model data. In fact, these operations are supported by the presented supported graphic systems . On the aspect of CAD programmers, they only need to know how to use STEP data form to develop the software, and no need to know how to use the supported graphic system. Through this framework the CAD/CAM developer can deal with their design in an uniform model without considering particular graphics system underneath.

2 Basic concept and design idea

STEP documents consist of three sets (see Figure 1),(1) Representation methods and implement methods;(2) Integrated resources;(3) Application protocols representation methods providing the data model description language[2], EXPRESS. EXPRESS is used to defined all STEP data. Implementation methods direct the developing work on Express data model. Integrated resources is composed of generic resources and application resources. They are used to describe product information. Generic resources are independent from any specific application and can be used by any applications. Application protocols add some necessary restrictions and modifications on generic resources to make them fit for particular application needs. Application protocols are defined for the sake of making STEP be suited to some particular environment. AP203 is one of the STEP application protocols. It is about 3-dimensional product shape representation and configuration controlled data definitions within the product design phase. Product shape representation is a kernel part in the product design. AP203 refers to lots of data definitions of ISO 10303-42[4] to describe 3D product shape.
In a CAD/CAM integrated system, the graphics information between CAD and CAPP or among various application systems can be transferred by STEP neutral file[1] or be transferred by sharing product model data. It is necessary for the developing of CAD/CAM integrated system to keep the unique representation of STEP semantic model and the complete property of this model. Also, it is convenient to display the semantic model in a visual graphics model during the developing of CAD/CAM systems. The graphic operation can be done directly in the visual mode. Therefore, it is very important for developing CAD/CAM integrated system directly to use the geometric representation graphics package which is identical to STEP semantic model or to adapt any graphics package (X) as the visual display and as the graphics

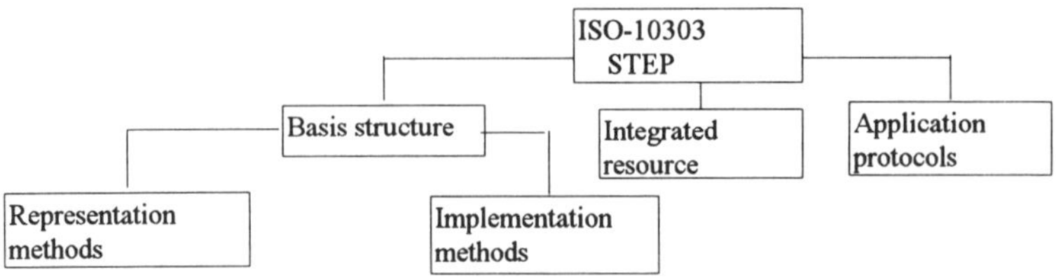

Fig. 1 The structure of STEP

operation for STEP geometric representation. Therefore the problems on graphics representation of CAD/CAM integrated system based on STEP is described as (take an example of AP203[3]):

How to make the STEP semantic sharing on geometric representation and how to give the graphics operation and visualization based on the sharing. In other words, this is the problem on graphic operation and visualization of STEP/AP203 semantic model or the problem on transformations between the two semantic models of STEP/AP203 and X graphics package.

Problem solving:

By using framework mechanism 203-FM to solve the problem. (see Fig. 2)

Objectives:

- CAD/CAM programs can do the sharing of STEP/AP203 semantic model.
- CAD/CAM developers need only concentrate their attentions on STEP semantic model system structure design.
- Through supported graphic system, graphic operations and displays for STEP/AP203 semantic models are provided.

3 System model

Let the framework model of geometry platform with heterodoxy architecture be 203-FM, then 203-FM consists of quintuple at least:

$$203\text{-FM}=\{Q, M, GS, I, R\} \quad (1)$$

where,

Q: a Set of all the data models and objects in 203-FM;

M: a Set of operations of 203-FM;

GS: a Set of architectures of various graphic systems;

I: an Interface model of users or on Interface of software components, providing various functions and I/O forms which meet component technology;

R: Integrated relation models, representing various inclusion and collection relation among Q, M. GS and I.

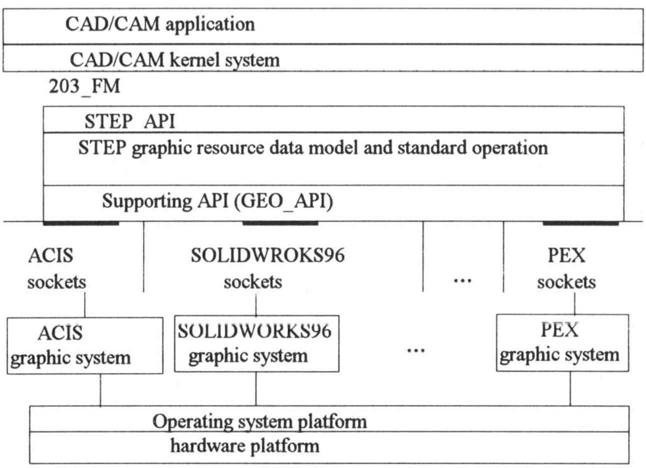

Fig.2 System structure

203-FM defines the semantic models which meet the STEP/AP203 requirements. It can be used as a geometric description platform for various graphic packages, providing an unified operating interface and being independent from any specific graphic systems. The inner structure of 203-FM is with the cutable feature and it provides reusable C++ library by using O-O technique.

4 System Architecture

As Fig. 2, the architecture shows hierarchical relationship among CAD application software, generic geometry platforms, particular graphic systems and operation systems.

The 203-FM provides the selection choice among the various commercial graphic package for CAD/CAM system developer (open property, flexibility and quality). The developing work is mainly as follows:

(1) To create a generic framework 203-FM with STEP model data definitions and geometric operations;

(2) To add standard operations upon STEP data model represented by C++ classes.

(3) To develop the application interface GEO-API which can recognize specific graphic system and can make it compatible with STEP standard data models.

Since 203-FM is built upon different graphic packages, therefore it has different selections on these graphics packages. GEO-API provides the interface sockets to recognize and automatically match different graphic system.

203-FM provides a group of standard logical operations for graphic application programmers. The objects operated by 203-FM are the data models of STEP geometric represented graphics resources. Since GEO-API is developed, these standard operations and the entities represented by STEP data model can be shown by using different graphic systems . (see Fig.3).

From the system structure (see Fig.2) it is shown that CAD application users work upon CAD/CAM system. The CAD/CAM system designers use 203-FM through STEP-API to operate particular graphic package. Considering the CAD/CAM system, all the defined product data conform with STEP requires.

5 Implementation Principles

As shown in Fig.4, STEP graphic resources are originally described by data description language , namely EXPRESS. EXPRESS provides only data description entities without any operations upon these data models. After Compiling EXPRESS model into C++ codes , operations can be applied to the model of C++ codes. C++ has the characteristic of a class enclosure which meets the STEP object-oriented requirement. It is the key work to translate the entities described in EXPRESS of STEP graphic resources data models into that represented by C++ classes .

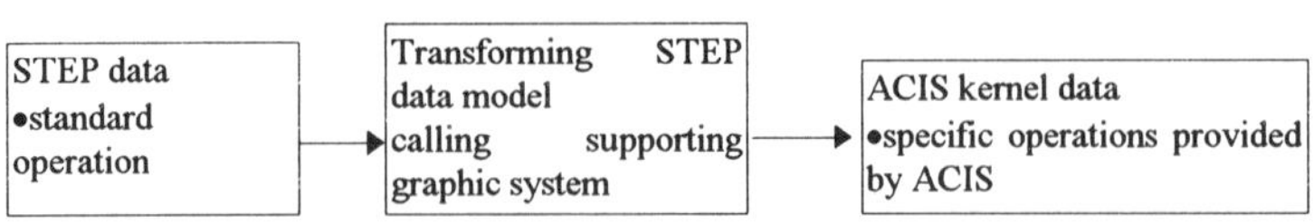

Fig.3 The steps for data model exchanges between STEP and ACIS

203-FM implementing steps are as following:

(1) To map and transform STEP graphic resources data models to the data models of any specific graphic system; First the supported graphic system must have object-oriented graphic data description, than the transformation from STEP graphic resources data models to the data models of the supported graphic system can be done. There are two key transforming methods(or operations): one is the attribute transformation from a STEP graphic resource object to a corresponding object of the supported graphic system. The other method does the opposite thing, which feedbacks the changing of the object made by the supported graphic system to STEP graphic resources, and the same change to STEP corresponding object will be made, i.e.

$$203\_FM\{Os\} \Longleftrightarrow GS\{Og\}.$$

(2) To add operations to graphic resources entities.

As shown in Fig.5, methods adding relies on the supported graphic system because STEP doesn't provide operations on data model. In the implementation of 203_FM, an operator(STEP_GEO_OPERATOR) is designed to call supported operations, such that

$$GS\{M\} \Rightarrow 203\_FM\{M\}$$

By designing a data transformation algorithm including specific graph calling algorithm and generic algorithm, 203_FM graphic operations can be realized. The security and enclosure of the original supported system is still kept. This operator can be represented as:

STEP_GEO_OPERATOR(DMT,INC)

Where,

DMT: Data model transformation algorithm;

INC: Algorithm of methods adding to 203_FM and methods calling to supported graphic system.

(3) To code C++ application programs, test the correctness of the expanded classes and verify the correctness and consistency of the transformation. This work is mainly to design the application program for the expanded classes and make the specific instances described by STEP graphic resources visible on screen so that the correctness of the expanded classes developing can be verified. (4)To build generic geometric platform framework named 203_FM to provide higher abstract level of C++ classes description.

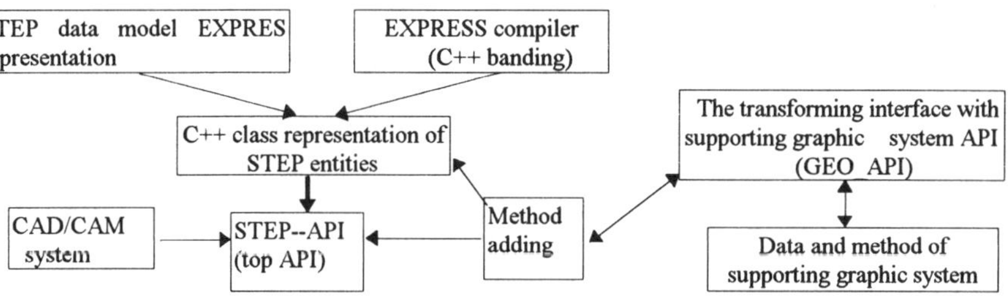

Fig. 4 The principle of 203_FM geometry platform developing

After developing a STEP transforming upon a solid modeling graphics system(e.g. ACIS), we can give a complete operation set on STEP data models, and make the abstract level further. Then we provide the operation set to users as a standard operation interface. This make the interface become one framework form and independent from any specific graphic systems. On the framework the operations is called STEP_API. When users select a specific graphic system, the interface software which transformates STEP operations to that specific graphic system is loaded. The software for these selection and allocation is called a socket of bottom API(GEO_API). For instance: the transformation for ACIS[5] is named ACIS sockets, the transformation for SolidWorks96[6] is named SolidWorks96 sockets, etc., namely 203_FM={O,M,GS,I,R};

Where GS={Sockets(GS1), Sockets(GS2), ...Sockets(GSn),}

(5) To integrate the sockets developed for various graphic subsystems to the generic geometric platform framework(203_FM). This kind of integration means that the management on STEP models transformation can be achieved during the executing of a socket selection on graphic packages. Then the different C++ libraries can be created and selected according to particular needs.

(6) To provide an opening and reusable C++ library. Because the framework is given in a C++ class library form, users can expand the classes directly on the library.

6 Conclusions

(1) In this paper a design project of generic STEP geometric semantic model platform has been given. In fact, it is a implementation plan for graphic resources of STEP/AP203[3]. It makes CAD/CAM programmers develop an application by using STEP data models directly. And in this way the different application programs can share STEP semantic model directly by reference.

(2) The object-oriented technique and basic software component methodology are used during the designing and implementing of the platform defined according to 203_FM structure and its API. This not only ensures to fit with STEP, but also enlarges the open property during the designing and developing of CAD/CAM systems.

(3) According to the conceptions, architectures and principles in this paper, the developed platform has been verified in a STEP based CAD/CAM integrated system.[8]

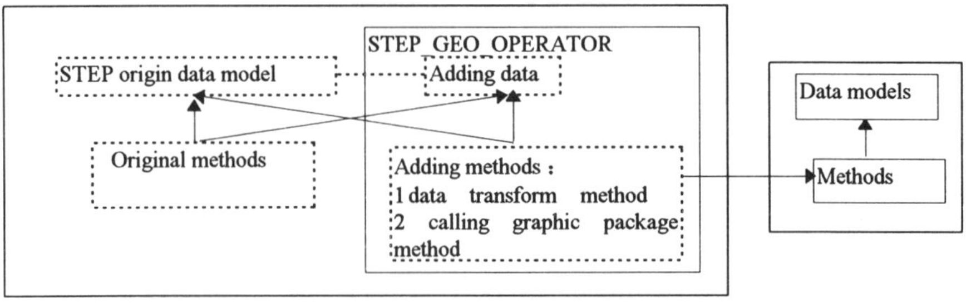

Fig. 5 The concept of method extending

References

1. ISO 10303--21 "Clear Test Encoding of the Exchange Structure"
2. ISO/DIS 10303--11 "Product Data Representation and Exchange --Part 11: The Express Language Reference Manual", 1992,8.
3. ISO/DIS 10303--203 "Product Data Representation and Exchange--Part 203: Application Protocol: Configuration Controlled Design"
4. ISO 10303--42 "Integrated Resources: Geometry and Topology Representation", 1992..9
5. "ACIS Geometric Modular Programmers Reference" SPATIAL TECHNOLOGY INC, 1995.
6. Alan Smith, "A Near look for Solid Modeling" Computer_Aided Engineering, 1996.4 P19--P24
7. "HP--PHIGS Graphics Techniques" Herlett--Packard Co., Vol.1, 1991.1
8. "LONICERA system" , 720 institute of Beijing University of Aeronautics and Astronautics, Beijing, China

A CAD-System for Dressmaking

Karl-Heinz Roediger

University of Bremen, Department of Mathematics & Computer Science
P.O. Box 33 04 40, D-28334 Bremen, Germany
e-mail: roediger@informatik.uni-bremen.de

Abstract. This paper reports about an R&D-project in the dressmakers' trade in Germany. A system for all tasks in small dressmakers' workshops is nearly completed. It covers office tasks as well as CAD of basic and model patterns. The special situation of the target group raised some problems: The final system should be a low cost one; it should also be handled very easily; and special attention should be paid to work organization. These constraints have been solved in a low cost system under Windows 95, which is also suitable as a system for those countries that will no longer be the workbench of the so-called first world.

1 Introduction

In Germany as well as in other - also developing - countries one hardly may find any information technology in small dressmakers' workshops. The only technical instrument apart from a - in the developing countries not everywhere electrical - sewing machine might somtimes be a rotary iron. Nevertheless there are a lot of repetitive tasks to be done in these workshops: Correspondence, bookkeeping, billing, maintaining of customer and supplier data as well as the construction of basic, model and production patterns. Especially the last task, often consisting of variant construction, is time-consuming. The fairly old production methods together with high wage costs have brought the dressmakers' trade in Germany into difficulties: They cannot realize the prices, they should ask for their hand-made clothes, because only few people are able to pay them. On the other hand, the dressmakers cannot pay the wages for their employees because of the prices they actually realize

They now become acquainted with the same problems Middle Europe's clothing industry had suffered from years ago: The production of ready-to wear clothing went over to Eastern Europe or - much more - to Asia; if they could not survive with design and pattern construction they went bankrupt. Now Germany's and also Middle Europe's dressmaking is in a dilemma, which is characterized by a fairly high demand for new clothing on one hand and production costs that are much too high on the other. The demand results from people wishing something different to the ready-to-wear clothes or having a so-called problem figure. The amount of the production costs is mainly resulting from the production methods. Nearly every new dress is constructed »from scratch«, that means, despite the fact that most of the new things are only variants of formerly made ones, the dressmakers start the process by designing the basic patterns followed by the model and the production patterns. This repetitive activities take a long time and are not the dressmakers' favorite ones. Therefore the dressmakers are looking for ways to modernize their old fashioned production methods.

In this situation the dressmakers' guild of Bremen, a county of Germany where - as all-over the country - the trades are organized in specific guilds, asked for help from the University of Bremen in 1993. Together with the guild we have defined a research and development project on »CAD in dressmaking«. In October 1994 this project sponsored by the program »Arbeit und Technik« (»Work and Technology«) of Bremen's county government was started. The objectives of the project are to look into the dressmakers' work organization and to propose new organizational forms if necessary, to investigate the requirements for a computer system and to compile a system that meets these requirements out of standard products or to develop a new one, and to examine the needs for training IT skills and to arrange adequate courses.

2 Work Organization

The project was started with a long phase of work analysis. Every dressmaker was observed and interviewed several hours, during which we got to know big differences in the work process even between the individual dressmakers. They all learned certain methods of pattern construction, mainly »Hohenstein« and »Mueller & Son«; during their working life they have modified these basic methods to their very own one. We - computer scientists - again learned what industrial science is postulating against F.W. Taylor's »Principles of Scientific Management« (1911): There does not exist only one best way for all dressmakers, there are many individual best ones. It was our first lecture from work organization.

The second lecture was: Whatever the individual design style looks like, all clothes can be traced back to basic patterns as they exist for dresses, skirts, suits, jackets, trousers, blouses, shirts etc. The way of constructing a new model pattern is one of varying a basic one by size, by appliqués, by odds and sods like collars, revers, pockets etc.

The third observation was one of a very specific division of labour: Because of the employees' education pattern construction is the privilege of the female masters in every workshop; everthing else is also done by the employees. Because all masters have their own specific method of creating new patterns, because it shows their hand and is often some sort of a trade mark for their workshop, they do not believe their employees being able to do it the same way. Nevertheless the resulting problem is that in cases of the master's absence or illness nobody is able to prepare the sewing of new clothes. If the journeywomen ever had this special skills they have lost it over the years they did not practice it. All participants in our project therefore decided that pattern construction should be no longer a privilege of a single person, because of the possible problems in work organization.

A fourth lecture when looking into work organization was: Whatever somebody is trying to do with a computer in a workshop like those under investigation, work with a computer is only a part of the whole working day of dressmakers. They are doing complete work along the production process of clothing. Except pattern construction they never divide this process neither horizontally nor vertically. Normally the whole process lies in one hand. The uniqueness of every piece might be the reason for such an organization of work.

These observations have several implications for the future work organization and for the requirement definition: First, a new software system must be open for all different methods of pattern construction as well as be dedicated to every individual process in order to give the right support. Second, a new system should contain a database with all basic patterns in order to support variant construction. Third, work with and at the computer must become an integral part of the whole production process. It must be easy to use and exactly match the work routine. Easy switches between pattern construction and office work like maintaining a customer's measures should be possible. These implications might be fullfilled by a new software system. Fourth, to enable the journeywomen for the construction of well fitting patterns is a matter of training pattern construction and IT skills. That means as a result to qualify the journeywomen as well as the masters for the use of computers.

3 COAT - A System for the Dressmakers' Workshops

After the investigation of work organization the existing CAD systems have been studied, especially those ones used in the textile industry, in order to look what features they are offering and how far they probably meet the requirements resulting from the specific work organization in small workshops. Pattern construction, grading and cutting cloth have been automated in the textile industry years ago. The proprietary systems used there are rather expensive, more than small workshops can pay, and they are not easy to use. They require a professional user being familiar with all features of a complete CAD system. The working situation of a dressmaker however is characterized by a high variety of totally different tasks during a day. Moreover, lot size in the clothing industry is always more than one; in dressmaking it is generally one; that means, having constructed the patterns for one piece of clothing a dressmaker will use the computer at the earliest when having finished the sewing for that piece. Because of these circumstances dressmakers are in terms of human-computer interaction research novices and occasional users by sometimes using a computer only a few minutes per day.

The industrial solutions to these problems are designed for a work place, where one specialist is practicing computer aided construction the whole day. This manageress is more or less familiar with the functionality and the user interface of the system. Nevertheless the industrially used systems often are so complicated that even the specialists cannot explain the functionality and the tools: They often do things without knowing their exact meaning, and they must keep in mind a lot of commands and actions in order to do the right things at the right time. Additionally, as in the small workshops most of the clothing enterprises use their own very special way of pattern construction and dressmaking and this method is sometimes implemented in the software. Some dressmakers involved in our project have tried to work with these systems. The only thing they could tell at first was: »Not that way!«.

In order to achieve all possible effects of rationalization by introducing computers we looked for a complete and simple software solution for all tasks in a dressmaker's workshop. That includes all office tasks as word processing, bookkeeping, billing, maintaining of customer and supplier data as well as a specific system for the construction and grading of new clothing. Because of the specific target group this project has some specific constraints.

First, the complete solution consisting of hard- and software must be extremely cheap because most of those workshops do not have any budget for information technology. Second, the solution must be extremely user friendly because nobody of the target group had any experience with computers and - even in the future - will work only a very short time at or with a visual display unit. The system also shall become part of the vocational training of the future dressmakers. Third, the solution must be work-oriented and should not affect their normal work organization, because the main task of a dressmaker is consulting, designing, cutting and sewing. Usually construction is only a small portion of the whole dressmaking process; for the expert women it is some kind of routine matter. Therefore the whole system must be embedded in the normal process of work organization.

Because there is no standard solution for these constraints, the project participants, computer scientists, dressmakers and teachers in dressmaking, have decided to develop a completely new system for a commercial PC environment, called COAT (CAD and Office Applications for the Taylors' Trade). A new training course in the use of a personal computer for administrative as well as for constructive tasks in dressmakers' workshops was developed. An evolutionary and participatory way of system development was chosen in order to fullfill the requirements of work organization and usability.

Concerning the first constraint, the upper limit for hard- and software investment in our project was about 5.500 USD. The hardware platform we use in our project is a Pentium PC with 16 MB RAM, 1 GB hard drive, a CD-ROM drive, a PCI graphics accelerator and a 17" or better 21" monitor; the operating system is Windows 95®. All together this may cost less than 3.000 USD today (Halfhill 1996). The difference between both sums should be more than enough for the purchase of standard software and the development of new dedicated CAD software, because the development is sponsored by the county government of Bremen.

3.1 Software Development

Looking around one may find a lot of standard software. But a complete and unified solution for all tasks in a dressmaker's workshop as mentioned above cannot be found. That means either to solve only parts of the problems or to force casual users to learn more about computers, operating systems and interfaces than necessary and opportune. Therefore it was at first decided to create a new unified system with a uniform graphical user interface. This decision is a great opportunity to create an adequate solution for the target group and a big challenge at the same time.

The main objectives of all software development are usability and flexibility. Concerning the frequency in which a PC tool is needed in a dressmaker's workshop an easy-to-use interface is of great importance. To create such an interface the whole system is implemented in an uniform way under Windows 95®. Most of the necessary interactions are of the types »drag and drop«, »cut and paste« or »point and click«. No commands must be learnt by heart; what can be done with the system is offered in menus, lists, and buttons. Whenever possible defaults are underlayed. All tools are self-explanatory - because of the terminology - for professional dressmakers; additionally they are explained in a help system. To enable the use of the system in other countries it has a language switch: any given language may be easily read in and used; until now it is realized in English, German and Italian. A Portuguese and a Spanish version are in progress. In one word: we have tried to realize the state of the art in human-computer interaction (Preece et al. 1994, Shneiderman 1992).

According to the working situation with alternating activities a very flexible system, which allows immediate changes between the different tasks, is necessary. Figure 1 presents the overall structure of COAT. It consists of two parts: an office and a workshop one. Correspondence, customer file, bookkeeping, stock, and supplier files are included in the office part; it is mainly designed out of standard packages like a word processing system, a database management system, spread sheets etc. together with connecting parts implemented via MS Access®. The central module of the total system is the database management system which is also serving the workshop part, the CAD module. The database consists of data about customers (adresses, measures, individual pecularities etc.), suppliers (materials, prices, discounts, lead time etc.), materials (stock, prices etc.), patterns etc.

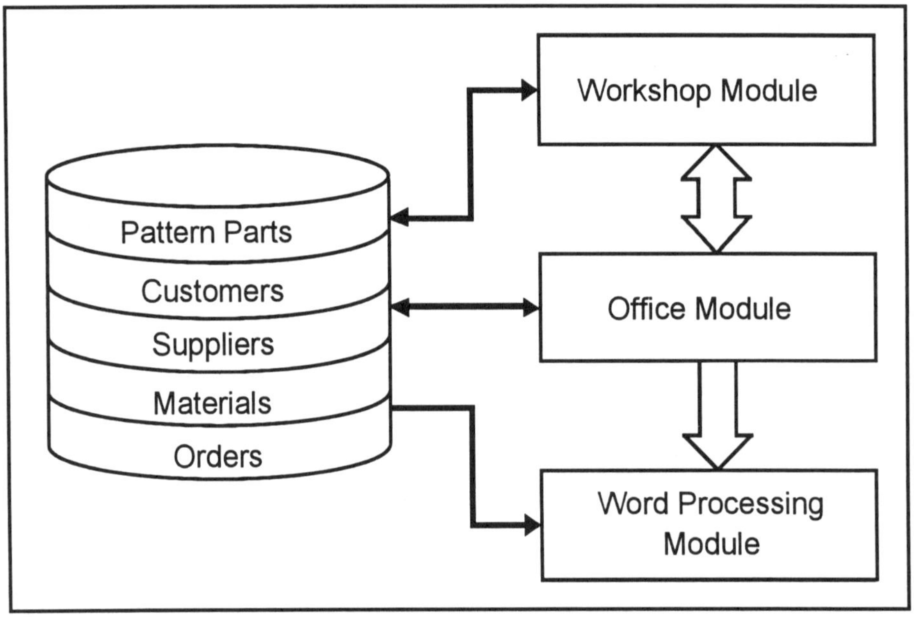

Fig. 1. Structure of COAT
The broad lines and arrows denote procedure calls between components of the overall system, the slim ones are showing data interfaces.

The workshop part consists of the CAD module together with the different data- bases for basic, model, and production patterns as well as for the bits and pieces. The database is the connecting link between workshop and office. Pattern construction can be supported automatically (size), partly automated by presenting databases for the small odds and ends, and by special construction tools with which the patterns can be modified just as dressmakers are doing it manually now. Nobody is forced to use the basic patterns, all can be done with the construction tools as well. The system also supports exploration and experimental action; because all construction steps are recorded in the background every state may be reconstructed by a sophisticated undo function.

Moreover, the dressmakers may store their newly created model patterns in a further database in order to use them for other customers or some years later for the same ones according to the first theorem of fashion »It has all been done before«. As well as in, for example, mechanical engineering, the basic idea of such a software design is that of variant construction. Most of what dressmakers are doing is in that way; designers however may additionally use the construction tools. As well as we know from the dressmakers examining the prototypes, the system is exactly fitting their problems. In most cases it reduces the time spent for pattern construction from hours to seconds.

Flexibility has a twofold meaning: The system must be flexible according to different styles of working as well as flexible in itself. The first kind of flexibility has been explained before. The second one is realized by multiple entries and multiple transitions offered by the system. Whatever one wants to do, she or he may start with or change to it. If during the construction process a dressmaker notices that material is missing, she may directly write an order; if a special basic pattern is missing in the pattern database, the dressmaker may construct it from scratch by using the construction tools.

3.2 Design Knowledge

The experiences with CAD systems in manufacturing have shown that geometric information about the final product design is not sufficient, especially in situations in which someone is reusing an old design, for example in variant construction. As mentioned before, variant construction is also widely spread in fashion design. In order to correctly use older designs or patterns dressmakers as well as manufacturing engineers need information about it at a semantically higher level than the pure geometric elements typically offered by CAD systems. Moreover, because dressmakers are not familiar with hyperbolic functions and splines, geometric data about patterns are of no use for them. Several attempts were made in CAD research to preserve the design knowledge in terms of knowledge-based systems (Kratz 1991), in terms of features of the product design (Mäntylä, Dana, Shah 1996) or in terms of macros, scripts and meta languages in which the design process can be documented.

In COAT we have chosen scripts, a meta language which can be read and understood by the dressmakers. The scripts have been especially designed and evaluated for this application. A script is automatically generated during every pattern construction process; it represents the process as an ASCII text. At the same time the scripts realize the undo function as they support the dressmakers by giving them that information about a pattern which they need to create a variant. With these scripts we also solve the grading problem: Instead of calculating the nonlinear process of grading with numerous tables of jump in values we take these scripts, feed them with the measures of the new size as parameters and reconstruct the model by interpreting the script. Dressmakers may complete or modify their constructions by using the script language, the elements of which are offered in drop down lists at every construction step. The only input they are forced to are parameters (length, width, height in cm).

3.3 Participatory Design

The whole process of software development is organized in a participatory, prototype-oriented and iterative way (Schuler and Namioka 1993). After an extensive phase of work and requirements analysis the description of work and the requirements definition were written by computer scientists and evaluated by dressmakers. The allover design of the new system was done in cooperation. The system specification was again elaborated by computer scientists with an object-oriented tool. This has formed the basis for the first prototype presented to the dressmakers. After having finished the first course in using computers they have evaluated this prototype in their workshops and criticized it in joint meetings.

The best results of these critique meetings are achieved when the discussions take place in small groups. This critique together with the scheme for iterative development again form the plan of action for the next phase. A new version of the total system is offered to the dressmakers approximately every two month. The office part is now stable; there is no further development. The part under iteration is now only the CAD system, which is implemented in Microsoft Visual C++®.

Participatory development is without doubt the best way to develop efficient and usable systems. But also without doubt, it is time-consuming for both sides, developers and users. Especially in small and medium-sized enterprises time is money: If one person is absent sometimes work stands still. Regarding this, the demands for time e.g. for project management, introductory courses, or participatory development should be handled very carefully.

4 Qualifying for the Forthcoming Challenges

A bigger problem was that no dressmaker in the project had ever worked with a computer before. They are familiar with sewing machines and rotary irons; personal computers however have been a totally new experience. All participants in the project are however convinced that the only possibility for the survival of the taylors' trade in Middle Europe is to renew the old fashioned production methods by introducing computers and software systems into the workshops as quick as possible. It is often argued that the introduction of information technology destroys workplaces. As it is done in this project no workplace is at risk; on the contrary we hope to create new workplaces because the situation concerning orders is very good. If the workshops can produce the clothes in a shorter time and therefore more economically some new workplaces may be realized. The central aspect of these expectations is that all dressmakers, masters as well as journeywomen, achieve as much IT skills as they need in order to rationalize their production methods. This in mind the first training was the most important step during the project. If it had failed we could have stopped the project.

So a special introductory course in using a computer had to be designed very carefully. In this training course a lot of time was spent with practicing the handling of mouse, cursor, and graphical user interface. Although the dressmakers are accustomed to handling needles and threads some of them have had problems with their micro motor avtivity. We succeeded in arousing and preserving the dressmakers' enthusiasm for the new media by working out the course strictly problem-oriented. We did not teach the use of an operating system, of a graphical interface or of a word processor. We took problems from their working life as designing sheets of paper with individual logos, maintaining the customer file, billing, advertising for new products, ordering of material, constructing simple patterns etc. The first course covered twelve evenings three hours each; presently we offer two evenings every second month in order to refresh the skills, to discuss problems and to teach new features of the incrementally developed CAD system. Regarding the capabilities of the dressmakers it seems to be sufficient.

Because pattern construction shall no longer be a privilege of the masters we also integrated the journeywomen of the workshops involved in the project into these training courses. They have passed them with the same, sometimes with more enthusiasm than their employers. Together with teachers we are now elaborating new elements for the primary vocational training of dressmakers concerning the use of a PC in administrative and constructive tasks of a workshop. The vocational school where the first training will be offered is well equipped with personal computers. The school will use COAT for primary training.

5 Outlook

As mentioned before, we are developing COAT in an evolutionary way: The dressmakers are using preliminary versions of the system during their normal work. They criticize these versions during our regular meetings and we, academic researchers, improve these versions in a spiral approach (Boehm 1988). The project ends in spring 1998; then the system will be completed and evaluated.

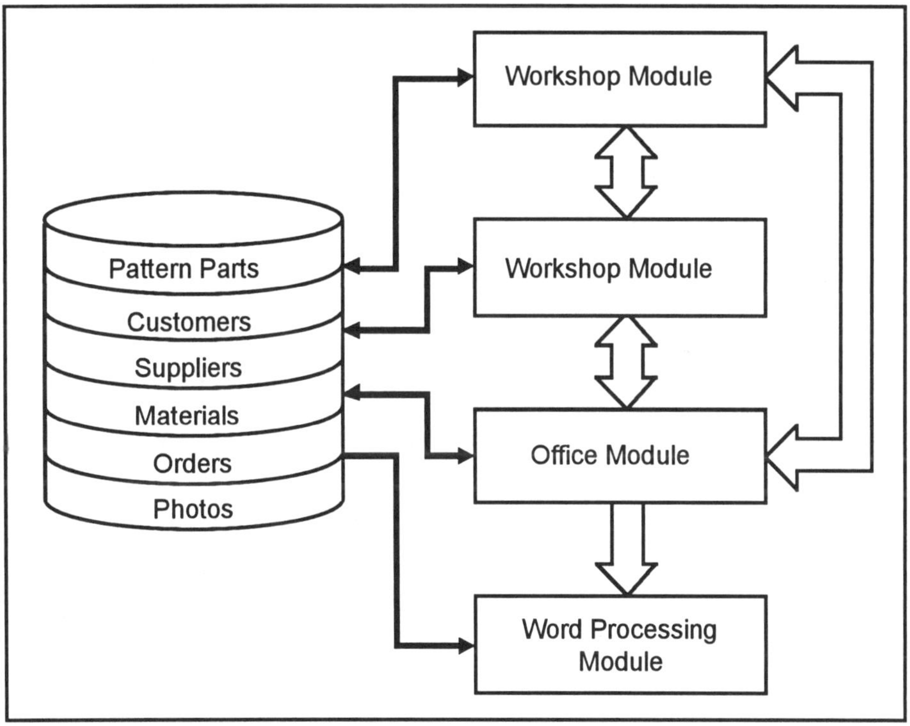

Fig. 2. Structure of COAT and CADwalk

In the meantime, we have started a new project in the same domain called CADwalk, a term which is derived from Computer Aided Design and catwalk. With this project we intend to support the consulting process between a dressmaker and her customer. We take photos of a customer with a digital camera, project them onto a monitor, and clothe the person with the new dress which she has just selected. In doing so, we provide a fast impression of the new garment before the first fitting. So the customer can see whether the new dress meets her requirements in shape, material, colour, and design. Because the computation or simulation of the material properties as for example the fall of the folds or of the material at all in reasonable time so far is an unsolved problem on PC platforms, this development will take a longer time. Figure 2 shows the structure of both systems, COAT and CADwalk. CADwalk is only a module operating on the same database which now is enlarged by photos of customers. The dressing is done by converting the two-dimensional patterns of COAT into three-dimensional clothes in CADwalk.

6 References

1. B.W. Boehm, A Spiral Model of Software Development and Enhancement, *IEEE Computer* **21** (1988) No. 5, pp. 61-72.
2. T.R. Halfhill, Inside the Web PC, *BYTE* **21** (1996) No. 3, pp. 44-56.
3. N. Kratz, Architektur eines wissensbasierten Systems zur Unterstützung der Konzeptionsphase in der Konstruktion, Dissertation, Universität Kaiserslautern, Kaiserslautern 1991.
4. M. Mänttylä, D. Nau, and J. Shah, Challenges in Feature-Based Manufacturing Research, *Communications of the ACM* **39** (1996) No. 2, pp. 77-85.
5. J. Preece, Y. Rogers, H. Sharp, D. Benyon, S. Holland, and T. Carey, Human-Computer Interaction, Addison-Wesley, Wokingham 1994.
6. K.-H. Roediger, Work Organization and IT Skills in Dressmaking, invited Paper for the International Conference on Information Technology for Competiti-viness, Florianopolis, Brazil, June 19-21, 1997.
7. K.-H. Roediger and U. Szczepanek, CAD in Dressmaking, in: R.J. Koubek and W. Karwowski (eds.), Manufacturing Agility and Hybrid Automation I, IEA Press, Louisville, KY 1996, pp. 620-623.
8. D. Schuler and A. Namioka, Participatory Design: Principles and Practices, L. Erlbaum Associates, Hillsdale, NJ 1993.
9. B. Shneiderman, Designing the User Interface, 2nd. ed., Addison-Wesley, Reading, MA 1992.
10. U. Szczepanek und T. Schriefer, Integrierte CAD-Lösungen für Klein- und Kleinstbetriebe, in: P. Brödner, H. Paul und I. Hamburg (Hrsg.), Kooperative Konstruktion und Entwicklung, R. Hampp Vlg., München 1996, S. 137-143.
11. F.W. Taylor, Principles of Scientific Management, Harper & Row, New York 1911.

A New Generation CAD System for Railway Weighing Apparatus and Its Implementation

Meng Xiangxu   Li Xueqing   Gong Bin

CAD Laboratory,
Institute of Computing Technology, Academia Sinica, P. R. China
Department of Computer Science, Shandong University, P. R. China

Abstract. Railway weighing apparatus (RWA) can be widely used in various plants, coal mines and railroads. The main dimensions of RWA often need to be changed to adapt to the application circumstance, but all of RWA has the similar topological structures. In this paper, we describe a new generation CAD system model for RWA based on knowledge representation and parametric geometry technology. The parametric geometry modeling can support complex drawing representations. The implementation of the system is also described.

1 Introduction

Railway weighing apparatus (RWA) can be widely used in various plants, coal mines and railroads. The main dimensions of RWA often need to be changed to adapt to the application circumstance, but all of RWA has the similar topological structures. The product design is expensive and error-prone tasks because of complex calculations and engineering drawings. In this paper, we describe a new generation CAD system model for RWA based on knowledge representation and parametric geometry technology. The new parametric geometry modeling based on directed hypergraph is a major component of the system. It can support the representation of dimensional constraints and geometric constraints. Moreover, the geometric structure changes and design constraints can also be represented. The parametric geometry modeling is finished while interactively constructing geometric objects. The constraint consistency is ensured by construction steps. Overconstrained or underconstrained sketches can not take place. The parametric geometry modeling can support complex drawing representations.

The CAD system for RWA is composed of four parts: *user interface*, *specification analysis and expert system*, *graph deduction and dimension parameter value transfer*, *product drawings output subsystem*.

User interface is used to accept user input (such as: product specifications and select design constraints) and provide a friendly interactive environment.

Specification analysis and expert system(SAES) is used to analyze product specifications and make a reasoning to calculate the main dimensions and basic structures of the RWA. A rule-based method is adopted to select product types and structural parameters. Experiential knowledge and experiential formulae can be easily incorporated into the subsystem. In the process of reasoning, user can also change and select the key value of product parameters interactively. So, it is fast and convenient to make a new product design in the system.

Graph deduction and dimension parameter value transfer(GDPT) is used to create correct engineering drawing instances from parametric drawings. It accepts initial main dimensions from the SAES and transfer to parametric assembly drawing. Then solve the parametric assembly drawing by graph deduction to make an instance of the parametric drawing, and transfer the new dimension derived from the drawing to the subassembly drawings. Finally all of the parametric drawings are calculated by the subsystem and create the engineering drawing for the product.

Product drawings output subsystem is to adjust the drawings and output it to plotter or display. User can also transfer the design results to other CAD software system (for example, AUTOCAD).

2 The Design Flowchart and System Architecture

2.1 The Traditional Design Flowchart

In the railway weighing apparatus (RWA) design fields, the engineering drawings of a family of products are similar and can be determined by several key parameters. Considering the structures of RWA, the structures of 100 tons RWA are similar to 30 ton's. The differences between them are the changes in specification for rails, steel plates, angle steel and I-steel, in the number of spandrel girders and rails. Figure 1 shows traditional design process.

According with the design flowchart, the parametric CAD system should satisfy with the following conditions:

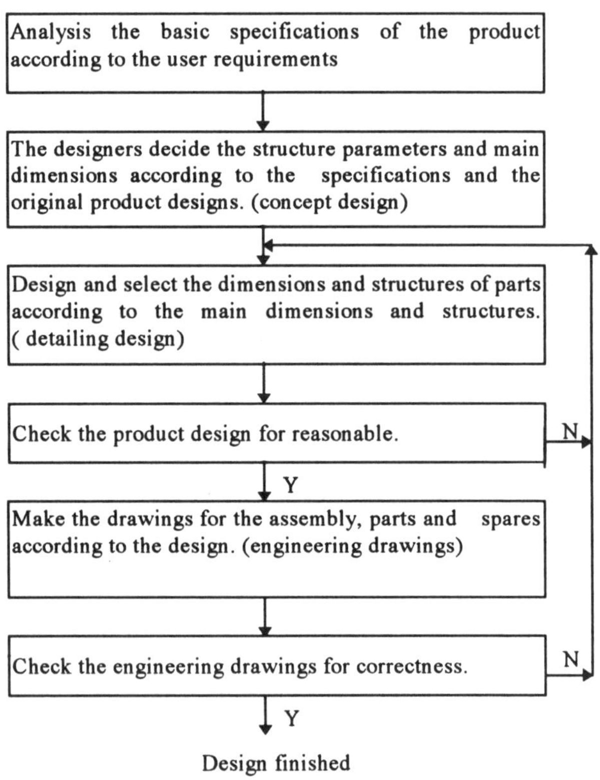

Figure 1

(a) The modeling should combine the traditional design calculation, optimization design and engineering drawings together.

(b) The parameterization of drawings can support the topological changes for the structures of products.

(c) The engineering drawing instance can be created automatically according to the given parameters.

(d) The modeling should ensure that the relative dimensions and parameters of spares, parts and assembly are consistent.

2.2 The Parametric CAD System Architecture For RWA

In view of above-mentioned reasons, the new generation parametric CAD modeling for RWA is as follows (as show in Figure 2):

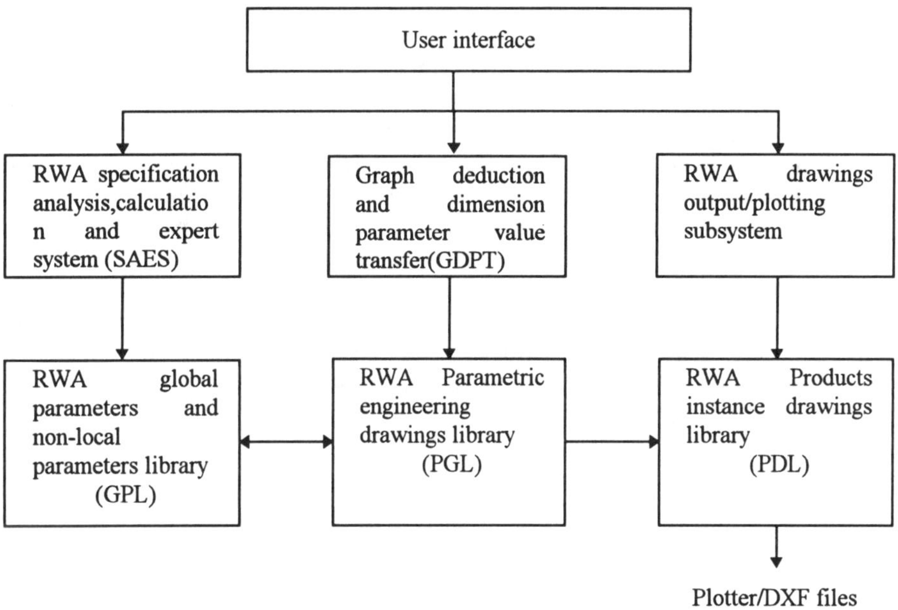

Figure 2

2.2.1 Specification Analysis, Calculation And Expert System

Specification analysis, calculation and expert system(SAES) is used to analyze product specifications and make a reasoning to calculate the main dimensions and basic structures of the RWA. A rule-based method is adopted to select product types and structural parameters. Experiential knowledge and experiential formulae can be easily incorporated into the subsystem. In the process of reasoning, user can also change and select the key value of product parameters interactively. So, it is fast and convenient to make a new product design in the system.

There are two kinds of design processes. One is certainty process that is basis on the numerical calculation, experiential formulae, FEM analysis and optimization. The other is concept design process that is basis on the design knowledge and logical reasoning. The design can be considered as a recursive process between synthesis and analysis.

The subsystem was written by PROLOG. The input data source can be specifications, dimensions and design constraint. By the execution of the system, the basic structures, main dimensions and other parameters can be gotten. The results were written to the RWA global parameter and non-local parameter library.

There are three kinds of knowledge in the knowledge library.

(a) RWA construction and protocol knowledge: to describe the RWA characteristic and the selection of basic structures.

(b) Numerical knowledge: the numerical and dimensional constraint among the parts.

(c) Design rules: the standards and rules for the RWA and the design experience.

2.2.2 Parametric Drawing Library And Parametric Drawings

A parametric drawing library (PGL) is used to store all the parametric drawings for a kind of RWA. It is the kernel for the parametric CAD system. The PGL library was build by the parametric drafting program generator (PDPG). Each parametric drawing can retrieve, read and write the global and non-local parameter library. It is easy to transfer and to share dimensional and structural parameter among parametric drawings. Because the expression calculation and variable assignment are supported in the drawing, the experienced formulae for design and dimensional deduction can be expressed directly in the drawing. The result dimension value can be passed to other drawings by the global and non-local parametric library. The initial parameter value of each drawing can be obtained from global parameter library or by user input interactively.

2.2.3 Graph Deduction And Dimension Parameter Value Transfer

Graph deduction and dimension parameter value transfer(GDPT) is used to create correct engineering drawing instances from parametric drawings. It accepts initial main dimensions from the SAES and transfer to parametric assembly drawing. Then solve the parametric assembly drawing by graph deduction to make an instance of the parametric drawing, and transfer the new dimension derived from the drawing to the subassembly drawings. Finally all of the parametric drawings are calculated by the subsystem and create the engineering drawing for the product.

2.2.4 Product Drawings Output/Plotting Subsystem

Product drawings output/plotting subsystem is to adjust the drawings and output it to plotter or display. User can also transfer the design results to other CAD software system (for example, AUTOCAD).

3. RWA Parameter Modeling and Dimensional Deduction

The aims to the RWA CAD system are:

For a group of given specifications and main dimensions, all the specified dimensional value of each drawing can be obtained automatically by the design calculation and constraint expression deductions based on the dependence relationship among parametric drawings.

For the RWA products design, the dependence relationship can be expressed by a tree structure. Figure 3 shows a typical structure for a RWA.

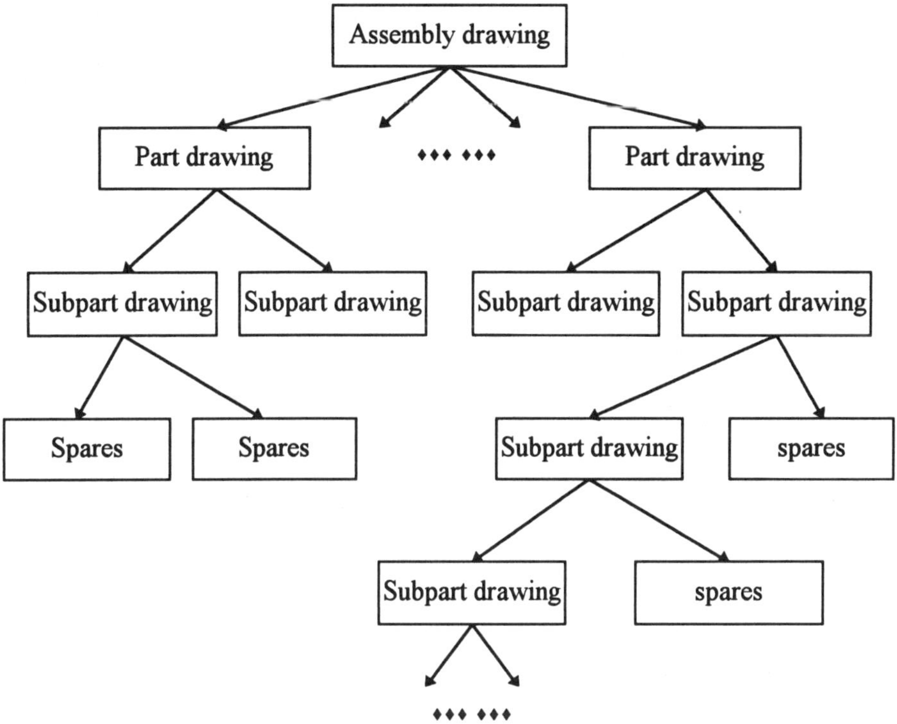

Figure 3 Typical structure for a RWA

Because the PDPG system supports expression deduction automatically, dimensional deduction can be realized as follow:

(a) Scanning and processing each drawing on the node of dependence relationship tree by breadth-first search.
(b) For each drawing, read each parameter initial value from the GPL and obtain freedom parameter from user interactively. Calculate the constraint expressions and make the graph deduction to create the instance of parameter drawing. Write new design parameters to the GPL and shared by other sub-drawings.
(c) Store the result instance drawing to the PDL for plotting. The drawing can also be converted to other system data format for further processing (Such as AUTOCAD, Pro-engineer).
(d) Repeat scanning from (b) to (d) until all the node was handled.

By the end of processing, the whole set of specific detailing design drawings for RWA was obtained. The whole calculation course was finished in several minutes.

4 The Implementation and Examples

A parametric RWA CAD prototype system based on the model has been developed by the researcher. It was written in C++ language and PROLOG. It can run on IBM PC compatible computer with DOS or WINDOWS 3.X operating system. It has been successfully used to create the parametric engineer drawings for various types of RWA. Figure 4 show the engineering drawings produced by the system.

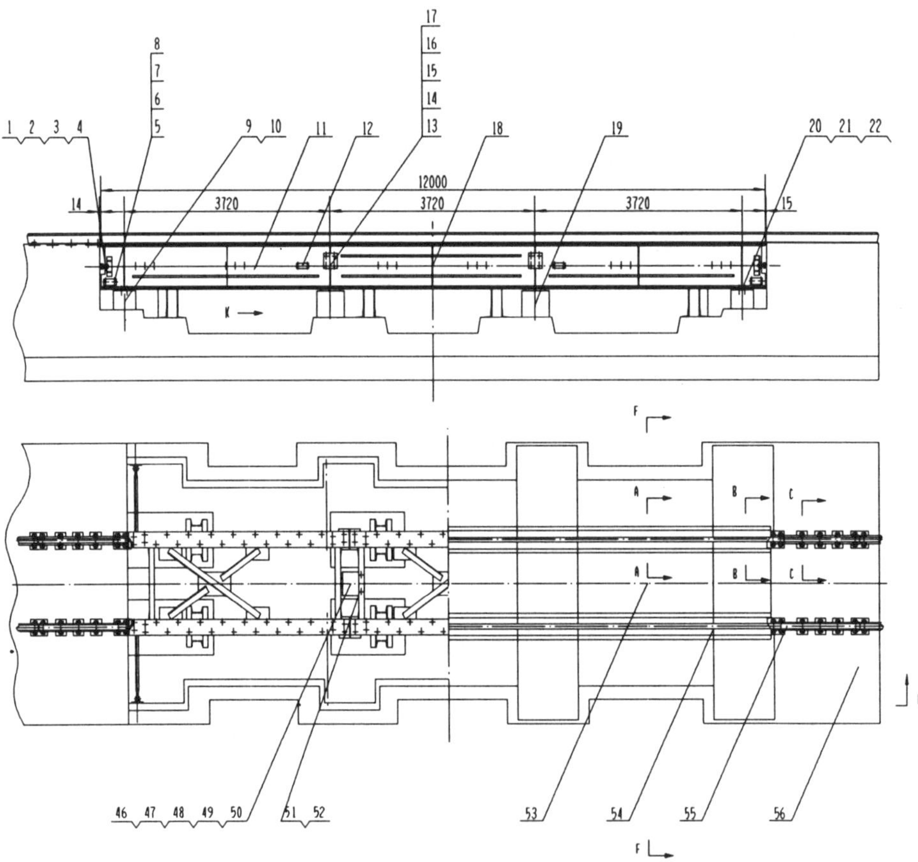

Figure 4 The assembly drawing of RWA produced by the system. When the main dimensions are changed by users, all of the drawings in a RWA will be regenerated to satisfy with the constraints.

5 Conclusion

In this paper, we describe a new generation CAD system model for RWA based on knowledge representation and parametric geometry technology. The new parametric geometry modeling based on directed hypergraph is a major component of the system. It can support the representation of dimensional constrains and geometric constrains. Moreover, the geometric structure changes and design constraints can also be represented. The parametric geometry modeling is finished while interactively constructing geometric objects. The constraint consistency is ensured by construction steps. Overconstrained or underconstrained sketches can not take place. The parametric geometry modeling can support complex drawing representations. The implementation suggests that the modeling is practical and convenient for parametric geometry design. It can be used to realize various products oriented parametric designs and drafting. The model will be extended to 3D Geometry to support parametric solid CAD modeling.

6 Acknowledgments

The project was supported by Shandong Natural Science Foundation. We gratefully acknowledge their support. The author also thanks all research fellows for many helpful discussions and suggestions.

7 References

1. V. C. Lin, D. C. Gossard, R. A. Light, Variational geometry in computer aided design, computer graphics, 15, 3 (1981) 171-177
2. R. A. Light, D. C. Gossard, Modification of geometric models through variational geometry, CAD, 14, 4, (1982) 209-214
3. B. Aldefeld, Variational geometries based on a geometric-reasoning method, CAD, 20, 3, (1988) 117-126
4. A. Verroust, F. Schonek, and D. Roller, Rule-Oriented Method For Parameterized Computer-Aided Design, CAD, 24, 3, (1992)
5. X. Meng, J. Wang, A New Generator For Parametric Drafting Program, CADDM, 2, 1, (1992) 38-45
6. D. Roller, An approach to computer-aided parametric design, CAD, 23, 5,(1991) 385-391
7. H. Suzuki, H.Ando, F. Kimura, Geometric constraints and reasoning for geometrical CAD system, Comput. & Graph, 14, 2, (1990) 211-224
8. K. Martini, Hierarchical geometric constraints for building design, CAD, 27, 3, (1995) 181-191

Designing a Computer-Aided Product Design Advisor

Tay Eng Hock

Department of Mechanical Engineering
National University of Singapore
10 Kent Ridge Crescent
Singapore 119260

Abstract. In this paper, the specifications and the customer requirements of a computer-aided product design advisor are identified and described. These specifications are derived from a survey and experiential interviews conducted in a design firm.

After identifying the needs of potential users, a proposed integrated architecture for design capture and analysis is described. The architecture will provide the development tools and methods to build and integrate design tools. A designer will be able to easily make enquiries to the common knowledge sources containing design rules and guidelines, manufacturing capabilities, existing products and parts, and a classification scheme. A common user interface will provide communication between the user and the underlying tools. The architecture will evolve over time and development will start with a common representation of the data and interfacing mechanisms.

1 Introduction

Two major goals of firms in the nineties are to significantly reduce product cost and time to market (TTM). To meet these goals, quality product designs will have to be developed that will minimize the impact on the manufacturing facilities allowing for rapid product introduction. The design community has recognized that today's design tools and environments are insufficient in meeting their future needs.

The typical design evaluation tool can only deal with single parts and do not consider the system aspects of the design. Further, much of the data to provide answers is not readily available from design capture systems, such as UG-II™ or Pro/Engineer ™, and therefore must be entered again into these evaluation tools. In many cases, this occurs too late into the design cycle to impact the design positively. If these questions and subsequent answers were asked earlier during the actual design process, then manufacturing considerations would be easier to incorporate into the product's design.

From the survey of user needs and subsequent reports, there exists a very large demand to implement a tool that will allow for quick location of standardized parts being able to locate using non-geometric attributes. Further, there is no single product/part design process among those surveyed. Different designers use Computer-Aided Design (CAD) tools in different ways: some to capture the design beginning with the conceptual design, others do the conceptual design on paper, and transfer it to a CAD system as the design becomes more detailed, while a third group does the conceptual and detailed design work on paper, and loads it to a CAD system only when the design has been detailed.

Therefore, the computer-aided product design advisor should have access to Product/Part Information stored in the CAD systems, to detailed manufacturing process information, and to a battery of design evaluation tools such as tolerancing and costing. A system for determining overall design "goodness" based on predetermined criteria will be included.

2 Design Tools

Design tools fall into the following categories:

1. Design capture - where the design is input into some system
2. Evaluation/analysis - where the design is measured for some variable(s)
3. Design rule checkers - where the design is checked for violations of design practices and standards
4. Advisory tools - where the design is analysed and suggestions are provided to produce a better design
5. Synthesis tools - where the design is generated by the tool, given a set of design specifications.

In the mechanical design space the technology is in the infancy stages of design capture and evaluation/analysis tools. In the electrical arena, where the problem is more constrained and perhaps better defined, tools have advanced to the point where design synthesis has been partially realised. This is a component of the future design system that will allow the designer to perform a sustained high level of performance creating a design and have the system perform the more mundane tasks automatically.

3 Design Survey Results

The survey was targeted at three groups within the company: product/part designers, manufacturing process designers, and design tool developers. Distribution of the survey was by hardcopy, electronic mail, and posting to the Design for Assembly (DFA) and Design For Manufacturing (DFM) notes files. Responses were collected by written response, electronic mail, and many personal interviews. Due to the complexity and length of the survey, and the workloads of the respondents, some difficulty was encountered in completing the surveys.

Some of the findings are categorised and summarised below:

The Product/Part Design Process

There is no single Product/Part Design Process within the company. Different designers use Computer-Aided Design (CAD) tools in different ways. Some to capture the design beginning with the conceptual design, others do the conceptual design on paper, and transfer it to a CAD system as the design becomes more detailed, while a third group does the conceptual and detailed design work on paper, and loads it to a CAD system only when the design has been detailed.

In many cases, the difficulty in finding parts to perform a function, cause designers to design a new part. Since most designs are targeted for a specific plant, even more proliferation can occur. Likewise, the difficulty in locating a standard design for a part feature, causes these to redesigned also. These duplications generate unnecessary costs.

The design function that an organisation performs varies widely within the company. Few perform only a single function such as new designs only or even Engineering Change Orders (ECOs) only.

Manufacturing capabilities are factored into designs by interaction between the designers and manufacturing engineering. This is a manual interaction. This works well for products that are introducing a new manufacturing process, but those using an existing process often receive attention too late in the design process to influence the design.

Design tradeoffs should be tracked, including the reason for a decision made. A design advisor tool could help the designer to evaluate the different tradeoffs.

Product/Part Design Parameters

The majority of those surveyed indicated that there were at least nine different design parameters. When asked to indicate which were critical design parameters, the majority could agree on only two parameters: cost and product introduction time.

The emphasis on different design parameters changes from design to design, and frequently changes during a design project. Designers are frequently trying to hit a moving target of design parameters.

Design Advisor Requirements

A design advisor could improve the design process in a number of ways. By reducing Time to First Revenue Shipment (FRS), reducing cost, increasing product quality, and improving the design, among other things.

The design advisor should have access to product/part information stored in the CAD systems, to the detailed manufacturing process information, and to a battery of design evaluation tools (EMI, Tolerancing, Costing, and others), including a system for determining overall design "goodness". It should include both a common parts library with detailed part information about existing parts (geometry, function, requirements, specifications, etc.), and a common feature library with details of common features that could be incorporated directly into designs in a CAD system.

The design advisor should be flexible enough to be used by a designer, no matter what their designing style, the type of design function they are performing, or the kind of product or part they are designing.

The interface to the design advisor should be flexible. The users should be able to indicate how much help and design assistance they want, and when their designs should be evaluated (in an ongoing mode, when saved, on demand, etc.).

Summary

the survey results clearly pointed to the need for a design advisor system that will check designs early for various considerations such as DFM. A design advisor system could improve the availability of manufacturing process information to design suggestions based on industry and corporate standards. This would result in a number of improvements in the design process: increased product quality, reduced time to FRS, reduced design time, reduced numbers of common mistakes, reduced design duplication at both the part and feature levels.

4 Proposed Design Advisor

A design advisor is required that offers design suggestions to the product designer that assists in attaining cost and other product goals/metrics. The design shell will integrate and utilise current evaluation tools to perform product analysis. Furthermore, tools will be developed that perform evaluation and estimation on incomplete designs, such as conceptual designs, and suggest design alternatives thereby assisting to produce a better final product design. In its final form the design advisor will greatly enhance the manufacturing to engineering relationship by providing the design engineer with specific manufacturing information at the time of design. It will provide design engineering with estimates of manufacturing costs, recommended manufacturing process, and provide manufacturing process engineers with product information. It will advise in areas of expertise such as safety, plastics, and sheet metal as well as methods to reduce process steps or improve manufacturing yield. Rules included will be analytic, comparative and deductive in nature. Included with each rule will be an impact assessment, both positive and negative, so that the designer can make an informed decision whether to apply the rule or not.

Furthermore, the design advisor should be an overall architecture which integrates CAD tools, point out design solutions, workbenches, and part databases. It provides and automates data translations/extractions, manages the search for data, and manages remote processing. It does this focused at the design attribute level and not at the tool or file level.

5 Design Methods

There are several approaches or methods that one could take for the design advisor. There re three methods chosen to allow for incremental development of the total design advisory system. The methods are: post-design, palette-design, and embedded-design. The post-design method utilises feature extraction from CAD files to generate input to evaluation modules. This is similar to the traditional CAD post processing methods. The pallette-design method utilises a context based menu it guide the designer in the selection of desired components. The embedded-design method utilises tight integration between the modeling tool(s) and evaluation tools by focusing at the design attribute level. The embedded-design method is the long term vision for implementation of design advisors.

6 Design Structure

There are several options for structuring advisors and rule bases. One option is to build a single advisor inference engine with a single rule base providing just one set of applicable rules. Another is to build several smaller domain specific advisor inference engines with their own separate rule base providing multiple sets of applicable rules. A hybrid approach is to develop the smaller domain specific advisors and couple them into a domain independent advisor each with their own rule base. Each of these advisory implementation options would operate on the current product model continued within the blackboard.

The first option single advisor with a single rule base has the following advantages and disadvantages:

- advantages
    1. one advisor to support
    2. one rule base to support
    3. able to deal with multiple domain tradeoffs through conflict resolution
    4. incremental knowledge
    5. intermediary between the product model and the rule base
- disadvantages
    1. potential size problem due to the number of different domains required
    2. must validate new knowledge using the complete advisor
    3. all domains may not be best represented in a single representation paradigm

4. may have to extract the rules from the many existing advisors, some of which may be unknown or embedded.

The second option, multiple advisors with independent rule base has the following:

- advantages
  1. independent advisors, one per domain area
  2. modular
  3. independent testing and validation
  4. able to utilise each advisor independently
  5. able to support multiple types of advisors, such as externally developed advisors that use a foreign structure
- disadvantages
  1. may have many duplicate rules
  2. no mechanism for resolving conflicts between competing domains
  3. no mechanism for supporting interaction between domains.

The third option or the hybrid approach:

- advantages
  1. one rationalisation advisor with several independent domain specific advisors
  2. all of the positive points of the second option
  3. the rationalisation advisor can be added later when conflict resolution and interactions are understood better, this gives potential for applying learning to a rationalisation advisors
  4. ease of expansion by adding in domain specific advisors
  5. could potentially build a hierarchy of advisors that fully segment the rule sources.
- disadvantage
  1. may be difficult to obtain the methodology for resolving conflicts between domains.

The third option has been selected as the structure for developing the advisor.

7 Conclusion

The survey has clearly indicated the need for a design advisor that will impact the market place in terms of product that are cost effective and timely. While the characteristics of such a design advisor are being worked out, the design structure and design method has been adopted to be the hybrid structure and the embedded-design method.

References

1. Finger, S. and Dixon, J.R., A Review of Research in Mechanical Engineering Design, Part 1: Descriptive, Prescriptive, and Computer-Based Models of Design Processes, Research in Engineering Design, Vol. 1, pp. 51-67, 1989.

2. Pugh, S., Total Design: Integrated Methods For Successful Product Engineering, Addison-Wesley Publishing Company, 1991.

3. Sriram, D. et al, An Object-Oriented Framework for Collaborative Engineering Design in Computer-aided Cooperative Product Development, MIT, Cambridge, USA, Springer-Verlag, 1989.

4. Tomiyama, T., Kiriyama, T. and Umeda, Y., Toward Knowledge Intensive Engineering in Proceedings of the 1994 Lancaster International Workshop on Engineering Design CACD '94, Lancaster, U.K., 1994.

Development of An Intelligent Grating CAD System----GRTCAD

Ma YongSheng

Gintic Institute of Manufacturing Technology, 71 Nanyang Drive, Singapore 638075

and

Gao HuiXing

SOE Dept., Ngee Ann Polytechnic, 535 Clementi Road, Singapore 599489

Abstract. This paper reports a CAD product developed for grating design. In this system, methods for designing different types of grating panels are modelled using an object oriented approach. Design rules can be defined interactively by using friendly dialogue boxes. Advanced generic 2D dividing algorithms with the consideration of grating area boundaries, openings, gap size, supporting structure and panel properties are coded. Drawing geometry, layout and bill of material are generated automatically. Industrial feedback has indicated improvement on design quality and productivity. **Key words: CAD, Grating.**

1. Terminology

Grating panel: A completed piece of grating, cut to size banded and notched as per cutting drawing.
Bearing bars: Load carrying bars of uniform section in rolled steel spanning between supporting steel members. It is also known as *Loading Bars*.
Crossing bars: Rods of uniform section in square twisted steel, electro-forged at right angle to the bearing bars.
Length of panel (L): The overall measurement of a panel when measured parallel with the bearing bars.
Width of panel (W): The overall measurement of a panel when measured parallel with the crossing bars.
Bearing bar pitch: The standard distance between two neighbouring bearing bars.
Crossing bar pitch: The standard distance between two neighbouring crossing bars.
Openings: The spaces cut out in grating areas to accommodate some obstacles, such as pipes, columns, tanks, etc.

Nominal circle: A circle defined by the user and used for circular platform to measure panel material size. Particularly, the panel width is measured along this circle.
Standard width: The width of a standard panel (W). For polar bearing bar direction, this parameter is measured along the circumference of the nominal circle.
Minimum width: The minimum size of a separated panel in the width direction. Similarly, for polar bearing bar direction, this parameter is measured along the circumference of the nominal circle.
Gross area of panel: The total area of flooring including openings, notches, etc; as used in material estimation (Area = L x W).
Net area of panel: The real geometrical area of a panel excluding openings, notches, etc, as used in weight calculation and certain accurate quotation.
Bill Of Material (BOM): A list of grating panels with series number, group number, gross length and width, gross or net area, cutting drawing number, etc. This document is used in production planning and invoicing.
Cutting list: A set of production assignment documents with cutting drawings for all grating panels, planned time schedule, and other technical requirements.
Packing list: Document contains information for a set of panel groups for delivery purpose. Panels are grouped according to panel size, total weight of each group, and other delivery instructions.
Start cutting point: A point indicating the distance between the first crossing bar in a panel and the parallel panel cutting edge measured along the bearing bar direction.

2. Introduction

Grating is necessary in shipbuilding, refinery construction, steel plant construction, etc. In this process, a platform with different layouts is covered by a set of panels so that the platform can be used for servicing or loading purpose. However, so far, there is no any specific CAD system for grating design. Grating design requires experienced engineers to calculate panel layout from structural layout.

Grating design includes selecting grating category according to the load range, and calculating gratings layout for a given area, which is constrained by the supporting structure and obstacles according to certain distributing patterns. Grating layout drawings are then produced. The total gross grating area is calculated so that quotation for estimating and invoicing can be obtained. Each panel's geometry is later used to produce the panel cutting drawing. This cutting drawing is used to cut the panel shape from standard grating panel. The net area of each panel is used to calculate the weight. This is used for planning and shipment packing. Throughout the process, the Bill Of Material (BOM) is required.

Grating areas can mainly have rectangular or circular shapes. Certain openings are necessary in order to accommodate pipe lines or any other obstacle. Loading bar alignment among panels is important in order to achieve consistent appearance. Because the panel material comes with standard width and length, dividing methods should be optimised to reduce material waste. Traditionally, the grating design and material estimation were done manually.

3. System description

The system is essentially a specially developed, automatic design and drawing generation system for the grating engineering industry. PC and AutoCAD were selected as the hardware platform and the basic graphic tool respectively. This CAD system can automatically produce design drawings for a grating area with a given supporting structure. The profiles of the supporting structures and the openings for obstacles consist of straight lines and circular arcs. After users click in the drawing contours of the outer support structure, openings, and inner supports in the grating area, grating layout drawings are automatically generated with the consideration of the technical rules for dividing the grating area into panels. Rules are selected by the user. It can also calculate the grating area, BOM, start cutting point, cutting list, packing list while panels are created. All operations are menu-driven. After the system generates and displays the drawing of panels, users can still modify these panels interactively.

3.1 Specifications

- The system can deal with an area with inner support lines, even when they have cross openings.
- Standard panel width is used whenever possible.
- Gap generation is automatic for a fixed gap value and floating gap values optimised in a range.
- If the panel width is smaller than the minimum width specified, the small panel is then merged with its neighbouring panel.
- Panels can be modified easily.
- Text in panel labelling and other automatically generated documents, such as bill of material, cutting points, can be modified easily.

3.2 Data Structure and Repository

For creating the relationship between a geometrical block and a panel data structure, a special user defined object type is used. Each grating piece is drawn as an AutoCAD block which has many attributes, such as a unique name and a grouping number. A separate data file is created when panels are generated. All panel information including segment definition, position and block name are written to this file. During an interactive operation process, such as modifying the position of a panel grouping number, when a panel is clicked, the system can return the panel's name. Thus, a relationship between the panel and the data file is established. If there is any modification on any panel, its data records are updated accordingly. The panels with the same dimensions, in the same project, can be assigned with the same cutting number if the user selects this option. The user can take a batch or several batches to process the layout in a design session.

4. Panel Generation Algorithm and General Rules

Two types of grating areas are considered, i.e., rectangular and circular platforms. Three panel distribution patterns are designed according to the bearing bar direction, i.e., horizontal, vertical, and polar patterns. Many intelligent judgements can be performed, such as standardising the width of the panels, selecting the suitable dividing line within an internal opening, setting the gap between panels, and labelling panels automatically. Certain technical requirements, such as maximum and minimum grating panel dimensions are observed.

4.1 Standard Panel Width and Sub-Standard Width

In order to minimise material waste, it should be appreciated that generated panels should have standard widths whenever possible. In Fig.1, the grating area is a simple rectangle. The bearing bar direction is vertical, so the panels created are in a row. In Fig.1(a), the gap value is set as a fixed value, 10 mm, and the standard panel width is 905mm. When dividing the grating area from right to left, the first eight identical panels are created with the standard width 905mm, while the last one has a width of 740mm.

For grating panel cutting, it is always preferred to cut next to a bearing bar so that the side edge of the panel is neat with a solid bar. Therefore, if the standard panel width cannot be achieved, a width that can accommodate certain number of bearing bars is preferred. There could be many width values that satisfy this condition. These values are referred as sub-standard panel widths. Sub-standard widths are calculated by this formula: *Bearing_bar_pitch X (No. of bearing bars - 1) + Bearing_ bar_width.*

In the examples shown in this paper, the bearing bar thickness is 5mm and the pitch is 30mm. If the minimum and the maximum numbers of bearing bars are 4 and 45 respectively, the sub-standard values are the integer series from 95mm to 1325mm with an increment of 30mm.

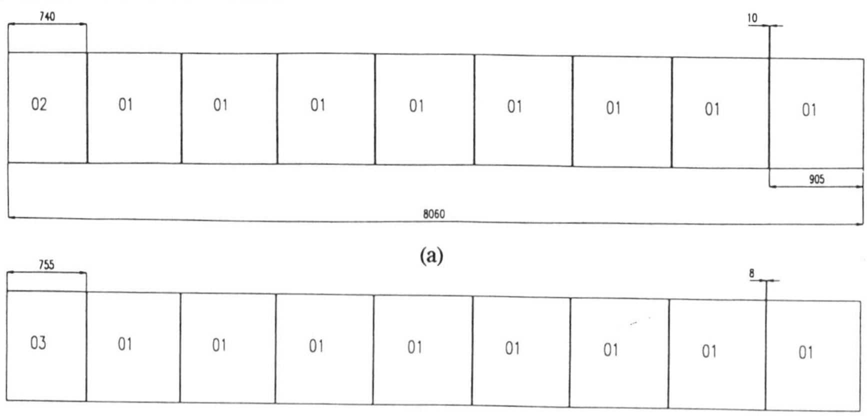

(a)

(b)

Fig.1 Simple rectangular platforms

To achieve the sub-standard, the gap value has to be adjustable. In Fig.1(b), the gap value is adjusted to 8mm. The width of the last panel becomes 755mm which is one of the sub-standard widths. When the user selected either horizontal or vertical loading bar direction, the rules in this section are applied.

When the user activates the "Auto-adjust gap" option, panel gaps will be dynamically adjusted in a range, between the minimum and the maximum values allowable, to ensure that the panel widths are equal to the standard width or sub-standard widths. Otherwise, gap values will be kept constant throughout the panel generation process.

4.2 Rules to Deal with Openings

Most of the openings are either circles or rectangles. There must be a dividing line passing through each opening. Otherwise, grating panels have to be installed from the top of the obstacle, and doing this is often not possible.

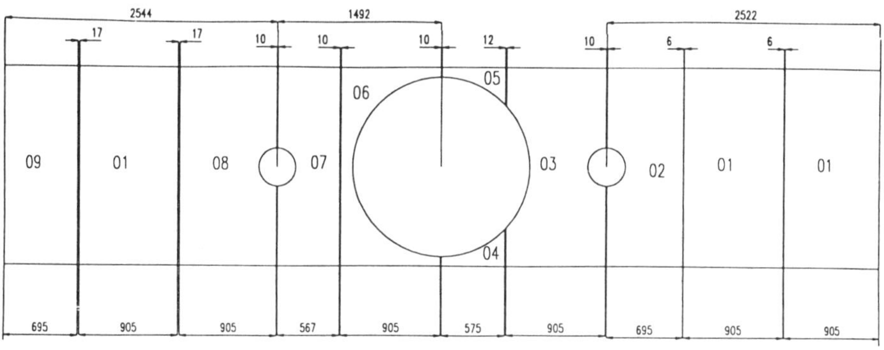

Fig.2 Grating area with circular openings

For a circular opening, the dividing line must go through its centre point. An example is shown in Fig.2. There are three circular openings. With the gap value varying from 6mm to 17mm, the panels created have either standard or sub-standard widths.

However, when the opening is not a circle, e.g. a rectangle, the dividing line can be at any position within the range in which the opening has the maximum height. This range is defined as the limit zone of the opening.

For example, when using the vertical bearing bar direction, opening (1) in Fig.3 can be divided at any horizontal position between A and B, while opening (2) can be between P and Q. Again, in order to generate panels with widths either equal to the standard width or sub-standard widths, dividing line positions are optimised with great effort. In this system, the limits of each opening in the horizontal and vertical (or in radial and angular for polar pattern) directions and their corresponding zones in which the opening reaches limits are calculated.

It should be noted that if the bearing bar direction is vertical, dividing lines to generate panels are also vertical. Then, the dividing line through an opening must fall in the common horizontal range of the corresponding zones for its upper and lower limits. Similarly, limits zones can be found if the bearing bar direction is horizontal. If the bearing bar direction is polar, dividing lines are in radial directions, the dividing line through an opening, must fall in the common range of the corresponding angular zones for the outer and inner limits related to the reference centre of the platform.

GRTCAD provides an option for the user to decide whether the dividing line positions can be away from the centre of their limit zones. This is controlled by the "Dividing at square centre" toggle button.

In Fig.3, all panels' widths are adjusted by changing the neighbouring gap values which vary from 10 to 14. Note that opening (1) and opening (2) are overlapping, and there is some overlapping limit zone. The system automatically identifies this type of overlapping and decides a dividing position within the common limit zone, hence positively avoids small panels being generated. It should also be noted that the first dividing line from right to left results in two panels on the right side, i.e., panel "01" and "02". Similar rules are applicable to a polar pattern although the restrictions for a polar pattern is much more stringent because the dividing line must be in the radial direction in order to have reasonable load distribution.

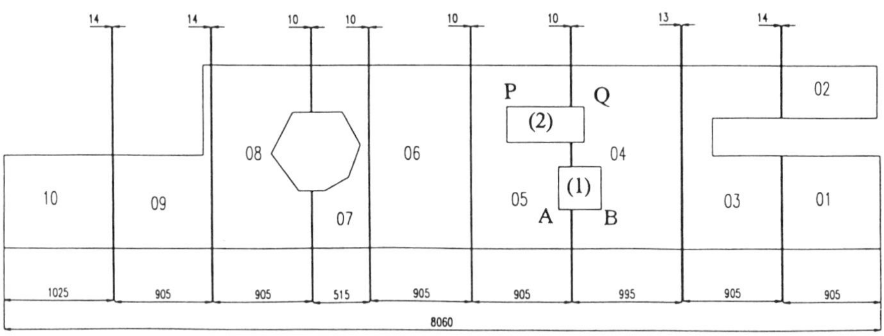

Fig.3 Rectangular grating area with irregular openings

4.3 Panel Group Numbering

The user can group identical panels into a group so that the machining of these panels can be carried out without repeating set-up processes. This system mainly identify those panels with simple shapes, such as rectangles and typical fan shapes. This is done automatically when the user toggle the "Group Numbering" switch on. Otherwise, the panel number is assigned in the sequence of their generation. The user is also allowed to insert, modify, or confirm the starting panel number for the current platform before starting to generate panels. For some special cases, where an opening is small rectangle, if gaps can be adjusted, the opening is divided at one end so that the panel manufacturer only needs to cut the opening on one side panel.

4.4 Rules to Deal with Inner Support Lines

Quite often, the grating area spans in a large space, inner support structure are therefore necessary to support grating panels to achieve satisfactory rigidity. The support structure design is out of the scope of this system. However they are considered as inner support lines in grating panel generation. For rectangular platforms, supporting lines are perpendicular to the bearing bar direction, while for polar platforms, the support lines are in radial directions.

In GRTCAD, the grating area is divided into sub-areas considering the gaps along the supporting lines, and the panels in each sub-area are then generated in order automatically. In Fig.4, it can be observed that there are two inner support lines in the horizontal direction (the bearing bar direction is vertical). Note the lower support line crosses the big opening and the system has created grating panels properly.

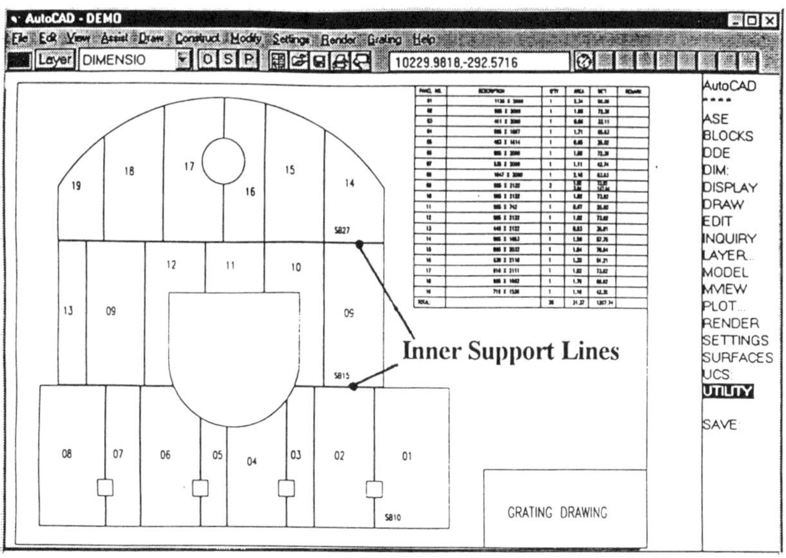

Fig.4 Grating area with inner support lines

4.5 Avoiding Small Panels

Panels generated cannot be too small because small panels are inconvenient to cut and install. In GRTCAD, in order to avoid small panels, the next generated panel size is predicted. If the panel width is smaller than the minimum width, the next small panel will not be generated. Instead, the current panel will include this small area. Hence, the current panel becomes bigger. For polar bearing bar direction, this minimum allowable width is measured along the circumference of the nominal circle, which is also used to measure the standard width in the dividing process.

5. Editing Capability

GRTCAD provides many user-friendly interactive modification tools, such as changing the shape and name of a panel, dividing one panel into two or merging two into one, etc. When panels are created or modified under the normal AutoCAD environment, they can be re-integrated into the grating database by running a "read-in" program. These modification functions make this system more flexible to meet customers' requirements and to cope with complicated geometry. A dialogue box showing panel modification options is shown in Fig.5.

6. User Interfaces

A customised menu group under "Grating" is inserted into AutoCAD's top menu bar. If the user clicks the grating command item, the GRTCAD is activated, and the first menu for selecting operations is then displayed. After clicking a grating operation button, a few AutoCAD-like dialogue boxes will be displayed for selecting operation options, and inputting data. For example, GRTCAD displays a dialogue box to allow the user to set initial parameters for panel generation. These parameters include:

- Bearing bar direction, the choice can be either Horizontal, or Vertical, or Polar.
- Standard panel width, minimum width, bearing bar pitch, thickness, and height; crossing bar pitch, starting panel number, etc.
- Dividing principles applied, such as whether the gap value can be floated, and the range for auto-adjustment; whether the dividing line for a rectangular opening must be at the centre or it can be adjusted; and whether panels are grouped with one panel number if they are identical, etc.

These dialogue boxes are defined by using AutoCAD Dialogue Control Language (DCL). A map with important menus is shown in Fig.5. Used items in dialogue boxes include *Radio buttons, Editing boxes*, and *Toggle switches*. The user's settings for each dialogue box are backed up in a data file and reloaded when the system is started. By doing so, the user will save a lot of time for inputting these values. This data file is updated instantly when the settings are modified. Hence, the system can "learn" from the user.

7. System Output

Grating drawings are the most important output. Fig.4 shows an example for a rectangular platform. In Fig.6, a polar platform is shown. All panels can be output individually to create cutting drawings. The system can automatically point out the position of cutting start point for each panel in order to align all the crossing bars of the neighbouring panels. These cutting start points are displayed on the

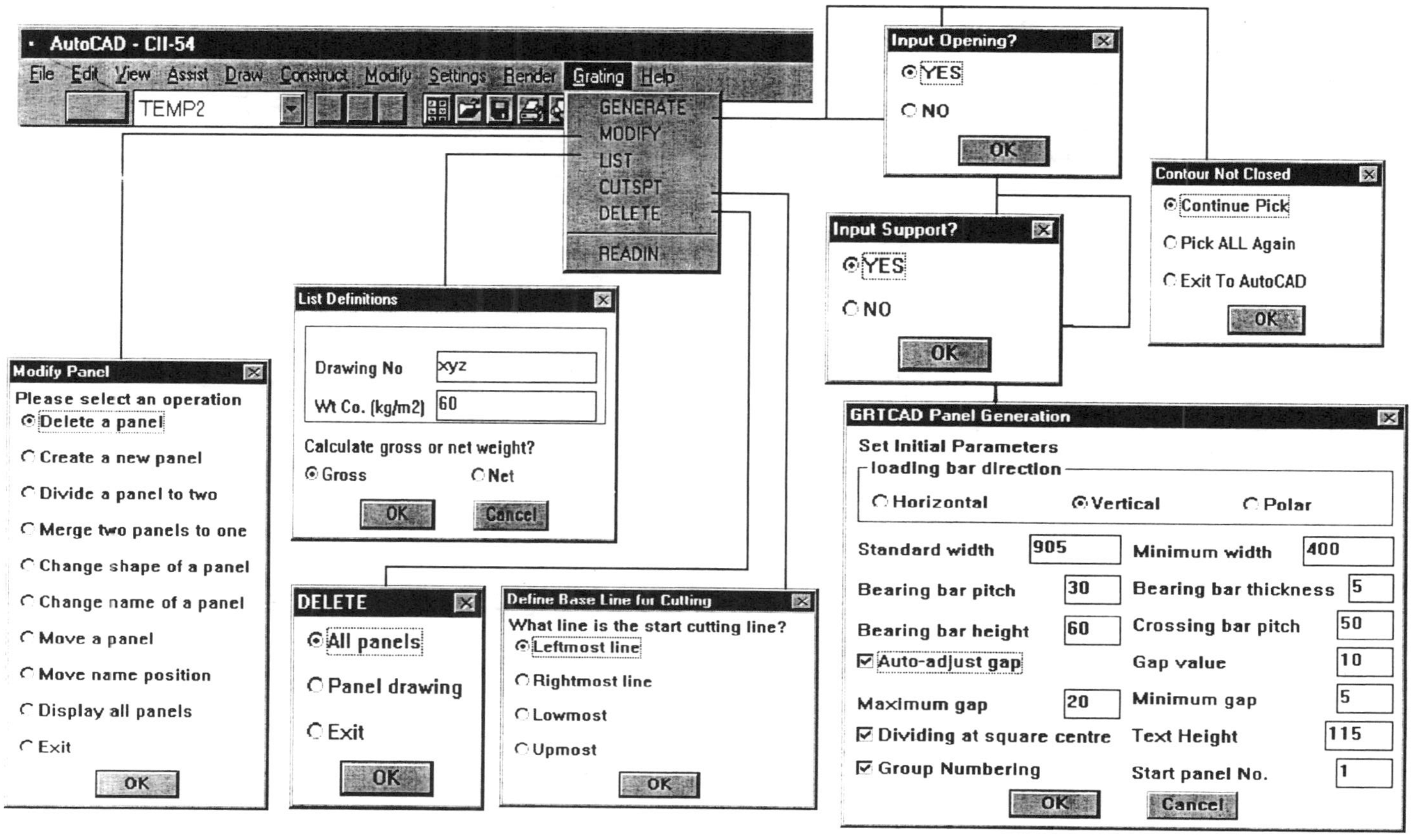

Fig.5 Menu Map in GRTCAD

manufacture drawings. All the output data documents, such as bill of material, cutting list and packing list, are generated automatically using the integrated data base. Hence, it eliminates human errors. The output data can also be exported to Lotus 1-2-3 spread sheets.

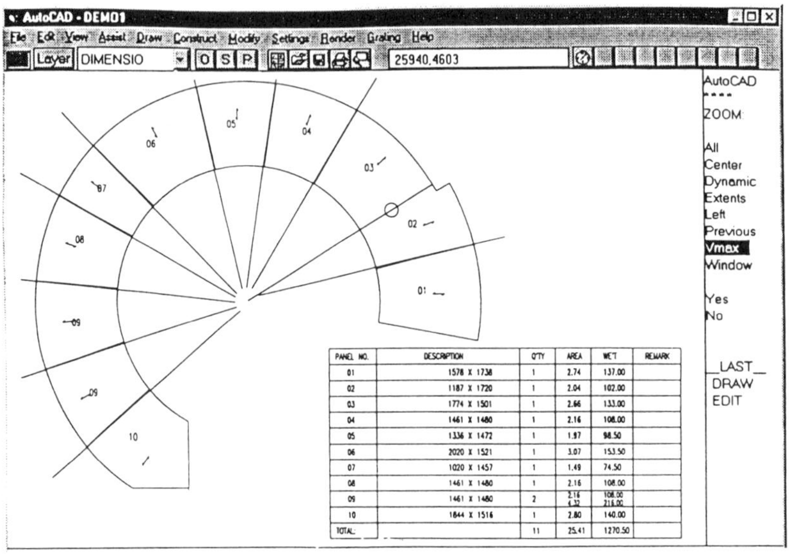

Fig.6 Circular grating area with openings

8. Conclusion

The intelligent features in this system allow the user to cut down operation and training time. The easy-to-use data entering format has shorten the learning curve. The user needs less man power to deal with grating panel design. Higher quality and more accurate grating drawings can be produced with high efficiency. Human errors are eliminated in the different output data documents. The approach used in this paper has been adopted by several other specialised CAD development projects such as generating scaffolding structures and design 3D metal form structures in construction engineering.

9. Acknowledgement

The authors would like to acknowledge Sangjin Dahe Industries Pte Ltd (S) and the EDB of Singapore government for their sponsorship, and Ngee Ann Polytechnic for permitting the relevant projects being carried out. AutoCAD ® is a trade mark of AutoDesk Inc.

References

1. Sangjin Dahe Industries Pte Ltd (S), Company Document, *Specifications for Grating Panel Cutting and Operational Documentation* (1994).
2. Y. Ma and H. Gao, *Grating CAD System --- GRTCAD User's Guide*, (1996).

Simultaneous Consideration of ecological & economical Parameters by using a feature-based design system

Dieter H. Müller, Kai Oestermann

Department of Mechanical Engineering, Bremen University
Badgasteiner Straße 1, 28359 Bremen, Germany
E-mail: bik@bik.uni-bremen.de

Abstract. This paper deals with a new method for the isochronous consideration of ecological and economical parameters in early stages of mechanical design. The method was transformed in a computer aided device for the ecological and economical optimisation of the product definition phase.

1. Introduction

Today's products do not only have to work satisfactorily. In addition to the functional requirements they have to comply with other - partly contradictory - demands /1/, /2/.

Due to the increasing responsibility of companies for the ecology and the increasing demand for environmentally conscious products, a sustainable industrial production is one of the challenges facing the industrialised world. Hence there is the need for an equal consideration of ecological and economical parameters within the process of product design.

One decision model for this is a complete eco balance consisting of the elements life cycle inventory and environmental impact assessment. Due to the complexity of a complete eco balance, this method is not practicable within the product development in small and medium sized enterprises (SME) which represent the backbone of German economic strength.

At present methods for product optimisation like „Design for Cost" and „Design for Environment" are only applicable in a sequential manner. In consequence of that the possibilities to obtain a comprehensive optimisation of products are strongly reduced (Fig. 1).

Therefore it is necessary to develop auxiliary means and methods for the simultaneous optimisation of functional, ecological and economic parameters, especially in the context of product development in SME.

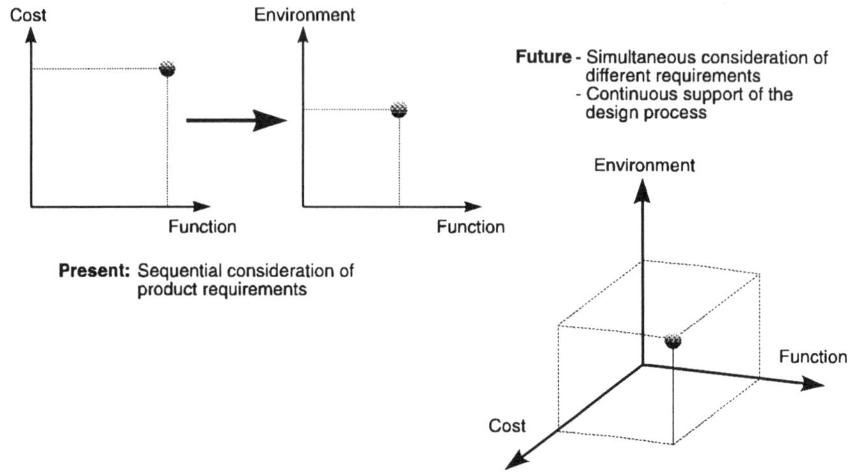

Fig 1: Differences between sequential and simultaneous consideration of different requirements during the process of product optimisation

2. Evaluation Method

In this paper a method will be presented which leads to a more transparent and more feasible evaluation concept for an ecologically oriented production. This method was developed within a project at the Bremen University.

Using this method it will be possible to assess the ecological effects of a certain way of product realisation at an early stage of product planning. This point is an important feature since approx. 70% of the product costs and all product features are predetermined by the first steps of product development (Fig. 2).

On the other hand it is important that the solution obtained fits equally-balanced to ecological and economical requirements.

For reasons of simplicity, our method is restricted to an eco balance at the level of a life cycle inventory of the production process (Fig. 3).

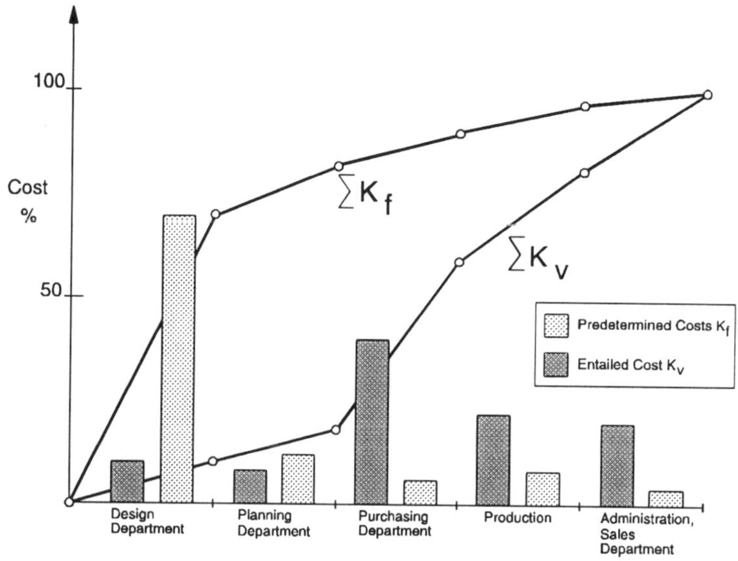

Fig. 2: Fixed and entailed costs during product realisation /3/.

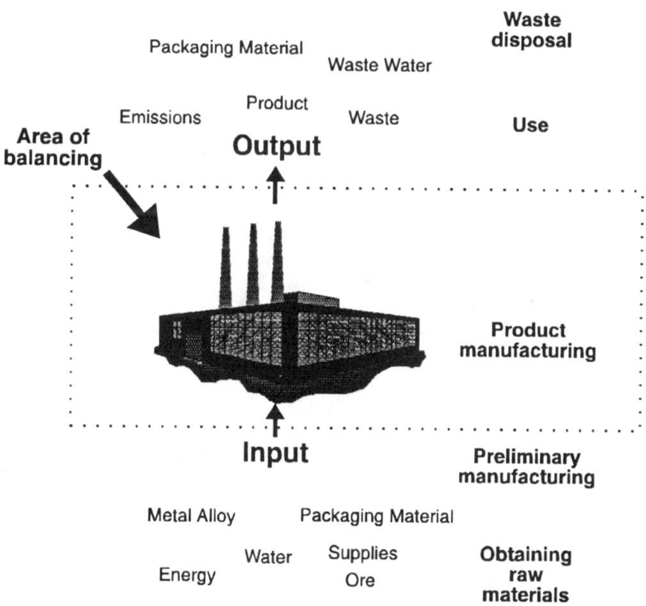

Fig. 3: Area of balancing for SME´s

The evaluation method used is the method of the eco-toxicological equivalent coefficient as used by the Department of Environmental Technology at the Bremen University /4/.

The project process is shown as follows: Using a machine manufactured product the life cycle inventory shown in Fig. 4 was carried out for different manufacturing methods and production sequences. The balance was built up in a modular way, where the smallest unit (production sequence) is one module which is connected with other modules linked to the border lines of the system by the mass and energy flows.

The environmentally relevant effects of a production sequence are described by an emission parameter, which is divided into three single parameters, one for gas, one for fluid and one for solid matter. It contains the finding of toxicological equivalents connected with the mass flows. Using this damage-profit-evaluation a simple and transparent description of the ecological effects for different manufacturing methods and production sequences becomes possible.

This method is a suitable device to uncover weak ecological points in manufacturing sequences.

In a second step, a computer based design system was developed for the ecological and economical assessment of technical products. Within this system a connection was established between manufacturing methods, product cost and functional requirements.

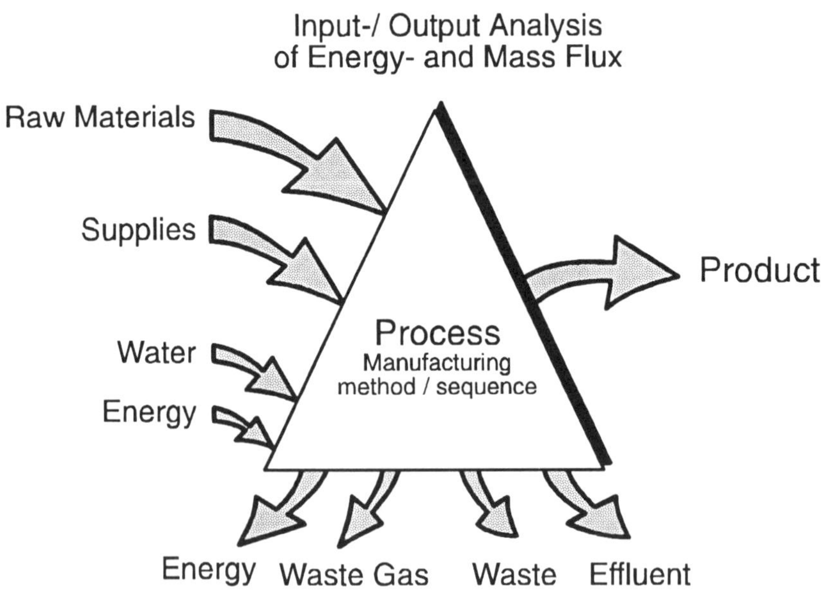

Fig. 4: Life cycle inventory /4/

3. Main results

The result of the project is a feature-based design system (Fig. 5) including ecological and economical decision parameters for the design of machine-manufactured, rotationally symmetrical products (e.g. shafts, bolts etc.). The term „feature" is defined as:

Feature = Form-Feature & Semantical Information /5/.

In addition to the geometrical information, features contain further information about the product, for example, possible manufacturing methods, production costs and eco-toxicological data.

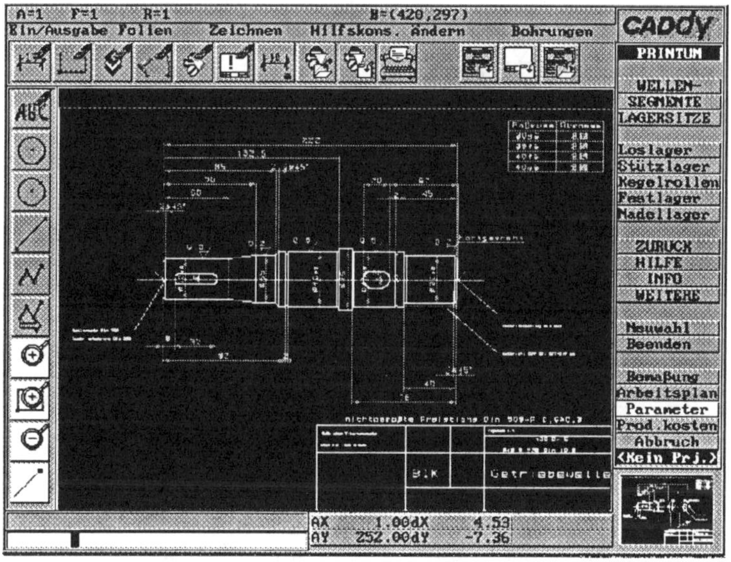

Fig. 5: Feature-based design system

Taking into account the required product functionality, alternative manufacturing methods (e.g. hard machining instead of grinding) or production sequences (e.g. dry machining) are suggested by the system during the design process. Also design actions like the renunciation of a definied surface quality, or the use of different materials, are implemented into the system.

At the same time the production costs are determined by the system. The basic evaluation method for this is the VDI-guideline 2225 /6/. Thus the opportunity of a more complete product optimisation is given to the designer by the simultaneous consideration of the product parameters function, cost and environment.

4. Example

The following example is meant to illustrate the feasibility of the evaluation method in general and the functionality of the developed feature-based design system. As an example we used a machine manufactured product, the shaft of an electromotor. The development task was the consideration of ecological and economical parameters during the design process.

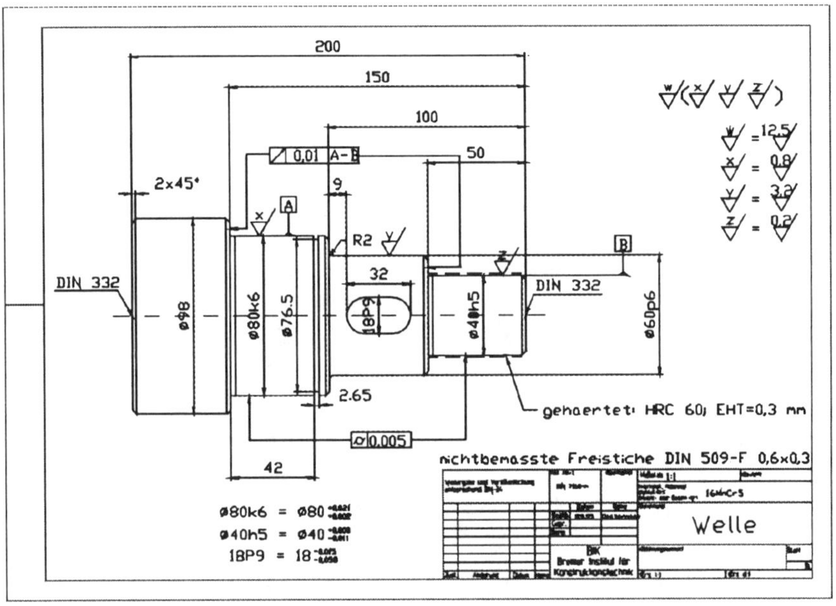

Fig. 6: Shaft of an electromotor

As a first step the preliminary design concept of the shaft was worked out by using the design system. After receiving the input data the process of product optimisation started. The decision of the final design concept was given by the comparison of the energy and ecological requirements of resources and the product costs.

The solution suggested by the system was the replacement of the manufacturing sequence „grinding of roller bearing seat“ by „hard machining of roller bearing seat“.

The application of the evaluation method lead to the conclusion that the demand for energy per piece could be reduced fr om 10.43 kWh to 9.24 kWh (Fig. 7).

As a result of the application of the evaluation method it was possible to illustrate the optimisation in money values. In this practical case a production cost reduction of about 6% has been achieved (Fig. 7).

Demand of energy		
Variant 1 [kWh]	Manufacturing sequence	Variant 2 [kWh]
0,030	Sawing	0,030
0,985	Roughturning	0,9 85
0,046	Smoothturning	0,046
8,168	Direct hardenening after cementation	8,168
1,203	Grinding of roller bearing sets	-
-	Hard machining of roller bearing sets	0,006
10,432	Total demand [kWh]	9,235

Production cost		
Variant 1 [DM]	Manufacturing sequence	Variant 2 [DM]
30,12	Material cost	30,12
2,56	Sawing	2,56
15,06	Roughturning and smoothturning	15,06
26,95	Direct hardenening after cementation	26,95
8,20	Grinding of roller bearing sets	-
-	Hard machining of roller bearing sets	3,20
82,91	Total Cost [DM]	78,41

Fig. 7: Demand of energy and Production cost

5. Conclusions

The sequential consideration of different requirements like function, cost and environment in early stages of mechanical design is necessary for today's products. We think that the method described in this paper is one step in this direction.

Mainly enterprises using metal-cutting manufacturing should be interested in the results. The basic idea of our simplified ecological balancing method and isochronous consideration of function, cost and environment is also applicable to other manufacturing technologies.

References

1. Müller, D.H and Oestermann, K. (1994). Wie können Marketing und Konstruktion gemeinsam das Produkt definieren ?. *Tagungsband des VDI-EKVIP Zirkel Führungskraftetreffens 9/94*, pp. 2.1-2.15. Düsseldorf: VDI-Verlag.

2. Müller, D.H and Oestermann, K. (1993). Integrative Zusammenarbeit von Konstruktion, Marketing und Design. *Proceedings of 9th International Conference on Engineering Design, Den Haag*, pp 1197-1201.

3. Vayna, S. (1991). Strategien zur Integration von Entwicklung, Konstruktion und Arbeitsvorbereitung. In *Die Konstruktion als entscheidender Wettbewerbsfaktor: Ziele, Strategien, Maßnahmen*; Tagung Bad Soden 7. und 8. März 1991, pp. 57-66. VDI-Gesellschaft Entwicklung, Konstruktion, Vertrieb. Düsseldorf: VDI-Verlag.

4. Räbiger, N. and Haase, C. (1995). Ein Bewertungsmaßstab für die ökotoxikologische Schwachstellenanalyse von Produktionsprozessen. *Tagungsband Colloqium Produktionsintegrierter Umweltschutz "Abwässer der Pharmazeutischen Industrie und Biozidherstellung" 9/95*, pp. 281-292. GVC - VDI-Gesellschaft Verfahrenstechnik und Chemieingenieurwesen und IUV-Institut für Umweltverfahrenstechnik, Universität Bremen (Hrsg.)

5. Krause, F.L., Ciesla, M. Reiger, E., Stephan, M., Ulbrich, A. (1994): Featureverarbeitung -Kernkomponente integrierter CAE-Systeme. In *Produktdatenmodellierung und Prozessmodellierung als Grundlage neuer CAD-Systeme*. Fachtagung der Gesellschaft für Informatik e.V. Paderborn CAD 94. Jürgen Gausemeier (Hrsg.). Müchen, Wien: Hanser.

6. VDI-Richtlinie 2225 (1996). Technisch-wirtschaftliches Konstruieren. VDI-Gesellschaft Entwicklung Konstruktion Vertrieb. Düsseldorf: VDI-Verlag

A Concurrent Integrated CAD (CICAD) Method for Ship and Sailing Yacht Design

Kok Thong Tan and Thomas P Bligh

Department of Engineering, Cambridge University, United Kingdom

Abstract. The objective of this paper is to introduce a new approach that eliminates the sequential nature of the ship design process. This new approach is called the Concurrent Integrated CAD (CICAD) System. It establishes a concurrent computer-aided design procedure, instead of following the traditional sequential design approach. It is modelled and analysed by utilising I-CAM Definition (IDEF) techniques. The paper starts with a brief review of the evolution of ship and yacht design processes from craftsmanship to computer-aided design methods. The authors point out the problems of the ship design process, and investigate the impacts and trends of applying CAD methods, in order to justify the new approach to an integrated CAD method. The paper demonstrates how the sailing yacht design process could be executed concurrently, and estimates its significance. Some preliminary findings of this new model are reported.

1 INTRODUCTION

The shipbuilding industry is actively involved in applying the computer technology since the 1960s [Hays et al, 1994]. Therefore, in the present information age, ship and yacht designers are able to enjoy the benefits of it, especially in reducing tedious calculations, and improving the design accuracy and quality. The industry perceived this transformation as a revolutionary one, for it had digitised traditional 2-D ship design drawings, carried out complex design calculations, and enhanced the productivity through computer numerical control machines [Ross, 1995]. In principle, computer technology does contribute to the ease of making changes throughout the ship design process. Thus designers may have relatively more time to explore many potential designs. Current innovation is concerned with the development of an Integrated Computer-aided Design (CAD) method for ship designs, which combines all stages of ship design programs into a single system, and even embraces the elements of the product's life cycle from conception to disposal [Mistree et al, 1990]. All of these exciting developments in the shipbuilding industry consolidate the belief that computer technology has revolutionised the ship design process. However, the principle of existing CAD methods capture the features of the Ship Design Spiral[1] concept, which is sequential, and therefore suffers from a number of problems, such as designers adversely alter other design characteristics while concentrating on a particular design aspect. For a design that employs the Ship Design Spiral concept, the sequential process is

[1] The Ship Design Spiral concept was introduced by Evans in 1959, and enables ship design problems to be solved systematically [Evans. 1959].

inevitable as each phase of design process requires the output from the preceding one. For example, a designer has to conduct a weight estimation, in order to determine the position of the centre of gravity, then stability studies can be carried out by using this value to obtain the righting arm, $\overline{GZ}$. As a result, this computer technology does not actually reform the ship design process. We, however, believe that it is possible to remodel the design process to avoid the ineffective sequential nature of the Ship Design Spiral. The objective of this paper, therefore, is to illustrate this new approach to an integrated CAD method for ship and sailing yacht design.

2 PROBLEMS OF THE SHIP DESIGN PROCESS

Ships, like aircraft, are examples of extremely complex systems. The design and building of ships was wholly a craft until the middle of the eighteenth century, before science affected ships design appreciably [Rawson et al, 1966]. For the next two centuries, the designer started with a number of assumptions and worked through the rules, which only imperfectly modelled the real situation, to see if the design satisfied the requirements, specified beforehand [Evans, 1959]. In 1959, Evans made a significant contribution to visualising and modelling the process of ship design, which is now known as the 'Ship Design Spiral' [Mistree et al, 1990]. The purpose of this attempt was to assist designers in organising the thought process, so as to enable ship design problems to be solved more effectively [Evans, 1959]. While some refinements have been made over time, the principle remains unchanged. Nowadays, this spiral model has become a widely accepted approach for both ship and sailing yacht design [Hassan et al, 1992, Larsson et al, 1994]. A typical design spiral for sailing yacht designs is shown in Figure 1. This spiral consists of eleven stages, where each stage corresponds to a design task. The design process keeps on iterating all these stages until it converges towards the centre, the final solution.

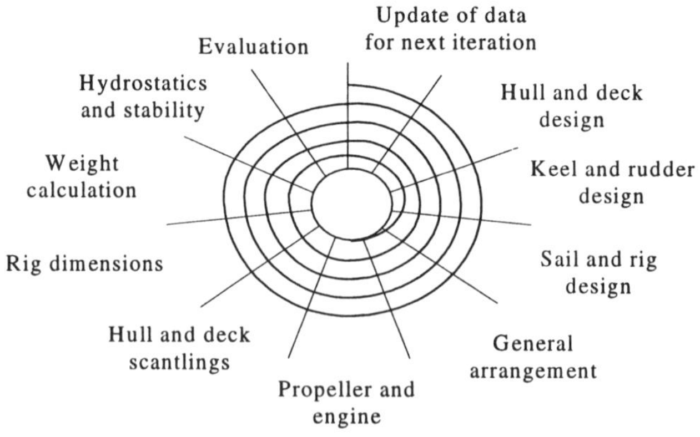

Fig. 1. The design spiral for sailing yacht designs [Larsson et al, 1994].

The major characteristic of the Ship Design Spiral concept is that the design process is sequential and iterative. Designers must satisfy a limited set of criteria at each design stage, however, having satisfied them, goes on to the next stage with little idea of how good the design actually is. Therefore the design is continuously altered by means of consecutive iterations until it reasonably fulfils all the requirements. As each design stage depends on the output of the preceding one, the process is arranged in a sequential manner. This leads to a design philosophy that based on a straightforward sequential and iterative procedure [Kroemker et al, 1996]. Moreover, for certain design stages, internal iterations are required, and this makes the design process even more time consuming.

Surprisingly, it is not easy for practising designers and naval architects to justify the effectiveness of the Ship Design Spiral approach, because by and large, it has been exercised for decades, and has become the routine procedure. Experienced designers may occasionally skip stages, as they are able to make a good guess, based on a rule of thumb [Condylis, 1997, Simpson, 1997]. These designers, however, have to comply with the design spiral, when subtle modification to the design takes place.

Despite the general acceptance of the spiral model, a few academics and operations researchers began to question its effectiveness. A number of general surveys were taken, and the results show that this method, by its very nature, may obstruct the exploration of optimum design [Kroemker, 1997, Kroemker et al, 1996, Mistree et al, 1990, Lyon et al, 1985]. Due to this sequential, iterative, tedious and time-consuming design process, most of the designers are still unable to achieve near optimal designs. This is because the sequential process requires a great deal of design time and thus designers have limited time to explore many potential designs [Kroemker et al, 1996]. Moreover, the longer the design time, the less competitive are designers in bidding the contract [Keane et al, 1996].

In addition to that, the spiral approach is unable to provide the designers with a global view of the designs, especially at an early stage, because the design information emerges at each stage as they work through the spiral. Consequently, they are incapable to recognise the effects of modifications on the design quickly. One drawback to this is that designers may adversely alter other design characteristics while concentrating on a particular design aspect [Condylis, 1997, Simpson, 1997].

3 THE EXISTING TOOLS FOR SHIP DESIGN

The advent of computer technology lead to the development of Computer-aided Design and Manufacturing (CAD/CAM) programs. The shipbuilding industry, likes others, extensively used this new technology from the 1960s [Hays et al, 1994]. Today, CAD/CAM is a key technology for ship design and construction [Hays et al, 1994]. Typical CAD programs for ship design are built around a geometry generation capability which allows the designer to create on the computer screen a mathematically defined hull surface. Furthermore, a number of distinct programs allow the post-processing of these surfaces in the areas of analyses and predictions, such as hydrostatics and stability calculation, resistant and powering, Velocity Prediction Program (VPP), Computational Fluid Dynamics (CFD) etc. Compared with the pre-computer era, these computer-based tools undoubtedly offer a number of advantages, which include replacing time-consuming laborious tasks, improving design accuracy, making changes easy [Ross, 1995].

The application of these tools have had great impact on ship designs. First, many of the design tasks are carried out at great speed by the computer, therefore, in an earlier stage more information about the design is available [Excell, 1997]. As the laborious tasks have been removed, it opens up greater opportunities for intellectual exploration, and eases the process of making changes. Second, as today's powerful hardware and software can effectively model and fair a hullform precisely, the manual lofting process can be eliminated [Hays et al, 1994, Larsson et al, 1994]. Third, it is possible to view a perspective 3-D plot of the hull prior to the construction. Compared with the manual approach, where only three standard views are employed, this visualisation is considerably important, because hulls that look good in lines plan may look quite ugly in reality [Larsson et al, 1994]. Fourth, the computer networking allows a multi-user environment. This forms a huge database that may provide new ideas or opportunities for designers, as well as ease the selection of components such as engines, deck equipment and fittings. Fifth, the data generated during the design process can be tailored in format and content so that they can support the ship production process [Ross, 1995]. From an integrated CAD/CAM feedback program, potentially costly mistakes can be ironed out before manufacture. Thus, this results in a production-oriented design, which may reduce the cost and time of manufacturing [Excell, 1997]. Sixth, the digitisation of design information, instead of traditional engineering drawings, promotes interdisciplinary co-operation [Mulligan, 1997].

Many believe that reusing data in other computerised tasks will maximise the benefits of computerisation, and this is called ***Integration*** [Hays et al, 1994]. Thus, they regard the integrated ship design program as a compelling concept, and one whose time has come [Ross, 1995, Hays et al, 1994]. This notion is driven by economic aspects. As it involves enormous sum of cost to computerise any given task in the design and manufacturing process, it is difficult to justify the computerisation if that task is taken as a standalone item. As a result, there is a definite trend that to integrate the computer-aided ship design programs [Ross, 1995]. In this case, the ship design process may be enhanced through individual programs sharing their results with each other, preferably from a common database, which may be extended to computer-aid manufacturing (CAM). A few state-of-the-art integrated ship design programs have been developed, the representative ones include HULLTECH, AutoSHIP System, FORAN, HICADEC, IMSA, TRIBON, NAPA and NAVSEA CAD-2. In general, these programs can be categorised into two groups, where it is either integrated among modules, or by means of a product model [Ross, 1995]. Integration among the modules of the ship design process is the most common one; it means that the various programs of design tasks are designed to communicate data with one another to at least some extent. The HULLTECH, AutoSHIP and IMSA are examples of this group. The second one, a product model, shares an integrated database of various ship design programs, which means there is no need for data conversion among the modules. This is a more advanced level of integration. However, it suffers from inflexibility of future growth [Ross, 1995]; recent developments involve object-oriented system design which exhibits the concept of abstraction, encapsulation and inheritance. This results in simpler structures which are easier to maintain [Wu, 1994]. Either group, however, captures the features of the Ship Design Spiral concept.

This trend towards integration seems to be a revolutionary concept for the shipbuilding industry, as it may reduce the lead time and cost, as well as improve the quality of the products in general. Computervision Corporation even anticipates that there should be saving of around 25% on shipbuilding costs [Pullin, 1997]. Hence, there cannot be serious opposition to the notion that an integrated system may enhance the shipbuilding industry. However, it would be interesting to find to what extend this new approach resolves the problems of the ship design process, and whether the remaining problems are of great significance.

The above integrated CAD approach, with its powerful computing capability, may minimise the problems of the ship design process, in terms of removing laborious tasks and making change easy. However, designers literally experience the same design process, as they did before the computer era, except that now they are provided with 3-D images and high speed calculations. In principle, they still adhere to the sequential nature of the Ship Design Spiral. Thus, this approach suffers from the consequences of the spiral approach, and therefore, still does not resolve the problems of the ship design process.

The significance of these problems is evidenced by various observations of designers. For a given specification, designers will require approximately the same number of iterations to converge to an acceptable design, no matter whether the programs are integrated or not. This is because in either case, the designers are lacking a global view of the design, thus the design is adjusted, on a trial and error basis, at respective design stages; this may well adversely alter other design characteristics, which is very annoying [Condylis, 1997, Simpson, 1997]. In fact, this can be avoided if they were provided with the global view of the design throughout the design process. Compared with previous computer-based tools, the above integrated one is unable to assist designers to attain at a solution much more rapidly, because the rate of converging towards a solution remains constant.

4 TOWARDS A NEW DESIGN APPROACH

As the existing approach does not resolve the problems of the ship design process, there is a need to establish a new approach to an integrated CAD system. The new approach must be able to provide the right information to the designers at the right time, in order to assist the designers to converge towards a solution quickly.

This approach is called the Concurrent Integrated CAD (CICAD) System. In general, it will be modelled and analysed by utilising I-CAM Definition (IDEF) techniques, developed by the US Air Force to perform modelling activities in support of enterprise integration [KBSI, 1995]. For the analysis of requirements, the IDEFØ technique, which is a structured approach, will be employed to capture a static view of the functions performed by designers. This is to explore the possibility of executing the design process concurrently, and to establish the function model. Besides, it helps focus on a specific aspect within the system. Subsequently, the IDEF4 technique, which is an object-oriented approach, will be used in the system development process in order to incorporate desirable life cycle qualities such as modularity, maintainability and reusability. This paper deals with the function model only.

In essence, CICAD is developed by remodelling the sequential ship design process in a concurrent manner. Since each task is highly dependent upon the output of the preceding one, it is crucial, at this point, to examine to what extend the process can be overlapped. This section demonstrates how the sailing yacht design process could be overlapped, and estimates its significance. The yacht design process is selected because it captures the basic tenets of the Ship Design Spiral, and has a moderate complexity.

To begin with, each design stage is modelled in a top-down fashion by adopting the IDEFØ technique, which minimises the need for elaborate descriptive text, and its graphical presentation provides a clear representation of a complex aspect of organisation. The model is supported by the inputs, controls, outputs and mechanisms (ICOMs). As this technique is designated to perform the process planning, it must be modified in order to represent the actual stages of the Ship Design Spiral. Two crucial modifications were made. First, a closed loop structure was incorporated to account for the nature of the iterations in the Ship Design Spiral. Second, the controls and mechanisms were omitted, because they do not play an important role in determining the interdependency, otherwise, they might complicate the entire model.

The overview of the design model is shown in Figure 2. The top level is known as the 'environment', and is given a node number L0 (Level 0), this is to identify the inputs and outputs that cross the boundary of the model. To clearly identify the subject and extent of the model, a 'context' diagram node number L n_i (Level n_i) is established. One level down is the 'viewpoint', node number L $n_i m_i$ (Level $n_i m_i$); this is used to establish each aspect of the 'context'.

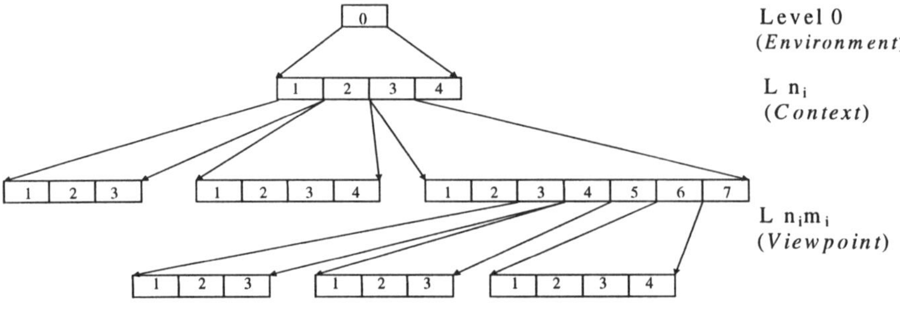

Fig. 2. Overview of the design model

The core diagram, L0 (see Figure 3) illustrates the generic description of sailing yacht design stages, which comprise of requirement development (L1), conceptual design (L2), preliminary design (L3) and detail design (L4). During the first stage, the design requirements are developed; it usually involves clients and designers, where the clients will try to elucidate their demands or wishes for the product. Those challenging requirements will then inspire the designers to generate many sketches in the conceptual design stage, where these initial solutions will be evaluated intuitively, in order to select suitable combinations. If the clients are satisfied with the combination of these potentially most promising solutions, then the preliminary design will be carried out to embody the ideas in a design model. The design is continuously adjusted to converge into a best possible solution. This is the part where the Ship Design Spiral concept is employed extensively, and many computer-based tools have been developed to support it. The result of the preliminary design will be used to bid the contract; therefore, it appears to be the most crucial part of the design process. The preliminary design is the key issue for a designer to be successful. If he does succeed in receiving the contact, then the detail design will be conducted. This involves the refinement of the preliminary design, and converts it into manufacturing instructions. For the requirement development, conceptual design and preliminary design stages, there are three closed loops respectively. The first two loops are intended to determine the viability of each corresponding stage, while the third loop decides the acquisition of the design contract.

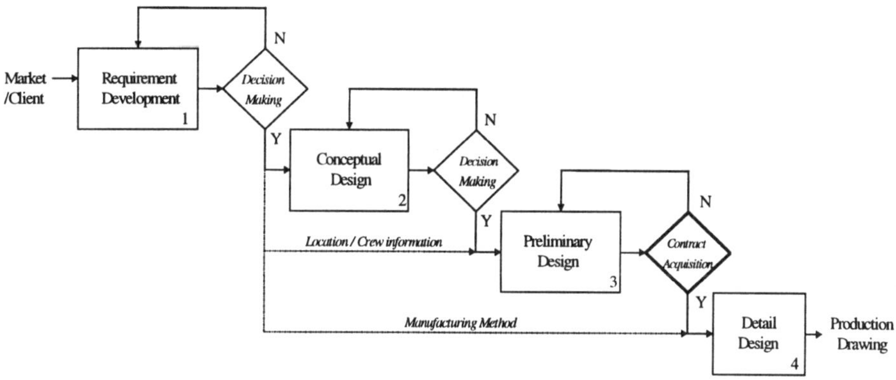

Fig. 3. Diagram Level 0, the sailing yacht design stages

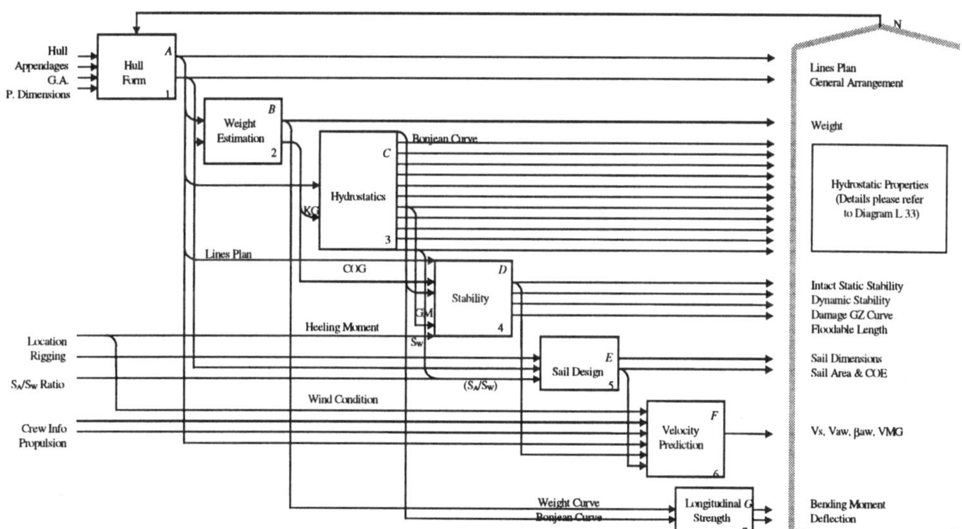

Fig. 4. Diagram L3, the preliminary design

The development of the new approach focus on the preliminary design stage, and its schematic diagram, L3 is shown in Figure 4. The reasons are twofold. First, this is the design stage that suffers most from the problems of the ship design spiral, whereas the conceptual and detail design stages do not. Second, the greatest chance of arriving at an optimum solution occurs during the preliminary design stage. This is because for sailing yacht designs, the conceptual designs do not have any significant meaning in term of performance, as they are usually in the form of sketches, which are approximate and immeasurable. Therefore, to achieve a best possible design, it requires the repetitive and subtle modifications throughout the preliminary design stage. We have concentrated on the preliminary design stage, because this will allow the designer to achieve an optimal design, and, this is the most effective way to improve the quality of a product, rather than through improving the organisational and production efficiency.

Diagram L3 indicates that the existing preliminary design is inevitable sequential, and it's order has been arranged in a fairly effective way. This can be ascertained by conducting the network analysis, which is known as the design structure matrix (DSM) [Smith et al, 1994]. The result shows that all the tasks can be completed sequentially without having to make any guess or assumption throughout the design process; thus, it appears to be a reasonably effective order. To a certain extend, this has justified the spiral concept of Evans, where he was attempting to assist designers in organising the thought process, so as to enable ship design problems to be solved more efficiently [Evans, 1959].

Tasks *C*, *D* and *F* (see Figure 4) have been expanded into Level $n_i m_j$, and are L33, L34 and L36 respectively (see Figures 5, 6 and 7). This is to provide an insight into elementary processes, in order to investigate the interdependency of each design task at a lower level. The diagrams L33, L34 and L36 show that certain design tasks do not actually require the output of the preceding one, but they do require some input from elsewhere. Moreover, a small numbers of tasks may even be carried out at once after the conceptual design. These two findings hint at the possibility of overlapping the whole design process.

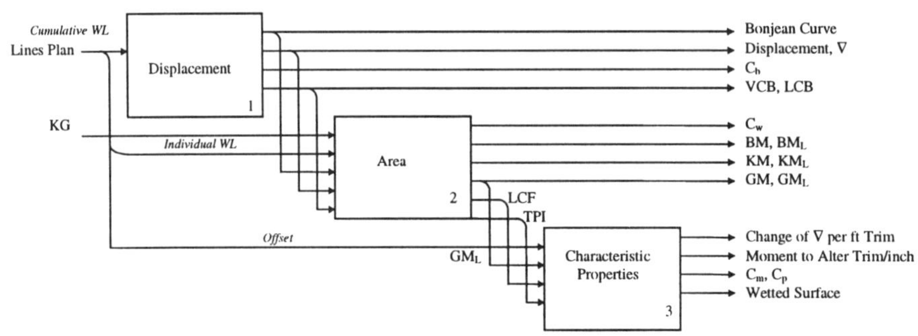

Fig. 5. Diagram L33 : Hydrostatics

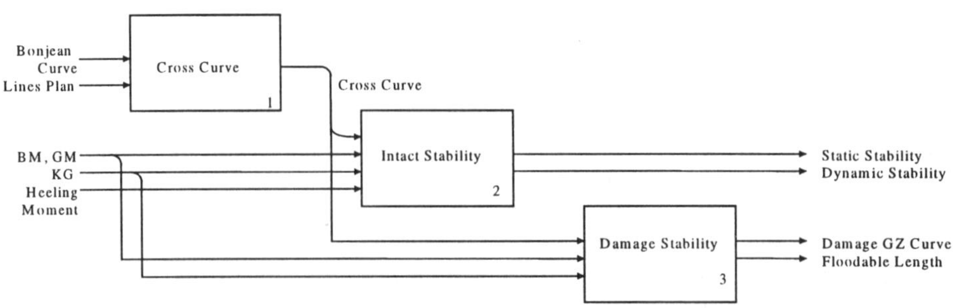

Fig. 6. Diagram L34 : Stability

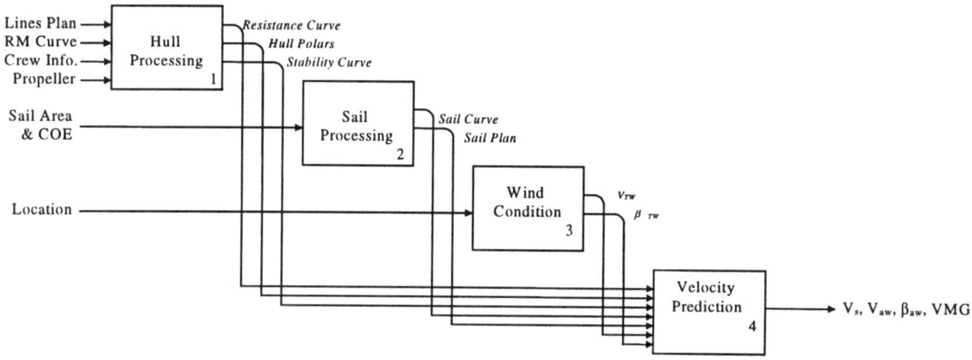

Fig. 7. Diagram L36 : Velocity Prediction

Assuming that the duration of task *A* is 6; task *B* is 2; task *C* is 2; task *D* is 3; task *E* is 2; task *F* is 3 and task *G* is 2 units of time, and the intervals to produce each output within a specified task are spread uniformly, an overlapped design process can be constructed (see Figure 8).

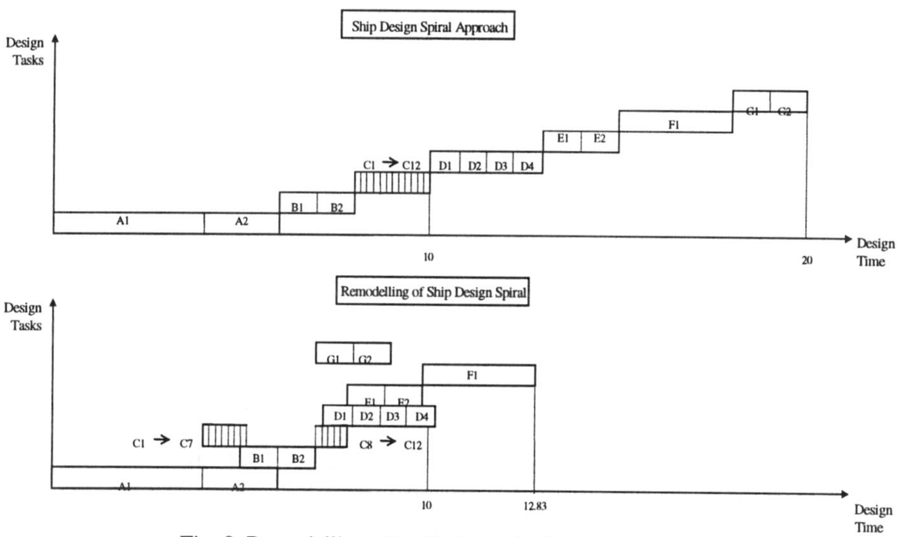

Fig. 8. Remodelling of preliminary design process

Figure 8 shows that the design process could be overlapped by up to 40%. Moreover, the right information becomes available to the designers at an average 30% earlier than in the spiral approach.

5 FUTURE WORK

Two aspects of the function model need to be investigated further prior to system development. First, the model requires refinement to ensure the acquisition of an accurate description of the problem situation. Second, the duration of each design task needs to be considered in terms of CPU and interactive times. This is to clarify where the significant improvements of the concurrent approach arise. Subsequently, the system development process will employ the IDEF4 technique to incorporate desirable life cycle qualities such as modularity, maintainability and reusability

6 CONCLUSION

This paper deduces that existing integrated CAD systems do not profoundly resolve the problems of the ship design process. Therefore, a new design approach, the CICAD, which overcomes the sequential nature of the Ship Design Spiral, is proposed. CICAD resolves the problems by integrating the design process concurrently; this advances the right information to the designers, although not exactly at the right time, it nevertheless would provide a better global view of the design.

REFERENCES

1. Condylis, T., 1997, Industrial collaboration : C & S Yacht Design, Southampton, UK.
2. Evans, J. H., 1959, Basic Design Concepts, Naval Engineers Journal, November, pp. 671 - 678.
3. Excell, J., 1997, Has CAD killed the designer ?, Design Engineering, UK, Vol. March, pp. 25.
4. Hassan, N. E. A., and Thoben, K., 1992, Practical Design of Ships and Mobile Units, Elsevier Applied Science, London, Vol. II, pp. 2.1284 - 2.1297.
5. Hays, B., and McNatt, T., 1994, The Integrated, Computer-based Ship Design Environment, Marine Engineering, Computational Mechanics Publications, Southampton, pp. 75 - 84.
6. KBSI, 1995, IDEF4 Object-oriented Design Method Report, Human Resources Directorate, Logistics Research Division, KBSI (Knowledge Based System Inc.).
7. Keane, R. G., and Tibbitts, B. F., 1996, A Revolution in Warship Design : Navy-Industry Integrated Product Teams, Journal of Ship Production, Vol. 12, No. 4, pp. 254 - 268.
8. Kroemker, M., 1997, Personal communication : Bremen Institute of Industrial Technology and Applied Work Science (BIBA), Germany, mk@biba.uni-bremen.de.
9. Kroemker, M., and Thoben, K. D., 1996, Re-engineering the Ship Pre-design Process, Computers In Industry, Elsevier Applied Science, London, Vol. 31, pp. 143 - 153.
10. Larsson, L., and Eliasson, R., 1994, Principles of Yacht Design, Adlard Coles Nautical, UK.
11. Lyon, T. M., and Mistree, F., 1985, A Computer-Based Method for the Preliminary Design of Ships, Journal of Ship Research, Vol. 29, No. 4, pp. 251 - 269.
12. Mistree, F., Smith, W., Bras, B. A., Allen, J. K., and Muster, D., 1990, Decision-based Design : A Contemporary Paradigm for Ship Design, SNAME Transactions, Vol. 98, pp. 565 - 597.
13. Mulligan, H., 1997, Computed Aided Drawing, Engineers Journal, Ireland, pp. 25 - 29.
14. Pullin, J., 1997, Shipshape and Computer Fashion, CIM Focus, London, Vol. Feb., pp. 15 - 16.
15. Rawson, K. J., and Tupper, E. C., 1966, Basic Ship Theory, Longman, London.
16. Ross, J. M., 1995, Integrated Ship Design and Its Role in Enhancing Ship Production, Journal of Ship Production, Vol. 11, No. 1, pp. 56 - 62.
17. Simpson, A., 1997, Industrial collaboration : Andrew Simpson & Associates, Poole, UK.
18. Smith, R., and Eppinger, S., 1994, A Predictive Model of Sequential Iteration in Engineering Design, MIT's Schools of Engineering and Management, Massachusetts.
19. Wu, B., 1994, Manufacturing Systems Design and Analysis, Chapman & Hall, London.

Product Analysis

Optimisation of Mould Design in Die Casting of Pewter Parts through Numerical Simulation

B.H. Hu, S.W. Hao, X.P. Niu, K.K. Tong and F.C. Yee

Gintic Institute of Manufacturing Technology
71 Nanyang Drive, Singapore 638075

Abstract. Mould design, especially the gating system design, plays a very important role in the die casting of high quality products. Numerical simulation was used for the evaluation and design of the gating and venting systems for the hot chamber die casting of a "merlion", a typical pewter part widely used for souvenirs in Singapore. It was found that the improper geometry and location of the gating and venting systems in the old design resulted in air being entrapped during both the earlier and later stages of mould filling which caused blisters in the casting. In addition, improper cavity numbers and plunger speed led to an excessive cavity filling time, and subsequently resulted in cold shut problems. The casting mould was re-designed with a new gating system, with the aid of CAE techniques. The mould filling process was simulated on the screen before physically cutting the die. The new design gave a more homogenous mould filling and an efficient air venting system was designed following the results of the simulation. Together with the reduction of the cavity filling time, the problems of blisters and cold shuts were solved and high quality pewter parts were achieved.

1 Introduction

Die casting is a high productivity process, in which molten metal is injected into a precisely dimensioned steel mould to produce a part by reproducing, with high accuracy, the finest detail of the impression of the mould cavity. Mould design, especially the gating system design, plays a very important role in the die casting of high quality products[1-4]. A gating system is a composite of the various channels cut in the die that guides the melt from the furnace to the cavities to form the final components. It dictates from which portion the part starts to be filled, the filling direction and speed, whether the flow is turbulent or non-turbulent, and where the overflows and vents should be cut. Finally, it contributes to the quality of the casting in areas such as gas porosity, shrinkage porosity, flow lines, cold shuts and surface finish [2,4,5].

Pewter, a tin based alloy, is typically suitable for die casting due to its low melting point and good fluidity. However, die casters still face problems in areas such as gas porosity and cold shuts during production.

2 Numerical Evaluation Based on an Old Mould

The "merlion", shown in Fig. 1, is a typical pewter product widely used for souvenirs in Singapore. Traditionally, it has been made by spin casting or cold pressing. However, due to the increasing demand for the products, a high productivity process is required. One such process that has been tried in the past was hot chamber die casting. However, it was found that air was entrapped and cold shuts existed within the castings produced(Fig. 2). Table 1 gives the general information for such a casting. The equipment used for this study was an old one phase hot chamber die casting machine. Some key parameters of the machine are shown in Table 2, and Table 3 shows the alloy and die information used for simulation.

The principle of the hot chamber die casting process is briefly illustrated in Fig. 3. The molten metal fills the shot sleeve chamber through the melt inlet. The plunger then moves downward to inject the melt through the gooseneck, nozzle and gating system into the casting cavity. The symmetrical half configuration and geometry of the old gating design and the venting locations of the mould, together with the castings, are shown in Fig. 4. The gate thickness is 0.8mm, about half the thickness of the "merlion" part. The total gate area is $48mm^2$ for 10 "merlion" parts.

Table 1 General information of the casting

Alloy	Casting weight	Mould Material	Mean wall thickness	Overall Dimensions
Pewter (2%wtCu, 1.6%wtSb)	8.5g	Hot work steel	1.5mm	53x24x1.6

Table 2 The machine information

Tonnage	Plunger diameter	Hydraulic pressure	Plunger velocity
50 ton	50	100 bar	0.2m/s

Table 3 The alloy and die information

Density	Specific heat	Latent heat	Liquidus	Solidus	Die conduct.	Pouring temp.	Die temp.
7300 kg/m^3	222 J/kg K	7080 J/kg	220^oC	180^oC	25.6 w/m ^oC	300 ^oC	101 ^oC

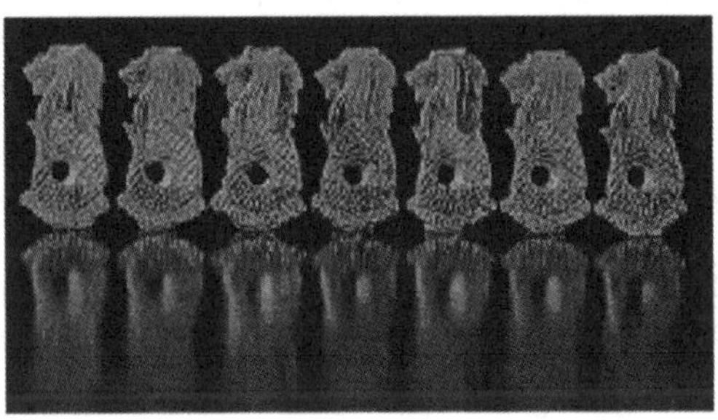

Fig. 1 "Merlion": A typical pewter part

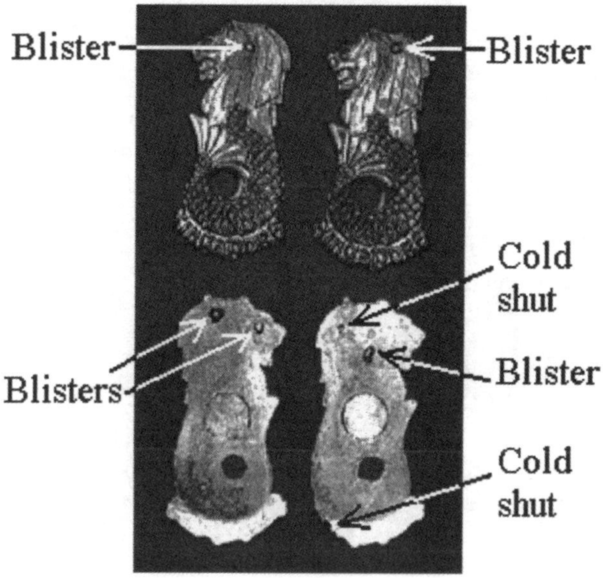

Fig. 2 Blisters in castings

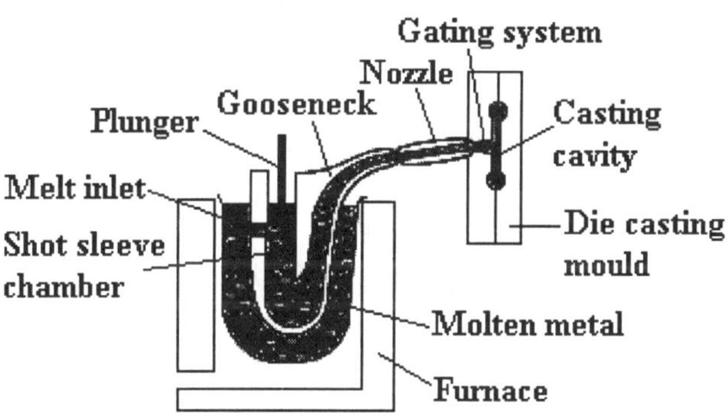

Fig.3 Principle of hot chamber die casting process

A commercial CAE package (MAGMAsoft) was used for numerical analysis. Figures 5 and 6 are selected simulation results (temperature distributions) of the mould filling at two different filling stages. The temperature of the melt is indicated by the grey scales. The grey colour at the bottom of the scale indicates the lowest temperature, and the white colour on the top of the scale the highest. The unfilled portion of the casting cavity is shown in light grey colour.

Fig. 5 shows the filling results at 30% cavity fill. It can be seen that as the liquid metal hits the basin of the sprue, air is entrapped in both the straight and the basin portions of the sprue. This is due to the increasing cross section of the sprue and the basin along the flow direction. Also, the fact that the air is entrapped inside the metal stream, and no possibility exists for it to be vented out of the cavity. Therefore, it will be brought further into the casting cavities during the later stage of filling. This will subsequently result in gas porosity or blisters in the casting.

Fig. 6 represents the filling pattern at 90% cavity fill. It can be seen that the melt fills the tail portion first and then back-fills the head of the "merlion". Since the overflows and air-vents in

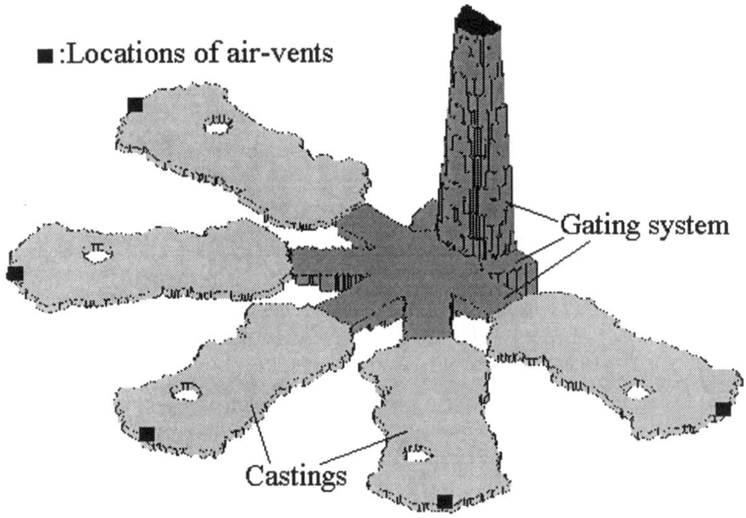

Fig. 4 Three dimensional layout of the casting gating system

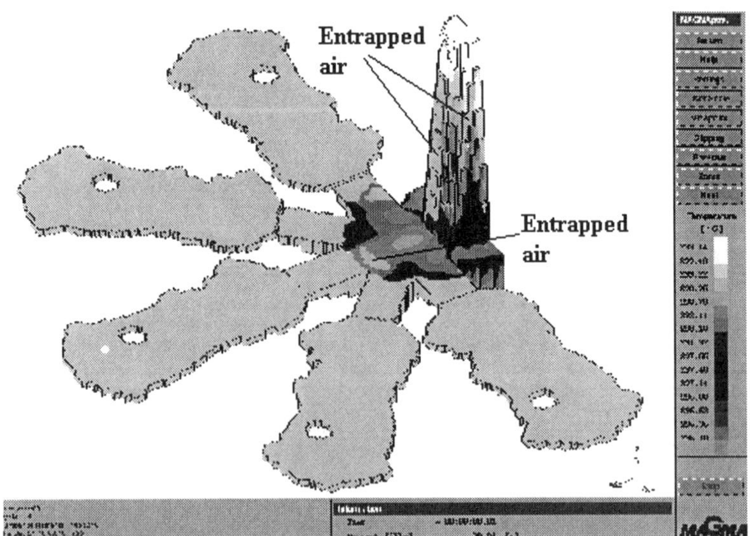

Fig. 5 a) Old die design: Mould filling (30% cavity fill) - Overall view

the old design are located at the tip of the tail instead of at the head of the "merlion", it is impossible to remove the entrapped air. This explains why blisters were often found at the head potion of the "merlion" (Fig. 2). The cavity filling time was calculated as 29.6ms, which seems to be excessive for such a small pewter part. Defects, such as cold shuts, were also found in the castings (Fig. 2).

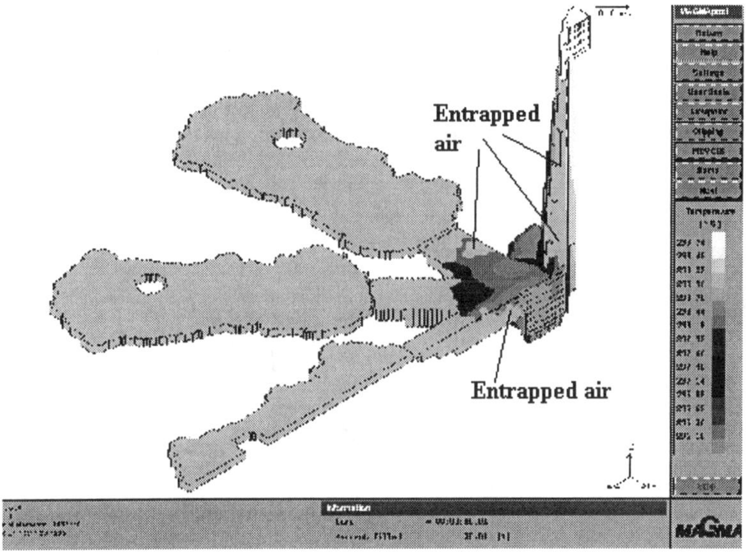

Fig. 5 b)Old die design: Mould filling (30% cavity fill) - Section view

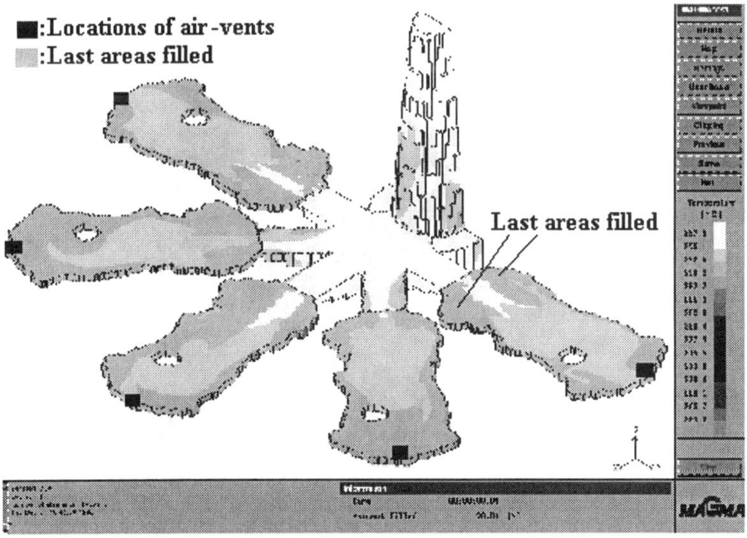

Fig. 6 Old die design: Mould filling (90% cavity fill)

3 New Design

Based on the above evaluation, the mould was re-designed with a new gating system with the aid of CASTFLOW, a computer aided gating design software. In order to get a more homogenous mould filling, a tapered runner system was introduced for this part. Considering the capability of the casting machine, the cavity numbers were reduced from ten to eight. A typical tapered runner is shown in Fig. 7. The layout (3-D) of the new symmetrical half gating system is shown in Fig. 8. The total gate area is 25.6mm^2 (4mm x 0.8mm x 8).

In order to demonstrate visibly the melt flow during the various filling stages and the melt solidification behaviour, to determine whether the new design needed further optimisation, a

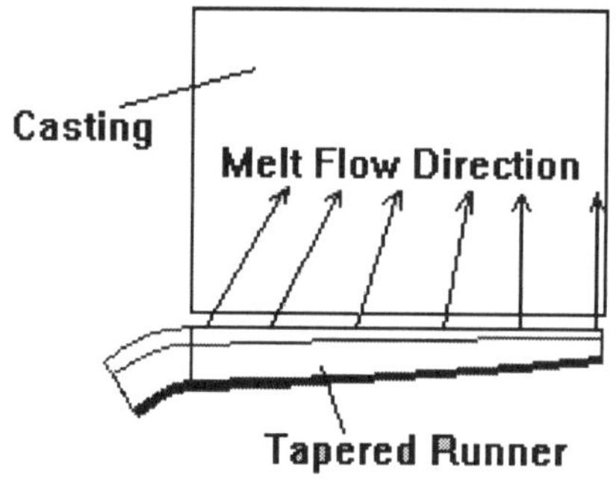

Fig. 7 Tapered runner

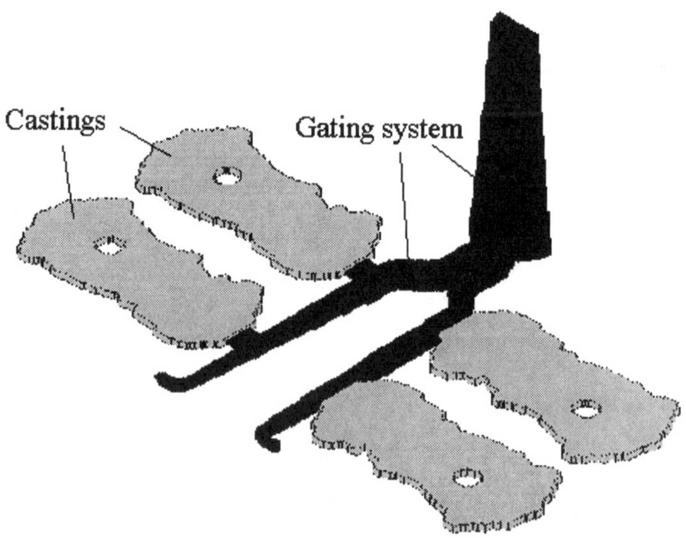

Fig 8. 3-D layout of the new runner and gating system for "merlion" parts

simulated "shot test" was conducted using MAGMAsoft. The simulation used the same alloy and process parameters as those used for the old design, except that the plunger speed was increased to 0.3m/s.

Figures 9 and 10 show the simulation results (temperature distributions) of the mould filling at 30% and 90% cavity fill respectively.

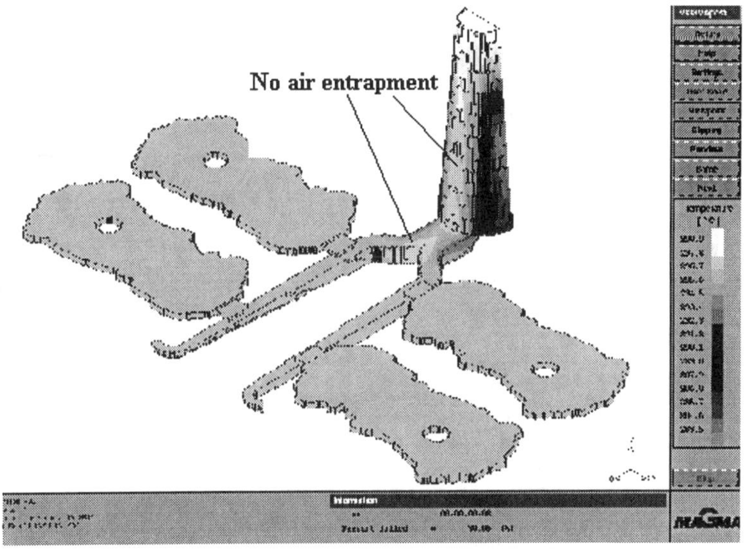

Fig. 9 a) New design: Mould filling (30% cavity full) - Overall view

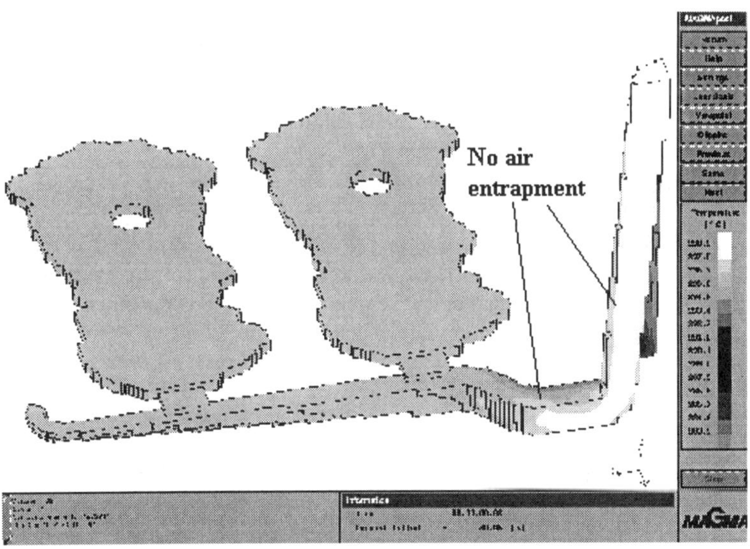

Fig. 9 b) New design: Mould filling (30% cavity full) - Section view

It can be seen from Figures 9 and 10 that, by using the new gating system, the melt steadily fills the sprue and runner without any gas entrapment.

With the new design, the melt also fills the tail-end of the "merlion" first, before back-filling the head portion. This is easily understood, from the geometry of the "merlion", irrespective of where the gate is located. Therefore, new air-venting channels are added to the last filled areas near the head of the "merlion" in the new mould design, to remove the air in these regions. However, in order to remove the air in the cavity during the earlier filling stages, the air venting channels at the tail end of each part remain. In addition, the calculated cavity filling time was reduced to 15.8ms. Further experiments verified that high quality castings without any blisters and/or cold shuts were achieved (Fig. 11).

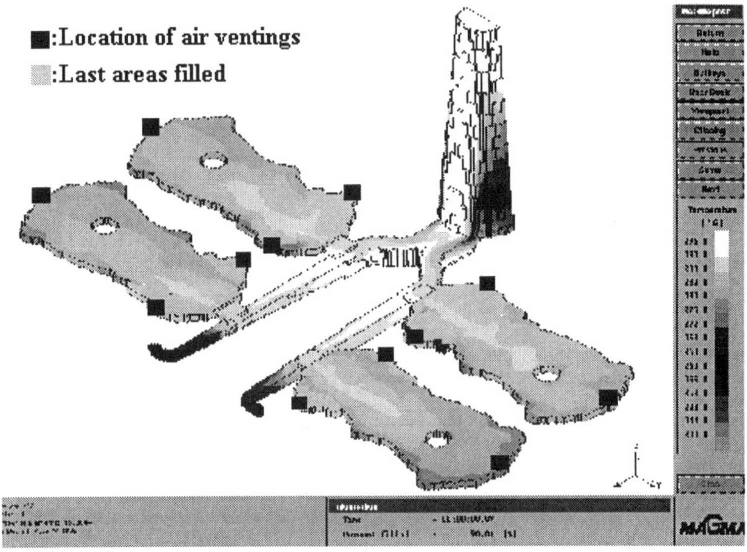

Fig. 10 New design: Mould filling (90% cavity full)

4 Conclusions

To secure good quality castings, it is important to have a proper gating and venting system in the mould to guarantee homogenous mould filling and efficient air venting. Numerical simulation was used for the evaluation and design of the gating and venting systems for hot chamber die casting of pewter parts, through which the mould filling process was made visible and predictable.

It was found that the cause of the blisters in the old design was the improper geometry and location of the gating and venting systems. This led to air entrapment during the earlier and later stages of mould filling. A long cavity filling time caused the cold shut problems.

With the aid of the CAE technique, a new gating system was designed. The mould filling process was simulated on the screen before physically cutting the die. The new design gave a more homogenous mould filling and an efficient air venting system was designed following the

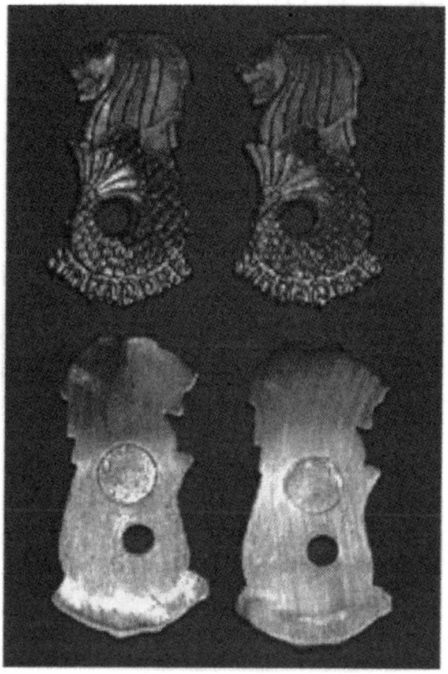

Fig. 11 Hot chamber die cast "Merlion" parts

results of the simulation. Together with the reduction of the cavity filling time, the problems of blisters and cold shuts were solved and high quality pewter parts were achieved.

Acknowledgements

This study is part of the work for the first phase of a project in collaboration with Phoenix Pewter-Phoenix Marketing Service, Singapore. The project was jointly funded by Phoenix Pewter and the Economic Development Board through the "Local Enterprise Technical Assistance Scheme". Special thanks are given to Mr. Kelly Tang, General Manager and Mr Tang Hon Wah, Technical Manager of Phoenix Pewter for their support and invaluable contributions to this project.

References

1. A. C. Street, *The Diecasting Book*, 2nd Edition, Portcullis Press Ltd (1986), 1986, pp.3-17
2. E.A. Herman, *Die Casting Dies: Design*, North American Die Casting Association (1992), pp.15-24
3. B.H.Hu, K.K.S.Tong, F.C. Yee, D. Sudheeran, and K.T. Tan, *Process Diagnosis and Optimisation in Pressure Die Casting*, Proceedings of the 3rd International Conference on Die & Mould Technology, Taipei, Taiwan, **A** (1995) 137-142
4. K.K.S.Tong, B.H.Hu and F.C. Yee, *Industrial Application of Computer Simulation in Mould Design for Pressure Die Casting*, Proceedings of International Conference on Mechanics of Solids and Materials Engineering, Singapore, **A** (1995) 235-240
5. T.H. Siauw and S. Tartaglia, *A Design Procedure for Feed Systems Incorporating Tapered Tangent Runners Used in Aluminium Pressure Die Casting*, Proceedings of the 2nd South Pacific Diecasting Congress, Sydney, (1983) 105-109

Fatigue Analysis based on FEA

Wolf-U. Zammert
Dept. Mechanical Engineering, Fachhochschule Technik Esslingen, Germany

Abstract. A PC software package is presented that enhances the performance of FEA-Systems to enable complete fatigue analysis. The program MAKE-FATIGUE is a software-system for general fatigue analysis of structural components, taking into account statistical effects in material behavior and load history. It incorporates the evaluation of fatigue life subject to deterministically or randomly distributed thermal and/or mechanical loads and the calculation of factors of safety due to periodically alternating loads with respect to the behavior in cases of overload. The several methods of fatigue analysis implemented in MAKE-FATIGUE are described and discussed by examples.

1 Introduction

It is the desire of every engineer to obtain a realistic view of the distribution of stresses in a component and to evaluate the consequences of fatigue life and factors of safety. The solution of this problem is the combination of FEA with numeric fatigue analysis.

FEA offers a complete view of the distribution of strains and stresses in a mechanical component. Few FEA-systems also offer a fatigue module for the calculation of usage factors. These fatigue modules however are not applicable for general analysis in the wide range of fatigue calculations. This task can be solved by using supplementary software like MAKE-FATIGUE. MAKE-FATIGUE is a PC software system for general fatigue analysis of structural components, taking into account statistical effects in material behavior and load history:

- evaluation of fatigue life subject to deterministically or randomly distributed mechanical loads,
- evaluation of fatigue life subject to deterministically or randomly distributed thermal loads,
- evaluation of fatigue life subject to a combination of deterministically or randomly distributed mechanical and thermal loads,
- calculation of factors of safety with respect to the behavior in cases of overload.

The input of stresses and load history can either be done manually or by using FEA results files. Structural elements like 2D-, 3D-solids and 3D-shells can be processed. Material properties are handled by files which are compatible with dBASE. The results of fatigue calculations due to an FEM calculation can be completely reviewed in the corresponding postprocessor. All features of the postprocessor can be applied. The MAKE-FATIGUE system consists of 2 modules: LIFETIME for evaluation of fatigue life and usage factors and DURABILITY for calculating factors of safety with respect to Goodman diagrams.

2 General Fatigue Analysis

2.1 Creep Fatigue Analysis

If structural components are being operated subject to elevated temperatures, the dependence of material properties upon temperatures has to be considered. With rising temperatures the strength decreases in general while the ductility increases. If the temperature stays below a limit T_K, which may be different from material to material, the temperature does not affect the microstructure of the material being deformed by the load. Temperatures exceeding the recrystallization temperature T_K induce a decrease of the load-dependent hardening of the microstructure: the material starts to creep. Material properties which have been measured in short-time tests at elevated temperatures e.g.

rupture strength and yield strength are only reliable, if the service temperatures are below the recristallization temperature of the corresponding material. Above the recristallization temperature the material strength not only depends upon temperatures it also depends on time. For fatigue design creep data are used, which show the dependence between temperature T_i, stress σ_i and time to rupture t_{Bi}, **Fig. 1**.

If a machine is operated at elevated temperatures subject to constant loads in a stationary working point, a constant creep fatigue acts on its components (e.g. thermal power plants). In general however, machines are working in different constant operating modes, e.g. full throttle - idle. In that case the components are exposed to a changing creep fatigue on different load levels. On every level a part of the initial creep fatigue strength is "consumed". **Fig. 2** shows as an example the load and temperature history of a pipe intersection that is stressed by a constant interior pressure. The superposition of the load and temperature history levels results in stress levels of the time t_i in which a constant temperature and a constant load act on the component. The collection of all stress levels of a component is called a stress spectrum. Stress spectra with different creep portions can occur in different ways. In general they result from a superposition of load and temperature history, but they can also occur by constant load and changing temperatures or constant temperatures and changing loads. In theses cases the stress spectrum consists of a load or temperature history only.

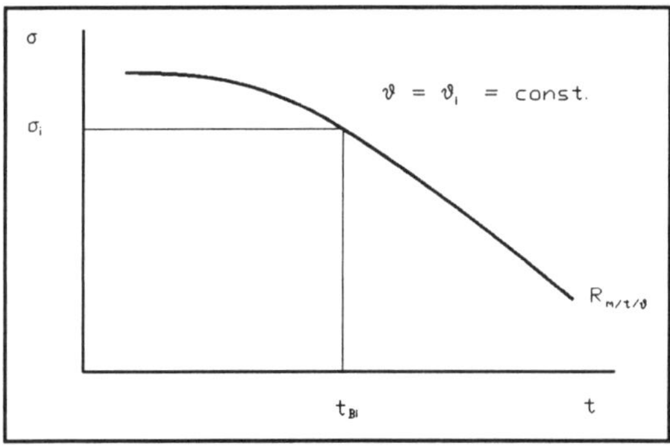

Fig. 1 Determination of time to rupture t_{Bi} of the component on stress level i due to service temperature ϑ_i

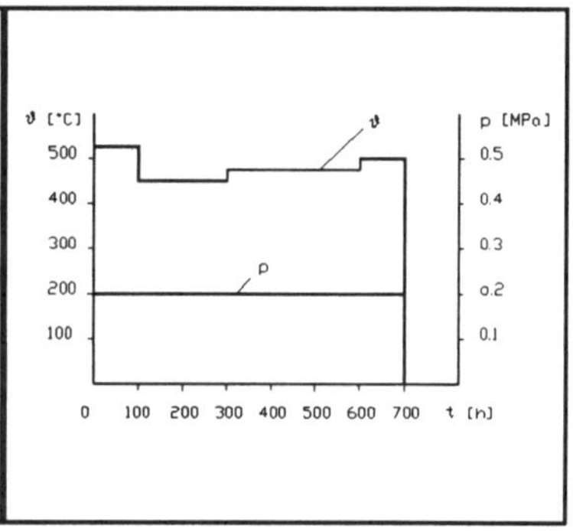

Fig. 2 Load and temperature history of a pipe intersection

An assumption for creep fatigue is however, that the hold time on any level lasts sufficiently long. If the stress levels change too quickly, a superposition of creep fatigue and fatigue due to cyclic stresses should be taken into consideration.

Robinson and Taira [1, 2] assume for their creep fatigue damage rule that the creep strength of a component subjected to varying loads and temperature is exhausted, if the accumulation of the creep fatigue ratio has reached its limit. The creep fatigue ratio $D_{\vartheta i}$ is calculated for every stress level as the ratio of hold time at the level t_i to time to rupture t_{Bi}, see Fig. 1. This model leads to the linear damage accumulation rule of Robinson and Taira:

$$\sum_{i=1}^{n} \frac{t_i}{t_{Bi}} = D_\vartheta \qquad (1)$$

D_ϑ	creep fatigue ratio
t_i	hold time on stress level i
t_{Bi}	time to rupture on level i
t_i/t_{Bi}	creep fatigue ratio on level i.

If the creep fatigue ratio D_ϑ reaches the limit $D_\vartheta = 1$, the creep fatigue strength of the component is exhausted. Creep fatigue ratios which do not reach this limit can be endured $1/D_\vartheta$-times. The estimated time t_B until the component breaks down is therefore

$$t_B = \frac{\sum_{i=1}^{n} t_i}{D_\vartheta} = \frac{t_{ges}}{D_\vartheta}$$

Fig. 3 and **Fig. 4** show the results of the creep fatigue analysis of a pipe intersection subjected to varying temperatures and constant internal pressure according to Fig. 2. The distribution of usage factors is shown in Fig. 3. Fig. 4 includes the distribution of the estimated lifetimes in the structure. For reasons of better representation the contour values have been set to specific data by the /cval-command.

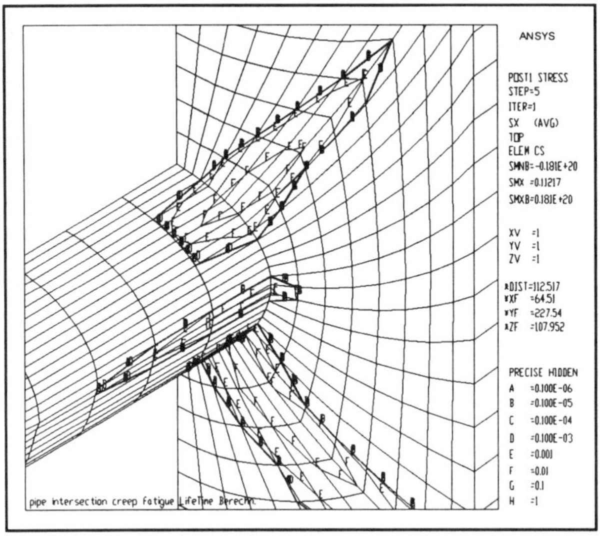

Fig. 3 Distribution of usage factors in the structure

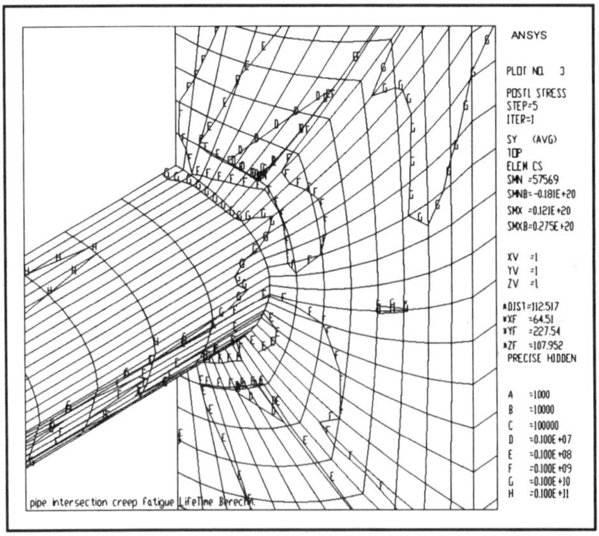

Fig. 4 Distribution of estimated lifetimes in the structure

2.2 Fatigue Analysis for Cyclic Loads

It is an experimental experience that σ-N-curves for steel and cast iron do not show any failure below a specific stress amplitude σ_A, even if the number of cycles is increased, **Fig. 5**. The stress amplitude $\sigma_{A(DFS)}$ at the failure limit of N_G cycles is the endurance stress amplitude. Plotting the endurance stress amplitude against the corresponding mean stress in the test leads to a Goodman diagram. In contrast to cubic volume-centered metals like steel or cast iron, face-centered metals like aluminum or magnesium alloys exhibit curves, which are apparently asymptotic to the N axis. For such materials, the stress $\sigma_{A(ZFS)}$ corresponding to some arbitrary number of cycles N_G is taken as the endurance limit.

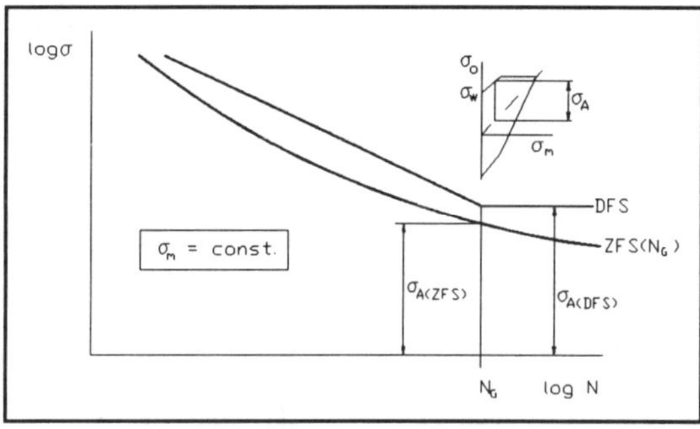

Fig. 5 σ-N-curves and Goodman diagram

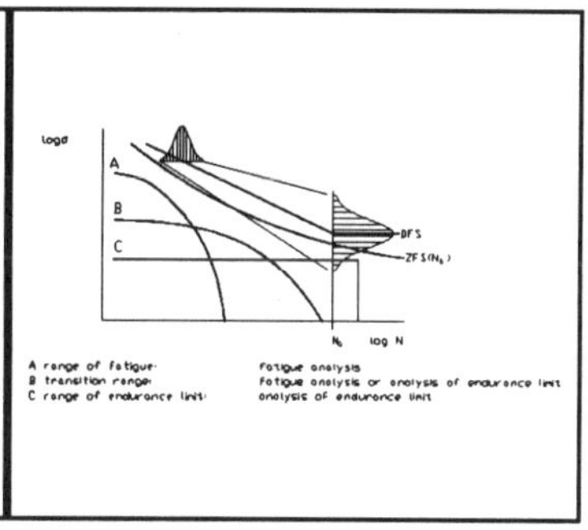

Fig. 6 Position of stress spectra to σ-N-curve

Subdividing the σ-N-curves into the range of fatigue and endurance strength, the position of the cyclic stresses to the σ-N-curve determines the method of fatigue analysis, **Fig. 6**. Structural stresses according to spectrum A, which are mainly in the range of fatigue, demand a fatigue analysis on the basis of σ-N-curves whereas components, mainly stressed periodically with constant amplitude in the range of the endurance limit (spectrum C) are dimensioned against the endurance limit with Goodman diagrams. At first glance it is not evident which is the more appropriate method of fatigue analysis for components with stresses in the transition range between fatigue and endurance, because both methods seem to be applicable. Due to the relatively extensive amount of stresses in the range of the endurance limit the use of Goodman diagrams seems to be more plausible. The actual stress spectrum must either be transferred into an equivalent stress spectrum with constant amplitude or a statistical mean stress amplitude has to be determined for fatigue. analysis [3-5]. Because the precise experimental investigation of Goodman diagrams needs a large amount of σ-N-curves and therefore is too expensive and time consuming, it is usual, to replace the Goodman diagram by straight lines, **Fig. 7**, which render an approximation by using only few characteristic material data. The approximation of the Goodman diagram can easily be transferred to the Haigh or Goodman diagram.

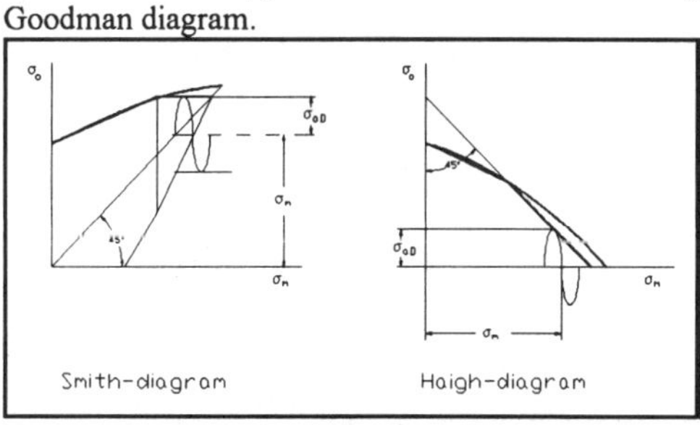

Fig. 7 Approximation of Goodman or Haigh diagrams by straight lines

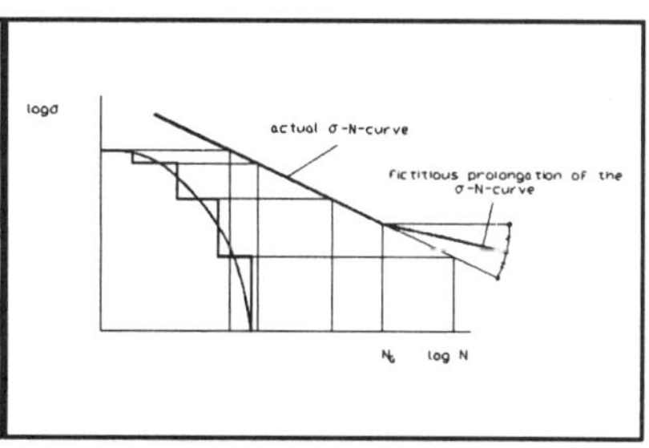

Fig. 8 Stepped stress spectrum and σ-N-curve for life estimation

2.2.1 Fatigue Analysis for Randomly Distributed Cyclic Loads

The basis of life estimation in a fatigue analysis is the precise investigation of the load history to be expected in service. The load history causes a stress spectrum, which usually consists of a mixture of randomly distributed high and low stresses. The range of time, in which the load history is recorded by a sample, must include every relevant service load condition in its characteristic frequencies. Numerous counting methods have been developed to transform the recorded load or stress history into a load or stress spectrum. The usually continuous spectrum is then split into approximating steps. This stepped spectrum and the σ-N-curve supplies the starting point for a numeric life estimation. In the range of the endurance limit the σ-N-curve is modified for life estimation according to [6], so that stress amplitudes below the endurance limit can be included, **Fig. 8**. The component has an initial resistance against fatigue, which is partially consumed on every step of the stress spectrum. Let N_i be the number of cycles to failure on a stress level i and n_i be the number of cycles of that level, the cyclic fatigue ratio $D_i = n_i/N_i$ determines the partial damage on level i. The linear accumulation of the cyclic fatigue ratios corresponds with the Palmgren-Miner rule [7, 8].

$$\sum_{i=1}^{k} \frac{n_i}{N_i} = D_\sigma \qquad (2)$$

D_σ	cyclic fatigue ratio
n_i	number of cycles on stress level i
N_i	number of cycles to failure on level i
n_i/N_i	cyclic fatigue ratio on level i.

According to the Palmgren-Miner rule the structure will fail, if $D_\sigma = 1$. Cyclic fatigue ratios which remain below the limit can be endured $1/D_\sigma$-times. The estimated time to failure N will be

$$N = \frac{\sum_{i=1}^{k} n_i}{D_\sigma} = \frac{N_{ges}}{D_\sigma}$$

Fig. 9 represents the stress amplitude spectrum of a pipe intersection. This load history can be due to deterministically or randomly distributed internal pressure. The resulting distribution of usage factors is shown in **Fig. 10**.

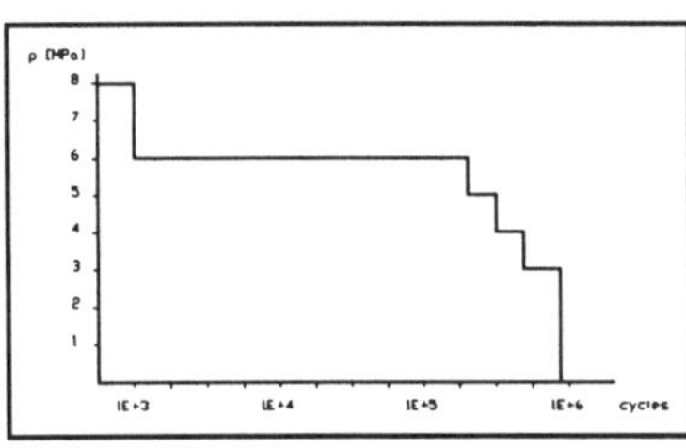

Fig. 9 Stress amplitude spectrum of a pipe intersection

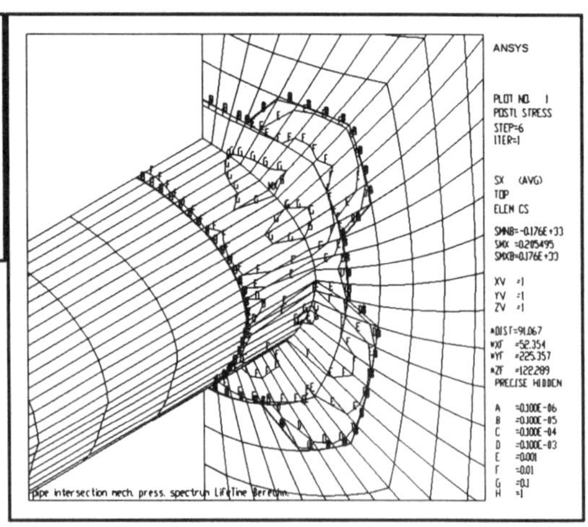

Fig. 10 Distribution of usage factors in a pipe intersection subject to varying pressure amplitudes

From a statistical point of view, fatigue data show large scatter and therefore fatigue material data usually are mean values (50%), because the mean tendency can already be determined with a relatively small number of specimen. As a rule the failure probabilities needed in life estimations of structural components are clearly lower than 50%, so that the life estimations calculated on a 50% basis have to be converted into values corresponding to the required failure probability. Conversion will be done with a safety factor S, which includes the statistical consideration of scatter and number of specimen representing the material data. If in Eq. (2) N_i is regarded as the mean value N_{i50}, the appropriate value for the cyclic fatigue ratio considering the required failure probability will be $N_i = N_{i50} / S$. Life estimations only make sense, if the statistical assumptions above are clearly quoted.

2.2.2 Analysis for Creep Fatigue Interaction

In process engineering, power plants or combustion engines etc., structural components are often operated in temperature ranges, in which creep fatigue has to be expected. If the resulting stresses include mean and cyclic stresses, the resulting damage will be produced by a combination of creep and cyclic fatigue. Both mechanisms of damage have to be considered in dimensioning the component. So the creep and cyclic fatigue stresses are not constant but vary according to the service point, they have to be recorded in appropriate spectra. The simplest method of accumulating creep and fatigue damage is the superposition of the Robinson-Taira and Palmgren-Miner rule. **Fig. 11** shows the scheme of linear damage accumulation using creep fatigue and σ-N-curves.

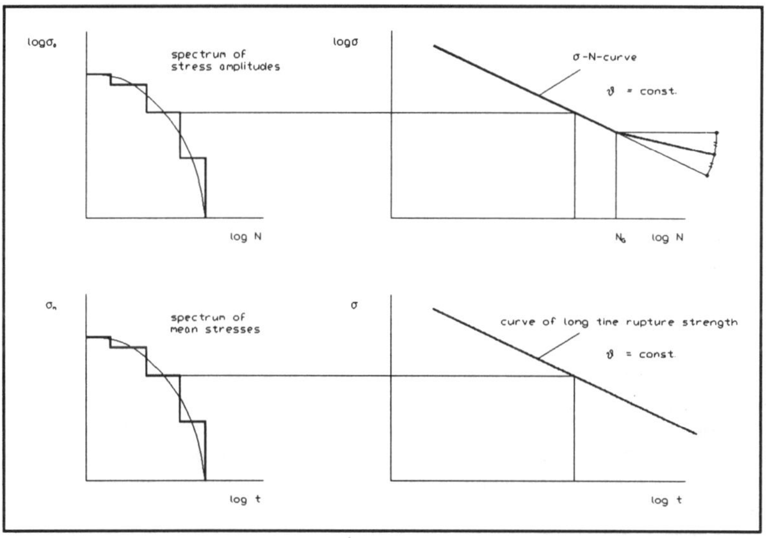

Fig. 11 Linear damage accumulation for creep fatigue interaction

The superposition of both linear accumulation rules leads to

$$\sum_{i=1}^{n} \frac{t_i}{T_{Bi}} + \sum_{i=1}^{k} \frac{n_i}{N_i} = D_{ges} \qquad (3)$$

Statistical aspects are now applied on the creep and fatigue data as mentioned above [9]. Fatigue design will be sufficient, if $D_{ges} < 1$. By using the method of linear superposition of creep and fatigue spectra it has to be kept in mind that the creep and fatigue ratios can only be compared with each other, if they are related to the same period of time. If the periods of time do not match, one fatigue ratio has to be converted to the period of time of the other. The creep fatigue interaction can then be endured $1/D_{ges}$-times.

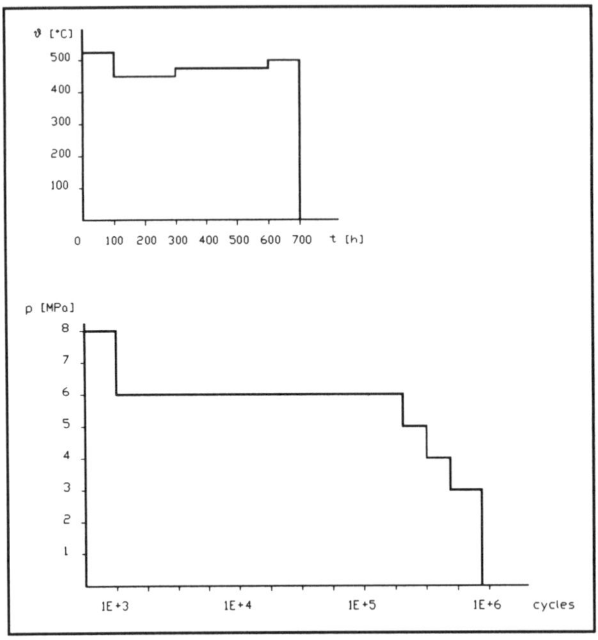

Fig. 12 represents the stress spectra for a combined creep and fatigue stress of the exemplary pipe intersection. The appropriate σ-N-curve will be selected for a mean temperature.

Fig. 12 Load spectra causing thermal and mechanical stresses

2.2.3 Analysis for Endurance Limit

The endurance limit of specimen and components is affected by numerous factors. The main factors are:

- type of load,
- mean stress,
- size.
- quality of surface,
- stress concentration,

Components, which are periodically stressed on a single load step are dimensioned against the endurance limit by reducing the Goodman diagram for unnotched specimen by reduction factors for size, quality of surface and stress concentration to an appropriate diagram for the corresponding cross section. The external loads of the component supply the nominal normal and shear stresses within the corresponding cross section, subdivided into nominal mean stresses and nominal stress amplitudes. According to the material's behavior of fracture the nominal mean stresses and nominal stress amplitudes are converted into nominal comparative mean stresses σ_{vm} and nominal comparative stress amplitude σ_{va}. The maximum comparative stress σ_{vo} is

$$\sigma_{vo} = \sigma_{vm} + \sigma_{va} \tag{4}$$

In FEA stress concentrations have already been calculated, therefore the consideration of a stress concentration factor will not be applied in the Goodman diagram. In FEA the comparative stresses σ_{vm}, σ_{va} and σ_{vo} correspond with the maximum stresses at the edge of a notch or discontinuity. The upper maximum comparative stresses σ_{vo} characterize the operating stresses of a component in the reduced Goodman diagram. Due to an overloading of the structure these operating stresses will be changed approaching the limitation of the Goodman diagram. The distance between σ_{vo} and the limitation constitutes a certain safety margin against failure. Overloading the structure can be performed by several ways. The different ways of overloading are shown in **Fig. 13**. They have to be selected by the designing engineer in correspondence with the operating conditions and the design of the component. If overloading of the structure results in an increase of the stress amplitudes while the mean stresses remain unaffected, case A will be applied. The safety margin will be exhausted, if the comparative stress σ_{va} reaches the limitation of the reduced Goodman diagram, which is called the upper endurance limit σ_{oG}. If the comparative mean stress σ_{vm} and the comparative stress amplitude σ_{va} increase, case B has to be selected. σ_{oG} will be determined by drawing a line from the origin through the upper maximum comparative stress σ_{vo}. If the lower maximum comparative stress σ_{vu} remains constant and σ_{vo} increases, case C will be applied. C is similar to B. Only the starting point of the line is not the origin but the intersection between the 45°-line and a horizontal line at the level of σ_{vu}. The last case D will be selected, if the stress amplitude remains constant while the mean stress increases. Fig. 13 shows the dependency between the endurance limit and the case of overload.

The upper endurance limit σ_{oG} consists of an endurance strength amplitude σ_{GA} and a corresponding mean stress. σ_{oG} and σ_{GA} are used to calculate factors of safety. S_o is the factor of safety against the upper endurance limit σ_{oG}. S_a is the factor of safety against the endurance strength amplitude σ_{GA}. The are calculated according to Eq. 5:

$$S_o = \frac{\sigma_{oG}}{\sigma_{vo}} \quad \text{and} \quad S_a = \frac{\sigma_{GA}}{\sigma_{va}} \tag{5}$$

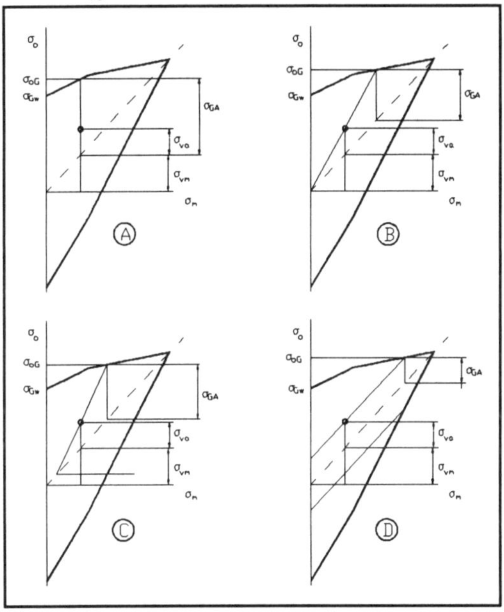

Fig. 13

Selection of case of over load for determining the upper endurance limit σ_{oG} using a Goodman diagram

Fig. 14 represents the distribution of safety factors against the upper endurance limit for a pipe intersection subjected to periodically alternating pressure between 3 and 9 MPa. The mean pressure therefore is 6 MPa and the amplitude pressure 3 MPa. The safety factors are shown at the top surface of the structure being modelled with shell elements.

Fig. 15 shows the definition of a path, along which the tendency of the safety factor against the upper endurance limit can be observed. The resulting values along the paths for the top- and bottom-position of the shell elements are represented in **Fig. 16**.

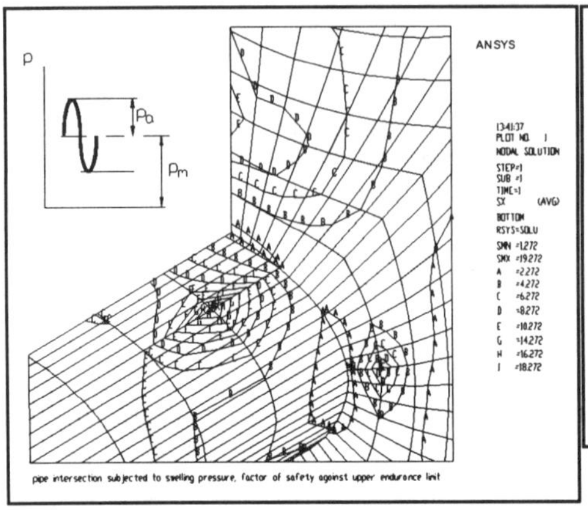

Fig. 14 Distribution of safety factors against the upper endurance limit

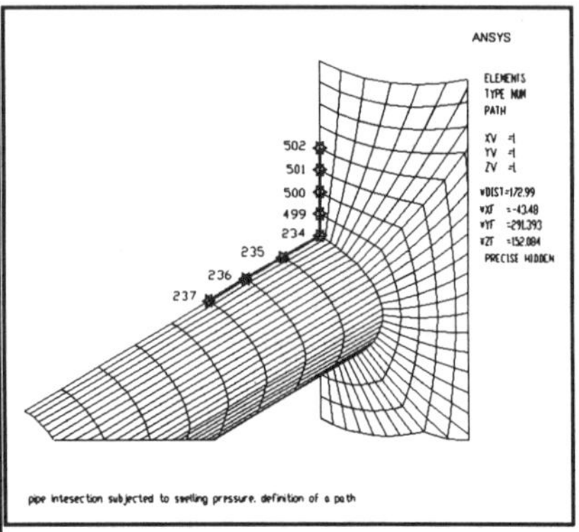

Fig. 15 Definition of a path for safety factors

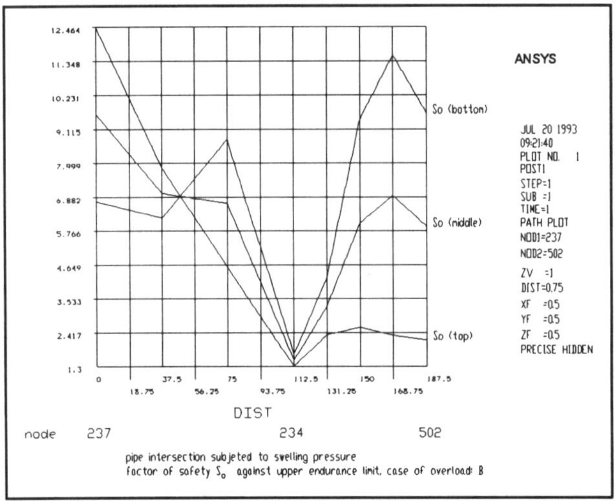

Fig. 16 Variation of safety factors along a path

3 Interfacing between an FEA-System and MAKE-FATIGUE

The stress spectra have to be defined as load steps in the FEA-system. The following remarks illustrate the interfacing between the FEA-system ANSYS® and MAKE-FATIGUE.

Interfacing between ANSYS Rev. 5.X and MAKE-FATIGUE uses temporary files, as shown in **Fig. 17** for an analysis of endurance limits. The structure will be modelled as usual and the load steps being defined in the solve-module. Starting the MAKE-FATIGUE system first initiates a preprocessor via a batch job which reads the stress tensor from file.rst and calculates the von Mises stresses at every node of the elements. These results are written to a temporary file from which MAKE-FATIGUE reads its input. MAKE-FATIGUE performs a fatigue analysis and writes the results to another temporary file from which a postprocessor makes the ANSYS Rev. 5.X pseudo-results file lifetime.rst for fatigue analysis or dauer.rst for the analysis of endurance limits.

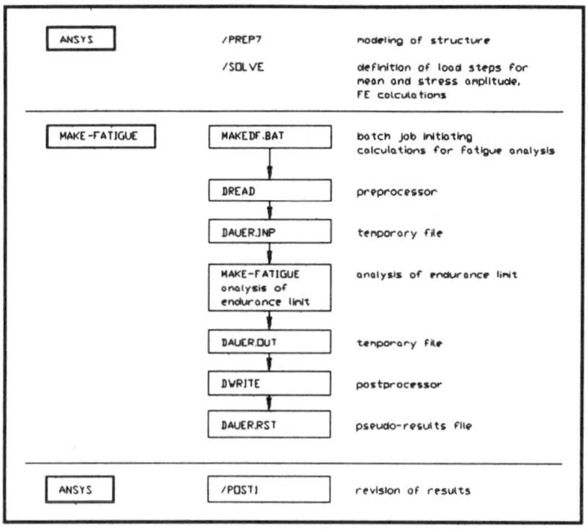

Fig. 17 Interfacing between ANSYS Rev. 5 and MAKE-FATIGUE

The results of the fatigue analysis are fatigue ratios (usage factors) and estimated lifetimes. They are treated as "stress-values" SX and SY in the pseudo-results file. The factors of safety S_0 against the upper endurance limit σ_{0G} and S_a against the endurance strength amplitude $\sigma_{GA}$are processed in the same way in an analysis of endurance limits for further reviewing in the ANSYS® postprocessor POST1.

4 Literature

[1] Robinson, C.L., Effect of Temperature Variations on the long-time Rupture Strength of Steels. Trans. ASME, 74 (1952), No. 5, p. 777-781.

[2] Taira, S., Lifetime of Structures Subjected to Varying Load and Temperature. Colloquium on Creep in Structures, Stanford University (California), Int. Union of Theoretical and Applied Mechanics, 1960.

[3] Kececioglu, D.; Haugen, E.B., A unified Look at Design Safety Factors, Margins and Measures of Reliability.Proceed. of 7th Reliability and Maintainability Conference 1968, p. 520 - 531.

[4] My Dao Thien, N.; Massoud, M., On the Relation between the Factor of Safety and Reliability. Trans. ASME, Series B, J. of Engng. for Industry 96 (1974) 3, S. 853 - 857.

[5] Haugen, E.B., Probabilistic Approaches to Design. London, New York, Sidney: J. Wiley & Sons 1968.

[6] Haibach, E., Beurteilung der Zuverlässigkeit schwingbeanspruchter Bauteile. Luftfahrttechnik-Raumfahrttechnik 13 (1967), No. 8, p. 188 - 193.

[7] Palmgren, A., Die Lebensdauer von Kugellagern. VDI-Z. 58 (1924), p. 339 - 341.

[8] Miner, M.A., Cumulative Damage in Fatigue. J. of Appl. Mech., Trans. ASME 12 (1945), p. A159 - A164.

[9] Zammert, W.-U., Betriebsfestigkeitsberechnung. Braunschweig, Wiesbaden: Vieweg 1985.

The Design and Implementation of a Visualization Tool for Structural Analysis

Yujin Hu, Jun Wu and Ji Zhou

CAD Center
Huazhong University of Science and Technology
Wuhan, Hubei, P. R. China

Abstract. This paper presents a finite element visualization facility, FEVS(Finite Element Visualization System), which has been implemented based upon PCs with Window95, NT and a graphic library --OpenGL. The main features of the FEVS are interactively easy operation, efficiency and robustness of algorithms, and powerful graphic processing functions. The design ideas, system structure, data management, interactive manner and various methods of finite element results processing and displaying are discussed in detail in this paper. Hierarchical data structure and algorithms for fast generating and displaying the contours, isosurface and cross sections using volume rendering are also proposed.

1. Introduction

The finite element(FE) analysis has become a very important tool in structural analysis as a result of the increase in computer power and the progress in numerical computation technologies. However, it produces a large amount of 3D unstructured data. To be able to interpret the data and understand the simulated process the scientists and engineers require visualization tools to be capable of extracting the interesting information from the data and displaying it in an easy comprehensive form.

FEVS is developed through analyzing the practical needs in engineering applications and adopting many new algorithms in visualization field. Since many advanced techniques in computer technology such as OpenGL and MFC programming are used, the system provides a user-friendly interface and many effective ways to display the analysis results. User can first use directly interface to analysis program to input the necessary data(including file input), then inquire the geometrical or topological or physical information, and can process static and dynamic data such as stress, strain and displacement etc., and can analyze or display interesting results in many easy comprehensive form. In this paper we will concentrate to discuss the design ideas, system structure, data management and main algorithms of FEVS. Figures shown in this paper were originally generated in color but are displayed in the paper as shades of gray.

2. System Function and Structure

FEVS is based upon Wondow95, NT and OpenGL environment, it provides a user-friendly graphical user interface and extensive graphical and numerical tools for gaining an understanding of analysis results. Users can be easy not only to operate by menu or commandlines but also to operate by multil-windows. Fig. 1 shows the interface of FEVS. In order to be maintained and further expanded in the future the system is designed into a modular structure. According to the functionality the system is divided into five main modules shown in Fig. 2. Each main module is also subdivided into many small modules.

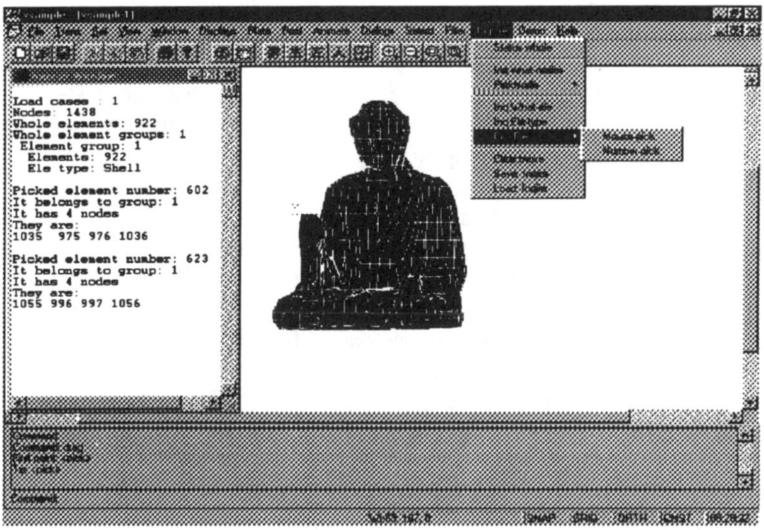

Fig. 1 The interface of FEVS

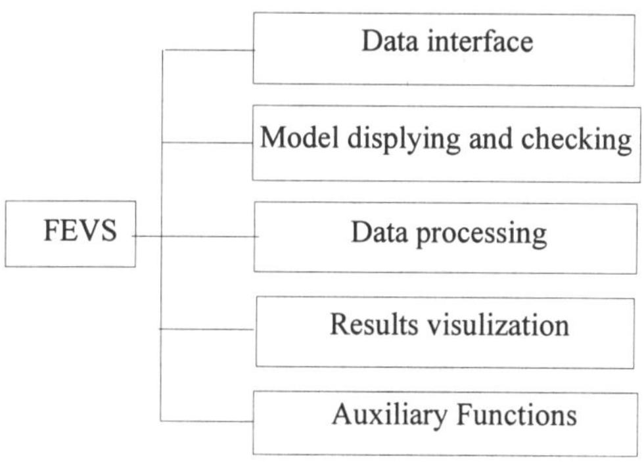

Fig. 2 System Structure for FEVS

A. Data interface module

Data interface module is data exchange module with popular FE analysis program. FEVS provides the interface to ADINA, ALGOR and ANSYS etc.. It supports many element types from FE analysis program. such as 2 to 3 nodes truss, 2 to 3 nodes beam, 3 nodes plate, 3 to 16 nodes shell, 4 to 27 nodes solid element and all plane element etc.

B. Model displaying and checking module

This module is used to display or check FE model in many easy comprehensive form , such as mesh, hidden, shrink, light shading and transparent model etc. users can rotate, pan and zoom model to make it at the best position and the best size.

C. Data processing module

This module is mainly used to extract the data that users interested from the basic output results of the FE analysis program. Also, it can compute new data types from the basic ones such as Von Mises stress and bending stress etc.

D. Results visualization module

This module is the core of the whole system. In order for user to easily understand the simulation results or make design decisions, this module provides many visualization methods for user, such as contour display with color or meshes, deformed geometry display and animation with meshes or shading colors, criterion display for given value and 2D function graphing etc..

E. Auxiliary functions module

It mainly consists of three sub-modules:

a. inquiring sub-module: This sub-module provides interactive tool to inquire the model's summary information(including number of sub-structures, number of element groups, number of elements and nodes etc.), geometrical and topological information of each node or element, physical attribute and the analysis results information.

b. filtering sub-module: Users can use this sub-module to filter out the uninterested elements, nodes or element groups etc. Filtering methods include color filtering, geometrical filtering, topological filtering, physical attribute filtering, analysis results range filtering etc.. After the filtering, users can concentrate on the most interested part of the model.

c. outputting sub-module: According to the results visualization, users can select some interested information to generate a report file and output it. Users can also output every virtual images displaying on the screen to a printer.

3. Data Structure and Management

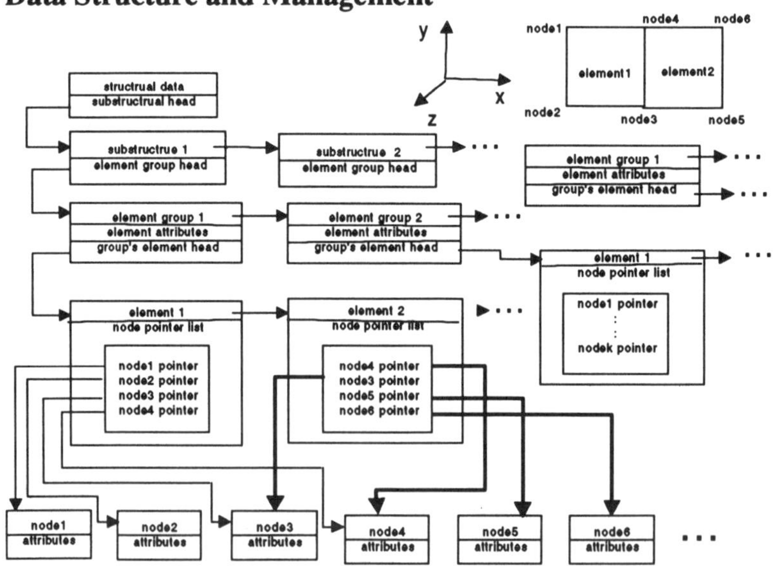

Fig. 3 Data Structure

In FEVS, the input information from FE analysis program can be classified into geometrical data(node coordinates etc.), topological data(element and element group etc.), physical attribute data(material, plate thickness etc.), load and constraint data, analysis results(node displacement, stress, temperature etc.) For storing and accessing these data more efficiently, FEVS uses Object-Oriented Data Base to manage them. In the computer the data relations are presented in Fig. 3(assume every element in the first element group have only four nodes like element1 and element2).

The table 1 shows the information included in the element group structure and node structure.

Table 1 structures for the element group and node

element group structure	node structure
element type	world coordinates x,y,z
number of elements	node displacement $\Delta x, \Delta y, \Delta z, \Delta\theta x, \Delta\theta y, \Delta\theta z$
node number in a element	node stress $\sigma x, \sigma y, \sigma z, \tau x, \tau y, \tau z$
material	node load Fx,Fy,Fz,Mx,My,Mz
layer	node constraints $Cx, Cy, Cz, C\theta x, C\theta y, C\theta z$
displaying color	node temperature t

4. Visualization Methods

4.1. Visualization of Displacement and Modes of Vibration

The method of displaying model's displacement is putting the undeformed model together with the deformed one in different colors. So users can see the model's deformed shape and each node's displacement clearly as shown Fig.4(left-window shows the hidden model, right-window shows deformed and undeformed meshes in different colors)

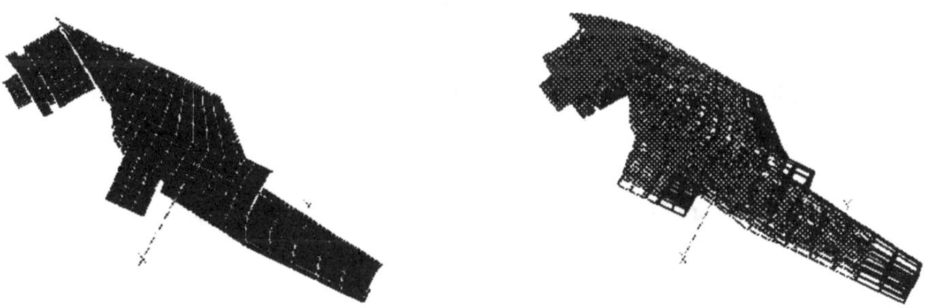

Fig.4 Deformed and undeformed meshes in different colors

FEVS can also simulate the model's deforming procedure in two ways according the result data provided by the FE analysis program. If the result data give only nodes' undeformed and deformed positions without intermediate positions, FEVS will select a suitable(For example sin(x)) interpolation function and a number of interpolation, then calculate out a series of nodes' intermediate positions from the undeformed position to deformed position, finally display the model according these intermediate positions in a sequence. Thus a simulation of the model's deforming procedure is achieved. In this method, all the intermediate positions are computed out according to the interpolation function. If the interpolation function does not fit the actual deforming procedure, the simulation is inappropriate. To get a more accurate simulation of the deforming procedure, it is necessary for analysis program to provide the intermediate positions in every given time interval instead of interpolation. This method is called deforming history animation, and it need more extra data.

There is vibration in the period of deforming. Vibration has many frequencies, and every frequency reflects a mode of vibration. FEVS uses static model or simulation by interpolation to view a mode of vibration as the same way described before.

4.2 Data Visualization

A. data processing

The FE analysis programs output basic data types such as displacement components ($\Delta x, \Delta y, \Delta z, \Delta\theta x, \Delta\theta y, \Delta\theta z$), stress components ($\sigma x, \sigma y, \sigma z, \tau x, \tau y, \tau z$), node temperature (t) etc. data processing is to extract one of them or compute a new data type from them shown as Fig. 5

processed data field

extracting processing

stress component | displacement component | temperature | Von Mises stress | Tresca stress | bending stress | accuracy stress

Fig. 5 Data Processing

Table 2. Methods of visualization

methods	implementation	part of data field described
2D function graphing	various ways of selecting nodes	node selected
Deformed display and animation	deformed geometry display and animation of displacement or of time varying results	part or whole geometry model
Criterion display		
isolines	number of isolines can be set	surface of the data field
contour	grade shading smooth shading	
isosurface	shown by colored mesh light shading on the model surface shown in transparency	whole data field
cross section through model	light shading in the model surface shown in transparency	part of interior data field
isosurface in volume rending	light integration	whole data field

After the data processing, every node has a correspondent scalar value, so the model become a unstructured scalar data field.

B. Visualization of the processed data field

As shown on table 2, FEVS provides many methods to visualize the data field from the surface of the model to the interior of the model, from the local data field to the whole data field.

X-Y plotting is used to show the data value variation over a series of nodes selected as shown in Fig. 6.

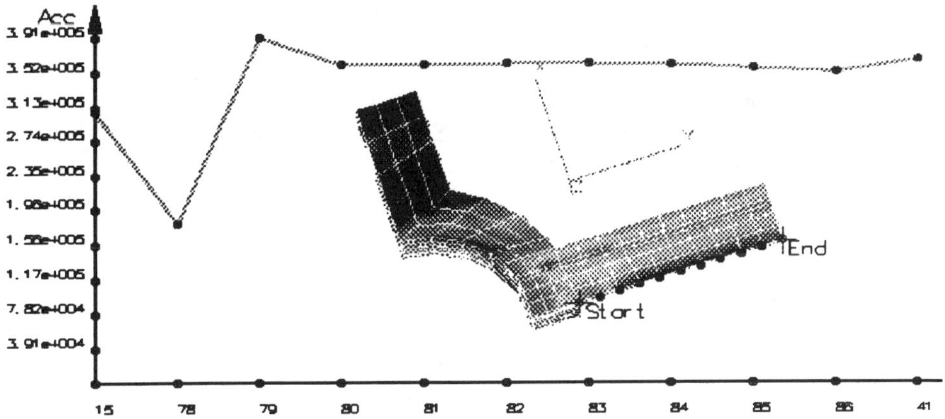

Fig. 6 X-Y plotting of nodes' Von Mises stress variation

In the same way, FEVS can use X-Y plotting to represent how a variable changes over time.

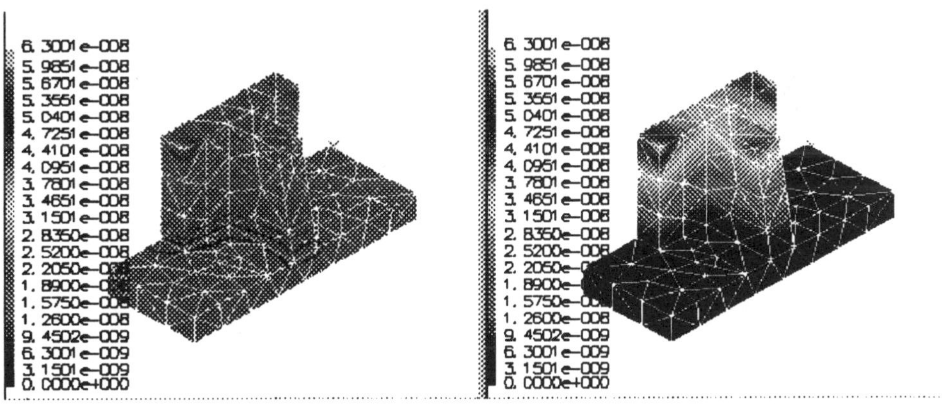

Isolines Grade shading

Fig. 7 Isolines and grade shading

Typically, FEVS uses isoline and grade shading color to present the data value on model's surface. The method to display the isoline and shading color has following steps: first find the data value range, then divide the range according to the isoline number set by the user and get many small value section $\Delta v_i = v_i - v_{i-1}$. We assign a uniform color to each Δv_i, the principle is the smaller the value, the darker the color, the bigger the value, the brighter the color. Then use the algorithm described in Section 4, we get a clear isoline or grade shading color representation of the data value on model surface. The area that has the brightest isoline or the color block is the location of the biggest data value. The relation between the color and the data value is also shown at the side of the window as shown in Fig. 7.

If we set the color number to a big one, for example, bigger than 100, the color blocks are so small that grade shading color become a smooth shading. The smooth shading is a more accurate representation of the continuous data field. Isosurfaces and cross section are used to present the data field in the interior of the model. Isosurface is a surface of constant value within a data field. It is a three-dimensional extension of two-dimensional isoline. What's more, FEVS can also use a block to represent a data value range which is between two isosurfaces. Cross section is a user defined cutting plane on which data value are shown by isoline or shading method. In FEVS, the plane can be defined in many ways and can be interactively oriented. Fig. 8 and Fig. 9 show these two methods.

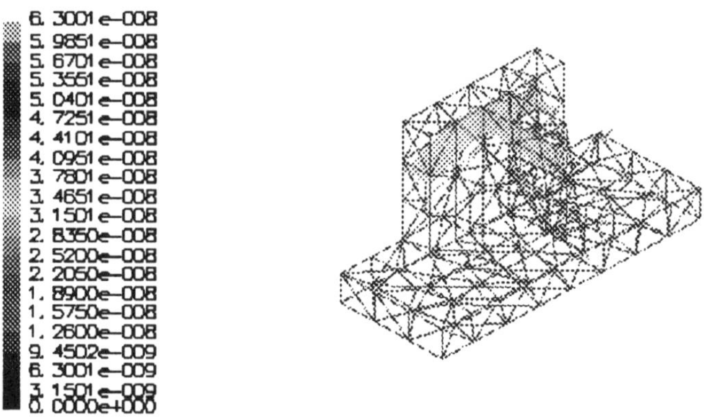

Fig. 8 Isosurface show in surface mesh(value=0.5*maxvalue)

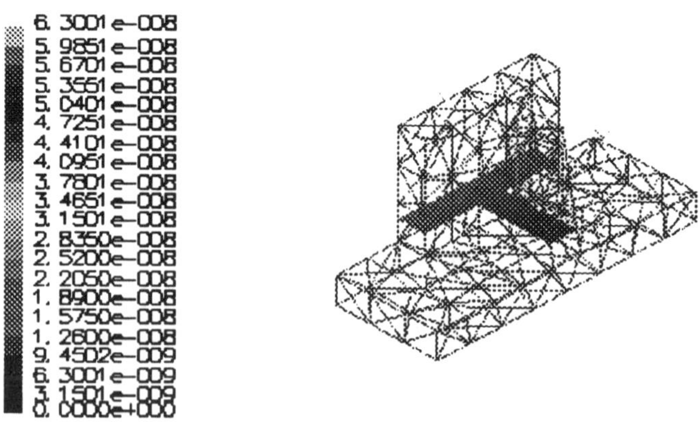

Fig. 9 Cross section shown in surface mesh

There are three ways of showing the isosurfaces(cross section). a. model and isosurface(or cross section) shown in different mesh colors. b. shading the isosurface(or cross section) in the model's surface shown in Fig.8(Fig.9). c. regard the model as a transparent object and isosurface(or cross section) a opaque one, show the isosurface in transparency.

Method of isosurfaces showing can only present the data value in the isosurface and the rest of the data field is still unknown. By using volume rendering technique we can get the whole data field's information at the same time. In this method, the isosurface is regard opaque, but the rest of the data field is semi-transparent, which emits light with a color reflect its value. By integrating the light emitted by them, we get a whole understanding of the data field. Fig. 10 is a example of this method. Readers can compare Fig.8 and Fig.10 to see the advantage of this method.

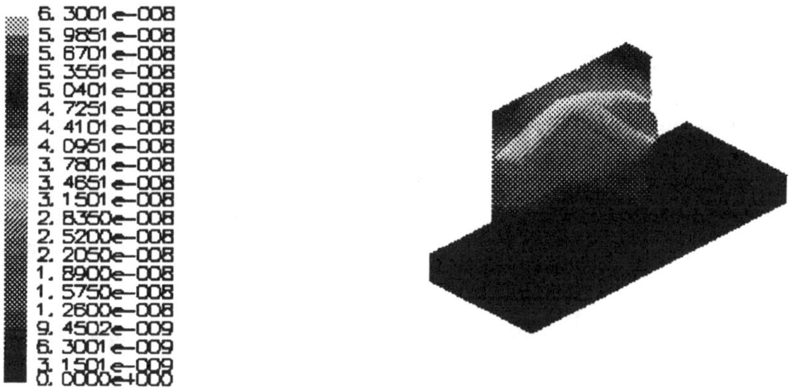

Fig.10 Volume rendered isosurface(value=0.5*maxvalue)

5. Main Algorithms of FEVS

In this section, we will discuss four main algorithms in the system. They are isoline generation algorithm, isosurface generation algorithm, cross section algorithm, isosurface generation using volume rendering method. In these algorithms, we assume the data value are linear in the edges of a element.

5.1 Isoline Generation Algorithm

This algorithm can be applied to the model consisting of solid elements(hexahedron, prism, tetrahedron), plate elements and shell elements. The algorithm has the following steps:

a. find the model's visible surfaces and translate these surfaces to triangles or quadrilaterals.

b. set the number of isolines n, find the data value range v_{min} and v_{max}. calculate the corresponding data value of isoline $v_i = v_{min} + (v_{max} - v_{min}) * i / n$, $(i = 1,2,3,...n)$ and assign each value a color according to the colormap defined before.

c. divide each quadrilateral into four triangles by interpolate a point in its shape center.

d. assume the current isoline value is v, current triangle is ΔABC and the values of

A,B,C have following relations: $v(A) \le v(B) \le v(C)$.Then if $v(A) \le v \le v(C)$, we can be certain that the v -valued isoline lies between the edge AC and one of the other two edges AB or BC. Similarly, if $v(A) \le v \le v(B)$, the other end of the isoline lies in edge AB. if $v(B) \le v \le v(C)$, it lies in edge BC. Calculate out the positions of the isoline ends by linear interpolation and connect them in the color correspondent to v.

e. for every isoline value v_i, do step d and draw the isolines in the triangle with correspondent colors.

f. for every visible surface triangle do step d and e, we get the model's all isolines .

5.2. Isosurface Generation Algorithm

Since isosurfaces exist only in three-dimensional solid elements(tetrahedrons, pentahedrons and hexahedrons), so this algorithm ignores the two-dimensional element types such as triangles, quadrilaterals and beam, truss elements. The algorithm has the following steps:

a. divide all the nonetetrahedral three-dimensional elements into tetrahedrons.

b. to maintain the correctness and consistency of the isosurface, adjust the diagonal orientation on ambiguous faces according to the isovalue using asymptotic decider method[1].

c. process each tetrahedron and get the isosurface patch in it[2].

d. calculate the light shading effect using Phone lighting method.

e. display the isosurface according to one of the three methods: colored mesh, isosurface in model's surface mesh and isosurface in transparency.

5.3 Cross Section Algorithm

This algorithm processes the model's three-dimensional elements and ignore other element types. The algorithm has the following steps:

a. get a plane defined by users.

b. get the surface of the model using hidden method.

c. construct the topological relations among the surface faces.

d. calculate the intersection points of the given plane and the surface faces.

e. connect the intersection points in a loop according to the topology of the surface faces and decide if the loop is a outer loop or inner loop.

f. calculate the intersection points of the given plane and the elements in the model.

g. triangularize the cross section plane using Delaunary method[3].

h. draw the isolines or shading colors in the cross section according the values in the intersection points.

5.4. Isosurface Generated by Volume Rendering

Similarly, this algorithm processes the model's three-dimensional elements and ignore other element types. Our algorithm is similar with the ray casting method [4] in which a ray is traced through the cells, the intersection of the ray with each cell is computed. The algorithm has following steps:

a. get a isosurface value input by users.

b. calculate the box of the model projection in the screen.

c. for every pixel in the box, we do[5]

1. a ray is cast to the model and if the ray intersect with a element a line section is obtained. Assume we get N line sections in the ray path.

2. calculate the color and opacity of every line section. If in the line section there is a point whose value equal to isosurface value, set its pacity=1.0;

3. sort the N ray line sections from near to far from the viewpoint.

4. integrate the N ray line section according to formulation below:

$$C_{acc} = (1 - O_{acc}) * C_{new} + C_{acc}$$

$$O_{acc} = (1 - O_{acc}) * O_{new} + O_{acc}$$

if $O_{acc} \geq 1.0$, stop integration.

5. calculate the shading effect on the pixel whose ray intersect with the isosurface.

6. output the color in the pixel.

5. Conclusion

In the present work, the visualization tool, FEVS, has been developed based on PC with window95, NT and OpenGL environment using MFC. FEVS provides a user-friendly interface and extensive graphical and numerical tools for gaining an understanding of analysis results. Data processing and data inquiring provides powerful data manipulation capabilities for user. Direct interfaces to more analysis program are available.

In the future, the visualization for vector and tensor fields as well as streamline will be added into FEVS.

References

1. Yu Hongqian, Hu Yujin, Zhou Ji, Isosurface Generation in Finite Element Models. in: Proc. of the First International Conference on Engineering Computation and Computer Simulation. Changsha.1995
2. Doi A, Koide A. An Efficient Method of Triangulation Equi-Valued Surface by Using Tetrahedral Cells. IEICE TRANSACTION, 1991,E74(1)
3. George P L.Automatic Mesh Generation: Applications to Finite Element Methods. New York: Willy, 1991
4. Levoy. M. Display of Surfaces from Volume Data, IEEE CG&A, 8(5),1988
5. Wen sili, Tang weiqing, Liu shenquan. Direct Integration in Volume Advances in Computer Aided Design & Computer Graphics,Vol(2),1993

Process Planning

CathViewer: An Interactive System For Catheter Design and Testing

Wang Yaoping, Chui Cheekong, Cai Yiyu, Lim Honglip

Institute of Systems Science
National University of Singapore
Heng Mui Keng Terrace, Kent Ridge 119597
Republic of Singapore
Tel: (65)7726724, Fax: (65)7744990
Email: {ypwang | cheekong | yycai | hllim} @iss.nus.sg

Abstract. This paper reports on a comprehensive, interactive simulation system of design and testing for catheter designers and manufacturers. The system features a physical computation model of finite element method built with specific biomedical knowledge. A 3D vascular structure which is segmented mostly from the Visible Human Data is used in the system as a realistic testing environment. In additional to conventional tools available in CAD systems, an easy-to-use catheter shape editor is also provided for manipulating catheter geometry. Design Engineers can freely design new catheters or modify the existed ones and then insert them into any position of the vascular structures for various purposes of testing. The navigation and deformation of the catheter and vascular structures in their interaction are computed real time using a customised Finite Element Modelling and Analysis module. The elastic property of catheter can be pre-defined or modified interactively during the navigation procedure.

1 Introduction

Minimally invasive therapeutic procedures, including surgery and interventional radiology, reduce patient discomfort, hospital stay, and medical costs. The socioeconomic impact of compensation for lost work time is also reduced. But interventional radiologists often face critical uncertainties in predicting the outcome of the procedures they plan to perform. More objective assessment of the equipment used in the procedures and better understanding of the treatment mechanism are also needed[1,2]. However, the evaluation of equipment relating to interventional procedures often depends on opinions of various operators. These may be accompanied by personal bias and experience, leading to differing views. In addition, the mechanism of treatment of vascular disease also remains unclear. Although several hypotheses have been put forward, majority of these studies were based on post-mortem examination. Different plaque morphology may respond differently. Moreover, patients also respond to balloon injury or various other devices differently. Generating a model to reproduce the varied scenarios will facilitate greater appreciation of these processes, thereby providing new information for risk assessment of patients and for the development of new devices. Developing a system to simulate the process not only provides a better understanding of the behaviour of these equipment but also allow assessment to be more objective. These factors will enhance modification and further development of the equipment.

However, conventional design systems/tools for designing catheter and other associated interventional devices are unable to address these issues properly. Engineers in catheter manufactures usually apply the existed CAD systems with no or very little direct medical knowledge for their catheter designs. They select the shapes for the catheter hooks and choose the material properties for the head, transition joint and body based on their past experience and some feedback from physicians practising in catheterization. Furthermore, they can very unlikely have an opportunity to validate the design concept themselves at a realistic checking environment. And the validation process is considered as the utmost crucial step in any design procedure[3].

Cathviewer is a joint R&D project involved with medical professionals, mechanical engineers and computer scientists to develop an interactive system for designing and testing catheter and other interventional devices in a medical knowledge based environment. An intuitive drawing tool is provided for engineers to draw, display and modify catheters very easily. A catheter library with all Cordis catheters [4] is also available for designer's reference. A comprehensive vasculature of human blood vessel from Visible Human data is employed in the system for providing the realistic testing bed for the newly designed catheters. An accurate biophysical model allow designers to realistically simulate, on the computer, the interaction between the vascular system and the interventional devices (the catheter, guidewire and various devices such as balloons, stents, etc.) through navigation manipulations, such as pushing, pulling and twisting the devices. The ultimate goal of this project is to develop an accurate and realistic Interventional Simulator [5] to simulate interventional procedures. **CathViewer** for Catheterising Viewer, as a part of this Interventional Simulator, emphasises on the development of new catheters. We will incorporate this model into the Interventional Simulator and deliver an invaluable tool for objective evaluation of the equipment, for development of new equipment, and for pretreatment planning in complex patient cases.

2 CathViewer Architecture

The design and implementation of CathViewer have been influenced by the following considerations.

- Incorporation of models. The model-based approach is followed. Physical models are used to represent the catheter and vascular systems.
- Real-time performance. The run-time system should avoid using time-consuming search and memory-management techniques and should provide predictable response times.
- Accurate vascular system. The credibility of the system relies heavily on the accuracy of the modelled vascular structure. Vascular structures have to be re-constructed from real human data, or professional model.
- An user-friendly drawing tool, good file management and comprehensive catheter library and material library.

The architecture of CathViewer is divided into several components, the Vascular Model Preparation Shell, Drawing Shell, Material Properties Manipulation Shell, and Finite Element Computation Shell. Fig. 1 shows the interaction between the various components.

Users can interact with the system through Vascular Model Preparation Shell and Equipment Manipulation Interface. The Vascular Model Preparation Shell represents a series of techniques we have exploited to develop accurate digital models to be used in the system. The digital model is refined in an iterative manner to improve its accuracy in modelling the required structure. The Equipment Manipulation Interface contains Drawing Shell, Material Properties Manipulation Shell and Navigation Shell. Drawing Shell is used to create and modify the shape of the interventional equipment through a graphic user interface. Material Properties Manipulation Shell deals with physical properties of the interventional equipment. Navigation allows the interventional equipment created/modified from the Drawing Shell and Material Properties Manipulation Shell to be tested on the vascular model from the Vascular Model Preparation Model. The 3D Rendering and Visualisation Shell provides the graphical and visualisation routines required to view the modelled interventional equipment and vascular structures. The Finite Element Computation Shell provides the numerical computation engine to analyse effect of forces on interventional equipment and simulate its navigation in the vascular structures.

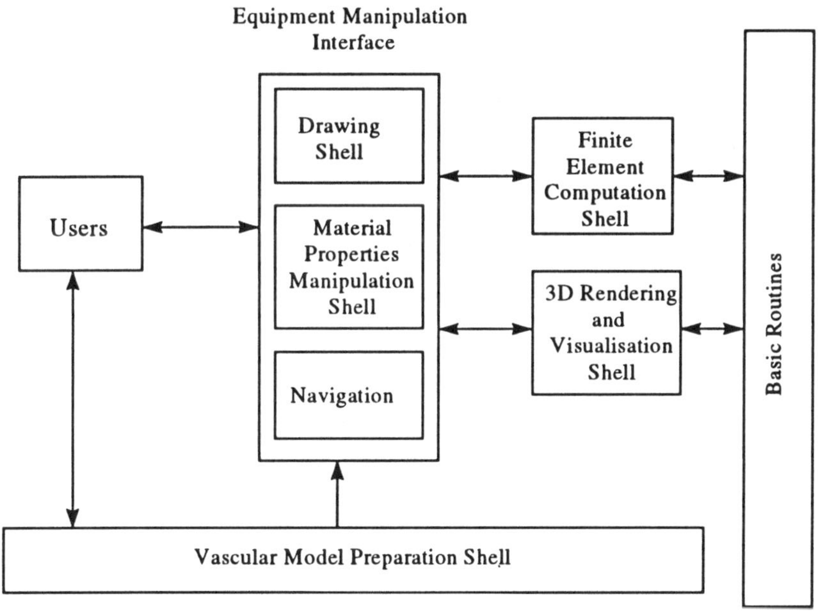

Fig. 1 Architecture of CathViewer.

3 Vascular Model Preparation Shell

Aiming to provide a realistic checking and validating environment for catheter design, we use human primary vasculature, cardiac and other secondary tertiary networks in Cathviewer as a virtual patient for the testing purpose. The primary arterial vasculature is from the VHD™ data, the secondary tertiary is from other scanned human vasculature data while the cardiac vessel network is from a professional steel model[6]. The vascular anatomy is represented by the central lines and their corresponding radii of the vessels. These vessels are assumed in the existing version of Cathviewer as cylindrical structures with varied radii. And these structures are treated as rigid.

The primary arterial vasculature and some associated secondary tertiary networks are shown in Figure 2. The extraction of the co-ordinates of the central lines and associated radii of the vessels is achived by applying semi-automatic and manual volume segmentation methods to the VH cyro-section data and the registered CT data[7]. The VHD™ data is stored in numerous 2D slices of uniform thickness of 1mm. We first manually segment the contours of the blood vessels concerned at each slice by using Photoshop, a commercial program. Then the centre of the projected cross-section of a vessel is calculated and the resultant co-ordinates of this centre is taken as the co-ordinates of the centre of the vessel at this slice. The corresponding radius of the vessel is also obtained by the equation: $R=A\cos\alpha/\pi$, where R denotes the radius, A denotes the area of the vessel at this slice, and α is the project angle which can be obtained by using the relationship between the co-ordinates of the central line at this and its two adjacent slices. The 3D model of the vessel is then built for the vessel by connecting all the co-ordinates at involved slices.

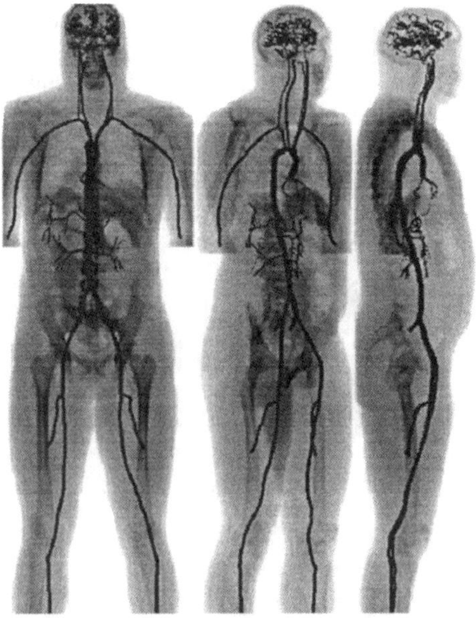

Fig. 2 Human primary vasculature and some secondary tertiary networks

Alternatively, we construct a 3-D model for the cardiac vessels from a professional steel model by using Coordinate Measuring Machine (CMM) Probing. A CMM equipped with an analogue probe or touch probe is a prominent example of the contact inspection method used to determine the 3D co-ordinates of the model. The determined co-ordinates is then fed into a process to generate the computational 3D model used in CathViewer. The related radii of the vessels are also found by using a mechanical measuring technique. The shape of the vascular system as shown in Fig. 3 has been examined and refined further by the medical professional using the Editor Tool.

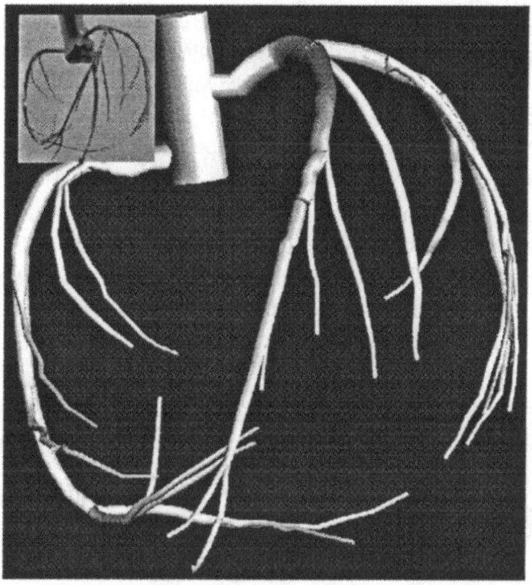

Fig. 3 Cardic networks used in CathViewer.

4 Drawing Shell

The shape of catheter determines how it is being used in the interventional procedure. The design of the catheter shape can be done using standard CAD tools. The Drawing Shell can import files created using these tools and convert them into the format used in CathViewer.

We have also develop a specific shape editor to design catheter. Assume that a catheter designer will start off with a straight rod in designing his/her new catheter. First consider creation and adjustment of the unstressed catheter shape. The UI begins by providing a straight starting form to work on, all the interactions except *new catheter* come under *adjustment*. The shape is built of rods, glued end-to-end without corners, starting from a fixed *proximal* end (where the design is to attach to the part of the catheter that joints the designed head to the end that never enters the patient), from which the parametrization by arc length s begins, and a free *distal* end which moves when, for instance, a point in the middle is bent. We will also call rod ends *proximal* and *distal*. There should be a user choice of marking the joints, or leaving them invisible.

Rods can be straight, flat arcs, or pieces of helix. The user does not need to select actively between them, since an arc and a straight line are special cases of a helix, and a straight line is a special case of an arc. Since computation is fastest in the straight case, and faster for arcs than helices, it makes sense to replace a *nearly straight* rod by a straight one, and a *nearly flat* helix by an arc, where *nearly* means that some number is less than a pre-set tolerance, say 0.01.

Adjusting the shape by changing one of the rods is like working with pliers on a wire, rather than adjusting control points on a curve whose ends stay fixed. It matches the way a doctor experiments with new shapes, by bending existing catheters. A rod can be lengthened, along its current straight/circular/helical path; it can be bent, by changing the arc or helix parameters; or it can be twisted around its joint with the previous rod. A rod r_i can be broken into two rods, or merged with the next rod r_{i+1} (which then disappears, while the length of r_i increases by the length of r_{i+1}).

Each rod is defined in its own coordinates, and mapped into the catheter coordinates by its position matrix. But it is convenient to note here that in each rod's internal coordinates it starts from (0,0,0) with tangent in the positive x-direction. In any interaction with the shape, the user needs to select a rod, a point on it, or a general point on the catheter. This selection will typically involve input of (x_{in}, y_{in}, z_{in}) values from a 3D stylus, or screen (x_{in}, y_{in}) values from a 2D mouse. As the input changes, different segments (at most one at any time: test the rods in order from proximal to distal) use highlighting to single that (x_{in}, y_{in}, z_{in}) is within their *touched* zone or that the line of $(x_{in}, y_{in}, z_{abitrary})$ meets it. Clicking the stylus or mouse button when a particular rod R is highlighted will start the interaction with R (such as twisting it around its start, or changing its length) that was picked by the most recent tool selection.

Many tool actions require widgets for user control. All the control functions of the classes involve changing one parameter at a time, so the 2D mouse interactions can all use a slider: for instance, in changing a length, display a slider with the current length, and let the user drag it to a different value.

The shape editor contains a set of tools to vary the length and cross sectional radius of this straight catheter; select reference point(s) along the catheter, bend and twist with respect to these points to create the shape of the catheter hook. We have yet to determine how well will this concept of shape editing be received by the industry. Not only do we have the ability to develop shape, we also have sophisticated rendering techniques to allow the catheter to be viewed in the material envisaged by the designer.

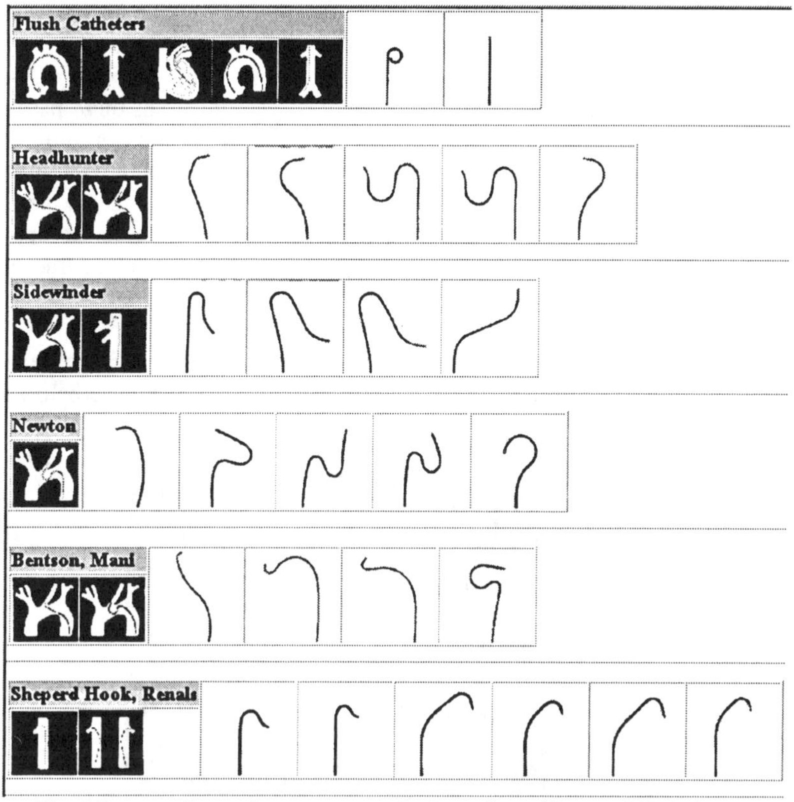

Fig. 4 Some catheters that are present in the catheter library of Cathviewer.

5 Manipulation of Material Properties

We are building a library of all the physical information for each material used in the manufacturing of catheters. Design engineer can assign varying physical properties to the elements defining the shape of catheter. The material models include rubber elasticity and plasticity for metals and foam, equation of state, and a graphical user interface for user material definition. The Young's modulus *E* and shear modulus *G* of *polyurethane*, which is commonly used to make various kinds of catheters, are used in cathviewer as the default material values for the newly designed catheter.

Along the length of the arc, there are distinct *physical bits* with assignable values of bendingstiffness (*E*), twiststiffness (*G*), and diameter (2*R*). Since these need not share endpoints with the geometric rod subdivisions of the shape, they are managed by a separate linked list. Each physicalbit in the list rccords the mechanical parameter values appropriate to it. The user is able to select an arbitrary pair of points p0 and p1 on the shape, create a new physicalbit and adjust the parameters by using sliders. The selection of the points is done by the geometric touch detection above; The rod(s) touched report the corresponding values of arc length (from distal end of the entire design), and the linked list of physicalbits is revised accordingly.

In the first version, the physical parameters of a physicalbit are to be treated as constants; later we can consider varying them along its length. One of the obvious schemes, mathematically, is interpolate linearly between values at p_0 and p_1, but this need not be the most easily manufactured variation. The impact of these parameters on the bending and twist stiffnesses will be taken into account in the function elemesh() (to be created by the FEMengine), which creates a model usable in simulations. This model must then be transferred to where an experimental simulation can use it: typically, the user will want to see it in 3D close up, going through the task it is designed for, rather than in the training simulation fluoroscopic shadow play.

Design engineer can test the effect of each variation of material properties by examining the effect of forces assigned at the given point. The Figure below shows how the catheter is bend under the impact of a constant force at a point.

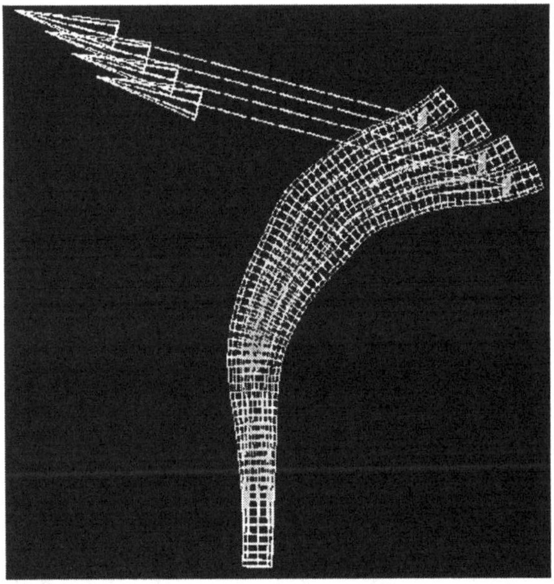

Fig. 5 Catheter bends under the impact of a constant force.

6 Finite Element Modelling and Analysis

With the goal of creating a realistic real time interactive environment for CathViewer, an incremental Finite Element Modelling (FEM) module has been developed for the analysis of catheters navigation in the vascular structure[8]. This FEM analysis uses a new algorithm to speed up the determination of interaction between the walls of vessels and the catheter, which is normally the most expensive portion in FEM calculation of contact problem. Note that the FEM nodes need not correspond to a point defining the hook shape of catheter.

In this FEM analysis, a catheter as a slender structure is discretized into a flexible multibody system with a finite number of 3D FEM beam elements. The blood vessels are assumed to be rigid circular tube-structure (a majority of the great vessels are of circular cross sections and they are well stretched by the surrounding muscles). The catheter navigation is considered as a sequence of movements such as push, pull and/or twist of the flexible multibody system inside the rigid tube-structure. The move of an element of the multibody system at each step is assumed to consist of a rigid-body displacement and a relative deformation. We first move the elements of the system as rigid bodies by applying multibody dynamics method at each step and then find the deformations at their equilibrium position by applying finite element method. Contact forces between the catheter and the walls of the vessels are introduced at the catheter nodes, which are outside the vessels during the rigid moves, to push the catheter back to the tubes. Those contact forces, however, are functions of the position of the catheter at each time step and their determination is very time consuming.

We have developed an efficient semi-implicit algorithm by using the hierarchy structure of the vasculature of the vessels for computing the contact forces involved. The computation of contact force is involved with two steps:

- checking whether catheter nodes are outside the vessels;
- calculate the contact force at the nodes if they are outside.

In order to achieve an efficient in/outside checking, all the branch segments of the vasculature are linked through a topology relation. As shown in Figure 6, a current vessel segment is connected with its parent, brothers and children segments. Each catheter node is associated with one branch segment and the in/outside checking for this node is conducted only with respect to its segment and parent-brother-child segment group, rather than to compare with all the branch segments of the vasculature. It therefore speeds up the checking process by a very good percentage. The contact force at the node outside is defined as $F=cdN$, where d is the distance between the node and the surface of the vessel, N is the unit vector pointing from the node to the surface along the distance direction, while c is the coefficient representing some physical characterisitc parameters of the contact. It is currently set up with an arbitrary number since the flexibility of the vessels has not been taken into consideration in the present study.

7 The Future

We plan to develop a hemodynamics model which will compute deformations of the associated arteries and fluid dynamics (pressure, velocity and temperature) of the blood flow in the arteries. In addition, the morphology, geometry and material properties of the plaque and diseased arteries will be taken into consideration. Careful computation of finite element analysis will be performed to take into account the interaction of the interventional devices with the diseased arterial wall and with the blood flow. The analysis of plaque cutting and removal will also be implemented.

Beyond the immediate goals to complete and refine the simulation and design system, there is a need to create additional patient models and a library of pathologies. Given the degree of variability of the sinus anatomy in the general population, it is important to have a variety of models representative of a diverse population. The models will help developers to test the newly designed catheters within a wide range of simulated conditions and could serve as an evaluation of the new catheters.

To facilitate and reduce the expense of creating patient models for simulation, better software tools are necessary. These same software tools would also facilitate creating patient-specific models from routine medical imaging examinations. This would enable clinical uses of CathViewer, allowing the design engineer to develop customised catheters; and the surgeon to preoperatively plan and rehearse a surgical procedure using these customised catheters. The design engineer and surgeon could experiment with competing design and surgical approaches respectively. The advent of medical digital imaging standards and the increasing appearance of medical image data networks have made it clinically practical to introduce new applications that use the patient's imaging data. In so doing, CathViewer can improve the efficiency and effectiveness as the catheter development system for interventional radiology procedures.

Acknowledgement: Support of this research by National Science and Technology Board of Singapore is gratefully acknowledgement. The generous help of Dr. T. Poston and B. T. Nguyen on the drawing shell of this research is sincerely appreciated.

References

[1] J.L. Berry, V.S. Newman, C.M. Ferrario, W.D. Routh, and R.H. Dean, A method to evaluate the elastic behaviour of vascular stents, *22nd Annual Scientific Meeting, Society of Cardiovasculatur & Interventional Radiology*, 1996, 7: 381-385.
[2] United States Food and Drug Administration guidance for the submission of research and marketing applications for interventional cardiology devices: PTCA catheters, atherectomy catheters, lasers, intravascular stents. Rockville, Md: FDA, May 1993.
[3] J. Thrner, J. Pegna, M. Wozny, *Product Modeling for Computer-aided Design and Manufacturing*, North-Holland, Amsterdam, 1991.
[4] CORDIS Corporation, Diagnostic and Interventional Radiology, *Technical Report*, Miami, USA 1992.
[5] J.H. Anderson, R. Raghavan, Y.P. Wang and C.K. Chui, daVinci – A vascular catheterization simulator, *Journal of Vascular and Interventional Radiology*, 1997, Vol.8, No.1, Part2, P261.
[6] Kyoto Scientific Specimen Co., Ltd, *Coronary Artery Model/key chart*, Kyoto, Japan.
[7] C.K. Chui, H.T. Nguyen, Y.P. Wang, R. Mullick, R. Raghavan, J.H. Anderson, Potential field of vascular anatomy for real-time computation of catheter navigation *First Visible Human Conference*, Bethesda, MD, USA, October, 1996.
[8] Y.P. Wang, C.K. Chui, Y.Y. Cai, R. Viswanathan and R. Raghavan, Potential field supported method for contact calculation in FEM analysis of catheter navigation, *XIXth International Congress of Theoretical and Applied Mechanics*, Kyoto, Japan, August, 1996.

A CAPP System Based on Concurrent Engineering System

Ying Sihong, Wang Rong, Ma Dengzhe

CIM Research Institute of Shanghai Jiao Tong University, Shanghai, 200030, China

ABSTRACT: This paper analyzes some PDM software, as PDM system is based on Concurrent Engineering and supports concurrent operation. The author points out that PDM is useful to manage CAD data, but is weekly effective to CAPP data. Now, researching on CAPP is emphasized on generation of process routings, not management of process routing, on part, not whole product, and most important thing is that how to share data resource is never considered.
The author puts forward thought of developing CAPP system based on Concurrent Engineering, and that single database, control mode of design process is the key to carry out concurrent operation in CAPP system. A functional system named SIPM has been developed after researching on methods of process planning, process routing data model and technical document management. It makes that all process planning engineers can share all data resource of their plant, and that is possible to established the interface between CAPP and MRPII. SIPM system has been used in many factories in China.
KEY WORDS: Concurrent Engineering, CAPP, CIM

1 Introduction

From 1995, most companies have many CAD seats to support product design, 41% drawings are in CAD format, and 60% of companies have two or more CAD systems. How to manage the design results and make design change becomes the most important problem. There are 42% companies have some kind of PDM system. PDM is an essential strategic technology underling a successful product development and support environment, and that provides a concurrent operating environment to design engineers. Now most companies are reported MRP (or MRPII or ERP) systems in use. Again, most of these are on different platforms than the design/document creation systems. MRP systems manage parts and materials through the scheduling and production process and many are beginning to offer elements of document management. Usually, PDM system outputs the BOM of a product in neutral format, and MRP system reads, that finishes the information integration, see figure 1.

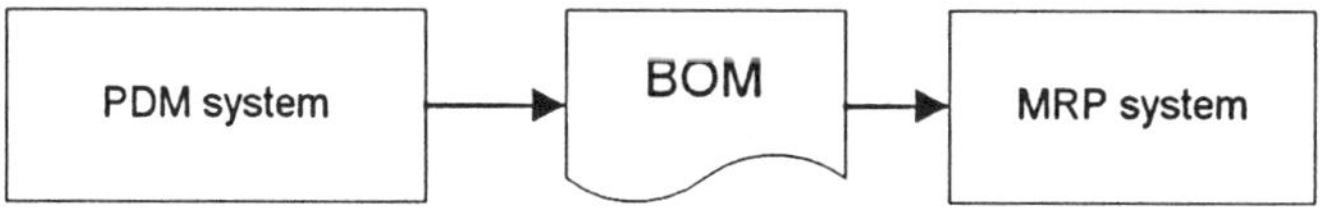

Fig.1 Information integration between PDM and MRP

PDM is useful to manage CAD data, but is weekly effective to CAPP data since PDM system is based on how to managing engineering documents such as drawing, specification, definition. So a total information integrated system should like figure 2. Process planning is that function within a manufacturing facility that establishes which processes and parameters are to be used to convert materials from an initial form to predetermined final form. Process planning may be defined as the act of preparing detailed work instructions to produce a part. For example, in the machined part domain, using a raw material shape as a starting point, the process planning engineer prepares a list of processes needed to convert this material into predetermined final shape. In addition to operation sequencing and operation selection, the selection of tooling jigs and fixtures is also a major part of the process planning function. And also a CAPP system should provide a concurrent environment for process planning and scheduling engineers.

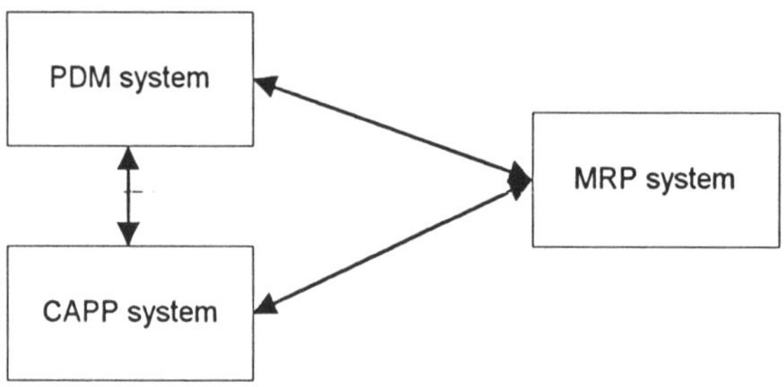

Fig.2 A full information integrated system

2 PDM System

The common problems are focused on PDM system are:

- ensuring everyone works to correct revision level
- distribution of drawings
- time to implement an engineering change
- time to access drawings
- control of variants/configurations
- exchange of drawings with suppliers, customers
- revising paper-based drawings

The author has learned some PDM software such as Iman, Metaphase, Pro/PDM and AutoManage as well. The main functions are similar, they are:

- request for engineering action
- engineering change order or engineering change
- document management including support for multiple versions and multiple formats
- product structure definition
- effectivity of products and occurrences by date, by product unit number, by customizable parameters

- configuration management
- test, maintenance and diagnostic information

PDM system is essentially document control and management, whose basic unit consists of controller and browser. The controller is used to coordinate operations among multiple users and provide a concurrent processing environment for them. The controller actually controls attributes of document. The browser is applied to identify document format and reinstate details of document. The more the browser supports document categories, the more its function is perfect. In general, electronic document is made up of one or a group files, and therefore file(s) and its(their) attributes constitute the substance of PDM management. On the other hand, process routing is a flow or a kind of graph. Especially in the case of non-linear process routing, it is a kind of data structure due to its various representation methods and depends on specific application system. Thus management attributes of process routings should be specific data structure and its attributes and can not be represented with one or a group files and then can not be joined into document structure managed by PDM system. Hence, how to provide a concurrent process planning environment must be accomplished by CAPP system.

3 Theory Model of Concurrent CAPP System

The aim of CAPP system is to solve problems of product process planning automation. Traditional CAPP system mostly is a kind of design mode based on serial works and design works at individual design stage are relatively independent, while prior design knows little about demand of subsequent design and subsequent design only has to accept passively results of prior design. Once problems are found appeared in prior design during subsequent design, the prior design has to restart. This causes the efficiency of product design to be reduced and the costs of product development to be increased. The more serious problem must be mentioned is that traditional CAPP system seldom takes into account constraints of production planning management resource and manufacturing resource from MRPII such as production orders, lot size and etc.. For this reason, it could hardly realize the optimization of manufacturing cycle and manufacturing costs of product.

Under concurrent environment, a CAPP system must be guided by theories on manufacturing- oriented, cost- oriented and product- oriented design. The procedure of process planing in CAPP system must be controlled by feedback information from MRPII. From these, it will optimize reasonability of process planing, economics of manufacturing process routings and availability of manufacturing resource, and integrate process planing with practical production and market requirement to the greatest extent.

The theory model of concurrent CAPP system oriented integration with MRPII is a multiple input/output closed-loop feedback system, which should have approximate stability in a wide range of balanced point and rapid time response. This can be best illustrated by Figure 3.

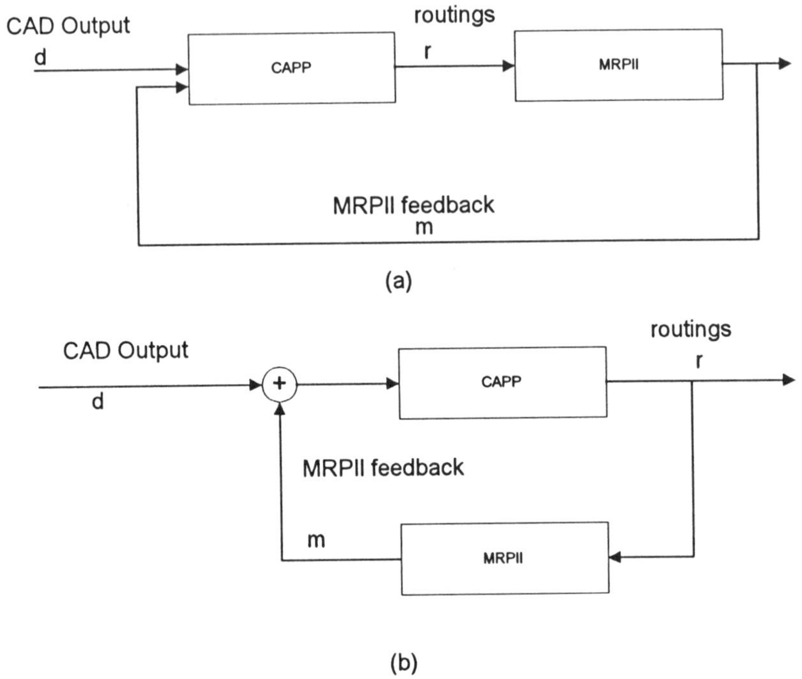

Fig.3 A closed-loop feedback model integrating concurrent CAPP system with MRPII system

In figure 3, figure 3(b) is transformed from figure 3(a), where d can be BOM, parts information and other design output information; r is results of process planning which can be expressed as r = $f_1(d, m)$; m is feedback information from MRPII including production lot size, historical production situations and so forth, and it is historical feedback of process routings and belongs to integral feedback in closed-loop system that can be expressed as follows:

$$m = \int_{t=0}^{T} f(r,t)$$

where t = 0 is start time of certain process routing and t = T is feedback time of certain process routing.

From the above analysis, when the system input is stepped function, the system time response performance can be shown as curve 1 in figure 5, which keeps stable by means of adjusting feedback continuously in a long time. To settle the stable problem of time response, a differential feedback may be inserted into the system for forecasting. Now the construction of system can be illustrated by figure 4, where p = f_2 (r') is forecasting feedback. The system time response performance(curve 2 in figure 5) is obviously more perfect than the former and therefore that makes CAPP system response more rapidly to changes of design and production demand, and realizcs concurrent engineering demands on the system. So, the author takes the concurrent CAPP system with forecasting as theory prototype.

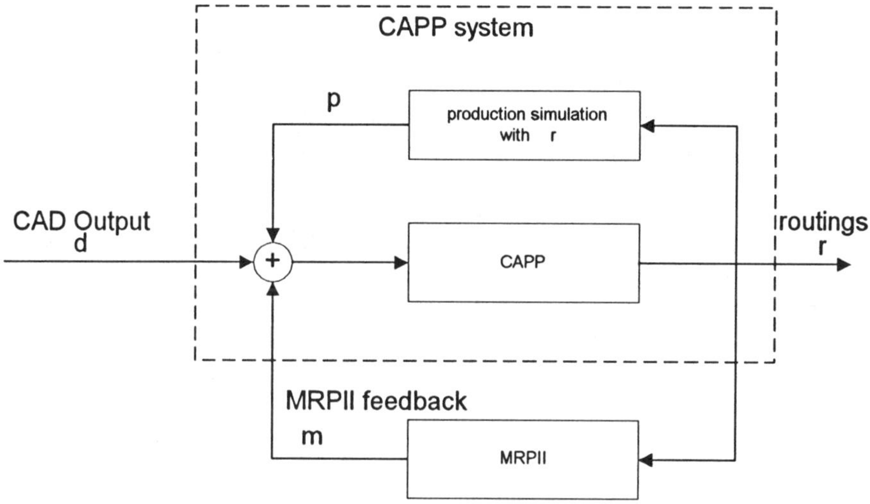

Fig.4 Theory model of concurrent CAPP system with forecasting

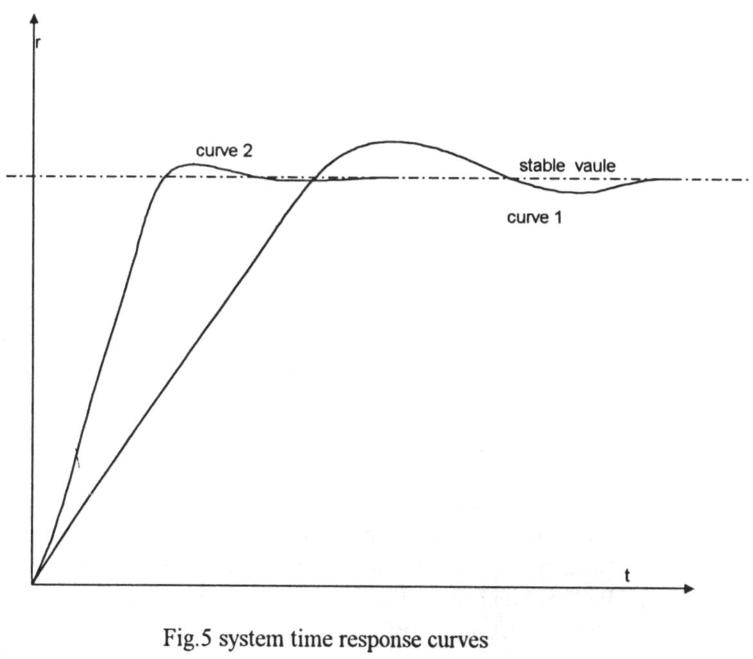

Fig.5 system time response curves

4 A Concurrent CAPP System with Forecasting - SIPM System

The Architecture of SIPM system, as illustrated in figure 6, consists of following functional modules:

. Manufacturing Resource Modeling

This functional module describes base data required for process planning including settings of plants, departments, workshops or sub-plants, situations of technical equipment, machines or workcenters, states of raw materials, standard terms of operations or sub-operations and their relations, user-defined GT rules and etc.. Whether the arrangement and entries of base data is scientific and intact will effect directly efficiency of process planning.

. Task of Process Planning Management

In SIPM system, this module is quite important because the whole process planning is task-driven. In this module, various function is supplied to handle tasks such as assignment, modification, completion status statistics, graphic analysis and so on. With the help of this module, SIPM system ensures data coordination and workflow management of process planning in the case of concurrent operations, that is, when the system is oriented to multiple users.

. Product Structure Modeling

Product structure is the base of process planning.

. Process Planning

In SIPM system, this module is the main module, which can finish process routings' design, process procedures' design and various specialization process routings. In the meantime, it supplies GT design method and the ability of typical routing reuse.

. Production Simulation With Current Routings

The results of process planning may be simulatively run in virtual production environment before they are carried out in practical production, the simulation results can be directly used to forecast the situation of the whole plant runs after these results of process planning have been issued. Through the simulation, some problems can be found out in time and the process plan can be adjusted respondingly in advance. In this way, it can integrate process plan well with demands of production and improve the system time response performance greatly.

. Input/Output

This module can carry out generating technical documents for management automatically and interface to MRPII or other management systems. It can reduce repetitive jobs of process planning engineers and improve speed and accuracy of process planning and achieve output results from CAD and feedback information from MRPII.

The SIPM system is analyzed and designed and developed in object-oriented method and based on single database and network communication technologies. The network system has been mainly implemented on different microcomputer platforms by means of task-driven. Its graphical user interfaces are very reliable and user-friendly and inherit the character of GUI of MS-Window. The SIPM database designed on the base of single database theory is central to all SIPM modules and used to exchange and synchronize information and avoid all unnecessary redundancies and ensure high flexibility for user specific customization.

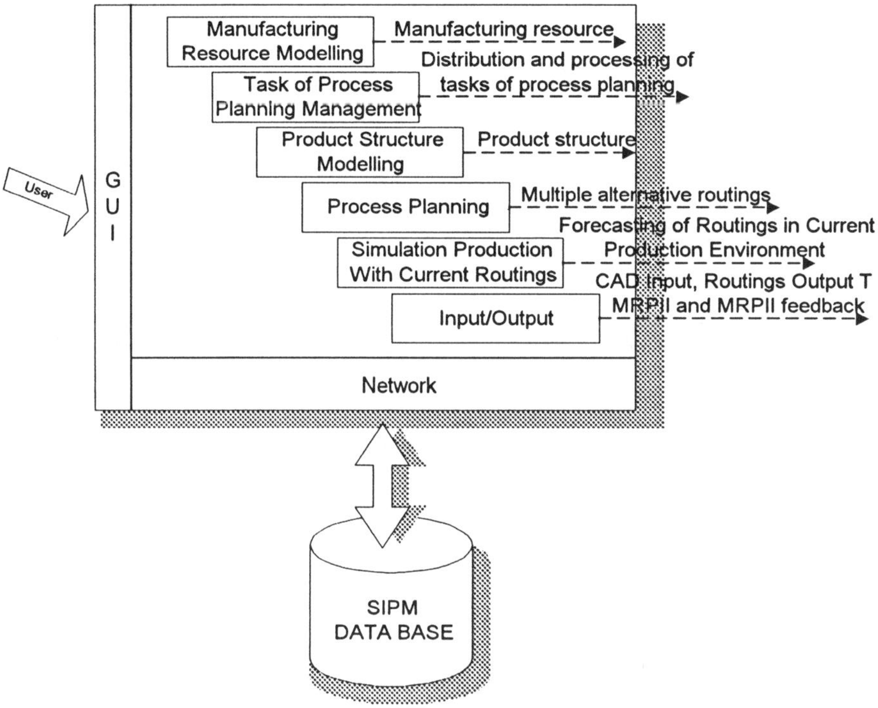

Fig.6 Architecture of the SIPM system

5 Conclusion

Due to commercial PDM system itself, PDM system has no way to concurrently manage process planning. While the SIPM system is analyzed and designed and developed on the base of practice and research in many factories or enterprises, this CAPP system gets over defects of traditional CAPP system, that is, traditional CAPP system can not be put into practice widely. And it is also shown that a CAPP system can be powerful and effective only by putting it into concurrent engineering integration environment.

Reference

1. B.Wu, Manufacturing System Design and Analysis, Chapman and Hall, 1992
2. S.Evans, The Implementation of CAPP in a Rapidly Developing Manufacturing Environment, Proc. 4th European Conf., Automated Manufacturing, 309-318, May 1987
3. Leo Alting and Hongchao Zhang, Computer Aided Process Planning: the state-of-the-art survey, Int. J. Prod. Res., 1989, Vol. 27, No. 4, 553-585

Designing and Planning Cable Harnesses using Immersive Virtual Reality

Foo Meng Ng, J.M. Ritchie and R.G. Dewar

Department of Mechanical and Chemical Engineering,
Heriot-Watt University,
Riccarton, Edinburgh EH14 4AS, UK.
Telephone: +44 (0)131 449 5111
Fax: +44 (0)131 451 3129
e-mail: f.m.ng@hw.ac.uk

Abstract. Currently the design, planning and routing of cables within electro-mechanical assemblies is usually carried out on physical prototypes towards the end of the product development process. The three dimensional harness is then transposed into two dimensions and drawn on to a board. This drawing is then used as a template for the manufacture of the cable harness. These characteristics of conventional harness design make the process more expensive and time consuming than it need be. A system has been developed which uses immersive virtual reality to allow engineers to design harnesses in a virtual chassis without the need for a physical prototype. The system incorporates novel tools to assist the user in performing the routing task. Design and planning information is generated on-line as the engineer plans the route.

1. Introduction

The design and planning of cable harnesses that are use to connect electrical modules within an electro-mechanical device is often considered a tail-end design activity. This process is, in fact, a complex design problem which is time-consuming and costly. The end connectors used will be dependent on the type of socket provided by individual modules, which in turn determines the types of wires to be used. These individual wires vary in size, and together, determine the stiffness and mass distribution of the entire cable harness. In turn, the stiffness, bend radii and mass of the cable harness then dictates the position and number of fasteners required.

Current practise often requires a physical prototype of an electro-mechanical device to be built to allow engineers to manually determine the cable lengths, paths and positions of fasteners. Once a suitable cable path is found, the components involved are selected from tables and the results entered into a database. This information is then converted into two dimensional drawings, part lists and instructions. Figure 1 shows a typical cable harness laid out in two dimensions.

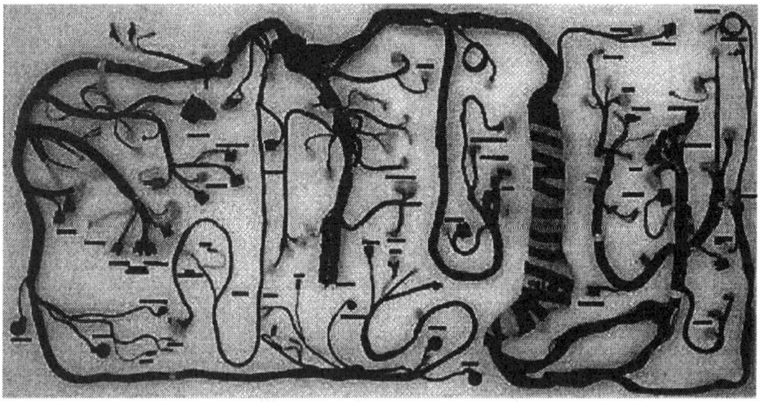

Fig. 1. An assembled cable harness

The routing problem is further complicated by the vulnerability of the cable harness to decisions made upstream. The cable harness may have to be reconfigured after only minor changes that affect the chassis and the individual modules within the prototype machine. The routing process can even result in the late and expensive re-design of the machine chassis to allow the cables to reach their terminal points.

2. Current State-of-the Art in cable harness routing systems

Proprietary software packages do exist to assist in the design of cable layouts, however these tend to be difficult to use and do not always come up with a suitable solution. A significant amount of manual 'tailoring' is often required. The authors are aware of at least one industrial application where this is the case.

There is little published literature on the subject of cable routing with the noticeable exception of the work being done at Stanford University's Centre for Design Research [1, 2, 3]. They have developed a system called First-Link which incorporates artificial intelligence to try and semi-automate the routing process.

However, the authors believe that these CAD-based cable routing systems are unable to entirely integrate the intuitive and creative nature of the expert human operator. By using immersive VR the user is presented with a richer environment. This environment excludes the need for any routing automation and allows users to experiment with alternative solutions.

3. Cablc Layout using Immersive Virtual Reality

The virtual design and planning cable routing system presented here is implemented on Hewlett-Packard workstation with additional VR hardware and software from Division Ltd. CAD models of a prototype assembly can be imported directly into the system which negates the need for any extra component modelling.

The virtual environment can be displayed on the computer monitor and the user can interact with virtual objects using a conventional mouse. This is know as "Desktop VR". However, the authors contend that it is necessary to use "Immersive VR" (that is, using an Head Mounted Display (HMD) which gives a stereo image of the virtual environment) to enable users to comprehend complex three dimensional assemblies. A three dimensional mouse is used to interact with the immersive virtual world.

This virtual cable routing process shown in Fig. 2 focuses on the expert user's creativity and intuition to complete the task rather than trying to automate the procedure. The system developed has three modes of operation - namely, Point-To-Point, Continuous Path and Rubber Banding. Point-To-Point and Continuous Path are creation modes, whereas Rubber Banding is an editing mode. Figure 3 shows the floating toolbox in a virtual world. This is used to select the required mode. This figure also shows a section of cable laid around some of the virtual objects in the environment. Notice the straight sections and nodes of the cable.

When the user is creating or editing sections of cable, if there is a collision between one of the sections and an object (or even another section of cable), visual and audio cues are used to highlight the occurrence and warn the user of the clash.

A Size Management VR tool has also been developed, which provides the user with the ability to alter their size with respect to the virtual objects. This allows the user to shrink to a manageable size to perform detail work or to enlarge themselves to create rough cable routes.

Internally, the system represents the network of nodes and cables as a multi-linked structure [4]. This provides a compact and flexible portrayal of the harness. As the cables are routed in real time, behind the scene, the system is generating a text file detailing the bill-of-materials and process planning information that is associated with the physical cable harness (for example, estimated length of the cables, types of fasteners and end connectors used).

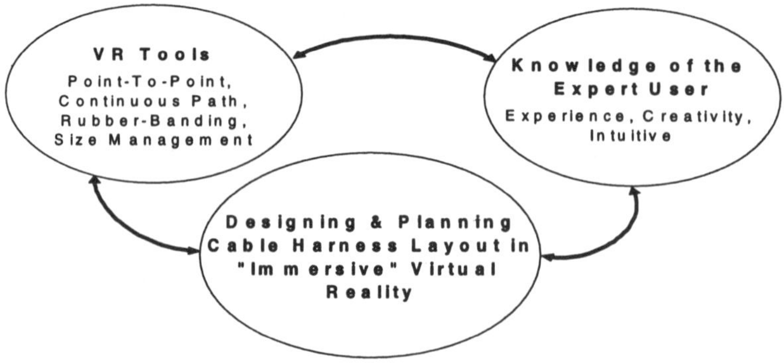

Fig. 2. Concept of designing and planning cable harness in immersive VR

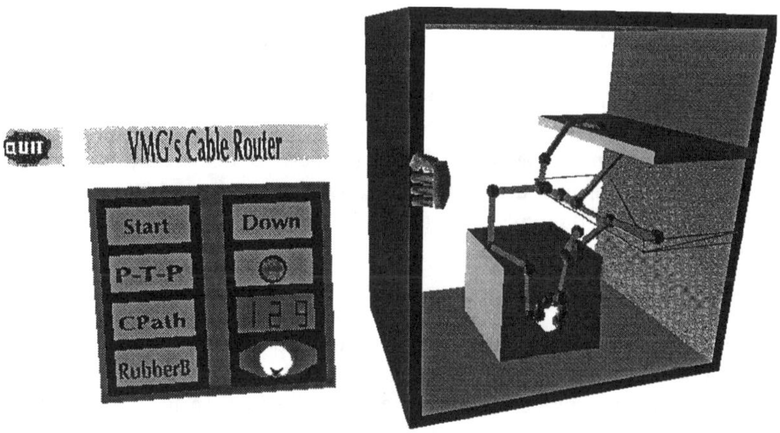

Fig. 3. A floating toolbox, some components and a cable harness in the virtual world

The following sections describe the three modes of operation of the system in more detail. They also detail how collisions are dealt with in the virtual world as well as mentioning the size management facility in the system.

3.1 Point-To-Point

The Point-To-Point technique of routing cables provides the capability to generate rough cable routes rapidly by picking positions or nodes in the virtual environment. The user simple probes a point in space and a section of cable appears between this point and the last node that was created. If the user picks an existing node, they can move this node about in three dimensions. This editing facility within point-to-point is effectively rubber banding which is described later. Additionally, this picking of another node makes that node active and any subsequent point chosen in space will create a section of cable between it and the active node. By choosing existing nodes, multiple branches can emanate from a single node. Some examples of this can be seen in Figure 3.

3.2 Continuous Path

The other technique developed for routing cables is known as the Continuous Path method. This technique routes a cable by extruding a new section from a user selected node. Thus, by picking an existing node, a new node is created and this is then attached to the virtual hand until the node is dropped. This method has rubber banding implicit in it as far as the new section changes in length and pose as the virtual hand moves.

This method is more suitable for detailed routes such as routing a cable around the curvature of the hole in the virtual object as shown in Figure 4. The user can see the collisions that the new section of cable is involved in and immediately take action to move the section and so avoid the clash. Again, this method allows for creating nodes with multiple branches.

3.3 Rubber Banding

Once the entire cable layout is produced, modifications may be required. A rubber banding facility allows the user to re-position entire sections of cable that are knitted together simply by holding on to and moving a node. Although this editing facility stands alone, it has already been mentioned that it is available in the point-to-point tool and, to some extent, in the continuous path tool.

3.4 Collision Detection

The ability to touch and feel objects in the real world is one that is taken for granted. However, this haptic feedback is still in its infancy in virtual environments [5, 6, 7]. For this reason, the system developed here has used alternative visual and audio cues to highlight collisions. A full polygonal collision detection algorithm is available in the Division software. Thus, when a collision occurs, the system utilises the messages sent from the algorithm to make objects in the virtual world turn to a wireframe representation to highlight to the user that something is amiss. This visual cue is accompanied by an audio cue which takes the form of a simple 'beep'.

When a cable section clashes with a virtual component in the chassis the chassis component turns to wireframe. When one cable touches another, both cables turn to wireframe. A section of cable is in collision with a component of the virtual chassis, in Figure 5. Here the chassis component has turned to wireframe. Using the rubber banding tool, the user is able to alter the position of the cable, thus removing the clash.

3.5 Size Management

Another VR tool that assists the human user in routing is the Size Management tool. This tool provides the user with the ability to enlarge or shrink themselves in the virtual environment. By increasing the size of the user with respect to the objects within the virtual environment, cables can be quickly routed using the point-to-point method. If detailed cable routes are required, users can shrink themselves. This provides the user with more information about valid paths and collisions and may be more suited to the continuous path method.

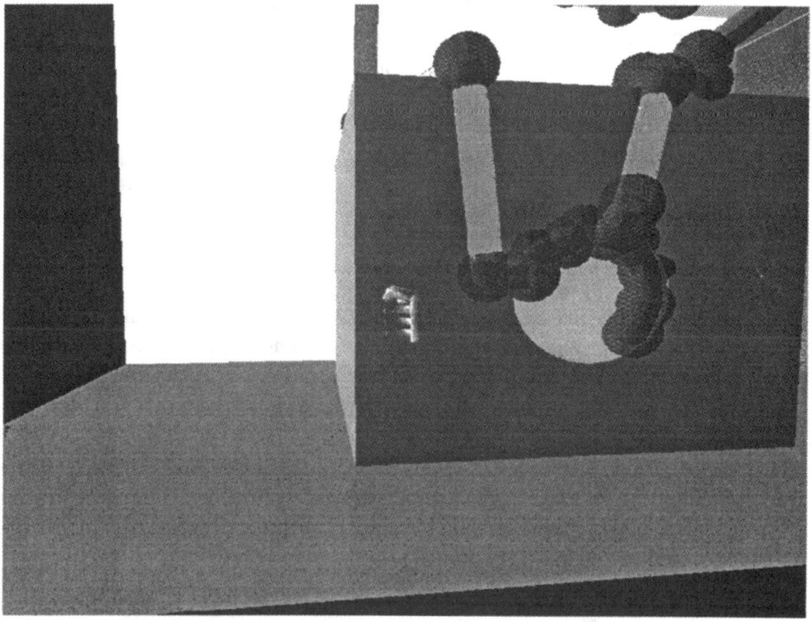

Fig. 4. Detail route using Continuous Path VR tool

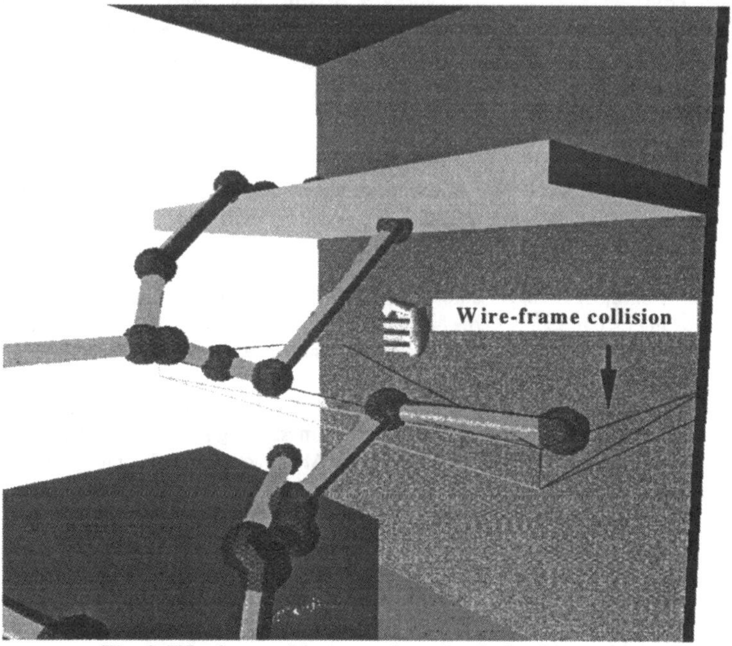

Fig. 5. Wireframe object warning of a clash with a cable

4. Conclusions

Current CAD cable harness routing systems are only able to provide users with two or three degrees of freedom of movement on a flat screen. This inhibits the user's ability to recognised and edit flaws in a cable layout design. Immersive VR, on the other hand, gives the user the ability to change their stereo viewpoint quickly and provides six degrees of freedom. In this paper novel VR tools have been presented to assist the user of virtual environments perform the routing process intuitively and efficiently. The ability also exists in this system to log the details of the routing process which can be used to create a bill-of-materials and two dimensional schematics of the harness.

5. Future Work

In the future the system will be extended to take account of bundle stiffness and mass to facilitate realistic bend radii and help with the positioning of harness fixtures. Also it would be advantageous to add new nodes bisecting existing straight lengths of cable to allow further path editing. Adding a catalogue database to the system will allow designers to choose suitable connectors, bundles and fasteners on-line. Another feature which may prove useful is an undo/redo option when creating nodes. Finally, it will be important to conduct industrial trials to evaluate the ergonomics and usability of the system.

Acknowledgements

The authors are grateful to NCR (Scotland) Ltd., Kinloch Electronics Ltd. and Division Ltd. for their support of this work and access to their expertise and knowledge. The support of the EPSRC, through access to the equipment provided under grant GR/K 41823, is also gratefully acknowledged.

References

1. Conru, A.B (1994), A genetic approach to the cable harness routing problem, Proceedings of the IEEE Conference on Evolutionary Computation, vol 1, pp 200-205.

2. Conru, A.B and Cutkosky, M.R. (1993), Computational support for interactive cable harness routing and design, Proceedings of the 19th Annual ASME Design Automation Conference, vol 65, pp 551-558.

3. Park, H., Cutkosky, M.R., Conru, A.B. and Lee, S. H. (1994), An agent-based approach to concurrent cable harness design, Artificial Intelligence for Engineering Design, Analysis and Manufacturing, vol 8, pp 45-61.

4. Langsam, Y., Augenstein, M.J. and Tenenbaum, A.M. (1996), Data structures using C and C++, 2nd Ed., Prentice Hall, New Jersey.

5. Taylor, P (1995), Tactile and kinaesthetic feedback in virtual environments, Transactions of the Institute of Measurement and Control, vol.17, no.5, pp.225-233.

6. Caldwell, D.G., Lawther, S., and Wardle, A. (1996), Tactile perception and its application to the design of multi-modal cutaneous feedback systems, Proceedings of the IEEE International Conference on Robotics and Automation, vol.4, pp.3215-3221.

7. Tateno, T, and Igoshi, M (1996), Hierarchical processing for presenting reaction forces in virtual assembly environments, Journal of the Japan Society for Precision Engineering, vol.62, no.8, pp.1182-1186.

A New Solution to the Integration of CAPP and PPC

Wang Liya
CIM Research Institute, Shanghai Jiao Tong University, P.R.China
Jiang Zhiping
CIM Research Group, Automation Department, Shanghai Jiao Tong University, P.R.China
Yan Junqi
CIM Research Institute, Shanghai Jiao Tong University, P.R.China
Wu Dezhong
CIM Research Institute, Shanghai Jiao Tong University, P.R.China

Abstract: The paper proposes a new architecture of multiple layer, concurrent and distributing integration of CAPP and PPC which supports the function integration and the overall optimization of their process plans and production plans. The architecture is further solidified with both the implementation of a system which is based on an integration platform and the realization for a prototype of a products production planning layer .

Keywords: Computer-aided process planning, production planning and control, functional integration, architecture, system realization

1 Introduction

The integration method for computer-aided process planning(CAPP) and production plan and control (PPC) has long been concerned in both research field and engineering field as well. Currently the mostly used, however, is an information integration method by CAPPs interfacing with PPCs. One of the widely known defects of such a method is that the CAPP and the PPC are still two functionally independent sub-systems[1,2]. With this architecture, the dynamic feature of manufacture resource and the feedback message from PPC is difficult to be fully considered in the stage of CAPP, and the PPC has to invariantly organize production based on whatever results come from CAPP. Therefore, it would be common that, in a real manufacture environment, a detailed process plan is infeasible or hard to be implemented. In order to solve these problems, we propose a new solution to support *function integration* (FI) of CAPP and PPC which takes into account, instead of merely information integration, the influence of the environment to a CAPP and a PPC, the mutual restriction to each other and the joint optimal objectives. In accordance with this criterion, we suggest in this paper a new architecture of multiple layer, concurrent and distributing integration, which support FI of CAPP and PPC. To demonstrate the instructive meaning of such an architecture to a class of job-shop enterprises featured with multiple products and small lot sizes, we put forward an implementation to a system based on an integration platform as well as an universal logic adaptable to the decision making of every layer of the system. A realization detail for the development of a prototype for a products layer is also given.

This work was supported by Natural Science Foundation of China

2 The new integration architecture

By analyzing their individual function and mutual relationship, we find that PPC is a layered procedure from planning down to real manufacturing, whereas the CAPP is only a technically preparatory procedure in advance to real manufacturing. Therefore, there is an obvious boundary line between these two procedures whether in terms of the time they apply or in terms of functions they take. We can thus be hinted that, in order to realize FI, the traditional function division among CAPP and PPC should be broken, and a new combination to their functions is necessary.

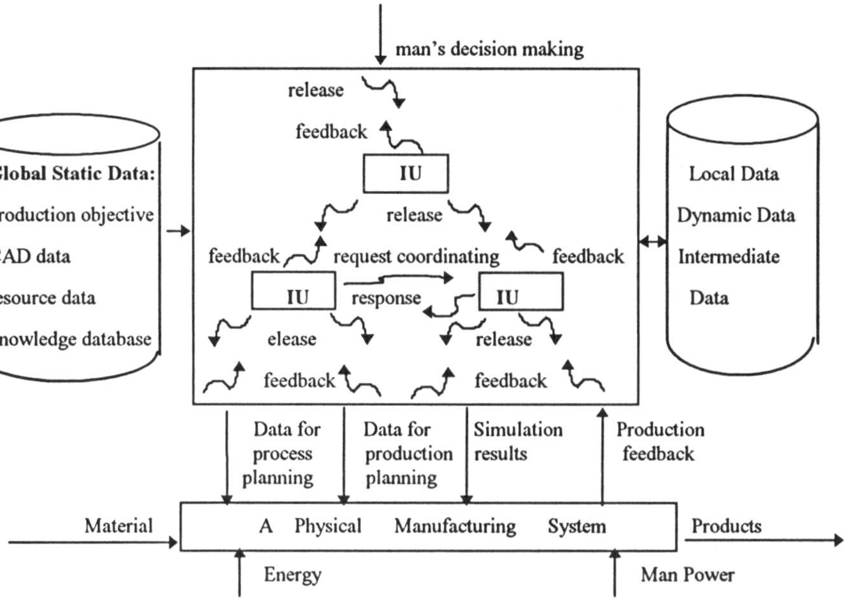

Figure 1 A structure for multiple-layer, concurrent and distributing integration

Figure 1 represents a new architecture of multiple layer, concurrent and distributing integration, in which multiple relationships among layers as well as among components of the same layer are expressed. It is easy to see that the hierarchy is such that all the components in the architecture are distributed in horizon and layered in vertical. Also depicted are the control relation from upper layers to lower layers, the feedback relation form lower layers up to upper layers, and the distributing relation among components of same layers, respectively. It is noted that such an architecture has an outstanding feature that the complexity for decision making is decomposed. For example, the people can make decision on different layers according to their diverse frequencies and/or diverse responses, which can be deemed as a heuristic process from macroscopic to microscopic. In this way, the decision-making process for an object on a layer can be done at its own block (unit) in distribution. Moreover, even a larger problem can also be dissolved to be processed in many small related blocks. It is seen that such an architecture is not only in accordance with the mode most enterprises organize their production but also the commonly used control structure in computer integrated manufacturing systems(CIMSs)[3].

From Figure 1, we can see that every node in the multiple layer distributing structure is an integration unit (IU). It is also the basic function unit for CAPP and PPC integration, composed of three fundamental components. One is a *CAPP function unit* (CAPPFU) which ensures to produce process plans which can manufacture required parts, components and products according to the original design; the second is a *PPC function unit* (PPCFU) which is responsible for working out the reasonable quantities of the parts, components and products of a specific period of time from the requirements; and the third is called an *integration decision function unit* (IDFU) which is separated from the original function used for time and/or cost control in traditional CAPP and PPC sub-systems, which realize the optimal control of process plans and production plans on a basis of timing or cost objectives. the CAPPFU and the PPCFU jointly fulfill the concurrent decision making procedure while the IDFU acts as a coordinator and participates cooperative decision making. Figure 2 shows the relations among these three units.

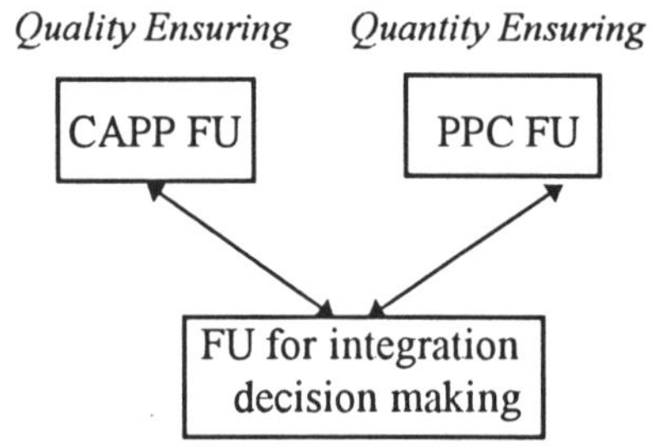

Figure 2 The Principle of an IU

It has been shown through real application that the architecture makes it possible to dissolve the complexity problem raised in CAPP and PPC integration, to solve the difficulty of how to determine the integration scale and to detail the way of information and functions coupling. What is more, the architecture also leads to a new model combining decision making in layer, in distribution and in concurrent cooperation as well. For a traditional CAPP and PPC integration problem, we can, according to the architecture, dissolve their original functions and then re-assign and finally put together those functions among CAPPFU, PPCFU and IDFU[2] Therefore the original mutually independent and centralized decision making procedure is now turned to be a concurrently interactive and distributively cooperative one. Further application shows that this kind of solution to CAPP and PPC integration can not only supports the overall optimization of the process plans and the production plans, but also fit the dynamic variation of the environment through coordination and feedback its mechanism.

3 Integration reference model

For a general class of enterprises featured with multiple products and small lot sizes, we here suggest a five-layer reference model for the system integration using the criterion outlined in the previous section. The model narrates the layers to which a CAPP or a PPC should belong (for detail see 2). In addition, for every layer, the functions to be fulfilled are also specified which include the division of an IU on shop floors, cells and equipment levels, respectively in addition to their function range on each layer, as is shown in Table 1 at the next page.

Table 1

Integration layer	*IU number*	*Functions of IUs*
products production planning layer	1	products process technique analysis, products production planning,process planning for products orders, construction of products process network, capacity checking ,adjustment of process plan and production plan , *etc.*
parts production planning layer	1	parts process technique analysis,parts production planning, process planning for parts orders, capacity checking,adjusting process and production plan for parts orders, feedback control, assigning and adjusting shopfloor tasks, *etc.*
Job-shop planning layer	depending on the division of the cells	planning process in shopfloors,shopfloor tasks decomposing, job shop scheduling ,capacity checking, process and tasks adjusting in a extend of shopfloors, coordinating shopfloor tasks the shopfloors among IUs, feedback control and adjusting cell tasks.
Job-cell planning layer	depending on the division of the cells	process planning in cells, cells tasks decomposing, scheduling and capacity checking, adaptivity adjusting(including process batching), coordinating cells tasks among IUs, feedback control, and adjusting equipment tasks.
Equipment planning layer	as needed	Job-equipment sorting, process adaptivity adjusting, coordinating tasks for equipment, timing and feedbacks among IUs.

From the model, we can see that the five layers are in a tight connection. For example, except the production plan layer, every other layer can make its own integration decision only based on the results of the decision made on its directly upper layer. And each layer can recursively solve the problem handed upward from its directly lower layer through feedback when the latter must but can not, due to its function limitation, the problem. In this way, the decision logic within each integration layer is similar though each has its own limitation of functions. As sketched in Figure 3, this arrangement of the decision process, we call it a *bottleneck-driven decision-making logic*, can efficiently optimize the process plan and the production plan in the integration of CAPP and PPC.

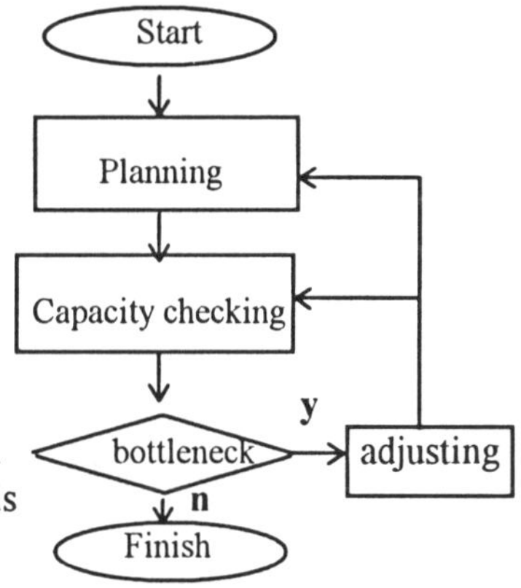

Figure 3 The bottleneck driven logic

4 Integration-platform-based system implementation

The above description defines only the basic framework of an integration system and its corresponding function domain for each of its IUs. For the system implementation, we propose a method based on an integration platform which is shown in Figure 4. The kernel parts of the integration platform is composed of the following

four parts:
(1) a product data model: includes the modeling of parts features, specification of the products structure and products data maintenance as well as data transformation, *etc.*;
(2) a manufacturing environment model[4]: includes the definition of the control structure of the factory, manufacturing resource modeling, process and manufacturing knowledge expressing as well as the function description and implementation of shop floor data management;
(3) an integration information model[5]: defines both the plans structure generated by the integration system and the information structure used for intermediate decision making; and
(4) a general data access and management: controls and administrates and maintains the fundamental databases integrity, completeness, security, *etc.*.

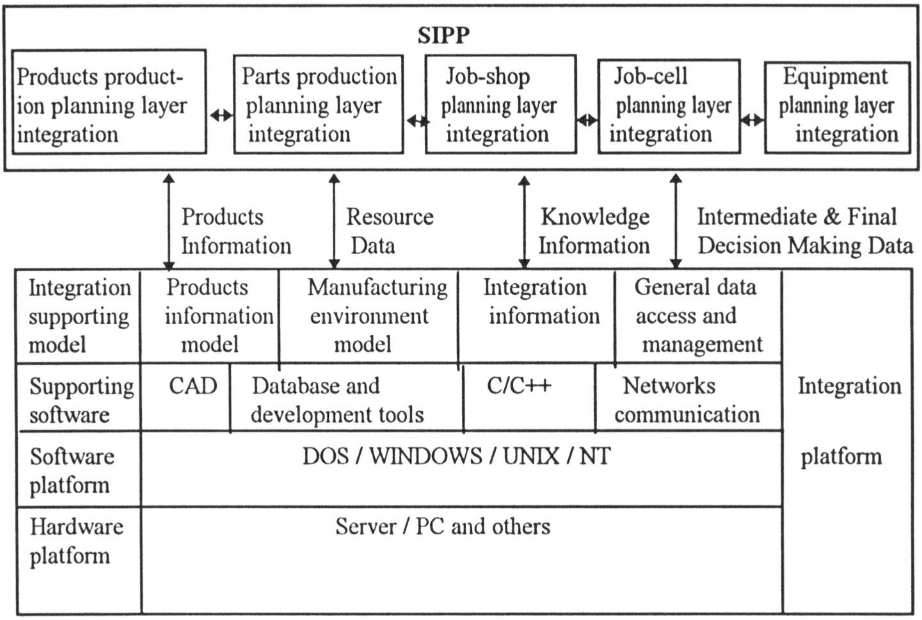

Figure 4 An Integration platform supporting SIPP

Such an integration platform will offer the following conveniences:

(1) Manage in a centralized way all the process decision-making information raised during CAPP and PPC integration. Since the efficient FI for CAPP and PPC should be assured through each of the CAPP and PPC making full use of other party's abundant procedure decision-making information, how to rapidly and correctly access all these information becomes compulsory. The procedure information structure of the integration information model in the integration platform provides us with a means to visit all the key message during the integration decision making on the current layer and all the results of the decision making on the non-current layers.

(2) Provide for CAPP and PPC a unified data environment. Traditionally, CAPP and PPC possess their own databases and data structures, which results in at least four drawbacks. One is that data would not be mutually sharable. Secondly the manufacturing environments data which comprises the basis for their own decision makings is usually inconsistent. The third is that each procedure is unable to make quick response to the variation of the data in the other procedure. And the fourth is that the relations between a process planning and an order can not be expressed in the data structure. That is why we establish an unified data model on the integration supporting layer of the integration platform. For example, in a manufacturing environment model, all the static and dynamic resource information is organically collected for CAPP and PPC sharing, while in the integration information model, the process plan and the production plan are tightly combined with each other through an integration plan structure, and thus forms a connection between process plans and orders, which supports the overall optimization.

In addition to the above merits, the platform technology has another distinctive feature that the information model can be efficiently separated from the function model[6], so the information model is highly re-usable, which reduces the hardness to develop or improve a system of integrated process planning and production planning and control (or briefly, SIPP) and also provides a handy integration interface between SIPP and other systems or functions .

5 System realization

As an example, here we give a concrete realization of an IU for a product planning layer (see Figure 5). The functions of all the blocks in the figure can be addressed below.

- Process technique analysis (P.T.A.): It is the first step in process planning. Since the accurate time when the process planning is being put into use is still not predictable at this stage, the purpose of process technique analysis is mainly to produce, in consideration of their techniques, the process scheme or scheme set for all involved products, components or even parts, and then assign each scheme a priority according to some criteria (e.g., the minimal machining cost, the shortest machining time, the least process number, *etc.*).

- Products production plan (P.P.P.): This is obtained by means of taking into consideration both the predicted and practical products orders. It is only a tentative plan reflecting the market or user demands because no process planning has been formed at the moment.

- Process planning for products production plan layer (P^5L.): This is based on the analysis results for process techniques and the preliminary products production plan. And then according to certain criterion, the first process planning for each order is selected, i.e., the process planning for products orders.

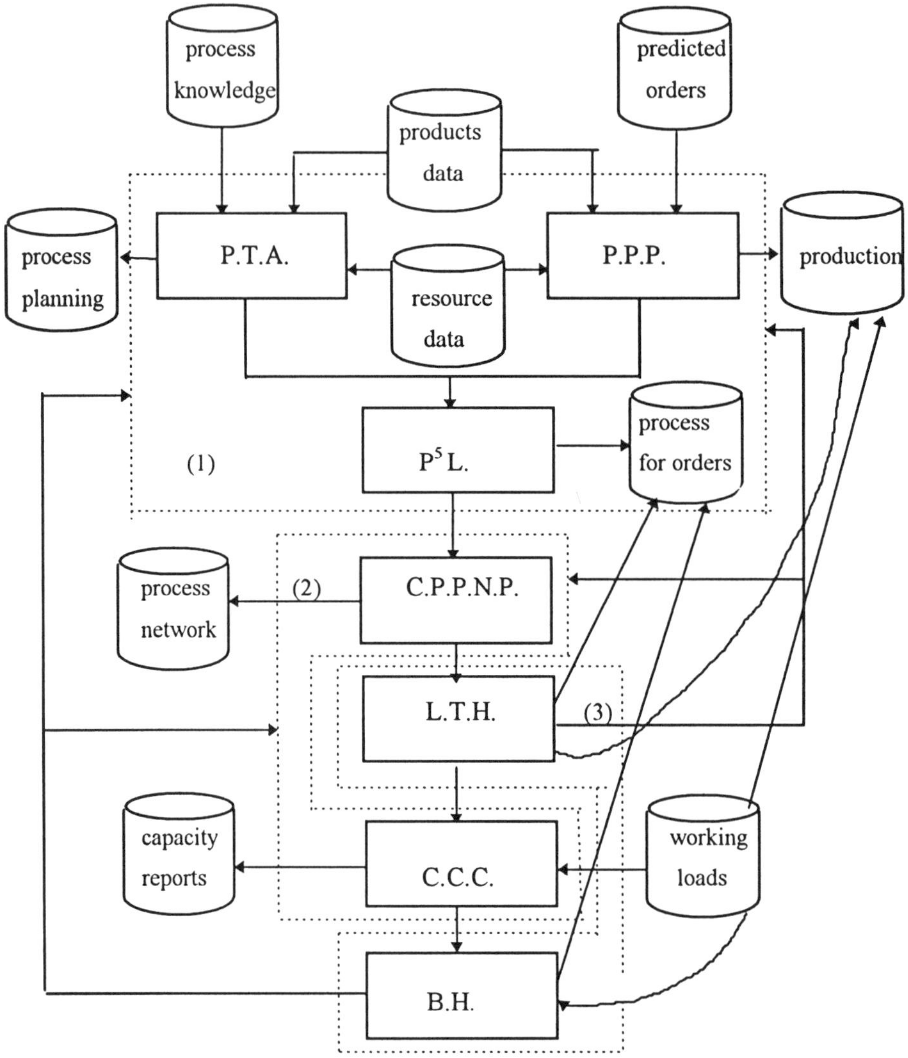

Figure 5 An Integration Model for product production layer

• Constructing progress plan network for products (C.P.P.N.P.): Since each the working load of each equipment might be different, it would be necessary to compute for each equipment its planned working load at different periods of time. To this end, the products production plan and the process plan for an order are decomposed and transformed into a progress plan network.

• Lag time handling (L.T.H.): According to the progress plan network , it would be easy to find out those processes in progress plan networks that will fall into the lagging time segments. The purpose for the lag time handling is to avoid this situation by means of adjusting the process planning for orders or modifying the due time in the production plan.

- Cycle capacity checking (C.C.C.): This is to get the status regarding the capacity of the equipment as well as the bottlenecks of resource. For this purpose, the progress plan network, the production plan and the structure tree of the products are utilized to calculate, in a specific period of time, the planned working loads of the equipment within various time segments.

- Bottleneck handling (B.H.): This is realized by scheduling processes (moving up some processes, for example), adjusting the process for the orders, and/or modifying the production plan.

It is noted that the integration decision making procedure of the production plan layer can further be divided into three levels: (1)the planning level cares the process technique analysis, products production planning and construction of the process plan for the products production planning layer; (2) the capacity checking, the second level, involves the establishment of the progress plan network for the products as well as the checking of the cycle capacity for the plan; and finally (3)the adjustment level handles lag time and bottleneck. It is also noted that there are feedbacks among the three levels. Through such feedbacks, the control of the execution frequency of the planning level and adjustment level becomes possible.

6 Conclusion

In this paper, we put forward a new solution to a the CAPP and PPC function integration, which takes into consideration not only the factors of dynamic variation of the manufacturing environment but also making reasonable use of the enterprises resource . Therefore, the overall optimization in the CAPP and PPC integration becomes realistic. To this end, a framework of a multiple-layer, concurrent and distribute integration architecture is constructed, in which the key is how to decompose and re-group the functions of CAPP and PPC by their multiple-layer distribution hierarchies. An integration reference model is henceforth proposed and it is enhanced with a small but illustrative system realization.

References

1 Larsen N.E., "Methods for Integration of Process Planning and Production Planning", *Int. J.* **Computer Integratein d Manufacturing, Vol.** 6,(1993) 152 ~ 162

2 Wu Dezhong, Yan Junqi and Wang Liya, "Research of CAPP and PPC Integration concurrency and distribution", *J.* **Shanghai Jiao Tong University, Vol.** 30, **No.** 12, (1996)1 ~ 6 (in Chinese)

3 Li Bohu, The Guide of the Convention, Standards and implementation of Computer Integrated Manufacturing Systems, **Armament Industry Press**, Beijing, (1994) (in Chinese)

4 Wang Liya, Yan Junqi, "Modeling of Manufacturing Environments in a Object Oriented Method", *J.* **Shanghai Jiao Tong University, Vol.** 30, **No.** 2, (1996)110 ~ 116 (in

Chinese)
5 Wang Liya and Yan Junqi, "An objet-oriented alternate process data model for CIMS", **Proceedings of International Conference on Intelligent Manufacturing**, Wuhan, China, June , (1995)149 ~ 160
6 Jin Zunhe and Wucheng, The Idea and Development of Integration Software Platform in Factory Automation, **CIMS-China'94**,(1994)6-44 ~ 6-49(in Chinese)

Product Renewal By Using IPD

Robert Bjärnemo

Department of Machine Design, Lund Institute of Technology, Sweden

Abstract. The origination of product ideas for new products is usually the result of a systematically executed product planning process. In smaller companies, most often with no access to such a product planning function, the product ideas are usually generated through less formal procedures. However, regardless of how these product ideas are generated, there is a strong demand for improving these activities in a way which contributes to an increased effectiveness and efficiency of the subsequent product development process and thus of the overall industrial innovation process. Here, the *product renewal procedure* is proposed as effective means for fullfilling these objectives, when the *Integrated Product Development procedure* is utilized for the subsequent product development process. A description of the background to and an example of the practical use of the product renewal process are also included in the paper, as well as a brief outline of the future development of both the product renewal procedure and of the IPD procedure.

1 Introduction

In order to successfully meet the demands originating from the rapid globalization of the economy, most manufacturing companies around the world have been forced to introduce major improvements in their approaches to the development of new products in order to maintain or to improve their competitiveness on the international market. The most significant of these improvements are undoubtedly those focusing on a combined increase of the efficiency as well as of the effectiveness of the *product development process*. The objectives set out in order to reach these improvements are often of a conflicting nature. One such conflict is created when the demand for a reduction of the development resources arises and there is a simultaneous demand for an increase of the quality of the product–to–be.

A prerequisite for creating more efficient and effective development procedures is the access to similarly effective and efficient methods and techniques to support these procedures. An often neglected fact is that the development of these methods or techniques is equally dependant on the existence of an overall development procedure model as the development of the product development procedure model is dependant on the access to, or the possibility to develop, suitable methods and techniques. By omitting this dependency it is more than likely that the method or technique selected or developed instead contributes to an increase instead of decrease of resource consumption, if not proven completely unsuitable for its purpose. The increase of resource consumption is most frequently the result of the need for an adaptation of product–related data to fit the demands from a subsequent method or technique in the development process, such as the frequent need for transformation of product geometry data from a CAD program to fit the demands on the geometry representation by a general purpose finite element program.

In other words, the *product development procedure model* provides the information on *which* problems should be solved during the process and *when* (in which order). The *methods* and *techniques,* on the other hand, provide the means for *how* a given problem, derived from the development procedure model, should be solved. As the main objective here is to present the use of an existing *product development procedure model* in *product renewal,* supporting methods and techniques will not be addressed in any detail, if at all.

In industry the product development process is in turn embedded in a more comprehensive process, the *industrial innovation process*. This innovation process encompasses all of the activities preceding the adoption of a new product on the market, such as basic and applied research, design and development, market research, marketing planning, production, distribution, sales and after sales service. In other words, the innovation process comprises more than the product development process – see Roozenburg and Eekels [13].

Preceding the product development process in the overall innovation process is the *product planning process.* In the perspective of developing new products, the main objective set out for this process is to generate and to decide upon which *product idea(s)* to be developed. For a successful realization of the product development process, the result obtained during the product planning process is of the utmost importance; especially with reference to the development and design of the product–to–be.

In order to facilitate an effective "link", or more precisely, an *integration* between the product planning process and the product development process, with the overall objective to contribute to an increased performance of both processes, the *product renewal process* is proposed here. The concept of product renewal is not confined to those, usually larger companies, in which product planning is an established function, but also to smaller companies lacking this function.

In spite of the technically oriented approach underlying the concept of product renewal, none of the additional aspects taken into account during a traditional product planning process are omitted in the concept.

Before elaborating on the concept of product renewal and the use of *Integrated Product Development (IPD)* in this process, it is convenient to start by introducing the concept of *(engineering) design methodology* as this forms the basis for the *IPD procedure* and thus for the product renewal process.

2 Design methodology and product development

In the late 60s, but more commonly during the 70s, many manufacturing companies introduced *design methodology* as the mean to improve the *(engineering) design process* as well as the product–to–be. In design methodology, the focus is set on the *technical realization* of the product, which means that virtually every phase of the technical realization process is covered – from *why* and *how* a design project is initiated until the completion of the documentation necessary for the materialization (manufacturing) of the product–to–be is achieved.

One of the most well–known of these early design methodologies is the one by the German Professors G. Pahl and W. Beitz. This methodology was first introduced in a series of articles in the German journal *"Konstruktion"* between 1972–74 and later, in 1977, published in a book. Since then, the methodology has been updated and enlarged, thus still holding a foremost position in the current choice of design methodologies – see G. Pahl and W. Beitz [14,15].

In order to give some insights into the contents of a design methodology, the design procedure model according to G. Pahl and W. Beitz is presented in Fig. 1.

As long as the technical performance of a product constituted the single most important competitiveness factor, design methodology alone represented the most effective and efficient means of increasing competitiveness. However, as additional factors (adaptation to business, design, market/consumer, production (primarily manufacturing), sales etc.) also became equally important in handling the competition on the global marketplace, new methodologies focusing on the

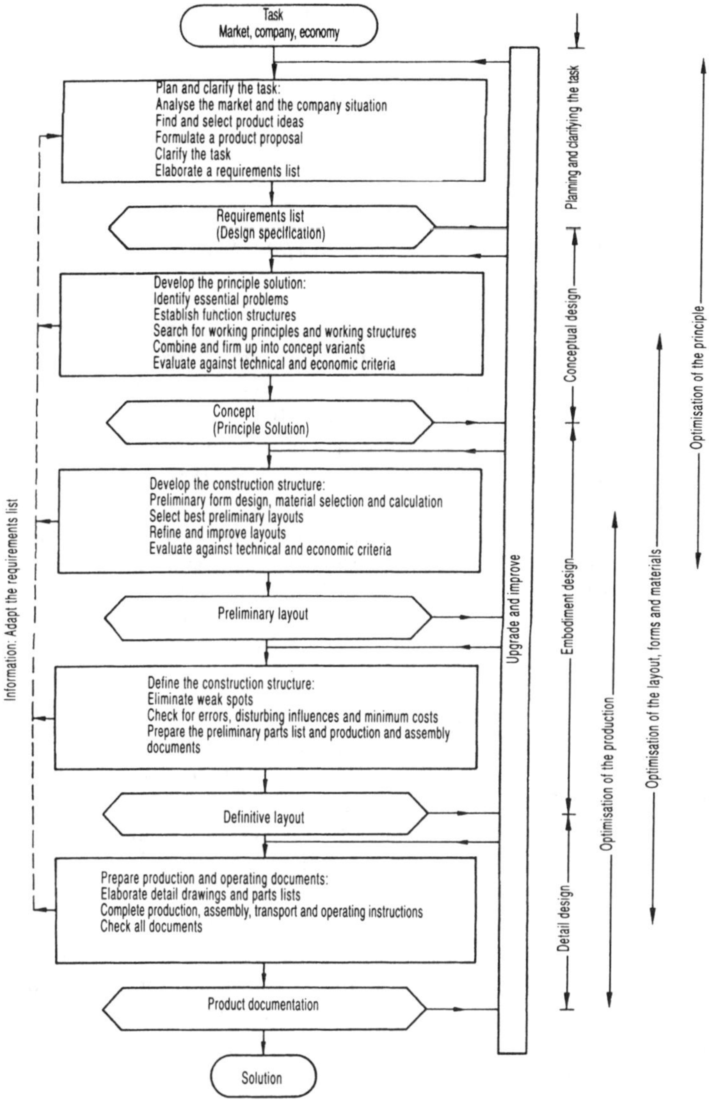

Fig. 1. The design procedure model according to G. Pahl and W. Beitz [15 p 66].

overall development process, the *product development process,* were developed. Note, that the creation of product development procedure models represents an *extension* of the models provided for within design methodology. The design procedure becomes, in other words, embedded into the product development procedure model.

The trends towards the development of *new,* more efficient and effective, development–/design–oriented, *product development approaches*, can be traced back to the beginning of the 80s. The most successful of these new approaches are undoubtedly *Concurrent Engineering (CE)* – see D. Clausing [3, 4], *Simultaneous Engineering (SE)*– see C. W. Allen [1], B. Evans [6] and F. Gordon and R. Isenhour [7] and *Integrated Product Development* – see L. Hein et al. [8], F. Olsson et al. [11], M. M. Andreasen and L. Hein [2] and K. Ehrlenspiel [5].

The common denominator in these new approaches is that the development process is carried out in close interaction between a number of the *functions* or *departments* of the company, such as *design, market, production* and *sales.* Regarding *which* functions to be included in a specific approach is dependent on both *type* (CE, SE or IPD) and *version* of the selected approach. Another characteristic of these procedures is that the work is performed by a *multifunctional product development team (MFPDT).* Note that these categories/types of product development procedures represent but a fraction of all currently available procedures. One example of an additional approach to and decomposition of the product development process is the managerially–oriented, sequential, *new–product development process* presented in P. Kotler et al. [9].

3 The IPD procedure

In 1969 the late Prof. F. Olsson introduced a procedure model for the product development process or, as he named it at the time, the *Product Origination Process*, [10]. In 1982, this model was revised and renamed. This revision was undertaken in close cooperation between Prof. Olsson and the two Danish researchers Prof. M. M. Andreasen and Technical Director L. Hein. Together they presented the new procedure model under the name *Integrated Product Development* or simply *IPD,* at the *Design Policy Conference* in London in 1982, [8]. Today, IPD is best known as a *general concept* for product development. From the original IPD procedure, two different versions have been derived – a Swedish one by F. Olsson in 1985 [11] and a Danish one by M. M. Andreasen and L. Hein in 1987 (originally presented in Danish in 1985), [2]. In Fig. 2, the overall, functional, structure of both procedure models is presented.

In both IPD procedure models the procedure model is decomposed into *five stages.* In the Swedish model the procedure integrates the *four "parallel" functions* or *subprocesses: market, production, (engineering) design* and *business/finance.* The main difference between the Danish and the Swedish model is that the *business/finance* function is excluded in the Danish version. The difference is due to Olsson's ambition to create a model well–suited for use in the setting up of new companies or subsidiaries, where finance and the establishment of new business are in focus. In the Danish model, on the other hand, primarily companies with an established product development facility and a well–established financial situation are addressed.

The main *integration* in the process takes place between the *four* or *three "parallel" functions/subprocesses.* When *integration* is addressed, it is usually this *"vertical"* integration which is referred to. Apart from this type of integration it is also necessary to mention the *"horizontal"* integration. This refers to the integration *along* the time–axis of the project and *within* each and every one of the subprocesses. One of the key subprocesses is *design and development,* which explains the close connections between design methodology and the IPD procedure models.

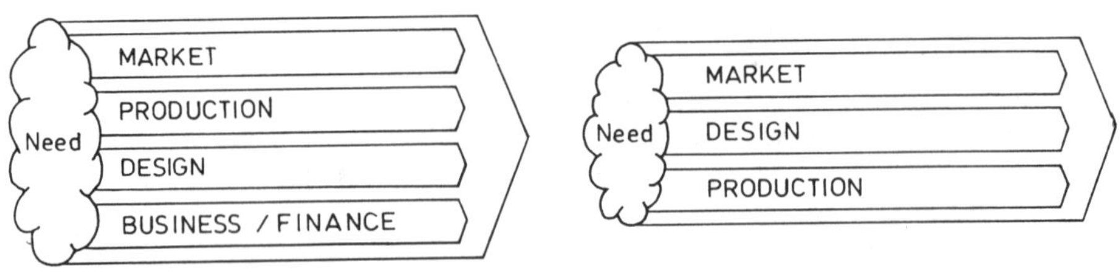

IPD procedure according to Olsson ***IPD procedure according to Andreasen and Hein***

Fig. 2. The overall structure of the IPD procedure models by Olsson [11] and Andreasen & Hein [2].

Finally, in order to give a more complete picture of the structure of the IPD procedure utilized in the *product renewal process,* a detailed procedure model is presented in Fig. 3.

4 Why product renewal?

Since the mid 60s industry–related product development projects have been carried out at the Department of Machine Design, Lund Institute of Technology, Sweden. In the majority of these projects, the focal point has been on the development of *new* products. We have progressed from the initial projects, with a duration of 7–14 weeks, to the one–year projects of today, representing not only an increased cooperation with Swedish industry, but also a continuous development of new procedures and methods/techniques to carry out these projects.

From the experience obtained during these and other industry projects in which the IPD procedure model by Olsson (see Fig. 3) has been used, the following observations represent some of the most important arguments in support of the product renewal procedure:

- The *product ideas* created during the product planning process are often provided in a form very close to a *conceptual solution* (= the working principle of the whole product and its most significant subsystems are determined). The product idea is in other words given in such a concrete form that the possibility to question the selected working principle(s) in the subsequent conceptual design phase is practically omitted. As the product ideas also frequently are created by non–technicians, this may result in major obstacles in the subsequent development in the IPD process.
- The product planning process is mostly dominated by or, in the worst cases, solely performed by non–technicians. This often results not only in situations in which technically insoluble problems are neglected or not identified, but also in situations in which solutions cannot be perceived.
- Product planning is often regarded as a "stand alone" function within the company, focusing on control rather than on participation in the overall innovation process. This is in no way in accordance with the overall objective of creating the best possible conditions for an effective and efficient organization to handle the development of new products.

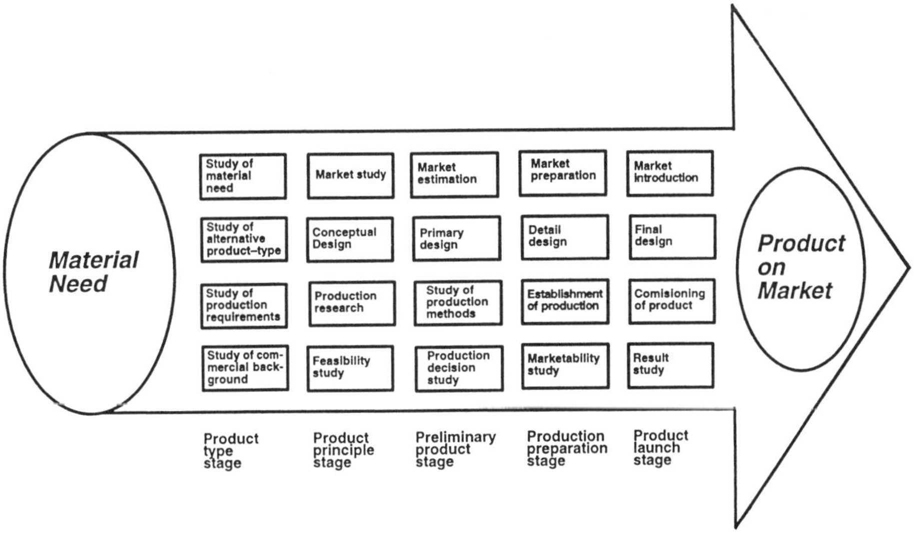

Fig. 3. The IPD procedure model according to F. Olsson [11].

Regardless of the size of the company, the decision to initialize a product development project must always be based upon a solid foundation of facts and figures. This implies that there is an equally important need for handling these issues in smaller companies, without the benefit of a product planning function, as there is in larger companies with access to such a function. As the original IPD procedure model is somewhat unclear as to how the procedure model is to be used in the absence of a product planning function, there is a demand for supplying additional procedural support in order to "substitute" the product planning process in smaller companies.

5 The product renewal procedure model

As the product renewal procedure model encompasses major parts of the activities constituting the product planning process, it is necessary to further elaborate on the contents of the product planning process, before presenting the product renewal procedure model.

The common denominator in the majority of the current product planning procedure models, is the existence of the two major activities *formulation of product policy* and *idea finding.*

The *formulation* of *product policy* consists of two parts, the proclamation of goals and the product market strategy by which the company stipulates which kinds of products are to be developed now and in the future.

The objective set out for the *idea finding* is to breed one or a number of promising ideas which later or immediately will be cultivated into detailed plans for new business activities.

As the product planning process is of a continuous nature, the activity of *controlling* the subsequent development process(s) is often also a part of the planning process. One such implementation is found in the Danish IPD procedure model, since this IPD procedure does not encompass the "built–in" control function provided in Olsson's model by the *business/ finance* function.

Like the IPD procedure, the product renewal process should, if possible, be performed by a *team* representing at least the following functions: *market, design, production* and *product planning*, if existing as a company function. Instead of resulting in a *product idea* as in the product planning process, the final results obtained during the renewal process should be a well–defined *product type.* Such a definition of the product type should only contain the minimal set of information necessary to understand the purpose of the product–to–be, thus avoiding situations in which the product idea has been given in the form of a complete product concept (conceptual solution).

As the product renewal procedure model is created with the objective to facilitate an easy integration to the IPD procedure, regardless of the existence of a product planning function, the IPD procedure model has been slightly revised. The revision consists of a replacement of the *business/finance* function with a *project management* function, as the business and financial aspects are already taken care of during the renewal project, but not the management of the development project. As the product renewal process results in product proposals, which are given in the form of product types, the first stage of the IPD procedure model is omitted. This gives a revised IPD procedure according to Fig. 4.

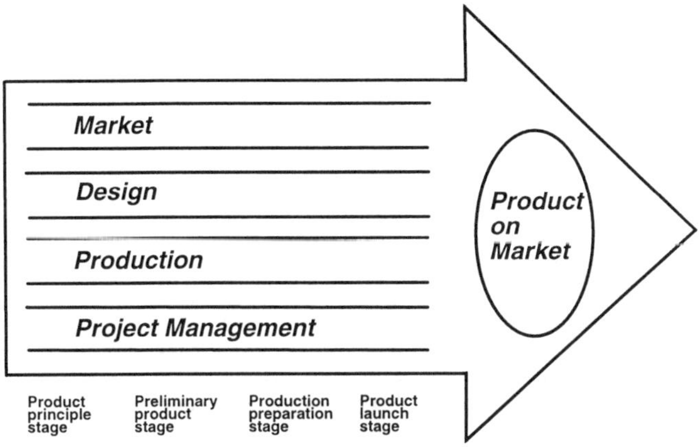

Fig. 4. The revised IPD procedure model.

The four constituent activities of the product renewal procedure model are as follows:

- **Preparation and setting the goals**
  During this activity the following subactivities are to be carried out:
  - □ Clarification of goals, inventory of resources available, critical examination of company "strengths and weaknesses"; the company history might also be utilized.
  - □ Study of existing product history, product assortment as well as the current market and technology status.
  - □ Establishment of product policy and goals and criteria (demands and wishes) for product generation and product evaluation and selection.
- **Initial product generation and primary product selection**
  For the initial product generation the objective is to come up with as many product proposals as possible. One way of achieving this is to utilize intuitive idea generation methods such as brainstorming etc. A search strategy based on *need, resources, situation* and *technology* might also be useful. After the generation of the product proposals, an evaluation of each and every one of the proposals is carried out as a preparation for the subsequent decision on which proposals to develop further.
- **Elaboration of product proposals and final selection of product proposals**
  The most promising of product proposals obtained during the previous activity (= selected proposals) are elaborated upon, thus creating the conditions necessary for the final evaluation and selection of product proposals to be developed (by the IPD procedure).
- **Establishment of product renewal programme**
  For the product proposals derived during the previous activity, a complete renewal programme(s) is to be established. In order to create this programme, time–plans, financial–plans etc. are derived for each and every one of the product proposals, after which an evaluation with reference to six factors such as risk involvement, resources needed, time available etc. is performed. Based on the outcome of this evaluations, a ranking of the proposals is made. This ranking is in turn utilized in creating the renewal programme, in which all of the product development projects are defined.

In Fig. 5, the interaction between the product renewal procedure model and the subsequent IPD procedure model is illustrated.

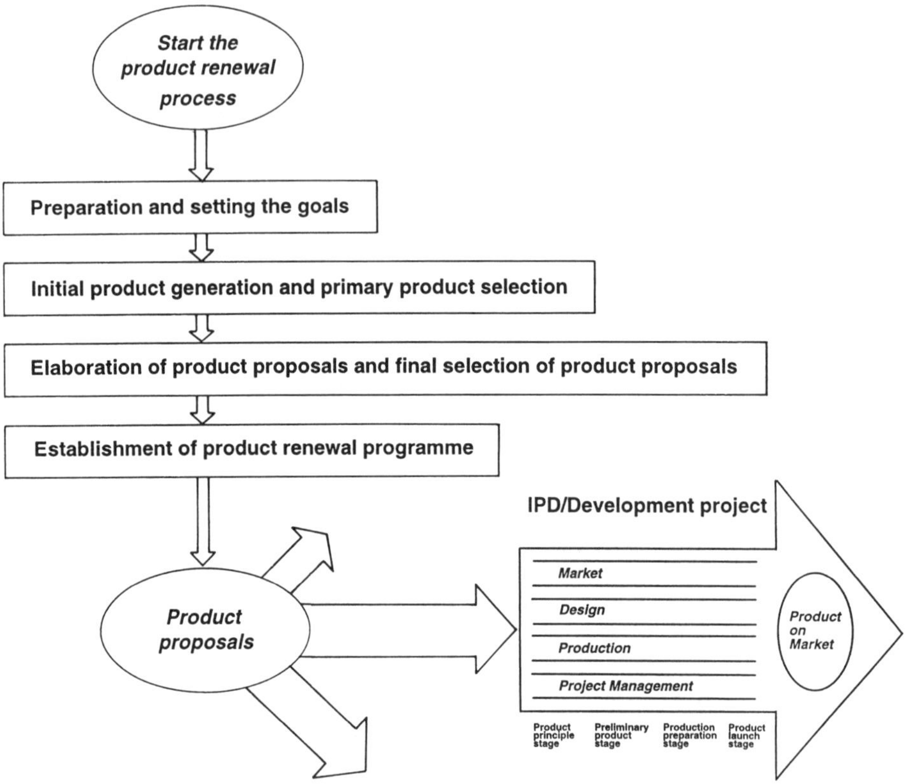

Fig. 5. The interaction between the product renewal and the IPD procedure models.

In a company with a well–established product planning function, the product renewal procedure is most efficiently used in those instances an urgent need for a renewal of the product assortment is at hand. However, this does not exclude the possibility to replace the current product planning function with a modified version of the product renewal procedure. If this concept is chosen, it is necessary to modify the renewal procedure and/or the organization to also address those issues handled by the present product planning procedure on a regular basis. Note that the need for this kind of modifications is only confined to those companies with a well–established product planning function.

6 Product renewal in practice

As previously described, the present state of the product renewal procedure model is the result of the practical experience obtained during a number of renewal projects in Swedish industry. The most important source of observation in this development activity is undoubtedly those projects performed by the students following the *Product Development and Design* curriculum at the Department of Machine Design, Lund Institute of Technology, Sweden.

To give a short insight into the nature of these projects and the feed–back from industry, the 1997 renewal project is briefly outlined. It should be noted that the renewal project forms the basis for a subsequent IPD project, during which only the early phases of the design process are performed. The objective is in other words not to perform a complete product development project, but to supply an insight into the practical use of the product renewal procedure as well as to illustrate the interaction between the renewal project and the subsequent IPD procedure.

This year's product renewal project was performed at *Granuldisk AB* in Malmö, Sweden. *Granuldisk AB* is one of 6 companies belonging to *GS Development AB. Granuldisk* is developing and selling a new type of dishwashers used for potwashing in restaurants and other commercial kitchens.

To remove food remains that have burnt on to trays or pots manually requires a great deal of hard work, patience and strong chemicals. The *Granuldisk* idea or working principle for their machines to remove food remains is copied from nature; the sea laps against stones and rocks and polishes them smooth, but instead of grains of sand *Granuldisk* utilizes specially designed pieces of plastic which scour kitchenware without scratching or damaging.

Granuldisk AB's business idea is based on: *"Improving efficiency and contributing to improving environment, hygiene and economics of all existing potwashing locations with customized systems solutions of the highest quality and reliability".*

An example of a dishwashing machine is given in Fig. 6.

The project to be undertaken by the students was primarily to generate new product ideas based on the working principle utilized in present dishwashers. No further demands or constraints were imposed by the company, which gave the students rather "free hands" to generate new and, hopefully, revolutionary ideas. In practice the students were also free to question and change the current product policy and the company business idea.

The students were divided up into 16 groups with 3 to 5 students per group. In all, 62 students participated in the project. The total time available to each group to perform the project was five weeks – part time.

The results obtained are rather promising. According to an evaluation of the project results performed by Product Manager T. Pehrsson [12], all of the groups have generated at least one product proposal which has the potential to become a future product of the company. This indicates that the students have successfully managed, in a very short time, to acquire the necessary facts and figures to generate and to evaluate and select promising product proposals. However, before a final evaluation of the project results can be performed, the results from the subsequent design activities are also to be considered. As these projects are currently in progress, a more complete evaluation is not currently available. Note that a more specified evaluation needs to involve additional company proprietary information, and is thus, for obvious reasons, not publishable.

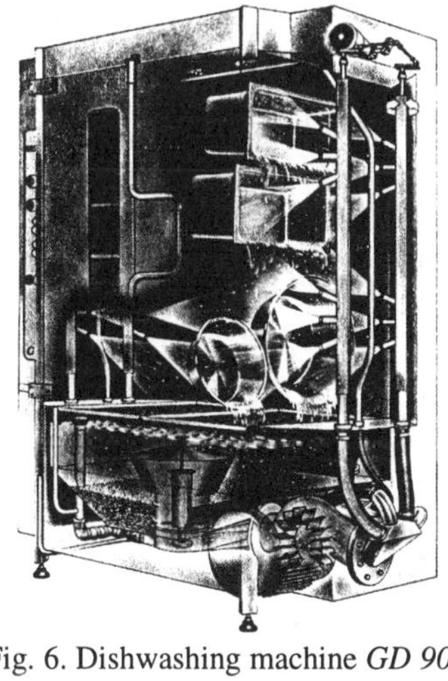

Fig. 6. Dishwashing machine *GD 90S.*

7 Conclusion

The results obtained from numerous projects utilizing product renewal in combination with the revised IPD procedure are very promising. However, this does not exclude the need for additional modifications of the present models, including both the renewal procedure and the IPD procedure.

At present the development of a new version of the IPD procedure model, the *A–IPD (Adaptive – IPD)* procedure is in progress and thus also an updated version of the product renewal procedure model. The main reason for these efforts is the need for a more flexible product development procedure model, which can more easily be adapted to the different forms of organizing a company now available. Another influencing factor is the rapid development of different tools to be used during the development process.

Especially the implementation of computer based tools calls for more flexibility to be effective and efficient, to continuously be able to introduce new methods and techniques in the process. An additional factor of interest is the possibility to spread the different functions of a company as well as to facilitate access to expert help by utilizing the possibilities provided by *Internet* and *Intranet.*

8 Acknowledgements

The author gratefully acknowledges the support given by Swedish industry in the development and evaluation of the product renewal process. A very special thanks to Managing Director Per Sandberg and Product Development Manager Lars Lundström at Granuldisk AB for allowing us to perform this year's product renewal projects at your company! I am also indebted to Product Manager Thomas Pehrsson, also at Granuldisk AB, for his help in evaluating the product renewal projects.

References

1. C. W. Allen, Simultaneous Engineering: What? Why? How?, in *Simultaneous Engineering – Integrating Manufacturing and Design,* edited by C. W. Allen, SME (1990), pp 63–68.
2. M. M. Andreasen and L. Hein, *Integrated Product Development*, IFS (Publications) (1987).
3. D. Clausing, World–Class Concurrent Engineering, in *Concurrent Engineering: Tools and Technologies for Mechanical Engineering,* edited by E. J. Haug, NATO ASI Series, Series F: *Computer and Systems Sciences,* Vol. 108, Springer–Verlag (1993), pp 3–40.
4. D. Clausing, *Total Quality Development – A Step–by–Step Guide to World–Class Concurrent Engineering*, ASME Press (1994).
5. K. Erhlenspiel, *Integrierte Produktentwicklung*, (in German), Carl Hanser Verlag (1995).
6. B. Evans, Simultaneous Engineering, in *Simultaneous Engineering – Integrating Manufacturing and Design,* edited by C. W. Allen, SME (1990), pp 3–4.
7. F. Gordon, F. and R. Isenhour, Simultaneous Engineering, *in Simultaneous Engineering – Integrating Manufacturing and Design,* edited by C. W. Allen, SME (1990), pp 14–16.
8. L. Hein, M. M. Andreasen and F. Olsson, Integrated Product Development, in the *Proceedings of the International Conference on Design Policy,* Vol. 2, edited by R. Langdon and S. Gregory, The Design Council (1984), pp 86–90.
9. P. Kotler et al, *Principles of Marketing*, Prentice Hall Europe (1996).
10. F. Olsson, *Produktframtagning*, (in Swedish), Lic. Eng. thesis, Institutionen för maskinkonstruktion, Lunds Tekniska Högskola (1969).
11. F. Olsson et al., *Integrerad Produktutveckling,* (in Swedish), Sveriges Mekanförbund (1985).
12. T. Pehrsson, *Private communication,* Granuldisk AB (1997).
13. N. F. M. Roozenburg and J. Eekels, *Product Design: Fundamentals and Methods*, Wiley, (1995).
14. G. Pahl und W. Beitz, *Konstruktionslehre – Methoden und Anwendung,* (in German), Springer–Verlag (1993).
15. G. Pahl and W. Beitz, *Engineering Design – A Systematic Approach,* Springer–Verlag, (1996).

Determination of Assembly Sequence for Hull Block in Shipbuilding

Hongtae Kim · Byeongse Yoo · Jaemin Lim

Shipbuilding System Department, Korea Research Institute of Ships Ocean Engineering, KIMM, 171, Jang-Dong, Yusong, Taejon, Korea
E-mail : kht@mailgw.kimm.re.kr

Abstract. When constructing a ship it is necessary to establish a process planning of shipbuilding, which includes fabrication, assembly, and erection of blocks, which are basic units in construction. The assembly process planning determines a production method, a production sequence, equipment and manpower to use, and working condition of block assembly. And the determination of an optimal assembly sequence using positional relationships among the components of a block is one of the most important parts in the assembly process planning. In this paper we developed a simulation program which adopts a neural network to determine a block assembly sequence for shipbuilding. The program derives for a block an assembly sequence based on the Hopfield network, which is created through precedence relationship and the quantification of difficulties in assembling the components of the block. In order to evaluate the performance and the effectiveness of the simulation program developed, the upper deck block of a container ship was selected, and the assembly sequence obtained shows a satisfactory result.

1. Introduction

In spite of rapid growing in the shipbuilding industry the process planning and the production control have been rather lagging behind in progress, resulting in low productivity and high production cost in shipyards. To solve this problem the industry has been concentrating on the research for CIM (Computer Integrated Manufacturing) for the total integration of the design, production and planning.

Currently, however, unnecessary and questionable works are being required in shipyards because of lack of correctly defined job sequence and order. This deficiency is primarily due to the fact that detailed and quality necessary for the production scheduling is not generated in the product design. As the design precedes the production scheduling, the former process naturally cannot produce enough and detailed information for the latter process.

To solve the problem, it is obvious to minimize the difference and it can be achieved through a so-called integrated system, which integrates both the design and production processes. And a considerable attention has been paid to automatic process planning called CAPP (Computer Aided Process Planning), which integrates CAD and CAM. Thus the automatic process planning in concerned with integration of entire production processes such as the production method, production sequence and resources. These are obtained based on the technical knowledge concerning with materials, part, and products through the entire production processes.

To realize automatic process planning, assembly planning is the most important field. Automation of assembly planning begins with analysis of assembly and part models of the product to assemble. Some constraints are frequently involved in automation of the assembly planning. These constraints are acquired from analysis of two models. Automation of assembly planning contains tool/fixture selection, assembly route selection and assembly sequence determination. Among them the determination of assembly sequence is a main part. To find an optimal assembly sequence is closely related to TSP (Travelling Salesman Problem).

This paper investigated an assembly process planning system, in which covers determination of assembly sequence is determined using Hopfield network and precedence-constraint function.

2. Assembly Process Planning

Up to these days, HBCM (Hull Block Construction Method) is widely used in shipyards. HBCM is a method which divides hull into many blocks and builds them up in dock (Storch and Bunch(1995)).

Usually one hull block is composed of several sub-assemblies or part-assemblies that also consist of several pieces or units of pieces.

Block division includes two stages. In the first stage, size of super blocks (pre-erection block) is determined based on the key plan (general arrangement, midship section, lines, construction profile and machinery arrangement). In the second stage, each super block is divided into assembly block considering assembly capacity.

In HBCM, it is practical to plan hull construction in seven levels as shown in Fig. 1. Starting with the block level, work is subdivided down to the part fabrication level to optimize workflow.

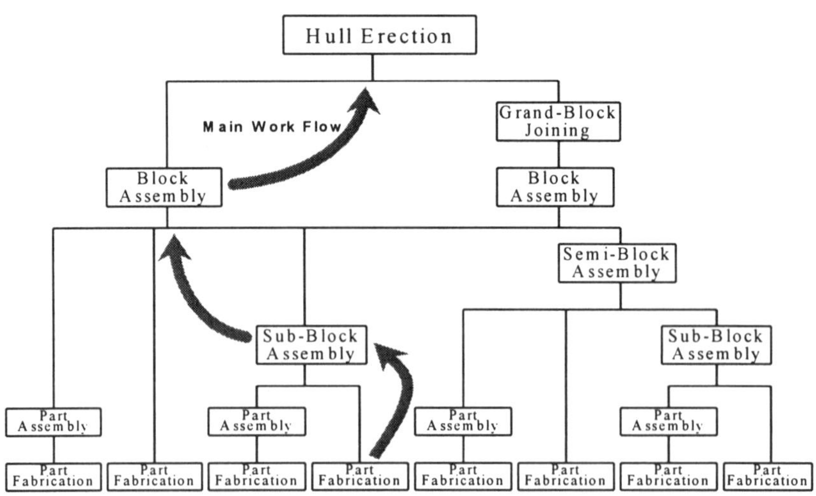

Fig. 1. HBCM Manufacturing Level

Information about block assembly to perform assembly process planning contains part relationship, cutting length, fitting length, welding length, welding posture, assembly area, facility and equipment.

Fig. 2 shows analysis of relationships between assemblies; it is a kind of BOM (Bill of Material). As shown in Fig. 2, there are two types of relationship in block analysis. The first, horizontal relationship is determined in design stage. The next, vertical relationship is determined considering economy of assembly in process planning stage.

```
Level 1         Level 2         Level 3          Level 4

1 B1101 ---- 2 LB12A/G --- 3 TB81A/S5 -- 4 BK54A/S1S5
             |            |               |- 4 BK55A/S1S5
             |            |               |- 4 BK56A/S1S5
             |            |               |- 4 BK57A/S1S5
             |            |               |- 4 BK58A/S1S5
             |            |               |- 4 BK59A/S1S5
             |            |               |- 4 BK60A/S1S5
             |            |- 3 FR82A/S5 -- 4 BK82B/S1S5
             |            |               |- 4 BK82C/S1S5
             |            |               |- 4 BK82D/S1S5
             |            |- 3 FR83A/S5 -- 4 BK83B/S1S5
             |            |               |- 4 BK83C/S1S5
             |            |               |- 4 BK83D/S1S5
             |            |- 3 FR84A/S5 -- 4 BK84B/S1S5
             |            |               |- 4 BK84C/S1S5
             |            |               |- 4 BK84D/S1S5
             |            |- 3 FR85A/S5 -- 4 BK85B/S1S5
             |                            |- 4 BK85C/S1S5
             |
             |- 2 LB12P/G ---
```

Fig. 2. Block Analysis

3. Determination of Block Assembly Sequence Using Neural Network

3.1 Determination of assembly sequence

Generally, assembly sequence represents transition from the initial state, in which all components are separated, to the goal state, in which all components are in consolidation. A state transition is related to an assembly task, and represents an assemblage of two sub-assemblies.

In assembly process planning, determination of assembly sequence is a very difficult problem, because there are numbers of feasible sequences as the number of component increases. And even an experienced assembly planner does not know all feasible sequences about complex assembly.

Five basic principles are considered to determine assembly sequence in shipbuilding area.

① Avoid a interference between part and sub-assembly by shape and position.
② Balance work load in production line.
③ Minimize transfer and change of direction in production process.
④ Maximize a stability of sub-assembly.
⑤ Maximize a fillet welding task.

To find an assembly sequence considering above principles, the assembly process planning is an essential process.

Various geometric informations have to be generated by computer in performing an automatic assembly process planning. But, up to these days, the assembly process is defined manually using part list and assembly drawings.

In this paper, we made an assembly tree using relationship between assemblies and determined assembly sequences in each level from the highest (level 1) to the lowest (level 4), as shown in Fig. 2.

There are many reviews of generating assembly sequence plans on robot assembly task. Bourjault (1984) considered the assembly of the product and proposed a method to build all the liaison precedence-constraints relative to an assembly. De Fazio and Whitney (1987) also took into account the assembly of the product and proposed a simplified method to elaborate the liaison precedence-constraints. They used Bourjault's concepts but made certain simplifications allowing the practical extensions of the technique to assemblies with more than Bourjault considered. Hommem de Mello and Sanderson (1990) suggested a decomposable production system in which the problem of disassembling one product is decomposed into distinct program. Ko and Lee (1987) presented a system that automatically generates an assembly procedure. They use a virtual link concept to represent relations between components of an assembly.

There are few reviews of application on shipbuilding area. Cho (1995) is reviewed a block assembly sequence planning using a CBR (Case Based Reasoning) method.

3.2 Neural network for assembly sequence

3.2.1 Hopfield network

It is the combinatorial optimization problem to find an optimal sequence among feasible sequences. And this problem is similar to TSP.

TSP is simple to state but very difficult to solve exactly. A set of N cities is given, and the shortest closed path, which passes through each city exactly once, must be found. If an initial arbitrary ordering of the ten cities A, B, C, D, ..., I, J is given, a solution to the TSP is represented as 10×10 permutation matrix.. As shown in Table 1, optimal path would be BAGFCJEDHI, and the total length of this path would be $d_{BA} + d_{AG} + d_{GF} + d_{FC} + d_{CJ} + d_{JE} + d_{ED} + d_{DH} + d_{HI} + d_{IB.}$

Table 1. TSP for 10-Cities

City \ Sequence	1	2	3	4	5	6	7	8	9	10
A	0	1	0	0	0	0	0	0	0	0
B	1	0	0	0	0	0	0	0	0	0
C	0	0	0	0	1	0	0	0	0	0
D	0	0	0	0	0	0	0	1	0	0
E	0	0	0	0	0	0	1	0	0	0
F	0	0	0	1	0	0	0	0	0	0
G	0	0	1	0	0	0	0	0	0	0
H	0	0	0	0	0	0	0	0	1	0
I	0	0	0	0	0	0	0	0	0	1
J	0	0	0	0	0	1	0	0	0	0

In Hopfield network, each entry*(X, i)*(where *X* denotes the city and *i* the position) in this permutation matrix is represented by the voltage output V_{xi} of a single neuronal amplifier. The output of each neuron *Xi* is connected to the input of each other neuron *Yj* through a resistance of value $T_{Xi,Yj}$. While neural voltages are allowed to vary continuously, the final result is interpreted by replacing each entry with either a 1 or a 0, depending on whether the neuron's value is high or low (Wilson and Pawley (1988), Hopfield and Tank (1988)).

To calculate Solutions to the TSP, network neurons are represented by an energy function considering the requirements. This requirement can be split into ensuring that stable states corresponding to a permutation matrix are favored and ensuring that matrices representing short paths are favored.

Energy function of TSP is

$$E = \frac{A}{2}\sum_{X=1}^{N}\sum_{i=1}^{N}\sum_{j=1,j\neq i}^{N} V_{Xi}V_{Xj} + \frac{B}{2}\sum_{i=1}^{N}\sum_{X=1}^{N}\sum_{Y=1,Y\neq X}^{N} V_{Xi}V_{Yi} + \frac{C}{2}\left(\sum_{X=1}^{N}\sum_{i=1}^{N} V_{Xi} - n\right)^2$$
$$+ \frac{D}{2}\sum_{X=1}^{N}\sum_{Y=1,Y\neq X}^{N}\sum_{i=1}^{N} d_{XY}V_{Xi}\left(V_{Y,i+1} + V_{Y,i-1}\right) \qquad (1)$$
$$X = successor, \quad Y = predecessor$$

where V_{xi} is the neuron state of the *X*th row and *i*th column, and d_{XY} the cost between two cities. A, B, C, and D are parameters which characterize the performance of the neural network.

The first term will be zero if each row corresponding to a city contains a single 1, with all the other values being zero. The second term is zero if each column corresponding to a position in the tour, contains a single 1. The third term is zero only when the total number of 1's in the network is *N*, the number of cities. The fourth term describes the length of the tour.

The connectivities $T_{Xi,Yj}$ of the Hopfield network are found to be

$$T_{Xi,Yi} = -Ad_{XY}(1 - d_{ij}) - Bd_{ij}(1 - d_{XY}) - C - Dd_{XY} \qquad (2)$$
$$d_{ij}, d_{XY}: \text{Delta Function}$$

3.2.2 Consideration of assembly precedence-constraints

There are precedence relationships between parts in assembly tasks. These relationships have to add into the energy function.

Chen (1990) proposed the concept of precedence-constraint assembly sequence. To solve precedence-constraint assembly sequences using Hopfield network, an additional term is added into Equation (1). An additional term is

$$E_P = \frac{F}{2}\sum_{i=1}^{N}\sum_{j=1,i<j}^{N} V_{Xi}V_{Yj} \qquad (3)$$
$$X = successor, Y = predecessor, F: \text{positive constant}$$

The successors and predecessor of precedence condition correspond, respectively, to the right and left sides of the "⇒" operator. That is, the mapping of the precedence-constraints to the Hopfield network can be done by setting the corresponding weights into negative values.

The investigation of parameter and initial conditions affecting the convergence of the Hopfield network has been widely studied.

3.2.3 Cost matrix for assembly task

While the cost matrix in TSP represents the distance between cities, the cost matrix in the assembly planning represents performance of the assembly.

There are two factors determining performance of assembly process. The first is qualitative factor (cost and man-hour), and the second is quantitative factor(stability of assembly, direction of welding). Welding posture is a very important because welding process has influence on both factors.

The types of welding posture are listed as follows:

① Down welding
② Horizontal welding
③ Vertical welding
④ Overhead welding

Assembly planning must try to minimize difficult welding postures like overhead or vertical welding.

In this paper, we developed a method for the cost matrix to quantify the cost for difficulty in assembling (DIA) when performing two tasks continuously. And assembly sequences are determined using assembly man-hour considering welding posture.

3.3 Simulation for determining assembly sequence

Performance of proposed model was tested on real ship. Double bottom block of container ship (D/W 7000 ton-under construction in shipyard A) was used in the simulation (see Figure 3). Construction of grand block "1101" includes part assembly, sub-block assembly and block assembly. In this paper, we concentrate on sub-block assembly and block assembly.

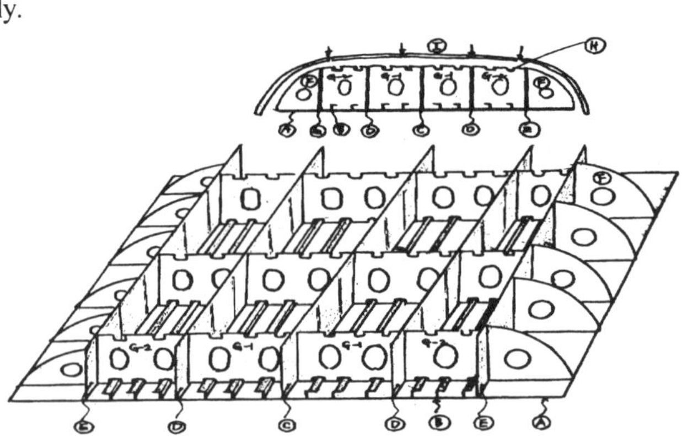

Fig. 3. Grand Block "1101"

As shown in Fig. 3, the drawing of grand block "1101" indicates a symmetry type centering on a center girder. And Table 2 shows a structure of grand block "1101".

Table 2. Structure of Block "1101"

No.	Sub-Assembly		Code
1	Tank Top Plate	101	A
2	Tank Top Longitudinal	102	B
3	Center Girder	103	C
4	No.1 Side Girder	104	D
5	No.2 Side Girder	105	E
6	Open Floor	106	F
7	No. 1 Solid Floor	107	G1
8	No. 2 Solid Floor	108	G2
9	Shell Longitudinal	109	H
10	Shell Plate	110	I

As shown in Table 2, structure of grand block "1101" is composed of ten sub-block assemblies. And Fig. 4 represents fourteen tasks for ten sub-block assemblies.

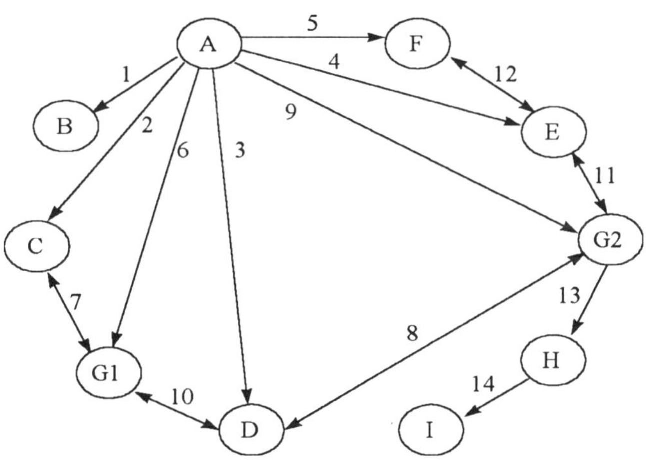

Fig. 4. Assembly Graph of Grand Block "1101"

Assembly sequences are generated by applying Hopfield network that contains precedence-constraints. And cost matrix can be obtained by using a concept of DIA. The actual assembly sequence for block "1101" is shown as follows:
(A) Tank Top Plate ⇒ (B) Tank Top Long. ⇒ (C) Center Girder ⇒ (G1) No.1 Solid Floor ⇒ (D) No.1 Side Girder ⇒ (G2) No.2 Solid Floor ⇒ (E) No.2 Side Girder ⇒ (F) Open Floor ⇒ (H) Shell Long. ⇒ (I) Shell Plate

Table 3 shows precedence-constraints of assembly based on the 14-assembly tasks. And Table 4 shows DIA matrix for block assembly.

Table 3. Precedence-Constraints of Assembly

No.	Preceding Task	Succeeding Task
1	1	5, 8, 9
2	2	5, 8, 9
3	3	5, 8, 9

Table 4. DIA Matrix for Block Assembly

Job	1	2	3	4	5	6	7	8	9	10	11	12	13	14
1	0.0	0.3	1.0	1.0	1.0	0.3	0.3	0.6	0.6	1.0	1.0	1.0	1.0	1.0
2	0.3	0.0	0.3	0.6	0.6	0.3	0.3	0.6	0.6	1.0	1.0	1.0	1.0	1.0
3	1.0	0.3	0.0	1.0	1.0	0.3	0.6	0.3	1.0	0.6	1.0	1.0	1.0	1.0
4	1.0	0.6	1.0	0.0	0.3	0.6	1.0	0.6	0.3	1.0	0.3	0.3	1.0	1.0
5	1.0	0.6	1.0	0.3	0.0	1.0	1.0	1.0	0.6	1.0	1.0	0.3	1.0	1.0
6	0.3	0.3	0.3	0.6	1.0	0.0	0.3	0.3	0.3	0.6	1.0	0.6	0.6	1.0
7	0.3	0.3	0.6	1.0	1.0	0.3	0.0	0.3	1.0	0.6	1.0	1.0	0.6	1.0
8	0.6	0.6	0.3	0.6	1.0	0.3	0.3	0.0	0.6	0.3	0.3	1.0	0.6	1.0
9	0.6	0.6	1.0	0.3	0.6	0.3	1.0	0.6	0.0	1.0	0.3	0.3	1.0	1.0
10	1.0	1.0	0.6	1.0	1.0	0.6	0.6	0.3	1.0	0.0	1.0	1.0	1.0	1.0
11	1.0	1.0	1.0	0.3	1.0	1.0	1.0	0.3	0.3	1.0	0.0	0.6	0.3	1.0
12	1.0	1.0	1.0	0.3	0.3	0.6	1.0	1.0	0.3	1.0	0.6	0.0	1.0	1.0
13	1.0	1.0	1.0	1.0	1.0	0.6	0.6	0.6	1.0	1.0	0.3	1.0	0.0	0.3
14	1.0	1.0	1.0	1.0	1.0	1.0	1.0	1.0	1.0	1.0	1.0	1.0	0.3	0.0

DIA values in Table 4 are divided into easy(0.3), moderate(0.6) and difficult(1.0) cases in performing two tasks continuously.

3.4 Experimental results

Based on the input data just described, simulation is performed. Simulation program is written in MS-C language, running on an IBM Pentium-PC which takes assembly model as input and generate the feasible assembly sequences.

A simulation program is carried out to generate 470 feasible assembly sequences for 10000 assembly sequences. Fig. 5 represents frequencies for sum of DIA.

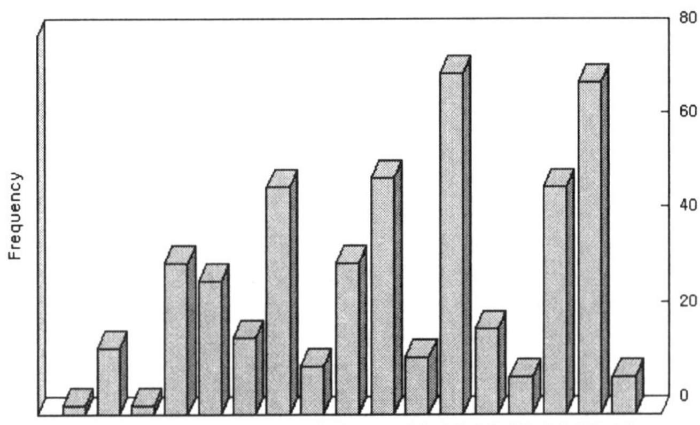

Fig. 5. Frequencies for Sum of DIA

As shown in Fig. 5, the result shows that two assembly sequences have minimum value for sum of DIA among all feasible sequences. Two assembly sequences are shown as follows:

① 1⇒2⇒7⇒6⇒3⇒8⇒9⇒4⇒5⇒12⇒10⇒11⇒13⇒14

② 2⇒1⇒7⇒6⇒3⇒8⇒9⇒4⇒5⇒12⇒10⇒11⇒13⇒14

Table 5. Assembly Man-hour and Welding Posture

Task	Assembly	MH	Welding Posture	Tas	Assembly	MH	Welding Posture
1	A ⇒ B	32	Down	8	D ⇒ G2	12	Down
					G2 ⇒ D	24	Vertical
2	A ⇒ C	4	Down	9	A ⇒ G2	24	Down
3	A ⇒ D	8	Down	10	D ⇒ G1	12	Down
					G1 ⇒ D	24	Vertical
4	A ⇒ E	8	Down	11	E ⇒ G2	20	Vertical
					G2 ⇒ E	20	Vertical
5	A ⇒ F	20	Down	12	E ⇒ F	12	Down
					F ⇒ E	20	Vertical
6	A ⇒ G1	24	Down	13	G2 ⇒ H	20	Vertical
7	C ⇒ G1	12	Down	14	H ⇒ I	32	Down
	G1 ⇒ C	20	Vertical				

To determine a final assembly sequence, calculation of assembly man-hour is performed for two sequences using assembly man-hour table in shipyard A (Table 5). As shown in Table 5, tasks(#1, #7, #8, #10, and #12) have a different man-hour depending on assembling orders, which determines welding posture.

Table 6 shows that two sequences (#1, #2), which have the same sum of assembly man-hour, are the final block assembly sequences.

Table 6. Assembly Man-hour for Sequences

No.	Assembly Sequences	Sum of Man-hour
1	A, B, C, G1, D, G2, E, F, G, H, I	252
2	A, C, B, G1, D, G2, E, F, G, H, I	252

In Table 6, #1 sequence is exactly same as the one used by a shipyard shown in Sect. 3.3, while #2 changes the order of B and C sub-blocks from #1 sequence. The best assembly sequence for grand block "1101" is determined that tank top longitudinal, center girder, #1 solid floor, #1 side girder, #2 solid floor, #2 side girder, open floor, shell longitudinal and shell plate, based on the tank top plate, are to be assembled sequentially.

4. Conclusion

In this paper we developed a simulation program which adopts a neural network to determine a block assembly sequence for shipbuilding. The program derives for a block an assembly sequence based on the Hopfield network which is created through the quantification of difficulties in assembling the components of the block.

In order to evaluate the performance and the effectiveness of the proposed model, the upper deck block of a container ship is tested. The result shows that the proposed model can provide a good planning model for the block assembly process. This simulation program is applied as sub-module of integrated hull process planning system that contains assembly planning.

Acknowledgement

The research reported in this paper has been supported by Ministry of Science and Technology of Korea.

References

1. R. L. Storch and H. M. Bunch, Ship Production 2nd edi., Cornell Maritime Press (1988) pp. 67-93.
2. A. Bourjault, Contribution a une approche méthodologique de 1' assemblage automatisé: Elaboration automatque des séquences opératiores, PhD thesis, L' Université de Franche-Comté (1984).
3. T.L. De Fazio and D.E. Whitney, Simplified Generation of all Mechanical Assembly Sequences, *IEEE J. Robotics and Automation*, **RA-3**, 6 (1987) 640-657
4. L. S. Homen de Mello and A. C. Sanderson, And/Or Graph Representation of Assembly Plan, *IEEE Trans. Robotics and Automation*, **6**, 2 (1990) 188-199.
5. H. Ko and K. Lee, Automatic Assembling Procedure generation from mating conditions Field Annealing, *CAD*, **19**, 1 (1987) 3-10.
6. K. Cho, Development of Computer Aided Process Planning System for Hull Assembly Shops, *IE Interface*, KIIE, **8**, 2 (1995) 41-53 (In Korean).
7. G. V. Wilson and G. S. Pawley, On the Stability of the TSP Algorithm of Hopfield and Tank, *Biological Cybernetics*, **58** (1988) 63-70.
8. J.J. Hopfield and D. W. Tank, Neural Computation of Decision in Optimization Problems, *Biological Cybernetics*, **52** (1985) 141-152.
9. C. L. Philip Chen, Neural Computation for Planning And/Or Precedence- Constraint Robot Assembly Sequences, *Proceedings of International Joint Conference on Neural Networks* (1990) pp. 127-142.

Computer Aided Systems for Industrial & Engineering Design - An Attempt to their Integration

Dieter H. Müller, Claus Aumund-Kopp

Department of Mechanical Engineering, Bremen University
Badgasteiner Str. 1, 28359 Bremen, Germany
e-mail: bik@bik.uni-bremen.de

Abstract. A method is presented which describes an integrative collaboration of both industrial and engineering designers.
A survey on requirements for both industrial and engineering design systems is given and the features and functions of common engineering CAD systems and those of „Industrial Design Systems" are compared.

1. Introduction

The question of industrial design nowadays has become more and more important not only for utility goods but also for industrial products. This is a result of technological developments and market trends which can be summarised as follows:

- due to the trend to miniaturisation (e. g. electronic control systems) and due to necessary coverings (safety) the functions of technical products are no longer visible from outside and therefore have to be made clear to the user by means of industrial design

- the heavy competition in supply industry and capital goods industry and the growing internationalisation of these industries are responsible for the fact that no longer only technical features of a product are responsible for the customers decision to buy the product. These days especially features of industrial design are taken into account more and more.

Although quite obvious - these facts have only been realised and put into action by a few companies yet. This understanding has already been used successfully in some industries, e. g. producers of machine tools. Within other branches industrial design is sometimes used as a subsequent improvement of an existing design only. Thus no Integration of engineering and industrial design takes place and therefore no product improvement is to be expected.

Doing this it has to be kept in mind that the design of a product is strongly effected by the following items /1/:

- technological and physical requirements
- ecological requirements
- manufacturing requirements
- economical requirements

Due to these demands an integration of engineering and industrial design will only be possible, if there is a close co-operation of both the engineer developing the product and the industrial designer determining its appearance. This co-operation should last during the whole phase of development already starting with defining the specifications of the product. Then a product can be expected which fulfils its requirements and which is not only given a functional but also an attractive form and appearance.

Following this procedure does not mean priority of industrial design but rather its integration. Main objective is still the fulfilment of specifications and perfect function of the product.

2. Integration of Product Development and Industrial Design

Qualitative Market Research and Demand Analysis

With the assistance of a method called "Conjoint-Measurement" (a further development by Chaselon & Henning, University of Bremen) an instrument for market research first used in the U. S. /2/ a prospective solution for future product development and product design was found in co-operation with technicians, designers and marketing experts (Fig. 1).

This methodology makes it possible to compare the benefit of use and the image of a product before it is actually brought to market.

A study having pilot character was carried out covering qualitative interviews with specialist customers and users with the help of assessment displays. These displays represent possible configurations of the future product influenced by the factors "brand", "price", "design" and "technology/material".

The evaluation of these assessment displays with the aid of the "Conjoint-Measurement" method and an additional examination of existing offers (products) of competitors then allowed a statement about the relatively most useful and most economical design of the product.

Results of this analysis were market demands, lists of faults of competitive products and a list of wanted requirements supplied by the asked costumers and users. This was used as a basis for the definition of further specifications.

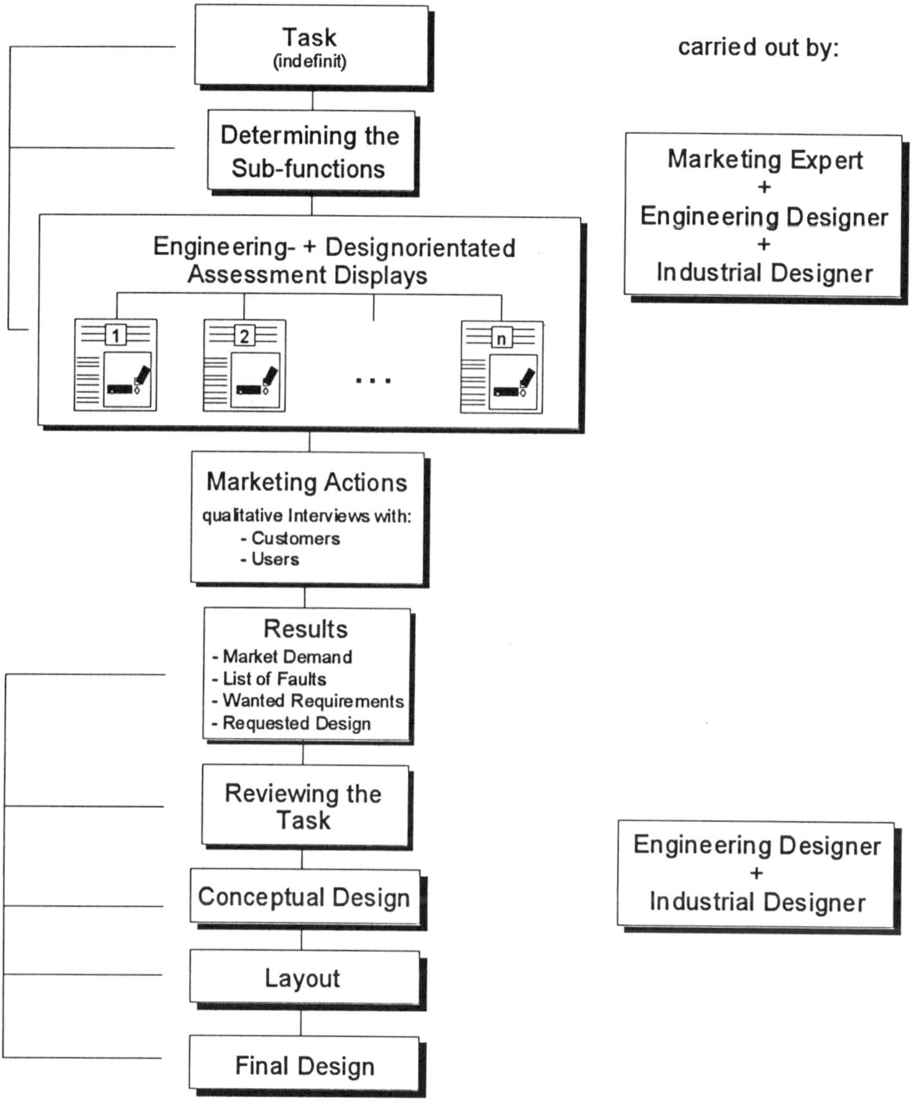

Fig. 1: Integrated Product Development /3/

Integrative Co-operation in Design

Starting from clear technical requirements thc conceptual solution for the task was worked out during an integrative co-operation of engineers and designers. According to the VDI-rule 2222 part 1 /4/ especially according to the "Procedure for the Creation of new Products" the design was made.

In Addition a phase had been inserted deliberately where engineers and designers separately dealt with certain problems and their solutions. Each of them using the classical procedures according to their discipline.

The engineers were looking for possible solutions of individual problems which then where morphologically listed and combined following the classical method. The designers likewise examined certain components whereas only the outer form was prescribed.

During the established meetings of the team the separated work was brought together again in the following way:

Industrial designers and engineers together investigated practicable variants for technical solutions which were designed from inside to outside (from sub-functions to overall function) with regard to the opportunity of industrial design and the possible integration into the overall design of the study.

Just as the engineers vice versa observed to what extend elements of form determined by the industrial designers were realisable.

Summarised this procedure can be described like this: both the results found by engineers and thus being designed from inside to outside (from sub-functions to overall function) and the solutions found by industrial designers and therefore essentially including the outer form only (designed from outside to inside) were examined concerning the adequate fulfilment of functions.

This step-by-step division of the work of both engineers and industrial designers and its integration afterwards lead to essential advantages:

- encouraged by prescribed designs of the outer form the engineers got new ideas and suggestions to fulfil required functions i. e. starting from an outer shape determined by the designers new engineering solutions were found
- industrial designers used pure functional relations to visualise these functions by an appropriate design of the outer shape
- mutual understanding of both disciplines (engineers and industrial designers)

This procedure was repeated several times and clearly showed that the finding of new ideas was not influenced but the engineers developed a certain understanding for the overall design of the outer shape and the designers understood technical necessities to provide functionality.

At the end production documents were generated and a prototype was manufactured and tested.

3. Technology and Design

As a rule design, the formal image of new products, nowadays is not yet a result of interdisciplinary teams and true co-operation of engineer and designer. Still to be found are industrial designers working more or less independently from the process of engineering design creating their own concepts and ideas of the shape and appearance of the new product. It is quite obvious that in this way no mutual enrichment among engineer and designer can come about. Thus it is not surprising that the task of industrial design is also not connected to the schedule of the usual engineering design process.

A direct result of the fact described above is that a formal image is prescribed from the start as a solution of the task of industrial design and engineers have to comply with the appearance of the product. Vice versa industrial design is adapted to a finished or almost finished technical solution.

Using the method of integrated product development this problem is taken into account by the fact that the task of industrial design is already situated at the very beginning the procedure (Fig. 2).

Because of that engineers for example benefit from representation techniques of industrial designers:

- new ideas can be translated visually
- the formal impression of new solutions can be judged early
- different 3-dimensional representations economise manufacture of samples and prototypes
- industrial designers bring in unconventional but visually ascertainable and therefore “understandable“ ideas

On the other hand designers are kept away from following unrealistic “dreams of styling“ too long.

Mutual criticism and help lead to a useful and optimised result with lowest waste of time and money.

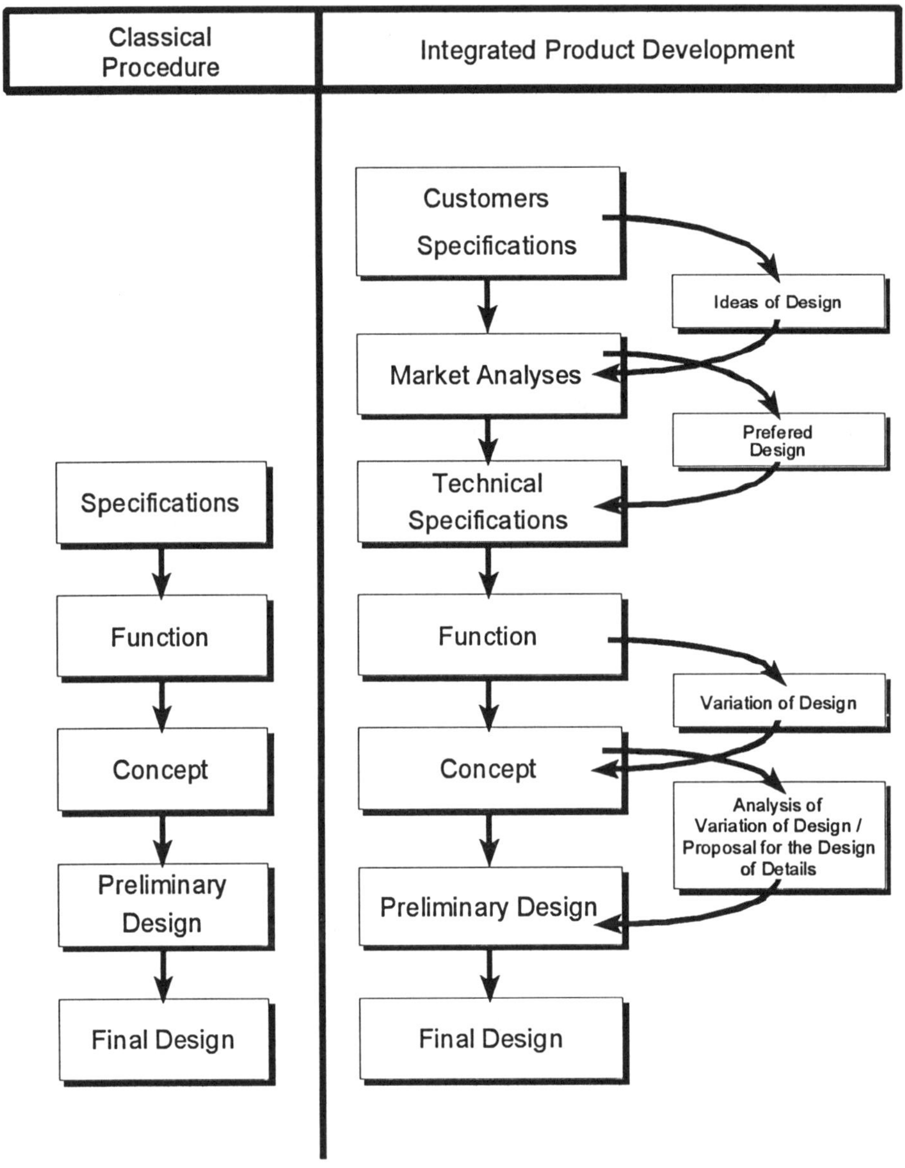

Fig. 2: Comparison of Classical Design Procedure and Integrated Product Development /3/

4. Demands upon a CAD System for Engineers and Industrial Designers

As we have seen engineers and designers can or rather must work together. Nevertheless they still work on different features of a product and follow different ways to work. Thus there are different demands upon the CAD systems they apply.

CAD systems for engineers have to enable their users to work precisely and they have to make clear representations possible.

Looking at the systematic way an engineer develops a product: from the determination of its function (requirement, specification) via finding an effect carrier (part, component) to the complete product, these kinds of systems also have to support their user by finding out further technological data. These additional system functions can be an FEM-module for example determining load and stress/strain behaviour of a part or it can be sub-routines to check the kinematic behaviour of a part in an early stage of design.

A system especially unable for industrial design has to include features concerning the development of the outer shape and appearance of a product because a major part of an industrial designers work deals with this „cover“. Doing this the usual way of an industrial designer is the description and representation of an overall function towards the elementary functions and their effect carriers.

Thus the external form and design of a product is divided into separate parts and developed one after another. At first such system for the purpose of industrial design can do without precision. However it principally has to possess extended functions to edit and adapt shape, colours, textures and graphic features of a product.

Software given the attribute „industrial design“ primarily means systems that exceed the ability of common CAD systems concerning the features mentioned before. These systems particularly provide:

- 3-dimensional modelling concepts to generate geometry, including freeform geometry
- facilities to generate and to work on graphic features
- modules for photorealistic presentation and visualisation

5. Comparison of Features of Common CAD Systems and a Computer Aided Industrial Design System

As an example we have taken a software bought as an „CAID“ system (Computer Aided Industrial Design System) which is capable of 3D-modelling and photorealistic visualisation /5/, /6/.

The procedure to generate the form and design of the product takes place in the following steps - analogous to the separate parts of the external form and design of a product (Fig. 3):

- Generation of geometry
- Shaping
- Colouring
- Texturing

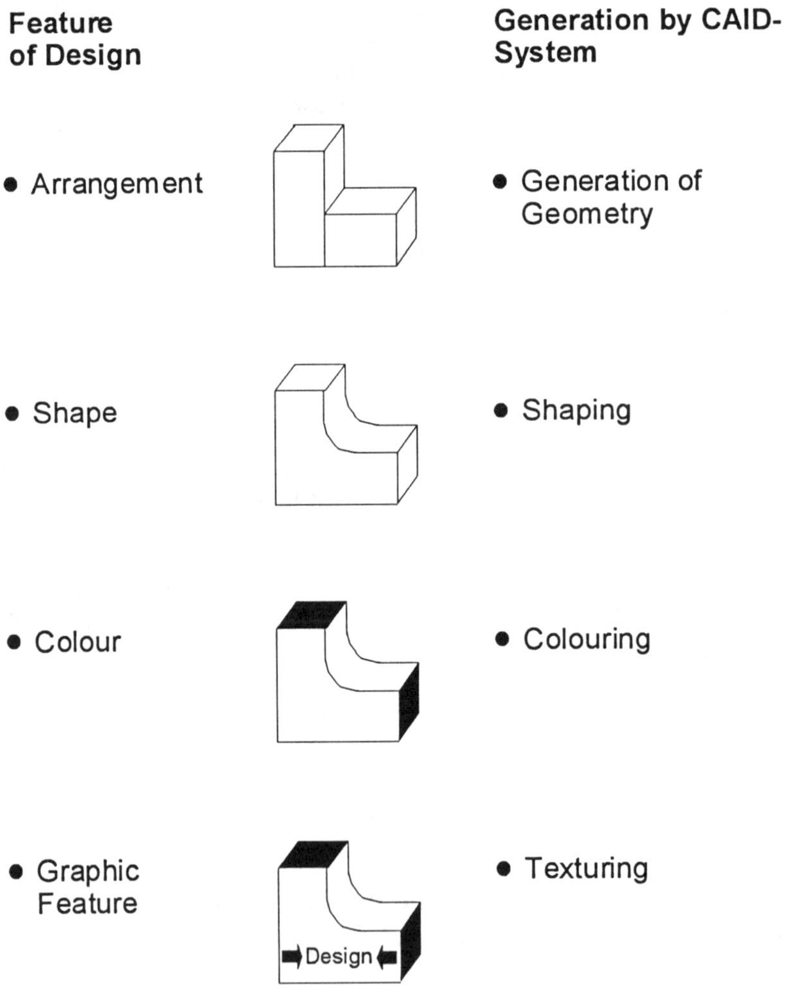

Fig. 3: Analogy of Product Design

In contrast with usual systems for engineers which functions are often limited and which only have predefined basic geometries like cylinders or rectangular solids to generate geometries the examined industrial design system enables its user to model freeform shapes intuitively. Doing this the designer can abstain from precision but he does not necessarily have to. The generation of freeform shapes here is done by rotation or extrusion of any cross section for example without being limited to simple circles or lines as basic elements.

On the other hand the abilities of the examined industrial design system concerning common features as measuring and hatching functions are quite restricted. There are also no special functions to determine technological data.

For the further forming of design the shapes of geometries generated as described before can be edited in multiple ways.

Formed geometries can be trimmed (when intersecting each other), they can be blended or they can simply be arranged in new positions. All this is supported by sophisticated preview modes which enable the designer to regard the geometry in different ways e. g. in illuminated scenes and surfaces or with varying colours where other systems often offer only rotated views. Thus the user can have a close look at his model and get a first impression of its overall appearance.

Of particular importance with systems used for industrial design are their functions for defining colours and the appearance of materials as well as functions for texturing and graphic processing. In these systems many more features of that kind are implemented than in „simple“ systems for mechanical engineering only. These are especially possibilities to mix colours or to define and assign physical effects like reflection, roughness or refraction and corresponding parameters to surfaces.

Functions to define 3-dimensional textures like „marble“ and to assign them to surfaces of geometries in that way that even edges look like having been cut off considerably reduce the expenditure which would usually be necessary to the get same effect.

Furthermore the system being tested here enables its user to import graphic features like drawings, pictures or patterns, to edit them and to attach them to one or more surface elements. For example as a tag in the centre of a surface or mapped around the surface element or elements in total.

Particularly for the creation of photorealistic images the tested software possesses modules to define the complete scene of a picture including bottom, fore- and background, perspective of the camera, lights and shades etc.

Comparing the features of common CAD systems and those of special systems for industrial design it can be seen that most of the different requirements have been taken into account during the system development.

Nevertheless and due to the modular architecture of present CAD systems there is no sharp boundary between systems merely for mechanical and engineering design used to produce drawings and technological data or systems for industrial design mainly visualising the outer form and appearance of a product. Many engineering systems nowadays can be upgraded by adding modules performing computation of kinematic data on the one hand and modules visualising products as photorealistic images on the other hand.

References

1. H. Seeger, Design technischer Produkte, Programme und Systeme: Anforderungen und Lösungen. Berlin, Heidelberg: Springer Verlag (1992)

2. J. Henning, F. Chaselon, Abschlußbericht „qualitative Marktstudie und Bedarfsanalyse für das HATEDE F & E Verbundprojekt", unveröffentlichter Bericht, Universität Bremen (1992)

3. D. H. Müller, K. Oestermann, Integrative Zusammenarbeit von Konstruktion, Marketing und Design, Proceedings of 9th International Conference on Engineering Design, The Hague (1993), 1197-1204

4. VDI-Richtlinie 2222, Blatt 1, Konzipieren technischer Produkte, Düsseldorf VDI-Verlag (1977)

5. C. Aumund-Kopp, Design Systeme - Nutzen und Anwendbarkeit für Ingenieure, - Bremen, Universität, Bremer Institut für Konstruktionstechnik, Studienarbeit (1994).

6. C. Aumund-Kopp, Integration des Designsystems "DeskArtes" in die Prozeßkette eines Yacht- und Schiffskonstruktionsbüros, Universität Bremen, Bremer Institut für Konstruktionstechnik, Diplomarbeit (1995)

Integrated System of CAIP and Quality Forecasting for Mold Manufacturing in CE Environment

Chen Kang Ning, Gao Guo Jun, Chen Shan Hong, Zhang Jun Mei

School of Mechanical Engineering
Xian Jiaotong University, Xian, P.R.China

Abstract. This paper describes a integrated system of the CAIP and quality forecasting for sculptured surface of dies and molds in ball-end milling process. In the system, the machining process planning, inspecting planning and machining error prediction are carried out concurrently. Five interrelated modules are being developed to construct the proposed integrated system. They are: (1) part accuracy feature module, (2) the platform for concurrent development, (3) automatic machining process planning module, (4) automatic inspecting planning module, (5) machining error predication module. The machining error predication module is established based on contributory factors of machining errors. The optimum matching method between the measurement data and CAD-base design data is presented. The machining accuracy evaluation of sculptured surface is described. The predicting results of the dimensional errors to determine the number of the measurement points and their distribution are discussed.

1. Introduction

The traditional approach for the mold manufacturing is in sequence and series fashion. The machining process planning, the NC code generated , the machining execution , the inspection planning and the evaluation of inspecting results are carried out sequentially. The modules are of separation and there is no information exchange between these processes. Once the inspection result is beyond the scope of specification, it is necessary to repeat the processes sequence again. In this paper , a integrated system of CAIP and quality forecasting for mold manufacturing in CE environment is presented. The system is consisted of the accuracy feature module, platform for concurrent development, automatic machining process planning module, automatic inspection planning module and machining error predication module. The client/server pattern is used in the system. On the basis of analysis major error source, the machining error predication model is established. In order to eliminate errors caused with the difference between the machining datum and the inspecting datum, the optimum iteration progress is presented. The variation of inspecting results associated with the change of sampling scheme is closely related to the curvature of surface profile and process capability. The predicting results of the dimensional errors to determine the number of the measurement points and their distribution are discussed.

2. Framework of the System

The Fig.1 shows the structure of the system.

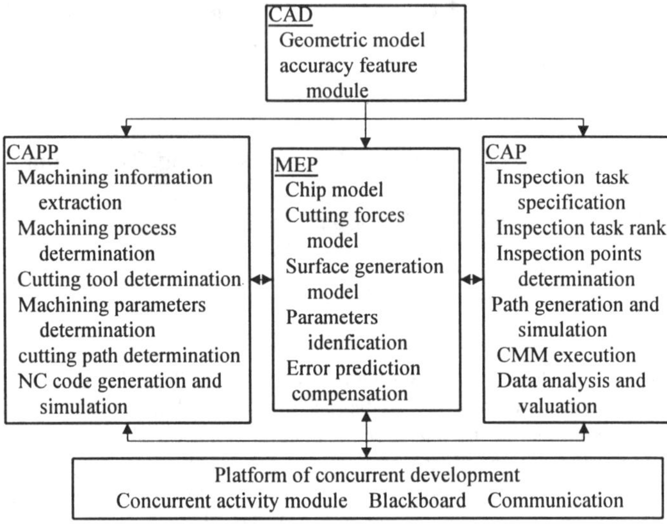

Fig.1 Structure of the system

• CAD module Based on the commercial software the accuracy feature model of the part was developed by use of the neutral file. The CAPP and CAIP accept the accuracy feature model of the part as input and extract concurrently the machining and inspecting information .

•CAPP module The following tasks are implemented in the module. They are: (1) selection machining fashion on the sculptured surface based on the code rules, (2) determination of cutting tools, (3) determination of machining parameters, (4) determination of cutting paths, (5) generation of NC code and the simulation.

• Machining error predication module According to the cutter information, machining parameters and cutting path from CAPP module , the chip geometry model, the cutting force model , the tool defection model and surface generation model are established. The predication result of machining error is fed back to the CAPP module to modify the machining paths. At the same time, the predication results are used to modify the number of inspection point and their distribution in CAIP module. The inspecting results is used to modify the parameters of the predication model of machining error. So that the machining error predication module is integrated with the CAPP module and the CAIP module.

• CAIP module According to the accuracy feature model of CAD and the predication results of machining error, the six functions are developed in the CAIP module. They are : (1) the definition of inspection information and information acquisition, (2) align part, selection probe and inspection sequence, (3) determination the number of inspection points and their distribution, (4) determination the inspection planning and simulation, (5) CMM execution , and (6) analysis and evaluation.

• Platform for concurrent development In order to support the resource management, control and eliminating the conflict between the tasks, the platform is consisted of concurrent activity model, blackboard and communication. The function of platform is shown in Fig. 2.

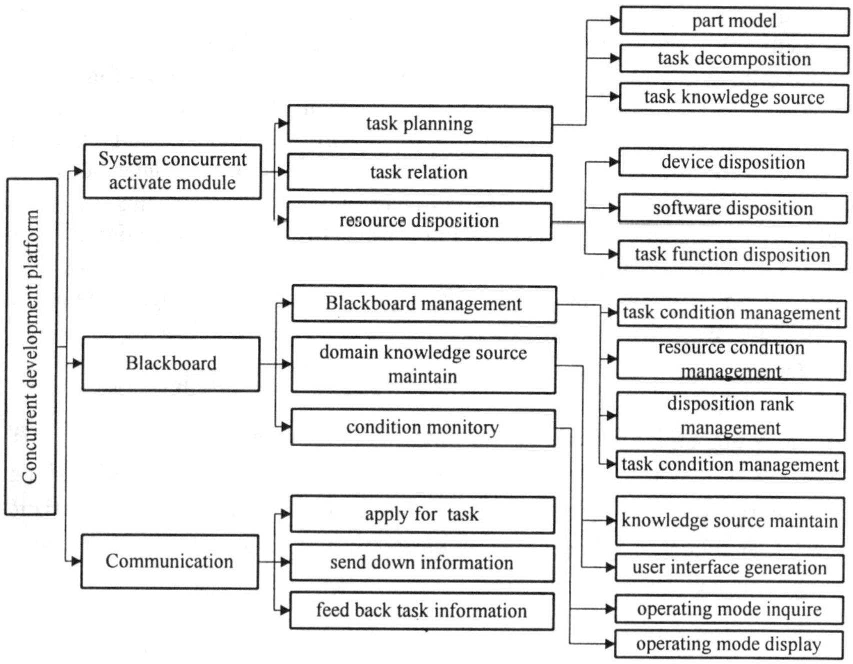

Fig. 2 The function of platform

concurrent activity module supports the functions, including the task planning ,task relation establishing and resource disposing. The concurrent activity model provides the relationship between the tasks and the task decomposition. When the tasks are accurately modeled , the optimum control of the system is realized.

Blackboard provides the system management and control, domain knowledge source maintain and system condition monitoring. To ensure the information insistent and safety, the communication between the tasks is completed through the blackboard. The condition monitoring shows various task condition in real time.

Communication To ensure effectively transfer information between various tasks, the double-action communication are built through the platform for concurrent development. The double-action function is as following: task number, disposition information, feed back information and operating condition.

3. Machining Error Prediction Module

The machining error prediction module accept the parameters of the cutter , machining condition and cutting paths from CAPP module and the number of inspection points and their distribution from CAIP module as inputs to predict machining errors, while CAPP module and CAIP module accepts the predication results of machining error as inputs to modify cutting paths, the number of inspection points and their distribution. It is very important to build up the accuracy prediction model of machining error.

Compared with the flat-end milling process the cutting process for sculptured surface using ball-end milling is more complicated due to the complicated and constantly changing cutting geometry. the predication model of machining error was presented based on the result that the tool deflection is the dominant factor in the ball-end milling process[1]. In order to obtain the instantaneous cutting force model, a series of carefully randomized model-building experiments was performed. The cutting forces are depend on the cutting conditions of the part-cutter system and in the X and Y directions were measured using a Kister model three-component piezoelectric force dynamometer. It is necessary to completed a large amount of experiments. An approach to identify the parameters of cutting force using the result of machining errors is presented and implemented.

the machining error of the surface normal vector at a designed point can be approximated as

$$e = \left(\sum_{z=z_{\min}}^{z_{\max}} \left(\sum_{i=1}^{n} \left(K_T(z)[t_i(\theta,z)]^{M_T}[-\cos\theta_i(\theta,z)] + K_R(z)[t_i(\theta,z)]^{M_R}[\sin\theta_i(\theta,z)]\right)\right) \cdot h\right) \cdot \sin\phi \Big/ k \quad (1)$$

where ϕ is the angle between the tool axis and the surface normal vector, k is the stiffness of the cutter, h is the width of the cut of the differential cutting edge along the z direction, $t_i(\theta, Z)$ is the radial chip thickness, M_t and M_r are the parameters characterizing the size effects of the part material, and $K_t(Z)$ and $K_r(Z)$ are the specific cutting force coefficients.

$$K_T(z) = a_0 + a_1 \cdot (\frac{z}{R}) + a_2 \cdot (\frac{z}{R})^2 + a_3 \cdot (\frac{z}{R})^3, \quad 0 \le z \le R$$

$$K_R(z) = c_0 + c_1 \cdot (\frac{z}{R}) + c_2 \cdot (\frac{z}{R})^2 + c_3 \cdot (\frac{z}{R})^3$$

$$K_T(z) = a_0 + a_1 + a_2 + a_3 \qquad Z>R \quad (2)$$

$$K_R(z) = c_0 + c_1 + c_2 + c_3$$

The variables in the function (1) can be divided to two categories. A part of the variable can be accurately calculated through geometric relation such as ϕ, $t_i(\theta, Z)$, $\theta_i(\theta, Z)$ and the others have be obtained through the method of parameter identification based on the machining errors which are measured by CMM.

We use the developed software to calculate the variables of first category and the variables of second category are determined to use the method of system identification. For exempla, based on the results of measured errors , a numerical fitting procedure is devised to evaluate the model parameter. The tests were carried out and details of the cutting conditions are shown as following.

Test model: One set slope model and one 2-D extended surface model;

Cutting strategy: Cutting along -X direction;

Machine tool: MAHO MC50 3 coordinate horizontal machining center;

Cutter: high speed steel ball-end mill with 5 mm diameter, two right-handed flutes with 35 helix angle;

Part material: aluminum;

Cutting depth: 2mm; Spindle speed: 1000rpm; Feed: 350mm/min;

Without coolant: The identification coefficients of cutting force $K_T(z)$ and $K_R(z)$ are:

$$K_T(z) = 1360 - 2082 \cdot (\frac{z}{R}) + 622 \cdot (\frac{z}{R})^2 + 278 \cdot (\frac{z}{R})^3$$

$$K_R(z) = 458 + 324 \cdot (\frac{z}{R}) - 1664 \cdot (\frac{z}{R})^2 + 1001 \cdot (\frac{z}{R})^3$$

$$0 \le z \le R$$

$$K_T(z) = 178 \quad , \quad z > R$$

$$K_R(z) = 119$$

$$M_T = 0.86$$

$$M_R = 0.85$$

In order to verify the prediction ability of the developed approach, a machining sculptured surfaces experiment was performed . the inspection results was listed in Table 1 and in Fig,3. It can be seen that the correspondence between the prediction data and measured data is well.

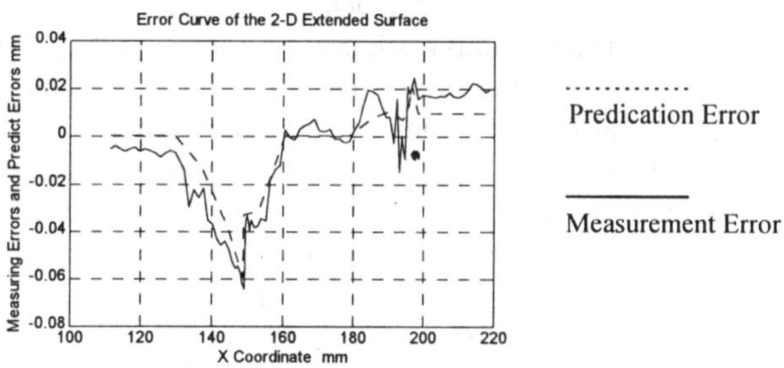

Fig. 3 Error curve

Table 1 Inspection result

	X Coordinate (mm)	Y Coordinate (mm)	Z Coordinate (mm)	Predict Error (mm)	Measuring Error (mm)
Predict Point 1	131.7077	21.2	-31.9336	0.000	-0.0056
Predict Point 2	138.2839	21.2	-30.3808	-0.0132	-0.0225
Predict Point 3	146.4617	21.2	-24.5949	-0.0387	-0.0468
Predict Point 4	151.1904	21.2	-15.9133	-0.0592	-0.0623
Predict Point 5	152.0111	21.2	-9.5795	-0.0326	-0.0402
Predict Point 6	154.7824	21.2	-3.9356	-0.0312	-0.0368
Predict Point 7	169.9032	21.2	-2.0000	0.000	0.0042
Predict Point 8	184.5253	21.2	-2.8131	0.008	0.018
Predict Point 9	188.7286	21.2	-5.3661	0.011	0.0084
Predict Point 10	191.7029	21.2	-9.3394	0.0072	-0.0023
Predict Point 11	193.0710	21.2	-15.9944	0.0116	0.021
Predict Point 12	194.4556	21.2	-19.5208	0.223	0.0224
Predict Point 13	203.2589	21.2	-24.2875	0.01	0.019

4. Inspection Planning Module

The inspection planning module aims at obtaining accuracy valuation for inspected part based on less measurement points. The number of the required inspection points, their distribution and the accuracy valuation of the measurement data are primarily concerned in this module. Owing to the complicated dimensional measurement , the number of required points for obtaining effective quality assurance is unknown (4). The accuracy of sculptured surface is defined as the profile error. It is necessary to use other datum of part when the part is machined and inspected. The errors which is caused the difference between the using the machining datum and inspecting datum are larger and it usually leads to exceed the tolerance limit. In this paper, the number of measurement points and their distribution in close relationship to the curvature of the surface profile and process capability is investigated, and for the accuracy valuation of the measured data an optimal match algorithm is discussed.

• The Number of Measurement Points and their Distribution

A surface profile is within tolerance if the deviation at any point on the surface is within the specified bound. Cases may occur when sampled deviations are within bound while non-sampled deviations are in fact out of tolerance. How to select a set of minimum number of measurement points for achieving the best possible accuracy it is necessary. It is well known that the machining errors of the sculptured surface mainly suffer tool defection and various random factors in the milling process. Through the prediction model of machining error, the machining paths of cutter are modified to unify with designed ideal profile. It means that the number of measurement points is mainly related to designed profile and process capability. The experiments of machining the sculptured surface with various curvature have carried out. The relationship are shown in Fig 4. The variance of the measured errors relates closely to the number of measurement points and the curvature of profile. In the same variance of measured errors, the greater the curvature is the more the number of the inspecting points needs. Thus the number of required measurement points for quality assurance of the surface profile not only depend on the process capability but also depend on the curvature of profile. In the Fig 5, curve(a) represents the result of using equally spaced sample and curve(b) represents the result of the sample interval to relate to curvature. It is shown in order to obtain equal variance of the measurement error , that the equally spaced sample needs more measurement points.

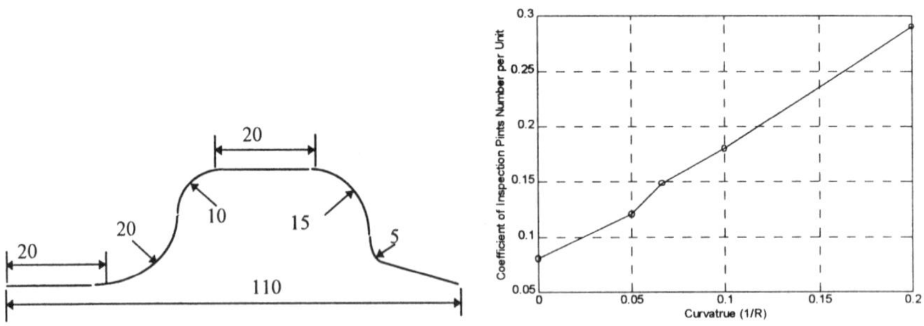

(a) (b)

Fig. 4 (a) Measured profile

(b) Relationship between curvature and number of inspection points

• Error Analysis and Valuation

Owing to the difference between the using the machining datum and inspecting datum and various factors in the machining process, two part errors were contained in the measurement data of the surface profile, that is systematic errors and random errors. In order to obtain the systematic error from total errors, the principle of used algorithm is minimization of the sum of .squared distance of the measured coordinates from the surface with respect to the six variables of a rigid body transformation. The program flowchart of algorithm is shown in Fig. 6. The measured data and transferred data after using the algorithm are shown in Fig 7. It is obvious that before the algorithm the greatest normal error and the object function F is 0.257mm and $1.225mm^2$ respectively, but after the algorithm the greatest becomes 0.059mm and $0.038mm^2$. This algorithm can be adopted for the use of locating design coordinate system and matching measurement data with design data as well.

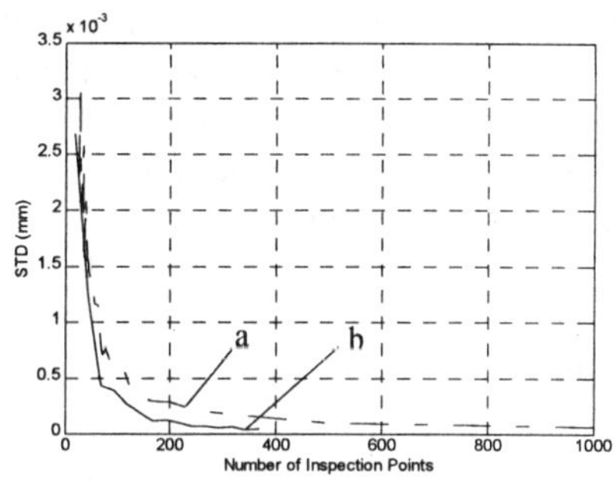

Fig 5 Result of deferent sampling scheme

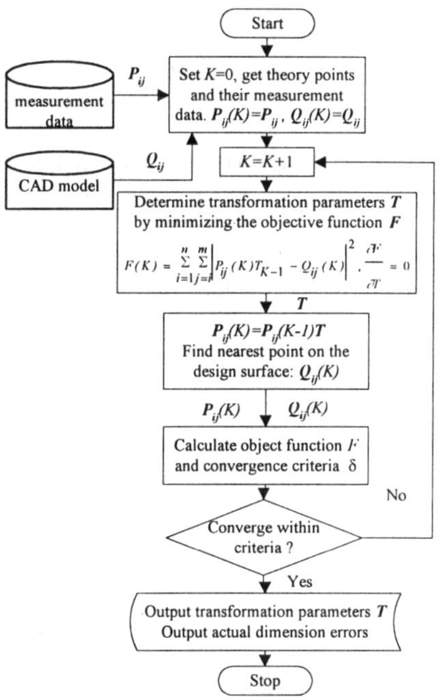

Fig. 6 Flow diagram of optimal matching algorithm

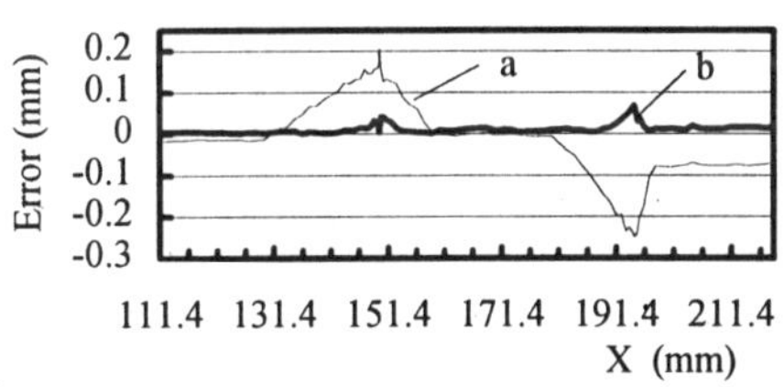

Fig.7 Measured result (a) and transferred result (b)

5. Conclusion

In this paper we construct a integrated system of the CAIP and quality forecasting for mold manufacturing in CE environment. The various information and resource in the system is shared. The concurrent operation and feed back control between modules is realized. In order to obtain conveniently accuracy prediction model of machining error, the method of parameters identification based on the machining errors which are measured by CMM is effect. Based on experiment and analysis, the number of required measurement points for quality assurance of the surface profile not only depend on the process capability but also depend on the curvature of profile. When the process capability is higher and the errors were compensated according to the prediction model of machining error , the number of required measurement points mainly depend on designed part profile.

6. Acknowledgment

The authors would like to thank the National Science Foundation of China and the Doctorate Academic Foundation of Education Committee for support to this work.

Reference

1. E.M.Lim, H.Y.Feng, C.H.Menq and Z.H.lin. The Prediction of Dimensional Error for Sculptured Surface Productions Using the Ball-End Milling Process. Int.J. Mach. Tools Manufact. Vol.35.No.8.1995
2. H.Y. Feng and C.H.Menq. The Prediction of Cutting Forces in the Ball-End Milling Process-1. Model Formulation and Model Building Procedure. Int. J. Mach. Tools e. Vol. 34. No.5. 1994
3. Z.C.Yan and C.H.Menq, Uncertainty Analysis for Coordinate Estimation Using Discrete Measurement Data. MED-Vol. 2-1/MH-Vol. 3-1, Manufacturing Science and Engineering ASME 1995
4. C.H.Menq, H.T.Yan and G.Y.Lai, Automated Precision Measurement of Surface in CAD Directed Inspection. Journal of Robotics and Automation . Feb. 1990.

Rapid Prototyping

Development of Complex Plastic Parts with Simulation and Rapid Prototyping

Michael Wilmsen

TransCAT GmbH, Karlsruhe, Germany

Abstract. As the development and construction of plastic parts specially for the automotive industrie needs to proceed faster, certain CAE-moduls have been implemented into 3D-CAD-Systems to optimise this proces. One possibility is the simulation of the production from parts for injection moulding. So it is possible to analyse the part during the construction. the best injection points can be chosen, welding lines and airtraps can be prevented. The following step is the production of a „Rapid Prototyping" part. With this model it is possible to produce an easy injection mould for first parts in thermoplastic material within 2 or 3 weeks. The following paper shows samples of products which have been produced in the way described.

1 Introduction

In the last two years, the most CAE software producers spread the slogan of "better - cheaper - faster", which should be a statement for the more efficient use of their programs. Thinking about these points means not only to buy a good CAD system for the development and construction of parts, but also the possibility of using the models for several following steps.

The production of plastic parts (e.g.: injection moulding) allows the engineer to design free surfaces for the best visual and technical effect. That means as well that complex models, which are produced in a 3D CAD have to be designed only once. After designing it should be possible, to take these data and use other CAE systems to optimise, simulate or produce prototypes.These following steps need a special system and a specialist to handle it. Figure 1 shows the possibilities in developing plastic parts and the required injection moulds. Because of this knowledge, that each module contains, it is not possible, to develop an universal CAE system. Some CAE developer have already recognised this circumstances and stopped their own research on special programs like for example the simulation of injection moulding.

The reason is the big knowledge that expert systems, for example simulation software needs, to copy the real part forming process. So we reach the most important points for an effective CAE system:

◆Using several expert-systems for the different steps of a product development.

◆Combining these special software products in one CAD system.

Only the attention of these two points gives the user a helpful system for developing and constructing parts perfectly. Paying attention to these points, we must recognise, that in the last years, the expert systems got better and better, but they had only few possibilities, to transform part information from a CAD system to an expert system without loosing important information of surfaces or anything else.

So the second step, after developing and during optimising expert systems, is the connection of these systems to the CAD. That means designing the part structure only once and using this geometry for all the following steps as shown in figure 1.

Let us have a look to two samples of these modules. At first we see the different possibilities in simulating a mould injected part and interpreting the results. The second module is the realisation of first parts with rapid prototyping and rapid tooling.

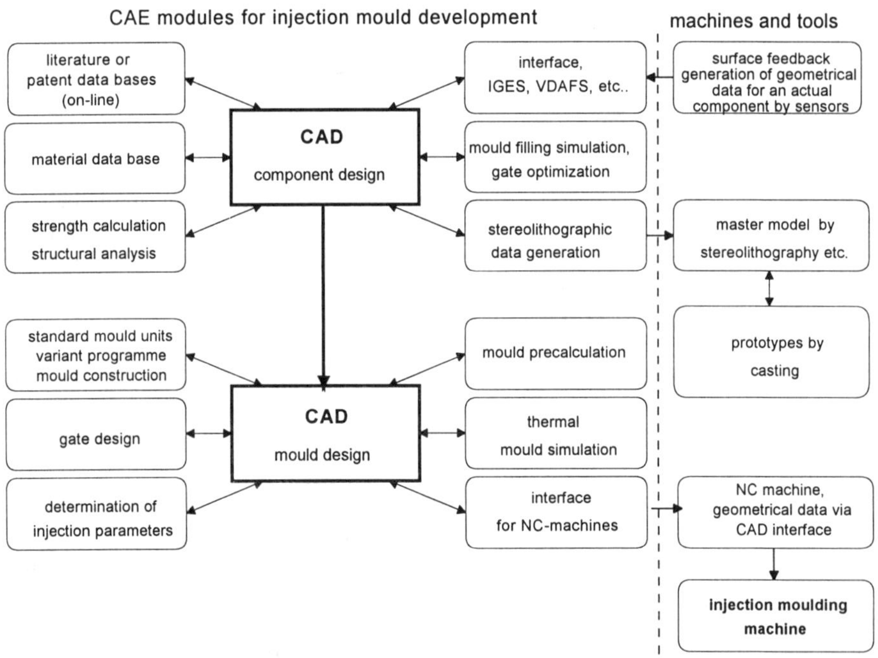

Fig. 1. CAE modules for developing plastic parts and injection moulds

2 Simulation

Complex plastic parts should be designed in 3D CAD systems (figure 2). Then you have a lot of surfaces or a solid part. Either, the first step is the transformation of the model in a part, which can be simulated. As we know that the filling process of an injection mould is a complicate rheological and thermal process, it is necessary to use simplifications for describing the process in a simulation. For example, we cannot consider the thermal conductivity in the direction of flow without raising the simulation time drastically. So it can happen, that the results have to be interpreted critically.

The most important simplification is the use of shell elements for simulating the flow path. That presents the problem of the preprocessing. Either, we take one of the part surfaces (outside or inside) or we need the middle plane. In each case we have to change the 3D part and make new surfaces, mesh it and optimise the mesh. (figure 3).

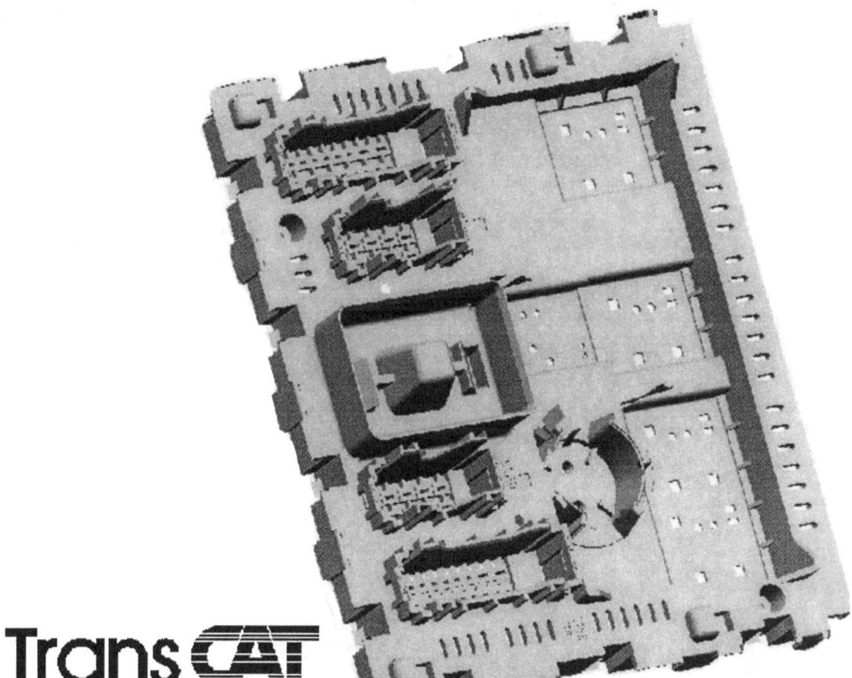

Fig. 2. 3D part (platine) before generating a shell mesh (CATIA; picture: TransCAT GmbH)

Meanwhile, designing this geometry can be realised in the connected CAD system. As we see the steps in figurc 2 and 3. As you can imagine, this additional work keep a lot of designers away from using this expert system. What are the possibilities in the future of optimising this procedure:

◆Simplification of the meshing (e.g. meshing the 3D modell and producing automatically the middle plane shell mesh)

◆Meshing the 3D part with shell and tetraeder elements (depending on the geometry).

◆Meshing the 3D part with a tetraeder mesh and simulating the formfilling in 3D.

The three points show the steps of developing the system. Meanwhile it is possible to mesh the part from a 3D mesh to a shell mesh. The other steps will be realised within the next years, depending on the performance of the hardware. These points stand for the commercial use of the system.

At this step another important point of the simulation has to be looked at: The interpretation of the results. This is still the work of the user, who needs not only a good knowledge of the system but also of the real production process.

For example taking the platine that you have seen on the further page. The geometry shows regions of only 0.7 mm thickness. The main region consists of 1.5 to 2 mm thickness. So we can imagine, that filling problems will occur in the thinner regions.

TransCAT

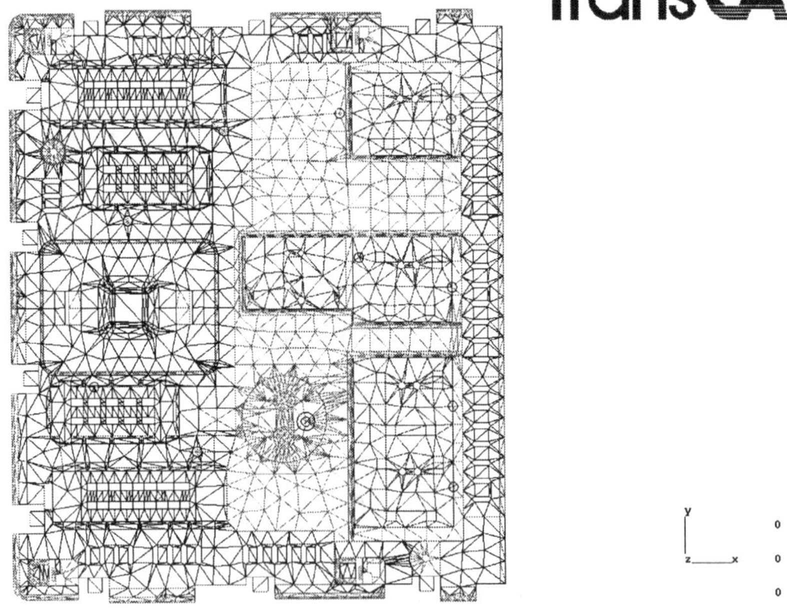

Fig. 3. Shell mesh of the platine (CATIA MESH; picture: TransCAT GmbH)

As the simulation shows (figure 4), it happens that the part cannot be filled completely, because the melted material, a glasfibre reinforced Polybutylentherephtalat (PBT), needs more pressure near the gate to fill the region. Further away of the gate, the flow front has nearly filled the thicker walls, the pressure in the part has been grown high enough and the other thin regions can be filled. But now the melt at the thin plate near the gate has already been frozen, the filling is no more possible. With this part we can see a typical case of the limit between simulation and reality. The result is realistic, as we can see in the part, which has been filled incompletely at first. But at the end of the optimisation at the injection moulding machine, it was possible to fill the part with high pressure.

This effect appears in few cases, because simulation software cannot consider the rheological fluidity in the regions of very slow filling, recognising that a very thin frozen layer at the flow front can be broken through with high pressure. With the right background knowledge it is possible to recognise this software problem in advance and give the right result to the client. Nevertheless the simulation of plastic parts is an important tool for developing and designing complex plastic parts in shorter times.

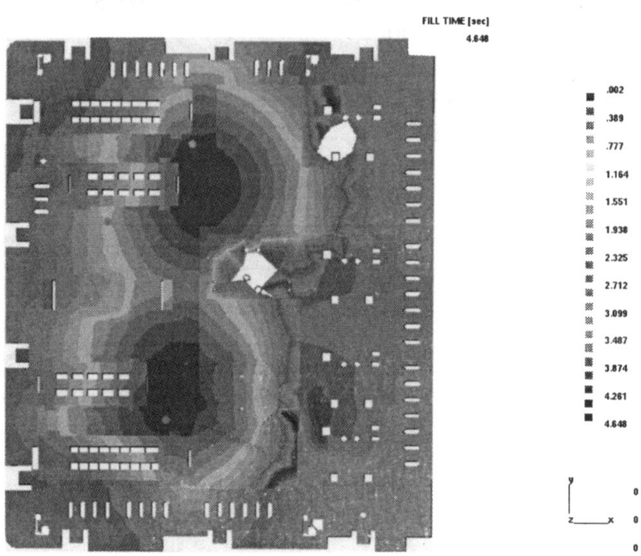

TransCAT

Fig. 4. Simulation result (flow front) of the platine.

3 Rapid Prototyping

The next developing step of a plastic part after the optimisation is the production of first prototypes. Here we have to distinguish between:

◆first prototype

◆design model

◆samples for first testing (similar material, e.g. thermosets)

◆samples for specified testing (original material)

Meanwhile for the first prototype, several "rapid prototyping" systems are existing on the market (e.g. stereolithographie). Herewith it is possible, to build a realistic part out of a 3D CAD system within some hours. If we look at the five best known prototyping systems, we recognise the pros and cons of these systems (table 1). All systems work with producing slice by slice of a part. The system work either with a bath of liquid thermosets (Stereolithographie), a small extruder for thermoplastic material (Fused Deposition Modeling), a thermoset on a solid ground (Solid Ground Curing; Cubital), thermoplastical pellets (Selective Laser Sintering) or paper sheets (Laminated Object Manufacturing).

Table 1. pros and cons of the different rapid prototyping systems.

Property	STL/SLA	Cubital	SLS	LOM	FDM
Thermoset	++	++	0	0	0
Thermoplastic material	0	0	++	0	++
Parts with high elasticity	+	-	-	+	++
Parts with high strength	+	-	+	++	++
Short production thin parts	++	+	-	+	+
Easy finishing	+	+	-	-	+
Short production thick parts	+	++	-	++	+
Production of moulds	+	-	-	++	-
Each geometrie	++	++	+	-	-
Design model	+	++	-	++	+
Test model	+	-	+	++	+
Simple using of the machine	-	-	-	+	++
Tolerance	++	++	+	(+)	-

0 not possible; - bad results; + good results; ++ best results

Regarding table 1, the possibilities of the rapid prototyping systems are in the first two, maximum three model variations (see points above). But what happens, when a part of origin thermoplastic material is needed in a very short time (e.g. the production of an injection mould takes mostly 2 months).

Therefore the following step is to produce an injection mould with these systems or with a rapid tooling material.

4 Rapid Tooling

Therefore a mould can be produced with a prototype (figures 5, 6). This happens with a aluminium filled thermoset in the following steps:

◆preparing the prototype as needed (e.g. coating or polishing)

◆preparing the different inserts for the mould (figure 5)

◆positioning the part in the mould to get a good ejection

◆preparing the separation zone

◆filling the different mould parts with thermoset around the prototype, positioned in an aluminium frame.

◆fitting the aluminium frame in a prepared injection mould out of steel (figure 6)

◆installing the ejectors.

Fig. 5. Prototype with prepared cores for the mould

Producing the mould takes about 2 weeks. It is possible to produce between 50 and 500 parts in origin material (e.g. Polyamid 6.6 with 30% glasfibre). Also in this process it is possible, to integrate the CAE modules. The contours of the mould inserts can be prepared with the milling mashine, taken directly from the system.

Fig. 6. Complex mould with several cores, with aluminium frame and steel mould

6 References

1. Mit Kunststoffen zu neuen Produkten; Symposium FhG Karlsruhe 1997

2. Michael Wilmsen: Seminar: Rapid Prototyping and Rapid Tooling 1996

3. TransCAT Firmenschriften: Kunststofftechnologie, 1996

4. Moldflow Firmenschriften

5. Gebhardt, A.: Rapid Prototyping, Carl Hanser Verlag München 1996

6. M. Wilmsen, u.a.: Bauteil-Entwicklungszeit mit ..., Kunststoffe 10, 1995

7. M. Wilmsen:
   Wettbewerbsvorteile durch CAE-Anwend., Kunststoffberater, Nov.1996

Reverse Engineering Physical Models For Generation Of Surface Models

Liang-Chia Chen, Grier C. I. Lin

Centre for Advanced Manufacturing Research
School of Engineering, University of South Australia
The Levels, Pooraka, SA 5095, Australia

Abstract. This paper presents an innovative approach to reverse engineering physical models for generation of subdivision surface models. The approach integrates stereo vision detection, Delaunay Triangulation, and smooth subdivision surface refinement scheme into an automatic surface reconstruction process. In the approach, a coordinate measurement machine equipped with a touch-trigger probe is integrated with a CCD camera, in order to enhance digitising efficiency while high accurate measurement is also obtained. The proposed self-driven digitising process guarantees that the subdivision surface generated reaches user-specified digitising accuracy. By integrating these capabilities, the approach is aimed to reconstruct smooth arbitrary topological surfaces efficiently and accurately. Consequently, the problems encountered in reverse engineering processes can be reduced to a minimum.

1. INTRODUCTION

Reverse engineering has played an important role in obtaining accurate CAD surface models and shortening product development lead time. The product design process often needs to apply physical models for initial conceptual and aesthetic design, product prototyping, performance testing and model modification to obtain an optimal product design. Reverse engineering is an extremely important tool in reconstructing physical models into mathematical CAD models, in terms of the speed of reconstruction and the accuracy of CAD models generated.

However, there are still many problems in effectively implementing reverse engineering. Optical digitisers such as laser range finders, stereo image detectors, moire interferometers and structured lighting devices are very popular due to their advantages in measuring speed and non-contact property. Unfortunately, their scanning accuracy to date cannot be guaranteed to meet strict requirements. For example, laser triangulation probes are poor in terms of surface edge measurement and the measured errors sometimes may be up to 50 μm. These kind of large digitised errors would be very difficult to eliminate in further data processing. Moreover, the huge amounts of data generated from these fast digitisers is irregular, unformatted, error prone, and require extensive processing for generating smooth surfaces[1]. Consequently, the surfaces reconstructed from these digitised data might lose their surface characteristics and become invalid for further product development.

Some engineers apply coordinate measurement machines (CMMs) with touch-triggered probes for their reverse engineering applications. CMMs with contact probes are excellent in measuring accuracy. The problem is that the scanning rate is much slower than 3D optical digitisers due to the inefficiency and limitation of the measuring principles. Although the automatic digitisation of a CMM can be easily achieved by scanning objects along rectangular patches, it is still time consuming and difficult to digitise a surface with an arbitrary topological type.

This paper proposes an integrated reverse engineering approach to reconstructing smooth subdivision surfaces in a coordinate measurement machine (CMM) equipped with a computer vision system and a touch-triggered probe (TTP). A smooth subdivision surface is defined as the limit of a subdivision process applied to a control triangular mesh. Subdivision surfaces can be used to model everywhere smooth surfaces of arbitrary topological types using numbers of unconstrained parameters[2]. This kind of surface representation is also capable of modelling surface discontinuities such as creases and corners which would cause difficulties in fitting Non-Uniform Rational B-Spline Surfaces (NURBS). Furthermore, the proposed strategy aims to integrate the data digitisation and surface modelling into a single self-driven surface reconstruction process so that surface reconstruction can be implemented accurately and automatically[3,4]. Therefore, the major problems encountered in reverse engineering processes can be solved effectively.

2. OVERVIEW OF THE PROPOSED APPROACH

The proposed approach integrates two steps, namely the initial vision-driven surface triangulation process (IVSTP) and the adaptive subdivision-surface refinement processing (ASSRP), into a single automatic process of surface reconstruction. The IVSTP generates an initial triangular patch by using data filtering, stereo image detection and constrained Delaunay triangulation. The ASSRP which integrates surface digitisation refinement and subdivision surface generation, is then used to reconstruct piecewise smooth subdivision surface and control the digitising accuracy within the required digitising tolerance.

The flow chart of the proposed approach is shown in Figure 1. In the IVSTP, a 3D stereo detection method is used to detect 3D surface boundaries[3]. Multiple images are taken by a CCD camera and appropriate image processing methods, such as noise filtering, are applied to reduce image noise to an acceptable level for further processing. A Laplacian edge detection operation is then used to extract image edges effectively. To obtain an accurate boundary measurement, a CCD camera calibration method is also used to establish a camera model which can adequately transform the image coordinates of the surface boundaries to the CMM world coordinates and reduce the transformed errors to a minimum. Furthermore, in order to save the computing time for further 3D stereo image detection, only characteristic points on the surface boundaries are extracted to represent their original contours by using a data filtering strategy developed by the authors[3]. These extracted points can then be constructed into an initial triangular patch by applying a constrained Delaunay Triangulation[3].

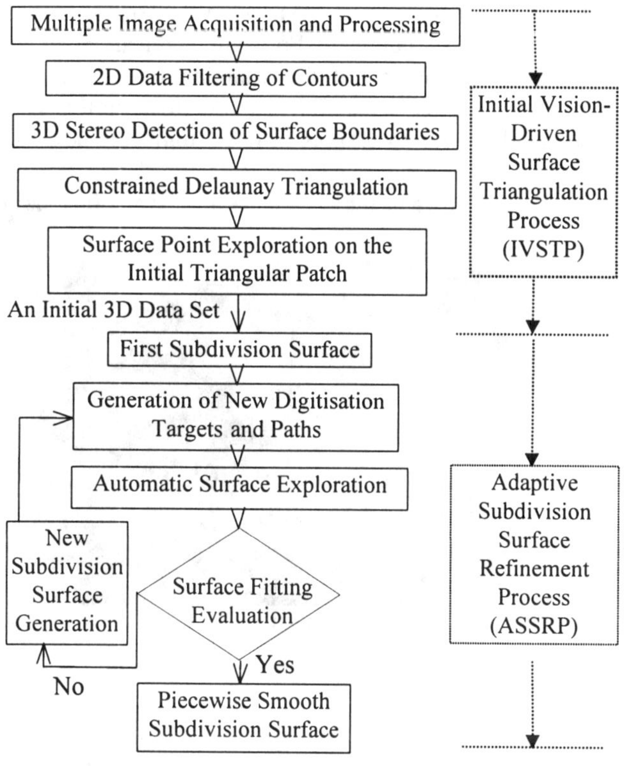

Figure 1 The flow chart of the surface reconstruction algorithm

Following the IVSTP, the ASSRP is then used to refine surface reconstruction by digitising more surface points along their collision-free digitising paths[4]. The ASSRP continues until the digitising accuracy satisfies the user-specified tolerance. As a result, piecewise smooth subdivision surface can be obtained. The detailed strategies will be discussed in the following sections.

3. INITIAL VISION-DRIVEN SURFACE TRIANGULATION PROCESS (IVSTP)

3D stereo vision detection is increasingly being used to enhance the efficiency of various manufacturing tasks due to excellent characteristics such as fast scanning capability and non-contact property. In order to facilitate a CMM with a visual capability for fast digitisation, a CCD camera is integrated with a TTP in a CMM for the purpose of rapidly detecting the object's position and measuring its surface boundary coordinates. The system hardware setup (shown in Figure 2) is designed to achieve the above aim. The system includes a Brown&Shape CMM with a Renishow touch trigger probe and a vision system from the Matrox company. For some applications in which the measuring accuracy is not a major concern but the scanning rate is more important, a laser triangulation probe can then replace a TTP as the digitiser in this system. This indicates that the approach has flexibility to adapt to different digitising devices.

The main aim in the IVSTP is to rapidly detect all surface boundaries, such as edges, corners and creases, and triangulate these surface features into initial surface triangular patches. The strategies involved in the IVSTP are camera calibration method, object's edge detection, data reduction of surface edges, three-dimensional stereo image detection, and initial surface Delaunay Triangulation[5].

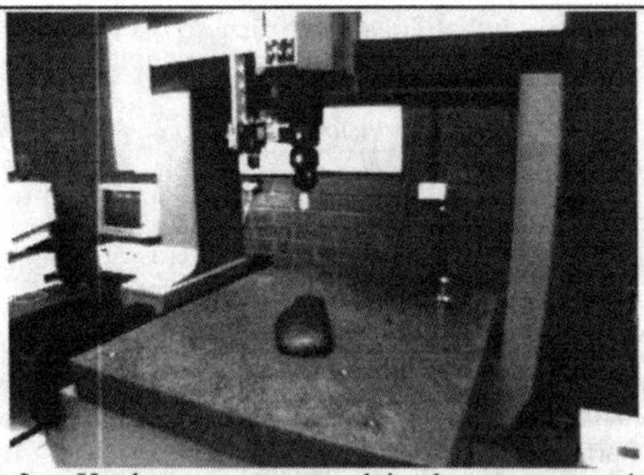

Figure 2 Hardware system used in the reverse engineering process

The flow chart of the IVSTP is shown in Figure 3. In the process, a vision system is used to detect 3D surface boundaries by using a 3D stereo detection method. Prior to the stereo image detection, multiple images are taken and appropriate image processing methods, such as noise filtering, are applied to reduce image noise to an acceptable level for further processing. A Laplacian edge detection operation is then used to extract 2D surface boundaries effectively. To obtain an accurate boundary measurement, a CCD camera calibration method is also used to establish a camera model which can adequately transform the image coordinates of the surface boundaries to the CMM world coordinates and reduce the image system errors to a minimum. Furthermore, in order to save the computing time for further 3D stereo image detection, only characteristic points on the surface boundaries are extracted to represent their original contours by using a data reduction strategy developed by the authors. These extracted points can then be constructed into an initial triangular patch by applying a constrained Delaunay Triangulation method. The detailed principles for the IVSTP have been described in references 3 and 5.

Figure 4 shows an example using the IVSTP to detect the initial surface triangular patches of a toy clay model. The 3D initial triangular patch of a reconstructed toy model was rapidly constructed by the IVSTP. In this case, 75 characteristic points were extracted to represent the shape of the model. The triangular patch was constructed by the Delaunay surface triangulation method. Following this, this initial patch is used in

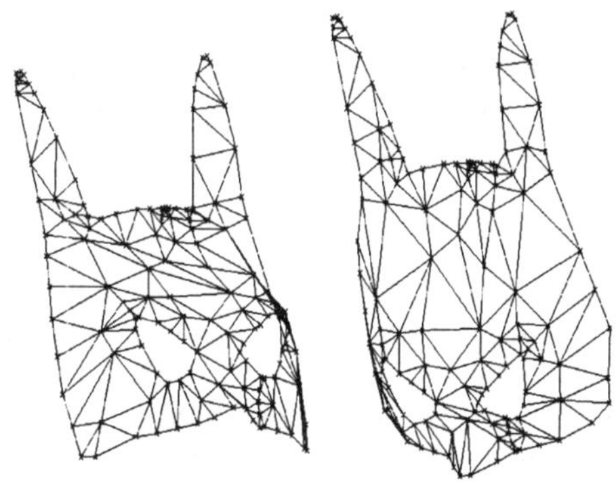

Figure 4 The 3D initial triangular patch of a reconstructed toy model (the left picture is the right ISO view and the right picture is the left ISO view)

the following surface refinement process.

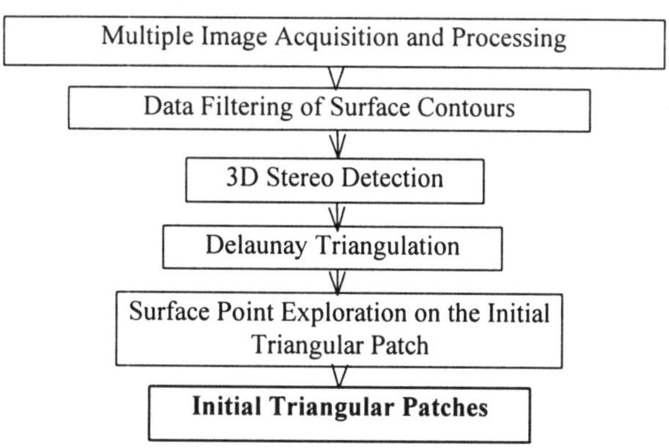

Figure 3 The Initial Vision-Aided Surface Triangulation Process

4 SMOOTH SUBDIVISION SURFACE REFINEMENT PROCESS

The main idea of the digitising strategy is to digitise the surface points following the surface local curvature trend and to automate the digitising process. At first, an initial subdivision surface model can be reconstructed by parameterising the currently available surface points. This model can then be provided to calculate the midpoints between the triangular edges in the initial triangular patch. Moreover, the calculations of the geometric centre of each triangular area and its normal direction provide probe exploration paths to the surface refinement process. The surface digitising process which explores the model-based midpoint estimates on the existing triangular patch can continue to refine the triangular patches whose digitised deviation exceeds the user-specified digitising tolerance. This iterative reconstruction repeats until all new digitised points have satisfactory deviation with their estimated digitising targets. The principles applied in the surface refinement process are discussed in details in the following sections.

4.1 Subdivision Surface

Surface modelling techniques are applied to parameterise digitised points into a surface model. In Computer Aided Geometric Design (CAGD), a variety of curve and surface modelling methods have been developed for different applications. Among these, the most influential one is Non-uniform Rational B-spline curves and surfaces (NURBS). Because the format of NURBS is the umbrella under which both the spline, Bezier, and conic format fit comfortably[6], NURBS has become the standard representation for complex smoothly surfaces.

However, the main disadvantage of using NURBS is that control points should be in a regular rectangular grid. Due to this limitation, NURBS can only be used to represent limited surface topologies rather than arbitrary topologies[2]. Therefore, this approach applied another surface representing form, called subdivision surfaces, to overcome the above problem. Subdivision surfaces are not only easy to implement, but also can parameterise surfaces with arbitrary topological type. Moreover, initial surface triangular patches generated from the IVSTP which detects surface features such as surface boundaries, creases and corners can be directly applied to this type of surface representation.

Subdivision Surfaces were first introduced by Catmull et al. [7,8]. A subdivision surface S(M) is defined as the limit of a subdivision process applied to a control mesh (M) (In our approach, the vertices of the surface triangular patches can be considered as M). There are many subdivision schemes to date. Among these, Loop's scheme using a triangular scheme provides a generalisation of the subdivision surface scheme to generate tangent plane continuous surfaces of arbitrary topological type[9]. Hoppe further modified Loop's subdivision rules to parameterise sharp surface features such as edges and creases[10]. Although subdivision surfaces lack a closed form, some recent research has contributed to the knowledge of how to calculate exact properties such as surface points and surface normals. These breakthroughs have made it possible for subdivision surfaces to be applied in this approach.

4.2 The Adaptive Subdivision-Surface Refinement Process (ASSRP)

The main idea of the ASSRP is to digitise the surface points following the local surface curvature trend and to automate the digitising process. The digitised vertices on the initial surface triangular patches can be used to construct the first subdivision surface by using Hoppe's subdivision rules. The subdivision surface S(M) associated with a control mesh M = (K, V) (shown in Figure 5) is defined as the limit of a refinement process applied to M [9,10]:

$$M,\ M^1 = R(M),\ M^2 = R(R(M)),\ \ldots. \tag{1}$$

where

K is a *simplicial complex* representing the connectivity of vertices, edges, and faces, thus determining the topological type of the mesh M;

$V = \{v_1, \ldots, v_m\}$, $v_i \in R^3$ is a set of vertex positions defining the shape of the mesh M in R^3.

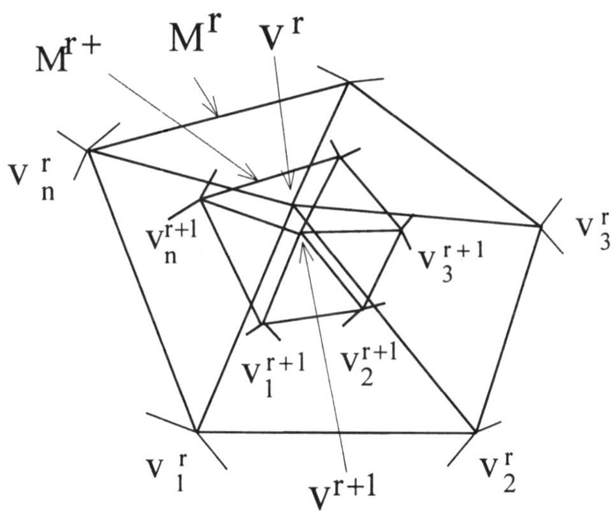

Figure 5 The neighbourhood around a vertex v^r

The refinement procedure R divides each triangle into four subtriangles. The vertices of the new generated mesh are determined by summing weighted averages of the vertices in the existing unrefined mesh as follows[10]:

The new vertex points and the new edge points are computed by the following subdivision rules[10]:

$$v^{r+1} = \frac{\alpha(n)v^r + v_1^r + \cdots + v_n^r}{\alpha(n)+n} \tag{2}$$

$$v_i^{r+1} = \frac{3\,v^r + 3\,v_i^r + v_{i-1}^r + v_{i+1}^r}{8}, \quad I= 1,\ldots,n \tag{3}$$

where $\alpha(n) = \frac{n(1-a(n))}{a(n)}$, with $a(n) = \frac{5}{8} - \frac{(3+2\cos(2\pi/n))^2}{64}$

The matrix form of Equations 1 and 2 can be expressed as[9,10]:

$$\begin{aligned}(v^{r+1}, v_1^{r+1}, \ldots, v_n^{r+1})^T &= S_n(v^r, v_1^r, \ldots, v_n^r)^T \\ &= S_n^{r+1}(v^0, v_1^0, \ldots, v_n^0)^T\end{aligned} \tag{4}$$

where
$$S_n^{loop} = \frac{1}{8}\begin{bmatrix} 8-8\,a(n) & \frac{8\,a(n)}{n} & \frac{8\,a(n)}{n} & \frac{8\,a(n)}{n} & \frac{8\,a(n)}{n} & \cdots & \frac{8\,a(n)}{n} \\ 3 & 3 & 1 & 0 & 0 & \cdots & 1 \\ 3 & 1 & 3 & 1 & 0 & \cdots & 0 \\ \cdots & & & & & \cdots & \cdots \\ 3 & 1 & 0 & 0 & \cdots & 1 & 3 \end{bmatrix}$$

Therefore, by using these subdivision rules, the first subdivision surface estimate can be generated from these vertices on the first triangular patch. Based on this initial estimate, new digitised target points can be determined by Equation 5 [9, 10]:

$$v^\infty = \frac{l_0 v^0 + l_1 v_1^0 + \cdots + l_n v_n^0}{l_0 + l_1 + \cdots + l_n} \tag{5}$$

where $(l_0, \ldots, l_n)$ is the dominant left eigenvectors of Sn

The normal vectors can also be computed by Equation 6:

$$\vec{n} = u_1 \times u_2 \tag{6}$$

where: $u_1 = c_1 v_1^0 + c_2 v_2^0 + \cdots + c_n v_n^0$
$u_2 = c_2 v_1^0 + c_3 v_2^0 + \cdots + c_1 v_n^0$
$c_i = \cos(2\pi\, i / n)$

The ASSRP then explores the target points which locate on the midpoints of each edge of each refining triangle. Following this, the adaptive process estimates how well the new digitised points fit with the corresponding target points and determines whether the next exploration procedure will be taken on each refining triangular area. The estimated deviation is computed as Equation 7.

$$E_S(\Delta) = \sum_{i=1}^{3} \| \vec{E P_i} \| \tag{7}$$

Where $\| \vec{E P_i} \|$: the length of the deviation vector of each vertex of an exploration triangle (Δ)

If the estimated deviation is larger than the user-specified tolerance, a new subdivision surface model is generated by parameterising all currently available digitised points located on the triangular patches. The new surface estimate can then be used to calculate the new exploration target points for the next loop. The ASSRP which explores the model-based midpoint estimates on the existing triangular patches, continues to refine the triangular patches in which digitised deviation exceeds the specified digitising accuracy. The ASSRP continues the automatic data exploration process until the digitising accuracy reaches the user's specified deviation tolerance. Consequently, the user-specified digitising accuracy can be satisfied in the surface reconstruction process.

5. SUMMARY AND FUTURE WORK

An innovative reverse engineering approach has been proposed to reconstruct piecewise smooth subdivision surfaces. The approach has many advantages in terms of high digitising accuracy, digitising efficiency enhancement and capability to digitise arbitrary topological surfaces. By integrating an accurate contact probe with a fast digitising CCD camera and the proposed digitising strategies, the deficiency of a touch-trigger probe, namely low digitising rate, can be minimised while high accurate measurement still remains. Moreover, tangent plane continuous surfaces (C^2 quartic triangular B-Splines) of arbitrary topological type can be generated. The drawbacks in using NURBS surface modelling such as limited surface topologies, lack of local refinement procedure and problems in modelling surface discontinuities can be avoided by applying subdivision surfaces. Therefore, the proposed approach provides a good solution to most problems encountered in reverse engineering processes.

The system for the IVSTP has been successfully implemented in our hardware system setup. The ASSRP will be implemented for the integration of the whole approach in the next step. The convergence of the ASSRP will also be investigated to ensure the stability of the whole approach. Some industrial examples will then be used to test the feasibility of the approach.

ACKNOWLEDGMENTS

The authors would like to convey their appreciation towards the Australian Federal Department of Employment, Education and Training for supporting this project under its Targeted Institutional Link Program. Assistance from Study advisers in University of South Australia is also appreciated.

REFERENCES

1. Milroy, M. J., Weir, D. J., Bradley, C., Vickers, G. W. : Reverse engineering employing a 3D laser scanner: case study. Int. J. Adv. Manuf. Technol. **12** (1996) 111-121.
2. DeRose, T., Duchamp, T., Hoppe, H. and McDonald, J.: Subdivision surfaces [Online, accessed on April,1996]. URL:http://www.cs.washington.edu/homes/derose/grail /Projects/ subdivision.html
3. Lin, Grier C. I. and Chen, L. C., "A Vision-Aided Reverse Engineering Approach to Reconstruct Free-Form Surfaces", In Int. Conf. of *CAD/CAM Robotics and Factories of the Future*, London, England, August 14-16, (1996), 854-859.
4. Lin, G. C. I. and Chen, L. C.: An integrated reverse engineering approach to reconstruct free-form surfaces in coordinate measurement machines. In Int. Conf. of *Automation '96*, Mechanical Industry Research Laboratories, Industrial Technology Research Institute, Hsinchu, Taiwan, R.O.C., July 8-11, (1996), 239-246.
5. Lin, G. C. and Chen, L. C.: An intelligent surface reconstruction approach for rapid prototyping manufacturing. Fourth Int. Conf. on Control, Automation, Robotics and Vision, Singapore, Dec. (1996), 43-47.
6. Ma, W: NURBS-based CAD modelling from measuring points of physical models. *Ph.D. thesis*, Faculteit der Toegepaste Wetenschappen, Departement Werktuigkunde, Katholieke Universiteit Leuven, (1995).
7. Catmull, E. and Clark. J. : Recursively generated B-spline surfaces on arbitrary topological meshes. Computer-Aided Design, **10** (1978) 350-355, Sep.
8. Doo, D. and Sabin, M. : Behaviour of recursive division surfaces near extraordinary points. Computer-Aided Design, **10,** 6 (1978) 356-360, Sep.
9. Loop, C. : Smooth subdivision surfaces based on triangles. Master's Thesis, Department of Mathematics, University of Utah, Aug. (1987).
10. Hoppe, H. : Surface Reconstruction from Unorganised points. Ph.D thesis, Department of Computer Science, University of Washington, (1994).

Model-enhanced Rapid Prototyping Using Range Imaging

YY Cai[1], AYC Nee[2], HT Loh[2] And T Miyazawa[3]

[1]Institute of Systems Science
[2]Department of Mechanical and Production Engineering
[3]Sony Precision Engineering Center (Singapore)
National University of Singapore
Kent Ridge Crescent, S119597, Republic of Singapore
Tel: (65)7726724 Fax (65)7744990 Email: yycai@iss.nus.sg

Abstract. Models are essential in computer-aided design and manufacturing (CAD/CAM). They are important for object representation, manipulation, and data communication, etc. With fast and explicit availability of 3D object coordinates, range imaging is increasingly used in various engineering applications such as measurement, inspection, etc. We use a laser digitiser to capture object geometry by rotation scanning. Modelling of the digitised range object is therefore one of the fundamental tasks in this work.

Two modelling approaches are presented here. A NURBS method is first developed to model the sculpture object by employing the NURBS theory. The smooth model generated is then converted to SLA model for stereolithographic rapid prototyping. We also propose a novel approach to modelling range objects using a feature-centred method. That is realized by range image understanding to extract qualitative geometric features of range objects. The feature information can be applied to rectify the errors introduced during the processes of object scanning and numerical calculation. The qualitative feature information is served to enhance various processes in rapid prototyping application including optimized model re-orientation, etc.

1. Introduction

Models are essential in CAD/CAM domain. In a computer integrated manufacturing (CIM) environment, scan-paths for rapid prototyping or probe-paths for coordinate measuring machine (CMM) inspection are traditionally derived from CAD models of the parts. Typically, CAD models are designed using CAD tools and are retrievable from CAD databases. They provide an avenue for object representation, manipulation, data communication, etc. However, there are complex industrial parts that may not be available in CAD databases. In car or shipbuilding industries, for instance, manufacturing of quite number of objects often starts with a crafted master model. With the advent of high quality scanning equipment, a master CAD model can now be generated and manipulated based on the digitisation of a physical master model. This allows engineers to modify or analyse the master model with the aid of CAD tools without the burden of having to recreate the master model. Rapid prototyping of laser scanned object can reap the benefits of shortening the life circle of the product/process development, and improving cost effectiveness, productibility and product quality.

Model reconstruction from digitised object is one of the important tasks in geometric modelling. Approaches to object reconstruction from scanned images for CAD/CAM applications have long been investigated. Triangulating and free-form surface fitting are commonly used in model reconstruction. Delaunay graphs were used by many researchers in order to generate more uniform triangular approximation for range objects [1-2]. Chen and Schmitt [3] discussed range surface modelling by constrained triangulation on discontinuity edges. [4] investigated the free-form surface reconstruction using interpolating techniques for scattered range data. With distinct characteristic of smooth interpolation, NURBS model is widely used in designing and manufacturing applications. It is therefore a basic requirement in this work to have free-surface modelling functionality.

Features are of paramount importance in design and manufacturing contexts, as they are a natural choice for capturing and describing the identity of individual objects, as well as establishing relationships between certain objects [5]. Objects in different levels of abstraction demonstrate different feature information. For rapid prototyping applications, certain feature information may enhance the rapid prototyping processes like error reducing, optimized model re-orienting and re-scaling, etc., to achieve high production quality. One of the common drawbacks of current approaches to object reconstruction from scanned digital data is the lack of the ability of geometric feature understanding. It is also difficult for any CAD-based feature recognisers to detect geometric features from such digital-originated reconstructed models. Form features, like cylindrical bosses, are likely to be represented by a set of polygons or control polygons in the reconstructed models. Besides, due to the nature of imaging, noise-free scanning is usually not possible. For manufacturing applications, the intolerance of down-stream processes to noise in the model makes the use of simply reconstructed models from scanning unsuitable.

In the next sections, we first present the 3D range object capturing using depth mapping. This is followed by the discussions on object reconstruction using NURBS approach and feature-centred approach. The model-based stereolithographic rapid prototyping is then developed.

2 3D Range Object Capturing Using Depth Mapping

To capture the depth information of an engineering part to be measured, a ray of light generated by a laser beam is projected onto the spatial object. With known directions of the source and a displaced detector, the range value of a point spotted by the light on an object can be deduced very quickly. By measuring sequential points and producing a standard list of XYZ coordinates, the laser scanner interfaces to vision and CAD systems with 3D depth maps. Using a unique triangulation technique instead of complicated imaging optics and sophisticated detector-array, the ranging system provides high resolution and accuracy while maintaining a small, fixed triangulation angle across a large work volume [6].

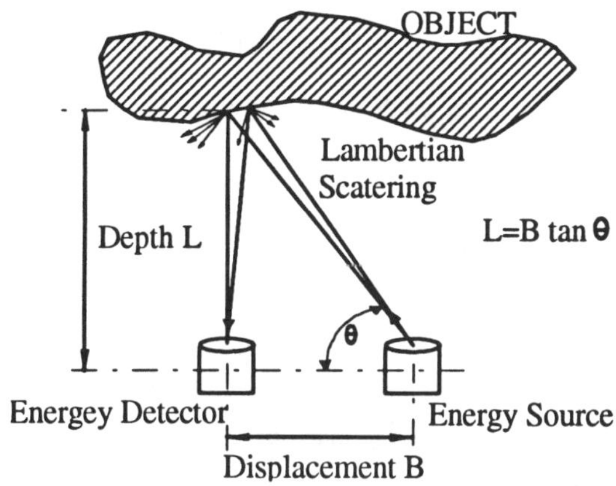

Figure 1. Depth measuring using laser scanner (source: [6])

3 NURBS Modelling For Range Objects

Many engineering applications require generating smooth model to represent objects. Several techniques in computer-aided geometry design (CAGD) have been developed to model the free-form surface objects [7]. In the following, we will apply NURBS theory to to reconstruct complicated free-form surfaces for range objects. If the XYZ coordinates of the data points $\{P_{ij} = (x_{ij}, y_{ij}, z_{ij}),\ i=0,\ldots,m,\ j=0,\ldots n\}$ are captured via laser scanning, a B-spline surface using the given points as the control net can be easily generated with the standard B-Spline surface function. However, since such a reconstructed surface usually does not pass through the measured points or the control net, the resultant free form surface is not accurate. To generate a more accurate free-form surface which passes through all the given points, it is therefore required to calculate a real control net $\{V_{ij},\ i=0,\ldots,m,\ j=0,\ldots n\}$ so that a B-spline surface based on the new control polygon is able to interpolate all the given points. A fast algorithm using the bi-cubic B-spline approach is developed here to get the desired surface with the constraint that the resultant surface will interpolate all the measured points. A linear system can be derived from the constraint by considering both boundary and internal conditions of the control nets. Our program is designed to solve the boundary and internal control vertices with the given scan data. Smooth NURBS model can then be obtained by solving the linear systems. Figure 2 shows a reconstructed smooth model from range image using a NURBS approach.

4 Feature-assisted Model Reconstruction From Range Images

Objects in range images are expressed by digital depth values. Homogeneous pixels can be clustered and grouped into object shells, which are features at a global level. We will concentrate on the quadratic features in this work. A fuzzy clustering approach has been developed to partition the shell features by range image understanding [8]. The technique developed can segment and fit shell features in range images simultaneously. In addition, different shells in range images can be partitioned concurrently. Clustering the quadric and planar shells in 3D space and extracting their prototypes from range images are of our particular interest as those information are important to our model-enhanced rapid prototyping application.

Given image points $\{(x_k, y_k, z_k), k = 0,1,\ldots,K-1\}$, the squared distance of the k-th point to l-th primitive can be defined as

$$d_{lk}^2 = d^2(x_k, y_k, z_k, \pi_l) = (U_l', V_l')M_k \begin{pmatrix} U_l \\ V_l \end{pmatrix} \quad (1)$$

where π_l: $S \cdot U_l + T \cdot V_l = 0$ (called the prototype of the l-th primitive),

$$S = \begin{cases} (x^2, y^2, z^2, \sqrt{2}yz, \sqrt{2}zx, \sqrt{2}xy), & \text{quadrics} \\ (x, y, z), & \text{plane} \end{cases}$$

$$T = \begin{cases} (x, y, z, 1), & \text{quadrics} \\ (1), & \text{plane} \end{cases}$$

$$U_l' = \begin{cases} (a_{00}^{(l)}, a_{11}^{(l)}, a_{22}^{(l)}, \sqrt{2}a_{12}^{(l)}, \sqrt{2}a_{20}^{(l)}, \sqrt{2}a_{01}^{(l)}), & \text{quadrics} \\ (a_0^{(l)}, a_1^{(l)}, a_2^{(l)}) & \text{, plane} \end{cases}$$

$$V_l' = \begin{cases} (2a_{03}^{(l)}, 2a_{13}^{(l)}, 2a_{23}^{(l)}, a_{33}^{(l)}), & \text{quadrics} \\ (a_3^{(l)}) & \text{, plane} \end{cases}$$

$$M_k = \begin{pmatrix} B_k & C_k \\ C_k' & D_k \end{pmatrix}$$

$$B_k = S_k' \cdot S_k, \quad C_k = S_k' \cdot T_k, \quad D_k = T_k' \cdot T_k$$

$$S_k = \begin{cases} (x_k^2, y_k^2, z_k^2, \sqrt{2}y_k z_k, \sqrt{2}z_k x_k, \sqrt{2}x_k y_k) \\ (x_k, y_k, z_k), \end{cases}$$

$$T_k = \begin{cases} (x_k, y_k, z_k, 1), \\ (1), \end{cases}$$

The notion of distance d_{lk} is a similarity measure to indicate how far the k-th point can be classified into the l-th cluster. The shell thickness $J(\pi_l, \mu_l)$ of the l-th primitive is the membership-weighted squared distances of all the points to the l-th particular cluster and the total shell thickness $J(\pi, \mu)$ can therefore be defined as:

$$J(\pi, \mu) \equiv \sum_{l=0}^{C-1} J(\pi_l, \mu_l) = \sum_{l=0}^{C-1}\sum_{k=0}^{K-1} (\mu_{lk})^m d_{lk}^2 \quad (2)$$

where const $m > 1$,

μ is a (m+1) x (n+1) fuzzy membership matrix, and

$$\sum_{l=0}^{C-1} \mu_{lk} = 1 \text{ (for all } k\text{) and } 0 < \sum_{k=0}^{K-1} \mu_{lk} < K, \ \mu_{lk} \in [0,1] \text{ (for all } l \text{ and k).}$$

Partitioning of a finite set of points in a three-dimensional Euclidean space into fixed C fuzzy clusters is to determine the membership grades μ_{lk}, (l=0,1,...,C-1, k=0,1,2,...,K-1) of each element. From the viewpoint of optimisation, the partition of K points into C clusters is equivalent to minimise the above total distance. In order to avoid trivial solution of optimal shell clustering, necessary constraints need to be introduced. A unit normal constraint in planar case and an invariant constraint of $\|U_l\|^2=1$, (l= 0,...,C-1) in quadric case are chosen here [2]. This leads to a constraint optimisation problem $\underset{\|U_l\|^2=1,\ 0\le l<C}{MIN} \left\{ \sum_{l=0}^{C-1} J(\pi_l, \mu_l) \right\}$. By applying the Lagrange multiplier method, the above constraint optimisation is transferred into a non-constraint problem

$$MIN\left\{\sum_{l=0}^{C-1}\left\{\begin{pmatrix}U_l' & V_l'\end{pmatrix}\begin{pmatrix}E_l & F_l\\ F_l' & G_l\end{pmatrix}\begin{pmatrix}U_l\\ V_l\end{pmatrix}+\lambda_l\left[1-\begin{pmatrix}U_l' & V_l'\end{pmatrix}\begin{pmatrix}I & 0\\ 0 & 0\end{pmatrix}\begin{pmatrix}U_l\\ V_l\end{pmatrix}\right]\right\}\right\} \qquad (3)$$

where $E_l=\sum_{k=0}^{K-1}(\mu_{lk})^m B_k$, $F_l=\sum_{k=0}^{K-1}(\mu_{lk})^m C_k$, $G_l=\sum_{k=0}^{K-1}(\mu_{lk})^m D_k$,

$$I=\begin{cases}\text{6x6 unit matrix, quadrics}\\ \text{3x3 unit matrix, plane}\end{cases}$$

Differentiating (3) leads to following equations:

$$\begin{cases}V_l=-G_l^{-1}F_l'U_l\\ A_lU_l=\lambda_lU_l\end{cases} \qquad (4)$$

where $A_l=\left(E_l-F_lG_l^{-1}F_l'\right)$, ($l$=0,1,...,$C$-1).

It is not difficult to prove A_l is positive semi-definite, and thus, the optimal shell clustering problem is to find the minimal eigenvalue and its corresponding eigenvector of each shell eigensystem of matrix A_l, (l=0,1,...,C-1). The eigensystem A_l has three eigenvalues for planar shells and six eigenvalues for quadric shells. A convergent membership matrix μ can be iteratively obtained using Bezdek's updating role [9].

5. Model-enhanced Rapid Prototyping

5.1 Invariant-based Qualitative Feature Identification

The intrinsic relationship between qualitative and quantitative analysis of shape feature can be explored by employing its geometric invariance property. The classical theory of invariance focused on polynomial invariants that are independent of certain admissible transformations. For any quadrics in algebraic form, there are basic independent to three-dimensional general transformations. Based on the basic invariants, the eigenvalue invariants, spherical invariant and revolutional invariant can be defined [8]. We have developed a mechanism to classify the quadric features and represent explicitly the feature parameters using the above invariant quantities of any quadratic shells. As a result, the spherical invariants and the revolutional invariant can be used for inspection or measurement of spherical and revolutional features, respectively. The quadric primitives can be classified according its invariants values [8]. Furthermore, these invariant quantities can be used to explicitly represent the parameters of quadric features such as cylinder radii, cone angles, etc.

The integration of the fuzzy clustering and invariant analysis makes the identification of the qualitative features very efficient. Figure 3 illustrates a range object with three quadratic features and Figure 4 shows three extracted shell from the range image using the partitioning algorithm (the top half part only due to symmetry). The individual shell features can be classified based on its invariant values.

Figure 2. A NURBS model reconstructed for a spiral range object

5.2 Feature-centred Model Rapid Prototyping

Using the NURBS approach we reconstruct the range object. We then utilise the quadratic and planar feature to help achieving better quality of RP manufacturing. In fact, in the presence of noise in 3D data, the geometrical features extracted by range image understanding usually have various shape variations. The designed spherical and revolutional invariants can be used to monitor the variations. By checking the values of the spherical invariant or the revolutional invariant, one can determine whether the variations of the spherical feature or revolutional feature are within allowable ranges. Tolerance information from range images can also be detected utilising invariants. For instance, a sphere is most likely deformed into an ellipsoid-like shape. The maximum and minimum spheres associated with the ellipsoid can be computed using the three calculated radii of the ellipsoid.

Once the model is available, CAD tools can be applied to manipulate the model (rotating, translating and scaling, etc.). Good orientation of the object, however, may lead to good quality of the model rapid prototyping. It is not easy to find a better orientation for the NURBS-based model by purely interactive mainuplation. We propose a novel method to determine its good orientations by referencing the partitioned qualitative features. This is indeed a localisation process for feature identification. It is not difficult to determine its axial or symmetric directions for all quadratic surfaces (Figure 5). These directions, together with the feature areas, determine the final orientation for better RP quality and a model re-orienting operation can then be performed before STL formating. For sculpture model without any quadatic or planar features, however, one has to use the other way to estimate a good orientation which may be highly relied on users' experience. Finally, we convert the NURBS model into a STL format. The layered STL model is then fabricated with a stereolithographic RP system. Figure 6 shows the spiral object manufactured using the SOLID CREATION system at the SONY Precision Engineering Centre (Singapore).

Conclusions. Addressing the needs of manufacturers moving towards variety-based and time-based competition, the newly emerged RP technology provides tools that can significantly streamline the design and product development process, while substantially reducing costs. The basic RP principle allows a conversion of CAD models into layered information used for process control in different layered. To reduce the relatively lengthy process of interactive design and manufacturing, this work discusses the rapid prototyping of range object captured by laser scanning. Range-based modeling techniques are developed to reconstruct physical models. A NURBS modelling method and a feature-centred modelling method are presented. With the reconstructed model, high quality rapid prototyping can be expected. Advanced vision-systems like CT or MRI can be employed to capture the full structure information of 3D objects. Extraction of the boundary surface from CT/MRI volume is one of the challenge tasks. Range-based modelling and rapid prototyping techniques developed may be further applied to this area.

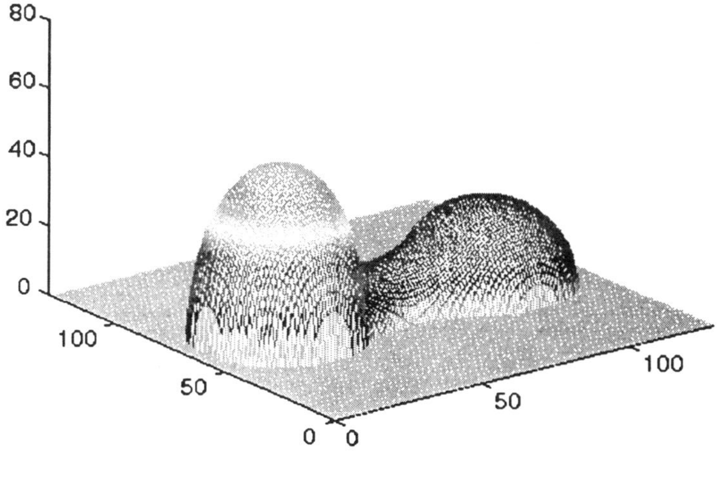

Figure 3. A top view of a range object

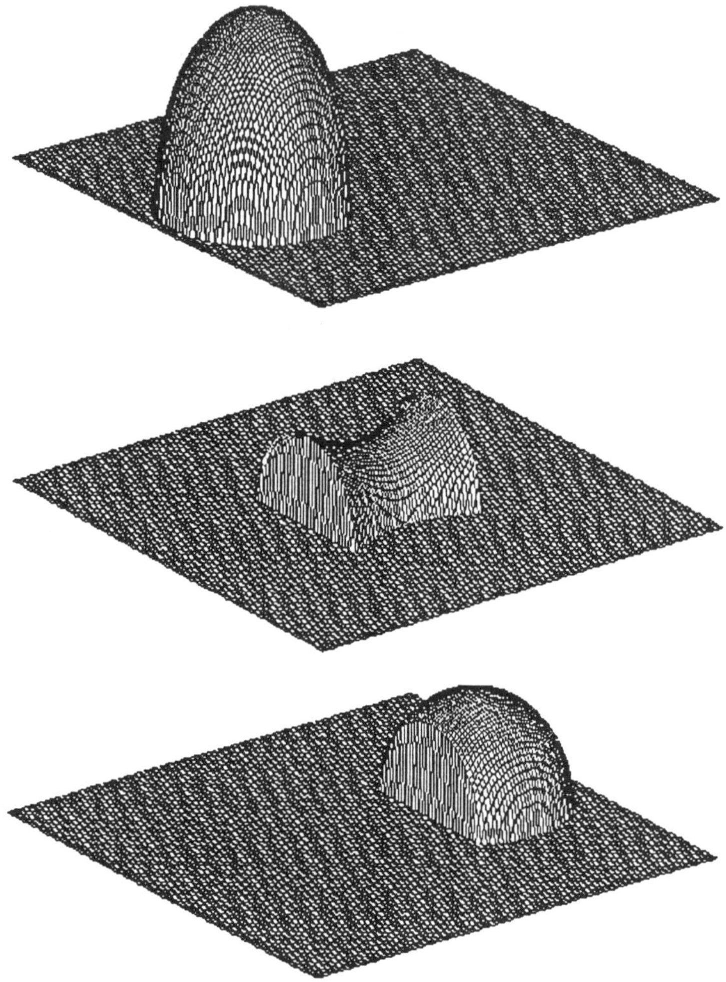

Figure 4. Three quadratic shell features extracted for the range object

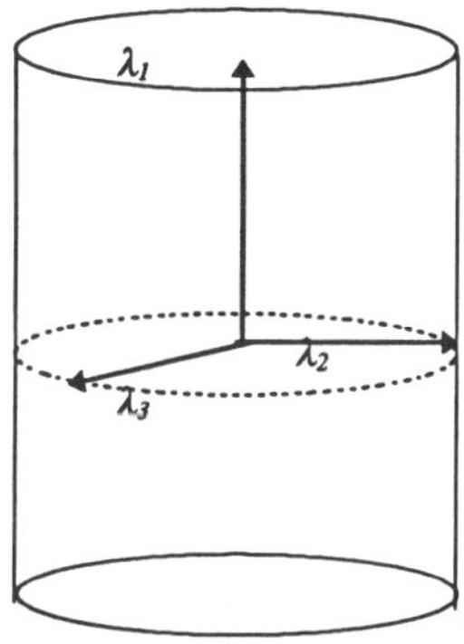

Figure 5. Eigenvectors and eigenvalues with a quadric invariant system

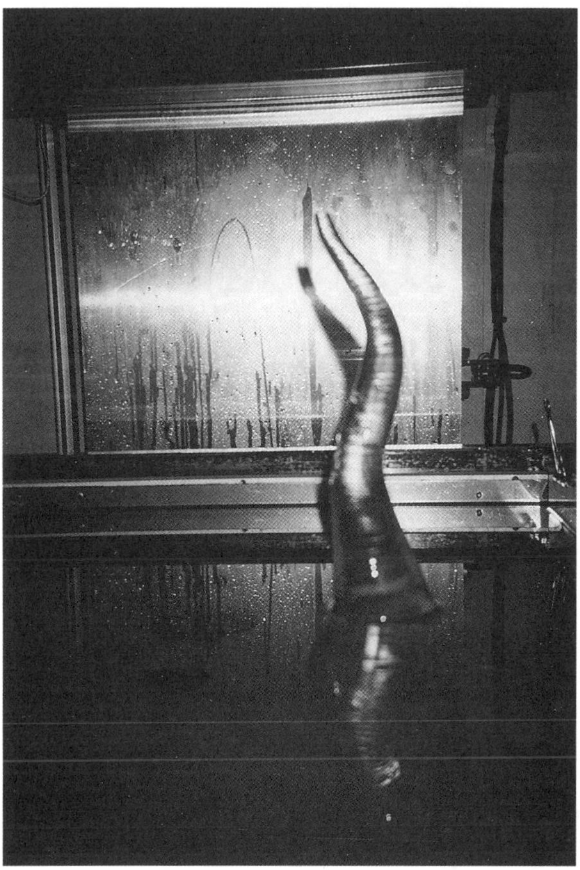

Figure 6. The stereolithographic rapid prototyping of the spiral range object

References

1. J. D. Boissonnat, Representing 2D and 3D Shape with the Delaunay Triangulation, *Proc. of IEEE International Conference on Pattern Recognition*, 1984, pp. 745-748.
2. O. D. Faugeras, *Three-dimensional Computer Vision*, MIT Press, Cambridge, 1994.
3. X. Chen and F. Schmitt, Surface Modelling of Range Data by Constrained Triangulation, *CAD,* **26**(8), 1994, pp. 632-645.
4. C. Bradley and G. W. Vickers, Free-form Surface Reconstruction for Machine Vision Rapid Prototyping, *Optical Engineering*, **32**(9), 1993, pp. 2191-2200.
5. M. J. Pratt, Aspects of form feature modelling, in *Geometric Modelling Methods and Applications* (Eds: H. Hagens and D. Roller), Springer-Verlag, New York, 1991, pp. 227-250.
6. R. Jarvis, Range Sensing for Computer Vision, in *Three-dimensional Object Recognition Systems* (Jain A. K. and Flynn P. J., Eds.), Elsevier Science Publishers, 1993, pp. 17-56.
7. G. Farin, *Curves and Surfaces for Computer Aided Geometric Design: A Practical Guide*, Academic Press, Boston, 1988.
8. Y. Y. Cai, H. T. Loh, and A. Y. C. Nee, Qualitative Primitive Identification Using Fuzzy Clustering and Invariant Analysis, *Image and Vision Computing*, **14**(1996), pp. 451-464.
9. J. C. Bezdek, *Pattern Recognition with Fuzzy Obective Function Algorithms*, Plenum Press, New York, 1981

Numerical Simulation of Cavity Filling of the Powder-Binder Mixture in a Rapid Prototyping Process

Lin Bilon and Gong Haiqing*

School of Mechanical and Production Engineering,
Nanyang Technological University
Nanyang Avenue, Singapore 639798

Abstract. The analysis and simulation of the cavity filling process of the powder-binder mixtures with shear-induced powder migration in a powder injection rapid prototyping process was presented in this paper. A fully-implicit time discretization scheme was applied and extended such that the velocity, pressure, temperature, powder and flow front location were iterated and solved simultaneously. The purpose of this work is to develop a simulation technique to help design engineers to understand the physical process, detect non-homogeneous powder distributions and improve the quality of the final parts.

1 Introduction

Rapid prototyping (RP) is a computer-controlled, layer by layer, additive process for fabrication of three-dimensional solid objects without tooling. Application of RP technology is becoming more popular due to its tremendous time saving in the building of complex 3D parts. There are many types of the rapid prototyping techniques, including a laser to print photopolymer layers, mask photopolymer exposure with wax support, laser sintering ceramic powder with a liquid binder, ballistic particle formation, and fused extruded filament formation. For RP technique using powder-binder mixture as building materials, the numerical simulation of cavity filling of the powder-binder mixture is important to the entire rapid prototyping process, in which, the effect of non-homogeneous powder distribution is a critical factor to the quality of the final products.

Flow simulation of the powder-binder mixtures in the cavity filling were carried out by many authors (Lee, et al 1989, Liaquat, et al 1991, Rhee, et al 1992 and Wang, et al, 1992, 1994). The rheological characteristics, thermal properties and mechanical properties of powder-binder mixture were utilized by Fox, et al (1992) to simulate two-dimensional flow through a thin cavity. However, the current flow simulations were all based on the homogeneous powder distribution.

In the present study, a new numerical scheme based on finite element and finite difference method was developed for modeling the cavity filling process with the non-isothermal, non-Newtonian powder-binder mixtures in the rapid prototyping process. Shear-induced migration of powder was modeled in terms of powder diffusion equation.

---

* To whom all correspondence should be addressed

2 Powder Injection Rapid Prototyping Process

A new rapid prototyping process called Powder Injection Rapid Prototyping (PIRP) process was developed by the authors. The working principles of the powder injection rapid prototyping can be summarized as follows:

- A single component resin is injected uniformly into the working platform by the injection head to form a uniform plate as a outer geometry of the model part.
- The powder-binder mixture is heated and then injected into the cavity formed by the resin at previous step through a computer-controlled nozzle by a cylinder-piston system to create a layer of the model part. The binder in the mixture solidifies when it comes into contact with the platform or the previous layer.
- The next layer of the resin solidifies and adheres to the previous layer, then the outer boundary of the model part is cutted and formed, and the second layer of the powder-binder mixture is injected into the new cavity to form the model part which adheres to the previous layer. The part is built layer by layer until it is complete. The outer support part (resin part) is then removed by heating

3 Governing Equations

A typical filling problem in the thin cavity can be reduced to the two-dimensional. The pressure is a function of x and y, and velocities u and v are the functions of x, y and z. The governing equations can be reduced to the following forms:

3.1 Continuity Equation for the Powder-Binder Mixture

$$\frac{\partial[h\bar{u}(x,y)]}{\partial x}+\frac{\partial[h\bar{v}(x,y)]}{\partial y}=0 \tag{1}$$

3.2 Momentum Equations for the Powder-Binder Mixture

$$\frac{\partial p(x,y)}{\partial x}-\frac{\partial}{\partial z}\left[\mu_s\frac{\partial u(x,y,z)}{\partial z}\right]=0 \tag{2}$$

$$\frac{\partial p(x,y)}{\partial y}-\frac{\partial}{\partial z}\left[\mu_s\frac{\partial v(x,y,z)}{\partial z}\right]=0 \tag{3}$$

where $h=h(x,y)$ is the gapwidth of the cavity, which depends upon the x and y co-ordinate, $u(x,y,z)$ and $v(x,y,z)$ are the velocity components in the x and y direction, and $\bar{u}(x,y)$ and $\bar{v}(x,y)$ are the mean velocities averaged over the cavity gap, which is given as,

$$\bar{u}(x,y)=\frac{1}{h}\int_0^h u(x,y,z)dz\,,\qquad \bar{v}(x,y)=\frac{1}{h}\int_0^h v(x,y,z)dz \tag{4}$$

3.3 Shear-induced Powder Diffusion Equation

The hydrodynamic interaction among neighboring powder particles often creates a non-homogeneous powder distribution in flowing suspension of the powder-binder mixture that were initially well mixed. In concentrated suspensions undergoing shear, a flux of particles is usually reduced from regions of high shear to low and from regions of high particle concentrations to low. The powder diffusion equation is obtained by balancing the fluxes due to bulk convection, gravitational sedimentation and shear-induced diffusion.

$$\frac{\partial\Phi}{\partial t}+u(x,y,z)\frac{\partial\Phi}{\partial x}+v(x,y,z)\frac{\partial\Phi}{\partial y}=\lambda\frac{\partial}{\partial z}\left[\hat{D}_c(\Phi)\dot{\gamma}\frac{\partial\Phi}{\partial z}+\hat{D}_s(\Phi)\frac{\partial\dot{\gamma}}{\partial z}+\frac{2}{9Fr}\Phi f(\Phi)\right] \tag{5}$$

where the scalars $\hat{D}_c(\Phi)$ and $\hat{D}_s(\Phi)$ are shear-induced diffusion coefficients, $\dot{\gamma}$ is the shear rate; the modified Froude number $Fr=\mu_b U/((\rho_p-\rho_b)gL^2)$; $\lambda=(a/L)^2$, the square of the ratio of the particle size a to the characteristic length scale L, ρ_b and ρ_p are the densities of the binder and powder, respectively. Based on dimensional analysis and the mechanisms of shear-induced diffusion, for the unidirectional flow of suspension, these two coefficients are given as:

$$\hat{D}_c(\Phi)=0.43\Phi+0.65\Phi^2\frac{1}{\mu}\frac{d\mu}{d\Phi}\quad,\quad\hat{D}_s(\Phi)=0.43\Phi^2 \tag{6}$$

The relative viscosity μ is modeled as,

$$\mu=\frac{\mu_s}{\mu_b}=\left(1-\frac{\Phi}{\Phi_m}\right)^{-1.82} \tag{7}$$

where Φ_m is the solid volume fraction beyond which the suspension can no longer flow, as in the studies by Leighton, et al (1986, 1987), Schaflinger, et al (1990), Phillips, et al (1992) and Raman, et al (1994). The value of Φ_m depends on the properties of the powder and binder. μ_b and μ_s are the viscosity of the pure binder system and the mixture, respectively. $f(\Phi)$ can be expressed as follows:

$$f(\Phi)=\frac{1-\Phi}{\mu} \tag{8}$$

3.4 Energy Equation for Powder-Binder Mixture

Viscosity of the filling process changes drastically with temperature and convection plays an important role in heat transfer during the filling process; therefore, flow and heat transfer are strongly coupled. Moreover, the energy equation remains strongly three-dimensional. Heat conduction is dominant in the gapwise direction, while the convection terms are dominant in x and y direction. The energy equation may be written as follows:

$$\rho C_p\left(\frac{\partial T}{\partial t}+u(x,y,z)\frac{\partial T}{\partial x}+v(x,y,z)\frac{\partial T}{\partial y}\right)=K\nabla^2T(x,y,z)+Q(x,y,z) \tag{9}$$

where $Q(x,y,z)$ represents the viscous dissipation, which is given as follows:

$$Q(x,y,z)=\mu_s\left[\left(\frac{\partial u(x,y,z)}{\partial z}\right)^2+\left(\frac{\partial v(x,y,z)}{\partial z}\right)^2\right] \tag{10}$$

4 Formulation of the Solution

4.1 Derivation of the Pressure Equation Using the Hele-Shaw Approximation

If the gapwidth $h(x,y)$ maintains constant along the flow direction, integrating Eq. (2) and Eq. (3) with respect to z and using the boundary conditions, velocities can be obtained as follows:

$$u(x,y,z) = -\frac{\partial p}{\partial x}\int_z^h \frac{\hat{z}d\hat{z}}{\mu_s} \tag{11}$$

$$v(x,y,z) = -\frac{\partial p}{\partial y}\int_z^h \frac{\hat{z}d\hat{z}}{\mu_s} \tag{12}$$

Integration of the continuity equation over the gapwidth with the assumption that the velocities vanish at $z=0$ and $z=h$.

Flow quantity $S(x,y)$ is introduced as follows:

$$S(x,y) = \int_0^h \frac{z^2}{\mu_s} dz \tag{13}$$

From the continuity equation, the mean velocities can be derived as follows:

$$\bar{u}(x,y) = -\frac{S}{h}\frac{\partial p(x,y)}{\partial x} \tag{14a}$$

$$\bar{v}(x,y) = -\frac{S}{h}\frac{\partial p(x,y)}{\partial y} \tag{14b}$$

Combining with continuity equation, we may derive a flow equation in terms of pressure.

$$\frac{\partial}{\partial x}\left(S\frac{\partial p}{\partial x}\right) + \frac{\partial}{\partial y}\left(S\frac{\partial p}{\partial y}\right) = 0 \tag{15}$$

4.2 Boundary Conditions

The pressure along the advancing flow front can be assumed as constant. For the sake of simplicity, this pressure at the flow front is set to zero. The heat loss from the flow front to the ambient is small in comparison with the other heat transfer process and is neglected. Powder diffusion between the flow front and the ambient is very small, therefore the volume fraction gradient of the powder can be considered as zero.

At the entrance into the cavity, the flow rate is specified as a function of time. The inlet temperature is held at a constant temperature across the gap and taken as either the melt temperature or the calculated bulk temperature at the end of the delivery system. The volume fraction of powder is also treated as a constant.

At any impermeable boundary, the pressure gradient in the normal direction to the boundary is zero. The powder fluxes into any impermeable boundaries, such as wall boundary and the symmetrical boundary are required to be equal to zero. The heat conduction in x and y directions may be neglected in the energy equation, so that the temperatures on the walls of the mold have to be given as a function of time.

5 Modeling of Material Properties

Without considering the elastic effect, the flow behavior of powder-binder mixture can be represented by the generalized Newtonian fluids. Viscosity of the binder is calculated in terms of a power law, i.e.,

$$\mu_b = K_0 \exp(\frac{E}{RT})\dot{\gamma}^{n-1} \tag{16}$$

where $\dot{\gamma}$ is the shear rate, n is the constant of the viscosity, K_0 is an empirical constant, E is the activation energy flow, R is the universal gas constant, and T is the absolute temperature. Thermal conductivity and heat capacity of the powder-binder mixture are calculated in terms of the mixture law according to the properties of the component.

6 Numerical Implementation

The computations were performed using the successive finite element meshes generated on the filled part of the mid-surface. A finite element and finite difference model was used for the non-Newtonian non-isothermal powder-migration flow analysis based on the generalized Hele-Shaw flow approximation. In the present approach, a single finite element mesh on the mid-surface of the cavity was used to solve for the moving flow front. The two-dimensional pressure field was solved by the finite element method. The temperature and powder distributions at each node along the gapwise direction were solved by the finite difference method.

An efficient front tracking algorithm based on a new FEM/VOF technique and a global/local co-ordinate transformation technique (Couniot, et al 1986, Subbiah, et al 1989, Friedrichs, et al 1993) was extended in this simulation. Based on the previous methods, a local re-meshing technique restricted to the area close to the free surface while keeping most of the elements unchanged was developed to treat the mesh generation and re-mesh of the moving solution domain.

The bilinear quadrilateral elements and local co-ordinate system are used in the solution of the pressure equation. Eq. (15) is multiplied by the shape function and integrated over the solution domain. After integration by part to eliminate the higher order derivatives, this will leads to the following equation:

$$\int_{\Omega} \nabla N \cdot S\nabla p d\Omega + \int_{\partial\Omega} NS\nabla p \cdot \vec{n} dl = 0 \tag{17}$$

The pressure can be expressed approximately as follows:

$$p(x,y) = \sum_{i=1}^{n} p_i N_i(x,y) \tag{18}$$

where $N_i(x,y)$ is the shape function

A system of non-linear algebraic equation is obtained:

$$[A]\cdot[p] = [R] \tag{19}$$

where $[A]$ is the stiffness matrix for the system, while $[p]$ represents the vector of unknown pressures and $[R]$ represents for forcing vector depending upon the given flow rate and the pressure boundary conditions, which are given as follows:

$$[A] = \sum_{i=1}^{n} A_{ij} = \sum_{i=1}^{n} \int_{\Omega} S\nabla N_i \cdot \nabla N_j d\Omega \quad \text{and} \quad [R] = \int_{\partial\Omega} N_i S\nabla p \cdot \vec{n} dl \tag{20}$$

Once the nodal pressure is solved, u and v can be computed by differentiating the interpolation function for pressure according to Eq. (11) and Eq. (12).

The diffusion and energy equations were discretized using the finite different method in the gap direction. The time derivative of powder distribution was expressed by forward finite difference. The space derivative at z-direction was expressed as by central difference. The grid points of the powder distribution in the cavity might be equally spaced. The shear-induced diffusion and convection terms were evaluated using the results from the previous time step. This method allows powder distribution as a function of gapwise position from node to node.

The successive finite element meshes were applied in the simulation, which were generated on the filled part of the midsurface. Starting from an initial finite element mesh developing from the initial fluid domain, the pressure equation was solved using an iterative procedure. Then, the velocities of the mixture were obtained using the Eq. (11) and Eq. (12). Then, the powder and temperature distributions can be iterated based on TDMA method. The flow front was moved to a new location with a small, stable time increment. A new mesh was then generated over the new flow domain using the previous mesh as a first guess. The mesh size was enlarged in the flow direction.

7 Results and Discussions

The simulation was performed on a simple cavity as shown in Figure 1. The material employed in this study is an iron/wax-based binder mixture, and the average diameter of the iron powder is 4 micron. The viscosity of which is given as in equation (16), in which $R = 8.3143J/(mol.K)$, $n = 0.90$ and $E = 14.5KJ/(mol.K)$. The inlet velocity of the powder-binder mixture is set constant at 0.142 m/s. Powder volume fraction is assumed as constant at the inlet section, i.e., $\Phi_{in} = 0.50$. The computation was performed under the simulation conditions given in Table 1.

Table: Input parameters for the simulation

Range of Melting Temperature	360 –400 K
Initial Mold Temperature	310 –330 K
Injection Temperature	380 K
Density	4500 kg/ m^3
Special Heat	1300 J/(kg K)
Thermal Conductivity	2.0 W/(m K)

In order to simplify the simulation and analysis, the mold thickness remains constant.

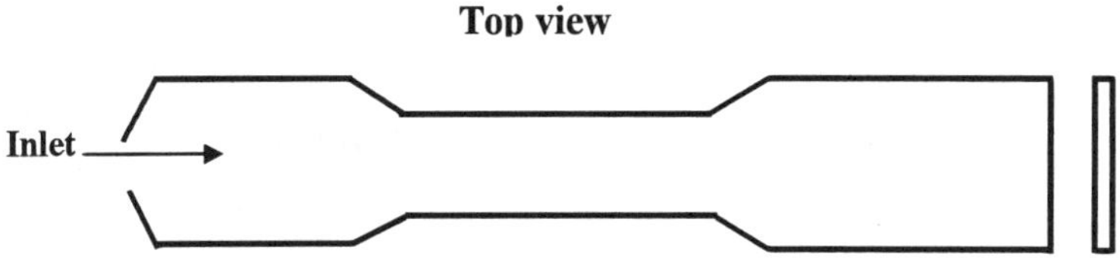

Figure 1 Schematic of the cross section of a simple cavity for simulation

Figure 2 shows the transient numerically generated meshes over the flow domain at some chosen time steps during the filling simulation. The number of the meshes increases as the flow domain expands, so that it can reduce the computational cost. Figure 3 shows the typical results for pressure profile near the completion of the filling stage. To represent effects of the temperature and powder distribution, the relative gapwidth Z is introduced. Z stands for the ratio of the z coordinate to the gapwidth. The three-dimensional temperature distributions are displayed in Figure 4 as a set of contour plots in the x-y plane for four different layers along the cavity gapwise direction: (a) the central plane Z=0.50, (b) Z=0.625, (c) Z=0.75, (d) Z=0.875. The temperature distributions along the centerline of the cavity in the gap-wise direction are shown in Figure 5. The three-dimensional powder distributions are displayed in Figure 6 as a set of contour plots in the x-y plane for five different layers in the mold gap-width: (a) the central plane Z=0.50, (b) Z=0.625, (c) Z=0.75, (d) Z=0.875, (e) the cavity surface Z=1.0. Figure 6 indicates that the powder distribution change greatly at the different layers of the cavity during the filling process. This non-homogeneous powder diffusion is mainly caused by the shear-induced powder diffusion and flow convection. The gapwise powder distributions along the centerline of the cavity are also shown in Figure 7, which shows that the powder volume fraction along the gapwise direction changes greatly.

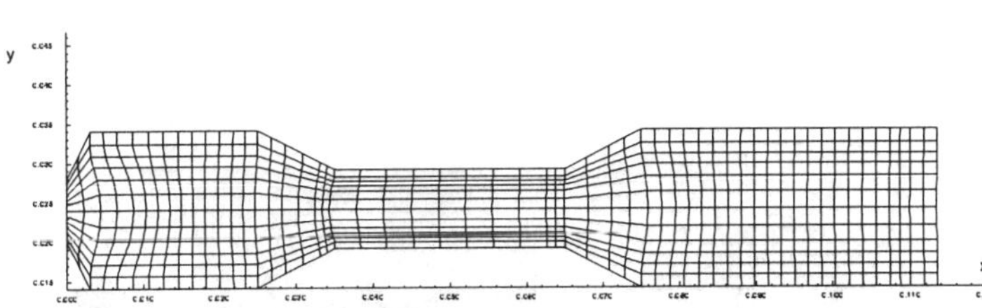

Figure 2 The transient numerically generated meshes over the flow domain at some chosen time steps during the filling simulation

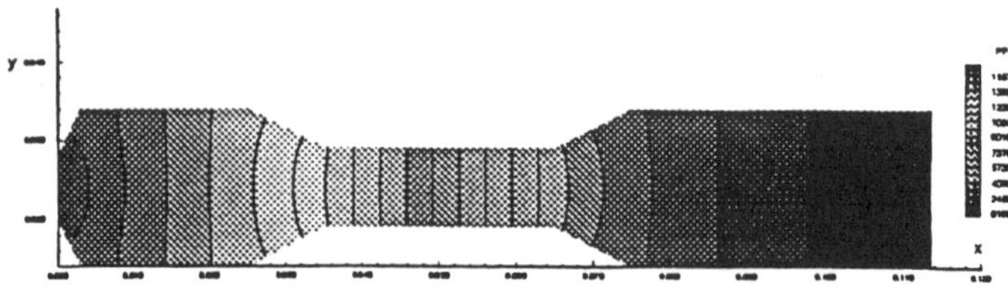

Figure 3 Pressure profiles near the completion of the filling stage

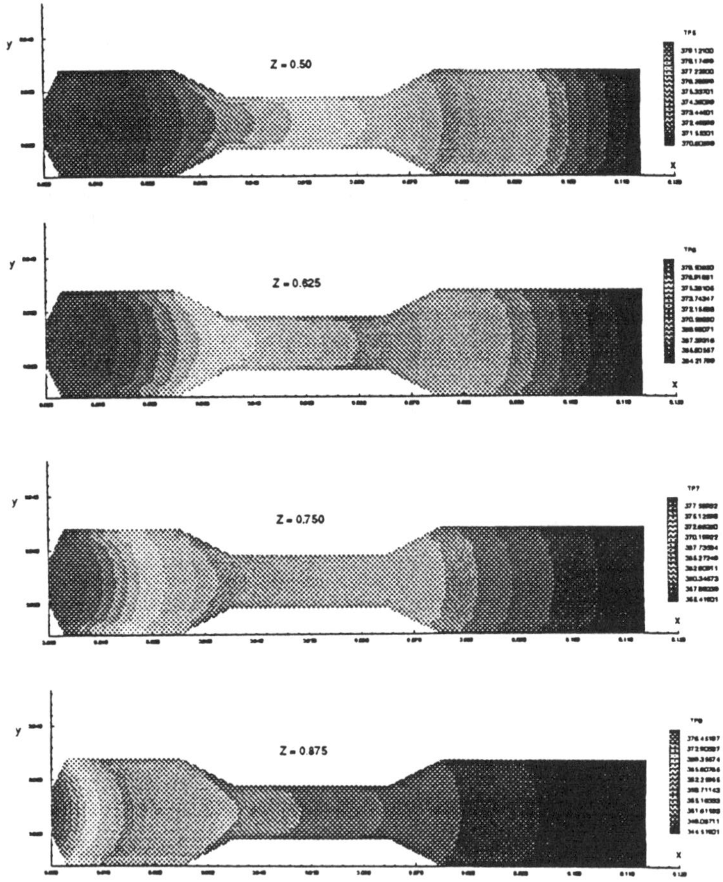

Figure 4 Temperature distributions of the powder-binder mixture at some selected planes: Z=0.50, Z=0.625, Z=0.75 and Z=0.875

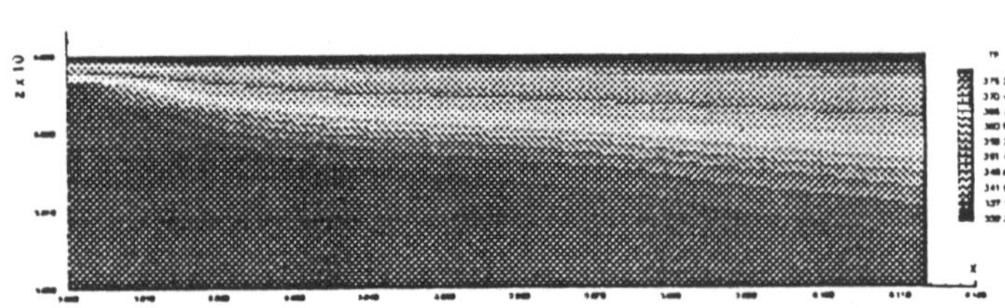

Figure 5 Temperature distribution along the centerline of the cavity in the gap-wise direction

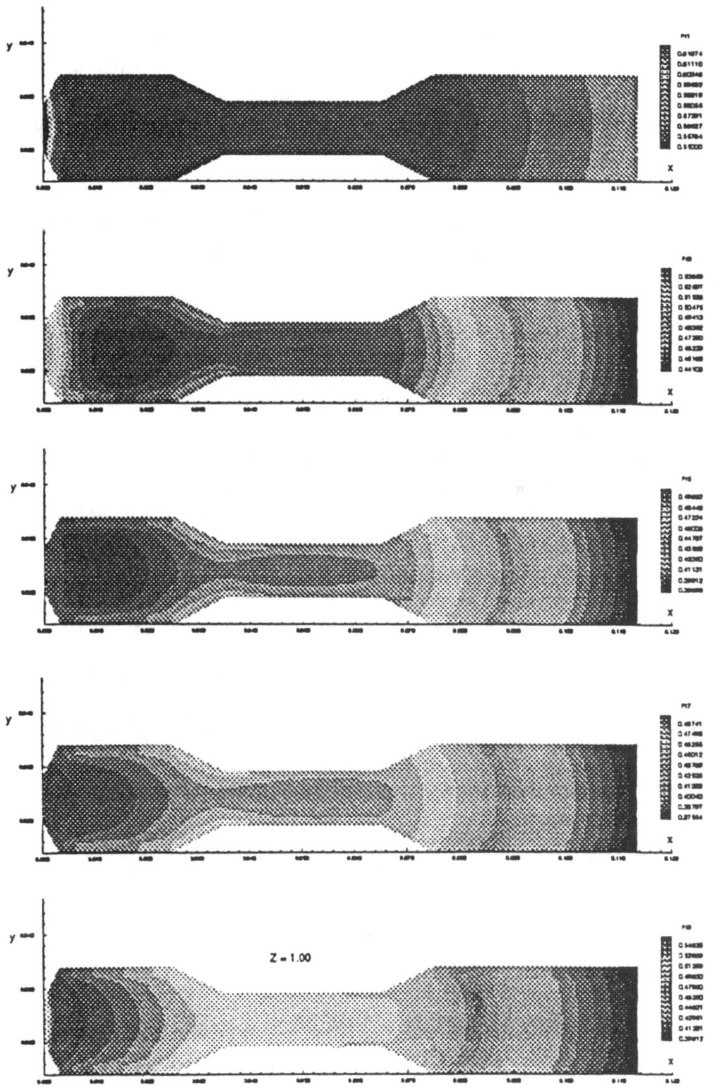

Figure 6 Powder distributions of the powder-binder mixture at some selected planes: Z=0.50, Z=0.625, Z=0.750 Z=0.875 and Z=1.0

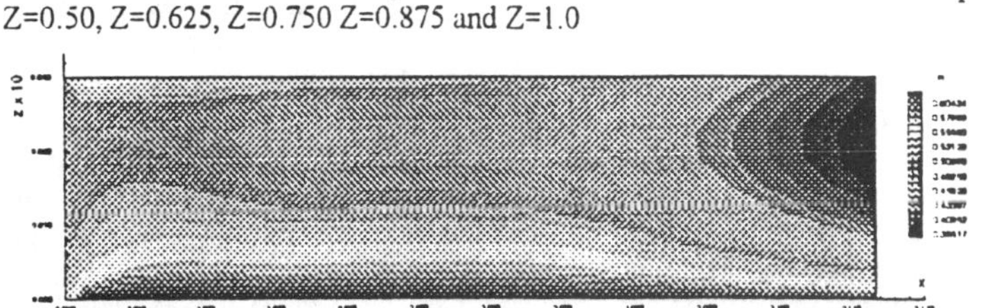

Figure 7 Powder distribution along the centerline of the mold in the gap-wise direction

8 Conclusions

A computational method based on the finite element and finite difference hybrid method was developed to simulate the cavity filling process of non-Newtonian powder-binder mixture in the rapid prototyping process. A new grid generation approach to the analysis of the cavity filling process was also presented to reduce the computational cost. Shear-induced diffusion under non-isothermal conditions was investigated in the simulation. This method is advantageous in analyzing and detecting the non-homogeneous powder distribution of the powder-binder mixture during the filling process, which may be useful to improvement of quality of the final product.

References

Chiang, H. H. and Himasekhar, K., "Integrated simulation of Fluid Flow and Heat Transfer in Injection Molding for the Prediction of Shrinkage and Warpage", HTD-Vol. 175/Md-Vol. 25, Heat and Mass Transfer in Solidification Processing, ASME, pp. 133-146.

Couniot, A. and Crochet, M. J., "Finite Element for the Numerical Simulation of Injection Molding", Proceedings of the NUMIFORM' 86, Gothenburg, Sweden, 1986.

Fauchon, D., Tanguy, P. A. and Dannelongue, H. H., " A Rigorous 2-D Mold Filling Model Using The Finite Element Method", Proceedings of the NUMIFORM' 89, Fort Collins, Colorado, USA, 1989, pp. 259-263.

Friedrichs, B. and Guceri, S. I., "A Novel Hybrid Numerical Technique to Model 3-D Fountain Flow in Injection Molding Processes", Journal of Non-Newtonian Fluid Mechanics, 49(1993), 141-173.

Leighton, D. and Acrivos, A., "Measurement of Shear -Induced Self-Diffusion in Concentrated Suspension of Spheres", Journal of Fluid Mechanics, Vol. 177, 1987, pp. 109 -131.

Leighton, D. and Acrivos, A., "The Shear - Induced Migration of Particles in Concentrated Suspensions", Journal of Fluid Mechanics, Vol. 181, 1987, pp. 415 - 439.

Liaquat Alinajmi and Lee, D., "Modeling of Mold Filling for Powder Injection Molding", Polymer Engineering and Science, Mid-August 1991, Vol. 31, No. 15, pp. 1137 -1148.

Phillips, R. J., et al, "A Constitutive Equation for Concentrated Suspension that Accounts for Shear-induced Particle Migration", Phys. Fluids A 4(1), January, 1992, pp. 30 - 40.

Raman, R. and German, R. M., "Mixing of Injection Molding Feedstock Using Batch and Continuous Mixers", Advances in Powder Metallurgy and Particulate Materials, Vol. 4, 1994, pp. 1 - 14.

Rhee, B. O., Cao, M. Y., Zhang, H. R., Streicher, E. and Chung, C. I., "Improved Wax-Based Binder Formulations for Powder Injection Molding", Advances in Powder Metallurgy and Particulate Materials, Vol. 2, 1991.

Schaflinger, U., Acrivos, A. and Zhang, K., "Viscous Resuspension of a Sediment within a Laminar and Stratified Flow", International Journal of Multiphase Flow, Vol. 16, pp. 567 - 578, 1990.

Subbiah, S., Trafford, D. L. and Guceri, S. I., "Non-isothermal Flow of Polymers into Two-dimensional, Thin cavity Molds: A Numerical Grid Generation Approach", International Journal of Heat and Mass Transfer, Vol. 32, No. 3, pp. 415-434, 1989.

Wang, C. M., Carr, K. E. and McCabe, T. J., "Computer Simulation of the Powder Injection Molding", Vol. 4, Advances in Powder Metallurgy and Particulate Materials, pp. 15-26, 1994.

Wang, C. M., Leonard, R. C., Posteraro, R. A. and McCabe, T. J., "A Finite-Difference Computing Model for 3-D Simulation of the Powder Injection Molding Process", Proceedings of the 1992 Powder Injection Molding Symposium, pp. 435-449, 1992.

Zhang, K. and Acrivos, A. "Viscous Resuspension in Fully Developed Laminar Pipe Flows", International Journal of Multiphase Flow, Vol. 20, No. 3, pp. 579-591, 1994.

Rotary Axis Compensation for a 5-axis Micro Milling Rapid Prototyping System

Ng Peow, Fang Wei, Ngooi Chan Siang, Li Hui, Li Bo, Zou Jiguo and Gong Haiqing*

School of Mechanical and Production Engineering, Nanyang Technological University, Singapore

Abstract. In view of the 'staircase or stepped effect' in the present commercial rapid prototyping (RP) systems, a new approach is proposed and is currently under development to produce true stepless parts. The new RP system is a combination of 5-axis motion control technology and state-of-the-art micro milling. As in any rapid prototyping system, accuracy is one of the many key issues. This paper presents the rotary axis compensation of the system to improve the overall machining accuracy.

1 Introduction

In today's highly competitive industries, product quality, time-to-market and cost are the critical factors for success. Faster development cycle requires the generation of highly accurate model directly from the CAD data. Rapid prototyping, or RP technologies offer the precise solution for this purpose.

RP technology was commercially introduced in 1987 with the presentation of StereoLithography (SLA) [1,2,3,4]. Several processes are now commercially available in the United States, Europe, and Asia. To name a few, like the Solid Ground Curing from Cubital [2,5], Fused Deposition Manufacturing (FDM) from Stratasys [2,6] and Laminated Object Manufacturing (LOM) from Helisys [2,7,8]. Although different techniques are used in these RP systems, there is a common drawback. Parts produced by the current commercial systems have *staircase*, or *stepped* slanted surfaces. This inspires the search for a new RP technology, which will produce *truly accurate stepless* parts. The system currently under development is a new 5-axis CNC micro milling system, which consists of three linear and two rotary axes. The two rotary axes enable tangent cutting, which eliminates the staircase effect.

2 5-Axis Micro Milling for Rapid Prototyping

Like the other RP systems, the 5-axis micro milling RP system 'decomposes' a CAD model into boundary contours of thin slices. Each slice is then 'cut out' by micro milling, and built one layer on top of the other, as in LOM. Figure 1 shows the picture of the system currently under development, it consists of a x-y machine table and a z-axis. At the end of the z-axis is a robot wrist (i.e. the roll and pitch axes). It is similar to a 3-axis CNC milling machine, but with the addition of two rotary axes to enable tangent cutting. The cutting method, i.e. micro milling, is similar to conventional slot milling, except the tool diameter is much smaller.

* To whom all correspondence should be addressed

The building of each layer in this system requires 4 processes as listed below:

1. Laying of photopolymer on the machine table.
2. Solidify the photopolymer with UV (the solidification process is beyond the scope of this paper).
3. Micro-milled the boundary contour of the model.
4. Final face milling for cosmetic purpose.

Figure 1: Picture of the 5-axis Micro Milling RP Machine

After the final face milling, it is ready to build the next layer.

Current commercial RP systems use a reasonably thin build layer to minimize the 'staircase effect'. Take LOM for example, each build layer thickness is around 0.0038" to 0.01" (0.0965 to 0.254mm), and a fixed thickness is used throughout the whole building process [7]. Consequently, to build a part of 15mm height will require 60 to 150 layers! The 5-axis micro milling RP system on the other hand, does not have a fixed build layer thickness. The thickness of each build layer depends on the complexity of the boundary contour, and is maximized whenever possible. This approach minimizes the number of build layers required, and subsequently speeds up the building process.

3 Layer Transfer Interface (LTI) Format

Presently, there are three main techniques in slicing the STL file, they are:

1. *Fixed slicing*: Uniform layer thickness is applied throughout the part. The drawback of this technique is that flat surfaces that lie in between two slicing planes may be missed out.
2. *Semi-fixed slicing*: This is a modified fixed slicing technique. Special attention is given to those 'problematic flat surfaces' to ensure that they are not missed out during the slicing operation.
3. *Adaptive slicing*: The layer thickness is not fixed. It depends on the complexity, or difference between the two adjacent slice contours.

Among these techniques, the adaptive slicing technique is the best as far as building speed is concerned. However, its implementation can be difficult. For instance in SLA, variable build layer thickness implies that the laser intensity must always vary accordingly. Furthermore, the benefits of adaptive slicing can only be fully utilized in a system that allows tangent cutting. For system that only allows normal cutting, to minimize the staircase effect, the system has to revert to thin layer thickness for slanted surfaces, as illustrated in figure 2. The STL slicer for the 5-axis micro milling RP machine adopts the adaptive slicing technique. The sliced data is presented in a new file format known as the LTI (Layer Transfer Interface) format[(1)] Unlike the other 2D sliced

---

[(1)] LTI format is a new data format specially developed for the 5-axis micro milling RP system by a group led by Dr. Thomas Gong Haiging in the Nanyang Technological University

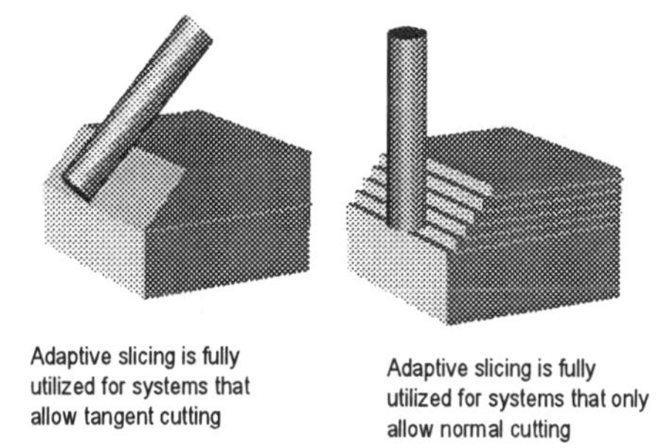

Figure 2: Utilizing Adaptive Slicing

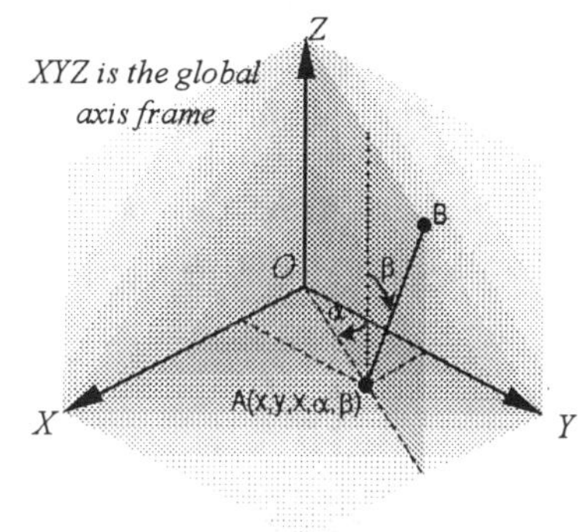

Figure 3: The LTI Format

data formats such as the SLC [9] and CLI [10], the LTI format includes additional data, like the slope information at every point, to orientate the cutter for tangent cutting. In this format, every point has 5 coordinates as shown in figure 3. 'x', 'y' and 'z' are the cartesian coordinates, while 'α' and 'β' are the roll and pitch coordinates respectively.
The LTI format is a 'two ways data format', the data can be transferred into the STL facetted object for modifications. This is a feature which neither the SLC nor CLI format can provide. Similarly, the LTI format can be converted into the SLC and CLI formats, but the reverse is not possible. Although the LTI format contains more information than the other two formats, its file size is considerably smaller as it uses the adaptive slicing technique to minimize the number of slices required to build a part. This reduces the storage size and speeds up the exchange between the computer and RP system.

4 Tool Diameter Offset

In CNC milling, it can be cumbersome and difficult for the NC programmer to program the cutter's path based on the size of the milling cutter being used. Also if the cutter size changes, it would be infeasible to change the program based on the new cutter size. For this reason, tool diameter offset allows the programmer to ignore the cutter size as the program is written. However, in this system, the approach is slightly different, instead of specifying the tool diameter offset at the tool path generation stage, the offset is specified during the STL file slicing stage. That is, the tool diameter is already set before the file is sent for tool path generation. The reasons for such approach are as follows,

1. During the STL to LTI format translation, some of the data are purposely ignored to keep the LTI file size small. Take a plane that is formed by any two consecutive segment points for example, to offset the tool diameter, we need to know which side of the plane should we place the cutter. Consequently, additional information must be added for each and every segment point. This will result in a much larger file size and eventually slow down the processing time.
2. The STL slicer, besides slicing the STL files, is also a off-line process simulation software. The tool diameter is an essential parameter for this application.
3. As mentioned earlier, the STL slicer adopts the adaptive slicing technique to minimize the number of build layers required to build a part. To do this, the STL slicer needs to know what is the maximum slicing thickness allowable. This slicing

thickness depends on the diameter of the tool since the maximum depth of cut allowable is two times its diameter. (i.e. depth of cut $_{max}$ = 2 x Tool Diameter)

Due to the additional information for the two rotary coordinates, the 'point' generated by the STL slicer describes more than just a point. It represents a line segment of undefined length. Therefore, from this point onwards, the five coordinates representation, i.e. $P(x,y,z,\alpha,\beta)$ will be referred to as *segment point*, while $P(x,y,z)$ will be referred to as *point*.

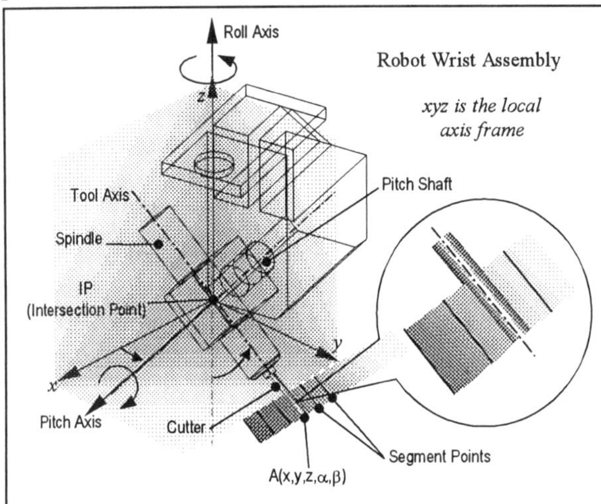

Figure 4: Tool Orientation

Figure 4 shows the schematic of the robot wrist assembly. Given a segment point $A(x,y,z,\alpha,\beta)$, (x,y,z) represents the position of the cutter tip along its centreline. 'α' is the orientation of the pitch axis, it is measured from the X-axis. 'β' is the orientation of tool axis, it is measured from the roll axis. Because the tool diameter offset is already specified in the LTI file, each segment point then represents the position whereby the tool axis must be aligned with. Therefore, if a tool measured L_{tool} from the tip of the micro mill along its centreline (or tool axis) to the IP position, by replacing |AB| with L_{tool} (refer to figure 3), point 'A' represents the actual cutting (milling) position, while point 'B' represents the IP position. The tool path is formed by joining these IP positions with *straight* line segments.

5 Motion Compensation

To build a RP machine with high accuracy and precision, compensation is always required. There are two different ways to compensate the deviations or errors in machines. For instance, the straightness error of a carriage guide can be compensated, either by purely *hardware techniques*, such as scraping the guide, or by *software techniques*, which is also known as the *motion compensation*. The open loop compensation is used in the current 5-axis micro milling RP system, however, the closed loop compensation technique will be employed in the future system(s).

Motion compensation is a three steps procedure:

Step 1: Describe the machine setup by a simplified mathematical model. Two models are required, one for the ideal setup, and the other for the actual setup, which includes all the possible geometrical errors. Because of the geometrical errors, the two models will have different mathematical representations for the same motion transformation. By comparing the difference between the two representations, the compensation function for the machine is established. (This step will be discussed in detail in the coming sections.)

Step 2: Establish an experimental procedure to measure the parameters (i.e. the geometrical errors) in the model for the actual setup. Due to the constraint of the physical setup of the machine itself, direct measurement can be difficult to carry out at times, in such cases, indirect measurement should be used [11].

Step 3: Conduct experiment to verify to effectiveness of the compensation.

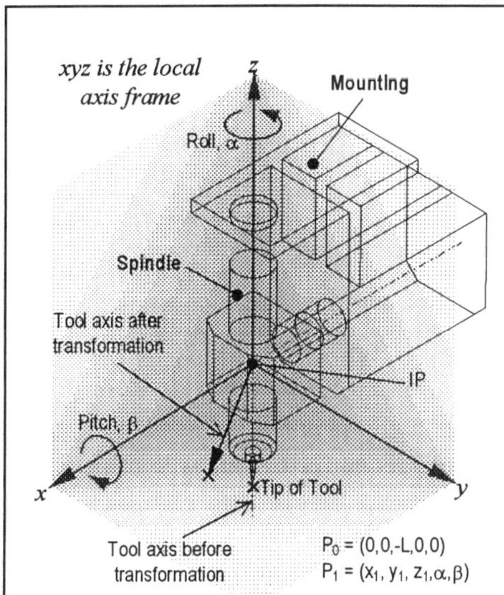

Figure 5: Ideal Coordinate System for the Robot Wrist

Figure 5 shows the ideal coordinate system of the robot wrist used in the RP system. It has the following characteristics:

1. The roll axis coincides with the z-axis;
2. The roll axis is perpendicular to the pitch axis; and
3. The tool, roll and pitch axes intersect at *one* point.

Consider an arbitrary transformation **[M]**, such that

$$[\mathbf{P1}] = [\mathbf{M}].[\mathbf{P0}]$$

where $[\mathbf{P_0}]$ is tip the of the tool when it is in the vertical position, and $[\mathbf{P_1}]$ is the tip of the tool after the transformation.

[M] consists of a pitch rotation of angle β, followed by a roll rotation of angle α z-axis.

i.e. $[\mathbf{M}] = [\mathbf{R}]_{roll,\alpha} \cdot [\mathbf{R}]_{pitch,\beta}$ ...eq.(5.1)

For the ideal system, the transformation $[\mathbf{R}]_{roll,\alpha}$ can be represented by a rotation about the z-axis with an angle α, $[\mathbf{R}]_{z,\alpha}$. Similarly, $[\mathbf{R}]_{pitch,\beta}$ can be represented by a rotation about the x-axis with an angle -β, $[\mathbf{R}]_{x,-\beta}$. Therefore, eq.(5.1) becomes

$$[\mathbf{M}]_{ideal} = [\mathbf{R}]_{z,\alpha} \cdot [\mathbf{R}]_{x,-\beta} \quad \text{...eq.(5.1a)}$$

Consider only the cartesian coordinates,

$$\begin{bmatrix} x_1 \\ y_1 \\ z_1 \\ 1 \end{bmatrix} = \begin{bmatrix} \cos\alpha & -\sin\alpha & 0 & 0 \\ \sin\alpha & \cos\alpha & 0 & 0 \\ 0 & 0 & 1 & 0 \\ 0 & 0 & 0 & 1 \end{bmatrix} \cdot \begin{bmatrix} 1 & 0 & 0 & 0 \\ 0 & \cos(-\beta) & -\sin(-\beta) & 0 \\ 0 & \sin(-\beta) & \cos(-\beta) & 0 \\ 0 & 0 & 0 & 1 \end{bmatrix} \cdot \begin{bmatrix} 0 \\ 0 \\ -L \\ 1 \end{bmatrix} = \begin{bmatrix} L\sin\alpha\sin\beta \\ -L\cos\alpha\sin\beta \\ -L\cos\beta \\ 1 \end{bmatrix}$$

where (x_1,y_1,z_1) is the cartesian coordinates of the tool tip after the transformation, L is the tool length (measured from the tool tip to the IP position) and the fourth parameter is the *scaling factor* which is usually included in robotics.

$$\text{Hence } P_1 = \begin{bmatrix} L\sin\alpha\sin\beta \\ -L\cos\alpha\sin\beta \\ -L\cos\beta \end{bmatrix} \quad \text{...eq.(5.2)}$$

5.1 Angular Offsets of the Roll and Pitch Axes

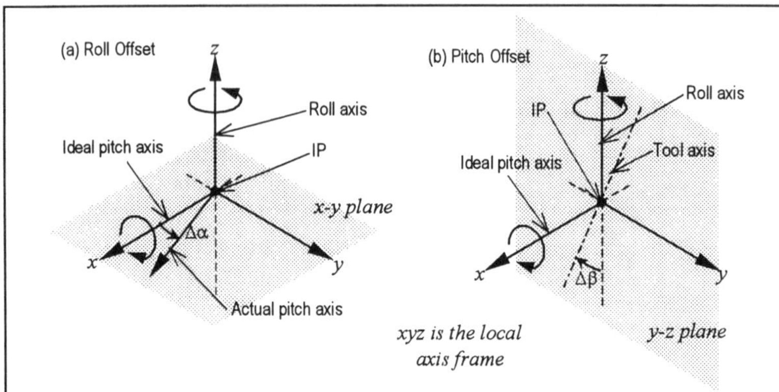

Figure 6: Angular Offsets of the Roll and Pitch Axes

Figure 6a shows the roll axis offset of the robot wrist. When the roll axis is reset to its origin position, ideally, the pitch axis should lie on the xz-plane, or $\Delta\alpha = 0$. However, due to the geometrical and dimensional tolerance of the mounting, deviation from the ideal pitch axis usually occurs. Hence $\Delta\alpha$, in general, is not zero. This type of deviation can be easily compensated as follow,

$$\alpha_{compensated} = \alpha_{nominal} + \Delta\alpha \qquad \text{...eq.(5.3a)}$$

Similarly, deviation $\Delta\beta$ (see figure 6b) can be compensated by eq.(5.3b)

$$\beta_{compensated} = \beta_{nominal} + \Delta\beta \qquad \text{...eq.(5.3b)}$$

5.2 Linear Offsets of the Roll and Pitch Axis

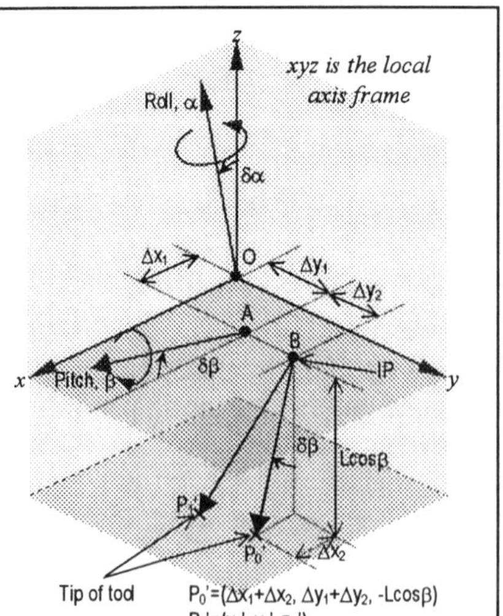

Figure 7: Actual Coordinate System for the Robot Wrist

Compensation of the roll and pitch axes only ensures that both the roll and pitch axes are parallel to the x-z plane. The pitch axis, for instance, may be some distance (Δy_1) away from the x-z plane and at an angle ($\delta\beta$) to the x-y plane as shown in figure 7. The actual IP (intersection point) position, instead of coinciding with O, which is the ideal IP position, may be Δx_1 from the y-axis and Δy_2 from the pitch axis. These are all the possible geometrical errors in the actual setup. Since there isn't any rotary axis along the y-axis, $\delta\alpha$ and $\delta\beta$ cannot be compensated by software as in the case of $\Delta\alpha$ and $\Delta\beta$. However, $\delta\alpha$ and $\delta\beta$ can be minimized through hardware compensation. Therefore, we may assume that $\delta\alpha \approx 0$ and $\delta\beta \approx 0$. If $\delta\beta \approx 0$, $L\cos(\delta\beta) \approx L$. However, the displacement of the tool tip along the x-direction (Δx_2) due to $\delta\beta$ can be

significant depending on L, hence, it must also be taken into consideration during motion compensation. With all these geometrical errors, the cartesian coordinates of the tool tip before transformation can be expressed as $P_0'(\Delta x_1+\Delta x_2,\ \Delta y_1+\Delta y_2,\ -L)$, where " ' " represents the actual coordinate system.
Consider the transformation **[M]** in eq.(5.1), we may either use the nominal or compensated values for 'α' and 'β' (refer to eq.(5.3a) and eq.(5.3b)). However, if the nominal values are used, then the positional error, or deviation becomes a combined effect of both the linear and angular offsets.

The latter approach is preferred and we shall see its advantages in section 5.3.1. After the angular compensation,

$$\alpha' = \alpha + \Delta\alpha \qquad \text{...eq.(5.3c)}$$

and $$\beta' = \beta + \Delta\beta \qquad \text{...eq.(5.3d)}$$

where α and β are the nominal values.
α' and β' are the compensated values.

With reference to eq.(5.1), for the actual system, the transformation $\mathbf{[R]}_{roll,\alpha}$ is equivalent to a rotation about the z-axis with an angle α', $\mathbf{[R]}_{z,\alpha'}$. Transformation $\mathbf{[R]}_{pitch,\beta}$ is more complicated in the actual system since the pitch axis does not coincide with the x-axis (refer to figure 7). The transformation $\mathbf{[M]}_{actual}$ can be expressed as,

$$\mathbf{[M]}_{actual} = \mathbf{[R]}_{z,\alpha'} \cdot \{\mathbf{[T]}_{O\to A} \cdot \mathbf{[R]}_{x,-\beta'} \cdot \mathbf{[T]}_{A\to O}\} \qquad \text{...(5.1b)}$$

Let $P_1'(x_1', y_1', z_1')$ be the coordinates of the tool tip after the transformation, we have

$$\begin{bmatrix} x_1' \\ y_1' \\ z_1' \\ 1 \end{bmatrix} = \begin{bmatrix} \cos\alpha' & -\sin\alpha' & 0 & 0 \\ \sin\alpha' & \cos\alpha' & 0 & 0 \\ 0 & 0 & 1 & 0 \\ 0 & 0 & 0 & 1 \end{bmatrix} \cdot \begin{bmatrix} 1 & 0 & 0 & \Delta x_1 \\ 0 & 1 & 0 & \Delta y_1 \\ 0 & 0 & 1 & 0 \\ 0 & 0 & 0 & 1 \end{bmatrix} \begin{bmatrix} 1 & 0 & 0 & 0 \\ 0 & \cos(-\beta') & -\sin(-\beta') & 0 \\ 0 & \sin(-\beta') & \cos(-\beta') & 0 \\ 0 & 0 & 0 & 1 \end{bmatrix} \cdot \begin{bmatrix} 1 & 0 & 0 & -\Delta x_1 \\ 0 & 1 & 0 & -\Delta y_1 \\ 0 & 0 & 1 & 0 \\ 0 & 0 & 0 & 1 \end{bmatrix} \cdot \begin{bmatrix} \Delta x_1 + \Delta x_2 \\ \Delta y_1 + \Delta y_2 \\ -L \\ 1 \end{bmatrix}$$

$$= \begin{bmatrix} (\Delta x_1 + \Delta x_2)\cos\alpha' - \Delta y_2 \sin\alpha' \cos\beta' + L \sin\alpha' \sin\beta' - \Delta y_1 \sin\alpha' \\ (\Delta x_1 + \Delta x_2)\sin\alpha' + \Delta y_2 \cos\alpha' \cos\beta' - L\cos\alpha' \sin\beta' + \Delta y_1 \cos\alpha' \\ -\Delta y_2 \sin\beta' - L\cos\beta' \\ 1 \end{bmatrix}$$

$$\text{Hence } P_1' = \begin{bmatrix} (\Delta x_1 + \Delta x_2)\cos\alpha' - \Delta y_2 \sin\alpha' \cos\beta' + L \sin\alpha' \sin\beta' - \Delta y_1 \sin\alpha' \\ (\Delta x_1 + \Delta x_2)\sin\alpha' + \Delta y_2 \cos\alpha' \cos\beta' - L\cos\alpha' \sin\beta' + \Delta y_1 \cos\alpha' \\ -\Delta y_2 \sin\beta' - L\cos\beta' \end{bmatrix} \qquad \text{...eq.(5.4)}$$

Comparing eq.(5.2) with eq.(5.4), the deviation in P_1 becomes

$$\Delta P_1 = -\begin{bmatrix} (\Delta x_1 + \Delta x_2)\cos\alpha' - \Delta y_2 \sin\alpha' \cos\beta' - \Delta y_1 \sin\alpha' + L(\sin\alpha' \sin\beta' - \sin\alpha \sin\beta) \\ (\Delta x_1 + \Delta x_2)\sin\alpha' + \Delta y_2 \cos\alpha' \cos\beta' + \Delta y_1 \cos\alpha' - L(\cos\alpha' \sin\beta' - \cos\alpha \sin\beta) \\ -\Delta y_2 \sin\beta' - L(\cos\beta' - \cos\beta) \end{bmatrix} \qquad \text{...eq.(5.5)}$$

such that $P_{1,compensated} = P_{1,nominal} + \Delta P_1$.

Therefore for every nominal segment point P(x,y,z,α,β), the corresponding compensated segment point is given by P'(x',y',z',α',β'), such that,

$$\begin{bmatrix} x' \\ y' \\ z' \\ \alpha' \\ \beta' \end{bmatrix}_{compensated} = \begin{bmatrix} x \\ y \\ z \\ \alpha \\ \beta \end{bmatrix}_{nominal} + \begin{bmatrix} -(\Delta x_1 + \Delta x_2)\cos\alpha' + \Delta y_2 \sin\alpha' \cos\beta' + \Delta y_1 \sin\alpha' - L(\sin\alpha' \sin\beta' - \sin\alpha \sin\beta) \\ -(\Delta x_1 + \Delta x_2)\sin\alpha' - \Delta y_2 \cos\alpha' \cos\beta' - \Delta y_1 \cos\alpha' + L(\cos\alpha' \sin\beta' - \cos\alpha \sin\beta) \\ \Delta y_2 \sin\beta' + L(\cos\beta' - \cos\beta) \\ \Delta\alpha \\ \Delta\beta \end{bmatrix}$$

...eq.(5.6)

5.2.1 Experiment to determine the linear offsets of the roll and pitch axis

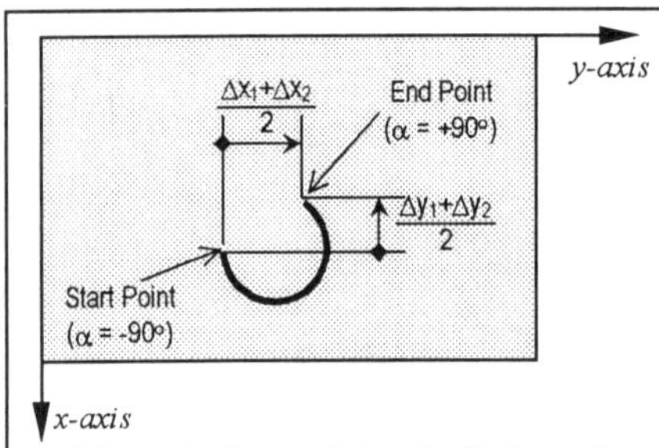

Figure 8: Determining the Values of ($\Delta x_1 + \Delta x_2$) and ($\Delta y_1 + \Delta y_2$)

From eq.(5.6), the geometrical errors required are ($\Delta x_1+\Delta x_2$), Δy_1 and Δy_2. Indirect measurements are used to determine these values. There are only two steps,

Step 1 : Do the necessary angular offsets in section 5.1.

Step 2 : Make a cut on a test surface by rotating the roll axis from -90° to +90°. Note that these are the compensated values.

For the ideal system, only a dot will be observed. However, if a semi-circle is observed (see figure 8), it implies that there are some deviations. Since, the angular offsets are already compensated, the deviation must be caused by the linear offsets of the roll and pitch axis.

The horizontal and vertical deviations are due to ($\Delta x_1+\Delta x_2$) and ($\Delta y_1+\Delta y_2$) respectively. Depending on the position of the start and end points, these deviations may have negative values. Unfortunately, from this experiment, the individual of values of Δy_1 and Δy_2 cannot be determined, but we shall see how they can be deduced in section 5.3.1.

5.3 Implementation of Motion Compensation

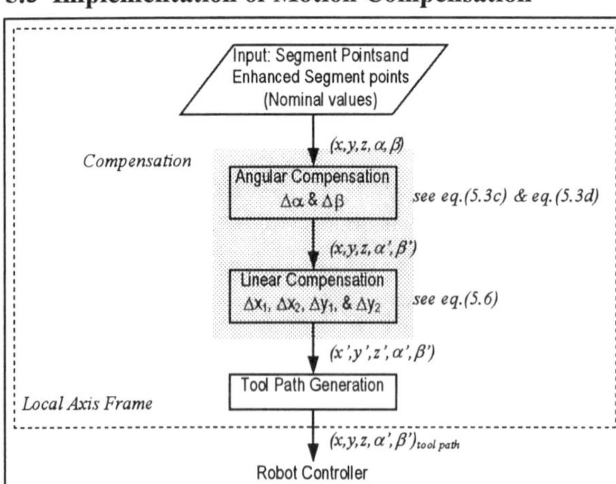

Figure 9: Implementation of the Motion Comp. Algorithm

The implementation of the motion compensation is shown in the flow chart (figure 9).

5.3.1 Verification of Concentricity Compensation

A test file was written to instruct the machine to cut a slot by rotating its roll axis from -90° to +90°. To ease investigation, the smallest micro-mill (∅0.2mm) was used. As milling (cutting) causes deflection in both the z-axis and the cutter itself, it is desirable to reduce these deflections by minimizing the cutting force during the experiment. This is achieved by:

1. Ensuring a shallow depth of cut;
2. maintaining a low feed rate; and
3. using a high spindle speed (at 100,000rpm) for cutting.

The pictures of the slot is captured and analyzed by an image processing software

Figure 10 shows the concentricity error before and after the compensation, different values of Δy_2 (refer to section 5.2.1) are tested until the minimum values of DX and DY are obtained.

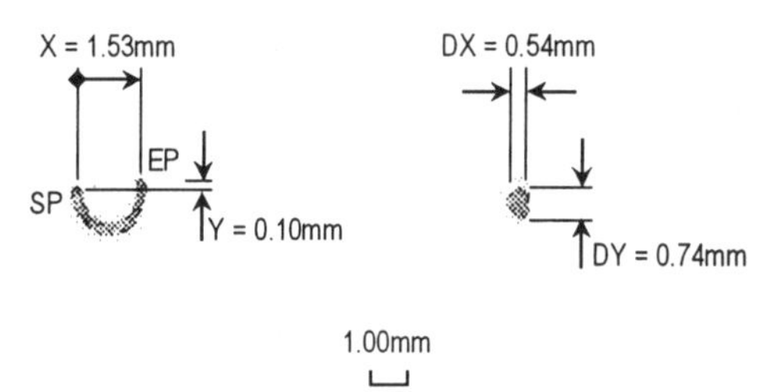

Figure 10: Concentricity Error before and after Compensation

6 Conclusions

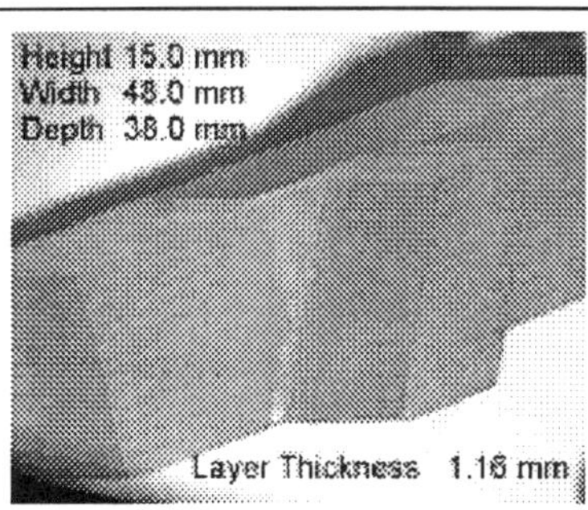

Figure 11: Picture of a 14 Layer Prototype

The concept of 5-axis micro milling for RP application was proposed to eliminate the staircase effect in current rapid prototyping. A prototype system has been built and has proven its feasibility and potential.

A motion compensation algorithm was also developed to improve the machining accuracy. Experiment were conducted to verify the effectiveness of motion compensation, and the outcome was very positive.

For the verifications of the tangent cutting and other building processes, a simplified 14 layers (15mm) prototype was built. The result was acceptable, the concept of tangent cutting to produce stepless part was proven feasible. However, there is still room for improvement in terms of building speed and accuracy.

References

1. Crowell, George. StereoLithography. 1989. Dearborn, Michigan: Society of Manufacturing Engineers.
2. Jacobs, P.F. 1996. StereoLithography and other RP&M technologies : from rapid prototyping to rapid tooling. Dearborn, Mich.: Society of Manufacturing Engineers in cooperation with the Rapid Prototyping Association of SME; New York: ASME Press.
3. Schweikhart, R. 1990. StereoLithography : applications for rapid prototyping. Dearborn, Mich. : Society of Manufacturing Engineers.
4. Welcome to 3D Systems. 1996. [web page]. http://www.3dsystem.com/.
5. Cubital: The Faster, the Better. 1996. [web page]. http://www.iquest.net/cubital/.
6. Welcome to Stratasys, Inc. 1997. [web page]. http://www.stratasys.com/.
7. Michael J.B. 1995. Tele-manufacturing rapid prototyping on the Internet. IEEE computer graphics and applications. IEEE Computer Society: National Computer Graphics Association.
8. Helisys: Laminated Object Manufacturing. 1996. [web page]. http://helisys.com/.
9. .SLC File Format. 1996. [web page]. http://cadserv.cadlab.vt.edu/bohn/rp/SLC.html.
10. Common Layer Interface (CLI). 1994. [web page]. http://www.cranfield.ac.uk/aero/rapid/CLI/cli_v20.html.
11. Sartori, S & Zhang G.X. 1995. Geometric error measurement and compensation of machines. Annals of the CIRP, 44, (2). 599-609.

Generation of Tooth Models for Ceramic Dental Restorations*

Sven Gürke

Department of Cognitive Computing & Medical Imaging, Fraunhofer Institute for Computer Graphics, Darmstadt, Germany

Abstract. Complete CAD/CIM-systems for dental restorations require user interaction for the construction of tooth inlays in ceramics. An automation is only possible if the typical geometry of the teeth is known by the system. One realizable way for the solution of this problem is the restoration of the occlusal surface by adaptation of an appropriate tooth model.

In this paper, we present a method for developing and generating of tooth models including dental medical knowledge about the occlusal surface. Furthermore, our approach includes an adaptation process which can be extended to an automatic tooth restoration system.

1 Introduction

1.1 Conventional Method

In the last few years, on the field of tooth restorations we can ascertain an increased acceptance of ceramic inlays besides plastic and gold inlays. Ceramic restorations possess an inconspicuous appearance like natural teeth and exhibit similar hardness. Furthermore, in contrast to amalgam ceramic inlays have no toxical pertinence and a long durability. Unfortunately, most ceramic inlays are produced in dental labs and therefore the patient must accept at least two visits at the dentist.

1.2 Alternative

One popular CAD/CIM system is the CEREC-device [8] (see figure 1) of the Siemens AG. In this system, the dentist has to take an optical impression of the prepared tooth with an intra-oral 3D-sensor. Then he designs the inlay interactively with an integrated CAD-system. Finally, an integrated milling machine produces the inlay from a block of ceramics. In the processing pipeline, the interactive construction process represents the difficult part of this method. Often the surface of the occlusal part of the inlay has to be modeled manually after the inlay is inserted into the cavity.

* Sponsored by the Bayerische Forschungsstiftung

1.3 New Technique

The aim of a research project is to minimize the complicated user interaction for the construction of an inlay. An automation of a CAD/CIM system is only possible if the typical geometry of teeth is known by the system. One realizable approach for the solution of this problem is the restoration of the occlusal surface by adaptation of an appropriate tooth model.

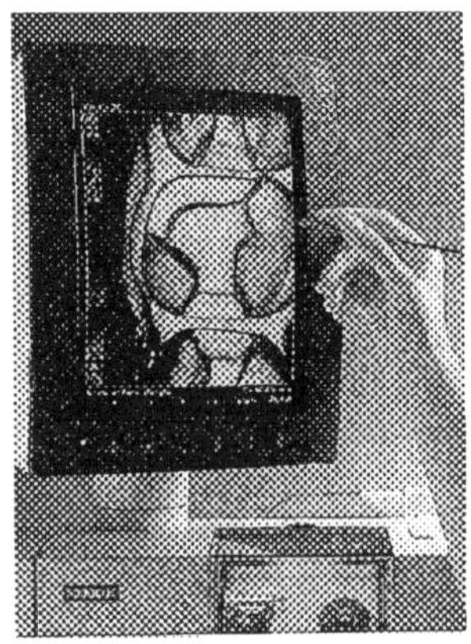

Fig. 1. CEREC-device for dental restorations

In this paper, we present a method for the generation of suitable tooth models. Input data are two-dimensional range images. They can be captured by the CEREC scanner or an another intra-oral 3D scanner. After generation of a range image a corresponding tooth model is chosen out of the tooth library. The development of such a tooth model is described in section 3. By a series of deformations and refinements the model adapts automatically to the scanned tooth. Result is a triangular representation of the occlusal surface. The processing pipeline is shown in figure 2.

This step is necessary for quality estimations and improvements of the developed tooth model, but also as a basis process for an automatic dental restoration system.

Due to the deformable 3D tooth model and the included adaptation process this approach differs from other methods as it is proposed in [9, 10] for instance.

2 Structure of the Tooth Library

There are a lot of dental medical analyses [1, 3, 12–14] dealing with the division of different teeth into only a few numbers of types. Siebert [12] shows that for each tooth group like the first upper or first lower molar only few types are needed to cover all forms occurring in the nature. Our tooth models make a great use of

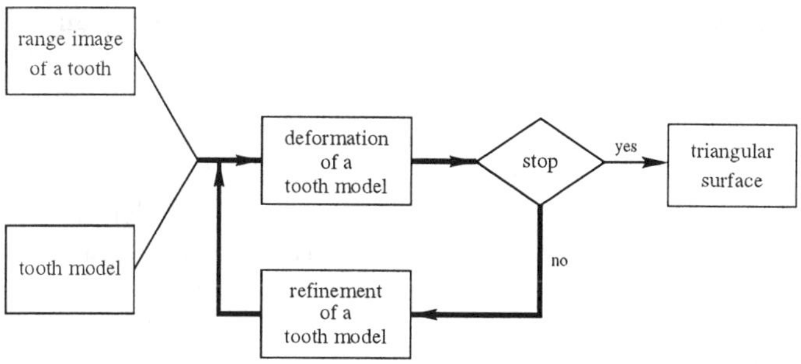

Fig. 2. Occlusal surface restoration using a tooth model (heavy arrows indicate design loop)

this analysis. For each tooth group at least one tooth model is designed. Then, it is guaranteed that our tooth library represents all typical forms occurring in the nature and the task of the adaption process is reduced to fashion the individual tooth anatomy of the patient.

3 The Tooth Model

All the tooth models within the library are built up on the same principle. In this section it is explained how we can derive these main principles and how this information is used for the generation of a tooth model.

The tooth model must fulfill the following demands:

1. It must contain the dental medical knowledge about this tooth type,
2. it must be deformable in an easy way, and
3. it must represent the tooth surface.

3.1 The Geometrically Deformable Model

A model which suffices in particular the above listed demands is the geometrically deformable model (GDM).

Geometrically deformable models were introduced by Miller et al. [2, 7] for the segmentation and visualization of two- and three-dimensional objects. D. Rückert [11] uses geometrically deformable models for the segmentation of two-dimensional medical images.

GDMs present a very intuitive approach to generate models of objects whose shape cannot be described in a simple mathematical form. They can operate on a sparse sampling of the data while still allowing the level of detail extracted to be set by the application. By altering the level of detail, quality models can be generated while still providing a compact representation. This fact is used in the provided adaptation process by reiterating a deformation and a refinement step.

A geometrically deformable model is created by placing an initial simple model in the data set which is then deformed by minimizing a set of constraints. The process starts with a simple non-self-intersecting triangulation of the occlusal surface. Then, the model is deformed based on the set of constraints, so that the model grows and shrinks to fit the scanned tooth. There is an energy function associated with every control point of the GDM determining the local deformation, the properties of the model, and the relationship between the construction features of the tooth. By minimizing these functions the model adapts the scanned tooth automatically.

The dental medical knowledge can leave its remark on the model twice. On the one hand the outer characteristic features of the tooth surface are included in the model by the configuration of the triangulation. On the other hand the relationships between the features are introduced by assigning special energy functions to the control points of the triangulation.

Hence, our proposed tooth model consists of a triangular surface and a set of specially chosen energy functions.

3.2 The Surface of the Model

The crowns of the teeth and especially the occlusal surface consist of hills and valleys. They are the basic elements of the tooth surface [6]. Also the classic technique for producing crowns bases on these elements.

These features are used for the generation of a triangulation of the occlusal surface by assigning each basic element of the tooth some points or lines. These geometrical elements build the frame of the triangulation. Figure 3 shows wireframe representations of the models of the first lower molar and of the first upper molar. Each control point and most edges of the triangulation represent at least one basic element of teeth.

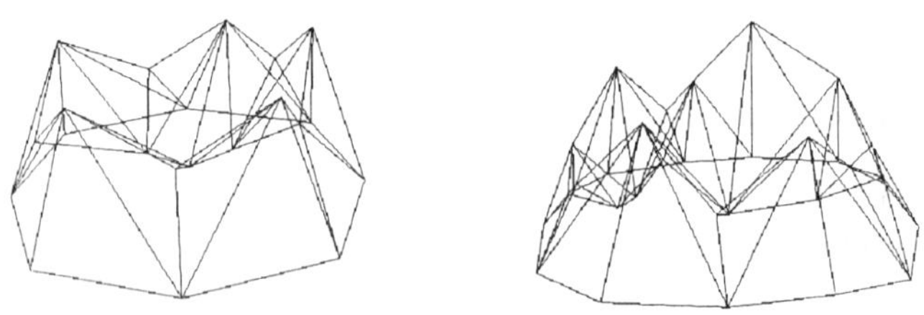

Fig. 3. Model of the first lower and the first upper molar representing main information of the occlusal surface

The extraction of the typical feature points is done by a semi-automatic process. For an undamaged typical tooth the maxima, minima and, saddle points are calculated and entered into the image. After an interactive check of the detected points we get the coordinates of the most important points on the tooth surface. The list is completed by searching a fixed number of points on the equatorial line of the tooth. This can be done by following some automatically detected lines. An example is shown in figure 4.

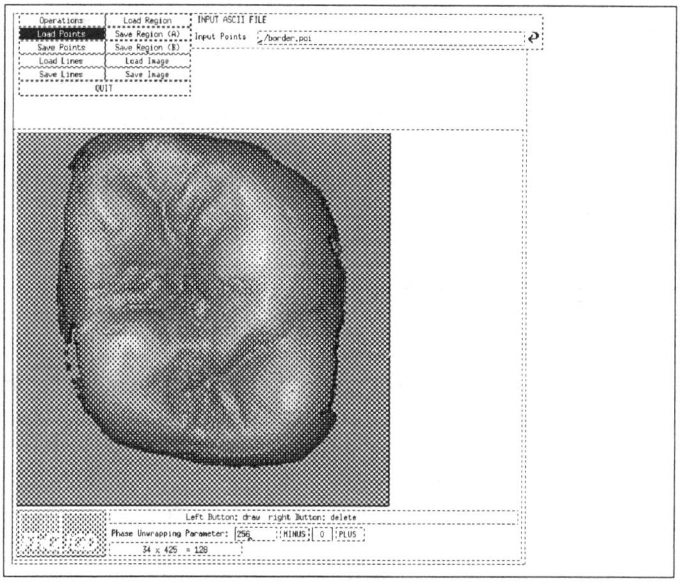

Fig. 4. Semi-automatic extraction of the most important surface points

3.3 The Energy Functions

The energy functions occupy a very important part within the model. Their quality determine the quality of the final restorations.

In contrast to Miller et al. [2, 7] and D. Rückert [11] in our approach every control point *cp* of the GDM gets its own energy function

$$E_{cp} : \mathbb{R}^3 \to \mathbb{R}.$$

They are defined as a weighted sum of given functions $E_1, \ldots, E_n$

$$E_{cp}(x) = \sum_{k \in I_{cp}} w_k E_k(x) \qquad \text{with} \qquad I_{cp} \subset \underline{n}.$$

Each of these functions E_k focuses on a specific point, curve, or surface characteristic. In combination, they represent a measurement for the quality of a point with regard to a particular tooth feature. As each energy function is minimized, the model will deform while searching for the surface of the tooth, and maintaining its topology.

In order to develop a new tooth model we search for each basic element of the teeth, as the cusp tips or the central fossa, a precise characterization in terms of the differential geometry and image processing. For example, the equatorial line is determined by a strong change of the gray values in the neighborhood of the line. Using a derivative filter, we are able to detect these changes. A function for the judgement of local maxima is suitable for the detection of the cusp tips. For the other basic elements it can be found further functions.

There are also functions measuring the distance between a model point and the corresponding image point or for the preservation of the surface topology or a fixed smoothness.

4 The Adaptation Process

To judge the quality of the model we apply it to a series of range images of undamaged teeth of the same type. Figures 5a) and c) show two examples. On the left side you can see the range images of a first lower molar and on the right side the image of a first upper molar. Wireframe representations of the corresponding tooth models are shown in b) and d). Now, the adaptation occurs by interactions between the model and the range image.

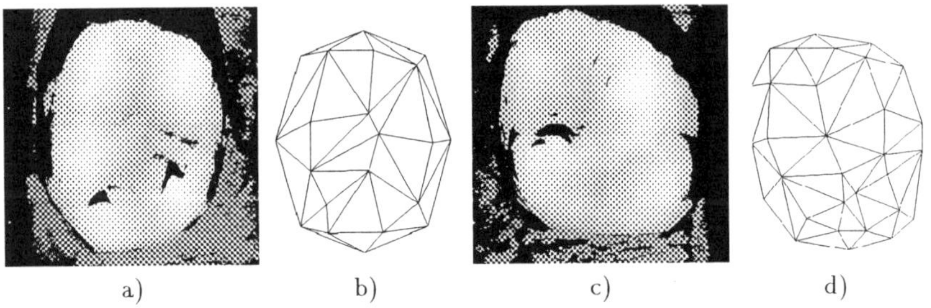

Fig. 5. a) and c) range images of the 1. lower molar and the 1. upper molar, b) and d) the triangular surfaces of the corresponding models

The optimal geometrical approach is realized by two partial processes (see also figure 2):

- In a first step the GDM adapts to the tooth surface by deformations using simulated annealing optimization.
- In a second step the GDM is refined by a subtriangulation process.

These two steps are repeated since a sufficient degree of refinement is reached. The quality of the deformation process is controlled by the above described energy functions. The real deformation takes place by minimizing the energy functions. It has been shown that the segmentation process of Miller [2] is very susceptible to noise and artifacts. The reason lies in the usage of the Hillclimbing process. Rückert provided as an improvement the simulated annealing optimization. It has been shown that this method in theory always finds the global minimum of a function. Due to the enormous complexity of our energy functions we decided the usage of simulated annlealing [4, 5] as our optimization method.

4.1 The Deformation Process

During the deformation process at each temperature T all control points are passed through. If a control point cp is moved to a randomly generated position p is decided by the condition

$$\exp\left(-\frac{E_{cp}(cp) - E_{cp}(p)}{T}\right) < r,$$

where $r \in [0, 1]$ is a random number. A violation of the topology is reduced by the usage of suitable energy functions. The process terminates, if a negligible impovement of the total error is reached.

After the deformation step the whole process stops if the triangulation has reached a sufficient degree of refinement.

4.2 The Refinement Process

Otherwise we subtriangulate the model. Therefore, we are able to reconstruct finer structures of the scanned tooth. We distinguish between four cases. They are shown in figure 6.

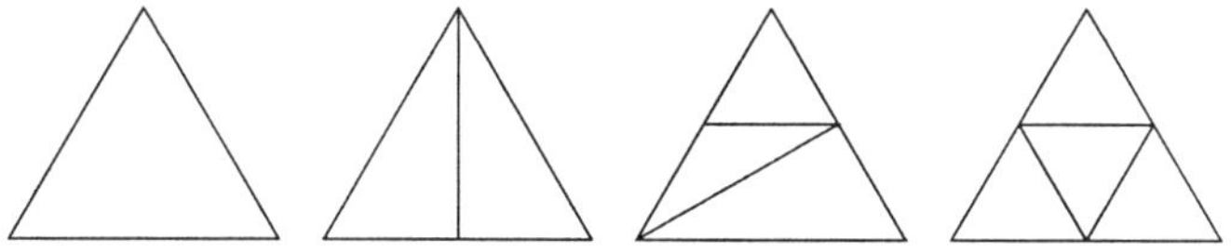

Fig. 6. Possible cases of subtriangulation

As described, the adaptation process is controlled by minimizing of energy functions. After subtriangulation we have to assign suitable energy functions to the new control points. Since the new control points are always in the middle of the old edges they have the common features of the two vertices. Hence, the new control points get the weighted sum of the common functions E_k of the two vertices.

5 Results

In this section we show the results of the adaptation process which judge our generated tooth model. Therefore we introduce two criteria:

- a numerical criterion which rests on the mean distance of a control point to the image and describes the precision,
- a visual criterion which shows the behavior of the construction features within the adaptation process.

The first criterion is a necessary condition of the reconstruction process. A very good adaptation of the occlusal construction features is requirement for a successful implementation of a tooth restoration. This can be judged with the second criterion.

The dental clinic in Munich has put a lot of tooth images at our disposal. We use these images for the development of the tooth models and for the tests of our algorithms. The used 3D-scanner provides range images with a resolution of 30x30x1 μm in x-, y- and z-direction.

Figure 7 shows the results of the deformations in different resolutions starting with the initial model and terminating after four refinement steps with a mean error of 1–2 μm calculated in the vertices of the final triangulation. The resulting triangulation is modeled by 4261 triangles. The whole process has taken 62 seconds on a SPARC 10 workstation. Typical computing times were 60–80 seconds for the calculation of four respectively five refinement steps.

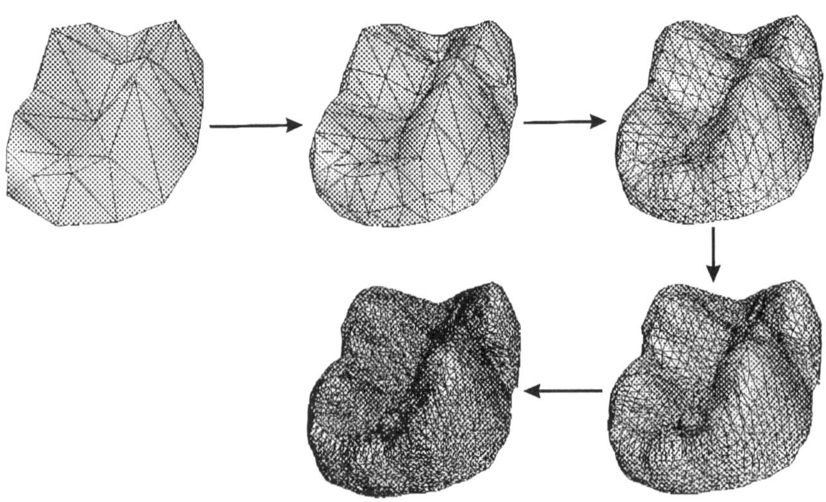

Fig. 7. Result of an adaptation process in form of triangular meshs

Figures 8 and 9 show the the corresponding changes of the range image in the different resolution steps. At the end of the chain we can see the original scanned tooth image for making a comparism. We recognize an effect of smoothing as it is visible particularly in the fissures.

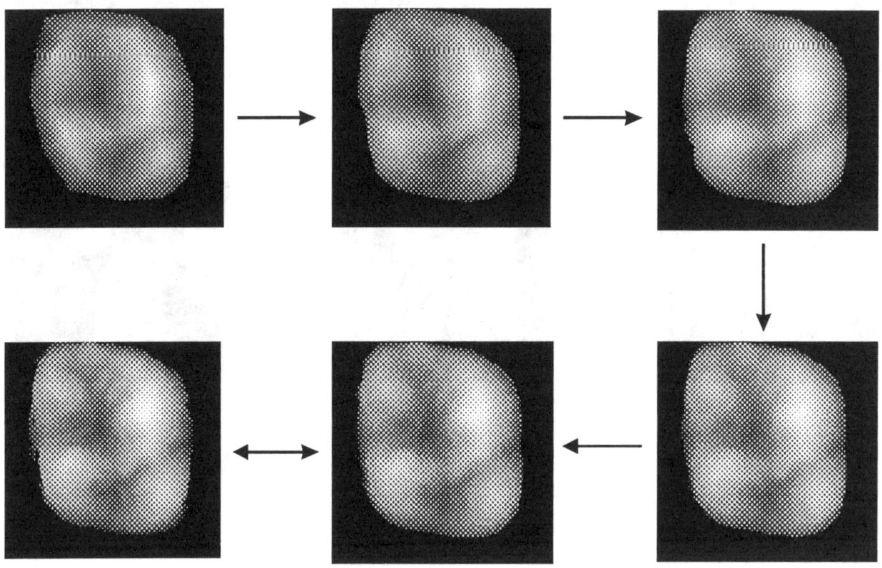

Fig. 8. Result of an adaptation process in form of range images

Figure 9 shows the difference images. Following the images counterclockwise we ascertain an decrease of the distance between the original range image (at the end of the series) and the range images in the different resolutions of the adaptation process. The higher distance values at the border of the tooth are caused by the given range images (see figure 5).

Figure 10 shows the use of the second criterion. The control points of the resulting triangulation which belong to a chosen construction feature as the equatorial line or the cusp tips are entered the rendered range image. Now we are able to judge the quality of the adaptation in the construction features of the tooth. An abnormal behavior as the slipping off the cusp line to the equatorial line is ascertainable in an easy way. However, at the same time the result of the reconstruction shows no losses of accuracy.

6 Conclusion

In this paper, we presented a method for the generation of suitable tooth models for dental restorations. The approach includes an adaptation process which can

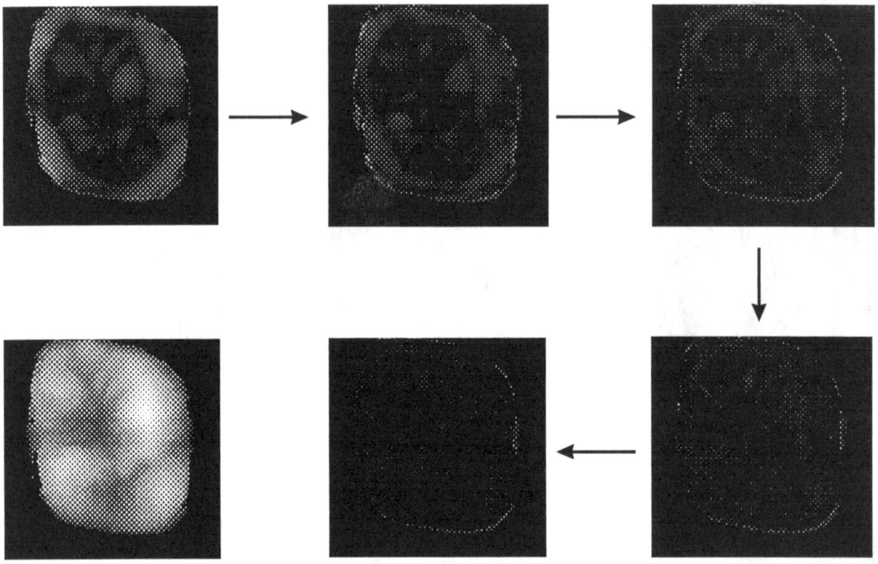

Fig. 9. Result of an adaption process in form of the difference images

be extended to an automatic tooth restoration system. An important topic for future research is the development of methods which lead to a restoration of the destroyed occlusal surface. Therefore, during the adaptation the points above the cavity are detected. A change of their behavior of adaptation will lead to a closure of the cavity according to the dental medical knowledge of the model.

Acknowledgement

I wish to express my thanks to G. Saliger from the Siemens AG in Bensheim, and K.-H. Kunzelmann from the dental clinics of the University Munich for their assistance and preparation of images. Special acknowledgement is due to P. Neugebauer and D. Barnickel, who have made a lot of useful ideas, comments, and suggestions.

References

1. L. Abrams, R.E. Jordan, and B.S. Kraus. *Dental anatomy and occlusion.* The Williams and Wilkins Company, Baltimore, 1969.
2. D.E. Breen, W.E. Lorensen, J.V. Miller, R.M. O'Bara, and M.J.Wozny. Geometrically deformed models: Method for extracting closed geometric models from volume data. *Computer Graphics*, 25(4):217–226, July 1991.
3. T.E. De Jonge Cohen. *Mühlreiters Anatomie des menschlichen Gebisses.* Verlag Arthur Felix, Leipzig, fourth edition, 1920.

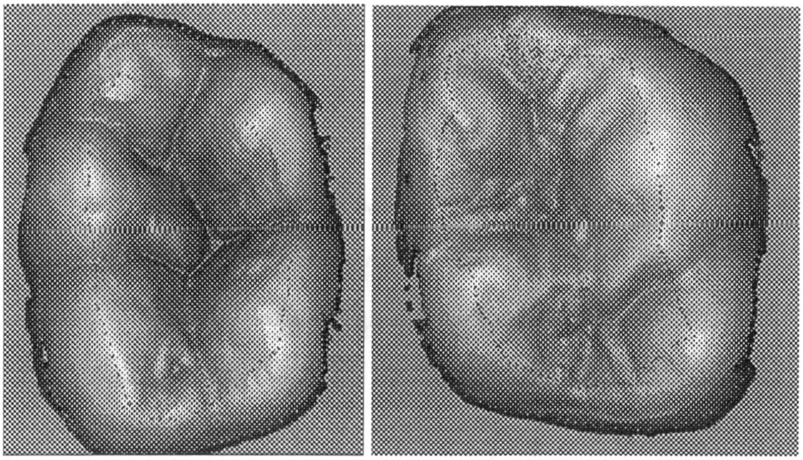

a) b)

Fig. 10. Construction features of the model after the adaption process: a) 1. lower molar, b) 1. upper molar

4. S. Geman and D.Geman. Stochastic relaxation, gibbs distribution, and the bayesian restoration of images. *IEEE Transaction on Pattern Analysis and Machine Intelligence*, 6(2):721–741, 1984.
5. L. Ingber. Simulated annealing: Practice versus theory. *Mathl. Compu. Modelling*, 18(11):29–57, 1993.
6. William H. McHorris. *Einführung in die Okklusionslehre.* Quintessenz Verlag, Berlin, 1983.
7. J.V. Miller. On gdm's: Geometrically deformed models for the extraction of closed shapes from volume data. Master's thesis, Rensselaer Polytechnic Institute, Troy, New York, December 1990.
8. W.H. Mörrmann and M. Brandestini. *Die CEREC Computer Rekonstruktion.* Quintessenz-Verlag, 1989.
9. K. Myszkowski, V.V. Savchenko, and T.L.Kunii. Computer modeling for the occlusal surface of teeth. In *Computer Graphics International'96*, Pohang, Korea, June 1996. IEEE Computer Society.
10. P.J. Neugebauer. Generating CAD-models of Teeth. In *Proceedings of the Computer Assisted Radiology CAR'96 Conference*, Paris, 26.-29. June 1996. Elsevier Science.
11. D. Rückert. Bildsegmentierung durch stochastisch optimierte Relaxation eines 'geometric deformable model'. Master's thesis, TU Berlin, 1993.
12. G. Siebert. Zur Morphologie der Zähne unter funktionellen Aspekten. *Österreichische Zeitschrift für Stomatologie*, 80:139–147, April 1983.
13. G. Wetzel. *Lehrbuch der Anatomie für Zahnärzte und Studierende der Zahnheilkunde.* Fischer Verlag, Jena, sixth edition, 1951.
14. R.C. Wheeler. *An atlas of tooth form.* W.B. Saunders Company, Philadelphia, fourth edition, 1969.

An Investigation on Parameter Optimization of Selective Laser Sintering Process for Rapid Prototyping Applications

Dr Zhu Chun Bao

CIM Center, German-Singapore Institute, Nanyang Polytechnic
Jurong East, 10 Science Center Rd, Singapore 609079
Email: zhucb@nyp.ac.sg

Leonard Loh, Ng Poey Leong & Jeffrey Chan H C

PE Center, German-Singapore Institute, Nanyang Polytechnic
Jurong East, 10 Science Center Rd, Singapore 609079

Abstract

The quality of solid prototype from the Selective Laser Sintering (SLS) process depends very much upon process parameters selected. The paper describes an investigation on optimal selection of SLS parameters for rapid prototyping application using Taguchi method. The effects of six important parameters, laser power, part temperature, temperature difference between part and piston heaters, laser scan space, layer thickness and powder material composition (e.g., ratio of virgin powder to recycled powder) on part geometrical properties, such as surface finish, dimension accuracy and profile distortion (roundness and curling), have been tested. An L_{16} (2^{15}) orthogonal array was used to accommodate the experiments and to analyze the experimental data.

The paper reports the experiment method used and benchmark designed for part property measurement. Based on statistical regression analysis of the experimental data obtained in the research, the paper indicates the optimal combination of SLS parameters which would result in the desired prototype with respect to several property parameters.

Keywords: Selective laser sintering, Taguchi method, Regression analysis, Rapid prototyping, Parameter optimization.

1 Introduction

Last few years have seen fast development of rapid prototyping (RP) applications. The techniques used in the RP industry are rather complex which involve various aspects of modern scientific and technological developments, such as laser scanning techniques, CAD modeling, CAM, CAD data conversion and transferring, electronic automatic control, etc. Selective Laser Sintering (SLS) is one of the new techniques commonly used today.

The number of factors involved in the SLS process is very large [1] (Zhu CB, Wong D S K, Loh L and Sue H Y, 1996) and some are not thoroughly studied. It is therefore difficult to develop a comprehensive mathematical model for parameter optimization which would result in desired parts. In RP practice, the input conditions for making prototypes are often not optimal. It is particularly true that the best rapid prototyping parts are normally built on a given machine only after multiple attempts, with some tuning or selected process parameters after each iteration. This learning process is very costly and the experience and production expertise gained this way is not systematic, which makes it very difficult to make use of the knowledge for RP industries.

There is a need to investigate the optimal parameter setting for a given rapid prototyping machine and known application requirements. Two major problems should be addressed in order to carry out experimental optimization for the known SLS machine, 1) the number of tests should be small as running the SLS Sinterstation can be very costly and 2) the limited tests should provide enough information for exploring the cause-effect of the SLS process. Taguchi method meets just the above requirements and therefore is used in this research.

2 SLS Part Geometrical Properties

The SLS part geometrical properties can be characteristicsed by a number of parameters, such as deviation of measured dimensions from the desired dimensions determined by a CAD model, dimensional tolerance and trueness of geometrical profile, such as sharpness of corners and edges [2][3], roundness, curling and creep distortions as well as growth. Surface quality of SLS parts can be measured using roughness (R_a), surface texture and aliasing [4]. Considering easy-to-measure nature of these parameters, in this investigation, following are selected for quantification of part geometrical properties.

1) ***Part surface finish R_a***, surface finish of SLS parts may range from 5 to 50 μm[5, 6].

2) ***Dimensions deviation (Δl_i),*** measurement of actual part size of L_i provides data for setting offset/scaling for part dimension compensation due to part shrinkage after or during the sintering process. In the investigation, deviation (Δl_i) of each dimension L_i is determined as follows :

$$\Delta l_i = (L_{io} - L_{im})/ L_{io} \tag{1}$$

Where, L_{io} is design size, L_{im} is the actual dimension measured, normally, it is a mean of multiple measurements on the same dimension.

3) ***Roundness (K),*** K typically represents the trueness of a geometrical feature (circularity). It is defined by

$$K = (\Phi_{max} - \Phi_{min})/ \Phi_m \tag{2}$$

Where, Φ_{max} and Φ_{min} are the maximum and minimum diameter measured on the round element, Φ_m is the mean of the multiple measurements of the diameter.

4) ***Curling (β),*** Curling is a phenomenon that the SLS part's edges or corners rise up (see Fig.1). During the selective laser sintering process, temperature differences can exit in different areas of the part being built. These temperature differences cause uneven shrinkage, which in turn causes curling.

There are two kinds of curling, i) in-build curling and ii) post-build curling. Severe in-build curling may cause the part to shift in the part bed when the roller passes over the part, while post-build curling generates permanent part profile distortion. In this investigation, curling (β) is quantified as follows:

$$\beta = \Delta h/L \quad (3)$$

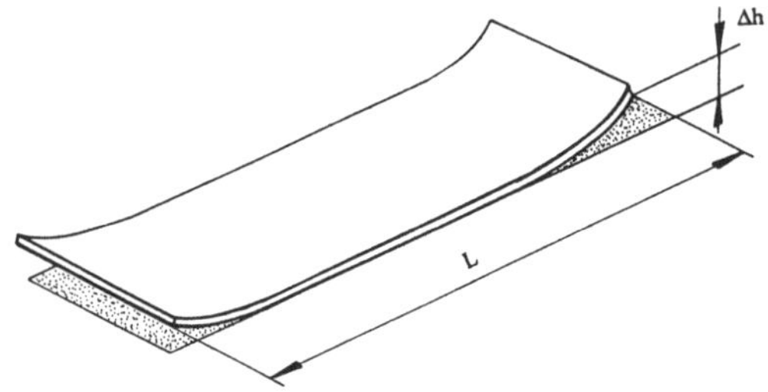

Fig.1. Curling and measurement

3 Description of Experiment Design and Implementation

3.1 Benchmark Design

A benchmark (shown in Fig.2) has been designed to provide a test piece with all the necessary geometrical elements that ease measurement of part geometrical properties, e.g., roundness, curling, dimension accuracy, surface finish and growth. The geometrical elements designed are distributed on and supported by a flat base (2 mm in thickness).

This part contains stepped boxes, the dimensional tolerances in X, Y and Z are characterized, as are other factors, such as cone and cylinder circularity, cone angle errors, edge straightness. Flat areas at different locations and in different directions provide the surfaces for roughness measurements. The long stripes in both X and Y direction are used for curling measurement.

During test, each benchmark is numbered which is correspondent to the trial number of the experiment and each geometrical measurement element is also named so that the data collected could be easily stored and analyzed accordingly.

Fig.2. The benchmark designed for testing SLS part properties

3.2 Control Factors

Although there are a number of parameters affecting SLS process [1], only six parameters (see Table 1) are selected as controllable factors in the research due to their effects on the SLS part geometrical properties. For each factor, two levels are considered in the experiment design.

Table 1 Factor & Level Specification

	Factors	Symbol	Units	Level 1	Level 2
A	Powder material composition (Virgin:recycled)	R	-	1:1	1:2
B	Laser power 9% (3W)	P	% (W)	10% (3.5)	9% (3.0)
C	Part Temperature	T	°C	185	186
D	Temperature difference btw part & piston heaters	ΔT	°C	82	80
E	Scan space	d	mm	0.17	0.15
F	Layer thickness	Δa	mm	0.10	0.11

In Table 1, R is the ratio of virgin powder materials to recycled ones mixed in the build. Factor P (laser power) takes two levels, 3.5 (w) and 3.0 (w) which covers most commonly used power setting range for building prototype with Nylon Composite (Laserite Powder, LNC7000). Two temperature factors (T and ΔT) have been selected to investigate their effects on part profile distortion and shrinkage. The scan space (d) and layer thickness (Δa) are used to test their influence on part geometric accuracy and surface finish.

Some other parameters are fixed, for instances, 1) temperatures for powder material pre-heating are set as 80 °C (Left bed), 80 °C (left bed), 80 °C (Part piston); 2) the slice thickness is set at 0.11 mm.

3.3 Experiment Plan

The L_{16} (2^{15}) array is selected to accommodate the experiment. The factor column design is shown in Table 2. As can be seen from Table 2, the factors A, B, C, D, E and F are assigned into Columns 1, 2, 4, 8, 11, 13 respectively in L_{16} (2^{15}) array. This design is of low resolution [7], in which the main factor effects can be estimated but two-factor-interactions are confounded each other. The interaction confounding that exists in this experiment is noted but ignored as the first round of the test is to try to investigate the main effects of the major factors on the output parameters, and the interactions will be investigated in further research.

Table 2 L_{16} (2^{15}) Array Factor Column Design

Col No	1	2	3	4	5	6	7	8	9	10	11	12	13	14	15
Factors	A	B	A×B	C	B×F	B×C	e_1	D	B×E	B×D	E	C×D	F	e_2	B×F

Table 3 is the layout of the experiment plan. As can be seen from Table 3, the powder material composition (the ratio of virgin powder to recycled powder) is assigned to Column 1 as this parameter is the most difficult one to alter. The rest factors are randomly assigned into main effect columns 2, 4, 8, 11, 13 accordingly. Column 7 and 14 are used for error evaluation [8, 9].

Table 3 Experiment Plan and Sample Data (Part Curling Test)

Trial No	Col No						β_{Xi}	β_{Yi}
	1 (A)	2 (B)	4 (C)	8 (D)	11 (E)	13 (F)		
	Factor							
	R (V:R)	P (W)	T (^oC)	ΔT (^oC)	d (W)	Δa (mm)		
1	1: 1:1	1: 10%	1: 185	1: 82	1: 0.17	1: 0.10	0.395	1.113
2	1: 1:1	1: 10%	1: 185	2: 80	2: 0.15	2: 0.11	0.017	0.042
3	1: 1:1	1: 10%	2: 186	1: 82	1: 0.17	2: 0.11	0.034	0.170
4	1: 1:1	1: 10%	2: 186	2: 80	2: 0.15	1: 0.10	0.387	0.622
5	1: 1:1	2: 9%	1: 185	1: 82	2: 0.15	1: 0.10	2.794	5.083
6	1: 1:1	2: 9%	1: 185	2: 80	1: 0.17	2: 0.11	0.034	0.085
7	1: 1:1	2: 9%	2: 186	1: 82	2: 0.15	2: 0.11	0.017	0.042
8	1: 1:1	2: 9%	2: 186	2: 80	1: 0.17	1: 0.10	1.535	3.525
9	2: 1:2	1: 10%	1: 185	1: 82	2: 0.15	2: 0.11	0.068	0.643
10	2: 1:2	1: 10%	1: 185	2: 80	1: 0.17	1: 0.10	0.051	0.556
11	2: 1:2	1: 10%	2: 186	1: 82	2: 0.15	1: 0.10	0.017	0.684
12	2: 1:2	1: 10%	2: 186	2: 80	1: 0.17	2: 0.11	0.102	0.512
13	2: 1:2	2: 9%	1: 185	1: 82	1: 0.17	2: 0.11	0.378	0.171
14	2: 1:2	2: 9%	1: 185	2: 80	2: 0.15	1: 0.10	0.017	0.001
15	2: 1:2	2: 9%	2: 186	1: 82	1: 0.17	1: 0.10	0.239	0.042
16	2: 1:2	2: 9%	2: 186	2: 80	2: 0.15	2: 0.11	0.342	0.042

3.4 Experiment Implementation

The experiments were carried out in Rapid Prototyping Laboratory of Precision Engineering Center at German-Singapore Institute, Nanyang Polytechnic. All experiments are implemented using the SinterStation^{TM} 2000 System to explore the effect of the control variables on part geometrical properties. The major characteristics of the system are listed as follows:

Laser	- 25- or 50 watt CO_2CW laser
Max build envelope	- 305 mm in diameter and 381 in height (for Wax Polycarbonate)
	- 203 ~ 254 mm in diameter and 381 in height (for Nylon Composite/Nylon)
Nominal slice thickness	- 0.13 ~ 0.18 mm
Tolerances	- 0.38 mm (for initial part build accuracy)
	- 0.13 (for optimized parameter settings)

4 Analysis of Experiment Data and Discussion of the Results

4.1 Experiment Data and Discussion

A large amount of data are collected from the investigation and systematically analyzed using orthogonal regression method [8, 9]. Here, as an example, two sets of measured data on the part curling distortion is listed in Table 3, where, β_{Xi} is the curling measured on the part in X direction, while β_{Yi} is measured in Y direction. β_{Xi} and β_{Yi} are mean values of multi-readings taken from the parts. Same methods have been used to take readings for other geometrical measurements, e.g., part surface finish R_a(μm), dimensions L_i (mm) and roundness (K). Typical graphical representation of level average analysis of the test results is given in Fig.3, where the effects of various SLS parameters on part geometrical properties can be clearly viewed. Within each effect analysis diagram (e.g., curling, roundness, dimensions and roughness), two sets of data are presented, and designations of the data are also given in the diagram.

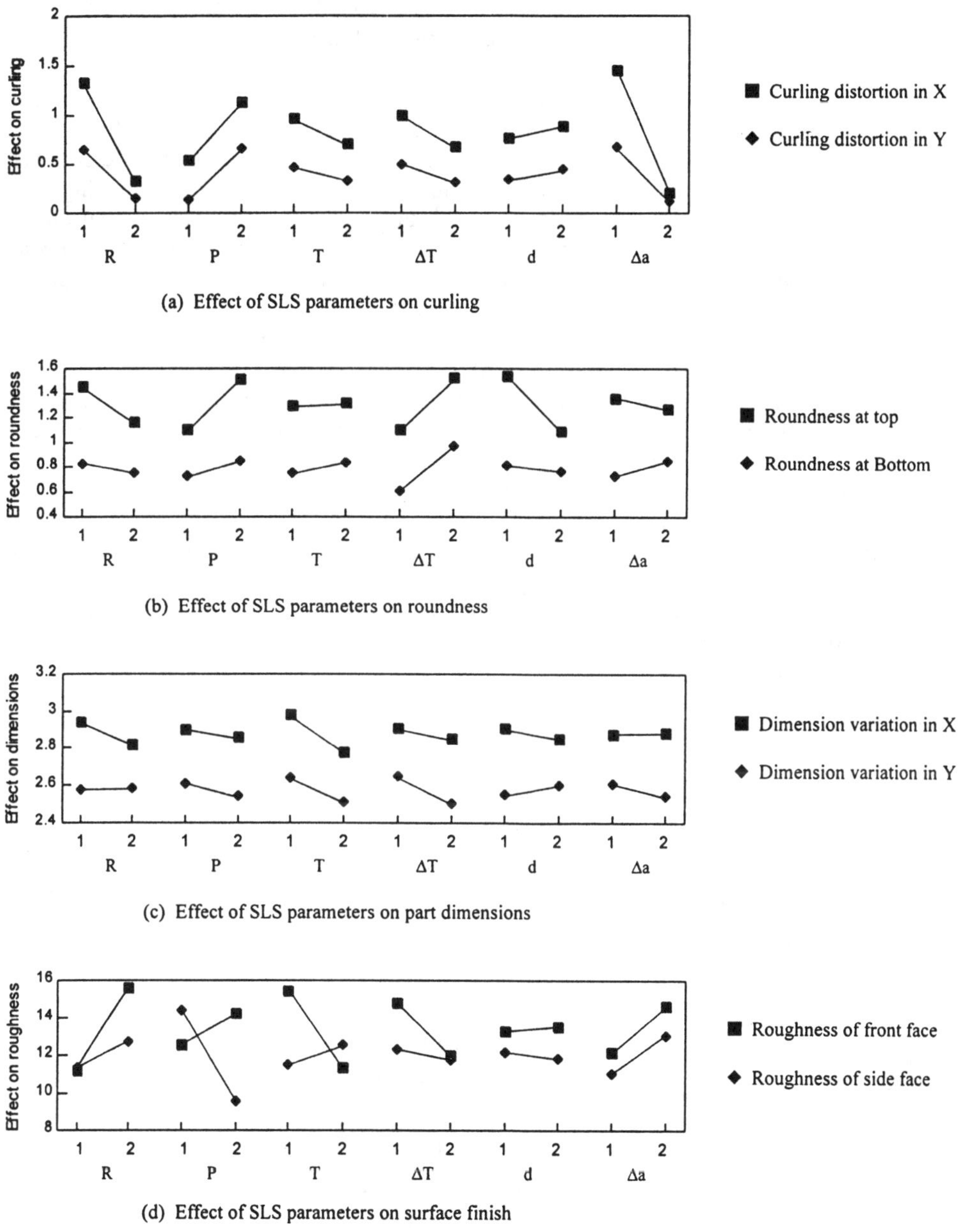

(a) Effect of SLS parameters on curling

(b) Effect of SLS parameters on roundness

(c) Effect of SLS parameters on part dimensions

(d) Effect of SLS parameters on surface finish

Fig.3. Response of part geometrical properties via the major process parameters

As can be seen from Fig.3, R at level 2 (e.g. 1:2, more recycled powder used) generates low curling distortion, better roundness and less dimension variation. But it results in poorer surface finish (Fig.3 (d)). Using higher laser power (Level I: 3.5 W) comparatively improves curling, roundness, and results in better surface finish on the front surface of the part, but it will generate poorer roughness on the side face due to the side-effect of growth which normally is caused by high temperature and high laser power. Higher laser power also makes part dimension variation larger (see Fig.3 (c)).

Temperature is always an important parameter to consider in SLS process. From Fig.3, it seems that the part temperature (T) and the temperature difference (ΔT) between part and piston heater have similar effects on various part property parameters. Refer to the factor and level specification in Table 1 with Fig.3, we can see that increasing part temperature will decrease the part curling and dimension distortion and improve front face's roughness, however, as mentioned earlier, higher T will cause poorer surface finish on the side surface. On the other hand, higher value of ΔT makes dimension and curling distortion high and front face roughness poorer, but improves roundness (Fig.3 (c)).

The effects of scan space (d) and layer thickness (Δa) on surface finish are very much similar, as can be seen from Fig.3(d), both larger scanning space (0.17 mm) and thicker layer thickness (0.11) result in poorer surface finish and roundness. Scan space d has no significant effect on curling distortion, but thin layer thickness may cause the heat from the previous layer(s) below to increase the temperature on the current layer and therefore generate higher curling distortion.

It should be noted that the benchmark pieces were built without any geometrical compensation and that the dimension variation (see Fig.3 (c)) caused by various factors can be compensated during machine set-up using offset for laser beam size and scaling for shrinkage. The results show that the X has an average dimensional inaccuracy of 2.75 mm and Y 0.76 mm. What needs to be determined is the effect of each parameter on the dimensional accuracy.

4.2 Analysis of Variance and Factor Significance

Following table shows the results of analysis of variance against curling (in X direction).

Table 4 Analysis of Variance

Source	f	S	V	S'	F	ρ (%)
A: Powder composition (R)	1	0.9995	0.9995	0.9258	13.567 **	12.2
B: Laser power (P)	1	1.1476	1.1476	1.0739	15.577 **	14.1
C: Part Temperature (T)	1	0.0730	0.0730			
D: Temperature difference (ΔT)	1	0.1327	0.1327	0.059	1.8009	0.7
E: Scan space (d)	1	0.0496	0.0496			
F: Layer thickness (Δa)	1	1.2338	1.2338	1.1601	16.747 **	15.3
CxD, AxF	1	1.9425	1.9425	1.8689	26.376 **	24.6
CxE, BxF	1	0.6340	0.6340	0.5603	8.6058 *	7.38
e - pooled	3	0.221	0.0737			
Total	15	7.5909	0.4744			100

Note: * At least 90 % confidence, ** At least 95 % confidence, *** At least 99 % confidence.

Table 5 is the summary of analysis of variance for all four part property parameters studied in the investigation, which shows how significant each factor affects the output.

Table 5 Summary of Analysis of Variance for Various Factors

Source	Curling		Roundness		Dimensions		Roughness	
	X	Y	Top	Bottom	X	Y	Front	Side
A: Powder composition (R)	**	**	**		**		***	*
B: Laser power (P)	**	**	***					***
C: Part Temperature (T)					***	**	***	
D: Temperature difference (ΔT)			***	**		**	**	
E: Scan space (d)			***					
F: Layer thickness (Δa)	**	**					**	**

Based on the investigation, for a given geometrical property criterion, optimal parameter settings can be obtained (see Table 6). Following are two examples:

(1) The optimal setting for good prototype profile and dimension accuracy is: A_2 (R =1:2), B_1 (P=3.5W), C_2 (T = 186 ^oC), D_2 (ΔT =82 ^oC) , E_2 (d = 0.15mm) , F_2 (Δa = 0.11).

(2) The optimal conditions for good part surface finish : A_1 (R = 1:1) , B_2 (P = 3.0W), C_2 (T = 186 ^oC), D_2 (ΔT =82 ^oC) , E_2 (d = 0.15mm) , F_1 (Δa = 0.10).

Table 6 Optimal Parameter Settings

Level / Goal	Curling		Roundness		Dimensions		Roughness	
Source	X	Y	Top	Bottom	X	Y	Front	Side
A: Powder composition (R)	A_2 ⊗	A_2	A_2	A_2 ⊗	A_2	A_2 ⊗	A_1	A_1
B: Laser power (P)	B_1	B_1	B_1	B_1 ⊗	B_2 ⊗	B_2 ⊗	B_1 ⊗	B_2
C: Part Temperature (T)	C_2 ⊗	C_2 ⊗	C_1 ⊗	C_1 ⊗	C_2	C_2	C_2	C_1 ⊗
D: Temperature difference(ΔT)	D_2 ⊗	D_2 ⊗	D_1	D_1	D_2 ⊗	D_2	D_2	D_2 ⊗
E: Scan space (d)	E_1 ⊗	E_1 ⊗	E_2	E_2 ⊗	E_2 ⊗	E_1 ⊗	E_1 ⊗	E_2 ⊗
F: Layer thickness (Δa)	F_2	F_2	F_2 ⊗	F_1 ⊗	F_1 ⊗	F_2 ⊗	F_1	F_1

⊗ Effect of the factor is not significant.

4.3 Regression Analysis of Experiment Data

The main effects of the SLS output parameters normally can be expressed as functions of input operation parameters in the form of empirical/regression formulae [8], namely, in following format:

$$Y = C_o[R]^{n1}[P]^{n2}[T]^{n3}[\Delta T]^{n4}[d]^{n5}[\Delta a]^{n6} \tag{4}$$

Equation (4) provides a basis for regression analysis of the experiment data. The data collected from the experiment investigation can also be expressed using following matrix:

$$Y = X\beta + \varepsilon \tag{5}$$

Where, Matrix X is constructed by the factors A, B, C, D, E and F, each varying with two levels. it can be coded using following equations:

$$x_{i,j} = \frac{2\left(Log\Psi_{ij} - Log\Psi_{imax}\right)}{Log\Psi_{imax} - Log\Psi_{imin}} \tag{6}$$

In Equation (6), $\Psi_{i,j}$ (i =1,2,3,4,5,6) represents the Factor A, B, C, D, E, F respectively, $x_{i,j}$ (i=1,2,3,4,5,6) represents the level code for each level. The subscript, j, can be either 1or 2. When j=1, $\Psi_{i,1}$ takes upper level (e.g., Ψ_{max}), when j=2, $\Psi_{i,2}$ takes lower level (e.g., Ψ_{min}).

Computing with the above Equation (6) and the factor's level specifications in Table 1 results in following level coding table (Table 7).

Table 7 Factor's Level Coding

	A: (R)	B: (P)	C: (T)	D: ΔT	E: (d)	F: (Δa)	Code
Samples	x_1	x_2	x_3	x_4	x_5	x_6	
Upper level	1:1	3.5	186	82	0.17	0.11	+1
Lower level	1:2	3.0	185	80	0.15	0.10	-1
Gap	0.5	0.5	1.0	2.0	0.02	0.01	

The coefficients matrix β in Equation (5) can be calculated using following regression analysis model:

$$\beta = (X^TX)^{-1} (X^TY) \tag{7}$$

As an example, the calculation results using Equation (7) for prediction of roundness (K) are listed in Table 8.

Table 8 Regression Analysis of Experiment Data (Roundness at top)

S/N \ Code	x_0	x_1 A: R	x_2 B: P	x_3 C: T	x_4 D:ΔT	x_5 E: d	x_6 F: Δa	Roundness k_i	y_i =log (k_i)
1	+1	+1	+1	-1	+1	+1	-1	1.391	0.1433
2	+1	+1	+1	-1	-1	-1	+1	0.604	-0.2190
3	+1	+1	+1	+1	+1	+1	+1	0.932	-0.0306
4	+1	+1	+1	+1	-1	-1	-1	1.764	0.2465
5	+1	+1	-1	-1	+1	-1	-1	1.232	0.0906
6	+1	+1	-1	-1	-1	+1	+1	1.875	0.2730
7	+1	+1	-1	+1	+1	-1	+1	1.103	0.0425
8	+1	+1	-1	+1	-1	+1	-1	2.760	0.4409
9	+1	-1	+1	-1	+1	-1	+1	1.217	0.0853
10	+1	-1	+1	-1	-1	+1	-1	1.556	0.1920
11	+1	-1	+1	+1	+1	-1	-1	0.288	-0.5406
12	+1	-1	+1	+1	-1	+1	+1	1.149	0.0603
13	+1	-1	-1	-1	+1	+1	+1	1.645	0.2162
14	+1	-1	-1	-1	-1	-1	-1	0.867	-0.0619
15	+1	-1	-1	+1	+1	+1	-1	1.003	0.0013
16	+1	1-	-1	+1	-1	-1	-1	1.656	0.2190
X^TY	1.1589	0.8158	-1.2843	-0.2799	-1.1428	1.4349	0.1348		
X^TX	16	16	16	16	16	16	16		
$(X^TX)^{-1}(X^TY)$	0.0724	0.0510	-0.0803	-0.0175	-0.0714	0.0896	0.0084		

Equation (4) can be expressed as its Log format,

$$\log(Y) = \log(C) + n_1\log(R) + n_2\log(P) + n_3\log(T) + n_4\log(\Delta T) + n_5\log(d) + n_6\log(\Delta a) \tag{8}$$

Equation (8) can be simplified as :

$$y = b_0 + b_1x_1 + b_2x_2 + b_3x_3 + b_4x_4 + b_5x_5 + b_6x_6 \tag{9}$$

The coefficients $(b_0, b_1, b_2, b_3, \ldots b_6)$ in the above equation (9) can be calculated using Equation (6) and experiment data (y_i) listed in Table 8, the result is shown below:

$$y = 0.0724 + 0.051x_1 - 0.08x_2 - 0.0175x_3 - 0.0714x_4 + 0.896x_5 + 0.0084x_6 \tag{10}$$

Using level coding equations (6), following regression equation can be generated finally

$$K = 2.556 \times 10^{4} R^{0.33876} P^{-0.14967} T^{-0.01543} \Delta T^{-0.075} d^{3.2928} \Delta a^{0.4071} \tag{11}$$

The same procedure has been followed to compute regression formulae for prediction of other geometrical properties, e.g., curling, dimension variation and surface finish. Following table (Table 9) lists all the constants C_0 and exponential coefficients n_i ($i = 1 \sim 6$, see Equation 4).

Table 9 Summary of Constants and Exponential Coefficients

Prediction / Coefficient	Curling		Roundness		Dimensions		Roughness	
	X	Y	Top	Bottom	X	Y	Front	Side
C_0	1.33E-8	1.61E-7	2556.51	17.041	4.1537	1.0995	388.24	166.72
n_1	0.76446	1.6289	0.3387	-0.05435	0.06068	-0.0029	-0.4174	-0.1193
n_2	-0.3950	0.5251	-0.14967	0.09185	0.00767	0.01034	-0.0501	0.15728
n_3	0.02947	0.04928	-0.01544	0.01059	-0.0137	-0.00896	-0.04831	0.01611
n_4	0.05301	0.20589	-0.07506	-0.09549	0.00417	0.01254	0.04038	0.00867
n_5	5.04136	7.0063	3.29278	0.19604	0.14434	-0.12360	-0.1322	0.2315
n_6	-11.738	-10.620	0.40708	0.98760	0.03937	-0.26563	1.5469	1.3344

5 Conclusions

An experimental investigation has been carried out to explore the relationship between the input parameter and the properties of the SLS parts with an attempt to optimize the input parameters for a given rapid prototyping application. As the behavior of the SLS process on output varies significantly based on what input is selected for building the SLS parts, this results in a huge number of variables that need to be considered in the experimental investigation. To minimize the amount of tests, the Taguchi Method was used to design the experimental plan and to analyze the experimental data. This provides more efficient and less time-consuming method of searching for the ideal parameter setting for rapid prototyping applications.

By doing an analysis of variance for each input property parameter, the significance of the parameter on the output parameters was estimated. From these estimations, an optimal combination of input parameter settings was obtained which resulted in desirable SLS parts. The optimal parameter combinations were found to conform to known trends in selecting parameters. It should be noted that the results are by no means the best setting possible, but rather the optimal setting for the input parameters chosen in the research.

The regression analysis presents the experiment data in the form of empirical formulae which would help to predict possible output with a satisfactory degree of accuracy although the formulae obtained may need to be improved in further research.

6 List of References

1. Zhu CB, Wong D S K, Loh L and Sue H Y, *Parameter interactive architectures and Fuzzy sets analysis of SLS process for optimising rapid prototyping applications,* the proceeding of the seminar of CAD/CAM for Rapid Product Development, 23-24, May, 1996, Singapore, pp102-114.

2. Christian Nelson, Kevin McAlea, Damien Gray, *Improvements in SLS part accuracy,* proceedings of Solid Freeform fabrication Symposium, 9-11 August, 1993, The University of Texas at Austin, Austin, Texas, USA.

3. Uday Lakshminarayan, Christian Nelson and Kevin McAlea, *Rapid prototyping accuracy: a view of recent results,* proceedings of Solid Freeform fabrication Symposium, 9-11 August, 1993, The University of Texas at Austin, Austin, Texas, USA.

4. David Thompson and Richard Crawford, *Optimizing part quality with orientation*, the proceedings of Solid Freeform fabrication Symposium, September, 1995, The University of Texas at Austin, Austin, Texas, USA, pp362-368.

5. Lakshminarayan U, McAlea K, and Booth R, *Manufacture of iron-copper composite parts using Selective Laser Sintering (SLS^{TM}),* the proceedings of Solid Freeform fabrication Symposium, 9-11 August, 1993, The University of Texas at Austin, Austin, Texas, USA.

6. Irem Tumer, David Thompson, Richard Crawford and Kristin Wood, *Surface characterization of polycaobonate parts from Selective Laser Sintering,* the proceedings of Solid Freeform fabrication Symposium, September, 1995, The University of Texas at Austin, Austin, Texas, USA, pp181-188.

7. Phillip J. Ross, *Taguchi Techniques for Quality Engineering, Loss Function, Orthogonal Experiments, Parameter and Tolerance Design,* McGraw-Hill Book Company, New York, USA.

8. Zhu CB, An investigation on grinding power in creep feed grinding using the method combining dimension analysis with orthogonal regression, *Journal of Northeast Univ. of Technology,* December, 1983, No.37, p67-76.

9. Genichi Taguchi, *System of Experimental Design, Engineering Methods to Optimise Quality and Minimise Costs* (Volume One, Volume Two), Unipub by Kraus International Publications, White Plains, New York, and American Supplier Institute, Inc., Dearborn, Michigan, USA, 1989.

CE Environments

Management and Scheduling of Design Activities in Concurrent Engineering*

Qian Zhong Yingping Zheng
The Institute of Automation, Chinese Academy of Sciences, Beijing, P.R.China

Abstract: The objective of management and scheduling of design activities in concurrent engineering (CE) is to improve the concurrence of design process and minimize the total design time. In this paper, firstly, some pecularities of design activities in CE are analyzed, and the limitations of conventional project management techniques are presented. Then a new feasible and systematic approach for management and scheduling of concurrent design activities is developed, and two kinds of scheduling problem are formulated and their resolutions are given.

1. Introduction

Concurrent Engineering has probed to be a valuable tool for maintaining competitiveness in today's ever changing and expanding world market by providing a systematic approach to the integrated, concurrent design of products and their related processes. The main objective of CE is to shorten the product life cycle, improve the product performances and reduce producation costs[1].

The design process involves a number of design activities and their inter-connections, a limited amount of available resources restrain the simultaneous execution of these activities. Scheduling of these design activities is a very important step to improve the concurrence of design process and shorten the total design time[2]. Some conventional techniques, such as critical path methods (CPM) and program evaluation and review technique(PERT)[3] have been successfully adopted for different types of projects. However, these techniques need to be extended to resolve the management and scheduling problem in CE due to the fact that the concurrent design activities have some special characteristics, such as uncertainty, randomness, concurrence and repetitiveness.

There have been some work of scheduling of concurrent manufacturing projects[4,5,6,7,8], however, it is still lack of a systematic formulation and techniques to this problem. The main purpose of this paper is to present a feasible and systematic approach for efficient management and scheduling of design activities in CE. According to our understanding of CE, we should use a global view to schedule the design activities. Specially, we emphasize the resource and information factors, because in concurrent design process, whether or not one activity can be executed depends on the available resources provided by the system and the available information delivered by the preceding activities.

In the following sections, on the base of new pecularities of design

* Supported by National 863 High-tech Project Foundation and National Natural Science Fund

activities in CE, an improved and systematic approach is proposed, including the extended method to establish the network of design activities, the more feasible assessment of time distributions, the formulation of scheduling problem and the global optimal scheduling rules.

2. New Pecularities of Design Activities in CE

CE differs from the traditional sequential product development by considering the whole concerns in product life cycle at early design stage and dispatching the activities that are performed according to precedence constraints to be executed simultaneously as many as possible[9]. There are some new pecularities of design activities in CE, such as:

1) CE allows the serial, overlaping or simultaneous execution of activities according to the technical constraints, procedural constraints and information availability. If represented in precedence relationships among activities, there are three types: lead, lag and overlap. But in CPM and PERT, there is only one type of precedence relationship among activities: lead relationship.

2) A main characteristic of CE is to evaluate the design results at any time and correct the design defects timely to obtain good performances, which will cause some activities to be repeated. And the results of evaluation are stochastic, so the processing time of them are not easy to be predicted in advance. Thus it is necessary to use the dynamic scheduling policies in CE. However, in CPM and PERT, the decision are taken in a static mode.

3) The available resources are limited in design process, the simultaneous execution of activities will cause resource conflicts, thus how to assign resources to activities is very important in CE. But, CPM and PERT assume that the relationship between the resources and duration of activities is known a priori for all activities.

4) In CE, the execution of succeeding design activities is based on the information provided by the preceding activities. The most important logic relationship among activities is information dependence relationship. But CPM and PERT methods do not consider the information dependence relationships.

Therefore, from the four respects mentioned above, the conventional techniques need to be extended to satisfy the specific and unusual project scenarios of CE.

3. Model of Design Activities

The basis for management and scheduling of activities is to build the model of design activities to express relationships among activities clearly and evaluate the time distribution of activities accurately.

3.1 Network of design activities

To represent the complex relationships among activities more clearly, such as time order and information dependence, we propose a simple method---extended activity-on-node(E-A-O-N) to establish the network of design activities.

E-A-O-N can be represented as a directed gragh G = (N, A, C), which is developed from A-O-N method. N is a set of nodes, each node indicates a design activity in design process. A is a set of arcs, the arrow of arc drawn from activity 'i' to activity 'j' indicates that the activity 'j' depends on the information provided by the activity 'i'. C is a set of weights of arcs, indicating the time order between the activities. $C_{ij} \in C$, $(i,j) \in N$, $C_{ij} \geq 0$ means that activity 'j' must be executed after activity 'i' finishes with the lag of C_{ij} , $C_{ij} < 0$ means that activity 'i' and activity 'j' has an overlap of $|C_{ij}|$ (see in figure 1).

Adding the weights of arcs in A-O-N can represent the time order and information dependence relationships among the design activities more clearly and accurately.

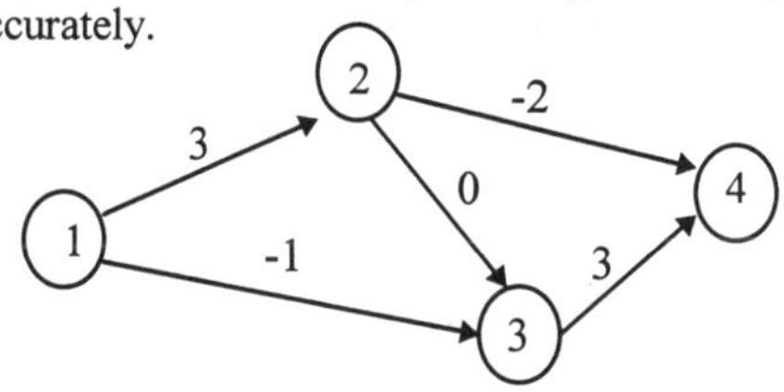

Fig. 1 A network of design activities in extended A-O-N

3.2 Assessment of Activity Time Distribution

It is a key factor in management and scheduling problem. Some design activities are new and unique in nature, it is not possible to predict the duration of them in advance, and the processing time of some design activities are uncertain due to the stochastic evaluation activities. It is a difficult problem to find a good approximation to assess the time distribution of design activities.

Keffer and Bodily[10] have explored the use of a simple k-point discrete approximation to estimate the processing time which appears robust to mis-specification yet make only modest demands upon the assessor. According to our experience, the processing time of one activity must be in an interval (A_1 , A_2), which could be divided into several time slots, and the processing time could be in one slot with estimated possibility. For example, the time distribution of activity 'i' could be approximated by the discrete random variables T_i. It is assumed that there are K possible time slots, according to the previous knowledge, the processing duration of 'i' could be in the time slot (A_{ir1}, , A_{ir2}) with probability p_{ir} (r=1,2,...K). So the time distribution of activity 'i' can be expressed as: $P\left(A_{ir1} \leq T_i \leq A_{ir2}\right) = p_{ir}, r = 1,2,\cdots,K$.

Each activity may have its own value of K. If K is chosen sufficiently large, it is evident that the approximation will approach the true distribution to any desired degree of accuracy. Normally, K–6 is ok. The estimated processing duration of activity 'i' is: $\bar{t_i} = \sum_{r=1}^{K}(\frac{A_{ir1} + A_{ir2}}{2})p_{ir}$, the estimation deviation σ_i is: $\sigma_i = \sum_{r=1}^{K} (\frac{A_{ir1} + A_{ir2}}{2} - \bar{t}_i)^2 p_{ir}$, and

the standard deviation S_i is: $S_i = \sqrt{\sigma_i}$.

Then the earliest/latest start time and earliest/latest finish time of each activity in the network of design activities can be obtained by forward and backward calculations. Firstly, we must derive distributions corresponding to convolutions and maximization. For two activities 1 and 2, activity 1 has K_1 possible duration and activity 2 has K_2 possible duration. This process gives rise to $L=K_1 \times K_2$. If this procedure is repeated at successive nodes throughout the network, the value of L rapidly becomes large. Ord[11] has proposed a simple and easy-to-implement procedure to reduce the computation by merging successive sets of K_2 , named "collapsing the distribution". Full details have been provided in the literature.

From this way, it can obtain an acceptable degree of approximation and require only minimal computing facilities.

4. Scheduling Problem

The scheduling problem in CE is a resources-constrained scheduling[12] . According to the different requirements in CE, we think there are two types of scheduling problem in design process:

1) Scheduling simultaneous execution of the combination of activities to minimize the sum of tardiness when design activities have different due dates.

2) Sequencing and scheduling design activities to be performed simultaneously such that the completion time of design process is minimized, given the processing time distribution and the resource requirements of each design activity.

Firstly, we give the formulation of such two types of scheduling problem.

4.1 Formulation

To formulate the scheduling problem, it is assumed that every design activity in the network of design activities should obey:

1) Every activity cannot be interrupted once begun.

2) Activities are permitted to wait for resources when the resource conflicts exist.

3) Resources are assumed to be available per period in constant amounts and are also demanded by an activity in constant amounts throughout the duration of activity.

4.1.1 Formulation of the first scheduling problem

The formulation is as below:

$$\min \sum_{i \in A} w_i T_i \tag{1}$$

subject to:

$$f_j - f_i \geq C_{ij} + d_j, (i,j) \in H \tag{2}$$

$$\sum_{S'} r_{ik}(t) \leq R_k(t), t = 1,2,\cdots,f_n \qquad k = 1,2,\cdots,m \tag{3}$$

$$T_i = \max(0, f_i - D_i) \quad \text{where} \tag{4}$$

A: set of activities in design process, i∈A, i=1,2, ···,n

w_i: : weight of activity 'i'.

T_i: : delaying time of activity 'i'

d_i : processing time of activity 'i'

f_i : finishing time of activity 'i'

D_i : due date of activity 'i'

H : set of pairs of activities indicating the precedence constraints

$r_{ik}(t)$: amount of resource type k required by activity i at time t

S' : set of activities executed simultaneously at time t

$R_k(t)$: total availability resource type k at time t.

The objective of scheduling problem is given in (1). Constraint (2) indicates the time relation of activity i and j. The resource constraints given in constraint (3), indicating that at each time and for each resource type k, the resource amounts required by the activities executed simultaneously cannot exceed the resource availability. Constraint (4) gives the computation of delay time T_i.

4.1.2 Formulation of second problem

To formulate the second scheduling problem, at first, let us consider two dummy nodes, namely 0 and n+1, such that their processing time and amount of resource requirement are 0. The problem can be formulated as:

$$\min f_{n+1} \tag{1}$$

subject to :

$$f_{n+1} \geq f_i, \forall i \in A \tag{2}$$

$$f_j - f_i \geq C_{ij} + d_j, (i,j) \in H \tag{3}$$

$$\sum_{S'} r_{ik}(t) \leq R_k(t), t = 1,2,\cdots,f_n \qquad k = 1,2,\cdots,m \tag{4}$$

The meanings of parameters are as above.

4.2 Scheduling Rules

The two types of scheduling problem above mentioned are usually NP-hard, so it is hardly to find any efficient algorithm to solve it. Normally, there are three kinds of methods to resolve them in the literature[13]. The most effectively method is to develop an heuristic solution procedure for satisfactory solutions[14]. In this paper, we only propose some scheduling rules and show a simple example of heuristic scheduling.

In concurrent design process, a simultaneous execution of a number of activities would cause resource conflicts. The scheduling rules should decide which activity will be delayed. We know, working on the wrong activity could delay the delivery of some important outcome (design information). Thus the critical design activities that influence the total design time should not be allowed to be idle. Proper dispatching at the upstream design activities is a feasible way to ensure an uninterrupted flow of information through whole design stages[5].

In concurrent design process, there are three key factors for successful and timely completion of the design process:

1) Using the limited resources effectively;

2) Scheduling simultaneously as many activities as possible without violating the constraints;

3) Optimizing the flow of information, that means to provide the information firstly to those activities whose information is needed by downstream activities mostly.

Therefore, the scheduling rules in CE should consist of the rules of allocation of the limited resources and optimization the flow of information.

4.2.1 Rules of allocation of resources

Firstly, we give the definition of *set of schedulable activities---S_t*. A set S_t is the set of activities which can be implemented at time t according to the information they need and their preceding design activities have been completed.

In order to deal with the resource conflicts, Demeulemeester and Herroelen[15] proposed the concept of *minimal delaying alternatives*, and proved that *it is sufficient to consider only the minimal delaying alternatives to resolve the resource conflicts.* To our problem, the supplementary sets of minimal delaying alternatives --- S^k,k=1,2,…,L are proposed, named *maximal scheduling alternatives*. A set S^k is a subset of S_t, which is defined as the such combination of design activities from the set S_t that can be performed simultaneously without violating the resource constraints, but the addition of any one more activity in S^k from set S_t would result in resource conflicts. All subsets S^k consist of *a scheduling set--- S(t)*. It can be proved that in order to resolve the resource conflicts, it is sufficient only to schedule the maximal scheduling alternatives.

The sets of S^k can be obtained by enumerative procedures.

4.2.2 Rules of information optimization

We propose the concept of *information weighted factor(IWF)* to deal with information priority problem. *IWF,* denoted by $\lambda(0 \leq \lambda \leq 1)$, is a measure of the information provided by one activity how to influence its succeeding activities. IWF of activity 'i' could be decided by the numbers of activities that depend on the information provided by activity 'i' . There are unified rules to obtain the value of IWF of each activity by analyzing the information dependence relationships among the activities in *network of activities*. The more activities that depend on the information delivered by activity 'i' ,the larger value of λ_i will be. So the critical activities in design process are those activities that have large value of IWF .

When a number of activities compete for the resources, the activities with larger values of IWF will be given higher priority and executed earlier, thus to avoid the resource conflicts and assure the smooth flow of information to the the critical activities.

Different scheduling problem will have different scheduling rules, but all feasible scheduling rules in CE should consist of the two rules mentioned above.

4. 3 A Simple Example

We use the example presented in Kusiak's paper[5] to illustrate how to use the rules of allocation of resources and information above mentioned.

Suppose that, the ongoing design phase is D and the set of schedulable activities consists of 5 activities. The design process under consideration consumes 4 types of resources. The status of these four resources is indicated by a status vector. Suppose that the resource status vector at time t is R= (5 5 4 0.9). The known condition are listed in table 1.

Table 1. Activities with their resource requirement, priorities, and estimated duration

Activities	Resource Requirement	weights	Estimated Duration	Latest StartTime	Latest Finish Time
5	(2 1 1 0.15)	0.10	2	9	11
7	(2 3 1 0.2)	0.20	7	6	13
8	(3 3 3 0.5)	0.10	3	11	14
10	(4 5 3 0.25)	0.30	6	13	19
11	(1 1 2 0.4)	0.10	3	13	16

The main intention is to schedule as many activities concurrently as possible. The scheduling rule we used in this paper is to minimize the sum of maximal waiting time.

The maximal waiting time---D_j is computed as:

$$t^* = t + \max_{i \in S^k} d_i$$

$$D_j = \max\,(t^* - LS(j)\,,\,0)$$

The heuristic scheduling rule is as:

$$Z = \min_{D}\,(\sum_{j \in D^k} (w_j \times D_j))$$

The scheduling procedures are:

Step 1: The schedulable set S_t ={5,7,8,10,11}. The resource vector at time t is R=(5 5 4 0.9).

Step 2: The maximal scheduling alternatives are S^1 = {5,7,11},S^2= {10}, S^3= {5,8}.

Step 3: t^*=5+max d_i =5+7=12. In S^1, Z_1=0.10×max(12-11,0)+0.30×max(12-13) = 0.10.

Perform next iteration, Z_2=1.20 , Z_3=0.40.

All possible subsets S^k and corresponding value of objective funtion are tabulated in Table 2.

Table 2. The subsets of schedulable activities and their weighted tadiness

S^k	Total Resource Utilized	Z_k
S^1={5,7,11}	(5 5 4 0.75)	10
S^2={10}	(4 5 3 0.25)	120
S^3={5, 8}	(5 4 4 0.65)	40

Step 4: The subset of schedulable activities with minimum weighted tadinesses is chosen, which is found to be S^1 = {5,7,11}, min Z=0.10. Hence, the activities 5, 7, 11 are scheduled simultaneously. The activities 8, 10 are delayed.

The details of scheduling algorithms to the two kinds of scheduling problem in CE will be discussed in other paper. We have developed a software prototype to support the decision making in concurrent engineering environment with Lotus Notes and Visual C++ language under Windows NT LAN, management and scheduling of design activities is a subsystem of it.

5. Conclusions

In this paper, to resolve the problem of management and scheduling of design activities in CE, a feasible and systematic approach is proposed in global view. On the base of new characteristics of design activities in CE, there are limitation of conventional project management techniques to be used in CE scenarios. We give a systematic approach to deal with this problem, including building a model of design activities, formulation of scheduling problem and providing some scheduling rules of allocation resources and information. A simple numerical example is given to illustrate the rules, and the details of concrete scheduling procedures will be discussed in other papers.

REFERENCES

1. Sohlenius, G.,"Concurrent Engineering", Annal of the CIRP, 41, 2 (1992), 645-655
2. Gu, P. & Kusiak, A. (editor), Concurrent Engineering: Methodology and Applications, Elsiver Science Publishers,1993.
3. Moder, J.J., Phillips, C.R. and Davis, E.W.,"Project Management with CPM, PERT and Precedence Diagramming (3rd edition) Van Nostrand Reinhold, New York,1993.
4. Adedeji, B.B.,"Scheduling of Concurrent Manufacturing Projects", Concurrent Engineering Contemporary Issues and Modern Design Tools, (1992) 93-109.
5. Kusiak, A.,"Scheduling Design Activities", Information and Collaboration Models of Integration, S.Y.Nof (ed.), Kluwer Academic Publishers, (1993) 43-60
6. Badiru,A.B.,"Project management in manufacturing and high technology operations", Wiley, New York,(1988,).
7. Badiru, A.B.,"Project Management Tools for Engineering and Management Professionals", Industrial Engineering & Management Press, Norcross, GA, (1991)
8. Badiru,A.B.,"Scheduling of Concurrent Manufacturing Projects", Concurrent Engineering : Contemporary Issues and Modern Design Tools,(1993) 446-464
9. Palmer, B.& Korbley, L. et.al., "Modeling the Concurrent Engineering Process", CERC Technical Report, West Virginia University,(1992).
10. Keefer, D.L., & Bodily, S.E.,, "Three-point Approximation for Continuous Radom Variables", Management Sci.,29(1983) 595-609
11. Ord, J.,"A Simple Approximation to the Completion Time Distribution for a PERT Network", Res. Soc. Vol. 42, 11(1992) 1011-1017
12. Christofides, N., Alvarez-Valdes,R. & Tamarit,J.M.,"Project scheduling to resource Constraints: A Brach and Bound Approach,European J. Open. Res.,29 (1987) 262-273
13. Davis, E.W.,"Project Scheduling Under Resource Constraints: Historical Review ad Categorization of procedures", AIIE Trans., 5, 4 (1973), 297-313
14. Kurtulus, L. & Davis, E.W.,"Multi-Project Scheduling: Categorization of Heuristic Rule Performance", Management Sci,,28, 2(1982), 161-172
15. Demeulemeester,E. & Herroelen W.," A Branch-And-Bound Procedure for the Multiple Resource-Constrained Project Scheduling Problem", Management Sci.,38,12(1992),1803-1818

PIKS: Product Information And Knowledge Servers For Concurrent Engineering Environments

D. Domazet., QZ Yang, YZ Zhao

Gintic Institute of Manufacturing Technology, Singapore

Abstract. In this paper we present the STEP information database server, known as PIKS, developed at Gintic as the result of the PIKS project. PIKS stores and manages STEP AP203 objects by using ObjectStore ODBMS, and exchanges product information with its users (e.g. CAD systems) with STEP Part-21 data files. PIKS provides STEP objects consistency and integrity, as well as concepts that are convenient for concurrent engineering.

1. Introduction

In concurrent engineering (CE) environments, product information needs to be shared among different computer systems/applications. In general, different data models are used for representing product information and problems occur when exchanging product data. STEP standard [1] specifies a standard product data model that can be used as a neutral product representation in heterogeneous computer environments. Standard STEP Part-21 data files [2] may be exchanged between different systems (such as CAD/CAM/CAE systems) if they are provided with adequate STEP translators that converts their proprietary data models into the STEP data model and vice versa.

Product data exchange is not very convenient in collaborative and concurrent engineering environments as it is difficult to control and manage the data consistency, integrity and propagation of design changes. Product Data Management (PDM) systems can distribute and manage STEP files, but do not have a control of their content as they usually do not interpret data files. Applications integrated by a PDM system have exclusive control and responsibility of their content. Therefore, PDM systems have limited control over application (or product) data and its integrity and consistency.

In this paper we present a solution for the management of STEP information that is based on the object-oriented database technology. Instead of data exchange, PIKS (Product Information and Knowledge Server) provides sharing of STEP AP203 [3] information. Main PIKS features are listed in Section 2, and its architecture is presented in Section 3. Database organization of STEP AP203 objects is presented in Section 4 and use cases supported by PIKS version 1.0 are described in Section 5. Section 6 summarizes the presentation of PIKS and implemented concepts.

2. Main Features of PIKS

PIKS version 1.0 provides the following functions:

- *Object persistent storage:* PIKS stores STEP AP203 objects in an ObjectStore object-oriented database [5] whose schema is compliant with STEP standard [3].
- *STEP file check-in/check-out:* STEP AP203 objects are transferred by STEP Part 21 data files [2] when they are checked-in and checked-out. STEP Part 21 files are used only for the transfer of data between PIKS and applications, and are not persistently stored. PIKS re-establishes external references of the checked-out objects with objects that were not checked-out, once they are checked-in back into the database.
- *UoF Check-out data granularity:* STEP data sets may be checked out according to specified Units of Functionality [3].
- *Data consistency and integrity:* Concurrent access to the data, and data consistency and integrity is managed by the ODBMS concurrency control based on the two-phase locking protocol.
- *Support of collaboration:* Support of collaborative group work by providing project-, group-, and personal workspaces. For instance, only users (individual or groups) of a group workspace may access (with write or read privileges) some design objects.
- *Event-driven notification:* PIKS notifies users of a shared (project- or group-) workspace on some user-specified events, such as check-in and check-out of specified objects, or modification of some data objects.

3. PIKS Architecture

PIKS uses client /server software architecture. Figure 1 presents the architecture of the server site of PIKS.

Schema Manager translates STEP AP203 EXPRESS constructs into C++ definitions, installs AP203 and PIKS class libraries to form a STEP-compliant product data model, and creates database sche-mas from this model. *STEP Object Manager* (SOM) analyses objects found in an imported STEP file, puts them into the database, and organizes the database according to the UoF (Units of Functionality) specifications of AP203 [3]. It also gets STEP objects from the database using requested product ID and UoF, and creates an export STEP file that is sent to the user who requested STEP objects. *PIKS Object Manager* (POM) puts or gets meta (control) objects needed for normal operation of PIKS, such as users, groups, projects, assignments, group- and project workspaces, available products etc. (see the Administration module later).It also maintains the data consistency among PIKS meta objects. *GUI* (Graphical User Interface) module provides an end-user interface to PIKS (forms and menus). Current version is based on X11/Motif windows tool-kit.

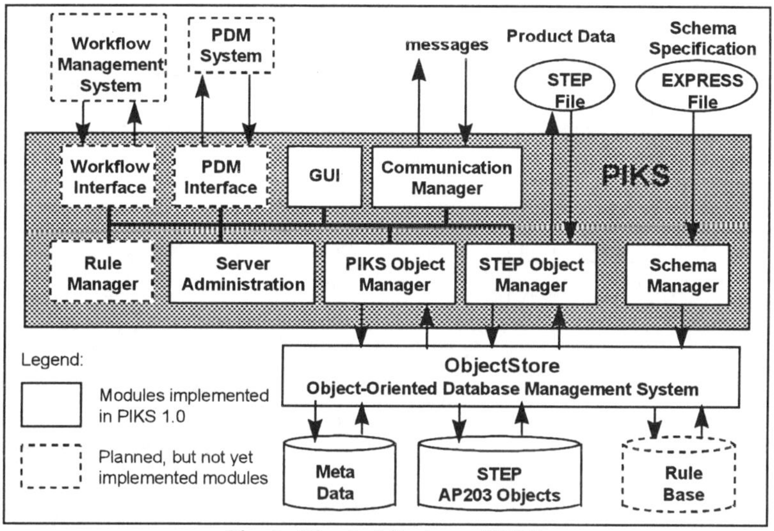

Fig. 1. The Architecture of PIKS

Communication Manager distributes automatically notification messages generated by the system and receives messages from the client site of PIKS. In future versions of PIKS, the *Rule Manager* will handle Event-Condition-Action (ECA) rules [4] in order to provide features of active databases (e.g. automatic database updates, design change propagation, notifications). *Workflow Interface* will provide an interface to a workflow system that will coordinate activities of PIKS users and their interaction with PIKS. *PDM Interface* will provide an integration of PIKS with a Product Data Management system.

PIKS is a C++ class library that uses the ObjectStore ODBMS for storing both application (STEP) and meta (PIKS) data [5] and ST Developer [6] to translate AP203 schema into C++ classes, and to convert data formats from STEP Part-21 to ObjectStore, and vice versa. ILOG View 2.1 is used as a GUI builder [7], and ILOG Broker [8) is used for object distribution, communication, and interaction between the server and client of PIKS in heterogeneous computer environments.

4. STEP AP203 Database Organization

PIKS is a specialized database server for the persistent storage and management of STEP AP203 objects [3]. The STEP application protocol 203 specifies 188 data entities, and many functions, local and global rules related to the configuration controlled design of mechanical products. STEP AP203 standard specifies these data sets in the form of 14 Units of Functionality (UoF). Three of them are shown in Figure 2, and Bill-of-Material (BOM) UoF with Application Interpreted Model (AIM) objects is presented in Figure 3. As it can be seen, the data model for STEP AP203 is very complex. In order to achieve higher data retrieval efficiency of such complex data models, we decided to use the object-oriented database technology for the server's back-end [5].

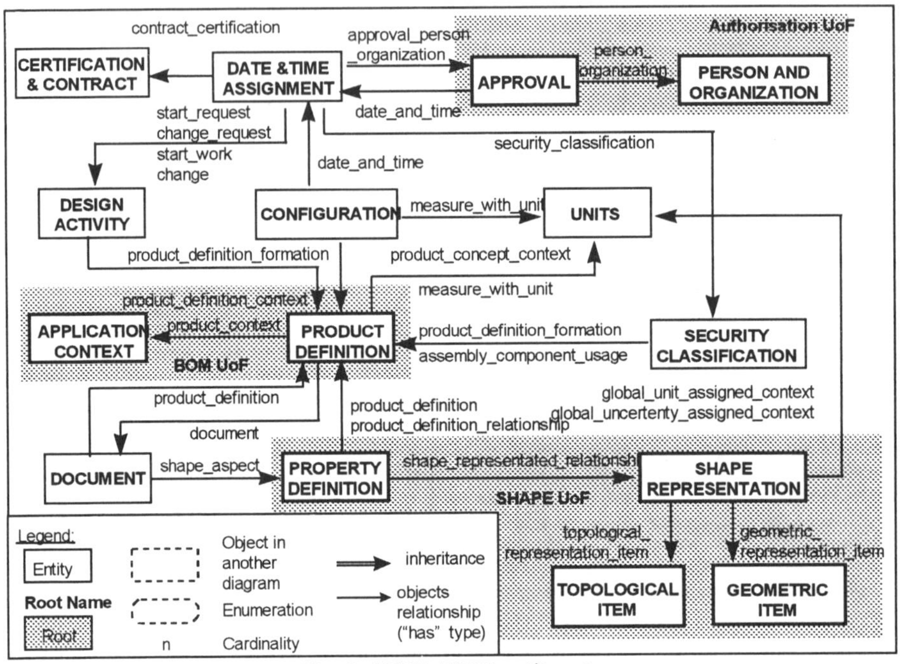

Fig. 2. STEP AP203 entity sets

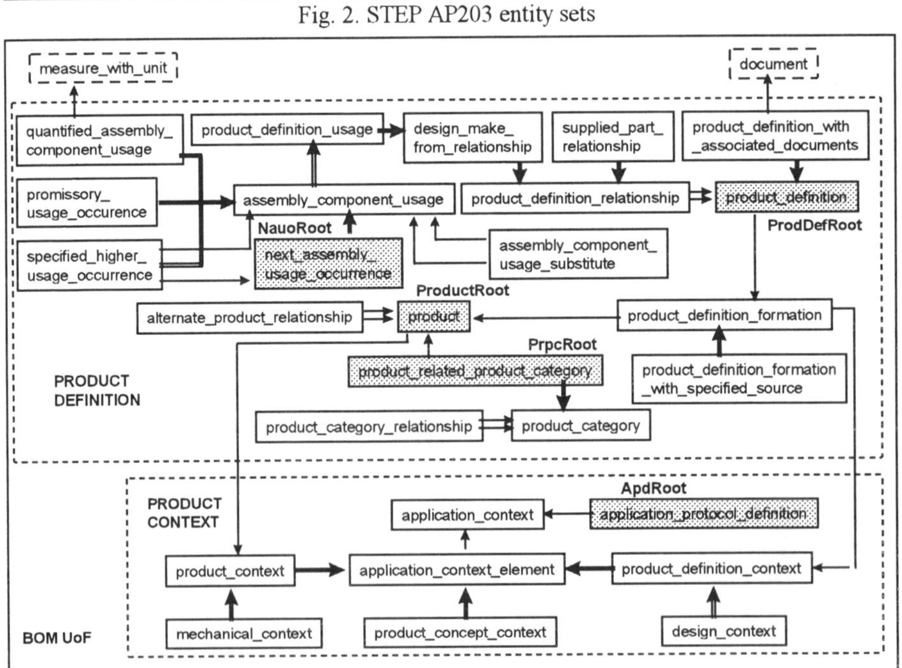

Fig. 3. Bill-of-Material UoF entities

Entities of the Application Interpreted Model (AIM) of STEP Application Protocol 203 are converted to C++ classes that specify the database schema of PIKS. Most of applications (e.g. CAD/CAM systems) that support STEP Part-21 data files, need STEP data sets, and not a single STEP data entity (object). PIKS 1.0 AP203 AIM database is organized according to UoFs, and provides needed UoF-based data sets when PIKS users specify UoFs required for their applications.

Navigational data access provides an efficient mechanism to collect or update needed data objects for each UoF required by a user. As this require identification of the root objects, from which the navigation starts, for each UoF one or more root entities are determined. We will explain this in the case of BOM UoF (Figure 3).

The root is an entry point for retrieval of all objects in the list. For BOM UoF, three database roots are created to ensure every object pertaining to this UoF can be reached by database applications through these root objects (NauoRoot, PrpcRoot, and ApdRoot). For instance, to create a root object pointing to the next_assembly_usage_occurrence OS list, the following OS API will be used:

db->create_root("NauoRoot")->set_value(next_assembly_usage_occurrence) (1)

We use the following API to do the root updating:

db->find_root("NauoRoot")->set_value(next_assembly_usage_occurrence) (2)

Through "NauoRoot" root all objects in the next_assembly_usage_occurrence list can be accessed/extracted by another OS API:

db->find_root("NauoRoot")->get_value() (3)

If a given product need to be extracted , its ID (PID) number is adjusted as a top level PID. Each of three OS lists holding BOM objects will be traversed by three OS cursors. An OS cursor keeps track of the iteration over an OS list to ensure every element in the list to be visited. All objects extracted from these three lists are then moved and saved as a STEP file by calling Rose API [6]: ROSE.saveDesign(). Actually, This API translated objects from the Rose internal data format into STEP Part 21 file format.

If a user requests only BOM objects related to an sub-assembly, an OS cursor will still be used to iterate over an OS list, such as the next_assembly_usage_occurrence list. For each list, a search path must be created to reach a part ID object from any one of objects in the list. Sometimes, this path may be very long and complex. Through this path, a part ID object will be retrieved. Then an evaluation will be performed to check whether an object in the list matches that one given by the user. If it is true, return this object. If it is not matched, next object in the list will be examined. Only requested objects will be moved and written into a STEP file and sent to the user.

5. PIKS Classes

as PIKS is a middleware system component between a STEP object-oriented database and an application or user. It is developed as a client/server system. Figure 4.a shows developed class categories of the server site of PIKS. We use Booch's notation [9] here, as we used Rational Rose OO development tool for design of the system [10]. In the design stage we tried to satisfy criteria for robust design, by separating all classes in three types of objects: interface objects, control objects, and entity objects [11]. We also tried to group classes that share similar properties, or are closer inter-related, into the same class categories with clear interfaces. For instance, if we want to change the underlying database system, we only need to replace DB_Access class category with the new one relevant to the new database system. Similar is valid for the GUI system (Server GUI class category), or for system administration (Administration class category).

Figure 4.b shows currently available classes in the Server Manager class category and their relationships with server's classes in other class categories. PiksServer class is the main control object of the server site of PIKS. For each specified use-case (explained later), a separate sub-class was created. New such classes will be added in future, for new use-cases. PiksAPI is a class with API methods that can be used for the integration of PIKS with other systems, and for communication with the client part of PIKS. Notifier class distribute notification messages to PIKS users.

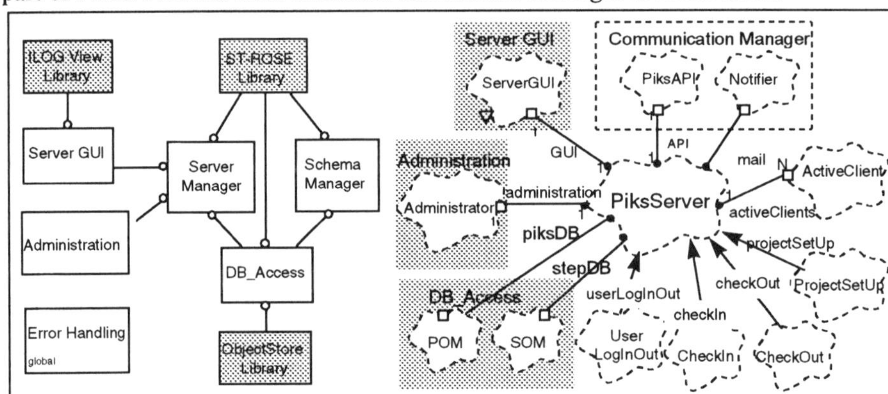

Fig. 4. PIKS server class categories and basic classes

ServerGUI is an abstract class. Its subclasses create different windows, dialog forms and menus. We used IlogViews in development of these classes [7]. The Administration class provides methods for specifying PIKS users, groups, data access privileges, roles, workspaces etc. Figure 5.a shows classes linked with the Administration class that are responsible for storing and managing these meta data. Figure 5.b shows classes of the DB_Access class category. SOM (STEP Object Manager) class manages all STEP persistent objects. For each UoF, a separate sub-class was created, responsible for putting, getting and updating STEP objects of the corresponding UoF. InBox and OutBox classes manage import and export STEP data files, respectively. POM (PIKS Object Manager) class stores and retrieves all PIKS meta objects (such as objects in the Administration class category), that are stored in the ObjectStore database.

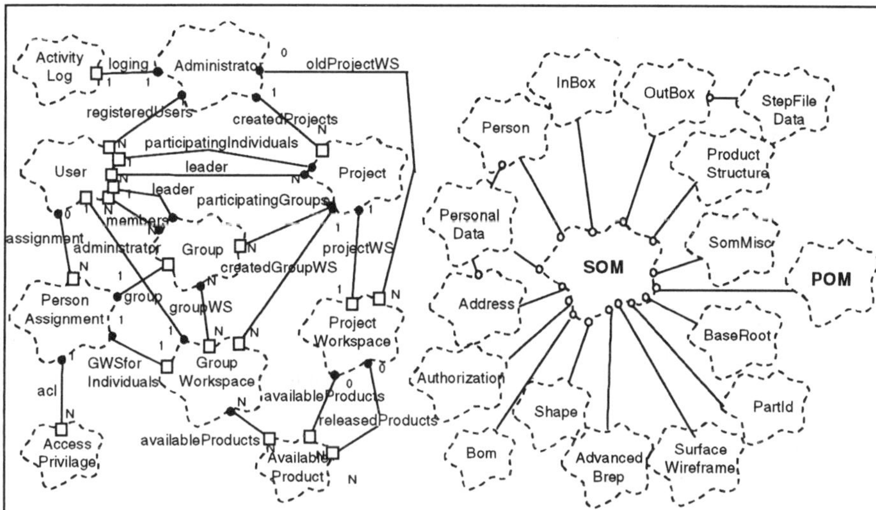

a) Classes of the Administration class category b) Classes of the STEP Object Manager

Fig. 5 Classes of the Administration and SOM class categories

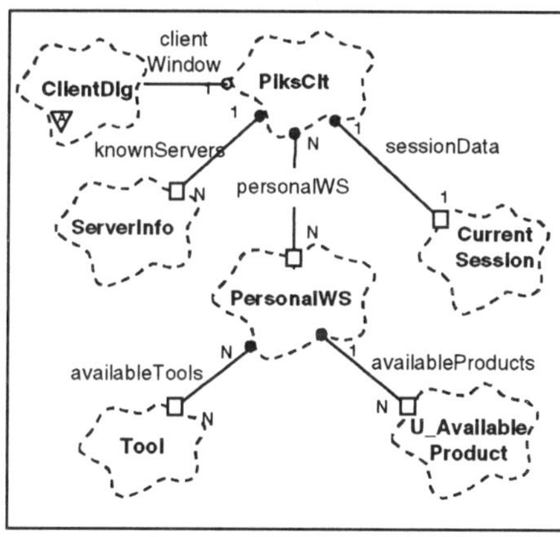

Fig.6. PIKS client classes

Figure 6 shows PIKS classes of the client site. PiksClt is the main control object. CltDlg is the abstract class. Its sub-classes create different windows, forms and menus that are used on the client site for the interaction with a user. Other classes are entity classes. ServerInfo class provides information about existing PIKS servers, and CurrentSession class - about current session data. PersonalWS class manages the personal workspace for a user. It manipulates with existing STEP files described by U-AvalableProduct objects, and software tools (applications, such as a CAD system) are described by the Tool objects.

6. Use Cases Supported by PIKS 1.0

According to Jacobson's use case analysis [11], we implemented the following use cases in PIKS 1.0:

- *Project setup*: All project related objects (see Fig. 5.a) are specified, such as PIKS users, groups, projects, project and group workspaces, users assignments and privileges, etc.

- *User log-in & log-out:* Provides users registration (log in), log out and open/close/switch a session with a chosen project or group workspace of PIKS.
- *STEP file check-in & STEP file check-out:* STEP objects found in an import STEP file are put in the object-oriented database (check in use case) or are read from the database and put into a created export STEP file sent to the user who requested STEP information (check out use case). These two use cases will be explained in more details in next two sections.

6.1 STEP File Check In

If an imported STEP file contains new STEP objects, the new objects are created in the database. Otherwise, old objects are updated with restored relationships with other objects of the STEP AP203 data model. Using appropriate GUI forms, the user chooses one STEP file from his personal workspace and sends the message fileChekIn to the PIKS server site, i.e. invokes fileCheckIn method of the PiksAPI class. This API method has the following signature:

fileCheckIn (char* userName, char* userRole, char* projectID, chargroupWSName, char* dbName, char* stepFileName, char* checkInPID)

Before checkIn method of the SOM class is invoked, CheckIn class execute some usual servers activities, such as: checks and verifies user's authorization (if rejected, notifies the user) and copies the named STEP files from the user's personal workspace into the directory specified for the InBox server's object. SOM class analyzes the imported STEP file (stepFileName) and updates the ObjectStore database (dbName) by using appropriate UoF-related classes. Previously created database roots are updated and objects found in the STEP file replace old objects. If some of them are new, they are stored as new objects. After completing the check in task, CheckIn class logs the checkIn activity (using the ActivityLog class) and informs all users of the group workspace (groupWSName) through Notifier class.

6.2 STEP File Check Out

When a user needs STEP objects, he/she needs to open appropriate check out GUI form and specify:

- product (by ID number) or a component of a product,
- a desired STEP file name (that will be created at the server's site),
- desired UoF data sets related to the specified product, and
- read or write privilege

Based on the specified data, PIKS client send a fileCheckOut message to PiksAPI class on the server machine (Figure 7). This check out API method has the following signature:

fileCheckOut (char* userName, char* userRole, char* projectID, char* groupWSName, char* dbName, char* productID, char privilage, UOF uof, char* stepFileName)

UOF is a composite object that specifies the list of UoF numbers/names needed for check out. The PiksAPI class invokes checkOut method of the CheckOut class (Figure 10). It verifies the user's, registers him as an active client and invokes checkOut method of the SOM class. This method gets all stored STEP objects related to the requested UoF and product ID, and creates an export STEP file with collected STEP objects. OutBox class sends this file to the user who sent the check-out request and notifies all users of the used group workspace..

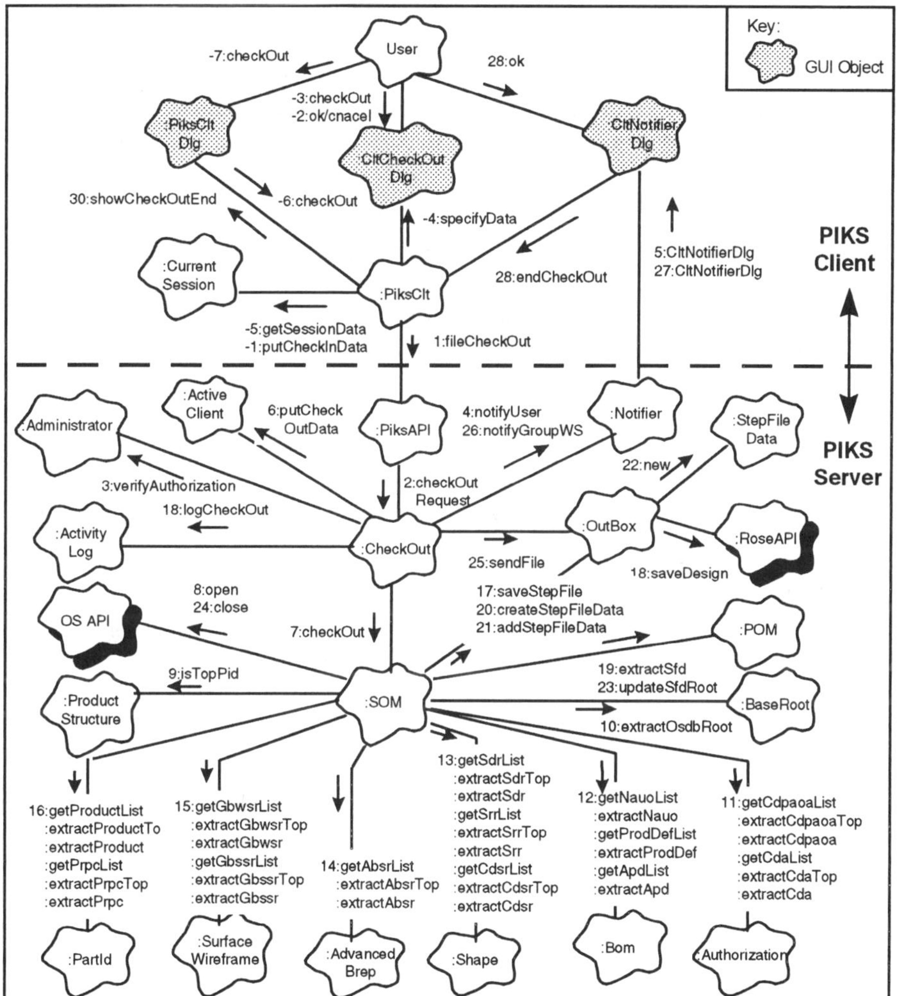

Fig.10 Object interaction diagram for STEP file check-out use case

7. Conclusion

PIKS 1.0 provides the basic functionality for persistence storage and management of STEP AP203 objects, such as project set-up, users log in and log out, and STEP file check in and check out. Its back-end object oriented database (ObjectStore) stores

STEP objects, and STEP files are used only for data transfer between the clients and server and vice versa. The developed query system uses a set of predefined (and programmed) UoF-based queries that the user can choose by using appropriate GUI forms and menus. PIKS can be integrated with other systems by using its API class (PiksAPI class) with the developed API methods.

For next versions of PIKS we plan to add also the following new features:

- Java interface to PIKS (PIKS clients as Java applets),
- use of more advanced and appropriate concurrency control mechanism (e.g. optimistic concurrency control, and new lock and transaction types),
- flexible data access granularity levels (from an attribute, to an UoF),
- STEP SDAI (Standard Data Access Interface) interface to the PIKS database,
- an interface to PDM and workflow systems
- implementation of other STEP application protocols(e.g. shipbuilding industry)

PIKS may be very suitable for inter-enterprise collaboration (virtual enterprise concept). Organizations involved in a joint product development and manufacture can share product related information stored and managed by PIKS, as a common STEP data server that supports heterogeneous computer environments.

Acknowledgment

The paper presents some of results of the PIKS project, the three-year in-house project of Gintic that started in October 1994. Authors wish to express their gratitude to the Gintic Institute of Manufacturing Technology for providing very good research conditions for this project.

References

1. ISO 10303 - 1: Industrial automation systems and integration - Product data representation and exchange - Part 1: Overview and Fundamental Principles, 1994
2. ISO 10303 - 21: Industrial automation systems and integration - Product data representation and exchange - Part 21: Implementation methods: Clear text encoding of the exchange structure, 1994
3. ISO 10303 - 203: Industrial automation systems and integration - Product data representation and exchange - Part 203: Application protocol: Configuration controlled design, 1994
4. Domazet D., Active data driven design using dynamic product models, *Annals of the CIRP* Vol. 44/1/1995, pp 109-112
5. ObjectStore Release 4, *Reference Manual*, Object Design Inc., USA, 1995
6. ST-Developer Version 1.4, *ROSE Library Reference Manual*, STEP Tools Inc, USA, 1995
7. Ilog Views 2.3, *Reference Manual*, Ilog, France 1996
8. Ilog Broker 2.1, *Reference Manual*, Ilog, France, 1996
9. Grady Booch, *Object-Oriented Analysis and Design*, Second Edition, The Benjamin/Cummings Publishing Company Inc., 1994
10. Rational Rose/C++ 2.0.14, Rational, USA, 1994
11. Objectory Version 3.0, Rational, USA, 1996

CONCURRENT ENGINEERING THROUGH DATA SUPPORT AND COMMUNICATION IN COMPUTER INTEGRATED MANUFACTURING ENVIRONMENT

P.K. Jain, S.C. Jain and Tafesse G-S.B

Dept. of Mechanical & Industrial Engineering. University of Roorkee
Roorkee - 247 667 (India)

Abstract. The current status of CE in CIM environment, its impact in manufacturing, its possible application in the future and the need of integrated databases for CE in CIM environment is discussed. A hybrid interfaced/distributed CIM database integration system with global and local data model for integration based on modular and functional approach is proposed. In parallel a KBS and a CE Blackboard module is included in the system architecture to enhance the coordination, cooperation, decision support, data support and communication of CE in CIM environment.

Key Words: Computer Integrated Manufacturing (CIM), Concurrent Engineering (CE), Data Base Management System (DBMS), Design For Manufacurability (DFM)

1. Introduction

Traditional design and Manufacturing of products is a sequential process and it depends on iterative design and manufacturing process. However, the iterative process is time consuming and since the design determines significant amount of the product cost (about 70 to 80% of the production cost), a new approach Concurrent Engineering (sometimes called simultaneous engineering or life-cycle engineering) is practiced [1]. Concurrent Engineering is the simultaneous consideration of production, function, design, materials, manufacturing process and cost, taking into account later-stage considerations such as testability, serviceability, quality and redesign [2]. The implementation of CE by considering the life-cycle factors avoids the possible resources spent for a redesign. CE is characterized by a focus on customers' requirement, moreover it embodies the belief that quality is in built in the product and quality is the result of continuous improvement of a process which, is supposed to be an integral part of CIM considerations from the order receipt till the product life cycle ends. The objectives of CE in CIM include:

- elimination of waste
- reduction in lead time for product development & product delivery
- improvement of quality
- reduction of cost
- continuous improvement

An ideal CIM system applies computer technology to all of the operational functions and information processing functions in manufacturing from order receipt, through design and production to product shipment [Groover M.P., 1987]. CE targets to produce a product which is economically sound, environment friendly, attractive in its design, produced with in shortest lead time etc.[4]

The marriage of CIM into CE is the manufacturing philosophy for futuristic industries, where life cycle considerations including the product quality, lead time, cost effectiveness and recyclability are the basic issues. The CIM environment which, is gradually getting reality through the advancement of CAD/CAM is on the verge of totally implementing the concepts laid by CE through the advancement of information technology in CIM framework. Where, integrated business and manufacturing activities link the need of the customers with the manufacturer.

2. Integrated Database for CE in CIM

The heart of CIM system is a management of information system which processes, handles and controls the shared data for administration, design, planning, scheduling and control. The efficiency of operating CIM depends on the quality and integration of a well-designed information management system[5].

Databases for life-cycle product design considerations in the CIM framework must be integrated to communicate with various interrelated functions of manufacturing activities to secure this target Different approaches for database models has been implemented including: a loose collection of independent databases, a central database, an interfaced database and a distributed database[6].

In CE environment manufacturing industry handles a lot of information needed for administration ,design, planning, scheduling & control. There is a close relationship between data which is being processed and used in various activities for example design information is needed for planning , scheduling, machine programming, quality control & so on. For this reason an information system must be designed as an entity which comprises all manufacturing activities in CE frame work. The essential features which are needed to be incorporated for CE in CIM environment, to assure the life cycle issue considerations includes distributed computer systems, a communication system for real time information exchange, a common data & knowledge base which can be reused for different manufacturing activities and decision support module to coordinate and communicate at each levels for strategic planning decision.

CE requires diverse heterogeneous data that addresses various needs of design, planning and manufacturing. Similarly CIM database is composed of CAD databases, accounting and administrative control databases and manufacturing requirement planning databases. However, handling efficiently these separate and heterogeneous databases are very difficult. To challenge this reality CIM database should be capable of sharing data, handling heterogeneous data and avoiding data inconsistency. Meanwhile, different database integration methods have been applied to handle the design and manufacturing activities including[7,8,9]:

A. Developing a single, new, unified homogeneous database system to store and manage the various types of design planning, manufacturing and accounting data, based on a uniform data model by discarding the individual functional database in CIM.
B. Developing an integrated data administration system to manage heterogeneous distributed databases.
C. Developing a frame-work for knowledge-based systems to integrate the heterogeneous multi-database of CIM.

3. Reinforcement of data support and communication for CE in CIM environment

CIM is highly dependent on data communication where, design considerations are weighted against manufacturing considerations, including material availability, manufacturing tolerances, cost considerations etc., within the context of the prescribed quality levels. Subsequently, the design data is transferred to manufacturing requirement planning system to verify the availability of manufacturing resources and to the accounting and administrative control system, to estimate the planned cost of manufacturing to test the feasibility of manufacturing the product[7]. Lastly, the system determines the possible cost trade-offs between such attributes as features, tolerances and quality and feeds this information back to the CAD system. Where the majority of these databases serve multiple applications to designer, system operators and other related personnel.

The CIM system gradually grows with the advancement of the technological and scientific knowledge, while the integration of the manufacturing & business activities need a revolutionary change to merge the concepts of CE in to CIM frame-work. CIM comprises many separate modules or subsystems , there is no generally agreed sub-system structure, not even a generally accepted list of subsystem, that was the main reason why the ESPRIT CIM project has addressed three separate but related goals[13]:

1. To modularize the total CIM into functionally discrete sub-systems.
2. To describe the minimum specifications of each sub-system.
3. To identify the interrelationships that exists between any one CIM sub-system and all other sub-systems .

The proposed CIM system which is targeted to further advancement of integration of CIM modules based on functionality of each module oriented to generalized activities. Communication of CIM needs to be built in such a way that all data communications to take place in real time and it should take into consideration factors like, different hardware configuration, software , mode of processing, vendor's attitude etc.

To solve these integration issues for effective information exchange a communication network protocol must be chosen in addition to the proposed architecture of CIM frame work in CE environment [fig.1]. For all modules to communicate conceptually over the same net, the proposed hybrid distributed/interfaced CIM model in concurrent engineering environment facilitates communication and data exchange at the local and global model.

4. Global and local data model in CIM

A grand scale integration and developing a series of data is the main concern of CIM. Coordinating the whole manufacturing activity for competitive production is the basic target but, because CIM is a complex in the sense that it involves accessing large databases for data storage and retrieval purpose, conventional data processing (i.e. numerical, alphanumeric computations and symbolic processing) and computer decision making. An alternative communication module which enhances seamless access and integration of manufacturing activities in the CIM framework is essential. The emergence of different knowledge-based systems, database management systems and data models can be incorporated for a CIM data model where, subdividing the whole manufacturing activity into specialized groups to form a global and local data model can be a basis for communication. These data models consists of interconnected complex computer controlled manufacturing activities including:

1. Computer aided planning and logistics (CAPL) data model.
2. Computer aided design (CAD) data model.
3. Computer aided manufacturing (CAM) data model.

The engineering information system in such a frame work assists efficient and fast handling of data often in real time. Since function based similar manufacturing data models can be easily handled by the same type of DBMS (for example a CAD data can be transferred using graphical & alphanumeric data model) for different functions in the sub-module.

5. Hybrid Interfaced/Distributed data communication in CIM environment

To integrate the basic concepts of CE, to manage complex data in CIM environment, to handle design, planning, scheduling, administration, controlling and other related data, CIM functions are classified in to CAD database, CAM database and CAPL databases. To simplify the organization of data-structure, handle interrelated heterogeneous data, and to enhance communication between them, a hybrid interfaced/distributed data communication system architecture is proposed for CE applications in CIM environment (Fig.1)The proposed system architecture considers the following facts:

1. To modularize the CIM system into discrete functional CIM subsystems where enhanced communication between each sub modules can be categorized.
2. To integrate different local data management system for communication and data exchange in a neutral data format for goal oriented specifically localized applications.
3. To provide coordination of the distributed data models for cooperative, hierarchical tasks through these modules and to strengthen the link and communication for collaborative works.

The distributed databases are integrated by a post and pre-processor for data exchange and communication. The KBS handles data extraction, manipulation and decision support for CE applications. Functional level communication is performed at the data level with out any additional data modification and the data structure can be custom designed to meet the specific requirement of the manufacturing needs. The CE blackboard system coordinates the collaboration and communication of each database group, where decision support system with a user interface can be embedded into it.

The control blackboard uses data interpreted by a KBS, an additional function of this module is evaluation & controlling of the system by formulating and retrieving hypothesis concerning the system. The decision support system embedded in the global control unit coordinates on line and off line planning, monitoring systems, diagnosis of abnormal statuses and intervention to modify system parameters.

Complex applications like design and manufacturing needs considerations of large collection of graphical and alpha numeric data concurrently. That is the reason why an object-oriented data base management systems, where object-orientation concept such as classes, inheritance, encapsulation and complex objects are applied to improve integrity, flexibility, reliability and scalability of these large and complex data. Additionally object models are semantically rich, and they provide a variety of type, abstraction mechanisms and relations for expressing the complex inter schema relationships and potential conflicts among data at each component system [11]. Based on the facts mentioned above, the object-oriented data modeling is selected for the proposed hybrid interfaced/distributed CIM architecture since, they are the best alternative to model data and to facilitate automation of design and manufacturing functions.

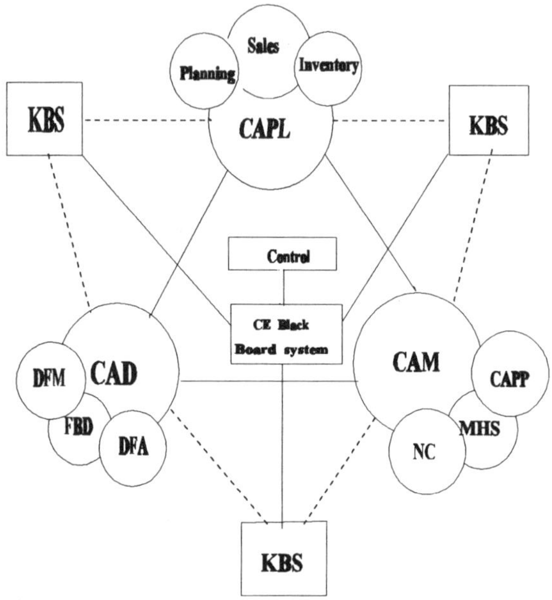

Fig. 1 A hybrid distributed/ interfaced CIM Database integration system for CE application

CAPL - Computer Aided Planning & Logistics
MHS- Material Handling and Servicing
CAPP- computer aided process planning
CAM -Computer Aided Manufacturing
DFM -Design for Manufacturability
NC- Numerical Control
DFA- Design for assembly,
FBD - Feature-Based Design
CAD- Computer Aided Design
KBS - Knowledge-Based System

The decision support system of the proposed model is typically controlled by the Blackboard system where, information is passed through the KBS. The KBS eases the communication between the local data modules and the distributed DBMS's. It also adds reasoning capabilities to the system which, in turn improves the efficiency of the CIM system. Additionally the KBS, combining knowledge and data driven systems and applying semantic integrity constraints it provides the system with intelligent answers at all levels of the manufacturing activities. The Blackboard system, based upon the dynamic data and knowledge available improves the communication, coordination, cooperation, conflict resolution and decision support for CE in CIM environment.

The proposed Hybrid interfaced/distributed data integration module can be used to solve problems of concurrent engineering. A subset of concurrent engineering approach, i.e. Design For Manufacturability (DFM) is taken into consideration to demonstrate the information flow and the communication net work by the methodology used in our research work in DFM.

6. Application of some concepts of CE in CIM environment

The result of applying the DFM (a subset of concurrent engineering) introduces the adoption of concurrent rather than serial approach to the various product design and process planning functions in the existing manufacturing facility. To disclose the soundness of the proposed CIM architecture, a DFM module for machined part manufacturability analysis is considered, where Knowledge-Based approach is implemented, for integrating multiple sources of knowledge (which are required for DFM evaluation) .
In order to fulfill the requirements set by the manufacturability analysis module of the proposed system the following points are considered as a basis in order to facilitate the DFM analysis of a machined-part in CIM environment

A modular problem decomposition approach, to capture different activities, to organize these modules hierarchialy, to reduce the complexity of the modules, and to enhance easy management and abstraction is used.

During product development, the product passes through different stages of design and manufacturing activities. These activities are by nature modular and a modular approach to capture the intents of each activities will give the best problem solving approach. Modularizing into discrete functional CIM subsystems simplifies the integration, and complexity of the entire system. Because functional level communication is performed at the data level (locally) and additional data modification can be easily custom designed to meet the specific requirement of the manufacturing needs. Modular approach organises design, manufacturing and planning activities as modules which are at the same level and represented by an object-oriented data model to pass message between each modules.

Integration by part representation scheme to enhance the design and manufacturing data exchange in real time and to ensure concurrency.

Designing for manufacturability requires the fusion of the designers intention for a part with a manufacturing requirements for the product . The designers intentions are a set of functions which the product will provide or require. These functions are related to the features of part. Feature technology, provides a better approach to integrate design and applications following design such as: engineering analysis, process planning , machining and inspection. Multiple representation of features depending upon their application, are used to model these different design and manufacturing activities. Combining these different views is difficult, but different techniques has been proposed including: feature refinement, feature conversion, feature mapping, feature translation, feature transformation, and feature relaxation (Solomon et.al.). To achieve the greatest possible flexibility with respect to application-related specification, geometrical and technological attributes (face relations, edge classification, dimensions, location/orientation/size parameters and other non-geometric attributes) are used to model and evaluate manufacturability (Appendix- 2).

Integration by database.

Various heterogenous dynamic data available in the data base of CIM information system can integrate DFM and KBS in CIM environment through a real time active database support. Additionally the data available in the database enhances integration of the design and manufacturing activities, enhances communication between different modules and facilitates integration. Database systems are active integration tools, by which extraction of data, storing of data, and generating rules from the data base is important for manufacturability analysis in CIM environment . As described above part representation using "features " is convenient for implementation of object-orientation concept, and different attributes can be modelled in the CAD database including: design rules for building rules , active values/demons and methods, to represent feature prototypes with objects , classes, methods and dynamic bindings, feature derivation for representing class hierarchy and inheritance. These CAD database can be used also to model manufacturing and planning of production, where it acts as an integration agent for DFM analysis in the design activities. Object-oriented concept that uniformly models any real world entity as an object is applied in manufacturing to model a part, operation process, machine tool fixture and operating parameters as entities, so the relationships between each entity are mostly many-to many and form a very complex graph. DFM analysis is enhanced in CIM environment using both local and global data models and integrating them by an object-oriented-database that views the world as entities defined by their functional characteristics, which supports collaborative works of the team of developers and the customers.

Integration by a communication software to realize the feed back from the customer at the design stage or after operation of the product for a possible redesign of the product or a part to enhance two-way communicability.

Co-ordination of design, manufacturing and planning parallel with a customer oriented production are the core demand for the CIM activities, designed to adapt in the changing product specification. In this particular case, at one end the customer sets his preferences and priorities while the manufacturers at the other end are prepared to change their product in respect to the demands of the customer, using the available material resources or by adapting the resources with out spending major capital expenditure which is the objective of DFM. Real time co-ordination of these factors is difficult, but their consideration is vital, in today's competitive market. That is the basic reason for searching a real time integration software on the top of the existing CAD\CAM software to ensure two way communication between the user and the manufacturer, to practically avoid redesign, give better service, maintenance, and to design custom tailored product for a specific user. Compatibility issues for establishing data format standards , communication network standards, computer architectural standards are among the difficulties hampering integration and real time communication, but the research in these areas are encouraging.

Implementation of the above mentioned integration strategies using features requires the modelling of the part using the object oriented concepts. The part model in our DFM analysis uses the object oriented paradigms objects, classes, instances of classes and etc. to model the part features in the feature hierarchy modeled in Table.1. The KBS are organised for different functions (e.g fixturability analysis, etc.). Other related activities are organised in a modular manner in a predefined hierarchical relationships, to make effective the whole manufacturing activity and to encourage real time communication of the design, manufacturing and planning, as to produce a product which is custom tailored to the needs of the customer.

1. To combine the two activities (Design and Manufacturing) a feature based modelling is implemented with a new class of features (Appendix -2) to capture the basic tasks of the two activities "function" of the part and "manufacturability" of the part.
2. The manufacturability analysis module uses the same feature-based part modelling to represent material charcteristics, tolerances, and surface finish indexes with additional attributes added to represent these product attributes.
3. The module considers a number of process plans which are of a realizable type among which to choose after the manufacturability analysis based on the manufacturability rating. Initially a good plan is considered as the current best plan, and then search of state space of possible choices are accomplished.
4. The fixtureability of the design is checked by the corresponding knowledge base module. Considering the part to be machined and the existing resource model.

The manufacturability of the part is analysed step by step by considering the features in each part, their shape , dimension, and location, the available resource, etc. The final evaluation of the part is based on the manufacturability rating index. The steps-wise description of the DFM task followed in the present work considers the procedures listed in the (Appendix - 1)

Conclusion

The communication and data support issue is vital for CE in CIM environment, that is why the proposed hybrid interfaced /distributed CIM communication module with local and global data model, the KBS and the CE Blackboard control module can be implemented to solve the problems of CIM which includes sharing of data among CIM subsystems and the existence of inconsistencies among these data. Local and global data models and CE Blackboard control module are included in the proposed hybrid interfaced/distributed CIM communication architecture to enhance communication, through data support and to make the decision support method more effective, by extending and improving data application in CIM. To disclose the soundness of the proposed CIM architecture, a DFM module for machined part manufacturability analysis is considered, where Knowledge-Based approach, object-oriented paradigm and distributed databases concepts are implemented, for integrating multiple sources of knowledge which are the important parts for the DFM evaluation methodology. There are still problems to be addressed in the future to solve the communication of information in the CE application in CIM environments and these problems includes modelling events, handling anomalous information, incomplete knowledge and exception handling.

Appendix-1

Step-wise description of the task

No	*Description of the task*
1.	product design from a Feature-Based CAD solid modeller.
2.	Feature based stock modelling for stock material representation (user editable) and selection of stock material or near-net shape workpieces by comparing the features of the part with the stock material.
3.	Feature based geometric and tolerance modelling and feature-feature relationship modelling
4.	Feature by Feature comparison of stock material versus part feature model for cut dimensions depth of cut, number of passes and total machining allowances (t, i, h) and preliminary cutting sequence selection (recommended cutting variables and preliminary cutting sequence selection from the best possible plan stored in the database).
5.	Input of the required resources or constraints (user editable) and representing them in the form of feature-cutting operation-resource model.
6.	Limitation imposed by the machine horse power, maximum spindle speed and maximum depth of cut will be imposed (user editable).
7.	Checking about accessability, rigidity, fixiturability and non-destruction of parts using the knowledge-based module for DFM analysis.
8.	Identifying the feature of the part with serious problems and assigning a penalty factor for manufacturability; the penalty factors considers: tolerance, shape(morphological), resource, cost, etc. analysis.
9.	Constraint checking for the limitation imposed using feature-based modelling applying feature-cutting operation -machinetool-tool model and feature-feature relationship model, and selection of operational sequences.
10.	Corrections for tool approach and over travel made, then calculation of machining time and result display.
11.	Amendment f or the values for the hardness of the work material, tool material and size of the cuts,operational sequences, etc...(cost, resource, quality or user defined selection criteria will be available for customization) and assigning penalty factors.
12.	Part design is accepted or rejected for manufacturing.

Appendix-2

Part modelling using a hierarchical feature-modelling of the part.

No	*Feature type*	*Description*	*Purpose*
1.	General features	cylindrical, plane, etc.	To represent stock shape, material, overall dimensions
2.	Basic features	External and internal cylindrical surfaces, conical, etc.	To represent the manufacturing attributes including shape, dimension, etc.
3.	Auxiliary features	holes, chamfers, threads, etc.	To represent the functional properties of the feature.
4.	Tolerance modelling features	edges, vertices, faces, chamfers, centre lines, datum features (among the auxiliary features)	To represent the geometry, topology, and tolerance of the part.

References

1. Abdella H.S. & J. Knight (1994), An Expert System for Concurrent product and process design of Mechanical parts, *Instn. Mech. Engrs*, part B, Vol. 208, pp. 167-172.
2. Young R.E.(1992), A Grief and P.O'Grady, An Artificial intelligence based network system for concurrent engineering, *Int. Journal of Prod. Res.*, Vol. 30(7), pp. 1715-1735.
3. Groover M.P. (1987), *Automation, production system, and Computer Integrated Manufacturing*, Prince-hall,India.
4. Bayliss D.C, R. A. Kuesum, R. Parkin and J.A.G.Knight(1995), Concurrent Engineering philosophy implemented using computer optimized design, *Proc. of Instn. Mech.* Engrs., Part. B, Vol. 209, pp. 193-199.
5. Andrew Kusiak(1988), *Artificial Intelligence for CIM*, IFS/Springer Verlag.
6. Remobold U., Nnaji B, Stor A.(1993), *Computer integrated manufacturing and engineering*, Addison Wesley.
7. Kaiser G.E. and et al.summer 1988, Database support for knowledge based Engineering Environments, *IEEE Expert*.
8. Dills D.H. and W. Wu, Using knowledge-based technology to integrate CIM databases, *IEEE Trans. on Knowledge and data Engg.* Vol. 3(2), 1991, pp. 237-245.
9. Beynon-Davis (1981), *Expert data base system*, McGraw Hill, N.Y.
10. Levent Orman (1988), Functional development of database application", *IEEE Trans. on Software Engg.*, Vol. 14(9).
11. Ahmed Elmagarmid and Evaqqelia Pitoura(Oct. 1995), " Using Object technology for databse system integration, *IEEE Computer*, pp. 67-68.
12. Subramanyam S. and S.C.-Y, Lu(1991), Computer aided Simultaneous engineering for components manufacturing in small and medium lot-sizes, Trans. of ASME, *Journal of Engineering for Industry*, pp. 450-464, .
13. Yemons R., Choudry A., Hagen P.J.W. Ten(1985), *Design rules for a CIM*, NHPC.
14. Don Ralstan & Tony Munton (1987), Computer integrated manufacturing, *Computer Aided Engg J.,Vol.4(4)*, pp 167-174.
15. Knox Charles S.(1987), *Organizing data for CIM application*, Marcel Dekker Inc.
16. Beach et al.(1988), Integration soft ware the key to CIM ,*IFS Springer Verlag*, pp 445-453.
17. Allen D.K(1986), Architecture for CIM,*Annals of CIRP Vol.35(1)*, pp351-355.
18. Martin R. & Dans J.E.(1985), The role of data in CAD/CAM integration, *Computer Aided Engg J. Vol.2 (3)*, pp97-101.
19. Gerzco J.M.& Buchman A.P. (1985), *An object-oriented language for CAD and requied data base capabilities , language for automation Chang S.(ed)*, Plenum press N.Y.
20. Nagel R.A. et al. (1986) Developing and implementing integration ststegies for manufacturing companies, *Annals of CIRP Vol.35(1)*, pp 313-316.
21. Wirt et al. (1989), Requirments on interfaces & data for NC data transfer in view of Computer integrated manufacturing, *Annals of CIRP Vol.38(1)*, pp 443-448.
22. Saleman Roy M. (1985), The Evolution from CAD/CAM to CIM: possibilities, problems and strategiesfrom the future, *Computer and Graphics, Vol.9(4)*, pp 435-439.
23. Salomons O.W., Van Houten F.J.A.M., Kals H.J.J.(1993) , Review of research in feature-based design, *Journal of manufacturing systems Vol. 12(2)*, pp 113-132.
24. Gupta S.K.(March 1997), Using manufacturing planning to generate manufacturability feedback, *ASME Journal of Mechanical Design Vol.119*, pp 73-80.
25. Arimoto S., Ohashi T., Ikeda M., and Miyakawa S.(1993), Development of machining-producability evaluation method (MEM), *Annals of the CIRP Vol . 42/1*, pp 119-122.
26. Juri A.H, Saia A., and De Pennington A.(1990), reasoning about machining operations using feature -based models, *Int J. Prod. Res. Vol. 28(1)*, PP 153-171.
27. Korde U.P., Bora B.C., Stelson K.A. and Riley D.R. 1992), Computer-aided process planning for turned parts using fundamental and heuristic principles, *ASME Journal of Engineering for Industry Vol. 114*, pp 31- 40

The Management of Parts for Complex Assembly System in PC Environment

Yin Wensheng, Wu Yongming, Wang Tongyang and Zhou Ji

CAD Center, HuaZhong University of Sciencc and Tcchnology, P.R.China

Abstract. It is important to manage the parts in complex assembly design system. It will be useful if an effective environment is provided to designers to manage the parts and subassemblies in this assembly or in other assemblies. In these environment designers can save, find and modify the parts or subassemblies hierarchically, orderly and conveniently. It is required for assembly design system to support the top-down design process and to let designers have a clear and complete realization about their own products. The management of assembly design system for parts now generally uses the method of managing files directly or by other database management systems, but it is hard or not convenient for complex assembly systems. A data structure for managing parts and subassemblies in complex assembly system is presented and with this structure one can establish a code rule easily in PC environment and can simply manage complex assembly system that has many parts. It can also be used for the management of other complex systems.

1 Introduction

One of the important purposes of computer design system for assembly is to provide an effective environment to designers. In these environments designers can specify the relative positions of all of the parts, tell the interactions between these parts, and analyse the kinematic properties of the assembly and assembly tolerances. It can also save, find and modify the parts or subassemblies hierarchically, orderly and conveniently. It is under the management of parts and assemblies that we can design a product according to the habit of real process of product design and realize the top-down design in the computer design system. We can make the designers understand their products from the start to the end completely and clearly and help them to realize their design easily and quickly.

Generally a part or a subassembly is saved and opened as a file. In order to help designers to understand the hierarchical relation of a assembly, a hierarchical tree such as featuremanager in SolidWorks is provided in some CAD systems. There are somc disadvantagcs of management of parts in these system and it is hard to manage the large scale assembly system. We use a code management of parts in our assembly system of CDS(Concurrent Design System).

2 Characteristics of complex assembly and management of parts

A real assembly may be a complex assembly and there are a large amount of parts and subassemblies in it. We must manage all the parts and subassemblies and their attributes such as the features of parts, the mates of parts, and etc. With the development of the computer technology, the management of production becomes more and more convenient. We can simply manage all parts and assemblies of a factory easily.

2.1 Assembly design representations

In the assembly system, the use of data structure must adopt a particular application. The first type of assembly structure is the virtual link structure that is a well-known hierarchical assembly representation (Lee and Gossard, 1985) and has influenced the data structure used in other parts. It is a tree that tells us the structure of the assembly and assembly sequence in Fig. 1.

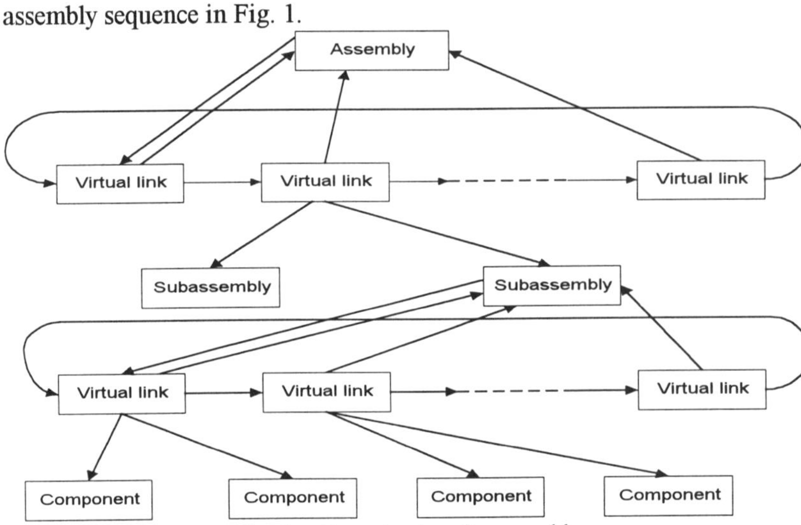

Fig. 1. Tree structure for assembly

The second type of assembly structure is the general graph structure that more explicitly represents part-to-part relationships than the tree and is particularly well-suited to tolerance and kinematic chain generation (Wang and Ozsoy, 1993) (Kim and Lee, 1989). The disadvantage of the general graph is that it lacks the hierarchical model. The best structure should be the combination of these two representations and will process not only the hierarchical relations but also the mate relations. Fig. 2 shows the relations of parts and subassemblies.

The feature-based part representation is in common use of assembly model and it has more advantages than the old ones. Generally, a feature is a geometric primitive consisting of a predefined pattern of low-level geometric entities, typically a volume or a set of surfaces, that have more information than part geometry and have significance to an application. The features of assembly are provided in some design systems such as in SolidWorks, which represents some assembly operations. Feature-based part representation is used in CDS.

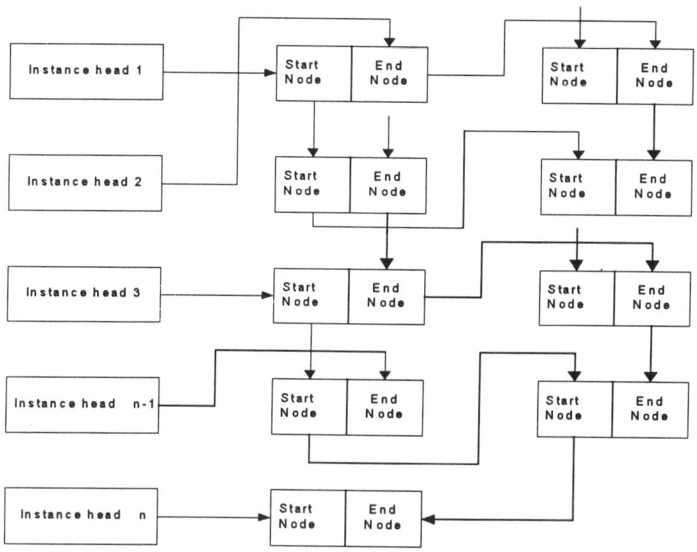

Fig. 2. The relations of parts and subassemblies

2.2 Simple assembly system

The simple assembly system is the system that has a few parts and the management of this system is very simple. In this system, we insert a file by a name to load the data of a part or a subassembly into the assembly system, and we save the file also by the name to write the data of part or an assembly to the disk. The file name we see is the name of the part or assembly. It is the way that manages parts and assemblies directly by file management system. It is useful and effective for simple assembly systems but it is required to remember all names that represent the parts and subassemblies.

2.3 Complex assembly system

Generally, a product consists of thousands of parts and subassemblies. If we use simple assembly system the disadvantages are:

It is difficult to inquire the parts or assemblies; There are too many files on the disk to find the one we need; Although we can make a lot of catalogues for the assembly, it is required that we know not only the structure of assembly but also the rule of saving file.

The file name will be different from the name of the part or assembly. If we save the file of part by the same name of the part, it will confuse the management of parts. For example, a part or a component can have only one name, but when it is used in an assembly system, it will have many instances which have the same data structure and different names. The name we see will be the name of the instance but not the name for saving. If we use Chinese name for representing the part in China, we will find it very hard to establish and inquire the Chinese file name for saving and loading the part.

In the real design process, a part has a name and a code. The name consists of English or Chinese word which has a very clear meaning and easy to remember, and the code consists of a series of alphabets which present a certain meaning. Each code of part must obey code rule and present a part exclusively. Designers and workers remember the parts by the name and look for the parts by the code.

2.4 Managing files of part directly

The simple management of parts is the method of using the files directly. Data of each part consists of data files. Although each part has different format of files, it can be read or written as a unit. The function of every computer operation system is the access of files, so it naturally uses the management of files directly. It is very effective when the assembly system is small.

2.5 Managing parts by other systems

In order to manage the complex assembly system, we can use other database management systems to manage the parts. For example, we can use Sybase to create a management system of part or we can use PowerBuild to manage the technological documents, including the part files. The parts are managed as a database unit like float, integer data type. They can be inserted in OLE way in Windows environment if the part and assembly system supports OLE way. This way is widely used in the management of parts and takes an very important role. But there are some disadvantages in it:

Higher cost

If we only have the assembly system, we can not manage the parts. We have to buy the database management system which is costly.

Lower speed

Use of other database management systems will lower the speed of assembly system. The database management systems will take up a lot of memory and take a lot of time. If the data of parts is processed as a block of OLE, many programs will be processed simultaneously and this will make the speed more slowly.

2.6 Management of featuremanager design tree

A method of management of featuremanager design tree has been proposed in the SolidWorks. It provides a featuremanager design tree to manage assembly, parts, features of parts and mates among parts and subassemblies. The featuremanager design tree allows designers to quickly make changes or evaluate the steps used to create the model and let designers understand the model construction. It is useful for designers to inquire the parts and subassemblies in the assembly but it is hard to know where the part or subassembly is.

2.7 The process of top-down

At the stage of conceptual design of a product development, the designers first generate a functional representation of the design and often take some methods to simplify the design. We often sketch a simple part to present a complex product, and we create a chart of processes to realize the functions of the product. We can get a real product by converting the chart of processes to the parts and subassemblies and detailing the simple parts. It is important for designers to construct a draft of product at the stage of conceptual design and change the parameters to get new designs.

Construction of parts and subassemblies will be determined at the stage of conceptual design and can be described as a code rule of product. Generally, this code rule will be active not only in this product development but also in the other products of the same series of products. It is useful to establish a code rule quickly for the development of products.

3 Managing parts by code

If we use the access of files of operating system, it is necessary for us to remember all the names of parts and subassemblies. If we use the other database systems, it will be costly. If we use the featuremanager design tree, we can find a part in this assembly easily, but we cannot find out the part which is not in this assembly. It is necessary to use a new method to solve this problem, which is not directly associated with the files or position of files. Because designers do not want to know the physical position of parts, their only concern is the logical position of files and this will be good for top-down design process and concurrent assembly. It is proposed that using codes to manage parts is a useful method.

3.1 Detaching of part and its file name

Detaching of part and its file name is very important. If we use part name as the file name for saving the part data, it will be hard to manage the versions of part. It is required that the part name is clear for us to understand the meaning of the part. The name will be long or express in English or other languages. For example, Chinese will be used in China to name the parts. It is not convenient to use these names to save the parts in file management of system. We must detach the name of part and the name of file for saving the data of part, so we can interact with assembly system by its Chinese name or English name which has a meaning for designers to remember. The physical position for saving the part will be determined by assembly system itself, and versions of part will be maintained by the system itself too.

3.2 Management of top-down and bottom-up

A product has been developed according to the following flowchart:

According to this chart the design is a repeating process and will be modified at any stage. It is required that the assembly design system can provide an ability to modify the design at any time. The top-down design is a design that designers make functional representations of product at first and then break down the complex functional representations into simpler functional representations and satisfy the functional constraints gradually. It is widely used in all industrial design. The bottom-up design is a design that designers design the parts or subassemblies at first and then assembly these parts or subassemblies into a assembly. It is hardly to find in traditional design, but it is used in computer aided design (CAD) because the top-down process is hard to realize while the bottom-up process is easily implemented.Computer aided design system is powerful to design a part for its extreme ability of 3D construction and can provide a realistic part which has no error for designers, so it is widely used in the development of product. In order to support the real design of product, the CAD system must provide the ability of top-down design. It is useful to use the code of product to manage the parts and subassemblies in top-down design and bottom-up design. Using code of product we can easily find the part we want and understand the progress of design. We can easily make functional representations of product and dispatch the tasks to other designers. The functional decomposition of product will be recorded automatically and design can be carried out by different designers simultaneously. The general engineer will know the progress from the messages among the designers. Each modification of part or subassembly can be known by the related designers. Because a CAD system has an ability of bottom-up design and by code of parts it can manage the parts for top-down and bottom-up design.

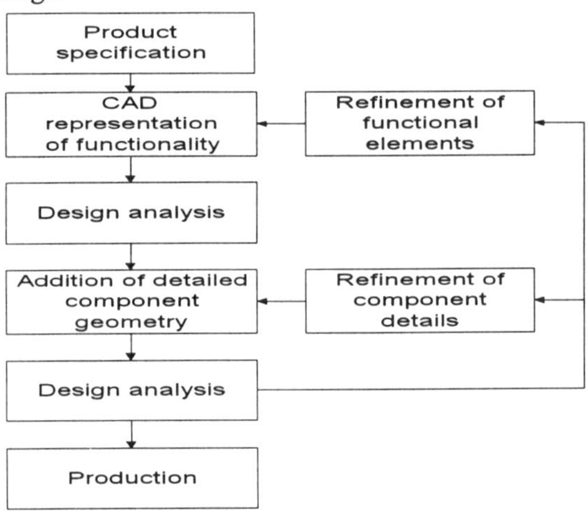

Fig. 3. Top-down process

3.3 Creating a code rule at run-time

In order to manage the development of a product, designers usually make a code rule for the product. The code rule is not only used in the development of this product but also used in the developments of other products. The code rule is usually designed to manage a series of products or even the whole products of a factory. We can easily make an interface of code rule for a factory, but it is hard to make interface for every factory and when the code rule is changed, it is required to modify the program code. It is proposed that we can make a code rule at run-time and when the code rule is changed we can update the code rule at run-time too. Designers are required to write a text file which describes the code rule by a certain regulation and lets the system to read in. The code rule written by text format is very easy for designers to read and modify.

When the code rule is read by system, the legality of code rule will be checked by the system. If there are some errors, the system will prompt the type of error and indicate the place where there is an error. The designers will not remember the code rule again because the system will give the definition of each code. This is a sample of text file describing the code rule:

```
// Code rule
{       // Code start
// The first code segment
        1, Machine tool, Describe the type of machine tool;
        {       @L,Lathe;
                P,Planer;
                ... ... }
// The second code segment
        1,Group, Describe the main constructs;
        {
                {               # Machine tool,0,1;
                        {       L; }
                        @6,Boring lathe;
                        ... ...
                          # Machine tool, 0,2;
                        {       P; }
                        @5,Hydraulic planer;
                        ... ...
                }
        }
... ...
// The fifth code segment
        1, Version, Describe the version of design , , T;
        {               @A,Z, Version;  }
... ...
// The twelfth code segment
        2, serial number of part, serial number of part, ,T;
        {               @00,99, serial number of part;      }
}                       // End of code
```

When the designers update the code rule it is necessary to ensure that the new code rule does not confuse with the old one.

The system supports code rules used in most of the factories and companies. It is required that the code rule has hierarchical and left to right structure. The code rule the system can solve is shown as Fig. 4

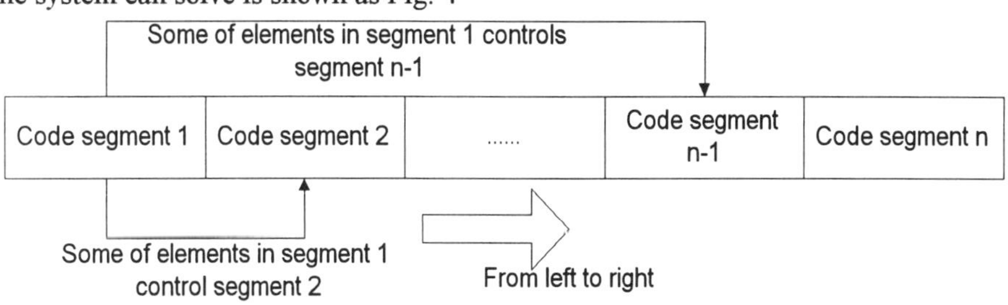

Fig. 4. Code rule of product that the system can solve

3.4 Various and intelligent requisition

Various and intelligent requisition is supported in this system. We can inquire a part in these ways:

Key in a code completely

Designers input the whole code and the system will match the code with code rule and find the demanding part. If there is not a part which has the same code as the designers input, the system will prompt an error message and provide some parts which have code similar to the input code and let designers select correct one.

Select the code under the navigation of the system

The system displays each code and its meaning from left to right, layer by layer. The designers select a code and come into the next code layer. When all codes of layers have been selected, the complete code is made. That is the code of part you want.

Key in the code incompletely

Sometimes we know the code partially for we know there is not other part which has the same code fraction, or the code has the default values, or we will search some parts which have the same regulation, we can input the code incompletely and the system will provide all of parts which have the same code we input.

Because some code rules are designed for people to read so it is hard for computers to recognize. This system is of some intelligence and can match most codes like designers.

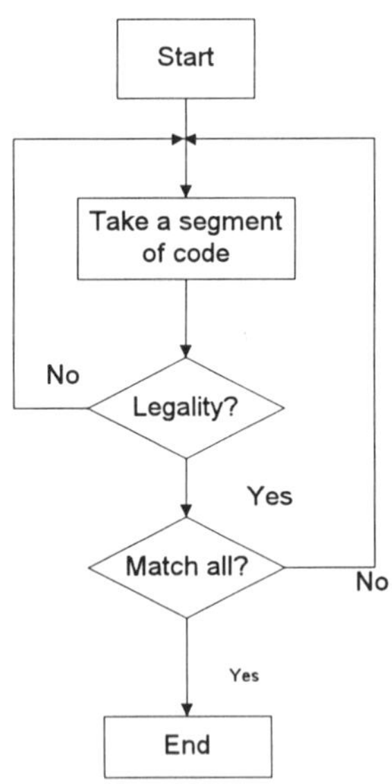

Fig. 5. the process of code matching

Key in a code which has no regulation

Sometimes we create a part only for temporary use and we save this part directly by file management system. We can input the part name and indicate the path of file without selecting the code rule.

3.5 Need no database
It will be costly for CAD system running in PC environment to have a database management system. Security and other properties of design system will be raised if we use database management, but it will decrease the speed of running of CAD system. Generally, if we use the database management system, the system will consume a lot of memory to run the database management system and the CAD system will run in the OLE way. It is required to increase the cost in PC environment. By the management of code of parts, the database management system is no longer required and this can be written into the CAD system. As part of assembly system, it takes little memory and has high speed to inquire the parts.

3.6 Management of assembly
An assembly has a hierarchical tree which is described as code in product to record the relation of parts and subassemblies. We not only deliver a hierarchical tree for designers to operate on the parts in this assembly but also deliver a way of code of parts to insert a part or subassembly which is not in this assembly before.
BOM is one of the contents of assembly design. This system takes little memory and can run in higher speed. It is easy to produce BOM tables for an assembly and also for all products of a factory.

3.7 Mini PDM system
Product Data Management(PDM) is an important technology. It stores and controls the data that is critical to product design, manufacturing and technical support, so it manages the process of product development. It is required to base on the database management system and becomes very complex. Although our system loses some properties, it is helpful to support conceptual design and top-down design because the code rule implies information of design process. This system is the mini system of PDM and the data structure is already used in the InteMAN system(the PDM system of CAD Center of HUST).

4 Data structure
The assembly system consists of various software components written in C++, and C++-based ACIS solid modeller is used. The structure of code of product has four classes shown in Fig. 6.
The classes of code of product are derived from class of ENTITY of ACIS which is the base class of model. They inherit the features of ENTITY of ACIS and are easy to access data of code. CODE_TREE is the class which describes the whole code rule; class CODE describes the value and meaning of each code segment; class CODE_GROUP describes the data of groups in each code segment; and class CODE_ELEMENT describes the value and meaning of each code element or records the value of code of current part.

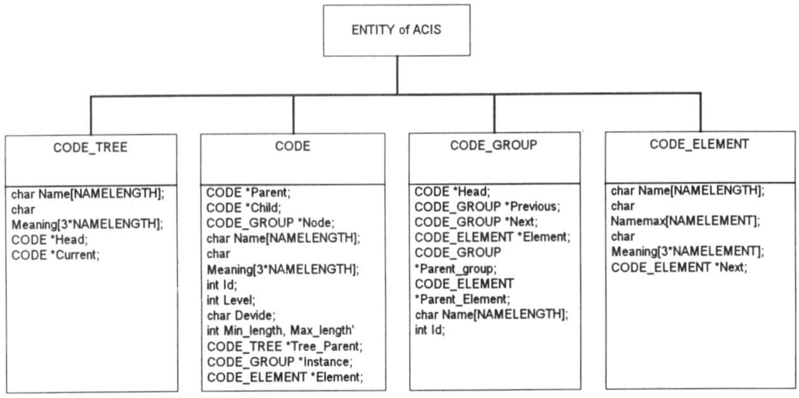

Fig. 6. the components of code of product

5 Conclusion

The data structure for managing parts and subassemblies in complex assembly system is used in CDS. In this system the designers need not load and save the parts or subassemblies directly by their file names. They can use Chinese words to represent the parts or subassemblies, and it is useful for Chinese engineers to design products.

In this system designers can create a code rule at run-time and inquire a part in various ways. One can key in a code completely, or select the code under the navigation of the system or key in the code incompletely. This system has some intelligence and can search and match the code. If one likes to use file management system, he can key in a code which has no regulation. The code data of product has been built into the system so it makes the modelling system effective.

This system will be helpful to top-down design process and concurrent assembly because the code of product can represent the process of design.

Reference

1. Kunwoo Lee and David C Gossard, A hierarchical data structure for representing assemblies: part 1 , Computer-Aided Design, Vol. 17, No. 1, pp. 15-19, 1985.
2. D.A Beach & D.C. Anderson, A Computer Environment for Realistic Assembly Design , Proceedings of The 1996 ASME design Engineering Technical Conferences and Computers in Engineering Conference August 8-12, 196, Irvine, California. 96-DETC/CIE-1336.
3. Gritt Ahrens, Oliver Tegel, Conceptual product development , Proceedings of The 1996 ASME design Engineering Technical Conferences and Computers in Engineering Conference August 8-12, 196, Irvine, California. 96-DETC/EIM-1402.
4. Rajneet sodhi and Joshua U Turner, Towards modelling of assemblies for product design, Computer-Aided Design Volume 26 Number 2 February 1994, pp.85-96
5. Kim, S. H. and Lee, K. An assembly modeling system for dynamic and kinematic analysis, Computer-Aided Design, Vol. 21, No. 1, pp.2-12, 1989.
6. Wang, N. and Qzsoy, T. M., Automatic Generation of Tolerance Chains From Mating Relations Represented in Assembly Models, J. of Mechanical Design, Vol. 115, No. 4, pp. 757-761, 1993.

An Intelligent Mechanism to Support DFX -abilities in Automated Designer's Environment

Wulanmuqi and Jiati Deng

The State 863/CIMS Design Automation Laboratory (DeAu Lab.)
Institute of Manufacturing Systems (720 Institute)
Department of Manufacturing Engineering
Beijing University of Aeronautics and Astronautics, Beijing, 100083, P.R.China

Abstract. The paper presents an intelligent DFX mechanism for product design. The authors solved this problem by taking the product life-cycle issues into consideration during design process from the designer-oriented perspective of view. At first the designer-oriented computer environment *DesignerSpace* is introduced for understanding the designer how to implement design activities and DFX method better. In order to integrate design knowledge from downstream aspects for the optimized decision-making, an intelligent DFX mechanism is developed to incorporate knowledge-base, algorithm-base and monitoring/debugging tools. *DesignerSpace* with DFX abilities is implemented and the further development is strongly intended.

1 Introduction

As we know, time, quality and cost have been three key criterion of product competitiveness in the market, and now the focus is placed on product development time reduction from which other metrics probably can be obtained. The whole product design and development process is carried out along the timeline of product life-cycle. At the stage of design decision making, 70% to 80% of the product life cycle value is formed at the expense of about 20% of product life cycle[1]. So it will be most effective means to get time advantage by continuously improving product design, especially at the early stage - the stage of conceptual design. Concurrent Engineering (CE) is a product life-cycle-concerned methodology. In addition to the functionality that product concerns, CE emphasizes the optimization of both product and product development process. Distinct from the traditionally sequential way, the design is largely depentent on the integrated product development process. So considering DFX(Design For X) into the early stage of design that could be one of the fundamental concerns will affect the optimization of product and process. DFX methodology embodies product design philosophy of CE, where X corresponds to manufacturing, assembly, serviceability, disassembly and recyclability , or other post-manufacturing issues. DFX takes the requirements into account from manufacturing, assembly, marketing, maintenance and recycling, etc. The final result is more efficient product development cycle: less development time, improved quality and less cost. The challenge drives research on design methods of product and product development process, and their implementations. According to S. Sivaloganathan et al.[2], an integrated design system "Design Function Deployment (DFD)" was proposed to incorporate Quality Functions and customer requirements throughout design. Feng and Kusiak[3], Hashemian and Gu[4], Schmitz and Desa[5], Hsu, Lee and Su[6], Boothroyd and Dewhurst[7], Makino, Barkan and Pfaff[8], and Ishii, Eubanks and Marks[9] have given their contributions. Topics on design for manufacturing (DFM) and design for assembly (DFA) are the most common areas.

The authors propose to establish a designer-oriented environment named *DesignerSpace* as an intelligent framework assisting in fulfilling DFX, and also it functions as an organizer to implement each unit activity in product design and development processes. All product design and development activities organized, managed and implemented by *DesignerSpace* are controllable and reconfigurable. *DesignerSpace* establishes the infrastructure for the design and development processes, and can be automated well. So *DesignerSpace* behaves as a knowledge-based and distributed intelligent problem-solving mechanism adopting software agent technology. DFX is treated as a kind of mechanism integrated in *DesignerSpace* for optimizing the product design at the designated activity stage under *DesignerSapce* management. The paper is organized as follows: after a brief introduction in the first section, *DesignerSpace* is described in Section two. Then the intelligent DFX mechanism and its components in *DesignerSpace* are explained for product design. Finally the implementations and conclusions are presented, and the future research is expected.

2 *DesignerSpace*: Designer-Oriented Environment

Product design and development processes[1] are performed by a series of concurrent or sequential engineering activities which have certain inputs, outputs, control and leverage mechanism[2] . So every design activity is organized and carried out within/by *DesignerSpace*, that is, a working space which can help the designer to accomplish the design activities[3] in the way of automation. Such a *DesignerSpace* is developed to incorporate utilities to decompose and coordinate design processes, provide design assistance tools, resources and optimized design methods, and carry out each design activity partitioned from the process under the constraint/control which is specified or configured from the higher level for management of the activities. The framework of *DesignerSpace* illustrated in Fig.1, has one *Management* module and three

---

[1] : Generally there are the following major processes in product design and development: customer requirement collection, planning, design, process planning, production, service and maintenance, etc.
[2] : Their definitions conform to those of IDEF0.
[3] : Design activity is a distinct stage of a design's development whose boundaries are usually characterized by changes in design fidelity[11].

fundamental modules: *Process* module, *Resource* module and *Mechanism* module.

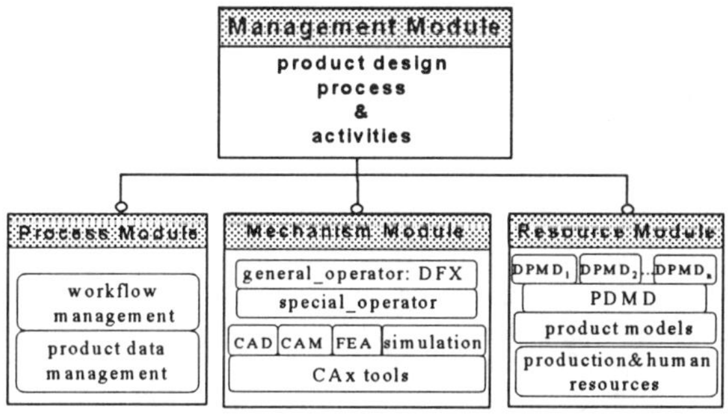

Fig. 1. The organization and management of *DesignerSpace*

- *Management* module, functions as a platform integrating *Process* module, *Resource* module and *Mechanism* module, and managing the operation procedure of deign activities. ① For certain activity a_i (i=1,...,n) starting at the time t_i, any creation, modification or improvement on product can be precisely described by the changes of Product Definition Model Data(PDMD) in *Resource* module. So the activity a_i is said only an operation on PDMD. ② It accesses to the existing PDMD created by the previous activity a_{i-1}, and outputs new instance of PDMD generated by current activity a_i within the timeframe from t_i to t_{i+1}, which is the beginning time of next activity a_{i+1}. ③ Using DFX and CAx tools from *Mechanism* module can provide the means to solve problems adaptively. These operations described above is for solving product design problems levered by a knowledge-based and distributed intelligent multi-agent system. ④ Friend user interface is easier for different disciplinary engineers to solve their specific engineering design works.
- *Process* module, like a leader organizing design process in higher level, is responsible for design process decomposition, workflow management and information management. It provides *DesignerSpace* facilities that transmit the designer concrete design task to be completed in the form of task order, and permit the designer to partition a design problem into manageable subproblems. The sample of a task order is shown in Table 1. The leader of a product design team is able to control and monitor the workflow of product development.

Table 1. Task order from *Process* Module

Description	Input	Output	Resource	Coordination
Task_#:	Item:	Item:	Hardware:	Perspectives:
Goal:	From_whom:	From_whom:	Tools:	
Designer(s):	Time:	Time:	Document:	
Constraints:				
Duration:				

- *Resource* module, offers the meta-data of product information models and data concerning product design processes and activities, and specifies the real production system components covering the technological and human resources referenced by product design and development processes. The physical and logical relations between them are defined. According to the workflow management in *Process* module, the resources assigned for each designer to finish his design task are reached from *Resource* module.
- *Mechanism* module, provides CAx design assistance tools required for the designer's easy and effective design activities throughout design processes. As for different design tasks, different product model data are required as inputs and outputs, and different CAx tools are used; for example, CAD/CAM system with 3D modeling, 2D draughting, assembly, NC codes generating and other functions. When a designer receives the definite design task and possesses CAx tools helping finish the task, *Mechanism* module also offers design knowledge about how to solve the problem with optimization. In the following it will be discussed in detail.

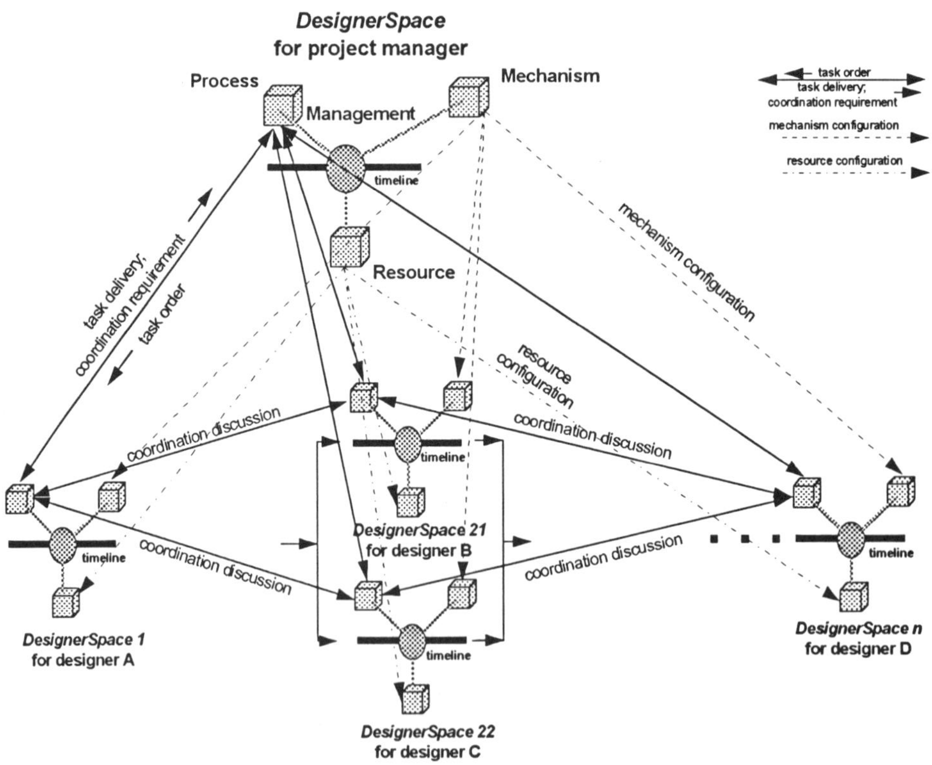

Fig.2. The operation of the design activities in *DesignerSpace*

When a design process is partitioned into various design activities and designated for different team members to finish, *DesignerSpace* is actually worked as a designer working space for fulfilling their activities. One *DesignerSpace* corresponds to a design activity. During the product design process, many *DesignerSpaces* are required to work cooperatively although the contents of different *DesignerSpace* from *Process* module, *Resource* module and *Mechanism* module are not the same at all. The activities of design process carried out within *DesignerSpace* are illustrated in Fig.2.

As seen in Fig.2, *DesignerSpace* is originally constructed for the team leader (such as the project manager) at top level with all configurable resources, mechanism and process timeline, and then is copied to low level *DesignerSpace* (numbered 1, 21, 22, or n) as a template for each team member (A,B,C or D) in specified design activity with configured resource and mechanism. Within each activity that a *DesignerSpace* manages is the operation procedure which requests the access to resources and mechanism. The multiple agents within *DesignerSpaces* cope with the fulfillment of design activities. Project leader at top level has a project-agent to help organizing team, configuring resources and mechanism, and coordinating team activities. At middle level, each team member has engineering-agent help to cooperate with other members and accomplish his own engineering activity. At low level (which is not shown in Fig.2.) the R-agent help each middle level *DesignerSpace* to manage and access its resoures adaptively.

3 Intelligent DFX Mechanism

From the information integration point of view, design activities are the processes of the information handling. *Mechanism* module in *DesignerSpace*, like an operator, operates on the information (which can be thought as the input or output) being processed during certain design activity, as shown in Fig.3. Therefore *Mechanism* is a comprehensive framework which intends to integrate more computational technologies along the whole product design process.

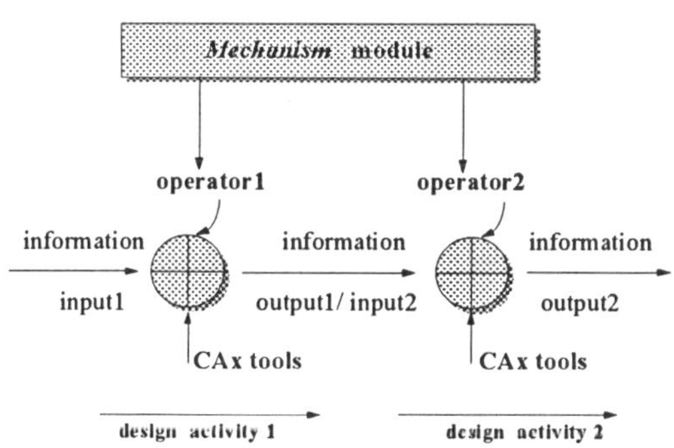

Fig.3. The role of *Mechanism* module in the design information processing

Because of different engineering application domains involved in solving design problem, *Mechanism* offers *general_operator* and *special_operator*. The former is proposed for general mechanical product design, such as DFX; and the latter has definite application scope and needs special design technologies, for example, blankdisc design in aircraft. Any mechanism operator is giving the constraints on design, but CAD system is needed to help construct PDMD under the constraints. *General_operator* or *special_operator* are made up of three basic components: *knowledge_base, algorithm_base* and *monitoring/debugging*. The components of DFX mechanism with *Process* and *Resource* modules in *DesignerSpace* are described in EXPRESS-G in Fig.4.

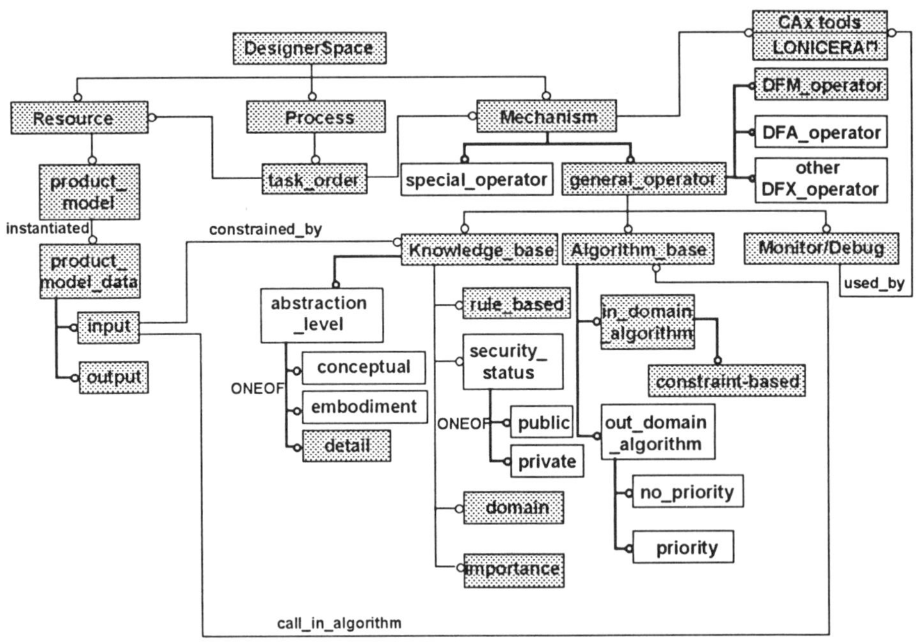

(*): LONICERA is a integrated CAD/CAPP/NCP system based on STEP, developed by DeAu Lab, BUAA.

Fig.4. The components of DFX mechanism in *DesignerSpace*

3.1 Product models and operated product model data

The DFX intelligent mechanism operates on the product model data for obtaining the optimized product model data which are of DFX abilities. The product models used in *DesignerSpace* are structured into two levels: geometric_model and feature_model[12].

- Geometric_model, is the representation of product geometry and topology using geometrical and topological elements based on AP203/STEP[4] .
- Seeing that the geometric_model lacks engineering semantics, feature_model describes the geometrical shape with product design and engineering knowledge

[4]: AP203/STEP: Part 203, Application Protocol: Configuration controlled design;

in features. Referencing AP214/STEP[5] , the feature-based product model called product_definition_model_data (PDMD) is established, and its transformed model data called domain_product_model_data (DPMD) with a particular domain application are derived from the feature_model.

The product model data operated by DFX mechanism are the instantiated data of the feature_model: PDMD. They are the basis of information integration in the product life cycle. The support of the geometric_model are indispensable.

3.2 Knowledge Base (KB)

In such a DFX intelligent mechanism, design knowledge is involved in multi-disciplines and multi-domains. So *Knowledge_base* is a collection of design requirements representing expertise from different domains covering the product life cycle, such as manufacturing, assembly, marketing, service and recycling, etc. These design requirements need to be classified, abstracted and represented as knowledge written and stored in *Knowledge_base*. For example, from manufacturing point of view, fabrication means (such as machining, casting, forging or moulding), equipment (such as machine tool, cutting tool, or die), and shopfloor production resources and conditions, constrain the product design. *Knowledge_base* is intended to support the design process from conceptual, embodiment to detail design. Therefore each piece of design knowledge is labeled which domain it belongs to, and is abstracted or refined which design level it is in. There are the conceptual level, the embodiment level and the detail level according to different design phases. Such classification of design knowledge is convenient and efficient for information management during design processes. But more difficulties exist in the knowledge representation during the conceptual design and the embodiment design than during the detail design.

Whenever some knowledge is added into or deleted from *Knowledge_base,* the consistency with other knowledge in either the same or different domain will be checked. If it is violated, changes in knowledge are required until the consistency is achieved and remained. The representation of knowledge in *Knowledge_base* is rule-based and object-oriented. The inheritance and poly-morphism support the design information management of *Knowledge_base.*

The general representation of knowledge is described in the following form by reference to [13].

```
{Knowledge:
    {form                   ;the equation or logical relation
     abstraction_level      ;name of the abstraction level
```

---

[5] : AP214/STEP: Part 214, Application Protocol: Core data for automative mechanical design processes.

```
            domain                    ;which domain offers this piece of knowledge
            constrains                ;product model's entities knowledge concerns
            importance                ;on a scale of 1 to 10
            parent                    ;supertype of this knowledge
            child                     ;subtype of the knowledge
            security_status           ;public or private?
            }
    };
```

3.3 Algorithm Base (AB)

After *Knowledge_base* about certain kind of product design is established, decision-making methods are needed. *Algorithm_base* in the DFX intelligent mechanism is responsible for it. Since one DFX concerned is not optimized at expense of the others, DFM, DFA, DFS(Design For Service), or DFR(Design For Recycling) should be considered concurrently. *Algorithm_base* provides two algorithms which are *in_domain_algorithm* and *out_domain_algorithm* respectively.

Considering only one DFX, such as design for manufacturing, usually there is an uniform representation in product model and knowledge. We can find a common algorithm. Constraint-based design is one of the effective methods to cope with this kind of DFX problems. So *in_domain_algorithm* offers inference mechanism and search algorithm inside certain domain for one DFX. While taking two or more DFXs into account at the same time, for example, DFM and DFA, the characteristics of DFM and DFA determine that different methods are used for different domain applications. Therefore *out_domain_algorithm* is required and gives compromising the conflicts among more than one DFX seeking for the optimized design result. If necessary, human interactions are in demand. There are *forward_processing* and *backward_processing* methods in *in_domain_algorithm* for domain knowledge inference. For *out_domain_algorithm,* the importance or priority of each rule in domain knowledge and that of each domain knowledge in *Knowledge_base* has to be defined.

3.4 Monitoring/Debugging

Monitoring/Debugging in fact is a visual tool for monitoring the design activity and debugging the design problems. The elements of *Knowledge_base* can be checked; for example, the relationship between rules is able to be visualized in network diagram. Any correction or change in rule network will be reflected in *Knowledge_base* written in text form. During inference processing, the operation of which rule is active or not are displayed. So the designer can set or clear breakpoints dynamically. It is very easy for the designer to operate the design activity. Then the satisfactory and optimized result about part or component can be displayed in CAD/CAM system from *Mechanism* module.

4 Implementation and Conclusions

The intelligent DFX mechanism is developed within the framework of *DesignerSpace* and partially implemented to achieve optimized mechanical part/component and its process design during the detailed design phase. A prototype system *Intelli-DFX* coded in C++ on HP workstation can be demonstrated now. The core part is a STEP based integrated CAD/CAM system-LONICERA (also the registered trademark), which is being developed in our institute. LONICERA system has integrated a 3D parametric feature based modeler on AP214/STEP and a STEP model data access environment , and has integrated workflow management module. The implementation of DFX mechanism in *DesignerSpace* framework is briefly illustrated in Fig.5.

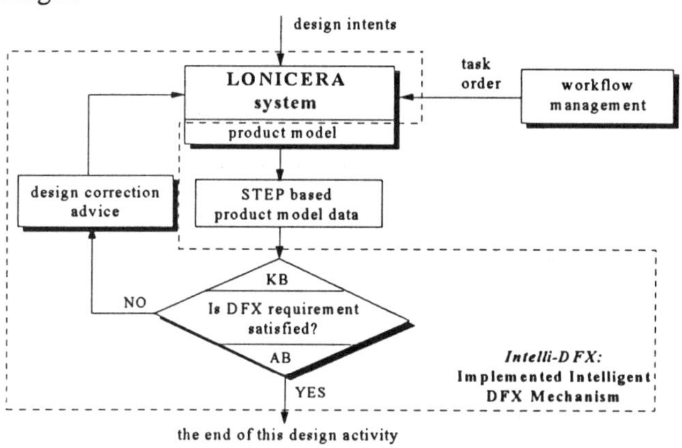

Fig.5. Implementing DFX mechanism within *DesignerSpace*

Designers write design requirements and knowledge concerning their design activities into the knowledge base in their own *DesignerSpace* during the process of requirement analysis. DFX mechanism works while their design tasks being in progress. Once the product model data, which could not be the end required output, are produced, *Intelli-DFX* is provided the designer to check the manufacturability, and informs him of design problems in the form of text or visualization for correction.

The main idea and implementation of the designer's design environment *DesignerSpace* with intelligent DFX abilities have been presented in this paper. *Process, Mechanism* and *Resource* modules integrated in *Management* module constitutes the computer architecture of *DesignerSpace. Mechanism* provides the part/ component/product design with intelligent abilities. The implementation helps us find out problems to be improved and promote the further development.

1. A single function of DFX, like only DFM or DFA, has the single objective besides satisfying the functional requirements. Research in the past was mostly involved in one separate DFX. In fact, a product design problem is always constrained by more than one objective at the same time. So we are developing

further to integrate DFM and DFA in simultaneous consideration of manufacturing and assembly.

2. High performance of DFX mechanism is tightly related with those of the other parts in *DesignerSpace*. So the fine integration of *Process, Resource* and *Mechanism* modules and development of product data management will increase the efficiency of design more.
3. Since the designers in a design team concern different design knowledge and may also be geographically distributed, a distributed intelligent design cooperative system is needed for the operation of *DesignerSpace* using multi-agent technology. Upon the developed *DesignerSpace*, the research supported by National 863/CIMS Hi-Technology Research and Development Project (865-511-706) is underway concerning about distributed artificial intelligence for virtual product design environment based on Internet/Infranet.

References

1. P.O'Crady, D.Ramer, J.Bolsen, Artificial intelligence constraint nets applied to design for economic manufacture, *Computer Integrated Manufacturing*, **1**, 4 (1988) 204-209.
2. S.Sivaloganathan, N.F.O.Evbuomwan, A.Jebb, The development of a design system for concurrent engineering, *Concurrent Engineering:Research and Applications*, **3**, 4 (1995) 257-269.
3. C.X.Feng., A.Kusiak, Constraint-based design of parts, *Computer-Aided Design*, **27**, 5 (1995) 343-352.
4. M.Hashemian, P.Gu, A constraint-based system for product design, *Concurrent Engineering:Research and Applications*, **3**, 3 (1995) 177-186.
5. J.Schmitz, S.Desa, The development of virtual concurrent engineering and its application to design for producibility, *Concurrent Engineering:Research and Applications*, **1**, 1 (1993) 159-169.
6. W.Hsu, G.Lee, S.F.Su, Feedback approach to design for assembly by evaluation of assembly plan, *Computer-Aided Design*, **25**,7 (1993) 395-410.
7. G.Boothrody, P.Dewburst, Product design for manufacturing and assembly, *Manufacturing Engineering*, 4 (1988) 42-46.
8. A.Makino, P.Barkan, R.Pfaff., Design for serviceability, *The 1989 ASME Winter Annual Meeting*, 1989, USA.
9. K.Ishii, C.F.Eubanks, M.Marks, Evaluation methodology for post-manufacturing issues in life-cycle design, *Concurrent Engineering:Research and Applications*, **1**, 1 (1993) 61-68.
10. J.T. Deng, Product design and development *technical report*, Institute of Manufacturing Systems, Beijing University of Aeronautics and Astronautics, 1996
11. M.A.Hale, J.I.Craig, F.Mistree, D.P.Schrage, DREAMS and IMAGE: a model and computer implementation for concurrent,life-cycle design for computer systems, *Concurrent Engineering:Research and Applications*, **4**, 2 (1996) 171-186.
12. J.T.Deng, LONICERA- a STEP based CAD/CAM integrated system, *Proceedings of Conference on CALS*, Nov. (1996), Singapore.
13. Navin chandra, Mark S. Fox, Eric S. Gardner, Constraint management in design fusion, *Concurrent Engineering: Methodology and Applications*, edited by P. Gu and A. Kusiak, Elsevier Science Publishers B.V. (1993) 1-30

FACTORY AUTOMATION

Shopfloor Integration

JAVA Implementation of Shopfloor Access and Monitoring of A Manufacturing System

Ong Chee Wee, Osamu Kaneko, Kang Siew Hwa
Japan-Singapore Institute, Nanyang Polytechnic

Abstract

The software developing environment has been changing very rapidly with the explosive growth of the Internet as well as the Intranet over the last two years. The development of the JAVA language by Sun Microsystems is one of the most significant event in the Internet world. In this paper, we would like to introduce the feasibility and possibility of using the JAVA language for implementing shopfloor access and monitoring of a manufacturing system.

1 Introduction

An automated manufacturing system as shown in figure 1 is currently being developed in Nanyang Polytechnic. Its major system building blocks are four robot assembly cells complete with conveyor systems, an automated guided vehicle (AGV) and an automated warehouse (AWH). A shopfloor control and monitoring network system is used to coordinate and schedule the activities among the various equipment in the automated assembly system. This 10BaseT Ethernet network system consists of a host computer, a network hub and a number of node computers. The various equipment in the automated system are linked to the node computers through serial communication interfaces. There is also a gateway node computer which is connected to the Internet to link the system to the external world. This setup will provide an Intranet for borderless access and monitoring of the automated manufacturing system.

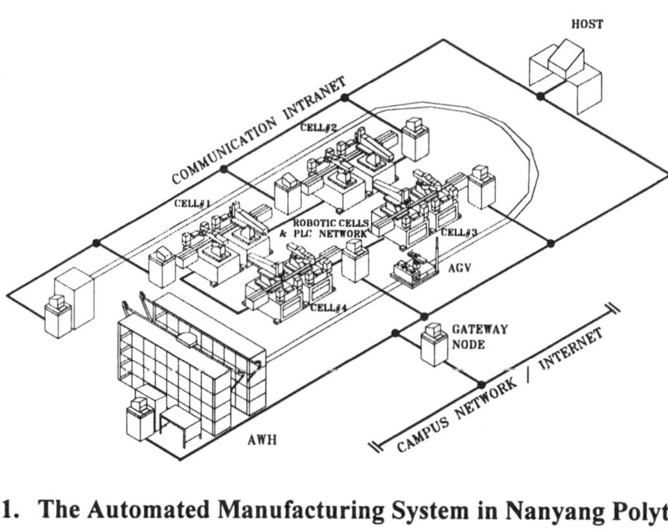

Figure 1. The Automated Manufacturing System in Nanyang Polytechnic

This borderless access and monitoring of the automated manufacturing system can be achieved in the following main domains as shown in figure 2.

i. **Automated Manufacturing System**
   - Flexible Assembly using 4 Robot Assembly Cells With Conveyor Systems
   - Automated Material Storage/ Retrieval Using An Automatic Warehouse
   - Intelligent Material Transportation Using An Automated Guided Vehicle

ii. **Host Computer**
   - Scheduling & Production Planning
   - Data logging & Statistical Data Processing
   - Database for Automatic Warehouse, Products and Production Output
   - Remote Control & Monitoring of Assembly Cells, Automatic Warehouse and Automated Guided Vehicle
   - Communication Error Detection
   - Contingency Procedures

iii. **4 Cell Node Computers With Vision Systems**
   - Uploading/Downloading of Robot Application Programs
   - Control & Monitoring of Robots (2x) during operation
   - Communication with Host Computer

iv. **PLC Network Node Computer**
   - Uploading/Downloading of PLC Application Programs
   - Control & Monitoring of PLCs (4x) during operation
   - Communication with Host Computer

v. **AWH Node Computer**
   - Control & Monitoring of the AWH through its controller unit
   - Editing, Updating & Maintenance of the AWH Database
   - Communication with Host Computer

vi. **AGV Node Computer**
   - Remote Control & Monitoring of the AGV through its controller unit
   - Scheduling of the AGV operations
   - Communication with Host Computer

vii. **Gateway Node Computer**
   - Remote access and monitoring terminal
   - Communication window to the external world via the Internet

viii. **Fast Ethernet**
   - Data communication highway in real-time (100Mbps)
   - Transmission of audio and video signals for remote monitoring

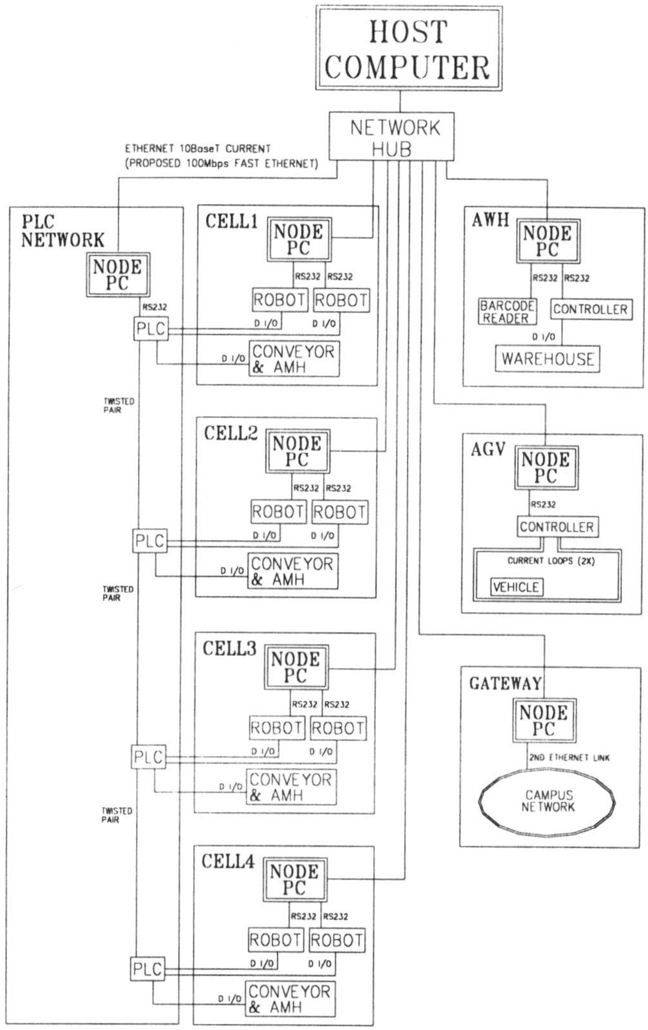

Figure 2. Block Diagram of The Automated Manufacturing System Incorporating A Borderless Access and Monitoring Through The Internet

The system would demonstrate the use of current networking technology in a factory automation environment providing shopfloor supervisory control and monitoring, including SCADA and MRP functions through an Intranet which is connected to the Internet. This method of access, control and monitoring of an automated system through a borderless network is the current trend in factory automation. It would also provide a platform for the development of solutions for industrial automation systems as well as research in the area of remote control, access and monitoring through the Internet.

JAVA language is growing in popularity at lightning speed for applications through the Internet. This is because the JAVA environment provides the development platform for independent stand-alone applications and protocol handlers for distributing dynamic data through the Internet. JAVA language is currently being improved, and many people are working on it through the Internet. In addition to this, JAVA language is easy to access and can be developed seamlessly across various platforms from small to large computer systems, including UNIX, Window NT and Window95 environment.

2 Features of The JAVA Language

The Java language allows platform-independent program code to be dynamically loaded and run through the Internet. This functionality provides revolutional changes to the passive nature of the Internet and the World Wide Web. The followings are some of the features that are incorporated in the Java language that make it the most promising protocol for the Internet in the near future.

Java is an Object-Oriented language that allows the inheritance and reuse of program code both statically and dynamically for run-time extensibility. Multithreading are provided at every level, allowing independent task control by each thread.

Java is platform-independent, therefore its program can run on any machine that has the Java interpreter ported to it. This allows application programs written in Java to be executed in an Open Network System through Internet.

Java language also support multimedia programming, where animation with sound can be incorporated into the application programs.

Java can also access SQL database. This allows a common database to be built within the network for data access and modification can be performed from each node.

3 The Experimental System

The experimental system, as shown in figure 3, has been setup to demonstrate the capabilities of borderless remote access and monitoring. It consists of a Sun workstation, which acts as a host computer, and is configured as a DNS server, as well as a SQL Database server using mSQL. There are two node computers which are used to link the Automatic Warehouse and the PLC network to the host computer. One PC has been assigned as the gateway to the external world through the Internet. Its Web's page shown in figure 4. The Host computer controls the production scheduling and coordinates the AGV's and AWH's schedules in parts/products transportation between the AWH and the Robot Assembly Cells. The hardware of the network is 10baseT Ethernet.

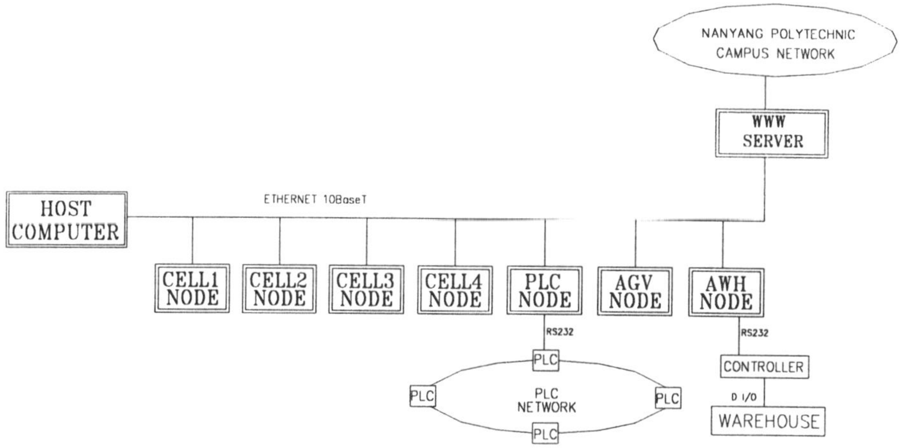

Figure 3. Block Diagram of The Experimental Borderless Access and Monitoring - Manufacturing System

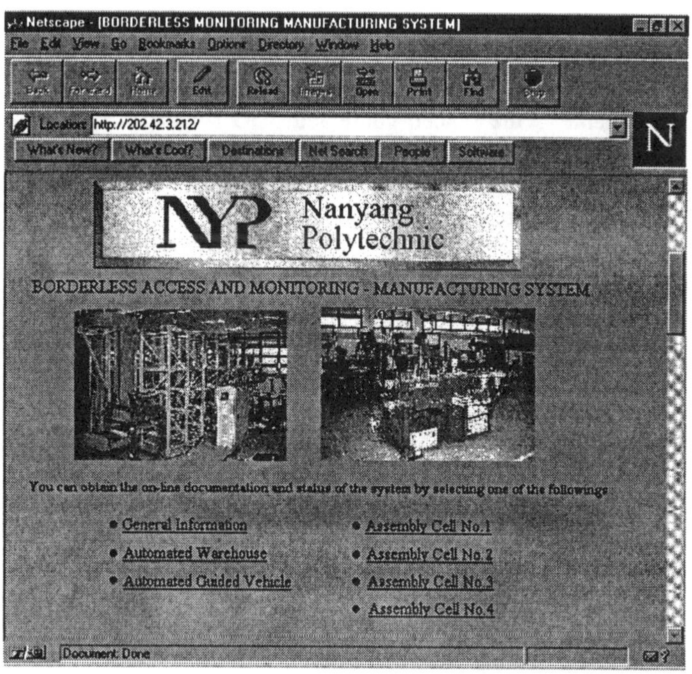

Figure 4. The Gateway Computer's Web Page

3.1 Device Level Control through JAVA Language

JAVA language is well known as a development tool for WWW home page, therefore it is strong in graphic user interface, animation and networking. However, for low level control of devices, such as PLC or Robot, the RS232C serial communication is commonly used. Here, we discuss how to incorporate the RS232C serial communication in JAVA application software.

We used Runtime and Process classes in "java.lang" package. First, prepare an executable file compiled in C language, for the basic RS232C control and communication. This executable file can then be called using Runtime and Process classes in the JAVA application program as shown below.

```
String command[]=new String[2];
try{
    Process process = Runtime.getRuntime().exec(command);
    inFromCommand = new DataInputStream(new
    BufferedInputStream(process.getInputStream()));
    String line = inFromCommand.readLine();
   }catch(IOException e){}
```

Here, "command" is String array and command[0] is executable file name and command[1] is outgoing data through RS232C. The response from the RS232C is stored in the "line" string. JAVA can pick up the response by using "process.getInputStream()", the related command in C language is printf(). If the device is "dead" or not connected, the program will check device status and report "NO POWER". Without this process, the JAVA application will hang the PC, because the C executable program will be out of control.

The following program segment can be used to detect the status of the local device that is linked to the node PC through its RS232C port after powered up.

```
if((0x10 & inportb(0x2fe)) == 0x00)
  { printf("%s",err);
    exit(1); }
```

Generally, the procedure to initialize the system is to turn on the Host then follow by the node PCs. The Host has to recognize the status of the PCs continuously. The "ping" command can be used to detect the most low level node PC conditions, for example, no power or no connection. We realize continuous monitoring of the node PCs by executing the getStatus() method of SyncClient class program every 5-10 seconds as shown in the following JAVA program listing.

```
class SyncClient{
DataInputStream in;
PrintStream out;
Socket client;
boolean flag=true;
Status status;
String command[]=new String[3];
String response;
int port;
String node;
public SyncClient(int p,String n,Status s){
      status=s;
      node=n;
      port=p;
      command[0]="/usr/sbin/ping/";
      command[1]=node;
      command[2]="1"; }
public boolean getStatus(){
try{
      Process process = Runtime.getRuntime().exec(command);
      DataInputStream dis = new DataInputStream(process.getInputStream());
      response = dis.readLine();
      }catch(IOException ioe){ System.out.println(ioe.getMessage());
                                    return false; }
if(response.substring(0,2).compareTo("se") == 0)
    { try
        { client = new Socket(node,port);
          out = newPrintStream(client.getOutputStream());
          in = new DataInputStream(client.getInputStream());
         }catch(IOException e){ flag = false;
                                    status.toNotReady(); }
      if(flag == true) { status.toReady();
                         status.onPower(); }
      else{ status.noPower(); }
      return true; }

public void close(){
try{
      out.close();
      in.close();
      client.close();
        }catch(IOException e){ System.out.println(ioe.getMessage());}
   }
}
```

3.2 SQL database

Database management is important for networking in an automated manufacturing system. We used mSQL, which is available over the Internet. The mSQL language offers a subset of the features provided by ANSI SQL. mSQL is a shareware product authored by David J. Hughes, and is free to educational and research organizations. Sun Microsystems provides JDBC Database Access API for JAVA. The JDBC drivers for mSQL are also available through the Internet.

The Main Databases for our system are as follows.

Autowarehouse	: Products/Parts in the warehouse
Product_part	: Products/Parts ID codes and Barcodes
Production	: Production Data, ie. Lot number; Yield; Time; etc.
Error Log	: Production Error, ie. Error code; Node number; Time; etc.

The Host controls the overall production schedules, and therefore, these databases are updated or modified by Host only, except the Error log. The node PCs can get information from these databases. They enter data into the Error log when error occurs. The two examples below, show the use of the SQL. The first example is to create table Product_PartDB, the second is the use of the SELECT command.

```
msql.Query("CREATE TABLE Product_PartDB
                    (ID int primary key,name char(20),barcode char(5))" );

result = msql.Query("select * from AWH ORDER BY position");
```

3.3 Typical JAVA Network Program Between Node PC and Host Computer

The network is based on socket connection of TCP/IP. In the case of this type of connection, the Server or Client status does not depend on whether it is a Host or Node PC, but depend on the network program. A server application program listens to a specific port waiting for connection requests from a client. When a connection request arrives, the client and server establish a dedicated connection. We prepare three connections between Host and each Node PC. The Node PC has two server programs and the Host has one server program. The Thread program is used for server application. The related instances are transferred to these communication programs to do some jobs in accordance with the information from remote side.

The followings are examples of the JAVA communication classes that have been developed in the Host Computer and the Node PC for Client/Server applications.

Cell Node PC Server Programs

SyncClientHandler : Receive the access from Host and send back the Node's status.

The status data for testing are random numbers between 1-50 generated to simulate production data. It is periodically accessed, for example every 5-10 sec. Using "Thread" to enable it to accept a call at any time.

AsycClientHandler : Receive the access from Host and execute some jobs.

In this case, get an instance of Content class(extends Applet), and do some jobs. Using "Thread" to enable it to accept a call at any time.

Host Server Program

ClientHandler : Receive the access from Node and execute some job.

In this case, issue a musical chime to call for attention. This has an instance, "audio"(instance of class Audio). Using "Thread" to enable it to accept a call at any time.

Host Client Programs

SyncClient : This has an instance, "status"(instance of class Status)

This is used to get the status information from Node and set its current status on the Host's graphic user interface.

AsyncClient : This has an instance, String "sendout"

This is used to send out data to Node. It has a function for sending out only.

Node Client Program

Client: This is used to call the attention of the Host.

In this case, send the help command, when the Help button is pushed.

Using these JAVA communication classes, the Client/Server applications that have been established between the Host and each of the Node PCs are illustrated below.

Host Computer		Node PC
SyncClient(status)	→	*SyncClientHandler( )*
ClientHandler(audio)	←	*Client( )*
AsynClient("updateDB")	→	*AsyncClientHandler( )*

3.4 Open Network

We prepare one Node PC (Windows NT) which serve as the gateway to the external world through the Internet. This Node PC has two network cards, one is connected to the local Intranet and another is connected to our campus network. The WWW server program that has been installed is the "Front Page" of Microsoft Internet Explorer. On this Gateway PC, The WWW server and one JAVA program are running concurrently. The JAVA program accesses the Host program every 150 sec. and update the file in "Front Page" of the WWW server directory. Therefore, it is possible to access the production data and status of the automated manufacturing system through Internet from anywhere by using an Internet browser.

In the JAVA applet program in the WWW server directory, the file can be read and stored in "is" instance as follows, and the content can be shown as graphical way. The HTML(Hyper Text Markup Language) refer the applet class in it.

```
is=DataInputStream;
try{ is=new DataInputStream(new BufferedInputStream(
    new URL(getDocumentBase(),"cell4.txt").openStream()));
    }catch(MalformedURLException e) {}
catch(IOException e){}
```

Conclusion

The experimental system for the shopfloor access and monitoring of the manufacturing system has demonstrated that it is possible to extend the applications of the Internet to the manufacturing industry through an Intranet for remote control and data acquisition. This will enable the production facility and the various departments within a manufacturing plant to be easily linked together to create an information network for Computer Integrated Manufacturing through the Internet using the JAVA language for developing the required Client/Server applications.

References

1. JAVA JDK and tutorial, http://www.sunsoft.com
2. mSQL code, ftp://bond.edu.au/pub/Minerva/msql
3. JDBC API, http://splash.javasoft.com/jdbc

Real-Time Supervisory Control of a Flexible Manufacturing System Using AI-based Method

Zhong Jiangsheng

Wuyi University, Mechanical Engineering Department, Jiangmen, Guangdong 529020
China

Wong Chun Chong

Ngee Ann Polytechnic, Mechanical Engineering Department, 535 Clementi Road
Singapore 599489

Abstract. To be able to achieve cost-performance benefits from a flexible manufacturing system (FMS), the main objective considered in supervisory control of FMS is to increase the utilization of all processing cells, e.g. machining cell, inspecting cell, and assembly cell etc. which a FMS consists of. This paper addresses the key issues involved in design and operation of real-time supervisory control for a FMS that consists of following major components, i.e. a machining cell including one CNC lathe and one CNC machining center, a conveyor, an inspecting and assembly cell, an automated storage and retrieval system (ASRS), and an automated guided vehicle (AGV), using artificial intelligent (AI)-based real-time software tool known as G2. The concept of virtual equipment is introduced. The proposed material tracking method is elucidated with an example, and the event-driven AGV dispatching strategy is addressed.

1 Introduction

Flexible manufacturing systems (FMSs) are collections of machines and related processing equipment linked together by an automated material handling system, and all under computer control. These systems are capable of producing a wide variety of items in small batches and in random order (Williams et al 1994).

A typical FMS consists of the following components:

- computer numerical control (CNC) machine tools, usually equipped with tool magazines and tool changers;
- a part load/unload station, and other stations such as inspection or part wash;
- a set of pallets and fixtures where parts are loaded into fixtures, which in turn are mounted on pallets;
- an automated material handling system (automated guided vehicles (AGVs), conveyors, tow carts, etc.);
- a tool storage area and tool transport system;
- part buffers at machines and a storage area for pallets.

Supervisory control is the most important function in a FMS. It includes the monitoring and control of machine tools, material handling equipment, work transportation devices, and auxiliary equipment, i.e. all components a FMS consists of (Hwang et al 1984). In this investigation an artificial intelligence (AI)-based software tool known as G2 is adopted to code the supervisory control of a FMS, which consists of a machining cell including one CNC lathe and one machine center, a conveyor, an inspecting and assembly cell, an automated storage and retrieval system (ASRS), and an AGV. The concept of virtual equipment is introduced. The proposed material tracking method is elucidated with an example, and the event-driven AGV dispatching strategy is addressed.

2 Description of the FMS

The layout of the FMS for the present investigation is shown in Fig. 1. The salient features of the system are described.

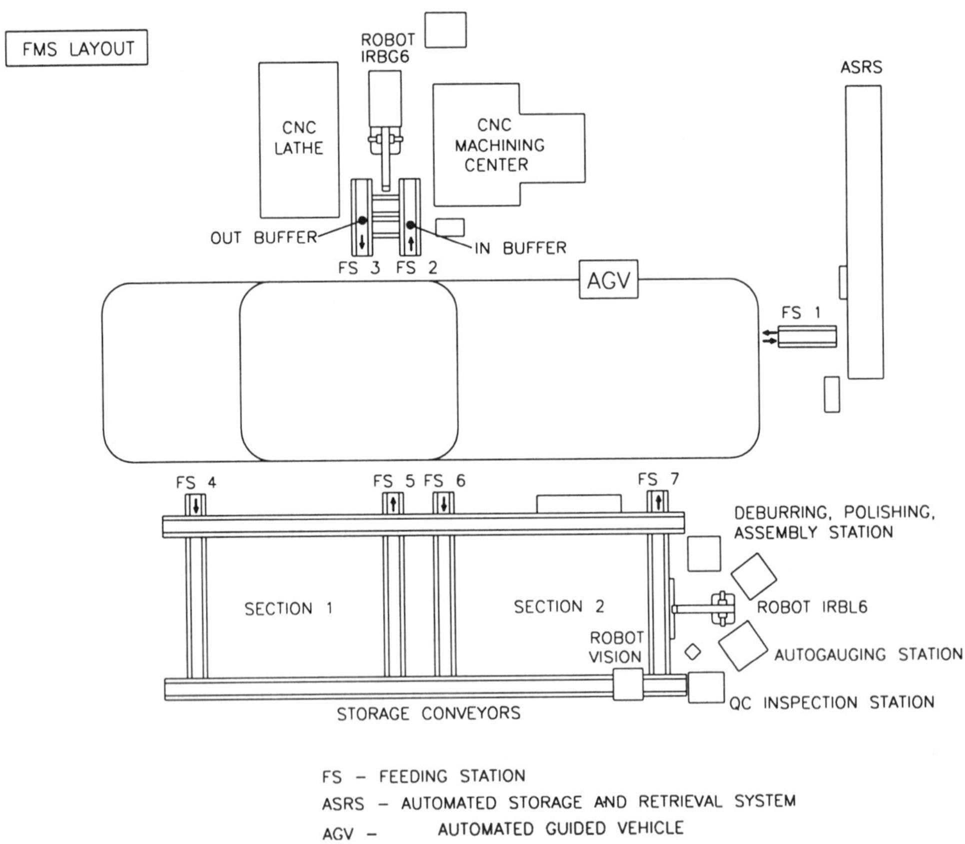

Fig. 1. Layout of a flexible manufacturing system

It consists of two cells and an automated storage and retrieval system (ASRS) connected by an AGV. In the cell 1 (machining cell) there are two machine tools, i.e. one CNC lathe and one CNC machining center, an input buffer and an output buffer, and a robot handling with tools/parts load/unload jobs. The cell 2 (inspecting and assembly cell) consists of an inspection station, an autogauging station, a deburring, polishing and assembly station, and a robot. The local storage, i.e. the section 1 of the storage conveyors is provided for storage parts which need another machining operation after turning/milling operation. At the cell 2 there is also an assembly local storage for temporally store parts to be assembled further. There are seven feeding stations, feeding station 1 to 7 (FS 1 to FS 7), as shown in Fig. 1, for AGV load/unload. Six types of part are processed simultaneously. The six part types are in three product groups, i.e.

group one: part type 1, part type 2, and part type 3;
group two: part type 4 and part type 5;
group three: part type 6.

In the FMS operation all the above parts are machined from the raw materials. The part types 1 and 2 are assembled onto the part type 3. In the group two the part type 4 is assembled onto the part type 5. The part type 6 is an independent part. The system is able to work on different part types as long as the features can be handled by the grippers, fixtures, pallets and other mechanical accessories provided. All workpieces (raw material and semi-finished/finished parts) are held by the pallet in operation. Two pieces of the same type are held in one pallet for the turned parts and one piece for the milled parts. Pallets are being used randomly for the six part types at different stages in the FMS operation. The processing sequences for all six part types are shown in Fig. 2.

The AGV track layout consisting of both unidirectional segments is shown also in Fig. 1. When the AGV completes a delivery task, it waits at home-station provided near the section 1 of the storage conveyors if it is unassigned. The AGV travels only by the shortest route between any two locations.

3 Virtual Equipment

The design of the supervisory control system (SCS) of the FMS begins with the development of classes. The physical equipment (PE) of the FMS in the shop floor are matched into virtual equipment (VE) as classes in SCS. Each VE class is defined with its attributes related to specification and characteristics of its PE and assigned responsibilities for certain tasks required for the supervisory control. As the classes are designed, relationship between classes are defined. The VE classes form tree hierarchy by which a class can inherit attribute(s) from its superior class. For example, for class *Feeding-station* the data structure is defined as:

Class name	*feeding-station*
Superior class	*plant-item*

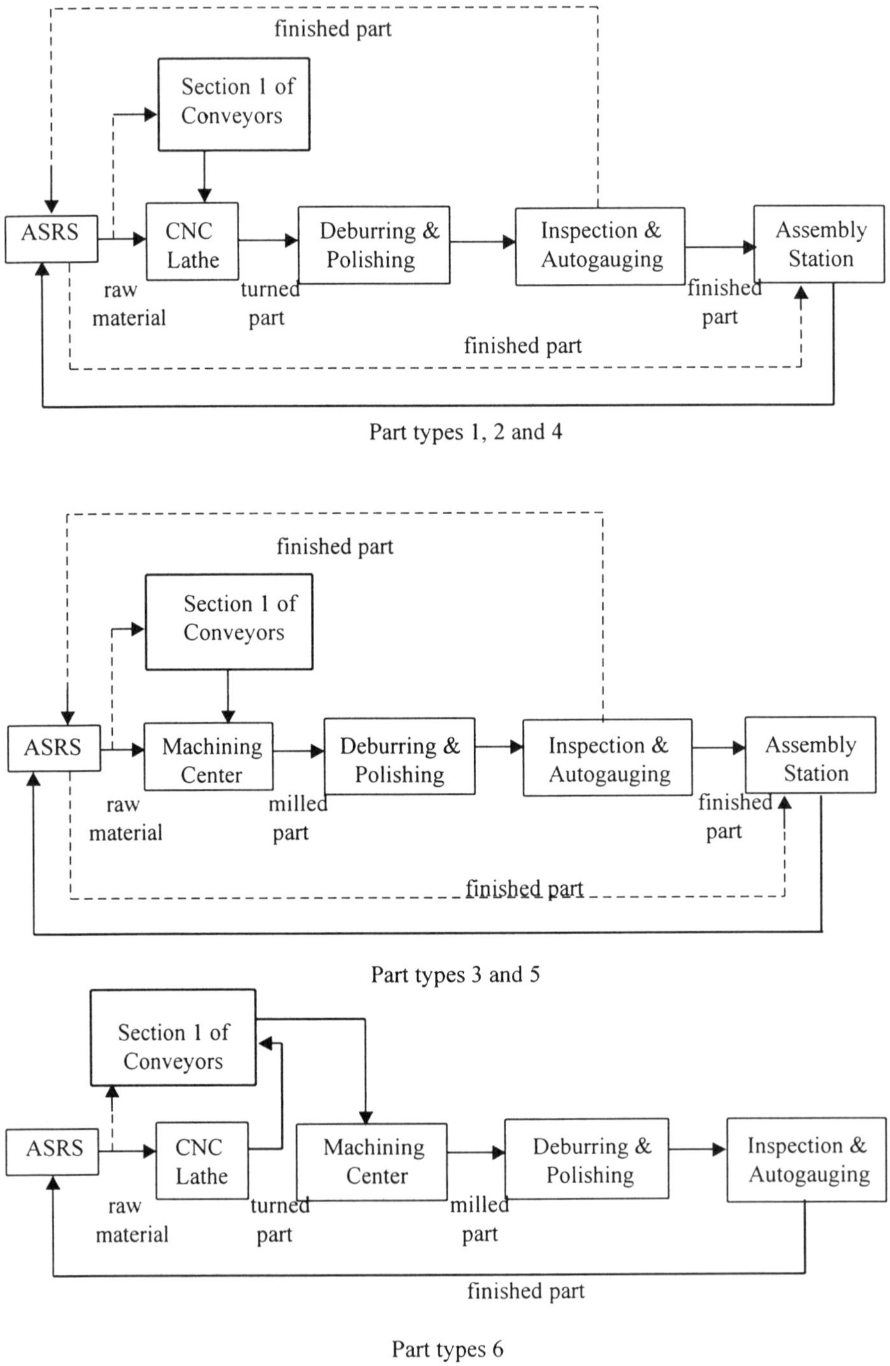

Fig. 2. Processing sequences for the six part types.

Attributes specific to class	*agv-arrival is given by a fms-value*
Inherited attributes	*identity;*
	current-action;
	current-event;
	current-pallet is 0;
	last-pallet is 0;
	start-time;
	expected-duration;
	pallet-sensor is given by a fms-value

In addition to the attribute *agv-arrival is given by a fms-value* specific to this class, the class *Feeding-station* inherits the attributes as above mentioned from the superior class *plant-item,* which is the defined top class for VE. The *fms-value* is another class that is responsible for catching sensor values from shop floor.

In relation to PE configuration and the number of plant items of the same type the instances of VE classes are connected together logically, as shown in Fig. 3 and Table 1, so that during SCS process data can be correctly and easily transformed bi-directionally, i.e. from SCS to shop floor and vice versa.

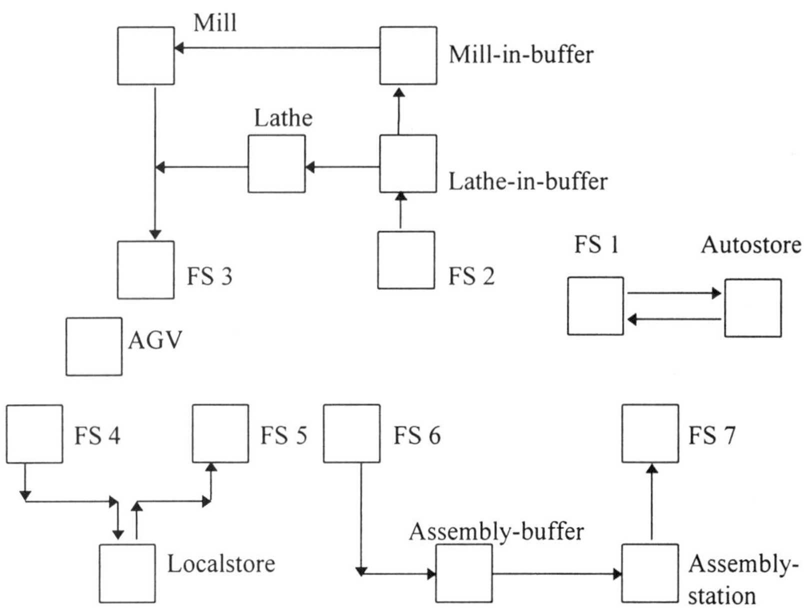

Fig. 3. The connection of virtual equipment

Table 1. Virtual equipment corresponding to physical equipment

Virtual Equipment (VE)	Physical Equipment (PE)
Mill	CNC machining center
Mill-in-buffer	Input buffer for milling operation
Lathe	CNC lathe
Lathe-in-buffer	Input buffer for turning operation
Autostore	Automated storage and retrieval system (ASRS)
AGV	Automated guided vehicle (AGV)
Localstore	Section 1 of storage conveyors
Assembly-buffer	Section 2 of storage conveyors
Assembly-station	Cell 2 including inspection, autogauging, deburring, polishing, and assembly stations
FS 1 to FS 7	feeding station 1 (FS 1) to feeding station 7 (FS 7)

4 On-line Material Tracking

As shown in Fig. 2, in the FMS part(s) can be machined on different machine tools, inspected and/or assembled with other part(s) on the other station(s), if necessary. In order to deliver the right part(s) in the right time at the specific feeding station, it is necessary to track material on-line in the SCS. *Material* (part(s)) placed on pallet is expressed in SCS as a class with an attribute of *pallet code* which will change upon a certain operation has completed. A class *activity* with its many instances is constructed with the attributes of *pre-code and post-code* corresponding to status of material before and after an operation. By means of identifying the specific processing plant item and its status, the status of all plant items are monitored through a set of rules which trigger the identification procedure(s) of work-in-process pallets. For instance, the code number of raw material of part type 6 is 3360 (see Fig. 2), and after turning and milling operations the code number is changed to 3361 and 3362, indicating semi-machined and machined part respectively. Then it should be transported to section 2 of the storage conveyors by the AGV. After deburring and polishing operation it will get code number 3363, 3364 or 3369, if two, one or no part(s) pass(es) through inspection and autogauging operations respectively.

5 Event-driven AGV Dispatching

In a FMS environment a typical part may visit several work cells, and a variety of parts may be processed simultaneously. AGVs perform a variety of delivery tasks to transfer the parts between the work cells. Upon completion of a delivery task, an AGV is reassigned if there is any unassigned handling task; otherwise the AGV is set idle and waits for a task to emerge. The AGV may have to be moved to an alternative position, however, before it is permitted to become idle. If the number of unassigned tasks exceeds that of free AGVs, prioritization of tasks is essential. Mahadevan and Narendran (1990) analyzed the problems arising from multi-AGV systems and strategies for resolving them using analytical and simulation models. Raju and Chetty

(1993) proposed a Petri net-based methodology for modeling and simulating AGVs for FMSs.

In this investigation, AGV dispatching job can be divided into two kind of circumstances during the FMS operation:

1. more than one feeding station request AGV to transport pallets at the same time or during the last AGV trip;
2. No requests for AGV when AGV returns its home after completing the last trip.

When the input buffer of cell 1 is idle or the output buffer is full, or the feeding station of the cell 2 request, AGV immediately takes its trip based on first-come first-served (FCFS) rule. If the requests occur at the same time, then the cell 1 will get higher priority because of its smaller buffer capacity. This event-driven AGV dispatching strategy ensures the machining or assembly process can be continued without interruption.

To maximize utilization of AGV (Dirne 1990), when AGV returns its home with no requests, SCS will assign AGV to retrieve raw material from ASRS, and it will be transported into the section 1 of the storage conveyors according to material priority.

6 Conclusions

The design and operational control strategies of real-time supervisory control for a FMS are presented, that consists of following major components, i.e. a machining cell including one CNC lathe and one CNC machining center, a conveyor, an inspecting and assembly cell, an ASRS, and an AGV. The concept of virtual equipment is introduced, and the proposed material tracking method is elucidated with an example. To this end, the event-driven AGV dispatching strategy is addressed. SCS of the FMS is being tested now with shop floor physical equipment together and the testing results show that SCS works successfully.

Acknowledgment

The authors are grateful to the colleagues in Mechanical Engineering Department and Advanced Manufacturing Workshop, Ngee Ann Polytechnic, who involved in this investigation for their assistance.

References

1. C. W. G. M. Dirne, The quasi-simultaneous finishing of work orders on a flexible automated manufacturing cell in a job shop, *International Journal of Production Research* **28** (1990) 1635-1655.
2. S. -L. Hwang, W. Barfield, T. -C. Chang, and G. Salvendy, Integration of humans and computers in the operation and control of flexible manufacturing systems, *International Journal of Production Research* **22** (1984) 841-856.
3. B. Mahadevan, and T. T. Narendran, Design of an automated guided vehicle-based material handling system for a flexible manufacturing system. *International Journal of Production Research* **28** (1990) 1611-1622.

4. K. R. Raju, and O. V. K. Chetty, Design and evaluation of automated guided vehicle systems for flexible manufacturing systems: an extended timed Petri net-based approach. *International Journal of Production Research* **31** (1993) 1069-1096.
5. T. J. Williams, J. P. Shewchuk, and C. L. Moodie, The role of CIM architectures in flexible manufacturing systems. In *Computer control of flexible manufacturing system,* S. B. Joshi and J. S. Smith (eds) Chapman & Hall (1994) London.

The Description、 Verification and Implementation of MMS Interconnection between Heterogeneous Networks

Xiang Fei Guanqun Gu

Department of Computer Science and Engineering, Southeast University , Nanjing 210096,People's Republic of China,E-mail:xfei@seu.edu.cn

Jieyi Wu Zheng Li

CIMS Research Center, Southeast University , Nanjing, People's Republic of China

Abstract. In this paper, the protocol specification and service definition of MMS (Manufacturing Message Specification) are briefly introduced. According to the state transition diagram of MMS, a gateway machine used for the interconnection of heterogeneous networks is defined and described. Then ,based on the Petri nets, the MMS interaction between heterogeneous networks is described and analyzed. At last, a MMS communication system in CIMS environment is implemented.

1 Introduction

The core of CIM(Computer Integrated manufacturing) is integration, the basis of which is computer network. The Manufacturing Message Specification(MMS) is the ISO standard communication protocol specific to manufacturing, the purpose of which is to meet communication requirements among various programmable devices in CIM environment. MMS is stemmed from MAP, but it is independent of specific network platform. In a real CIM environment, there exist heterogeneous networks, such as Ethernet and Fieldbus, because of various communication requirements. Therefore it is necessary to study the MMS interconnection in such a case. Fig 1 shows the framework of heterogeneous networks in a typical CIM environment.

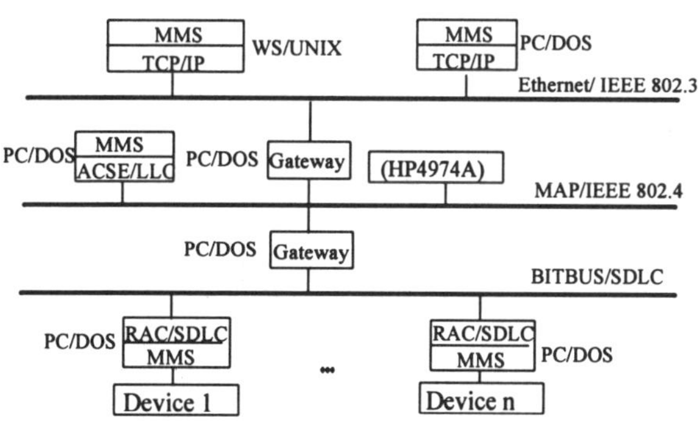

Fig 1. The Framework of MMS Communication System

In order to correctly understand and implement MMS services and MMS interconnection between heterogeneous networks, it is essential to develop a formal methodology to describe the various service units in MMS, based on which the MMS protocol machine can be implemented conformed to international standard. Such a formal methodology is further used to design the structure and functions of MMS gateway , describe the behavior of MMS interactions and test the correctness of MMS interconnections, so that the MMS integrated communication system is implemented successfully with reliability and consistence.

2 System Framework and MMS Overview

According to the communication requirements in today's CIM environment, a framework of *MMS Integrated Communication System* is shown in Fig 1. In this framework there exist three kinds of network products: Ethernet、 MAP and BITBUS. Two gateways are used to interconnect Ethernet/MAP and MAP/BITBUS respectively. HP7974A is a MMS instrument for testing protocol.

MMS mainly consists of two parts: MMS service definition and MMS protocol specification. In MMS service definition, client/server model is used to describe its service interactions with client issuing requests for services to be performed by the server. Object model is used to abstractly describe the services provided by a manufacturing device to a controlling device. A class of object consists of attributes of an object, the operations that can be performed on the object and the behavior of the object when receiving operation requests. Generally there are 10 classes of services in MMS. MMS protocol specification specifies the procedure for the transfer of data and control information in the MMS context, the means of selecting the services to be used by the application entities, and the structure of the MMS PDU used for the transfer of data and control information.

MMPM(Manufacturing Message Protocol Machine) is an abstract machine that carries out the procedures specified in MMS protocol specification.The state transition diagrams are applied in MMPM to describe the protocol. CS(Confirmed Service) is the basis of MMPM and almost all the MMS services have to be handled by CS unit. So in this paper , we will focus on the analysis of CS, the design of the STD (state transition diagram) of CS in MMS gateway, and the description of the CS interactions between heterogeneous network application entities.

3 Description of MMS-GWPM

The main functions of gateway are interconnecting different types of network, providing data associations between application entities and converting heterogeneous protocols, i.e. converting the information constructed by one network protocol to what could be identified by another network protocol, so that two application processes in heterogeneous networks can communicate with each other.

Gateway is the mean to interconnect heterogeneous networks at high level according to the 7-layer reference model. Because there exist apparent differences among the framework and protocols of MAP、 Ethernet and BITBUS, and MMS lies in application layer, we decide to design the gateway at application layer.

According to the general requirements of gateway, the main functions of MMS gateway should be:

1. Providing associations between different networks at application layer;

2. Converting MMS service primitives and protocols between heterogeneous networks.

3. Having no change on hardware、 protocols and application software of each network;

4. Providing necessary control and management functions such as flow control and route selection.

Referring to the states transition diagrams of CS in MMS protocol specification, the MMS-GWPM(Gateway Protocol Machine) is designed and shown in figure 2.

Where function 2 and 4 are implemented in the state "Converting in Gateway". In order not to change the STD of CS, this MMS-GWPM includes the responses to all the state transitions of MMS requester and MMS responder. For special MMS-PDU such as cancel-PDU, GWPM will response it directly so that the process delay is reduced. Because MMS is connection-oriented, function 1 is accomplished in Environment and General Management Service.

4 Description and Analysis of the Interaction of MMS CS through Gateway

Having designed the STD of MMS-GWPM, we should further consider the description and analysis of the dynamic process when interactions of MMS CS take place through MMS gateway. Compared with the formal technologies such as Estelle

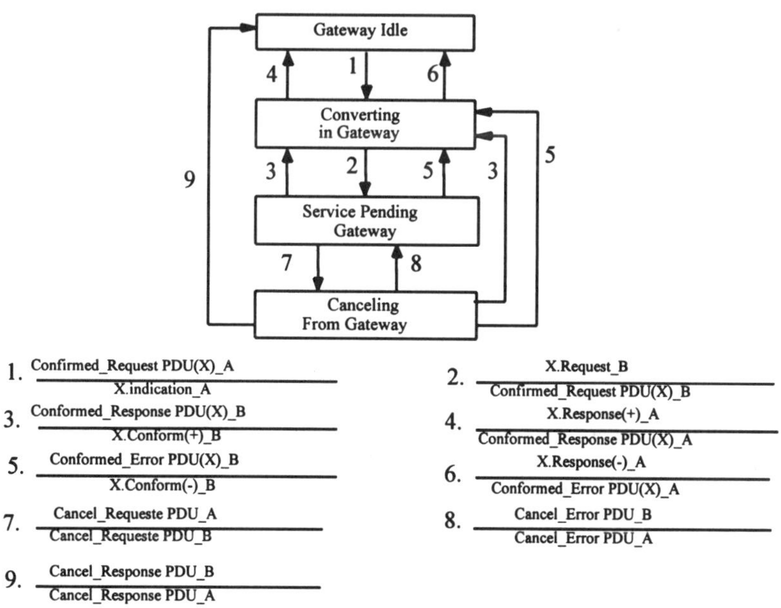

Fig. 2. State Transition Diagram of MMS-GWMP

and Lotos, Petri nets have several advantages as protocol engineering: Petri nets can describe asynchronous and concurrent system using the form of figure; it has rigorous mathematical theory which can be used to analyze and verify the behavior of protocol.

4.1 Description of the Interaction of MMS CS through Gateway Using Petri Nets

According to the STD of CS and GWPM, the *Petri Net for the Interaction of CS through Gateway* could be established following the principles mentioned below:

1 . State is expressed by the state place, the message is expressed by the token, the protocol action is expressed by transition;

2 . The service provider of lower layer is characterized by two appositional FIFO queues with reverse direction. Each queue is represented by one place.

3 . The resources in the network are expressed by the number of tokens, including the arrived request, of the outstanding services, i.e., not yet confirmed services, and etc. .

Figure 3 presents the *Petri Net for MMS CS through Gateway*. For clarity, we assume the length of the arriving requests and the number of outstanding services are all one. The actual length of the arriving requests and the number of outstanding services will be shown in section 4.2. At the same time, for simplifying the complex of the Petri net, the net is shown without place *channel*. Place *channel* has an ingoing arc to each of transitions ni (i=1 ~ 5, 6 ~ 10) and an outgoing arc to the same transition ni . The net consists of five parts: the service requester(place Ps1, P1, P8), the responder(Ps3, P4, P5), the gateway(Ps2, P2, P3, P6, P7), the service request queue(Q1, Q2), and the communication network(all other places). t1 ~ t19 present the transition corresponding to the STD of requster, responder and gateway. n1 ~ n5, n6 ~ n10 present the process and transmision at lower layer, tQ presents the arriving of the service requet.

4.2 Analysis of Petri Net for the Interaction of MMS CS through Gateway

According to the viewpoint of protocol engineering, after being described by formal methodology, the model should be analysed and verified . The properties of Petri nets are boudness 、 liveness and etc. . Two techniques are used to analyze Petri nets: metric analysis(invariant analysis) and reachability tree analysis.

By using the package PESIM, P-invariants and T-invariants of the net are determined.

P-invariants: The following are the four minimum P-invariants.

P1: m(Q1)+m(Q2)=K

P2: m(Ps1)+m(P1)+m(P8)=S

P3: m(Ps2)+m(P2)+m(P6)+m(P3)+m(P7)=S

P4: m(Ps3)+m(P4)+m(P5)=S

Where K represents the maximun number of services allowed to enter the CS. The number of outstanding services allowed in MMS is indicated by S. In this Petri net, K=S=1.

Analysis: The meaning of P1 indicates that the maximun of service requests is no more than K, if the request number is more than K, it can not be accepted. P2, P3 and

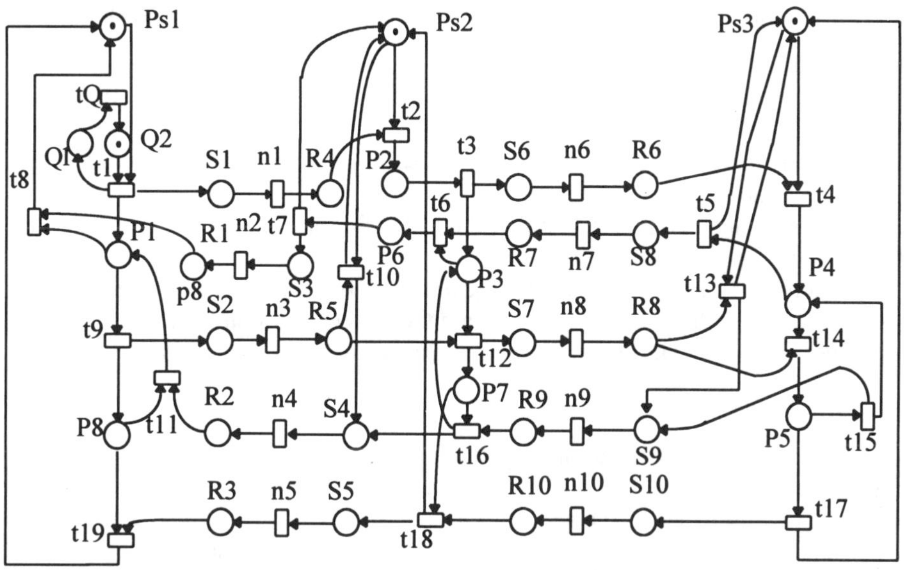

Fig. 3. Petri Net for the Interaction of MMS CS through Gateway

P4 indicate that the subnets of petri net corresponding to the service requester、gateway and responder are strictly conservative. Furthermore, since all the places are covered by some P-invariants, the petri net is bounded.

T-invariants: T-invariants can be used to determine the normal service operations. A normal service cycle is defined to be any transition sequence which leads the Petri Nets from the initial marking and returns it to the initial marking. There are five T-invariants:

T1: t1,tQ,n1,t2,t3,n6,t4,t5,n7,t6,t7,,n2,t8;
T2: t1,tQ,n1,t2,t3,n6,t4,t9,n3,t12,n8,t14,t15,n9,t16,n4,t11,t5,n7,t6,t7,n2,t8;
T3: t1;tQ,n1,t2,t3,n6,t4,t5,t9,n3,t12,n8,t13,n9,t16,n4,t11,n7,t6,t7,n2,t8;
T4: t1,tQ1,n1,t2,t3,n6,t4,t5,n7,t6,t7,t9,n3,t10,n4,t11,n2,t8;
T5: t1,tQ,n1,t2,t3,n6,t4,t9,n3,t12,n8,t14,t17,n10,t18,t19;

Analysis: The T-invariants mentioned above represent five types of service cycle. T1 specifies the type of standard service executions in which no requests of service canceling have been issued by the requester. T2 ~ T4 give the types of unsuccessful canceling service. In t2, canceling requests have been issued by the requester but denied by the responder for some reasons. T3 is cancel error executions in which requests of service canceling from the requester are received by the responder after the services requested have been completed. T4 is also cancel error because the service requested have been fulfilled by both responder and gateway before requests of service canceling are received. Finally, the service executions defined by T5 are successful canceling service execution in which canceling requests have been issued and accepted.

By using the package PESIM, we could get the reachable tree for this petri net, from which we can see that this petri net is live, have no deadlock marking.

5 Implementation of MMS Integrated Communication System

Based on the STD of GWPM and Petri net for the interaction of MMS CS, we can implement the MMS integrated communication system showed in figure 1 by applying the following steps:

Implementing MMS over MAP 、 Ethernet and BITBUS respectively using protocol migration principles.

Implementing GWPM in terms of the STD after the correct configuration of PC for gateway.

Finally , as to the *Petri Net for the Interaction of MMS CS through Gateway*, debugging the whole system, so as to ensure the correctness of the implementation.

The model of MMS interconnection is showen in figure 4.

6 Conclusion

In order to design and implement a reliable and efficient MMS integrated communication system, it is necessary to use formal methodology for design and analysis. In this paper, STD and Petri net is applied to describe and analyze the interaction of MMS confirmed service between heterogeneous networks . The results show that the principles of protocol software development mentioned above are effective and efficient. The work of this paper provides a basis for further analysis, which will improve the system with higher performance.

7 References

1 *ISO/IEC 9506 Manufacturing Message Specification*, Part 1 and Part 2, (1990).

2 Fei_Yue Wang, Performance Analysis of MMS Using GSPN, *Proceedings of the IEEE Conference on Robotics and Automation*, (1991) 1573 — 1578.

3 Guanqun Gu, MMS Communication System in Heterogeneous Network, *Computer Research and Development,* No.3, (1995) 1 — 7.

4 P.Pleinevaux, An Analyis of the MMS Object Model, *IEEE Transaction on Industrial Electronics*, Vol. 41, No. 3, (1994) 265 — 268.

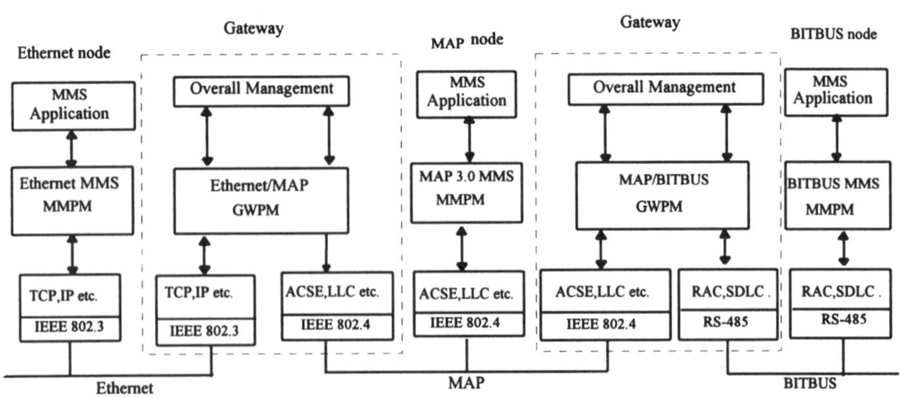

Fig. 4. Model for MMS Interconnection

Assembly Automation

Analysis Method For Automation of A Socket Assembly System - A Case Study

Ng Kok Weng

B.Eng. (Mech & System Eng.), M.Sc (Manufacturing System Eng.)
Production Systems Group, National CAD/CAM Centre
SIRIM Berhad, 1, Persiaran Dato' Menteri, P.O. Box 7035,
Section 2, 40911 Shah Alam, Malaysia

Hamiraj Fahry Abdul Hamid

B.Sc (Industrial Engineering)
Production Systems Group, National CAD/CAM Centre
SIRIM Berhad, 1, Persiaran Dato' Menteri, P.O. Box 7035,
Section 2, 40911 Shah Alam, Malaysia

Abstract. There is a need to automate assembly systems to enable increase in productivity and reduction of manual operation processes. Labour intensive industries are currently facing severe problems of maintaining their skilled labour force as the labour market gets more competitive. Though automation maybe the solution to the current shortages of skilled labour force problem, investments on automated assembly systems require detail and careful study and planning to ensure the well returns of the investments. This means automation of assembly systems must also lead to significant improvement in rate of assembly and a reduction of manpower. Currently there are very few guidelines and assistance to industry in automating their assembly systems. Hence this paper look into the analysis method to be used as a reference platform for Malaysian industry in automating their assembly systems especially on manual intensive assembly processes. Though the technology of this method presented is not something new but the method used is applicable in the process of making decision on automation. Currently, this method was applied to a domestic electrical appliances manufacturer in Malaysia and the results are discussed in this paper.

1.0 Introduction

In the process of automating an assembly system which involves various assembly processes such as inserting of parts, screwing, positioning of parts, etc., deciding which process to be automated is critical. A well analysed and study on the assembly systems is necessary and automating the critical process will lead to significant improvement on productivity and reduction of manpower. In the process of analysing the assembly system, usually weaknesses such as bottleneck stations, high reject processes, manual intensive processes, starvation stations, delicate and stressful processes, hazardous processes, high speed processes, etc. are the main targets of automation. However in most cases, a combination of all these weaknesses occurs. Depending on the severity of the weaknesses, different companies will have different critical weaknesses. Some companies will have significant amount of manual intensive processes and some may have severe bottlenecks' problems but all of them have limited amount of allocation budget. Hence it is vital that the investment in automation will give the biggest impact in improving their productivity.

2.0 Assembly Systems Overview

Basically an assembly system has various activities of assembly. The activities may involve a combination of manual operation, semi-automated operation and fully automated operation [5]. The product involved will be of variants or totally different type. Hence inventories, work in progress, finished goods, etc. will be of variety types too.

Currently in Malaysia, assembly systems that are manually intensive will have operators positioned at various places of the production flow. Each operator will either assemble the complete product or assemble a portion of the product (assemble in modules[6]). In the case of assembling a portion of the product (the assembly is divided into smaller process). Generally the idle assembly process flow would be the cycle time for the work of each operator is equally divided to ensure a balanced assembly line [1]. The line balancing or workstation balancing of any assembly systems will require balancing manual operation processes and machine operation processes cycle time. The balancing between machine and human operators work cycle time requires simulation analysis and is a very difficult process [2]. The machine has the ability to work 24 hours a day (non-stop) with preventive maintenance stoppages occasionally which is planned. The machine will work much faster than a human operator and without varying in their performances.

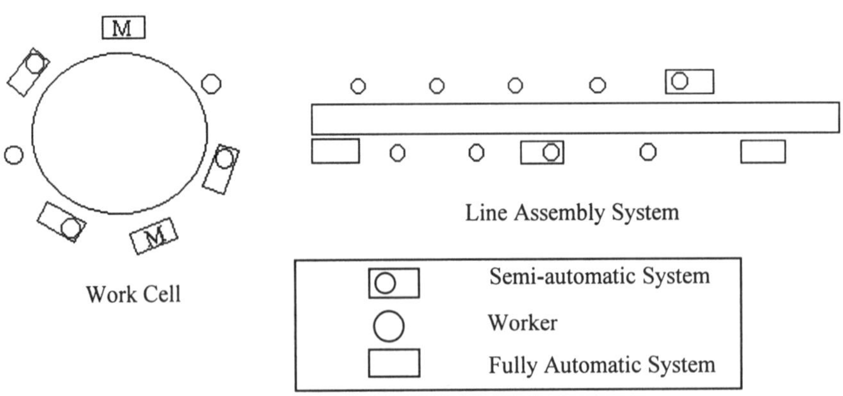

Figure 1 : The Common Layout of Assembly Systems

Based upon this, to balance these two categories of assembly approach, it is preferable to match the human operators' rate to the machine working rate. This is done because, for cost effective systems, machine utilisation must be maximum or 100% (for ideal case) . Figure 1 shows 2 of the common assembly systems layout for a company in Malaysia.

Hence, increase of human operators is carried out in process of matching the work rate of the machine. In local industries, balancing work load is usually neglected hence problems like bottlenecks and starvation occur. Most local (Malaysian) and small industries are interested in automating their assembly processes. However their budget is very tight and assembly systems are usually not well planned and organised. Therefore to automate industries that have these problems and gradual introduction of automation with detail analysis on balancing the system as well as partitioning the system for the process of step-by-step automation. The final aim is to automate the entire system with minimal effect on the output and spread out the cost to avoid sudden financial burden due to heavy investment. The further advantage of this automation approach is as technology and market favour changes in time, the system can adapt to this changes because of the gradual introduction of automation. This lead to a significant reduction of risks involved on the capital investment for the company concerned.

3.0 Analysing Assembly Systems

This automation approach starts with an observation and identification of assembly process which is "generic". "Generic" in this context means applicable to all the product variants that are being produced by the company or can be adapted to other product variants with minimal modification and cost. The "common" process usually involved similar parts or required similar assembly process such as screwing, riveting, etc. Hence the automation process will also be eased or enhanced by application of group technology, design for assembly, etc. onto the products or part of the product itself.

With this in mind, the analysis system will require the follow up of processes like design for assembly, simplification of assembly process, probable redesigning of layout, group technology, etc. Hence after identifying common processes, a work study is required to detail out the movement of an operator's arm and body in the process of assembling the product. Generally each worker has to do a specified movement of hands and body (sometimes even legs) to assemble the product parts. All this micro study on operators in performing their work will be listed down for analysis purposes. This study process is part of a well-known approach known as Method Time Measurement (MTM) where micro motion of worker is used to analyse a particular operation. Hence with MTM, the cycle time for each micro motion of individual worker is obtained.

The time and operation data obtained via observation and MTM must be counterchecked with the real cycle time (obtained by carrying out work study). Appendix 1 illustrates an example of a complete assembly process for a socket (all data are obtained from Industry) and Appendix 2 illustrates the cycle time obtained from MTM analysis in comparison to the actual cycle time and shows how categorisations are made to decide on which process to automate and what to automate. Using process categorisation, the decision on the level of automation can also be made easier and the overall assembly system can be simplified. With this process categorisation, further analysis like cost analysis, automation decision and implementation, stages of automation and various consideration can be made to ensure the company will have a gradual, optimal, efficient and effective application of automation processes over their existing manual intensive system.

The entire approach is very dependent on the accuracy of the data obtained and the observations made. However consideration must also be made on the transfer the product of a non precise transfer system to a highly precise transfer system during planning for automation. The transferring methods are can be various and should be looked into in details.[3, 4]

In short, the entire process is summarised in Figure 2.

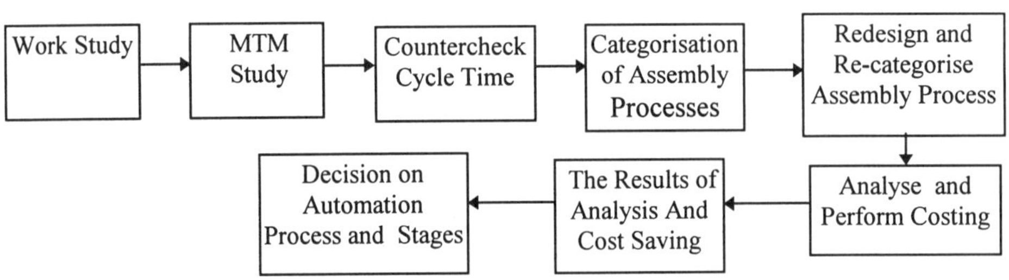

Figure 2 : The flow process of analysis system for automating assembly systems

In case study of socket assembly, based upon the categorisation process (refer Appendix 3), the categories of assembly process for a socket can be listed as below :

1. Base Positioning
2. Metal Parts Assembly
3. Dolly Switch Assembly
4. Plate Positioning
5. Shutter Assembly
6. Assembling of Plate to Base
7. Screwing Process
8. Final Assembly

Hence by accumulating all the assembly sub-processes under the above mentioned category of processes, the cycle time for each category can be listed as below :

Activities	Cycle Time(s)	% of Total Time
1. Base Positioning	3.10	5.69
2. Metal Parts Assembly	5.84	10.70
3. Dolly Switch Assembly	13.32	24.41
4. Plate Positioning	2.19	4.01
5. Shutter Assembly	5.29	9.70
6. Assembling of Plate to Base	2.55	4.68
7. Screwing Process	16.24	29.77
8. Final Assembly	6.02	11.04

It is important to know that the cycle times of these categories of assembly processe are obtained by adding all the sub-processes under each category. All sub-processe cycle time is obtained via MTM and hence though there maybe some slight error or inaccuracy, the ratio of cycle time among the processes is consistent. Therefore the percentage of total time each category of processes consumed is accurate.

With this data, a chart (refer Figure 3) can be plotted to analyse the assembling process situation more clearly.

Title: Assembly Activities vs Percentage of Assembly Time Chart For Socket E15

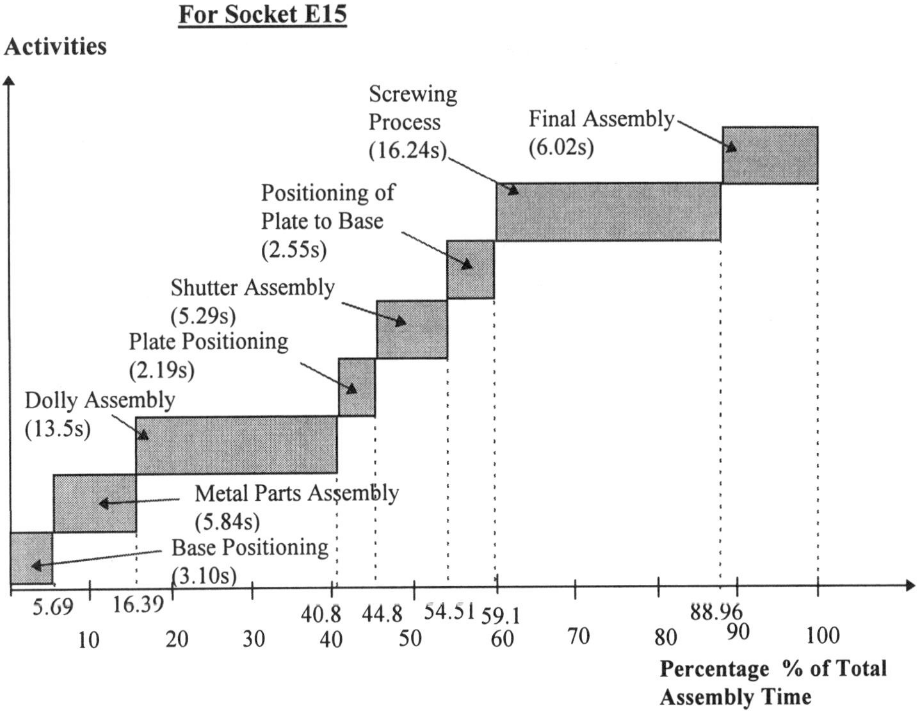

Figure 3 : The Categorising of Assembly Processes For Automation

By referring to the chart above, it is very vivid that screwing process is the most critical and time consuming processes among all the processes involved. The dolly assembling process is the second most time consuming process. It is very important to note that this case study is carried out on an assembly process where one worker assembles a complete socket. So all process above is done by a single worker and the entire line consists of several workers. Hence if there are any processes in this particular system that are to be automated, the most time consuming processes should be the priority. This is because the amount of cycle time for the overall processes will be significant reduced if the slow processes are removed and automated. When these slow processes are automated, the process can be perform in parallel. In this way, with the output of the system will be significantly increase. However it is vital to acknowledge that not all slow processes or bottlenecks can be automated. Automation is influenced by the cost involves and to the complexity of the assembly process concerned.

4.0 Decision and Planning for Automation

All analyse results lead to a clear indication of which assembly processes should to be the priority whenever a company wanted to automate. However the final decision is still dependent on the cost and return (pay back) evaluation. Only cost feasible processes should be automated. The method of automation is to start small and in stages. With reference to the socket case study above, screwing process seems to be the main bottleneck for the socket assembly. If the automation of screwing processes is feasible in terms of cost, then considerations on the cycle time for the prospective automated screwing process must be determined. The key point is to automate with the assembly line well balanced. Hence it is very critical to ensure the objectives of the automation process are met with, because some may want to automate to reduce operator and some may want to increase assembly output , etc. Therefore in the process of planning for automation, a chart to indicate the cycle time of the existing assembly system and the future assembly system must be developed side by side to allow comparison and frequent counter checking as shown in Figure 4 for the case study of socket assembly. Figure 4 shows the entire assembly system and it is vivid that the bottleneck for the overall system is at the testing and label sticking. In this case study, the stress is on the main assembly where 18 workers are working. However it is also important to balance the overall system after improving the main assembly.

5.0 Conclusions

This approach assists significantly in making decision to automate a particular assembly and in this case we tried it on domestic electrical appliance assembly systems. Currently there is no clear assistance, references or guidelines for the industry in Malaysia to evaluate their assembly systems and to decide how and which processes to automate. In fact, during analysis, any processes that can be removed or replaced with a better process that is designed for assembly must be take into account seriously. The product itself should be analysed using Design For Assembly (DFA). The current method mentioned here should not be a main consideration point for automation but should be a complement to other various existing method such as DFA, etc.

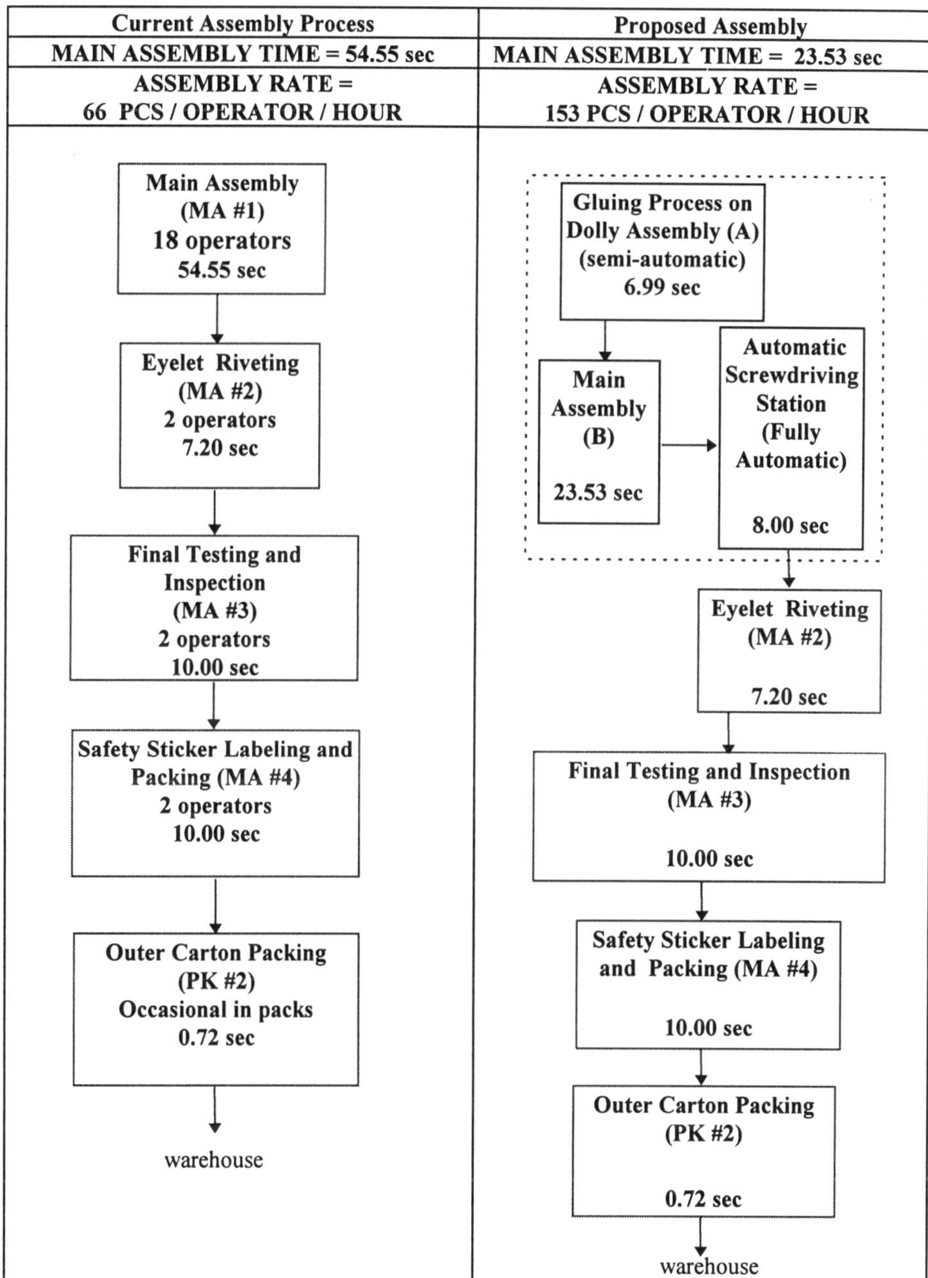

Figure 4 : Comparison Chart For Automation

Appendix 1

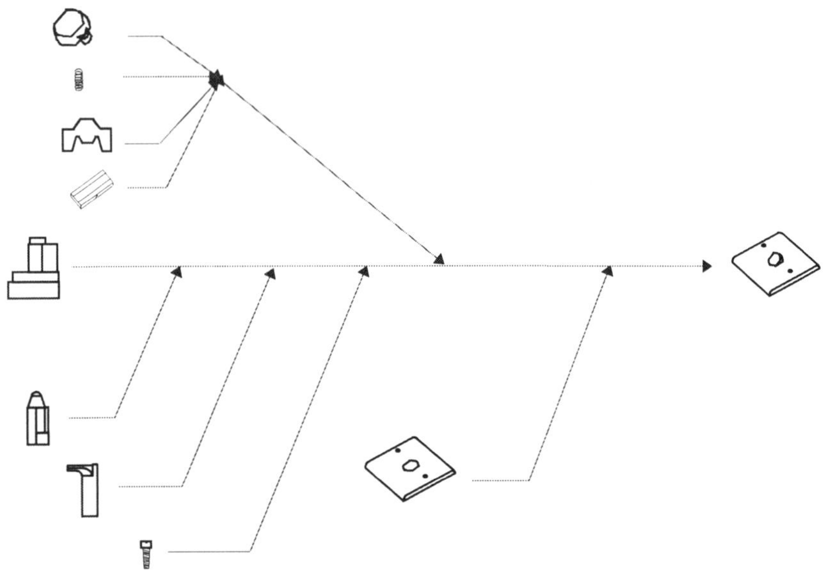

Complete assembly process for a socket

Appendix 2

Main Activities	MTM Activities	Total Estimated Time	% of Total	Actual Time
Base Positioning	1. Pick and Place Base (45 T.M.U.) 2. Inspect Base (15 T.M.U.) 3. Turn Base (5 T.M.U.) 4. Place Base Onto Jig (20 T.M.U.)	85 T.M.U. (3.06 sec)	5.69	86.15 T.M.U. (3.10 s)
Metal Parts Assembly	5. Pick and Place SA #1 (55 T.M.U.) 6. Pick and Place SA #2 30 T.M.U.) 7. Pick and Place SA #3 (55 T.M.U.) 8. Pick and Place SA #4 (20.T.M.U.)	160 T.M.U (5.76 sec)	10.70	162.16 T.M.U. (5.84 s)
Dolly Switch Assembly	9. Pick and Hold Bridge (45 T.M.U.) 10. Get Stick With Grease (40 T.M.U.) 11. Applying Grease (25 T.M.U.) 12. Place Bridge Onto Base (20 T.M.U.) 13. Pick and Hold Dolly (45 T.M.U.) 14. Pick and Hold Spring (20 T.M.U.) 15. Insert Spring (30 T.M.U.) 16. Pick and Place Yoke (40 T.M.U.) 17. Visual Control (15 T.M.U) 18. Grip Dolly Tightly (10 T.M.U.) 19. Turn Upside Down (5 T.M.U.) 20. Place Onto Base (30 T.M.U.) 21. Test Dolly (10x4 T.M.U.)	365 T.M.U. (13.5 sec)	24.41	369.93 T.M.U. (13.32 s)
Plate Positioning	22. Pick and Place Plate (45 T.M.U.) 23. Visual Inspection (15 T.M.U.)	60 T.M.U. (2.16 sec)	4.01	60.81 T.M.U. (2.19 s)
Shutter Assembly	24. Pick Shutter (45 T.M.U) 25. Pick Shutter Spring (20 T.M.U) 26. Insert Spring Into Shutter (20 T.M.U.) 27. Assemble Onto Plate (30 T.M.U.) 28. Control Motion (30 T.M.U.)	145 T.M.U. (5.22 sec)	9.70	146.96 T.M.U. (5.29 s)
Positioning of Plate to Base	1. Place Plate onto Base (40 T.M.U.) 2. Visual Control (15 T.M.U.) 3. Flip Jig (15 T.M.U.)	70 T.M.U. (2.52 sec)	4.68	70.95 T.M.U. (2.55 s)
Screwing Process	4. Pick and Place Screws (65 T.M.U.) 5. Fasten Screws into Holes (5x4T..M.U.) 6. Go to Next Hole (20x3 T.M.U.) 7. Return Unused Screws (20 T.M.U.) 8. Get Screwdriver (70 T.M.U.) 9. Fasten Screws (30x4 T.M.U.) 10. Transport Screwdriver (30x3 T.M.U.)	445 T.M.U. (16.02 sec)	29.77	451.01 T.M.U. (16.24 s)
Final Assembly	11. Flip Back Jig (15T.M.U.) 12. Take out Assembled Socket(20 T.M.U.) 13. Test Dolly (10x4 T.M.U) 14. Pick and Place Rivet (55 T.M.U.) 15. Move Socket to Conveyor (25 T.M.U.) 16. Press Counter (10 T.M.U.)	165 T.M.U. (5.94 sec)	11.04	167.23 T.M.U. (6.02 s)

Categorisation of Socket E 15 Assembly Processes

References :

1. Boothroyd G., and A.H. Redford, (1982), Mechanized Assembly, McGraw Hill Publishing Co. Ltd.

2. Buzacott, J.A., L.E. Hanifin, (December 1978), "Transfer Line Design and Analysis" - An Overview," Proceedings, 1978 Fall Industrial Engineering Conference of AIIE.

3. Buzacott, J.A., (1967), "Automatic Transfer Lines with Buffer Stocks", International Journal of Production Research, Vol.5, No.3, pp.183-200.

4. Groover, M.P., (1987), "Automation, Production Systems and Computer Integrated Manufacturing"

5. Groover, M.P., (1975), "Analyzing Automatic Transfer Machines", Industrial Engineering, Vol. 7, No.11, pp. 26-31.

6. Riley, F.J., (1983), "Assembly Automation", Industrial Press, Inc., New York.

Design and Analysis of a Flexible Workcell for Automated Assembly of Electric Motors

Shamsudin H. M. Amin, Mohamad Noh Ahmad, Zaharuddin Mohamed,
Jameel Mukred, Mohammad Ghulam Rahman
Center for Artificial Intelligence and Robotics, Faculty of Electrical Engineering.
University Technology Malaysia, Locked Bag 791,80990 Johor Bahru, Malaysia
Tel: +607-5505119 Fax: +607-5566272
sham@fkeserv.fke.utm.my

Abstract. Current manual assembly process of electric motors has been found to be incapable of keeping pace with the requirements of the manufacturing industry which aims to be globally competitive. The automation of the assembly process is the way to be in the competitive market. In this paper we propose the design of a flexible workcell for the assembly of different electric motors. An analysis of the design has been performed. The Robot Time and Motion (RTM) methodology has been implemented to compute the overall workcell cycle time. This study will be the basis of the implementation of an automated electric motor assembly factory.

1. Introduction

Traditionally the assembly of a composite product (like motors of automobiles) is mostly done through human operation. This type of job is highly repetitive, extremely boring and sometimes tedious. Usually the quality of the job cannot be constantly maintained due to human fatigue after long hours of work. So in order to be in the competitive global market it is necessary to maintain product quality, lower the cost and reduce the lead time ahead of the competitors. This can be achieved by automation of the assembly system. Automated assembly of composite products has been a goal of robotics researchers since the beginning of the field. In this paper a layout design of a flexible workcell for automated assembly of motors is presented.

A great deal of research have been done to develop high-level automatic generation of the assembly sequence plan([1] -[7]), where mating and insertion constraints are considered from only the product component side and assembly sequence has been considered as the reverse of the disassembly sequence of that product ([8], [9]). But in designing a workcell layout for assembling a product task assignment [10], part routing [11], fixturing or gripping, material transferring system, sequencing of the workstations, scheduling of the operations of different machines and other production related criteria are also to be considered.

The following alternative conditions relating to a design process which concern the relations between the product and the system may be identified.

A: Existing product - New system (system design after the product).

B: New product - New system.

C: New product-Existing system(product design adapted to the existing system).

The methods concerning assembly system design may be briefly classified into the following general categories. The relations to the conditions above are shown (A and/or B):

a) Simultaneous design of product and system (B).

b) Methods for total system design (A, B).

c) General problem solving methods relating to assembly system design (A, B).

d) Methods for specific types of assembly systems-mainly flexible automatic assembly system(A,B).

e) Expert system models, often related to the methods in category d (A).

f) Case studies of design and development of assembly systems by companies (A).

g) Human oriented methods often based on technical ideas where the assembly system is regarded as a combination of social and technical systems (A).

A system design can be considered as an integration of two or more criteria. Like System Design Method [12], our design method is a combination of the principle category of b, c, d and e.

An assembly system proposed at Draper Laboratory ([13], [14]) applies only in linear assembly plans (where only one component is added in a single operation to the assembly). An improved design was proposed in [15] where works were being done on different subassemblies in workstations, as shown in Fig.1. Two subassemblies were built and then assembled .

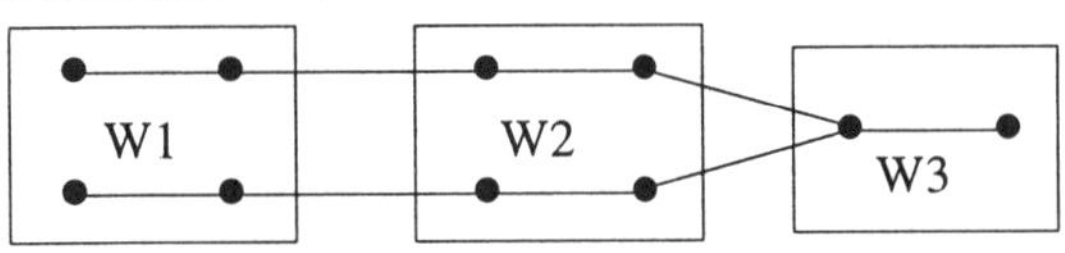

Fig.1. Assembly system with subassemblies

For the transfer of the assembly product, conventional linear assembly line had been well investigated ([20], [21]). A cyclic assembly line where a central conveyor with workstations situated in circular form at outside of the conveyor was proposed in [16] to reduce the work-in-process inventory accumulation as shown in Fig.2. In this paper an assembly workcell design with transfer systems resembling an α-shape with rotary indexing conveyor is chosen.

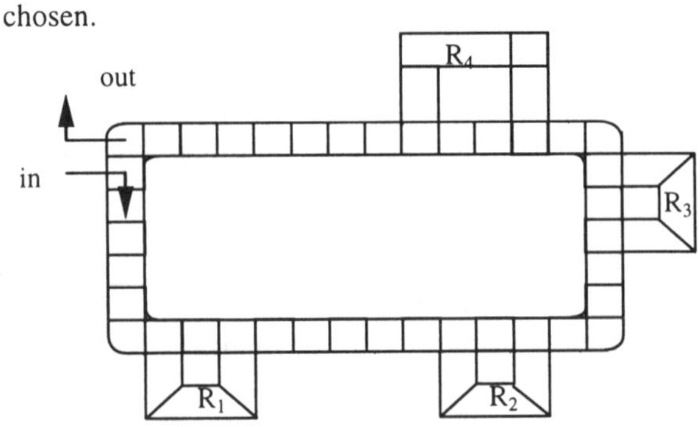

Fig.2 . Cyclic assembly line with robots R1,...,R4

From the exploded view of different motors, process flow for each motor was decided and based on that the required workstations of the workcell were designed. Accordingly a layout design and analysis were performed. The analysis and cycle time computation were performed with alternator. The procedure to achieve our target of study is shown in Fig.3.

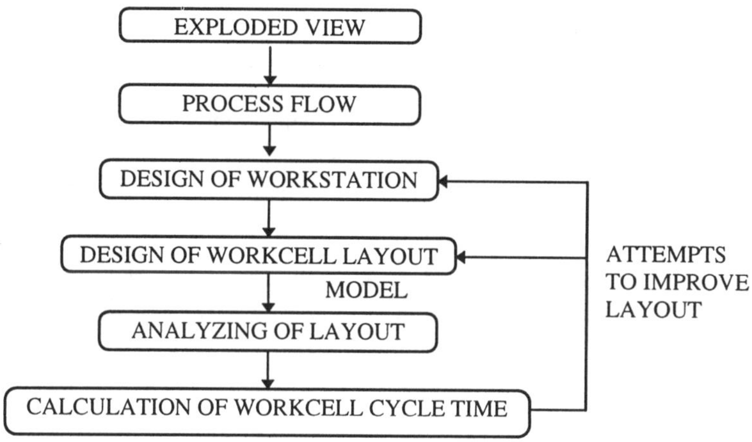

Fig.3. Work Procedure

Robot time and motion (RTM) ([17], [18], [22]) method had been developed for estimating the time that a given robot manipulator model needs to complete a given task cycle. In order to determine the time required to assemble one alternator, RTM methodology had been used. In this paper, design of the workcell is explained in section two, analysis of the workcell design is in section three and results are discussed in section four.

2 . Workcell Design

The main aim is to design a flexible assembly workcell that can be used to assemble different motors used in automobiles. We proposed a workcell that can be used to assemble alternator, starter and wiper motors of automobiles.

The proposed layout of the workcell is based upon the following motors:

1. Ford autolite alternator [23].
2. Typical Ford or American Motors two speed wiper motor [24].
3. Typical positive two engagement movable pole shoe starter.

Following a careful study of each of the above motor's exploded view, preassembled parts were identified and process flow was recognized. According to the motor part's dimension, weight and precedence constraints, a workcell with six workstations was proposed. The number of workstations was estimated from the need of the assembly tasks of the motor that has the largest number of parts. The proposed workcell consists of four robot workstations and two press stations distributed in sequential layout [19] basis.

At the beginning, the designed workcell was a line-shaped assembly process, but it is found that in case of wiper motor, workstation 1 is lightly loaded whereas workstation 6 is overloaded. By following the procedure of Fig.2, ∝-shaped conveyor is designed to perform the assembly process, so that workstation 1 can share the load with workstation 6 through the rotary intersection conveyor part, which indexed during the pallet forwarding from workstation 1 to workstation 2 and during pallet forwarding from workstation 6 to workstation 1. Pallet with fixtures were used to transfer the work-in-process on the conveyor. Complete workcell layout is shown in Fig.4. Design of workstation 1 and workstation 2 are shown in Fig.5 and Fig.6 respectively.

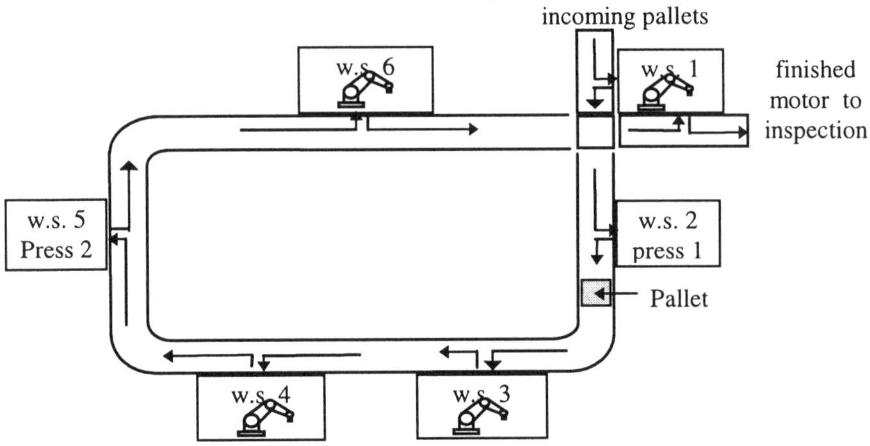

Fig. 4. Workcell Layout

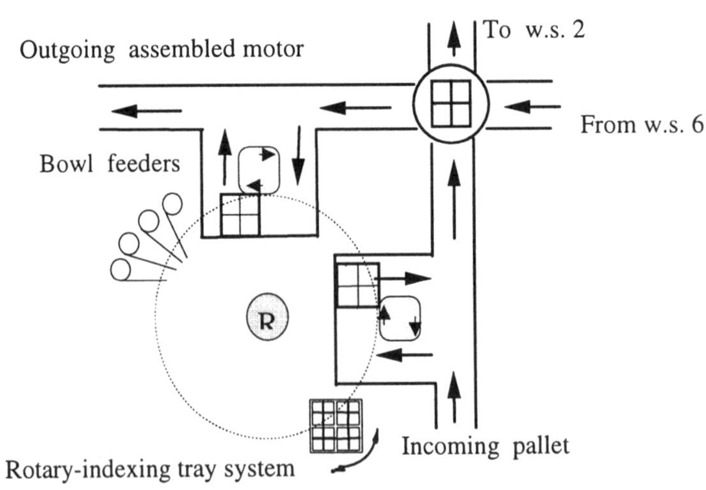

Fig. 5. Workstation 1 layout

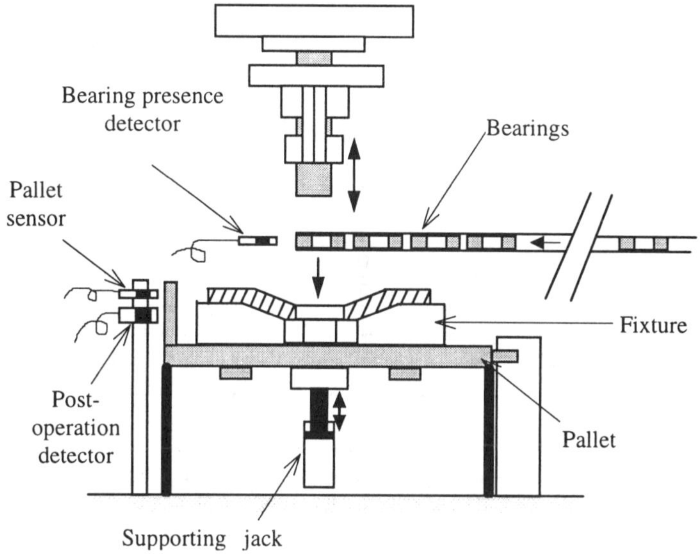

Fig.6. Work station 2 (press station 1)

Trays, bowl feeders(B.F.) and linear vibratory feeder were used in the workstations to supply the different parts to the robots and press units. Large and weighty parts were supplied by trays and smaller parts were supplied by the bowl feeders. Rotary-indexing tray system is used in the workstation to place the trays at certain position within the work envelope of the robot. The linear vibratory feeder placed the bearing at a predetermined position below the press. From the analysis of the exploded views it is found that alternator is made of 14 different types of parts, starter motor is of 26 different parts and wiper motor is made of 36 different parts. The number of trays and bowl feeders required for each motor are shown in Table-1.

Table 1 . Trays and bowl feeders(B.F.) for each workstation(w.s.)

	Alternator		Starter motor		Wiper motor	
	Tray	B.F.	Tray	B.F.	Tray	B.F.
w.s. 1	3	0	4	0	3	4
w.s. 2	0	2*	0	0	0	0
w.s. 3	0	4	1	3	3	5
w.s. 4	1	0	3	4	0	6
w.s. 5	0	0	0	0	0	0
w.s. 6	1	3	2	3	2	4

* linear vibratory feeder

In order to assemble all the three types of motors through the same assembly line, maximum number of trays and bowl feeders for each work station were selected as shown in Table 2.

Table 2. Max. number of trays and bowl feeders(B.F.)

	Tray	B.F.
w.s. 1	4	4
w.s. 2	0	2*
w.s. 3	3	5
w.s. 4	3	6
w.s. 5	0	0
w.s. 6	2	4

Workstation 2 i.e. first press station is to press and insert bearings inside the front and rear housings of the motors. The bearings are positioned exactly below the press tool by passing them through an extended tunnel to the correct position. The other press station (workstation 5) is used to press and fit heavy parts to the shaft of the motors.

In each robot workstation Staubli Unimation RX90 industrial robot with multi-gripper end-effector was used to handle the different parts. The Staubli Unimation RX90 industrial robot was selected for its following characteristics:

— Six axes revolute.
— Work-envelope: reach at wrist center 900 mm max. (between joint 2 and joint 5)
reach at wrist center 290 mm min. (between joint 2 and joint 5)
— Load capacity: 6 kg at nominal speed and 9 kg at low speed.
— Maximum Cartesian speed : 1.5 m/s.

3. Analysis

The proposed layout can be used to assemble any or all of the three types of motors. But analysis was made for only the assembly of alternator. For the assembly of alternator, a pallet with four fixtures was devised to transfer the work-in-process on the conveyor system. All the six work stations were used to assemble alternator. In the robot work station multi-gripper end-effector robot was used to assemble different parts (light parts, heavy parts, nuts, screws etc.). Also turning over of assembly-in-process was done in workstation 4 and workstation 6. In the first press station front bearing was pressed into the front housing and rear bearing was pressed into the rear housing. In workstation 5, fan and pulley assembly was pressed onto the rotor shaft over the assembled parts.

The amount of time required for the work cycle is an important consideration in planning of the workcell. The cycle time determines the production rate of the job which is a significant factor in determining the economic success of replacing the existing system with the new one.

The time taken by the robot to complete an assembly task in a specified workstation was calculated by the RTM methodology. The total time is that consumed in all the workstations and that taken by the pallet to transfer from incoming to the outgoing on the conveyor. Total conveyor length and part transfer time were also determined.

Then the total cycle time for a single alternator was computed. The production rate is calculated with the assumption that the parts transferred in a synchronous way. The RTM calculation for workstation 4 (robot 3) is shown in Table.3.

Table 3. Calculation of cycle time using R.T.M. methodology.

Home position at a height	=	217 mm/s
Robot velocity without load (in arc)	=	800 mm/s
Robot velocity with load (in arc)	=	700 mm/s
Robot velocity without load (in straight line)	=	100 mm/s
Robot velocity with load (in straight line)	=	80 mm/s

Seq.	RTM symbol	Weight	DISTANCE							ELEMENT TIME		Description
			ARC LENGTH			STRAIGHT LENGTH			velocity	for (AL)	for (SL)	
		(lb)	angle	radius	(mm)	1ST	2ND	(mm)	mm/s	sec.	sec.	
1	R1		45	770	604.5				800	1.2		Move in arc from home position to fixture I
2	R1					770	1110	340	100		3.8	Move straight to fixture I
3	R1					217	22.5	194.5	100		2.3	Move down
4	GR1									0.1		Grasp assembled parts [(3)+(4)+(5)+(2)*3] of fixture I
5	M1d	1.3				22.5	217	194.5		3.0		Unload assembled parts from fixture I >1lb
6	M1	1.3				1110	900	210	80		3.2	Move back straight
7	M1t	1.3								2.0		Turn over the assembled parts
8	M1	1.3	13	900	204.1				700	0.6		Move in arc to fixture III
9	M1d					217	104	113		3.0		Load on assembled parts on fixture III
10	R1					104	217	113	100		1.5	Move up
11	R1					900	770	130	100		1.7	Move back straight
12	R1		108	770	1450.7				800	2.2		Move in arc to the tray system
13	R1					217	30	187	100		2.3	Move down
14	GR1									0.1		Grasp part (1) >1lb
15	M1	2				30	217	187	80		2.9	Move up
16	M1	2	108	770	1450.7				700	2.7		Move in arc to fixture III
17	M1	2				770	900	130	80		2.2	Move straight to fixture III
18	M1d					217	135	82		3.0		Load on assembled parts on fixture III
19	R1					135	217	82	100		1.2	Move up
20	R1					900	770	130	100		1.7	Move back straight
21	R1		32	770	429.8				800	0.9		Move in arc to home position
				Total time for movement in arc					=	18.8		
				Total time for movement in straight line						=	23.0	
				Total cycle time for work station # 4							=	41.7

4. Results and Discussion

Total time computed for the production of a single alternator is 10.9 minutes. By the use of RTM method, cycle time for each robot work station was computed. It is found that each workstation has different cycle time and workstation 6 (robot 4) took the highest time (187.1 sec) to complete the assigned assembly work as shown in Table.4. So the transfer mechanism was considered synchronous in order to avoid work-in-process accumulation in other workstations.

If 26 working days and 7 active hours per shift per day are considered, and the time span between adjacent two motors is 187.1sec., then:
production per day =1+(420-10.9)/3.118=132.2 units, and monthly production rate is 3437 units of alternator for a single assembly line. If three shifts per day are considered then the production rate will become 10312 units of alternator per assembly line.

Table 4. Cycle time of the workcell using (R.T.M.) method.

	Cycle Time (second)
1st work station (robot 1)	65.1
2nd work station (press station 1)	4.0
3rd workstation (robot 2)	111.1
4th workstation (robot 3)	41.7
5th work station (press station 2)	3.0
6th workstation (robot 4)	187.1
total transfer time over conveyor	244.7

5. Conclusion

An assembly workcell to assemble different types of motor was proposed in this paper. Selection of robot, workstations were also described in this paper. An analysis of the workcell design was also performed. Even though analysis of the workcell was done for the assembly of alternator, the same line can efficiently be used to assemble starter and wiper motor as well. By the RTM method, the time required to complete the assembly of one alternator was computed. Although this was not done on-line, this investigation can be the basis of the implementation of an automated assembly factory.

References

1. J. D. Wolter, A Combinatorial Analysis of Enumerative Data Structures for Assembly Planning, *Proceedings of IEEE International Conference on Robotics and Automation*, 1991, pp. 611-618.
2. S. Chakrabarty and J. Wolter, A Structure Oriented Approach to Assembly Sequence Planning, *IEEE Transactions on Robotics and Automation*, January 1995.

3. D. F. Baldwin, T. E. Abell, M. C. M. Lui, T. L. De Fazio and D. E. Whitney, An Integrated Computer Aid for Generating and Evaluating Assembly Sequences for Mechanical Products, *IEEE Transactions on Robotics and Automation*, Vol.7, No.1, 1991, pp.79-94.
4. B. Romney, C. Godard, M. Goldwasser and G. Ramkumar, An Efficient System for Geometric Assembly Sequence Generation and Evaluation, *Proceedings of 1995 ASME International Conference on Computers in Engineering*, pp. 699-712.
5. R. Hoffman, Automated Assembly in a CSG Domain, *Proceedings of IEEE International Conference on Robotics and Automation*, 1989, pp. 210-215.
6. M. Goldwasser, J. C. Latombe and R. Motwani, Complexity Measures for Assembly Sequences, *Proceedings of IEEE International Conference on Robotics and Automation*, 1996, pp.1581-1587.
7. J. Wolter, S. Chakrabarty and J. Tsao, Mating Constraint Languages for Assembly Sequence Planning, *Proceedings of IEEE International Conference on Robotics and Automation*, 1992, pp. 2367-2374.
8. L S. H. Mello and A. C. Sanderson, A Correct and Complete Algorithm for the Generation of Mechanical Assembly Sequences, *IEEE Transactions on Robotics and Automation*, , Vol. 7, No. 2, 1991, pp. 228-240.
9. R. H. Wilson and J. C. Latombe, Geometric Reasoning About Mechanical Assembly, *Artificial Intelligence* 71(2), 1994 .
10. A. Agnetis, F. Nicol`o, C. Arbib and M. Lucertini, Task Assignment in Pipeline Assembly Systems, *Proceedings of IEEE International Conference on Robotics and Automation*, 1992, pp. 1133-1138.
11. A. Agnetis, C. Arbib, M. Lucertini and F. Nicol`o, Part Routing in Flexible assembly Systems, *IEEE Transactions on Robotics and Automation*, Vol. 6, No. 6, 1990, pp. 697-705.
12. M. Lundstrom, M. Bjorkman and C. Johansson, A Method for Assembly System Design Including an Integrated Computerized Design Support, *CompEuro Proceedings on Computers in Design, Manufacturing and Production*, 1993, pp. 52-61.
13. D. E. Whitney, T. L. De Fazio, R. E. Gustavson, S. C. Graves, C. Holmes and J. C. Klein, *Computer Aided Design of Flexible Assembly Systems*, Report No CSDL-1947, Cambridge, Massachusetts, August 1986.
14. D. E. Whitney, T. L. De Fazio, R. E. Gustavson, S. C. Graves, K. Cooprider, J. C. Klein, M. Lui and S. Pappu, *Computer Aided Design of Flexible Assembly Systems,* Report No CSDL-R-2033, Cambridge, Massachusetts, January 1988.
15. V. Minzu, J. M. Henrioud, Systematic Method for the Design of Flexible Assembly Systems, *Proceedings of IEEE International Conference on Robotics and Automation*, 1993, pp. 56-62.
16. G. Finke and L. Dupont, Combinatorics of Cyclic Assembly Systems, *Proceedings of IEEE International Conference on Systems, Man and Cybernetics*, 1993, pp. 297-301.
17. S. Y. Nof and R. L. Paul, A Method for Advanced Planning of Assembly by Robots, *Proceedings of SME Autofact*, 1980, pp. 425-435.
18. S. Y. Nof and H. Lachtmann, The RTM Method of Analyzing Robot Work, *Industrial Engineering*, April 1982, pp. 38-48.
19. G. Marlin, A Study Case of Critical Analysis, Relating to an Automatic Assembly System, *Proceedings of 6th International Conference on Flexible Manufacturing Systems*, November 1987, pp. 169-176.
20. A. C. Hax and D. Candea, *Production and Inventory Management*, Prentice-Hall, Englewood Cliffs, New Jersey, 1984.

21. A. Kusiak, *Intelligent Manufacturing Systems*, Prentice-Hall, Englewood Cliffs, New Jersey, 1990.
22. M. P. Groover, M. Weiss, R. N. Nagel and N. G. Odrey, *Industrial Robotics Technology, Programming and Applications*, McGraw-Hill Book Company, 1988.
23. W. K Toboldt and L. Johnson, *Automobile Encyclopedia: Fundamental Principles Operation, Construction, Service and Repair*, Goodheart-Wilcox Company Inc. 1972.
24. W. K Toboldt and L. Johnson, *Automobile Encyclopedia, Fundamental Principles Operation, Construction, Service and Repair*, Goodheart-Wilcox Company Inc. 1975.

Modeling and Estimation of a Pneumatic Positioning System For Lumber Processing

Xiaochun George Wang and Dorothy Wong

National Research Council of Canada
Integrated Manufacturing Technologies Institute, Western Lab,
3250 East Mall, Vancouver, B.C. Canada V6T 1W5
Tel.: 01-604-221-3061, Fax: 01-604-221-3001
e-mail: george.wang@nrc.ca

Key words: wood industry, lumber positioning, pneumatic system, position, force compliance, control, modeling, Kalman filtering, adaptive control, self-tuning control, singular pencil models.

Abstract. This paper presents results of a research project on the modeling, estimation and control of a pneumatic lumber positioning system. Such systems are used in machinery in saw mills and other applications. Control of such systems present a major challenge due to the nonlinearity and variation in dynamics. A dynamic model and the simultaneous on-line recursive estimation of the model parameters and internal variables (state variables) are developed, using the singular pencil models. This approach has shown superior performance on the real system than conventional state space model based approaches, and has show potential for improved self-tuning control based on estimated dynamic parameters as well as the state variables.

1 Introduction

This paper is concerned with the position and force control of heavy rollers of a log feeding machine for lumber production. Typically, such a machine employs from 20 to 60 such rollers to position the log for sawing based on laser scanned information about the logs. Both hydraulic and pneumatic actuators are used, often in series connection, for fast positioning and soft handling of lumber. This research focuses on the use of pneumatic actuators to achieve both force compliance and fast and accurate positioning, for reasons of reliability, efficiency and cost reduction. The accurate control of the position, however, is a challenging problem due to the nonlinearity induced by the pneumatic system and the variation of dynamics. A test facility has been built where part of the real machine is integrated with the most difficult roller mounted. The project has undergone through the first phase successfully and is proceeding to the second phase of implementation. The first phase of the project focused on the modeling and system identification of the system. Many set of data was taken under different operating conditions with different input signals such as step, pseudo-random signals etc. Measurement of positions and pressure at different condition was then used for estimating the parameters of a linearized system model. The model constructed from the estimated parameters are then used with the Kalman filter to give the optimal estimation of the state variables. Since the parameters are time varying for such a system, it presents a problem for state estimation and later on the control based on state feedback, since one set of parameters is only good for certain operating condition, not good for other operating condition. All this is complexed with the nonlinearity exist the system.

The problem of parameter and state estimation is fundamentally nonlinear in the state space model and the common input-output model. A new class of model, the Singular Pencil (SP) model is introduced. The SP model contains the input-output and state space model as subsets, but has certain advantages. One of the advantage is shown in the simultaneous state and parameter estimation where, when formulated in the SP model format, the simultaneous state and parameter estimation problem can be formulated into a linear problem. An Ordinary Kalman filter is used for this problem, giving good convergence. Since the parameters and state are estimated at the same time, the estimated states are valid over a wider range of the operating conditions. The optimal adaptive control can be achieved using the estimated parameters and states, in the state feedback optimal design.

2 The Test Facility

A test facility for development of the modeling, estimation and control was setup, as shown in Figure 1. The function of this pneumatic system is to position wood handling rollers at high speed, yet maintaining accuracy and force compliance. It consists of the cylinder, and two valves that control the air flow to and from the top chamber and bottom chamber of the cylinder. The position of the cylinder, read from a position sensor, can vary from 0 inch at the bottom to 12 inches at the top, corresponding to 0.0 V to 10.0 V. This cylinder goes upward when the bottom valve voltage increases and goes downward when the top valve voltage increases, with the two valve voltages ranging from 0.0 to 5.0 V sent by the computer. Due to the non-linearity of the system, there is a delay between the input voltages and the movement of the cylinder, which is present in a form of hysterisis.

Fig. 1. The test rig and the control system

3. Modeling Of The System

A Singular Pencil Model is used to describe the relationship between the inputs (top and bottom voltages) and output (position of the roller) of the system. It is of the form

$$[E-FD, G]\begin{bmatrix} x \\ w \end{bmatrix} = P(D)\begin{bmatrix} x \\ w \end{bmatrix} = 0, \tag{1a}$$

where $x \in R^n$ is the *internal (auxiliary) variable* vector, and $w \in R^{p+m}$, with known p and m, is the *external variable* vector, usually consisting of input/output variables of the system. E and F are $(n+p)\times n$ matrices, G is $(n+p)\times(p+m)$, and D is a linear operator. Equation (1a), representing a multi-input, multi-output system, also includes state-space models as a special case. When D is replaced by z, it is written as

$$\left\{\begin{bmatrix} A \\ C \end{bmatrix} - \begin{bmatrix} I_n \\ 0 \end{bmatrix} z\right\} x_k + \begin{bmatrix} 0 & B \\ -I & D \end{bmatrix}\begin{bmatrix} y_k \\ u_k \end{bmatrix} = 0, \tag{1b}$$

where $u_k \in R^m$ is the input vector, $y_k \in R^p$ is the output vector, and *(A,B,C,D)* are the state-space matrices. Another special case is the class of matrix fraction or ARMA (Auto-Regression Moving Average) models of the form

$$\sum_{i=0}^{n} A_i z^{n-i} y_k = \sum_{i=0}^{m} B_i z^{m-i} u_k \tag{2a}$$

the A_i and B_i are matrices and z is the shift operator. This representation can be written in first-order form as

$$\begin{bmatrix} I & & \\ & \ddots & \\ & & I \\ & & 0 \end{bmatrix} z x_k = \begin{bmatrix} 0 & & \\ I & & \\ & \ddots & \\ & & I \end{bmatrix} x_k + \begin{bmatrix} -A_n & B_n \\ \vdots & \vdots \\ -A_o & B_o \end{bmatrix}\begin{bmatrix} y_k \\ u_k \end{bmatrix} \tag{2b}$$

In general, singular pencil models have the following characteristics which can be some of the reasons to use them:

1. Unification: SP models include as special cases the two types of most commonly used models: state space models and input-output (or MFDs, ARMA) representations.
2. Generality: unlike state space representations, SP models can describe improper systems. PID controllers, tachometers, and inverses of strictly proper systems are some improper systems that are useful both in concept and in fixed stages of design. Futhermore, SP models admit non-uniqueness of the output y and constraints on the input u of the internal variable x of the system.
3. Simplicity: Useful transformations can be simply implemented as sequences of elementary (for sparsity) or orthogonal (for stability) row and column operations on the constituent matrices E, F, G of (1a).

For details of the realization, as well as design and analysis of control systems using such models, the reader is referred to [1,2,3]. The simultaneous estimation of states and system parameters was developed by Salut, Aquilar Martin and Lefebvre[4,5], and Chen, Aplevich and Wilson[6]. The distinction between inputs and outputs in w

is generally unimportant for identification since both types of variables must be measured. For the present development it can even be assumed that the definition of variables in w is unknown or is not unique, since it is not explicitly required. The auxiliary (internal) vector $x \in R^n$ is necessary in order to write the system as a set of simultaneous first-order difference equations. To determine the model structure, the input/output data collected from the open-loop system with sufficient excitation , as shown in Fig 2, is used to fit an ARMAX (Auto-Regression Moving Average with eXogeneous input signal) model of different orders (model structures). It is of the form

$$\sum_{i=0}^{n} A_i z^{n-i} y_k = \sum_{i=0}^{m} B_i z^{m-i} u_k + \sum_{i=0}^{n} C_i z^{n-i} e_k \, , \tag{3}$$

where A_i and B_i are matrices and z is the shift operator.

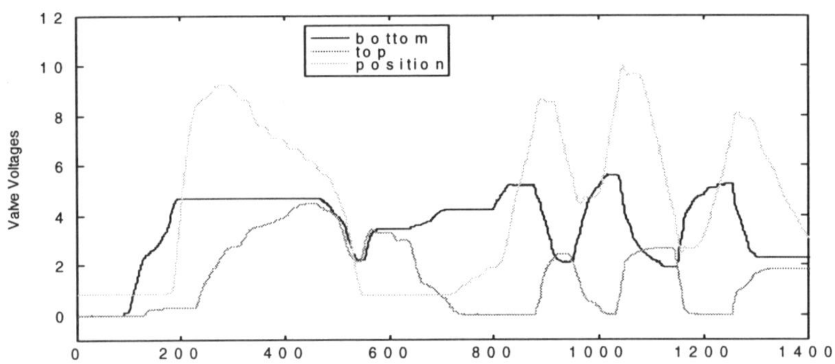

Fig. 2. Input voltages and position

For the pneumatic system discussed in this paper, there are two inputs and one output. If error is accounted for, the ARMAX model is of the form

$$\sum_{i=0}^{n} a_i z^{n-i} y_k = \sum_{i=0}^{m} b_i^1 z^{m-i} u_k^1 + \sum_{i=0}^{m} b_i^2 z^{m-i} u_k^2 + \sum_{i=0}^{n} c_i z^{n-i} e_k \, , \tag{4}$$

where y_k is the roller position detected by the position sensor, u_k^1 is the bottom valve voltage, u_k^2 is the top valve voltage, e_k is the equation error, accounting for the noise in the process and measurement, and is assumed to be an i.i.d.(independent identically-distributed) Gaussian process, a_i, b_i^1, b_i^2 and c_i are the parameters of the model, and n and m are the orders of the system. To find the optimal model structure, the performance of the model, judged by many factors such as the variance of the prediction error curve (e_k = *actual position - predicted position*) and the time necessary for the estimation of parameters and states to settle, can be used. Without losing generality, m is made equal to n when comparing models of increasing orders. It was observed that the quantity J_N , the summation of error, didn't improve much beyond n=4 and thus the fourth order will be used.

For a fourth order system, the ARMAX model is of the form

$$(z^4 + a_1 z^3 + a_2 z^2 + a_3 z + a_4) y_k = (b_1^1 z^3 + b_2^1 z^2 + b_3^1 z + b_4^1) u_k^1 + (b_1^2 z^3 + b_2^2 z^2 + b_3^2 z + b_4^2) u_k^2$$
$$+ (z^4 + c_1 z^3 + c_2 z^2 + c_3 z + c_4) e_k \tag{5}$$

This representation can be written in SP form as

$$\begin{bmatrix} 1 & 0 & 0 & 0 \\ 0 & 1 & 0 & 0 \\ 0 & 0 & 1 & 0 \\ 0 & 0 & 0 & 1 \\ 0 & 0 & 0 & 0 \end{bmatrix} zx_k = \begin{bmatrix} 0 & 0 & 0 & 0 \\ 1 & 0 & 0 & 0 \\ 0 & 1 & 0 & 0 \\ 0 & 0 & 1 & 0 \\ 0 & 0 & 0 & 1 \end{bmatrix} x_k + \begin{bmatrix} -a_4 & b_4^1 & b_4^2 & c_4 \\ -a_3 & b_3^1 & b_3^2 & c_3 \\ -a_2 & b_2^1 & b_2^2 & c_2 \\ -a_1 & b_1^1 & b_1^2 & c_1 \\ -1 & 0 & 0 & 1 \end{bmatrix} \begin{bmatrix} y_k \\ u_k^1 \\ u_k^2 \\ e_k \end{bmatrix}. \tag{6}$$

By simple matrix manipulation, equation (6) can be formulated as

$$\begin{aligned} x_{k+1} &= E_* x_k + G_* w_k + C_* e_k \\ 0 &= E_o x_k + G_o w_k + e_k \end{aligned}, \tag{7}$$

with the matrices being

$$E_* = \begin{bmatrix} 0 & 0 & 0 & 0 \\ 1 & 0 & 0 & 0 \\ 0 & 1 & 0 & 0 \\ 0 & 0 & 1 & 0 \end{bmatrix}, E_o = \begin{bmatrix} 0 & 0 & 0 & 1 \end{bmatrix}, G_* = \begin{bmatrix} -a_4 & b_4^1 & b_4^2 \\ -a_4 & b_3^1 & b_3^2 \\ -a_4 & b_2^1 & b_2^2 \\ -a_4 & b_1^1 & b_1^2 \end{bmatrix}, G_o = \begin{bmatrix} -1 & 0 & 0 \end{bmatrix}, \tag{8}$$

and

$$C_* = \begin{bmatrix} c_4 & c_3 & c_2 & c_1 \end{bmatrix}^T, w = \begin{bmatrix} y_k & u_k^1 & u_k^2 \end{bmatrix}^T. \tag{9}$$

Furthermore, the expression $G_* w_k$ can be rewritten as

$$G_* w_k = \begin{bmatrix} -a_4 & b_4^1 & b_4^2 \\ -a_3 & b_3^1 & b_3^2 \\ -a_2 & b_2^1 & b_2^2 \\ -a_1 & b_1^1 & b_1^2 \end{bmatrix} \begin{bmatrix} y_k \\ u_k^1 \\ u_k^2 \end{bmatrix} = \begin{bmatrix} -y_k & 0 & 0 & 0 & u_k^1 & 0 & 0 & 0 & u_k^2 & 0 & 0 & 0 \\ 0 & -y_k & 0 & 0 & 0 & u_k^1 & 0 & 0 & 0 & u_k^2 & 0 & 0 \\ 0 & 0 & -y_k & 0 & 0 & 0 & u_k^1 & 0 & 0 & 0 & u_k^2 & 0 \\ 0 & 0 & 0 & -y_k & 0 & 0 & 0 & u_k^1 & 0 & 0 & 0 & u_k^2 \end{bmatrix} r_k$$

$$= \tilde{G}_*(w_k) r_k . \tag{10}$$

where $r_k = \begin{bmatrix} a_4 & a_3 & a_2 & a_1 & b_4^1 & b_3^1 & b_2^1 & b_1^1 & b_4^2 & b_3^2 & b_2^2 & b_1^2 \end{bmatrix}^T$. If noise is ignored, the system can be described by

$$\begin{aligned} x_{k+1} &= E_* x_k + \tilde{G}_*(w_k) r_k \\ y_k &= E_o x_k + [G_o + J] w_k \end{aligned} \tag{11}$$

with the matrices being: $G_o = \begin{bmatrix} -1 & 0 & 0 \end{bmatrix}, J = \begin{bmatrix} 1 & 0 & 0 \end{bmatrix}, G_o + J = \begin{bmatrix} 0 & 0 & 0 \end{bmatrix}$ (12)

In the presence of noise, we have
$$\begin{cases} x_{k+1} = E_* x_k + \tilde{G}_*(w_k) r_k + C_* e_k \\ y_k = E_o x_k + e_k \\ r_{k+1} = r_k + \pi_k \end{cases} \tag{13}$$

letting s_k hold the states and parameters, i.e. $s_k = \begin{bmatrix} x_k & r_k \end{bmatrix}^T$, the following equations are obtained:

$$s_{k+1} = F_k s_k + \begin{bmatrix} C_* e_k & \pi_k \end{bmatrix}^T, \tag{14}$$

$$y_k = H_k s_k + e_k, \tag{15}$$

where $$F_k = \begin{bmatrix} E_* & \tilde{G}_* w_k \\ 0_{12\times 4} & I_{12\times 12} \end{bmatrix}, \tag{16}$$

and $$H_k = [E_0 \quad 0_{1\times 12}] = [0 \ 0 \ 0 \ 1 \ 0 \ 0 \ 0 \ 0 \ 0 \ 0 \ 0 \ 0 \ 0 \ 0 \ 0 \ 0]. \tag{17}$$

The time-varying matrices F_k and H_k are known because E_* and E_0 are determined from the $\{k_i\}$,and $\tilde{G}_*(w_k)$ and $\tilde{G}_0(w_k)$ are constructed from the measure-ment w_k .

4. Estimation of States and Parameters

Having developed a model to describe the system, the next step is to actually estimate the states x_k and parameters r_k of the model, in other words, the vector s_k . A Kalman filter can be applied directly to give the optimal linear estimate of this vector, by using the recursive equations

$$\hat{s}_{k+1} = F_k \hat{s}_k + K_k (y_k - H_k \hat{s}_k), \tag{18}$$

$$K_k = (F_k P_k H_k^T + S)(H_k P_k H_k^T + R)^{-1}, \tag{19}$$

$$P_{k+1} = F_k P_k F_k^T + Q - K_k (H_k P_k H_k^T + R) K_k^T, \tag{20}$$

with initial conditions

$$\hat{s}_0 = E\{s_0\}, P_0 = E\{(s_0 - E\{s_0\})(s_0 - E\{s_0\})^T\}, \tag{21}$$

where *E{x}* denotes the mathematical expectation of *x*, and

$$E\left\{\begin{bmatrix}\begin{bmatrix} C_* e_k \\ \pi_k \end{bmatrix} \\ e_k \end{bmatrix}\begin{bmatrix}\begin{bmatrix} C_* e_k \\ \pi_k \end{bmatrix} \\ e_k \end{bmatrix}^T\right\} = \begin{bmatrix} Q & S \\ S^T & R \end{bmatrix} \tag{22}$$

where *Q, S,* and *R* are noise covariance matrices of the model. Thus, provided the structural integers, the dynamic noise parameters C_* , and *Q, S,* and *R* are known, the system auxiliary vector and non-structural parameters can be estimated using an ordinary Kalman filter, which converges under well-defined conditions.

Considerations are taken when selecting the initial values for the above matrices. The state covariance matrix P_k should be symmetrical and large to allow for error at initial conditions. It is chosen as $10000 \times I_{(n+n+2m)\times(n+n+2m)}$ ($I_{a\times b}$ denotes identity matrix with *a* rows and *b* columns). Based on measurement of signals in the system and tuning of the matrices for optimal performance, the matrices *Q* and *S* are chosen to be identity matrices $I_{(n+n+2m)\times(n+n+2m)}$ and $I_{(n+n+2m)\times 1}$, respectively, and *R* is selected as 0.1 of the range of deviation in the position, which is around 0.97 when the cylinder is operating in the middle region. Matrix C_* , a measurement of the noise parameters, is chosen to be very small assuming negligible error. It is chosen as

$$C_* = [0 \quad 0 \quad 0 \quad 0.1]^T . \tag{23}$$

The matrix of variance of the parameters, π_k , is set as 0.5 after a few trials.

Next, performance of the model using the matrices mentioned above is compared with that when matrices Q and S are selected according to equation 22. It can be shown that the estimation error improves slightly with the new Q and S. With R=0.97, the values of the matrices used in calculation are summarized as follows:

$$\pi_k = \begin{bmatrix} 0.5 & 0.5 & 0.5 & 0.5 & 0.5 & 0.5 & 0.5 & 0.5 & 0.5 & 0.5 & 0.5 & 0.5 \end{bmatrix}^T \qquad (24)$$

$$Q = \begin{bmatrix} 0 & 0 & 0 & 0 & 0.25_{1\times12} \\ 0 & 0 & 0 & 0 & 0.25_{1\times12} \\ 0 & 0 & 0 & 0 & 0.25_{1\times12} \\ 0 & 0 & 0 & 0.097 & 0.25_{1\times12} \\ 0.25_{12\times1} & 0.25_{12\times1} & 0.25_{12\times1} & 0.25_{12\times1} & 0.25_{12\times12} \end{bmatrix} \qquad (25)$$

$$S = \begin{bmatrix} 0 & 0 & 0 & 0.09655 & 0 & 0 & 0 & 0 & 0 & 0 & 0 & 0 & 0 & 0 & 0 \end{bmatrix}^T \qquad (26)$$

$$Pk = 10000 * I_{16}; \qquad (27)$$

where $0.25_{a\times b}$ is a matrix of a rows and b columns with every entry equal to 0.25. In this section, it has been shown that the two steps of estimating the states and parameters can be done simultaneously by using a linear Kalman filter with the SP model (equation 18-20). For the input/output data shown in Fig 2, the states and parameters $x_4 \ldots x_1, a_4 \ldots a_1, b_1^1 \cdots b_4^1$, and $b_1^2 \cdots b_4^2$ estimated using this model are plotted in the following figures, where $b_1^1 \cdots b_4^1$ and $b_1^2 \cdots b_4^2$ are denoted as b11...b14, and b21...b24, respectively.

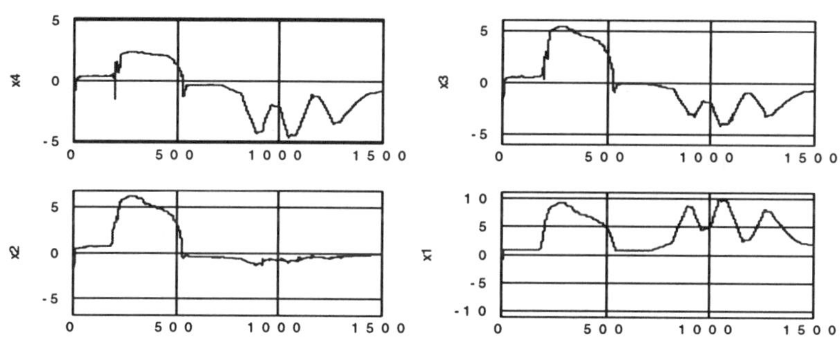

Fig. 3. Estimated states ($x_4 \ldots x_1$)

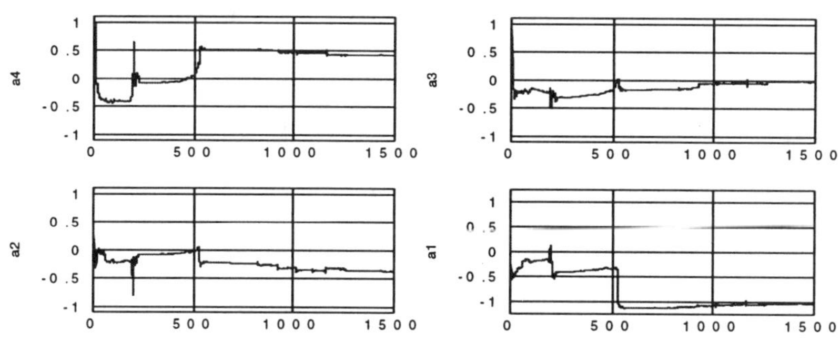

Fig. 4. Estimated parameters ($a_4 \ldots a_1$)

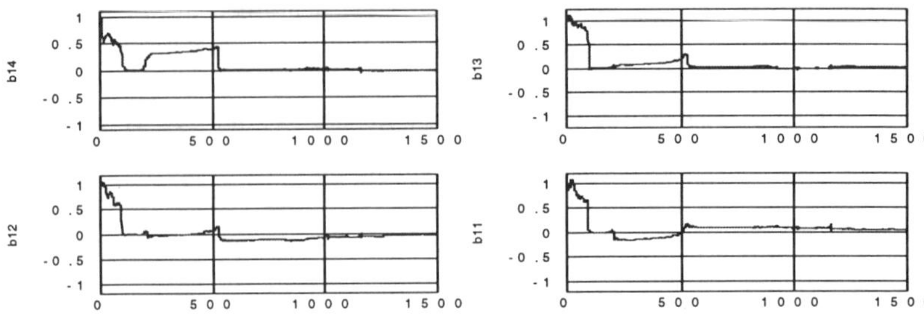

Fig. 5. Estimated parameters (b14...b11)

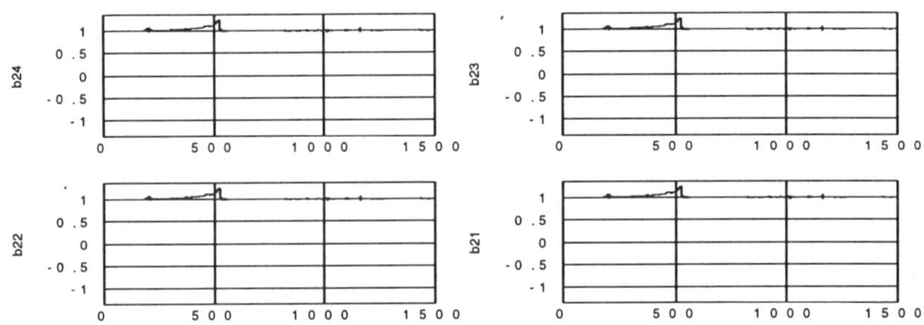

Fig. 6. Estimated parameters (b24...b21)

From the above figures, it can be observed that the parameters settle to constant values after a short period of time. No matter what values they are initialized at, they should settle to their true values in the end.

5. Validation

To confirm that the SP model developed in the previous sections gives correct results, an output is simulated using the regression model with input u_k and the estimated parameters r_k, and is compared with the actual position y_k. The simulated output y_{sim_k} is

$$y_{sim\_k} = \phi_k \hat{r}_k \tag{28}$$

where $\phi_k = \left[-y_{sim\_k}(k-n) \quad \cdots \quad -y_{sim\_k}(k-1) \quad u_k{}^1(k-n) \quad \cdots \quad u_k{}^1(k-1)\right]$. (29)

The following plot shows the simulation compared to the real position. It can be observed that the simulated position is very close to the actual one and thus the estimations are accurate.

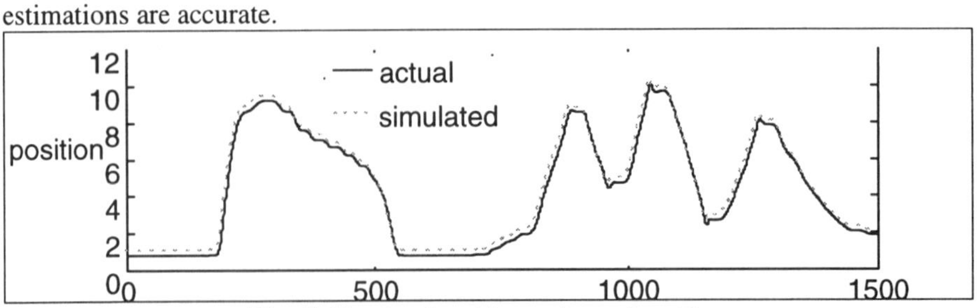

Fig. 7. Simulated position and actual position

Besides being able to provide a way to estimate the states and parameters in one single step as mentioned in the last section, it can be shown that the SP model gives a smaller estimation error ($e_k = y_k - \hat{y}_k$) and takes much less time for the estimation ($\hat{y}_k$) to converge to the real value (y_k) comparing to the State Space Model usually used. Shown in figures 9 and 10 are plots of position estimation and estimation error for the SP model and State Space Model.

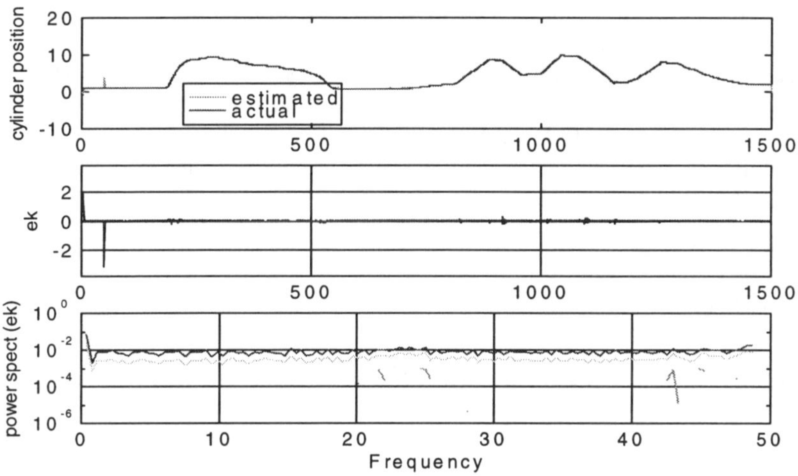

Fig. 8. Performance of SP Model (Top: actual position and estimation, middle: prediction error, bottom: power spectrum of prediction error)

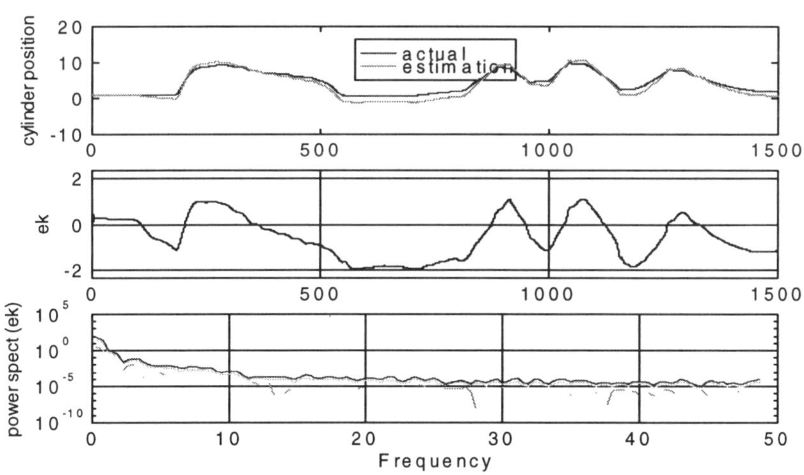

Fig. 9. Performance of State Space Model (top: actual Position and estimation, middle: Prediction error, bottom: Power spectrum of prediction error)

6. Conclusion and Further Work

This paper presents results on the modeling and simultaneous state and parameter estimation of an industrial pneumatic positioning system for lumber processing machines using the singular pencil model. The dynamics of such systems changes dramatically with load, speed, force, air supply, environment etc. thus a control scheme using the estimated parameters and state variables to maintain optimal performance would be preferable. Such schemes can be the ones discussed in [8,9,10]. Good convergence property is achieved by the use of the linear Kalman filter estimating the state and parameters. A self-tuning control scheme, based on state feedback with the estimated state variables, and using the estimated parameters in the controller design, can promise improved performance over schemes that uses estimated states based on a state space model constructed with pre-estimated model parameters or boot-strap methods. Major challenges, however, still exist in that significant level of nonlinearity is observed in the system, in the form of hysteresis between the input voltage to the valves and the output (the position) of the actuator. This nonlinearity is being compensated and helps in improving the estimation as well as in control (results to be published later).

7. References

1. J.D.Aplevich, Time-Domain input-output representations of linear systems, *Automatica,* vol. 17, no. 3 (1981) 509-521.
2. Aplevich, Minimal representations of implicit linear systems, *Automatica,* vol. 21, no. 3 (1985) 259-269.
3. Aplevich, *Implicit Linear Systems, Lecture Notes in Control and Information Sciences,* vol. 152, Heidelberg: Springer-Verlag (1991).
4. Salut, J. Aquilar-Martin, and S. Lefebvre, Canonical input-output representation of linear multivariable stochastic systems and joint optimal parameter and state estimation, *Stochastica,* vol. III, no. 1 (1979) 17-38.
5. Salut, J. Aquilar-Martin, and S. Lefebvre, New results on optimal joint parameter and state estimation of linear stochastic systems, *ASME Trans. J. Dyn. Syst. Meas. and Control,* vol.l02 (1980) 28-34.
6. Chen, J.D. Aplevich and W. Wilson, Simultaneous estimation of state and parameters for multivariable linear systems with singular pencil models, *IKE Proceedings,* vol. 133 (March 1986) Pt. D(2):65-72.
7. Xiaochun George Wang, *Model Structure Selection for On-Line System Identification Using Overlapping Singular pencil Models.* Ph.D. thesis, University of Waterloo, Waterloo, Ontario, Canada (1991).
8. Xiaochun George Wang, Guy Dumont and Michael Davies, Modeling and identification of basis weight variations in paper machines, *IEEE trans. Control Systems Technology,* Vol. 1, No. 4 (1993) 230-237.
9. Xiaochun George Wang, Adaptive pole placement of paper machines with singular pencil models, Proceedings *of the International Conference On Intelligent Manufacturing' 95,* Wuhan, China (June 14-17, 1995) 770-777.
10. Xiaochun George Wang, Self-tuning control with Singular Pencil Models, Proceedings of *1995 IEEE International Conference on Systems, Man and Cybernetics* ,Waterfront Center Hotel, Vancouver, B.C., Canada (Oct 22-25, 1995) 2604-2610.

Inspection and Diagnostic Systems

A Fuzzy Expert System for Monitoring Dimensional Accuracy of Components

[1]Henry Lau, [2]Ralph W.L. Ip [3]Felix T.S. Chan

[1]Advanced Technology Education Centre, Regency Institute, Australia
[2]Dept. of Manuf. Engg. & Engg. Mgt., City University of Hong Kong
[3]Dept. of Ind. & Manuf. Systems Engg., University of Hong Kong

Abstract. Fuzzy expert system uses fuzzy logic control[1] which is based on a "superset" of Boolean logic that has been extended to handle the concept of "partial truth" and it replaces the role of mathematical model with another that is built from a number of rules with fuzzy variables such as output temperature and fuzzy terms such as hot, fairly cold, probably correct. A fuzzy expert system has been implemented in a aluminium die casting company for monitoring dimensional quality of output products. This fuzzy expert system was developed based on the experience of a similar approach for quality control of injection molding components. The die casting parts are supplied to a local car-manufacturing plant and used as assembly components for the motor engines and therefore the dimensional accuracy of them is of utmost importance. This paper presents the implementation of this monitoring system using fuzzy logic theory. Practical examples with descriptions of how the fuzzy rules are fired and the operations of the fuzzy inference engine are also covered.

1. Introduction

The principle of aluminium die casting is to force molten aluminium under certain pressure into a permanent steel mold to produce the required shaped parts. Many intricately-shaped components of various sizes are produced in these molds ranging from small knobs, handles to engine blocks, large appliance castings, automotive crankcase and transmission housings. During operation, high temperature of up to 1150°F is required to melt the aluminium alloy which is injected into the mold. Studies show that dimensional quality of parts from aluminium die casting operations depends largely on the parameters of the moulding conditions. There are four parameters which may affect the surface finish, mechanical properties and dimensional stability of output components and these include mold temperature, molten alloy temperature during injection, injection pressure, operation time.

To monitor the quality of the products, a monitoring system based on fuzzy logic has been developed and implemented in a car-component supplier in Australia. Fuzzy rules with linguistic terms are set based on alloy die-casting theory, past experience and trial results. New rules can be added anytime whenever some new results are found. Instead of specifying exact numeric value, fuzzy rules support the use of statements with linguistic terms. Just like a classical expert system[2,3] fuzzy expert system consists of knowledge base, rules and inference engine, but instead of dealing with crisp (exact) information, it handles fuzzy (uncertain, imprecise) data as well[4].

2. Traditional Technique

Quality of die cast parts depends largely on the parameters of the moulding conditions as described in the above context. Although these parameters can be set before moulding process starts, the actual conditions may alter due to other factors e.g. blocking of passage for cooling water, unforeseen mechanical or electronic failures. Actual states of moulding conditions can be obtained by using sensing devices and the actual readings can be recorded by instruments or displayed on computer screen if the sensor devices are connected to a computer via some analogue/digital hardware.

Traditionally, if there is suspected dimensional discrepancy occurred in certain components, experts are called in to provide advice for correctional action. In most cases, the parameter settings such as temperature, pressure are to be adjusted in order to rectify the dimensional problem. However, this is not an efficient way to handle the situation and in addition, expert advice is normally based on personal opinion and knowledge.

3. Fuzzy Logic System

A dimensional monitoring system based on fuzzy logic concept has been tested and implemented in a company. This system is designed in such a way that it is able to store knowledge and experience from previous tests and production, with the important feature that expert advice is available whenever it is needed.

Specific components, which require high dimensional and strength quality, are first chosen to be monitored by this system. Take for example the component as shown in Figure 1, it is important to specify which dimensions of the product are to be monitored. As shown in the Figure, the dimensions "a", "b", "c", "d" and "e" of the component are the key dimensions that require high accuracy. Normally, the dimensions of a component depend on two key factors which are the design of the mold and the molding conditions during the process. During the pre-production stage, the mold design (such as draft, runners and gates, side slides) as well as the molding conditions (such as temperature, pressure) are finalised before actual production. However, due to some unexpected reasons, the quality of output components may not be sustained.

A fuzzy expert system is implemented to deal with this problem. The first step of designing the fuzzy expert system is to determine the fuzzy sets[5,6] for the monitored dimensions. Each dimension has to be represented by a fuzzy set which is designed in compliance with the fuzzy logic principle. In this example, the required dimension for "a" is, say, 110.50 ± 0.02 mm. This dimension is then "fuzzified" and represented as fuzzy set shown below :
Undersized - represented by Z function with data "d-0.1" and "d-0.01" (d = 110.50)

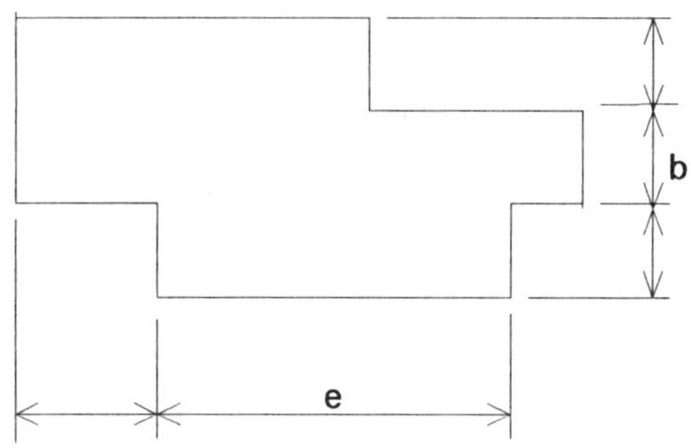

Figure 1 : Sample Component

OK - represented by Π function with data "d" and "0.02"
oversized - represented by S function with data "d+0.01" and "d+0.1"

Figure 2 shows the fuzzy representation of dimension "a". In FuzzyCLIPS[6,7], the above functions can be represented by a template and in this case, it is called Dimension_a:

(deftemplate Dimension_a
d-0.01 d+0.01 mm
(undersized (z d-0.1 d-0.01)
(OK (pi 0.02 d)
(oversized (s d+0.01 d+0.1)))))

For example, the dimension "a" of the component is measured and found to be equal to 110.56 or "d + 0.06" which is obviously oversized. The result obtained is a "crisp" value and needs to be "fuzzified" for fuzzy inference by the fuzzy expert system. In FuzzyCLIPS, the fuzzy set of the actual dimension can be represented as :

(assert (Actual_DimA (- ?Actual_DimA 0.005) 0.0) (?Actual_DimA 1.0)
(+ ?Actual_DimA 0.05) 0.0)

As shown in Fig. 3 , the fuzzy set is :

(Actual_DimA (110.555 0.0) (110.56 1.0) (110.565 0.0))

For the component as shown in Fig. 1, there are five dimensions to be controlled and assuming that there are four parameters to be considered including mold temperature (P1), molten alloy temperature during injection (P2), injection pressure (P3) and operation time (P4), the table for the adjustments of parameters is as shown in Table 1.

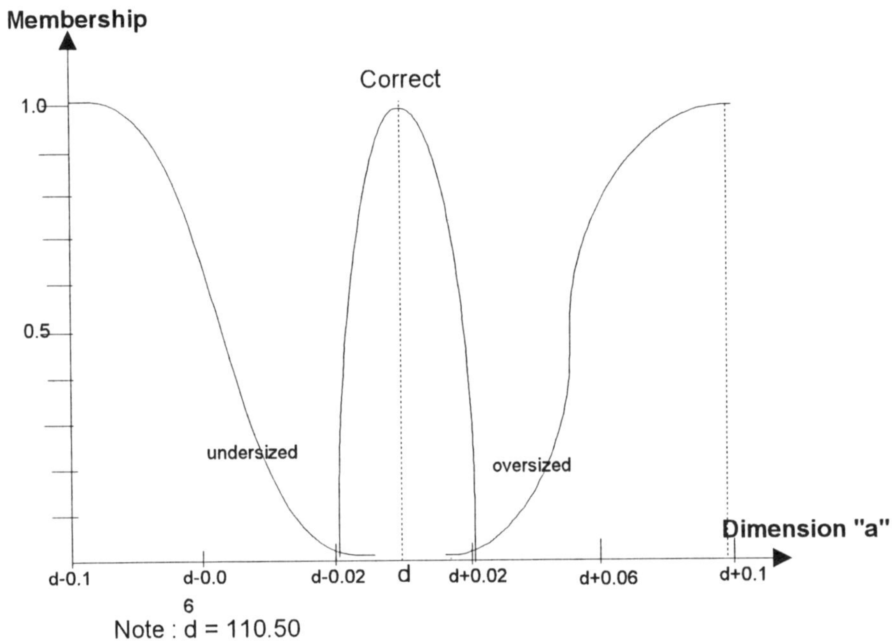

Figure 2 : Fuzzy Set for Actual Dimension of "a"

4. Modifiers and Linguistic Expressions

A modifier may be used to further enhance the ability to describe the fuzzy sets. Modifiers (very, slightly) used in phrases such as very hot or slightly cold modify (fine tune) the shape of fuzzy sets in a way that suits the meaning of the word used. FuzzyCLIPS has a set of predefined modifiers that can be used at any time to describe fuzzy concepts when fuzzy terms are described in fuzzy deftemplates, fuzzy rule patterns are written, or fuzzy facts or fuzzy slots are asserted.

5. Structure of Rules

A number of rules are established based on the data from Table 1. The antecedents and consequents of the rules contain fuzzy data as shown in the following example.

<If dimension A is oversized and dimension B is undersized>
Then
<Increase operation time slightly and reduce the molten alloy temperature considerably>

According to Table 1, the recommended action for an oversized dimension "a" is a considerable decrease (CD) of mold temperature (P1). As shown in Fig. 3, the fuzzy pattern of Actual Dimension of "a" cuts the oversized fuzzy pattern at 0.79 (maximum point). According to Compositional Rule for Multiple Antecedents, the result of fuzzy reference is referring to the membership grade 0.79 as shown in Fig. 4.

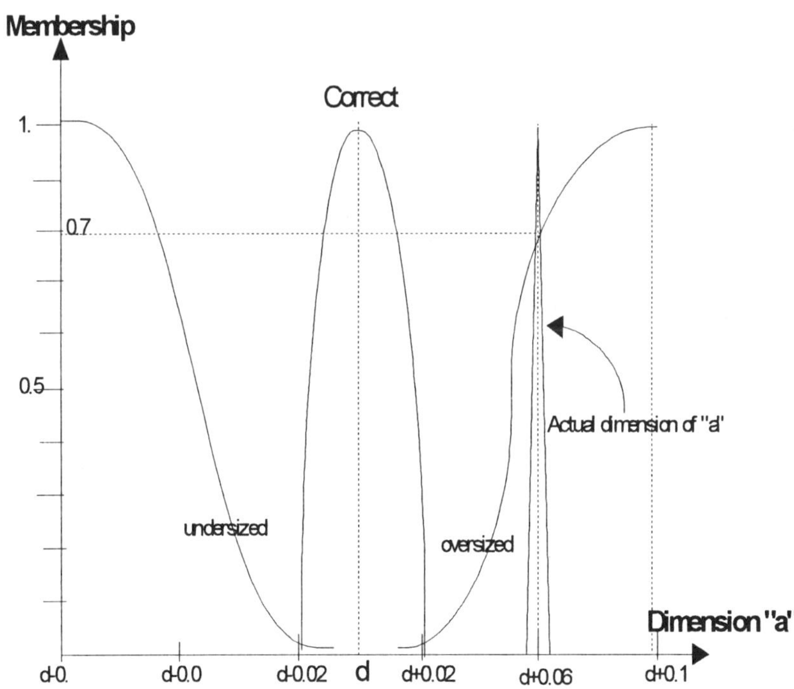

Figure 3 : Fuzzy Set for Actual Dimension of "a"

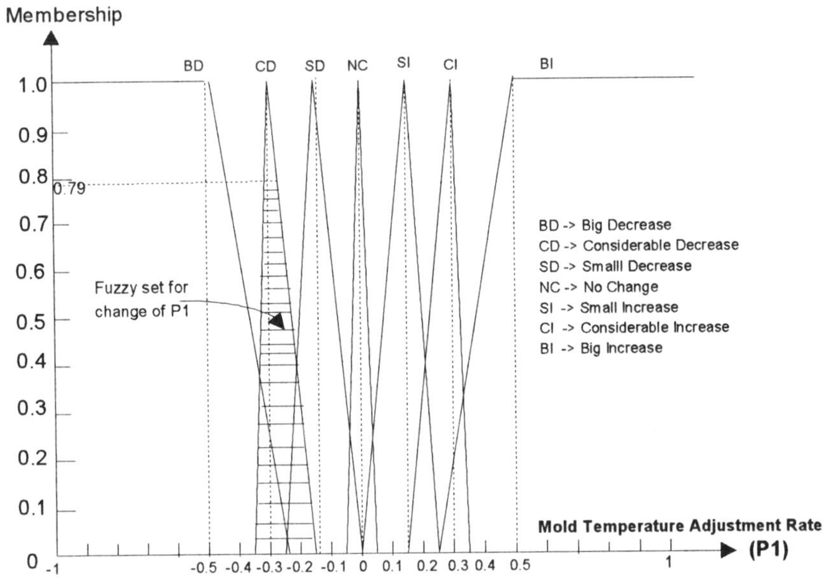

Figure 4 :Change of Mold Temperature Recommended

Table 1. Parameter Change Based on Dimensional Results

OS = OverSized
US = UnderSized
SI = Small Increase
CI = Considerable Increase
BI = Big Increase

NC = No Change
SD = Small Decrease
CD=Considerable Decrease
BD = Big Decrease

Dim. a	**Dim. b**	**Dim. c**	**Dim. d**	**Dim. e**	**P1**	**P2**	**P3**	**P4**
OS					CD			
OS	US					CD		SI
		OS					CI	
			US					
				OS		BI		
	US		OS				SD	
OS				US		BD		SD
	OS	OS					SI	
OS		US		US		CD		

The fuzzy set for the change of mold temperature is determined and shown in Fig. 4 (shaded area). It should be noted that the outcome of the fuzzy inference process is another fuzzy set[8]. However, in this case, it is essential that only a single discrete action is applied and so a single point that reflects the value of the set needs to be determined. The processing of reducing a fuzzy set to a single point is known as defuzzification[6,9]. FuzzyCLIPS allows users to choose either one by specifying in the codes. In this example, the crisp value is -0.252 i.e. 25.2% reduction of existing value. With this approach, the value recommended for adjustment of one of the important parameter is obtained. Each dimension (totally five in this example) and each parameter (totally four) have its own fuzzy pattern and should be designed based on fuzzy logic principle. The FuzzyCLIPS solution can handle a large number of fuzzy sets and produce the required results to users.

6. Conclusion

This paper presents the application of fuzzy logic in an expert system for the monitoring of dimensional quality of die cast components. Instead of using the traditional technique based on individual's experience and knowledge, the fuzzy expert system provides another approach, which is based on fuzzy control principle, to tackle the problem In this case study, fuzzy sets are used to represent the dimensions of the components. Modifiers can also be used to modify the shape of the fuzzy sets in a way that suits the meaning of the word (very, slightly) used. This system has been used in a company and proved to be quite successful and cost-effective. Full automatic control is under development with the PC interfacing to the automatic sensing equipment and the controlling hardware of the automatic die-casting machines so that whenever there is any dimensional discrepancy, the fuzzy expert system will respond and determine the adjustment of related molding parameters. This is possible due to the fact that FuzzyCLIPS is written in C with source codes provided. Modification of the source codes with some additional interfacing functions allows the direct control of the parameter settings of molding conditions.

References

1. L.A. Zadah, Fuzzy Sets, *Information and Control*, Vol. 8, 1965, pp.338-383.
2. M. Stefik, J. Aikins, R. Balzer, J. Benoit, L. Birnbaum, F. Hayes-Roth, F and E. Sacerdoti, E, *Basic Concepts for Building Expert Systems*. Addison-Wesley, 1983.
3. B.G. Buchanan and E.H. Shortliffe, *Rule-Based Expert Systems*, Addison-Wesley, 1984.
4. J. Giarratano and G. Riley, *Expert Systems: Principles and Programming,* International Thompson Publishing, 1993.
5. K.S. Leung and W. Lam, Fuzzy Concepts in Expert Systems. *IEEE*. September: 43-56, 1988.
6. R.A. Orchard, *FuzzyCLIPS Version 6.02A User's Guide*, National Research Council Canada, 1994.
7. CLIPS Reference Manual, Version 6.0, 1993.
8. J. Buckley and W. Siler, Fuzzy Operators for possibility Interval Sets, *Fuzzy Sets and Systems*, 1987, Vol.22, pp 215-227.
9. T. Chiueh, Optimization of Fuzzy Logic Inference Architecture, *Computer*. May, 1992, pp 67-71.

Fuzzy Classification of Fault Diagnosis Using ANNs

Qian Huang Dan Pan GanYing Luo S. K. Tso

Department of Electronics & Communications Engineering,
South China University of Technology, China

Abstract In recent years, fuzzy logic and neural networks have been gradually introduced in the field of fault diagnosis. However, both fuzzy logic based methods and conversational neural networks have their own disadvantages. One practical way is to integrate fuzzy logic with neural networks. We have presented the feasible method of fuzzy classification of fault diagnosis patterns. In the neural networks using back propagation algorithm, the input vector consists of membership values to fault symptoms while the output vector is defined in terms of fuzzy fault reason class membership values. The effectiveness of the algorithm is demonstrated with the fault diagnosis of a radar equipment.

1 Introduction

Failure is a nearly unavoidable phenomenon with technological products and systems. By fault we mean a dynamic (system) state which deviates from the desired system state. The tasks of fault diagnosis may include diagnosing where the fault occurred and what are the type and kind of the fault, assessing the damage of the fault, and reconfiguring the system to accommodate the fault. In a narrow sense, fault diagnosis partially answers one of the basic issues in system failure engineering: why system failure. Obviously, various symptoms of a system during its operation are essential to implement tasks of fault diagnosis. However, vague symptoms frequently emerge. Fuzzy methodology is a natural tool to incorporate symptoms of this kind and can be also used to deal with vagueness in system models and in human perceptions. Moreover, because fault diagnosis can be thought of as pattern recognition, artificial neural networks (ANNs) are well suited to this task[1]. ANNs using the back propagation algorithm can be used for character recognition. Several studies that examine the use of ANNs for fault diagnosis appear in the literature. Whereas, conventional two-state neural net models generally deal with the ideal condition, where an input feature is either present or absent and each pattern belongs to either one class or another. They do not consider cases where an input feature may belong to more than one class with a finite degree of "belongingness"[2]. While in fault diagnosis, the fault symptoms which we based to implement tasks of fault diagnosis are often vague symptoms[3].However, one practical way is to integrate fuzzy logic with neural networks[4]. The proposed neural network model using the gradient-descent-based back propagation algorithm incorporates concepts from fuzzy sets and is capable of classification of fuzzy patterns.

2 Multilayer Perceptron Using Back Propagation Of Error

The multilayer perceptron (MLP) consists of multiple layers of simple, two-state sigmoid processing elements (nodes) or neurons that interact with weighted connections. The strong functions of the MLP are mainly brought about with the nonlinear property of hidden neurons.[5] At the present, there are many effective studying algorithms for the multilayer percetron, especially back propagation algorithm. The Back Propagation (BP) learning model of MLP converts the problems of input and output in a set of patterns to the problems of the nonlinear optimization, using the most popular least mean square (LMS) algorithm. It is because of the hidden layer that the adjustable parameters of the optimal problem increase so that the accurate results can be gained. If the MLP using BP algorithm is considered as the mapping from the input space to the output space, the mapping is highly nonlinear and can reflect the complex phenomena in real life.

Now consider the network with a hidden layer. Set x_i (i=1, 2, ... ,n_0) and $w_{ij}^{(l)}$ (l=1, 2) as the input vectors and the weights of the network respectively. So for respective layer, the inputs are:

$$I_i^{(0)} = x_i \qquad i=1, 2, \ldots, n_0 \tag{1}$$

$$I_i^{(1)} = \sum_{j=1}^{n_0} w_{ij}^{(1)} o_j^{(0)} - \theta_i^{(1)} \qquad i=1, 2, \ldots, n_1 \tag{2}$$

$$I_i^{(2)} = \sum_{j=1}^{n_1} w_{ij}^{(2)} o_j^{(1)} - \theta_i^{(2)} \qquad i=1, 2, \ldots, n_2 \tag{3}$$

Where $w_{ij}^{(1)}$ is the weight of the connection from the jth neuron in the input layer to the ith neuron in the hidden layer, $w_{ij}^{(2)}$ is the weight of the connection from the jth neuron in the hidden layer to the ith neuron in the output layer, $o_j^{(0)}$ is the output of the jth neuron in the input layer, $o_j^{(1)}$ is the output of the jth neuron in the hidden layer, $\theta_i^{(1)}$ is the threshold of the ith neuron in the hidden layer, $\theta_i^{(2)}$ is the threshold of the ith neuron in the output layer, n_0 , n_1 , n_2 is the number of neuron in the input, hidden output layer respectively.

For respective layer, the output are

$$o_i^{(0)}=x_i \qquad i=1, 2, \ldots, n_0 \tag{4}$$

$$o_i^{(1)}=g(I_i^{(1)}) \qquad i=1, 2, \ldots, n_1 \tag{5}$$

$$o_i^{(2)}=g(I_i^{(2)}) \qquad i=1, 2, \ldots, n_2 \tag{6}$$

where $g(I)=\dfrac{1}{1+e^{-I}}$ is the sigmoid function.

The least mean square error in output vectors, for a given network weight vector **W** is defined as

$$E(w)=\frac{1}{2}\sum_{\mu=1}^{n_1}\sum_{i=1}^{n_2}(\zeta_{\mu_i} - o_{\mu_i}^{(2)}(w)) \tag{7}$$

where ζ_{μ_i} is the desired output of the ith neuron in the output layer in input-output case μ specified by the teacher and $o^{(2)}_{\mu_i}(w)$ is its actual output. The error $E(w)$ is minimized by the BP algorithm using gradient descent. We start with any set of weights and repeatedly update each weight by an amount

$$\Delta w_{ij}^{(l)}(n+1) = -\alpha \frac{\partial E(n)}{\partial w_{ij}^{(l)}(n)} + \varepsilon \Delta w_{ij}^{(l)}(n), \qquad l=1, 2. \tag{8}$$

where the learning rate $\alpha > 0$ controls the descent, $0 \le \varepsilon \le 1$ is the damping coefficient or momentum and n denotes the number of the iteration currently in progress.

3 Pattern Representation With Fuzzy Concept

In the fault diagnosis, many fault features can not simply be divide by crisp two-value logic but have a certain extent, for example weak, weaker, ordinary, strong and stronger. Therefore, fault feathers should be fuzzed in order to make them closer to the fact in the fault diagnosis .

Now, there is an example about the fault diagnosis of a radar equipment. In the radar equipment, a kind of fault reason usually leads to many of fault features and different fault features contribute differently to a certain kind of fault reason .

Fig. 1. is a typical single stage electrical circuit diagram. Since the parameters of components such as capacitance, resistance value have tolerable errors (usually the tolerable error of ordinary resistance is $\pm 10\%$) and may give rise to the drifts with the change of environmental temperature and time. Hence, the relations between the features and the reasons are not fully decided. Moreover, the human's diagnosis experiences are usually very fuzzy. So the diagnosis models should be based on the fuzzy system theory instead of the classical mathematics. The voltage of every measurement spot is the continuous variable and all of three fault features are fuzzy values.

There are n fault reasons and m fault features, which are denoted with $X=\{x_1, x_2, \ldots, x_n\}$ and $U=\{u_1, u_2, \ldots, u_m\}$ respectively.

Define $R: X \times U \rightarrow [0, 1]$, $r_{ij}=R(x_i, u_j) \in [0, 1]$, R is the membership matrix.
The membership functions r_{11} - r_{53} are shown as below.

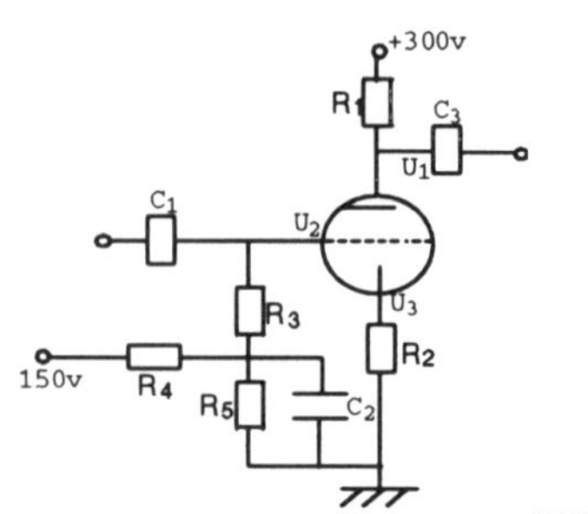

Fig1. the single stage electric

$\tilde{R}_1$	r_{11}	r_{12}	r_{13}
$\tilde{R}_2$	r_{21}	r_{22}	r_{32}
$\tilde{R}_3$	r_{31}	r_{32}	r_{33}
$\tilde{R}_4$	r_{41}	r_{42}	r_{43}
$\tilde{G}$	r_{51}	r_{52}	r_{53}

$$r_{12} = r_{22} = r_{52} = \begin{cases} \frac{1}{2} + \frac{1}{2}\sin\frac{\pi}{3}(x - \frac{-39}{2}), & -21 \le x \le -18 \\ 1, & -18 \le x \le -12 \\ \frac{1}{2} - \frac{1}{2}\sin\frac{\pi}{3}(x - \frac{-21}{2}), & -12 < x < -9 \\ 0, & other \end{cases} \tag{9}$$

$$r_{11} = \begin{cases} 1, & 0 \le x \le 2 \\ \frac{1}{2} - \frac{1}{2}\sin\frac{\pi}{2}(x - \frac{6}{2}), & 2 < x \le 4 \\ 0, & other \end{cases} \tag{10}$$

$$r_{13} = \begin{cases} \frac{1}{2} - \frac{1}{2}\sin\frac{\pi}{0.8}(x - \frac{0.8}{2}), & 0 \le x \le 0.8 \\ 0, & other \end{cases} \tag{11}$$

$$r_{21} = r_{41} = r_{51} = \begin{cases} \frac{1}{2} + \frac{1}{2}\sin\frac{\pi}{6}(x - \frac{594}{2}), & 294 \le x \le 300 \\ 1, & x > 300 \\ 0, & x < 294 \end{cases} \tag{12}$$

$$r_{31} = \begin{cases} \frac{1}{2} + \frac{1}{2}\sin\frac{\pi}{15}(x - \frac{235}{2}), & 110 \le x \le 125 \\ 1, & 125 < x < 140 \\ \frac{1}{2} - \frac{1}{2}\sin\frac{\pi}{16}(x - \frac{290}{2}), & 140 \le x \le 150 \\ 0, & other \end{cases} \tag{13}$$

$$r_{32} = \begin{cases} \frac{1}{2} + \frac{1}{2}\sin\frac{\pi}{1}(x - \frac{-1}{2}), & -1 \le x \le 0 \\ 0, & other \end{cases} \tag{14}$$

$$r_{42} = \begin{cases} 1, & x < -150 \\ \frac{1}{2} - \frac{1}{2}\sin\frac{\pi}{5}(x - \frac{-295}{2}), & -150 \le x < -145 \\ 0, & other \end{cases} \tag{15}$$

$$r_{23} = \begin{cases} \frac{1}{2} + \frac{1}{2}\sin\frac{\pi}{2.5}(x - \frac{9.5}{2}), & 3.5 \le x < 6 \\ 1, & 6 \le x \le 10 \\ \frac{1}{2} - \frac{1}{2}\sin\frac{\pi}{2.5}(x - \frac{22.5}{2}), & 10 < x \le 12.5 \end{cases} \tag{16}$$

$$r_{43} = r_{53} = \begin{cases} \frac{1}{2} - \frac{1}{2}\sin\frac{\pi}{0.5}(x - \frac{-0.5}{2}), & -0.5 \le x \le 0 \\ 0, & other \end{cases} \quad (17)$$

$$r_{33} = \begin{cases} \frac{1}{2} + \frac{1}{2}\sin\frac{\pi}{1}(x - \frac{9}{2}), & 3 \le x < 4 \\ 1, & 4 \le x \le 5 \\ \frac{1}{2} - \frac{1}{2}\sin\frac{\pi}{1}(x - \frac{11}{2}), & 5 < x \le 6 \\ 0, & other \end{cases} \quad (18)$$

In this example, the fault features U_1 ,U_2 , U_3 have respectively 3 , 3 , 4 different membership functions because of $r_{21} = r_{41} = r_{51}$,$r_{12} = r_{22} = r_{52}$ and $r_{43} = r_{53}$. So each input feature U_{ji} , ($i=1, 2, 3$) in the jth input feature pattern can be expressed as the membership values according to different membership functions of the ith feature attaching to different fault reason. Therefore an input pattern $\vec{U}_j = [u_{j1}, u_{j2}, u_{j3}]$ may be represented as :

$$\vec{F}_j = [\mu_{1(u_1)}(u_{j1}),\dots, \mu_{3(u_1)}(u_{j1}), \mu_{1(u_2)}(u_{j2}),\dots, \mu_{3(u_1)}(u_{j2}), \mu_{1(u_3)}(u_{j3}),\dots, \mu_{4(u_3)}(u_{j3})] \quad (19)$$

acting as actual input of ANNs corresponding to the jth input feature pattern $\vec{U}_j$.

According to experts' experiences, decide the continuos fuzzy membership values lying in the interval [0,1] of a input pattern (fault features pattern) belonging to every fault reasons and these values are considered as actual desired output pattern corresponding to the input pattern. Thus a learning pattern set can be gained.

Then, the neural network which includes input layer, a hidden layer and output layer is constructed. The numbers of nodes in the input, hidden and output layer are 10 , 8 , and 5 ,$\alpha = 0.3$, $\varepsilon = 0.5$, learning time =1500, then E=0.002. The result of learning is shown in the Table 1.

Table 1. Input ,desired output, and actual output vectors for a set of sample patterns presented to the three-layer neural network

Input Feature		Output Vector	
u_1, u_2 u_3	Input Vector	Desired	Actual
1 0 1	1 0 0 0 0 0 1 0 0 0	1 0 0 0 0	.97 .02 .01 .01 .00
294 0 8	0 0 0 0 0 0 0 1 0 0	0 .7 0 0 0	.01 .69 .00 .01 .01
132 0 3.5	0 0 1 0 1 0 0 0 .5 .0	0 0 .95 0 0	.01 .01 .94 .01 .00
294 -150 -1	0 0 0 0 0 1 0 0 0 0	0 0 0 .8 0	.01 .02 .01 .79 .01
300 0 -0.5	0 1 0 0 0 0 0 0 0 0	0 .3 0 .1 1	.00 .03 .01 .09 .97

In practice, different fault features reflect a certain kind of fault reasons to different extent. In other words, different fault features have different weights for a certain kind of fault reasons. Now, ANN is utilized to gain these weights from the learning patterns and stores them in itself weights. When a new test input pattern (fault features pattern) is inputted, the output value of every neuron in the output layer can be considered as its belongingness attaching to every fault reason respectively.

4. Conclusion

Now, the procedure of the fault diagnosis with Fuzzy ANNs is concluded.

(1) Decide the membership matrix R. According to the experiences, membership function r_{ij} is introduced, namely:

$$r_{ij} = \mu_{x_i}(u_j) \subset [0,1]$$

where $\mu_{x_i}(u_j)$ is the membership function of the fault feature u_j attaching to the fault reason x_i .

(2) Choose the input patterns for learning. In view of every fault reasons, the input patters are typical as soon as possible.

(3) According to experts' opinions, decide the desired output pattern corresponding to every input pattern.

(4) Train the ANN with the learning patterns till success.

(5) Make fault diagnosis utilizing the ANN.

The above-all is an extension to the conventional MLP model that deals with actual input and *crisp* (binary) output values. The proposed model is found to classify fuzzy data with overlapping class boundaries in a more efficient manner. Therefore, it has a promising application of prediction diagnosis.

REFERENCES

1. Won-Yong Lee, Jokn M. House, Cheol Park, et. Fault Diagnosis of on Air-Handling Unit Using Artificial Neural Networks, ASHRAE Technical Data Bulletin, *Fault Detection For HVAC Systems*, (1996): 23-32
2. Sankar K. Pal, Sushmita Mitra, Multilayer Perceptron, Fuzzy Sets, and Classification, *IEEE Trans. on Neural Networks*, Vol3, No.5, Sep.(1992): 683-696
3. Kai-Yuan Cai, System failure engineering and fuzzy methodology an introductory overview, *Fuzzy Sets and Systems*, 83 (1996):135-141
4. Sushmita Mitra, Sankar K. Pal, Fuzzy Multi-Layer Perceptron, Inferencing and Rule Generation, *IEEE Trans. on Neural Networks*, Vol. 6, No. 1, Jan. (1995): 51-63
5. David E. Rumelhart, James L.Mcclelland, And the PDP Research Group, *Parallel Distributed Processing*, The MIT Press, London (1987) pp. 319-328.

Intelligent Visual Evaluation System for Mirror Finished Surfaces

Kazuhiko Kato*, Toshihiro Ioi*, Masahisa Matsunaga** and Nobuaki Iguchi***

* Department of Project Management, Chiba Institute of Technology, Narashino, Chiba 275 Japan
** Institution, Chiba Institute of Technology
*** R & D Department, Kuroda Precision Industries LTD, Kawasaki, Kanagawa, 239 Japan

Abstract. This paper describes an automated visual evaluation system for mirror finished surfaces of heat treated steel. The neural network is applied to predict the surface roughness of various kinds of mirror finished surfaces. The surface integrity will be clarified by the learning system if the correlation between the characteristics of each surface image and some optical properties of mirror finished surfaces can be analyzed. It is found that the neural network system is very effective means for evaluating the finished surfaces when image features of mirror finished surfaces are used as the input data and surface roughness and/or kinds of finishing process are derived as the output data.

1 Introduction

The mirror finished surfaces of heat treated steel as block gauge have been made through grinding and lapping processes. These iterative finishing operations and the evaluations of the surfaces after finishing have been performed by expert workers. If the features of mirror finished surfaces are recognized by a vision system and a neural network, the degree of surface roughness and/or kinds of finishing process will be clarified in relation to the image data of surfaces.

The researches(1)-(6) of visual evaluation system for metal surfaces have been reported. These are related to evaluation systems concerning the cross section surfaces(1),(2) after metal cutting, the friction surfaces(3), the surfaces(4) after blast processing, and the surfaces(5) of different metal materials. However, these researches using image processing methods are applied mainly to the evaluation of rough metal surfaces. The visual evaluation systems which can predict the surface roughness for mirror finished surfaces after grinding and lapping processes have not yet been developed, because of the difficulty of measuring methods and the complexity of image processing analysis.

The objectives of this paper are to develop the intelligent visual evaluation system for mirror finished surfaces(6). It is found that the features of image data of mirror finished surfaces can be decided by texture analyses. The neural network systems are also applied to predict the surface roughness and/or kinds of finishing processes. As the results, it is possible to construct the neural network system which is composed of the features of surface image as input data and the degree of surface roughness and/or kinds of finishing processes as output data.

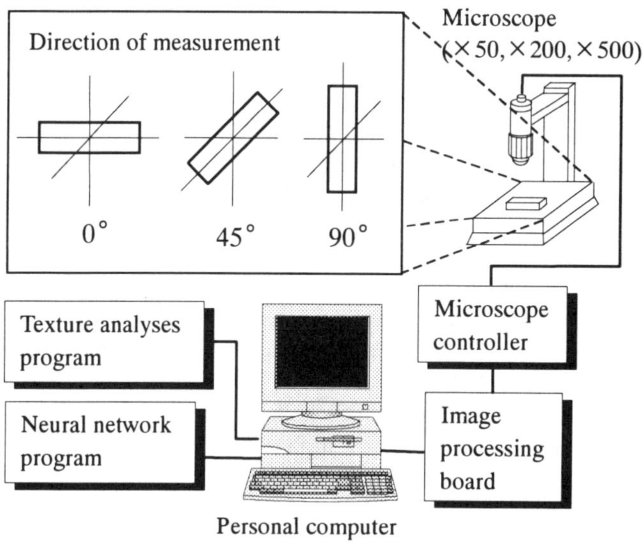

Fig.1. Experimental system

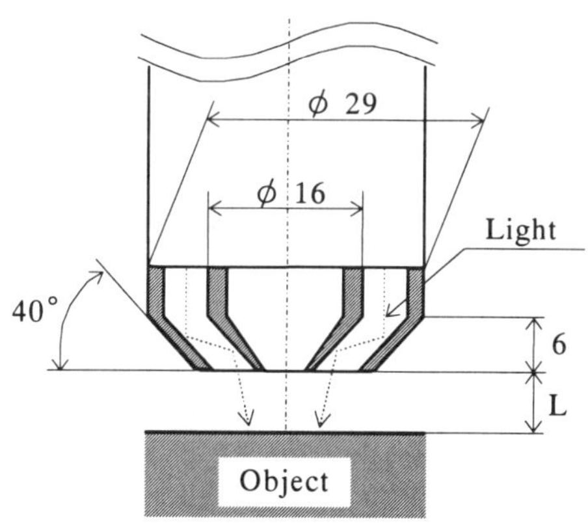

Magnification	Distance L (mm)
× 50	12.8
× 200	3.5
× 500	3.5

Fig.2. Lighting mechanism of a microscope

2 Experimental System and Procedure of Experiments

Fig.1 shows the experimental system which is used in the study. The system is composed of a microscope, a controller, an image processing board and a microcomputer. The mirror finished surfaces which are used in the experiment are the testpiece of standard surface roughness finished by grinding and lapping processes. The light reflection data from a microscope can be analyzed using an image processing board in a microcomputer. The image data are displayed by the multiplication of the magnification of optical measuring system and the digital processing (expansion of pixel data) by microcomputer. The mirror finished surfaces are measured by three measuring directions of 0, 45 and 90 degree for the light axis of microscope as shown in figure 1, and also three magnifications of 50, 200 and 500 are used as the measuring conditions. Fig.2 shows the non-contact lighting head (x 50) of a microscope. A lamp in lighting system is set in a microscope controller, and the mirror finished surfaces can be illuminated by a photo fiber cable of circle type. The uniform illumination for the surfaces is possible using this lighting system. Table 1 shows the surface roughness of each testpiece and Table 2 is the experimental conditions. Fig.3 shows the mirror finished surface images in measuring condition of 200 magnifications of microscope. It is well understood that the different image are obtained by each finishing process; grinding, lapping process 1, lapping process 2 and final lapping. As the result, it is possible to construct the neural network which is compose of the features of surface image as input data and the degree of surface roughness and/or kinds of finishing processes as output data.

Table 1. Finishing process and surface roughness

Finishing process	Number of parts	Surface roughness μm Ra
Grinding process	No.1	0.462
	No.2	0.432
Lapping process 1	No.3	0.016
Lapping process 2	No.4	0.007
	No.5	0.008
Final lapping	No.6	0.003

Table 2. Experimental conditions

Image size	128pixel×128pixel
Concentration level	256Gradation
Element for shot	CCD image sensor
Light source	100W halogen lamp
Lighting	Oblique irradiation from multi-direction
Resolution (×500)	1 μ m/pixel

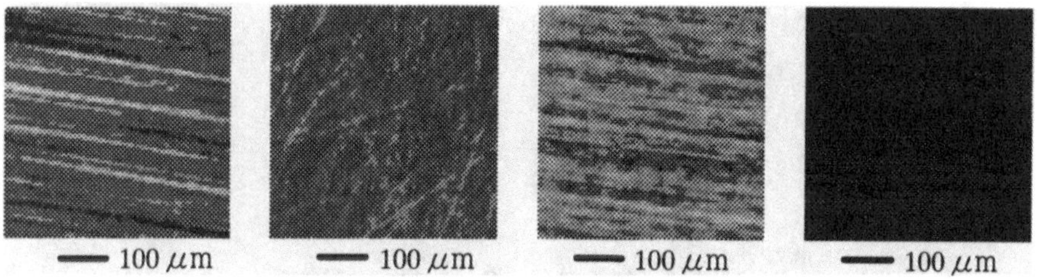

(a) Grinding process 0.432 μm Ra (b) Lapping process 1 0.016 μm Ra (c) Lapping process 2 0.008 μm Ra (d) Final lapping 0.003 μm Ra

Fig. 3. Mirror finished surface images (×200)

3 Texture Analyses for Mirror Finished Surfaces

3.1 Texture Analysis

In general, texture analysis is applied to the statistical treatment of gray level distribution in local area of image processing. The gray level co-occurrence matrix and the gray level run length matrix in the texture analysis(7) are applied to the image processing of mirror finished surfaces. The former analysis is classified into an angular second moment (equation (1)), a contrast (eq. (2)), an energy (eq.(3)) and an entropy (eq.(4)). The others are classified into a short runs emphasis (eq.(5)), a long runs emphasis (eq.(6)), a gray level uniformity (eq.(7)) and a run length uniformity (eq.(8)). Where, M(f,l) is the gray level co-occurrence joint probability, L(f,g) is the distance to each matrix element from the principal diagonal axis of M(f,l), and f and g are the elements of the gray level. M(f,l) is gray level run length probability, and l is the continuous length of the elements. The maximum unit number of the square matrix is 128, and the maximum gray level is set to the value of 256 in the study. It is important in the image processing analysis to decide the scanning direction of texture analyses for mirror surfaces. It is necessary to get the same image features even if the measuring point of microscope, which is faced with mirror surfaces, is rotated for the light axis. If the texture analysis algorithm which is not depended on the scanning directions for surface is developed, it is possible to get the same image features not related to the measuring direction. However, it is difficult to develop the software algorithm. The scanning direction for texture analysis in the research is set to one direction (0 degree) considering the image processing speed, as the results, the measuring condition for CCD camera is set to three directions of 0, 45 and 90 degree for the light axis of microscope as shown in Fig.1. The correlation characteristics between the geometrical properties of mirror surfaces and the image features of texture analyses are investigated in the study. The image features of mirror finished surfaces are indicated by the non-dimensional quantities of gradation. The equation for normalizing is shown in equation (9). Where, X_1 is normalized data, X is gradation value by image feature, X_{max} is maximum value of image feature and X_{min} is minimum value of image feature.

$$I = \sum_f \sum_g \left[\{L(f,g)\}^2 \cdot P(f,g) \right] \tag{1}$$

$$C = \sum_f \sum_g \left\{ (f - g)^2 \cdot P(f,g) \right\} \tag{2}$$

$$E = \sum_f \sum_g \{P(f,g)\}^2 \tag{3}$$

$$ENT = \sum_f \sum_g \left[P(f,g) \cdot ln\{P(f,g)\} \right] \tag{4}$$

$$S_1 = \frac{\sum_f \sum_l \{M(f,l)/l^2\}}{\sum_f \sum_l M(f,l)} \tag{5}$$

$$S_2 = \frac{\sum_f \sum_l \{M(f,l) \cdot l^2\}}{\sum_f \sum_l M(f,l)} \tag{6}$$

$$S_3 = \frac{\sum_f \left\{ \sum_l M(f,l) \right\}^2}{\sum_f \sum_l M(f,l)} \tag{7}$$

$$S_4 = \frac{\sum_l \left\{ \sum_f M(f,l) \right\}^2}{\sum_f \sum_l M(f,l)} \tag{8}$$

$$X_1 = \frac{X - X_{\min}}{X_{\max} - X_{\min}} \cdot 65535 \tag{9}$$

3.2 Results by Texture Analysis for Mirror Finished Surfaces

3.2.1 Characteristics of Image Features by Luminous Intensity

The difference between surface roughness in each finishing process and image features of mirror surfaces must be recognized distinctly as image features to evaluate the surface roughness. The streaks on the surface by grinding processing are recognized as shown in figure 3(a), and the light reflection disperses strongly at the streaks. The other side, the surface after lapping process 2 seems to be smooth as shown figure 3(c), and the image picture is dark. The optimum lighting condition, that the different image features by texture analyses are recognized clearly in each finishing process, is investigated. Table 3 shows the luminous intensity in each lighting level. Fig.4 and 5 show the calculated results of entropy and gray level uniformity by texture analyses. The parameter of the graph is lighting level as shown in Table 3, and 200 magnifications of microscope is applied to the experiment. As the surface roughness increases, these texture features decrease except the surface of lapping process 2. The same feature is obtained in other texture analyses. As the analytical results, it is found that the suitable luminous intensity for the measurement of mirror finished surfaces is from level 2 to level5 at 50 magnifications of microscope, from level 3 to level 6 at 200 magnifications, and from level 5 to level 8 at 500 magnification. The luminous intensity of level 4 at 50 magnification, and level 7 at 200 magnification are set in the experiment.

Table 3. Luminous intensity in each lighting level

Lighting level	Luminous intensity lux		
	×50	×200	×500
Level 1	4580	11700	12600
Level 2	7780	18000	20600
Level 3	12000	27700	32700
Level 4	15800	39100	46900
Level 5	21500	51600	64500
Level 6	28300	69800	82600
Level 7	34700	86000	–
Level 8	47300	–	–

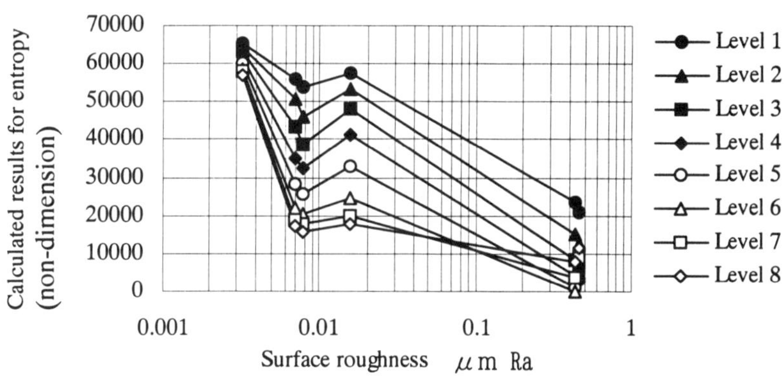

Fig.4. Effect of surface roughness on entropy (×200)

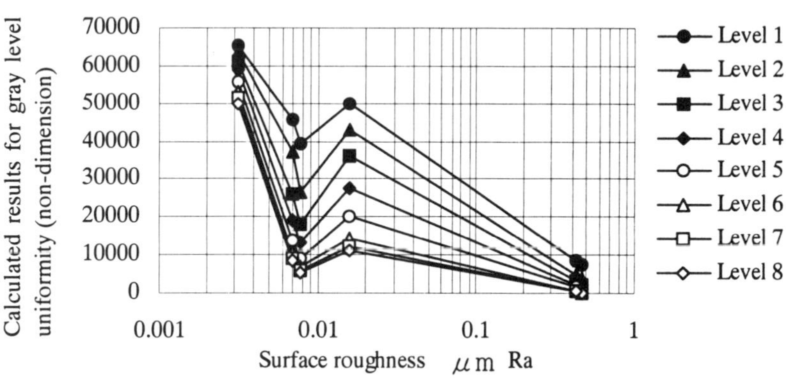

Fig.5. Effect of surface roughness on gray level uniformity (×200)

3.2.2 Characteristics of Image Features by Magnification of Microscope and Geometrical Property of Mirror Finished Surface

The correlation characteristics between the geometrical property of mirror finished surfaces and the image features of texture analyses are investigated in the chapter. Fig.6 shows the radar charts for the characteristics of image features by texture analyses at 200 magnification of microscope. The parameter is measuring direction of CCD camera shown in Fig.1. Fig.6(a) is the image features by texture analyses for the surface after grinding process, and it is clear that the chart pattern differs slightly by the measuring direction for the regular streaks on the surface. Especially, it is well understood that angular second moment, contrast and run length uniformity depend on the influence of surface geometry. Fig.6 (b) shows the image features for the surface after lapping process 1. The irregular lapping streaks on the surface are recognized,

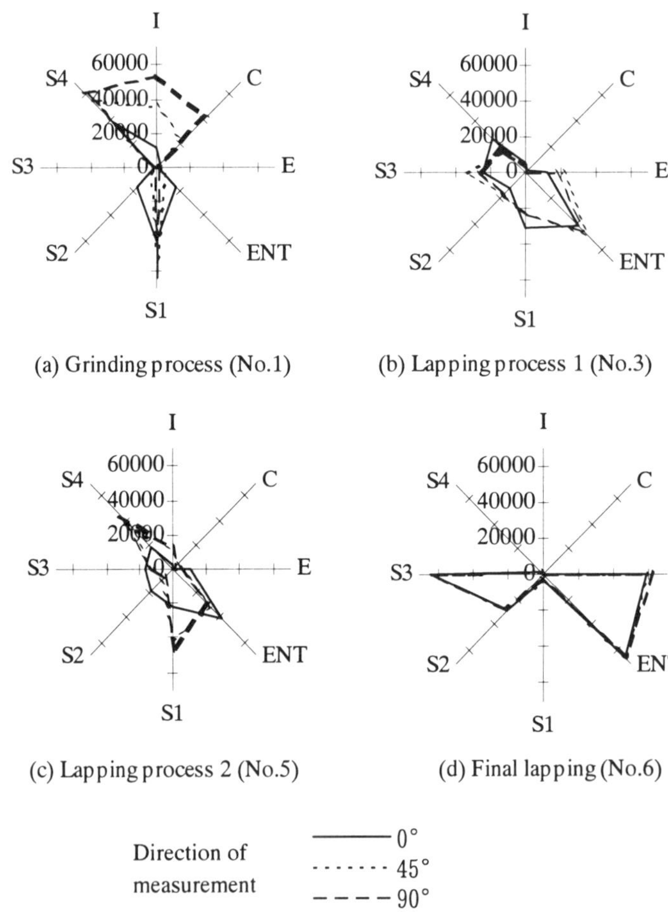

Fig.6. Characteristics of texture analyses by radar chart

and the streaks on the surface are not shown. Therefore, the pattern of radar chart is almost same. Fig.6 (c) shows the image features for the surface after lapping process 2, and it is clear that the chart pattern differs by the measuring direction for regular streaks in longitudinal direction of testpiece. Fig.6 (d) shows the image features for the surface after final lapping process, and it is well understood that the pattern of radar chart is almost same, and the image features do not depend on the measuring directions, because the surface after final lapping is very smooth and there are no thin lapping streaks in condition of 200 magnifications of microscope. The almost same results are obtained for other experimental conditions.

4 Construction of Neural Network and the Verification Results

Fig.7 shows the neural network system developed in the study. The back propagation learning rule of three layers is applied to the system. The input data is the value of eight image features by texture analyses which are shown from equation (1) to (8). The output data is four kinds of surface roughness and/or four kinds of finishing processes. The neural network systems are classified to two types. The one is the system that the measuring direction is set to one direction, 0 or 45 or 90 degree, and as the results, the three kinds of the neural networks are constructed. The another is the system that the measuring direction is set to the total sum of 0, 45 and 90 degree, and the system is learned using these measured data. The magnification of microscope is set respectively to three of 50, 200 and 500. As the results, twelve neural networks are constructed in the study. It is found that the good convergence of square error at the 1,000 to 1,200 iterations in the learning operation was obtained in each network system in case the number of elements of hidden layer is set from 8 to 12. Table 4 shows the verification results for the prediction of surface roughness. Table 4 shows the recognition ratio for the number of testpieces for verification when the microscope

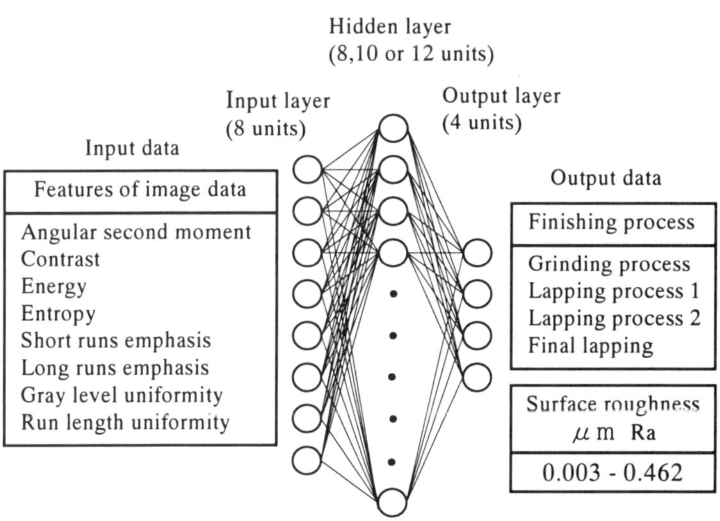

Fig.7. Neural network system for mirror finished surfaces

Table 4. Verification results of neural network for mirror finished surfaces

		Recognition ratio					
Finishing process		Grinding process		Lapping process 1	Lapping process 2		Final lapping
Number of parts		No.1	No.2	No.3	No.4	No.5	No.6
Surface roughness μm Ra		0.462	0.432	0.016	0.007	0.008	0.003
0°		100% (45/45)	100% (45/45)	100% (40/40)	100% (45/45)	100% (45/45)	100% (40/40)
45°		100% (45/45)	100% (45/45)	100% (40/40)	100% (45/45)	100% (45/45)	100% (40/40)
90°		100% (45/45)	100% (45/45)	100% (40/40)	100% (45/45)	100% (45/45)	100% (40/40)
Learning of three directions	0°	100% (45/45)	100% (45/45)	98% (39/40)	100% (45/45)	100% (45/45)	100% (40/40)
	45°	100% (45/45)	100% (45/45)	100% (40/40)	100% (45/45)	100% (45/45)	100% (40/40)
	90°	100% (45/45)	100% (45/45)	100% (40/40)	98% (44/45)	100% (45/45)	100% (40/40)

Threshold = 0.75

Table 5. Verification results of neural network for mirror finished surfaces

		Recognition ratio					
Finishing process		Grinding process		Lapping process 1	Lapping process 2		Final lapping
Number of parts		No.1	No.2	No.3	No.4	No.5	No.6
Surface roughness μm Ra		0.462	0.432	0.016	0.007	0.008	0.003
Learning of three directions	23°	100% (50/50)	100% (50/50)	100% (50/50)	100% (50/50)	100% (50/50)	100% (50/50)
	68°	100% (50/50)	100% (50/50)	100% (50/50)	100% (50/50)	100% (50/50)	100% (50/50)

Threshold = 0.75

of 200 magnifications is used. Upper column shows the results for measuring direction of 0, 45 and 90 degree. Lower column shows the results for learning of three directions. The recognition results in each measuring condition is very high (greater than 98%) in condition of threshold value of 0.75. Table 5 shows the verification results for the neural network system that the mirror surfaces are measured in condition of three directions of 0, 45 and 90 degree. It is found that the recognition results of the neural network for non-learning measuring directions of 23 and 68 degree is also very high if the features of three measuring directions data are used to construct the neural network.

5 Conclusions

The conclusions drawn from the research are as follows ;

(1) It is found that features of image data of mirror finished surfaces after grinding and lapping processes can be decided by texture analyses.

(2) It is understood that the optimum measuring condition can be decided by the correlation analyses between the measuring conditions and the characteristics of image features of texture analyses. Additionally, the correlation between the influence between geometrical direction of mirror finished surfaces and the image features is cleared.

(3) It is possible to construct the neural network system, which is composed of image features of mirror finished surfaces as input data, and surface roughness and/or kinds of finishing processes as output data.

(4) It is shown that the neural network system has rapid convergence characteristic, and the recognition results of surface roughness of each mirror finished surface are fairly good. The vision system which is developed in the study has great advantages as non-contact, in-line, high speed and free camera positioning system.

References

1. K. Minoshima : Automated distinction cross section surface after cutting using image processing, Journal (A) of JSME, **56**, 525 (1990) 1317. (in Japanese)
2. A. Ishii et al : Automated distinction of surface integrity using a neural network, 6th destruction dynamics symposium, Japan Society of Materials, (1991) 12. (in Japanese)
3. K. Odo et al : Texture analyses for frictional surfaces by image processing, Proceedings of Kyushu branch of JSME, **46** (1993) 136. (in Japanese)
4. Y. Kawaguchi : Application of image processing for evaluation of blasted surfaces, Report of Chugoku industrial technology laboratory, **34** (1990) 21. (in Japanese)
5. Huynh.V.M et al : Texture analysis of rough surface using optical fourier transform, Meas. Sci. Technol., **2**, 9 (1991) 831.
6. K. Kato et al : Development of visual evaluation system for mirror finished surfaces, Proceedings of JSPE, (1995) 253. (in Japanese)
7. For example, H. Ozaki : Image processing - From it's basic to application - , the second edition, (1988) (in Japanese)

A SURVEY OF NON-CONTACT 3D SURFACE IMAGING TECHNIQUES FOR CRANIOFACIAL AND RECONSTRUCTIVE SURGERY

* B T Sim, C K Chua, S M Chou, W S Ng
# S T Lee, S C Aung

* School of Mechanical and Production Engineering
Nanyang Technological University, Singapore

Department of Plastic Surgery
Singapore General Hospital

Abstract. Traditionally, in the planning of cosmetic or reconstructive surgery, the surgeon has to rely heavily on manual measurement of the human anatomy, which is laborious and error prone, and the photographs of the patient. With the advanced development in computer technology, simulation of the surgery is now possible by manipulating the 3D model of the human parts, which is digitised using the non-contact, optical imaging technique. Other benefits include surface measurement, monitoring of pre- and post-intervention results, design and fabrication of prosthesis, etc. A collaborative project between the Nanyang Technological University and Singapore General Hospital is carried out to design a new portable and mobile 3D surface imaging system. This paper presents the results of part of the project: the review of the current state-of-the-art 3D imaging techniques. The laser stripe scanning, the moiré fringe contouring and the holographic method are studied and compared for their feasibility in the digitisation of the human body.

1 Introduction

3D optical surface imaging techniques have been widely used in a variety of applications including apparel design, quality inspection, robot navigation, robotics assembly, gauging, etc. [1]. These non-contact measurement techniques, which offer the ability to measure the shape of the objects accurately in a short period of time, have also successfully found its niche in the biomedical field. Craniofacial surgeons can use 3D surface scanning to obtain a facial model of the patient before surgery, modify the model, and show the patient the anticipated outcome of the surgery. The digitised image will also aid the surgeons in the pre-planning of the reconstructive surgery by making critical anthropometric measurement. In addition, breast augmentation, and the design and fabrication of prosthesis will also benefit form this technology.

The current state-of-the-art optical surface imaging techniques are: laser stripe scanning and moiré fringe contouring. Linney et al have developed a laser-based scanning system for the measurement of the human anatomy [2-4]. Furthermore, active developers like Cyberware Inc. [5], 3D Scanners [6], Laser Design Inc. and Digibotics have commercial laser systems to produce 3D models for clinical application as well as animation industry. Marshall et al [7-9] have developed a 3D clinical facial imager, which utilises the projection moiré fringe contouring method to acquire accurate 3D coordinates from the surface of a patient's face. The Cencit system developed by Vannier et al [10-11] has employed 6 sets of cameras and projectors for whole body surface imaging. For works in apparel design and human engineering, a moiré-based system known as Phase Measuring Profilometry [12] is being developed by the Textile and Clothing Technology Corporation in Cary, North Carolina. Several research groups have also developed their own 3D optical active imaging systems for research and industrial applications [13-18].

This paper describes part of the collaborative project between Nanyang Technological University and Singapore General Hospital, and reviews the current state of the optical based 3D imaging techniques. Three non-contact optical techniques: laser stripe scanning, moiré fringe contouring and holographic method are studied for their feasibility in the digitisation of the human body. These techniques are then compared in detail with reference to parameters such as the acquisition time, accuracy and resolution.

2 System Performance Parameters

Before going into the description of various imaging technique, one must understand the common parameters that are used to specify a 3D imaging system. Based on these parameters, one can define the requirements of a imaging system, and decide the suitability of a particular imaging technique for certain application, such as biomedical imaging. Thus, the application or problem must be properly defined before a solution is determined. The various parameters are depicted in Figure 1.

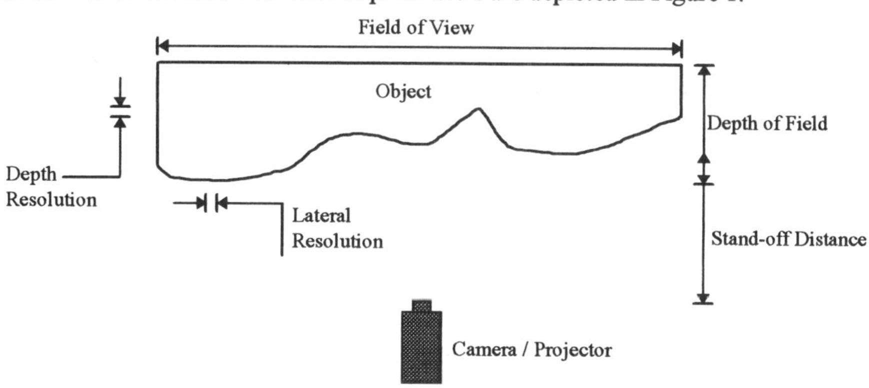

Figure 1. Performance parameters

Another critical parameter, which is especially concerned in the imaging of live subjects, is the cycle time required to capture and display the 3D model. The total time required is usually divided into two categories: image acquisition time and data processing time. The image acquisition time is the time taken to capture the full image of an object. The captured image is then digitised and fed into the computer. The three-dimensional coordinates are then computed by triangulation and the data is processed to display in a different format selected by the user. These processes will contribute to the data processing time. In order to minimise the total imaging time, both the image acquisition process and data processing can work simultaneously by overlapping the data capturing for one measurement with the processing of data for previous measurement. It should be noted that the accuracy of an imaging system can be enhanced by slowing down the acquisition process. Thus, there is a trade off between accuracy and data capturing time. For data processing, the required time will be determined by the data size, processing algorithm and the system hardware (such as the use of a transputer).

In addition to the above parameters, one should also consider the ambient lighting condition. Since most commercial scanners are active systems which project structured illumination onto the object, intense ambient light will increase the noise level and impair the accuracy of the results. Thus, the ambient light should be carefully modulated or the system should be designed to eliminate the unwanted light. Three methods are suggested:

a.) By subtracting the image of the object with a reference image, the projected stripe or other structured pattern will be highlighted. In this case, two images are captured, one with no illumination, and the other with illumination.

b.) By fitting an optical filter in front of the camera, the ambient light can be rejected and allow only the structured light to pass through.

c.) By enclosing the object and the imaging system in a dark chamber.

3 3D Surface Imaging Techniques

In the development of an active optical non-contact imaging system, most of the works have employed the structured light techniques to illuminate and extract the geometric information of the object. Structured light is a method in which a regular geometric pattern (dots, single stripe, multiple lines) is projected onto the object's surface, and viewed by one or more cameras at different positions. In this section, three techniques are discussed: laser stripe scanning, moiré fringe contouring and holography. The first two methods use the triangulation method to compute the 3D data.

3.1 Laser Stripe Scanning

The set-up of the laser stripe scanning system is depicted in Figure 2. A low intensity, eye-safe laser beam is passed through a cylindrical lens so as to project a vertical stripe onto the scanning surface. A charged-coupled device (CCD) camera is used to record the image of the projected stripe. Viewing at an angle to the illuminating source, the laser stripe is distorted according to the object's shape. The laser stripe is then recorded by the CCD camera, and digitised by a frame grabber. The frame grabber is a device that interprets the analogue modulated signals output from the camera as a frame of pixels. The digitised data is stored either in the frame buffer or read directly into the computer memory. To get the full view of the object, the object is rotated on a platform under computer control to obtain a 360° coverage, while the image capturing unit (source + camera) remains stationary. Alternatively, the scanner can rotate around a stationary object. As a result, a complete three-dimensional image of the object is captured. In general, the capturing time varies from 15 to 30 sec, depending on the type of application, working range, data size and the required accuracy.

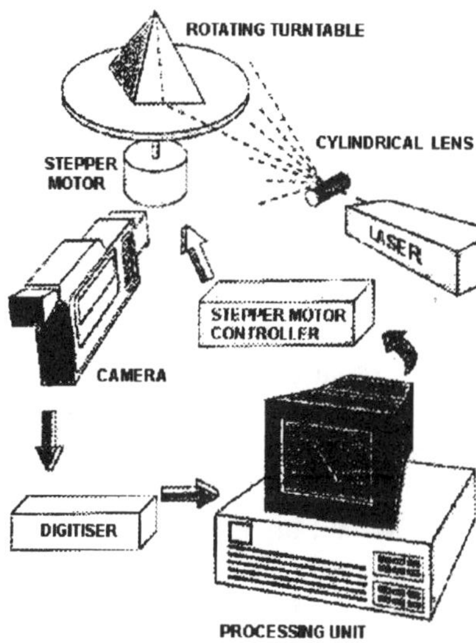

Figure 2 . Laser stripe scanning system. From S Sunthankar, SPIE, vol. 1826, pp 495, (1992).

Next, the image data is processed to obtain the 3D coordinates of the surface points lying along the projected laser stripe. The x and y dimensions can be readily determined by spatial measurement in the image plane, while the third dimension (depth) is computed using the triangulation principle. In addition, a calibration factor is to be included to transform the pixel values to real world coordinates. Thus, a calibration program must be carefully designed and performed to calculate accurate and reliable results.

Finally, the image data file is displayed in surface model, wireframe, etc. depending on the user's selection. Subsequently, for clinical application, further development of more comprehensive software is necessary to manipulate the data. The software will benefit the surgeon in pre- and post-intervention assessment, and the planning of surgical procedures for reconstructive surgery. Besides displaying the 3D model in any desired viewpoint, other graphics facilities should include distance, area and volume measurements, editing of surface shape, and the anatomical registration of images. In addition, the data file can be imported to the CNC machine or rapid prototyping system for the fabrication of 3D models and prostheses. These 3D models will enable the patient to visualise the results of the intervention.

To ensure the accuracy of the scanned data, the projected laser stripe must coincide with the axis of rotation of the turntable. A disadvantage of the laser stripe scanning is the shadowing effect of surface concavities, such as the area under the chin, around the eyelid, the inside of the nostrils, and the wounded surface. Surface infoldings such as the surface behind the ears are still unattainable by all optical scanning system. Furthermore, with an image capturing time of more than 10 sec, this technique is prone to error due to movement artifact, especially with children or infants.

3.2 Moiré Fringe Contouring

The word "moiré" is the French name for a fabric known as watered skill, which exhibits patterns of light and dark bands [19-20]. This moiré effect occurs when two grid patterns of similar period are superimposed. For example, this effect can be seen in two overlapping nets or when two layers of window screen are in contact.

Early work in this field required subjective visual evaluation, or laborious manual analysis of the moiré fringe pattern. However, with the advanced development in computer and peripherals, this moiré technique is revived. In the industry, there are a number of moiré methods developed to fit the need of a particular application, like the shadow moiré, projection moiré, reflection moiré etc. These moiré methods differ in how the deformed grating is produced. Among these, the projection moiré method is found to be more suitable for the application in 3D imaging of live human subjects. The schematic diagram of the projection moiré is shown in Figure 3. In this case, a line grating, which is an array of dark and bright lines, is projected onto the object. The deformed grating is also detected by a CCD camera linked to a computer. By superpositioning the projected grating lines with a reference grating, the moiré fringes are generated. Since the projected grating is deformed by the irregularities on the object's surface, the resulting moiré pattern describes the surface contours, resembling how a topographic map delineates the contours of the land (see Figure 4).

The moiré pattern can only provide a qualitative visualisation, although it contains the depth information of the object represented by the contour interval. In order to obtain the quantitative results (3D coordinates), a technique known as phase-shifting is selected with the aim of automating the analytical process, and improving the system resolution. Phase-shifting involves shifting of the projection grating or the reference grating by predetermined steps (fraction of the grating's pitch), and determining the intensity of the resulting fringes. A minimum of 3 shifted images are sufficient for the evaluation of the moiré fringe pattern. The fringe pattern is then transformed into a phase distribution, which automatically detects the convexity and concavity of the object's surface. The difference in phase together with a calibration factor are used to calculate the depth of the object with respect to a specific plane. The calibration factor is initially obtained by performing a calibration of an object of known depth. Thus, a set of 3D coordinates are derived and displayed in surface models as described in section 3.1.

Using the phase-shifting technique, a precise mechanism is required to shift the grating by a proportion of its pitch. For example, the Auto-MATE system [7] makes use of the piezo-electric translator which is coupled to a precision translation stage onto which the projection grating is fixed. Alternatively, avoiding the use of a mechanical translator, Asundi [21-23] has described a computer-aided method, which shifts a computer generated grating precisely with no moving parts. The computer grating is made up by an array of dark and bright pixels. The dark pixel has a binary value of 0, whereas the bright pixel has a binary value of 1. As a result, the pitch of the grating, which is the sum of the dark and bright line width, can be easily created for a particular application. Moreover, the grating can have any profile like sinusoidal, square or triangular profile. With the help of a liquid crystal display (LCD) projector, the computer grating can be projected, shifted and analysed using the phase-shifting method. The LCD projector, linked to the computer, is able to display the contents on the computer monitor via a suitable graphics adapter. Furthermore, Asundi has also introduced another method to obtain the moiré patterns by superpositioning of the computer gratings using logical operations. For example, the logical AND of two one-bit binary digital gratings will have the same results equivalent to that obtained by physical superposition of two gratings. More detail on this operation is presented in [23].

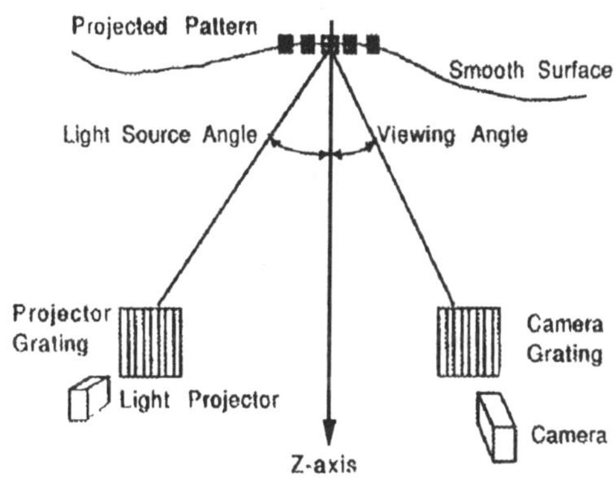

Figure 3. Projection moiré configuration. From P J Besl [1].

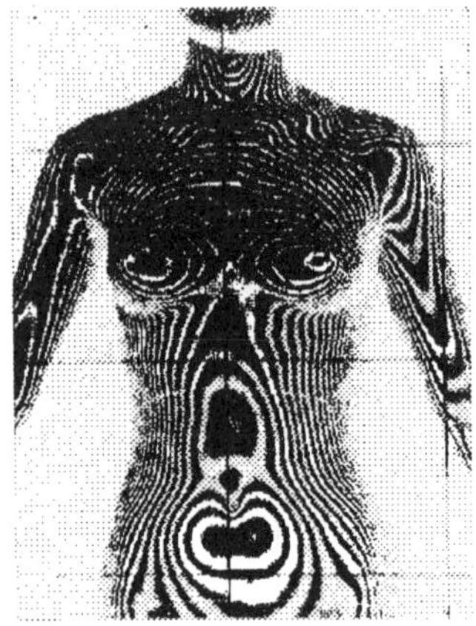

Figure 4. Infinite fringe moiré pattern of a human body.
From H. Takasaki, Appl. Optics 9, 1469 (1970).

3.3 Holography

Holography was discovered by Dennis Gabor in 1947 [24], in connection with 3D viewing of x-ray images before the development of a laser. It has become a practical technique only with availability of the coherent monochromatic light which is obtained from a laser. Its use though growing, is still in its infancy. The lack of maturity of this technique provides several areas which are of interest to researchers for exploration. To most people, holography is like 3D photography. However, besides 3D art work, applications such as the storage of optical information, are probably more important from the scientific point of view. Using the holographic method, a special dark room, which has minimised air flow and sufficient sound proofing, is required to produce the holograms. Furthermore, it is very sensitive to small disturbances and requires a strong frame or isolation platform to eliminate external vibrations. As a result, if it is used to scan the human body, the patient has to remain still throughout the scanning process, even breathing may seriously affect the accuracy of the result. In medical applications, holographic techniques have been used with a certain degree of success in dentistry, urology, otology and orthopaedics [26]. Thus, due to the limitations mentioned above, the holographic method is rejected as the imaging technique for a mobile and portable scanning system.

4 Evaluation of Imaging Techniques

The performance parameters of laser stripe scanning and the moiré topography are summarised in table 1. More detailed comparison of the performance parameters is described the following sections.

Table 1. Laser stripe scanning versus Moiré topography

Performance parameters \ Imaging techniques	Laser stripe scanning	Moiré topography
Accuracy \ resolution	1 μm	1 μm
Depth of field	100 m	10 m
Acquisition time	7 - 15 sec or more	less than 4 sec
Data processing time	30 sec or more	10 sec to few min
Lighting condition	Dim lighting condition	Dim lighting condition
Safety	Eye safe for Class I to Class IIIa laser	Ordinary white light
Field of view	Scanning mechanism required	Vary

4.1 Acquisition time

From table 1, both techniques satisfy the design requirements, and the results are relatively similar. Their suitability for a specific application will depend on the importance of each design parameter. In fact, both the laser stripe scanning and the moiré topography employ the same triangulation principle to obtain the coordinates of the surface points. The laser stripe scanning projects line by line onto the subject during the scanning process while the moiré method projects a series of lines at one time. As a result, the acquisition time for laser stripe scanning is greater than the moiré system due to the horizontal scanning of the laser beam. By adding moving components, the cost of the system will increase significantly for the expensive hardware. In addition, the control of the motion mechanism constitutes another problem, especially in the synchronisation of the moving parts and the rate of data capturing. On the other hand, the acquisition time for moiré system is far superior because there are no moving parts involved. The short capturing time will allow the patient to be scanned during a single breath hold, and motion artifact can be avoided. This will result in a better accuracy for breast scanning, and enable the scanning of young children to be possible. Furthermore, without the mechanical motion required, the errors resulting from vibration, backlash, wear and tear etc. can be eliminated.

4.2 Data processing time

Data processing time for both techniques depends on a few factors such as data size, processing algorithm, types of frame grabber etc. Quick processing time to display the scanned image of the patient will enable the surgeon to visualise the quality of the scan. If the image is distorted due to noise or patient's movement, another scan can be made immediately, instead of asking the patient to come back after a few days, and the scan area can be selected to focus on the critical features. A data processing time of about 1 minute is practical according to the surgeon in SGH. For laser stripe scanning, this requirement is achievable. For example, the facial scanner developed by the University College of London [4] is able to produce the 3D surface model in less than 30 sec with the help of a transputer based processing board, while for moiré topography or patterned light projection, the processing time will depend on the type of fringe analysis algorithm and the method to obtain the interference images or the projected pattern. In general, there is an additional step to identify and locate each stripe of projected light or grid intersection, before applying triangulation to compute the 3D data sets. Marshall et. al. [7] had developed a clinical facial imager, which employs the projection moiré fringe contouring method and has a data capture time of 1.8 sec but a data processing time of about 600 sec. However, the long processing time may be due to the merging of different data sets (from different views) to obtain a full 3D model. On the other hand, the moiré method can produce the instantaneous moiré pattern for qualitative assessment.

4.3 Accuracy and resolution of laser stripe scanning

From table 1, the accuracy and resolution of the laser scanning system can be as fine as 1 μm, which is far superior than the design requirement. However, in 3D surface imaging of the human body, the accuracy and resolution of commercial systems of Cyberware Inc. and 3D Scanners Ltd are claimed to be around 0.5 mm. Keep in mind that high accuracy and resolution can be achieved by increasing the acquisition time and reducing the working range. Thus, a trade-off between these parameters is made to suit the requirements of the application. The resolution of the laser scanning system depends on the size of the CCD array. The standard number of points measured along the laser stripe is either 256 or 512. By reducing the laser spot size or the width of the laser stripe, better resolution can also be obtained. Unfortunately, there are practical limits to the possible reduction of the laser beam's footprint, i.e. the corresponding reduction of the working range. In addition, the laser beam is easy to focus and its intensity is higher than the normal white light employed in moiré topography. Thus, superior quality images can be obtained. Despite the high resolution, a problem arises when there is a sharp change in the surface height (such as at the edge) that occur between two successive laser stripes. The sensor will fail to detect the abrupt change, and the resulting data will be inaccurate. Besides these, the laser stripe scanners tend to be more accurate at the centre of each sweep and less accurate at the fringes.

4.4 Accuracy and resolution of moiré fringe contouring

The moiré system is capable of achieving the same resolution and accuracy of 1μm as in the laser-based system. However, in the 3D digitisation of human body, problems such as the light scattering properties of the human skin, the skin colour and the complexity in the body shape, like the face, will affect the results. The performance of an optical sensor depends greatly on the detection of light scattered from the surface of the object [8]. Human skin is not completely opaque, which allows a proportion of the projected light to be scattered directly from the surface, while a proportion will penetrate through the skin and is scattered back form the flesh immediately beneath [8,9]. This scattered light will reduce the contrast of the projected grating lines, as well as the contrast of the laser stripe for the laser system. To improve the image contrast, one solution will be the use of a more powerful light source. Another solution will require makeup to apply on the surface to enhance the quality of the image captured. Nevertheless, the facial imager developed by Marshall et al [7] has achieved a resolution and accuracy of approximately 0.5 mm. Besides the illuminating source, the sensitivity of the moiré method can be improved by varying the pitch of the grating. With the help of the computer generated grating, the pitch can be changed rapidly and easily, and a minimum pitch can go to as fine as 2 pixels.

4.5 Field-of-view

Another parameter to be considered for a moiré system is the Field-Of-View (FOV) of the light detector. If the sensor has a small FOV, more than one scan will be needed to capture the whole surface area. Consequently, this will cause a registration problem to merge a few sets of data to produce a whole image, and the processing time will increase significantly. Although it is possible to merge the measurement data from two measurements at different views using one set of the projector-camera system, possible movement of the subject between the measurements can introduce serious errors in the merged data. The solution adopted is to make simultaneous measurements using two camera-projector units. The Cencit system [10], for example, has six projector-camera sets. Since the laser scanning system uses motion mechanism to scan the whole area of interest, there will be no problem concerning FOV. However, merging of data sets is still required for whole body scanning as the length of a laser stripe can only cover a certain part of the body. The Cyberware WB4 system [5] requires four scanning units to capture the whole human body.

4.6 Lighting condition

The laser scanning system needs to be operated in a low levels of ambient light because the intensity of the visible laser beam is between 1 to 5 mW. The intensity of the laser beam can be increased to enhance the light detection by the camera, but this may endanger the patient. For eye safe application, the laser intensity must be kept below 5 mW, i.e. Class I to Class IIIa laser. For a moiré system, although the light source is absolutely safe to the human eyes, the intensity of the white light is lower than that of the laser beam, the ambient light must therefore be modulated to increase the signal-to-noise ratio. Other methods commonly used to reduce the stray light problem are described in section 2.

4.7 Multiple view point

In general, for any triangulation system, occlusion is unavoidable. It occurs when the camera's view of the illuminated stripe on the object is blocked by an extrusion part of the object. Another problem of a single view system is that of obscuration, since the camera can only detect the scene that is visible to it. These problem can be overcome by using more than one sensor to capture the image multiple view point. For a moiré system, multiple sensors are required to resolve the FOV problem described above.

5 Conclusion

The current state-of-the-art optical sensing techniques are reviewed and discussed for their application in 3D digitisation of the human body. The three techniques are evaluated with respect to the performance requirements of the new scanner to-be-developed, such as the accuracy and resolution, acquisition time, data processing time, and the scan area. Holography is found to be unsuitable for the designated application. Both the laser and moiré techniques are comparable in performance, and the major differences are image capturing time and the cost of the system. For the commercial systems, the Cyberware WB4 system costs about $400,000, while the PMP moiré-based system is targeted at a price of $50,000[12]. Based on these findings, further development work on the conceptual design of the new scanner, which utilises the selected imaging techniques, can then proceed.

Acknowledgements
The authors wish to thank Miss Fiona Khong Lai Meng for her help in the research work. This project is funded by the National Medical Research Council (NMRC), Singapore.

References

1. P J Besl, "Active Optical Range Imaging Sensors", Machine Vision and Applications, 1: 127-152, 1988.
2. A D Linney, A C Tan, R Richards, J Gardener, S Grindrod and J P Moss, " Three-dimensional visualization of data on human anatomy: diagnosis and surgical planning", Journal of Audiovisual Media in 1993, 16: 4-10, 1993.
3. S Arridge, J P Moss, A D Linney and D R James, "Three-dimensional digitisation of the Face and Skull", J. max. -fac. Surg., vol. 13, pp. 136-143, 1985.
4. J P Moss, A D Linney, S R Grindrod and C A Mosse, "A Laser Scanning System for the Measurement of Facial Surface Morphology", Optics and Lasers in Engineering, vol. 10, pp. 179-190, 1989.
5. Cyberware Inc., "WB4: Whole Body Scanner", http://wwww.cyberware.com
6. 3D Scanners Ltd, "Products", http://www.3dscanners.com
7. S J Marshall, R C Rixon, D N Whiteford and J T Cumming, "Development of a 3-D clinical Facial imager", Proc. SPIE, vol. 1889, pp. 255-262, 1993.
8. S J Marshall, G T Reid, S J Powell, J F Towers and P J Wells, "Data capture techniques for 3-D facial imaging", Computer Vision and Image Processing, ed. A N Barrett, pp 248-275, Chapman and Hall, London, 1991
9. S J Marshall, R C Rixon, D N Whiteford, P J Wells and S J Powell, "The development of a 3-D data acquisition system for human facial imaging", Proc. SPIE Medical Imaging IV, vol. 1231, pp 61-74, 1990
10. Internet document: Surface Imaging Group, "Surface Imaging of the Human Body", Mallinckrodt Institute of Radiology, Washington University School of Medicine.
11. M W Vannier, G Bhatia, P Commean, T K Pilgram and B Brunsden, "Medical facial surface scanner", Proc. SPIE Image Capture, Formatting, and Display, vol. 1653, pp. 177-184, 1992.
12. S Paquette, "3D Scanning in Apparel Design and Human Engineering", http://www.computer.org/pubs/cg&a/apps/g50011.html.
13. K C Ng, B F Alexander, S H Boey, S Daly, J C Kent, D Q Huynh, R A Owens and P E Hartmann, "Biostereometrics - A Noncontact Noninvasive Shape Measurement Technique for Bioengineering Applications", Australasian Physical & Engineering Sciences in Medicine, vol. 17, no. 3, pp. 124-130, 1994.
14. C B Cutting, J C McCarthy and D B Karron, "Three-dimensional Input of Body Surface Data Using a Laser Light Scanner", Annals of Plastic Surgery, vol. 21, no. 1, pp. 38-45,

1988.
15. T T Wohlers, "Reverse Engineering Systems: From Product to CAD and Back again", Cadence, January, pp 45-57, 1993.
16. S M Dunn, R L Keizer and J Yu, "Measuring the Area and Volume of the Human Body with Structured Light", IEEE Transactions on Systems, Man, and Cybernetics, vol. 19, no. 6, pp. 1350-1364, 1989.
17. K Patorski, M Rafalowski and M Kujawinska, "Computer aided postural deforming studies using moiré and grid projection method", Proc. SPIE, vol. 2340, pp. 442-448, 1994.
18. S T Young, S W Yip, H C Cheng and D B Shieh, "Three-Dimensional Surface Digitizer for Facial Contour Capture", IEEE Engineering in Medicine and Biology, pp 125-128, 1994.
19. O Kafri, I Glatt, The Physics of Moiré Metrology, Wiley, New York, 1991.
20. J W Dally, W F Riley, Experimental Stress Analysis, McGraw-Hill, 3rd ed., New York, 1991.
21. A Asundi, "Computer Aided Moiré Methods" Optics and Lasers in Engineering, vol. 18, pp. 213-238, 1993.
22. A Asundi, "Moiré Methods using computer-generated gratings", Optical Engineering, vol. 32, no. 1, pp. 107-116, 1993.
23. A Asundi and K H Yung, "Phase-shifting and logical moiré", Journal of the Optical Society of American A, vol. 8, no. 10, pp 1591-1600, 1991.
24. W Hawkes, OptoElectronics An introduction, 2nd edition, pp. 243-253.
25. Chaimowizz, J.C.A, Holography Technology, pp. 259-270.
26. R Jones and C Wykes, Holographic & Speckle Interferometry, 2nd edition, Cambridge University Press, Cambridge, 1989.
27. P C Mehta and V V Rampal, Lasers and Holography, World Science, Singapore, 1993.

A Condition Monitoring and Fault Diagnosis System for Modern Manufacturing Equipment

Wenbin Hu, Zude Zhou and Youping Chen

School of Mechanical Science and Engineering,
Huazhong University of Science and Technology,
Wuhan,Hubei,430074,P.R.China
Email: wbhu@hotmail.com

Abstract. After analysis of the monitoring and diagnostic tasks, a condition monitoring and fault diagnosis system is designed and implemented, which covers the major aspects in the condition monitoring and fault diagnosis of modern manufacturing equipment. The overall hardware and software design of the system, approach to some functional sub-systems or modules are presented. In particular the condition monitoring, data acquisition, knowledge acquisition and representation, as well as the integrated diagnostic reasoning strategy of the integrated expert system are presented in detail.

1 Introduction

The large-scale automation and integration of modern manufacturing equipment, which has become possible with the development of low-cost digital computers and communication networks, permits more efficiency and flexibility in meeting production schedules and can potentially lead to lower-cost and higher-quality products. This kind of manufacturing equipment such as FMS and CIMS, is being more and more widely used because of its potential to improve the strategic and competitive position of firms.

However, such a manufacturing equipment is very dependent upon the trouble-free operation of all its component parts. When a fault occurs, it is critical to isolate the causes as rapidly as possible and to take appropriate maintenance action. For this reason, corresponding diagnostic techniques and systems are studied extensively with the application and dissemination of modern manufacturing equipment. The condition monitoring and fault diagnosis system for modern manufacturing equipment, presented in the following, is designed to meet this need.

2 Monitoring and Diagnostic Tasks Analyzed

Condition monitoring and fault diagnosis of modern manufacturing equipment means a group of special control functions, namely:

- observation of the machine and process condition by sensors;
- recognition of any incorrect event;

- decision on the necessary control action;
- analyzing the diagnostic information;
- displaying and storing the fault information for maintenance planning.

The fundamental tasks of modern manufacturing equipment is to realize the relative motion of the work-piece and the tool and the auxiliary motions setting out the geometric and technological data of the part program. These motions are to be previously and deterministically planned by the programmer. An unexpected and catastrophic change of the condition may cause the machine or process faults.

In general, the correct operational behavior of manufacturing equipment may be characterized by a series of state transitions of the equipment used during the manufacturing of a product. These state transitions occur because of the proper functioning of causal agents responsible for the transitions. The state is characterized by discrete and continuous variables.

Discrete state variables are not only binary or digital control signals but also the signals of switching sensors observing the motion of auxiliary and positioning mechanisms. The error-free cycle of these mechanisms is the foundation of automatic operation. The tasks of monitoring and diagnosis based on the discrete variables are:

- diagnosis of the CNC functions;
- diagnosis of the integrated PLC's;
- supervision of the right schedule of the cyclical operating mechanisms.

Continuous state variables are the sensor signals measuring the physical state of the machine or process. The tasks of monitoring and diagnosis based on the continuous state variables are:

- monitoring and diagnosis of machining operation;
- indirect monitoring and diagnosis of the most important mechanisms, gearboxes, and functional elements of the manufacturing equipment;
- monitoring and diagnosis of the cutting features of the tools.

3 Overall Hardware and Software Design

Taking into consideration the rapid development of hardware and software tools and the continuously increasing demands of the industry, the design of new, more complex and powerful monitoring and diagnosis systems is necessary.

The vast majority of modern manufacturing equipment have automatic monitoring available for faults characterized by discrete state signals in their controllers. These discrete state signals indicate the machine operating state, by which further diagnosis can be carried out. These signals can be obtained directly by using an on-line linkage between the controllers and the monitoring and diagnostic system computer (This can also be carried out via several information-technical levels using LAN).

However, there is little evidence that the machine and process conditions are continuously monitored inside the manufacturing equipment. Those continuous state variables must be acquired by designing an external data acquisition system. Hence the hardware of the integrated monitoring and diagnosis system with a modular structure is designed as shown in Fig. 1.

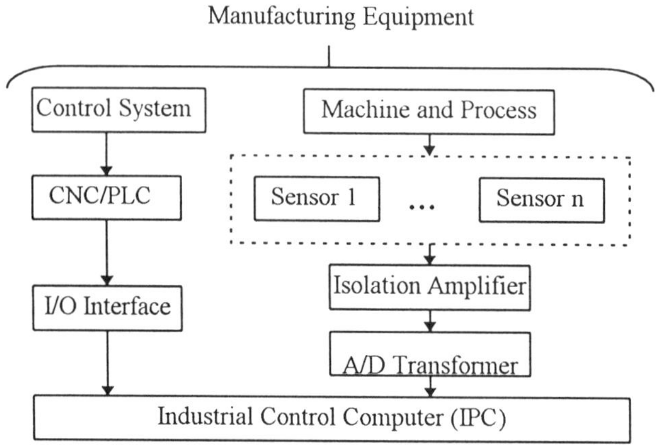

Fig. 1. The hardware block diagram of the system

In order to automate the condition monitoring and fault diagnosis, artificial intelligence in particular expert system is often adopted. However, a highly automatic and integrated manufacturing equipment is a combination of very complex machinery integrated through an equally complex computer system. It is not only limited to just mechanical parts, but will also have electronic, hydraulic, software and human elements, each having different fault distributions. Traditional expert system is far unequal to the task. Therefore, an integrated expert system is designed.

Fig. 2 shows the software block diagram of the integrated monitoring and diagnosis system. This is an integrated expert system with also a modular structure. It is the integration of several functional sub-systems or modules, the integration of numerical computation and symbolic reasoning, etc. All these together form a distributed large expert system. Besides these sub-systems or modules, there is a special sub-system called metal-system, which manages, coordinates and controls the whole system, calls relevant sub-systems or modules to complete corresponding tasks and provides a good environment for man-machine interaction.

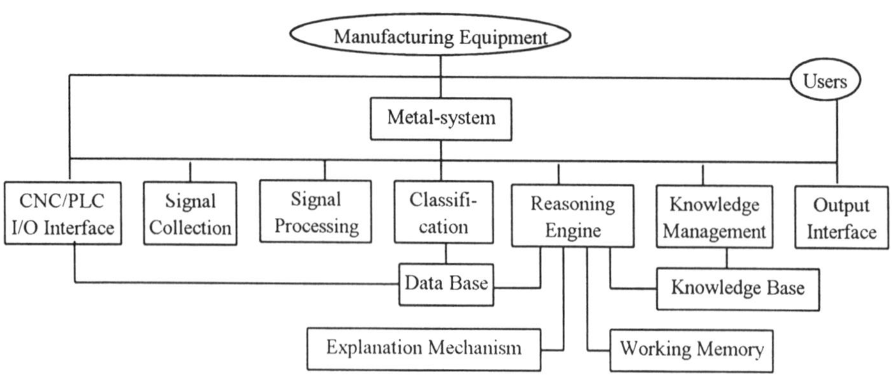

Fig. 2. Integrated monitoring and diagnosis expert system

4 Approach to Some Functional Sub-systems or Modules

In the integrated condition monitoring and fault diagnosis expert system, condition monitoring, data base, knowledge base and reasoning engine are three major functional sub-systems or modules. A more detailed description is given below.

4.1 Condition Monitoring

Early condition monitoring systems relied on the sensing and processing of a single parameter by a single sensor. This kind of monitoring strategy is simple and has poor usability and often brings about false or incomplete diagnosis. This single-sensor and single-parameter strategy is only suited for the monitoring of simple process with a single condition or not frequently changed conditions.

The processes of modern manufacturing equipment are complex and changeable. All parts associated with these processes are closely related to each other. The processes involve a large number factors and the relationship between these factors and processes is very complex and to some extent is fuzzy. To these situations, traditional single-sensor and single-parameter monitoring strategies are powerless. Therefore, a fuzzy hybrid strategy with multiple sensors and multiple parameters must be used to extract multiple integrated parameters from those important parts of the manufacturing equipment, and then hybrid analysis and judgment are carried out so as to reach a significant conclusion from multiple associated parameters.

Parameters Monitored

Considering the sensitivity and easy acquisition, parameters including power, vibration, temperature and pressure are chosen and the monitored objects are spindle, feed axes in three directions (X, Y, Z), hydraulic oil and pneumatic gas (Table 1).

- Power parameters: voltage (U), spindle drive current (I_s), X-axis drive motor current (I_x), Y-axis drive motor current (I_y), Z-axis drive motor current (I_z). Power (P) is calculated by P=UI.
- Vibration parameters: accelerations at the spindle and the three feed axes (X,Y,Z).
- Temperature parameters: temperature of the spindle motor (T_s), oil temperature in the spindle box (T_b), temperature of three axis drive motors (T_x, T_y, T_z).
- Pressure parameters: pressure of the pneumatic gas for claming devices (P_c), pressure of the hydraulic oil for revolving devices (P_r), pressure of the hydraulic oil for feed drives (P_f).

Table 1. Monitored objects and parameters

Objects	Parameters							
	Voltage	Current	Power	Vibration			Temperature	Pressure
				X	Y	Z		
Spindle	U	I_s	P_s	a_{xs}	a_{ys}	a_{zs}	T_s, T_b	P_c
X-axis	U	I_x	P_x	a_{xx}	a_{yx}	a_{zx}	T_x	P_r
Y-axis	U	I_y	P_y	a_{xy}	a_{yy}	a_{zy}	T_y	P_f
Z-axis	U	I_z	P_z	a_{xz}	a_{yz}	a_{zz}	T_z	

Feature Extraction
The following two kinds of features are extracted from above parameters:

- Feature that indicates the total average changing trend of the machine or process status at the moment, defined as $\Sigma(k)$. It also represents the deviation of current status relative to its initial status. If the deviation goes beyond a preset limit, the condition associated with the feature is considered to be abnormal. This feature is suited for the situations where faults occur gradually.
- Feature that indicates the instantaneous changing rate of the machine or process status at the moment, defined as $\Delta\phi$, $\Delta\phi=\phi(k)-\phi(k-1)$. If the deviation goes beyond a preset limit, the condition associated with the feature is considered to be abnormal. This feature extraction strategy is suited for the situations where faults occur abruptly.

Thus for each monitored object, extracted features include: U, $\Sigma(P)$, $\Delta\phi(P)$, $\Sigma(I)$, $\Delta\phi(I)$, $\Sigma(a_x)$, $\Delta\phi(a_x)$, $\Sigma(a_y)$, $\Delta\phi(a_y)$, $\Sigma(a_z)$, $\Delta\phi(a_z)$, T. Besides, there is a temperature feature - T_{b},, and three pressure features - P_c, P_r, P_f,. Totally there are 52 features.

4.2 Data acquisition

The data base is a dynamic base which is generated and used by both monitoring and diagnosis. There are three kinds of fault data in the data base, they are: signals in the controller, condition monitoring results and the observed symptom. Fig. 3 shows the acquisition of these data.

4.3 Knowledge acquisition and representation

The knowledge base is another important part of the integrated system. There are two kinds of knowledge: the deep knowledge and shallow knowledge. The former consists of knowledge based on system structure and behavior, based on fault tree analysis and parameter/state estimation; the latter are mainly from engineers' and experts' experience, fault statistics and process history.
In the design of knowledge base, one of the key problem is knowledge representation. In this system, deep knowledge is represented in rule or object-oriented frame, while shallow knowledge is represented as rules or independent facts.
The retailed knowledge acquisition and representation process is described in Fig. 4.

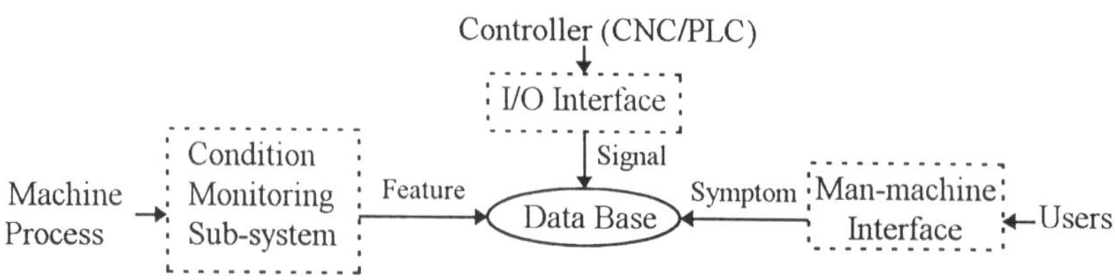

Fig. 3. Data acquisition

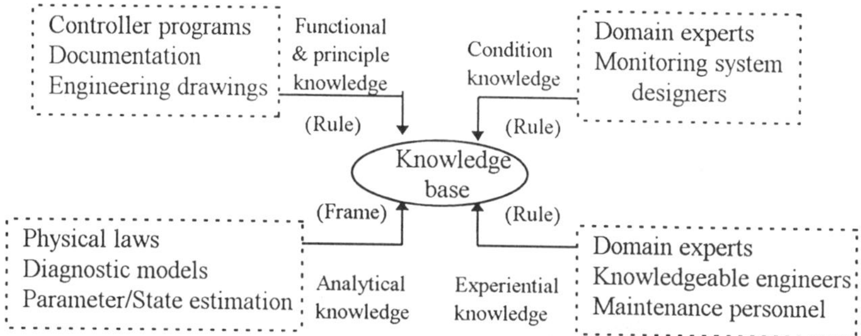

Fig. 4. Diagnostic knowledge acquisition and representation

4.4 Integrated diagnostic reasoning

The reasoning engine is the kernel of the integrated condition monitoring and fault diagnosis expert system. In order to realize integrated diagnosis and imitate the usual fault propagation process of manufacturing equipment, diagnostic knowledge in knowledge bases is also divided into three different levels, that is, functional knowledge (functional decomposition), principle knowledge (decomposition according to the principle of operation) and experiential knowledge (domain experts' or knowledgeable engineers' experience). Meanwhile, all the knowledge bases are in the form of a fault tree, respectively, functional fault tree, principle fault tree and experiential fault tree or rule tree. During the diagnosis of a fault, a human expert or maintenance personnel usually locates the faulty functional modules at first by using of functional fault trees, and then uses the faulty functional modules related principle fault trees to find the rough fault causes, at last obtains the final and more accurate fault causes with the help of corresponding rule trees. Therefore, similar diagnostic reasoning process in the integrated system is carried out as described in Fig. 5.

In Fig. 5, diagnostic reasoning based on the functional fault tree and principle fault tree is performed by the strategy of breadth-first search combined with failure probability of each node in the fault trees. Whether a node is faulty or not is determined by various signals in controllers and the logic relationship among these signals. Reasoning based on experiential knowledge or rule tree is more complicated. For rules associated with machine or process condition, reasoning is performed with the help of condition monitoring results, while for other rules, reasoning is performed as a sequential hypothesize-test cycle. In the cycle, a cost-weighted entropy criterion is used to choose the next part of the rule tree to activate. This entropy criterion helps to select the rule that gives the maximum fault discernment per unit cost in cases where multiple tests might be performed.

Supposing R_{kj} is a node in the kth level of the rule tree and $R_{k+1,1}$, $R_{k+1,2}$, $\cdots$, $R_{k+1,m}$ are nodes in the $(k+1)$th level, cost-weighted entropy of node R_{kj} can be calculated by the following equations:

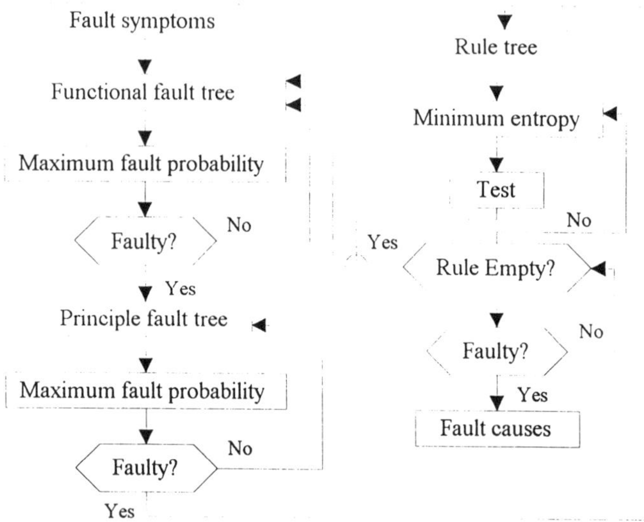

Fig. 5. Integrated diagnostic reasoning procedure

$$H=-\sum_{j=1}^{m} w_j P_j \ln P_j \qquad m \geq 2 \tag{1}$$

where

$$1 \geq w_j \geq 0, \qquad \sum_{j=1}^{m} P_j = 1$$

The weighting factor, w, is a normalized cost, determined by dividing the actual cost of a measurement operation, by the maximum of the set of measurement costs for all components at the current rule level. P_j is the probability with that $R_{k+1,j}$ is the cause of R_{ki}, under the condition that the test result of node R_{ki} is known.

Cost-weighted entropy is used to select and activate a part of experiential knowledge base or rule tree. It selects the measurement that will give the most discernment at the lowest cost. That is to say, the next rule to be tested, its entropy must be the minimum.

5 Case study

The above integrated system has been implemented on FFS-1500-2 FMS installed at Zhengzhou Textile Machinery Plant in China. FFS-1500-2 FMS consists of a PFZ 1500 FMC, a KBNG85 MC and an automatically guided vehicle (AGV). PFZ1500 FMC is made up of some functional modules such as tool change, adapter change, axis drive and hydraulic drive, etc. Axis drive can also elaborately be divided into spindle drive, X-axis drive, Y-axis drive and Z-axis drive. This kind of decomposition is based on fault tree analysis method. In case a process fault is detected in PFZ1500 FMC, the diagnostic research is shown in Fig. 6 and the faults corresponding to each node are:

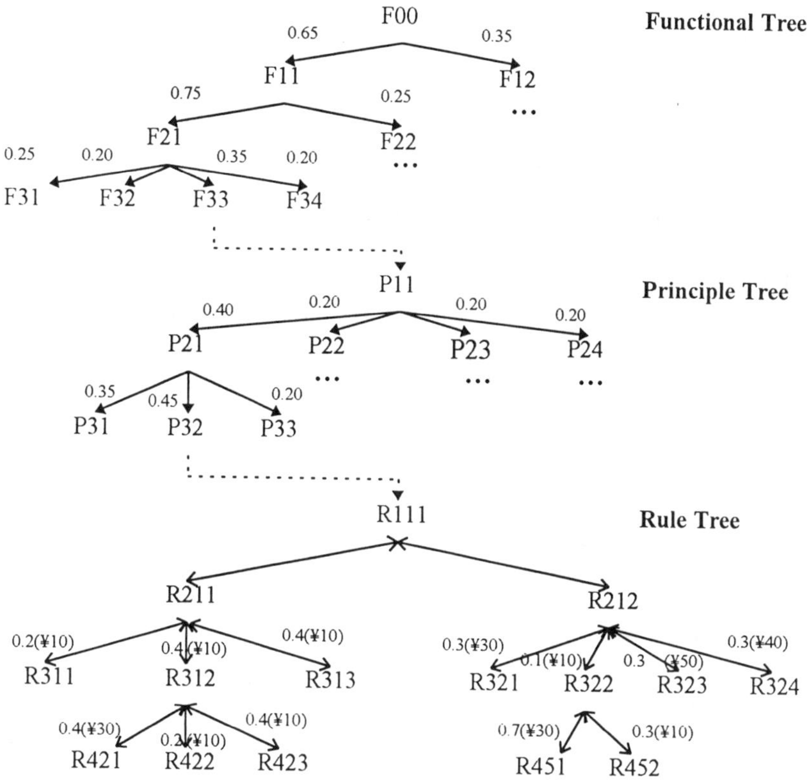

Fig. 6. Sample integrated diagnostic reasoning

F00: FFS-1500-2 FMS
F11: Machine Tools
F12: AGV
F21: PFZ1500 FMC
F22: KBNG85 MC
F31: Tool Change
F32: Adapter Change
F33: Machining Process
F34: Hydraulic Drive
P11: Machining Process
P21: Spindle
P22: X-axis
P23: Y-axis
P24: Z-axis
P31: Spindle Feed
P32: Spindle Motor
P33: Spindle Transmission
R111: Spindle Motor
R211: Motor Drive
R212: Control Circuit
R311: Mechanical Parts
R312: Motor Temperature
R313: Motor Connection
R321: Circuit Connection
R322: Power Connection
R323: Control Amplifier
R324: Power Amplifier
R421: Tool Edge
R422: Cooling System
R423: Feed Load
R451: Fuse
R452: Power Switch

A first search through functional knowledge base (functional fault tree) leads to F33, which is a terminal functional knowledge node. Principle knowledge base P11 is activated at this point and another search through principle knowledge base (principle fault tree) leads to P32, which is a terminal principle knowledge node. Whether a node is faulty or not depends on signals in the controllers and condition

monitoring results.
Then experiential knowledge base (rule tree) R111 is then activated and the cost-weighted entropy is computed for R211 and R212 groups. The rule with minimum entropy is tested first and maintenance personnel are instructed to check the corresponding part of the equipment. If it is faulty, then diagnosis terminates, otherwise the system will check other observed symptoms before backtracking within the functional and principle knowledge bases.

6 Conclusions and future work

The research described in this paper has introduced a systematic methodology for the design of an integrated condition monitoring and fault diagnosis system for modern manufacturing equipment. This is an intelligently integrated system with a modular and reconfigurable structure, and the functions of condition monitoring, fault diagnosis, maintenance planning and so on. The system has been implemented on FFS-1500-2 FMS and can be applied to other manufacturing equipment with little change.
Further work on this research will focus on refinement of monitoring and diagnostic algorithm, improvement of the system design and implementation, and investigation of the potential of a learning strategy to make the integrated system adaptable to changes in the monitoring and diagnostic environment. Furthermore, in order to improve the efficiency of the system, it is significant to identify a generic strategy for hosting condition monitoring and fault diagnosis expert system for manufacturing equipment on their controllers. This kind of system is called embedded system.

References

1. E. Ferenc and S. Csongor, Monitoring tasks on boring and milling production cells, *Computers in Industry*, 7(1986) 65-71.
2. M. Weck and M. B. Hummels, Diagnosis of flexible manufacturing systems (FMS) — aspects of on-line error detection and knowledge processing, *IFAC Fault Detection, Supervision and Safety for Technical Processes*, Baden-Baden,Germany (1991) 361-367.
3. H. Tang and B. K. Xu. Computer Aided Diagnosis of Mechanical Equipment, Tianjin: Tianjin University Press, 1992.
4. W. B. Hu. Research on a Quality-Control-Based Fault Diagnosis System in a Flexible Manufacturing Environment. *Ph.D. Dissertation*, Huazhong University of Science and Technology, 1995.
5. S. M. Alexander, C. M. Vaidya and J. H. Graham, Model for the diagnosis of CIM equipment, *Computer - Elect. Eng.* 19(1993) 175-183.
6. J. H. Graham, S. M. Alexander and W. Y. Lee. Knowledge-based software for diagnosis of manufacturing systems. *Intelligent Manufacturing — Programming Environment for CIM*, Spinger - Verlarg, 1993.

An Indirect Machine Tool Condition Monitoring System

P. Oksanen, P. H. Andersson

Institute of Production Engineering, Tampere University of Technology, Finland

Abstract. Quality systems according to ISO 9000 standard preconceive that production machinery is to be inspected regularly. Conventional inspection methods are time consuming and the tests need to be performed by skilful personnel. The inspections have also been departed from workpiece measurements. In the *Eureka Maine FMS Maint* project machine tools were measured regularly with the aid of a double ball bar (DBB) device, test pieces and actual workpieces. The Quality level of machining was analyzed by DBB results database and *Polar Check Analysis* software [1]. Correlation was found with temperature conditions, NC compensation parameters, test pieces and actual workpiece measurements. The Introduced maintenance control method is an useful aid for maintenance planning and it gives vital information about state of individual machine tools.

1 Introduction

The key questions in machine tool maintenance are: *when to carry out the needed preventive maintenance actions* and *how to measure and adjust machine's geometry or compensation parameters* [2]. It is evident that maintenance based on calendars only is not the best solution, because machines can be run different shifts or variety of products. The dispersion of the sizes in the final product features is a function of the characteristics of the workpiece, fig 1.

The concept for the integrated machine tool condition monitoring system is shown in figure 2. In order to be able to draw conclusions of the production quality, a large amount of information was gathered and stored in a common database. This database includes information of the basic machine tool structures, characteristics, tools, workpieces and parameters. An important part of the complete system was an automatic data collection (ADC) system. All this information was necessary in order to be able to interprete the measuring data. Information system with databases forms a central and important role of the concept. Measuring methods can be divided to machine tool measurements with conventional devices or quick test methods and production measurements monitoring test pieces or actual workpieces [4]. This work concentrates machine tool measurements with DBB method and result analysis has been made with the aid of the other parts of the concept.

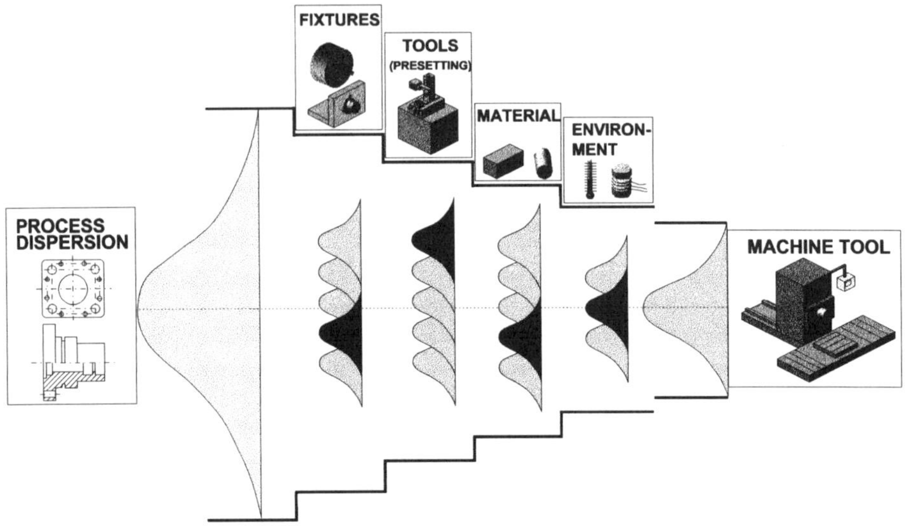

Fig 1. Process dispersion vs. deviation caused by the machine tool.[3]

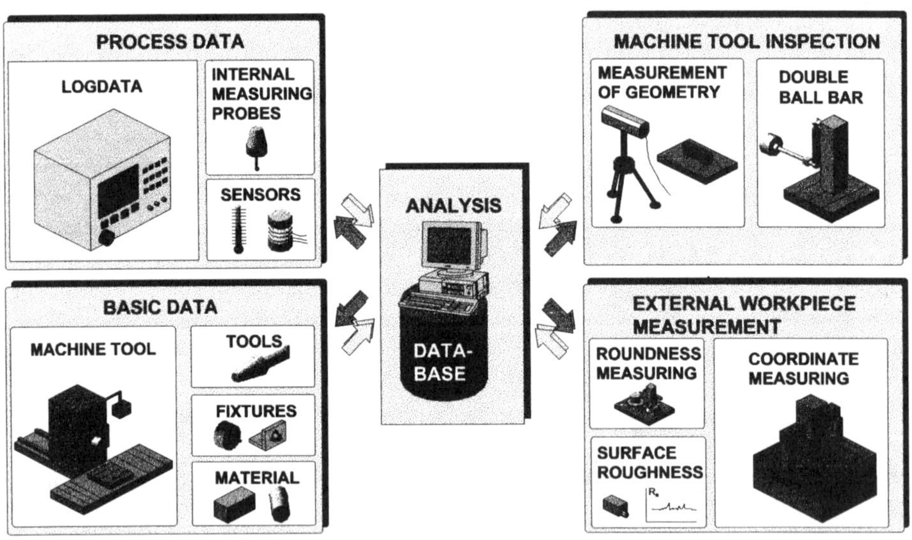

Fig 2. Integrated maintenance system using process control and measurements.[3]

2. Machine Tool Maintenance Control by DBB Method

Before the measurement phase, a comprehensive training session for the pilot factory personnel was made. The initial state of the machinery was examined in a thorough inspection using conventional measuring equipment. Controller parameter changes were made to adjust the machine tools as accurate as possible.

DBB tests were carried out once a month in connection with a test piece run. Workpieces to be measured were selected among those of the highest volume. The features measured were choosed in accordance with indicating the production accuracy. All work and test piece measurements were performed in a coordinate measuring machine.

2.1. Analysis of DBB Measurements

The accuracy of three identical machine tools were evaluated regularly, the time span of the whole project was about three years. All machines were measured with the same NC-program, the position of the measurement was the same. The measuring data and the trends were analyzed with the *Polar Check Analyser* software.

The most important indicator for the quality of machining is the change of the circularity on xy-plane of an individual machine tool. The upper limit of the circularity deviation was set to 15 µm. The trend of measurements were quite similar for all the machines. However, the error sources that caused the circularity deviations varied. Figure 3 shows an example of trend analysis.

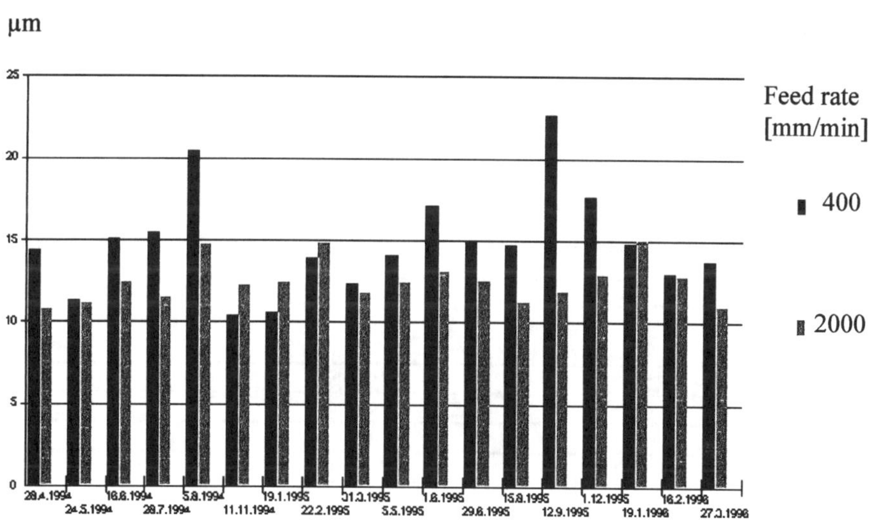

Fig 3. Trend analysis of machine tool errors: xy-plane circularity deviation with DBB method. [5]

The trend analysis shows that the circularity values of the machine tools were generally between 10 and 15 µm. The upper limit was exceeded occasionally. The measurement results were reanalyzed using individual error parameters. This analysis showed that the greatest circularity deviations were caused by the backlash of the linear axes (Fig. 4) .

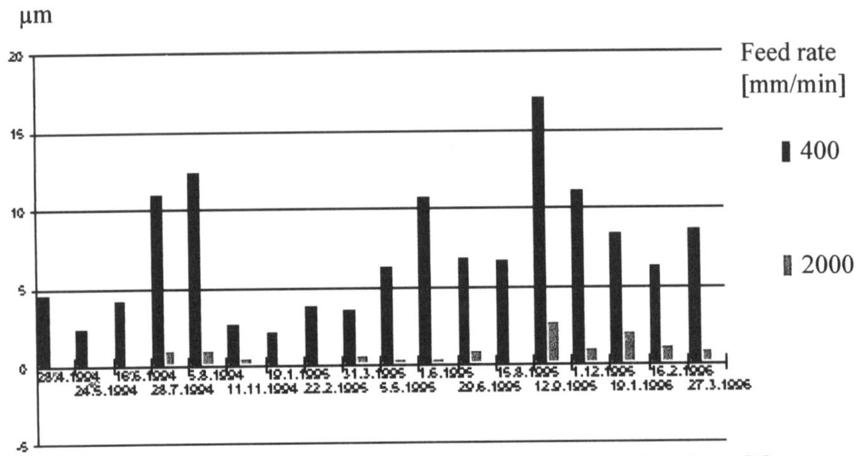

Fig 4. Backlash deviations detected with DBB trend analysis. [5]

As no compensation parameters were changed or major services performed, the reason for circularity error had to be found elsewhere. The temperature analysis of the machining centres showed a relationship between the circularity error and the temperature. Graph of temperature values during DBB measurements can be seen in figure 5. The key observation is that the greatest backlash errors and the highest temperatures occurred the same day. When the temperature changes back to it's normal level, the backlash error reduced without any adjustments. The temperature diversion of the machining centres were about 5... 6°C. The exceptional warm periods of summers '94 and '95 in Finland can be seen from the graphs as well. The effect could be detected on each of the three machining centres. If a warm period lasts at least one week, an effect can be detected by analysing the backlash values.

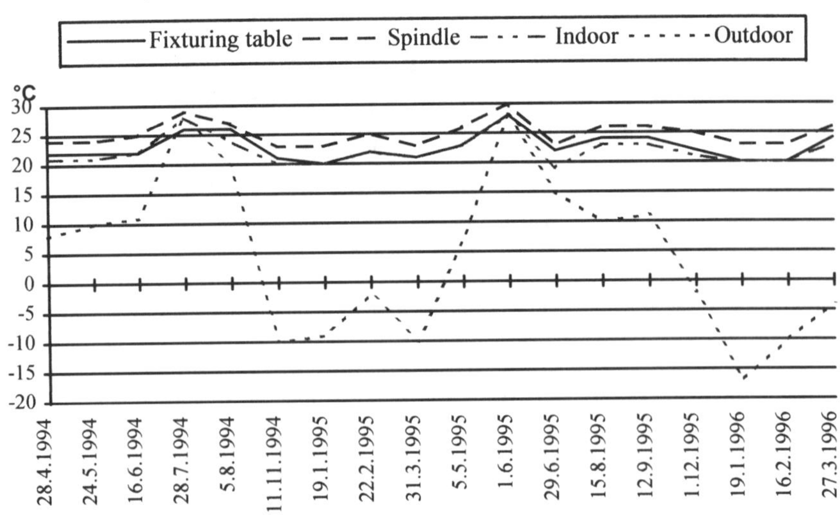

Fig 5. Machine and environment temperatures during the DBB-measurement dates. [5]

Another remarkable observation from the trend analysis graphs was the effect of the NC-controller compensation parameters on the amount of backlash. Also the traverse speed of the machine tool slides has a significant impact on the actual reversal error. The magnitude of the error showed to be greater when using slower feed rates and vice versa.

Conventionally the backlash compensation values are evaluated with laser-interferometer measurements using rapid speed as the positionings during the machining operations are usually performed the same way. The backlash is then properly compensated only in these circumstances. When contour milling using circular interpolation and a slow feed rate, the inertia of the moving masses affects dissimilarly and the effective backlash will be remarkably greater than measured and compensated.

In this project a laser-interferometer and rapid speed was used in parameter compensation in the first phase, later on the method was changed to the DBB and normal machining feed rates. Figure 6 shows the effect of this change.

Only after a major adjustment the deviations could be set to acceptable limits during the slow feed rate operations (overcompensation at higher speeds are however unavoidable). The same pattern could be detected from the test piece measurements.

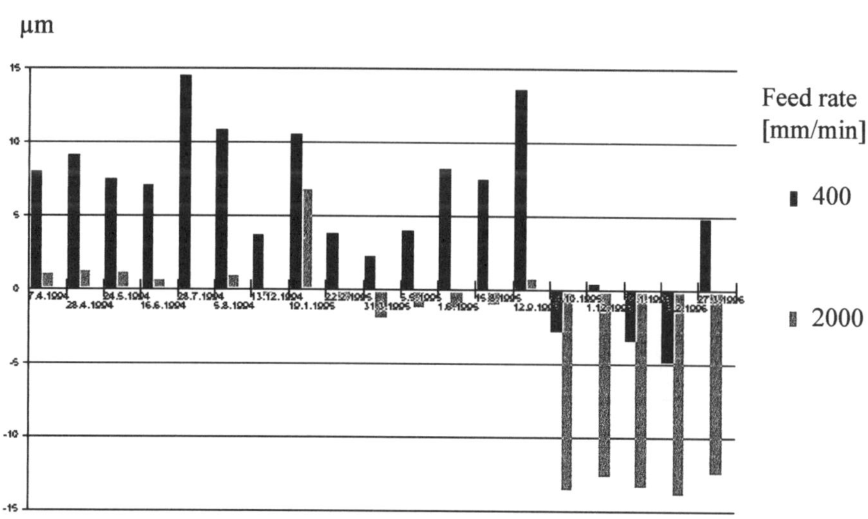

Fig 6. Effect of changes in compensation parameters on DBB trend analysis. [5]

2.2. DBB Analysis Compared with Actual Workpiece Measurements

Actual workpieces produced by the pilot company were measured in the project and the results were analyzed statistically. The results were compared with DBB analysis and the object was to find out how the individual error parameters could be detected with both methods. Especially importance was to know the influence of cutting forces, because DBB measurements are carried out in no load condition.

A comprehensive example of the comparison material is the influence of the backlash error of two different machining centres. Both machines were measured identically. An identical workpiece, with a circular interpolation feature, was machined with both machine tools as well. The result of this comparison is presented in figure 7.

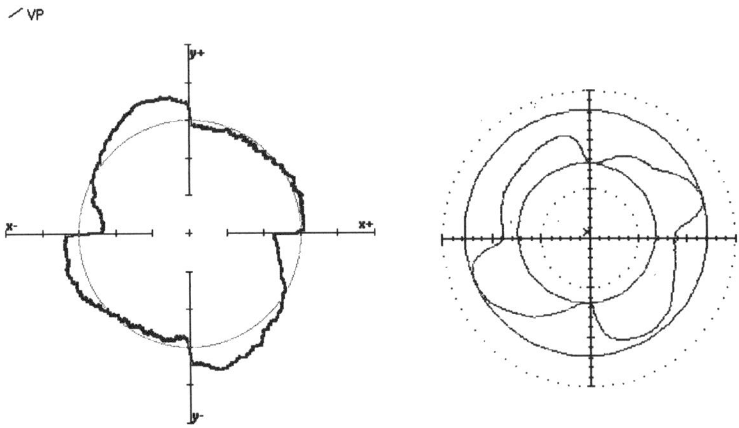

Fig 7. DBB measurement of machining centre A, xy-plane (left) compared with actual workpiece measurement (right). [5]

As can be seen from the graph, the results of both methods are quite similar. As the tool radius acts like a filter, the effect of backlash is somewhat smoother on the workpiece and the amount of error is a bit smaller. The similar comparison at the machining centre B gives however totally different results. Those paths are shown in figure 8. When comparing the DBB results from the both machine tools, the cause of major errors seems to be the same - the axis backlash. However, the workpiece measurement graph from the machining centre B shows great squareness error between the linear axes. The DBB measurement being a static method could not detect the specific deviation. This is due to the cutting forces bending the frame of the machine tool.

By comparison of several test methods could be found that a particular machining centre is individual. Same interpretations from one machine tool can not be generalised to all machine tools, even if they were same type. The DBB method is an useful aid in maintenance control if it is used together with test and workpiece measurements. The DBB could then be a constant control method and other measurements are used as supporting actions when needed.

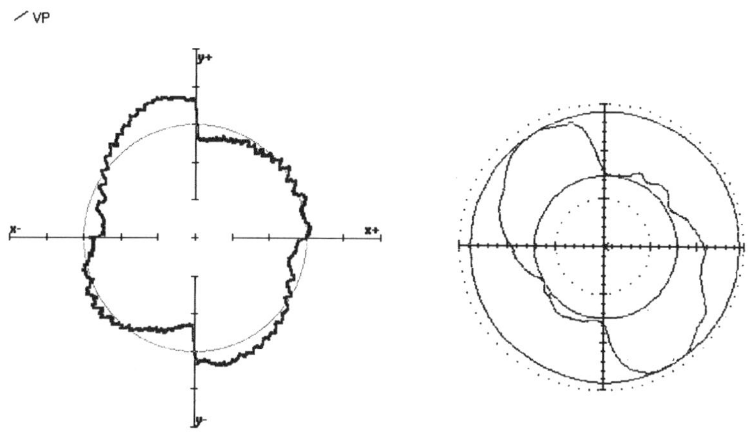

Fig 8. DBB measurement of machining centre B, xy-plane (left) compared with actual workpiece measurement (right). [5]

3. Experiences of the DBB as Part of a Maintenance Control System

As a result of project the DBB can be concluded to be a very effective and quick method indicating trends in machine tool accuracy. This was partly because of very thorough project planning which included static measurement methods. DBB is really a quick test - average measuring time for one machining centre with pallet changes was one hour. Before an effective analysis of measuring results machine tools' dynamic behaviour must be known. Test pieces give a great support for interpretation. The overall concept is visualised in figure 9.

Actual statistical analysis of workpiece measurements should be divided to Statistical Process Control (SPC) of normal production and machine tool condition dependent SPC. If normal SPC gives an error signal, there is no need to make any measurements on the machine tool, but to try to found the error causes from i.q. tooling or materials. Machine tool measurements are not needed until error signals from machine tool condition dependent SPC occurs. Test could then be made by the DBB with the aid of dynamic behaviour information from test piece measurements. Some parameters can be changed by DBB and conventional methods should be used only when major geometrical adjustments or positioning error compensation is needed

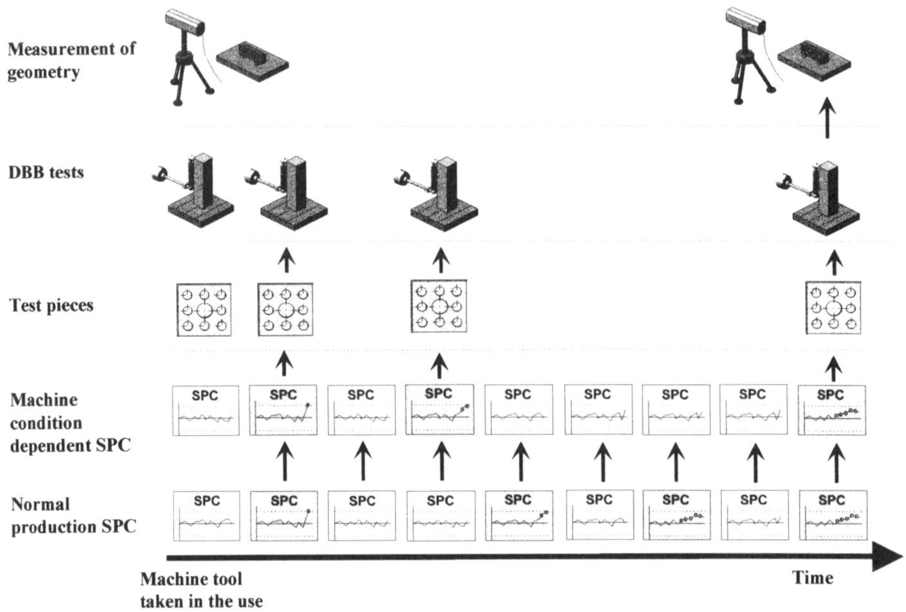

Fig 9. Maintenance control with machine tool, test piece and workpiece measurements. [3, 5]

4. Conclusions

Carefully prepared machine tool maintenance control with the DBB method assisted by the statistical control data from the test and workpieces serve effectively predicting machine tools maintenance and service needs. Almost all the necessary changes in control parameters could made according to the DBB data without any other measurement aid. In some cases the need of a maintenance action was predicted before any parts were broken. This had an positive influence on the machine tool productivity. One of the most important results of the project was establishing the connection between the DBB results and the machine tool temperature behaviour.

References

1. J. Hölsä, Analyse and Simulation of Double Ball Bar Method. Tampere: Tampere University of Technology, (1995). M.Sc. Thesis. 101 p. (Finnish)
2. S. Torvinen, P.H. Andersson, J. Vihinen and R. Milne, Monitoring the accuracy characteristics of the machinery by using a dynamic measurement approach.. In: Winsor, J., Sivakumar, A.I., Gay, R. (eds.) Proceeding of the 3rd international conference computer integrated manufacturing., (1995). Vol 2, 1435-1442.
3. P.H. Andersson, S. Torvinen and J. Vihinen, Indirect maintenance control of machine tools. In: Konepajamies **6** (1994), 54-56. (Finnish)
4. S. Torvinen, P.H. Andersson, J. Vihinen and J. Hölsä, An off-line condition monitoring system for machine tools. In: Hope, A.D., Smith, G.T., Blackshaw, D.M.S. (eds.) Second

International Conference on Laser metrology and Machine Perfomance "LAMDAMAP 95", (1995). 237-250.

5. P. Oksanen, A Machine Tool Inspection and Maintenance Control by Double Ball Bar Method. Tampere: Tampere University of Technology, (1996). Lic.Techn. Thesis. 101 p. (Finnish)

Sensor System for Disassembly of Electrical Motors

Björn Karlsson, Nils Karlsson and Alexander Lauber

Department of Physic and Measurement Technology, Linköping University, Sweden

Abstract The role of reuse and recycling has become more and more important during the last years. There are a lot indications that the producers in the future will have to take care of their own products when they are worn out. During the lifetime of a product it will change to a large extent. To make industrial robots able to perform disassembly in a proper way measurement and sensors will play an important role [1].

In this paper the sensor support of a robotized work station for the disassembly of industrial asynchronous motors is presented. In an industrial motor the copper is situated in the stator windings and in the junction box. There are three steps in the proposed disassembly work. In the first step the functionality of the motor is checked. The second is a manual disassembly part where the terminal box, the shields and the rotor are removed. The third is the robotized automatic part.

In the automatic part the stator winding made of copper will be removed from the stator. There are four main steps in this activity. The first step is to place the motor in the correct position, which is done with a robot guided by a vision system. The second step is to cut off the stator winding at one end of the motor. This is achieved by means of a cutting tool guided by an eddy current sensor and a vision system. The third step is to pull out the stator winding from the stator, which is done by a hydraulic tractive system. After removal of the stator winding a check has to be done to guarantee that all copper has been removed. This test will be performed by an eddy current probe mounted on the robot.

1 Introduction

From a recycling view copper is toxic for steel. The amount of copper in the steel leaving the shredder is typically 0.25-0.3% [2]. If the amount of copper in the steel scrap exceeds 0.02%, the steel scrap can not be used for the production of steel sheets [2]. In electrical motors the copper amount is about 10% of the motor weight. Today small electrical motors when worn out are put in a shredder, due to disassembly problems. If the copper could be removed from the motor before shredding, the usability of the materials out of the shredder would increase significantly. To make the removal of the copper acceptable from economical and ergonomic point of views automatic disassembly supported by sensors should be used.

When a product reaches the scrap dealer a classification has to be done. From the scrap dealer's point of view it is impossible to have knowledge of all products reaching his shop. To guide the scrap dealer an advance sensor and classification system is needed. For an electrical motor, the construction principle is simple but there are a lot of variations in design depending of branch and the field of applications. The classification separates motors in two parts, one for repair and one for disassembly.

The motors classified for repair are sent to repair and the motors classified for disassembly are going to the next step at the disassembly work station. In this step manual disassembly of the shields and the junction box is performed, figure 1. After that it is time for the last step in the chain, the automatic disassembly, which is in focus in this paper. Figure 2 illustrates the test and disassembly system.

2 Test of Electrical Motors

When a motor reaches the disassembly station a condition test of the motor will be done. The different test methods are:

- Vibration measurement for detecting bearing wear, defective rotors and electrical unbalance. The measurements are preferably done with a triaxial accelerometer, threaded or mounted with temporary adhesive, on a horizontal (vertical) machine surface. Vibration measurement should be recorded during start-up and when the motor works without load at nominal r.p.m.[3].
- Acoustic measurement can be used mainly to discover mechanical problems with the electrical motor. Acoustic measurement are to be seen as a complement to vibration measurement.
- Electrical performance [4],[5]
    - Measure of the start-up current characteristics during the first 0.5 s give a lot of information about the condition of the motor, such as stator winding status and unbalance.
    - Measuring stator resistance will give information about broken windings.
    - Perform a high voltage test to get information on the condition of the stator winding.

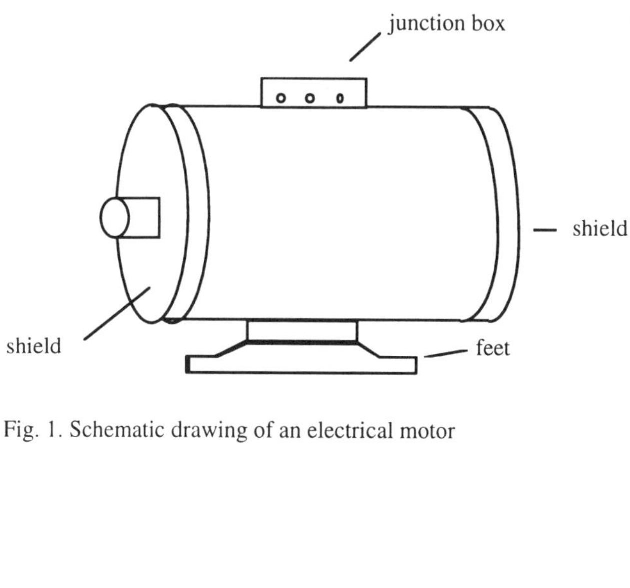

Fig. 1. Schematic drawing of an electrical motor

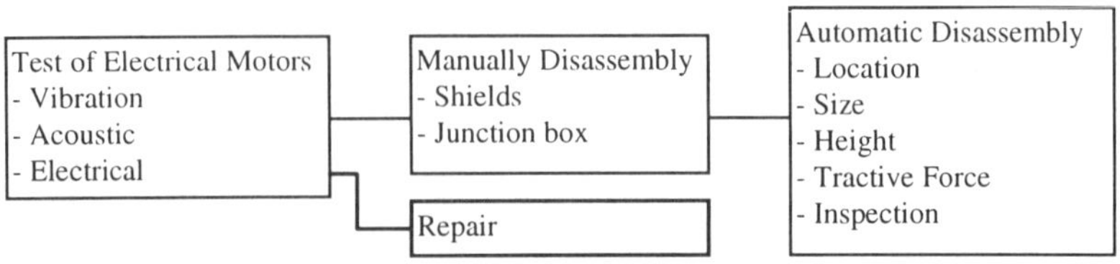

Fig. 2. Block diagram of the disassembly work station

3 Automatic Disassembly

After manual disassembly it is time for the automatic one. The automatic disassembly will be carried out with an industrial robot and for the removal of the stator winding a hydraulic jack will be used. Both the robot and the hydraulic jack will be guided by fused high level information from a sensor system.

3.1 Vision for Location

For location vision is used. One of the bigger problems is how to fuse the information from the vision system with the information from the other sensor systems. To clarify things, the number of parameters detected should be reduced early in the process. The wanted vision parameters are the centre point of the cylindrical part of the motor house, the diameter of the motor house and the location of the feet. These parameters are calculated and the motor is placed in the desired location.

3.2 Eddy Current Measurement for Slot Inspection and Determination of Height

The electric and magnetic properties of the copper winding differ from those of the iron where they are mounted. To measure the height of the motor (figure 3) and to inspect that the winding is successfully removed from the stator after disassembly, eddy current measurements are used, figure 3.

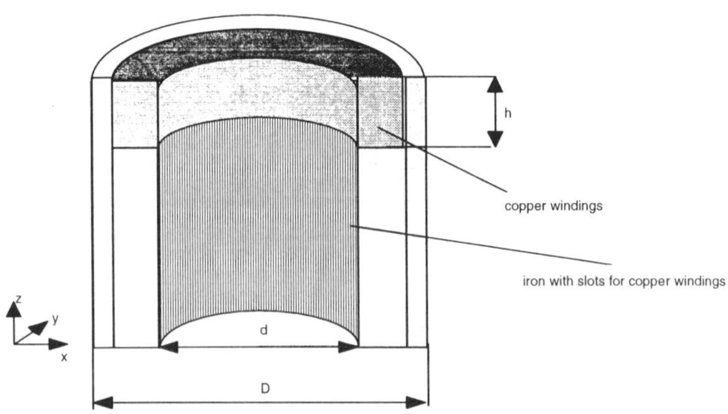

Fig. 3. Schematic picture of the upper part of the electric motor cut in halves as seen from the side.

3.2.1 Principle for Inductive Testing

The inductive testing also called eddy current testing, is based on the law of induction [6]. When a coil is placed in a time-varying magnetic field an electric voltage is induced. The induced voltage can be calculated by Faraday's Law of induction.

$$e_0 = -N \frac{d\Phi}{dt} \tag{1}$$

N = number of turns of the coil.
Φ = the magnetic flux through the coil.
e_0 = the voltage induced in the coil.
The law of induction is also valid when a solid electrically conducting object is placed in time-varying magnetic field. The solid object can be seen as one short-circuited turn of a coil. The induced voltage now causes currents called eddy currents. Figure 4 shows how a magnetic field is generated by a coil connected to an AC-voltage source. This magnetic field H_p, is the primary field which induces eddy currents in the object. The eddy currents produces in their turn a secondary magnetic field H_s. The magnitude and phase shift of H_s compared to H_p depend on the electric and magnetic properties of the object.

It can be seen that the eddy currents in the test object are concentrated more or less strongly to the surface of the test object, adjacent to the coil causing the primary field. This is called the skin effect phenomenon. This effect increases with increased frequency, test object conductivity and magnetic permeability. The standard depth of penetration of eddy currents, δ, is defined as that depth below the surface of the test object where the current density is 1/e of the current density at the surface. The standard depth of penetration of eddy currents, δ, is determined by:

$$\delta = \frac{1}{\sqrt{\pi f \mu_r \mu_0 \sigma}} \tag{2}$$

f = the frequency of the magnetic field
μ_r = the relative permeability of the object
μ_0 = permeability of free space ($4\pi * 10^{-7}$ H/m)
σ = the conductivity of the test object

With knowledge of the conductivity and permeability of the material of the test object and the thickness of the object a suitable test frequency can be chosen.

3.2.2 Determination of the Height for the Cutting of the Stator Winding

In the disassembly procedure the copper windings have to be cut. The upper part of the windings has to be removed before the copper windings in the slots can be removed. The windings ought to be cut just above the slots figure 3. To be able to find this location in the x/y-plane the vision system is used, and to get the information about the z-position the inductive probe can be used.

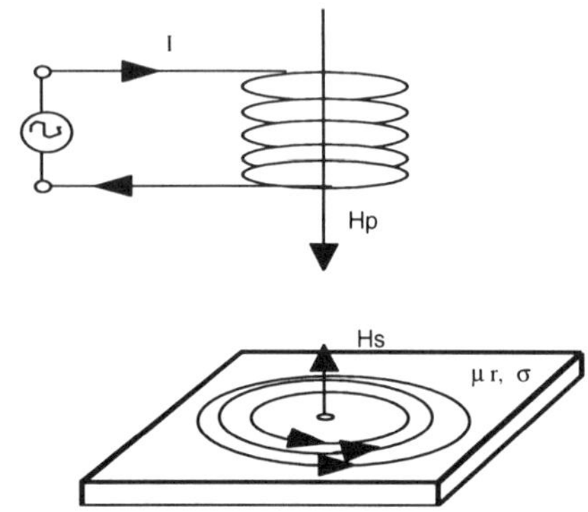

Fig. 4. The production of eddy currents in an object of conductive material placed in the proximity of a coil connected to an AC-voltage [6].

The robot will place the inductive probe, e.g. the coil, just above the motor. The position in the x/y-plane is obtained by the vision system. The edge of the sensor must be near the projected inner edge of the motor. The sensed material at the edge of the sensor is air. The robot lowers the sensor (the negative z-direction in figure 3) until the sensed material is copper windings. The position of the inductive sensor in the x/y-plane might now have to be checked by the vision system (the projection error is smaller at shorter distances) and perhaps adjusted. The coil is lowered until it senses the iron. Then the sensor is moved in the positive z-direction until the sensor indicates copper windings again. This z position is the position to cut the copper windings.
The signal of the sensor that corresponds to iron and the copper windings has to be tested out; it can not be calculated analytically. An approximate result might be simulated by computer programs based on FEM calculations.

3.2.3 Inspection of Stator Slot by Inductive Testing

To check that the removal of the copper windings in the stator slots is successful an inductive sensor will be used. The coil of the inductive sensor will be placed inside the motor with its axis perpendicular to the inner wall of the motor, figure 5. The coil will then be moved (rotated) one full turn around the inner wall.

The sensor output for the coil placed near solid steel is different than the output from the coil when the sensor is placed near an empty slot. As copper has a much higher conductivity than the air in an empty slot the sensor output for copper windings in a slot will be different from the sensor output when all copper has been successfully removed from the slot. If all copper is removed the output of the sensor ought to be the same when passing all slots. The measurement will cause a pattern. If this pattern is periodic the removal of the copper out of the slots has been successful. The expected sensor output can not be analytically calculated. Tests therefore have to be made to get reference signals. Information about the sensor output can however be achieved approximately by computer simulations. The problem of the limited case: which is the maximum possible undetectable amount of copper remaining in a slot must be solved by practical testing.

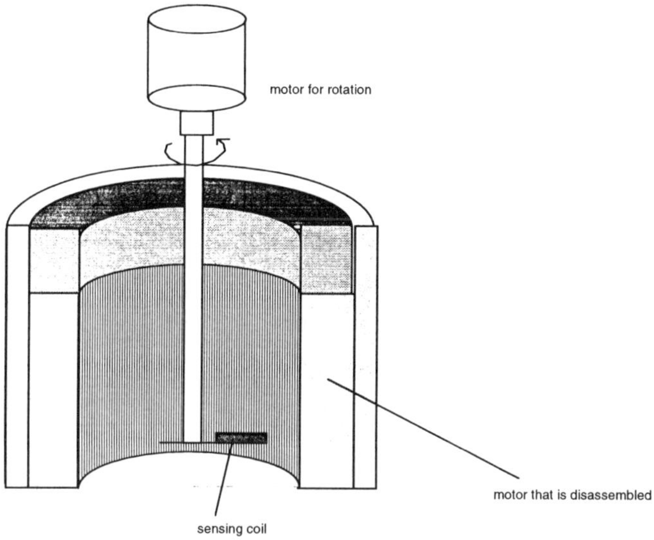

Fig. 5. Laboratory test set-up for eddy current measurement

To test the measuring principle a test setup has been made in the laboratory, see figure 5.

4 Sensor Data Fusion

To make it possible to fuse information from the different sensors in the system, pre-processing of the data is needed. The great amount of information from the vision system needs to be simplified. The human brain chooses to handle only one part at a time, i.e. in vision, location and shape are handled separately[7]. If we continue to compare with the human a distributed data reduction is performed before the information reach the brain. Every second the human body get more than 40 Mbits of information, but the amount that reaches the brain is only 10-40 bits [8].

The eddy current measurement to detect if all copper is removed or not is made by three vertical placed probes. If all copper is removed the output of the three probes will be the same. Due to disturbances it is important to classify the output. A possible approach is to use a fuzzy set, figure 6 [9].

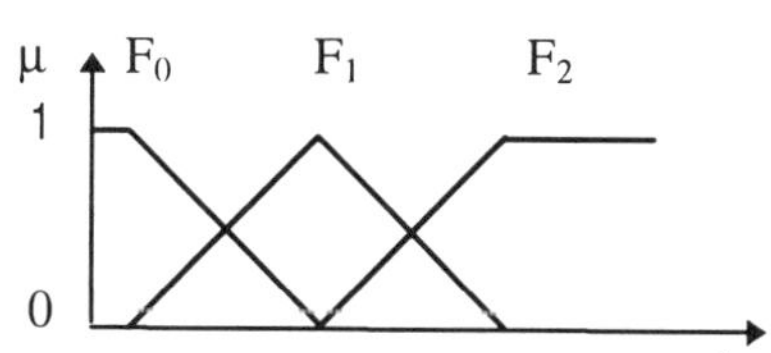

Fig. 6. The fuzzy sets for a sensor

Each fuzzy set is defined as a triangular membership function $F: U \rightarrow [0,1]$ given as

$$F(u;\alpha,\beta,\gamma)=\begin{cases} 0 & u<\alpha \\ \dfrac{u-\alpha}{\beta-\alpha} & \alpha\le u\le\beta \\ \dfrac{\gamma-u}{\beta-\alpha} & \beta\le u\le\gamma \\ 0 & u>\gamma \end{cases} \tag{3}$$

where u belongs to the overall range of the sensor output and α and γ are these values of u for which the degree of membership of α and γ to F is zero, and β is this value of u for which the membership of β to F is one. The output from the probes is classified with (3) in three different classes:

F_0 = Air with iron on either side of it
F_1 = Copper with iron on either side of it, or iron with air in the slots on either side of it
F_2 = Iron with copper in the slots on either side of it

When measuring both near the slot and near the iron between, for a motor with 24 slots it will be 48 measurements. In figure 7 measurements on three different positions as well as the fused values f(μ) of the three probes are shown. Equation (4) is used to fuse the measured data, In (4) a value of K=2 is used and N=3 due to that three sensors are used. The K value determine the weighting function, with K=1 we get the usual arithmetic mean value and a bigger K value gives a more unlinear weighting function.

$$f(\mu_1,\ldots,\mu_N)=\frac{1}{K^N-1}\,\frac{-1+K^N\prod_{i=1}^{N}\dfrac{1+(K-1)\mu_i}{K-(K-1)\mu_i}}{1+\prod_{i=1}^{N}\dfrac{1+(K-1)\mu_i}{K-(K-1)\mu_i}} \tag{4}$$

The result out of figure 7 is that in position 1 it is a slot with air, position 2 is between two slots and position 3 is a slot with copper left. The conclusion is that it is still copper left in the motor which has to be removed manually.

5 Results of the inductive measurements

To determine the proper height to cut the copper windings an inductive probe is used. The inductive probe is guided by a vision system in the x/y-plane and positioned by the robot near the inner edge of the sensor. Figure 8 shows the output from the sensor as it is lowered by the robot. Point A is inside the motor, the current through the coil decreases. Point B is the border between the upper winding and the stator iron. The current through the sensor still decreases until point C. The current is stable i. e. the surroundings of the coil do not change with the height. This tells us that the sensor must be in the stator iron region. We turn the direction of the coil and moves it upwards. Point D is the same as B. The increase of the current through the sensor indicates that the region of stator iron is left. This is the height to cut the copper windings. The z-coordinate is got by the use of the robots z-position.

	Position 1			Position 2			Position 3		
	F_0	F_1	F_2	F_0	F_1	F_2	F_0	F_1	F_2
Probe 1:	0.7	0.3	0	0	0.3	0.7	0	0.8	0.2
Probe 2:	0.6	0.4	0	0	0.3	0.7	0.7	0.3	0
Probe 3:	0.4	0.6	0	0	0.1	0.9	0.5	0.5	0
Fused value:	0.67	0.41	0	0	0.18	0.82	0.37	0.54	0.04

Fig. 7. Classified sensor data from eddy current measurement

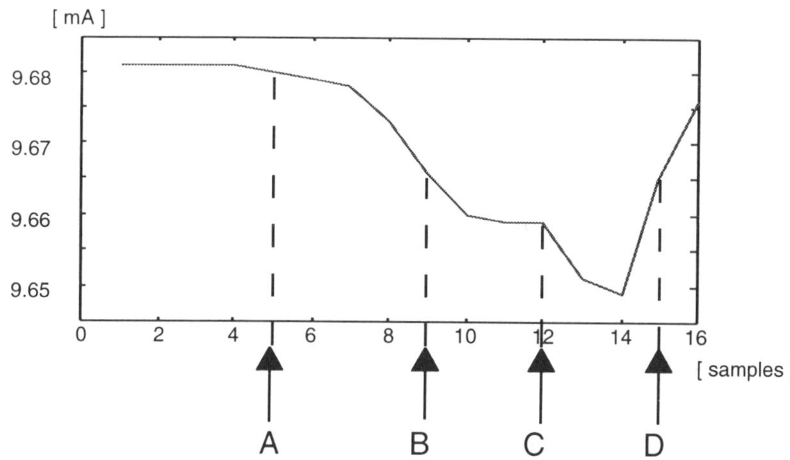

Fig. 8. Output for the inductive sensor for the determination of the height to cut the copper winding. Point A is the height for the upper winding, B is the height for the stator iron, C marks that the sensor has switched directions and D is the height for cutting.

To check that the removal of the copper windings in the stator slots is successfully performed an inductive probe is used. The inductance for the coil used in our experiments has an inductance of 342.4 μH as it is placed near the iron of the stator slot, with copper in the slots on either side of it. As the coil is placed near a slot with copper inside the inductance is 341.6 μH. When copper is removed and the sensor is placed near the stator iron the inductance is 341.7 μH and as it is placed near an empty slot the inductance is 338.6 μH. The output from the sensor for a test of a badly disassembled motor can look as the signal in figure 9. Point A indicates the output for an empty slot, the sensor is moved to the iron point B and so on until point C where the expected notch for an empty slot is exchanged by the output of the copper.

The figures in the result section show that the inductive probe can be used both to determine the correct height to cut the winding and to check that the copper is properly removed from the stator slots. One however has to comment that the sensor signal is rather small and that the sensor has to be carefully positioned to get correct signals. More advanced signal handling can take care of some of these drawbacks but the user should hold in mind that the position and orientation of the inductive probes are important.

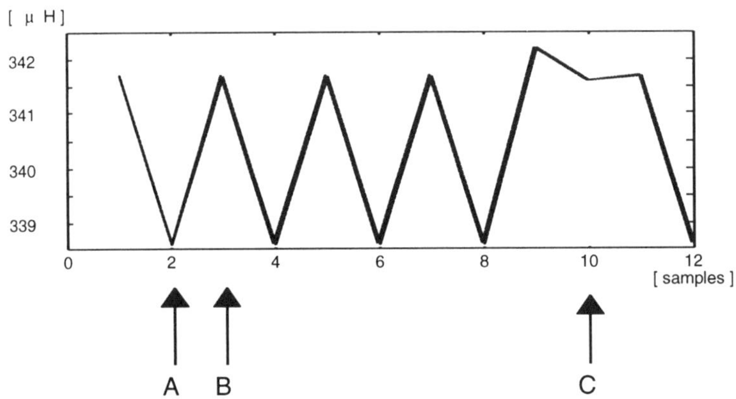

Fig. 9. Output from the inductive probe when it is placed at the iron (point A) and at a slot with air (point B). Point C indicates that it is copper in the slot instead of air.

Acknowledgement

This work has been performed at the Laboratory for Measurement Technology at Linköping University. We take this opportunity to thank technician I. Grahn for fruitful discussions and for building the sensor system. The work is a part of a research project "REMPRODUSE-Cu" in the Programme Environmental and Climate 1994-1998, sponsored by EU Commission, whose support is gratefully acknowledged.

References

1. For a review see B. Karlsson and A. Lauber, The role of measurement in industrial recycling, Proc IEEE-IMTC/96 - IMEKO TC-7, Brussels, Belgium, June 4-6, 1996 pp 1491-1494.
2. P Norrthon, En studie i kretslopp. Hinder och möjligheter för ökad material återvinning (A Study of Recycling), Ecocycle Commission of the Swedish Ministry of Environment, February, 1996.
3. International standard CEI/IEC 34-2: 1972
4. C. Rossi, A. Tonelli, On-line estimation technique for electrical and mechanical parameters of DC and AC motors, *EPE 91/4th European conference on power electronics and applications*, Firenze, September 1991.
5. A. Pouliezos, G. Stavrakakis, C. Lefas, Fault detection using parameter estimation, *Quality and reliability engineering international*, Vol 5, 283-290, 1989.
6. D. H. Libby, Introduction to electromagnetic nondestructive test methods, John Whiley & Sons, Inc. 1971.
7. D H. Ballard et al Deictic Codes for the Embodiment of Cognition, Computer Science Department, University of Rochester, accepted for publication (Copyright 1996: Cambridge University Press)
8. M. Zimmermann, The Nervous System in the Context of Information Theory, *R.F Schmidt & G. Thews (eds.): Human Physiology, 2nd ed.*, Springer-Verlag, Berlin, 1989, pp 166-173.
9. P. Wide and D. Driankov, A fuzzy approach to multi-sensor data fusion for quality profile classification, *Proc IEEE/SICE/RSJ*, Washington D.C., U.S.A., December 8-11, 1996.

AUTOMATED VISUAL INSPECTION APPARATUS FOR CIM SYSTEMS

Dr G Bright
brightg@eng.und.ac.za

Mr R Mayor
mayorj@eng.und.ac.za

Department of Mechanical Engineering
University of Natal, Durban
Private Bag X 10, Dalbridge, 4014, South Africa

Abstract. Automated visual inspection is vital in the production of manufactured products acceptable to world market standards. The volatile nature of the consumer markets is placing increased emphasis on manufacturers to produce cheaper products of higher quality. This paper proposes an unique, cost effective automated visual inspection apparatus suitable for application in the computer integrated manufacturing environment. The inspection apparatus utilises specifically designed camera positioning and part manipulation systems to position a single digital camera at known view points within its inspection envelope. View planning algorithms can then be implemented to perform three dimensional part inspection utilising inspection by multiple views techniques.

1 Introduction

The development of automation technologies has been highlighted as an area of research since the late 1960's [1]. The evolution of group technology (GT), flexible manufacturing systems (FMS), computer aided design and manufacturing (CAD / CAM) and computer integrated manufacturing (CIM) has delivered the technology of manufacturing to the brink of the fully automated factory of the future.

A major retarding factor in the development of the fully automated manufacturing environment has been the markedly slower automation of the quality control and inspection tasks [2]. This paper proposes an innovative, cost effective automated visual inspection apparatus that could be suitable for application in the automated CIM environment. Newman *et. al.* [1] report that the implementation of *in-process verification* techniques, i.e the inspection of the part at multiple steps in the process to assure the operation has been done and done correctly, dramatically reduces the reworking and scrapping costs typical to most manufacturing processes. The inspection apparatus should, therefore, be able to integrate simply and easily with the CIM environment thereby monitoring the development and quality of the part throughout the entire manufacturing process.

This paper proposes the design of unique, dedicated camera positioning and part manipulation subsystems based on established, cost effective mechatronic principles.

Through the coordinated interaction of these two subsystems, the inspection apparatus should be able to position a single digital camera at any arbitrary view point within its *inspection envelope.* This ability to position the camera at known view points should then allow, with the appropriate image processing, the development of reliable three dimensional models of the inspected parts [3].

2 Design Criteria For The Inspection Apparatus

The design of the inspection apparatus was focused by a set of predetermined design criteria. The criteria were adopted to ensure that the inspection apparatus would be able to interact with a flexible manufacturing system that is currently under development at the University of Natal, Mechanical engineering's Mechatronic department. The criteria are listed below:

- The apparatus must be capable of acquiring an image of all 6 major faces of a manufactured part using a single CCD camera. The major faces of a manufactured part (ref Fig 2.1) are defined as top, left side, front side, right side, rear side and bottom, in order of inspection.
- The apparatus must be able to accept a pallet from an entry conveyor at height 0.8 m, manipulate the pallet in any manner necessary for a complete inspection and then return the pallet to an exit conveyor at height 0.8 m. The pallet must be able to accept a pallet of at least 200 x 200 mm in base, but no greater that 325 x 325 mm in base.

3 Conceptual Development Of The Apparatus

The final design concept for the inspection apparatus evolved through an iterative process of conceptual design and critical analysis. This section discusses the evolution of the final design for the inspection apparatus, concentrating on the respective disadvantages and failings of the preliminary design concepts and how these were improved upon in the next concept.

The Gantry System. The initial design concept (ref Fig 3.1 (a)) for the inspection apparatus station was based on a gantry type system similar to those used in coordinate measuring

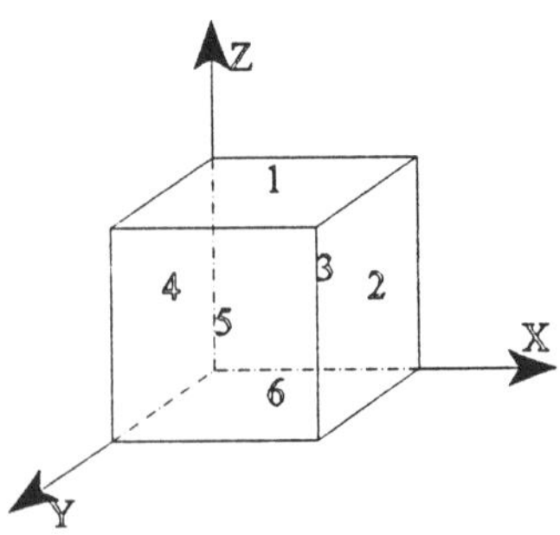

Figure 2.1. A figure defining the six major faces of a 3D part. The coordinate system, XYZ, is also defined where 1 is the top face and 5 is the front face (in the direction of part flow)

machines (CMM's). A large gantry construction was suggested to achieve the required three dimensional positioning of the camera. The disadvantages of the gantry based system are listed below.

- The gantry system could not position the camera below the table rendering the system incapable of acquiring an image of the bottom face of the part.
- The camera positioning system required the movement of a significant structural element, the gantry. The overall mass of the components of the gantry system would cause significant inertial loading on the drive motor and therefore the degree of precision required to correctly position and orientated the camera could not be achieved.
- The risk of damaging the part after the inspection had occurred could not be removed.

The Bridge System. The revised concept (ref Fig 3.1 (b)) replaced the gantry system with a bridge construction. The bridge construction was rigidly attached to the frame of the system and was vertically aligned with a glass / perspex viewing window introduced to the inspection table. The viewing window allowed for the acquisition of an image of the bottom face of the inspected part. A system of flat belt conveyors was introduced in order to control the flow of the inspected parts through the apparatus. The conveyor system was capable of accurately positioning the inspected part in the plane of the camera and therefore the camera positioning system was rationalised to vertical (Z) and horizontal (X) placement in the XZ plane. The improved bridge based system did however have two major failings.

- The bridge system was not capable of capturing all six views. Faces 3 and 5 were not accessible (ref Fig 2.1)
- The movement of the camera along the inner face of the bridge would have required a complex drive mechanism.

The Rotating Cylinder System. This concept (ref Fig 3.1 (c)) considered the replacing of the bridge structure with a cylinder structure to which the camera could be rigidly mounted. With the camera rigidly attached to the cylinder, the camera could be positioned at the four stations indicated in figure 3.1 (c) simply by rotating the cylinder. This rotation of the cylinder could be easily achieved using a motor and a gearing system. A pinion gear could be mounted on the shaft of the motor which would engage the appropriate gear mounted on the outer surface of the cylinder. The costs involved in the fabrication of the rotating cylinder and the machining of the outer gear would have been prohibitively expensive.

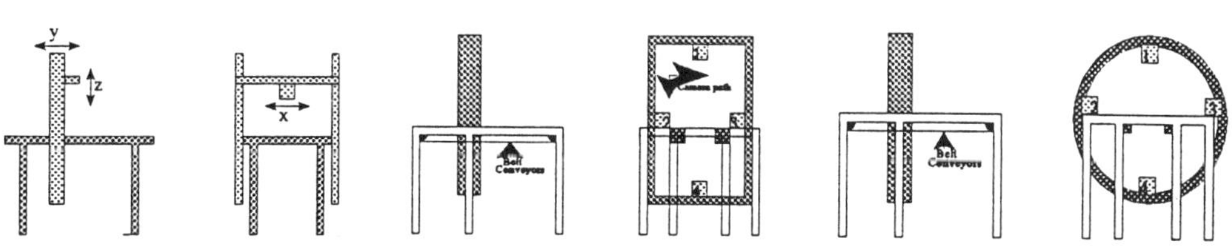

(a) (b) (c)

Figure 3.1. Schematic illustrations of the Inspection apparatus design concepts. (a) the Gantry concept, (b) the Bridge concept, (c) the Rotating Cylinder concept

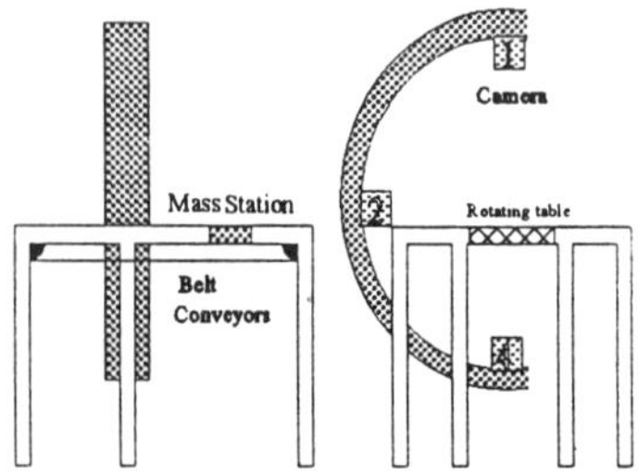

Figure 3.2. The final concept for the inspection apparatus unit, showing the incorporation of the rotary table and the mass station.

The Final Design. The final design concept (ref Fig 3.2) introduced a rotary actuator to the inspection table that enabled the inspected part to be rotated through 360° in the plane of the table. The ability to rotate the part through any angular displacement in the XY plane enabled the acquisition of images of the previously obscured faces of the inspected part; faces 2, 3 and 4. Through the interaction of the rotary inspection table and the camera positioning system (explained in more detail in section 5), the inspection apparatus was now able to fully satisfy the design criteria as presented in section 2.

4 Component Systems Of The Inspection Apparatus

The structure of the inspection apparatus comprised of three interdependent systems; the part positioning system, the part manipulation system, and the camera positioning system.

4.1 The Base Frame

The base frame of the apparatus was designed to provide a stable foundation on which the three principle sub-systems could be mounted. The major focus in the design of the frame was to enhance the reliability and validity of the data collected during the image acquisition process. Key aspects considered were the deformation of the inspection table under typical loading conditions and overall vibration absorption characteristics of the frame. The finite element analysis package, *COMOSM*, was used to investigate the deflections of the frame and modifications to the design were made as required. A list of design parameters used in the design of the base frame for the inspection apparatus is presented below.

- Max. pallet size : 325 x 325 mm
- Max. pallet load = 20 kg
- Max. permissible deflection under max pallet load = 0.001 m
- Mass of the trolley track = 30 kg

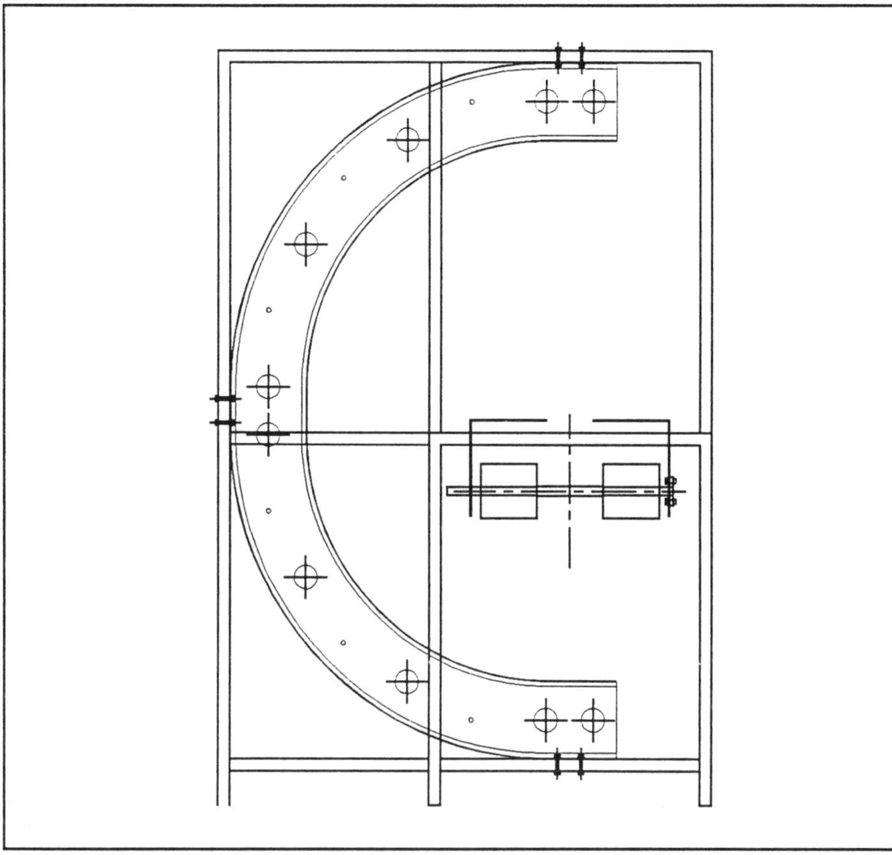

Figure 4.1. A side elevation of the inspection apparatus.

4.2 The Camera Positioning System

Any part recognition and inspection processes performed by the inspection apparatus would be dependent on the information contained in digital images of the six principle views of a part. The acquisition of these images required the development and design of a camera positioning system capable of positioning the digital camera at three principle positions (ref Fig 3.2); top dead centre (TDC), side dead centre (SDC) and bottom dead centre (BDC). This necessitated the design of an unique trolley track system in which the camera is rigidly mounted onto a trolley running in an I-beam type track of constant curvature (ref Fig 4.2). The trolley is positioned in the track using a chain drive system powered by a 12 V DC motor delivering 20 Nm of torque at 75 rpm, capable moving the camera from station to station (eg. TDC to BDC) in a traverse time of 5.4 seconds.

The trolley was designed to be flexible with respect to the type of digital camera used. The maximum permissible trolley load was set at 10 kg. The deflection of the camera focal axis had to be restricted to ensure that the variance in the focal point of the camera was minimised. An upper limit of 2° was set for the allowable angle of deflection, ensuring that the focal point of the camera remained within 17 mm of the true centre of the part (ref Fig 4.3). The image recognition system was then able to operate in an non-fuzzy environment in which the position of the part with respect to the camera was precisely known.

4.3. The Part Positioning And Manipulation System

The part positioning system. The pallet transport conveyor controlled and regulated the movement of the inspected parts through the inspection apparatus. The conveyor was driven by a 12 V DC motor running through a 2:1 reduction pulley system that presented the conveyor shaft with 40 Nm of torque at 35 rpm. The pallet transport conveyor was then able to perform a full traverse, from entrance to exit, in 4.8 seconds.

The part manipulation system. The rotary inspection table allowed for the rotation of the pallet required to acquire images of the three obscured sides. The table was designed in order to enable the rotation of the inspected part about a vertical axis passing through the centre of the inspection area. The design of the rotating table necessitated the design of a pallet that could interact with the table and prevent slippage of the inspected part during rotation. A chamfered step was machined into the top surface of the rotating table that introduced a self-locating capability that countered small variances in pallet delivery inherent in the pallet transport conveyor system.

4.4 Lighting Technique

The inspection apparatus utilised a compound lighting format which included peripheral lighting and high intensity, directional front lighting. The peripheral lighting was provided by four corner mounted spot lights. The spot lights were arranged such that the individual incident beams converged at the geometric centre of the inspection area in order to reduce the degree of shadow casting [4]. The high intensity front lighting was provided by a circular tube florescent lamp mounted on the camera positioning trolley. The camera was mounted concentrically (with respect to the circular tube) on the trolley. The directional capability of the high intensity front lighting system ensured a constant incident light at every camera position.

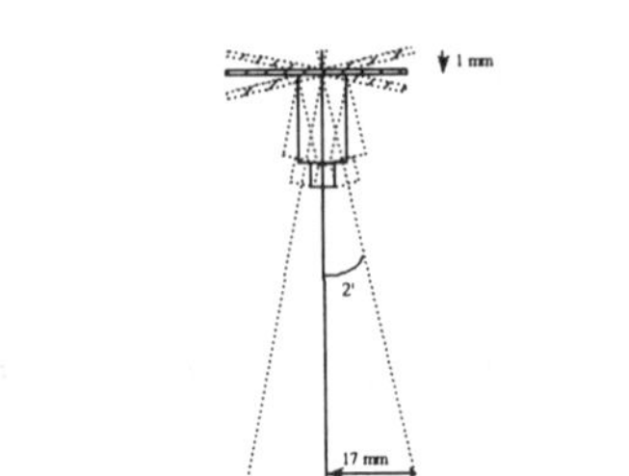

Figure 4.3. A schematic diagram illustrating the defection of the camera's focal axis resulting from a 1mm defection of the camera mounting plate.

5 Typical functioning of the Inspection Apparatus

This section details the typical functioning of the inspection apparatus, explicitly defining the interaction of the camera positioning system and the part manipulation system. A typical inspection routine would begin with the transportation of the inspected part to the rotary inspection table by the pallet transport conveyor. The part, at this stage, would then be positioned at the geometric centre of the *inspection envelope*. The inspection envelope is a complex geometric surface that defines all possible camera viewing positions with respect to the inspected part. The envelope is described mathematically in equation 5.1, where r is the radius of curvature, ϕ is the angle of rotation from the Y axis and φ is the angle of declination from the Z axis (ref Fig 5.1). The radius of curvature, r, is a constant defined by the geometry of the I-beam camera track. However, both the angle of rotation, ϕ, and the angle of declination, φ, are variable within the limits indicated in equation 5.1.

$$\begin{aligned} r &= const \\ 0 < \phi &< 360 \\ 0 < \varphi &< 90 \end{aligned} \qquad \ldots\ 5.1$$

The camera positioning system was capable of positioning the camera at any specified angle of declination, φ. Similarly the part manipulation system was capable of positioning the part at any specified angle of rotation, ϕ. The coordinated interaction of the camera positioning system and the part manipulation system could therefore position the camera at any viewing point within the inspection envelope as defined by equation 5.1.

6 The Apparatus Control System

The design of the inspection apparatus control system was focused by the identification of two principle objectives; the control system needed to be computer based and the control system had to ensure the smooth and coordinated functioning of the inspection apparatus. This necessitated the control of three drive motors and the collection of data from the respective three position feedback circuits.

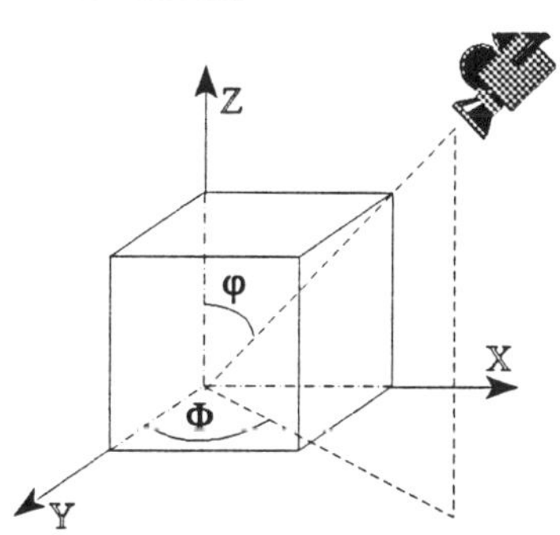

Figure 5.1. A figure defining the angles of declination, φ, and rotation, Φ, for an arbitrary camera viewing point.

The Camera Positioning System. In order to ensure the required degree of precision in the positioning of the camera, a two speed traverse system was adopted. For the last 50 mm on approach and the first 50 mm on exit from any of the three stations, the systems DC motor, MTR1, was driven with half its rated voltage (ie 6 Vdc). Outside of these limits the motor was driven with its full rated voltage (12 Vdc). Through this voltage switching, a rudimentary ramping of the motor speed was achieved and consequently the degree of overshoot in the trolley positioning was reduced. Position feedback information was gathered using optical sensor based electronic circuitry incorporated on the camera track.

The Part Positioning System. The pallet transport conveyor system was driven by a DC motor, MTR2, and the position of the pallet along the conveyor was determined by a limit switch feedback circuit. The rudimentary speed ramping system used in the camera positioning system was again adopted in the part positioning system to improve the accuracy and precision of the pallet positioning.

The Part Manipulation System. Once the inspected part had been positioned in the plane of the camera, part rotation was still required to present the obscured faces of the inspected part to the camera. The part manipulation system used a 12 Vdc DC motor, MTR3, to rotate the inspection table. Angular position of the table was monitored using a rotary optical encoder mounted on the shaft of the motor. The feedback signal, PRF1, was used to monitor the angular position of the table.

A further set of control signals was required to control the Inspection apparatus's pneumatic circuits, responsible for raising and lowering either the mass station or the inspection table. Two control signals, PNM1 and PNM2, were used to activate the pneumatic circuits. Table 5.1 summarises the control signals and feedback signals used to control the inspection apparatus.

CONTROL AND FEEDBACK SIGNAL CHART		
SYSTEM	CONTROL ACTION	SIGNAL
Camera Positioning System	MTR1 speed 1	CPC1
	MTR1 speed 2	CPC2
	MTR1 direction	CPC3
	Camera position feedback	CPF1
Part Positioning System	MTR2 speed 1	PPC1
	MTR2 speed 2	PPC2
	Pallet position feedback	PTF1
Part Manipulation System	MTR3 speed 1	PMC1
	Pallet Rotation feedback	PRF1
Pneumatic Circuits	Activate circuit 1	PNM1

CONTROL AND FEEDBACK SIGNAL CHART		
SYSTEM	CONTROL ACTION	SIGNAL
	Activate circuit 2	PNM2

Table 5.1. Control signal and feedback signal chart listing all the signals used in the Inspection apparatus control system. A definition of the 4 bit code words used in the table follows:

MTR	:	DC motor number
CPC	:	Camera positioning system Control signal
CPF	:	Camera positioning system Feedback signal
PPC	:	Part positioning system control signal
PTF	:	Pallet transport conveyor feedback signal
PMC	:	Pallet manipulation system control signal
PRF	:	Pallet rotation feedback signal
PNM	:	Pneumatic circuit control signal

The control system has been coded using Microsoft Visual C++ , version 4.1. Communication between the processor and the Inspection apparatus was achieved using the standard Input / Output card via D-type 25 pin parallel port. The DC motors were controlled from the computer using an application specific driver circuit [6]. The circuit was designed to use control signals sourced from the parallel port to switch relays in the driver circuit to change voltage supply and polarity to the motor. The driver circuit enabled the computer to select direction and perform crude, two step speed control.

7 Conclusion

An automated visual inspection apparatus was developed using cost effective, off-the-shelf technologies based on mechatronic principles. The inspection apparatus was developed as a computer based technology and therefore was suitable for integration into a CIM environment. The inspection apparatus was incorporated into a low cost CIM system, still under development at the University of Natal, Durban, and has enhanced the effectiveness of that system.

Specifically designed camera positioning and part manipulation systems were developed that enabled the inspection apparatus to position the single digital camera at known view points within the apparatus' inspection envelope. The control resolution for both the camera positioning system and the part manipulation system was such that both the angle of declination, φ, and the angle of rotation, Φ, could be incremented in steps of $5°$. The computer based control system was able to smoothly coordinate the functioning of the two systems, accurately and precisely positioning the camera at specified view points within the limits imposed by the control resolution of the motors.

Through this interaction of the camera positioning system and part manipulation system, the inspection apparatus was suitable for use as a test bed for the implementation of view planning algorithms and inspection by multiple view techniques. Further research is being conducted in order to further investigate the development of reliable and accurate 3D models of inspected parts by implementing view planning algorithms. Research will include investigations into different methods of archiving such data and sharing it with the CIM environment.

8 References

1 T.S. Newman, and A.K. Jain, A Survey of Automated Visual Inspection, *Computer Vision and Image Understanding*, Vol. 61, No. 2, March 1995, pp 231-262

2 M.P. Groover, Automation, Production Systems, and Computer Integrated Manufacturing, *Prentice Hall International*, 1987.

3 W. Brent Seales and O.D. Faugeras, Building Three Dimensional Object Models From Image Sequences, *Computer vision and Image Understanding*, Vol. 61, No. 3, May 1995, pp 308-324

4 L. Van Gool, P. Wounbacq and A. Oosterlinck, Intelligent Robotic Vision Systems, *Intelligent Robotic Vision Systems* (S.G. Tzafestas, Ed.), Dekker, New York, 1991, pp 457-507

5 A.D. Marshall, The Automatic Inspection Of Machined Parts Using Three-dimensional Range And Model-based Matching Techniques, *Ph.D. dissertation,* Department of Computing Mathematics, University of Wales, College of Cardiff, 1989.

6 G. Kaplan, Industrial Electronics, *IEEE Spectrum*, Vol. 31, No. 1, 1994, pp 74-76

Non - Destructive Testing for Thin - Film Magnetic Rigid Disks

A. Wichainchai, P. Nabokin, S. Rukvichai, and Ch. Puprichitkul*

Magnetic Research Lab. , Applied Physics Department, Faculty of Science, King Mongkut's Institute of Technology Ladkrabang, Bangkok 10520, Thailand.

*Failure analysis Department, Seagate Teparuk Thailand Ltd. Co., Thailand

Abstract. An inexpensive non-destructive testing system was designed and built to support magnetic characterization of longitudinal recording thin-film rigid disks for hard disk drive application. The instrument's principle is magneto-optic Kerr-effect measurements of the magnetic film M-H hysteresis loop. These measurements being combining with ordinary testing by reading/writing head provide the data on rigid disk quality and the sources of read/writing failure. The system operates in AC mode with magnetic field amplitude up to 4 kOe and rise/fall times more than 1 ms. The magnetic field parallel to the plane of disk, is produced by a pair of "big" magnetic heads (small electromagnets) supplied with triangular current pulses. The testing program, disk positioning and data processing are computer controlled. The main system's features are: 2% - relative magnetization resolution, 1% and 2% - repeatability and accuracy of coercive force H_c determination for H_c values (500 - 3500) Oe, comparing with commercial systems' (1 - 5)% absolute accuracy for H_c.

1 Introduction

Magnetization - field, M-H hysteresis loop is the most important characteristic curve for magnetic recording thin-film media, providing the major parameters of the media: coercive force, remanent magnetization, squareness' S and S^* [1]. Magnetic recording has reached to the density of 1 Gbits/inch^2 and is expected to grow to higher densities [2]. Writing difficulties due to the high coercivity of disks and the effect of thermal relaxation for high density longitudinal recording media result in vital necessity of testing the thin-film disks magnetic parameters: coercivity, magnetization and S^*. Non-destructive measurements of M-H loops are needed both at thin- film disks manufacturing and the hard disk components testing for assembling and read/write failure analysis.

Some instruments are available now for non-destructive M-H measurements. For rigid disks of any diameters the most interesting commercial proposals are of Digital Measurements Systems, Inc. [3] and Innovative Instrumentation Incorporated [4]. The first is the magneto-optic Kerr effect utilizing machine operating in DC modc (3 seconds field sweep time) with high power 4 inches pole pieces electromagnet producing magnetic field up to 13 kOe in the 5.25 inches gap. The second is the magnetic head instrument utilizing scaling principle : the recording a strip of oppositely directed magnetization by head with head gap and head - media separation four orders of magnitude bigger that of an actual recording head, then examining the transition region at the ends of this strip by rotating the disk at low speed (typically 12 rpm) past a sensor. Both systems have fine technical specifications but they are very expensive.

We will describe here AC-mode operating system which we have built for non-destructive testing of rigid disks. The setup is based on the magneto-optic Kerr-effect examining of the film magnetization at checked point (region) of the disk under remagnetizing of this region with AC magnetic field produced by symmetrical pair of big magnetic heads.

2 Technique

Schematic diagram of the test set is shown on Fig.1. Planopolarized laser's light beam is splittered by a beamsplitter onto two : one - to the tested sample, then through absorbing filter (or polarizer, for equalizing light beams intensities) to #1 photodetector, another - to #2 photodetector directly. The photodetectors signals are amplified by differential amplifier with subtracting of low frequency additions due to the laser noise and the system mechanical vibrations (splitting of the beam and differential detection doesn't principle, it depends on : 1) desirable ΔM resolution, 2) how large is the laser noise and 3) how well, about vibrations, the system is constructed and placed). The amplifier's output signal proportional to the sample's magnetization in the region of the light beam spot is fed into Y input of a digital oscilloscope operating in average XY mode with X input signal, from magnetic field sensor, proportional to the magnetic field acting on the sample at the checked point. The averaged M - H loop is stored by digital oscilloscope and / or computer processed or copied by a printer.

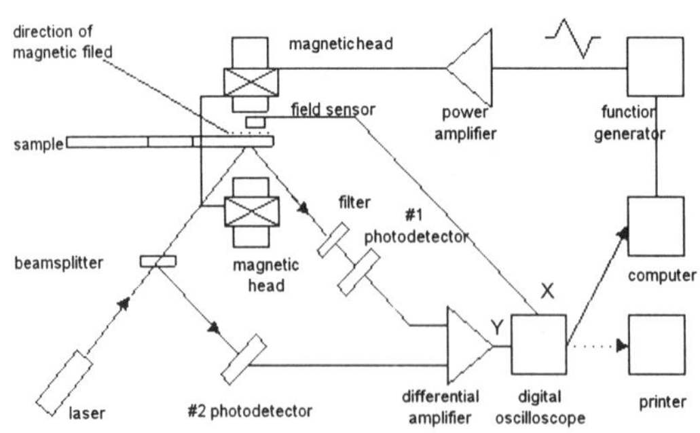

Fig. 1. Schematic diagram of the test set

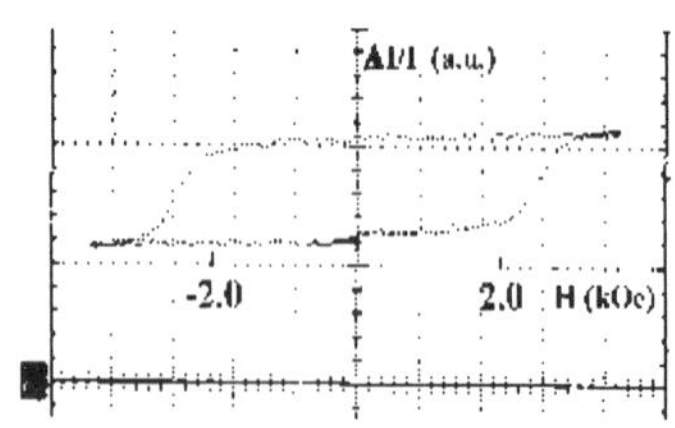

Fig. 2. Transverse Kerr-effect M-H loop recording of rigid disk

Transverse Kerr effect is used for M-H curves recordings : the change of the light beam intensity reflected by magnetic surface is proportional to the component of magnetization perpendicular to the plane of the light incidence. The incident angles $\theta \geq 50°$. Light beam spot at the reflecting disk surface is elliptical in shape with main axes about 1.5 mm x 1.0 mm. Longitudinal Kerr-effect geometry can also be used in our system.

The most serious problem for nondestructive magneto-optic testing is the high amplitude magnetic field creation. The problem is solved for DC-mode systems by use of big electromagnet with placing of the testing rigid disk in the inter-poles gap [3]. Such Kerr-effect system for longitudinal recording disks is needed in very high power and high stable programmable DC source for the electromagnet (it is not the ordinary problem to create (5 - 10)kOe in the gap 10cm) and is needed in good dampers for mechanical vibrations' insulation (because single M - H loop is recorded, without averaging). Conventional recording magnetic heads which are used for data writing can not be used with optical testing systems because of their extremely small gaps and heights of flights (about 0.3 μm and 0.03 μm respectively).

We have built symmetrical system of big magnetic heads ≡ small electromagnets, with soft steel (transformer sheets) cores of about 20 mm^2 cross section area, 2 mm head gap and 4 mm inter-heads poles separation. The coils of the heads system, 4 Ohms total resistance in series, are supplied with alternating current from function generator through 300 W power audio amplifier. Magnetic fields up to 4 kOe in amplitude, triangular pulses with rise/fall times (10-1) ms, were produced by the system at the coils currents less than 10 A in amplitude. Repetition rates about $(100 \text{ x pulse width})^{-1}$ were used to operate without the coils overheating.

A small sensitive coil located at the inter-poles gap, with linear integration circuit, is used as AC magnetic field sensor. Correction for the checked point, which is the light beam spot at the surface of the disk, was obtained during calibrating procedure with additional sensitive coil located at the checked point without disk and calibrated by use of standard disks. The magnetic field homogeneity in the operating volume, being specified better than 2 % , was checked by high precision moving of the additional coil through operating volume (steps : 10 μm in Z direction, perpendicular to the disk plane, and 100 -150 μm in X-Y directions).

3 Results

An example of M-H loop recording for rigid disk with Co-Cr-Pt magnetic layer (300 Å thickness), covered by diamond like carbon film (200 Å thickness) and lubricant film (25 Å thickness) is shown on Fig. 2. The recording was made in transverse KE geometry, $\theta = 50°$, H parallel to the easy axis , one light beam optical scheme, with laser diode light source (λ = 670 nm, beam power 2 mW, low frequency noise $\simeq$ 0.5 %), field rise time 1.5 ms, 256 sequence averaging mode, at ordinary third floor room with undamped equipment, without additional antivibration efforts. The total change of the reflected light intensity corresponding to the magnetization reversal (from $-M_s$ to $+M_s$) is $2\,\Delta I / I_{refl} = 2 \times 10^{-3}$, M_s is the saturation magnetization.

The main systems specifications (for the two-beam optical scheme) are as following :

$\Delta M / M_s$	resolution	2%
coercive force :	repeatability	1%
	accuracy	2% relative to standard disk
magnetic field amplitude		up to 4 kOe
test time :		1 sec for 1 point
		30 sec for disk change
measurement area		1.0mm x 1.5mm
disks diameters		90 , 65mm (larger and smaller sizes possible)
capability for measurements		both H ll. easy axis and H ⊥ easy axis
determined parameters		$M_{remanent} / M_s$, coercive force, S*

The system has similar to [3], [4] magnetometers features (for example, 2% accuracy of coercive force determination compared to 1% [3] and <5% [4]) with total cost of the equipment which the setup consists of about one order less.

References

1. J. E. Manson , Recording measurements, in *Magnetic Recording Handbook: Technology and Applications*, edited by C. D. Mee and E. Daniel, Mc Graw-Hill, Inc.(1990), pp. 396-399.
2. D. N. Lambeth, E. M. Velue, G. H. Bellesis, L. L. Lee, and D. E. Laughlin, Media for 10 Gb/in.^{2} hard disk storage: Issues and status, *J. Appl. Phys.* **79**, 8 (1996) 4496-4501.
3. Kerr - Effect Magnetometer System, Model 660 L - KEM, *Rev. Sci. Instr.* **63**, 4 (1992) p 2347.
4. R. M. Josephs, Remanent Moment Magnetometer, *US Patent* # 5 136 239 (1993).

Digital Filtering of Images Using Compact Optical Processors

Yu-hua Li, Yi-mo Zhang, Wen-yao Liu and Hong Hui

The Faculty of Precision Instruments and Photoelectrons Engineering
TianJin University, TianJin 300072,P.R.China
Tel: (086)022 27404459, Fax: (086) 022 27404547

Abstract. Based on the optical 4-f system, designing the compact automatic 2-D target classification implementation. The length of the system can be reduced to about 1/10 of original system length by using two Fourier transform telephoto lenses and an adding phase compensating lens in a 4-f optical system. The focal length of the phase compensating lens is calculated using optical theory. Lastly, the result of a computer simulation with four - kinds of aircraft classification is given.

1 Introduction

Automatic on-line 2-D multi-target classification, which is an important branch of pattern recognition, has been researched in many practical application fields for nearly two decades[1-4]. However, a satisfactory result has not been obtained yet. Its application is limited by four main factors: 1) the precision of classification; 2) the abilities of error-tolerance; 3) the requirement of real-time classification; 4) the size of system. The techniques of traditional Computer target classification can meet the factors of (1) and (4) [5-6], but its consuming time will be much longer than our expecting, especially when the shapes of the target are very complicated. Compared with Computer 2-D data processing, optical image processing has many advantages due to optical features, such as high speed, amass-parallel processing model, large capacity etc.. Therefore, Combining them with the hybrid opto-electronic hybrid processors have been proposed. It will have a future in industry automation, missile guidance, fingerprint identification, and security applications etc[7-8].

The initial efforts and developments are directed at using it for large space-bandwidth product systems, with emphasis on the large information-handling capacities of such systems[9-10]. Its typical structure is shown in Fig.1. It is based on an optical 4-f system that is used for optical information processing. In Fig.1 f is the focal length of the Fourier transforming lenses. The outside object is collected by CCD and as an input pattern f(x,y) is placed in plane P_1. The 2-D Fourier transform F(u,v) of f(x,y) is taken by the Fourier lens L_1 and is presented in plane P_2. The matched spatial filter, which represents the 2-D Fourier transform G*(u,v) of a test pattern g(x,y), is placed in plane P_2. The transmitted light through plane P_2 consists of the product F(u,v)G*(u,v) and is Fourier transformed by the lens L_3. The correlation of f(x,y) and g(x,y) is then formed in plane P_3, and the outside object will be classificated — the basic method of optical pattern recognition.

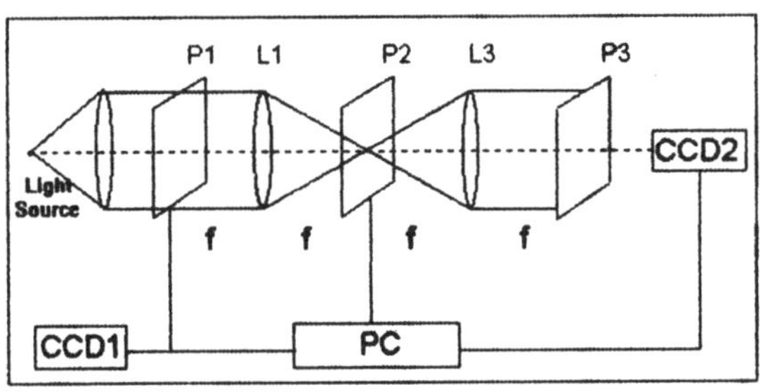

Fig.1 The typical structure of opto-electric hybrid 2-D target classification system

This optical system has a physical length equal 4f, and this physical length is often unacceptably large in its many practical application fields. Meanwhile, these opto-electric hybrid processors required high quality optical components with system stability requirements that are best met only in a laboratory environment. These requirements limit the development of hybrid opto-electronic element in its application fields.

In this paper, we propose the new-style optical construction of real-time compact optical processors to be used for automatic 2-D target classification. It is based on an optical 4-f system, and in order to reduce the system length, a five-lens optical system is designed. As a result, total optical system length will be compressed about 10 times. This system will serve as a test bed to further determine applications for which processors could be useful.

2. The Theory of Compact Optical Processors Designing

2.1 Fourier Transform Basic Properties of Lenses and Optical 4-f System

It is very useful that a 2-D Fourier transformation may be obtained from a positive lens[11-13]. The simple optical system is shown in Fig.2. In Fig.2, P_1 as a planar light source, P_2 as output plane of Fourier transformation lens. Let us assume that the complex light field at P_1 is f(x,y). Then the complex light distribution at P_2 $g(\alpha,\beta)$ may be determined by means of Huygen's principle:

$$g(\alpha,\beta) = C\iint_{S_1} f(x,y)\exp(ikr)dxdy \qquad (1)$$

where S_1 denotes the P_1 surface integral, C is an arbitrary complex constant, and

$$r = \left[L^2 + (\alpha - x)^2 + (\beta - y)^2\right]^{1/2} \approx L + \frac{1}{2L}\left[(\alpha - x)^2 + (\beta - y)^2\right] \qquad (2)$$

At last, we can obtain:

$$g(\alpha,\beta) = C_1 \exp\left[-i\frac{\pi}{\lambda f}\left(\frac{1-v}{v}\right)\left(\alpha^2 + \beta^2\right)\right] \times \iint_{S_1} f(x,y)\exp\left[-i\frac{2\pi}{\lambda f}(\alpha x + \beta y)\right]dxdy \qquad (3)$$

where v = f/L, C_1 is another arbitrary complex constant.

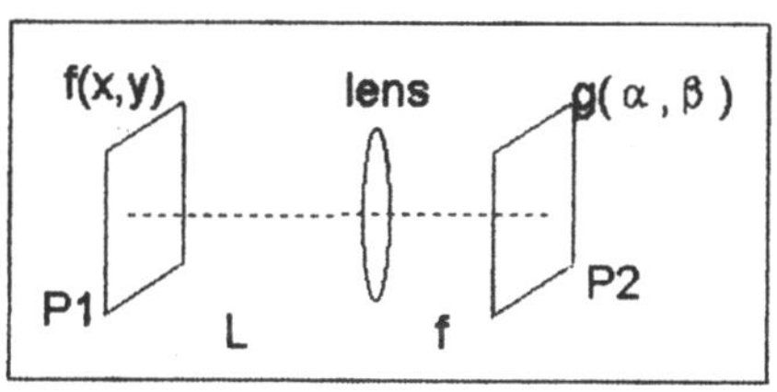

Fig.2 Geometry for determination of the optical Fourier transformation

From equation (3), clearly, except for a spatial quadratic phase variation, g(α , β) is the Fourier transform of f(x,y). In fact, the quadratic phase factor vanishes if L = f. Furthermore, it can easily be shown that a quadratic phase factor also result if the signal plane P1 is placed behind the lens, and the quadratic phase factor vanishes too if f = L. It is a typical optical 4-f system.

2.2 The Phase Transformation of Thin Lenses and the Theory of Designing Shortened Optical 4-f System

Optical 4-f system has a physical length 4f, and the sizes of the system seem very long for its application. Therefore, one of the important issues is the miniaturization of the dimension of optical processors to meet the requirement of available space in a particular application. Meanwhile, the sizes of the 2-D Fourier transform of an input object that is produced optically must exactly match the dimension of the 2-D Fourier transform which is stored on spatial filter. The focal length f of a Fourier lens must satisfy:

$$f = \frac{N_2 d_1 d_2}{\lambda} \tag{4}$$

where d_1 and d_2 are the pixel spacing of the two planes, N_2 is the number of pixels in the second plane, and λ is the wavelength used.

In the equation (4), the focal length f has been fixed. The choice of lenses and input, output plane spacing has a significant effect on processor's size. The consequence is that the exact 2-D Fourier transform is performed by the lens but is multiplied by a quadratic phase front according to Fourier transform basic properties of lens. Thus, another method of designing shortened processors should be considered.

We know, lenses are made of glass or some other transparent material. When a light ray passing lens, accord the theory of optical information processing, the phase transformation can be shown to be:

$$T(x, y) = C_2 \exp\left[\ i\frac{2\pi}{\lambda f}(x^2 + y^2)\ \right] \tag{5}$$

where C_2 is a complex constant.

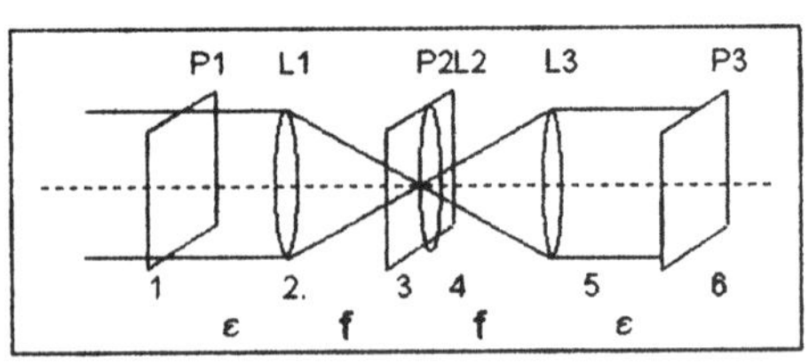

Fig.3 The structure of optical 4-f system with adding an phase compensating lens

We consider reduction sizes of the optical processors. Therefore, a new-style optical 4-f system which adding a phase compensating lens has been designed , as shown in Fig.3. In Fig.3, L_2 is an adding phase compensating lens, it immediately adjacent to behind of plane P_2. At first, let us assume that Fourier lenses L_1 and L_3 are same, its focal length is f, and the focal length of phase compensating lens L_2 is f_2. According to the theory of optical information processing, we can obtain:

$$g(x_6,y_6)=C\{\ \Big\{\Big\{\Big\{\big[f(x_1,y_1)*h(x_1,y_1)\big]T(x_2,y_2)\Big\}*h(x_2,y_2)\Big\}F(x_3,y_3)T(x_4,y_4)\Big\}*h(x_4,y_4)\Big\}\times T(x_5,y_5)\ \}*h(x_5,y_5) \qquad (6)$$

where $f(x_1,y_1)$ and $g(x_6,y_6)$ correspond to the complex light field at input plane P_1 and output plane P_3, h(x,y) is the spatial impulse response, and T(x,y) is the phase transformation of the lens. C is an arbitrary complex constant, and' * ' denotes the spatial convolution.

In the equation, let us assume that all lenses and planes are so thin that can be negligible in their thickness, and the spatial filter G*(u,v)=1. Then equation (6) may be written in the integral form:

$$g(x_6,y_6)=\iint\limits_{S1}\iint\limits_{S2}\iint\limits_{S4}\left[\iint\limits_{S5}\exp(i\frac{k}{2}\Delta)dx_5dy_5\right]f(x_1,y_1)dx_4dy_4\ldots dx_1dy_1 \qquad (7)$$

where, $\iint\limits_{S1}\iint\limits_{..}\iint\limits_{S5}$ denote the surface integrals in plane S_1, S_2, S_4, and S_5, and

$$\Delta\equiv\{\frac{1}{\varepsilon}\left[(x_2-x_1)^2+(y_2-y_1)^2\right]+\frac{1}{f}\left[(x_4-x_2)^2+(y_4-y_2)^2\right]+\frac{1}{f}\left[(x_5-x_4)^2+(y_5-y_4)^2\right]$$
$$+\frac{1}{\varepsilon}\left[(x_6-x_5)^2+(y_6-y_5)^2\right]-\frac{1}{f}\left[(x_2^2+y_2^2)+(x_5^2+y_5^2\right]-\frac{1}{f_2}(x_4^2+y_4^2)\} \qquad (8)$$

Through calculating, we can get, when $\varepsilon\neq f$, the additional quadratic phase fronts of optical system will remove too using a phase compensating lens L_2, and its focal length of L_2 satisfies equation (9)

$$f_2=\frac{f^2}{2(f-\varepsilon)} \qquad (9)$$

where ε is the distance between plane P_1 and lens L_1 (or L_3 and P_3).

When $\varepsilon=0$, $f_2=f/2$. In this condition, the plane P_1 and plane P_2 are immediately adjacent Fourier lenses L_1 and L_3. So that, the size of total optical system is compressed 2 times, and its physical length only equal to 2f.

2.3 The Telephoto Fourier Lens and Compact System Designing

Applying the basic theory of geometrical optics[14-15], we know that a telephoto Fourier lens can reduce its working distance. The telephoto lens is consisted by two lenses (a positive lens L'_1 and a negative lens L'_2), and the working distance L' of telephoto lens (the distance between first lens and the focal plane of the telephoto lens) is less than its focal length f'. We define that k'= L' / f' is its compressing ratio, and $k' \leq 1$ for the telephone lens. Therefore, we can farther shorten the size of optical system in condition of keeping the focal length of Fourier lens invariability. We can get (10) using the Principe of geometrical optics

$$\frac{1}{f'} = \frac{1}{f_1'} + \frac{1}{f_2'} - d' \frac{1}{f_1' f_2'}$$

$$x_H' = f' \frac{d'}{f_2'} \tag{10}$$

where f'_1 and f'_2 are separately the focal length of L'_1 and L'_2, and d' is the distance between them, X'_H is the distance between object main plane and lens L'_1, and f' is the focal length of telephoto lens.

As a result, we design a new-style compact optical processor, it is shown in Fig.4. Comparing with Fig.3, we replace two single Fourier lenses (L_1 and L_3) in optical system with two combinations of the positive and negative lenses (L_1 and L_4, L_3 and L_5) in order to shorten size of the 4-f optical system. In Fig.4, an input pattern f(x,y) is placed in $plane_1$ (LCD) (Liquid Crystal Display), and the matched spatial filter which represents the 2-D Fourier transform G*(u,v) of a test pattern g(x,y) is placed in $plane_2$ (LCLV) (Liquid Crystal Light Valve). This one acts as a filter by displaying a stored image that has been transformed electronically. The correlation result of f(x,y) and g(x,y) is then formed in plane P_3, where correlation peaks appear, the images match.

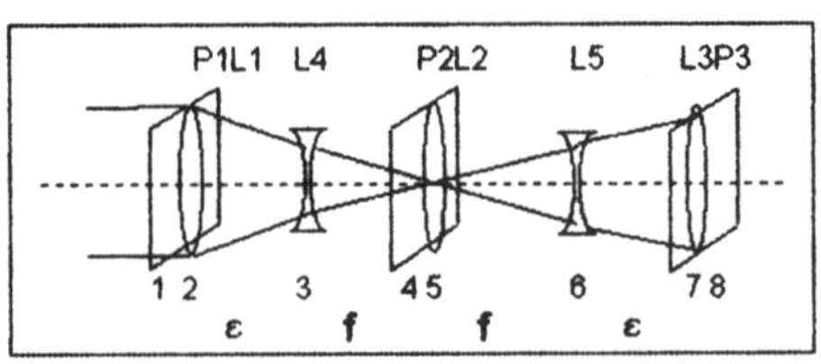

Fig.4 The farther structures of compact optical 4-f system with the telephoto Fourier lenses

In Fig.4, we design that the lens L_1 and L_3, and L_4 and L_5 are same, i.e. $f_1 = f_3$, $f_4 = f_5$, and the optical system is symmetry about the axis of the P_2 and L_2. One of the reasons for this is to reduce cost of designing and manufacturing lenses, and another, to reduce some image aberration in this optical system.

Keeping this in mind, we calculate the focal length f_2 of the phase compensating lens L_2 when the plane P_1 and plane P_3 are immediately adjacent lenses L_1 and L_5. $\varepsilon = X_H^{(1)} = X'_H{}^{(2)}$, applying equation (9) and (10), we get,

$$f_2 = \frac{f \times f_4}{2(f_4 - d)} \tag{11}$$

where, f is total focal length of combination lenses L_1 and L_4(or L_3 and L_5), f_2 is focal length of lens L_4 (or L_5), and d is the distance between lenses L_1 and L_4 (or L_3 and L_5).

2.4 The Example of Compact Optical Processors Designing

In input plane P_1, we selected LCD is PVG161501PYN model LCD. The panels have 150 × 160 pixels, and each measures about 270 × 270um with a center-to-center spacing of about 300 × 300 um. In the spatial filter plane P_2, we selected LCLV is 40 lines / mm, the number of pixels in each row of the filter $N_2 = 128$. The model of the diode laser is TOLD9412(S), its wavelength is 0.65 um. Through calculating, the focal length of Fourier lens f = 1477 mm, and 4 f = 5908 mm and 2 f =2954 mm. It will be very large and unacceptable length.

We design compact optical processors according to above method. We select: 1) the focal length of the positive lens L_1 and L_3 is 170 mm, $f_1 = f_3 = 170$mm; 2) the focal length of the negative lens L_4 and L_5 is -20mm, $f_4 = f_5 = -20$mm; 3) the distance between L_1 and L_4 (or L_3 and L_5) is 152.3mm, i.e. d=152.3mm. Through calculation, the total focal length of Fourier lens that is combined by two lenses L_1 and L_4 (or L_3 and L_5) is 1477 mm, and the focal length of the phase compensating lens L_2 is 9.87 mm, i.e. f_2=9.87mm according equal (11). The compressing ratio is k = 1/5. Thus, the total length of the compact 4-f optical system that we design is only 609 mm, and the system length is compressed about 10 times.

3 Simulation Result of Using Compact Optical 4-f System in Target Classification

As an example which the compact 4-f optical system application, taking four classes of aircraft(bomber, rocket, airliner and fighter as shown in Fig. 5) as the targets to be classificated, we have taken the bomber as test pattern. In the match spatial composite filter is BPOF (binary phase-only filter)[16-17]. The result of a computer simulation is shown in Fig.6.

4 Summary

A scaled shortened processor with a total system length equal to about f/3 is proposed, and the system length is compressed about 10 times. In future, it will be small enough to fit the PCI (Peripheral Component Interconnect) slot of a personal computer with processing up 30 flames per second. It will be acted a part of computer and accelerated image processing speed of the computer.

5 Acknowledgment

This work was supported by the research fund of the National Nature Science Foundation of P.R.China and Post-Doctoral Foundation of P. R. China.

References

1. D.Casasent, "High capacity pattern recognition associative processors," N.N. Vol.5, 687-698(1992)
2. Qing Tang, "Multiple-object detection with a chirp-encoded joint transform correlator," App. Opt., Vol.32,No.26, 5079(1993)
3. J.A.Davis, "Compact optical correlator design," App.Opt. Vol.28, No.1, 10(1989)
4. D.L.Flannery, "Real-time coherent correlator using binary magnetooptic spatial light modulators at input and Fourier planes,"App. Opt., Vol.25, No.4, 466(1986)
5. A.J.Katz, "Generating image filters for target recognition by genetic learning," IEEE Trans. PAMI, Vol.16, No.9, 906(1994)
6. T.Poggio, "Computational vision and regularization theory," Nature, No.317, 314(1985)
7. Bahram Javidi, "Position-invariant two-dimensional image correlation using a one-dimensional space integrating optical processor : application to security verification", Opt. Eng. Vol.35, No.9, 2479(1996)
8. Thomas J.Grycewicz, "Experimental comparison of binary joint transform correlators used for fingerprint identification," Opt. Eng., Vol.35, No.9, 2519(1996)
9. Juris Upatnieks, "Portable real-time coherent optical correlator," App. Opt., Vol.22, No.18, 2798(1983)
10. X.J. Lu, "Basic parameters for miniature optical correlators employing spatial light modulators," Opt. Eng. Vol.35, No.2, 429(1996)
11. Francis T.S.Yu, 《Optical Information Processing》, A Whiley-Interscience Publication, 1983
12. Guo-guang Mu, Yuan-ling Zhan, 《Optics》, The People Educational Publication, P.R.China, 1978
13. Zhi-jiang Wang, 《The handbook of Optical Technology》,The Mechanic Publication of China, 1987
14. R. Kingslake, 《Lens Design Fundamentals》 ACADEMIC PRESS, INC. 1978
15. S.G.Lipson & H.Lipson, 《Optical Physics》, Second Edition, Cambridge University Press, 1981
16. Danny Roberge, Yunlong Sheng, "Optical real-time correlator for implementation of phase-only composite filters," Opt. Eng. Vol. 35, No. 9, 2541(1996).
17. B.V.K.Vijaya Kumar, "Tutorial survey of composite filter designs for optical correlators,"App. Opt. Vol. 31, No. 23, 4773 (1992).

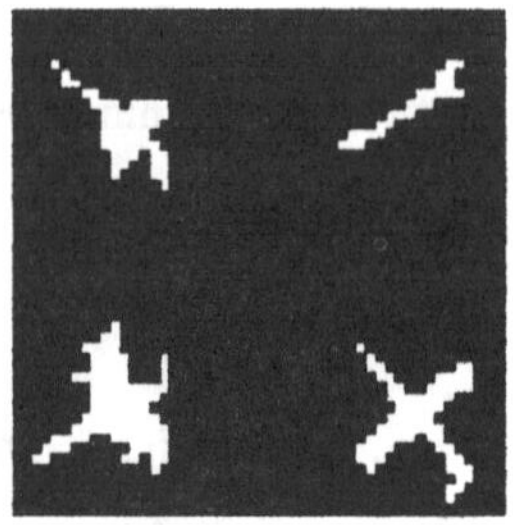

Fig.5(a) The four kinds of aircraft
(fighter, rocket, bomber, airliner)

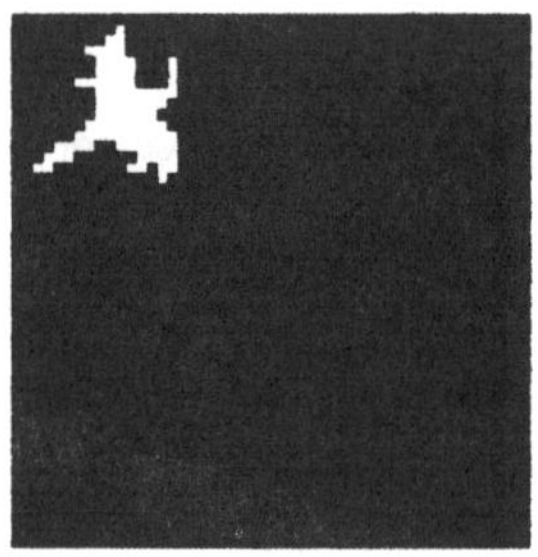

Fig.5(b) The test pattern (bomber)

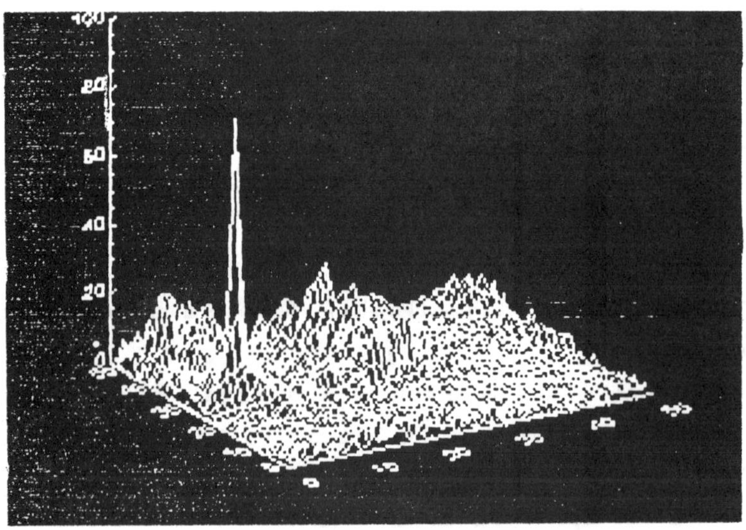

Fig.6 The result of computer simulation on BPOF correlation intensity pattern

Robotics and AGV Systems

An Easy-to-Use Robotic Welding System for Shipyards[1]

Marcelo H. Ang Jr.♣, Lin Wei*, and Lim Ser Yong*

Centre for Intelligent Products and Manufacturing Systems
Department of Mechanical and Production Engineering
National University of Singapore 119260

*Gintic Institute of Manufacturing Technology
71 Nanyang Drive
Singapore 638075

Abstract. Welding is a fundamental task in ship building and repair. The welding process involves joining stiffeners with large steel plates which may be flat or curved, and fastening webs, T-bars, and brackets to the steel plates. Harsh working conditions coupled with the complexity of the welding patterns represent a major challenge for designing robotic welding systems. Welding webs, T-bars, brackets and sections with curvature is not currently feasible using automated systems. In this paper we describe a robotic welding system designed welding stiffeners, webs, T-bars, brackets, and flat steel panels for shipyards. The central component of the system is a robotic manipulator that can be programmed in a *walkthrough* mode. The welder programs the robot by guiding it once through the actual welding process as though he is doing it himself. Critical positions and orientations are recorded during teaching and the robot thereafter does the unpleasant welding job. Welding is further improved with arc-sensing technology. An intelligent welding subsystem is incorporated to automatically select the welding and other parameters. We also present an easy-to-use interface for the welding operators. We describe the details of the robotic welding system for shipyards and highlight system integration aspects.

1. Introduction

Welding is fundamental task in shipyards and marine/offshore companies. It is used in building and repairing structures. The welding process involves joining plates together and joining stiffeners with large steel plates which may be flat or curved, and fastening webs, T-bars, and brackets to the steel plates. In ship building, and more so in ship repair, welding is done in restricted and confined spaces with a very harsh environment. These environmental factors coupled with the complexity of the welding patterns represent a major challenge for designing robotic welding systems.

Welding steel plates together and fastening the first set of stiffeners can be done almost fully automatically together with the handling and transport of plates and stiffeners. These operations are shown in Fig. 1. Automatic and semi-automatic welding systems that exist today are confined to relatively simple welding operations.

---

[1] Presented at the International Conference on Computer Integrated Manufacturing, 21-24 October, 1997. Project supported by NSTB Grant 17/3/16 and Far East Levingston Shipbuilding Ltd. For more information, contact Marcelo H. Ang Jr. (mpeangh@nus.sg). The support of the Singapore National Science and Technology Board through research grant NSTB/17/3/16 is gratefully acknowledged.

These operations include straight welds (Fig. 1) or smoothly-curved weld paths wherein a flexible track can be laid to guide the welding head. The challenge is the more complicated welding paths as shown in Fig. 2. Fig. 2 shows the welding paths (in bold) required to weld "cross-stiffeners" onto the plates in Fig. 1. Welding the second set of stiffeners ("cross-stiffeners) or webs is done manually due to the complexity of the welding paths required. The cross-stiffeners run perpendicular to the first set of stiffeners. Fig. 2 shows the motions required.

The welding operations require horizontal and vertical motions as well as rounding motions around the corner. The size of the plates and stiffeners may vary. For example, the stiffeners can be as high as 4 m. Currently, the process is done manually with the operator having to stop and start the welding. This affects the quality of the weld. A robotic solution is found the be the most feasible approach to complicated welding paths, as shown in Fig. 2. With a robotic system, the starts and stops during welding operations can be minimised.

It can be observed that all shipyards involve welding operations with motions that shown in Fig. 2. The variation is in the size of the webs, stiffeners and plates. It should also be emphasised that unlike manufacturing environments, the parts and workpieces in shipyards are rarely standardised. Each part has a unique shape and dimension with some similarities. This makes traditional robotic programming of detailed teaching not feasible since the time spent in robotic teaching is not worth the effort since only one piece will be handled. The use of the robotic system cannot be justified. This emphasises the importance of having an **easy** teaching method, i.e., much easier than the actual welding.

We have developed a Ship Welding Robotic System (SWERS) to address this complicated welding. SWERS is currently in use by Keppel Far East Levingston Shibuilding in Singapore. In this paper, we describe the SWERS and highlight important issues. We discuss the current status and future plans for the SWERS.

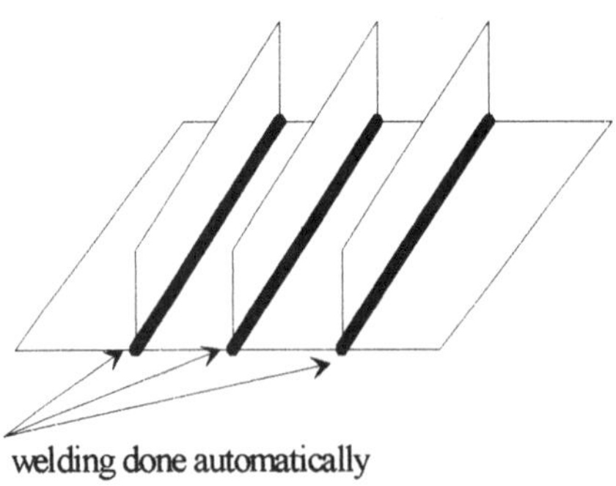

Fig. 1 Welding Plates and Stiffeners

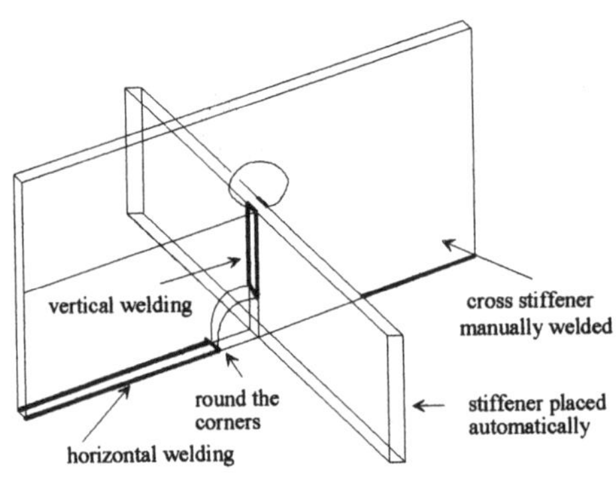

Fig. 2: Manual Welding of Cross-Stiffeners

2. Methodology

Robotic welding has been recognized as the next step in technological advancement of shipyards[1]. There many commercially available robotic welding systems such as the robotic systems of Odense Steel Shipyard Ltd in Denmark, but almost all are not attractive to the Singaporean shipyards. The are many important reasons for this. Robotic welding system are very complicated to use, they require a robot programmer and/or application engineer which shipyards do not have. The workforce is welding operators and supervisors whose training is up to the high school level at best. Also, CAD data of plates, webs, stiffeners are not available. Furthermore, the workplace is very unstructured and locations of work pieces are always uncertain. CAD-based robotic solutions such as those in [2] are not suitable. Offline robot programming systems [3] are also not feasible because of the uncertainty and unstructured nature of the environment.

The need for improved productivity and quality is however crucial to the survival of shipyards. A very important requirement of our robotic solution is its ease of use, that is, it must be "idiot-proof" such that an unskilled or untrained worker can operate it out of his/her own "common sense". This is one of the most important design targets in our development of SWERS. Our approach is to develop a robotic system as a super tool for humans to use, rather than developing a completely autonomous system. The importance of human operators in man/machine teams has been recently recognized to be more useful that purely autonomous systems, especially so in welding [4].

Our approach to developing SWERS to exploit currently available technology and build on top of it to make it useable and attractive for shipyards to use. A very important consideration in deciding which robotic welding system to use is the "openness" of its architecture. The system must allow us low level access to its controller so that we add our own features to the system. This is crucial because we do not want the robotic system to restrict our ability to implement advanced algorithms to make the robotic system usable by shipyards. For example, we need to be able to give current and/or torque commands to each robot joint motor, thereby by-passing the robot controller.

Although many robotic systems claim to provide full access, almost all of them were built without the "openness" feature we require. They did not expect that the target customers would do R&D on a robot, and hence would require low level access. We have however found a robot manufacturer, REIS robotics, who is willing to work with us on this. They have provided a customised version of an EPROM in their robot controller that allows us a clean and multi-level access to their controllers, the lowest level of which is complete by-pass of their robot controller through software.
Another important criterion in our design is the use of industrially proven hardware for implementing our algorithms and for smooth integration into the robot controller. We have decided on the use of the industrially proven VME bus system, which is the same bus used by the robot controller. Our algorithms implementing the additional features run on a single board computer (the HyperSparc 125 MHz board) that plugs

into the same VME bus of the robot controller. Our HyperSparc works "hand-in-hand" with the robot CPU, which is a 68030 based computer.
The SWERS has the following main features:

- A VMEbus-based open architecture robot controller.
- Off-the-shelf robot and welding equipment.
- A new robot teaching method called *Walk-Through Teaching* (WTT).
- Custom designed *Man-Machine Interface* (MMI).

3. SWERS

SWERS consists of a six-axis articulated welding robot (REIS SRV6) that is mounted upside down on a 3-axis gantry system. A schematic diagram of SWERS is illustrated in Fig. 3. Fig. 4 shows a picture of SWERS at FELS. The workspace of the robot is about 3 meters in diameter and the gantry moves the base of the robot in a work volume of 12 m x 12 m x 2 m. The gantry was designed to accommodate the 12m x 12 m panel size requirements in FELS. The whole gantry moves on a pair of tracks with length 16 m.

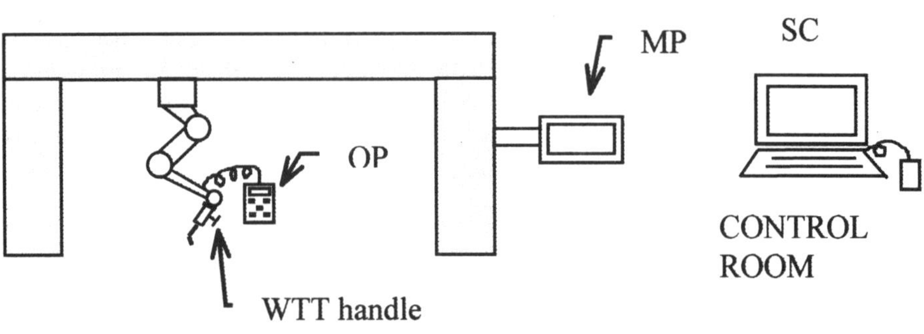

Fig. 3: Schematic Diagram of SWERS

Fig. 4: SWERS in the Shipyard Work Area

A key feature of the system is the walk-through-teaching (WTT) handle. An operator using the WTT handle specifies welding positions and motions. Advanced algorithms are employed in the robot controller to allow the operator to grab the WTT handle and move the robot as if the robot is operating in a zero-gravity environment. Parameters can be adjusted to change the dynamic response of the robot from the forces exerted by the operator. The robot can be made to appear very light of fast, or to behave as if it's underwater or in a very viscous environment.

Fig. 5 shows the various subsystems of SWERS and their components. The REIS robot controller is a 68030-based computer on a VME bus. We have installed a single board HyperSparc 125 MHz computer on the same VME bus to run our enhancements in conjunction with the 68030-based robot controller. The standard teach pendant of the REIS robot has been supplemented with the OP, MP, SC, and WTT handle to allow a welding operator to easily use the system without the need for technical knowledge on robot programming. The HyperSparc acts as the supervisory controller that allows high level and low level access to the robot controller.

3.1 Robot Subsystem

The Reis SRV6 robot used is a 6 degree-of-freedoms articulated configuration robot. It has a work envelop of 3m diameter and a repeatability of ±0.05mm. The main feature of the robot is the VME-based open architecture controller. The robot subsystem incorporates our walk-through algorithms and the robot controller. The welding controller sits on the same VME bus as the robot controller and information is shared between the robot and welding subsystems via shared memory. Communications between the HyperSparc and the robot and welding subsystems are via shared memory, while communications with the other five subsystems are via serial communications.

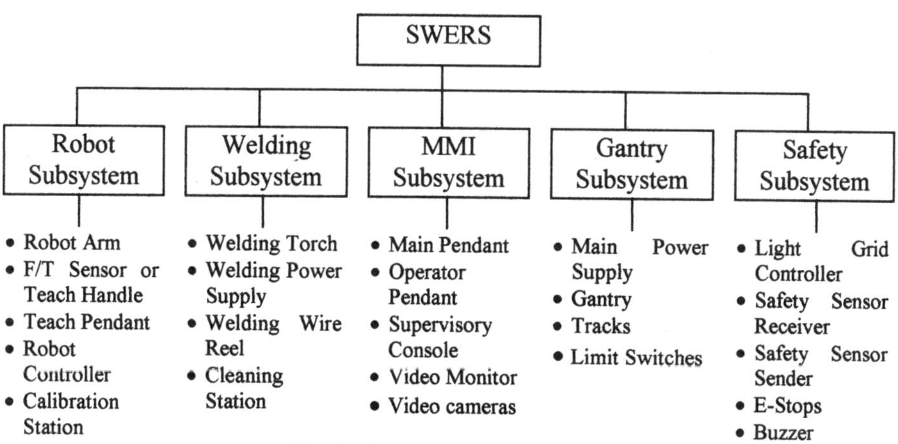

Fig. 5: Components of the SWERS

Mounted together with the robot are the welding accessories such as the wire reel drum, the wire feeder, a torch cleaning station. The welding torch and a WTT Handle (Walk-Through Teach Handle) are equipped at the end of the robot arm. The welding power source is placed on top of the gantry. The robot controller, Supervisory Console and the gantry controllers are housed in the air-conditioned Control Room at one end of the gantry. The control Room moves with the gantry.

3.2 Welding Subsystem

The welding system consists of a power source, a wire feed unit, a welding torch and a torch nozzle cleaning unit. It also comes with a built-in seam tracking sensor for following of welding path. It compensations for errors in welding taught paths due to workpiece dimensional errors (due to heating) or positioning errors.

Table 1 shows a list of important welding process parameters. We have done a thorough study of the welding process and arc sensor characteristics. Many of the welding process parameters are coupled with arc sensor parameters. This study together with the current welding code resulted in an intelligent database of values for the different welding process and arc sensor parameters. This database consists of recommended values for the welding process (Table 1) and arc sensor parameters. The user, however, is allowed to select and adjust parameter values before welding. Typically adjustments are done by a welding engineer in the control room; and welders need not adjust the values. The recommended values serves as the default values for the parameters if the user does not select or input values. An improved feature we are currently working on is the provision for online adjustment of welding current and voltage during welding. The welding position (vertical, overhang, or horizontal) can also be automatically determined from the taught positions.

3.3 Man-Machine Interface (MMI)

To facilitate the walk-through teaching and to make the system easier to operate, custom-designed MMI software and hardware are developed.

The hardware of the MMI consists of a MAIN PENDANT (MP), an OPERATOR PENDANT (OP, Fig. 6) and a SUPERVISORY CONSOLE (SC, Fig 7). Important welding and other positions are recorded through the operator pendant (OP). The main pendant (MP) consists of a touch screen LCD that allows the operator to move the gantry or robot base. Through the MP, the robot can be set to WTT mode or to welding mode. In the welding mode, the safety subsystem is activated and the robot does the actual welding using the positions previously taught in WTT mode.

The Supervisor Console (SC) provides a higher level user interface for the welding supervisor. In the SC, important welding parameters, such as welding voltage and current can be set or default parameters from an intelligent database can be used. This database contains important parameters ideal for different welding configurations. The database has been developed through a detailed welding process study and in consultation with current welding practices and codes. From the taught positions using the WTT handle, the system can automatically select the correct welding parameters. As in any intelligent system, the user can always over-ride these parameters through the SC.

The OP is attached on the robot near the welding torch. It is a control pendant for WTT only. The MP and the OP are located outside the gantry. They are for overall control of the SWERS and data/parameters entry.

The handle located at the welding torch is for WTT. The handle is designed under the considerations of ease in teaching and no interference during welding.

The SUPERVISORY CONSOLE is located inside the control room of the SWERS. It consists of a monitor, a keyboard and a mouse. Upon start off the SWERS, the user firstly uses the SC to select and adjust the welding parameters according to the workpiece he is working on.

A Graphical User Interface (GUI) is developed to facilitate the operation of the system. Fig. 7 illustrates an example of the menus and dialog boxes of the GUI. The GUI allows the user simply to click the mouse button to select the desired welding data such as leg length, wire type, welding mode and no. of passes .The selected data is matched with the default and optimized values of the welding parameters such as current, voltage, speed, oscillation etc. The user may adjust the values of the parameters if he/she so desired. The default values of the parameters are based on a welding study and are consistent with the welding specifications. The user can, however, add, delete, and change the records in the database. This ensures that the database grows and improves with experience, The user updates the data using the mouse and the keyboard.

Table 1: Important Welding Process Parameters

• Welding Mode (Position)
• Leg Length
• Wire Type
• No. Of Passes
• Welding Current
• Welding Voltage
• Welding Speed
• Stick out Length
• Oscillation Frequency
• Oscillation Amplitude
• Oscillation Hold Time

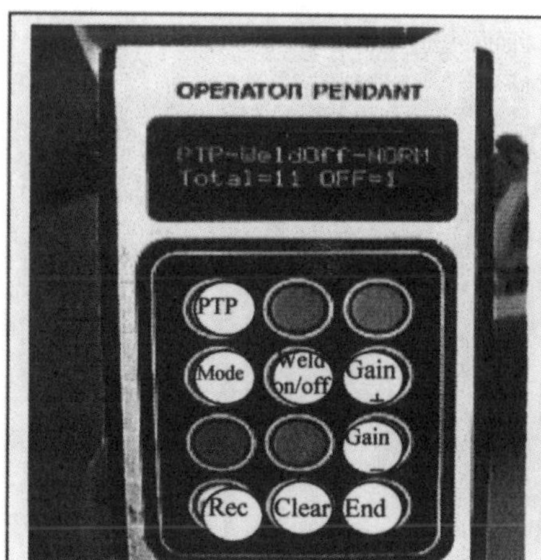

Fig. 6: Operator Pendant

The MP is a device for overall control of the whole welding system. It is the only device in the commands and controls movements of the system. Using the MP located near the control room at the side of the gantry, the user can move the gantry, transfer the robot on the gantry, clean the welding torch, and start the teach mode and welding mode. For safety purpose, it can only be operated when the OP is hanged up on the robot arm.

The MP consists of a touch screen terminal. The touch screen has the advantages of ease of use and programming flexibility. Fig. 8 illustrates some screen pages available in the MP. The screen displays the desirable buttons for the user to press at different states. It also displays the robot and welding status.

The OP is a hand-held device for the operator to perform walk-through teaching. It is activated by the MAIN PENDANT located remotely from the robot. The OP is connected on the robot at a position near the welding torch. To begin the teaching, the operator firstly removes the OP from the robot arm and a sensor detects that the OP is detached. After the teaching procedure, he/she attaches the OP back to the robot and all the buttons are then disabled. The operator then leaves the work envelope for the MP. The feature is to ensure that no operator is around the robot when it is moved by the MAIN PENDANT.

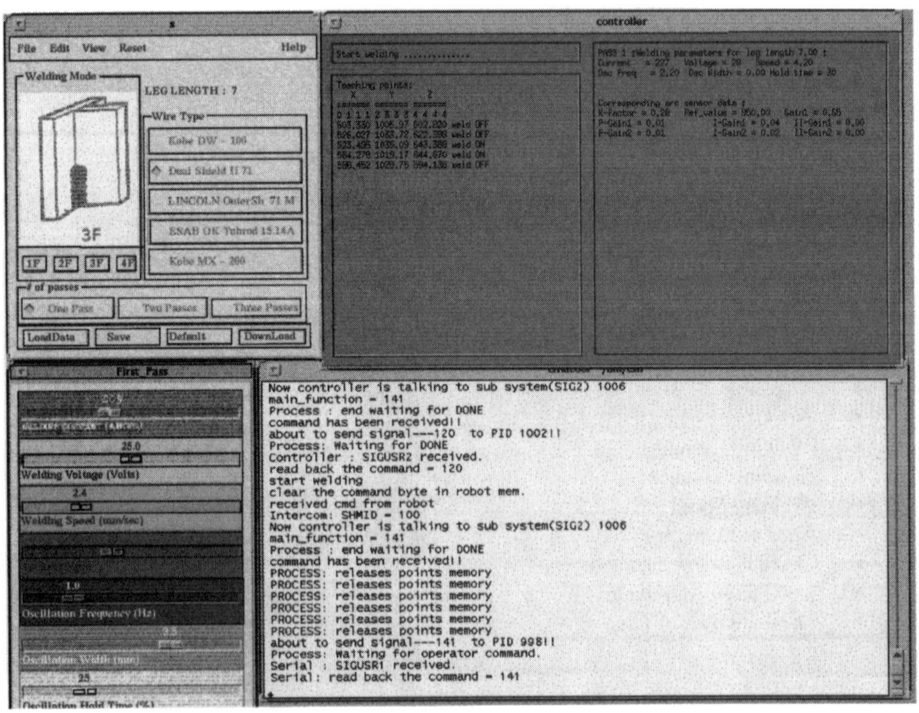

Fig.7 : Supervisory Console GUI (Welding Parameters)

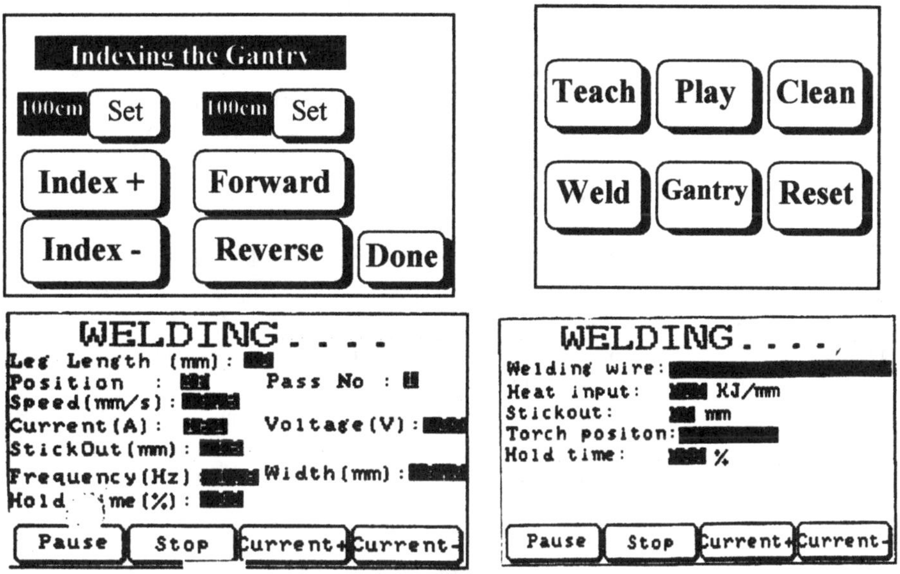

Fig 8: Sample MP (Touch Screen) Pages

As shown in Fig. 6, there are eight buttons and a 2x16 LCD on the OP. The LCD displays the current teaching state. The *PTP* button allows Point-to-Point for walk-through teaching. As the operator leads the torch to the point to be recorded, he/she can select the mode of robot movement by pressing the *Mode* button. Three modes can be select with the toggle button, namely the Normal, Rotate and Translation modes. The Normal Mode allows the operator to rotate and shift the position of the torch. This mode involves more computation time thus the torch is slower to follow the hand movement. On the other hand, the Rotate Mode and the Translate Mode decouple the two modes. It allows a easier movement either by rotating the torch about the TCP (Tool Center Point) Point or just moving the torch position. The *Gain +* and *Gain -* buttons increases or decreases the gain of the robot speed. The operator can adjust the gain value for the easiness of moving the robot.

There are two states, namely *Weld On* and *Weld Off*. The operator can teach the robot for the desired state by pressing the corresponding buttons as he/she tracing the robot path. The point just recorded can be deleted if he/she so desired. Table 2 lists the description of each buttons.

Table 2 *Description of the OP Buttons*

PTP	• Pressed to put robot in walk-through Point-to-Point mode • State after pressed: PTP (WELD OFF)

		• Enable: Walk-through' program
Mode	:	• Toggle button • Pressed to put robot in the mode of NORM, ORIENTATION, or TRANSLATION
Gain +		• Pressed to increase the speed of robot
Gain -		• Pressed to decrease the speed of robot movement
WELD ON/OFF	:	• Toggle button • Toggle between PTP (WELD OFF) state and PTP (WELD ON) states • The appropriate sets of welding parameters are determined from the taught motion data base which indicates if it is horizontal motion (2F) or vertical motion (3F) etc.
RECORD	:	• Pressed to record current position (all joint coordinates) • Pressed to record welding parameters under PTP (WELD ON) state. • Pressed to change welding path (for PTP state) and welding modes. • Used in the following states: PTP (WELD ON); PTP (WELD OFF)
CLEAR	:	• Pressed to CLEAR the previous pressed data entered.
END	:	• Pressed to get into IDLE state • States can be terminated: PTP (WELD ON); PTP (WELD OFF)
LCD display	:	• show the following - states [PTP(WELD ON); PTP(WELD OFF) - Number of points recorded

3.4 Gantry Subsystem

The 3-axis gantry is driven by a pair of AC induction motors along the flow of the panel (X drive). The robot is transferred across the panel by an AC servo motor (Y drive). Due to the height of the webs and limited size of the robot work envelope, the robot has to be raised up for the next welding zone. (Z drive). The communication between the robot controller and the gantry controller is through simple RS232 interface.

The robot controller is housed inside an air-conditioned control room. The room is mounted together with the welding source at one end of the gantry. Due to the long span of the gantry, the cables lines for controller and welding equipment are up to 30m. The drum for wire feed unit and the nozzle cleaning station are located on top of the gantry and moved together with the robot.

The gantry is to index the robot at a number of points along the webs, The locations of these points are determined such that they ensure the robot to cover the entire required welding area. The fine positioning of the arm compensates position errors contributed by the gantry.

The MP is a touch-screen panel and serves as the user interface for the operator to move the gantry and the robot mounting base. A video monitor is also available so that the operator can see the work area as seen from the top by the robot.

3.5 Safety System

Generally, the overall system is designed and constructed under standard safety guidelines of industrial robots [5]. Some of the safety features are as follows:

- Emergency Stops are available on the Reis Robot Controller, MAIN PENDANT and OPERATOR PENDANT.

- Software module of each subsystem includes *exception handling routines* which ensure that the software will not be aborted abnormally leaving the system in an unknown state.

- In "TEACH mode", the velocity of the robot is restricted at a safe speed and the robot cannot be moved on the gantry.

- The MAIN PENDANT is operational only when the TEACHING PENDANT and all other control stations are deactivated.

- A sensor is implemented to detect whether the OPERATOR PENDANT is detached from the robot.

- The program ensures that the robot is in up position when the robot moves in X or Y direction.

- Light curtain is installed for vertical safe-guarding of the robot work envelope.

4. Performance Evaluation

The SWERS have been certified and performance has been good. Certification includes extensive laboratory tests of welded specimens. These include hardness tests and micro-edge inspections, and macro-etch tests for cracks and porosity. Extensive experimental runs and demonstrations have been conducted. Typical runs involving welding paths that are U-shaped (100 mm down, 600mm right, 100mm up). The tests have shown that the quality of the weld by the SWERS is better than the manual FCAW and SMAW in terms of uniformity, shape and consistency. Macro-etching inspection and micro hardness tests have shown that the samples are free of porosity and cracks. For the cycle time, the SWERS also performed better than the manual processes (Table 3)

Table 3: Typical Cycle Times

Time	SWERS	FCAW	SMAW
Set-up	N/A	60 sec	30 sec
Teaching	150 sec	N/A	N/A
Welding	234 sec	306 sec	608 sec
Total	**384 sec**	**366 sec**	**638 sec**

The cycle time is expected to be even faster in actual workpieces where it has longer welding seams and repeated patterns. For such job, the PTP mode and "Teach-Weld-Weld" mode can be used so that the teaching time can be significantly reduced.

The flexibility of SWERS has also been demonstrated in pipe welding operations. Although not originally designed for this application, the general applicability of the walk-through teaching methods has made the application to pipe welding very straightforward with minimal effort.

5. Conclusions

We have described the SWERS that we have developed for Far East Levingston Shipbuilding Ltd. The system demonstrates an attractive and feasible approach to robotics applications wherein state-of-the-art technology is used and additional features are added to make the system attractive to industry. An important requirement is the "openness" of commercially available systems to allow further R&D work to improve and/or add features.

A unique feature of SWERS is the walk-through programming capability, which is very natural and easy to use. The difficulty and tedium involved in robot programming has always been a disadvantage in choosing a robotic solution. Walk-through programming addresses this problem and is applicable not only in robot welding but in general robot programming as well.

An increasing trend in robotic applications is the use of the robot as a "super-duper" tool for humans to use. This opposes the original thinking in which robots would replace humans. Indeed the SWERS demonstrates how the robot can be used as an effective tool. The concept of "de-skilling" is demonstrated by SWERS. The welding skill is delegated to the robot, while the programming and decision making tasks are retained by the human.

SWERS is continuously evolving through improved features. Current work hopes to improve the productivity of SWERS by incorporating additional features. These features include the welding of similar parts by teaching only reference positions, and the closer coordination of gantry and robot motion during welding.

Acknowledgements. The contributions of the following members of the project team have made this project a success. They include Ng Teck Chew from Gintic Institute of Manufacturing Technology; Fung Mok Wing, Patrick Yang, Danny Tan, and Zeng Xiao Ming from Singapore Productivity and Standards Board; Roy Lim Weng Chin, Lim Tow Koon, Wee Teck Guan, Lee Weng Kee and Sam Weng Kuan from Far East Levingston Shipbuilding Ltd; Ng Joon Leong from University of Western Australia, Yang Jiandong, Lee Kim Seng, Kang Yanjun, and Pubudu N. Pathirana from National University of Singapore. The support of the National Science and Technology Board (Grant NSTB/17/3/16), Gintic Institute of Manufacturing Technology, Singapore Productivity and Standards Board, and Keppel Far East Levingston Shiobuilding Ltd are also acknowledged.

References

[1] C E Skjolstrup and S Ostergaard, Shipbuilding using automated welding processes, *Welding International Review*, February 1994, pp. 213-217.

[2] T. Kangsanant, and R.G. Wang, CAD-based robotic welding system with enhanced intelligence, in *Proceedings of the Third International Conference on Computer Integrated Manufacturing*, 11-14 July 1995, World Scientific Publishing, Singapore.

[3] R.O. Buchal, et.al., Simulated off-line programming of welding robots, *International Journal of Robotics Research,* Vol 8, No 3, June 1989, pp. 31-43.

[4] Arc welding is still very much a craft, *ABB Robotics Review, No. 1, 1993.*

[5] *Singapore Standard CP53: 1990: The Safe Use of Industrial Robot.*

Fixturing 2D Concave Polygonal Objects Using a Three Fingered Flexible Fixturing System

H. Du and Prof. Grier C I Lin

Centre for Advanced Manufacturing Research
School of Engineering
University of South Australia
The Levels, SA 5095 Australia

Abstract. This paper presents the development of an algorithm for fixturing 2D concave polygonal objects using a three fingered flexible fixturing system. The three fingered flexible fixture system includes two computer numerically controlled (CNC) modules and an auxiliary mechanism. One module has two fingers which can be positioned along two circumferences of two circles, whereas the other module has one finger which can be adjusted along a slot. The auxiliary mechanism applies the clamping force to the object and the control of the clamping force can be achieved automatically.

The algorithm presented in this paper is the implementation of the concept of the maximal inscribed circle of a polygon, and it is an extension of the work conducted by the authors. The algorithm mainly consists of determining clamping points on the concave polygonal object and the optimal configuration of the three fingers in the fixturing system. The algorithm is implemented by a program written in C language for Windows. Examples of the application of the algorithm are described in this paper. The results show that the algorithm can guarantee a secured fixture in a trial with a concave polygonal object.

1 INTRODUCTION

A fixture is a device that locates and holds a part for machining, assembly, inspection and other processes in manufacturing. In order for a Flexible Manufacturing System (FMS) to be really flexible, every process needs to be as flexible as possible. It is thus desirable that a fixture is flexible to accommodate geometries and shapes of a variety of parts or assemblies in manufacturing.

Research has been conducted into the minimum number of fingers needed to immobilise an object in robot hand grasping; this has provided a new approach to flexible fixturing. Markenscoff et al [1] have established that four contact points suffice to immobilise a generic 2D object, and seven suffice to immobilise a generic 3D object. On the other hand, Czyzowicz et al [2] have shown that generic 2D and 3D polygonal objects can be immobilised respectively by three and four frictionless point contacts. There are two advantages to using the minimum number of fingers in a fixture. One advantage is that it may lead to more adaptable or more efficient fixturing techniques and fixturing planning algorithms. The other advantage is that the possibility of using fewer fixtures may lead to more efficient re-fixturing methods in machining process involving frequent repositioning of fixtures. Therefore, a need has occurred to implement the concept of the minimum number of fingers in fixturing for further development.

Based on the number of contact points for a planar object, a three fingered fixturing system is proposed by the authors and is currently being developed. Figure 1 shows a schematic diagram of the system, while Figure 2 shows a prototype of the system. The three fingered fixturing system consists of two computer numerically controlled (CNC) modules, one fixed and the other movable. The fixed module has two fingers positioned along two circumferences of two adjacent circles, while the movable module has one finger which can be adjusted along a slot. A detailed mechanical design of the system can be found from the paper published by Lin and Du [3].

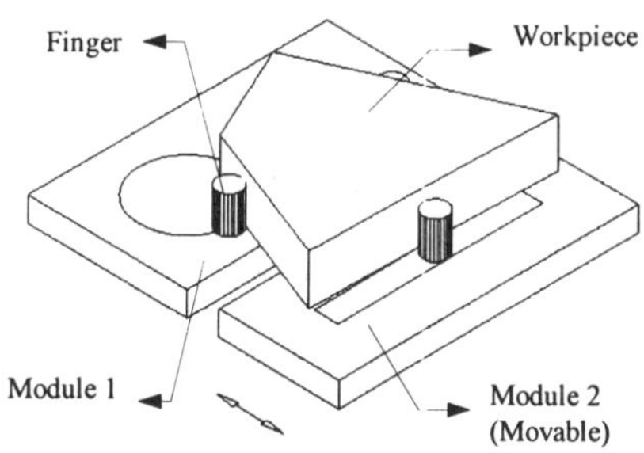

Figure 1 A three fingered fixturing system

Figure 2 Prototype of the system

The motivation behind this research is that an algorithm needs to be developed for the three fingered flexible fixturing system. The authors have developed an algorithm for fixturing planar objects [4], but it can only deal with convex polygons. This paper presents the development of an algorithm for fixturing a 2D concave polygonal object, based on the concept of the maximal inscribed circle in a polygon. In comparison with the previous one, this algorithm is faster and more efficient. It will be tested in the three fingered fixturing system in the future. In this paper, after reviewing some previous research in fixturing algorithms, the proposed algorithm is presented and explained, followed by an example of its application. Finally, conclusions are given as to the efficiency of this algorithm.

2 LITERATURE REVIEW

In recent years considerable progress has been made on fixturing research, and substantial literature exists on automated fixture design and flexible fixturing, as reviewed by Hazen and Wright [5] and Trappey and Liu[6]. Two major geometric algorithms proposed by Brost and Goldberg, Wallack and Canny, along with the concept of the maximal inscribed circle, are discussed as follows.

Brost and Goldberg [7] have proposed an algorithm for fixturing prismatic polyhedral objects using a modular fixturing system, consisting of fixture elements and a translating clamp, all mounted on a fixture plate. Wallack and Canny [8] have also proposed an algorithm which is similar to that developed by Brost and Goldberg. Their algorithm synthesises a class of modular fixtures with four round locators to be in form closure on a split lattice that can be closed like a vice. Due to the characteristics of modular fixturing, these algorithms generate all the possible configurations for a polygonal workpiece. However, they can only be applied in modular fixturing.

The concept of the maximal inscribed circle has been proposed by many researchers to secure stable grasps for both a three fingered robot hand in robotics and fixturing system. Rimon and Burdick [9] have shown that three contact points, where a circle maximally inscribed meets the sides of a polygon, can determine a feasible equilibrium grasp, and thus these three points secure a stable fixturing configuration. Even though a few schemes for determining the maximal inscribed circle have been suggested by these authors, no details are given regarding their implementation. Czyzowicz et al [2] have explored the idea of using the maximal inscribed circle with a Voronoi diagram to immobilise a polytope. But similarly, no practical implementation has been given. The Voronoi diagram is a very useful tool in computational geometry to determine the maximal inscribed circle, but it is complicated and difficult to implement in fixturing. Preparata and Shamos [10] have proposed converting the nonlinear problem of the maximal inscribed circle into a linear program, but the implementation of the idea is not straightforward.

The algorithm previously developed by Lin and Du [4] was the implementation of the concept of the maximal inscribed circle in a polygon for fixturing in manufacturing. However, it was not very efficient and it can only deal with convex polygons.. The algorithm developed here is faster and more efficient.

3 THE ALGORITHM

The determination of the maximal inscribed circle in a polygon lies the principal part of the algorithm. Once the maximal inscribed circle is obtained, three clamping points for fixturing are chosen either on or close to tangential points of the circle maximally inscribed within the boundary of the object. These three tangential points secure the fixturing configuration. In this section, an overview of the algorithm is followed by an explanation.

3.1 Overview of the Algorithm

Figure 3 shows a flow diagram of the algorithm. Once the geometry of the data of a polygon is extracted from a CAD file, the algorithm checks whether it is the convex polygon or not. The algorithm then applies the downhill simplex method (Press et al [11]) to determine the maximal inscribed circle in the polygon. After the circle is found, the Newton-Raphson method (Faux and Pratt [12]) is used to obtain three tangential points where the maximal inscribed circle meets the sides of the polygon. Configuration of the three fingers in the three fingered system can then be determined from these three tangential points. Optionally, the fixturing configuration can be verified and simulated by using two software modules developed by Du and Lin [13]. The algorithm has been implemented in a program written in C language for Windows.

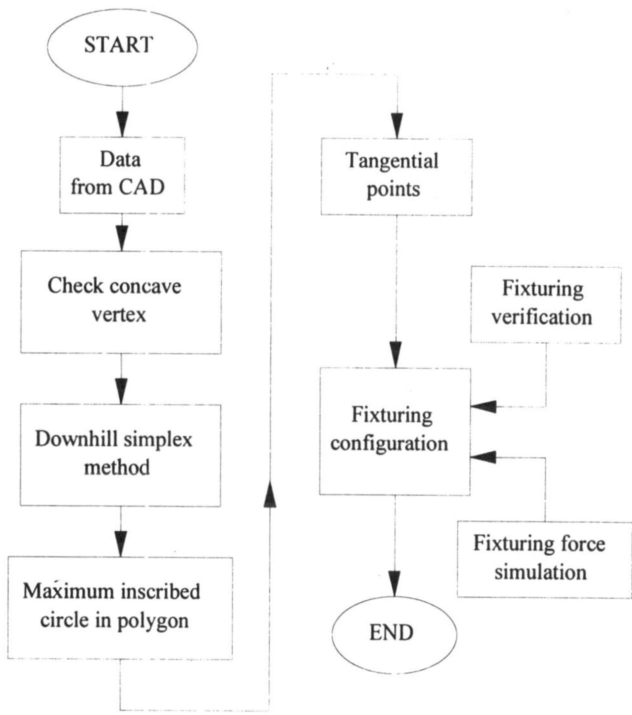

Figure 3 Algorithm flow diagram

3.2 Algorithm for Checking Concave Polygons

Vertices of a polygon are referred to by their indices, which increase counter clockwise. It is assumed that the first index of the polygon be known, and the linear constraint equations for all edges be determined.

The concave vertex can be found by using the following procedure:

1) Start from the first vertex of the polygon, for example, vertex 1 in Figure 4, obtain the middle point $P(x, y)$ of the line segment formed by the first vertex and the third vertex.

2) Check if the point is within the polygon, by comparing its position relative to the two edges of the second vertex. The lines formed by the two edges provide enough information to determine if the point is within the polygon.

3) If the point is within the polygon, replace the first vertex with the third one, and return to step 1. If it is not, then the second vertex is a concave vertex. Next set flags to the two edges of the vertices. Replace the first vertex with the third one, return to step 1.

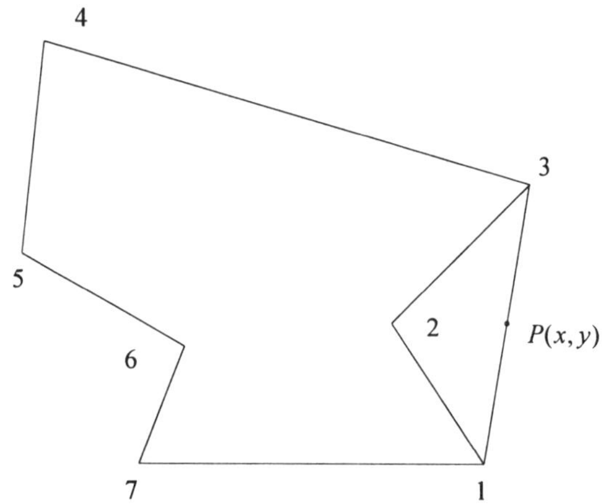

Figure 4 Finding the concave vertex

Repeat the above procedure, until all polygon vertices have been checked. If there is any vertex which is a concave vertex, then the polygon is concave. The concave vertices will provide constraints in the determination of the maximal inscribed circle in the polygon and this will be discussed in the following section.

3.3 Determination of the Maximal Inscribed Circle in a Polygon

The penalty function of nonlinear programming techniques [14], has been used by Lin and Du [4] to determine the maximal inscribed circle. However, this process is not very efficient, so in this section, a different approach is used to determine the maximal inscribed circle in a polygon.

The maximal inscribed circle in a polygon is formulated as an optimisation problem, listed as follows:

$$\begin{aligned} &\text{Maximise} && r \\ &\text{Subject to} && \mathbf{x} \in \mathbf{X} \end{aligned} \tag{1}$$

where $\mathbf{x}$ is the coordinates of the points on the circumference of an inscribed circle in the polygon, and r is the radius of the inscribed circle in the polygon.

The above unconstrained optimisation problem is solved by using the downhill simplex method for multiple dimensions [11]. The downhill simplex method requires only function evaluation, not derivatives. This is the main reason why it is used to optimise the maximal inscribed circle. The method can be briefly explained as follows:

The method must start with N+1 points, defining an initial simplex. N is the dimension of the optimisation problem. If one of these points is used as an initial point $\mathbf{P}_0$, then the other N points are

$$\mathbf{P}_i = \mathbf{P}_0 + \lambda \mathbf{e}_i \quad (2)$$

where the $\mathbf{e}_i$'s are N unit vectors, and where λ is a constant which is an estimate of the problem's characteristic scale of length.

This method involves many steps, and most steps only move the point of the simplex where the function is largest through the opposite face of the simplex to a lower point. These steps are constructed to conserve the volume of the simplex. If this can be done, large steps will be taken to expand the simplex in one direction or another. When the simplex reaches a valley floor, the method contracts in the transverse direction and tries to moves rapidly down the valley. Where the simplex tries to pass through a very small hole, it contracts itself in all directions, pulling itself in around its lowest point. The optimised points have thus been found.

Having discussed the downhill simplex method, the procedure of the unconstrained optimisation is explained as follows. With a given point in the polygon as the centre of a circle, the maximal radius of the circle within the polygon is obtained by comparing the distances between the centre point and all edges of the polygon, and the distance between the central point and the concave vertex. The maximal radius is the minimum of all distances. As shown in Figure 5, for example, r_1 is the maximal radius of the feasible circle within the polygon, with a given point o as the centre of a circle in the polygon.

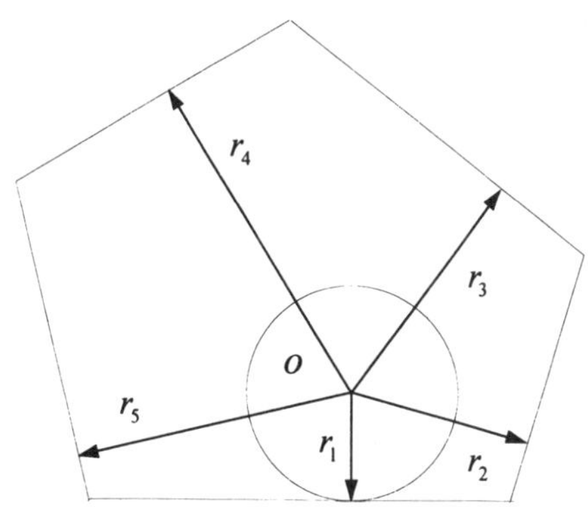

Figure 5 Maximal radius of an inscribed circle

Once the maximal radius of the circle has been obtained from a given point as the centre of the circle, the downhill simplex method generates the search direction of the maximal radius of the circles within the polygon. The initial points of the simplex can be generated by a random function to achieve the automation of calculation. A new centre point is then generated by the algorithm. This process repeats until the optimisation method converges to the maximal point. The final result of the optimisation is a maximal inscribed circle in the polygon.

An alternative geometric method for determining the maximal inscribed circle is to locate the maximal radius of the circle according to any three edges of a polygon. By searching all three edge combinations and the circles based on them, the maximal inscribed circle can be found. However, this method requires an exhaustive search of all possible combinations. If the number of edges of the polygon is large, the computation time will be extensive. By comparison, the proposed algorithm only searches the circle maximally inscribed in the polygon, with an initial estimation of the centres of the circles. Hence, it is efficient and its computation time is small.

The key issue in the above algorithm lies in determining the maximal inscribed circle in a polygon. Once the three clamping points are obtained, configuration of the three fingered system can be determined accordingly [4]. Optionally, the fixturing configuration can be verified and simulated by using the software modules developed by Du and Lin [13]. In the next section, an example is given of an application of the algorithm.

4 EXAMPLE OF APPLICATION OF THE ALGORITHM

In this section, a concave polygonal object is used as an example to show the effectiveness of the algorithm and the flexibility of the three fingered system developed.

A program of implementation of the algorithm has been written in C language for Windows. An input interface has been developed to read the CAD data in DXF file format. The CAD data for a concave polygonal object with five edges is generated by a CAD package as an input for the algorithm. The coordinates of the vertices of the polygon are (0,0), (20, 90), (100, 140), (140, 50) and (200, 0). Figure 6 shows the maximal inscribed circle in the concave polygon and the results of the calculation. Figure 7 shows the results of the three tangential points where the maximal inscribed circle meets the sides of the polygon.

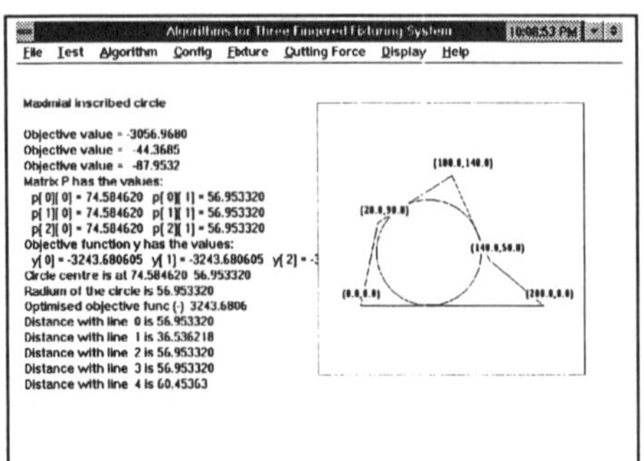

Figure 6 Maximal inscribed circle

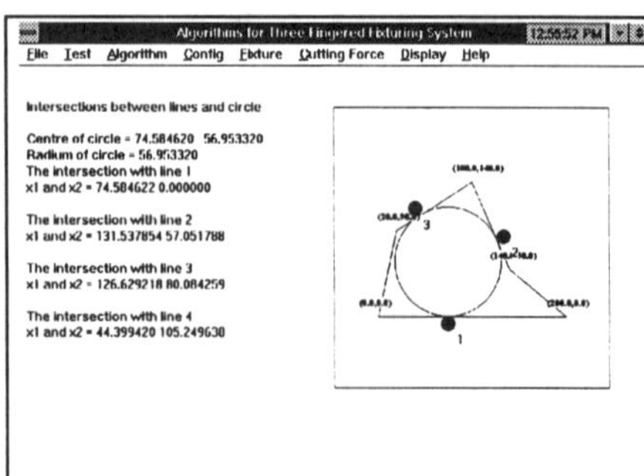

Figure 7 Results of tangential points

After the three fixturing points having been obtained, the fixturing verification module can be used to verify whether the fixturing configuration is valid. The results in Figure 8 show that the fixturing configuration is valid for this example.

The configuration of the three fingers in the three fingered system can then be obtained from the three clamping points. Figure 9 shows one of the fixturing configurations of the three fingers and the results of the calculation. The results include the coordinates of the vertices of the object and the three fingers, and the distance between the two modules. The results of the fixturing configuration are used to configure the three fingers in the three fingered system. It shows that the algorithm can guarantee a secured fixturing configuration for the three fingered fixturing system in a machining process.

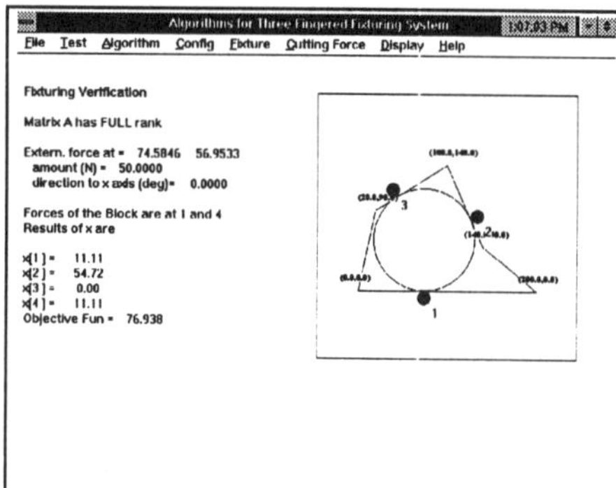

Figure 8 Results of fixturing verification

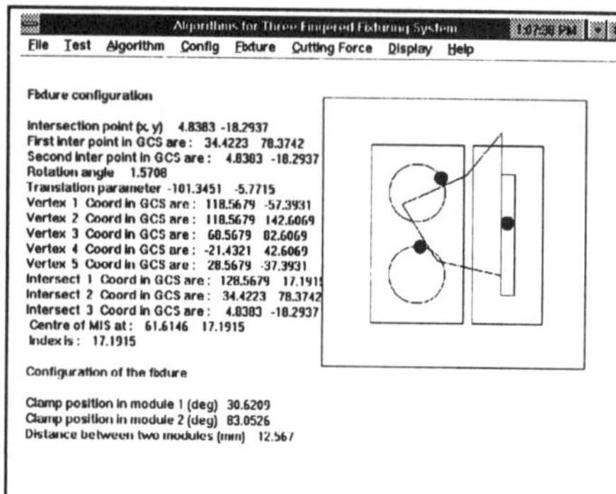

Figure 9 Results of fixturing configuration

5 CONCLUSIONS

An algorithm has been developed for fixturing concave polygonal objects by using the concept of the maximal inscribed circle for a three fingered fixturing system. It is an extension of the research conducted previously by Lin and Du [4]. The difference between them is that this algorithm can deal with concave polygonal objects, and is more general than the previous one. In addition, by considering the distance between the centre point and the concave vertices of a polygon, the algorithm tests the suitability of the concave vertices as clamping points.

The results of an application of the algorithm show that the algorithm guarantees a secured fixture in a trial with a concave planar polygonal part. The results also show that the algorithm is faster and more efficient than the previous one. The algorithm will be tested in the future in a real manufacturing operation for the three fingered fixturing system.

6 REFERENCES

1. X. Markenscoff, L. Ni, and C.H., Papadimitriou, The Geometry of Grasping, *International Journal of Robotics Research*, **9**, 1(1990) 61-74.
2. J. Czyzowicz, I. Stojmenovic, and J. Urate, Immobilising a Polytope, *Lecture notes in Computer Science*, **519** (1991) 214-227.
3. G.C.I. Lin and H. Du, On Developing a Flexible Fixture for Planar Objects, Proc of the International Conf. on Manufacturing Automation, Hong Kong, 28 - 30 April 1997, **2** (1997) 629-634.
4. G.C.I. Lin and H. Du, Fixturing Planar Objects Using the Concept of the Maximal Inscribed Circle, Australasian Conference on Manufacturing Technology, New Zealand, February 12-13, (1997).
5. F.B. Hazen and P.K. Wright, Workholding Automation: Innovations in Analysis, Design, and Planning, *Manufacturing Review*, American Society of Mechanical Engineers, **3**, 4 (1990) 224-237.
6. J.C. Trappey and C.R. Liu, Literature survey of fixture-design automation, *International Journal of Advanced Manufacturing Technology*, **5**, 3 (1990) 240-255.
7. R.C. Brost and K.Y. Goldberg, A Complete Algorithm for Synthesizing Modular Fixtures for Polygonal Parts, IEEE International Conference on Robotics and Automation, (1994) 535-542.
8. A. Wallack and J.F. Canny, Planning for Modular and Hybrid Fixtures, IEEE International Conference on Robotics and Automation, (1994) 520-527.
9. E. Rimon and J. Burdick, New Bounds on the Number of Frictionless Fingers Required to Immobilise Planar Objects, *Journal of Robotic Systems*, **12**, 6 (1995) 433-451.
10. F.P. Preparata and M.I. Shamos, *Computational Geometry: an Introduction*, Springer-Verlag (1988)
11. W. Press, B. Flannery, S.Teukolsky and W. Vetterling, *Numerical Recipes in C*, Cambridge University Press, (1979) 305-309.
12. I.D. Faux and M.J. Pratt, *Computational Geometry for Design and Manufacture*, Halsted Press, (1979) 297-300.
13. H. Du, and G.C. I. Lin, Clamp Location Optimisation in Fixturing Configuration, Proceedings of the 12th International Conference on CAD/CAM Robotics and Factories of the Future, Middlesex, UK, (1996) 1054-1060.
14. Bazaraa, M.S. and Shetty, C. M. 1979, *Nonlinear Programming: Theory and Algorithms*, John Wiley & Sons.

Study of Self-Learning Control for Special Shape Robots

Sui Qing Xie Ming

School of Mechanical & Production Engineering
Nanyang Technological University, Singapore 639798

Abstract. In this paper, self-learning control for special shape robots is studied thoroughly by a case that a special shape robot (i.e., having the shape of a Greek cross "+"), which has a perception system for sensing environment, passes through a crack whose width is less than the width of the robot body. This means that the robot cannot absolutely pass through the narrow crack unless it knows how to take advantage of its self special shape intelligently. The objective of this work is to investigate how to automatically generate a trajectory for the robot based on its own behavior. The advantage of self-learning control for special shape robots is demonstrated and verified from principle and experiment views.

1 Introduction

With the development of robotic technique, special shape robots and associated control technique are studied widely recently. The special shape robots are useful systems in the military or defense industry. Their applications can also be found in manufacturing industry or practical life for the purpose of performing some particular tasks (e.g., snake robots exam and repair pipelines). In this paper, a self-learning control approach for special shape robots is studied. The study case considered is a cross mobile robot (i.e., having the shape of a Greek cross "+"), which has a perception system for sensing environment, passing through a crack which width is less than the robot body's width or length. This means that the robot cannot pass through the narrow crack unless it knows how to take advantage of its self special shape intelligently. Normally, it is difficult to plan their trajectories and program them in advance for the special shape robots like the cross mobile robot. It should be admitted that the case to be studied here is not an easy problem, so choosing such a case is without lose of generality for studying behavior-based cognitive control for special shape robots. The objective of our work is to investigate how to automatically generate a trajectory for the robot based on its own behavior.

In the paper we first discuss the state and action space of the special mobile robot, and give some useful definitions and conclusions. Then we investigate a self-learning algorithm using reinforcement learning idea [1][2][3](more precisely, Q-learning) for the special robot. The presented algorithm, however, is somewhat different from the traditional learning method in that 1) in the traditional algorithm the space for choosing a random action is fixed, now we change this static space to a dynamic space so that its size reduces gradually with the training time increasing; 2) For the states whose corresponding Q values are never updated before, we modify the mapping from state space to action space so that the Q values of the most similar states will updated when no unique action for those states exists. Finally, some simulation results for the cross mobile robot are given in order to prove the validity and efficiency of the developed self-learning control strategy for special shape robots.

2 The State and Action Space of the Mobile Robot

In fact, teaching a robot to perform a task essentially reduces to learning a mapping from perceived state space to desired action space. The preferred mapping is one that maximizes the robot's performance at the particular task. In this section, we will discuss the state and action space in accordance with our problem considered in this paper, and give some useful definitions and conclusions.

2.1 Action space

First let's give the following definitions and conditions for an action:

Definition 1. Supposing that T_1 and T_2 are two motion modes for a moving robot, if their step lengths are equal but their motion directions are opposite (e.g., translate along positive and negative x axes or rotate in a clockwise and counterclockwise directions), then the T_1 and T_2 are called a pair of symmetric motion modes;
Definition 2. The action is invalid if it includes two adjacent steps which are symmetric.
Condition 1. The robot has n modes for each motion step, i.e., T_1, T_2, ..., T_n (n is an even number). Where T_1 and $T_{n/2+1}$, T_2 and $T_{n/2+2}$, ..., $T_{n/2}$ and T_n are symmetric respectively.
Condition 2. Each action of the robot includes N motion steps;

According to the above definitions and conditions, we have the following conclusion:
Conclusion. If an action is denoted as $A_1 \ldots A_i A_j \ldots A_N$ ($N \geq 2$, $j=i+1$), where $A_i A_j$ ($1 \leq i,j \leq N$) corresponds to the robot motion modes $T_u T_v$ ($1 \leq u$, $v \leq n$), and $|u-v| = n/2$, then this action is invalid.

Let's illustrate the above conclusion by giving an example. Suppose that a robot moves in 2D space, more specifically, it moves along positive X, negative X, positive Y, negative Y axes of its own coordinate system, or rotates about the origin of the coordinate system in the clockwise and counterclockwise directions. And again suppose that each action includes 2 steps, then we have 36 different types of actions totally. In fact, however, there are only 30 reasonable actions in the action space after reducing 6 invalid actions according to the conditions and conclusion given above.

2.2 State space

Assume that the robot is equipped with proximity sensors so that the robot can detect if some obstacles exist around. The state of the mobile robot in a given environment at a time is determined by the following factors 1) the robot self orientation, i.e., relationship between the robot coordinate system and world coordinate system; 2) the goal orientation, i.e., the direction from the robot coordinate origin to the goal point; 3) the environment information which can be detected by the robot sensors. Now we give the following general conclusion.

Conclusion.
IF
1) A robot self orientation is represented by N_s quadrants approximately (detected by one sensor), and any self orientations of the robot which belong to $[(2p/N_s)\cdot(i-1),(2p/N_s)\cdot i]$ (*i*=*1,2,3,...,* N_s) are regarded as identical; 2) the direction of the goal is represented by N_g quadrants approximately (detected by one sensor), and any directions of the goal which are in $[(2p/N_g)\cdot(i-1),(2p/N_g)\cdot i]$ (*i*=*1,2,3,...,* N_g) are regarded as identical; 3) the environment information around the robot is detected by n sensors, i.e., the environment states N_e is represented with n bits ($N_e = 2^n$)
THEN
States in the state space are $N_s \cdot N_g \cdot N_e$.

Now let's give an example. Suppose a robot self orientation is represented with 8 quadrants (N_s=8). The goal direction is represented with 8 quadrants (N_g=8). The robot is equipped with 4 sensors which are used to detect the information of the environment. Therefore, the all states in state space are: $N_s \cdot N_g \cdot N_e = 8\cdot 8\cdot 2^4$.

3 The Learning Algorithm

Here we wouldn't plan to describe the basics of reinforcement learning due to the space limitation. For a review of reinforcement learning in general, see [2][3][4]. In the light of the problem concerned in this paper, we give the following specific algorithm using reinforcement learning idea for the special shape robot cognitive control . This algorithm is somewhat different from the traditional learning method in that 1) in the traditional reinforcement algorithm the space for choosing a random action is fixed, now we change this static space to the dynamic space so that its size reduces gradually with the training time increasing; 2) For those states whose corresponding Q values are never updated before, we modify the mapping from state space to action space so that the Q values of the most similar states will updated when no unique action for those states exists.

THE LEARNING ALGORITHM

1) Define Step: *X_STEP, Y_STEP, Θ_STEP*

2) Initialize *Q[s][a]* whose initial entries are 0;

3) For ()

(1) Call function *WORLD()* to observe the current world state *s*

**WORLD () {*

Read the sensors on the robot to get the state of the robot in the environment.

}

(2) If *rand()* $\in$ *(0, rd)* (Let *rd* $\in$ *(0,1)*)

Choose an action that maximizes *Q(s,a)*

Else

Choose a random action

(3) Call function *ACTION()* to carry out the action *a*

**ACTION() {*

Actuate the robot and check collision.

}

(4) Call function *WORLD()* to observe the next state s' and *CRITERION(),* let the reward $r = F_r(J)$ and $E(s') = \max_a(s',a)$

**CRITERION() {*

$J = \alpha_1 \cdot d_1 + \alpha_2 \cdot d_2 + \alpha_3 \cdot d_3$

where d_1 : the distance from the robot to the goal;

d_2 : the angle from the self orientation of robot to the desired orientation;

d_3 : the complexity of the environment;

$\alpha_1, \alpha_2, \alpha_3$ are weight factors and $\alpha_1 + \alpha_2 + \alpha_3 = 1$.

}

(5) Choose parameter β and γ, and update *Q(s, a)*

$$Q(s,a) = Q(s,a) + \beta[r + \gamma E(s') - Q[s,a)]$$

End For()

4 Simulation

In order to testify efficiency and feasibility of the self-learning algorithm for the control of the special shape mobile robot, we simulated some typical cases. Due to the limit of space, we present here only one case, i.e., a mobile robot with cross shape passes through a narrow crack. The width of the crack is less than width and length of the robot body. This means that the robot cannot absolutely pass through the arrow crack unless it takes advantage of its self special shape intelligently.

Simulation description:

Mobile robot: cross shape; the width is Lr. (the length is equal to the width).

Environment: one narrow crack formed by several broken lines; the width of the crack is Le.

Goal: at the lower left corner of the environment.

Conditions: 1) the width of the crack is less than the width of robot body, i.e., Le<Lr.

2) the robot can move along X axis and y axis of its own coordinate system, or rotate in clockwise and counterclockwise directions.

3) each action includes 2 steps.

From above description, we know that the robot body's length is greater than the crack's width. So in order to pass through the crack, the robot has to rotate properly near the crack, then translate, after that, rotate and translate again, maybe repeat such procedures a lot of times. That means that no collision will happen only if the robot understands how to take advantage of its own special shape. Fig.4.1---Fig.4.2 show the results of training. Fig.4.1 is the training result for first time. Fig.4.2 is for 70th time. From these Figures, we can see that the robot moved at random basically at the beginning, and could pass through the crack very easily and smoothly at last.

5 Conclusion

In this paper, the technique of self-learning control for special shape robots is studied thoroughly by a case. The main contribution includes some useful definitions and conclusions for the state and action space, a self-learning algorithm for special shape robots, and a lot of simulation results. In addition, the advantage of self-learning control for special shape robots is demonstrated and verified from principle and experiment views.

References

1. Jonathan H. Connell, Sridhar Mahadevan, "Robot Learning", Kluwer Academic Publishers, 1993.
2. Richard S. Sutton, "Learning to Predict by Method of Temporal Defferences", The Journal of Machine Learning, 3(1), 1988.
3. Richard S. Sutton, "Reinforcement Learning", Kluwer Academic Publishers, 1992.
4. Leslie Pack Kaelbling and Stanley J. Rosenschein, "Action and Planning in Embeded Agents" , Robotics and Automation, (6) 1990.
5. Meystel, Alex, "Autonomous Mobile Robots Vehicles with Cognitive Control", Word Scientific Publishing Co. Pte. Ltd 1991.

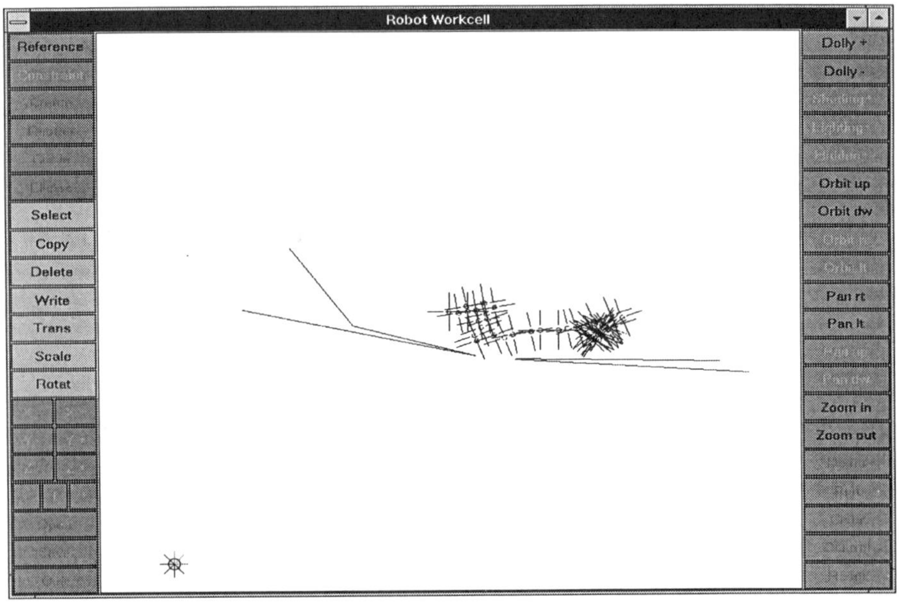

Fig.4.1 The training result for the first time

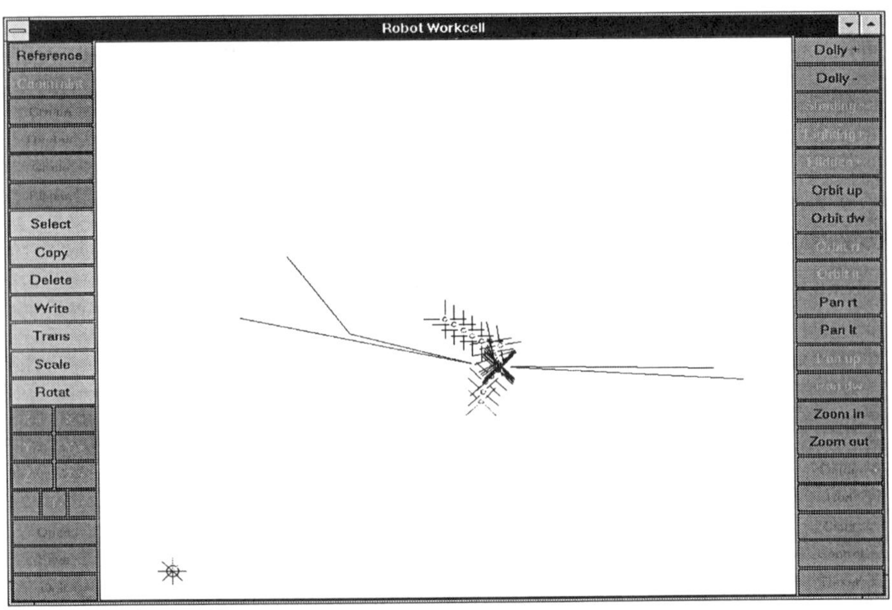

Fig.4.2 The training result for the seventieth time

DEVLOPMENT OF AN HIGHLY FLEXIBLE TWO-LEGGED WALKING DEVICE

PD Prof. h. c. Dr.-Ing. Josef Schlattmann

University of Paderborn, Germany

Abstract. The paper addresses the development of a static steady walking two legged totally modulized robot.

1 Mobile Robots

Mobile robots are often used in environments that are dangerous for humans, or where humans cannot reasonably be expected to work. In these environments they carry out, for example inspections, maintenance work or are used as observers in potentially dangerous situations. Typical applications are in irradiated areas in nuclear power stations, for manipulating tasks in laboratories or for sealing leaks in chemical plants as well as for fires in potentially explosive environments. These robots usually move using tracks or rollers. Only a few are constructed as walking robots and thus move around on feet. The construction of the movement unit is based on the conditions under which the robot has to fulfil its tasks /1/.

1.1 Robot Concepts

Wheel and track drives are suitable for movement on flat and uneven ground. However, application limits are set by the potential to overcome obstacles, to move on stairs or curbs and to manoeuvre in narrow, winding passages. Special constructions are suitable to overcome certain obstacles by having devices that enable them to place the tracks or driven wheels on steps or curbs and then to move onto them. However these devices considerably increase the machine's weight and are usually only suitable for mounting straight stairs. The size of these extra machine parts further reduces the machine's ability to manoeuvre in narrow spaces.
Walking robots use another principle of movement. Movement in these machines is not realised by a rolling movement, but by moving the body relative to the feet. As this movement is limited by the length of the legs, it is necessary for the feet to change their position to enable it to travel long distances. Nature shows us many examples of solutions for this problem.

2 Walking Forms

When considering these examples from nature we should remember that machines at the moment, are not yet capable of dynamic movement over long distances, as, for example, a galloping horse would be. This is because of the machines relatively low power capacity compared to its body weight as well as the complexity of control necessary. "Dynamic" movement means that movement is realised in conditions in which it cannot be stopped because the existing conditions are not stable. In contrast we should name, for example, the stable form of movement of insects, which is characterised by always having at least three legs touching the floor at any time. They describe a surface on the ground, i.e. the positions where the legs touch the ground are not in one line /2/.

2.1 The Arangement of Legs Provides Stability

That a surface is exactly defined by three points, or, in other words, a body with three legs on a surface is stable, is the reason for animals to usually have at least four legs. Also, animals usually prefer stable forms of movement where at least three legs describe a surface on the ground while the other leg(s) move to a new position. The energy required to co-ordinate the leg movement increases with the number of legs, while at the same time the energy needed to keep up stability decreases. During the process of leg movement in a quadruped (horse, dog, cat, etc.) leg movement must be exactly co-ordinated. However, in a six-legged being (crabs, insects, etc.) there are a multitude of possible stable leg positions due to the 2x3 arrangement of the legs /3/. Thus, the fewer legs an animal has the more likely it is to get into an unstable state. For this reason the energy necessary to regulate stability must be greater. Only creatures with highly developed brains are capable of moving continually on two legs. The high amount of energy needed to stabilise movement is justified by optimal mobility and the ability to move in any direction from any position.

The information about the movement of creatures has been transferred to walking robots. Most of these robots have six legs and copy the movement of insects. These walking robots can be used on even surfaces and open terrain. With the relevant co-ordination it is even possible to overcome obstacles. The long body form is a distinct disadvantage as it prevents manoeuvring in narrow passages and movement up spiral stairs. Additionally, it requires a greater amount of energy to co-ordinate leg movement and to calculate the individual leg load necessary because the positioning of the feet in uneven terrain cannot be predicted.

Principally, two-legged robots have the same flexibility of movement as humans, however, due to the technical difficulties in controlling dynamic movement, robots are not yet able to make an unlimited amount of controlled steps. Continual improvements in sensors and information processing make definite improvements possible in the near future.

3 The New Robot Concept

At the University of Paderborn, a concept for a two-legged robot has been developed using a systematic process. This combines the advantages of a six-legged creature with the mobility of a two-legged one. The aim was to develop a modular robot that would be capable of offering maximal mobility with minimal control technology. Also, the weight should be low enough to make it suitable for use by mobile units of the fire brigade.

3.1 The Feet

The invention of a new foot form that enables a two-legged walking machine to be able to move in a stable way was the basis of the design. Each foot is shaped so that it forms a standing surface within which the machine's centre of gravity falls, and can be placed inside the area covered by the other foot. The foot shape enables the robot to stand stable on each foot and to move its body relative to the foot inside an area limited sideways by the expansion of the foot and forwards and backwards by the maximal extension of the leg. Furthermore, by continually transmitting both foot surfaces the robot can be moved in any direction. At no time during this movement does the robot become unstable. As the foot shape allows movement in any direction, it is necessary to have a leg construction that enables equal movement in any direction, i.e. it has a rotation-symmetrical working space and enables the foot to be positioned in any of six degrees of freedom.

Figure 1 shows by example of a "one-legged-structure" the stabile area. This area is limited by projection of the foot-area along the direction of gravity force. The area is limited by the stability border. Transgressing this border would cause the robot to be in an unstabilized mode.

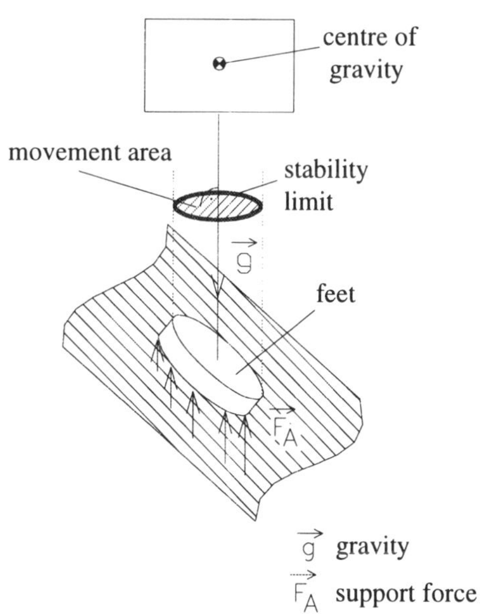

Fig. 1. project movement area with stability limit

Fig. 2. Form of the feet

3.2 The Legs

All these criteria are covered by constructing the legs as hexapods. The hexapod, also known as the "Steward platform" shown in figure 3, has been used successfully for many years in the movement devices of flight simulators /4/. It consists of six legs each of which can be altered in length. The legs join two platforms so that by altering the length of the legs the position and orientation of the platforms to one another inside the construction limits can be varied in any way. Further characteristics are the high rigidity of the construction, the simple rotationally symmetrical construction and the high load capacity relative to the low weight of the hexapod. These characteristics

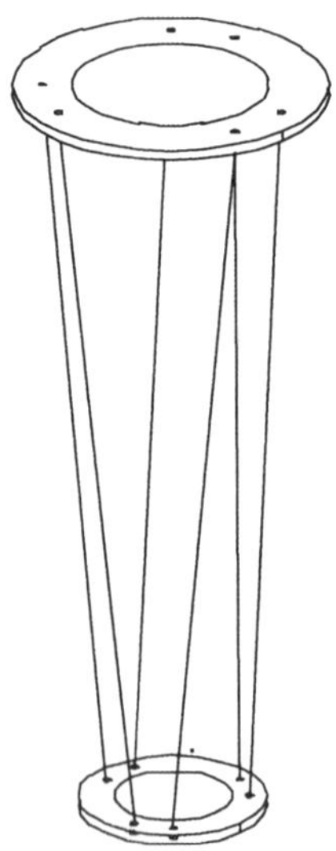

Fig. 3. Hexapod

are the reasons for the system having recently been used in the construction of various manipulating systems.

The construction of the leg as hexapod combined with the foot shape described above enable the robot's body to be moved inside the constructive limits in any direction in relation to the foot on the floor as well as the other foot. This enables the robot to perform manipulating tasks and walking operations. A robot of this type is capable of moving on slopes, straight and spiral staircases and, while moving and stationary, of performing exactly controlled activities.

The figure 4 shows the complete assembly. Table 1 is an overview of the important design parameters.

Table 1. Technical Dates of the Two-Legged-Robot

overall width	700 mm
max. overall height	1400 mm
min. overall height	900 mm
max. step length	400 mm
max. stroke length of a leg	500 mm
motion speed	150 mm/sec
weight of the mechanical unit	80 kg
max. balancing moment of the MVS	25 Nm
max. allowable tilting moment	80 Nm

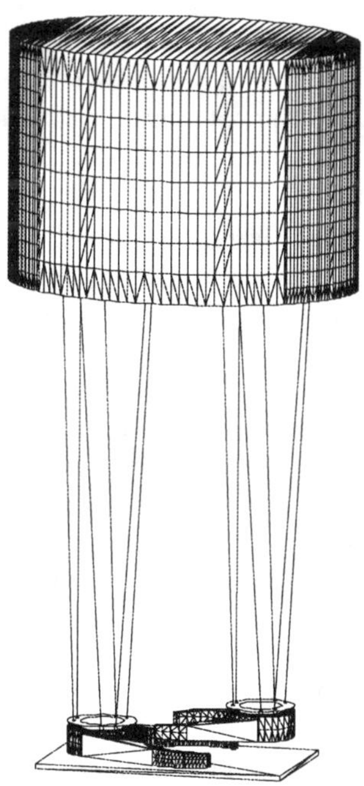

Fig. 4. Two-Legged-Robot

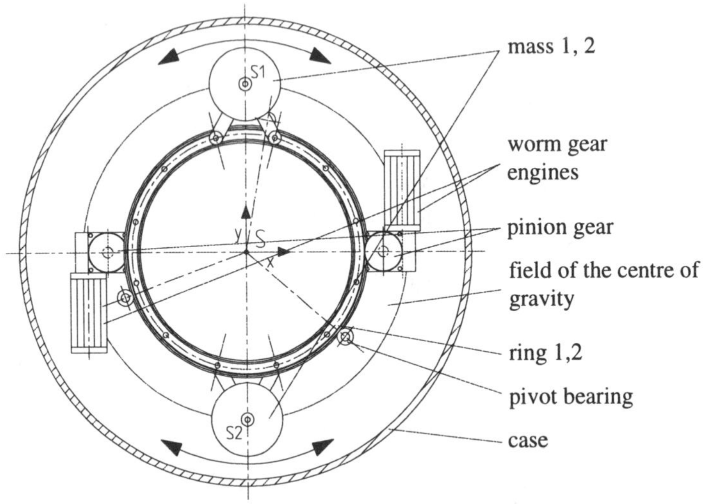

Fig. 5. The Mass Displacement System

3.3 The Mass Displacement System

To extend the movement of the machine's centre of gravity outside the foot area a special device that enables the machine's centre of gravity to be moved within the construction limits is used. At a further stage of development this mass displacement system (*MassenVerlagerungsSystem MVS*) will also be used for dynamic impact stabilisation.

This function is fulfilled by two excentric co-axial masses that are capable of rotating on a ring guide and that are driven by a servo drive. The positioning of these masses can translate the common centre of gravity inside the shaded area shown in figure 5.

3.4 The Control System

Because of the modular construction the control tasks can be based on solutions for individual systems with low complexity. The co-ordination of these systems takes place through a higher ordered unit. The legs are completely identical in their construction and control systems. The control task for the legs is limited to position control of its linear units.

The desired position of the foot platforms at the end of the leg and thus the position and orientation of the foot surface are given by a separate system that is responsible for the route planning. Load cells contained in the linear unit of the leg give the route planner the required information about the foot load.

With the aid of the foot geometry and the measured foot load, the area can be calculated in which the body may be moved relative to the feet with a given acceleration without the system becoming unstable. If movement is planned outside this area, one leg is moved while the robot is stationary, in such a way to enable this foot to be newly positioned. Because the foot surfaces can be placed inside one another they form a redundant system in view of the stability of the body. As long as the foot surfaces do not need to be re-positioned the robot stands on one foot and the other foot is parallel to and above it. The stressed leg can be re-positioned by moving the appropriate linear units into the new position. Control of stability is guaranteed through the foot on which the machine is standing. If the calculated load moment is too high the moving foot is stopped and lowered. Because the dynamic force on the moving foot is relatively small compared to the static forces that act during a collision with an obstacle the moving foot is used to measure the load during movement. This measurement is used as an input quantity for collision recognition.

3.5 The Concept for Collision Recognition

A sudden change in the load is interpreted as a collision of the foot with an obstacle and leads to a stop in movement. When lowering the foot, the collision means that the foot has been placed on the surface. This starts an algorithm that, under consideration of the measured moment in the foot surface centre of gravity, places the foot squarely on the surface. If the foot collides during movement to a desired position the movement is stopped and then moved back a given distance in the opposite direction. After moving back, the foot is then lowered as described above.

4 Module Oriented Optimisation through a Structural Construction

The algorithm for calculating the position of the foot relative to the surface and the positioning of the foot as well as decision making in case of a collision or stability loss can be tested with simple test equipment. These factors can be optimised independently of each other. Errors in movement behaviour can be relatively easily allocated to the modules and corrected. The development of the control tasks can take place independent of the whole construction subject to the interface. After a short development period, a basic version can be implemented that can be immediately employed in a limited task area. It can be further developed by information gained while in operation and can be adapted to take on new tasks.

The advantage of the modular design of the construction offers itself to the development of the robot as a project in the field of simultaneous development at university. At the beginning of the project the development team were working on the following projects:

- construction of the feet
- development of a prototype of the "Stewart platform"
- design of a device to help the robot back on its feet, without outside help, after falling
- construction of the mass displacement system (MVS)
- development of an alternative solution for driving the "Stewart" platform with the aim of making the legs lighter and more dynamic.

The aim during the first stage was to simultaneously develop a prototype in a short amount of time that could be used as a functional model for the development of the modules used for planning routes, recognising collisions and measuring the foot forces. The prototype has been designed so that further development is possible by merely changing the individual parts. Basic units such as the MVS, feet and body remain the same. To a great extent, the information about the characteristics of this unit gained through the prototype can thus be transferred to the new models. The number of possible source errors can be lessened by optimisation. In a later development stage these components will be adapted to suit the specific requirements. Further details will be given in the oral presentation.

5 Conclusions

Future development is aimed at the optimization of control algorithm in order to improve system stability during locomotion.

A future focal point in research is developing strategies towards coping with obstacles. Development is enhanced by employing virtual reality (VR) and simulation of the mechanical components. The gained results are to be directly transfered towards the mechanical unit. This allows for a highly flexible layout of the controls, which can be swiftly adapted to changing environment data.

6 References

/1/ *Alexandre, P.; Preumont, A.* Walking Machines: A State of the Art in Europe. Zeitschriftenaufsatz: European Journal of Mechanical Engineering, Band 40, Heft 1, Seite 27-33, 1995

/2/ *Pfeiffer, F.; Cruse, H.* Bionik des Laufens - technische Umsetzung biologischen Wissens. Zeitschriftenaufsatz: Konstruktion, Band 46, 1994, Heft 7/8, Seite 261-266

/3/ *Ferrell,C.:* A Comparsion of three Insect - Inspired Locomotion Comtrollers. Zeitschriftenaufsatz: Robotics and Autonomous Systems, Band 16 (1995) Heft 2-4, Seite 135-139

/4/ *Fichter, E. F.* A Stewart Platform-Based Manipulator: General Theory and Practical Construktion. Zeitschriftenaufsatz: The International Journal of Robotics Research, Band 5, Heft 2 Sommer 1986. Massachusetts Institute of Technology.

Design of Active Links for Modular Robot[1]

Wei Tech Ang and Ming Xie
(techang@post1.com) (mmxie@ntu.edu.sg)

School of Mechanical and Production Engineering
Nanyang Technological University, Singapore 639798

Abstract. This paper presents a new design concept of modular robot. Our design of modular robot features two distinctive characteristics (i) we do not separate links and joints, the modules are *active links*[2]; (ii) each module has two degrees-of-freedom: a rotational DOF and an angular DOF. The active links are serially connected to form a modular robot manipulator. Making use of the design and geometric advantages of the active links, we have also developed a closed-form geometric inverse kinematics method that is more efficient than the conventional analytical and numerical methods. This allows more computational resources to be allocated to the vision system that is computationally intensive so as to achieve real-time guidance.

1. Introduction

The conventional applications of robotic system involve manufacturing and assembly tasks. In recent years, much effort has been made in developing new field of non-manufacturing applications for robots. The perceivable trend of technology is moving towards the direction of service robot, i.e. a robot that can provide services for both human and machine [1,2,3]. Saturation in the industrial robot market, the increasing numbers of elderly people in most developed countries, the shortage of manpower in hospitals, the involvement of human in hazardous or harsh working environment etc. are some of the important factors that are providing the necessary background for the development of service robots.

Here, we are developing a general purpose vision guided mobile service robot for structured indoor environment. The robot is targeted to perform a large variety of tasks ranging from house-keeping to mail delivery to machine maintenance. In order for the robot to adapt to tasks of different nature, we want it to possess both software and hardware flexibility. The re-programmability nature of software systems has made software flexibility easily attainable. Hardware flexibility, on the other hand, can be achieved by adopting a modular design for our robot arm. *Modular robot* introduces a new dimension to hardware flexibility and ensure an individual global optimal arm geometry or degree-of-freedom for each task in hand.

Research in the area of modular robot design has been addressed by both the robotics system developers/ manufacturers (e.g. TOMMS by Toshiba Corp. [4]), as

[1] This project is supported by the Ministry of Education, Singapore under the grant RG72/96
[2] "Active" in the context of modular robotics refers to the module that is the motion provider.

well as academic institutions (e.g. the "Modular Robot System" by University of Stuttgart [5], the "Reconfigurable Modular Manipulator System (RMMS) by Carnegie Mellon University [6, 7] and a nameless modular robot system by University of Toronto [8] and RISC modular robotic system by Nanyang Technological University [9]). Some other literature in this area had presented pneumatic and electro-mechanical modular robots for small parts assembly [10, 11], for spot welding [12], and other material handling applications [13]. Modular robotics has also been of interest in the context of space stations as reported in [14].

The TOshiba Modular Manipulator System (TOMMS) in [4] consists of one type of joint module (active) and one type of link module (passive) with one degree-of-freedom and the control unit. The system allowed two types of configuration, namely the horizontal type and the vertical type. A single control software was developed such that it could control any configuration of assembled manipulators without special handling.

The Modular Robot System by University of Stuttgart [5] consists of a variety of rotational joints actuated by AC motors in conjunction with differential gears and links of square cross-section. The two modular units manufactured are : a two DOF "main axis" (large module) joint and a 2 DOF "wrist" (small module). Due to its weight the main axis can only be used at the base, it would have to be substantially reduced in weight (and torque) for use at a location such as the elbow of the arm.

The "Reconfigurable Modular Manipulator System" (RMMS) by CMU in [6,7] consists of "rotate" (roll) and "pivot" (pitch) type one DOF rotary joints actuated by DC motors in conjunction with harmonic-drive transmissions, and links of circular cross-section. Each module is self-contained with sensor interfaces, power amplifiers and control processors. The joint designs were accomplished by using V-band flanges. It was commented in [8] that this joint design were more suitable to modular robotics than those reported in [5].

The modular robot by University of Toronto in [8] consists of rotary joints (active with one DOF) of two sizes, prismatic joints, a Roll-Pitch-Yaw end-effector and passive connectors. To avoid difficulties and efficiently connecting a vast number of wires, a serial communication method was selected. The design of the connector made use of a 45° connection scheme which enable the arm to be reconfigured into in-plane or out-of-plane configuration.

The RISC modular robot system by Nanyang Technological University, Singapore in [9] once again consists of two types of one DOF active joint modules: revolute and prismatic, and two types of passive link modules with "connecting ports". The "connecting ports" were the places where the joint modules were attached, a cubic box link had six ports while a square prismatic links had ten.

The objective of our research work is to propose a new concept of modular robot that made use of our two DOF active links. The two main research interests in modular robotics have been mechanical design and motion control. The following sections describe the proposed design of the active links and the geometric inverse kinematics method.

2. Design of Modular Active Links

Our design of modular robot features two main differences as compared to those as reported in [4,5,6,7,8,9]:

(i) There is only one type of module, i.e. we do not separate links and joints. Therefore, the modules are *active links*.

(ii) Each module has two degrees-of-freedom: a rotational DOF with a limit-to-limit range of 360° and an angular DOF with a limit-to-limit range of 180° as shown in Fig. 1.

To cater for different mechanical stiffness and torque requirements, three sizes of modular active links are designed namely Type L, M and S. Every module houses two YASKAWA Σ series AC servo motors, one for each DOF. Harmonic Drive transmission is proposed in our design to provide the necessary torque because of its following features:

(i) High positional accuracy and repeatability, since 10% of the teeth are always meshed;

(ii) Low backlash, (≈1 arc minute) due to pre-loading and radial mesh of the teeth;

(iii) High efficiency (approximately 79 percent for 100:1 reduction ratio) because of the single stage speed reduction.

The design of a modular link is presented in Fig. 2. The modules are to be serially assembled to form a manipulator. The rotational DOF of a module is connected to the angular DOF of its predecessor. The rotational motor is housed in a hollow cylindrical motor housing, whose base flange is connecting to the base flange of the previous module. The motor output shaft is directly coupled to the Harmonic Drive, and the output spline of the Harmonic Drive is attached to a rotating plate which connects to the upper angular DOF section. For the angular DOF, the motor power is fed into the horizontally oriented Harmonic Drive via timing belt and pulleys. A output shaft extension is bolted to the output spline of the Harmonic Drive which drives the connector flange and hence the next module.

To ensure modularity and compatibility, (see Fig, 3) the dimensions of the connector and base flanges, the bolt holes size and positions are standardised throughout all types of modules. When Module i is removed, Module $i+1$ can be connected to Module $i-1$ without problem. The size of the hollow cylinder and the distance between the flange legs, however, are different in order to suit the different sizes of motors and Harmonics Drives. The slot in the connector flange enables the signal and power cables to run through without difficulty.

The central control unit (CCU) and power source are separated from the arm. In our design, the power and communication cables are run internally, the signal and power cables have to run through the modules to reach the motors and sensors. To avoid interference problem, the signal wires and power cables of each module are in separate strands. The strands of signal wires through a module are bundled into a single strand, and so do the power cables. Fig. 4 shows the distribution of the wires running internally to the four modules, two bundles of four strands run through the base, two bundles of three through the Module 1-2 interface, two bundles of two through Module 2-3 interface and two strands through Module 3-4 interface.

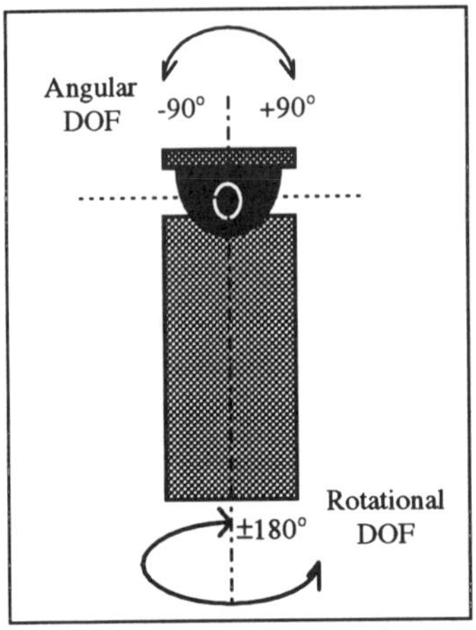

Fig. 1. The two DOF active link

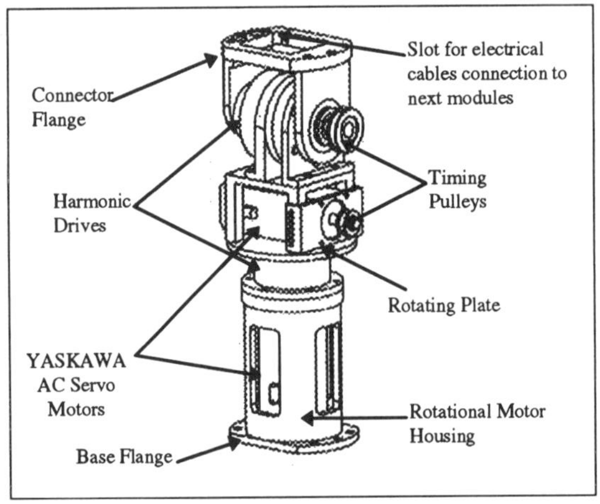

Fig. 2. Design of a modular link

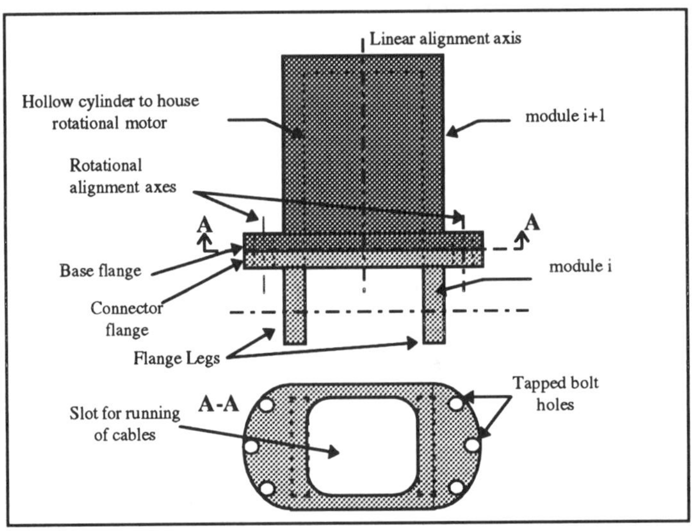

Fig.3. Design of connecting flanges

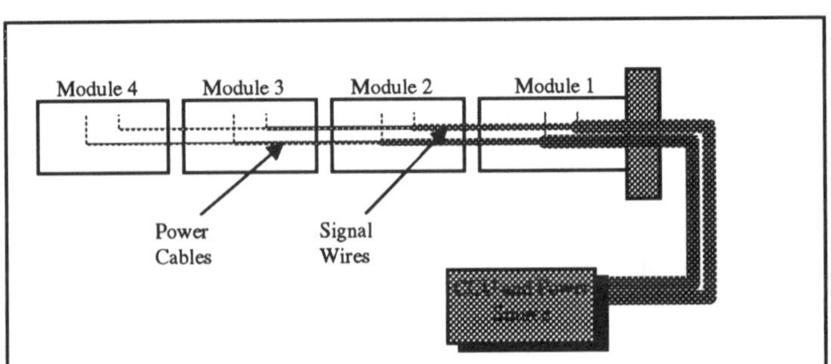

Fig. 4. Distribution of internal signal wires and power cables

Each module is self-contained with its wiring networks and communicate with its predecessor and successor through wire connectors at the base and connector flanges. In a module, there are male type wire connectors at the base flange, evenly spaced around the internal circumference of the hollow rotational motor housing. These male connectors are connected to the female type wire connectors fixed to the four sides of the rectangular slot at the connector flange of the predecessor module.

Fig. 5 explains the arrangement of wires within a module, the wires are constrained at various appropriate positions and enough slacks (additional length) are given to ensure limit-to-limit operations.

The physical and technical specifications of the modular active links are summarised in Table 1.

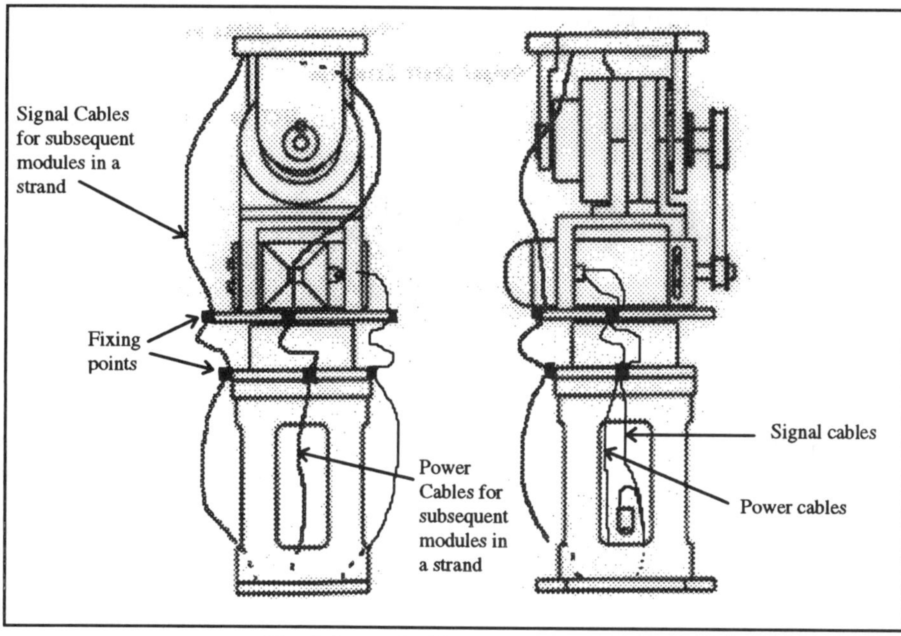

Fig. 5. Internal wiring network in module

Table 1. Technical specifications of modular active links

Item		Active Link Type L	Active Link Type M	Active Link Type S
Dimension l x d (mm)		340 x Ø140	310 x Ø130	250 x Ø110
Weight (kg)		4.4	3.5	2.0
Motor Power Rating (W)	rotational	200	100	30
	angular	100	100	30
Rated Output Torque (Nm)	rotational	60.7	25.3	7.5
	angular	40.5	30.3	9.6
Maximum Output Torque (Nm)	rotational	150.9	75.1	22.1
	angular	120.1	91.0	28.2
Rated Output Speed (rpm)	rotational	30	30	30
	angular	15	20	20

3. Modular Arm

3.1 Mechanical Design

Fig. 6 shows a 4-links manipulator made up by four serially connected active links. The degree-of-freedom of the manipulator is scaleable. If an application requires higher flexibility and mobility, additional modules can be added to provide more degrees-of- freedom. Conversely, the DOF of the manipulator can be scaled down when the application is simple. A number of configurations are possible with different combination of the three types of modules (i.e. Type L, M and S), depending on the requirement of the task in hand. Fig. 7 shows a modular arm with a 4-links M-M-S-S configuration.

The technical and physical specifications of three different configurations of modular robot arm (not including the end-effector), namely 4-links configurations L-M-M-S and M-M-S-S, and 3-links configuration L-M-S are summarised in Table 2.

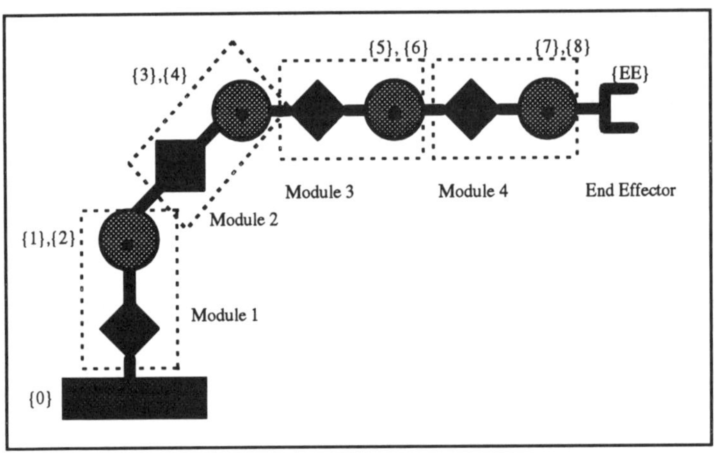

Fig. 6. Serially connected modular arm of 4

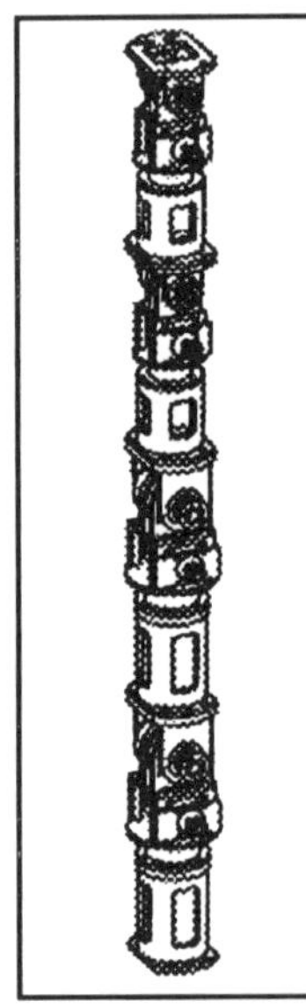

Fig.7. Modular arm with M-M-S-S configuration

Table 2. Technical specifications of different configuration of modular arm

Item	Configurations		
	L-M-M-S	M-M-S-S	L-M-S
Design Payload (kg)	4.5	2.0	9.0
Tip Velocity (m/s)	2.73	2.54	1.76
Reach (m)	0.87	0.81	0.56
Total Weight (kg)	13.4	12.8	9.9
Overall length (m)	1.21	1.14	0.9

3.2 Control

3.2.1 System Hardware

The modular arm is to be guided by real-time vision system. The central control unit consists of a Pentium 200 PC with 32MB RAM, vision system (2 CCD cameras and a SMT303 frame grabber card), a TMS 320C81 DSP based image processing system by LSI and a NextMove PC 8-axes motion controller card.

3.2.2 The Geometric Inverse Kinematics Method

Instead of getting the inverse Jacobian matrix (or the Pseudoinverse matrix in cases where there is redundancy in the number of DOF) from pure mathematical manipulations by the conventional analytical [15] and numerical [16] methods, we are proposing a geometric inverse kinematics method that exploits the modular characteristic and the design advantages of the two DOF active links. The objective of the geometric inverse kinematics method is to achieve a good solution (not necessary the best solution) by injecting geometric insight into the analysis. Due to the constraint in the number of pages, only a brief description of the method will be discussed.

Consider a desired position $\boldsymbol{P_0}$ in 3D space and orientation $\boldsymbol{V_0} = [\hat{n}_0 \quad \hat{s}_0 \quad \hat{a}_0]$ with respect to the base co-ordinate system {0} of the modular arm (refers to Fig.6 for assignment of co-ordinate frames). Refer to Fig. 8, now we offset Frame {0} in the direction of Z_0 such that Frame {0}, {1} and {2} coincides.

Let $\boldsymbol{T_0}$ be the target position offset from $\boldsymbol{P_0}$ in the direction of $-\boldsymbol{a_0}$ with an magnitude equal to l_e the length of the end-effector, $\boldsymbol{T_0} = [\,{}^0T_x \; {}^0T_y \; {}^0T_z\,]^T = \boldsymbol{P_0} - l_e\,\boldsymbol{a_0}$ (Refer to Fig.8). If we rotate Frame {0} by $\theta_1 = \tan^{-1}({}^0T_y / {}^0T_x)$ about the Z_0 axis such that $\boldsymbol{T_0}$ is housed in the X_1-Z_1 plane, we transform the problem into a 2-dimensional.

Once the problem is simplified to 2D, the inverse kinematics or joint angles for subsequent DOF can be solved by simple geometric manipulations. The solution in the case of 3-links configuration is trivial as shown in Fig. 9. The solution of a 4-links configuration, on other hand, can be further broken down into two different implementation schemes as shown in Fig. 10 depending on the distance from $\boldsymbol{T_0}$ to {0}. Details in dealing with different orientations of the tooltip will not be discussed.

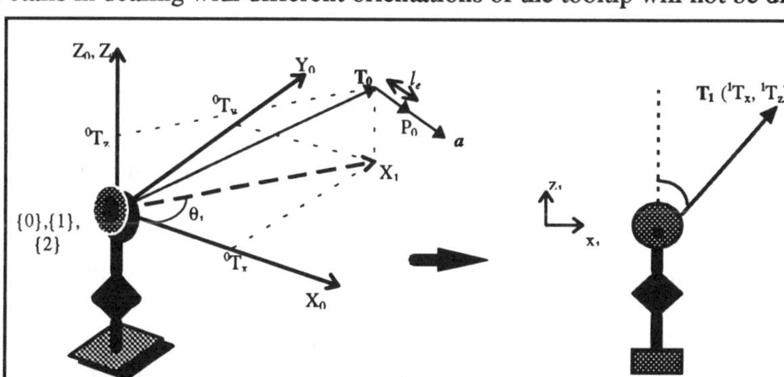

Fig. 8. Transformation into a 2D problem

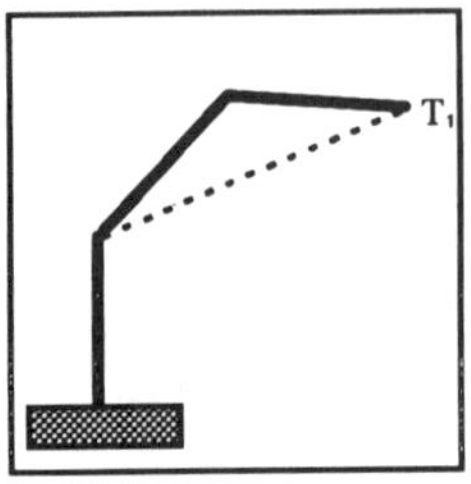

Fig.9. Inverse kinematics solution of 3-link configuration

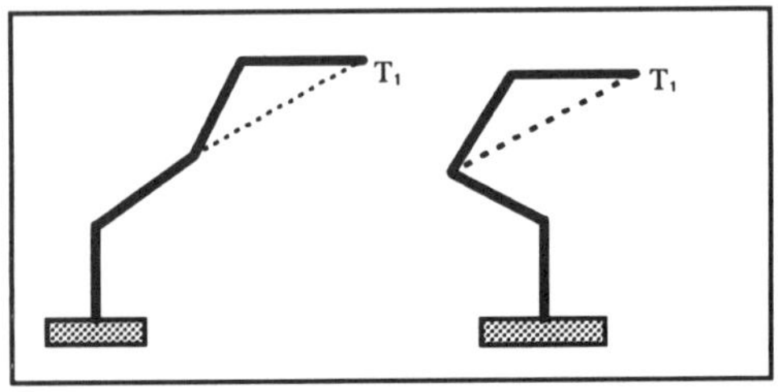

Fig.10. Two implementation schemes of inverse kinematics solution of 4-links configuration

The geometric inverse kinematics method of a 4-links M-M-S-S configuration is formulated using Mathematica Ver.2.2, together with the closed-form polynomial method proposed by Raghavan et. al. [15] and the numerical method proposed by Kholsa [16]. The efficiencies of the three algorithms are evaluated and simulation results have shown that the geometric inverse kinematics algorithm consumes 7.1 and 20.6 times less CPU time than the numerical and the closed form polynomial algorithm respectively. This would allow more computational resources to be allocated to the vision system that is computationally intensive so as to achieve real-time guidance.

4. Modular Service Robot

The development of modular active links and modular arm is part of the effort to develop a general purpose vision guided service robot for structured indoor environment (Fig.11). The main components that made up our service robot consist of a mobile base of 3 DOF (2 translational DOF in the X-Y or ground plane and 1 rotational about the Z-axis), one or two modular robot arms (depending on the application) and a robot head with a pair of CCD cameras mounted on a modular neck. The computing system, image processing system and the motor amplifiers are all housed in the mobile base.

4.1 System Advantages

The fundamental aim of the design schemes adopted by the other modular robot developers is to achieve an optimum hardware configuration for each task in hand. Most of them had chosen to have a variety of links (passive) and joints (active) of different sizes, and hardware flexibility is achieved by different combinations of the stock of modules. The fact that there might be numerous possible configurations for a given task has exerted tremendous pressure on the flexibility of the control software. The software has to decide on the optimum DOF, the best combination of hardware, check the singularity points with respect to the nature of the given task, perform calibration, analyse the dynamics and re-evaluate the kinematics of the new configuration.

One major draw back of these design schemes is that they are built around the concept of achieving an optimum configuration for a particular task. This means that the optimal configuration for a grasping operation might not be able to tighten screws due to singularity problem. For a robot that is required to perform a variety of tasks,

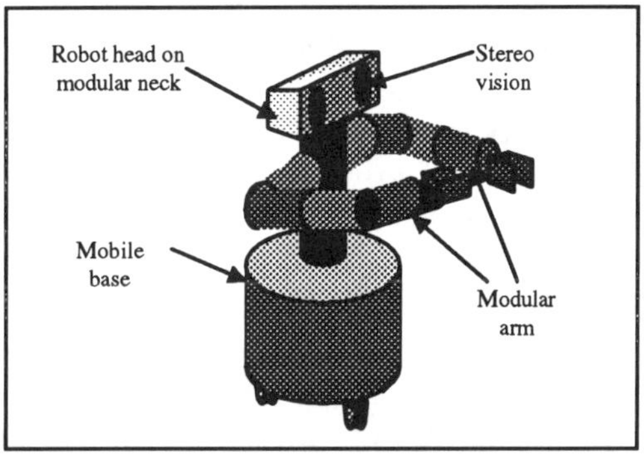

Fig. 11. A schematic view of the service robot

the engineers would have to reconfigure the robot for almost every task. This situation is definitely intolerable for our application as a general purpose service robot. This again comes back to the problem of limited hardware flexibility, which means that the fundamental problem has not yet been solved.

In addition, most researchers has been spending a lot of effort in achieving hardware flexibility and neglected system simplicity and efficiency which is utmost important for real-time control. In other words, they are heading towards the direction of developing a complicated system with ideal flexibility and adaptability in both software and hardware which requires a great deal of engineering effort and money.

The essences of our proposal are design simplicity, control efficiency and hardware flexibility through control strategy. Firstly, instead of *the optimum* configuration for a given task, our emphasis is placed on achieving a robot with *an optimal* DOF for a variety of tasks. With the additional 3 DOF from the mobile base, we are able to adopt a control strategy that is inspired by the way human being perform tasks. When the robot arm is unable to reach certain positions or orientations along the planned trajectory, the mobile base will re-position the robot so that the desired positions and orientations are within its reach and out of the singularity zones. In this way, nearly all positions and orientations in the workspace can be reached without difficulties. This is our proposed solution to "hardware flexibility".

Secondly, since there is only one type of module in our design, and all modules are serially connected, the tasks of dynamics analysis, forward kinematics and calibration of a new configuration is greatly simplified. Hence, much engineering effort and resources can be saved in the software development.

Last but not least, by taking the design and geometric advantage of the active links we are able to develop a simple and efficient inverse kinematics scheme. The significant reduction of CPU consumption time of the modified geometric method over both the conventional closed-form and numerical methods has made more computational resources available to the real-time visual servoing system.

5. Conclusion

We have presented our research work on the design and control of a new concept of modular robot arm. The modular robot arm is made up of serially connected active links with 2 DOF. Three sizes of active links have been design to meet different requirement of torque and material stiffness depending on the task in hand. Exploiting the design and geometric advantage of the active links, we have developed a simple and efficient geometric inverse kinematics scheme. The modular arm is to be guided by real-time vision system, and this new inverse kinematics scheme allows more computational resources to be allocated to the computational intensive real-time visual servoing.

References

[1] Hasegawa Y., "A New Wave of Japanese Robotization", *Proceedings of the 4th International Conference on Control, Automation, Robotics and Vision*, Singapore, Dec.1996, pp.1640-1645.

[2] Xie M., "Towards visual Intelligence of Service Robot", *Proceedings of the 4th International Conference on Control, Automation, Robotics and Vision*, Singapore, Dec 1996, pp. 2217-2221.

[3] Dario P., Guglielmenlli E., Genovese V. and Toro M., "Robot assistants: Applications and evolution", *Robotics and Autonomous Systems*, Vol.18, 1996, pp. 225-234.

[4] Matsumaru T., "Design and Control of the Modular Robot System: TOMMS", *IEEE International Conference on Robotics and Automation*, 1995, pp. 2125-2131.

[5] Wurst K.H., "The Conception and Construction of a Modular Manipulator System", *Proceedings of the International Symposium Industrial Robotics*, 1986, pp.37-44.

[6] Schmitz D., Khosla P. and Kanade T., "The CMU Reconfiguarble Modular Manipulator System", *Proceedings of the 19th International Symposium on Experimental Robotics*, Sydney, Australia. 6-10 Nov, 1988, pp. 473-488.

[7] Paredis C.J.J., Brown H.B., Casciola R.W., Moody J.E. and Khosla P.K., "A Rapidly Deployable Manipulator System", *In International Workshop on Some Critical Issues on Robotics*, Singapore, 1995, pp. 175-185.

[8] Cohen R., Lipton M.G., Dai M.Q. and Benhabib B., "Conceptual Design of a Modular Robot", *ASME J. Mechanical Design*, 114: pp. 117-125, Mar 1992.

[9] Chen I-M., "On Optimal Configuration of Modular Reconfigurable Robots", *Proceedings of the 4th International Conference on Control, Automation, Robotics and Vision*, Singapore, Dec.96, pp.1855-1859.

[10]Gruzdev, Korytko and Yurevich, "Modular Electro-Mechanical Industrial Robots", *Electrotekhnika*, Vol.55, No.4, 1984, pp. 4-7.

[11]Harrison, Weston, Moore and Thatcher, "Industrial Applications of Pneumatic Servo-Controlled Modular Robots", *Proceedings of the 1st National Conference on Production Research (UK)*, 1986, pp.229-236.

[12]Smith and Crazes, "Modularity in Robots-Technical Aspects and Applications", *Proceedings of the International Conference on Robotics and Automotive Industry (UK)*, 1982, pp.115-122.

[13]Muck and Mammern, "Modular Mechanical Engineering", *Proceedings of the International Conference on Advances in Manufacturing*, 1984, pp. 271-282.

[14]Jenkins L.M., "Telerobotic Work System-Space Robotics Applications", *IEEE International Conference on Robotics and Automation*, 1986, pp. 804-806.

[15]Raghavan M. and Roth B., "Inverse Kinematics of the General 6R Manipulator and Related Linkages", *Transactions of the ASME*, Vol. 115, 1993.

[16]Khosla P.K. and Neuman C.P., "An Algorithm for Seam Tracking Applications", *The International Journal of Robotics Research*, Vol.4, No.1, MIT, 1985.

[17]Xie M., "Robotic Hand-Eye Co-ordination: Solutions with uncalibrated stereo cameras", *Journal of The Institution of Engineers, Singapore*, Vol.37, No.1, 1997.

An Adaptive Tracking Controller for Robots Including Motor Dynamics

Yin Guang and Song Qing

School of EEE, Nanyang Tech Univ, Singapore

Abstract. This paper presents a direct adaptive tracking control scheme for robotic applications with motor dynamics. It is proved that the tracking error is semi-globally asymptotically stable and all parameters are bounded. Simulation results are provided to illustrate the performance.

1 Introduction

The main focus of this paper is to develop an observer-based adaptive control scheme which considers both robotic manipulator and motor dynamics without using the measurement of acceleration. We attack the same problem posed in [1], that is, the tracking control of rigid-link electrically-driven robots with the same assumption regarding actuator parameters as in [2], but use a different approach, *i.e.* the adaptive observer-based control algorithm. We first construct a velocity estimator and then design a stable-feedback embedded controller which uses only position measurements. The voltage input control signal,therefore, does not depend on the acceleration with the estimated velocity variable. This adaptive control law is designed to achieve semi-globally asymptotical stability of the link position tracking error in the face of parameter uncertainties. The result is an extension of the work of Slotine and Li [3], adding the non-negligible motor dynamics. In this way, we combine the approaches of embedded control as in [1] with the adaptive observer-based control concept to guarantee semi-globally asymptotical stability.

2 Preliminaries

Assume that the robot under consideration is driven by armature-controlled DC motors with voltages as inputs. The model of a robot manipulator including motor dynamics and some physical properties of the robot is known, and can be used in the control synthesis procedure. Following [4], a model of the robot manipulator, including the dynamics of DC motors, can be expressed as:

$$D(q)\ddot{q} + C(q,\dot{q})\dot{q} + g(q) = K_t I \tag{1}$$

$$L\dot{I} + RI + K_b\dot{q} = u \tag{2}$$

where $q \in \Re^n$ represent link angles; $D(q) \in \Re^{n\times n}$ is the link inertia matrix, $C(q,\dot{q})\dot{q}$ represents Coriolis and centrifugal force; $g(q)$ represents gravitation terms. $K_t \in \Re^{n\times n}$ is a diagonal positive definite matrix. The vectors u and $I \in \Re^n$ represent the armature input voltage and current. $R, L, K_b \in \Re^{n\times n}$ are diagonal positive definite matrices of the resistance, inductance and back EMF of the motor respectively.

In this study $\|x\|$ denotes the Euclidean norm of an arbitrary vector $x \in \Re^n$ [5]. This norm induces a norm $\|\cdot\|_i$ on the space $L(\Re^n, \Re^n)$ of linear transformation such that for $A \in L(\Re^n, \Re^n)$, we define:

$$\|A\|_i := \sqrt{\lambda_M(A^T A)} \tag{3}$$

where $\lambda_M(A^T A)$ is the largest eigenvalue of the (positive semi-definite) symmetric transformation $A^T A$. Finally, assuming that $\|A(t)\|_i$ is bounded, we define:

$$\alpha_m := \sup_{x\in\Re^n} \|A(t)\|_i, \quad \alpha_M := \inf_{x\in\Re^n} \|A(t)\|_i \tag{4}$$

Property 1 — Inertia [6]: The inertia matrix $D(q)$, defined in (1), is symmetric, positive definite and uniformly bounded by the following condition:

$$\lambda_m\|x\|^2 \le x^T D(q)x \le \lambda_M\|x\|^2 \tag{5}$$

It is easy to see then that $D^{-1}(q)$ is also uniformly bounded since:

$$\frac{1}{\lambda_M}\|x\|^2 \le x^T D^{-1}(q)x \le \frac{1}{\lambda_m}\|x\|^2 \tag{6}$$

where λ_m and λ_M are positive scalar constants and $x \in \Re^n$ is an arbitrary vector. It should be noted that the above bounds only hold for revolute robots.

Property 2 — Skew Symmetry [6]: The inertia matrix $D(q)$ and the Coriolis and centrifugal matrix $C(q,\dot{q})$ are not independent, and the matrix $\dot{D} - 2C$ is skew symmetric by selecting a suitable definition of the matrix $C(q, \dot{q})$. However, the following quadratic form is always zero regardless of the choice of $C(q, \dot{q})$.

$$x^T(\dot{D}(q) - 2C(q, \dot{q}))x = 0 \tag{7}$$

for an arbitrary vector $x \in \Re^n$.

Property 3 — Coriolis and centrifugal Matrix: Given arbitrary vector v_1 and v_2, there exists:

$$C(q, v_1)v_2 = C(q, v_2)v_1 \tag{8}$$

Property 4 — Boundedness on Coriolis and centrifugal Matrix: The norm of $C(q, \dot{q})$ satisfied the inequality:

$$\|C(q, \dot{q})\| \leq k_c\|\dot{q}\| \tag{9}$$

where k_c is a constant.

Property 5 — Linear in Parameters: The manipulator dynamic equation (1) is linear in its parameters:

$$D(q)\ddot{q} + C(q, \dot{q})\dot{q} + g(q) = Y(q, \dot{q}, \ddot{q})\theta \tag{10}$$
$$D(x_1)\phi_1 + C(x_1, \phi_2)\phi_3 + g(x_1) = Y(x_1, \phi_1, \phi_2, \phi_3)\theta \tag{11}$$

where $\theta \in \Re^p$ is a vector of unknown constant parameters and $\phi_1, \phi_2, \phi_3 \in \Re^n$. Matrix $Y \in \Re^{n \times p}$ is usually known as the manipulator regressor [6].

In this paper, the parameter θ is assumed to be unknown, while the structure of the regressor matrix Y is assumed to be available. The target is to design a control law to ensure asymptotic tracking of the desired trajectory q_d. Let $q_d(t) \in C^3$ denote a desired link trajectory which is continuously differentiable up to the third order, and the desired trajectory and its first two time derivatives are bounded. The $3n$-dimensional state-space Equation (1) and (3) for the model of the robot manipulator and motor are given by:

$$\dot{x}_1 = x_2 \tag{12}$$
$$\dot{x}_2 = D^{-1}(x_1)[-C(x_1, x_2)x_2 - g(x_1) + K_t x_c] \tag{13}$$
$$\dot{x}_c = L^{-1}(-Rx_c - K_b x_2 + u) \tag{14}$$

where $x_1 := q$, $x_2 := \dot{q}$ and $x_c := I$. The starting point is to define a tracking error as the following:

$$e = x_1 - q_d \tag{15}$$
$$x_r = \dot{q}_d - (x_1 - q_d) \tag{16}$$
$$\dot{x}_r = \ddot{q}_d - (x_2 - \dot{q}_d) \tag{17}$$
$$s = x_2 - x_r = \dot{e} + e \tag{18}$$

Now, with the definitions:

$$\dot{\hat{x}}_r = \ddot{q}_d - (\hat{x}_2 - \dot{q}_d) = \dot{x}_r + \tilde{x}_2 \tag{19}$$

$$\hat{s} = \hat{x}_2 - x_r = s - \tilde{x}_2 \tag{20}$$

where $\hat{x}_2$ determined later is the estimate of x_2 and $\tilde{x}_2 = x_2 - \hat{x}_2$ is the observation error vector. Now, we can rewrite (13) in terms of Equation (18) to yield:

$$\begin{aligned} D(x_1)\dot{s} &= D(x_1)[\dot{x}_2 - \ddot{q}_d + (x_2 - \dot{q}_d)] \\ &= -C(x_1, x_2)x_2 - g(x_1) + K_t x_c - D(x_1)\dot{\hat{x}}_r + D(x_1)\tilde{x}_2 \\ &= -C(x_1, x_2)s - D(x_1)\dot{\hat{x}}_r - C(x_1, \hat{x}_2)x_r - g(x_1) + K_t\tau \\ &\quad -K_t\eta + D(x_1)\tilde{x}_2 - C(x_1, x_r)\tilde{x}_2 \end{aligned} \tag{21}$$

where $\tau \in \Re^n$ is an embedded control input vector to be defined. This embedded control signal can be viewed as an adaptive control variable that specifies a desired link torque. As shown later, this control signal τ is actually embedded inside the overall control strategy which is designed at the voltage input level for each motor. This control strategy can be best described in terms of the perturbation term:

$$\eta = \tau - x_c \tag{22}$$

If η equals to zero, we can see from (21) that the desired control torque is effectively applied to each joint of the robot. In general, η will not be equal to zero, so we must design our overall control strategy to make η "small" enough so that we can still achieve the desired stability result of the link position tracking error. To do this, the dynamic characteristics of the perturbation term is needed. Therefore, we differentiate equation (22) to yield:

$$\dot{\eta} = \dot{\tau} - \dot{x}_c \tag{23}$$

And multiply (23) by L to give

$$L\dot{\eta} = L(\dot{\tau} - \dot{x}_c) = L\dot{\tau} + Rx_c + K_b x_2 - u \tag{24}$$

where the vector $u \in \Re^n$ is the final control signal that determines the input voltage for each motor.

It is reasonable to assume that the matrices K_t, L, R, and K_b are accurately known , because they are basically motor related parameters. Hence the final control signal u can be described from the torque signal τ.

Now, we select the control torque τ as:

$$\tau = K_t^{-1}[\hat{D}(x_1)\dot{\hat{x}}_r + \hat{C}(x_1, \hat{x}_2)x_r + \hat{g}(x_1) - K_D\hat{s}] \tag{25}$$

where K_D is a positive definite matrix and define:

$$d_m = \lambda_m(K_D) \quad \text{and} \quad d_M = \lambda_M(K_D) \tag{26}$$

Therefore, the final voltage input is given by:

$$u = L\dot{\tau} + Rx_c + K_b x_2 - K_t^T(s + \tilde{x}_2) \tag{27}$$

Hence, using (25) we obtain:

$$\begin{aligned} D(x_1)\dot{s} &= -C(x_1,x_2)s + Y_k\tilde{\theta} - K_D\hat{s} \\ &\quad -K_t\eta + D(x_1)\tilde{x}_2 - C(x_1,x_r)\tilde{x}_2 \\ &= -C(x_1,x_2)s + Y_k\tilde{\theta} - K_D s - K_t\eta \\ &\quad +[D(x_1) - C(x_1,x_r) + K_D]\tilde{x}_2 \end{aligned} \tag{28}$$

where:

$$Y_k\tilde{\theta} = Y(x_1,\hat{x}_2,x_r,\dot{\hat{x}}_r)\tilde{\theta} = \tilde{D}(x_1)\dot{\hat{x}}_r + \tilde{C}(x_1,\hat{x}_2)x_r + \tilde{g}(x_1) \tag{29}$$

$$\tilde{D}(x_1) = \hat{D}(x_1) - D(x_1) \tag{30}$$

$$\tilde{C}(x_1,\hat{x}_2) = \hat{C}(x_1,\hat{x}_2) - C(x_1,\hat{x}_2) \tag{31}$$

$$\tilde{g}(x_1) = \hat{g}(x_1) - g(x_1) \tag{32}$$

The following Lemma is useful for the stability analysis:

Lemma 2.1. *Consider the function* $f(\cdot,\cdot) : \Re \times \Re \to \Re$

$$f(w_1,w_2) = 2\alpha_1 w_1 w_2 - \alpha_2 w_1^2, \quad w_1,w_2 \in \Re^+ \tag{33}$$

where α_1 *and* α_2 *are positive constants, then we have:*

$$f(w_1,w_2) \le \frac{\alpha_1^2}{\alpha_2} w_2^2 \tag{34}$$

3 Adaptive Velocity Estimator

In order to remove the acceleration measurements in the final control signal, we design a velocity estimator to reconstruct the velocity state. In addition, we use these estimated states together with the adaptive control signal to get semi-globally asymptotical stability.

Based on the above model, the proposed velocity estimator is described by:

$$\dot{\hat{x}}_1 = \hat{x}_2 + \tilde{x}_1 \tag{35}$$

$$\hat{x}_2 = z + k\tilde{x}_1 \tag{36}$$

$$\dot{z} = \dot{\hat{x}}_r + k\tilde{x}_1 \tag{37}$$

where $\hat{x}_1$ and $\hat{x}_2$ are the estimates of x_1 and x_2 respectively, and $\tilde{x}_1 = x_1 - \hat{x}_1$ and $\tilde{x}_2 = x_2 - \hat{x}_2$ are the observation error vectors. And z is an auxiliary variable, and k is a positive constant.

Comparing equations (12) and (35), (13), (36) and (37), we can derive the following error equation:

$$\dot{\tilde{x}}_1 = \dot{x}_1 - \dot{\hat{x}}_1 = x_2 - \hat{x}_2 - \tilde{x}_1 = \tilde{x}_2 - \tilde{x}_1 \tag{38}$$

From this we can obtain:

$$\begin{aligned} \dot{\tilde{x}}_2 &= \dot{x}_2 - \dot{\hat{x}}_2 = \dot{x}_2 - \dot{z} - k\dot{\tilde{x}}_1 = \dot{x}_2 - \dot{\hat{x}}_r - k\tilde{x}_2 \\ &= D^{-1}(x_1)[-C(x_1, x_2)x_2 - g(x_1) + K_t\tau - K_t\eta - D(x_1)\dot{\hat{x}}_r] - k\tilde{x}_2 \end{aligned} \tag{39}$$

In addition, using the embedded control law (25), we get:

$$\begin{aligned} D(x_1)\dot{\tilde{x}}_2 &= -C(x_1, x_2)x_2 - g(x_1) + K_t\tau \\ &\quad -K_t\eta + D(x_1)\dot{\hat{x}}_r - kD(x_1)\tilde{x}_2 \\ &= -C(x_1, x_2)\tilde{x}_2 - C(x_1, x_2)\hat{x}_2 - g(x_1) + \hat{D}(x_1)\dot{\hat{x}}_r - K_t\eta \\ &\quad +\hat{C}(x_1, \hat{x}_2)x_r + \hat{g}(x_1) - K_D\hat{s} - D(x_1)\dot{\hat{x}}_r - kD(x_1)\tilde{x}_2 \\ &= -C(x_1, x_2)\tilde{x}_2 + Y_k\tilde{\theta} - C(x_1, \hat{x}_2)s - K_D\hat{s} \\ &\quad -K_t\eta - kD(x_1)\tilde{x}_2 \\ &= -C(x_1, x_2)\tilde{x}_2 + Y_k\tilde{\theta} - C(x_1, \hat{x}_2)s - K_Ds \\ &\quad -K_t\eta - [kD(x_1) - K_D]\tilde{x}_2 \end{aligned} \tag{40}$$

To simplify the stability analysis, k is now explicitly defined in terms of bounding coefficients and a nonlinear damping gain [7] as the following:

$$k = \frac{1}{\lambda_m}(d_m + d_M + \sigma_1^2\frac{k_0}{d_m} + \sigma_2^2\frac{k_0}{d_m}) \tag{41}$$

where k_0 is a positive, nonlinear damping gain, and σ_1, σ_2 are positive "bounding" constants which will be determined later.

The parameter update law is provided by:

$$\dot{\hat{\theta}} = -\Gamma^{-1}Y_k^T(s + \tilde{x}_2) \tag{42}$$

where the $\Gamma \in \Re^{p\times p}$ is a diagonal positive definite matrix

4 Stability Analysis

We now show that observation errors converge to zero if the initial error is within a compact set. In other words the equilibrium point is Lyapunov semi-global stable. To simplify the proof of the following theorem, we define $x_s := [s^T\ \tilde{x}_2^T]^T$, $x_f := [s^T\ \tilde{x}_2^T\ \tilde{\theta}^T\ \eta^T]^T$, and $F := diag\{D(x_1)\ D(x_1)\ \Gamma\ L\}$.

Theorem 4.1. *Assume that for any $t \geq 0$, $\|\dot{q}(t)\| \leq k_q$, and for the control signal in (25) and (27) as well as adaptive law given by (42), if the control gain k_0 of (41) is chosen such that:*

$$k_0 > 1 \tag{43}$$

Then

$$\lim_{t\to\infty} e = \lim_{t\to\infty} \dot{e} = \lim_{t\to\infty} \tilde{x}_2 = 0 \tag{44}$$

Moreover, a region of attraction is given by:

$$B = \left\{x_f \in \Re^{3n+p}, \|x_f\| < \sqrt{\frac{\varrho_m}{\varrho_M}(k_0 - 1)}\right\} \tag{45}$$

where ϱ_m and ϱ_M are positive constants defined as $\varrho_m = min\{\lambda_m,\ \lambda_m(L),\ \lambda_m(\Gamma)\}$, $\varrho_M = max\{\lambda_M,\ \lambda_M(L),\ \lambda_M(\Gamma)\}$.

Proof: Omitted due to the page limit and available upon request.

5 Simulation

The performance of the adaptive controller and velocity estimator have been applied to the simulation plant, referred to a two-link robot arm with DC motor. The dynamic model is described by (1), (2) and a parameterization scheme of this robot is given in [3]:

$$\theta_1\ddot{x}_{11} + (\theta_3 c_{21} + \theta_4 s_{21})\ddot{x}_{12} - \theta_3 s_{21}\dot{x}_{12}^2 + \theta_4 c_{21}\dot{x}_{12}^2 = k_{t1}x_{c1} \tag{46}$$
$$(\theta_3 c_{21} + \theta_4 s_{21})\ddot{x}_{11} + \theta_2\ddot{x}_{12} + \theta_3 s_{21}\dot{x}_{11}^2 - \theta_4 c_{21}\dot{x}_{11}^2 = k_{t2}x_{c2} \tag{47}$$

where $x_1 = [x_{11}^T\ x_{12}^T]$, $\theta = [\theta_1\ \theta_2\ \theta_3\ \theta_4]$ $c_{21} = \cos(x_{12} - x_{11})$ and $s_{21} = \sin(x_{12} - x_{11})$. It is clearly linear in terms of the four parameter $\theta_1, \theta_2, \theta_3$ and θ_4, which are related to the physical parameter of the links with the load treated as part of the second link. Defining the component of matrix C as:

$$C(1,1) = C(2,2) = 0 \tag{48}$$
$$C(1,2) = (\theta_4 c_{21} - \theta_3 s_{21})\dot{x}_{12} \tag{49}$$
$$C(2,1) = (\theta_3 c_{21} - \theta_4 c_{21})\dot{x}_{11} \tag{50}$$

the skew-symmetry of the $\dot{D} - 2C$ can also be satisfied.

6 Conclusion

In this paper, we have proposed an adaptive control law with a velocity estimator for the robot link. Using this estimated speed, the controller is capable of tracking a given desired trajectory with guaranteed stability The dynamic part of the controller consists of a nonlinear velocity estimator which guaranteed semiglobal uniformly asymptotic stability of the error dynamics.

References

1. M. M. Bridges, D. M. Dawson, and X. Gao, "Adaptive control of rigid-link electrically-driven robots," Proc. IEEE Conference on Decision and Control, vol. 1, pp. 159–165, 1993.
2. S. S. Ge and I. Postlethwaite, "Nonlinear adaptive control of robots including motor dynamics," Proc. American Control Conference, pp. 1423–1427, 1993.
3. J.-J. E. Slotine and W. Li, "Adaptive manipulator control: A case study," IEEE Transactions on Automatic Control, vol. 33, pp. 995–1003, 1988.
4. M. W. Spong and M. Vidyasagar, Robot Dynamics and Control, New York: John Wiley & Sons, 1989.
5. M. Vidyasagar, Nonlinear System Analysis, New Jersey: Prentice Hall Inc., 1978.
6. R. Ortega and M. W. Spong, "Adaptive motion control of rigid robot: A tutorial," Automatica, vol. 25, pp. 877–888, 1989.
7. T. Turg, D. M. Dawson, and P. Vedagarbha, "A redesigned dcal controller without velocity measurements: Theory and experimentation," Proc. IEEE Conference on Decision and Control, vol. 1, pp. 824–828, 1994.
8. S. Sastry and M. Bodson, Adaptive Control, New York: Prentice Hall, 1979.

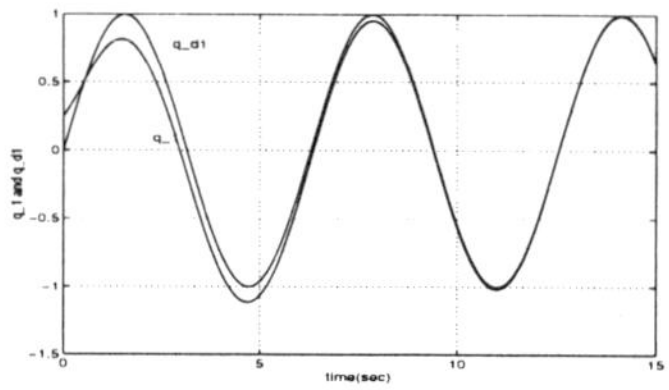

Figure 1: Output tracking q_1 and its desired trajectory q_{1d} for first link.

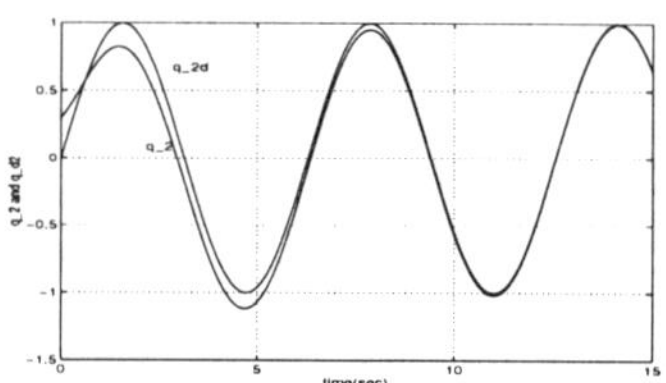

Figure 2: Output tracking q_2 and its desired trajectory q_{2d} for second link.

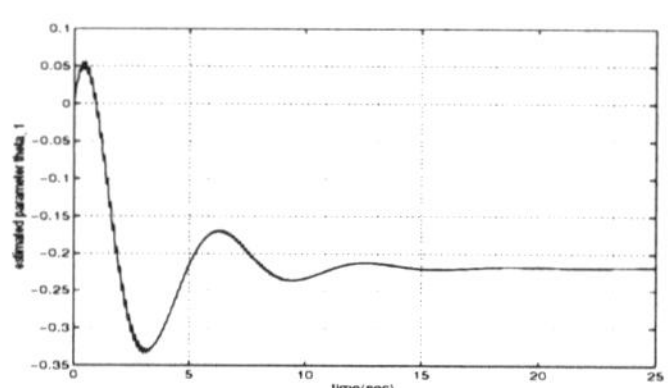

Figure 3: The estimated parameter value $\hat{\theta}_1$.

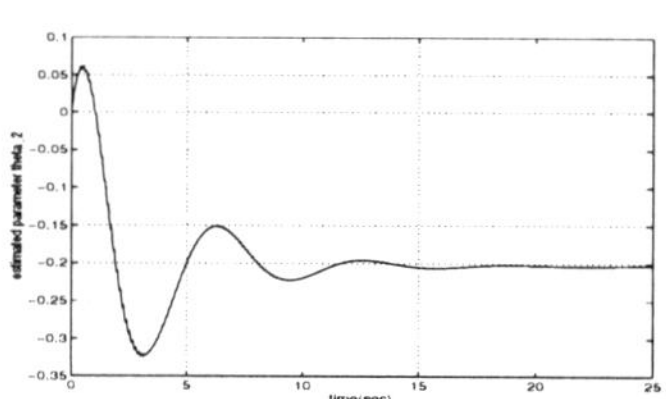

Figure 4: The estimated parameter value $\hat{\theta}_2$.

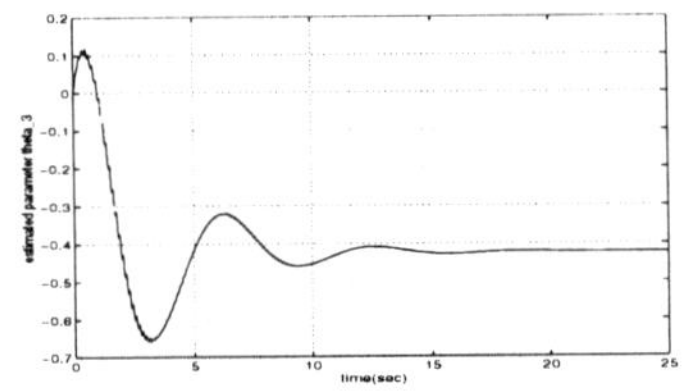

Figure 5: The estimated parameter value $\hat{\theta}_3$.

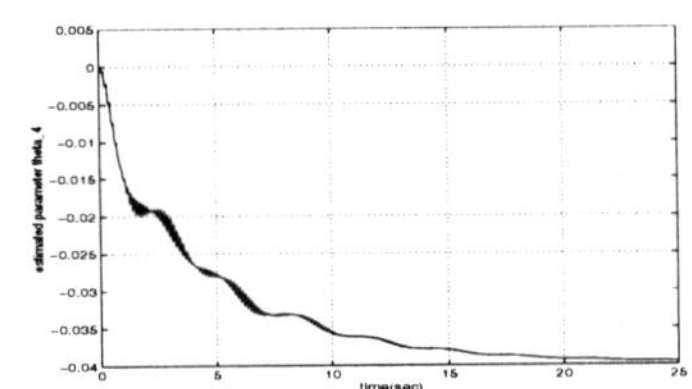

Figure 6: The estimated parameter value $\hat{\theta}_4$.

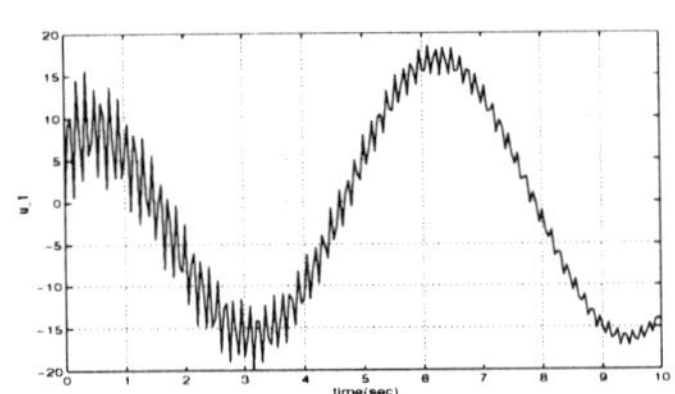

Figure 7: Control signal u_1 of the first link.

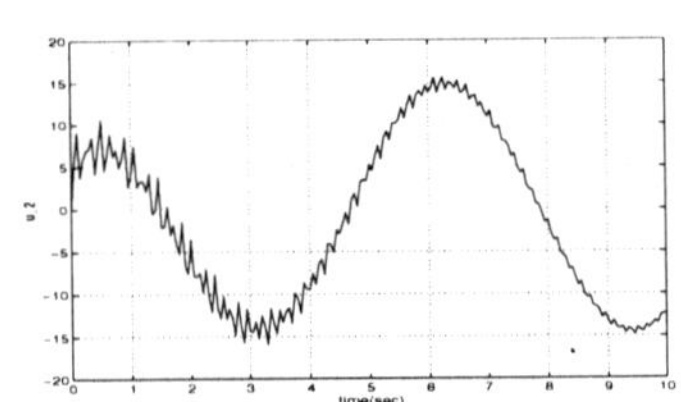

Figure 8: Control signal u_2 of the second link.

Development of A New Industrial Robot

Hong-You Lee and W. Wälischmiller

Hans Wälischmiller GmbH, Tele Robot Engineering
Klingleweg 8, D-88709 Meersburg, Germany
E-Mail: hongyou@mee.hwm.com and ww@mee.hwm.com

Abstract. This paper presents a new general purpose and modular industrial robot, trade named TELBOT, developed by Hans Wälischmiller GmbH. The robot has six revolution joints. A unique drive system of the robot contains all of the electronics, cables, motors, reduction gears and encoders. The transmission of motions of the arm are translated from the unique drive system by sets of concentric tubes that make up to the links and sets of bevel gears that make up the joints. There is no electrical parts in the arm itself. This allows unrestricted rotation of each joint. That means, all the revolution joints can rotate over 360 degrees continuously. The arm itself is relatively compact and lightweight. The inverse kinematics problem of this robot has still up to 16 solutions, which correspond to the different robot configurations. The Jacobian matrix of the robot is very simple when it is formulated with respect to a selected special coordinate system, so that it is possible to be inverted symbolically. Even more, the singularities of the robot can be identified easily, because the determinant of the Jacobian is very concise. The robot has been used for steam generator maintenance in Canada and for decommissioning glove boxes in Japan.

1 Introduction

A general purpose and modular industrial robot, trade named TELBOT, has been developed by Hans Wälischmiller GmbH (Wälischmiller and Frager, 1995). The TELBOT robot has been successfully used for cleaning steam generator tubes at the primary side in CANADA (Bains, Majarais, and Scott, 1995). The robot has been also used for decommissioning glove boxes in Japan. The robot has six degrees of freedom and a unique drive system (Fig. 1). All the motors, measuring system and the transmission gear box of the manipulator are located in the robot base. There is not any electrical cables from the base to the end-effector. The transmission of the motion from the motors to the end-effector is realized via concentric tubes and bevel gears, so that all the revolution joints can rotate over 360 degrees continuously. Very recently, Prof. Jacques M. Hervé of Ecole Centrale Paris suggested the first author a reference (Kersten, 1977) with a conceptual design of the similar manipulator kinematics. Because of the novel design and motion transmission, the end-effector or the tool can reach its work position in any orientation. Furthermore, due to there is no electrical cable in the whole arm, the manipulator can be easily protected from the dirty environment, and also some of the parts can be easily changed if necessary.

The present paper will focus on the novel mechanical design, the inverse kinematics, singularity identification and the application in hazardous environments.

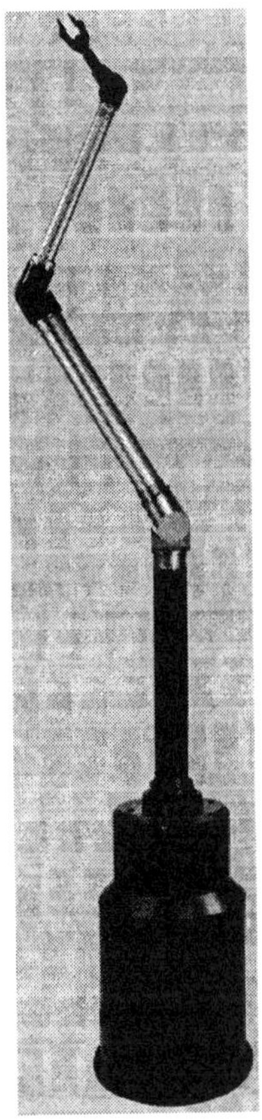

Fig. 1 TELBOT Robot

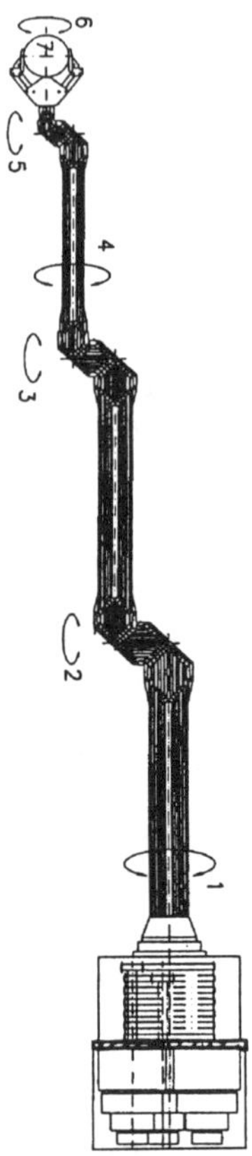

Fig. 2 Transmission Principle

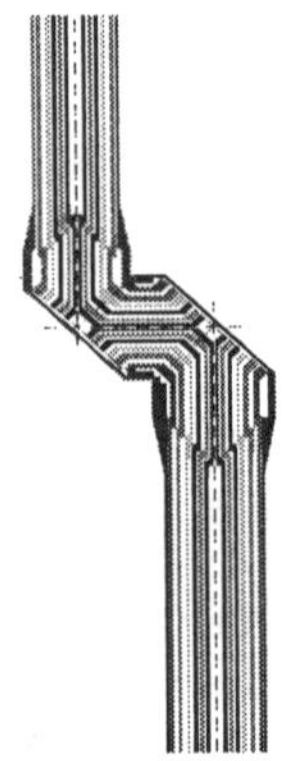

Fig. 3 Joint Design

Fig. 4 Robot Base

2 Mechanical Design

The TELBOT robot has six degrees of freedom and a unique drive system that allows unrestricted motion of each joint (Fig. 1 and Fig. 2). The rotation of all the revolution joints and the opening and closing movement of the gripper are transmitted with the concentric tubes in the arm and bevel gears in the joints. The mechanical system possesses many novel properties and advantages:

- **Unique Drive System**

A unique drive system is located at the robot base. All of the motors, electronics and cables are contained within the base (Fig. 4). There is no electrical parts in the arm itself.

- **Motion Transmission**

The transmission of motions of the arm is translated from the unique drive system by sets of concentric tubes that make up the links and sets of bevel gears that make up the joints (Fig. 2 and Fig. 3).

- **Kinematic Skeleton**

The first three joint axes of the TELBOT robot are in a standard configuration with one vertical and two parallel horizontal axes. The three wrist joint axes have an offset between the fourth and sixth joint axis, so that all the three wrist joints can also be rotated over 360 degrees continuously (Fig. 5).

- **Modular Design**

The modular design of the gear block allows for adaptation to various transmission ratios and moments. Tubes and joints with different sizes are designed for meeting the requirements of different payloads. The modular parts can be quickly changed when it is needed.

- **Unrestricted Joint Rotation**

The novel mechanical system allows unrestricted rotation of each joint. All the revolution joints can rotate over 360 degrees continuously.

- **Compact Arm and Lightweight**

The arm itself is relatively compact and lightweight. This facilitates increasing the workspace and improving the dynamic properties.

3 Geometrical Modeling

A kinematic diagram of the TELBOT manipulator showing all link parameters is given in Fig. 5. The directions of the joint axes are sequentially labeled with the unit vectors $\vec{s}_i$, $i = 1, 2, \ldots, 6$. The directions of the common normals between two successive joint axes $\vec{s}_i$ and $\vec{s}_j$ are labeled with the unit vectors $\vec{a}_{ij}$, $ij = 12, 23, \ldots, 56$, and the lengths of the common normals (link lengths) with a_{ij}. The mutual perpendicular distances between pairs of successive links $\vec{a}_{ij}$ and $\vec{a}_{jk}$ along the joint axis $\vec{s}_j$ are denoted by S_j, $j = 1, 2, \ldots, 6$, and are called joint offsets. The joint angles θ_i and the twist angles α_{ij} are measured by right-hand rotations about $\vec{s}_i$ and $\vec{a}_{ij}$, respectively.

A robot base frame is located with $\vec{z}$ axis coaxial with $\vec{s}_1$, the axis of the first joint. The position of the end-effector is specified in the robot base frame by $\vec{R}_H$ while the orientation is defined by the pair of orthogonal unit vectors $\vec{a}_{67}$ and $\vec{s}_7$.

It is important to note that only the joint angles (θ_j) are unknown quantities; all other Denavit-Hartenberg parameters, i.e., twist angles (α_{ij}), offsets (S_j), and link lengths (a_{ij}) have constant values that are listed in Table 1.

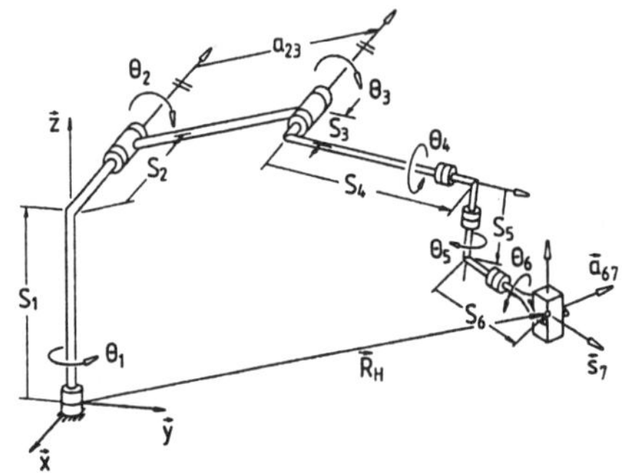

Fig. 5 Robot Kinematic Skeleton

Table 1: Link Parameters

i	S_i	$a_{i(i+1)}$	θ_i	$\alpha_{i(i+1)}$
1	S_1	0.0	θ_1	$\pi/2$
2	S_2	a_{23}	θ_2	0.0
3	S_3	0.0	θ_3	$\pi/2$
4	S_4	0.0	θ_4	$\pi/2$
5	S_5	0.0	θ_5	$\pi/2$
6	S_6		θ_6	

4 Inverse Kinematics

It was shown by Lee (Li) and Liang (1988) and Lee (Li) and Reinholtz (1996), that there are up to 16 solutions of the inverse kinematics for robots with 6R joints. The TELBOT robot with non-zero offset of the 5th joint has no closed-form solution of the inverse kinematics either, although it still has special geometry. It will be shown that the inverse kinematics problem of the TELBOT manipulator can be reduced to a 16th degree polynomial in one unknown and five linear equations in the other five unknown joint angles. The univariate polynomial is derived from only three constraint equations. Once the Polynomial is solved for the unknown joint angle numerically, all the other five unknown joint angles are determined by solving linear equations. The real solutions correspond to the different robot configurations. The optimal configuration for a given task and limited workspace can be chosen once all the possible solutions are found. Because all the joint rotations are transmitted by concentric tubes in the arm and concentric bevel gears in the joints, the joint angles must be then converted in the rotation angles of the tubes and finally rotation of the motors.

It is well known that a 16th degree polynomial for the general 6R manipulator can be derived from 14 closure equations (Lee and Liang, 1988), while the derivation of such a polynomial for the TELBOT robot, because of its special geometry, requires only three closure equations. The basic approach is to disconnect the manipulator into two subchains at two joints and then formulate scalar products of the vectors representing the joints and links for both subchains. These scalar equations are then solved simultaneously to yield a single polynomial equation in one unknown joint angle. Once the polynomial is obtained and solved for the joint angle, all the remaining joint variables follow from linear equations.

4.1 Univariate Polynomial

The 16th degree polynomial in one unknown is derived by using the following three equations:

$$[\vec{s}_4 \cdot \vec{s}_1]_A = [\vec{s}_4 \cdot \vec{s}_1]_B \tag{1}$$

$$\left[\vec{R} \cdot (\vec{s}_1 \times \vec{s}_4)\right]_A = \left[\vec{R} \cdot (\vec{s}_1 \times \vec{s}_4)\right]_B \tag{2}$$

$$\left[\vec{R} \cdot \vec{R}\right]_A = \left[\vec{R} \cdot \vec{R}\right]_B \tag{3}$$

where

$$\left[\vec{R}\right]_A = [(S_2 + S_3)\vec{s}_2 + a_{23}\vec{a}_{23}] \tag{4}$$

$$\left[\vec{R}\right]_B = -\left[S_4\vec{s}_4 + S_5\vec{s}_5 + S_6\vec{s}_6 - (\vec{R}_H - S_1\vec{s}_1)\right] \tag{5}$$

The three closure equations are derived by disconnecting the TELBOT manipulator at joints 1 and 4. This divides the manipulator into two subchains A and B and the vector sum of subchain A equals the sum of subchain B. Note that the subscripts A and B on a bracketed quantity indicates that all vectors within the brackets must be expressed using the vectors and rotations associated with that subchain.

Although it may not be immediately apparent, these three scalar equations contain only θ_5, θ_6 and $(\theta_2 + \theta_3)$, as unknowns. It should, therefore, be possible to solve this system. A key to efficient formulation is to express vectors in the most convenient coordinate frame before the scalar and triple scalar product operations are performed. A change in reference frame is easily accomplished by using standard joint and twist rotation matrices.

The given end-effector position vector and the orientation matrix are defined in the robot base frame as follows:

$$\vec{R}_{H0} = \vec{R}_H - S_1\vec{s}_1 = \begin{bmatrix} r_x \\ r_y \\ r_z - S_1 \end{bmatrix} \qquad M_H = \begin{bmatrix} \vec{a}_{67} \\ \vec{s}_7 \times \vec{a}_{67} \\ \vec{s}_7 \end{bmatrix} = \begin{bmatrix} h_{00}\ h_{10}\ h_{20} \\ h_{01}\ h_{11}\ h_{21} \\ h_{02}\ h_{12}\ h_{22} \end{bmatrix} \tag{6}$$

The three eqs. (1)-(3) can be evaluated and may be written as:

$$c_5h_{22} + s_5(-h_{20}c_6 + h_{21}s_6) = \cos(\theta_2 + \theta_3) \tag{7}$$

$$\begin{aligned} &c_5\left[c_6(h_{20}S_5) - s_6(h_{21}S_5) + (-h_{12}r_x + h_{02}r_y)\right] + \\ &s_5\left[c_6(h_{10}r_x - h_{00}r_y - h_{21}S_6) + s_6(-h_{11}r_x + h_{01}r_y - h_{20}S_6) + (h_{22}S_5)\right] \\ &= (S_2 + S_3)\sin(\theta_2 + \theta_3) \end{aligned} \tag{8}$$

$$\begin{aligned} &c_5S_4(H_z - S_6) + s_5(-c_6S_4H_x + s_6S_4H_y) - c_6S_5H_y + s_6S_5H_x + \\ &\left[(1/2)(S_4^2 + S_5^2 + S_6^2 + R^2 - (S_2 + S_3)^2 - a_{23}^2) - S_6H_z\right] = 0 \end{aligned} \tag{9}$$

where S_i and a_{23} are constant parameters defined in the Table 1, $c_i = \cos\theta_i$, $s_i = \sin\theta_i$ (the abbreviations c_i and s_i will be used throughout this paper), and

$$[H_x,\ H_y,\ H_z]^T = M_H\vec{R}_{H0} \tag{10}$$

$$R^2 = r_x^2 + r_y^2 + (r_z - S_1)^2 \tag{11}$$

The three closure equations (7)-(9), which contain all the information needed to derive the 16th degree polynomial, now can be written in the following form:

$$F_1(\theta_5, \theta_6) = \cos(\theta_2 + \theta_3) \quad (12)$$
$$F_2(\theta_5, \theta_6) = (S_2 + S_3)\sin(\theta_2 + \theta_3) \quad (13)$$
$$F_3(\theta_5, \theta_6) = 0 \quad (14)$$

Where the functions $F_1(\theta_5, \theta_6)$, $F_2(\theta_5, \theta_6)$ and $F_3(\theta_5, \theta_6)$ are linear combinations of the terms (c_5c_6, c_5s_6, c_5, s_5c_6, s_5s_6, s_5, c_6, s_6).

Eliminating $(\theta_2 + \theta_3)$ from Eqs. (12) and (13) gives

$$F_1^2 + [F_2/(S_2 + S_3)]^2 = 1 \quad (15)$$

The next step is to eliminate θ_5 (or θ_6) from Eqs. (14) and (15). For convenience, the two equations can be expressed as polynomials in $x_5 = \tan(\theta_5/2)$ and $x_6 = \tan(\theta_6/2)$ as follows:

$$A_0 + A_1x_5 + A_2x_5^2 = 0 \quad (16)$$
$$B_0 + B_1x_5 + B_2x_5^2 + B_3x_5^3 + B_4x_5^4 = 0 \quad (17)$$

where A_0, A_1 and A_2 are quadratic polynomials in x_6, and B_0, B_1, B_2, B_3 and B_4 are quartic polynomials in x_6.

Eliminating x_5 from the two eqs. (16) and (17) yields the desired 16th degree univariate polynomial in x_6 with constant coefficients e_i:

$$\sum_{i=0}^{16} e_i x_6^i = 0 \quad (18)$$

4.2 Determination of All the Joint Variables

Once the desired hand position and orientation (6) is specified, all e_i coefficients of Eq. (18) can be calculated. The roots of the polynomial give the values of θ_6 which correspond 16 possible configurations of the manipulator. Each value of θ_6 will give a unique value of x_5 by solving the two equations (16) and (17). For each pair of θ_5 and θ_6, a value of $(\theta_2 + \theta_3)$ can be easily obtained from equations (12) and (13). Then for each set of $(\theta_2 + \theta_3)$, θ_5 and θ_6, the joint angles θ_1 and θ_4 can be determined from the following two pairs of equations:

$$[\vec{s}_1 \cdot \vec{x}]_A = [\vec{s}_1 \cdot \vec{x}]_B \quad (19)$$
$$[\vec{s}_1 \cdot \vec{y}]_A = [\vec{s}_1 \cdot \vec{y}]_B \quad (20)$$

$$[\vec{s}_4 \cdot \vec{a}_{45}]_A = [\vec{s}_4 \cdot \vec{a}_{45}]_B \quad (21)$$
$$[\vec{s}_4 \cdot \vec{s}_5 \times \vec{a}_{45}]_A = [\vec{s}_4 \cdot \vec{s}_5 \times \vec{a}_{45}]_B \quad (22)$$

The equations (19)-(22) can be formulated so that they are linear in terms of sines and cosines of θ_1 and θ_4 respectively, thus can be solved for directly.

The joint angles θ_3 can be easily determined from the following equations, respectively:

$$\left[\vec{R}\cdot\vec{a}_{34}\right]_A = \left[\vec{R}\cdot\vec{a}_{34}\right]_B \tag{23}$$

$$\left[\vec{R}\cdot(\vec{s}_3\times\vec{a}_{34})\right]_A = \left[\vec{R}\cdot(\vec{s}_3\times\vec{a}_{34})\right]_B \tag{24}$$

The two equations (23) and (24) are linear in terms of sines and cosines of θ_3. If θ_3 is found, the joint angle θ_2 can be determined using the value of $(\theta_2+\theta_3)$.

It should be noted that if $(\theta_2+\theta_3)$ near to zero or 180 degrees, the proposed equations (19) and (22) cannot be used. The joint angles θ_1 and θ_4 must be calculated by using some other closure equations.

5 Singularity Identification

5.1 Jacobian Matrix

It is well known that the Jacobian matrix of a 6R robot is defined by the following matrix:

$$J_{acobian} = \begin{bmatrix} \vec{s}_1 & \vec{s}_2 & \vec{s}_3 & \vec{s}_4 & \vec{s}_5 & \vec{s}_6 \\ \vec{R}_{P1}\times\vec{s}_1 & \vec{R}_{P2}\times\vec{s}_2 & \vec{R}_{P3}\times\vec{s}_3 & \vec{R}_{P4}\times\vec{s}_4 & \vec{R}_{P5}\times\vec{s}_5 & \vec{R}_{P6}\times\vec{s}_6 \end{bmatrix} \tag{25}$$

where $\vec{R}_{Pi}$ with $i=1,2,3,4,5,6$ are the position vectors of the point P to an arbitrary point of the axis of i_{th} joint. If the point P is selected at the cross point of the axes of joints 3 and 4, for example, the six position vectors $\vec{R}_{Pi}$ can be defined as following:

$$\vec{R}_{P1} = -(S_2+S_3)\vec{s}_3 - a_{23}\vec{a}_{23} \tag{26}$$

$$\vec{R}_{P2} = -S_3\vec{s}_3 - a_{23}\vec{a}_{23} \tag{27}$$

$$\vec{R}_{P3} = \vec{0} \tag{28}$$

$$\vec{R}_{P4} = \vec{0} \tag{29}$$

$$\vec{R}_{P5} = S_4\vec{s}_4 \tag{30}$$

$$\vec{R}_{P6} = S_4\vec{s}_4 + S_5\vec{s}_5 \tag{31}$$

Evaluating the Jacobian matrix with respect to the coordinate system ($\vec{a}_{34}$, $\vec{s}_3$, $\vec{s}_4$) which is attached to the joints 3 and 4, results in the following:

$$J_{acobian} = \begin{bmatrix} s_{23} & 0 & 0 & 0 & s_4 & c_4s_5 \\ 0 & 1 & 1 & 0 & -c_4 & s_4s_5 \\ -c_{23} & 0 & 0 & 1 & 0 & -c_5 \\ c_{23}(S_2+S_3) & s_3a_{23} & 0 & 0 & c_4S_4 & c_4c_5S_5 - s_4s_5S_4 \\ -c_2a_{23} & 0 & 0 & 0 & s_4S_4 & s_4c_5S_5 + c_4s_5S_4 \\ s_{23}(S_2+S_3) & -c_3a_{23} & 0 & 0 & 0 & s_5S_5 \end{bmatrix} \tag{32}$$

Note that $c_{23} = cos(\theta_2 + \theta_3)$ and $s_{23} = sin(\theta_2 + \theta_3)$.

The lengths of the axes S_2, a_{23}, S_3, S_4 and S_5 are scaled in the Jacobian matrix so that the maximum value of S_2, a_{23} S_3, S_4 and S_5 is equal to 1. This reduces the influence of the dimension of lengths and also makes the Jacobian matrix dimensionless. Actually, all the lengths and the end-effector position $\vec{R}_H$ in the controller algorithm are divided by the maximum value of the original values of S_2, a_{23}, S_3, S_4 and S_5.

5.2 Determinant of the Jacobian Matrix and Singularities

The determinant of the Jacobian matrix of the robot is evaluated

$$D_{jacobi} = c_2 s_4 c_5 S_5 [c_3 c_4 a_{23} + s_4(S_2 + S_3)] - s_5(s_{23} S_4 + c_2 a_{23})(c_3 S_4 - s_3 s_4 S_5) \tag{33}$$

Obviously, the robot is in singularity configurations when the joint angles meet one of the following conditions:

1. $\theta_2 = \pm 90$ degrees, and at the same time, $\theta_3 = \pm 90$ degrees, it means that the first joint axis $\vec{s}_1$ is parallel to the upper arm $\vec{a}_{23}$, and at the same time, the upper arm $\vec{a}_{23}$ to the fourth joint (lower arm) axis $\vec{s}_4$;

2. $\theta_3 = \pm 90$ degrees, and at the same time, $\theta_4 = 0$ or 180 degrees, the upper arm $\vec{a}_{23}$ is parallel to the fourth joint axis $\vec{s}_4$ and at the same time, the third joint axis $\vec{s}_3$ to the fifth joint axis $\vec{s}_5$;

3. $\theta_4 = 0$ or 180 degrees, and at the same time, $\theta_5 = 0$ or 180 degrees, the third joint axis $\vec{s}_3$ is parallel to the fifth joint axis $\vec{s}_5$, and at the same time, the fourth joint axis $\vec{s}_4$ to the sixth joint axis $\vec{s}_6$;

4. $\theta_2 = \pm 90$ degrees, and at the same time, $\theta_5 = 0$ or 180 degrees, the first joint axis $\vec{s}_1$ is parallel to the upper arm $\vec{a}_{23}$, and at the same time, the fourth joint axis $\vec{s}_4$ to the sixth joint axis $\vec{s}_6$.

Finally, it should be noted that the Jacobian matrix, which is expressed with respect to the coordinate system ($\vec{a}_{34}$, $\vec{s}_3$, $\vec{s}_4$) attached to the joints 3 and 4, can be inverted symbolically, and even more the inverse matrix is not very complex. This greatly facilitates the analysis of velocity and statics.

6 Application

An important application of the TELBOT system is for nuclear steam generator maintenance. Fig. 7 shows the steam generator primary head tube cleaning system block diagram. The steam generator contains a shell-and-tube type heat exchanger with thousands of tubes. The plurality of the tubes is encased in a vertical cylindrical pressure vessel with a hemispherical lower end (bowl). The tube ends are inserted through a plate (tube-sheet) and welded and forms a planar cap over the bowl. The tubes are bent into an upside down "U" shape. These tubes must be inspected and repaired from within the bowl quarter spheres. The steam generator bowl is a high radiation area. Some robotic systems were

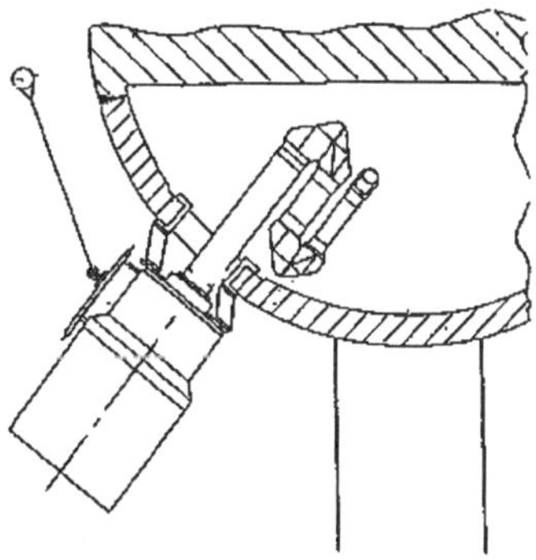

Fig. 6: Man Way Mounting Position

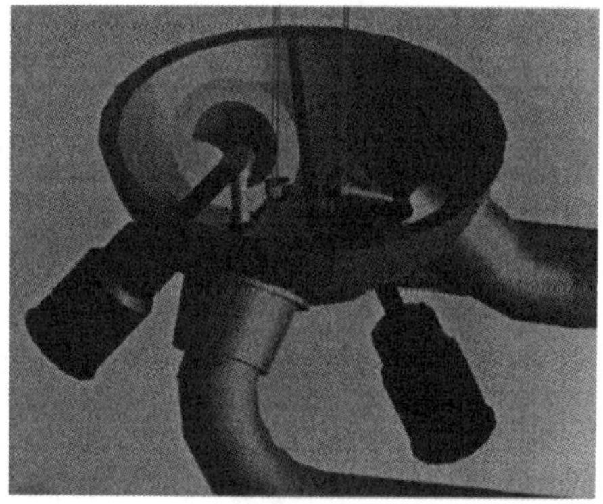

Fig. 7: Steam Generator Maintenance

developed and used for cleaning the tubes, for example, COBRA robotic system developed by B&WNT (Tidwell and Glass, et al, 1991). An advantage by using TELBOT system is that the motors, gear block and all the electrical cables are OUTSIDE of the bowl (Fig. 6 and Fig. 7). The robot arm parts which are inside the bowl can be very easily protected from dirty radioactive materials. Some arm parts can be quickly changed if necessary, because of the modular design. This can greatly reduce the cost of nuclear steam generator maintenance.

Once the manipulators were installed at the man-ways, the operator could initiate the calibration program. An initializing routine was used to calibrate the manipulator base coordinates with respect to the overall tube-sheet reference coordinates. The manipulator automatically moved to a pre-programmed location at the tube-sheet underneath a tie-rod location which was convenient to use for calibration as its position was known and was visually different from the tubes. They were also well spaced. For this initial calibration the stand-off distance of the tool to the tube-sheet was greater than the working distance. This allowed for any misplacement of the manipulators during the manual setting up operation. In order to ensure the manipulators moved in a plane parallel to the tube-sheet, the tool had a metallic "finger" that was used to touch the tube-sheet in three places each time the "finger" touched the tube-sheet a circuit was closed and the operator was alerted that the manipulator was in contact with the tube-sheet. From the three locations the manipulator base frame was calculated through the software and referenced to the tube-sheet plane. The manipulator could now be operated in straight line interpolation mode with respect to the tube-sheet. This made the tele-operation of the manipulator much easier than using the individual joint control. Once the manipulator was in this mode it was referenced to the actual tubes. Again the manipulators were initiated to move to the tie-rod locations. The manipulator was then tele-operated so that the center of the tube adjacent to the tie rod location was in the center of the field of view of the cameras. This was simply carried out using a set of cross hairs over-layed on the monitors. Once this was achieved for two or more tie rod locations the manipulator was referenced with respect to the tubes. During the cleaning process this type of calibration was repeated several times.

This means that the robot base frame is determined with respect to the world frame (attached to the bowl) in two steps. First, only the plane of tube-sheet is calibrated to the robot base frame, and then the position of the tubes with respect to the robot base frame. This greatly simplified the process of the calibration of the robot base frame with respect to the world frame, because there are not three or more sharp points in the bowl whose position is known exactly and could be touched by tool tip of the robot. The problem to plug the finger into the tubes is that the depth of the tool finger in the tube is very difficult to be exactly measured by using normal camera. Therefore, the plane of the tube-sheet is first determined , and then the position of tubes.

Another important application of the TELBOT robot is for decommissioning glove boxes in Japan. The robot is used for dismantling boxes in radioactive environment and for loosing screws on the boxes and then to cut the boxes. The novel properties of the robot, such as no limit continuous rotation of the tool, greatly facilitate reducing the working time and the cost.

7 Conclusions

The TELBOT system is a universal industrial robotic system. Its novel kinematic structure, control system and simulation system allow itself to be able to perform complicated tasks in hazardous environments and in manufacture automation. The TELBOT manipulator without a doubt has proven itself in the field and its revolutionary design coupled with its flexibility and proven reliability will impact the advanced manipulator field for years to come.

References

1. Bains, N., Majarais, B. and Scott, D.A., 1995, "Cleaning of Steam Generator Tubes at the Primary Side Using a General Purpose Manipulator", ANS conference, San Francisco, October.
2. Kersten, L., 1977, "The Lemma Concept: A New Manipulator", International Journal of Mechanism and Machine Theory, Vol. 12, pp.77-84.
3. Lee (Li), H.-Y. and Liang, C.-G., 1988, "Displacement Analysis of the General Spatial 7-Link 7R Mechanism", International Journal of Mechanism and Machine Theory, Vol. 23, pp.219-226.
4. Lee (Li), H.-Y. and Reinholtz, C.F., 1996, "Inverse Kinematics of Serial-Chain Manipulators", ASME Journal of Mechanical Design Vol.118 (September), pp.396-404.
5. Tidwell, P. H., Glass, S. W. and et al, 1991, "COBRA - Design and Development of a Manipulator for Nuclear Steam Generator Maintenance", Proceedings of the 2nd National Applied Mechanisms and Robotics Conference, USA, Vol. 1, pp. IVB1-1 to IVB1-10.
6. Wälischmiller, W. and Frager, O., 1995, "TELBOT - A Modular Tele-Robot System for Hostile Environments with Unlimited Revolutionary Joints and Motors, Drive Sensors and Cables in the Base", Proceedings of the ANS 6th Topical Meeting on Robotics and Remote Systems, Monterey, California, USA, Vol. 1, pp.274-280.

Adaptive Gait Planning for Multi-Legged Robots Using a Stability Compensation Method

Wenjie Chen, K. H. Low and S. H. Yeo

School of Mechanical and Production Engineering
Nanyang Technological University
Republic of Singapore 639798

Email: m94013h61 | mkhlow | myeosh@ntuvax.ntu.ac.sg

Abstract. Adaptive gait planning is an important consideration in the development of the control system for any multi-legged robot applied to rough terrain. The problem of adaptive gait generation is to find a sequence of suitable foothold on rough terrain so that the legged system maintains static stability and walking continuity. Due to the limit of static stability, deadlock situation during finding suitable foothold may occur if terrain contains a large amount of forbidden zone. This paper develops an improved method for adaptive gait planning through active compensation of stability margin, through the adjustment of center of gravity (CG) in longitudinal and lateral axes. Simulation results show that the presented method provides the legged machine with a much better terrain adaptivity and deadlock-avoidance ability.

Key words: *Multi-legged Robot, Gait Planning, Motion Planning.*

1 Introduction

The problem of gait planning in multi-legged system can be generally formulated as how to coordinate the motion of leg and body to make the machine traverse over a particular terrain with static stability manner from the starting position to the given objective location along a trajectory specified in advance [6]. The task of an adaptive gait planning in control includes determining the optimal schedule for the lifting and placing of the legs, and finding suitable foothold as supporting point(s) for transfer leg(s) through the whole trajectory [3, 6]. To achieve an effective adaptive gait, some limits, namely, terrain condition, kinematic (or geometry) constraint and static stability, must be taken into account in planning [4, 10].

A considerable amount of prior work has been devoted to the study of adaptive gait over past decades [1-4, 6, 9]. Representative of the these works is *free gait* algorithm which was first recognized and formalized by Kugushev and Jaroshevskij [3] and subsequently improved and developed for a hexapod robot by McGhee and Iswandhi [6]. The general strategy of free gait is to find a sequence of support points so that there is an overlap of the *existence segment* of each support state with that of the proceeding support state. To reduce the complexity of the problem, the algorithm determines the support pattern sequence only one stage forward rather then over the whole trajectory. However, such an approach may lead to a situation of deadlock if the terrain contains a large number forbidden zones. For a four-legged robot, due to fewer choices about selecting its steps, free gait algorithm may become more susceptible to deadlock situation. Hirose [2] presented a hierarchical algorithm to overcome the deadlock problem of quadruped robot. In his algorithm, the selection of new support points was conducted through the three reflexes from sensors, which trend to restrict the search area of new foothold. More recently, an improved free gait algorithm based on the approach of graph search over whole trajectory [9] was also presented to avoid deadlock.

Careful examination of the methodology of all algorithms for free gait shows that the trade-off between stability and adaptability is critical for successful generation of free gait. Deadlock will occur if a contradiction appears between stability and adaptability. It is clear that to enforce or compensate the minimum value of stability margin would decrease the likelihood of deadlock [1, 5, 8].

In this paper, a method to improve adaptive gait planning through the compensation of stability margin is proposed. The compensation of stability margin here is implemented by using CG adjustment in longitudinal and lateral axes. The significant new feature of this method is that it can provide machine a much greater adaptivity over rough terrain with a reasonable desired walking speed. The remainder of the paper is organized as follows. The scheme of compensation of stability margin for a quadruped prototype is described in section 2. Some necessary terminology about adaptive gait is introduced in section 3. The method to improve adaptive gait planning under consideration of stability margin compensation is addressed in section 4 and the strategy of CG adjustment is also included in this part. The simulation result is given in section 5 to show the promising behavior of the proposed method. Finally, a summary and the conclusions are made in section 6.

2 Description of stability margin compensation for a quadruped robot

For the study of adaptive locomotion of legged system, a quadruped robot, referred as NTU-Q1, is at present under construction at the Nanyang Technological University. A preliminary simulation prototype of NTU-Q1 is shown in Fig. 1. When completed, NTU-Q1 will be a self-contained

Fig. 1. Computer model of NTU-Q1

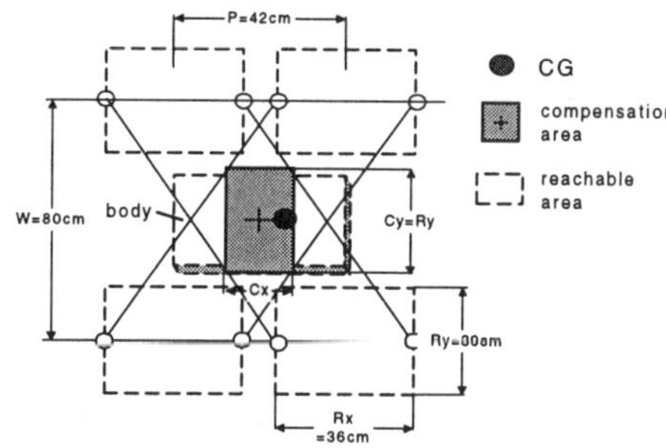

Fig. 2. The compensation area of NTU-Q1

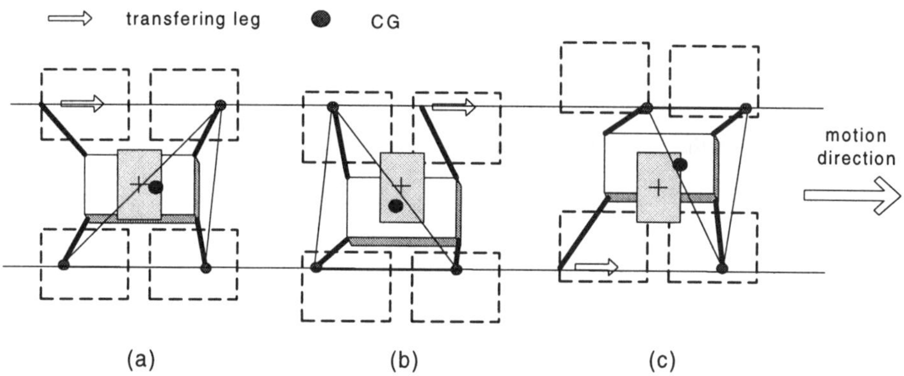

Fig. **3**. Top view of CG adjustment of NTU-Q1 in some support pattern to maintain static stability. (a) CG adjusted forward only in longitudinal direction (b) CG is adjusted only in lateral direction (c) CG is adjustment in both longitudinal and lateral directions.

walking robot intended for material handling on irregular terrain environment. Each of its four legs will has three independent rotary joints arranged in an arthropod configuration. Two terrain scanners located in the front of body will be used to provide terrain preview data which can then be processed by the control computer to predict suitable foothold position. To achieve superior terrain adaptivity, a device for the compensation of stability has also been incorporated into the design of this machine.

The stability margin compensation scheme adopted for NTU-Q1 involved CG adjustment relative to body frame along longitudinal axis and body frame translation in lateral axis. Such an arrangement will allow the machine to enhance the minimum value of stability margin, while independent of machine's locomotion in longitudinal direction. To adjust CG position in longitudinal axis effectively, a moveable weight which may consist of batteries and control parts is considered in the design of the machine. The moveable weight can move forward and backward along the longitudinal axis so as to shift CG to the desired location. On the other hand, due to arthropod configuration of each leg, the machine possesses a large lateral stroke which will allow the body frame to translate laterally in a sufficient large range. The adjustment of CG position in lateral direction is thus obtained.

The region with which the CG can be adjusted through the motion of movable weight in longitudinal axis and that of body frame along lateral axis composes a *compensation area.* Clearly, the compensation area will be a rectangle if the body is not allowed to rotate. The length of compensation area in lateral direction depends on the position of support legs in lateral stroke, while the length in longitudinal direction is determined by the maximum distance covered by moveable weight. The size of compensation area may affect the extent of compensation to the stability margin. A reasonable size for compensation area is needed to maintain machine's static stability on any support pattern permitted by the reach of support leg. For NTU-Q1, if legs are assumed to have only longitudinal motion and all support points on the same side are on central line of lateral stroke, then the required compensation area is determined by (see Fig. 2)

$$\begin{aligned} C_y &= R_y \\ C_x &= R_x - \frac{C_y}{W} P. \end{aligned} \tag{1}$$

According to the parameter value given in Fig. 2, the adjusting distance of CG in longitudinal axis, C_x, is about 20.2 cm. Some preliminary simulations have shown that such an adjusting distance of CG is acceptable in the mechanism design aspect.

As long as appropriately adjusting CG position in this compensation area, the machine will be able to keep its static stability at any instance. Figure 3 shows several examples of how to adjust CG position in compensation area to maintain the static stability of machine.

3 Terminology

The *support state* [7] of a *K*-legged locomotion system is actually the state of all legs touching or lifting the ground. A support state is called *F-state* if its all legs are supporting leg, otherwise, called *T-state.* The shifting of support state from one to another is caused by the occurring of at least one event. An *event* [3] of a gait is the placement or lift of a leg. An entire gait always consists of a finite number of support states corresponding to the events occurring. To take the sequence of a support state into account, an integer number *n* is used for each support state according to the time order. If the location of a support state in the series formed by all support states is *n*, then this support state will be denoted as S(n). The time length occupied by a support state is called the *state period.* A state period associated with S(n) can be denoted as $\nabla t(n)$. The moment of shifting from S(n-1) to S(n) is denoted as $t^-(n)$, likewise, the moment from S(n) to S(n+1) as $t^+(n)$, both of which are measured from the origin time.

Three frame-systems, namely, *global-frame, body-frame* and *trace-frame* are used to describe the machine's motion. The global frame is fixed on the ground, used to defined the motion of body-frame. The body frame is attached on the body chassis and used to depict leg motion relative to body. While trace-frame, which is fixed on the compensation area, is used to describe CG adjustment relative to compensation area. During the period of a support state, the body of machine may move or rest. A *body motion state* [11] is defined as 1 if the body has moved, and as 0 if the body has not moved. The *locomotion state* of a machine in any time is possibly one of four cases, (T, 0), (T,1), (F,1), and (F,0). If the locomotion state of S(n) is (F,0), such a S(n) is called *transitional state* which will be used in our research for CG adjustment in some cases. The *foot condition* of a state is the position of the support feet respect to the body-frame, which can also be presented by the *kinematic margins* [6] of the feet. Kinematic margin of a leg is the length of a vector opposite motion direction from current support point to the intersection with the reachable area boundary. In this paper, $K_i(n)$ is used to denote the kinematic margin of leg i in S(n). It is clear that foot condition in different instance of this state will be different if body motion state is 1 in a support state. The foot condition at the moment of the beginning of a support state is called *initial foot condition*, which determines the locomotion characteristics of the support state.

Both body speed and swing speed of transfer leg have direct effect on the shifting of support states. To consider these effect, two parameters $\beta(n)$ and $\alpha(n)$, also called β-factor and α-factor respectively, are defined by

$$\beta(n) = \frac{v_s}{v_s + v_b(n)}, \qquad \alpha(n) = \frac{v_b(n)}{v_s + v_b(n)} \tag{2}$$

where v_s is swing speed which is given by system specification, $v_b(n)$ is body speed in S(n). Swing speed here is assumed a constant specified in advance while body speed may be changed in different support state. For a given task, it is often useful to specify a desired β-factor in advance, this desired β-factor can be denoted by β. Since it is assumed $v_b(n) \geq 0$, therefore $v_s + v_b(n) \geq 0$. The following relation about β(n) and α(n) are always hold in a support state

$$1 \geq \beta(n) > 0, \quad 1 > \alpha(n) \geq 0, \quad \alpha(n) + \beta(n) = 1\,. \tag{3}$$

It can be proven that, if the sequence of support state is periodic, $\beta(n)$-factor in quantity will equate to the duty factor defined in periodic gait.

The following terminology is also helpful for understanding of this paper. The *support pattern* [7] associated with a given support state is the convex hull (minimum area convex polygon) of the point set in a horizontal plane which contains the vertical projections of the feet of all supporting legs. *Stroke of a leg* [11], denoted by R_x, is a length of reachable area of this leg in the motion direction. *Foothold* [6] is a permitted cell which lies in the reachable area of a swing leg.

4 Improved adaptive gait planning

The problem of how to use the concept of stability compensation to plan the adaptive gait for NTU-Q1 to travel over rough terrain will be addressed in this part, which includes the rule of selection of transfer leg and locomotion state, the way to select foothold and the strategy of CG adjustment.

4.1 Rules of selecting transfer leg and support state

The selection of swing leg is based on the initial foot condition, and selecting process follows the rules:

Rule 1: the leg with minimum kinematic margin will always first be selected as transfer leg.

Rule 2: a leg can't be selected twice successively

Rule 3: if there are two or more legs with minimum kinematic margin, the leg that, when lifted, machine will have the largest stability margin should be selected as swing leg.

Rule 4: if above rule can't still determine swing leg, then the sequence of swing leg, 1-4-2-3, should be used, which possess largest stability margin in period gait [7].

Rule 1, also referred as minimum kinematic margin principle in [1], means that if only one leg has minimum kinematic margin, then this leg will be selected as transfer leg. Rules 2 through 4 give the way of how to determine transfer leg when there are two or more legs with minimum kinematic margin.

Having determined the transfer leg, it is still needed to consider whether this swing leg is lifted immediately, which result in a T-state, or this leg is lifted after a period, which result in an F-state. The way for generation of support state here is given by following rules.

Rule 5: if support state tends to be static unstable when selected leg is lifted, then an F-state should be used.

Such an F-state is actually a transitional state. During the period of this transitional state, CG of the machine will be adjusted to a suitable position so as to guaranty the selected swing leg can be lifted.

Rule 6: if the minimum value of kinematic margin among all legs is zero, then a T-state should be used.

For the case which minimum kinematic margin is greater than zero, both T-state and F-state can be used for machine locomotion. However, it is often desired to suitably use the characteristics of two types of support state for the purpose of keeping $\beta(n)$ of desired β so as to make the motion of machine more "smooth". If the kinematic margins of all legs at instance $t^{'}(n)$ are arranged in order from the smallest to the biggest and denoted by $K(1,n)$, $K(2,n)$, $K(3,n)$ and $K(4,n)$, following rule can be used for the above purpose.

Rule 7: If $K(2,n) > \dfrac{[R_x - K(1,n)]\alpha(n)}{\beta(n)}$ and $K(1, n) > 0$, then an F-state should be used, otherwise a T-state should be used.

As will be seen in part 6, above rules can generally result in machine achieving a better locomotion behavior when walking over rough terrain.

4.2 Foothold Selecting

The underlying logic for foothold selection here is to find a region in which all feasible footholds satisfy the constrains requirement. After the region is determined, the suitable foothold will be selected from this region according to some optimum rules.

4.2.1 *Geometry constraint*

Geometry constraint requires that all feet of the machine must be placed within respective reachable area. To determine the foothold region satisfying the geometry constraint, the maximum distance covered by the transfer leg should be determined. Clearly, if without the motion of body, the maximum distance covered by swing leg in S(n), $C_g(n)$, is R_x-$K(1,n)$, where R_x is the stroke of the swing leg. $C_g(n)$ may increase if having the aid of body motion. Denote the maximum distance body can translate in S(n) as $D(n)$, this distance, in geometry, should be equal to the minimum kinematic margin of support legs of S(n), i.e., $D(n)=K(2, n)$. Therefore, with the aid of body motion the maximum distance the swing leg can cover should be $C_g(n) = R_x - K(1, n)+K(2, n)$. Notice that $K(1, n)$ and $K(2,n)$ should be measured at the instance $t^{'}(n)$.

4.2.2 *Kinematic constraint*

When the legged system walk at a desired β specified in advance, the following cases may appear: 1) Swing leg reaches the boundary of its reachable area, while the support legs don't. 2) Support leg with kinematic margin $K(2,n)$ reaches the bounder of its reachable area, while transfer leg doesn't. The time through which transfer leg reaches its boundary is $(R_x-K(1,n))/v_s$ and the time body covers the distance of $D(n)$ is $D(n)/v_b$, which is also the time that the support leg with $K(2, n)$ reaches its boundary. It is obvious that if $D(n)/v_b > (R_x-K(1, n))/v_s$, then transfer leg reaches boundary earlier than support leg, thus first case appears; if $D(n)/v_b < (R_y-K(1, n))/v_s$, then the support leg reaches boundary earlier than the transfer leg, second case appear; if $D(n)/v_b = (R_y-K(1, n))/v_s$, both of transfer leg and support leg will reach their boundary at the same time. In any case, to satisfy the kinematic constraint, the maximum distance the swing leg can cover in S(n) is determined in the form

$$C_k(n) = \min\{(R_x - K(1,n) / \beta(n), D(n) / (1 - \beta(n))\} \tag{4}$$

Obviously, $C_k(n) \le C_g(n)$. if appropriately adjusting the value of $\beta(n)$, $C_k(n)$ will be able to get its maximum value $C_g(n)$. It can be found in equation (4) that if let $(R_x - K(1,n)) / \beta(n) = D(n) / (1 - \beta(n))$,

$$\text{or } \beta(n) = \beta_m(n) = \frac{K(2,n)}{R_x - K(1,n) + K(2,n)}, \tag{5}$$

then $C_k(n)$ will be equal to $C_g(n)$, i.e., R_y-$K(1, n)$+$K(2,n)$. At this point both the support leg and the swing leg will reach their boundary simultaneously. It means that as long as $\beta(n)$ is adjusted to $\beta_m(n)$ in S(n), the machine will achieve maximum $C_k(n)$.

4.2.3 *Stability constraint*

Static stability requires the CG of the machine located within support polygon or the stability margin of the machine positive at any time. Since the backward motion of machine is forbidden, only the *front stability margin* (S_l) needs to be considered in foothold selection. It has been proven that the front stability margin in the period of a support state is monotone function and the minimum value occurs at the end of that support state [4]. It implies that if the front stability margin of S(n) is S_l, then under the consideration of static stability, the maximum distance body can translate in S(n) will be S_l. Thus the maximum distance transfer leg can cover under the condition of the stability constraint, $C_s(n)$, can be determined using the same way as that in section 4.2.2 in the form

$$C_s(n) = \min\{(R_x - K(1,n)) / \beta(n), S_l / (1 - \beta(n))\}. \tag{6}$$

Comparing $C_s(n)$ with $C_k(n)$, the maximum distance that transfer leg can cover under kinematic constrain and stability constrain, $C(n)$, should be the lesser one which can be expressed as

$$C(n) = \min\{(R_x - K(1,n)) / \beta(n), \min(S_l, D(n)) / (1 - \beta(n))\}. \tag{7}$$

Likewise, the value $\beta_m(n)$ can also be obtained by

$$\beta_m(n) = \frac{\min(S_l, D(n))}{R_x - K(1,n) + \min(S_l, D(n))}. \tag{8}$$

If without the compensation of stability margin, S_l will passively depend on the support polygon. It is possible that S_l is very small in some support states if terrain is very rough. At that case, $C(n)$ may decrease even become zero, thus terrain adaptivity of machine will be reduced, the deadlock case may occur. If there is compensation of stability margin, S_l will be independent on the support polygon. The value of S_l can be increased through adjusting CG position within the compensatior area.

C(n) defines a range (or region) in which any a feasible foothold can be selected as a suppor point for next state S(n+1). Such a defined range is also called *foot range.*

4.2.4 *Foothold Selection*

In general, we select suitable foothold based on the such a rule: the feasible foothold wit maximum distance from last support point will always be selected as next support point fc transfer leg. It means if forbidden zone totally locates inside of foot range, then the foothold wit a distance of *C(n)* will be selected as support point, if the forbidden zone crosses the fror boundary of foot range, then the foothold nearest to forbidden zone will be selected as suppo point. This rule tends to make the swing leg get maximum ability of obstacle avoidance.

Once the support point is selected based on the foot range, the real distance covered by body can be finally determined. If the distance from the last support point to selected foothold is *c(n),* then the state-period of S(n) will be

$$\nabla t(n) = c(n) / (v_s + v_b(n)) = c(n)\beta(n) / v_s. \tag{9}$$

The real distance covered by body is $d(n)=c(n)\alpha(n)$ or $c(n)(1-\beta(n))$, and the kinematic margin of the transfer leg in S(n+1) is $c(n)\beta(n)$.

After all parameters of S(n) are determined, the beginning time, $t^-(n+1)$, can be determined by $t^-(n+1)= t^-(n)+ \nabla t(n)$ and the kinematic margin of all leg of S(n+1) at the instance $t^-(n+1)$, i.e., state foot condition of S(n+1), is also obtained by.

$$\begin{aligned} K_i(n+1) &= K_i(n) - d(n) && \text{if leg } i \text{ is support leg in S(n)} \\ K_i(n+1) &= K(1,n) + c(n)\beta(n) && \text{if leg } i \text{ is transfer leg in S(n).} \end{aligned} \tag{10}$$

4.3 Strategy of CG adjustment

Static stability of a S(n) is heavily associated with its support pattern. To clearly address the strategy of CG adjustment, represent the support pattern at the beginning of S(n) as $P^-(n)$ and the support pattern at the end of S(n) as $P^+(n)$. Here $P^-(n)$ and $P^+(n)$ also represent the area covered by them when observed from body frame. Denote the location of CG projection relative to the trace frame at instance $t^-(n)$ as $g^-(n)$, at $t^+(n)$ as $g^+(n)$, and assume the support state at $t^-(n)$ is stable. In addition, denote the intersection area of support pattern $P^+(n)$ and $P^-(n+1)$ as *A*, the intersection area between $P^+(n)$ and compensation area, $\Re$, as *B*, and the intersection area between $P^-(n+1)$ and $\Re$ as *C*, the relationship of which can be written in mathematics by

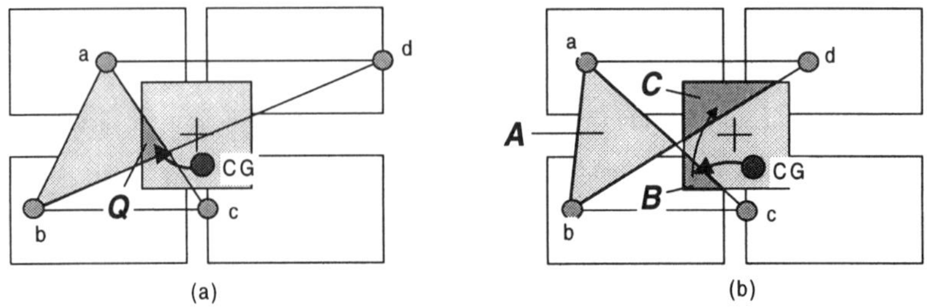

Fig. 4. Schematic diagram for the strategy of CG adjustment, where a-b-c is support pattern at $t^+(n)$ and a-b-d is support pattern at $t^-(n+1)$. (a) the case in which $Q \neq 0$, (b) the case in which $Q = 0$.

$$
\begin{aligned}
A &= P^+(n) \cap P^-(n+1) \\
B &= P^+(n) \cap \Re \\
C &= P^-(n+1) \cap \Re .
\end{aligned}
\tag{11}
$$

It is clear that if $g^+(n) \in B$, then support state S(n) is stable at $t^+(n)$, if $g^+(n) \in C$, then S(n+1) at instance $t^-(n+1)$ is stable, if $g^+(n) \in A$, then the shifting from S(n) to S(n+1) is stable. Therefore to guaranty static stability of S(n) and the shifting from S(n) to S(n+1), the CG location at $t^+(n)$ should satisfy

$$
g^+(n) \in Q = A \cap B \cap C = P^+(n) \cap P^+(n+1) \cap \Re \tag{12}
$$

where Q is the intersection area among A, B and C. Since we have assumed S(n) at $t^-(n)$ stable, if $g^-(n) \in Q$ and let $g^+(n) = g^-(n)$, then CG can satisfy stability requirement during period of S(n) without any adjustment. However, if $g^-(n) \notin Q$, CG location must be adjusted to $g^+(n)$, a new position satisfying equation (12).

It should be noted that, in the terrain containing a large amount of forbidden zone, it is possible that $Q = 0$. In that case, it is necessary to insert a *transitional support state* S([n]), Since S([n]) is with full support legs, the support pattern $P([n])$ covers $P^+(n)$, $P^-(n+1)$ and $\Re$. Thus we can get

$$
\begin{aligned}
g^+(n) &= g^-([n]) \\
&= P^-([n]) \cap P^+(n) \cap \Re = P^+(n) \cap \Re \\
g^+([n]) &= g^-(n+1) \\
&= P^+([n]) \cap P^-(n+1) \cap \Re = P^-(n+1) \cap \Re .
\end{aligned}
\tag{13}
$$

It means that if $Q = 0$, CG should be first moved to a position $g^+(n)$ within B, then through transitional state S([n]) to final position $g^+([n])$ within C. The state period of S([n]) should be determined according to the moving speed of movable weight.

The strategy of CG adjustment can be summered as follows. First look for the area Q, B and C. Second, if $Q = 0$, judge whether $g^-(n)$ is located inside the interior of Q, If true, then let $g^+(n) = g^-$

(n), that means no any adjustment is needed; otherwise find a point as $g^+(n)$ in Q, and then mov CG from $g^-(n)$ to $g^+(n)$. Third, if $Q = 0$, then judge whether $g^-(n)$ is inside B, if true, $g^+(n) = g^-(n$, otherwise, find a point as $g^+(n)$ within B, move CG from $g^-(n)$ to $g^+(n)$, and then add a transition state following S(n), find a point in C as $g^-(n+1)$, move CG from $g^+(n)$ to $g^-(n+1)$ durin transitional state. Obviously, $g^+(n)$ and/or $g^-(n+1)$ may be not unique. Figure 4 shows th schematic picture of CG adjustment strategy.

5 Simulation Results

The program for adaptive gait generation has been written in Maple V Language (release 3) an implemented on a PC486 computer. The animation is conducted using UNIX system an displayed on SUN-station through a commercial software package DADS (release 8), which also helps to do dynamic and kinematic analysis of the prototype. The same simulating model as that of NTU-Q1 shown in Figure 1 is used and only the locomotion with zero crab angle is considered in the simulation. This program allows the user to set up the terrain data, select desired β-factor and/or vary initial foot condition according to requirement to test the terrain adaptivity of prototype.

Figure 5(a)-(j) shows ten representative phases of a simulation experiment with swing speed of transfer leg, $v_s = 25$ *cm/s,* and desired β-factor $\beta = 0.8$. Here, the transfer leg is indicated with an arrow near that leg. It can be seen from Figure 5(c), (e), and (f) that the negative stability margin or deadlock situation would occur at time t_1 (11.9s), t_2(15.3s) and t_3(19.5s) if without compensation of stability margin. The trajectories of CG adjustment relative to float-frame in longitudinal and lateral direction are shown in Figure 6.

Many different terrain conditions have been investigated with the program. It is noteworthy that as long as the maximum width of forbidden zone is less than stroke of reachable area, no deadlock cases have been found even although the machine traverses such a continuous forbidden zone as ditch. It is also been found that if reducing the size of forbidden zone or increasing the value of β-factor in the program, the continuity of body motion will be improved. If the terrain becomes perfect, the program will automatically produce a periodic gait for locomotion of the machine. These features are highly desired in the control of system.

6 Conclusions

This paper has presented an alternative method for the improving of the terrain adaptivity of a multi-legged machine traveling over rough terrain. It is done through 2-D CG adjustment intending to compensate stability margin according to the terrain condition. This method determines foothold based on the current support pattern rather then over the whole trajectory.

Due to the aid of stability compensation, such determined foothold is capable of maintaining the machine static stability during current support state and the shifting from current support state to the proceeding support state. Therefore this way can easily help machine to avoid deadlock case unless the forbidden zone is larger than its reachable area. Some simulation works have been implemented to test the effectiveness of the proposed method.

The concept of stability margin compensation has been applied to the design of the quadruped robot NTU-Q1. Since the moveable weight provides a way to control stability margin actively, it is believed that NTU-Q1 will possess a larger degree to accommodate rough terrain during locomotion.

It should be pointed out that the trajectory of CG adjustment is not unique. Although a feasible CG trajectory has been given in this paper, apparently, it may not be the most optimum one. Further research is therefore necessary in finding the strategy to achieve the most optimum trajectory of CG adjustment based on the given criteria such as minimum adjusting distance and smoothness.

References

1. Y. Ding, and E. Scharf, "Deadlock avoidance for a quadruped with free gait," Proc. of IEEE Conf. Automation and Robotics, (1994), pp. 143-148.
2. S. Hirose et al., "A study of design and control of a quadruped walking vehicle," Int. J. of Robotics Res., vol. 3, no. 2, (1984), pp. 113-133.
3. E. I. Kugushev and V. S. Jaroshevskij, "Problems of selecting a gait for an integrated locomotion robot," Proc. Fourth Int. Conf. Artificial Intelligence Tbilisi, Georgian SSR, USSR, (1975), pp. 789-793.
4. Y. J. Lee and Z. Bien, "A hierarchical strategy for planning crab gaits of a quadruped walking robot," Robotica (1994) vol. 12, pp. 23-31.
5. D. A. Messuri and C. A. klein, "Automatic body regulation for maintaining stability of a legged vehicle during rough-terrain locomotion," IEEE Journal of Robotics and Automation. vol. RA-1, no. 3, Sept. (1985), pp. 132-141.
6. R. B. McGhee and G. I. Iswandhi, "Adaptive locomotion of a multilegged robot over rough terrain," IEEE Trans. System, Man and Cybernetics, vol. SMC-9, no. 4, (April 1979), pp. 176-182.
7. R. B. McGhee and A. A. Frank, "On the stability properties of quadruped creeping gaits," Math. Biosci., 3, 331-351, (1968).

8. P. V. Nagy and S. Desa, "Energy-based stability measures for reliable locomotion of statically stable walkers: theory and application," The Int. Journal of Robotics Research, vol. 13, no. 3, (June 1994), pp. 272-287.
9. P. K. Pal and K. Jayarajan, "Generation of free gait--a graph search approach," IEEE Tran. Robotics. Autom. vol. 7, no. 3, (June 1991), pp. 299-305.
10. J. Stoner and R. H. Davis, "A gait simulator for a quadruped walking robot," Robotica (1992) vol. 10, pp. 57-64.
11. S. M. Song and K. J Waldron, *Machines That Walk: The Adaptive Suspension Vehicle*, The MIT Press, Cambridge, MA, (1989).

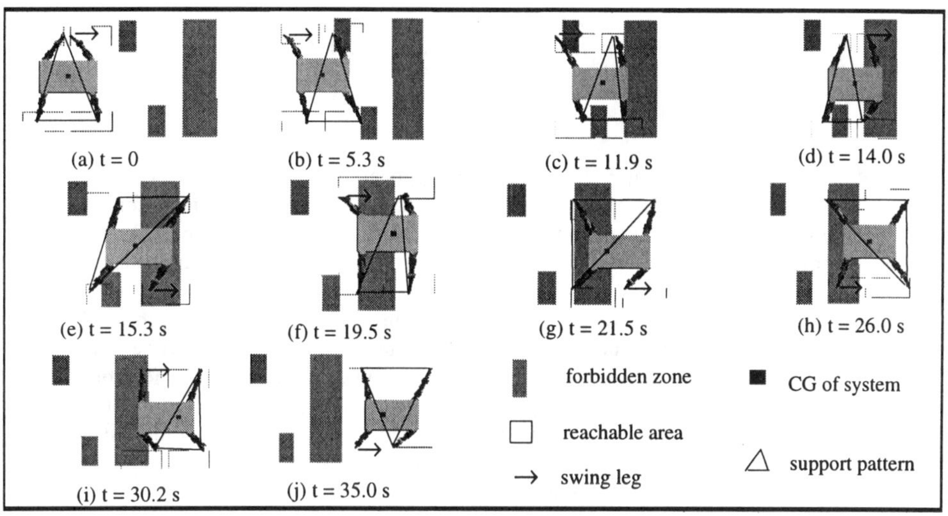

Fig. 5. Some representative phases when simulation model of NTUQ-1 walks over a typical rough terrain

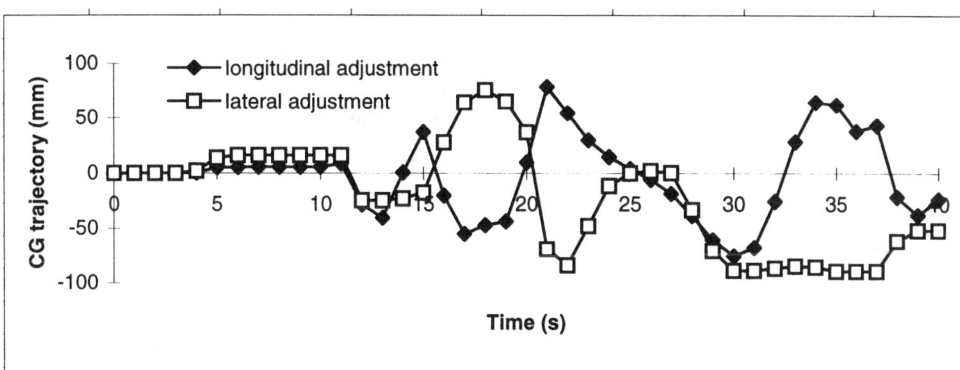

Fig. 6. CG adjustment trajectory in longitudinal and lateral direction

Application of Pareto-Based Multiobjectives Genetic Algorithm in Minimum Time Motion Planning

Ang Mei Choo

B.Sc (Maths.), M.Sc (Eng. in Control Systems)
Dept. of Industrial Computing, Faculty of Information, Science and Technology
National University of Malaysia, 43600 Bangi, Malaysia.

Dr. A.M.S. Zalzala

BEng, PhD, MIEEE, AMIEE
Dept. of Automatic Control and Systems Engineering
University of Sheffield, S1 3JD, UK

Abstract. This paper looks into the application of Pareto-based Genetic Algorithm (GA) in obtaining the minimum time motion planning for an industrial robot. A common practice in multiobjective optimisation GAs for minimum time motion planning is to apply classical aggregation approach to the objective formulation while in this study, Pareto-ranking method is used. A suitable objective vector is organised to include the total travel time and the two joint constraints: velocity and acceleration limits. The objective is to obtain a optimal motion with minimum travel time and within the kinematics limitations. The optimisation process involves first producing a fixed number of joint displacements using the genetic operators, and then scaling the travel time such that it is not violating the kinematics constraints. The feasibility of this method is shown by simulation results with an RTX SCARA robot. Cubic spline functions are used in the construction of the joint trajectory.

1. Introduction

Minimum time motion planning for a industrial manipulator is an important subject in the area of robotics. However, most multiobjective optimisation GAs used in minimum time motion planning apply classical aggregation approach to the objective formulation (see for e.g., [8, 9, 11]). This paper is intended to develop a Pareto-based multiobjective genetic algorithm, which has recently being proposed [4, 5], for the minimum time motion planning of an RTX robot (with six joints). The travel distance and kinematics characteristics of each joint are varied. Thus, to obtain a feasible minimum motion time, the combination of the kinematics characteristics for their joints such as velocities, accelerations and jerks have to be optimised. The objective vector in this project includes the total travel time, the criticality to joint velocity and acceleration limits. The GA motion planning is using the cubic spline theory [7] to construct the joint trajectories. The method allows manipulator motion time, velocity, acceleration and jerk to be scaled such that the kinematics constraints are met. Note that the path planning is carried out in joint space and only kinematics constraints (joint velocities, accelerations and jerks) are considered in this studies. This paper is organised as follows: Section 2 explains the application of cubic spline joint trajectories and the time scaling method. Section 3 incorporates the parameters obtained in Section 2, i.e. optimal motion time, the criticality to velocity and acceleration limits, into the objective function formulation. Section 4 gives the Pareto-based GA motion planning. Section 5 shows the results when applying this method. Section 6 gives the discussions and conclusions.

2. Cubic Spline Joint Trajectories and Time Scaling Method

Each joint trajectory is fitted to a number of joint displacements at a sequence of time instants by using piecewise cubic polynomials. Let $t_1 < t_2 < t_3 < t_4 < ... < t_{n-2} < t_{n-1} < t_n$ be an ordered time sequence and the position (or knot) of the jth joint at time $t = t_i$ is $\theta_{ji}(t_i)$. Thus, the vector of knots for the jth joint along the path is given as $[\theta_{j1}(t_1), \theta_{j2}(t_1), ..., \theta_{jn}(t_n)]$. The interval time is defined as $h_i = t_{i+1} - t_i$ $(i = 1,2,...,n-1)$ whereas the velocity and acceleration of joint j at knot i are denoted as v_{ji} and w_{ji} respectively. The cubic spline joint trajectories allow the joint velocities, accelerations and jerks to be evaluated at every instant of motion time [7]. Every initial motion time will be tested such that it is optimal and the motion will not violate the kinematics constraints. The intervals must be scaled up or down depending on the scaling factor in equation (4). The knot velocities, accelerations and jerks should be compared with their own limits to obtain the time-optimal path satisfying the joint constraints. Let the absolute values of the jth joint velocity, acceleration and jerk limits denoted as VC_j, WC_j, and JC_j respectively. The scaling factor λ can be obtained as follows:

$$\lambda_1 = \max_j \left[\max_{i=1,...,n} \left\{ |v_{ji}|, |v_{ji}(\bar{t}_i)| \right\} / VC_j \right] \tag{1}$$

where $\bar{t}_i$ satisfies $w_{ji}(\bar{t}_i)=0$ and is in the interval $[t_i, t_{i+1}]$,

$$\lambda_2 = \max_j \left[\max_i |w_{ji}| / WC_j \right] \tag{2}$$

$$\lambda_3 = \max_j \left[\max_i \left| \frac{w_{j,i+1} - w_{ji}}{h_i} \right| / JC_j \right] \tag{3}$$

$$\text{and } \lambda = \max\left(\lambda_1, \sqrt[2]{\lambda_2}, \sqrt[3]{\lambda_3}\right). \tag{4}$$

If the time interval h_i is replaced by λh_i for $i = 1,2,..., n-1$, then the velocity, acceleration and jerk will be replaced by factors of $1/\lambda, 1/\lambda^2, 1/\lambda^3$, respectively. These changes assure the satisfaction of constraints on velocities, accelerations and jerks [7].

3. Objective Formulation

The objective vector following from the definition of Pareto optimality [13] would look like this:

$$\text{Minimise: } \left(\sum_{i=1}^{n-1} h_i \,,\ 1\text{-}\lambda_1,\ 1\text{-}\lambda_2\right) \tag{5}$$

$$\text{subjects to constraints: } \lambda_3 \leq 1,$$

where h_i is the time interval i, 1-λ_1 is the criticality to the velocity constraints, 1-λ_2 is the criticality to acceleration constraints. The parameters: λ_1, λ_2 and λ_3 are computed by equations (1), (2) and (3) respectively. Criticality is a measurement of a trajectory on how close it is to the joints' velocity and acceleration limits.

4. Pareto-based GA Motion Planning

The schematic diagram for GA minimum time motion planning is illustrated as follows:

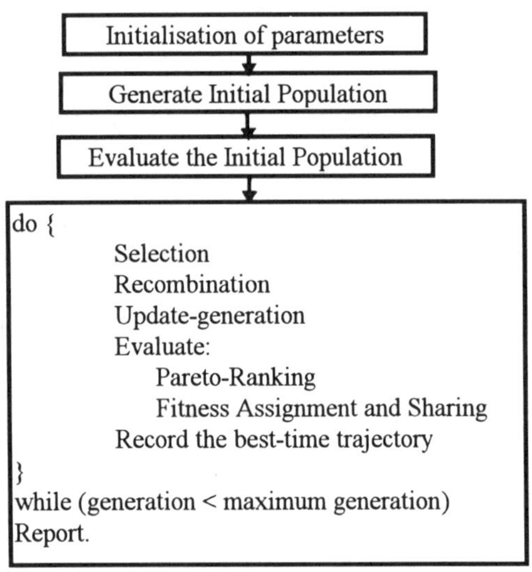

4.1 Generate Initial Population

From the knots vector described in Section 2, the path is encoded directly as string of floating point to be used by the GA as [θ_{11}, θ_{12},..., θ_{1n}; θ_{21}, θ_{22},..., θ_{2n}; ...; θ_{m1}, θ_{m2},..., θ_{mn}]. The knot values of each joint are generated randomly using a transition scheme. The joint is restricted to move in only 5 directions from any knot. To create the initial population, two trajectories, one begins from start position and the other one begins from end position will be generated. The trajectories start from the end position will have higher tendency to move downward whereas the other one will have higher possibility to move upward. A valid trajectory will be created once they meet at a particular point. The travel time between intermediate knots is initially set to their lower bounds for each joint, according to the formula as follows:

$$(h_1, h_2, h_3, \ldots, h_{n-1}) = \left(\max_j \frac{|\theta_{j2} - \theta_{j1}|}{VC_j}, \max_j \frac{|\theta_{j3} - \theta_{j2}|}{VC_j}, \ldots, \max_j \frac{|\theta_{j,n} - \theta_{j,n-1}|}{VC_j} \right) \quad (6)$$

and then the time scaling method will convert the time into a feasible one so that the trajectories will not exceed the limits on all the six joints

4.2 Evaluate The Population

All trajectories are ranked based on its total travel time, the criticality to the joints' velocity and acceleration limits (see Section 3) based on Pareto ranking. According to the procedure proposed by [4], an individual in a population can be ranked by counting the number of individuals that dominate it. When all individuals are ranked, the fitness values will be assigned to them according to their rank. In this paper, the fitness assignment is done by interpolating a linear function [12] from the best individual (rank = 0) to the worst individual (rank<maxpop). Following that, same rank individuals will receive the same fitness values by averaging the total values assigned to them. Fitness sharing uses a sharing parameter h to control the extend of sharing, which is a measurement of the maximum distance between individuals that could form niches. It is computed as follows [4]:

$$h = [8C_n^{-1}(n+4)(2\sqrt{n})^n / N]^{1/(n+4)} \quad (7)$$

where n is the number of decision variables, and N is the population size. It is also known as smoothing parameter for the Epanechnikov kernel [10]. Each trajectory's share count is set to zero initially, and then it is incremented by a certain amount for every trajectory in the population, including the trajectory itself. Raw fitness value is then recalculated by dividing the values with the total share count. The share count can be obtained by using the Epanechnikov kernel [4]. However, the calculation of Epanechnikov kernel function will require the inverse of the sample covariance matrix and also the determinant of the covariance matrix [10]. To avoid the problem of matrix singularity, an approximation is implemented and the share count function will look like the following [6]:

$$sh(d^*) = \begin{cases} 1 - (d^* / h)^2 & \text{if } d^* / h < 1 \\ 0 & \text{otherwise} \end{cases} \quad (8)$$

where d^* is computed such that each decision variable x_i is weighted by its variance. The variance is calculated as follows:

$$\sigma_i^2 = \frac{N\left(\sum_{j=1}^{N} x_{ij}^2\right) - \left(\sum_{i=1}^{N} x_{ij}\right)^2}{N(N-1)} \tag{9}$$

where N is the number of observations or in other word the population size and d^* is computed as:

$$\left(d^*\right)^2 = \sum_{i=1}^{n} \frac{\left(x_i - y_i\right)^2}{\sigma_i^2}. \tag{10}$$

4.3 Selection Scheme And Mating Restriction

Selection scheme is a process to determine the number of trials a particular individual is chosen for reproduction. The selection technique adopted in this project is based on stochastic universal sampling (SUS) introduced by [1]. This method uses a single spin and N equally spaced pointers, where N is the number of population size. Arbitrary recombining pairs of trajectories may produce new offspring that do not represent any niche [3]. It is therefore desirable to reduce crossover between trajectories of different niches. A simple mating restriction scheme is implemented by setting the mating parameter to be equalled to the sharing parameter [3]. The mating parameter is actually the measurement of maximum distance between individuals (trajectories) that allow them to be paired for recombination. Thus, if an individual (trajectory) is within a distance of mating parameter, then a mating companion is found and mating can be performed, otherwise another individual (trajectory) is tried.

4.4 Genetic Recombination

The selected trajectories will be paired up for crossover or recombination subject to their mating distance (i.e. mating restriction) and cross-over probability. A robot trajectory consists of joint angles that may produce large position jump in the offspring strings after conventional crossover. To tackle this problem, a customised genetic operator named as path redistribution and relaxation operator is used [9]. The technique involves fitting cubic splines onto the offspring's knots with each time interval set to one. The path length is then computed as the Euclidean distance between the start and end knots along the splines. Each joint knots are then 'redistributed' evenly over these splines at equal intervals. The paths are then relaxed by moving each knots with a small step towards the point that will bisect the line between its neighbouring knots. Single point cross-over is applied in this GA motion planning. A special purpose mutation operator [9] named as injection is implemented where a fixed number of new trajectories are injected into the population and randomly replace selected trajectory. The number of new trajectories is kept low but it has the effect of preventing premature convergence and creating new search space. After the recombination, the parent trajectory motion time is passed on to the child trajectory as the initial time. The child path together with the motion time will be tested by time scaling method to obtain an optimal motion which will not violate the kinematics constraints. The resulted optimal time and the criticality values to joint velocity and acceleration limits will be used to access the performance of the child trajectory.

5. Results

The multiobjective optimisation GA incorporating all the techniques described in the sections above is implemented. In this simulation, the robot is assumed at rest initially, and comes to a full stop at the end of the time interval. In other words, at the initial time $t = t_1$ and the terminal time $t = t_n$, the joint velocity, and joint acceleration are given as: $v_{j1} = 0 = v_{jn}$, and $w_{j1} = 0 = w_{jn}$. The initial and final configurations of the path planning are shown in Table 1. The limits of the velocities, accelerations and jerks are given in Table 2 [2]. The path redistribution-relaxation operator is experimented with different population size, and 0.9 crossover probability [9]. The injection rate is 2.5% of the population size. The results for 100 generations with different population size are shown in Table 3. The motion profiles for 200 population size are given in Figures 1-4.

Table 1. Initial and final configurations for the path planning of RTX robot.

	Column (m)	Shoulder (rad)	Elbow (rad)	Yaw (rad)	Pitch (rad)	Roll (rad)
Initial Configuration	0.4	$-\pi/6$	$-\pi/3$	$-\pi/2$	0	$-\pi/4$
Final Configuration	0.8	$\pi/6$	$\pi/3$	$\pi/2$	$-\pi/6$	$\pi/4$

Table 2. Velocity, acceleration and jerk constraints for RTX robot.

	Column	Shoulder	Elbow	Yaw	Pitch	Roll
Velocity	0.1116	0.1654	1.2092	1.9715	1.3780	1.2412
Acceleration	1.7755	6.2018	14.081	31.055	28.063	26.180
Jerk	297.59	894.67	3718.9	3377.6	3933.1	4172.7

Note: The zed velocity, acceleration, and jerk are in m/s, m/s^2, and m/s^3, respectively. The other joint angle velocities, accelerations and jerks are in rad/s, rad/s^2, and rad/s^3, respectively.

Table 3. Results From GA Minimum Time Motion Planning

Population Size	Minimum Time(sec)
100	3.9530
200	3.9335
300	3.9049

6. Discussions And Conclusions

Pareto-based multiobjective GA involves ranking process of non-dominated solution to generate the joints trajectories using cubic spline function can produce a optimum path which results in minimum time. This result of minimum motion time improves as the population size increases. However the processing time in obtaining results also increases. Therefore an appropriate population size has to be selected to obtain a reasonably minimum motion time with acceptable length of processing time.

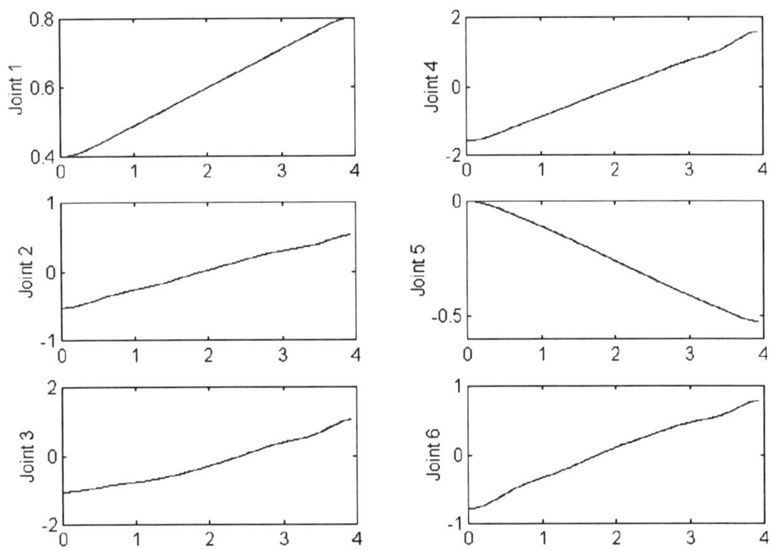

Fig. 1. Position vs. Time

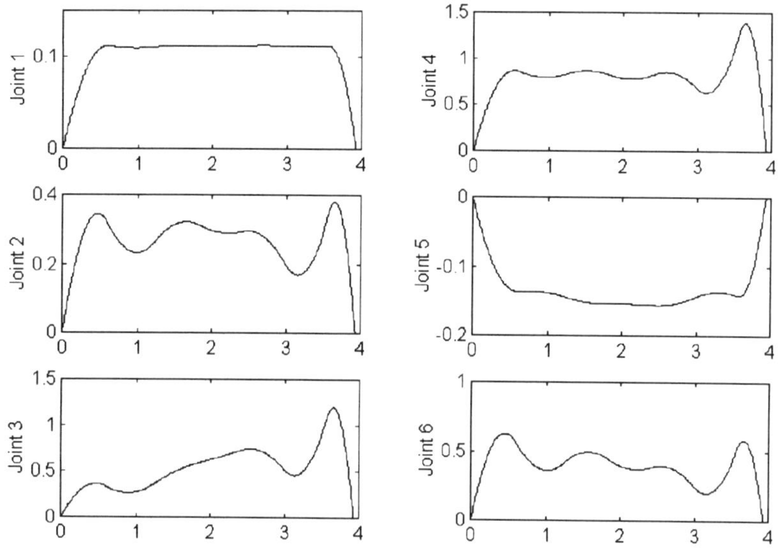

Fig. 2. Velocity vs. Time

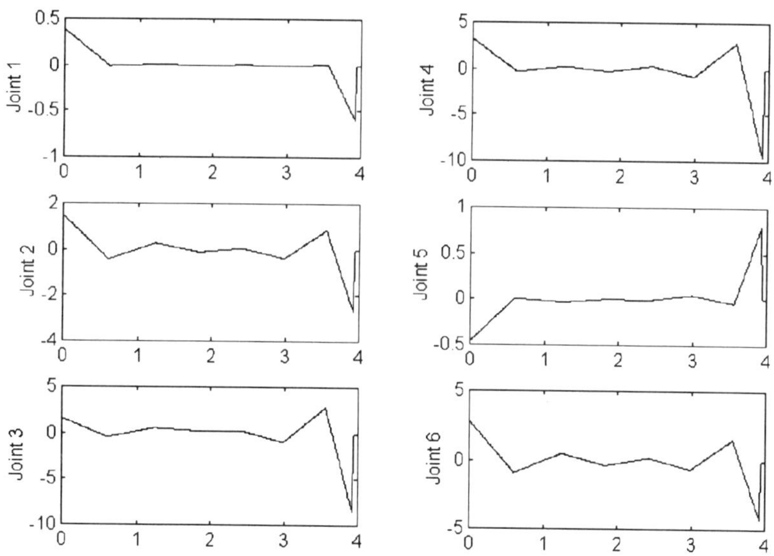

Fig. 3. Acceleration vs. Time

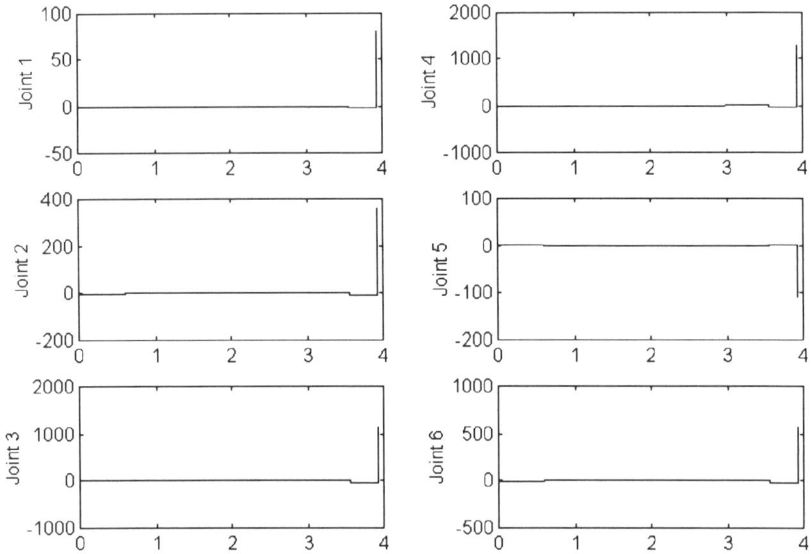

Fig. 4. Jerk vs. Time

References

1. Baker, J. E. (1987). Reducing bias and inefficiency in the selection algorithm, *Proceedings of the Second International Conference on Genetic Algorithms*, pp. 14-21 Grefenstette, J. J. ed., Lawrence Erlbaum Associates, Publishers.

2. Cao, B., and Dodds, G. I. (1994). Time-Optimal and smooth joint path generation for robot manipulators. *Int. Conf. CONTROL'94,* (Coventry, UK), pp. 1122-1127.

3. Deb, K and Goldberg, D. E. (1989). An Investigation of Niche and Species Formation in Genetic Function Optimization. *Genetic algorithms: Proceedings of the Third International Conference on Genetic Algorithms,* pp. 42-50. Schaffer, J. D. ed., Morgan Kaufmann.

4. Fonseca, C. M. and Fleming, P. J. (1995a). Multiobjective genetic algorithms made easy: selection, sharing, and mating restriction. *First International Conference on 'Genetic Algorithms in Engineering Systems: Innovations and application', GALESIA '95*, (Sheffield, UK, 1995), pp. 45-52.

5. Fonseca, C. M. and Fleming, P. J. (1995b). An Overview of evolutionary algorithms in multiobjective optimisation. *Evolutionary Computation.*, **3**, No. 1, pp. 1-16.

6. Fonseca, C. M. (1996). Personal communication.

7. Lin, C.-S., Chang, P.-R. and Luh, J. Y. S. (1983). Formulation and optimisation of cubic polynomial trajectories for industrial robots. *IEEE Trans. Automatic Control*, **AC-28**, No. 12, pp. 1066-1073.

8. Rana, A. S. and Zalzala, A. M. S. *An Evolutionary Algorithm for the Collision Free Motion of Multi-Arm Robots.* Research Report 570, Depart. of ACSE, University of Sheffield, UK, April, 1995.

9. Rana, A. S. and Zalzala, A. M. S. (1996). *Minimum Time Motion Planning of the RTX Robot Using an Evolutionary Algorithm.* Research Report 620, Depart. of ACSE, University of Sheffield, UK, April 1996.

10. Silverman, B. W. (1986). *Density Estimation for Statistics and Data Analysis.* London: Chapman and Hall.

11. Wang, Q. and Zalzala, A. M. S.(1995). *Genetic Optimisation and Experimentation for the PUMA 560 Manipulator.* Research Report 567, Depart. of ACSE, University of Sheffield, UK, April 1995.

12. Whitley, D. (1989). The GENITOR algorithm and selection pressure: why rank-based allocation of reproductive trials is best. *Genetic algorithms: Proceedings of the Third International Conference on Genetic Algorithms,* pp. 116-121. Schaffer, J. D. ed., Morgan Kaufmann.

13. Yacov, Y. H., Tarvainen, K, Shima, T. and Thadathil, J. (1990). *Hierarchical Multiobjective Analysis of Large-Scale Systems.* New York: Hemisphere Publishing Corporation.

NC Programming

A Curvilinear Grid Generation Technique Applied to Tool-Path Planning of a Five-Axis Milling Machine

D. Batanov*, E.Bohez*, S.S. Makhanov**,
K. Sonthipaumpoon** and M. Tabucanon*

*School of Advanced Technologies,
Asian Institute of Technology, Bangkok 12120, Thailand
**Faculty of Information Technology,
King Mongkut's Institute of Technology, Bangkok 10520, Thailand

Abstract. We propose a new algorithm to correct tool-paths of a five-axis milling machine based on elliptic grid generation. Our weighting function is represented in terms of errors related to nonlinear kinematics of the milling machine. Numerical experiments reveal that the proposed technique provides a significant increase in the accuracy of milling.

1 Introduction

Tool-path planning and optimization of a five-axis milling machine has received considerable attention [1-9]. Recent results have displayed a number of sophisticated methods to optimize a "zigzag", spiral or window frame pattern[1-3]. These methods are often combined with techniques dealing with geometric complexity of a work-piece [4,5]. There exists also a variety of off-line techniques to generate a non-uniform tool-path, for instance, methods based on the Voronoi diagram [6] or on neural network modeling[5]. However, a robust algorithm to generate such complicated patterns is still an open problem. On the other hand, the "structured" zigzag pattern is simple and robust and therefore it has been used in conventional CAD/CAM systems.

Adaptive grid generation techniques constitute an essential tool in various CFD and CAD/CAM applications. (See[10,11] for fairly comprehensive reviews on this topic).

In this paper, we present a technique to optimize the zigzag tool-path based on an adaptive grid refinement. We generate the path in a curvilinear "computational" space adapted to milling errors expressed in terms of some global spatial estimates. The resulting curvilinear grid determines the zigzag tool motion adapted to the error variations.

Finally, numerical experiments demonstrate a significant increase in the accuracy of milling.

2 Mathematical formulation

Let $S \equiv S(u,v)=(x(u,v),y(u,v),z(u,v))$ be a machined surface, where u and v are independent variables. Consider a set of points $\{(u,v)_{i,j}\}$,$0 \le i \le N_\xi$, $0 \le j \le N_\eta$, being a discrete analogy of a mapping from the "computational region" $\Delta=\{0 \le \xi \le N_\xi, 0 \le \eta \le N_\eta,\}$ into the "physical region" $D=\{0 \le u \le 1, 0 \le v \le 1\}$, i.e. the points $(u,v)_{i,j}$ belong to some curvilinear structured grid .

Next, consider the tool-paths $T_{i+0.5,j}(\xi)$, $T_{i,j+0.5}(\eta)$ of a five-axis milling machine between the points $(u,v)_{i,j}$,$(u,v)_{i+1,j}$ and $(u,v)_{i,j}$, $(u,v)_{i,j+1}$, respectively, and let us define the following surfaces:

$$T_\xi(\xi,\eta) = L_\eta(T_{i+0.5,j}, T_{i+0.5,j+1}), \qquad \text{if} \quad i \le \xi \le i+1 \quad \text{and} \quad j \le \eta \le j+1,$$

$$T_\eta(\xi,\eta) = L_\xi(T_{i,j+0.5}, T_{i+1,j+0.5}), \qquad \text{if} \quad i \le \xi \le i+1 \quad \text{and} \quad j \le \eta \le j+1,$$

where L_τ corresponds to a linear approximation in the τ-direction, $\tau=\xi$ or $\tau=\eta$.

A quality criterion of a global tool path is formulated in terms of the spatial estimate

$$\omega_\tau = |S(\xi,\eta) - T_\tau(\xi,\eta)|,$$

corresponding to the zigzag motion in the τ-direction.

Let $N_\xi \times N_\eta$ be an arbitrary but fixed number of grid points. We call "an optimal tool path in the τ-direction", a solution of the following optimization problem:

$$\underset{(u,v)_{i,j}}{\text{minimize}} \; E_\tau, \; i=1,\dots,N_\xi, \; j=1,\dots,N_\eta,$$

where

$$E_\tau = \max_{\xi,\eta}\{\omega_\tau\}\cdot$$

Note that the above estimate could be modified in order to involve some additional technological effects such as machining techniques, tool inclination, etc.[8]. However, such modifications do not affect the main results presented in this paper.

Furthermore, observe that the tool path depends on a prescribed policy to calculate the orientation of the tool. Such a policy can be formulated in terms of some supplementary relations between components of the normal vector and the rotation angles.

Therefore, in this section we formulate the problem of tool-path planning in terms of a grid adaptation. As a matter of fact, we have to redistribute a given number of points belonging to a structured curvilinear grid in order to find the best approximation with regard to the maximum-norm. However, the tool path is not linear with respect to ξ or η. Moreover, the derivatives $\partial T_\tau / \partial\xi$,$\partial T_\tau / \partial\eta$ could be discontinuous. Hence, prior error estimates, for instance [12,14], can not be used to optimize the grid.

3 Non-linear kinematics of a five-axis milling machine

In order to calculate the tool-path between successive positions, we first transform part-surface coordinates into machine coordinates. Secondly, the rotation angles and the corresponding machine coordinates are assumed to change linearly between the prescribed points. Finally, we invoke an inverse kinematic equation technique[3], [8]. Observe that inverse kinematics depend upon the structure of a machine. For the machine configuration presented in this paper (Fig.1) the resulting kinematics equations, involving two rotations and three translations, are given by

$$M=B\,(A(P+R)+T)+C,$$

where M, P, R, T, C are respectively the machine, the part-surface, the rotary table, the tilt table and the cutter center coordinates. A, B are respectively the matrices of rotation about the primary and the secondary axes.

The rotation angles a and b are given by

$$a=\arctan(i/j),\quad b=\arctan((i^2+j^2)^{1/2}/k),$$

where (i, j, k) denotes the normal vector.

The five variables (X, Y, Z, a, b) (where $(X, Y, Z)\equiv M$) provide reference inputs for servo-controllers of the milling machine.

Remarks

A procedure to compute the rotation angles becomes "ill-conditioned" in the regions where $|i + j|< \varepsilon$, or $|k|< \varepsilon$ (ε is a small number), for instance for "flat" surfaces[15]. As a matter of fact, small variations of the normal vector in the above mentioned regions can lead to sharp variations of the rotation angles and to numerical instabilities. Therefore, in order to provide a "continuous" change of the rotation angles, special interpolation procedures are implemented in the regions where the normal vector degenerates. Note also that if the a -angle "jumps" from a small value to 360^0 or vice versa, then the sequence of the angles should be "smoothed" in order to minimize the difference between the successive values.

4 Grid generation technique

Next, we propose to apply a modification of an elliptic grid generator [12,14], [16]. A grid generation problem is formulated in terms of Euler equations for a linear combination of functionals characterizing some desirable properties of the grid, namely

$$I = I_s + \lambda I_v ,$$

where

$$I_s = \iint_\Delta [(\nabla \xi)^2 + (\nabla \eta)^2]\, d\xi\, d\eta ,$$

$$I_v = \iint_\Delta J \omega_\tau\, d\xi\, d\eta ,$$

and where J is the Jacobian of the mapping. λ is a calibration parameter. Observe that I_s is related to the smoothness of the grid (Winslow-functional[17]), whereas I_v is related to an adaptation of the grid to regions of large milling errors. Fig.2 illustrates a grid adapted to an elliptical region.

In order to compute E_τ for a fixed λ, we implement the following iterative algorithm:

1) Generate the initial grid G.
2) Calculate $\omega_{\tau,i,j}\equiv\omega_\tau(\xi_i, \eta_j)$ and E_τ on G.
3) Scale and smooth the grid-function $\omega_{\tau,i,j}$ [16].

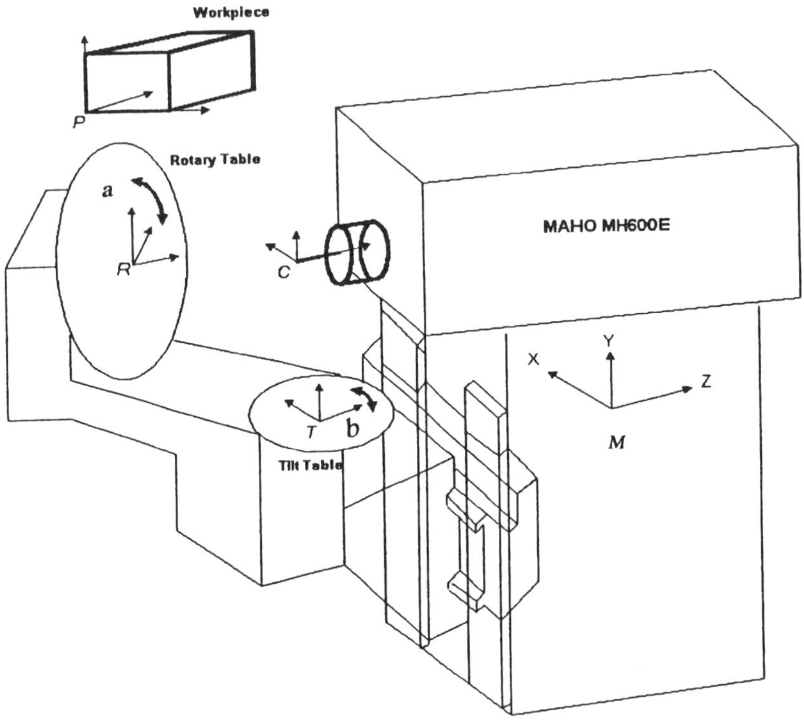

Fig.1 Five-axis milling machine with rotary axes on the table

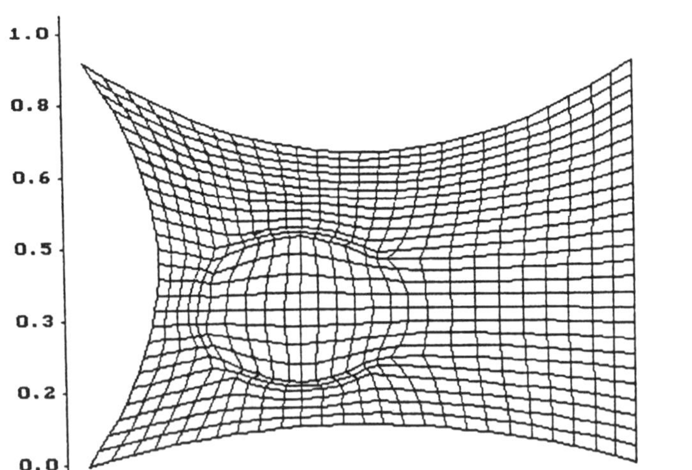

Fig.2 Grid adapted to an ellipse-shaped zone

4) Generate the adapted grid G^{new}, i.e. solve numerically the corresponding Euler equations.
5) Calculate $\omega_{\tau,i,j}{}^{new}$, $E_\tau{}^{new}$ on G^{new}.
6) If $| E_\tau{}^{new} - E_\tau | > \delta$ (where δ is some prescribed value) then $E_\tau := E_\tau{}^{new}$, $\omega_{\tau,i,j} := \omega_{\tau,i,j}{}^{new}$ return to step 3.
Note that there exist various numerical strategies to increase the convergence and the stability of the grid generation algorithm(see, for instance, [14]).
Finally, in order to find an "optimal grid", we apply an appropriate modification of the standard one-dimensional search[18]. But observe that the proposed procedure only searches a local minimum. The existence of a global minimum is still an open problem.

5 Application

A large series of surfaces has been tested by means of the proposed algorithm. Our techniques provide a significant increase in accuracy of about 40%.
A typical testing surface is depicted in Fig.3. The corresponding error distribution on the grid 25x25 and the zones of large milling errors are depicted in Fig.4. The grid adapted by means of the weighting function ω_τ, is shown in Fig.5. Clearly, the grid adaptation corresponds to error variations. The convergence of our proposed algorithm requires 4 iterations, for a prescribed δ=0.01.The resulting λ-value is 180.5. Table 1, presenting errors for the sequence of nested rectangular and curvilinear grids, demonstrates that an increase of grid-points in 144 times leads to an accuracy increase of about 140 times for the uniform grid and of about 235 times for the adaptive grid.

Conclusions

A new algorithm to correct tool-paths of a 5-axis milling machine based on elliptic grid adaptation has been proposed and tested. Numerical experiments reveal that the proposed technique provides a significant increase in the accuracy of milling.

Finally, we remark that the proposed approach is in particular suitable to be embedded in a conventional CAD/CAM software in so far as the algorithm deals with a curvilinear version of the standard zigzag pattern. Such modifications only require an additional mapping obtained by means of the grid adaptation procedures.

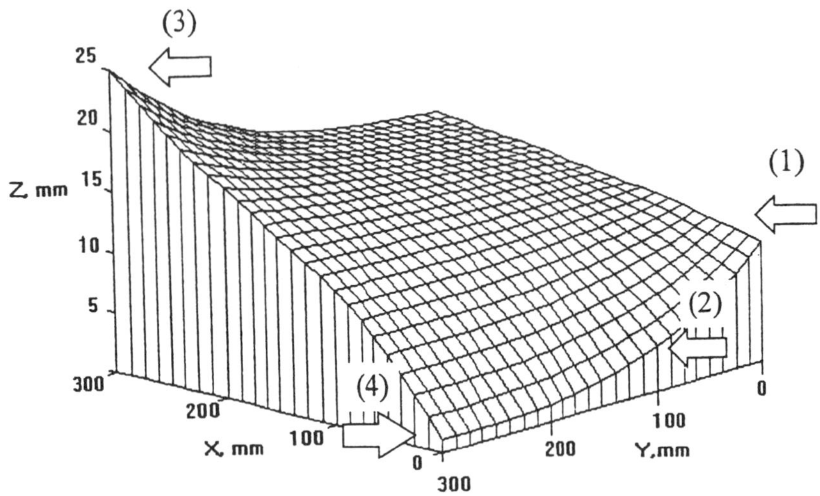

Fig.3 Machined surface, (1),(2),(3),(4) regions of large milling errors

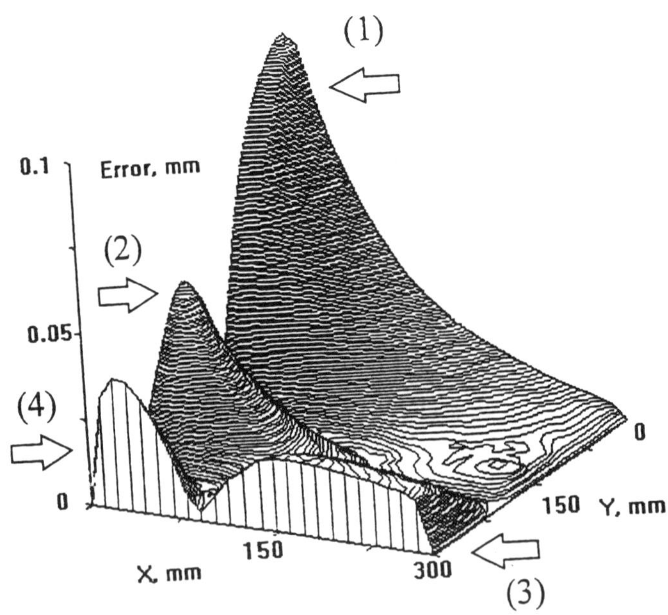

Fig.4 Spatial error distribution

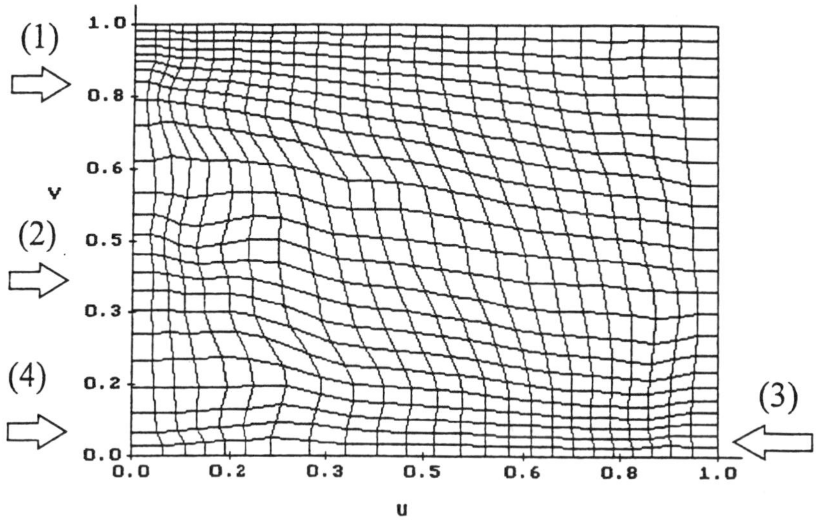

Fig.5 Grid adapted to milling errors

Table 1. Error on uniform and adaptive grids

Number of grid-points	Step for uniform grids, mm	Max. error on uniform grids, mm	Max. error on adaptive grids, mm
100	30.00000	0.14233	0.09538
400	15.00000	0.05692	0.02845
1600	7.50000	0.01594	0.00715
3600	3.75000	0.00752	0.00480
6400	1.87500	0.00421	0.00268
10000	0.93750	0.00161	0.00096
14400	0.46875	0.00099	0.00041

References

1.M.I. Kamien and L.Li, Subcontracting, Coordination, Flexibility and Production Smoothing in Aggregate Planning, *Management Science*, **36**, 11 (1990) 1352-1363.

2. J.Chou and D. Yang, On the Generation of Coordinate Motion of Five-Axis CNC/CMM Machines, *Journal of Engineering for Industry*, **114** (1992) 15-22.

3. Y. Koren, Five-Axis Interpolators, *Annals of CIPR*, **44**, 1 (1995) 379-382.

4. K. Kim and J. Jeong , Finding Feasible Tool-Approach Directions for Sculptured Surface Manufacture, *IIE Transactions*, **28**, 10 (1996) 829-836.

5. S.-H. Suh and Y.-S. Shin, Neural Network Modeling for Tool Path Planning of the Rough Cut in Complex Pocket Milling, *Journal of Manufacturing Systems*, **15**,5 (1996) 295-324.

6. R. Ahmadi and H. Matsuo, The Line Segmentation Problem, *Operation Research*, **39**, 1 (1991) 42-55.

7. R. Lin and Y. Koren, Real Time Interpolator for Machining Ruled Surfaces, *Proc. ASME Annual Meeting*, DSC-**55**, 2, (1994), 951-960.

8. E.A. Gani, B.Lauwer, P.Klewais and J. Detand, An Investigation of the Surface Characteristics in Five-Axis Milling with Thoroid Cutters, *Proc. Pacific Conference on Manufacturing* (1994) 53-60.

9. Y. Huang and J.H. Oliver, Integrated Simulation, Error Assessment, and Tool Path Correction for Five-Axis NC Milling, *Journal of Manufacturing*, **14**,5 (1995) 331-344.

10. V.D. Liseikin, The construction of structured adaptive grids-a review, *Journal of Computational Mathematics and Mathematical Physics*, **36**, 1 (1996) 1-32.

11. J.F. Thompson and N.P. Weatherill, Aspects of Numerical Grid Generation: Current Science and Art, *AIAA-93-3539-CP* (1993) 1029-1070.

12. S.A. Ivanenko, Construction of Curvilinear Meshes for Computing Flows in Basins, in *Modern Problems of Computational Aerodynamics,* CRC Press, London (1992), pp.211-250.

14. S.A. Ivanenko, Adaptive Grids and Grids on Surfaces, *Journal of Computational Mathematics and Mathematical Physics*, **33**, 9 (1993) 1179-1194.

15. M. Daniel, A Note on Degenerate Normal Vectors, *Computer Aided Geometric Design*, 12 (1995) 857-860.

16. J.U. Brackbill and J.S. Saltzman, Adaptive Zoning for Singular Problems in Two Dimensions, *Journal of Computational Physics*, **46**, 3 (1982) 342-368.

17. A.M Winslow, Numerical Solution of Quasilinear Poisson Equation on Non-Uniform Triangle Mesh, *Journal of Computational Physics*, **1**, 2 (1967) 149-172.

18. W. T. Vetterling, S.A. Teukolsky, W. H. Press and B.P. Flannery, *Numerical Recipes in Pascal*, Cambridge, University Press (1990).

Intelligent, Distributed NC Systems

Sally M. Chan
Terence L. Lammers

Boeing Commercial Airplane Group - Information Systems
P.O. Box 3707, MS 6C-MT
Seattle, WA, USA 98124
{sally.m.chan, terence.l.lammers}@boeing.com

Abstract This paper presents the integration of Artificial Intelligence Technologies and Object-Oriented Technology into a domain engineering process, Object-Oriented Domain Engineering (OODE). This process is demonstrated in the Numerical Control (NC) domain. Specifically, the paper discusses how the integration of four techniques can be used to produce intelligent, distributed NC applications. These four techniques are: Knowledge Acquisition (KA), the Knowledge Based Systems (KBS) approach, Object-Oriented Analysis and Design (OOA&D), and architecture based on a distributed, heterogeneous object bus.

The first part of the paper discusses Knowledge Acquisition methods, and how these methods capture requirements and expert knowledge for KBS development and for OOA&D. Here the paper describes what a KBS is, its purpose, benefits, and application to the NC domain. The second part of the paper talks about the application of Object-Oriented Analysis and Design to NC for domain engineering. This section describes the OODE process and deliverables in the NC domain, and discusses how these support the rapid development of intelligent distributed NC applications. The third part of the paper presents an overview of a preliminary heterogeneous, distributed object architecture for the NC domain. This architecture then could be used as the framework for developing intelligent, distributed NC applications. Finally, the paper presents the future direction of AI and OOT as seen by the authors.

1. Introduction

This paper discusses the integration of Artificial Intelligence (AI) Technologies and Object-Oriented Technology (OOT)[2][3][6] to achieve rapid development of intelligent Numerical Control (NC) systems based on a distributed architecture. The AI Technologies discussed are Knowledge Acquisition (KA)[5] and Knowledge Based Systems (KBS). The paper shows the use of KA to build a KBS application. In addition it shows that KA is needed to extend and complement Object-Oriented Analysis and Design (OOA&D) for Object-Oriented Domain Engineering (OODE)[1]. Both AI and OOT are important to form an NC framework[4].

Communication by applications across heterogeneous environments (such as the one created by integrating AI and OOT) is enabled by an object bus architecture with its Object Request Broker (ORB). This architecture encourages reuse, open frameworks, and standard object interfaces, all of which make Rapid Application Development (RAD) possible. Rapidly developing applications by extending a framework defines a new software development paradigm.

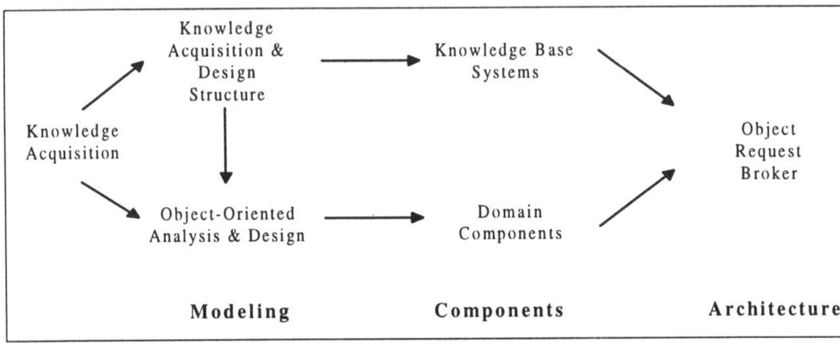

Figure 1. Information Map

Figure 1 is an information map of the content of this paper. It shows the use of KA by both KBS development and OOA&D. It shows how the products of the two approaches, KBS and domain components, are integrated by the object bus architecture, represented in the diagram by the Object Request Broker.

2. Artificial Intelligence Technologies

The goal of Artificial Intelligence Technologies is to develop computer systems which can mimic human intelligence. In order to mimic human intelligence, knowledge must be acquired and stored. This section discusses how Knowledge Acquisition (KA) and Knowledge Based Systems (KBS) accomplish this.

2.1 Knowledge Acquisition (KA)

The Knowledge Engineers (KEs) use the KA process to gather a domain expert's complex problem solving knowledge to develop a KBS. Knowledge Acquisition and Design Structure (KADS) is a methodology widely used by KEs. KADS approaches KBS development as a modeling activity. It uses cognitive models and Problem Solving Templates (PSTs) to illustrate the reasoning pattern domain experts follow when solving complex business problems. KADS consists of four modeling steps: the Process Model, the Conceptual Model, the Problem Solving Template, and the Strategic Model. The Process Model structures the behavior used to solve problems and complete a process. The Conceptual Model structures the information needed to solve problems and perform a process. The PST illustrates the reasoning patterns inherent in business and system processes. The Strategic Model identifies the high level workflow and core processes of the system to define the architecture of the KBS.

Figure 2 illustrates the application of KADS modeling techniques to the NC domain to develop a manufacturing KBS. The models in the Figure represent preliminary work which has not been implemented in any specific application.

KADS models are highly abstracted descriptions of typical problem solving tasks. They are easy to understand because their cognitive structure represents high level collaborations and because they identify clear boundaries and linkages for business processes. KADS models help KEs identify the information necessary for KBS development effectively, therefore shortening the system development cycle.

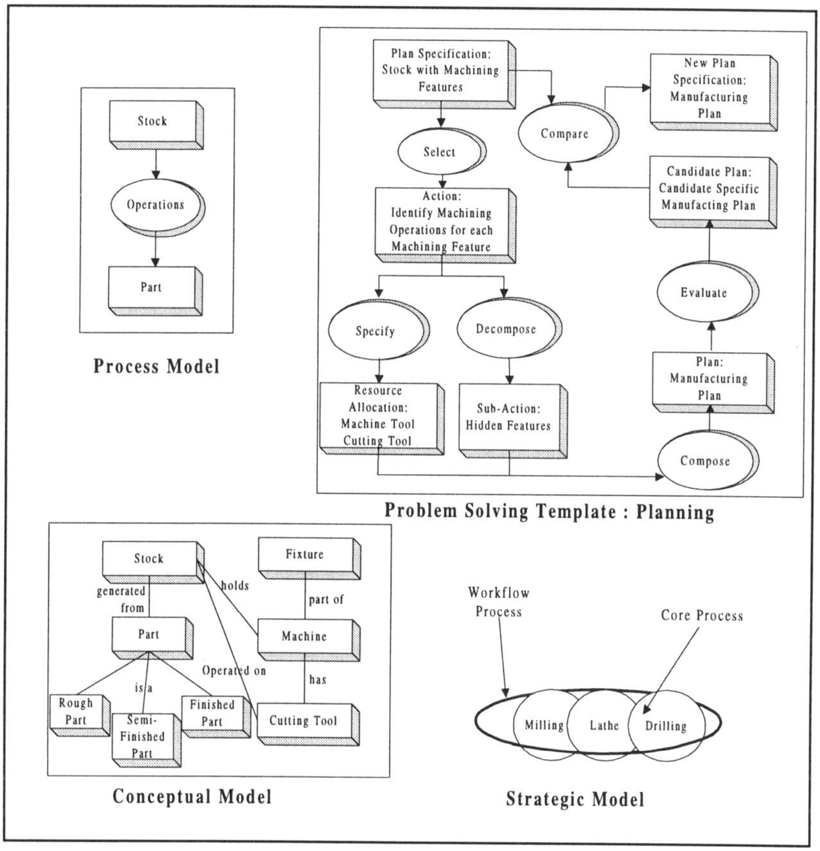

Figure 2. Examples of KADS Models for Knowledge Acquisition

In KBS development, the modeling focus is on "how" the domain expert solves a complex business problem; in OOA&D, the modeling focus is on "what" the interactions between objects are in the system. Despite these different focuses, KA can benefit OOA&D as a process for gathering systems requirements. Before any application is developed, the requirements need to be understood, collected, and documented. In OOA&D, user requirements are defined in Use Cases, which are textual descriptions of an actor's interaction with the application. From the Use Case, one can identify the system or domain objects, their collaborations, relationships, and subsystems. Applying KADS models increase understanding of the problem and this enhances the application of OOA&D. For example, the Process Model can graphically capture user requirements and interaction in Use Cases. The Problem Solving Templates (PSTs) add understanding of how something is done, and the Process Model and Conceptual Model aid the abstraction of application-level Use Cases into Domain

Use Cases for framework development. Figure 3 shows the integration of KA models and OOA&D models; the un-shaded boxes represent the KA models.

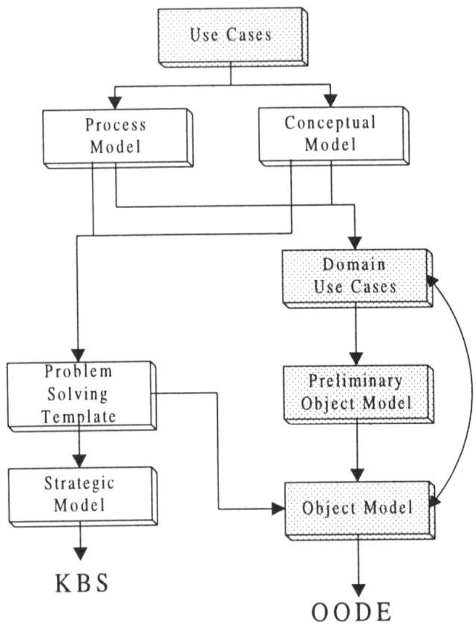

Figure 3. Integration of KADS and OO Models

```
Priority : 9
Rule Type:       forward, backward
Rule Analysis:   forward, onrequest, singlefire

Rule Definition:

If
   hole(->hole_instance).machine is 'PW01' and
   operation_type(in001).rout_finish is 'defined' and
   hole(->hole_instance).hole_diameter < rmaxhold and
   hole(->hole_instance).hole_diameter >= rminhold
then
   operation_typw(in001).insert_holddown_bolts is 'defined'
   message (insert_holddown_bolts)
end
```

Figure 4. Sample KBS Rule

2.2 Knowledge Based Systems (KBS)

KA is an input to KBS development. The KBS processes subject matter expert's knowledge for solving complex problems within a domain. It is designed to solve diagnostic, monitoring, configuration, planning, design, or scheduling problems. A KBS consists of an Inference Engine and a Knowledge Base.

The Knowledge Base is usually implemented in rules, such as IF_THEN rules, meta rules, and/or methods. Figure 4 shows an example of a manufacturing rule in a KBS. Knowledge Engineers collect these rules in KA sessions from subject matter experts. The Inference Engine processes the rules in the Knowledge Base by forward or backward chaining, agendas, truth maintenance, confidence factors, conflict resolution, and by using daemons. Other functionality like explanation and pattern matching are additional capabilities which facilitate KBS development. A KBS can provide realistic conclusions to actual problems as well as the steps and materials involved in the reasoning process.

Using a KBS adds to an application the complex decision making capability which usually requires human expertise. The KBS Inference Engine separates the decision-making logic from the process control logic. It does so by not embedding business decision rules within application code, as in conventional programming. Separation accommodates ever-changing business requirements and provides maintainability. NC is a good area for KBS application because NC programming requires experience and knowledge.

3. Object-Oriented Domain Engineering

3.1 Background

A manufacturing KBS is only one component of the NC domain framework. To develop complex systems such as NC applications, it is crucial to also apply OOT to maximize reuse, provide plug-and-play capability, and enable RAD. Object-Oriented Domain Engineering (OODE) supports these goals by creating a framework of collaborating components. OODE extends the techniques and concepts of Object-Oriented methodologies and applies them to framework development.

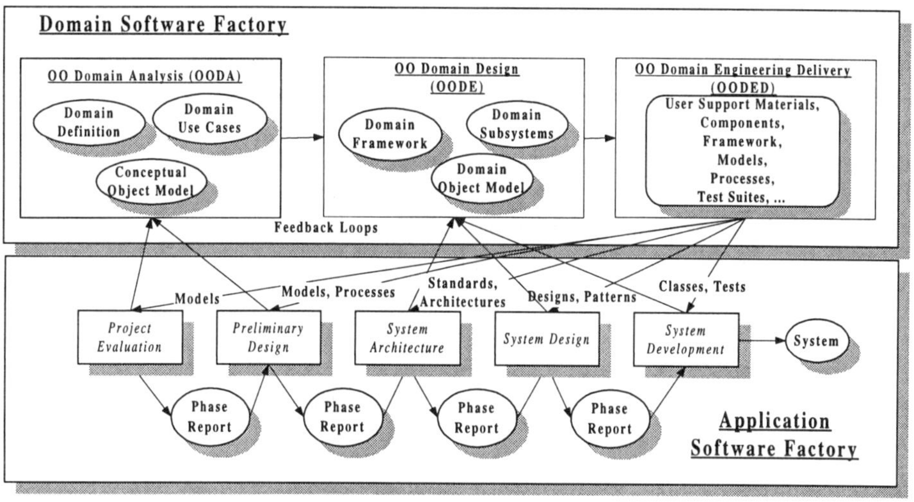

Figure 5. Application and Domain Engineering Dual Lifecycles

OODE is one of the two lifecycles in a dual lifecycle software development method. The OODE lifecycle consists of three phases: Object-Oriented Domain Analysis (OODA), Object-Oriented Domain Design (OODD), and Object-Oriented Domain Engineering Delivery (OODED). These phases run parallel to the Application Engineering (AE) lifecycle. Figure 5 shows the interrelations, including feedback loops, between the OODE and AE lifecycles. The major OODE end user deliverables are a Domain Object Model, a Domain Framework, and User Support Materials.

3.2 Object-Oriented Domain Engineering Lifecycle Phases

Object-Oriented Domain Analysis (OODA) is the first phase of OODE. It has three key deliverables: 1) the Domain Definition; 2) the Domain Use Cases; and 3) the Conceptual Object Model. The Definition focuses and constrains the OODE process. The Domain Use Cases capture the domain requirements, and validate the completeness of the Model. The Conceptual Object Model captures the domain classes, their relationships, and their responsibilities. The Conceptual Object Model and Domain Use Cases span multiple domain applications and the domain through time.

The second phase of OODE is OODD. Its deliverables are: 1) the Domain Framework; 2) the Domain Subsystems; and 3) the Domain Object Model. These are elaborated from the OODA deliverables. Once the Domain Object Model is complete, a make or buy study can determine the best approach for implementing the framework.

OODED is the third phase of OODE. The major risk OODE faces is that once a framework is delivered, projects will not use it. From the application project's point of view, the framework could be viewed as an unknown entity. This critical issue is addressed by the OODE Delivery Phase which supplies training, documentation, and support to transition the framework into the application project. The major deliverables in the OODE delivery phase are the Test Suites and User Support Materials. Test Suites are derived from the Domain Use Cases. Thus the Domain Use Cases play a role in the OODE process from modeling through testing, ensuring the robustness and correctness of the framework.

4. OODE Deliverables for the Numerical Control (NC) Domain

The previous section of the paper presented the OODE process. This section describes the deliverables produced by that process for the NC domain. The deliverables discussed in this section are preliminary and meant to illustrate OODE concepts. They support Rapid Application Development (RAD) in the NC domain or any domain, because they are in place before an application development project begins. The project only needs to customize the deliverables for its special needs and then can rapidly assemble them into an application.

4.1 Domain Definition

After a domain has been identified for OODE, the first step is to define it. Combinations of events and actors within the domain define the domain boundary. Domain subject matter experts contribute their expertise to identifying domain actors and events and to drawing the boundary. Experience with OODE has shown that the key to good domain definitions is effective Knowledge Acquisition techniques. Thus KADS Process Models complement the Domain Definition Model. Figure 6 is a sample domain boundary diagram, showing an NC Programmer as an initiating actor.

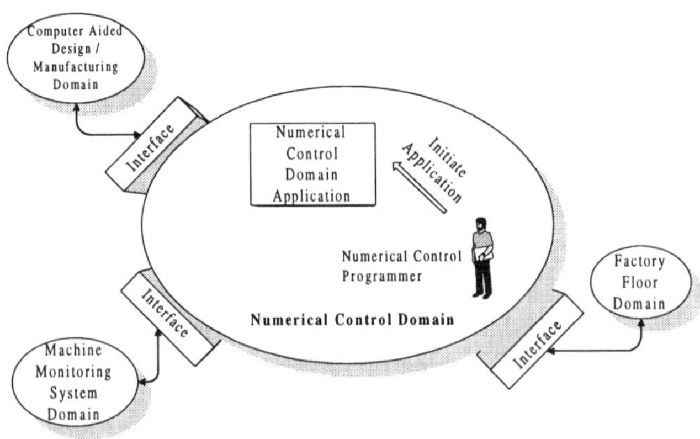

Figure 6. Numerical Control Domain Boundary

4.2 Domain Use Cases

Domain Use Cases capture abstract business processes which span an entire domain. The KADS Process and Conceptual Models make the process of abstracting application-specific Use Cases to domain level Use Cases easier. They are applicable to all the applications in the domain. It is the goal of the Domain Use Cases to not only capture what exists in a domain, but to describe "to be" processes as well. "To be" processes are defined by reengineering the existing processes in light of future business requirements and technological advancements. By capturing these processes OODE fulfills its responsibility for having components ready when they are needed by an application project, even a future project.

A Domain Use Case contains the actors, the intent of the Use Case, assumptions which underlie the Use Case, and the results achieved by it. The actor initiates the Use Case and it is carried forward to completion by the interaction of the system and the actor. System responses to actor events are mapped to domain classes which provide them. From the Domain Use Case the domain's classes are identified and later formalized in the Domain models.

4.3 Domain Models

Domain Use Cases capture essential business processes, and the Domain Models build on these to define generic domain classes and their relationships. The OODE process produces and elaborates three models: the static object model, the dynamic object model, and the subsystem model. These three make up the Domain Object Model.

The static object model presents the domain classes, their inheritance and composition relationships, responsibilities, services, and attributes. Figure 7 portrays only one NC subsystem from the entire NC static object model. The model in Figure 7 is preliminary and does not represent a framework which has yet been implemented.

Class collaboration relationships are modeled in the OODE dynamic domain class model and the subsystem models. They indicate how the domain objects interact over time and demonstrate whether the class responsibilities have been well distributed.

4.4 Domain Subsystems

A domain subsystem is both a container for the classes which compose it, and a controller for their activities. It presents a public interface, or contract, to the world. The contract in Figure 8 is the half circle on the right edge of the subsystem boundary. To fulfill its contract the subsystem sequences the services of its internal objects. This sequence is indicated by the numbers on the lines. The lines themselves represent service requests from the subsystem to its members to support its contract.

The domain subsystem models continue the process of elaborating the object models, but their primary purpose is to aggregate classes into larger entities. Subsystems increase the granularity of the domain components, as seen in Figure 8, where the NC Strategy Planner Subsystem aggregates a number of classes. This is a goal of OODE: to enable Rapid Application Development by reducing the number of components needed to assemble an application.

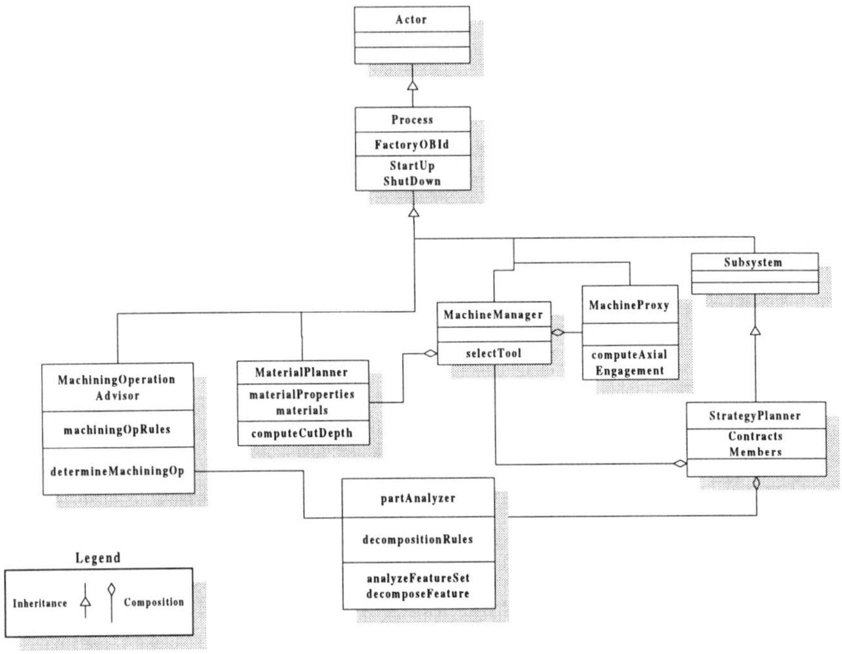

Figure 7. NC Strategy Planner Subsystem Static Object Model

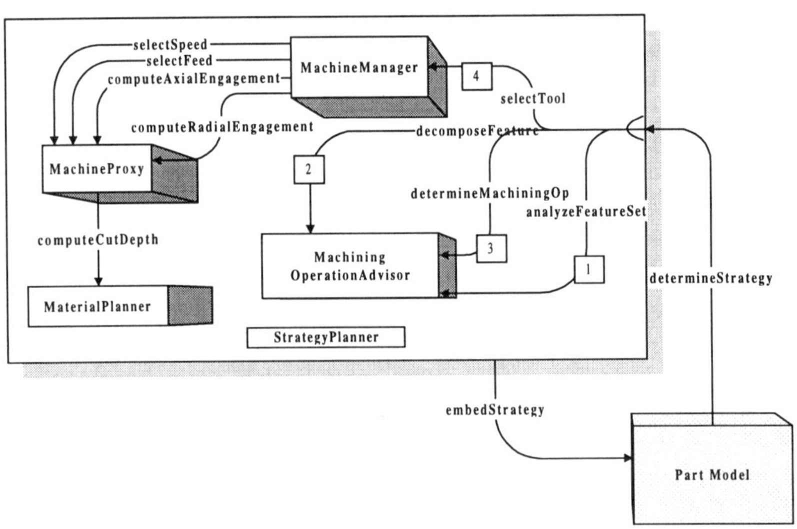

Figure 8. NC Strategy Planner Subsystem

5. Distributed Object Architecture

This part of the paper presents a preliminary heterogeneous distributed object architecture for the NC domain. The architecture is the culmination of the first two phases of the OODE process. Figure 9 presents the architecture in terms of subsystems and an object bus. These are the building blocks for application development.

The architecture is heterogeneous because components built with a variety of languages and for different platforms can be integrated into it. It is distributed because components execute on independent hosts and communicate via an object bus. The Inter Bus Gateway manages communication between different object bus standards.

The architecture is the foundation of Rapid Application Development, because components plug quickly into the object bus with minimal effort through standard interfaces such as the CORBA Interface Definition Language (IDL). OODE supports this approach because components are specifically designed for assembly and plugability.

5.1 Benefits

Domain architectures are beneficial because they enable comparisons among domains. Common, reusable components which exist in more than one domain are easily identified. This leads to cross-domain reuse and greater return on investment.

Distributed architectures make the best use of available resources. For example, in Figure 9 the StrategyPlanner is seen hosted on a separate machine, because it is computationally intensive. Other components are located on one host because they need fewer resources or in order to communicate locally. By distributing components, data movement on the network is localized and therefore minimized overall.

Heterogeneous architectures maximize the reuse of existing applications and components. Applications need not be rewritten, but only made compatible with a standard object bus interface.

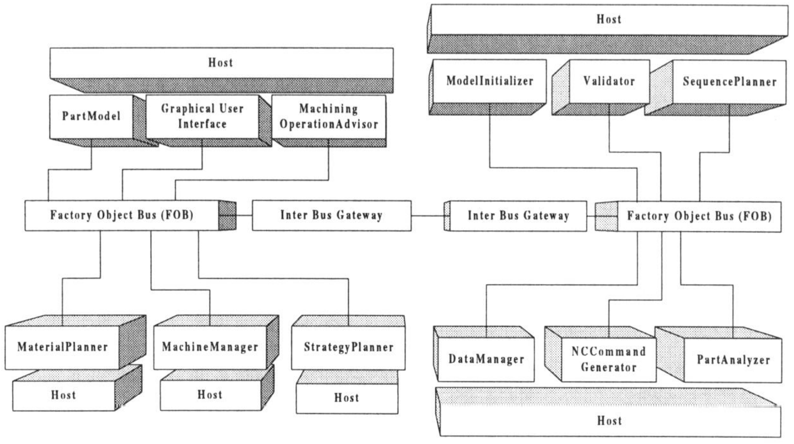

Figure 9. Distributed, Heterogeneous OO Numerical Control Architecture

6. Conclusion

This paper concludes that the integration of Artificial Intelligence (AI) and Object-Oriented Technology (OOT) will lead to the rapid development of intelligent, distributed Numerical Control applications. AI and OOT complement each other. KA methods are strong where OOT is weak. Their integration leads to better framework design. Frameworks in turn enable the rapid development of applications.

At the architectural level, the object bus allows domain subsystems to communicate with each other through standard interfaces. Heterogeneous components such as a KBS and domain components are integrated by a distributed object architecture. These lead to rapid assembly of applications and software reuse.

7. Future Direction of OO Technology and Artificial Intelligence

This paper describes how OOT and AI complement each other. In the future there will be a more complete merger of the two. The State Object is the most common type in the current practice of OOT. State Objects change state but are not aware it. Aware Objects know their state and make use of it to perform their responsibilities. At this point AI begins to be deeply integrated into OOT. Intelligent Objects have integrated Knowledge Based Systems which give them the capability to perform complex, unstructured services. Artificial Life Objects are intelligent and have the ability to learn and evolve. This means that Artificial Life NC systems can rapidly adapt to new parts, new tools, new materials, and new business requirements. Living Systems are systems of Artificial Life Objects. These Systems demonstrate behavior which emerges from the interaction of their members. Because its objects evolve, the entire System evolves to respond to changes in its environment. Some of the enabling AI technologies which contribute to this are machining learning, genetic programming, neural nets, and adaptive programming.

References

[1] Chan, S., Lammers, T. , *Object-Oriented Domain Engineering*, Reuse'97 Workshop Tutorial, 1997

[2] Jacobson, I. *et al.*, *Object-Oriented Software Engineering - A Use Case Driven Approach*, Addison Wesley, 1992

[3] Orfali, R., Harkey, D., Edwards, J., *The Essential Distributed Objects Survival Guide*, Katherine Schowalter, 1996

[4] Schmid, H. A., "Creating Applications from Components: A Manufacturing Framework Design", IEEE Software, November 1996, Vol. 13, No. 6, pp. 67-75

[5] Tansley, D., and Hayball, C., *Knowledge-Based Systems Analysis and Design: A KADS Developer's Handbook.* Prentice Hall, 1993

[6] Wirfs-Brock, R., Wilkerson, B., and Wiener, L., *Designing Object-Oriented Software*, Prentice Hall, 1990

Automation Components

DEVELOPMENT OF AN AUTOMATIC AND FLEXIBLE FEEDING SYSTEM FOR DELICATE CYLINDRICAL PARTS WITH ASYMMETRICAL FEATURE

Patrick S.K.Chua
School of Mechanical and Production Engineering
Nanyang Technological University
Nanyang Avenue
Singapore 639798

ABSTRACT

This paper concerns the development of a flexible and automatic feeding system which is capable of feeding cylindrical parts which are delicate or powdery in nature and possess asymmetrical features such as a groove near to one end. Controlled by a programmable logic controller and electropneumatics, the system has active orientating capability. System evaluation results showed that the average jam rate is below 10 % and the percentage of correctly orientated parts is above 90 %. With fine tuning, the system could become a very useful feeder for industry in the future.

1 INTRODUCTION

In industries where assembly costs form a large portion of the production costs, manufacturers are increasingly looking towards assembly automation to improve efficiency as well as productivity. Special purpose automatic assembly machines are however only economically feasible in very large volume production and they are usually dedicated only to that particular product, thus lacking flexibility.

Recently, there is an increase in demand for customised products with special features and options to meet the particular needs of the customer. To meet this market needs, manufacturers must be able to reduce manufacturing lead times, accommodate for product changeovers and part design changes in order to stay competitive. It is envisaged that flexibility of the assembly systems is the key to survival in the competitive manufacturing market. Therefore there is now a growing interest in the development of a flexible parts presentation system to achieve a total flexible manufacturing system

Conventional vibratory feeders commonly seen in industry today have the following shortcomings or limitations in scope:

(a) They are usually dedicated types and therefore unsuitable for productions in which there are frequent batch changes or in which the product life cycle is short.

(b) They are also not suitable for fragile, delicate or powdery parts because of the vibration of the feeder.

(c) The orientating devices used in the vibratory bowl feeders are usually of the passive type, i.e. by rejecting parts which travel on the track in the undesired orientation. Thus there is a reduction in efficiency.

(d) They are incapable of feeding and orientating cylindrical parts which have unsymmetrical features at the circumference e.g. a groove near one end as shown in Fig. 1.

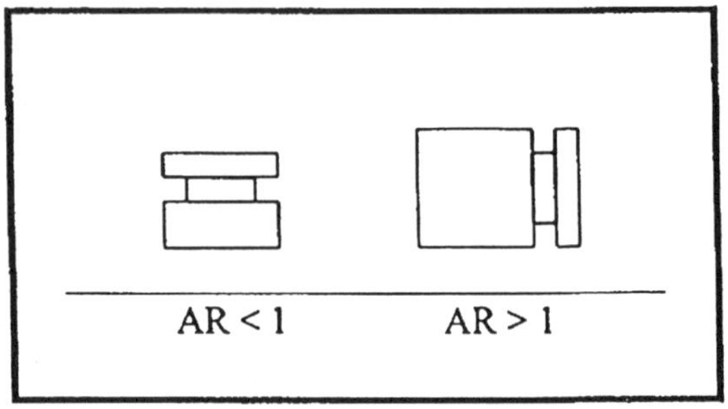

Fig. 1 :Asymmetrical Parts

Moreover, existing designs of non-vibratory feeders have similar shortcomings as mentioned in items (a), (c) and (d) above. These limitations in scope of existing feeding systems call for a flexible, non-vibratory feeding system with active re-orientating devices capable of orientating cylindrical parts even if these parts have asymmetrical features such as a groove near one end.

2 OBJECTIVE AND SCOPE

The work involved the conceptualisation, experimental evaluation of the conceptualised design, design, construction and evaluation of a suitable feeding system for feeding and orientating fragile asymmetrical cylindrical parts of various aspect ratios.

3 LITERATURE REVIEW

When it comes to the feeding of delicate powdery parts, vibratory feeders are certainly unsuitable for use because cracking and dust accummulation can occur as a result of the rigorous vibration. Many non-vibratory feeders have been developed over the years but the objective of their developments was as a low cost alternative to the more expensive vibratory bowl feeders and not because delicate parts were to be fed. Several such non-vibratory feeders can be found in Boothroyd et al [1, 2], Denhamer [3] and Chua and Lim [4].

Another reason for developing non-vibratory feeders was flexibility which necessitated the use of feature recognition sensors such as infra-red fibre-optic sensors and cameras. These feature recognition devices require that the surface that the part is travelling be non-vibratory in nature inorder to ensure reliability in identification of the orientation of the moving part. Examples of such flexibile non-vibratory feeders can be found in Pherson et al [5] and Chua and Sung [6].

The works of Pherson et al [5] and Chua and Sung [6] concentrated on the feeding of non-rotational parts with asymmetrical feature such as a blind hole at one corner of a rectangular part. No work was carried out on the feeding of rotational parts with an asymmetrical feature such as a groove near to one end. The reason for this lack of work on rotational part is probably due to the additional difficulty of the part's ability to roll freely. To be able to convey it without vibratory means while at the same time orientating it to the desired orientation for pick-up by the manipulator is not an easy task.

4 DESIGN SPECIFICATIONS AND REQUIREMENTS

The feeding system to be developed should be able to feed cylindrical parts as follows:

Part Diameter (mm)	Aspect Ratio (AR)
13 - 25	0.5 - 0.8 1.2 - 1.6

Table 1 : Cylindrical Parts of Interest

The part has a groove of 2 mm (width and depth) near to one end as shown in Fig. 1. The feed rate should not be less than 10 parts per minute.

5 CONCEPTUAL DESIGN

The components required of the feeding system is shown in Fig. 2.

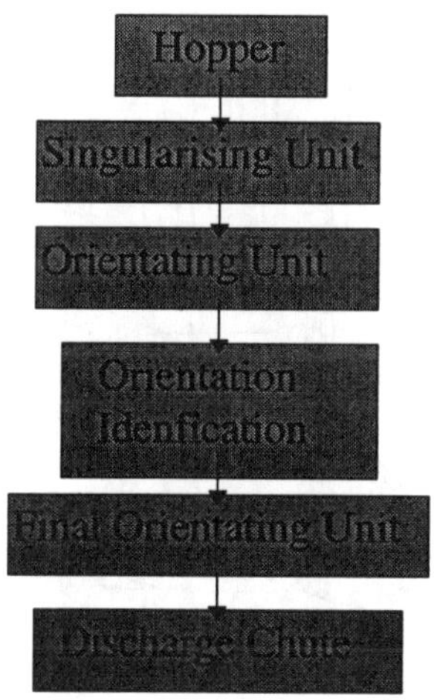

Fig. 2 : Sub-systems of the Feeding System

Various conceptual designs were generated and the final design that emerged as the most suitable one is as shown in Fig. 3.

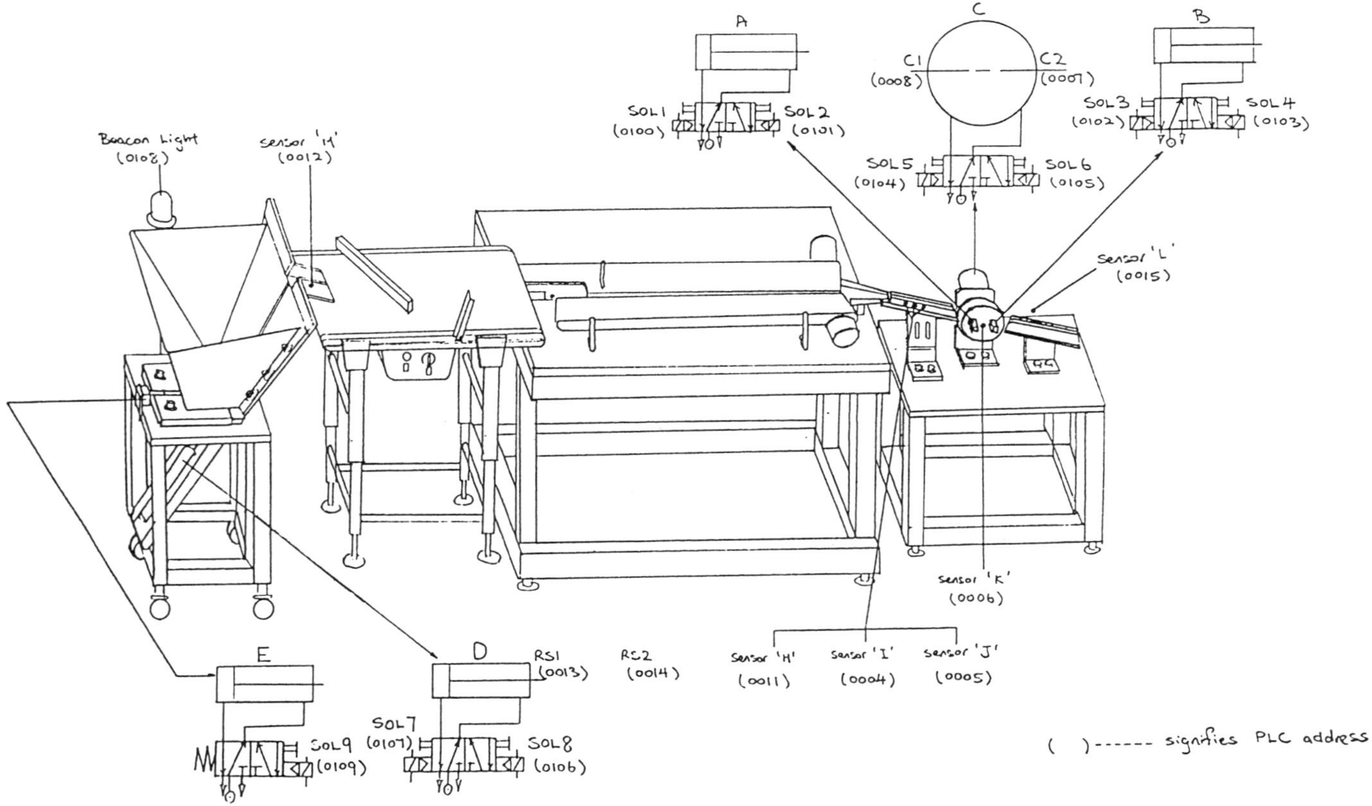

Fig. 3 : The Final Design of the Feeding System

DESCRIPTION OF FEEDING SYSTEM DEVELOPED

6.1 Loading Module

The loading module consists of a hopper for feeding the cylindrical parts. It is flexible in the sense that minimum adjustment/alignment is made when there is a change of part size. Fig. 4 illustrates the fabricated design of the hopper unit.

The mechanism of the hopper basically uses the reciprocating action of a pneumatic linear actuator. When the actuator is activated, the piston rod extends and pushes those parts that fall on the track up to the gate. The opening gap of the gate can be adjusted to cater for any change in part size so as to control the feedrate.

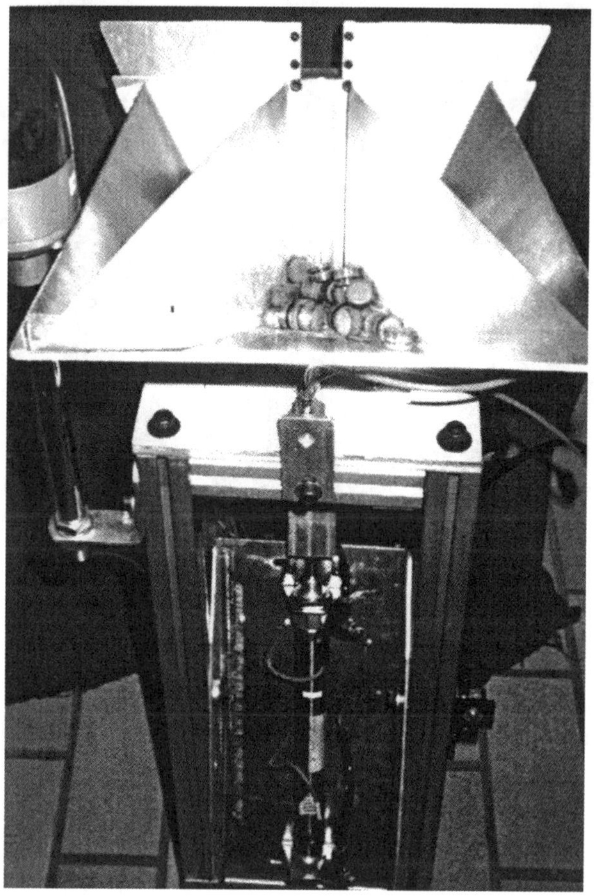

Fig.4 : Front View of Loading Module (Hopper)

6.2 Singularising Module

The singularising module consists of a conveyor belt system which comes after the hopper unit and its requirement is to separate those parts that come close to each other before proceeding to the next station. Figure 5 illustrates the fabricated design of the singularising conveyor unit.

The unit consists of a rough conveyor belt which gives a good grip of the parts (without rolling off) after having been deflected from the deflector blades.

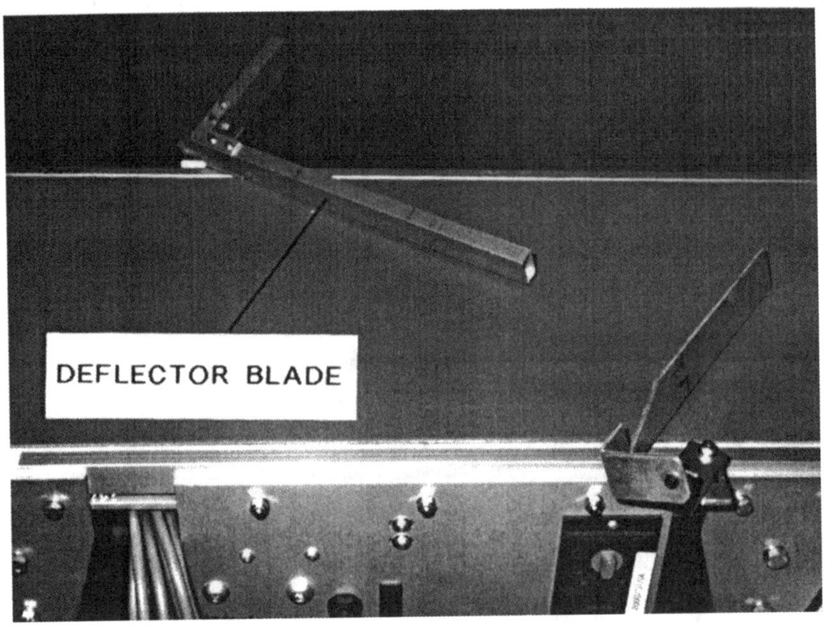

Fig. 5 : Singularising Conveyor Unit

6.3 V-Belt Orientator

The V-belt orientator (Fig. 6) serves to orientate cylindrical parts of different aspect ratios (length/diameter). It consists of two inclined conveyor belts of different roughness and moving at two different speeds. Parts travelling on the conveyor are orientated till they reach a stable position by the means of rotation caused by the differential speed of the moving V-belt. The tilting angle of the V-belt was experimented and analysed for various aspect ratios of the parts and it was found that the optimum inclination is 30° [7].

Evaluation of the orientator showed that parts with an aspect ratio less than one will stabilize in resting aspect 1, whereas parts with an aspect ratio greater than one will stabilize in resting aspect 2 (Fig. 7).

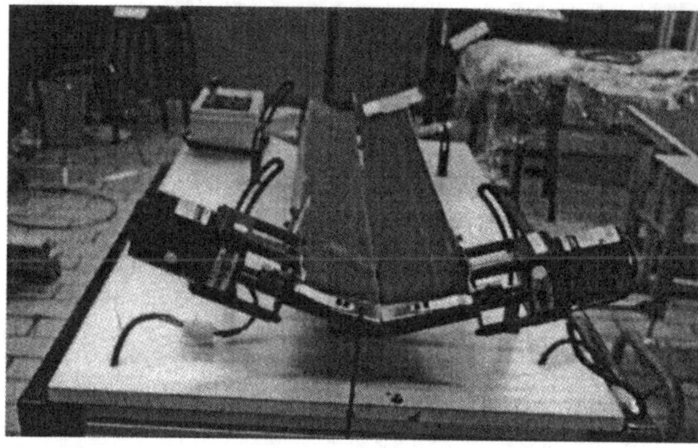

Fig. 6 : V-belt Orientator

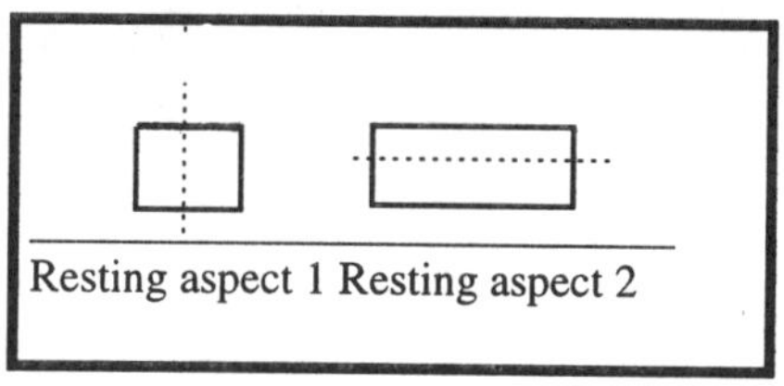

Fig. 7 : Resting Aspect of Parts

Parts with an aspect ratio not equal to one do not have any orientation problems. All of them managed to be orientated within the length of the V-belt. Whereas, parts that have aspect ratio of one or very close to one do not have one stable position. They usually assume the position of either resting aspect 1 or 2.

6.4 Transfer Mechanism

The transfer mechanism basically consists of thin aluminium plate and is installed just after the V-belt unit to transfer the orientated parts to the next station. Fig. 8 illustrates the fabricated design of the transfer mechanism.

From the figure, the two metal strips located by the side are to induce a springing effect to any part that comes in contact so as to assist the part to drop down the slope of the transfer mechanism smoothly. They can be adjusted in the gap opening to suit the different part sizes. The slope of the transfer mechanism is set between 25° to 30°. To avoid any part getting stuck, the entering edge of the transfer mechanism before the V-belt is chamfered.

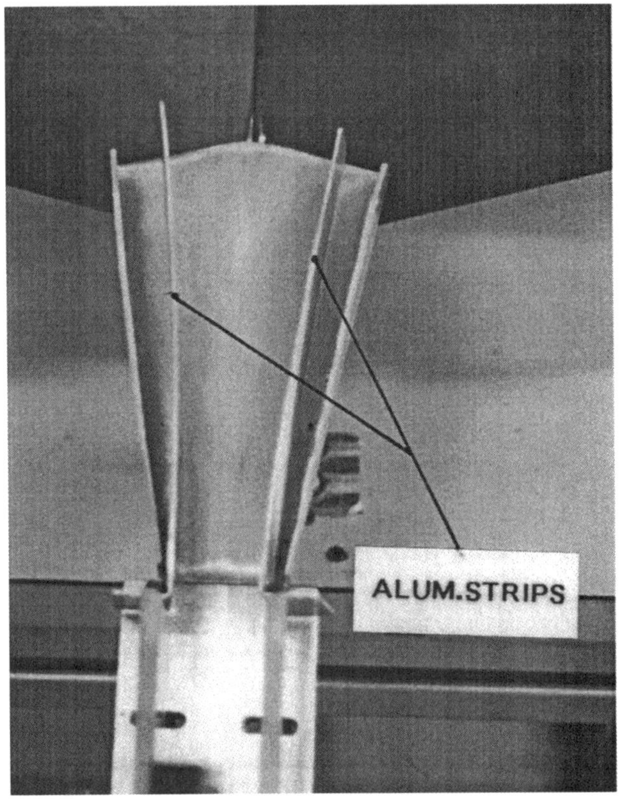

Fig. 8 : Transfer Mechanism

6.5 Unloading Module

The unloading module basically consists of upper delivery chute, holding device (re-orientator) and lower delivery chute. The holding device will re-orientate any part that is in an undesired orientation to the desired orientation. In other words, the principle of active orientation is employed. Fig. 9 illustrates the fabricated unloading unit.

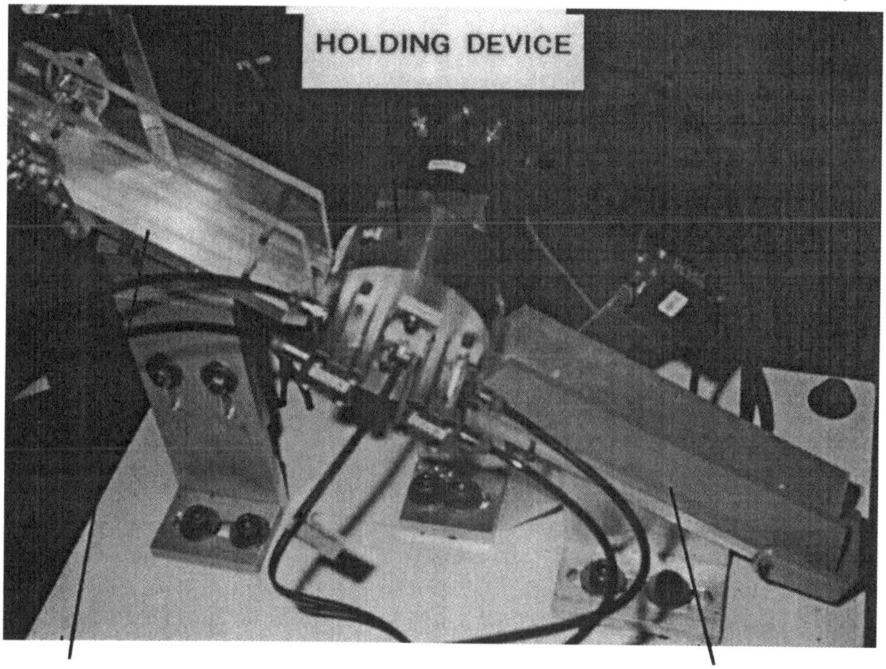

Upper Delivery Chute *Lower Delivery Chute*

Fig. 9 : Unloading Unit

6.5.1 Upper Delivery Chute

The upper delivery chute is tilted at an angle of 30° with respect to the horizontal after evaluating the smooth sliding of parts at several angles and the height is made adjustable to provide flexibility. The whole section here is important because this is the area where the orientated parts are sensed for the desired orientation. The desired orientation for a part having an aspect ratio less than one is with the groove at the top whereas for a part having an aspect ratio more than one is the groove at the bottom (Fig. 10).

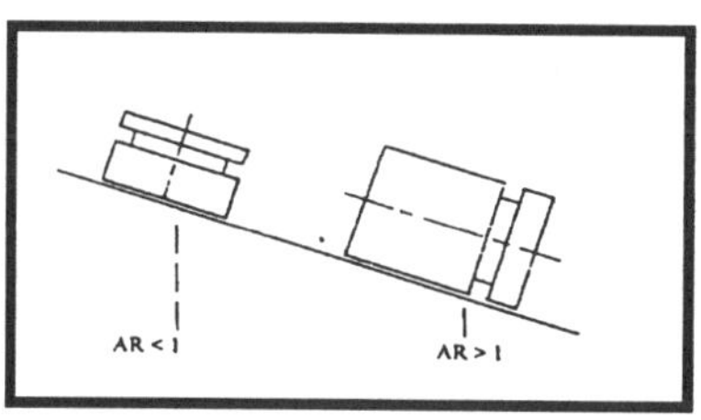

Fig. 10 : Desired Orientation for Range of Aspect Ratio

As for wrong orientation, the parts will be sensed by fine fibre-optic sensors. A pattern of on/off signals depending on the stage of the part is created as parts move past the strategically located sensors. This technique is relatively low in cost compared to the expensive vision system. Fig. 11 illustrates the positions of the sensors for wrong orientations.

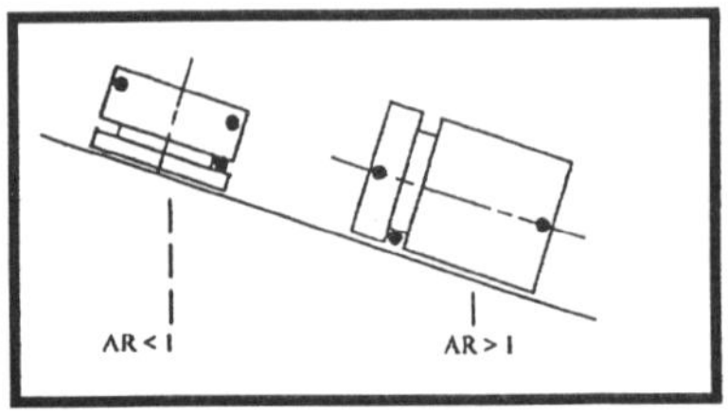

Fig 11 : Position of Sensors

Through-beam sensors are used for this application to detect grooves. The sensors shown in the figure above are all in the 'on' stage to activate the holding device.

6.5.2 Holding Device

The holding device re-orientates any part that is in an undesired orientation to the desired orientation through rotation of 180°. A rotary pneumatic actuator is installed to operate the rotation. Speed of rotation can be controlled through the throttle valves attached on the actuator. Linear compact pneumatic actuators are used to act as stoppers in the holding device. The holding device is also made flexible for small parts by attaching a rectangular slab to close up the gap.

6.5.3 Lower Delivery Chute

The lower delivery chute is also tilted at an angle of 30° with respect to the horizontal. This is the area where all completed parts in their desired orientation are collected. A photoelectric sensor is installed at the upstream end of the chute to alert the operator that excess parts collected along the chute are not being removed for packing.

6.6 Overall System

The individual modules of the fabricated designs are integrated together to form a full flexible feeding system as illustrated in Fig. 12. It is designed to operate continuously with minimal human intervention. Control of the whole system is achieved using programmable logic controller and electropneumatics.

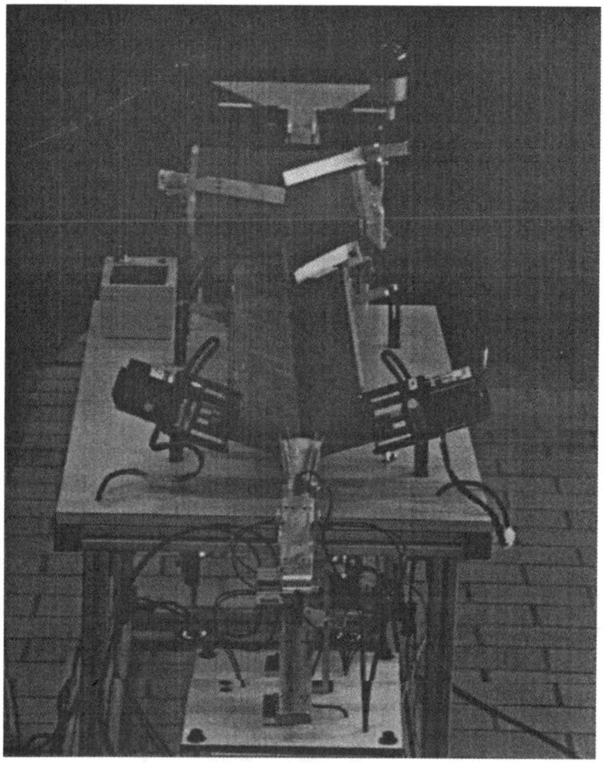

Fig. 12 : The Complete Feeding System

6.6.1 Operation Sequence of the New System

The whole system now consists of the feeder, singularising conveyor, V-belt orientator, re-orientator, chute and the controller unit.

First, the gate opening gap of the hopper, V-belt tilting angle, sensors mounting position and gap opening of the holding device have to be set for the particular part to be processed. The start button can be initiated to start the sequence.

Once there is a part/s ejected out from the hopper (detectable by parts sensor), the hopper will pause for a couple of seconds before the next action begins so as to prevent parts congestion at the next station. Parts arriving at the singularising conveyor will be separated by the deflector blades before proceeding to the V-belt conveyor. The V-belt will orientate the parts as they travel along the belt after which they will slide down the transfer mechansim to the upper delivery chute. Here, the sensors will detect the position of the groove to check if the part is in the correct orientation. Correctly orientated parts will be allowed to slide towards the lower delivery chute for collection.

If the part is in the wrong orientation, the stopper pins will be activated. They will stop the part in the holding device as it arrives from the chute. Once the part is in the holding device, the rotary actuator will rotate to turn the part to its correct orientation. Once this active re-orientation has been achieved, the stopper pins will retract to allow the part to be collected at the lower delivery chute. The rotary actuator will then rotate back to its original position and waits for the next part.

7 SYSTEM EVALUATION AND DISCUSSION

Evaluations were carried out to test the workability of the new system developed. They are :

(a) Sub-evaluation at the transfer mechanism

(b) Evaluation on the performance of the whole system for AR<1 and AR>1.

7.1 Sub-Evaluation at the Transfer Mechanism

An evaluation was done to analyse the orientation of the parts on the transfer mechanism after emerging from the V-belt. Fig. 13 illustrates the results obtained for four categories of parts.

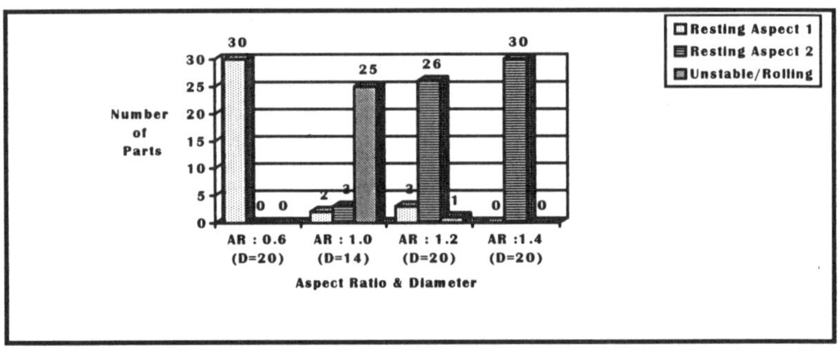

Fig. 13 : Orientation of Parts on Transfer Mechanism

It is observed that parts with an 'AR' less than or greater than 1 maintain the orientation from the V-belt well on the transfer mechanism. Whereas parts with an 'AR' very close to 1 do not have a stable orientation.

7.2 Evaluation Results for AR<1

Parts with a diameter of 20 mm but with two different aspect ratios (0.6and 1.2), were run to test on the efficiency of the feeding system for an input size of 30 parts. The parameter settings were as follows :

Speed of Singularising Conveyor	*Speed of V-belt*
85 mm/s	Smoother belt : 65 mm/s Rougher belt : 110 mm/s

Table 2 : Parameter Settings

Data was collected on parts completed per minute at the last station. A total of 6 runs were conducted for the same part size. The results are illustrated in Figs. 14 and 15 respectively.

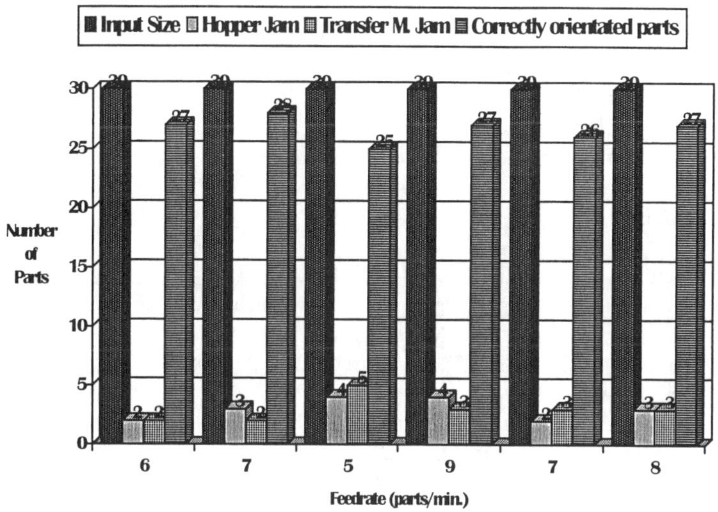

Fig. 14 : Feeder Performance for AR<1

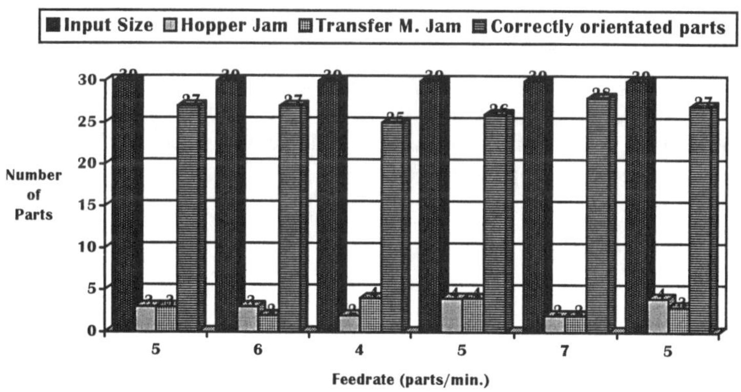

Fig. 15 : Feeder Performance for AR>1

The results showed that for both aspect ratios, the average percentage of correctly orientated parts is above 90 % and that jam rates are typically in the region of 10 %. Such results are indeed encouraging and indicate that the performance of the feeding system can be improved through fine tuning.

8 CONCLUSION

A flexible non-vibratory feeder with active orientating capability for fragile or powdery cylinderical parts with asymmetrical feature has been developed. Delicate and/or powdery cylindrical parts (including those with asymmetrical features) with aspect ratios ranging from 0.5 - 0.8 and 1.2 - 1.6 can be fed with this feeding system. It represents the first of its kind in providing for such feeding needs. Employing active orientation technique, it has a high feeding efficiency due to the absence of rejection. Fine tuning will certainly improve the performance of the feeding system.

REFERENCES

[1] Boothroyd, G., Poli, C.R.and Murch, L.E., Handbook of Feeding and Orienting Techniques for Small Parts, Dept. of Mechanical Engineering, University of Massachusetts, Amherst, Massachusetts.

[2] Boothroyd, G., Poli, C.R.and Murch, L.E., Automatic Assembly, Marcel Dekker Inc., USA, 1982.

[3] Denhamer, H.E., Interordering : A New Method of Component Orientation, Elsevier Scientific Publishing Co. , 1980.

[4] Chua, P.S.K. and Lim,L.E.N., "Development of an Automated Feeding, Orientating and Joint-Straightening System for Desk-Top Pen Holder", 10th International Conference on Production Research (10th ICPR), 14 - 17 August 1989, Nottingham, U.K.

[5] Pherson, D., Boothroyd G. and Dewhurst, P., “Programmable Feeder for Non-Rotational Parts”, Manufacturing System, Vol. 13, No. 2.

[6] Chua, P.S.K. and Sung, E.,"Development of a Programmable Parts Feeder for Non-rotational Parts", 10th International Conference on Production Research (10th ICPR), 14 - 17 August 1989,Nottingham, U.K.

[7] Justin Quak, Development Of A Flexible Feeder For Cylindrical Parts, Internal Report, School of Mechanical & Production Engineering, Nanyang Technological University, Singapore, 1993

Parametric Parts Feeder in Assembly Workcell for Mass Customization

Banna G. Rao

Dept. of Industrial Engineering and Engineering Management
Hong Kong University of Science and Technology
Clearwater Bay, Kowloon, Hong Kong

Abstract. Assembly automation is well established in the manufacturing industry, where economic advantages are proven in mass production. However, modern industries are following the concept of mass customization to meet business competition. The main challenge is to produce an increasing variety of products without any significant trade-off in production costs or lead time. Conventional mass production tooling in the assembly work cells are not flexible enough to handle the product variations and the present flexible assembly tooling is not giving better throughput rates. This research is focused on the issue of redesigning the tooling in assembly workcells to impart flexibility, while retaining the capability of high throughput rates. In this paper we have presented the detailed steps when redesigning a conventional parts feeder into a parametric parts feeder.

Key words: Mass customization, Assembly workcell, Parametric parts feeder

1 Introduction

1.1 Literature review

Modern industries are adopting the concept of mass customization in both manufacturing and service sectors to meet today's business competition. Mass customization is a new way of viewing business competition, one that identifies and fulfills customer demands without sacrificing efficiency, effectiveness and cost benefits [Pine 93]. The main idea of mass customization is the production and distribution of customized goods and services on a mass basis. The mass production environment is premised on high volume production in less variety at lower costs, but in the customized manufacturing environment, products are being produced in small batches and in a wide variety. Mass customization is aimed at uniting these two opposing strategies. In order to meet these challenges, today's business is looking for new approaches to combine mass production efficiency with customized production flexibility and variety. Research efforts were made to reduce the product design time and thereby achieve a short product realization cycle. Some researchers have applied axiomatic design methods to formulate a product family architecture (PFA) which allows for optimization of commonality in product design and process selection [Tseng 96]. However, there is also need for better tooling in manufacturing to suit mass customization requirements. Modular and flexible tooling is already in use in machining automation but is not well developed in assembly automation. Many attempts have been made to increase the flexibility of the tooling in the assembly workcells. Modular Orienting Devices (MODs) were developed to replace the fixed orienting devices on the conventional vibratory bowl feeders [Lim 93]. MODs offer greater flexibility in sequencing the orienting devices on the bowl feeder for the given features of the part to be fed. Recently modular and parametric orienting devices have been designed to further enhance the flexibility of the dedicated vibratory bowl feeders [Joneja 97]. These devices were designed as a subset of a Modular and Parametric Assembly Tool Set (MPATS) that is being developed by our group. The main idea is to impart modularity and parametric variability in the dedicated assembly tooling. This research explores the methods needed to redesign some special conventional parts feeders into modular and/or parametric parts feeders.

1.2 Problem Description

Automated assembly work cells are typically equipped with parts feeders, assembly fixtures and assembling tools. Redesigning these assembly tools to suit the mass customization needs is the main objective of our research, however in this paper our focus is limited to a subset of assembly tooling namely parts feeders. Feeding parts in the correct attitude/orientation and at a desired feed rate is an important requirement in automated assembly operations. Parts feeders are the devices used to segregate the parts and feed them into assembly machines. Existing parts feeders are designed for the particular dimensions of the part to be fed, and hence the same feeders can not be extended to feed the same type of part with varying dimensions. This inability has become an obstacle to meet diverse customer needs in a mass customization environment. We propose an approach to redesign these conventional feeders into modular and/or parametric feeding and orienting devices. Parts feeders in general can be classified as vibratory feeders or non vibratory feeders. The selection of parts feeder depends on the type of part to be fed. Conventional vibratory bowl feeders have been redesigned to equip with modular orienting devices [Lim 93] and further they have been modified and designed with modular and parametric orienting devices [Joneja 97]. However the same ideas have not been applied on non vibratory parts feeders. The utility of these feeders can be greatly extended by redesigning them into modular and/or parametric feeders. This paper presents the detailed steps involved when redesigning a Rotary Centerboard Hopper Feeder into a parametric parts feeder.

2. Approach

2.1 Redesign considerations

A good redesign should result in a feeder that performs its original function efficiently and economically within the imposed constraints.

1. New designs should provide the flexibility for manipulating all the critical operating parameters independently.
2. New functional modules/devices in the redesigned feeder should be easily reconfigured and also easily assembled and disassembled.

3. New modules/devices should fit together seemlessly and should not obstruct the flow of the parts during feeding.

Original designs of the non vibratory feeders were extensively developed and reported in a catalogue [Boothroyd 82]. For each feeder, catalogue contains drawings with description, general design data, details of the range of applications and operating speeds. The redesigning process can be approached in one of the following ways:

- Development of existing designs by introducing modifications to its functional modules/devices to suit the new design requirements. This type of work represents a large proportion of the design effort in industry.
- Generation of new designs from scratch by taking part features into consideration. It often requires a solution for problems which have not been encountered before and could require the addition of new resources to the organization. The introduction of a totally new design usually involves a large amount of experimental and prototype work [Farag 89].

We have followed the first alternative when developing the new designs for the parts feeders by modifying the existing designs into modular and/or parametric designs. All major steps of the redesign process are depicted in the flow chart shown below.

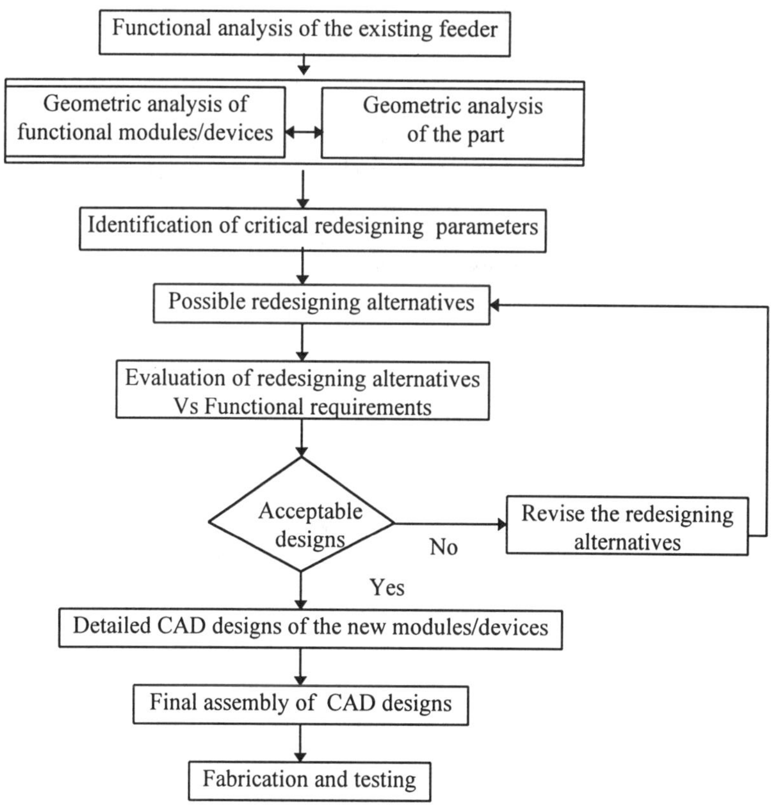

Fig. 1. Major steps of redesign process

3. Implementation

3.1 Example:

The Rotary centerboard hopper feeder is a non vibratory parts feeder. It is basically used to feed U - shaped parts to automated assembly machines. The rotary centerboard hopper feeder mainly consists of the following parts: a hopper, a hopper groove and a rotary bladed wheel. The hopper is a conical bowl designed to carry the parts when placed at random as a heap. It acts as a housing for the parts during agitation by the bladed wheel. It collects the rejected parts and returns them to the heap. It has a slot on its inclined wall to make a path for wheel movement. Rubber or cork may be placed along the inside wall of the hopper to reduce noise and damage to parts. A groove in the hopper bottom has dimensions such that the parts in the hopper may be accepted in the desired orientation only. The groove acts like an inclined track, allowing the parts to slide down and retaining them at the bottom. The upper end of the groove is curved so that the parts can be recirculated back into the hopper. The rotary bladed wheel rotates across the hopper and has blades whose dimensions are tailor-made to suit the parts to be lifted. The edges of the blades should be profiled to collect parts in the desired attitude and lift them clear of the bulk of the parts.

3.1.1 Principle of operation

The rotary wheel agitates the parts causing them to fall into the shaped groove in different orientations. Some of these orientations are suitable to fit into the groove and others are not. Once the blade sweeps up through the slot some of the unsuitable orientations are brought to suitable orientations and the remaining are rejected by the blade. Finally parts that fall with desired orientations are accepted by the groove for feeding. As the wheel continues to rotate further, these parts are collected by the blades and raised clear of the batch of parts. These correctly oriented parts slide along the blade onto the delivery track when the blade and the track are aligned.

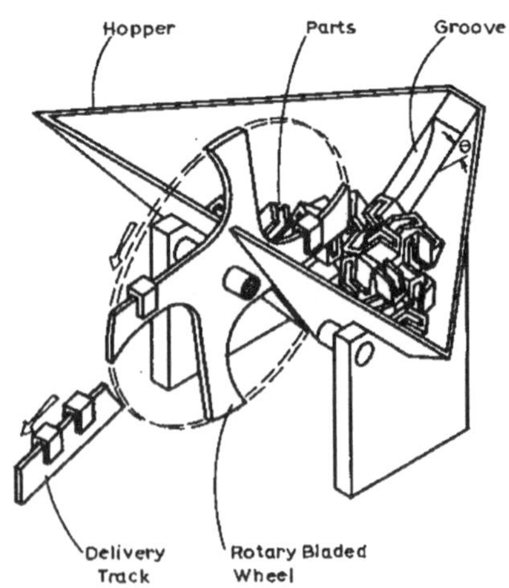

Fig. 2. Rotary centerboard hopper feeder [Boothroyd 91]

3.2 Redesigning process

3.2.1 Functional analysis of the feeder

The functional mechanism of the rotary centerboard hopper feeder performs the following basic operations:

1. Collection of parts as a heap at the bottom of the feeder.
2. Continuous agitation of these parts and allowing them to orient in correct attitude.
3. Accepting the parts that are in correct orientation and rejecting the rest into the heap to make further agitation until they attain the desired orientation.
4. Smooth conveyance of these parts to the feed track without disturbing their orientation.

In order to perform these operations we can identify the following components in the original design:

- A Hopper to house the parts and direct the parts always towards the zone of agitation which further helps to orient the parts into right attitude.
- A Hopper groove to accept the parts only in correct orientation.
- A Bladed wheel that aids in both agitating the parts continuously until they attain correct orientation and also conveying the correctly oriented parts to the feed track without disturbing their orientation.

3.2.2 Geometric analysis of the part

Geometry of the part is important in two aspects in the redesign process:

1. *To determine the possible orientations of the part*: Since the parts are filled into the feeder as a heap, it is the function of the orienting device to reject all but one orientation of the part. Hence a knowledge of the possible orientations of the part (fig. 3a) is important when designing the orienting device to maximize the occurrence of correct orientation. The orienting device restricts the resting poses of the part and attempts to orient every part in the desired attitude. This is an essential requirement to increase the feed rate of the feeder.
2. *To determine the geometric constraints on the redesign:* The geometry (fig. 3b) of a typical U-shaped part can be represented by its parameters such as length of the part (L), height of the part (H), width of the part (W) and thickness of the part (T). The part has a central slot whose width and height are dependent on the thickness of the part. Among all the possible orientations shown in the fig.3a, the orientation 'g' is suitable for feeding and hence it dictates the parameters to be considered to fit the parts into the hopper groove in this orientation. Therefore the variations in width and height of the part play a vital role when redesigning the hopper groove. Since correctly oriented parts exactly sit on the blade to convey them smoothly onto the feed track, the variations in the dimensions of the central slot of the part are important to decide the redesign parameters of the blade cross section. The performance of the feeder is measured as the number of parts delivered per blade (N_P) per unit time. The number of parts that can be conveyed per blade may vary depending on the availability of space in the groove. The variations in the length of the part affect the available space for the correctly oriented parts to fit in the hopper groove. Therefore the variations in the part length alter the length of the groove at the hopper bottom.

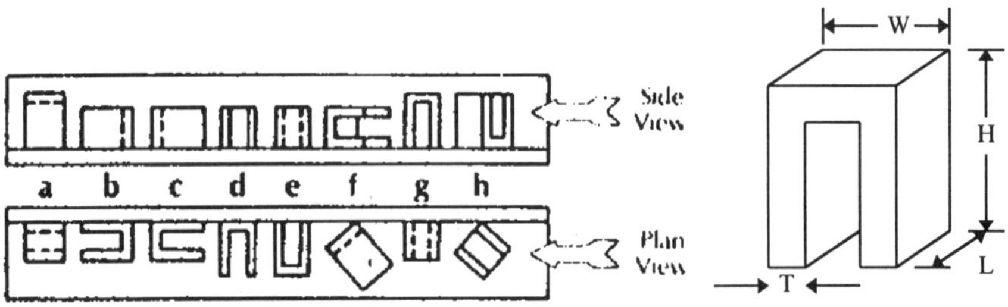

(a) Part orientations

(b) Part geometry

Fig. 3. Part geometry and orientations [Boothroyd 82]

3.2.3 Geometric analysis of the modules/devices

The geometry of the hopper cross section is very important for collecting the parts as a heap at the bottom of the hopper during agitation. Since it has a reducing cross section towards the bottom, parts can easily fall either into the hopper groove in correct orientation for acceptance or rejoin the heap. The dimensions of the hopper generally depend on the volume of the parts to be handled and the diameter of the bladed wheel. The important parameter of the hopper is the hopper wall inclination (θ) which helps the correctly oriented parts to slide down the groove and line up at the bottom of the hopper. If the parts are not correctly oriented, the inclined wall helps them to join the heap at the bottom to make further attempts. The geometry of the hopper groove depends on the parameters of the part in the correct orientation. In order to accept the correctly oriented parts into the groove, the width (W_g) and the height (H_g) of the groove should match the variations in the width and height of the part. The number of correctly oriented parts that can line-up in the groove depends on the length of the part and length of the groove (L_g). Hence 'L_g' should be enough to handle the variations in the part length to maintain optimum number of parts in the groove. The diameter of the wheel and the cross section parameters of the blade are important in the geometry of the bladed wheel. Wheel diameter (d) is specified in terms of part width (W) as the ratio of 'd/W' [Boothroyd 82] and hence the wheel diameter can be decided according to the variations in the width of the blade. In order to convey the correctly oriented parts onto the feed track without disturbing their attitude, there should be an exact fit between the central slot parameters of the part and the cross section parameters of the blade. Hence the width (W_b) and the depth (D_b) of the blade cross section are to be in agreement with the central slot parameters of the part when redesigning the blade.

3.2.4 Critical redesigning parameters

From the geometric analysis of functional modules/devices and the part, the following parameters are identified as the critical parameters for the redesign process: hopper wall inclination (θ), width of the hopper groove (W_g), height of the hopper groove (H_g), length of the hopper groove (L_g), width of the blade (W_b), depth of the blade (D_b), and diameter of the bladed wheel (d).

3.2.5 Solution alternatives and evaluation

We have come up with the following design as an alternative solution to parameterize the feeder modules/devices. The new design has a main body upon which all the other components can be assembled. The body itself is mounted on a base with an angular tilt adjustment. The hopper design is modified so that it has two halves, allowing for the groove width to be manipulated independently. The hopper wall inclination can be varied using the angular tilt mechanism on the base. The hopper groove is designed as a separate device with independent height adjustment and can be easily assembled to the hopper. The original design of the bladed wheel has been modified into two halves with a hollow expandable cover between the halves. The expandable cover is made up of flexible material and is fastened to the two blade halves. This cover fills the gap between the two halves and can prevent the parts from jamming. The parameterization of the wheel diameter is a difficult task because of its circular cross section. Therefore the ratio of 'd/W' as specified by Boothroyd *et al.* is used and the diameter is set to the maximum possible diameter for maximum width of the part that can be handled by the feeder. As mentioned in the geometric analysis of the part, the width (W_b) and the depth (D_b) of the blade are dependent on variations in the central slot dimensions of the part. Since we have set the wheel diameter suitable for the maximum part width, the corresponding depth of the blade can be fixed to handle the maximum part height. Since the manipulation of the parameters in the new design is done by adjusting the moving parts in the devices, the range of parameterization in the linear parameters is less than twice the dimension of the original part. For example, the height of the hopper groove (H_g) can be manipulated by the two sliding sections of the groove as shown in the simplified figure 4. However the maximum movement is restricted by the dimensions of the sections.

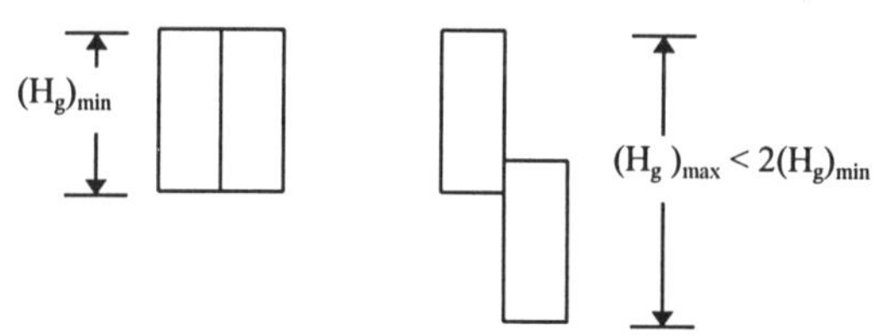

Fig. 4. Hopper groove height adjustment

3.2.6 Detailed designs & Final assembly

Detailed designs of all the components in the functional modules/devices are developed using the Pro/Engineer CAD package. A typical U- shaped part is selected and the designs of the feeder functional modules/devices are developed to suit a range of variations in dimensions of the part. These new designs are shown in the figures 5a & b. The final assembly of these modules/devices is developed to realize the CAD model of the assembled design and are shown in the figure 6.

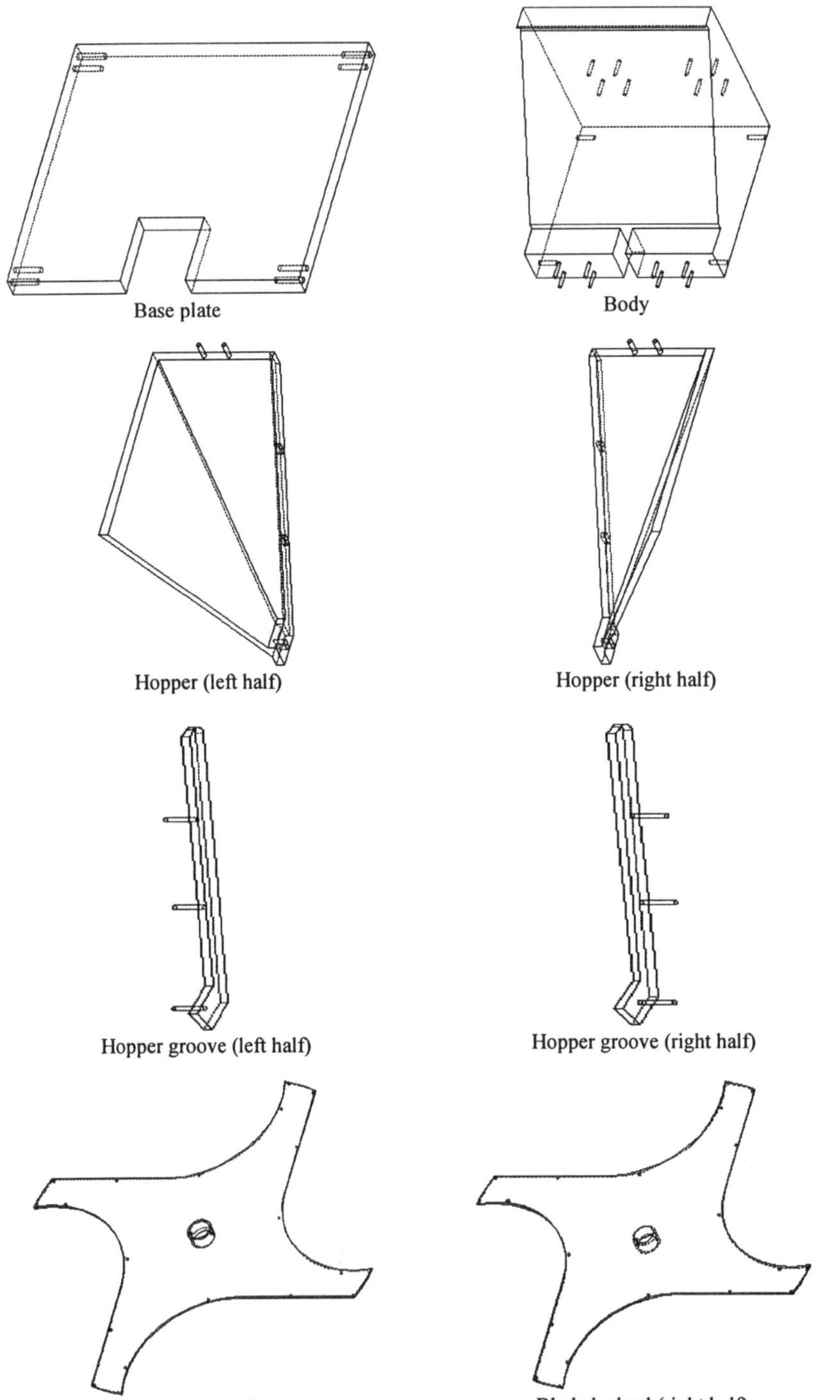

Fig. 5a. New designs of the feeder components

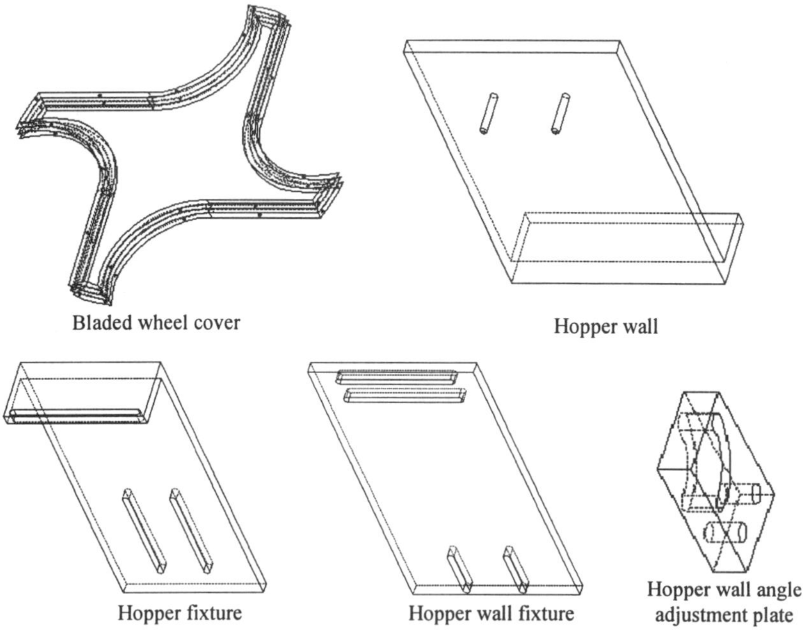

Fig. 5b. New designs of the feeder components

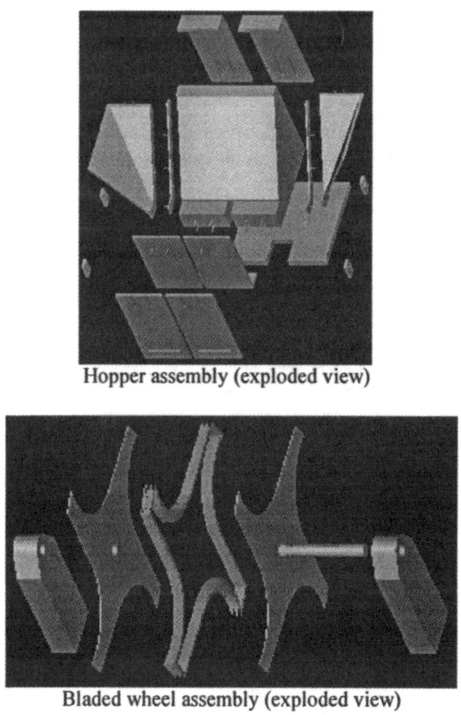

Fig. 6. Final assembly of the parametric parts feeder

3.2.7 Fabrication and testing

Since the designs are available, the devices can be made and tested using the U-shaped part for which it is designed. A set of U-shaped parts with varying dimensions can be used to test the feeding characteristics of the feeder. After analyzing the performance of the feeder, we can fine tune the device designs for better results.

Conclusions. In order to meet the needs of the mass customization in a manufacturing environment, an approach has been proposed to redesign the tooling in the assembly work cell. We have focused our attention on a subset of hardware namely parts feeders in assembly work cell. Our approach is based on adopting the existing parts feeder designs and performing modifications to develop the new designs into modular and/or parametric parts feeder. Detailed steps of the redesigning process are explained using a rotary centerboard hopper feeder as an example. Though the new designs meet the basic functions of the feeder, there is enough scope for refinement. However our basic aim in this paper is to emphasize the need for modular and/or parametric tools in the assembly workcell which suit the needs of mass customization. One limitation to the validity of these designs is that we have not made the prototype and tested it to examine the feeder's performance characteristics. Hence our future work will focus in this direction.

4 References

1. Boothroyd, G., Poli, R. C., and Murch, L. E., Handbook of Feeding and Orienting Techniques for Small Parts, Dept. of Mechanical Engineering, University of Massachusetts, 1982.
2. Boothroyd, G., Assembly Automation and Product Design, Marcel Dekker, 1991.
3. Pine II, B. J., Mass customization: The New Frontier in Business Competition, Harvard Business School Press, Boston, 1993.
4. Farag, M. M., Selection of Materials and Manufacturing Processes for Engineering Design, Prentice Hall International (UK) Ltd., 1989.
5. Lim, L. E. N., Ngoi, B. K. A., Lee., S. S. G., Lye., S. W., and Tan., P. S., "Flexible Vibratory Bowl Feeding Using Modular Orienting Devices", Journal of The Institution of Engineers, Singapore, vol. 33, no. 4, June 1993.
6. Joneja, A., and Lee, N., "MPATS: A Modular, Parametric Assembly Tool Set for Mass Customization", Technical Report in Department of Industrial Engineering and Engineering Management, Hong Kong University of Science and Technology, Hong Kong, 1997.
7. Tseng, M. M., and Jiao, J., "Design for Mass Customization", CIRP Annals, vol. 45, no 1, 1996.
8. Pro/Engineer on-line manual, Version 15.

Advanced Mechatronic System for Turbine Blade Manufacturing and Repair

U. Berger, R. Janssen, E. Brinksmeier

Institute for Material Science (IWT), Division of Manufacturing Technologies
Badgasteiner Str. 3, 28359 Bremen, Germany

Abstract: Aspects of advanced technology for individual jet engine components repair are discussed. The technological solutions lie in the domain of cutting path planning and machining of difficult-to-cut materials like titanium alloys. The paper presents an integrated process chain for multi-axis high speed milling of TiAl6V4. The main development items are based on a simulation and verification system and a milling reference model. The milling reference model is established by milling experiments performed in order to enable safe machining processes and supports the machine operators in selection of cutting data for repair processes on turbine blades. The results of these research developments are applied to an European project in the field of Intelligent Equipment and Control.

1 Introduction

Jet engine components are an important field of modern aircraft technology. Especially the manufacturing of turbine blades and vanes due to safety and functional features will lead to higher engine efficiency, reduced fuel consumption and noise emission. On the other side, are turbine blades requiring regular and certified repair and maintenance cycles. The problem related to the manufacturing and repair processes is twofold:

- Due to optimised aerodynamic features modern turbine blades and vanes combine complex free form surface geometry and high finishing accuracy. Thus the machining processes require geometric flexibility and ensurance of precise cutting paths.
- Use of materials with high specific strength and heat resistance, mainly difficult-to-machine materials such as titanium alloys like TiAl6V4. The low heat conductivity and high chemical reactivity causes increased tool wear, high cutting temperatures and strong adhesion between tools and chips. Increasing the cutting speed leads to high thermal load of tool and workpiece, which affects the material microstructure. This transformation has to be avoided due to the functional behaviour of titanium parts under loading.

Titanium alloys are used mainly for the stationary vanes and the rotating blades in the "cold" compressor stages of a turbine. Since original manufacturing of turbine blades and vanes is done by contour milling under mainly automated manufacturing conditions, the repair processes are done by manual operation. The repair of turbine blades starts with the removal of the damaged area and replacement by raw block-material in a welding process. Finally the welded material has to be machined back due to the original shape. This operation is highly time and cost consuming and has a lack of sufficient process monitoring in view of traceability demands.

Automation of the machining process has not progressed significantly, due to the great difficulties encountered when trying to control specific aspects of a component's necessary repair characteristics:

- follow a precise path and adjust certain varying part parameters
- program the machining path to follow the complex geometric part surface in the repair sector.

Due to their exposure in the jet engine environment, airfoils can vary from other airfoils or new ones. It is therefore difficult to automate the machining processes required, because a unique repair process with its own parameters is necessary for each airfoil. Fig. 1 shows a typical turbine blade before repair takes place.

An advanced machining system based on a multi-axis high-speed milling center enables adaptive reshaping of blades with individual geometries due to the functional requirements. The RTD work performed so far to reach these goals are:

- Conceptualisation of a technology model combining geometrical and high speed milling features.
- Development of a milling reference model, based on analogous field test experiments, for selection of tool and coolant supply and setting of machining parameters. The milling reference model leads to a manufacturing database, which can be connected directly to the machine tool controller.
- Laboratory set-up of a 4-axis high speed milling center including tool and pallet changer and special fixtures such as experimental field test demonstrator, equipped with force/torque measurement sensors for in-process monitoring and evaluation of machining results.
- Set-up of a 3D shape measurement sensor system for integration in the machining platform, based on coded light and phase shifting principles.

The investigations and verification experiments performed so far contribute to increasing of process stability and tool life, improvement of surface roughness and avoidance of tensile residual stresses. The reference model implemented in the manufacturing demonstrator supports the staff involved in repair work preparation and the machine operators in selection of tools, coolant supply and machining parameters. The throughput time and machining costs can be substantially reduced. Furthermore, the part quality will be ensured at the required level.

Fig. 1. Jet engine components in repair status

2 Process Chain for Repairing Jet Engine Components

The machining of individual jet engine components as turbine blades and vanes require different technological issues than existing ones deliver. Applied systems should combine adaption features, easy and transparent programming and shorten the throughput cycles due to cost and delivery time demands.

To set up a multi-axis milling cell in view of reverse engineering and quality inspection, has been developed a process chain from geometry design to quality control of machined parts. The process chain is shown in Fig. 2 and consists mainly of two loops.

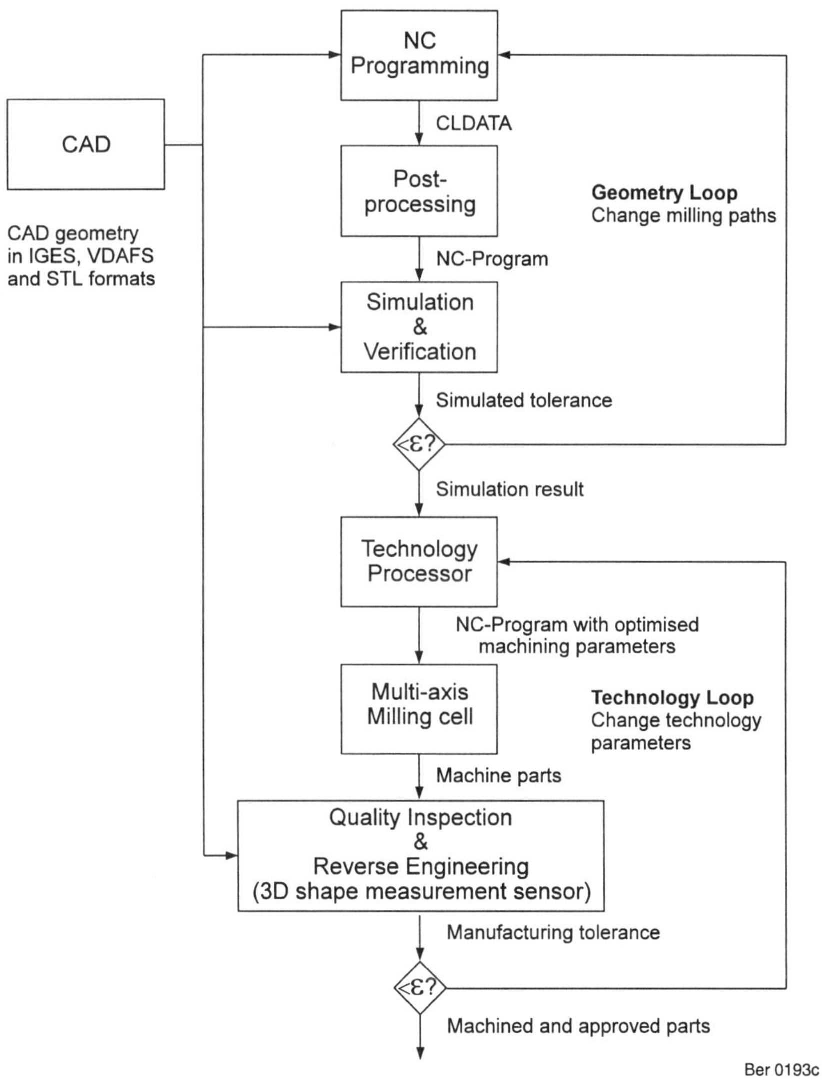

Fig. 2. Process chain for multi-axis milling cell

In the first loop a multi-axis NC programming system generates the milling path for repairing a turbine blade based on its geometry description. The geometry of the blade is constructed by a CAD system and then transferred into the NC programming system via neutral data exchange formats (e.g. IGES and VDAFS). The tool paths generated by the NC programming system describes the tool location of a pre-selected tool in CLDATA format. Then, a multi-axis postprocessor produces the necessary machine control codes for a specific NC-machine. These machine control codes are checked for their correctness by a realistic NC simulation system. In the simulation only geometric aspects of machining processes are considered. These include the detection of collisions in the machining environment, the visualization of material removal of the workpiece, handling of fixtures and jigs as well as the numerical verification of the workpiece surface quality. If deviations between the original CAD geometry and the simulated geometry exceed a specified machining tolerance, the NC program must be corrected in the NC programming system and checked again by the simulation system. Otherwise, the simulation result is put into the technology processor in the second loop that determines necessary process parameters for the NC programme. After calibration of the multi-axis machine tool center and set-up of the fixture, tool and workpiece, a test part is machined. The machined part is finally measured and a geometric model is generated by a reverse engineering software [1], [2]. If the geometry model is within the allowable tolerance zone, the real part can be machined. Otherwise, machining parameters must be adaptively changed. The machining technology processor is based on a milling reference model, which contains operational parameter settings.

3 Simulation and Verification of Multi-Axis Milling Processes

Due to complex geometry of turbine blades multi-axis milling machines are needed for the repair purpose. However, the generation of error-free NC programs for these machines is still a difficult and time-consuming task despite computer-assisted tools. An NC program consists of a series of cutter tool movements which remove material from a piece of raw stock to create a prototype part, mould or stamping die. In this process one can use a high level language like APT or a CAD system to define the geometry and cutter sequence, calculate cutter offsets and produce a cutter location data (CLDATA) file. The CLDATA file must then be postprocessed into a machine control data file which contains the instructions to control a specific machine tool. The next step is the verification of the NC programme to eliminate any errors.

Generally, verification of NC programs is a time-consuming process. Modern CAD systems provide a limited form of NC simulation by displaying machining paths with respect to workpiece. But it is difficult and often even impossible to imagine how the machine axes move just from looking at the cutter location data. This complicated process can be significantly simplified by using computer simulation technology. In principle, the motion of machine axes can be displayed on a graphical screen by compiling the NC programs. The milling process (removing material away from workpiece) can be simulated by the Boolean operation of subtracting the tool swept volume from the workpiece. From a kinematic point of view, a NC-machine is mainly built up of translational and rotational segments which can be divided into two types: one is workpiece carrying and the other is the tool carrying. These are represented by two kinematic chains, whose end segments are workpiece and tool respectively. Both chains start with the machine base as shown in Fig. 3.

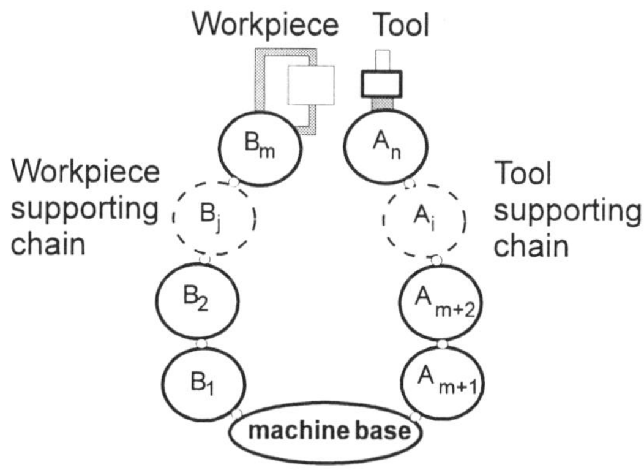

Fig. 3. Kinematic configuration of NC-machine

To describe the kinematic relationship between the segments the coordinate frames are embedded in each segment [3]. Hence, the kinematic configuration of these two chains can be represented in the model:

$$\begin{cases} T_m = B_1 * B_2 * \ldots * B_m & \text{for workpiece} \\ T_n = A_{m+1} * A_{m+2} * \ldots * A_n & \text{for tool} \end{cases} \tag{1}$$

In this model the (4x4) homogeneous transformation matrices T_n and T_m define the relative position and orientation of the tool with respect to the machine base respectively. Conceptually, the matrices A_i and B_j describe the spatial relationships between coordinate frames fixed relative to each of the two segments in the kinematic chains. These transformation matrices can be expressed as:

$$A_i = A(q_i) = \begin{cases} \text{Rot}\ (q_i) & \text{for rotat. sgmts} \\ \text{Trans}\ (q_i) & \text{for trans. sgmts} \end{cases} \tag{2}$$

where the motion axis is defined by the origin and motion direction. The input parameters to this model are the machine axis moving coordinate q_i. According to this kinematic model the transformation from the workpiece coordinate system to the machine coordinate system can also be easily derived for postprocessor development. Since an NC-machine consists mainly of a set of segments, one can model the geometry of these segments separately. The geometric construction of the segments is done interactively by combining simple geometric primitives. The primitives currently implemented in the system are box, cylinder,sphere, cone and torus. They are geometrically defined by a few parameters, e.g. a box can be specified by length, height, width and a corner point. A solid modelling programme based on boundary representation is used to create and manipulate these primitives. To simplify implementation of the Boolean set operations the solid modelling programme is designed to handle only planar faced objects. It is, however, easily possible to approximate analytical surfaces as multiply-connected planar faces. For user convenience, the system supports also basic geometric operations of the object manipulations in the modelling space, such as translation, rotation and scaling etc. Once the segments have been geometrically created, they can be assigned to the corresponding machine components to build the whole machine geometry. This begins with the machine base component and completes at the tool and workpiece carrying components respectively. Each segment is specified with machine axis type, moving axis and working range. In order to build the simulation environment easily and quickly a library with a variety of geometric segments is also provided. A number of milling machines have been modelled by the system in [4].

4 Milling Reference Model and Manufacturing Experiments

The repair process for turbine components requires high demands on dimensional accuracy, surface roughness and material microstructure. To close the gap between simulation and verification of multi-axis milling processes, there will be established several milling reference processes according to material specifications in an experimental environment. This reference processes gain the required machining parameters; tool and workpiece behaviour and transformation of material properties under real machining conditions. Further experiments were performed in regard to increasing process stability and tool life, improvement of surface roughness and avoidance of tensile residual stresses. The reference processes will lead to an overall milling reference model, which supports both, the work preparation staff and the machine operators in choosing of machining parameters. The titanium alloy TiAl6V4 is in aerospace industrial use to a great extent and was also chosen for the first machining experiments. Titanium alloys perform high specific strength and heat resistance (strength-to-density ratio), but in any cases they are difficult to machine. The low thermal conductivity and high chemical reactivity of titanium alloys causes increased tool wear, high cutting temperatures and strong adhesion between tool and part surface [5],[6].

Accordingly, practical milling investigations on the titanium alloy TiAl6V4 were carried out on a multi-axis CNC-machining center. For this a representative series of uncoated and coated end mills of carbide metal and high speed steel were selected for the experiments. Experimental research focusses on cutting tool performance (e.g. breakage of cutting edges under high speed milling conditions and wear mechanisms while using different lubricants and cooling supply strategies). To characterize the performances of different tools and parameters, the cutting forces, the tool wear, the surface quality and the chip forming were analyzed.

The effect of the cutting speed on the tool life travel of uncoated HSS tools is shown in Fig. 4. It can be established that an increasing cutting speed reduces the tool life of HSS tools rapidly. An increase of the cutting speed to 90 m/min shortens the tool life down to 0,50 m. A further increase to 100 m/min is almost impossible. In this case, the tool failed after a milling path of 0,20 m. It can be established that milling processes under high speed conditions with cutting speeds over 100 m per minute are impossible with the HSS tools. For additional investigations, the experiments were extended with carbide tools. Compared with HSS tools the carbide tools have shown clearly better performance under high speed cutting conditions. Increasing cutting speeds also lead to reduced tool life travel, but compared with HSS tools higher cutting speed can be achieved. The effect of cooling lubrication on the tool wear is shown in Fig. 5. It illustrates the progress of the width of wear land VB_{max} in relation to the tool life travel. Generally, the correct use of coolants during machining operations greatly extents tool life. Especially in the case of milling the use of minimum quantity lubrication is suitable to reduce the tool wear [7]. In this case a special device atomizes the coolant and supplies it with very low flow rates.

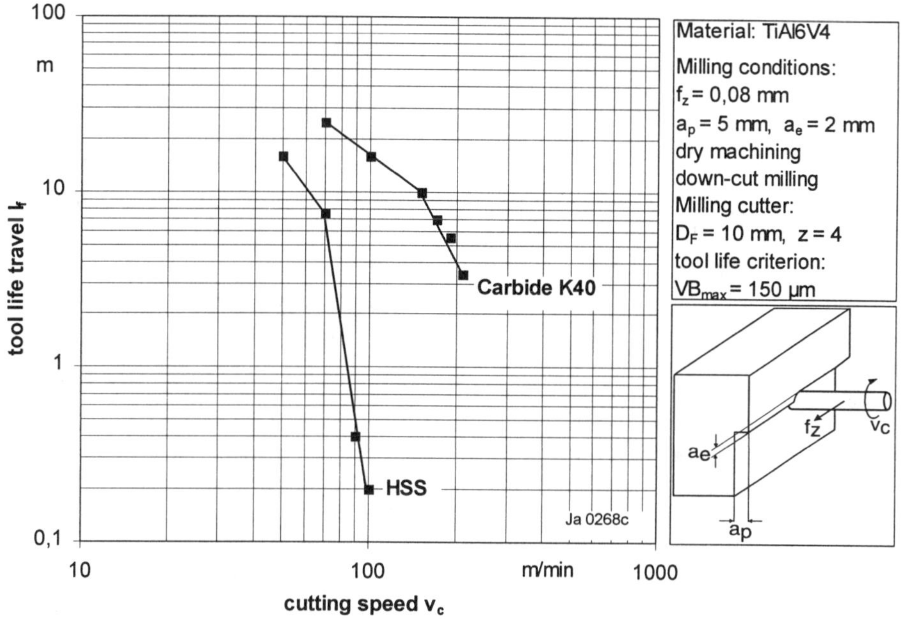

Fig. 4. Relation between tool life travel and cutting speed

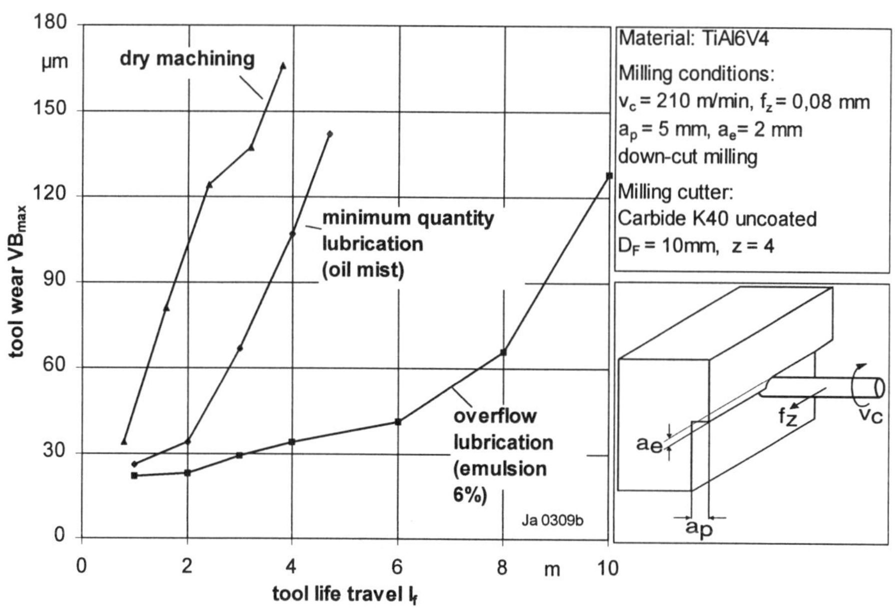

Fig. 5. Effect of cooling lubrication on the tool wear

The investigations have shown that the use of overflow lubrication with a 6% emulsion reduces the tool wear visible. The reduction of the heat generated by the process reduces also the tendency of welded chips on the tool surface. The use of minimum quantity lubrication which has a lower cooling effect is also shown in the figure. It can be established that also the use of minimum quantity lubrication decreases the tool wear clearly. For the use of HSS tools, it can be established that the tendency of the titanium to pressure weld to the tool flank can almost completely avoided by the use of oil mist lubrication.

5 Three Dimensional Shape Measurement System

The three dimensional shape measurement system is based on Coded Light Approach and phase shifting techniques [8]. The measurement process results are largely influenced by tolerances (e.g. workpiece, jig and fixtures and the system itself). These tolerances can be compensated for, by scanning of the 3D-geometry of the workpieces and the manufacturing environment as well as for intermediate and final inspection of the workpiece. The selected measurement principle is very robust in respect to changing surface qualities (coating, reflection etc.). Also the system is designed to comply with the rough conditions of the machining environment in respect to mist, vibrations and even shock.

The main feature of the Coded Light Process in comparison with Triangulation processes is the mathematical connection between space and time in a topometric scene, which is realised by the use of a sequential projection of n grey coded stripe patterns in the selected field of view.

The stripe patterns enable a differentiation of 2^n variable projection directions x_p, which can be securely identified by a characteristic bright/dark sequence and the specific code of the projection. A sequence of n = 7 stripe pattern enables therefore a differentiation of $2^7 = 128$ directions of projection.

Fig. 6a shows the physical principle of topometric measurement by use of Coded Light Process. The stripe patterns on the object of interest are generated by a transparent Liquid Crystal Display (LCD). The images are received and collected by a CCD camera, which is positioned in a defined direction to the projector. Now the point coordinates of a three dimensional object can be calculated by matching the plane generated from the stripe pattern and a line with direction through the centre of the CCD-camera. This line is based on the image plane of the camera and the location of the object point in camera coordinates {K}. The stripe projector defines for each of the different stripes a specific plane. This plane can be calculated by the projector centre and each line, which is projected in the scene. The existing planes are then according to the constraint (λ and μ are current parameters):

$$P_E = \lambda \begin{pmatrix} 0 \\ 1 \\ 0 \end{pmatrix} + \mu \begin{pmatrix} x_p \\ 0 \\ 1 \end{pmatrix} \qquad (3)$$

For each image point can be a specific code defined through binarisation of the image and check, for which of the projections the image point is illuminated (value = 1) or not (value = 0).

The available binary images can be collected to a bit plane stack. Each pixel of this stack contains for the specific image point a n-ranging bit-sequence, which describes the projector direction to this scenery point. This code defines the projection direction x_p and the constraint of the plane. Therefore the measurement result of this process is singular and absolute. With the bit plane stack and the defined location of camera and projector the three dimensional coordinates of the object can be calculated. The main advantage of the Coded Light Process in comparison with other techniques is the fast image processing of topometric data. This is related to the fact, that regarding to the resolution of the CCD-camera only a small number of patterns has to be projected in the scene. Additionally, the analysis of the bit plane stack as well as the calculation of three dimensional coordinates can be done by simple image analysis operations. So a whole topometric scene can be measured within seconds. Also the Coded Light Process serves advantages at changing illumination conditions and different surface reflexions in comparison to other techniques.

Due to inaccuracies in scanning different views of the workpiece, points in the overlapping areas do not exactly occupy the same positions. This causes problems for constructing a solid model from the acquired data. Then, the data often includes points of environment elements such as fixtures etc. These points must be eliminated, otherwise the reconstructed model would be incorrect or invalid. Therefore, there has been developed a special-purpose algorithm overcoming the above mentioned difficulties. Fig. 6b shows a modelling example for a turbine vane. The model is scanned in six different views and obtained contains 23152 points.

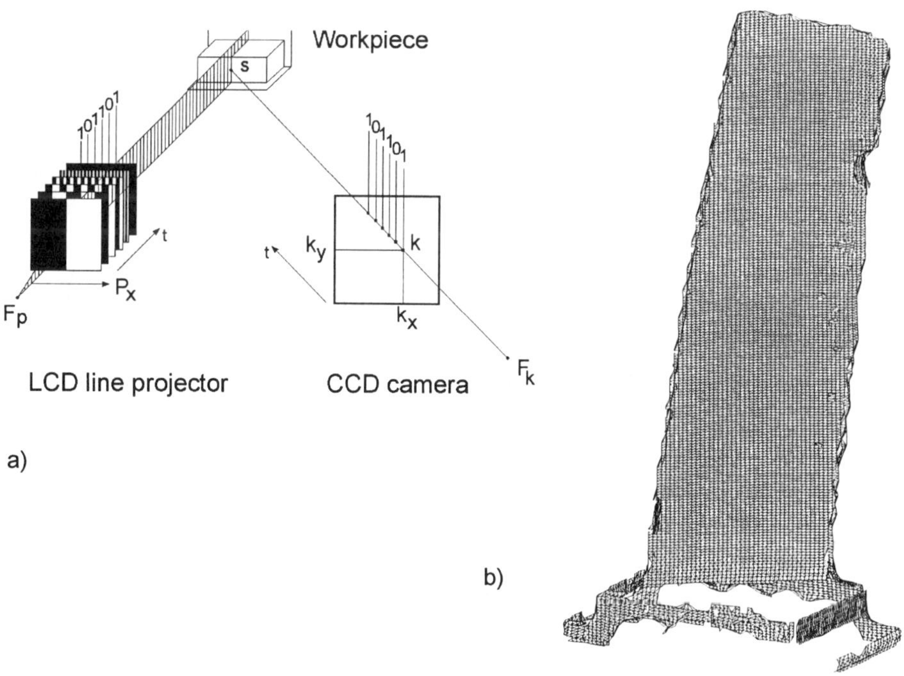

Fig. 6. Measurement of the Coded Light Approach and modelling example for a turbine vane (length appr. 150 mm)

6 Summary

The basic concept of the described developments is the integrated process chain for multi-axis milling. Due to the requirements in machining of individual jet engine components as turbine blades and vanes specific attention is paid to high-speed milling aspects. The geometrical loop concerning cutting path planning and adaptation is based on a mathematical model for a generic machine tool system. Features like collision avoidance, jig and fixture modelling and universal postprocessing are included. For quality inspection and reverse engineering purposes a 3D shape measurement system based on Coded Light Approach and phase shifting techniques is installed. The presented results of the carried out investigations show an overview of the possibilities for influencing the milling process of titanium alloys. Due to the investigations it can be established that higher cutting speeds on the machining of TiAl6V4 can be realized. The appropriate selection of tool materials, cutting parameters and coolant lubrication constitutes the basis for a safe machining process. Further investigations will be performed to achieve a wide range of machining data for suitable milling conditions to ensure increasing process stability. The pursued parameter database supports the machine operators involved in repair work preparation and selection of cutting data.

7 Acknowledgement

The presented paper is based on an RTD project funded by the the European Union in the ESPRIT Program with the title: Advanced Mechatronics Technology For Turbine Blades Repair (AMATEUR). The other project partners are ZENON S.A. (GR), Hellenic Aerospace Industry (GR), APS Gesellschaft für Automatisierung, Prozeßsteuerung, Schweißtechnik m.b.H. (DE), Bremen Institute of Industrial Technology and Applied Work Science at the University of Bremen (DE), Staubli France S.A. (FR).

8 References

1. U. Berger, Entwicklung eines Sensorgestützten Planungs- und Programmiersystems für den Industrierobotereinsatz in der Unikat- Einzel- und Kleinserienfertigung. Dissertation, Universität Bremen, Verlag Mainz, Aachen, 1995.
2. U. Berger, X. Sheng and A. Walter, Simulation and Verification of 5-Axis Milling Processes for Repairing Turbine Blades, *SURFAIR XI*, 11ème Journées Internationales d'Etude sur les Traitements de Surfaces dans l'Industrie Aéronautique et Spatiale, 12/13/14 Juin 1996, Cannes.
3. R. P. Paul, Robot Manipulators: Mathematics, Programming and Control, MIT Press, Cambridge, MA, 1981.
4. X. Sheng, Solid Modelling Based Geometric and Graphical Simulation of Multi-axis Milling Processes, PhD Dissertation, University of Bremen, 1992.
5. M. Eckstein et.al., Schaftfräsen von Titanlegierungen mit hohen Schnittgeschwindig-keiten, *VDI-Z* **133** (1991) 12, 28-34
6. H. Schulz, High-Speed Machining, Carl Hanser Verlag, Munich, 1996
7. E. Brinksmeier; A. Walter: Examples of the Employment of Reduced Amount of Coolant and Dry Machining, *Tribology - Solving Friction and Wear Problems*, 10th International Colloquium, 09. - 11. January 1996, Technische Akademie Esslingen.
8. Berger and A. Schmidt, Active Vision System for Planning and Programming of Industrial Robots in One-of-a-kind Manufacturing. *Proc. SPIE*, Automated 3D and 2D Vision, Casasent, D.P.; Hall, E.L.(eds.), VOL. 2588, SPIE Bellingham, WA, U.S.A., 1995.

The Development of an Intelligent Flexible and Programmable Vibratory Bowl Feeder Incorporating Neural Network

Patrick Soon-Keong Chua, Meng-LeongTay and Siang-Kok Sim

School of Mechanical and Production Engineering
Nanyang Technological University
Nanyang Avenue
Singapore 639798

Abstract This paper is concerned with the development of a vibratory bowl feeder which is not only flexible and programmable but also intelligent with neural network incorporated. Controlled by a microcomputer and driven by electropneumatics and mini-motors, the feeding system is capable of identifying the orientation of non-rotational parts with external and internal (e.g. blind holes of different depths and diameters) features as well as having the ability to actively re-orientate the parts so as to feed them in the desired orientation. The system developed was evaluated and the test results showed that with fine tuning, the feeding system would be an invaluable asset to industry.

1 Introduction

A parts feeder can be defined as a device that receives randomly oriented parts at its input and delivers these parts in the desired orientation at its output. The vibratory bowl feeder is a very common parts feeder in industry. The industry has accepted the use of the vibratory bowl feeder for parts feeding for many decades now and it appears that the industry is unlikely to turn away from it in the forseeable future. The reasons are:

(a) Its compactness.

(b) Its proven track record in its ability and reliability to convey parts smoothly along the vibrating track.

However, two of its main draw backs are its mainly passive orientating system and its inflexibility in feeding different parts with the same feeder. In today's highly competitive manufacturing environment, the ability to rapidly adjust to product changes due to changing market demand is a key factor in survival. To make automated workpiece orientation feasible in batch production, it requires the feeders to be flexible, reliable, able to accommodate a wide range of parts and deliver them at the required feeding rate.

A programmable parts feeding system should have the flexibility to orientate most of the parts from one or more part families, with down time to changeover from one part to the next not requiring more than a few minutes. Attempts at flexible assembly system such as G.M.s' Programmable Universal Machine for Assembly (PUMA), Westinghouse's Adaptable Programmable Assembly System (APAS), and Olivetti's Programmable Assembly System (SIGMAS) have emphasized the need for such flexible feeding devices. Salmon and d'Auria [1] states that".......the lack of flexible feeding devices causes the main limits of programmable assembly". Beecher [2] states that "Whether or not programmable assembly devices succeed could very well be determined by the success or failure of parts feeding and orienting mechanisms".

2 Objective

The objective of this work is to develop a computer-controlled, flexible and intelligent programmable vibratory bowl feeder with the following salient features:

(a) Ability to reliably identify the orientation of moving parts on the vibrating track of the bowl feeder
(b) Application of neural network in the feature recognition process
(c) Ability to flip inverted moving parts travelling on the vibrating track
(d) Ability to perform active re-orientation of parts to the desired orientation
(e) Ability to reject parts that are faulty or unrecognisable

3 Literature Review

Several researchers such as Maul and Goodrich [3] had attempted to transform the dedicated vibratory bowl feeder into a flexible and programmable feeding system. They converted the vibratory bowl feeder into a flexible and manually programmable feeder which could be used for batch as well as mass production. The orientating devices such as wiper, pressure break, narrow track, slotted track, etc. of the bowl feeder were made adjustable by means of lead screws and dials were used to record the positions of the tools which are located along the bowl track near to the exit. The feeder could be programmed by placing a part in the desired orientation at each tool whose position was then adjusted and recorded.

Other attempts had also been made to increase the versatility of automatic parts feeders. One approach was to restrict the feeder's ability to only parts singularisation, leaving parts recognition to sensors at the exit of the feeder and the subsequent orientation by a manipulator [4]. This unfortunately slowed down the manipulator. The ideal situation was to have the parts already correctly orientated and waiting for the manipulator at the pick-up point.

Suzuki and Kohno [5] had developed a multi-parts feeding system using a multi-storey bowl feeder, separating unit and chutes and a parts holding/reversing unit in conjunction with a camera. The bowl feeder singularised the different types of thin parts in either one of two possible orientations via the chute to the holding/reversing unit where its orientation was identified by the camera system. Flipping then took place if the part was inverted. It could then be picked up by the manipulator for assembly. An important feature of their system was that the bowls did not need to be configured to perform part orientation and hence were easily adapted to accommodate different although limited part types.

Similar developments using the vibratory bowl feeder and camera system for image recognition were carried out by Cronshaw [6], Uno [7] and Warnecke et al [8]. The image recognition system was first taught to recognise the various possible input orientations of a part in the teach mode and the results stored in a computer as reference signatures. In the run mode, the image acquired was compared to the reference data for identification of orientation.

The above camera vision techniques, although achieving the objective of providing flexibility and programmability, were unable to detect the orientation of parts with only blind features and differentiate between blind features of different geometry. In an attempt to avoid the problems associated with camera vision systems (e.g. high costs, lighting, other environmental problems), Park [9] used fibre optic sensors to create a pattern of on/off signals as parts moved past the strategically located sensors. This silhouette detection technique would still be unable to detect blind features on a part.

4 Description of the Vibratory Bowl Feeder System Developed

The vibratory bowl feeder is designed to be flexible, programmable and intelligent. Instead of welded passive orienting devices such as wiper blades and permanently constructed track width, mini-motor controlled wiper blades and adjustable track width are incorporated into the system to make it programmable to suit parts of different sizes. As shown in Figure 1, the vibratory bowl feeder consists of nine specially designed stations along its track for feeding of non-rotational parts. These stations are controlled by a micro-computer.

4.1 Wiper Blade Station

Basically, this is a passive orienting device commonly used in the vibratory bowl feeder to reject or wipe off components that are stacked on top of one another and also those components that are higher than the set height limit of the wiper blade (i.e. specimen resting on its side). In this feeder, in order to achieve flexibility of such device, the wiper blade was attached to a ballscrew assembly driven by a miniature stepper motor, mounted vertically above the bowl track. This allows the wiper blade to be set at any desired height controlled by the computer.

4.2 Programmable Track Width Station

This is also a passive orienting device commonly used in the vibratory bow feeder to ensure that all the components would travel in single file longitudinally on the track, rejecting components that are travelling abreast to another part. In this feeder, the programmability of the track width is achieved by constructing a hinged wall pair that guides the components into a narrowing track. This hinged wall pair is also attached to a ballscrew assembly driven by a miniature stepper motor and therefore allows the narrow track to be set at any desired width controlled by the computer.

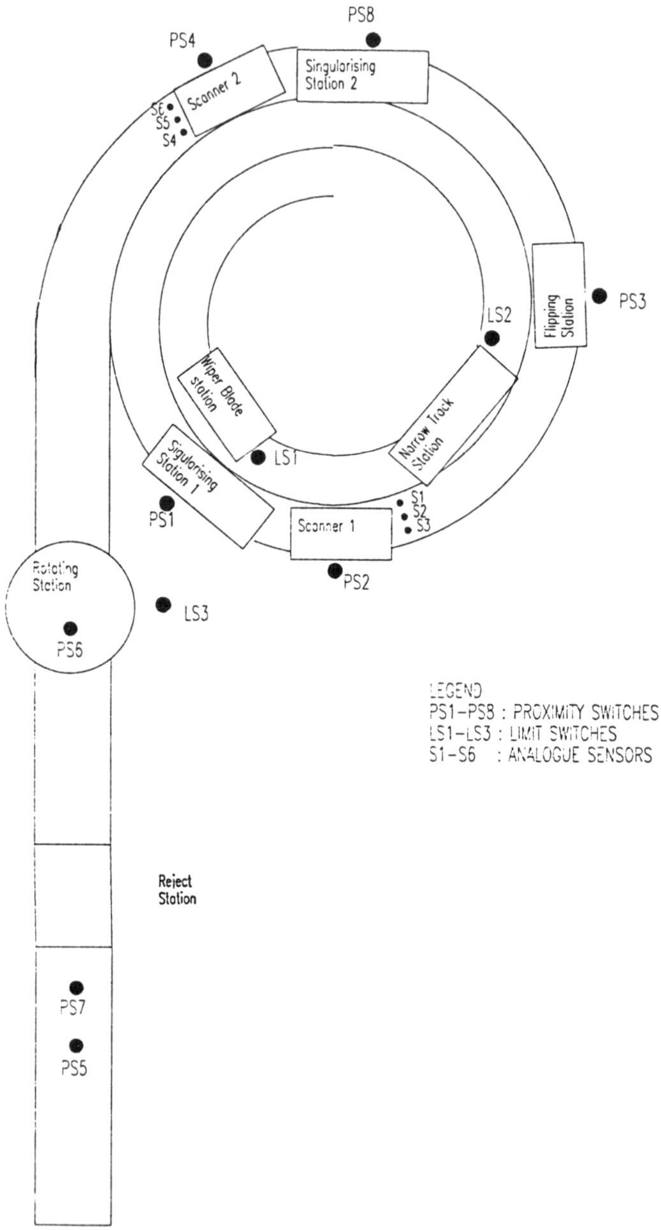

Figure 1. Schematic Layout of the Vibratory Bowl Feeder

4.4 First Scanning Station

This station serves to determine the orientation of a feeding component with an array of analog fibre optic sensors (scanners -- OMRON E3XA). The technique of sensing will be discussed in section 5. When the orientation of the feeding component has been determined, the computer would decide on one of the following:

(a) no action if component is in the desired orientation or 180° reversed.
(b) request flipping operation if the component is upside-down.
(c) activated scanning station #2 if a poor scan or identification is encountered.

4.5 Flipping Station

This station is designed to flip the component which is identified to be upside-down by the first scanning station. The orientating device consists of a 90° vee-track, an air-jet and a fibre-optic sensor. The flipping operation is activated upon request by the computer, after determining that the component is upside-down. Therefore, as the feeding component arrived and is sensed by the fibre-optic sensor, the air-jet would blow the component to rest on the other side of the vee-track, and thereby flipping it 180°.

4.6 Second Scanning Station (Scanner No. 2)

This station would make a second identification to determine the orientation of the feeding component. This station uses the multi-beam sensor which is modulated for ambient light rejection. In the same way, the computer would decide the appropriate actions to be taken after scanning such as:

(a) send "accept" signal when component is in desired orientation.
(b) request rotating operation if the component is 180° reversed.
(c) reject the component if a poor scan or identification is encountered
(d) reject an unflipped upside-down component.

In addition, as a precautionary measure, it would compare the scanning result with that of the first scanning station and would also attempt to use the successful first scan result whenever possible if a poor scan or identification is encountered.

4.7 Reject Station

This stage would reject any component upon sensing it unless an "accept" signal is received. This is to ensure that only parts determined to be in the desired orientation will enter the discharge chute. The reject station consists of a fibre-optic sensor and a diverting chute. Upon sensing of feeding parts by the fibre-optic sensor, the chute would open to divert the components out of the track. This reject operation would be inhibited when an "accept" signal is received from the computer.

4.8 Rotation Station

This is the last station of the whole system and is used to re-orientate a component that is 180° reversed. It consists of a 180° pneumatic rotary actuator which rotates a short chute by 180°, two pneumatic actuators acting as stoppers at the entrance and exit of the chute and two fibre-optic sensors as part sensor. The overall sequencing of the rotation operation is controlled using a relay logic circuit.

The computer would signal the relay logic circuit to perform a rotation operation when required. The exit of the orientator would first be closed by the pneumatic actuator to catch the component. Upon sensing of feeding component in the orientator, the entrance would be closed by another pneumatic actuator and the rotary actuator would perform a 180° rotation before releasing the component out from the orientator. The feeding component would then slide out and activate another fibre-optic sensor to signal "end of cycle" and return the orientator to its initial condition. If the component is in the desired orientation, no signal would be generated by the computer and it would be allowed to slide through the station as it moves down the chute.

5 Pattern Recognition of Bowl Feeder System

A pattern/feature is a description of an object. It contains discriminatory information and characteristics pertaining to that object. A good pattern will be one that is as unique to the object as possible. Hence, pattern recognition is defined as the categorisation of input data or information into the appropriate pre-defined classes of information via the extraction of significant attributes or features from a vast amount of details.

In general, there are three main phases in pattern recognition. The first is the data acquisition phase, whereby a pattern representation for the objects has to be found. The pattern is normally digitised by a vision system. Digitisation is the conversion of the image of a scene or picture into discrete/digital form. The second phase is the data preprocessing and feature extraction phase, whereby the image is being processed to extract salient features. Lastly, the classifier which is in the form of a set of decision functions is applied to the extracted feature. This is the recognition stage where the input pattern will be recognised. The most commonly used techniques for pattern recognition can be categorised as follows:

(a) Template matching
(b) Structural and syntactic techniques.
(c) Statistical pattern recognition techniques
(d) Neural network.

The first three methods rely on pre-stored patterns (e.g. template patterns, ordered composition of simple subpatterns or primitive patterns, or features extracted from patterns) as basis for comparison with input pattern to perform the recognition task. These methods are not easily amendable to changes in pattern configurations that one would expect in large variety small batches of parts required in today's dynamic manufacturing environment. By virtue of its ability to learn and hence adapt to changing patterns/features of different batches of parts required in part orientation system, neural network is chosen for the task of feature recognition of parts. Furthermore neural network is less susceptible to variation in lighting conditions which might result in noisy or incomplete digitised input after pre-processing of images of the vision system.

5.1 Feature Recognition using Neural Network

The feature recognition ability of the bowl feeder system is achieved by four modules, namely:

(a) Part singularising station.
(b) Collection of scan signatures.
(c) Preprocessing and image extraction.
(d) Neural network module.

5.2 Singularising Station

This station is used to control the flow of part into the scanner. The singularising station which consists of a hinge, a pneumatic cylinder, and a fiber optic proximity sensor, will release one part at a time for scanning. The computer will first send a signal to singulate or release a part. The cylinder will then retract and a part will move forward until the part reach the proximity switch. The proximity switch will then send a signal to cause the cylinder to extend again and stop the subsequent part from moving forward.

5.3 Collection of Scan Signatures

The scanners consist of three analog fiber optical sensors arranged in a row across the track. The line image sensing technique employed in this system. Three signatures of the part profile will be captured by the three analogue sensors in the scanner block after the part has passed through it.

For full analogue conversion, it was found that the pacer rate of the data acquisition card must not be set above 100 kHz. A memory for an array of 600 floating point numbers was allocated for the purpose of storing the raw pattern. Since each row in the raw pattern set has three readings, the scan signature must not exceed 200 rows ($200 \times 3 = 600$). Hence, the pacer was set at around 1-2 kHz to get around 100 scan rows of readings. A sample of the scan signature at pacer setting of 1.6 kHz and bowl speed of 8 is given in Figure 2. The collected signature has to be preprocessed before it is passed into the input of the neural network.

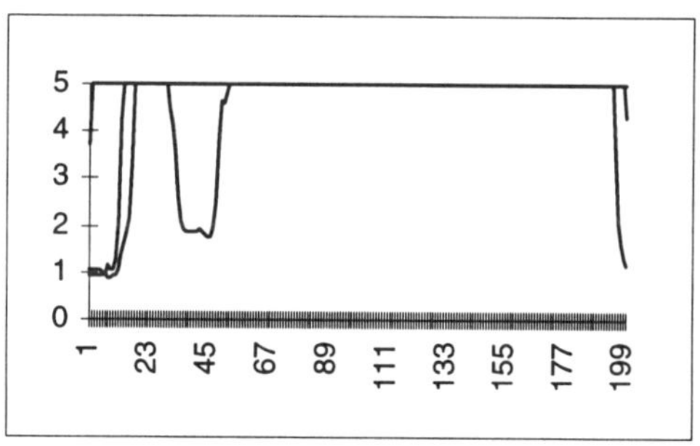

Figure 2: Sample of the scan signature

5.4 Preprocessing and Image Extraction

The collected signatures will be passed through a preprocessing module to extract the input pattern file as input into the neural network module. For example, if the input nodes of the network is specified as 30, then the preprocessing module will compress the raw pattern collected into 3×10 numerical values to be input to the network. The raw pattern collected usually consists of over 100 rows of values or $3 \times$ (over 100 rows) of numerical values (See Figure 3).

5.5 Neural Network Module in Bowl Feeder System.

A control program has been developed to operate the training and operation of the neural network in this feeder system through a user interface menu. A brief description of the operation of the neural network module is as follows:

5.5.1 Configuring the Neural Network

The architecture of the feedforward backpropagation neural network [10] in terms of the number of input nodes, number of hidden nodes and number of output nodes can be re-configured through the settings in the configuration file. In general, the selection of the number of input nodes would depend on image resolution selected in relation to the part feature that could be discriminated.

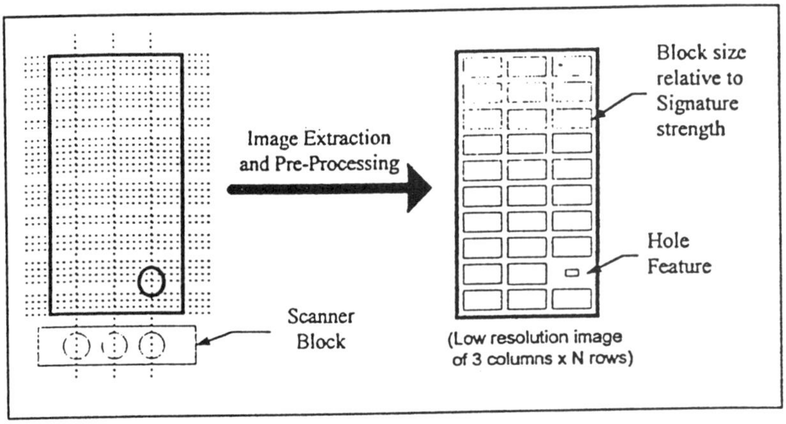

Figure 3. Preprocessing and Image Extraction

The selection of hidden nodes bears a relationship to the number of statistically significant factors that exist in the input pattern. A reasonable value to start with is given by ((input nodes + output nodes)$^{1/2}$ + a few nodes) [11]. If too few of the hidden nodes are used, the network will not train. If too many are chosen, the network beside taking forever to train, tends to develop a network that memorise everything and will not recognise a new pattern it has never seen before, in other words, the network does not generalize.

The number of output nodes assigned should be equal to the number of classifications or orientations involved. For example, in the feeding operation of the mixed specimens of A and B (shown in Figure 4), possible number of orientations of the parts are seven and hence the number of output nodes is set to seven when configuring the network.

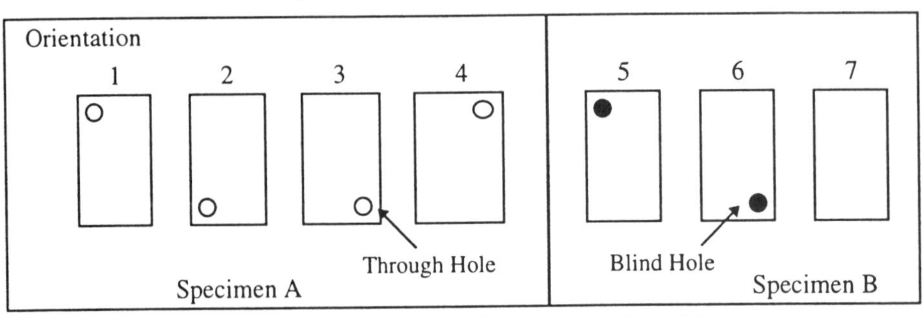

Figure 4. Orientations of Specimens A and B

5.5.2 Training the Neural Network

Two methods can be used in the training of the neural network. One method is tc complete both weight adjustments of feedforward and backward activities in a singl(pass for each pattern presentation. The error is propagated back and the connectin; weights are adjusted after each pattern is presented to the network. This is known a: online training. The other method is to accumulate the changes of weights for al training sets in a batch, and then propagate the error back based on the grand tota change in weight. This is called batch or epoch training.

The advantage of online training is that less memory is needed during training whe compared with batch training. However, the disadvantage is that the network ma 'unlearn' what it has learned in the previous pattern when it is trained by the preser pattern. The batch training method is adopted for training the network in this project.

In this experiment, the neural network will be trained using the following training pattern sets collected:

For scanning station #1 - S1SET1.PAT, S1SET2.PAT, S1SET3.PAT.
For scanning station #2 - S2SET1.PAT, S2SET2.PAT, S2SET3.PAT.

The raw scan signatures of S1SET1.PAT collected are plotted and a sample of which is shown in Figure 2. The training pattern sets consist of all the seven orientations of the mixed specimens A and B. The neural network will be trained to recognise the seven orientations of the two specimens as shown in Figure 4. The training of the neural network consists of the preparation of the training configuration file (TRAIN.CFG), the preparation of the training pattern files (*.PAT) as inputs to the neural network module. After the training, weight files obtained are then used to operate the feeding action.

6 Testing the Accuracy of the Feature Recognition

6.1 Flipping Station Performance: Test on Recognition Ability of Scanner #1

The experimental procedure to test the scanner 1 (OMRON EX3A) recognition accuracy on the seven orientations of specimens A and B is described as follows:

Step 1 : The parts are arranged in orientation 1 on the bowl track.

Step 2 : Change the classification action of the TEST.CFG file to 1 0 0 0 0 0 0 0 0 0. This will cause the flipping station to flip part in orientation 1 and perform no action on the other orientations.

Step 3 : Record the number of parts in orientation 1 not flipped by the flipping station.

Step 4 : Go to Steps 1 to 3 but with the parts arranged in orientation 2 on the track and the classification action of the TEST.CFG file changed to 0 1 0 0 0 0 0 0 0 0. This procedure is repeated for all the other orientations (i.e. 3 to 7).

The results of the test are summarised in Table 1 below.

Table 1. Test Results on Flipping Station

Orientation	1	2	3	4	5	6	7	Average error (%)
Unflipped parts /100 parts	12	3	26	5	50	52	13	23

From Table 1 above, it can be observed that various orientations of specimen A (with through holes) gives a better set of results than that of specimen B (with Blind holes) Orientations 2 and 4 of specimen have better results when compared with the rest of the orientations. For the recognition of the blind holes specimens (orientations 5 and 6), the error is approximately 51%. This implies that either the sensors cannot differentiate the depth feature well or the neural network was not trained properly to recognise the depth variation.

6.2 Rotation and Reject Stations : Test on Recognition Ability of Scanner # 2

Similar to the flipping station test, the experimental procedure for the test on rotation and reject stations is as follows:

Step 1 : Arrange the parts in orientation 1 on the track.

Step 2 : Change the classification to 2 0 0 0 0 0 0 0 0 0 in the TEST.CFG file. This will cause the parts in orientation 1 to be rotated.

Step 3 : Record the error.

The above steps are repeated for the rest of the orientations with the appropriate classifications settings. The results of the performance of rotation and diverting chute stations are summarised in Table 2.

From Table 2, it is noted that the recognition ability of scanner 2 is very good.

Table 2. Test Results on Rotation/Diverting Chute Stations

Orientations	1	2	3	4	5	6	7	Average % error
No of mistakes per 100 parts	0	0	0	0	2	0	0	0.03

6.3 Testing on the Accuracy of the Overall Feeder System

The experimental procedure is as follows:

Step 1: Mix an equal number of specimen A and specimen B (20 from each group) and place the mixed parts in the bowl.

Step 2: Set the classification of the TEST.CFG file (0 1 2 1 3 3 3 0 0 0) such that the system orientates parts to orientation 1 at its output.

Step 3: Start the feeding operation of the feeder system.

Step 4: Record the number of parts in wrong orientations collected in the delivery chute and the number of specimen A rejected.

The results of the test is summarised as followed :

Table 3: Test results of the Overall Feeder System

No. of parts in wrong orientations at the delivery chute per 100 parts processed by the system.	0
No. of specimen A rejected by the system per 100 parts processed by the system	6

From the results in Tables 1 and 2, it is observed that there is a significant difference in recognition accuracy between scanner 1 and scanner 2. Scanner 1 has an accuracy of approximately 77% and scanner 2 an accuracy of approximately 99.97 %. Since both the scanners use the same neural network, the different in the accuracy is most likely due to the scanning ability of the sensors used.

From observations on the scan signatures collected from both stations, it is noticed that scanner 1 signatures are more ambiguous than scanner 2 signatures. This is caused by the inferior recognition ability of scanner 1 when compared with scanner 2. From the data collected in Table 3, out of 100 parts processed by the system, 6 parts of specimen A is rejected by the system. However, from Table 2, the scanner 2 can recognise specimen A almost perfectly, thus the rejected specimen A are probably due to the unflipped parts caused by the errors of scanner 1 and flipping station.

7 Discussion

From the results of the experiments described in Section 6, the main factor that impairs the recognition capability of the system is the sensor's reliability at scanner 1. Although various parameters of the neural network has an effect on its classification ability, the performance of the network is dependent on the input pattern used for training. Since real patterns are used for the training, the reliability of the sensor has a direct effect on the performance of the neural network. Hence, by replacing the sensors at scanner 1 with better sensors (e.g. multi-beam), the recognition ability can be enhanced because the latter is modulated for ambient light rejection..

Another problem of the feature recognition may be caused by irregular movement of part during the scanning process. This may cause the scanner to read in a pattern that is quite different from that used in the training set. Although the neural network can be trained to recognise as many patterns as possible by including all those patterns near the 'boundaries', it may be too time consuming to collect all the possible 'boundaries' pattern.

Another problem observed is that the sensors used in the scanner is not spaced close enough. This could cause some important feature on the part to be missed by the sensors. The present data acquisition card allows up to 16 single ended or 8 differential analogue inputs. Hence additional analogue sensor can be added to the scanner for a bigger and closer scan.

The classification ability of the feedforward backpropagation neural network is dependent on the number of different features of the similar parts used during training. Should different parts with different features be required in the manufacturing process, there is a need to re-train the neural network each time a different batch of parts are used. For the intelligent bowl feeder to be highly flexible, one approach is to used the Adaptive Resonance Theory (ART) neural network [10, 12]. ART is a two-layer nearest-neighbour classifier that stores an arbitrary number of spatial patterns using competitive learning. The neural network architecture is chosen because it updates its memory with any new input pattern without corrupting the existing patterns. The training algorithm conducts a search through the encoded patterns associated with each class of components trying to find a sufficiently close match with the input pattern. If no class exists, a new class of pattern is learned.

8 Conclusion

A flexible, programmable and intelligent vibratory bowl feeder with neural network incorporated has been developed. It represents one of the first attempts, if not the first, to perform feature identificaton of the feeding parts on the vibrating track as well as using neural network. It converts the vibratory bowl into a more versatile parts feeding equipment and also retains the compactness of the vibratory bowl. Evaluation results have been encouraging and with further fine tuning, the feeding system will be an important productive asset to industry.

References

[1] Salmon, M. and d'Auria, Programmable Assembly System, *Computer Vision and Sensor-based Robots*, Plenum Press, London, 1979.

[2] Beecher, R.C., "PUMA: Programmable Universal Machine for Assembly", *Computer Vision and Sensor-based Robots*, Plenum Press, London, 1979.

[3] Maul, G.P. and Goodrich, J.L., "Programmable Parts Feeder", *Manufacturing Systems*, Vol. 13, No. 2, pp. 75 - 81.

[4] Parham, R.J., "Automatic Parts Handling Systems", *Proceedings, of the 2nd International Conference on Assembly Automation*, U.K., May 18 - 21, 1981, pp.161 - 166.

[5] Suzuki, T. and Kohno, M., "The Flexible Parts Feeder which Helps a Robot Assemble Automatically", *Assembly Automation*, Vol. 1 No. 2, Feb. 1981, pp. 86 - 92.

[6] Cronshaw, A.J., "A Practical Vision System for Use with Bowl Feeders", *Assembly*, Sept. 1990, pp. 265 - 274.

[7] Uno, M., "Multiple Parts Assembly Robot Station with Visual Sensor", *5th International Conference on Assembly Automation*, 22 -24 May 1984, Paris, France, pp. 55 - 64.

[8] Warnecke, H.J., Schraft, R.D., Lindner, H., Schmid, S. and Sieger, E., "Components and Modules for Automatic Assembly", *Proceedings of the Automation-in-Manufacturing'91* (AiM'91) International Conference, 1991, Singapore.

[9] Park, I. O., "Development of a Programmable Bowl Feeder Using Fiber Optic Sensor", *10th International Conference on Assembly Automation*, Japan, 1989, pp. 191 - 198.

[10] Fausett, L., *Fundamentals of Neural Networks*: Architecture, Algorithms, and Applications, Prentice Hall International, 1994.

[11] Russell C. Eberhart and Roy W. Dobbins. *Neural Network PC Tools.-A practical guide*. Academic Press, 1990.

[12] Vellani, M and Dagli, C. H., Artificial Neural Network Approach in Printed Circuit Board Assembly, *Journal of Intelligent Manufacturing*, 1993, Vol. 4, No. 4, 109, 119, Chapman & Hall.

Enhanced Intelligent Control for Strip Thickness in a Rolling Mill

Shaowei Wang and A. Kiet Tieu

Department of Mechanical Engineering, University of Wollongong, NSW 2522, Australia

Abstract This article addresses two enhanced intelligent control schemes for rolling process. A mathematical model is formulated to simulate a rolling mill thickness control process; then two intelligent control schemes, an integral neural control and an enhanced fuzzy control are developed for the process. The integral neural control is composed of an inverse neural network controller, a PID controller and a process modelling neural network. The enhanced fuzzy control consists of a fuzzy inference controller and an adaptive neurofuzzy model parallel to the rolling process. The simulation results show that the integral neural control could significantly improve the strip thickness, and more excellent performance is obtained with the enhanced fuzzy control.

1 Introduction

The conventional PID control could achieve good performance for linear systems, but has a limitation when dealing with non-linear plants, whereas neural networks and fuzzy inference systems are finding increasing use in complex process control, due to their non-linear capabilities.

The application of neural networks for rolling process mainly focused on the parameter estimation[1-5]. The use of neural models for direct control had been investigated only by D. Sbarbaro-Hofer *et al* [6], in which static neural networks were created with data obtained by solving several equations of an analytical model of rolling mill. But the motivation of using neural networks to replace mathematical models (the equations of analytical models of the plants) is that neural network based controllers generally depend less on *a priori* knowledge of the process to be controlled but more on the input-output observations. And the input-output observations are the data which could be collected from the dynamic process, rather than those from the static equations. D. Sbarbaro-Hofer *et al* did not contribute a strategy of real time process application value. Fuzzy inference system has been successfully used for gain tuning in thickness control of rolling mill [7], but fuzzy application in rolling process has not been discussed widely in the literature.

In this article, a dynamic mathematical model for the rolling process is formulated which is represented by the connection of multiple transfer function blocks. Based on the simulation results of this dynamic model, a rolling process modelling neural network and an inverse neural network are developed, which, together with a PID controller, compose an integral neural control. Within the fuzzy control scheme, a fuzzy inference controller is implemented with rolling technology expertise, and based on the mathematical model simulation results, an adaptive neurofuzzy model is incorporated to enhance the control structure. The methodology proposed in this article could be verified by and applied to real time process.

2 Control Problem

The rolling process of a four-high mill could basically be described as shown in Figure 1, where *f* is rolling force, *H* strip input thickness, *h* output thickness, *L* and *l* the distances of input and output thickness sensors from rolling mill respectively.

Suppose S_0 is the roll gap with zero rolling force ($f = 0$), *K* is the rolling mill elastic modulus, the strip output thickness is given by

$$h = S_0 + f / K \tag{1}$$

where the rolling force is [8]

$$f = \sigma_c \left[\sqrt{\frac{DH\gamma}{2}} + \alpha \sqrt{\frac{fD}{E}} + \frac{\alpha^2 \mu D f (2-\gamma)}{2EH(1-\gamma)} \right] \tag{2}$$

where σ_c is compressive yield stress of the material being rolled($\mathrm{N/mm^2}$), *D* diameter of work rolls(mm), γ fractional reduction ($\gamma = \frac{\mathrm{H-h}}{\mathrm{H}}$), α dimensionless constant of value 1.08, *E* elastic modulus of the work rolls($\mathrm{N/mm^2}$), and μ effective coefficient of friction in the bite.

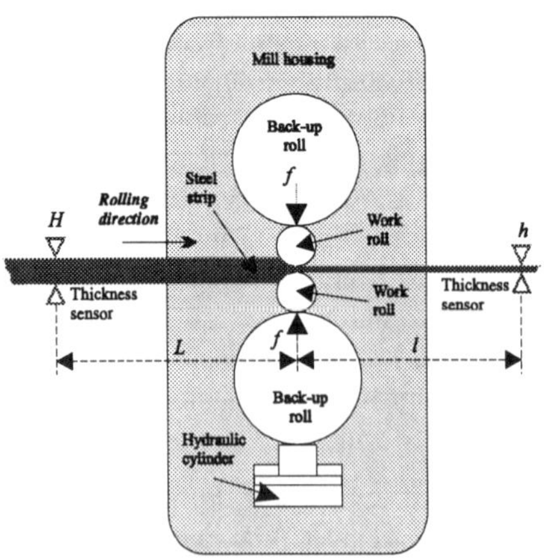

Fig. 1. General sketch of a four-high mill rolling process

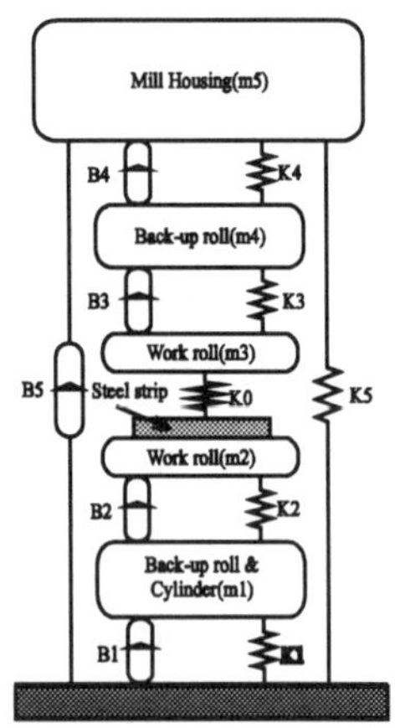

Fig. 2. Mathematical model of a four-high rolling mill

The control objective is to maintain the output thickness as close as possible to the desired value h_{ref}. Equations 1 and 2 indicate that the input thickness variation ΔH will result in the output thickness deviation Δh, and their relationship is non-linear. The basic control concept is adjusting the work roll gap by ΔS to compensate the output thickness deviation Δh incurred by input thickness variation ΔH.

$$\Delta S = (M / K)\Delta H \tag{3}$$

where M is plastic modulus of steel strip, the value of which varies as to different input thickness values and different reduction values.

For this kind of non-linear process, the combination of pre-compensation control (feedforward control) and monitor control (feedback control) is conventionally a representative control scheme, as shown in Figure 3, in which the hydraulic valve and cylinder are omitted for simplicity. In order to increase the control performance to the maximum, it is necessary to set the parameters (e.g. M, K etc.) accurately at every operational point (corresponding to every different input thickness and reduction value) for pre-compensation control, and to obtain the accurate and prompt value of output thickness for feedback control. Unfortunately, it is difficult to obtain the values of M and K with sufficient accuracy, and the output thickness transport delay exists inevitably due to the distance between the output thickness sensor and the roll bite. As a result, an integral neural control and an enhanced fuzzy control are approached.

The control performance is indexed by the integral of the square error between output thickness reference value and measured value over an unit time, which is defined as

$$I = \frac{\int_{t=t_0}^{t_{\max}} \left(h_{ref} - h\right) dt}{t_{\max} - t_0} \tag{4}$$

The test input thickness profile is shown in Figure 4, and the desired output thickness is 3.5 mm.

3 Mathematical Model of Rolling Mill

In order to simulate the rolling process control and obtain data for neural network training and neurofuzzy model adapting, a mathematical model for the rolling mill is formulated.

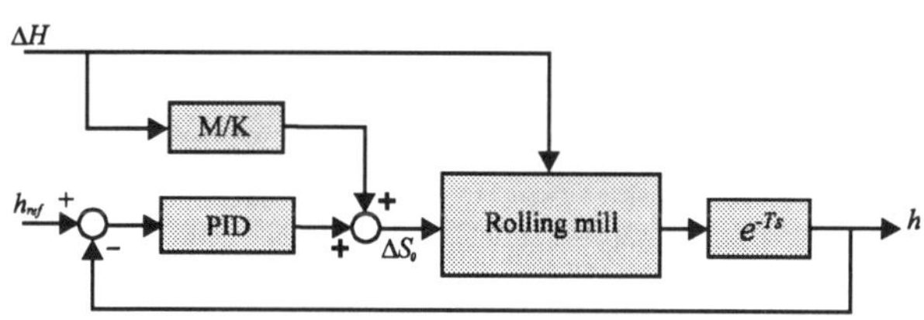

Fig. 3. Feedback linear control and feed forward control for strip thickness

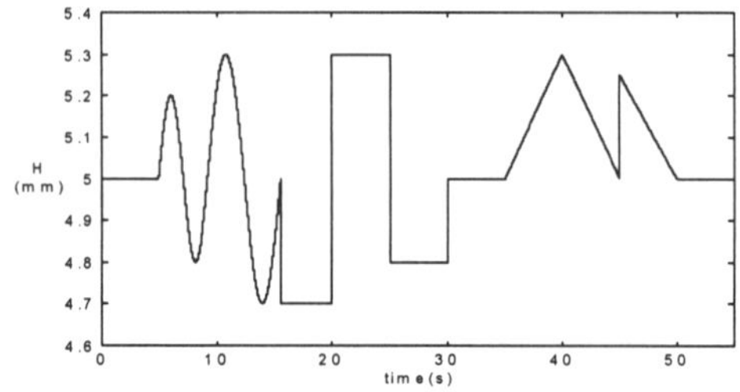

Fig. 4. Input thickness profile

Normally, a mechanical element of translational motion can be described as a mass-spring-damper system[9], in which the spring represents the elastic behaviour, and the viscous friction is represented by the damper. Therefore, the four-high rolling mill is modelled by a five-degree-of-freedom mass-spring-friction (damper) system, as shown in Figure 2, in which the roll force and hydraulic screw-down cylinder are considered as external forces to the system.

The force equation of the system is

$$\{[M]s^2+[B]s+[K]\}[X]=[F] \tag{5}$$

where $[M]$ is mass matrix (kg), $[B]$ damping friction matrix (N/m/s), $[K]$ spring constant matrix (N/m), $[X]$ mass displacement vector (m) and $[F]$ external force vector (N). In block diagram, the Equation 5 is represented by connection of multiple transfer function blocks[10].

4 Integral Neural Control

As neural networks have great capabilities in solving non-linear problems by learning from the observations of input-output performance of the process[11], they are used to model the rolling process dynamics and the process inverse, with which the process modelling network and the inverse network are obtained.

There are several types of neural networks for non-linear system modelling, such as multiple-layer perceptron, radial basis function (RBF) network and back-propagation (BP) network. Here the most popular network BP network is used. To overcome the conventional BP network limitation of slow training, an alternative method to gradient descent, an approximation of Newton's method called Levenberg-Marquardt optimisation is used. The Levenberg-Marquardt weight update rule is

$$\Delta W=\left(J^T J+\lambda J\right)^{-1} J^T e \tag{6}$$

where ΔW is weight change matrix, J the Jacobian matrix of derivatives of each error to each weight, λ a scalar, and e an error vector. When the scalar is very large, the above expression approximates gradient descent, whereas when it is small the above expression becomes the Gauss-Newton method. The Gauss-Newton method is faster and more accurate near an error minimum, so the aim is to shift towards the Gauss-Newton method as quickly as possible. Thus, λ is decreased after each successful step and increased only when a step increases the error.

4.1 Feedforward Control with an Inverse Neural Network

With an inverse neural network, the feedforward control to compensate the output thickness deviation could be formulated as shown in Fig. 5, in which the hydraulic valve and cylinder are omitted for simplicity. The neural network for the rolling process inverse is trained with data from the simulation results of PID feedback control, taking input thickness and output thickness as inputs and roll gap as output.

4.2 Integral Neural Control

The result of feedforward control with an inverse neural network is much better than that of linear PID feedback control. But simulation results also indicate that the transient response of this control is still not quite satisfactory. Obviously that is the inevitable outcome of open loop control system, due to the dynamic behaviour of mill stand. In this sense, a feedback PID linear controller parallel to feedforward inverse neural network controller is implemented.

To overcome the output signal delay problem, a process modelling neural network is applied to estimate output thickness, which is trained with the data collected from PID control simulation results. The estimated output thickness is used to replace the delayed process output thickness as the feedback signal. The error between the process network output (after delayed) and rolling process output (only delayed value available) is also used as a feedback signal and passed to PID controller, taking into account the uncertainty of the process and inaccuracy of the process network.

The control structure is as shown in Figure 6, in which the hydraulic valve and cylinder are omitted. In this way the control performance is improved significantly. Figure 9b is the response of this control to the input given by Figure 4.

5 Enhanced Fuzzy Inference Control Structure

Fuzzy inference systems are able to successfully control non-linear processes[12] when the membership functions and rules are properly determined. For the rolling

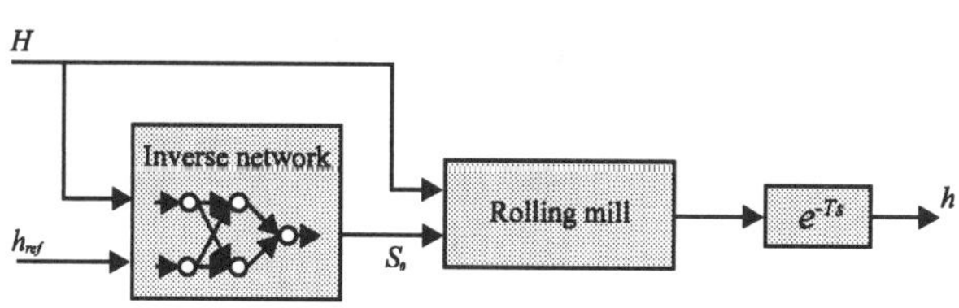

Fig. 5. Feedforward control with an inverse neural network

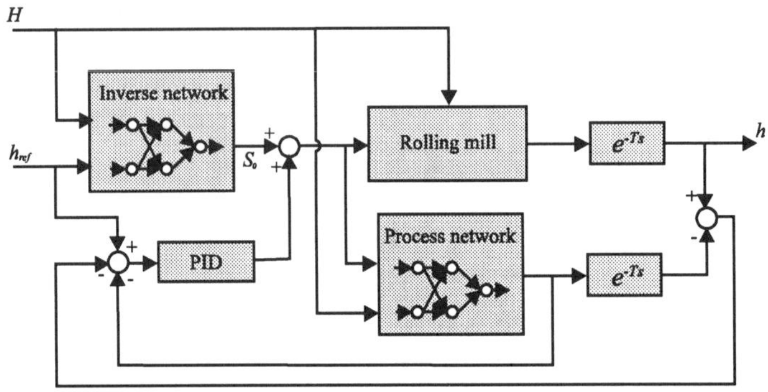

Fig. 6. Integral neural control

process, the deviation of output thickness Δh and its derivative $\frac{d\Delta h}{dt}$ are used as the inputs of the fuzzy controller.

To enhance the fuzzy control scheme, a rolling mill model is built by an adaptive neurofuzzy inference system to estimate the prompt process output thickness. The estimation result, instead of the delayed process output signal, is adopted as the input of fuzzy inference controller. And the error between the real process output (only delayed value available) and neurofuzzy model estimation (after delayed) is fed back to account for the real process uncertainty and neurofuzzy model inaccuracy. This enhanced fuzzy control structure is described as shown in Figure 7.

5.1 Fuzzy Inference Controller

The fuzzy inference controller has two input variables Δh and $\frac{d\Delta h}{dt}$, as stated above, and one output variable *u*, the control signal to hydraulic servo-valve.

The two inputs and one output are assigned different fuzzy sets as shown in Figure 8, based on rolling theory and rolling technology expertise. The linguistic variables are defined as follows:

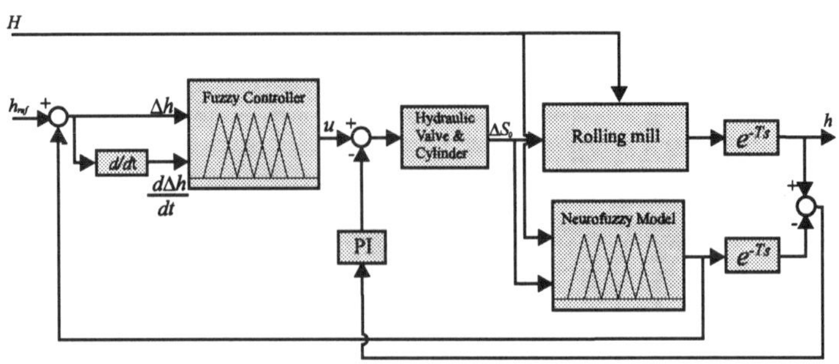

Fig. 7. The enhanced fuzzy control scheme

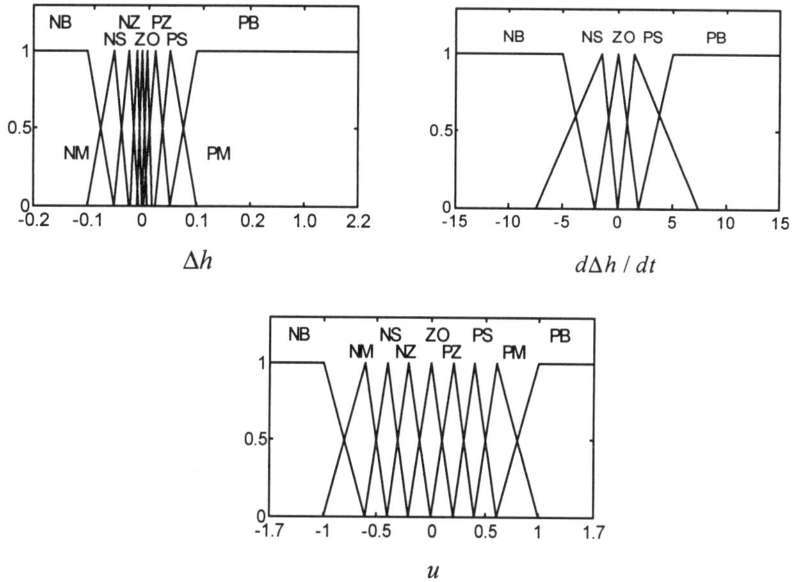

Fig. 8. Membership functions of fuzzy inference controller

PB: positive big;	PM: positive medium;	PS: positive small;
PZ: positive zero;	ZO: zero;	NZ: negative zero;
NS: negative small;	NM: negative medium;	NB: negative big.

With these fuzzy sets, 37 rules are used with Mamdani's fuzzy inference method[13].

5.2 Neurofuzzy Inference Modelling

The rolling process is modelled by a fuzzy inference system, and its parameters are turned by a back-propagation neural network structure based on the rolling process input-output data. The neural network structure minimises the fuzzy inference model output error. As the fuzzy inference model is optimised by neural learning, it is called adaptive neurofuzzy inference modelling.

For fuzzy process modelling, a more efficient fuzzy inference method, Sugeno method is employed[14]. Sugeno method[15] adopts singleton spikes to find the weighted average of a few data points, greatly simplifies the computation required to find the centroid of a two-dimensional shape, thus enhances the efficiency of the defuzzification process and speed up the adaptive learning.

For rolling process modelling, 6 membership functions are assigned to each input of the fuzzy inference system, S_0 and H. Therefore, there are 36 membership functions for the output of the fuzzy inference system, represented in the form of the following equation

$$u_{ij} = a_{ij} S_0 + b_{ij} H + c_{ij} \qquad i,\ j{=}1, 2, \ldots\ldots 6. \qquad (7)$$

where a_{ij}, b_{ij} and c_{ij} are parameters optimised by the neural network structure.

Generalised bell curve membership functions and triangular membership functions are separately employed for the fuzzy inputs. After training for 40 epochs, the neurofuzzy model with generalised bell curve membership functions estimates more accurate rolling process output than that with triangular membership functions.

The enhanced fuzzy control performs excellently, after several times of parameter regulation according the simulation results. The response of this control scheme indicates that the output thickness is almost perfect, as shown in Figure 9c, except for the several spikes, which are caused by the sudden changes of input thickness.

6 Results and Discussions

Figure 9 and Table 1 summarise the results of integral neural control and enhanced fuzzy control with PID linear control. The integral neural control improves the control performance greatly. This is due to the control philosophy that the non-linear relation between the input variables and output variable is compensated by the inverse neural network, and the PID controller is used to further minimise the error, especially the steady error.

Table 1. Comparison of the different control schemes

Control strategy	Index(t_0=0)	Index(t_0=5)
PID linear control	0.0082	6.2634e-005
Integral neural control	0.0024	2.1529e-006
Enhanced fuzzy control	0.0023	9.6746e-007

Note: the index is defined by Equation 4 as t_{max} =55.

More excellent performance obtained with enhanced fuzzy control is due to the fact that fuzzy inference controller is a non-linear controller, which could seize on the non-linear characteristics of the rolling process by itself. It is worth noting that the output thickness steady error could be controlled within a very small limit by these two enhanced intelligent control schemes. For the input thickness given by Figure 4, the output thickness error (except the spikes caused by sudden input changes) could be controlled within $\pm\,0.0005$ mm with the integral neural control, and the enhanced fuzzy control reduces the output thickness tolerance to $\pm^{0.0003}_{0.0002}$ mm.

7 Conclusion

From this work, it has been shown that rolling process and its process inverse can be successfully modelled by neural networks and the rolling process could be modelled also by neurofuzzy inference system. The integral neural control with process modelling network and inverse network improves the rolling process performance significantly. The enhanced fuzzy control with a fuzzy inference controller and an adaptive neurofuzzy model parallel to rolling process can obtain excellent performance, better than integral neural control.

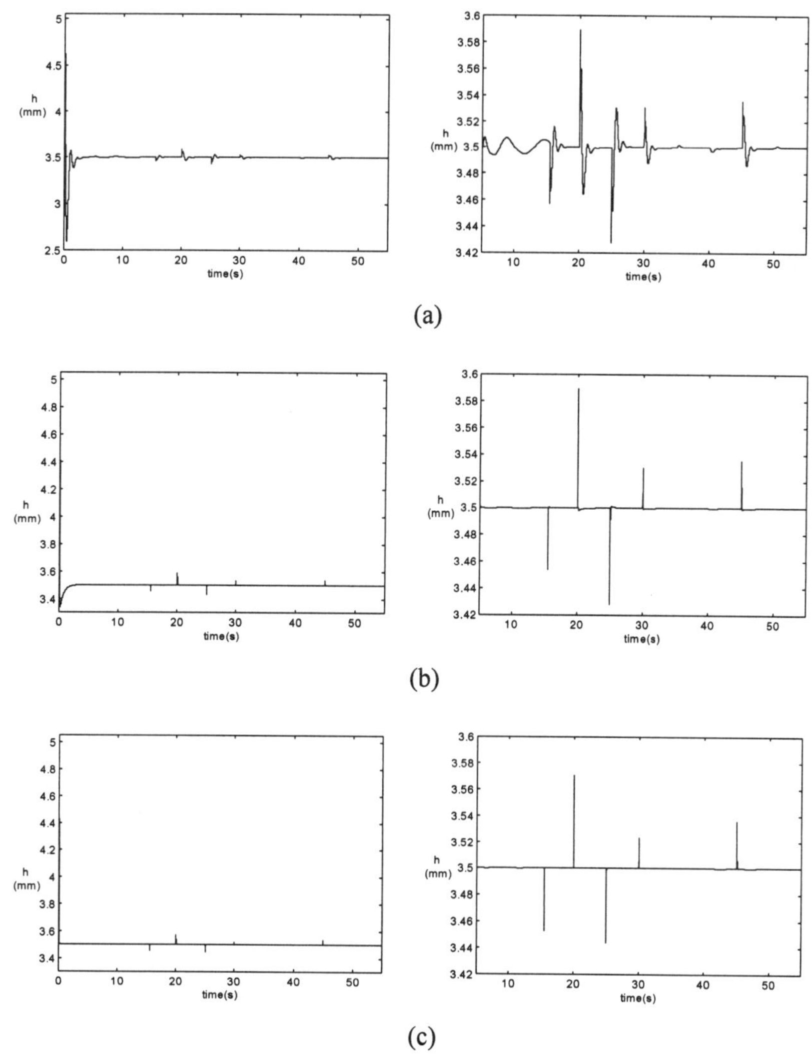

Fig. 9. Response of the rolling thickness control system with difference control schemes

(a) PID feedback control (b) integral neural control (c) enhanced fuzzy control

Note: The graphs in the right hand part are the magnification of those in left hand part with starting time of 5th second

References

1. S. Wang and A. K. Tieu, Application of Neural Network to Width Estimation in Flat Rolling, *Proceedings of Second Biennial Australian Engineering Mathematics Conference '96*, The Institute of Engineer, Australia, 1996, pp.231-235.

2. N. F. Portnann, *et al*, Application of Neural Networks in Rolling Mill Automation, *Iron and Steel Engineer*, Vol. 72, February 1995, pp.33-36.

3. T. Fechner, *et al*, Adaptive Neural Network Filter for Steel Rolling, *Proceedings of the 1994 IEEE International Conference on Neural Networks*, Vol. 6, 1994, IEEE, Pitscataway, pp.3915-3920.

4. M. Roscheisen, *et al*, Neural Control for Rolling Mills: Incorporating Domain Theories to Overcome Data Deficiency, *Advances in Neural Information Processing*, Vol. 4, Hupson and Lippman, Eds, Morgan Kaufman, 1992, pp.659.

5. A. C. Tsoi, Application of Neural Network Methodology to the Modelling of the Yield Strength in a Steel Plate Mill, *Advances in Neural Information Processing*, Vol. 4, Hupson and Lippman, Eds, Morgan Kaufman, 1992, pp.698.

6. D. Sbarbaro-Hofer, *et al*, Neural Control of a Steel Rolling Mill, *IEEE Control Systems*, June 1993, pp.69-75.

7. K. Nose, *et al*, Application of Fuzzy Inference to Gain Tuning in Thickness Control of Multi-high Rolling Mills, *12th World Congress International Federation of Automation Control: Prints of Paper*, The Institute of Engineer, Australia, Vol. 6, pp.135-138.

8. W. L. Roberts, A Simplified Cold Rolling Model, *Iron and Steel Engineer*, October 1965, pp.75.

9. B. C. Kuo, *Automatic Control System*, Prentice Hall International, Inc. 1991, pp.128.

10. S. Wang, *et al*, A Mathematical Model for a Rolling Mill Thickness Control System and its Digital Simulation, *Proceedings of Second Biennial Australian Engineering Mathematics Conference '96*, The Institute of Engineer, Australia, 1996, pp.125-131.

11. A. G. Barto, Connectionist Learning for Control: An Overview, *Neural Networks for Control*, W.T. Miller, III, R. S. Sutton, and P.J. Werbos, Eds. Cambridge, MA: MIT Press, 1990. pp.5.

12. M. Brown and C. Harris, *Neurofuzzy Adaptive Modelling and Control*, Prentice Hall International, 1994.

13. E. H. Mamdani and S. Assilian, An Experiment in Linguistic Synthesis with a Fuzzy Logic Controller, *International Journal of Man-Machine Studies*, Vol.7, No.1, 1975, pp.1-13.

14. T. Terano, *et al*, *Fuzzy Systems Theory and its Application*, Academic Press, Inc., Harcourt Brace Jovanovich, Publishers, 1991.

15. M. Sugeno, *Industrial Applications of Fuzzy control*, Elsevier Science Pub. Co., 1985.

ADVANCED MANUFACTURING TECHNOLOGY

A Modular Head-Eye Platform for Real-time Active Vision [1]

Chee-Kong FONG[2]
jfong@cyberway.com.sg

Ming XIE
mmxie@ntu.edu.sg

Robotics Research Center
School of Mechanical & Production Engineering
Nanyang Technological University, Singapore 639798

Abstract. This paper describes the development of a high speed, controlled stereo head-eye platform which facilitates the rapid redirection of gaze in response to visual input. It details the mechanical device, which is based around geared DC motors, and describes the hardware aspects of the motion controller and vision system, which are implemented on a modular structured mobile robot. To arrive at an active head-eye relationship, two adaptations of head-eye coordination algorithms are implemented.
Our **N**TU **A**ctive **S**tereo **H**ead (NASH) is a 6 DOF camera head. That includes independent auto-focus and independent control of pan for each of the two cameras. A combined set of pan and tilt motors are used for emulation of the neck movements. Servo control of all the axes is accomplished by a programmable multi-axis controller (PMAC) card.
To achieve real-time and robustness in our head-eye coodination problem, we adopted two methods. The first method involves using planar primitive, which requires to iteratively converge observed image towards that of the reference image. The second method is a closed-for solution, it uses two new images to determine where the position of the reference image has been taken.

1. Introduction

The ability to move one's head and eyes is imperative in order to perform even simple tasks. Imagine trying to catch a ball without moving either your head or eyes. At certain points during its flight, the ball may not even be in your field of view, so how are you expected to align yourself to catch it? Because animals and robot exist in a complex and dynamic world, one may argue that head and eye movements are necessary to function in the environment. In addition, one may also argue that moving one's head or eyes makes many visual task simpler. That is, movement can enhance the sensory acquisition process. For example, the ability to move our eyes allows us to track a moving object and thus we can predict it's position and/or velocity (such as in catching an object). Or, we can maintain the image of a moving object within our field of view to perform further processing. We can also move our eyes to view portions of the scene, to examine areas of interest or areas where we are uncertain of what we see.

---

[1] This research work is supported by Ministry of Education of Singapore (MOE) under the grant RG72/96.
[2] Author is a graduate student pursuing his Master degree.

In nature, heads are expected to be on shoulders and not on arms. This playing with words is actually referring to two different approaches in building head-eye systems. Some researchers are, rationally, interested in putting head-eye systems on a robot arms or have chosen other approaches [2,3,4]. NASH that is presented in this paper, on the contrary will be put on shoulders, that is a platform. Having a head and not just a pair of eyes on a robot arm (manipulator) is also more convenient for mounting on a moving platform. When talking about a 'head', the neck movements are to be included as part of the head. That is, having the three basic degrees of freedom (roll, pitch and yaw) in the manipulator will already do the job of the neck would do. In this sense, most of the head-eye systems implemented up to now are actually a pair of eyes and not head-eye systems.

Active vision research was initiated at the University of Pennsylvania in the early eighties [1]. This led to construction of robotic head-eye systems at a number of US universities including the University of Pennsylvania [5], Harvard University [6], and the University of Rochester [7,8]. These heads can be characterized as first generation heads, which to a large extent are based on home-made mechanisms etc. As active vision research has gained more wide-spread interest, there has also been an increased activity related to the construction of head-eye systems. Examples are those presented by Crowley [9], Prelove [10], TRISH [11], and KTH [12]. There seem to be two different approaches to the construction of head-eye systems, (1) general purpose, fast and high accuracy heads, e.g. the KTH head, and (2) heads which are flexible enough for a number of different tasks in computer vision, e.g. the Harvard head.

Under the Strategic Research Program: Vision and Control, Nanyang Technological University (NTU), there is also a considerable interest in the use of active vision to emulate an intelligent human behaviour, i.e. head-eye coordination. To enable further research works on robotic head-eye relationship, **N**TU **A**ctive **S**tereo **H**ead (NASH) has been developed. At the same time, a customised mobile platform has been designed and it will be integrated with NASH to form a mobile head-eye system. To gain insight into the design and use of a head-eye system, the effort our research program was directed towards the construction of a head in category (2) as that mentioned in the later part of the above paragraph. Based on the experience from construction of such a prototype, subsequent new and improved designs would be specified and implemented. This paper presents the work carried out in order to construct the first prototype.

2. Requirements For a Head

Generally, a robotic head consists of a mechanical eye and a neck mechanism. There have been intensive studies on eye movements in human being. Less attention has been directed towards the neck movements and body movements. This might be the major source of confusion in the design of head-eye systems, because the eye-movements in present head-eye systems are actually a hybrid of eye and neck movements.

Eye movements are different from other movements that affect the quality of visual cues. This is true from the fact that the fovea centralis is occupied by images from the different parts of the scene, so, no geometrical distortion is generated in the image by normal eye movements. This is not the case with neck movements and other body movements. On the contrary, neck movements are supposed to change the perspective of the images and generate parallaxes, this is caused by the eccentric position of the spine. In a sense, body movements and neck movements are serving the same tasks, although there are still functional differences between them.

In the process of designing NASH, we have identified a number of different sets of degrees of freedom that are potentially of interest. There are grouped into (1) intrinsic parameters, (2) extrinsic parameters of individual cameras, and (3) head/neck parameters as shown below:

(1) Intrinsic camera parameters (eye configuration);
- Focus,
- Aperture,
- Zoom/Focal length.

(2) Extrinsic camera parameters (eye motion / position);
- Vergence and version.

(3) Head parameters (neck motion / position);
- Roll angle,
- Pitch angle,
- Yaw angle,
- Length of baseline - distance between cameras.

2.1 Intrinsic Camera Parameters

In the above list, though intrinsic camera parameters were mentioned, they are somehow not critical in our intended applications. Firstly, focusing can be provided by auto-focus Iris lens. Secondly, for the time being, NASH will only work in a structured environment with constant light conditions, thus camera aperture setting is not needed. And finally, since NASH will be mounted on a mobile platform, it is not crucial to have a zooming facility just to have a larger field of view.

2.2 Extrinsic Camera Parameters

By panning the cameras, it is possible to change the position where the optical axes intersect (the gaze or fixation point).

For change of gaze point, a symmetrical or independent control of the vergence angles θ_1 and θ_2 may be used. Most of the first generation head-eye systems used an arrangement where two cameras are coupled through use of a spindle to provide symmetric angles. In such a configuration, the gaze point may be changed only to a position along a line perpendicular to the baseline (line between the focal points). This is different in human vision, where the gaze point may move around in the plane without any motion by the neck. If the goal of active vision is to gain insight into biological vision, it is desirable to have independent vergence angle control for the two cameras.

In human vision, there is limited ability to perform independent tilt of the eyes. In general the use of the tilt will complicate the stereo reconstruction and it will not be considered, besides, tilt of the entire head can allow change of the gaze point to any position in the 3-D space much more appropriately and easily.

2.3 Head Parameters

As mentioned, the best possible accuracy for vergence stereo is achieved with symmetric camera vergence angles. To achieve symmetric camera vergence angles for any point in 3-D space, it is necessary to be able to perform panning for the head. Therefore the field of view is much larger and the specified accuracy will allow positioning of the point of interest close to the optical axis where a minimum of aberration is necessary. This consideration applies likewise for the tilt action of the head.

The human visual system baseline is about 6 cm,. This is much smaller than it will be possible to realize due to physical limitations (size of cameras, etc) for a binocular robotic head. If the cameras are to verge, then there has to be somewhere for the cameras and lens to move, thus the smallest realistic baseline has to be about 32 cm.

3. Head-Eye Coordination Algorithm

Having provided the mechanical structure for a head-eye system. The next step is to write the instruction sets in achieving a head-eye coordination solution, which sometimes also known as image based visual servoing or even mis-appropriately called hand-eye coordination.

Apparently, almost all the head-eye coordination algorithms proposed thus far used some form of convergent methods to servo the cameras to a position where the observed image is equal to the reference image.

At a particular position (called Reference View), the vision system can capture an image as reference (called Reference Image). Figure 1 shows one example that a camera is at a reference view position in which a reference image is taken.

In actual situation, the cameras may be at any position and orientation (as in Figure 2) other than the reference view position. Therefore, the observed image is different from the reference image.

Now, the problem is how to control the head-eye system which will bring the cameras to a position where the observed image is equal to the reference image? This problem has been extensively investigated so far.

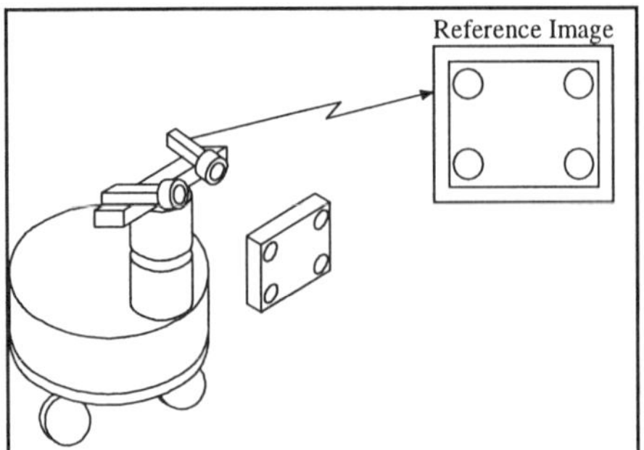

Figure 1 Reference View and Reference Image

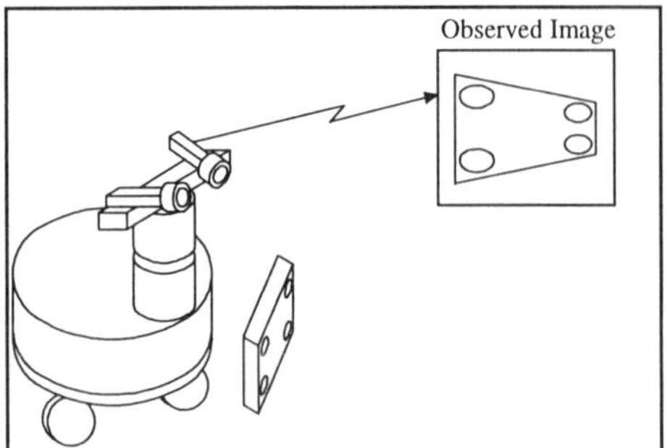

Figure 2 An Arbitrary View and Observe Image

There are many iterative approaches based on minimization process of a cost function [18], while there is only one deterministic method presented [20].

In the following sections, we will demonstrate two head-eye coordination solutions, one is a iterative method extracted from [19] while the second is a deterministic method developed by us.

3.1 Iterative Methods using Planar Primitive

Iterative approaches are based on the use of specially designed minimization process to generate motion control signal which guides head to a position where the image seen by eyes (the observation) looks like a reference image (the goal).

Planar primitive method was formed by Corke [19], in this method, it calculates a series of motions (translations and rotations) in order to iteratively converge the observed image towards the reference image (see Figure 3). The equations to work out the translation and rotation motions are as follows.

Translation:

$$Tx = k_1 * (x - x_c^r)$$

$$Ty = k_2 * (y - y_c^r)$$

$$Tz = k_3 * (L_0 - S_0)$$

Rotation:

$$\theta x = k_4 * [(S_0/(L_0-1)) + (L_2/(S_2-1))]$$

$$\theta y = k_5 * [(S_1/(L_1-1)) + (L_3/(S_3-1))]$$

$$\theta z = k_6 * \pm * a\cos(S_0 * L_0)$$

where $k_{1,\ldots 6}$ are control gains.

The main advantage iterative approaches has is that cameras do not have to be calibrated because 'self-calibration' will be achieved after the first few consecutive iterations.

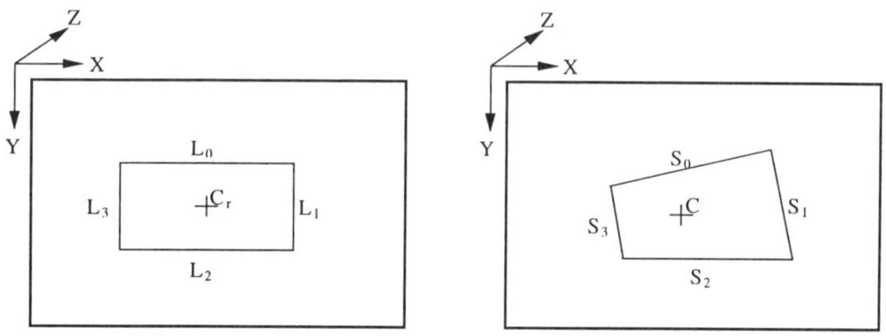

Figure 3 Iterative Approach

3.2 Deterministic Method

In our deterministic approach, the strategy for a closed-form solution is to estimate the position where the reference image has been taken. With two cameras and thus two known views, it is possible to (1) determine a third unknown view from 6 line segments by a combined least-squares estimation or (2) determine the third unknown view from 8 line segments by two separate least-squares estimations, where the first estimation is for rotation motion while the second one is for translation motion (see Figure 4).

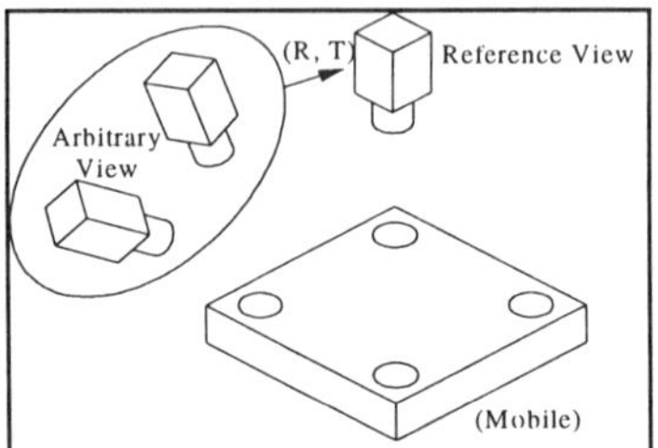

Figure 4 Deterministic Approach

The steps in arriving at the third unknown view from two known views can be demonstrated as follows:

1) Camera model: Perspective view.

$$x = X/Z$$

and $y = Y/Z$

2) Input views: A total of three views are needed, they are reference view, left view and right view each being represented by $V3$, $V2$ and $V1$ respectively.
3) Input primitives: From left and right views, extract at least 6 or 8 line segments.
4) ***Step 1.*** At reference view ($V3$),
   - A 2D line segment is described by,

     $$a_{v3}x_{v3} + b_{v3}y_{v3} + c_{v3} = 0$$

   - The corresponding projection plane is,

     $$a_{v3}X_{v3} + b_{v3}Y_{v3} + c_{v3}Z_{v3} = 0$$

   - If $S_{v3} = (a_{v3}, b_{v3}, c_{v3})$ and $P_{v3} = (X_{v3}, Y_{v3}, Z_{v3})$, then,

     $$S_{v3} * P_{v3}^{t} = 0 \qquad \ldots\ldots\ldots \qquad \textbf{(1)}$$

5) ***Step 2.*** Relationship between left/right view and reference view,

   $$P_{v3}^{t} = R_{v2v3} * P_{v2}^{t} + T_{v2v3} \qquad \ldots\ldots\ldots \qquad \textbf{(2)}$$

6) ***Step 3.*** General equation 1, substitute equation (2) into (1),

   $$S_{v3} * (R_{v2v3} * P_{v2}^{t} + T_{v2v3}) = 0 \qquad \ldots\ldots\ldots \qquad \textbf{(3)}$$

7) ***Step 4.*** At left/right view,
   - From a pair of points (P'_{v2}, P''_{v2}) on a 3D line segment,

     $$S_{v3} * (R_{v2v3} * P'^{t}_{v2} + T_{v2v3}) = 0$$

     $$S_{v3} * (R_{v2v3} * P''^{t}_{v2} + T_{v2v3}) = 0$$

   - Subtract above two equations,

     $$S_{v3} * R_{v2v3} * (P'^{t}_{v2} - P''^{t}_{v2}) = 0$$

8) ***Step 5.*** General equation 2,

   define $D_{v2} = (P'_{v2} - P''_{v2}) / \| P'_{v2} - P''_{v2} \|$

   then $S_{v3} * R_{v2v3} * D^{t}_{v2} = 0 \qquad \ldots\ldots\ldots \qquad$ **(4)**

The direction vector of a 3D line segment can be directly determined from the two known views [20]. It should be noted that equation (4) does not involve the unknown T_{v2v3}. Therefore, we can first estimate the rotation matrix and then the translation vector of the motion transformation between left/right view and reference view.

There are three independent unknown parameters in T_{v2v3}. Hence, a set of three line segments are enough to fully determine R_{v2v3} and T_{v2v3} from equations (3) and (4). In practice, it may be preferred to work with linear system and use Least-Squares estimation method. In order to do so, we can consider that there are eight unknown parameters in R_{v2v3} (factorize out, for instance, the parameter of row 3 and column 3 from R_{v2v3}, and recover it later by using the orthogonality constraint that the matrix R_{v2v3} has to meet). In this way, a set of eight line segments allow to establish a linear system for the estimation of the matrix R_{v2v3}. Once we know R_{v2v3}, T_{v2v3} can be estimated from a linear system with a set of three line segments only.

Contrary to iterative approach, our proposed deterministic approach needs the cameras to be calibrated. Since our approach uses 'one-move' from an unknown cameras position to a reference image position, therefore the perspectives of cameras have to be found.

4. Design of a Mobile Head-Eye Coordination Robot

Based on the considerations in Section 2, NASH was constructed. At the same time, a customized mobile platform was also designed to add locomotion to the head. The intended purpose for our robot is to serve as an experimental testbed for our future research works on active vision and also on autonomous robot. Mechanically, the robot is modular in design. Currently, it has 3 separate modules, and they are (1) the NASH module which consists of a pair of active 'eyeballs' (cameras), a neck (manipulator unit) and a baseline rig, this head module is mounted on top of a (2) command-center module (CCM) where it's backplane houses a Pentium computer running at 200 MHz, a DSP card and a motion controller card, in addition, the CCM also holds the battery units as well as all the servo motor amplifiers, below the CCM is the (3) mobile base module (MBM), this module has 3 independent wheels and each wheel has independent driving and steering actions. A schematic view of the mobile robot is shown in Figure 5.

4.1 Design of the NTU Active Stereo Head (NASH) Module

NASH is made up of a pair of eye units, the baseline rig, and a manipulator unit.

Lately, there has been a keen interest and rather novel concept in the design of robotic arm, that is modularity. Research works presented in [13,14,15,16,17] focused on complete modularity in which joints and links are classified as separate modules. In such a design, one joint provides one motion, therefore a range of different joints having different degree-of-freedoms is probable, so, when coupled serially together with the links, a customized manipulator is formed based on its intended application. Such an idea offers a higher flexibility in manipulator selection and configuration.

In the design of the neck, we work on similar concept, that is, the neck which is constructed like a manipulator, should be made up of modules. But in our design, we do not separate the link with the joints. In fact, we have integrated two joints (with the same or different degree-of-freedom) with a common link between them to form an 'active link'. Our active link is simpler in design and construction as compared to our counterparts. Two of such units are serially stacked together to form the neck. A schematic drawing on such an active link is depicted in Figure 6.

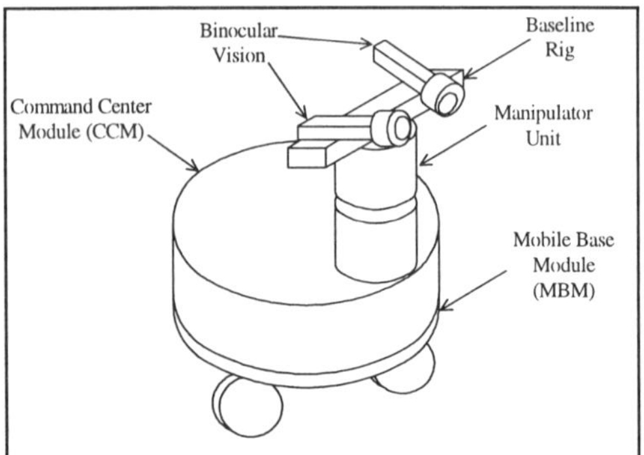

Figure 5 Schematic of Mobile

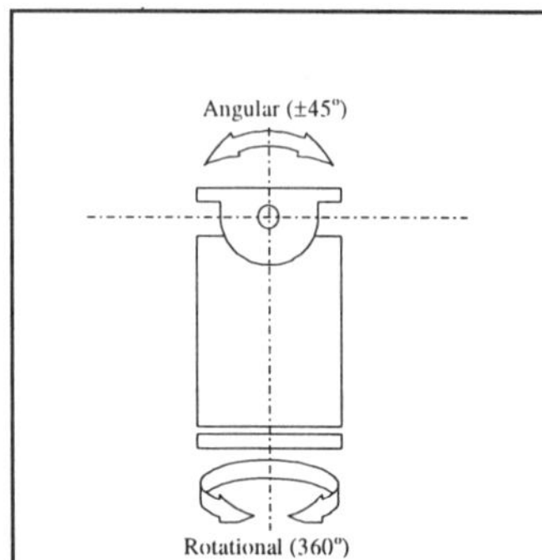

Figure 6 Schematic of an Active Link

The motion actuators in each active link include a pair of 100W MAXON motor (one for each DOF), each motor is coupled to a Harmonic Drive reduction gear via a set of timing pulley and belt.

To ensure compatibility, mounting hole dimensions on both the mating flange ends are the same. Electrical connectors of respective match are also located on each end of the flange. Each link has its own wiring network, and each network includes wires for its own motion actuators as well as wires for the neighbouring modules.

The technical specifications of an active link is summarized in Table 1.

The baseline rig provides the ability to change the length of the baseline dynamically. The rig containing the baseline mechanism represents also the connection between the eye units and the neck. This rig is purely for experimental reasons.

The only eye movement available on the eye unit is the panning action that allows both that cameras to verge and fixate at the point of interest.

4.2 Design of the Mobile Base Module (MBM)

To add mobility to NASH so that it can move around in a structured environment, a mobile base module (MBM) has to be designed. This module consists of a base plate with three wheels mounted underneath it. Each wheel has independent steering as well as driving, thus providing holonomic capability. Our mobile platform is capable of 3 DOF, two translational dof along the X-Y direction and the third is a rotational DOF about Z-axis.

Three wheels are mounted around and inside the circumference of a ϕ160mm base plate at an angle of 120° apart. Steering and driving actions from each wheel are realised by 2 separate sets of geared DC motor. Currently, the designed weight it can support is 115 kg. A schematic view of a wheel unit is shown in Figure 7.

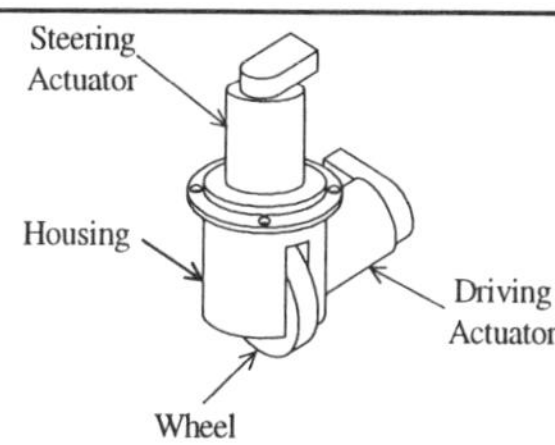

Figure 7 Isometric View of a Wheel Unit

As shown in the diagram, the wheel unit can be regarded as one independent unit. If future modification arises, the wheel can be removed easily and a new design can be fitted, without having to disturb the rest of the structure.

Technical specifications of the wheel unit can be summarized in Table 2.

Table 1 Active Link Specification Summary

Active Link Design Specification					
Active Link (Pan/Tilt)	1	Operating angular velocity	Pan	rpm	44
			Tilt	rpm	44
	2	Operating torque	Pan	Nm	0.12
			Tilt	Nm	4.12
	3	Operating power	Pan	W	0.55
			Tilt	W	18.98
	4	Acceleration	Steer	sec	1
			Drive	sec	1
DC Servo Motor (Electronic Commutation)	1	Assigned power rating		W	100
	2	Nominal voltage		Volt	30
	3	No load speed		rpm	4800
	4	Max. permissible speed		rpm	15000
	5	Designed operating speed	Pan	rpm	4400
			Tilt	rpm	4400
	6	Designed operating torque	Pan	Nm	0.22
			Tilt	Nm	0.22
Harmonic Drive Reduction Gear	1	Reduction ratio		-	1:100
	2	Max. input speed	Oil	rpm	14000
			Grease	rpm	5000
	3	Max. output torque		Nm	28
	4	Rated output torque at 2000rpm		Nm	7.8
	5	Designed input speed	Pan	rpm	4400
			Tilt	rpm	4400
	6	Designed output speed	Pan	rpm	44
			Tilt	rpm	44
	7	Rated output torque at 44rpm		Nm	27.84

Table 2 Wheel Unit Specification Summary

Wheel Unit Design Specification					
Wheel Unit	1	Operating angular velocity	Steer	rpm	44
			Drive	rpm	44
	2	Operating torque	Steer	Nm	1.2
			Drive	Nm	20.59
	3	Operating power	Steer	W	5.5
			Drive	W	94.92
	4	Acceleration	Steer	sec	0.3
			Drive	sec	0.3
DC Servo Motor (Electronic Commutation)	1	Assigned power rating		W	100
	2	Nominal voltage		Volt	30
	3	No load speed		rpm	4800
	4	Max. permissible speed		rpm	15000
	5	Designed operating speed	Steer	rpm	4400
			Drive	rpm	4400
	6	Designed operating torque	Steer	Nm	0.22
			Drive	Nm	0.22
Harmonic Drive Reduction Gear	1	Reduction ratio		-	1:100
	2	Max. input speed	Oil	rpm	14000
			Grease	rpm	5000
	3	Max. output torque		Nm	28
	4	Rated output torque at 2000rpm		Nm	7.8
	5	Designed input speed	Steer	rpm	4400
			Drive	rpm	4400
	6	Designed output speed	Steer	rpm	44
			Drive	rpm	44
	7	Rated output torque at 44rpm		Nm	27.84

5. Conclusion

We have presented a conceptual layout of a modular robot. The robot consisits of a stereo camera head system (NASH) mounted on a mobile platform. The mechanical parts and head-eye coordination algorithms were described in order to achieve active vision.

We presented a new concept of modular manipulator made up of a series of active links. Each active link has 2 DOF. By stacking two of the active links serially together, the neck of NASH is formed.

We have also presented two head-eye coordination approaches. The first is a iterative method using planar primitives, while the second method is a deterministic approach, which may be the first in the world, developed by our group. Both methods complement each other, future tests are needed to validate the real-time responses from each of the methods.

Acknowledgement. This project is funded by Ministry of Education, Singapore. It is part of the effort to develop one complete general purpose vision guided modular robot.

References

[1] R. Bajcsy, 'Active Perception', *Proc. of IEEE 76*, pp. 996-1005, 1988.

[2] D. H. Ballard, 'Animated Vision', *Proc. of 11th International Conference on Artificial Intelligence*, 1989.

[3] E. P. Krotkov, 'Exploratory Visual Sensing for Determining Spatial Layout with an Agile Stereo System', *Ph.D Thesis*, 1987.

[4] J. J. Clark, N. J. Ferrier, 'Modal Control of an Attentive Vision System', *Proc. of 2nd International Conference on Computer Vision*, 1988.

[5] E. P. Krotkov, K. Henriksen, R. Korries, 'Stereo Ranging with Verging Cameras', *Transcript of IEEE Pattern Analysis and Machine Intelligence 12*, pp. 1200-1205, 1990.

[6] N. J. Ferrier, 'The Harvard Binocular Head', *Technical Report 91-09*, Harvard Robotics Laboratory, University of Harvard, 1991.

[7] D. H. Ballard, A. Ozcandarli, 'Eye Fixation and Early Vision: Kinematic Depth', *Proc. of 2nd International Conference on Computer Vision*, 1988.

[8] T. J. Olson, D. J. Coombs, 'Real-Time Vergence Control for Binocular Robots', *International Journal on Computer Vision 7*, pp. 67-89, 1991.

[9] J. L. Crowley, P. Bobet, M. Mesrabi, 'Gaze Control for a Binocular Camera Head', *Proc. of SPIE Application of Artificial Intelligence 5: Machine Vision and Robotics*, pp. 47-61, 1992.

[10] J. R. G. Pretlove, G. A. Parker, 'A Lightweight Camera Head for robotic-Based Binocular Stereo Vision: An Integrated Engineering Approach', *Proc. of SPIE Application of Artificial Intelligence 5: Machine Vision and Robotics*, pp. 62-67, 1992.

[11] M. Jenkin, E. Milios, J. Tsotsos, 'TRISH: A Binocular Robot Head with Torsional Eye Movements' *Proc. of SPIE Application of Artificial Intelligence 5: Machine Vision and Robotics*, pp. 36-46, 1992.

[12] K. Pahlavan, J. O. Eklundh, 'Heads, Eyes and Head-Eye Systems', *Proc. of SPIE Application of Artificial Intelligence 5: Machine Vision and Robotics*, pp. 14-25, 1992.

[13] D. Schmitz, P. K. Khosla, T. Kanade, ' The CMU Reconfigurable Modular Manipulator System', *Proc. of 19th International Symposium on Experimental Robotics,* pp. 473-488, 1988.

[14] K. H. Wurst, 'The Conceptual and Construction of a Modular Robot System', *Proc. of 16th International Symposium on Industrial Robotics,* pp. 37-44, 1986.

[15] T. Matsumaru, 'Design and Control of the Modular Robot System: TOMMS', *Proc. of IEEE International Conference on Robotics and Automation*, pp. 2125-2131, 1995.

[16] R. Cohen, M. G. Lipton, M. Q. Dai, B. Benhabib, 'Conceptual, Design of a Modular Robot', *ASME J. of Mechanical Design*, vol. 114, pp. 117-125, 1992.

[17] T. Fukuda, S. Nakagawa, 'Dynamically Reconfigurable Robotic System', *Proc. of IEEE International Conference on Robotics and Automation*, pp. 1581-1586, 1988.

[18] K. Hashimoto, 'Visual Servoing', World Scientific, 1994.

[19] P. I. Corke, 'Visual Control of Robot Manipulator', *Visual Servoing*, Editor: K. Hashimoto, World Scientific, pp. 1-31, 1994.

[20] M. Xie, 'Head-Eye Coordination: A Closed-Form Solution', *Proc. of 2nd World Automation Congress*, 1996.

Third Millennium Manufacturing: Healing the Executive/Production Schism by Computers or Conversation?

Duley K W and Maj S P

Department of Computer Science, Edith Cowan University,
Mt Lawley, Perth, Western Australia

Abstract. Since the turn of the century there has been a perceptible dichotomy between management and workshop in discrete manufacturing. Production-line Manufacturing, developed by Henry Ford, and Scientific Management, developed by Frederick Winslow Taylor, left an enduring legacy in the division between executive and production, This dichotomy has manifested itself demographically (in a new class system), sociologically (in labour unrest), technologically (in separate automation systems) and philosophically (in political division). Australian companies are seeking to close this gap - this paper looks at the two major solutions.

Computer Integrated Manufacturing and, more recently, Manufacturing Execution Systems, seeks to establish a digital link between the Management Information and Decision Support systems at executive level and the processes, control systems and general automation on the shop floor, creating a replacement for the intimate knowledge of workshop processes and progress which were the province of the owner/ manager but which vanished with the development of the multi-divisional (arid then multi-national) companies. Japanese success with a person to person solution and with modified processes rather than automated (remote) control provides an interesting alternative.

This paper follows the development and impact of the dichotomy, looks at the two solutions and seeks an appropriate Australian answer.

Keywords : Computer Integrated Manufacturing, Continuous Improvement, Cultural Differences, Discrete Manufacturing, Just-In-Time, Manufacturing Execution Systems, Pull Manufacturing, Push Manufacturing, Third Millennium Manufacturing, Zero Defects Quality

1 Evolution of the Division between Executive and Production

Discrete manufacturing has been around in some form since before man learnt to practice agriculture and could, thereby, settle and establish a defined civilisation - club makers and flint chippers produced a manufactured product. As the agricultural civilisations of Europe and the Middle East developed, a range of arts and crafts evolved becoming increasingly codified and sophisticated, culminating in the self-regulating Guild structure of Europe's Middle Ages, Renaissance and Reformation. Steam Power, providing the means to aggregate a large workforce around many machines in one workplace and revolutionising communication and trade by fast, cheap rail and sea transport, changed the techniques of manufacturing in the Industrial Revolution.

While the establishment of large factory manufacturing, the "*dark, satanic mills*" of the mid-1800s, was a significant change in the way that items were produced, little changed in the way in which production was managed. Managers tended to be Owners and also to have an understanding of the processes of production. While the crafts and trades had advanced their abilities and technologies, the economic structure which underpinned this, essentially *cottage*, style of industry remained unchanged. However, the economic foundation of production had changed. For example, prior to the Industrial Revolution the products of shearing, spinning, weaving and finishing changed ownership between the farmer and the various craftspeople involved. Each was, therefore, able to compute profit and loss by simple comparison of outlay and income (costs being defused by market forces). Under this regime, businesses used accounts primarily to record the results of these market exchanges.

By the start of the nineteenth century, textile merchant/entrepreneurs were taking control of spinning, weaving and finishing within a single enterprise. This change necessitated an emphasis on accounting for interests within the company and on the use of accounting records in administrative control of the enterprise. Despite this subtle change, the essential point is that management still understood the processes on the shop floor and were able to maintain an overview of the progress of product through a plant by personal inspection and direct, verbal reports from supervisors on the floor. This direct, almost *hands-on*, management style remained common until after the turn of the century.

Henry Ford, in moving the production of the Model 'T' to a purpose-built factory at Highland Park (IL, USA) in 1910, initiated the most significant change in the methods of discrete manufacturing since the industrial revolution itself. While increasing his staff from 1,908 to 4,110 (+215%), Ford's innovation in implementing production-line manufacturing resulted in an increase in output from 1,708 Model 'A's to 34,528 Model 'T's (+2,021%). (Ford, 1991) This, better than nine-fold, increase in *per capita* productivity attracted widespread interest.

It should be understood that this revolution in manufacturing took place in a period of US history where population, personal wealth and, consequently, market was booming. Ford's factory was designed to produce one item as well as possible, as cheaply as possible and as fast as possible. This was genuine *push* manufacturing in which parts and raw materials are thrust in one end of a factory and forced out the other as product at the greatest possible rate. Emphasis on market research did not emerge until the late 1920s (amongst Ford's competitors) since all that could be produced would find a ready market.

"No manufacturer anywhere in the world was able to exactly repeat Henry Ford's extraordinary success with mass production. This is hardly surprising. In the history of commerce there have been few opportunities to exploit such an enormous, untapped and reasonably homogeneous market. Yet mass production - modified to accommodate

different markets and labour conditions... was... widely initiated." (Batchelor, 1994, p. 66)

Despite the success of production-line manufacturing, Ford faced considerable criticism on the grounds that his practices alienated his workers, treating them as machines enslaved to the greater machine (the production line itself) and forcing him to pay extraordinarily high wage rates in an effort to stem the flow of staff turnover. This alienation was exacerbated by the concurrent work of Frederick Winslow Taylor on the concept of *Scientific Management*, the idea that the application of scientific principles to management could significantly enhance productivity.

Taylor discerned a tendency for all members of a workforce to slow their output to the pace of the slowest.

"*When a naturally energetic man works for a few days beside a lazy one, the logic of the situation is unanswerable. 'Why should I work hard when that lazy fellow gets the same pay that I do and does only half as much work?*" (Taylor, 1967, p. 19)

Taylor's counteraction to this tendency which he termed 'soldiering' was the study of time and motion, a process which identifies the atomic parts of a task and then discovers the most efficient way to perform that atom of work and the minimum time that efficient atom should take a suitable person without injuring that person's health or well being. Employing carefully chosen work teams, at significantly higher than average rates of pay and on the basis that the workers followed his prescription for the job to the letter, Taylor was able to obtain exceptional improvements in productivity - it is worth noting, however, that his attitude to the people in his workforce might not be seen as politically correct today.

"*The [Frederick W.] Taylor method does not recognise the hidden abilities workers possess. It ignores humanity and treats workers like machines. It is no wonder that workers resent being treated that way and show no interest in their work*". (Ishikawa & Lu, 1985, p. 25)

While it would be impossible to refute Taylor's success in productivity improvement, it might be said that his more enduring legacy was his enhancement of the alienation for which Ford was being criticised. Believing planning work to be clerical in nature and that every detail of a job should be thought out in a '*planning department*' Taylor developed an extensive hierarchy of middle management under the planning department to ensure that each task was carried out precisely according to plan including '*Gang Bosses*', '*Speed Bosses*', '*Inspectors*' and a '*Shop Disciplinarian*'. Each of these, in turn, had his foreman over him. (Taylor, 1993) As this hierarchy became institutionalised, so too did the dichotomy between white collar and blue, between executive and production. Management would manage and workers merely had to do as they were told. This dichotomy became virtually universal, persisting for more than five decades. In terms of the communications breakdown between executive and production, however, worse was on the way.

With the wall between the office and the workshop now a significant obstacle, management were entirety dependent on the accounting records as the basis for administrative control. In this they were assisted by developments in accounting theory, notably those in concert with the American Society of Mechanical Engineers (ASME), which produced information on efficiency and throughput. Two further, subtle changes rendered this unworkable. Firstly, management themselves changed - the giant, multi-divisional companies which were appearing tended to appoint accountancy and business school graduates to executive positions. Management now lacked an intuitive, practical understanding of the processes of production and could be said to have had to '*manage by numbers*'. Secondly, the ownership of the company were normally '*absentee industrialists*', shareholders who might never set foot in the factory and were only interested in the return on their investment. For these people, and, of course, for the taxation department, the accountants now had to produce financial reports and these reports quickly became the central focus of the accountant's employment. Communication of strategic information between production and executive levels was now virtually impossible since managers who needed relevant 'numbers' no longer received them.

"*In [the place of procedures for computing managerially relevant product costs] appeared the costing procedures that twentieth-century accountants developed to value inventories for financial reports. While those procedures yield cost information that apparently aids financial reporting, the same information is generally misleading and irrelevant for strategic product decisions.*" (Johnson & Kaplan, 1991, p. 126)

Disconnection is a term which could from this time be applied to the relationship between executive and production. This disconnection has manifested itself demographically (in a new class system). sociologically (in labour unrest), technologically (in separate automation systems) and philosophically (in political division). Its effects in manufacturing management linger:

"*A survey of information preparers and users in an automated manufacturing environment indicated that 54 percent of preparers were dissatisfied with their costing methodologies, and 62 percent of users were similarly dissatisfied. Performance measurement systems were perceived as even worse, with 57 percent of preparers and 69 percent of users dissatisfied.*" (Linnegar, 1988, p. 39)

Next we consider how companies survived this cervical severance.

2 Surviving the Communication Breakdown

Bearing in mind the market context in which manufacturing thrived, apart from the Great Depression of the '30s, through the expansion of the US, the impetus of two World Wars and the post-WWII global boom, corporate survival was relatively simple despite executive isolation from the shop floor. Inability to predict, monitor and control the passage of components and raw materials through a plant towards the finished goods inventory could be disguised and compensated for by high levels of work in progress inventory. Conventional managerial thinking of the time required all machines and processes to be running at maximum rate whenever possible - this is *push* style manufacturing - leading to rising inventories particularly at bottleneck processes. (If the mean arrival rate of batches at a workstation is 'λ' and the mean service rate is 'μ', and if any random variations exist in either, then where $\mu = \lambda$ the length of the queue (size of the inventory) will build to infinity. Where a batch arrival may occur at a process at any random time interval then the length of the queue 'L_Q' is found by:

$$L_Q = \lambda^2 / \mu(\mu - \lambda)$$

Equation 1 : Growth of Inventory

Where $\mu = \lambda$ then $\mu(\mu - \lambda) = 0$ so $L_Q \rightarrow \infty$ (Gibson, Greenhalgh, et al, 1995, p. 75)

Raw Materials and Finished Goods Inventories (RMI and FGI) also tended to rise throughout this period, RMI as an hedge against disruption of supply and FGI as a buffer against unexpected surge in demand. Conventional cost accounting permitted this rise in inventory through recognition of inventory as an asset (Mitchell & Granof, 1981, p. 195) (Kirkwood, Ryan, et al., 1989, p. 13) (Mathews, Perera, et al., 1991, p. 160) yet the results of this artificial imbalance are obvious. Company stockholders, accepting financial reports showing inflated company assets, demanded better return on their *share* of what they perceived to be their *investment*. By the 1960s, manufacturing management were seeking ways to improve that return and two major streams of development arose - one in the Western World (dominated by the USA) and the other in Japan.

3 Western Industry Embraces Computerisation

Calculation of manufacturing schedules requires the asking of four fundamental questions:

1.	What are we going to make?	Master Production Schedule (MPS)
2.	What does it take to make it?	Bill of Materials
3.	What do we have in stock?	Inventory Records
4.	What do we need to get?	Materials Requirement Plan (MRP)

(Storey, 1994, p. 159)

Time Phased Ordering (Orlicky, 1975) sought to resolve these questions by determining ordering rates and sequences (an MRP) which would meet the MPS while minimising RMI, IBM's "Production, Inventory and Control System" (PICS),which was to support sales forecasting, requirements planning, capacity planning, engineering data control, shop floor control, operations scheduling, purchasing and inventory control, lead in the mid 1960s to the development of computerised MRP

"*[in] an attempt to use the 'number-crunching 'power of the computer to develop a production plan for an entire plant in which the production of each individual item is coordinated with a master production schedule for the production of end products.*" (Kerr, 1991, p. 17)

Although the move to MRP failed, largely due to its inability to respond to unexpected contingencies such as delivery failure or plant breakdown, PICS and the subsequent development of the 'Communications Oriented Production Information and Control System' (COPICS) by IBM in the early 1970s set Western manufacturing firmly on the road to computerisation as the solution to the executive / production communication failure. Evolution of MRP and interaction with the COPICS concept lead to MRP II (the acronym now referring to Manufacturing Resources Planning) which covered labour, machine, capital, purchasing, marketing and shipping requirements as well as the areas covered by MRP. Computerisation, however, was moving into every facet of manufacturing.

In the office, systems from word processors and spreadsheets to complex company-specific Management Information Systems (MIS) and Decision Support Systems (DSS) are, these days, ubiquitous; Computer Aided Design and Drafting (CADD) has almost completely replaced the traditional drawing board and pencil; support software is available for purchasing, quality control, supply chain, management, training, transportation and distribution and warehousing as well as the traditional ledgers and payroll. Meanwhile, on the shop floor, computerised numerical control (CNC) of production machinery, robotics, automated storage and retrieval systems, automated guided vehicles for product transfer, coordinate measuring machines and automatic data collection for quality control are typical moves towards full computerisation. Computer Integrated Manufacturing (CIM) and Manufacturing Execution Systems (MES) are not single application programs but the drawing together of a huge range of software, hardware, programming languages, operating systems, data exchange protocols, graphics display systems, databases etc. into one organism which encompasses the whole plant from reception desk to scrapyard.

"*MES systems are arguably one of the more complex Computer Information Systems in existence. This is not to say that you can't think of several other systems that you personally consider more complex, perhaps you can and it's not that the code is necessarily complex either, but rather that the whole system itself is. The interaction of all the components is critical to the net result.*" (McDonough, 1995)

Complexity leads to expense, and until recent moves towards Open Systems development of CIM / MES software few Australian companies would have been able to support the expenditure of millions of dollars on powerful centralised computer systems and large, expensive proprietary software to run on them. Today, scores of software companies offer management support packages which run on widely used interface systems and operating systems such as DOS, OS/2, MS Windows, Windows95, WindowsNT and UNIX. System integration can start with packages which are needed and affordable, then develop as the needs of the company and the sophistication of the users grow. Reductions in inventory and lead times, smoother flows through the plant and improvements in quality are achieved yet the systems are imposed top-down and, while improving executive monitoring and control of production, do not address the fundamental problem of executive / production dichotomy. Neither, regardless of the method of implementation, does CIM / MES change the underlying processes of production.

Where two solutions to any problem exist, it is common for them to be substantially different in emphasis. For example, it is said that the NASA (when the US first started putting men into space) spent millions of dollars devising a ballpoint pen which would work in weightless conditions while the USSR solved the weightless writing problem by issuing its cosmonauts with pencils. Similarly, the Japanese response to the pressure for improved return on investment was fundamentally different from that in the West.

4 Japan's Solution (Just-In-Time) is People-based

"*After World War II, Japanese planners developed strategies for competing with the United States and Western Europe... Growing international price competition forced continual reductions in manufacturing and marketing costs. When Japan's competitive advantage derived from low-cost labour become exhausted, purchasing and inventory management systems became a focus for cost reductions. This led to JIT, although it did not become widespread throughout Japan until the OPEC embargo in 1973.*" (Meredith, Ristroph, et al., 1991, p. 448)

Japan's economy collapsed to a state of zero growth under the effects of the 1973 oil crisis yet Toyota's earnings increased and the widening gap between it and other companies, many of whom had continued to use the conventional American-style system of mass production, generated interest in the production system developed in the 1960s and 70s by Toyota under the leadership of Vice President Taiichi Ohno. (Robinson, 199 1, p. 133) Although commonly referred to as '*Just-In-Time*' (or '*Kanban*' after the inventory control cards used) the system is not the application of a single idea but the integration of a range of concepts which continue to develop under the umbrella of Continuous Improvement ('*Kaizen*').

Ohno initiated the solution of the executive/production communication problem by means of simple, human communication. By opening dialogue between workers and management, establishing *bone fides* by acting on the suggestions of shop floor staff, Toyota's middle management (with Ohno's enthusiastic support) unleashed a *bottom-up* restructuring of the actual processes of production themselves, a situation which persists within the company - from a workforce of 60,000 Toyota received 2.6

million process improvement suggestions in the calendar year 1986. Of these 96% were implemented. (Hutchins, 1988, p. 10) In this, the Toyota system not only re-opened the communications cut off at the turn of the century but also rejected Taylor's philosophies, identifying them with low motivation, low job interest and absenteeism. (Hutchins 1990, p. 49) By the time of the oil crisis in 1973, Toyota had instituted a new system of production which enabled it to weather the economic storm which threatened to wreck conventional industry worldwide. Process changes included:

- Elimination of Waste which JIT defines as any activity which doesn't add value to the product (Hernandez, 1993, p. 10) including transport time, inspection time and waiting time in inventory buffers as well as the traditional rework and scrap.

- Reorganisation of the Shop Floor including a) movement of machines into process related groups which reduces materials transfer time and permits robotic handling (Kerr, 1991) and b) the establishment of worker movement patterns to avoid workers getting in each other's way and to equitably share the workload (Sartori, 1988)

- Zero Defects Quality utilising the mistake-proofing ('Poka-Yoke') techniques devised by Shigeo Shingo in which simple changes to processes and machines, made immediately on the production of a defect, prevent the recurrence of the defect. This may be seen to supersede Total Quality Management as it is understood in the West which merely seeks to reduce the production of defects to an acceptable level. (Shingo, 1986)

- Autonomation ('*Jidoka*') which precipitates immediate attention by operators and supervisors to the production of a defect by stopping the production line at a signal from workers or the automatic machines themselves. Immediate abnormality identification and process improvement compensate for the down-time with a subsequent lack of defects.
  (Robinson, 199 1) (Yoshikawa & Burbridge (eds), 1991)

- Single Minute Exchange of Dies (SMED) reducing set-up times and permitting economic production in small lot sizes. (Dyck, Varzandeh, et al., 1991)

- Reduction of Inventory to the minimum required to allow the plant to operate. This is made possible only by the achievement of Zero Defects Quality (since there is little allowance to cover for mistakes). Toyota seeks to keep inter-process inventory below one tenth of a day's production. (Graham, 1988)

Just-In-Time Process Control in which movement of batches is controlled by authorisation cards (kanbans) which replace the *'work order'* in the conventional '*push*' system factory and from which they differ in three ways.

1. a kanban is issued when needed and not procedurally and at a pre-determined time, thereby eliminating traditional order states and their administrative overhead

2. kanbans are issued by the downstream process and until an empty palette or bin, with its kanban, arrives at a workstation that workstation remains idle regardless of any theoretical loss of productivity

3. each kanban represents an immediate requirement to be filled in *real time* thereby directly, connecting the adjacent processes and making superfluous the holding of buffer stocks of materials between departments. (Sartori, 1988)

Inventory level in any area is controlled by the number of kanban cards which Toyota calculates according to:

$$y = D(T_W + T_P)(1 + n)a$$

Equation 2 : Number Of Kanban Cards

where 'y' the number of kanban, 'D' is the demand per unit time, 'T_w' is the waiting time for the kanban to be returned to the supplying area, 'T_p' is the process time, 'a' is the container capacity and 'n' is the policy variable which allows some excess stock to accommodate disruptions and variations in usage rate. (Graham, 1988)

Throughout the restructuring process, Toyota strenuously resisted the idea of control by management through massive, centralised computer systems. Osamu Kimura, a General Manager of Toyota Motor Corporation's Transport Administration Office, illustrated information flows in the two types of system:

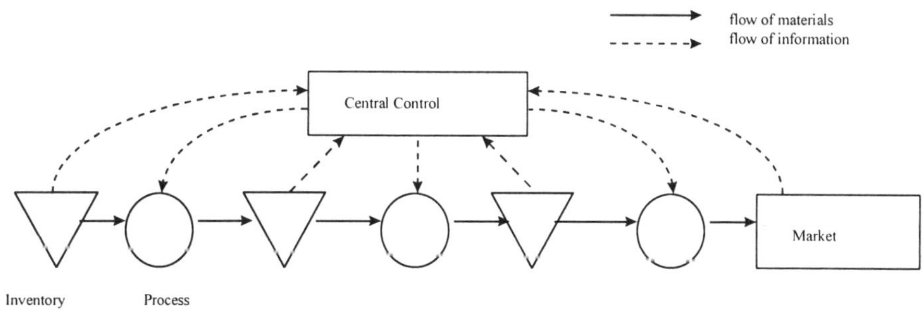

Fig 1 : Control In a Centralised System

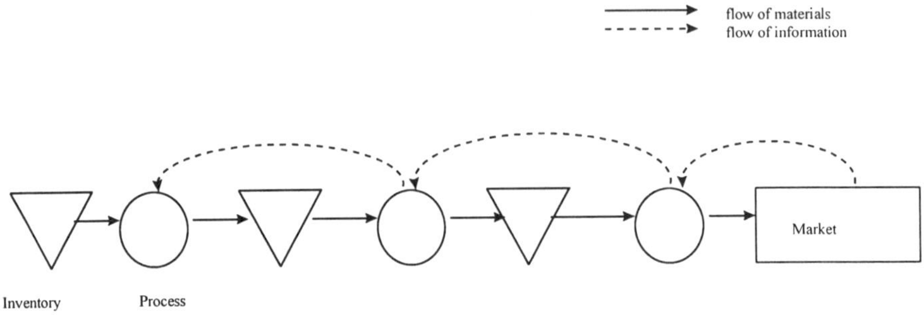

Fig 2 : Control in a Kanban System

"We should be careful not to centralise the system," he said, "by means of mammoth computers and information networks which may only lead to death by strangulation." (Yoshikawa & Burbridge, 1987, p. 18)

5 Solution to Schism a Question of Choice, not of Culture

Westerners sometimes correlate the success of the Toyota system, now widespread in Japan, with supposed cultural differences between Japanese workers and their Western counterparts - loyalty, long service, dedication and enthusiasm are seen to be parts of the Japanese culture and their absence in the West, therefore, to preclude the transplantation of the Japanese methodology. While some cultural differences are seen to exist, many commentators are adamant that management methodologies are exportable. (Fanuc, the Japanese CNC giant, is renowned for its high level of CIM and robotics implementation.)

Hofstede (1984) surveyed 50 countries ranking them comparatively on a variety of character traits and demonstrating quite clearly the existence of cultural differences. He concludes that the differences do not preclude the importation/exportation of systems across cultural boundaries but rather that:

"Effectiveness within a given culture, and judged according to the values of that culture, asks for management skills adapted to the local culture." (Hofstede, 1984. p. 98)

Translocation of JIT into the USA has frequently been highly successful, notably in that US bike icon Harley-Davidson. Other cases listed may include Apple, GM, General Electric, Gillette, Xerox and Hewlett Packard. (Meredith, Ristroph, et al., 1991,p.448)

"The attitude that Toyota's management and workers bring to each of these issues is not inherent to their "culture". In fact, a sizeable number of

American, firms have both the same philosophy and similar results." (Bignell, Dooner, et al., 1985, p. 154)

6 Appropriate Option may be a Mixture

Proposal 1: That clearly demonstrable benefits may accrue from the implementation of management support software.

Progressive implementation of currently available software in support of strategic management decision making, coupled with appropriate staff training and incentive, regularly produces significant improvements in efficiency, lead-time reduction, throughput and profit while reducing overhead and inventory. Despite the recent move to Open Systems implementation with reduced software costs and vendor dependence, however, the complete MES/CIM implementation remains a complex, expensive, long-term project best suited to major industrial operations.

Proposal 2: That realignment of management attitudes and restructuring of production processes may produce at least comparable benefits.

Just-In-Time manufacturing, including its associated systems and techniques, provided the impetus for the outstanding success of Japanese industry in the '80's at low cost and without massive computerisation. Management commitment to human-level communication and the consequent elimination of artificial executive / production barriers rekindled craftsmanship, pride in work, motivation and involvement at all staff levels. Resurgence of enterprise-wide focus, teamwork and willingness to make appropriate changes to fundamental processes cost little and achieved much.

Proposal 3: That Third Millennium Manufacturing may, to be successful, require a mixture of both solutions.

Manufacturing environments tend to be individually unique - JIT implementation at Nippon Steel must vary in practice from that at Sony and practice at Sony from that at Harley-Davidson; CIM implementation at Du Pont must vary from that at Fanuc and Fanuc's implementation from that at the wastewater treatment plant in Lexington, KY (USA). Australian industry is typified by small size and short production runs, consequently solutions applicable to multinational companies, directly transferred, are likely to be inappropriate. Careful blending of concepts from both mainstream solutions appears to offer the most applicable options.

Proposal 4: That reassessment of production processes should precede automation of control.

Logic demands that the best control of a flawed system can never produce the best results. Survival of Australian industry into the third millennium may depend on management accepting the challenge, and pain, of reformation far more than on management willingness to accept the cost of computerisation.

References

1. Batchelor, Ray, *Henry Ford: Mass Production, Modernism and Design.* New York, NY (USA): Manchester University Press (1994).
2. Bignell, Victor, Dooner, Mike, Hughes, John, Pym, Chris, Stone, Sheila (eds), *Manufacturing Systems: Context, Applications and Techniques.* Oxford (UK): Basil Blackwell Ltd (1985).
3. Dyck, Harold, Varzandeh, Jay, McDonnell, Jack, *Quality Impacts on JIT Performance Measures: A Factory Simulation*, Technology Management - The New International Language (Cat No 91CH3048-6) (October 27-31, 1991).
4. Ford, Henry, *Ford on Management, Harnessing, the American Spirit.* Oxford (UK): Basil Blackwell Ltd (1991)
5. Gibson, P., Greenhalgh, G., Kerr, R. *Manufacturing Management.* London, (UK): Chapman and Hall (1995)
6. Graham, Ian R, *'Just-in-Tiine' Management of Manufacturing.* Oxford (UK) : Technical Communications and Elsevier Advanced Technology Productions (1988)
7. Hernandez, Arnaldo. *Just-in-Tirne Ouality.* Englewood Cliffs, NJ (USA): PTR Prentice Hall (1993)
8. Hofstede, Geert. *Cultural Dimensions in Management and Planning.* Asia Pacific Journal of Management. 1 (2) 81-99 (Jan, 1984)
9. Hutchins, David. *Just In Time.* Aldershot, Hants. (UK) : Gower Technical Press Ltd. (1988)
10. Hutchins, David. *In Pursuit of Quality: Participative Techniques for Quality Improvement.* London (UK): Pitman Press (1990)
11. Ishikawa. Kaoru, Lu, David J. *What is Total Quality Control? The Japanese Way.* Englewood Cliffs, *NJ,* (USA): Prentice-Hall Inc. (1985)
12. Johnson, H. Thomas, Kaplan, Robert S, *Relevance, Lost : The Rise and Fall of Management Accounting.* Boston, MA (USA): Harvard Business School Press (1991)
13. Kerr, Roger, *Knowledge-Based Manufacturing Management: Applications of Artificial Intelligence to the Effective Management of Manufacturing Companies.* Sydney, NSW: Addison-Wesley Publishing Company (1991)
14. Kirkwood, L., Ryan, C., Falt, J., Stanley, T. *Accounting: An Introductory Perspective.* Melbourne, VIC: Longman Cheshire Pty Ltd (1989)
15. Linnegar, Gary J. *An Investigation into the Management of Change in Cost Accounting System Requirements During the Transition from Traditional Manufacturing to Just In Time Concepts.* Mississippi (US): Mississippi State University (1988)
16. Mathews, M. R., Perera, M.H. B., Ng, L.W., Chua,F.C., *Accounting Theory and Development.* South Melbourne, VIC. Thomas Nelson Australia (1991)
17. McDonough, Brian., *Considerations for Manufacturing Execution System Implementation.* [on -line] Available: http://www.smartdocs.com/~bmcd/accufacts9000/ art3.html (1995)
18. Meredith, Paul H, Ristroph, John H., Lee, Jim. *Implementing JIT: The Dimensions of Culture, Management, and Human Resources*, Technology Management - The New International Language (October 27-31, 1991)
19. Mitchell, Geoffrey B., Granof, Michael H., *Principles of Accounting.* Sydney,NSW.. Prentice-Hall of Australia (1981)
20. Orlicky, Joseph., *Material Requirements Planning : The New Way of Life in Production and Inventory Management.* New York, NY (USA): McGraw-Hill, Inc, (1975).

21. Robinson, Alan (ed.)., *Continuous Improvement in Operations : A Systematic Approach to Waste Reduction.* Cambridge, MT (USA), Productivity Press (1991).
22. Sartori, Luca G., *Manufacturing Information Systems.* Wokingham (UK), Addison-Wesley Publishers Ltd(1988)
23. Shingo, Shigeo, *Zero Quality Control : Source Inspection and Poka-Yoke System.* Norwalk, CT (USA), Productivity Press (1986)
24. Storey, J (ed.), *New Wave Manufacturing Strategies: Organisational and Human Resource Management Dimensions.* London (UK), Paul Chapman Publishing (1994)
25. Taylor, Frederick Winslow., *The Principles of Scientific Management.* New York, NY (USA): W. W. Norton & Company Inc. (1967)
26. Taylor, Frederick Winslow. *Principles of Scientific and Shop Management.* London (UK), Routledge/Thoemrnes Press(1993)
27. Yoshikawa, H., Burbridge, J, L. (eds). *New Technologies for Production Systems.* Amsterdam (ND): Elsevier Science Publishers B. V. (1987)

AUTHOR INDEX